Date Due

TREES, SHRUBS, AND WOODY VINES OF THE SOUTHWEST

THIS BOOK IS PUBLISHED WITH THE ASSISTANCE OF THE *Dan Danciger Publication Fund*

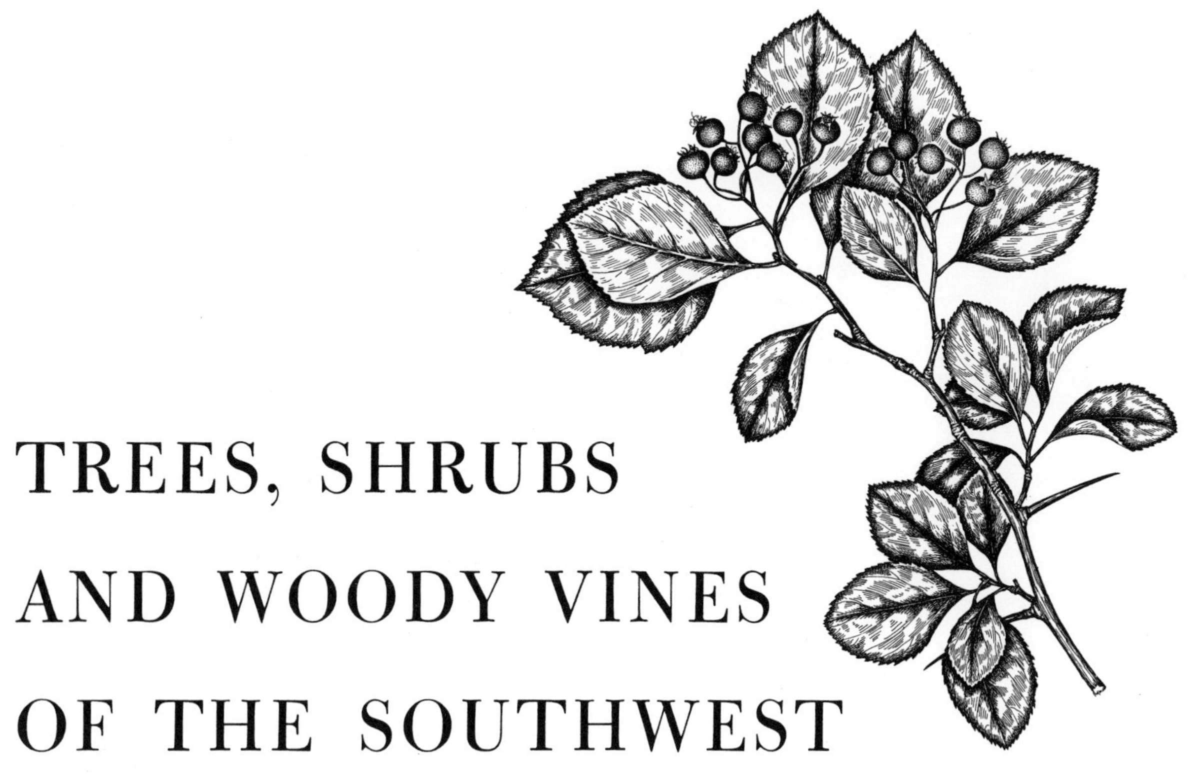

TREES, SHRUBS AND WOODY VINES OF THE SOUTHWEST

By Robert A. Vines

With drawings by Sarah Kahlden Arendale

UNIVERSITY OF TEXAS PRESS AUSTIN, 1960

Library of Congress Catalog Card Number 59–8129

Manufactured in the United States of America

Second Printing, 1969

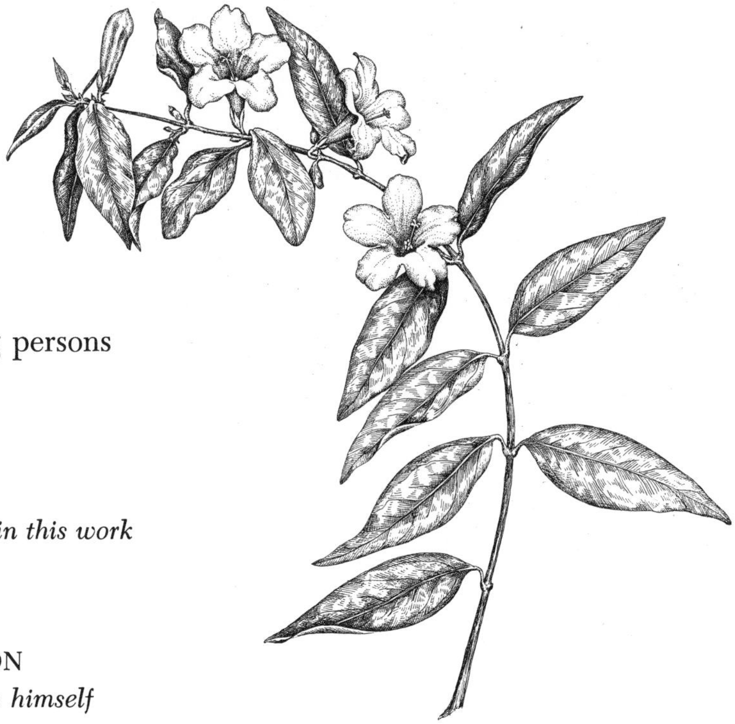

This book is dedicated to the following persons

TO MY WIFE
RUBY VALRIE VINES
whose love and encouragement sustained me in this work

TO MY PASTOR
THE REVEREND FERRIS NORTON
who taught me that man is not sufficient unto himself

TO MY FRIEND
PAUL WINKLER
who shared with me the pleasures of field work and the disciplines of the botanical sciences

TO MY BOYHOOD TEACHER
MABEL CASSELL
who instilled in me a love of the outdoors and an appreciation of its beauties

TO MY FRIEND AND PATRON
MRS. T. S. MAFFITT, JR.
who loyally rallied the support of many friends

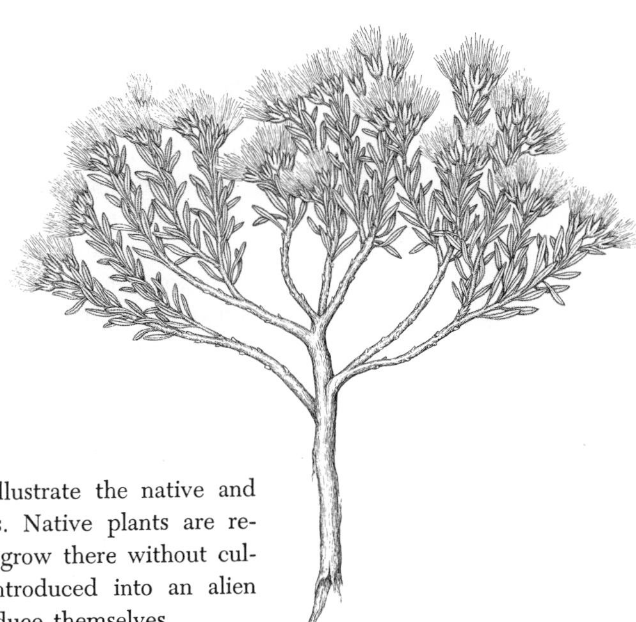

PREFACE

This book represents an effort to describe and illustrate the native and naturalized woody plants of the southwestern United States. Native plants are regarded as those which are indigenous to an area and which grow there without cultivation. Naturalized plants are those which, after being introduced into an alien region, have escaped cultivation and have continued to reproduce themselves.

The term "woody plants" is herein taken to include trees, shrubs, and woody vines, and also certain low and partially woody plants. The term "the Southwest," here used in a somewhat restricted and modified sense, includes Texas, New Mexico, Oklahoma, Arkansas, and Louisiana. Although this book is primarily intended to cover only these states, the extensive ranges of the majority of the species treated will make it quite useful in adjacent areas of surrounding states as well. The dates of earliest cultivation, unless otherwise indicated, refer to the United States.

In the vast area included in the five states taken to be the province of the book, a great variety of plant habitats is represented, ranging from the subtropical lower Gulf Coast through various temperate environments to the montane areas of the West. The many species are grouped into 102 chapters, each chapter representing a different plant family. Some chapters, such as those dealing with the rose, legume, and composite families, are quite large, while certain others contain only one or a few species.

Some of the problems encountered in the preparation of the book may be of interest to the reader. In order to gain a firsthand knowledge of the plants in the field, the author has traveled more than 250,000 miles by automobile. Areas inaccessible by automobile were reached by foot and on horseback. Collecting plants in the dry regions of the Southwest is not always an easy matter, for much of the area is desert-like and mountainous. Owing to the failure of some plants to produce flowers and fruit in especially dry years, collecting trips to certain areas had to be deferred for as long as several years, until rainfall restored the depleted ground moisture.

A large amount of literature, some of it quite rare and difficult to locate, was borrowed from libraries and botanical institutions; but, because of their incompleteness or vagueness, some of the older publications were nearly useless. Plants known only from such fragmentary descriptive matter have been completely redescribed from field collections.

The inadequacy of label data lessened the usefulness of many of the borrowed herbarium specimens of Southwestern plants. Specimens described by early field collectors and botanists frequently bore no indication of place of collection, and field notes regarding such matters as flower color, ecological preference, and frequency were the exception. Because dry pressed specimens often lack a third dimension, their usefulness in preparing illustrations was rather limited. Many of the very old original type-collections of some species were found to be badly fragmented and decomposed. These difficulties made it necessary to collect fresh material, not only to ensure drawings that are lifelike and as typical as possible, but also to make possible the verification and, where necessary, the amplification of previously published descriptions.

The determination of correct names, common and scientific, for each genus, species, variety, and form was sometimes perplexing. Simply because a name is in common use does not always mean it is the best name or the most nearly correct one.

For some of the more widespread species many common names exist, each having currency in a more or less local area. Some plants have English names as well as Mexican and Indian names. Several approaches were followed by the author in selecting a single or preferred common name. Usually, the common names listed by H. P. Kelsey and W. A. Dayton in *Standardized Plant Names,* 1942 edition, were adopted. For the common names of plants not treated in this work, the name most general in other authoritative sources has been used. Where no common name has been published, the author has drawn upon personal knowledge and conversations, and in some few instances he has used his discretion in coining a common name. It will be noted that the preferred or most general common name is given at the heading of each species description. Secondary names are listed in the paragraph entitled "Remarks." As a rule, a name in general use has been chosen over those employed locally. For some readers, this procedure may result in the joining of an unfamiliar common name to an otherwise familiar plant, but, as all known common names occur in the Index, there should be no real problem.

The problem of selecting correct scientific names required a different procedure. In this case *Standardized Plant Names* could not be used. Instead, the appropriate rules and recommendations laid down in the *International Rules of Botanical Nomenclature,* 1956 edition, were adhered to. For more specific guidance the most recent monographic treatments and botanical manuals were consulted. In cases where discrepancies occurred, recognized authorities on certain genera and families were followed where possible; in some few instances it was necessary to make decisions at variance with the prevailing ones.

The author feels that twenty-five years of study has given him a firsthand knowledge of the woody plants of the Southwest, their peculiarities, their preferences, and their variability. However confident one may feel in possession of such information, it is clear that botanical science never has been, nor can be, the work of any individual. Each worker co-operates with his contemporaries, building on the work of others and knowing his own work will be added to by his successors. It is fully expected and hoped that other workers will refine, expand, correct, and otherwise modify the content of this book. Grateful mention of the many institutions and individuals that have assisted the author, spiritually and materially, is made in the Acknowledgments section which, because of its length, has been placed at the end of this book.

Robert A. Vines

September, 1959

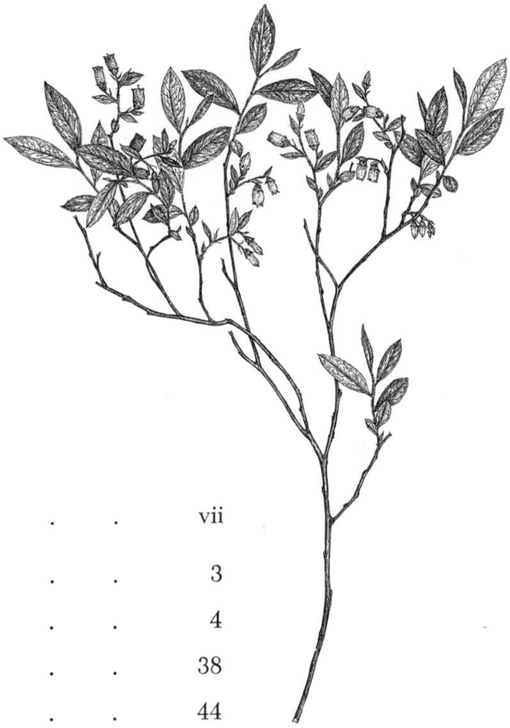

CONTENTS

TREES, SHRUBS, AND WOODY VINES OF THE SOUTHWEST

GINKGO FAMILY (*Ginkgoaceae*)

GINKGO
Ginkgo biloba L.

GINKGO

Ginkgo biloba L.

FIELD IDENTIFICATION. An introduced tree known to attain a height of 100 ft and a diameter of 6 ft. Trunk slender and straight with an ovoid crown.

FLOWERS. In late spring, small; staminate and pistillate on different trees; staminate flowers in slender, loose catkins, anthers borne in stalked pairs on a slender axis; pistillate flowers solitary or paired, long-stalked with two ovules, one ovule abortive, the other naked ovule developing, embryo with two cotyledons.

FRUIT. A drupe usually solitary, obovoid or ellipsoid, about 1 in. in diameter, flesh pale yellow at maturity, acid but sweet, odor disagreeable; seed solitary, large, angular-oval, cream-colored, thin-shelled, sweet.

LEAVES. Deciduous, alternate or 2–6-clustered on short spurs, 2–4 in. across, wedge-shaped, apex cleft and wavy, base gradually narrowed, texture thickened, surfaces striate with numerous parallel veins, glabrous; petioles 1–4 in. long, slender, green to yellowish, glabrous.

TWIGS. Stout, smooth, green to yellow or brown, lustrous.

BARK. Dark brown or gray to black, smooth at first, roughened by fissures later.

RANGE. A native of China, cultivated for ornament in the United States. Much planted in Washington, D.C. Grows well in north and east Texas, but not extensively planted. Persistent about abandoned garden sites. A number of trees have been planted on the campus of Baylor University at Waco, Texas, and seem to be growing well.

PROPAGATION. Ginkgo can be propagated by seed stratified in autumn, and varieties can be obtained by budding and grafting.

HORTICULTURAL FORMS. A number of horticultural forms of *Gingko* are known.

Cut-leaf Ginkgo, *G. laciniata*, has deeply incised and divided leaves.

Large-leaf Ginkgo, *G. macrophylla*, has larger leaves than other ginkgoes.

Sentry Ginkgo, *G. fastigiata*, has ascending branches and a columnar head.

Weeping Ginkgo, *G. pendula*, has pendulous branches.

Variegated Ginkgo, *G. variegata*, has yellow-variegated leaves.

Yellow-leaf Ginkgo, *G. aurea*, has bright yellow leaves.

REMARKS. The genus name, *Ginkgo*, is from the Chinese word meaning "silver fruit." The species name, *biloba*, refers to the 2-lobed leaves. The tree was introduced to Japan from China by the Buddhist priests, who planted it by their temples. It was brought to America in 1784. The tree is the sole survivor of a family that was widespread in early geologic times. Fossil specimens are found in both America and Asia. The kernels of the fruit are sweetish and resinous, and are highly esteemed as food in China and Japan. The flesh is made into a preserve, or is baked and eaten between meals to aid digestion. The fruit has a disagreeable odor, and for this reason the pistillate trees are not planted as often as the staminate.

Ginkgo is relatively free of insect damage and is hardy as far north as Massachusetts and Michigan.

PINE FAMILY (*Pinaceae*)

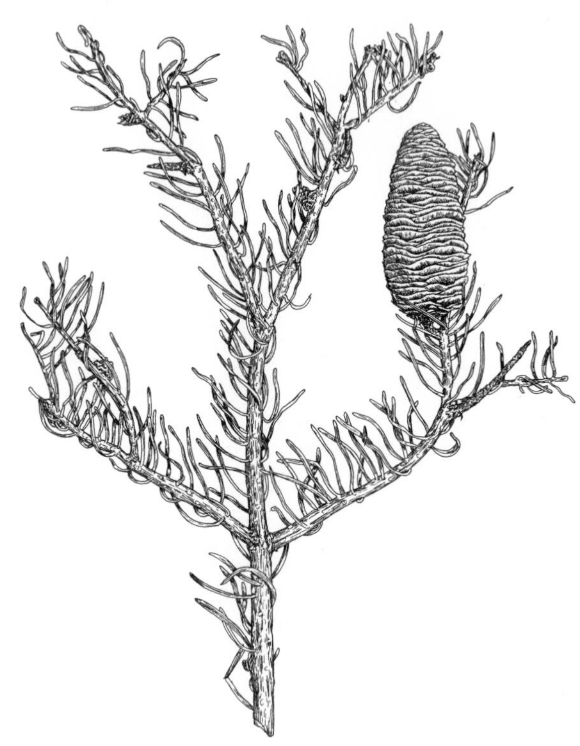

White Fir
Abies concolor (Gord. & Glend.) Lindl.

White Fir

Abies concolor (Gord. & Glend.) Lindl.

FIELD IDENTIFICATION. Tree of the Rocky Mountain region attaining a height of 100–250 ft, diameter of 3–6 ft. Young trees generally dense, symmetrical and conic; older specimens in the open irregular or rounded, with stout branches almost to the ground. When in crowded stands, trunk often clear one half to two thirds its height. Branches regularly in whorls of 4–5, the whorls forming flat masses of foliage. Root system shallow.

FLOWERS. May–June, axillary, pendent, from buds formed the previous season on branches of that year, involucres of enlarged bud scales at base; staminate abundant on lower side of branches above middle of tree, oval to oblong, dark red or rose-colored, anthers yellow or red, surmounted by short knoblike projections; pistillate ovoid to oblong, with numerous 2-ovuled imbricate scales; bracts oblong, obcordate, laciniate above, short-pointed, strongly reflexed.

FRUIT. Ripening September–October, soon breaking apart to expose a persistent axis; cone ovoid to oblong, cylindric, green to yellow or purplish, puberulous; apex rounded or obtuse; length 3–5 in.; scales closely imbricate, nearly twice as wide as high, apex rounded; bracts emarginate or truncate with a spikelike tip, shorter or longer than the scales; seeds dark brown, ⅓–½ in. long, seed coat with a thick outer layer and thin membranous inner one, seed bearing lustrous rose-colored wings with a truncate apex.

LEAVES. Crowded, spreading at all angles but showing a definite tendency to be erect, more or less 2-ranked by the twist of their bases, sessile, linear; ¾–3 in. long, 1/16–⅛ in. wide, pale bluish green or silvery blue, glaucous, eventually dull green, flat, straight, apex acute to obtuse or rounded, 2 resin ducts along the lower epidermal surface, more or less stomatiferous above; leaves on flowering branches rather thick, keeled, often curved, apex usually acute; leaf scars circular.

TWIGS. Rather stout, yellowish or orange at first, later yellowish green to gray or reddish brown, glabrous and lustrous; some with conspicuous resin blisters; winter buds subglobose, resinous, ⅛–¼ in. thick, puberulent to glabrous, yellowish green to brown.

BARK. Smooth on young trunks and branches, gray, thin, sometimes with numerous resin vesicles; on old trunks 5–7 in. thick, gray to reddish brown, deeply fissured with ridges broad and broken into irregularly shaped platelike scales.

WOOD. Pale brown to buff or nearly white, heartwood not very distinct, grain coarse but rather straight and even, specific gravity when dry 0.42, soft, brittle, not durable, moderately weak in bending, strong in endwise compression, rather stiff, moderately low in shock resistance, easy to work, paints and glues well, nailholding capacity low, moderate shrinkage, easily cured, subject to surface checking if dried too rapidly. Used for general construction, boxes, crates, refrigerators, planing-mill products, pulpwood, woodenware, and mine timbers.

RANGE. Usually in gravelly soil or sandy loam at altitudes of 3,000–11,000 ft and upward, usually with a northern exposure. Widest distribution of any of the Western firs. In the Sandia, Mogollon, White, Sacramento, and Capitan mountains; north to Wyoming, west to Washington, Oregon, and California, and south to Baja California and Sonora, Mexico.

PROPAGATION. Optimum seed-bearing age is 50–100 years, with good crops every 2–4 years. Tree starts bearing annually about the 40th year. Cones are gathered before the scales separate, and are spread out to dry in the sun or in ventilating sheds. Seeds may be rubbed out of the cone by hand and the chaff separated by a fanning mill. The seeds are soft and easily damaged by dewinging. A bu of cones yields 48–82 oz of seed with an average of 10,000–15,000 seeds per lb. The commercial purity is about 91 per cent and soundness 60 per cent. Seeds lose vitality quickly and should not be stored more than 1 or 2 years. Embryo dormancy can be reduced by stratification in sand at 40°F. for 60–90 days, but untreated seeds germinate fairly well. Stratified seed is sown in spring, or unstratified in fall, broadcast or in drills using 60–80 seeds per sq ft and covered about ⅛ in. with sandy loam. Screen protection from birds and mammals usually required. Beds may be treated with sulphur to prevent damping-off. Half-shade is needed for the seedlings. Some seed lots may be worthless owing to infestation by the chalcid fly *(Megastigmus strobilobius)*. Older trees are sometimes severely damaged by the fir engraver beetle *(Scolytus ventralis)*.

VARIETIES. The following horticultural varieties are recognized:

Golden White Fir, *A. concolor* forma *aurea* Beissn., has golden yellow leaves.

Short-leaf White Fir, *A. concolor* forma *brevifolia* Beissn., has broad, short, obtuse leaves.

Conical White Fir, *A. concolor* forma *conica* Slavin, has a dwarf pyramidal form and leaves ¾–1½ in. long.

Globe White Fir, *A. concolor* forma *globosa* Niemetz, has a globose habit and white branches.

Pacific White Fir, *A. concolor* forma *lowiana* (A. Murr.) Lemm., has smaller buds, leaves pectinately arranged, 2–2¾ in. long, rounded or bifid at apex, shallowly grooved above.

Weeping White Fir, *A. concolor* forma *pendula* Beissn., has a narrow columnar head and strong pendulous branches.

Purple-cone White Fir, *A. concolor* forma *violacea* (Roesl.) Beissn., has bluish white leaves and purplish cones.

Wattez White Fir, *A. concolor* forma *wattezii* Beissn., has pale yellow to silvery white leaves.

REMARKS. The genus name, *Abies*, is the classical name of the fir-tree, and the species name, *concolor*, refers to the uniform color of the needles. It also has the vernacular names of White Balsam, Balsam Fir, Silver Fir, Colorado Fir, and Pino Real Blanco.

The tree is handsome in cultivation and has been grown since 1851. It attains an age of 350 years. At first it is rapid-growing, but slower later, and is resistant to fire, heat, and drought. The seeds are eaten by sooty grouse, Mount Pinos grouse, Inyo chipmunk, and California pine squirrel, and browsed by black-tailed deer.

Alpine Fir

Abies lasiocarpa (Hook.) Nutt.

FIELD IDENTIFICATION. Evergreen tree of the Rocky Mountain area. Under the severe conditions at timberline sometimes a prostrate shrub; at lower elevations a tree to 175 ft, with a symmetrical, moderately tapering trunk to 4 ft in diameter. Branches rather short and crowded, extending almost to the base on old trees, forming a dense narrowly conic or spirelike crown.

FLOWERS. In June, axillary, with large involucres of bud scales at base; staminate ones pendent, dark blue or violet at maturity, anthers yellow or red; pistillate ones erect, globose, scales imbricate in many series, obovate, rounded above, cuneate below, much shorter than the acute, mucronate bracts, 2-ovuled.

FRUIT. Maturing September–October, erect, oblong-cylindric, dark brown to purple, length 2½–4 in., width 1–1¼ in.; scales closely imbricate, thin, reddish brown, apex rounded to almost truncate, ½–⅘ in. wide or long, scales covering the much shorter bracts which are

Alpine Fir
Abies lasiocarpa (Hook.) Nutt.

laciniately cut with slender black tips; seeds about ¼ in. long, with conspicuous resin vesicles, wing broad and lustrous; seed coat thin and in 2 layers, the outer coriaceous, the inner membranous; scales usually deciduous, and intact cones hard to find on the ground.

LEAVES. Crowded, spreading, on young twigs appressed and directed forward, length ½–1¾ in., about 1/12 in. wide, sessile, linear; apex rounded to acute or occasionally notched (young leaves sometimes with long slender, rigid points), flat and grooved above, marked usually above the middle with 4–5 rows of stomata, lower surface marked by 2 broad bands of 7–8 rows of stomata.

TWIGS. Stout, reddish and pubescent at first, later orange-brown to gray or whitish; leaf scars conspicuous and circular; winter buds subglobose to ovoid, ⅛–¼ in. long, resinous, covered with closely imbricate light orange-brown scales.

BARK. At first smooth and gray to white, sometimes with numerous horizontally elongate resin pockets. Later ¾–1½ in. thick, breaking into shallow furrows, gray to grayish brown or reddish, scales thick and closely appressed.

WOOD. Pale brown to almost white, sapwood pale, light, soft, not strong.

RANGE. In the spruce and alpine belts, 2,700–8,000 ft and upward. From the Sandia, Mogollon, and Tunitcha mountains of northern and western New Mexico north through the mountains to Alberta; also in the Cascade Mountains of Oregon, Washington, and British Columbia.

PROPAGATION. Optimum seed-bearing age of the tree is 150–200 years. Seed crops are abundant every 2 years. The cones are spread out to dry in the sun or in well-ventilated sheds. The chaff is separated by a fanning mill. The seeds average about 37,500 per lb, with a purity of 94 per cent and a soundness of 53 per cent. The seed vitality is not long, and germination averages about 24 per cent. About 5,000 usable plants may be obtained per lb of seed. Embryo dormancy may be broken by stratification in sand at 41°F. for 60 days. Seeds are drilled or broadcast in well-drained sandy soil for 60–80 seedlings per sq ft and covered about ⅛–⅜ in. with soil. Should be protected from the sun by a half-shade. Sulphur treatment of the beds is advisable for damping-off protection.

VARIETY AND FORMS. Alpine Cork-bark Fir, *A. lasiocarpa* var. *arizonica* (Merriam) Lemmon, has a creamy white, thick and corky bark with leaves whitish beneath, emarginate at apex. Occurs from central Colorado to southeastern New Mexico and northern Arizona.

Dwarf Alpine Fir, *A. lasiocarpa* forma *compacta* (Beissn.) Rehd., is a dwarf compact form.

REMARKS. The genus name, *Abies,* is the classical Latin name of Silver Fir, *A. alba* Mill. The species name, *lasiocarpa,* means "hairy-fruited." The tree is also known under the vernacular names of Rocky Mountain Fir, Subalpine Fir, Balsam Fir, White Fir, Balsam, White Balsam, Pino Real, and Blanco de las Sierras.

The tree is slow-growing and moderately long-lived. A tree 15 in. in trunk diameter may be 175 years old. It was introduced into cultivation in 1863. The wood is little used except for fuel and pulpwood. It has some value for wildlife, the seeds being eaten by at least 6 species of birds and various squirrels; mountain sheep and black-tailed deer browse the foliage and young shoots.

ENGELMANN SPRUCE
Picea engelmannii Parry

ENGELMANN SPRUCE

Picea engelmannii Parry

FIELD IDENTIFICATION. Western evergreen shrub, at very high elevations often prostrate as a result of wind pressure, or at lower elevations a large tree to 120 ft high, with a trunk 3 ft in diameter. The trunk straight and tapering, and the whorled, short, lateral branches forming a compact pyramidal head. Root system rather shallow.

FLOWERS. June–July, male and female separate on the same tree; staminate catkin-like cones, long-stalked, axillary, dark purple; anthers numerous, spirally arranged with connectives produced into broad, nearly circular, toothed crests; pistillate cone terminal, scarlet, with numerous 2-ovuled bracted scales; the bracts obovate-oblong, acute to acuminate, apex denticulate.

FRUIT. Cone maturing August–September, deciduous during the winter, horizontal or pendulous, cylindric ovoid to oblong, ends narrowed, sessile or short-stalked, about 1–2½ in. long, green to reddish at first, later light brown and shiny; scales rhombic-oblong, thin, tough, papery, apex truncate or acute, margin erose or almost entire; seeds 2 under each scale, shed by October, small, about ⅛ in. long, obtuse, compressed, nearly black, with a much longer broad, oblique wing.

LEAVES. Borne on prominent leaf cushions, spirally disposed, sometimes appressed and pointed toward the apex of the twig, linear, soft and flexible, flattened, straight or slightly incurved, apex acute or blunt, length ¾–1¼ in., marked on each face with 3–5 rows of stomata, at first with a glaucous bloom but eventually dull steel-blue or dark bluish green, producing a disagreeable odor when bruised.

TWIGS. Slender, minutely glandular-pubescent, younger dark yellow to orange-brown, or gray tinged with brown.

BARK. Light reddish or cinnamon-red to grayish, separating into large, thin, loose scales; winter buds brown, obtuse, scarious, appressed or only slightly reflexed at apex.

WOOD. Yellowish to reddish or brown, sapwood thick and hardly distinguishable, rather close and straight grained, lustrous, light, soft, specific gravity when dry about 0.37, moderately stiff, rather low in shock resistance, moderately weak in endwise compression, does not split easily, glues well, low in nail-holding ability, average in paint-holding, works easily, shrinks moderately, easy to kiln-dry, when seasoned stays in place well, not durable under decay conditions.

RANGE. At altitudes of 5,000–15,000 ft. New Mexico, Arizona, Colorado, Utah, Wyoming, Nevada, Montana, California, Oregon. Occasionally planted for ornament in the Northeastern states and in Europe.

PROPAGATION. The optimum seed-bearing age is 150–200 years. The yield is about .05 lb of seed per bu of cones. The seed averages about 135,000 per lb, with a commercial purity of 90 per cent and a soundness of 67 per cent. Germination tests are variable, seed lots running 34–68 per cent. Pretreatment of the seed is usually not required. Tests have indicated that with proper storage the seed viability was as high as 69 per cent after 5 years. Trees are subject to the attack of the Engelmann spruce beetle, *Dendroctonus engelmannii* Hopk., in the central Rocky Mountain states.

FORMS. It has been cultivated since 1862, and has been known to reach an age of 500–600 years. It is desirable because of its dense foliage and pyramidal habit. A number of horticultural forms have been developed:

Blueleaf *(glauca)*, glaucous leaves; Dwarf Globe *(microphylla)*, small-leaved dwarf globose form; Silver *(argentea)*, silvery-gray foliage; and Fendler *(fendleri)*, pendulous branches and slender leaves with 2 bands of 4 rows of stomata above and half as many beneath.

REMARKS. The genus name, *Picea*, is from the ancient Latin *pix, picis* ("pitch"), and the species name *engelmannii*, is in honor of George Engelmann (1809–1884), German-born physician and botanist of St. Louis, an authority on the conifers. It is also known as White Spruce. It is generally distinguishable from the Colorado Spruce, *Picea pungens*, by its pubescent twigs and soft flexible leaves. The bark is used for tanning leather, and the wood for many purposes including paper pulp, fuel, charcoal, cooperage, general construction, planing-mill products, car construction, shipbuilding, musical instruments, furniture, cabinets, ladder rails, oars, and woodenware.

The tree is also valuable for watershed protection and for wildlife food and cover. The seeds are known to be eaten by at least 10 species of birds, and the plant foliage, or bark, is browsed by bighorn sheep, black-tailed deer, white-tailed deer, and porcupine.

COLORADO SPRUCE

Picea pungens Engelm.

FIELD IDENTIFICATION. Evergreen tree of the Rocky Mountain region, attaining a height of 150 ft and a diameter of 3 ft. Branches somewhat rigid, stout, spreading or whorled, forming a broad, regular, pyramidal or conical crown, sometimes extending to the ground. The root system is shallow in wet soils and deeper in dry soils.

FLOWERS. In April or May, staminate ones reddish, axillary, in catkin-like cones, ½–1½ in. long; anthers numerous, spirally arranged, with connectives produced into broad, nearly circular, toothed crests; pistillate ones terminal, in the axis of accrescent bracts, shorter than the scale at maturity, each bract 2-ovuled.

FRUIT. Cone maturing in August, drooping, sessile or short-stalked, oblong-cylindric, usually 2¾–3½ in. long, at first green or reddish, later light chestnut-brown and lustrous; scales oval to rhombic, flat, thin, flexible, apex rounded or truncate, margins erose; 2 seeds under each scale, seed about ⅛ in. long, compressed, winged. Optimum seed-bearing age of tree 50–150 years.

LEAVES. Spirally disposed on all sides of the twig, linear, entire, spiny-acuminate, rigid, incurved, ½–1⅛ in. long, with 4–7 rows of stomata on each side, with 2 resin canals, varying in color from silvery white to light or dark bluish green; buds brownish yellow, usually with reflexed scales.

TWIGS. Stout, rigid, glabrous, orange or yellowish brown to grayish brown, at first glaucous green, with prominent leaf cushions, pointing apically and forming flat-topped masses of foliage; winter buds stout, obtuse to acute, pale brown, ¼–½ in. long, margin reflexed.

COLORADO SPRUCE
Picea pungens Engelm.

BARK. On young trees pale gray and smooth, on older ones gray to reddish brown and deeply divided into broad ridges and with small closely appressed scales.

WOOD. Pale brown, sapwood paler or sometimes hardly distinguishable, close-grained, light, soft, with numerous resin canals.

RANGE. Along water courses on hillsides or canyons at altitudes of 6,500–11,000 ft. New Mexico, Arizona, Colorado, Utah, Idaho, and Wyoming.

PROPAGATION. Good seed crops are borne every 2 or 3 years. The cones are collected in fall before they open, and are spread out to dry in the sun or in well-ventilated drying sheds. The seeds may then be shaken out of the cones and cleaned by a blower. About 12–20 ounces of seed come from a bu of cones. The seed averages about 106,000 seeds per lb, with commercial seed having a purity of 88 per cent and a soundness of 93 per cent. Some lots have considerable embryo dormancy. Stratification in moist sand at 32°–41°F. for 30–90 days is helpful for spring sowing. Sowing in drills or broadcasting for about 75 seedlings per sq ft is a good ratio. The seedbed should be of well-drained sandy loam and seeds covered about ¼ in. Fall beds should be mulched over winter and protected from birds and rodents by screens.

HORTICULTURAL FORMS. The following horticultural forms have been developed:

Green *(viridis)*, deep green foliage; Yellow *(flavescens)*, whitish-yellow foliage; Yellow *(lutea)*, yellow foliage; Blue *(glauca)*, bluish green foliage; Cerulean *(coerulea)*, bluish white foliage; Dwarf *(compacta)*, dwarf, compact, flat-topped form with the branches in almost horizontal layers and with rigid dark green leaves; Golden *(aurea)*, golden yellow foliage; Hunnewell *(hunnewelliana)*, dwarf form with a dense pyramidal habit, with rather flexible leaves; Koster *(kosteriana)*, pendulous branches and bluish foliage; Moerheim *(moerheimii)*, compact habit and deep blue foliage; Silver *(argentea)*, silvery white foliage; and Weeping *(pendula)*, pendulous branches.

REMARKS. The genus name, *Picea*, is the classical name of the spruce, and the species name, *pungens*, means "sharp-pointed" and refers to the leaves. Also known under the vernacular names of Silver Spruce and Blue Spruce. The wood is used for posts, poles, and fuel, but is the least valuable of any of the spruces. It is sometimes planted for wildlife cover, windbreaks, and watershed protection. Because of the variation in shape and foliage color, it is often planted as an ornamental in the eastern and northern states and in western and northern Europe. It was introduced into cultivation in 1862. Old trees, however, become ragged in appearance as a result of the loss of the lower branches. The tree is rather long-lived, slow-growing, and wind-firm. The seeds are eaten by numerous species of birds and browsed somewhat by black-tailed deer.

WHITE CEDAR

Chamaecyparis thyoides (L.) B. S. P.

FIELD IDENTIFICATION. Tree 70–80 ft, with a diameter 3–4 ft. Slender, horizontal or pendulous branches forming a narrow conical head. Trunks of trees growing close together are limbless for a considerable distance upward.

FLOWERS. March–April, monoecious, minute, solitary, terminal, staminate and pistillate on separate branches; staminate oblong, terminal, sessile or nearly so, 1/12–⅛ in. long, stamens 4–6 pairs, decussate, filaments short with ovate connectives; anther cells 2, globose in sides of a shieldlike scale; pistillate terminal, subglobose, pink; scales ovate, acute, spreading, with 1–4 basal, erect, bottle-shaped, blackish ovules.

FRUIT. Ripe in autumn, persistent, cone erect, globose, maturing at end of first season, ⅕–⅓ in. diameter, lower sterile scales of flowers persistent at base, fruit formed of ovule-bearing scales; scales swollen at apex, flattened with short prominent points or knobs, green to reddish brown and glaucous, fleshy, sessile; seeds 1–3, under each fertile scale, ovoid, acute, grayish brown, about ⅛ in. long, wings dark brown.

LEAVES. Buds minute; leaves scalelike, closely appressed, 4-ranked, overlapping, ovate, acute or acuminate, 1/25–⅛ in. long, bluish green, keeled and glandular or glandular punctate dorsally, becoming russet brown, persistent; leaves on vigorous shoots often spreading, needle-shaped or linear-lanceolate.

TWIGS. Two-ranked in 1 horizontal plane, compressed, forming a fan-shaped spray, later more terete, green to light or dark brown, the thin bark separating into papery scales.

BARK. Gray to light reddish brown, with narrow fissures and flat, confluent, sometimes twisted, ridges, exfoliating into shaggy strips.

WOOD. Light reddish brown, sapwood nearly white and narrow, straight-grained, fine-textured, somewhat oily and fragrant, moderately soft, compact, not strong, weak in bending and endwise compression, low shock-resistance, splits easily, works readily with tools, finishes smoothly, paints well, shrinks little, weighs 20.70 lb per cu ft, durable in contact with soil or water.

RANGE. Usually in cold, acid, sandy swamps. Possibly to be found, but not definitely recorded, from eastern Louisiana, eastward in Mississippi, Alabama, and Florida.

PROPAGATION. Cultivated for ornament since 1727, and occasionally grown in Europe. Rather large crops of seed produced, beginning at about 5 years of age. Fruit matures September–October and is persistent 2–3 years. Seeds are collected by hand and dried. Dewinging should be avoided because the seeds damage easily. About 4–10 lb of seed can be obtained from 100 lb of fruit. Commercially-cleaned seed averages 460,000 seeds per lb. Dry storage in sealed containers at near-freezing temperatures seems adequate for a year. Germinative capacity 70–90 per cent. Germination occurs best on moist peat or mucky mineral soil. Seed should be covered with about ¼ in. of soil, and seedlings given part shade.

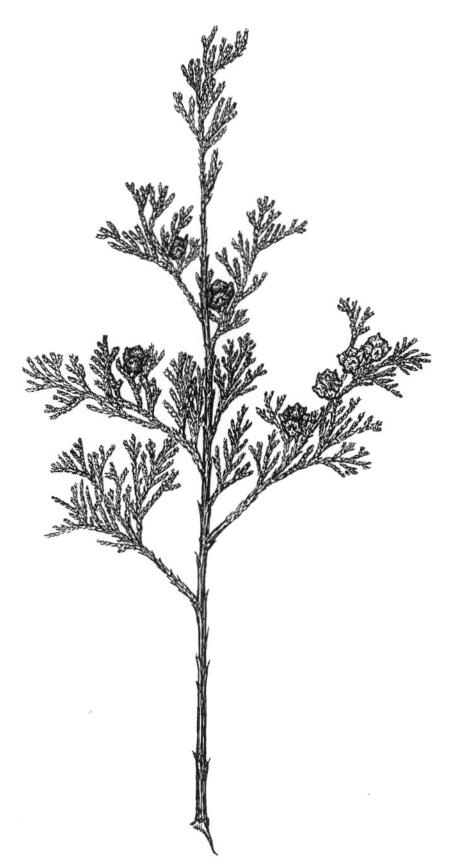

WHITE CEDAR
Chamaecyparis thyoides (L.) B. S. P.

VARIETIES. A number of horticultural forms have been listed by Rehder (1940) as follows:

Blue White Cedar, *C. thyoides* forma *glauca* (Endl.) Sudw., is a compact shrub with glaucous or nearly silvery white leaves.

Variegated White Cedar, *C. thyoides* forma *variegata* Sudw., has branches variegated with yellow.

Heath White Cedar, *C. thyoides* forma *ericoides* (Carr.) Sudw., is a low compact pyramidal bush; leaves linear, spreading, with 2 glaucous lines beneath, usually changing to reddish brown in winter.

Andley White Cedar, *C. thyoides* var. *andelyensis* Schneid., is of upright habit with loose, appressed subulate leaves or integrating with Heath White Cedar.

REMARKS. The genus name, *Chamaecyparis*, is the Greek name for the Ground Cypress of the Old World, and the species name, *thyoides*, means "Thuya-like," or like Arbor Vitae. Other vernacular names are Northern White Cedar, Atlantic White Cedar, Southern White Cedar, Swamp Cedar, Swamp Juniper, Juniper-tree, and White Cypress. The wood is used for boxes, crates, fencing, cooperage, boatbuilding, shingles, mine timbers, posts, poles, furniture, lumber, ties, finish for houses, and charcoal. It has little value as a wildlife food. The tree is easily fire-damaged, somewhat subject to storm damage, comparatively free of insects, and is attacked by only a few fungi.

COMMON DOUGLAS FIR

Pseudotsuga menziesii (Mirb.) Franco

FIELD IDENTIFICATION. Tree attaining a height of 200 ft, with a diameter of 3–4 ft. The branches somewhat slender, elongate and pendulous, forming a pyramidal crown. The older trees usually have a narrower crown, and when crowded the lower branches shed.

FLOWERS. Spring–summer, on branchlets of the previous year, terminal, or in the upper leaf axils, cones solitary; pistillate orange-red; scales spiraled, rounded, ovate, with the bracts longer, 2-lobed and slender-tipped; staminate scattered, axillary, solitary, oblong-cylindric; bracts slender, elongate, reddish-tinged; anthers numerous, globose, their connectives with short, spurlike apices.

FRUIT. Cone ripening August–September, maturing in one season, pendent, ovoid-oblong, 2–4½ in. long, dark reddish brown; scales closely appressed, semiconcave, rigid, rounded, longer than broad, bluish green to purplish or reddish; bracts much exserted, 2-lobed, 1/5–1/4 in. wide, terminally free about 1/2 in. beyond the scales, the midribs terminating in rigid, woody awns, basal ones lacking the scales; seeds 2 under each scale, semitriangular or rounded, reddish brown, somewhat lighter-spotted, shiny, about 1/4 in. long and 1/8 in. wide; seed wings oblong, broadest above the middle, oblique and rounded at apices; seeds dispersed by the winds.

LEAVES. Long-persistent, spreading nearly at right angles, petiolate, linear, flattened, entire, straight or semicurved, apex obtuse or rounded (or acute on vigorous shoots), bluish green or yellowish green, upper surface grooved, lower surface with rows of stomata on each side of midrib.

TWIGS. Slender, for 3–4 years, retaining pubescence which is highly variable in amount, orange to reddish brown, somewhat shiny, with age dark grayish brown and dull; terminal bud fusiform, lustrous, acute, about 1/4 in. long, much longer than the lateral ones.

BARK. On young trees and branches of older trees rather smooth, thin, dark reddish to grayish brown, often with resin blisters; on old trunks 4–24 in. thick, separated by large, irregular rounded ridges into oblong plates which are in turn covered with small, appressed, dark, reddish brown scales; fissures deep and irregular; bark corky on some mountain forms.

WOOD. The following description of the wood characteristics are from Brown, Panshin, and Forsaith (1949–1952, pp. 473–74):

"Sapwood whitish to pale yellowish or reddish white, narrow (Rocky Mountain type) to several inches in width (Pacific Coast type); heartwood ranging from yellowish or pale reddish-yellow (slow-grown stock) to orange-red or deep-red (fast grown stock), the color varying greatly in different samples; wood with a characteristic resinous odor when fresh (different from that of pine), without characteristic taste, usually straight and even or uneven grained, medium to fairly coarse-textured, moderately light (sp. gr. approx. 0.40 green, 0.45 ovendry—Rocky Mountain type) to moderately heavy (sp. gr. approx. 0.45 green, 0.51 ovendry—Pacific Coast type), moderately hard, moderately strong in bending, moderately strong (Rocky Mountain type) to very strong (Pacific Coast type) in endwise compression, moderately stiff and moderately low in shock resistance (Rocky Mountain type) to stiff and moderately high in shock resistance (Pacific Coast type), difficult to work with hand tools, machines satisfactorily, splits easily, glues satisfactorily, takes and holds paint poorly, difficult to treat with preservatives (material to be treated is usually incised to allow better penetration of preservatives), shrinks moderately (Rocky Mountain type shrinks the least), easy to kiln-dry if proper methods are used, moderately durable to durable (dense stock) when exposed to conditions favorable to decay. Growth rings very distinct, frequently wavy, delineated by a pronounced band of darker summer wood."

COMMON DOUGLAS FIR
Pseudotsuga menziesii (Mirb.) Franco

RANGE. The form here taken to be the species ranges from the Pacific Coast region of southwestern British Columbia south through western Washington and western Oregon to central coastal California, east to the Cascade Mountains in Oregon and the Sierra Nevada Mountains in California and western Nevada (see range of the Mountain form under VARIETIES).

PROPAGATION. The commercial seed-bearing age is 9 years at a minimum, optimum 100–200 years, maximum 600 years. The average tree produces about 2½ bushels of cones, with an average bushel containing about 1,000 cones. The unopened cones are gathered from standing or felled trees August–September. The following four methods have been recommended for extraction of the seed:

1. Spread the freshly collected cones out to dry in the sun or place them in a simple convection type cone kiln for 10–15 hours at 104°F. (for fresh cones) or 110° (for precured cones) and gradually increase the temperature to 130°.

2. Run the cones through a shaker or flail to remove the seed.

3. Dewing the seeds by running them through a dewinger, by kneading them in a sack, or by rubbing them over a 1/16 in. mesh screen.

4. Clean the seed by fanning or blowing off the chaff. The seed averages about 42,000 seeds per lb, but shows considerable variation from different localities. Commercial seed has a purity of 93 per cent and

a soundness of 72 per cent. Seed stored in sealed containers at 40°F. still retained their original viability at the end of 4 years. Dormancy appears to vary with different seed lots. Some lots are nondormant and some will not begin to germinate after more than 60 days unless pretreated. For dormant lots stratification in moist sand at 41°F. for 30–60 days is recommended. The seed should be sown in drills in the fall, or nondormant or stratified seed in spring, using 5–10 oz per 50 sq ft of seedbed to produce 80–120 usable seedlings per sq ft of bed. The seed is covered about ¼ in. with soil and the seedbed mulched until germination begins. Half shade is used for seedbeds the first season.

VARIETIES AND FORMS. Because of the great variability of Douglas Fir over its wide range considerable confusion has arisen over the nomenclature. The two forms now generally accepted as being distinctive are the Rocky Mountain form, considered to be classified herewith under the name Mountain Douglas Fir, *P. menziesii* var. *glauca* (Mayr) Franco, and the Pacific Coast form given the species rank of Common Douglas Fir, *P. menziesii* (Mirb.) Franco (as described above). Apparently the chief differences are that the former has smaller cones (rarely 3 in. long), with more exserted and strongly reflexed bracts; the latter in comparison has cones often 4 in. long, and more or less appressed bracts. Also the leaves of Mountain Douglas Fir show a greater abundance of leaves with glaucous (bluish green) color as against the yellowish green color of the Common Douglas Fir. However, the two forms seem to intergrade considerably. The Mountain Douglas Fir ranges through the Rocky Mountain region. In the area of our manual it occurs in Trans-Pecos Texas, New Mexico, and Arizona; northward in Colorado, Utah, Wyoming, Nevada, Montana, Oregon, Washington, British Columbia, and Alberta. Southward it occurs in Mexico in Sonora, Chihuahua, Durango, Zacatecas, Coahuila, Nuevo León, Hidalgo, and Puebla.

For a list of the scientific synonyms of the two forms discussed the reader is referred to Little (1953, pp. 305–307).

The following horticultural forms of Douglas Fir are listed by Kelsey and Dayton (1942, p. 504):

Blue *(glauca)*, Columnar *(fastigiata)*, Compact *(compacta)*, Dwarf *(pumila)*, Frets *(fretsi)*, Globe *(globosa)*, Gray *(caesia)*, Green *(viridis)*, Silverleaf *(argentea)*, and Weeping *(pendula)*.

REMARKS. The genus name, *Pseudotsuga*, is from the Greek word *pseudo*, "false," and the Japanese *tsuga*, referring to its relationship with *Tsuga* (Endl.) Carr. The species name, *menziesii*, is in honor of Archibald Menzies (1754–1842), Scotch physician and naturalist, who discovered it in 1791 at Nootka Sound in Vancouver Island, British Columbia. Vernacular names are Yellow Fir, Red Fir, Coast Douglas Fir, and Oregon Douglas Fir. The seeds are eaten by dusky and Franklin's grouse, Douglas squirrel, sierra chickaree, red spruce squirrel, and Magdalena chipmunk. It is occasionally browsed by mule deer and Olympic wapiti. The wood is used for many kinds of construction, theatrical scenery, railway ties, fuel, cooperage, boxes, crates, shipbuilding, furniture, piles, posts, poles, pulpwood, veneer, sash work. The bark is used for tanning leather. Douglas Fir comprises approximately 60 per cent of our standing timber in the West. It furnishes more timber than any other American species. The tree requires a high relative humidity for best growth; cannot stand severe cold or drying winds; grows best on well-drained sandy loam; growth rate rapid; often a pioneer in burns; sometimes planted for erosion control and windbreaks. It is often planted (principally *P. menziesii* var. *glauca*) in the eastern United States and Europe for ornament. It has been in cultivation since 1827.

Arizona Cypress

Cupressus arizonica Greene

FIELD IDENTIFICATION. Evergreen tree attaining a height of 90 ft, with a diameter of 5 ft, but generally less than half that size. The short, stout, horizontal branches form a dense conical or narrowly pyramidal crown, though sometimes broad and flat.

FLOWERS. Borne in spring, staminate and pistillate on different twigs of same tree; staminate cones borne on end of branchlets, ovoid-oblong, cylindrical; stamens 3–8 with broad-ovate, acute, yellow connectives; anther cells pendulous and globose; pistillate cones small, erect, greenish, with ligneous peltate scales; ovules numerous, erect, in several rows at base of scales.

FRUIT. Cone on stout pedicels ¼–½ in. long, long-persistent, maturing at end of second year, subglobose, dry, woody, ¾–1 in. long, dark reddish brown, often glaucous, with 6–8 shield-shaped woody scales having stout, flattened, incurved prominent bosses, the scales tardily spreading; seeds numerous, dark reddish brown, oblong to triangular, compressed, narrowly winged, about 1/16–⅛ in. long.

LEAVES. Minute, scalelike, about 1/16 in. long, thick, appressed, imbricate, acute or obtuse, rounded, keeled, with or without a resin gland on the back, fetid when bruised, pale green, turning reddish brown later, very persistent.

TWIGS. Stout, stiff, dark gray after leaves fall.

BARK. On young branches breaking into thin, large, irregular scales; on old trunks and branches, longitudinally furrowed and fibrous-shreddy, dark reddish brown to grayish.

WOOD. Gray or yellowish-streaked, light, moderately soft, close-grained, specific gravity 0.48, durable when seasoned.

RANGE. Dry, well-drained soil in the sun at altitudes of 3,000–8,000 ft. On gravelly slopes or cuts with northern exposure. Trans-Pecos Texas (Chisos Mountains in Brewster County), west through New Mexico and Arizona to California. In Mexico in Sonora, Coahui-

ARIZONA CYPRESS
Cupressus arizonica Greene

la, and Durango. Planted for ornament in Europe and to some extent in the United States.

PROPAGATION. Fruit borne rather abundantly each year, long persistent on the branches, and seeds of unopened cones viable for several years. Mature cones brown, older ones gray. Cones are opened by sun-drying or in a cone kiln. Storage for 2–3 years in sealed containers at about 40°F. is practiced. Cones contain 48–112 seeds each. A lb of cleaned seed contains about 40,000 seeds, with a commercial purity of 82 per cent and a soundness of 55 per cent. Seed may be planted without pretreatment, but stratification in moist sand for 60 days at 41°F. will give a germination to 30 per cent on good lots of seed. Stratified seed is planted in spring in drills and covered about ¼ in. with soil. Germination occurs 2–3 weeks after spring planting. Plants grow best in deep sandy loam. Propagation by cuttings or veneer grafting is also practiced.

VARIETIES AND HORTICULTURAL FORMS. Beautiful Arizona Cypress, *C. arizonica* var. *bonita* Lemmon, has larger cones, very glaucous when young, and a conspicuous resin gland on the leaves. However, intergrading forms with the species often occur.

Blue Arizona Cypress, *C. arizonica* forma *glauca*, has the silvery gray juvenile foliage.

Compact Arizona Cypress, *C. arizonica* forma *compacta*, has a low rounded shape.

Dwarf Arizona Cypress, *C. arizonica* forma *nana*, is of dwarf habit.

Pyramid Arizona Cypress, *C. arizonica* forma *pyramidalis*, is pyramidal in shape.

REMARKS. The genus name, *Cupressus*, is from the name of the Italian Cypress, *C. sempervirens* L. The species name *arizonica*, refers to the state of Arizona, where it grows. Also known under the vernacular names of Cedro, Cedro de la Sierra, Pinobete, Cuyamaca Cypress, Pinte Cypress, and Arizona Smooth Cypress. Sometimes planted for ornament, erosion control, or for windbreaks. It was first cultivated in 1882. It is attractive when young because of its silvery gray bloom. The tree is rapid-growing in good soil, slow-growing in poorer soils, long-lived, susceptible to fire damage. The largest trees are seldom more than 700 years old.

MONTEZUMA BALD CYPRESS

Taxodium mucronatum Ten.

FIELD IDENTIFICATION. Tree reaching a height of 60–90 ft, with a straight trunk and an enlarged base. The very robust branches are continually subdivided into slender twigs. Roots large and spreading, bearing conical, upright "kneelike" projections.

FLOWERS. In February, staminate and pistillate separate on the same tree. The staminate are numerous, borne in slender racemes, these composed externally of sharp, peltate, ovoid-triangular bracts spiraling around a central axis, each one of which represents the connective of an anther provided at its inner base by 4–10 biserrate and 1-loculed globules, opening by a longitudinal fissure; pistillate flowers much fewer than the staminate, usually on thicker branchlets of the previous year, globose and sessile, with about 20 spreading scales, each scale with 2 internal ovules at base.

FRUIT. Cones subglobose, to 1 in. in diameter, scales spirally imbricate, trapezoid, thick, adhering to the ovuliferous lamina at the inferior half; the apex free, incurved, often mucronate on the median dorsal surface; surfaces often resinous, green, roughened, becoming woody in mature cones, seeds 2 together at a scale.

LEAVES. Leaves, and many of the young branches, deciduous, alternate but crowded, 2-ranked in the same plane, ¼–½ in. long, about 1/12–1/25 in. wide, green, linear, straight or somewhat incurved, terminated by a minute point, only the main vein conspicuous.

BARK. Reddish brown, rather smooth but shreddy.

WOOD. Light to dark brown or yellowish, takes a good polish.

RANGE. Along the lower Rio Grande River in Texas. At one time more common, but now rare in the United States. Recorded from Havana, Texas (Clover, 1937). In Mexico widespread from Nuevo León, Tamaulipas, Coahuila, Sinaloa, and southward to Guatemala.

MEDICINAL USES. "The tree furnishes an acrid resin which was used in pre-Conquest times for the cure of wounds, ulcers, cutaneous diseases, toothache, gout, etc., and which is still used extensively in popular practice. The bark is employed as an emmenagogue and diuretic, and the leaves are applied as a resolutive and as a cure for itch. Chips of the wood are placed in an excavation in the ground, covered with earth, and fired, and as a result there is obtained a kind of pitch which is used commonly as a cure for bronchitis and other chest affections." (Standley, 1920–1926, pp. 60–61)

REMARKS. The genus name, *Taxodium*, is from *Taxus*, the genus including the yews, referring to the yewlike leaves. The species name, *mucronatum*, means "a short, sharp point," with reference to the mucronations on the conelet. It is very closely related to *T. distichum* (L.) Rich., to which trees from south Texas were referred. First distinguished in the United States by Britton (1926). Also by Cory and Parks (1937).

The tree is known under many vernacular names in Mexico. Perhaps the most popular name is Ahuehuete, from the Nahuatl name of Ahuehuetl. Others are Pentamu, Pentamón, Ciprés, Cipreso, Sabino, Ciprés de Montezuma, Tnuyucu, Yucuchichicino, or Yaga-quichi xiña.

Standley (1920–1926, p. 60) makes the following comments concerning the tree and its uses:

"This bald cypress is one of the best-known trees of Mexico, being noted especially for its size. The largest individual reported is the famous tree at Santa María del Tule, Oaxaca, near the city of Oaxaca, which has a height of 38.6 meters and a trunk circumference of 51.8 meters; the greatest diameter of its trunk is 12 meters, and the spread of its branches about 42 meters. The Cypress of Montezuma, in the gardens of Chapultepec, has a height of 51 meters and a trunk circumference of 15 meters. It was a noted tree four centuries ago, and has been estimated to be about 700 years old. Other trees have been estimated to have attained a much greater age. A third famous tree is the 'Arbol de la Noche Triste,' in the village of Popatela, near the City of Mexico, which is noted for its association with Cortés."

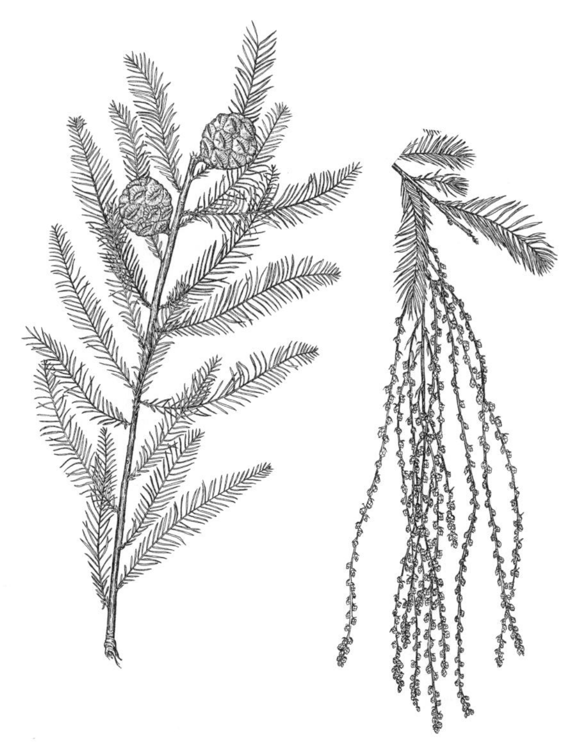

MONTEZUMA BALD CYPRESS
Taxodium mucronatum Ten.

COMMON BALD CYPRESS
Taxodium distichum Rich.

COMMON BALD CYPRESS

Taxodium distichum Rich.

FIELD IDENTIFICATION. Deciduous conifer growing in swampy grounds, attaining a height of 130 ft and a diameter of 8 ft. It is reported that some trees have reached an age of 800–1,200 years. The trunk is swollen at the base and separated into narrow ridges. Curious cone-shaped, erect structures called "knees" grow upward from roots of trees growing in particularly wet situations. Branches horizontal or drooping.

FLOWERS. March–April, staminate cones brownish, 3–5 in. long; stamens 6–8, with filaments enlarged and anthers opening lengthwise; pistillate cones solitary, or 2–3 together, clustered in the leaf axils, scaly and subglobose; scales shield-shaped, with 2 ovules at the base of each.

FRUIT. Ripening October–December, cone globose,

closed, rugose, ¾–1 in. in diameter, formed by the enlargement of the spirally arranged pistillate flower scales; scales yellowish brown, angular, rugose, horny, thick; seeds 2-winged, erect, borne under each scale, dispersed by water or wind; large crops occur every 3 to 5 years with lighter crops between.

LEAVES. Deciduous, alternate, 2-ranked, ½–¾ in. long, flat, sessile, entire, linear, acute, apiculate, light green, lustrous, flowering branches sometimes bear awl-shaped leaves, deciduous habit unusual for a conifer.

TWIGS. Green to brown, glabrous, slender, flexible, often deciduous.

BARK. Gray to cinnamon-brown, thin, closely appressed, fairly smooth, finely divided by longitudinal shallow fissures.

WOOD. Light or dark brown, sapwood whitish, straight-grained, moderately hard, not strong, very durable, weighing 28 lb per cu ft, not given to excessive warping or shrinking, easily worked; heart often attacked by a fungus, the disease being known as "peck."

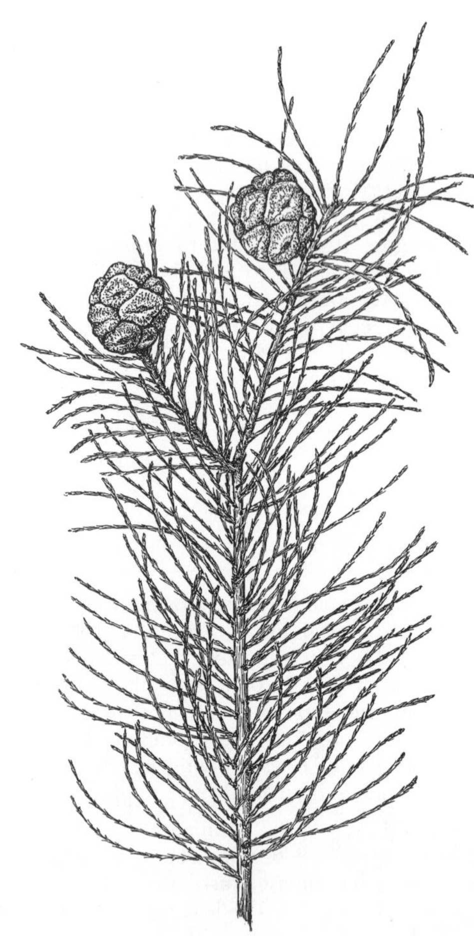

POND BALD CYPRESS
Taxodium distichum var. *nutans* (Ait.) Sweet

RANGE. Texas, Oklahoma, Arkansas, and Louisiana; eastward to Florida, northward to Massachusetts, and west to Missouri.

PROPAGATION. The seed cones are hand-picked when ripe and spread out to dry. The cones may be flailed in a bag to break them up. About half the weight of the cones is seed. Seed average about 4,800 seeds per lb. Commercial seed averages about 45 per cent in purity and 40 per cent in soundness. Several methods of pretreatment for sowing are: (1) store dry and stratify 30 days at 41°F.; (2) store dry and soak in water 4–8 weeks before sowing; and (3) sow immediately after collection. The stratification method seems the best and speeds up germination of the otherwise dormant embryos. Wet muck forms a good natural seed bed. Seeds are sown broadcast, or planted in drills, covered about ½ in. with sand, loam, or peat moss, and mulched with leaves. The beds should be watered well. Germination takes 15–90 days. Semishade is sometimes needed. Usable seedlings usually average about 50 per cent of nursery stock. No special protection from rodents or birds is needed.

VARIETIES AND HORTICULTURAL FORMS. Pond Bald Cypress, once described as a species, *T. ascendens* Brongn., has been recently suggested as a variety under the name of *T. distichum* var. *nutans* (Ait.) Sweet, since the differences between it and Common Bald Cypress are minor. It occurs on the coastal plain from Virginia south to Florida and west to southeastern Louisiana. Horticultural forms of Common Bald Cypress include Columnar *(fastigiatum)*, Dwarf *(nanum)*, Pyramid *(pyramidale)*, and Weeping *(pendens)*.

REMARKS. The genus name, *Taxodium*, is from the Greek and means "yewlike," in reference to the leaves, and the species name, *distichum*, means "two-ranked," and also refers to the leaves. Other vernacular names are White Cypress, Gulf Cypress, Southern Cypress, Tidewater Red Cypress, Yellow Cypress, Red Cypress, Black Cypress, Swamp Cypress, and Sabino-tree. Cypress wood is used for boatbuilding, ties, docks, bridges, tanks, silos, cooperage, posts, shingles, interior finishing, car construction, patterns, flasks, greenhouses, cooling towers, stadium seats, etc. It is very durable in contact with soil and water. It is easily worked and takes a good polish. The knees are sometimes made into souvenirs, and the cone resin used as an analgesic for wounds. The conical erect knees serve as a mechanical device for anchoring the tree in soft mud, and some authorities believe that the knees also aerate the roots. The seeds are eaten by a number of species of birds, including wild ducks. Common Bald Cypress is often planted for ornament and has been in cultivation in Europe since about 1640. Fossil ancestors of Bald Cypress, at one time, covered the greater part of North America in company with the ginkgoes, sequoias, and incense-cedars. At present it is concentrated in the swamps of the Southern states and middle to lower Mississippi Valley. Florida has about one third of the total amount of acreage of Common Bald Cypress.

Mexican Pinyon Pine

Pinus cembroides Zucc.

FIELD IDENTIFICATION. Conifer to 50 ft, with a rounded or pyramidal shape. The trunk is short and the lower branches often wide-spreading.

FLOWERS. In unisexual cones, perianth absent; staminate clusters short, dense, stamens spirally disposed; pistillate clusters red, scales spirally disposed with 1–2 ovules at base.

LEAVES. Thickly covering the twigs, persistent, bluish green, glaucous, somewhat curved, ¾–1¾ in. long, slender, in clusters of 2–3, mostly threes.

FRUIT. Cones mature August–September, soon falling, ovoid to globose, 1–1¾ in. long; scales large, thick and fleshy when green, reddish brown later, irregularly pyramidal, strongly keeled and resinous; seed borne in cavities usually at the base of the middle scales which spread widely at maturity, diversely shaped, triangular to rounded, somewhat flattened or rounded at the base, ½–¾ in. long, oily, brown to black, wing rudimentary, edible.

Mexican Pinyon Pine
Pinus cembroides Zucc.

TWIGS. Branches rough and scaly, young twigs very smooth, gray, leaf scars and lenticels numerous.

BARK. Reddish brown to almost black on old trunks, broken into thick broad plates with smaller thin scales and deep fissures.

WOOD. Yellow, soft, light, close-grained, somewhat fragrant when burned, specific gravity 0.65.

RANGE. Mexican Pinyon Pine is found usually at higher elevations of 4,000–7,000 ft in west Texas, New Mexico, and Arizona to California. South into Mexico in Chihuahua to Baja California and Hidalgo.

VARIETIES. A closely related variety of the Mexican Pinyon, and overlapping in distribution, is Colorado Pinyon Pine, *P. cembroides* (Engelm.) var. *edulis* Voss, which differs in such characters as the leaves being somewhat less crowded, yellowish green in color, and mostly in bundles of twos instead of threes. The cone is broad-ovoid, with minute incurved lips on the scales. Also the Colorado Pinyon Pine has a tendency to be a rather small, scraggy tree seldom exceeding a height of 35 ft. However, upon examination of numerous specimens in the field, the author has found many intergrades between the Mexican Pinyon Pine and the Colorado Pinyon Pine. With this in mind, it appears that the latter is a rather weak variety, and is certainly not distinct enough for a species rank. It is distributed in western Texas, New Mexico, Colorado, Utah, Wyoming, Arizona, and California.

More distinct, as a variety of the Mexican Pinyon Pine is the Single-leaf Pinyon Pine, *P. cembroides* (Torr. and Frem.) var. *monophylla* Voss, which has solitary, green-glaucous, rigid, aromatic leaves ¾–1½ in. long. The cone is broad-ovoid and 1½–2 in. long, scale tips with a small incurved prickle. It is found in Colorado, Utah, Nevada, California, and Arizona.

Parry Pinyon Pine, *P. cembroides* var. *parryana* Voss, is another variety of Mexican Pinyon Pine, with 3–5 (usually 4) leaves, 1¼–1½ in. long and rigid; the cone is subglobose and the scales have recurved prickles. The distribution is apparently confined to Lower California.

REMARKS. The genus name, *Pinus*, is the classical name, and the species name, *cembroides*, denotes its superficial resemblance to *P. cembra*, Swiss Stone Pine. Other vernacular names are Nut Pine, Rocky Mountain Piñon, Pino, and Ocote. The wood is used for fuel and posts, but rarely for lumber. The tree is valued for its edible seeds which are gathered and sold in great quantities by the Mexican and Indian people. The seeds are eaten raw or roasted and have an excellent flavor. The seeds are also consumed by many species of ground squirrels, chipmunks, porcupine, black bear, Mearns's quail, Merriam's turkey, and thick-billed parrot. Goats and mule deer browse the foliage. A resin obtained from the tree is used as a waterproofing material and cement for pots and baskets, and for mending articles of jewelry. The tree is usually of slow growth.

Apache Pine

Pinus engelmannii Carr.

FIELD IDENTIFICATION. A pine attaining a height of 70 ft and a massive diameter of 2 ft. The branches are few, rather stout and tortuous, and develop into a broad, open, rounded crown. It is very variable in the length of the leaves and the size of the cone. The very long leaves on saplings are unlike those of any other Western pine, most closely resembling those of the Eastern Longleaf Pine. The older trees resemble Ponderosa Pine, but have fewer, stouter twigs, longer needles, and slightly larger cones. The wood is similar to that of Ponderosa Pine.

FLOWERS. Borne in separate staminate and pistillate cones; staminate borne in numerous oblong clusters, each catkin ¾–1¾ in. long and ⅜–½ in. broad, reddish brown; pistillate flowers not seen.

FRUIT. Cone 4–5½ in. long, reddish brown, apices of the scales with a short sharp prickle, several scales remaining on the twig when the cones shed; seed pale brown, ovoid to ellipsoid, asymmetrical, flattened, about ¼ in. long; wing thin and papery, ¾–1 in. long, one side curved more than the other.

LEAVES. Usually in bundles of 3 (sometimes 4–5), 8–12 in. long, or on saplings 15 in. or more, rather stout and stiff, pale green or dark green, one side slightly convex, the other side with a longitudinal ridge, margin obscurely and finely serrulate under magnification; sheaths persistent, dark reddish brown, ½–1¼ in. long.

TWIGS. Rather few, stout, reddish brown, surfaces very scaly and roughened by the persistent leaf sheaths, older ones more glabrous.

BARK. Cinnamon-red or dark brown, thick, broken into large plates with deep furrows.

WOOD. Yellowish or reddish brown, the sapwood thick and pale, hard, heavy, containing dark bands of small very resinous summer cells and faint medullary rays, weighing about 30.96 lb per cu ft, used for the same purposes as Ponderosa Pine.

RANGE. Not common anywhere, on mountainsides at altitudes of 5,000–8,200 ft. In extreme southwestern New Mexico. In Arizona in the southeastern corner in Cochise and Pima counties, also in adjacent Mexico.

REMARKS. The genus name, *Pinus*, is the classical name for the pine. The species name, *engelmannii*, is in honor of George Engelmann (1809-1884), German-born physician and botanist of St. Louis, Missouri. The tree has been given various names, such as *P. macrophylla* Engelm., *P. latifolia* Sarg., *P. apacheca* Lemmon, *P. mayriana* Sudw., and *P. ponderosa* var. *mayriana* (Sudw.) Sarg. The tree was discovered in the autumn of 1877 on the southern slopes of the Santa Rita Mountains in southern Arizona, growing with *Quercus hypoleucoides*, by Heinrich Mayr of the Bavarian Forest Department. It is also known under the vernacular name of Arizona Longleaf Pine.

Limber Pine

Pinus flexilis James

FIELD IDENTIFICATION. A low shrub at high altitudes, or a long-lived evergreen tree 40–70 ft, and 2–5 ft in diameter. Branches plumelike, drooping, trunk noticeably tapering into a wide, round-topped crown. Growth not rapid, the trees often occurring in pure stands or in small groves mixed with other conifers.

FLOWERS. In June, staminate cones reddish; pistillate cones clustered, bright reddish purple, scales becoming thickened and woody.

FRUIT. Cones maturing in September, good crops about every 3 years, light crops in intervening years, cone subterminal, more or less stalked, horizontal or slightly declined, subcylindric to oval, green or purplish at first, later light brown, length 2½–10 in.; scales narrow, more or less reflexed, slightly thickened and somewhat squarrose at apex, spines absent, opening at maturity; seeds dispersed September–October, large, flattened, ⅓–½ in. long, shell thick, slightly winged or wing obsolete.

LEAVES. Sheaths loose and deciduous, leaves needlelike, 5 in a cluster, terminal, stout, rigid, 1½–3 in. long,

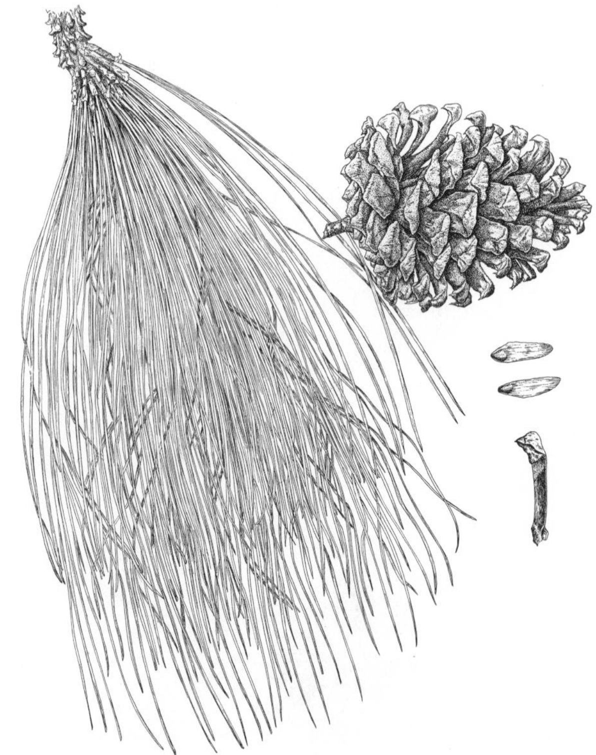

Apache Pine
Pinus engelmannii Carr.

LIMBER PINE
Pinus flexilis James

marked by 1–4 rows of stomata on each side, persistent 3–6 years.

TWIGS. Often somewhat drooping, orange-green to brown or gray, at first puberulent but later glabrous.

BARK. Young branches gray to silvery white, old trunks gray to brown or almost black, divided into deep furrows separated by broad, flat ridges bearing small, close scales.

WOOD. Yellow to reddish, close-grained, light, soft, specific gravity 0.43.

RANGE. The species is found at altitudes of 3,500–12,000 ft and is evidently adjustable to many soil types. Often on ridges or rocky foothills on dry, rocky, shallow soil. Trans-Pecos Texas; west through New Mexico to California, northward to British Columbia and Alberta, south in Mexico to Coahuila and Nuevo León.

PROPAGATION. The most cones are produced by trees in open stands at low altitudes. The cleaned seed averages about 4,400 seeds per lb, with a commercial purity of 99 per cent and a soundness of 80 per cent. The seed may be stored at ordinary room temperature for several years, or stratified in moist sand or peat at 41°F. for 30–90 days. A combination of stratification for 90 days plus scarification has been found advantageous. The average germinative capacity is about 36 per cent. The seeds may be broadcast or sown in drills 4–6 in. apart and covered about ¼ in. with sandy loam. Seedlings have rather large taproots and are susceptible to blister rust. Older trees are hard to transplant.

VARIETY. A variety, *P. flexilis* var. *reflexa* Engelm. (*P. strobiformis* Engelm.), is known under the various names of Mexican White Pine, Border Limber Pine, White Pine, Border White Pine, and Southwestern White Pine. It differs from the species by having branches not as close to the ground, leaves bluish green with stomata absent from the back, and by the apex of the cone scales being narrowed and not as truncate. However, intermediate forms between the species and variety occur. The variety occurs in the mountains of southwestern New Mexico and southeastern Arizona; and in Mexico in Sonora and Chihuahua; south to Sinaloa, Durango, and Zacatecas.

REMARKS. The genus name, *Pinus,* is the classical Latin name, and the species names, *flexilis,* refers to the pliant branches. It is also known under the vernacular names of White Pine and Rocky Mountain Pine. The tree is long-lived, rather slow-growing, occasionally used for rough lumber, and has possibilities for erosion control. It has been in cultivation since 1851.

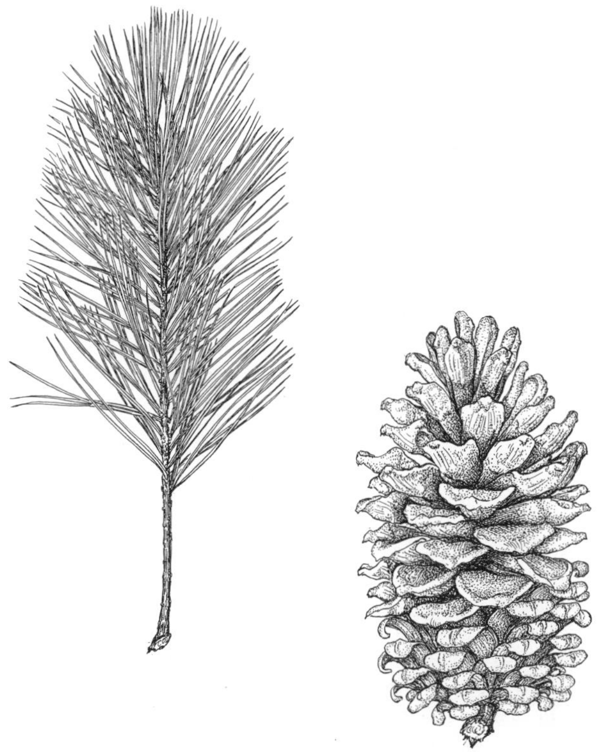

BORDER LIMBER PINE
Pinus flexilis var. *reflexa* Engelm.

PONDEROSA PINE

Pinus ponderosa Laws.

FIELD IDENTIFICATION. Western pine known to attain a maximum height of 230 ft and a diameter of 5–8 ft. Branches stout, thick, somewhat drooping, but upcurved terminally. The crown varying from short-conical to rounded or flat-topped. Trunk symmetrical and usually devoid of branches for one half or more of its length. Very variable in length of leaves, size of cone, and color of bark over its wide distribution.

FLOWERS. April–June, staminate inflorescences yellow, elongate, pistillate cones paired or clustered, dark red.

FRUIT. August–September, cone ellipsoid to ovoid, solitary or clustered, sessile or short-stalked, horizontal or slightly declining, length 3–6 in., when mature light reddish brown, scales narrow with a thickened apex which is somewhat diamond-shaped and transversely ridged; umbo with a slender prickle or deciduous; cone deciduous soon after opening; seeds ovoid, about ¼ in. long, base rounded, apex slightly flattened, shell mottled brown or purplish, wings ¾–1¼ in. long, apex narrowed and oblique.

LEAVES. Borne terminally on naked branchlets, mostly in 2- or 3-leaved clusters, rather stout, yellowish green to dark green, bearing stomata on 3 faces, length 5–11 in., deciduous the second or third year.

TWIGS. Aromatic when bruised, stout, orange to dark gray or black.

BARK. Rather thick with deep irregular fissures; separating into large, elongate plates, which break into small, appressed, cinnamon-red to brown or black scales.

WOOD. Heartwood reddish to orange or yellowish brown, sapwood pale yellow to white, growth rings distinct, summerwood darker, springwood zone narrow to wide, transition to summerwood abrupt, odor resinous, grain variable and even to uneven, specific gravity when dry about 0.42, moderately soft, moderately weak in endwise compression and bending, moderately low in shock resistance, fairly easy to work, average in paint holding, glues well, nail-holding ability average, moderately low in shrinking, when thoroughly dry does not warp, seasons very well, not durable under decay conditions.

RANGE. The species tolerates many types of soil and grows generally at elevations of 2,000–10,000 ft. Occurs in the Davis, Guadalupe, and Chisos mountains of west Texas; north through northwestern Oklahoma and west through New Mexico and Arizona to California, eastern Washington, and British Columbia.

PROPAGATION. The minimum seed-bearing age is about 20 years and the optimum about 150. Good crops are produced every 2–5 years and lighter crops every 2–3 years. The cones may be collected September–October and average about 250 closed cones per bu. They may be sun-dried 96–144 hours. One hundred lb of fresh cones yields 2–7 lb of seed, and average about 12,000 seeds per lb. Commercial seed has a purity of 96 per cent and a soundness of 81 per cent. Viability of seed at 50 per cent or better for 3–4 years has been maintained by storage in sealed containers at 32°–41°F. Some lots of seed are seemingly more dormant than others as a result of seed coat and embryo dormancy. Stratification in moist sand or peat 30–60 days at 41°F. is helpful. The seed may be sown in fall or spring, either broadcast or in drills, usually 4–6 in. apart and covered about ¼ in. with well-drained sandy loam. Damping-off may be controlled by use of aluminum sulphate in the beds before sowing. Winter blister rust *(Cronartium filamentosum)* is controlled by clearing out infested stock and destroying host plants such as *Castilleja* and *Orthocarpus;* likewise, sweet-fern blister rust *(Cronartium comptoniae)* by destroying the host plants *Comptonia* and *Myrica.* Dwarf Mistletoe *(Arceuthobium)* is controlled by cutting out infected plants.

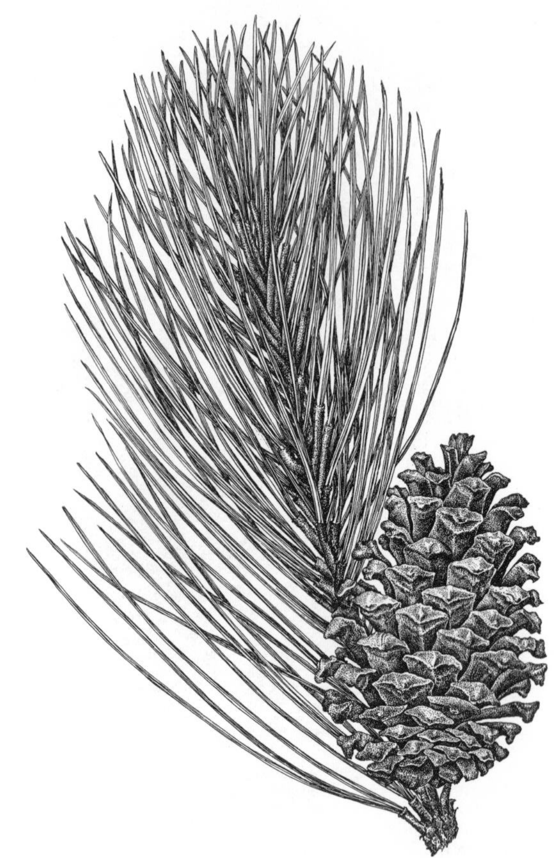

PONDEROSA PINE
Pinus ponderosa Laws.

VARIETIES. At least 4 distinct races of Ponderosa Pine occur. However, since there are so many intergrades, and variations in young and older trees, only 2 varieties seem constant enough to merit segregation. Rocky Mountain Ponderosa Pine, *P. ponderosa* var. *scopulorum* Engelm., formerly known as *P. brachyptera* Engelm., has furrowed grayish or nearly black bark; leaves 2–3 in a cluster and 3–6 in. long; cone smaller, scales with slender backward-hooked prickles. It intergrades with the species. A common variety of the Rocky Mountains from Nebraska, Wyoming, and Montana; south to Colorado, Arizona, New Mexico, and western Texas.

A form with drooping branches is known as *P. ponderosa* var. *scopulorum* forma *pendula.*

Arizona Ponderosa Pine, *P. ponderosa* var. *arizonica* Shaw, has more slender needles 5–7 in. long and 3–5 in a bundle; cone ovoid, 2–3½ in. long, spines on scale slender and recurved, seed about ⅛ in. long with wings broadest above the middle. It is sometimes listed as a distinct species (*P. arizonica* Engelm.). Usually at altitudes of 6,000–9,000 ft on high cool slopes in New Mexico and Arizona, and in Mexico in Sonora, Chihuahua, Nuevo León, and Tamaulipas.

REMARKS. The genus name, *Pinus,* is the classical Latin name. The species name, *ponderosa,* means "heavy," perhaps in reference to the wood. Other vernacular names are Rocky Mountain Ponderosa, Western Yellow Pine, Yellow Pine, Bull Pine, Blackjack Pine, Rock Pine, California White Pine, Arizona White Pine, Pino Real, and Pinobete. The tree is rapid-growing at first but slows down considerably when older. It is rather drought- and fire-resistant, but not very resistant to insects, and is subject to red heart-rot disease. It is used somewhat in shelter-belt planting, and has been cultivated since 1827.

The seeds of Ponderosa Pine and its varieties are eaten by many species of birds and mammals, including California quail, Richardson's grouse, sooty grouse, porcupine, Columbian gray squirrel, Sierra chickaree, California gray squirrel, white-tailed squirrel, Richardson's pine squirrel, Abert's squirrel, tassel-eared squirrel, spruce squirrel, Say's ground squirrel, and chestnut-mantled squirrel. It is also browsed by mule deer, white-tailed deer, and mountain sheep. The wood of Ponderosa Pine has a great variety of uses including lumber, general construction, mine timbers, boxboards, railway crossties, fencing, and fuel. It is probably the most important pine in western North America, furnishing more timber than any other American pine.

Bristle-cone Pine

Pinus aristata Engelm.

FIELD IDENTIFICATION. Western shrub, at high elevation sometimes prostrate as a result of winds, or in favorable locations a tree to 50 ft, a trunk 2–3 ft in diameter. When young the crown pyramidal and densely bushy. Later the branches irregular with the lower pendulous and the upper more erect.

FLOWERS. In July, staminate orange to red; pistillate dark purple.

FRUIT. Cone maturing September–October, cone cylindric, ovoid to oblong, 2–3½ in. long, resinous, violet to reddish brown; scales with thickened rhombic ends and armed with lanceolate-subulate, brittle, often recurved prickles; discharging the seed in about October; seeds nearly oval, compressed, mottled-brown and black, about ¼ in. long; wings about ⅓ in. long and ¼ in. wide, broadest at about the middle; seed averages about 23,000 seeds per lb, with a commercial purity of 91 per cent and a soundness of 65 per cent, germination average about 66 per cent.

LEAVES. Five in a cluster, somewhat appressed, 1–2 in. long, stout, slender, straight or curved, submucronate, dark green, lustrous, marked by numerous rows of stomata on the ventral faces, white-lined, some with copious exudations of resin, very persistent over a number of years.

TWIGS. At first glabrous or puberulent, orange-colored to dark gray, brown, or black later, foliage in tufts.

BARK. At first thin, smooth and whitish, later reddish brown and thick, ridges flat with small close scales.

WOOD. Light red, light, soft, not strong.

Bristle-cone Pine
Pinus aristata Engelm.

RANGE. Mostly in exposed situations of dry rocky soil, on gravelly slopes, at altitudes of 8,000–12,000 ft, at the upper limit of tree growth. In New Mexico, Arizona, Colorado, Utah, Nevada, and California. Introduced into cultivation throughout the eastern United States and in western Europe.

REMARKS. The genus name, *Pinus,* is the classical Latin name, and the species name, *aristata,* refers to the bristles on the cones. Other vernacular names are Fox-tail Pine and Hickory Pine. The wood is sometimes used for fuel and mine timbers. The tree is very cold- and drought-resistant and is hardy in cultivation as far north as Massachusetts. It has been cultivated since 1861, and is mostly shrubby, with branches and leaves with white resinous grains. It is rather slow-growing and is subject to blister rust. The seeds are eaten by rodents and porcupine. Owing to embryo dormancy, the seed should be stratified in moist sand at 41°F. for 60 days or more before planting.

CHIHUAHUA PINE

Pinus leiophylla var. *chihuahuana* (Engelm.) Shaw

FIELD IDENTIFICATION. Western evergreen tree to 50 ft, 2 ft in diameter. Branches rather stout and ascending to form a narrow round-topped crown.

FLOWERS. May–June, staminate yellow, clustered at base of leafy growing shoots of the year with basal scalelike bracts; anthers numerous, sessile, imbricate in many ranks; pistillate yellowish green, subterminal or lateral, scales in the axils of nonaccrescent bracts.

FRUIT. Ripe in November, cone woody, very persistent, horizontal or somewhat declining, long-stalked, ovoid, ripening the third year, 1½–2¼ in. long, light brown and lustrous; scales with a broad base and abruptly narrowed toward the apex, umbos pale and with recurved, gradually deciduous prickles; seeds oval, rounded or pointed terminally, about ⅛ in. long, dark brown, wings about ⅓ in. long.

LEAVES. In clusters of 3, slender, pale glaucous-green, 2½–4 in. long, about 1/25 in. wide, marked by 6–8 rows of stomata on each face, irregularly deciduous at 3–4 years with the sheaths.

BARK. Dark brown to reddish brown or almost black, ¾–1½ in. thick, plates broad and flat, scales thin and closely appressed.

TWIGS. Slender, orange-brown to dull reddish brown later.

WOOD. Light orange to yellowish, sapwood lighter colored, light, soft, not strong but durable, specific gravity 0.54.

RANGE. At altitudes of 5,000–7,000 ft on dry rocky slopes, not common. In the mountains of New Mexico and Arizona. More abundant in Mexico in Chihuahua, Sonora, Durango, Nayarit, and Jalisco.

CHIHUAHUA PINE
Pinus leiophylla var. *chihuahuana* (Engelm.) Shaw

SYNONYMS AND VARIETY. Chihuahua Pine was once known under the name of *P. chihuahua* Engelm. However, it now has the status of a variety under *P. leiophylla* Schneid. & Deppe, which has 5 needles in a cluster instead of 3, and is found in Mexico from Chihuahua to Michoacán, Veracruz, and Oaxaca.

REMARKS. The genus name, *Pinus,* is the classical name; the species name, *leiophylla,* means "flexible"; the variety name, *chihuahuana,* refers to the state of Chihuahua, Mexico, where it was first discovered.

The closed cones of Chihuahua Pine number about 1,300 cones per bu and can be collected in November. A bushel of cones will produce 12–14 oz of seed which averages 40,000 seeds per lb.

SLASH PINE

Pinus elliottii Engelm.

FIELD IDENTIFICATION. A rapid-growing tree attaining a height of 100 ft, with a diameter of 3 ft. The trunk is clean, straight, and symmetrical, and the branches heavy, horizontal, or ascending, to form a handsome round-topped head. The root system reaches a depth of 9–15 ft.

FLOWERS. Borne January–February before the new leaves. Staminate inflorescences in short many-flowered, purplish brown clusters, young pistillate cones cylindrical and ¾–1½ in. long; pistillate cones solitary, lateral, pinkish, ovate, long-peduncled.

FRUIT. Cone borne on a peduncle ¾–1 in. long, persistent until the following summer, lateral, symmetrical, variable in size, 2–6 in. long, 2–3½ in. wide when open, ovoid to ovoid-conic, rich brown, shiny; scales flat, thin, flexible, apex lustrous-varnished, armed with a minute recurved prickle; seeds about ¼ in. long with a wing ¾–1 in. long; seed body almost triangular, sides full and rounded, dorsally ridged, gray to black or mottled, shell thin and brittle; wing rather thin and transparent; cotyledons 5–9.

LEAVES. Stout, stiff, in 2–3-leaved clusters (mostly 2), enclosed at base in a persistent, membranous sheath, 5–10 in. long (rarely longer or shorter), rather dark, lustrous green, marked with numerous bands of stomata on each face, deciduous at the end of the second year. The heavy, shiny green foliage differs from the associated Longleaf Pine with its coarser, rather bluish green foliage. Loblolly Pine, like Longleaf Pine, has 3 bluish green leaves in a cluster, but has smaller leaves than Longleaf Pine or Slash Pine.

TWIGS. Stout, orange to brown, leaves tufted at the ends; buds with silvery brown scales, loose at the apex; terminal bud large, reddish brown, elongating in spring into a straight, stout, light gray "candle," about as thick as a pencil. (Longleaf Pine has a larger bud and candle 1 in. or more in diameter.)

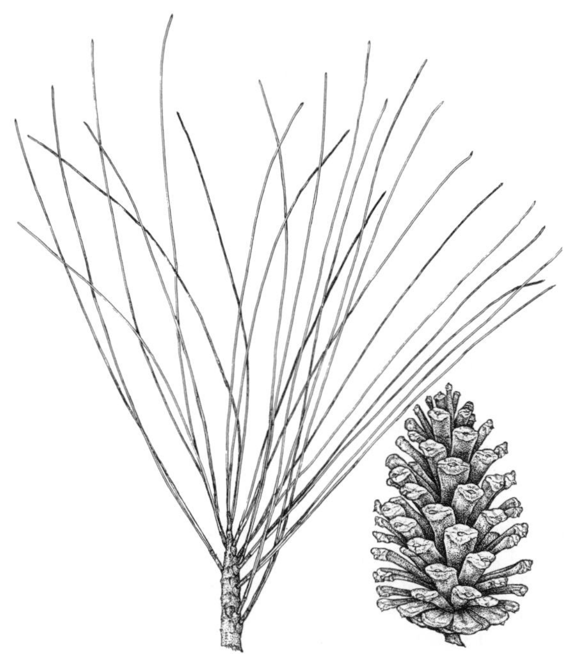

SLASH PINE
Pinus elliottii Engelm.

BARK. On mature trees ¾–2 in. thick, separating into large flat plates deeply furrowed between, surfaces breaking into scales which are thin, papery, and reddish to orange or silvery. Fairly similar to the Longleaf Pine. Differing from Loblolly Pine in being less deeply furrowed.

WOOD. Yellow to dark orange or brown, sapwood nearly white; coarse-grained, strong, tough, stiff, hard, durable, weighing about 45 lb per cu ft, seasons easily, strong in endwise compression, moderately high in shock resistance, holds nails and screws well, glues well, low in ability to take and hold paint. Considered to be the hardest, strongest, and heaviest of all commercial conifers in the United States.

RANGE. Rolling lands of the southern coastal plains. Replacing Longleaf Pine of the lowlands. Sandy-loam soils of swamps and along streams. In Louisiana in St. Tammany, Washington, Tangipahoa, and Livingston parishes; eastward to Florida and Georgia.

PROPAGATION. Trees 15–20 years of age start producing seed. Good crops occur every 2–3 years with light crops between. In close stands the crops are reduced. Seed trees should be left in the open. Slash Pine reseeds naturally with ease, if not crowded, too heavily cut, or fire damaged. A bushel contains about 215 ripe closed cones. A bushel yields about 1⅕ lb of seed, or an average of 87 seeds to the cone. The seed averages about 15,500 per lb with about 90 per cent fertile seeds. The wings are easily removed. If the seeds are to be used soon they may be dried and stored in ordinary jars at room temperature. They can be sown in early November in wet lands, germination is in February or March. The best results are obtained from nursery sown seed in sandy loam in beds about 4 ft wide and sides 1–4 in. high. The seed may be broadcast on beds measuring 144 sq ft (4 x 36 ft); or drilled in rows 6 in. apart at the rate of 60 seeds per linear ft of drill; 1 lb of seed will drill a bed 4 x 32 ft. About 40 seedlings result per sq ft broadcast, and 20 seedlings per linear ft drill. After seeding the seeds are pressed in with a roller, and covered lightly with a layer of sand or pine straw litter, or burlap. The mulch should be removed as soon as the seeds germinate within 2 weeks or so. Shade for the seeding is not necessary. Seedlings are ready for transplanting when about 8–10 in. high. Seedlings are set tightly in the soil, with the roots spread and not balled up. If for turpentining later, they may be set 5 ft apart in furrows 10 ft apart, or 870 to the acre. Otherwise a planting of 8 x 8 ft apart requiring about 680 seedlings is used in commercial plantings. More can be used if thinning is to be practiced extensively later. Fire protection is very essential to Slash Pine during its life, but particularly the first 3–5 years of age.

VARIETY AND SYNONYM. There has been considerable confusion in the use of the scientific names. Slash Pine is often listed in the literature under the name of *P.*

caribaea Morelet. However, it is now thought that the latter name should best apply to a closely related species of the Bahama Islands, Isle of Pines, Western Cuba, and Central America from British Honduras, eastern Guatemala, and northeastern Nicaragua. (See Little, 1953a, pp. 264–265).

A variety has been segregated as the South Florida Slash Pine, *P. elliottii* var. *densa* Little & Dorman, and is described as having dense, very heavy, hard wood with very thick summerwood, also grasslike seedlings with crowded needles, very thick hypocotyl, and thick taproot, and thick hypoderm of the needles. In South Florida and the Keys.

TURPENTINING. Slash Pine yields more turpentine than Longleaf Pine and a higher grade of resin. The following data on Slash Pine (Mattoon, 1939) gives the amounts of gum which can be obtained. Slash Pine growing in an open stand with 32 streaks during the season and scraped, will yield as follows: if 9 in. in diameter (breast high), about 123 ounces; 10 in., about 140 ounces; and 12 in., about 180 ounces. In units of naval stores the yield of Slash Pine may be illustrated by the average 9 in. tree, which produces a little under 8 lbs (123 ounces) of gum, which will make about one fifth of a gallon of spirits of turpentine and about 5 lbs of resin. A 12 in. tree will yield 11.2 lb of gum, which will make about one third of a gallon of spirits of turpentine and about 7 lbs of rosin.

WOOD VALUE. Well-stocked stands of Slash Pine on average land at ages of 15–25 years produce wood for pulp wood, stave blocks, fuel, etc. at a rate of ½–1 cord an acre a year, or sawlogs at the rate of 100–300 board ft an acre a year. On good quality land well-stocked stands are growing at the rate of 1–2 cords, or 300–600 board ft per acre yearly.

INSECTS AND DISEASES. Slash Pine is subject to damage by the southern pine beetle grub which girdles the inner bark, especially in dry years. The pine sawyer beetle also bores in the sapwood of dead or felled timber. It may be controlled by peeling, or drying the timber, or immersing in water. Some occasional damage is done by the Nantucket tip moth on young shoots. Rust canker or Woodgate canker may cause deformations on stems or trunk. Red heart-rot is also damaging, especially northward.

REMARKS. The genus name, *Pinus,* is the classical name, and the species name, *elliottii,* honors its discoverer, Stephen Elliott (1771–1830), botanist and banker of South Carolina.

Slash Pine is often planted for ornament along highways in the South. The wood has a wide variety of uses, such as crossties, posts, poles, joists, flooring, railroad cars, veneers, tanks, silos, boxes, baskets, crates, planing-mill products, excelsior, agricultural implements, paving blocks, woodenware and novelties, shipbuilding and boatbuilding, fuel, paper pulp, mine timbers, bridges and trestles, and general rough construction of warehouses and factories. Large amounts of resin and turpentine are obtained from the gum for use as naval stores.

LOBLOLLY PINE

Pinus taeda L.

FIELD IDENTIFICATION. Handsome conifer attaining a height of 170 ft and a diameter of 6 ft.

FLOWERS. Staminate flowers yellowish green or violet, spirally arranged in slender inflorescences about 2 in. long; involucral scales 10–13, overlapping; stamens almost sessile, anthers 2, opening lengthwise; pistillate cone ovate, about ½ in. long, yellowish green, also spirally arranged in scaly catkins, each scale with 2 ovules at the base.

FRUIT. Ripening September–November, the spirally arranged imbricate scales of the pistillate inflorescences harden to form a cone; cone persistent, ripening in 2 years, 3–5 in. long, elongate-oblong or ovoid, reddish brown, sessile, scales thickened at the apex and bearing short incurved or straight spines; seeds 2 on each scale, rhomboid, mottled-brown, about ½ in. long, attached to a thin wing about ¾ in. long.

LEAVES. Persistent, in clusters of 3, rarely a few in pairs, glaucous, light to dark green, rigid, slender, 3-sided, 5–10 in. long, sheath about ½ in. long.

TWIGS. Stout, reddish brown, scaly.

BARK. Reddish brown, rough, thick, deeply furrowed, scaling into coarse, large segments with large, appressed, papery scales.

WOOD. Streaky yellow and brown, coarse-grained, resinous, soft, brittle, weighing 34 lb per cu ft.

RANGE. Over wide areas on low grounds. East Texas and Louisiana; eastward to Florida and north into southeastern Oklahoma, southern half of Arkansas, southern Tennessee, Georgia, South Carolina, North Carolina, Virginia, Maryland, and Delaware.

PROPAGATION. The seed is dispersed from fall to spring. The commercial seed-bearing age is 12–15 years, with an optimum of 35–60 years and a maximum of 150 years. Good crops are borne every 3–10 years, with light crops in intervening years. Cones may be collected from standing trees or recently felled trees October–November. The closed cones average 400 to 1,100 per bu. The cones may be air dried, or dried and the seeds extracted in convection kilns, of which there are several types. The extraction schedule for cured cones is 6–48 hours at a temperature of 120° to 140°F., at a relative humidity of 20 to 30 per cent. The seeds average 2 to 3 lb per 100 lb of fresh cones, with an average of 18,400 cleaned seed per lb. Commercial seed has a purity of 97 per cent and a soundness of 90 per cent. Seed may be used fresh, or can maintain viability 7–9 years stored in sealed containers at a temperature of 32°–41°F. A moisture content of 7–9 per cent is recommended. Seed dormancy is quite general and stratification may be practiced in moist acid peat or sand at a temperature of 35°–38°F. for a period of 30–90 days. The longer periods are required for large seed lots of 50 lb or more. Planting may be made in well-drained sandy-loam seedbeds for best results. The seeds may be either broadcast or planted in drills 4–6 in. apart. A depth of ⅜ in. is adequate. From 6–10 oz of seed per 100 sq ft of seed bed is sufficient. Seedlings may be protected from damping-off by acidifying the beds with aluminum sulphate prior to sowing. Sowing should produce 80–100 good seedlings per sq ft of seed bed. Seed beds should be mulched with leaves, cloth, or burlap and protected from rodents with a wire screen.

REMARKS. The genus name, *Pinus,* is the old Latin name. The species name, *taeda,* is for the resinous wood. Other vernacular names are Frankincense Pine, Black Pine, Lowland Shortleaf Pine, Torch Pine, Slack Pine, Sap Pine, Swamp Pine, Bastard Pine, Long-straw Pine, Indian Pine, Long-shucks Pine, Fox-tail Pine, Shortleaf Pine, Rosemary Pine, and Old-field Pine. The date of earliest cultivation is 1713. The wood is used for lumber, cooperage, pulp, boxes, crossties, posts, and fuel. Loblolly Pine is not usually worked for turpentine, because the flow of gum is checked quickly and labor costs are too high. Razorbacks do not injure the saplings of Loblolly Pine as much as those of Longleaf Pine.

Loblolly Pine grows faster than Longleaf Pine or Shortleaf Pine. Slash Pine grows fastest to 20 years of age, but from then on Loblolly Pine is the fastest grower of all Southern pines. Slash Pine somewhat resembles Loblolly Pine, but Slash Pine has longer, glossier, greener leaves. Although growth varies considerably according to good or poor soil, or to condition of the stand, the average growth of Loblolly Pine on average soil is reported to be, for a 30-year-old tree, a height of 61 ft and a diameter of 10.6 in.; for a 70-year-old tree, a height of 96 ft and a diameter of 18.4 in. An acre of Loblolly Pine will average 300–1,000 board ft of saw timber annually, depending, of course, on soil, age of trees, and density of stand. Second-growth Loblolly Pine produces more cut timber than all other Southern pines added together. Loblolly Pine growing in the open assumes a bushy form with widespread branches. This produces an inferior, knotty lumber. For best lumber, young Loblolly Pine should be grown in denser stands to produce tall, straight, limbless trunks. For natural reseeding, 3–6 large-topped trees per acre should be left on cut-over land. If artificial planting is done, year-old seedlings can usually be obtained from federal, state or private nurseries at small cost. Planting should be done in late winter or early spring while trees are still dormant. Planting should be made about every 6 ft apart, or 1,210 plants per acre. Pine straw on the ground holds the moisture and improves growing conditions. Fire is the worst enemy of pines and every effort should be made to protect trees by making fire lanes, and destroying litter and tops after cutting. Trees weakened by fire are more subject to attacks of southern pine beetles. Dead trees or felled timber become infected with the pine sawyer, which often can be checked by peeling, drying, or immersing timber. Young shoots are sometimes injured by the Nantucket tip moth in early summer. Such branches are often retarded and deformed in growth. Older trees are sometimes attacked by red heart-rot fungus. Symptoms of the disease are "bumps," and all such diseased trees should be cut. Loblolly Pine is also subject to the attack of *Cronartium* cankers in some areas. Bulletins and leaflets with detailed accounts of the growing and harvesting of Loblolly Pine may be obtained from the U. S. Forest Service, various state forest services, U. S. Department of Agriculture, and forest experiment stations. A particularly good bulletin is Mattoon's *Loblolly Pine Primer* (1926).

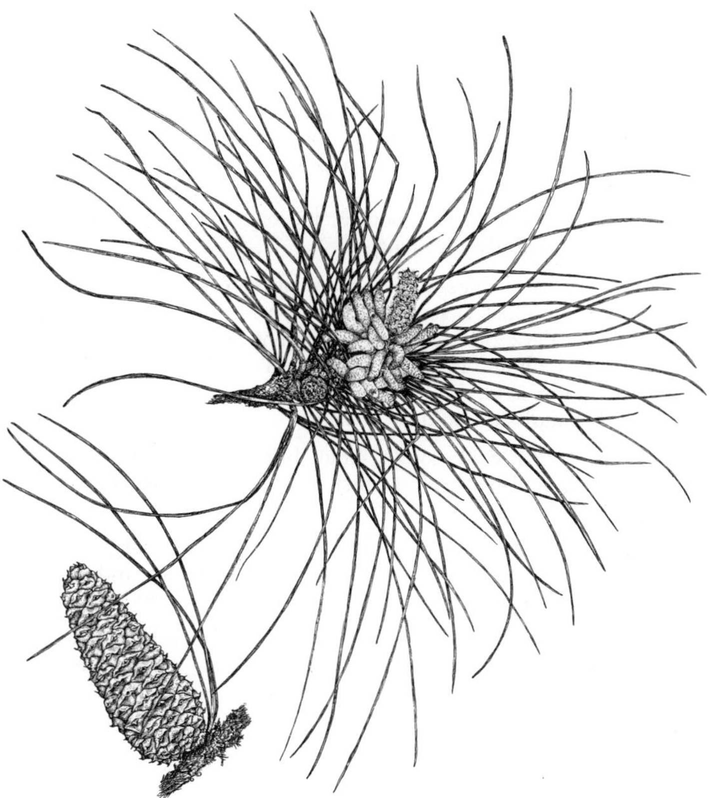

LOBLOLLY PINE
Pinus taeda L.

SHORTLEAF PINE

Pinus echinata Mill.

FIELD IDENTIFICATION. Valuable coniferous tree to 110 ft, and 2 ft in diameter. The branches are whorled and the crown is rather short and pyramidal to oblong.

FLOWERS. March–April, staminate inflorescences borne in larger clusters, sessile, about ¾ in. long, yellowish brown to purple; pistillate in small clusters of 2–3 cones with stout peduncles, about ¼ in. long, rosy pink, with scaly bracts and inverted ovules.

FRUIT. Maturing the second year October–November, persistent on the branches, cone solitary or a few together, sessile or short-stalked, borne laterally, reddish brown, 1½–2½ in. long, ovoid to oblong-conic; scales separating at maturity, woody, thickened at apex, armed with sharp but weak spines which are often deciduous; seeds 2 on each scale, brown marked with black, triangular, about ¾ in. long, bearing an oblique wing about ½ in. long.

LEAVES. Needle-like, from persistent sheaths, usually 2 in a cluster, sometimes 3, 3–6 in. long, slender, flexible, dark bluish green, persistent. A ton of straw contains about 14.2 lb of nitrogen and 5.4 lb of phosphoric acid. Pine straw protects the soil and increases the moisture-holding capacity.

TWIGS. Stiff, rough, stout, brittle, glaucous, brownish to greenish purple at first, later dark reddish brown to purple.

BARK. Thick, brownish red, broken into large, angular, scaly plates with small appressed scales and coarse fissures.

WOOD. Variable in color and quality, yellow, orange, or yellowish brown, sapwood lighter, coarse-grained, fairly heavy, medium-hard, not as resinous as other yellow pines, weighing 38 lb per cu ft.

RANGE. Forming dense stands, doing best on uplands or foothills. East Texas, Oklahoma, Arkansas, and Louisiana; eastward to Florida, northward to New York, and west to Illinois.

PROPAGATION. A good crop is borne every third year, years in between light to medium. Commercial seed-bearing begins at about 16 years of age, with an optimum of 40–50 years of age and a maximum of 280 years in between light to medium. Commercial seed-standing or felled trees. There are approximately 1,400–2,500 closed cones per bu. They may be sun-dried, or dried by forced draft, 6–8 hours at a room temperature of 130°F. The yield is 2–3 lb of seed per 100 lb of fresh cones. The seed averages 48,000 seeds per lb, with an average purity of 96 per cent and soundness of 95 per cent. The seeds may be sown fresh or sealed, stored at 32°–41°F. for 7–9 years. A maintained moisture content of 7–9 per cent is recommended. The seed may be stratified for spring planting in moist acid peat or sand at a temperature of 35°–38°F. for 30–45 days. Dormancy is encountered in both fresh and stored seed. Planting may be made in well-drained sandy loam for best results. The seeds may be either broadcast or planted in drills 4–6 in. apart. A depth of ⅜ in. is recommended. From 6–10 oz of seed per 100 sq ft of seedbed is adequate. The seedlings may be protected from damping-off by acidifying the beds with aluminum sulphate prior to sowing. They are sometimes subject to attack by the rust-gall diseases caused by *Cronartium cerebrum* and *Cronartium fusiforme*. Infected stock may be culled or sprayed with Bordeaux mixture for control.

It is a profitable crop, but the growth is commensurate with the type of soil. The richer the soil the faster

SHORTLEAF PINE
Pinus echinata Mill.

the growth. The average size of the tree is small, but it grows naturally in dense stands and hence the yield is high per acre. It will average a height of 34 ft and a diameter of 4.5 in. in 20 years, and in 70 years a height of 82 ft and a diameter of 12 in. When grown in dense stands and protected from fire, it will show growth on an average of a cord a year per acre, or 300–500 board ft. It also has the ability to sprout from the stump, but such growth seldom yields first-class timber. An abundance of seed is produced, averaging a large crop every third year. Tapping the tree for resin and turpentine is not considered very profitable because of the rapid checking of the gum flow. The tree grows rapidly at first and is fairly fire-resistant. It is also fairly resistant to disease and insects, but in dry seasons suffers some from the attacks of the southern pine beetle. Dead or felled trees are sometimes attacked by the pine sawyer, and young shoots occasionally by the

Nantucket tip moth. Old trees are susceptible to red heart-rot. Shortleaf Pine is sometimes planted for erosion control.

It has been found that careful thinning produces more rapid growth. The first thinning should be done when the trees are 15–20 years old. It is called "lower thinning" because undesirable, small, or weak trees are removed. These can often be sold for pulp, poles, staves, heading, or excelsior to take care of the thinning costs. "Higher thinning" is sometimes done in a more mature stand. This consists of cutting out older, larger trees to make room for the smaller ones to grow. Such a method generally insures a profit every 10 years or so. Three to five large-topped trees should be left on every acre for seed trees. It is advisable not to practice thinning April–September because of the increased activity of wood beetles during that period. Shortleaf Pine is eighth in importance in the United States in reforestation use.

REMARKS. The genus name, *Pinus,* is the ancient Latin name, and the species name, *echinata,* refers to the hedgehog-like, or echinate, bristly needles. Vernacular names are Yellow Pine, Rosemary Pine, Forest Pine, Old-field Pine, Bull Pine, Pitch Pine, Slash Pine, and Carolina Pine. The wood is valuable because of its softer and less resinous character. It is used for general construction, exterior and interior finishing, planing-mill products, veneer, cooperage, excelsior, boxes, crates, agricultural instruments, low-grade furniture, posts, poles, woodenware, toys, etc. A number of species of birds and rodents feed on the seeds.

Longleaf Pine

Pinus palustris Mill.

FIELD IDENTIFICATION. Coniferous tree to 125 ft, and a diameter of 4 ft. Trunk straight and tall with few branches.

FLOWERS. Cones February–April, staminate cones purple, 2–3½ in. long, borne on conspicuous scaly, clustered inflorescences; stamens short-filamented, anthers 2, opening lengthwise; pistillate cones reddish purple, bearing spirally arranged scales, each scale with 2 ovules at base.

FRUIT. Ripening September–October, cone large, dry, reddish brown, subsessile, conical-oblong or cylindrical, slightly curved, 6–12 in. long; the scales much thickened and bearing at the apex a short, recurved spine; seeds mostly triangular, blotched, ridged, ¼–½ in. long with an oblique, thin wing about 1½ in. long.

LEAVES. Borne in terminal, plumelike clusters, needles in bundles of 3, flexible, slender, shiny, dark green, 3-sided, 10–15 in. long, sheaths long. Plants 3–12 years of age almost grasslike in leaf appearance.

TWIGS. Stout, scaly, orange-brown, buds long, white-silvery, scaly.

BARK. Smooth, thin-scaled, separating into large reddish brown plates with coarse fissures.

WOOD. Very desirable, resinous, heavy, hard, strong, tough, durable, coarse-grained, yellow, yellowish brown or orange, sapwood whitish, weighing about 44 lb per cu ft.

RANGE. Mostly in pure stands on deep sandy land. Texas and Louisiana, eastward to Florida and northward to Virginia. In Texas west to the valley of the Trinity River. Rare, if at all, in Oklahoma and Arkansas.

PROPAGATION. Large seed crops are produced every 3–7 years with light crops between. Minimum seed-bearing age 20 years, optimum 40–60 years, maximum 350 years. The cones may be collected in October and average 60–120 closed cones per bu. The seeds may be extracted by forced draft and convection by curing for 8–16 hours at a temperature of 115°F. at a relative humidity of 20–30 per cent. One to three lb of seed are obtained from 100 lb of fresh cones. Cleaned com-

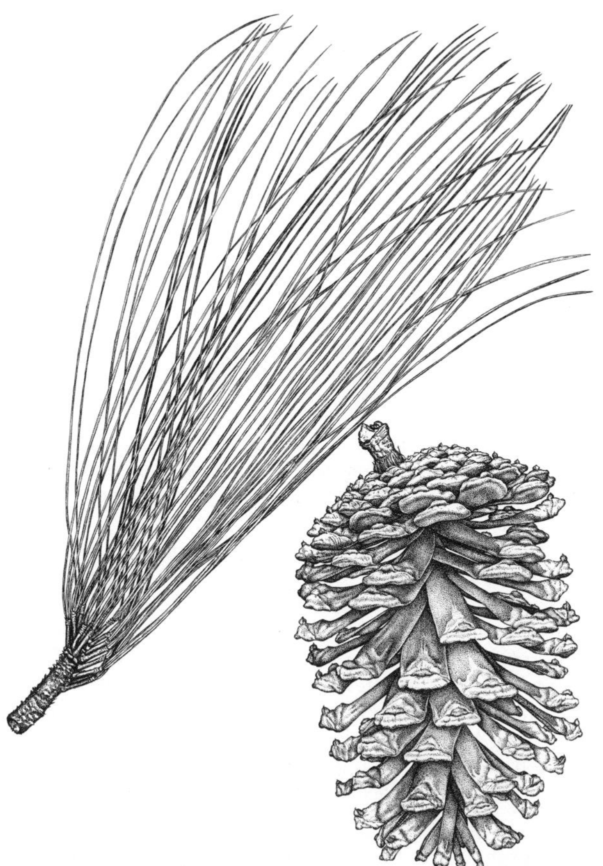

Longleaf Pine
Pinus palustris Mill.

mercial seed averages about 4,200 seeds per lb, with a purity of 92 per cent and a soundness of 90 per cent. The seed spoils easily and should be sown or stored promptly after collection. Vitality may be maintained for at least 2 years by sealed storage at 32°–41°F. A moisture content of 7–9 per cent is recommended. Seed dormancy is not great but occasionally occurs in dried or stored seed. Stratification is practiced in moist, acid soil at a temperature of 35°–40° F. for 15–30 days. Germination may be improved by soaking seed in water 6–12 hours at 80°F. The seed may be either broadcast or planted in drills 4–6 in. apart. From 6–10 ounces of seed per 100 ft of seedbed is sufficient. Planting at a depth of ⅜ in. is recommended. Seedbeds may be mulched with litter or leaves and protected from rodents by a mesh screen. Seedlings are sometimes attacked by brown-spot needle blight, *Scirrhia acicola* (Dearn.) Siggers. The recommended control is spraying with 4–4–50 Bordeaux between late May or early June and the latter part of December.

Longleaf Pine is a little slower growing than Shortleaf, Loblolly, or Slash Pine at first, but the growth continues somewhat longer. The superior wood compensates for the difference, and also, because of its long taproot, it can grow on deep sandy soils not suitable for the other pines. Records show that under average conditions it reaches a height of 36 ft and a diameter of 5 in. in 20 years, and in 70 years, a height of 82 ft and a diameter of 11 in. The growth is most rapid to 25 years of age. Longleaf Pine 40 years old, on average soil, will yield 39 cords of unpeeled wood, or 10,000 board ft of saw timber per acre. It is rather difficult to transplant older trees because of the long taproot, but seedlings one year old are transplanted easily January–March. State, federal, and private nurseries now make these seedlings available to the public, either free or at a nominal fee. The species ranks fifth in reforestation use in the United States.

HYBRID. Longleaf Pine is known to hybridize with Loblolly Pine to produce *P.* × *sondereggeri* H. H. Chapm.

TURPENTINE YIELD. The yield of turpentine varies considerably according to the size of tree and conditions of growth. Mattoon (1940a) has reported on the yield as follows:

"A 'crop' consists of 10,000 cups. In Florida the average yield per crop is 31 barrels (50 gallons each) of turpentine, while in the South as a whole the average runs about 35 barrels. About 3⅓ barrels of rosin are produced to each barrel of turpentine. Longleaf pine trees 7 in. in diameter (at breast height) and worked 32 streaks a season, should give an average yield of about 14 barrels of turpentine; 8 inch trees, 21 barrels; 9 inch trees, 27 barrels; 10 inch trees, 34 barrels; 11 inch trees, 40 barrels; 12 inch trees, 46 barrels.

. . . .

"Turpentining pine trees does not lower the strength or amount of resin in the wood. . . . Heavy working greatly checks the rate of growth. Conservative turpentining on second-growth trees, with one face per tree, checks the growth about one third of normal; working two faces reduces growth nearly one half its former rate. A natural healing-over growth takes place following the working of the tree, faster in the more vigorous and healthy-topped trees."

Turpentine and rosin are used in varnish, paint, soap, shoe polish, wax, lubricants, pitch, linoleum, and sweeping compounds. Although turpentine is of some medicinal value it is rarely used because of inconvenience of administration for internal use. It is occasionally used externally for local irritant effects or to stiffen ointments where a stimulating effect is not undesirable.

REMARKS. The genus name, *Pinus,* is the classical Latin name, and the species name, *palustris,* refers to the marshy habitat. Vernacular names are Georgia Pine, White Rosin-tree, Pine Broom, Southern Pine, Yellow Pine, Hard Pine, Texas Yellow Pine, Pitch Pine, Fat Pine, Heart Pine, Turpentine Pine, and Florida Pine.

The seeds of Longleaf Pine are eaten by at least 10 species of birds and by many rodents. Razorbacks are very destructive to Longleaf Pine, eating the seeds and the thick, succulent root bark of young seedlings and saplings. Goats also do considerable damage to young plants. After cuttings, the new seedling should be allowed 5 years or so to become established. When the saplings are 5–8 ft high moderate grazing in the area by horses, cattle, or mules can be maintained without appreciable damage. Longleaf Pine is resistant to fire. During hot dry years southern pine beetles cause damage, and pine sawyers sometimes infest dead or felled trees. The wood is very desirable because of its strength and durability. It is used for a great variety of purposes, especially for interior finishing, flooring, fencing, piling, paper pulp, bridges, ties, heavy construction timbers, fuel, charcoal, and shipbuilding. The date of its earliest cultivation is 1727.

Spruce Pine

Pinus glabra Walt.

FIELD IDENTIFICATION. Tree growing singly or in small groves. Rather rapid-growing and attaining a height of 120 ft and a diameter of 2–3½ ft. Branches usually small and horizontal or drooping.

FLOWERS. In March, staminate inflorescences about ¼ in. long, in crowded clusters, yellow, scales erose at apex; pistillate cones raised on slightly ascending peduncles, subglobose; scales bracted, at length woody and spreading, ovules inverted.

FRUIT. Cone persistent, solitary or in clusters of 2–3, reflexed on short, stout stalks, length ½–2¼ in., width to 2 in., ovoid to oblong, at first reddish brown and lustrous, later gray and dull, scales slightly concave, armed with very small and delicate spines or spineless; seeds samara-like, deltoid, sides rounded, length

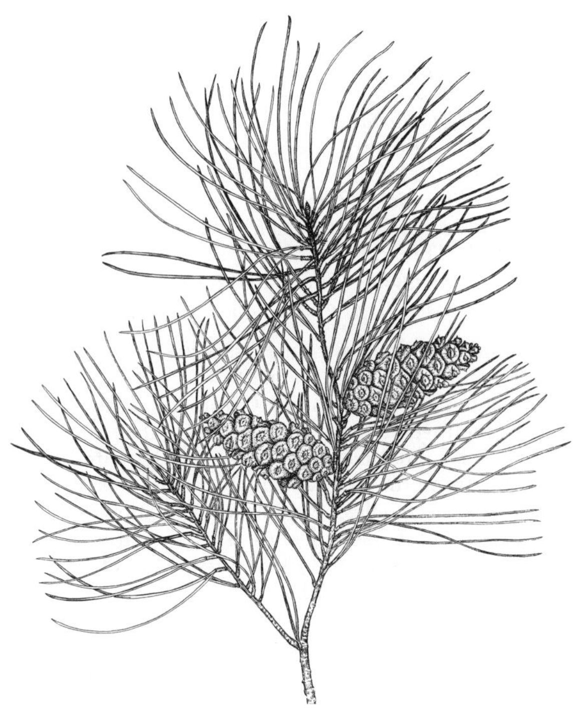

SPRUCE PINE
Pinus glabra Walt.

1/8–1/6 in., shell mottled-gray to black; wings broadest below the middle, 5/8 in. long, 1/4 in. wide.

LEAVES. Buds brown, about 1/10 in. long, sheath 1/5–1/3 in. long, the needle-like leaves numerous, somewhat crowded, in clusters of 2, soft, flattened, very slender, dark green, glaucous, length 1½–4 in., marked by numerous rows of stomata, deciduous in the second or third year.

TWIGS. Slender, smooth, flexible, reddish brown or purplish.

BARK. On young branches pale gray, on older trunks gray to almost black, irregularly divided into shallow furrows and flat ridges, over-all pattern fairly smooth, somewhat resembling the bark of certain oaks.

WOOD. Light brown with thick pale sapwood, close-grained, not strong, soft, brittle.

RANGE. Nowhere very abundant, in sandy-loam soil of river swamps, bottom lands, or on bluffs. Southeastern Louisiana (Livingston Parish), Mississippi, Alabama, and northwestern Florida, northward to South Carolina.

PROPAGATION. Propagation is usually by seeds. The cones are gathered in October and curing may be done at ordinary room temperature for 264 hours. The cleaned seed averages about 70,000 seeds per lb. Storage is usually done in sealed containers at temperatures slightly above freezing. Or seeds may be sown in fall, either broadcast or in drills 4–6 in. apart, and covered 1/4 in. with sandy loam.

REMARKS. The genus name, *Pinus*, is the classical Latin name. The species name, *glabra*, refers to the smoothish bark. This species is also known under the name of White Pine, Cedar Pine, and Walter's Pine. It was given the name Spruce Pine by John Bartram who saw it for the first time on September 25, 1765, near Savannah, Georgia. The cones resemble those of Shortleaf Pine somewhat, but are more globose, often smaller, and the prickles are minute and weak.

EASTERN RED-CEDAR

Juniperus virginiana L.

FIELD IDENTIFICATION. Evergreen tree of variable shape, attaining a height of 50 ft or rarely more. Leaves of two kinds, either scalelike and appressed, or awl-shaped and spreading.

FLOWERS. March–May, dioecious, catkins small and terminal; staminate catkins oblong or ovoid; stamens 10–12, golden brown, pollen sacs 4; female cones globular; scales spreading, fleshy, purplish, bearing 1–2 basal ovules.

FRUIT. Ripening September–December, cone berry-like, on straight peduncles, pale blue, glaucous, subglobose, 1/4–1/3 in. in diameter, sweet, resinous; seeds 1–2, ovoid, acute, 1/6–1/8 in. long, smooth, shining.

LEAVES. Of two kinds; one kind scalelike, appressed, glandular, dark green, acute or obtuse, about 1/16 in. long, 4-ranked; the other awl-shaped, sharp-pointed, glandless, glaucous, 1/2–3/4 in. long; some of the leaves are intermediate between the two forms.

TWIGS. Reddish brown, round or angled.

BARK. Light reddish brown, separating into long fibrous strips; trunk more or less fluted and basally buttressed.

WOOD. Red, sapwood white, knotty, light, brittle, soft, even-textured, compact, weighing about 30 lb per cu ft, shrinks little, very resistant to decay.

RANGE. Growing in all types of soil, on hilltops or in swamps. Almost throughout the eastern United States. West into Texas, Oklahoma, Arkansas, Kansas, Nebraska, and North and South Dakota.

PROPAGATION. The cleaned seed averages about 43,200 seeds per lb, with a commercial purity of 89 per cent and a soundness of 70 per cent. Most of the seeds germinate naturally the second year, and some the third, only a few germinate the first year. The dormancy is probably caused by both the embryo and seed coat. Stratification is best accomplished in sand or peat 100–120 days at 41°F. Because of the thick seed coat, treatment with diluted sulphuric acid or mechanical scarification is sometimes resorted to before stratification. The germinative capacity averages about 32 per cent. Various methods of planting are recommended in *Woody-Plant Seed Manual* (U.S.D.A., 1948, p. 209) as follows: (1) store seed in fruit 1 year, then clean, scarify, and sow in the fall; (2) store seed in fruit 1 year, then clean and stratify in peat for 100 days at 41°F., and sow in the spring; or (3) stratify outdoors in the shade from May until sowing time in the fall. Germination in all 3 cases will occur the following spring. The fruit can also be stratified outdoors or sown in the nursery in the fall and kept mulched until sown in the spring; this should be done early enough so that germination will be practically complete before air temperatures go higher than 70°F. Seed is sown 6–8 in. apart and covered about ¼ in. with firmed soil. Mulching with straw or burlap and screen protection helps. Nursery germination is about 30 per cent. Seedlings are sometimes injured by cedar blight, caused by *Phomopsis juniperivora*. Spraying in early spring with Bordeaux mixture at 2–3-week periods will help. Red-cedar can also be grown from cuttings of nearly ripened wood in fall under glass.

VARIETIES AND FORMS. Northern Eastern Red-cedar, *J. virginiana* var. *crebra*, is a variety with an erect symmetrical form. Kelsey and Dayton (1942), list the following horticultural forms:

Burk *(burkii)*, Canaert *(canaertii)*, Chamberlayn *(chamberlaynii)*, Column *(cylindrica)*, Creeping *(horizontalis; reptans)*, Cypress *(cupressifolia)*, *Elegans*, Feather-white *(plumosa-alba)*, Fountain *(tripartita)*, Globe *(globosa)*, Gold-tip *(elegantissima)*, Green Pyramid *(venusta)*, Hill Dundee *(hillii)*, Keteleer *(keteleeri)*, Koster *(kosteri; kosteriana)*, Lebreton *(lebretonii)*, Maccabe *(maccabei)*, Purple Pyramid *(pyramidiformis)*, Pyramidal *(pyramidalis)*, Reeves *(reevesiana)*, Schott *(schottii)*, Silver *(glauca)*, Smith *(smithii)*, Thread-leaf *(filifera)*, Triomphe D'Angers, Upright *(fastigiata)*, Variegated *(albo-variegata; variegata)*, Weeping *(pendula)*, and Whitetip *(albo-spica)*.

REMARKS. *Juniperus* is the classical name, and the species name, *virginiana*, refers to the state of Virginia. Other names are Red Savin, Carolina Cedar, Juniper-bush, Pencil-wood, and Red Juniper. The capital of the state of Louisiana, Baton Rouge (Red Stick), gets its name from the red wood. The wood is used for novelties, posts, poles, woodenware, millwork, paneling, closets, chests, and pencils. The aromatic character of the wood is considered to be a good insect repellent. The extract of cedar oil has various commercial uses. The tree is host to a gall-like rust which in certain stages attacks the leaves of apple trees. Twig-laden bagworm cocoons are also frequent on the branches. A few borers attack the tree, and it suffers greatly from fire damage. It is sometimes used in shelter-belt planting, and has been cultivated since 1664. The fruit is eaten by at least 20 species of birds and the opossum.

EASTERN RED-CEDAR
Juniperus virginiana L.

ROCKY MOUNTAIN JUNIPER

Juniperus virginiana var. *scopulorum* (Sarg.) Lemmon

FIELD IDENTIFICATION. Shrub on high, dry slopes, or a tree to 50 ft and 15–30 in. in diameter. Trunk usually short and dividing into stout branches to form an irregular round-topped crown.

FLOWERS. In spring, cones small, inconspicuous, yellowish, staminate and pistillate borne separately; staminate with about 6 stamens, connectives entire, anther sacs 4–5; pistillate scales spreading, acute or acuminate, obscure on the mature fruit.

FRUIT. Scales of pistillate cones gradually becoming fleshy and uniting into a berry-like, indehiscent conelet, November–December, ripening at the end of the second season, subglobose, ¼–⅓ in. in diameter, bright blue and very glaucous, flesh resinous and sweet; seeds 1–2, about $\frac{3}{16}$ in. long, grooved and angled, acute; outer coat thick and bony, inner coat thin and membranous, hilum 2-lobed, endosperm fleshy, embryo straight.

LEAVES. Scale-like, $\frac{1}{25}$–⅙ in. long, closely appressed, opposite or in threes, rhombic-ovate, acute or acuminate, entire, obscurely glandular dorsally, dark green

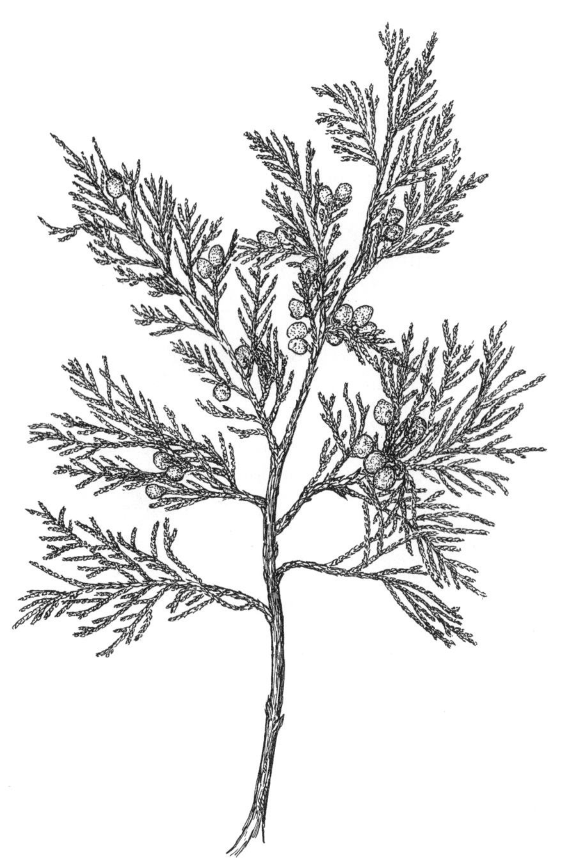

ROCKY MOUNTAIN JUNIPER
Juniperus virginiana var. *scopulorum* (Sarg.) Lemmon

to yellowish green, often pale or very glaucous, leaves of young shoots often awl-shaped and sharply pointed.

TWIGS. Slender, angular and flattened at first, rounded later, brown or gray, scaly with age.

BARK. Reddish brown to gray, fibrous, thin, fissures shallow, ridges flat and interlacing, breaking into shreddy scales.

WOOD. Generally the same as that of Eastern Red-cedar, knotty, heartwood in varying shades of brown to red, or streaked lighter, sapwood thin and nearly white, texture rather fine and even (except at knots), specific gravity about 0.49, moderately weak in bending, fairly strong in endwise compression, works rather easily, shock resistance high, shrinks very little in drying, resistant to decay.

RANGE. Gravelly or rocky soils, dry ridges or bluffs. At altitudes from sea level to 9,000 ft. The most widely distributed juniper of the West. Western Texas, New Mexico, Arizona, Colorado, Nebraska, Nevada, South Dakota, Oregon, Washington, British Columbia, and Alberta.

PROPAGATION. Good crops of fruit are borne every 2–5 years with lighter crops in intervening years. The minimum fruit-bearing age is 10 years and the maximum 300 years. The conelets number about 4,200–8,600 per lb, with 100 lb of fruit yielding about 22–25 lb of seed. The seed averages about 28,600 seeds per lb. Commercial seed averages about 94 per cent in purity and 59 per cent soundness. The conelets are hand-picked or flailed into a canvas. Care should be taken to obtain 2-year fruit instead of half-developed 1-year fruit. They may be soaked in a lye solution (one teaspoon per gallon of water) for a day or two to soften the pulp. They may then be run through a macerator and the residue floated off. The seed may then be sown, stratified, or stored. Dry storage in sealed containers at low temperatures should maintain at least a 50 per cent viability to 3½ years. The seed has both embryo dormancy and seed-coat impermeability. This can be broken by stratification for about 120 days at 68°–86°F., alternated diurnally. The sowing is generally in drills and rows 6–8 in. apart. Mulch and screen protection is usually used and removed just prior to germination. The beds are generally kept moist at first. Light to medium shade for seedlings is helpful the first year. The seedlings may be subject to attack by cedar blight, caused by *Phomopsis juniperivora*. Spraying in spring at 2–3-week intervals with 4–4–50 Bordeaux mixture is helpful. Propagation is also practiced with cuttings.

VARIETIES AND FORMS. Some of the forms in horticulture are as follows:

Silver Column *(argentea)* is a pyramidal form with silvery white leaves.

White *(glauca)* has more glaucous foliage.

Fraser *(fraseri)* has silvery leaves and grows to 20 ft.

Hill *(hillii)* has light bluish foliage.

Weeping *(pendula)* has a weeping shape to 15 ft and silvery leaves.

Roller *(rollerensis)* is erect and compact with bluish green foliage.

Green Pyramid *(viridifolia)* is pyramidal with bright green leaves.

Columnar *(columnaris)* is columnar in shape.

Horizontal *(horizontalis)* has upright spreading branches and very glaucous leaves.

Other horticultural forms are Chandler Blue *(chandleri)*, Cypress-leaf *(cypressifolia)*, Round *(globosa)*, Creeping *(prostrata)*, Marshall, and Moonlight.

REMARKS. *Juniperus* is the classical Latin name; the species name, *virginiana*, is for the state of Virginia; and the varietal name, *scopulorum*, refers to its habitat of rocky cliffs and crags. Also known under the vernacular names of Western Juniper, River Juniper, Western Red-cedar, Red-cedar, Mountain Red-cedar, Colorado Red-cedar, and Cedro Rojo. The tree is slow-growing and long-lived. It is used for reforestation to some extent, and for shelter-belt planting on prairies and plains. It also has value in ornamental planting and has been cultivated since 1836. The wood is used for chests, closets, millwork, interior finish, posts, poles, pencils, water buckets, woodenware, novelties, and fuel, and is reputed to have insect repellent properties. The fruit is eaten by a number of species of birds and grazed by bighorn sheep.

Southern Red-cedar

Juniperus silicicola (Small) Bailey

FIELD IDENTIFICATION. Pungent evergreen tree sometimes attaining a height of 50 ft, with a trunk to 2 ft in diameter. Branches spreading when the tree grows in the open to form a broad irregular crown. The upper branches usually erect and the lower pendulous. When crowded by other trees the branches form a more symmetrical, narrower, pointed crown. The root system is rather shallow.

FLOWERS. Generally opening in March, minute, dioecious, axillary or terminal; staminate cones usually terminal, solitary, oblong-ovoid, ⅛–¼ in. long; stamens 10–12, filaments enlarged with yellow scalelike rounded connectives, usually with 8 pollen sacs at the base; pistillate cones ovoid, scales at base persistent, acute, some ovulate with 1–2 ovules opposite the scales, scales later fusing with the fleshy fruit.

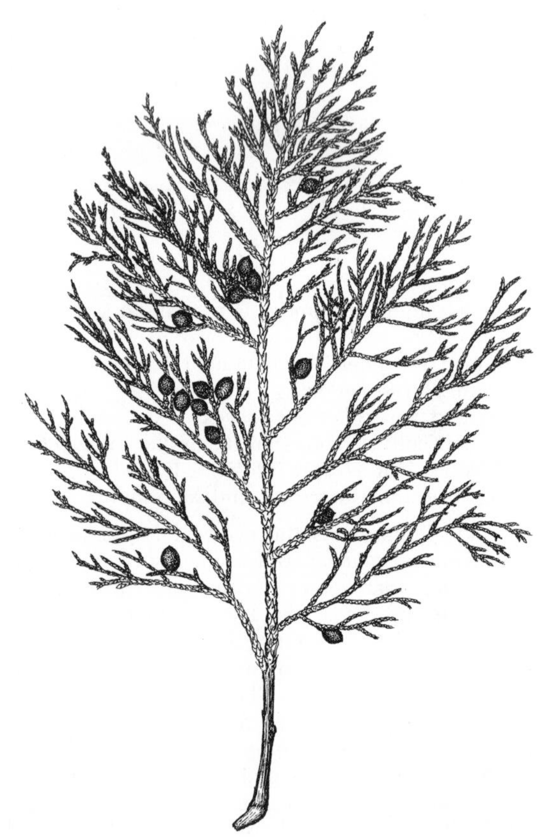

Southern Red-cedar
Juniperus silicicola (Small) Bailey

FRUIT. Berry-like, persistent, succulent, formed by the coalescence of the flower scales, subglobose to short-oblong, 1/12–1/6 in. in diameter, dark blue, glaucous when ripe, skin thin, flesh sweet and resinous; seeds 1–2, ovate, acute, prominently ridged.

LEAVES. Minute, sessile, persistent, opposite, or in whorls of 3, acute or acuminate, dorsal surface with a conspicuous oblong gland; leaves sometimes of 2 kinds; either linear-subulate and sharp-pointed, or scalelike, ovate, imbricate and appressed.

TWIGS. Slender, flexuous or curved on the lower branches, young ones green, soon light reddish brown or eventually gray, the older leaves becoming brown, woody, and persistent.

BARK. On the older branches and trunk reddish brown, thin, separating into long fibrous strips.

WOOD. Dull red, sapwood lighter, close-grained, light, soft, fragrant, very durable.

RANGE. River swamps, streams and creek margins in low woods. Eastern and southeastern Texas and Louisiana; eastward to Georgia and Florida and northward to South Carolina.

RELATED SPECIES. Southern Red-cedar is distinguished from Eastern Red-cedar by the former having a usually irregular-shaped crown, lower pendulous branches, by the much smaller fruit (1/12–1/6 in. in diameter), and by its growth in low wet grounds. However, some authors maintain that these differences are not distinct enough to merit a specific rank for it. Southern Red-cedar has been listed by some authors under the names of *J. barbadensis* L. and *J. lucayana* Britt.

REMARKS. The genus name, *Juniperus*, is the classical name. The species name, *silicicola*, refers to its growing in sandy soils. Also known as Red-cedar and Eastern Red-cedar. The tree is slow-growing, long-lived, and very desirable for ornamental purposes. The wood is commercially important, but the supply is considerably diminished.

Alligator Juniper

Juniperus deppeana Steud.

FIELD IDENTIFICATION. Tree large, evergreen, pyramidal or round-topped, attaining a height of 65 ft. The trunk is stout and short, and an unusual feature for a juniper is the conspicuous squarish scales of the bark, which resemble somewhat the scales of an alligator.

FLOWERS. January–March, terminal on short lateral branches, dioecious, small; staminate solitary, fleshy, stout, about ⅛ in. long; stamens 10–12, scalelike, connectives ovate, pollen sacs near the base; pistillate ovoid, scales ovate and acuminate, ovules attached to the scales.

FRUIT. Maturing the second year September–Decem-

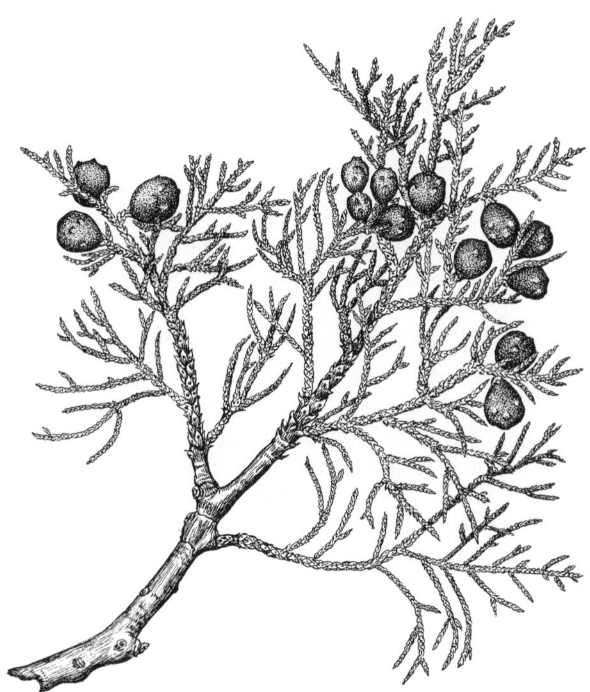

Alligator Juniper
Juniperus deppeana Steud.

ber, a berry-like cone, ¼–½ in. long, subglobose or short-oblong, reddish brown, resinous, aromatic, glaucous, roughened by the minute scale excrescences; flesh dry, mealy, thin; seeds 2–4, about 3/16–¼ in. long, ovoid, obtuse, smooth, shiny, somewhat flattened on one or both sides.

LEAVES. Bluish green, resinous, glaucous, scalelike, 1/25–⅛ in. long, in threes or pairs, closely appressed, imbricate, thickened, ovate, obscurely keeled, acute or obtuse or apiculate; acicular leaves longer, ¼–⅜ in. long, stout, oblong, abruptly spinose.

TWIGS. Stout, reddish brown, resinous, young twigs clothed with the large, scalelike acicular leaves.

BARK. Gray, divided into squarish plates 1–2½ in. long with deep fissures between, reddish brown under the plates.

WOOD. Light red, sapwood lighter, close-grained, light, soft, brittle, specific gravity of 0.58.

RANGE. In high mountains of Trans-Pecos Texas. At altitudes of 4,000–8,000 ft in the Chisos, Davis, Eagle, and Limpia mountains of Texas. Also in New Mexico and Arizona; southward into the Mexican states of Chihuahua and Sonora to Zacatecas and Puebla.

SYNONYM AND FORM. Although this tree has been listed under the name of *J. pachyphloea* Torr. (Kelsey and Dayton, 1942) later research has established the older name of *J. deppeana* Steud. (Little, 1948).

REMARKS. *Juniperus* is the classical name, and the species name, *deppeana,* honors Ferdinand Deppe (d. 1861), a German botanist.

The thick square-plated bark has been compared to the rough plates on an alligator's back and the vernacular name, Alligator Juniper, has been given the tree. Other names are Checker-bark Juniper and Tascate. The tree is slow-growing, long-lived, and occasionally sprouts from the stump. Its leaves are browsed by livestock and deer, and the fruit is eaten by gray fox, black bear, rock squirrel, and wild turkey. The leaves are also reputed as being used as a local remedy for rheumatism and neuralgia in Chihuahua, Mexico. The wood is used for fuel and occasionally for posts, although posts from other junipers seem to be more durable.

Red-berry Juniper

Juniperus pinchotii Sudw.

FIELD IDENTIFICATION. Scraggly shrub or evergreen tree rarely over 25 ft. The numerous branches widespreading and the lower ones often close to the ground, irregular or pyramidal and clump-forming.

FLOWERS. Cones small, dioecious, terminal or axillary on short branchlets; stamens numerous with filaments enlarged into scalelike connections with 2–6 pollen sacs at the base; pistillate cones oblong-ovoid,

Red-berry Juniper
Juniperus pinchotii Sudw.

solitary, with small persistent bracts, minute scales at the base with 1–2 ovules attached to the inner face, scales later enclosed by the fleshy, berry-like cone.

FRUIT. Maturing within the year, a berry-like cone ¼–⅜ in. long, sessile or short-peduncled, subglobose, roughened by the scale remnants, red, somewhat glaucous, thin-skinned, mealy, resinous; seeds solitary, ⅛–3/16 in. long, ovoid, obtuse at one end and rounded at the other, somewhat grooved, brown, lustrous; hilum large, bilobed.

LEAVES. Aromatic, scalelike, yellowish green, appressed, imbricate, disposed in ranks of twos or threes, broad-ovate, obtuse or acute, keeled, dorsally glandular-pitted, 1/16–⅛ in. long, those on young shoots ¼–½ in. long, linear to lanceolate, acuminate, sharp-pointed, buds with a conspicuous resinous gland.

TWIGS. Young twigs rather rigid, greenish, older twigs red.

BARK. Gray to reddish brown, peeling off in shaggy longitudinal strips.

WOOD. Reddish brown to white, soft, moderately durable in contact with the soil, used locally for fuel and fence posts.

RANGE. Dry hillsides and canyons of western Texas and the Texas Panhandle area. Recorded in the following Texas counties: Kimble, Val Verde, Menard, Hood, Briscoe, Tom Green, Randall, Armstrong, Potter, and Hartley. Abundant in the vicinity of Sanderson, Texas. The type specimen is from the Palo Duro Canyon in the Texas Panhandle.

REMARKS. *Juniperus* is the classical name of the Red-berry Juniper, and the species name, *pinchotii,* is for the botanist Gifford Pinchot (b. 1865). Red-berry Juniper seems to be rather hardy, especially as to fire damage, and will sprout from the stump. This trait should make the tree valuable for reforestation of burned-over areas.

ASHE JUNIPER

Juniperus ashei Buchholz

FIELD IDENTIFICATION. Shrub or evergreen tree rarely over 30 ft. Usually irregular, leaning, low-branched, with a fluted and twisted trunk.

FLOWERS. Minute, dioecious, terminal; staminate oblong-ovoid, about ⅙ in. long, stamens 12–18; filaments enlarged into connectives which are ovoid, obtuse, or somewhat cuspidate, pollen sacs near the base; pistillate 1/12–⅛ in. long with ovate, acute spreading scales which are 1–2-ovuled at the base. The fruit is formed by the cohesion of the enlarged fleshy scales.

FRUIT. August–September, sessile or short-peduncled, a fleshy, berry-like cone about ¼ in. long, bluish green, glaucous, ovoid to subglobose, skin thin and roughened somewhat by the scale remnants, flesh sweetish but resinous; seed solitary (rarely 2), ovoid, acute or obtuse at apex, rounded at base, somewhat grooved on 2 sides, lustrous, light to dark brown, about 3/16 in. long.

LEAVES. Usually at the ends of the twigs, 1/25–1/16 in. long, scalelike, opposite in 2–4 ranks, appressed, imbricate, ovate, acute, keeled, minutely denticulate or fringed on the margin, usually nonglandular but resinous and aromatic; on young shoots the awl-like or acicular leaves are ¼–½ in. long, lanceolate, rigid, apex long and sharp-pointed.

TWIGS. Gray to reddish, scaly, aromatic, rather stiff.

BARK. Gray to reddish brown, reddish brown beneath, shredding into shaggy, longitudinal strips.

WOOD. Streaked reddish brown, sapwood lighter colored, close-grained, hard, light, not strong, but durable in contact with the earth, specific gravity 0.59, somewhat aromatic.

RANGE. Mostly on limestone hills. In Texas, Arkansas, Oklahoma, and Missouri. The common juniper of central Texas. Southward and westward into Mexico and Guatemala.

VARIETY AND SYNONYM. A complex taxonomical problem arises concerning the relationship of *J. ashei* Buchholz of the Arkansas and Oklahoma mountains with *J. mexicana* Spreng. of Texas. Some authorities hold that the two are different trees, but others contend that the former is merely a more northern and isolated

ASHE JUNIPER
Juniperus ashei Buchholz

variation of J. *mexicana.* If they are the same, then it serves the best purpose of nomenclature standards to refer to the Texas tree as *J. ashei* Buchholz also. See Johnston (1943c, p. 337).

Some authorities have relegated the One-seed Juniper, *J. monosperma* (Engelm.) Sarg., of west Texas to the status of a variety of Ashe Juniper. The two are very similar, and more field study is needed to determine the exact relationship.

REMARKS. The genus name, *Juniperus,* is the classical name, and the species name, *ashei,* is in honor of William Willard Ashe (1872–1932), American botanist. Names used are Mountain Cedar, Cedar Brake, Texas Cedar, Sabino, Enebro, Tascate, Taxate, and Cedro. The wood is used for fuel, poles, posts, crossties, and small woodenware articles. The foliage is occasionally browsed by goats and deer, and the sweet fruit eaten by a number of species of birds and mammals, including the bobwhite quail, robin, Gambel's quail, cedar waxwing, curved-bill thrasher, gray fox, raccoon, and thirteen-lined ground squirrel. The tree is occasionally cultivated for ornament and is apparently resistant to the cedar-apple rust.

UTAH JUNIPER
Juniperus osteosperma (Torr.) Little

UTAH JUNIPER

Juniperus osteosperma (Torr.) Little

FIELD IDENTIFICATION. Usually an evergreen many-stemmed shrub, but sometimes a small tree to 25 ft, with a trunk 2 ft in diameter. The trunk short and the branches erect to form a rounded open head.

FLOWERS. Borne in short, separate, staminate and pistillate cones terminally borne on short axillary twigs; staminate solitary, oblong-ovoid; stamens 18–24, opposite or in threes, connectives rounded; pistillate flower scales ovate, acute, spreading, with the ovules in their axils.

FRUIT. A small berry-like succulent cone, ripening during the autumn of the second year, very persistent, subglobose to short-oblong or broadly ellipsoid, some showing the tips of the fleshy flower scales, reddish brown but appearing bluish because of the thick glaucous coating, length ¼–⅓ in.; flesh thin, dry, sweet; seed solitary or rarely 2, ovoid, acute at apex, base rounded, obtusely or acutely angled, marked centrally by the hilum, outer seed coat hard and bony, inner thin, cotyledons 4–6.

LEAVES. Scalelike, closely appressed, crowded, ⅛–1/12 in. long or less, sessile, rhombic-ovate, apex acute or acuminate, concave above, convex below, usually not glandular dorsally, yellowish green, mostly opposite, rarely in threes, very persistent; acicular leaves sometimes in threes.

TWIGS. Slender, light yellowish green, old leaves persistent, eventually light reddish brown and scaly.

BARK. On older branches and trunk light gray to almost white, ridges broad and rounded, fissures deep and irregular, breaking into long, shreddy, persistent scales.

WOOD. Light brown, sapwood thicker and paler, fragrant.

RANGE. Often in pure stands at altitudes of 3,000–8,000 ft on arid slopes or in valleys. In New Mexico, Arizona, Colorado, Utah, Nevada, Idaho, Wyoming, and California.

PROPAGATION. Good crops of fruit are borne about every two years. The cones average about 800 per lb, and the cleaned seed averages 5,000 seeds per lb, with a soundness of 20–78 per cent. The fruit may be dried and stored for 1 year, then cleaned, scarified, and stratified in peat for 100 days at 41°F., and then sown in spring. Planting is done in drills 6–8 in. apart and covered ¼ in. with firmed soil. The beds are kept moist and mulched before germination.

SYNONYMS AND VARIETIES. The scientific name used by some authors is *J. utahensis* (Engelm.) Lemmon, but the older name is given preference (Little, 1948; Van Melle, 1950).

Some authorities also list Utah Juniper as a variety of California Juniper, *J. californica* var. *utahensis* Engelm.

A variety with a large fruit to ¾ in. in diameter, a single tall trunk, and leaves in threes, has been separated as *J. utahensis* var. *megalocarpa* (Sudw.) Sarg.;

however, it appears to be only a form rather than a distinct variety.

REMARKS. The genus name, *Juniperus*, is the classical Latin name, and the species name, *osteosperma*, means "bone-seeded."

Utah Juniper was first cultivated about 1900, but is seldom grown and has no great ornamental value. The fruit is eaten by various species of chipmunk and ground squirrel, and the leaves are occasionally browsed by black-tailed deer. The wood is occasionally used for fencing, fuel, and interior finishing. Indians eat the fruit raw or bake it into cakes.

ONE-SEED JUNIPER

Juniperus monosperma (Engelm.) Sarg.

FIELD IDENTIFICATION. Evergreen tree sometimes attaining a height of 50 ft, with a trunk to 3 ft in diameter, often with several trunks. The branches are stout and form a rather irregular crown. Often reduced to a low much-branched shrub when at high altitudes or on sterile soils.

FLOWERS. March–April, dioecious, terminal or axillary, borne on branches of the previous year; staminate solitary, oblong-ovoid; stamens 8–10, filaments enlarged into entire or erose, ovate, rounded, pointed, scalelike connectives, anther sacs near the base; pistillate flowers ovoid, bearing many very small, pointed scales, some scales bearing 1–2 ovules inwardly at the base.

FRUIT. In September, fleshy, dark blue to brownish, often somewhat glaucous, subglobose, ⅛–¼ in. long, flesh thin; seeds usually 1–2, sometimes extruded from the fruit apex, ovoid, apex obtuse, often angled.

LEAVES. Minute, grayish green, ¹⁄₁₆–⅛ in. long, opposite or in threes, apex acute or acuminate, fleshy-thickened, dorsally rounded, often glandular, margin minutely and denticulately fringed; leaves on young shoots often ovate, acutely and rigidly pointed, glandular, ¼–½ in. long.

TWIGS. Slender, old leaves reddish brown, the thin reddish brown bark loosely spreading.

BARK. On trunks gray, ridges rather flattened and irregular to separate into elongate, loose shreddy scales.

WOOD. Heartwood light reddish brown, sapwood yellowish or white, heavy, slightly fragrant.

RANGE. Usually on lower hills approaching taller mountains at altitudes of 3,000–7,000 ft. New Mexico, western Texas, and western Oklahoma; northward in Colorado, Utah, Wyoming, and Nevada, westward into northern Arizona, where it is reported to reach its maximum size, and southward into Mexico.

PROPAGATION. The fruit persists 1–2 years. It is usually obtained by hand-picking or by shaking or flailing into a canvas. Care should be taken not to collect unripe fruit, but collection should be made as soon as ripe to prevent loss by birds and mammals. Since some trees bear sterile fruit, cutting tests should be made before collection. The fruit is then run through a macerator or a hammer mill and the pulp and empty seed floated off. Seeds of dry fruit may be extracted by rubbing on screens. One hundred lb of fruit yield about 15 lb of seed. The cleaned seed averages 18,000 seeds per lb. It has a commercial purity of 89 per cent and a soundness of 51 per cent. Germination of seed occurs in the early spring. Dormancy may be the result of both embryo and seed coat, but some seeds are not dormant. Dormancy may be broken by stratification in sand or peat for 60 days at 41°F. However, more study is needed to determine the true dormancy periods of this species. Sowing should be in fall, or the seed stratified. They should be planted in drills in rows 6–8 in. apart and covered with about ¼ in. of firmed soil or sand. A mulch of straw or burlap can be used until germination begins. Planted seeds should be kept moist. Germination should start in 6–10 days if stratified seed are used and be completed in 4–5 weeks. Nursery germination is usually 65–70 per cent.

RELATED SPECIES. Some authors consider that the name *J. monosperma* (Engelm.) Sarg. is a synonym of *J. ashei* Buchholz (*J. mexicana* Spreng.). Apparently the two are closely related, but more study is needed before a combination can be definitely made.

Cory (1936) states that *J. monosperma* should be

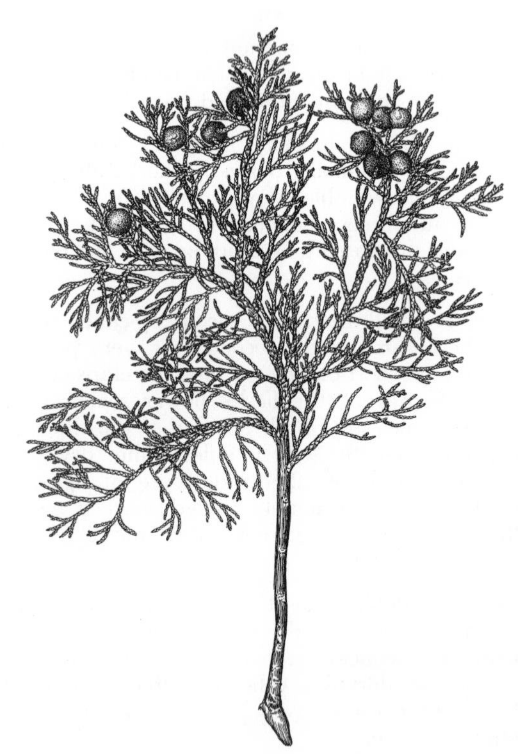

ONE-SEED JUNIPER
Juniperus monosperma (Engelm.) Sarg.

relegated to a variety under the name of *J. mexicana* Spreng. var. *monosperma* (Engelm.) Cory.

Cory has also described two new species from the mountains of southwestern Texas, which are apparently closely related to *J. monosperma*. These are *J. gymnocarpa* (Lemmon) Cory and *J. erythrocarpa* Cory. These descriptions are also set forth in the paper cited above. Since the matter is one of very close taxonomical distinctions, the reader is referred to the original paper. The matter is also discussed by Van Melle (1952).

REMARKS. The genus name, *Juniperus*, is the classical Latin name. The species name, *monosperma*, refers to the solitary seed. Other vernacular names are Cherrystone Juniper, Red-berry Juniper, West Texas Juniper, and Sabina. The fruit is sometimes ground into flour and made into bread by the Indians. The fibrous bark is used for mats, saddles, and breechcloths. The tree grows rapidly for a juniper and is long-lived. It has been cultivated since 1900. The wood is used locally for posts and fuel. The fruit is known to be eaten by at least four species of songbirds and by Gambel's quail. It is also eaten by coyote, fox, raccoon, rock squirrel, and Hopi chipmunk. It is occasionally browsed by goats.

MEXICAN DROOPING JUNIPER
Juniperus flaccida Schlecht.

MEXICAN DROOPING JUNIPER

Juniperus flaccida Schlecht.

FIELD IDENTIFICATION. Evergreen shrub or tree attaining a height of 30 ft. Easily recognized by the drooping branchlets and leaves which give the tree a wilted appearance.

FLOWERS. Dioecious, terminal; staminate slender, stamens 16–20; connectives ovate, dorsally keeled; pistillate scales acute or acuminate.

FRUIT. Maturing the second season September–October, persistent, a berry-like cone, ¼–½ in. in diameter, subglobose, reddish brown, glaucous, roughened, resinous, flesh dry; seeds 4–12 (usually 6–8), angular or distorted, somewhat flattened, ⅛–¼ in. long, acute or obtuse.

LEAVES. Scalelike, 1/16–⅛ in. long, yellowish green, imbricate, opposite, spreading, ovate to lanceolate, acuminate, acute, or mucronate at the apex, dorsally glandular or glandless.

TWIGS. Slender, reddish, fibrous, drooping and graceful.

BARK. Reddish brown, deeply furrowed, and shredding in long, thin, fibrous strips.

RANGE. A Mexican species found at altitudes of 4,000–8,000 ft in the states of Chihuahua and Sonora southward. In the United States found only in the Chisos Mountains in Brewster County, Texas, where it is abundant.

REMARKS. The genus name, *Juniperus*, is the classical name, and the species name, *flaccida*, refers to the flaccid drooping habit of the branchlets. Also known under the vernacular names of Weeping Cedar, Weeping Juniper, Cedro, and Cedro Colorado. The tree is slow-growing and long-lived. The wood is considered to be durable, and is used locally for posts. It is cultivated in gardens of south Europe and north Africa.

COMMON JUNIPER

Juniperus communis L.

FIELD IDENTIFICATION. Varying greatly in size and form. The typical species form is a small tree with a short crooked trunk, attaining a height of 10–40 ft, the crown open and irregular. However, mostly inclined in its many variations to be low and shrubby, or even prostrate, with no well-defined main trunk. The shrubby varieties often forming broad dense patches of growth.

FLOWERS. April–May, axillary on branches of the year, minute, dioecious or monoecious; staminate solitary, ⅛–¼ in. long, oblong-ovoid, composed of 5–6 whorls of 3 decussate stamens, connectives ovate, acute, short-pointed, the 3–4 globose anther cells at the base; pistillate ovoid, surrounded by 5–6 whorls of ternate scales and 3 minute fleshy scales; ovules 3, enlarged and apically opened.

FRUIT. Maturing August–October, the fleshy strobile berry-like, persistent, maturing the third season, sessile or very short-peduncled, ¼–⅓ in. in diameter, globose to subglobose or short-oblong, bright blue, glaucous, formed by the coalescence of the flower scales, fleshy, succulent, soft, mealy, resinous, sweet, aromatic, somewhat terebinthinate to the taste when dry; seeds 1–3, ovoid, acute, about ⅛ in. long, brown, warty, flattened or obtusely 3-angled, with numerous deep resin glands, outer seed coat thick, inner membranous.

LEAVES. Very persistent, usually in whorls of 3, ascending or spreading at nearly right angles to the stem, length ⅓–⅗ in., 1/32–1/12 in. wide, linear-lanceolate, apex sharp-pointed, lower surface green and keeled or strongly convex, upper surface deeply grooved with a broad white band of stomata, stiff and firm, dark yellowish green, in autumn turning bronze, mature leaves often with small ones in their axis; also small scalelike leaves found on some shoots.

TWIGS. Slender, smooth, shiny, yellowish to reddish and finally rather dark reddish brown with small thin scales, at first somewhat 3-angled but later terete; buds ⅙–⅛ in. long, ovoid, acute, scaly.

BARK. Young bark yellowish to purplish or reddish, later dark reddish brown to gray, rather thin, separating irregularly and perpendicularly into loose papery scales.

WOOD. Light brown, sapwood paler, very close-grained, hard, light, and easily worked, very durable in contact with the soil.

RANGE. Usually dry well-drained soil on sunny sites, at high altitudes of 3,000–15,000 ft, or coastal variations lower. Widespread in the mountainous areas of both eastern and western North America. New Mexico in the Sangre de Cristo Mountains and Sandia Mountains; north to northern Canada, west to Alaska, British Columbia, Washington, Oregon, and California, east to Greenland, and extending southward along the Allegheny Mountains to Georgia. Also in the boreal regions of Europe and Asia.

MEDICINAL USES. All parts of the plant contain a volatile oil. A terebinthinate juice exudes from the tree and hardens on the bark. Juniper berries owe their medicinal virtues to a volatile oil known as "oleum juniperi." The proportion of volatile oil is highest in the green, but fully grown, fruit. The fruit contains about 1 per cent oil. In the overripe black fruit the oil is changed to resin. The oil is chiefly used as an adjuvant to the so-called alternative diuretics in treatment of cystitis and similar complaints. For most purposes the oil is preferred. An ounce of bruised berries in a pint of boiling water makes an infusion which is sometimes used. (See Wood and Osol, 1943, p. 583)

PROPAGATION. The small berry-like strobiles are picked from the branches when mature. The cleaned seed averages about 35,000 seeds per lb, with a purity of 95 per cent and a soundness of 78 per cent. Because of embryo dormancy, and perhaps also the impermeable seed coat, the seed should be stratified in sand or peat for 120 days at 41°F. plus 60–90 days at 68°–86°F. Stratified seed are planted in spring in drills and in rows 6–8 in. apart and covered with about ¼ in. soil. They are kept moist and mulched until germination.

COMMON JUNIPER
Juniperus communis L.

The plant, with its many shrubby forms, is very desirable for planting in rock gardens, along walls, or where a low evergreen is desired. The plants do best in a cool moist situation and are hardy when fully established.

VARIETIES AND HORTICULTURAL FORMS. A number of ornamental dwarf and pyramidal forms are planted in the eastern United States and in countries of west-central and northern Europe. They grow on almost any type of soil and are rather slow-growing, but long-lived. The principal ones are listed below.

Big-fruited Common Juniper, *J. communis* var. *megistocarpa* Fern. & St. John, has large fruit ⅓–½ in. in diameter, sweet, pulpy, edible; seeds ⅕–⅓ in. long.

Old-field Common Juniper, *J. communis* var. *depressa* Pursh, has branches ascending from a procumbent base, lower branches decumbent, forming mats to 5 ft high, rather flat-topped and broad-spreading; leaves ⅓–¾ in. long, straight or slightly curved, sharp-pointed, white-striped on the upper surface; fruits ¼–⅖ in. in diameter.

Golden-head Common Juniper, *J. communis* var. *depressa* forma *aurea-spicata*, has drooping branches and golden yellow young growth.

Erect Common Juniper, *J. communis* var. *erecta*, has an upright stature.

Bloomy Common Juniper, *J. communis* var. *glaucescens*, has glaucous-white leaves.

Irish Common Juniper, *J. communis* var. *hibernica*,

has leaves shorter and less spreading, dark green, tips of branches upright, form narrow and columnar.

Blue Irish Common Juniper, *J. communis* var. *hibernica* forma *glauca,* has bluish leaves.

Dwarf Irish Common Juniper, *J. communis* var. *hibernica* forma *nana,* is a dwarf form.

Jack Common Juniper, *J. communis* var. *jackii,* is a prostrate shrub with trailing branches often 3 ft long, leaves linear-lanceolate and incurved.

Nippon Common Juniper, *J. communis* var. *nipponica,* has leaves deeply sulcate and keeled below; high mountains of Japan.

Mountain Common Juniper, *J. communis* var. *saxatilis,* is completely decumbent or prostrate, leaves crowded, linear-oblong, 1/6–1/3 in. long, 1/25–1/12 in. wide, curved and very concave above; fruit globose and glaucous (syn. *J. communis* var. *sibirica, J. communis* var. *montana, J. communis* var. *alpina).*

Swedish Common Juniper, *J. communis* var. *suecia,* has a columnar form, rather spreading and light bluish green leaves; branches with nodding tips (syn. *J. communis* var. *fastigiata).*

Dwarf Swedish Common Juniper, *J. communis* var. *suecia* forma *nana,* is a dwarf form.

Ashford Swedish Common Juniper, *J. communis* var. *suecia* forma *ashfordii,* has leaves darker and branches partly upright.

Broad Weeping Swedish Common Juniper, *J. communis* var. *suecia* forma *oblongopendula,* has a columnar form with drooping branches.

Dense-ball Swedish Common Juniper, *J. communis* var. *suecia* forma *hemisphaerica,* has a low, dense, rounded shape, usually under 3 ft high, leaves straight, stiff, and short.

Tight Swedish Common Juniper, *J. communis* var. *suecia* forma *compressa,* has a dwarf form and short crowded branchlets.

Golden Swedish Common Juniper, *J. communis* var. *suecia* forma *aurea,* has golden yellow young growth.

Flat Swedish Common Juniper, *J. communis* var. *suecia* forma *prostrata,* is a prostrate form (syn. *J. communis* var. *horizontalis).*

Golden-flat Swedish Common Juniper, *J. communis* var. *suecia* forma *prostrata-aurea,* is prostrate with golden yellow growth.

Gray Swedish Common Juniper, *J. communis* var. *suecia* forma *grayi,* is a pyramidal, quick-growing sport.

Hedgehog Swedish Common Juniper, *J. communis* var. *suecia* forma *echiniformis,* has a growth to 2 ft, with densely crowded branchlets.

Polish Swedish Common Juniper, *J. communis* var. *suecia* forma *cracovia,* has dropping branchlets.

Upright Swedish Common Juniper, *J. communis* var. *suecia,* forma *columnaris,* has a columnar shape.

Weeping Swedish Common Juniper, *J. communis* var. *suecia* forma *pendula,* has drooping branchlets.

Day's Swedish Common Juniper, *J. communis* var. *suecia* forma *dayi,* is of columnar shape.

Other forms listed in the literature include *J. communis* var. *hibernica* forma *keyonoi, J. communis* var. *koraiensis,* and *J. communis* var. *pyramidalis.*

REMARKS. The genus name, *Juniperus,* is the classical name, and the species name, *communis,* means "common" or "widespread." Vernacular names are Horse Savin, Aiten, Hackmatack, Ground Cedar, Dwarf Juniper, Prostrate Juniper, Melmot-tree, Low Savin, and Gin Juniper. The wood is used for fuel in Europe, burned for incense in India, and used as a fumigant elsewhere. The plant is valuable for wildlife food, the fruit being eaten by over 30 species of birds, including the bobwhite quail, ruffed grouse, sharp-tailed grouse, Hungarian partridge, and ring-necked pheasant. The fruit is also eaten by many small mammals, and the foliage is browsed by white-tailed deer and moose. The fruit is used in Europe as a common constituent of Holland gin. They have also been known to be roasted and used as a very poor substitute for coffee.

EPHEDRA FAMILY (*Ephedraceae*)

ROUGH EPHEDRA

Ephedra nevadensis var. *aspera* (Engelm.) Benson

FIELD IDENTIFICATION. Erect shrub attaining a height of 3½ ft, the branches opposite or whorled.

FLOWERS. April; dioecious, spikes of the staminate flowers usually in pairs at the nodes, 1/6–1/3 in. long, sessile or short-pedunculate, obovate; bracts in 6–10 whorls, obovate, about 1/8 in. long and 1/12 in. broad, thin, yellow or brown; perianth as long as the subtending bract or longer; staminal column 1/6–1/5 in. long, exserted about one third its length; anthers 4–6, sessile or short-stipitate; pistillate spikes paired, or sometimes whorled or solitary, 1/4–2/5 in., sessile or short-pedunculate; bracts in whorls of 5–8, oval, 1/12–1/5 in. long, 1/12–1/5 in. broad, red to brown, margins thin.

FRUIT. Seed solitary, narrowly ovoid, rounded to somewhat 3-angled, light to dark brown, 1/5–1/3 in. long, 1/12–1/6 in. broad, about one third longer than the bracts; tubillus about the same length as bracts or slightly exserted, the apex conical and twisted.

LEAVES. Opposite or occasionally ternate, 1/25–1/10 in. long, apex obtuse and thickened below, connate to the stem from one half to almost the complete length, sheath generally splitting.

STEMS. Rather stiff, terete, opposite or whorled, about 1/8 in. thick on branches; internodes 1/25–2¼ in.; young stems yellowish green to dark green, noticeably rough to the touch (because of the minute papillae), or some smooth and glaucous; older stems thickened, with longitudinal fissures.

RANGE. Dry hills, rocky ravines, flats, grassy slopes to an altitude of 5,000 ft. Trans-Pecos Texas and New Mexico; west through Arizona to California. In Mexico in Tamaulipas, San Luis Potosí, Coahuila, Zacatecas, Chihuahua, and Baja California.

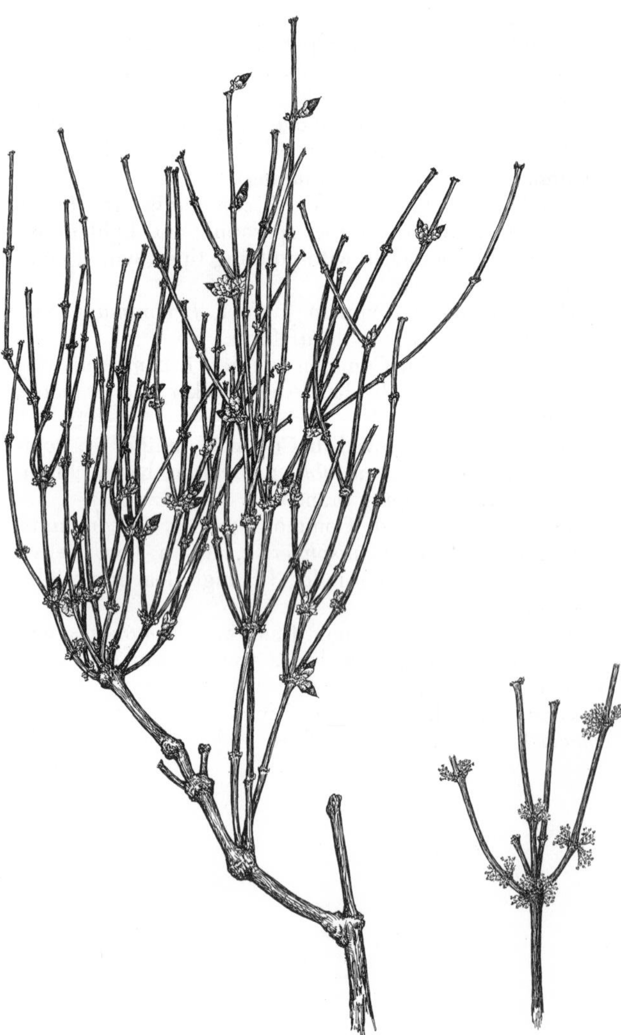

ROUGH EPHEDRA
Ephedra nevadensis var. *aspera* (Engelm.) Benson

REMARKS. The genus name, *Ephedra*, is from the Greek for "horsetail." The species name, *nevadensis*, is for the state of Nevada, where it is abundant, and the variety name, *aspera*, refers to the rough stems.

EPHEDRA FAMILY

Green Ephedra

Ephedra viridis Cov.

FIELD IDENTIFICATION. Dioecious shrub 1–4 ft, the numerous branches dense, rigid, jointed, and broomlike.

FLOWERS. March–June, staminate spikes in pairs or numerous at the nodes, about ¼ in. long, obovate to nearly spherical, short-stalked, usually with 2 flowers and 3–5 pairs of bracts; bracts opposite, ovate, acute, 1/12–1/6 in. long, thin, yellow or reddish; staminal column one fourth to one half exserted, anthers large, 5–8, sessile or nearly so; pistillate spikes with stalks 1/12–¼ in. long, usually several or opposite at the nodes, obovate, ¼–⅖ in. long; bracts opposite, ovate, in whorls of 3–5 pairs, green near the center, 1/12–⅕ in. long.

FRUIT. Seeds usually 2, nutlike, hard, varying from brown to green or grayish, smooth, sharply angled, flattened inwardly, ⅕–⅓ in. long, about 1/10 in. broad, about one fourth exserted from the bracts; tubillus exserted, straight or recurved.

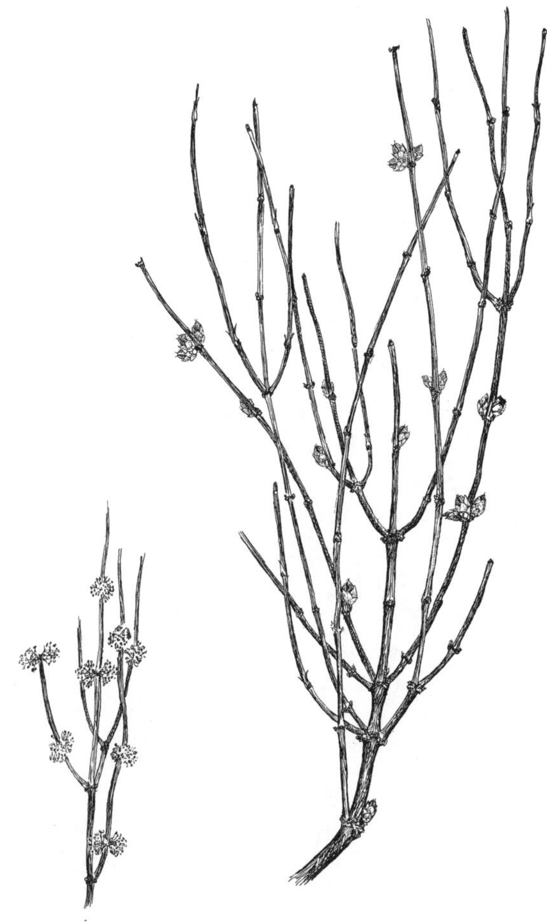

Green Ephedra
Ephedra viridis Cov.

LEAVES. Opposite, 1/15–1/6 in. long, awl-like, subulate, base dark brown, connate to three fourths of its length, apex obtuse to setaceous, upper part of leaf sheath deciduous with age.

TWIGS. Slender, 1/16–⅛ in. in diameter, opposite or numerous at the nodes, internodes ⅜–1¾ in.; young ones finely striate and somewhat scabrous, yellowish green to dark green; older ones stout, reddish brown to gray, bark cracked and fissured; terminal buds obtuse and conical, 1/25–1/12 in. long.

RANGE. The species is found on rocky or sandy slopes at altitudes of 3,000–7,500 ft. New Mexico; west through Arizona to California and northward to Colorado, Utah, and Nevada.

VARIETY. Sticky Green Ephedra, *E. viridis* var. *viscida* (Cutler) L. Benson, (*E. cutleri* Peebles) is a variety with stems somewhat stouter and shorter than in the typical species, and somewhat sticky; ovulate spikes with peduncles ⅕–⅖ in. It occurs on sandy plains, sagebrush deserts, short-grass prairies, or in juniper-pinyon woodlands at altitudes of 4,000–6,000 ft. In northwestern New Mexico, Arizona, Utah, and Colorado.

REMARKS. The genus name, *Ephedra*, is the ancient Greek name, and the species name, *viridis*, refers to the green stems. Vernacular names for the plant are Joint-fir, Mormon Tea, and Brigham Tea. A tonic was made from the dried stems and flowers by the early settlers and Indians. The plant is browsed by black-tailed deer and mule deer, and is important for livestock browse.

Torrey Ephedra

Ephedra torreyana Wats.

FIELD IDENTIFICATION. Erect shrub 1–3½ ft. Branches usually in threes, spreading, flexuous, averaging about 1/12 in. in diameter, jointed, the internodes ¾–2 in. long.

FLOWERS. In May; dioecious, axillary, solitary or in a whorl; staminate spikes ¼–⅓ in. long, sessile or nearly so, spherical or ovoid with 6–8 whorls of flowers; bracts ternate (in 6–9 whorls), broadly ovate or almost oval, erose-margined, base rounded or truncate, 1/12–⅛ in. in diameter, the yellowish perianth longer than the bracts; stamens solitary at base of each bract, filaments united into a clavate staminal column; anthers sessile or short-stipitate, with a pore at apex; pistillate spikes 1–3, ovate, sessile or short-peduncled, ⅖–½ in. long, erect, terminated by a stylelike process which is ¼–⅖ in. long; bracts in many cycles of 3, very membranous, ¼–⅓ in. long, about ⅛ in. broad.

FRUIT. Equal to or exceeding the bracts, solitary or in twos or threes, oblong-lanceolate to narrowly ovoid or tetragonal, light colored, scabrous, beaked; nutlets

TORREY EPHEDRA
Ephedra torreyana Wats.

usually 1–2, ⅓–⅖ in. long, leathery-skinned, light brown to yellowish green; tubillus noticeably exserted, the contorted limb about the same length as the nutlet body or shorter, enfolded in the bracts.

TWIGS. Stout, 1/12–⅛ in. in diameter, bluish green to yellowish green, finely furrowed; terminal buds conical, about ⅙ in. long; bark on old stems cinereous, cracked, irregularly furrowed.

LEAVES. In whorls of 3 at the nodes, reduced to sheathing scales, brown, tardily deciduous, 1/12–⅕ in. long, connate one third to two thirds of length but later somewhat spreading, apices acute to obtuse, usually erect in age, margin thin and membranous.

RANGE. Dry areas of sandy or gravelly plains, mesas, canyons, hillsides, at altitudes of 2,000–6,000 ft. In western Texas and New Mexico; west through Arizona to California and northward into Colorado, Utah, and Nevada.

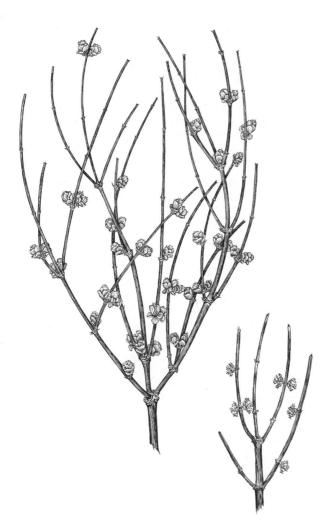

HYBRID. *E. torreyana* hybridizes with *E. viridis* var. *viscida* to produce *E.* × *arenicola* Cutler.

REMARKS. The genus name, *Ephedra,* is from the Greek word used for "horsetail." The species name is for John Torrey (1796–1875), American botanist. The plant is also known under the vernacular names of Mormon Tea and Torrey Joint-fir. It is sometimes grazed by cattle during the winter. Records indicate that it was first grown in cultivation in 1912.

LONG-LEAF EPHEDRA
Ephedra trifurca Torr.

FIELD IDENTIFICATION. Erect dioecious shrub attaining a height of 1½–6 ft. Branches solitary or whorled at the nodes, rather rigid, terete, internodes 1¼–3½ in.

FLOWERS. March–April, at the nodes, staminate spikes solitary or numerous, obovate, ¼–⅖ in. long; peduncles short and many-scaled; bracts in threes, in

LONG-LEAF EPHEDRA
Ephedra trifurca Torr.

8–12 whorls, 1/8–1/6 in. long, 1/12–1/8 in. broad, reddish brown, membranaceous; perianth about as long as the subtending bract; staminate cones broadly ovoid to spherical, 1/5–1/4 in. long; staminate column about one fourth exserted, 1/6–1/5 in. long; anthers 4–5 and short-stipitate; pistillate spikes solitary or many at the nodes, obovate, 2/5–3/5 in. long, not stalked, or short-pedunculate and scaly; bracts in cycles of 3 in 6–9 whorls, orbicular, clawed at base, 1/3–1/2 in. long, 1/3–1/2 in. broad, oval or nearly so, margins entire and membranous, bases notched, reddish brown and translucent.

FRUIT. Cones maturing April–May, solitary, 2–3, narrowly ovoid to tetragonal, smooth, reddish brown, 1/3–1/2 in. long, 1/15–1/8 in. wide, membranous-margined, equaling the bracts; tubillus (beak) noticeably exserted, 1/6–1/4 in. long; seed germination of at least 34 per cent.

LEAVES. In whorls of 3, 1/5–1/2 in. long, united for one half to three fourths their length, later becoming frayed and gray, apex slender-pointed and erect, dorsal side thickened along the center; sheath persistent, thin.

TWIGS. Young ones slender, about 1/15 in. in diameter, pale green to yellow or gray, smooth or nearly so, finely furrowed; older ones cinereous, with irregular fissures and cracks, usually with weak apical spines.

RANGE. From sea level to an altitude of 5,000 ft. Semidesert foothills, dry creek beds, sandy or gravelly places in Texas and New Mexico; west through Arizona to California; in Mexico in Coahuila, Chihuahua, Sonora, and Baja California.

HYBRID. A hybrid of *E. trifurca* and *E. torreyana* has been named *E* × *intermixta* and is intermediate to the parents by the following characters: branches numerous to few at the nodes; angle of branch divergence 35°–40°; bracts from brown and entire to yellow and erose; seeds smooth and light brown. It occurs in New Mexico; the type was collected 3 miles west of Elephant Butte, Sierra County.

REMARKS. The genus name, *Ephedra*, is from the Greek word for "horsetail." The species name, *trifurca*, refers to the 3-forked branches and leaves. The seeds are known to be eaten by scaled quail. The plant is browsed by mountain sheep, Mexican bighorn, and jack rabbit; and is also grazed by cattle, especially in winter.

Cory Ephedra

Ephedra coryi Reed

FIELD IDENTIFICATION. Slender jointed shrub attaining a height of 1/2–3 ft. Branches opposite or whorled, terete, 1/12–1/8 in. in diameter, internodes 3/4–1 7/8 in. long.

FLOWERS. Staminate spikes in April, paired or numerous at nodes, obovate, sessile or short-pedunculate, 1/6–1/3 in. long; bracts opposite, slightly connate at the base, in 5–9 whorls, ovate, 1/12–1/6 in. long, 1/12–1/8 in. broad, membranous, light yellow, perianth slightly larger than the bracts; staminal column 1/12–1/6 in. long, about one fourth exserted; anthers 5–7, sessile or short-stipitate; pistillate spikes in pairs or several together at the nodes, obovate to spherical, 1/4–3/5 in., peduncle 1/8–4/5 in. (with a pair of bracts); bracts opposite in 3–4 whorls, ovate to orbicular, apex acute.

Cory Ephedra
Ephedra coryi Reed

FRUIT. Spring or early summer, seeds 2, trigonal, brownish, about 1/4 in. long, as long as the bracts or slightly longer; tubillus straight, limb recurved.

LEAVES. Opposite, scalelike, connate one third to three fourths their length; apex acute, rather thickened dorsally; sheath early deciduous, but the dry base persistent.

TWIGS. Young ones green, finely furrowed, older ones reddish brown with irregular furrows; terminal buds conical, obtuse, 1/12–1/8 in. long; underground stems with an extensive rhizome system.

RANGE. Cory Ephedra is known from west-central Texas. It has been collected in Gaines, Martin, and Howard counties, in most instances in loose sandy soil of dry well-drained sites.

REMARKS. The genus name, *Ephedra,* is from the Greek word used for "horsetail." The species name, *coryi,* is in honor of V. L. Cory, former range botanist for Texas Agricultural and Mechanical College.

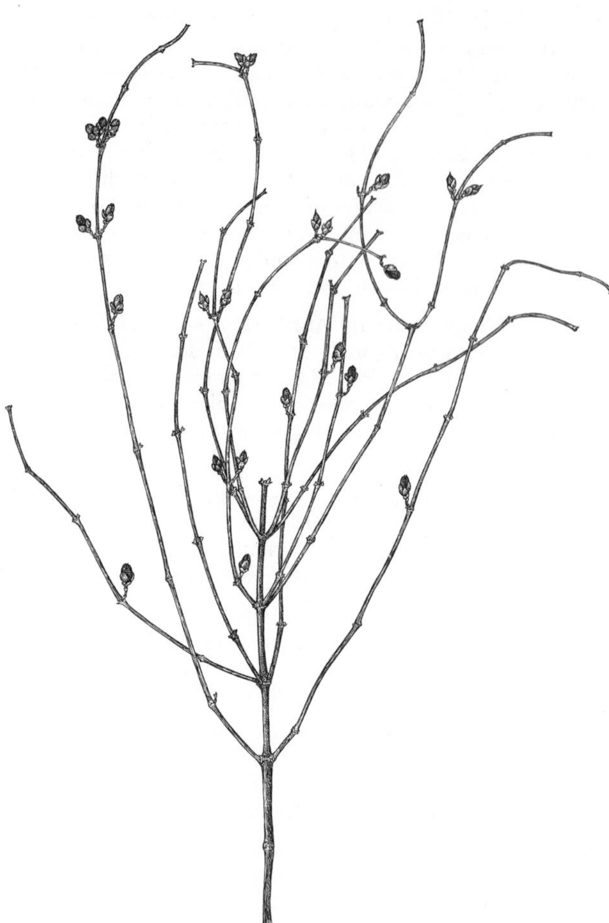

VINE EPHEDRA
Ephedra pedunculata Engelm.

ERECT EPHEDRA

Ephedra antisyphilitica Berl.

FIELD IDENTIFICATION. Shrub with an erect or spreading habit and attaining a height of 1–3 ft. The branches are opposite or whorled at the nodes.

FLOWERS. Dioecious, the staminate ones borne in solitary or paired spikes at the upper nodes, lanceolate to elliptic, 1/5–1/3 in. long, sessile or nearly so; bracts in 5–8 pairs, obovate, short-clasping at the base, 1/12–1/7 in. long and 1/12–1/8 in. broad, margin thin, color green to yellowish or reddish; perianth as long as the subtending bract or slightly longer; staminal column 1/6–1/5 in. long; anthers 4–6, sessile or short-stipitate; pistillate flowers solitary or paired, occasionally a few more, elliptic, 1/4–1/2 in. long, sessile or nearly so; bracts usually 4–6 pairs, from slightly to almost entirely connate, ovate, reddish and fleshy when ripe.

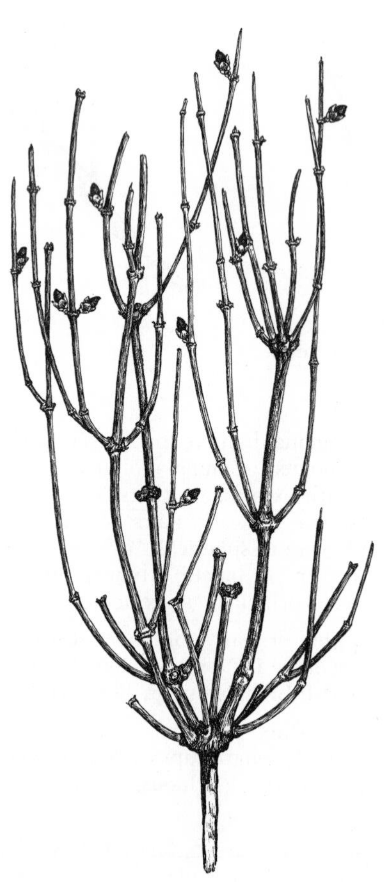

ERECT EPHEDRA
Ephedra antisyphilitica Berl.

FRUIT. Seed solitary, somewhat 3- or 4-angled, light to dark brown, smooth, 1/4–1/3 in. long, 1/12–1/7 in. broad, noticeably prolonged beyond the bracts; tubillus barely exserted, apex straight or a little twisted.

LEAVES. Leaves opposite, 1/25–1/8 in. long, dorsally thickened and thinning to an obtuse apex, usually connate for more than half of their length; sheath thin, splitting, deciduous; terminal buds 1/12–1/8 in. long.

STEMS. Stiff, terete, younger ones green to yellowish, glaucous, smooth or nearly so, longitudinally lightly furrowed, about 1/6 in. thick; older ones darker, cinereous, cracked, and fissured.

RANGE. The species is found on dry soils of gravelly plains, rocky hillsides, old fields and pastures, and calcareous slopes. Central and western Texas, southwestern Oklahoma, and Mexico in Nuevo León and San Luis Potosí.

VARIETY. A variety has been named *E. antisyphilitica* var. *brachycarpa* Cory, with shorter pistillate spikes, $\frac{1}{4}$ in. long and broad or less; seed about $\frac{1}{8}$ in. wide, distinctly 3-angled and included. It occurs in Kent and Bexar counties, Texas.

REMARKS. The genus name, *Ephedra*, is from a Greek word for "horsetail." The species name, *antisyphilitica*, refers to the plant's use in domestic medicine in treatment of syphilis. However, related species are officially listed for this use. The plant is not to be confused with *E. pedunculata*, which has a clambering habit and long-stipitate anthers. A plant described as *E. texana* Reed (Reed, 1935) is considered to be a synonym of *E. antisyphilitica*.

Vine Ephedra

Ephedra pedunculata Engelm.

FIELD IDENTIFICATION. Shrub with vinelike habit, clambering on other shrubs or trailing on the ground. Stems attaining a length of 20 ft. Branches alternate, or sometimes whorled at the nodes, rather lax, terete, to $\frac{1}{8}$ in. thick; internodes $\frac{3}{8}$–$2\frac{3}{4}$ in.

FLOWERS. Staminate spikes borne at the nodes, solitary or paired, lanceolate to elliptic, $\frac{1}{6}$–$\frac{1}{3}$ in. long; peduncles to $\frac{1}{2}$ in.; bracts 6–12 pairs, $\frac{1}{15}$–$\frac{1}{7}$ in. long, $\frac{1}{5}$–$\frac{1}{8}$ in. broad, membranaceous, yellowish to reddish; perianth slightly longer than bracts; staminal column about one half exserted, $\frac{1}{8}$–$\frac{1}{5}$ in. long, anthers 4–6; pistillate spikes solitary or paired at the nodes, elliptic, $\frac{1}{4}$–$\frac{2}{5}$ in.; peduncles $\frac{1}{25}$–$\frac{4}{5}$ in.; bracts usually in 3–6 pairs, ovate, inner ones fleshy and reddish.

FRUIT. Seeds in pairs, conspicuously exserted, trigonous, light brown to chestnut, smooth, $\frac{1}{6}$–$\frac{2}{5}$ in. long, $\frac{1}{12}$–$\frac{1}{7}$ in. wide; tubillus with a contorted limb.

LEAVES. Binate, connate the greater part of their length, $\frac{1}{25}$–$\frac{1}{8}$ in. long, apex obtuse, dorsally thickened; sheath membranaceous, eventually splitting.

TWIGS. Young ones grayish green, glabrous, with a few longitudinal furrows; older ones green to yellowish green, somewhat fissured and cracked; terminal buds $\frac{1}{25}$–$\frac{1}{8}$ in. long, apex gradually narrowed.

RANGE. In Texas known from Zapata, Sutton, Val Verde, Uvalde, and Brown counties. In Mexico in Tamaulipas, Nuevo León, Coahuila, Chihuahua, San Luis Potosí, Zacatecas, and Durango.

REMARKS. The genus name, *Ephedra*, is from the Greek word for "horsetail." The species name, *pedunculata*, refers to the pedunculate flowers.

GRASS FAMILY (*Gramineae*)

Giant Cane
Arundinaria gigantea (Walt.) Chapm.

Giant Cane

Arundinaria gigantea (Walt.) Chapm.

FIELD IDENTIFICATION. Canes arborescent, woody, branched above, to 30 ft.

FLOWERS. At long, indefinite periods of time. Borne in panicles composed of large, flattened spikelets from branches on the old wood, or, in one form, with spikelets on nearly leafless shoots directly from the rhizome; spikelets 1½–2½ in. long, mostly 5–15-flowered; glumes 1 or 2, distant, the first sometimes absent; lemmas pubescent or glabrous, acute or acuminate, many-nerved, about ¾ in. long; paleae scarcely shorter than the lemmas, prominently 2-keeled; lodicules 3; stamens 3; styles 2 or 3, stigmas plumose.

FRUIT. Grain furrowed, free, enclosed in flattened spikelets.

LEAVES. Smallest somewhat crowded at ends of branches, oblong-lanceolate to linear, acuminate at apex, rounded or cuneate at base, serrulate, blades 4–12 in. long, ⅚–1⅓ in. wide.

SHEATHS. Ciliate on margin, fimbriate at apex, otherwise glabrous.

RANGE. In low grounds, or in waters of ponds, rivers, and swamps. Texas, Oklahoma, Arkansas, and Louisiana; also eastward to Florida, north to southern New Jersey and west to southern Missouri.

VARIETIES. Some authors segregate this species into two forms and give the name of *A. tecta* (Walt.) Muhl. to plants which bear the spikelets on short, leafless, or almost leafless, shoots directly from the rhizome. The typical form bears the spikelets on branches from old wood. Seemingly these differences are only the result of environmental conditions. (Gleason, 1952, Vol. 1, p. 101)

REMARKS. The genus name, *Arundinaria*, is derived from the word *arundo*, the Latin name for "reed." The species name, *gigantea*, refers to the "gigantic" stature of the plant. Another vernacular name is Southern Cane. Growing in colonies to form the canebrakes of the South. Indians and early settlers used the seeds for food. The young shoots were used as a pot herb in the same manner as the bamboo shoots of tropical countries. Livestock eat the shoots and fruit. Canes also used for fishing poles and for mats.

PALM FAMILY (*Palmae*)

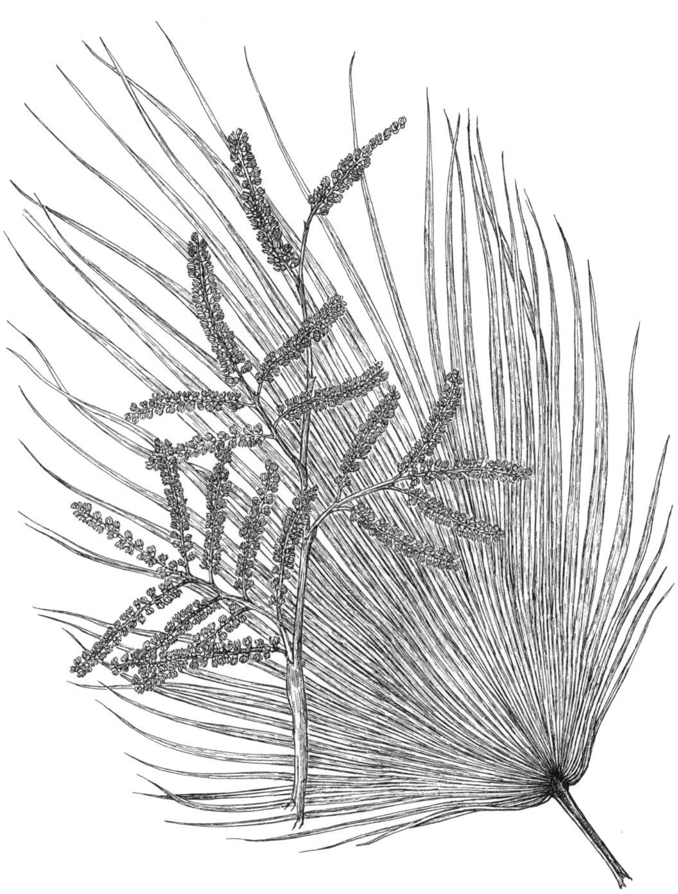

Texas Palm (Palma de Micharos)
Sabal texana (Cook) Becc.

Texas Palm (Palma de Micharos)
Sabal texana (Cook) Becc.

field identification. Tree attaining a height of 20–48 ft, with a trunk diameter of 12–32 in. The large fan-shaped leaves form a dense rounded crown, with dead leaves often pendent and persistent on the trunk.

flowers. March–April, spadix commonly drooping or spreading, 7–8 ft long, decompound, flattened, the branchlets 2–4 in. long, evenly and rather densely flowered, borne from numerous acuminate spathes which become brown, woody, and persistent; flowers perfect, white or greenish, fragrant at maturity, glabrous, sessile, ⅙–⅕ in. long; calyx cup-shaped, striate; sepals 3; petals 3, valvate in the bud, much exceeding the calyx, nearly united, narrow ovate-oblong, acute, concave, erect; stamens 6, those opposite the corolla-lobes generally longest, filaments united below into a shallow cup and adnate to the corolla; anthers ovoid, introrse, 2-celled, opening longitudinally; ovary superior, 3-carpellate, 3-lobed, 3-celled; style elongate, 3-lobed, truncate and stigmatic at apex; ovule basilar and erect.

fruit. Berry maturing in summer, short-peduncled, globose, or oblate, lobed, about ½ in. by ¼ in. (or sometimes ¾–1 in. across), sides rounded, dull black; flesh thin, dry, sweet; seeds chocolate-brown, shiny, bottom truncate, apex convex, ½–⅗ in. across, seed coat thin, hilum conspicuous and orange-colored, micropyle lateral and prominent.

leaves. Persistent, alternate, blade fan-shaped, parallel-veined, lustrous, yellowish green, coriaceous, 4–6 ft long, and 4–7 ft wide, palmately divided almost to the middle into narrow folded segments, each segment 2-cleft at apex, 1¾–2 in. broad at base, scarcely, if at all, hanging, margins filamentose; ligule adnate to the rachis, acute and concave; petiole stout, straight or curved, usually as long as the blade or longer, dorsally ridged, ventrally rounded and concave, margin acute and entire, with sheaths of brown fiber.

bark. Gray to reddish brown, marked with the prominent leaf scars, and with long persistent sheaths, often with pendent dead leaves toward upper part of trunk; old leaf bases (boots) are either shed or retained.

wood. Inner wood pale, reddish brown, outer rind lighter colored, soft, light. Old trees are generally smooth.

RANGE. Arthur Carl Victor Schott, who was a surveyor in the United States–Mexico boundary survey, spent 1852 on or near the lower Rio Grande River of Texas and remarked (Emory, 1857, Vol. 1, p. 44), "It [Texas Palm] extends along the Rio Bravo [Rio Grande] up to about 80 miles from the Gulf." Since Schott's explorations much of the lower Rio Grande area has been converted to agriculture and the clearing has destroyed most of the palms.

In 1925 Robert Runyon, well-known botanist of Brownsville commented: "The Frank Rabb Ranch contains the remaining largest and most beautiful group of palms of about 100 acres. Above the Rabb Ranch only a very few scattered palms are found on the Texas side of the Rio Grande, and there are none above Santa Maria, which is about 60 miles from the Gulf of Mexico.

"On the Mexican side of the Rio Grande the distribution is confined to a narrow strip along the river. A very pretty grove is located on the Santa Rosa Ranch six or seven miles below Matamoros, and a few scattered palms grow along the banks of the river as far up as the Huasteca Ranch. There is a large grove extending about a mile along the river. With the exception of some trees in Matamoros the above lines indicate about all the palm growth on the Mexican side of the Rio Grande." (Small, 1927, pp. 140–142)

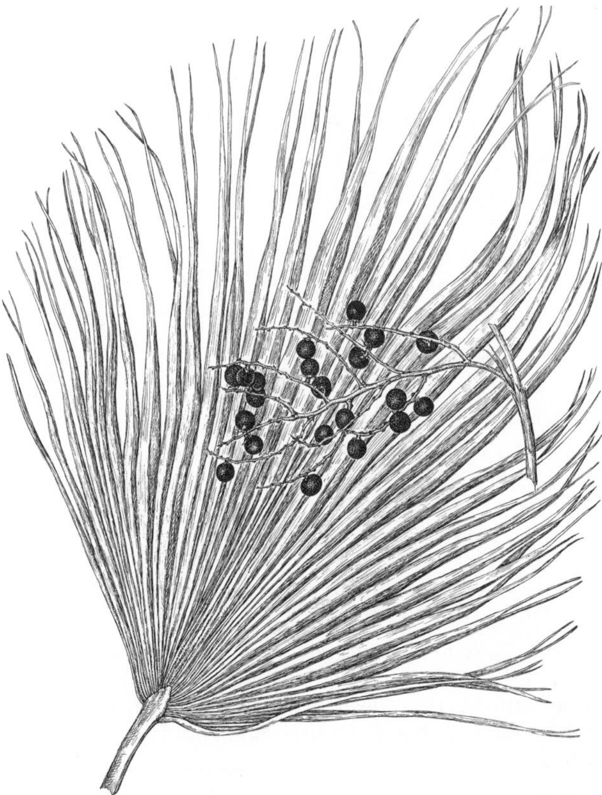

LOUISIANA PALM
Sabal louisiana (Darby) Bomhard

The author has had the privilege of visiting the Rabb Ranch with Mr. Runyon in 1954. A much smaller part of the palm grove remained at that time. It would be most unfortunate if all of these beautiful groves of picturesque palms were destroyed entirely. Their preservation would make an area of great value and interest to visitors. Such an area could readily be converted into a tropical botanical garden which would greatly benefit the people of Brownsville and vicinity. Besides the palms, the following trees and shrubs are found there: *Pithecellobium flexicaule, Pithecellobium pallens, Celtis pallida, Fraxinus berlandieri, Mimosa berlandieri, Leucaena pulverulenta, Leucaena glauca, Malpighia glabra, Zanthoxylum fagara, Condalia obtusifolia, Bumelia angustifolia, Diospyros texana,* and *Salvia greggii.*

REMARKS. The genus name, *Sabal,* is of uncertain origin. The species name, *texana,* refers to its habitat in Texas. Some venacular names for the tree are Rio Grande Palmetto, Palma Real, and Palma de Micharos. The fruit, known as "micharos," is occasionally sold in the markets of Brownsville and Matamoros. Chair seats and thatching for roofs are made from the leaves. The trunks are sometimes used as posts or for wharf piles. Many southwestern Texas towns use the tree as an ornamental in parks and along streets. The growth is usually slow, a tree reaching maturity in 20–35 years, and becoming rather hardy when mature. Propagation from seed is rather easy, but transplanting is difficult.

LOUISIANA PALM

Sabal louisiana (Darby) Bomhard

FIELD IDENTIFICATION. An arborescent palm often confused with *S. minor* when in juvenile form. The trunk averages 3–6 ft tall (rarely, to 18 ft), usually exhibiting three zones: (1) a region of roots at the base, but occasionally an additional root development occurs fairly high on the trunk, indicating some previous high water level; (2) a narrow girdle of grayish brown, rough bark; (3) a boot area below the leaf crown. Trunk diameter (bark only) rarely more than 27 in., and usually less.

FLOWERS. June–July, or even delayed until the following spring, borne in 4–6 spadices, stiff, erect; spathes 20 or more, long-pointed, tubular, overlapping, covering the length of the axis of the inflorescence, upper ones sterile, basal ones fertile; inflorescence thrice compound, some as long as 3 ft; some of the lower panicles may flower but the rest may be abortive.

Individual flowers white, sessile, 1/5–1/4 in. high, spirally placed above the rachillae at rather regular intervals, spaced several millimeters apart. Subtended by two unequal bracteoles, the base of the smaller one being partially enclosed by the larger; calyx about 1/12 in. high, cylindric and thick below, 3-angled, with 3 short, triangular, unequal, slightly carinate, thin, nerved

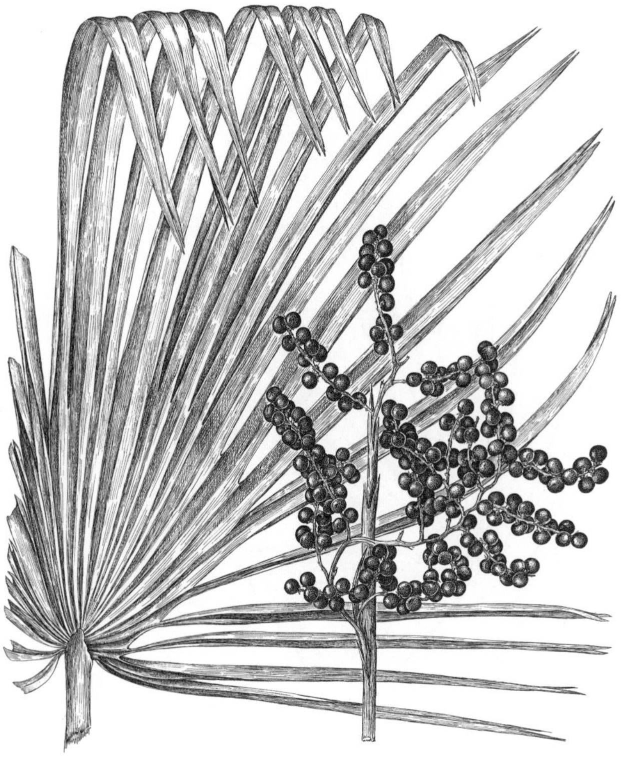

Dwarf Palm
Sabal minor (Jacq.) Pers.

the two boot halves erect or ascending, persistent; an interesting character of this species is the peculiar collapse of the dying leaves at the juncture of the petiole and blade, giving a half-closed-umbrella effect.

RANGE. In Louisiana along Bayou Sauvage, north of Chef Road; Frenier Beach, west shore of Lake Pontchartrain; Bayou Bienvenue; Bayou Vermilion; Bayou des Allemands; east of Berwick Bay on bayous Black and Chacahoula, and south of it on Bayou Shaffer. In Texas a stand was discovered by the author containing about 20 plants, one with a trunk 18 ft in height. This stand is located 8 miles west of Brazoria on the Brazoria–Cedar Lane cutoff road; other locations of Texas stands are on the bottom lands of the Lavaca River west of Lolita; 4 miles south of Cleveland within sight of the bridge (U. S. Highway 59) that crosses the East Fork of the San Jacinto River; north of Rockport, eastern shore of Copano Bay; on the Blanco River south of Blanco; on Hog Bayou, about 8 miles south of Port Lavaca; isolated stands are also known from Alabama and Florida.

REMARKS. The genus name, *Sabal,* is of uncertain origin. The species name *louisiana,* is for the state of Louisiana, where it is found in greatest abundance. This palm has long been unrecognized as a distinct species and has been linked historically with both *S. minor and S. palmetto*. Diverse opinions still exist as to its relationship. The fullest discussion of the history of the nomenclature is given by Bomhard (1935).

lobes; corolla more or less united with the stamens into a short pseudotube at base; petals 3, broadly ovate, 1/8–1/7 in. high, 1/2 in. broad at base, thin, involute, minutely serrulate, thickened and hooded at apex, auricled at base, 5–7-nerved; stamens 6, the alternate shorter than the opposite that are adnate to the petals; filaments subulate-lanceolate, dorsiventrally flattened; anthers bright yellow, introrse, short-saggitate, 1/25 in. long, anther sacs somewhat unequal; pistil comprised of 3 carpels, 1/8–1/6 in. long, about 1/25 in. or more broad at the enlarged ovary base, stylar portion 3-angled, apex truncate.

FRUIT. Ripening in November, brown to black, suborbicular, averaging about 2/5 in. in diameter; seeds reddish brown, sublustrous, enclosed in a thin integument, micropyle lateral.

LEAVES. Palmate, comparatively thin, bluish green, attaining a width of 80 in.; rachis winged below and supporting the lower one third of the blade, then deeply split into 2 halves beyond; segments 36–50, splitting the outer half or two thirds of the blade, acuminate, rather stiff, the apices usually bifid and flaccid, margins of younger segments often with threadlike fibers; hastula flat, platelike, asymmetrical, averaging about 1 1/2 in. long; petiole unarmed, longer than the blades, upper surface concave, lower surface rounded, margin very sharp and faintly denticulate toward the base;

Dwarf Palm

Sabal minor (Jacq.) Pers.

FIELD IDENTIFICATION. A palm without a trunk, the leaves fanlike and arising in a crown from a subterranean rootstock.

FLOWERS. May–June, on the coast of the Gulf of Mexico, spadix 2–8 ft high from a long spathe, erect or ascending; secondary flowering panicles 4–11 in. long from sheaths 1–5 in. long which are split on one side, striate and acuminate; ultimate flowering divisions 4–20, 3–6 in. long; flowers small, numerous, perfect, sessile or nearly so, 2/16–3/16 in. long, subtended by one or more minute bracts; calyx-lobes 3, lobes unequal, about 1/25 in. long, rounded; petals 3, erect, white, elliptic, concave, apex obtuse or rounded, about 1/12 in. long; stamens 6 (sometimes fewer); filaments flattened, broadened at base and adnate to the corolla, slightly longer than the petals; anthers yellow, ovate-saggitate, about 1/25 in.; pistil included, columnar, stigma small and truncate; gynoecium of 3 carpels, usually developing only one.

FRUIT. Drupe on a peduncle 1/8 in. long or less, subglobose, slightly broader than long, 1/4–1/3 in. in diameter, black, remnants of the short style basal; seed coat thin and membranous; seed solitary, white, lustrous, hard, bony, flattened at base.

LEAVES. Clustered from the base, flabellate (fanlike), suborbicular, 2–8 ft in diameter, pale green or glaucous; segments shallowly cleft at apex, almost as long as the leaf or much shorter, mostly entire on the margin or more rarely with filiferous threads, midribs very short, petioles shorter than the blades, ligules ⅜–¾ in.

RANGE. Wet alluvial ground in Texas and Louisiana; north to Oklahoma, Arkansas, and North Carolina. Abundant in the river bottoms and swamps of southern Louisiana, and in Texas west to the valley of the Colorado River.

REMARKS. The genus name, *Sabal*, is of obscure meaning, and the species name, *minor*, refers to this palm's dwarf, or trunkless, stature. Also known locally under the names of Dwarf Palmetto, Blue Palm, Blue Stem, and Swamp Palm. It is occasionally browsed by cattle. Although rather attractive, it is usually not cultivated for ornament.

Much botanical controversy has been waged concerning the relationship between the Dwarf Palm and the Louisiana Palm, S. *louisiana* (Darby) Bomhard. For a discussion of this problem the reader may refer to the description of the Louisiana Palm.

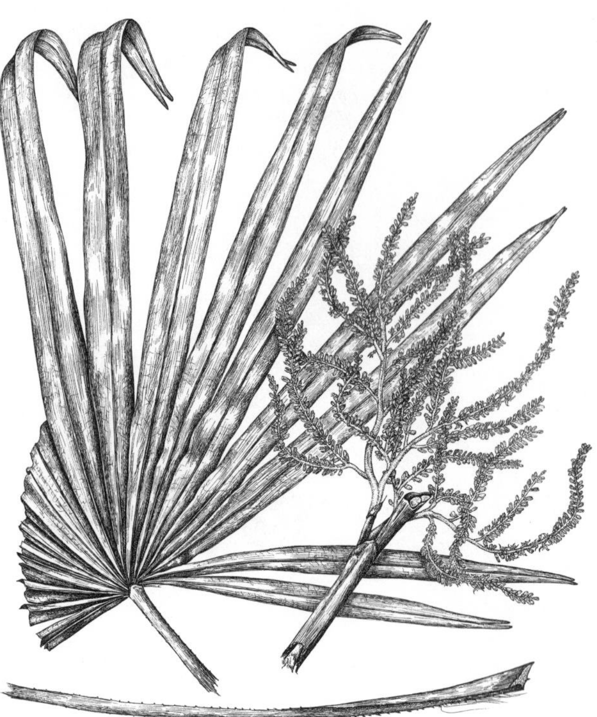

SAW PALMETTO
Serenoa repens (Bartr.) Small

SAW PALMETTO

Serenoa repens (Bartr.) Small

FIELD IDENTIFICATION. Usually 3–7 ft, with creeping or horizontal stems often rooting beneath, occasionally becoming a small tree, attaining a height of 25 ft, with an erect, simple, or branched trunk.

FLOWERS. Appearing from spring to early summer. Spadix erect or spreading, elongate, zigzag, densely pubescent, paniculately branched, shorter than the leaves; flowers usually numerous, each subtended by a bract and 2 bractlets, perfect, white, fragrant, ⅙–¼ in. long; calyx cup-shaped, about 1/25 in. high; sepals 3, partially united; petals 3, ⅙–⅛ in. long, valvate, keeled within, oblong; stamens 6, unequal, filaments adnate to the corolla below, anthers introrse; carpels 3, free at the base, 3-angled, each narrowed into a slender style with a minute stigma.

FRUIT. Drupe ovoid-oblong, black to dark brown, ⅗–1 in. long, pericarp thin and fleshy; seeds subglobose, slightly flattened on the ventral side, erect, solitary; fruiting panicles sometimes weighing as much as 9 lb.

LEAVES. Clustered, erect or ascending, fan-shaped, suborbicular, 1–2½ ft broad, base cordate, rather stiff, green or glaucous, much cleft, the segments each 2-cleft; petioles usually longer than the blades, armed with sharp, rigid, recurved teeth.

RANGE. Usually on sandy pinelands, prairies, hammocks, or dunes, seemingly adjusting to acid or alkaline, fresh or brackish waters in the soil. Eastern Louisiana; eastward to Florida and northward to South Carolina.

MEDICINAL USES. The following is a description of the drug *Fluidextractum Serenoae* N.F. quoted from Wood and Osol (1943, pp. 971-972):

"'Serenoa is the partially dried, ripe fruit of *Serenoa repens* (Bartram) Small (Fam. *Palmae*). Serenoa contains not more than 1 per cent of foreign organic matter, and yields not more than 2 per cent of acid-insoluble ash. Serenoa contains not more than 15 per cent of moisture nor more than 10 per cent of immature fruits which are not well filled, and whose surfaces are not creased or wrinkled.' (*National Formulary*, 7th ed.)

. . . .

"Constituents.—By pressure saw palmetto berries yield about 1.5 per cent of a brownish-yellow to dark-red oil. Shermann and Briggs (*Ph. Archiv.*, 1899, p. 101, and *Ph. Rev.*, 1900, p. 217) found this to be composed of about 63 per cent of free fatty acids and 37 per cent of ethyl esters and of these acids; there was no alkaloid present. Mann (*Amer. J. Pharmacy*, 138:517, 1916) concluded that the so-called volatile oil was a mixture of the ethyl esters of the various fatty acids present due to the combination with the ethyl alcohol which is used as a preservative in shipping them. The fatty acids are caproic, caprylic, capric, lauric, palmitic and oleic. . . .

"Uses.—Saw palmetto appears to exert a stimulant action upon the mucous membrane of the genito-urinary tract, similar to, but milder and less irritant, than cubeb or copaiba. It is used chiefly in chronic and subacute cystitis. It has been especially recommended in cases of enlarged prostrate of old men; it is not probable that it has a direct influence upon the prostatic gland itself but there is clinical testimony as to its value and it probably acts by reducing the catarrhal irritation and the relaxed condition of the mucous membrane of the bladder and urethra, which are often present in prostatic hypertrophy."

REMARKS. The genus name, *Serenoa,* is in honor of Sereno Watson (1826–1892), distinguished American botanist. The species name, *repens,* refers to the creeping habit of this species. The plant is relatively hardy and will withstand a temperature of 10°F. The leaves are shipped North for Christmas decorations in considerable amount. The flowers are a significant source of honey, and the stems are a source of tannic acid extract.

LILY FAMILY (*Liliaceae*)

Aloe Yucca
Yucca aloifolia L.

Aloe Yucca
Yucca aloifolia L.

FIELD IDENTIFICATION. A plant to 10 ft, with a simple or branched stem.

FLOWERS. In a stout, conic, showy, white panicle 1–1½ ft high and 6–10 in. wide; panicle usually set ¼–½ within the upper leaves; peduncles and pedicels minutely puberulent or glabrous; individual flower pedicels ½–1 in. long, mostly flexed downward; sheathing bracts ¼–½ in. long, lanceolate-ovate, acuminate; mature flowers not spreading, petals 3 and sepals 3, approximately the same color, shape, and size, semiconcave, oblong-elliptic, apex acute or obtuse, base slightly rounded or cuneate, many-veined, waxy-white, sometimes purplish at base, 1–1¼ in. long, ⅓–½ in. wide; pistil with lower style stout, columnar, greenish, ½–¾ in. long, of 3 united carpels; stigmatic portion about ⅛ in. long, whitish, 3-angled; stamens 6, erect or nearly so, surrounding the pistil, about ½ in. long, white, flattened, puberulent; anther sessile, yellow, falcate, 1/25–1/16 in. long.

FRUIT. Capsule 1½–2 in. long, ¾–1 in. wide, oblong or slightly larger toward the apex, apex abruptly obtuse or rounded, 3-parted by broad, flat troughs between the valves, glabrous, light green at first, black to purplish at maturity, pulpy, indehiscent.

LEAVES. Dagger-like, closely spiraled, stiff, crowded, spreading, deflexed and usually deciduous below but sometimes persistent, flattened or slightly concave, light green, 12–20 in. long, 1–1½ in. wide; widest at the middle or slightly below; gradually narrowed toward the apex and base; apex stiff, ending in a stiff, sharp, dark brown spine; leaf margin sharp with minute, erose teeth.

STEMS. Simple or branched, usually 3–6 ft high, rarely to 10 ft, leaves on upper stems spreading and stiff, on older plants lower leaves persistent, but reflexed and brown.

RANGE. Usually on sand dunes or shell mounds close to the coast. Louisiana; east to Florida, and north to Virginia. Also in Mexico and the West Indies. Escaped from cultivation on the Texas coast.

VARIETIES AND FORMS. A number of horticultural varieties and forms are known as follows:

Margined Aloe Yucca, *Y. aloifolia* forma *marginata* Bommer, has yellow-margined leaves.

Stripleaf Aloe Yucca, *Y. aloifolia* forma *tricolor* Bommer, has yellow or white stripes in the leaf center and yellow leaf margins.

Menard Aloe Yucca, *Y. aloifolia* forma *menardii* Trel., has very narrow leaves with a yellow or white stripe in the leaf.

Branched Aloe Yucca, *Y. aloifolia* forma *draconis* (L.) Engelm., has a trunk branched above, leaves to 2 in. across, more flexible, recurved, and not rigid.

REMARKS. *Yuca (Yucca)* is the native Haitian name for the *Manihot* genus, erroneously applied to this group of plants. The species name, *aloifolia,* refers to the *Aloe*-like leaves.

Propagation of Aloe Yucca may be made from seeds, offsets, or cuttings from stems or rhizomes. The leaves were used in pioneer days for string or twine to hang up cured meats.

Faxon Yucca

Yucca faxoniana (Trel.) Sarg.

FIELD IDENTIFICATION. Tree 6–40 ft, with a diameter 8–24 in. Either unbranched or with a few short branches at the top. The long leaves are borne in a dense head 4–7½ ft broad, eventually contributing to a dense pendent thatch which clothes the flank below.

FLOWERS. Maturing in April, borne in glabrous densely flowered panicles 3–4 ft long, exceeding foliage one third to one half its length; branchlets of panicle 35–40; pedicels thin, glabrous, drooping, ⅓–1¾ in. long, bracts very variable in size, basal ones 6–12 in. long, decreasing in size along the branchlets upward, with those at the pedicel base ½–3 in. long, shape of bracts varying from broad-triangular to lanceolate, with apices acuminate to acute; texture eventually thin, papery, dry, and brittle; flowers 2–3⅓ in. long, the 6 segments white or greenish, concave, thin, widest above the middle, the 3 segments of the outer rank narrower, segments united from ⅛–½ in. at base; stamens 6, usually shorter than the ovary, filaments hairy above the middle, anthers ⅙–⅕ in. long; pistil 1¾–2½ in. long; ovary four to five times longer than broad, conspicuously ridged, usually fusiform, sometimes asymmetrical, style ⅙–⅕ in., slender, elongate, stigmatic openings wide.

FRUIT. Maturing in early summer, 3–4 in. long, 1–1-½ in. wide, at first brown or orange-colored, but later blackening, somewhat angled or very smooth, apex usually tapered and ending in a hooked beak, base often with remains of persistent perianth, flesh thick and succulent, rather bitter, 3-celled; seeds about ¼ in. long and ⅛ in. thick.

LEAVES. Borne in a radiating mass, length 2–4 ft, width 2–3 in., rather abruptly narrowed at the base, then gradually widened to above the middle, and then tapered to the apex which bears a short, stout, dark spine; margins entire with gray or brown fibres which sometimes form a mass at base; surfaces concavo-convex with the upper surface somewhat flattened and the lower rounded, rigid, smooth, dark green.

BARK. Often covered with dead reflexed leaves, dark reddish brown, breaking into small loose scales.

RANGE. On the high desert plateau of western Texas. Known from Jeff Davis, Presidio, Culberson, and Hudspeth counties. In the vicinity of Sierra Blanca and Van Horn, Texas. Also in Chihuahua, Mexico.

REMARKS. The genus name, *Yucca,* is from the native Haitian name. The species name, *faxoniana,* honors Charles Edward Faxon (1846–1918), artist of Sargent's *Silva of North America* (1891–1902). Known under the vernacular names of Spanish Bayonet and Spanish Dagger.

Carneros Giant Yucca

Yucca carnerosana Trel.

FIELD IDENTIFICATION. Large robust plant attaining a height of 6–15 ft and a trunk diameter of 6–15 in. The trunk usually solitary, but sometimes with a few

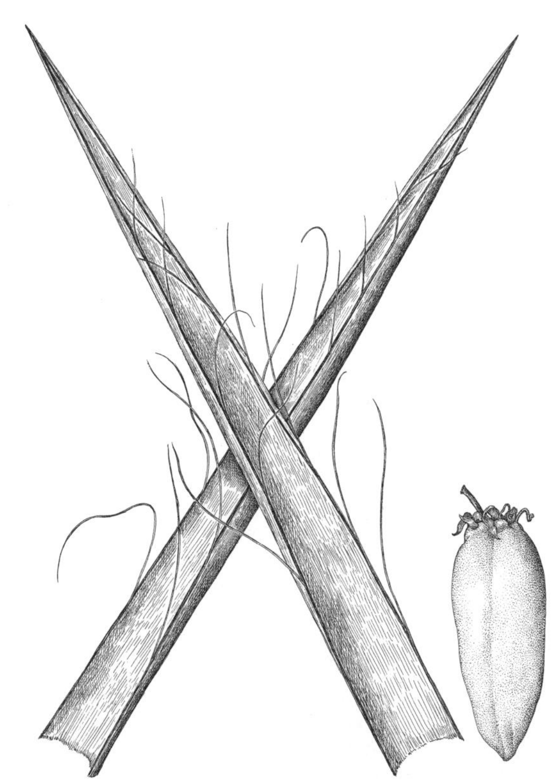

FAXON YUCCA
Yucca faxoniana (Trel.) Sarg.

Carneros Giant Yucca
Yucca carnerosana Trel.

short branches. The dead reflexed leaves form a thatch often to the ground. The heads of living leaves are very handsome and symmetrical. The flower panicle is ball-like and set at a distance above the topmost leaves. This is in contrast to the Torrey Yucca, which has the flower panicle set down among the leaves, and the heads of leaves usually untidy and unsymmetrical. Also the Torrey Yucca is generally found on the desert flats, and the Carneros Giant Yucca on higher hillsides.

FLOWERS. Inflorescence, including scape, 4½–6 ft tall, flower panicle nearly spherical, usually set about 12 in. above the upper leaves; branchlets 15–30, erect to ascending; pedicels ¾ in. or more in length; bracts variable in size, those near base of scape 9–12 in. long, 2–2½ in. broad at base, those on middle part of scape 6–10 in. long, graduated to smaller ones at the apex of the scape and finally those at base of branchlets and pedicels only ⅜–1 in. long, triangular to oblong, at the apices acute to short-acuminate and often spinescent, texture of bracts leathery to thin, dry, brittle, and paper-like. Flowers proper 2–3¾ in. long, segments 6 (3 petals and 3 sepals), united at base, stamens 6, anthers ⅕–⅓ in.; pistil 2⅓–2¾ in.; ovary about five times as long as broad, about ⅓–½ in. in diameter, symmetrical, style about ⅓ in. long, stigmatic openings variable in size.

FRUIT. Capsule reddish brown to black, 3–4½ in. long, about 1½ in. in diameter, apices abruptly or gradually narrowed, ovoid to oblong, 3-celled, seeds numerous.

LEAVES. Heads of leaves 3–4½ ft high, usually considerably wider, leaf blade 1½–3¾ ft long, averaging about 2½ ft, at widest point 2–3 in. across, gradually widened just above the base to near the middle, and then gradually narrowed to a stout spine ⅜–½ in. long, concavo-convex most of the length.

RANGE. Usually on limestone foothills approaching higher mountains. In Texas confined to Brewster County at altitudes of 2,700–6,300 ft. Best developed in the area between the Chisos and Del Carmen mountains. In Mexico in Zacatecas, Coahuila, and San Luis Potosí.

REMARKS. The genus name, *Yucca*, is from the native Haitian name, and the species name, *carnerosana*, is from Carneros Pass, Coahuila, Mexico, where the type specimen was found. Other vernacular names are Giant Dagger and Palma Samandoca. In Mexico the trunks are used for palisade construction, fences, or walls of huts, and the leaves for roof thatch. The leaves are sometimes used for the production of "palma istle fiber." Various parts of the plant are eaten. It is reported that the Mexicans and Indians split open the trunks so that cattle can eat the soft interior. Cattle also eat the flowers and fruit. The Indians are known to have eaten the immature flowers boiled or roasted, as well as the bittersweet fruit. Often planted along highways in Trans-Pecos Texas for ornament.

Datil Yucca

Yucca baccata Torr.

FIELD IDENTIFICATION. Plants acaulescent, solitary or clump-forming, less often with a short prostrate or ascending trunk. The leaves in spreading rosettes.

FLOWERS. March–June, panicle dense, upright, 1–3 ft in length, usually only 4–8 in. above the uppermost leaves, scape stout and heavy, glabrous or somewhat pubescent, green to reddish or purplish, becoming woody; flowering branches 11–18 in number; pedicels ¼–1⅔ in. long, bracts varying from 1–10 in. long, those on scape the longest, shape mostly lanceolate-oblong or triangular-ovate, usually spinescent, papery and brittle at maturity; flowers pendulous, very variable in size on individual plants, 2–5¼ in. long, usually open-campanulate, sometimes barely open, segments somewhat united at base, white; petals 3, sepals 3, lanceolate to oblanceolate, somewhat thickened; stamens 6, ¾–1¾ in. long, about equaling the ovary; filaments fleshy, pubescent, somewhat flattened toward the base; anthers short, sagittate, ⅕–¼ in. long; ovary oblong, sometimes curved or distorted, 3-lobed and 3-fissured (or 6-lobed); style about ¼ in. long, ending in 3 short and broad emarginate stigmas.

FRUIT. Capsule pendent, fleshy, indehiscent, very

variable in size, 4–9 in. long, 2–3 in. in diameter, conic-ovoid to oblong, occasionally oval, cylindric, surfaces smooth and rounded or sometimes with constrictions or lobes, green to dark purple, often long-beaked, edible, early deciduous; seeds numerous, thick, rugose.

LEAVES. Leaf pale bluish green, rosettes 1½–2¾ ft high, 3–3¼ ft wide, usually on the ground, or occasionally on short unbranched stems, thick and rigid, spreading, sometimes twisted, length of mature blade 20–40 in., width 1¼–2½ in., concave and convex for entire length, surfaces roughened; margins with recurved, broad, coarse fibers ⅝–4 in. long, apex with a short, stout, stiff, brown spine.

RANGE. Open dry plains and mesas at altitudes of 2,000–8,000 ft. Trans-Pecos Texas and New Mexico; west to California, north to Colorado, Utah, and Nevada, and south to Chihuahua, Mexico.

VARIETIES. Thornber Yucca, *Yucca baccata* var. *brevifolia* (Schott) Benson & Darrow, has been distinguished from the species by the stems being solitary, or several, the longer ones 3–6 ft and often branched; leaf fibers finer and more threadlike; inflorescence stalkless or on a stalk to 12 in. In New Mexico in Hidalgo County, also in Arizona and Sonora, Mexico.

REMARKS. The genus name, *Yucca,* is from the native Haitian name. The species name, *baccata,* refers to the berry-like fruit. Vernacular names are Banana Yucca, Blue Yucca, Dátil, Spanish Bayonet, Palma, Palma Criolla, and Hosh-Kawn. The fibers of the leaves are used by the Indians to make baskets. The fruit is eaten raw, dried, or roasted. A soap substitute made from the roots is known as "amole."

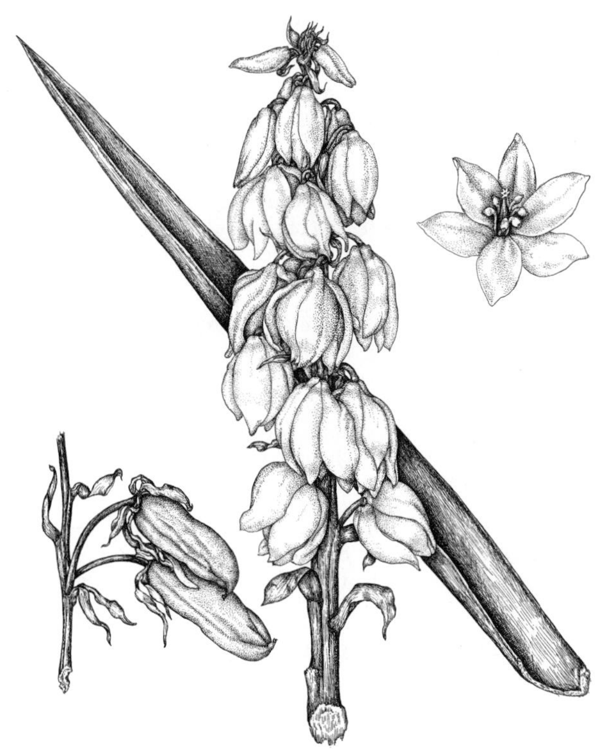

TRECUL YUCCA
Yucca treculeana Carr.

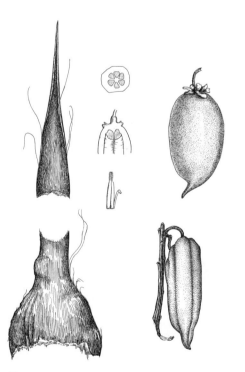

DATIL YUCCA
Yucca baccata Torr.

TRECUL YUCCA

Yucca treculeana Carr.

FIELD IDENTIFICATION. Tree 5–25 ft, with a simple trunk or with a few stout spreading branches at the top, crowned by large symmetrical heads of radiating sharp-pointed leaves. The plant sometimes occurs as a thicket-forming shrub.

FLOWERS. Maturing December–April, borne in a large dense, showy glabrous or puberulous panicle 1½–4 ft long; pedicels ½–3 in.; bracts ovate to lanceolate, often spinescent at apex, varying from 1 in. at base of pedicels to 1 ft at base of main stem, becoming dry, thin, and papery; flowers creamy-white, rather globose, later expanding broadly, the 6 segments ovate to ovate-lanceolate, acute to acuminate at apex, waxy, brittle, thin, 1–2 in. long; stamens 6, filaments slightly papillose above, usually finely and shortly pubescent below, about as long as the pistil; pistil ¾–1⅓ in. long, ovary slender and oblong-cylindric; style very short (⅛–⅖ in.); stigmas 3, abruptly spreading, nearly horizontal at anthesis, deeply lobed.

FRUIT. Capsule indehiscent, 2–4½ in. long, about 1 in. thick, reddish brown or later black, oblong-cylindric, rather abruptly contracted at the acute or acuminate apex, surfaces often with fissures or deeply cleft, filaments and perianth often persisting, heavy and thick-walled, 3-celled, flesh sweetish and succulent; seeds numerous, flat, about 1/16 in. thick, ⅛–¼ in. broad, with a narrow border to the rim.

LEAVES. In large radiating clusters, bluish green, length 2½–4 ft, 1–3½ in. wide, usually straight, concavo-convex, apex acute to short-acuminate, with a brown or black, short, sharp spine, margin entire, rigid, inner surface rather smooth, outer surface scabrous to the touch; dead leaves hanging below the crown and long-persistent.

BARK. Dark reddish brown, on older trunks ¼–½ in. thick, with shallow or deep irregular fissures. The intervening ridges broken into thin oblong plates with small appressed scales.

WOOD. Light brown, spongy, fibrous, heavy, not easily cut.

RANGE. Well-drained hillsides, chaparral regions, or open flats near the Gulf of Mexico. From the shores of Matagorda Bay, Texas, westward and southward along the coast to Brownsville. From San Antonio in Bexar County, westward to the Rio Grande and Pecos rivers. In Mexico in Nuevo León, Tamaulipas, Durango, and Coahuila.

VARIETY. Succulent Trecul Yucca, *Y. treculeana* var. *succulenta* McKelvey, has been described by McKelvey (1938, pp. 80–81), as follows: "Having the habit and foliage of the typical plant. Inflorescence fleshy, succulent throughout, inflorescence proper narrowed below and above, somewhat slender, at anthesis well filled but not crowded; flowers campanulate, especially those of outer row, often much thickened, brittle and concave; fruit unknown. Range confined mainly to Bexar County, Texas, and adjacent regions."

REMARKS. The genus name, *Yucca,* is from a native Haitian name. The species name, *treculeana,* is in honor of A. A. L. Trecul (1818–1896), who in 1850 took the plant to France from Texas. Also known under the vernacular names of Spanish Dagger, Spanish Bayonet, Don Quixote Lance, Pita, Palma Pita, Palma de Dátiles, and Palma Loca. The plant is a handsome ornamental for use in central or coastal Texas or Louisiana, and is sometimes grown in southern Europe. The leaves are very tough and were used in frontier days for making twine or rope. The blossoms were made into pickles or cooked like cabbage. The spines on the leaves are used by the Mexican people to jab the wound of a snake bite and induce bleeding. In this manner much of the poison is carried away. The Chihuahua Indians fermented the fruit of various species of *Yucca* to make an intoxicating beverage. The trunks are sometimes used for posts, and the leaves for thatch, in making huts. It is also reported that the seeds have purgative qualities.

SCHOTT YUCCA

Yucca schottii Engelm.

FIELD IDENTIFICATION. An arborescent plant attaining a height of 6–20 ft, with a diameter of 8–12 in. The trunk may be solitary or branched several times, and is generally clothed with reflexed dead leaves. Where no leaves are present the trunk is dark brown and roughened by horizontal leaf scars.

FLOWERS. Maturing July–September; panicle large, somewhat overtopping the leaves, erect, tomentose, 1–2¾ ft long; branches of the panicle 8–25, erect, ascending or horizontal; pedicels short and decurved, bracts of pedicels ⅝–1 in. long, narrow, bracts of the scape to 1 ft long; flowers perfect, white, globose when immature, cup-shaped when mature, declined, 1–2 in. long, petals and sepals 6, segments broadest near the middle, broadly lanceolate or elliptic to oblong or obovate, apex acute, pubescent; stamens 6, about two thirds as long as the ovary, pilose at base, anthers about 1/10 in.; pistil ¾–1 in., ovary 3-celled, oblong-cylindric, green, abruptly terminated in a style about ⅛ in., stigmas 3, bifid.

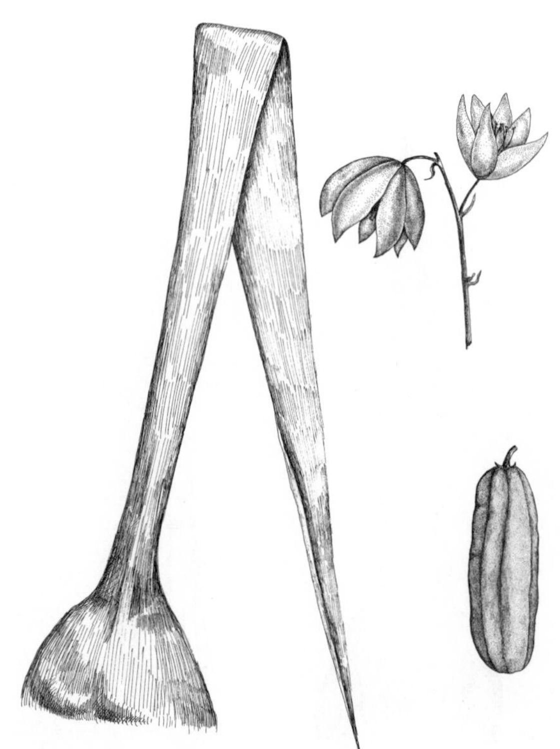

SCHOTT YUCCA
Yucca schottii Engelm.

FRUIT. Maturing October–November, capsule fleshy, succulent, indehiscent, green at first, black or brown later, oblong or less often oval, sometimes constricted at the center with upper part larger, rounded at base, apex tapering then abruptly ending, obscurely angled, sometimes distorted or unsymmetrical, length 3–5½ in., 1–1½ in. in diameter, style and stigmas persistent; seeds about ¼ in. wide and ⅛ in. thick, with a thin marginal rim.

LEAVES. Radiate, length 1½–3 ft, width about 1½ in., straight or occasionally somewhat curved, flattened at the center, apex long-acuminate and inrolled on margins; apical spines brown to reddish, sharp, strong, 1/12–⅙ in. long; base green to brown or reddish, margins sometimes with a red line and generally with no free fibers, surfaces gray to yellowish green, shiny, finely striate.

RANGE. Dry plains, upper grasslands, and woodlands at altitudes of 4,000–7,000 ft. New Mexico in the San Luis and Animas mountains. Also in southern Arizona and Sonora, Mexico.

REMARKS. The genus name, *Yucca*, is from the native Haitian name. The species name, *schottii*, is in honor of Arthur Schott, who helped prepare reports on the vegetation and geology of the United States–Mexico boundary survey under Major William Emory. Vernacular names are Hoary Yucca and Mountain Yucca. A handsome species often cultivated for ornament.

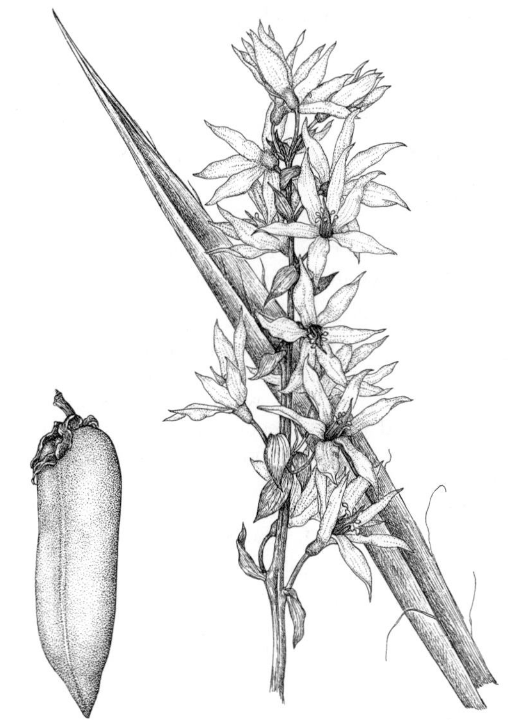

TORREY YUCCA
Yucca torreyi Shafer

TORREY YUCCA
Yucca torreyi Shafer

FIELD IDENTIFICATION. An arborescent plant 3–24 ft, with a simple or few-branched trunk. The crowded radiating leaves usually untidy in appearance, this in part because of the persistent thatch of dead, reflexed leaves on the trunk below.

FLOWERS. Panicle densely flowered, 3–3½ ft long including the scape, extending above the leaves one fourth to one half of its length; branchlets of the panicle 20–35, erect to ascending, ridged and somewhat flattened; pedicels ¾–2 in. long, terete or nearly so, some curved; bracts variable in size, mostly short-acuminate to acute at apex, those on lower part of scape ½–2 ft long, 1¾–2½ in. wide, those on upper parts of scape 4–7 in. long, those at base of pedicels small, narrowly ovate; flowers 1¾–4 in. long, perianth somewhat united at base, creamy-white, or tinged purplish, waxy, segments 6 (3 petals and 3 sepals), concave, apex pubescent, inner series sometimes shorter and broader; stamens 6, papillose to pubescent, rather stout, anthers ⅛–⅕ in. in length; pistil 1–1½ in., style ⅕–⅓ in., gradually broadened into the ovary; stigmas spreading, longer than broad, emarginate.

FRUIT. Capsule 4–5½ in. long, 1¼–2 in. wide, dark brown to black, gradually narrowed toward the apex which is abruptly acute or short-acuminate; surfaces with the primary divisions rather deep, secondary divisions less so, remnants of the perianth and filaments persistent at base; 3-celled; seeds numerous.

LEAVES. Borne in a radiating head, dark yellowish green, straight, rigid, rarely curved, concavo-convex, but less so near base, gradually tapering from just above base to an acuminate apex bearing a rigid, sharp spine ¼–½ in. long, margin with curly to straight persistent fibers, surfaces usually scabrous, length of blade 2–4½ ft.

RANGE. In Texas from the Devil's River area westward, northwest into New Mexico and southwest into Mexico.

FORMS. Small-flower Torrey Yucca, *Y. torreyi* forma *parviflora* McKelvey, has been described by McKelvey (1938, pp. 112–113), as follows: "Similar in habit, foliage and form of inflorescence to the typical plant. Inflorescence few-branched and soon ligneous. Flowers globose, small, 1⅝–2⅜ in. long, greenish-cream, much colored with reds and purple at anthesis, the color effect of the cluster somewhat sordid. Although less common, found practically throughout the range of the typical plant."

REMARKS. The genus name, *Yucca*, is from the Haitian name. The species name, *torreyi*, is in honor of John Torrey (1796–1873), American botanist of Columbia

University. The plant is closely related to the Trecul Yucca.

Moundlily Yucca

Yucca gloriosa L.

FIELD IDENTIFICATION. A low caespitose, simple or few-branched, tree-like dune plant, 6–15 ft and 4–6 in. in diameter, usually with dead leaves to the base.

FLOWERS. In late summer to autumn, panicles showy, narrow, on scapelike simple stalks 2–4 ft long; flowers pendulous, large, perfect, white or occasionally purplish; sepals and petals 3 each, oblong-lanceolate, apex obtuse or acute, 1½–2 in. long, slightly united at base, deciduous; stamens 6, hypogynous, about as long as the ovary; ovary slightly lobed, 6-sided, with 3 spreading stigmatic lobes, 3-celled, ovules numerous.

FRUIT. Stipe short and stout; capsule pendulous, 2–3¼ in. long, about 1 in. in diameter, oblong-ovoid, 6-ribbed, constricted at or near the middle, cuspidate; outer coat black, thin and leathery, indehiscent, pulpy; seeds ¼–⅓ in. long, lustrous, black, thin, flattened, slightly margined.

LEAVES. Numerous, stiff, straight, firm, erect, spreading, broadly linear, apex rigidly spine-tipped, base somewhat constricted, gradually narrowed upward, nearly flat or concave near apex, margin with a few minute teeth when young and a few threads when old, glaucous green, sometimes reddish-tinged.

RANGE. On coastal dunes, known from Breton Island, Louisiana, and Mississippi. Also cultivated in its many forms in south Texas and other Gulf Coast states. Persistent about old gardens and waste places. Probably only indigenous as a species on the coastal dunes of South Carolina, Georgia, and northeastern Florida.

HORTICULTURAL FORMS. The following forms are listed by Trelease (1902):

Y. gloriosa forma *minor* Carr. is a small form.

Y. gloriosa forma *obliqua* (Haworth) Baker has glaucous leaves somewhat twisted to one side.

Y. gloriosa forma *medio-striata* Planchon is a garden sport with a median white stripe on the leaves.

Y. gloriosa forma *robusta* Carr. is rather intermediate between *Y. gloriosa* and *Y. recurvifolia,* with the outermost of the evanescently glaucous, usually slightly plicate, leaves somewhat stiffly recurved.

Y. gloriosa forma *nobilis* Carr. has leaves scarcely plicate, glaucous, the outer ones recurved, sometimes twisted to one side. A small-flowered variation of this form is also known.

Y. gloriosa forma *longifolia* Carr. is very similar to forma *nobilis* except the young leaves are much narrower.

Y. gloriosa forma *plicata* Carr. has more permanently glaucous, usually shorter and hence relatively broader, concave leaves plicate toward the apex.

Y. gloriosa forma *superba* (Haworth) Baker is a cultivated form of forma *plicata,* becoming 1–1½ ft high, with greener leaves.

Y. gloriosa forma *maculata* Carr. is a low garden form, with plicate dark green leaves slightly roughened on the margin. The varietal name refers to a mottled variation of the usual red-tinged flowers. The leaves are more elongate and recurved than in the species.

FERTILIZATION IN YUCCAS. The flowers of the yuccas have their anthers located at a lower level than the stigmas, in such a manner that the pollen cannot be placed on them without being artificially transported.

It is known that certain insects visit some flowers for nectar. But, in the case of the yuccas, the principal visitor is not seeking nectar but is a small moth, *Pronuba yuccasella,* which bores a hole in the ovary of the yucca flower to deposit her eggs. The ovules will not develop and furnish food for her larvae unless fertilized, so the moth gathers pollen in a mass with her palps and introduces it on the stigmas. In this manner the insect benefits by the developing ovules producing food for her larvae. Likewise the plant is assured of production of seeds because only a certain number of ovules serve as food for the larvae.

However, some Yuccas may not be pollinated as described above because the maturity of the flowers does not agree with the life cycle of the adult moth.

REMARKS. The genus name, *Yucca,* is from a native

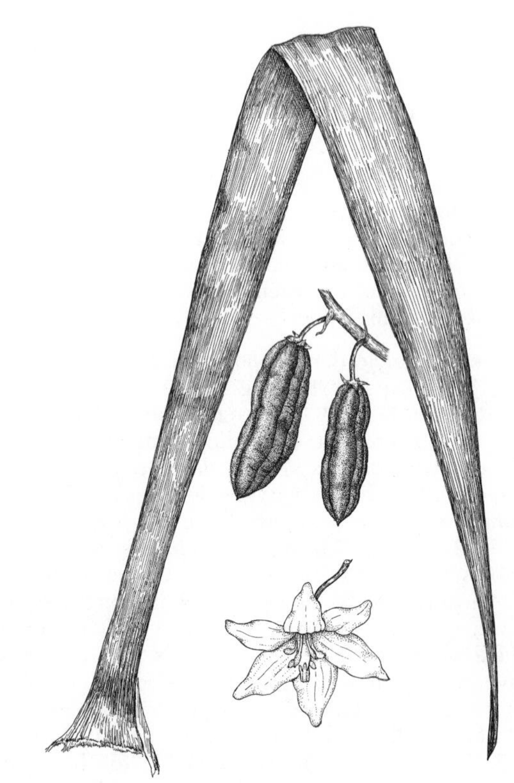

Moundlily Yucca
Yucca gloriosa L.

Haitian name. The species name, *gloriosa,* refers to the beauty of the flowers. Known also under the names of Spanish Bayonet and Sea Island Yucca.

Curve-leaf Yucca

Yucca recurvifolia Salisb.

FIELD IDENTIFICATION. Plant with a leafy stem 3–6 ft, simple or branched.

FLOWERS. Borne in a narrow panicle which is elevated only slightly above the leaves on a scapelike stem; perianth 6-parted; calyx of 3 white or greenish sepals; corolla of 3 similar petals; stamens 6, shorter than the perianth, filaments enlarged above, anthers sagittate; gynoecium of 3 united carpels; style stout and shouldered; ovules numerous in the cavity of each carpel.

FRUIT. Capsule oblong, indehiscent, erect, 6-ribbed or -winged (wings mostly infolded over the nectarial grooves), 1–1¾ in. long; seeds numerous, thin, margined, dull, about ¼–⅓ in. long, albumen not ruminated.

LEAVES. Numerous, crowded, closely alternate, green, pliable, recurved, surface nearly plane, often slightly plicate above, about 2 in. wide, margin narrowly yellow or brown, often with a few microscopic teeth, at maturity entire or slightly filiferous.

RANGE. Usually in sandy soil of the Gulf Coast Plain from eastern Louisiana eastward to Florida and Georgia.

RELATED SPECIES. *Y. recurvifolia* is very closely related to *Y. gloriosa* L. It has been listed by various authors under the names of *Y. gloriosa* forma *recurvifolia* Engelm., *Y. gloriosa* forma *planifolia* Engelm., *Y. recurva* Haw, and *Y. pendula* Groenl.

VARIETIES AND FORMS. Trelease (1902, pp. 46, 47, 84) has listed the following forms of *Y. recurvifolia:*

Dull Curve-leaf Yucca, *Y. recurvifolia* forma *tristis* (Carr.) Trel., has blackish-purple bracts.

Girdled Curve-leaf Yucca, *Y. recurvifolia* forma *rufocincta* Baker, Gard., is a low form with rather pronounced accentuation of the reddish brown leaf margin.

Margined Curve-leaf Yucca, *Y. recurvifolia* forma *marginata* (Carr.) Trel., has leaves bordered with yellow, and often also rosy tinted. A garden form.

Variegated Curve-leaf Yucca, *Y. recurvifolia* forma *variegata* (Carr.) Trel., is a garden sport with median yellow stripes.

Elegant Curve-leaf Yucca, *Y. recurvifolia* forma *elegans* Trel., differs in having the median stripe reddish.

REMARKS. The genus name, *Yucca,* is an old Haitian name, probably incorrectly applied to this plant. The species name, *recurvifolia,* refers to the recurved, flaccid leaves. The plant may be propagated from seeds, offsets, stem or rhizome cuttings.

Pale-leaf Yucca

Yucca pallida McKelvey

FIELD IDENTIFICATION. Plants without stems forming clumps with orderly few-leaved heads. Recognized by the rather erect, short, flat, broad, sage-green leaves.

FLOWERS. Panicle and subjacent scape together 3–7½ ft tall, the racemose apex 6–8 in. long; branchlets of inflorescence 6–8 in. long toward the base, 1–4 in. long toward the apex; pedicels slender, somewhat flattened or terete; bracts near base of scape 5–10 in. long; bracts broadened above base and then long-acuminate–pointed; gradually shortened upward to become only 2–3 in. long, long-triangular, papery and fragile, with an acute apex, margin usually denticulate; bracts at pedicel base only ¼–¾ in. long and ⅜ in. broad at base; sometimes a bractlet about ⅜ in. long is present; inflorescence with numerous flowers (as many as 100, but generally much fewer); flowers campanulate, pendent, 2–2⅔ in. long; perianth segments 6, varying from narrow to broad, elliptic or ovate to slightly obovate, inner 3 somewhat broader, very concave, thin, veiny, margin entire or somewhat toothed, center greenish, but white toward margins; stamens 6, filaments ¾–1⅓ in., papillose at apex, pubescent at base, anthers sagittate and about ⅜ in.; pistil 1¼–1⅝ in.; oblong-cylindric toward the base, and yellowish green with the neck paler to white; style oblong-cylindric to ovoid, white, ⅜–¾ in., basal tips extending down on ovary about halfway; stigmas 3, mostly erect, papillose, about ⅜ in. long with central apex blunt and thickened, and lateral lobes more noticeable.

FRUIT. Capsule oblong-cylindric, 1½–2 in. long, ¼–¾ in. in diameter, most often symmetrical, apical beaks about ⅜ in. long, wall of capsule fairly thick, smooth or roughened, dehiscent to the base on the primary fissures, not splitting (or only a short distance) on the secondary fissures; carpels 3, dorsally rounded or sometimes semiflattened or indented; old segments of corolla sometimes persistent and basal section about ⅜ in. long; seeds in capsule small.

LEAVES. At maturity 6–16 in. long, widest point at or above the middle of blade ¾–1¼ in., gradually narrowed to an acuminate, somewhat concave, apex from which is produced a rather blunt yellowish spine ⅝–1¼ in.; narrowed to just above the base where it is ¼–⅜ in. wide, but the base itself triangular and 1½–2⅜ in. wide; margins horny, finely denticulate, yellowish; surfaces flat, flexible, striate, bluish green or sage-green with a glaucescent appearance.

RANGE. Pale-leaf Yucca is mostly confined to the blackland prairie region of Texas. From Palo Pinto, Dallas, and Tarrant counties on the north, southward to Travis County.

RELATED SPECIES. *Y. pallida* may possibly be confused with *Y. rupicola,* but the distinctions between the two species have been well described by McKelvey (1947, p. 61) as follows:

"*Yucca pallida*—Ovary sturdy, lower portion oblong-cylindric with ill-defined shoulders, upper neck stoutish; style generally moderately thick, the 3 stigmas erect or only slightly spreading at anthesis; each stigma has, at apex, a thickened central portion (the upper end of the basal lobes of the style) which is shorter and less conspicuous than two round-tipped lobes which flank it, these giving to the stigma a blunt semiobtuse appearance.

"*Yucca rupicola*—Ovary slender, at times slightly stouter below than above, with well-defined shoulders and slender, tapered neck; style slender, stigmas widespread at anthesis; each stigma has, at apex, a thickened central portion which appears to extend beyond the two flanking lobes (these soon inroll) so that each stigma looks slender and acuminate. Also the style is very often tightly twisted in anthesis, as if to close the stigmatic tube (this latter character is only occasionally noted in *Y. pallida,* and then merely at the top of the style)."

REMARKS. The genus name, *Yucca,* is from the native Haitian name, and the species name, *pallida,* refers to the pale glaucous leaves.

TWIST-LEAF YUCCA
Yucca rupicola Scheele

TWIST-LEAF YUCCA

Yucca rupicola Scheele

FIELD IDENTIFICATION. A low plant without visible stem, appearing as single heads of leaves or sometimes a few heads together. Recognized usually by the twisted, rather arching leaves, which are rather few in number.

FLOWERS. Usually opening in May, the scape attaining a height of as much as 9 ft (but usually only 1½–5 ft), the flowering portion limited to the upper quarter or half; inflorescence panicled or semiracemose, branchlets 8–24; pedicels ⅜–1 in., terete or somewhat flattened; lower bracts of scape 4–10 in. long, long-acuminate at apex, thin and membranous, denticulate on margins; bracts on upper part of scape 1–2½ in. long, broadly triangular, apex acute, margins entire or denticulate; bracts at base of pedicels ⅜–¾ in. long or bractlets even smaller; flowers at base of inflorescence opening first, campanulate, pendent, 2–3¼ in. long; segments 6, white or greenish white or sometimes also purplish, either narrowly or broadly ovate, concave, veined, short-acuminate, ¾–1⅓ in. broad, inner 3 segments often broader than the outer 3; stamens 6, appressed to the ovary, as long as or above the shoulders of the ovary, pubescent, erect or ascending at apex; pistil 1–1⅔ in.; ovary oblong, cylindric toward base, apex slender and tapering to noticeable shoulders, yellowish green; style slender, white, apex with 3 spreading stigmas about ⅜ in. long.

FRUIT. Capsule brown, maturing in August, 2–3 in. long, about 1 in. wide before opening, ovoid to obovoid or often variously distorted or asymmetrical, apex long-beaked, carpel walls moderately thick, splitting to the base along the primary fissures at dehiscence, but only part way on the secondary fissures; carpels 3, dorsally rounded or occasionally flat, sometimes with a raised median rib; remains of perianth often persistent at base.

LEAVES. Mature leaf blade 8–24 in. long, narrowed toward base and toward apex, at the greatest width measuring ½–2 in., stiff and rigid toward base and apex, where widest usually flexible, usually curved, twisted or lopsided, concave, revolute toward the terminal spine which is ⅛–⅙ in. long, stout, blunt and yellowish brown; surfaces olive-green or marked with reddish brown, glabrous, coarsely striate, leaf margins with minute sharp teeth and usually yellow, orange, or red.

RANGE. On rocky limestone hillsides, edges of ravines, in canyons, or on grassy flats at an average elevation of 1350 ft. Confined mostly to the southeastern part of the Edwards Plateau area of Texas, with the greatest concentration in Gillespie, Kerr, Kendall, Travis, Hays, and Comal counties.

REMARKS. The genus name, *Yucca,* is from the native Haitian name. The species name, *rupicola,* is from the Latin and means "lover of rocks." The plant was first introduced to cultivation in Europe by A. A. L. Trecul, the French botanist.

Thompson Yucca

Yucca thompsoniana Trel.

FIELD IDENTIFICATION. A small tree 4–7½ ft, with a solitary trunk or with a few branches near the top. Leaves borne in a dense radiating cluster 1–1½ ft long or wide. Trunk covered with dead reflexed leaves often to the ground.

FLOWERS. Sometimes appearing on young and stemless plants; panicle 2–3½ ft long, with about 25 branchlets which are erect or ascending to horizontal below, green to purplish, pubescent at first but glabrous later, somewhat angular; pedicels ½–⅝ in., rather slender; bracts numerous and variable in size, those at base of scape 4–6 in. long, those of the inflorescence proper 1½–2 in. long, those at base of pedicels about ½ in., shape of bracts from ovate to oblanceolate with acute or acuminate apices, margins entire, or on the larger ones corneous or intermittently denticulate, texture of bracts (especially the smaller ones of the pedicels) tending to become papery, crisp, and curling with age; flowers numerous, globose to campanulate, 1¼–2½ in. long, the 6 segments ovate to obovate, concave to flattened, thin, apex acute to short-acuminate, margins entire or toothed, white or greenish; stamens 6, ¾–1 in., pubescent above, anthers sagittate and less than ⅕ in. in length; pistil 1¼–1¾ in., ovary oblong-cylindric, greenish yellow, style slender, about ½ in. long, white, apex bearing 3 stigmas with thick rounded lobes.

FRUIT. Abundant, persistent, reddish brown, length 1¼–2½ in., to ¾ in. wide, symmetrical, occasionally constricted, at apex abruptly narrowed with short, reflexed tips ¼–½ in. long, walls thin and strong, dehiscent the length of the primary fissures, carpels 3; remnants of the corolla segments long-persistent.

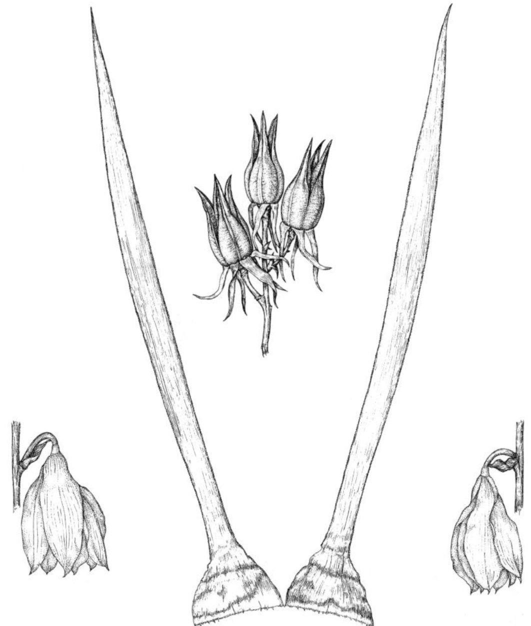

Thompson Yucca
Yucca thompsoniana Trel.

LEAVES. Borne in a radiating mass, yellowish green or bluish green to somewhat glaucous, 10–18 in. long, averaging ⅜–½ in. wide at the greatest width at about the center, gradually narrowed to a long acuminate apex terminating in a sharp, slender, fragile spine; gradually narrowed also to the base and becoming as narrow as ⅙ in. just above the base; leaves usually straight, or occasionally when falcate broader and shorter, surfaces scabrous, plano-convex, flattened, or becoming concavo-convex near the apex, thin and flexible, margins with numerous, minute, regular teeth.

RANGE. Rocky slopes and foothills at altitudes of 700–3,400 ft. In Texas in southeastern Val Verde, Terrell, and Brewster counties. In Mexico known from Chihuahua and Coahuila.

REMARKS. The genus name, *Yucca,* is from the native Haitian name. The species name, *thompsoniana,* is in honor of the botanist C. H. Thompson.

Beaked Yucca

Yucca rostrata Engelm.

FIELD IDENTIFICATION. An arborescent plant attaining a height of 6–12 ft, with a trunk 5–8 in. in diameter. Trunk simple, or with a few short branches, bearing a rather symmetrical head of crowded leaves. With age the leaves become shaggy and reflexed, often clothing the trunk to the ground.

FLOWERS. Inflorescence a glabrous panicle becoming racemose at the apex, scarcely exceeding the ends of the uppermost branchlets, 1½–4½ ft long, the flowering part about half the length of the naked scape below; branchlets 20–45, erect to ascending or older ones more horizontal, somewhat angled; bracts very variable in size and shape, largest at base of scape; flowers large, 2–2⅔ in. long, white or sometimes purplish; segments 6 (3 petals and 3 sepals) varying from broad-ovate to elliptic, apices acute to acuminate, those of the inner row about 1 in. broad, those of outer row narrower; stamens 6, filaments about as long as the ovary and style or longer; pistil 1½–2 in.; ovary oblong-cylindric, narrowed into the style, stigmas erect or somewhat spreading.

FRUIT. Pedicels about ½ in. long, slender, capsules numerous, erect, persistent, 2–3¼ in. long, ½–1 in. broad, occasionally constricted, symmetrical, gradually attenuate at apex into 3 erect beaks; dehiscent at maturity along primary sutures, otherwise entire or only slightly split at apex, valves light to dark brown, hard, dorsally rounded, median-ribbed, 3-celled, old perianth remaining persistent; seeds numerous, small, black, smooth.

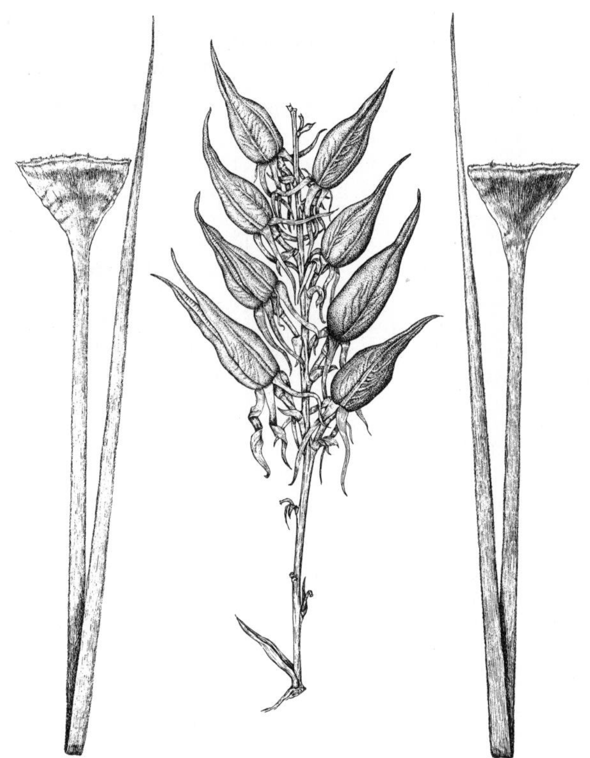

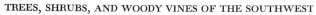

Beaked Yucca
Yucca rostrata Engelm.

LEAVES. Borne in a radiating mass, stiff, younger ones cream-colored to yellowish green, later light brown, striate, glaucous, blade length 8–24 in., width ⅜–½ in., gradually widened from the base to just above the middle, beyond attenuated to an acuminate apex which becomes a needle-like, strong, light-colored spine; margins corneous, finely denticulate or smooth, yellow or reddish; surfaces flattened to slightly convex or keeled but near apex more concavo-convex.

WOOD. White to brownish, spongy, soft.

RANGE. In Texas in the southeastern tip of Brewster County. More widespread in Coahuila and Chihuahua, Mexico.

REMARKS. The genus name, *Yucca,* is from the native Haitian name. The species name, *rostrata,* refers to the "beaked" fruit. The plant is also known by the vernacular names of Palmita and Soyate.

Soaptree Yucca

Yucca elata Engelm.

FIELD IDENTIFICATION. Tree attaining a height of 6–20 ft, or rarely as high as 30 ft, with a trunk diameter of 6–12 in. The 2–3 short branches (or more in older specimens) bear radiating heads of leaves nearly hemispherical on young plants, but on old ones less so with the older dead leaves pendent to form a thick thatch on the trunk and branches.

FLOWERS. Maturing May–June, in glabrous panicles 1½–6 ft long, on naked stems 3–7 ft; branchlets of panicle 25–30, soon woody, erect to ascending or spreading; pedicels slender, spreading, often recurved, ⅜–1 in. long; bracts variable in size, those at base of scape 6–8 in. long, acuminate, on scape proper reduced in size upward to become small and inconspicuous; bracts at base of pedicels ⅜–½ in. long, long-triangular, generally thin, papery, whitened, fragile, a small bractlet often present also; flowers numerous, in bud ovoid and acute, white or greenish, campanulate later, segments 6 (sepals 3, petals 3), slightly united at base, 1⅓–1¾ in. long, ovate to narrowly elliptic or slightly obovate, acute, those of the outer rank generally narrower; stamens 6, about ¾ in. long, as long as the ovary or longer, filaments slender, pale green, pubescent below, papillose at apex, anthers ⅛–⅕ in.; ovary oblong-cylindric, about twice as long as the white or greenish abruptly tapering style; stigmas 3, erect or semispreading, 2-lobed.

FRUIT. Capsule dry, persistent, brown, oblong-cylindric, less often obovoid or ovoid, 1½–2 in. long, 1–1½

Soaptree Yucca
Yucca elata Engelm.

in. wide, symmetrical or sometimes constricted, apices obtuse, rounded, short-pointed, 3-ribbed, carpels with rounded ridges dorsally, dehiscent longitudinally from apex to base between the carpels, but only dorsally at the apex; epidermal layer yellowish or light tan; seeds about ⅓ in. wide, about 1/32 in. thick, narrowly wing-margined.

LEAVES. Borne in rather symmetrical radiating clusters, pale yellowish green, narrowly linear, blades 10–30 in. long, ⅛–½ in. broad, 1/32–1/16 in. thick, usually straight, upper surface flattened, lower surface slightly rounded or plano-keeled, smooth or slightly scabrous and striate, tapering very gradually to the apex and ending in a needle-like straw-colored or brownish spine about ⅛ in.; margins when young white, thin and papery; when older with slender curly, white fibers attached to the papery tissue.

BARK. About ¼ in. thick, dark to light brown, irregularly fissured with thin plates.

WOOD. Brown to yellowish, light, soft, spongy.

RANGE. High plains, arid grasslands, desert washes, and dry gravelly mesas at altitudes of 1,500–6,000 ft. Trans-Pecos Texas; northwest to central New Mexico and western or central Arizona; south in Mexico to Sonora, Chihuahua, and Coahuila.

REMARKS. The genus name, *Yucca,* is from the native Haitian name, and the species name, *elata,* refers to the tall, or elevated, size of the plant. Some of the vernacular names used for the plant are Palmilla, Soapweed, Amole, and Jabonilla. The roots and stems contain a mucilaginous substance used for a soap substitute. Indians eat the young flower stalks, which are also eaten by cattle in time of stress. The leaves are also chopped up and mixed with other feeds for cattle. The tough leaf fiber is used for twine and rope.

Narrow-leaf Yucca

Yucca angustissima Engelm.

FIELD IDENTIFICATION. A usually stemless plant, or occasionally with short procumbent stems. Heads mostly solitary and rounded, the stem if present covered with persistent dead leaves.

FLOWERS. The scape and raceme together 2½–4½ ft tall; inflorescence branches slender and well spaced; pedicels ½–1 in., terete; bracts toward base of scape 4½–6½ in. long, usually widest at the middle, taper-pointed; bracts toward the upper part of scape about 3 in. long, long-triangular and acute at the apex, rather papery and fragile; bracts becoming progressively smaller up the inflorescence until some 1 in. long or less and broadly triangular with an acute point, green to purplish; bractlets very small or sometimes absent; flower globose or campanulate, white, or tinged greenish or purplish, 1½–2⅓ in. long; perianth segments 6, inner 3 broad-elliptic, outer 3 narrow-elliptic, rather thin, margins entire or wavy and sometimes pubescent; stamens 6, filaments erect or ascending, about 1 in. long, about half the length of the style, pubescent toward the base, apex papillose; ovary oblong-cylindric, abruptly necked, about ½ in. long, pale green, papillose, basally extended into 3 tips to the middle of the ovary; stigmas 3, erect or nearly so, white, the lobes obtuse or rounded.

FRUIT. Capsule tan, 2–3½ in. long, ¾–1 in. wide, usually constricted at the center, ending in mucronate beaks ⅛ in. long or less, surface smooth or roughened; capsule walls rather thin but tough, at dehiscence splitting to the base on primary fissues, but only part way on the secondary fissures; carpels dorsally rounded, sometimes with a rib; flower base becoming about ⅜ in. long and somewhat tubular, with the perianth remaining persistent; seeds dull black.

LEAVES. Mature leaves yellowish green or bluish green, straight, barely broadest at the middle or slightly above (about ⅜ in.), gradually narrowed to an acuminate apex, and also gradually narrowed almost to the base where widened 1–1¾ in; outer leaf surface rounded or slightly keeled, inner surface convex or somewhat flattened at widest point, surface smooth, margins with white curly thread-like fibers, especially near the base of leaf, with age the upper margins of leaves showing fewer fibers, apex with a sharp spine ⅛–⅓ in. long.

RANGE. In New Mexico in Valencia, Rio Arriba, and McKinley counties, extending to Arizona and Utah.

REMARKS. The genus name, *Yucca,* is from the old Haitian name. The species name, *angustissima,* refers to the very narrow leaves.

Standley Yucca

Yucca standleyi McKelvey

FIELD IDENTIFICATION. Plant stemless, or with a stem 6–8 in. long. Heads of leaves solitary or to 15.

FLOWERS. Borne in a raceme 1¼–3 ft high, inflorescence crowded down among the leaves or just above them; perianth 2–2½ in. long, the united portion about ¼ in. long; segments white, 6, ovate or obovate, 1⅛–1¼ in. long, about 1 in. broad, concave, outer series of 3 narrower than the inner; stamens 6, filaments about ¾ in., below top of ovary, pubescent toward base, papillose; ovary oblong-cylindric, about ⅜ in. in diameter, green, with 3 short basal lobes and 3 deeply cleft stigmas at apex and continuous to form a rib.

FRUIT. Capsule light brown, smooth or roughened, oblong-cylindric to obovoid, entire or occasionally constricted, mucronate at the apex, splitting to base on primary fissures or only part way on the secondary fissures; carpels 3, erect or spreading somewhat, dorsally rounded; remains of filaments and corolla persistent at base.

LEAVES. Heads of leaves varying from rather loose to densely crowded, central ones semierect or ascending, lower ones spreading; blade 6–12 in. long, gradually narrowed from base to apex, or slightly narrowed just above base to become slightly wider toward the center and then tapering to the apex, stiff, straight or falcate, plano-convex, or near the tip concavo-convex or concavo-keeled, yellowish green, glossy and smooth or slightly roughened, margins with numerous white fibers, spine at apex of blade short and sharp.

RANGE. Usually in sandy or rocky soils on hillsides at altitudes of 6,000–7,500 ft. In New Mexico, Arizona, Colorado, and Utah.

RELATED SPECIES. Since *Y. standleyi* and *Y. angustissima* are apt to be confused, the more significant differences are hereby quoted from McKelvey, (1947, pp. 112–113).

"Both are acaulescent, with racemose inflorescence (*Y. angustissima* very occasionally produces one or two basal branchlets); the inflorescence of *Y. angustissima* is taller (by 0.3–0.4 m.) or at times more, more slender for its length and at anthesis more open, the flowers less congested; the flowers of both (as in all the *Elatae*) have a white to greenish style, paler than the ovary at anthesis, which varies from nearly oblong-cylindric to slender-ovoid; the whole pistil is slightly stouter for its size in *Y. standleyi* than the *Y. angustissima;* the fruit of the last shows more tendency to constriction, perhaps because of its more fragile wall, and the epidermis is smoother; the leaves of *Y. angustissima* are less numerous in one head, average larger, are more supple, and except at the juvenile stage, less filiferous. Standley attributed 'much smaller flowers to *Y. angustissima* than to *Y. standleyi (Y. baileyi),* in which they are 60–65 mm. long.' Flowers in all species vary considerably in respect to size; the dimensions noted by the writer were 4.5–5.7 cm. *(Y. angustissima),* 5–6.5 cm. *(Y. standleyi).* It may be that the flowers expand first at different positions on the axis in the two species; in *Y. angustissima* they appear always to open first near the base of the flowering portion and thence upward; in *Y. standleyi* they sometimes open first near the center, last at the base and apex of the axis; how often this happens is uncertain."

REMARKS. The genus name, *Yucca,* is from the native Haitian name, and the species name, *standleyi,* is in honor of Paul C. Standley, American botanist.

Intermediate Yucca

Yucca intermedia McKelvey

FIELD IDENTIFICATION. Small, stemless, or more rarely with a very short stem, the heads of leaves solitary or 3–4 heads in a clump.

FLOWERS. Borne in a raceme (occasionally paniculate at base), usually 2–4½ ft tall; branchlets numerous, erect or ascending, glabrous, green or sometimes with purplish markings; flower-bearing portion of the scape usually beginning in the leaf head; flowers rather crowded at maturity, pedicels averaging about ½ in. long, with small bracts ¼–1⅔ in. long; bracts at base of the inflorescence proper 2–4 in. long, but those at base of the scape 4–12 in. long; bracts generally papery and fragile, concave; perianth campanulate to globose, 2–2⅔ in. long; segments 6, concavo-convex, white or greenish, sometimes tinged purple or brown, the segments about equal in length, but the 3 inner ones broader (1–1¼ in. wide); stamens 6, about 1 in. long, filaments slender and flattened, anthers ⅛–⅕ in.; pistil 1–1¼ in. long; ovary about 1 in. long, oblong-cylindric; style about ⅜ in. long, yellowish green to white.

FRUIT. Capsule oblong-cylindric, usually symmetrical, entire or sometimes constricted near the middle, apex mucronate; wall of capsule at first brown but later turning black, somewhat roughened; dehiscent along the primary fissures to the base, and along the secondary fissures only part way; the carpels usually erect or ascending, dorsally rounded below and flattened toward the apex, median ribs present or absent; remains of the perianth persistent, ⅜–½ in. long; seeds numerous, flattened, dull black or lustrous.

LEAVES. Mature leaf 8–24 in. long, 1¼–1½ in. wide at base, narrowest point just above base ¼–⅜ in. wide, and then gradually tapering to a long-acuminate apex, coarse and stiff, margins whitened with few or no threadlike fibers, surfaces usually plano-convex but near the base sometimes plano-keeled.

RANGE. Intermediate Yucca occurs in New Mexico in Socorro, Bernalillo, Sandoval, Mora, and Colfax counties.

VARIETY. Branched Intermediate Yucca, *Y. intermedia* var. *ramosa* McKelvey, is a variety which has been segregated owing to its large broad panicle, branched for the greater part of its length; but it is connected by intermediate forms to those plants with a simple raceme. Plant generally with a more vigorous form, foliage, and inflorescence. Forms large caespitose clumps in sandy areas. Occurs in grasslands and desert areas in Torrance, De Baca, and Guadalupe counties of New Mexico.

REMARKS. The genus name, *Yucca,* is from the native Haitian name. The species name, *intermedia,* was given because it occurs in an area "intermediate" between *Y. glauca, Y. angustissima,* and *Y. standleyi.* The following quotation from McKelvey (1947, pp. 117–118) explains the taxonomic position of *Y. intermedia* in relation to other species:

"Although *Y. intermedia* closely resembles in certain aspects each of the three species mentioned above, it also differs from each in one important aspect at least. In all these species the habit is acaulescent, any stems short and inconspicuous; in all the leaf-blade is wiry; in all the typical inflorescence is a raceme. *Y. intermedia* (like *Y. angustissima* and *Y. standleyi*) differs from *Y. glauca* in form and color of style at anthesis; because of this they are placed in different series; fruit of these two yuccas is very similar (as is that of *Y. standleyi*); the clumps are much the same size although those of *Y. angustissima* in capsule—its form and size, strength dead leaves intermixed with the living; both species demonstrate the tendency to depart from the simple raceme and each has produced a localized, panicled variety, the form of the flowering portion differing considerably however. *Y. intermedia* differs radically from *Y. angustissima* in capsule,—its form and size, strength and thickness of wall, texture and color of epidermis, manner of expansion of the valves after dehiscence, etc.; the inflorescence attains a greater height (1.5 as against 1.3 m. or usually less) and appears more vigorous with a greater tendency to a panicled form; the clumps are somewhat larger; the pistil is slightly longer and somewhat 'heavier' in all parts. *Y. intermedia* differs from *Y. standleyi* primarily in foliage; the leaves are larger, less rigid, do not approach the triquetrous and have less filiferous margins; the inflorescence is considerably longer (that of *Y. standleyi* does not exceed and rarely attains 1 m.) although it is not more vigorous; no tendency to depart from the simple raceme has been noted in *Y. standleyi;* they are nearly hemispherical and are more densely crowded, and the spine is especially unpleasant to encounter; fruit is much the same in both; the stigmas do not appear to produce on the interior the thickened, central ribs which are usual in those of *L. standleyi*."

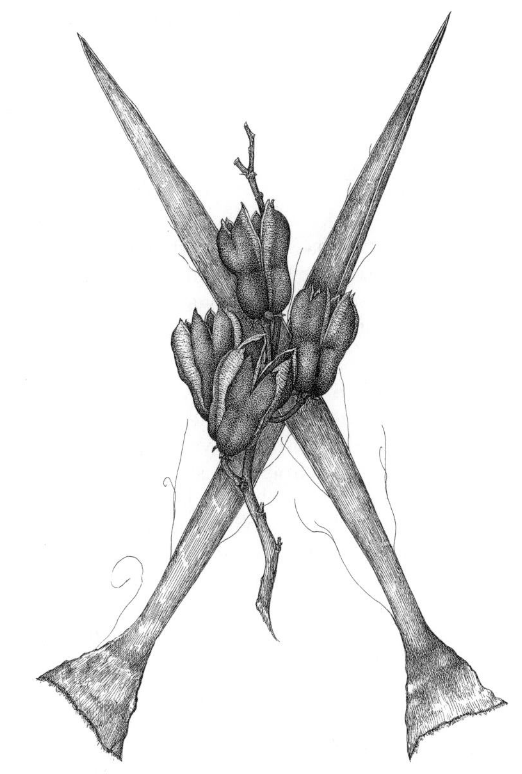

Buckley Yucca
Yucca constricta Buckl.

Buckley Yucca

Yucca constricta Buckl.

FIELD IDENTIFICATION. Plant usually stemless, but sometimes developing weak, erect or semiprostrate stems to 5 ft long. Plants occuring either singly or in clumps.

FLOWERS. Scape as long as the inflorescence proper or longer, which is an ellipsoid to ovoid panicle produced a considerable distance above the leaves. Branches of inflorescence not crowded, 6–10 in. long near the base and 1–3 in. long at the apex; bracts varying in size, those at the base of the pedicels ⅜–1 in. long, rather broad and papery; bracts near base of scape 5–7 in. long, broadest near middle and taper-pointed; bracts near apex of scape 3–4 in. long; flowers varying from globose to semicylindric; segments 6, broadly or narrowly elliptic, the 3 outer ones narrower than the 3 inner ones; stamens 6, filaments slender, as long as the base of the style or shorter; pistil 1–1¼ in.; ovary oblong-cylindric, about ¾ in., neck short; style rather slender, ¼–⅜ in., basal lobes passing into neck of ovary, pale green or white; stigmas white, erect or spreading.

FRUIT. Capsule oblong-cylindric, at maturity dark brown or black, variable in size, length 1⅔–2⅓ in., width ½–1 in., symmetrical, or on some plants considerably constricted, surfaces smooth or slightly roughened, dehiscent to the base on the primary fissures but only part way on the secondary fissures, the 3 carpels dorsally rounded or flattened; remains of perianth persistent; seeds glossy black.

LEAVES. Heads symmetrical or asymmetrical, blade at maturity 8–20 in. long, broadest somewhat above the middle, ¼–½ in. wide, from there long-tapering with a stout, short spine at the apex; blade narrowed toward base but the base proper expanding and nearly triangular, width of the insertion area ⅜–1¾ in.; surfaces flattened and bluish green (rather grasslike); margins with long curly filiform threads.

RANGE. In Texas from the eastern margin of the Edwards Plateau southwest to the Gulf of Mexico, westward to the Rio Grande River, and across the Pecos River to the lower plains country.

RELATED SPECIES. The species formerly described as *Y. tenuistyla* Trel. is considered to be partially included in the above concept of *Y. constricta*.

REMARKS. The genus name, *Yucca*, is from the native Haitian name. The meaning of the species name, *constricta*, is obscure and may possibly refer either to the constrictions of some capsules or to the heads of leaves which are somewhat constricted below and spreading above.

Harriman Yucca

Yucca harrimaniae Trel.

FIELD IDENTIFICATION. Plant short-stemmed or stemless, forming one or several clumps. Often flowering when the leaf rosette is not over 4–5 in. wide, the small, broadly spatulate leaves unlike that of any other mature yucca.

FLOWERS. Scape 8 in.–3 ft tall, usually about one half overtopping the leaves, simple, glabrous, flowering from close to the base; bracts varying ½–9 in. long, those at the base of scape largest, purplish to white, thin and papery with age; flowers large, 2–2½ in. long, rounded, greenish white; sepals 3 and petals 3, ovate to elliptic, somewhat concave, acute or obtuse; stamens 6, filaments about ¾ in. long and pubescent, anthers about ⅛ in.; pistil ¾–1⅓ in., ovary ⅝–¾ in., with 6 dorsally rounded lobes; style slender, green, ⅜–⅝ in., terminating in 3 white lobed stigmas.

FRUIT. Capsule brown, broad-oblong, 1¼–2½ in. long, apices with short recurved mucros, constricted, dehiscent into 3 dorsally rounded, erect or ascending carpels; seeds ⅙–¼ in. long, dull black; pedicels short, ⅜–1 in., stout, and terete.

LEAVES. Spreading, rigid, older ones more flexible, blades 4–18 in. long, width ⅜–1 in., or more rarely to 1⅗ in., linear to spatulate-lanceolate, flattened at the middle sections, becoming convex where narrowed, apex narrowed and acuminate; margins thin, papery, narrowly brown-bordered or whitish, with coarse white fibers; surfaces smooth and finely striate, light to yellowish green or glaucous.

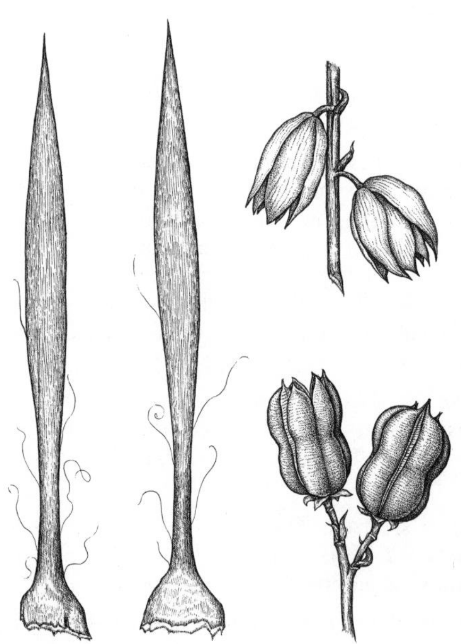

Harriman Yucca
Yucca harrimaniae Trel.

RANGE. Rocky hillsides at altitudes of 7,500–8,500 ft, in New Mexico, Arizona, Utah, and Colorado.

REMARKS. The genus name, *Yucca*, is from the native Haitian name. The species name, *harrimaniae*, is in honor of Mrs. Edward H. Harriman, hostess to William Trelease on the Harriman Alaska expedition.

Arkansas Yucca

Yucca arkansana Trel.

FIELD IDENTIFICATION. Plants stemless and usually with a solitary head of leaves, or more rarely in few-headed clumps. Usually inconspicuous because of its small size.

FLOWERS. In racemose inflorescences 1½–3 ft tall overall, the scape 10–24 in.; flowering portion with the basal part included in the leaves or starting immediately above the leaves, greenish or purplish; bracts on scape numbering 6–12, the lower ones larger, 1½–2¾ in. long with an acuminate or acute apex, upper bracts of scape ¼–1¾ in. long, broader than the lower bracts, thin and fragile; pedicels ⅜–1 in., slender, terete, erect at fruiting period; flowers borne singly or in pairs, globose in outline, 1¼–2⅝ in. long, united a short distance at base; perianth segments 6, inner 3 and outer 3 of the same length, but inner ones broader and ellipsoid to nearly orbicular, outer row narrower and oblong to lanceolate, rather thin, apices usually obtusish, margins pubescent or papillose; stamens 6, pubescent toward the base, at anthesis erect or erect-ascending, anthers sagittate; pistil 1–1½ in.; ovary oblong-cylindric, neck short, shoulders well defined, pale green; style dark green at anthesis, ⅜–½ in., papillose, shortly 3-lobed at base; stigmas 3, erect and stout.

FRUIT. Capsule dark brown or black at maturity, sometimes shortly mucronate, sometimes asymmetrical or contorted near the middle; wall fairly thick but weak, surfaces smooth or roughened, dehiscent to the base on the primary fissures but only part way on the secondary fissures; carpels 3, dorsally flattened or rounded toward the base, sometimes concave toward the apex; perianth segments and stamen remnants long-persistent.

LEAVES. Mature blade bluish green or yellowish green, 8–24 in. long, finely striate, greatest width varying ⅜–1 in., narrowed to 3/16–⅜ in. just above base, tapering to a long-acuminate apex, surfaces flat above but revolute at apex or plano-convex at base, usually thin and flexible (grasslike), margin bearing white fibers, often papery near base; apical spine 1/15–⅛ in. long, sharp but often broken.

RANGE. In chalky or gravelly soil on rocky hillsides

ARKANSAS YUCCA
Yucca arkansana Trel.

and prairies. In Texas from Uvalde County northward and eastward to northwestern Louisiana; northward to central and eastern Oklahoma and western Arkansas.

REMARKS. The genus name, *Yucca,* is from the native Haitian name. The species name, *arkansana,* is for the state of Arkansas.

LOUISIANA YUCCA

Yucca louisianensis Trel.

FIELD IDENTIFICATION. A stemless plant with an inflorescential height of 4–8 ft. The branchlets of the inflorescence and the scape are densely pubescent.

FLOWERS. Usually appearing April–May, borne in a panicle 2–3½ ft long, often of pyramidal form, or narrowed toward the base and attenuate toward the apex; branches of panicle 5–18 in. long, stout, green to reddish, pubescent, flattened somewhat at the nodes; individual flower pedicels ⅛–⅜ in., pubescent; bracts at base of pedicels ⅛–¼ in. long, ovate to lanceolate, apex acute to acuminate, pubescent, striate, clasping at base; larger bracts of inflorescence long-acuminate, lanceolate, convex, puberulent, striate, pale green to brownish, 1–3½ in. long; flower buds narrowly conic, pale green; young flowers greenish, pendent, semiglobose to ovoid; older flowers greenish to white, semi-spreading, with 6 segments; outer 3 segments concave, elliptic or oval, apex acute, parallel-veined, somewhat pubescent toward the base; inner 3 segments somewhat broader than the outer 3; stamens 6, filaments stout, white, ⅓–⅝ in., apex thickened and curved outward, densely puberulent, anthers somewhat mucronate; pistil green, stout, ⅔–1 in., angled, abruptly narrowed to shoulders above the middle and terminating into green styles with 3-lobed stigmas; lower half of pistil with a number of short grooves from the base and each terminating in pairs of narrowly elliptic depressions.

FRUIT. Capsule stout, brown, 1½–2 in. long, at first somewhat angular, symmetrical or somewhat obovoid, usually much constricted near the middle, dehiscent along the fissures; walls of capsule rather fragile, smooth or roughened on the outer surface.

LEAVES. Numerous, 1–3½ ft long, ¼–1⅗ in. wide at the widest part, flexible, grasslike, the older ones sometimes drooping and twisted near the apex, dull green, margin with white filiform threads, or some threadless, apex long-acuminate and spinose, blade thicker, flatter and narrower toward the base.

RANGE. Usually in sandy soils. In Texas near Dallas, Fort Worth, Texarkana, Huntsville, and Houston. In Louisiana near Alexandria, Alden Bridge, and Minden. In Oklahoma near Otoka, Standley, and Poteau. Also in southern Arkansas.

RELATED SPECIES. Louisiana Yucca has been variously interpreted by different authors. McKelvey (1947, p. 148) states:

"Examination of the material cited by Trelease for his species (as well as similarly determined but uncited collections) indicate that his *Y. louisianensis* was an aggregate concept, some plants described (in his text) by a 'pale and oblong', some (in his key and in his text) by a 'swollen, green' style,—both are illustrated in his Pl. 34. For reasons stated later the last-named form has been considered typical *Y. louisianensis*—it is not believed that the style of any one species normally assumes two such different forms and colors. The 'swollen, green' style, the grass-like foliage, the fruit, the panicled form of inflorescence, all suggest that Trelease's species may prove to be an eastward extension of the variety distinguished here as *Y. arkansana* var. *paniculata* McKelvey; but the two differ in certain respects. Until the more eastern species can be compared in the field, it seems reasonable to leave their identity open to question."

The original descriptive matter on *Y. louisianensis* Trel. may be consulted in Trelease (1902, pp. 64–65).

Shinners, (1951c, p. 170), has presented a key for the segregation of *Y. louisianensis* Trel. and *Y. arkansana* Trel., together with a new species designated as *Y. freemanii* Shinners. This key is reproduced under the description of *Y. freemanii.*

REMARKS. The genus name, *Yucca,* is from the ancient

Haitian name. The species name, *louisianensis*, is for the state of Louisiana, where it is most abundant.

Freeman Yucca

Yucca freemanii Shinners

FIELD IDENTIFICATION. A stemless plant usually with a solitary head of leaves. Long confused with *Y. louisianensis* and *Y. arkansana*, this plant has been described as a new species by Shinners (1951c). The following description is based on that of Shinners except that the metric measurements have been changed to the linear to conform with the remainder of the text:

FLOWERS. "Scape slender (1/6–1/5 in. thick at base of inflorescence), 3–3 2/3 ft high, flower-bearing in terminal 1/4–1/3. Panicle 16 in.–2 ft long, rather narrowly ovoid, the branches spreading-ascending, 8–16 in. long, glabrous; bracts of axis oblong-ovate, 2/5–3/5 in. long. Perianth apparently funnelform-globose (only pressed specimens seen), pendulous, on glabrous pedicels 1/5–1/2 in. long. Sepals [perianth segments] rather thin, yellowish green, about equal in length, 1 1/4–1 1/2 in. long; outer elliptic-oblanceolate, with median vein dorsally thickened toward apex, mucronate, 3/8–1/2 in. wide; inner broadly elliptic, obtuse or barely mucronate, 5/8–7/8 in. wide. Filaments thick, densely papillose-pubescent, 1/2–3/4 in. long; anthers 1/10–1/8 mm. long. Pistil about 1 in. long; ovary narrowly ovoid or ovoid-cylindric, 3/8–3/4 in. long, about 1/4 in. thick; style about 1/4–1/3 in. long." (pp. 168–169)

FRUIT. "Capsule oblong-obpyriform (sometimes obpyramidal), rounded-trigonous, slightly constricted just below equator, glabrous, about 1 3/8 in. long by 5/8 in. thick, tipped by the persistent style, and with persistent, reflexed sepals at base; seeds flat, approximately deltoid, 1/3 in. long by 1/5 in. wide, black." (p. 169)

LEAVES. "Narrowly to broadly lanceolate, 3/8–1 5/8 in. wide, 20 in.–2 ft long with narrow petiole-like section of 1 1/2 in. above the short, widely clasping base (base 3/8–3/4 in. long, wider than long), soft and rather limp, the apex broadly acute to subacuminate (inrolled in drying and apparently narrow acuminate), the margins with few and mostly long white threads." (p. 168)

RANGE. The type specimen of the new species was collected 2.5 miles southwest of Redwater, Bowie County, Texas, in red sandy-clay soils by H. A. Freeman. Deposited at Southern Methodist University.

SPECIES. Lloyd Shinners has also published the following key on the species of *Yucca* of northeastern and north central Texas (1951c, p. 170):

Leaf margins minutely toothed or smooth, not shredding .... *Y. pallida* McKelvey
Leaf margins shredding into prominent white threads (these often largely disappearing late in the year).
Inflorescence panicled, with branches in more than half its length, elevated well above the leaves; plant 3–12 ft high, flowering late May–July.
Panicle branches glabrous; scape 3–3 1/2 ft high, slender (1/6–1/5 in. thick at base of inflorescence); leaves 3/8–1 5/8 in. wide, broadly acute to subacuminate, rather soft and limp; flowers mid-June–July .... *Y. freemanii* Shinners
Panicle branches densely pubescent; scape 4 1/2–12 ft high, stout (1/5–2/5 in. or more thick at base of inflorescence); leaves 1/4–3/4 in. wide, narrowly acute or acuminate, rather firm and stiff; flowering late May–June (sporadically later) .... *Y. louisianensis* Trel.
Inflorescence spicate-racemose, occasionally with 1–2 branches near base, borne close to or even partly below the summit of the leaves; plant 21 3/5–36 3/5 in. high, flowering late April–mid-May .... *Y. arkansana* Trel.

REMARKS. The genus name, *Yucca*, is from the native Haitian name, and the species name, *freemanii*, is given in honor of H. A. Freeman of Garland, Texas, former biology instructor of Southern Methodist University.

Navajo Yucca

Yucca navajoa Webber

FIELD IDENTIFICATION. The following description closely follows that of Webber (1945): Plant forming at length a very dense, compact mass of rosettes ranging 1–44 but averaging 10 rosettes per plant with 0.75–1.47 rosettes per sq ft of soil; caudex from soil level to center of rosette 0–40 in., 2–3 1/4 in. in diameter, simple or with 2–4 short branches, each terminating in 1–3 rosettes.

FLOWERS. Racemes densely flowered, to 1 1/2–3 ft long, usually about 2 ft long, peduncles short, rarely extending above the leaf rosette; flowers short and opening widely, or long, narrow, and remaining closed; sepals generally purple-tinged, averaging about 1/5 in. wide and 1 3/4 in. long; petals white or slightly purple-tinged, 1–1 1/2 in. wide and 1 1/2–2 1/4 in. long; ovary light to dark green, thick, ending abruptly in the style; style white to pale green, gradually tapering into the stigma, 1/4–2/5 in. long; filaments generally abruptly bending outward at or near base of style, 3/5–1 1/8 in.

FRUIT. Capsule when 3–4 weeks old broadly oblong, 1 1/12–1 2/5 in. wide, 2 1/2–2 4/5 in. long, from deeply constricted to not constricted.

LEAVES. Thin but firm, rigid, leathery, dagger-like, linear to oblanceolate, narrowly white-margined, 1/5–2/5 in. wide and 4 1/2 in.–1 ft 4 1/2 in. long, averaging about 9 in. long.

RANGE. Sandy and rocky soil at an elevation of 6,000–6,500 ft. Type locality: west side of U. S. Highway 666,

4.9 miles northeast of Tohatchi, eastern Navajo Indian Reservation, McKinley County, New Mexico.

REMARKS. The genus name, *Yucca,* is from a native Haitian name, and the species name, *navajoa,* refers to the Navajo Indian Reservation near Tohatchi, New Mexico, where this species is found. Webber (1945) considers that *Yucca navajoa* belongs to the *Y. glauca* complex which also includes *Y. baileyi* Woot. & Standl., *Y. angustissima* Engelm. *ex* Trel., and *Y. constricta* Buckl. It is most closely related to *Y. baileyi,* and differs more from *Y. glauca* than any other species or forms of the complex. *Y. navajoa* is unique in that it is the only species in the group with dense clumping mainly due to the branching of an aerial caudex. The formation of a caudex itself is rather unusual, as elongate aerial stems are rarely formed within the *Y. glauca* group, and all the other species of the group have been described as acaulescent. A third distinction of the species is the characteristic smallness of the leaves, these being mainly linear, but often oblanceolate.

Aside from the preceding major distinctions *Y. navajoa* differs from *Y. glauca* in that the inflorescence is strictly racemose and densely flowered, the peduncle is short and never extends above the leaves. It also differs from some of the various *Y. glauca* forms in flower color, pistil color and shape, and in the prominence of carpel sutures and of anther depression in the ovary. *Y. navajoa* has many more leaf rosettes per plant than *Y. baileyi* and slightly smaller flowers and flower parts. Additional distinctions of the species from *Y. angustissima* and *Y. constricta* are the comparatively large capsules and short peduncles of *Y. navajoa.*

SMALL SOAPWEED YUCCA
Yucca glauca Nutt.

SMALL SOAPWEED YUCCA

Yucca glauca Nutt.

FIELD IDENTIFICATION. Plant usually acaulescent, or with a short stem to 12 in. high, or decumbent. The heads of leaves very variable in size and number.

FLOWERS. May–July, fragrant, in a racemose panicle 1–3 ft high, simple or short-branched near the base, only slightly overtopping the leaves; flowers opening from the base upward; pedicels terete, variable in length; bracts variable in size, the upper ones on the scape very small and the lower larger and elongate to short-triangular, green to purplish or brownish, when mature drying to become fragile, papery, and white; flowers 1½–2½ in. long, campanulate to globose, petals 6, sepals 6, greenish white, concave, ovate to lanceolate, apices acute; stamens 6, filaments stout, erect-ascending, anthers sagittate and about ⅙ in. long; ovary about ¾–1 in. long, oblong-cylindric to obovoid; style ovoid, green, pubescent; stigmas 3, erect, short, thick, white.

FRUIT. Oblong-cylindric to obovoid, length 1¾–2¾ in., width ¾–2 in., 6-sided, rough, loculicidally 3-valved from the apex, carpels dorsally rounded near base but flattened toward the apex; seeds black, flat, semiobovate, about ⅖ in. long.

LEAVES. Heads widely radiating, often asymmetrical, blades 8 in.–3 ft long, ⅕–½ in. broad, usually straight, occasionally curved, stiff, linear, tapering to an acuminate, concavo-convex apex bearing a spine of varying length, most of the leaf surface plano-convex, margin inrolled, whitened and sometimes with fine white fibers, especially near the base, surfaces smooth.

RANGE. Badlands, dry plains, and sandhills. In Texas; west through New Mexico to Arizona and north through Oklahoma to Montana and North Dakota.

VARIETIES. Soft Soapweed Yucca, *Y. glauca* var. *mollis* Engelm., has flatter, more flexible leaves to ⅝ in. wide.

Pink Soapweed Yucca, *Y. glauca* var. *rosea* D. M. Andrews, has flowers tinted rose-pink outside.

Erect Soapweed Yucca, *Y. glauca* var. *stricta* (Sims) Trel., has a distinct erect stem with a longer branched inflorescence.

Y. × *karlsruhensis* Graeb. is a hybrid developed in 1899 between *Y. glauca* and *Y. filamentosa* with inflorescence branched below the middle and leaves about ¾ in. wide.

REMARKS. The genus name, *Yucca*, is from the native Haitian name. The species name, *glauca*, refers to the bluish green, glaucous leaves. Also known under the vernacular names of Adam's-needle and Palmillo. The short crowns were sometimes used as a soap substitute by Indians.

PLAINS YUCCA

Yucca campestris McKelvey

FIELD IDENTIFICATION. Plant with a solitary head or with numerous heads in thicket-like clumps. Either stemless or with stems to 3 ft. Rather unsymmetrical when mature. Often with a hill of sand around the base.

FLOWERS. Borne in a dense, slender, or broadly ellipsoidal panicle; the panicle and scape together 1–3 ft tall; the base of the panicle usually among the leaves or sometimes above; branches of panicle numerous, rather slender, 4–5 in. long toward the base, 2–3½ in. long toward the apex; flowers proper greenish or sometimes pinkish, usually 3–4½ long. usually globose, tubular or slightly spreading above; segments 6, about of equal length, the 3 inner concave, about 1 in. wide and broader than the 3 outer ones; stamens 6, filaments rather slender, erect to spreading; anthers about ⅛ in. long or less; pistil stout, 1–1¼ in.; ovary ½–¾ in. long, ovoid or obovoid, indented by anthers; style ovoid, green, papillose, enlarged at base; stigmas erect at anthesis.

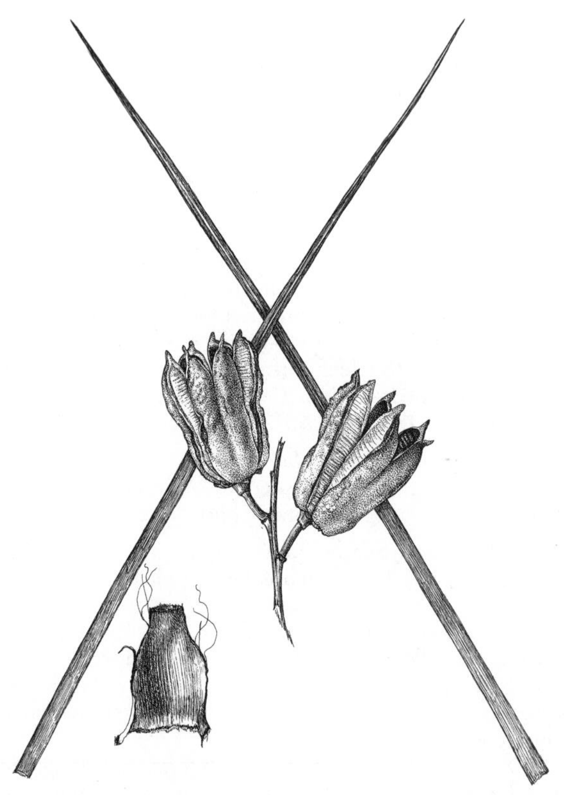

PLAINS YUCCA
Yucca campestris McKelvey

FRUIT. Capsule subglobose or short-oblong, brown to gray, 2–4½ in. long, 1–2 in. broad, symmetrical but sometimes constricted, mucros present or absent; dehiscent to base on primary fissures, on secondary fissures only part way; carpels 3, dorsally rounded or flattened, apex sometimes flaring, median rib present or absent; old floral base tubular and 6-parted; perianth segments persistent; seeds ⅜–½ in. long or broad, thin, and glossy.

LEAVES. Blade slender and wiry, bluish green, sometimes as long as 3 ft, but generally much shorter, about ⅜ in. at widest point, this usually being above the middle; sometimes the blade tapering the entire length and not noticeably widened near the center; surfaces plano-convex or plano-keeled; margins of young leaves whitened with fine threads, threads often absent with age; apices tipped with a slender spine about ⅜ in. long.

RANGE. Usually in very sandy soil in Ward, Ector, Midland, and Howard counties of Texas.

REMARKS. The genus name, *Yucca*, is from the ancient Haitian name. The species name, *campestris*, refers to the plains or prairies where it grows.

SMOOTH-LEAF SOTOL

Dasylirion leiophyllum Engelm.

FIELD IDENTIFICATION. Plant with a clump of slender, spiny-edged basal leaves arising from a short trunk.

FLOWERS. Inflorescence a stout, narrow, spicate panicle borne on a rather tall scape; the very short pedicels articulated close to the dioecious flowers; segments of the perianth scarcely over 1/12 in. long, rather obovate, apices obtuse, margin denticulate, thin, greenish or white; stamens 6; ovary 1-celled, style short, stigmas 3.

FRUIT. Capsule coriaceous, indehiscent, 3-sided and with 3 papery wings; obovate to subelliptic, ⅙–⅕ in. wide, ¼–⅓ in. long; style thick and about equaling the moderately open and deep apical notch, or exserted if wings are not fully developed, 1-celled; seed solitary, 1/12–⅛ in. long, bluntly triangular.

LEAVES. Clumped at the base, rigid, linear, from ⅗–⅘ in. wide, hardly 3 ft long, somewhat brush-tipped, green or glaucous, smooth, margin serrulate-roughened with prickles ⅖–⅗ in. apart, ⅛–⅙ in. long, and yellow to orange or red.

RANGE. In southern Texas in the Rio Grande region; northward into New Mexico and southward to central Chihuahua, Mexico. In Texas at Presidio, Eagle Pass,

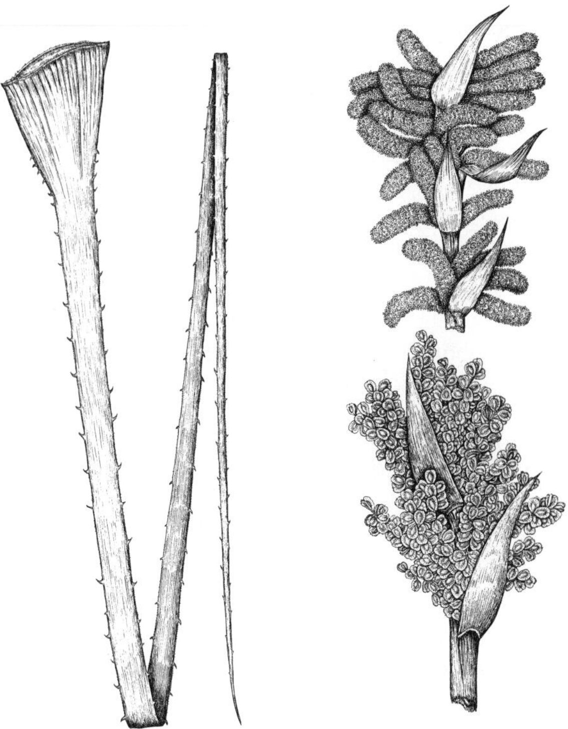

SMOOTH-LEAF SOTOL
Dasylirion leiophyllum Engelm.

and Van Horn; in New Mexico in the Florida Mountains.

REMARKS. The genus name, *Dasylirion*, means "tufted lily." The species name, *leiophyllum*, refers to the smoothness of the leaf surfaces. The leaves are used to make mats, baskets, and paper.

WHEELER SOTOL

Dasylirion wheeleri Wats.

FIELD IDENTIFICATION. Plant with slender leaves basally clumped. The trunk sometimes to 3 ft, either buried or above ground.

FLOWERS. In early summer, dioecious, scape 9–15 ft high, inflorescences spikelike, arranged in the axils of leaflike bracts; pedicels short, articulated close to the flowers; perianth campanulate, the segments (petals and sepals) alike, about 1/10 in. long, white or greenish, obovate, obtuse, denticulate; stamens 6, exserted, filaments slender; style short, stigmas 3.

FRUIT. Capsule coriaceous, circular to obovate, about 1/4 in. by 1/3 in., 3-winged, style normally equaling the open, moderately deep apical notch, 1-celled; seed solitary, about 1/8 in., obtusely triangular.

LEAVES. Clumped, rather rigid, linear, to 3 ft long, 3/5–1 in. wide, surfaces somewhat glaucous, nearly smooth, dull; marginal prickles directed apically, 1/5–2/5 in. apart, 1/12–1/8 in. long, yellow or becoming brown.

RANGE. On rocky and gravelly hillsides or slopes at altitudes of 3,000–5,000 ft. In Texas at El Paso; west to the Organ, Burro, and Pinal mountains; and south in Mexico to Chihuahua.

PROPAGATION. The plant will stand several degrees of frost and will easily grow from seed. It can be grown in rock gardens in pots. The leaves are inverted in floral arrangements to display the white, shiny, broad, spoonlike bases. The plant is sometimes known in the floral trade as Spoon-flower or Spoon-leaf.

VARIETY. The following variety has been described by Trelease (1911a, p.443): "*D. wheeleri* var. *wislizenii* Trel. has a short trunk; leaves 3/5–4/5 in. wide, scarcely 3 ft long, green or slightly glaucous, typically smooth and rather glossy; prickles 1/5–2/5 in. apart, 1/12–1/8 in. long, reddish brown or yellow; inflorescence ample; fruit triangular-obcordate, 1/4–1/3 in.; style thick and about equaling the open, moderately deep notch. Immature seeds about 1/8 in. long. Known from north to central Mexico and adjacent Texas—apparently grading into *D. wheeleri*. Adjoining or overlapping the area of *D. wheeleri* and *Nolina erumpens*. El Paso, Texas (Franklin Mountains), Chihuahua, Paso del Norte."

REMARKS. The genus name, *Dasylirion*, means "tufted lily." The species name, *wheeleri*, is in honor of William Archie Wheeler (b.1876). In times of stress ranchers burn the leaves and split the short round heads for cat-

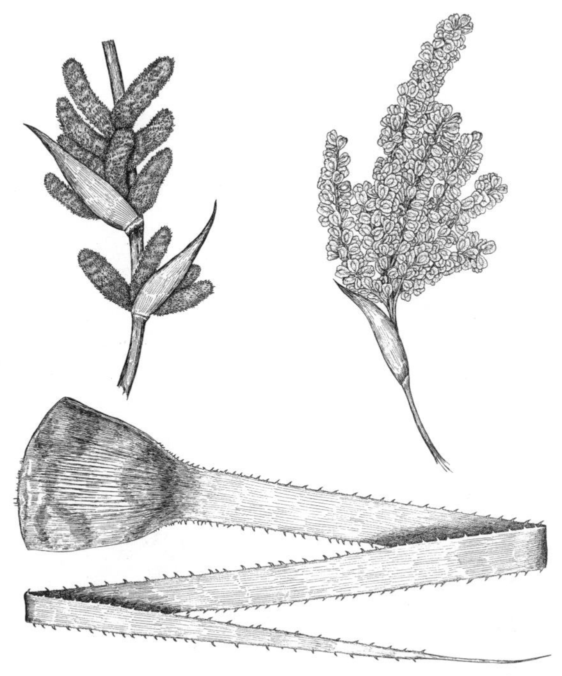

WHEELER SOTOL
Dasylirion wheeleri Wats.

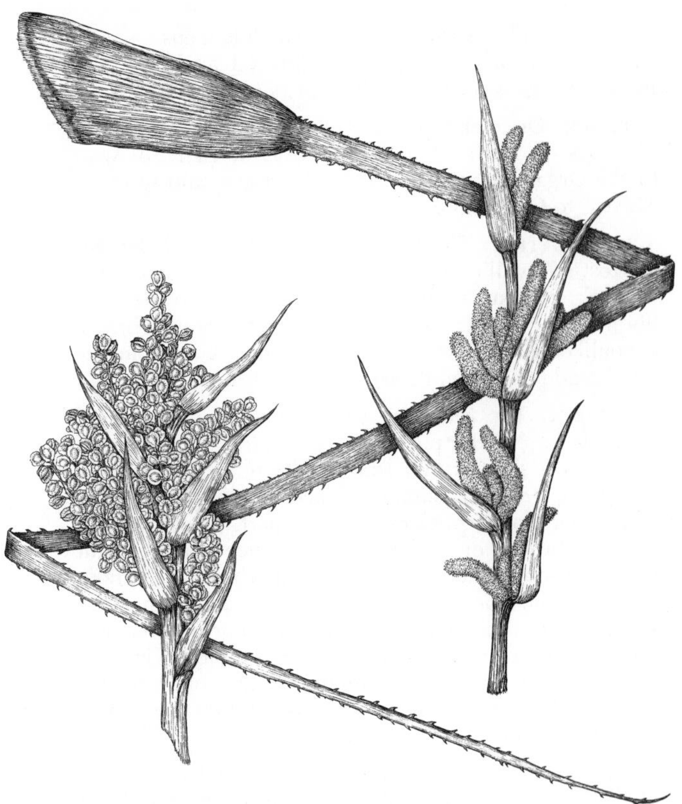

Texas Sotol
Dasylirion texanum Scheele

tle feed. The Indians and Mexican people prepared an alcoholic drink known as "sotol" by roasting the heads in a pit for 24 hours and then distilling the expressed juice. The leaves are used to make mats, baskets, thatch, and paper.

Texas Sotol

Dasylirion texanum Scheele

FIELD IDENTIFICATION. Plant with slender, basal, clumped leaves. Trunk woody and short, often buried in the ground.

FLOWERS. Dioecious, scape 9–15 ft high, the narrow inflorescence in a spikelike panicle, 2–3 ft long; the short pedicels articulated close to the flowers; floral bracts broadly ovate and acute, lacerately toothed; perianth 1/12–1/8 in. long, segments white or green, obovate, obtuse, the margins toothed; stamens 6, exserted, filaments slender, style short, stigmas 3.

FRUIT. Capsule short-pedicellate, coriaceous, 1/6–1/4 in. wide, 1/4–1/3 in. long, 3-sided, 3-winged, elliptical, 1-celled, 1-seeded, the seed obtusely triangular; style very short and equaling or surpassing the open shallow apical notch.

LEAVES. Clumped at the base, linear, attenuate, stiff, 2/5–3/5 in. wide, scarcely 3 ft long, somewhat brush-tipped, base dilated and entire; margin with prickles 1/5–2/5 in. apart, 1/12–1/8 in. long, yellow or brownish; surfaces green, glossy, smooth or rough-keeled.

RANGE. In dry, arid, stony soil. Known mostly from south central Texas, occurring in the vicinity of New Braunfels, Blanco, Kerrville, Sanderson, Marathon, and Fort Davis.

VARIETY. A Mexican variety with dull, somewhat glaucous leaves about 3/5 in. wide has been named *D. texanum* var. *aberrans* Trel. It is found in Nuevo León and Coahuila, Mexico.

REMARKS. The genus name, *Dasylirion,* means "tufted lily." The species name, *texanum,* refers to the state of Texas, to which this species is restricted. It is easily propagated from seed. The plant has high ornamental value and can be grown in large pots. Cattle feed on the round heads when the leaves are burned off.

SMILAX FAMILY (*Smilacaceae*)

REDBEAD GREENBRIER
Smilax walteri Pursh

REDBEAD GREENBRIER
Smilax walteri Pursh

FIELD IDENTIFICATION. Slender, spiny, or almost spineless vine climbing to a height of 5–20 ft by numerous tendrils. Bearing long underground runners, but not tuberous.

FLOWERS. Borne April–June in separate staminate and pistillate umbels of 6–15 flowers; peduncles shorter or sometimes a little longer than petioles; staminate peduncles ⅛–¼ in., and pedicels about the same length; sepals 3 and petals 3, very similar, greenish or bronze, linear, recurved, about ¼ in.; stamens 6, ⅕–¼ in.; pistillate peduncles flat, ⅛–1 in., pedicels ⅛–⅓ in., segments not as recurved as in the staminate flower; stamens 1–6, usually abortive; ovary oval, styles 2–3, stigmas 2–3 and spreading.

FRUIT. Maturing September–October of the first year, but rather persistent during the winter; peduncle flattened, berry about ⅓ in. in diameter, red to orange (rarely white), globose to obovate or more rarely blunt-pointed; seeds usually 1–2, occasionally 3, dull reddish brown, ⅕–¼ in. in diameter.

LEAVES. Deciduous, alternate, ovate to triangular-ovate or lanceolate, apex obtuse or abruptly acute, mucronate, margin entire, base rounded to truncate or subcordate; thin, smooth, coriaceous; upper surface green, glabrous, lustrous; lower surface also green, glabrous, prominently reticulate, main veins 5–7; length of blade 1¾–5 in., width ¾–3¼ in.; petioles ⅕–½ in.

STEMS. Canes slender, glabrous, terete or obscurely angled; young ones green to reddish brown, spineless or nearly so; older ones dark brown, spines more numerous, flattened, 1/15–⅙ in.

RANGE. Not common anywhere, on margins of sandy acid swamps, or along streams, often growing in the water. In eastern Texas, Arkansas, and Louisiana; eastward to Florida and northward to New Jersey and Maryland.

REMARKS. The genus name, *Smilax*, is from the Greek word *smilax*, of obscure meaning. The species name, *walteri*, is in honor of Thomas Walter (1740–1789), author of *Flora Caroliniana*. The plant is also known under the vernacular names of Red-berry Bamboo, Coral Greenbrier, and Sarsaparilla. The fruit is known to be eaten by at least two species of birds, and is nibbled by marsh rabbit.

Dwarf Greenbrier

Smilax pumila Walt.

FIELD IDENTIFICATION. Weak, slender, unarmed, usually prostrate vine, occasionally with a few tendrils and clambering over low objects. The upper stems are herbaceous and softly hairy, but the lower ones are woody near the base.

FLOWERS. Borne in staminate and pistillate axillary umbels; peduncles densely woolly and usually shorter than the petioles, 1/12–⅗ in.; pedicels 6–30, 1/25–1/12 in.; petals 3 and sepals 3, yellowish green; staminate flowers with segments recurved, lanceolate to linear, ⅙–⅕ in.; stamens 6, anthers erect and shorter than filaments; usually with no ovary; pistillate flower with segments about ⅛ in., usually with 1–6 abortive stamens; ovary 1-celled, 1-ovuled, style short, stigmas 1–2.

FRUIT. Borne in umbellate clusters, persistent, red to orange, ovate to oval, blunt-pointed, shiny, glabrous, ⅕–⅓ in.; seed solitary, oval, hard, basal disk black, about ⅛ in.; peduncle hairy.

LEAVES. Deciduous to persistent southward, varying from ovate to oval or broadly heart-shaped, sometimes oblong, apex obtuse to acute or abruptly mucronulate, base cordate, margin entire, thin but firm, reticulate-veined; main veins 3–5, reddish and prominent on the lower leaf surface, which is also clothed with dense gray to brown hairs; upper surface lustrous, sparingly pubescent to almost glabrous (young leaves with both surfaces densely woolly-tomentose); length of leaf blade 2–4 in., width 1¾–3¼ in.; petiole brown-tomentose, ⅛–¾ in., or on vigorous shoots 1–3 in.; stipular sheath one fourth to one half as long as the petiole; tendrils on petiole, if any, 1/12–⅕ in. from base.

STEMS. Young stems weak, slender, densely brown-hairy, herbaceous; tendrils solitary or in pairs at nodes; no thorns as in other *Smilax* species; usually prostrate or creeping.

RANGE. Usually in open sandy woods. Eastern Texas, Arkansas, and Louisiana; eastward to Florida and northward to South Carolina.

REMARKS. The genus name, *Smilax*, is from the Greek word *smilax*, of obscure meaning. The species name, *pumila*, is for its dwarf habit. Vernacular names are Hairy Greenbrier, Prostrate Greenbrier, Brier-vine, Ground-brier, and Wild Sarsaparilla. It differs from most greenbriers by its thornless stems, prostrate habit, densely tomentose leaves, and fruit in the spring.

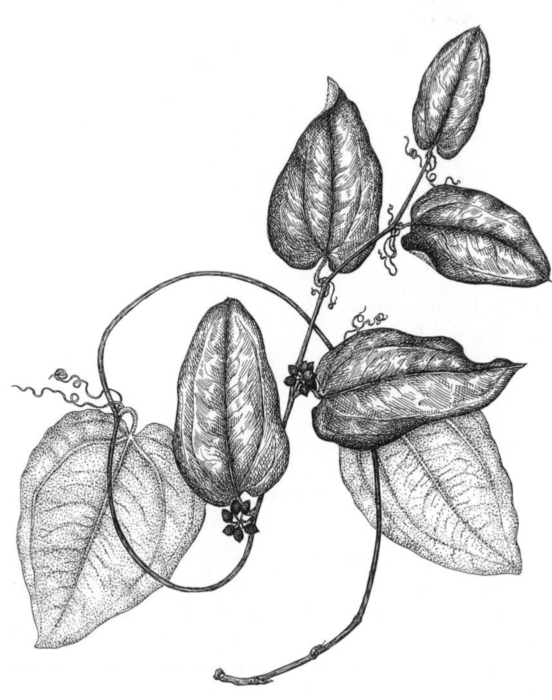

Dwarf Greenbrier
Smilax pumila Walt.

Saw Greenbrier

Smilax bona-nox L.

FIELD IDENTIFICATION. Stout spiny vine with 4-angled branches, either low clambering or extensively climbing by tendrils. The underground tubers are unusual among the greenbriers in being densely set with single or branched spines, which are also on the runners.

FLOWERS. April–June, in axillary, staminate and pistillate umbels; peduncles usually much longer than petioles; staminate peduncles ⅝–1¼ in., flat, slender; pedicels ⅙–⅓ in.; sepals 3 and petals 3; in the staminate flowers the segments linear or nearly so, ⅙–¼ in.; stamens 6, filaments ⅛–⅙ in.; pistillate flowers with flat peduncles ⅜–1⅜ in. (rarely to 2 in.), pedicels 1/7–⅖ in.; pistillate flowers 8–20 in the umbel, segments of the flower linear, ⅛–⅙ in.; stamens 1–6, usually abortive; ovary short-clavate, 1-celled, 1-ovuled (but showing evidence of being aborted from 2–3 cells), styles short.

FRUIT. Maturing September–November, berry globular or ovate, about ¼ in. thick, shining or dull black, sometimes a little glaucous; seed solitary (rarely 2) ellipsoid-obovoid to subglobose, reddish brown, with a black basal disk.

LEAVES. Tardily deciduous especially southward, greatly variable in shape, ovate, cordate, broadly lanceolate, deltoid, panduriform or hastate, apex acute or abruptly cuspidate; base rounded, truncate, cordate, sometimes with strong basal lobes, margin entire or set with prickles, stiff and coriaceous; surfaces green and glabrous, sometimes blotched white, more or less reticulate, the 5–9 larger veins sometimes with tiny spicules; blade length 1½–4½ in., width ⅔–3 in.; petioles ¼–⅗ in., often prickly.

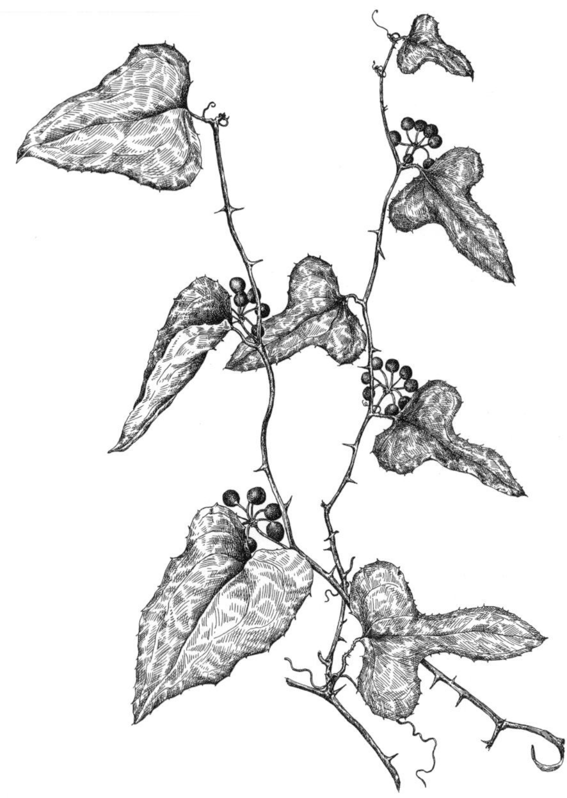

SAW GREENBRIER
Smilax bona-nox L.

STEMS. Canes stout, strongly 4-angled above, more terete toward base, with warty patches of whitish spinules or "hairs." Also set with spines at nodes and internodes, spines stout, green to brown or black, to ⅖ in., more numerous toward base of canes, those at nodes often 2 with several smaller ones on each side. Flowering stems noticeably flexuous.

RANGE. In thickets, dry woods, roadsides, and fields. Texas, Oklahoma, Arkansas, and Louisiana; eastward to Florida, northward to Massachusetts, and westward to Nebraska, Kansas, and Missouri. Also in Mexico and the Bahama Islands.

VARIETIES AND SYNONYMS. Since the plant is so variable a number of species and varieties have been described, but most of these have been reduced to synonyms. These include S. *tamnoides, sensu* Chapm. & A. Gray, not L., S. *renifolia* Small, and S. *variegata* Walt.

Beach-saw Greenbrier, S. *bona-nox* var. *littoralis* Coker, has gray tubers without spines, stolons without or with a few spines; aerial canes with no scurfy "hairs"; otherwise as in the species. Type collected at Myrtle Beach, South Carolina.

Hastate-saw Greenbrier, S. *bona-nox* var. *hastata* (Willd.) DC., has leaves with horizontally divergent, oblong, basal lobes which are perpendicular to the narrowly oblong to linear-lanceolate summit. Known from Texas to Florida and northward to Virginia.

Earless-saw Greenbrier, S. *bona-nox* var. *exauriculata* Fern., has leaves without basal ear-lobes, narrowly lanceolate, base cordate. Known from Virginia.

Ivyleaf-saw Greenbrier, S. *bona-nox* var. *hederaefolia* (Beyrich) Fern., has leaves broadly cordate-ovate or deltoid-ovate to shallowly panduriform, the broad basal lobes not prolonged. Known from Texas to Florida; northward to Delaware and westward to Kansas and Missouri.

REMARKS. The genus name, *Smilax*, is from the Greek word *smilax*, of obscure meaning. The species name, *bona-nox*, means "good night." Vernacular names used are Fiddle-leaf Greenbrier, China-brier, Bull-brier, Tramp's-trouble, Stretch-berry, Cat-brier, Greenbrier, and Fringed Greenbrier. The fruit is known to be eaten by thirteen species of birds, including the ruffed grouse.

BRISTLY GREENBRIER
Smilax hispida Muhl.

FIELD IDENTIFICATION. Stout, woody, high-climbing vine, bearing tendrils at the nodes. Mostly thornless above, but lower stem densely covered with straight needle-like spines. Underground stems without tubers or runners of any length.

FLOWERS. Dioecious, in separate staminate and pistillate axillary umbels, 5–26-flowered; peduncles flattened, always longer than the petioles; flowers small, green, sepals and petals in threes, distinct and deciduous; staminate peduncles ½–⅘ in.; pedicels ⅕–⅖ in.; petals and sepals linear to linear-elliptic, about ¼ in.; stamens 6, filaments about ¼ in.; anthers erect and about 1/12 in.; pistillate flowers numerous, peduncles ⅜–2 in. (averaging 1–1¼ in.), slender and flattened; pedicels ⅙–⅓ in.; pistillate flowers smaller than the staminate, petals about 1/7 in.; stamens rudimentary in pistillate flowers; ovary oval, with 3 cavities, ovules 3, but usually with 1 or 2 abortive; styles 3; stigmas 3, rather thick and spreading.

FRUIT. Maturing August–November of first year, persistent, on flattened peduncles 1–2 in., longer than the petioles; individual pedicels in clusters of 5–26, about ¼–⅓ in.; berry bluish black, usually without bloom, subglobose or slightly longer than wide, ¼–⅓ in. in diameter; seed usually solitary (rarely 2–3), ⅙–3/16 in., rounded to ovoid, somewhat flattened, reddish brown to dark brown.

LEAVES. Alternate, variable in shape, oval, ovate, heart-shaped, elliptic-ovate, orbicular-ovate, ovate-lanceolate, base rounded or cordate, margin usually entire but sometimes denticulate, apex acute to obtuse or abruptly cuspidate, thin and coriaceous, upper surface green, glabrous, lustrous, with the 5–7 main veins

Bristly Greenbrier
Smilax hispida Muhl.

from the base impressed, lower surface paler green and glabrous, sometimes with minute, toothlike nodules scattered along the conspicuous veins, length of blades 2–6 in., width 2–5½ in.; petiole ⅓–¾ in., gradually expanding into the blade.

STEMS. Rather high climbing, woody, stout, tendrils at nodes; young stems green, terete, finely grooved, usually spineless; older stems green to brown, lower part often densely beset with slender, straight, brown to black spines 1/25–⅓ in.

RANGE. Rich woods along streams or low grounds; eastern Texas, Oklahoma, Arkansas, and Louisiana; eastward to Florida, northward to Connecticut, and westward to Kansas, Nebraska, and Missouri.

VARIETIES. Fiddle-leaf Bristly Greenbrier, S. *hispida* var. *australis* Norton, has pandurate (fiddle-shaped) leaves. The basal lobes are broad and rounded and blade is narrowed to the apex. In other characteristics it is like the species.

Mountain Bristly Greenbrier, S. *hispida* var. *montana* Coker, has the primary rootstock which is short, black, thickly covered with long stout roots, but differing from the typical in giving rise to 1 or several long (to 8–10 ft) runners. These runners are stout or slender, densely to sparsely armed, at times unarmed, usually without, but sometimes with, enlarging rooting joints often with aerial shoots; fruit large, few in a cluster, round or oblate-flattened, many containing 2–3 seeds; peduncles averaging about ¾ in. Aerial stems and leaves as in typical *hispida.* Known from mountainsides at altitudes of 4,400–4,500 ft, in Macon County, North Carolina (Shortoff and Satulah mountains); also between Buck Creek and Perry Gap, Clay County, North Carolina.

Fernald (1950, p. 450), has reduced S. *hispida* to a variety of S. *tamnoides* L.

REMARKS. The genus name, *Smilax,* is from the Greek word *smilax,* of obscure meaning. The species name, *hispida,* is for the hispid, or thorny, stems. Vernacular names are Hagbrier, Devil Greenbrier, Hellfetter, and Wild Sarsaparilla. Bristly Greenbrier can be trained into a hedge plant with proper pruning.

Common Greenbrier

Smilax rotundifolia L.

FIELD IDENTIFICATION. Vine climbing by spirally coiled tendrils. Sometimes forming tangled thickets. The stems perennial, woody, evergreen, and erratic. The thickened underground stems have no runners.

FLOWERS. Borne April–June, in staminate and pistillate axillary umbels; peduncles shorter or a little longer than petioles; flowers small, regular, greenish yellow; staminate flower with pedicels ¼–⅖ in.; with 3 sepals and 3 petals, the segments strap-shaped, pointed, recurved above the middle; stamens 6, distinct; filaments two to three times as long as the anthers which are 2-celled and introrse; pistillate flowers generally smaller than the staminate, sepals and petals about 1/25 in., oblong, acute; stamens often reduced and abortive; ovary 3-celled, styles 3 and prominent, stigmas thick and spreading.

FRUIT. Ripening the first year, persistent, borne in rounded umbellate clusters; peduncles ½–1½ in., glabrous; pedicels ⅛–¼ in. in clusters of 3–25; berry ovoid to subglobose, black, 3/16–¼ in., skin thin, flesh scanty

Common Greenbrier
Smilax rotundifolia L.

and tough; seed mostly solitary, occasionally 2–3, ovoid, rounded at ends.

LEAVES. Alternate, deciduous or sometimes tardily so southward, shape very variable, from ovate to orbicular or heart-shaped, more rarely lanceolate; margin entire or occasionally erose-denticulate; apex from obtuse to acute or rounded, often apiculate; base cordate; thin and leathery; reticulate–veined with the 5 main veins from the base, often erose; upper surface dark green and glabrous; lower surface slightly paler, glabrous; blade length 2–6 in., width 1–6 in.; petiole ¼–¾ in., somewhat winged at blade junction, sometimes erose.

STEMS. Woody, high-climbing to 15–20 ft, erratic, terete (or 4-angled in some varieties), bright green to brown, glabrous or nearly so; spines rather scattered, stout, usually not at nodes, green to brown or black, ⅛–½ in.

RANGE. In low woods, thickets, fence rows. Widespread in Texas except in the Trans-Pecos and Panhandle areas; Oklahoma, Arkansas, and Louisiana; eastward to Florida, northward to Nova Scotia, and westward into Ontario, Minnesota, Iowa, Kansas, South Dakota, and Missouri.

VARIETIES. Square-stem Common Greenbrier, S. *rotundifolia* var. *quadrangulata* Gray, has squarish stems. However, since intermediate forms are sometimes found, the distinction is not always clear.

Crenate-leaf Common Greenbrier, S. *rotundifolia* var. *crenulata* Small & Heller, has lanceolate to ovate-lanceolate leaves with strongly nodular and crenate margins.

REMARKS. The genus name, *Smilax*, is from the Greek word *smilax*, of obscure meaning. The species name, *rotundifolia*, is for rounded shape of leaves. Vernacular names for the plant are Biscuit-leaves, Wait-a-bit, Niggerhead, Bamboo-brier, Devil's-hop-vine, Bread-and-butter, Hungry-vine, Greenbrier, Cat-brier, Horse-brier, and Sow-brier. The fruit is eaten by 15 species of birds, and by rabbit, opossum, and raccoon.

LAUREL GREENBRIER
Smilax laurifolia L.

LAUREL GREENBRIER
Smilax laurifolia L.

FIELD IDENTIFICATION. Stout evergreen vine climbing to a height of 5–15 ft by paired tendrils, and bearing large rigid spines. The underground stems become tubers, which are very large, thick, half-woody, and reddish. There are no true stolons.

FLOWERS. July–August, borne in peduncled umbels, peduncles shorter or a little longer than the petioles (1/25–⅓ in.); pedicels 1/12–¼ in.; staminate and pistillate flowers separate; flowers numerous, rather crowded, regular, greenish white, petals 3 and sepals 3; segments of the staminate flowers longer than those of the pistillate flowers; segments of pistillate flower linear-lanceolate, ⅕–⅙ in.; stamens 6, erect, about as long as the petals, anthers about one third shorter than filaments; (stamens of pistillate flowers sometimes abortive); ovary 2–3-carpellate, usually with only one fertile ovule because of abortion, styles 2–3, stigmas 2–3, short, thick, fused below, early deciduous.

FRUIT. Ripening in October, in clusters of 5–25 (usually 6–12), maturing the second year, persistent, black, young somewhat glaucous, later nonglaucous, rather shiny, subglobose to ovoid, ⅕–⅓ in. in diameter; seed solitary, rarely 2.

LEAVES. Evergreen, alternate, blades more or less conduplicate, broadly to narrowly elliptic, oblong, lanceolate, or in some forms linear; apex from acute to obtuse or acuminate, cuspidate; base cuneate to rounded or subcordate, length 3–5 in., width ½–1½ in., thick and coriaceous; surfaces pale green, smooth, glabrous, sometimes glaucous; lower surface with 3 prominent veins, the others inconspicuous; petiole ⅛–¼ in., essentially glabrous.

STEMS. Perennial, woody, green, finely grooved; spines usually not at nodes, stout, flattened, straight or slightly curved, green to black, ⅛–½ in., sometimes spineless; fruiting canes sometimes bear more than 1 crop of fruit.

RANGE. In sandy, acid swamps or wet woods. Eastern

Texas and Arkansas; eastward to Tennessee, Kentucky, Georgia, and Florida, and northward to New Jersey.

VARIETY. Linear-leaf Laurel Greenbrier, S. *laurifolia* var. *bupleurifolia* Delile, has very narrow linear leaves.

REMARKS. The genus name, *Smilax*, is from the Greek word *smilax*, of obscure meaning. The species name, *laurifolia*, is for the smooth, thick, laurel-like leaves. Vernacular names are Bamboo-vine, China-brier, Laurel-brier, Blaspheme-vine, and Bay-leaf Smilax. The leaves are conspicuous because of their pale green color and the upright angle at which they are set on the stem.

At least 5 species of birds are known to eat the fruit. It has been reported that the large semiwoody tubers were once used by the Indians and early settlers for food. They were beaten to a pulp, strained to remove the woody fibers, and then ground to make a meal which was made into a dough or mush, or used in a cooling drink. However, it seems most probable that the more tender tubers of S. *lanceolata* and S. *auriculata* were more often used for the purpose.

LANCELEAF GREENBRIER
Smilax lanceolata L.

LANCELEAF GREENBRIER
Smilax lanceolata L.

FIELD IDENTIFICATION. Evergreen vine, vigorous, stout, usually spineless, climbing to heights of 5–40 ft by bifid tendrils. Underground tubers very large, thick, and jointed.

FLOWERS. Borne in axillary, umbellate clusters of separate staminate and pistillate flowers; individual flowers somewhat fragrant, spreading, yellowish green, regular, petals 3, sepals 3; peduncle of staminate flowers 1/8–1/3 in., pedicels the same or longer; segments of staminate flowers narrowly lanceolate, recurved, 1/6–1/4 in.; stamens 6, nearly as long as the petals, filaments longer than anthers and inserted at base of segments, anthers introrse; peduncle of pistillate flower 1/8–2/5 in. (sometimes to 2/3 in.), pedicels about same length; pistillate flowers smaller than staminate, sepals slightly broader than petals on some plants, sometimes with 1–6 abortive stamens; ovary 3-celled, ovules 3, but by abortion only 1–2 maturing; styles 2–3, stigmas 2–3.

FRUIT. Ripening in June of the summer of second year, berries in umbellate clusters of 4–10 (rarely as many as 40), dull red or reddish brown at maturity, with or without bloom, subglobose or somewhat flattened, 1/6–1/3 in. in diameter; seeds 1–3, mostly 2, brown.

LEAVES. Evergreen, alternate, often conduplicate, shape varying from lanceolate to ovate-lanceolate or elliptic (on some plants more or less hastate); margins entire or with tiny cuplike nodules; apex acute or acuminate; base broadly cuneate to almost truncate or acute (more rarely subcordate); thin but coriaceous; upper surface deep green, glabrous, shiny, sometimes with mottled paler areas; lower surface paler, glabrous and sometimes glaucous, inconspicuously 5–7-veined; blade length 2–5 in.; width 3/4–2 in.; petioles 1/12–1/2 in., glabrous, their sheaths about one half as long, with entire or ciliolate margins; some leaves with paired coiled tendrils as appendages of the petiole base.

STEMS. Vigorous, tough, stout, woody, green to reddish brown, glabrous, glaucous, terete or slightly angled; usually spineless; but sometimes with a few spines toward the lower parts of the old stems (fruiting canes always spineless); spines, if present, either at nodes or internodes, flattened, slightly recurved, 1/16–3/8 in.

RANGE. In rich grounds, thickets, fields, edges of ditches or streams. Eastern Texas, Arkansas, and Louisiana; eastward to Florida and northward to Virginia.

REMARKS. The genus name, *Smilax*, is from the Greek word *smilax*, of obscure meaning. The species name, *lanceolata*, is for the lanceolate leaves. Also known under the vernacular names of Coral Greenbrier, Jackson-brier, Bamboo-vine, and Thornless Smilax. The branches are much used for winter decorations, and the vine itself is of ornamental value for planting on porches, trellises, or arbors. The large tubers were ground into a flour and made into bread or a cooling drink by the Indians and early settlers. It attains a weight of as much as 16 lb and a length of 2 ft or more.

Auricled Greenbrier

Smilax auriculata Walt.

FIELD IDENTIFICATION. A vine, either low and clambering over bushes or climbing to a height of 20–30 ft. The slender zigzag stems are either spiny or spineless. The thickened underground stems become larger, reddish, and tuberous, with no stolons.

FLOWERS. May–June, in separate axillary clusters of staminate and pistillate flowers; peduncles and pedicels very variable in length on different plants; staminate peduncles usually 1/25–2/5 in.; pedicels 1/12–1/3 in.; flowers yellowish green, small, regular, petals 3, sepals 3; staminate flowers usually with no ovary, segments narrow and about 1/5 in.; stamens 6, with erect, 2-celled, introrse anthers which are shorter than filaments; pistillate flowers with segments varying from linear to elliptic to oblanceolate, 1/6–1/5 in., sometimes with 1–6 abortive stamens; ovary oval, ovules 3, but sometimes only 1–2 mature because of abortion; styles 3, stigmas 3, and broadly spreading.

FRUIT. Abundant, maturing second year, borne in axillary umbels on stout peduncles 1/25–3/5 in. (usually 1/8–1/3 in.), pedicels 1/12–1/3 in.; berries usually 4–20 in a cluster (occasionally as many as 40), 1/4–1/2 in. in diameter, subglobose, flattened, or sometimes slightly 3-lobed, reddish to purplish or finally black, with or without bloom; seeds 1–3, short-oblong to rounded, somewhat flattened, dark brown, lustrous.

Auricled Greenbrier
Smilax auriculata Walt.

LEAVES. Alternate, variously shaped, oval, ovate, elliptic, lanceolate or hastate, the more typical form broadly extended into 2 lateral auriculate lobes, some with bases broadly cuneate or rounded; apex abruptly pointed and mucronate or rounded to notched; margin entire, if examined with a lens from the underside the marginal thickening is double with a narrow groove between (a good way to distinguish it from species such as *S. bona-nox*); thin, leathery, prominently reticulate-veined; primary veins usually 5 in larger leaves or fewer in smaller leaves; surfaces glabrous and lustrous green above and beneath, or slightly glaucous beneath; blade length 3/4–4 3/4 in., width 1/2–3 in.; stipular sheath at base of petiole and about half its length; petiole about 1/2 in., often with a bifid coiled tendril as an appendage.

STEMS. Perennial, evergreen, slender, zigzag, rounded or angled, finely lined, climbing by coiled tendrils from petiole base; unarmed or with spines, spines mostly toward base if present, usually not at nodes, hardly more than 1/6 in., stout and dark.

RANGE. In sandy lands of flat woods or on dunes. Louisiana; eastward to Florida and Georgia and northward to North Carolina.

REMARKS. The genus name, *Smilax,* is from the Greek word *smilax,* of obscure meaning. The species name, *auriculata,* is for the earlike leaf bases. Vernacular names are Bamboo-vine, Horse-brier, and Arrow-leaf Smilax.

Cat Greenbrier

Smilax glauca Walt.

FIELD IDENTIFICATION. Slender, spiny vine climbing by coiled tendrils at the nodes. The underground tubers are borne either in potato-like strings or in knotty thickened masses.

FLOWERS. May–June, small, regular, greenish, borne in separate staminate and pistillate umbels; peduncles much longer than petioles, 1/3–1 3/8 in., slender, flat; pedicels averaging 6–12, 1/6–2/5 in.; petals 3 and sepals 3; segments in the staminate flower linear to oblong, about 1/5 in., stamens 6; pistillate flower segments narrowly lanceolate, about 1/8–1/6 in.; ovary ovate, glaucous, about 1/12 in., 3-carpellate, styles 3, stigmas 3.

FRUIT. Ripening the first year, September–October, berry subglobose to oblate, 1/4–1/3 in. in diameter, black with a glaucous bloom, with maturity somewhat shiny; seeds 2–3, somewhat flattened, brown.

LEAVES. Deciduous, or long persistent in protected

CAT GREENBRIER
Smilax glauca Walt.

places, alternate, very variable in shape, mostly broadly ovate, but also sometimes oval, orbicular-ovate, elliptic-ovate, orbicular-reniform, lanceolate or panduriform-lanceolate; apex acute to acuminate and cuspidate; base rounded to truncate or semicordate; upper surface dark green or sometimes with lighter blotches; lower surface glabrous, smooth and conspicuously white-glaucous or somewhat roughened (if roughened, minute papillae show under magnification); main veins at base 3 (rarely 5); blade length 1½–4 in., width 1¼–3 in.; petiole ¼–½ in., sometimes with a spine on each side at the base.

STEMS. Slender, climbing by delicate tendrils; young ones glabrous or somewhat glaucous, spineless or with a few spines; older stems green to dark brown, glabrous; spines sometimes abundant, green to reddish brown or black, ¼–⅜ in., straight or slightly curved, borne either on the nodes or internodes (sometimes the nodes have 3–6 spines instead of 1–2).

RANGE. Usually in loose dry sandy soil. Eastern Texas, Oklahoma, Arkansas, and Louisiana; eastward to Florida, northward to Massachusetts, and westward to Nebraska, Kansas, and Missouri.

VARIETY. Smooth Cat Greenbrier, S. *glauca* var. *leucophylla* Blake, has leaves smooth and glabrous beneath.

REMARKS. The genus name, *Smilax*, is from the Greek word *smilax*, of obscure meaning. The species name, *glauca*, refers to the whitened undersurface of the leaf. Vernacular names in use are Glaucous-leaf Greenbrier, Sow-brier, and Sarsaparilla-vine. The plant is persistent, often forming thickets which have decided resistance to fire damage. The fruit is eaten by a number of species of birds, including the cardinal, bobwhite quail, wild turkey, and ruffed grouse. The large tubers are eaten by hogs.

AMARYLLIS FAMILY (*Amaryllidaceae*)

PARRY AGAVE

Agave parryi Engelm.

FIELD IDENTIFICATION. Plants large, acaulescent, often in colonies expanding by means of underground offsets.

FLOWERS. Scape 6–15 ft high, the branches spreading; perianth 2–3 in. long, perfect, constricted above into a neck and then expanding into 6 greenish yellow or reddish semierect, linear segments (3 petals and 3 sepals); stamens 6, long-exserted and more than twice as long as the perianth segments, yellowish green; anthers yellow, somewhat falcate; style light green to whitish, 2–3 in.; ovary almost cylindrical, 1–1⅗ in. long, about ⅓ in. in diameter, 3-celled.

FRUIT. Primary rachis flattened and grooved, stout, divergent into short, terminal and lateral, finger-like peduncles ¼–¾ in. Capsules ½ in. long, ⅘–1 in. in diameter, oblong to broadly cylindroidal, apex abruptly acute to obtuse, stigma remnants persistent, light green, dehiscent into 3 stiff, coriaceous valves; seeds numerous, 3/16–¼ in. long, black, lustrous, flattened, lunate.

LEAVES. Blades numerous, crowded, 10–16 in. long, 2–4 in. wide, curving upward, oblong or obovate, centrally concave, grayish green, gradually constricted toward the base and then broadened to clasp the stem, overlapping each other; apex abrupt or acutely pointed, apical spine stout, purple to black or gray, ⅘–1⅕ in. long, grooved dorsally; marginal spines ⅛–¼ in. long, straight or recurved, black or gray.

RANGE. The common agave of western New Mexico. Known from the Mogollon, Bear, Big Hatchet, Burro, Las Cruces, and Organ mountains. Also in eastern Arizona; southward in Chihuahua on slopes of the Sierra Madre Occidental.

REMARKS. The genus name, *Agave,* is from the Greek word *agaue,* meaning "noble." The species name, *parryi,* is in honor of C. C. Parry, who prepared reports on the United States–Mexico boundary survey in 1854.

The Indians roasted the crowns of Parry Agave in rock-lined ovens to prepare the food known as "mescal."

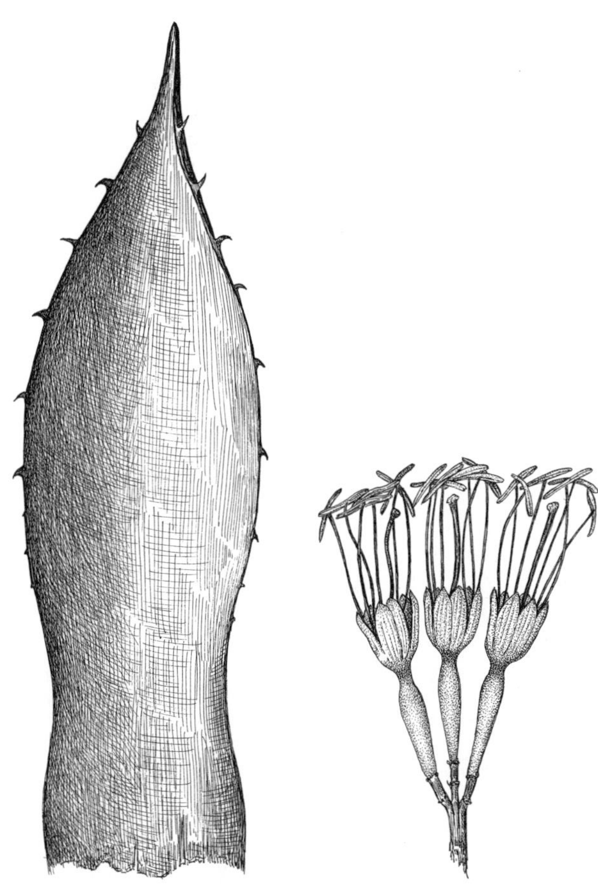

PARRY AGAVE
Agave parryi Engelm.

Havard Agave

Agave havardiana Trel.

FIELD IDENTIFICATION. An acaulescent plant with large fleshy basal leaves arranged in a rosette.

FLOWERS. Plant attaining a height of 12 ft or more, including the scape and loose yellow-flowered panicle toward the apex; perianth 2–2¼ in. long, tubular-funnelform, the 6 divisions (3 petals and 3 sepals) linear and almost equal; stamens 6, long-exserted, adnate to about the middle (or sometimes lower) of tube; ovary about equaling the lobes and tube of the perianth, inferior, 6-celled, style slender with an obscurely 3-lobed stigma.

FRUIT. Capsule slender, prismatic, coriaceous, apically acute; seeds numerous, flattened.

LEAVES. Blades 8–15 in. long, 4–6 in. wide, dull gray, rigid, smooth, concave, broadly ovate to suboblong, narrowed above the base, broader above the middle, apex acute or acuminate; terminal spine blackish, stout, subflexuous, round-grooved, decurrent; marginal spines rigid, the upper longest and straight, the lower smaller and subdeflexed, usually ⅜–⅘ in. long, the margin between slightly hollowed.

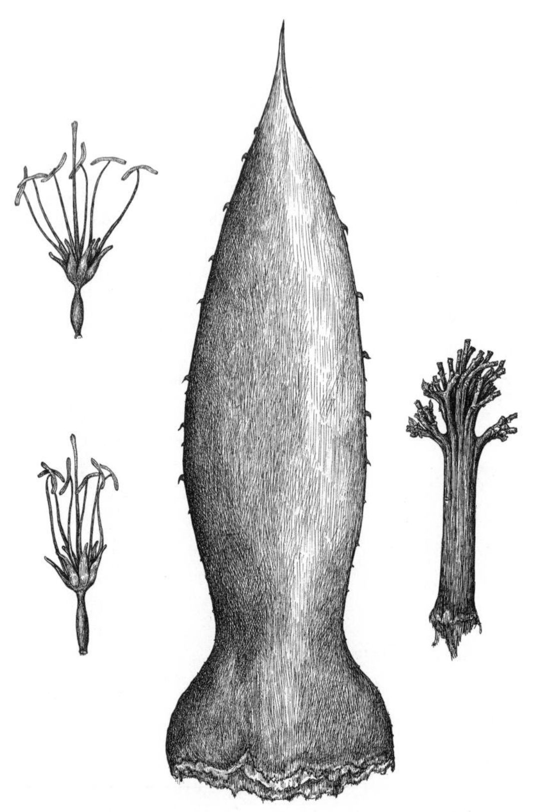

Havard Agave
Agave havardiana Trel.

RANGE. Havard Agave is scattered over the foothills and slopes of western Texas in the Chisos, Davis, and Sierra Blanca mountains, and perhaps in other ranges of that region.

NOMENCLATURE PROBLEMS. *A. havardiana* Trel. is listed by some authors under the names of *A. wislizenii* Engelm. and *A. scabra* Salm-Dyck. The name *A. havardiana* now seems most acceptable and was applied by Trelease (1911b).

The agaves of western Texas need a more careful study and revision. What is now known as *A. havardiana* seems to be very variable over a wide range. No doubt other species occur within the area which have not been properly classified. The following names have been applied by various authors to west Texas Agaves: *A. applanata* L., *A. heterocantha* Zucc., *A. huachucensis* Baker, *A. parryi* Engelm., *A. asperrima* Jacobi, and *A. chisosensis* Mull. The more herbaceous species of agaves in Texas have not been included here, such as *A. maculosa* Hook., *A. tigrina* (Engelm.) Cory, *A. variegata* Jacobi, and *A. virginica* L.

REMARKS. The genus name, *Agave*, is from the Greek word *agaue*, meaning "noble." The species name, *havardiana*, is in honor of Valery Havard (1846–1927), United States army surgeon of French birth, who collected plants in Texas and other states while stationed at army posts.

The baked crowns of agaves of western Texas were abundantly used by the Indians for food (mescal). The procedure seemed to vary between the pit-baking method (see under *A. neomexicana*), and that involving use of specialized refuse-heap ovens aboveground. However, it is often difficult to ascertain whether *A. havardiana* or *A. lechuguilla* was the species used. Some species of *Yucca* and *Dasylirion* were also roasted and the chewed portions, known as "sotol quids," are found at many sites.

Lechuguilla Agave

Agave lechuguilla Torr.

FIELD IDENTIFICATION. A very variable plant, perennial, long-lived, often occurring in extensive colonies. With a tall flower stalk rising from a cluster of numerous thick basal leaves.

FLOWERS. Blooms only once when 3–4 years old. The old plant dies, but young plants are produced at the base. The scape is 3–12 ft tall, the spikelike panicle at the apex with very short branches having flowers usually in clusters of 2–3 or more; bracts lanceolate from a broad and clasping base; perianth tubular-funnelform, green or yellow but sometimes almost white, often tinged purple; segments 6, narrowly oblong, nearly equal, ⅗–⅘ in. long; stamens inserted at the top of floral tube but long-exserted, filaments about 1⅗ in.; anthers about ⅗ in.; style 1–1⅗ in.; ovary ⅖–⅗ in. long.

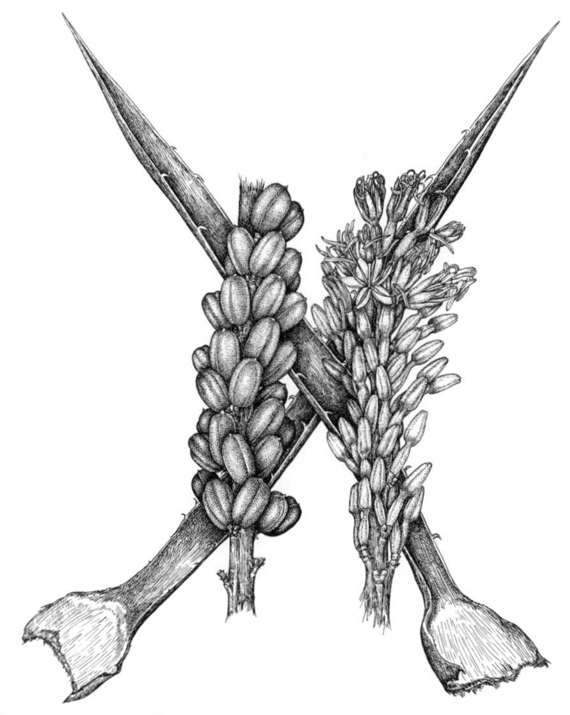

LECHUGUILLA AGAVE
Agave lechuguilla Torr.

FRUIT. Capsule brown to black, coriaceous, ⅗–1 in. long, ½–⅗ in. in diameter, oblong, almost cylindrical to obtusely triangular, with a short, broad acumination, 3-celled; seeds numerous, flat, black, shiny, semiorbicular, ⅙–⅕ in. long, about ⅛ in. broad.

LEAVES. From 10–30, grayish green or yellowish green, 8–24 in. long, upper face sometimes with a paler stripe, 1–1½ in. broad, at least six times longer than broad, rounded beneath and somewhat channeled above, usually falcate, apex terminating in a stout spine; marginal prickles hooked downward, rather distant, triangular, ⅕–⅖ in. long, the border between the prickles horny and easily detachable.

RANGE. Of wide distribution, covering an area about 100 miles wide and 700 miles long, from southern New Mexico through western Texas; and in Mexico southward in Chihuahua, Tamaulipas, Zacatecas, and San Luis Potosí.

REMARKS. The genus name, *Agave*, is from the Greek word *agaue*, meaning "noble." The species name, *lechuguilla*, is the Mexican common name, meaning "little lettuce." The root, under the name "amole," is extensively used as a substitute for soap. The plant is considered to be poisonous to livestock. When eaten by sheep and goats in time of an extreme dearth of food, it produces a feverish condition known as "swellhead." Its leaves, with other species of *Agave*, are used to make twine and rope, called "istle" or "tampico fiber." The plant is also often cultivated for ornament in gardens of succulent desert plants.

PALMER AGAVE

Agave palmeri Engelm.

FIELD IDENTIFICATION. An agave varying considerably in size, with the scape and inflorescence together, 8–23 ft tall. The leaves are basal, spiny-toothed, to 3 ft long and 2–4 in. wide.

FLOWERS. June–August, in spreading open panicles, the branches horizontal; perianth-tube 1–1½ in. long, green to purplish; tube segments 6, as long as the tube or shorter, linear or nearly so, apex obtuse; stamens 6, exserted, filaments purplish; anthers linear, yellow, attached near their center; ovary about 1¼ in. long, ¼–⅓ in. broad, green; style exserted, purplish, longer or shorter than the stamens.

FRUIT. Brown, cylindrical or slightly widened toward the apex, base gradually narrowed, length 1⅗–2⅖ in., width ⅗–¾ in., or sometimes much less at the base; seeds ⅕–¼ in. long, about ⅙ in. broad.

LEAVES. Basal cluster 2–5 ft across, blade linear-lanceolate, stiff, thickened, concave; marginal spines ¼–2 in. apart, apical spine ½–1⅔ in. long.

RANGE. At altitudes of 3,000–7,800 ft on rocky hillsides or desert grasslands. Southwestern New Mexico, Arizona, and Sonora, Mexico.

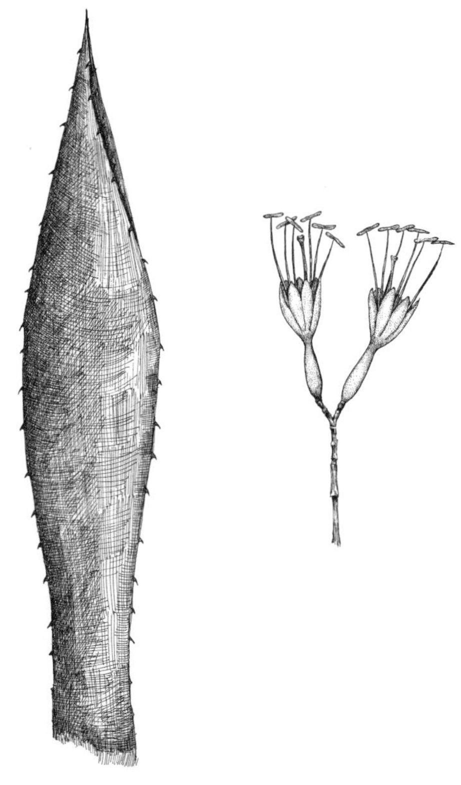

PALMER AGAVE
Agave palmeri Engelm.

VARIETY. Golden-flowered Agave, *A. palmeri* var. *chrysantha* (Peebles) Little, is a variety distinguished by the golden yellow congested flowers; leaf spines more distant and surface glaucous; flower stalks shorter. Known to occur only in Arizona.

REMARKS. The genus name, *Agave,* is from the Greek *agaue,* meaning "noble." The species name, *palmeri,* is in honor of Edward Palmer, who collected extensively in New Mexico and Arizona until 1890, and a large number of plants bear his name. The plant is also known as the Century Plant and Mescal. The Indians of the Southwest bake the crown to obtain the mescal for food. The leaves also are used for making twine and rope.

New Mexico Agave

Agave neomexicana Woot. & Standl.

FIELD IDENTIFICATION. Plant acaulescent, with numerous basal leaves crowded together and forming a compact, almost globose, rosette 1½–2 ft in diameter when mature.

FLOWERS. Panicle, including scape, 8–15 ft high with 10–15 divergent lateral branches bearing subumbellate clusters of flowers; flowers dull brownish to reddish outside, deep yellow to orange within, 2–2½ in. long, perianth complete and composed of 6 segments (3 petals and 3 sepals); stamens 6, filaments attached by an expanded base to the base of the perianth segments in a saccate depression, about ¾ in. long, yellow, anthers versatile and about ⅗ in. long; ovary inferior, 3-celled; style 1, slender, somewhat exceeding the stamens, the stigma obscurely 3-lobed and capitate.

FRUIT. Capsule oblong to elliptic, light brown, 1–1½ in. long and ⅘ in. in diameter.

LEAVES. From 4–12 in. long, ovate-lanceolate, bluish green, glaucous, tipped with a very sharp spine, this being brownish black when young, grayish in age, decurrent into a horny, toothed margin bearing 3–4 brown or gray retrorsely hooked teeth on the upper half of the leaf, the lower part armed with fewer smaller spines.

RANGE. In southern New Mexico in the Tortugas, Organ, San Andrés, Las Cornudas, Sacramento, and Guadalupe mountains. In Texas in the vicinity of Pine Springs, just below the New Mexico line, on slopes of the Guadalupe Mountains.

REMARKS. The genus name, *Agave,* is from the Greek word *agaue,* meaning "noble." The species name *neomexicana,* refers to the plant's habitat in New Mexico. Some authors list *A. neomexicana* as a synonym of *A. parryi* Engelm.

The Mescalero and Chiricahua Apache baked the crown of the plant for food, and also used it as a source of beverage. The following method of preparing the mescal is given by Castetter, Bell, and Grove (1938, pp. 28–29):

"The crowns of the mescal plants were dug out with three-foot sticks cut from oak branches and flattened at the end. This end, when pounded with a rock into the stem of the plant just below the crown, permitted the crown to be removed readily. A broad stone knife was used to chop off the leaves, two being left for tying the crowns together, thus making transportation more convenient. The naked crowns were bulbous, white in color, and from one to two feet in circumference.

"Pits in which crowns were baked were about ten to twelve feet in diameter and three or four feet deep, lined with large flat rocks. On the largest rock, which was placed in the center, a cross was made with black ashes. Rocks were piled on the flat stones, but care was always taken that the top should be level. Upon this, oak and juniper wood was placed, and before the sun came up was set on fire. By noon the fire had died down, and on these hot stones was laid moist grass, such as bunchgrass *(Sporobolus airoides),* side-oats grama *(Bouteloua curtipendula),* Texan crabgrass *(Schedonardus paniculatus),* big blue-stem *(Andropogon furcatus),* mesquite grass *(Muhlenbergia wrightii),* marsh foxtail *(Alopecurus aristulatus),* [New Mexican muhly] *(Muhlenbergia neomexicana),* or the leaves of bear grass *(Nolina microcarpa),* but bear grass was usually preferred since it did not burn readily. The largest mescal crown was selected and a cross made on it with tule or cat-tail pollen *(Typha latifolia),* when this was available, the pollen always being placed on the crown from east to west and from north to south. The Indians then prayed. Extending the large crown toward the opening of the pit four times, they tossed it in and threw the other crowns after it. Next the youngest child present stood at the east of the pit and threw four stones into it. It should be made clear that this little ceremony, held at the time of baking, varied among different family groups and the above description should be regarded as one of a type rather than a fixed performance. After the mescal had been covered with the long leaves of bear grass and the whole with earth to a depth sufficient to prevent steam from escaping, the crowns were allowed to bake the rest of the day and all night. Early in the morning the pit was opened and a crown examined and eaten. The pit was again closed and the Indians refrained from drinking until noon of this day so as to prevent rain. The following morning the mescal was removed."

. . . .

"In addition to the food product known as mescal, certain Indian tribes prepared from agave an alcoholic beverage known also as mescal. Among the Mescalero and Chiricahua Apache, the agave crowns were roasted as described above, then the inner portion cut into pieces, pounded until soft, and the pulp placed in a pouch made of animal hide. This was buried in the ground where it was usually allowed to remain for two days, although the Indians claimed that the longer it was buried the better it became. When removed from the ground, the juice was squeezed from the pulp into a container and allowed to ferment for two or three days, when it was ready for use. Thus prepared, the drink was quite potent, and occasionally an Apache would allow the juice to spoil (change to a high percentage of acetic acid) before getting drunk on it. Cremony [*Life Among the Apaches,* San Francisco, 1868, p. 217], in 1868, reported that the Apache residing on the headwaters of the Gila River, and the Mescalero and Jicarilla Apache made a drink by macerating the mescal root (probably the crown) in water and allowing the mixture to ferment for several days, after which the liquid was boiled down to produce a strong intoxicating beverage." (p. 60)

Century Plant Agave

Agave americana L.

FIELD IDENTIFICATION. An acaulescent plant with a basal rosette of large leaves. Not woody, but persisting for many years and dying after blooming. New plants are produced as offsets at the base.

FLOWERS. Borne in a panicle 3–4½ ft long and the scape two to six times its length; perianth yellow, perfect, 2–2¾ in. long, with the 6 segments (3 petals and 3 sepals) about 1 in. long; stamens 6, long-exserted, filaments inserted near the throat of the tube, anthers versatile; ovary about 1¼ in. long, inferior, 3-celled; style 1, stigma obscurely 3-lobed.

FRUIT. Capsule brown, ellipsoid, loculicidal, without bulbils; seeds numerous, black, flat, thin.

LEAVES. Ascending, the outer ones recurved usually above the middle, length 60–80 in., width 6–9 in., lanceolate or oblanceolate, stiff, fleshy, smooth, somewhat concave, gray-glaucous; apical spine brown or gray, ⅙–⅘ in., stout, straight or recurved, round-grooved, very shortly decurrent; marginal teeth raised on horny ridges, gray, ⅕–⅖ in. long.

RANGE. A native of Mexico. Often cultivated for ornament from Texas to Florida. Sometimes escaping cultivation in western Texas. Cultivated extensively in the West Indies and in Europe along the Mediterranean.

VARIETIES AND FORMS. The Miller Century Plant, *A. americana* var. *milleri* Baker, has narrower leaves with the spines more hooked and scarcely decurrent. This variety is often seen in cultivation in the following forms:

A. americana var. *milleri* forma *marginata* has leaf margins yellow to white.

A. americana var. *milleri* forma *striata* is variously lined with yellow or white.

A. americana var. *milleri* forma *medio-picta* has a broad median yellow band.

REMARKS. The genus name, *Agave,* is from the Greek word *agaue,* meaning "noble." The species name *americana,* refers to the habitat of the plant in the Americas. Vernacular names in use for the plant are Maguey, American Aloe, Flowering Aloe, and Spiked Aloe.

The following discussion of uses of agaves is quoted from Wood and Osol (1943, p. 1247):

"Although the *Agave americana* is the best-known form, botanists have described fifty species of the genus which are indigenous to South America and the southern portions of North America, and many of which contribute to the economic products produced in that country from the agave plant. The number of these products is very great. *Sisal grass* or *sisal hemp* and *Tampico hemp,* also known as *Pita hemp* or *Pita fiber,* are the most important of the various fibers obtained from the agave leaves, although a number of other forms are locally known in Mexico. From a number of species of Agave are produced in Mexico large quantities of a fermented liquor, known as *pulque,* and distilled liquors known as *mescal* or *tequila.* All of the pulque agaves have thick leaves. When they are about to bloom the central bud is cut out, leaving a large cavity into which the sap *(aguamiel* or *honey water)* exudes rapidly. At first clear green, yellowish or whitish, the sap soon by fermentation becomes milky and acquires a cider-like taste or, if the process is allowed to go on, is rapidly converted into vinegar. Pulque is said to contain about 7 per cent of alcohol, and is very largely used as a beverage by the Mexicans. The juice has in it an optically inactive reducing sugar, *agavose.* The juice as well as the herbage of the agave contain a saponin and are used in Mexico in the place of soap.

"The fresh juice is said to be laxative and diuretic. The leaves are said to be used as counterirritants, and Lenoble has found in them an acrid volatile oil.

"Jones *(J. Pharmacol. and Exper. Therapeutics,* 48:1, 1933) obtained results from aguamiel in experimental nephritis of dogs which would seem to suggest a possible usefulness as a remedial agent in inflammations of the kidney.

"Agave gum has been compared to gum arabic, but differs in containing a much larger proportion of calcium, and in being only partially soluble in water, the soluble portions resembling arabin, but the larger insoluble portions having all the characteristics of bassorin."

Schott Agave

Agave schottii Engelm.

FIELD IDENTIFICATION. An acaulescent plant, smaller than most other agaves, but spreading by underground stems to form crowded colonies, and thus often covering areas of considerable extent.

FLOWERS. Usually maturing May–August, scapes and flowering portion together 5–9 ft tall, slender and unbranched, spikelike; flowers numerous, light yellow, rather crowded, in pairs, somewhat scented, usually curved, 1⅕–2 in. long; tube of perianth ⅓–½ in. long, constricted proximally, broadly funnelform, expanded into 6 slender perianth segments (3 petals and 3 sepals); stamens 6, long-exserted, inserted in upper part of the perianth-tube; filaments whitish or pale yellow, ¾ in., anthers also yellowish and ¼–½ in.; style pale, 1⅓–1¾ in.; ovary green, ⅓–⅖ in. long, about ⅕ in. in diameter, 3-celled, style elongate, stigma capitate and obscurely 3-lobed.

FRUIT. Globose or short-oblong, brown to black, almost cylindrical, ⅖–1 in. long, ⅓–¾ in. in diameter; seeds numerous, black, flattened, ⅙ in. long, about ⅛ in. wide.

LEAVES. Linear, yellowish green, falcate, blades 8–12 in. long, ¼–½ in. wide near the middle, concave above, gradually narrowed to the apex, margins usually entire and shreddy, or sometimes with a few small teeth at the base, surfaces sometimes marked with white from adjoining leaf margins.

RANGE. On dry rocky slopes at altitudes of 3,000–7,000 ft. Found in southwestern New Mexico in Hidalgo County. In Arizona known from Gila, Cochise, Pima, and Santa Cruz counties. Also in northern Sonora, Mexico.

VARIETY. The Trelease Agave, *A. schottii* var. *treleasei* (Toumey) Kearney & Peebles, has dark green leaves 10–16 in. long and about ⅝ in. wide at the middle, nearly flat on the upper surface. It is known from the Santa Catalina Mountains near Tucson, Arizona.

REMARKS. The genus name, *Agave,* is from the Greek word *agaue,* meaning "noble." The species name *schottii,* is in honor of Arthur Schott, who assisted Major William H. Emory on the United States–Mexico boundary survey in 1854. The plant is sometimes known as "Amole," and the crown is used for soap by the Mexicans and Indians. The crown was also roasted to make mescal, but to a lesser extent than other species.

WILLOW FAMILY (*Salicaceae*)

WHITE POPLAR
Populus alba L.

FIELD IDENTIFICATION. Tree attaining a height of 100 ft, with a trunk diameter of 3–4 ft. Sometimes spreading by root-suckers to form thickets in old fields or about abandoned dwelling sites. Recognized by the conspicuous white-tomentose undersurface of the leaves.

FLOWERS. Borne in pendulous catkins, pistillate about 2 in., slender, stigmas 2, each deeply 2-parted; staminate 1½–4 in.; scales dentate, fringed with long hairs; stamens 6–10 (usually about 8).

FRUIT. Capsule narrowly ovoid, ⅛–⅕ in. long, tomentose, 2-valved; seeds minute, numerous, with a tuft of long silky, white hairs.

LEAVES. Simple, alternate, rather variable, on vigorous shoots palmately 3–5-lobed, the lobes also coarsely toothed or with additional small lobes, base rounded to subcordate, blades 2⅓–5 in. long, upper surface dark green, lower surfaces conspicuously white-tomentose; on older branches leaves often smaller, ovate to elliptic-oblong; margin sinuate-dentate; petioles terete, densely tomentose; young twigs and branches also white-tomentose.

BARK. Greenish gray to white, usually smooth on branches or young trunks, toward the base of old trunks roughened into firm dark ridges.

WOOD. Reddish to yellowish, sapwood nearly white, tough but light and soft.

RANGE. Adapted to many soil types, on dry, well-drained sites in the sun. Grown for ornament in Texas, New Mexico, Oklahoma, Arkansas, Louisiana, and more or less throughout the United States. A native of central and southern Europe to western Siberia and central Asia.

VARIETIES. A number of horticultural variations are known as follows:

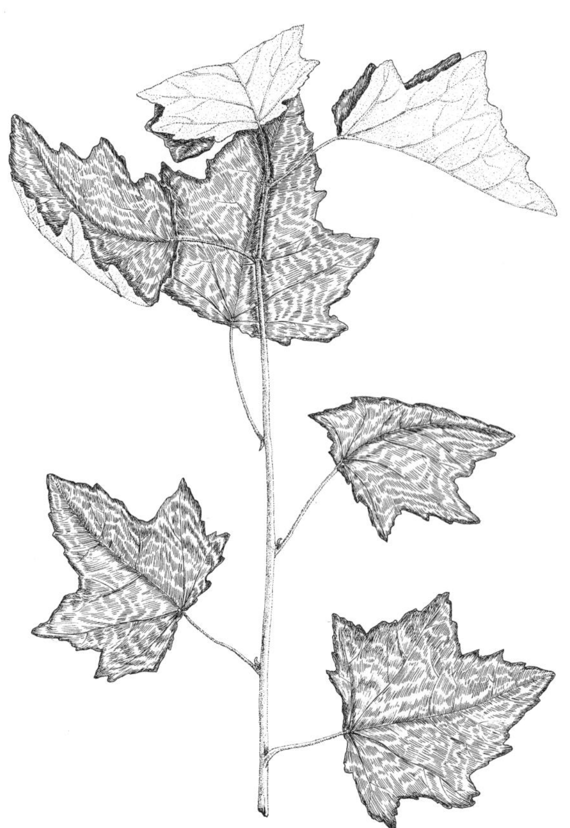

WHITE POPLAR
Populus alba L.

Silvery-white Poplar, *P. alba* forma *nivea* Ait., has lobed leaves densely silvery white-tomentose beneath. The most commonly cultivated form.

Richard White Poplar, *P. alba* forma *richardii* Henry, has leaves yellow above.

Bolle's White Poplar, *P. alba* forma *pyramidalis* Bge., is a columnar tree; leaves of the short branches orbicular, coarsely toothed, glabrescent and green above.

Globose White Poplar, *P. alba* forma *globosa* Spaeth, is a shrub or small tree with a dense oval head; leaves slightly lobed, lower surface grayish-tomentose, pinkish when young.

Weeping White Poplar, *P. alba* forma *pendula* Loud., has pendulous branches.

REMARKS. The genus name, *Populus,* is the ancient Latin name, and the species name, *alba,* refers to the white undersurface of the leaves. The tree is also known under the vernacular name of Abele. White Poplar has long been grown for ornament. It is very conspicuous because of the contrasting white and green leaf surfaces. However, the white undersurfaces catch soot and dust easily and become unsightly in some localities. The tree grows rapidly, transplants easily, prunes well, and has few insects or fungus pests, but is short-lived.

Fremont Cottonwood

Populus fremontii S. Wats.

FIELD IDENTIFICATION. Tree attaining a height of 100 ft, a diameter of 4–6 ft. The stout, spreading or pendulous branches form a broad, open crown.

FLOWERS. In separate staminate and pistillate catkins; staminate catkins 1¼–3¼ in., peduncles glabrous; disk broad, oblique, thickened, revolute; stamens 50–60, exserted at anthesis, anthers large and red; pistillate catkins 4–5 in. at maturity, peduncles stout and glabrous, flowers remote; pedicels variable but mostly 1/25–⅛ in.; scales thin and cut into filiform lobes at the apex; disk cup-shaped and variable in size, 1/25–⅕ in. in diameter; ovary ovoid or ovoid-oblong, glabrous; stigmas 3–4, irregularly lobed.

FRUIT. Capsule brown, ovoid to ellipsoid or nearly globose, acute or obtuse, ⅙–½ in. long, dehiscing into 2–4 thick-walled valves; seeds minute, ovoid, acute, light brown, with copious silky hairs.

LEAVES. Simple, alternate, deciduous, deltoid or reniform, apex acute or abruptly short-acuminate, occasionally rounded or emarginate, base truncate to subcordate or abruptly cuneate, margin crenate-serrate with the teeth incurved, blade length 2–2½ in., width 2–3 in., young ones somewhat pubescent, older ones bright green on both sides as well as glabrous and lustrous, firm, midrib yellowish with 4–5 pairs of slender veins; petioles 1½–3½ in., flattened, yellowish, glabrous; lateral buds resinous, about ⅙ in. long, ovoid, sharp-pointed, glabrous, terminal bud about ⅓ in. long.

TWIGS. Stout, green to yellow and later gray; leaf scars 3-lobed.

BARK. On young branches or trunk light gray or brownish, thin, smooth or somewhat fissured; on old trunks ¼–2 in. thick, light gray to dark brown or reddish-tinged, deeply broken into irregular rounded ridges and furrows.

WOOD. Sapwood nearly white; heartwood brown, light, tough, not durable.

RANGE. Moist soil along streams or water holes from sea level to an altitude of 7,000 ft. Southwestern New Mexico, Colorado, Arizona, Utah, Nevada, and California; in Mexico in Lower California and Sonora.

VARIETIES. The following varieties have been segregated:

The Silver City Poplar, *P. fremontii* var. *macrodisca* Sarg., has a broad disk nearly enclosing the ellipsoidal fruit.

San Diego Poplar, *P. fremontii* var. *pubescens* Sarg., has the foliage and young twigs pubescent.

Thornber Poplar, *P. fremontii* var. *thornberi* Sarg., has more numerous serrations on leaf margins; capsules ellipsoid; smaller disks and shorter pedicels.

Toumey Poplar, *P. fremontii* var. *toumeyi* Sarg., has a large disk and subcordate leaves narrowed and cuneate to the petiole insertion.

MacDougal Poplar, *P. fremontii* var. *macdougalii* (Rose) Jepson, has dense pubescense on young parts and a bluish green leaf color.

Parry Poplar, *P.* × *parryi* Sarg. is a hybrid between *P. fremontii* and *P. trichocarpa.*

Rio Grande Cottonwood, *P. fremontii* var. *wislizenii* (Torr.) S. Wats., is probably the most abundant of the varieties and is described in full on another page.

REMARKS. The genus name, *Populus,* is the ancient name given by Pliny. The species name, *fremontii,* is in honor of General John Charles Fremont (1813–

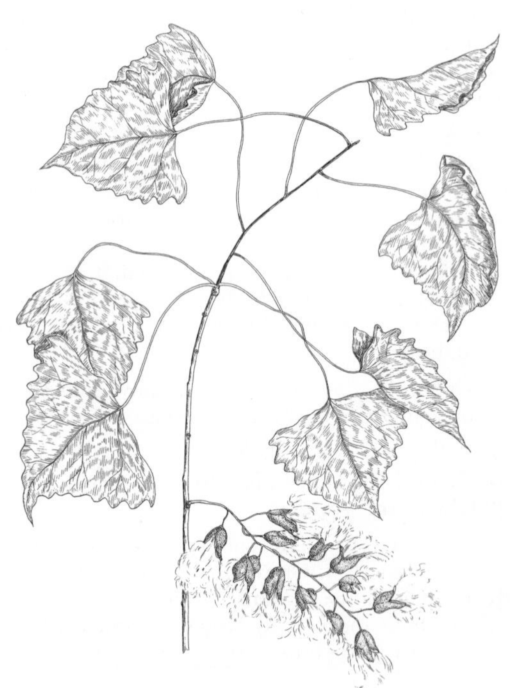

FREMONT COTTONWOOD
Populus fremontii S. Wats.

1890), politician, soldier, and explorer of the western United States. The tree is also known under the vernacular names of Fremont Poplar, Arizona Cottonwood, MacDougal Cottonwood, and Alamo. It was first introduced into cultivation in 1904, and is much planted as a shade tree in the Southwest. It is easily propagated by cuttings, rapid-growing, short-lived, subject to wind damage, and is much attacked by mistletoe. The wood is used locally for fuel and posts. The Indians ate the uncooked catkins, used the inner bark as an antiscorbutic, and made baskets from the young twigs. The foliage is browsed by mule deer and cattle.

Rio Grande Cottonwood

Populus fremontii var. *wislizenii* (Torr.) S. Wats.

FIELD IDENTIFICATION. A large tree with a thick trunk, attaining a height of 90 ft. Branches wide-spreading to form a large crown.

FLOWERS. Staminate and pistillate catkins slender and elongate; pistillate catkins very slender, 3–6 in. long, longer than the staminate catkins; scales of catkins reddish, roughened, slender-lobed; staminate flower disk broad and oblique; stamens numerous, anthers large, filaments short; disk of pistillate flower cup-shaped, margin dentate, ovary rounded and long-stalked.

FRUIT. Fruiting pedicels ½–¾ in., usually as long as or much longer than the capsules; capsules scattered on the glabrous peduncle, about ¼ in. long, ovoid to narrowly ellipsoidal, apex acute, slightly ridged, pale, dehiscent into 3 or 4 valves.

LEAVES. Simple, alternate, deciduous, broadly ovate to triangular, apex abruptly acute or long-acuminate, base subcordate to rounded or sometimes truncate, margin coarsely crenate except at apex, length or width of blades 2–4 in., surfaces yellowish green, shiny, glabrous; veins delicate, yellowish, veinlets reticulate; petioles slender and glabrous, 1¼–2 in., flattened laterally.

TWIGS. Stout, yellowish to orange or gray, glabrous; buds with acute apex, resinous, shiny and puberulous; bark of trunk pale gray or light brown, fissures deep and irregular, ridges broad and flat.

WOOD. Brownish, soft, specific gravity 0.46.

RANGE. Stream banks and valleys at altitudes of 2,500–7,000 ft. Trans-Pecos Texas along the valley of the Rio Grande River; northward into New Mexico, southern Colorado, and southern Utah. In Mexico in Chihuahua and Sonora.

VARIETY. Rio Grande Cottonwood is closely related to Fremont Cottonwood, *P. fremontii* S. Wats., but the former is distinguished by the very slender, elongate pistillate and staminate pedicels, and the fruit is more ellipsoid. The size and shape of the leaves are so variable that they serve as no criteria for separating the species from the variety.

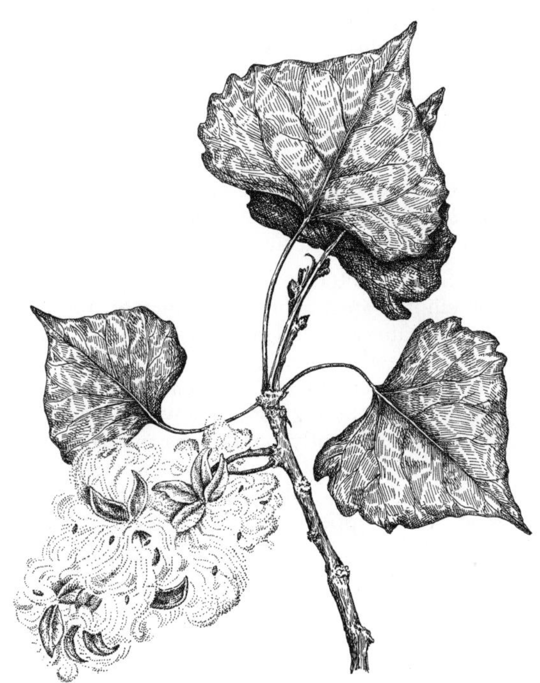

Rio Grande Cottonwood
Populus fremontii var. *wislizenii* (Torr.) S. Wats.

REMARKS. The genus name, *Populus*, is the classical Latin name. The species name, *fremontii*, honors General John Charles Fremont (1813–1890), politician, soldier, and explorer of the western United States. The variety name, *wislizenii*, is for Friedrich Adolph Wislizenus (1810–1889), German-born physician of St. Louis who collected plants in Mexico and the Southwestern United States in 1846–1847. The tree is sometimes listed under the species name of *P. wislizenii* (Wats.) Sarg. Also known under the vernacular names of Wislizenus Cottonwood, Valley Cottonwood, Big Bend Poplar, Alamo, and Güerigo.

The tree is rapid-growing on moist sites, is short-lived, and occasionally browsed by cattle. It is used for posts, fuel, rafters, and rough lumber.

Narrow-leaf Cottonwood

Populus angustifolia James

FIELD IDENTIFICATION. Western tree to 60 ft tall, 2 ft in trunk diameter. Branches slender, erect, or ascending, to form a narrowly pyramidal head.

FLOWERS. In drooping staminate and pistillate catkins before the leaves, length 1–4 in., pistillate longest, densely flowered, glabrous; scales brown, obovate, margins thin and fimbriate; disk shallow and irregularly lobed; ovary ovoid, style short, stigmas 2, divergent, irregularly lobed; staminate catkins 1–2½ in., disk cup-shaped; stamens 10–20, filaments short, anthers red.

FRUIT. Capsule borne on pedicels about ⅓ in. long, ovoid, narrowed toward the short-pointed apex, dehiscent into 2 valves; seeds numerous, brown, ovoid to obovoid, apex obtuse, bearing tufts of apical hairs.

LEAVES. Simple, alternate, deciduous, ovate to lanceolate or elliptic, apex acute to acuminate, rarely obtuse, base cuneate or rounded, margin bluntly glandular-serrulate, or teeth acute, blade length 2–4 in., width ½–2 in. (some longer or wider on young shoots); upper surface light green and glabrous, lower surface paler green, glabrous or slightly puberulous, midrib yellowish, secondary veins slender; petiole slender, less than one third as long as the blade, mostly terete or slightly flattened on the upper surface.

TWIGS. Slender, terete, glabrous or puberulous, green to yellow or orange, eventually gray; buds resinous and aromatic, slender, ovoid, ¼–½ in. long, long-pointed, glabrous, scales thin and brown.

LANCE-LEAF COTTONWOOD
Populus × *acuminata* Rydb.
(Populus angustifolia and *Populus sargentii)*

NARROW-LEAF COTTONWOOD
Populus angustifolia James

BARK. Green to yellowish or gray, branches and upper trunk rather smooth, lower trunk with irregular shallow fissures divided by broad flat ridges.

WOOD. Pale brown, sapwood lighter colored, weak, soft, not durable, specific gravity about 0.39.

RANGE. Usually along streams in mountainous regions at altitudes of 2,000–7,000 ft. Trans-Pecos Texas; west to Arizona, north to British Columbia, Alberta, and Saskatchewan, and south to Chihuahua, Mexico.

HYBRID. *P. angustifolia* and *P. sargentii* hybridize where they contact to produce *P.* × *acuminata* Rydb., which was long considered to be a distinct species.

REMARKS. The genus name, *Populus,* is the classical Latin name. The species name, *angustifolia,* refers to the narrow leaves. Also known under the vernacular names of Black Cottonwood, Narrow-leaf Poplar, Mountain Cottonwood, and Alamo. It is a rather fast-growing tree and subject to storm damage. The wood is sometimes used for fuel or fence posts. The bark is eaten by beaver and young branches sometimes browsed by deer and livestock. Paul C. Standley reports that the Gosiute Indians of Utah use the young shoots for basket-making. A kind of honeydew produced on the under side of the leaves by aphids is gathered and used in much the same way as sugar. The tree was introduced into cultivation about 1893. It is sometimes planted in Western cities for ornament.

Great Plains Cottonwood

Populus sargentii Dode

FIELD IDENTIFICATION. Tree attaining a height of 90 ft, a diameter of 6 ft. The branches erect and spreading to form a broad crown.

FLOWERS. In smooth staminate and pistillate catkins; staminate catkins 2–2½ in., scales light brown, roughened, apex fimbriate; disk broad, oblique, margin thickened; stamens 20–25, filaments short, anthers yellow; pistillate catkins 4–8 in., disk cup-shaped, margin slightly lobed; stigmas laciniately 3–4-lobed, ovary subglobose.

FRUIT. Capsule maturing June–August, about ⅖ in. long, oblong-ovoid, apex obtuse, three to four times longer than the pedicel; seeds oblong to obovoid, apex rounded, about 1/16 in. long.

LEAVES. Simple, alternate, deciduous, ovate to broadly deltoid; apex mostly long-acuminate, base truncate or subcordate, margin crenate-serrate; leaves slightly hairy at first; when mature glabrous, shiny, light green to yellowish green, veins rather slender and delicate; blade length 3–3½ in., width 3½–4 in.; petiole slender, laterally flattened, 2–3½ in. long, often with 2 small glands at the apex.

TWIGS. Stout, glabrous, green to yellowish, somewhat angular or rounded, leaf scars rather large; buds ovoid, acute, resinous, scales puberulous and brownish.

BARK. Smooth on young trees and branches, pale gray, later with deep fissures, rounded ridges and irregular scales closely appressed.

RANGE. In stream beds and on hillsides at altitudes of 3,500–7,000 ft. In the Texas Panhandle, Oklahoma, and New Mexico; northward into southern Saskatchewan.

HYBRIDS AND RELATED PLANTS. Some botanists consider the Great Plains Cottonwood to be only a xerophytic Western form of the Eastern Cottonwood, *P. deltoides*. However, it is usually smaller in stature, with glabrous and yellowish green leaves bearing coarser and fewer teeth, the flower pedicels shorter and fruit more obtuse-pointed.

P. sargentii is known to hybridize with *P. angustifolia* to produce *P.* × *acuminata* Rydb. At one time *P. acuminata* was considered to be a species instead of a hybrid. It is known from Trans-Pecos Texas, New Mexico, Colorado, Nebraska, Wyoming, South Dakota, and Montana.

REMARKS. The genus name, *Populus*, is the classical Latin name. The species name, *sargentii*, honors Charles Sprague Sargent (1841–1927), American dendrologist and at one time director of the Arnold Arboretum. The wood of Great Plains Cottonwood is used for fuel, posts, veneer, and baskets.

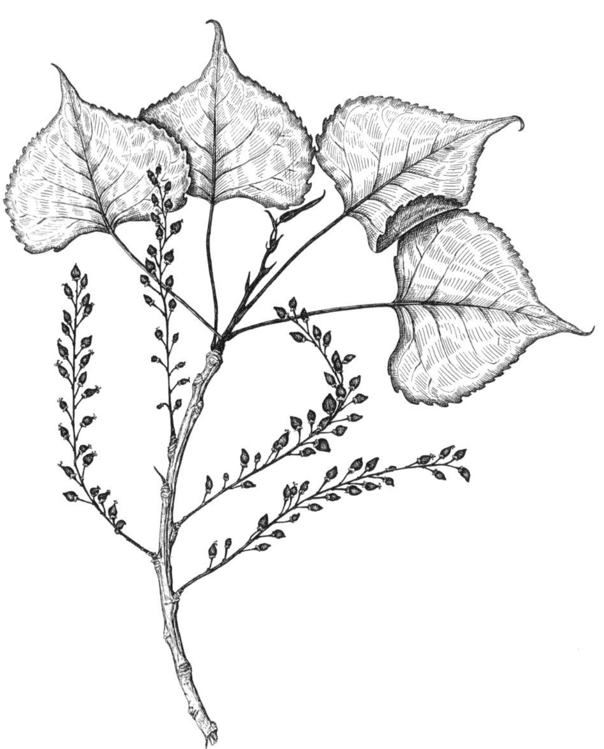

Great Plains Cottonwood
Populus sargentii Dode

Palmer Cottonwood

Populus palmeri Sarg.

FIELD IDENTIFICATION. A Texas tree 30–60 ft tall, with a straight trunk 1–3 ft in diameter. The branches are smooth, pale, and erect, forming an open pyramidal head.

FLOWERS. In pendulous, stalked catkins, the pistillate longer than the staminate; scales membranous, ovate to obovate, more crowded on the staminate.

FRUIT. Fruiting catkins glabrous, 3–6 in. long, capsules borne on slender pedicels, ¼–⅓ in., ovoid, obtuse, ¼–⅓ in., puberulous to glabrate, slightly pitted, dehiscent into 3–4 thin valves, disk deeply lobed.

LEAVES. Involute in the bud, blades 2½–5 in. long, 2–4¾ in. wide, ovate to broadly ovate; apex entire, long-acuminate or abruptly pointed; base truncate or abruptly narrowed; margin serrate with incurved teeth, glabrous (ciliate on the margin when young, but later glabrous); petiole slender, glabrous, tan to reddish brown, laterally compressed, 1½–4 in. long, with glands at apex of the petiole; leaf scars with 3 fibrovascular bundles; stipules caducous.

TWIGS. Pale green to tan or light reddish brown at first, later pale grayish brown.

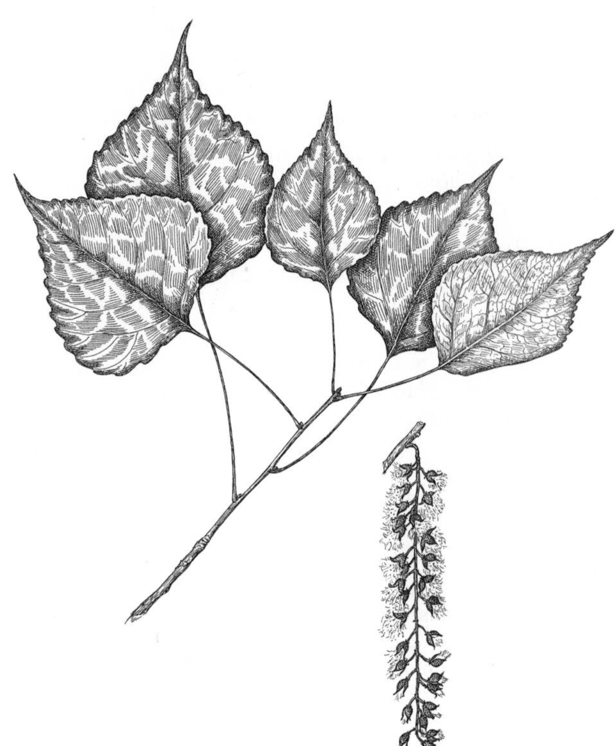

PALMER COTTONWOOD
Populus palmeri Sarg.

BARK. Pale, 2–4 in. thick, roughened by narrow ridges which are divided by wide fissures.

RANGE. Known only from the moist alluvial valley of the Nueces River to Uvalde County, Texas.

REMARKS. The genus name, *Populus*, is the classical name, and the species name *palmeri*, honors E. J. Palmer, American botanist.

SWAMP COTTONWOOD

Populus heterophylla L.

FIELD IDENTIFICATION. Tree attaining a height of 90 ft, with a trunk diameter of 3 ft. It has a narrow, round-topped head with slender branches.

FLOWERS. Usually April–May; staminate and pistillate catkins borne separately; staminate catkins rather stiffly pendent, oblong, cylindric, 2–4 in. long, ⅓–½ in. wide, densely flowered; scales brown, oblong-obovate, glabrous, lobes reddish brown and elongate; disk asymmetrical; stamens 12–20; anthers red, linear, ¹⁄₁₆–⅛ in., filaments filiform; pistillate catkins pendulous, slender, raceme-like, few-flowered, 2–6 in. long; scales concave, oval to obovate, brown, fimbriate, deciduous early; disk long-stalked, irregularly triangular-toothed; ovary ovoid, style short with a thickened 2–3-lobed stigma.

FRUIT. Maturing in May, borne on elongate pendent racemes 3–6 in. long, usually glabrous, capsule peduncles ¼–½ in. long; capsule shorter than the peduncles or the same length, ovoid-oblong, pointed, reddish brown, thick-walled, 2–3-valved; seeds very small, reddish brown, obovoid, tufted with hairs, dispersed by wind.

LEAVES. Alternate, simple, deciduous, 3-nerved, ovate to oval, margin crenate-serrate, apex obtuse or rounded, base broadly cuneate, rounded or cordate; surface dark green and glabrous above or with tomentose veins; lower surface often white-tomentose, especially on younger leaves; blades 2½–7 in. long, width 1¼–5 in.; petiole sometimes as long as the blade, slender, tomentose to glabrous; stipules, when present, lanceolate.

TWIGS. Stout, varying shades of gray, reddish, or brown, young densely white-tomentose or glaucous, later glabrous; leaf scars large and triangular; lenticels small, elongate.

BARK. Brown to reddish, ridges broad and flattened, separating into long, loose flakes, fissures shallow.

WOOD. Dull brown, sapwood lighter, close-grained, moderately soft when dry, weighing about 26 lb per cu ft, low in durability, weak in bending, low in shock resistance, warps somewhat, resistant to splitting, shrinks considerably, does not hold nails well, moderately hard to work, glues well, takes paint moderately well; rays very fine, scarcely visible.

RANGE. Along the margins of swamps and streams. Best developed in the lower Mississippi Valley and in Southeastern swamps. Louisiana; eastward to Florida; north into Arkansas, Missouri, Kentucky, Tennessee,

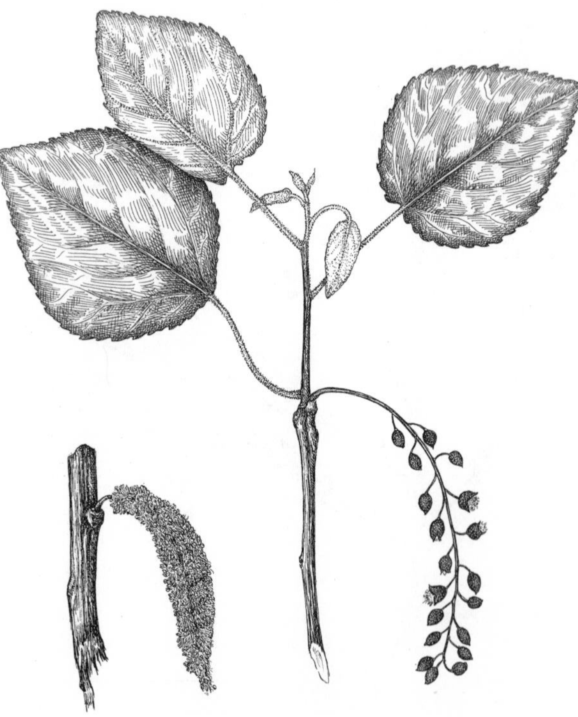

SWAMP COTTONWOOD
Populus heterophylla L.

Illinois, and Indiana; and along the Atlantic coastal plain to Connecticut. Not in the Appalachian region.

REMARKS. *Populus* is the classical name of the Poplar, and the species name, *heterophylla,* refers to the variously shaped leaves. Other vernacular names are Black Cottonwood, Black Poplar, Swamp Poplar, and Downy Poplar. The wood is often used under the name of Black Poplar for the same general purposes as that of other poplars, including excelsior, veneer, musical instruments, boxes and crates, poultry coops and brooders, ironing and washboards, tubs and pails, and concealed parts of furniture. Swamp Poplar is rapid-growing, short-lived, and subject to heart rot. It is rarely cultivated for ornament. Distinguished from other poplars by the round, instead of flattened, petiole.

EASTERN COTTONWOOD

Populus deltoides Marsh.

FIELD IDENTIFICATION. Tree to 100 ft high and 8 ft in diameter. The trunk is often rather short, the branches massive, the top rounded, and the root system spreading and shallow.

FLOWERS. February–May, borne in separate staminate and pistillate catkins; staminate catkins densely flowered, 1½–2 in. long, ½–¾ in. wide, disk oblique and revolute; stamens 30–60, filaments short, anthers large and red; pistillate catkins at first 3–3½ in. long, loosely flowered; bracts brown, glabrous, apex fimbriate; disk ovoid, obtuse, enclosing about one third of ovary, ovary sessile, style short, stigmas 2–4, large, spreading, laciniately lobed.

FRUIT. Ripening May–June, racemose, 8–12 in. long at maturity, capsules on slender pedicels ⅛–⅖ in., ovoid to conical, acute, about ¼ in. long, 1-celled, 3–4-valved; seeds numerous, small, brown, oblong-obovoid, buoyant with cottony hairs when the capsule ruptures. The minimum seed-bearing age about 10 years and maximum about 125. The seed averages 10–30 seeds per capsule, with an average of 3,032,000 seeds per lb. The commercial purity is about 40 per cent, with a soundness of 95 per cent. Germination averages 60–90 per cent.

LEAVES. Simple, alternate, deciduous, broadly deltoid-ovate, margin crenate-serrate, apex abruptly acute or acuminate, base truncate to heart-shaped or abruptly cuneate, blades 3–7 in. long, about as broad; upper surface light green, glabrous, lustrous, main vein stout, yellow to reddish; lower surface paler and glabrous with primary veins conspicuous; petiole smooth and glabrous, flattened, stipules linear.

TWIGS. Yellowish to brown or gray, stout, angular, lenticels prominent; buds ovoid, acute, resinous, brown, about ½ in. long or less, laterals much flattened; leaf scars triangular or lunate, with 3 bundle marks, pith star-shaped.

EASTERN COTTONWOOD
Populus deltoides Marsh.

BARK. Thin and smooth on young branches or trunks, green to yellow; older trunks gray to almost black with flattened, confluent broad ridges broken into closely appressed scales.

WOOD. Dark brown, sapwood white, weak, soft, weighing about 24 lb per cu ft, moderately weak in bending, weak in endwise compression, low in shock resistance, moderately easy to work with tools, takes paint well, easy to glue, warps and shrinks considerably, low in durability, below average in ability to stay in place, nails easily, does not split easily.

RANGE. Eastern Cottonwood is found in rich, moist soil, mostly along streams. It and its varieties occur over practically the entire United States east of the Rocky Mountains. Texas, New Mexico, Oklahoma, Arkansas, and Louisiana; eastward to Florida and northward into Canada.

PROPAGATION. The seeds should be collected as the capsules start to burst open. It is not necessary to remove the fuzzy hairs before planting. Since the seeds lose viability quickly it is advisable to plant soon after gathering. The seeds may be sown broadcast, or in drills, and need not be covered, but can be pressed down. The seed beds should be kept moist and shaded, but good aeration is essential because of susceptibility to heat or fungus damage. Cottonwood seedlings are subject to attack by leaf rusts, bark beetles, and poplar leaf beetles. Most cottonwoods can be easily propagated from hardwood cuttings and can also be grafted. Eastern Cottonwood has been cultivated since 1750.

VARIETIES. Two Southern varieties of the cottonwood are generally recognized. The Southern Cottonwood, *P. deltoides* var. *missouriensis* Henry, has branches more strongly angled, leaves oval to broad-ovate, usually longer than broad, margins more finely crenate-serrate; petioles 3–4-glandular at apex. It occurs from Florida and Mississippi north to Missouri, Ohio, and Vermont.

Hairy Cottonwood, *P. deltoides* var. *pilosa* (Sarg.) Sudw., is a variety with petioles and leaves pubescent when young and retaining some pubescence on the veins beneath. It occurs in Georgia, Louisiana, Oklahoma, and Kansas.

The original name for Northern Cottonwood, *P. deltoides* var. *virginiana* (Castiglioni) Sudw., is now considered to be a synonym of *P. deltoides* Marsh.

P. × *eugenei* Limon-Louis is a hybrid of *P. deltoides* × *P. nigra* var. *italica.*

P. × *jackii* Sarg. is a hybrid of *P. deltoides* × *P. balsamifera.*

REMARKS. The genus name, *Populus,* is the ancient name given by Pliny, and the species name, *deltoides,* refers to the triangular shape of the leaf. Other vernacular names are Carolina Poplar, Necklace Poplar, Water Poplar, Southern Cottonwood, Yellow Cottonwood, and Alamo. Alamo is the Spanish name for the tree, and was also the name given the famous Texas fort which was surrounded by the trees. "Remember the Alamo" was the battle slogan of the Texas-Mexico War. The tree is often planted for ornament and for erosion control in dune-fixing. The air-borne cottony seeds are undesirable at fruiting season. The leaves flutter rapidly and make a rustling sound in the wind because of the flexible flattened petiole. The tree sprouts from the stumps and roots, is easily storm damaged, easily fire damaged when young, is much attacked by fungi, and grows rapidly. The foliage is known to be browsed by cattle, black-tailed deer, and cottontail. The seeds are eaten by rose-breasted grosbeak and evening grosbeak. The wood is used for paper pulp, cases and crates, tubs and pails, excelsior, veneer for plywood, musical instruments, dairy and poultry supplies, laundry appliances, and fuel.

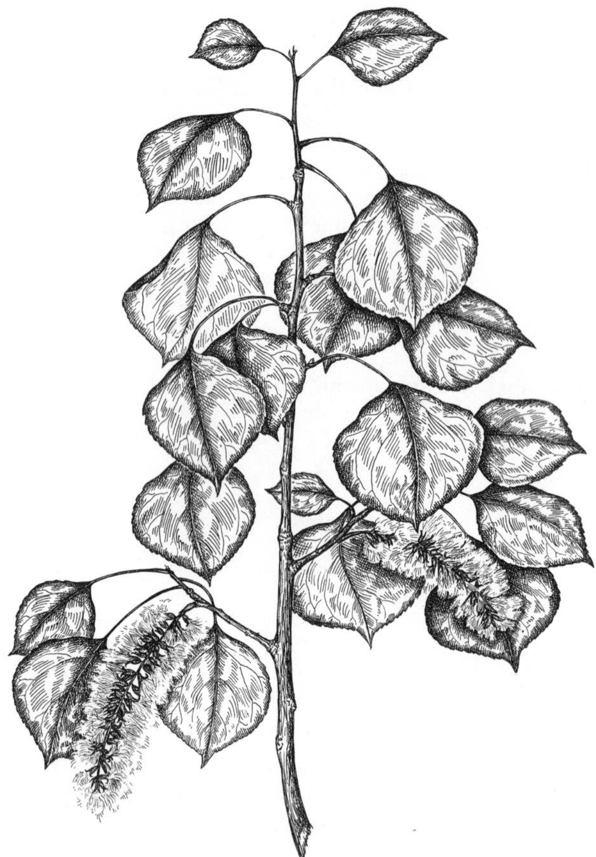

QUAKING ASPEN
Populus tremuloides Michx.

QUAKING ASPEN

Populus tremuloides Michx.

FIELD IDENTIFICATION. Tree 20–40 ft high, with a trunk diameter of 15–20 in. Rarely to 100 ft high and 3 ft in diameter. Usually with a long, slender trunk and a narrow, rounded top. Bark conspicuously whitened.

FLOWERS. April–May, on drooping catkins, staminate 1½–2½ in. long, ¼–⅓ in. in diameter; disk oblique, entire; stamens 6–12, on short filaments; scales divided into 3–5 triangular-lanceolate, acute to acuminate, hairy-fringed lobes; pistillate catkins about 4 in. long at maturity, disk slightly crenate; ovary conic, style short and thickened, stigmas 2, erect, thick, with linear, spreading lobes.

FRUIT. Ripening May–June, capsule borne spirally on stalks 1/25 in. long, body ⅕–⅓ in. long, oblong-conic, light green to brown, thin-walled, 2-valved; seeds obovoid, light brown, about 1/32 in. long.

LEAVES. Rather variable, those on young shoots much larger than leaves of older trees. Simple, alternate, deciduous, ovate to broad-ovate or reniform, apex short-pointed or acuminate, base usually truncate or rounded, margin closely crenate-serrate with mostly rounded teeth, thin but firm, yellowish green to bluish green, semidull or lustrous and glabrous above, dull green and glabrous beneath, main vein conspicuous and whitish, primary veins and veinlets minutely reticulate, blade 1–4 in. long; petiole slender, flattened, 1–3 in.

TWIGS. Slender, reddish brown to gray, lustrous and

glabrous or slightly hairy, roughened by elevated leaf scars; lenticels scattered, oblong, orange to yellow; buds both terminal and lateral, about ¼ in. long, somewhat resinous, glabrous or ciliate, reddish brown, shiny, conic, apex acute; flower buds with 6–7 scarious scales; pith star-shaped.

BARK. Thin, smooth, greenish to gray or white, marked with large excrescences and rows of leaf scars, lower part of trunk on old trees becoming dark brown and scaly with deep furrows and ridges.

WOOD. Light brown, sapwood nearly white, weighing about 25 lb per cu ft, straight-grained, soft, weak, odorless when dry, disagreeable odor when wet, moderately shock resistant, low in stiffness, tools moderately hard, easy to glue, takes paint well, low in nail-holding ability, tending to warp in seasoning, shrinks considerably, not durable on exposure.

RANGE. From sea level to an altitude of 10,000 ft. One of the most widely distributed trees in North America. Found in New Mexico and Trans-Pecos Texas, west to southern California, eastward to West Virginia and Pennsylvania, and north to Canada. Extending southward in the mountains of Mexico.

MEDICINAL USES. The bark of Quaking Aspen was used extensively by the pioneers and Indians as a remedy for fevers and as an antiscorbutic. It contains salicin and populin. Salicin occurs as colorless, silky, shining needles or prisms, or as a white crystalline powder. It is odorless and has a very bitter taste. Populin occurs as white crystals insoluble in cold water, but soluble in 70 parts of boiling water and in alcohol. When anhydrous it melts at 180°C.

Populin and salicin are also found in the buds of other poplars such as *P. candicans* and *P. tacamahacca;* and in the bark of some willows as *S. helix, S. fragilis,* and *S. purpurea.* The official drug is known as "syrupus pini albae compositus." Its action is somewhat analogous to turpentine. It is sometimes used externally as a counterirritant in muscular rheumatism. Internally it is of value chiefly as a stimulating expectorant in subacute or chronic bronchitis. It is comparatively little employed except in combination with other drugs. (Wood and Osol, 1943, pp. 881–882)

PROPAGATION. The tree spreads rapidly by root sprouts and is valuable in erosion control because of its ability to cover denuded and burned areas quickly, but is usually followed by other more permanent trees, such as conifers. The tree is short-lived, rapid-growing, and easily damaged by fire, wind, fungus, and insects. It has been in cultivation since 1811. The seed-bearing age is 50–70 years, with heavy crops every 4–5 years and lighter crops in intervening years. The capsules should be gathered when they first show signs of rupturing. The cottony hairs on the seed need not be removed before planting. The cleaned seed averages about 3,600,000 seeds per lb, with a purity of about 50 per cent. In test lots the seed is very variable, some averaging about 40, and others to 83 per cent germination. The seed may be sown immediately after gathering, or dried for 3–4 days and stored in sealed containers at 41°F. Good viability can be maintained for at least a year. No pretreatment of seeds is necessary. The seeds may be broadcast or planted in drills, or the capsules spread over a prepared seed bed and the seeds will disperse naturally. No soil covering is required, or at most only a very light one. During germination the seedlings are shaded and good aeration provided. They are susceptible to fungi, heat, and drought. Root cuttings can also be made for propagation.

VARIETIES. Rocky Mountain Quaking Aspen, *P. tremuloides* var. *aurea* Daniels, is the common form of the Rocky Mountain area, with leaves often broadest near the middle, apex with a short, slender point, margin remotely and irregularly serrate, turning orange to vivid yellow in fall.

Vancouver Quaking Aspen, *P. tremuloides* var. *vancouveriana* (Trel.) Sarg., has branches pubescent or puberulous, and leaves tomentose at first and glabrate later with margins coarsely crenate-serrate. Introduced into cultivation in 1922.

Big-leaf Quaking Aspen, *P. tremuloides* var. *magnifica* Vict., has large leaves.

Weeping Quaking Aspen, *P. tremuloides* var. *pendula* Jaeg., has drooping or pendulous branches. Also known under the name of Parasol de St. Julien.

REMARKS. *Populus* is the classical Latin name, and the species name, *tremuloides,* is from the Latin word meaning "trembling," referring to the movement of the leaf in the wind. Some of the vernacular names in use are Alamo Blanco, Popple, Poplar, White Poplar, Trembling Poplar, Aspen, Golden Aspen, American Aspen, Mountain Aspen, Trembling Aspen, Vancouver Aspen, Quiver Leaf, Noisy Leaf, and Woman's Tongue.

The wood is used for pulp, boxboards, matchwood, tubs, pails, clothespins, laundry, dairy and apiary supplies, excelsior, and occasionally for rough lumber not in contact with the soil. The flowers are considered a good bee food. The seeds are eaten by a number of species of birds, including the ruffed and sharp-tailed grouse. The twigs, bark, or buds are browsed by beaver, pika, white-tailed and mule deer, mountain sheep, moose, black bear, red squirrel, cottontail, snowshoe rabbit, and porcupine. The leaves are palatable to cattle, sheep, and goats.

Blue-stem Willow

Salix irrorata Anderss.

FIELD IDENTIFICATION. Western shrub to 12 ft high, but only rarely a small tree. Recognized by the reddish brown but very glaucous stems, and the oblong-lanceolate leaves which are white-glaucous beneath and remotely toothed.

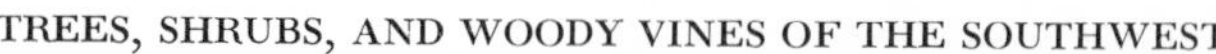

Blue-stem Willow
Salix irrorata Anderss.

Flowers. Buds rather large, catkins before the leaves, rather short and dense, subsessile; scales ovate to obovate, dark, densely white-villous; staminate catkins ⅝–1 in. long, about ⅜ in. thick, oblong, rather closely appressed to the twig; stamens 2, filaments glabrous, united at base, anthers yellow or reddish; pistillate catkins ⅝–1¾ in. long, with or without basal bracts, pedicels very short or almost sessile; ovary glabrous, perianth with a single scale, style less than 1/25 in. long, stigmas short, thick, and entire or marginate.

Fruit. Capsule subsessile, ovoid-conic, glabrous, woolly-hairy at base, ⅛–⅙ in. long; seeds numerous, very small.

Leaves. Simple, alternate, deciduous, narrowly oblong to lanceolate or elliptic, apex acute to short-acuminate, base cuneate or gradually narrowed, margin entire or remotely low-serrate, slightly revolute; upper surface dark green, lustrous, glabrous; lower surface whitened-glaucous, veins delicate, yellow to reddish, blades 2½–4 in. long (maximum of 6 in.); petiole ⅛–⅖ in., yellow to brownish.

Twigs. Slender, rather straight, young ones somewhat angled and reddish brown to purplish, covered with a conspicuous glaucous-white coating; buds stout, broadly ovoid, light brown to black.

Range. In sunny sites along streams, usually at altitudes of 5,500–7,500 ft. In western Texas, New Mexico, Arizona, and eastern Colorado.

Remarks. *Salix* is the classical Latin name, and the species name, *irrorata*, means "unwanted." The plant is conspicuous in winter because of the very glaucous stems, which characteristic also distinguishes it from the closely related *S. lasiolepis* Benth. It was first introduced into cultivation about 1898.

Gulf Black Willow

Salix nigra Marsh.

Field Identification. A rapidly growing tree sometimes attaining a height of 125 ft.

Flowers. April–May, dioecious, borne in many-flowered catkins preceding the leaves or with them; staminate catkins cylindrical, slender, 1–2 in. long; bracts obtuse, yellow, hairy below; stamens 3–7, filaments hairy below, anthers yellow; pistillate catkins 1½–3 in. long; bracts deciduous; pistil solitary, style short, the 2 stigmas thickened.

Fruit. May–June, borne on slender spreading pedicels, capsule light brown, conic-ovoid, sharp-pointed, glabrous, ¼–⅛ in. long, splitting into two valves; seeds minute, green, pilose with long hairs.

Leaves. Simple, alternate, deciduous, blades 3–6 in. long, ¼–¾ in. wide, narrowly lanceolate; apex long-attenuate and sometimes falcate; base rounded, acute; margin finely glandular-serrate; green above and paler below, glabrous or puberulent along the veins, or pubescent when young; petiole short, puberulent; stipules variable, either large, persistent, semicordate, pointed, and foliaceous, or small, ovate, and deciduous.

Twigs. Slender, brittle, reddish brown.

Bark. Light brown to black, rough, deeply fissured, ridges dividing into thick shaggy scales, rich in tannin.

Wood. Light brown, soft, light, weak, not durable, weighing 27 lb per cu ft.

Range. In wet soil, Texas, Louisiana, Oklahoma, and Arkansas. Reaching its largest size on the banks of the Brazos, San Bernard, and Colorado rivers in Texas; east to North and South Carolina, north to New Brunswick, and west to North Dakota.

Propagation. The date of the earliest cultivation of Black Willow was 1809. The minimum seed-bearing age is 10 years, the optimum is 25–75 years, and the maximum 125 years. The ripe capsules are collected in bags. The tiny seeds need not be separated from the capsule. There are 2 million seeds per lb. Commercial seed is not available because of short viability. No dormancy occurs, seeds germinate 12–24 hours after falling, with a germination of from 75–95 per cent.

Seeds are broadcast and the beds firmed and kept moist. Seedlings should be well shaded at first, and transplanted at 3–4 weeks. Field planting is done with one-year-old plants. Leaf rusts caused by *Melampsora* should be sprayed. Nursery stock also infected by black canker and fungus scab should not be field-planted, nor should plants with dead bark caused by *Cytospora*. Good results can be had with Black Willow by stem or root cuttings.

VARIETIES. The species *Salix nigra* Marsh is said to be replaced in the Texas gulf coast areas by the variety *altissima*. However, the characters which separate the two, such as pubescence of leaves and twigs, larger petioles, and acute angles of the leaf bases, are so variable that clear-cut distinctions between the species and variety are questionable.

S. nigra var. *lindheimeri* Schneid. is a glabrous variety known as Texas Black Willow and occurs in central Texas.

REMARKS. The genus name, *Salix,* is the classical Latin name, and the species name, *nigra,* refers to the black bark. Vernacular names are Scythe-leaved Willow, Swamp Willow, and Pussy Willow. The bark was formerly used as a home remedy for fever ailments. The wood is used for artificial limbs, charcoal, toys, doors, fuel, cheap furniture, boxwood, and excelsior.

WESTERN BLACK WILLOW
Salix nigra var. *vallicola* Dudley

GULF BLACK WILLOW
Salix nigra Marsh.

WESTERN BLACK WILLOW

Salix nigra var. *vallicola* Dudley

FIELD IDENTIFICATION. Western tree to 45 ft, and 3 ft in diameter.

FLOWERS. Dioecious, staminate catkins 1–2 in. long, cylindric; scales linear-oblanceolate, apex acute, yellow, with dense long soft hairs; stamens 3–6, filaments villose; pistillate catkins 2–3½ in. long at maturity, scales early deciduous; ovary conic, apex acuminate, pubescent to glabrous; stigmas 2, short and broad.

FRUIT. Capsules spread well apart on mature catkins, pedicels 1/12–1/8 in. long; capsule body 1/5–1/4 in. long, conic-ovoid to lanceolate, apex acute, reddish brown, pubescent at first but more glabrous later.

LEAVES. Simple, alternate, deciduous, elliptic to

lanceolate, some curved, long-acuminate at apex, base narrowly cuneate or attenuate, margin finely serrate, young with pale hairs, mature leaves glabrous and dull green on both sides, blade length 1½–3 in., width ¼–½ in., on young shoots considerably larger; petioles about ¼ in. long or less, pubescent to glabrous; stipules orbicular, base cordate, margin serrate, surface pubescent.

TWIGS. Slender, yellow to orange or grayish, pubescent or glabrous when older.

BARK. On trunk dark gray to black, rather deeply furrowed, ridges breaking into thickish plates.

WOOD. Reddish brown to pale brown, soft, weak, specific gravity 0.44.

RANGE. In clumps or patches along water courses at altitudes of 300–4,000 ft. In western Texas and New Mexico; west to California, north to Nevada and Utah, and south in Mexico to Chihuahua, Sonora, Sinaloa, and Baja California.

REMARKS. The genus name, *Salix,* is the classical Latin name, and the species name, *nigra,* refers to the dark bark. The variety name, *vallicola,* means inhabitant of low places or valleys. Some authorities consider the tree to be a separate species instead of a variety of Black Willow, and list it under the name of Gooding Willow, *S. goodingii* Ball. Leslie Newton Gooding was a botanist of the U. S. Department of Agriculture and collected the type specimen. However, the differences between the plants are so slight that a varietal name under *S. nigra* seems most fitting. The twigs of the variety are yellow to gray as compared to reddish brown twigs of *S. nigra,* also the leaves of the former are elliptic-lanceolate as compared to narrow-lanceolate. Also known as the Dudley Willow. The young shoots are sometimes browsed by livestock and mule deer. A decoction of the leaves is reported as being used in Chihuahua for fevers.

Sand-bar Willow

Salix interior Rowlee

FIELD IDENTIFICATION. A slender, upright shrub forming thickets by stolons, or a small tree to 30 ft.

FLOWERS. Dioecious, April–May on leafy twigs. Staminate and pistillate catkins slender, cylindric, linear, borne on different plants; staminate catkins terminal or axillary, dense, ¾–2 in. long, about ⅓–⅜ in. broad; stamens 2, exserted, filaments distinct, hairy at base; pistillate catkins loosely flowered, 2–3 in. long and about ¼ in. broad; scales light yellow, hairy, ovate to obovate, entire or erose; ovary short-stalked, oblong-cylindric, silky-hairy when young, less hairy or glabrous later; stigmas 2, subsessile, lobed; young capsule with long white silky hairs.

Sand-bar Willow
Salix interior Rowlee

FRUIT. Capsule matures in April, sessile or short-peduncled, narrowly ovoid-conic, gradually narrowed to a blunt apex, ⅙–¼ in. long, brownish, glabrous or villous, 1-celled, splitting into 2 reflexed valves; seeds minute, attached to long white hairs, buoyant in the wind.

LEAVES. Deciduous, alternate, blades 2–6 in. long, ⅛–⅓ in. wide, linear-lanceolate, sometimes falcate, thin, apex acuminate, base gradually narrowed into a short petiole, margin with remote, denticulate, glandular teeth, main vein prominent; upper surface dark green and glabrous or puberulent along the main vein; paler and pubescent beneath; petioles ⅛–³⁄₁₆ in., pubescent; stipules small or absent. Young leaves silky-hairy beneath.

TWIGS. Slender, erect, green to brown or red, glabrous or puberulent and sometimes glaucescent.

BARK. Green to gray or brown, smooth; on older trunks furrowed and broken into closely appressed scales; lenticels sometimes large and abundant.

WOOD. Soft, light, reddish brown, sapwood pale brown, weighing 31 lb per cu ft, little used except for fuel or charcoal.

RANGE. The species is found in alluvial soil along streams and lakes over a wide area; Texas and Louisiana coast, and north through Arkansas and Oklahoma to Canada and Alaska. Also in northern Mexico in the states of Nuevo León, Tamaulipas, and Coahuila.

VARIETIES AND CLOSELY RELATED SPECIES. Texas Sand-bar Willow, S. *interior* var. *angustissima* (Anderss.) Dayton, is separated from the species by a silky pubescent ovary and no dorsal gland in the male flowers. However, intergrades between it and the species are found. It occurs in northwest Texas, New Mexico, and northeastern Mexico.

Narrow-leaf Sand-bar Willow, S. *interior* var. *pedicellata* (Anderss.) Ball, of the Northwestern states, has narrower, linear leaves with larger fruit pedicels and glabrous ovaries.

Wheeler's Sand-bar Willow, S. *interior* var. *wheeleri* Schneid., is of shrubby habit with densely silky-pubescent leaves. It is found from Oklahoma north to New Brunswick, Nebraska, and North Dakota.

Coyote Willow, S. *exigua* Nutt., is considered to be closely related to the Sand-bar Willow. It is a Western willow to 15 ft, with leaves linear-lanceolate, subsessile, almost entire, yellowish green and silky-hairy.

Silver-leaf Willow, S. *argophylla* Nutt., is also closely related to the Sand-bar Willow. It is a shrub or small tree with leaves narrowly lanceolate, subsessile, usually entire, silky-hairy and silvery-white beneath, 2–4 in. long; ovary densely pubescent. It is found from Texas to California, and north to Montana and Washington.

REMARKS. The genus name, *Salix,* is the classical Latin name, and the species name, *interior,* refers to the plant's inland distribution along water courses. It was formerly listed under the scientific names of S. *longifolia* Muehl. and S. *fluviatilis* Sarg. Known under the vernacular names of Riverbank Willow, Osier Willow, Shrub Willow, Long-leaf Willow, Narrow-leaf Willow, Red Willow, and White Willow.

SILVER-LEAF WILLOW
Salix argophylla Nutt.

SILVER-LEAF WILLOW
Salix argophylla Nutt.

FIELD IDENTIFICATION. Thicket-forming shrub or small tree 3–18 ft tall.

FLOWERS. Borne in May, in separate staminate and pistillate catkins, terminating leafy shoots; catkins 1–3, ¾–1¾ in. long, cylindric, rather narrow; scales oblong and obtuse in the staminate catkin, in the pistillate narrow-lanceolate and more acute, villous; stamens 2, lower half of filament densely hairy; pistil single, ovary densely white-villous, stigmas sessile, short, thick.

FRUIT. Capsule on pedicels about 1/50 in. long, lanceolate, ⅕–¼ in. long, silky-villous, 2-valved; seeds numerous, very small, hairy.

LEAVES. Simple, alternate, narrowly lanceolate to linear, acute at both ends or some acuminate at the apex, margin entire or remotely denticulate, both sides with lustrous silvery-white hairs, veins obscure because of the hairs, length of blades 1½–3¼ in., width ⅓–½ in., short-petiolate or subsessile.

TWIGS. Young ones with white villous hairs, yellow to orange, older ones reddish brown and glabrous.

RANGE. Along stream beds or irrigation ditches in the artemisia or yellow pine belt. In Trans-Pecos Texas and New Mexico; west to California, north to Montana, and west to Washington and Oregon.

REMARKS. The genus name, *Salix,* is the classical Latin name, and the species name, *argophylla,* refers to the silvery leaf caused by the silky hairs. Also known under the vernacular name of Coyote Willow. Sometimes sheep and cattle browse the foliage.

SCOULER WILLOW
Salix scouleriana Barratt

FIELD IDENTIFICATION. A thicket-forming Western shrub, or tree attaining a height of 30 ft and a diameter of 10 in. Branches slender and pendulous, forming a round-topped crown.

SCOULER WILLOW
Salix scouleriana Barratt

FLOWERS. Dioecious catkins appearing before the leaves, pedicels short and villous; staminate catkin oblong-cylindric, ¾–1¼ in. long, about ½ in. or more thick, erect, sessile or nearly so on short scaly, hairy branchlets; stamens 2, with free filaments exceeding the scale in length; filaments pilose at base, anthers yellowish red; pistillate catkin short-peduncled, ¼–2 in. long and ½–⅝ in. thick; scales oblong to obovate or oval, ends narrowed, blackened, white-hairy; ovary short-stalked, single, cylindric, style distinct with broad emarginate stigmas, carpels 2; pedicels hairy and less than half the length of the scale.

FRUIT. Pedicels 1/25–1/12 in., pubescent; capsule oblong-ovoid or rostrate, acuminate, ¼–⅓ in. long, reddish brown, pale-pubescent, 1-celled, splitting into 2 recurved valves later, the 2 stigmas persistent and about 1/25 in. long; seeds minute, dark brown, narrowed.

LEAVES. Simple, alternate, deciduous, shape very variable, oblong-obovate to oval or elliptic, apex acute to obtuse or abruptly acuminate, base gradually narrowed and cuneate or unsymmetrical, margin entire or remotely crenate-serrate, thin but firm, blades 1¼–4 in. long, ½–1½ in. wide, upper surface dark green, shiny and glabrous, lower surface glaucous and often densely hairy (more rarely glabrate), veinlets rather strongly reticulate; petioles slender, glabrous or puberulous, ⅜–¾ in.; stipules falling early, ⅛–¼ in. long, either minute and entire, or foliaceous, half-curved and glandular-serrate.

TWIGS. Stout, terete, tough, at first pubescent, yellow to orange, later glabrous but roughened and dark reddish brown; winter buds ovoid, terete or somewhat flattened, apex acute, yellow or orange, glabrous or hairy, about ¼ in. long; leaf scars with 3 small fibrovascular-bundle marks.

BARK. Dark reddish brown, rather thin, divided into broad, flat, scaly ridges, with a peculiar fetid odor.

WOOD. Reddish brown, sapwood nearly white, close-grained, light, soft.

RANGE. In the yellow pine, spruce, and aspen belts at altitudes of 3,000–11,000 ft. New Mexico and Arizona; northward through South Dakota and northern California to Saskatchewan and Alaska.

VARIETIES. A number of varieties are recognized:

Yellow Scouler Willow, *S. scouleriana* var. *flavescens* Schneid., is a shrub or small tree with obovate, rounded, yellowish leaves and branchlets. It occurs in California, Wyoming, South Dakota, and northern New Mexico.

Rough Scouler Willow, *S. scouleriana* var. *crassijulis* Andr., has shorter, broader obovate leaves, apex rounded, pubescent below; twigs tomentose; winter buds larger and pubescent; catkins short and stout; becomes a tree 70 ft high and 2½ ft in diameter. From central California to San Bernardino County.

REMARKS. The genus name, *Salix*, is the classical Latin name, and the species name, *scouleriana*, honors John Scouler, Scotch naturalist and physician who collected plants on the northwest coast of North America 1825–1827. Vernacular names are Black Willow, Fire Willow, Mountain Willow, and Nuttall Willow. It is fair browse for mule deer and livestock. It recovers well from grazing and comes back quickly on burned-over areas. It is rather rapid-growing and was first introduced into cultivation in 1918.

BABYLON WEEPING WILLOW

Salix babylonica L.

FIELD IDENTIFICATION. Cultivated tree attaining a height of 50 ft. The drooping twigs give the tree its name.

FLOWERS. April–May, catkins small, appearing on short lateral leafy branches; staminate catkins to 1⅝ in. long and ¼–⅓ in. wide on peduncles ⅜–⅝ in. long (pistillate catkins smaller); bracts ovate-lanceolate, yellowish, obtuse, deciduous; stamens 3–5, free, pubescent at base; style almost none.

FRUIT. Capsule ovoid-conic, sessile or nearly so, glabrous, style almost absent, stigmas minute.

LEAVES. Alternate, narrowly lanceolate, apex long-acuminate, base narrowed, margin serrulate, at first somewhat silky pubescent, glabrous with maturity,

BABYLON WEEPING WILLOW
Salix babylonica L.

lower surface glaucous, blade length 3–7 in., width $\frac{1}{4}$–$\frac{1}{2}$ in., sometimes curling; stipules wanting or if present lanceolate and $\frac{1}{12}$–$\frac{1}{3}$ in. long.

TWIGS. Slender, glabrous, elongate, drooping, green at first, later yellowish or brownish.

RANGE. Grows best in damp sandy soils near water courses. In Texas, Arkansas, Oklahoma, and Louisiana; eastward to Florida, northward to Virginia and Connecticut, and westward to Michigan. A native of North China, cultivated throughout North America below an altitude of 3,500 feet.

HYBRID. It is known to hybridize with the Crack Willow, *S. fragilis* L., and the White Willow, *S. alba* L.

REMARKS. The genus name, *Salix*, is the classical name. The species name, *babylonica*, refers to its once-presumed West Asiatic origin, but this is a misnomer because it is a native of China.

The tree of the Biblical reference (Psalms 127: 1–2) is now known to be the willow-like Euphrates Poplar, *Populus euphratica* Oliv. Babylon Weeping Willow sometimes escapes cultivation by the distribution of its twigs. It is known also as the Garb Willow, Napoleon Willow, and Weeping Willow.

BEBB WILLOW
Salix bebbiana Sarg.

FIELD IDENTIFICATION. Shrub or small tree with a single trunk 3–24 ft tall and a bushy top.

FLOWERS. April–June, dioecious, catkins expanding with the young leaves, lax and open; staminate catkins rather small, $\frac{3}{8}$–$\frac{5}{8}$ in. long; each flower with a basal gland; bracts yellow to brown, sparingly hairy; stamens 2, filaments free and slender, a few hairs at the base; pistillate catkins $\frac{3}{4}$–$2\frac{3}{4}$ in. long, very lax, on bracted peduncles $\frac{1}{5}$–$\frac{4}{5}$ in.; scales narrowly oblong to lanceolate or ovate or oval, acute, $\frac{1}{25}$–$\frac{1}{12}$ in. long, greenish yellow, apices pilose and reddish; ovary silky-hairy, stigmas entire or divided, style very short or absent.

FRUIT. Pedicels of fruit three or four times the length of the subtending bract, slender, finely pubescent; capsule body lanceolate-rostrate, somewhat pubescent.

LEAVES. Simple, alternate, deciduous, oblong to el-

BEBB WILLOW
Salix bebbiana Sarg.

liptic or oblanceolate, sometimes ovate, apex acute or short acuminate, base acute or rounded, margin entire or crenate-toothed, length of blade 1½–2¾ in. (rarely to 4 in.), width ⅝–1¼ in., surfaces of young ones with gray to white tomentum, mature ones pubescent or glabrescent above and lower surface tomentose and rugose-veiny; petioles 1/16–¼ in.

TWIGS. Slender, divaricate, brownish, with gray tomentum, older ones reddish brown, pubescent to glabrous and lustrous.

RANGE. A widespread species. Often on riverbanks or in moist soil at altitudes of 4,000–9,000 ft. New Mexico; west to California, north to Washington, Canada, and Alaska; also to South Dakota, east to Maryland, Pennsylvania, New Jersey, and New England.

VARIETIES. Smooth Bebb Willow, S. *bebbiana* var. *perrostrata* (Rydb.) Schneid., has looser catkins; leaves more obovate, entire and scarcely reticulate beneath. It is found in New Mexico and Arizona, northward to Alaska and Newfoundland.

A number of other Northern varieties have been recorded from Newfoundland to Quebec and along the St. Lawrence River.

REMARKS. The genus name, *Salix*, is the classical name, and the species name, *bebbiana*, is in honor of Michael Schuck Bebb (1833–1895), American specialist on willows. Vernacular names for the plant are Beaked Willow, Long-beaked Willow, Chaton, and Petit Minou. The young plants are browsed by livestock and the cottontail.

PRAIRIE WILLOW
Salix humilis Marsh.

PRAIRIE WILLOW

Salix humilis Marsh.

FIELD IDENTIFICATION. Shrub 3–9 ft tall with wandlike branches. Very variable in characters of inflorescence and leaf, and hence divided into varieties or species by some authors.

FLOWERS. Catkins precocious, sessile or almost so; staminate catkins ¼–1¼ in. long and ⅜–⅞ in. wide, oval to obovoid; stamens 2, filaments long, free, glabrous; anthers reddish purple; pistillate catkins at maturity ⅜–3¼ in.; styles about 1/50 in., equaling the divided short, thick stigmas; scales oblanceolate, 1/25–1/12 in. long, dark brown or black, long-villous.

FRUIT. Pedicels 1/25–1/12 in., capsules narrowly lanceolate-rostrate, ⅙–⅓ in. long, gray-cinereous, persistent styles less than 1/50 in., equaling the divided stigmas.

LEAVES. Alternate, blade oblanceolate to narrowly obovate or subelliptic, 1¼–4 in. long (rarely to 6 in.), ⅜–¾ in. wide (rarely to 1¼ in.), apex obtuse to acute, narrowed to the base, margin entire or undulate-dentate, upper surface grayish green and glabrous or puberulent, lower surface more or less gray-pubescent but becoming glabrescent; stipules lanceolate, acute, dentate, often deciduous.

TWIGS. Slender, wandlike, young ones pubescent, later glabrate, yellowish to brown.

RANGE. In open woodlands, dry prairies, or barrens. Arkansas, Oklahoma, east Texas, and Louisiana; eastward to Florida, northward to Newfoundland and Quebec, and westward to North Dakota.

VARIETIES AND HYBRIDS. Prairie Willow is known to hybridize with S. *bebbiana* Sarg., S. *discolor* Muhl., S. *gracilis* Anderss., and S. *sericea* Marsh.

Wrinkled Prairie Willow, S. *humilis* var. *hyporhysa* Fern., is a variety with the leaves glabrate or barely puberulent beneath, the prominent veinlets becoming conspicuous beneath; mature blades ¼–¾ in. broad. It occurs in Oklahoma, Arkansas, east Texas, and Louisiana; eastward to Florida, northward to West Virginia, New Jersey, New York, and Connecticut; and westward to Michigan, Wisconsin, and Iowa.

Some authors list S. *tristis* Ait. as a small-leaved variety of Prairie Willow under the name of S. *humilis* var. *microphylla* (Anderss.) Fern., but to this author the plant seems sufficiently distinct to warrant its retention as a species. (See description of S. *tristis* Ait.)

Several other Northern varieties and forms not occurring in the Southwest have been described by various authors.

REMARKS. The genus name, *Salix*, is the classical name, the species name, *humilis*, refers to its low stature. It is also known as the Small Pussy Willow, Upland Willow, and Gray Willow.

Coyote Willow

Salix exigua Nutt.

FIELD IDENTIFICATION. Shrub with spreading stems to 12 ft tall, or sometimes a tree to 25 ft and 5–6 in. in trunk diameter. The branches spreading and forming a rounded crown.

FLOWERS. Appearing after the leaves on glabrous twigs 1–2 in. long, catkins terminal or axillary; staminate catkins 5/8–3/4 in.; pistillate catkins 5/8–1 5/8 in. Scales hoary-pubescent, varying from lanceolate to obovate with an acute or rounded apex; stamens 2, with the filaments hairy below; ovary sessile, villose, stigma bifid, sessile, style absent.

Coyote Willow
Salix exigua Nutt.

Nevada Coyote Willow
Salix exigua var. *nevadensis* (Wats.) Schneid.

FRUIT. Capsule ovoid-lanceolate, apex acuminate, glabrous or nearly so, 1/5 in. or less long, sessile or with pedicels shorter than the gland.

LEAVES. Alternate, blade linear to lanceolate or oblanceolate, apex acuminate, some curved, base gradually narrowed, margin glandular-serrate, often entire below the middle, upper surface grayish green and glabrous, lower surface with white appressed hairs (or in some forms only puberulent), length 1 1/2–4 in., 1/8–1/4 in. wide (on young growth sometimes 4 1/2 in. long and 1 1/2 in. wide); stipules absent or minute; sessile or nearly so.

TWIGS. Slender, glabrous, reddish brown to gray.

BARK. Grayish brown, thin, longitudinally fissured, older trunks more furrowed.

RANGE. Along streams in mountain woodlands or desert grasslands at altitudes of 4,000–7,000 ft. Oklahoma, western Texas, and New Mexico; westward to California and northward to British Columbia. In Mexico from Chihuahua to Baja California.

VARIETIES. Narrow-leaf Coyote Willow, S. *exigua* var.

stenophylla (Rydb.) Schneid., has very narrow leaves. It occurs in western Texas and New Mexico.

Nevada Coyote Willow, S. *exigua* var. *nevadensis* (Wats.) Schneid., is segregated by having 2 glands on the glabrous ovary. From Arizona to Nevada.

REMARKS. The genus name, *Salix,* is the classical name. The species name, *exigua,* refers to the small-sized leaves. Also known under the vernacular names of Basket Willow, Gray Willow, Sand-bar Willow, Narrow-leaf Willow, Slender Willow, and Acequia Willow.

Peach-leaf Willow

Salix amygdaloides Anderss.

FIELD IDENTIFICATION. Tree to 40 ft, or rarely larger. The branches are ascending, or sometimes drooping to form an open irregular head.

FLOWERS. April–May, appearing with the leaves, borne on the ends of short lateral branches; staminate and pistillate catkins borne separately, glands of both elongate and free; staminate catkins 1–2 in., slender, elongate, cylindric, hairy; scales deciduous, yellowish green, hairy, ovate, apex rounded; stamens 5–9 (usually 5), filaments distinct, hairy at base, anthers yellow; pistillate catkins 1½–3 in., lax in fruit; scales narrowly oblong to obovate, early deciduous; style short, stigmas nearly sessile, bifid; ovary oblong-conic, glabrous.

Peach-leaf Willow
Salix amygdaloides Anderss.

FRUIT. Capsules borne on the elongate catkins, about as long as the glabrous pedicels or shorter, lanceolate to narrowly ovoid or conic, ⅙–⅕ in. long, light yellow or reddish, 1-celled, valves 2–3 and recurved when ripe; seeds numerous and very small.

LEAVES. Revolute in the bud, simple, alternate, deciduous, pubescent when young, lanceolate to ovate-lanceolate, some falcate, apex long-acuminate, base broad-cuneate to rounded, some inequilateral, margin finely and sharply serrate, thin, upper surface yellowish green and glabrous, lower surface paler and slightly glaucous and glabrous, veins prominent and yellowish, length of blade 2½–6 in., width ¾–1¼ in.; petiole slender, ¼–¾ in., twisted, glandless; stipules falling early, reniform, to ½ in.

TWIGS. Long, slender, terete, flexible, drooping or erect, glabrous, lustrous, varying from yellow or orange to reddish brown or gray later; lenticels scattered, small and pale; buds ⅛–⅙ in. long, ovoid, shiny, light or dark brown.

BARK. Reddish brown to dark brown, fissures irregular, ridges breaking into thick flat scales.

WOOD. Heartwood light brown, sapwood whitish, close-grained, soft, weak, weighing about 28 lb per cu ft.

RANGE. Sunny sites along margins of lakes or streams, in Western areas at altitudes of 3,000–7,000 ft. In Trans-Pecos and northwestern Texas and New Mexico; east through Kentucky to Pennsylvania, north to Quebec, and west to Manitoba, Washington, and Oregon.

VARIETIES AND FORMS. Wright Peach-leaf Willow, S. *amygdaloides* var. *wrightii* (Anderss.) Schneid., has yellow or yellowish brown, glabrous twigs; leaves 1½–2 in. long, ¼–⅓ in. wide and narrowly-lanceolate. It has been recorded in Texas from Ward, El Paso, and Potter counties; also in southern New Mexico. Some botanists do not consider it as distinct from the species, but only a narrow-leaved form.

Two Western forms have been described as follows:

Hairy Peach-leaf Willow, S. *amygdaloides* forma *pilosciuscula* Schneid., has young leaves and twigs hairy; catkins denser and fruits closer together.

Narrow Peach-leaf Willow, S. *amygdaloides* forma *angustissima* E. C. Smith, has linear-lanceolate leaves, at maturity about ⅜ in. wide. Near Cheyenne, Wyoming. This may be the same as Wright Peach-leaf Willow, S. *amygdaloides* var. *wrightii* (Anderss.) Schneid.

Peach-leaf Willow and Black Willow sometimes hybridize to produce the Glatfelter Willow, S. × *glatfelteri* Schneid.

REMARKS. The genus name, *Salix,* is the classical Latin name. The species name, *amygdaloides,* is from *amygdalus* ("Peach") and refers to the peachlike leaves.

Heart-leaf Willow
Salix cordata Muhl.

Also known under the vernacular names of Almond Willow, Almond-leaf Willow, Peach Willow, Southwestern Peach Willow, Wright Willow, and Wright Peach-leaf Willow. The tree has been in cultivation since 1895 and is rapid-growing and short-lived. The wood is used mostly for fuel and charcoal.

Heart-leaf Willow

Salix cordata Muhl.

Field identification. Shrub attaining a height of 6–12 ft.

Flowers. Dioecious, appearing May–June, catkins 2–3¼ in.; peduncles ⅜–1 in., with or without foliaceous bracts, sometimes the catkins subsessile; scales oblong-lanceolate, brown, densely villous; stamens 2, filaments glabrous; pistillate catkins with bracts narrowly ovate, brownish; ovary glabrous, style short.

Fruit. Capsule narrowly conic to lanceolate, acute, glabrous, 2-valved, ⅙–¼ in. long; styles at apex 1/15 in. long, stigmas 1/50 in., entire to somewhat bifid, gland short; seeds numerous, minute, hairy.

Leaves. Simple, alternate, deciduous, 1½–5 in. long, ⅝–1¾ in. wide, ovate-lanceolate to broadly ovate, apex acuminate, sometimes abruptly so, margin glandular-dentate or serrate, base rounded to narrowed to subcordate, surfaces green on both sides, or paler and more or less lanate and strongly veined beneath (young leaves densely gray-villous on both sides); stipules persistent, cordate-ovate or subreniform, oblique, dentate, blades ¼–⅗ in. long; petioles stout, ⅙–½ in., villous, somewhat clasping.

Twigs. Stout, young ones green to brown and gray, pubescent, older ones more glabrous.

Range. Wet sandy or alluvial soils of beaches and dunes. Arkansas, Missouri, and Colorado; west to California and northward into Wisconsin, Michigan, Illinois, Quebec, Nova Scotia, and Newfoundland.

Varieties. Purple Heart-leaf Willow, S. *cordata* forma *purpurascens* (Dieck) Schneid., is a form with the leaves purple when young.

Smooth Heart-leaf Willow, S. *cordata* var. *abrasa* Fern., has twigs, petioles, and blades (except the midrib) more or less glabrous.

Remarks. The genus name, *Salix*, is the classical Latin name, and the species name, *cordata*, refers to the cordate base of the leaves. However, the base is not cordate on some leaves. The plant was first introduced into cultivation in 1812.

Yew-leaf Willow

Salix taxifolia H. B. K.

Field identification. Large, Western shrub, or tree to 50 ft, and 18 in. in diameter. Branches short and divaricate, forming a broad open crown.

Flowers. Borne March–May. Catkins on leafy branches, terminal, or axillary, usually clustered, densely flowered, staminate and pistillate catkins separate; shape of catkins subglobose to cylindric-oblong, ¼–½ in.; scales deciduous, yellowish, oblong or obovate, apex acute to rounded or apiculate, outer surface densely tomentose, inner surface glabrous or pubescent, margins ciliate; stamens 2, filaments free and pubescent below; ovary sessile or short-stalked, ovoid-conic, hairy; stigmas 2, deeply emarginate.

Fruit. Capsule sessile or short-stalked, about ¼ in. long, cylindric, ovate-conical, acuminate-pointed, reddish brown, silky-hairy at first but more glabrate later; seeds numerous, small, hairy.

Leaves. Alternate, deciduous, linear-lanceolate or oblanceolate, graduated to apices, acute and mucronate, slightly curved, margin slightly revolute, obscurely and remotely denticulate above the middle or entire, soft white-hairy when young, later grayish green and finely pubescent above, lower surface paler and somewhat more hairy, blade length ⅓–1⅓ in., width 1/12–⅛ in.; leaves sessile, or petioles puberulous and less than 1/12 in.; stipules ovate, acute, minute, early deciduous.

Bark. Light gray to brown, on the trunk of mature

Yew-leaf Willow
Salix taxifolia H. B. K.

trees ⅓–1 in. thick, fissures deep and irregularly longitudinal, with intervening broad, flat ridges breaking into small, close scales.

TWIGS. Slender, reddish brown, densely tomentose at first but glabrous later except for roughened leaf scars; buds brown, 1/16–⅛ in. long, ovoid, puberulous.

RANGE. On creek banks and in canyons at altitudes of 3,000–6,000 ft. In Trans-Pecos Texas, southwestern New Mexico, and southern Arizona. In Mexico it is widespread, especially in Sonora, Chihuahua, Coahuila, and Baja California. Also in Guatemala.

REMARKS. The genus name, *Salix*, is the classical Latin name. The species name, *taxifolia*, refers to the small, narrow, yewlike leaves. Also known in the Southwest and Mexico under the vernacular names of Sauz, Sauce, Jaray, Taray, Taray de Rio, and Tarais. It is considered to be a good browse for livestock. The branches are used for brooms and the bark as a remedy for malaria by the Mexican Indians.

Summit Willow
Salix saximontana Rydb.

FIELD IDENTIFICATION. Low or prostrate alpine shrub, stems buried or rooting where they touch the earth. Forming dense, matted, turflike clumps, usually less than 6 in. high but with thick woody bases.

FLOWERS. The small flowers in separate staminate and pistillate catkins, ⅜–¾ in., 10–20-flowered, scales broadly obovate to rounded, yellowish to reddish, almost glabrous; staminate catkins ovoid to nearly globular, terminating leafy glabrous twigs; stamens 2, exserted; pistillate with ovary superior, solitary, 1-celled and with parietal placentation; style very short, stigma short, bifid.

FRUIT. Capsules ⅛–⅙ in. long, sessile, densely white-tomentose, 2-valved; seeds numerous, with tufts of apical hairs.

LEAVES. Clustered at the ends of short branches, elliptic-oblong to suborbicular, apex obtuse to acute, base rounded to acute, margin entire and often revolute, glabrous, upper surface light green, lower surface glaucous and reticulate-veiny, blade length ⅝–1⅓ in., width ⅜–¾ in.; petioles 1/12–½ in.

STEMS. Prostrate or ascending terminally, stout, often gnarled and crooked, very short and low to the ground, younger ones reddish brown, older ones gray to almost black with small broken scales, often rooting where they touch the ground.

RANGE. Boreal zones of high elevations, 10,500–12,500 ft. New Mexico; west to California and north to Alberta and British Columbia.

REMARKS. The genus name, *Salix*, is the classical Latin name, and the species name, *saximontana*, refers to its rocky habitat in mountains. Also known as Rocky

Summit Willow
Salix saximontana Rydb.

Dwarf Pussy Willow
Salix tristis Ait.

Mountain Willow. Snow Willow, S. *nivalis* Hook., is a closely related species which has orbicular, oblong-obovate to elliptic leaves not over ½ in. long and pistillate catkins only 3–6-flowered. Some botanists feel that the two are so closely related that Summit Willow should be a variety of Snow Willow and be given the name of S. *nivalis* var. *saximontana* (Rydb.) Schneid. Other botanists maintain the two as separate species. They are both found in New Mexico.

Dwarf Pussy Willow

Salix tristis Ait.

FIELD IDENTIFICATION. Shrub attaining a height of 1–3 ft. Branches numerous, leafy, erect or diffuse. Roots long and slender.

FLOWERS. March–April, catkins precocious, dioecious, ⅝–¾ in., subglobose to oval; scales oblanceolate, obtuse, blackish, villous, 1/25–1/15 in. long; stamens 2, orange-red, filaments free and glabrous.

FRUIT. Capsules narrowly lanceolate or oblanceolate, ⅕–¼ in. long, gray pubescent; styles equaling the divided stigmas; pedicels 1/25–1/12 in., pubescent.

LEAVES. Simple, alternate, deciduous, crowded toward the ends of twigs, narrowly oblanceolate to elliptic, blades ½–2 in. long, ⅓–½ in. wide, apex acute, base gradually narrowed, margin revolute and entire or nearly so, upper surface dark green, densely tomentose when young, becoming glabrate with age, lower surface persistently gray-tomentose and glaucous; petioles 1/12–¼ in., pubescent.

TWIGS. Slender, brown to gray, pubescent at first, glabrous later.

RANGE. Usually on sandy or gravelly banks in acid soil. Louisiana, Oklahoma, and Arkansas; eastward to Florida, north to Massachusetts and Vermont, and west to North Dakota, South Dakota, Nebraska, and Missouri.

REMARKS. The genus name, *Salix*, is the classical Latin name, and the species name, *tristis*, refers to the dull green leaves. Also known under the vernacular names of Dwarf Gray Willow and Dwarf Upland Willow. The plant is closely related to S. *humilis*, but is smaller. In fact, it has been given a varietal name of S. *humilis* var. *microphyllus* (Anderss.) Fern.

Arroyo Willow

Salix lasiolepis Benth.

FIELD IDENTIFICATION. Low Western shrub with many stems from the base, or a small tree 15–35 ft with a trunk 3–7 in. in diameter, the branches rather slender and erect to form an open head.

FLOWERS. Dioecious, borne February–April, mostly before the leaves, densely silky-tomentose in the bud, catkins sessile or nearly so, rather erect, cylindric, densely flowered; staminate catkins ¾–1½ in. long, about ½ in. thick, much thicker than the pistillate; stamens 2, filaments elongate and glabrous, somewhat united at the base; pistillate catkins ¾–1¼ in. long; ovary cylindric, long-pointed, glabrous (or puberulent in some varieties); stigmas spreading, nearly sessile; scales obovate or oblong, apex acute, dark brown, persistent, white-hairy.

FRUIT. Fruiting catkins 1–2¼ in., scales brown or black, densely hairy; pedicel of capsule about 1/25 in. long, capsule body ⅕–¼ in. long, oblong-cylindric, reddish brown, glabrous or pubescent, stigma remnants persistent.

LEAVES. Simple, alternate, deciduous, rather variable in shape, mostly oblanceolate to oblong, sometimes obovate, or linear on the other extreme, blades 2–6 in. long, ⅜–1 in. wide, apex acute to acuminate, base cuneate or less often rounded; leaf sometimes inequilateral and falcate, margin obscurely serrulate or seemingly entire, often somewhat revolute; young leaves densely hairy, eventually with the upper surface dark green and glabrous; lower surface paler and glaucous

Arroyo Willow
Salix lasiolepis Benth.

or puberulent, reticulate venulose; thick and coriaceous; petioles slender, ⅛–½ in.; stipules dark green above, densely tomentose, paler beneath, ovate to semilunar, margin entire or toothed, size very small or large and foliaceous.

TWIGS. Stout, yellowish to dark red or brown, pubescent at first, brown to black and glabrous later; winter buds ovoid, acute, flattened, yellow to brownish, puberulous to glabrous.

BARK. On younger branches and upper trunks brown to gray, thin, smooth, later becoming darker with broad flattened ridges and shallow fissures.

WOOD. Light brown, sapwood whitish, close-grained, light, soft, weak, specific gravity about 0.56.

RANGE. Usually in wet grounds from sea level to an altitude of 7,500 ft. Texas and New Mexico; west to California, north to British Columbia, and south to Chihuahua and Coahuila, Mexico.

VARIETIES. A number of varieties of Arroyo Willow have been named as follows:

Baker Arroyo Willow, *S. lasiolepis* var. *bakeri* (V. Seem) Ball, is very similiar to the species except the capsule is pubescent and slightly larger. It occurs in the San Francisco Bay area.

Sandberg Arroyo Willow, *S. lasiolepis* var. *sandbergii* (Rydb.) Ball, has leaves more broadly oblanceolate to obovate and young parts with a denser, more persistent pubescence. It occurs in northern California, Oregon, Washington, and western Idaho.

Bigelow Arroyo Willow, *S. lasiolepis* var. *bigelovii* (Torr.) Bebb, has leaves broadly cuneate-oblanceolate to obovate, entire on the margin and more densely pubescent beneath. It occurs with the species in the San Francisco Bay area.

REMARKS. The genus name, *Salix,* is the classical name, and the species name, *lasiolepis,* refers to the white-hairy scales of the flowers. The plant is also known under the vernacular names of White Willow and Ahuejote. The wood is occasionally used for fuel, and the foliage is sometimes browsed by mule deer.

Gray-leaf Willow

Salix glauca L.

FIELD IDENTIFICATION. Subalpine shrub to 3–6 ft.

FLOWERS. Staminate catkins ⅜–1¼ in. long, and ⅜–⅝ in. wide; scales oval to oblong, dark brown toward the apex, villous; filaments pilose at base; pistillate catkins 1¼–2 in. long, about ⅝ in. wide; ovary solitary, short-stalked, silky-hairy; style fuscous or yellowish,

Gray-leaf Willow
Salix glauca L.

about 1/25 in., with oblong bifid stigmas; ventral gland truncate, as long as the pedicel or longer.

FRUIT. Pedicels stout, 1/25–1/15 in.; capsule 1/5–1/3 in. long, conic, gray-tomentose, 2-valved, 1-celled; seeds numerous.

LEAVES. Simple, alternate, deciduous, lanceolate to obovate or oblanceolate to oblong, apex acute or obtuse, base acute to obtuse or subcordate, blades 1¼–3¼ in. long, margin entire; when young the leaves are silky or white-tomentose, upper surface becoming glabrate with maturity and dark green, lower surface grayish green with silky hairs or white tomentum, rarely green; petioles yellowish, 1/5–5/8 in.; stipules small, ovate or semicordate, or sometimes absent.

TWIGS. Short and stout, lustrous, chestnut-brown or yellowish, older ones gray and often flaky.

RANGE. In boggy meadows, in high mountain regions, at altitudes of 9,000–12,000 ft. In New Mexico on the Taos Mountain summits; west to California; and northward through British Columbia and Alberta to Alaska.

VARIETIES. Smooth Gray-leaf Willow, S. *glauca* var. *acutifolia* (Hook.) Schneid., has well-developed stipules, longer pedicels, and glabrescent filaments.

Blue Gray-leaf Willow, S. *glauca* var. *glabrescens* (Anders.) Schneid., is less pubescent throughout than the species.

REMARKS. The genus name, *Salix*, is the ancient Latin name. The species name, *glauca*, refers to the whitened lower surface of the leaves. Gray-leaf Willow is considered to be important as a livestock browse plant in the high mountain meadows. It was first cultivated for ornament in 1813.

MOUNTAIN WILLOW
Salix monticola Bebb

MOUNTAIN WILLOW
Salix monticola Bebb

FIELD IDENTIFICATION. Shrub attaining a height of 18 ft.

FLOWERS. Dioecious, staminate catkins ¾–1¾ in. long, 3/8–5/8 in. thick, sessile or short pedunculate, with or without basal bracts; stamens with filaments not pilose at base, anthers yellow; pistillate catkins 1¼–2 in. long, 3/8–5/8 in. thick, subsessile or stalked; scales dark, obtuse, lanceolate to obovate or oval, somewhat narrower in the staminate catkins, densely villous; stigmas short, bifid or entire, styles 1/25–1/15 in.

FRUIT. Pedicels about 1/25 in. or less, pilose, capsule rostrate to ovoid, glabrous, 1/5–¼ in. long; seeds numerous and minute.

LEAVES. Simple, alternate, deciduous, blades 1¼–2¾ in. long, ¾–1½ in. wide (or larger on vigorous shoots), blade elliptic, oblanceolate, obovate or oval, apex acute or short-acuminate, base usually narrowed or rounded or sometimes cuneate-subcordate, margin crenate-serrate, upper surface dark green and glabrous, lower surface subglaucous to glaucous, puberulent to glabrate, strongly reticulate; petioles 1/8–2/5 in.; stipules large, ovate or semicordate, serrate, on fertile branches small or absent.

TWIGS. Stout and divaricate, glabrous, lustrous, at first yellow or greenish, later reddish brown.

RANGE. Along streams, or in boggy meadows at altitudes of 3,500–11,500 ft. New Mexico, Colorado, Utah, and Wyoming; north to Alberta.

REMARKS. The genus name, *Salix*, is the classical Latin name, and the species name, *monticola*, means "mountain-dweller." The plant was introduced into cultivation in 1898.

SKYLAND WILLOW
Salix petrophila Rydb.

FIELD IDENTIFICATION. Very low plant hardly over 6 in. high, the stems creeping either above or below the ground, producing short erect branches.

FLOWERS. Dioecious, the catkins terminating short leafy branches; staminate ⅜–¾ in. long, stamens 2, filaments glabrous and free, anthers red to purplish; pistillate catkins ¾–1¾ in., slender, somewhat drooping; scales broadly lanceolate to obovate or oval, apex often rounded, margin entire or somewhat erose, brown to black, clothed with long hairs within, sparsely so externally; styles about 1/25 in., red; stigmas about 1/50 in., entire or deeply cleft.

FRUIT. Capsule ⅙–¼ in. long, lanceolate, sessile or nearly so, gray-tomentose or villous; seeds numerous, small, hairy.

LEAVES. Simple, alternate, deciduous, blade elliptic to obovate, apex abruptly acute to obtuse and somewhat apiculate, base acute, margin entire, blade length ⅝–1½ in., width ⅜–⅝ in., upper surface deep-green and glabrous, lower surface paler to subglaucous and rather strongly veined, more or less hairy when young, especially on the midrib; petioles slender, yellow; stipules absent or very small.

TWIGS. Erect, hairy at first, glabrous later, brown to yellowish.

RANGE. On alpine summits in the sun at altitudes of 7,500–14,500 ft. New Mexico; west to California and north to British Columbia and Alberta.

SKYLAND WILLOW
Salix petrophila Rydb.

VARIETY. A variety named Alpine Willow, S. *petrophila* var. *caespitosa* (Kennedy) Schneid., has leaves acute at both ends, or apex subacuminate, densely pilose-tomentose, and scales narower.

REMARKS. The genus name, *Salix,* is the classical Latin name. The species name, *petrophila,* refers to its rock-loving habit.

PLANE-LEAF WILLOW

Salix planifolia Pursh

FIELD IDENTIFICATION. Shrub 3–10 ft and divaricately much-branched.

FLOWERS. Dioecious catkins June–August, with or before the leaves, sometimes with a few small leaves at the base, rather dense, sessile or with pedicels as long as the gland or slightly exceeding it in length; scales ovate to obovate, or lanceolate, acute to obtuse, dark, 1/12–⅛ in. long, surfaces with long white hairs; staminate catkins ⅜–¾ in. long and ⅜ in. thick, sessile; stamens 2, filaments glabrous or slightly hairy at base, anthers small and reddish purple to yellowish; pistillate catkins dense, ¾–1¾ in. long and ⅜ in. thick; style 1/25–1/15 in., base somewhat pubescent, stigmas thick, almost entire or cleft.

FRUIT. Capsule sessile or short-peduncled, ovate to conic or rostrate, ⅙–¼ in. long; seeds numerous and small.

LEAVES. Simple, alternate, deciduous, subleathery, blade broadly elliptic to ovate or obovate, blade length ¾–2½ in., width ⅜–¾ in., apex acute or obtuse, sometimes abruptly so and apiculate, base acute to narrowed or slightly rounded, margin entire or occasionally somewhat glandular-serrate, upper surface bright green, lustrous and glabrous, lower surface paler or glaucous (young leaves short silky-pilose beneath); petioles 1/12–½ in., glabrous; stipules absent or very minute; buds large, chestnut-brown.

TWIGS. Short, divaricate, glabrous, various shades of brown or purplish, occasionally glaucous.

RANGE. Usually in thickets above timber line in mountain bogs at altitudes of 6,000–13,500 ft. New Mexico; west to California, north to Alaska, and eastward to Labrador.

VARIETIES. Two varieties have been segregated from Plane-leaf Willow as follows:

Mono Plane-leaf Willow, S. *planifolia* var. *monica* (Bebb) Schneid., has leaves broadly elliptic or oval and apex acute; staminate catkins ⅜–¾ in. long, ⅜–⅝ in. wide, with bracts usually absent; twigs slightly pruinose.

Nelson Plane-leaf Willow, S. *planifolia* var. *nelsoni* (Ball) Ball, has leaves narrowly elliptic, oblong or ob-

Plane-leaf Willow
Salix planifolia Pursh

lanceolate and apex acute or acuminate; staminate catkins elliptic-cylindrical, ⅝–1¼ in. long, ¼–⅜ in. wide, leaves present on tips of twigs which are slightly pruinose.

REMARKS. The genus name, *Salix*, is the classical Latin name, and the species name, *planifolia*, means "flat-leaved." The plant is occasionally browsed by mule deer.

Slender Willow

Salix petiolaris J. E. Sm.

FIELD IDENTIFICATION. Clumped shrub with slender erect branches, attaining a height of 3–9 ft. More rarely a small tree to 22 ft in Northern areas.

FLOWERS. May–June, catkins lax, appearing with the leaves, sessile or on bracted peduncles ⅜–¾ in.; bracts pale brown, linear-lanceolate to narrowly oblanceolate or spatulate, glabrous or slightly pubescent, 1/25–1/12 in. long; staminate catkins ellipsoid-obovoid, ⅜–¾ in. long, ⅜–½ in. thick; pistillate catkins in fruit ⅝–1⅝ in. long, on leafy peduncles about ¾ in.; style very short, entire or divided; stamens usually 2, filaments free and glabrous or pubescent at base; flowers each with one basal gland.

FRUIT. Capsule borne on pubescent pedicels 1/12–⅙ in., divergent, lanceolate-rostrate, obtuse or acute, silky-hairy.

LEAVES. Young ones thinly silky, mature ones dark green above, glaucous and glabrous beneath, narrowly lanceolate or narrowly oblanceolate, blades 2–4 in. long (or occasionally to 6 in.), ⅜–¾ in. wide (or occasionally to 1¼), apex acute or acuminate, base acute, margin minutely glandular-serrate or entire; petioles slender, ⅕–⅗ in.; stipules absent or minute and caducous.

TWIGS. Slender, tough, young ones puberulent and yellowish, older ones becoming glabrous and dark or reddish brown. Sometimes the twigs are glaucous.

RANGE. Occasionally in Oklahoma and Colorado; east to Virginia, north to New Brunswick, and west to Alberta and Montana.

Slender Willow
Salix petiolaris J. E. Sm.

SYNONYM. Fernald (1950, p. 516), adopted the later name S. *gracilis* Anderss. for this species interpreting S. *petiolaris* as a European Willow. However, Ball (1948) showed that S. *petiolaris* is a native of America, though described from plants introduced in Scotland, and should be retained.

REMARKS. The genus name, *Salix*, is the classical name. The species name, *petiolaris*, refers to the leaf petioles, longer than those of most willows. Also known under the vernacular name of Meadow Willow.

Bonpland Willow
Salix bonplandiana H. B. K.

FIELD IDENTIFICATION. Small to large tree attaining a height of 50 ft and a diameter of 4–20 in. The slender, ascending, or somewhat drooping branches, form a round-topped crown.

FLOWERS. Borne on leafy branches, catkins dioecious, cylindric, slender; staminate catkins 1¼–2¼ in., somewhat longer than the pistillate which average ¾–1½ in.; scales obovate, apex rounded, outer surface yellow or grayish and villose, inner surface slightly villose or subglabrous; stamens 3, slightly hairy at base of filaments; ovary slender, oblong-conic, short-stalked, glabrous, stigmas much-thickened, club-shaped, nearly sessile.

Bonpland Willow
Salix bonplandiana H. B. K.

FRUIT. Capsule short stipitate, ovoid-conic, base rounded, yellowish to light reddish, seeds numerous and very hairy.

LEAVES. Simple, alternate, deciduous, narrowly lanceolate to oblong or broadly linear, apex long-acuminate, base narrowly or broadly cuneate, margin finely serrulate, blade length to 5 in., width ⅝–¾ in., upper surface green and glabrous, lower surface silvery white and glabrous or pubescent when younger; midrib broad and yellow; petioles ⅛–½ in., stout, grooved, glabrous or slightly pubescent; stipules ovate, entire or undulate, thin and scarious, ⅛–¼ in. broad.

TWIGS. Slender, yellowish at first, later brown to reddish, slightly hairy to glabrous later.

BARK. Dark brown to black, ridges broad and flat with appressed scales, fissures narrow.

RANGE. At altitudes of 2,500–5,000 ft in well-drained soil in the sun. Southeastern New Mexico and central Arizona; south in Mexico from Sonora, Chihuahua, and Coahuila to Oaxaca and Guatemala.

REMARKS. The genus name, *Salix*, is the classical name, and the species name, *bonplandiana*, is in honor of Aimé Bonpland (1773–1858), a French botanist. Also known under the vernacular names of Toumey Willow, Sauce, and Sauz. Cattle browse the young plants.

Pacific Willow
Salix lasiandra Benth.

FIELD IDENTIFICATION. Large shrub or small to large tree to 60 ft, with a trunk diameter of 2–3 ft. The branches are rather straight and erect or ascending.

FLOWERS. Catkins borne terminally with the leaves, dioecious, narrow-cylindric, staminate catkins ¾–2½ in. long, ⅜–½ in. wide; scales varying from obovate to lanceolate, entire to dentate; stamens 5–9, filaments hairy toward the base; pistillate catkins 1¼–4 in. long, ⅜–¾ in. wide, scales obovate to ovate and dentate toward the apex, sparsely hairy; ovary long-stalked, abruptly narrowed to an acuminate apex, sometimes glandular, style short and thickened.

FRUIT. Capsules borne on pedicels $\frac{1}{25}$–$\frac{1}{12}$ in., four to six times as long as the gland; body light brown, lanceolate, ⅙–¼ in. long, dehiscent to release small seeds with pilose hairs.

Pacific Willow
Salix lasiandra Benth.

LEAVES. Simple, alternate, deciduous, lanceolate to broadly lanceolate or sometimes oblanceolate, apex acuminate, base acute to rounded, margin finely serrate-glandular, blade length to 4½ in., width ⅝–1½ in., upper surface dark green and lustrous; lower surface glaucous to glaucescent; petioles to ½ in., with yellowish glands near the base of the blade; stipules small, mostly less than 3⅟₁₆ in., acute, glandular-dentate.

TWIGS. Rather stout, lustrous, young ones pubescent to tomentose, greenish yellow to reddish brown or purplish, often with a glaucous bloom; older ones reddish brown to orange or purple and glabrous; buds about ¼ in. long, ovate, acute, scales brown and lustrous.

BARK. On trunks dark brown or somewhat reddish, ½–¾ in. thick, ridges flat and scaly, fissures shallow.

RANGE. Sandy soil along streams at altitudes of 5,000–6,500 ft. New Mexico; west to California and north to Alberta, British Columbia, and Alaska.

VARIETIES. Lance-leaf Pacific Willow, S. *lasiandra* var. *lancifolia* (Anderss.) Bebb, has the twigs and lower surface of young leaves densely covered with gray or rusty pubescence. Big-leaf Pacific Willow, S. *lasiandra* var. *macrophylla* (Anderss.) Little, has generally larger leaves than those of the species.

REMARKS. The genus name, *Salix*, is the classical name, and the species name, *lasiandra*, means "shaggy-hairy," the stamens being pubescent at the base. Also known under the vernacular names of Black Willow, Red Willow, Western Black Willow, and Yellow Willow. Mule deer browse the foliage.

Yellow Willow
Salix lutea Nutt.

FIELD IDENTIFICATION. Clumped shrub or small tree 9–21 ft.

FLOWERS. Borne in separate staminate and pistillate catkins which are sessile or on very short-bracted peduncles; staminate 1–1¾ in. long, about ⅜ in. wide; stamens 2, united at base, anthers reddish; pistillate catkins ¾–2¼ in. long, about ⅜ in. thick or less, terminal on short leafy twigs; scales bicolored, light base and reddish brown tip, darker later, thin, crisp, thinly pilose; styles about ⅟₅₀ in.; stigmas short, thick, notched or entire, reddish.

Yellow Willow
Salix lutea Nutt.

FRUIT. Capsules on peduncles 1/35–1/12 in., body ovate-conic, glabrous, 1/6–1/5 in. long; seeds numerous, small.

LEAVES. Simple, alternate, deciduous, narrowly to broadly lanceolate or ovate-lanceolate, often widest at the middle, apex acute to short-acuminate, base rounded to subcordate, margin serrulate to entire, blade length 2–4 in., width 3/4–1 3/4 in., upper surface yellowish green and glabrous, glaucous beneath; petioles 1/8–2/3 in.; stipules usually small, or on vigorous shoots to 3/8 in., mostly ovate to reniform, margin entire to glandular-dentate or glandular-serrate.

TWIGS. Yellow, turning reddish later, puberulent at first, glabrous later.

RANGE. Along streams or ditches in the sun, at altitudes of 3,800–8,000 ft. New Mexico; west to California and north to Alberta and Manitoba.

VARIETIES. Two varieties of Yellow Willow have been named as follows:

Tongue-leaf Yellow Willow, *S. lutea* var. *ligulifolia* Ball, has twigs mostly dark, sometimes yellowish, and often very pubescent; leaves narrowly lanceolate to oblong, straplike, 2–4 in. long, 3/8–3/4 in. wide, subentire, dark green; pedicel about 1/25 in., capsule 1/6–1/5 in. long.

Flat-leaf Yellow Willow, *S. lutea* var. *platyphylla* Ball, has longer, more slender petioles 1/3–3/5 in.; leaves broader and shorter, elliptic-obovate to ovate, 1 3/4–3 1/4 in. long, about 1 1/3 in. wide, on shoots even larger.

The species and varieties are often browsed by livestock.

REMARKS. The genus name, *Salix*, is the classical Latin name, and the species name, *lutea*, refers to the yellow twigs or yellowish green leaves.

Blue Willow

Salix subcoerulea Piper

FIELD IDENTIFICATION. Shrub attaining a height of 3–9 ft.

FLOWERS. Dioecious catkins densely flowered, short-pedunculate or sessile; staminate catkins stout, precocious, 3/8–1 1/4 in. long, about 3/8 in. thick; stamens 2, filaments united at base, anthers reddish; pistillate catkins subsessile, often with bracts at the base, 3/4–1 1/2 in. long and 3/8 in. thick; style about 1/25 in., yellow or brown; stigmas thick, 2-lobed or entire, red; scales black, ovate to obovate, acute to obtuse, sparsely pilose.

FRUIT. Pedicels about 1/25 in. or less, about half as long as the gland, capsule body about 1/8 in. long, ovoid or lance-ovoid, silvery-pubescent with short appressed, silvery hairs.

LEAVES. Simple, alternate, deciduous, blade oblong-lanceolate to elliptic or oblanceolate, apex acute or acuminate, base acute or cuneate, margin entire or rarely obscurely crenulate, blade length 3/4–3 1/4 in., width 3/8–5/8 in., sometimes on young shoots to 4 in. long and

Blue Willow
Salix subcoerulea Piper

1 1/4 in. wide, upper surface green and minutely pubescent, glabrescent with age, lower surface with appressed silvery hairs, midrib yellow; petioles 1/12–1/5 in., on vigorous shoots 1/2 in. or longer.

TWIGS. Young ones slightly pubescent, glabrous later, sublustrous, very pruinose, purplish to brown or black.

RANGE. Moist soil in alpine meadows in the aspen and spruce belt, at altitudes of 3,000–12,000 ft. New Mexico; west to California and north to British Columbia and Alberta.

REMARKS. The genus name, *Salix*, is the classical Latin name. The species name, *subcoerulea*, refers to the somewhat bluish color of the twigs, as a result of the pruinose covering. This bluish color gives a handsome appearance to the branches in winter. Blue Willow has been cultivated since 1890. Livestock sometimes browse it.

Whiplash Willow

Salix caudata (Nutt.) Heller

FIELD IDENTIFICATION. Clumped shrub attaining a height of 6–20 ft.

FLOWERS. Dioecious, borne on leafy pubescent twigs; staminate catkins 1 1/4–1 3/4 in. long; stamens 5, filaments somewhat pubescent at base, anthers yellow; pistillate catkins 1¼–2½ in. long, ⅜–⅝ in. thick; scales lanceolate, oblanceolate, or obovate, green to yellowish, somewhat pilose to glabrous externally, deciduous; styles about 1/50 in., stigmas notched.

FRUIT. Borne on pedicels 1/25–1/15 in., capsules ⅕–¼ in. long, lanceolate, straw-colored to brownish, dehiscent; seeds numerous, small, hairy.

LEAVES. Simple, alternate, deciduous, lanceolate to oblanceolate (on young shoots sometimes ovate-lanceolate or even ovate), apex long-acuminate or caudate and almost tail-like, base acute to rounded, margin glandular-serrate, upper surface green and glabrous, lower surface paler but not glaucous, on flowering branches blades ¾–1½ in. long, on young sterile shoots to 5 in. long and 1½ in. wide; petioles 1/12–¼ in.; stipules caducous, small, lanceolate to reniform or semicordate, margin glandular-serrate.

TWIGS. Elongate, pubescent at first, glabrous later, lustrous, reddish brown to chestnut-brown.

RANGE. In moist soil along mountain streams or meadows, usually in the conifer belt at altitudes of 5,500–10,000 ft. New Mexico, north to British Columbia.

VARIETIES. Several varieties have been named:

Small-leaf Whiplash Willow, S. *caudata* var. *parvifolia* Ball, has smaller leaves 2–3¼ in. long and ⅓–½ in. wide.

COASTAL PLAIN WILLOW
Salix caroliniana Michx.

WHIPLASH WILLOW
Salix caudata (Nutt.) Heller

Bryant Whiplash Willow, S. *caudata* var. *bryantiana* Ball & Bracelin, has the branchlets and peduncles glabrous, or nearly so, from the first, and by the mature leaves having stipules with margins very glandular.

REMARKS. The genus name, *Salix*, is the classical Latin name. The species name, *caudata*, refers to the whiplike (literally, "tail-like") apex of the leaves. It is also given the vernacular name of Caudate Willow.

COASTAL PLAIN WILLOW

Salix caroliniana Michx.

FIELD IDENTIFICATION. Shrub or small tree to 30 ft, and 18 in. in diameter. Branches spreading or drooping to form an open irregular crown. Closely related to Black Willow, S. *nigra* Marsh, and known to hybridize with it. However, Black Willow leaves are green beneath and Coastal Plain Willow leaves are very glaucous.

FLOWERS. May–June, buds single-scaled, expanding with the leaves; catkins terminal, slender, lax, narrow-cylindric, to 4 in. long; scales yellow, ovate to obovate, apex rounded or obtuse, densely villose-pubescent; glands of staminate flowers lobulate or forming a false disk; stamens 3-12, exserted, separate, filaments hairy at base, anthers yellow; gland of pistillate flower clasping the base of the pedicel; ovary stipitate, ovoid-conic, acute; style short, 2-lobed.

FRUIT. Capsule ovoid-conic, 1/6–1/4 in. long, granular-roughened, abruptly long-pointed, remnants of the 2 persistent stigmatic lobes almost sessile, base with pedicel to 1/4 in.; seeds numerous, silky-hairy.

LEAVES. Involute in the bud, simple, alternate, deciduous, length 2–7 in., width 3/8–1 1/3 in., lanceolate to lanceolate-ovate, sometimes falcate, apex acuminate or acute, base gradually narrowed on young leaves, on older ones often rounded; upper surface bright green and glabrous; lower surface whitened or glaucous, somewhat puberulent when young, glabrous later, veins yellowish, delicate, margin finely serrate; petioles 1/8–1/2 in. long, densely hairy, glandless; stipules usually small on normal leaves, on vigorous shoots large (to 3/4 in. wide), foliaceous, conspicuous, ovate to reniform, mostly serrate above the middle.

TWIGS. Slender, yellowish to reddish brown or grayish, more or less pubescent, eventually glabrous; winter buds small, brown, lustrous.

BARK. Reddish brown to gray, ridges broad, fissures deep, conspicuously checkered, breaking into closely appressed scales.

WOOD. Dark reddish brown, sapwood nearly white, light, soft, not strong.

RANGE. Mostly along gravelly banks and shores of streams or lakes or in low woods. Texas, western Arkansas, eastern Oklahoma, and Louisiana; east to Florida, north to Maryland, and west to Kansas.

REMARKS. The genus name, *Salix,* is the classical Latin name, and the species name, *caroliniana,* refers to the states of Carolina. Also known under the vernacular names of Ward Willow and Carolina Willow. The scientific terminology of this willow has been very confused. The following names have been applied from time to time: S. *occidentalis* Bosc. *ex* Koch, S. *longipes* Shuttl., S. *nigra* var. *wardii* Bebb, S. *occidentalis* var. *longipes* (Anderss.) Bebb., S. *wardii* Bebb, S. *marginata* Wimm. *ex* Small, S. *amphibia* Small, S. *longipes* var. *venulosa* (Anderss.) Schneid., S. *longipes* var. *wardii* (Bebb) Schneid., S. *harbisonii* Schneid., S. *chapmanii* Small, and S. *floridana* Chapm.

GARRYA FAMILY (*Garryaceae*)

MEXICAN SILKTASSEL
Garrya ovata Benth.

MEXICAN SILKTASSEL

Garrya ovata Benth.

FIELD IDENTIFICATION. Western evergreen shrub or small tree 2–18 ft, all parts densely pubescent with gray or brown curled hairs.

FLOWERS. Dioecious, catkins axillary; staminate catkins shorter than the pistillate, bracts small; calyx with 4 linear sepals; petals absent; stamens 4, filaments distinct, anthers linear, ovary none; pistillate catkins 1–3 in., bracts large and leaflike; calyx limb abbreviated, stamens absent; styles 2, inwardly stigmatic, persistent on fruit; ovary 1-celled with 2 pendulous ovules.

FRUIT. Drupe 1/6–1/3 in., sessile or short-pedicellate, globose to ovoid, bluish purple, sometimes drying brownish, eventually glabrous, stigma persistent, flesh thin; seeds 1–2, subglobose to short-oblong, one side more flattened than the other, 3/16–1/4 in. long.

LEAVES. Persistent, blades 1–2½ in., opposite, leathery, variable in shape, either narrowly lanceolate to ovate, or oblong to oval or obovate, apex acute, obtuse, or rounded, usually mucronate, base broadly to narrowly cuneate or in some rounded, margin entire and sometimes irregularly thickened and muricate; young leaves densely curly-hairy on both sides; at maturity pale and densely curly-hairy beneath; becoming glabrous, or nearly so above and either dull or semilustrous, veins finely reticulate; petioles connate at base, 1/8–3/4 in., at first densely curly-hairy but more glabrous later.

TWIGS. Stout, 4-angled to terete later, densely curly-hairy, but eventually glabrous, reddish brown to gray; lenticels small, slitlike or elliptic.

RANGE. Rocky limestone hills and canyons of central Texas, and westward into the Trans-Pecos counties. In Mexico in Chihuahua, San Luis Potosí, and Puebla. The type specimens are from Guanajuato.

REMARKS. The genus name, *Garrya*, is in honor of Nicholas Garry, secretary of the Hudson's Bay Company, and the species name *ovata*, refers to the ovate leaves. A variety known as the Lindheimer Silktassel, *G. ovata* var. *lindheimeri* Coult. & Evans, has been named as occurring in central and western Texas. However, the characters by which it is described—oblong or obovate, acute, nonthickened or muricate-margined, and curly-haired leaves—do not appear distinct enough to warrant the segregation of the variety.

Wright Silktassel, *G. wrightii*, is a closely related, more glabrous species of extreme west Texas, New Mexico, and Arizona.

Wright Silktassel
Garrya wrightii Torr.

FIELD IDENTIFICATION. Western evergreen shrub or small tree 2–15 ft.

FLOWERS. Dioecious, catkins axillary, more or less branching, loosely flowered; staminate ones in clusters of 2–3, or solitary, with small but distinct bracts; petals absent; sepals 4, valvate, linear; stamens 4, filaments distinct, anthers linear, ovary absent; pistillate catkins 1–3½ in., subtended by leaflike bracts which become smaller above, calyx almost obsolete; no petals; stamens absent or rudimentary; styles 2, stigmatic within, persistent on fruit; ovary 1-celled, the 2 ovules pendulous.

FRUIT. Sessile, baccate, globose, bluish purple, sometimes glaucous, becoming glabrous, ⅙–⅓ in. in diameter, flesh thin; seeds 1–2, brown, short-oblong, ends rounded, somewhat flattened, about ¼ in. or less.

LEAVES. Opposite, leathery, persistent, length of blades ¾–2 in., light green but dull, drying bluish green to grayish, elliptic to oblong or lanceolate, ends mostly acute, mucronate, margin entire but sometimes thickened and muricate, young with sparse scattered white hairs, at maturity surfaces glabrous or nearly so, main vein yellowish and lateral veins more delicate and obscure; petioles connate at base, ⅛–½ in., pubescent at first, glabrous later.

TWIGS. Stout, 4-angled, older terete, at first pubescent but glabrous later, green to reddish brown or gray.

RANGE. Rocky banks and canyons at altitudes of 3,000–8,000 ft, in extreme western Texas, New Mexico, and Arizona; in Mexico in Chihuahua and Sonora.

REMARKS. The genus name, *Garrya*, is in honor of Nicholas Garry, secretary of the Hudson's Bay Company, and the species name, *wrightii*, is for Charles Wright, American botanist who was attached to the United States–Mexico boundary survey in 1851. The plant is also known under the vernacular name of Quinine-bush and Coffeeberry-bush. Some species contain an alkaloid, garryin, which is used medicinally. It is occasionally browsed by deer and cattle. *G. ovata* Benth., of Texas and Mexico, is a closely related but more hairy plant.

Wright Silktassel
Garrya wrightii Torr.

WAX-MYRTLE FAMILY (*Myricaceae*)

Dwarf Wax-myrtle
Myrica pusilla Raf.

FIELD IDENTIFICATION. Shrub very similar to Southern Wax-myrtle, and considered by some botanists as a dwarf variety of it. Forming small clumps to 2½ ft from spreading horizontal rootstocks.

FLOWERS. Staminate and pistillate catkins borne on separate plants; staminate in short-oblong axillary catkins, ¼–⅓ in.; bracts loosely imbricate, reniform; stamens 2–8 on a bract; filaments short; anthers erect, ovoid, extrorse, 2-celled; pistillate catkins axillary, smaller than the staminate, short-ovoid; subtending bracts suborbicular or broader than long; gynoecium generally of 2 united carpels on a bract; styles short, bifid, inwardly stigmatic; ovary solitary, sessile, 1-celled, subtended by short scales.

FRUIT. Nutlet on short peduncles or sessile, persistent, May–October, ⅛–⅙ in. in diameter, subglobose, epicarp white-waxy and granular.

LEAVES. Aromatic, simple, alternate, evergreen, coriaceous, sessile or nearly so, narrowly lanceolate, obovate or linear-spatulate; margin usually coarsely toothed above the middle, somewhat revolute; apex acute or mucronate; base gradually narrowed; surface minutely and abundantly glandular pellucid-punctate above and beneath; veins inconspicuous except the main veins; dark green above, pubescent or glabrous, sometimes brownish beneath; blade length 1½–2¼ in., width 3/16–⅜ in.

STEMS. Often clumped from horizontal rootstocks, usually not over 2½ ft, when young rather hairy, older stems smooth and shiny. Twigs smooth, slender, brownish gray.

RANGE. In sandy pinelands or on prairies. Eastern Texas and Louisiana; eastward to Florida, and north to North Carolina.

Dwarf Wax-myrtle
Myrica pusilla Raf.

REMARKS. The genus name, *Myrica,* is from an ancient Greek word for tamarisk, and the species name, *pusilla,* refers to the dwarf size of the plant. Vernacular names are Dwarf Candle-berry, Bayberry, Waxberry, and Wax-myrtle. The Dwarf Wax-myrtle is so similar to the Southern Wax-myrtle, *M. cerifera,* that the only distinct difference appears to be the dwarf form and subterranean stems which produce clumps or patches of growth.

Southern Wax-myrtle
Myrica cerifera L.

FIELD IDENTIFICATION. Crooked evergreen shrub, but sometimes a tree to 40 ft.

FLOWERS. Borne March–April, with staminate and pistillate catkins on different plants. Staminate catkins oblong, cylindric, ¼–¾ in.; scales acute, ovate and ciliate; stamens 2–8, yellow, anthers 2-celled, reddish yellow; pistillate catkins short, ovoid; ovary 2–4-scaled at base; stigmas 2, slender and spreading.

FRUIT. Drupe maturing September–October, spikes short; bracts deciduous; drupes persistent, about ⅛ in., globose, light green, covered with granules of bluish white wax; seed pale, minute, solitary.

LEAVES. Simple, alternate, tardily deciduous; 1½–5 in. long, ¼–¾ in. wide, oblanceolate to elliptic, apex acute or rounded, cuneate or narrowed at base and decurrent on the short stout petiole, margin entire or coarsely serrate above the middle, shining above, resinous with orange-colored glands beneath, aromatic when crushed.

TWIGS. Reddish brown to gray, young parts with early-deciduous orange-colored glands.

BARK. Gray, or grayish green, smooth, compact, astringent.

WOOD. Light, brittle, soft, fine-grained, dark brown, sapwood lighter, weighing 35 lb per cu ft.

RANGE. Sandy swamps or low acid prairies. Eastern Texas, Oklahoma, Arkansas, and Louisiana; eastward to Florida and north to New Jersey.

PROPAGATION. The fruit may be gathered by hand and the waxy surface removed in a stirring machine. One lb of fruit yields about 53,000 seeds, of about 98 per cent purity in commercial lots. Clean stratified seeds have an average germination of 70 per cent. Seeds may be sown in fall or spring, by drilling in rows 8–12 in. apart, and covering with about ¼ in. of firmed soil. If sown in fall they should be mulched. Stratified seeds should be sown in spring. The plant can also be propagated by cuttings, layers, or suckers.

REMARKS. The genus name, *Myrica*, is the ancient name of the tamarisk, and the species name, *cerifera*, refers to the waxy fruit. Vernacular names are Waxberry, Spice-bush, Candleberry, Bayberry, Sweet-oak, and Tallow-shrub. The fruit is eaten by about 40 species of birds, especially bobwhite quail and turkey. It was first cultivated in 1699, and makes a desirable ornamental. Candles were formerly made by boiling the waxy-blue berries. The bark and leaves are reputed to have medicinal properties. Southern Wax-myrtle is closely related to the Dwarf Wax-myrtle, *M. pusilla*, which is smaller, and spreads by underground runners.

SOUTHERN WAX-MYRTLE
Myrica cerifera L.

ODORLESS WAX-MYRTLE
Myrica inodora Bartr.

FIELD IDENTIFICATION. Evergreen shrub with erect branches, or sometimes a small tree to 20 ft.

FLOWERS. Small, greenish, borne in separate staminate and pistillate catkins ⅓–1 in., bearing scaly bracts; staminate catkins short-oblong, bracts ovate-orbicular and about as broad as long; stamens numerous, inserted on base of bracts, anthers oblong, yellow; pistillate catkins ¼–⅓ in. long, bracts loosely imbricate, usually broader than long, gynoecium usually in pairs, styles 1–2, bifid, red, stigmatic on inner surface; ovary sessile, 1-celled.

FRUIT. Nutlets produced sparingly on short spikes ¼–½ in.; sessile or nearly so, globose, black, papillose, waxy-scaled, about 3/16 in. in diameter; seed oblong-oval, apex acute, base rounded, brownish, about ⅛ in.

LEAVES. Simple, alternate, evergreen, more abundant terminally on the twigs; oblong-obovate to ovate; apex obtuse or rounded and often apiculate; base cuneate and decurrent on the petiole; margin entire or rarely a few teeth toward the apex and slightly revolute; thick and coriaceous, veins rather obscure except somewhat glandular midrib; upper surface dark green, lustrous, glabrous; lower surface paler green, glabrous, or with a few scattered hairs on veins, blade length 2–4 in., width ¾–1½ in.; petiole about ¼ in., pubescent or glabrous and somewhat glandular, gradually expanded into blade, petiole and blade non-odorous.

STEMS. Young branches erect, rather smooth, reddish brown, lenticels longitudinal; older branches grayish brown to gray eventually.

TWIGS. Young ones slender, reddish brown, scurfy, tomentose at first; older twigs gray, glabrous or pubescent; young leaves with pale yellow glands.

ODORLESS WAX-MYRTLE
Myrica inodora Bartr.

RANGE. Usually in pineland swamps near the Gulf Coast, but not common. In Washington Parish in eastern Louisiana; near Poplarville, Mississippi; near Mobile, Alabama; and near Apalachicola, Florida.

REMARKS. The genus name, *Myrica,* is the ancient name of the tamarisk, and the species name, *inodora,* refers to the odorless leaves, which is unusual for a southeastern Wax-myrtle. The plant was named by William Bartram, who found it near Taensa, Alabama, in 1778. He mentions it as an evergreen shrub, growing in wet sandy soil in swamps, having erect branches, dark green, entire, lanceolate, and odorless leaves. He also stated that the French inhabitants of the area used the wax of the berries for making candles. Also known under the vernacular names of Waxberry, Candleberry, and Wax-tree.

BAYBERRY WAX-MYRTLE

Myrica pensylvanica Lois.

FIELD IDENTIFICATION. Half-evergreen, divaricately branched shrub to 9 ft.

FLOWERS. Dioecious, the catkins appearing after the leaves; staminate catkins ¼–¾ in., oblong-cylindric; stamens 2–4 on short filaments about 1/25 in., anthers erect, 2-celled, base of stamens attached to a hairy, ovate, acute or rounded bract about 1/16 in.; pistillate catkins axillary, short-oblong, smaller than the staminate catkins, not bristly, bracts ovate and acute.

FRUIT. Drupe persistent, globose, about ¼ in. in diameter, waxy-granular, white, resinous, in crowded clusters.

LEAVES. Alternate, half-evergreen, blades 2–4 in. long, ½–1½ in. broad; oblong, oblanceolate, elliptic or lanceolate, generally of broad type; margin mostly entire, slightly revolute, occasionally with a few low, coarse, remote teeth near the apex; apex obtuse, acute or rounded; base gradually narrowed; young leaves with white hairs above and below; older leaves dark green and semilustrous above, glabrous or finely pubescent along the veins; lower surface pale and dull green, inconspicuously white-hairy along the veins with minute, scattered, brown, peltate scales (scales more noticeable under magnification); leaves not conspicuously reduced in size toward branch tips (as in *M. cerifera*); petioles 1/5–½ in.

TWIGS. Gray, stout, glabrous, somewhat glandular-scaly.

BARK. Grayish brown, smooth, tight.

RANGE. Usually in sandy, boggy soils. East Texas and Louisiana; eastward to Florida, north to Connecticut and Pennsylvania, and on the shores of the Great Lakes. Rare in east Texas, specimens collected in Angelina County State Park at Boykin Springs, also on the banks of the San Jacinto River near Humble, Texas.

MEDICINAL USES. The bark of both Southern Wax-

BAYBERRY WAX-MYRTLE
Myrica pensylvanica Lois.

myrtle and Bayberry Wax-myrtle contains a saponin-like principle known as "myricine acid." Wood and Osol (1943, p. 1451) give the following medicinal uses of Wax-myrtle bark: "Wax myrtle bark appears to be moderately stimulating and astringent, and in large doses emetic. It was used by the eclectics in diarrhea, jaundice, scrofula, etc. Externally the powdered bark was used as a stimulant to indolent ulcers, and the decoction as a gargle and injection in chronic inflammation of the throat and leukorrhea."

RELATED SPECIES AND FORMS. Bayberry Wax-myrtle was at one time listed by some botanical authors under the name of *M. carolinensis* Mill. Research proved that this name was a synonym for *M. cerifera* L., so the name was changed to *M. pensylvanica* Lois., as is described above. The true *M. cerifera* L. and *M. pensylvanica* Lois., as described, hydridize and produce a plant known under the name of MacFarlane's Wax-myrtle, *M.* × *macfarlanei* Youngken.

Also, the areas of contact of *M. pensylvanica* and *M. cerifera* produce some small-fruited, broad-leaved forms to which some authors have given the name of *M. heterophylla* Raf.

The Bayberry Wax-myrtle can be distinguished from the Southern Wax-myrtle by the large, broader leaves which are not conspicuously reduced in size toward the ends of the branches; by the leaves being mostly toothless except a few toothed near the apex; and by the under surface having inconspicuous, brown, scattered scales. Southern Wax-myrtle has much smaller, serrate leaves, which are so conspicuously brown, glandular-scaly underneath that they appear golden brown.

REMARKS. The genus name, *Myrica,* is the ancient name of the tamarisk, and the species name, *pensylvanica,* refers to the state of Pennsylvania, where it grows abundantly. Vernacular names in use are Waxberry, Tallow Bayberry, Small Waxberry, Northern Bayberry, Tallow-shrub, Swamp Candleberry, Candlewood, Candle-tree, and Tallow-tree.

Bayberry Wax-myrtle has some value for erosion control in dry, sterile soil. However, it is most widely known as a source of bayberry wax. The wax occurs in fine granules on the white fruit, and is obtained by steeping the fruit in boiling water and skimming off the wax for candlemaking. The wax is also used for the manufacture of soap in Europe. It contains a substance known as "palmitin." The leaves and berries are sometimes substituted for bay leaves in flavoring stews and soups.

The seed of Bayberry Wax-myrtle number about 55,500 seeds per lb. They are known to be eaten by about 40 species of birds, particularly by the ruffed grouse and bobwhite quail.

CORKWOOD FAMILY (*Leitneriaceae*)

Corkwood

Leitneria floridana Chapm.

FIELD IDENTIFICATION. Swamp-loving shrub or tree attaining a height of 20 ft.

FLOWERS. Dioecious, staminate catkins clustered, many-flowered, 1–2 in. long; stamens 3–12, subtended by triangular to ovate, scalelike bracts; filaments incurved, slender, anthers oblong, 2-celled; pistillate catkins few-flowered, shorter than the staminate; pistil surrounded by 3–4 glandular-fringed bractlets; style flattened, recurved, inwardly stigmatic; ovary pubescent, 1-celled, ovule solitary.

FRUIT. In clusters of 2–4, flattened, about ¾ in. long, ¼–⅓ in. wide, oblong, pointed at apex, rounded or acute at base, rugose, reticulate, dry, brown; seed light brown, flattened.

LEAVES. Alternate, entire, firm, oblong or elliptic-lanceolate, apically acute, obtuse or acuminate, basally narrowed, blades 3–6 in. long, 1–3 in. wide, when young pubescent above and densely tomentose below, when mature bright green, glabrous or pubescent above, villose-pubescent below; petioles ⅓–1¼ in. long, villose-pubescent.

BARK. Gray to brown, ridges narrow, fissures shallow.

TWIGS. Reddish brown to gray, glabrous, finely furrowed, lenticels numerous, leaf scars semilunate.

WOOD. Pale yellow, soft, close-grained, exceedingly light, about 12½ lb per cu ft.

RANGE. Rare in Texas. Reported from Velasco and West Columbia. Also on railroad embankments in the vicinity of Port Arthur and High Island. Eastward to Florida and north to Georgia and Missouri.

CORKWOOD
Leitneria floridana Chapm.

REMARKS. The genus name, *Leitneria,* is in honor of the German naturalist E. F. Leitner, and the species name, *floridana,* refers to the early so-called "Floridian Provinces" of the southeastern states. The exceedingly light wood is used to float fishing nets.

BEEFWOOD FAMILY (*Casuarinaceae*)

HORSETAIL BEEFWOOD
Casuarina equisetaefolia Forst.

HORSETAIL BEEFWOOD
Casuarina equisetaefolia Forst.

FIELD IDENTIFICATION. Tree attaining a height of 70 ft or more, and a diameter of 3 ft. Peculiar because of the very slender, drooping, jointed branches.

FLOWERS. Monoecious, staminate spikes long-stalked, cylindric, terminal, ⅜–1⅔ in. long, subtended by bracts which are imbricate, appressed, and slender; stamens exserted, with elongate filaments and short anthers, filaments with scales attached; pistillate flowers on short lateral peduncles, in dense globose heads, perianth absent, style branches slender and united at the base, ovary 1-celled, ovules 1–2 in a cavity and subtended by bracts.

FRUIT. Spike globular to oblong, compact, ⅜–¾ in. in diameter, becoming an aggregate of achenes with pubescent to glabrous valves; seeds 1–2, about 1/12 in. long, wing about ⅕ in. long, testa membranous; seed scattered by wind or water.

LEAVES. No true leaves, but 6–8 lanceolate sheathing teeth 1/12–⅛ in. long, in a whorl at the nodes, with the vein decurrent on the internode, appressed or later recurved, acute, ciliate.

TWIGS. Numerous, loosely jointed (resembling stems of *Equisetum*), drooping, pale green, slender, more or less whorled, internodes rather short, older bark gray.

WOOD. Red to brown (hence the name Beefwood), very hard, strong, durable, specific gravity about 0.93, burns quickly.

RANGE. Planted in the Gulf Coast states from Texas eastward to Florida; also in southern California, escaping cultivation in the two latter states. Apparently grows well in brackish or saline soil. A native of Asia and Australia, but widespread in cultivation in Mexico, West Indies, and Central and South America.

REMARKS. The genus name, *Casuarina*, is from *casuarius* ("the cassowary bird"), because of the featherlike branches. The species name, *equisetaefolia*, means "horse-tailed leafed," referring to the resemblance of the branches to those of the *Equisetum* plant. Also known under the vernacular names of Australian-pine, Horsetail-tree, She-oak, Pino, Pino de Australia, Sauce, and Cipres. The tree is of some value as a sand binder on the seacoast, because it grows well near salt water. It is rapid-growing and is propagated by seeds or by cuttings of half-ripened wood. In Mexico the bark has been reported as having been used for tanning and dyeing, producing a bluish black or reddish color. The bark is also used in domestic medicine for its tonic and astringent properties.

WALNUT FAMILY (*Juglandaceae*)

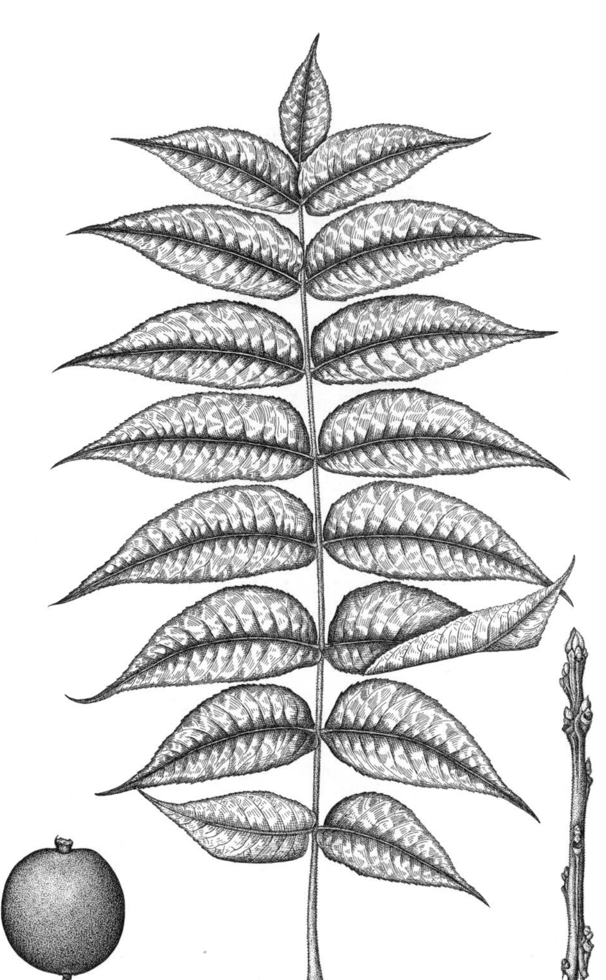

EASTERN BLACK WALNUT
Juglans nigra L.

EASTERN BLACK WALNUT

Juglans nigra L.

FIELD IDENTIFICATION. Tree to 125 ft with a rounded crown.

FLOWERS. May–June, staminate and pistillate flowers on same tree; staminate catkins 2–5 in. long, stout, stamens 20–30, sessile, calyx 6-lobed, lobes oval and pubescent; bracts triangular, brown-tomentose; pistillate flowers in 2–5-flowered spikes, about ¼ in. long; stigmas 2, plumose, yellowish green; style short on a subglobose ovary; calyx-lobes acute, ovate, green, pubescent.

FRUIT. Ripening September–October, solitary or clustered, subglobose; husk yellowish green, thick, papillose, indehiscent, 1½–2½ in. in diameter; nutshell hard and bony; nut dark brown to black, compressed, corrugated, 4-lobed at base, oily, sweet, edible.

LEAVES. Pinnately compound, 1–2 ft long, yellowish green, deciduous; petioles puberulent; leaflets 11–23, sessile or short-stalked, ovate-lanceolate, acute or acuminate, rounded or subcordate at base, inequilateral, serrate, glabrous above, pubescent below, 3–5 in. long, 1–2 in. wide.

BARK. Grayish brown, black, or reddish, fissures deep, ridges broad, rounded, and broken into close scales.

WOOD. Very beautiful, dark rich brown, sapwood white, durable, strong, heavy, hard, close-grained, is easily worked, glues well, does not warp, shrink, or swell much, takes a good polish, weighs 38 lb per cu ft. The whitish sapwood is sometimes stained to match the color of the heartwood to bring a better price.

RANGE. Oklahoma, Arkansas, Louisiana, and Texas; east to Florida, north to Minnesota, New York, and Ontario; and west to Nebraska.

PROPAGATION. The minimum commercial seed-bearing age of Black Walnut is about 12 years, an optimum of 30 years and a maximum of 80 years. Good crops are borne every 2–3 years with intervening light crops. Clean seed averages about 40 seeds per lb, with a commercial purity of 99 per cent and a soundness of 87 per cent. The fruits may be collected in fall when mature from the ground or shaken from the branches. The green husks are easier to remove than the dry ones.

They are removed by hand or by running through a corn sheller. The fruits should be dried to store if the husk is not removed. A good tree will produce several bushels of nuts. They can be sown in the fall soon after collection, or stratified in sand or peat for spring planting at a temperature of 33°–50°F. for 60–120 days. The sowing is generally in drills 8–12 in. apart, with about 15 nuts per linear ft, and covered with 1–2 in. of sandy soil. Protection with screens against rodents is desirable. Fall plantings are mulched with leaves or straw for protection. Root-pruning is helpful at a depth of 8–10 in. to insure development of a good root system. A good, well-drained, loamy soil is best for transplanting 1–2-year-old stock or stratified seed in spring. The seedlings are subject to root rot by the fungus, *Phytophthora.*

REMARKS. The genus name, *Juglans,* is from the Latin *Jovis glans,* meaning "acorn [or any nut of similar shape] of Jove," and the species name, *nigra,* refers to the dark wood. The wood is used in making superior furniture, cabinets, veneers, musical instruments, interior finish, sewing machines, caskets, coffins, posts, railroad crossties, and fuel. Large amounts were used for gunstocks during the Civil War and World War I. Trees about 12 years old begin to bear nuts. Confections and cakes are made from the nuts, which were also a favorite with the American Indians. Squirrels are fond of the large nuts. The tree makes a fine ornamental because of its shape and beautiful large leaves. Black Walnut and English Walnut are known to hybridize. Eastern Black Walnut is sometimes planted in shelter belts, and has been cultivated since 1686. It is a rapid grower and is usually found mixed with other hardwoods. Over 69 horticultural clones have been listed by Kelsey and Dayton (1942).

TEXAS BLACK WALNUT
Juglans microcarpa Berl.

TEXAS BLACK WALNUT

Juglans microcarpa Berl.

FIELD IDENTIFICATION. Strong-scented, many-stemmed shrub or small tree attaining a height of 30 ft, often with a number of leaning trunks from the base.

FLOWERS. Borne March–April in separate catkins on same tree. Staminate catkins from twigs of previous year, solitary, simple, long, slender, pendulous, pubescent at first, more glabrous later, 2–4 in. long; calyx short-stalked, greenish, puberulous, 3–6-lobed, lobes rounded; bracts densely hairy, ovate lanceolate; stamens 20–30 in several rows; anthers almost sessile, yellow, connectives somewhat lobed; pistillate catkins solitary, or several together, at the end of branches of the current year, oblong, rufous-tomentose; calyx-lobes usually 4, ovate and acute; bracts and bractlets laciniate; ovary inferior; stigmas spreading, plumose, greenish red.

FRUIT. Borne solitarily, or several together, globose, ½–¾ in. in diameter; hull with persistent calyx remnants at apex, rusty-pubescent at first but more glabrous later, thin, indehiscent; nut subglobose or ovoid, deeply grooved longitudinally, 4-celled toward base, shell thick; kernel small, oily, sweet.

LEAVES. Alternate, 9–12 in. long, odd-pinnately compound of 11–25 leaflets; the leaflets lanceolate to narrowly lanceolate, often scythe-shaped; margin entire or serrate with appressed teeth; apex acuminate; base cuneate or rounded, inequilateral, petiolules short; surface pubescent when young, becoming glabrous later; light green; aromatic when crushed; petiole and rachis pubescent.

TWIGS. Slender, orange-reddish or gray, usually pubescent, lenticels pale, pith in plates.

BARK. Gray to dark brown, fissures deep.

WOOD. Dark brown, sapwood white, heavy, hard, not strong.

RANGE. In valleys and rocky stream beds, Texas, Oklahoma, New Mexico, and northern Mexico. In Texas from the valley of the Colorado River west into the mountains of the Trans-Pecos, passing into the closely related Arizona Black Walnut in the far West.

PROPAGATION. In nursery planting the nuts may be planted, hulled or unhulled, in the fall. A sandy soil with an inch or two of firmed soil is preferable. Planting is done in drills 8–12 in. apart, with about 15 nuts per linear ft of row. A straw mulch will prevent winter freezing. The seeds average about 75 per lb. The young plant is a fairly rapid grower with a deep taproot.

CLOSELY RELATED SPECIES AND SYNONYMS. Johnston (1944c, p. 436) points out that the scientific name for Texas Black Walnut should be *J. microcarpa* Berlandier, instead of the *J. rupestris* Engelmann of some authors. The Berlandier collection was made from the upper Uvalde Canyon, and notes were published December 7, 1828. This date precedes the Engelmann date. In fact, Mr. Johnston points out that the later collection attributed to Engelmann was really made by Bigelow on the Devils River, Val Verde County, and was described and illustrated by Torrey. A later name of *J. nana* Engelmann also is not valid.

Texas Black Walnut is closely related to Arizona Black Walnut, and the two seem to intergrade where the distributions overlap. However, typical Arizona Black Walnut is more arborescent in habit, with larger nuts, fewer and broader leaflets, and generally seems to prefer higher altitudes of extreme west Texas, New Mexico, and Arizona.

REMARKS. The genus name, *Juglans*, is from the Latin *Jovis glans*, meaning "acorn [or any nut of similar shape] of Jove," and the species name, *microcarpa*, refers to the small fruit. Vernacular names used are Dwarf Walnut, Little Walnut, Nogal, and Nogalillo.

The wood is used for cabinet work, furniture, paneling, and veneers. The tree is sometimes cultivated for ornament in the United States and in Europe. It is also used in shelter-belt planting, and has some value as a wildlife food, especially for squirrels.

Arizona Black Walnut

Juglans major (Torr.) Heller

FIELD IDENTIFICATION. Tree attaining a height of 50 ft, with a diameter to 3–4 ft.

FLOWERS. Borne in spring in separate staminate and pistillate catkins; staminate catkins long, slender, pendulous, pubescent, 4–8 in. long; calyx long-stalked, 4–6-lobed, lobes greenish, ovate, acute, densely hairy; stamens 30–40, anthers yellow, almost sessile, about 1/12 in. long, with connectives; pistillate catkins shorter than the staminate catkins, solitary or a few in a cluster; ovary inferior, pistil solitary and green, stigmas 2, spreading.

FRUIT. Subglobose or somewhat ovoid, 1–1½ in. in diameter; husk thin, densely rufous-tomentose; nut brown or black, ovoid, apex rounded or acute, base rounded or flattened, grooves broad and deep, shell thick; seed solitary, large, sweet.

LEAVES. Alternate, 7–12 in. long, odd-pinnately compound of 9–15 (rarely 19) leaflets; leaflets ovate to lanceolate, somewhat scythe-shaped, margin coarsely serrate, apex acuminate, base cuneate or rounded or asymmetrical; surface pubescent when young, later glabrous above with scattered pubescence on midrib beneath, yellowish green, length 1½–4 in.; width 1–1½ in.; petiole and rachis slender and pubescent.

TWIGS. Slender, rufous-pubescent at first, becoming reddish brown to gray later and almost glabrous, lenticels small and pale.

BARK. Gray to brownish black, deeply fissured with prominent ridges.

RANGE. At elevations of 2,000–7,000 ft, along rocky stream beds and canyons of west Texas, New Mexico, and Arizona; also north to Colorado. In Mexico in the states of Chihuahua and Sonora to Durango.

CLOSELY RELATED SPECIES AND VARIETIES. Arizona Black Walnut is closely related to Texas Black Walnut, and is considered by some botanists as a variety of it. It appears that the distributions of the two walnuts overlap in the mountains of the Trans-Pecos and many intermediate forms are found. Typical Arizona Black Walnut is generally found at higher altitudes, is more arborescent, has larger nuts, and the leaflets are fewer, broader, and with strongly oblique leaf bases.

An intermediate form, between the Texas and Arizona Walnut, with arborescent habit, large fruit, and numerous elongate leaflets has recently been described as Stewart Walnut, *J. major* var. *stewartii* I. M. Johnston. It occurs in Cañón Indio Felipe in the Sierra Hechiceros, Coahuila, Mexico.

Arizona Black Walnut
Juglans major (Torr.) Heller

REMARKS. The genus name, *Juglans*, is from the Latin *Jovis glans*, meaning "acorn [or any nut of similar shape] of Jove," and the species name, *major*, refers to the larger size. Vernacular names are River Walnut, Mountain Walnut, Nogal, and Nogalillo. The tree grows rapidly when young, has a well-developed taproot and is long-lived. The wood is durable in contact with the soil, and is used for posts and occasionally for furniture or veneer. A good crop of fruit is borne every 2–3 years, there being about 45 seeds per lb. The germination rate is usually less than half. It has considerable wildlife value, especially for squirrels and rodents.

BUTTERNUT

Juglans cinerea L.

FIELD IDENTIFICATION. Tree attaining a height of 100 ft, diameter of 4 ft. A tree in St. Joseph County, Michigan, has been reported with a circumference of 11 ft, a height of 75 ft, and a spread of 70 ft. The trunk is short, and the wide-spreading ascending and horizontal branches form an irregular or round-topped crown. The roots are rather deep, and there are both lateral roots and a taproot.

FLOWERS. Borne April–June, monoecious; staminate catkins 1½–5 in. long, appearing when the leaves are ⅓–½ grown from axillary buds of previous year; calyx about ¼ in. long, 4-lobed, yellowish green, puberulent; bract adhering to the calyx, reddish-pubescent, apex acute; stamens 7–15, filaments free but very short, anthers dark brown and glabrous; pistillate spikes borne on shoots of current season, solitary or several together, ⅓–½ in. long, 5–8-flowered, constricted above; calyx minute, lobes 4, linear-lanceolate; bracts brownish, 3-lobed, cup-shaped, acute, glandular-hairy; style short, red, inwardly stigmatic, ovary 1-ovulate, ovule erect or orthotropous.

FRUIT. Maturing September–November of the first year, racemes drooping, 1–5-fruited, odor pungent, 1½–3 in. long, ovoid-oblong, obsurely 2–4 ridged, densely hairy with reddish matted, sticky hairs; hull thick, semi-fleshy, indehiscent; nut light brown, ovoid, apex acute or acuminate, ribs 4, thin, sharp, sometimes with 4 fainter ribs, interspaces deeply sculptured, basally 2-celled, 1-celled above; seed sweet, oily, edible.

LEAVES. Alternate, deciduous, 10–25 in. long, odd-pinnately compound with a glandular-hairy rachis; leaflets 11–19, oblong-lanceolate to elliptic, apex acute or acuminate, base rounded and some unsymmetrical, margin finely and shallowly appressed-serrate except near the base, length 2–5 in., width 1½–2 in.; young leaves sticky-glandular, older leaves yellowish green and finely pubescent above, paler and glandular-pubescent beneath, eventually turning yellowish and early deciduous.

TWIGS. At first greenish orange or reddish with clammy pubescence, later reddish brown to gray and only puberulent; lenticels pale and abundant; leaf scars large, gray, 3-lobed, raised, hairy-fringed on the upper margin, bundle scars in 3 U-shaped groups; terminal buds ½–¾ in. long, depressed, obtuse at apex, oblique at base, pubescent, scales entire or lobed; axillary buds, ovoid, depressed, apex rounded, rusty-pubescent, about ⅛ in long; pith thin-plated, diaphragms thickened.

BUTTERNUT
Juglans cinerea L.

BARK. On trunk gray to light brown, fissures fairly deep; ridges broad, flat, short, confluent into roughly diamond-shaped patterns, scales small and appressed; bark of branches smooth and gray, inner bark bright green.

WOOD. Heart light brown, sapwood white to grayish brown, grain coarse but straight, rather weak in bending and endwise compression, moderately high in shock resistance, rather soft and works well, stains well, lustrous and takes a good polish, shrinks moderately, not very durable, weighs about 25 lb per cu ft.

RANGE. Seldom in pure stands, usually in mixed forests, mostly in moist rich soils of wooded hillsides and stream banks. Northeastern Oklahoma and Arkansas; east to Georgia, north to New Brunswick, and west to Minnesota, the Dakotas, Nebraska, and Kansas. Not found on the lower Atlantic and Gulf Coast plain.

PROPAGATION. The fruit should be gathered when brown and dehulled before drying. The seeds can be spread out to dry, and have been known to retain vitality 4–5 years, stored in sealed containers at temperatures a little above freezing. One hundred lb of fruit yields 20–30 lb of seed, with an average of 30 cleaned seeds per lb. The seed has a commercial purity of 100 per cent and a soundness of 96 per cent. Dormancy may be broken by stratification in sand and peat, or sandy loam at 35°–45°F. for 90–120 days. Fresh nuts are sown in the fall, or stratified seed in spring, in drills 8–12 in. apart, with 51 seeds per linear ft, and 1–2 in. deep. Germination averages about 80 per cent. Screening provides protection from rodents, and fall-sown seed are mulched over winter. Two-year-old seedlings are transplanted and young plants sometimes root-pruned. Butternut can be grafted on other species of *Juglans* potted stock in the greenhouse or old trees can be top-grafted.

HYBRIDS. Butternut hybridizes with the English Walnut, *J. regia* L., to produce *J.* × *quadrangulata* (Carr.) Rehd.

It also crosses with the Japanese Walnut, *J. sieboldiana* Maxim., to produce *J.* × *bixbyi* Rehd.

REMARKS. The genus name, *Juglans,* is from the Latin *Jovis glans,* meaning "acorn [or any nut of similar shape] of Jove." The species name, *cinerea,* refers to the ashy gray hairs of the young parts. Vernacular names in use are White Walnut and Oil Nut. The tree grows rapidly at first and has a maximum age of about 75 years. The foliage is susceptible to insect attack. It is easily damaged by fire and storm, is wind firm, and rather difficult to transplant. The tree has been cultivated since 1633.

The wood of Butternut is used for furniture, cabinet work, interior trim of houses, instrument cases, boxes and crates, woodenware, toys, and novelties. A decoction of the inner bark of the root was once used as a cathartic. It is collected in autumn and comes chiefly from North Carolina, Virginia, and Michigan. The drug is officially known as "Underground Juglans." It has a faintly aromatic odor and a bitter, somewhat acid taste. The fruit hull yields a yellow or orange dye which was much used during Civil War days for soldiers' uniforms. The soft green fruit is sometimes pickled with vinegar, sugar, and spices, and was much used by the pioneers. The ripe kernel is also eaten. Sugar has been extracted from the sap in a manner similar to that of extracting maple sugar. Various species of animals feed on Butternut, including a number of species of squirrel, white-tailed deer, and cottontail.

PECAN

Carya illinoensis (Wangh.) K. Koch

FIELD IDENTIFICATION. Tree to 150 ft, with a broad rounded crown. The largest of all hickories.

FLOWERS. March–May, borne in staminate and pistillate catkins on same tree, subject to frost damage; staminate in slender, fascicled, sessile catkins 3–6 in. long; calyx 2–3-lobed, center lobe longer than lateral lobes; stamens 5–6, yellowish; pistillate catkins fewer, hairy, yellow, stigmas 2–4.

FRUIT. Ripening September–October, in clusters of 2–10, persistent; husk thin, aromatic, splitting along its grooved sutures into 4 valves at maturity; nut oblong to ellipsoid, cylindric, acute, bony, smooth, reddish brown, irregularly marked with darker brown, 1½–3½ in. long; seed deeply 2-grooved, convoluted on surface.

LEAVES. Alternate, deciduous, odd-pinnately compound, 9–20 in. long; leaflets 9–17, sessile or short-stalked, oblong-lanceolate, falcate, acuminate at apex, rounded to cuneate and inequilateral at base, doubly serrate on margin, 4–8 in. long, 1–2 in. wide; aromatic when crushed, dark green and glabrous above, paler and glabrous or pubescent beneath; rachis slender, glabrous or pubescent.

TWIGS. Reddish brown, stout, pubescent, lenticels orange-brown.

BARK. Grayish brown to light brown under scales; ridges flattened, narrow, broken, scaly; fissures narrow, irregular.

WOOD. Reddish brown, sapwood lighter, coarse-grained, heavy, hard, brittle, not strong, weighing 45 lb per cu ft, inferior to other hickories.

PECAN
Carya illinoensis (Wangh.) K. Koch

RANGE. Rich river-bottom soils. Texas, Oklahoma, Arkansas, and Louisiana; eastward to Alabama, and north to Kansas, Iowa, Indiana, and Tennessee.

PROPAGATION. The minimum seed-bearing age is about 20 years and the maximum 300 years. The optimum is 75–200 years. The nuts average about 100 per lb, but this varies considerably because of the increased nut-size in some of the improved varieties. Commercial seed averages 100 per cent purity and 95 per cent soundness. Preplanting treatment to break seed dormancy consists of using various acids and hydroxides. The most generally used method is stratification in sand, peat, or loamy soil at 35°–45°F. for 30–90 days. The germinative capacity is about 50 per cent. Planting can be done in fall or spring. For spring planting only stratified seed should be used. The seeds are planted in drills using 6–8 per linear ft in rows 8–12 in. apart and covered with ¾–1 in. of firmed soil. A mulch of straw or leaves should be provided until germination. Screening sometimes helps keep the mulch in place and protects against rodents. Pecan is also propagated from cuttings, and by budding or grafting. Some natural hybrids of pecan have also been listed as follows:

HYBRIDS. Brown's Hickory, *Carya* × *brownii* Sarg., is a cross of *C. cordiformis* and *C. illinoensis,* named after its discoverer, George M. Brown. It occurs in Mississippi, Arkansas, Oklahoma, and Kansas.

Bitter Pecan, *Carya* × *lecontei* Little, is a cross of *C. aquatica* and *C. illinoensis,* named after Major John Eaton Le Conte (1784–1860), American naturalist. It is found in Missouri, Arkansas, east Texas, Louisiana, and Mississippi.

Nussbaumer Hickory, *Carya* × *nussbaumeri* Sarg., is a cross of *C. illinoensis* and *C. laciniosa,* named for its discoverer, J. J. Nussbaumer, of Okawville, Illinois. It also is found in Iowa and Missouri.

Schneck Hickory, *Carya* × *schnecki* Sarg., is a cross of *C. illinoensis* and *C. tomentosa,* named for its discoverer, Jacob Schneck (1843–1906), of Mount Carmel, Illinois. It occurs in Illinois and Iowa.

REMARKS. The genus name, *Carya,* is the ancient name for walnut, and the species name, *illinoensis,* refers to the state of Illinois, the tree at one time having once been called Illinois Nut. It is widely planted as an ornamental, and for its sweet edible nuts. The wood is not important commercially, but is occasionally used for furniture, flooring, agricultural implements, and fuel. The nut is valuable to wildlife, being eaten by a number of species of birds, fox squirrel, gray squirrel, opossum, raccoon, and peccary. The bark and leaves are sometimes used medicinally as an astringent. Pecan is a rather rapid grower for a hickory, and is long-lived, but is subject to bark beetle attacks. It has been cultivated since 1766.

Over 116 horticultural clones of pecan have been listed by various authors.

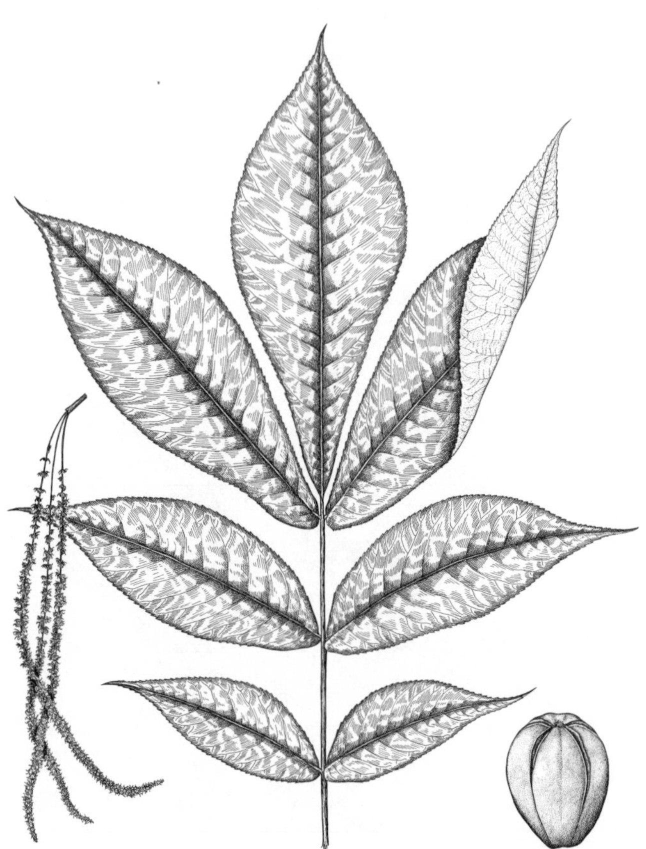

SHELLBARK HICKORY
Carya laciniosa (Michx. f.) Loud.

SHELLBARK HICKORY

Carya laciniosa (Michx. f.) Loud.

FIELD IDENTIFICATION. Tree attaining a height of 120 ft, with short, stout limbs and a narrow crown. A specimen at Big Tree State Park, Missouri, has been reported with a circumference of 12 ft 9 in., a height of 128 ft, and a spread of 70 ft.

FLOWERS. April–June, borne in separate staminate and pistillate catkins on the same tree; staminate catkins in threes, 5–8 in.; bracts linear-lanceolate, acute, scurfy-tomentose; calyx 3-lobed, central lobe longer than lateral ones; stamens 3–10, hairy, yellow, anthers emarginate; pistillate in 2–5-flowered spikes, tomentose; bracts longer than calyx-lobes; stigmas short.

FRUIT. Ripening September–November, solitary or 2–3 together, ellipsoid to globular, depressed at apex, 1–3 in. long; hull light to dark orange-brown, hard, woody, ¼–½ in. thick, dehiscent along the 4 ribs; nut yellowish white, ellipsoid to obovoid or globose, usually rounded and flattened at ends but sometimes pointed at apex, somewhat compressed, bony, hard, thick-shelled; kernel sweet. Largest of all hickory nuts.

LEAVES. Large, 1–2 ft long, alternate, deciduous, odd-pinnately compound of 5–9, usually 7, leaflets; leaflets ovate to oblong-lanceolate or acute at apex, cuneate or rounded and inequilateral at base, finely serrate on margin; dark green, glabrous and shiny above; pale green beneath and velvety pubescent when young, glabrous later; terminal leaflet petioled, lateral leaflets sessile or nearly so; petiole and rachis pubescent or glabrous with age.

TWIGS. Stout, dark brown to reddish orange, lenticels elongate.

BARK. Gray, separating into long, thin, shaggy plates hanging loosely.

WOOD. Dark brown, sapwood lighter, close-grained, hard, strong, tough, heavy, very flexible, weighing 50 lb per cu ft.

RANGE. Northeast Texas, Oklahoma, Arkansas, and Louisiana; east to Alabama, and north to Nebraska, Minnesota, Iowa, Kansas, Delaware, and Ontario.

PROPAGATION. Shellbark Hickory has been cultivated since 1800. The minimum commercial seed-bearing age is 40 years, and the maximum 350, with an optimum of 75–200 years. Healthy trees can produce 2–3 bushels of nuts. The cleaned seed averages about 30 seeds per lb because of the large size. The commercial seed purity is 100 per cent, with a soundness of 95 per cent. Planting can be done in spring or fall. For spring planting seed should be stratified in sand, peat, or sandy loam at 32°–45°F. for 120–150 days (30–60 days is sufficient if seed has been dry-stored for a year). Seeds should be planted in drills with 6–8 nuts per linear ft in rows 8–12 in. apart. They should be covered with firmed soil about 1½–2½ in. deep and covered with a mulch of straw or leaves. A screen will help hold the mulch in place and protect seeds from rodents. A little shade is helpful to the seedlings (statistics from U.S.D.A., 1948).

HORTICULTURAL HYBRIDS, NATURAL HYBRIDS, AND CLONES:
Dreppard (Drep., 1934)
Eversman (Snyd., 1926)
Hage (Hage, 1926)
Landis (Red., 1934)
Stanley (Jones, 1916)
Tama Queen (Snyd., 1920)
Beecher, *C. laciniosa* × *C. ovata* (Zim., 1933)
Berger, *C. laciniosa* × *C. tomentosa* (Ber., 1934)
Burlington, *C. laciniosa* × *C. illinoensis* (Jones, 1915)
Des Moines, *C. laciniosa* × *C. illinoensis* (Snyd., 1924)
Dintleman, *C. laciniosa* × *C. illinoensis* (Din., 1922)
Gerardi, *C. laciniosa* × *C. illinoensis* (Ger., 1930)
Joy, *C. laciniosa* × *C. ovata* (Joy, 1934)
McCallister, *C. laciniosa* × *C. illinoensis* (McCal., 1890)
Nussbaumer, *C. laciniosa* × *C. illinoensis* (S. Mil., 1888)
Rockville, *C. laciniosa* × *C. illinoensis* (Jones, 1900)
Sande, *C. laciniosa* × *C. ovata* (Sande, 1925)
Weiker, *C. laciniosa* × *C. ovata* (Jones, 1903)

A natural hybrid of *C. laciniosa* × *C. ovata* has been listed as *C. dunbari* Sarg., named in honor of John Dunbar, at one time assistant superintendent of parks of Rochester, New York. It is found in New York State.

REMARKS. The genus name, *Carya*, is the ancient name for walnut, and the species name, *laciniosa*, refers to the deep furrowing and splitting of the bark. Vernacular names are King Nut, Bottom Shellbark, Big Shellbark, and Thick Shellbark. The wood is used as that of other hickories, particularly for tool handles, baskets, and fuel. The large nuts are edible and sweet, but are considered to lack the flavor of the Shagbark Hickory. The tree is long-lived, slow-growing, hard to transplant, and subject to insect damage. It is rather similar in appearance to the Shagbark Hickory, but has larger leaves and nuts, and the bark is somewhat less shaggy in appearance. However, the two species are known to hybridize.

Shagbark Hickory

Carya ovata (Mill.) K. Koch

FIELD IDENTIFICATION. Tree attaining a height of 100 ft, with an oblong crown and shaggy bark. A specimen in Turkey Run State Park, Indiana, has been reported with a trunk diameter of 9 ft 8 in., a height of 127 ft, and a spread of 50 ft.

FLOWERS. Appearing March–June, borne in separate staminate and pistillate catkins; staminate catkins in threes after the leaves appear, 4–5 in. long, slender, green, hairy-glandular; bract ovate-lanceolate and longer than the ovate and acute calyx-lobes; stamens 4, reddish yellow, hairy; pistillate catkins 2–5-flowered, rusty-tomentose.

FRUIT. Ripening September–October, very variable in size and shape. Borne 1–3 together, 1–2½ in. long, oval to subglobose or obovoid, depressed at apex; hull blackish to reddish brown, glabrous or hairy, ¼–½ in. thick, splitting freely to the base into 4 valves along the grooved sutures; nut light brownish white, oblong-obovate, somewhat compressed, usually prominently 4-angled, barely acute, or rounded, or truncate at apex; rounded at base, shell thin; kernel light brown, aromatic, sweet, edible.

LEAVES. Alternate, deciduous, 8–17 in. long, odd-pinnately compound of 3–5 (rarely 7) leaflets; lateral leaflets sessile or nearly so, ovate-obovate or elliptic-lanceolate, acuminate at apex, cuneate and unequal at base, serrate on margin, yellowish green and glabrous above, paler and glabrous or hairy beneath, 4–7 in. long, 2–3 in. wide; terminal leaflet stalked, terminal leaflet and the upper pair of leaflets considerably larger than the lower pairs; rachis and petiole glabrous or pubescent.

TWIGS. Orange-brown, stout, glabrous or pubescent.

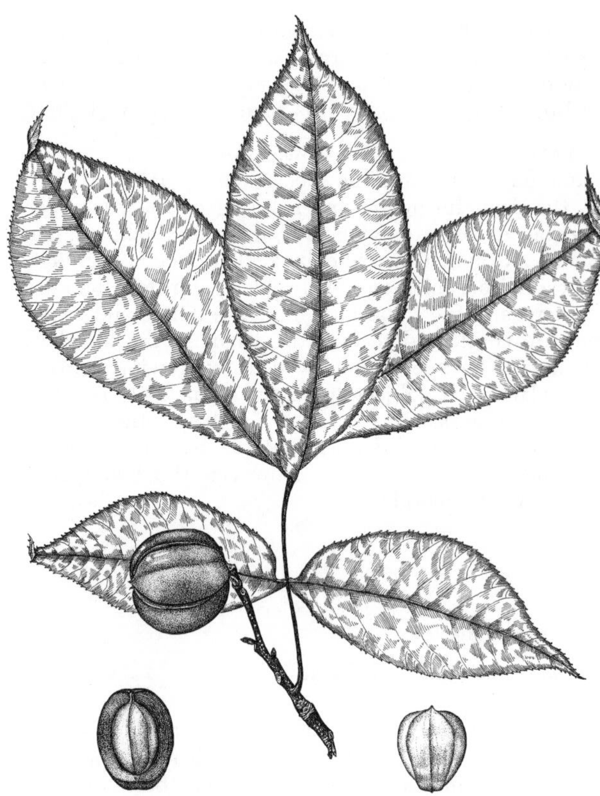

SHAGBARK HICKORY
Carya ovata (Mill.) K. Koch

BARK. Gray, conspicuously exfoliating into long shaggy strips.

WOOD. Light brown, sapwood lighter, close-grained, heavy, hard, strong, tough, flexible, weighing 52 lb per cu ft.

RANGE. East Texas, Oklahoma, Arkansas, and Louisiana; east to Alabama, north to Maine and Quebec, and west to Minnesota, Michigan, and Nebraska.

PROPAGATION. Shagbark Hickory has been cultivated since 1911. The minimum commercial seed-bearing age is 40 years, the optimum 60–200 years, and the maximum 300 years. The seeds average about 100 per lb cleaned. Commercial seed has a purity of 100 per cent and a soundness of 95 per cent. The seed may be sown in fall or spring. If sown in spring the seed should be stratified in sand, peat, or sandy loam at 35°–45°F. for 90–150 days (only 30–60 days' stratification if seeds have been dry-stored for a year). They should be drilled in beds with 6–8 nuts per linear ft in rows 8–12 in. apart, and covered with firmed soil ¾–1½ in. thick. A mulch of straw or leaves helps until germination is complete. A screen helps to hold the mulch in place and keep away rodents. No shading is necessary. The germinative capacity averages 80 per cent. (Statistics from U.S.D.A., 1948).

NATURAL VARIETIES, NATURAL HYBRIDS, HORTICULTURAL HYBRIDS. The natural varieties are as follows:

Missouri Shagbark, *C. ovata* var. *ellipsoidalis* Sarg., has a decided ellipsoidal nut.

Ashleaf Shagbark, *C. ovata* var. *fraxinifolia* Sarg., has leaves resembling those of White Ash.

Littlenut Shagbark, *C. ovata* var. *nuttallii* Sarg., has smaller, obcordate, compressed, angled fruit.

Southern Shagbark, *C. ovata* var. *pubescens* Sarg., has more pubescent leaves.

The natural hybrids are as follows:

Dunbar Hickory, *C.* × *dunbarii,* is a cross of *C. laciniosa* and *C. ovata,* and is named after John Dunbar, at one time assistant superintendent of parks, Rochester, New York, who discovered it. It is found in New York.

Laney Hickory, *C.* × *laneyi* Sarg., is a cross of *C. cordiformis* and *C. ovata,* and is named after C. C. Laney, at one time superintendent of parks, Rochester New York, where it was discovered. It occurs in southern Quebec, southern Ontario, Vermont, New York, and Pennsylvania.

Horticultural hybrids:
Barnes, *C. tomentosa* × *C. ovata* (H. Bar., 1919)
Beaver, *C. cordiformis* × *C. ovata* (Jones, 1916)
Beecher, *C. laciniosa* × *C. ovata* (Zim., 1933)
Burton, *C. illinoensis* × *C. ovata* (E. Burt., 1880)
Creager, *C. cordiformis* × *C. ovata* (Snyd., 1925)
Fairbanks, *C. cordiformis* × *C. ovata* (Jones, 1916)
Joy, *C. laciniosa* × *C. ovata* (Joy, 1934)
Laney, *C. cordiformis* × *C. ovata* (Dunb., 1916)
Peck, *C. cordiformis* × *C. ovata* (G. L. Smi., 1934)
Sande, *C. laciniosa* × *C. ovata* (Sande, 1925)
Stocking, *C. cordiformis* × *C. ovata* (E. Sto., 1932)
Weiker, *C. laciniosa* × *C. ovata* (Jones, 1903)
Woods, *C. laciniosa* × *C. ovata* (Zim., 1927)

Besides the above, about 63 horticultural clones have been listed by Kelsey and Dayton (1942).

REMARKS. The genus name, *Carya,* is the ancient name for walnut, and the species name, *ovata,* refers to the ovate-shaped leaflets. Vernacular names are Scaly Bark Hickory, White Hickory, Upland Hickory, Red Heart Hickory, Sweet-walnut, and White-walnut. The tree is long-lived, slow-growing, and subject to insect attacks. The wood is used for fuel, tool handles, baskets, wagons, and other general uses. This nut, next to the pecan, is the best of any American tree, and is the common hickory nut of commerce.

WATER HICKORY

Carya aquatica (Michx. f.) Nutt.

FIELD IDENTIFICATION. Water-loving tree attaining a height of 100 ft, with an irregular, narrow crown. A specimen near Camden, South Carolina, has been reported to be 120 ft high, with a trunk circumference of 10 ft 7 in., and a spread of 68 ft.

FLOWERS. In separate staminate and pistillate catkins

on the same tree, staminate catkins hairy, 2½–3 in., solitary or in threes, pubescence yellow-glandular; stamens usually 6, anthers yellow-pubescent; calyx-lobes almost equal in length; pistillate catkins 2–10-flowered.

FRUIT. Often clustered, ovoid or obovoid, conspicuously flattened in comparison with other hickories, 1–1½ in. long, about 1 in. wide; hull thin, shallowly 4-winged, splitting from the base, yellow-pubescent; nut very flat, reddish, angled and corrugated, thin-shelled; seed bitter, dark reddish brown.

LEAVES. Alternate, deciduous, odd-pinnately compound, 9–15 in. long, composed of 7–15 leaflets; leaflets lanceolate-ovate, often falcate, long-acuminate, the laterals unequally cuneate at base, serrate, 3–5 in. long, ½–1 in. wide, dark green and glabrous above, pubescent or glabrous beneath; rachis and petiole puberulent to pubescent.

TWIGS. Reddish brown or gray, tomentose at first, glabrous later, lenticels pale.

BARK. Grayish brown, tinged with red; scales small, thin, and brittle.

WATER HICKORY
Carya aquatica (Michx. f.) Nutt.

WOOD. Dark brown, sapwood lighter, heavy, hard, brittle, close-grained, weighing 46 lb per cu ft.

RANGE. East Texas, Oklahoma, Arkansas, and Louisiana; also eastward to Florida and north to Illinois, Missouri, and Virginia.

PROPAGATION. Cleaned seed averages 199 seeds per lb. For planting, the seed should be stratified in sand or peat at 30°–35°F. for 90–150 days. Seeds that have been dry-stored for about a year may need only 30–60 days' stratification. The maximum germinative percentage is 81. Only stratified seeds are used in spring planting. Six to eight seeds are drilled per linear ft in rows 8–12 in. apart and covered about 1 in. with firmed soil. Generally no shading is required. The tree was first cultivated in 1800.

VARIETY AND HYBRIDS. A variety known as the Narrow-leaf Water Hickory, *C. aquatica* var. *australis* Sarg., has very narrow leaflets, small ellipsoidal fruit with pale brown nuts, fewer wrinkles, and a less scaly bark.

Bitter Pecan, *Carya* × *lecontei* Small, is a hybrid of *C. aquatica* and *C. illinoensis*, named in honor of Major John Eaton Le Conte (1784–1860), American naturalist. It is found in Arkansas, eastern Texas, and Louisiana; also north to Missouri, and east to Mississippi.

Louisiana Hickory, *Carya* × *ludoviciana* (Ashe) Little, is a hybrid of *C. aquatica* and *C. texana*, and the hybrid name means "delusive."

REMARKS. The genus name, *Carya*, is the ancient name for walnut, and *aquatica* refers to the tree's wet habitat. Vernacular names are Bitter Pecan, Swamp Hickory, and Water Pignut. The wood, rather inferior to that of other hickories, is hard to work, and is used in small amounts for fuel, posts, and props. The nuts have been found in the stomachs of mallard duck and wood duck.

MOCKERNUT HICKORY

Carya tomentosa Nutt.

FIELD IDENTIFICATION. Tree to 100 ft, with rather short limbs and a broad or oblong crown. A specimen has been reported from Turkey Run State Park, Indiana, with a circumference of 9 ft 6 in., a height of 146 ft, and a spread of 52 ft.

FLOWERS. April–May, borne in separate staminate or pistillate catkins; staminate catkins 3-branched, 4–5 in. long, yellowish green, hairy; bracts ovate to lanceolate, hairy, much longer than calyx-lobes; stamens 4, with red, hairy anthers; pistillate in 2–5-flowered hairy spikes; bracts ovate and acute, longer than bractlets and calyx-lobes; stigmas dark red.

FRUIT. Ripening September–October, solitary or paired, very variable in size and shape, usually obovoid, globose, or ellipsoid, 1–3½ in. long, acute at apex,

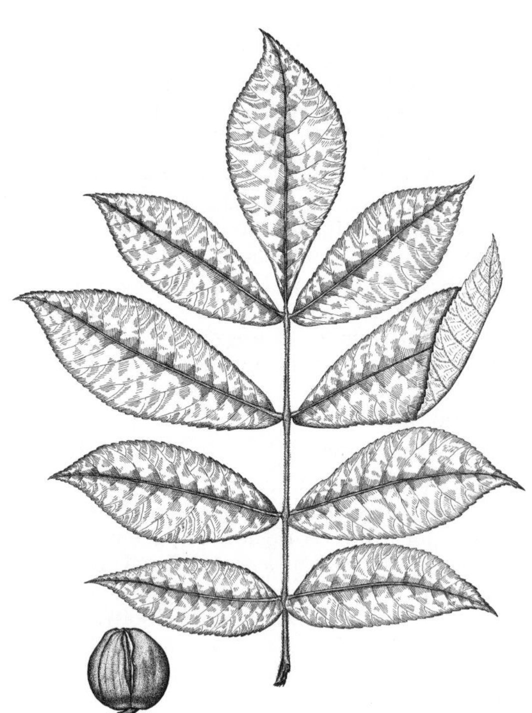

MOCKERNUT HICKORY
Carya tomentosa Nutt.

rounded or rarely with a short necklike base; hull dark reddish brown, woody, hairy or nearly glabrous with yellow resinous dots, 4-ribbed, dehiscent down the deep ribs to the middle or near the base, about ⅛–¼ in. thick; nut variable in shape, obovoid-oblong to globose or ovoid, rounded at base, acute or acuminate at apex, slightly flattened, noticeably or obscurely 4-ridged, brownish white or reddish, shell thick and hard; kernel dark brown, small, shiny, sweet, edible.

LEAVES. Alternate, deciduous, 8–24 in. long, odd-pinnately compound of 5–9 (usually 7) leaflets; lateral leaflets sessile or nearly so, oblong to lanceolate or obovate, acute to acuminate at apex, rounded or broadly cuneate at base and inequilateral, serrate on margin; shiny yellowish green above, paler beneath and clothed with brownish orange hairs, glandular and resinous, fragrant when crushed, 5–8 in. long, 2–5 in. wide, terminal leaflets and upper pairs generally larger than lower pairs; rachis and petiole glandular-hairy.

TWIGS. Stout, grayish brown to reddish, hairy at first but more glabrous later; buds distinctively large and tomentose.

BARK. Gray, close and rough but never shaggy, ridges rounded and netted, separated by shallow fissures.

WOOD. Dark brown, sapwood lighter, close-grained, heavy, hard, strong, tough, flexible, weighing 51 lb per cu ft.

RANGE. Texas, Louisiana, Arkansas, and Oklahoma; east to Florida, and north to Nebraska, Ontario, Iowa, Illinois, Michigan, and Maine.

PROPAGATION. Mockernut Hickory was first cultivated in 1766. The cleaned seed averages about 90 seeds per lb, with a commercial purity of 100 per cent and a soundness of 95 per cent. The seed may be planted in fall or spring. For spring planting the seed should be prestratified in sand, peat, or loamy soil at a temperature of 30°–35°F. for 90–150 days (30–60 days will suffice for seeds dry-stored a year). The average germinative capacity is 66 per cent. The seeds are planted in drills with 6–8 every linear ft in rows 8–12 in. apart, and covered with firmed soil ¾–1½ in. deep. A mulch of straw or leaves should remain until germination. A screen helps hold the mulch in place and protects the seeds from rodents.

VARIETIES AND HYBRIDS. Gulf Mockernut Hickory, *C. tomentosa* var. *subcoriacea* (Sarg.) Palm. & Steyermark, is a variety with thicker, more pubescent leaves, and a fruit more prominently angled with thicker husk and larger nuts.

Fig Mockernut Hickory, *C. tomentosa* var. *ficoides* Sarg., is a variety with a stipelike base to the fruit.

Ovoid Mockernut Hickory, *C. tomentosa* var. *ovoidea* Sarg., is a variety with long-acuminate, ovoid fruit.

Horticultural hybrids:

Barnes, *C. tomentosa* × *C. ovata* (H. Bar., 1919)
Berger, *C. tomentosa* × *C. laciniosa* (Berg., 1934)
Siers, *C. tomentosa* × *C. cordiformis* (Siers, 1916)

A natural hybrid has been listed as Schneck Hickory, *C.* × *schneckii* Sarg., a hybrid of *C. tomentosa* × *C. illinoensis,* and named after Jacob Schneck (1843–1906), who discovered it in Iowa and Illinois.

REMARKS. *Carya* is the ancient name for walnut, and the species names *tomentosa,* refers to the tomentose hairs of the leaves. Vernacular names are Whiteheart Hickory, White Hickory, Red Hickory, Black Hickory, Whitebark Hickory, Hardbark Hickory, Bigbud Hickory, Bullnut Hickory, and Fragrant Hickory. It is long-lived, a rapid-grower when young, will sprout from the stump, and is subject to insect damage. The foliage is occasionally browsed by white-tailed deer. The wood is important commercially and is used for vehicle parts, handles, fuel, and agricultural implements. The nut is sweet and edible but used in smaller quantities than other edible species.

BITTERNUT HICKORY

Carya cordiformis (Wangh.) K. Koch

FIELD IDENTIFICATION. Tree attaining a height of 100 ft, with stout limbs and a broad spreading head. Tree in West Feliciana Parish, Louisiana, has been reported to be 171 ft high, with a trunk 12 ft 6 in. in circumference.

FLOWERS. April–May, staminate and pistillate catkins

separate but on the same tree; staminate catkins in threes, 3–4 in. long, reddish-hairy, bracts ovate and acute; calyx 2–3-lobed; stamens 4, anthers yellow and hairy; pistillate catkins mostly in ones or twos, sessile.

FRUIT. September–October, solitary or paired, obovoid to subglobose, about ¾–1 in. long; husk yellow-scaled, thin, 4-winged above the middle and splitting to somewhat below the middle; nut globose to ovate, small, smooth, white, thin-shelled, abruptly pointed into a conical beak, slightly flattened; kernel convoluted and bitter to taste.

LEAVES. Alternate, deciduous, 6–12 in. long, odd-pinnately compound of 7–11 leaflets on a pubescent or glabrous rachis; leaflets ovate-lanceolate, some falcate, sessile or nearly sessile, terminal leaflets petioled, acuminate at apex, cuneate or rounded at base, serrate on margin, shiny green and glabrous above, paler and glabrous or pubescent beneath, 3–6 in. long, 1–2 in. wide; rachis slender and pubescent; leaflets are generally smaller and more slender than those of other species of hickories.

TWIGS. Winter buds scurfy, bright yellow; twigs greenish brown to reddish brown, lustrous, pubescent at first and glabrous later, stout; lenticels numerous and pale.

BARK. Grayish brown, smooth for a hickory, with thin scales in the flattened ridges, furrows shallow.

WOOD. Reddish brown, heavy, hard, strong, tough, weighing 47 lb per cu ft. Used for the same purposes as other hickories, but wood considered to be somewhat inferior.

RANGE. In moist woods of bottom lands; east Texas, Oklahoma, Arkansas, and Louisiana; east to Florida, north to Minnesota, Maine, and Ontario.

PROPAGATION. The cleaned seed averages about 155 seeds per lb, with a commerial purity of 100 per cent and a soundness of 95 per cent. The seed is stratified before planting in sand, peat, or loamy soil at a temperature of 32°–45°F. for 90–120 days (30–60 days' stratification for year-old dry seed). The germinative capacity is about 55 per cent. Planting is done in spring with stratified seed in drills with 6–8 seeds per linear ft and rows 8–12 in. apart. They are covered with firmed soil about 1 in. thick. Bed should be mulched until germination is complete. No shading is generally needed. Bitternut Hickory stock is much-used in budding and grafting.

NATURAL VARIETIES, NATURAL HYBRIDS, HORTICULTURAL HYBRIDS. Two known natural varieties are:

Ashe Bitternut Hickory, *C. cordiformis* var. *elongata* Ashe, which has elongate fruit.

Broad-leaf Bitternut Hickory, *C. cordiformis* var. *latifolia* Sarg., which has broader leaves.

Natural hybrids are as follows:

Brown's Hickory, *Carya* × *brownii* Sarg., is a hybrid of *C. cordiformis* and *C. illinoensis,* and is named for its discoverer George M. Brown. It is found in Oklahoma, Arkansas, Mississippi, and Kansas.

Demaree Hickory, *C.* × *demareei* Palmer, is a hybrid of *C. cordiformis* and *C. illinoensis,* and is named after Delzie Demaree, botanist of Arkansas.

Laney Hickory, *C.* × *laneyi* Sarg., is a cross of *C. cordiformis* and *C. ovata,* and is named for C. C. Laney of Rochester, New York, where it was discovered. It is recorded from southern Quebec, southern Ontario, Vermont, New York, and Pennsylvania.

Horticultural hybrids:

Beaver, *C. cordiformis* × *C. ovata* (Jones, 1916)
Creager, *C. cordiformis* × *C. ovata* (Snyd., 1925)
Fairbanks, *C. cordiformis* × *C. ovata* (Jones, 1916)
Galloway, *C. cordiformis* × *C. illinoensis* (Gal., 1894)
Laney, *C. cordiformis* × *C. ovata* (Dunb., 1916)
Mall, *C. cordiformis* × *C. illinoensis* (Mall, 1934)
Peck, *C. cordiformis* × *C. ovata* (G. L. Smi., 1934)
Pleas, *C. cordiformis* × *C. illinoensis* (Pleas, 1916)
Siers, *C. cordiformis* × *C. tomentosa* (Siers, 1916)
Stocking, *C. cordiformis* × *C. ovata* (E. Sto., 1932)

REMARKS. The genus name, *Carya,* is the ancient name for walnut, which is close kin to the hickories, and the species name, *cordiformis,* means "heart-shaped," in reference either to the fruit or the base of the leaflets. Vernacular names are Swamp Hickory, Pig Hickory, White Hickory, Pignut Hickory, and Red Hickory. The wood is used for wheel stock, handles, and fuel. It was first cultivated in the year 1689. It is per-

BITTERNUT HICKORY
Carya cordiformis (Wangh.) K. Koch

haps the most rapid grower of all hickories and sprouts from the stump. It is suitable as a good park and shade tree, and is often used as a potted stock for grafting pecan varieties.

Black Hickory

Carya texana Buckl.

FIELD IDENTIFICATION. Tree attaining a height of 80 ft, with short, crooked branches forming a narrow crown.

FLOWERS. Borne in separate staminate and pistillate catkins, staminate catkins 2–3 in. long with acuminate bracts considerably longer than the calyx-lobes; stamens 4–5 with somewhat hairy anthers; pistillate catkins 1–2-flowered and red-hairy on all parts.

FRUIT. Hull of nut 1¼–2 in. in diameter, puberulent, subglobose to obovoid, or sometimes with a short basal neck, splitting at sutures to the base with valves 1/12–1/6 in. thick; nut globose, or obovoid, somewhat compressed, rounded at base, suddenly narrowed to an acute apex, 4-angled along upper part especially, reddish brown, reticulate-veined, shell about ⅛ in. thick; kernel small, rounded, sweet.

LEAVES. Alternate, deciduous, 8–12 in. long, odd-pinnately compound of 5–7 (usually 7) leaflets; leaflets sessile or nearly so, 4–6 in. long, lanceolate to oblanceolate or obovate, acuminate or acute at apex, cuneate and somewhat inequilateral at base, serrate on margin; dark green, shiny, and usually glabrous above; paler and rusty-pubescent below, especially when young, later becoming more glabrous; petioles rusty-hairy when young. The occurrence of rusty hairs and white scales on buds and young parts is a conspicuous feature.

TWIGS. Rusty-pubescent and reddish brown at first, grayish brown and glabrous later.

BARK. Dark gray to black, ridges irregular, broken into deep fissures.

WOOD. Brown, sapwood paler, hard, tough, brittle, used chiefly for fuel.

RANGE. The species, *texana,* and its variety, Arkansas Black Hickory, are distributed through east and south-central Texas and Louisiana; north to Oklahoma, Arkansas, Indiana, Illinois, Missouri, and Kansas.

PROPAGATION. The nuts are picked from the ground, or from the tree. The husks are removed by hand or by flailing. A good tree will produce 2–3 bushels of nuts. The nuts may be stored 3–5 years in closed containers at about 41°F., with a relative humidity of 90 per cent. For spring planting, the nuts can be stored indoors by stratifying in sandy loam at a temperature of about 40°F. For outdoor storage they can be covered with mulch, soil, or leaves for about 2 ft to protect from freezing in colder regions. Planting can also be done in the fall. The seedbed should be a moist, fertile loam, and the nuts drilled in rows 8–12 in. apart and covered about 1½ in. with firmed soil. Only stratified seed is used in spring beds. Beds are mulched until germination is complete. No shading is generally required.

Black Hickory
Carya texana Buckl.

VARIETIES, HYBRIDS. The most common variety, especially in Arkansas, Louisiana, and East Texas, is Arkansas Black Hickory, *C. texana* var. *arkansana* (Sarg.) Little, with an obovoid or pear-shaped hull, and a pale brown, compressed, slightly 4-angled nut; kernel small and sweet.

Louisiana Hickory, *C.* × *ludoviciana* (Ashe) Little, has been described as a hybrid of *C. texana* and *C. aquatica.* It is found in northern and central Louisiana.

Black Hickory was at one time listed under the name of *Hicoria buckleyi* (Durand) Ashe and *C. buckleyi* Durand.

REMARKS. The genus name, *Carya,* is the ancient name for walnut, and the species name, *texana,* refers to the state of Texas. Some of the vernacular names for it are Buckley Hickory and Pignut Hickory. The thick shell makes the nut almost impossible to extract, but hogs sometimes crack them. Wood used for fuel.

Nutmeg Hickory

Carya myristicaeformis (Michx. f.) Nutt.

FIELD IDENTIFICATION. Tree attaining a height of 100 ft, with a narrow open crown.

FLOWERS. Borne in separate staminate and pistillate catkins, staminate catkins 3–4 in., covered with brown scurfy pubescence; bracts ovate to oblong; stamens 6, with yellow anthers; pistillate catkins oblong, also with brown pubescence.

FRUIT. Usually solitary, 1–1⅜ in. long, ellipsoid to obovoid, covered with yellowish brown scurfy pubescence; hull very thin, distinctly 4-ridged at the sutures, splitting nearly to the base; nut small, bony, ellipsoid, acute or rounded at ends, smooth, reddish brown, often marked with blotches or bands, shell thick, resembling a nutmeg; kernel sweet but astringent.

LEAVES. Alternate, deciduous, odd-pinnately compound, 7–14 in. long, composed of 5–9 leaflets; leaflets oblong, lanceolate to ovate-lanceolate, the lateral ones sometimes ovate to obovate, acute or acuminate, unequally rounded or cuneate at the base, sharply serrate, dark green and nearly glabrous above, pubescent to glabrous and lustrous silvery-white beneath, thin and firm, short-stalked or almost sessile, 4–5 in. long, 1–1½ in. wide; petioles slender and pubescent.

TWIGS. Grayish brown to reddish brown, covered with small golden brown scales.

BARK. Dark brown to reddish; scales small, thin, appressed.

WOOD. Light brown, sapwood lighter, close-grained, tough, strong.

RANGE. Nowhere abundant, but scattered from east Texas to Louisiana and Alabama; north to South Carolina and west to Oklahoma and Arkansas.

REMARKS. *Carya* is the ancient name for walnut, and the species name, *myristicaeformis,* means "shaped like myristicae or nutmegs." Another vernacular name is Bitter Waternut. The tree is difficult to transplant, and is seldom seen in cultivation. The wood is considered inferior to that of other hickories and is of little commercial importance. Although very few data on the methods of propagation by seeds are obtainable for this species, the general methods employed for other hickories will suffice.

NUTMEG HICKORY
Carya myristicaeformis (Michx. f.) Nutt.

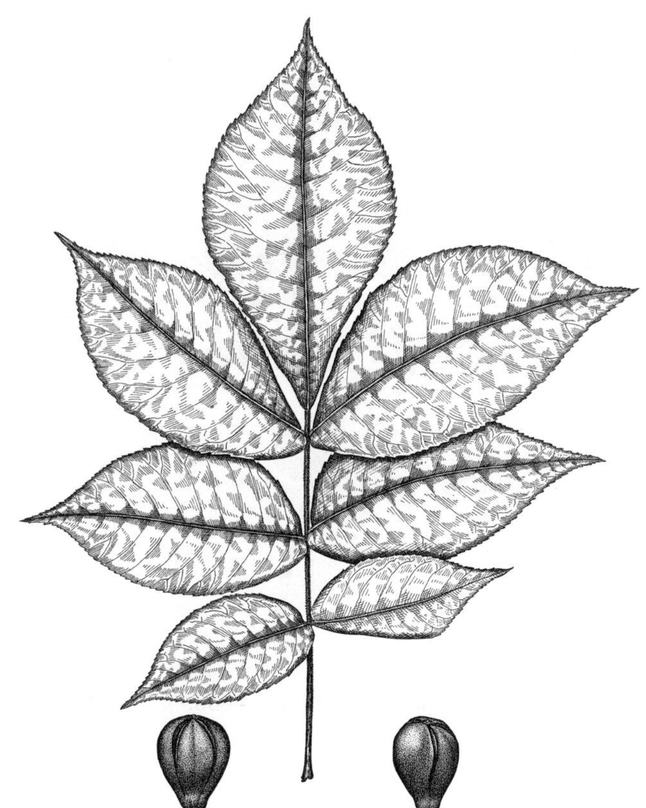

SWAMP HICKORY
Carya leiodermis Sarg.

SWAMP HICKORY

Carya leiodermis Sarg.

FIELD IDENTIFICATION. Tree attaining a height of 80 ft, with a rounded crown.

FLOWERS. Opening after the leaves, in staminate and pistillate catkins 4–5 in. long; staminate catkins pubescent; bracts lanceolate-ovate, glandular-hairy, longer than the ciliate calyx-lobes; stamens 4, hairy, anthers red; pistillate spikes few-flowered, shorter, pubescent, stigmas short.

FRUIT. Solitary or clustered, obovoid or globose, 1½–2 in. long, about 1¼ in. in diameter, little-compressed, or occasionally depressed, at apex; husk fairly thick

SAND HICKORY
Carya pallida (Ashe) Engelm. & Graebn.

(about ¼ in.), white-scaly, splitting along the 4 sutures to base; nut obovoid or ellipsoid, rounded at ends, reddish or brown, smooth, thin-shelled; kernel small and sweet.

LEAVES. Alternate, deciduous, 12–14 in. long, composed of 7 (rarely 5) leaflets; leaflets oblong-obovate to lanceolate, acuminate at apex, cuneate and inequilateral at base, serrate on margin; dark green, glabrous and shiny above; paler and slightly pubescent beneath; 4–5 in. long, and 2–2½ in. wide; lateral leaflets sessile or nearly so, terminal one short-stalked; distal leaflets larger than proximal ones; petiole and rachis pubescent at first and glabrous later.

TWIGS. Slender, grayish brown.

BARK. Grayish brown, close, rather smooth.

WOOD. Light reddish brown, sapwood paler, hard, strong, used for same purposes as that of other hickories.

RANGE. The species, *leiodermis*, is distributed from Mississippi to Louisiana and east Texas, and north into Arkansas.

VARIETY. Ruddy Swamp Hickory, *C. leiodermis* var. *callicoma* Sarg., is a variety having bright red young leaves and a thinner husk to the fruit. It occurs with the species, and has been found on the Neches River near Beaumont, Texas.

REMARKS. The genus name, *Carya*, is the ancient name for walnut. The species name, *leiodermis*, refers to the smooth leaf surfaces.

SAND HICKORY

Carya pallida (Ashe) Engelm. & Graebn.

FIELD IDENTIFICATION. Trees usually 30–40 ft (rarely to 100 ft), with a diameter of 1–3 ft. Branches erect or pendulous into an open, spreading crown.

FLOWERS. In separate staminate and pistillate catkins March–April; staminate catkins 3–8 in., slender, branched, peduncle hairy and scaly; bracts 3-lobed, center lobe longest; stamens 4, anthers ovate-lanceolate, hairy; pistillate catkins 2–6-flowered; flowers usually sessile; apetalous; involucre perianth-like, 4–5-lobed.

FRUIT. Nut subglobose, ellipsoid or obovoid, ½–1½ in., apex bluntly pointed, base rounded, dark brown, when young finely pubescent and covered with minute rufous-brown scales, with age less so, tardily dehiscent, hull 1/10–1/6 in. thick; nut oval, flattened, nearly white, sometimes tipped with the style remnants, smoothish or ridged to the base, shell ⅛–1/12 in. thick, kernel small and sweet.

LEAVES. Pinnately compound, 6–11 in. long, rachis and petiole densely fascicled, hairy; leaflets 5–7 (rarely 9); oblong, lanceolate, oblanceolate, or ovate; short-stalked or sessile; terminal leaflet 4–6 in., usually larger than lateral leaflets; apex acuminate; base cuneate, rounded or asymmetrical in lateral leaflets; margin finely serrate, sometimes entire at base; upper surface light green and lustrous, glabrous or hairy along the veins; lower surface much paler with minute punctations and densely hairy, especially along the veins; young leaves scurfy, or with yellowish scales.

TWIGS. Slender, reddish brown when young, pubescent or glabrous; older twigs gray to almost black, glabrous.

BARK. Gray to black, ridges rough, furrows deep.

WOOD. Brown, sapwood nearly white, tough, moderately strong.

RANGE. Usually in dry, sandy or gravelly soils. Occurs in Louisiana (Tangipahoa Parish), Mississippi, and commonly in Alabama; also northward to New Jersey, Maryland, and Tennessee.

REMARKS. The genus name, *Carya*, is the ancient name for walnut, and the species name, *pallida*, refers to the pale lower surface of the leaves. Vernacular names are Pale Hickory and Brown Hickory. The wood is little used except for posts and fuel. The methods of seed propagation are similar to those of other hickories.

PIGNUT HICKORY

Carya glabra (Mill.) Sweet

FIELD IDENTIFICATION. Tree with a narrow, oblong crown, and somewhat pendulous branches. Known to attain a height of 120 ft. A specimen from near Cross-wicks, New Jersey, has been reported with a circumference of 14 ft 9 in.

FLOWERS. April–May, borne in separate staminate and pistillate catkins on the same tree, staminate catkins 3-branched, 2–2½ in., yellowish green; bracts hairy; stamens 4, anthers yellow; pistillate few-flowered, calyx unequally 4-lobed; stigmas short.

FRUIT. Ripening September–October, variable in size and shape, but usually pear-shaped to ovoid, often with a necklike base, somewhat compressed, about 1¼ in. long and ¾ in. wide; hull tardily dehiscent along 2–4 sutures, or sometimes not at all, about 1/16 in. thick; nut obovate, short-beaked, compressed, not ridged; bony, brownish, thick-shelled; kernel small, usually sweetish but astringent, sometimes bitter.

LEAVES. Alternate, deciduous, 8–12 in. long, odd-pinnately compound of 5 (rarely 3–7) leaflets; leaflets sessile or nearly so, lanceolate-obovate or ovate, acute or acuminate at apex, rounded or cuneate at base, serrate on margin, yellowish green and glabrous above, paler and glabrous or pubescent on veins beneath, terminal leaflet and upper pair larger than lower pairs; rachis and petiole usually glabrous.

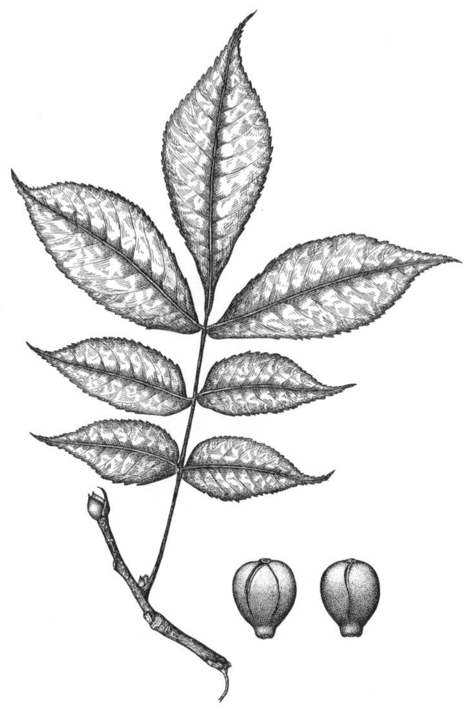

PIGNUT HICKORY
Carya glabra (Mill.) Sweet

TWIGS. Reddish brown, glabrous or nearly so, dotted with pale lenticels.

BARK. Gray, close, ridges narrow and scaly.

WOOD. Brown, sapwood lighter, heavy, hard, strong, tough, elastic, 51 lb per cu ft.

RANGE. Eastern Texas, Louisiana, and Arkansas; eastward to Florida, north to Maine, and west to Ontario and Michigan.

PROPAGATION. The tree has been cultivated since 1750. The cleaned seed numbers about 200 seeds per lb, with a commercial purity of 100 per cent and a soundness of 95 per cent. Preconditioning for planting includes stratification in sand, peat, or loam at 32°–45°F. for 90–120 days (if dry-stored for a year, only 30–60 days' stratification needed). Stratified seeds are planted in spring with 6–8 nuts drilled in linear rows 8–12 in. apart. Mulching with straw or leaves helps keep the seed moist through germination.

VARIETY. A variety has been described as Coast Pignut Hickory, *C. glabra* var. *megacarpa* (Sarg.) Sudw., which has larger leaves, larger fruit (1–2 in. long), thick husk, and thick-shelled nut. It occurs with the species.

The great variations in size and shape of Pignut Hickory fruit have caused a number of other varieties to be described by some botanists, but most have been reduced to the status of synonyms in later treatments.

A species of hickory formerly listed in the literature as Red Hickory, *C. ovalis* Sargent, and which also has numerous variations, has now been combined with *C. glabra* as a synonym. (Little, 1953a, pp. 88–89).

Demaree Hickory, *C.* × *demareei* Palmer, is a hybrid of *C. glabra* and *C. cordiformis,* and is named after Delzie Demaree, botanist of Arkansas, who first collected it.

REMARKS. The genus name, *Carya,* is the ancient name of walnut, and the species name, *glabra,* refers to the smooth character of the leaves and petioles. Vernacular names are Broom Hickory, Switch Hickory, Black Hickory, Red Hickory, White Hickory, Brown Hickory, and Switchbud Hickory. The wood is used for fuel, tool handles, wagons, and agricultural implements. Settlers made brooms from the tough, flexible wood. The tree is slow-growing and hard to transplant.

BATIS FAMILY (*Batidaceae*)

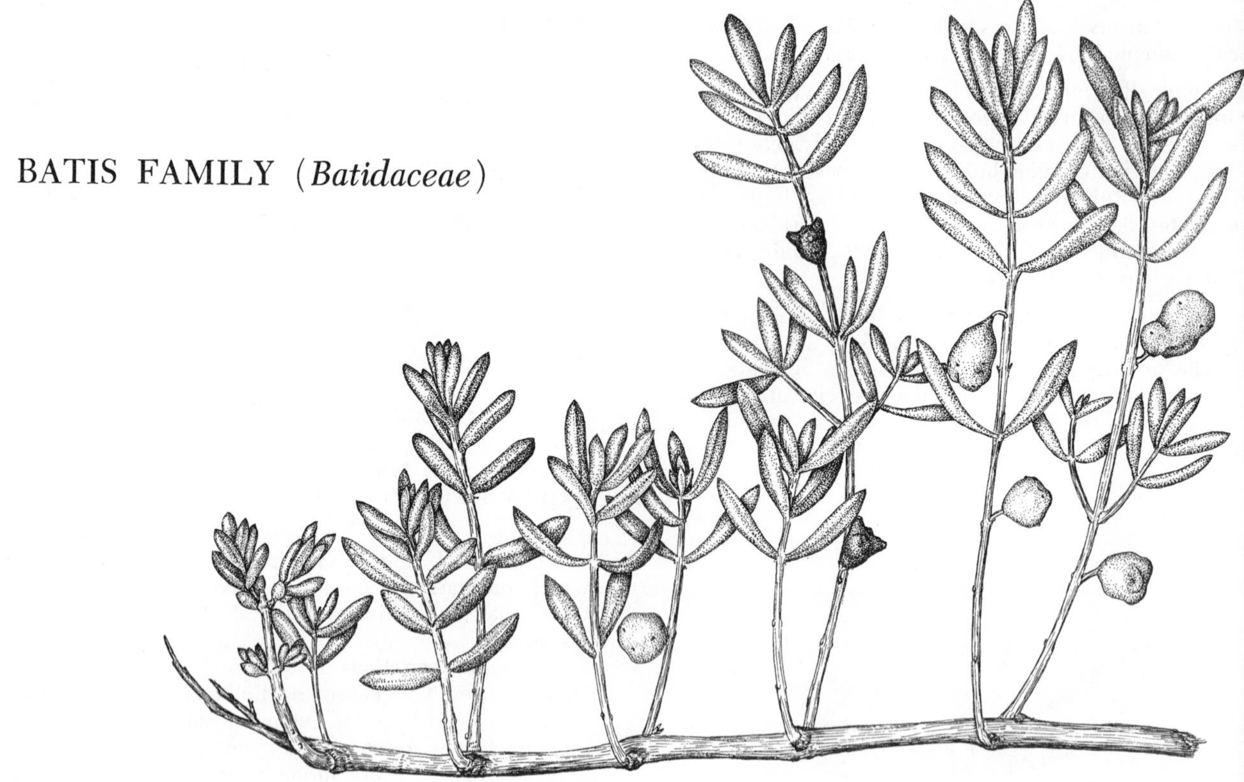

MARITIME SALTWORT *Batis maritima* L.

MARITIME SALTWORT
Batis maritima L.

FIELD IDENTIFICATION. Low prostrate shrub of sandy beaches, mud flats, and salt marshes. Stems usually creeping, but occasionally sending up shoots as much as 6 ft in height.

FLOWERS. Small, in separate staminate and pistillate conelike spikes 1/5–2/5 in.; staminate sessile, ovoid, composed of imbricate scales which have a subtending flower; calyx cup-shaped, 2-lipped; stamens 4–5, alternating with the triangular staminodia, and longer than the staminodia, filaments thick, anthers introrse; pistillate spikes on peduncles 1/12–1/5 in., 4–12-flowered, no calyx or corolla, scales deciduous; ovary 4-celled, with a sessile, spongy, somewhat 2-lobed stigma.

FRUIT. Often abundantly borne, short-stalked, drooping, a fleshy aggregate from the pistillate inflorescence, waxy-green, glabrous, lumpy, oblong, ovoid or obovate, often angular, 3/8–3/4 in. long, about 1/3 in. wide; seeds 4, erect, clavate, falcate, seed coat thin.

LEAVES. Decussate-opposite, sessile, waxy green, glabrous, thick, succulent, half-terete or somewhat boat-shaped, rounded or bluntly keeled below, somewhat flattened above, often falcate, asymmetrical, entire, apex obtuse, acute or minutely mucronulate, base gradually narrowed, 3/8–1 in. long, 1/8–3/16 in. thick.

STEMS. Green or brown, somewhat shreddy, rooting at the nodes, creeping or sending up stems from the nodes.

RANGE. On sandy beaches, mud flats, and saline marshes near the sea. From Texas eastward to Florida and north to North Carolina; also in Mexico, Central and South America, the West Indies, and Hawaii. Collected by the author on Galveston Island, Matagorda Island, and at Corpus Christi, Texas.

REMARKS. The genus name, *Batis*, is from a Greek word meaning "prickly roach," and refers to the fruit. The species name, *maritima*, refers to its habitat close to the sea. Vernacular names are Sea Sapphire, Beachwort, Lechuga de Mar, and Barrilla. The succulent leaves crunch when trodden upon. They have a salty flavor and have been used in salads, in domestic medicine for treating ulcers, and as a diuretic. The ashes have been used in the West Indies in the manufacture of soap and glass.

BIRCH FAMILY (*Betulaceae*)

New Mexican Alder
Alnus oblongifolia Torr.

New Mexican Alder

Alnus oblongifolia Torr.

FIELD IDENTIFICATION. Western shrub, or small tree to 30 ft, and 3–8 in. in diameter, sometimes larger, with erect or spreading branches to form a rounded crown.

FLOWERS. Staminate borne in short, stout-peduncled catkins, yellow-orange or brown, at first about ½ in. long but at maturity 1¾–2½ in.; scales ovate, acute; calyx 4-lobed; stamens 2–3 with reddish yellow anthers; pistillate catkins ⅛–¼ in. during the winter; scales brown, ovate, rounded; stigmas bright red with a 2-celled ovary.

FRUIT. Strobile ½–1 in. long, scales thickened and apex truncate; nutlet solitary, broadly ovoid, border narrow and membranaceous.

LEAVES. Simple, alternate, deciduous, elliptic to oblong or lanceolate, or more rarely ovate, with few to many black glands, apex acute to obtuse or occasionally rounded, base gradually to broadly cuneate, margin sharply and doubly serrate, infrequently shallowly lobed, upper surface yellowish green and glabrous or puberulent, lower surface paler and glabrous to puberulent and also with brown axillary hair tufts; blades 2–3 in. long, ¾–1½ in. wide; petioles slender, pubescent, grooved, ½–¾ in.; stipules about ¼ in., ovate to lanceolate, brown, caducous.

TWIGS. Slender, puberulent at first but glabrous later, shiny, orange-red to reddish brown or gray; leaf scars raised and roughened; winter buds ¼–½ in. long, acute, red, shiny, glabrous.

BARK. Thin, smooth, light brown tinged with red, or grayish brown on older trunks.

RANGE. Along streams in mountains at elevations of 5,000–7,500 ft. In New Mexico in the Magdalena, Mogollon, and Black ranges; west to Arizona, and south into Sonora, Mexico.

REMARKS. The genus name, *Alnus,* is the classical Latin name of the alder, and the species name, *oblongifolia,* refers to the oblong leaves. Vernacular names in use are Mexican Alder and Arizona Alder.

Seaside Alder

Alnus maritima (Marsh.) Muhl.

FIELD IDENTIFICATION. Thicket-forming shrub to 15 or 20 ft, with numerous, slender, spreading stems and a round-topped crown. Or sometimes a small tree to 30 ft and 4–6 in. in diameter.

FLOWERS. Maturing long after the leaves August–September; staminate catkins pendulous, peduncles slender, usually less than 1 in. long, catkin 1½–2½ in. long, width ¼–½ in., pubescent, glandular; bracts ovate, acute, lustrous, ciliate, green to brown, each bract subtending 3 or more flowers; stamens 4, filaments short, orange-colored; pistillate catkins ⅛–⅙ in. long at an-

SEASIDE ALDER
Alnus maritima (Marsh.) Muhl.

thesis, borne on stout peduncles, green to red, ovoid to oblong, solitary, or in clusters of 2–5; bracts ovate, acute, ciliate, each one with 2–3 flowers in its axil; ovary sessile, 2-celled, styles 2, bracts eventually woody, with a deeply erose margin.

FRUIT. Maturing the following year; peduncles stout; ⅝–1 in. long, ½–¾ in. wide, reddish brown to black, ovoid to oblong, apex obtuse or rounded, base rounded or depressed; bracts woody, thickened, lobed and erose at apex; nut oblong-obovoid, apex narrowed and crowned with the persistent styles, margin acute but wingless.

LEAVES. Simple, alternate, deciduous, elliptic to oblong or ovate, occasionally obovate, length of blade 2–3½ in., width ¾–1¼, apex acute to acuminate or rounded, base narrowed or cuneate, margin sharply crenate-serrate, teeth incurved and glandular, surfaces at first pale pubescent; later the upper surface dark green, shiny and glabrous or nearly so; lower surface paler, pubescent to glabrous, sometimes glandular-dotted on the prominent veins; petioles ⅓–1 in. long, slender, yellow to reddish, glandular, or glandless, flattened and grooved above; stipules about ⅛ in. long, oblong, acute, reddish brown, early deciduous.

TWIGS. Slender, greenish orange or reddish brown, eventually gray, puberulent at first, lustrous and glabrous later; lenticels small, yellow to orange, glandular-dotted at first, pith 3-angled; winter buds about ¼ in. long, acute, pale-hairy; bark of trunk brown to gray, smooth, about ⅛ in. thick.

WOOD. Light brown, sapwood slightly paler, close-grained, weighing about 31 lb per cu ft.

RANGE. In wet soil or on granite outcrops. In southern Oklahoma in Johnston and Bryan counties; east to Delaware and eastern Maryland.

REMARKS. The genus name, *Alnus*, is the classical Latin name of the alder. The species name, *maritima*, refers to its occurrence on the coast. Also known as Oklahoma Alder and Brook Alder.

THINLEAF ALDER
Alnus tenuifolia Nutt.

FIELD IDENTIFICATION. Often a shrub with spreading stems from the base to form clumps, or in favorable situations a tree to 30 ft and 6–8 in. in diameter, the slender branches forming a narrow or ovoid rounded head.

FLOWERS. Opening in the spring before the leaves, staminate catkins on stout peduncles about ½ in. long or nearly sessile, 3 or 4 borne together in racemes; catkin body at first purplish and about ¾ in. long, at maturity 2 in. long and ¼ in. thick; bractlets 4–5 under each short-stalked, shield-shaped scale; calyx 3–5-parted, shorter than stamens; stamens 4, filaments short and simple; pistillate catkins ovoid to oblong, usually erect, reddish brown, about ¼ in. long, scales imbricate, acute, apiculate, ovary 2-celled.

FRUIT. Strobiles persistent, 3–9, on peduncles ⅛–⅓ in. long, body obovoid-oblong, ⅓–½ in. long; scales much-thickened and woody, apex 3-lobed and truncate; nutlet solitary, small, flattened, oval to obovoid, circled by a thin, membranous border, seeds averaging about 675,000 per lb.

THINLEAF ALDER
Alnus tenuifolia Nutt.

LEAVES. Simple, alternate, deciduous, ovate to oblong, apex obtuse to acute or short-acuminate, base rounded or cordate, or sometimes suddenly narrowed and cuneate, margin doubly serrate and often cut-lobed as well, young leaves densely hairy or tomentose, mature leaves thin, upper surface dark green and somewhat pubescent or glabrous, lower surface yellowish green and slightly pubescent to glabrous, blades 1¼–4 in. long, 1½–2½ in. wide, veins prominent and impressed above; petioles stout, ½–1 in. long, orange-yellow; stipules ovate to lanceolate, acute, thin, pubescent, about ½ in. long or less, caducous.

TWIGS. Slender, at first with brown pubescence, later more glabrous and brown to reddish brown or grayish; lenticels few, large, scattered, yellow to orange; winter buds ¼–⅓ in. long, enveloped by a pair of stipular scales, reddish and puberulent.

BARK. Rather thin, reddish brown to gray, dark brown on old trunks, scales small and closely appressed.

RANGE. Along creeks and canyons of the pinyon and aspen belts at altitudes of 5,000–10,000 ft. New Mexico, Arizona, and northern Mexico; north to Yukon, and west to Alaska, British Columbia, Washington, Oregon, and California.

VARIETY. A more northern form with larger, more sharply toothed leaves has been segregated under the name of Western Thin-leaf Alder, *A. tenuifolia* var. *occidentalis* (Dipp.) Collier.

REMARKS. The genus name, *Alnus*, is the classical Latin name of the alder, and the species name, *tenuifolia*, means "thin-leaved." Also known under the vernacular names of Mountain Alder, River Alder, and Barana.

Thin-leaf Alder has a limited value for livestock grazing. Because of its clump-forming habit it is of some use on the edge of streams for erosion control. The Indians made a red dye by mixing its powdered bark with ashes of *Juniperus monosperma* and *Cercocarpus montanus*. It also furnishes some food and cover for game.

Hazel Alder

Alnus serrulata (Ait.) Willd.

FIELD IDENTIFICATION. Irregularly shaped shrub or slender tree to 20 ft.

FLOWERS. Borne in separate staminate and pistillate catkins; staminate catkins in clusters of 2–5, 2–4 in. long, cylindric, drooping, bracts subtending the flowers; stamens 3–6; pistillate catkins in clusters of 2–3, about ¼ in. long, green to purple; bracts 3-lobed, each subtending 2–3 pistils.

FRUIT. About ¾ in. long, ovoid, a conelike aggregation of woody bracts, each subtending a nutlet; nutlet small, ovate, flattened, sharp-margined, less than ⅛ in. long.

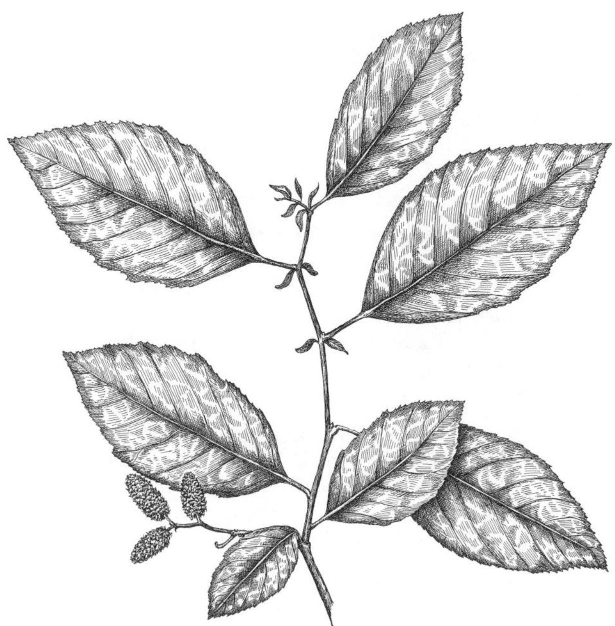

Hazel Alder
Alnus serrulata (Ait.) Willd.

LEAVES. Alternate, simple, deciduous, thick, blades 1–5 in. long, obovate to oval, obtuse or rounded at apex, acute or cuneate at base, sharply serrulate on margin, rather dark green on both sides, veiny and somewhat pubescent or glabrous beneath; petioles glabrous or pubescent, ⅓–½ in. long; stipules oval, deciduous.

TWIGS. Reddish brown or orange, pubescent at first, glabrous later, slender.

BARK. Gray, smooth, thin, slightly roughened with age.

WOOD. Light brown, soft, brittle, weighing 29 lb per cu ft.

RANGE. In wet soil along streams. East Texas, Louisiana, Oklahoma, and Arkansas; east to Florida, north to Maine, and west to Minnesota.

REMARKS. The genus name, *Alnus*, is the classical name of the alder, and the species name, *serrulata*, refers to the finely toothed leaves. This species was formerly known as *A. rugosa* (Du Roi) Spreng., which is a name better applied to *A. incana* (L.) Moench, a European species, as suggested by Fernald, (1945d, pp. 333–361). Vernacular names are Common Alder, Smooth Alder, Tag Alder, Green Alder, Red Alder, Speckled Alder, and American Alder. The bark yields tannic acid and is an astringent. It was formerly used in the treatment of intermittent fever. The fruit is eaten by a number of species of birds. Hazel Alder is sometimes planted to prevent erosion on stream banks.

Water Birch

Betula occidentalis Hook.

FIELD IDENTIFICATION. Shrub with spreading stems and forming clumps, or sometimes a tree to 35 ft high, and 12–14 in. in diameter, the slender branches ascending or drooping to form a broad open crown.

FLOWERS. Staminate catkins bibracteate, in ones to threes, length eventually 1¾–2½ in., about ⅛ in. thick; scales ovate, acute, ciliate, brown; filaments 2, each 2-forked; pistillate catkins short-peduncled, hardly over ¾ in. long; scales acute at apex and greenish; styles 2, bright red, stigma terminal.

FRUIT. In a conelike strobile, erect or pendent, on glandular peduncles ¼–¾ in. long; body cylindric, 1–1½ in. long, ⅕–½ in. thick, puberulent to glabrous; scales 3-lobed, center lobe the longest, ciliate, acute; nutlet ovoid to obovoid, puberulent, wing wider than the seed body, seeds averaging about 1,134,000 per lb.

LEAVES. Simple, alternate, deciduous, thin, broadly ovate to almost orbicular in some, apex acute to acuminate, base rounded or abruptly cuneate, margin with sharp single or double teeth, sometimes shallowly lobed, mature leaves dull green above, lower surface paler and glandular-dotted, blade length 1–2 in., width ¾–1 in., with 3–5 pairs of glandular veins and reticulate veinlets; petioles stout, green to yellow or reddish, glandular-dotted, flattened above, length ⅓–½ in.; stipules ovate, acute, or rounded, margin somewhat ciliate, bright green at first, paler later.

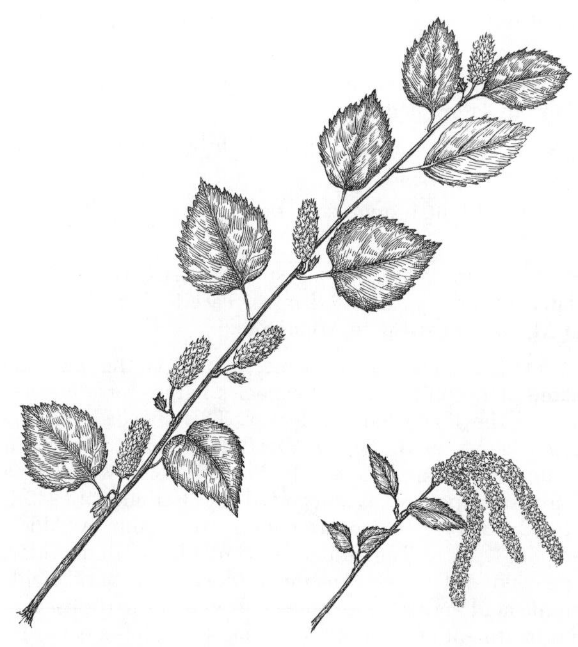

Water Birch
Betula occidentalis Hook.

TWIGS. Slender, green to dark brown, puberulous to glabrous later, shiny, resinous-glandular; winter buds about ¼ in. long, with several scales which are ovoid, acute, brown, resinous.

BARK. Rather thin, gray to reddish brown or dark bronze, lenticels pale brown, horizontal, becoming long and conspicuous on old trunks.

WOOD. Heart light brown, sapwood thick and lighter colored, soft but rather strong.

RANGE. In moist canyons of the pinyon, yellow pine, and spruce belts at altitudes of 5,000–9,000 ft. New Mexico and Arizona; north to North Dakota, and west to Washington, Oregon, and California.

RELATED SPECIES AND VARIETY. The plant was formerly listed as *B. fontinalis* Sarg., but it has been shown that *B. occidentalis* Hook. is the correct name (Fernald, 1945c, pp. 312–317).

A variety, described as the Fertile Water Birch, *B. occidentalis* var. *fecunda* Fern., becomes a tree to 60 ft and 18 in. in diameter; strobiles longer and narrower with lateral lobes of bracts broad, spreading, and auricled at base. It was formerly listed under the name of *B. fontinalis* var. *piperi* (Britt.) Sarg., and occurs in Montana and eastern Washington.

REMARKS. The genus name, *Betula,* is the classical name of the birch, and the species name, *occidentalis,* means "western," and refers to its westerly distribution in North America. The Water Birch is also known under the vernacular names of Black Birch and Red Birch. The wood is used for fuel and fencing, and the foliage is fair-to-good browse for sheep and goats.

River Birch

Betula nigra L.

FIELD IDENTIFICATION. Tree commonly of wet ground to 90 ft high, with dull reddish brown bark peeling off in curly thin flakes.

FLOWERS. Staminate catkins clustered, sessile, 1–3½ in. long; scales brown and shining; stamens 2, bifid; pistillate catkins cylindric, about ½ in. long; scales ovate, green, pubescent, ciliate; ovary sessile, with 2 spreading styles.

FRUIT. Strobile oblong-cylindric, about 1½ in. long and ½ in. thick, erect, borne on peduncles ½ in. long; scales 3-lobed; nutlet about ⅛ in. long, ovoid-obovate with a thin reniform wing, ripening April–June.

LEAVES. Simple, alternate, deciduous, rhombic-ovate, acute, cuneate at base, doubly serrate or some lobed with double serrations, blades 1½–3½ in. long, 1–2 in. wide, lustrous dark green above, tomentose beneath; petioles tomentose, slender, averaging about ½ in. long.

BARK. Reddish brown or grayish, marked by darker

River Birch
Betula nigra L.

elongate lenticels, peeling into conspicuous papery strips.

WOOD. Light brown, hard, strong, close-grained, weighing 36 lb per cu ft.

RANGE. Texas, Oklahoma, Arkansas, and Louisiana; north to Massachusetts and west to Minnesota and Kansas.

PROPAGATION. Birch seed is gathered into bags by picking the strobiles, and then broken up by flailing and screening or fanning. The seed are very small, averaging 374,000 seeds per lb, with a commercial purity of 58 per cent and a soundness of 42 per cent. The average germination capacity is about 34 per cent. The seeds may be sown in spring or early summer without treatment. They can be sown broadcast and covered about 1/16 in. with nursery soil and kept moist.

REMARKS. *Betula* is the ancient classical name of the birch, and the species name, *nigra,* means "black," but no reason can be found for the name. Vernacular names are Red Birch, Water Birch, and Black Birch. The wood is used for furniture, woodenware, wagon hubs, and fuel. The seeds are sometimes eaten by birds and the foliage is browsed by white-tailed deer. Sometimes cultivated as an ornamental along streams or ponds; and is also used for erosion control. It has been cultivated since 1736.

American Hornbeam
Carpinus caroliniana Walt.

FIELD IDENTIFICATION. Small crooked tree attaining a height of 35 ft, with a fluted gray trunk and pendulous branches.

FLOWERS. April–June, staminate and pistillate flowers separate but on same tree; staminate flowers green, borne in linear-cylindric catkins, 1–1½ in.; scales of catkin triangular-ovate, acute, green below, red above; stamens numerous, filaments short and 2-cleft; pistillate catkins about ½ in.; flowers with hastate bracts which develop into a 3-lobed green involucre; styles slender, stigmas 2.

FRUIT. August–October, nutlet about ⅓ in. long, ovoid, acute, nerved, borne at base of a 3-lobed foliaceous bract, many together forming loose pendent clusters 3–6 in. long. The middle lobe of the bract is lanceolate and entire or dentate, and much longer than the lateral lobes which are usually incised-dentate on one side.

LEAVES. Simple, alternate, deciduous, sometimes falcate, acute or acuminate at apex; base rounded, wedge-shaped, or heart-shaped, often somewhat inequilateral; margin sharply double-serrate, teeth glandular except at base; dull bluish green and glabrous above, paler and hairy in axils of the veins below; petioles about ⅓ in. long, slender, hairy; stipules ovate-lanceolate, hairy, reddish green.

American Hornbeam
Carpinus caroliniana Walt.

TWIGS. Slender, zigzag, gray or red.

BARK. Smooth, tight, thin, bluish gray, sometimes blotched with darker or lighter gray (some gray blotches may be due to crustose lichens), trunk fluted into muscle-like separations.

WOOD. Light brown, sapwood lighter, strong, hard, tough, heavy, close-grained, weighing about 45 lb per cu ft.

RANGE. East Texas, Oklahoma, Arkansas, and Louisiana; east to Florida, north to Virginia, and west to Illinois.

PROPAGATION. Good crops of seed are produced every 3–5 years, with light intervening years. Fruit is handpicked before shattering and placed in a stirring machine, or beater, to separate the seed from the bracts. Clean seed averages about 30,000 seeds per lb, with a commercial purity of 96 per cent. The germination rate is rather low for stored seed, but if collected slightly green and planted in August, germination rates are high. The seeds should be drilled in rich, moist loamy soil for best results.

VARIETY. The above description applies to the Southern form which is distinguished from the Northern variety, *C. caroliniana* var. *virginiana* (Marsh.) Fern., by narrower and smaller leaves with shorter teeth, and by the bracts which have fewer and blunter teeth or none. However, intermediate forms seem to occur between the Northern and Southern forms.

REMARKS. The genus name, *Carpinus*, is the classical name for hornbeam, and the species name, *caroliniana*, refers to the states of Carolina. Vernacular names are Blue-beech, Water-beech, Lean-tree, and Ironwood. The wood is used for golf clubs, handles, fuel, mallets, cogs, levers, and wedges. The seed is eaten by at least nine species of birds. The tree has been cultivated since 1812.

Knowlton Hop-hornbeam

Ostrya knowltonii Coville

FIELD IDENTIFICATION. Western tree to 30 ft and 18 in. in diameter. Trunks usually short and divided into a number of slender, crooked branches to form a round-topped crown.

FLOWERS. Usually before the leaves in spring, staminate catkins 1–3 together at the tips of the branches, slender, cylindrical, pendulous, ½–1¼ in. long, yellowish green to reddish or brown; bracts ovate, concave, several stamens in the axil of each bract; pistillate catkins about ¼ in. long; bracts yellowish green, puberulous, ciliate; generally with 2 flowers in the axil of each bract; ovary 2-ovuled, apex with 2 slender stigmas and inclosed in a tubular involucre.

FRUIT. Peduncles about ½ in. long, the inflated involucres loosely imbricate to form a strobile resembling a hop, clusters about ¾ in. broad and 1½ in. long; nutlets about ¼ in. long, solitary, sessile at base of bract, their involucres puberulent or glabrous, green or reddish.

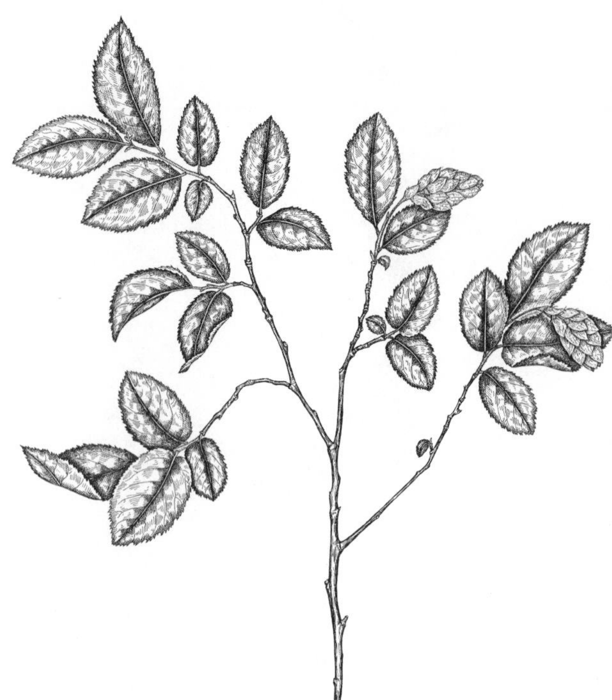

Knowlton Hop-hornbeam
Ostrya knowltonii Coville

LEAVES. Simple, alternate, deciduous, concave in the bud, blades 1–2 in. long, 1–1½ in. wide, ovate to obovate or elliptic, apex acute, obtuse or occasionally rounded, base cuneate to rounded or cordate, and sometimes unequal, margin sharply and doubly serrate; upper surface at maturity dark yellowish green, pubescent to glabrous; lower surface paler and pubescent, main vein and primary veins slender; petiole ⅛–¼ in. long; stipules ⅛–¼ in. long, green to reddish, deciduous.

TWIGS. Slender, tomentose at first to glabrous and lustrous later, dark green to reddish brown or gray; buds reddish brown, ovoid, about ⅛ in. long.

BARK. Light gray to reddish, about ⅛ in. thick, shallowly furrowed and breaking into loose, small scales 1–2 in. long.

WOOD. Light reddish brown, sapwood thin, hard, tough, durable, used for fuel or posts occasionally.

RANGE. Usually in canyons of the pinyon belt at altitudes of 5,000–7,000 ft. Trans-Pecos Texas in the Guadalupe, Davis, and Chisos mountains; west through New Mexico to Arizona, and north to Utah.

PROPAGATION. Unless seeds are sown soon after maturity or are stratified, they will not germinate until the second year. Rarer kinds may be grafted.

REMARKS. The genus name, *Ostrya*, is from the Greek word *ostrua*, designating a tree with very hard wood. The species name, *knowltonii*, is in honor of Frank Hall Knowlton (1860–1926), American botanist. It is sometimes listed under the name of *O. baileyi* Rose. Some vernacular names are Western Hornbeam and Ironwood.

Woolly American Hop-hornbeam

Ostrya virginiana (Mill.) K. Koch. var. *lasia* Fern.

FIELD IDENTIFICATION. Tree with grayish brown bark, attaining a height of 60 ft.

FLOWERS. Monoecious, staminate catkins 1–3 at ends of branches, 1½–3 in. long; scales triangular-ovate, acuminate, nerved, ciliate, green to red; stamens 3–14, filaments short and forked, anthers villous; pistillate catkins small, usually solitary, slender, about ¼ in. long; scales lanceolate, acute, ciliate, hirsute and red above, developing into pubescent-nerved bladdery sacs; ovary 2-celled.

FRUIT. In conelike imbricate clusters 1½–2 in. long; peduncles hairy, about 1 in. long; each papery sac about ¾ in. long, ⅔–1 in. wide, ellipsoid, strongly tomentose at apex; nuts enclosed in the sac, small, ovoid, brown, faintly ribbed, about ¼ in. long.

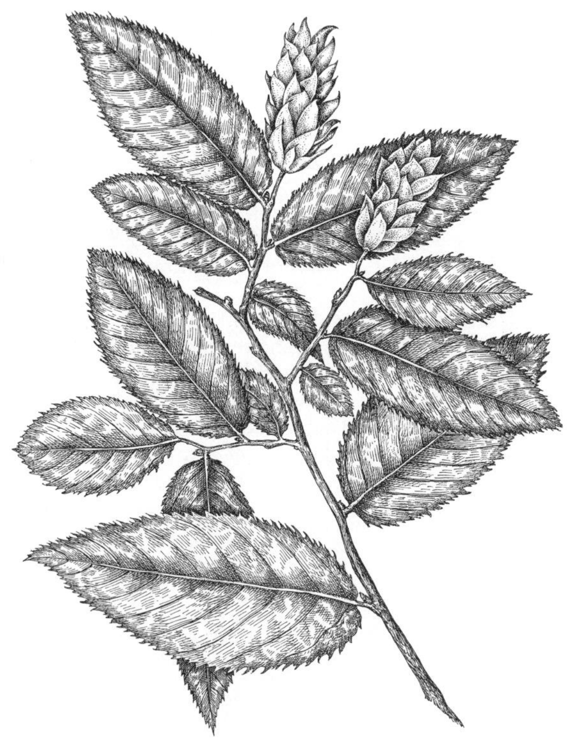

Woolly American Hop-hornbeam
Ostrya virginiana (Mill.) K. Koch var. *lasia* Fern.

LEAVES. Simple, alternate, deciduous, blades 2½–4½ in. long, 1½–2½ in. wide, ovate, or oblong-lanceolate, apex acute or acuminate, base rounded, heart-shaped, or wedge-shaped, often inequilateral, margin sharply and doubly serrate, glabrous and yellowish green above, hairy and paler below, turning yellow in autumn; petioles about ¼ in. long, hairy; stipules acute, rounded, ciliate, hairy, about ½ in. long.

TWIGS. Terete, crooked, slender, yellow-orange to brown, pubescent.

BARK. Grayish brown, broken into small, narrow, oblong, shreddy scales.

WOOD. Hard, strong, tough, close-grained, durable, light brown to white, sapwood lighter, weighing 51 lb per cu ft.

RANGE. In rich, moist woods. East Texas, Oklahoma, Arkansas, and Louisiana; eastward to Florida; and the Northern variety to Ontario, west to Minnesota and Nebraska. The exact overlapping distribution of the Southern and Northern varieties is not known.

PROPAGATION. The seed clusters are picked by hand before they shatter. The seeds may be separated from the involucre by beating and fanning. The cleaned seed averages about 30,000 seeds per lb. Commercial seed average 97 per cent purity and 80 per cent soundness. The seeds may be sown in either fall or spring. Seed can be planted immediately in fall in mulched beds using straw or leaves. For spring sowing the seed may be stratified over winter in sand or peat, at 41°F. Planting is usually done at a depth of ¼ in. in well-drained loamy soil. There is a potential germination of 85 to 90 per cent.

REMARKS. The genus name, *Ostrya*, is the ancient Greek name, and the species name, *virginiana*, refers to the state of Virginia. Vernacular names are Ironwood, Leverwood, Deerwood, Hardhock, and Indian-cedar. The wood is used for posts, golf clubs, tool handles, mallets, and woodenware. The fruit is eaten by at least 5 species of birds. The tree is rather slow-growing but has possibilities as an ornamental; has been cultivated since 1690.

The east Texas form appears to be the Southern variety, *lasia*, recently described by Merritt Lyndon Fernald, and varying from the Northern variety mostly by pubescence of leaves, petioles, and twigs. The Northern variety *O. virginiana* var. *glandulosa*, has glandular hairs on young parts.

American Filbert (Hazelnut)

Corylus americana Walt.

FIELD IDENTIFICATION. Thicket-forming, spreading shrub, attaining a height of 3–10 ft.

FLOWERS. March–April, monoecious, staminate catkins mostly solitary, 3–4 in. long, elongate, cylindric;

scales and bractlets small, subtending a single naked flower, calyx absent; stamens 4, filaments deeply divided, each division bearing half an anther, sac villous above; pistillate catkins clustered at the ends of short branches of the season, each in the axil of a bract, the imbricate bracts concealing all except the protruding stigmas, 2 bractlets becoming greatly enlarged at maturity and enclosing the nut; ovary inferior, incompletely 2-celled, adnate to a minute calyx, with a short stipe and the 2 stigmas slender.

FRUIT. Maturing July–August, in clusters of 2–6, rarely solitary, involucral bracts foliaceous, large, reddish brown, pubescent, or sometimes basally glandular, compressed, striate, apex laciniately cut into triangular lobes, extending beyond and closely investing the nut, ½–1½ in. long; nut ⅜–⅝ in. long, ovoid to subglobose, usually wider than long, light brown, sweet, edible.

LEAVES. Simple, alternate, deciduous, ovate to oval or obovate, apex acute or abruptly acuminate, base broadly rounded or subcordate, margin finely double-serrate, thin, upper surface dark green, somewhat rough and glabrous or nearly so, lower surface paler and finely tomentose to more or less pubescent, length 3–6 in., width 2–4½ in.; petiole ⅙–⅓ in. long, often with stalked, reddish, glandular hairs, or long white hairs.

AMERICAN FILBERT (HAZELNUT)
Corylus americana Walt.

TWIGS. Slender, strongly ascending, reddish brown at first with stipitate, glandular hairs, gray and glabrous later.

RANGE. Hedgerows, thickets, and woods. Oklahoma and Arkansas; east to Georgia and Florida, north to Maine, and west to Ontario, Minnesota, and Kansas.

PROPAGATION. Good crops are borne every 2–3 years. The fruit should be gathered when half dry, and drying completed by spreading out. The seed is separated from the involucre by flailing. One hundred lb of fruit yields 25–30 lb of seed. The cleaned seeds average 250–400 per lb, with a commercial purity of 96 per cent and a soundness of 87 per cent. Because of a dormant embryo, pretreatment is needed to hasten germination. Stratification is practiced for 60 days at 41°F., plus 67 days at 65°F., plus 30 days at 41°F. The stratified seed is sown in drills in spring in rich, well-drained soil. Germination is about 80 per cent, with 60 usable plants per lb of seed. Propagation may also be practiced by suckers or layers, in fall or spring. Budding and grafting are also practiced. Cuttings of mature wood taken in fall may be held over winter in sand, and planted in spring. The tree coppices freely after cuttings.

VARIETIES AND FORMS. Several varieties and forms of American Filbert are known:

Indehiscent American Filbert, *C. americana* var. *indehiscens* Palmer & Steyermark, has the fruiting involucral bracts united on one side.

Missouri American Filbert, *C. americana* forma *missouriensis* (DC.) Fern., is a form with no stipitate glands on foliage or twigs.

REMARKS. The genus name, *Corylus*, is the classical Greek name, which probably originated from the hood-like or helmet-like appearance of the involucre. The species name, *americana,* denotes its origin. The name Filbert is of French origin, the European Filbert becoming ripe about the time of St. Philbert's Day (August 22).

American Filbert has been cultivated for ornament since 1798 and is used for shelter-belt planting and as wildlife food and cover. It is known to be eaten by bobwhite quail, ruffed grouse, ring-necked pheasant, blue jays, squirrels and white-tailed deer; and is sometimes grazed by sheep. The sweet nuts are sold on the market and may be eaten raw, or ground and made into a cakelike bread.

BEECH FAMILY (*Fagaceae*)

Overcup Oak
Quercus lyrata Walt.

Overcup Oak
Quercus lyrata Walt.

FIELD IDENTIFICATION. Tree attaining a height of 100 ft, and a diameter of 2–3 ft. Branches small, crooked, often drooping, forming an open, irregular head. When growing in swamps the base often buttressed.

FLOWERS. March–April, staminate and pistillate catkins on the same tree; staminate clusters loosely-flowered, 3–6 in. long, slender, hairy; calyx yellow, hairy, irregularly lobed, lobes acute; pistillate catkins sessile, or on peduncles ⅛–¾ in., solitary or a few together; peduncles and bracts tomentose; style short, stigmas recurved.

FRUIT. Acorn annual, sessile or short-peduncled, solitary or in pairs, ovoid or depressed-globose, base broadly flattened, chestnut-brown, ½–1 in. high, about 1 in. broad, nearly or quite covered by the cup; cup hemispheric or spheroid, depressed, ⅗–1⅕ in. broad, ⅖–⅘ in. deep, tomentose, thin, ragged, often splitting at the apex, thicker toward the base; scales reddish brown, ovate, acute, lower coarse and thickened, marginal ones small, thin, and appressed.

LEAVES. Alternate, simple, deciduous, thin, obovate-oblong, blades 3–10 in. long, 1–4 in. wide, apex acute to acuminate or rounded, base cuneate or attenuate, margin with lobes very variable, usually 7–9, rounded, acute or acuminate, middle lobes generally broadest, sinuses deep and broadly rounded; upper surface dark green, glabrous and lustrous; lower surface white-tomentose or glabrate later; petioles ⅓–1 in., glabrous or pubescent.

TWIGS. Slender, green and pubescent at first, later grayish brown and glabrous; buds about ⅛ in. long, ovoid, obtuse, chestnut-brown; stipules deciduous and subulate.

BARK. Gray to brown or reddish, broken into irregular ridges with thin flattened scales.

WOOD. Dark brown, sapwood lighter, durable, hard, strong, tough, close-grained, weighing about 51 lb per cu ft.

RANGE. On wet, poorly drained clay soils. East Texas, Oklahoma, and Arkansas; east to Florida, north to New Jersey, and west to Missouri.

HYBRIDS. A number of hybrids have been listed as follows:

Compton Oak, *Q.* × *comptoniae* Sarg., is a cross between *Q. lyrata* and *Q. virginiana.* It is named for (Miss) C. C. Compton of Natchez, Mississippi, who aided C. S. Sargent in field study of this hybrid. It occurs in southeastern Virginia, Alabama, Mississippi, Louisiana, and southeast Texas.

Mounds Oak, *Q.* × *humidicola* Palmer, is a cross of *Q. bicolor* and *Q. lyrata;* the name refers to its moisture-loving habitat. It occurs in Illinois and Missouri.

Sterrett Oak, *Q.* × *sterrettii* Trel., is a hybrid of *Q. stellata* and *Q. lyrata,* named for its discoverer, William Dent Sterrett, forester with the U. S. Forest Service, and is found in Arkansas.

REMARKS. The genus name, *Quercus,* is the ancient

classical name, and the species name, *lyrata*, refers to the lyrate-pinnatifid leaves. Other vernacular names are Water White Oak, Swamp White Oak, Swamp Post Oak, and White Oak. The wood is used for the same purpose as the true White Oak, *Q. alba*. The trees are slow-growing, long-lived, generally free from insects and disease, and resistant to disease. Overcup Oak was first introduced into cultivation about 1786. The young plants are browsed by deer and cattle.

Bur Oak

Quercus macrocarpa Michx.

FIELD IDENTIFICATION. Tree to 150 ft, with heavy spreading limbs and a broad crown.

FLOWERS. Borne in staminate and pistillate catkins; staminate catkins 4–6 in. long, yellowish green; calyx deeply 4–6-lobed, hairy; pistillate flowers sessile or nearly so, solitary or a few together; involucral scales ovate, red, tomentose; stigmas red.

FRUIT. Acorn very large, variable in size and shape, sessile or short-stalked, solitary or paired; ¾–2 in. long, ellipsoid to ovoid, apex pubescent; cup subglobose or hemispheric, thick, woody, tomentose, enclosing one half to three fourths of nut; scales imbricate, thick, upper scales with awnlike tips to produce a fringed border on the cup, giving a mossy appearance.

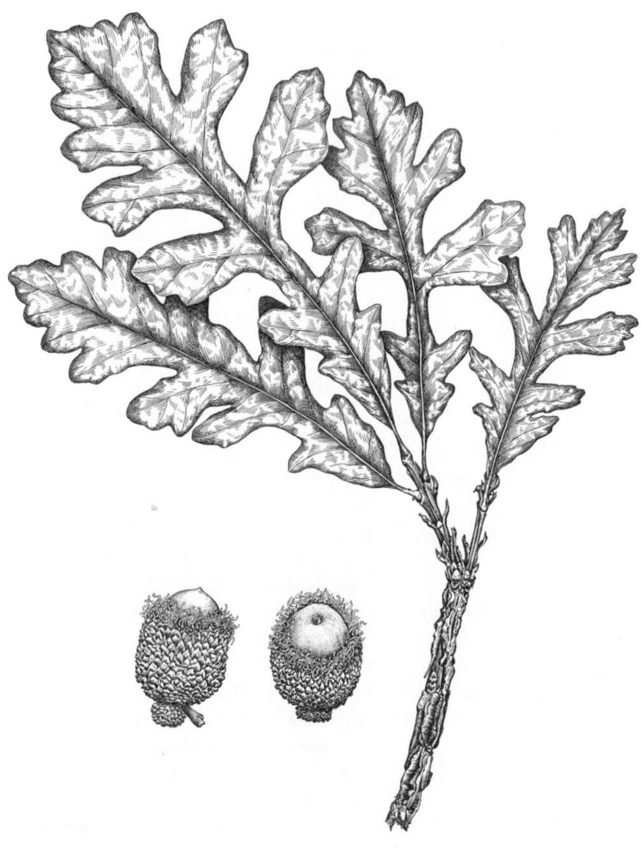

Bur Oak
Quercus macrocarpa Michx.

LEAVES. Simple, alternate, deciduous, obovate-oblong, 5–9-lobed, lobes separated by very deep sinuses; terminal lobe usually largest and obovate with smaller lobes or coarse teeth, apex usually rounded; base cuneate from smaller lobes; blades 6–12 in. long, 3–6 in. wide; dark green, lustrous and glabrous above, paler and pubescent beneath; petioles stout, pubescent, ⅓–1 in.

TWIGS. Light brown and pubescent, later becoming dark brown and glabrous, and sometimes with corky ridges; terminal buds reddish brown, pubescent, ovoid, obtuse, about ¼ in. long.

BARK. Light gray, or reddish brown, thick, deeply fissured and broken into irregular narrow flakes.

WOOD. Dark or light brown, close-grained, heavy, hard, strong, tough, durable, weighing about 46 lb per cu ft.

RANGE. Not apparently at home on the Atlantic and Gulf Coast plains, but on higher grounds. Central and east Texas, Oklahoma, Arkansas, and Louisiana; east to Georgia, north to Nova Scotia, and west to Manitoba, Kansas, and Wyoming.

PROPAGATION. Bur Oak has been cultivated since 1811. The flowers appear April–May, and the fruit ripens August–September. The minimum commercial seed-bearing age is 35 years, the maximum 400 years. Good seed crops are produced every 2–3 years. The cleaned seed averages about 75 seeds per lb, with a purity of 86 per cent and a soundness of 85 per cent. The seed may be stored in cool cellars or in dry sealed containers at 32°–36°F., although it is best not to let the acorns dry. Immediate planting is best or stratification at 41°F. for 30–60 days for spring sowing. The average germination capacity is about 45 per cent.

HYBRIDS. A number of hybrids of Bur Oak have been listed by Little (1953a, p. 339):

Bebb Oak, *Q.* × *bebbiana* Schneid., is a hybrid of *Q. macrocarpa* and *Q. alba*, and is named for its discoverer Michael Schuck Bebb (1833–1895), American botanist. It occurs in Quebec, Vermont, Indiana, Illinois, and Missouri.

Covington Oak, *Q.* × *byarsii* Sudw., is a cross of *Q. macrocarpa* and *Q. michauxii*, and was named for James Byars, who discovered it in 1888. It occurs in Tennessee (Covington, Tipton County).

Deam Oak, *Q.* × *deamii* Trel., is a cross of *Q. macrocarpa* and *Q. muehlenbergii*, named after its discoverer, Charles Clemon Deam, botanist of Indiana. It occurs in Indiana (Wells County).

Schuette Oak, *Q.* × *schuettei* Trel., is a hybrid of *Q. macrocarpa* and *Q. bicolor*, and is named for J. H. Schuette, of Green Bay, Wisconsin, who collected in

1893. It occurs in southern Quebec, New York, Michigan, Wisconsin, Minnesota, and Indiana.

REMARKS. The genus name, *Quercus*, is the ancient classical name, and the species name, *macrocarpa*, refers to the large acorn. Vernacular names are Mossycup Oak and Overcup Oak. The wood is similar to that of White Oak and is used for baskets, lumber, ties, fences, cabinets, ships, and fuel. The acorns are greedily eaten by squirrels and white-tailed deer, and young plants are browsed by livestock. A narrow-leaved and small-fruited variety of Bur Oak has been given the name of *Q. macrocarpa* var. *olivaeformis* Gray. A dwarf form of Bur Oak known as Bur Scrub Oak, *Q. macrocarpa* var. *depressa* (Nutt.) Engelm., is found in Minnesota, South Dakota, and Nebraska and has acorn cups about ⅗ in. wide, slightly fringed, innermost scales caudate-attenuate; acorns ovoid, about ⅗ in. long. It is usually 3–8 ft high with corky branches. However, many intermediate forms occur between it and the species.

WHITE OAK
Quercus alba L.

WHITE OAK

Quercus alba L.

FIELD IDENTIFICATION. Large tree to 150 ft, with a broad open head.

FLOWERS. Appearing with the leaves April–May, staminate and pistillate on the same tree; staminate catkins solitary, hairy, about 3 in. long; calyx yellow, pubescent, with acute lobes; stamens 6–8; pistillate catkins usually solitary, 2–3-flowered, about ½ in. long, red; involucral scales hairy, ovate; calyx-lobes acute, ovate; styles erect, short.

FRUIT. Ripening September–October. Acorn sessile or short-stalked, solitary or in pairs; nut ellipsoid-ovoid, light brown, lustrous, ¾–1 in. long, enclosed to one fourth its length in cup; cup bowl-shaped, scales woody-tuberculate, thickened, somewhat fused, closely appressed, acorn maturing the first season. Minimum commercial seed-bearing age 30 years, optimum 50–100, maximum 150. Good crops about every 3 years, with light crops intervening.

LEAVES. Alternate, simple, deciduous, oblong-obovate, 5–9 in. long, 7–11-lobed; lobes oblique, rounded, elongate, the terminal lobe usually shallowed, 3-parted; leaf base cuneate; bright green and glabrous above, paler or glaucous below.

TWIGS. Slender, reddish brown to gray, glabrous; terminal buds about 3/16 in. long, subglobose, glabrous, brown; leaf scars half-moon–shaped, pith stellate in cross section.

BARK. Light gray, or reddish brown beneath the flat loose ridges, which are separated by shallow fissures.

WOOD. Light brown, hard, strong, heavy, close-grained, durable, weighing about 46 lb per cu ft.

RANGE. On bottom lands, rich uplands and on gravelly ridges. East Texas, Oklahoma, Arkansas, and Louisiana, east to Florida, north to Maine, Ontario, and Minnesota, and west to Nebraska.

PROPAGATION. One hundred lb of acorns yield 60–90 lb of usable seed. In cleaned seed the average number of seeds per lb is 150, with a commercial purity of 96–100 per cent and a soundness of 92 per cent. White Oak acorns should be gathered and stored immediately after falling to retard early germination. They may be stored in dry sealed containers at 32°–36°F., or in cool humid cellars for shorter periods. Seeds of the White Oak lose vitality in storage sooner than those of Red Oak. No pretreatment is required to break dormancy. However, fall planting is best for the White Oak generally. Planting is done in drills 8–12 in. apart, and covered ½–1 in. with soil which is kept moist until germination. The bed should be mulched with leaves or straw held in place by covers for rodent protection. Germination of White Oak acorns averages about 66 per cent. Direct seeding is highly desirable, instead of use of nursery stock; providing rodent control is practiced. Diseases of seedlings include collar rot which may be controlled by treating each sq ft of bed with a solution of 1½ fluid oz of formaldehyde in 2 pt of water. Powdery mildew may be controlled by dusting with sulphur.

FORMS. White Oak occurs in several forms, such as:

Louisiana White Oak, *Q. alba* forma *repanda* (Michx.) Rehd., differs from the species by shallow leaf sinuses and short-stalked fruit.

Forked White Oak, *Q. alba* forma *pinnatifida* (Michx.) Rehd., has leaves deeply pinnatifid, with narrow, often deeply toothed lobes, and a slender peduncled acorn.

HYBRIDS. A number of hybrids of White Oak follow:

Beadle Oak, *Q.* × *beadlei* Trel., is a hybrid of *Q. alba* and *Q. michauxii*. It is known from New Jersey, Delaware, Indiana, Virginia, North Carolina, Florida, and eastern Texas. Named after Chauncey Delos Beadle, American botanist.

Bebb Oak, *Q.* × *bebbiana* Schneid., is a hybrid of *Q. alba* and *Q. macrocarpa*. Known from Missouri, Illinois, Indiana, Ohio, and Vermont. Named after Michael Schuck Bebb (1833–1895).

Faxon Oak, *Q.* × *faxonii* Trel., is a hybrid of *Q. alba* and *Q. prinoides*. Known from Michigan, New York, and Massachusetts. Named after Charles Edward Faxon (1846–1918), American botanical artist.

Fernow Oak, *Q.* × *fernowii* Trel., is a hybrid of *Q. alba* and *Q. stellata*. Known from East Texas and Alabama; north to Missouri, Illinois, Indiana, Virginia, District of Columbia, Maryland, Delaware, and New Jersey. Named after Bernhard Eduard Fernow (1851–1923), at one time chief of the U. S. Forest Service.

Jack Oak, *Q.* × *jackiana* Schneid., is a hybrid of *Q. alba* and *Q. bicolor*. It is found in Illinois, Indiana, New York, Rhode Island, Massachusetts, and southern Quebec. Named for John George Jack (1861–1949), American dendrologist.

Saul Oak, *Q.* × *saulii* Schneid., is a hybrid of *Q. alba* and *Q. prinus*. It is found from Maryland, District of Columbia, Virginia, North Carolina, Alabama, Kentucky, Ohio, Pennsylvania, New Jersey, Rhode Island, Massachusetts, and New York. Named for John Saul (1819–1897), nurseryman of Washington, D. C.

REMARKS. The genus name, *Quercus*, is the classical name, and the species name, *alba*, refers to the white bark. Vernacular names are Stave Oak, Fork-leaf White Oak, and Ridge White Oak. The wood is used for fuel, ties, baskets, cabinets, barrels, tools, furniture, and construction work. The dried, powdered inner bark of this and other oaks has some medicinal value because of the quercitanic acid it contains. It is used almost solely as an astringent wash, or occasionally as an injection in leucorrhea or hemorrhoids. Indians ground the acorns into meal and poured water through it to leach out the tannin before baking into bread. Squirrel, white-tailed deer, wild turkey, and bobwhite quail eat the acorns, and livestock browse the foliage. White Oak is very desirable for park and street planting and is rather free of insect pests, but is somewhat difficult to transplant. It has been in cultivation since 1724.

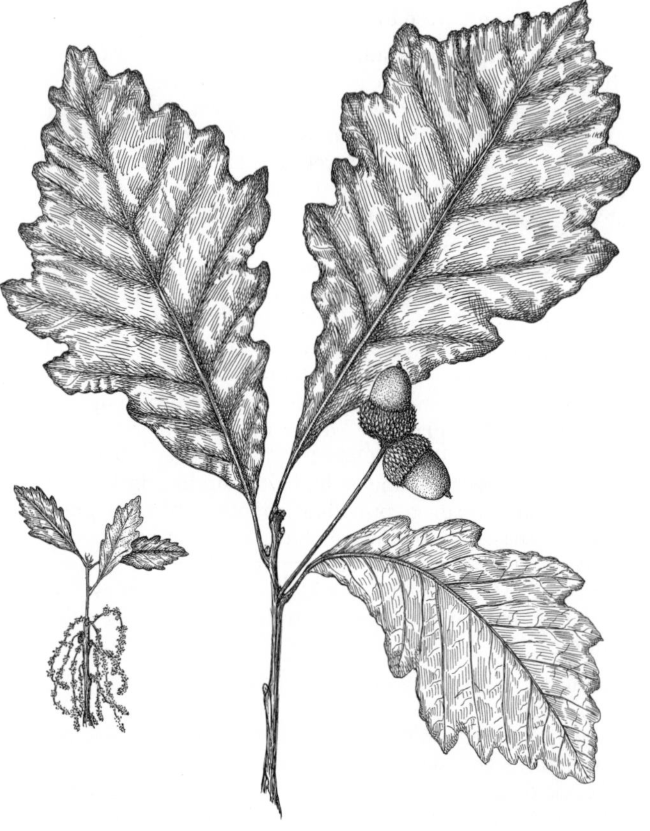

SWAMP WHITE OAK
Quercus bicolor Willd.

SWAMP WHITE OAK
Quercus bicolor Willd.

FIELD IDENTIFICATION. Large, majestic, narrow, round-topped tree attaining a height of 110 ft, and a diameter of 8 ft. Lower branches rather numerous and pendulous.

FLOWERS. In separate staminate and pistillate catkins, appearing after the leaves are about one third grown; staminate catkins 2–4 in. long, hairy; calyx 5–9-lobed, lobes lanceolate, acute, yellowish green, hairy; pistillate catkins few-flowered, borne on hairy peduncles ¾–3½ in., much longer than the leaf petioles; stigmas short, spreading, red.

FRUIT. Maturing annually September–October, borne in clusters of 1–3 (usually 2) on slender, dark brown peduncles two to six times as long as the leaf petioles; acorn oblong to ovoid, base wide, apex rounded to acute and pubescent, length ¾–1½ in., width ½–¾ in., color chestnut brown, meat rather sweet; cup hemispheric to bowl-shaped, covering one third to one half of acorn, ½–¾ in. high, ½–1 in. wide, light brown and tomentose; scales appressed, often tuberculate, lower scales more ovate and obtuse, upper lanceolate, acute and acuminate, sometimes forming a short fringe on the border.

LEAVES. Alternate, simple, deciduous, when young bronze green above and silvery-tomentose beneath; mature leaves varying from obovate to oblong or elliptic, margin coarsely sinuate-toothed or sometimes lobed nearly to the center, slightly revolute or flattened, apex obtuse or rounded, base cuneate or rounded, firm and leathery, upper surface dull green and glabrous, lower surface densely white or tawny tomentose, blade length 4–7 in., width 3½–4½ in.; petiole pubescent or glabrous, ¼–¾ in.

TWIGS. Rather stout, when young green, lustrous and pubescent, becoming orange-brown, reddish brown to light brown, and glabrous.

BARK. Brownish on young trunks, changing to various shades of gray to dark brown on older trunks, fissures deep, ridges flattened and loosely curling back at the ends to give a rough appearance.

WOOD. Light to dark brown, sapwood lighter, close-grained, straight-grained, strong, hard, tough, weighing about 48 lb per cu ft, very durable on contact with the soil, equal to the true White Oak, *Q. alba,* in all respects and not separated from it commercially.

RANGE. In moist or swampy soil. Oklahoma, Arkansas, Missouri, Tennessee, Georgia, and Kentucky; north to Maine, Quebec, and Ontario and west to Minnesota, Michigan, and Nebraska.

PROPAGATION. The acorns should be gathered immediately when fallen because of their quick germination rate. Fall sowing is better than spring. Planting is done in rows 8–12 in. apart, the acorns covered ½–1 in. When planting in the fall a straw or leaf mulch is desirable. If protection against rodents is desired, a screen of hardware cloth can be used. A mixture of 1½ oz of formaldehyde and 2 pt of water for each sq ft of soil will help control collar rot. Powdery mildew can be controlled by dusting with sulphur.

HYBRIDS. The following hybrids are known for Swamp White Oak:

Mounds Oak, *Q.* × *humidicola* Palmer, is a cross of *Q. bicolor* and *Q. lyrata,* named from its moisture-dwelling habit in low woods. It occurs in Illinois and Missouri.

Jack Oak, *Q.* × *jackiana* Schneid., is a cross of *Q. bicolor* and *Q. alba,* named after John George Jack (1861–1949) American dendrologist, who discovered it in 1894. It occurs in southern Quebec, Massachusetts, Rhode Island, New York, Indiana, and Illinois.

Schuette Oak, *Q.* × *schuettei* Trel., is a cross of *Q. bicolor* and *Q. macrocarpa,* named for J. H. Schuette of Green Bay, Wisconsin, who collected it in 1893. It occurs in southern Quebec, New York, Michigan, Wisconsin, Minnesota, and Indiana.

Substellate Oak, *Q.* × *substellata* Trel., is named for its resemblance to *Q. stellata,* one of the parents with *Q. bicolor.* It occurs in New Jersey.

REMARKS. The genus name, *Quercus,* is the ancient classical name, and the species name, *bicolor,* refers to the contrasting upper and lower leaf surfaces. Also known as Swamp Oak and White Oak.

Swamp White Oak sprouts from the stump, is moderately resistant to fire damage, and is rather free from insects and disease. The leaves turn yellowish brown to brown or reddish in autumn. It was introduced into cultivation about 1800. The long fruiting peduncles distinguish it readily from the Swamp Chestnut Oak, the Chinquapin Oak, and the true White Oak.

The wood of the Swamp White Oak is used for the following purposes: agricultural implements, boxes, crates, caskets, refrigerators, sewing machines, cabinets, furniture, planing mill products, vehicles, flooring, firewood, paneling, furniture, veneer, piling, poles, timber, and cooperage. Along with White Oak it possesses the following characteristics: smooth and attractive in appearance, shows resistance to abrasion, hardness, strength, durability, impermeable to liquids, resistant to shock, glues well, and holds nails well. However, it has a tendency to shrink considerably and to split, and it is somewhat difficult to dry.

CHINQUAPIN OAK
Quercus muhlenbergii Engelm.

CHINQUAPIN OAK

Quercus muhlenbergii Engelm.

FIELD IDENTIFICATION. Narrow, round-topped tree rarely over 60 ft.

FLOWERS. In separate staminate and pistillate catkins on the same tree; staminate catkins 3–4 in. long, hairy; calyx 5–6-lobed, yellow, hairy, ciliate, lanceolate; stamens 4–6, filaments exserted; pistillate catkins sessile, short, tomentose; calyx 5–6-lobed; stigmas red.

FRUIT. Acorns mostly solitary or in pairs, sessile or short-peduncled, broadly ovoid, brown, shiny, ½–¾ in.

long, enclosed about half its length in the cup; cup thin, bowl-shaped, brown, tomentose; scales of cup appressed, obtuse to acute or cuspidate; kernel sweet, edible.

LEAVES. Alternate, simple, deciduous, 4–6 in. long, 1–3½ in. wide, oblong to lanceolate or obovate, acute or acuminate at apex, cuneate to rounded or cordate at base; margin with coarse, large, acute, mucronate, often recurved teeth; dark green, lustrous and glabrous above; paler gray-tomentulose and conspicuously veined beneath; petiole slender, ½–1 in.

TWIGS. Slender, hairy or glabrous, reddish brown to gray, terminal buds orange to reddish brown, ovoid, acute.

BARK. Light grayish brown, broken into narrow, loose plates.

WOOD. Reddish brown, sapwood lighter, close-grained, durable, hard, heavy, strong, weighing 53 lb per cu ft.

RANGE. Well-drained uplands. Texas, Louisiana, Oklahoma, and Arkansas; east to Florida, north to Maine, and west to Ontario, Michigan, Wisconsin, and Nebraska. In Mexico in Coahuila and Nuevo León.

REMARKS. The genus name, *Quercus,* is the classical name of the oak-tree, and the species name, *muhlenbergii,* is in honor of G. H. E. Muhlenberg (1753–1815), botanist and minister in Pennsylvania. Other vernacular names are Pin Oak, Shrub Oak, Scrub Oak, Yellow Oak, Chestnut Oak, Rock Oak, and Chinkapin Oak.

Chinquapin Oak sprouts from the stump, grows rather rapidly, and is fairly free of insects and disease. The wood is used for posts, ties, cooperage, furniture, and farm implements. Bray Chiquapin Oak, *Q. muhlenbergii* var. *brayi* (Small) Sarg., is somewhat similar and is found on the Edwards Plateau and into west Texas, and south into Mexico. It has nuts sometimes to 1¼ in. long, and deeper cups to 1 in. in diameter. However, these differences are not distinct, and it is now considered by some botanists as only a form instead of a variety.

Deam Oak, *Q.* × *deamii* Trel., is a cross between *Q. muhlenbergii* and *Q. macrocarpa.* It is named after Charles Clemon Deam (b. 1865), botanist of Indiana, where it occurs in Wells County.

SWAMP CHESTNUT OAK.

Quercus prinus L.

FIELD IDENTIFICATION. Long-lived tree attaining a height of 100 ft, with a compact narrow head.

FLOWERS. Borne April–May. Arranged in separate staminate and pistillate catkins; staminate catkins slender, hairy, 3–4 in.; calyx 4–7-lobed, hairy, green; pistillate catkins few-flowered, involucral scales tomentose; stigmas red.

FRUIT. Ripening September–October. Acorn sessile, or stalked, solitary or paired, ovoid-oblong, ¾–1½ in. long, shiny brown, set one third to one half its length in the cup; cup bowl-shaped, thick; scales hard, stout, ovate, acute, reddish brown.

LEAVES. Simple, alternate, deciduous, obovate-oblong, acute or acuminate at apex, cuneate or rounded at base, undulately crenate on margin, glabrous and lustrous green above, paler and pubescent beneath, blades 4–8 in. long, 1½–4 in. wide; petioles about ¾ in.

TWIGS. Smooth, reddish brown to gray later.

BARK. Pale gray, broken into shaggy strips which are reddish beneath, furrows deep.

WOOD. Light brown, heavy, tough, hard, strong, close-grained, durable.

RANGE. In greatest abundance and size on the coastal plain, generally in moist soil. Texas, Arkansas, and Oklahoma; east to Florida, north to Delaware, and west through Indiana and Illinois to Missouri.

PROPAGATION. The minimum commercial bearing age is about 20 years, and optimum 50–200 years. The cleaned seed averages about 100 seeds per lb, with a commercial purity of about 98 per cent and a soundness of 60 per cent. The acorns should be gathered as soon as they fall, to prevent drying, and stored in sealed containers at 32°–36°F., or stored in cool cellars. Usually no pretreatment is required to break dormancy. The germinative capacity is about 87 per cent.

HYBRIDS. Some botanical authors list this tree under

SWAMP CHESTNUT OAK
Quercus prinus L.

the name of *Q. michauxii* Nutt. and preserve the name *Q. prinus* L. for the more northern oak once known as *Q. montana* Willd. It is known to hybridize with other oaks as follows:

Beadle Oak, *Q.* × *beadlei* Trel., is a hybrid of Swamp Chestnut Oak and White Oak. It is named for Chauncey Delos Beadle, American botanist. It is found in east Texas and Florida; north to New Jersey, Delaware, Virginia, and North Carolina.

Covington Oak, *Q.* × *byarsii* Sudw., is a hybrid of Swamp Chestnut Oak and Bur Oak. It was discovered by James Byars in 1888, and grows near Covington in Tipton County, Tennessee.

REMARKS. The genus name, *Quercus*, and the species name, *prinus*, are both classical names applied to certain oaks of Europe. Vernacular names are Cow Oak, Basket Oak, Michaux Oak, Swamp White Oak, and Swamp Oak. The tree has been in cultivation since 1737. The wood is used for posts, tools, baskets, splints, cooperage, boards, veneer, and fuel. The acorns are eaten by mourning dove, wild turkey, and white-tailed deer, and the leaves frequently browsed by livestock.

DWARF CHINQUAPIN OAK
Quercus prinoides Willd.

DWARF CHINQUAPIN OAK

Quercus prinoides Willd.

FIELD IDENTIFICATION. Slender shrub, or more rarely a small tree to 15 ft, with a trunk diameter of 1–4 in. Usually growing in thicket-forming clumps on rocky hillsides. Closely related to the Chinquapin Oak, *Q. muhlenbergii*, but differing in its shrubby stature, smaller leaves with generally shorter petioles, deeper acorn cups, with thicker scales, and shorter stamens.

FLOWERS. Borne April–May in separate staminate catkins and pistillate catkins; staminate catkins 1–2½ in. long, pendent, cylindric, loosely and remotely flowered, perianth densely hairy, lobes thin, scarious, ovate to oblong; stamens numerous, exserted, filaments short, anthers large and short-oblong; pistillate flowers solitary or paired, sessile, stigmas yellowish red.

FRUIT. Ripening September–October, maturing the first season, often abundant, acorn covered one half or more by the cup, chestnut-brown, ovoid to ellipsoid, apex obtuse or rounded, ⅖–1 in. long; cup sessile or nearly so, hemispheric, thin, deep, ½–¾ in. wide; scales appressed, pale brown, densely hairy, finely tuberculate, triangular-ovate to oblong-lanceolate, apex obtuse to truncate, acorns average about 400 per lb; sweet and edible.

LEAVES. Alternate, simple, deciduous, blades 2–6 in. long, width 2–3 in., obovate to oblanceolate or elliptic, apex acute or short-acuminate, base cuneate, margin undulate-serrate; teeth 3–7 to each side, large, short, acute or obtuse; upper surface olive-green to bright green, lustrous, glabrous; lower surface much paler, finely gray-pubescent to tomentulose; veins conspicuous and rather straight; petiole slender, channeled, ¼–¾ in., sparsely puberulent or glabrous; leaves brilliant red in autumn.

TWIGS. Young reddish brown and puberulent to glabrous, older ones gray and glabrous; bark of older limbs and trunk dark brown to gray, broken into flat, scaly, checkered ridges and shallow furrows.

RANGE. Dwarf Chinquapin Oak is found in sunny sites, often in rocky or acid sandy soil. Northeast Texas, Oklahoma, and Arkansas; east to Alabama, north to Vermont, and west to Minnesota and Kansas.

VARIETIES AND HYBRIDS. Brown Dwarf Chinquapin Oak, *Q. prinoides* var. *rufescens* Rehd., has mixed white and brownish hairs on young twigs and the lower leaf surfaces. It is found in North Carolina, New Jersey, and Massachusetts.

Stelloides Oak, *Q.* × *stelloides* Palmer, is a hybrid of *Q. prinoides* × *Q. stellata*, named for one of the parents. It is found in Massachusetts, New Jersey, Missouri, Kansas, and Oklahoma.

Faxon Oak, *Q.* × *faxonii* Trel., is a hybrid of *Q. prinoides* and *Q. alba*. It was named for Charles Edward Faxon (1846–1918), American botanical artist. It occurs in New York, Massachusetts, and Michigan.

REMARKS. The genus name, *Quercus*, is the ancient classical name. The species name, *prinoides*, refers to its resemblance, especially in the leaves, to *Q. prinus*. It also has the vernacular names of Scrub Chestnut Oak, Dwarf Chestnut Oak, Running White Oak, and Chinquapin Oak. It was introduced into cultivation about

1730. The acorn is known to be eaten by a number of species of birds and mammals, including the gray squirrel and ruffed grouse, and the foliage is browsed by cottontail.

Post Oak

Quercus stellata Wangh.

FIELD IDENTIFICATION. Shrub or tree to 75 ft, with stout limbs and a dense rounded head.

FLOWERS. Appearing with the leaves March–May, borne on the same tree in separate catkins; staminate in pendent catkins 2–4 in. long; calyx yellow, hairy, 5-lobed; lobes acute, laciniately segmented; stamens 4–6, anthers hairy; pistillate catkins short-stalked or sessile, inconspicuous; scales of involucre broadly ovate and hairy; stigmas red, short, enlarged.

FRUIT. Ripening September–November. Acorns maturing the first season, sessile or short-stalked, borne solitary, in pairs, or clustered; acorn oval or ovoid-oblong, broad at base, ½–¾ in. long, striate, set in cup one third to one half its length; cup bowl-shaped, pale and often pubescent within, hoary-tomentose externally; scales of cup reddish brown, rounded or acute at apex, closely appressed.

LEAVES. Simple, alternate, deciduous, oblong-obovate, blades 4–7 in. long, 3–4 in. wide, 5-lobed with deep rounded sinuses; lobes usually short and wide, obtuse or truncate at apex; middle lobes almost square and opposite giving a crosslike appearance to the leaf; terminal lobe often 1–3-notched; base of leaf cuneate or rounded; dark green, rough and glabrous above; paler and tomentose beneath; leathery and thick; petioles short, usually ½–1 in., pubescent.

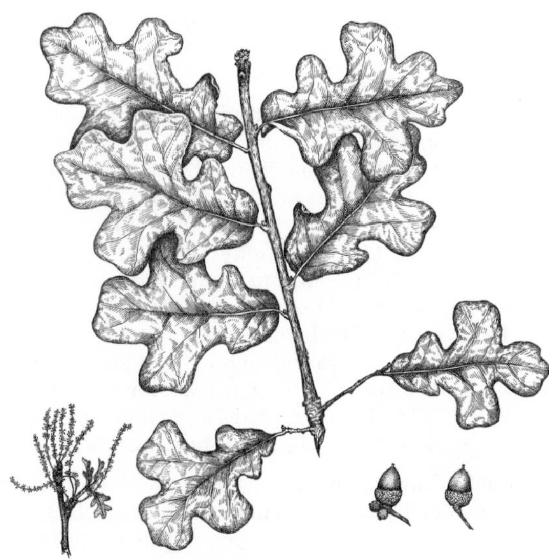

Post Oak
Quercus stellata Wangh.

TWIGS. Brown, stout, pubescent to tomentulose, or becoming glabrous later; buds ¹⁄₁₆–⅛ in. long, subglobose, brown.

BARK. Gray to reddish brown, thick, divided into irregular fissures with platelike scales.

WOOD. Light to dark brown, durable, heavy, hard, close-grained, difficult to cure, weighing about 52 lb per cu ft.

RANGE. Post Oak is distributed in the Edwards Plateau of Texas, adjacent Oklahoma, and Arkansas; east to Florida, north to New England, and west to Iowa and Kansas.

PROPAGATION. Post Oak has been cultivated since 1819. The minimum commercial seed-bearing age is 25 years and the maximum 250 years, with good crops every 2–3 years. The cleaned acorns average about 400 per lb, with a commercial purity of 98 per cent and a soundness of 68 per cent. The acorns require no pretreatment to break dormancy and can be planted when gathered in the fall, or stored in a cool moist cellar or in sealed containers at 32°–36°F., with care taken to prevent their becoming too dry. The germinative capacity of test lots averages about 81 per cent. The planting is done in drills, in rows 8–12 in. apart, and covered 1 in. with firm moist soil.

VARIETIES, HYBRIDS AND SYNONYMS. The extreme variation of Post Oak and its tendency to hybridize freely have given rise to a multitude of names for the varieties and hybrids as follows:

Sand Post Oak, *Q. stellata* var. *margaretta* (Ashe) Sarg., is a variety named for Margaret Henry Wilcox, late Mrs. W. W. Ashe. It is usually scrubby, often sending out underground stolons, and has smaller leaves rounded at the tips and woolly twigs. Usually in sandy lands in eastern and central Texas, Louisiana, and eastern Oklahoma; north to Missouri and Virginia.

Delta Post Oak, *Q. stellata* var. *mississippiensis* (Ashe) Little, was named after the state of Mississippi, and occurs in the Mississippi bottom lands, southeastern Arkansas, and eastern Louisiana.

Drummond Post Oak, *Q. drummondii* (Liebm.) Muller (Muller, 1951, pp. 51-52), is herewith considered to be a synonym only of Sand Post Oak, *Q. stellata* var. *margaretta* (Ashe) Sarg.

Swamp Post Oak, *Q. stellata* var. *similis* Ashe (Sudworth, 1927, p. 107), is herewith considered to be a synonym of *Q. stellata* Wangh.

Brownwood Post Oak, *Q. stellata* var. *parviloba* Sarg. (Sargent, 1918c), is now listed as a synonym only of *Q. stellata* Wangh. (Muller, 1951, p. 47).

East Texas Post Oak, *Q. stellata* var. *araniosa* Sarg., has now been relegated to a synonym of Sand Post Oak, *Q. stellata* var. *margaretta* (Ashe) Sarg. (Little, 1953a, p. 354).

The following hybrids of Post Oak, *Q. stellata* Wangh., are listed by Little (1953a) and Muller (1951):

Bernard Oak, *Q.* × *bernardiensis* W. Wolf, is a

SAND POST OAK
Quercus stellata var. *margaretta* (Ashe) Sarg.

hybrid of *Q. stellata* and *Q. prinus,* named after Saint Bernard College in Cullman County, Alabama, where it was discovered.

Nameless Post Oak, *Q. stellata* var. *anomala* Sarg. (Sargent, 1918b), has now been relegated to the status of a hybrid of *Q. stellata* and *Q. havardii* by Muller (1951, p. 55). It is recorded from the deep sands of hills and plains throughout the eastern range of *Q. havardii* in the southeastern counties of the Texas Panhandle into Oklahoma, and southwestward into the lower plains.

Ruddy Post Oak, *Q. stellata* var. *rufescens* Sarg. (Sargent, 1918d), has now been named a hybrid of *Q. stellata* and *Q. havardii* (Muller, 1951, p. 55).

A hybrid of Post Oak, *Q. stellata* Wangh., and *Q. margaretta* Ashe (Muller, 1951, pp. 50–51) is not followed here because *Q. margaretta* has been relegated to a variety of *Q. stellata* (as *Q. stellata* var. *margaretta* (Ashe) Sarg.), and the hybrid of the species and its variety is a dubious separation (Little, 1953b, pp. 305–306).

Post Oak, *Q. stellata,* and Mohr Oak, *Q. mohriana,* have been described as producing a hybrid in Nolan and Taylor counties, Texas (Muller, 1951, p. 49). We quote Muller's description as follows: "Twigs 1/25–1/12 in. thick, densely buff- or gray-tomentulose. Buds about 1/12 in. long, ovoid, acute, reddish brown, sparsely pubescent at the apex. Leaves 2–2¾ in. long, ⅝–1⅜ in. broad, cuneate and acute or rounded at both ends, undulate or shallowly few-lobed, glabrate and shiny above, densely gray-tomentulose or sparsely so beneath. Acorns similar to those of *Q. stellata.*"

Fernow Oak, *Q.* × *fernowii* Trel., is a hybrid of *Q. alba* and *Q. stellata,* named after Bernhard Eduard Fernow (1851–1923), first chief of the Division of Forestry, U. S. Department of Agriculture. It occurs in New Jersey, Delaware, Maryland, District of Columbia, Virginia, Indiana, Illinois, Missouri, Alabama, and east Texas.

Harbison Oak, *Q.* × *harbisonii* Sarg., is a hybrid of *Q. stellata* and *Q. virginiana,* named after its discoverer, Thomas Grant Harbison (1862–1936), American botanist. It occurs in Jacksonville, Florida, and eastern and central Texas.

McNab Oak, *Q.* × *macnabiana* Sudw., is a hybrid of *Q. stellata* and *Q. durandii,* and is named from McNab, Arkansas, where it was discovered. It occurs in southwestern Arkansas, southern Oklahoma, and central Texas. Hybrids are also formed with *Q. durandii* var. *breviloba* (Torr.) Palmer, in Eastland and Palo Pinto counties, Texas (Muller, 1951, p. 67).

Tharp Live Oak, *Q.* × *neo-tharpii* A. Camus, is a hybrid of *Q. stellata* and *Q. oleoides* var. *quaterna* (Muller, 1951, p. 76). It is named for Benjamin C. Tharp, Texas botanist. It occurs in Texas in Calhoun County near the coast.

Stelloides Oak, *Q.* × *stelloides* Palmer, is a hybrid of *Q. stellata* and *Q. prinoides,* named after its relationship with *Q. stellata,* Post Oak. It occurs in Massachusetts, New Jersey, Missouri, Kansas, and Oklahoma.

Sterrett Oak, *Q.* × *sterrettii* Trel., is a hybrid of *Q. stellata* and *Q. lyrata,* and is named in honor of William Dent Sterrett of the U. S. Forest Service. It occurs in Arkansas.

Substellata Oak, *Q.* × *substellata* Trel., is a hybrid of *Q. stellata* and *Q. bicolor,* and is named for *Q. stellata,* one of the parents. It occurs in New Jersey.

REMARKS. The genus name, *Quercus,* is the classical name; the species name, *stellata,* refers to the stellate hairs of the leaves and petioles. Vernacular names are Iron Oak, Cross Oak, Branch Oak, Rough Oak, and Box Oak. The wood is used for railroad crossties, fuel, fence posts, furniture, and lumber; and the acorns are eaten by deer and wild turkey.

DELTA POST OAK

Quercus stellata var. *mississippiensis* (Ashe) Little

FIELD IDENTIFICATION. Slender tree to 65 ft, with a trunk diameter of 16–28 in. Branches rather short and spreading, forming an oval to oblong head.

FLOWERS. Similar to those of the species *Q. stellata* Wangh., in separate staminate and pistillate catkins ¾–1 in. long and pubescent.

FRUIT. Maturing October–November, borne on gray-canescent peduncles 1/12–⅗ in. long, solitary or in clusters of 2–4; acorn ovate to oblong, ½–¾ in. long, ⅖–½ in. thick, about one third enclosed in the cup, strongly beaked, at first canescent, later dark brown to tan; cup ⅖–½ in. wide, ⅓–½ in. long, turbinate; scales ovate,

DELTA POST OAK
Quercus stellata var. *mississippiensis* (Ashe) Little

appressed, gray-canescent, apices truncate; interior gray-canescent.

LEAVES. Simple, alternate, deciduous, 3¼–5¼ in. long, 2½–4 in. wide (larger on upper branches), broadly obovate, base cuneate to narrowly round, margin with 1–3 pairs of short lobes, some lobes dilated, prominent, notched at apex, rounded, or retuse; upper surface dark green, shiny, sparingly stellate-pubescent; lower surface paler and dull, densely stellate-pubescent; midrib prominent, often arcuate, lateral veins usually deliquescent before reaching the margin; petiole ⅝–1 in., sparingly stellate-pubescent.

TWIGS. Slender, tan to brown, 1/12–⅙ in. thick, at first stellate-hairy, later glabrous; buds 1/12–⅙ in. long, ovate, obtuse or acute; scales 20–25, reddish brown, obtuse, pubescent.

RANGE. From Chico County, Arkansas, southward in the Mississippi River Valley to St. Landry Parish, Louisiana. Also in Richland, Ouachita, Morehouse, West Carroll, and Caldwell parishes, and in western Mississippi.

REMARKS. The genus name, *Quercus*, is the classical name, and the species name, *stellata*, refers to the starlike hairs of the leaves. The variety name, *mississippiensis*, refers to its habitat in the lower Mississippi River Valley. Ashe (1931a, pp. 39–41) makes the following observation concerning it: "This is one of the common oaks on the intermediate or better drained classes of alluvial lands of the lower Mississippi River Valley proper. It is a timber tree of importance not being separated in marketing from that of upland white oak." Also known as Mississippi Valley Post Oak and Yellow Oak.

DURAND OAK

Quercus durandii Buckl.

FIELD IDENTIFICATION. Handsome tree with a dense round top, attaining a height of 90 ft and a diameter of 2½ ft.

FLOWERS. Staminate catkins hairy, 3–4 in. long; lobes of calyx hairy and acute; stamens 3–12; pistillate catkins 1–2-flowered, about ⅕–¼ in. long, peduncles stellate-tomentose to slightly pubescent; bracts linear and acuminate, stigmas red, reflected.

FRUIT. Annual, on peduncles subsessile or to ⅜ in. long, solitary or paired, ovoid to ellipsoid or subrotund, chestnut-brown, glabrous or minutely puberulent, ½–⅔ in. long, one fourth or less included in the cup; cup about ½–⅘ in. broad, ⅙–⅓ in. high, saucer-shaped to shallowly cup-shaped, margin not enrolled, base flat or abruptly rounded; scales ovate, acute, thin, appressed, silvery-pubescent, margins reddish.

LEAVES. Simple, alternate, deciduous; variable, often in three distinct forms; either obovate or oblong-elliptical; from either rounded at apex or somewhat 2–3-lobed, to either entire on the margin or irregularly lobed laterally; length 2½–7 in., width ½–3½ in., leathery; somewhat stellate-pubescent when young, upper surface later dark lustrous green and glabrous; lower surface white-silvery, stellate-tomentose, or more glabrate and green on shade leaves; petioles stellate-pubescent or glabrous, ¼–⅓ in.

TWIGS. Slender, pale gray to brown, stellate-pubescent to glabrous later, lenticels small; buds 1/12–⅛ in. long, subglobose to ovoid, obtuse, scales reddish brown, glabrous to puberulent, ciliate; stipules falling early, linear to spatulate, about ⅕ in. long, pubescent.

WOOD. Reddish brown, sapwood pale, heavy, hard, of no economic value except as posts or fuel.

RANGE. Usually in alluvial soil of the coastal plain, though nowhere very abundant. Calhoun and Milam counties, Texas, central and eastern Texas, southwestern Oklahoma, Arkansas, and Louisiana; east to Florida and north to North Carolina.

VARIETIES AND HYBRIDS. A variety of Durand Oak is Bigelow Oak, *Q. durandii* var. *breviloba* (Torr.) Palmer, found on the limestone hills of central Texas and western Oklahoma. It compares with Durand Oak by being generally more scrubby in size and form, the leaves smaller and shorter lobed, and the deeper cups barely constricted. However, intermediate forms are found where the species and variety overlap at the edge of the Texas Edwards Plateau, and they are difficult to distinguish.

Another variety is known as Bluff Oak, *Q. durandii* var. *austrina* (Small) Palmer. The leaves are larger than those of Durand Oak and more regularly and deeply lobed, and are green beneath (not silvery or whitened). In general appearance and in habit as well as in bark characters, it closely resembles Durand Oak.

Durand Oak is known to hybridize with Post Oak,

Q. stellata, to produce McNab Oak, *Q.* × *macnabiana* Sudw. It is named after McNab, Arkansas, where it was found. It occurs also in southern Oklahoma and central Texas.

An excellent paper, "The Durand Oak and Its Close Relatives," by Ernest J. Palmer (1945) offers the following key to separate the varieties:

1. Leaves entire or variously and often symmetrically lobed, pale and silvery or green beneath; cups of acorns enclosing ¼ or less of well developed nuts.
   a. Mature leaves thin; cup of acorn very small and often disc-like, enclosing ⅕ or less of well developed nuts, the scales thin and never corky thickened at base. A tree usually in alluvial soil of coastal plain. . . . . *Q. durandii.*
   b. Mature leaves firm to subcoriaceous; cup of acorns rounded and enclosing about ¼ of well developed nuts, the scales sometimes slightly thickened at base. A shrub on limestone hills, or rarely a small tree in protected places. . . . . *Q. durandii var. breviloba.*
2. Leaves normally all lobed at least toward apex, green beneath; cups of acorns rounded or turbinate enclosing about ⅓ of well developed nuts. A tree along streams or on low hills of the southeastern coastal plain. . . . . *Q. durandii var. austrina.*

Bigelow Oak
Quercus durandii var. *breviloba* (Torr.) Palmer

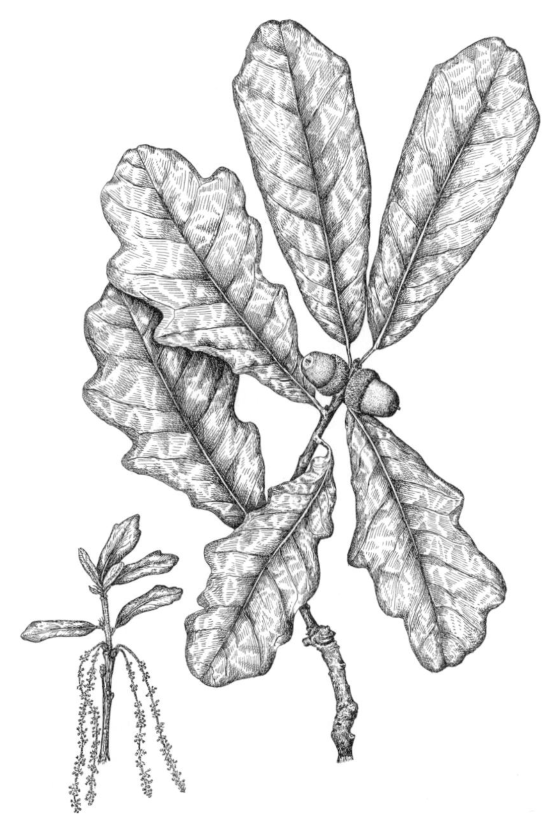

Durand Oak
Quercus durandii Buckl.

REMARKS. The genus name, *Quercus,* is the classical name for the oak-tree, and the species name, *durandii,* is in honor of Elias Magloire Durand (1794–1873), American botanist. Also known by the vernacular names of Bluff Oak, Pin Oak, White Oak, Durand White Oak, Bastard White Oak, and Basket Oak. Wood used for making splint cotton baskets in early times.

Bigelow Oak

Quercus durandii var. *breviloba* (Torr.) Palmer

FIELD IDENTIFICATION. Usually a dwarfed, thicket-forming shrub, or a small round-topped, scrubby tree rarely to 25 ft.

FLOWERS. Staminate and pistillate catkins separate; staminate catkins 1–2 in. long, tomentose, loosely flowered; calyx-lobes rounded, hairy; stamens barely exserted, anthers red; pistillate catkins 1–3-flowered, tomentose, ⅛–⅓ in. long, stigmas red.

FRUIT. Acorn annual, solitary or in clusters of 2–3, sessile or on short pubescent peduncles 1/12–⅓ in.; ovoid-oblong or ellipsoid, length ⅜–1 in., width ¼–½ in., about one fourth to one third enclosed in a shallow cup, apex rounded or abruptly apiculate (sometimes depressed), base truncate; light yellowish brown and shiny, glabrous; cups shallowly cup-shaped or turbinate, ⅖–½ in. broad and ⅕–⅓ in. high, base usually rounded or sometimes flattened, margin of cup thin and smooth; lower scales closely appressed, upper ones loosely; scales ovate to oblong, thickened and tomentose at the base; apex obtuse or acute, reddish brown, puberulent or glabrous; acorns very variable, sometimes very small, greatly flattened and hardly protruding from the cup; on other specimens very elongate and narrow, cup sometimes enclosing only one fifth of base.

LEAVES. Deciduous, alternate; very variable in shape, obovate, oblong or elliptic, usually broadest above the middle; margin entire, undulate, or lobed with shallow rounded lobes, apex broadly rounded or sometimes 3-lobed; base cuneate to rounded or occasionally attenuate; 1–3 in. long, width ½–1¼ in.; rather firm; upper surface glabrous, grayish green and lustrous or with scant, minute, fasciculate hairs; lower surface much paler, almost whitened with dense, close, grayish tomentum (sometimes lower surface green, with sparse hairs or glabrous); petioles 1/16–½ in., fasciculate-hairy, more glabrous later.

TWIGS. Terete or rounded, young twigs greenish gray or brown, stellate-tomentose; older twigs darker gray, glabrous; lenticels minute and numerous.

BARK. Gray, rough, flaking off in strips.

RANGE. Dry, limestone uplands of central and west Texas, Oklahoma, and Mexico.

REMARKS. The genus name, *Quercus,* is the ancient classical name. The species name, *durandii,* is in honor of E. M. Durand (1794–1873), and the variety name, *breviloba,* refers to the short lobes of the leaves. Some botanists list this tree under the names of *Q. annulata* Buckl. or *Q. breviloba* Sarg. Some hold the opinion that Bigelow Oak is only a xerophytic, or dry land, Durand Oak which differs only by dwarf stature, smaller leaves, and smaller fruit with deeper cups. The author believes that Bigelow Oak is so closely related to Durand Oak that it should be maintained as a variety of Durand Oak. Durand Oak and Bigelow Oak produce many intermediate forms where the ranges of the two overlap. Either species or variety hybridizes with Post Oak to produce McNab Oak, *Q.* × *macnabiana* Sudw., which has intermediate characters. It occurs in southwestern Arkansas, southern Oklahoma, and central Texas.

BLUFF OAK

Quercus durandii var. *austrina* (Small) Palmer

FIELD IDENTIFICATION. Tree attaining a height of 80 ft, with a diameter of 2–3 ft. The spreading or ascending branches form a broad open crown.

FLOWERS. Borne in separate staminate and pistillate catkins; staminate 1¼–3 in., interrupted on a slender tomentose rachis; calyx-lobes hairy, obtuse or rounded, about 1/25 in. long or less; stamens 3–12; pistillate catkins 1–2-flowered, pubescent, bracts reflexed, stigmas yellowish red.

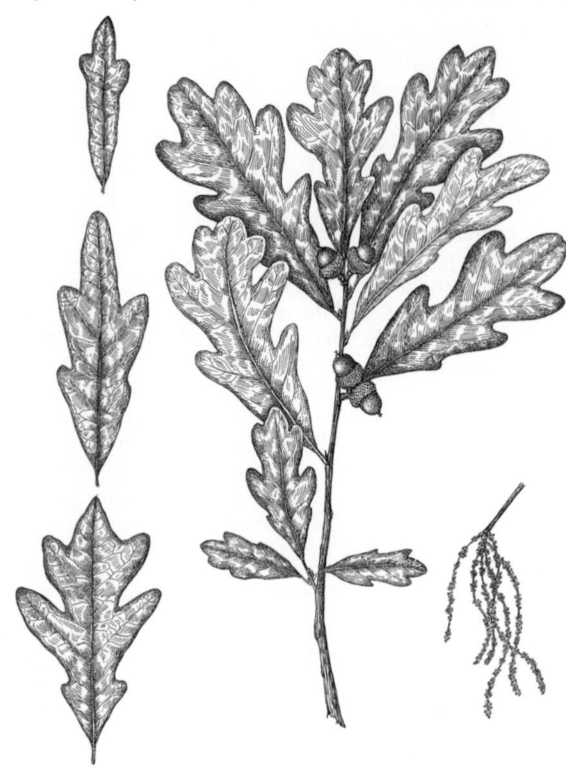

BLUFF OAK
Quercus durandii var. *austrina* (Small) Palmer

FRUIT. Solitary or paired, with a stout peduncle to ½ in. long or sessile; acorn ½–¾ in. long, one third to one half enclosed in the cup, apex narrowed, rounded and pubescent at first; acorn cup deep and cup-shaped, pale-tomentose within, scales loosely appressed, narrow, thin, obtuse, tomentose.

LEAVES. Simple, alternate, deciduous, blades 3–8 in. long, 1–4 in. wide, apex rounded or obtuse to acute, base cuneate to rounded, or gradually narrowed, upper lateral lobes surpassing the lower pair, sometimes 3-lobed at the apex or only somewhat undulate; young leaves somewhat hairy, later the upper surface dark green, glabrous and lustrous, green beneath; main vein prominent, lateral veins slender; petioles ¼–⅓ in., slender, pubescent at first, glabrous later.

TWIGS. Slender, eventually glabrous, reddish brown or gray; winter buds ellipsoid to ovoid, ⅛–¼ in. long; scales imbricate, brown, acute, ciliate, puberulous.

BARK. Pale gray, broken into broad ridges and irregular fissures, scales loosening but clinging persistently.

RANGE. Mostly on limestone soils. To be looked for in southeastern Louisiana because it is found in adjacent southern Mississippi; eastward to Florida and north to North Carolina.

REMARKS. The genus name, *Quercus,* is the ancient classical name of the oak; the species name, *durandii,* is in honor of E. M. Durand (1794–1873); and the variety name, *austrina*, refers to its Southern distribution.

Bluff Oak is similar to Durand Oak in general appearance and habit, and has a close intergrade of leaf form. However, the leaves are larger than those of Durand Oak, and more regularly and deeply lobed, and green beneath instead of white or silvery.

Hinckley Oak

Quercus hinckleyi C. H. Muller

FIELD IDENTIFICATION. Dwarf, Western, thicket-forming shrub to 3 ft. Branches intricate and divaricate.

FLOWERS. Staminate catkins usually less than 1/4 in. long, peduncle tomentose, loosely few-flowered, calyx densely lanate-hairy, stamens with rounded anthers somewhat exserted; pistillate catkins densely hairy.

FRUIT. Annual, solitary or paired, sessile or on peduncles 1/5 in. long or less, acorns ovoid, 1/4–1/2 in. broad, brown, glabrous; cup basal, shallow, saucer-shaped, 1/2–3/5 in. broad, 1/25–1/8 in. deep, margins undulate; scales closely appressed, thickened toward the base, ovate, glabrous or slightly puberulent, margins ciliate.

Hinckley Oak
Quercus hinckleyi C. H. Muller

LEAVES. Persistent, 1/5–3/5 in. long and broad, oval or sometimes broader than long, apex acute to obtuse and spiniferous, base cordate or auriculate, marginal teeth large, coarse, spinescent (2–3 on each side), thickened, with the margins undulate-crisped, both surfaces glabrous and glaucous, veins obscure; petioles 1/25–1/12 in., glabrous, greenish red; stipules about 1/12 in. long; ligulate, hairy.

TWIGS. Brown, 1/25–1/12 in. thick, glabrous, or remotely and minutely stellate-hairy, later glaucous, lenticels obscure; buds about 1/25 in. long, reddish brown, glabrous, ciliate.

RANGE. Known only from the dry slopes of the neighborhood of Solitario Peak, at an altitude of about 4,500 ft, in Presidio County, Texas.

REMARKS. The genus name, *Quercus*, is the ancient classical name, and the species name, *hinckleyi*, is in honor of L. C. Hinckley, botanist at Sul Ross Teachers College, Alpine, Texas, who made outstanding collections of the Trans-Pecos flora. Evidently a very localized species. Named by Muller (1951, p. 40).

Lacey Oak

Quercus laceyi Small

FIELD IDENTIFICATION. Tree to 45 ft, and 1 1/2 ft in diameter, with stout, erect, spreading branches, but sometimes only a clumpy shrub. Easily identified by the grayish green leaves which lend a rather smoky appearance when seen at a distance.

FLOWERS. Borne in spring in separate staminate and pistillate catkins; staminate catkins loosely flowered, 2–2 1/2 in. long; perianth deeply divided into 4 or 5 ovate lobes which are acuminate at the apex; stamens exserted, longer than the perianth lobes, anthers subglobose; pistillate catkins 1–3-flowered distally, 1/5–3/5 in. long.

FRUIT. Acorns borne annually, 1–3 in a cluster, sessile or short-peduncled, short-oblong, ovoid or ellipsoid, apex truncate or rounded, shiny chestnut-brown, enclosed in cup one fourth to one half its length, or at the base only; cup stout, cup-shaped or saucer-shaped, about 1/4–1/2 in. wide, reddish brown, rather corky; scales closely appressed, apex rounded to obtuse, margin ciliate, tomentose, more glabrous later.

LEAVES. Alternate, tardily deciduous, blades 2–5 in. long, 3/4–2 in. wide, leathery, reticulate-veined, upper surface conspicuously grayish green and glabrous; lower surface paler, glabrous to glaucous or pubescent on the veins; oblong, elliptic or obovate, mostly shallowly

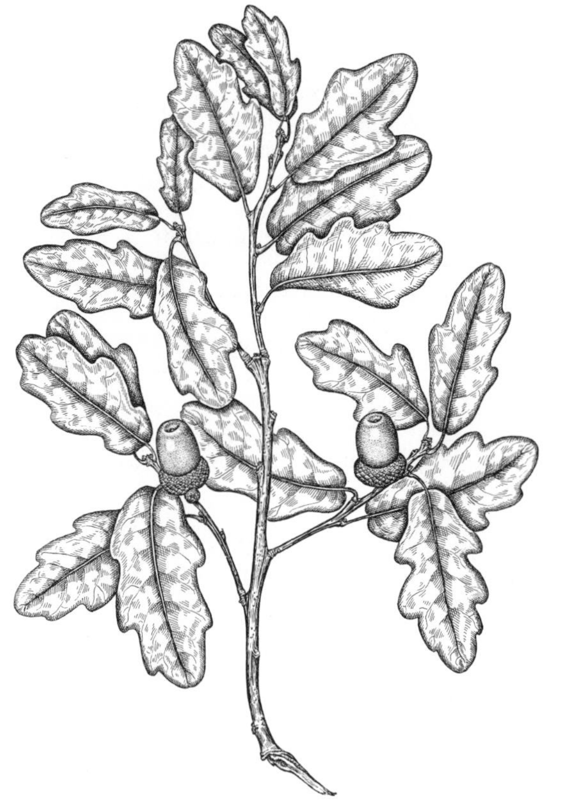

LACEY OAK
Quercus laceyi Small

sinuate, 2-lobed on each margin, apex rounded or occasionally 3-lobed; base rounded to cuneate or rarely cordate; petiole slender, glabrous or sparingly pubescent, 1/4–1/3 in.; stipules caducous, about 1/25 in. long, linear to subulate.

TWIGS. Young twigs green to light brown, pubescent at first, glabrous later; older ones gray, smooth, tight, marked with small, light, orbicular lenticels.

BARK. Gray, thick, broken into flat, narrow ridges and deep fissures.

RANGE. Rocky soils of bluffs and riverbanks, apparently confined to the limestone Edwards Plateau of Texas. In addition to the type locality, the Lacey Oak is also found in Real, Terrell, Edwards, Kimble, Uvalde, Menard, Bandera, Medina, and Kendall counties of Texas. Also in the mountains of Nuevo León, Tamaulipas, Coahuila, and San Luis Potosí, Mexico.

REMARKS. The genus name, *Quercus,* is an ancient classical name, and the species name, *laceyi,* is for Howard Lacey, who first collected it on his Kerrville ranch. Vernacular names are Rock Oak, Canyon Oak, Mountain Oak, Smoky Oak, and Bastard Oak. The wood is occasionally used for fuel and posts. Lacey Oak is sometimes confused with Bigelow Oak, *Q. durandii* var. *breviloba* Sarg., but the former has grayish green leaves as a distinguishing feature.

MEXICAN BLUE OAK
Quercus oblongifolia Torr.

FIELD IDENTIFICATION. Tree attaining a height of 35 ft, with a diameter of 20 in. The branches are rather heavy and contorted, forming a round-topped crown.

FLOWERS. Borne in the typically separate staminate and pistillate catkins on the same tree; staminate catkins densely tomentose, calyx yellowish red, pilose, laciniately cut into 5–6 acute sepals; pistillate catkins sessile or on short tomentose peduncles; stigmas red.

FRUIT. Acorn usually short-peduncled, solitary or sessile, ovoid to ellipsoidal, apex usually rounded and often with a pubescent ring, light to dark brown, striate, shiny, 1/2–3/4 in. long, about 1/3 in. thick, enclosed for one third its length in the cup; cup turbinate or cup-shaped, pubescent within; scales of cup ovate-oblong, acute and reddish at the apex, margin ciliate, surface densely tomentose.

LEAVES. Simple, alternate, deciduous, young leaves densely tomentose and reddish, mature ones bluish green and shiny above, paler beneath, elliptic-oblong or ovate to slightly obovate, apex rounded to acute or sometimes emarginate, base cordate to rounded or cuneate; margin entire or somewhat undulate and thickened, or with small remote teeth, or sometimes with 3

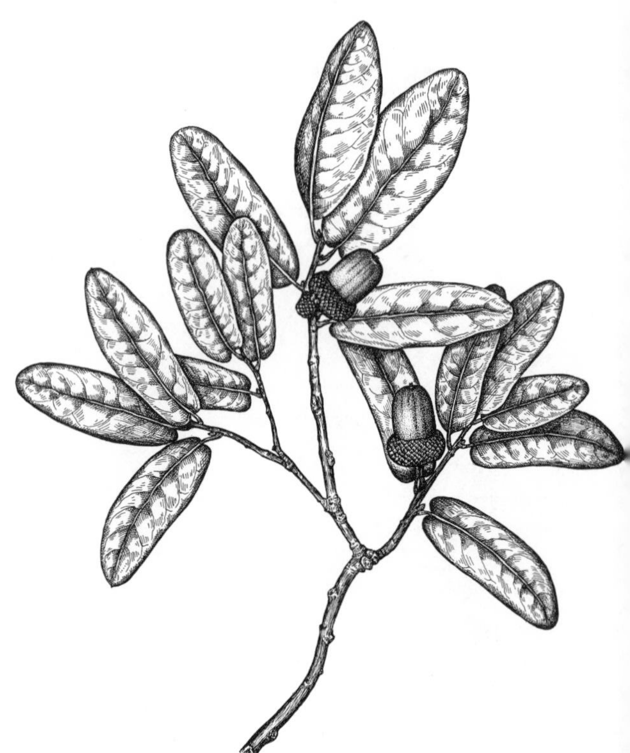

MEXICAN BLUE OAK
Quercus oblongifolia Torr.

teeth at the apex, length 1–2¼ in., width ½–¾ in. (larger on young shoots), veins conspicuous with reticulate veinlets; petioles about ¼ in. long.

TWIGS. Slender, rigid, young ones with gray or brownish tomentum, light to dark reddish brown, later gray and glabrous; buds subglobose 1/16–⅛ in. long with brown scales.

BARK. Gray, with small rectangular or square, thickened scales.

WOOD. Dark brown to black, sapwood lighter, strong, hard, brittle, very heavy.

RANGE. Canyons of mountain slopes, upper mesa coves, at altitudes of 4,000–6,000 ft. Southwestern New Mexico and Arizona; and in Sonora, Chihuahua, and Coahuila, Mexico.

REMARKS. The genus name, *Quercus*, is the classical name, and the species name, *oblongifolia*, refers to the oblong leaves. The wood is sometimes used for fuel or posts.

Mexican Dwarf Oak

Quercus depressipes Trel.

FIELD IDENTIFICATION. Thicket-forming shrub to 3 ft, from underground stolons.

FLOWERS. In staminate and pistillate catkins; staminate to ⅘ in. long, loosely flowered, pubescent, stamens barely exserted, anthers red; pistillate catkins solitary or 2–3-flowered distally, peduncle with stellate hairs or glabrous later.

FRUIT. Acorn annual, solitary or paired, peduncles ⅓–⅗ in.; acorn elliptic to ovoid, ½–⅗ in. long, ⅓–⅖ in. broad, apex rounded, light brown, at first glaucous, later glabrous, one fourth to one half included in the cup; cups ⅓–⅖ in. broad, ⅙–¼ in. high, bowl-shaped, base rounded or constricted; scales closely imbricate and appressed, base densely gray-tomentose, apex reddish brown, dorsal surface glabrous to pubescent, margins ciliate.

LEAVES. Simple, alternate, half-evergreen, length ⅕–1⅕ in., ⅙–½ in. broad, oblong to elliptic, apex rounded or obtuse to acute, base cordate or rounded, margin somewhat revolute, entire, or with a few remote teeth toward the apex, thick and leathery; upper surface and lower surface dull grayish green, somewhat glaucous, glabrous or slightly hairy on the veins beneath, veins 5–6 on each side; petioles 1/25–1/12 in., flattened at the distal end, reddish and glaucous, glabrous or slightly hairy; stipules falling away early, ⅛–⅕ in., ligulate, reddish tan, ciliate.

TWIGS. Reddish brown or gray, at first brown-hairy but later glabrate, hardly over 1/15 in. thick, somewhat fluted; lenticels scattered, inconspicuous; buds subglobose, 1/25–1/15 in. long, brown, glabrate to ciliate.

Mexican Dwarf Oak
Quercus depressipes Trel.

RANGE. Known from the summit of Mount Livermore in the Davis Mountains of Texas, and from Chihuahua and Durango, Mexico.

REMARKS. The genus name, *Quercus*, is the ancient classical name, and the species name, *depressipes*, refers to the depressed or dwarf stature of the plant. This species was first recorded in Texas by C. H. Muller (1951, pp. 39–40).

Gambel Oak

Quercus gambelii Nutt.

FIELD IDENTIFICATION. Very variable and widespread Western species. Sometimes only a thicket-forming shrub, or in favorable locations a tree to 50 ft, and 2 ft in diameter, with a rounded head.

FLOWERS. In May with the young leaves in separate staminate and pistillate catkins; staminate catkins 1–1½ in., loosely flowered, borne from thin, brown, hairy, oblong to spatulate, obtuse to rounded scales; calyx scarious, narrowly 5–6-lobed; anthers exserted, reddish brown; pistillate flowers inconspicuous, 1–3-flowered at the apex, ⅛–⅓ in. long.

FRUIT. Acorn borne annually, solitary or several together, sessile or on tomentose peduncles ⅜–⅗ in.; ovoid, oblong, or ellipsoid, apex narrowed and rounded, base rounded, light brown and glabrous or nearly so, ½–¾ in. long, ⅓–⅖ in. broad, about one third to one half included in the cup; cups hemispheric or deeply cup-shaped, bases rounded, margins thin and smooth, ⅕–⅓ in. high, ⅓–⅗ in. broad; scales of cup appressed, ovate-acuminate, tomentose at base; apex glabrous or pubescent, obtuse or narrowly rounded.

LEAVES. Young leaves brownish gray, tomentose; older ones deciduous, oblong, obovate, oval or elliptic; 2½–6 in. long (usually 2–4), 1½–3 in. wide (usually ¾–2); margin 5–9-lobed, middle lobes often the largest, terminal lobe rounded or sometimes 3-parted; lobes mostly oblong and entire, but some notched or with secondary lobes; sinuses either shallow or almost reaching to the midrib, base rounded or acute; base of leaf cuneate or truncate; upper surface shiny green with minute fascicled hairs; lower surface duller, varying from almost glabrous or pubescent to densely velvety-tomentose, on some trees glaucous, veins conspicuous; petiole slender, pubescent or glabrous, ¼–1 in. long. The leaves show a wide range of variation in size, shape, number of lobes, and hairiness, however intergrading forms demonstrate the close relationship of these races.

TWIGS. Rather slender, young twigs brown to reddish brown, pubescent to glabrous; older twigs grayish brown, stout; lenticels not abundant.

BARK. Gray, thick, fissures deep, scales small and appressed.

RANGE. At altitudes of 4,000–8,000 ft in western Texas; northward into New Mexico, Arizona, Colorado, Utah, Nevada, and Wyoming. Southward in Mexico in Coahuila and Chihuahua.

REMARKS. The genus name, *Quercus,* is the ancient name, and the species name, *gambelii,* refers to the botanist William Gambel, who collected in 1844 in the Rocky Mountains. The tree is so variable in size, leaf, and fruit characters that it has been described by botanists under many different names, but the recent tendency is to consider the variants as isolated geographical races rather than as species. In this category may be placed New Mexican Oak, *Q. novo-mexicana* (A. DC.) Rydb.; Gunnison Oak, *Q. gunnisonii* (Torr. and Gray) Rydb.; the Socorro Oak, *Q. leptophylla* Rydb. Vernacular names used for the tree are White Oak and Rocky Mountain White Oak. The foliage is sometimes browsed by livestock, black-tailed deer, and porcupine. The sweet acorn is eaten by swine. The wood is hard, strong, and used for fuel or posts.

GAMBEL OAK
Quercus gambelii Nutt.

SHRUB LIVE OAK

Quercus turbinella Greene

FIELD IDENTIFICATION. Stiffly branched shrub or small tree sometimes attaining a height of 15 ft.

FLOWERS. Staminate catkins ¼–¾ in. long, yellowish green, bracts densely brown-tomentose, lobes rounded or obtuse, stamens exserted, anthers large, yellow, filaments short and densely hairy; pistillate catkins sessile or peduncled, very small, ⅛–¼ in. long, solitary or clustered, scales brown-tomentose.

FRUIT. Borne on tomentose peduncles ⅜–1¼ in., solitary or several at the distal end, acorn annual, length about ⅗ in. or less, width about ½ in., ovoid to short-oblong, light brown, pubescent at first, less so later, one fourth to one half included in the cup; cup shallowly cup-shaped, or turbinate, ⅓–½ in. broad, ⅙–¼ in. high, base rounded, scales ovate, closely appressed, reddish brown, densely tomentose to pubescent to semiglabrous, apex obtuse to rounded and thin, base thickened.

LEAVES. Evergreen, stiff, thick, leathery, sometimes crinkled, broadly elliptic or ovate, blade length ⅖–1⅗ in., width ⅕–⅘ in., apex obtuse to acute or rounded, spinose, base subcordate or rounded, each margin coarsely 3–5-spinose-toothed, flattened or subrevolute and thickened (rarely entire), upper surface dull or semilustrous to glaucous, grayish green, lower surface densely stellate-tomentose, veins fairly prominent; petioles tomentose at first, but less so later, 1/25–⅙ in.

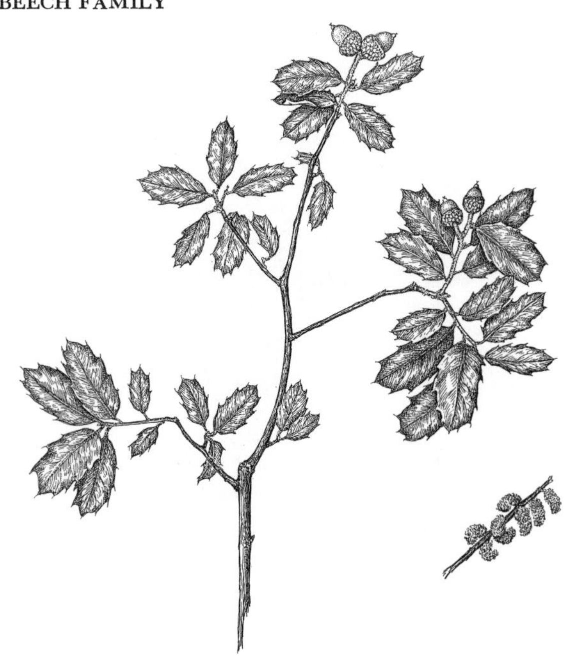

Shrub Live Oak
Quercus turbinella Greene

TWIGS. Brownish gray, tomentulose to glabrate later, 1/25–1/5 in. thick, almost rounded or fluted, rather rigid; buds ovoid, 1/25–1/12 in. long, brown-tomentose; stipules persistent, 1/12–1/6 in., linear-filiform, hairy.

RANGE. In Trans-Pecos Texas, New Mexico, Arizona, California, Southern Colorado, Utah, and Nevada. Usually at altitudes of 3,500–8,000 ft.

REMARKS. The genus name, *Quercus*, is the ancient classical name, and the species name, *turbinella*, refers to the turbinate cup of the acorn. It is usually known as Scrub Oak locally and is very variable and comprises several races. Occasionally it resembles toothed forms of Gray Oak, *Q. grisea*.

Net-leaf Oak

Quercus reticulata H. & B.

FIELD IDENTIFICATION. Shrub or tree to 40 ft, and 1 ft in diameter.

FLOWERS. Staminate and pistillate flowers borne separately on the same tree; staminate catkins in axils of leaves of the year, 1¼–2 in. long, loosely flowered, short-tomentose, anthers slightly exserted, calyx greenish yellow, pale-hairy, 5–7-lobed, the lobes ovate and acute; pistillate catkins slender, ¾–2½ in. long, densely tomentose, distally flowered, involucre scales also tomentose, stigmas bifid, dark red.

FRUIT. Acorn annual, solitary, or in twos or threes, peduncles slender, ¾–2 in. long, rarely longer; acorn 2/5–4/5 in. long, ⅓–½ in. broad, rarely longer, ovoid to oblong or elliptic, apex rounded or acute, base broad, light brown, finely pubescent or glabrous later, one fourth to one half included in the cup; cups saucer- or cup-shaped, 2/5–½ in. broad, 1/5–2/5 in. high, base constricted or rounded; scales appressed or somewhat spreading, ovate, acute, apex thin, base thickened, brown, pubescent to canescent-tomentose.

LEAVES. Simple, alternate, subevergreen, blades 1¼–2½ in. long (occasionally to 5 in.), ⅝–1½ in. wide (occasionally to 4 in.), broadly to narrowly obovate or elliptic, apex mostly rounded to obtuse or almost acute, base cordate or rounded, margin entire or remotely toothed toward the apex, revolute; firm, rigid, thickened, flat or concave, sometimes undulate; upper surface dark green to grayish green, shiny, glabrate to somewhat stellate-pubescent, veins impressed; lower surface dull green, varying from glabrate to glaucous, or brownish, pubescent to tomentulose, veins conspicuously raised and netted; petiole stout, ⅛–⅓ in., at first densely tomentose, later glabrate.

TWIGS. Younger ones brown to orange and tomentose, later brownish gray and glabrous; buds ovoid to oval, about ⅛ in. long, scales reddish brown, ciliate.

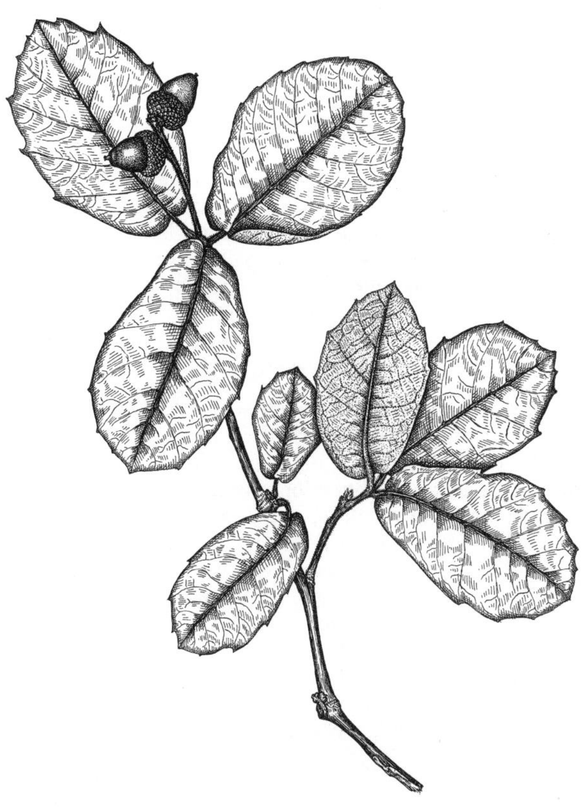

Net-leaf Oak
Quercus reticulata H. & B.

BARK. Dark to light brown or gray, scales small, thin, and tight.

WOOD. Dark brown, sapwood lighter, close-grained, heavy, and hard.

RANGE. Not abundant, on wooded slopes at altitudes of 4,000–6,000 ft on the high mountains of Trans-Pecos Texas, New Mexico, Arizona, and Mexico.

REMARKS. The genus name, *Quercus,* is the ancient classical name for the oak, and the species name, *reticulata,* refers to the net-veined lower leaf surface. The tree is listed by some botanists under the name of *Q. diversicolor* Trel. It has been cultivated since 1883. The young leaves are fair browse for cattle, and the acorns are eaten by the thick-billed parrot.

Arizona White Oak

Quercus arizonica Sarg.

FIELD IDENTIFICATION. Shrub at high elevations or on lower mountain slopes, or a tree to 40 ft, with wide-spreading branches and a rounded top.

FLOWERS. Borne in separate staminate and pistillate catkins; staminate catkins with a tomentose rachis ⅔–3 in. long, flowers fairly dense, calyx pale yellow, 4–7-lobed, lobes acute and ciliate, anthers included; pistillate catkins with a tomentose rachis ⅕–⅘ in. long, 2–6-flowered distally, scales tomentose.

FRUIT. Maturing September-November, acorn annual, solitary or paired, subsessile or short-peduncled, ⅓–⅔ in. long, about ½ in. thick, light to dark brown, lustrous, often striate, ovoid to oblong, apex obtuse to rounded, puberulent at first, glabrous later, about one half included in the cup; cup hemispheric or cup-shaped, ⅕–¼ in. high, ⅖–⅗ in. broad, base rounded; scales ovate, closely appressed, pale to brown tomentose, apex reddish brown, thin, obtuse to rounded, base rounded and roughly thickened.

LEAVES. Evergreen, or deciduous just before the new leaves in spring, firm and rigid, blade length 1¼–4 in., width ⅜–2 in., elliptic to oblong or obovate to oblanceolate, apex acute to obtuse or rounded, base cordate or rounded; margin revolute, entire or toothed with large or small teeth, especially toward the apex; upper leaf surfaces dull bluish green or semilustrous, glabrous, or with scattered stellate hairs, veins impressed or slightly prominent; lower surfaces duller, with pale to brown stellate-pubescence or tomentose, sometimes glaucous; veins rather prominent and reticulate, midrib broad and yellow, lateral veins united near the revolute margin; petioles stout, somewhat stellate-pubescent to tomentose, 1/12–⅖ in.

TWIGS. At first pale gray or fulvous with dense stellate tomentum, later becoming reddish brown and pubescent to glabrate, fluted, 1/15–1/10 in. thick; buds chestnut-brown, subglobose, about ⅛ in. long, pubescent to subglabrous, scales loosely imbricate, ciliate on margin; stipules moderately persistent, threadlike, ⅕–⅓ in. long, pubescent.

Arizona White Oak
Quercus arizonica Sarg.

BARK. Pale, thin, scaly on young branchlets, later with deep fissures and broad ridges broken into plate-like, gray scales.

WOOD. Heartwood dark brown to black, sapwood paler, close-grained, heavy, hard, strong.

RANGE. Widely distributed at altitudes of 5,000–10,000 ft in Trans-Pecos Texas, New Mexico, Arizona, and Mexico. In Texas reported from El Paso, Culberson, and Brewster counties. In Mexico in Sonora, Coahuila, and Durango.

REMARKS. The genus name, *Quercus,* is the ancient classical name for the oak, and the species name, *arizonica,* refers to the state of Arizona, where it is abundant. A rather puzzling oak, very variable over its distribution, and in the presence of many forms which probably represent hybrid origins. Evidently closely related to *Q. reticulata,* but has narrower leaves, shorter fruiting peduncles, cup scales thickened and tightly appressed, and grows at lower, drier altitudes.

Organ Mountains Oak, *Q.* × *organensis* Trel., is

considered to be a hybrid of *Q. arizonica* and *Q. grisea.* It was first discovered in the Organ Mountains, Dona Ana County, New Mexico.

Coahuila Scrub Oak

Quercus intricata Trel.

FIELD IDENTIFICATION. Intricately branched, thicket-forming, low, shrubby oak.

FLOWERS. Staminate rachis tomentose, catkins ¾–1¼ in., loosely flowered, perianth glabrous or slightly pubescent, anthers oval, barely exserted; pistillate rachis tomentose, ⅛–⅖ in., 1–5-flowered distally.

FRUIT. Acorns on tomentose peduncles to ⅝ in. long, or subsessile, borne annually, solitary or paired; acorn ⅓–½ in. long, ⅓–⅖ in. wide, ovoid, light brown, slightly hairy or glabrous, one third to one half included in the cup; cup ¼–⅓ in. high, ⅓–⅖ in. wide, deeply cup-shaped, rounded at the base, sparsely pubescent or glabrate, and thinner toward the apex.

LEAVES. Simple, alternate, evergreen, thick and coriaceous, ovate to oblong, apex obtuse, base cordate or cuneate, margin entire or remotely toothed, revolute and sometimes wavy; upper surface semilustrous, glabrate or sparsely stellate-pubescent; veins impressed; lower surface gray or brownish-tomentose, becoming glabrous with age, veins raised and rather prominent; petioles 1/12–⅛ in., tomentose-hairy.

TWIGS. Covered with gray or brownish tomentum, gray to darker later, about 1/12 in. thick; buds 1/25–1/12 in. long, reddish brown, pubescent to glabrate later; stipules persistent, filiform to linear, pubescent, about 1/12 in. long.

Coahuila Scrub Oak
Quercus intricata Trel.

RANGE. Known in the United States only from the Chisos Mountains in Brewster County, Texas, and from the Eagle Mountains in Hudspeth County, Texas. Collected by the author in the Chisos Mountains on the ridge above Green Gulch, at the summit of the Lost Pine Trail. Also in Mexico in Coahuila, Chihuahua, Zacatecas, and Nuevo León. A characteristic species in the Coahuilan chaparral.

REMARKS. The genus name, *Quercus,* is the ancient classical name, and the species name, *intricata,* refers to the intricate, stiff twigs.

Mohr Oak

Quercus mohriana Buckl.

FIELD IDENTIFICATION. Usually, a thicket-forming shrub, but sometimes a small, round-topped tree to 20 ft.

FLOWERS. Borne in separate staminate and pistillate catkins; staminate catkins ¾–1½ in. long, loosely flowered, from sparsely to densely hairy, anthers barely exserted, red; calyx hairy, divided into ovate lobes; pistillate catkins 1–3-flowered toward the apex, hairy, 1/12–⅓ in.; calyx-lobes and bracts hairy.

FRUIT. Acorns borne annually, solitary or 2–3, sessile, or on densely pubescent peduncles ¼–¾ in. (usually about ⅜ in.); ovoid to ellipsoid or oblong; apex abruptly rounded and apiculate; young acorns with fascicled hairs, older ones brown and lustrous; length ⅓–⅗ in., width ¼–⅓ in., enclosed one half to two thirds of length in cup; cup turbinate or cup-shaped, ⅕–½ in. high and ⅓–¾ in. broad, color reddish brown and tomentose; margin thin and smooth; base flattened or rounded; scales closely appressed, ovate to oblong, thickened and more tomentose toward the cup base; smaller, thinner, and more glabrous toward the cup rim; apices elongate, obtuse or acute.

LEAVES. Alternate, persistent, coriaceous, oblong to elliptic to lanceolate or obovate; margin entire, or with a few, large, coarse, apiculate teeth, sometimes with a few rounded lobes, the plane surface of the margin either undulate or flattened and slightly revolute; apex acute, rounded or acuminate; base rounded or cuneate, sometimes unequal-sided; upper surface usually dark green and shiny, sparsely and minutely stellate-pubescent; lower surface usually densely gray or white-tomentose; blade length ¾–4 in., width ½–1½ in.; petiole 1/12–¼ in., tomentose; stipules caducous, subulate, about ⅛ in. long.

TWIGS. Young parts brownish gray, tomentose, and

MOHR OAK
Quercus mohriana Buckl.

fluted; older ones gray, smooth and glabrous; buds reddish brown, smooth or pubescent.

BARK. Grayish brown, thin, pale, deeply furrowed.

RANGE. In dry, well-drained, preferably limestone soils of the West. In west-central Texas, southwestern Oklahoma, and Coahuila, Mexico.

REMARKS. The genus name, *Quercus,* is the classical name, and the species name, *mohriana,* refers to Charles Mohr (1824–1901), German-born druggist and botanist of Alabama. Vernacular names are Scrub Oak, Shin Oak, and Limestone Oak. Mohr Oak is known to hybridize with Post Oak, Havard Oak, and Gray Oak where the contact is made. Gray Oak and Mohr Oak, especially, produce a very varied assemblage of hybrid forms. Muller (1951, p.59) states, "*Quercus mohriana* is rather strictly confined to limestone outcrops ranging from the Del Norte Mountains and the Guadalupe Mountains eastward. *Quercus grisea* occurs principally on igneous slopes but may also range onto limestone locally. In the Del Norte Mountains igneous rock and limestone are contiguous. At their line of contact a great number of apparent hybrids is encountered."

The wood of Mohr Oak, or its hybrids, has little value except as fuel or posts.

HAVARD SHIN OAK

Quercus havardii Rydb.

FIELD IDENTIFICATION. Low shrubs, hardly over 3 ft, forming thickets by underground rhizomes in deep sands. Rarely a small tree.

FLOWERS. Borne in separate staminate and pistillate catkins; staminate catkins pubescent, heavily flowered, ½–1½ in.; anthers pubescent, moderately exserted; pistillate catkins, ⅛–⅓ in., 1–5-flowered toward the apex.

FRUIT. Acorn rather large, annual, very variable in size and shape, solitary or 2–3 in a cluster, sessile or short-peduncled, enclosed one third to two thirds in the cup, length ½–1 in., ½–¾ in. wide, ovoid to short-oblong, color chestnut-brown, lustrous and glabrous or slightly glaucescent; cup deeply bowl-shaped to goblet-shaped, very variable in size, ½–1 in. broad and ⅖–½ in. high, base mostly rounded, margins either thin or thick; scales reddish brown, pubescent, ovate-oblong, apex long-acuminate, blunt, thinner than the base.

LEAVES. Alternate, deciduous, leathery, very variable in size and shape, blades ¾–4 in. long, ¾–1½ in. wide, oblong, elliptic, lanceolate, or oblanceolate, ovate or obovate; margin entire or variously undulate, coarsely toothed or lobed, sometimes margins falcate or asym-

HAVARD SHIN OAK
Quercus havardii Rydb.

metrical, revolute or flattened; apices broadly rounded to obtuse or acute; base cuneate or rounded; upper surface bright green, lustrous, glabrous or with minute fascicled hairs; lower surface densely brown to gray-tomentose; petioles 1/12–1/4 in., pubescent, about 1/5 in. long.

TWIGS. Rounded or sulcate, young ones densely brownish yellow-tomentose, older twigs gray to reddish brown, glabrous or nearly so. Bark gray, smooth or scaly.

RANGE. Across the sandy plains of the lower Texas Panhandle area into eastern New Mexico.

REMARKS. The genus name, *Quercus*, is the ancient name. The species name, *havardii*, honors the botanist Valery Havard (1846–1927), whose work is recorded in the Proceedings of the U. S. National Museum (Havard, 1885). Vernacular names are Shinnery Oak, Sand Oak, Panhandle Shinnery, and Sand Scrub. Also known to hybridize with Mohr Oak and with Post Oak on the Eastern contact of the species. The acorns and leaves of these hybrids vary greatly in size and shape. The acorn of Havard Oak has some value to wildlife, being eaten by peccary, prairie chicken, and bobwhite. It is reported to cause some stock poisoning.

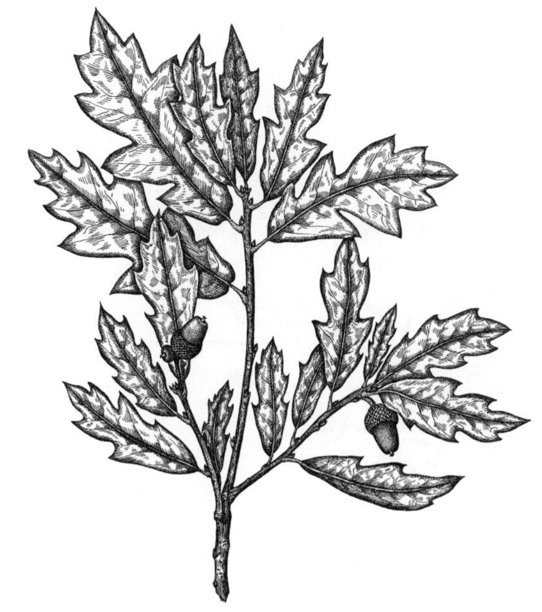

WAVY-LEAF OAK
Quercus undulata Torr.

WAVY-LEAF OAK

Quercus undulata Torr.

FIELD IDENTIFICATION. Mostly shrubby, or occasionally a small tree.

FLOWERS. Flowers in separate male and female catkins.

FRUIT. Acorn borne annually, solitary or paired, subsessile or on a tomentose peduncle to 3/8 in., pale brown, smooth, ovate to cylindric, about 2/5 in. broad and 3/5 in. long, about one third included in the cup; cup shallow or rather deep, 1/5–1/3 in. high, 2/5–1/2 in. broad, base rounded; scales triangular-ovate, closely appressed, thickened below and densely tomentose, apex obtuse, thin, and almost glabrous.

LEAVES. Deciduous or subevergreen, thick and leathery, blade length 3/4–2 1/4 in., width 3/8–1 1/4 in., elliptic to oblong, apex obtuse to subacute, base truncate to rounded or cuneate, margin coarsely toothed or irregularly and shallowly lobed and mucronate; upper surface dark green, with scattered stellate hairs, sometimes roughened; lower surface paler, usually densely soft-hairy, in age less so, veins more prominent below than above; petiole tomentose, 1/12–1/3 in.

TWIGS. Gray, lightly or densely tomentose, with age almost glabrous; buds ovoid, reddish brown, 1/25–1/12 in. long, pubescent to glabrous; stipules early-shedding, subulate, pubescent, 1/8–1/5 in.

RANGE. In Texas in Culberson and Jeff Davis counties, as well as in others. Also in New Mexico and Arizona; northward to Utah, Colorado, and Nevada. Also in Coahuila, Mexico, in the Sierra del Carmen, Cañón Sentenela, on high northwest slopes.

REMARKS. The genus name, *Quercus*, is the classical name, and the species name, *undulata*, refers to the wavy leaf margin. Also known as the Scrub Oak, Shin Oak, and Switch Oak.

This oak has been very much confused in botanical nomenclature and has been interpreted in several ways. Also it exhibits a number of intermediate forms, a fact which has added to the problem.

GRAY OAK

Quercus grisea Liebm.

FIELD IDENTIFICATION. Sometimes only a shrub on exposed mountain slopes, or becoming a tree to 65 ft in alluvial canyons.

FLOWERS. Monoecious, staminate catkins 2/3–1 1/2 in., or rarely longer, hairy or tomentose; anthers exserted, glabrous; pistillate catkins tomentose, 1/5–1 3/5 in., 1–6-flowered.

FRUIT. Acorn borne annually, sessile, or peduncle to 1 1/3 in. long, solitary or paired, 1/2–3/4 in. long, 1/3–1/2 in. high, ellipsoid to ovoid, light brown, glabrous or puberulent, one third to one half included in cup; cup 1/3–1/2 in. broad, 1/6–2/5 in. high, cup-shaped or goblet-shaped;

Gray Oak
Quercus grisea Liebm.

scales appressed, reddish brown, ovate to oblong, acute, pubescent to tomentose.

LEAVES. Grayish green, entire or toothed, variable in size and toothing, thick and leathery, blades ¾–3 in. long, ⅓–1½ in. broad, ovate, oblong, or elliptic; apex acute, obtuse or rounded, mucronulate; base rounded or cordate; margin entire or with small mucronate-tipped teeth; upper surface dull grayish green, minutely stellate-pubescent; lower surface stellate-pubescent or tomentulose and conspicuously veined; petioles 1/12–⅖ in., tomentose.

TWIGS. Slender, grayish brown, all parts (leaves, petioles, and twigs) with a dense grayish green to buff tomentum; stipules filiform, persistent; buds small, reddish.

BARK. Dark gray, with narrow ridges and rather straight furrows.

RANGE. Over wide areas in the Trans-Pecos region of Texas at altitudes to 7,800 ft, also in southern New Mexico, and is less abundant in Arizona. Southward it occurs in Chihuahua and Coahuila, Mexico.

HYBRIDS. Gray Oak is known to hybridize with Mohr Oak, *Q. mohriana* Buckl. *ex* Rydb., to produce a polybrid series, some with predominant Gray Oak characters and some with Mohr Oak features. The two species are very similiar in appearance, and many of the hybrids are not clearly defined. Muller (1951, p. 59) has pointed out that Mohr Oak is partial to limestone areas and Gray Oak to igneous regions. Overlapping of these geological areas bring the two into close proximity, and facilitates their production of a nonuniform polybrid series. Probably the most distinctive comparative character of the foliage of Gray Oak and Mohr Oak is the fact that the former is grayish green and tomentose, both above and below, while the latter is dark green above and whitish with tomentum below.

Gray Oak has also been reported under the name of *Q. endemica* (Muller, 1940a, p. 706).

In the northern part of its range Gray Oak also hybridizes with Arizona White Oak, *Q. arizonica*, to produce the Organ Mountains Oak, *Q.* × *organensis* Trel., named from a single tree in the Organ Mountains, Dona Ana County, New Mexico. Shrub Live Oak, *Q. turbinella* Greene, is also thought to hybridize with Gray Oak and produce a vague series of strains.

REMARKS. The genus name, *Quercus*, is the classical name, and the species name, *grisea*, refers to the gray appearance of the leaves. Vernacular names are Shin Oak, Scrub Oak, Encina, Prieta, and Encina Blanca. The wood is used for fuel and posts, and the leaves are browsed by livestock, deer, and porcupine. Many species of ground squirrels feed on the nut, also some birds, including Viosca's pigeon and the thick-billed parrot.

Sandpaper Oak

Quercus pungens Liebm.

FIELD IDENTIFICATION. A shrub or small tree.

FLOWERS. Staminate, loosely flowered, ½–1½ in.; pistillate catkins ⅛–⅕ in., 1–3-flowered.

FRUIT. Acorn annual, solitary or paired, peduncle to ⅛ in. long, sometimes sessile, length ⅓–⅖ in., ⅕–⅖ in. broad, ovoid to subcylindric, apex rounded to subacute, light brown, glabrous, one fourth included in the cup; cup ⅙–⅓ in. high, ¼–½ in. broad, shallowly to deeply cup-shaped, base rounded; scales ovate to oblong, closely apressed, apex thin, base thickened, reddish brown, gray-tomentulose to pubescent.

LEAVES. Evergreen, simple, alternate, coriaceous, stiff, blades ⅜–1½ in. long (rarely to 3½ in.), width ⅜–¾ (rarely to 1½ in. wide), elliptic to oblong, apices acute to obtuse and mucronate, bases rounded or cordate, margin coarsely toothed or incised, teeth and lobes acute and mucronate, margins crisped-wavy on the plane surface; upper surface shiny, stellate-hairy and rather scabrous to the touch; lower surfaces densely hairy with simple or stellate and appressed or spreading hairs, veins rather conspicuous, becoming less hairy

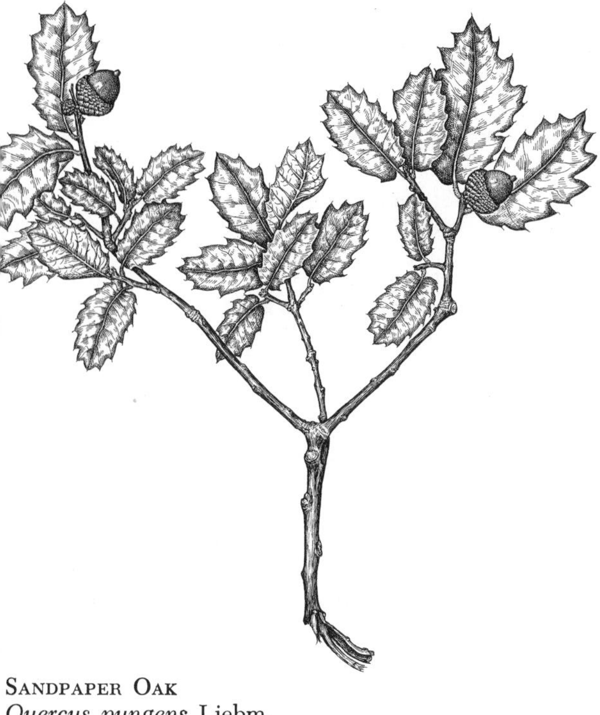

SANDPAPER OAK
Quercus pungens Liebm.

than the inter-vein spaces with age; petioles densely hairy, 1/12–1/5 (or to 1/2) in. long.

TWIGS. Gray, densely hairy, later glabrate, fluted, 1/25–1/12 in. thick, lenticels small and scattered; buds ovoid, reddish brown, slightly pubescent, about 1/12 in. long; stipules caducous, about 1/6 in., subulate, pubescent.

RANGE. Usually on dry limestone slopes and along arroyos at lower elevations in the mountains. In Texas in Brewster, Culberson, El Paso, and Hudspeth counties. Also in adjacent New Mexico and Arizona. In Mexico in Coahuila, Chihuahua, Nuevo León, and Tamaulipas.

REMARKS. The genus name, *Quercus,* is the ancient classical name, and the species name, *pungens,* refers to the sharply toothed leaves. The rough feel of the leaf surface, because of the stiff, short hairs, has given it the name of Sandpaper Oak. Also known as Scrub Oak and Encino.

VASEY OAK

Quercus pungens var. *vaseyana* (Buckl.) C. H. Muller

FIELD IDENTIFICATION. Usually a shrub 1–5 ft, forming thickets over wide areas, or under favorable conditions a small tree to 20 ft.

FLOWERS. Staminate and pistillate catkins borne separately; staminate 1–1¼ in. long, loosely flowered, hairy; stamens exserted; calyx 4–5-lobed, lobes ovate, scarious, apex obtuse to rounded; pistillate 1–3-flowered on tomentose peduncles 1/8–1/5 in., subtended by ovate, pubescent scales shorter than the pubescent calyx, stigmas red, reflexed.

FRUIT. Annual, solitary or in pairs, sessile or peduncle 1/12–1/8 in.; acorns ellipsoid to oblong or subcylindric; light brown, shiny and glabrous; one fifth to one third included in the cup, ½–¾ in. long, ¼–½ in. broad; cup saucer-shaped or cup-shaped, thin, base rounded, puberulous within, ¼–½ in. broad, 1/8–1/6 in. high; scales closely appressed, ovate; apex glabrous or nearly so, thin, acute, reddish; base thickened and tomentose.

LEAVES. Half-evergreen, alternate, coriaceous; oblong, obovate or lanceolate; margin with short, mucronate lobes, coarsely serrate to entire; apex acute, obtuse or rounded; base cuneate to rounded; blade length ¾–2½ in., ½–¾ in. wide; upper surface grayish green to dark green, shiny, glabrous or nearly so; lower surface paler, densely pubescent or varying to glabrous, veins prominent; petiole pubescent at first, glabrous later, 1/25–¼ in.; stipules about 1/8 in., subulate, caducous, pubescent.

TWIGS. Young slender, fluted, reddish brown to gray, tomentulose, later gray and glabrous; buds 1/25–1/8 in. long, ovoid to obovoid, reddish brown, pubescent or glabrous.

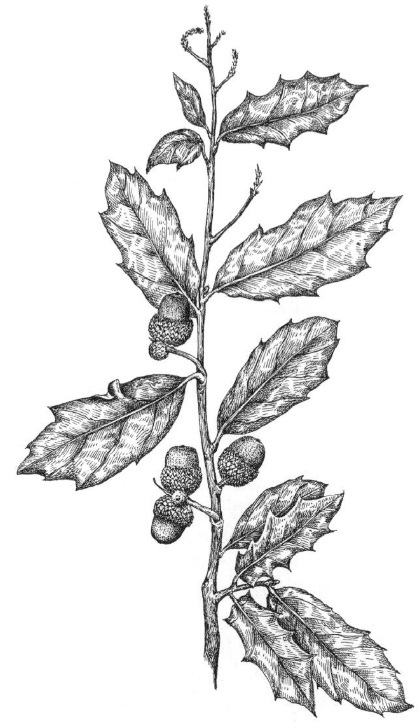

VASEY OAK
Quercus pungens var. *vaseyana* (Buckl.) C. H. Muller

LIVE OAK
Quercus virginiana Mill.

BARK. Grayish brown, rough, furrows deep, ridges scaly.

RANGE. Covering wide areas on dry limestone hills of central west Texas. Also in the Mexican states of Nuevo León, Tamaulipas, and Coahuila. In Texas in the counties of Edwards, Crockett, Kimble, Kinney, Kendall, Kerr, Menard, Pecos, Real, Schleicher, Sutton, Terrell, Uvalde, and Val Verde. The type was collected in Val Verde County.

REMARKS. The genus name, *Quercus,* is the ancient classical name. The species name, *pungens,* means "prickly," referring to the mucronate teeth of the leaf, and the variety name, *vaseyana,* is in honor of the botanist G. R. Vasey, who made collections in Texas, New Mexico, and Arizona in 1881. Vernacular names are Scrub Oak and Shin Oak. Various species of ground squirrels eat the acorn. The wood is scarcely used except as fuel. This variety has been reported on by some authors as *Q. vaseyana* Buckl. and *Q. undulata* var. *vaseyana* (Buckl.) Rydb. Cornelius H. Muller has set up a new combination for the nomenclature. (Muller, 1951, pp. 70-71)

LIVE OAK

Quercus virginiana Mill.

FIELD IDENTIFICATION. Evergreen tree to 60 ft, with a wide-spreading crown and massive limbs close to the ground.

FLOWERS. Staminate and pistillate borne in separate catkins on same tree; staminate catkins hairy, 2–3 in. long; calyx yellow, with 4–7 ovate lobes; stamens 6–12, filaments short, anthers hairy; pistillate catkins fewer, on pubescent peduncles 1–3 in. long; scales and calyx-lobes hairy; stigmas 3, red.

FRUIT. Acorn on peduncles ¼–4 in. long, in clusters of 3–5; nut ellipsoid-obovoid, brownish black, shiny, ⅓–½ in. long; enclosed about one half its length in the cup; cup turbinate, light reddish brown, hoary-tomentose; scales of cup ovate, acute, thin, appressed.

LEAVES. Simple, alternate, persistent, coriaceous, dark green and lustrous above, paler and glabrous to pubescent beneath; very variable in size and shape, 2–5 in. long, ½–2½ in. wide, oblong or elliptic or obovate; apex rounded or acute, base cuneate, rounded or cordate, margin entire and often revolute, sometimes sharp-dentate, especially toward the apex; petioles about ¼ in. long, stout, glabrous or puberulent.

TWIGS. Grayish brown, glabrous, slender, rigid; terminal buds ovate to subglobose, about ⅙ in. long, light brown; leaf scars half-moon–shaped.

BARK. Dark brown to black (in some varieties gray), furrows narrow and interlacing, scales closely appressed, small.

WOOD. Light brown, sapwood lighter, close-grained, tough, hard, strong, heavy, weighing about 59 lb per cu ft, difficult to work.

RANGE. The species is usually found in sandy-loam soils but may also occur in heavier clays. In Texas, Oklahoma, and Louisiana; east to Florida and north to Virginia.

VARIETIES AND HYBRIDS. On the Edwards Plateau area of Texas the species passes into the shrubby Texas Live Oak, *Q. virginiana* var. *fusiformis* (Small) Sarg., differing by oblong-ovate leaves with an acute apex, rounded or cuneate at base, entire or dentate margin, and a smaller acorn. However, intermediate forms occur between the variety and the species and the distinctions are often difficult.

Bay Live Oak, *Q. virginiana* var. *maritima* (Michx.) Sarg., is a shrub, or smaller tree than the species, found in sandy soil, usually near the coast, from eastern Louisiana, eastward to Florida and northward to North Carolina. In Florida it is not only found on the coastal sand dunes but also in the scrub lands of the central part of the peninsula. Trunk often leaning, with light to dark gray bark which is shallowly and irregularly fissured, on the coast where subjected to strong winds assuming an irregular, dwarfed, and contorted shape; leaves 1½–3 in. long, ⅓–1 in. wide, oblong to elliptic

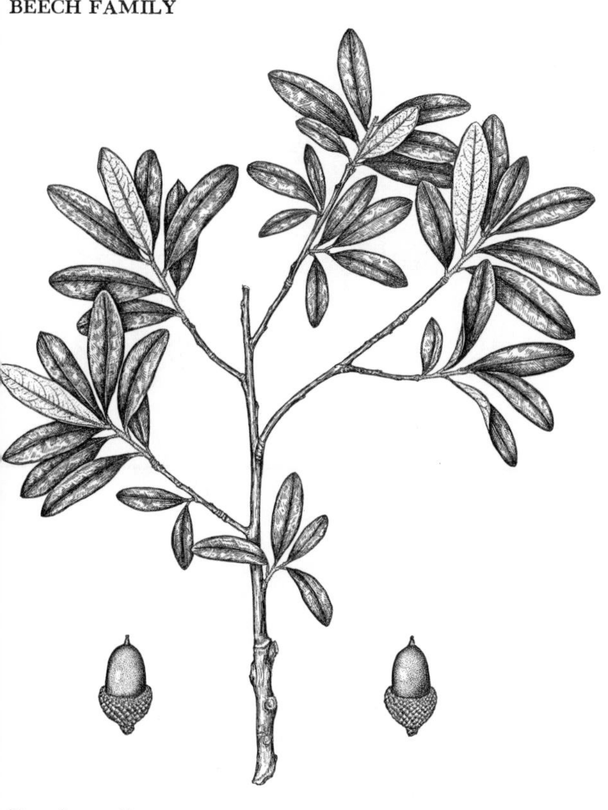

Bay Live Oak
Quercus virginiana var. *maritima* (Michx.) Sarg.

or somewhat narrowly obovate, apex acute, base narrowed gradually or rounded, margin conspicuously rolled back and thickened, surface reticulate-venulose, upper surface shiny and more or less glabrous, lower surface conspicuously hoary-tomentose, often somewhat curved and boat-shaped in appearance; fruit and flowers the same as in the species. This variety has been listed by some authors under the names of *Q. virginiana* var. *geminata* (Small) Sarg., *Q. geminata* Small, *Q. virginiana* var. *eximia* Sarg., *Q. phellos* var. *maritima* (Michx.) Sarg., *Q. andromeda* Ridd., and *Q. virens* var. *maritima* (Michx.) Chapm.

Compton Oak, *Q.* × *comptoniae* Sarg., is a hybrid of *Q. lyrata* × *Q. virginiana.* It is named for (Miss) C. C. Compton of Natchez, Mississippi. It occurs in southeastern Texas.

Harbison Oak, *Q.* × *harbisonii* Sarg., is a hybrid of *Q. stellata* × *Q. virginiana* and was named for Thomas Grant Harbison (1862–1936), American botanist. It occurs in Florida and in eastern and central Texas.

REMARKS. The genus name, *Quercus,* is the classical name, and the species name, *virginiana,* refers to the state of Virginia. The wood is used for hubs, cogs, shipbuilding, or for any other purpose where a hard and strong wood is required. The tree is often planted in the Southern states for ornament along avenues. It is comparatively free of insect pests and diseases, and can stand considerable salinity, often growing close to the sea. The bark was formerly much used in production of tannin, and acorn oil was used in cooking by the Indians. Live Oak seems to be susceptible to soil types, and produces dwarf varieties and diverse leaf forms under certain conditions. The fruit seems to vary least in the varieties.

Dwarf Live Oak

Quercus oleoides var. *quaterna* C. H. Muller

FIELD IDENTIFICATION. Shrubs 1½–9 ft, forming dense thickets by spreading rhizomes.

FLOWERS. Not seen.

FRUIT. Annual, solitary or paired, or variously grouped on peduncles, ⅜–⅔ in. long, cups ½–⅗ in. in diameter, turbinate, at base strongly constricted, the scales usually markedly vertical-seriate. Acorns elliptic to narrowly ovoid, glossy brown, usually less than one half included.

Dwarf Live Oak
Quercus oleoides var. *quaterna* C. H. Muller

LEAVES. Evergreen or subevergreen, ⅝–1½ in. long or rarely longer, ¼–¾ in. broad, entire, dentate, or only undulate, oblong or oblanceolate to obovate; acute or rounded at apex; the bases obtuse or narrowly cuneate; margins markedly revolute; upper surfaces glabrate and shiny, becoming yellow or yellowish brown upon drying, lower surfaces canescent with closely appressed minute stellate hairs.

TWIGS. Densely canescent, with minute appressed stellate hairs.

RANGE. On heavy soil on the coastal plain and on sand ridges or dunes near the beaches, probably following the Gulf Coast and merging with the species in Veracruz, Mexico. In Texas in Aransas, Brazoria, Calhoun, Jackson, Jefferson, Nueces, and Victoria counties.

HYBRID. Dwarf Live Oak is known to hybridize with Post Oak, *Q. stellata,* along the Texas coast.

REMARKS. The genus name, *Quercus,* is the classical name, and the species name, *oleoides,* means "olive-like," because of the leaf color.

There has long been considerable question concerning the identity of the so-called Dwarf Oaks which occur in the sandy soils close to the Gulf of Mexico from Texas to Florida. Muller (1951, p. 76) has described the new Texas variety, *quaterna,* which he reasons has its closest relative *(Q. oleoides)* in the Gulf region of Mexico and Central America. The notes on this new variety closely follow Muller's original description.

Toumey Oak

Quercus toumeyi Sarg.

FIELD IDENTIFICATION. Shrub 3–6 ft, or sometimes a tree to 30 ft, with a trunk diameter of 6–8 in. The branches form a broad irregular crown.

FLOWERS. Staminate and pistillate in separate catkins; staminate yellowish green, usually less than 1 in. long; pistillate sessile.

FRUIT. Maturing June–July, solitary or in pairs, acorn enclosed about one half its length in the cup, oval to ovoid, ½–⅔ in. long, about ¼–⅓ in. in diameter, light brown, shiny, apex sometimes puberulent; cup saucer-shaped, shallow, thin; scales closely imbricate, reddish brown, ovate, apex obtuse or rounded, surface pale-tomentose.

LEAVES. Simple, alternate, dense; ovate, elliptic, oblong or oval, ⅜–1½ in. long, ¼–¾ in. wide, margin entire or few-toothed, apex acute, mucronulate or sometimes minutely 3-toothed, base rounded or cordate, thin and firm, upper surface light bluish green and glabrous, lower surface paler and sparsely pubescent, conspicuously reticulate-venulose; petioles stout, pubescent, about 1/16 in. long or longer.

Toumey Oak
Quercus toumeyi Sarg.

TWIGS. Slender, young ones reddish brown and tomentose, later darker and glabrous, with small thin scales; bark on trunk scaly and flaky.

RANGE. In dry well-drained soil at altitudes of 3,500–6,000 ft. Southwestern New Mexico, southern Arizona, and Mexico in Sonora and Chihuahua.

REMARKS. The genus name, *Quercus,* is the classical name, and the species name, *toumeyi,* is in honor of James William Toumey (1865–1932), American forester and botanist.

Canyon Live Oak

Quercus chrysolepis Liebm.

FIELD IDENTIFICATION. Sometimes an evergreen shrub to 9 ft under unfavorable conditions, but usually becoming a large tree to 50 ft and 2–5 ft in diameter. The large heavy limbs are often horizontal. This tree is represented in extreme southwestern New Mexico by the variety known as Palmer Canyon Live Oak. *Q. chrysolepis* var. *palmeri* (Engelm.) Sarg. The variety is described separately.

FLOWERS. May–June, staminate and pistillate in separate catkins; staminate slender, 2–4 in. long, tomentose; calyx 5–7-lobed, each lobe broad-ovate, acute, ciliate, surfaces yellow to reddish; stamens short; pistillate flowers sessile or nearly so, solitary or occasionally in 2–3-flowered spikes, scales broadly ovate, brown-tomentose; stigmas bifid, red.

FRUIT. Maturing September–October, usually solitary, short-stalked or sessile; cup basal, thin and hemispheric to thick-turbinate, the rim thickened, internally reddish brown, externally with scales deltoid, closely appressed, clothed with light or dense brown tomentum.

LEAVES. Simple, alternate, persistent, elliptic to oblong or ovate, apex acute or cuspidate, base rounded to cuneate or cordate, varying from entire to remotely stiff-spinescently toothed, somewhat revolute and thickened; young leaves with dense brown tomentum; mature leaves thick and stiff, upper surface yellowish green and glabrous, lower surface brown-tomentose at first, gradually becoming whitish or pale and glabrous, usually persistent for 2–4 years; blades 1–4 in. long, ½–2 in. wide; petioles slender, yellowish, ⅕–½ in. long.

TWIGS. Slender, rigid, at first brown-tomentose, later reddish brown, and eventually gray, gradually less hairy to glabrous; winter buds ⅙–⅛ in. long, ovoid, acute, scales imbricate, brown, puberulous to pubescent.

RANGE. The species Canyon Live Oak is distributed from Oregon through California on the coast ranges and the slopes of the Sierra Nevada.

REMARKS. The genus name, *Quercus*, is the classical name. The species name, *chrysolepis*, refers to the golden yellow acorn-cup scales. Locally the tree is sometimes known as Maul Oak. The wood is used for furniture, agricultural implements, pack saddles, and fuel. It is planted for ornament, watershed protection, and soil stabilization.

CANYON LIVE OAK
Quercus chrysolepis Liebm.

PALMER CANYON LIVE OAK

Quercus chrysolepis var. *palmeri* (Engelm.) Sarg.

FIELD IDENTIFICATION. A shrub or small tree attaining a height of 25 ft, forming thickets on mountainsides.

FLOWERS. May–June, staminate and pistillate in separate catkins; staminate slender, 2–4 in. long, tomentose; calyx 5–7-lobed, each lobe broad-ovate, acute, ciliate, surfaces yellow to reddish; stamens short; pistillate flowers sessile or nearly so, occasionally in 2–3-flowered spikes, scales broadly ovate, brown-tomentose; stigmas bifid, red.

FRUIT. Short-peduncled or almost sessile; acorn cups turbinate to saucer-shaped, ⅖–½ in. broad, thin-walled, inner surface of shell pale-tomentose, externally the scales appressed, basally thickened, golden-tomentose; acorn fitting loosely, ovoid to oblong, acute, ¾–1½ in. long.

LEAVES. Evergreen, varying from elliptic to lanceolate or oblong to ovate; margins crisped, set with long mucronate or aristate teeth, body somewhat concave, rigid, very coriaceous; upper surface dull green; lower surface usually yellowish and resinous-pubescent when young, later glabrous and whitish, length ¾–1⅓ in.; petiole tomentulose.

BARK. With small checked, grayish brown scales.

RANGE. In the grassland belts or in desert areas, canyons, moist ridges, flats. Southwestern New Mexico, Arizona, and California; southward into Mexico.

REMARKS. The variety name, *palmeri*, is in honor of Edward Palmer, English-born American physician and collector. This variety seems to intergrade very closely with the species *Q. chrysolepis* Liebm. However, some authors have considered it distinct enough to merit a species standing, and have given it the name of *Q. palmeri* Engelm. or *Q. wilcoxii* Rydb.

Silver-leaf Oak

Quercus hypoleucoides A. Camus

FIELD IDENTIFICATION. Shrub, or moderate-sized tree, attaining a height of 35 ft and a diameter of 15 in., the slender branches forming a narrow head.

FLOWERS. Staminate and pistillate catkins borne separately on same tree; staminate 1⅕–5 in. long, loosely flowered, slender, white-hairy and reddish green; calyx with 4–5 ovate, rounded lobes; anthers exserted, red or yellow, ellipsoid, acute, glabrous; pistillate catkins ⅕–⅖ in. long, 1–2-flowered terminally, peduncle stellate-tomentose; scales thin, scarious, soft-hairy; stigmas red.

FRUIT. Annual, solitary or paired, sessile or on hairy peduncles ⅕–⅗ in. long; acorn about one third included in the cup, oblong to ovoid, apex acute to rounded, densely hairy to glabrescent later, at first green and sometimes striate, later light brown, length ½–⅗ in., ¼–⅖ in. broad; cup turbinate or hemispheric, thick, about ⅖ in. in diameter, interior of cup white-pubescent, exterior of cup with scales which are appressed, thin, broadly ovate, obtuse or rounded at apex, pale brown, and densely silvery white-pubescent.

LEAVES. Simple, alternate, persistent, blades 2–4 in. long, ⅗–1⅓ in. wide, lanceolate to elliptic or oblong, sometimes falcate, apex acute and apiculate, base rounded to cuneate or cordate, margin entire or occasionally with small, rigid, spinose teeth or on young shoots with oblique, acute serrations; strongly revolute, thick and leathery; upper surface in mature leaves dark yellow or bluish green and lustrous, glabrous or minutely puberulent; lower surface densely silvery or fulvous canescent-tomentose, prominently veined; petiole stout, ⅛–⅗ in. long, pubescent or tomentose.

TWIGS. Light brown or reddish brown, younger gray-tomentose, older smoother and glaucous to gray or black; buds about ⅛ in. long, ovoid, obtuse, reddish brown, scales pubescent to glabrescent, ciliate.

BARK. Grayish black, fissures deep, ridges broad, dividing into thick scales.

WOOD. Dark brown, sapwood lighter, close-grained, heavy, hard.

RANGE. In Trans-Pecos Texas, New Mexico, Arizona, and Mexico, usually at altitudes of 4,000–9,000 ft. Nowhere abundant, but scattered in pine forests in canyons or on slopes and high ridges. In Texas collected by the author in the upper reaches of Madera Canyon on the slopes of Mount Livermore, in Jeff Davis County, Texas. In New Mexico in the Mogollon and Black mountains. In Mexico in Chihuahua, Coahuila, and Sonora.

REMARKS. The genus name, *Quercus,* is the classical name of the oak, and the species name, *hypoleucoides,* refers to the very white undersurface of the leaf. It is also known as White-leaf Oak. It is a rather attractive tree worthy of cultivation. The acorns are eaten by the Apache squirrel and thick-billed parrot. It is known to hybridize with Graves Oak, *Q. gravesii,* to produce Livermore Oak, *Q.* × *inconstans* Palmer. The hybrid is very variable and occurs where the two species make contact in the Davis Mountains of Jeff Davis County, Texas.

Silver-leaf Oak
Quercus hypoleucoides A. Camus

Emory Oak

Quercus emoryi Torr.

FIELD IDENTIFICATION. Shrub or tree to 60 ft, with a rounded shape and a black trunk.

FLOWERS. Monoecious; staminate catkins 1–2 in. long, hoary-tomentose; perianth yellow, hairy, lobes 5–7, ovate and acute, anthers exserted; pistillate sessile or short-stalked, solitary or 2 together, stellate-tomentose.

FRUIT. June–September, acorn solitary or paired, sessile, or on short, stellate-tomentose peduncle, oblong or broad-ovoid, apex rounded or acute, hardly over ½–¾ in. long, one half to three fourths its length extruded from the cup, dark brown to black when mature; cup ⅓–½ in. broad, ⅙–⅖ in. high, cup-shaped or hemispheric, scales brown, appressed, ovate, ciliate, acute, pubescent or glabrous later.

LEAVES. Half-evergreen, varying considerably in size on different trees, ¾–3½ in. long, ⅓–1⅓ in. broad; margins entire, or with remote, coarse, spinulose teeth, oblong to lanceolate or narrowly ovate, acute to obtuse and spinulose at the apex, rounded to cordate or trun-

Emory Oak
Quercus emoryi Torr.

cate at base, dark green and lustrous above, paler beneath, stiff and leathery, surface with scattered stellate hairs especially along the midrib beneath; petioles ⅛–½ in. long, flattened, brownish stellate-hairy, more glabrous later.

TWIGS. Reddish brown to gray, fluted, pubescent with stellate hairs, becoming glabrous later.

BARK. Dark brown or black, on old trunks roughly broken into thick plates with small scales, the fissures deep.

WOOD. Dark brown, close-grained, heavy, strong, rather brittle, specific gravity about 0.93.

RANGE. At altitudes of over 4,000 ft. Common in the Chisos, Davis, and Limpia mountains of Texas, and in southern New Mexico and Arizona. Also in Mexico in the states of Nuevo León, Chihuahua, and Sonora.

CLOSELY RELATED FORMS. Muller (1951, p. 79) has described a hybrid between *Q. emoryi* and *Q. graciliformis* as occurring in Blue Creek Canyon of the Chisos Mountains in Brewster County, Texas. It is now recorded as Tharp Oak, *Q.* × *tharpii* C. H. Muller.

Benjamin C. Tharp is a Texas botanist, for many years chairman of the Botany Department of the University of Texas.

Formerly Muller had thought this to be a distinct species (Muller, 1938, p. 586). It is known by narrow, glabrous, oblanceolate leaves with 4–5 spiny teeth on each side and entire below the middle; the apex is acute and the base cuneate or rounded.

Muller has also described a hybrid between *Q. emoryi* and *Q. gravesii* as occurring in the Chisos Mountains of Brewster County, Texas (Muller, 1951, p. 80). Muller originally presented this hybrid as a species (Muller, 1934, pp. 119–120). "The pubescence of the fruit, and the frequently elongated terminal leaf lobes indicate a close relationship to *Quercus gravesii*, while the coriaceous blades and failure to become crimson in fall are characteristics of *Quercus emoryi*. A study of a large number of collections in Oak Canyon, the type locality of *Quercus robusta*, has revealed a great number of intermediates between the proposed species and the two parents." The hybrid is therefore now known as Robust Oak, *Q.* × *robusta* C. H. Muller.

It is also believed by some botanists that *Q. emoryi* hybridizes with *Q. pungens* Liebm. (Wooton and Standley, 1915, p. 169). *Q. emoryi* is also thought to hybridize with *Q. hypoleucoides* A. Camus.

REMARKS. The genus name, *Quercus*, is the classical name, and the species name, *emoryi*, is in honor of William H. Emory (1811–1887), who was a member of the commission which conducted a survey of the United States–Mexico boundary in 1857. Vernacular names are Western Black Oak, Blackjack Oak, and Bellota. The leaves are browsed extensively by mule deer. The

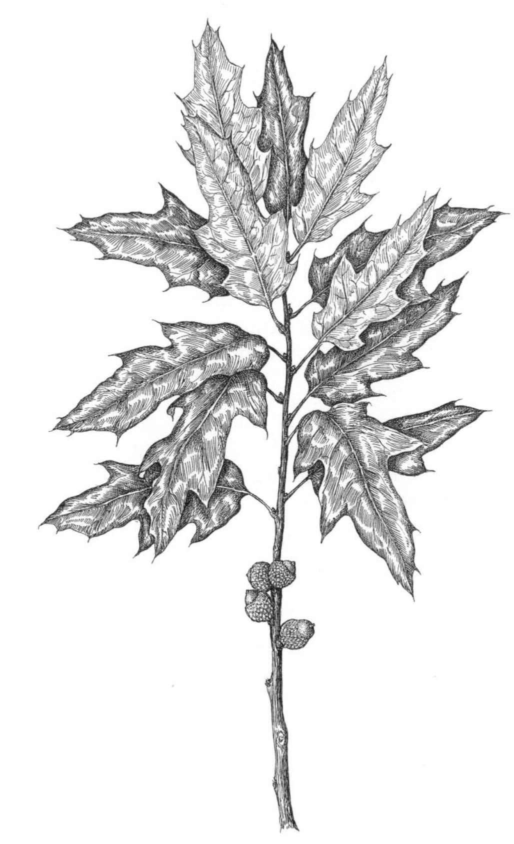

Chisos Hybrid Oak
Quercus × *robusta* C. H. Muller
(Quercus emoryi and *Q. gravesii)*

acorns are eaten by the Gambel's, Mearns's, and scaled quail, and are also greedily consumed by many species of chipmunk and ground squirrel. They are often gathered for food by the Indian and Mexican people.

Willow Oak
Quercus phellos L.

FIELD IDENTIFICATION. Tree to 130 ft, and 6 ft in diameter, with a rounded or broad-oblong crown.

FLOWERS. Borne in spring in separate staminate and pistillate catkins; moderately close-flowered, hairy, yellowish green, with 4–5 acute calyx-lobes, anthers oval and exserted; pistillate flowers usually solitary, occasionally in pairs, glabrous peduncles 1/25–1/8 in. long, 1–3-flowered; scales and calyx hairy; stigmas red, slender, recurved.

FRUIT. Maturing in 2 years, solitary or in pairs, sessile, or on short peduncles to 1/5 in. long; acorn subglobose to ovoid, 2/5–3/5 in. long, nearly as broad as long, densely puberulent or glabrate later, yellowish to dull brown, sometimes striate, about one fourth of base enclosed in cup; cup 2/5–3/5 in. broad, 1/6–1/3 in. high, saucer-shaped, shallow, margin not inrolled, enveloping only the base of the acorn; scales closely appressed, imbricate, small, thin, ovate, greenish brown, finely tomentose. Acorns averaging about 600 per lb.

LEAVES. Revolute in the bud, alternate, simple, deciduous, linear-lanceolate to elliptic, entire on margin, apex acute and bristle-tipped, base cuneate or narrowly rounded; light to dark green and glabrous above, or slightly pubescent on the midrib; lower surface paler and glabrous or pubescent along the midrib, blade length 2–5 in., width 1/3–1 in.; petioles stout, 1/25–1/4 in. long, tomentose at first and glabrate later.

TWIGS. Reddish brown and pubescent at first, gray and glabrous later, slender, fluted; buds brown, ovoid to lanceolate, apex acute, 1/12–1/6 in. long, scales mostly glabrous and ciliate; stipules caducous, 1/4–1/3 in. long, filiform, villous.

WOOD. Light brown, close-grained, soft, moderately strong, not durable, weighing 46 lb per cu ft, of rather low quality in comparison with that of other oaks.

RANGE. Willow Oak grows usually on rich, wet bottom lands of clays or loams. Eastern Texas, Oklahoma, Arkansas, and Louisiana; eastward to Florida, north to New York, and west to Illinois.

HYBRIDS. The following hybrids have been recorded by Little (1953a, pp. 345–346).

Capesius Oak, *Q.* × *capesii* W. Wolf, is a cross of *Q. phellos* and *Q. nigra*. It was named after A. Capesius who discovered the hybrid in seedlings raised from a cultivated tree at St. Bernard, Alabama. It also occurs in New Jersey, South Carolina, and southeast Texas (in Hermann Park along ravines, Houston, Harris County, Texas).

Willow Oak
Quercus phellos L.

Filial Oak, *Q.* × *filialis* Little, is a cross of *Q. phellos* and *Q. velutina*. The hybrid name is from the filial generation of a cross. It occurs in New Jersey, Delaware, Missouri, Arkansas, and Louisiana.

Gifford Oak, *Q.* × *giffordii* Trel., is a cross of *Q. phellos* and *Q. ilicifolia*. It was named after John Clayton Gifford (1870–1949), American forester. It occurs in New Jersey and Delaware.

Bartram Oak, *Q.* × *heterophylla* Michx. f. is a hybrid of *Q. phellos* and *Q. rubra*. The name means "various-leaved." It occurs in New York, New Jersey, Pennsylvania, Delaware, District of Columbia, Virginia, Tennessee, Missouri, Arkansas, southeastern Oklahoma, Alabama, and North Carolina.

St. Landry Oak, *Q.* × *ludoviciana* Sarg., is a hybrid of *Q. phellos* and *Q. falcata*. It was named for the state of Louisiana where it was discovered. It is also found in Virginia, Kentucky, Missouri, Arkansas, southeast Texas, Mississippi, Alabama, and Georgia. (In Brazoria County and Hardin County, Texas.)

Moulton Oak, *Q.* × *moultonensis* Ashe, is named from Moulton Valley of the Tennessee River, Lawrence County, Alabama. It occurs also in Virginia, Tennessee, and Arkansas. It is a hybrid of *Q. phellos* and *Q. shumardii.*

Rudkin Oak, *Q.* × *rudkinii* Britton, is a hybrid of *Q. phellos* and *Q. marilandica.* It was named for William H. Rudkin. It occurs in New York, New Jersey, Delaware, Virginia, North Carolina, South Carolina, Georgia, Florida, Louisiana, and Arkansas.

Schoch Oak, *Q.* × *schochiana* Dieck, is a hybrid of *Q. phellos* and *Q. palustris.* It was named for State Garden Director Schoch, who discovered it in a park at Worlitz, Germany. It occurs in Virginia, Kentucky, Illinois, and Arkansas.

REMARKS. The genus name, *Quercus,* is the classical Latin name of the oaks, and the species name, *phellos* ("cork"), is the ancient Greek name of *Q. suber* L., Cork Oak. Other local names are Water Oak, Peach Oak, Sandjacks Oak, Red Oak, Swamp Oak, Swamp Willow Oak, and Pin Oak. Willow Oak is often called Pin Oak in many sections of the South, but that name should apply to the true Pin Oak, *Q. palustris* Muenchh. The author has never seen the true Pin Oak in Texas outside of cultivation. However, there is a remote possibility that the true Pin Oak may extend far enough southward to reach into the northeastern corner of the state. Other oaks which are apt to be confused with the Willow Oak are Laurel Oak, Bluejack Oak, Water Oak, and Diamond-leaf Oak. Willow Oak makes an exceedingly handsome ornamental tree and has been cultivated since 1723. The wood is somewhat inferior to that of other commercial oaks but is used for fuel, charcoal, ties, shingles, sills, planks, and general construction. The acorns are eaten by wild turkey, bobwhite quail, dove, jay, gray fox, and squirrel.

Bluejack Oak

Quercus incana Bartr.

FIELD IDENTIFICATION. Shrub or tree to 35 ft, with stout crooked branches.

FLOWERS. In spring, in staminate and pistillate catkins; staminate catkins clustered, 2–3 in. long, hairy; calyx-lobes 4–5 in. ovate, acute, red to yellowish green; stamens 4–5, yellow; anthers apiculate; pistillate catkins on stout, tomentose, short peduncles; scales and calyx-lobes of pistillate flowers tomentose, stigmas dark red.

FRUIT. Maturing the second season, sessile or short-stalked, globose to ovoid, sometimes flattened, brown with grayish pubescence, often striate, about ½ in. long, set in a shallow cup one third to one half its length, kernel bitter; cup shallow, saucer-shaped; scales imbricate, thin, ovate, tomentose, reddish brown.

LEAVES. Alternate, simple, deciduous, entire (or rarely 3-dentate at the apex), oblong-lanceolate to elliptic, distinctly grayish green, densely tomentose beneath, smoother above, cuneate or rounded at base, acute or rounded at the apex, apiculate, 2–5 in. long, ½–1½ in. wide; petiole ¼–½ in. long, stout.

TWIGS. Gray to dark brown, slender, smooth.

BARK. Grayish brown to black, broken into small blocklike plates.

WOOD. Reddish brown, close-grained, hard, strong.

RANGE. Usually in dry sandy pinelands of east Texas, Louisiana, Oklahoma, and Arkansas; north and east to North Carolina and Virginia.

HYBRIDS AND SYNONYMS. Considerable diversity of opinion has existed concerning the scientific name of the Bluejack Oak. At one time it was listed as *Q. cinerea* Michx., but the validity of the name of William Bartram has been pointed out by Francis Harper (1943) and E. D. Merrill (1945). It should now be known as *Q. incana* Bartram.

The following hybrids have been listed by Little (1953a, p. 333):

Ashe Oak, *Q.* × *asheana* Little, is a cross of *Q. incana* and *Q. laevis,* and is named after William Willard Ashe (1872–1932), pioneer forester of the U. S. Forest Service. It occurs in central Florida and Mississippi.

Albemarle Oak, *Q.* × *atlantica* Ashe, is a cross of *Q. incana* and *Q. laurifolia,* and is named after the Atlantic. It occurs in South Carolina and Georgia, to central and northwestern Florida and Alabama.

Bluejack Oak
Quercus incana Bartr.

Caduca Oak, *Q.* × *caduca* Trel., is a cross of *Q. incana* and *Q. nigra,* and means "promptly falling." It occurs in Virginia, Georgia, Florida, Alabama, Mississippi, and eastern Texas.

Carolina Oak, *Q.* × *cravenensis* Little, is a cross of *Q. incana* and *Q. marilandica,* and is named for Craven County, North Carolina. It occurs in southeastern Virginia, North Carolina, South Carolina, Georgia, Alabama, and East Texas.

Footleaf Oak, *Q.* × *podophylla* Trel., is a cross of *Q. incana* and *Q. velutina,* its Greek species name meaning "foot and leaf," in reference to the stout petiole.

The Subentire-leaf Oak, *Quercus* × *subintegra* Trel., is a cross of *Q. incana* and *Q. falcata,* and the name applies to the almost-entire leaf margin. It is found in Maryland, Virginia, South Carolina, Georgia, Florida, and Alabama.

A variety of Bluejack Oak was at one time listed as *Q. cinerea* var. *dentato-lobata* A. DC., with three-lobed leaves, but this variety has been reduced to a synonym with the species.

REMARKS. The genus name, *Quercus,* is the ancient classical name, and the species name, *incana,* refers to the grayish green tomentum of the leaves. Vernacular names are Upland Willow Oak, High-ground Willow Oak, Sandjack Oak, Turkey Oak, and Cinnamon Oak. The trunk is generally too small to be of much value except for fuel or posts.

SHINGLE OAK
Quercus imbricaria Michx.

SHINGLE OAK

Quercus imbricaria Michx.

FIELD IDENTIFICATION. Tree usually to 60 ft in height, with a narrow rounded crown and a trunk diameter of 2–4 ft. Rarely attaining heights of 90–100 ft.

FLOWERS. April–May, borne in separate staminate and pistillate catkins; staminate numerous, slender, tomentose, 2–3½ in. long, calyx 4–lobed, yellow, pubescent; pistillate usually solitary on short, slender, tomentose peduncles; scales acute, shorter than the calyx-lobes or as long; stigmas greenish yellow, recurved.

FRUIT. Acorn biennial, mature September–October, solitary or paired; on short, stout peduncles to ½ in. long; ovoid to subglobose, ⅜–⅝ in. long; light to dark chestnut-brown, shiny, often striate, covered one third to one half its length by the cup, about 450–800 acorns per lb; cup thin, turbinate, about ½ in. across; cup scales small, appressed, imbricate, reddish brown, ovate-oblong, pubescent, margin ciliate, apex obtuse to truncate or rounded.

LEAVES. Alternate, simple, deciduous; oblong, elliptic, or oblong-lanceolate; apex acute, rounded or apiculate; base cuneate or rounded; margin entire, somewhat revolute, undulate or rarely 3-lobed; surface lustrous, dark green and glabrous above; paler and brownish pubescent beneath; midrib stout and yellow; petiole short, stout, pubescent, ⅕–⅗ in. long.

TWIGS. Slender, glabrous at maturity, green to reddish brown or gray, younger somewhat pubescent.

BARK. On young branches brownish gray and smooth, on older trunks grayish brown, close, eventually with low scaly ridges separated by shallow, perpendicular furrows.

WOOD. Brownish red, sapwood lighter, coarse-grained, hard, heavy, strong, weighing about 47 lb per cu ft.

RANGE. Hillsides and riverbanks. North Louisiana, Oklahoma, and Arkansas; east to Georgia, north to Massachusetts, and west to Kansas.

HYBRIDS. Shingle Oak is known to produce a number of hybrids. The following are listed by Little (1953a, p. 333):

Two-edge Oak, *Q.* × *anceps* Palmer, is a hybrid of *Q. imbricaria* and *Q. falcata.* It occurs in southeastern Illinois.

Eggleston Oak, *Q.* × *egglestonii* Trel., is a hybrid of *Q. imbricaria* and *Q. shumardii,* and was named for William Webster Eggleston (1863–1935), botanist of the U.S. Department of Agriculture. It occurs in Kentucky and Missouri.

Shinglepin Oak, *Q.* × *exacta* Trel., is a hybrid of *Q. imbricaria* and *Q. palustris,* named after the uniform spreading teeth. It occurs in Pennsylvania, Indiana, Illinois, and Missouri.

Lea Oak, *Q.* × *leana* Nutt., is a hybrid of *Q. imbricaria* and *Q. velutina,* and named for its discoverer Thomas Gibson Lea (1785–1844). It occurs in Pennsylvania, the District of Columbia, Ohio, Michigan, Indiana, Illinois, Iowa, Missouri, and North Carolina.

Bottom Oak, *Q.* × *runcinata* (A. DC.) Engelm., is a hybrid of *Q. imbricaria* and *Q. rubra,* and is named for the sharp, backward-pointing lobes of the leaves. It occurs in Pennsylvania, Ohio, Indiana, Missouri, Kansas, Kentucky, and Maryland.

St. Louis Oak, *Q.* × *tridentata* (A. DC.) Engelm., is a hybrid of *Q. imbricaria* and *Q. marilandica.* The name refers to the 3-toothed leaves. It occurs in Pennsylvania, Michigan, Illinois, Missouri, and Virginia.

REMARKS. The genus name, *Quercus,* is the ancient classical name, and the species name, *imbricaria,* means "overlapping," as with shingles. Michaux gave the name to the plant when he found early settlers making hand-hewn shingles from the tree. Other names are Lea Oak, Jack Oak, and Northern Laurel Oak. It is also used in small amounts for construction and fuel. It is generally marketed as Red Oak, but is considered most valuable as a shade tree in the North because of its dense, lustrous, dark green foliage. It has been cultivated since 1724. The acorn is eaten by squirrel and woodpecker.

Laurel Oak

Quercus laurifolia Michx.

FIELD IDENTIFICATION. Dense, round-topped tree attaining a height of 100 ft.

FLOWERS. Staminate and pistillate catkins borne separately on same tree in spring. Staminate catkins clustered, red, hairy, 2–3 in. long; calyx 4-lobed, pubescent; pistillate catkins short-stalked with brown-hairy involucral scales; calyx-lobes acute; stigmas red with short spreading styles.

FRUIT. Acorn usually solitary, sessile or subsessile, ovoid to hemispheric, dark brown, about ½ in. long, enclosed about one fourth its length in the cup; cup thin, saucer-shaped, reddish brown, with appressed ovate scales.

LEAVES. Alternate, simple, deciduous in the North, half-evergreen in the South, elliptic or oblong or occasionally obovate and lobed, sometimes falcate; leaves on young shoots or young plants often variously cut and lobed. Apex acute and apiculate, base narrowed, deep shiny green and glabrous above, paler and lustrous beneath, blades 2–6 in. long, ½–1 in. wide; petioles yellow, stout, rarely more than ¼ in. long.

TWIGS. Reddish brown to gray, glabrous, slender.

BARK. When young nearly smooth, dark brown tinged with red, when older dark gray to black, furrows separated by flat ridges.

WOOD. Reddish brown, coarse-grained, heavy, hard, strong, weighing 48 lb per cu ft, warps easily.

RANGE. In low grounds, eastern and coastal Texas and Louisiana; east to Florida and north to South Carolina and Virginia.

VARIETY AND HYBRIDS. A variety, *Q. laurifolia* var. *rhombica* (Sarg.) Trel., occurs in swamps and low wet woods, principally near the coast, from the Neches River Valley in eastern Texas and Red River Valley in Louisiana eastward to Florida and Virginia. The variety is distinguished by having leaves generally larger, rhombic in form, and apically rounded.

Beaumont Oak, *Q.* × *beaumontiana* Sarg., is a hybrid of *Q. laurifolia* var. *rhombica* and *Q. rubra,* and is named from the type locality, Beaumont, Texas.

The following hybrids of the Laurel Oak are listed by Little (1953a, pp. 335–336):

Albermarle Oak, *Q.* × *atlantica* Ashe, is a hybrid of *Q. laurifolia* and *Q. incana,* its species name meaning "Atlantic." It occurs in South Carolina and Georgia, to central and northwestern Florida and Alabama.

Mellichamp Oak, *Q.* × *mellichampii* Trel., is a hybrid of *Q. laurifolia* and *Q. laevis,* named after its discoverer, Joseph Hinson Mellichamp (1829–1903), physician and amateur botanist of South Carolina. It occurs in South Carolina and northern and central Florida.

Laurel Oak forms part of a very complex group including the Willow Oak, Water Oak, and Diamond-leaf Oak. Botanical authors do not all agree on the exact relationship of these species.

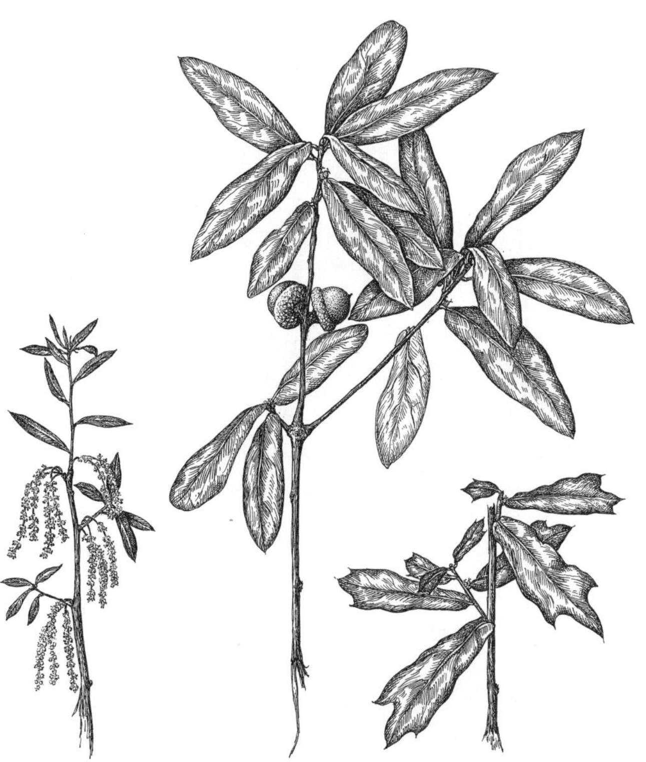

LAUREL OAK
Quercus laurifolia Michx.

REMARKS. The genus name, *Quercus,* is the classical name of the oak-tree, and the species name, *laurifolia,* refers to the laurel-like foliage. Vernacular names are Water Oak, Willow Oak, and Live Oak. The tree is often cultivated for ornament in the South. The wood is mostly used for fuel and charcoal.

MYRTLE OAK

Quercus myrtifolia Willd.

FIELD IDENTIFICATION. Often an intricately branched shrub on exposed dunes near the coast, but under more favorable conditions a tree to 40 ft or more. The trunk usually 3–8 in. in diameter, but more rarely to 15 in. or more. The slender spreading branches form a rounded crown.

FLOWERS. Staminate and pistillate borne separately on same plant; staminate catkins drooping, hoary-pubescent, 1–1½ in. long; calyx campanulate, usually with 5 ovate acute segments; stamens 2–5, filaments filiform, longer than the calyx; pistillate flowers sessile or short-pedunculate, solitary or several together, involucral scales reddish and tomentose; styles short, usually 3, with 3 dilated stigmas; ovary 3-celled, ovules 2 in each cavity but usually only 1 maturing.

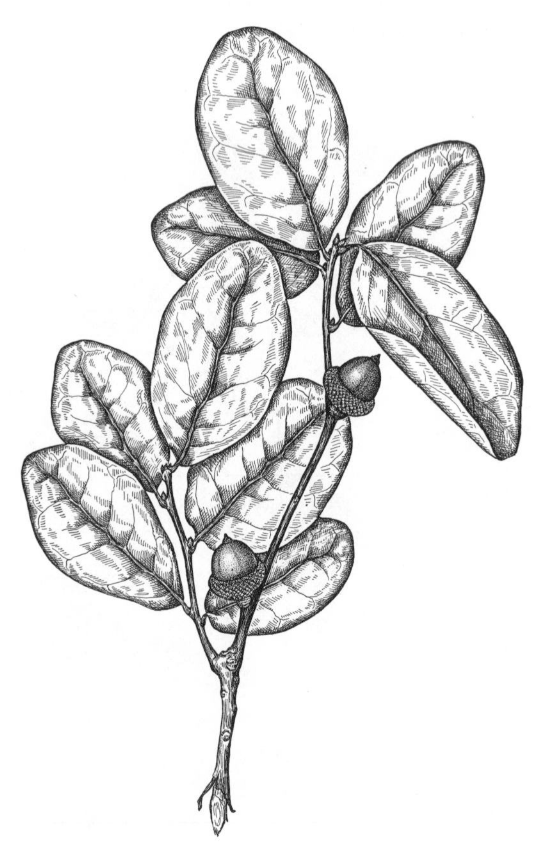

MYRTLE OAK
Quercus myrtifolia Willd.

FRUIT. Acorn solitary or paired, sessile or short-stalked; cup saucer-shaped or hemispheric, ⅖–½ in. wide, closely imbricated with scales which are brown, pubescent, ciliate and apically rounded; acorn ovoid to ellipsoid-ovoid to subglobose, ¼–½ in. long, dark brown, shiny, often striate, apex pubescent.

LEAVES. Simple, alternate, somewhat persistent, or even evergreen in protected sites, blade oval to oblong or obovate, ½–2 in. long and ¼–1¼ in. wide, apex rounded or acute, apiculate, base rounded or cordate; margin entire, thickened and revolute, or on shoots sinuate-toothed or lobed; young leaves reddish pubescent; mature leaves thick and coriaceous; upper surface shiny, dark green, glabrous, reticulate-venulose; lower surface paler green, glabrous or pubescent, sometimes with rufous hairs in the vein axils; petioles 1/25–⅛ in., stout, yellow, pubescent.

TWIGS. At first pale reddish or gray, with reddish tomentum, later becoming darker and pubescent and eventually glabrate.

BARK. Brown to gray, smooth on the upper trunk and branches, slightly furrowed near the base.

RANGE. Mostly on dry sandy ridges near the coast and on islands, sometimes forming extensive low shrubby thickets. Eastern Louisiana; eastward along the coast of Mississippi and Alabama to Florida and northward to South Carolina.

REMARKS. The genus name, *Quercus,* is the classical name of the oak, and the species name, *myrtifolia,* refers to the myrtle-like leaves. It is also known as Scrub Oak or Sand Oak.

WATER OAK

Quercus nigra L.

FIELD IDENTIFICATION. Tree attaining a height of 80 ft, with a round top and grayish black bark. Leaves variously shaped, wedge-shaped and entire at the apex, or 3-lobed at apex, or variously cut into oblique bristle-tipped lobes.

FLOWERS. Appearing with the leaves in spring in separate staminate and pistillate catkins; staminate catkins clustered, 2–3 in. long; calyx 4–5-lobed, pubescent; pistillate catkins short-peduncled, scales rusty-hairy; stigmas red.

FRUIT. Ripening September–October. Acorn sessile or short-peduncled, solitary or paired, globose-ovoid, ⅓–⅔ in. high, light yellowish brown, often striate, usually somewhat pubescent, flat at base, enveloped in cup one third to one half its length; cup shallow, saucer-shaped, thin, reddish brown, pubescent; scales small, thin, closely appressed, imbricate.

WATER OAK
Quercus nigra L.

LEAVES. Simple, alternate, persistent, variously shaped; typically entire, obovate or spatulate; with a rounded or 3-lobed apex; margins on some leaves often with deep, oblique, bristle-tipped lobes, a variety of shapes frequently appearing on the same twig or on different twigs; blades 2–4 in. long, 1–2 in. wide, upper surface lustrous green and glabrous, lower surface lighter and glabrous, or pubescent in vein axils; petioles short and stout. Leaves half-evergreen in the southern Gulf Coast area.

TWIGS. Slender, glabrous, reddish gray; buds ovoid, acute, angled, reddish brown, ⅛–¼ in. long.

BARK. Grayish black to light brown, bark so tightly appressed as to appear almost smooth, ridges flattened and thin.

WOOD. Light brown, sapwood lighter, close-grained, heavy, hard, strong.

RANGE. In low woods, or borders of streams or swamps. From the Colorado River of Texas eastward through Louisiana to Florida, northward into Oklahoma, Arkansas, and Missouri, and on the Atlantic Coastal Plain to New Jersey.

PROPAGATION. Water Oak has been cultivated since 1723. The commercial minimum acorn-bearing age is about 25 years and the maximum is 175 years. The acorns average about 400 per lb, with a commercial purity of 100 per cent and a soundness of 95 per cent. For spring sowing the acorns are stratified from collection time at 30°–35°F. for 30–60 days in moist sand or peat. The acorns may be stored in sealed containers at 32°–36°F., or storage in cool humid cellars is often adequate. The planting is in drills, the rows 8–12 in. apart, and covered 1 in. with firmed moist soil.

HYBRIDS. The following hybrids have been recognized by Little (1953a, p. 343):

Caduca Oak, *Q.* × *caduca* Trel., is a hybrid of *Q. nigra* and *Q. incana,* the name meaning "promptly deciduous." It occurs in southeastern Virginia, Georgia, Florida, Alabama, Mississippi, and eastern Texas.

Capesius Oak, *Q.* × *capesii* W. Wolf, is a cross of *Q. nigra* and *Q. phellos,* and is named after A. Capesius, who discovered the hybrid in seedlings raised from a cultivated tree at St. Bernard, Alabama. It occurs in New Jersey, South Carolina, and southeastern Texas.

Demaree Oak, *Q.* × *demareei* Ashe, is a cross of *Q. nigra* and *Q. velutina,* and is named for Delzie Demaree, botanist of Arkansas. It occurs in Arkansas and Louisiana.

Garland Oak, *Q.* × *garlandensis* Palmer, is a cross of *Q. nigra* and *Q. falcata,* and is named after Garland County, Arkansas, where it was discovered. It occurs in Virginia, Alabama, Arkansas, and Louisiana.

Shumagra Oak, *Q.* × *neopalmeri* Sudw., is a cross of *Q. nigra* and *Q. shumardii,* and is named for Ernest Jesse Palmer, American botanist, who discovered it. It occurs in Georgia, central Florida, Alabama, and Arkansas.

Blackwater Oak, *Q.* × *sterilis* Trel., meaning "sterile" or "nonfruiting," and occurring in North Carolina, Georgia, and eastern Texas, is a cross of *Q. nigra* and *Q. marilandica.*

Walter Oak, *Q.* × *walteriana* Ashe, is a cross of *Q. nigra* and *Q. laevis,* and is named for Thomas Walter (1740–1789), English-born planter and botanist of South Carolina. Known to occur in North Carolina, South Carolina, Georgia, Florida, and Alabama.

REMARKS. The genus name, *Quercus,* is the classical name and the species name, *nigra,* refers to the black bark. Vernacular names are Bluejack Oak, Duck Oak, Pin Oak, Spotted Oak, Barren Oak, Punk Oak, and Possum Oak. The wood is used for fuel, crossties, and poles. Water Oak is extensively planted as a street shade tree in the South and is subject to attack by mistletoe. Trident Water Oak, *Q. nigra* var. *tridentifera* Sarg., is a variety described as having leaves more acute at the apex, but this character does not appear to be constant, and it has been relegated to the status of a synonym of Water Oak.

DIAMOND-LEAF OAK
Quercus obtusa (Willd.) Ashe

FIELD IDENTIFICATION. Tree attaining a height of 150 ft and a trunk diameter of 4½ ft, but usually only one fourth to one half that size, or even scrubby in ap-

DIAMOND-LEAF OAK
Quercus obtusa (Willd.) Ashe

pearance. The branches are smooth, stout, and form a broad open crown.

FLOWERS. Similar to those of Laurel Oak, *Q. laurifolia,* to which it is closely related. Staminate and pistillate catkins borne separately on same tree.

FRUIT. Acorn sessile or short-stalked, ovoid, pubescent, ⅖–½ in. long, ½–⅗ in. thick; cup only at the base, saucer-shaped, pubescent within, scales reddish brown, appressed, apices free, somewhat pubescent, margins ciliate.

LEAVES. Very persistent, simple, alternate, deciduous, oblong-obovate to lanceolate, in the most typical form broadest at the middle and tapering evenly to the base or apex (diamond-shaped), apex acute to rounded and apiculate, base cuneate, margins entire or slightly wavy (or short-lobed in young leaves), young leaves pubescent and with axillary tufts of hairs beneath; mature leaves 2½–4 in. long, 1½–2 in. wide, upper surface dark green, shiny and glabrous, lower surface paler and glabrous, yellow in autumn; petioles ⅕–½ in., stout, yellow.

TWIGS. Slender, glabrous, reddish brown, older ones dark gray.

BARK. Tight, thin, gray, or greenish, smooth or very shallowly furrowed, scales small and appressed, frequently marked with warts, white lichens, and various excrescences.

WOOD. Reddish brown to brown, sapwood cream-colored and usually wide, straight-grained, heavy, hard, checks badly in drying.

RANGE. Wooded swamps of the Gulf Coast plain. Eastern Texas (Neches River); eastward through Louisiana, Mississippi, and Alabama and north to the Red River Valley. Also known from a few localities in Florida.

RELATED SPECIES, VARIETIES. Diamond-leaf Oak, Water Oak, and Laurel Oak form a very complex group. Botanists do not agree as to the exact nomenclature. The following statement is from Little (1953a, p. 336): "In the 1927 Check List *Quercus obtusa* was accepted as a distinct species, though Trelease (1924, p. 157) had regarded it as a variety. Fernald (1950, pp. 549-550, figs. 931, 935) used *Q. laurifolia* Michx. in place of *Q. obtusa* and adopted *Q. hemisphaerica* Bartr. for *Q. laurifolia,* as explained in the reference cited above. However, Palmer (1948) did not accept those changes. Muller's union of all these variations under a single species, *Q. laurifolia,* is a simplified solution to the confused nomenclature and is adopted here (Muller, 1951, pp. 85-86)."

Diamond-leaf Oak is sometimes confused with Water Oak, but the latter has leaves tapering evenly from the apex to the base. Diamond-leaf Oak leaves are widest at the middle and taper to each end.

A variety of Diamond-leaf Oak has been described as *Q. obtusa* var. *obovatifolia* (Sarg.) Sudw., with leaves obovate and rounded and slightly 3-lobed or undulate at the apex. This variety is possibly only named from leaf forms found on young vigorous shoots of the species.

REMARKS. The genus name, *Quercus,* is the classical name of the oak-tree. The species name, *obtusa,* refers to the blunt apex of the leaf. It is also known as Obtusa Oak, Water Oak, Swamp Laurel Oak, Pine Oak, Spotted Oak, and Laurel-leaved Oak.

The lumbering possibilities of Diamond-leaf Oak are minor because it is not common, easily accessible, or of high average quality. Trees containing 2 good logs are uncommon. It is generally relegated to lower-than-average grades and is used mostly for flooring and planking. The tree is of average susceptibility to fire injury, but more than average to mineral stain, grub, and bird peck. Insect and fungus damage are usually not serious.

BLACKJACK OAK

Quercus marilandica Muenchh.

FIELD IDENTIFICATION. Shrub, or round-topped symmetrical tree attaining a height of 60 ft and a diameter of 2 ft.

FLOWERS. With the leaves in spring, in staminate or pistillate catkins; staminate catkins clustered, loosely flowered, slender, hairy, yellowish green, 2–4 in. long; stamens 3–12, filaments filiform, anthers exserted; calyx thin, pubescent, reddish green, 4–5-lobed; pistillate flowers solitary or paired, ⅛–⅕ in. long, pubescent to glabrate, peduncles rusty-tomentose and short; styles recurved, stigmas red.

FRUIT. Acorn ripening in 2 years, solitary or in pairs, sessile or on peduncles 1/8–2/5 in. long; light brown, enclosed one third to two thirds in cup, 3/5–4/5 in. long, 1/2–3/4 in. high, often striate, ovoid-oblong to subglobose, pubescent; cup thick, turbinate, 3/5–4/5 in. broad, bases rounded or suddenly constricted; scales of cup imbricate, loose, obtuse, ovate to oblong, thin.

LEAVES. Simple, alternate, deciduous, stiff, coriaceous, broadly obovate to clavate, margin revolute; apex 3-lobed to entire or dentate, bristle-tipped, base rounded, cordate or cuneate; upper surface dark green, glossy and glabrous (or young leaves tomentose and hairy along the veins); lower surface semiglabrate or scurfy and yellow-hairy, veins conspicuous, length 3–7 in., width 2–5 in.; petioles 1/2–3/4 in., stout, glabrous or pubescent; stipules caducous, 1/4–1/3 in., glabrous or pubescent.

TWIGS. Grayish brown, stout, stiff, densely tomentose at first, glabrous later; buds 1/6–1/3 in. long, ovoid to lanceolate, apex acute, reddish brown, slightly or densely tomentose.

BARK. Dark brown or black, broken into rough, block-like plates.

WOOD. Dark brown, sapwood lighter, heavy, hard, strong, weighing 46 lb per cu ft.

RANGE. Usually on dry, sandy, sterile soils. Central Texas, Oklahoma, and Arkansas; eastward through Louisiana to Florida, north to New York, and west to Minnesota, Michigan, Illinois, and Kansas.

HYBRIDS. The following hybrids are listed by Little (1953a, p. 339):

Britton Oak, *Q.* × *brittonii* W. T. Davis, is a cross of *Q. marilandica* and *Q. ilicifolia,* named for its discoverer, Nathaniel Lord Britton (1859–1936), at one time director of the New York Botanical Garden. It occurs in New York, New Jersey, and southeastern Pennsylvania.

Bush's Oak, *Q.* × *bushii* Sarg., is a cross of *Q. marilandica* and *Q. velutina,* named after its discoverer, B. F. Bush (1858–1937), botanist of Missouri. It occurs in east Texas, Alabama, Mississippi, Arkansas, Oklahoma, Missouri, Pennsylvania, Indiana, Iowa, Nebraska, Kansas, Tennessee, Georgia, North Carolina, South Carolina, Virginia, and Maryland.

Carolina Oak, *Q.* × *cravenensis* Little, is a hybrid of *Q. marilandica* and *Q. incana,* named after Craven County, North Carolina, where it was found. It occurs in Virginia, North Carolina, South Carolina, Georgia, Alabama, and east Texas.

Hastings Oak, *Q.* × *hastingsii* Sarg., is a cross of *Q. marilandica* and *Q. shumardii,* named after its discoverer, Stephen Harold Hastings, agronomist of the U. S. Department of Agriculture.

Rudkin Oak, *Q.* × *rudkinii* Britt., is a cross of *Q. marilandica* and *Q. phellos,* named after its discoverer, William H. Rudkin. It occurs in New York, New Jersey, Delaware, Virginia, North Carolina, South Carolina, Georgia, Florida, Louisiana, and Arkansas.

Small Oak, *Q.* × *smallii* Trel., is a cross of *Q. marilandica* and *Q. georgiana,* named after its discoverer, J. K. Small (1869–1938), American botanist and authority on plants of the Southeastern United States.

Blackwater Oak, *Q.* × *sterilis* Trel., means sterile, or not fruiting. It is a cross of *Q. marilandica* and *Q. nigra* and occurs in North Carolina, Georgia, and eastern Texas.

St. Louis Oak, *Q.* × *tridentata* (A. DC.) Engelm., is a hybrid of *Q. marilandica* and *Q. imbricaria.* The name refers to the 3-toothed leaves. It occurs in Pennsylvania, Michigan, Illinois, Missouri, and Virginia.

Blackjack Oak presents a number of small-leaved variations on the high dry lands of central Texas and Oklahoma, but these may be considered as only forms.

REMARKS. The genus name, *Quercus,* is of classical origin, and the species name, *marilandica,* refers to the state of Maryland. Also known by the vernacular names of Iron Oak, Black Oak, Jack Oak, Barren Oak, and Scrub Oak. The wood is used mostly for posts, fuel, and charcoal, and the acorns sought by wild turkey and white-tailed deer.

BLACKJACK OAK
Quercus marilandica Muenchh.

CHISOS OAK

Quercus graciliformis C. H. Muller

FIELD IDENTIFICATION. Small oak to 25 ft, with slender graceful branches.

FLOWERS. Borne in separate staminate and pistillate catkins. Staminate catkins unknown. Pistillate catkins 1–3-flowered distally, about 1/6 in. long, glabrous, or nearly so, reddish brown.

FRUIT. Acorns biennial, solitary or paired, sessile or very short-peduncled, narrow-ovoid, acute, length about ¾ in., width about ⅖ in., finely pubescent, striate, basally enclosed only by the cup; cup hardly more than ⅛ in. deep, about ⅖ in. in diameter; scales ovate, closely appressed, brown, thin, hairy to glabrate later, somewhat ciliate.

LEAVES. Partially evergreen, thin but leathery, blades 3–4 in. long, ¾–1¼ in. wide, narrowly lanceolate, apex long-acuminate and bristle-pointed, base cuneate, margin 8–10-toothed or lobed and bristle-pointed, lobes with shallow, or deep, rounded sinuses between; upper surfaces lustrous green, mostly glabrous; lower surface paler and duller, glabrous, or occasionally a tuft of hairs in the leaf axils; petiole flexible to cause a pendent leaf, flattened above, glabrous, greenish red or yellowish, length ⅗–⅘ in.

TWIGS. Lustrous, reddish brown and gray later, somewhat hairy at first but soon glabrate, fluted, the small, pale lenticels inconspicuous; buds shiny brown, ovoid, apex acute to rounded, 1/25–1/12 in. long, scales ciliate; stipules early deciduous, ⅛–¼ in. long, slightly hairy, setaceous to spatulate.

RANGE. The species, variety, and hybrid, are known only from the Chisos Mountains of Brewster County in western Texas. Its closest relatives are in Mexico.

HYBRID AND VARIETY. Chisos Oak is reported as hybridizing with Emory Oak to produce Tharp Oak, *Q. × tharpii* C. H. Muller.

A variety of Chisos Oak has been named *Q. graciliformis* var. *parvilobata* C. H. Muller. It has fewer leaf lobes than the species.

FEW-LOBED CHISOS OAK
Quercus graciliformis var. *parvilobata* C. H. Muller

REMARKS. The genus name, *Quercus*, is the ancient classical name, and the species name, *graciliformis*, refers to the slender, drooping, graceful branches.

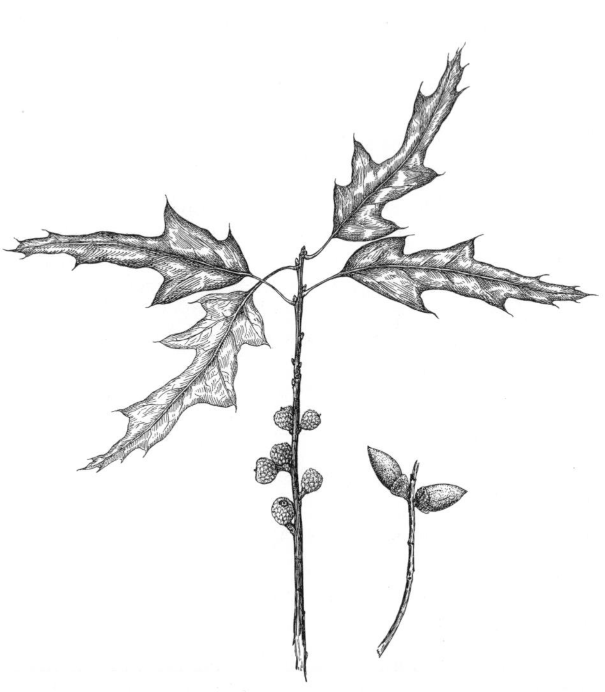

CHISOS OAK
Quercus graciliformis C. H. Muller

PIN OAK

Quercus palustris Muenchh.

FIELD IDENTIFICATION. Beautiful tree attaining a height of 120 ft, with a diameter of 5 ft. The crown oblong or broadly pyramidal. Branches numerous and spreading, the lower often pendulous. The short tough branchlets from the trunk produce a "pin-like" appearance in winter.

FLOWERS. Borne April–May, in separate staminate and pistillate catkins; staminate catkins slender, pubescent, 2–3 in. long, calyx-lobes 4–5, oblong, rounded at apex, denticulate on margin, shorter than the stamens, stamens exserted; pistillate flowers 1–3, on slender tomentose peduncles, scales of involucre ovate, tomentose, shorter than the acuminate calyx-lobes, styles slender, spreading, stigmas red.

FRUIT. September–October, binennial, solitary or

clustered, sessile or on stalks ⅖–⅗ in.; nut hemispheric or sub-globose, about ½ in. in diameter, light brown, often striate, bitter, enclosed only at the base in a shallow, saucer-shaped cup; cup ⅓–½ in. broad, thin, reddish brown, scales closely appressed, ovate, acute or obtuse, puberulous, margins darker.

LEAVES. Simple, alternate, deciduous, 4–6 in. long, 2–5 in. wide, 5–9-lobed, lobes with rounded sinuses cut two thirds or more to the midrib, lobes oblong to lanceolate or triangular with apices 2–4-toothed and bristle-tipped; general shape of leaf ovate, to obovate to broadly oblong; apex acute to acuminate; base truncate to broadly cuneate; thin and firm; upper surface dark green and lustrous; lower surface paler and glabrous except for axillary tufts of hairs, scarlet in autumn; petioles slender, yellow, puberulent to glabrous, ⅝–2½ in. long.

TWIGS. At first slender, tough, puberulent, green to reddish brown or orange; later brown to gray and glabrous. Older dead twigs tardily deciduous and resembling pins thrust into the tree, hence the name of Pin Oak. The twigs are subject to the attack of gall-producing insects. The small buds are about ⅛ in. long, ovoid, and reddish brown.

BARK. On young trunks and branches light brown, smooth and lustrous, later brownish gray and slightly roughened by shallow fissures and small, closely appressed scales.

WOOD. Light brown, coarse-grained, heavy, hard, strong, often knotty owing to persistence of many small limbs, weighing 43 lb per cu ft.

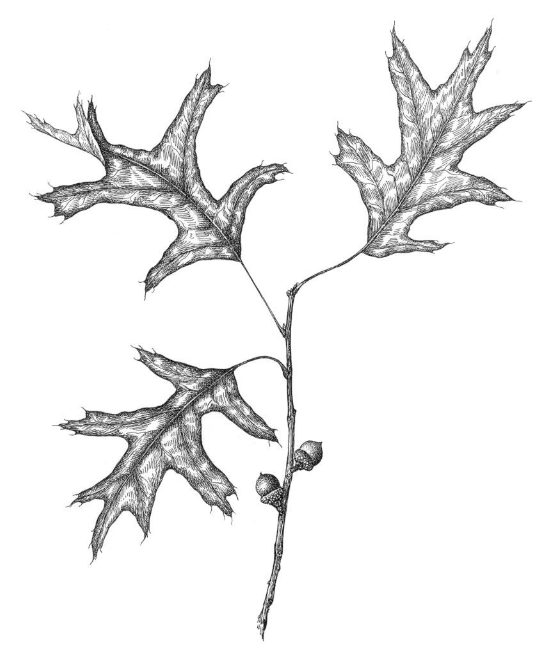

PIN OAK
Quercus palustris Muenchh.

RANGE. Deep rich soil of bottom lands. Oklahoma, Arkansas, and Mississippi; east through Tennessee and Kentucky to Delaware, north to Massachusetts, and west to Ontario, Wisconsin, Iowa, and Kansas.

HYBRIDS. The following hybrids are listed for Pin Oak by Little (1953a, p. 345):

Shingle Oak, *Q.* × *exacta* Trel., is a cross of *Q. palustris* and *Q. imbricaria,* and is named after the uniform teeth of the leaves. It occurs in Pennsylvania, Indiana, Illinois, and Missouri.

Variable Oak, *Q.* × *mutabilis* Palmer & Steyerm., is a cross of *Q. palustris* and *Q. shumardii,* named after the variable leaves. It occurs in Missouri.

Schoch Oak, *Q.* × *schochiana* Dieck, is a cross of *Q. palustris* and *Q. phellos,* named after State Garden Director Schoch, who discovered it in a park at Worlitz, Germany. It occurs in Virginia, Kentucky, Illinois, and Arkansas.

Wandering Oak, *Q.* × *vaga* Palmer & Steyerm., is a cross of *Q. palustris* and *Q. velutina.* The name means "of uncertain distribution." It is known to occur in Missouri.

REMARKS. The genus name, *Quercus,* is an ancient classical name, and the species name, *palustris,* means "of low grounds," referring to the tree's habitat. Vernacular names are Swamp Spanish Oak, Swamp Oak, and Spanish Oak. The wood of Pin Oak is used for fuel, interior finish, crossties, construction, shingles, clapboards, and cooperage. It is similar to Northern Red Oak but inferior to it. The wood was formerly much used as wooden pins to hold together squared timbers. The trees are generally free of insects and disease.

Pin Oak is used extensively for street and yard planting in the Northeastern states and in Europe, and was first cultivated in 1770. Its rapid growth, slender form, and often pendulous branches make it highly desirable. It apparently can also stand considerable amounts of dust and smoke, and coppices fairly well. The acorns average about 400 per lb and direct seeding in spring is better than nursery planting. They should be stratified over the winter, but not permitted to dry out. Storing in a cool humid place is necessary. They may be planted in rows about 12 in. apart and ½–1 in. deep. The average germination is 68 per cent. Fall beds should have a mulch of straw or leaves for protection and also be covered with hardware-cloth as a protection against rodents.

NORTHERN RED OAK
Quercus rubra L.

FIELD IDENTIFICATION. Massive tree attaining a height of 150 ft and a diameter of 4 ft. Branches large, spreading, forming a round-topped crown.

FLOWERS. May–June, in separate staminate and pistillate catkins; staminate catkins 3–5 in. long, pubescent;

NORTHERN RED OAK
Quercus rubra L.

calyx-lobes 4–5, shorter than the stamens, ovate, rounded at apex; stamens 4–6; pistillate catkins on glabrous or pubescent peduncles in clusters of 1–3; scales of the involucre reddish, ovate; bracts linear and acute; calyx-lobes lanceolate and acute; stigmas green, slender, elongate, spreading.

FRUIT. Maturing October–November, solitary or paired, sessile or short-stalked, ovoid or slightly obovoid, apex rounded to acute and sometimes tomentose, pale brown, lustrous, 1–1¼ in. long, ½–¾ in. thick, bitter, usually two to four times as long as the cup or enclosing cup only at the base; cups saucer-shaped, reddish brown, puberulous; ⅝–¾ in. across; scales small, closely appressed, ovate, acute, darker colored on tip or margins.

LEAVES. Simple, alternate, deciduous, blades 5–9 in. long, 4–6 in. wide, oblong to obovate or oval, apex acute or acuminate, base cuneate or rounded, margin 7–11-lobed; lobes extending halfway to the middle with apices 1–3-toothed and bristle-tipped; sinuses oblique and rounded at the base; upper surface dull, dark green, smooth, glabrous; lower surface paler and glabrous (rarely puberulous) except for occasional tufts of axillary hairs; dark red in autumn; petiole stout, reddish or yellow, 1½–2 in. long.

TWIGS. Green to reddish brown, lustrous, glabrous, moderately stout, pith star-shaped; terminal buds brown, ovoid or narrowly conical, acute, glabrous or somewhat pubescent, about ¼ in. long.

BARK. At first smooth and greenish brown; later dark brown to reddish or almost black, with shallow fissures and broad flat ridges with small scales.

WOOD. Reddish brown, heavy, hard, strong, close-grained, weighing 41 lb per cu ft.

RANGE. Oklahoma and Arkansas; east through northern Mississippi and Alabama to northern Florida, north along the Appalachian Mountains to Delaware, thence along the coast to New England and Quebec, and west to Ontario, Wisconsin, Minnesota, Nebraska, and Missouri, reaching its largest size in the Ohio Valley.

VARIETIES AND HYBRIDS. The nomenclature of Red Oak has been rather confused. For a time the scientific names of *Q. borealis* Michx. f. and *Q. borealis* var. *maxima* (Marsh.) Sarg., were used by some botanists instead of *Q. rubra* L. However, the latter name has now been restored (Svenson, 1939).

Many hybrids of Red Oak have been recognized by Little (1953a, p. 350):

Fernald Oak, *Q.* × *fernaldii* Trel., is a cross between *Q. rubra* and *Q. ilicifolia,* named for its discoverer, Merritt Lyndon Fernald (1873–1950), botanist of Harvard University, who prepared two revisions of Gray's *Manual of Botany* (1908, 1950). It occurs in Massachusetts and Virginia.

Hawkins Oak, *Q.* × *hawkinsiae* Sudw., is a cross between *Q. velutina* and *Q. rubra,* and is named after Mrs. Eugene Hawkins. It occurs in Maine, Massachusetts, New York, Pennsylvania, Ohio, Tennessee, and Missouri.

Bartram Oak, *Q.* × *heterophylla* Michx. f. is a cross between *Q. rubra* and *Q. phellos,* and means "various-leaved." It is found in southeastern New York, New Jersey, Pennsylvania, Delaware, District of Columbia, Virginia, Tennessee, Missouri, Arkansas, southeastern Oklahoma, Alabama, and North Carolina.

Bottom Oak, *Q.* × *runcinata* (A. DC.) Engelm., is a cross between *Q. rubra* and *Q. imbricaria,* named from the deeply cut, or runcinate, leaves. It occurs in Pennsylvania, Ohio, Indiana, Illinois, Missouri, Kansas, Kentucky, and Maryland.

Beaumont Oak, *Q.* × *beaumontiana* Sarg., is a cross of *Q. rubra* and *Q. laurifolia* var. *rhombica,* and is named from the type locality, Beaumont, Texas.

REMARKS. The genus name, *Quercus,* is an ancient classical name, and the species name, *rubra,* refers to its reddish wood. Vernacular names are Leopard Oak, Gray Oak, Mountain Red Oak, Champion Oak, Black Oak, and Spanish Oak. The wood is considered to be better than that of other Red Oaks, but is not quite as good as White Oak. It is used for furniture, interior finishing, crossties, posts, mine timbers, and general construction. It has a smooth finish, machines well, glues moderately well, is shock-resistant, and resists bending. On the other hand, it has a tendency to check, split, and shrink, and is difficult to dry. The tree grows rapidly, sprouts from the stump freely, is long-lived, but is easily fire damaged and subject to beetle and fungus attack, especially heart rot. The acorns average about 90 per lb, with an average germination of 58 per cent. They are eaten by bobwhite quail and ruffed grouse, and the foliage browsed by white-tailed deer and cottontail. However, the bitter acorns do not seem to be eaten by wildlife as much as those of other oaks. The tree casts dense shade, and is often planted in the Northeastern states and in Europe as an ornamental. It was first introduced into cultivation about 1880.

Southern Red Oak

Quercus falcata Michx.

FIELD IDENTIFICATION. An open, rounded tree, forming a broad top and attaining a height of 80 ft. Leaves very variable, 3–7-lobed, (or 5–13 lobes in some varieties), lobes often falcate, brownish white-tomentose beneath.

FLOWERS. March–May, staminate and pistillate catkins borne separately on same tree; staminate catkins clustered, tomentose, 3–5 in. long; calyx-lobes 4–5, round, thin, hairy; pistillate flowers solitary or several together, borne on downy peduncles; scales of involucre with reddish tomentum; calyx-lobes acute.

FRUIT. Solitary or in pairs, sessile or short-peduncled, small, globular or hemispheric, orange-brown; often striate, pubescent, about ½ in. long, enclosed to one third its length in the shallow, saucer-shaped, thin cup; cup scales reddish, pubescent, ovate-oblong, acute or rounded at apex; matures during the second season.

Southern Red Oak
Quercus falcata Michx.

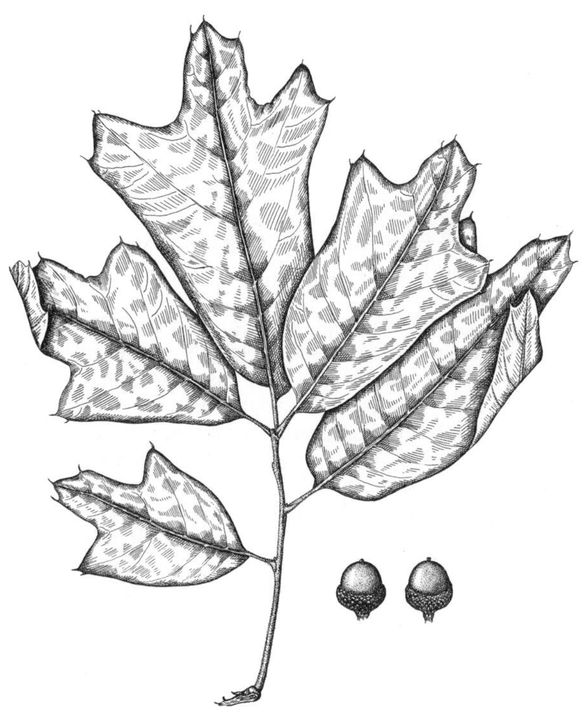

Three-lobe Red Oak
Quercus falcata var. *triloba* (Michx.) Nutt.

LEAVES. Simple, alternate, deciduous, very variable in shape and lobing, ovate-oblong to obovate, with usually 3–7 bristle-tipped lobes (usually 3-lobed in the variety *triloba,* but numerous lobes in the varieties *pagodaefolia* and *leucophylla);* lobes often falcate, slender, narrowed, rounded or cuneate at the base; dark green and lustrous above, paler with brown to grayish white tomentum beneath; leaf blades 6–7 in. long, 4–5 in. wide; petioles slender, flattened, 1–2 in.

TWIGS. Reddish brown, stout, pubescent at first, later glabrous; terminal buds ⅛–¼ in. long, ovoid, acute, reddish brown.

BARK. Grayish black, broken into deep fissures and appressed scales.

WOOD. Light red, sapwood lighter, coarse-grained, durable, heavy, hard, strong, weighing 43 lb per cu ft.

RANGE. The typical species of Southern Red Oak occur from the Brazos River of Texas eastward through Louisiana to Florida, Oklahoma, and Arkansas and northward to New York, Pennsylvania, Ohio, and Illinois.

PROPAGATION. The minimum commercial seed-bearing age is 25 years and the maximum 125 years. The acorns average about 595 per lb, with a commercial purity of 98 per cent and a soundness of 60 per cent. The acorns may be stored in sealed containers for a year or two at 32°–36°F., but it is best to plant immediately when gathered or to stratify at 32°–38°F.

for 30–45 days before spring planting. They should never dry out. They are planted in drills 8–12 in. apart and covered about ½ in. with moist firmed soil.

VARIETIES. Red Oak Complex. The former name of *Q. rubra* L., meaning "Red Oak," has for many years been a vague and confusing term applying to a complex of Northern and Southern oaks. It was suggested by Rehder, Palmer, and Croizat (1938, pp. 283–284) that the name be replaced by *Q. falcata* Michx., as applying to Southern Red Oak and *Q. borealis* Michx., as applying to Northern Red Oak.

Southern Red Oak, *Q. falcata,* is in itself a very variable species, and not all botanists agree as to the exact status of some of its variations. The difficulty lies in the unstable character of the species and its tendency to produce intergrading forms over wide areas.

The first variety, known as the Three-lobe Red Oak, *Q. falcata* var. *triloba* (Michx.) Nutt., is a form in which the leaves are 3–lobed at the apex. Other differences between it and the species are negligible. However, the number of 3–lobed leaves on different trees varies considerably. Some trees have practically all leaves 3–lobed, some trees perhaps half, and others very few. Some botanists conclude, therefore, that this 3–lobed form is not constant enough to warrant a special varietal name.

A second variety, now known as Swamp Red Oak, *Q. falcata* var. *pagodaefolia* (Ell.) Ashe, is described by some botanists as a separate species (Pagoda Oak, *Q. pagoda* Raf.). However, a varietal standing does seem to fit the tree better than species rank. The fruit and flowers are similar to the typical species, but the leaves are distinctly pagoda-shaped, with 5–13 lobes, and have a cuneate or truncate base. Also, the bark has a tendency to be tighter and resemble the bark of Wild Cherry, hence giving it the name of Cherry-bark Oak in some areas.

SWAMP RED OAK
Quercus falcata var. *pagodaefolia* (Ell.) Ashe

CHERRY-BARK RED OAK
Quercus falcata var. *leucophylla* (Ashe) Palmer & Steyerm.

The third variety, known as the true Cherry-bark Red Oak, *Quercus falcata* var. *leucophylla* (Ashe) Palmer & Steyerm., is very closely allied to the Swamp Red Oak. In fact, if the Swamp Red Oak were given a species name, then the Cherry-bark Red Oak would fall into a varietal classification under it. It has similar acorns and cherry-like bark, but the leaves have somewhat more irregular lobes and the lower surface is more or less white-tomentose. The lower leaves on Cherry-bark Oak-trees resemble those of Black Oak leaves in shape.

HYBRIDS. Southern Red Oak has many hybrids which are listed herewith, following Little (1953a, p. 325):

Two-edge Oak, *Q.* × *anceps* Palmer, is a hybrid of *Q. falcata* and *Q. imbricaria.* It occurs in southeastern Illinois.

Bluffton Oak, *Q.* × *blufftonensis* Trel., is named for Bluffton, South Carolina, where it was discovered. It is a cross of *Q. falcata* and *Q. laevis.*

Garland Oak, *Q.* × *garlandensis* Palmer., is a cross of *Q. falcata* and *Q. nigra,* and is named for Garland County, Arkansas. It occurs in Virginia, Alabama, Arkansas, and Louisiana.

Joor Oak, *Q.* × *joorii* Trel., is named for Joseph F. Joor (1848–1892), who collected in 1884. It was found at Galveston, Texas. It is considered to be a hybrid of *Q. falcata* and *Q. shumardii.*

St. Landry Oak, *Q.* × *ludoviciana* Sarg., is a hybrid between *Q. falcata* and *Q. phellos,* and named for the state of Louisiana. It occurs also in Virginia, Kentucky, Missouri, Arkansas, southeastern Texas, Mississippi, Alabama, and Georgia.

Subentire-leaf Oak, *Q.* × *subintegra* Trel., is a cross of *Q. falcata* and *Q. incana,* and is named from the almost-entire leaves. It occurs in Maryland, Virginia, South Carolina, Georgia, Florida, and Alabama.

Willdenow Oak, *Q.* × *willdenowiana* (Dippel) Zabel, is a cross of *Q. falcata* and *Q. velutina,* and is named for Karl Ludwig Willdenow (1765–1812), German botanist. It occurs in North Carolina, Tennessee, Arkansas, and Georgia.

REMARKS. *Quercus* is an old classical name, and the species name, *falcata,* refers to the scythe-shaped leaves. Vernacular names are Spanish Oak, Turkey Oak, Pagoda Oak, and Cherry-bark Oak. The tree is often planted for ornament, and the wood is used for general purposes, rough lumber, and furniture (chairs and tables). The bark is excellent for tanning and is used as an astringent in medicine.

Turkey Oak

Quercus laevis Walt.

FIELD IDENTIFICATION. Tree attaining a height of 20–60 ft and a trunk diameter of 2 ft, the branches ascending and spreading to form an open, rounded crown. The root system is rather deep. Sometimes on poor soil a shrub rather than a tree.

FLOWERS. Staminate catkins slender, reddish, hairy, 4–5 in. long; calyx-lobes 4–5, ovate, acute, puberulent; pistillate peduncles short and tomentose, scales red, pubescent; stigmas red.

FRUIT. Acorn in twos and threes or solitary, short-peduncled, ¾–1 in. long, and ½–¾ in. broad, ovoid to oval, ends rounded, light brown, apex woolly, inserted about one third its length in the turbinate and thin cup; scales of cups ovate-oblong, densely pubescent, covering the rim and extending inside part way.

LEAVES. Simple, alternate, nearly deltoid or obovate to oblong, margin deeply cut with 3–5 or sometimes 7 lobes; base wedge-shaped; terminal lobe elongate, entire or repand-dentate on the margin, sometimes 3-toothed at apex; lateral lobes spreading, usually curved, 3-lobed or 3-toothed at acute apices, margin otherwise entire; thick and coriaceous, upper surface yellowish green and shiny; lower surface pale, glabrous, with axillary tufts of reddish hairs, blades 2¾–12 in. long, 1–10 in. wide (usually about 5 in. long and wide); petioles ¼–1 in., stout.

TWIGS. Bearing fascicled hairs at first, red, eventually reddish brown and glabrous; winter buds elongate, acute, brown-pubescent, about ½ in. long.

BARK. Dark gray to black or sometimes reddish, broken into ridges with irregular furrows and small appressed scales.

WOOD. Reddish brown, sapwood paler, close-grained, hard, heavy, strong.

RANGE. Usually in dry sandy sterile soil, often in the Longleaf Pine belt, southeastern Louisiana; eastward to Florida and north to Virginia.

HYBRIDS. A number of hybrids are as follows:

Ashe Oak, *Q.* × *asheana* Little, is a hybrid of *Q. laevis* and *Q. incana.* It is named for William Willard Ashe (1872–1932) of the U. S. Forest Service. It is found in Mississippi, east to Georgia and Florida.

Bluffton Oak, *Q.* × *blufftonensis* Trel., is a hybrid of *Q. laevis* and *Q. falcata.* It is named for Bluffton, South Carolina, where it was first found. It is known to occur only in South Carolina.

Mellichamp Oak, *Q.* × *mellichampii* Trel., is a hybrid of *Q. laevis* and *Q. laurifolia.* It was named in honor of Joseph Hinson Mellichamp (1829–1903), physician and botanist of South Carolina and northern and central Florida.

Walter Oak, *Q.* × *walteriana* Ashe, is a hybrid of *Q. laevis* and *Q. nigra.* It was named in honor of Thomas Walter, English planter and botanist of South Carolina. It is known to occur in North Carolina, South Carolina, Georgia, Florida, and Alabama.

REMARKS. The genus name, *Quercus,* is the ancient classical name, and the species name, *laevis,* refers to

Turkey Oak
Quercus laevis Walt.

the smooth leaves. Vernacular names are Catesby Oak and Scrub Oak. It is generally short-lived, rapid-growing, and usually free of insects and disease. The wood is sometimes used for fuel.

BLACK OAK
Quercus velutina Lam.

FIELD IDENTIFICATION. Stout tree attaining a height of 90 ft, with a spreading open crown. Bark dark brownish black. Leaves cut into usually 7 oblique lobes with sinuses of different depths.

FLOWERS. April–May, appearing with the leaves in staminate and pistillate catkins; staminate catkins clustered, tomentose, 3–6 in. long; calyx hairy with acute lobes; stamens 4–12; pistillate catkins a few together on short tomentose peduncles; bracts ovate; stigmas red.

FRUIT. Maturing September–October, acorn solitary or paired, sessile or short-stalked, ovoid-oblong or hemispheric, brown, often striate, pubescent, ½–1 in. long, one half to three fourths of length enclosed in cup; cup turbinate, light brown, pubescent or glabrous, ¾–1 in. broad; scales closely appressed toward base of cup but loose and spreading near the rim—a diagnostic feature distinguishing it from other similar oaks.

LEAVES. Simple, alternate, deciduous, oval-obovate, usually with 7 oblique bristle-tipped lobes, middle lobes longest, apex acuminate or acute, base cuneate or truncate, surface dark green and lustrous; paler below and either pubescent or glabrous, tufts of hairs in axils of veins; 4–10 in. long, 3–7 in. broad; petioles 3–6 in., stout, yellow, glabrous or puberulous.

TWIGS. Reddish brown, stout, tomentose at first, glabrous later.

BARK. Dark brown to black, ridges flattened with platelike scales between deep fissures; inner bark orange-yellow, bitter.

WOOD. Reddish brown, sapwood paler, coarse-grained, strong, heavy, hard, weighing about 43 lb per cu ft, not commercially distinguished from other Red Oaks.

RANGE. Apparently not at home on the Gulf Coast plain. Often on poor, dry, sandy, heavy clay, or gravelly soils. Eastern Texas, Louisiana, Oklahoma, and Arkansas; eastward to Florida, north to Maine, and west to Ontario, Wisconsin, and Iowa.

PROPAGATION. The minimum seed-bearing age is 20 years, optimum 40–75, and the maximum 100. Good seed crops are borne every 2–3 years with light crops on intervening years. The cleaned seed averages about 250 seeds per lb, with a commercial purity of 100 per cent and a soundness of 85 per cent. For best germination the fruit should be stratified from collection time to seeding in spring at 33°–40°F. for 30–60 days. Good seed lots should germinate about 90 per cent. The planting is done in drills 8–12 in. apart on well-prepared soil and covered ½–1 in. with firmed soil.

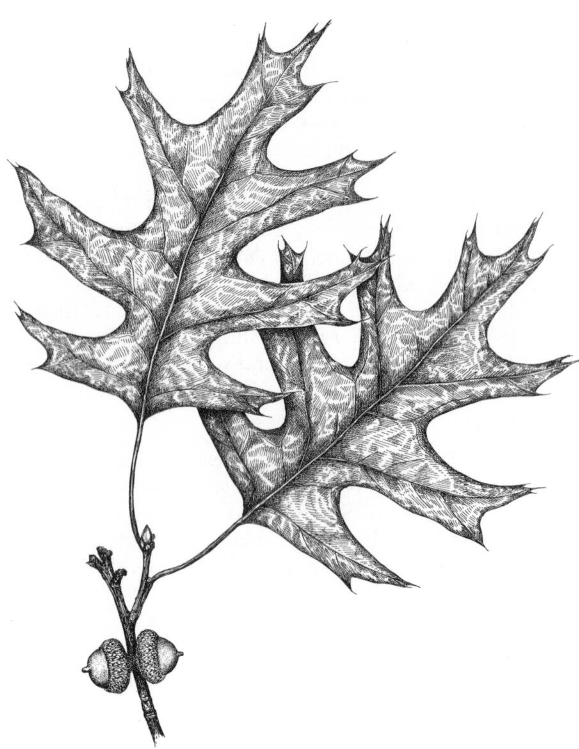

BLACK OAK
Quercus velutina Lam.

VARIETIES AND HYBRIDS. Missouri Black Oak, *Q. velutina* var. *missouriensis* Sarg., is a variety with leaves permanently rusty pubescent beneath. The bark is smoother and the cup scales more loosely imbricate. It occurs in Arkansas and Missouri.

Bush's Oak, *Q.* × *bushii* Sarg., is a hybrid of *Q. velutina* and *Q. marilandica* and is named for its discoverer, Benjamin Franklin Bush (1858–1937), botanist of Missouri. It has been found in eastern Texas, Oklahoma, Arkansas, Mississippi, Alabama, and Georgia; north to North Carolina, Virginia and Maryland, Missouri, Kansas, Nebraska, Iowa, Illinois, Indiana, and Pennsylvania.

Demaree Oak, *Q.* × *demareei* Ashe, is a hybrid of *Q. velutina* and *Q. nigra*. It is named for its discoverer, Delzie Demaree, botanist of Arkansas. It occurs in Arkansas and Louisiana.

Cross Oak, *Q.* × *filialis* Little, is a hybrid of *Q. velutina* and *Q. phellos*. The name is from a filial generation of a cross. It occurs in Louisiana, Arkansas, Missouri, Delaware, and New Jersey.

Hawkins Oak, *Q.* × *hawkinsiae* Sudw., is a hybrid of *Q. velutina* and *Q. rubra*, and is named for Mrs. Eugene Hawkins, who discovered it. It occurs in Missouri, Tennessee, Ohio, Pennsylvania, New York, Massachusetts, and Maine.

Lea Oak, *Q.* × *leana* Nutt., is a hybrid of *Q. velutina* and *Q. imbricaria.* It is named for its discoverer, Thomas Gibson Lea (1785–1844). It occurs in Missouri, Iowa, Ohio, Indiana, Michigan, Pennsylvania, North Carolina, Maryland, and District of Columbia.

Fink Oak, *Q.* × *palaeolithicola* Trel., is a hybrid of *Q. velutina* and *Q. ellipsoidalis,* so named because it grows in soil on Paleozoic rock, and occurs in Illinois, Iowa, Indiana, and Wisconsin.

Foot-leaf Oak, *Q.* × *podophylla* Trel., is a hybrid of *Q. velutina* and *Q. incana.* It is named for the Greek words meaning "foot" and "leaf," referring to the stout petiole. Reported from Virginia and North and South Carolina.

Rehder Oak, *Q.* × *rehderi* Trel., is a hybrid of *Q. velutina* and *Q. ilicifolia.* It is named for Alfred Rehder (1863–1949), American dendrologist of German birth, who first described and illustrated this hybrid. It occurs in Pennsylvania, Maine, Massachusetts, and Rhode Island.

Wandering Oak, *Q.* × *vaga* Palmer & Steyerm., is a hybrid of *Q. velutina* and *Q. palustris.* It occurs in Missouri.

Willdenow Oak, *Q.* × *willdenowiana,* is a hybrid of *Q. velutina* and *Q. falcata.* Named for Karl Ludwig Willdenow (1765–1812), a German botanist.

REMARKS. The genus name, *Quercus,* is the classical name, and the species name, *velutina,* refers to the velvety pubescence of the lower leaf surface. Vernacular names are Dyers Oak, Spotted Oak, Yellow-bark Oak, Yellow Oak, and Quercitron. The inner bark yields a tannin and a yellow dye for woolen goods. It is also a source of quercitannic acid, which has medicinal uses. However, because of the large amount of tannin, the bark is less often used in medical practice than White Oak bark. The drug is officially known as "Quercus Cortex" and is used mostly as a mild astringent.

The wood of Black Oak is used for rough lumber, crossties, and fuel, or generally for the same purposes as Red Oak, not generally being separated from it in the lumber trade. Black Oak is seldom used for ornamental planting because it lacks the brilliant fall coloring of some of the other oaks. It is rather slow-growing and trees over 200 years of age are seldom seen. It has been cultivated since 1802.

SHUMARD OAK

Quercus shumardii Buckl.

FIELD IDENTIFICATION. Tree attaining a height of 120 ft, with an open head and stout spreading branches. Leaves 5–9-lobed, usually 7-lobed, green and glabrous on both sides except for tufts of hairs in the axils of the veins beneath. Distinguished from Southern Red Oak by the smoothness and lobing, the Southern Red Oak having leaves densely pubescent beneath and irregularly lobed. Also, Southern Red Oak has much smaller acorns than Shumard Oak.

FLOWERS. Borne in spring on separate staminate and pistillate catkins; staminate catkins slender, 6–7 in. long, usually clustered; calyx-lobes 4–5, hairy; stamens 4–12; pistillate flowers solitary or paired, peduncles pubescent; involucral scales ovate, pubescent, brown or greenish; stigmas red.

FRUIT. Acorn sessile or short-stalked, solitary or paired, ovoid to oblong-ovoid, pubescent or glabrous, sometimes striate, ¾–1 in. long, ½–1 in. wide, set only at base in shallow, thick cups; cup covering one fourth the length of acorn; scales appressed, imbricate, thin or tuberculate, acuminate.

LEAVES. Simple, alternate, deciduous, obovate, or elliptic-oblong, cut into 7–9 more or less symmetrical lobes, lobes sometimes lobulate and bristle-tipped, sinuses broad and varying in depth. The leaves of upper and lower branches often vary considerably in the number and length of lobes. Upper surface dark green, glabrous and lustrous, lower surface paler and glabrous with tufts of axillary hairs; petioles glabrous, grayish brown.

TWIGS. Grayish brown, glabrous; branches smooth.

BARK. Gray to reddish brown, smooth or broken into small tight interlacing ridges.

WOOD. Light reddish brown, close-grained, hard, strong, weighing 57 lb per cu ft.

RANGE. Moist hillsides or bottom lands in clay soils. Central Texas, Oklahoma, and Arkansas; eastward

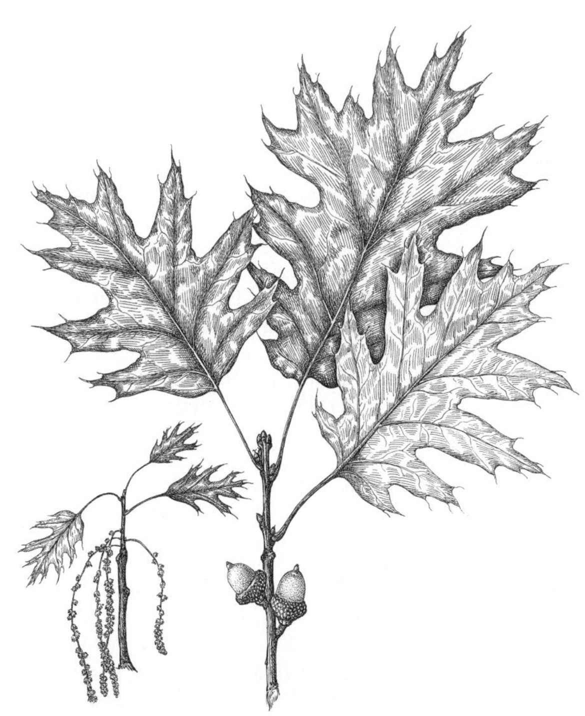

SHUMARD OAK
Quercus shumardii Buckl.

through Louisiana to Florida, northward to Pennsylvania and west to Kansas.

VARIETIES. Shumard Complex: Shumard Oak over its range occurs in a number of varieties. Schneck Oak, *Q. shumardii* var. *schneckii* Sarg., is very similar to the species but has deeper turbinate or hemispheric acorn cups. It is sometimes found associated with the species, but on the Texas Gulf Coast plain the predominant form appears to be the species. However, intermediate forms between the two often occur.

Texas Oak, *Q. texana* Buckley, of the central Texas limestone region has a close affinity with Shumard Oak, and is considered by some botanists as a variety. It has smaller acorns with turbinate cups. It also appears that some of the oaks of the high mountains of the Texas Trans-Pecos region may also have close kinship with the group, particularly the Chisos Mountains Red Oak, *Q. gravesii.*

Nuttall Oak, *Q. nuttallii* Palmer, is another tree closely related to Shumard Oak and grows in the valleys of the Mississippi and Red rivers from Missouri down through Mississippi, Louisiana, and eastern Texas. The leaves are 5–9-lobed with deep wide sinuses. The distinctive acorns are 1–1½ in. long, oblong-ovoid to short-cylindric, and enclosed one third to one half of their length in the cup; the cup is prolonged into a nipple-like base.

Mapleleaf Oak, *Q. shumardii* var. *acerifolia* Palmer, is a variety of the Shumard Oak found in Arkansas. The leaves are often as broad as long and deeply 3-lobed; the acorns are ⅕–⅓ in. long and enclosed ⅕–⅓ in. in the shallow cup.

HYBRIDS. Shumard Oak has a number of hybrids which have been listed by Little (1953a, p. 352).

Eggleston Oak, *Q.* × *egglestonii* Trel., is a hybrid of *Q. shumardii* and *Q. imbricaria,* named after William Webster Eggleston (1863–1935), botanist of the U. S. Department of Agriculture. It occurs in Kentucky and Missouri.

Hastings Oak, *Q.* × *hastingsii* Sarg., is a hybrid of *Q. shumardii* and *Q. marilandica,* named after Stephen Harold Hastings, agronomist of the U. S. Department of Agriculture. It occurs in central Texas.

Joor Oak, *Q.* × *joorii* Trel., is a hybrid of *Q. shumardii* and *Q. falcata,* named for Joseph F. Joor (1848–1892), who collected in 1884. It occurs at Galveston, Texas.

Moulton Oak, *Q.* × *moultonensis* Ashe, is named from Moulton Valley of the Tennessee River, Lawrence County, Alabama. It occurs in southeastern Virginia, Tennessee, Arkansas, and Alabama. It is a cross of *Q. shumardii* and *Q. phellos.*

Variable Oak, *Q.* × *mutabilis* Palmer & Steyermark, is a cross between *Q. shumardii* and *Q. palustris,* and the name means "variable." It occurs in Missouri.

Shumagra Oak, *Q.* × *neopalmeri* Sudw., is a cross of *Q. shumardii* and *Q. nigra,* named after Ernest J. Palmer, American botanist and authority on woody plants, who discovered it. It occurs in Georgia, central Florida, Alabama, and Arkansas.

REMARKS. The genus name, *Quercus,* is the classical name, and the species name, *shumardii,* refers to Benjamin Franklin Shumard (1820–1869), state geologist of Texas. Vernacular names are Spotted Oak, Leopard Oak, and Spanish Oak. The wood is not commercially distinguished from that of the other Red Oaks, and is used for veneer, cabinets, furniture, flooring, interior trim, and lumber. It is a beautiful tree with a symmetrical leaf design, and, being rather free from insects and diseases, could be more extensively cultivated for ornament.

Texas Oak

Quercus texana Buckl.

FIELD IDENTIFICATION. Shrub or small tree rarely over 35 ft in height, with spreading branches. Confined mostly to the high dry uplands of central Texas. The leaves are small, averaging about 3½ in. long.

FLOWERS. In catkins borne separately on same tree; staminate catkins 1⅓–3½ in. long, loosely flowered, hairy; calyx hairy-fimbriate, divided into 4–5 acute lobes, shorter than the stamens, anthers glabrous; pistillate catkins with short tomentose peduncles, catkins 1½–3½ in. long, 1–3-flowered, involucral scales reddish brown, stigmas red.

FRUIT. Acorn biennial, sessile or on short peduncles 1/12–⅓ in. long, solitary or paired, short-oblong or ovoid, ¼–¾ in. long (mostly about ½), reddish brown, pubescent, often striate, narrowed or rounded at the apex or abruptly apiculate, rounded at the base, one fourth to one half included in the cup; cup turbinate, abruptly constricted at the base, reddish, densely pubescent, ⅜–½ in. high and wide; scales ovate, appressed, thin, pubescent, apex obtuse, rounded or truncate, margins darker brown and thin.

LEAVES. Deciduous, turning red in autumn, 2½–5½ in. long, 2–3½ in. wide, ovate to obovate or rounded, margin with 3–7 (mostly 5) aristate lobes with intervening broad or rounded sinuses; terminal lobe often longest, entire or 2–3-lobed at the apex, acute; upper lateral lobes sometimes broad and entire or divided at the apex into lesser acuminate lobes; lower lateral lobes often much smaller (except where entire leaf is only 3-lobed); margin thin and slightly puberulent below along the veins, occasionally with tufts of hairs in the vein angles; petioles ⅓–1½ in. long, slender, glabrous, reddish yellow; lenticels small and pale; buds ⅛–⅕ in. long, lanceolate-ovoid, acute, reddish brown, tomentose; young unfolding leaves densely pubescent and reddish.

TWIGS. Slender, younger ones reddish brown, mostly glabrous; older ones gray and glabrous.

BARK. Dark gray to black with thick short ridges and platelike scales, fissures deep.

RANGE. Common on the dry uplands of central and

TEXAS OAK
Quercus texana Buckl.

west Texas (Edwards Plateau), but not known to occur beyond the Pecos River. Also in southern Oklahoma in the Arbuckle Mountains.

CLOSELY RELATED SPECIES, VARIETIES, AND HYBRIDS. The close relationship of Texas Oak and Shumard Oak has been a botanical classification problem. Some botanists prefer to classify the two as separate species, but others prefer to reduce Texas Oak to a varietal standing under Shumard Oak and give it the name of *Q. shumardii* var. *texana* (Buckl.) Ashe. The two oaks overlap in distribution on the eastern margin of the Edwards Plateau, and in that area trees with characteristics of both are in evidence. Whether most of this is a result of hybridization, or is an adjustment from alluvial coastal plain soils to the dry xerophytic conditions of the Texas hill country, is problematic. However, outside of these intermediates, Shumard and Texas oaks show distinct differences when growing in their own typical habitats.

As early as 1918 Charles Sprague Sargent noted the differences and relationships of Texas Oak and Shumard Oak (Sargent, 1918, p. 425). Sargent says: "The leaves of the sterile branches of Shumard Oak are often difficult to distinguish from those of the Texas Oak, and the best characters by which these oaks can be distinguished are found in the red-brown, more or less pubescent buds and reddish branchlets of Texas Oak and its variations, and the usually glabrous buds and grayish branches of the Shumard Oak and its varieties. The close relationship of these trees is shown, however, in the occasional occurrence in Missouri of trees of Shumard Oak with reddish, slightly pubescent buds and reddish branchlets.

"Descending sometimes from the hills in better soil the Texas Oak grows taller and produces fruit 1 in. in length with a turbinate cup comparably less deep than that of the smaller fruit produced on the neighboring hills. On the Edwards Plateau in western Texas trees occur with acorns acute at the apex, about ¾ in. long, and only ⅓ in. in diameter. On some trees the leaves are 5-lobed with broad and shallow sinuses."

Texas Oak is also very closely related to Graves Oak, which occurs in the high mountains of the Trans-Pecos Texas region. See the description of Graves Oak for a discussion of this matter.

Texas Oak also is thought to hybridize with Black Oak and Blackjack Oak.

REMARKS. The genus name, *Quercus,* is the classical name, and the species name, *texana,* refers to the state of Texas, its native habitat. The first specimen described in the literature was found on limestone hills near Austin, Texas. Also known under the names of Texas Red Oak, Rock Oak, Hill Oak, Spotted Oak, Red Oak, and Spanish Oak. The wood is used locally for fuel and posts, but the tree is generally too small for lumber.

GRAVES OAK

Quercus gravesii Sudw.

FIELD IDENTIFICATION. An oak attaining a height of 40 ft.

FLOWERS. Borne in separate staminate and pistillate catkins; staminate catkins loosely flowered, stellate-hairy, about ¼ in. long, perianth red, ciliate; pistillate catkins borne on reddish brown peduncles, ⅛–⅗ in. long, 1–3-flowered.

FRUIT. Acorn mature in September, solitary, or several together on peduncles 1/12–½ in. long, ovoid, ½–⅗ in. long, ⅓–⅖ in. broad, when green covered with dense tomentum, later glabrate or nearly so, light brown, one fourth to one half included in the cup; cup light brown, ⅖–⅗ in. broad, ⅕–⅓ in. high, deeply turbinate or hemispheric; scales ovate, obtuse at apex, tomentose, appressed.

LEAVES. Alternate, deciduous, red or yellow in autumn, 2–4 in. long, or rarely longer, 1½–3½ in. wide or rarely wider, ovate, obovate or oblong, obtuse or long-acute at apex, cuneate or rounded at base, variously 3–7-lobed with bristle-tipped teeth, the terminal lobe often somewhat elongate, sinuses rounded; dark green above, shiny, at first with fascicled hairs but later glabrate, lower surface paler green, later scattered stellate-hairy or glabrous, occasionally with dense tufts of hairs

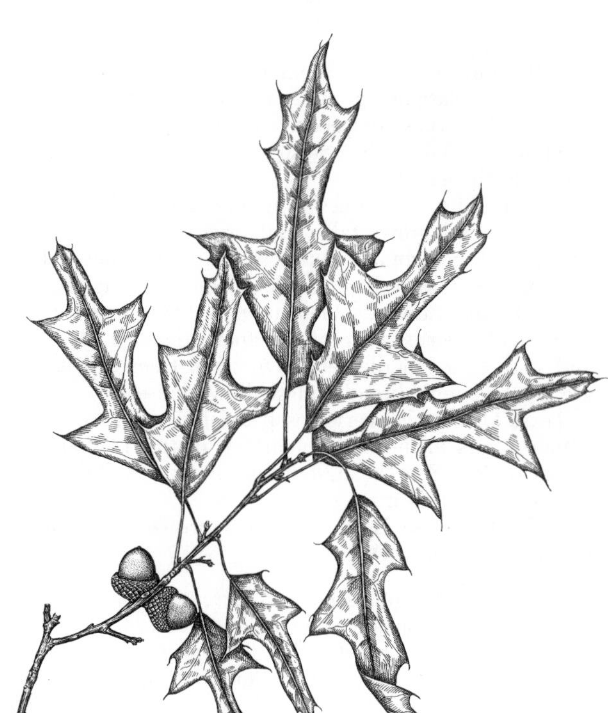

Graves Oak
Quercus gravesii Sudw.

in the vein angles above or below; petioles ⅓–2 in. or longer, slender, yellowish green to reddish, glabrous.

TWIGS. Rather slender, fluted, grayish green or reddish brown, glabrous or with stellate hairs, lenticels pale, inconspicuous.

BARK. Dark grayish black or black, on old trees roughened into flat ridges and narrow fissures, branches smooth and gray.

RANGE. Graves Oak is apparently confined to the high mountains of the Trans-Pecos Texas area, at altitudes of 4300–7800 ft in the Glass, Davis, and Chisos ranges. It is abundant in moist canyons in the Chisos Mountains particularly. Also in Coahuila, Mexico.

CLOSELY RELATED SPECIES, VARIETIES, FORMS, AND HYBRIDS. Botanical authorities are rather divided concerning the exact status of Graves Oak. Some classify it as the same tree as the Texas Oak, with minor variations such as small acorns, elongate terminal leaf lobes and occasional axillary hair tufts on the leaves.

However, Texas Oak is also in a controversial position and some consider it only a Western, highland variety of Shumard Oak, *Q. shumardii* var. *texana* (Buckl.) Ashe, instead of a separate species. If this position is taken then it is only logical to also reduce Graves Oak to a varietal standing under Shumard Oak. This question has been carefully weighed by Muller (1951, p. 99), who points out that variations between Graves Oak and Texas Oak, because of geographical overlapping, do not occur, and that Texas Oak rarely extends west of the Pecos River, while Graves Oak is confined to isolated high mountains of the Trans-Pecos area, such as the Chisos and Davis ranges. He notes, however, that the two trees do exhibit enough variation within themselves to form a nongeographical basis for an undeniable close relationship of the two.

The writer feels that at the present stage in this botanical investigation the differences are sufficient between Texas Oak and Graves Oak to maintain them as separate species. The relationships of Texas Oak and Shumard Oak are discussed under Texas Oak.

A form of Graves Oak with tufts of fascicled hairs on the leaf surfaces and branchlets has been found in the Davis mountains above Fort Davis. Although variously described by authors as a separate species, or as a variety of Texas Oak, or as a variety of Graves Oak, it seems most appropriate to consider it as only a hairy form of Graves Oak with no specific title.

Hybridization of Graves Oak with various oaks has been suggested. A Graves Oak and Emory Oak hybrid has been described. At one time this hybrid was considered as a possible species (Muller, 1934, pp. 119–120). It is now known as Robust Oak, *Q.* × *robusta* Muller, and is a cross of *Q. emoryi* and *Q. gravesii*, and is found in the Chisos Mountains, in Brewster County, Texas.

Another hybrid is known as Livermore Oak, *Q.* × *inconstans* Palmer, because of its inconstant character. It is found in the Davis Mountains, Jeff Davis County, Texas. It is a hybrid of *Q. gravesii* and *Q. hypoleucoides*.

REMARKS. *Quercus* is the classical name, and the species name, *gravesii*, is for the botanist H. S. Graves (1871–1950), dean of Yale University School of Forestry. Vernacular names are Chisos Oak, Chisos Red Oak, Texas Red Oak, Mountain Red Oak, Rock Oak, and Encina. The acorn is eaten by various species of ground squirrels and the foliage occasionally browsed by deer.

Lateleaf Oak

Quercus tardifolia C. H. Muller

A very rare oak known only from the Chisos Mountains of western Texas, described by Muller (1936b, pp. 154–155). The tree is inadequately known and the following description is adopted from Muller's account:

FIELD IDENTIFICATION. Small erect trees with short stiff branches and hard furrowed bark.

FLOWERS. Unknown.

FRUIT. Acorn biennial, solitary or paired, subsessile, young cups with scales thin and closely appressed, short apices truncate, glabrous and brown, tomentose basally. Mature fruit unknown.

Lateleaf Oak
Quercus tardifolia C. H. Muller

LEAVES. Evergreen, the new ones appearing with dense tomentum about the first of July, rather thick and chartaceous, blades 2–2½ in. long, 1–2¼ in. wide, sometimes to 4 in. long and 3 in. wide, oblong-ovate to subobovate, 3–4-lobed on each side with shallow or moderate sinuses, the lobes aristate-tipped and entire or rarely 2-toothed, short and rather broadly acute like the apices, basally subequilateral and slightly cordate, the upper surface dull bluish green, at length glabrate, lower surfaces detachably stellate-tomentose; veins 6–8 on each side, alternately passing into the teeth, hardly evident above but prominent beneath; petioles 3/5–4/5 in. long, 1/2–1/15 in. thick, red at base, glabrate with the leaves; buds 1/8–1/6 in. thick, acutely fusiform, hairy at the apices, the broadly truncate scales usually split at the end, slightly pubescent and often ciliate; stipules at length deciduous, 1/5–1/4 in. long, setiform to broad-ligulate, pubescent at apex.

TWIGS. About 1/12 in. thick, somewhat fluted, densely fulvous, with stellate tomentum the first season, glabrate or nearly so the second and reddish brown with minute inconspicuous lenticels, finally becoming gray.

RANGE. C. H. Muller states that only 2 clumps of this species have been found in the Chisos Mountains of western Texas in woodlands at an altitude of about 7,000 ft.

REMARKS. The genus name, *Quercus*, is the classical name, and the species name, *tardifolia*, refers to the late appearance of the leaves. The flowers and mature fruit are unknown. More study is needed to determine the plant's place in taxonomy.

Arkansas Oak
Quercus arkansana Sarg.

FIELD IDENTIFICATION. Tree to 70 ft high, and 1 ft in diameter, the stout ascending branches forming a narrow head.

FLOWERS. Staminate catkins 2–2½ in. long, white-hairy; calyx 3–4-lobed, hairy; stamens 4; anthers ovoid-oblong, apiculate, dark red; pistillate on stout peduncles with tomentose involucre scales; stigmas dark red.

FRUIT. Acorn solitary or paired, sessile or on short, glabrous peduncles; body ovoid, apex rounded, at first pubescent with fascicled hairs, later more glabrous, light brown, slightly striate, length 1/4–1/3 in., width 1/2–5/8 in.; cup saucer-shaped, enclosing the acorn only at the base, scales closely appressed, apex obtuse, pubescent.

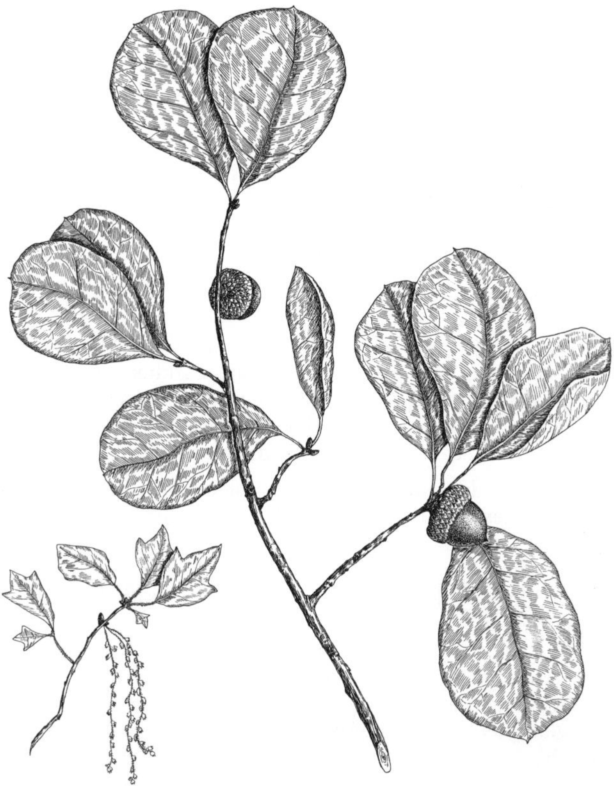

Arkansas Oak
Quercus arkansana Sarg.

LEAVES. Simple, alternate, deciduous, broadly obovate, shallowly 3-lobed in some at the apex, most rounded or acute, often slender-mucronate; 2–2¾ in. long and broad, on vigorous shoots 4½–5½ in. long and 2½–2¾ in. wide, upper surface glabrous at maturity and yellowish green; lower surface paler and glabrous or pubescent in the axils of the veins; petioles slender.

TWIGS. Slender, brownish gray, pubescent at first, glabrous later; buds brown, ovoid, acute, pubescent or semiglabrous.

BARK. Dark to grayish brown or black, fissures deep, ridges long and narrow, scales thick and tight.

RANGE. Sterile clays, gravels, and sandy hills. In Arkansas in Hempstead and Clark counties. On Yellow Creek, Bridge Creek, and Sandy Bois d'Arc Creek and other small tributaries of the Little River and Red River. In southeastern Alabama, northern Florida, and southwestern Georgia.

REMARKS. The genus name, *Quercus,* is the classical name, and the species name, *arkansana,* refers to the state of Arkansas, where it occurs in most abundance. It was at first considered a hybrid of *Q. marilandica* and *Q. nigra,* but Palmer (1925b) has set forth evidence to sustain it as a distinct species of ancient lineage, formerly of greater abundance than at present. Also known locally as Water Oak and Arkansas Water Oak.

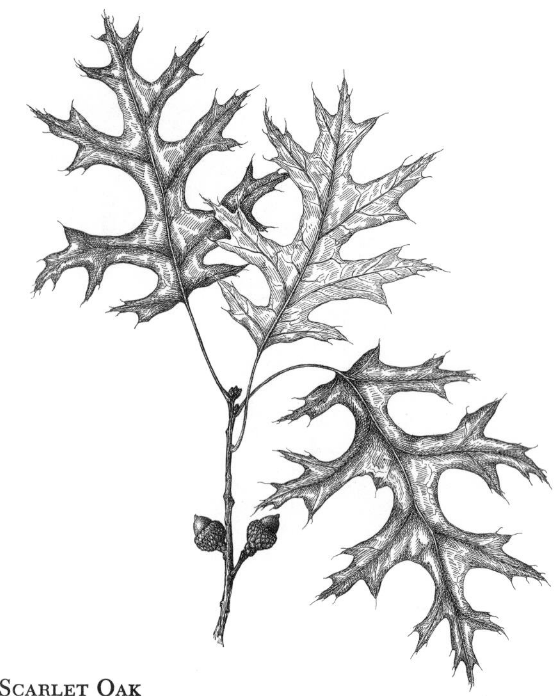

SCARLET OAK
Quercus coccinea Muenchh.

SCARLET OAK

Quercus coccinea Muenchh.

FIELD IDENTIFICATION. A tree attaining a height of 100 ft, with a diameter of 4 ft. The crown is rounded to narrowly oblong and rather open with irregular branches.

FLOWERS. Catkins in April, staminate 3–4 in. long, slender, glabrous or pubescent, loosely flowered, red at first; calyx-lobes 4–5, ovate and acute, shorter than the stamens, stamens 3–5; pistillate catkins on pubescent peduncles about ½ in. long, few-flowered, red; involucral scales shorter than the acute calyx-lobes; ovate, pubescent; styles slender, spreading, recurved.

FRUIT. September–October, biennial, solitary or paired, sessile or short-stalked, oval to ovate or hemispheric, apex rounded, base rounded or truncate, brown and sometimes striate, occasionally with concentric rings close to the apex, length ⅜–1 in., width ⅓–⅔ in., enclosed in the cup one third to one half its length, kernel white and bitter; cup turbinate or hemispheric, ½–1¼ in. in diameter, more or less contracted at base; scales of cup rather large, closely imbricate, shiny, mostly glabrous, thin, triangular to lanceolate or ovate, apex obtuse.

LEAVES. Simple, alternate, deciduous, blades 3–7 in. long, 2–5 in. wide, oblong to elliptic or obovate, apex acute or acuminate, base truncate or broadly cuneate; lobes 7–9 (rarely 5) extending more than halfway to the midrib, divergent or ascending, oblong to obovate or sometimes falcate, the apices with smaller repand bristle-tipped lobes; sinuses wide and rounded; young leaves densely pubescent; mature leaves thin, firm; upper surface bright green, glabrous, smooth, lustrous; lower surface paler, often with axillary tufts of rusty hairs; leaves brilliant scarlet in autumn; petioles long, slender, terete, 1–3 in. long.

TWIGS. Slender, greenish at first, later orange-red or brown, glabrous or pubescent, pith star-shaped in cross section; buds ovoid to ellipsoid, glabrous or pubescent, ⅛–¼ in. long.

BARK. Brownish black, broken by shallow fissures and irregular ridges, later roughly corrugated, flaking into small light brown to reddish scales.

WOOD. Light to reddish brown, coarse-grained, heavy, hard, strong, weighing 47 lb per cu ft, inferior to some of the other Red Oaks of commerce, but not officially distinguished from them.

RANGE. In dry sterile, sandy soil. Eastern Oklahoma and Mississippi; east to Florida, north to Massachusetts, Maine, and Ontario, and west to South Dakota, Wisconsin, Michigan, and Illinois.

VARIETIES AND HYBRIDS. Sargent Scarlet Oak, *Q. coccinea* var. *tuberculata* Sarg., is a Southern variety with larger acorns, thicker scales on the cup, and the axillary tufts of hairs are often absent. It occurs from Georgia and Alabama, north to Massachusetts, Indiana, and Missouri.

Scarlet Oak hybridizes with Northern Red Oak, *Q. rubra,* to produce Bender's Oak, *Q.* × *benderi* Balnitz.

Scarlet Oak also hybridizes with Scrub Oak, *Q. ilicifolia* to produce Robbins Oak, *Q.* × *robbinsii* Trel.

REMARKS. The genus name, *Quercus,* is the ancient classical name, and the species name, *coccinea,* refers to the scarlet leaves in autumn. Vernacular names are Spanish Oak and Red Oak. For later planting the acorns may be stored in a cool, humid place. They should be stratified at collection time if spring sowing is practiced. The seeds average 156–400 seeds per lb, with a germination average of 75 per cent. The tree is beautiful because of its deeply cut leaves and scarlet coloring in autumn. Scarlet Oak was introduced into cultivation in 1691 and is still much planted for ornament. However, it is not favored for row planting because the shape is often rather irregular. It is rapid-growing, short-lived, shallow-rooted, rather free of insects and fungus diseases, subject to dry rot, not very drought resistant, and coppices from young trees only. The foliage is occasionally browsed by white-tailed deer, and the fruit eaten by red and gray squirrel, blue jay, and sharp-tailed grouse.

NUTTALL OAK

Quercus nuttallii Palmer

FIELD IDENTIFICATION. Tree attaining a height of 120 ft, with a diameter of 3 ft. Small trees have a rather narrow pyramidal crown which becomes broad, open and wide-spreading with age. The upper branches are ascending and the lower horizontal or drooping. The trunk is often branched close to the ground, and older trees are almost always strongly buttressed.

FLOWERS. Borne in separate staminate and pistillate catkins similar to those of Red Oak.

FRUIT. Acorn and cup variable in size and shape; sessile or short-stalked, oblong-obovoid, length ¾–1¼ in., width ½–1 in., light to dark reddish brown, usually striate, apex rounded and narrowed, at first scurfy but later shiny, enclosed about one fourth to one half of length in a hemispheric or turbinate cup; cup about ⅝ in. broad, gray-puberulent, abruptly continued into a conspicuous (1⁄12–3⁄16 in. long) neck at the base, which is an important diagnostic feature.

LEAVES. Very variable in size and shape, simple, alternate, deciduous, rather symmetrically 5–7-lobed, the lobes horizontal or slightly ascending, the central pair usually longer and broader, usually terminating in 2–5 bristly teeth and commonly abruptly squared or angular; terminal lobe either entire, acuminate and bristle-tipped, or 3-toothed with the middle tooth much larger than the others and all bristle-tipped; smaller and lower lobes triangular, acuminate, and more or less entire (particularly in the small leaves in the upper part of the crown), sinuses wide, deep, obtuse or angularly rounded, base cuneate to truncate, length of blade 4–8 in., width 2–5 in., upper surface dull dark green and glabrous, lower surface paler and glabrous or with tufts of pale hair in the axils of the main veins, thin but firm; petioles slender, glabrous, ¾–2 in. long; winter buds pubescent to glabrous, ciliate, ovoid, acute to acuminate, about ¼ in. long; autumnal leaves yellow or dull light brown, rarely red.

TWIGS. Young ones glabrous, olive-green to reddish brown, later brown to gray.

BARK. On young trees greenish brown, tight, smooth, often shiny; on older trees very close, hard, firm, ½–1 in. thick; fissures shallow, irregular and narrow; ridges broad, flat and gray-scaly; light to dark gray or brownish to black, often with small excrescences, warts, or burls.

WOOD. Reddish brown, sapwood yellowish brown, hard, heavy, mineral stains common, checks excessively. The short trunk and many horizontal or drooping branches result in many pin knots in the logs.

RANGE. Usually in low, poorly drained clay, silty clay, or occasionally loamy flats in both first and second bottom lands. Also sometimes on fairly well-drained clay ridges of first bottoms. Not usual in permanent swamps

NUTTALL OAK
Quercus nuttallii Palmer

but more common in shallow swags and drains. On the Gulf Coast plain or adjacent provinces. Eastern Texas, Louisiana, Mississippi, Alabama, Oklahoma, Arkansas, and southeastern Missouri.

RELATIONSHIP TO PIN OAK. Nuttall Oak is easily confused with the Northern Pin Oak, *Q. palustris* Muenchh., especially in Arkansas where the ranges of the two overlap. The following account of the differences is given by Putnam and Bull (1932, pp. 117–118):

"In Arkansas, however, where both species are common, and possibly in central and northern Mississippi where the two ranges are imperfectly known, there will frequently be difficulty in making positive identification. The habitat and bark are often practically identical. The form of the tree and the leaves are so nearly the same, in the usual absence of a uniform set of perfectly typical characteristics, that they should not be relied upon to separate the two. What slight differences there seem to be between typical leaves of the two species can be summarized briefly as follows: *nuttallii* leaves are rather dull, or at least not very glossy or lustrous on the upper surface; *palustris* leaves are ordinarily distinctly lustrous above. *Nuttallii* leaves are hardly ever 9-lobed; *palustris* leaves are rather frequently 9-lobed. *Nuttallii* leaves (except from the tops of large crowns or the tips of large branches) are generally broader and less acuminate at the tip than are *palustris* leaves and have slightly shorter petioles and more angular, more widely divergent, broader, lateral lobes with more sharply truncate ends; they are also much more commonly broadest just above the middle, with a sharp taper toward a narrow cuneate base, than are *palustris* leaves.

"The fruit or the winter buds, or both, must therefore be observed. These are both typically very different in the two species and, although variations will be found that will occasionally make separation difficult, in the vast majority of cases they can be depended upon to point out the correct identification. The fruit of Nuttall oak is an oblong-ovoid acorn, ½–1 in. wide by ¾–1¼ in. long, that is enclosed for from one fourth to one half of its length in a thick-based deep cup, the base of which is drawn out into a distinct, stout, short stalk. The fruit of pin oak is an almost hemispherical acorn, rarely over ½ in. in length and breadth, that is enclosed for about one fourth (sometimes more) of its length in a thin, shallow, saucer-like cup. The size and shape of the acorn and the thickness, shape and depth of the cup are the important points to be noted. The color of each acorn is much the same (light to dark lustrous reddish-brown) and each is usually striped.

"The winter buds of Nuttall oak are usually just a shade less than ¼ in. long, plump, ovoid, acute at the apex, inconspicuously angled, light medium or grayish-brown or dull straw color, and entirely smooth or only slightly downy. The winter buds of pin oak are rarely over ⅛ in. long, plump, ovate, acute at the apex, rarely angled, light chestnut brown and more or less smooth. Nuttall oak buds strongly resemble and are hardly distinguishable from those of Shumard red oak. Their greater size and usually distinctive color are the important characteristics separating them from the buds of pin oak."

REMARKS. The genus name, *Quercus*, is the classical name for the oak. The species name, *nuttallii*, is in honor of Thomas Nuttall (1786–1859), an English-born American botanist and ornithologist. It is also known under the name of Red Oak, is not segregated from it in the lumber trade, and is of considerable commercial value. Other names for it are Smooth-bark Red Oak, Tight-bark Red Oak, Yellow-butt Oak, Striped Oak, Red River Oak, and Pin Oak. The tree has more than average susceptibility to fire injury and fungus damage. It is also subject to leaf beetle and grub infestation in damaged or waning specimens. A variety of Nuttall Oak has been given the name of *Q. nuttallii* var. *cachensis* Palmer. It has smaller, short-oblong or depressed conic acorns included about ⅓ in. in the cup, length ⅗–¾ in., width ½–⅖ in. In Arkansas in the bottoms of the Cache River.

AMERICAN BEECH

Fagus grandifolia Ehrh.

FIELD IDENTIFICATION. Beautiful tree attaining a height of 120 ft, with a rounded top and spreading branches. Easily sprouting from the roots to form thickets.

FLOWERS. April–May after the leaves unfold in separate staminate and pistillate clusters; staminate in globose heads, about 1 in. in diameter, pendent on hairy peduncles 1–2 in. long; stamens 8–10 with green anthers; pistillate flowers in clusters of 2–4 borne on short hoary peduncles ½–1 in. long; calyx campanulate, 4–5-lobed, hairy; pistil composed of a 3-celled ovary and 3 inwardly spreading stigmatic styles.

FRUIT. Ripening September–November, borne on stout hairy peduncles, composed of burlike involucres ½–¾ in. long with straight or recurved prickles, full-grown at midsummer but becoming brown and persistent on the branches, splitting into 4 valves to release a pair of small brown, 3-angled, sweet nuts, dispersed after first frost, good seed crops every 2–3 years.

LEAVES. Simple, alternate, deciduous, straight-veined, ovate-oblong; acuminate at apex, cuneate to rounded or cordate at base, coarsely serrate on margin; when mature glabrous and dark green above; paler and pubescent, especially in the axils of veins beneath; 3–6 in. long; petioles ⅙–½ in., hairy; stipules ovate-lanceolate to linear.

TWIGS. Slender, zigzag, green and hairy at first, later glabrous and orange to yellow or reddish brown; lenticels oblong and orange-colored.

BARK. Light gray, often mottled, smooth.

WOOD. Varying shades of red, sapwood lighter, close-

American Beech
Fagus grandifolia Ehrh.

grained, hard, strong, tough, difficult to cure, not durable, weighing 43 lb per cu ft.

RANGE. Eastern Texas, Louisiana, Arkansas, and Oklahoma; eastward to Florida, north to Nova Scotia, and west to Wisconsin, Michigan, Illinois, and Missouri.

PROPAGATION. The seeds may be thrashed from the trees and raked up on the ground after the first heavy frost. They can be spread out to dry and shaken to remove the dried burs. Cleaned seed averages about 1,600 seeds per lb, with a commercial seed purity of 97 per cent and a soundness of 88 per cent. It can be stored over winter at 41°F. in sealed containers. The seed can be sown immediately in fall, or stratified for spring sowing in sand at 41°F. for 60–90 days. The germinative capacity of stratified seed averages about 85 per cent. A seed bed of sandy-loam soil should be prepared, and the seeds may be broadcast or drilled and covered with soil about ½ in. deep. Rodent protection is necessary, and partial shade required for most of the first year. Transplanting is usually done with 1–3-year-old plants on sandy-loam or limestone soils.

VARIETIES. Some botanical authorities hold that the Northern and Southern beeches vary, and have described the Southern form as Carolina Beech, *F. grandifolia* var. *caroliniana* Fern. & Rehd. Carolina Beech has leaves which are darker green, thicker, short-ovate, not as coarsely toothed, and the involucre is smaller with fewer and shorter prickles. Also a form has been named as *F. grandifolia* var. *pubescens* Fern. & Rehd., which has leaves soft-pubescent beneath.

REMARKS. The genus name, *Fagus*, is from an old Greek word referring to the edible nuts, and the species name, *grandifolia*, refers to the large leaves. Vernacular names are Red Beech, Ridge Beech, and White Beech, also Beechnut. The tree is a very desirable one for ornamental planting, but grass has a difficult time growing under the dense foliage. It was first cultivated in the year 1800. It is long-lived, free of disease, and sprouts easily from the roots to form thickets. The wood is sold commercially for chairs, tool handles, shoe lasts, flooring, cooperage, crates, spools, brush backs, toys, and fuel. The small sweet edible nuts are sometimes gathered and sold on the markets in the Northern states and Canada, and are a source of vegetable oil and swine feed. They are eaten by many species of birds and by raccoon, opossum, porcupine, gray fox, red fox, and white-tailed deer.

Ozark Chinquapin

Castanea ozarkensis Ashe

FIELD IDENTIFICATION. Tree attaining a height of 50 ft, with a diameter of 2 ft.

FLOWERS. Staminate and pistillate catkins borne separately on same tree; staminate catkins 2–8 in. long, pendent, linear, ¼–⅜ in. wide at anthesis, peduncle densely white-hairy; stamens numerous in dense, close clusters; filaments filiform with very small, oval anthers; bracts linear-lanceolate, acuminate, white-hairy, about

Ozark Chinquapin
Castanea ozarkensis Ashe

1/16 in. long; pistillate flowers much smaller and in clusters at the base of the catkins.

FRUIT. Spiny burs often in large, heavy clusters, ½–1½ in. across, brown at maturity; spines sharp, elongate, in dense stellate clusters; seed edible, solitary, subglobose to ovoid, dark brown, sweet.

LEAVES. Simple, alternate, deciduous, 3–10 in. long, 1½–3 in. wide, elliptic-oblong, apex acute to acuminate, base cuneate or some rounded or semicordate, veins conspicuously straight; upper surface dark green, smooth and glabrous, lower surface paler and velutinous, margin very coarsely toothed and teeth elongate; petioles ¼–½ in., glabrous or nearly so.

TWIGS. Slender, young ones light brown, covered with dense pubescence, older ones gray and glabrous.

BARK. Light brown to reddish brown or grayish, the flat ridges breaking into platelike scales.

WOOD. Dark brown, sapwood rather narrow, coarse-grained, hard, rather brittle, weighing about 31 lb per cu ft.

RANGE. Rocky slopes and moist hillsides. In southern Missouri, Arkansas, eastern Oklahoma, and northeastern Louisiana. Most abundant on the slopes of the Ozark and Ouachita mountains.

VARIETY. Smooth Ozark Chinquapin, *C. ozarkensis* var. *arkansana* Ashe, is a variety with scant pubescence on sun-exposed leaves which are glaucescent on their lower surface. It is confined mostly to the Ozark Mountain region, but extending westward south of the Arkansas River as far as Mena. Intermediate forms between it and the species occur.

REMARKS. The genus name, *Castanea,* is for a town in Thessaly, and the species name, *ozarkensis,* is for the Ozark Mountains of Arkansas, where it is most abundant. The tree often sprouts from stumps. The wood is cut and marketed with Chestnut for fence posts and railroad crossties. Dwight M. Moore, of the Botany Department of the University of Arkansas, reports that a very large specimen 52 ft high and 10 ft 4 in. in diameter grows in Washington County, 5 miles north of Fayetteville, and ¾ miles east of Johnston, Arkansas.

ALLEGHENY CHINQUAPIN

Castanea pumila (L.) Mill.

FIELD IDENTIFICATION. Thicket-forming shrub or tree with slender spreading branches and a round top, sometimes attaining a height of 50 ft.

FLOWERS. Both staminate and pistillate catkins on same tree; some catkins all staminate, and some with both staminate and pistillate flowers; staminate catkins cylindric, slender, hoary-tomentose, 2½–6 in. long, ¼–⅓ in. in diameter; calyx small, 6-lobed; stamens 8–20 with 2-celled anthers; pistillate flowers generally in threes, or scattered toward base of catkins, and staminate flowers toward the tip; involucre prickly, sessile or short-stalked; ovary imperfectly 6-celled; styles linear, exserted with small stigmas.

ALLEGHENY CHINQUAPIN
Castanea pumila (L.) Mill.

FRUIT. In spikelike clusters; burs formed by the prickly involucre and 1–1½ in. in diameter; spines of bur in crowded clusters, slender and basally tomentose or glabrous; bur opening by 2–3 valves to expose the nutlet; nutlet solitary, small, shiny brown, round-ovoid, pointed and somewhat pubescent at the apex; kernel sweet and edible.

LEAVES. Alternate, simple, deciduous, blades 3–4 in. long, 1½–2 in. wide, elliptic-oblong to oblong-obovate, acute at the apex, unequal and rounded or broadly cuneate at the base; margins coarsely serrate with pointed teeth; upper surface glabrous, yellowish green, lower velvety-white pubescent; petioles short, stout, flattened, pubescent at first, glabrous later, ¼–½ in. long; stipules yellowish green, ovate to lanceolate or linear, pubescent.

TWIGS. Green to reddish brown or orange-brown, pubescent at first, glabrous later.

BARK. Smoothish, reddish brown, furrows shallow, ridges flat with platelike scales.

WOOD. Dark brown, coarse-grained, light, hard, strong, durable, weighing 37 lb per cu. ft.

RANGE. East Texas, Oklahoma, and Louisiana; eastward to Florida, north to New Jersey, and west to Missouri.

HYBRID. Chinknut, *C.* × *neglecta* Dode, is a hybrid of *C. pumila* × *C. dentata.*

REMARKS. The genus name, *Castanea,* is for a town in Thessaly, and the species name, *pumila,* for the tree's small stature. The tree is generally too small for commercial use but is occasionally used for posts, railroad crossties, and fuel. The sweet nuts are sometimes gathered for the market and are eaten by a number of birds and mammals.

Ashe Chinquapin

Castanea ashei Sudw.

FIELD IDENTIFICATION. Shrub or small tree to 30 ft high, with usually several trunks from the base, and a broad round-topped head.

FLOWERS. In slender cylindric catkins; staminate catkins about 5 in. long; stamens 8–20, conspicuous, exserted on long filaments with small yellow anthers; calyx bell-shaped, 6-lobed, yellowish green, pubescent; pistillate flowers on the lower part of bisexual catkins in little clusters of 2 or 3 with spiny involucres; sterile stamens often present; ovary 6-celled with spreading linear styles.

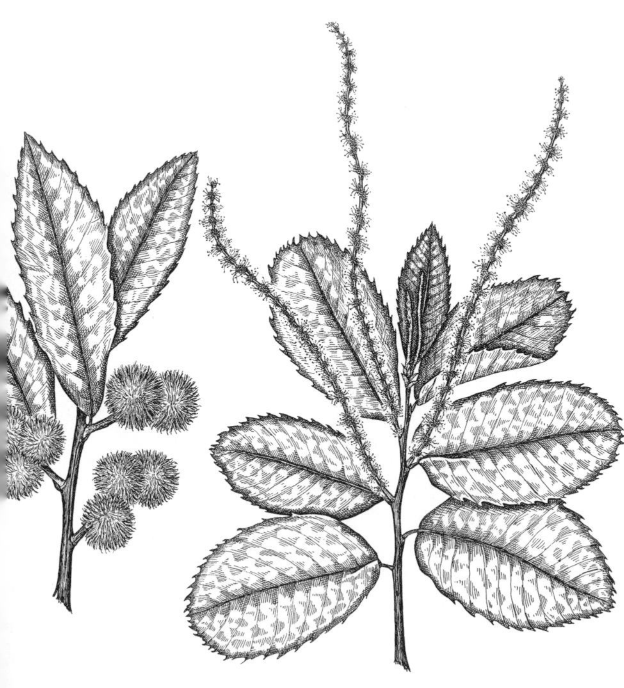

Ashe Chinquapin
Castanea ashei Sudw.

FRUIT. Borne in subglobose brown burs about 1 in. long; spines of bur short, stubby, pubescent, in rather distant clusters; bur dehiscent into 2–4 valves to expose the nuts; nuts solitary, ovate, lustrous brown, point pubescent and often stellate; kernel sweet; edible.

LEAVES. Alternate, simple, deciduous, firm, elliptic to narrowly obovate, acute or rounded at apex, narrowly rounded at base, coarsely serrate on margin, dark green and glabrous above, gray-downy beneath, about 3 in. long and 1½ in. wide, but sometimes larger on young twigs.

TWIGS. Yellowish brown, slender, gray-tomentose, especially when young.

BARK. Rather smooth, brownish gray, ridges flat and broad, furrows shallow.

WOOD. Hard, strong, tough, not large enough for commercial use.

RANGE. Thickets on hillsides. East Texas, Arkansas, and Oklahoma; east to Florida and north to Virginia.

REMARKS. The genus name, *Castanea,* is for a town in Thessaly, and the species name, *ashei,* is in honor of W. W. Ashe (1872–1932), a dendrologist of the U. S. Forestry Service. This species may be confused with the Allegheny Chinquapin, but the leaves are usually smaller and narrower, and the burs have shorter, stubbier pubescent spines in less crowded clusters. However, some authors feel that Ashe Chinquapin is not a species but only a variety of Allegheny Chinquapin, and have accepted the name of *C. pumila* var. *ashei* Sudw. Ashe Chinquapin may be distinguished from Florida Chinquapin by the gray-tomentose undersurface of the leaf. Florida Chinquapin is lustrous and smooth, and only occasionally thinly tomentose.

Florida Chinquapin

Castanea alnifolia var. *floridana* Sarg.

FIELD IDENTIFICATION. Tree to 40 ft high, with spreading branches forming a narrow crown; often with many trunks from the base.

FLOWERS. Borne in 2 kinds of catkins on same tree, one kind bearing all staminate flowers, and the other bearing staminate flowers on the distal part and pistillate on the proximal; staminate, catkins 4–5 in. long, cylindric, pubescent; stamens 8–20, exserted with 2-celled anthers; calyx 6-lobed; androgynous catkins with pistillate flowers below the middle, bearing prickly involucres and a 6-celled ovary terminating in linear exserted styles.

FRUIT. A bur formed from the spiny involucre, ¾–1¼ in. in diameter, globose or short-oblong, tomentose;

FLORIDA CHINQUAPIN
Castanea alnifolia var. *floridana* Sarg.

spines stout, pubescent, fascicles somewhat scattered with bald spots between; bur splitting into 2–3 valves to expose the nut; nut ovoid, shiny brown, acute, ½–¾ in. long; kernel sweet, edible.

LEAVES. Alternate, simple, deciduous, oblong-obovate to elliptic, acute at apex, rounded or cuneate at base, shallowly bristle-toothed on margin, thin, dark green and glabrous above, lighter green and glabrous below, or some showing a thin tomentum, 3–4 in. long, 1–1¾ in. wide; petiole stout and glabrous.

TWIGS. Reddish brown, slender, pubescent or glabrous.

BARK. Smoothish, ridges flat, furrows shallow.

WOOD. Hard, strong, durable, brownish.

RANGE. Rich moist soil of thickets or roadsides. Eastern Texas, Louisiana, and Oklahoma; eastward to Florida and northward to North Carolina.

OTHER SPECIES AND VARIETIES. Florida Chinquapin is considered to be a variety of Trailing Chinquapin, *C. alnifolia* Nutt., which is a thicket-forming shrub with creeping rootstocks, broader obovate leaves, and somewhat puberulous beneath. It occupies about the same range as the species.

Chinquapins are rather similar in appearance but may be distinguished chiefly by the leaf pubescence and spines of the fruit. The Allegheny Chinquapin has no rootstocks, leaves with velvety-white tomentum beneath, and the bur spines are rather long, slender, and glabrous, or sparsely pubescent. Florida Chinquapin has bur spines which are sparser and shorter, and the leaves glabrous and lustrous, or only thinly tomentose beneath. Ashe Chinquapin has even shorter, stubbier, and more pubescent spines than Florida Chinquapin, and the leaves are gray-downy beneath.

C. × *alabamensis* Ashe is a hybrid of *C. alnifolia* and *C. dentata.*

REMARKS. The genus name, *Castanea,* is for a town in Thessaly; the species name, *alnifolia,* refers to the alder-like leaves; and the variety name, *floridana,* refers to the Floridian habitat.

ELM FAMILY (*Ulmaceae*)

LINDHEIMER HACKBERRY

Celtis lindheimeri K. Koch

FIELD IDENTIFICATION. Tree to 40 ft, trunk 12–18 in. in diameter. The stout spreading branches form a broad irregular head. Recognized by the decidedly grayish green leaves.

FLOWERS. Borne March–April on branches of the year, minute polygamo-dioecious or rarely monoecious, pedicels pubescent; staminate fascicled; pistillate solitary or few-flowered; calyx greenish yellow, deciduous, the 5 lobes oblong, scarious, apex narrowed and rounded; stamens inserted on the margin of a tomentose torus, filaments subulate, anthers ovoid; ovary ovoid, sessile, green; style short and sessile (rudimentary in the staminate flowers), lobes divergent and stigmatic within.

FRUIT. Drupe maturing in September, persistent, peduncles ¼–⅔ in., tomentose; body subglobose, about ¼ in. in diameter, ellipsoid, reddish brown, shiny, flesh thin; seed a bony nutlet.

LEAVES. Simple, alternate, blade length ¾–3 in., width ¾–2 in., blade ovate to lanceolate or ovate to oblong, apex acute to acuminate, base cordate or rounded, often asymmetrical, margin entire, or crenate-serrate on young growth, upper surface grayish green and scabrous, lower surface paler and conspicuously reticulate-veiny, clothed with white hairs when young, less so or almost glabrous at maturity, 3-veined; petioles ¼–½ in., densely villous-pubescent.

TWIGS. Slender, gray, pubescent at first, glabrous later, lenticels numerous and rough.

BARK. Gray to brown, roughened by thick, wartlike excrescences and broken into narrow ridges and shallow fissures.

WOOD. White to brownish, not strong or durable, of little commercial value, used occasionally for fuel.

LINDHEIMER HACKBERRY
Celtis lindheimeri K. Koch

RANGE. Rich bottom lands adjacent to limestone hills. The central Texas Edwards Plateau area—at Austin, San Marcos, New Braunfels, Kerrville, and San Antonio.

REMARKS. The genus name, *Celtis*, is the ancient classical name, and the species name, *lindheimeri*, in honor of Ferdinand Lindheimer (1801–1879), German botanist, who resided more than 30 years in Texas and made a large collection of plants. These were reported on by Gray and Engelmann. A vernacular name is Palo Blanco.

SUGAR HACKBERRY

Celtis laevigata Willd.

FIELD IDENTIFICATION. Tree attaining a height of 100 ft, with a spreading round-topped or oblong crown.

FLOWERS. In spring, monoecious-polygamous, small, inconspicuous, greenish, borne on slender glabrous

Sugar Hackberry
Celtis laevigata Willd.

pedicels; staminate fascicled; calyx 4–6-lobed, (usually 5), lobes ovate-lanceolate, glabrous or pubescent; stamens 4–6; pistillate flowers solitary or 2 together, peduncled; ovary 1-celled, surmounted by 2 stigmas.

FRUIT. Drupe ripening in late summer, pedicel ¼–½ in., subglobose-obovoid, orange-red to black, about ¼ in. in diameter, flesh thin and dry, sweetish; seed solitary, pale brown, roughened. Fruit pedicel often longer than the leaf petiole.

LEAVES. Simple, alternate, deciduous, ovate-lanceolate, often falcate, long-acuminate at apex, rounded or wedge-shaped and inequilateral at base, entire or a few teeth near apex, thin, light green and glabrous above, paler and smooth beneath, conspicuously 3-veined at base beneath, 2½–4 in. long, 1–2½ in. wide.

TWIGS. Light green to reddish brown, somewhat divaricate, lustrous, glabrous or pubescent.

BARK. Pale gray, thin, smooth or cracked, with prominent warty excrescences.

WOOD. Yellowish, close-grained, soft, weak, weighing 49 lb per cu ft.

RANGE. The species is found in Texas, Arkansas, Oklahoma, and Louisiana; east to Florida, and north to Missouri, Kansas, Indiana, and Virginia. Also in Nuevo León, Mexico.

PROPAGATION. Sugar Hackberry has been cultivated since 1811. The fruit may be gathered by hand, or stripped into canvas, when it turns reddish brown, after the leaves fall. The fruit need not be cleaned to remove the pulp from the seed, but if desired the seeds can be macerated and run through a hammer mill. The seeds average about 4,400 per lb, with a commercial purity of 98 per cent and a soundness of 94 per cent. The dried seed may be stored in sealed containers at 35°–40°F. Seed may be sown in fall, or stratified seed may be sown in the spring. The stratification method is in moist sand at 41°F. for 60–90 days. Fruits with macerated pulps respond more evenly to stratification than nonmacerated pulps. The seeds are sown in rows 8–10 in. apart with about three times the amount of seeds for seedlings desired. Fall beds should be mulched with straw or leaves and held down with screens for bird or rodent control. The surface should be kept moist. Sugar Hackberry can also be propagated by cuttings.

VARIETIES. This species seems to present a considerable number of local variations which have caused some botanists to name a number of varieties, while other botanists feel that the distinctions are too slight. Some of these are as follows:

Texas Sugar Hackberry, *C. laevigata* var. *texana* (Scheele) Sarg., is scattered in Texas and extends over the Edwards Plateau limestone area to the west. It has leaves which are ovate-lanceolate, acuminate, mostly entire, rounded or cordate at the base, glabrous above and pubescent beneath with axillary hairs; fruit orange-red with pedicels longer than the petioles; branches gray to reddish brown and pubescent.

Uvalde Sugar Hackberry, *C. laevigata* var. *brachyphylla* Sarg., is a form with thicker and shorter leaves, found on the rocky banks of the Nueces River, Texas.

Scrub Sugar Hackberry, *C. laevigata* var. *anomala* Sarg., is a sandy-land shrub of Callahan County, Texas, having oblong-ovate, cordate leaves and dark purple, glaucous fruit.

Small Sugar Hackberry, *C. laevigata* var. *smallii* (Beadle) Sarg., is a small tree with sharply serrate,

Small Sugar Hackberry
Celtis laevigata var. *smallii* (Beadle) Sarg.

Texas Sugar Hackberry
Celtis laevigata var. *texana* (Scheele) Sarg.

acuminate, somewhat smaller leaves. It occurs from the Gulf Coast plain north to North Carolina and Tennessee.

Arizona Sugar Hackberry, *C. laevigata* var. *brevipes* Sarg., is an Arizona variety with ovate, mostly entire leaves 1½–2 in. long, yellow fruit, and glabrous reddish brown branchlets.

Net-leaf Sugar Hackberry, *C. laevigata* var. *reticulata* (Torr.) L. Benson, is considered by some botanists as a distinct species, but others feel that it has such close affinities it should be classed as a variety of the Sugar Hackberry with xerophytic tendencies. The flowers and fruit are similar except for more pubescence on the fruit pedicel. The leaves are smaller, broadly ovate, yellowish green, stiff, coriaceous, entire or serrate, conspicuously reticulate-veined beneath, the veins are pubescent beneath, and the leaf petiole is densely pubescent. As a tree it is rarely over 30 ft and quite often it is only a large shrub. Subsequently it may be found that Arizona Sugar Hackberry is a synonym of Net-leaf Sugar Hackberry. West Texas to California; north to Colorado, Utah, Washington, and Oklahoma; and south into Mexico. In Texas generally on limestone hills west of the Colorado River. Occasionally on shell banks near the Gulf as far east as Houston.

REMARKS. *Celtis* is a name given by Pliny to a sweet-fruited African lotus. The species name, *laevigata*, means "smooth." The wood is used to a limited extent for furniture, flooring, crating, fuel, cooperage, and posts. The dry sweet fruit is eaten by at least 10 species of birds. The tree is often used for street planting in the lower South.

Common Hackberry

Celtis occidentalis L.

FIELD IDENTIFICATION. Tree attaining a height of 120 ft, with a rounded crown. The gray bark bears corky warts and ridges.

FLOWERS. In spring with the leaves, perfect or imperfect, small, green, borne in axillary, slender-peduncled fascicles, or solitary; staminate fascicles few-flowered; calyx 4–6-lobed, lobes linear-oblong; stamens 4–6 (mostly 5); no petals; pistillate flowers usually solitary or in pairs; ovary sessile, ovoid, with 2 hairy reflexed stigmas.

FRUIT. Drupe variable in size and color, globose or subglobose to ovoid, orange-red turning dark purple, ¼–½ in. long, persistent; flesh thin, yellow, sweetish, edible; seed bony, light brown, smooth or somewhat pitted, pedicels longer than leaf petioles.

LEAVES. Simple, alternate, deciduous, ovate to elliptic-ovate, often falcate, short acuminate or acute, oblique at base with one side rounded to cuneate and the other somewhat cordate, usually coarsely serrate but less so near the base, 3-nerved at base, light green and glabrous above (or rough in var. *crassifolia*), paler green and soft pubescent or glabrous beneath, blades 2½–4 in. long, 1½–2 in. wide; petioles slender, glabrous, about ½ in. long. The leaves are broader in proportion to width, not as long taper-pointed, more serrate on margin, and drupes larger than in Sugar Hackberry.

TWIGS. Green to reddish brown, slender, somewhat divaricate, glabrous or pubescent.

BARK. Gray to grayish brown, smooth except for wartlike protuberances.

Net-leaf Sugar Hackberry
Celtis laevigata var. *reticulata* (Torr.) L. Benson

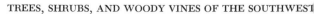

Common Hackberry
Celtis occidentalis L.

WOOD. Yellowish white, coarse-grained, heavy, soft, weak, weighing 45 lb per cu ft.

RANGE. Texas, western Oklahoma, Arkansas, and Louisiana; eastward to Georgia, north to Quebec, and west to Manitoba, North Dakota, Nebraska, and Kansas.

PROPAGATION. Common Hackberry has been cultivated since 1656. The fruit may be gathered by hand, or stripped into canvas, when it turns purplish black after the leaves have fallen. The fruit need not be cleaned to remove the pulp. The seed averages about 4,300 seeds per lb, and commercial seed is 99 per cent pure and 95 per cent sound. The dry seed may be stored in sealed containers at 35°–40°F. Seed can be sown in fall, or stratified seed sown in the spring. The stratification method is in moist sand at 41°F. for 60–90 days. Fruits with macerated pulps respond more evenly to stratification than nonmacerated. The seeds are sown in rows 8–10 in. apart, with about three times the number of seeds for seedlings desired. Fall seed beds should be mulched with leaves or straw and held down with screens for bird or rodent control. The surface should be kept moist. Common Hackberry can also be propagated by cuttings.

VARIETIES. Common Hackberry is rather variable in size and shape of leaves and fruit, and botanists have described a number of varieties to fit these differences.

Big-leaf Common Hackberry, *C. occidentalis* var. *crassifolia* Gray, is perhaps the most common variety in east Texas. The leaves are large, very coarsely serrate, cordate at base, and rough to the touch above, and the hairy petioles are shorter than the fruit pedicels. It occurs mixed with the species, and intergrading forms appear to be rather common.

Dog Hackberry, *C. occidentalis* var. *canina* Sarg., is a variety with oblong-ovate acuminate leaves, abruptly cuneate at base, finely serrate and glabrous or hairy on veins beneath; petioles glabrous or rarely pubescent. Not known to occur in Texas, but common in Oklahoma and northward.

REMARKS. *Celtis* was a name given by Pliny to a sweet-fruited lotus, and *occidentalis* means "western." Vernacular names are Nettle-tree, False-elm, Bastard-elm, Beaverwood, Juniper-tree, Rim-ash, Hoop-ash, and One-berry. The tree is drought resistant and often planted for shade in the South and for shelter-belt planting. The wood is occasionally used commercially for fuel, furniture, veneer, and agricultural implements. The fruit is known to be eaten by 25 species of birds, especially the gallinaceous birds. It was also eaten by the Indians.

Small Hackberry
Celtis tenuifolia Nutt.

FIELD IDENTIFICATION. Eastern shrub or small tree to 24 ft.

FLOWERS. In spring, small, axillary, 5–6-parted; staminate in small pedunculate clusters near the base of twigs of the year, calyx 5-lobed; stamens 5, opposite the calyx-lobes, exserted, usually no ovary in staminate flowers; pistillate flowers appearing with leaves, pediceled, solitary or paired from upper axils, stamens present or absent; ovary ovoid, 1-celled, 1-ovuled, ovule single and suspended; style very short, stigmas 2, elongate, subulate, recurved, divergent.

FRUIT. Drupe September–October, subglobose, ⅕–⅓ in. in diameter, orange to brown or red, thin-fleshed, sweet; stone ⅕–¼ in. long, subglobose, light to dark brown, seed coat somewhat granular; peduncles ⅛–⅖ in., shorter than the subtending petioles, or longer.

LEAVES. Simple, alternate, deciduous, usually broad-ovate to deltoid, apex blunt, acute or short-acuminate, base oblique and 3-nerved, mostly entire on margin (on young shoots sometimes few-toothed), blade length ¾–4 in., width ½–1¾ in., thin and smooth, surfaces grayish green but darker above, lower surfaces more or less pubescent and veiny, petioles pubescent.

TWIGS. Slender, reddish brown, pubescent at first, later glabrous and darker brown to gray; bark often with corky ridges.

RANGE. On dry and rocky foothills. Southern to eastern Oklahoma, Arkansas, and Louisiana; eastward to northern Florida, and northward to Pennsylvania, Indiana, Kansas, and Missouri.

VARIETY. Georgia Hackberry, *C. tenuifolia* var. *georgiana* (Small) Fern. & Schub., is a variety with leathery, pubescent, scabrous leaves. It was formerly listed as *C. georgiana* Small and *C. pumila* var. *georgiana* Sarg. It occurs in Arkansas, Oklahoma, Louisiana, Georgia, Missouri, Virginia, and District of Columbia.

REMARKS. *Celtis* is the name given by Pliny to a sweet-fruited African lotus. The species name, *tenuifolia,* refers to the thin leaves. It is listed as *C. pumila* by some authors. Vernacular names are Sugarberry, Nettleberry, and Nettle-tree.

Spiny Hackberry

Celtis pallida Torr.

FIELD IDENTIFICATION. Spiny, spreading, densely branched evergreen shrub attaining a height of 18 ft.

FLOWERS. Axillary, inconspicuous, in 2-branched, 3–5-flowered cymes, pedicels about 1/12 in. long; flowers greenish white, polygamous or monoecious; corolla absent; calyx-lobes 4 or 5; stamens as many as the calyx-lobes; style absent; stigmas 2, each 2-cleft and spreading; ovary sessile, 1-celled.

FRUIT. Drupe subglobose or ovoid, yellow or orange, thin-fleshed, mealy, acid, edible, 1/5–1/3 in. in diameter; stone ovoid, oval or obovoid, reticulate, acute, about 1/4 in. long and 3/8 in. wide.

LEAVES. Alternate, simple, small, 3-nerved, deep green, scabrous and puberulent, elliptic to oblong-ovate; rounded, acute or obtuse at the apex; oblique and somewhat semicordate at base; coarsely toothed on margin, or entire; blades 1/2–2 1/4 in. long, 1/2–1 in. wide; petioles pubescent, 1/16–3/16 in., or longer.

Small Hackberry
Celtis tenuifolia Nutt.

Spiny Hackberry
Celtis pallida Torr.

BARK. Mottled gray to reddish brown, rather smooth and tight, sometimes with long, stout, gray or brown spines. Bark only rough at base of very old trunks.

TWIGS. Divaricate, flexuous, spreading, smooth, gray or reddish brown, glabrous or puberulent; with stipular spines 1/4–1 in., stout, straight, single or paired, often at ends of shoots; lenticels small, pale, and usually numerous.

RANGE. Central, western, and southern Texas, New Mexico, Arizona, and Mexico. In Mexico from Chihuahua to Baja California, and south to Oaxaca.

REMARKS. The genus name, *Celtis,* is the classical name for a species of lotus, and the species name *pallida,* refers to paleness of the branches. Commonly used vernacular names in Texas and Mexico are Desert Hackberry, Chaparral, Granjeno, Granjeno Huasteco, Capul, and Garabata.

The Indians of the Southwest are reported to have ground the fruit and eaten it with fat or parched corn. It is reported that the larvae of the snout butterfly feed upon the foliage. Spiny Hackberry is also considered a good honey plant. Many birds consume it, particularly the cactus wren, cardinal, pyrrholuxia, towhee, mockingbird, thrasher, scaled quail, Gambel's quail, and green jay. The raccoon, deer, and jack rabbit eat it occasionally. The wood is used for fence posts and fuel, and the plant is of some value in erosion control.

Slippery Elm
Ulmus rubra Muhl.

FIELD IDENTIFICATION. Tree attaining a height of 75 ft, with spreading branches and a broad open head.

FLOWERS. February–April, before the leaves, borne in dense fascicles on short pedicels; perfect; no petals; calyx campanulate, green, hairy; calyx-lobes 5–9, lanceolate and acute; stamens 5–9, with elongate yellow filaments and reddish purple anthers; pistil reddish, compressed, divided into a 2-celled ovary and 2 exserted, spreading, reddish purple stigmas.

FRUIT. Ripening April–June. Samara short-stalked, green, oval to orbicular, apex entire or shallowly notched, ¼–¾ in. long; seed flattened with the surrounding wing reticulate-veined; seed area reddish brown and hairy; and wing area glabrous. Minimum seed-bearing age of tree 15 years, optimum 25–125, and maximum 200 years. Good crops every 2–4 years and light crops intervening.

LEAVES. Buds densely rusty-tomentose; leaves simple, alternate, deciduous, blades 4–8 in. long, 2–3 in. wide, obovate, ovate to oblong, acuminate at apex; rounded, cordate to cuneate, and inequilateral at base; margin coarsely and sharply double-serrate; dark green and very rough above because of tiny pointed tubercles, also pubescent when young and later more glabrous; paler and soft-pubescent beneath, often with axillary hairs; petioles ⅓–½ in., stout; leaves fragrant when dry.

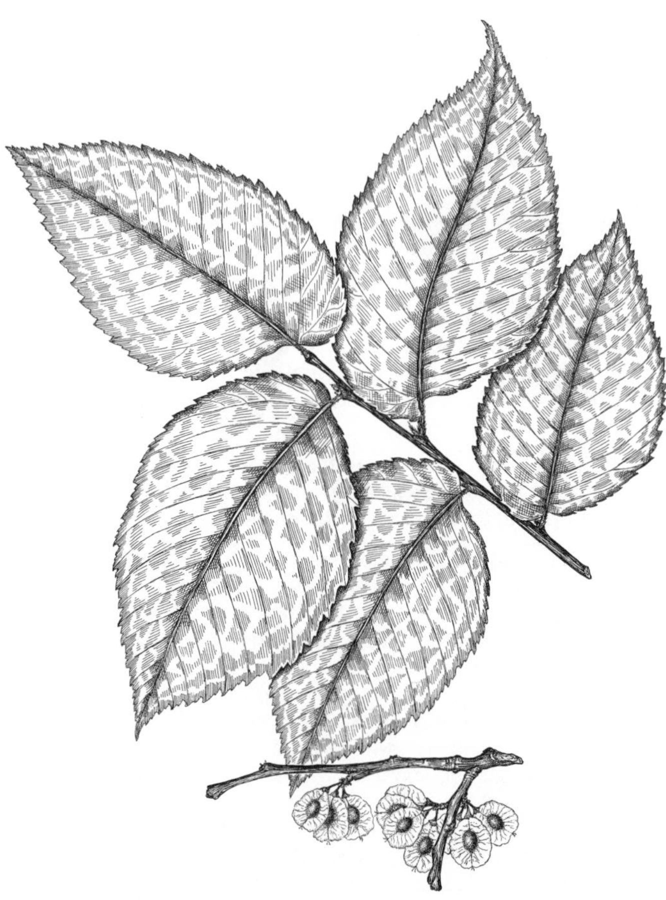

Slippery Elm
Ulmus rubra Muhl.

TWIGS. Gray, stout, roughened and densely pubescent when young, more glabrous later.

BARK. Gray to reddish brown, ridges flattened, fissures shallow, inner bark mucilaginous and fragrant.

WOOD. Reddish brown, tough, strong, heavy, hard, compact, durable, weighing 43 lb per cu ft.

RANGE. Texas, Oklahoma, Arkansas, and Louisiana; eastward to Florida, north to Maine and Quebec, and west to Ontario, Minnesota, Wisconsin, and Nebraska.

MEDICINAL USES. Elm bark, known medicinally as "Ulmi Cortex," contains a mucilaginous substance forming a jelly when mixed with water. A warm infusion was used as a treatment of coughs and diarrhea by the early settlers. It was prepared by stirring an ounce of powdered bark in a pint of hot water. Sometimes the strips of elm bark prevented fat from becoming rancid, and settlers sometimes cooked bear fat with the bark.

PROPAGATION. Elm seeds are collected by sweeping them up on the ground or beating the branches. The fruit should be air-dried for a few days, but not excessively. The wings are not usually removed, but sometimes it is done by placing in bags and beating. The cleaned seed averages 41,000 seeds per lb, with a commercial purity of 79–99 per cent and a soundness of 96 per cent. For pretreatment dry-stored seed may be stratified for 60–90 days at 41°F. for best results, though some lots of seed show more dormancy than others. Some growers prefer to sow the seed immediately after ripening in the fall. Planting is done in drills 8–10 in. apart, with about 15 seeds per linear ft, covered with about ½ in. of firmed soil. Beds are kept moist until germination is complete. No shade is necessary. Seedlings are sometimes affected by damping-off which can be controlled by treating the soil with formaldehyde before seeding. One-year-old seedlings can be set out in the field.

REMARKS. *Ulmus* is the ancient Latin name for elm, and the species name, *rubra,* refers to the reddish wood. Fernald (1945b, pp. 203–204), notes that the older name *U. rubra* Muhl. has precedence over the name *U. fulva* Michx., as used in most books. Vernacular names are Rock Elm, Red Elm, Sweet Elm, Indian Elm, Moose Elm, Gray Elm, and Soft Elm. The wood is used for making posts, ties, sills, boats, hubs, agricultural implements, furniture, slack cooperage, veneer, and sporting goods. In frontier days the bark was often chewed as a thirst-quencher. White-tailed deer, rabbit, porcupine, and moose browse the twigs and foliage. The tree is rather short-lived and subject to insect damage. It has been cultivated since 1830.

American Elm

Ulmus americana L.

FIELD IDENTIFICATION. Much-loved and famous American tree, admired for graceful vaselike shape. Attaining a height of 120 ft, but generally under 70 ft, and often buttressed at base. Known to reach an age of 300 years or more.

FLOWERS. February–April, before the leaves, in axillary, 3–4-flowered (occasionally to 12) fascicles; pedicels slender, jointed, drooping, nearly sessile or to 1 in. long; individual flowers perfect, petals absent; calyx campanulate, red to green, lobes 7–9 and short; stamens 5–8, exserted, with slender filaments and red anthers; pistil pale green, compressed, composed of a 2-celled ovary and 2 spreading inwardly stigmatic styles.

FRUIT. Ripening March–June. A samara, about ½ in. long, red to green, oval-obovate, consisting of a central flattened seed surrounded by a membraneous wing; wing reticulate-veiny, glabrous but ciliate on margin, a deep terminal incision reaching the nutlet. The minimum seed-bearing age is 15 years and the maximum 300 years. Good crops occur most years.

LEAVES. Simple, alternate, deciduous, 4–6 in. long, 2–3 in. wide, oval, obovate or oblong, acute or abruptly acuminate at apex; veins conspicuous to the serrations; somewhat cordate on one side at base and rounder or cuneate on the other side, giving an inequilateral shape; margin coarsely and doubly serrate, upper surface dark green and mostly smooth (occasionally scabrous on vigorous shoots); lower surface pubescent at first, but glabrate later; petioles ⅕–⅓ in., rather stout.

TWIGS. Slender, varying shades of brown, pubescent at first, glabrous later.

BARK. Light to dark gray, ridges flattened and scaly, fissures deep.

WOOD. Light to dark brown, sapwood whitish, coarse-grained, tough, heavy, hard, strong, weighing 40 lb per cu ft, difficult to split, durable, bends well, shrinks moderately, tends to warp and twist.

RANGE. Moist soils of bottom lands and upland flats. Texas, Oklahoma, Arkansas, and Louisiana; eastward to Georgia and Florida, northward to Newfoundland, and west to Ontario, North Dakota, Montana, and Nebraska.

PROPAGATION. The winged seeds are collected by sweeping them from the ground or beating them from the branches. They may be spread out to dry. Dewinging is accomplished by flailing the seed in bags. The cleaned seed averages about 68,000 seeds per lb, with a commercial purity of 92 per cent and a soundness of 96 per cent. The seeds may be stored for as long as 2 years in sealed jars at 40°F. Embryo-dormancy can be overcome by stratifying in sand for 60 days at 41°F. prior to sowing. Seed of elm ripening in spring can be sown that same spring if desired. About 15 seeds per linear ft are sown 8–10 in. apart, and covered with about ¼ in. of firmed moist soil. Damping-off can be controlled with formaldehyde treatment of the soil before seeding.

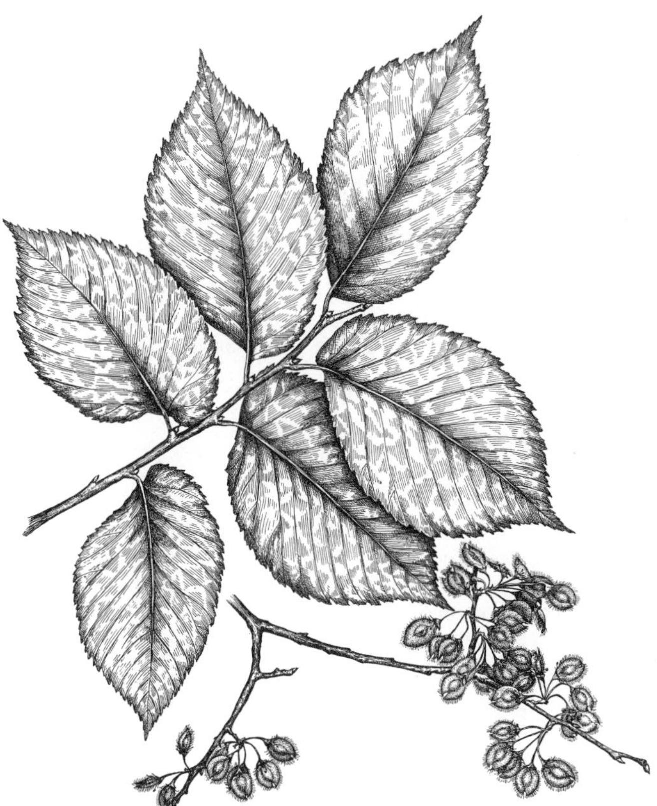

American Elm
Ulmus americana L.

FORMS. American Elm has been known in cultivation since 1752. A number of horticultural forms of the American Elm have been developed by propagation through cuttings, and are classified as the Ascendens, Column, Golden, Lake City, Littleford, Moline, Princeton, Vase, and Weeping forms.

REMARKS. *Ulmus* is the ancient Latin name for elm, and the species name, *americana,* refers to its native home. Vernacular names are Rock Elm, Common Elm, Soft Elm, Swamp Elm, White Elm, and Water Elm. Known to the lumber trade as White Elm and makes up the greater part of elm lumber and logs. In most cases the trade does not distinguish between the elm species. A very desirable ornamental tree for street and park planting, attaining large size and much admired for the graceful upsweep of the branches. It is sometimes used for shelter-belt planting in the prairie states, but *U. pumila* is considered superior for that purpose. The wood is used for woodenware, vehicles, baskets, flooring, veneer, furniture, cooperage, cabinets, sporting goods, stock staves, boxes, crates, framework, agricultural implements, trunks, handles, toys, car construction, shipbuilding and boatbuilding, and fuel. It is reported that the Indians used the wood for canoes and the bast fiber for ropes. The fruit is often eaten by gallinaceous birds, and the young twigs and leaves are browsed by white-tailed deer, opossum, and cottontail. In the northern part of its range it appears to be subject to the attack of the elm-leaf beetle, *Galerucella xanthomelaena*, and in some areas large numbers of trees are killed by the Dutch elm disease, caused by a fungus, *Graphium ulmi*, and by phloem necrosis, caused by a virus. American Elm is distinguished from Slippery Elm by the former having leaves less scabrous above and the samara being ciliate on the margin.

Cedar Elm

Ulmus crassifolia Nutt.

FIELD IDENTIFICATION. Tree attaining a height of 90 ft, with slender, somewhat drooping branches and a narrow or rounded crown. Twigs or branches often with lateral corky wings.

FLOWERS. Borne usually in July in small, 3–5-flowered fascicles; pedicels slender, ⅓–½ in.; calyx campanulate, hairy, red to green, 6–9-lobed, lobes hairy and acute; no petals; stamens 5–6, with slender filaments and reddish purple anthers; pistil green, flattened, pubescent, composed of a 2-celled ovary and 2 exserted spreading styles.

FRUIT. Samara borne in late summer, small, ¼–½ in. long, oval-elliptic or oblong, green, flattened, pubescent; composed of a central seed surrounded by a wing which is deeply notched at apex and ciliate on margin.

LEAVES. Simple, alternate, somewhat persistent, blades 1–2 in. long, ¾–1 in. wide, elliptic to ovate, acute or obtuse at apex, rounded or cuneate to oblique at base, doubly serrate on margin; dark green, stiff and very rough to the touch above, pubescent beneath; petiole about ⅓ in., stout, hairy.

TWIGS. Reddish brown, pubescent, often with brown, thin, lateral, corky wings. The only other Texas elm with corky wings is the Winged Elm.

BARK. Brown to reddish, or gray, ridges flattened and broken into thin, loose scales.

WOOD. Reddish brown, sapwood lighter, brittle, heavy, hard.

RANGE. Often in limestone soils. Texas, Oklahoma, Arkansas, and Louisiana; east to South Carolina, north to New York, and west to Kansas.

REMARKS. *Ulmus* is the ancient Latin name, and *crassifolia* refers to the rough, thick leaves. Vernacular names are Scrub Elm, Lime Elm, Texas Elm, Basket Elm, Red Elm, and Southern Rock Elm. It is often planted as a shade tree, but the wood is considered inferior to other elms because of its brittle and knotty character. It is sometimes used for hubs, furniture, and posts.

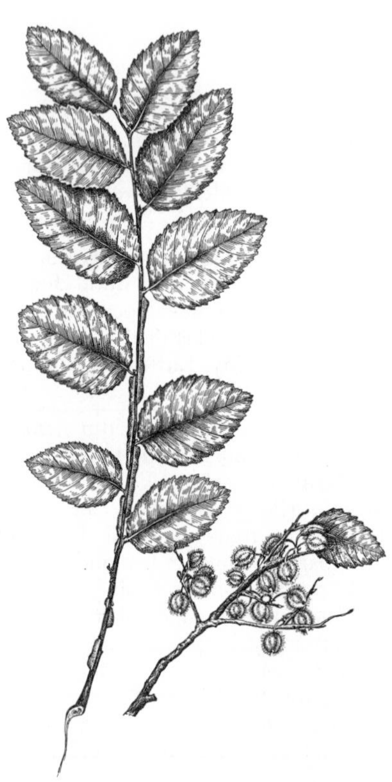

Cedar Elm
Ulmus crassifolia Nutt.

Winged Elm

Ulmus alata Michx.

FIELD IDENTIFICATION. Tree attaining a height of 60 ft, with slender branches and a rounded or oblong crown. Often with conspicuous corky wings on the twigs and branches.

FLOWERS. Before the leaves in spring, borne in few-flowered drooping fascicles on filiform pedicels; flowers perfect, petals absent; calyx campanulate, red to yellow, the 5 lobes obovate and rounded; stamens 5, with long, slender filaments and reddish anthers; pistil green, hairy, flattened, composed of a tomentose 2-celled ovary tipped by 2 spreading styles.

FRUIT. Samara reddish or greenish, long-stipitate, ovate to elliptic or oblong, ¼–⅓ in. long; seed solitary, flattened, ovoid; wing flat, thin, narrow, prolonged into divergent, apically incurved beaks; seed and wing hairy, especially on margin; the reddish samaras giving the tree a reddish appearance when in fruit.

LEAVES. Simple, alternate, deciduous, ovate-oblong

WINGED ELM
Ulmus alata Michx.

to oblong-lanceolate, occasionally somewhat falcate, blades ½–3 in. long, coarsely and doubly serrate on margin, acute or acuminate at apex, wedge-shaped or subcordate at base, pale-pubescent or glabrous beneath, with axillary hairs and prominent veins; petioles about ⅓ in. long, stout, pubescent.

TWIGS. Reddish brown, slender, pubescent at first, glabrous later, often with conspicuous, opposite, thin, corky wings. The only other elms having corky wings are the Cedar Elm, September Elm, and Rock Elm.

BARK. Reddish brown to gray, ridges flat with closely appressed scales, fissures irregular and shallow.

WOOD. Brown, close-grained, compact, heavy, hard, difficult to split, weighing 46 lb per cu ft, not considered as strong as other elms.

RANGE. Texas, Oklahoma, Arkansas, and Louisiana; eastward to Florida, north to Virginia, and west through Ohio and Indiana to Kansas and Missouri.

REMARKS. *Ulmus* is the ancient Latin name, and *alata* refers to the corky wings on the twigs. Vernacular names are Cork Elm, Water Elm, Wahoo Elm, Red Elm, and Witch Elm. The Winged Elm is a favorite shade and ornamental tree. It is easily transplanted, sprouts readily from seed, is a rapid-grower, and is rather free of disease and insects. Directions for the handling and sowing of seeds is similar to that practiced with other spring-fruiting elms. The wood is generally used for the same purposes as other elms, such as tool handles, vehicle parts, and agricultural implements. Formerly the bark was used in some localities for baling twine.

SEPTEMBER ELM

Ulmus serotina Sarg.

FIELD IDENTIFICATION. Tree to 60 ft and 3 ft in diameter. The branches are rather short and pendulous forming a rounded crown. Some branches bear corky wings.

FLOWERS. Borne in September, in the axils of leaves of the season; racemes ½–1½ in. long, many-flowered, pendulous, glabrous; pedicels conspicuously jointed, slender, 1/12–⅙ in.; petals absent; calyx 5–6-parted nearly to base, reddish brown, lobes oblanceolate; stamens 6, exserted, filaments slender, elongate; anthers yellow to reddish, erect, apex emarginate, base subcordate; ovary sessile, narrowed below and pubescent, superior, flattened, 1–2-celled, ovule solitary and suspended; styles 2, short, divergent, stigmatic inwardly.

FRUIT. Samara ripening in November, stipitate, winged, about ½ in. long, rhombic to ovate or elliptic,

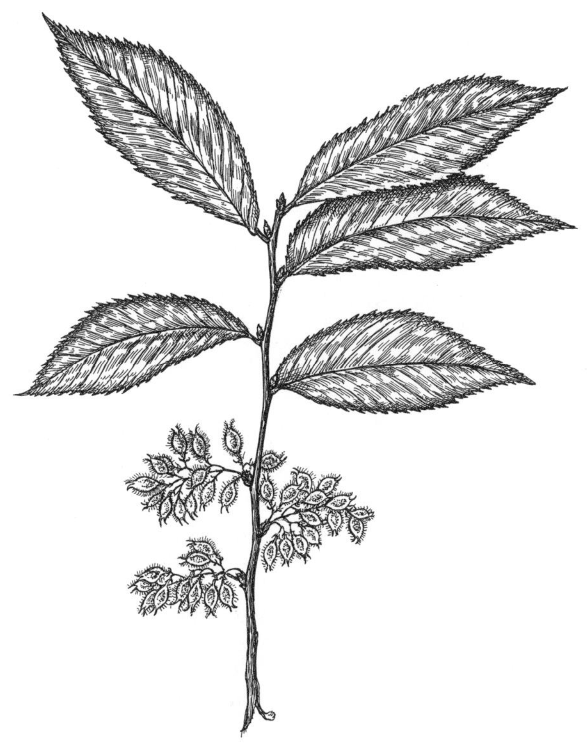

SEPTEMBER ELM
Ulmus serotina Sarg.

divided at apex into 2 hornlike divisions and notched; body of fruit flattened, pubescent, margin densely white to brown-hairy, 1-celled and 1-seeded, albumen none.

LEAVES. Simple, alternate, deciduous, oblong to obovate, apex acuminate or acute, base oblique, margin coarsely double-serrate, thin but firm, upper surface yellowish green, smooth and shiny, lower surface paler and pubescent, especially on the veins, main veins strongly straight and often forked near the leaf apex; petioles about ¼ in., stout; stipules about ¼ in., linear-lanceolate, base broad and clasping, persistent.

TWIGS. Slender, pendulous, brown to gray, pubescent to glabrous, often corky-winged later, lenticels pale and elliptic to oblong; buds about ¼ in. long, ovoid, acute, dark brown; leaf scars semicircular, bundle scars 3.

BARK. Reddish brown to light brown, rather thin, fissures shallow and separated by broad, flat ridges with thin close scales.

WOOD. Light reddish brown, yellowish sapwood, close-grained, strong, tough, hard.

RANGE. Usually in limestone soils of hills or riverbanks. Northwestern Arkansas and eastern Oklahoma; east to Georgia, and north to Kentucky and Illinois.

REMARKS. The genus name, *Ulmus*, is the classical Latin name, and the species name, *serotina*, refers to the autumnal flowers. Also known as the Red Elm. It resembles the American Elm somewhat in shape of the tree, but is not as large; the leaves are smaller; corky wings sometimes form on the twigs; and the flowers are borne in fall instead of spring. The wood is used for fuel and veneer occasionally. It is sometimes planted as a shade tree in the southeastern Gulf states.

SIBERIAN ELM
Ulmus pumila L.

FIELD IDENTIFICATION. Graceful cultivated shrub or small tree with slender drooping branches.

FLOWERS. March–April, appearing with or before the leaves, axillary, inconspicuous, greenish, clustered, short pediceled, perfect or rarely polygamous, petals absent; calyx campanulate, 4–5-lobed; stamens 4–5, with green to violet anthers; style 2-lobed, ovary flattened and 1-celled.

FRUIT. Samara April–May, clustered, ¼–½ in. long and broad, rarely more; oval to obovate, composed of a central, dry, compressed nutlet surrounded by a wing which is thin, reticulate-veined, membranous, semitransparent, apex with a notch sometimes reaching one third to one half way to the nutlet; pedicel 1/25–⅛ in.

LEAVES. Simple, deciduous, alternate, oval to ovate or elliptic, blade length 1–2 in., width ½–1 in., margin doubly serrate, apex acute, base cuneate or somewhat asymmetrical, leathery and firm; upper surface olive-green to dark green, glabrous, veins impressed; lowe surface paler and glabrous, or somewhat pubescen when young or with axillary tufts of hair; turning yellow in autumn; petiole glabrous or pubescent, ⅙–⅓ in., stipules caducous.

TWIGS. Slender; when young brownish and pubescent; when older brown to gray and glabrous; bark of trunk gray to brownish.

RANGE. A native of Asia, extensively cultivated in the United States. In its typical form a small-leaved shrub or tree from Turkistan to Siberia, Mongolia, and North China.

PROPAGATION. It is propagated by seed as soon a ripe. From 12 to 20 seeds are sown per linear ft in drills 10 in. apart, and covered ¼ in. with firmed soil The beds should be kept moist, but not particularly shaded. One-year-old seedlings are old enough to plant The germination rate averages about 60 per cent, but only 12–20 per cent of the seedlings are usable. The cleaned seed averages about 65,000 seeds per lb. Dwarf forms are sometimes grown from greenwood cuttings under glass.

VARIETIES AND HYBRIDS. Weeping Siberian Elm, *U pumila* var. *pendula* (Kirchn.) Rehd., has slender pendulous branches and is the form commonly cultivated

SIBERIAN ELM
Ulmus pumila L.

Narrow Siberian Elm, *U. pumila* var. *arborea* Litvinov, is pyramidal in shape with pinnately-branched shoots; leaves ovate to lanceolate, doubly serrate, lustrous, blades 1½–2¾ in., petioles pubescent. It apparently originated in Turkistan, and came into cultivation about 1894.

Androssow Siberian Elm, *U. pumila* forma *androssowi* (Litv.) Rehd., is a cultivated form with a dense head, spreading corky branches, and has been cultivated since 1934.

Siberian Elm is known to hybridize with Scotch Elm, *U. glabra* Huds., to produce *Ulmus* × *arbuscula* Wolf., a shrubby tree with elliptic to oblong leaves ¾–2¾ in., doubly serrate and nearly equal at base. It originated in 1902.

REMARKS. The genus name, *Ulmus,* is the ancient Latin name for elm, and the species name, *pumila,* refers to its shrubby habit in some forms. Often wrongly called Chinese Elm, but that name should properly apply to *U. parvifolia* Jacq. Both the Siberian and Chinese Elm are cultivated in the Gulf Coast states for ornament. The Siberian Elm is being extensively planted in the prairie-plains region as shelter belts, and has some use as a game cover. It is rather drought resistant, and seems to be less susceptible to the Dutch elm disease, *Graphium ulmi,* than the native elm species. The wood is hard, heavy, tough, rather difficult to split, and is used in China for agricultural implements, boatbuilding, and wagon wheels. The inner bark was once made into coarse cloth.

ROCK ELM
Ulmus thomasi Sarg.

ROCK ELM

Ulmus thomasi Sarg.

FIELD IDENTIFICATION. An elm confined mostly to the Eastern and Northern states, but found in northern Arkansas also. Attaining a height of 100 ft and a diameter of 4 ft. Trunks clean and straight for a considerable distance. Branches slender and spreading into a rather rounded-topped crown. When growing in the open the top is oblong, with the lower branches drooping. Closely resembling the American Elm in habit.

FLOWERS. March–May, before the leaves, axillary on the twigs of last season, borne in slender, drooping, 2–4-flowered cymes which elongate to become racemose and to 2 in.; calyx campanulate, green to reddish, 7–8-lobed, lobes oblong and rounded; stamens 5–8, filaments erect, slender, exserted, anthers purple; styles 2, greenish, spreading, inwardly stigmatic; ovary compressed, 1–2-celled, ovule solitary and suspended.

FRUIT. Maturing May–June, samara about ½ in. long. Oval to elliptic or obovate, apex shallowly notched, styles persistent and incurved, base oblique, surface pubescent and obscurely veined, margins densely ciliate and very narrow, obscurely passing into the solitary seed. Good crops borne every 3–4 years, with lighter crops in intervening years. The seed averages about 7,500 seeds per lb, with a commercial purity of 95 per cent and a soundness of 97 per cent.

LEAVES. Simple, alternate, deciduous, oval to oblong or obovate, apex abruptly narrowed or short-acuminate, occasionally rounded, base cuneate to subcordate and slightly inequilateral, margin coarsely double-serrate, blade length 2–6 in., width 1⅓–3½ in., thick and firm; upper surface dark green, shiny and smooth; lower surface paler and pubescent with main veins nearly straight; petioles ¼–⅓ in., pubescent; stipules deciduous early, ovate-lanceolate, base oblique-cordate and clasping, margin coarsely dentate.

TWIGS. Slender, pale brown and pubescent at first, later dark brown to gray and puberulous to glabrous; lenticels scattered, linear to oblong; buds large, acute, pubescent and cilate; leaf scars rather large and orbicular; some older twigs with corky wings which are dark gray to brown, to ½ in. wide, and irregularly interrupted.

BARK. On the trunk broken into broad flat ridges with deep irregular fissures, gray to reddish-tinged, ¾–1 in. thick.

WOOD. Heart reddish brown, sapwood lighter, grain close and straight or interlocking, strong in bending, tough and flexible, very high in shock resistance, heavy, hard, difficult to work with tools, weighing about 45 lb per cu ft, very durable but difficult to season because of its tendency to warp and check, moderately resistant to decay.

RANGE. Rocky riverbanks, gravelly slopes, or calcareous uplands. Northwestern Arkansas; east to Tennessee, north to Quebec, and west to South Dakota, Minnesota, Wisconsin, Nebraska, and Kansas.

REMARKS. The genus name, *Ulmus,* is the classical name for the elm. The species name, *thomasi,* honors its discoverer, David Thomas (1776–1859), American civil engineer and horticulturist. It is listed in some literature under the name of *U. racemosa* Sarg. It is also known under the names of Cork Elm, Cliff Elm, Hickory Elm, Swamp Elm, Racemed Elm, Corky White Elm, and Wahoo Elm. The tree has been planted for ornament since 1875. It has a wide-spreading root system, is long-lived, slow-growing, wind-firm, severely injured by leaf insects, and fairly free from fungi. It is rather easy to transplant. The young leaves are nibbled by cottontail.

WATER ELM

Planera aquatica (Walt.) Gmel.

FIELD IDENTIFICATION. Contorted shrub or tree to 40 ft, growing in swampy ground.

FLOWERS. Three kinds of flowers on the same tree—male, female, and perfect. Staminate flowers fascicled, 2–5-flowered; petals none; calyx 4–5-lobed, bell-shaped, greenish yellow, lobes ovate and obtuse; stamens 4–5, filaments filiform, exserted; pistillate flowers 1–3 together perfect; ovary ovoid, stalked, tubercular, 1-celled; styles 2, reflexed, stigmatic along inner side.

FRUIT. Peculiar, covered with irregular warty excrescences, leathery, oblong-ovoid, compressed, ridged, about ⅓ in. long, short-stalked; seed ovoid.

LEAVES. Elmlike, alternate, deciduous, blades 2–4 in. long, ½–1 in. wide, ovate or oblong-lanceolate, crenulate-serrate, acute to obtuse at apex, cordate or oblique at base, dark green, paler below; petioles stout, puberulent, about ¼ in. long; stipules lanceolate, caducous.

BARK. Light reddish brown or gray, dividing into large shreddy scales.

WATER ELM
Planera aquatica (Walt.) Gmel.

WOOD. Soft, weak, light, close-grained, light brown, weighing 33 lb per cu ft.

RANGE. In swamps or river-bottom lands. Texas, Oklahoma, Arkansas, and Louisiana; eastward to Florida, northward to North Carolina, and west through Kentucky and Illinois to Missouri.

REMARKS. The genus name, *Planera,* is in honor of the German botanist Johann Jakob Planer (1743–1789), a professor at the University of Erfurt, and the species name, *aquatica,* refers to the swampy habitat of the tree. The peculiar little warty fruit is considered to be an important duck food in swampland areas. Squirrels also eat the fruit. The wood has no commercial importance.

MULBERRY FAMILY (*Moraceae*)

COMMON PAPER MULBERRY
Broussonetia papyrifera (L.) Vent.

COMMON PAPER MULBERRY
Broussonetia papyrifera (L.) Vent.

FIELD IDENTIFICATION. Small tree, rarely to 50 ft, with irregular spreading branches.

FLOWERS. Dioecious, staminate catkins peduncled, cylindric, pendulous, 2½–3½ in. long; no petals; stamens 4; calyx 4-lobed; pistillate in globose heads with a tubular perianth; ovary stipitate, stigma filiform and slender.

FRUIT. Globose, about ¾ in. across, a multiple fruit composed of many 1-seeded drupelets which are reddish orange, and which protrude from the persistent calyx.

LEAVES. Alternate, deciduous, long-petiolate, blades 3–8 in. long, ovate, margin coarsely dentate and often deeply lobed; apex acuminate; base cordate or rounded; rough above, conspicuously veined and velvety-pubescent beneath; stipules ovate-lanceolate, deciduous.

TWIGS. Stout, hirsute, tomentose.

BARK. Smooth, tight, reticulate, green to yellow.

WOOD. Coarse-grained, soft, light, easily worked.

RANGE. Texas, New Mexico, Oklahoma, Arkansas, and Louisiana; eastward to Florida, and northward to Missouri and New York. A native of Asia. Cultivated and escaping to grow wild in some areas in the United States.

VARIETIES. A number of varieties are used in horticulture as follows:

Curled-leaf Paper Mulberry, *B. papyrifera* var. *cucculata* Ser., has small leaves curled upward.

Large-leaf Paper Mulberry, *B. papyrifera* var. *macrophylla* Ser., has large usually undivided leaves.

Cut-leaf Paper Mulberry, *B. papyrifera* var. *laciniata* Ser., has finely dissected leaves, sometimes reduced to only the veins, and a low shrubby habit of growth.

Variegated Paper Mulberry, *B. papyrifera* var. *variegata* Ser., has leaves variegated white and yellow.

White-fruit Paper Mulberry, *B. papyrifera* var. *leucocarpa*, Audib., has white fruit.

REMARKS. The genus, *Broussonetia,* is named in honor of Auguste Broussonet, a French naturalist; and the species name, *papyrifera,* refers to the use of the inner bark in papermaking. The inner bark is also used for making cloth in the tree's native home of Japan and China. The famous tapa cloth of the South Pacific Islands is also made from the bark by macerating it and pounding with a wooden mallet. It is often planted for ornament in the United States, being drought-resistant, a rapid grower, and sprouting freely from the root. The fruit is also eaten by a number of species of birds.

Red Mulberry

Morus rubra L.

FIELD IDENTIFICATION. Handsome tree to 70 ft, with a rather broad, spreading crown.

FLOWERS. With the leaves in spring, green; petals absent; staminate spikes cylindric, 2–3 in. long; stamens 4, green; filaments flattened; calyx with 4 ovate lobes; pistillate spikes about 1 in., cylindric, sessile; calyx 4-lobed; styles 2; ovary ovoid, flat, 2-celled, 1 cell generally atrophies.

FRUIT. Ripening May–August, a cylindric syncarp ¾–1¼ in. long, resembling a blackberry, red at first, becoming purplish black, juicy, edible; achene ovoid, acute, light brown, covered by the succulent calyx. Minimum seed-bearing age of the tree 10 years, optimum 30–85 years, maximum about 125 years.

LEAVES. Simple, alternate, deciduous, 3–9 in. long, ovate or oval, or 3–7-lobed, doubly serrate, rough and glabrous above, soft pubescent beneath, very veiny, acute or acuminate at apex, cordate or truncate at base, turning yellow in autumn; petiole 1–2 in.; stipules lanceolate and hairy. The lobing of the leaves varies considerably on different trees or even on the same tree; some are only serrate, while others have numerous lobes.

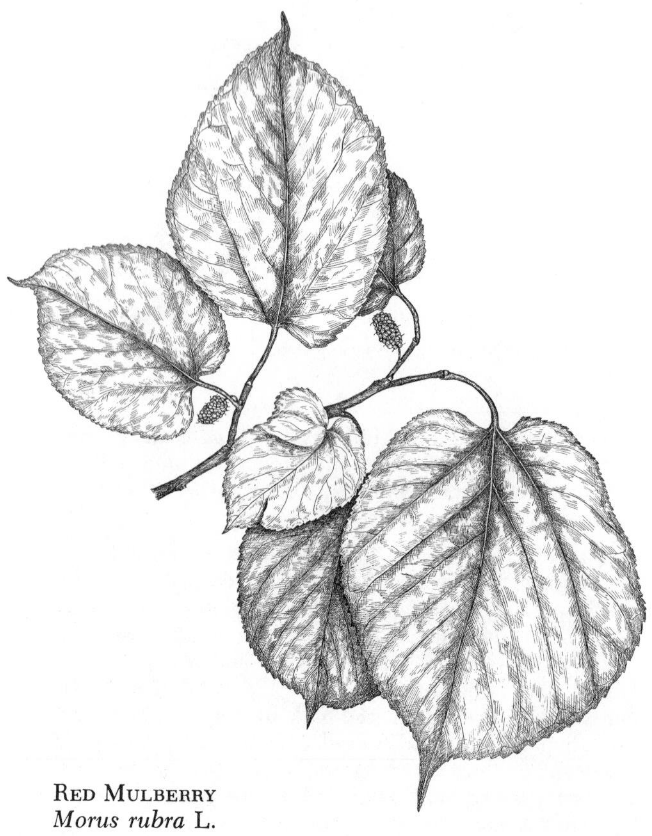

Red Mulberry
Morus rubra L.

BARK. Dark brown to gray, ½–¾ in. thick, divided into irregular, elongate plates separating into appressed flakes.

WOOD. Light orange, sapwood lighter, durable, close-grained, light, soft, weak, weighing about 45 lb per cu ft, used for boats, fencing, cooperage, and railroad crossties.

RANGE. Usually in rich moist soil. Does not grow well on thin, poor soil. Texas, Oklahoma, Arkansas, and Louisiana; eastward to Florida, north to Vermont, and west to Ontario, Wisconsin, Michigan, Minnesota, Nebraska, and Kansas.

PROPAGATION. The following methods of handling seed are recommended: The ripe fruit is collected by flailing or shaking the fruit into canvas. The seed is extracted by mashing and soaking the berries in water for 24 hours and then running them wet through a macerator, skimming or floating off the pulp. A lye solution of 1 per cent is reported to aid the extraction and cleaning process. The cleaned seed should be air-dried in the shade. One hundred lb of fruit yield 2–3 lb of seed. The seed numbers about 360,000 seeds per lb. The commercial seed purity is 85 per cent and soundness 90 per cent. For best germination results the seed may be stratified in moist sand 90–120 days at 41°F. The seed may be broadcast or sown in drills in the spring, 8–12 in. apart, using about 50 seeds to the linear ft of the row. It is best to mulch with burlap and keep moist until germination. If sown in fall the beds should be mulched with straw or leaves and protected by screens against birds or rodents. Partial shade is helpful for a few weeks after germination. Germination usually occurs in 1–2 weeks in spring sowing. The various species of mulberry are sometimes attacked by leaf-spot disease caused by the fungi *Cercospora* and *Phleospora maculans*. The disease may be controlled by early applications of 4-4-50 Bordeaux mixture. For protection against insects the seeds are soaked in water mixed with camphor and the beds sprinkled with a mixture of lime, ashes, and white arsenic. Propagation by budding is sometimes practiced. The root-knot nematodes sometimes attack the roots of both seedlings and older trees. If infestation occurs, the nursery beds for the mulberry should be changed to fresh sites. Occasionally mulberry heartwood is known as "waterwood" because of excessive infiltration by water moving upwards from dead roots or branches by capillary force. External symptoms are grayish bark, dead limbs below the crown, thin bark plates, and thin crowns. Occasionally there are no specific external symptoms. Waterwood has a much higher water content than other wood and cures more slowly.

VARIETY. Woolly Red Mulberry, *M. rubra* var. *tomentosa* (Raf.) Bur., is a large-fruited Texas variety with leaves white-tomentose beneath. Most of the varieties of mulberries in cultivation are varieties of the White Mulberry, *M. alba*.

REMARKS. The genus name, *Morus*, is the classical name of the mulberry, and the species name, *rubra*, a

BLACK MULBERRY
Morus nigra L.

Latin word for "red," has reference to the red, immature fruit. The fibrous bark was used to make cloth by early Indians. The fruit is known to be eaten by at least twenty-one kinds of birds, fox squirrels, and human beings. Although the fruit is sweet, it does not seem to be very much in demand for culinary uses. For fruit-bearing purposes the trees may be planted 20–40 ft apart. Trees should not be planted next to walks because the abundant ripe fruit mashes readily underfoot. Often planted for ornament, and known in cultivation since 1629.

BLACK MULBERRY
Morus nigra L.

FIELD IDENTIFICATION. Cultivated shrub or tree attaining a height of 30 ft, or occasionally larger. The trunk is short and the wide-spreading branches form a broad, rounded, or irregularly shaped crown.

FLOWERS. Staminate flowers in cylindrical spikes ⅓–1 in., longer than the peduncles; stamens 4, inserted opposite the sepals under the ovary, filaments filiform; sepals 4; pistillate spikes cylindric-oval, ⅓–¾ in., shorter than the pubescent peduncles; sepals 4, lateral ones largest, sepals enclosing the fruit at maturity and becoming succulent; ovary sessile, 1-celled, style terminal and short, stigmas 2 and ascending.

FRUIT. Syncarp dark red or black, fleshy, oval-oblong, ⅓–1 in. long, achenes included in calyx and tipped by persistent stigmas.

LEAVES. Simple, alternate, deciduous, thin, ovate to oval, blades 1½–6 in. long, apex acute or short-acuminate, margin coarsely toothed or sometimes with one or more lobes, base rounded, cordate or semitruncate, upper surface dull dark green, usually rough and becoming glabrous, lower surface paler and sparingly pubescent on the veins to glabrous; young foliage pubescent; petioles usually shorter than the blade, one fourth to one half as long.

TWIGS. Young ones green to brown and pubescent, older ones darker brown to gray or glabrous.

RANGE. Old gardens, roadsides, thickets, and waste grounds. Cultivated in Texas, Louisiana, Oklahoma, and Arkansas; eastward to Florida and north to New York. A native of western Asia.

REMARKS. The genus name, *Morus*, is the classical Latin name of mulberry, and the species name, *nigra*, refers to the black color of the fruit. It is sometimes grown for fruit or for shade. Although Black Mulberry is sometimes reported as being cultivated in the Southwest, many times incorrect identifications are made because of its close resemblance to a black-fruited race of White Mulberry, *M. alba* var. *tatarica* (L.) Ser. (See *M. alba* var. *tatarica* description.)

WHITE MULBERRY
Morus alba L.

FIELD IDENTIFICATION. An introduced tree to 40 ft, and attaining a diameter of 3 ft.

FLOWERS. Staminate and pistillate catkins axillary, borne on the same tree or on different trees; staminate catkins ⅜–1 in., cylindric, slender, drooping; calyx 4-parted, lobes ovate; stamens 4, elastically expanding; pistillate catkins drooping, oblong or oval to subglobose, cylindric, ½–⅔ in. long, about ¼ in. in diameter; calyx 4-parted, lateral sepals largest, calyx greatly enlarging to envelop the achene at maturity; ovary sessile, 2-celled, 1 cell atrophies; styles 2, linear, stigmatic down the inner side.

FRUIT. Borne June–August on slender, glabrous or pubescent peduncles ¼–⅔ in. long, pendent, subglobose to oval or oblong, white to pink (rarely black), ½–¾ in. long and about ¼ in. wide, sweet; fruit a syncarp, or an aggregation of ovate, compressed achenes each covered by the succulent, thickened calyx, the whole fruit as a unit thus juicy and elongate.

LEAVES. Alternate, deciduous, ovate to oval or asym-

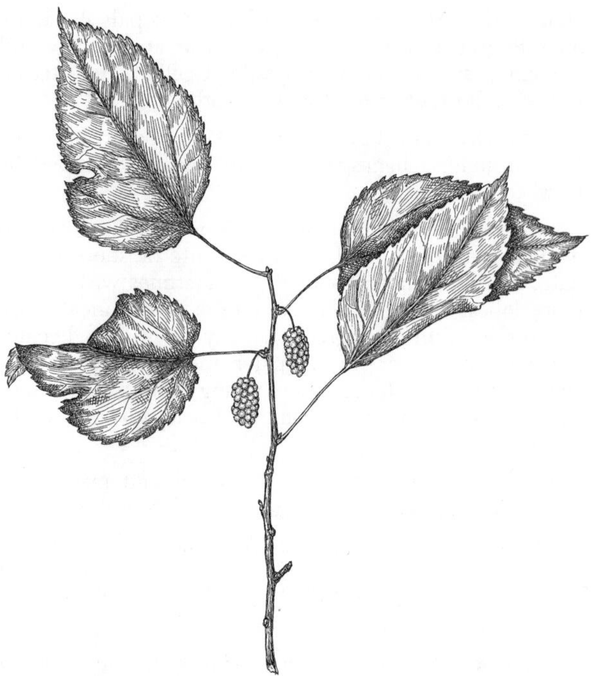

WHITE MULBERRY
Morus alba L.

metrical, heart-shaped, blades 2½–8 in. long, 1–4½ in. wide; margin with blunt, crenate teeth or often also 1–6-lobed, apex acute or short-acuminate; base semicordate, rounded or truncate, 3-veined; thin and smooth; upper surface olive-green, lustrous and glabrous, paler and glabrous beneath; petiole ½–1½ in., shorter than the blade, slender, glabrous or slightly pubesecent.

TWIGS. When young reddish brown, glabrous to slightly pubescent, when older slender, glabrous, gray.

BARK. Light to dark gray, broken into narrow furrows and irregular, often twisted, ridges.

RANGE. Escaping cultivation to roadsides, fields, and thickets. Naturalized Texas, Oklahoma, Arkansas, and Louisiana; north to Maine, and west to Minnesota and Wisconsin. Native home not positively known, either Europe or China, but most authorities cite an Asiatic origin. A cosmopolitan plant, known in nearly all parts of the world.

PROPAGATION. The following quotation gives methods of collection, extraction, storage, and planting of the seed (U. S. D. A., 1948): "[The] fruits may be collected by stripping, shaking or flailing them from the trees into a tarpaulin. If collection is not made as soon as they are ripe, they may be destroyed by birds or other animals. To extract the seed, mash and soak the berries in water for 24 hours and run them wet through a macerator, skimming or floating off the pulp. [The seed averages about 300,000 per lb.] Fermentation of the fruits for one day before extraction improves germination of *Morus alba* var. *tatarica* seed. A one per cent lye solution is reported to aid the extraction and cleaning process. Small samples may be pulped by rubbing them gently through a No. 6 screen and then floating off the pulp and empty seed. The cleaned seed should be air-dried in the shade for a short period before storage or use. The size, yield, purity, soundness, and cost of commercial cleaned seed vary by species. . . .

. . . .

"Mulberry seed mixed with sand or sawdust may be sown broadcast or in drills in the fall; or seed which has been stratified or water-soaked for 1 week may be sown in the spring and covered with ¼ in. of soil. Drills 8 to 12 in. apart, using about 50 seed to the linear ft of row, are satisfactory. Fall-sown beds should be mulched with straw or leaves and protected by bird or shade screens until germination begins in the spring. Spring-sown beds should be mulched with burlap and kept moist until germination begins. Seedbeds should be given half-shade for a few weeks after germination. Germination usually takes place 1 to 2 weeks after spring sowing. In the case of *Morus alba* var. *tatarica* about 12 per cent of the viable seed sown produce usable seedlings."

The bacterial canker disease which sometimes attacks the tree may be controlled by weekly spraying of 0.06 per cent corrosive sublimate plus 1.5 per cent slaked lime. Some fungi leafspot diseases of *Cercospora* and *Phleospora* attack the plants also.

It has been said that White Mulberry is cultivated in more countries than any other tree, and that more has been written about it than about any other tree. One reason is that it is the favorite food of the silkworm caterpillar, and silk-making gives employment to large numbers of people. At one time an effort was made to introduce the tree into the United States to start the industry, and thousands of trees were planted, but the high cost of labor made the project impossible. The tree was introduced into Texas about 1880 and planted for windbreaks in the Texas Panhandle area, but is perhaps of most use only in the sandy South Plains where few other trees will grow.

VARIETIES AND HORTICULTURAL FORMS. White Mulberry is very variable in leaf size and shape, fruit characters, and habit of growth. Some of the best known variants are as follows:

Russian Mulberry, *M. alba* var. *tatarica* (L.) Ser., is a small bushy-topped tree, with small leaves 2–4 in. long which are lobed or undivided, and fruit white to dark red, small, and insipid. A hardy variety often planted in rigorous climates, it is reported to have been first introduced into the Western states in 1875 by Russian Mennonites.

Large-leaf White Mulberry, *M. alba* var. *macrophylla* Loud., has large undivided leaves 6–8½ in. long and red fruit.

White-vein White Mulberry, *M. alba* var. *venosa* Delile, has strongly marked white veins and rhombic-ovate leaves, irregularly serrate and acute, acuminate or rounded at the apex.

Silkworm White Mulberry, *M. alba* var. *multicaulis* (Perrottet) Loud., is shrubby, or a small tree, with large, undivided leaves 3–6 in. long with apices obtuse, and a large, nearly black fruit.

Skeleton-leaf White Mulberry, *M. alba* var. *skeletoniana* Schneid., has regularly and deeply divided leaves with long-pointed lobes.

Weeping White Mulberry, *M. alba* forma *pendula* Dipp., has long, slender, pendulous branches.

Pyramid White Mulberry, *M. alba* forma *pyramidata* Ser., has a pyramidal form with usually lobed, acute leaves.

REMARKS. The genus name, *Morus*, is the classical Latin name, and the species name, *alba*, refers to the white fruit. It is also known under the names of Silkworm Mulberry, Russian Mulberry, Morera, and Morea. The fruit is not as juicy as the native Red Mulberry, and is somewhat smaller. It also seems to vary as to sweetness, on some trees being very sweet and on others so insipid and dry as to be hardly edible. Although the fruit of the species is most commonly white or pink, some varieties produce red or black fruit. The fruit has some wildlife value, being eaten readily by a number of species of birds, opossum, and raccoon, as well as by poultry and hogs. The wood is hard and durable, and is used for furniture, utensils, and boatbuilding.

Texas Mulberry
Morus microphylla Buckl.

FIELD IDENTIFICATION. Shrub, or sometimes a small, scraggy tree to 20 ft.

FLOWERS. Dioecious, small, green, inconspicuous, borne in ament-like spikes; staminate spikes on short pedicels, many-flowered, ½–¾ in. long; petals absent; calyx hairy, 4-lobed, lobes rounded, green to reddish; stamens 4, filaments filiform, anthers yellow with dark green connectives; pistillate sessile, drooping in short-oblong, few-flowered spikes rarely over ½ in. long; calyx hairy, 4-lobed, lobes thick, rounded, 2 larger than the others; ovary green and glabrous, 2-celled at first with one soon atrophying; stigmas 2, short, spreading.

FRUIT. A syncarp (multiple fruit) ripening in May, subglobose or short-ovoid, red at first to black later, sweet or sour, with scant juice, edible. Syncarp composed of numerous small, 1-seeded drupes; drupes about ⅙ in. long, ovoid, rounded at ends, containing a thick-walled, crustaceous, brown nutlet; seed pendulous, ovoid, pointed, pale yellow.

LEAVES. Simple, alternate, petioled, blades 1½–2½ in. long, ¾–1 in. wide; ovate to oval; margin coarsely serrate, sometimes 3-lobed; base truncate, rounded or semicordate; apex acute to short-acuminate; thin but firm; upper surface dull green, somewhat pubescent, tubercular-roughened, veins inconspicuous; lower surface paler, glabrous to somewhat hairy, veins delicate-reticulate, 3-veined at base; petioles slender, pubescent, about ⅓–¾ in., stipules linear-lanceolate, somewhat falcate, apex acute, white-tomentose, about ½ in.

TWIGS. Slender, white-hairy at first, glabrous later, light reddish brown to gray; lenticels small, round, pale.

BARK. Light gray or tinged with red, smooth, tight, shallowly furrowed and broken on the surface into narrow ridges and broad, flat fissures with slightly appressed scales.

WOOD. Dark orange or brown, sapwood lighter, heavy, hard, elastic, close-grained, specific gravity 0.77.

RANGE. Texas, New Mexico, Arizona, and Mexico. In Texas generally west of the Colorado River on dry limestone hills. In Mexico in the states of Chihuahua to Durango.

REMARKS. The genus name, *Morus*, is the classical name, and the species name, *microphylla*, refers to the small leaves. Vernacular names are Mexican Mulberry, Dwarf Mulberry, Wild Mulberry, Mountain Mulberry, Tzitzi, Hamdek-kiup, and Mora. The Indians of Arizona and New Mexico are reported to have grown the tree for its fruit and made bows from the wood. In Mexico the wood is used occasionally in carpentry. The fruit is rather small and dry, but is edible. It is consumed by a number of species of birds, including mockingbird, cardinal, mourning dove, Mearns's quail, scaled quail, and Gambel's quail, and is sometimes browsed by the white-tailed deer. Texas Mulberry is easily distinguished from Red Mulberry by much smaller leaves and

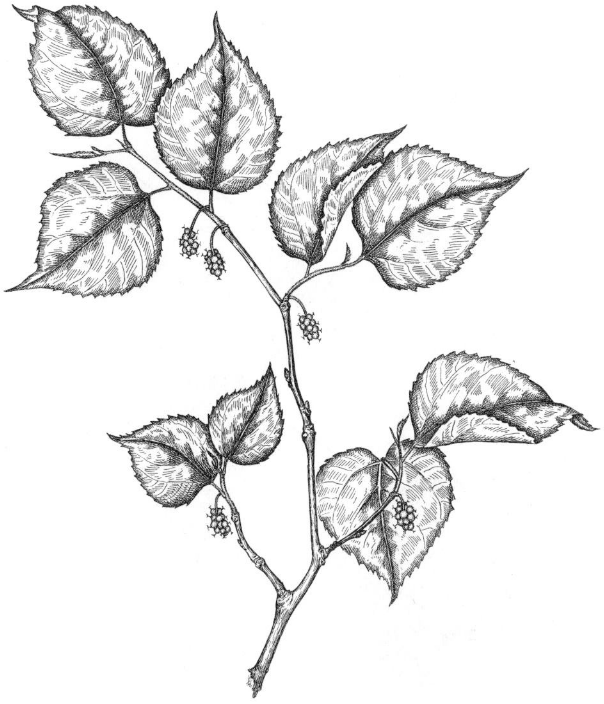

TEXAS MULBERRY
Morus microphylla Buckl.

fruit. Texas Mulberry is generally found on dry, limestone soils. At one time the botanist E. L. Greene split the species into a number of segregates. However, most of these divisions are not constant in character, so the recent tendency is to regard them as geographical forms of the same species, *M. microphylla* Buckl.

Osage-orange

Maclura pomifera (Raf.) Schneid.

FIELD IDENTIFICATION. Tree attaining a height of 60 ft, with a milky sap and bearing stout thorns.

FLOWERS. April–June, dioecious, green, staminate in long-peduncled axillary racemes, 1–1½ in. long; petals none; stamens 4, exserted; calyx 4-lobed; pistillate in globose dense heads about 1 in. in diameter; calyx 4-lobed, thick, enclosing the ovary; ovary ovoid, compressed, 1-celled; style filiform, long, exserted.

FRUIT. September–October, a syncarp, or aggregation of 1-seeded drupelets, globose, yellowish green, 4–5 in. in diameter; achenes surrounded by enlarged fleshy calyx; juice of fruit milky and acid.

LEAVES. Deciduous, alternate, entire, broad-ovate to ovate-lanceolate, rounded or subcordate at base, or broadly cuneate, acuminate at apex, 3–6 in. long, tomentose at first, lustrous later, yellow in autumn, petioles ½–2 in., stipules triangular, small, early deciduous.

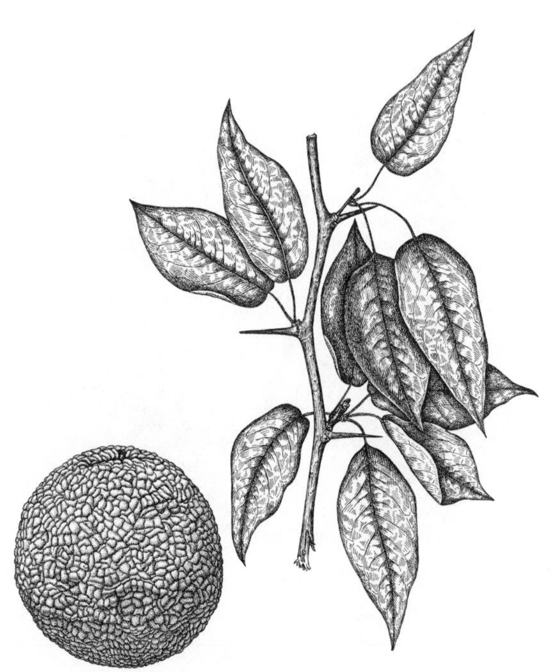

Osage-orange
Maclura pomifera (Raf.) Schneid.

BARK. Brown to orange, deeply furrowed, ridges rounded and interlacing.

WOOD. Bright orange or yellow, heavy, hard, durable, strong, weighing 48 lb per cu ft.

RANGE. Arkansas, Oklahoma, Louisiana, Missouri, and south into Texas. Well developed in the Oklahoma Red River Valley. Also escaping cultivation throughout Eastern United States.

VARIETY. *M. pomifera* var. *inermis* (André) Schneid. is a thornless variety.

PROPAGATION. The achene may be removed by crushing and macerating the fruit. A bushel of fruit yields about 24,500 seeds or about 2 lb. If stored at 41° F., seeds retain viability for 3 or more years. The achene is soaked in water, or stratified in sand or peat for 30 days. It is sown the following spring. Average germination is about 58 per cent. The seeds should be drilled in rows 8–12 in. apart for cultivation. About ¼ in. of firmed moist soil over the seed will suffice.

REMARKS. The genus name, *Maclura,* is in memory of William Maclure, an early American geologist, and the species name, *pomifera,* means "fruit-bearing." The name Bois d'Arc was given to it by the French, meaning "bow-wood," with reference to the fact that the Osage Indians made bows from the wood. Vernacular names are Hedge-apple, Horse-apple, Mock-orange, and Yellow-wood. Yellow dye was formerly made from the root bark. Also the bark of the trunk was used for tanning leather. Squirrels feed on the little achenes buried in the pulpy fruit, and black-tail deer browse the leaves. The tree was formerly much planted as windbreaks or hedgerows, and has been cultivated since 1818.

Common Fig

Ficus carica L.

FIELD IDENTIFICATION. Deciduous spreading shrub or tree to 30 ft. Branches numerous, stout, glabrous, spreading or ascending, forming a rounded or flattened crown. Sap thick and milky.

FLOWERS. Borne inside a hollow pear-shaped receptacle with a narrow orifice, axillary, solitary, greenish or brown to violet, length 1½–3⅓ in., staminate flowers nearly sessile, with 2–6 sepals and 1–3 stamens; pistillate flowers short-stalked; style lateral and elongate; ovary sessile, 1-celled.

FRUIT. Synconium obovoid to ellipsoid, fleshy, achenes small, numerous, included in fruit; taste mild, sweet, mucilaginous; firm and leathery.

LEAVES. Simple, alternate, ovate to oval, usually 3–7-lobed, some leaves lobed a second time; lobes obovate or obtuse at apex and irregularly dentate, blades 4–8 in. long and about as broad, base cordate or truncate, scabrous above and below with stout, stiff hairs, vena-

COMMON FIG
Ficus carica L.

tion palmate; petioles ¾–2 in., usually one half to two thirds as long as the blade, lightly to densely pubescent.

TWIGS. Smooth, stout, gray, pubescent at first, branches glabrous later.

RANGE. Cultivated in Texas, Oklahoma, Arkansas, and Louisiana; eastward to Florida and Tennessee, and northward along the Atlantic coastal plain as far north as New York. Not escaping cultivation readily because of frost kill. In old gardens, fields, and along roadsides. A native of western Asia. Cultivated since ancient times.

MEDICINAL USES. Ripe figs are considered to be nutritious, laxative, and demulcent, and have been used for food for many centuries. The unripe fig is known to yield an irritant juice which inflames the skin of some people. Dried figs are much more saccharine than the fresh fruit. The absolutely dry material averages 5.75 per cent nitrogenous matter and 72.26 per cent sugar. Both ripe and dried figs contain vitamins A and C (but lose about 30 per cent of vitamin A in drying). The latex of the fig tree has been used as an anthelmintic against internal parasites. The chief medicinal use of the fig is as a laxative article of diet.

PROPAGATION. Common Fig stands up to 10°–20° frost under favorable conditions. Two or more crops per year are borne, when grown under glass. Usually cuttings are taken in the fall and severed just below the bud or through the nodes. Only the ends are allowed to protrude barely above the ground. Plants begin to bear in 2–4 years. New varieties are obtained from seed. Because of nematode infestation of sandy soils figs apparently do best in clay soils. For orchard planting, trees are usually placed 10 by 15 or 12 by 20 ft apart in well-cultivated ground. Most planting is done in January or February. Usually the plants are cut back to several trunks for winter protection.

VARIETIES AND HORTICULTURAL FORMS. The botanical varieties are generally listed as *F. carica* var. *sylvestris* (the wild fig of Asia), *F. carica* var. *smyrniaca*, *F. carica* var. *hortensis*, and *F. carica* var. *intermedia*. The first of these varieties is the host of the fig wasp, *Blastophaga grossorum*, which pollinates the Smyrna fig. The *F. carica* var. *hortensis* group is self-fertilized. The *F. carica* var. *intermedia* group is partly so.

Many horticultural forms are known, the chief of which are as follows:

Adriatic, Black Douro (Turkey), Black Ischia, Brown Turkey (Turkey), Brunswick (Magnolia), Celeste (Celestial), Croisic (Cordelia), Dauphine (Ronde Violette), Doree, Dottato (Kadota), Endich (Kadota), Genoa (Genoa White), Green Ischia, Ischia (White Ischia), Lemon (Marseilles White), Leon, Lob Ingir (Lob Injur, Sari), Markarian 1 (Samson), Milco (Caprifig var.), Mission, Osborn (Osborn Prolific, Ronde Noire), Panoche, Ramsey, Rolding No. 1 (Caprifig var.), Rolding No. 2 (Caprifig var.), Rolding No. 3 (Caprifig var.), Rolding No. 4 (Caprifig var.), San Pedro Black (Turkey), San Pedro White, Sari Lob (Lob Ingir), Stanford (Caprifig var.), Verdal Longue.

REMARKS. The genus name, *Ficus*, is the ancient Latin name. The species name, *carica*, is from Caria in Asia Minor.

MISTLETOE FAMILY (*Loranthaceae*)

CHRISTMAS AMERICAN MISTLETOE
Phoradendron flavescens (Pursh) Nutt.

FIELD IDENTIFICATION. Semiparasitic shrub on deciduous trees. Branches yellowish green, glabrous or slightly pubescent.

FLOWERS. Borne May–July, axillary, dioecious, bracts fleshy; spikes ⅜–2 in. long, solitary, or 2–3 together, linear, jointed, shorter than leaves; joints usually about 4, some joints 6-flowered in pistillate inflorescences, others 12-flowered when staminate; petals absent; calyx of staminate flowers globose, deeply 3-lobed, each lobe with a sessile anther at base which opens transversely; calyx of pistillate flowers also 3-lobed and adnate to inferior ovoid ovary; stigma 1, subsessile, obtuse.

FRUIT. Berry sessile, white or creamy, translucent, subglobose, about ⅙ in. long, fleshy, 1-seeded.

LEAVES. Opposite, oblong or obovate, flattened, blades 1–2 in. long, ½–⅚ in. wide, margin entire, apex rounded, base cuneate, surfaces dark green to yellowish green, glabrous, leathery, 3–5-nerved; petioles 1/12–⅓ in.

TWIGS. Yellowish green to dark green, stout, thick, terete, brittle at base, at first minutely puberulent, becoming glabrous.

RANGE. On deciduous trees of many species. Texas, New Mexico, Oklahoma, Arkansas, and Louisiana; eastward to Florida, north to New Jersey, Pennsylvania, and Ontario, and west to Missouri and Kansas.

VARIETIES. *P. flavescens* var. *orbiculatum* Engelm. has orbicular leaves.

P. flavescens var. *macrophyllum* Engelm. has larger glabrous leaves.

P. flavescens var. *pubescens* Engelm. has leaves stellate-pubescent or short-hirtellous, branches tufted. This variety was once known under the name of *P. engelmannii* Trel.

P. flavescens var. *villosum* (Nutt.) Engelm. has dense velvety pubescence with simple hairs.

REMARKS. The genus name, *Phoradendron*, is from the Greek and means "tree-thief," from its parasitic habit. The species name, *flavescens*, refers to the yellowish herbage. The plant is much used for Christmas decorations.

CHRISTMAS AMERICAN MISTLETOE
Phoradendron flavescens (Pursh) Nutt.

Cory American Mistletoe

Phoradendron coryae Trel.

FIELD IDENTIFICATION. Semiparasitic plant found attached to various species of oaks. The jointed branches are rather long and stout, with internodes densely short stellate-tomentose. No cataphylls are present.

FLOWERS. Spikes short and clustered, 3/5–4/5 in.; peduncles 1/12–1/8 in., with scales tomentose; joints 3–4, swollen, tomentose, flowers small, sessile, inconspicuous; staminate spikes 18-36-flowered, petals absent, perianth calyx-like with 2–4 teeth; stamens inserted at base of sepals with anthers sessile or nearly so and 2-celled; pistillate spikes 12-flowered, style short, stigma barely capitate, ovary inferior and 1-ovuled.

FRUIT. In close whorls, white, small, about 1/8 in. in diameter, round-ovoid, pubescent above; seed solitary, albuminous, viscid; sepals persistent, deltoid, erect, widely parted.

LEAVES. Opposite, round-ovate, broadly elliptic or obovate, broadly obtuse, blade length 1–1½ in., width 5/8–1¼ in., surfaces closely puberulent; petioles stout, about 1/8 in.

RANGE. At altitudes of 3,500–8,500 ft. Western Texas, New Mexico, Arizona; Baja California and Sonora, Mexico.

Cory American Mistletoe
Phoradendron coryae Trel.

REMARKS. The genus name, *Phoradendron*, is from the Greek *phor* ("a thief"), and *dendron* ("tree"), from its parasitic habit. The species name, *coryae*, is in honor of V. L. Cory, former range botanist for Texas Agricultural and Mechanical College.

Hairy Mistletoe
Phoradendron flavescens var. *villosum* (Nutt.) Engelm.

Shortspike Mexican Mistletoe

Phoradendron brachystachyum Nutt.

FIELD IDENTIFICATION. Semiparasitic, semiwoody plant on trees. Not forked, the long slender branches without cataphylls. The following description is based on that of William Trelease (1916, p. 47).

FLOWERS. Dioecious in axillary, mostly solitary, spikes, about 2/5 in. long (scarcely reaching 3/5 in. in fruit), sparingly puberulent with 2–3 swollen joints; peduncles 1/5–1/8 in., scales nearly glabrous; flowers sessile, small, inconspicuous, usually sunken in the joints; petals absent; sepals distinct, deltoid, valvate, persistent; joints of staminate inflorescences 18–30-flowered, stamens inserted on base of sepals, anthers nearly sessile and 2-celled; joints of pistillate inflorescences 8–12-flowered, ovary inferior, 1-celled, 1-ovuled, stigma terminal and scarcely dilated.

FRUIT. Baccate, white, globose, umbonate, glabrous, about 1/8 in. in diameter; seed solitary, endocarp fibrous, mesocarp viscid; calyx persistent.

Shortspike Mexican Mistletoe
Phoradendron brachystachyum Nutt.

LEAVES. Very variable even on same shoot, from typically oblong-lanceolate or oblanceolate to obovate or orbicular, the narrower forms more or less falcate, obtuse, margin entire, blades either 1⅜ in. long and ¼ in. wide, 2 in. long and ⅝–¾ in. wide, or ⅝ in. long and ⅝ in. wide; base cuneately narrowed rather than petioled for ⅛–⅕ in.

STEMS. Long and slender, internodes short, about ⅘ in. long and 1⁄12 in. in diameter, or some to 2⅘ in. long and ⅛ in. in diameter, somewhat varnished, evanescently hispid like the foliage.

RANGE. Growing close to the border of the United States and Mexico in the state of Tamaulipas, and to be looked for in southwestern Texas. Southward it occurs in Mexico in Veracruz, Oaxaca, Sonora, Mexico D.F., Jalisco, and Colima.

REMARKS. The genus name, *Phoradendron*, is from the Greek and means "tree-thief," from its parasitic habit. The species name, *brachystachyum*, refers to the short spikes. It is known to use as host plants the *Arbutus* and *Quercus* species.

Havard American Mistletoe

Phoradendron havardianum Trel.

FIELD IDENTIFICATION. Semiparasitic, semiwoody plant on trees. Not forked, the rather short but stout branches without cataphylls. The following description is based on that of Trelease (1916, p. 44).

FLOWERS. Spikes closely tomentose, solitary, axillary, short (scarcely ⅖ in.), with 2 or 3 joints; peduncle about 1⁄12 in. long, scales puberulous; flowers sessile, small, inconspicuous, usually sunken in the joints; petals absent; sepals distinct, deltoid, valvate, persistent; joints of staminate inflorescences 8–12-flowered, stamens inserted on base of sepals, anthers nearly sessile and 2-celled; pistillate flowers with ovary inferior, 1-celled, 1-ovuled, stigma terminal and scarcely dilated.

FRUIT. Baccate, white or greenish, subglobose, glabrous, about ⅛ in. in diameter, sepals scarcely 1⁄50 in. long, persistent, suberect, pubescent, at least at base; seed solitary, endocarp fibrous, mesocarp viscid.

LEAVES. Simple, opposite, small, blades scarcely ⅜ in. wide and ⅝–¾ in. long, obovate-orbicular or elliptic (prevailingly rounded), broadly obtuse at apex, margin entire, base rather abruptly subpetioled for about 1⁄12 in., surfaces minutely stellate-tomentose.

STEMS. Short, stout, cataphylls absent; internodes short, ⅘ in. long, 1⁄12–⅛ in. in diameter, rather minutely stellate-tomentose like the foliage.

RANGE. The type locality is listed as the Guadalupe Mountains of Texas. It is also found in the Cornudas Mountains, the Hueco Tanks area near El Paso, and adjacent New Mexico.

REMARKS. The genus name, *Phoradendron*, is from the Greek and means "tree-thief," from the parasitic habit. The species name, *havardianum*, is in honor of Valery Havard, who made extensive collections along the valley of the Rio Grande and in adjacent territory. An account of Havard's work is given by Winkler (1915). The plant is known to be parasitic on species of oaks such as *Quercus arizonica, Q. chrysolepis, Q. emoryi, Q. gambelii, Q. hypoleuca, Q. oblongifolia, Q. reticulata, and Q. toumeyi.*

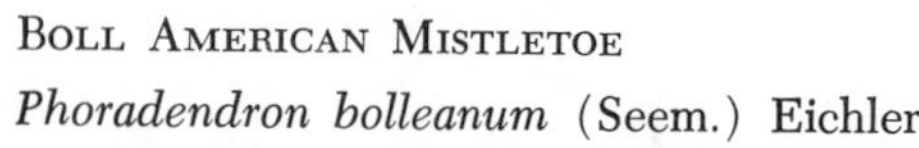

Boll American Mistletoe

Phoradendron bolleanum (Seem.) Eichler

FIELD IDENTIFICATION. Yellowish green plant with jointed much-branched stems, semiparasitic on *Arbutus* and *Juniperus* species.

FLOWERS. Axillary, in monoecious jointed spikes; staminate of 2 6–12-flowered joints, calyx 3-parted, anthers sessile on base of calyx and opening by a pore; pistillate spikes with 2 flowers at each node, spike hardly more than ⅕ in., calyx 3-lobed; stigma sessile and obtuse.

FRUIT. About ⅛–⅓ in. in diameter, white, pulpy, semitranslucent, 1-seeded.

LEAVES. Persistent, opposite, linear to spatulate or oblong, blades ¼–1 in. long, about ⅙ in. wide, margin entire, thick and firm, apex obtuse, base narrowed and blade sessile, surface papillate or hispid.

TWIGS. Much-branched, jointed, woody, brittle, greenish yellow or brown, at first puberulent, later glabrous.

RANGE. Within the *Juniperus* and *Arbutus* range at altitudes of 3,700–8,000 ft. Western Texas, New Mexico, Arizona, and California; north to Oregon and south into Mexico.

VARIETY. A variety with stellate-tomentose foliage has been named *P. bolleanum* var. *capitellatum* (Torr.) Kearney & Peebles.

REMARKS. The genus name, *Phoradendron,* is from the Greek *phor* ("a thief"), and *dendron* ("tree"), from its parasitic habit.

Juniper American Mistletoe

Phoradendron juniperinum Engelm.

FIELD IDENTIFICATION. Shrubby, semiparasitic plant growing on species of juniper, and more rarely on pines. The woody branches stout, rather dense, glabrous, 4–12 in.

FLOWERS. Small, inconspicuous, in jointed, axillary, opposite, few-flowered spikes; flowers usually 1-sexed on separate plants; petals absent; both staminate and pistillate calyx 3-lobed; staminate spikes terminating in a single 6–8-flowered joint; anthers 3, sessile at base of calyx-lobes, opening by a pore; spike of pistillate inflorescences 2-flowered, ovary adnate to calyx below, inferior, 1-celled, stigma sessile and obtuse.

Boll American Mistletoe
Phoradendron bolleanum (Seem.) Eichler

Juniper American Mistletoe
Phoradendron juniperinum Engelm.

FRUIT. Subglobose, about ⅛ in. in diameter, semitransparent, white to light red, crowned by the persistent sepals, 1-seeded.

LEAVES. Reduced to small opposite scales 1/25–1/12 in., spreading, ovate or deltoid, obtuse, base strongly connate, margin ciliate.

TWIGS. Usually 2-forked, crowded, stout, stiff, jointed, terete or somewhat quadrangular, woody.

RANGE. At altitudes of 4,000–7,500 ft in western Texas, New Mexico, Arizona, Colorado, Utah, Oregon, and Washington; also in Chihuahua, Mexico.

REMARKS. The genus name, *Phoradendron,* is from the Greek *phor* ("a thief") and dendron ("tree"), from the parasitic habit. The species name, *juniperinum,* refers to its habit of growing on juniper trees.

Mesquite American Mistletoe

Phoradendron californicum Nutt.

FIELD IDENTIFICATION. Plant semiparasitic on various trees and shrubs. The leaves reduced in size and scale-like.

FLOWERS. Spikes dioecious, canescent, short, axillary, mostly solitary, ⅕–⅖ in. long, joints about 4 and short, usually 2– but sometimes 4–6-flowered in 2–4 series; peduncles 1/25–1/8 in., with acute scales; flowers small, inconspicuous; petals absent; sepals valvate; pistillate spikes with 3 or more nodes and usually 2 flowers at each node; style short, stigma scarcely dilated; ovary inferior, 1-celled; staminate flowers variable in number; stamens as many as the sepals and inserted at their base, anthers 2-celled and sessile or nearly so.

FRUIT. Berry red (more rarely white in a variety), subglobose, sometimes rostrate, about 1/8 in. in diameter, smooth, glabrous; seed solitary, mesocarp viscid; sepals persistent and nearly or quite meeting.

LEAVES. Scalelike, simple, opposite, entire, short, about 1/12 in. long or less, triangular, acute, spreading, only slightly connate, not disarticulating from stem.

STEMS. Slender, terete, flexuous, canescent or glabrous later, without cataphylls, internodes 2/5–4/5 in. long and 1/25–1/8 in. in diameter.

RANGE. In the Mohave and Colorado deserts area; as well as in west Texas and New Mexico; also in Sonora, Mexico.

VARIETIES AND FORMS. *P. californicum* var. *distans* Trel. has the fruiting spikes elongate and the whorls of fruit somewhat separate.

P. californicum forma *leucocarpa* Trel. has white fruit.

REMARKS. The genus name, *Phoradendron,* comes from the Greek *phor* ("a thief") and *dendron* ("a tree"), in reference to the semiparasitic growth on the trees. The species name, *californicum,* is for the state of California, a vernacular name for it being California Mistletoe.

It is known to be semiparasitic on species of *Prosopis, Acacia, Olneya, Condalia, Cercidium, Dalea,* and *Parkinsonia.* It is never found on conifers.

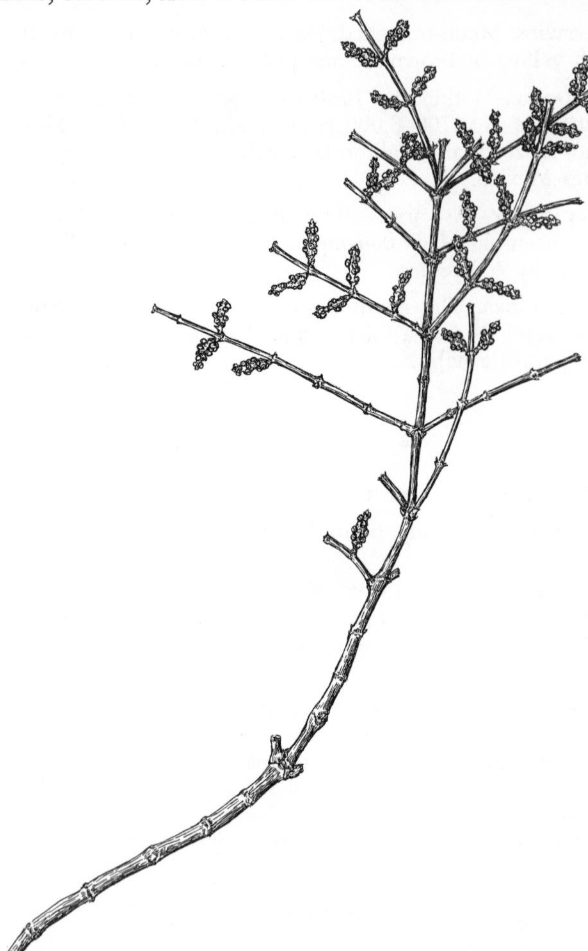

MESQUITE AMERICAN MISTLETOE
Phoradendron californicum Nutt.

BIRTHWORT FAMILY (*Aristolochiaceae*)

Woolly Pipevine

Aristolochia tomentosa Sims

FIELD IDENTIFICATION. Twining, high climbing, woody vine to 75 ft.

FLOWERS. May–June, on stout hairy pedicels; petaloid perianth yellow or greenish yellow, 1–2 in. long, densely tomentose, tube sharply curved, throat almost closed; limb 3-lobed, reflexed, wrinkled, dark purple; stamens 6, anthers sessile, adnate to stigma; styles united into a 4–6-angled column, ovules numerous in each cavity.

FRUIT. September–October, capsule pendulous, naked, oblong-cylindric, septicidally 6-valved; ovary 6-seeded; seeds flat, testa crustaceous.

LEAVES. Alternate, entire, persistent, suborbicular to broad-ovate, apex obtuse to rounded, base deeply cordate, densely tomentose beneath, less so above, reticulate-veined, blades 3–6 in. long, 2–3½ in. broad; petiole rather stout and densely tomentose, one half to two thirds the length of blade.

STEMS. Gray to brown or black, downy when young, more glabrate and somewhat grooved when older.

RANGE. In rich woods, eastern Texas, Oklahoma, Arkansas, and Louisiana; eastward to Florida, northward to North Carolina, and westward to Missouri, Kansas, and Illinois.

REMARKS. The genus name, *Aristolochia*, is from its supposed medicinal properties. The species name, *tomentosa*, refers to the tomentose leaves and stems. Another name is Dutchman's-pipe, because of the curious shape of the flowers. This plant, and others of the same family, are often cultivated for ornament. Most are greenhouse-grown as plant oddities, but the native species described is a rapid-growing vine suitable for growing outdoors in the South.

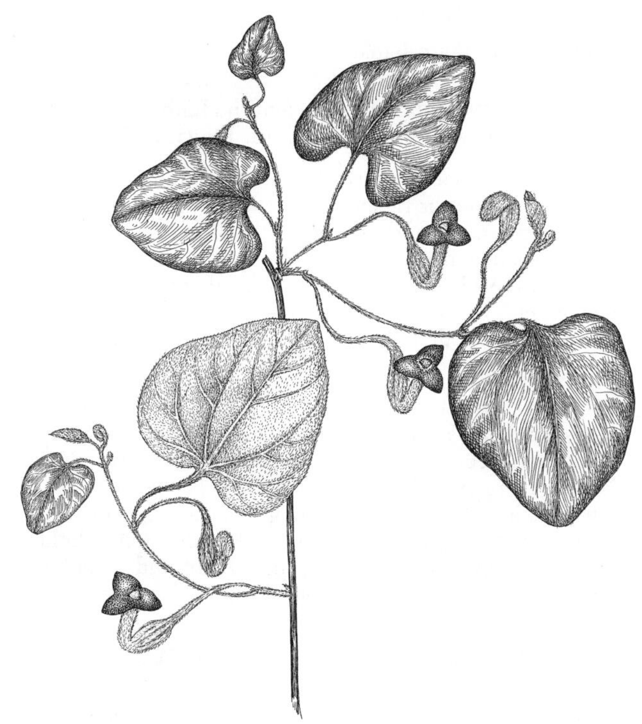

WOOLLY PIPEVINE
Aristolochia tomentosa Sims

BUCKWHEAT FAMILY (*Polygonaceae*)

AMERICAN BUCKWHEAT-VINE

Brunnichia cirrhosa Banks

FIELD IDENTIFICATION. A tendril-climbing, woody vine to 40 ft, with green to reddish brown stems.

FLOWERS. June–August, borne in terminal tendril-bearing panicles or racemes 2–10 in. long; perianth ¾–1½ in. long, composed of a petal-like, 5-parted, greenish calyx which is borne on a slender pedicel flattened into a wing ⅕–⅙ in. wide; sepals 5, about ¼ in. long, convergent, linear-oblong, acute or obtuse at apex; stamens included, 7–10 (usually 8), filaments filiform, enlarged below; ovary 3-celled, 3-sided; stigmas 3.

FRUIT. Pedicels filiform, ⅛–¼ in.; a fruiting body ¾–1½ in. long, somewhat falcate, enclosed by the hardened coriaceous calyx which becomes brown and leathery, upper half with the remains of the 5 sepals, lower half flattened into a wing with a ridge on one side; achene tightly enclosed, ¼–15/16 in. long, light brown, smooth, ovoid-oblong, triangular and rounded on the angles or distinctly grooved, base truncate or rounded, apex acute or apiculate.

LEAVES. Alternate, deciduous, entire, blades 1¼–6 in. long, ½–3 in. wide, ovate to lanceolate, petioled; base truncate, subcordate or broadly cuneate; apex acute or acuminate; thin, upper surface light green and glabrous, lower surface glabrous or slightly pubescent; petioles ¼–1 in. long, glabrous or puberulent.

STEMS. High-climbing, sometimes as long as 40 ft, green to reddish brown, finely grooved; tendrils terminal, delicate, on bare or leafy lateral shoots.

RANGE. Usually along streams in rich moist soil on the coastal plain. Texas, Oklahoma, Arkansas, and Louisiana; eastward to Florida, north to South Carolina, and west to Missouri.

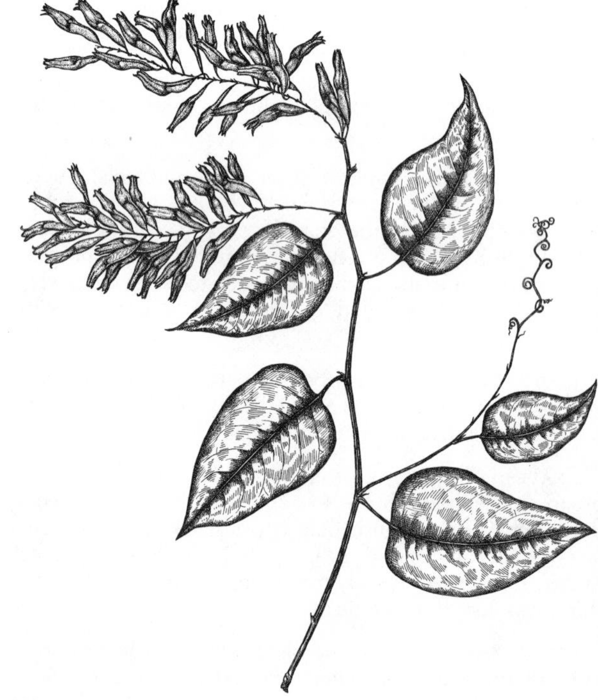

AMERICAN BUCKWHEAT-VINE
Brunnichia cirrhosa Banks

REMARKS. The genus name, *Brunnichia,* is in honor of M. T. Brunnich, a Norwegian naturalist of the eighteenth century. The species name, *cirrhosa,* means "having tendrils." The fruit somewhat resembles ear pendants and the vine is known as Eardrop-vine in some localities. It could be more extensively cultivated.

Slender-leaf Eriogonum

Eriogonum leptophyllum (Torr.) Woot. & Standl.

FIELD IDENTIFICATION. A perennial plant, definitely woody at the base, attaining a height of 6 in.–2 ft.

FLOWERS. In rather dense broomlike cymes averaging about 2½ in. high or less, the numerous branches erect or ascending; bracts small and triangular; peduncles rather short; petals absent; involucre sessile, 1⁄12–⅛ in. long, cylindric-turbinate, glabrous or sparingly ciliolate, lobes much shorter than the tube; sepaloid perianth segments 6, white or pink, 1⁄12–⅛ in. long, glabrous, obovate or oblong; stamens 9, inserted on the base of the perianth; styles 3 with 3 capitate stigmas; ovary superior, 3-carpellate and 1-celled.

FRUIT. A 3-angled achene, invested by the perianth.

LEAVES. Extending upward on flowering branches to the inflorescence; linear, apex acute or obtuse, margin entire and so strongly revolute as to nearly conceal the white-lanate lower surface, upper surface glabrous, length of blade usually ¾–2 in.

Slender-leaf Eriogonum
Eriogonum leptophyllum (Torr.) Woot. & Standl.

Wright Eriogonum
Eriogonum wrightii Torr.

STEMS. Rather slender, leafy, glabrous or obscurely puberulent, at base much thickened and woody.

RANGE. On dry hills at altitudes of 5,000–6,000 ft. New Mexico and Arizona, northward to Colorado.

REMARKS. The genus name, *Eriogonum,* is from the Greek words *erion* ("wool") and *gonu* ("knee"), for the swollen woolly nodes of some species.

Wright Eriogonum

Eriogonum wrightii Torr.

FIELD IDENTIFICATION. A white-tomentose, leafy perennial 6 in.–2 ft, often woody toward the base.

FLOWERS. June–September, peduncles seminaked, solitary or irregularly 2–3-forked, slender, ascending, tomentose; involucres loosely spicate, sessile, 1⁄12–⅛ in. long, teeth rigid and acute; petals absent; sepaloid segments of the perianth 6, 1⁄12–⅛ in. long, glabrous, white or pink; stamens 9; styles 3, stigmas capitate, ovary 1-celled.

FRUIT. Achene solitary, erect, scabrous, base acute, 3-angled.

LEAVES. Mostly on the lower half of the plant, scattered, alternate or clustered, oval to oblong or linear to lanceolate, apex acute, base tapering, margin entire, less than 1 in. long (usually ¼–½), both surfaces densely white-tomentose above and beneath.

STEMS. Erect and slender, woody toward the base, tomentose above, bark reddish brown, thin-scaly.

RANGE. In foothills, mountains, and canyons at altitudes of 2,500–7,200 ft. Trans-Pecos Texas; west through New Mexico to California, and south in Mexico from Chihuahua and Sonora to San Luis Potosí.

VARIETY (SUBSPECIES). The species has also been known under the name of *E. wrightii* subsp. *typicum* Stokes, and a low, spreading, mat-forming variety has been named *E. wrightii* subsp. *subscaposum* Stokes (Stokes, 1936).

REMARKS. The genus name, *Eriogonum*, is from the Greek words *erion* ("wool") and *gonu* ("knee"), for the swollen woolly nodes. The species name, *wrightii*, honors Charles Wright, botanist attached to the United States–Mexico boundary survey commission. The plant is regarded as important cattle and deer browse in some areas of Arizona and New Mexico. It is also reported as a good bee food. About 25 per cent of sown seed will germinate.

ROSEMARY FLAT-TOP ERIOGONUM
Eriogonum fasciculatum var. *polifolium* (Benth.) Torr. & Gray

FLAT-TOP ERIOGONUM

Eriogonum fasciculatum Benth.

FIELD IDENTIFICATION. A semierect shrubby plant attaining a height of 1–3 ft.

FLOWERS. May–October, perfect, inflorescence capitate to cymosely compound, simple or divided; peduncles 3–10 in.; rays 3–9, unequal; bracts linear to lanceolate, subtending the umbel or umbellets; involucres sessile or short-stalked, with short acute teeth; petals absent; perianth cone-shaped with the 6 sepaloid segments somewhat united at the base, not stipelike, white or pink, glabrous or hairy; stamens 9, anthers pink; pistil 1, styles 3 and separate to the base, stigmas 3, capitate; ovary superior, 3-carpellate and 1-celled.

FRUIT. A 3-angled achene with an adherent calyx.

LEAVES. Alternately fascicled in the axils of bracts, evergreen, ⅜–¾ in. long, about ¼ in. wide or less, linear to oblong or oblanceolate, base sessile or short-petioled, apex gradually narrowed, margins entire but usually revolute, rolled, upper surface glabrous (or tomentose in var. *polifolium*), lower surface white-tomentose.

STEMS. Numerous, slender, flexible, older stems glabrous or nearly so, bark thin and exfoliating.

RANGE. Usually in sand or dry rocky places at altitudes of 1,000–4,500 ft. In Arizona and southern California, north to Nevada and Utah.

VARIETIES. Rosemary Flat-top Eriogonum, *E. fasciculatum* var. *polifolium* (Benth.) Torr. & Gray, has leaves short cinereous-tomentose both above and beneath; involucres in dense headlike clusters, truncate or shortly toothed; calyx campanulate, pubescent, white or pink. This variety occurs in New Mexico, Arizona, Utah, Nevada, and California. It is apparently connected with the species by intermediate forms. The plant has possibilities for cultivation in rock gardens.

Leafy Flat-top Eriogonum, *E. fasciculatum* var. *foliolosum* (Benth.) Stokes, is a variety with leaves more revolute, green and pubescent beneath; peduncles 5–10 in. long. In southern California.

REMARKS. The genus name, *Eriogonum*, is from the Greek words *erion* ("wool") and *gonu* ("knee"), from the swollen woolly stem joints. The species name, *fasciculatum*, refers to the fascicled leaves.

SPREADING ERIOGONUM

Eriogonum effusum Nutt.

FIELD IDENTIFICATION. Plant low, leafy, shrubby, effusely branched, woody toward the base. A loosely defined species passing into a number of intergrading varieties within its range.

FLOWERS. Stems bearing inflorescences 8–18 in. tall, somewhat scapose, tomentose to floccose, cymes open, widening to become corymbose and branched 2–3 times or more; bracts usually small and triangular; involucre narrowly campanulate, sessile or pedunculate, 1/15–1/10 in. long, lobes shorter than the tube, tomentulose; petals absent; perianth shortly cone-shaped, 1/12–1/10 in. long, limb spreading; sepaloid segments of perianth 6, obovate to elliptic, mostly united near the base, rose-colored or white, glabrous; stamens 9, inserted on the base of the perianth; styles 3, each ending in a capitate stigma; ovary superior, 3-carpellate, 1-celled.

FRUIT. A 3-angled achene invested by the perianth.

LEAVES. Alternate, linear to oblanceolate or narrowly oblong, apex obtuse or acute, base narrowed or cuneate to a short petiole, margin plane or undulate, blade length 3/4–1 5/8 in., when mature rather coriaceous, pale to deep green.

STEMS. Slender, effusely branched, tomentose when young, later gray and glabrous but scaling away in strips, nodes swollen.

RANGE. On dry hills or plains at altitudes of 4,000–9,500 ft. In New Mexico; northward to Montana and westward into Arizona.

SPREADING ERIOGONUM
Eriogonum effusum Nutt.

VARIETIES AND SUBSPECIES. The species has been given the name of *E. effusum* subsp. *typicum* Stokes. A number of subspecies have been segregated. (Stokes, 1936).

Corymbed Eriogonum, *E. effusum* subsp. *corymbosum* (Benth.) Stokes, is suffrutescent, tomentose, pale; leaves numerous and alternate, short-petiolate, elliptic to narrowly ovate, undulate or partly revolute, slightly stiff, 1 1/4–2 in. long; inflorescence divisions umbellately corymbose, rays slightly rounded, U–shaped, bracts foliaceous; involucres solitary or sparingly clustered, not racemose, 1/8–1/5 in. long, sessile or short-pedunculate in the forks, firm, perianth white or rose. In New Mexico, Arizona, Colorado, and Nevada.

Fendler Eriogonum, *E. effusum* subsp. *fendlerianum* (Benth.) Stokes, has stout stems with leaves 1 1/2–3 1/3 in. long and 1/4–1/3 in. wide, blades rolled or undulate, petiole short; inflorescence stout, glabrate, branching, U–shaped, corymbose; involucres solitary or slightly congested, coriaceous, turbinate-campanulate, 1/6–1/5 in. high; perianth pale, segments 6, narrow. In New Mexico and Colorado in the grass and pinyon belts.

Thin-stem Eriogonum, *E. effusum* subsp. *leptocladon* (Torr. & Gray) Stokes, is shrubby with leaves rather numerous, lanceolate to linear, nearly plane or undulate; inflorescence primarily cymose, but final involucre racemosely disposed, campanulate, glabrate, 1/12–1/8 in. long, perianth pale to pink. In New Mexico, northern Arizona, and Utah.

Simpson Eriogonum, *E. effusum* subsp. *simpsoni* (Benth.) Stokes, has an openly branched base and leafy stem; inflorescence diffuse with slender branches, lightly tomentose; perianth pale, narrowly campanulate, shortly pedunculate in the lower axils, sessile above; leaves alternate. New Mexico, Arizona, Colorado, and Nevada.

REMARKS. The genus name, *Eriogonum*, is from the Greek words *erion* ("wool") and *gonu* ("knee"), in reference to the swollen, woolly stem joints.

SLENDER-BRUSH ERIOGONUM

Eriogonum microthecum Nutt.

FIELD IDENTIFICATION. Slender leafy plant with woody branches, attaining a height of 4 in.–2 ft.

FLOWERS. In cymose inflorescences on peduncles averaging about 4 in. long, 2–3–forked, the cymes V–shaped, often flat-topped, sometimes corymbose; bracts small and triangular; involucres sessile or peduncled in the axils of the inflorescence branches, 1/10–1/8 in. long, turbinate-campanulate, teeth short, more or less tomentose; petals absent; perianth 1/12–1/8 in. long, yellow or white; the sepaloid perianth segments 6, glabrous, oval to obovate, base attenuate; stamens 9, inserted on the base of perianth segments; pistil 1, with

SLENDER-BRUSH ERIOGONUM
Eriogonum microthecum Nutt.

3 styles, 3 stigmas and a superior 3-carpellate, 1-celled ovary.

FRUIT. Achene 3-angled, pointed, invested by the perianth.

LEAVES. Usually scattered and alternate, 3/8–3/4 in. long, oblanceolate or lanceolate, apex acute, or obtuse, base long-cuneate to a short petiole, margin often revolute, stiffish, upper surface glabrate to floccose, lower surface tomentose.

STEMS. Leafy, the lower vegetative stem portion sometimes longer than the inflorescence, tomentose to floccose; older stem near base more glabrate.

RANGE. The species is found on dry plains or foothills, of the larrea and grass belts, upward to the spruce belt, usually at altitudes of 4,500–8,000 ft. In New Mexico, Arizona, California, Utah, Montana, Nebraska, and Washington.

VARIETIES (SUBSPECIES). Golden Slender-brush Eriogonum, *E. microthecum* subsp. *aureum* (Jones) Stokes, has been named by Susan G. Stokes (1936). It has a woody base and is intricately branched, with the alternate leaves mostly terminal, oval to oblong, 3/8–3/4 in. long, base with a short petiole, lower surface with pale or rufous tomentum; peduncles to 2½ in. long, inflorescence borne in a flat-topped corymb with rays usually divaricate; involucres solitary or somewhat clustered, oblong; perianth yellow or pale, limb spreading; sepaloid perianth segments 6, outer somewhat oval, inner more narrow, 1/12–1/8 in. long. Known from New Mexico, Arizona, Colorado, Utah, and Nevada.

REMARKS. The genus name, *Eriogonum,* is from the Greek words *erion* ("wool") and *gonu* ("knee"), from the swollen woolly joints. The species name, *microthecum,* is for the small sheathlike flowers.

NAKED ERIOGONUM

Eriogonum nudicaule (Torr.) Small

FIELD IDENTIFICATION. A depressed or suberect perennial attaining a height of 4 in.–1 ft, the stems somewhat woody at the base with the inflorescences usually surpassing the leafy portion.

FLOWERS. Usually cymosely corymbose, two to three times divided; bracts scalelike; involucres sessile, solitary or subcapitate, glabrous, 1/8–1/6 in. long, campanulate, the lobes about one fourth as long as the tube, white or purplish; corolla absent; perianth 1/10–1/7 in. long, white, sometimes veined purple, glabrous, the 6 sepaloid perianth segments obovate or oblong; stamens 9, borne on the base of the perianth; styles 3, stigmas 3; ovary superior, 3-carpellate and 1-celled.

FRUIT. Achenes glabrous, angled, not winged, invested by the perianth.

LEAVES. Rosulate or apparently alternate, leaf-bearing portions of stem short, close-set at base of peduncles, erect, linear to narrowly oblanceolate, base long-cuneate to a short petiole, apex obtuse or acute, margin revolute, length 3/4–2 in., upper surface floccose or glabrate, lower surface tomentose.

RANGE. On dry hills and plains at altitudes of about 6,500 ft.; New Mexico, Arizona, Colorado, and Utah.

VARIETIES (SUBSPECIES). A taxon of considerable variation; some of its variants have been segregated as species by various authors and designated varietal status by others. The species described has been given the name of *E. nudicaule* subsp. *typicum* Stokes, and the following subspecies segregates recognized (Stokes, 1936):

E. nudicaule subsp. *scoparium* (Small) Stokes has long leaves, strongly revolute; involucres distinctly campanulate, about as wide as long, flowers pale. In Colorado.

E. nudicaule subsp. *tristichum* (Small) Stokes, has the involucres turbinate, about 1/6 in. long; flowers large for the group, pale. In New Mexico and Colorado.

REMARKS. The genus name, *Eriogonum,* is from the Greek words *erion* ("wool") and *gonu* ("knee"), for the woolly nodes of some species.

James Eriogonum

Eriogonum jamesii Benth.

FIELD IDENTIFICATION. A stout low perennial, with branches varying from tufted and matted to suberect and exposed. The caudices are woody toward the base, but become herbaceous above.

FLOWERS. In dichotomous or trichotomous cymes on stems 4–12 in. high, the peduncles villous-tomentose; bracts ternate and leaflike; central involucre sessile, lateral ones sessile or short-pedunculate, involucres usually turbinate-campanulate with short erect teeth, 1/5–2/5 in. long, tomentose; petals absent; perianth 1/8–1/4 in. long, with a stipelike base, cream or yellow, externally pubescent; the sepaloid perianth segments 6, broadly oblong or obovate, the inner ones exceeding the outer; stamens 9, inserted on the base of the sepals; styles 3, stigmas 3 and capitate, ovary superior, 3-carpellate, 1-celled.

FRUIT. A 3-angled achene invested by the perianth.

LEAVES. Rosulate at base or slightly scattered, length 5/8–2 in. Oblong to elliptic-lanceolate or ovate to spatulate, apex obtuse or subacute, base narrow or broadened, margins often undulate, upper surface thinly tomentose to glabrate, lower surface densely white-tomentose.

STEMS. The lower perennial section of stem usually short, stout, and tufted.

RANGE. On hills and mountains at altitudes of 5,000–11,500 ft. In western Texas; west through New Mexico to Arizona, northward to Wyoming, and south into Mexico.

VARIETY AND SYNONYMS. Yellow James Eriogonum, *E. jamesii* var. *flavescens* Wats., has stout scapes and branches divergent at a rather wide angle; the flowers are large and bright yellow. Known from Arizona, Colorado, and Wyoming.

The synonymy of this very variable species is much confused. The following names have been assigned as species synonyms by some authors or as varieties by others: *E. arcuatum* Greene, *E. bakeri* Greene, *E. vegitivus* A. Nels., and *E. jamesii* subsp. *bakeri* Stokes.

REMARKS. The genus name, *Eriogonum*, is from the Greek words *erion* ("wool") and *gonu* ("knee"), from the swollen woolly stem joints of some species. Also known by the vernacular name of Antelope-brush. The root is reported to be used medicinally by certain Indians.

Matted Eriogonum

Eriogonum tenellum Torr.

FIELD IDENTIFICATION. A perennial plant usually growing in depressed mats. The rather short, densely branched stems are woody at the base. Although usually growing in flattened tufts, occasionally the herbaceous upper stems are erect and rarely attain a height of 1½ ft.

FLOWERS. In open rather sparingly branched dichotomous cymes, the floriferous portion of the axis usually more extensive than the naked portion below; bracts scalelike, in twos or threes; involucre sessile or on peduncles 3/8–3/4 in. long in the axils of the branches; involucre turbinate, about 1/8 in. long, glabrous, lobes ovate-triangular; petals absent; perianth 1/10–1/8 in. long, glabrous, white or pink; sepaloid perianth segments 6, obovate to orbicular, often obtuse, inner ones linear-oblong; stamens 9, filaments slender and inserted on the base of the perianth; pistil 1 with 3 styles, 3 capitate stigmas and a 3-carpellate, 1-celled, glabrous ovary.

FRUIT. A 3-angled achene invested by the perianth.

LEAVES. Usually crowded on the caudex, ovate to orbicular, base cuneate, apex obtuse, length 1/4–5/8 in., both surfaces permanently white-tomentose, petiole more or less tomentose but sometimes glabrous.

STEMS. Usually short, crowded and depressed from a woody caudex, sometimes elongate.

RANGE. Dry plains or rocky canyons of grass belts, at altitudes of 4,000–5,000 ft. West Texas and New Mexico; north to Colorado and Utah and south in Mexico.

VARIETY (SUBSPECIES). A variety has been assigned the name of *E. tenellum* subsp. *platyphyllum* (Torr.) Stokes (Stokes, 1936). It has more or less ascending leafy branches 4–10 in. long; the inflorescence is tall and freely branched with larger flowers; leaves averaging somewhat longer (to 5/8 in. long). Known from central and southwestern Texas.

REMARKS. The genus name, *Eriogonum*, is from the Greek words *erion* ("wool") and *gonu* ('knee"), from the swollen woolly nodes of some species. The species name, *tenellum*, means "slender."

Southern Joint-weed

Polygonella americana (F. & M.) Small

FIELD IDENTIFICATION. A somewhat branched, half-evergreen, stout perennial to 4 ft; sometimes woody toward the base and long-rooted.

FLOWERS. From midsummer to fall in stout racemose panicles 1–3 in. long. Perianth white or pink, solitary, borne on rigid, divergent, jointed pedicels; base of pedicels subtended by densely imbricate, asymmetrical, attenuate-apiculate ocreae; perianth petaloid, segments 5, persistent, the 3 inner about 1/12 in. long, accrescent, loosely enveloping the achene to form orbicular wings; the 2 outer reflexed in fruit; stamens 8, included; styles 3, stigmas capitate, ovary 1-celled and 3-angled, ovule solitary, embryo slender and straight.

FRUIT. Achene stout, elliptic to ovoid, 3-angled, pointed at both ends, about 1/12–1/8 in. long, invested by the loose calyx, chestnut-brown, smooth, shiny, endosperm mealy.

LEAVES. Numerous, alternate, often fascicled, jointed to the ocreae, persistent, 1/3–1½ in. long, linear to spatulate, sessile, apex obtuse and sometimes revolute, fleshy; ocreae cylindric, scarious on margin, split on one side.

STEMS. Light to dark brown, erect to ascending, somewhat flexuous, jointed, ridged, and scaly.

RANGE. Sandy soils of the coastal plain. Texas, New Mexico, Oklahoma, Arkansas, and Louisiana; eastward to Florida, and northward to South Carolina.

REMARKS. The genus name, *Polygonella,* is a diminutive of *polygonum,* which is a word from the Greek meaning "many-knees" and refers to the swollen joints of some species. The meaning of the species name, *americana,* is obvious. A rather attractive species in bloom and worthy of cultivation in loose sandy soils.

SOUTHERN JOINT-WEED
Polygonella americana (F. & M.) Small

GOOSEFOOT FAMILY (*Chenopodiaceae*)

PICKLEWEED

Allenrolfea occidentalis (Watson) Ktze.

FIELD IDENTIFICATION. Erect shrub with succulent upper branches and woody toward the base, attaining a height of 2–5 ft. Almost leafless, or leaves reduced to small broad scales.

FLOWERS. June–August, spikes numerous, cylindric, ⅜–1 in. long, slightly thicker than the branches, sessile or nearly so, densely flowered; flowers small, slightly exserted, crowded, arranged spirally in threes in the axils of bracts; bracts subsessile, persistent, crowded, rhomboidal; petals absent; calyx of 4 (rarely more) sepals, imbricate, more or less united, keeled, concave, becoming spongy and covering the fruit; stamens 1–2, filaments slender, exserted from between the bracts; ovary superior, 1-celled, styles 2–3, distinct.

FRUIT. Achene naked, membranous, surrounded by the fleshy calyx, the 1 seed less than 1/25 in. long.

LEAVES. Sometimes almost lacking; leaves, when present, very small, slightly raised at the nodes, scalelike, alternate, broadly triangular, acute.

TWIGS. Much-branched, alternate, slightly and regularly constricted to appear jointlike, swollen, green and chorophyll-bearing, upper branches succulent, lower woody.

RANGE. An indicator of alkaline soil, in flats and valleys at altitudes to 5,000 ft. West Texas, New Mexico and Arizona; north through Colorado, Utah, and Nevada to eastern Washington and southern Idaho.

REMARKS, The genus name, *Allenrolfea*, honors Allen Rolfe, botanist at the Kew Gardens in England, and the species name, *occidentalis*, means "western," because of its distribution in the western United States. Also known by the vernacular names of Bush Pickleweed, Kern-greasewood, Iodine-bush, and Hierba del Burro. The crushed stems yield a dark, iodine-like stain. It is browsed by livestock only in times of great want. The abundant pollen is said to be the source of some hay fever.

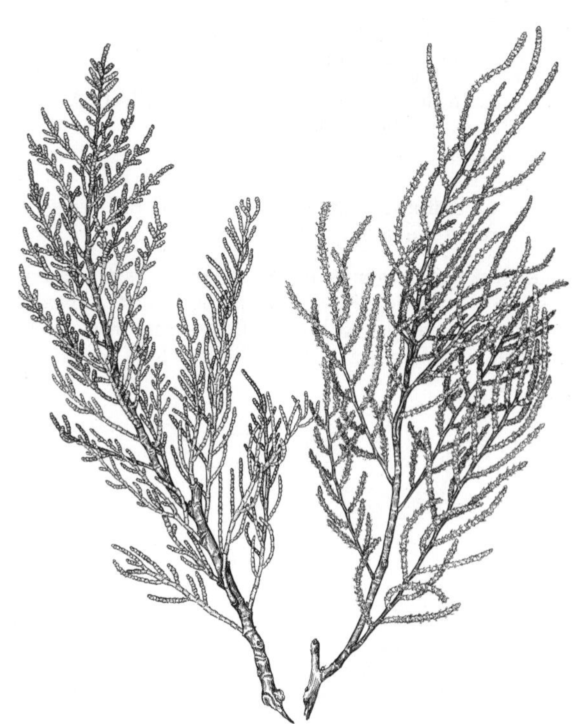

PICKLEWEED
Allenrolfea occidentalis (Watson) Ktze.

TUBERCLED SALTBUSH

Atriplex acanthocarpa (Torr.) Wats.

FIELD IDENTIFICATION. Western subshrub, herbaceous above, woody and branched from the base. Bracts of the fruit conspicuous with numerous flattened tubercles.

FLOWERS. Staminate and pistillate flowers on different plants; staminate in seminaked glomerules in elongate panicles, perianth 5-cleft, stamens 3–5; pistillate panicles axillary, few-flowered, more leafy than the staminate, perianth absent, stigmas 2.

FRUIT. Bracts subsessile or on peduncles 1/12–4/5 in. long, thick, united near the apex, rounded or broadly elliptic, 1/3–3/5 in. long or broad, apex beaklike; plane surfaces with numerous, flattened, often toothed appendages or tubercles, commonly longer than the body; seeds 1/25–1/12 in. long, brown, radicle superior.

LEAVES. Alternate, or opposite below, petiole winged, blade shape varying from lanceolate to oblong-elliptic or obovate, base cuneate, apex obtuse, margin sinuate-dentate or entire, blade length 3/4–2 in., width 3/8–1 in., thickish, surfaces white and densely scurfy.

TWIGS. Stout, angled to rounded, densely white-scurfy, glabrous later and old bark exfoliating in layers.

RANGE. In dry soil, to an altitude of 3,500 ft. Western Texas, New Mexico, and Arizona; south into Mexico in Chihuahua, Nuevo León, and San Luis Potosí. In Texas in the Franklin Mountains near El Paso. In Arizona near Safford, Graham County, and the type from the Burro Mountains in New Mexico.

REMARKS. The genus name, *Atriplex*, is the ancient classical name, and the species name, *acanthocarpa*, refers to the appendaged bracts of the fruit. It is of minor importance, being grazed to a slight extent by cattle.

TUBERCLED SALTBUSH
Atriplex acanthocarpa (Torr.) Wats.

FOUR-WING SALTBUSH

Atriplex canescens (Pursh) Nutt.

FIELD IDENTIFICATION. Shrub erect, evergreen, diffusely branched, variable in shape, to 8 ft tall, deep-rooted. Flowers borne in separate male and female spikes.

FLOWERS. In June–September, dioecious (more rarely monoecious); staminate heads densely spicate from terminal panicles which are leafy toward the base; perianth 3–5 cleft, stamens 3–5, inserted on base of perianth, anthers 2-celled, pollen abundant; pistillate spikes and panicles densely leafy-bracted, perianth absent, stigmas 2, base subtended by 2 connivent bractlets, enlarged in fruit.

FRUIT. Borne August–September, often an abundant crop every year, fruit varying in size and shape in different regions; bracts sessile or short-peduncled, 1/6–1/2 in. long (rarely more) developing 2 pairs of wings; margin of wings entire to laciniate; plane surfaces smooth or with small excrescences between the wings, veiny, base sometimes decurrent, apex bifid.

LEAVES. Numerous, evergreen, alternate, sessile or short-petioled, linear to elliptic or oblong to spatulate, apex usually obtuse, base narrowed, margin entire, blade length 3/8–2 in., width 1/8–1/2 in., 1-nerved, thick, surface densely gray-scurfy above and below.

TWIGS. Stout, terete, brittle, smooth, gray-scurfy, older bark gray and exfoliating in thin layers.

RANGE. Widely distributed; western Texas and Oklahoma; west to southern California, north to Canada, and south into Mexico in Coahuila and San Luis Potosí. Adaptable to many soil types, grassy uplands to sandy deserts, or salt or alkali flats.

PROPAGATION. The abundant seed is easily gathered by hand when ripe. No cleaning is required, the seed averages 22,500 seeds per lb, with about 85 per cent purity and about 50 per cent or more soundness. It can be stored for several years in a dry place without too much loss in viability. The seeds may be sown at almost any time of the year by broadcasting in sandy loam and covering about 1/8 in. with soil. Shading of the seedlings is not necessary. The seedlings are susceptible to damping-off during the first 2 weeks. Germination averages 30–60 per cent.

VARIATION. The plant is found widely in the western United States, and displays considerable variation in certain regions, particularly as regards the leaf shape and size, and fruiting bract shape and size. These minor variations have been recorded by Hall and Clements (1923, p. 343).

REMARKS. The genus name, *Atriplex*, is the ancient name, and the species name, *canescens* refers to the canescent or silvery scurf of the leaves. It is known also by the vernacular names of Wing-scale, Shad-scale, Orache, Sage-brush, Cenizo, Chamiso, Chamiza, and Costillas de Vaca. The plant is considered a valu-

FOUR-WING SALTBUSH
Atriplex canescens (Pursh) Nutt.

able, palatable, and nutritious feed for cattle, sheep, and goats. In a few isolated instances it has been known to kill goats and cause scours in cattle by concentrated feeding. The fruit is eaten by scaled quail, porcupine, rock squirrel, gray spotted squirrel, and jack rabbit. The plant is deep-rooted, is useful for erosion control, and is very drought-resistant. The pollen has been known to cause hay fever, and is used in immunization extracts. The Southwest Indians were known to grind the seeds to use as a baking powder in breadmaking.

SHADSCALE SALTBUSH

Atriplex confertifolia (Torr. & Frem.) Wats.

FIELD IDENTIFICATION. Western clump-forming shrub 1/2–3 ft high, outline normally rounded, branches numerous, erect, rigid, spiny, woody.

FLOWERS. Dioecious, staminate spikes with leafy bracts subtending the small, dense glomerules, perianth 5-parted; pistillate flowers solitary or a few together in the upper leaf axils, perianth absent.

FRUIT. Achenes maturing in September; bracts sessile, 1/5–3/5 in. long, 1/5–2/5 in. broad, the pairs convex, facing and united over the achenes, flat and free above, orbicular to broadly elliptic, surfaces smooth, margins mostly entire except in some variations, achenes average about 15,000 per lb.

LEAVES. Rather crowded, but later deciduous to expose spiny twigs, alternate, shape variable from orbicular to ovate, obovate or elliptic, base rounded or cuneate, apex obtuse, margin entire, blade length 3/8–3/4 in., width 1/25–1/2 in., firm, surfaces grayish green scurfy, obscurely 1–3 nerved, short-petioled.

TWIGS. Stout, young light brown and scurfy, later light to dark gray and glabrous, ending in spines, bark exfoliating on old stems.

RANGE. On hard, stony, alkaline soils, a characteristic dominant of eroded desert areas. At altitudes of 4,000–6,000 ft. From extreme west Texas and New Mexico to southern California, north to Oregon and Wyoming.

REMARKS. *Atriplex* is the ancient Latin name, and the species name, *confertifolia,* refers to the crowded leaves. Also known under the vernacular names of Spiny Saltbush, Round-leaf Saltbush, Hop-sage, and Sheep Fat. It is valuable for winter sheep browse, but less so for cattle because of the spines. It is also browsed by mule deer, and is resistant to overgrazing. The seeds are known to be eaten by at least 6 species of birds. It is also said to be a cause of some hay fever.

SHADSCALE SALTBUSH
Atriplex confertifolia (Torr. & Frem.) Wats.

Mat Saltbush

Atriplex corrugata Wats.

FIELD IDENTIFICATION. Shrub forming dense matlike mounds 4–12 in. high. Stems decumbent or erect, closely branched, densely furfuraceous, woody at base.

FLOWERS. Usually dioecious, occasionally monoecious, stems bearing the inflorescence mostly erect and slender; staminate spikes with sessile, nearly naked glomerules, perianth 5-cleft, stamens as many as the perianth segments; pistillate in elongate terminal spikes, pistil with 2 appressed bracts.

FRUIT. Utricle with the enveloping fruiting bracts ⅙–¼ in. long, ⅛–⅙ in. wide, sessile or subsessile, united nearly to the summit, obovate or narrowly fan-shaped, widest above the middle, surfaces flattened, thick, tubercled or appendaged dorsally, reddish brown, apical margins entire or denticulate.

LEAVES. Crowded, opposite or alternate, linear to oblong, or linear-spatulate, length ¼–1 in., apex obtuse to rounded, base cuneate or gradually narrowed, margin entire, surfaces densely white-mealy or furfuraceous.

TWIGS. Young ones green to gray-mealy or furfuraceous, older ones light to dark gray, woody, sometimes with shreddy bark.

RANGE. Saline or alkaline soil on the plains, at altitudes of 4,000–6,500 ft. New Mexico, Arizona, Colorado, and Utah.

Big Saltbush
Atriplex lentiformis (Torr.) Wats.

Mat Saltbush
Atriplex corrugata Wats.

REMARKS. The genus name, *Atriplex*, is the ancient classical name, and the species name, *corrugata*, refers to the wrinkled or wartlike appendages of the bracts. The plant is also known under the vernacular name of Poison Clover. It is occasionally grazed by sheep, but some ranchers consider it to be poisonous.

Big Saltbush

Atriplex lentiformis (Torr.) Wats.

FIELD IDENTIFICATION. An erect rapid-growing shrub forming a rounded or globular outline, and ½–5 ft in height (rarely, to 10 ft). The branches numerous, slender and rather flexuous.

FLOWERS. Maturing June–August, dioecious or more rarely monoecious, borne in paniculate spikes which are rather crowded and sometimes pendent; staminate spikes 1–3 in., perianth 3–5-cleft; stamens as many as the perianth lobes; pistillate flower with no perianth, pistil solitary, stigmas 2, ovary 1-celled, each subtended by 2 compressed bracts.

FRUIT. Maturing September–October, utricle enclosed in the 2 bracts. Bracts sessile, flattened, united at the base or half way to the apex, oval to orbicular or broadly elliptic, 1/12–1/4 in. long or broad, margins entire or crenulate, surfaces smooth; seed about 1/25–1/15 in. long, brown, about 412,000 per lb.

LEAVES. Numerous, alternate, persistent in moist situations, early deciduous in desert environments, blade ovate to rhombic or ovate to oblong, apex obtuse to rounded or short-mucronate, base cuneate to truncate, thickish, length 3/8–2 in., width 1/4–1 1/2 in., margin entire or subhastate, surfaces gray to silvery with a fine scurf, 1-nerved; sessile or short-petioled.

TWIGS. Wide-spreading, terete or angled, young ones gray-scurfy, later glabrous, white to straw-colored, bark rough and gray on old woody trunks.

RANGE. In strongly alkaline dry soils of flats or depressions in the sun, at altitudes of 300–4,500 ft. New Mexico, Arizona, and southern California; north to Colorado, Utah, and Nevada, and south into Sonora, Mexico.

VARIETIES. The following varieties of Big Saltbush have been segregated by Hall and Clements (1923, pp. 334–336):

Brewer's Big Saltbush, *A. lentiformis* var. *breweri* (S. Wats.) McMinn, has twigs not sharply angled; bracts entire or nearly so, 1/8–1/3 in. broad.

Torrey's Big Saltbush, *A. lentiformis* var. *torreyi* (S. Wats.) McMinn, has twigs sharply angled; mature bracts orbicular; leaves deltoid to ovate, rarely oblong.

Griffith's Big Saltbush, *A. lentiformis* var. *griffithsii* (Standley) Benson, has twigs sharply angled, mature bracts broader than long; leaves ovate to narrowly oblong.

REMARKS. The genus name, *Atriplex*, is the ancient classical name, and the species name, *lentiformis*, refers to the lenslike fruiting bracts. Also known under the names of Lens-scale Saltbush and Quail-bush. It is suspected to be a hay fever plant. Growing in dense stands it affords excellent cover for wildlife, and is browsed by mule deer. During winter or early spring it affords some browse for livestock. The plant is used as a hedge in coastal California. The Cahuilla Indians made a flour from the seeds.

Matamoros Saltbush

Atriplex matamorensis Nelson

FIELD IDENTIFICATION. Erect, or partially decumbent, to 2 ft. Branched at or above the base and woody, the upper stems herbaceous.

FLOWERS. Staminate and pistillate flowers on different plants, axillary, glomerules solitary or several together, spikes rigid and leafy; perianth 3–5-parted in staminate flowers, bractless, stamens 2–5, inserted on the base of the perianth, anthers 2-celled; pistillate flower subtended by 2 bracts becoming larger and enclosing the fruit, stigmas 2.

FRUIT. Utricle small, 1-seeded; fruiting bracts sessile, flattened, united to the middle or higher, orbicular, apex obtuse or short-mucronate, margin sharply toothed, faces 1–3-nerved and smooth, length and width 1/12–1/8 in., seed about 1/25 in. long, yellow, pericarp sheathlike, radicle superior, seed enclosed within the bracts.

LEAVES. Mostly opposite, sessile at base, lanceolate to oblong, base broadened, apex obtuse to acute, margin entire, length 1/12–1/5 in., width 1/25–1/8 in., thick, 1-nerved, surfaces densely scurfy-hairy.

TWIGS. Often crooked, slender, terete, young sparsely scurfy-hairy, older gray and exfoliating in thin scales.

RANGE. Dry, saline, or sandy soil, mostly near the coast. On the Texas coast at Sabine Pass, Galveston Island, Matagorda, Rockport, Corpus Christi, and Boca Chica. In Mexico in Tamaulipas.

Matamoros Saltbush
Atriplex matamorensis Nelson

REMARKS. The genus name, *Atriplex*, is the ancient classical name, and the species name, *matamorensis*, is for the Mexican city of Matamoros. Of no value, except of minor importance in erosion control.

Nuttall Saltbush

Atriplex nuttallii Watson

FIELD IDENTIFICATION. Western, much-branched perennial, the stems erect from a decumbent woody base.

FLOWERS. Dioecious or rarely monoecious; staminate in terminal glomerate spikes which are leafy below, staminate perianth 3–5-parted; pistillate spikes terminal, dense and leafy below, perianth absent, each flower subtended by 2 bracts which are enlarged and persistent later.

FRUIT. Bracts sessile or short-stalked, ⅙–¼ in. long, thickened, generally ovate to rounded, terminating in a lanceolate or cuneate-oblong beak more or less toothed at summit, margins often appendaged or tuberculate, faces smooth or tuberculate; seed 1/25–1/12 in. long, radicle superior.

LEAVES. Alternate, or lower sometimes opposite, blade length 1–2 in., width 3/16–⅜ in., oblong–linear to ovate or obovate, base narrowed or cuneate, apex obtuse or rounded, margin entire, thick and firm, surfaces grayish green with a dense mealy scurf.

Nuttall Saltbush
Atriplex nuttallii Watson

Obovate-leaf Saltbush
Atriplex obovata Moquin

TWIGS. Slender, herbaceous above, woody below, scurfy grayish green or whitened at first, later glabrous to dark brown or gray and roughened below.

RANGE. Usually in alkaline clay soils of northwestern prairie regions. New Mexico, west to California, and north to Canada.

VARIATIONS. The plant is very variable over its great range and many forms and subspecies have been suggested by various botanists. However, most of these are connected by numerous intergradations. The various segregates are listed by Hall and Clements (1923, p. 328).

REMARKS. The genus name, *Atriplex*, is the ancient Latin name, and the species name, *nuttallii*, honors the botanist Thomas Nuttall. The plant is considered to be a very important browse, being good for cattle and sheep on the Western ranges and forming a large part of winter forage. It easily reseeds and maintains itself. Indians formerly used the parched seed to make a pinole flour.

Obovate-leaf Saltbush

Atriplex obovata Moquin

FIELD IDENTIFICATION. Western subshrub, somewhat woody at the base, hardly over 2 ft.

FLOWERS. Dioecious, staminate borne in small, sessile heads in spikes forming part of terminal, nearly leafless panicles; perianth 5-cleft, stamens 3–5; pistillate flowers in leafy elongate terminal spikes; perianth absent, stigmas 2.

FRUIT. Bracts sessile or short-pedunculate, united at the truncate or subcordate base or up to the middle, obovate to orbicular, length 1/6–1/5 in., width 1/5–1/4 in., margins sharply toothed, plane surface smooth or with a few small tubercles; seed 1/12–1/8 in. long, brown, radicle superior.

LEAVES. Alternate or opposite, short-petioled, obovate or broadly elliptic, base cuneate, apex obtuse or retuse, margin entire or somewhat undulate, blade length 3/8–1½ in., width 3/16–3/4 in., thick and firm, surfaces white with a smooth permanent scurf.

TWIGS. Rigid, erect, gray-scurfy, older with gray exfoliating strips of bark.

RANGE. Dry soil, at altitudes to 5,500 ft. In western Texas, New Mexico, Arizona, and Colorado. In Mexico in Chihuahua and Zacatecas.

REMARKS. The genus name, *Atriplex*, is the old classical name, and the species name, *obovata*, refers to the obovate leaf shape. The plant is of no great commercial value, forming only a minor part of desert forage for cattle.

Australian Saltbush

Atriplex semibaccata R. Br.

FIELD IDENTIFICATION. A diffuse, much-branched perennial to 3 ft. The stems prostrate, or nearly so, and somewhat woody near the base.

FLOWERS. Monoecious, staminate in terminal glomerules with bractlike leaflets, perianth 3–5-cleft, stamens as many as the perianth lobes; pistillate solitary or a few together in the upper leaf axils; pistil naked, subtended by 2 appressed bractlets, perianth absent, ovary 1-locular.

FRUIT. An achene, the fruiting bracts sessile or short-stalked, 1/6–1/5 in. long, red at maturity, fleshy-thickened and convex at first, flattened and dry later, united only below the middle, rhombic to deltoid-cuneate, margins entire or coarsely few-toothed; seed vertical, about 1/12 in. long, brown or black, margin grooved.

LEAVES. Numerous, alternate, blade oblong to elliptic or slightly spatulate, length 5/8–2 in., width 1/4–1/2 in., apex obtuse or acute, base cuneate or gradually narrowed, margin entire or shallowly dentate, upper surface glabrate, lower surface white-scurfy, short-petioled.

TWIGS. Terete or nearly so, wiry, at first mealy-scurfy, soon glabrate and green to straw-colored, older stems gray and rough near the base.

RANGE. Escaping cultivation along irrigation canals, sandy fields, and flats. Originally from Australia and Tasmania. Naturalized extreme west Texas, New Mexico, Arizona, California, and Utah.

REMARKS. The genus name, *Atriplex*, is the ancient classical name. The species name, *semibaccata*, refers to the baccate-like, or thickened, fruiting bracts. The plant was brought into the United States in 1888 for use as a forage and hay. Now it is sometimes used for sheep forage, but it is also browsed by horses, cattle, and hogs. The seeds are eaten by chickens. It is reported that heavy yields are made under the right conditions. It is said to be rich in protein. Sometimes it is used as a soil-binder.

Australian Saltbush
Atriplex semibaccata R. Br.

Common Winter Fat

Eurotia lanata (Pursh) Moq.

FIELD IDENTIFICATION. Plant woody toward the base, 8 in.–3 ft high, ascending or spreading, many slender branches from the base. Root system deep and spreading. All parts with simple or star-shaped, conspicuous, long hairs.

FLOWERS. April–August, staminate and pistillate flowers on same plant, less often on different plants, axillary, or in spikelike clusters on upper part of the stems, small and greenish; staminate flowers without petals or bracts; calyx 4-lobed, lobes hairy, acute; stamens 4, opposite the lobes, exserted; pistillate flowers branched and often solitary, sepals absent, each flower enclosed by 2 bracts which are partly united and bear 2 short horns at the apex; styles 2, reflexed, hairy.

FRUIT. Utricle ripening August–October, enclosed by 2 bracts, bracts lanceolate, 1/5–1/4 in. long, 1/8–1/5 in. wide, 2-horned above, covered with 4 dense tufts of long white hairs to 3/16 in.; seed solitary, vertical, endosperm mealy, embryo coiled.

Common Winter Fat
Eurotia lanata (Pursh) Moq.

LEAVES. Alternate, or somewhat fascicled, sessile or short-petioled, linear to narrowly lanceolate, apex obtuse, base narrowed, margins entire and revolute, blade length 1/2–1 3/4 in., width 1/12–1/8 in., surfaces densely hairy with simple or stellate hairs, midrib prominent, lateral veins few. Smaller leaves less than 1/2 in. long, sometimes in axils of larger ones.

TWIGS. Stout, ascending or spreading, very leafy, densely hairy with simple or stellate hairs which are white or reddish brown later, subspinescent with age.

RANGE. Dry subalkaline soils of mesas and plains at altitudes of 2,000–8,000 ft. From Texas and New Mexico northward to Canada.

PROPAGATION. Plants one year old bear seed, and good crops occur almost every year. The seed may be gathered by hand in late summer. It may be cleaned, or sown immediately as is, but drying is necessary if it is to be stored any length of time. The cleaned seed averages about 90,000 seeds per lb, averaging about 50 per cent purity and 95 per cent soundness. If stored in sealed containers at room temperature it loses 10–50 per cent in viability the first year, and more the second. The average germinating capacity of fresh seeds is 58 per cent. Germination occurs in 2–5 days after sowing in the fall, and the seed should be covered about 1/8 in. with sifted sand. They can be sown directly in the field using 3–4 lb of seed per acre.

VARIETY. Winter Fat has a variety known as *Eurotia lanata* var. *subspinosa* (Rydb.) Kearney & Peebles, which is more woody, branches more spreading, is somewhat spinescent, and has hairs with few or no elongate rays. However, intermediate forms occur between the variety and species.

REMARKS. The genus name, *Eurotia*, is from the Greek word *euros* ("mould"), referring to the hairy covering of the plant. The species name, *lanata*, also is for the woolly interwoven hairs. Local names are Roemeria, Lamb's Tail, Sweet-sage, White-sage, and Feather-sage. Besides being of great value as a winter livestock forage, it provides food for elk, California mule deer, and rabbit. It is also reported as used by the Indians, in the form of a powder from the roots, for burns, and also a decoction of the leaves for treating fever. It also has some value in erosion control, is drought-resistant, and has been cultivated since 1895.

Greenmolly Summer-cypress

Kochia americana S. Wats.

FIELD IDENTIFICATION. Perhaps best considered as a perennial herb, only lignescent toward the base. Attaining a height of 1/2–2 ft with stems numerous, simple, slender and villous, but becoming glabrous with age.

FLOWERS. Very small, solitary or in twos or threes

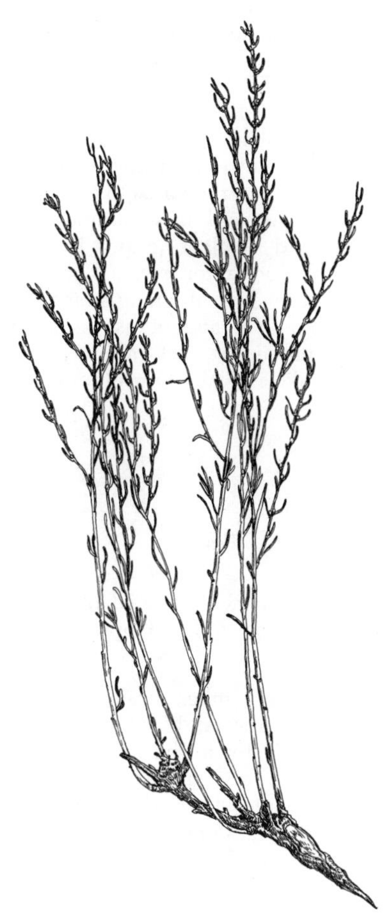

GREENMOLLY SUMMER-CYPRESS
Kochia americana S. Wats.

in the leaf axils, mostly perfect, but sometimes with a few pistillate ones, clothed with persistent tomentum, calyx with the 5 lobes incurved; stamens 3–5, exserted, sometimes with abortive ones and then the flower pistillate; ovary very pubescent, nearly equaling or equaling the calyx; styles 2 and filiform.

FRUIT. Utricle depressed-globose, about 1/12 in. across, the fruiting calyx forming 5 wedge-shaped, or cuneate, scarious-margined, horizontal wings; pericarp membranous and free from the horizontal seed; embryo annular, surrounding scanty endosperm.

LEAVES. Alternate or opposite, erect or ascending, sessile, narrowly linear and subterete, somewhat fleshy, 1/5–1¼ in. long, about 1/25 in. wide, apex acute, surfaces silky-pubescent to glabrate.

RANGE. Usually on alkaline soils at altitudes of 4,000–5,500 ft. New Mexico, Arizona, and California; north through Colorado and Nevada to Wyoming, southern Idaho, and eastern Oregon.

VARIETY. The American species also has a variety which has been given the name of Gray Summer-cypress, *K. americana* var. *vestita* (S. Wats.) Rydb., and is characterized by being more permanently villous-tomentose with an oblong ovary nearly equaling the calyx.

REMARKS. The genus name, *Kochia,* honors W. D. J. Koch (1771–1849), professor of botany at Erlangen, who wrote a flora of Germany and Switzerland. The species name, *americana,* refers to the native home of the species. The Garden Summer-cypress. *K. scoparia,* is a European species, often cultivated in gardens.

BLACK GREASEWOOD

Sarcobatus vermiculatus (Hook.) Torr.

FIELD IDENTIFICATION. Shrub 2–10 ft, with the branches divaricate, rigid, somewhat spinescent.

FLOWERS. Small, borne in separate staminate and pistillate inflorescences on the same plant, or on different plants; staminate ones in terminal ament-like spikes ¼–1 in. long, 1/12–1/6 in. in diameter, cylindric,

BLACK GREASEWOOD
Sarcobatus vermiculatus (Hook.) Torr.

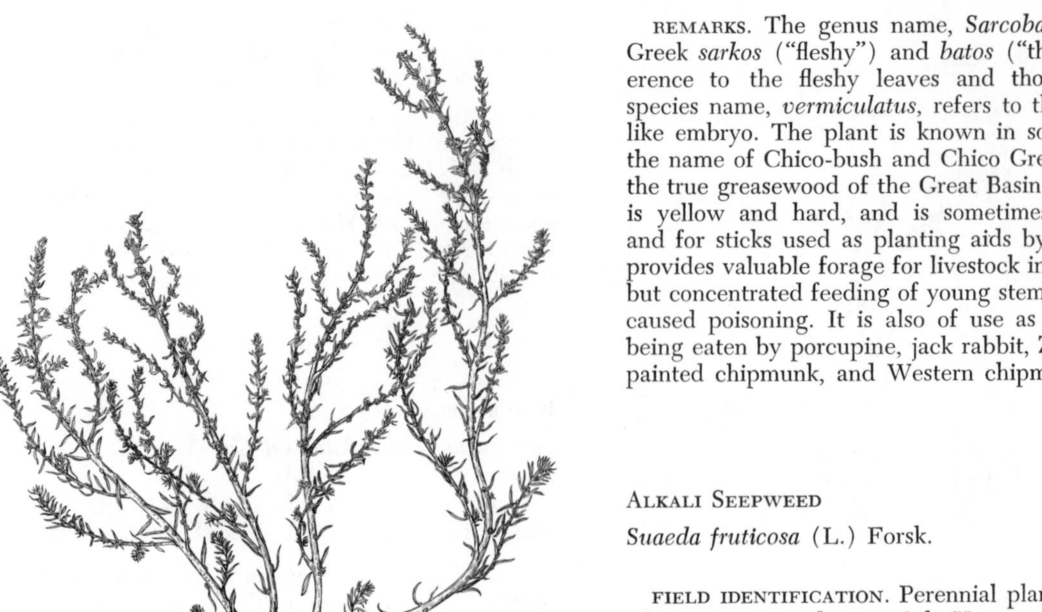

ALKALI SEEPWEED
Suaeda fruticosa (L.) Forsk.

short-peduncled or sessile; stamens 2–5, filaments short with fleshy anthers grouped together under scales which are spirally arranged, stipitate, peltate or rhombic acute; pistillate flowers in the upper leaf axils, solitary or a few together, sessile or nearly so; calyx compressed, saccate, oblong to ovoid; calyx margined by a narrow wing which is circular, horizontal, membranous, veiny, glabrous; stigmas 2, subulate, papillate, exserted; ovary superior, 1-celled.

FRUIT. Maturing September–October, wing ¼–½ in. broad, green to tan or reddish, achene ⅙–⅕ in. long, vertical, testa translucent, double.

LEAVES. Simple, alternate, deciduous, blade length ½–1½ in., width about 1/12 in., sessile, linear to filiform, flat above, rounded beneath, margin entire, apex obtuse or subacute, base narrowed, succulent, at first puberulent, glabrous later.

TWIGS. Leafy, spreading, dense, rigid, slightly angled, whitish to tan, some leafless and spinescent terminally.

RANGE. On alkaline or saline soils of dry plains and slopes of grasslands. Northwestern Texas and New Mexico; west to California and north to Canada.

VARIETY. Bailey Greasewood, S. *vermiculatus* var. *baileyi* (Torr.) Coville & Jepson, is a smaller, more spiny plant with gray bark; hairy leaves ⅓–⅗ in. long; staminate inflorescences ¼–⅜ in. long.

REMARKS. The genus name, *Sarcobatus*, is from the Greek *sarkos* ("fleshy") and *batos* ("thorn"), with reference to the fleshy leaves and thorny stems. The species name, *vermiculatus*, refers to the spiral, worm-like embryo. The plant is known in some areas under the name of Chico-bush and Chico Greasewood, and is the true greasewood of the Great Basin area. The wood is yellow and hard, and is sometimes used for fuel, and for sticks used as planting aids by the Indians. It provides valuable forage for livestock in fall and winter, but concentrated feeding of young stems and leaves has caused poisoning. It is also of use as a wildlife food, being eaten by porcupine, jack rabbit, Zuni prairie dog, painted chipmunk, and Western chipmunk.

ALKALI SEEPWEED

Suaeda fruticosa (L.) Forsk.

FIELD IDENTIFICATION. Perennial plant much-branched, erect or ascending, to 3 ft. Upper stems herbaceous, woody below.

FLOWERS. July–September, branches of inflorescence ascending or erect, flowers less than ⅛ in. broad, solitary, or 2–5 in a cluster in the upper leaf axils, sessile, bisexual ones mixed with unisexuals; perianth 5-lobed, lobes obtuse or acute, back rounded, one lobe often larger than the other; stamens 5, filaments short; ovary superior, 1-celled and 1-ovuled, stigmas 2–3; bracts 2, minute.

FRUIT. Utricle dry, ovoid to ellipsoid, enclosed in the fleshy perianth; seed 1/25–1/12 in. long, shiny black, embryo spirally coiled, cotyledons long and narrow.

LEAVES. Crowded, alternate, fleshy, narrowly linear to subterete, obtuse to acute, ⅜–1¼ in. long, glabrous and glaucous, grayish green, sessile at base.

TWIGS. Ascending or erect, green and glaucous to tan or whitened.

RANGE. Seacoasts and alkaline flats inland, at altitudes to 5,500 ft. In western Texas to southern California; north to Canada, and south into Mexico (Coahuila). Also in Europe, Asia, and Africa. Type locality on the coast of France. Introduced into North America and closely related to S. *torreyana* (S. Wats.) Standl., to which it is referred by some authors.

REMARKS. The genus name, *Suaeda*, is from the Arabian *Suidah*, name of an Asiatic species. The species name, *fruticosa*, means "shrubby." Also known under the name of Alkali Blight. The heathlike appearance gives it some value for planting in gardens, especially in saline or sandy soils. It has been cultivated since 1789. About 40 species of *Suaeda* are known, mostly from the seacoasts of all continents, and inland in saline soil. The plants are sometimes burned to secure ashes for soapmaking. The juices were used by the Coahuila Indians to obtain a black dye for baskets.

Desert Seepweed

Suaeda suffrutescens S. Wats.

FIELD IDENTIFICATION. Plant suffrutescent, fleshy above, woody toward the base, branches narrow, slender, diffuse, ascending at a narrow angle, tomentulose, attaining a height of 3 ft or more.

FLOWERS. Borne in the axils of much-reduced leaves, solitary or clustered, unisexual or bisexual; small, perfect, greenish, sessile, petals absent, perianth 5-lobed, lobes short and fleshy; stamens 3–5; ovary superior, 1-celled, stigmas 2–3.

FRUIT. Utricle enclosed in the 5 fleshy lobes of the perianth, seed black, less than 1/25 in. long.

LEAVES. Simple, alternate, evergreen, fleshy, 3/8–5/8 in. long, subterete, linear or narrowly oblong, apex obtuse to acute, base narrowed, surface pubescent to tomentulose, those of flowering branches much smaller.

RANGE. On saline or alkaline soils to an altitude of 5,000 ft. Western Texas, New Mexico, Arizona, and California; south into Chihuahua, Mexico.

REMARKS. The genus name, *Suaeda* (Suidah), is the Arabian name for a related species, and the species name, *suffrutescens*, refers to its shrubby habit. It is known in New Mexico as Yerba de Burro. It is browsed by cattle only in times of starvation. The Indians cooked the young leaves for greens, and made pinole flour from the seeds. The dried leaves were also applied to sores. The Desert Seepweed is very similar to the Alkali Seepweed, *S. fruticosa* (L.) Forsk., but has smaller, pubescent leaves and stems.

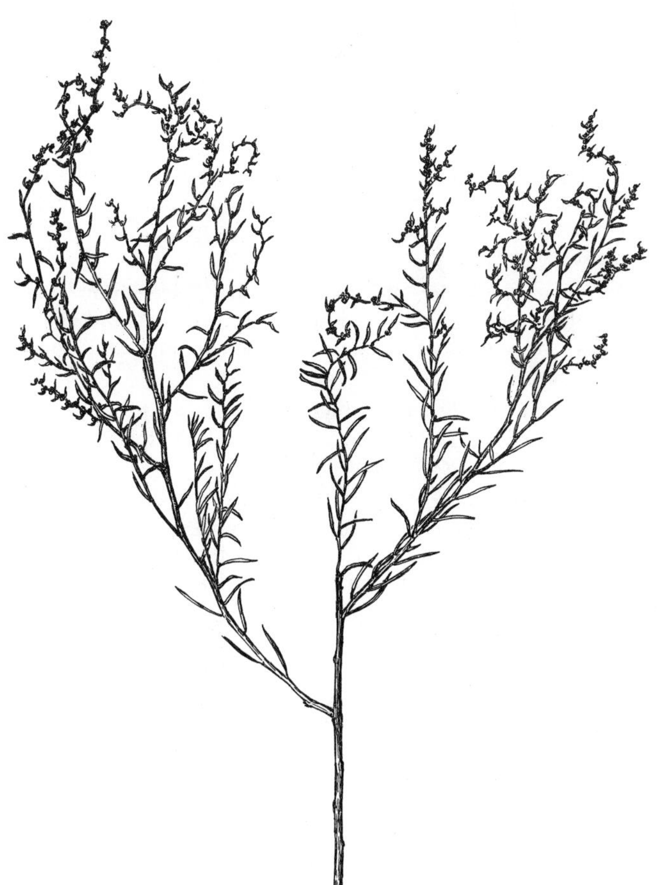

Torrey Seepweed
Suaeda torreyana (S. Wats.) Standl.

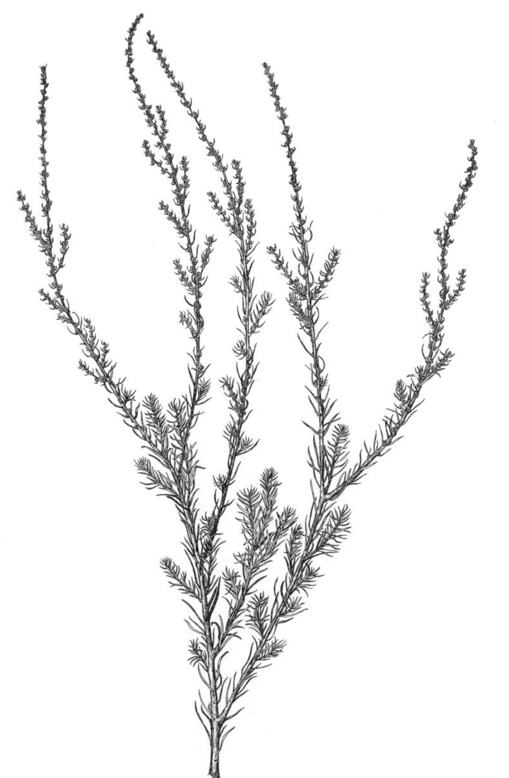

Desert Seepweed
Suaeda suffrutescens S. Wats.

Torrey Seepweed

Suaeda torreyana (S. Wats.) Standl.

FIELD IDENTIFICATION. Western perennial plant with a woody base. Attaining a height of 1–3 ft. rarely to 7 ft, with erect or spreading, green to light brown stems.

FLOWERS. May–September, branches of inflorescence slender, lax and open; flower axillary, sessile, less than 1/8 in. broad, clustered or solitary in the upper axils (1–6), perfect or 1–sexed; petals absent; calyx 5-lobed, lobes nearly equal and parted to below the middle; stamens 5, filaments short; styles usually 2; ovary 1-celled, superior.

FRUIT. Utricle enclosed in the fleshy calyx, 1/25–1/15 in. broad; embryo flattened and spirally coiled, endosperm scanty.

LEAVES. Simple, alternate, persistent, 3/8–1 1/3 in. long (those of inflorescence much reduced in size), fleshy, sessile, green, linear to narrowly lanceolate, subterete to strongly flattened, margin entire, apex acute to acuminate, surface glabrous.

TWIGS. Slender, flexuous, elongate, ascending and wide-spreading, tan to straw-colored, smooth and slick, bark on old wood at base roughened and dark gray to black.

RANGE. Saline or alkaline valleys, at altitudes of 4,000–7,000 ft. Texas, New Mexico, and California; north to Colorado, Oregon, and Alberta.

VARIETY. A variety of Torrey Seepweed has been named the Branched Torrey Seepweed, S. *torreyana* var. *ramosissima* (Standl.) Munz. It is very branchy and finely pubescent on stem and leaves. It is found from Arizona to Baja California.

REMARKS. The genus name, *Suaeda,* is the ancient Arabic name of an Asiatic species. The species name, *torreyana,* honors the early botanist, John Torrey. The plant is closely related to S. *fruticosa* (L.) Forsk., but Torrey Seepweed is not glaucous, the leaves are strongly flattened and very small in the inflorescence.

Vernacular names for Torrey Seepweed are Desert Blite and Sea Blite. The young plants are used as greens by the Southwestern Indians, and the dried leaves applied to sores.

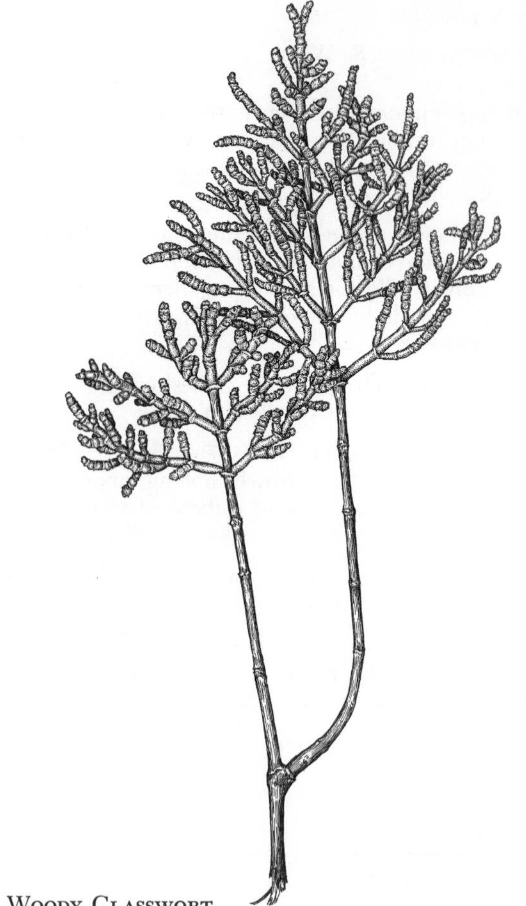

WOODY GLASSWORT
Salicornia virginica L.

WOODY GLASSWORT

Salicornia virginica L.

FIELD IDENTIFICATION. Often forming a colony with matlike, trailing or ascending subwoody perennial stems, rooting at the nodes, and rhizomes creeping. Often found on sandy beaches.

FLOWERS. August–October, borne in terminal cylindric spikes 5/8–2 1/2 in. long, about 1/12–1/5 in. thick, with age more loose, the scales becoming horizontally divergent on the margin; joints of the spike wider than high; flowers 3 together in cavities of the upper joints, lateral flowers separated by the central one, central flower cuneate-obovate, apex truncate, almost equal in height, perfect; stamens 1 or 2, basally united, exserted, filaments filiform, anthers exserted and opening lengthwise; ovary ovoid, laterally compressed; styles 2, short; calyx obpyramidal, opening by a terminal slit, fleshy at maturity.

FRUIT. Enclosed in the spongy calyx, sunken in the rachis, pericarp thin and free; seed minute, 1/50–1/25 in. long, vertical, endosperm none.

LEAVES. Reduced to opposite minute scales, upper scales forming a terminal spike.

TWIGS. Often depressed or spreading, green, gray, or brownish; flowering stems solitary, opposite or tufted, erect or ascending, 4–12 in. long.

RANGE. Sandy beaches or saline marshes, Texas and Louisiana; eastward to Florida and northward to Virginia and Massachusetts. Also from California to Alaska. West Indies, Europe, and South Africa.

REMARKS. The genus name, *Salicornia,* is from *sali* ("salt") and *cornu* ("horn"), referring to the hornlike branches. The species name, *virginica,* is for the state of Virginia. It is also known under the name of Sea Sapphire, Lead-grass, and Perennial Saltwort.

AMARANTH FAMILY (*Amaranthaceae*)

THIN-BRUSH

Dicraurus leptocladus Hook.

FIELD IDENTIFICATION. Western, white-pubescent, slender shrub to 3 ft, usually with many stems from the base.

FLOWERS. Very small, disposed in terminal panicles 1–5 in.; flowers dioecious, bracted, densely woolly; bracts oblong to rounded, thin, scarious, translucent, shiny; calyx 5-parted, segments linear and obtuse; stamens 5, hypogynous, filaments united at base, usually only 2 anther-bearing; stigmas 2, sessile, subulate, recurved.

FRUIT. Utricle broadly oval, indehiscent.

LEAVES. Simple, alternate, a few sometimes opposite, scattered, 1/4–1 in. long, 1/8–1/3 in. wide, ovate to lanceolate, apex acute, base with petiole 1/25–1/12 in. or leaf gradually sessile, margin entire, dull yellowish green above and beneath with densely appressed sericeous hairs, veins obscure.

TWIGS. Very slender, green and densely fine-hairy, when older gray to brown and becoming glabrous.

RANGE. Rocky slopes in the sun. Western Texas, and Northern Mexico in Chihuahua to San Luis Potosí.

REMARKS. The genus name, *Dicraurus,* is from *di* ("two") and *oura* ("tail"), for the 2-forked panicles. The species name, *leptocladus,* refers to the slender stems.

THIN-BRUSH
Dicraurus leptocladus Hook.

PALMER AMARANTH

Iresine palmeri (S. Wats.) Standl.

FIELD IDENTIFICATION. Plant slender, climbing and twining or scandent to decumbent, attaining a length of 15 ft. Stem angled and mostly herbaceous, but becoming brittle and woody near the base.

FLOWERS. Borne in short axillary and terminal racemes 1–3 in., branches pubescent to glabrous later; bracts glabrous, scarious, rounded, concave, subcoriaceous; flowers globose, small, dioecious, basal hairs very copious; perianth segments somewhat villous, ovate; filaments very short; ovary glabrous; stigmas with 2–3 subulate lobes, stout, nearly sessile.

Palmer Amaranth
Iresine palmeri (S. Wats.) Standl.

FRUIT. Subglobose, 1-seeded, cotyledons broad, the long woolly hairs from the calyx very conspicuous in fruit.

LEAVES. Opposite, blade ovate to oblong or lanceolate, apex acuminate to acute (rarely obtuse), base cuneate or gradually narrowed, margin entire, blades ¾–3 in. long, sparingly pubescent to glabrous later; lateral veins obscure; petioles ⅛–⅜ in.

STEMS. Slender, weak, often twining, young ones sparingly pubescent with rufous-brown hairs, later becoming glabrous and grayish brown.

RANGE. Collected by the author in Cameron County, Texas, in a palm grove *(Sabal texana)* on Rabb Ranch near Brownsville. In Mexico from Nuevo León to Veracruz.

REMARKS. The genus name, *Iresine,* is from the Greek word *eiresione* ("wool-bearing"), in reference to a wreath, or here applying to the long woolly hairs of the calyx. The species name, *palmeri,* is in honor of E. J. Palmer, American botanist.

Upright Tidestromia

Tidestromia suffruticosa (Torr.) Standl.

FIELD IDENTIFICATION. Erect, diffusely branched, suffruticose plant 4–12 in. Herbaceous above and woody toward the thickened base.

FLOWERS. Clustered in leaf axils, solitary or a few together, very small, yellowish white, perfect; bracts 3, shorter than calyx, concave, thin; calyx persistent, 1/12–1/10 in.; sepals 5, equal, erect, rigid, membranous, oblong to lanceolate, densely stellate-hairy; petals none; stamens 5, hypogynous, filaments united at base, anthers large and oblong, staminodia very short; ovary subglobose, style short, ovule solitary.

FRUIT. Utricle ovate to subglobose, indehiscent, seed smooth and enclosed.

LEAVES. Opposite or some alternate, blades 3/16–½ in., oval to ovate, apex obtuse, base rounded or truncate, margin entire or slightly undulate, both surfaces densely gray or brownish tomentose with branched hairs, veins obscure except the thickened main vein below; petioles 1/16–3/16 in.

STEMS. Slender, erect, ascending, much-forked, densely stellate-hairy, lower stems less so and light to dark brown, base woody.

RANGE. In Texas from the Pecos and Limpia rivers to the Rio Grande. In dry hills of the larrea belt. In New Mexico near Bishop's Cap in the Tortugas Mountains.

REMARKS. The genus name, *Tidestromia,* honors Ivar Tidestrom, Swedish-born American botanist, senior author of *A Flora of Arizona and New Mexico* (Tidestrom and Kittell, 1941). The species name, *suffruticosa,* refers to the suffruticose or semiwoody character of the plant.

Upright Tidestromia
Tidestromia suffruticosa (Torr.) Standl.

FOUR O'CLOCK FAMILY (*Nyctaginaceae*)

BERLANDIER ACLEISANTHES
Acleisanthes berlandieri Gray

BERLANDIER ACLEISANTHES

Acleisanthes berlandieri Gray

FIELD IDENTIFICATION. Diffusely branched, prostrate or semi-scandent vinelike plant, attaining a length of 2–10 ft, herbaceous above, woody near the base.

FLOWERS. May–June, opening late in the afternoon or in the early morning, fragrant, the slender pinkish, puberulent tube ¾–1¼ in. long, gradually expanding into a 5-lobed, white or pink limb about ½ in. across; lobes shallowly rounded, thin, delicate, margin erose, faintly marked within with 5 faint greenish lines, stamens 5, long-exserted; filaments slender, filiform, reddish, united into a cup at the base; anthers red, 2-lobed; style long-exserted, reddish, filiform, stigma minute and capitate; calyx 5-ribbed with several subulate, filiform bracts at the base.

FRUIT. Oblong, cylindric, 5-angled, truncate, constricted under the apex; seed filling the pericarp.

LEAVES. Opposite, blades ⅓–1¼ in. long, ovate, apex obtuse or acute, base cordate, margin entire, upper surface dull green and glabrous or slightly puberulent, lower surface paler and more densely puberulent; petiole ⅛–¾ in. long, pubescent.

TWIGS. Young ones delicate, green to reddish and pubescent, older ones more glabrous and gray to white.

RANGE. Lower Rio Grande Valley area of Texas. In Cameron County at Olmito, near San Benito and Brownsville. Also in Mexico.

REMARKS. The genus name, *Acleisanthes,* is from the Greek meaning "not-closed flowers"; however the flowers are open only in the early morning or late afternoon. The species name, *berlandieri,* is in honor of Luis Berlandier, Belgian botanist for the Mexican Government who assisted in the United States–Mexico boundary survey.

BOUGAINVILLEA

Bougainvillea Spp.

FIELD IDENTIFICATION. Cultivated shrubs or woody climbers in many horticultural forms and species. Recognized by the highly colored flower bracts. The commonly grown species are *B. rosea* Choisy and *B. spectabilis* Willd., both having many variations.

FLOWERS. Small, inconspicuous, terminal, tubular, margin 5–6-lobed; stamens 7–8 on unequal capillary filaments; ovary stipitate; bracts enclosing the flowers usually in threes and very conspicuous, highly colored—red, purple, mauve, yellow, or white—leaflike in size and texture, ovate to cordate, acute to obtuse at the apex.

LEAVES. Simple, alternate, entire, ovate to elliptic or lanceolate, apex acute to acuminate, bright green, glabrous or pubescent, petioled.

RANGE. Grown in a wide variety of soils in the

BOUGAINVILLEA
Bougainvillea Spp.

sun. Outdoors in the South and as greenhouse plants in the North. Natives of Brazil.

PROPAGATION. The plant may be used for porch and fence covers and trellis or specimen plantings on lawns where plenty of room is available for the arching stems. Also good for pot plants and summer bedding. Cuttings of half-ripened wood may be taken April–June, and grown in moist sandy soil at a temperature of 65°–70°F., and transplanted later to pots; or they may be grown in fields and moved later.

REMARKS. The genus name, *Bougainvillea,* is in honor of Louis Antoine de Bougainville (1729–1811), a French navigator.

CLIMBING SPIDERLING

Commicarpus scandens (L.) Standl.

FIELD IDENTIFICATION. Long-stemmed, prostrate or scandent, herbaceous or semiwoody perennial.

FLOWERS. June–September, opening in the early morning or late afternoon, very small, borne in 4–8-flowered umbels; peduncle axillary, slender, glabrous, ½–2 in. long; individual flower pedicels 1/16–⅛ in. long; flowers regular and perfect; calyx linear, cylindric, ribbed, about ⅛ in. long, with numerous prominent black glands toward the apex; corolla greenish yellow, tube slender, ⅛–⅓ in. long, gradually expanded into a 5-lobed, rotate or semiopen limb; lobes about 1/16 in. long, acute; stamens apparently only 2, barely exserted, filaments somewhat coiled; stigma simple, short; ovary 1-celled, superior, ovules solitary and erect.

FRUIT. Anthocarp indehiscent, ¼–⅜ in. long, dry, narrowly linear, apex truncate, obscurely 10-ribbed, glabrous, bearing viscid, black stipitate glands.

LEAVES. Opposite, blades ½–2¼ in. long, ovate to heart-shaped, apex acute to obtuse, minutely mucronulate, base cordate, margin entire; upper surface dull, dark green, glabrous to thinly pubescent; lower surface paler and glabrous to pubescent, veins obscure; petioles ¼–½ in. long, green to reddish, puberulent.

STEMS. Scandent or prostrate, long-stemmed, weak, slender, pale green, glabrous to slightly puberulent; older stems glabrous, gray to white, brittle, soon leafless.

RANGE. In dry soil of fence rows, waste grounds, edges of ravines, gravelly valleys, chaparral thickets, to an altitude of 4,500 ft. Widely scattered in south-

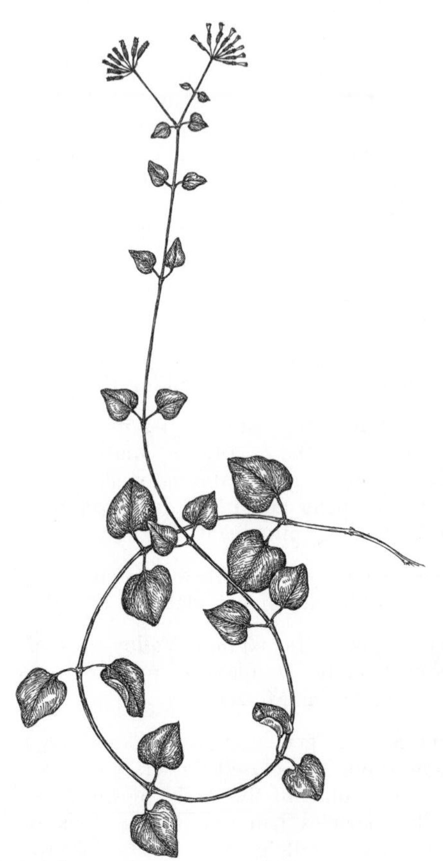

CLIMBING SPIDERLING
Commicarpus scandens (L.) Standl.

west and west Texas, New Mexico, Arizona, and southward through Central America to Peru. Also in the West Indies.

REMARKS. The genus name, *Commicarpus,* refers to the calyx compounded into a tube and enclosing the seed. The species name, *scandens,* refers to the plant's habit of supporting itself on other objects or plants. It is known in Central and South American countries under the names of Bejuco de Purgación and Sonorito. A decoction of the leaves is reported to be used for venereal diseases.

Narrow-leafed Moonpod

Selinocarpus angustifolius Torr.

FIELD IDENTIFICATION. A dichotomously branched plant, 6 in.–1½ ft, distinctly woody toward the base.

FLOWERS. Usually solitary, or sometimes 2–3 together in the leaf axils, becoming pendent with age, perfect; calyx corolla-like, funnelform, viscid-pubescent, tube about ⅜ in. long (more rarely to ½ in.); stamens 5, slender, elongate, didymous, hypogynous; stigma peltate, ovary 1-celled and 1-seeded.

FRUIT. An indurate, nutlike pericarp, ¼–⅓ in. long, elongate, grooved, bearing 3–5 broad, veinless, membranous wings, the dried calyx persistent at the apex; peduncle ⅛–¼ in. long, white-pubescent.

LEAVES. Opposite, pairs rather distant, narrowly elliptic or linear to linear-oblong, ¼–¾ in. long, 1/25–⅛ in. wide, apex acute or obtuse, base rather abruptly narrowed to a pubescent petiole 1/25–1/16 in. long, margin entire, both surfaces bearing short, white, appressed hairs, also subvicid.

STEMS. Young stems herbaceous, slender, finely striate, whitened with short, scabrous, appressed hairs; older stems becoming woody, less hairy and eventually glabrate with gray shreddy bark.

RANGE. Western Texas along the Rio Grande River, also in New Mexico, southward into Coahuila, Mexico. Type collected at Presidio Del Norte, Texas.

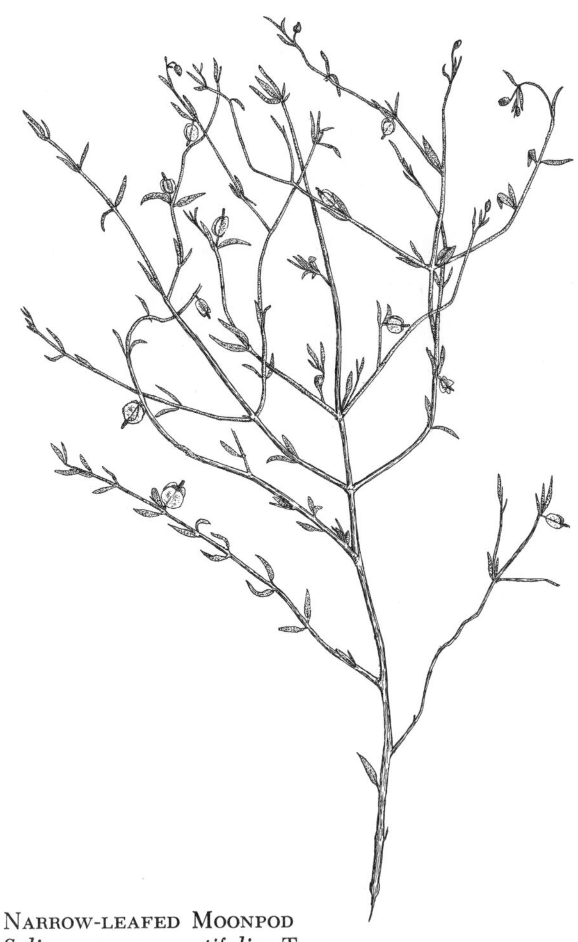

Narrow-leafed Moonpod
Selinocarpus angustifolius Torr.

REMARKS. The genus name, *Selinocarpus,* is from the Greek words *selinum* ("parsley") and *carpus* ("seed"), with probable reference to the winged seeds of some members of the parsley family.

PISONIA FAMILY (*Pisoniaceae*)

Devils-claw Pisonia
Pisonia aculeata L.

FIELD IDENTIFICATION. Spiny, semierect or scandent vine climbing to 20 ft, to 4 in. in diameter.

FLOWERS. In many-flowered compound cymes, pedicels slender, pubescent; corolla absent; flower consists of a rotate, yellow-to-purple calyx; lobes 5, mostly longer than wide, united below, ovate to triangular, apex acute, margin ciliate; stamens 5, exserted; ovary elongate-sessile, 1-celled, style terminal, stigma lobed.

FRUIT. Utricle slender-pediceled, club-shaped, dry, 10-ridged, with 5 rows of glands from base to apex.

LEAVES. Opposite, sometimes with smaller ones in the axils, elliptic to oval or ovate, apex acute or short-acuminate, surface glabrous, short-petioled.

TWIGS. Often with opposite, short, stout, lateral branchlets; green to gray or brownish, with pale scattered lenticels, and stout hooked axillary spines ⅛–¾ in.

RANGE. Texas Rio Grande Valley area. Usually along old resaca beds in Cameron County, Texas. In southern Florida, also in Mexico and the West Indies.

REMARKS. The genus name, *Pisonia,* is in honor of William Pison, a physician and naturalist of Amsterdam, who died in 1678. The species name, *aculeata,* refers to the spines. Vernacular names are Garabato Prieto, Garabato Blanco, Cockspur, Old Hook, and Pull-and-hold-back. About 40 species of the Pisonia Family are known, mostly from tropical regions.

Devils-claw Pisonia
Pisonia aculeata L.

POKEWEED FAMILY (*Phytolaccaceae*)

BLOODBERRY ROUGE-PLANT
Rivina humilis L.

FIELD IDENTIFICATION. Low slender plants with spreading branches, or sometimes vinelike. Rarely over 3 ft, mostly herbaceous or occasionally somewhat woody near the base. Usually under shade of other plants.

FLOWERS. Small, perfect, borne in loose, many-flowered, slender, narrowly oblong racemes. Racemes 1½–6 in. Individual flowers solitary, borne on glabrous or puberulent pedicels, white or pink, about ¼ in. across, spreading at maturity; petals absent; sepals 4, oblong to linear or cuneate, apex rounded and somewhat erose, occasionally mucronulate, 1/12–⅛ in.; calyx-tube 1/16–⅛ in., pubescent within, calyx subtending fruit base; stamens 4, erect or ascending, shorter than sepals, borne below the ovary, filaments filiform, anthers ovoid; ovary ovate to oblong, abruptly contracted into a single slender, curved style with a capitate stigma.

FRUIT. Borne on a pedicel ⅛–⅔ in., red or orange, translucent, shiny, subglobose or slightly 2-lobed, or slightly flattened, apiculate with the stigma remnants ⅛–3/16 in. long, pericarp thin and fleshy; seed lenticular, testa crustaceous.

LEAVES. Petiolate, alternate, 1–6 in. long, ovate to lanceolate or elliptic, margin entire, slightly wavy on the plane surface edge, apex acute to acuminate or obtuse, base acute or truncate, dark green, glabrous or pubescent; petioles about one third as long as the blades; stipules minute and caducous.

RANGE. Thickets and sandy lands. New Mexico, Oklahoma, Arkansas, Texas, and Louisiana; and eastward to Florida. Also Mexico, Central America, South America, and the West Indies.

REMARKS. The genus name, *Rivina*, is in honor of A. Q. Rivinus, professor of botany at Leipzig (1652–1723). The species name, *humilis*, refers to the low habit of growth. Vernacular names are Baby-pepper, Pokeberry, and Pigeon-berry. It can be planted as a summer annual, and propagated by seeds or cuttings in spring over heat, or by root divisions. Good use can be made of it for borders or backgrounds, and it is able to withstand considerable shade. The fruit yields a red dye, or ink, and the leaves were once used in domestic medicine for catarrh and for treating wounds.

BLOODBERRY ROUGE-PLANT
Rivina humilis L.

Snake-eyes

Phaulothamnus spinescens Gray

FIELD IDENTIFICATION. Rather rare shrub to 10 ft. The branches are dense, very divaricate and spiny.

FLOWERS. Solitary, or in small racemes on short pedicels, dioecious; perianth 4–5-parted, lobes imbricate, scarious-margined, puberulent, ovate to oval, apex rounded, length 1/25–1/16 in., stamens 12, some rudimentary, filaments distinct, slender; anthers linear-oblong, basifixed; ovary ovoid and 1-loculed; stigmas 2, filiform.

FRUIT. Borne on short peduncles about 1/8 in. or less; body subglobose, about 3/16 in., white, transparent, fleshy, indehiscent; seed solitary and black, easily seen within and giving the appearance of a small eye, about 1/16 in. long, subglobose or slightly flattened, granular and rugose.

LEAVES. Simple, alternate, or clustered at the nodes, length 1/2–1 in., width 1/8–1/4 in., linear to narrowly spatulate or oblong, margin entire, apex rounded or obtuse, minutely mucronulate, base narrowed into a short petiole, or sessile, veins obscure except the main vein, surfaces grayish green and glabrous or puberulent, often glaucous on young leaves or stems, thickish.

TWIGS. Slender, elongate, often with many short axillary branchlets, brittle, ending in straight gray or brown spines, also with axillary leafless spines; young twigs dark green or glaucous, smooth; when older gray, smooth or striate; bark on trunk mottled gray, smooth at first, with age breaking into small, thin scales.

RANGE. Not common anywhere, but in scattered locations in the lower Rio Grande Valley area of Texas—in Cameron, Kenedy, Starr, Jim Wells, Willacy, Nueces, and Kleberg counties.

REMARKS. The genus name, *Phaulothamnus,* is from the Greek and indicates an uncomely shrub, ill to handle. The species name, *spinescens,* refers to the spines. In south Texas it is known by the Mexican people under the name of "Putia."

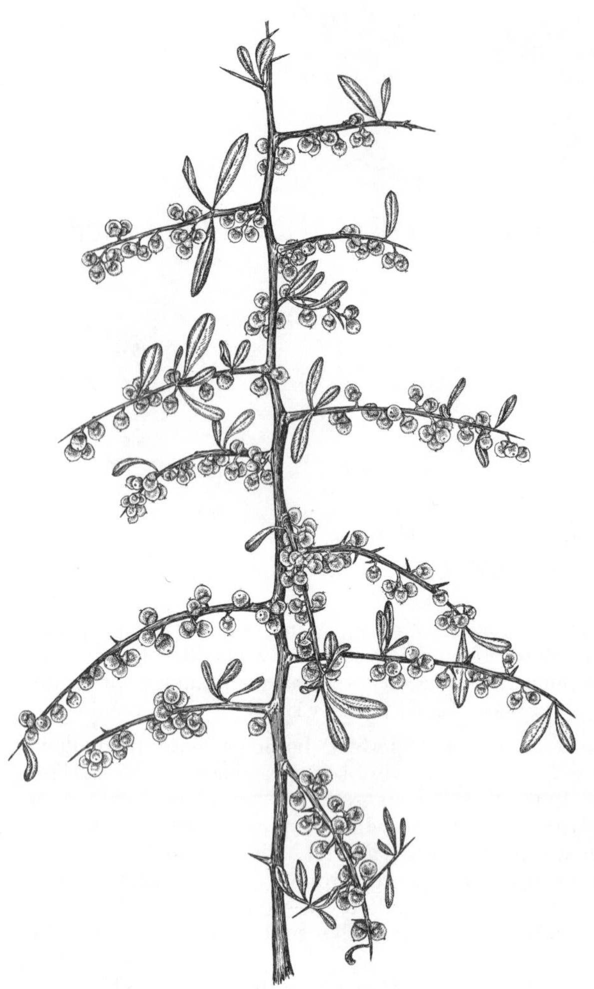

Snake-eyes
Phaulothamnus spinescens Gray

PETIVERIA FAMILY (*Petiveriaceae*)

GARLIC GUINEAHEN-WEED
Petiveria alliacea L.

GARLIC GUINEAHEN-WEED
Petiveria alliacea L.

FIELD IDENTIFICATION. An ill-smelling subshrub 1–4 ft, from creeping rhizomes. Stems tall and wandlike, closely pubescent.

FLOWERS. Spikes virgate, 4–12 in. long, flowers small, perfect; corolla absent; calyx short-pediceled or sessile, herbaceous, greenish white, conical at base; sepals 4, spreading in flower, erect in fruit, nearly equal, persistent, ⅛–⅙ in., linear to linear-lanceolate, obtuse; stamens 4–8, shorter than sepals, unequal in length, inserted at base of perianth on a disk; anthers linear, sagittate; ovary 1-celled, elongate, flattened, style absent, stigma 2-lobed.

FRUIT. Achene appressed to the narrow, spikelike rachis, linear-cuneate, ribbed, about ¼ in. long, twice as long as the persistent sepals, and surrounded by the sepals, apex with 1–6 subulate reflexed spines about half as long as the achene body; seed narrow and erect, testa membranous.

LEAVES. Alternate, thickish, elliptic to oblanceolate or obovate, apex acute to short-acuminate, base narrowed, margin entire, blade length 1½–4¾ in., surfaces somewhat pubescent (particularly on the veins); petioles ⅛–¾ in., pubescent to glabrous, stipules small and narrow.

RANGE. In woods and thickets. In southwest Texas and Florida. Common in tropical America.

MEDICINAL USES. The root, known under the name of "Pipi," attracted some attention as a medicinal plant in the latter part of the nineteenth century. It was used to stimulate expectoration and diaphoresis.

REMARKS. The genus name, *Petiveria*, is in honor of James Petiver (1665–1718), an apothecary and botanist of London. The species name, *alliacea*, is the Latin word for "garlic," and refers to the plant's disagreeable odor when crushed. It is known in Central American countries under the name of Hierba de las Gallinitas. The plant is sometimes grown in greenhouses in the North, and outdoors in the South.

MADEIRA-VINE FAMILY (*Basellaceae*)

MIGNONETTE MADEIRA-VINE
Boussingaultia baselloides H. B. K.

MIGNONETTE MADEIRA-VINE
Boussingaultia baselloides H. B. K.

FIELD IDENTIFICATION. Delicate twining vine to 20 ft. Often with tubers attached to the rootstocks. Essentially an herbaceous plant which freezes down easily, but in mild winters in south Texas the leaves persist, and new shoots spring up quickly from the roots before the old leaves are lost.

FLOWERS. September–November, in terminal leafy panicles composed of dense racemes; total panicle to 1 ft long, racemes 2–3 in.; individual flower pedicels 1/16–1/12 in., much shorter than petals, ovate to oblong, acute; petals 5, distinct almost to base, about 1/10 in., oval to elliptic-oval, greenish white, membranous, somewhat glutinous; stamens 5, borne opposite the petals, long-exserted; ovary superior and 1-celled, styles 3 and distinct.

FRUIT. A small utricle, seed solitary and erect.

LEAVES. Simple, alternate, deciduous, fleshy, oval to ovate or broadly elliptic, apex acute or acuminate, sometimes abruptly so, base cuneate to abruptly narrowed or semicordate, margin entire, both surfaces bright green, smooth, glabrous or lower surface slightly paler; petiole green to reddish, very short, or blade sessile.

STEMS. Twining and branching, glabrous and smooth, green to dark red, rather delicate, clambering on other plants for support.

RANGE. Cultivated for ornament, escaping to grow in waste places in south Texas and Florida. Collected by the author at Rockport and Brownsville, Texas. Climbing over chaparral shrubs in alluvial soil. Considered to be a native of tropical America.

PROPAGATION. The vine is propagated by seeds, by rootstocks, or from tubercles produced in the leaf axils. It is popularly grown for porches and arbors. The roots may be stored in winter and planted after frost, or grown in greenhouses in the North. In the South it may be planted outdoors, but cannot stand freezes.

REMARKS. The genus name, *Boussingaultia*, is in honor of J. B. Boussingault, a famous agricultural chemist, born in 1802. The species name, *baselloides*, denotes its similarity to members of the genus *Basella*.

CROWFOOT FAMILY (*Ranunculaceae*)

LEATHER-FLOWER CLEMATIS
Clematis viorna L.

FIELD IDENTIFICATION. Vine climbing by means of petioles of the upper stem, attaining a length of 10 ft or more. Generally herbaceous above but somewhat woody close to base.

FLOWERS. May–August, borne on long, axillary, naked or 2-bracted peduncles, 1–7-flowered, nodding, ovoid-campanulate or pitcher-shaped; sepals 4, valvate, purple or reddish purple, ⅝–1¼ in. long, ovate to oblong-lanceolate, apices acuminate, woolly, recurved, very thick and leathery, short appressed-pubescent, margins white-tomentose, not expanded, about as long as the stamens or slightly longer; petals absent; stamens numerous, adnate, long and linear; pistils numerous in a head.

FRUIT. Peduncles erect in fruit, achenes ovate, suborbicular or elliptic, ⅙–¼ in. wide, margins thickened and appressed-pubescent, 1-seeded; styles tail-like, about 1 in. long, very plumose, spreading or coiled, tan to yellowish or white.

LEAVES. Numerous, opposite, simple and entire, or 3-lobed, or ternate, or pinnately divided or lobed; stem leaves usually pinnate with 3–4 pairs of divisions; terminal leaflets very small with slender tendril-like filaments; floral leaves very simple and short-petioled or subsessile; petiolules generally ⅓–½ in. long; leaflets 1–3½ in. long, bright green, membranaceous, ovate to lanceolate or oblong, apices acute to acuminate or obtuse, base cuneate to subcordate, margin mostly entire (unless lobed); upper surface 3–5-nerved, not reticulate-veined, glabrous; lower surface usually pubescent; petioles slender.

STEMS. Slender, 6-angled and corrugated in drying, pubescent along the nodes.

RANGE. The species is known from eastern Texas, Arkansas, and Louisiana; east to Georgia, northward to Missouri and Pennsylvania, and west to Illinois.

LEATHER-FLOWER CLEMATIS
Clematis viorna L.

VARIETY. Weak Leather-flower Clematis, *C. viorna* var. *flaccida* (L.) Erickson, is a variety with very thin, velvety-pubescent, more simple leaves (or rarely 2–3-lobed); sepals lavender at base with apices more greenish. It was originally known only from Warren County, Kentucky.

C. × *divaricata* is a hybrid of *C. viorna* and *C. integrifolia.*

REMARKS. The genus name, *Clematis,* is from the Greek word for a climbing vine. The species name, *viorna,* is the old generic name of the European *C. vitalba* L.

Glaucous Leather-flower Clematis
Clematis glaucophylla Small

FIELD IDENTIFICATION. Slender, reddish vine, climbing by leaf petioles and attaining a height of 15 ft.

FLOWERS. Borne May–June, peduncles 2-bracted or naked, often solitary or several flowers in a leafy corymb; petals absent; calyx reddish purple, conic-ovoid, about 1 in. long; sepals ovate-lanceolate, thick, dorsally glabrous, glossy, slightly spreading at the caudate-acuminate apex, margin not expanded, white-tomentose; anthers about as long as the filaments.

FRUIT. Achenes nearly orbicular, inequilateral, ¼–⅓ in. wide, rim moderately wide, strigose; achene tails to 2 in. long at maturity, plumose, tawny, spreading or intertwined.

LEAVES. Simple or compound, opposite; some cauline leaves with 3–5 pairs of leaflets; basal leaflets entire, or trifoliate, or variously lobed; inflorescential leaves usually simple and smaller; leaves or leaflets usually ovate to broadly cordate or some lanceolate; apices acute to obtuse or acuminate, also usually apiculate; bases mostly cordate, subcordate, or rounded, petioled or subsessile; surfaces glabrous but the lower surface glaucous at maturity, rather strongly nerved but not prominently reticulate; leaflet blades 1–4 in. long.

STEMS. Slender, climbing, glabrous (some stems glaucous), ribbed, young ones tan to brown, older ones reddish.

RANGE. Moist woods, thickets, and riverbanks. Oklahoma and Arkansas; eastward to Alabama, Georgia, and northern Florida, and northward in Kentucky, Tennessee, Virginia, and North and South Carolina.

REMARKS. The genus name, *Clematis*, is a name given by Dioscorides to a slender climbing plant. The species name, *glaucophylla*, refers to the pale undersurface of the leaves. Erickson (1943) makes the following comments on the relationships of this species: "*C. addisoni* is closest to this species. In other relationships it appears to be with *C. viorna*, on the one hand, which it closely resembles in leaf characters, and with which it may be found to intergrade; and on the other hand, with *C. texensis* and *C. versicolor*. With the latter two species it shares a very similar flower structure and glaucous leaves; it differs from them in the degree of reticulation of the leaves."

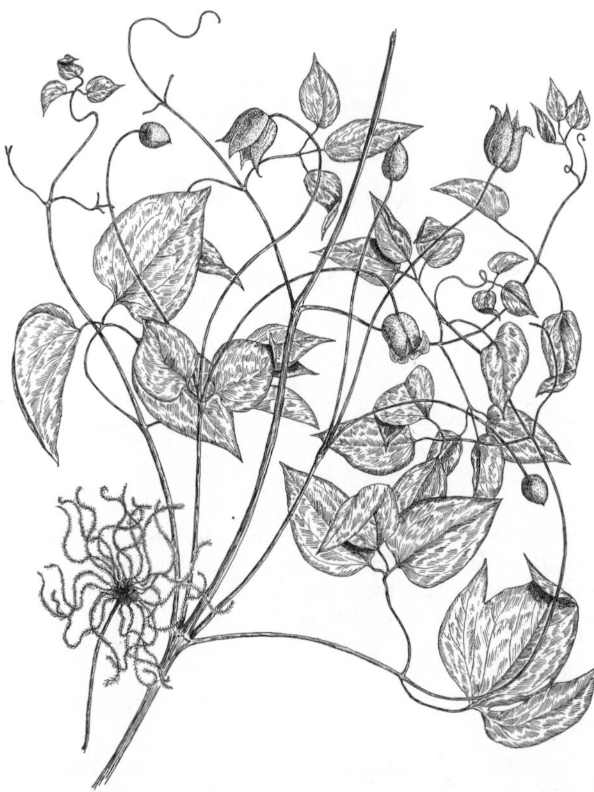

Glaucous Leather-flower Clematis
Clematis glaucophylla Small

Pitcher Clematis
Clematis pitcheri Torr. & Gray

FIELD IDENTIFICATION. High-climbing vine, simple or somewhat branched, climbing by means of the twisting leaf petioles. Stems herbaceous above but slightly woody near base.

FLOWERS. Appearing June–August, solitary or as many as 7 on axillary peduncles; petals absent; flower nodding, about 1 in. long, campanulate or ovoid to urceolate, externally dull purple to brick-red, ribbed and pubescent, internally darker purple or greenish; sepals 4, thick, as long as the stamens, or twice as long, somewhat recurved or spreading at apex, margin not expanded or only slightly expanded and white-tomentose; stamens numerous, anthers erect and long-tipped; pistils numerous in a globose head, style elongate.

FRUIT. Achene ¼–⅓ in. wide, orbicular or slightly quadrangular, somewhat inequilateral, rim broad and thick, surfaces appressed-pubescent; tail of the achene ¾–1⅓ in. long, filiform and naked or short-appressed silky-hairy (not plumose).

LEAVES. Those of the floral stems usually smaller, and with shorter petioles, but similar to the cauline leaves otherwise; cauline leaves sometimes stiffly erect, usually with 3–5 leaflets (occasionally to 9); rachis usually ending in a slender tendril-like filament; leaflets opposite, simple and entire, or 2–5-lobed or 3-foliate, apex obtuse to acute and mucronate, base acute to rounded or cordate, shape of leaflet very variable, from ovate to cordate or elliptical, the larger ones on the lower part of the stem 1⅓–4 in. long; surfaces usually glabrous above when mature and pubescent

PITCHER CLEMATIS
Clematis pitcheri Torr. & Gray

below, noticeably reticulate-veined but not as closely so as *C. reticulata,* coriaceous; petioles distinct but variable in length.

STEMS. Slender, green to reddish brown, ribbed or with 6 angles, pubescent, especially so at nodes, later glabrous.

RANGE. The species is found in thickets or borders of woods in low grounds. Texas, Louisiana, Oklahoma, and Arkansas; north to Indiana and west to Nebraska.

VARIETIES. A variety has been given the name of Cutleaf Pitcher Clematis, *C. pitcheri* var. *filifera* (Benth.) Robins. The leaves are small, much-divided, apices acute or acuminate, leathery, strongly net-veined, lower surface pubescent; sepals usually with expanded margins. Found in Texas in Brewster, Culberson, Jeff Davis, and Presidio counties; also in New Mexico.

Sargent Pitcher Clematis, *C. pitcheri* var. *sargentii* Rehd., has smaller paler flowers and leaflets rarely lobed.

REMARKS. The genus name, *Clematis,* is a name given by Dioscorides to a slender climbing plant. The species name, *pitcheri,* is in honor of its discoverer Zina Pitcher (1797–1872). This plant is sometimes confused with *C. reticulata,* but the latter has plumose achene tails and a greater degree of reticulation, and the leaflets are of a different shape.

BIGELOW CLEMATIS
Clematis bigelovii Torr.

FIELD IDENTIFICATION. Plant attaining a height of 20 in. Usually erect, simple or few-branched. The following description is based on that of Erickson (1943).

FLOWERS. Solitary, on long peduncles, terminal, petals none, calyx nodding, subcampanulate, ⅝–1 in. long; sepals 4–5, broadly lanceolate, thickish, finely pubescent externally, margins tomentose; stamens numerous, erect, as long as the sepals or shorter; pistils numerous, style elongate.

FRUIT. Borne in globose heads; achene bodies ⅙–⅕ in. broad, suborbicular or obovate, rim slight, surfaces appressed-pubescent; styles becoming tail-like, about 1¼ in. long, glabrous or pubescent near the base (not plumose).

LEAVES. With 3–4 pairs on the primary stem, pinnate with 7–11 primary divisions deeply 3-lobed or 3-foliate; leaflets usually 2–5-lobed, mostly broadly ovate, some lanceolate, apex blunt or rounded, mucronate, ⅝–1 in. long, surfaces glabrous and sometimes glaucous beneath, not conspicuously veined, rather thin.

RANGE. In New Mexico in the Tunitcha and Sandia

BIGELOW CLEMATIS
Clematis bigelovii Torr.

mountains. In eastern Arizona near Fort Apache, Navajo County.

RELATED SPECIES. Erickson points out: "The position of the species is problematical. In habit and vegetative characters it closely resembles *C. palmeri*. However, it resembles *C. pitcheri* in the appearance of the flowers, and in its non-plumose and fragile tails, and was regarded as a variety of *C. pitcheri* by Robinson."

REMARKS. The genus name, *Clematis*, is the ancient name of a climbing vine given by Dioscorides. The species name, *bigelovii*, is in honor of Jacob Bigelow (1787-1879), American botanist.

PALMER CLEMATIS
Clematis palmeri Rose

FIELD IDENTIFICATION. Plant climbing or scandent with an erect, simple or much-branched stem, rather woody below, attaining a length of 40 in.

FLOWERS. Usually solitary, on long peduncles, terminal or axillary, 1¼–1⅝ in. long; petals none; sepals thick, erect, dull brown or purplish, conspicuously white-lanate externally; stamens numerous; pistils numerous.

FRUIT. In globose heads, achenes obovate, ⅙–⅕ in. long, slightly rimmed, appressed-hairy; tail-like styles 1⅝–2 in. long, loosely intertwined, the plumose hairs brown.

LEAVES. Borne in 5–6 pairs, pinnate with 5–11 divisions which are 3-lobed or 3-foliate; leaflets 3–5-lobed or coarsely toothed, 1¼–2¾ in. long, ovate to oblong, apices obtuse or acute, surfaces with a few hairs or mostly glabrous, slightly glaucous, thin.

RANGE. Western New Mexico and eastern Arizona.

RELATED SPECIES. *C. palmeri* greatly resembles *C. hirsutissima* Pursh in characters of flowers and fruit, but has been separated from it on the bases of its more robust habit, broad leaflets, and nearly complete lack of pubescence.

REMARKS. The genus name, *Clematis*, is a name given by Dioscorides for a climbing vine. The species name, *palmeri*, is in honor of E. J. Palmer (b. 1875) American botanist.

PALMER CLEMATIS
Clematis palmeri Rose

CURLY CLEMATIS
Clematis crispa L.

FIELD IDENTIFICATION. Slender vine climbing to 10 ft or more by means of twisting leaf petioles. The stems herbaceous above but sometimes slightly woody near the base.

FLOWERS. Borne April–August, solitary and terminal at the ends of long naked peduncles; flowers valvate in the bud, nodding, fragrant; petals absent; calyx bluish purple, thickened, tubulate below, the upper half with 4 sepals 1–2 in. long, thin, spreading, lanceolate, margin widely expanded, crisp or undulate above the middle, glabrous except the white-tomentose margin; stamens numerous, anthers adnate, long and narrow; pistils numerous, in a head.

FRUIT. Achenes ¼–⅓ in. broad, suborbicular to quadrangular, inequilateral, moderately rimmed, finely pubescent; tails of fruit finely appressed-pubescent or glabrate (not plumose), spreading or coiled near the tip, ¾–1⅓ in. long.

LEAVES. Opposite, either simple or pinnately compound with 2–5 pairs of leaflets, 1¼–3¼ in. long, terminating in a tendril-like filament or occasionally simple or 3-foliate; leaflets most often lanceolate, but sometimes linear or ovate to cordate, margin either entire or 2–3-lobed, apex acute or acuminate, surfaces glabrous with 3 conspicuous veins from the base but not reticulate, thin.

STEMS. Slender, elongate, 6–12-angled or ribbed, often pubescent along the nodes, green to brown or gray.

CURLY CLEMATIS
Clematis crispa L.

RANGE. In swamps or low grounds. Texas, Arkansas, and Louisiana; east to Florida, north to Virginia, and west to Illinois.

VARIETY. Walter's Curly Clematis, *C. crispa* var. *walteri*, is a variety with very narrow linear-lanceolate leaves. However, there are so many intermediate forms between it and the species that it is a somewhat doubtful segregate.

REMARKS. The genus name, *Clematis*, is the name given by Dioscorides to a slender climbing plant. The species name, *crispa*, is for the crinkled margin of the sepals. Also known under the vernacular names of Curl Flower, March Clematis, and Blue-jasmine.

SMALL'S CLEMATIS

Clematis obliqua Schneid.

FIELD IDENTIFICATION. Vine climbing by twisting leaf petioles, herbaceous above and half-woody near the base.

FLOWERS. Peduncles bractless, petals absent, calyx valvate in the bud, conic, about 1½ in. long; sepals 4, lanceolate, thickened, leathery, margins cottony, apices shortly recurved; stamens numerous and erect, anthers long and narrow, pistils numerous.

FRUIT. Achenes usually elliptic, about ¼ in. wide, 1-seeded; achene tails long and plumose.

LEAVES. Opposite, usually simple, blade ovate to ovate-lanceolate, 1¼–3¼ in. long, apex acute or subacuminate, margins erose-denticulate and undulate, base truncate and inequilateral, long-petioled, surfaces veiny (but not reticulate) and glabrous.

STEMS. Slender, glabrous or somewhat pubescent along the nodes.

RANGE. In pinelands or in swamps; eastern Texas, eastward to Florida and Georgia.

REMARKS. The genus name, *Clematis*, is the name given by Dioscorides to a climbing vine. The species name, *obliqua*, refers to the inequilateral leaves. Also known as Leather-flower.

NET-LEAF CLEMATIS

Clematis reticulata Walt.

FIELD IDENTIFICATION. Slender-stemmed vine climbing by leafstalks, and attaining a length of 9 ft. Recognized by very rigid and conspicuously net-veined leaflets.

FLOWERS. In July, peduncles axillary, 1–3-flowered, floral leaves usually less than 1½ in. long, otherwise similar to the cauline leaflets; flowers dull purple, drooping, ⅝–1 in. long; petals absent; sepals 4, leathery, valvate in the bud, equaling or longer than the stamens, varying from velvety canescent to nearly glabrous, apices acute, recurved, margins narrowly expanded and tomentose; stamens numerous and erect, anthers long and narrow; pistils numerous.

FRUIT. Achenes borne in a head, symmetrical, suborbicular to elliptic, about ⅕–¼ in. long, rim prominent, surfaces appressed-pubescent, 1-seeded; achene tails persistent, loosely intertwined, plumose, pale yellowish brown, 1–2 in.

LEAVES. Opposite, numerous, rachis ending in a tendril-like filament or minute leaflet; cauline leaves long-petioled; leaflets usually 4 pairs, entire or less often 2–3-lobed, blade elliptic to ovate or even lanceolate to oblong, apex rounded or acute and mucronate, length ¾–2½ in. or more, both surfaces very strongly reticulated, rigid and coriaceous, long-petioluled.

STEMS. Slender, reddish, 6-angled, nodes somewhat pubescent.

RANGE. In sandy soils, fence rows, fields, and thickets. East Texas, Louisiana, and Arkansas; east to Florida, and northward to South Carolina and Tennessee.

SYNONYM. Since the species, *reticulata*, has considerable variation, especially as to the amount of pubescence, the species known as *Viorna (Clematis) subreticulata* has been combined with it. *C. subreticulata* was described as having somewhat thinner leaflets and lack of canescent pubescence on the outside of sepals.

REMARKS. The genus name, *Clematis*, is the Greek name for a climbing vine. The species name, *reticulata*, refers to the strongly net-veined leaves.

Beadle Clematis

Clematis beadlei (Small) Erickson

FIELD IDENTIFICATION. Slender vine climbing by twisting petioles. The following description is based on that of Erickson (1943):

FLOWERS. Peduncles axillary, 1-flowered, floral leaves about 1 in. long, simple and entire, short-petioled; uppermost bractlike, sessile or petiolate; sepals 4, ⅗–1 in. long, thick, petal-like, valvate; true petals absent; stamens erect and numerous; pistils numerous.

FRUIT. Borne in globose fruiting heads about 2 in. in diameter; achenes ovate, inequilateral, ⅙–¼ in. broad, compressed, moderately rimmed, appressed pubescent; tails of achenes 2–2½ in., plumose, loosely intertwined.

Net-leaf Clematis
Clematis reticulata Walt.

LEAVES. Opposite, cauline leaves with 3–4 pairs of leaflets; rachis slender, finely pubescent, geniculate; leaflets ovate to lanceolate, apex acute or acuminate, simple or entire, or the basal ones 2–3-lobed, 1–3¼ in. long, texture thin and chartaceous, moderately and distinctly reticulate with age.

STEMS. Slender, reddish brown, 6-angled, mostly glabrous or somewhat pubescent at nodes.

RANGE. Southeast Texas (Harris County), Arkansas, Mississippi, Tennessee, and Georgia.

RELATED SPECIES. Evidently closely related to *C. reticulata* but clearly outside the range of variation of that species.

REMARKS. The genus name, *Clematis*, is an ancient Greek name for a climbing vine. The species name, *beadlei*, is in honor of C. D. Beadle, at one time director of the Biltmore Herbarium.

Scarlet Clematis

Clematis texensis Buckl.

FIELD IDENTIFICATION. Vine with slender stems to 10 ft, and climbing by means of bending leafstalks. Upper stems herbaceous, partly woody close to the base.

FLOWERS. Sometimes hidden by the foliage, July–September; peduncles axillary, elongate, 1–7-flowered, lower pair of floral leaves about same size and character as cauline leaves but short-petioled; flower valvate in bud, ovoid to globose-ovoid or bell-shaped, scarlet to reddish purple, ¾–1¼ in. long, large, perfect; petals absent; sepals 4, petal-like, thick and leathery; apex recurved, spreading, or wholly connivent; glabrous, glaucous, lustrous externally, felty internally, margin scarcely expanded, white-tomentose; stamens numerous, erect, anthers long and narrow; pistils numerous.

FRUIT. Achenes borne in a head, ¼–⅓ in. broad, rather symmetrical, with a prominent rim, appressed-pubescent, 1-seeded; achene tails very plumose, loosely intertwined or spreading, 1–2¾ in.

LEAVES. Rachis slender, geniculate, terminated by a tendril-like filament, leaves opposite; cauline leaves with 4–5 pairs of leaflets; leaflets simple and entire or 2–3-lobed, or 2–3-foliate, ovate to ovate-lanceolate to orbicular, apex obtuse to rounded or mucronate, sometimes notched, base truncate or subcordate, lower blades 1–3½ in. long, glabrous, glaucous beneath, usually strongly reticulate, coriaceous, long-petiolate.

STEMS. Slender, green to gray or brown, ribbed, glabrous, glaucous.

RANGE. On rich soils, or on shady limestone ledges, often along streams. Mostly confined to the Texas Edwards Plateau area from the Colorado to the Rio Grande rivers. In Bexar, Blanco, Comal, Edwards, Hays,

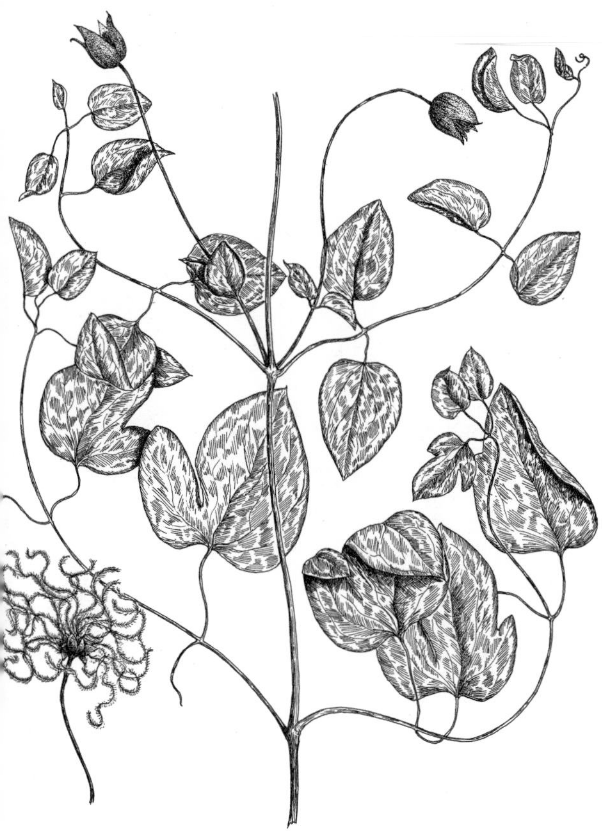

Scarlet Clematis
Clematis texensis Buckl.

Gillespie, Kerr, Lampasas, Travis, Dallas, and Uvalde counties.

REMARKS. The genus name, *Clematis*, is the Greek name for a climbing vine. The species name, *texensis*, refers to the state of Texas, where it grows most abundantly. Also known under the vernacular names of Leather-flower and Pipe-vine.

Many-color Clematis

Clematis versicolor Small

FIELD IDENTIFICATION. Slender, simple, or much-branched vine, climbing by leaf petioles, and attaining a length of 3–12 ft. Herbaceous above, somewhat woody toward the base.

FLOWERS. May–June, peduncles axillary, 3–7-flowered, lowest pair of inflorescential leaves similar in shape and character to the cauline leaves but smaller and short-petioled; flowers subglobose, ⅝–2 in. long, blue to lavender or purple, greenish toward apex; petals absent; sepals 4, petaloid, narrowly ovate to lanceolate, short-acuminate, tips scarcely recurved, glabrous and glaucous, margins unexpanded, white-tomentose; stamens numerous, spreading; pistils numerous, styles densely pubescent at anthesis.

FRUIT. Achenes orbicular or asymmetrical, ⅕–¼ in. broad, rimmed, surfaces appressed-pubescent; achene tails 1¼–1¾ in., plumose, pale yellow to whitish, loosely intertwined or spreading.

LEAVES. Numerous, rachis slender, geniculate, terminating in a tendril-like filament; leaflets simple or rarely 2–3-lobed, ovate to ovate-oblong or cordate-ovate, apex obtuse to rounded, mucronate, base obtuse to subcordate, glabrous on both sides, upper surface glaucescent, moderately reticulate, lower surface glaucous and conspicuously reticulate, coriaceous, length ¾–2¾ in.

STEMS. Slender, ribbed, glabrous, glaucous, brown to gray.

RANGE. Dry stony woods or barrens, or on rocky ledges. In Oklahoma, Arkansas (possibly in Texas), Mississippi, Missouri, Tennessee, and Kentucky.

RELATED SPECIES. Erickson (1943) makes the following remarks concerning the relationships of *C. versicolor*:

"Although this species is very closely related to *C. texensis*, it can be readily distinguished by its flower color if living material is seen. In herbarium material the distinction may be made on the basis of its somewhat more slender habit and smaller size throughout, its less compound leaves and less pointed leaflet tips. *C. versicolor* is a less variable species than *C. texensis*. Further study is desirable, especially in the regions between the Ozarks and the Edwards Plateau, to determine whether there are connecting forms between the two species as here understood. The writer has seen comparatively few specimens which are not either from the Edwards Plateau or the Ozarks. Those from southeastern Oklahoma and southwestern Arkansas have rather arbitrarily been assigned to *C. versicolor*, and those from northeastern Texas (Smith County) to *C. texensis*. Field study may change this interpretation. The specimens from Kentucky and Tennessee, also, suggest intergrades between *C. glaucophylla* and *C. versicolor*."

REMARKS. The genus name, *Clematis*, is the Greek name for a climbing vine. The species name, *versicolor*, refers to the various colors of the flowers. It is also known as the Leather-flower.

Rocky Mountain Clematis

Clematis pseudoalpina (Kuntze) A. Nelson

FIELD IDENTIFICATION. Plant with stems short and trailing or sometimes vinelike and climbing. Upper stems herbaceous, lower somewhat woody near base.

FLOWERS. Usually April–June, solitary, or a few together in leaf axils; petals absent; sepals 4, petaloid, valvate, thin, spreading, violet or purple, more rarely white, lanceolate, acute, margin entire or few-toothed, ¾–2 in. long, surfaces glabrous or slightly pubescent; stamens numerous, spreading; staminodia with petaloid filaments hardly longer than normal stamens, more or less antheriferous; pistils numerous, styles elongate.

FRUIT. In a globose cluster, achene densely pubescent or glabrous, terminating in a tail 1–2 in. long, and either plumose, pubescent, or rarely glabrous.

LEAVES. Opposite, biternately or ternately compound, 2–4 in. long, petiole channeled; petiolules usually less than ¼ in.; leaflet blade ¼–1 in. long, lanceolate or ovate, apex acute, margin more or less lobed and incisely 3–7-toothed, or cleft, or sometimes pinnately so, upper surface dull green, lower surface paler and glabrous.

STEMS. Slender, mostly prostrate, occasionally climbing, young ones green and glabrous, older ones tan to brown.

RANGE. Well-drained soil in thickets, or open wooded hillsides, at elevations of 6,000–10,000 ft. New Mexico and Arizona, northward to South Dakota.

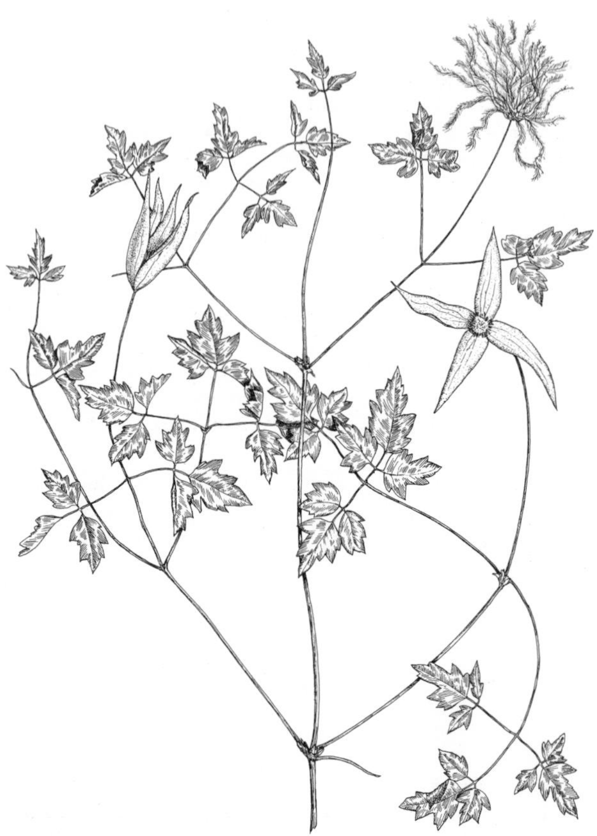

Rocky Mountain Clematis
Clematis pseudoalpina (Kuntze) A. Nelson

Oriental Clematis
Clematis orientalis L.

REMARKS. The genus name, *Clematis*, is the Greek name for a climbing vine. The species name, *pseudoalpina*, means "false-alpine," as apposed to *C. alpina* (L.) Mill. from Europe and Asia.

Oriental Clematis

Clematis orientalis L.

FIELD IDENTIFICATION. Slender cultivated vine climbing by twisting petioles, attaining a length of 9–20 ft. Upper stems herbaceous, lower somewhat woody near the base.

FLOWERS. Usually August–September, solitary, or in few-flowered cymes in leaf axils, on slender pedicels 1½–4 in.; dioecious or polygamo-dioecious; petals absent; sepals petaloid, valvate, erect or spreading, 4–5 in number, yellow or tinged with green, elliptic-ovate or elliptic-oblong, ½–2 in. long, surfaces usually pubescent, stamens and pistils numerous.

FRUIT. A conspicuous head of achenes with elongate plumose styles ¾-2 in.

LEAVES. Opposite, biternately or triternately compound, to 5 in. long including petiole; leaflets ovate to oblong-ovate to lanceolate, ½–2½ in. long, the terminal leaf usually largest; margins entire, cut-toothed, or lobed; upper surface dull green and glabrous, lower surface paler and glabrous (sometimes glaucescent) or minutely pubescent; petiolules ¼–¾ in.

STEMS. Young ones slender, green to reddish brown, glabrous or with sparse white hairs, older ones brown and glabrous.

RANGE. Moist river valleys or canyons at altitudes of 6,000–8,000 ft. A native of the Asiatic Himalaya region to Persia. Escaping cultivation in New Mexico, Colorado, and other Western states.

REMARKS. The genus name, *Clematis*, is the Greek name for a climbing vine. The species name, *orientalis*, is for the plant's East Asiatic origin. The plant was introduced into cultivation in 1731.

Columbia Rock Clematis

Clematis columbiana (Nutt.) Torr. & Gray

FIELD IDENTIFICATION. Plant a half-woody, slender vine, climbing by leaf petioles.

FLOWERS. May–August, usually solitary on elongate bractless peduncles; sepals valvate, 4–5, petaloid (true petals absent), blue to purple, more rarely white, 1–2 in. long, lanceolate to oblong, thin, spreading; stamens numerous, spreading, the outer filaments sometimes sterile and somewhat petaloid; pistils numerous, styles elongate.

FRUIT. Achenes borne in a head, densely pubescent, styles persistent and tail-like, plumose, 1–2 in. long.

LEAVES. Opposite, 3-foliate, leaflets narrowly to broadly ovate, apex acute or acuminate, base obliquely cordate, margins usually entire but occasionally toothed or cleft, length of blades 1¼–1⅝ in., surface sparsely hairy to glabrous later, slender-petioled.

RANGE. Usually climbing over bushes in dense mountain woods at altitudes of 6,000–10,000 ft. In New Mexico, west to California, and northward through Colorado to Alberta and British Columbia.

SYNONYMS. Some writers have listed the plant under the names of *C. verticillata* var. *columbiana* Gray; *Atragene columbiana* Nutt., *A. occidentalis* Horn., *A. grosseserrata* Rydb., *A. diversiloba* Rydb.; and *A. tenuiloba* (Gray) Britt.

REMARKS. The genus name, *Clematis*, is the ancient name given by Dioscorides to a climbing vine. The species name, *columbiana*, refers to its growth in the Columbian provinces of the northwestern United States and Canada. Also known as the Virgins-bower.

Drummond Clematis

Clematis drummondii Torr. & Gray

FIELD IDENTIFICATION. Perennial, clambering or straggling, delicately branched vine. Younger stems herbaceous, sometimes woody near base. Climbing by means of twisting leaf petioles. Very conspicuous in fruit because of long plumose styles.

FLOWERS. Usually March–September, flowers solitary, or in panicles of 10 or fewer in leaf axils of upper stems; dioecious or polygamo-dioecious; peduncles with simple or branched hairs; flowers white, about ¾ in. across; petals absent; sepals petaloid, ⅖–½ in. long, spreading, narrowly oblong, some slightly widened toward apex, silky-villous externally; stamens numerous, exserted, spreading, with short blunt anthers; pistils flattened and numerous.

FRUIT. Most abundant August–October, very conspicuous, borne in globose heads; achenes 1⁄12–⅛ in. long, brown, flattened, elongate, asymmetrical, pubescent, 1-seeded; styles very slender, 2–4 in. long, densely silky-plumose.

LEAVES. Opposite, remote on the stem, pinnately

Columbia Rock Clematis
Clematis columbiana (Nutt.) Torr. & Gray

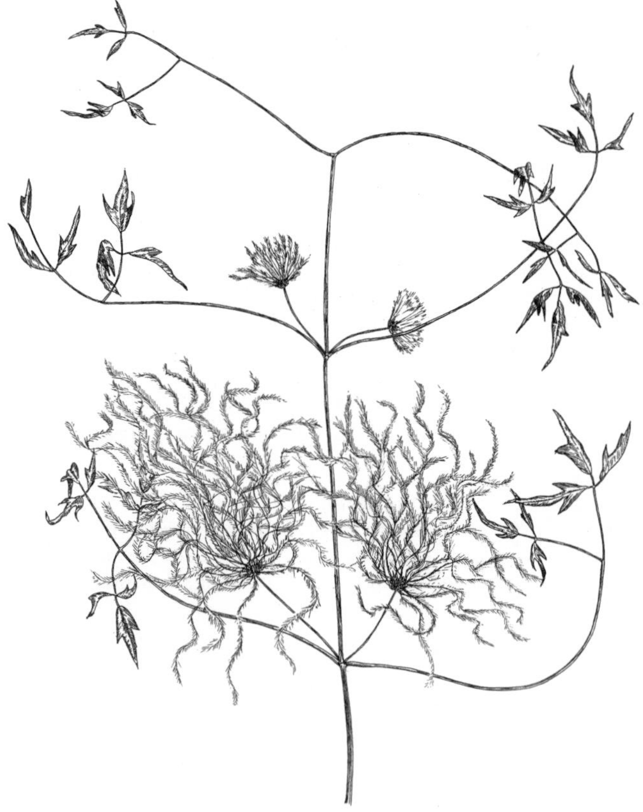

DRUMMOND CLEMATIS
Clematis drummondii Torr. & Gray

compound, 3–7-foliate; leaflets ½–2 in. long, lanceolate to ovate, margin coarsely toothed, lobed, or parted (often 3-lobed), the divisions more or less flaring, apex acute to acuminate or attenuate; surface usually cinereous-pubescent to gray pubescent, more glabrous with age; petioles long and slender, pubescent, twisting to assist the plant in climbing.

STEMS. Very slender, elongate and twisting, green, pubescent, angled; older stems tan to brown or gray, angled, glabrous, woody near base.

RANGE. Dry, well-drained soil in the sun; roadsides, fence rows, thickets, hillsides, and canyons. Central and western Texas, New Mexico, and Arizona. In Mexico in Tamaulipas, San Luis Potosí, and Sinaloa.

REMARKS. The genus name, *Clematis*, is the Greek name of a climbing vine. The species name, *drummondii*, is in honor of Thomas Drummond, an early botanist sent to Texas by the Glasgow Botanical Society. Vernacular names for the plant are Texas Virgins-bower, Graybeard, Grandad Beard, Old Man's Beard, Love-in-the-mist, Goat Beard, Barbas de Chivato, and Hierba de los Averos. Plants intermediate between Drummond Clematis and Western Virgins-bower are thought to be of hybrid origin.

WESTERN VIRGINS-BOWER CLEMATIS
Clematis ligusticifolia Nutt.

FIELD IDENTIFICATION. Vine climbing over bushes and rocks by means of tortuous petioles, sometimes attaining a length of 20–40 ft. Herbaceous above but somewhat woody near base.

FLOWERS. Often abundant, usually borne March–September, in terminal or axillary long-stalked cymes, staminate and pistillate flowers on different plants; peduncles 1–4 in. long, angled, pubescent; petals absent; sepals 4–5, white, petaloid, oblanceolate or oblong, spreading, about ⅜ in. long, tomentulose; stamens numerous, equaling or shorter than the sepals, anthers adnate (pistillate flowers with sterile stamens), carpels numerous, styles long-hairy.

FRUIT. Maturing August–November, achenes borne in large heads, ovoid, densely pubescent; mature styles elongate, 1–2 in. long, plumose with long straight hairs, 1-seeded.

LEAVES. Very variable in shape and size, opposite, pinnately compound, pinnae remote; leaflets 3–7 (usually 5) blade ovate to oblong or lanceolate, 1–3 in. long, apex acute or acuminate, base rounded to truncate or cuneate to cordate, margin coarsely toothed and often 3-lobed, thickish, yellowish green to bright green upper surface glabrous or slightly pubescent, lower surface somewhat more pubescent to glabrate; petioles strigose and twining like tendrils.

STEMS. Young ones green to tan or brown, angled, pubescent or strigose; older ones brown to gray and glabrous with stringy bark.

WESTERN VIRGINS-BOWER CLEMATIS
Clematis ligusticifolia Nutt.

RANGE. Prefers fertile, light, loamy soil. In thickets, along roadsides, or in moist canyons at altitudes of 3,000–8,500 ft. In western Texas; west through New Mexico and Arizona to California, and north to North Dakota, Montana, and British Columbia.

CLOSELY RELATED SPECIES. It is closely related to the Drummond Clematis, and is thought to hybridize with it in areas of contact.

VARIETIES. A California form has been given the name of *C. ligusticifolia* var. *californica* Wats., with leaves silky-tomentose beneath, but since so many intergrading plants are found many authors do not recognize the variety.

A plant with more pubescent leaves and more rounded teeth or lobes has been classified under the name of *C. neomexicana* Woot. & Standl. Intergrades between it and the species make this segregation dubious also.

C. ligusticifolia var. *brevifolia* Nutt. has leaves ovate in outline, usually cordate at base, nearly or quite glabrous, petiolules shorter than those of the species (hence the name *brevifolia*); branches of the inflorescence more spreading and shorter; sepals somewhat spatulate. A variety common in Washington and Oregon.

REMARKS. The genus name, *Clematis*, is the Greek name of a climbing vine. The species name, *ligusticifolia*, means "*Ligusticum*-like leaves." *Ligusticum* is a genus of the Umbelliferae. The vine is sometimes grown for ornament in mass effects for its plumose fruit. It has been cultivated since 1880. May be propagated by layers, divisions, or cuttings. Black-tailed deer and mule deer sometimes browse the foliage. It is reported that the Indians used the plant for treatment of colds and sore throat. The leaves were also used in an infusion for healing cuts and sores on horses.

New Mexican Clematis

Clematis neomexicana Woot. & Standl.

FIELD IDENTIFICATION. Woody climber from 3–6 ft, the stems striate and finely pubescent. Closely resembling *C. ligusticifolia.*

FLOWERS. Borne in a loose, few-flowered panicle on a peduncle about 2 in. long; pedicels about 1 in.; sepals 4, spreading, white, much exceeding the stamens, oblong-spatulate, apex obtuse, finely pubescent, about ½ in.; stamens erect, numerous, pistils numerous.

FRUIT. Borne in a globose head, the achenes orbicular or elliptic, flattened, densely hairy; styles tail-like, persistent, ⅗–1⅗ in. long, plumose.

LEAVES. Pinnately 5-foliate, on petioles 1½–2½ in.; leaflets ovate, 1⅗–2⅗ in. long, 1–1¾ in. wide, shallowly 3-lobed, never long-attenuate, the lobes entire or coarsely crenate with obtuse teeth; both surfaces finely pubescent, lower surface slightly paler.

Sweet-autumn Clematis
Clematis paniculata Thunb.

RANGE. To an altitude of 6,000 ft in the San Luis and Organ mountains of New Mexico. In Arizona in the Pinal and Chiricahua mountains.

RELATED SPECIES. The plant is closely related to *C. ligusticifolia,* but differs in the pubescent leaflets and stem, different form of leaflets, shorter tails of carpels, and shape of carpels. However, intermediate forms do occur, and some authors consider that they are synonymous.

REMARKS. The genus name, *Clematis*, is the ancient Greek name for a climbing vine. The species name, *neomexicana*, refers to its first being found in New Mexico.

Sweet-autumn Clematis

Clematis paniculata Thunb.

FIELD IDENTIFICATION. Vigorous woody climber attaining a length of 30 ft.

FLOWERS. Appearing in September, borne in many-flowered, axillary or terminal, fragrant panicles; each flower ½–1½ in. across, dull white; petals absent; sepals 4, oblong, nearly glabrous externally; stamens spreading, filaments longer than the blunt, linear-oblong anthers; pistils several in a globose head.

FRUIT. Achenes ⅕–¼ in. long, somewhat asymmetrical, margined, thinly appressed-pubescent, styles plumose.

LEAVES. Ternate or pinnate with 3–5 leaflets, the divisions often lobed or entire, varying from ovate to elliptic or orbicular, apex acuminate or acute, base subcordate or rounded, surfaces glabrous; coriaceous; leaflet blades 1–4 in. long.

STEMS. Slender, green to brown, glabrous or nearly so.

RANGE. Japan and Korea, introduced into cultivation in the United States, and escaping cultivation to appear along roadsides and in thickets. Arkansas and Tennessee; eastward to Florida and northward to Canada.

VARIETY. Yam-leaf Sweet-autumn Clematis, *C. paniculata* var. *dioscoreaefolia* (Levl. & Vant.) Rehd., has leaflets generally ovate to cordate, apex rounded to mucronate or emarginate, thickened; sepals broader, obovate-oblong to oblong.

REMARKS. The genus name, *Clematis*, is a name given by Dioscorides for a plant with slender branches, from *clema* ("a shoot"). The species name, *paniculata*, refers to the flower panicles. The vine was first introduced into American gardens from Korea in 1864. It and its varieties are the most common of the fall-blooming clematis in gardens, and will stand severe pruning.

Catesby Clematis

Clematis catesbyana Pursh

FIELD IDENTIFICATION. Slender vine attaining a length of 6 ft or more. Essentially a perennial herbaceous vine, but sometimes confused with other more woody species of *Clematis*.

FLOWERS. A few borne in a panicle, dioecious; petals absent; calyx petaloid, about ¾ in. wide, sepals 4, linear to cuneate or oblong to spatulate, ¼–½ in. long, obtuse, pubescent, thin, white, spreading; stamens numerous and spreading, anthers short and blunt (pistillate flower with sterile stamens); pistils several, carpels with a single ovule each, style elongate.

FRUIT. Achene body about ⅙ in. long, orbicular or somewhat inequilateral, flattened, 1-seeded; style persistent on the achene as a plumose tail 1–1⅝ in. long at maturity.

LEAVES. Numerous, opposite, two to three times compound; leaflet blades suborbicular, ovate or lanceolate, apex acute or acuminate, margins usually 3-lobed, less commonly toothed or entire, surfaces usually finely pubescent but with age subglabrous; membranous.

RANGE. In sandy woods; Louisiana, eastward to Florida and northward to South Carolina.

CLOSELY RELATED SPECIES. The plant is sometimes listed as a variety under the name of *C. virginiana* var. *catesbyana* Britt.

REMARKS. The genus name, *Clematis*, is a name given by Dioscorides for a climbing vine. The species name, *catesbyana*, is in honor of Mark Catesby (1679–1749), author of natural histories of parts of North America. It is also known as Leather-flower.

Eastern Virgins-bower Clematis

Clematis virginiana L.

FIELD IDENTIFICATION. Half-woody vine climbing to a height of 20 ft by twisting leaf petioles.

FLOWERS. Maturing July–September in axillary, leafy-bracted panicles or corymbs; flowers numerous, creamy-white; petals absent; sepals 4, valvate in the bud, wide-spreading, thin, oblong-lanceolate or spatulate, ¼–½ in. long, pubescent externally (flower ⅗–1¼ in. broad when expanded); stamens numerous, spreading in a head, anthers blunt and about $\frac{1}{25}$ in. long, filaments glabrous; pistillate flowers with sterile stamens, pistils numerous in a head, styles elongate.

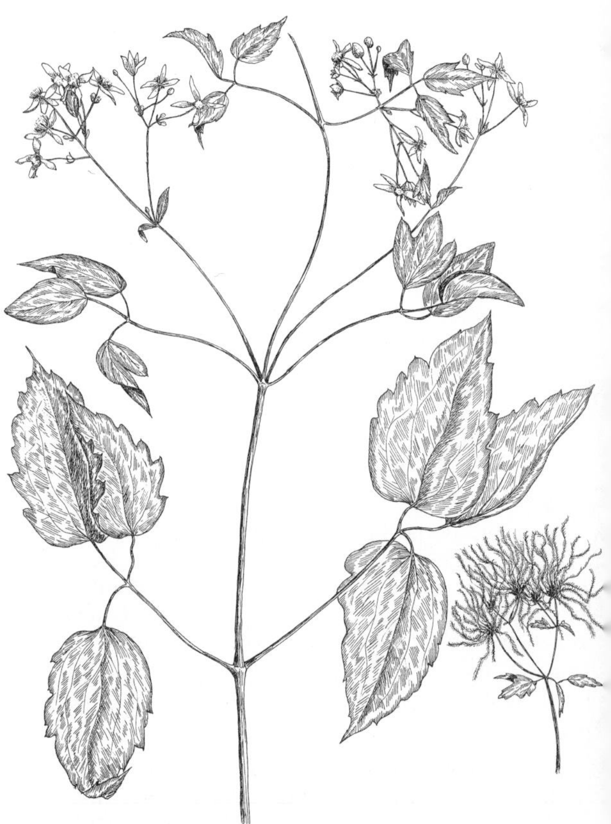

Eastern Virgins-bower Clematis
Clematis virginiana L.

FRUIT. Maturing August–September, in large fruiting heads; achenes 1/6–1/5 in. long, very inequilateral, brown, villous-hirsute, styles plumose and about 1 in. long.

LEAVES. Opposite, 2–3½ in. long, mainly 3-foliate, lower sometimes pinnately 5-foliate; leaflets thin, broad-ovate, apex acute or acuminate, base subcordate or rounded, margin entire, or incisely few-toothed and somewhat lobed; surfaces sparingly hairy to almost glabrous when mature; petioles green to brown, twisting on supports to pull the plant upward.

STEMS. Slender, green to brown, sparingly pubescent or glabrate.

RANGE. Low grounds along streams, woodlands, and thickets. Arkansas, Louisiana, Mississippi, Alabama, and Georgia; northward to Nova Scotia and westward to Manitoba and Kansas.

VARIETY. Missouri Virgins-bower, *C. virginiana* var. *missouriensis* (Rydb.) Palmer & Steyerm., has 5 leaflets 2–3 in. long which are densely pilose beneath.

REMARKS. The genus name, *Clematis*, is a name given by Dioscorides to a slender climbing plant. The species name, *virginiana*, is given for the state of Virginia, or the colonial territory once assigned broadly under that name. Vernacular names are Devil's Darning Needle, Herbe aux Gueux, and Gander-vine. The plant was first introduced into cultivation in 1720.

YELLOW-ROOT
Xanthorhiza simplicissima Marsh.

YELLOW-ROOT

Xanthorhiza simplicissima Marsh.

FIELD IDENTIFICATION. Weak diffuse shrubby plant 4–24 in., with long yellow roots. Often sending up suckers in the spring.

FLOWERS. Borne in spring in slender drooping panicles 2–6 in., pediceled, solitary or several together, petals absent; sepals 5, petalloid, spreading, about 1/8 in. long, brownish purple, ovate-lanceolate, apex acute, deciduous; staminodia 5, clawed, 2-lobed; stamens 5–10, filaments stout; pistils 5–10, style subulate; ovary sessile, 2-ovuled, one ovule suppressed at maturity.

FRUIT. Follicles sessile, whorled, divergent, about 1/8–1/6 in. long, obliquely oblong or elliptic, curved, minutely beaked, inflated, light yellow to brownish, thin-walled; seed solitary and distal.

LEAVES. In a crowded cluster, pinnate or bipinnate, slender-petioled, 6 in.–1 ft long; leaflets 3–5, 1–3 in. long, blade ovate to elliptic or oblong, apex acute to acuminate; base cuneate, sessile or nearly so, margin incised, toothed, cleft or divided, thin, shining, sparingly pubescent to glabrous.

STEMS. Short, brittle, glabrous, brown to gray, leaf scars narrow and numerous, inner bark yellow and bitter.

RANGE. Sandy or rocky soil in moist shady woods, banks, or on ledges. Seemingly escaped from cultivation only in Louisiana. Native from Alabama and Tennessee to Florida, and northward to New York.

VARIETY. A horticultural variation has been given the name of *X. simplicissima* var. *ternata* Huth. It has more deeply lobed leaflets, usually trifoliate, turning golden yellow in fall. It is sometimes used for ground cover in shady places.

REMARKS. The genus name, *Xanthorhiza*, is from the Greek word *xanthos*, meaning "yellow," and *rhiza*, meaning "root," which refers to the yellow roots. The species name, *simplicissima*, refers to its unbranched habit. Some vernacular names for the plant are Brook-feather and Parsley-leaf Yellow-root. The plant is propagated by seed or by root divisions in autumn or early spring.

BARBERRY FAMILY (*Berberidaceae*)

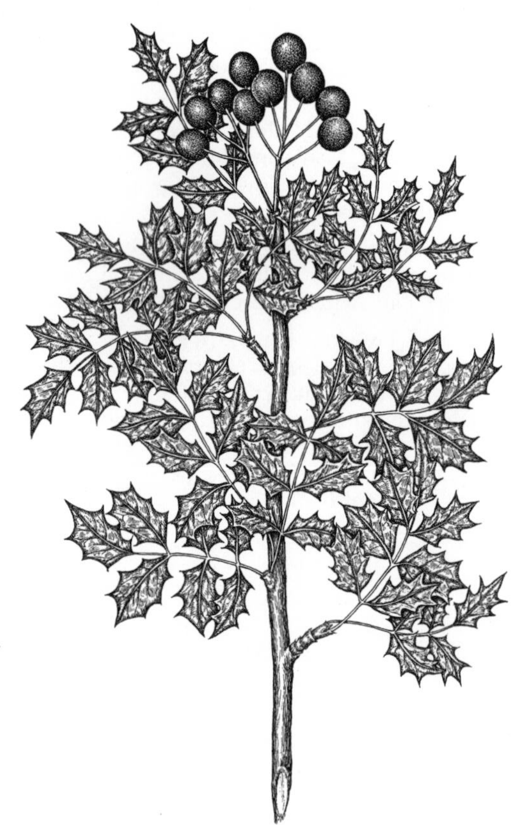

FREMONT MAHONIA
Mahonia fremontii (Torr.) Fedde

FREMONT MAHONIA

Mahonia fremontii (Torr.) Fedde

FIELD IDENTIFICATION. Western evergreen upright shrub, often in dense rounded clumps with many rigid stems from the base, 3–8 ft. Occasionally treelike in appearance and to 15 ft.

FLOWERS. May–July, racemes 3–8-flowered, fascicled terminating short stubby branches, from axis of bud scales, erect, 1–1½ in. long; peduncles ½–3 in., with pedicels ¼–¾ in.; flowers yellow, regular, perfect, hypogynous; petals and sepals imbricate and deciduous; petals 6, in 2 series, obovate, concave, glandular at base; sepals 6, in 2 series; stamens 6, anthers opening by uplifting valves; ovary superior, 1-celled, ovules few, style short, stigma peltate and depressed.

FRUIT. Maturing August–September, berry ⅓–½ in. in diameter, globose to ovate, bluish black, dry and inflated at maturity; 1–3 seeded.

LEAVES. Alternate, evergreen, 1½–3 in. long, odd-pinnately compound of 3–7 (usually 5) leaflets; leaflets ovate-oblong to lanceolate in outline, length ½–1½ in., width about ½ in., stiff and rigid; marginal teeth repand, large, remote, usually less than 5 to a side; surface bluish green, moderately glaucous, reticulate; lateral leaflets generally very short-petioled, the terminal ones longer; terminal buds with numerous persistent pointed scales.

RANGE. Well-drained sunny slopes, canyons, and mesas, at altitudes of 4,000–7,000 ft. Extreme western Texas, New Mexico, Arizona, California, Colorado, Utah, and Nevada; in Mexico in Baja California and Sonora.

REMARKS. The genus name, *Mahonia*, honors Bernard M'Mahon, American horticulturist (1775–1816). The species name, *fremontii*, is for General Charles Fremont (1813–1890), politician, soldier, and explorer of the western United States. The plant is also listed in the literature under the names of *Odostemon fremontii* (Torr.) Abrams and *Berberis fremontii* Torr. Also known under the vernacular names of Holly-grape, Desert Barberry, and Desert Mahonia. *M. haematocarpa* is a closely related species with red fruit about ⅓ in. across. The wood of Fremont Mahonia is used by the Indians for making small woodenware articles. From the root was prepared a tonic and also a yellow dye. The leaves and fruit are eaten by mule deer, black-tailed deer, and New Mexico black bear. In common with other Mahonias, the plant contains the alkaloid berberine. Because of its being a host for the black stem-rust (wheat rust), it is generally considered a menace in cereal-raising areas. It is rarely cultivated for ornament, but may be propagated by seeds, layers, or cuttings of half-ripe wood under glass. Germination about 90 per cent.

Red Mahonia

Mahonia haematocarpa (Woot.) Fedde

FIELD IDENTIFICATION. Western spiny-leaved shrub 3–12 ft.

FLOWERS. Racemes 1½–3 in. long, few-flowered, springing from the axils of numerous crowded bud scales; pedicels slender, mostly opposite; flowers ¼–⅜ in. across, perfect, regular, yellow, perianth segments free; sepals and petals in 2 series; petals 6, smaller than sepals, nectaries at base; sepals 6–9; stamens 6, included, filaments distinct, anthers opening by 2 uplifting apical valves; ovary 1-celled, ovules few.

FRUIT. Berry June–August, blood-red, globose or nearly so, about ⅓ in. across, juicy, acidulous, few-seeded, seeds oblong with a crustaceous testa. Fruit-bearing age about 4 years. Seed averages 103,000 seeds per lb.

LEAVES. Numerous, evergreen, alternate, some clustered, odd-pinnately compound, leaflets 3–7, (mostly 5), ovate to lanceolate-oblong, some falcate, apex long-acuminate, margin sinuate-dentate with spiny teeth, base acute to broadly cuneate and sessile in lateral leaflets; stiff, coriaceous, some twisted; surface grayish green, reticulate-veined, some glaucous, blade length 1–2½ in., terminal leaflet long-attenuate, usually longer than laterals and short-petioled; stipules minute and subulate; terminal buds with numerous pointed scales.

Creeping Mahonia
Mahonia repens (Lindl.) G. Don

Red Mahonia
Mahonia haematocarpa (Woot.) Fedde

TWIGS. Stout, rather stiff, rough with numerous imbricate scales, older twigs gray to reddish brown, some striate. Inner bark and wood yellow.

RANGE. On dry sunny sites at altitudes to 4,400 ft. Trans-Pecos Texas, New Mexico, Arizona, and adjacent Mexico.

REMARKS. The genus name, *Mahonia,* honors Bernard M'Mahon, American horticulturist (1775–1816). The species name, *haematocarpa,* refers to the blood-red berries which are pleasantly acid to the taste, and are used for making jellies. The fruit is also eaten by many species of birds including Gambel's quail. The shrub is evergreen and worthy of cultivation. It is susceptible to the black stem rust (wheat rust) of cereals.

Creeping Mahonia

Mahonia repens (Lindl.) G. Don

FIELD IDENTIFICATION. Evergreen stoloniferous shrub 4–12 in. Stems ascending from prostrate stolons, very short or almost absent.

FLOWERS. Borne April–June, yellow racemes 1¼–3 in. long, many-flowered, at ends of branches, pedicels opposite or alternate; sepals petal-like; petals 6, obovate, concavo-convex, in 2 series, each bearing 2 small glands at base; stamens 6, opposite petals, anthers opening by 2 uplifting valves; stigma circular, depressed, sessile; ovary superior, 1-celled.

FRUIT. Pedicels ¼–½ in., glabrous; berry ovoid to spheroid, black with a glaucous bloom, ¼–⅓ in. thick, rather dry at maturity; 1–few-seeded.

LEAVES. Evergreen, alternate, blades 3–10 in. long, pinnately compound of 3–7 (usually 5) leaflets, leaflets 1–2½ in. long, ⅝–1¾ in. wide, ovate to oblong, apex acute to obtuse or rounded, spinulose-pointed, base oblique or slightly obcordate, margin spinulose-dentate, coriaceous, dull bluish green, glabrous, minutely reticulate-veined, lower surface paler and minutely papillose; lateral leaflets sessile, terminal petioled; petioles ¾–2 in., somewhat clasping; stipules minute and subulate.

TWIGS. Tan to brown, glabrous, roughened by old leaf scars, inner bark yellowish.

RANGE. Shaded hillsides, often in coniferous woods, at altitudes of 4,500–10,000 ft. Western Texas (Guadalupe Mountains), New Mexico, Arizona, and California; and north to Nebraska, Wyoming, Washington, and British Columbia.

VARIETIES. A number of varieties of Creeping Mahonia have been described as follows:

Big-berry Creeping Mahonia, *M. repens* var. *macrocarpa* Jouin., has upright stems to 3 ft and the 5–7 leaflets are spiny-dentate and slightly lustrous; fruit somewhat larger also.

Round-leaf Creeping Mahonia, *M. repens* var. *rotundifolia* (May) Jouin., has leaflets 5–7, broad-ovate, sparingly serrate or nearly entire.

Lap-leaf Creeping Mahonia, *M. repens* forma *subcordata* Rehd., has leaflets 5–7, crowded and overlapping, broad-ovate, base subcordate, margin few-toothed.

REMARKS. The genus name, *Mahonia,* honors Bernard M'Mahon (1775–1816), a prominent American horticulturist. The species name, *repens,* refers to its creeping habit. It has been cultivated since 1822, and is resistant to black stem rust. It is sometimes used as a ground cover for erosion control. The fruit is made into jelly, and the fruit is eaten by a number of species of birds. The foliage is browsed by black-tailed and white-tailed deer, but is reported to be poisonous to livestock. It is propagated by seeds, suckers, cuttings, and layering.

Texas Mahonia

Mahonia swaseyi Buckl.

FIELD IDENTIFICATION. Rather rare shrub to 3 ft, with erect or spreading branches.

FLOWERS. Usually in May, solitary, or in peduncled racemes ¾–3 in. long; flower heads about ⅜ in. across, perfect, yellow; sepals 6, also with 2–3 bractlets, the inner petaloid; petals 6, obovate, concave, smaller than sepals, imbricate in 2 rows, nectaries at base; bracts broad-ovate to suborbicular, about ⅕ in. long; stamens 6, filaments distinct, anthers extrorse and opening by 2 apical pores; ovary 1-celled, sessile, ovules few; stigma peltate and depressed.

FRUIT. Usually borne abundantly, berries subglobose, ⅖–½ in. in diameter, translucent, yellowish white or reddish-tinged, pulpy, pleasantly acid; seed solitary or a few, oblong, slightly flattened, black, lustrous, about ⅛ in. long.

LEAVES. Persistent, alternate or fascicled, odd-pinnately compound of 3–11 (mostly 5–9) leaflets; leaflets oval to oblong or elliptic to lanceolate, some falcate, mostly ⅜–1½ in. long (to 4 in.), basal leaflets usually smaller, margin repand-dentate with spiniferous teeth, acuminate or acute, base broadly acute to rounded or unequally truncate; stiff, leathery, thin, paler beneath and reticulate-veiny; lateral leaflets sessile, terminal short-petioluled.

TWIGS. When young very scaly and rough, reddish brown; when older light to dark gray, with confluent shallow grooves.

TEXAS MAHONIA
Mahonia swaseyi Buckl.

RANGE. Dry, well-drained soil in the sun. Known from the limestone hill region of central Texas. Specimens examined from the United States National Herbarium, E. J. Palmer, No. 12199, along rocky streams, "Devil's Backbone," Comal County, Texas. B. C. Tharp, No. 1437, Edwards Plateau near Austin, Texas.

REMARKS. The genus name, *Mahonia*, honors Bernard M'Mahon (1775–1816), American horticulturist, and the species name, *Swaseyi*, for the botanist Swazey. The plant was introduced into cultivation in 1917. It makes a very good ornamental hedge plant and is resistant to cold and drought. It is susceptible to black stem rust of wheat. First discovered along the Pedernales River in Texas.

Laredo Mahonia

Mahonia trifoliolata (Moric) Fedde

FIELD IDENTIFICATION. Evergreen shrub to 10 ft, with stiff, spiny, holly-like, trifoliolate leaves.

FLOWERS. In racemes, on pedicels ⅓–¾ in.; yellow, perfect, corolla ⅜–½ in. across; sepals 6, petal-like, greenish yellow, in 2 series, 3 long and 3 short, obovate; petals 6, in 2 series, outer 3 oval-obovate, concave, clawed, yellow with reddish streaks, inner 3 erect, ovate-oval; stamens 6, opposite the petals, filaments sensitive to touch, anthers opening by valves; pistil single, stigma circular, depressed.

FRUIT. Berry ripening in June, on short peduncles ⅓–¾ in. long, obovate-oval, somewhat flattened, red, lustrous, aromatic, about ⅓ in. long, pulpy, acid; seeds 1 to several, crustaceous.

LEAVES. Alternate, blades 2–4 in. long, 3-foliolate; leaflets sessile, lanceolate to oblong or elliptic, apex acuminate-spiny, margin 3–7-lobed and spiny-pointed, 1–2½ in. long, ½–1½ wide, thick, rigid, coriaceous, pale green, upper surface glabrous and reticulate white-veined, lower surface paler and less prominently reticulate; petiole reddish green, glabrous, 1–3 in.

TWIGS. Young shoots smooth and reddish green, older twigs gray to reddish brown.

BARK. On older stems gray to reddish brown, yellow within, broken into thin, small scales. Wood yellow.

RANGE. On dry, stony hillsides over most of Texas except in the east and southeast portions; west through New Mexico to Arizona, and south into Mexico in the states of Chihuahua, Coahuila, Nuevo León, and San Luis Potosí.

MEDICINAL USES. A decoction of the roots was used in frontier times as a remedy for toothache. The root contains 1.3 per cent berberine and 0.1 per cent other alkaloids. In official medicine the physiological action of the alkaloid berberine has been found to be relatively feeble, although in sufficient quantities it will cause pronounced and even fatal poisoning. Berberine

Laredo Mahonia
Mahonia trifoliolata (Moric) Fedde

also is toxic in its effect on protozoa and other lower forms of life, and it is sometimes used for the treating of certain tropical sores. In the United States it is little used medicinally except as a stomachic. (See Wood and Osol, 1943, pp. 203-205.)

REMARKS. The genus name, *Mahonia*, is in honor of Bernard M'Mahon (1775–1816), and the species name, *trifoliolata*, refers to the trifoliolate leaves. Vernacular names are Wild Currant, Chaparral Berry, Agarita, Agrito, Algerita, Agrillo, and Palo Amarillo. The red, acid berries make excellent jellies and wine, and are gathered by threshing the thorny bush with a stick. The roasted seed has been reported as used for a coffee substitute. Birds eagerly devour the berries. In some areas the plant is used as a hedge, and the flowers are considered a good bee food. However, it is susceptible to black stem rust. A yellow dye is made from the wood and roots.

Wilcox Mahonia

Mahonia wilcoxii (Kearney) Heller

FIELD IDENTIFICATION. Straggling spiny-leaved shrub to 6 ft.

FLOWERS. Inflorescence elongate, racemose, fragrant, yellow; pedicels opposite or alternate, about ¼ in. long; flowers perfect, regular, petals and sepals 6, distinct and free, each in 2 series; anthers opening by 2 apical pores; pistil simple, style 1, stigma dilated, ovary 1-celled.

FRUIT. Berry ovoid to ellipsoid, about ⅜ in. in diameter, bluish black, glaucous, seeds numerous.

LEAVES. Rachis 3–8 in. long, green to straw-colored

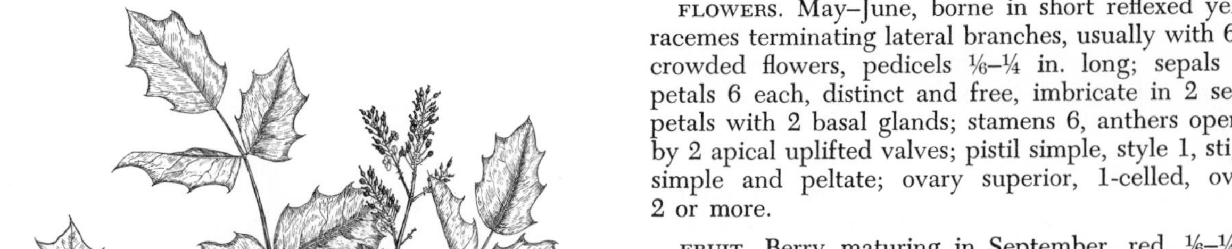

WILCOX MAHONIA
Mahonia wilcoxii (Kearney) Heller

or purplish, the compound leaves composed of 5–7 leaflets (rarely 3 or 9), blade oblong to ovate or rounded, margin set with coarse, remote, spiny teeth, apex acute to obtuse and spine-tipped, base rounded, asymmetrical or obliquely truncate, blade length 1–2¼ in., upper surface olive-green, glabrous, semilustrous, finely reticulate; lower surface paler green, dull, more strongly reticulate-veined; lateral leaflets sessile, and terminal rather long-petiolulate.

TWIGS. Stout, roughened by abundant leaf and inflorescence scars, young ones brown, older light to dark gray and glabrous, inner bark and wood yellow.

RANGE. Usually in dry, well-drained soil of the pinyon belt at altitudes of 5,000–8,500 ft. New Mexico, Arizona, and Sonora, Mexico.

REMARKS. The genus name, *Mahonia*, is in honor of Bernard M'Mahon (1775–1816), American horticulturist. The species name, *wilcoxii*, is named for T. E. Wilcox, the botanist who collected it in the Huachuca Mountains of Arizona. The plant may be propagated by cuttings, by layering, by grafting, or from seed.

FENDLER MAHONIA

Mahonia fendleri (Gray) W. & S.

FIELD IDENTIFICATION. Erect spiny shrub 1–6½ ft.

FLOWERS. May–June, borne in short reflexed yellow racemes terminating lateral branches, usually with 6–10 crowded flowers, pedicels ⅙–¼ in. long; sepals and petals 6 each, distinct and free, imbricate in 2 series, petals with 2 basal glands; stamens 6, anthers opening by 2 apical uplifted valves; pistil simple, style 1, stigma simple and peltate; ovary superior, 1-celled, ovules 2 or more.

FRUIT. Berry maturing in September, red, ⅙–¼ in. long, oval to short-ellipsoid, 1–3 seeded.

LEAVES. Alternate, deciduous, apparently simple and fasciculate, but really compound unifoliate, the lower leaflets becoming rigid spines ⅓–2½ in., blade ½–2½ in., elliptic to oblanceolate or spatulate, apex acute to obtuse, base gradually narrowed, margin entire or spiny-toothed; petiole short or leaflet sessile.

TWIGS. Smooth, lustrous, appearing varnished, reddish to purplish brown, later gray, spines 3-parted, present only at nodes.

RANGE. Open hillsides and valleys at altitudes of 4,500–8,500 ft. In New Mexico and Colorado.

REMARKS. The genus name, *Mahonia*, is in honor of Bernard M'Mahon, American horticulturist (1775–1816). The species name, *fendleri*, is in honor of Augustus Fendler, German naturalist and explorer of New Mexico. The shrub is worthy of cultivation as a hedge plant, but is susceptible to the black stem rust. The wood is yellow and rather weak.

FENDLER MAHONIA
Mahonia fendleri (Gray) W. & S.

MOONSEED FAMILY (*Menispermaceae*)

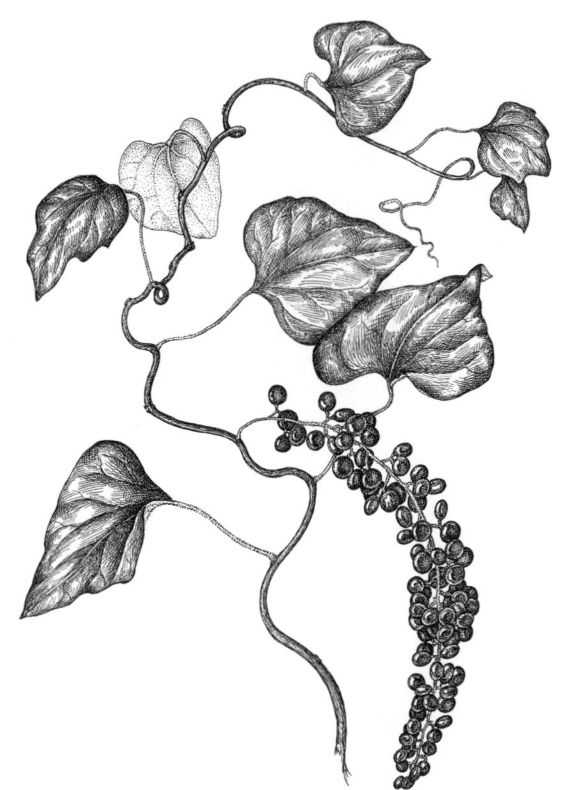

CAROLINA SNAILSEED-VINE
Cocculus carolinus (L.) DC.

CAROLINA SNAILSEED-VINE
Cocculus carolinus (L.) DC.

FIELD IDENTIFICATION. Slender, twining vine bearing racemes of brilliant red, somewhat flattened drupes.

FLOWERS. In loose racemes or panicles, 1–6 in., staminate in compound racemes, pistillate in simple racemes; corolla small, about 1/12 in. broad, greenish white; petals 6, in 2 series, smaller than the inner sepals, about 1/25 in., concave, erose, the outer 3 oblong, the inner 3 oval; stamens 6, clublike, anthers 4-celled, filaments distinct; carpels 3–6, erect, stigma entire and subulate; stamens more or less reduced, or lacking in pistillate flowers.

FRUIT. Drupe brilliant red, very conspicuous, in racemes 1–6 in. long, and hardly over 2 in. wide, primary pedicels ½–1 in., secondary pedicels about 1/16–¼ in.; body of drupe shiny, flattened laterally, about ¼ in., pulpy; stone solitary, flattened, curled into a spiral, crested on back and sides.

LEAVES. Simple, alternate, blades 2–4 in., broadly ovate to cordate, margin entire or 3–5 lobed; apex obtuse to acute or rounded, base truncate or cordate, 3–5 nerved; upper surface dark green and glabrous or pubescent, lower surface paler and densely pubescent, petioles slender, pubescent, ¾–4 in.

STEMS. Slender, twining, green, and pubescent, indistinctly and finely grooved, older stems green to brown, glabrous, somewhat woody toward base.

RANGE. Usually climbing on fences, hedgerows, and low shrubs, in well-drained moist soil in the sun. Texas and Louisiana; eastward to Florida, and north through Oklahoma, Arkansas, and Missouri to Kansas, Illinois, and Virginia. Also in Tamaulipas and Nuevo León, Mexico.

REMARKS. The genus name, *Cocculus*, refers to the curled, snail-like seed, and the species name, *carolinus*, is for Carolina. The author has been unable to ascertain whether the reference refers to North Carolina or South Carolina, or both. Vernacular names are Coral-bead, Coral-seed, Coral-vine, Moonseed, Red-berry Moonseed, Carolina Moonseed, Red Moonseed, Wild-sarsaparilla, Margil, and Hierba del Ojo.

This climbing vine could be more extensively cultivated for ornament because of the brilliant red fruit. It is consumed by a number of species of birds. The vine is sometimes mistaken for a kind of Smilax-vine, but most of the species of *Smilax* have much more leathery and stiff foliage, and thorny stems. The red fruit also distinguishes it from closely related species of the Moonseed Family.

Diverse-leaf Snailseed-vine

Cocculus diversifolius DC.

FIELD IDENTIFICATION. Slender, semievergreen, woody vine. Distinguished from other members of the Moonseed Family by the color of the fruit and the narrow leaves.

FLOWERS. In short axillary racemes ½–2 in.; pedicels 1/25–¼ in., glabrous; flowers dioecious, regular, small, about 1/16–⅛ in. across, cup-shaped; petals 3–6, white, about 1/16 in., ovate to oval, concave, margin erose, apex obtuse or rounded; calyx with 6 sepals in 2 sets (3 long and 3 short), petaloid, mostly greenish, elliptic; stamens 6, barely included, filaments erect or incurved, short, anthers 4-celled; style short and subulate, stigma minute; ovary incurved in fruiting so that seed and embryo form a coiled ring.

FRUIT. Drupe solitary, or in clusters of 2–6, on individual peduncles 1/16–¼ in., subglobose, bluish black, glaucous, somewhat flattened, to ¼ in., fleshy with purple juice; seed solitary, about 3/16 in. long, flattened, coiled like a snail, dorsal side sutured, ventral side with rugose transverse ridges, embryo curved, with flat narrow cotyledons.

LEAVES. Simple, alternate, deciduous, very variable in shape, cordate, ovate, oblong, or linear; apex obtuse to rounded, often notched and minutely mucronulate; base cordate or rounded, margin entire and slightly revolute, blade 1–3 in. long, ⅓–2½ in. wide, firm and leathery; upper surface dull green, glabrous, semilustrous, finely reticulate-veined; lower surface paler, glabrous or occasionally with slight puberulence on the veins toward the base, delicately reticulate-veiny; 3 main veins from the base, the 2 outer marginal; petiole ⅛–¾ in., pubescent with short, fine, appressed hairs.

STEMS. Twining on fences or other plants, older stems gray or brown, somewhat roughened by short, confluent ridges; younger stems green to gray or brown, finely grooved, pubescent with fine, white, appressed hairs.

RANGE. Western and southwestern Texas, New Mexico, and Arizona; and southward into Mexico. Rather common in the lower Rio Grande Valley of Texas in Cameron County. Ascending to altitudes of 5,000 ft in the mountains of New Mexico and Arizona. In Mexico from Tamaulipas to Sonora and Oaxaca.

REMARKS. The genus name, *Cocculus*, refers to the curled, snail-like seed, and the species name, *diversifolius*, is for the diverse shape of the leaves. Also known under the vernacular name of Sarsaparilla, and the Mexican people of southwest Texas call it Correhuela. Only a few species of *Cocculus* are cultivated for ornament, although some are very attractive. They are usually propagated by seeds, or by half-ripened wood in summer under glass.

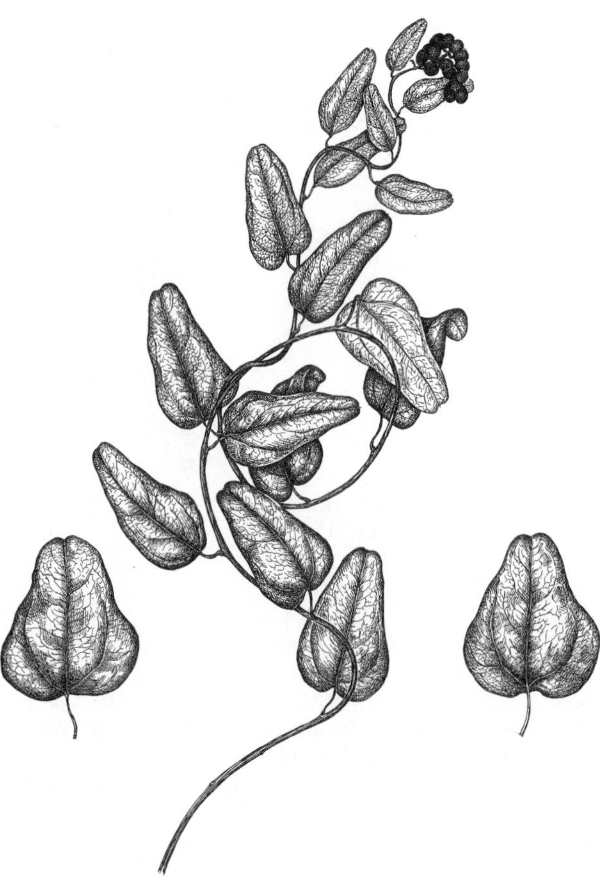

Diverse-leaf Snailseed-vine
Cocculus diversifolius DC.

Common Moonseed-vine

Menispermum canadense L.

FIELD IDENTIFICATION. Climbing vine, 6–12 ft, herbaceous above, somewhat woody near the base, fast growing.

FLOWERS. June–July, in loose, axillary, bracteolate panicles; individual flowers imbricate in the bud, hypogynous, in separate staminate and pistillate inflorescences; sepals 4–8, about 1/15 in., in 2 series, elliptic to oval, longer than petals; petals 4–8, in rows, about 1/8 in. wide, somewhat fan-shaped, sides involute, clawed, greenish white; stamens 9–24 (often reduced to 5 staminodia in the pistillate flower), filaments distinct, anthers 4-celled; pistils 2–4, raised on a short common receptacle, stigma broad and flat, ovule solitary.

FRUIT. Maturing September–October, in grapelike clusters, drupes ¼–1/6 in. in diameter, globose to ovoid, laterally somewhat flattened, black with a bloom; the solitary seed spirally curled or crescent-shaped.

LEAVES. Simple, alternate, broadly ovate to oval, margin entire and rounded or 3–7 angulate-lobed, apex acuminate to acute or obtuse, base cordate to truncate, barely peltate at the base of some, blade length 2–8 in., upper surface dull green and glabrate, much paler green beneath, membranous; petioles slender and elongate, about as long as the blades.

STEMS. Slender, twining, herbaceous toward the ends, green to brown or straw-colored, finely pubescent to glabrous later, finely grooved.

RANGE. Rich woods, thickets, and stream banks; Oklahoma and Arkansas; eastward to Alabama and Georgia; north to Missouri, Nebraska, Manitoba, and Quebec.

REMARKS. The genus name, *Menispermum*, is a combination of Greek words *meni,* ("moon") and *spermum* ("seed"). The species name, *canadense,* denotes its growth in Canada. Some vernacular names for the vine are Texas Sarsaparilla, Yellow Sarsaparilla, Yellow Parilla, Canadian Moonseed, Vine-maple, and Raisin de Couleuvre.

The rhizome, now thought to be inert, was at one time used as substitute for Sarsaparilla. The fluidextract was recognized by the *U. S. Pharmacopoeia,* 1890 (dose ½–1 fluid drachm). Barber (1884, p. 401) obtained a white amorphous alkaloid for which Maisch proposed the name "menispine." (See Wood and Osol, 1943, p. 1438.)

The fruit is considered poisonous to human beings, and is seldom eaten by livestock. It is occasionally grown for ornament, either from seed or ripened wood.

COMMON MOONSEED-VINE
Menispermum canadense L.

MAGNOLIA FAMILY (*Magnoliaceae*)

FLORIDA ANISE
Illicium floridanum Ellis

FLORIDA ANISE
Illicium floridanum Ellis

FIELD IDENTIFICATION. Strongly odorous, evergreen shrub or tree to 25 ft, with a trunk attaining about 4 in. in diameter. The trunk is often leaning or crooked, and the branches slender, forming an open round top.

FLOWERS. Peduncles long, nodding, ⅝–4 in. long, somewhat flattened, green to brown, glabrous; corolla 1–1½ in. wide, solitary or a few together in the leaf axils, unpleasantly scented, perfect; petals 20–30, deciduous, imbricate in rows, linear to narrowly lanceolate (inner ones narrower), ⅗–⅘ in., dark crimson and purplish; sepals 3–6, petaloid, over ⅖ in., ovate to oblong, membranous; stamens numerous, in several series, filaments mostly linear, thickened upward, longer than anthers; style recurved, ovary solitary and ascending; carpels in a single series forming dry follicles.

FRUIT. On elongate, drooping peduncles, ¾–4 in. Forming a star-shaped whorl of dry, crustaceous 1-seeded follicles, 1–1⅓ in. broad; each follicle sessile, flattened, ovate, acuminate, ⅜–⅝ in. long, splitting at the top to release the seeds which are oval, solitary, brown, lustrous and about ¼ in.

LEAVES. Alternate, mostly clustered terminally, simple, exstipulate, evergreen, unpleasantly scented, elliptic to oblong or lanceolate, apex acute or acuminate, base acute or cuneate, margin entire and slightly revolute, dark green and glabrous above, paler and glabrous beneath and minutely glandular-dotted, veins obscure, leathery and somewhat fleshy, blade length 2½–6 in.; petiole ⅜–1 in., grooved, glabrous.

TWIGS. Straight, ascending, smooth, green at first, later reddish brown to gray.

BARK. Dark brown to black or gray, surface broken into small flakes with shallow fissures.

RANGE. Moist, sandy soils along streams or swamps. In Louisiana in Washington Parish near Varnado, also at the crossing of the Tangipahoa River and Highway No. 35 in Tangipahoa Parish; eastward to Florida.

CHEMICAL PROPERTIES. The volatile oil from the leaves contains about 90 per cent safrol. A decoction of the seeds is said to produce violent gastrointestinal irritation, followed by motor and sensory paralysis with convulsions and death if the dose is sufficient. A poisonous glycoside from the kernel has been obtained, and a crystaline alkaloid from the leaves.

REMARKS. The genus name, *Illicium*, means "allurement," in reference to the flowers, and the species name, *floridanum*, means "of florida." Other vernacular names are Polecat-tree, Poison Bay, Sweet-laurel, Swamp-anise, Star-anise, Star-bush, Stink-bush, Purple Anise-tree, Matacaballos (horsekiller), and Ixcapantl. The illiciums may be propagated by seeds or by cuttings of ripened wood. The leaves are reported as being poisonous to stock.

Tulip-tree

Liriodendron tulipifera L.

FIELD IDENTIFICATION. Large, handsome tree capable of attaining a height of 100–200 ft, and a diameter of 4–10 ft. Lower part of trunk generally devoid of branches, upper branches forming a small, pyramidal to oblong crown. Trunks notable for columnar grandeur and straight, limbless uniformity.

FLOWERS. April–May, conspicuous, on green and glabrous peduncles 1–2½ in.; perfect, 3–4 in. across, cup-shaped; petals 6, erect, yellowish green, orange-banded at base, obovate to elliptic or oblong, apex rounded or truncate, 1½–2½ in.; sepals 3, greenish and thinly glaucous, elliptic to spatulate or ovate, rounded or truncate at apex, spreading or reflexed; stamens numerous, long, conspicuous; anthers linear, 1–1¼ in., as long as the filaments or longer; pistils flat, narrow, scalelike, imbricate into a spindle-shaped cone, stigmas short and recurved above.

FRUIT. A samara-bearing cone, ripening September–November, composed of closely imbricate, woody, brown, dry indehiscent carpels, about 3 in. long; seeds 1–2 (1 seed often abortive), testa thin and leathery.

LEAVES. Simple, alternate, deciduous, 4–6 in. long and broad, apex deeply notched, with 2 large acute lobes on each side of notch; lateral notches separating one large upper lobe and a smaller lower lobe, base of leaf truncate or cordate or sometimes rounded; upper surface dark green, glabrous, shiny; lower surface pale, glabrous, and somewhat glaucous, turning yellow in autumn; petioles 1–6 in., slender, glabrous; stipules persistent, foliaceous on young shoots.

TWIGS. When young reddish brown to greenish, shiny, smooth, glabrous, sometimes glaucous-purple; leaf scars rather large, semi-orbicular; older twigs slender, smooth, mottled gray, brittle.

BARK. At first gray, thin, tight, later gray to brown with rounded ridges and deep, confluent furrows.

WOOD. Growth rings distinct, heartwood variable in color, generally yellow, tan, or brown, sometimes with shaded mixtures of green, blue, or black; sapwood creamy white; straight-grained, light, soft, fairly stiff, moderately brittle, easily worked, split resistant, seasons well, except under decay conditions, takes glue readily, paint and varnish absorption good, takes a high polish, shrinkage considerable, nail-holding ability rather low, weighs about 26 lb per cu ft. The wood of Tulip-tree has a wide variety of uses including veneer, boxes, crates, plywood, pulp, hat blocks, furniture, cabinet-work, millwork, musical instruments, toys, novelties, cigar boxes, barrel bungs, and others.

RANGE. Seldom in pure stands, but mixed with a variety of trees. In rich, moist soil. Louisiana and Arkansas; eastward to Florida, and north to Massachusetts, Vermont, Rhode Island, Michigan, Wisconsin, and Ontario. Most abundant on mountain slopes west of the Blue Ridge.

MEDICINAL USES. A tonic prepared from the bark, principally root bark, was formerly much used in domestic medicine. The principal alkaloid ingredient, hydrochlorate of tulipiferine, was considered to be a heart stimulant and was also used in chronic rheumatism and dyspepsia. At one time it was believed to have antimalarial virtues.

PROPAGATION. Tulip-tree has been cultivated since 1663 and is a very desirable ornamental as well as a valuable timber tree. It has a deep, wide-spreading root system, is not readily transplanted (except in spring), grows rather rapidly when young, lives to a great age, is readily fire-damaged, rather free of diseases and insects, and coppices freely. About 7–13 lb of seed are obtained from a bushel of cones. The seeds are sown at a rate of 50-75 seeds per linear ft in drills of 8–12 in. apart and covered by soil about ¼ in. The germination rate is rarely over 15 per cent. A good, moist, rich soil is necessary and shade for young seedlings essential. Varieties are obtained by budding, grafting, and layering.

VARIETIES AND FORMS. Several varieties and forms are known.

Yellow-leaf Tulip-tree, *L. tulipifera* forma *aureo-marginatum* Schwerin, has leaves edged with yellow.

Whole-leaf Tulip-tree, *L. tulipifera* forma *integrifolium* Kirchn., has leaves rounded at base, without lobes.

Round-lobe Tulip-tree, *L. tulipifera* var. *obtusilobum* (L.) Michx., has a leaf with one rounded lobe on each side of base.

Tulip-tree
Liriodendron tulipifera L.

SOUTHERN MAGNOLIA *Magnolia grandiflora* L.

Pyramid Tulip-tree, *L. tulipifera* var. *fastigiatum* (L.) Jaeg., has branches forming a narrow-pyramidal head.

REMARKS. The genus name, *Liriodendron*, means "lily-tree," and the species name, *tulipifera*, may be translated "tulip-bearing." Other vernacular names in use are Yellow-poplar, Yellow Wood, White Wood, Tulip-poplar, Saddleleaf, Canoe Wood, Cucumber-tree, Blue-poplar, White-poplar, Lynn-tree, Saddle-tree, and Hickory-poplar. The use of the name "Poplar" is misleading, because the tree is not related to the genus *Populus* which contains the cottonwood trees.

The plant is known to be eaten by at least ten species of birds, including the bobwhite quail, and also by red squirrel, white-tailed deer, and cottontail. The birds and squirrels eat the seed, and the deer and rabbit browse the leaves.

SOUTHERN MAGNOLIA

Magnolia grandiflora L.

FIELD IDENTIFICATION. Large evergreen tree attaining a maximum height of 135 ft, but usually not over 50 ft. Possibly one of the largest trees known has been recorded from Pascagoula, Mississippi, with a trunk circumference of 13 ft 7 in., height of 52 ft, and spread of 92 ft.

FLOWERS. April–August, solitary on short pedicels, terminal, cup-shaped, 6–9 in. across; petals 6–18, rounded or obovate, white, often purple at base, arranged in a series; sepals 3, petal-like; stamens numerous, short, filaments purple; anthers linear, opening on the inner side; pistils numerous, coherent along a prolonged receptacle to form a fleshy cone, ovules 2.

FRUIT. Ripening July–October, cone ovoid to cylindric, rose-colored, fleshy, rusty-hairy, imbricate, 2–4 in., 1½–2 in. wide. Each carpel splits dorsally to expose 1 or 2 red, obovoid seeds suspended on thin threads. From 40–60 seeds per cone.

LEAVES. Very variable, alternate, simple, evergreen, coriaceous, blades 4–9 in. long, 2–3 in. wide, elliptic or oval, acute or obtuse at apex, cuneate at base, margin entire, shiny and dark green above, rusty-tomentose beneath; petioles stout, tomentose, about ¾ in.; stipules foliaceous, deciduous.

TWIGS. Green to olive, stout, hairy, or glabrous later.

BARK. Aromatic, bitter, grayish brown, breaking into small, thin scales.

WOOD. Creamy white, hard, weak, not durable, fairly heavy.

RANGE. In rich, moist soil; Texas, Oklahoma, Arkansas, and Louisiana; eastward to Florida and north to North Carolina. Cultivated for ornament as far north as Washington, D.C.

PROPAGATION. The pods are gathered when red by means of pruners or by hand. They can be spread out to dry and the seeds shaken out. If immediate planting is desired, the outer fleshy seed coat can be removed by maceration in water. Seed to be planted later can be dry-stored with the pulp left on. The cleaned seed average about 5,800 seeds per lb. Commercial seed has a purity of 94–100 and a soundness of 95 per cent. Stored seed can be carried over for several years if stored in sealed containers at 32-41°

F. If not stored in sealed containers at air temperature, they suffer some loss of viability. For spring planting seeds can be stratified in peat at 41°F. for 90–120 days. The germinative capacity average is 50 per cent. For fall sowing the cleaned seed, or seed in the pulp, are drilled in rows 8–12 in. apart and covered with about ¼ in. of rich, moist soil. A leaf or straw mulch affords protection from freezing. Nursery germination is 60–70 per cent. Protection from rodents by a wire screen is necessary in some areas. Propagation by grafting and cuttings is also practiced.

REMARKS. The genus name, *Magnolia*, is in honor of Pierre Magnol, professor of botany at Montpellier, and the species name, *grandiflora*, refers to the large flowers. Vernacular names are Bull-bay, Great Laurel Magnolia, and Loblolly Magnolia. The seeds are eaten by at least 5 species of birds and by squirrels. The wood is used for fuel, baskets, crates, woodenware, furniture, and shades. This species and others of the same genus are widely cultivated for their beautiful flowers and showy leaves, both in the United States and in Europe. It has been cultivated since 1734. Horticulturists have also developed a number of clones from the Southern Magnolia, among which are the Exmouth, Gallisson, Glorious, Goliath, Narrow-leaf, and Round-leaf.

ASHE MAGNOLIA

Magnolia ashei Weatherby

FIELD IDENTIFICATION. Mostly a shrub, but sometimes a small slender tree to 15 ft, with a broad round top. Easily recognized by the obovate to elliptic-lanceolate leaves, which are light green above and silvery white beneath, and from 1–2 ft long.

FLOWERS. In spring, terminal on stout pubescent or glabrous pedicels, fragrant, creamy white, very large, solitary, perfect, cup-shaped, often 12 in. in diameter; petals 6–8, 4–6 in. long, 2–3 in. broad, elliptic to oval; anthers opening inwardly; pistils coherent to form a fleshy cone.

FRUIT. In summer, cone cylindric-ovoid, tomentose, fleshy, green to rose-colored, 2–4½ in. long, 1–2 in. broad, carpels opening on the back and discharging red seeds attached by slender threads; seeds oval, flattened, conspicuous.

LEAVES. Simple, alternate, deciduous, thin, very large, sometimes to 2 ft long and 1 ft wide; obovate-spatulate or elliptic-lanceolate; cordate-auriculate at base; acute or rounded at apex; margin wavy, entire; light green and lustrous above; silvery white glaucous beneath and somewhat pubescent along main vein.

TWIGS. Reddish brown, smooth, slender, pubescent at first, later glabrous.

BARK. Gray to reddish brown, mostly smooth or slightly roughened.

WOOD. Soft, durable, easily worked, weighing about 30 lb per cu ft.

RANGE. Only western Florida and eastern Texas. Very rare in Texas, found possibly only in 2 counties.

REMARKS. The genus name, *Magnolia*, is in honor of Pierre Magnol, professor of botany at Montpellier, and the species name, *ashei*, is in honor of William Willard Ashe, American botanist (1872–1932). This tree is very similar to the Big-leaf Magnolia, *M. macrophylla*, but the fruit is ovoid-elliptic instead of globose-ovoid, the petals are smaller (4–6 in. long), the leaves are shorter and less pubescent below.

BIG-LEAF MAGNOLIA

Magnolia macrophylla Michx.

FIELD IDENTIFICATION. Tree to 50 ft and a trunk 20 in. in diameter, with a symmetrical broad crown. Probably the largest specimen ever recorded was from Baltimore, Maryland, with a trunk circumference of 6 ft 7 in., a height of 59 ft, and a spread of 64 ft.

ASHE MAGNOLIA
Magnolia ashei Weatherby

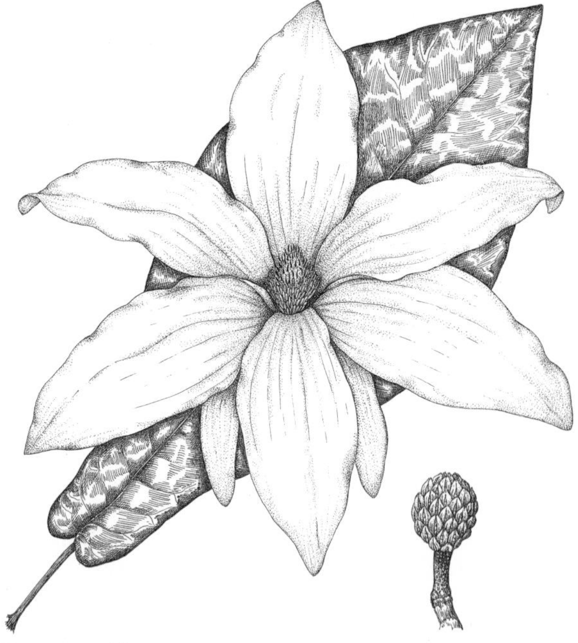

BIG-LEAF MAGNOLIA
Magnolia macrophylla Michx.

FLOWERS. April–May, flower buds to 9 in. long, borne on stout tomentose pedicels 1–1½ in.; flowers showy, white, fragrant, cup-shaped, perfect, 10–18 in. across; petals 6, the 3 inner ones narrower than the 3 outer, thick, ovate, acute or acuminate, concave, reflexed above the middle, 6–9 in. long, 3–4 in. wide, a purple spot at the base of the 3 inner petals; calyx membranaceous, sepals 3, green, turning dull yellow, ovate to obovate or oblong, apex rounded or obtuse, to 6 in. long and 1¾ in. broad at the center, narrower than the petals; stamens numerous, imbricate, in many series at the base of the receptacle, filaments short and stout; anthers introrse, adnate to the inner face of the filament; carpels obtuse or acute, imbricate in many series on an elongate receptacle, style persistent, ovules 2 in each cavity.

FRUIT. Conelike, ripening September–October, ovoid to nearly globose, composed of imbricate follicles, pubescent, 2½–3 in. long, rose-colored when mature; seeds obovoid, flattened, orange to red, ⅓–⅔ in. long, about ¼ in. broad, dangling on threads.

LEAVES. Buds tomentose, leaves terminal, alternate or in false umbrella-like whorls, obovate to oblong, apex acute, obtuse or rounded, base cordate with earlike lobes, upper surface green and glabrous, lower surface silvery white and pubescent, main vein prominent, stout and densely tomentose, membranous, length 20–30 in., width 9–10 in.; petioles stout, from gray-tomentose to pubescent, 2½–6 in.

TWIGS. Stout, brittle, yellowish green and tomentose at first, later reddish brown to tan or gray and less hairy, leaf scars large, irregularly shaped, oval to ovate.

BARK. Reddish brown to pale gray, younger branches and bark smooth, older ones with surface broken into small thin scales.

WOOD. Light brown, sapwood yellowish, satiny, close-grained, weak, hard, dry weighing about 33 lb per cu ft.

RANGE. Usually along streams in pinelands. Eastern Louisiana and Arkansas. Seemingly most abundant in south-central Mississippi and northern Alabama; eastward to Florida, northward to New York, and west to Kentucky and Tennessee.

MEDICINAL USES. A glycoside known as "magnolin" has been found in the bark, as well as a volatile oil. At one time Magnolia bark was used in domestic medicine for malaria and rheumatism.

PROPAGATION. The tree is conspicuous because of its very large leaves which are easily wind-damaged. It is hardy as far north as Massachusetts, and is sometimes planted for ornament in the Eastern states and Europe. It was first introduced into cultivation about 1800. Young trees will sprout from the stump. Transplanting should be done during the dormant period. The seed may be sown in fall or stratified for spring sowing. Green cuttings may be taken in spring.

REMARKS. The genus name, *Magnolia,* honors Pierre Magnol (1638–1715), professor of medicine and director of the botanical garden at Montpeller, France. The species name, *macrophylla,* means "large-leaved." Vernacular names are Large-leaf Magnolia, Large-leaved Umbrella-tree, Big-bloom Cucumber-tree, Silver-leaf, and Elk-bark. This tree was first discovered by Michaux in 1789 near Charlotte, North Carolina. It is closely related to Ashe Magnolia, *M. ashei* Weatherby.

UMBRELLA MAGNOLIA

Magnolia tripetala L.

FIELD IDENTIFICATION. Sometimes shrubby by forming basal shoots around the main trunk, or a well-defined tree to 40 ft, with a trunk 18 in. in diameter. Branches rather stout and contorted to form an open and irregular crown.

FLOWERS. Usually April–June, borne on slender, terminal, glabrous and glaucous pedicels 2–3½ in.; corolla large, white, solitary, cup-shaped, 8–10 in. in diameter, somewhat odorous; sepals 5–6 in. long and ½–1½ in. broad, reflexed later, deciduous, narrowly-obovate, pale green; petals 6–9, oblong-lanceolate to oblong-obovate, apex acute, mostly erect at first but later drooping, largest on the outer row are 4–5 in. long and 1–2 in. broad; stamens numerous, imbricate, deciduous; filaments purple, shorter than the adnate, introrse, 2-celled anthers; pistils numerous, coherent, crowded into a fleshy, conelike receptacle; ovary sessile, 1-celled, style short and recurved, ovules horizontal.

FRUIT. Maturing September–October, conelike, rose-colored at maturity, ovate to oblong, glabrous, length 2½–6 in., carpels producing follicles dehiscent on the back, with persistent styles prolonged into slender beaks, empty follicles wide-spreading, light brown and lustrous within, exterior dark brown; seed reddish, obovoid, about ½ in. long, suspended by a thread at maturity, outer coat fleshy, inner crustaceous.

LEAVES. Alternate, deciduous, crowded at ends of branches (umbrella-like), below the flowers, blades to 2 ft long and 10 in. wide, obovate-lanceolate to oblong-lanceolate, apex acute or abruptly short-acuminate, base tapering or cuneate, usually broadest above the middle, margin entire, thin and membranaceous, at first silky-hairy, at maturity green and glabrous above, lower surface paler and glabrous to slightly pubescent, primary veins slender, midrib thick and prominent; petiole stout, 1–3 in.

TWIGS. Stout, green to reddish brown or gray; lenticels small, scattered, pale; stipular scars circular at the nodes; leaf scars large, oval to broadly elliptic horizontally; terminal bud up to 1 in. long, obtuse or acute, smooth, purplish or reddish, glaucous; axillary buds more rounded, reddish brown.

BARK. Light gray, thin, smooth, marked by peculiar blister-like excrescences, bitter, aromatic.

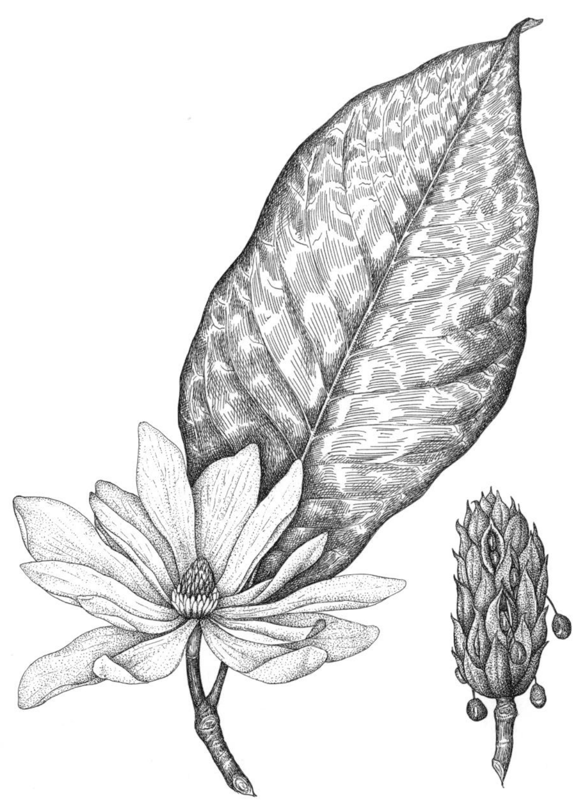

UMBRELLA MAGNOLIA
Magnolia tripetala L.

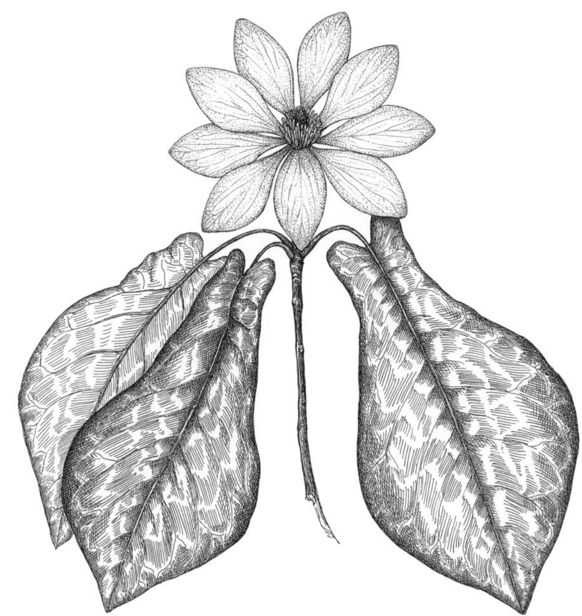

PYRAMID MAGNOLIA
Magnolia pyramidata Bartr.

WOOD. Heart light brown, sapwood creamy white, close-grained, not strong, soft, weighing 28 lb per cu ft.

RANGE. In rich, moist soil of swamps or wooded mountain slopes. In the Appalachians at altitudes to 2,000 ft. Central and southwestern Arkansas and southeastern Oklahoma; eastward to Alabama and Georgia, northward to Pennyslvania, and west to Missouri. Hardy as far north as New York (Long Island) and New Jersey.

REMARKS. The genus name, *Magnolia*, honors Pierre Magnol (1638–1715), director of the botanical garden of Montpellier, France. The species name, *tripetala*, refers to the 3 petaloid sepals. Also known under the names of Umbrella-tree, Cucumber-tree, and Elkwood. The name Umbrella-tree refers to the spreading umbrella-like leaves at ends of branches. The tree is rapid growing and is sometimes used as a stock for grafting. It is often cultivated for ornament in the United States and in Europe. The largest size ever recorded was a specimen from Lawrenceville, New Jersey, with a trunk circumference of 4 ft 3 in.

PYRAMID MAGNOLIA

Magnolia pyramidata Bartr.

FIELD IDENTIFICATION. Slender tree attaining a height of 30 ft, with ascending branches. Recognized by the almost whorled leaves with conspicuous earlike lobes at the base.

FLOWERS. Solitary on stout, terminal peduncles, some-

Cucumber-tree Magnolia
Magnolia acuminata L.

what fragrant, 3½–4 in. across at maturity, perfect; petals creamy white, imbricate, oblong to lanceolate, acute or acuminate, clawed, veiny, soon drooping; sepals 3, spreading or recurved, oblong to obovate, apex short-pointed, much shorter than petals; stamens distinct, numerous, ⅙–¼ in., imbricate in many series at base of receptacle, filaments short and distinct, anthers introrse and adnate to the inner side of the filament; styles and stigmas stout, forming a head of imbricate follicles on the elongate receptacle, ovules 2 in each cavity.

FRUIT. A conelike echinate fruit composed of numerous, rose-colored follicles, apices with short, incurved, persistent beaks; oblong to ellipsoid, length 2–2½ in.; seeds 2 in each follicle, red, ovoid, pendulous on a thread.

LEAVES. Deciduous, alternate, or almost whorled terminally, oblong to spatulate, apex obtuse or acute, base contracted with 2 spreading lobes, margin entire, thin, upper surface bright green, lower surface paler, glabrous and glaucous, especially when young, blade length 5½–8½ in., width 3½–4½ in., midrib slender and yellow, veinlets reticulate; petioles green to tan or brown, ⅓–2¾ in.

TWIGS. Slender, green to reddish brown, later gray, leaf scars oval or half-round, lenticels small and pale.

BARK. Smooth, tight, gray to reddish brown.

RANGE. Low rich woods near streams. In southeastern Louisiana in Washington Parish, and rarely in east Texas; eastward to Florida and Georgia, and northward to South Carolina.

REMARKS. The genus name, *Magnolia*, is in honor of Pierre Magnol (1638–1715), professor of medicine and director of the botanical garden at Montpellier, France. The species name, *pyramidata,* refers to the pyramidal fruit. Vernacular names are Southern Cucumber-tree and Mountain Magnolia. Pyramid Magnolia is closely related to the Fraser Magnolia, *M. fraseri,* but has smaller flowers, and a shorter fruit. Fraser Magnolia is confined mostly to the Appalachian Mountains and Piedmont region, while Pyramid Magnolia is a tree of low grounds of the coastal plain area. The latter tree was first cultivated in 1825, and is occasionally grown for ornament in the United States and in Europe.

Cucumber-tree Magnolia

Magnolia acuminata L.

FIELD IDENTIFICATION. Tree attaining a height of 90 ft and a diameter of 4 ft. The branches are rather slender, and the crown dense, pyramidal or rounded. The leaves, fruit, and flowers are very variable as to size and amount of hairiness.

FLOWERS. April–June, after the leaves appear, solitary, bell-shaped, greenish yellow, glaucous, perfect; borne on pedicels ½–¾ in., which are densely silky-tomentose at first and less so later; sepals 3, membranous, greenish, soon reflexed, deciduous, oblong-lanceolate or elliptic, acute, ¾–1½ in.; petals 6, longer than sepals, ovate to obovate or elliptic, concave, acute, veins apparent, 2–3 in., erect, outer petals wider than inner ones; stamens numerous, anthers long, linear, introrse; pistils numerous, spirally arranged on the receptacle in many series, styles filiform and later deciduous from the fruit follicle, ovules 1–2 in each cavity.

FRUIT. Cucumber-shaped cone, oblong to ovoid, sometimes knotty and distorted, almost cylindrical, red to brown, 2–3 in. long, ¾–1¼ in. wide, often curved, glabrous, composed of numerous rounded follicles; thickened valves of follicles dehiscent August–October to expose a reddish orange pendent seed on a filamentous thread; seeds 5–60 in fruit, obovoid to subglobose, oblique, flattened, ⅜–½ in. long, outer seed coat soft and fleshy, inner seed coat thin and membranous, endosperm large and fleshy.

LEAVES. Simple, alternate, deciduous, scattered, blades 5–12 in. long, 2½–6 in. wide, elliptic to oblong or ovate to oval; apex acute to obtuse, sometimes abruptly so; base rounded to broadly cuneate, rarely slightly cordate; margins entire, often undulate on the plane surface; young leaves densely-hairy when they unfold; mature leaves olive-green and glabrous on upper surface or slightly pubescent along the impressed veins (reticulate veiny under magnification); lower surface paler, semiglabrous or pubescent, secondary veins rather straight; leaves turning yellow or brown in autumn; petioles ½–1½ in., densely hairy.

TWIGS. Rather stout, red to brown, pubescent and lustrous at first, later glabrous and brown to gray; lenticels numerous, small, pale; winter buds terminal, ovoid to ellipisoid, acuminate, ½–⅗ in., densely silky hairy, lateral buds smaller and flattened; leaf scars narrow, elevated, with 5–9 bundle scars.

BARK. Dark brown to gray, ⅓–½ in. thick, furrowed with narrow confluent ridges, breaking on old trunks into thin scales.

WOOD. Ocher to yellowish brown, sapwood white, close-grained, satiny, weighing about 30 lb per cu ft, moderately hard, moderately strong in endwise compression, fairly stiff, moderately high in shock resistance, fairly difficult to tool but finishes smooth, glues satisfactorily, takes and holds paint, has a beautiful surface and stains well, average in nail-holding ability, shrinks considerably, stays in place well when properly seasoned, but not durable under decay conditions. The wood often mixed with that of Tulip-tree and sold under the name of Yellow-poplar. The wood is used for cabinets, flooring, interior finish, pump logs and troughs, Venetian blinds, boxes, crates, sidings, and general millwork. (The information on the character of the wood is from Brown, Panshin, and Forsaith, 1949, pp. 560–562.)

RANGE. In moist rich soil, banks of streams and rocky slopes, Louisiana, Oklahoma, and Arkansas; east to Georgia, and northward in Missouri, Tennessee, Kentucky, South Carolina, North Carolina, Ohio, Indiana, Illinois, New York, and Ontario.

PROPAGATION. Good crops of seed are borne every 4–5 years, commercial seed-bearing age starting at about 30 years. The cones are collected by hand when they turn red and are spread out to dry. The seeds are then shaken out of the cone. The fleshy seed coat can be removed by maceration in water or by rubbing on hardware cloth. Sometimes the seed is dried without removing the seed coat. It can be stored dry in sealed containers at 35° F. and seems to retain viability for at least 2 years.

The seed averages about 4,600 seeds per lb, with a commercial purity of 97–100 and an average soundness of 91 per cent. Germination is best assured by stratification in moist peat and sand at 41 degrees for 150–180 days. The germinative capacity averages 55 per cent. The seed may be planted in rich, moist seedbeds in drills and rows 8–12 in. apart, and covered with ¼ in. of firmed soil. The beds are kept moist until germination is finished. Half-shade is required for the young seedlings. If planted in fall a protective mulch protects from freezing temperatures. Screens may have to be used for rodent protection. Propagation is also practiced from green cuttings with a heel under glass. Cucumber-tree Magnolia is also used as a stock for grafting other magnolias.

VARIETIES AND FORMS. Cucumber-tree Magnolia is very variable in size and shape of leaf, flower, and fruit, and the amount of hairiness of the parts. One of these segregates is still in question, as to whether it should be listed as a distinct species or as a variety. It is known as the Yellow Cucumber-tree Magnolia, and has canary-yellow flowers, branches pubescent, and smaller more broadly ovate leaves. It has been listed under both the species name of *M. cordata* Michx. and the varietal name of *M. acuminata* var. *cordata* (Michx.) Sarg. It is a rare and local tree of North and South Carolina and Georgia. To the author it seems sufficiently distinct to remain as a species.

William Willard Ashe (1872–1932), of the United States Forest Service, published the following key suggesting the relationship of *M. acuminata* and its varieties (Ashe, 1931a, pp. 37–38):

Sun leaves pubescent and pale beneath; no leaves on vigorous shoots of a broadly ovate or broadly obovate type; flowers less than 7 cm. long; twigs glabrous.

Flowers yellow. . . . . *M. acuminata* var. *aurea*

Flowers green or yellow-green *M. acuminata typica*

Sun leaves pubescent; leaves on vigorous shoots often of a broadly ovate or obovate type; flowers green, 7 cm or more long, petals broad.

Twigs pubescent; leaves soft pubescent beneath; flowers 7.5 cm. long . . . . *M. acuminata* var. *alabamensis*

Twigs glabrous; leaves pubescent beneath; flowers 8-10 cm. long. . . . *M. acuminata* var. *ludoviciana*

Sun leaves glabrate or green beneath; twigs glabrous; flowers less than 6.5 cm. long, green or purplish, petals narrow. . . . . *M. acuminata* var. *ozarkensis*

The variety *alabamensis* Ashe is at one extreme in respect to copiousness of pubescence.

The variety *ozarkensis* Ashe is essentially glabrous at the other extreme.

The varieties *aurea* and *ludoviciana* and *acuminata typica* are intermediate.

The varieties *ludoviciana* and *alabamensis* have the largest flowers.

REMARKS. The genus name, *Magnolia,* is in honor of Pierre Magnol (1638–1715), director of the botanical garden at Montpellier, France. The species name, *acuminata,* refers to the pointed leaves. Other vernacular names for the tree are Mountain Magnolia, Cucumber-tree, Yellow Linn, and Black Linn. The Cucumber-tree Magnolia has been in cultivation since 1736 and is often planted as a street and park tree in the Eastern states and in Europe. It is the hardiest of our magnolias, is rapid growing, coppices freely, and is easily injured by fire.

Sweet-bay Magnolia

Magnolia virginiana L.

FIELD IDENTIFICATION. Swamp-loving shrub or tree attaining a height of 30 ft. The largest specimen ever recorded is in Cambden, South Carolina. This tree has a height of 67 ft and a circumference of 6 ft. One of the field marks for identification is the conspicuous white undersurface of the leaves.

FLOWERS. Borne on short slender pedicels, very fra-

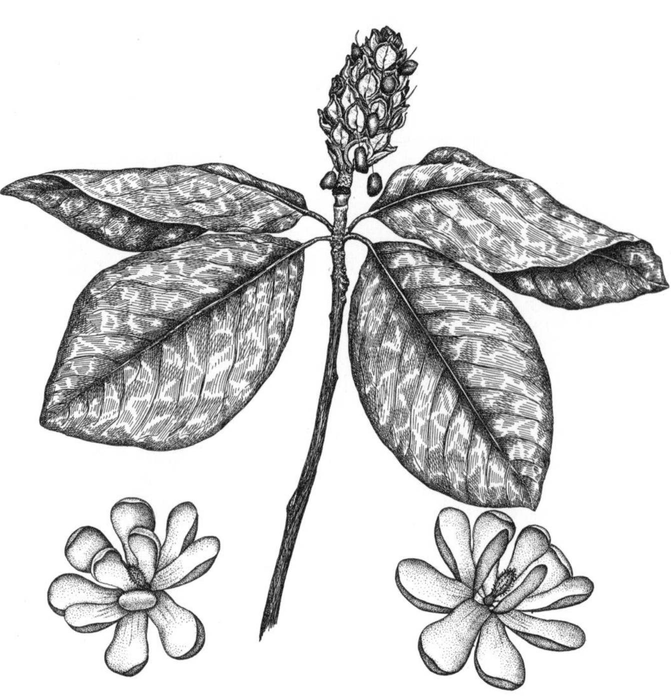

SWEET-BAY MAGNOLIA
Magnolia virginiana L.

grant, depressed-globose, 2–3 in. broad; petals 8–12, elliptic, obovate or oval, obtuse; sepals 3, obtuse, shorter than petals, obovate or oblong; stamens short-filamented, numerous on a prolonged receptacle.

FRUIT. Pistils coherent to form an ellipsoidal, imbricate, fleshy, red cone 1–2 in. long; cone splitting at maturity to discharge from each carpel 1–2 red, oval seeds suspended on thin threads.

LEAVES. Scattered, alternate, simple, undulate, oblong, elliptic or oval, obtuse at apex, broad-cuneate at base, green above, white or pale glaucous beneath, blades 3–6 in. long, 1–2½ in. wide; petioles slender, smooth, about 1 in.

TWIGS. Slender, bright green, glabrous or hairy.

BARK. Pale gray to brown, smooth, aromatic, bitter.

WOOD. Pale brown, sapwood white, soft, weighing 31 lb per cu ft.

RANGE. Usually in low, wet, acid, sandy soil. Texas, Oklahoma, Arkansas, and Louisiana; eastward to Florida, northward to Pennyslvania, and in isolated stations to Massachusetts.

MEDICINAL USES. Sweet-bay Magnolia bark was at one time the source of an official drug for the treatment of malaria and rheumatism, but its use has been abandoned.

PROPAGATION. The seeds can be sown in the fall by drills in rows 8–12 in. apart, and covered with firmed soil about ¼ in. deep. A rich, moist soil is required and the seeds should be mulched during the winter to protect from freezing. Partial shade is helpful. The germination varies from 50–75 per cent. Sometimes a screen wire over the bed is needed for rodent control. It may also be propagated by root sprouts, cuttings or grafting.

VARIETY. Southern Sweet-bay Magnolia, *M. virginiana* var. *australis* Sarg., is a variety bearing heavy white pubescence on the twigs and pedicels, and it appears to be the typical Southern form.

REMARKS. The genus name, *Magnolia,* is in honor of the botanist Pierre Magnol, and the species name is for the state of Virginia. Vernacular names are Swamp-bay, Beaver-tree, White-bay, White-laurel, Swamp Magnolia, Swamp-sassafras, and Indian-bark. This species is easily identified by the white undersurface of the leaves. Although deciduous in the North, the leaves tend to be evergreen in the South. It is reported that the flowers have been used in perfume manufacture, and the leaves as a flavoring for meats. The tree is

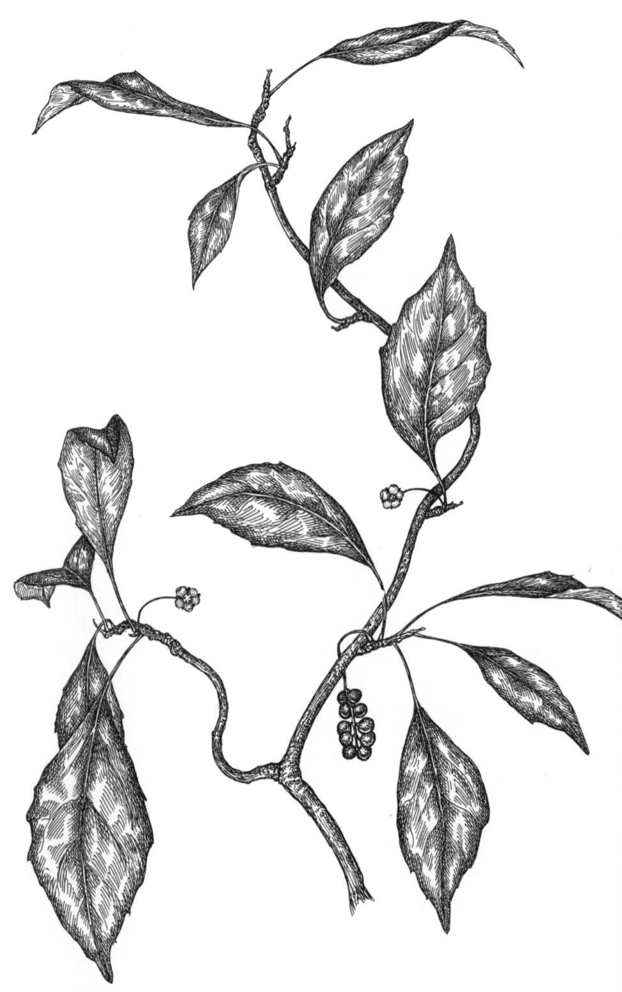

CAROLINA MAGNOLIA-VINE
Schizandra coccinea Michx.

sometimes confused with the Red Bay, *Persea borbonia*, which also has leaves used for flavoring purposes. Sweet-bay Magnolia is occasionally cultivated for ornament, and is thicket forming by sprouts from the roots. The flowers are small, but have a penetrating fragrance. The wood is used for making light woodenware articles.

Carolina Magnolia-vine

Schizandra coccinea Michx.

FIELD IDENTIFICATION. High-climbing woody, slender vine.

FLOWERS. Solitary, or in axillary drooping clusters, on slender, essentially glabrous, naked peduncles ½–2¾ in. Flowers monoecious, crimson-purple, ¼–½ in. across; petals 5, larger than sepals, deciduous, imbricate, oval to obovate-cuneate, apex rounded or truncate, base with a thickened claw; stamens 5, connate into a 5-lobed broad disk; filaments very short and flat; anther cells widely separated, short, introrse; carpels numerous, imbricate in several series, disposed on the elongate filiform receptacle, forming a drooping raceme, style very short; sepals 5, imbricate, petaloid, oval to ovate, thickened at base, ⅛–¼ in.

FRUIT. Ripe August–September, forming a loose raceme 1½–3 in., fruit orange-red, subglobose, 1/16–¼ in., 2-lobed when dry; seeds 2 in each berry, reniform, testa crustaceous.

LEAVES. Alternate, or somewhat clustered at nodes, deciduous, ovate to oval or elliptic, margin entire or with a few remote denticulate teeth; apex acute to acuminate and often abruptly so; base rounded to broadly or narrowly cuneate, sometimes attenuate on the petiole into a short wing; thin and membranous, blade length 2–6 in., width 1–3 in.; upper surface bright green and essentially glabrous, veins delicate; lower surface paler and glabrous or slightly pubescent; petioles ½–1 in., slender, channeled, glabrous.

RANGE. In rich woods; Texas, Louisiana, and Arkansas; eastward to Florida and north to South Carolina.

REMARKS. The genus name, *Schizandra*, is from the Greek *schizein* ("to cleave") and *aner-ados* ("man-stamen"), referring to the cleft anther cells. The species name, *coccinea*, refers to the red fruit. Also known under the vernacular names of Bay Star-vine and Wild Sarsaparilla. The plant thrives best in moist, sandy soil in partial shade. Propagation is by seeds, greenwood cuttings under glass, root cuttings, layers or suckers.

STRAWBERRY-SHRUB FAMILY (*Calycanthaceae*)

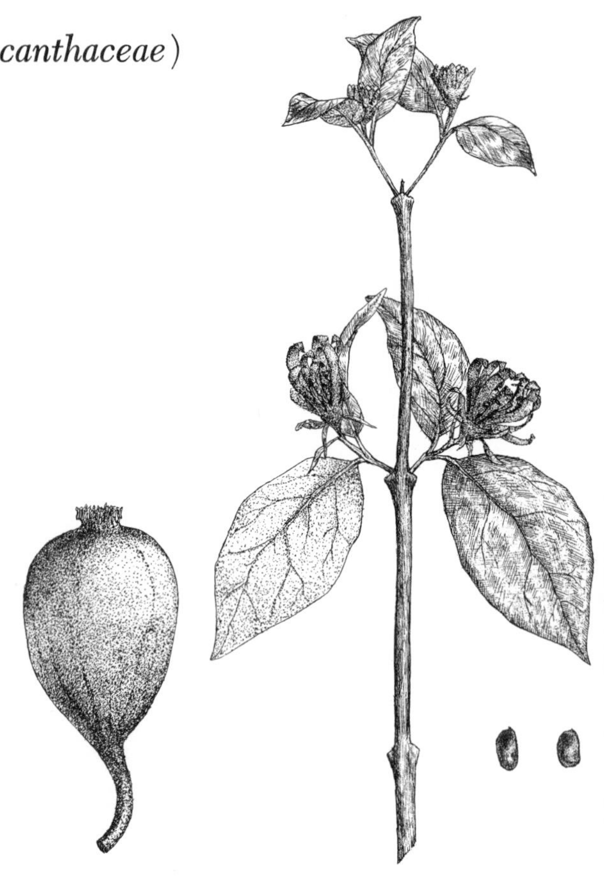

Common Sweet-shrub
Calycanthus floridus L.

Common Sweet-shrub

Calycanthus floridus L.

FIELD IDENTIFICATION. Shrub to 9 ft, with aromatic, deciduous leaves and pubescent twigs.

FLOWERS. April–June, terminal on short leafy branches, fragrant, short-peduncled, perfect, perigynous; petals and sepals imbricate, resembling each other, reddish brown, linear or nearly so, apex obtuse or acute, margin ciliate, surface pubescent, length ⅝–1 in., borne on edge of hypanthium; calyx united below into a fleshy inverted conical cup; androecium of many stamens, the outer 12 usually fertile, the inner reduced to staminodia, anthers adnate and extrorse; gynoecium of numerous distinct carpels borne on inside of hypanthium and inserted on its base or face, styles filiform.

FRUIT. Capsule obovoid to oblong, 1–2½ in., nodding, dry, leathery, finely tomentose at first, more glabrous later, constricted toward apex, hollow within; achenes numerous, obovoid, somewhat hairy, 1-seeded.

LEAVES. Opposite, deciduous, often only 1 or 2 pairs on short lateral branchlets, oval to elliptic, apex acute to acuminate, base rounded or broadly cuneate, margin entire, upper surface paler green and densely soft-tomentose; blade length 1½–5 in., width ¾–1¼ in.; petioles mostly less than ¾ in., densely tomentose.

TWIGS. Slender, rather straight, brown to gray, younger ones finely pubescent, older ones more glabrous, lateral branchlets rather remote and short, nodes somewhat flanged below; bark aromatic.

RANGE. Rich woods, hillsides, or stream banks, mostly on the Atlantic and Gulf Coast plains. Louisiana; eastward to Florida and northward to Virginia.

PROPAGATION. The various sweet-shrubs grow in either sun or shade in rich, moist soil. They can be propagated by seeds sown in spring, by layers put down in summer, or from divisions of older plants.

VARIETIES. Broadleaf Common Sweet-shrub, *C. floridus* var. *ovatus* Lav., has ovate to ovate-oblong leaves which are rounded or subcordate at base.

Pale Sweet-shrub, *C. fertilis* Walt., has a variety, Carolina Sweet-shrub, *C. fertilis* var. *nanus*, which has been recorded from Louisiana. Carolina Sweet-shrub has glabrous instead of tomentose-hairy leaves, as does Common Sweet-shrub.

REMARKS. The genus name, *Calycanthus*, is from the Greek *kalyx* ("calyx") and *anthos* ("flower"), the calyx being large and conspicuous. The species name, *floridus*, refers to the plant's conspicuous "flowering" habit. Vernacular names are Carolina Allspice, Sweet-scented-shrub, Bubby-bush, and Bubby-blossoms. It is often cultivated for its very fragrant flowers.

CUSTARD-APPLE FAMILY (*Annonaceae*)

COMMON PAWPAW
Asimina triloba (L.) Dunal

COMMON PAWPAW

Asimina triloba (L.) Dunal

FIELD IDENTIFICATION. Spreading shrub or broad-crowned tree attaining a height of 40 ft.

FLOWERS. Axillary, solitary, perfect, on stout rusty-hairy pedicels, 1–2 in. across, appearing with or before the leaves; petals 6, purplish green, veiny, the 3 outer ovate-obovate or orbicular, larger than the 3-pointed, erect, glandular inner ones; stamens many, short; pistils few to many; style inwardly stigmatic; ovary 1-celled, ovules numerous; calyx of 3 ovate, acuminate, pale green sepals, much smaller than petals.

FRUIT. Banana-like, borne singly, or in oblique clusters of 2–4, oblong-cylindric, often falcate, apex and base pointed or rounded, 2–7 in. long, 1–2½ in. thick, green when young, brown or black when mature; pulp sweet, white or yellow, aromatic, edible; seeds several, dark brown, large, bony, rounded, flat, horizontal, about 1 in. long and ½ in. broad.

LEAVES. Deciduous, alternate, simple, oblong-obovate abruptly pointed or acute at apex, obtuse or cuneate at base, entire, thin, rusty-pubescent when young, globous later, blades 4–11 in. long, 2–6 in. broad, odorous when bruised; petioles ⅓–1 in., stout.

TWIGS. Slender, olive-brown, often blotched, smooth, rougher when older, and often with warty excrescences.

BARK. Dark brown, thin, smooth, later with shallow fissures.

WOOD. Pale yellow, coarse-grained, soft, weak, weighing 24 lb per cu ft.

RANGE. Rich soil of bottom lands. East Texas, Arkansas, and Louisiana; eastward to Florida, and north to New York, Michigan, and Nebraska.

PROPAGATION. The fruit is picked from the plant, or from the ground after falling. The seeds can be removed by macerating and floating off the pulp, or the whole fruit can be planted. Cleaned seed averages about 697 seeds per lb, with a commercial purity of 100 per cent and a soundness of 96 per cent. The seeds are slow in germinating until the following spring. Stratification for 100 days at 50° F. seems to hasten germination. The seeds are sown in rich, moist soil and covered about ¾ in.

Zimmerman (1941, pp. 85–90) uses the following method of planting:

The seeds are placed in moist sand in a can and the can closed and placed in a cellar. After February they begin to sprout and when the weather is warm enough the sprouted seeds are picked out and planted in nursery rows a ft apart. The sprouted seed will grow about 6 in. the first year, a little more the second, and rapidly afterwards. They are transplanted to a permanent site when about a ft high. When they are established a little pruning from time to time forces new growth. Pawpaw can also be propagated by layers or root cuttings, by grafting and by hybridization.

Grafting is easily done using any of the native stock of *A. parviflora, A. obovata, A. incana, A. angustifolia, A. reticulata, A. pygmaea,* and *A. ruglei.* The grafting methods are the same as that used in nut culture, but a little earlier in the season. Interesting combinations may be had by hybridizing the Pawpaws with each other. Good crops of fruit are not the rule with wild plants due to the fact that the stigma often matures before the stamens shed their pollen. Hand pollination can correct this problem and nearly all species can be cross-pollinated to produce heavier fruit crops.

Zimmerman has listed the following horticultural clones and hybrids of the Common Pawpaw: Dr. Potter; Duck; Earlygold (Zimmerman, 1924); Fairchild (Fairchild, 1917); Gable (Joseph Gable); Holtwood (W. Hoopes); Hopes August (Benjamin Buckman); Ketter (Mrs. Frank Ketter, 1917); Longjohn (Benjamin Buckman); Martin; Osborne (A. Osborne); and Reese. *Trigustifolia* is a cross of *A. angustifolia* and *A. triloba. Trilobovata* is a cross of *A. triloba* and *A. obovata.*

REMARKS. The genus name, *Asimina,* is from the early French name *Asiminier,* which in turn was derived from the Indian *Arsimin.* The species name, *triloba,* refers to the petals which are in sets of 3. Vernacular names are Fetid-shrub and Custard-apple. Pawpaw fruit falls to the ground in autumn and must be stored until ripe. It may be baked into pies, made into dessert, or eaten raw with cream as a breakfast food. When eaten raw it is cloyingly sweet with a custard-like flavor. Seemingly a taste for it must be cultivated because some consider it nauseating. The food value is largely carbohydrate. The fruit varies greatly in size and flavor. Some are large, yellow-fleshed, highly flavored, and early-ripening. Others are white-fleshed, mildly flavored, and late-ripening. Handling the fruit is known to produce a skin rash on some people. The rough bark is sometimes used as a rope substitute. Birds are fond of the fruit, and it is also eaten by gray fox, opossum, raccoon, and squirrel. The seeds of the Pawpaw contain an alkaloid, asiminine, which is reported to have emetic properties. The bark was once used as a medicine, and contains the alkaloid analobine.

SMALL-FLOWER PAWPAW
Asimina parviflora (Michx.) Dunal

FIELD IDENTIFICATION. Irregular, straggling shrub to 12 ft.

FLOWERS. Imbricate in the bud, solitary, small, inconspicuous, green to dark purple; pedicels very short, 1/16–1/5 in., with reddish brown tomentum; sepals 3, broadly oval or ovate, 1/5–1/4 in., green or yellowish, hairy; corolla about 5/8 in. wide or less, composed of 6 petals (3 spreading and 3 erect), outer 3 petals reddish brown or purple, oblong to ovate, apex obtuse or rounded, 1/8–3/8 in., densely puberulent; inner 3 petals erect, recurved at apex, shorter than the outer 3, about 1/8 in., reddish brown, puberulent; stamens numerous, crowded on a somewhat convex receptacle; pistils protruding from the mass of stamens, few or many, ovaries 1-celled.

FRUIT. August–September, berry solitary or 2–4 together, oblong, ellipsoid or oval, often asymmetrical, sparingly pubescent; immature fruit greenish yellow, and mature turning black, 3/4–2 in. long, pulpy, edible, but cloyingly insipid; seeds several, bony, turgid.

LEAVES. Leaves simple, alternate, deciduous, heavy-scented, thin, pinnate-veined, shape variable, oblong to elliptic or obovate, mostly broadest above the middle, apex acute or short-acuminate, base narrowed; thin, dull green and glabrous above, lower side with more or less reddish brown tomentum, especially on the vein; blade length 2 1/4–6 3/4 in.; petioles 1/16–1/2 in., with densely reddish brown tomentum.

SMALL-FLOWER PAWPAW
Asimina parviflora (Michx.) Dunal

TWIGS. Young twigs with dense reddish brown tomentum; older twigs smooth, glabrous, gray to reddish brown, strong-scented.

RANGE. In rich, moist soil, often along streams. In eastern Texas in Hardin, Newton, Jasper, Montgomery, and Orange counties; eastward through Louisiana to Florida, and northward to North Carolina.

PROPAGATION. The plant has the same general growth characters as the larger Common Pawpaw, *A. triloba*, and the propagation methods are the same. The fruit is picked from the bush when soft. The seeds can be removed by macerating and floating off the pulp, or the whole fruit may be planted. Cleaned seed averages about 1,300 seeds per lb, with a purity of 98 per cent and a soundness of 94 per cent. The seeds usually do not germinate until the following spring. Stratification for 100 days at 50° F. seems to help germination. The seeds are covered about ¾ in. with moist, rich soil in spring planting. The growth of seedlings is slow. The seeds may also be sown in autumn, or the plant propagated by layers or root cuttings.

REMARKS. The genus name, *Asimina*, is from an old French name, *Asiminier*, and the species name, *parviflora*, refers to the small flowers. Vernacular names are Small-fruited Pawpaw, Small Custard-apple, and Custard-banana. The fruit is considered to have the same nutritive value, weight for weight, as the banana. The bark was at one time used in domestic medicine as a bitter.

LAUREL FAMILY (*Lauraceae*)

CAMPHOR-TREE

Cinnamomum camphora Nees & Eberm.

FIELD IDENTIFICATION. Handsome cultivated tree to 40 ft, and 2 ft in diameter, evergreen, dense, stout, round-topped.

FLOWERS. Borne in axillary, slender-peduncled, green and glabrous panicles 1¾–3 in.; individual flower pedicels ⅛–¼ in.; glabrous; petals absent; perianth-tube short, segments 5–6, early deciduous to leave a truncate cup-shaped receptacle about ¼ in. long which is rugose, glabrous, and granular; stamens in 2–4 dissimilar groups, some rows gland-appendaged and some abortive; ovary 1-celled, with a slender style and minute stigma.

FRUIT. Drupe solitary or a few together on the ends of the long, slender peduncles, globose, black, lustrous, fleshy, ¼–⅓ in. in diameter; seed solitary, globose, black, ridged around the center.

LEAVES. Persistent, essentially alternate but often opposite, elliptic to ovate, apex acute to acuminate, base acute, blade length 1¼–4½ in., width ¾–1½ in., margin entire and somewhat wavy on the plane surface, upper surface dark green, lustrous, and glabrous, veins yellowish green, lower surface paler and lightly glaucous, aromatic when bruised; petioles essentially glabrous, ¾–1½ in.

TWIGS. Slender, elongate, green, and glabrous.

BARK. On young branches smooth, green to reddish brown or gray, on old trunks light to dark gray, broken into deep fissures and flattened, confluent, wider ridges.

RANGE. A native of Japan, China, and Malaya. Planted as an ornamental in Texas, Oklahoma, Arkansas, and Louisiana; east to Flordia. Sometimes escaping cultivation. Often planted on the streets of Houston, Texas. Does well in cultivation in southern California.

CAMPHOR-TREE
Cinnamomum camphora Nees & Eberm.

MEDICINAL USES. Camphor is used for many purposes in the Orient, but in the United States its main use is as a mild counterirritant in muscular strains and in rheumatic conditions. It is sometimes used in conjunction with phenol or menthol for local anesthetic effects to relieve skin itchings or acute rhinitis.

PROPAGATION. The Camphor-tree is planted extensively as a shade tree in the lower Gulf Coast areas but cannot stand sustained below-freezing temperatures. It grows best in sandy loam, but can adjust to lighter clays. Propagation is usually by seeds. The seedlings are shaded until large enough to pot. Transplanting is usually done from these potted specimens. The tree can also be grown from cuttings or half-ripened wood rooted in coarse sand in spring.

REMARKS. The genus name, *Cinnamomum*, is the ancient Greek name, and the species name, *camphora*, refers to the camphorous aromatic resin which the tree contains. The wood yields the camphor of commerce. The camphor is obtained by passing a current of steam through the chips, and the volatilized camphor is then condensed into crystals and oil. Later the crystals and oil are separated by filtration under pressure. About 30 lb of chips is required for 1 lb of camphor.

POND-SPICE

Glabraria geniculata (Walt.) Britton

FIELD IDENTIFICATION. Rare spreading shrub to 9 ft, with zigzag branches.

FLOWERS. March–April before the leaves, 2–4 together in umbel-like involucrate cymes, with separate staminate and pistillate flowers, corolla absent; calyx yellow, about ⅛ in. across, sepals 6, spreading, oval to elliptic, apex obtuse, deciduous; stamens or staminodia 9, in 3 series, those of the first 2 rows glandless, the third row with glands; filaments filiform, glabrous, shorter than the sepals; anthers ovoid, 4-celled, 4-valved, sacs introrse; ovary ovoid, 1-celled, style elongate.

FRUIT. In summer, on the enlarged perianth-tube, drupe red, subglobose, ⅕–¼ in. long.

LEAVES. Simple, alternate, deciduous, elliptic to oblong, apex and base acute to obtuse, margin entire, length ⅝–2½ in., firm, surface dark green and glabrous above, lower paler and slightly reticulated; petioles short.

RANGE. In shallow ponds or swamps; Florida and Georgia; westward to Louisiana, and northward to Tennessee. Rather rare.

REMARKS. The genus name, *Glabraria* ("smooth"), refers to the glabrous character of the plant. The species name, *geniculata*, is the Latin for the jointed twigs. It is also listed by some authors under the name of *Malapoena geniculata* (Walt.) Coulter, *Litsea geniculata* Benth. & Hook, and *Laurus geniculata* Walt. Common names include Pond-bush and Crooked Swamp-bush.

POND-SPICE
Glabraria geniculata (Walt.) Britton

COMMON SPICE-BUSH

Lindera benzoin (L.) Blume

FIELD IDENTIFICATION. Stout, glabrous, aromatic shrub of damp woods. Attaining a height of 20 ft, with usually several stems from the base.

FLOWERS. Appearing before the leaves, polygamo-dioecious, yellow, fragrant, ¼–⅓ in. broad, in lateral, almost sessile, dense, umbel-like clusters of 3–6 flowers; involucre of 4 deciduous scales; petals absent; sepals 6, thin, obovate to elliptic, apex obtuse to retuse or truncate; staminate flowers with 9 stamens (in 3 series), some filaments glandular at base, anthers introrse, 2-celled and 2-valved; pistillate flowers with 12–18 rudimentary stamens in 2 forms (glandular and glandless); ovary globose, style slender and columnar.

FRUIT. Ripening August–September, drupes solitary or in small clusters on pedicels 1/12–⅕ in., orbicular to obovoid, elongate, about ⅖ in. long, red, fleshy, spicy; 1-seeded, seeds light brown, speckled darker brown.

LEAVES. Leaf buds scaly, leaves simple, alternate, deciduous, obovate to oval or elliptic, apex acute or short-acuminate, base acute or acuminate, margin entire, thin, bright green above, glaucous, beneath glabrous, more rarely pubescent, blade length 2–4¾ in., width 1–2½ in.; petioles 3/16–¾ in. Twigs often with 2 leaf sizes, much smaller ones sometimes at base of larger ones.

TWIGS. Slender, glabrous, smooth, brittle, bark with corky lenticels, spicy to the taste.

RANGE. Sandy or peaty soils in low woods or swamps. Central Texas, Oklahoma, Arkansas, and Louisiana;

eastward to Florida, north to Maine, and west to Ontario, Michigan, and Kansas.

PROPAGATION. The small yellow flowers in spring, followed by the red fruit in autumn, give it some value as an ornamental. This shrub has been in cultivation since 1683. The red drupes may be collected by hand September–October. The drupes may be depulped by rubbing on a screen, the pulp floated off in water, and then spread out to dry. Sometimes the entire fruit is dried with the pulp on. The cleaned seed averages about 4,300 seeds per lb, with a purity of 95 per cent. It is best to sow the seed soon after gathering because it soon loses its vitality. Pretreatment consists of stratification in moist sand or peat for 120 days at 41° F. The germinative capacity of pretreated seeds in test plots averages 85 per cent. If planted in the fall a mulch should be used and removed after danger of freezing in spring. The seedlings should be planted on moist sandy or peaty soils. Common Spice-bush may also be reproduced by layers in sand or peat, or by greenwood cuttings under glass.

VARIETIES. The Yellow-berry Common Spice-bush, *L. benzoin* var. *xanthocarpa* (Torr.) Rehd., has yellow fruit.

Common Spice-bush
Lindera benzoin (L.) Blume

Hairy Common Spice-bush
Lindera benzoin var. *pubescens* (Palmer & Steyermark) Rehder

Hairy Common Spice-bush, *L. benzoin* var. *pubescens* (Palm. & Steyerm.) Rehd., has pubescent and ciliate leaves and petioles.

Some authorities list the Common Spice-bush under the scientific name of *Benzoin aestivale* (L.) Nees.

REMARKS. The genus name, *Lindera,* is for John Linder, a Swedish physician (1676–1723). The species name, *benzoin,* denotes its similarity, in odor, to the true balsamic resin of *Styrax benzoin,* an Asiatic tree. Vernacular names are Benjamin-bush, Spice-wood, Fever-bush, Snap-bush, and Wild Allspice. There are also a few varieties of Common Spice-bush.

The leaves, twigs, bark, and fruit contain an aromatic oil which was made into a fragrant tea by the pioneers. The bark is aromatic, tonic, astringent, stimulant, and pleasant to chew. A substitute for allspice was once made from the dry, powdered drupes. Twenty-four species of birds are known to feed upon the fruit, also rabbit and white-tailed deer nibble the leaves.

Southern Spice-bush

Lindera melissaefolia (Walt.) Blume

FIELD IDENTIFICATION. Strong-scented shrub 1½–6 ft. Plant much less developed and smaller than Common Spice-bush. The branches and inflorescence pubescent.

FLOWERS. February–March, very showy, yellow, dioecious; clusters dense, lateral, almost sessile, umbel-like, 4–6-flowered, surrounded by an involucre of 4 deciduous scales; pedicels equaling the sepals or longer; staminate flower clusters larger and more abundant than pistillate clusters; calyx segments 6 (rarely 7–9), spreading, thin; stamens distinctly exserted, usually 9 in 3 series (first and second series usually glandless); anthers introrse, 2-celled, opening by uplifted valves; petals absent; gynoecium a single pistil; style simple, slender, columnar; ovary globular, 1-celled; ovules anatropous, pendulous; staminate calyx segments 1/25–1/15 in. wide; staminate pedicels pilose, filaments slender, narrower, about 1/50 in. wide, not dilated at base, almost 1/12 in. long; pistillate calyx segments 1/15–1/12 in. long, 1/25–1/15 in. wide; pistillate pedicels 1/10 in.

FRUIT. Maturing September–October, old fruiting pedicels persistent from previous year and lasting until time of anthesis the following year; fruiting pedicels stout, 1/3–1/2 in., conspicuously enlarged at summit, 1/12–1/8 in. wide; drupe seated on the accrescent calyx, somewhat elongate, elliptic-obovoid, red, 1/4–1/2 in. long, 1/5–1/3 in. wide, pulpy.

LEAVES. Drooping, aromatic when crushed, simple, alternate, deciduous, oblong to narrowly-elliptic or oval, apex acute to short-acuminate, base rounded or semi-cordate, margin entire, thin and membranaceous, length 2–6 in., width 3/4–2 1/4 in., surfaces concolorous, dark green, more or less densely pubescent on both sides, later more glabrous, the venation prominent below, lowest 2 pairs of lateral veins not parallel to ones above, conspicuously more ascending from midrib at 45°–50° angle; petiole stout, 1/12–1/4 in., pubescent.

TWIGS. Slender, reddish brown to gray, pubescent at first, glabrous later; buds pubescent; bark aromatic when bruised.

RANGE. Rare and local, along swamps and streams; Louisiana to Florida, north to North Carolina, and west to Illinois and Missouri.

REMARKS. The genus name, *Lindera,* honors John Linder (1676–1723), an early Swedish botanist, and the species name, *melissaefolia,* means melissa-like leaves. Vernacular names are Jove's-fruit, Spice-wood, Benjamin-bush, Hairy Spice-bush, and Pond-berry. The twigs are sometimes brewed into tea and used as a spring tonic. The plant is sometimes nibbled by marsh rabbit, and the fruit eaten by a number of species of birds.

Red Bay

Persea borbonia (L.) Spreng.

FIELD IDENTIFICATION. An evergreen tree to 70 ft, with a trunk diameter of 1–3 ft. The stout erect branches form a dense symmetrical head.

FLOWERS. Small, borne in few–several-flowered axillary panicles; peduncles pubescent to glabrous, 1/2–2 in., often reddish because of pubescence; corolla absent; calyx bell-shaped, pale yellow, 6-parted, in 2 series; outer series broadly ovate and puberulous; inner series oblong-lanceolate, acute, hairy; stamens 12, in 3–4 series, anthers opening by uplifted valves (anthers of 3 stamens turned down, others introrse); filaments flattened; pistil sessile, ovary 1-celled, style slender, stigma disklike.

FRUIT. Drupe about 1/2 in. long, globose or obovoid, lustrous, bright blue to black; seed solitary, ovoid, flesh thin and dry.

LEAVES. Simple, alternate, persistent, aromatic, elliptic-oblong, margin entire (more rarely remotely serrate), apex acute or acuminate, base wedge-shaped, bright green, lustrous above, glaucous beneath, 3–4 in. long, 1–1 1/2 in. broad, thickened, veins rather obscure; petiole 1/2–1 in. long, stout, grooved above, rigid, reddish brown; leaf scars circular, bundle scar solitary.

TWIGS. Slender, angled, gray- or brown-hairy, dark green at first, later light brown.

BARK. Reddish brown to grayish brown, fissures deep, ridges broad and scaly, 1/4–3/4 in. thick.

SOUTHERN SPICE-BUSH
Lindera melissaefolia (Walt.) Blume

RED BAY
Persea borbonia (L.) Spreng.

WOOD. Reddish brown, sapwood lighter, close-grained, fairly hard and strong, weighing about 40 lb per cu ft.

RANGE. Rich sandy soils of river-bottom lands or swamps, usually near the coast. Texas, Oklahoma, Arkansas, and Louisiana; eastward to Florida, and northward to eastern Virginia.

CLOSELY RELATED SPECIES AND VARIETIES. A closely related tree to the Red Bay is the Swamp Bay, *P. palustris* (Raf.) Sarg. It is so listed in Kelsey and Dayton (1942). It is chiefly characterized as having the petioles, peduncles, and flower parts more pubescent than in Red Bay. Fernald (1945a, 149–151) considered the differences as slight and has reduced it to a form of Red Bay, *P. borbonia* forma *pubescens* (Pursh) Fern.

Also a plant known as Shore Bay, *P. littoralis* Small, found on coastal dunes from Rockport, Texas, eastward to Florida, shows little difference when compared to the Red Bay. The fruit averages slightly larger and the plant somewhat more shrubby in habit, but many intergrading forms are found. It should be relegated to the status of a synonym of the Red Bay. It was originally described as a species by Small (1932).

Also a plant known as Silk Bay, *P. humilis* Nash, reported from Texas by Cory and Parks (1937, p. 47), may be considered as a synonym of Red Bay, as far as Texas material is concerned.

Therefore, for the purposes of this manual, the Red Bay, *P. borbonia* (L.) Spreng., appears to be the common species of the Gulf Coast region from Texas to Florida, with minor variations.

REMARKS. The genus name, *Persea,* is the ancient Greek name of an unidentified Egyptian tree. The species name, *borbonia,* is the old generic name of *Persea.* Vernacular names are Sweet Bay, Florida-mahogany, Tiss-wood, Laurel-tree, and Isabella-wood. The fruit is eaten by at least two species of birds. Soups and meats are often flavored with the leaves. The wood is used for furniture, boatbuilding, interior finish, and cabinetmaking, The tree is worth cultivation for its evergreen leaves and ornamental fruit. It is sometimes confused with Sweet-bay Magnolia, *Magnolia virginiana* L.

AMERICAN AVOCADO

Persea americana Mill.

FIELD IDENTIFICATION. Cultivated tropical tree. Evergreen and much-branched, attaining a height of 60 ft with a slender trunk.

FLOWERS. In subterminal, pubescent, naked panicles at the base of the current season's shoots; individual flowers small, perfect, yellowish green; corolla absent; calyx of 6 sepals in an inner and outer series of 3, ⅙–¼ in. long, about equal in length or the inner slightly longer, oblong to lanceolate, apex acute, united at base and remaining so when falling; stamens usually 12, in 4 series, but the inner series sometimes only staminodial; filaments slender, fine-hairy, inserted near the base of the calyx, sometimes with basal glands; anthers erect, oblong-ovate, flattened, 4-celled, the 4 valves hinged distally; ovary free, ovate-elliptic, tapering into a slender, simple style, 1-celled, ovule pendulous.

FRUIT. Drupe pyriform to elongate-elliptic or oval, length 3–8 in., outer skin green to black or violet, membranous to thick and woody in different varieties; pulp butter-like, white to green or yellowish, edible; seed conical to oblate, solitary; seed coats 2, often distinct, the outer usually thin and papery, endosperm none.

LEAVES. Persistent, alternate, leathery, blade length 3–10 in., margin entire, apex acute or short-pointed, base rounded or broadly cuneate; upper surface dark green and glabrous; lower surface paler and sometimes glaucous with veins more prominent; elliptic to oval; petiole ¾–2 in., puberulent to glabrous.

TWIGS. Rather stout, green, and puberulent to glabrous; younger branches and trunk rather smooth, light to dark brown or grayish, often marked with numerous paler lenticels, older bark becoming roughened longitudinally; wood light brown, specific gravity of 0.65.

RANGE. Native of tropical America and extensively cultivated. Cultivated in southern Florida, Central America, and South America. Sometimes naturalized in 'lorida. Sensitive to freezes in southern Texas. If ·lanted in a protected situation in gardens sometimes esistant to cold over several years, from Houston along he coast to Brownsville, Texas.

MEDICINAL USES. Various parts of the tree are used n domestic medicine in Mexico. The pulp of the fruit s said to be a stimulant and is applied to boils to asten their development. The skin of the fruit is used s a vermifuge especially for intestinal parasites. For his purpose, 8–10 grams are macerated in a glass of vater overnight and taken the following day. The seed as a bitter and disagreeable taste, and the milky juice lackens and can be used as ink. The fluid has also een used to cure certain neuralgias. A decoction of he seed macerated in alcohol is said to be a cure for heumatism. The seed is sometimes ground and mixed vith cheese or lard to poison rats.

PROPAGATION. The tree is propagated by seeds, or by elected cuttings of improved varieties. The growth is ather rapid and trees 5–6 years of age start producing. Maximum yield is at about 50 years. The fruit is gathered slightly before ripening and before falling. Both fruit and bark are sometimes attacked by a borer beetle, *Helipus lauri,* and the growing larva is so destructive as often to ruin the fruit and even kill the tree. The tree is also often grown as an ornamental.

AMERICAN AVOCADO
Persea americana Mill.

COMMON SASSAFRAS
Sassafras albidum (Nutt.) Nees

REMARKS. The genus name, *Persea,* is the ancient name, and the species name, *americana,* refers to its American origin. The tree is also listed under the scientific names of *P. gratissima* Gaertn. and *P. persea* (L.) Cockerell. It has many closely related forms. A number of these have been named, and have created considerable taxonomical confusion. Over 525 horticultural forms have been listed. Vernacular names applied to the American Avocado are Alligator-pear, Avocado-pear, Avocat, Abacate, Aguacate, Ahuacate, Ahuacaquahuitl, and Cuponda. The tree is probably one of the best known of tropical fruit trees, and is grown for its many varieties of fruit. It varies greatly in size and shape, from oval to pear-shaped, and in quality, the pulp sometimes being fibrous and disagreeable. The hard dry seed is sometimes used to carve tops and miniature figures. It also contains tannin used in leatherwork. The pulp of the fruit is mostly eaten uncooked, or a salad or sauce, known as guacamole, is prepared from it with lime juice, tomato, and condiments.

COMMON SASSAFRAS

Sassafras albidum (Nutt.) Nees

FIELD IDENTIFICATION. Tree attaining a height of 90 ft, with a flattened oblong crown, and short, crooked branches.

FLOWERS. March–April, dioecious, axillary, in racemes about 2 in. long; calyx yellowish green, of 6 spreading sepals; corolla absent; stamens in 3 sets of 3 each, the inner set glandular at the base; anthers 4-celled, with flattened, elongate filaments; pistillate flowers with an erect columnar style and depressed stigma, also 6 sterile stamens.

FRUIT. Drupaceous, blue, lustrous, ½ in. long, oblong or spherical, borne on a thickened red pedicel, pulpy; stone solitary, light brown, dispersed chiefly by birds.

LEAVES. Alternate, simple, deciduous, thin, aromatic, blades 3–5 in. long, ovate or elliptic, entire on the margin, or divided into 2–3 mitten-shaped lobes, lobes acute or obtuse; cuneate at base, bright green above, glabrous and glaucous beneath, often hairy on the veins; petioles about 1 in.

TWIGS. Yellowish green, mucilaginous, pubescent at first, turning glabrous and orange-red later.

BARK. Reddish brown to gray, aromatic, irregularly broken into broad flat ridges.

WOOD. Orange-colored, aromatic, durable, close-grained, soft, weak, brittle, weighing 31 lb per cu ft.

RANGE. Texas, Oklahoma, Arkansas, and Louisiana; eastward to Florida, north to Maine, and west to Ontario, Michigan, and Iowa.

MEDICINAL USES. The inner bark contains a mucilaginous, aromatic, somewhat pungent, volatile oil which is used medicinally. Mucilage of sassafras is prepared from the pith of the root, and was formerly used to soothe eye inflammations. Oil of sassafras, which is a volatile oil distilled from the root, is used as a flavoring material and as an antiseptic and disinfectant. If taken in large amounts the oil is said to be narcotic. A hot tea, prepared from various parts of the plant, is som times taken as a sudorific, and in weaker potions as pleasant beverage. The dried leaves, known as "gumb file," are used to give flavor to soups.

PROPAGATION. Sassafras has been cultivated sinc 1630. The fruit is picked by hand when ripe, and th pulp removed by rubbing on screens. The seed ave ages about 5,000 seeds per lb. The average purity c commercial seed is 98 per cent and soundness 85 pe cent. The natural dormancy may be broken by stratific tion in sand for 30 days at 41°F. for seeds from South ern trees. Germination averages about 85 per cen Stratified seeds should be sown in drills with rows 8–1 in. apart, and about ¼ in. of firmed soil over them The surface should be kept moist. Seeds planted in th fall should be mulched with leaves or straw. Fall-sow seeds should be planted late to avoid premature sprou ing.

VARIETY. Silky Sassafras, S. *albidum* var. *molle* (Raf. Fern., is a variety with buds and twigs pubescen leaves glaucescent and silky pubescent beneath, a least while young. Texas to Florida; north to Ontari and Michigan. In Texas west to the Brazos River.

REMARKS. The genus name, *Sassafras*, is a popula one derived from the word *salsafras*, which was give by early French settlers, with reference to its medicina properties; and the species name, *albidum*, refers to light-colored condition of the wood. Other commonl used vernacular names are Ague-tree, Cinnamon-woo Smelling-stick, Saloop, and Gumbo-file. The tree i long-lived and rather free of diseases. The fruit i known to be eaten by 28 species of birds, and th leaves browsed by woodchuck, white-tailed deer, mars rabbit, and black bear. The wood is used for posts rails, buckets, cabinets, and interior finish.

JUNCO FAMILY (*Koeberliniaceae*)

SPINY ALLTHORN

Koeberlinia spinosa Zucc.

FIELD IDENTIFICATION. Much-branched, usually leafless shrub or tree attaining a height of 24 ft, and consisting of a tangled mass of stiff green spines.

FLOWERS. Borne on slender pedicels 1/12–1/3 in. long in lateral racemes, each flower small, perfect, and about 1/4 in. across; petals 4, greenish white, linear to oblong, apex obtuse or sometimes notched, longer than sepals; stamens 8, filaments flattened in the middle and somewhat petaloid, anthers large, sagittate and deciduous; ovary 2–5 united carpels, styles united; calyx of 4 ovate, deciduous sepals about 1/25 in. long. In some flowers it is difficult to distinguish between the petals and petaloid stamens with deciduous anthers.

FRUIT. Borne in clusters about 1 in. long, peduncles clavate and about 1/3 in. long; berry black, subglobose, apiculate, 3/16–1/4 in. in diameter, fleshy, 2-celled; seeds 1–4, about 1/8 in. long, curled, wrinkled, and striate.

LEAVES. Alternate, consisting of minute scales which are early-deciduous, thus leaving the plant barren most of the year.

TWIGS. Green, smooth, stout, stiff, divaricate, all ending in large, sharp thorns.

BARK. Smooth, green to brown or gray on young trunks, older with small scales and shallow fissures.

WOOD. Black or brown, hard, resinous, close-grained, with a specific gravity of 1.12, emitting a disagreeable odor when burned.

RANGE. Arid places in western and southwestern Texas; west through New Mexico to Arizona, and south into Mexico in Sonora to Tamaulipas and Hidalgo.

VARIETY. E. R. Bogusch (1931) has described a variety as *K. spinosa* var. *verniflora* Bogusch. This variety is reported as flowering in March and April, and is described as a thorny, leafless shrub to 6 ft; twigs bright green, attenuated, 1¼–2 in., sometimes only 5/8 in. in extreme cases, tapering gradually in the typical form; vesture of unicellular puberulence, more common near the inflorescence; flowers in open fascicles; fruit as in the species. The type specimen was found in a pasture near Weslaco, Texas. Bogusch states that many intergrades between it and the species are to be found.

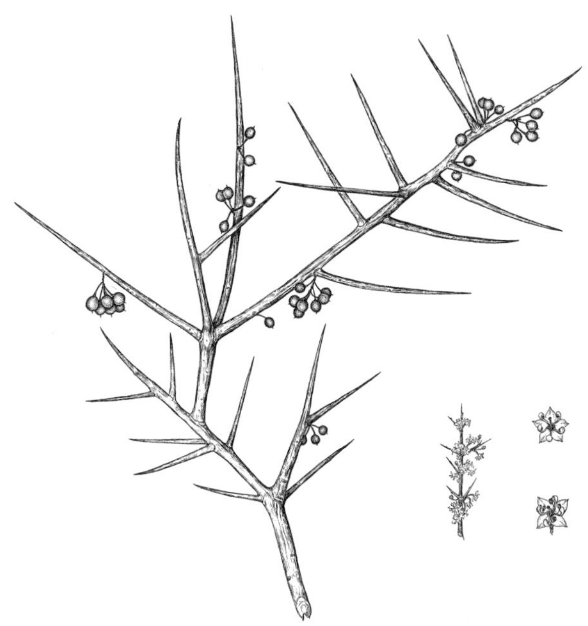

SPINY ALLTHORN
Koeberlinia spinosa Zucc.

REMARKS. The genus name, *Koeberlinia,* is in honor of C. L. Koeberlin, German clergyman and amateur botanist. The species name, *spinosa*, refers to the abundant spines. Common names for the plant are Junco, Corona de Cristo, and Crucifixion Thorn. Scaled quail have been seen to eat the fruit, as has the jack rabbit. The plant is perhaps of some value in erosion control. It is a perfect example of adjustment to desert conditions, with the green thorns and twigs carrying on the photosynthetic process. Being thicket-forming it presents an impenetrable green mass of thorns to any intruder.

MUSTARD FAMILY (*Cruciferae*)

DESERT PRINCESPLUME

Stanleya pinnata (Pursh) Britt.

FIELD IDENTIFICATION. Generally a stout, herbaceous perennial, but sometimes with a rather woody base. Simple or branched above and attaining a height of 2–5 ft.

FLOWERS. May–June, pedicels ascending or horizontal, about ⅜ in., buds elongate, crowded, borne in long terminal spike-like racemes 3 in.–2 ft; petals 4, limb golden yellow, ⅖–⅗ in., linear-oblong, one half to three fourths as long as the densely hairy, slender claw; calyx with 4 sepals ⅖–⅗ in. long, reflexed with age, slender, distinct; stamens 6, long-exserted, anthers long and narrow; style elongate, exserted, ovary superior, of 2 united carpels.

FRUIT. Capsule stipe ⅖–1 in., body 2–4 in., linear, curved, subterete.

LEAVES. Upper usually entire and broadly lanceolate to oblanceolate, ¼–2½ in., short-petioled or sessile; lower nearly entire to pinnately divided into oblong, elliptic or lanceolate segments, 2½–8 in., short-petioled or sessile, slightly puberulent to glabrous.

TWIGS. Light green, glaucous and glabrous, older woody parts at base light to dark gray.

RANGE. On plains and dry slopes, often in seleniferous soil, at altitudes of 1,000–8,000 ft. Trans-Pecos Texas; north into New Mexico, Colorado, Nebraska, South Dakota, and Wyoming; west into Arizona, California, Utah, and Idaho.

VARIETIES. A variety has been segregated as the Whole-leaf Desert Princesplume, S. *pinnata* var. *integrifolia*, which has leaves ovate to elliptic or ovate to lanceolate with entire margins.

DESERT PRINCESPLUME
Stanleya pinnata (Pursh) Britt.

Another variety is known as the Cut-leaf Desert Princesplume, S. *pinnata* var. *bipinnata,* with leaves mostly bipinnate and somewhat pubescent.

REMARKS. The genus name, *Stanleya,* is in honor of Lord Stanley, English ornithologist, and the species name, *pinnata,* refers to the pinnate leaves. Also known under the vernacular names of Desert Plume and Golden Princesplume.

HORSERADISH-TREE FAMILY (*Moringaceae*)

HORSERADISH-TREE

Moringa oleifera Lam.

FIELD IDENTIFICATION. Cultivated tree attaining a height of 33 ft, with open spreading branches.

FLOWERS. Fragrant, panicles reddish or white, 2–6 in. long, axillary or terminal, long-peduncled, many-flowered, densely pubescent; flowers perfect, slightly irregular, about 1 in. across; calyx borne on a short cuplike hypanthium which is 5-lobed; lobes petal-like, white or pink, linear, obtuse or acute, densely puberulent, ¼–½ in. long; petals resembling the sepals, linear to elliptic, white or pink toward the base, apex rounded or obtuse, about same length as sepals except one petal slightly longer than the others and tending to be erect; stamens 5, exserted, curved, reddish and hairy toward the base, ¼–⅜ in. long, alternating with the 2–5 shorter staminodia, anthers attached to the back, 1-celled, opening lengthwise; pistil simple, swollen below, ovary stipitate, stigma minute, ovules numerous in 2 series on each placenta.

FRUIT. Capsule ½–1½ ft long, 4–9 ribbed, slender, somewhat torulose, beaked, 1-loculed, 3-valved; seeds immersed in the spongy tissue, ⅜–¼ in. long, 3-winged.

LEAVES. Alternate, deciduous, blades ¼–2 ft long, rachis densely pubescent, reddish, swollen at the base; odd-pinnately compound (often 3-foliate), divisions opposite, primary divisions with 6–9 pairs of pinnae, each 2–3 in. long and a terminal pinna; secondary pinnae with 3–5 leaflets; leaflets ⅜–¾ in. long, oval to elliptic, apex rounded or slightly notched, base rounded to cuneate, margin entire, upper surface dull green with minute scattered hairs, lower surface paler and finely puberulent; petioles 1/25–1/12 in. long, puberulent, bearing a solitary stipule-like appendage at the base.

TWIGS. Stout, grayish green, puberulent to glabrous.

HORSERADISH-TREE
Moringa oleifera Lam.

BARK. Smooth, tight, grayish green, corky; leaf scars large, oval or horseshoe-shaped, some depressed.

RANGE. Usually in sandy soil; a native of South Asia, and cultivated in the United States in Florida, California, and the lower Rio Grande Valley area of Texas. Perhaps not naturalized in Texas, but spontaneous in adjacent Mexico. Planted on the streets and in parks in Brownsville, Texas.

REMARKS. The genus name, *Moringa*, is derived from the native Malabar name, and the species name, *oleifera*, means "oil-bearing." The roots smell like horseradish and are sometimes eaten, as well as the young fruit. The seeds produce an oil used in the arts and known as "behen oil." It is used in cosmetics and cooking, and also as a lubricant for delicate machinery.

SAXIFRAGE FAMILY (*Saxifragaceae*)

DECUMARIA-VINE

Decumaria barbara L.

FIELD IDENTIFICATION. Slender, smooth, woody vine climbing by aerial rootlets.

FLOWERS. May–June, inflorescence terminal, corymbose, compound, 1½–4 in. wide; flowers fragrant; regular, small, white, perfect, epigynous, ¼–⅙ in. broad, 7–10-merous; calyx top-shaped, adnate to the ovary, limb with 7–10 minute teeth; petals oblong, spreading; stamens 20–30 with elongate, subulate filaments inserted on the disk; ovary inferior, 10–15-ribbed, 5–10-celled; styles united, short, thick, with a 7–10-lobed stigma.

FRUIT. Capsule ⅙–¼ in. high, top-shaped, fragile, apex depressed conic with the persistent style, many-ribbed, dehiscent longitudinally between the ribs, persistent after the numerous seeds fall; seeds with membranous and reticulated testa, bearing a club-shaped appendage.

LEAVES. Opposite, deciduous, petioled, blades 2–4 in., ovate to ovate-oblong, apex acute to subacuminate, base narrowed to rounded, margin entire or somewhat denticulate, upper surface glabrous and lustrous, lower surface pubescent to glabrous later; petioles ½–1 in.

RANGE. In low rich ground or swamps. Louisiana; eastward to Florida and northward to Virginia.

PROPAGATION. The vine is sometimes cultivated for ornament, especially for covering tree trunks, rocks, or trellis work. Propagation is by greenwood cuttings in summer under glass, rarely by seeds. It is tender, but in protected places survives as far north as Massachusetts.

REMARKS. The genus name, *Decumaria*, is from *decum* ("ten"), referring to the sometimes 10-merous flowers. The species name, *barbara* ("of Barbary"), comes from the fact that the species was originally thought to be African.

DECUMARIA-VINE
Decumaria barbara L.

CLIFF FENDLER-BUSH

Fendlera rupicola Gray

FIELD IDENTIFICATION. Widely-branched, deciduous shrub to 6 ft.

FLOWERS. May–June, solitary, perfect, borne on lateral pubescent pedicels ⅛–¼ in.; corolla about ¾ in. across, white or pink-tinged; petals 4, spreading, ovate to rounded, puberulent, abruptly contracted into a long claw; stamens 8, erect, about half as long as the spreading petals, forked at apex; anthers cuspidate, borne between the lobes; calyx cup-shaped, 4-lobed, lobes longer than the cup, lanceolate to ovate, acute or acuminate at apex, pubescent to glabrous; ovary half superior, 4-celled, many-ovuled.

FRUIT. Capsule July–August, light brown, conical to an attenuate point, about ½ in. long, dehiscent at maturity into 4–5 woody valves; seeds compressed, oblong to triangular, ⅕–¼ in. long; calyx ribbed at maturity.

LEAVES. Opposite, or borne at distant nodes in small clusters, deciduous, thickish, sessile or short-petiolate, ½–1¼ in. long, lanceolate to linear, margin entire, often ciliolate, revolute, apex acute to obtuse; upper surface dark green, glabrous, or with a few scattered white hairs; lower surface paler and white-hairy, 3-nerved.

TWIGS. Gray to tan, elongate, arching, terete, striate, puberulent when young, when older fibrous.

RANGE. The species is distributed from the Sabinal River to the Pecos River in scattered locations in Texas, becoming more common in the higher mountains of the Trans-Pecos region. Occurs in wooded canyons of the Davis, Chisos, and Guadalupe mountains; northward and westward into New Mexico, Colorado, Utah, and Arizona, and southward into Mexico. It shows preference for canyon cliffs or rocky ledges at altitudes of 3,000–7,000 ft. The type locality is recorded as rocky ledges of the Guadalupe River above New Braunfels, Texas.

VARIETIES. Sickle-leaf Fendler-bush, *F. rupicola* Gray var. *falcata* Gray, is a variety with leaves narrow-elliptic, often falcate, strongly revolute, green and glabrate on both surfaces. It is distributed through New Mexico, Arizona, Colorado, and sparsely in Texas in the Trans-Pecos mountains. The type specimen was collected on Tunitcha Mountain, New Mexico.

Another variety is Wright Fendler-bush, *F. rupicola* var. *wrightii* Gray. The leaves are narrow, strongly revolute, more densely pubescent and white beneath; the flowers are smaller, and the pedicels and calyx are densely pubescent; the capsule averages about ⅜ in. long. It is distributed through west Texas, New Mexico, Arizona, and Chihuahua, Mexico.

REMARKS. The genus name, *Fendlera,* is for August Fendler, an early German botanical explorer of New Mexico (1813–1883). The species name, *rupicola,* means "lover of rocks," because the plant grows in stony ground. The Cliff Fendler-bush is suitable for rock gardens in well-drained, sunny situations, and has been found hardy as far north as New England. It is drought resistant, and is propagated by seeds or greenwood cuttings under glass. It is rather commonly grazed by deer and goats, and by cattle usually only in time of stress.

UTAH FENDLERELLA

Fendlerella utahensis (Wats.) Heller

FIELD IDENTIFICATION. Shrub 1–3 ft, upright, densely branched.

FLOWERS. May–August, cymes terminating leafy branches on rather short, naked or bracted peduncles, 3-7-flowered; flowers ⅛–⅕ in. across, regular, perfect; calyx united to the ovary base, tube elongate-turbinate, strigillose; sepals 5, shorter than the tube (1/25–1/15 in. long), lanceolate to oblong, ciliate; petals 5, inserted on the calyx-tube, distinct, white, 1/12–⅙ in. long, oblong to elliptic; stamens 10, unequal, filaments somewhat dilated distally; ovary half-inferior, conical, 3-celled, styles 3.

FRUIT. A capsule, ⅛–⅙ in. long, cylindric, cartilaginous, narrowed at base, styles persistent and spreading, septicidally dehiscent into 3 1-seeded valves.

LEAVES. Leaves numerous, simple, opposite or fascicled, deciduous or semipersistent, sessile or shortly petioled, blade length ¼–¾ in., width 1/16–3/16 in., elliptic or lanceolate to linear-oblong, or occasionally ovate, apex acute or obtuse, base gradually narrowed, margin entire, thickish, surface inconspicuously 3-veined, strigillose-hairy to subglabrous.

TWIGS. Slender, much-branched, green to gray, strigose-hairy, somewhat shreddy, internodes ⅓–¾ in. long.

RANGE. Dry mountain slopes and pine-clad canyons to an altitude of 8,000 ft. In Texas in the mountains west of the Pecos River, also in New Mexico, Arizona, Utah,

CLIFF FENDLER-BUSH
Fendlera rupicola Gray

Utah Fendlerella
Fendlerella utahensis (Wats.) Heller

and Southern California. In Mexico in Coahuila and Chihuahua.

VARIETY. A variety has been named the Arizona Fendlerella, *F. utahensis* var. *cymosa* (Greene) Kearney & Peebles, and has narrower and more pointed leaves than the species. It is found in southern Arizona, southern New Mexico, and northern Mexico. The type specimen was collected by G. Pringle in 1884, in the Santa Rita Mountains, Pima County, Arizona.

REMARKS. The genus name, *Fendlerella,* honors Augustus Fendler, German naturalist and explorer of New Mexico, and the species name, *utahensis,* refers to the state of Utah, where it was first collected at Kanab by E. P. Thompson. It is browsed by black-tailed deer and goats, but by cattle only when other plants are unavailable.

Oak-leaf Hydrangea

Hydrangea quercifolia Bartr.

FIELD IDENTIFICATION. Spreading, woolly shrub to 6 ft, with reddish brown exfoliating bark.

FLOWERS. May–June, borne in a large terminal showy panicle 4–12 in. long; rachis stout, elongate, clothed with dense, reddish brown or gray hairs; peduncles erect or ascending, 2–4 in., slender, hairs less dense; individual flower pedicels slender, 1/16–1/8 in. long, white-hairy; flowers of 2 types, seed-bearing and sterile; the sterile ones on the outer fringe of the panicle; sterile rays 4, petal-like, 1/2–1 1/2 in. across, orbicular to obovate, apex obtuse or rounded, base cuneate or abruptly short-clawed, white at first, turning reddish or purplish later, veiny; fertile flowers with 5 petals, 1/25–1/12 in. long, oblong, slightly concave, greenish white; stamens 8–10, conspicuously exserted, inserted at the base of a disk, filaments filiform; ovary inferior, completely or partially 2–4-celled, ovules numerous; styles absent, stigmas 2 and divergent; hypanthium urn-shaped, 1/8–1/6 in. wide at maturity, sepals 5, ovate, about one third as long as the hypanthium.

FRUIT. Capsule 1/8–1/5 in. wide, membranous, urn-shaped, ribbed, calyx teeth apparent, tipped by the 2 divergent stigmas with an opening between them for the escape of the numerous minute seeds; 2–4-celled.

LEAVES. Deciduous, opposite, blades 4–8 in. long and broad, ovate to suborbicular or oblong, deeply 3–7-lobed, the lobes again with smaller lobes or serrate with sharp teeth; apex of lobes acute or acuminate, sinuses either rounded or angled; leaf base somewhat rounded or truncate, and often decurrent into the petiole; upper surface dull green, somewhat glabrous except for long white hairs on the impressed veins; lower surface paler, densely woolly-hairy, thin; petioles one half to one fifth as long as the blade, densely woolly with white or reddish brown hairs.

Oak-leaf Hydrangea
Hydrangea quercifolia Bartr.

TWIGS. Younger ones densely reddish brown, or with white-woolly hairs; older with thin reddish brown bark exfoliating in shaggy strips; lenticels small, raised, numerous, pith ample.

RANGE. Most often in moist rich soil of shaded ravines or riverbanks. Eastern Louisiana; eastward to Georgia and Florida and northward into Kentucky and Tennessee.

PROPAGATION. This is a very beautiful plant which could be more widely cultivated for its handsome, large, white panicles of flowers. It can be propagated by cuttings of young wood in late winter, also by means of the root suckers. The tiny seed are seldom used. Pruning back considerably in fall or spring is beneficial. A moist, rich, porous soil is needed for the plant, either in partial shade or full sun.

REMARKS. The genus name, *Hydrangea,* is from the Greeks words *hydor* ("water") and *aggeion* ("a vessel"), referring to the cup-shaped fruit. The species name, *quercifolia,* refers to the oaklike leaves. Vernacular names are Sevenbark, Gray Beard, and Old Man's Beard.

Ashy Hydrangea

Hydrangea cinerea Small

FIELD IDENTIFICATION. Spreading shrub 3–6 ft.

FLOWERS. Maturing June–July in corymbs 2–8 in. broad, rather round-topped; a pair of small, ovate to lanceolate, pubescent bracts at base of the corymb; pedicels densely pubescent; flowers usually perfect except those on outer fringe; hypanthium campanulate, about 1/12 in. wide, ribbed; sepals minute, triangular, acute; petals 5, white, valvate in the bud, ovate, boat-shaped, broadest at apex, 1/15 in. long; stamens 8–10, conspicuously exserted, filaments filiform, inserted on the edge of an epigynous disk; ovary inferior, 2–4-celled; sterile flowers on the perimeter of the corymb, usually with 4–5 spreading, petal-like lobes.

FRUIT. Maturing September–November, capsule urn-shaped, usually higher than broad, about 1/12 in. in diameter, opening at the top between 2–4 spreading styles, strongly ribbed, 2–4-celled; seed minute and numerous.

LEAVES. Opposite, blade oval-elliptic, narrowly ovate or orbicular-ovate, length 2½–6 in., apex acute or acuminate, base rounded or cordate, margin serrate, veins rather conspicuous, upper surface bright green, lower surface light grayish and tomentose, especially along the veins; petioles ¾–3 in.

TWIGS. Slender, copiously pubescent and brown, when older glabrous and brown.

RANGE. Well-drained, moist soil in partial shade. Arkansas, Tennessee, Georgia, Alabama, and north to North Carolina.

FORM. A form with the sepals of the sterile flowers

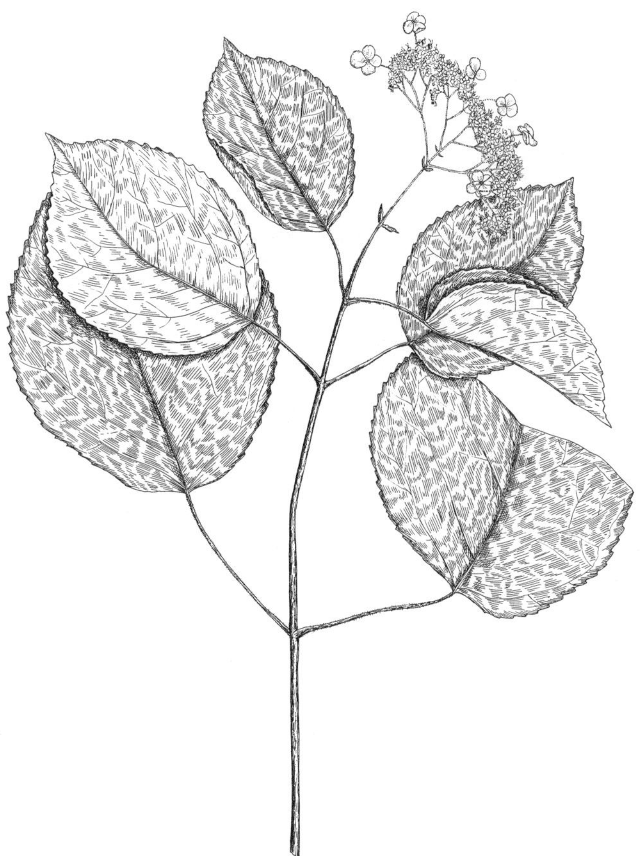

ASHY HYDRANGEA
Hydrangea cinerea Small

broad-ovate and obtuse has been listed as *H. cinerea* forma *sterilis* Rehd.

REMARKS. The genus name, *Hydrangea,* is from the Greek words *hydor* ("water") and *aggeion* ("a vessel"), from the shape of the capsule. The species name, *cinerea,* refers to the ash-colored hairs of the leaf. It was introduced into cultivation in 1906.

Ashe Hydrangea

Hydrangea ashei Harbison

FIELD IDENTIFICATION. Shrub 16–32 in., with clustered stems. The description follows that of Harbison (1928).

FLOWERS. Inflorescence a compound cyme, 1½–4¾ in. wide and 1¼–2½ in. high, more or less pubescent, the peduncle eventually nearly glabrous; flowers appearing the last of June to the middle of July, or occasionally later, cream-colored, the perfect ones about 1/10 in. long, including the stamens, with fugacious orbicular petals about 1/25 in. in diameter, anthers cream-colored; calyx a crown with pointed teeth; the usually few marginal sterile flowers about ½ in. wide, the small segments orbicular-ovate to orbicular.

FRUIT. Maturing September or later, prominently 10-ribbed, barely 1/12 in. long, including the short persistent spreading styles, and about the same width.

LEAVES. Blade ovate, broadly ovate to orbicular-ovate, or rarely oblong-ovate, 2¾–4⅜ in. long, 1⅘–3½ in. broad, apex abruptly acuminate-pointed, base subcordate or cordate, margin coarsely and sharply serrate with straight teeth nearly to the base; lower surface permanently pubescent with short, simple hairs on the surface of the blade as well as on the veins; petioles flattened and pubescent on the upper surface, rounded and glabrous on the lower.

TWIGS. Winter twigs of the year angled with smooth and tan-colored bark; bark on the older stems exfoliating in thin layers, the inner bark greenish gray.

RANGE. The proposed species seems to be essentially Ozarkian in its distribution, extending according to Ashe's specimens, from Montgomery County, Arkansas, to Stone County in the same state. He reports it as being not uncommon along rocky banks of small streams, but also as entering moist, stony woods as near the summit of the north slope of Boston Mountain in Newton County, Arkansas.

RELATED SPECIES. Harbison (1928, p. 256) makes the following comments on this plant:

"The affinities of this proposed species, in its broadly ovate and cordate leaves, in its pubescence and in its small marginal, sterile flowers seem to be with *Hydrangea cinerea* Small, rather than with *H. arborescens* L., which it superficially resembles. *H. arborescens* rarely has cordate leaves whereas at least the lower leaves of this plant are prevailingly cordate. The lower surfaces of the leaves of *H. arborescens* are essentially glabrous except on the veins, but the leaves of this species on their entire lower surface are invariably pubescent. The sterile flowers of *H. arborescens* as a rule are about twice the size of those of the proposed species, and both fertile flowers and capsules of *H. arborescens* seem to be consistently larger. In comparison with *H. cinerea* its pubescence is much less copious, never sufficient to whiten the lower surface of the leaf."

REMARKS. The genus name, *Hydrangea*, is from the Greek words *hydor* ("water") and *aggeion* ("a vessel"), from the shape of the capsule. The species name, *ashei*, is in honor of William Willard Ashe (1872–1932), of the United States Forest Service.

Smooth Hydrangea

Hydrangea arborescens L.

FIELD IDENTIFICATION. Widely-branched, straggly shrub growing in clumps and attaining a height of 2–10 ft.

FLOWERS. June–July, borne in rather flat, pubescent cymes 2–6 in. broad; staminate flowers generally marginal and radiant, slender-pediceled, some heads with a few sterile flowers or none at all; sterile flowers ¾ in. across, on pedicels about ¾ in., formed by a 4-parted, widespread, flat, dilated, colored calyx; pistillate flowers small, sometimes a cyme has all pistillate flowers; calyx-tube 8–10-ribbed, campanulate, adherent to the ovary, sepals minute and triangular, glabrous; petals valvate in the bud, deciduous, ovate, concave, about 1/25 in. long; stamens 8–10, conspicuously exserted, inserted on the disk, filaments slender and filiform, stigmas reddish, styles 2, ovary 2-celled, ovules numerous.

FRUIT. Ripening October–December on pedicels 1/12–⅛ in.; capsule membranous, 1/12–⅛ in. long or wide, prominently 8–10-ribbed, 2-horned, many-seeded, dehiscent at the base of the horny styles by a perforation.

LEAVES. Opposite, broadly ovate or orbicular to elliptic, apex acute or acuminate, base cordate or broadly cuneate, margin coarsely and sharply toothed, length 2–6 in., dark green above and usually glabrous, paler and more veiny and more or less pubescent beneath, petiole 1–4½ in., slender, more or less pubescent.

TWIGS. Slender, light brown, glabrous or pubescent, smooth on young stems, old bark sometimes shreddy.

RANGE. Usually on rocky, moist slopes of ravines to an altitude of 4,200 ft in Arkansas, Oklahoma, and Louisiana; eastward to Florida, north to New York, and west to Missouri and Iowa.

MEDICINAL USES. The official drug fluidextractum hydrangeae is made from the powdered, cylindrical, reddish yellow rhizome. Analysis has found it to contain a glycoside known as "hydrangin," a fixed oil and a volatile oil, several resins, together with saponin and sugar. No tannin is apparently present. At one time

Smooth Hydrangea
Hydrangea arborescens L.

hydrangin was used in the treatment of bladder stones and cystitis, but it is now seldom used. It is reported to have been a favorite medicine of the Cherokee Indians.

PROPAGATION. Smooth Hydrangea was introduced into cultivation in 1736. The plant may be grown from seed stratified and sown in the spring. It may also be started by tender cuttings taken in late winter or by means of green cuttings in summer under glass. It should be pruned back rather severely in fall or early spring. Apparently it blooms well in sun or partial shade.

VARIETIES. The following horticultural varieties and forms are listed by Rehder (1940):

Snowhill Hydrangea, *H. arborescens* forma *grandiflora* Rehd., has all sterile flowers forming subglobose heads 4–7 in. across with sepals ovate and acute. The large, white heads are conspicuous. It was introduced into cultivation about 1900 in Ohio.

Oblong-leaf Hydrangea, *H. arborescens* var. *oblonga* Torr. & Gray, has oblong, elliptic to ovate leaves. It was introduced into cultivation before 1800.

Sterile Smooth Hydrangea, *H. arborescens* var. *sterilis* Torr. & Gray, is considered by some authorities as only a form of the Oblong-leaf Hydrangea, with all sterile flowers and sepals oval to rounded and mucronate. It was found in cultivation in Pennyslvania before 1840.

West Virginia Smooth Hydrangea, *H. arborescens* var. *australis* Harbison, has petioles ⅜–2 in., with cordate, coarsely serrate leaves. It was first cultivated in West Virginia about 1888.

Besides the above, a variety known as Deam's Hydrangea, *H. arborescens* var. *deamii* St. John, has the entire lower leaf surface either lightly or densely hairy. It is reported as being found in Indiana, Ohio, and Illinois; south to Missouri, Oklahoma, and Georgia.

REMARKS. The genus name, *Hydrangea*, is from *hydor* ("water") and *aggeion* ("a vessel"), from the cuplike form of the capsular fruit. The species name, *aborescens*, refers to its likeness to a small tree. Vernacular names are Hill-of-Snow, Mountain Hydrangea, and Sevenbark. The name Sevenbark is given because the bark peels off in layers of different colors.

The flowers and fruit are eaten by some wildlife species, including wild turkey and white-tailed deer. The foliage is reported to be poisonous to livestock.

Virginia Sweetspire
Itea virginica L.

FIELD IDENTIFICATION. Slender-branched shrub to 10 ft, sometimes almost reclining.

FLOWERS. April–June in terminal racemes which are narrow, elongate, 2–5 in. long, somewhat drooping, pubescent to glabrous later, individual flowers perfect, fragrant, on puberulent pedicels ¹⁄₁₆–⅛ in.; calyx of 5 pubescent sepals, each sepal about ¹⁄₁₆ in. long, lanceolate to long-acuminate; petals white or pink, linear to lanceolate, erect or spreading, about ⅙ in. long, much longer than the calyx; stamens 5, inserted under the disk edge, about as long as the petals, anthers oblong; pistil puberulent below, style slender, stigma 2-lobed; ovary 2-celled, superior, ovules numerous.

Virginia Sweetspire
Itea virginica L.

FRUIT. Capsule oblong, ¼–⅓ in. long, brown at maturity, pubescent at first but more glabrous later, septicidally 2-grooved and 2-celled, acuminate-pointed by the hornlike remains of the united carpels.

LEAVES. Simple, alternate, deciduous, oblong to elliptic or oval, apex acute or abruptly acuminate, base cuneate, margin fine and sharply apiculate-serrulate; rather thin; upper surface dull green and mostly glabrous, lower surface pubescent to glabrous, red in autumn; blade 1–3 in. long, ½–1⅓ in. wide; petiole pubescent, ⅛–⅖ in.

TWIGS. Slender, smooth, virgate, pubescent at first, glabrous later, green to brown, leaves mostly borne toward the distal end.

BARK. Brown to gray, smooth or broken into small, thin scales on older stems.

RANGE. Usually in sandy, acid soils of swamps, or along streams. East Texas and Louisiana; eastward to Florida, north to New Jersey, and west to Missouri.

REMARKS. The genus name, *Itea*, is the Greek name for willow, and the species name, *virginica*, refers to the state of Virginia. Other names are Virginia-willow, Sweet-spine and Tassel-white. It has ornamental possibilities because of the long tassel-like racemes of white flowers, but since solitary plants are straggling in shape, mass planting brings the best results. The plants may be propagated by spring sowing in greenhouse flats. This is the only species occurring in North America, the other species being natives of Asia.

Cliff Jamesia

Jamesia americana Torr. & Gray

FIELD IDENTIFICATION. Western shrub 1–6½ ft, with erect diffuse branches, the bark exfoliating in thin shreddy flakes.

FLOWERS. May–June, borne in white-hairy cymes ½–1¾ in. high, pedicels strigose-hairy; individual flowers showy, slightly fragrant, ½–⅝ in. across, perfect; calyx turbinate to hemispheric, ⅙–¼ in. long, white-canescent; sepals 5, 1/12–1/10 in. long, triangular to lanceolate, acute; petals 5, ¼–½ in. long, convolute, narrowly oblong to obovate, pubescent within, white; stamens 10, filaments broad and flat, some shorter; styles 3–5, elongate after anthesis, becoming twice the length of the calyx, stigma terminal; ovary conical, imperfectly 3–5-celled, ovules numerous.

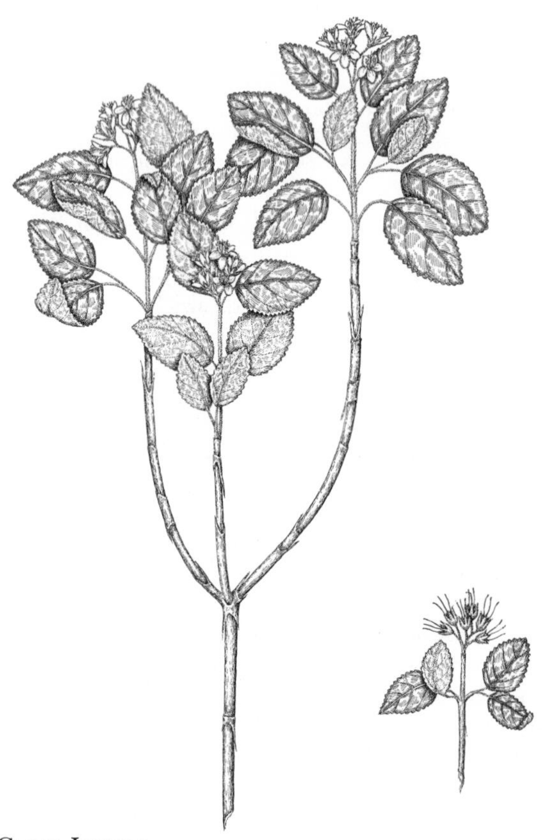

Cliff Jamesia
Jamesia americana Torr. & Gray

FRUIT. Capsule about ⅙ in. long, half-inferior, conic beaked by the persistent styles; 1-celled, but partially 3–5-valved; seeds numerous, striate-reticulate, small endosperm abundant.

LEAVES. Simple, opposite, deciduous, thin, ⅗–2½ in. long, oval to ovate, prominently crenate-mucronulate toothed, apex acute to obtuse; upper surface with impressed veins, sparsely hairy to glabrate and dark green lower surface densely whitened-tomentulose; petiole 1/12–⅗ in., tomentose.

TWIGS. When young short-villose, pith solid, outer scales 1–2, densely pubescent; bark gray to reddish brown, conspicuously shreddy.

RANGE. Well-drained, dry soil in sunny canyons, or along streams, at altitudes of 7,000–10,000 ft. In New Mexico in the Jemez, Sangre de Cristo, Manzano Sandia, Magdalena, Mogollon, and Sacramento mountains; west to California and north through Colorado and Utah to Wyoming.

VARIETIES. Two varieties have been segregated, the California Cliff Jamesia, *J. americana* var. *californica* (Small) Jepson, has a dwarf habit and rose-colored flowers.

Rose Cliff Jamesia, *J. americana* var. *rosea* Purpus, has pink flowers and has been cultivated since 1905.

REMARKS. The genus name, *Jamesia,* honors Edwin James (1797–1861), botanist for Long's expedition to the Rocky Mountains, and the species name, *americana,* refers to North America, the home of the only species. It has some value as an ornamental, is attractive but not showy, and has been cultivated since 1862. The leaves turn orange or scarlet in autumn. It can be used for borders of shrubbery or on rocky slopes. It is generally propagated from seed or ripened wood cuttings.

Thyme-leaf Mockorange

Philadelphus serpyllifolius Gray

FIELD IDENTIFICATION. Small, rigid, straggly Western shrub 1–5 ft. The branches rather crowded and arching.

FLOWERS. May–July, fragrant, rather numerous, usually solitary (rarely 2–3), and nearly sessile on short spurlike leafy branches, about ½ in. across, regular, bisexual; petals 4, white, ovate to rounded, convolute in the bud, about ⅖ in. long; stamens numerous; ovary inferior, 4-celled, styles 4, usually distinct above, much shorter than the stamens; receptacle and calyx silky-hispid, sepals 4–5, ovate-oblong, persistent.

FRUIT. Capsule small, globular, seeds numerous.

LEAVES. Simple, opposite, deciduous, blade narrowly ovate to elliptic or elliptic-lanceolate, length ¼–¾ in.,

Thyme-leaf Mockorange
Philadelphus serpyllifolius Gray

width 3/16–3/8 in., margin entire, apex obtuse or acute, base narrowed, green and pubescent or puberulent above, lower surface white appressed-hairy, 3-nerved from the base; leaf almost sessile, or with a strigose-hairy petiole 1/25–1/8 in.

TWIGS. Slender, rigid, brittle, finely grooved, when young puberulent and gray to brown, when older gray and glabrous.

RANGE. Edges of rocky ravines and limestone ledges. In Texas from the Sabinal River westward through the mountains of the Trans-Pecos region to New Mexico, Arizona, and California; northward to Colorado and Nevada.

RELATED SPECIES. Thyme-leaf Mockorange is closely related to the Small-leaved Mockorange, *P. microphyllus* Gray, but has smaller leaves and flowers. Perhaps it could be better segregated as a variety of the latter because intermediate forms are found. Small-leaved Mockorange has been assigned a number of varieties also (see description for data). Synonyms of *P. serpyllifolius* are *P. pumilus* Rydb. and *P. stramineus* Rydb.

REMARKS. The genus name, *Philadelphus*, is from Ptolemy Philadelphus, an ancient Egyptian king (283–247 B.C.). The species name, *serpyllifolius*, is for the thymelike leaves. The plant is also known under the names of Wild Syringa and Mirto. It was first cultivated in 1917. It may be grown from seeds, layers, suckers, and cuttings of mature wood, or from cuttings of softwood in summer in frames.

Mearns's Mockorange

Philadelphus mearnsii Evans

FIELD IDENTIFICATION. Low shrub, seldom over 3 ft, with more or less spinescent or rigid branches.

FLOWERS. Mostly solitary and terminal on short leafy branches; hypanthium gray-hairy, 1/25–1/10 in. long (about 1/8 in. long in fruit); sepals about 1/8 in. long, apex acuminate, upper surface strigose, lower surface woolly; petals white or yellowish, oblong-lanceolate to ovate-lanceolate, apex acute or sharply 2-toothed; stamens 16–20, rarely to 24, filaments short and free; styles less than 1/25 in., stigmas 1/25–1/12 in., almost completely united.

FRUIT. Capsule short-ovoid, about 1/8 in. high, flat-

Mearns's Mockorange
Philadelphus mearnsii Evans

SMALL-LEAVED MOCKORANGE
Philadelphus microphyllus Gray

tened, dark brown, glabrous, dehiscent into 4 spreading valves.

LEAVES. Elliptic to lanceolate, blades ¼–½ in. long, margin entire, apex acute, base acute, surfaces coarsely gray-hairy, thickened; petioles 1/25–⅛ in., densely coarse-hairy.

TWIGS. Slender, young ones brown to light gray and hairy; older ones darker gray, more glabrous and flaky.

RANGE. Mearns's Mockorange is found in New Mexico in Eddy County in the Guadalupe Mountains. Also in Culberson County, Texas, in the Guadalupe Mountains above McKittrick Canyon.

VARIETIES. A variety has been named *P. mearnsii* var. *bifidus* C. L. Hitchcock. The leaves are slender and longer than in the species; hairs of upper leaf surface slender, hardly at all appressed; petioles 1/12–⅛ in. long; petals about ⅖–½ in. long, the apices distinctly sharply 2-toothed; stamens about 24, filaments free. Known from the Sierra Madre at Monterrey, Nuevo León, Mexico.

REMARKS. The genus name, *Philadelphus,* is said to be named for Ptolemy Philadelphus, king of Egypt (283–247 BC.). The species name, *Mearnsii,* is in honor of E. A. Mearns, who in 1888 collected many plants in Yavapai and Coconino counties in Arizona. Also two years later, he was a naturalist of the second United States–Mexico boundary survey.

SMALL-LEAVED MOCKORANGE
Philadelphus microphyllus Gray

FIELD IDENTIFICATION. Western shrub to 4 ft, with delicate, leafy, ascending, or sprawling stems.

FLOWERS. June–August, very fragrant, usually solitary (rarely 2–3-flowered); pedicels 1/25–⅕ in., strigose-hairy; corolla ¾–1 in. across, perfect; petals 4, white, oval to oblong or obovate, apex obtuse to retuse, about ⅗ in. long; stamens numerous; styles united almost to the apex, stigmas distinct, ovary two thirds inferior, 4-celled; calyx sparsely pubescent to glabrous, lobes 4, 1/12–⅛ in. long, elliptic, short-acuminate, pubescent within on the margin.

FRUIT. Capsule globular, more or less leathery, ¼–⅜ in. long, ⅛–¼ in. wide, seeds numerous.

LEAVES. Simple, opposite, deciduous, blade varying from elliptic to ovate or oblong to lanceolate, apex and base broadly cuneate to acute, length ½–1¼ in., margin entire, upper surface green, glabrous and lustrous, lower surface paler and glaucescent and strigose-hairy; petiole very short, 1/25–⅛ in., or leaf subsessile.

TWIGS. Slender, rigid, pubescent and brownish at first, later glabrous, lustrous and tan to reddish brown or gray, bark exfoliating the second year.

RANGE. Well-drained rocky sites in the sun, at altitudes of 4,000–8,000 ft. New Mexico and Arizona; west to California and north to Colorado and Utah.

VARIETIES. A number of varieties (or subspecies) have been assigned the Small-leaved Mockorange by some botanists. The two most distinct are given as follows (Kearney and Peebles, 1951, pp. 366–367):

Leaves commonly more than ⅗ in. long:

Upper leaf surface strigose-pubescent; hairs of the calyx mostly appressed, not matted............ ......*P. microphyllus* subsp. *argenteus* (Rydb.) C. L. Hitchcock (Silver-leaf Mockorange).

Upper leaf surface hirsute, the hairs erect or nearly so; hairs of the calyx either matted and of two kinds, or not matted and all alike............ .....*P. microphyllus* subsp. *agyrocalyx* (Wooton) C. L. Hitchcock (Silver-calyx Mockorange).

REMARKS. The genus name, *Philadelphus,* is for Ptolemy Philadelphus, an ancient Egyptian king (283–247 B.C.).The species name, *microphyllus,* refers to the small leaves. The plant was first introduced into cultivation in 1883. It is known to be browsed by the Mexican bighorn sheep.

Hoary Mockorange

Philadelphus pubescens Lois.

FIELD IDENTIFICATION. Vigorous shrub 3–15 ft, with close gray bark.

FLOWERS. June–July, in conspicuous leafy racemes, of 5–9 flowers, 1–1½ in. across, perfect, regular, epigynous; petals 4, white, obovate to oval; stamens numerous, filaments elongate; ovary inferior, 4-celled, ovules numerous; styles elongate, united about two thirds of their length, stigma linear; sepals valvate, persistent, spreading, about ⅕ in. long, triangular-ovate, acuminate, pubescent.

FRUIT. Capsule obovoid, indurate, loculicidal, dehiscent into 3–4 valves, seeds numerous with membranous coats.

LEAVES. Simple, opposite, ovate to elliptic, margin entire or sharply and remotely denticulate, apex acuminate, base rounded to cuneate, sometimes subcordate on young shoots, length 2–3¼ in., upper surface dark green and glabrous, lower surface grayish pubescent; petiole ¼–½ in. long, pubescent.

TWIGS. Young ones green to yellowish or reddish brown, angled, later gray and glabrous, cracking but not shreddy.

RANGE. Well-drained limestone, or sandstone bluffs, often on riverbanks. Arkansas, east to Alabama and north to Illinois.

Hairy Mockorange
Philadelphus hirsutus Nutt.

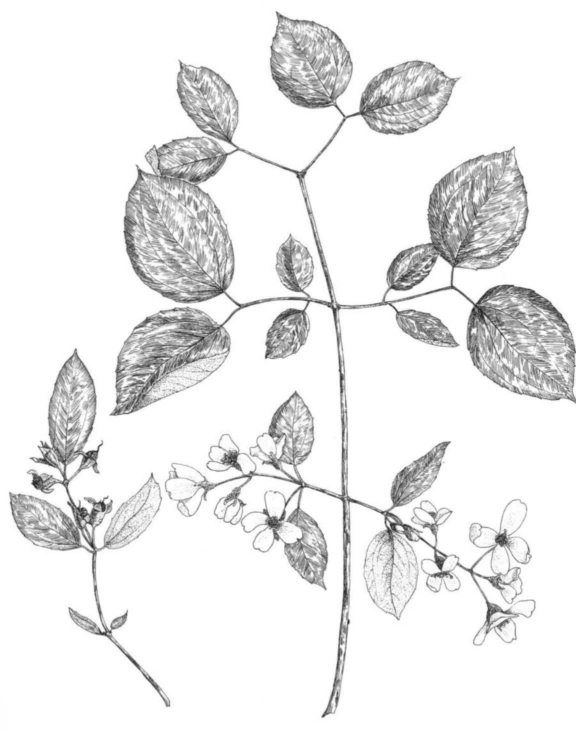

Hoary Mockorange
Philadelphus pubescens Lois.

VARIETY. A variety which has pedicels, hypanthium, and outer side of sepals glabrous has been named *P. pubescens* var. *intectus* (Beadle) Moore.

REMARKS. The genus name, *Philadelphus*, is in honor of Ptolemy Philadelphus (283–247 B.C.), king of Egypt. The species name, *pubescens*, is for the short hairs on the foliage. Introduced into cultivation in 1800.

Hairy Mockorange

Philadelphus hirsutus Nutt.

FIELD IDENTIFICATION. Straggly shrub 3–6 ft, with shreddy brown bark.

FLOWERS. May–June, axillary, solitary or in twos or threes, flowering branches often with only 1 pair of leaves; pedicels pubescent, the lower ⅕–⅘ in. long;

flowers regular, perfect, epigynous, about ¾ in. across; petals large, white, obovate to oval; stamens 15–20 or more, filaments elongate; ovary inferior, 4-celled, ovules numerous, styles glabrous and elongate; sepals valvate, persistent, divisions divergent and triangular-ovate.

FRUIT. Capsule obovoid, ¼–⅜ in. long, hairy when young, dehiscent into 3–4 valves and then glabrate, stigmas united until dehiscent, seeds numerous, seed coat prolonged terminally.

LEAVES. Simple, opposite, ovate-lanceolate to oblong or broadly ovate, margin sharply serrate, apex acuminate, base rounded to cuneate, 1–3¼ in. long, upper surface puberulent or eventually glabrescent, lower surface densely gray-puberulent, strongly 3-veined; petioles ¼–½ in., puberulent.

TWIGS. Younger ones brown and appressed-hairy, later gray, glabrous and exfoliating.

RANGE. In moist soil in sun or shade, on riverbanks or rocky slopes. Arkansas; east to Georgia and north to North Carolina.

REMARKS. The genus name, *Philadelphus*, is in honor of Ptolemy Philadelphus (283–247 B.C.), king of Egypt. The species name, *hirsutus*, refers to the hairs of the foliage. The plant was introduced into cultivation in 1820. It is not as ornamental as some of the other native species.

Pasture Gooseberry

Ribes cynosbati L.

FIELD IDENTIFICATION. Low straggly shrub with rigid, spreading or trailing branches. Generally armed with simple, reddish or black, slender epidermal spines usually subtending the leaves.

FLOWERS. April–June, racemes loose, 1–4-flowered, peduncles and pedicels slender; peduncles ⅕–1⅕ in. long to the first bractlets which are sheathlike, glandular-ciliate, and about 1/12 in. long; pedicels ⅕–½ in., glandular-ciliate or glabrous; calyx green, broadly campanulate, tube adherent to the ovary, 5-lobed; lobe much shorter than the tube, oblong to elliptic, obtuse, glabrous; petals 5, inserted in the calyx-throat, small, 1/25–1/12 in. long, reniform, truncate and cuneate; stamens 5, not exserted, alternate with petals, subulate, as long as the anthers or one half as long; style undivided, ovary bristly, 1-locular, with 2 parietal placentae.

FRUIT. Maturing July–September, reddish purple, globose, ⅓–½ in. in diameter, armed with long stiff prickles, or rarely smooth, crowned by the shriveled calyx remains, many-seeded.

LEAVES. Pleated in the bud, simple, alternate or clustered, deciduous, orbicular to ovate, base truncate to subcordate or rounded, margin deeply 3–5-lobed with the lobes crenate-dentate or incised, terminal lobe mostly acute and laterals rounded, surfaces soft-pubescent at first, more glabrate later, 1–2 in. long or wide; petioles slender, ½–1 in., pubescent, less so with age.

TWIGS. Slender, at first tan to brown and pubescent, later gray to almost black and glabrous; nodal spines slender, solitary or 2–3 together, straight, to ¾ in. long, young twigs often with numerous, slender, reddish or black spines between the nodes.

RANGE. Usually in open rocky woods. Arkansas; east to Alabama and North Carolina, north to New Brunswick and Quebec, and west to Missouri.

PROPAGATION. The fruit is generally stripped from the branches when ripe to prevent consumption by birds. The seed is extracted by macerating in water and floating off the residue, and also by fanning when dry. The cleaned seed averages about 200,000 seeds per lb, with a purity of 99 per cent and a soundness of 86 per cent. The dormancy of the embroyo may be broken by stratification in sand for 90 days or more at 41° F. Stratified seed is planted in spring, or dried seed may be stored in sealed containers for 2–3 years at low temperatures. About 40 viable seeds may be sown per linear ft of nursery row and covered to a

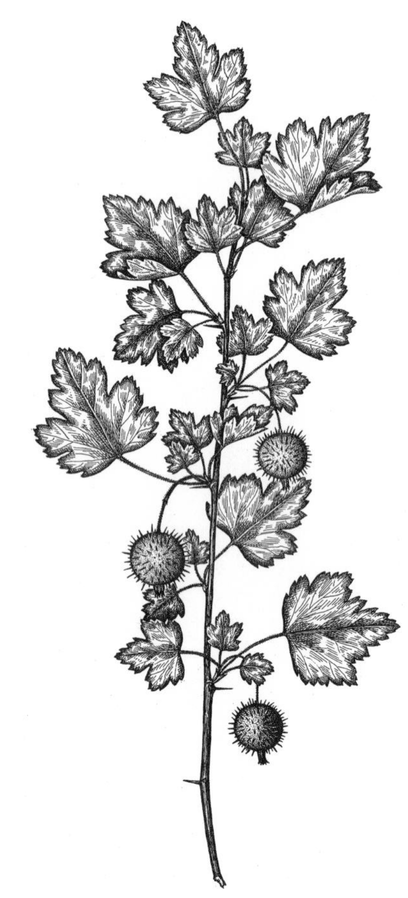

Pasture Gooseberry
Ribes cynosbati L.

ORANGE GOOSEBERRY
Ribes pinetorum Greene

depth of ⅛–¼ in. Propagation is also practiced with hardwood cuttings taken in the autumn.

VARIETIES, FORMS, AND HYBRIDS. Smooth Pasture Gooseberry, *R. cynosbati* var. *glabratum* Fern., has leaves glabrous or slightly hairy on the veins beneath. It occurs from Virginia to North Carolina, north to Michigan and Ontario.

Cruel Pasture Gooseberry, *R. cynosbati* var. *atrox* Fern., has the fruiting stems and young cones densely set with dark, stiff bristles.

Spineless Pasture Gooseberry, *R. cynosbati* forma *inerme* Rehd., is a form with no spines.

R. cynosbati is known to hybridize with *R. grossularia* L. to produce *R.* × *utile* Jancz. The leaves are similar to those of *R. grossularia,* with the flowers pale and slightly pubescent, style as long as the stamens, fruit red and sometimes with few spines.

REMARKS. The genus name, *Ribes,* is from the old Danish word *ribs,* the word for "red currant." The species name, *cynosbati,* translated means "dog-berry." Vernacular names are Dog-bramble and Prickly Gooseberry. Pasture Gooseberry has been in cultivation since 1759. It is of some use for wildlife food and cover, being eaten by a number of birds and by red squirrel and cottontail. It was sometimes used by the early pioneers for making jelly. Since it is known to be an alternate host of the white pine blister rust, its cultivation is to be discouraged.

ORANGE GOOSEBERRY

Ribes pinetorum Greene

FIELD IDENTIFICATION. Small or large, glabrous, spiny shrub to 6 ft.

FLOWERS. April–September, on short lateral branchlets, bracts very small; peduncles puberulent, shorter than the petioles; perfect, regular, orange-red, solitary or a few; petals 5, shorter than the sepals; sepals 5, linear-spatulate, hairy, twice as long as the receptacle; stamens 5; ovary inferior, 1-celled, many-ovuled; styles 2, glabrous, or more or less crenate.

FRUIT. Berry borne singly, reddish or purplish, globose to ovoid, ⅜–⅝ in. long, densely set with sharp, stout, yellow prickles, crowned by the persistent calyx-lobes, juicy; seeds albuminous, with a minute embryo.

LEAVES. Simple, alternate, but appearing clustered at the nodes, thin; blade rotund, cordate, margin usually cleft into 5 lobes which are irregularly dentate, upper surface dull green and glabrous, palmately nerved, lower surface sparingly pubescent, ¾–1½ in. long or wide; petioles as long as the blades or shorter, pubescent.

TWIGS. Stems spreading or reclining, young ones puberulent, green to straw color, older ones reddish brown to gray; spines 1–3 at the nodes, about ⅜ in. long or less; stem often with short spurlike lateral branchlets.

RANGE. Well-drained soil in coniferous forests at altitudes of 7,000–10,000 ft. New Mexico and Arizona.

REMARKS. The genus name, *Ribes,* is from an old Danish word *ribs,* for the "red currant." The species name, *pinetorum,* refers to its growth in pine forests. The plant has been cultivated since 1898. The orange-colored flowers make it very ornamental. The berry has a pleasant taste, but the spines are hard to remove.

GOOSEBERRY CURRANT

Ribes montigenum McClelland

FIELD IDENTIFICATION. Western shrub, low, straggly, the flexuous branches erect, ascending or procumbent, attaining a height of 1–2½ ft. The whole plant more or less densely pubescent or glandular, also with 1–3 spines at the nodes, bristly or naked internodes.

GOOSEBERRY CURRANT
Ribes montigenum McClelland

FLOWERS. Borne June–August, green to reddish purple, racemes short and 3–8-flowered; pedicels 1/12–2/5 in., sometimes glandular, bracts small; calyx-tube 1/25–1/12 in. long, saucer-shaped, glandular-hairy, about ¼ in. broad; calyx-lobes 5, reddish or purplish, 1/12–⅙ in. long, rather broad, veiny; petals 5, reddish, one half to one third as long as the calyx-lobes and alternate with them; stamens 5, filaments short; ovary inferior, 1-celled, with 2 parietal placentae; styles 2, stigmas terminal.

FRUIT. Berry globose, ¼–⅖ in. in diameter, red, glandular-bristly, withered flower remnants at apex, several-seeded.

LEAVES. Simple, alternate, some appearing clustered, deciduous, blades ⅝–1⅓ in. wide or long, suborbicular, margin usually 5-lobed or cleft, the divisions also incised or toothed, base cordate, surfaces varying from densely glandular-pubescent to almost glabrous; petioles glandular-hairy, usually shorter than the blades.

STEMS. Young ones tan to brown or straw-colored, older ones brown to gray, often with 1–3 stout sharp spines at the nodes (sometimes clustered), and with bristly or almost naked internodes.

RANGE. Exposed slopes and mountain summits at altitudes of 7,500–12,500 ft. In New Mexico; west through Arizona to California and north to Montan and British Columbia.

REMARKS. The genus name, *Ribes*, is an ancient Dan ish name, and the species name, *montigenum*, refe to the mountainous habitat of this species. It is als known under the vernacular names of Alpine Prickl Currant and Mountain Gooseberry. It was first cult vated in 1905. The berry is edible, and the plant i somewhat palatable to livestock.

ROTHROCK CURRANT
Ribes wolfii Rothr.

FIELD IDENTIFICATION. Upright shrub 1½–10 ft tall.

FLOWERS. Racemes 4–8-flowered, or more rarely a many as 12-flowered, dense, 1–1½ in. long; pedicel and bracts glandular-pubescent, pedicels ⅕–⅓ in. long about as long as the bracts; sepals 5, spreading, ovate oblong, greenish white or reddish, about ⅙ in. long three to four times the length of the puberulous re ceptacle; petals 5, white or pinkish-tinged, shorter tha

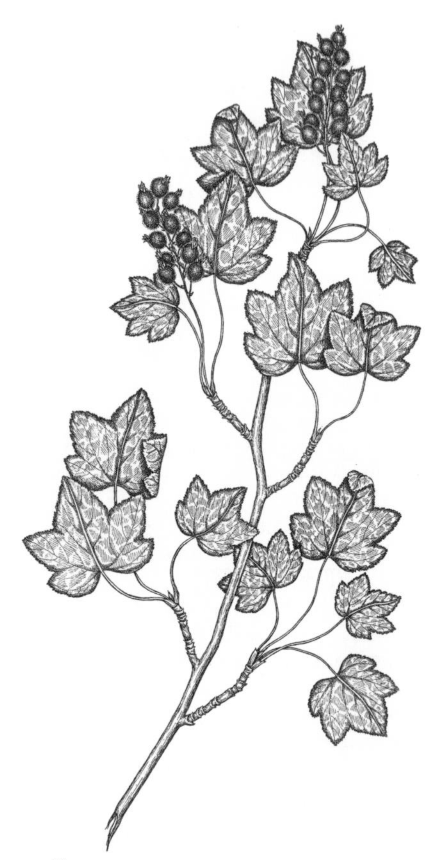

ROTHROCK CURRANT
Ribes wolfii Rothr.

Sticky Currant
Ribes viscosissimum Pursh

the sepals, inserted on the calyx-throat; stamens 5, half as long as the sepals; ovary inferior, glandular, bristly, 1-celled with 2 parietal placentae.

FRUIT. Maturing August, berry black, with or without a bloom, glandular-hairy, smoother with age, globose, ⅜–1½ in. in diameter, crowned by the calyx remains, juicy, several-seeded.

LEAVES. Simple, alternate, or in fascicles on short lateral spurlike branchlets, blade suborbicular, base cordate, margin 3–5-lobed with the lobes obtuse to acute and serrate-dentate, 1½–3½ in. long or wide, thickish, palmately veined, upper surface green and glabrous, lower surface paler green and glandular-pubescent to glabrous; petioles slender, ½–1½ in., glandular-pubescent or glabrate with age.

TWIGS. Young ones puberulous and sometimes also glandular, older ones reddish brown to tan or gray and glabrous.

RANGE. In damp woods in the aspen and conifer belt at altitudes of 6,000–12,000 ft. Trans-Pecos Texas, New Mexico, and Arizona; north through Colorado and Utah to Wyoming.

REMARKS. The genus name, *Ribes*, is from an old Danish word *ribs*, for the "red currant." The plant is of little value as an ornamental. The fruit is edible but not tasty. In Utah it is considered to be a fairly good cattle browse.

Sticky Currant

Ribes viscosissimum Pursh

FIELD IDENTIFICATION. Small unarmed shrub 1–4 ft, with a root system as much as 4 ft deep. The young branches glandular-pubescent.

FLOWERS. April–July, racemes spreading or ascending, 3–12-flowered, pedicels not over ¼–½ in. long, glandular-hairy; bracts spatulate or oblanceolate, often toothed at apex, glandular-hairy, flowers ½–¾ in. long from the base of the hypanthium, which is cylindric-campanulate, less than twice as long as wide; calyx-tube adnate to the ovary, ⅙–¼ in. long, ⅒–⅐ in. wide, cylindrical to almost campanulate, greenish to pink; calyx-lobes 5, spreading or reflexed, nearly or quite as long as the tube (rarely not much over one half as long), greenish white to pink; petals 5, erect, white, half as long as the calyx-lobes and alternate to them; ovary glandular-hairy; inferior, 1-celled with 2 parietal placentae; styles 2, united, stigma terminal; stamens 5, inserted on the corolla throat.

FRUIT. About ⅜–½ in. in diameter, black, glandular-pubescent, almost dry; crowned by the persistent calyx-lobes; seeds numerous.

LEAVES. Somewhat fragrant, alternate or clustered, ¾–3 in. wide, blade reniform-orbicular, base cordate, margin shallowly or deeply 3–5-lobed, with crenate-dentate margins and rounded apices, both surfaces glandular-pubescent, somewhat paler beneath; palmately 3–5-veined from the base; petioles shorter than the blade, dilated at base, glandular-hairy.

RANGE. On shaded mountainsides at altitudes of 6,000–6,700 ft. In west Texas and New Mexico; west through Arizona to California and north to Montana and British Columbia.

VARIETY. Hall's Sticky Currant, *R. viscosissimum* var. *hallii* Jancz., has a glabrous ovary and purple calyx.

REMARKS. The genus name, *Ribes*, is the ancient Danish name. The species name, *viscosissimum*, refers to the sticky viscid hairs. The fruit is known to be eaten by ruffed grouse and Richardson's grouse, and the foliage is browsed by cattle.

Mescalero Currant

Ribes mescalerium Cov.

FIELD IDENTIFICATION. An unarmed shrub to 6 ft, the erect or ascending branchlets glandular-pubescent.

FLOWERS. On short axillary branches in several-flow-

Mescalero Currant
Ribes mescalerium Cov.

ered racemes; pedicels jointed, each subtended by a bract and 2 bractlets; hypanthium well developed, nearly twice as long as broad; sepals 5, greenish white, much shorter than the hypanthium, calyx-tube adnate to the globose ovary; petals 5, erect, smaller than the sepals and alternate to them; stamens 5, anthers with a conspicuous cup-shaped apical gland; ovary greenish white, inferior, 1-celled, with 2 parietal placentae; styles 2 and united, stigma terminal.

FRUIT. Breaking from the pedicel, to 1/2 in. in diameter, black, globose, glandular-pubescent, calyx remnants persistent.

LEAVES. Alternate or clustered, deciduous, suborbicular, palmately 3–5-lobed, the lobes crenate-dentate on the margin, base cuneate to cordate, palmately veined, surfaces glandular-hairy.

RANGE. In dry, well-drained soil of yellow pine forests. In the mountains of western Texas and in the White and Sacramento mountains of New Mexico.

REMARKS. The genus name, *Ribes*, is the ancient Danish name. The species name, *mescalerium*, refers to its use by the Mescalero Apache Indians, who also made a beverage from the Mescal plant *(Dasylirion)*.

American Black Currant

Ribes americanum Mill.

FIELD IDENTIFICATION. Erect or spreading shrub attaining a height of 5 ft.

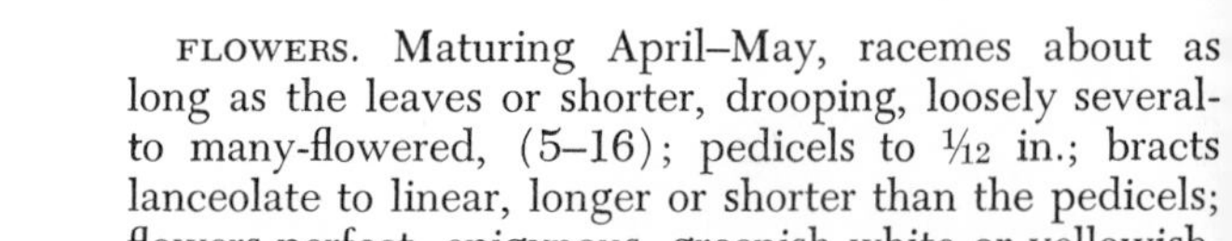

FLOWERS. Maturing April–May, racemes about as long as the leaves or shorter, drooping, loosely several- to many-flowered, (5–16); pedicels to 1/12 in.; bracts lanceolate to linear, longer or shorter than the pedicels; flowers perfect, epigynous, greenish white or yellowish, 1/3–1/2 in. long; calyx tubular, sepals 5, oblong-obovate, 1/6–1/5 in. long; petals 5, smaller than the sepals, oblong, very blunt, erect or nearly so, 1/12–1/8 in. long; stamens 5, included, opposite the sepals; ovary wholly inferior, glabrous, 1-celled with 2 parietal placentae, styles erect and bifid.

FRUIT. Maturing June–September, globose-ovoid, 1/4–1/2 in. in diameter, black, glabrous, smooth; seeds numerous, horizontal, obscurely angled; outer coat gelatinous, inner coat crustaceous; embryo small, terete, in fleshy endosperm.

LEAVES. Alternate or fascicled, blades 1–3 in. long or wide, outline suborbicular; lobes 3–5, sharply acute to obtuse, serrations single or double; leaf base broadly truncate to shallowly cordate, upper surface usually glabrous at maturity, lower surface somewhat pubescent and dotted with minute orange glands; petioles slender, shorter than the blade or longer, pubescent to glabrous.

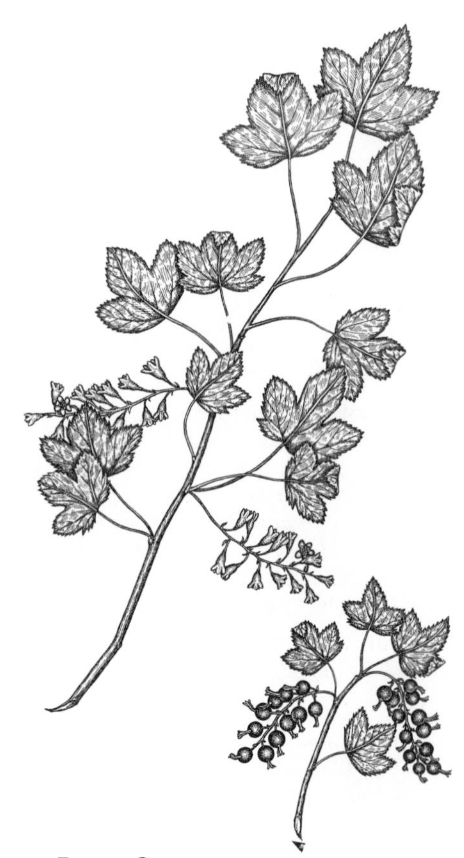

American Black Currant
Ribes americanum Mill.

STEMS. Upright to spreading, unarmed, slender, pubescent at first, glabrous later, occasionally dotted with orange glands.

RANGE. In moist woods at altitudes of 3,000–5,500 ft. Northern New Mexico and Colorado; west to California and north to Manitoba, Nova Scotia, New Brunswick, and Quebec; also eastward to Kentucky.

PROPAGATION. The mature fruit is picked by hand as soon as possible to prevent consumption by birds. It is spread out to dry to prevent heating. The seed is removed by macerating in water and the pulp floated off. The vitality of the seed declines considerably after a year or two of storage. Cleaned seed averages about 300,000 seeds per lb, with a purity of 97 per cent. Dormancy may be partially overcome by stratification in sand or peat for 60 days at 68–86° F. plus 90–120 days at 41°. About 40 viable seeds are sown per linear ft of nursery row and covered about ⅛ in. with soil. Propagation is also done with hardwood cuttings taken in autumn or by mound layers. The plant is considered to be a host to the white pine blister rust.

VARIETY. A variety has been described as the Indiana Black Currant, *R. americanum* var. *meschora* Nieuw., with flowers 2 weeks later than the species on longer racemes; sepals very narrow and acute; leaves larger, more deeply lobed and serrate; plant more slender and with longer straggly branches; twigs grayish white; reported from Mineral Springs, Porter County, Indiana.

REMARKS. The genus name, *Ribes*, is from an old Danish name, and the species name, *americanum*, refers to its native home of America. It is also known under the vernacular names of Wild Black Currant and Quinsyberry. The plant was first cultivated in 1727. The racemes of greenish white flowers and yellow-to-red leaves in the fall are attractive. The fruit is eagerly eaten by a number of species of birds, including the ringnecked pheasant.

Wax Currant

Ribes cereum Dougl.

FIELD IDENTIFICATION. Western, upright, much-branched shrub attaining a height of 2–4 ft, the young twigs unarmed and glandular-pubescent.

FLOWERS. May–June, regular and perfect, mostly on short axillary branches, small, inconspicuous, solitary, or in few-flowered drooping racemes; bracts cuneate-obovate, apex truncate to rounded and toothed, glandular-pubescent, longer than the very short pedicels; receptacle cylindric, ¼–⅓ in. long; sepals 5, ovate, erect, about 1/12 in. long; petals 5, minute, orbicular, erect, alternate with the calyx-lobes; stamens 5, anthers with a cup-shaped apical gland; ovary inferior, globose, glabrous or glandular, 1-celled, many-ovuled; styles 2, united, usually hairy above, stigma terminal.

FRUIT. Ripening July–August, berry bright red to

Wax Currant
Ribes cereum Dougl.

yellowish red, ¼–⅓ in. across, glandular or smooth, crowned by the withered remnants of the flower; parietal placentae 2, seeds numerous, albuminous, embryo minute.

LEAVES. Simple, deciduous, fragrant, alternate, but so crowded as to appear fascicled, reniform to orbicular, base broad-cuneate to subcordate, margin palmately 3–5-lobed with obtuse crenulate lobes, length or width of blades ⅜–1½ in.; upper surface glabrous or nearly so, lower surface light grayish green and somewhat glandular-pubescent; palmately veined; some of the petioles almost as long as the blade.

TWIGS. Tan to reddish brown and glandular-pubescent at first, eventually glabrous and light to dark gray.

RANGE. In dry sunny sites at altitudes of 3,500–8,500 ft, often in pine forest. Oklahoma and New Mexico; west to California and north to Montana and British Columbia.

VARIETY. Pale-leaf Wax Currant, *R. cereum* var. *farinosum* Jancz., is a variety with branches violet-colored and whitish puberulent with pink flowers.

REMARKS. The genus name, *Ribes*, is from the old Danish name, *ribs*, for the "red currant," and the species name, *cereum*, is for the waxy fruit. It is sometimes grown for ornament and is propagated by seeds, mound layers, and cuttings. The fruit is valuable for wildlife, being eaten by many species of song and game birds, especially grouse. The foliage has some limited value as browse for mule deer and livestock.

Georgia Gooseberry

Ribes curvatum Small

FIELD IDENTIFICATION. Diffusely branched, spiny shrub to 3 ft. Branches slender, graceful, and glabrate.

FLOWERS. Borne in spring, 1–5-flowered (usually one), on peduncles ⅓–½ in. long, jointed near the base; bractlets 2–3-lobed, ciliate; hypanthium glandular-pubescent, receptacle broad-campanulate; sepals whitish, ¼–⅖ in. long (longer than the free hypanthium), linear to linear-spatulate, reflexed and revolute, obtuse; petals 5, white, inserted in the calyx-throat, 1/25–1/12 in. long, obtuse with 2 lateral teeth near the apex, 1–2-nerved; stamens 5, as long as the sepals, alternate with the sepals, conspicuous, erect, about ⅛ in. long, filaments hairy; style 2-lobed, ovary 1-celled and inferior, ovules numerous.

Georgia Gooseberry
Ribes curvatum Small

FRUIT. In July, berry globose, ¼–⅓ in. in diameter, crowned by the persistent styles, greenish, glabrous, smooth, pulpy, acid; seeds somewhat angled, endosperm fleshy.

LEAVES. Solitary or a few together, suborbicular, palmately veined, ⅜–1 in. long, 3-lobed, the lobes obtusely toothed, terminal lobe often mucronate, base subcordate or broadly cuneate, somewhat pubescent; petioles slender, as long as the blade or shorter, usually somewhat hairy.

TWIGS. Slender, delicate, recurved or drooping; bearing slender, red, sharp spines ⅙–¼ in.; bark purplish to reddish brown, loose and exfoliating.

RANGE. In dry rocky soil in the sun, Georgia west to Louisiana and Texas. Rare in the latter two states.

REMARKS. The genus name, *Ribes*, is from an old Danish name, and the species name, *curvatum*, refers to the drooping twigs. Other names are Southern Currant and Slender Gooseberry. The plant was introduced into cultivation about 1898. The fruit is sometimes used for preserves, and is relished by many birds and small mammals. It is apparently only browsed lightly by livestock. Gooseberries generally serve as an alternate host to the white pine blister rust, *Cronartium ribicola*, and should be eliminated in areas where White Pine grows.

Colorado Currant

Ribes coloradense Coville

FIELD IDENTIFICATION. Small, glabrous or puberulent, unarmed shrub with branches decumbent or prostrate.

FLOWERS. Racemes erect, 4–12-flowered, pedicels ⅙–⅗ in. long, glandular and puberulent; bracts minute, linear to lanceolate, flowers perfect and regular; calyx with the 5 sepals ⅛–⅙ in. long, greenish or purplish, glandular externally, ovate, obtuse, the tube about 1/25 in. long, saucer-shaped; petals 5, shorter than the calyx-lobes, usually broader than long, purplish; stamens 5; ovary glandular-pubescent, 1-celled, with 2 parietal placentae, styles 2 with stigmas terminal.

FRUIT. Berry black, globose, about ⅖ in. in diameter, sparingly glandular, crowned by the flower remnants, breaking from the pedicel.

LEAVES. Simple, alternate, outline reniform, base cordate, margin with 5 ovate-triangular lobes and each lobe crenate to dentate or incised, length or width of blades 1–3½ in., both sides glabrous or sometimes also sparingly pubescent beneath and minutely glandular, palmately veined; petioles rather stout, puberulent, about equal to blade length.

TWIGS. Rather stout, young ones green to light brown or tan, puberulous, glandular; older ones brown to gray and glabrous.

RANGE. Well-drained sunny sites on high mountains in the spruce or subalpine belt at altitudes of 8,000–11,500 ft. In New Mexico, Colorado, and Utah.

REMARKS. The genus name, *Ribes*, is from an old Danish word, *ribs*, for "red currant." The species name, *coloradense*, refers to the state of Colorado where it grows. It was introduced into cultivation in 1905.

White-stem Gooseberry

Ribes inerme Rydb.

FIELD IDENTIFICATION. Shrub 3–15 ft, erect or semi-trailing, simple or branched, spiny or unarmed.

FLOWERS. April–May, borne on peduncles mostly shorter than the petioles; bracts very small, much shorter than the pedicels; peduncles and bracts glabrate or minutely glandular and puberulent; petals one third to one half as long as the sepals; rhombic-cuneate; calyx-tube adnate to the globose ovary and more or less produced above it, usually glabrous (rarely slightly hairy), 1/12–1/6 in. long, bell-shaped; sepals oblong, greenish to purplish, glabrous or slightly pilose, about as long as the tube, reflexed during anthesis; stamens shorter than the sepals or almost equaling them; ovary glabrous, 1-celled, with 2 parietal placentae, styles 2, stigmas terminal.

FRUIT. Berry maturing August–September, crowned with the floral remnants, purplish red, smooth, 1/3–1/4 in. long, edible, of good flavor.

LEAVES. Numerous, deciduous, thin, orbicular to reniform, margin 3–5-lobed, the lobes crenate-dentate and obtusish, base cordate or truncate, blades 3/8–2½ in. long or wide, upper surface glabrous and lustrous, lower surface pubescent and glandular to glabrous later; petioles about as long as the blade.

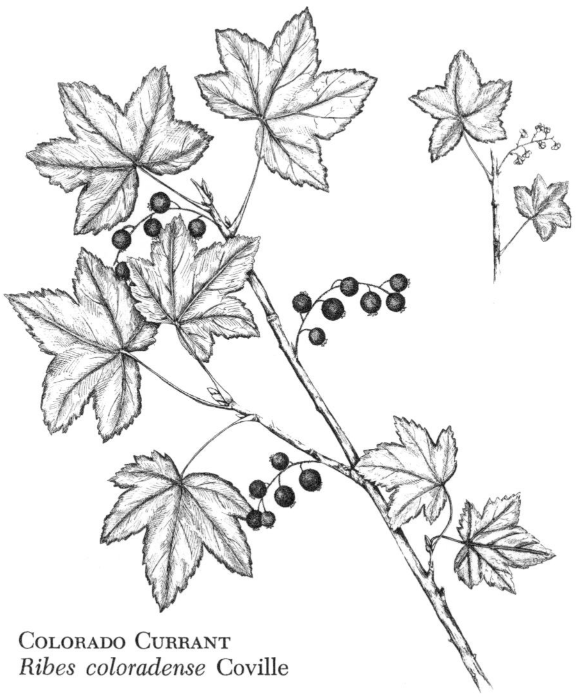

Colorado Currant
Ribes coloradense Coville

White-stem Gooseberry
Ribes inerme Rydb.

TWIGS. Young ones slender, white and glabrous, later dark reddish brown or grayish, splitting into thin flakes, unarmed, or with a few small spines (1–3 nodal), less than 3/8 in. long, rarely with a few internodal bristles.

RANGE. Shady mountain stream banks in the conifer belt, at altitudes of 5,000–11,000 ft. New Mexico; west through Arizona to California, and north to Montana and British Columbia.

VARIETY. Hairy White-stem Gooseberry, *R. inerme* var. *pubescens* Berger, is a variety with pubescent petioles, leaf blades, peduncles, and bracts.

REMARKS. The genus name, *Ribes*, is from an old Danish word, *ribs*, for "red currant." The species name, *inermis*, means "unarmed," but this is not wholly true. The plant has been cultivated since 1899. It is also of fair forage value for sheep browse. The fruit is agreeable to taste and is used for sauces.

Golden Currant

Ribes aureum Pursh

FIELD IDENTIFICATION. An unarmed, deciduous shrub attaining a height of 3–8 ft. The branches are erect or ascending.

FLOWERS. March–June, terminal on short, lateral branches, in 1–15-flowered racemes 1–2½ in. long; pedicels jointed beneath the ovary, about ⅛ in. long; bracts foliaceous, oblong to obovate, much longer than the pedicels; flowers sometimes fragrant, yellow, very showy, perfect, regular; hypanthium slender, smooth, ¼–⅖ in. long, about 1⁄15 in. wide; sepals 5, one fourth to one half as long as the cylindric tube which is ¼–½ in. long, erect or spreading, oval; petals 5, yellow, oblong, margins erose, reduced in size (1⁄12–⅛ in. long), inserted in the calyx-throat; stamens 5, alternate; ovary 1-celled, inferior, glabrous.

FRUIT. Maturing July–September, breaking from the pedicel; berry smooth, glabrous, not spiny, globose, juicy, ¼–⅓ in. across, crowned by the withered, persistent style, varying in color from red to yellow or black to brown; several-seeded.

LEAVES. Alternate or clustered, blade orbicular-reniform to obovate, base cuneate to subcordate, usually deeply 3-lobed, the lobes round and entire, or with few to several coarse or crenate teeth, ¾–2 in. wide, palmately and inconspicuously 3–5-veined, upper surface light green and glossy, lower surface from puberulous when young to almost glabrous later, firm and leathery; petioles ½–1½ in. long, glabrous.

GOLDEN CURRANT
Ribes aureum Pursh

TWIGS. Young ones puberulous, older ones glabrous, bark gray to brown or reddish.

RANGE. In moist canyons or shady ravines, often along streams of the grass, pinyon, and yellow pine belts, at altitudes of 3,500–8,000 ft. In New Mexico and Arizona; west to California and north to Montana, Washington, and adjacent Canada.

PROPAGATION. The general method of planting and seed extraction from the fruit is the same as for that of *R. odoratum.* The cleaned seed averages about 217,000 seeds per lb, with a purity of 97 per cent and a soundness of 98 per cent. Embryo dormancy may be overcome by stratifying in sand for 90 days at 41°F. Propagation is also practiced by hardwood cuttings taken in autumn.

FORMS AND VARIETIES. *R. aureum* forma *chrysococcum* Rydb., has orange-yellow fruit.

R. aureum var. *gracillimum* (Cov. & Britt.) Jepson, has been segregated on the bases of the calyx-tube being two to three times as long as the lobes. However, considerable variation exists between it and the species.

REMARKS. The genus name, *Ribes*, is from an ancient Danish name. The species name, *aureum*, refers to the golden yellow flowers. It is also known under the vernacular name of Buffalo Currant. The plant is handsome and is often grown for ornament, the varied color of the fruit making a striking contrast with the handsome foliage. It is also planted for erosion control and for wildlife food and cover.

MISSOURI GOOSEBERRY

Ribes missouriense Nutt.

FIELD IDENTIFICATION. Small or large shrub to 6 ft. The stems are thornless or thorny.

FLOWERS. April–May, peduncles ⅜–¾ in. long, drooping, long and slender, 2–3 flowers on peduncles much longer than the bracts; petals cuneate-obovate, pale green to white, about ⅛ in. long, inserted in the calyx-throat; stamens 5, alternate with the petals, ⅖–½ in. long; style erect, ⅖–½ in. long, bifid; ovary inferior, 1-locular with 2 parietal placentae and 2 styles, ovules numerous; sepals 5, each oblong-linear, two to three times as long as the cylindric pubescent receptacle.

FRUIT. Maturing June–September, berry smooth, blackish purple, subglobose, ⅓–⅗ in. across, subacid, crowned with the calyx remains, many-seeded.

LEAVES. Simple, alternate, outline rounded, palmately cut into 3–5 lobes, which are in turn coarsely toothed or obtusely lobed, base broad-cuneate to subcordate, palmately veined, blades ¾–2½ in. long or wide, upper surface green and glabrous, lower surface soft pubescent or glabrous later; petioles ½–1½ in.

MISSOURI GOOSEBERRY
Ribes missouriense Nutt.

TWIGS. Slender, gray to whitish, older ones darker gray to almost black, lateral ones short and spurlike; spines solitary or 2–3 together at the nodes, sometimes ¼–¾ in.

RANGE. Well-drained moist soil in the shade or sun. Arkansas, Tennessee, and Missouri; north and east to Connecticut and west to Minnesota and South Dakota.

VARIETY. Arkansas Gooseberry, *R. missouriense* var. *ozarkanum* Fassett, is a smooth-leaved variety occurring with the typical form in Arkansas and central Missouri.

REMARKS. The genus name, *Ribes*, is from an old Danish word, *ribs*, for "red currant." The species name, *missouriense*, is for the state of Missouri. Also known under the vernacular names of Slender Gooseberry and Illinois Gooseberry. It has been cultivated since 1907 and is valuable for wildlife food and cover.

CLOVE CURRANT
Ribes odoratum Wendl.

FIELD IDENTIFICATION. Erect spineless shrub attaining a height of 6 ft. The young branches pubescent.

FLOWERS. April–June, racemes short, nodding, 5–10-flowered, golden yellow, fragrant; bracts ovate to oval, foliaceous; rachis pubescent; calyx salverform, tube attached to the ovary, three to four times longer than the sepals; sepals 5, revolute and recurved, oblong-obovate, apex broadly rounded, ⅕–¼ in. long; petals 5, reddish, inserted in the calyx-throat, ⅒–⅐ in. long, apex erose; stamens 5, short, alternate with the petals; ovary 1-locular, with 2 parietal placentae, style erect.

FRUIT. Ripening June–August, berries black (yellow in some forms), smooth, globose to ellipsoid, ⅓–⅖ in. long, crowned by the shriveled calyx remains.

LEAVES. Alternate or fascicled, convolute in the bud, blade orbicular to cuneate-ovate, sometimes cordate or cuneate to truncate, usually 3–5-lobed, lobes entire or toothed toward the apex, usually entire toward the base, margin ciliate, glabrate above, lower surfaces puberulent to glabrate, 1¼–3¼ in. wide; petioles ⅜–1 in.

RANGE. Cliffs or rocky hillsides. East Texas, Oklahoma, Arkansas, and Louisiana; northward to South Dakota and eastward to Minnesota and Missouri. East of the Rocky Mountains. Escaping cultivation in the Eastern states.

PROPAGATION. The fruit should be gathered as soon as ripe to prevent consumption by birds. After picking the fruit should be planted immediately, or spread out to dry to prevent heating. For storing or stratification the seeds may be extracted from the fruit by macerating in water, allowing the empty seeds and pulp to float away. Other debris can be removed by a

CLOVE CURRANT
Ribes odoratum Wendl.

fanning mill. The seed averages about 125,000 seeds per lb, with a purity of 98 per cent and a soundness of 97 per cent. Usually the seeds have an embryo dormancy, and stratification in sand or peat is practiced for 60 days at 68°–86° F. plus 60–90 days at 41° F. However, conclusive tests have not been made and perhaps acid treatment could be substituted for warm temperature stratification. The fresh seeds should be sown in fall, or stratified seeds in spring. About 40 viable seeds are sown per linear ft of nursery row and covered to a depth of ⅛–¼ in. Propagation from hardwood cuttings taken in autumn is also practiced.

VARIETIES AND FORMS. *R. odoratum* forma *xanthocarpum* Rehd., has orange-yellow fruit.

R. odoratum var. *intermedius* (Spach) Berger, has racemes spreading; receptacle about ½ in. long; sepals about ⅓ in. long, spreading; fruit orange-brown.

R. odoratum var. *leiobotrys* (Koehne) Berger, has young branches and rachis glabrous, glandular; sepals recurved, not revolute; fruit black.

R. sanguineum × *R. odoratum* produces a hybrid known as *R.* × *gardonianum* Lem. It has the habit of *R. odoratum*; leaves usually truncate, 3-lobed, glabrate; racemes about 20-flowered, flowers yellow, tinged red outside, somewhat glandular, sterile.

REMARKS. The genus name, *Ribes*, is an ancient Danish name. The species name, *odoratum*, refers to the fragrant flowers. The plant has also been listed under the names of *R. aureum* Gray (not *R. aureum* Pursh) and *Chrysobotrya odorata* Rydb. Also known under the vernacular names of Buffalo Currant, Golden Currant, and Missouri Currant. The plant is widely used as an ornamental and was introduced into cultivation in 1812. Use has been made of it extensively in shelter-belt planting and as a wildlife food and cover.

TRUMPET GOOSEBERRY

Ribes leptanthum Gray

FIELD IDENTIFICATION. Slender Western shrub to 6 ft, with erect very spiny branches.

FLOWERS. Borne in 1–3-flowered racemes, peduncles short, pedicels very short or absent, bracts small, flowers perfect, greenish white; calyx-tube adnate to the globose ovary and more or less produced above it; ⅙–⅓ in. long, cylindric, pubescent; sepals 5, oblong, shorter or about as long as the tube; petals 5, erect, shorter than the calyx-lobes, white to pink; stamens 5, inserted on the calyx-throat; styles 2, glabrous, stigmas terminal; ovary inferior, 1-celled with 2 parietal placentae.

FRUIT. Berry black, lustrous, ¼–⅓ in. in diameter, globose to elongate, glabrous or more rarely glandular-

TRUMPET GOOSEBERRY
Ribes leptanthum Gray

hispid, apex bearing the remains of the withered flower, juicy; several-seeded, seeds albuminous, embryo minute and terete.

LEAVES. Alternate, often clustered and crowded, deciduous, orbicular to reniform, palmately veined and lobed on the margin, the lobes rather deep and often with 3 smaller lobules at the apices or dentate, leaf base truncate or cordate, surfaces pubescent to glabrous, upper surface olive-green to dark green, lower surface paler, sometimes glandular; petioles as long as the leaves or shorter.

TWIGS. Tan to light brown or reddish, mostly smooth, occasionally bristly; spines ⅕–⅗ in. long, single or triple, rather stout, divergent and slightly curved or straight, yellowish to reddish brown or grayish.

RANGE. Canyons and woodlands at altitudes of 5,500–12,000 ft. Utah, Colorado, and Arizona; in New Mexico in the Sangre de Cristo, Raton, Sandia, Mogollon, Santa Rita, and Organ mountains.

REMARKS. The genus name, *Ribes*, is of old Danish origin, and the species name, *leptanthum*, means "thin-flowered." It is also known as the Alpine Gooseberry. The fruit is edible, and the plant has been cultivated since 1893. The foliage is browsed somewhat by black-tailed deer.

WITCH HAZEL FAMILY (*Hamamelidaceae*)

Vernal Witch Hazel
Hamamelis vernalis Sarg.

FIELD IDENTIFICATION. Shrub attaining a height of 9 ft, and often sending up sprouts from the base. Differs from the Common Witch Hazel in that the flowers are produced in winter or early spring.

FLOWERS. Borne January–April, fragrant, sessile or short-peduncled, solitary or clustered; calyx 4-lobed, lobes dark red within; bractlets 2–3 at base of the calyx; petals 4, ⅜–⅝ in. long, yellow or reddish at base, in some forms all red, long, narrow; stamens usually 8, very short, alternate with the petals, 4 perfect and anther-bearing, the remainder imperfect; styles 2, short.

FRUIT. Capsule woody, dehiscent from a 2-beaked apex, 2-coated, the inner and outer coat separating; seeds 1–2, large and bony, discharged elastically; calyx large and persistent.

LEAVES. Simple, alternate, deciduous, blade length 2–4 ¾ in., obovate to elliptic or oval, apex rounded to obtuse, base cuneate to rounded, margin sinuate-wavy; upper surface dull green, with veins impressed; lower surface paler green, sometimes glaucous or somewhat hairy, veins straight and prominent beneath. Some forms with rather permanent stellate pubescence.

TWIGS. Rather stout, light brown to reddish brown or gray, densely stellate-tomentose, later smooth and glabrous, light or dark gray.

RANGE. Rocky shores and stream banks. East Texas, Louisiana, Alabama, Oklahoma, Arkansas, and Missouri.

VARIETIES. Two varieties have been segregated. The Red-petal Vernal Witch Hazel, *Hamamelis vernalis* forma *carnea* Rehd., has the petals inside of calyx, and stamens dark red. The Woolly-leaf Vernal Witch Hazel, *Hamamelis vernalis* forma *tomentella* Rehd., has leaves densely hairy on the lower surface.

Vernal Witch Hazel
Hamamelis vernalis Sarg.

REMARKS. The genus name, *Hamamelis,* is from the Greek words *hama* ("at the same time") and *melon* ("apple"), possibly because of the presence of both fruit and flower simultaneously. The species name, *vernalis* ("spring"), refers to its early-blooming habit, from midwinter to spring. It is also known as Ozark Witch Hazel.

Common Witch Hazel
Hamamelis virginiana L.

FIELD IDENTIFICATION. Tall shrub or tree to 30 ft.

FLOWERS. In axillary or terminal perfect clusters, bright yellow, petals 4, crisped; sepals 4, recurved, brown to yellow within, subtended by 2 or 3 bracts; stamens 8, awl-shaped, 4 normal and 4 staminodial; styles 2, awl-shaped, stigma very small; ovary formed of 2 basally coherent 1-celled carpels.

FRUIT. Capsule woody, about ½ in. long, 2-beaked, 2-coated, surmounted by the 2 styles, dehiscent to discharge forcibly 2 oblong, bony, shiny, black seeds; calyx large and persistent.

LEAVES. Alternate, simple, deciduous, oval or obovate, wavy-toothed, straight-veined, acuminate or broadly acute at apex, oblique at base, usually glabrous above, somewhat pubescent beneath, blade 6 in. long and 3 in. broad on the average, or on older branches much smaller; stipules lanceolate, deciduous; petioles ⅕–⅗ in.

TWIGS. Slender, zigzag, reddish or orange, hairy or glabrous; lenticels pale, small; buds brownish, hairy, about ½ in. long; leaf scars lunate.

Common Witch Hazel
Hamamelis virginiana L.

BARK. Brown, thin, smooth when immature, and scaly when mature.

WOOD. Reddish brown, hard, close-grained, sapwood almost white, weighing 43 lb per cu ft.

RANGE. Usually in rich moist, sandy soil in semishade. Texas, Oklahoma, Arkansas, and Louisiana; eastward to Florida, and north to Ontario and Nova Scotia.

PROPAGATION. The capsules of Common Witch Hazel are picked by hand before splitting open, are dried, and the seeds removed by screening. The seed averages about 9,000 seeds per lb, with a commercial purity of 99 per cent and a soundness of 95 per cent. They may be stored for a year in sealed containers at 41° F., or stratified in sand and peat at 41° F. for 90 days. Soaking the seeds in hot water prior to stratification is said to shorten the stratification period. The germinative capacity of pretreated seed is 17–25 per cent. Some seeds do not germinate until the second year. The seed may be drilled, with 12 in. between rows. Fall sowing as soon as gathered is advisable, with a winter mulch of straw or leaves. The beds should be kept moist until germination time. Propagation by layering is also practiced.

VARIETIES AND RELATED FORMS. Although some botanists designate the Texas species as the Southern Witch Hazel, *H. macrophylla* Pursh, and record the Common Witch Hazel, *H. virginiana* L., as occurring farther north, the difference between the two seems vague and inconstant. Both are here considered the same plant, with minor variations in the Southern species.

Pink Common Witch Hazel, *H. virginiana* var. *rubescens* Rehd., is a variety with reddish petals, sepals yellow or greenish within.

Small-leaf Common Witch Hazel, *H. virginiana* var. *parvifolia* Nutt., has leaves 1½–4 in. long, thick and leathery, densely stellate-tomentose, and whitened to brownish beneath. It is known from Louisiana northward to Nova Scotia and Ontario, and from South Carolina, Pennyslvania, and Virginia.

REMARKS. The genus name, *Hamamelis,* is from the Greek words *hama* ("at the same time") and *melon* ("apple"), possibly because of the presence of both fruit and flower simultaneously; and the species name, *virginiana,* refers to the state of Virginia. The common name was applied because of the hazel-like straight veins of the leaves. Vernacular names are Winter Bloom, Spotted-alder, Tobacco-wood, Pistacio, Snappy Hazel, and Witch-elm.

Witch Hazel flowers are conspicuous because they bloom in the autumn when the leaves are about to be shed. The mechanically forcible discharging of the seeds is another unusual feature. In superstitious lore the twigs are used as divining rods to locate water or mineral deposits. The rods are held in the palm in a certain manner and will presumably turn and point toward the earth at the desired spot. Medicinal extracts, lotions, and salves are prepared from the leaves, twigs, and bark, mostly by distillation process. The distillate contains tannic acid, and is used to lessen inflammation, stop bleeding, and check secretions of the mucous membranes. The remedial qualities of the drugs are probably overestimated. The fruit and young twigs are eaten by at least 5 species of birds, cottontail, white-tailed deer, and beaver. It is reported that the Indians prepared a tea from the leaves. Seemingly it is gaining in popularity as an ornamental plant in the United States and in Europe. It was introduced into cultivation in 1736.

American Sweetgum

Liquidambar styraciflua L.

FIELD IDENTIFICATION. Large tree, attaining a height of 150 ft, with palmately lobed, serrate, alternate leaves. The branches and twigs are corky-winged, or wingless on some trees.

FLOWERS. March–May, monoecious, very small, greenish; perianth none; staminate flowers in terminal, erect, tomentose racemes 2–3 in. long; stamens numerous, set among tiny scales, filaments slender and short; pistillate flowers in axillary, globose, long-peduncled, drooping heads; styles 2, inwardly stigmatic, sterile stamens 4.

FRUIT. September–November, persistent, globular, spinose, lustrous, 1–1½ in. in diameter, long-peduncled, resulting from the aggregation of the many 2-celled ovaries which are tipped by the 2-beaked or hornlike styles; ovules many, but maturing only 1–2 flat-winged seeds, the rest abortive, light brown; good seed crops every three years, light years in between.

LEAVES. Simple, alternate, deciduous, petioled, broader than long, blades 3–9 in. wide, with 3–7 acuminate lobes; lobes oblong-triangular, glandular-serrate; slightly cordate or truncate at base; glabrous and glossy above, pubescent along the veins beneath, aromatic when bruised; petioles 2–4¾ in., slender, stipules falling away early.

TWIGS. At first with rusty-red tomentum, later glabrous and with wide corky wings, or some trees without wings.

BARK. Very rough, deeply furrowed, ridges rounded, brown to gray.

WOOD. Fine-grained, fairly hard, not strong, heartwood reddish brown, takes a high polish, sapwood white or pinkish, weighing about 37 lb per cu ft.

RANGE. Usually in low bottom-land woods. In east Texas, Oklahoma, Arkansas, and Louisiana; eastward to Florida, north to New York and Connecticut, and west to Illinois and Missouri; also in mountains of Mexico.

PROPAGATION. The fruit heads are collected when yellow, before the seeds fall. When dry, the seeds can be shaken loose. Trash or abortive seeds can be floated off in water. The seed averages 82,000 seeds per lb. The commercial purity is about 95 per cent, and soundness about 85 per cent. The germination average is 70 per cent, but seed should be stratified 30–90 days, or water-soaked 15–20 days before sowing in spring.

American Sweetgum
Liquidambar styraciflua L.

VARIETIES. Two forms of sweetgum are recognized in horticulture. The Round-lobed American Sweetgum, *L. styraciflua* forma *rotundiloba* Rehd., is a form with 3–5 short rounded lobes on the leaves. Weeping American Sweetgum, *L. styraciflua* forma *pendula* Rehd., has pendulous branches forming an almost columnar head.

REMARKS. The genus name, *liquidambar,* refers to the amber-colored liquid sap, and the species name, *styraciflua,* is from *styraci* ("storax") and *flua,* ("fluidus"). Vernacular names are White Gum, Alligator-tree, Opossum-tree, Red Gum, Bilsted, Satin-walnut, Gum-wood, California Red Gum, and Star-leaf Gum. Medicinally the tree is known as "copalm balsam," and the resinous gum is used extensively in Mexico and Europe, especially as a substitute for storax. Various ointments and syrups are prepared from it and are used in the treatment of dysentery and diarrhea. The gum is sometimes chewed by children. It is also used as a perfuming agent in soap, and as an adhesive. It is reported as excellent for healing wounds. The reddish brown wood is used for flooring, furniture, veneers, woodenware, general construction, boxes, crossties, barrels, sewing machines, cabinets, molding, vehicle parts, conveyors, musical instruments, tobacco boxes, and other articles. At least 25 species of birds are known to feed upon the fruit, as well as the gray squirrel and Eastern chipmunk. The autumn foliage is conspicuous because of its beautiful color variations of red and yellow. It has been cultivated since 1681, and is highly ornamental. It is rapid-growing, long-lived, and relatively free from insects and disease damage. Perhaps it could be used more extensively in reforestation projects because of its rapid growth in cutover lands.

SYCAMORE FAMILY (*Platanaceae*)

AMERICAN PLANE-TREE (SYCAMORE)
Platanus occidentalis L.

AMERICAN PLANE-TREE (SYCAMORE)
Platanus occidentalis L.

FIELD IDENTIFICATION. Tree attaining a height of 170 ft, with reddish brown bark which scales off to expose the white, smooth, new bark.

FLOWERS. April–May, monoecious, the separate heads globose and peduncled; staminate head red, with 3–8 short-filamented stamens accompanied by tiny, club-shaped scales; pistillate heads solitary, green at first, brown when mature, composed of angular ovaries set among tiny scales; ovary linear, 1-celled; style elongate, threadlike; carpels mingled with staminodia.

FRUIT. Ripe September–October, borne on peduncles 3–6 in. long, usually solitary, persistent, globose, 1–1½ in. in diameter, light brown; achenes numerous, obovoid, small, leathery, obtuse at the apex, hairy at base, 1-seeded.

LEAVES. Simple, alternate, deciduous, thin, broadly ovate, 4–12 in. across; margin usually set with 5 short, sinuate, acuminate lobes with large teeth between; truncate or heart-shaped at the base; bright green above, paler and densely pubescent along the veins beneath; stipules sheathing, conspicuous, toothed, 1–1½ in.; petiole stout, woolly, shorter than the blade, 3–5 in.

TWIGS. Slender, shiny, tomentose at first, glabrous later, orange-brown to gray.

BARK. Reddish brown, scaling off in thin plates to expose the conspicuous white, or greenish, new bark.

WOOD. Light brown, rather weak, close-grained, hard, weighing 33 lbs per cu ft, difficult to work.

RANGE. In rich bottomland soils, mostly along streams. Texas, Oklahoma, Arkansas, and Louisiana; east to Florida, and north to Maine, Minnesota, Nebraska, and Ontario.

PROPAGATION. For planting the seed the following method is recommended: The globose fruit heads may be hand picked, or obtained from the ground. Afte drying the seed can be easily rubbed from the heads; a screen will help remove the hairs or other debris The seed averages about 175,000 seeds per lb. Th commercial seed averages about 85 per cent in purit and 66 per cent in soundness. The seed may be pre treated before planting by stratification in sand or pea at 35°–41° F. for 45–60 days. Under test condition the average germination capacity is 35 per cent. Th seeds are sown in rows 6–8 in. apart and should b covered with about ¼ in. of firmed soil. A burla covering is beneficial until germination, and partia shade is helpful.

VARIETY. Smooth American Plane-tree, *P. occidentali* var. *glabrata* (Fern.) Sarg., is a variety with less nu merous and more angular teeth to the leaves. It occur in Texas on limestone soils from the Colorado River westward to the Devils and Rio Grande rivers. Also in Coahuila and Nuevo León, Mexico.

REMARKS. The genus name, *Platanus*, is the classical name of the Plane-tree, and the species name, *occidentalis*, means "western." Vernacular names are Buttonwood, Buttonball-tree, and Water-beech. The wood is used for crates, interior finishing, furniture, cooperage, rollers, butcher blocks, and tobacco boxes. It attains the largest size of any deciduous tree in the United States, and is often planted for ornament. It is slow-growing, but long-lived, and old trees are often hollow with decay. It was first cultivated in 1640. The seeds are eaten by a number of species of birds and sometimes by muskrat.

Arizona Plane-tree (Arizona Sycamore)

Platanus wrightii S. Wats.

FIELD IDENTIFICATION. Western tree attaining a height of 80 ft and a diameter of 5 ft. The branches spreading or erect and forming a broad, open crown. Conspicuous because of the bark, which exfoliates in large, thin plates to expose the smooth white inner layer.

Arizona Plane-tree (Arizona Sycamore)
Platanus wrightii S. Wats.

Smooth American Plane-tree
Platanus occidentalis var. *glabrata* (Fernald) Sarg.

FLOWERS. Monoecious, appearing with the unfolding of the leaves in 1–4 dense, pendent, spicate, tomentose heads. Individual flowers minute and in great numbers in the heads. The calyx and corolla reduced to minute scales. Staminate heads on axillary peduncles, dark red; calyx composed of 3–6 tiny scalelike sepals; petals somewhat longer than the sepals, cuneiform, sulcate, scarious; stamens as numerous as the sepals, opposite them, elongate-ovoid, filaments very short or lacking, anthers 2-celled, cells longitudinally dehiscent, the apices with capitate, truncate, hairy connectives; pistillate flowers on terminal peduncles, green to red, calyx with 4 (occasionally 3 or 6) rounded sepals, petals longer than the sepals; ovary superior, sessile, oblong, persistent-hairy at base; styles red, papillose, stigmatic to below the middle; ovules 1–2, suspended laterally.

FRUIT. A syncarp, persistent in winter, borne in 1–4 heads on slender glabrous peduncles 4–8 in. long, heads about ¾ in. in diameter; achene ⅕–¼ in. long, elongate, obovoid, rounded, truncate, styles persistent at the apex, surrounded by long hairs.

LEAVES. Simple, alternate, deciduous, broadly ovate, divided into 3–7 elongate, acute or acuminate lobes, sinuses between narrow, and often extending more than halfway to the base, lobes entire or occasionally with 1–2 large teeth, leaf base cordate or truncate, length or width 4–10 in., densely tomentose when young; with age upper surface light green and somewhat pubescent to glabrous, lower surface paler and pubescent; veinlets reticulate and united near the margin; petioles 1½–3 in., enlarged at base, rather stout; stipules foliaceous, large on vigorous shoots, sheathing, acute, entire or dentate.

TWIGS. Terete, somewhat divaricate, young pale-tomentose, older puberulous to glabrous, brown to reddish or gray, later darker; axillary buds conic, large, smooth, shiny, leaf scars narrow.

BARK. On old trunks at base with broad ridges and deep furrows with small close scales; on younger trunks and with branches shedding in large thin scales to expose thin, smooth, white or greenish bark.

WOOD. Light reddish brown, hard, heavy, not strong, with broad conspicuous medullary rays.

RANGE. On stream banks and in mountain canyons in valleys of the grass and pinyon belt at altitudes of 3,000–6,000 ft. Southwestern New Mexico, southern Arizona, and southward into Mexico.

REMARKS. The genus name, *Platanus*, is the classical name of the plane-tree of Europe, and the species name, *wrightii*, is in honor of Charles Wright, who collected plants in the Southwest 1847–1848, and in 1851, during the time he was a member of the United States–Mexico boundary survey. The tree has been cultivated since 1900. It is propagated by seeds, layers, or cuttings under glass.

ROSE FAMILY (*Rosaceae*)

Genus Crataegus

The Crataegus Problem

The genus *Crataegus* with its numerous variations and hybrids represents a very difficult taxonomic complex. Elbert L. Little (1953, pp. 125–126) states:

"*Crataegus* is regarded as an unstable genus characteristic of openings and exposed areas, which has expanded and evolved rapidly following the clearing of forests, and the origin of vast new areas suitable for colonization. The variable, expanding populations probably produced numerous hybrids. Progeny tests have shown that many variations are perpetuated, or true breeding. However, cytological evidence indicates that a large number, perhaps a majority, of the supposed species, are 'asexual apomictic triploids'; that is, they are clonal populations of hybrid origin with one and one-half the normal number of chromosomes but form viable seeds vegetatively without benefit of pollination, and thus perpetuate their characters the same as if they were propagated by grafting."

One of the difficulties appears to be in the separation of the so-called "successful hybrid misfits" from the normal, well-distributed, sexual, diploid species. To make the matter more complicated, many of the nonhybrid good species cannot be determined with certainty by any one set of parts or characters. Ernest J. Palmer (1946, pp. 471–472) makes the following remarks:

"Variations in characters in *Crataegus* must be relied upon for defining species, but it has to be recognized that most of them may vary greatly within the limits of what must be taken as a specific unit. Species can, therefore, only be defined and recognized on the basis of a synthesis of characters, and seldom upon any constant or invariable one. In such a situation descriptions and keys, though necessary and indispensable, definitely have their limits and must be taken with the proverbial grain of salt."

It is evident that too many species have been described. Most of the 1,100 specific names given during the last 25 years were applied by the authors C. S. Sargent, W. W. Ashe, and C. D. Beadle. Many of these have been, and more probably will be, reduced to synonymy as a better knowledge of the group is achieved.

Faced with the very difficult problem of choosing the tree and shrub species of the southwestern United States, the author has turned to Dr. Ernest J. Palmer of Webb City, Missouri, for advice. Dr. Palmer is the leading authority on this group and has done much to clarify many problems concerning its species. He has studied the group closely for more than thirty years and has contributed a monographic treatment in the 1950 edition of *Gray's Manual of Botany* by M. L. Fernald.

Dr. Palmer has graciously provided the author with a list of those southwestern species which he considers to be valid and also has contributed a key to both the series and species.

Using these keys and list as a basis of approach, the author has carefully reviewed all of the original descriptions of Palmer, Sargent, Ashe, and Beadle. This was supplemented by inspection of all of the type material available at the Missouri Botanical Garden, the New York Botanical Garden, the Smithsonian Institution, and the Arnold Arboretum.

Although five years were spent in the study of this complex group the author makes no claim to having clarified all of the problems concerning the southwestern species. The last word has certainly not been said, and the material presented is only broadly interpreted. Changes and corrections will undoubtedly have to be made when the species become better known.

Key

(Contributed by E. J. Palmer)

How to use this key:

The key is in two parts: a Key to the Series of Crataegus and a Key to the Species under the

SERIES. In the key to the series, the characters of the series are outlined in contrasting pairs of statements, or, occasionally, groups of three contrasting statements. Always choose the statement which most closely describes the plant you have to identify. For example, first compare the two statements designated as (a). If the first (a) is the one which seems to apply to the plant in question, then choose between the two (b) characters. It will be noted that the first (b) requires a further choice between (c) and (c), whereas the second (b) leads to the number and name of a series. Whenever a sequence of choices leads to the name of a series, turn to the key to the species under the series and proceed in the same manner.

KEY TO THE SERIES OF CRATAEGUS

(a) Veins of the leaves running to the sinuses as well as to the points of the lobes.
  (b) Leaves thin but firm, early deciduous; fruit ⅛–⅕ in. thick; nutlets 3–5. Native species.
    (c) Leaves mostly 1¼–1⅞ in. wide, ovate or deltoid in outline; flowers opening in May; fruit with deciduous calyx exposing tips of nutlets . . . . 1. **Cordatae** Beadle
    (c) Leaves mostly ⅝–1½ in. wide, narrowly obovate to deltoid in outline; flowers opening in March or April; fruit with persistent calyx . . . . 2. **Microcarpae** Loud.
  (b) Leaves thick, persistent until late in the season; fruit ⅜–¼ in. thick. Introduced species. . . . . 3. **Oxyacanthae** Loud.
(a) Veins of the leaves running only to the points of the lobes.
  (d) Fruit red or yellow or remaining green at maturity; thorns usually long and slender, to 1¾–2½ in. long, or rarely thornless.
    (e) Flowers single or 2–5 in simple clusters; stamens 20–25.
      (f) Leaves mostly 1⅜–2½ in. long; petioles slender, ⅜–⅝ in. long; sepals not foliaceous, entire or serrate; fruit ½–¾ in. thick, becoming mellow or succulent, edible. Arborescent shrubs or small trees in wet or swampy ground . . . . . . 5. **Aestivales** Sarg.
      (f) Leaves mostly ⅝–1¼ in. long; petioles stout, ⅛–⅕ in. long; sepals foliaceous, pectinate or deeply glandular-serrate; fruit ⅜–⅞ in. thick, remaining firm or hard, scarcely edible; slender shrubs 1½–6 ft tall, in dry or sandy ground . . . . 6. **Parvifoliae** Loud.
    (e) Flowers more numerous, usually 5–20 in simple or compound cymes or corymbs; stamens 5–20.
      (g) Flowers opening in late March through April according to latitude; nutlets plane on ventral surfaces.
        (h) Foliage and inflorescence glandular, usually conspicuously so.
          (i) Leaves mostly narrowly obovate or spatulate, broadest above the middle except at the ends of branchlets where sometimes broadly oval or suborbicular; fruit red or orange-red, becoming mellow; arborescent shrubs or small trees . . . . 7. **Flavae** Loud.
          (i) Leaves mostly ovate, oblong-ovate, or rhombic in outline, broadest at or below the middle, gradually or abruptly narrowed at base, usually lobed, especially at the ends of branchlets; fruit bronze-green or dull red, remaining firm or hard; shrubs usually less than 9 ft tall . . . . 8. **Intricatae** Sarg.
        (h) Foliage and inflorescence eglandular, or if slightly glandular, the glands small and soon deciduous.
          (j) Leaves mostly narrowly obovate, cuneate or oblong-obovate, unlobed or very obscurely lobed except at the ends of branchlets, where sometimes broadly obovate to oval or suborbicular.
            (k) Leaves thick or firm, glossy above in most species; fruit remaining hard and often green at maturity; nutlets 1–3 (or rarely 2–5 in a few species) . . . . 10. **Crus-galli** Loud.
            (k) Leaves thin to firm, not coriaceous, dull green above; fruit becoming soft or mellow; nutlets usually 3–5.
              (l) Leaves relatively thin, the veins obscure, mostly 1¾–2½ in. long, quite variable in shape, often oblong-obovate or rhombic, unlobed or slightly lobed except at the ends of branchlets where broadly oval or ovate and more deeply lobed; fruit ¼–7/16 in. thick; bark thin, exfoliating from orange-brown inner bark. . . . . 9. **Virides** Beadle
              (l) Leaves firm, more uniform in shape, mostly oblong-obovate, unlobed or with small shallow lobes above the middle; veins distinctly impressed above, mostly 1¼–2 in. long; fruit usually ⅜–⅝ in. thick; bark gray, thick, slightly scaly or ridged . . . . . 11. **Punctatae** Loud.
          (j) Leaves mostly oblong-ovate to rhombic, or broadly ovate to suborbicular at the ends of branchlets, all sharply lobed. New Mexico. . . . . 12. **Rotundifoliae** Eggleston
          (j) Leaves mostly ovate or deltoid in outline, broadest below the middle, often rounded, truncate, or subcordate at base. Arkansas and eastward, except some species in 16 and 19.
            (m) Sepals entire or serrate, not pectinate or deeply glandular-serrate; filaments as long or nearly as long

as the petals; nutlets 3–5, usually less than 5.

(n) Leaves thin, glabrous except for short pilose hairs above while young; stamens about 10; fruit less than 7/16 in. thick, becoming succulent . . 13. **Tenuifoliae** Sarg.

(n) Leaves firm or thick; fruit usually 7/16 in. or more thick, remaining firm or hard.

(o) Young leaves scabrate with sparse hairs above, becoming glabrous; fruiting calyx small and sessile . . . . 14. **Silvicolae** Beadle

(o) Young leaves glabrous above, glabrous or rarely pubescent beneath; fruiting calyx elevated and usually prominent . . . . 15. **Pruinosae** Sarg.

(m) Sepals conspicuously glandular-serrate or pectinate; filaments distinctly shorter than the petals; nutlets usually 5.

(p) Foliage and inflorescence pubescent; flowers 3/4–7/8 in. wide; fruit pubescent at least while young, ripening in August or early September . . . . 16. **Molles** Sarg.

(p) Foliage and inflorescence glabrous; flowers 3/4–1 in. wide; fruit glabrous, ripening in October. . . . . 17. **Dilatatae** Sarg.

(d) Fruit blue or black at maturity; thorns short and stout, usually less than 3/4 in. long.

(q) Leaves mostly abruptly pointed or rounded at the apex, lustrous above; fruit blue at maturity (except in rare form), glaucous. Eastern Texas and eastward . . . . 4. **Brevispinae** Beadle

(q) Leaves mostly acute or acuminate at the apex, dull green above; fruit turning from purple to black, lustrous but not glaucous. New Mexico and westward . . . . 18. **Douglasianae** Egglest.

(g) Flowers opening late, April or May according to latitute; nutlets pitted on ventral surfaces . . . . 19. **Macracanthae** Loud.

KEY TO THE SPECIES UNDER THE SERIES

1. **Cordatae** (Only one species in this area.) . . . . 1. *C. phaenopyrum.*

2. **Microcarpae**

a. Leaves mostly broadly ovate in outline, often as broad or broader than long, deeply incised, rounded to cordate at base; anthers red; fruit oblong. . . . . 2. *C. marshallii.*

a. Leaves mostly narrowly obovate or spatulate, unlobed or nearly so except at the ends of branchlets, cuneate or attenuate at base; anthers pale yellow; fruit subglobose . . . . . . 3. *C. spathulata.*

3. **Oxyacanthae** (Only one species in this area.) . . . . 4. *C. monogyna.*

4. **Brevispinae** (Only one species in this area.) . . . . 5. *C. brachyacantha.*

5. **Aestivales**

a. Pubescence on the under surface of the leaves rusty brown, mainly along the veins; fruit ripening in May. . . . . 6. *C. opaca* (typical).

a. Pubescence on the under surface of the leaves gray, mainly in the axils of the veins; fruit ripening in June. . . . . . 6a. *C. opaca* var. *dormanae.*

6. **Parvifoliae** (Only one species in this area.) . . . . 7. *C. uniflora.*

7. **Flavae** (Only one species in this area.) . . . . 8. *C. pearsonii.*

8. **Intricatae**

a. Foliage and inflorescence glabrous or essentially so.

b. Leaves mostly 1 5/8–2 1/2 in. long, 1–2 in. wide; flowers 5/8–3/4 in. wide.

c. Fruit remaining dry and hard; sepals glandular-serrate.

d. Terminal leaves often ovate and deeply lobed near the base; fruit remaining green or yellowish green.

e. Anthers white or pale yellow (rarely pink). . . 9. *C. intricata* var. *straminea.*

e. Anthers pink or red (rarely white); fruit subglobose or short-oblong; nutlets 2–4, usually 2–3. . . . . . . 9a. *C. neobushii.*

d. Terminal leaves usually oblong-ovate or broadly elliptic, not deeply lobed; fruit becoming dull red.

f. Fruit subglobose; nutlets 3–5. . . . . 10. *C. buckleyi.*

f. Fruit obovoid or oblong; nutlets 2–5, usually less than 5. . . . . 11. *C. rubella.*

c. Fruit becoming mellow or juicy; sepals entire or finely glandular-serrate. 12. *C. padifolia* var. *incarnata.*

b. Leaves mostly 1–1 5/8 in. wide; flowers 1/2–5/8 in. wide. 13. *C. pagensis.*

a. Foliage and inflorescence pubescent, at least while young. Young leaves and inflorescence sparsely pilose, becoming glabrous or nearly so; stamens about 20; anthers red; fruit usually less than 7/16 in. thick, glabrous.

g. Leaves mostly 1 5/8–2 in. long, 1 1/4–1 5/8 in. wide; flowers mostly 6–12 in corymb; fruit subglobose, about 7/16 in. thick, orange-colored. Arborescent shrub or small tree. . . . . . 16. *C. harveyana.*

g. Leaves mostly 3/4–1 5/8 in. long and wide; flowers mostly 3–8 in corymb; fruit oblong or pyriform, about 1/3 in. thick, dull red. Widely branching shrub 3–6 ft tall. . . . . 14. *C. ouachitensis.*

a. Foliage and inflorescence pubescent throughout the season; stamens about 10; anthers cream-white or pale yellow; fruit pubescent while young. . . . . 15. *C. biltmoreana.*

9. **Virides**

a. Mature leaves and inflorescence glabrous or essentially so (except in variety of no. 17).

b. Leaves firm but comparatively thin at maturity, dull green above; nutlets normally 5.
c. Leaves variable in shape, mostly oblong-ovate or oblong-elliptic, glabrous (except in variety); anthers pale yellow or rarely pink. . . . . 17. *C. viridis.*
c. Leaves more uniform in shape, mostly ovate or oblong-ovate, pubescent above as they unfold, soon glabrous; anthers pink. . . . . 22. *C. sutherlandensis.*
b. Leaves thick or subcoriaceous at maturity, glossy above, nutlets 3–5.
d. Leaves mostly 2–2¾ in. long; terminal leaves broadly ovate and sharply lobed; fruit ¼–7/16 in. thick. . . . . 18. *C. nitida.*
d. Leaves mostly 1⅝–2½ in. long; terminal leaves broadly ovate to suborbicular; fruit about ⅓ in. thick. . . . . 23. *C. glabriuscula* forma *desertorum.*
a. Foliage and inflorescence conspicuously pubescent while young, the leaves more or less pubescent throughout the season.
e. Mature leaves comparatively thin; flowers mostly 8–15 in corymb; fruit subglobose.
f. Leaves pubescent beneath throughout the season; flowers ½–⅝ in. wide . . . . *C. viridis* var. *velutina.*
f. Leaves strongly pubescent while young, becoming nearly glabrous; flowers about ¾ in. wide. . . . . 19. *C. anamesa.*
e. Mature leaves thick or subcoriaceous, pubescent while young, becoming glabrous and glossy above and slightly hairy along the veins beneath.
g. Flowers ¾ in. or more wide, mostly 10–20 in corymb; sepals narrowly lanceolate, long-acuminate. . . . . 20. *C. stenosepala.*
g. Flowers ⅝–¾ in. wide, mostly 5–15 in corymb; sepals lanceolate or deltoid-lanceolate, broad based.
h. Leaves mostly 1¼–1¾ in. long; fruit subglobose or ovoid, orange-red, becoming mellow. . . . . 21. *C. poliophylla.*
h. Leaves mostly 1–1¼ in. long; fruit subglobose, dull red, remaining hard and dry. . . . . 24. *C. amicalis.*

10. **Crus-galli**
a. Foliage and inflorescence glabrous or essentially so, except in no. 30 and in var. of no. 32, in which the young leaves are more or less pubescent.
b. Mature leaves thick or subcoriaceous and glossy above (except sometimes in shade).
c. Leaves mostly obovate or spathulate, distinctly longer than broad, broadest above the middle, except sometimes at the ends of branchlets.
d. Serration of the leaves sharp with acute teeth; fruit usually ⅓–½ in. thick; nutlets 1–3, usually 1 or 2.
e. Terminal shoot leaves unlobed or rarely very obscurely lobed; flowers 7/16–⅝ in wide; stamens about 10 (except in var *leptophylla)* . . . . 25. *C. crus-galli*
e. Terminal shoot leaves often slightly lobed; flowers about ¾ in. wide; nutlets usually 2. . . . . 26. *C. bushii*
d. Serration of the leaves shallow or crenate; fruit ¼–⅖ in. thick; nutlets usually 2. . . . . 34 and 34a. *C. pyracanthoides* vars.
c. Leaves broader, mostly broadly obovate, oblong-obovate or oval, only slightly longer than broad or often as broad as long at the ends of shoots.
f. Young leaves quite glabrous; terminal shoot leaves usually broadly ovate to suborbicular.
g. Flowers ½–⅝ in. wide; fruit 7/16 in. or less thick; nutlets usually 3; terminal leaves broadly ovate or oblong-ovate, sometimes slightly lobed toward the base. . . . . 27. *C. palmeri.*
g. Flowers ⅝–¾ in. wide; fruit 7/16–½ in. thick (or smaller in varieties); terminal leaves broadly oval or suborbicular, often with several small shallow lobes. . . . . 29. *C. reverchonii.*
g. Flowers about ½ in. wide; corymbs glabrous; stamens 10, anthers red or pink; fruit ellipsoidal; leaves oblong. . . . . 29a. *C. cherokeensis.*
f. Young leaves sometimes slightly villous, soon glabrous (except in var. of no. 32 where they are permanently pubescent).
h. Leaves mostly obovate or oblong-obovate; terminal leaves incisely lobed; flowers usually 8–15 in lax corymbs.
i. Leaves sharply and deeply serrate; terminal shoot leaves mostly oval with 2–3 pairs of small shallow lobes; stamens about 10; fruit 7/16–½ in. thick, dull red; nutlets 2–3. . . . . 32. *C. regalis.*
i. Leaves with sharp but shallow serrations; terminal leaves mostly elliptic, sometimes slightly lobed toward the apex; flowers about ¾ in. wide; stamens about 10; fruit about 7/16 in. thick, bright orange or orange-red; nutlets usually 3. . . . . 30. *C. mohrii.*
h. Terminal shoot leaves broad-obovate to elliptic, glabrous at maturity, unlobed but deeply and irregularly serrate; flowers mostly 5–6 in compact corymbs; stamens 20, anthers pink. . . . . 36. *C. sublobulata.*
h. Terminal shoot leaves ovate to oval or obovate, pale villose below at maturity. . . . . 36a. *C. warneri.*
b. Mature leaves comparatively thin, not subcoriaceous, yellowish green, slightly lustrous but not glossy above.

j. Leaves mostly elliptic or oblong-obovate, longer than wide except sometimes at the ends of shoots, the veins obscure; fruit subglobose or slightly obovoid, dull red at maturity. . . . . 28. *C. acutifolia.*

j. Leaves mostly broadly obovate or rhombic, nearly or sometimes quite as broad as long, the veins slightly impressed above; fruit oblong, green or yellowish flushed with red at maturity. . . . . 33. *C. sabineana.*

a. Foliage and inflorescence pubescent while young and usually throughout the season.

k. Leaves mostly obovate or oblong-obovate, broadest above the middle except sometimes at the ends of shoots.

l. Fruit 7⁄16 in. or less thick, remaining dry and hard.

m. Flowers 7⁄16–5⁄8 in. wide; fruit red or orange at maturity, not lustrous.

n. Flowers mostly 4–5 in compact corymbs; stamens about 20; anthers pale yellow. . . . . 38. *C. berberifolia.*

n. Flowers mostly 8–12 in loose corymbs; stamens about 10; anthers usually pink, rarely white. . . . . 39. *C. engelmannii.*

l. Fruit ½–¾ in. thick, becoming mellow or succulent; flowers 5⁄8–¾ in. wide. . . . . 31. *C. palliata.*

k. Leaves broader, mostly oblong-obovate, oval or elliptic, usually broadest about the middle.

a. Leaves pubescent beneath throughout the season; flowers flattish, not noticeably cup-shaped.

o. Leaves mostly broadly obovate or oval, those at the ends of shoots similar but larger and relatively broader; sepals entire or minutely serrate; anthers yellow.

p. Flowers about 5⁄8 in. wide; stamens about 20; fruit subglobose or short-oblong. . . . . 40. *C. subpilosa.*

p. Flowers about ¾ in. wide; stamens about 10; fruit ovoid. . . 32a. *C. regalis* var. *paradoxa.*

o. Leaves mostly rhombic or oval, those at the ends of shoots broadly oval to suborbicular; sepals conspicuously glandular-serrate; anthers pink or red. . . . . 41. *C. traceyi.*

a. Leaves slightly pubescent on both sides while young, becoming glabrous at maturity; flowers cup-shaped. . . . . 30. *C. mohrii.*

11. **Punctatae**

a. Leaves mostly obovate or oblong-obovate, or at the ends of shoots elliptic or oval; flowers 5–12 in villose corymbs.

b. Flowers mostly 5–8 in corymbs; usually less than ¾ in. wide; stamens 10–20, usually 10–15; anthers white or pale yellow. . . 42. *C. collina.*

b. Flowers mostly 8–12 in corymbs, usually ¾ in. or more wide; stamens about 20; anthers pink or rose. . . . . 44. *C. verruculosa.*

a. Leaves mostly broadly oval or ovate; flowers mostly 8–15 in glabrous corymbs. . . . . 43. *C. fastosa.*

a. Leaves oval to obovate, acute or acuminate at apex; fruit often rather longer than broad, bright canary yellow; flowers in broad 7–8-flowered, slightly villose corymbs. . . . . . 43a. *C. brazoria.*

12. **Rotundifoliae**

a. Leaves elliptic, oval or suborbicular, usually slightly lobed, more or less pubescent while young; fruit about 7⁄16 in. thick, dark red or rarely dull yellow at maturity. . . . . 45. *C. chrysocarpa.*

a. Leaves mostly ovate or obovate, glabrous; fruit about ⅓ in. thick, orange-red or reddish orange at maturity. . . . . 46. *C. erythropoda.*

13. **Tenuifoliae** (Only one species in this area.) . . . . 47. *C. macrosperma.*

14. **Silvicolae** (Only one species in this area.) . . . . 48. *C. iracunda* var. *silvicola.*

15. **Pruinosae**

a. Flowers ½–¾ in. wide; fruit 7⁄16–5⁄8 in. thick with prominent elevated calyx.

b. Leaves of flowering spurs mostly 1–1¾ in. wide; terminal shoot leaves larger, ovate or deltoid, sharply lobed.

c. Leaves mostly abruptly narrowed or rounded at the base; fruit usually pruinose. . . . . 49. *C. pruinosa.*

c. Leaves mostly rounded, truncate or subcordate at base; fruit not pruinose.

b. Leaves of flowering spurs mostly 1–1⅜ in. wide, the terminal lobe often conspicuously elongate especially at the ends of shoots. . . . . 51. *C. gattingeri.*

d. Leaves with shallow or obscure lobes, mostly rounded or abruptly narrowed at base; fruit with a narrow slightly elevated calyx. . . . . 51a. *C. disjuncta.*

b. Leaves of flowering spurs mostly 1⅜–1¾ in. wide; terminal shoot leaves sometimes as broad as long or broader, the terminal lobe not conspicuously elongate.

e. Leaves glabrous or essentially so from the first. . . . . 50. *C. mackenzii.*

e. Leaves short villose above while young, and pubescent along the veins beneath throughout the season. . . . . 50a. *C. mackenzii* var. *aspera.*

a. Flowers ¾–1 in. wide; fruit 5⁄8–¾ in. thick, subglobose or depressed-globose, often wider than long, with a broad, slightly elevated, calyx. . . . . 53. *C. platycarpa.*

16. **Molles**

a. Leaves of flowering spurs mostly oval or ovate, rounded at base; terminal shoot leaves broadly ovate, often truncate or subcordate at base.

b. Fruit bright red at maturity.

c. Leaves longer than broad except rarely at the ends of shoots.

d. Mature leaves firm but comparatively thin; flowers numerous, to 15–20 in corymb.

e. Fruit ripening in August or September; flesh succulent and edible; nutlets 4–5, usually 5. . . . . 54. *C. mollis.*

e. Fruit ripening in October; flesh dry and mealy; nutlets 3–5. .... 57. *C. limaria.*

d. Mature leaves thick or subcoriaceous; flowers mostly 5–12 in corymb.

f. Leaves bluish green; flowers mostly 5–12 in compound corymbs; stamens about 20. .... 60. *C. lanuginosa.*

f. Leaves dull yellowish green; flowers mostly 5–8 in simple corymbs. .... 64. *C. greggiana.*

c. Leaves often as broad as long, comparatively small; terminal shoot leaves sometimes broader than long. .... 62. *C. brachyphylla.*

b. Fruit bright yellow at maturity. .... 63. *C. viburnifolia.*

a. Leaves of flowering spurs mostly elliptic or oblong-ovate, noticeably longer than broad, gradually or abruptly narrowed at base; terminal shoot leaves broader, usually rounded or rarely truncate at base. Fruit red at maturity; sepals glandular-serrate.

b. Stamens 10 or less; nutlets 3–5. .... 56. *C. noelensis.*

b. Stamens about 20; nutlets 4–5, usually 5, except in no. 58.

c. Mature leaves thick; sepals foliaceous, deeply glandular-serrate; anthers large, dark red; fruit with thick mellow flesh, edible. .... 55. *C. texana.*

c. Mature leaves relatively thin; sepals not foliaceous, more or less glandular-serrate; fruit with thin dry or mealy flesh, scarcely edible.

d. Anthers white or pale yellow; sepals laciniately glandular-serrate; nutlets 3–5. ... . 58. *C. invisa.*

d. Anthers pink or rose, or sometimes white in no. 59; nutlets 4–5, usually 5.

e. Flowers about 1 in. wide; sepals glandular-serrate; fruit bright red or crimson and lustrous at maturity. .... 59. *C. dispessa.*

e. Flowers about ¾ in. wide; sepals sparingly and irregularly glandular-serrate; fruit dull dark red at maturity. .... 61. *C. dallasiana.*

17. **Dilatatae** (Only one species in this area.) .... 66. *C. coccinioides.*

18. **Douglasianae** (Only one species in this area.) .... 67. *C. rivularis.*

19. **Macracanthae**

a. Leaves relatively large, those of the flowering spurs mostly 2–4 in. long, 1¾–3¼ in. wide; flowers mostly 8–20 in loose compound corymbs.

b. Mature leaves dull yellowish green above; anthers pink or rarely pale yellow in variety. Arborescent shrubs or small trees to 18–24 ft. .... 68. *C. calpodendron.*

b. Mature leaves bright green, glossy above; anthers pale yellow; diffuse shrubs 6–9 ft tall. .... 70. *C. carrollensis.*

a. Leaves relatively small, those of the flowering spurs mostly 1–1¾ in. long, ¾–1¼ in. wide; flowers mostly 5–8 in compact corymbs; low branching shrubs 3–4½ ft. ...... 69. *C. thermopegaea*

Washington Hawthorn

Crataegus phaenopyrum (L.f.) Medic.—
Series **Cordatae** (1)

FIELD IDENTIFICATION. Shrub or tree to 30 ft, with a trunk to 10–12 in. in diameter, the trunk rather short with the ascending branches forming a rounded crown.

FLOWERS. Opening in May, corymbs glabrous and many-flowered, corolla white, composed of 5 rounded petals; calyx broadly obconic, smooth and glabrous, sepals 5, deltoid, margin entire and ciliate, externally glabrous, inner surface pubescent; stamens 20, anthers rose-colored; styles 2–5, densely hairy at base.

FRUIT. Maturing in September, persistent, globose with ends somewhat flattened, shiny, red, about ¼ in. in diameter, calyx deciduous; nutlets 3–5, apex rounded, base acute, about ⅛ in. long.

LEAVES. Simple, alternate, deciduous, broad-ovate to

Washington Hawthorn
Crataegus phaenopyrum (L. f.) Medic.

PARSLEY HAWTHORN
Crataegus marshallii Egglest.

triangular, apex acute to acuminate, base truncate, rounded or broadly cuneate, margin both coarsely serrate and mostly 3-lobed; when young pale-tomentose, at maturity dark green, shiny and glabrous above, lower surface paler and pubescent, rather thin, length 1½–2 in., width 1–1½ in.; petioles ¾–1½ in., slender and hairless.

TWIGS. Slender, divaricate, glabrous, brown to gray, armed with short slender spines 1½–2 in. long.

RANGE. In rich moist soil. Arkansas and Missouri, eastward to Virginia and Florida.

REMARKS. The genus name, *Crataegus*, is the classical name for the hawthorn. The species name, *phaenopyrum*, means "pear-like." Sometimes cultivated for ornament in the United States and in Europe.

PARSLEY HAWTHORN

Crataegus marshallii Egglest.—Series **Microcarpae** (2)

FIELD IDENTIFICATION. Shrub or small tree attaining a height of 20 ft, with smooth gray bark and spreading crooked branches.

FLOWERS. Borne in villose corymbs of 3–12; petals 5, white, spreading, rounded, inserted on the disk margin; calyx 5-lobed, lobes lanceolate, acuminate, often glandular-serrate; stamens about 20, with red anthers; styles 1–3; nearly the whole of the inflorescence white-pubescent.

FRUIT. Pome oblong or ovoid, about ⅓ in. long, bright red, shiny, slightly pubescent; flesh thin, yellow, edible; nutlets 1–3 usually 2, smooth, rounded.

LEAVES. Simple, alternate, deciduous, ovate to orbicular, ¾–1½ in. long; acute at the apex; truncate, cuneate, or subcordate at base; incised into 5–7 deep clefts and serrate on the margin; pubescent on both faces when young, when older more glabrous above but hairy along veins beneath; petioles 1–2 in., slender, tomentose.

TWIGS. Brown to gray, pubescent when young, smooth later, crooked; bearing stout, straight, brown, scattered spines 1–2 in. long.

BARK. Gray to brown, smooth, scaling off in large thin plates to expose reddish brown inner bark.

WOOD. Reddish brown, heavy, hard, strong, weighing 46 lb per cu ft, of no particular commercial value.

RANGE. Texas to Florida; north to Oklahoma, Arkansas, Missouri, and Virginia.

HYBRIDS. *Crataegus* × *notha* is a hybrid between *C. marshallii* and *C. brachyphylla*.
Crataegus × *lacera* is a hybrid between *C. marshallii* and *C. invisa*.

REMARKS. *Crataegus* is a Greek word meaning "strong," in reference to the tough wood, and the species name, *marshallii*, is in honor of the botanist Humphrey Marshall. Parsley Haw could be more extensively cultivated for its beautiful foliage, white flowers, and scarlet fruit.

LITTLE-HIP HAWTHORN

Crataegus spathulata Michx.—Series **Microcarpae** (2)

FIELD IDENTIFICATION. Shrub or small tree to 25 ft, with a broad open head and bearing sparse straight spines.

FLOWERS. March–May, borne in glabrous, many-flowered corymbs; individual flowers about ½ in. in diameter on slender pedicels; petals 5, white, rounded, spreading, inserted on the margin of the disk; stamens about 20. Calyx-tube obconic, glabrous, with 5 lobes; lobes deltoid, entire, minutely glandular at apex.

FRUIT. Pome ripe in October, globose or nearly so, bright red, ¼ in. or less in diameter, tipped with the persistent reflexed calyx-lobes; flesh dry, thin, and mealy; nutlets 3–5, slightly ridged or smooth on back.

LEAVES. Simple, alternate, deciduous; spatulate to oblanceolate, sometimes 3–5-lobed at the apex; crenate-serrate on the margin, the cuneate base entire and tapering to a winged petiole; apex acute or rounded; hairy when young on both sides, at maturity becoming firm, glabrous, shiny and dark green above, and glabrous or villose on the veins beneath; 1–2 in. long, 1–1½ in. wide; terminal leaves with stipules stalked, foliaceous, falcate, serrate, and to ½ in. wide.

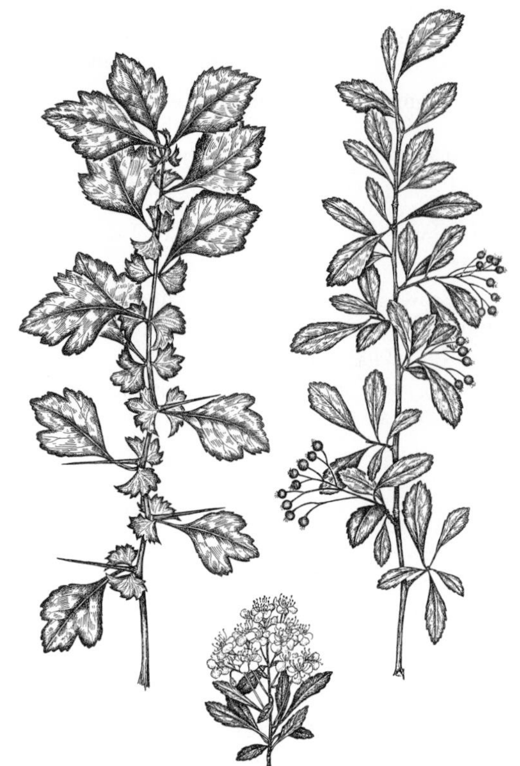

LITTLE-HIP HAWTHORN
Crataegus spathulata Michx.

TWIGS. Reddish brown, glabrous, crooked, armed or unarmed; spines sparse, slender, more or less straight, brown, 1–1½ in. long.

BARK. Light brown to gray, smooth, flaking off.

WOOD. Reddish brown, heavy, hard, strong, weighing 45 lb per cu ft, not large enough for commercial use.

RANGE. Oklahoma, Arkansas, and eastern Texas; eastward to Florida and northward to South Carolina and Virginia.

REMARKS. *Crataegus* is a Greek word meaning "strong," in reference to its tough wood, and the species name, *spathulata,* refers to the spathe- or spoon-shaped leaves.

SINGLE-SEED HAWTHORN

Crataegus monogyna Jacq.—Series **Oxyacanthae** (3)

FIELD IDENTIFICATION. An introduced shrub or tree to 40 ft, with a trunk 1½ ft in diameter, the branches slender, ascending, and armed with numerous thorns.

FLOWERS. May–June, in many-flowered glabrous, or thinly hairy, corymbs 1½–2¼ in. across; peduncles averaging ¾–1 in. long, with scattered pubescence, often glabrous later; pedicels averaging about ¼ in. long, scattered hairy at first; corolla ⅜–⅝ in. across, white to pink, 5-petaled; petals about ¼ in. long, oval to broadly obovate, deciduous; stamens about 20, exserted, anthers red and oblong, filaments filiform; style solitary, calyx 5-lobed, lobes deltoid, obtuse to acute, entire, pubescent at first, eventually glabrous.

FRUIT. Maturing September–October, pome globose to ellipsoidal, ¼–½ in. long, deep red, flesh firm and thin, nutlet solitary, calyx deciduous or sometimes persistent.

LEAVES. Persistent in mild winters, ovate to obovate or some almost deltoid, blade length ½–2 in., width ½-1⅘ in., apex acute, base cuneate to truncate, margin usually coarsely serrate or with 2–3 pairs of oblong lateral lobes, firm, deep-green, glabrous or nearly so; petioles slender, glabrous, from one third the length of blade to as long.

TWIGS. Slender, reddish brown to gray, slightly puberulent to glabrous; spines axillary or ending branches, slender, straight, lustrous, reddish brown, length ⅜-1 in.

WOOD. Yellowish white, fine-grained, heavy, hard, weighing 50 lb per cu ft.

RANGE. Introduced from Eurasia, escaping from cultivation in some areas in Arkansas and Oklahoma, and northward and eastward.

VARIETIES AND FORMS. A number of varieties and

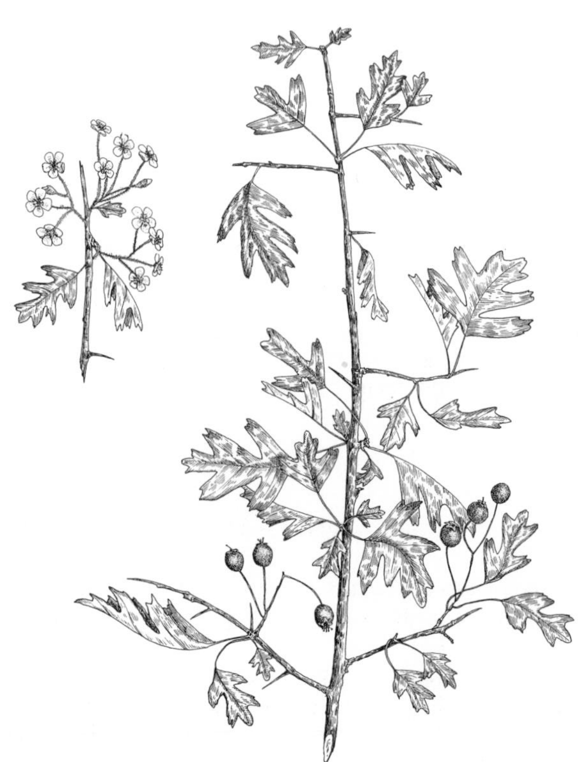

SINGLE-SEED HAWTHORN
Crataegus monogyna Jacq.

forms of the Single-seed Hawthorn are known in cultivation:

Wing-leaf Single-seed Hawthorn, *C. monogyna* var. *pteridifolia* (Loud.) Rehd., has leaves deeply lobed and closely incised-serrate.

Pyramidal Single-seed Hawthorn, *C. monogyna* var. *stricta* (Loud.) Rehd., has a fastigiate habit with erect branches.

Grisebach Single-seed Hawthorn, *C. monogyna* var. *azarella* (Griseb.) Koehne, has young branches, leaves, and inflorescence villous.

Weeping-white Single-seed Hawthorn, *C. monogyna* forma *pendula* (Loud.) Dipp., has pendulous branches.

Everblooming Single-seed Hawthorn, *C. monogyna* forma *semperflorens* (Andre) Dipp., is a low shrubby form flowering continuously until August.

Glastonburg Single-seed Hawthorn, *C. monogyna* forma *biflora* (West.) Rehd., blooms in mild seasons in midwinter, producing a second crop of flowers in spring.

Single-seed Hawthorn is sometimes confused with *C. oxycantha,* another Eurasian species, which has a number of horticultural varieties.

REMARKS. The genus name, *Crataegus*, refers to the hard wood. The species name, *monogyna*, refers to the solitary ovule. It is also known under the vernacular names of Whitethorn, Maythorn, English Hawthorn, Hedgethorn, Haw-tree, May-bush, Quickthorn, Quickset, Quick, Wick, and Wicken.

BLUEBERRY HAWTHORN
Crataegus brachyacantha Engelm. & Sarg.

BLUEBERRY HAWTHORN

Crataegus brachyacantha Engelm. & Sarg.—
Series Brevispinae (4)

FIELD IDENTIFICATION. Beautiful round-topped tree attaining a height of 40 ft, armed with short curved spines.

FLOWERS. Borne in many-flowered glabrous corymbs; flowers about 1/3 in. across, white at first and orange with age; petals 5, borne on the edge of the disk, rounded, spreading; stamens 15–20, anthers yellow; styles 3–5; calyx-tube obconic, glabrous, 5-lobed; lobes triangular to lanceolate, gland-tipped, entire.

FRUIT. Ripe pomes borne a few in a cluster on erect pedicels in August; depressed, subglobose or obovoid, bright blue or black and glaucous, 1/3–1/2 in. long, flesh thin; nutlets 3–5, round at apex, acute at base, rounded or with 2 slight grooves on the back, about 1/4 in. long, light brown.

LEAVES. Simple, alternate, deciduous, those on vigorous shoots often quite distinct from those on slow-growing spurs, oblong-obovate or oblong-lanceolate or ovate-rhombic, leathery; acute or obtuse at apex; cuneate to truncate or cordate at the base; crenate-serrate on margins and lobed on some; when lobed usually 3-lobed with the middle lobe longest and somewhat shallowly cleft again, or only serrate; dark green, glabrous and lustrous above; paler, and glabrous or pubescent beneath, 3/4–3 1/2 in. long, about 1/2–2 in. wide; petiole 1/2–3/4 in., slender, sometimes winged; stipules foliaceous, ovate-triangular, asymmetrical, broader than long, elongate and acuminate pointed on one side, the other side shortened and coarsely toothed also across the apex, to 1 in. long.

TWIGS. Green and minutely pubescent at first, glabrous and reddish brown to gray later; spines short, 1/3–2/3 in. long, stout, usually curved but some straight, gray to brown.

BARK. Smooth and gray on young trunks, on old trunks gray to brown and divided into narrow flattened ridges and shallow furrows, loosening with age into thin scales to expose reddish brown inner bark.

WOOD. Hard, strong, heavy, not large enough to be commercially usable.

RANGE. Margins of streams and swamps, Texas and Louisiana; east to Georgia and north to Arkansas. In Texas west to the Trinity River. Locally abundant between Hull and Saratoga, Texas.

FORM. A form with white or straw-colored fruit, of rare occurrence, is known as *C. brachycantha* forma *leucocarpa* Sarg. From Natchitoches Parish near Natchitoches, Louisiana.

REMARKS. *Crataegus* is from an ancient Greek word and means strength in reference to the tough wood, and the species name, *brachyacantha,* refers to the short thorns. The French gave the tree the name of *Pomette Bleu* in reference to the blue fruit which is unusual for a hawthorn.

May Hawthorn

Crataegus opaca Hook. & Arn.—Series **Aestivales** (5)

FIELD IDENTIFICATION. Shrub or small tree up to 30 ft, with slender erect branches and a rounded head.

FLOWERS. Borne before the leaves in February or March, in 2–5-flowered glabrous corymbs; corolla about ¾ in. across; petals 5, white, spreading, rounded, inserted on the margin of the disk; stamens about 20 with rose-purple anthers; styles 3–5; calyx-tube obconic, glabrous, 5-lobed; the calyx-lobes triangular, acute and gland-tipped, entire or serrulate on margin, often reddish-colored.

FRUIT. Pome borne in May, large, ½–¾ in. across, red, dotted, fragrant, globose, somewhat depressed, calyx persistent; flesh juicy, sweet-acid; nutlets 3–5, rounded.

LEAVES. Simple, alternate, deciduous, oblong to obovate or elliptic, acute or rounded at the apex, cuneate at the base, margin crenate-serrate or often 3-lobed, 1–2½ in. long, ½–1⅓ in. wide; dark green and usually glabrous above; lower surface clothed with dense rusty-brown pubescence, especially on the veins; petioles slender, rusty-pubescent.

TWIGS. Brown to gray, hairy at first, glabrous later; unarmed, or bearing stout, straight, brown spines ½–1 in.

BARK. Dark reddish brown, deeply fissured into persistent scales.

May Hawthorn
Crataegus opaca Hook. & Arn.

WOOD. Heavy, hard, strong, not large enough for commercial use.

RANGE. The species is found in wet soil in Texas, Arkansas, Louisiana, and north to South Carolina.

VARIETY. June Hawthorn, *C. opaca* var. *dormonae* (Hook. & Arn.) (*C. aestivalis* var. *dormonae* Ashe), differs from the species as follows:

The flowers are ⅗–⅘ in. wide, anthers deep rose-colored; calyx-lobes entire, short-triangular, glabrous, either spreading or with ascending tips from a keeled base and coloring with the fruit.

The fruit ripens and falls the last of June with the pedicels attached, bright glossy red, usually slightly depressed, rarely oblong, ⅖–⅗ in. long; flesh pale yellow, soft, and pleasantly acid when mature.

The leaves are similar to the species but thicker and dentate, appressed pubescent above as they unfold with the flowers early in March, eventually glabrate above except for the midrib, lower surface permanently pale gray-pubescent along the midrib and with prominent tufts in the axils of the veins; petiole permanently pubescent.

Named for Caroline Dormon from specimens collected at Black Lake, Natchitoches Parish, Louisiana. Sheets dated March and June, 1927, and March 9 and June 24, 1928 (type herbarium of W. W. Ashe).

REMARKS. *Crataegus* is from an ancient Greek word meaning "strength," in reference to the wood, and the species name, *opaca*, refers to the dull fruit. It was formerly listed under the name of *C. aestivalis* (Walt.) Torr. & Gray, but this name now applies to another species. Also known as the Riverflat Hawthorn. This is the famous May Haw of the South, from which preserves are made. The large size and acid character of the pomes make it particularly desirable for that purpose.

One-flower Hawthorn

Crataegus uniflora Muenchh.—Series **Parvifoliae** (6)

FIELD IDENTIFICATION. Slender shrub 3–12 ft, with crooked, thorny branchlets. Usually with a solitary flower or fruit.

FLOWERS. Opening in May, pedicels short, tomentose, single, or rarely 2–3 together; corolla ⅜–⅝ in. wide, petals 5, white, rounded, deciduous; calyx-lobes 5, foliaceous, lanceolate, pectinate-laciniate or deeply glandular-serrate; stamens 20 or more, filaments filiform; anthers small, oblong, white or pale yellow; styles 5–7.

FRUIT. Pome maturing in October, calyx persistent on the fruit with prominent reflexed lobes, fruit body ⅜–½ in. thick, pubescent, subglobose or slightly pyriform, greenish yellow to dull red, flesh firm, dry and mealy; nutlets 3–5 (usually 5), about ⅓ in. long, grooved dorsally, bony.

LEAVES. Simple, alternate, deciduous, subcoriaceous,

ONE-FLOWER HAWTHORN
Crataegus uniflora Muenchh.

apex obtuse to rounded or acute, base cuneate, margin sharply or crenately serrate, unlobed or sometimes obscurely lobed above the middle, leaf varying in shape from obovate to spatulate or cuneate, sometimes oblong or elliptic, upper leaf surface dark green and lustrous, with scattered hairs to glabrous; lower leaf surface pubescent, especially along the veins which run to the points of the marginal teeth; petioles very short or leaf almost sessile, pubescent.

TWIGS. Slender, stiffened, when young reddish brown to gray and densely pubescent, when older gray and glabrous; spines slender, straight or slightly curved, gray to black, ½–2¼ in. long; bark of trunks gray or dark brown.

RANGE. In woods or on sandy or rocky banks. Eastern Texas; eastward to Florida and Georgia, northward to Pennsylvania and New York, and westward to the Ozark region of Arkansas and Missouri.

REMARKS. The genus name, *Crataegus*, refers to the Greek word for hard wood. The species name, *uniflora*, means "one-flowered." It is also known as the Dwarf Thorn.

PEARSON HAWTHORN

Crataegus pearsonii Ashe—Series **Flavae** (7)

FIELD IDENTIFICATION. Small spreading tree to 15 ft, with rough bark and crooked recurved branches.

FLOWERS. Borne in simple, or compound, 3–5-flowered corymbs; pedicels white-hairy, ½–1 in.; corolla ⅘–1 in. wide; petals 5, white, oval to obovate, narrowed into a short claw; calyx more or less hairy at first; sepals 5, about ¼ in. long, oblong-linear, acuminate, serrulate, glandular; stamens 20, anthers yellow.

FRUIT. Early deciduous, on peduncles ½–¾ in. long, subglobose, ⅓–½ in. in diameter, bright red, flesh soft, juicy, acid; seeds 3–5, about ⅓ in. long, dorsally rounded, ventral side straight.

LEAVES. Of 2 types, smaller leaves spatulate to obovate or cuneate, apex rounded or abruptly acute, base gradually narrowed or cuneate, singly or doubly toothed on margin, teeth often glandular and glands often extending down on petiole which is ¼–⅔ in.; leaf surfaces glabrous at maturity above, or slightly puberulous beneath; leaves of young vigorous shoots often larger and almost oval, to 2¼ in. long and wide, apex rounded or abruptly acute, margin doubly serrate or some with short acute lobes as well as serrations above the middle, base almost rounded in some or abruptly narrowed, base of blade slightly decurrent on the petiole; petiole ¼–½ in., shorter in proportion to the blade

PEARSON HAWTHORN
Crataegus pearsonii Ashe

than in the smaller leaves, black glands on petiole as well as on teeth; veins impressed above and surface glabrous, lower surface with veins more prominent and mostly ending in the teeth, glabrous or slightly pubescent when mature.

TWIGS. Slender, light to dark gray or brown, mostly glabrous.

RANGE. Low sandy soil in Louisiana and Mississippi.

SYNONYM. *C. florens* Beadle.

REMARKS. The genus name, *Crataegus,* is the classical Greek name of the hawthorn.

Thicket Hawthorn

Crataegus intricata var. *straminea* (Beadle) Palmer—Series **Intricatae** (8)

FIELD IDENTIFICATION. An irregularly branched shrub to 10 ft. Branches spineless or with scattered spines 1–2 in. long. Foliage, inflorescence, and fruits glabrous.

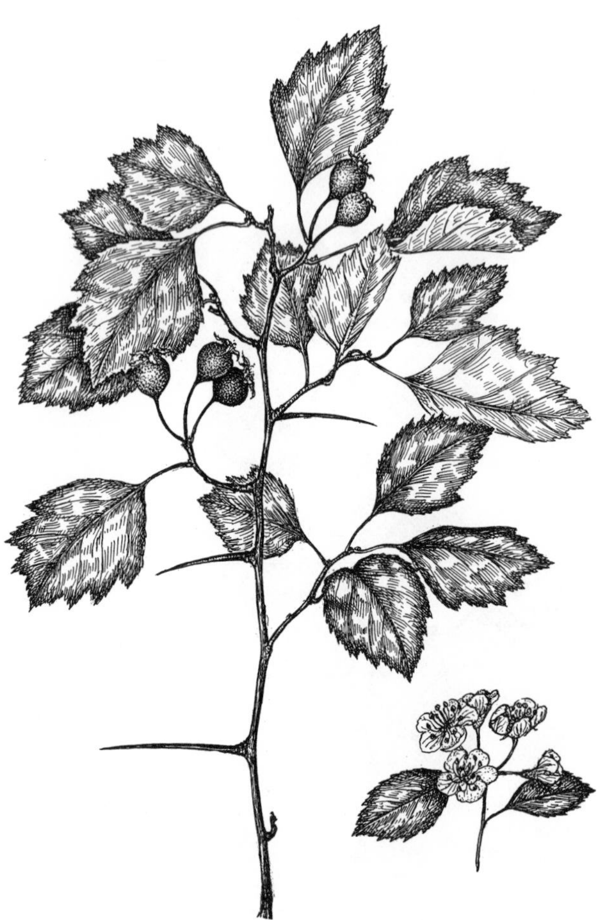

Thicket Hawthorn
Crataegus intricata var. *straminea* (Beadle) Palmer

FLOWERS. In 3–7-flowered glabrous corymbs; corolla white, 5-petaled, spreading, ⅓–⅖ in. broad, rounded, inserted on the disk margin; stamens usually 10, exserted, filaments filiform; anthers small, oblong, yellow or pink; styles 3–4, stigma terminal; ovules 1 in each carpel; calyx-tube prominent, sepals 5, lanceolate-acuminate, serrate toward the apex; bracts glandular, early deciduous.

FRUIT. Maturing in October, yellowish green, about ½ in. or less in diameter, pyriform to ellipsoid, rounded or angular, flesh hard and thick; seeds erect, flattish, testa membranous; calyx-lobes deciduous or reflexed.

LEAVES. Simple, alternate, deciduous, elliptic-ovate, ¾–2¾ in. long, ½–2 in. wide, firm, apex acute, base cuneate, margin singly or doubly serrate with teeth gland-tipped, often with 3 or 4 pairs of acute lobes toward the apex, lower pairs, or those on young shoots, more deeply cut, surfaces yellowish green.

TWIGS. Slender, rather stiff, divaricate, reddish brown to gray later, spines absent, or long and slender, straight or slightly curved, sharp, dark reddish brown to almost black, to 1½ in. long.

RANGE. On rocky hills; Arkansas and Missouri; eastward to Alabama, north to Vermont, and west to Michigan.

REMARKS. The genus name, *Crataegus,* is of Greek origin and refers to the strong wood. The species name, *intricata,* is for the intricate branching, and the variety name, *straminea,* is for the straw-colored or greenish yellow fruit. It was known at one time under the name of *C. straminea* Beadle. Also known under the vernacular name of Allegheny Thorn.

Monteer Hawthorn

Crataegus neobushii Sarg.—Series **Intricatae** (8)

FIELD IDENTIFICATION. Shrub 3–6 ft, unarmed or with short slender spines.

FLOWERS. About the middle of May, corymbs 4–5-flowered, pedicels slender, flowers about ¾ in. in diameter; bracts obovate to linear, glandular-hispid; bractlets reddish and semipersistent; calyx-tube narrowly obconic, lobes broad and abruptly acuminate-pointed with the apex entire and laciniately glandular-serrate toward the base, reflexed at maturity; petals 5, rounded; stamens 10, anthers rose-colored; styles 2–3, basally hairy.

FRUIT. Pomes borne on erect pedicels, corymbs 2–3-fruited, subglobose or short-oblong, orange-red or greenish red with dark dots, ⅜–½ in. in diameter, flesh scanty, yellowish green, hard, dry; nutlets 2–3, mostly obtuse at the ends, dorsally broad and deep-grooved, about ¼–⅓ in. long and ¼–⅕ in. wide; calyx prominent with spreading persistent lobes.

LEAVES. Simple, alternate, deciduous, ovate to rhombic, apex acuminate, base gradually narrowed or concave-cuneate and sometimes glandular; margin doubly serrate with straight or incurved glandular teeth, also usually with 3–4 pairs of broad acuminate lobes above the middle; surfaces at first soft white-hairy on the midrib and veins; at maturity dark yellowish green, semilustrous, glabrous or nearly so, lower surface paler, rather thin, blade length 1¾–2¾ in., width 1½–2 in., veins and midrib thin and yellow; petioles ⅝–1 in., slender, apex wing-margined, glabrous later, minutely glandular, green to rose-colored in autumn; leaves on young shoots varying from narrowly to broadly ovate, base cordate or less often truncate, margin coarsely serrate and some deeply 3-lobed, some as large as 2¾–3⅓ in. long and 2–2¾ in. wide, petioles broad-winged; stipules leaflike, lunate, coarsely serrate and deciduous.

TWIGS. Slender, almost straight, orange-red to brown or purplish, lustrous at first, dull later, lenticels pale, unarmed or with short slender spines.

RANGE. Arkansas and Missouri; north to Illinois and east to Pennsylvania.

REMARKS. The genus name, *Crataegus*, is the classical name of the hawthorn, and the species name, *neobushii*, is for its discoverer B. N. Bush (1858–1937), botanist of Missouri.

BUCKLEY HAWTHORN
Crataegus buckleyi Beadle

BUCKLEY HAWTHORN

Crataegus buckleyi Beadle—Series **Intricatae** (8)

FIELD IDENTIFICATION. Tree to 25 ft, with a trunk 4–7 in. in diameter. Bark gray or often dark brown, scaly. The branches are stout, spreading or ascending.

FLOWERS. About ¾ in. in diameter, on slender glabrous pedicels, in compact 3–7-flowered simple corymbs, with conspicuously glandular bracts and bractlets; petals 5, white, rounded; calyx-tube broadly obconic, glabrous, the 5 lobes broad, acuminate, laciniately cut toward the apex, and glandular with stipitate glands; stamens 10, anthers rose; styles 3–5, surrounded at base by tufts of pale hairs.

FRUIT. Maturing September–October, subglobose, usually angled, red, about ½ in. in diameter; calyx little-enlarged, with spreading or reflexed lobes; flesh thin, dry, mealy; nutlets 3–5, broad and rounded at base, rounded at the slightly narrowed apex, prominently ridged on the back, with a broad grooved ridge.

LEAVES. Broad-ovate or oval, acute, rounded or subcordate or narrowed and concave-cuneate at the entire base, coarsely and often doubly serrate above with straight glandular teeth, and more or less incisely lobed with acuminate lateral lobes, more than half-grown when the flowers open the middle of May, and then pale green and glabrous, with the exception of a few caducous hairs on the upper side of the base of the midrib, and at maturity dark green above, paler beneath, 1½–2 in. long and 1½–2 in. wide; petioles stout, conspicuously glandular above the base, wing-margined at the apex, glabrous, ½–¾ in. long.

TWIGS. Stout, glabrous, reddish brown, armed with thin, straight, shining spines ½ in. long, becoming much longer and branched on the trunk and large branches.

RANGE. On wooded slopes at altitudes of 2,000–3,000 ft. Arkansas and Tennessee, eastward to North Carolina and Virginia.

REMARKS. *Crataegus* is from a Greek word referring to the strong wood. The species name, *buckleyi*, is named in honor of Samuel Botsford Buckley (1809–1884), American botanist.

LOOKOUT MOUNTAIN HAWTHORN

Crataegus rubella Beadle—Series **Intricatae** (8)

FIELD IDENTIFICATION. Shrub 3–12 ft. Main stems solitary or several from the base, much-branched, bearing straight or slightly curved spines.

FLOWERS. April through May, appearing when the leaves are nearly grown; corymbs 3–6-flowered, glandular-bracteate; pedicels ⅜–¾ in. long; bractlets 1–3, pectinately glandular, deciduous, calyx obconic, 5-

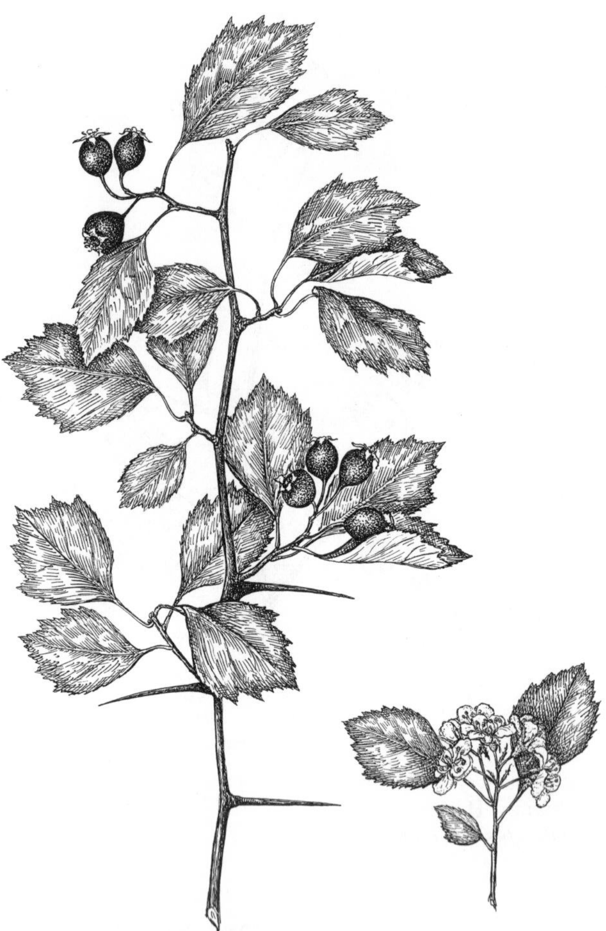

LOOKOUT MOUNTAIN HAWTHORN
Crataegus rubella Beadle

lobed, lobes 1/6–1/4 in. long, glandular-serrate; petals 5, white, rounded, rather broader than long, 1/3–1/2 in. wide, 1/4–2/5 in. long, short-clawed; stamens usually 10, 1/5–1/4 in. long, anthers light purplish; styles 2–4, occasionally 5, basally hairy.

FRUIT. Pome maturing in September, red or orange-red, pyriform or oval, 1/2–3/5 in. long, 2/5–1/2 in. wide; nutlets 2–3 (less often 4–5), hard, bony, 1/4–1/3 in. long, 1/8–1/6 in. wide, dorsally ridged and grooved, lateral faces plane.

LEAVES. Alternate, deciduous; oval, ovate, or obovate; blades 1 1/3–3 1/2 in. long, 5/8–1 3/4 in. broad, thin but firm, apex acute, base narrowed into a margined petiole, margin sharply and doubly serrate except near base, incisely lobed above the middle; young leaves reddish green or purplish at first and sparingly pubescent above, later glabrous, lower surface paler, yellowish in autumn; stipules early deciduous, oblong to linear-lanceolate, pectinately glandular; petioles margined, sparsely glandular, 3/8–1 in. long.

TWIGS. Numerous, slender, gray or reddish brown, younger smooth, older branches fissured slightly and scaly, spineless or armed with slender, straight or slightly curved chestnut-brown spines 1–1 3/4 in. long.

RANGE. The plant is also known to occur in Arkansas, Tennessee, and northern Mississippi, as well as in Alabama and North Carolina.

REMARKS. The genus name, *Crataegus,* is the classical name for the hawthorn, and the species name, *rubella,* refers to the reddish fruit. Beadle (1900) makes the following remarks regarding the plant: "*Crataegus rubella* is abundant on Lookout Mountain above Valley Head, Alabama, growing in the shade of oaks and pines, and has been collected in similar situations in eastern Tennessee and western North Carolina. It has been customary to refer this species to *C. flava, C. rotundifolia,* and *C. coccinea,* but I am inclined to place it near and compare it with *C. boyntonii* Beadle, from which it differs conspicuously in the outline of the leaves, shape and color of the fruit, and the purple color of the anthers. The type material is preserved in the Biltmore Herbarium."

CHERRY-LEAF HAWTHORN

Crataegus padifolia var. *incarnata* Sarg.—
Series **Intricatae** (8)

FIELD IDENTIFICATION. Usually a shrub or small tree 8–18 ft, with stout branches covered with dark corky bark.

FLOWERS. About 5/8 in. in diameter, petals 5, white, in small mostly 5–6-flowered corymbs, furnished with broad conspicuously glandular-serrate deciduous bracts; calyx glabrous with narrow acuminate glandular-serrate lobes often persistent on the ripe fruit; stamens 7–10; anthers pale pink; styles 2–4.

FRUIT. Ripening in early October, subglobose, dull crimson, punctate, about 5/8 in. in diameter; nutlets 3–4, rounded on the back and ends, slightly broader at the apex than at the base, 1/5–1/4 in. long.

LEAVES. Ovate, acute and short-pointed at the apex, rounded or abruptly narrowed at the base, frequently with acute double serrations, and often slightly divided into short acute lateral lobes, glabrous with the exception of a few caducous hairs on the upper side of the midrib early in the season, thin, dark green above, slightly paler below, blades 1 1/4–2 1/4 in. long and 1 1/4–2 in. wide, with a slender midrib deeply impressed on the upper side, and usually 5–6 pairs of slender primary veins; petioles slender, more or less glandular, with glands generally persistent during the season, usually almost 5/8 in. long; stipules oblanceolate, glandular-serrate, caducous.

TWIGS. Slender, erect, ascending or spreading, armed with numerous slender, straight or slightly curved, dark purple spines 1 1/2–2 in. long.

RANGE. In Arkansas near Cotter, Marion County, and

Magnet Cove, Hot Spring County. In Missouri, near Galena, Stone County.

REMARKS. The genus name, *Crataegus*, is the classical name for the hawthorn. The species name, *padifolia*, means "cherry leaf." The description is cited from Sargent's original notes (Sargent, 1925). Sargent also remarks, "The specimens of Mr. Palmer's sheet No. 3 can probably best be regarded as a variety of *C. padifolia* of the Intricatae group which occurs in this region, and from which it differs in its generally broader and usually slightly lobed leaves and its larger and softer crimson fruit."

PAGE HAWTHORN
Crataegus pagensis Sarg.

PAGE HAWTHORN

Crataegus pagensis Sarg.—Series **Intricatae** (8)

FIELD IDENTIFICATION. Thorny spreading shrub attaining a height of 9 ft. Branches slender, divaricate, and glabrous.

FLOWERS. Borne in April on slender glabrous pedicels in 4–5-flowered corymbs; bracts and bractlets oblong-ovate, acuminate, glandular, early deciduous; calyx-tube broadly obconic, glabrous, the 5 sepals gradually narrowed from a wide base; short, acute, laciniately glandular-serrate, glabrous externally, sparingly hairy on the inner surface; stamens 10, anthers rose-colored; styles 3–4; petals 5, white, rounded to obovate, flower when fully spread about 5/8 in. in diameter.

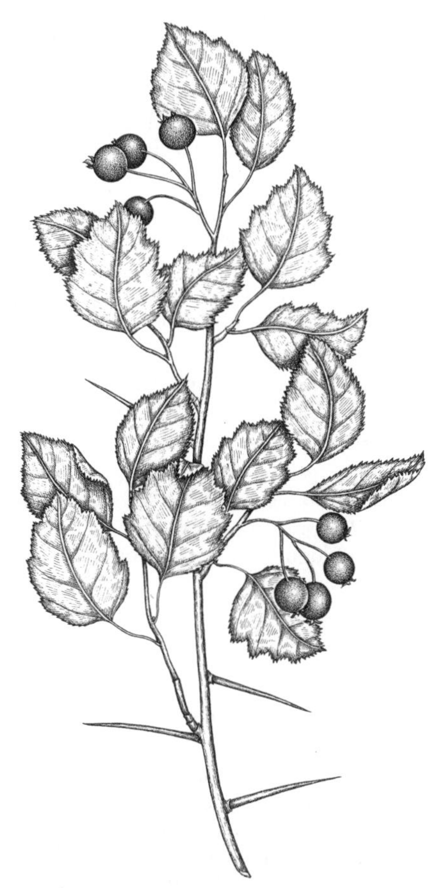

CHERRY-LEAF HAWTHORN
Crataegus padifolia var. *incarnata* Sarg.

FRUIT. Maturing in September, on short erect or spreading pedicels 1/4–3/4 in. long, clusters few-fruited, short-oblong to subglobose to slightly obovoid, dull orange-brown, about 3/8–1/2 in. long and broad, the calyx little-enlarged, with spreading lobes, and a deep terminal cavity; flesh thin and dry; nutlets 3–4, gradually narrowed and rounded at the ends, somewhat broader at the base than at apex, dorsally only slightly ridged, about 1/3 in. long and 1/5–1/4 in. wide.

LEAVES. Simple, alternate, deciduous, tinged reddish when young, slightly villose on the upper side of the midrib early in the season; mature leaves rather thin, glabrous, dark yellowish green above, lower surface paler, blades 1 1/2–2 in. long, 1 1/3–1 3/4 in. wide, midrib prominent and primary veins slender, ovate to slightly obovate or oval, apex rounded to acute, base concave-cuneate, occasionally and slightly lobed above the middle and doubly serrate with broad acute teeth; petioles 1/4–1/2 in., partially winged by the base.

TWIGS. Slender, stiff, somewhat divaricate, young ones

chestnut-brown to gray, lustrous, later gray to brown or almost black; spines 1–1½ in. long, numerous, stout or slender, nearly straight, gray to black.

RANGE. Rocky slopes and open hillsides near Page, Le Flore County, Oklahoma.

REMARKS. The genus name, *Crataegus,* refers to the hard wood of some species. The species name, *pagensis,* refers to the locality where it was found, near Page, Oklahoma. Charles Sargent noted in his original description (Sargent, 1923) that this species resembles *C. padifolia* Sarg. from Swan, Christian County, Missouri, which differs from it in its long ovate leaves and yellow anthers slightly tinged with pink, and its smaller subglobose fruit.

Harvey Hawthorn

Crataegus harveyana Sarg.—Series **Intricatae** (8)

FIELD IDENTIFICATION. Shrub 6–9 ft, with stems covered with smooth dark bark.

FLOWERS. Opening the middle of May, on stout pedicels thickly covered with long white hairs, in many-flowered villose corymbs, with large oblong-obovate sparingly villose glandular bracts and bractlets; calyx-tube broad-obconic, covered with short matted white hairs, the lobes separated by wide sinuses, acuminate, lanciniately glandular-serrate above the middle, sparingly villose on the outer surface, glabrous on the inner surface; stamens 20, anthers deep pink; styles 2 or 3.

Harvey Hawthorn
Crataegus harveyana Sarg.

FRUIT. Ripening the middle of September, subglobose, orange, about ⅜ in. in diameter, the little-enlarged calyx with a deep narrow cavity pointed in the bottom and spreading, often deciduous lobes; flesh thin and dry; nutlets usually 3, gradually narrowed and rounded at the ends, only slightly ridged on the back, ¼–⅓ in. long and ⅕ in. wide, the narrow dark hypostyle nearly to the middle.

LEAVES. Ovate, acuminate at apex, cuneate and often unsymmetrical at base, usually divided above the middle into short acute lobes, and finely double-serrate with straight gland-tipped teeth; early in the season covered above by short white hairs, and pale and villous below along the slender midrib and primary veins, and at maturity dull yellowish green and smooth on the upper surface and paler and nearly glabrous on the lower surface, 1½–2 in. long and 1¼–1½ in. wide; petioles slender, narrowly wing-margined at apex, often glandular, densely villose early in the season, becoming nearly glabrous, ⅜–¾ in. in length; leaves at end of vigorous shoots, smooth, ovate, acute, or acuminate at apex, rounded or cuneate at base, deeply lobed, coarsely serrate, 2⅓–3 in. long, and 1½–2½ in. wide, their petioles stout with broader margins and more numerous glands.

TWIGS. Slender nearly straight branchlets thinly covered early in their first season with matted pale hairs, light reddish brown and lustrous in their second season, becoming ashy gray or dark brown the following year, and armed with slender, straight, chestnut-brown spines 1⅓–2⅓ in. long.

RANGE. Near Eureka Springs, Carroll County, Arkansas. Also on road to Arkadelphia 10 miles out of Hot Springs, Arkansas.

REMARKS. The genus name, *Crataegus,* is the classical name of the hawthorn, and the species name, *harveyana,* honors Professor Le Roy Harvey, who, in 1883, published a catalogue of the trees of Arkansas.

Ouachita Hawthorn

Crataegus ouachitensis Palmer—Series **Intricatae** (8)

FIELD IDENTIFICATION. Shrub or small tree 12–18 ft, with slender ascending branches and dark gray bark, smooth on the branches and becoming scaly on old trunks. Abundantly armed with moderately stout, dark, purplish, straight or slightly curved spines ¾–1⅔ in. long.

FLOWERS. From ⅝–¾ in. in diameter, petals 5, white,

Ouachita Hawthorn
Crataegus ouachitensis Palmer

in 3–8-flowered corymbs; pedicels slender, ⅜–¾ in.; calyx-lobes lanceolate, serrate or laciniate; stamens about 20; anthers rose-colored or rose-purple; styles 3–4.

FRUIT. Obovoid or pyriform, ⅓–⅖ in. long, about ⅓ in. thick, remaining hard and green until late in the season, and finally flushed with red; calyx broad and prominent, with short tube and persistent lobes; nutlets 3–4 with shallow grooves and broad ridges on the dorsal surface, flattened or slightly concave ventrally, and with a distinct hypostyle.

LEAVES. Ovate or deltoid in outline, ¾–1¾ in. long and about as wide, sharply and deeply serrate with narrow acuminate teeth, and with 1–3 pairs of usually obscure lobes, acute or abruptly pointed at the apex, rounded or abruptly contracted at the base, tapering into the very slender petioles, ⅝–1 in.; glabrous at flowering time or with a few villose hairs on the upper surface and on the petioles; those on vigorous shoots sometimes larger, truncate or subcordate at base and more deeply lobed.

TWIGS. Slender, young ones olive-brown, older ones gray and glabrous.

RANGE. In the Ouachita Mountains in Arkansas.

VARIETY. A variety is named *C. ouachitensis* var. *minor* Palmer. The variety differs from the type in its smaller, relatively shorter leaves, which are sometimes broader than long, varying from short-ovate to broadly rhombic or elliptic in outline, less lobed, more deeply and finely serrate, and more copiously pubescent on the veins and on the extremely slender petioles, 1–1½ in. long, in the villous flowering corymbs and subglobose or slightly oblong fruit.

REMARKS. The genus name, *Crataegus,* is the classical name for the hawthorn. The species name, *ouachitensis,* is for the Ouachita Mountains of Arkansas.

Biltmore Hawthorn

Crataegus biltmoreana Beadle—Series **Intricatae** (8)

FIELD IDENTIFICATION. Branching shrub 3–15 ft.

FLOWERS. In May, when leaves are almost fully grown, corymbs simple, pubescent, 3–7-flowered; bracts caducous, lanceolate, pectinately glandular; pedicels pubescent, ⅓–1 in.; corolla 5-petaled, white, ¾–1 in. in diameter; calyx obconic, externally pubescent or tomentose; lobes 5, about ⅕ in. long, reflexed after anthesis, lanceolate, dentate or pectinate, glandular; petals broadly obovate or nearly orbicular, ⅓–½ in. long, contracted near the base into short claws; stamens

Biltmore Hawthorn
Crataegus biltmoreana Beadle

10, shorter than the petals, anthers pale yellow; styles 3–5, shorter than the stamens, basally pale-hairy.

FRUIT. Ripening in September or early October, greenish yellow, yellow or orange, sometimes with a red cheek; nutlets 3–5, ⅖–⅗ in. broad, ⅖–½ in. high, depressed-globose, bluntly angled, the cavity broad, surrounded by the calyx-lobes and portions of the filaments; nutlets dorsally grooved with a prominent ridge near the middle, the inner faces being nearly plane.

LEAVES. Simple, alternate, deciduous, ¾–2½ in. wide, 1¼–4 in. long including the petiole, or occasionally longer, ovate or round-ovate, apex acute, base usually wedge-shaped but some rounded, truncate or subcordate, margin sharply and irregularly serrate to acutely incised or slightly 5–9-lobed, rather thin at first but firmer later, young leaves pubescent above and beneath, mature leaves bright green above, paler beneath with 4–6 pairs of veins.

TWIGS. Reddish brown to gray, lenticels small and pale; buds globular, bright reddish brown; spines stout, ¾–2 in. long, slightly curved, brown to gray.

BARK. On trunk reddish brown, slightly fissured and broken into many small, persistent, ashy gray scales.

RANGE. Dry or rocky woodlands. Arkansas and Missouri; eastward to North Carolina and north to Vermont.

REMARKS. The genus name, *Crataegus*, is the classical name for hawthorn. The species name, *biltmoreana*, is for Biltmore, North Carolina. The type material was deposited in the herbarium at Biltmore University.

The description above cited in part from C. D. Beadle's descriptive original notes (Beadle, 1899). Professor Beadle notes that the Biltmore Hawthorn may be distinguished from *C. mollis* (Torr. & Gray) Scheele by its smaller size, simple corymbs, later time of blossoming, and by the color and texture of the fruit. It is sometimes listed in the literature under the name of *C. villicarpa* Sarg.

SUTHERLAND HAWTHORN

Crataegus sutherlandensis Sarg.—Series **Virides** (9)

FIELD IDENTIFICATION. Slender tree attaining a height of 15 ft. Bark dark gray, breaking into long, thin, oblong flakes to expose the reddish brown inner bark. The branches spreading, smooth and gray.

FLOWERS. Opening in March in lax 7–10-flowered corymbs, the pedicels slender and glabrous; calyx-tube narrow-obconic, glabrous, the 5 sepals slender, acuminate, often laciniately divided near the base into glandular teeth, glabrous on the outer surface, villose-pubescent inwardly; stamens 20, anthers faintly tinged with pink; styles 5; petals 5, white, rounded, entire corolla about ¾ in. across.

FRUIT. Maturing September–October, subglobose, base often truncate, orange-red, about ⅓ in. in diameter, calyx prominent, the 5 sepals erect or spreading and a narrow deep cavity; nutlets 5, apex rounded, base acute, obscurely grooved on dorsal side, ⅕–¼ in. long, ⅛–⅙ in. wide.

LEAVES. Simple, alternate, deciduous, the young ones coated with pale pubescence, maturing when the flowers open and then thin, upper surface dull yellowish green and glabrous, lower slightly paler, blades 1¼–1½ in. long, ¾–1 in. wide, midrib thin; blades ovate, apex acute, base gradually or abruptly narrowed and concave-cuneate, margin serrate with straight or incurved acuminate teeth; petioles slender, at first villose, later glabrous, ⅓–⅗ in.; leaves on young shoots irregularly divided into short wide lateral lobes, 1½–2 in. long.

TWIGS. Slender, slightly divaricate, green to orange, sparingly hairy at first, older ones reddish brown to gray and glabrous, set with occasional straight slender spines to 2 in. long.

RANGE. The species is found in south central Texas,

SUTHERLAND HAWTHORN
Crataegus sutherlandensis Sarg.

SPINY SUTHERLAND HAWTHORN
Crataegus sutherlandensis var. *spinescens* Sarg.

in Wilson County, on the Cibolo River near Sutherland Springs.

VARIETY. Spiny Sutherland Hawthorn, *C. sutherlandensis* var. *spinescens* Sarg., differs by the rather smaller leaves more pubescent early in the season, often with tufts of white axillary hairs, and spines larger and more numerous. It is a shrub or small tree to 15 ft, forming thickets of slender stems covered with dark scaly bark separating into narrow scales, slender zigzag twigs armed with many slender straight spines 1¼–2½ in. Found in the same range as the species.

REMARKS. The genus name, *Crataegus*, is an ancient name and refers to the hard wood. The species name, *sutherlandensis,* is named for Sutherland Springs, Texas, where this species was discovered.

GLOSSY HAWTHORN

Crataegus nitida Sarg.—Series **Virides** (9)

FIELD IDENTIFICATION. Tree attaining a height of 30 ft, with a rather tall trunk to 1½ ft in diameter. The branches stout and irregularly spreading to form a broad crown.

FLOWERS. Appearing in May, corymbs compound, broad, glabrous; corolla white, 5-petaled, about ¾ in. in diameter; calyx-tube narrowly obconic, lobes 5, glabrous, apex long-acuminate, margin entire or sparingly glandular-serrate; stamens 15–20, anthers pale yellow; styles 2–5.

FRUIT. Maturing in October, corymbs drooping, fruit numerous, short-oblong, heavily glaucous, often dark-dotted, ½–⅝ in. long and about ⅓ in. broad; fruiting calyx with lobes green or reddish, often caducous; nutlets 2–5, dorsally ridged, hardly over ¼ in. long.

LEAVES. Simple, alternate, deciduous, lanceolate to oblong or obovate, apex acuminate, base cuneate, margin coarsely serrate with glandular teeth except near the entire base, sometimes with 2–3 acute lobes also; young leaves reddish and hairy along the veins at first, mature leaves thick and leathery, upper surface dark green and shiny, lower surface paler and duller, in autumn turning red or orange, length 2–3 in., width 1–1½ in.; leaves on young shoots often larger and more conspicuously lobed; petiole ¼–¾ in., somewhat glandular, hairy at first, glabrous later.

TWIGS. Slender, smooth, glabrous, shiny orange-brown to reddish brown, at maturity gray and bearing shiny brown spines 1–1½ in.

RANGE. Rich bottom lands of the Mississippi drainage; Arkansas, Missouri, Illinois and Ohio.

REMARKS. The genus name, *Crataegus,* is the classical name, and the species name, *nitida,* refers to the glossy shining leaves. It is also known as the Shining Thorn.

GLOSSY HAWTHORN
Crataegus nitida Sarg.

Smooth Western Hawthorn

Crataegus glabriuscula forma *desertorum* Sarg.—
Series **Virides** (9)

FIELD IDENTIFICATION. Shrub to 9 ft. Bark on the branches thin, pale, and flaky.

FLOWERS. Appearing in April in small glabrous, 4–5-flowered corymbs; petals 5, white, oval to obovate; calyx broad-obconic, slightly villose, with the 5 sepals slender, acuminate, obscurely serrate, externally glabrous, inwardly villose; stamens 20, anthers pale yellow; styles 4–5.

FRUIT. Maturing in October, orange-red, subglobose, 1/6–1/5 in. in diameter, flesh dry and thin; calyx-lobes with 5 sepals erect or spreading, with a wide shallow cavity in the bottom; nutlets 4–5, rounded and rather broader at apex than at base, dorsally only slightly grooved, about 1/6 in. long, 1/10–1/8 in. wide.

LEAVES. Simple, alternate, deciduous, young with shiny white hairs, mature leaves thin, upper surface yellowish green and slightly roughened above by short white hairs, lower surface pale and glabrous or slightly villose toward the base of the prominent midrib below, with 3–4 pairs of primary veins extending to the points of the lobes, or 3-nerved from the base, ovate to slightly obovate or suborbicular, apex acute, acuminate, or rounded, base gradually or abruptly narrowed and cuneate, margin finely double-serrate usually above the middle, with blunt glandular teeth, and often divided into short acute lobes, 5/8–7/8 in. long, 3/8–5/8 in. wide; leaves on young shoots ovate, base broad and rounded, apex acute, often deeply lobed, 3/4–1 in. long and wide.

Smooth Western Hawthorn
Crataegus glabriuscula forma *desertorum* Sarg.

TWIGS. Slender, usually divaricate, at first reddish brown and slightly villose, later pale gray, armed with numerous thorns which are slender, straight or slightly curved, brown to gray, 1–2 in. long.

RANGE. In the rocky bed of a creek usually dry, but flooded during a few hours two or three times a year, near Uvalde, Texas.

REMARKS. The genus name, *Crataegus,* is the ancient name referring to the hard wood. The variety name, *glabriuscula,* means "somewhat smooth," and the form name, *desertorum,* is for its dry habitat. The following comments concerning it have been made by Sargent (1922c): "In its unusually zigzag branches, numerous long slender spines and minute fruit this is perhaps the most distinct hawthorn of the Virides group. The fact that it inhabits a region of rare rainfall where the soil in which it grows is only thoroughly wet two or three times in the year would be remarkable for any species of *Crataegus;* it is most remarkable for a species of this group, for the Virides, growing usually in low ground, are moisture loving plants. It is unfortunate that E. J. Palmer has been able to find only a single plant."

Since Sargent wrote the above the plant has been relegated to the status of a form only, under *C. glabriuscula,* instead of a species listed as *C. desertorum* Sarg.

Green Hawthorn

Crataegus viridis L.—Series **Virides** (9)

FIELD IDENTIFICATION. Tree attaining a height of 35 ft and forming a broad rounded crown. The trunk is often fluted, and the twigs are sparsely spined, or not at all.

FLOWERS. Opening March–April. Borne in many-flowered, glabrous corymbs; individual flowers on slender pedicels, about 3/4 in. in diameter; petals 5, rounded, spreading, inserted on the disk margin in the calyx-throat; stamens 15–20, anthers yellow; styles 2–5, hairy at the base; calyx-tube obconic, glabrous, with lobes entire, lanceolate, glabrous or puberulent, or pubescent to villose in some varieties.

FRUIT. Pome globose or depressed-globose, in drooping clusters, red to orange and often glaucous (sometimes remaining greenish), 1/6–1/4 in. in diameter; calyx-lobes 5, small, often dropping away early from the fruit; nutlets 4–5, obscurely ridged or grooved on the back, 1/6–1/8 in. long.

LEAVES. Simple, alternate, deciduous, ovate-oblong, acute to acuminate or rarely obtuse at the apex, cuneate or rounded at the base; serrate to doubly serrate and often shallowly lobed toward the apex, teeth usu-

GREEN HAWTHORN
Crataegus viridis L.

ally few or none at the cuneate base; dark green and shiny above and becoming glabrous later, paler beneath with axillary tufts of white hairs, blades ¾–3½ in. long, ½–2 in. wide; petiole slender and glabrous.

TWIGS. Gray or reddish, with or without spines; spines when present slender, pale, sharp, ¼–1 in.

BARK. Gray to reddish brown, shedding in small scales.

WOOD. Reddish brown, heavy, hard, tough, weighing about 40 lb per cu ft, not large enough to be commercially valuable.

RANGE. The species is known from Texas in the eastern and upper coastal regions; Louisiana, Arkansas, and eastern Oklahoma; eastward to Florida, northward to southeastern Virginia, and westward to southwestern Indiana, Kansas, and Missouri.

VARIETIES. Green Haw varies considerably over its wide range. The following varieties have been segregated:

Lance-leaf Green Hawthorn, *C. viridis* var. *lanceolata* (Sarg.) Palmer, has leaves of flowering branches oblong-elliptic to lance-elliptic, and on vegetative twigs ovate to oblong-ovate and sharply lobed. Known in Louisiana, Arkansas, Missouri, Mississippi, and Alabama.

Ovate-leaf Green Hawthorn, *C. viridis* var. *ovata* (Sarg.) Palmer, has ovate to oblong-ovate leaves with apices obtuse or rounded, base rounded or abruptly narrowed; those on vegetative shoots sometimes orbicular and only slightly lobed. Known from Arkansas and Missouri to North Carolina.

Mudbank Green Hawthorn, *C. viridis* var. *lutensis* (Sarg.) Palmer, grows on muddy margins of streams and swamps. Foliage of flowering branches oblong-obovate or elliptic, sharply and irregularly serrate almost to the base, terminal leaves on young shoots often broadly ovate with 2–3 pairs of spreading lobes. Known from Oklahoma, western Missouri, and Kansas.

Interior Green Hawthorn, *C. viridis* var. *interior* Beadle, is a small tree to 25 ft, with dark gray flaky bark. Leaf blades ovate, ovate-lanceolate or oblong, ¾–2½ in. long, ⅜–1¾ in. wide, glabrous or glabrate when mature, apices acute or acuminate, base rounded or abruptly contracted, margins serrate and deeply incised; petiole margined, slightly hairy on the upper surface, at least when young; corymbs compound, many-flowered, glabrous; pedicels, hypanthium and exterior surface of the sepals glabrous; corolla about ⅗ in. wide; stamens usually 20, the anthers yellow; fruit globose, bright red at maturity, about ⅓ in. wide; nutlets 3–5, about ⅕ in. long, the hypostyle occupying about two thirds of the ventral angle. Known from flat woodlands in Arkansas and Tennessee.

REMARKS. The genus name, *Crataegus*, is from an ancient Greek word meaning "strength," in reference to the tough wood, and the species name, *viridis*, means "green." Vernacular names are Green Thorn and Southern Thorn.

VELVETY GREEN HAWTHORN

Crataegus viridis var. *velutina* (Sarg.) Palmer—
Series **Virides** (9)

FIELD IDENTIFICATION. Tree attaining a height of 20–25 ft, with a trunk 8–10 in. in diameter covered with rough, dark, scaly bark.

FLOWERS. Opening the latter part of April, borne in hairy 7–12-flowered corymbs with slender villose pedicels; calyx-tube narrowly obconic, villose, the 5 lobes gradually narrowed from a broad base, short, acute, entire, slightly villose; stamens 20, anthers yellow; styles 5; petals 5, rounded, white.

FRUIT. Borne in slender, glabrous, few-fruited clusters, fruit body subglobose, orange-red, pale-dotted, about ¼ in. in diameter; calyx prominent, with a deep narrow cavity pointed inward from the bottom, and closely appressed lobes; flesh thin, dry and mealy; nutlets 5, acute at base, rounded, dorsally ridged and grooved, about ⅙ in. long and ⅛ in. wide.

LEAVES. Simple, alternate, deciduous, blade ovate to obovate, apex acute or rounded, base entire and gradually narrowed and cuneate, margin sharply and often doubly serrate with straight glandular teeth; half-grown when the flowers open, young ones covered with short white hairs above, hoary-pubescent below, and often with tufts of white tomentum; mature ones glabrous, smooth and lustrous above, and lower surface with pale matted hairs, length 1¾ in., width 1½–2 in.; petioles slender, thickly covered at first with matted hairs, later glabrous, length ½–1 in.

TWIGS. Slender, at first hoary-tomentose and light reddish brown; later becoming marked with lenticels and pubescent to glabrous, becoming gray, bearing straight chestnut-brown spines ¾–1½ in. long.

RANGE. Usually in dry sandy soils, but also on clays. In Texas in the valley of the lower Brazos River (Brazoria County). Also from eastern Texas to Louisiana, and Arkansas. In Arkansas in Hempstead and Bowie counties.

REMARKS. The genus name, *Crataegus,* refers to the strong wood, and the variety name, *velutina,* is for the velvety tomentum on young leaves.

Small-leaf Green Hawthorn

Crataegus viridis forma *abbreviata* (Sarg.) Palmer—Series **Virides** (9)

FIELD IDENTIFICATION. Shrub or small tree to 18 ft, with a slender trunk.

FLOWERS. Opening in April, corymbs slightly villose, compact, 10–15-flowered; petals 5, white, rounded; calyx-tube broad-obconic, slightly villose, the 5 sepals short, entire, often slightly villose or externally glabrous, villose within; stamens 20, anthers yellow, styles 4–5.

FRUIT. Maturing in October, in lax drooping clusters with slightly villose or glabrous pedicels, subglobose, dark red, ¼–⅓ in. in diameter, flesh thin and succulent, calyx little enlarged, with a deep cavity in the bottom; nutlets 4–5, rounded at the ends, broader at apex than at base, slightly grooved dorsally, ⅛–⅙ in. long.

LEAVES. Simple, alternate, deciduous, at first with short white hairs and dense tomentum below, expanding when the flowers open and then glabrous or slightly hairy along the midrib, or with a few axillary tufts below, yellowish green, thin, ovate to obovate, elliptic or suborbicular, apex acute or rounded and abruptly short-pointed, base narrowed and cuneate or rounded, margin often glandular-serrate, usually only above the middle, and often slightly divided usually toward the apex into short acute lobes, blade length 1–1⅓ in., ¾–1 in. wide; petiole slender, slightly villose-pubescent early in the season, soon becoming glabrous; leaves on young shoots broad-ovate to semiorbicular or elliptic, apex rounded or acuminate, base rounded or cuneate, often laterally lobed, 1½–2 in. long and broad.

TWIGS. Slightly villose at first, later glabrous and orange-brown, eventually gray to brown; unarmed, or with spines slender, straight, 1–1¾ in.

RANGE. Low woods on the Brazos River, near Brazoria, Brazos County, Texas.

REMARKS. The genus name, *Crataegus,* is the ancient name referring to the hard wood. The species name *viridis,* means "green," and the form name, *abbreviata* is for the short leaf. Sargent (1922b) remarks as follows:

"Although there is little in the flowers and fruit or in the habit of this plant to distinguish it from some of the other *Virides* species which grow in the valley of the lower Brazos River, where this group is represented by its greatest diversity of forms, the short leaves are so distinct in shape that until the Texas species are better known it appears necessary to treat it as a species."

Since the above was written by Sargent the species standing has been converted to that of a form by E. J. Palmer (*C. viridis* forma *abbreviata* (Sarg.) Palmer).

Thornless Green Hawthorn

Crataegus viridis forma *micracantha* (Sarg.) Palmer—Series **Virides** (9)

FIELD IDENTIFICATION. Tree with no spines or, rarely, a few small spines, attaining a height of 25 ft, with a diameter of 8–12 in. The spreading stout branches form a broad flat-topped head.

FLOWERS. Opening about the middle of May when the leaves are half-grown, cup-shaped, only about ¼ in. in diameter; petals 5, white, rounded; pedicels slender, with white matted hairs; corymbs broad, lax, many-flowered: calyx-tube narrowly obconic, villose, tipped with minute dark glands; stamens usually 10, occasionally 12, 15, or 20; anthers small, deep bright red; styles 5.

FRUIT. Maturing in October, on slender pubescent pedicels, in drooping many-fruited clusters, subglobose to short-oblong, full and rounded at the ends, bright orange-red, lustrous, about ¼ in. long, marked with large pale dots; calyx prominent with a short villose tube, the 5 hairy lobes often deciduous; nutlets 5, thin, ends acute, dorsally rounded and slightly grooved, about 3/16 in. long.

LEAVES. Simple, alternate, deciduous, blade outline oblong to obovate or oval, apex acute to acuminate or rarely rounded, gradually narrowed from about the middle to the entire cuneate base, coarsely crenulate-serrate, and occasionally 3-lobed above with short, broad, acute, lateral lobes; young leaves villose above and hoary-tomentose beneath; mature leaves thin but firm, dark yellowish green, lustrous and smooth above, paler below and tomentose on the midrib and venation, blade length 2–2½ in., width 1–1¼ in.; petiole slender, tomentose at first but becoming glabrous or pubescent; leaves on vigorous shoots often broadly rhombic to obovate, apex acuminate, frequently deeply 3-lobed or divided into 2–3 pairs of short lateral lobes, 2½–3 in. long.

TWIGS. Slender, straight, coated until after the flowering time with thick hoary tomentum, later becoming reddish brown and puberulent, eventually light to dark, dull reddish brown.

RANGE. Known from low rich woods in moist soil near Fulton, Hempstead County, Arkansas.

REMARKS. The genus name, *Crataegus*, refers to the strong wood, and the species name, *viridis*, means "green." The variety name, *micracantha*, is for "small thorns," but usually the tree is thornless.

COAST HAWTHORN

Crataegus anamesa Sarg.—Series **Virides** (9)

FIELD IDENTIFICATION. Shrub 12–15 ft, with trunks covered with gray slightly scaly bark. Branches erect, smooth, dark gray.

FLOWERS. Appearing at the end of March or early in April, ¾–1 in. in diameter, in compact mostly 10–15-flowered densely villose corymbs crowded on the branches; petals 5, white, rounded; calyx-tube narrow-obconic, glabrous except for occasional short white hairs, the lobes gradually narrowed from the base, entire or rarely minute-dentate, glabrous on the outer surface, villose on the inner surface, mostly deciduous from the ripe fruit; stamens 20, anthers pale yellow; styles 4 or 5.

FRUIT. Ripening early in October, on nearly glabrous pedicels, in few-fruited drooping clusters, subglobose to short-oblong or slightly obovoid, dark red, ⅓–⅖ in. in diameter, the calyx little enlarged, with a deep narrow cavity directed inward from the apex; nutlets 4 or 5, narrowed and rounded at the ends, only slightly grooved on the back, about ⅕ in. long, the dark narrow hypostyle extending to below the middle.

LEAVES. Elliptic to broad-ovate or slightly obovate, acute at apex, gradually or abruptly narrowed and cuneate at base, finely serrate with short broad teeth, and often slightly divided above the middle into broad rounded lobes; tinged with red as they unfold, and villose above and thickly covered below with matted white hairs. Not more than half-grown when the flowers open, and then glabrous or nearly glabrous above, sparingly villose and conspicuous below by the broad snow-white to dark gray band; apparently spineless. White hairs along the lower part of the midrib, and at maturity subcoriaceous, nearly glabrous, dark green and lustrous on the upper surface, pale on the lower surface, 1¼–2 in. long or wide; petioles slender, densely villose early in the season, becoming glabrous, ⅝–¾ in. long; leaves on vigorous shoots broad-ovate, rounded or acute at apex, rounded or broad-cuneate at base, finely serrate, slightly lobed with short broad lobes, to ¼ in. long and wide, their petioles stout, slightly wing-margined at apex, often furnished with occasional glands, 1–1⅓ in. in length.

TWIGS. Slightly zigzag, young ones densely white-hairy, later glabrous and reddish brown to dark gray, usually spineless, or occasionally a few spines.

RANGE. Fort Bend County, Texas.

REMARKS. The genus name, *Crataegus*, is the classical name of the hawthorn. The species name, *anamesa*, means "middle" or "intermediate" and refers to the fruit size. The size of the fruit of this species is intermediate between that of typical *C. viridis* L. and that of a small group of species with fruit ⅝–¾ in. in diameter of which *C. nitida* Sargent is the best known. Although much more pubescent, this Texas shrub resembles in the shape of its leaves another of the large-fruited Virides species, *C. atrorubens* Ashe of East St. Louis, Illinois. *C. anamesa*, has also been listed under the name of *C. antiplasta* Sargent.

COAST HAWTHORN
Crataegus anamesa Sarg.

NARROW-SEPAL HAWTHORN

Crataegus stenosepala Sarg.—Series **Virides** (9)

FIELD IDENTIFICATION. Shrub or small tree 12–15 ft, with stems forming large thickets, and covered with dark slightly scaly bark.

FLOWERS. Opening toward the end of March, ¾ in. in diameter, in wide loose 10–20-flowered, slightly villose corymbs; petals 5, white, rounded; calyx-tube broad-obconic, sparingly covered with long white ridged hairs, the lobes gradually narrowed from the base, slender, long-acuminate, minutely and irregularly serrate,

Narrow-sepal Hawthorn
Crataegus stenosepala Sarg.

glandular ciliate, glabrous on the outer surface, obscurely ciliate on the inner surface, ¼–⅓ in. long; stamens 20, anthers pale yellow; styles 5.

FRUIT. Ripening early in October, ellipsoidal to slightly obovoid, on slender glabrous pedicels in drooping clusters, orange-red, ¼–⅓ in. long, ⅕–¼ in. thick, with thin dry flesh, the calyx with a distinct tube, spreading lobes and a deep narrow cavity pointed in from the apex; nutlets 4 or 5, rounded at base, acute at apex, only slightly grooved on the back, ¼–⅓ in. long and ⅛–⅙ in. wide, the pale broad hypostyle extending to the middle.

LEAVES. Elliptic to oblong-elliptic or obovate, acute or acuminate at apex, gradually narrowed and cuneate at base, sharply and coarsely serrate above the middle with straight teeth, and often divided toward the apex into short lobes; when they unfold deeply tinged with red and slightly pubescent, nearly half-grown when the flowers open and then roughened above by short white hairs and conspicuous below by the thick snow-white pubescence along the midrib and on the petioles, the villose primary veins, and the axillary clusters of white hairs; and at maturity glabrous, yellowish green and lustrous on the upper surface, paler on the lower surface, blades 1½–2¼ in. long and ⅝–1¼ in. wide with a thin midrib and slender primary veins impressed above; petioles ¼–⅓ in. in length; leaves on vigorous shoots oblong-obovate, thicker, acuminate, cuneate at base, more coarsely serrate, more deeply lobed, and to 2¾ in. in length and 1½ in. wide.

TWIGS. Slender, slightly zigzag, white-hairy at first, becoming glabrous; at first reddish brown, later ashy gray and glabrous; spines numerous, nearly straight, slender, ¾–1⅔ in. long.

RANGE. Specimens collected by E. J. Palmer in Fort Bend County, Texas. Type specimen deposited in the herbarium of the Missouri Botanical Garden.

SYNONYM. *C. stenosepala* is sometimes listed under the name of *C. antemima* Sargent.

REMARKS. The genus name, *Crataegus,* is the classical name of the hawthorn, and the species name, *stenosepala,* refers to the remarkably long slender calyx-lobes. The leaves also are conspicuous when the tree flowers, owing to the broad band of snow-white tomentum covering the under side of the lower half of the midrib.

Rosemary Hawthorn
Crataegus poliophylla Sarg.—Series **Virides** (9)

FIELD IDENTIFICATION. Tree occasionally 12–15 ft, with a trunk 3–4 in. in diameter, covered with dark rough bark, smooth ashy gray branches.

FLOWERS. Opening late in March or early in April, about ⅝ in. in diameter, in wide lax 7–15-flowered densely villose corymbs; calyx-tube broad-obconic, villose, the lobes short, gradually narrowed from the base, glandular-serrate or nearly entire, glabrous on the outer surface, slightly villose on the inner surface; stamens 20, anthers yellow; styles 4–5; petioles 5, white, rounded.

FRUIT. Ripening late in September, in pendent clusters, globose to short-oblong or ovoid, orange-red, ¼–⅓ in. in diameter, the calyx prominent, with a short tube, reflexed lobes and a wide shallow cavity at the apex; nutlets 4 or 5, rounded at apex, gradually narrowed at base, slightly grooved on the back, ⅛–⅙ in. long, 1/12–⅛ in. wide, the narrow hypostyle extending to the middle.

LEAVES. Oblong-obovate to elliptic, acute to acuminate at apex, gradually narrowed and cuneate at base, finely double-serrate above the middle with straight teeth and usually irregularly divided toward the apex into short acute lobes; thickly covered when they unfold with white hairs longer and more abundant on the lower than on the upper surface, nearly glabrous above when the flowers open and more or less pubescent below, and at maturity subcoriaceous, glabrous, yellowish green and lustrous on the upper surface, paler on the lower surface, blades 1¼–1½ in. long and 1–1¼ in. wide, with a prominent midrib and slender veins deeply impressed above; on leading shoots to 2½ in. long and 1¾ in. wide; petioles slender, deeply grooved, narrowly wing-margined toward the apex, densely villose-pubescent early in the season, becoming glabrous, ⅜–¾ in. long.

TWIGS. Slender, thickly covered early in the season with long matted white hairs, becoming glabrous and ashy gray, and armed with slender straight spines ⅝–1 in. in length.

RANGE. Rare and local, Brazoria and Ft. Bend counties, Texas. Usually in thickets in drained soil.

REMARKS. The genus name, *Crataegus*, is the classical name of the hawthorn, and the species name, *poliophylla*, is from the "rosemary-like," or gray and woolly leaves. Sargent (1921) remarks as follows: "Distinct in the shape of the coriaceous leaves and in their villose covering while young, and in the villose corymbs. Two specimens collected at West Columbia, Texas, by B. F. Bush (No. 971), October 3, 1901, with thicker and more lustrous broad ovate leaves up to 3 in. long and 2½ in. wide, rather larger fruited, and more zigzag branchlets, probably represent an extreme form of this species. Much land has been cleared in the neighborhood of Columbia in recent years and this tree has probably disappeared as various attempts to rediscover it have failed."

ROSEMARY HAWTHORN
Crataegus poliophylla Sarg.

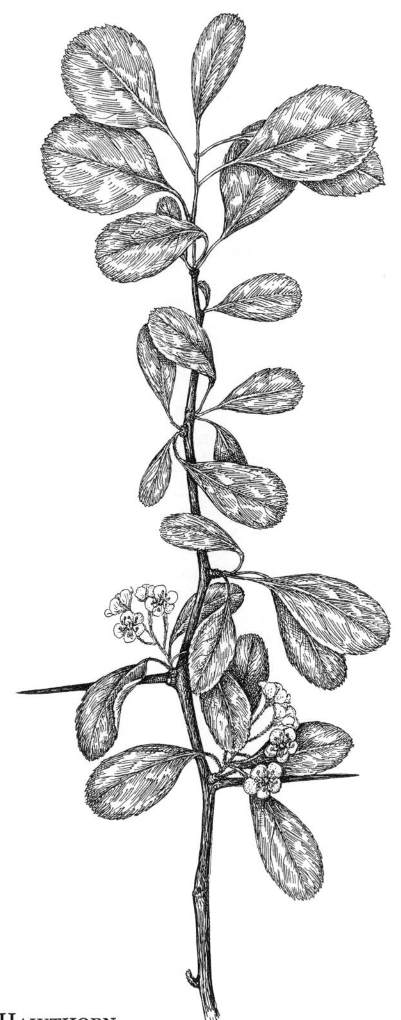

FULTON HAWTHORN
Crataegus amicalis Sarg.

FULTON HAWTHORN

Crataegus amicalis Sarg.—Series **Virides** (9)

FIELD IDENTIFICATION. Tree 15–25 ft, with a trunk to 10 in. in diameter.

FLOWERS. Borne in April, after the leaves are almost fully grown, in pubescent, wide, 5–12-flowered, slender-pediceled corymbs; flowers ⅝–¾ in. in diameter, with 5 white rounded petals; calyx-tube broadly obconic, hairy at base, the 5 lobes short, gradually acuminate at apex, margin entire or minutely glandular-dentate, slightly hairy outwardly, villose on the inner surface, reflexed after anthesis; stamens 20, anthers yellow; styles 5, basally hairy.

FRUIT. Maturing in October, on slender drooping pedicels, subglobose, dull red, ¼–⅓ in. in diameter; flesh dry and hard; nutlets 5, apices rounded, dorsally ridged with a narrow, low deeply grooved ridge, ⅕–¼ in. long and about ⅛ in. wide; fruiting calyx prominent, with spreading lobes and a deep broad cavity at the apex.

LEAVES. Simple, alternate, deciduous, oblong to obovate, apex rounded or abruptly acute, base cuneate and often unequal, margin sharply and doubly serrate above the middle with straight glandular teeth, when young pale yellowish green with short white hairs, glaucescent and villose beneath; when mature yellowish green, with a thick and firm texture; upper surface smooth, glabrous and lustrous; lower surface paler and somewhat hairy on the venation; blade length 1–1⅓ in., width ¾–1 in.; petioles about ⅜ in. long, stout, wing-margined to below the middle, densely hairy at first, becoming glabrous; leaves on young shoots oval to oblong-obovate, apex acute, some lobed above the middle and more coarsely serrate, to 2 in. long and 1⅓ in. wide.

RANGE. Known from upland woods near Fulton, Hempstead County, Arkansas.

REMARKS. The genus name, *Crataegus*, is the classical name for the hawthorn; the meaning of the species name, *amicalis*, is obscure.

COCK'S-SPUR HAWTHORN
Crataegus crus-galli L.

COCK'S-SPUR HAWTHORN

Crataegus crus-galli L.—Series **Crus-galli** (10)

FIELD IDENTIFICATION. Shrub or tree to 30 ft, and 6–12 in. in diameter. The branches are stout, rigid, horizontal or drooping, forming a round-topped or broadly depressed crown.

FLOWERS. Opening May–June after the leaves, in lax, many-flowered, glabrous corymbs; pedicels slender and glabrous, corolla ½–⅗ in. broad, petals 5, white, reflexed after anthesis; calyx-tube narrow and obconic, glabrous, sepals 5, ⅛–⅕ in. long, linear-lanceolate, entire or glandular-serrate; stamens 10, anthers pink or white; styles 2, hairy at the base.

FRUIT. Maturing in October, persistent over winter, short-oblong to subglobose or ovoid (occasionally slightly 5-angled), dull red, ⅓–½ in. long, with a terminal depression, flesh thin and dry; nutlets 2 (rarely 1 or 3), ridged dorsally, ends rounded, about ¼ in. long.

LEAVES. Simple, alternate, deciduous, thick and leathery at maturity, mostly obovate to oblanceolate, apex obtuse to rounded or acute, base gradually cuneate, margin sharply and minutely toothed above the middle, teeth often glandular; upper surface dark green and lustrous, lower surface paler and reticulate-veined, blade length ⅗–4 in., width ½–1⅓ in., turning yellow, orange, or red in the fall; petiole ½–¾ in. long, stout, winged above; leaves on young shoots often longer and apex acute or acuminate.

TWIGS. Stout, reddish brown to gray, glabrous, armed with sharp, straight or slightly curved spines 2–8 in. long, sometimes with lateral spines; bark of trunk dark brown to gray, breaking into small scales with irregular, moderately deep fissures.

WOOD. Heavy, hard, fine-grained, suitable for tool handles.

RANGE. Fence rows, woods, and thickets. The species and varieties are widespread. East Texas, Oklahoma, Arkansas, and Louisiana; east to Georgia, north to Michigan, Kansas, and southern Quebec, and west to Ontario.

VARIETIES. The tree is very variable over its wide area of distribution and has been separated by botanists into many varieties as follows:

C. crus-galli var. *bellica* (Sarg.) Palmer has very thorny branches and thickly crowded leaves less than 1¼ in. long and ¾ in. wide. It is known from southern Missouri, Arkansas, and Texas.

C. crus-galli var. *exigua* (Sarg.) Egglest. has fruit bright red, oblong or elipsoidal, ⅜–⅝ in. long, nutlets usually 1; leaves on vegetative shoots sometimes slightly lobed. Known from Connecticut, Georgia, Missouri, and Arkansas.

C. crus-galli var. *macra* (Beadle) Palmer has fruit about ⅓ in. thick, subglobose or short-oblong, nutlets 1–2. Known from Georgia, southern Missouri, and Arkansas.

C. crus-galli var. *barrettiana* (Sarg.) Palmer has leaves of flowering spurs duller, apex more rounded, and texture less coriaceous. Known from Illinois, Indiana, Missouri, and perhaps Arkansas also.

C. crus-galli var. *leptophylla* (Sarg.) Palmer has inflorescential leaves 1¼–2½ in. long, ⅔–1¼ in. wide; stamens more numerous than in the species; nutlets 3–4, rarely 5. Known from southern Missouri and Arkansas.

REMARKS. The genus name, *Crataegus*, is the classical Greek name for hawthorn. The species name, *crus-galli*, refers to the long thorns which resemble the spurs of a fowl. Cock's-spur Hawthorn is very desirable for cultivation because of the rounded crown, shiny leaves, and conspicuous flowers. It is perhaps the most widely planted hawthorn in the United States and Europe.

Bush's Hawthorn

Crataegus bushii Sarg.—Series **Crus-galli** (10)

FIELD IDENTIFICATION. Large shrub or tree to 25 feet, 10 in. in trunk diameter. The spreading branches forming a rounded crown.

FLOWERS. April–May, borne in broad, many-flowered, slender-pediceled, glabrous corymbs; corolla ¾–1 in. across; petals 5, white, oval, veiny; calyx-tube broadly obconic, sepals 5, ⅙–⅕ in. long, glabrous, linear-lanceolate, entire or obscurely toothed; stamens 20, anthers large and pink; styles 2–3, basally hairy.

Bush's Hawthorn
Crataegus bushii Sarg.

FRUIT. Maturing October–November, clusters drooping and few-fruited, short-oblong, green to reddish, length ⅓–⅗ in., sepals often deciduous; flesh thin, firm, dry; nutlets 2–3, about ¼ in. long.

LEAVES. Simple, alternate or clustered, deciduous, obovate or sometimes oval or elliptic, apex rounded to obtuse or acute, base cuneate or abruptly contracted, margin coarsely serrate above the middle, surfaces white-hairy at first, later firm and leathery, upper surface shiny, dark green and glabrous, lower surface paler and usually with scattered white hairs on the veins and midrib, veinlets reticulate, blade length 1¼–1½ in., width ½–1½ in.; petioles ⅕–⅗ in., white-hairy, or with age less so, somewhat winged by the gradually narrowed leaf bases; leaves on vigorous shoots usually somewhat larger and tending to be more elliptic.

TWIGS. Slender, elongate, nearly straight, essentially glabrous, chestnut-brown at first, gray later, spineless, or spines 1½–1¾ in. long, straight, brown, lustrous, some spines bearing reduced leaves; bark or trunk dark brown or grayish brown with irregular fissures and breaking into small scales.

RANGE. In Texas near Marshall, Harrison County; in Louisiana in Winn Parish, Calcasieu Parish, and Natchitoches Parish. Specimen examined from type tree, collected by B. F. Bush, No. 1368, Fulton, Arkansas, April 16, 1902. Deposited in the New York Botanical Garden Herbarium.

SYNONYM. *C. triumphalis* Sarg.

REMARKS. The genus name, *Crataegus*, is the classical Greek name for hawthorn. The species name, *bushii*, is in honor of its discoverer, Benjamin Franklin Bush (1858–1937), botanist of Missouri.

Pyracantha Hawthorn

Crataegus pyracanthoides Beadle—Series **Crus-galli** (10)

FIELD IDENTIFICATION. Shrub or small tree 6–15 ft, the trunk or main stems clothed with ashy gray or brownish, either smooth or scaly, bark.

FLOWERS. Open early in April when the leaves are almost fully grown, produced in compound, glabrous, many-flowered corymbs; pedicels and hypanthium glabrous; petals 5, white, rounded, spreading; sepals 1–1½ in. long, entire or remotely serrate, spreading or reflexed after anthesis; stamens 7–12, the anthers purplish; styles 2–3.

FRUIT. Ripens in September, globose or nearly so, ⅕–⅓ in. in diameter, bright red at maturity; nutlets mostly 2, ⅕–¼ in. long, the ventral surface nearly plane; hypostyle about half the length of the nutlet.

LEAVES. Obovate, oblanceolate or elliptic, the blades ⅝–2 in. long, ⅓–1⅓ in. wide, glabrous, acute or rounded

PYRACANTHA HAWTHORN
Crataegus pyracanthoides Beadle

at the apex, cuneate at the base, the margins serrate above the middle; they are glabrous, or when young a few bear weak hairs along the midrib on the upper surface, bright green and lustrous above, pale green beneath, eventually firm or subcoriaceous in texture, fading to tones of yellow, orange, and brown; petioles 1/12–2/5 in. long, margined.

TWIGS. Slender, brown to gray, armed with chestnut-brown or gray spines 5/8–1½ in. long.

RANGE. Louisiana, Alabama, and northern Florida; to Virginia and up the Mississippi Valley to Missouri and Indiana.

VARIETY. A variety has been described under the name of Tree Pyracantha Hawthorn, *C. pyracanthoides* var. *arborea* (Beadle) Palmer. It attains a height of 30 ft, either unarmed or with slender thorns; leaves oblanceolate to obovate, apex acute to obtuse or rounded, margin serrate except base usually entire, surfaces dark green and shiny, glabrous and firm; flowers in April, 3/8–5/8 in. wide, corymbs with 5–8 lax glabrous pedicels; stamens 10–20, anthers yellow to white or pink; fruit in October, subglobose or short-oblong, about 3/8 in. in diameter, red in color, flesh thin, nutlets 2–3. Usually in moist rich soil along streams. Indiana to Missouri and South Carolina; southward to Florida and east Texas.

REMARKS. The genus name, *Crataegus*, is the classical name of the hawthorn. The species name, *pyracanthoides*, refers to the pyracantha-like fruit. Also known as Montgomery Hawthorn.

UNIQUE PYRACANTHA HAWTHORN

Crataegus pyracanthoides var. *uniqua* (Sarg.) Palmer—Series **Crus-galli** (10)

FIELD IDENTIFICATION. Slender tree to 20 ft, with wide-spreading branches forming a flat head.

FLOWERS. Opening in April, in 5–8-flowered, glabrous corymbs, pedicels slender; calyx-tube narrowly obconic, the 5 lobes short, broad and acuminate, margin entire or slightly dentate, somewhat hairy within, later reflexed; stamens 20, anthers white; styles 2–3.

FRUIT. Ripening in October, on slender drooping pedicels, short-oblong, rounded at the ends, dull red, 3/8–½ in. long and about 1/3 in. thick; calyx conspicuous with a deep wide cavity broad in the bottom; sepals 5, reflexed, appressed, slightly hairy within, persistent, flesh thin, dry, hard; nutlets 2 or 3, broad and rounded at the base, keeled on the back with a high wide grooved ridge, ¼–1/3 in. long and about 1/8–1/6 in. wide, hypostyle conspicuous and broad, extending to below the middle of the nutlet.

LEAVES. About half-grown when the flowers open, simple, alternate, deciduous, oblong-obovate, apex acute or occasionally rounded, base gradually narrowed or cuneate; teeth on margin straight, incurved and glandular, usually above the middle, length ¾–1¾ in., width 3/8–¾ in., on vigorous shoots broadly obovate, acute or rounded at base, more closely serrate, 2–2½ in. long and 1–1 1/3 in. wide; surfaces when young glabrous except along the veins; at maturity glabrous, thin but firm, upper surface dark green and shiny, paler beneath, midrib and veins rather slender.

TWIGS. Slender, slightly divaricate, yellow to orange or reddish brown, armed or unarmed, spines straight or slightly curved, shiny brown, 3/8–¾ in.

RANGE. In southwestern Arkansas, eastern Texas, and northwestern Louisiana. This tree is found in Texas in low rich woods near Marshall, Harrison County; Louisiana at the marble quarry near Winnfield, Winn Parish.

SYNONYMS. This hawthorn has been variously described under the synonyms of *C. uniqua* Sarg., *C. arioclada* Sarg., and *C. cocksii* Sarg.

REMARKS. The genus name, *Crataegus*, is the classical name of the hawthorn. The variety name, *uniqua*, means "unique." Sargent (1913, p. 237) remarks that "the plant is interesting (or unique) as the only Hawthorn, of the *Crus-galli* group, with 20 white or yellow anthers and glabrous corymbs from the region west of the Mississippi. *Crataegus pyracanthoides* var. *arborea*, from Alabama has a glabrous corymb with 20 yellow anthers."

Palmer Hawthorn

Crataegus palmeri Sarg.—Series **Crus-galli** (10)

FIELD IDENTIFICATION. Tree attaining a height of 25 ft and a diameter of 12 in. The stout spreading branches shaped into a broad rounded crown.

FLOWERS. Opening in May, corolla of 5 white rounded petals, about ½ in. across, borne in many-flowered corymbs composed of slender pedicels; calyx-tube narrowly obconic, lobes 5, elongate-acuminate, margin entire or serrate, a few dark glands sometimes borne terminally; stamens 10 with yellow anthers; styles 3, pale-hairy at base.

FRUIT. Ripening in October, in drooping and few-fruited clusters, subglobose, about ⅓ in. in diameter, green or red, often pale-dotted, mature calyx-lobes persistent, erect or incurved; nutlets 3, acute terminally, dorsally ridged, ¼–⅓ in. long.

Palmer Hawthorn
Crataegus palmeri Sarg.

LEAVES. Simple, alternate, deciduous, oval to oblong, blade length 1½–2 in., width 1¼–1¾ in., apex acute or rounded, base narrowed or cuneate, margin coarsely serrate except toward the base, when mature leathery, upper surface shiny and dark green, paler beneath, primary veins 4–5 pairs; leaves on young shoots larger, oblong-ovate to elliptic, apex acute, margin serrate and often lobed, glandular toward the base.

TWIGS. Slender, shiny brown, glabrous, spineless, or with lustrous, reddish brown, straight thorns ¾–3 in. long.

RANGE. Oklahoma, Arkansas, Missouri, and southeastern Kansas.

REMARKS. The genus name, *Crataegus*, is the classical name of the hawthorn, and the species name, *palmeri*, honors the American botanist, Ernest Jesse Palmer, authority on the *Crataegus*.

Cherokee Hawthorn

Crataegus cherokeensis Sarg.—Series **Crus-galli** (10)

FIELD IDENTIFICATION. Small tree with slightly scaly bark.

FLOWERS. Opening early in April, ⅖–½ in. in diameter, on slender glabrous pedicels, in 5–7-flowered globose glabrous corymbs; petals 5, white, rounded, calyx-tube narrow-obconic, glabrous, the lobes slender, gradually narrowed from the base, long acuminate, entire or slightly and irregularly toothed above the middle, glabrous; stamens 10, anthers red; styles 1–3, usually 2.

FRUIT. Ripening late in September, ellipsoidal, dull orange-red, ⅖ in. long, about ¼ in. thick, the persistent calyx sessile or raised on a short tube; nutlets narrowed and rounded at the ends, only slightly ridged on the back, ¼–⅓ in. long, ⅛–⅕ in. wide, the narrow hypostyle extending to below the middle.

LEAVES. Oblong-obovate, apex rounded or acute, gradually narrowed and cuneate at base, finely, often doubly, serrate, usually only to the middle with acute teeth thickened at apex, glabrous with the exception of a few hairs on the upper side of the midrib early in the season, thin, dark green and lustrous above, paler below, blades 1½–1¾ in. long, ⅜–¾ in. wide, with a slender midrib and thin obscure primary veins, on vigorous leading shoots usually acute at apex, often acutely lobed above the middle, 1½–2 in. long, 1½–1¾ in. wide; petioles slender, often wing-margined nearly to the base, glabrous, on vigorous shoots ⅕–½ in. long.

TWIGS. Slender, glabrous, often zigzag, reddish brown to orange-brown at first, later dark grayish brown; spines numerous, straight, slender, 1–1½ in. long.

RANGE. In upland thickets, near Larissa, Cherokee County, Texas.

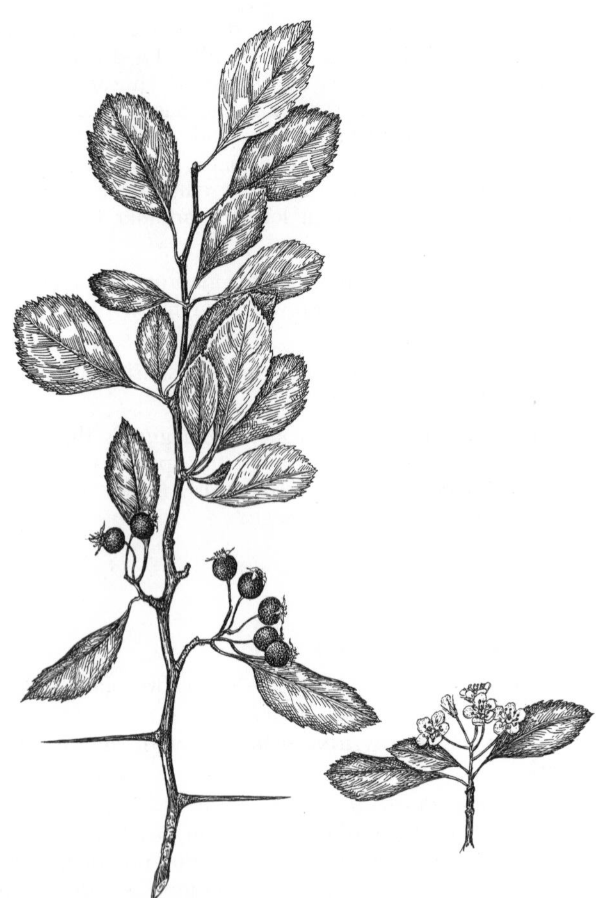

CHEROKEE HAWTHORN
Crataegus cherokeensis Sarg.

REMARKS. The genus name, *Crataegus*, is the classical Greek name of hawthorn. The species name, *cherokeensis*, refers to its being in Cherokee County, Texas. Sargent (1921a) notes: "the narrow oblong-obovate finely serrate leaves and ellipsoidal fruit distinguish this species from *C. Reverchonii* Sargent of the Dallas region of Texas, with nearly orbicular coarsely serrate leaves and smaller globose fruit, the only Crus-galli species with glabrous corymbs, 10 stamens and red or pink anthers which has been found before in the Arkansas, western Louisiana and Texas region."

REVERCHON HAWTHORN

Crataegus reverchonii Sarg.—Series **Crus-galli** (10)

FIELD IDENTIFICATION. Shrub 3–9 ft, usually with many stems from the base, or occasionally a small thorny tree to 26 ft.

FLOWERS. April–May, borne in slender pediceled, compact, glabrous, few-flowered corymbs; each flower about ⅓ in. in diameter, petals 5, white, rounded; calyx-tube narrowly obconic, 5-lobed, lobes slender, apices acuminate with a small red gland, margins entire or obscurely serrate; stamens 10–15, styles 3–5, usually 5.

FRUIT. Maturing in September on drooping few-flowered corymbs, with slender pedicels; subglobose, some slightly broader than long, light scarlet, shiny, marked by occasional large dots, about ⅓ in. in diameter; flesh sweet, juicy, yellow, thick; nutlets usually 4, dorsally and prominently ridged, about ¼ in. in diameter; calyx-lobes slender and deciduous.

LEAVES. Simple, alternate, deciduous, oval to obovate, margin finely crenate-serrate with gland-tipped teeth above the base; at first yellowish green and mostly glabrous above, lower surface slightly hairy on the midrib and veins; mature leaves leathery, dark green and shiny above, lower surface paler; blade length 1¼–1½ in., width ¾–1 in.; midrib stout, yellow, primary veins 5–6 pairs; petioles stout, about ⅓ in., grooved, wing-margined almost to the base; stipules minute, linear, reddish later, falling early; leaves on vigorous shoots rounded, less often ovate or elliptic, margin coarsely serrate or slightly lobed, about 1½ in. in diameter, petioles short-glandular, broad-winged.

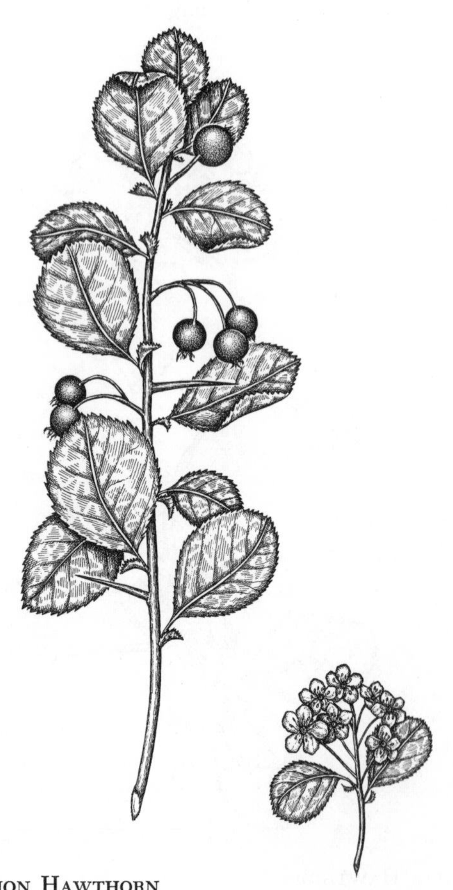

REVERCHON HAWTHORN
Crataegus reverchonii Sarg.

TWIGS. Erect, divaricate, glabrous, lustrous, orange-brown to reddish brown, lenticels small, pale, and numerous; older twigs gray, armed wtih spines 1½–3 in. long (occasionally to 5 in.), slender, nearly straight, lustrous, reddish brown or purple.

RANGE. Southwestern Oklahoma, Arkansas, northern and central Texas (Dallas County), Missouri, and eastern Kansas.

VARIETY. *C. reverchonii* Sarg. var. *discolor* (Sarg.) Palmer is a tree to 24 ft, round-topped, the branches slender and thorny or thornless; bark scaly in thin flakes over orange-brown inner bark. Flowers ½–¾ in. wide, usually 6–15 in glabrous corymbs; stamens 18–20 with pink or pale yellow anthers. Sepals entire or minutely glandular-serrate. Fruit ripening in October, subglobose or short-oblong, ⅜–½ in. thick, dull red or orange-red, with thin mellow flesh and 3–4 or rarely 5 nutlets. Leaves ovate, oblong-ovate or nearly orbicular on shoots, dark green, glossy, firm or subcoriaceous at maturity, ¾–1¼ in. long, ⅝–¾ in. wide, or larger and deeply serrate or dentate on vegetative shoots. Petioles slender, ¼–½ in. Thickets and borders of woods along small streams. In Arkansas, southern Missouri, and southern Kansas.

REMARKS. The genus name, *Crataegus,* is the classical name of the hawthorn, and the species name, *reverchonii,* honors Julien Reverchon (1837–1905), a Texas plant collector of French birth. Charles Sprague Sargent remarks that the Reverchon Hawthorn is one of the most typical species of the Crus-galli group.

Stevens Hawthorn

Crataegus reverchonii var. *stevensiana* (Sarg.) Palmer —Series **Crus-galli** (10)

FIELD IDENTIFICATION. An intricately branched shrub to 9 ft. Bark dark and slightly scaly.

FLOWERS. Opening in May in 10–15-flowered glabrous corymbs, pedicels slender, bracts and bractlets linear, scarious, caducous; petals 5, white, oval to obovate, corolla about ½ in. in diameter; calyx-tube narrow obconic, glabrous, the 5 sepals gradually narrowed from a wide base, acuminate, coarsely serrate below the middle, externally glabrous, inner surface slightly villose; stamens 10, anthers yellow; styles 2–3.

FRUIT. Ripening in late September, on slender pedicels, short-oblong to slightly obovoid, greenish red, ⅕–⅓ in. long ⅙–⅕ in. in diameter; calyx little enlarged, with a wide cavity pointed in from its base; flesh thin and dry; nutlets 2 or 3, ends rounded, dorsally only slightly ridged, ⅙–⅕ in. long, ⅛–⅙ in. wide.

LEAVES. Simple, alternate, deciduous, blade obovate, apex rounded, base gradually narrowed and cuneate, margin doubly serrate above the middle with acute teeth, subcoriaceous, upper surface dark green and shiny, lower surface dull and paler, blades 1¼–1⅔ in. long, ¾–1 in. wide, midrib thin, primary veins obscure; petioles slender, glabrous, narrowly winged to the middle by the decurrent blade, ⅓–⅖ in. long; leaves on vigorous shoots suborbicular, apex rounded or minutely pointed, base broad-cuneate, finely serrate above the middle, 1¼–1½ in. long and wide, their petioles broadly wing-margined at apex, ⅙–¼ in. long.

TWIGS. Slender, straight or divaricate, glabrous, young ones brown and shiny, later becoming dull grayish brown, spines 1–2 in. long.

RANGE. Thickets along streams or on rocky hillsides. Oklahoma in Snyder and Greer counties and in Kansas in Wilson County.

REMARKS. The genus name, *Crataegus,* is an ancient name and refers to the hard wood. The variety name, *stevensiana,* is in honor of G. W. Stevens, a botanist who collected it near Granite, Greer County, Oklahoma, in 1913. The species name, *reverchonii,* is in honor of Julien Reverchon (1837–1905), a Texan of French birth who collected plants in Texas.

Piedmont Hawthorn

Crataegus regalis Beadle—Series **Crus-galli** (10)

FIELD IDENTIFICATION. Tree to 30 ft, with a trunk 8–12 in. in diameter, covered with ashy gray or brownish scaly thin bark. Branches ascending or spreading.

FLOWERS. From ½–⅗ in. wide, petals 5, white, rounded, spreading; opening the last of April when the leaves are almost fully grown, produced in glabrous, compound, many-flowered corymbs, the lowest branches of which arise from the axils of leaves; pedicels and hypanthium glabrous; sepals ⅙–⅕ in. long, linear-lanceolate, entire or remotely serrate, spreading or reflexed after anthesis; stamens about 10, the anthers yellow; styles 1–2.

FRUIT. Ripening September–October, oblong, about ⅓ in. thick and ⅜ in. long; nutlets mostly 2–3, about ⅓ in. long, the lateral and ventral surfaces nearly plane; hypostyle ⅕–⅙ in. long.

LEAVES. Oval to elliptic, the blades 1¼–3 in. long, ⅝–2 in. wide, acute or acuminate at the apex, contracted or broadly cuneate at the base, the borders serrate and on leading shoots often incised; they are glabrous, or when young bear a few weak hairs along the midrib and on the upper surface, bright green and lustrous above, pale beneath, firm to subcoriaceous, fading in autumn to tones of yellow, orange, and brown; petioles ⅕–⅗ in., margined.

TWIGS. Gray or brown, bearing chestnut-brown or gray spines.

RANGE. In low woods. Arkansas; east to Georgia, north to North Carolina, and west to Indiana and Kansas.

VARIETY. A variety of Piedmont Hawthorn has been

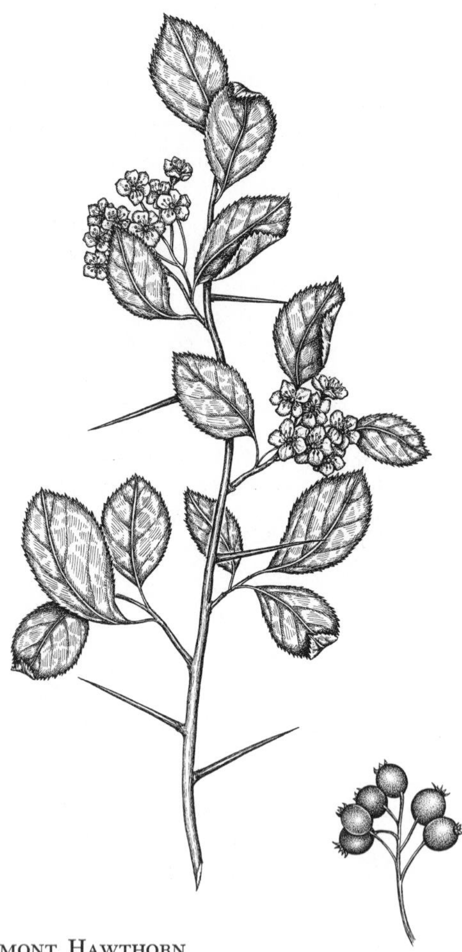

Piedmont Hawthorn
Crataegus regalis Beadle

described as *C. regalis* var. *paradoxa* (Sarg.) Palmer. It has the young branchlets, foliage, and inflorescence slightly pubescent; fruit ellipsoid or oblong. It occurs in Arkansas, Missouri, and Kansas.

REMARKS. The genus name, *Crataegus*, is the classical name for the hawthorn, and the species name, *regalis*, means "royal." A handsome symmetrical tree which has broad, lustrous foliage and good horticultural possibilities.

Mohr Hawthorn

Crataegus mohrii Beadle—Series **Crus-galli** (10)

FIELD IDENTIFICATION. Tree attaining a height of 30 ft, the trunk usually less than 8 in. in diameter. Branches ascending or pendulous, shaped into a loose and rounded crown.

FLOWERS. Opening in April or May, corymbs composed of numerous slender elongate pedicels which are hairy at first but more glabrous later; corolla white, 5-petaled, cup-shaped, hardly over ¾ in. across; calyx-tube narrowly obconic, glabrous or somewhat hairy; lobes 5, narrowly lanceolate, margin glandular-serrulate; stamens 20, anthers small, light yellow; styles 3–5, hairy at the base.

FRUIT. Maturing in October, corymbs many-fruited, drooping, reddish, globose to short-oblong, orange-red, about ⅓ in. thick; fruiting calyx conspicuous with erect, often early-deciduous lobes; nutlets 2–3, about ¼ in. long.

LEAVES. Simple, alternate, deciduous, obovate to rounded, apex acute or acuminate, base cuneate, margin doubly serrate; young leaves glabrous and hairy along the veins; mature leaves with the upper surface dark green and shiny, paler and glabrous or puberulous beneath, blade length 1–1½ in., width ⅔–1 in.; leaves on young shoots often larger, broadly elliptic, apex acute or acuminate, sometimes with short, acute lobes as well as serrations; petioles ⅓–½ in. long, green or reddish.

TWIGS. Slender, reddish brown to gray, furnished with shiny brown, straight, slender spines ¾–2½ in.

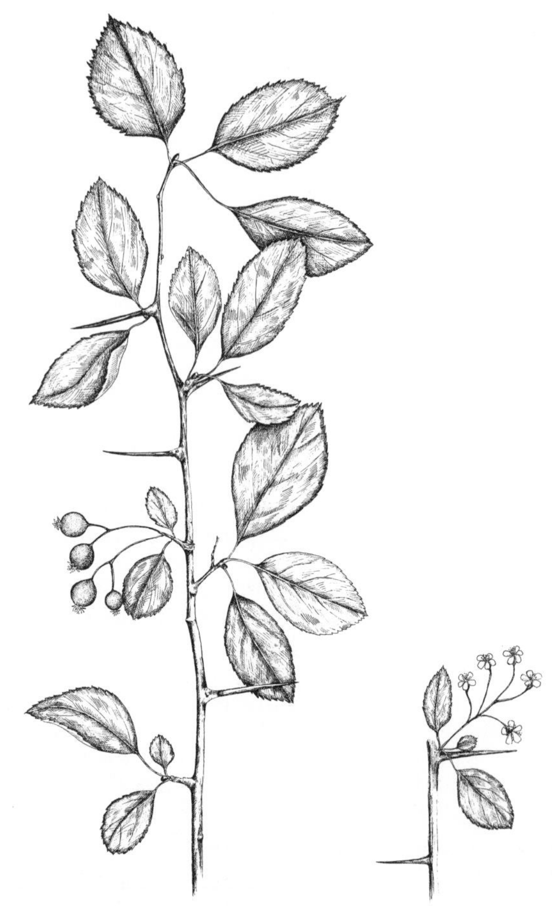

Mohr Hawthorn
Crataegus mohrii Beadle

San Augustine Hawthorn
Crataegus sublobulata Sarg.

RANGE. Arkansas, Alabama, Kentucky, and Tennessee.

REMARKS. The genus name, *Crataegus* is the classical name for the hawthorn. The species name, *mohrii,* honors Charles Theodore Mohr (1824–1901), German-born manufacturing druggist and botanist of Alabama.

San Augustine Hawthorn

Crataegus sublobulata Sarg.—Series **Crus-galli** (10)

FIELD IDENTIFICATION. A round-topped tree 24–30 ft high, with a short trunk covered with soft, corky, slightly ridged bark, the branches wide-spreading.

FLOWERS. Opening late in March and early in April, about ⅝ in. in diameter, on slender glabrous pedicels in compact 5–6-flowered corymbs; petals 5, white, rounded; calyx-tube narrow-obconic, glabrous, the lobes slender, acuminate, entire or furnished above the middle with occasional slender teeth, glabrous on the outer surface, sparingly villose-pubescent on the inner surface; stamens 20, anthers pink; styles 2–5.

FRUIT. Mature August–September, short-oblong to subglobose or rarely to obovoid, orange-red, ⅓–⅖ in. long and about ¼ in. broad; nutlets usually broader at apex than at base, prominently ridged on the back, ⅕–⅙ in. long and ⅛–⅕ in. wide, the broad hypostyle extending to below the middle and often nearly to the base.

LEAVES. Broad-obovate to elliptic, acute or rounded at apex, abruptly or gradually narrowed and cuneate at base, coarsely, deeply, and irregularly serrate with acuminate teeth, and often sublobulate with acuminate lobules, tinged with red and covered above with short white hairs, deciduous before the flowers open except from the upper side of the midrib, glabrous at maturity, thick, dark green and lustrous on the upper surface, paler on the lower surface, 1¼–1½ in. long, 1–1⅓ in. wide with a stout midrib and prominent primary veins, or on vigorous shoots often 1½–2 in. long and 1⅓ in. wide; petioles stout, wing-margined often nearly to the middle, grooved and villose-pubescent on the upper side, ⅜–⅝ in. long.

TWIGS. Slender, slightly zigzag, when young reddish brown, later gray to brown and glabrous; spines numerous, slender, straight, lustrous, chestnut-brown to grayish brown, ½–1½ in. long.

RANGE. Known from borders of woods near San Augustine, San Augustine County, Texas.

REMARKS. The genus name, *Crataegus,* is the classical Greek name of the hawthorn. The species name, *sublobulata,* refers to the deep nearly lobelike leaf serrations, unusual in plants of the *Crus-galli* group. It is well distinguished from *C. bushii* Sargent (the species of southern Arkansas, western Louisiana and eastern Texas, with 20 stamens and rose-colored anthers) by the shape of the leaves.

Warner Hawthorn

Crataegus warneri Sarg.—Series **Crus-galli** (10)

FIELD IDENTIFICATION. Shrub or tree to 24 ft, with dark bark scaly near the base. The erect branches form a narrow head.

FLOWERS. Opening from the 10th to the middle of April, ⅜–⅝ in. in diameter, on stout densely villose pedicels in compact many-flowered villose corymbs; petals 5, white, spreading; calyx-tube narrowly obconic, thickly covered with matted pale hairs, the lobes narrowed from a broad base, slender, acuminate, glandular-serrate, slightly villose on the outer surface, puberulous on the inner surface; stamens 10, anthers reddish purple; styles 2, rarely 3.

FRUIT. Ripening in September, on slightly villose pedicels; ellipsoidal to subglobose, orange-red, about ⅓ in. long, the calyx little enlarged with a short tube and a wide shallow cavity in the bottom, and with spreading, often deciduous, lobes; nutlets 2 or 3, rounded at the ends, ridged on the back wtih a broad deeply grooved ridge, ⅙ in. long and about ⅛ in. wide, the narrow hypostyle extending to the middle.

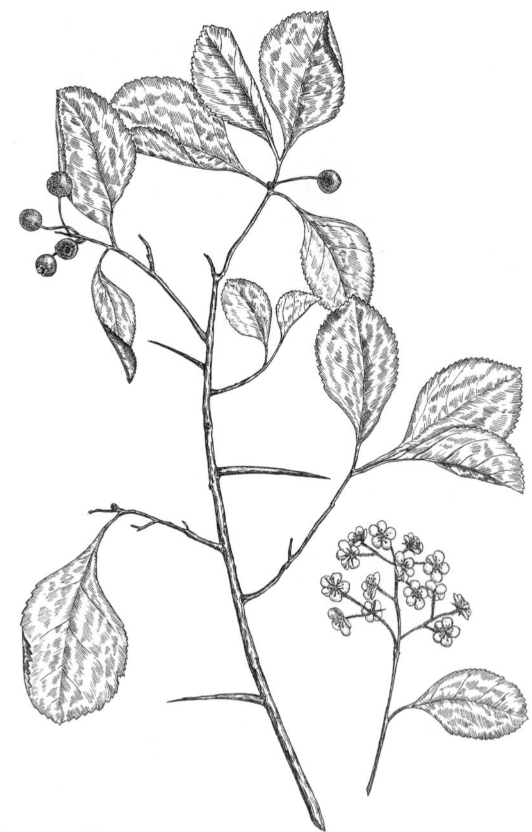

Warner Hawthorn
Crataegus warneri Sarg.

LEAVES. Ovate to oval or obovate, rounded or acute and short-pointed at apex, gradually or abruptly narrowed and cuneate at base, and coarsely serrate above the middle with straight gland-tipped teeth, nearly full-grown when the flowers open, and then covered above with short white hairs and villose below along the midrib and primary veins; at maturity thin, dark green and glabrous or occasionally still villose on the midrib above, pale and still villose below along the slender midrib and primary veins, 1¼–2 in. long and 1–1½ in. wide; petioles stout, wing-margined to the base, densely villose at maturity, ⅕–⅓ in. in length; leaves on vigorous shoots broad-ovate to semiorbicular, short-pointed at the rounded or acute apex, rounded and gradually narrowed below into a broad wing extending nearly to the base of the short petiole, more coarsely serrate, subcoriaceous, roughened above, 1¾–2⅓ in. long and broad, with a stout midrib and primary veins villose below.

TWIGS. Slender, pale hairy, reddish brown, glabrous with maturity; spines stout or slender, chestnut-brown, 1–2 in. long.

RANGE. In Texas at Huntsville in Walker County, near Larissa in Cherokee County, and near Palestine in Anderson County.

REMARKS. The genus name, *Crataegus*, is the classical name for the hawthorn, and the species name *warneri*, honors S. R. Warner, professor of botany at Sam Houston State Teachers College, Huntsville, Texas. Charles Sargent (1922) makes the following remarks concerning the species: "Extremely rare in the three stations where it is found, this thorn is arborescent only at Palestine, Texas where it grows in woods. In Walker and Cherokee counties where it grows on dry banks it is a shrub, with several stems not more than 9 ft high. Until the stones of the fruit are examined this species might pass for one of the Macracanthae group, although the entire absence of lobes from the leaves and the rather compact corymbs are unusual in plants of that group. Although it is an extreme form it is now referred with some doubt to the Crus-galli group. The species of that group which it most closely resembles is *C. sublobulata* Sargent from San Augustine, Texas, which differs in its slightly lobed, glabrous, thicker leaves, its broader corymbs, and its 20 stamens with pink anthers."

St. Louis Hawthorn

Crataegus acutifolia Sarg.—Series **Crus-galli** (10)

FIELD IDENTIFICATION. Tree attaining a height of 30 ft, with a diameter of 18 in. A rounded crown formed by the stout spreading and ascending branches.

FLOWERS. Borne in glabrous, compact, many-flowered corymbs in May; petals 5, white, rounded, when spreading the flower about ½ in. in diameter; stamens usually 10 (rarely to 13), anthers yellow; styles 2–3; calyx narrowly obconic, sepals 5, lanceolate, acuminate, margin entire or slightly glandular-serrate.

FRUIT. Maturing August–September, in lax, open drooping clusters; red, subglobose to short-oblong ½–¾ in. long, nutlets 2–3, dorsally with a rounded ridge, about 3/16 in. long; calyx-lobes usually deciduous early.

LEAVES. Simple, alternate, deciduous, oval to oblong obovate, apex acute or acuminate or rarely rounded, base cuneate and usually entire, margin finely serrulate, sometimes often only on the upper half, blade length about 1½ in., width about 1 in., thin and firm, upper surface dark green and lustrous, glabrous or with a few hairs on the principal veins; lower surface paler and slightly hairy but eventually glabrous; petioles ¼–½ in., at first minutely glandular. Leaves on young shoots to 3 in. long and 2 in. wide with a few acute short lobes.

TWIGS. Stout, gray or brown, spineless, or with spines lustrous brown, straight, slender and ½–2 in. long.

RANGE. In hilly woodlands. Arkansas, Missouri, Tennessee, Illinois, and Indiana. Particularly abundant in East St. Louis.

St. Louis Hawthorn
Crataegus acutifolia Sarg.

REMARKS. The genus name, *Crataegus*, is the classical name for the hawthorn, from an ancient Greek word referring to the hardness of the wood. The species name, *acutifolia*, is for the acute leaves. A variety has been given the name of *C. acutifolia* var. *insignis* (Sarg.) Palmer. It has longer and firmer leaves; flowers ⅝–¾ in. in diameter; pedicels somewhat slightly hairy; fruit ⅜–⅝ in. thick. Known from East St. Louis and Kahokia, Illinois.

Sabine Hawthorn

Crataegus sabineana Ashe—Series **Crus-galli** (10)

FIELD IDENTIFICATION. Tree 12–30 ft.

FLOWERS. Borne in few-flowered corymbs, petals 5, white, rounded; stamens 10–20, stigmas usually 3.

FRUIT. Corymbs simple, 3–5-fruited, glabrous, pedicels slender, 1–2 in. long; fruit body oblong, ½–⅗ in. long, base narrowed, greenish yellow with a red cheek; flesh thin and green; calyx-lobes 5, reflexed, linear, apex acuminate, margin sharply serrate, later brown and deciduous, cavity conical and shallow.

LEAVES. Simple, alternate, deciduous, broadly obovate or rhombic, apex acute, base cuneate, margin coarsely and sharply serrate, blade length 1–1¾ in.; petioles 1/12–⅛ in. long, winged at the blade base.

TWIGS. Stout, glabrous, tan to reddish.

RANGE. Eastern Texas and western Louisiana.

REMARKS. The genus name, *Crataegus*, is the classical name, and the species name, *sabineana*, is from the Sabine River. W. W. Ashe collected it on October 19, 1921, in Sabine Parish, Louisiana. He remarks that it is found on the loose, red Orangeburg soils, and that he considered it most closely related to *C. uniqua* Sarg. (now known as *C. pyracanthoides* var. *uniqua* (Sarg.) Palmer), of northeastern Texas from which it "differs by having much longer and differently shaped leaves and larger fruit mostly green, and not wholly red in color." (Ashe, 1925)

Barberry Hawthorn

Crataegus berberifolia Torr. & Gray—
Series **Crus-galli** (10)

FIELD IDENTIFICATION. Spiny tree attaining a height of 25 ft and a diameter of 10 in.

FLOWERS. Borne March–April, corymbs ample, 4–6-

Barberry Hawthorn
Crataegus berberifolia Torr. & Gray

flowered, simple or compound, villose; pedicels slender and hairy; calyx-tube obconic, hirsute-pubescent, sepals 5, ⅛–⅕ in. long, slender, acuminate, entire or serrate, villose when young and pubescent or glabrous with age; corolla ⅖–⅗ in. wide, petals 5, white, oval to obovate, abruptly narrowed into a short slender claw; stamens 16–20, anthers yellow; styles 2–3, hairy at base.

FRUIT. Maturing September–October, clusters few-fruited and drooping, subglobose or nearly so, ⅖–½ in. in diameter, yellowish orange; flesh thin, yellow, juicy, acid, of pleasant taste; nutlets 2–3, ¼–⅓ in. long, slightly ridged dorsally, hypostyle about ⅕ in. long.

LEAVES. Simple, alternate, deciduous, oblong-cuneiform or spatulate to obovate-cuneiform, apex rounded or obtuse, gradually narrowed to the cuneate base, margins serrate above the middle, teeth rather coarse, mature leaves firm, upper surface dark green and somewhat roughened but later glabrous and lustrous, lower surface paler and pubescent, especially on the veins, length 1½–2½ in., width ¾–1¼ in.; petioles 1/12–½ in., densely villose at first, later less so or almost glabrous.

TWIGS. Slender, brown and white-hairy when young, gray to brown and more glabrous later, spineless or with slender reddish brown to black spines 1–1½ in. long; bark on trunk dark gray and breaking into small scales.

RANGE. In low grounds of Arkansas, Oklahoma, and Louisiana near Opelousas, St. Landry Parish.

REMARKS. The genus name, *Crataegus*, is the classical name for the hawthorn, and the species name, *berberifolia*, refers to the barberry-like leaves. The species is considered to be closely related to *C. crus-galli* and perhaps could be described as a Southern pubescent variety of it. In fact at one time it was listed in the literature as *C. crus-galli* var. *berberifolia* (Torr. & Gray) Sarg. (Sargent, 1889, p. 424). Other species which are now regarded as synonyms of *C. berberifolia* are *C. crocina* Beadle, *C. edura* Beadle, *C. fera* Beadle, and *C. tersa* Beadle.

Marshall Hawthorn

Crataegus berberifolia var. *edita* (Sarg.) Palmer—Series **Crus-galli** (10)

FIELD IDENTIFICATION. Tree attaining a height of 35 ft and a diameter of ¾–1 ft. The stout branches ascending or spreading to form a rounded crown.

FLOWERS. Usually opening in April, corymbs narrow, few-flowered, compound, pedicels ½–¾ in., villose; petals 5, rounded, corolla when open ½–⅔ in. across; calyx-tube narrowly obconic, glabrous or somewhat hairy toward the base, sepals 5, linear-lanceolate, green to reddish, entire or obscurely glandular-toothed, glabrous or puberulous; stamens 20, anthers purplish; styles 2–3.

Marshall Hawthorn
Crataegus berberifolia var. *edita* (Sarg.) Palmer

FRUIT. Maturing October–November, clusters drooping and few-fruited, pedicels about ½ in. long, somewhat hairy or glabrous, fruit body short-oblong, greenish red, ¼–⅓ in. long, sepals deciduous or semipersistent; flesh green, dry, thin; nutlets 2 or 3 with broad low ridges, about ¼ in. long, hypostyle ⅙–⅕ in. long.

LEAVES. Simple, alternate or somewhat clustered, deciduous, oblong-obovate or oval to elliptic, apex acute or rounded, base cuneate, margin singly or doubly serrate (on younger shoots sometimes lobed and larger), dark green and lustrous, glabrous or slightly roughened, paler and scabrate beneath with some persistent pubescence along the midrib, blade length 1¼–2 in., width ½–1 in.; petiole ⅓–½ in., stout, white-hairy at first, sparsely pubescent to glabrous later.

TWIGS. Slender, rather straight, brown at first to gray later, white-hairy at first but eventually glabrous; spines slender, straight, chestnut-brown to gray, ¾–2 in. long; trunk with dark scaly bark.

RANGE. Swampy woods near Shreveport, Louisiana; Marshall, Texas; and Camden, Arkansas. The specimen examined was collected by B. F. Bush, Oct. 9,

1901, at Marshall, Texas, sheet No. 1009. Deposited in the New York Botanical Garden Herbarium.

REMARKS. The genus name, *Crataegus*, is the classical Greek name for the hawthorn. The species name, *berberifolia*, is for the barberry-like leaves, and the variety name, *edita*, means "standing out" or "elevated."

ENGELMANN HAWTHORN

Crataegus engelmannii Sarg.—Series **Crus-galli** (10)

FIELD IDENTIFICATION. Tree 15–20 ft, with a diameter of 5–6 in. Branches usually wide-spreading and usually horizontal, forming a low flat-topped or rounded head.

FLOWERS. April–May, flowers white, 5-petaled, about ¾ in. in diameter, in wide 8–12-flowered, slender-branched cymes thickly coated with long pale hairs; bracts about ½ in. long, linear-lanceolate, tomentose or villous; calyx tomentose, villous, or nearly glabrous, the lobes narrow, acuminate, entire, glabrous on the outer surface, and usually more or less pubescent on the inner surface, reflexed after anthesis, often deciduous before the ripening of the fruit; stamens 10; filaments slender; anthers small, rose-colored; styles two or three.

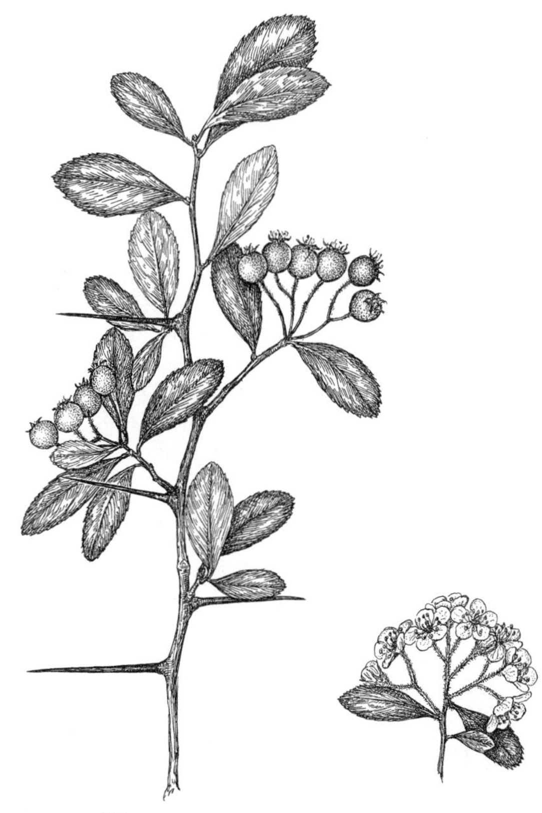

ENGELMANN HAWTHORN
Crataegus engelmannii Sarg.

FRUIT. Ripening early in November, globose, about ⅓ in. in diameter, bright orange-red with a yellow cheek and thin dry green flesh; tube of the calyx prominent, the cavity broad in proportion to the size of the fruit, shallow, nutlets 2 or 3, thick, prominently ribbed on the back with high rounded ridges, ¼ in. long.

LEAVES. Simple, alternate, deciduous, broadly obovate or rarely elliptic, apex rounded or short-pointed, gradually narrowed below into short glandular pilose petioles, coarsely glandular-serrate with incurved teeth usually only above the middle and generally only at the apex, coriaceous, dark green, lustrous and roughened on the upper surface with short rigid pale hairs, pale on the lower surface, pilose above and below on the slender midribs and on the thin obscure primary veins and veinlets, 1–1½ in. long, ½–1 in. broad; stipules linear-lanceolate, light red, ⅓ in. long, caducous.

TWIGS. Slightly zigzag, marked with large scattered white lenticels, at first clothed with pale hairs, becoming nearly glabrous and reddish brown during the first season, and lighter colored and gray or gray tinged with red during their second year, and armed with remote slender straight or slightly curved chestnut-brown spines 1½–2½ in. long.

RANGE. Dry hillsides and slopes of limestone soil. Oklahoma, Kansas, Missouri, Arkansas, Tennessee, Kentucky, Alabama, Mississippi, and Illinois.

VARIETY AND SYNONYMS. A variety of Engelmann's Hawthorn has been named by E. J. Palmer as *C. engelmannii* var. *sinistra* (Beadle) Palmer (*C. sinistra* Beadle, *C. villiflora* Sarg.), and differs in the smaller oval fruit, the smaller flowers and leaves, and more densely and harshly pubescent corymbs and foliage.

C. engelmannii forma *nuda* Palmer has been named as a form with the twigs, foliage, and inflorescence being glabrous or nearly so.

Synonyms of *C. engelmannii* are *C. pilifera* Sarg., *C. torva* Beadle, and *C. berberifolia* var. *engelmannii* (Sarg.) Egglest.

REMARKS. The genus name, *Crataegus*, is the classical name of the hawthorn. The species name, *engelmannii*, honors George Engelmann (1809–1884), a German-born American physician and botanist of St. Louis, Missouri, who first collected it.

PALE HAWTHORN

Crataegus palliata Sarg.—Series **Crus-galli** (10)

FIELD IDENTIFICATION. Tree to 25 ft, armed with stout spines 1½–3 in. long.

FLOWERS. April–May, borne in villose, 10–12-flowered corymbs; pedicels slender, flowers about ¾ in. wide or less; calyx-tube narrowly obconic, white densely appressed-hairy, 5-lobed, lobes gradually narrowed toward the acuminate apex, margin finely glandular-serrate below the middle; stamens 10, anthers red to purple; styles 2–5, basally hairy.

FRUIT. Ripening in October, pome subglobose, dark red, lustrous, pale-dotted, about ⅝ in. in diameter, flesh thick and succulent; nutlets 3–5, base rounded, apex acute, dorsally ridged and grooved, about ⅓ in. long and ⅙ in. wide; hypostyle narrow, light-colored, extending to the center of the nutlet.

LEAVES. Simple, alternate, deciduous, oval to obovate, apex acute or acuminate, base entire and cuneate-concave, margin coarsely double-serrate with straight glandular teeth, some on young shoots slightly lobed, at first light yellowish green with short white hairs, later shiny and glabrous above and villose-pubescent beneath, length 2⅓–3 in., 1½–2¼ in. wide, ribs stout, primary veins prominent; petioles ¼–⅓ in., stout, winged to below the middle, densely hairy at first, glabrous later.

TWIGS. Stout, divaricate, at first light chestnut-brown and shiny, bearing long matted white hairs, later pale grayish brown; spines chestnut-brown, lustrous, stout, nearly straight, 1½–3 in. long.

RANGE. In Arkansas, rich moist bottom lands near Fulton, Hempstead County, and in Miller County.

REMARKS. The genus name, *Crataegus*, is the classical name, and the species name, *palliata*, refers to the covering of pale or white hairs on young parts.

EUREKA SPRINGS HAWTHORN
Crataegus subpilosa Sarg.

EUREKA SPRINGS HAWTHORN

Crataegus subpilosa Sarg.—Series **Crus-galli** (10)

FIELD IDENTIFICATION. Tree 12–15 ft, with a trunk 8 in. in diameter, covered with pale gray scaly bark, the wide-spreading branches forming a round-topped head.

FLOWERS. Appearing early in May, about ½ in. in diameter, on slender pedicels in compact mostly 12–15-flowered corymbs, like the pedicels slightly pilose; petals 5, white, rounded, spreading; calyx-tube narrow-obconic, glabrous, the lobes slender, long-acuminate, entire, glabrous on the outer surface, sparingly pubescent on the inner surface; stamens 20, anthers pale yellow, styles 2 or 3.

FRUIT. Ripening the end of September, in small drooping glabrous clusters, short-oblong to subglobose, dull orange-red, ⅓–⅖ in. in diameter, with thin dry flesh, calyx little enlarged, with a narrow cavity pointed in the bottom; nutlets 2 or 3, narrowed at the rounded ends, ridged on the back with a narrow ridge, ¼–⅓ in. long and ⅙–⅕ in. wide.

LEAVES. Broad-obovate, rounded at apex, gradually narrowed and cuneate at base, finely and often doubly serrate to below the middle, with straight or slightly incurved callous-tipped teeth, slightly pilose on the upper side of the midrib early in the season, otherwise glabrous, thick, dark green and lustrous above, paler below, 1½–2½ in. long and ¾–1¼ in. wide; petioles slender, wing-margined to below the middle, sparingly pilose in May, becoming glabrous, ⅕–⅓ in. long.

TWIGS. Stout, slightly pilose at first, glabrous later, orange-brown or dark reddish brown; spines numerous, straight, purplish, 1⅓–2 in. long, some compound on older branches.

RANGE. Rocky hillsides in Arkansas; near Eureka Springs, Carroll County; Magnet Springs, Hot Spring County; and Faulkner County.

REMARKS. The genus name, *Crataegus*, is the classical Greek name for hawthorn. The species name, *subpilosa*, refers to the scant hairs on the upper midrib.

TRACY HAWTHORN

Crataegus tracyi Ashe—Series **Crus-galli** (10)

FIELD IDENTIFICATION. Bushy tree rarely more than

12–15 ft, with a trunk 8–12 in. in diameter, the branches erect and spreading.

FLOWERS. Opening late in April, about ⅝ in. in diameter, on villose pedicels ¼–½ in., in compact mostly 7–10-flowered villose corymbs, their bracts and bractlets linear-obovate, glandular-serrate; calyx-tube broadly obconic, glabrous or with occasional hairs near the base, the lobes gradually narrowed from a wide base, glandular-serrate, sometimes laciniate near the acuminate apex, glabrous on the outer surface, villose on the inner surface; petals 5, white, rounded to obovate; stamens 10–15, usually 10; anthers pink; styles 2–3.

FRUIT. Ripening in September or early October, on erect nearly glabrous or villose pedicels, short-oblong to ellipsoidal, orange-red, about ⅓ in. long, the calyx much enlarged; flesh thin, dry and mealy, yellowish green; nutlets 2–3, rounded at apex, ridged on the back with a low broad rounded ridge, about ¼ in. long.

LEAVES. Simple, alternate, deciduous, obovate to oval, rhombic or suborbicular; rounded, acute or acuminate, or abruptly short-pointed at apex, concave-cuneate at base, sharply serrate to below the middle with straight, acuminate, glandular teeth; dark green, lustrous and scabrate above, pale yellowish green below, blades 1–1⅓ in. long, ½–1 in. wide, with a thin midrib and prominent primary veins; petioles slender, wing-margined at apex, villose early in the spring, becoming glabrous later, about ¼ in. long.

TWIGS. Slender and nearly straight, the young ones covered with long scattered pale hairs, becoming dull brown or reddish and glabrous, eventually gray.

RANGE. Rocky banks of streams of the Texas Edwards Plateau area and westward. In Kendall, Comal, Bandera, Edwards, Callahan, Brown, Brewster, and Jeff Davis counties in Texas.

REMARKS. The genus name, *Crataegus*, is the classical name for hawthorn. The species name, *tracyi*, is in honor of C. Tracy, a Texas botanist who collected the type. The description cited is based on C. S. Sargent's original one of *C. montigava*, which is now considered a synonym of *C. tracyi* Ashe (Sargent, 1920).

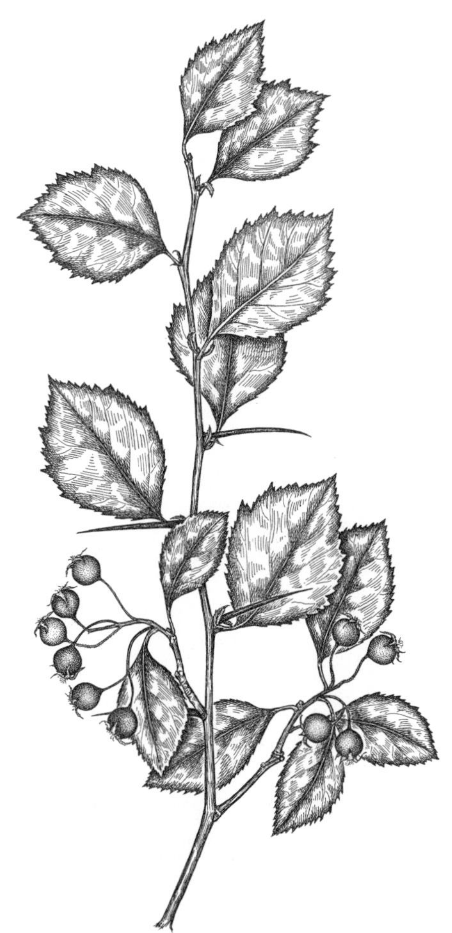

TRACY HAWTHORN
Crataegus tracyi Ashe

SANDHILL HAWTHORN

Crataegus collina Chapm.—Series **Punctatae** (11)

FIELD IDENTIFICATION. Tree attaining a height of 25 ft, with a well-shaped crown formed by stout spreading branches.

FLOWERS. Borne in many-flowered corymbs, the pedicels stout and villose; corolla about ¾ in. across with 5 rounded white petals; calyx-tube obconic, the 5 lobes gradually acuminate, margin finely glandular-serrate, glabrous externally, hairy within; stamens usually 20, pale yellow; styles 5.

FRUIT. Ripening in September, clusters few-fruited, erect or drooping, subglobose, ⅓–½ in. long or broad, dull red, often pale-dotted; mature calyx persistent, usually appressed, somewhat enlarged, flesh yellow; nutlets 5, ends rounded, dorsal surface grooved, about ¼ in. long.

LEAVES. Simple, alternate, deciduous, oval to obovate, apex acute, base broadly cuneate, margin irregularly serrate with straight or curved teeth except toward the base; young leaves red and pale-hairy, when mature upper surface yellowish green and essentially glabrous, lower surface glabrous or hairy along the midrib and 4–5 pairs of primary veins, blades 1½–2 in. long, 1–1¼ in. wide; leaves on young shoots larger and often with broad acute lobes; petioles ¼–½ in., slender, hairy at first, more glabrous later, yellow or reddish, somewhat winged by the leaf base.

TWIGS. Reddish and silky-hairy at first, later only puberulous and reddish brown, eventually gray with shiny spines 2–3 in. long, or even to 8 in. long and branched.

RANGE. Sandhill Hawthorn occurs in rich soil in Oklahoma, Mississippi, and Georgia; northward into Missouri, Indiana, and Virginia.

SANDHILL HAWTHORN
Crataegus collina Chapm.

FLOWERS. Opening early in May, borne in villose, 6–12-flowered, compound corymbs; corolla white, 5-petaled, about ¾ in. across; calyx-tube broadly obconic, densely hairy, lobes 5, gradually acute, margin glandular-serrate, surface green to reddish and pubescent; stamens 20; anthers pale rose-colored; styles 3–5, hairy at base.

FRUIT. Maturing in October, corymbs drooping, few-fruited, pedicels pubescent, subglobose, slightly flattened terminally, dark red; lobes of calyx semipersistent or deciduous; nutlets 3–5, ends acute, dorsally rounded and obscurely grooved, about ¼ in. long.

LEAVES. Simple, alternate, deciduous, rounded to obovate, apex acute or rounded, base cuneate, margin coarsely and sharply serrate except near the base, blade length 1½–2 in., width 1–1¼ in., young leaves reddish and densely white-hairy, upper surface dark green and roughened, lower surface paler and pubescent; when mature upper surface shiny and smoother, lower surface pubescent with conspicuous oblique veins; petioles about ¼–½ in. long, at first densely hairy, later only pubescent; leaves on young shoots larger, doubly and sharply serrate or some lobed.

TWIGS. Rather straight and stout, at first reddish

VARIETIES. A number of varieties of Sandhill Hawthorn are known:

C. collina var. *succincta* (Sarg.) Palmer has leaves relatively narrow, and mostly obovate, with a slight pubescence; stamens about 20; anthers yellow; nutlets 5. It occurs in Arkansas and southern Missouri.

C. collina var. *secta* (Sarg.) Palmer has leaves oblong-obovate or rhomic with small shallow lobes on the upper half, pubescent while young; stamens 5–10, usually 5. It occurs in Arkansas and southern Missouri.

HYBRIDS. *C.* × *incaedua* Sarg. is a hybrid of *C. collina* and *C. calpodendron.*

REMARKS. The genus name, *Crataegus,* is the classical name, and the species name, *collina,* means "of the hills." It is also known as the Chapman Hill Thorn.

WARTY HAWTHORN

Crataegus verruculosa Sarg.—Series **Punctatae** (11)

FIELD IDENTIFICATION. Tree attaining a height of 25 ft, with a trunk 6–12 in. in diameter, branches forming a compact rounded crown.

WARTY HAWTHORN
Crataegus verruculosa Sarg.

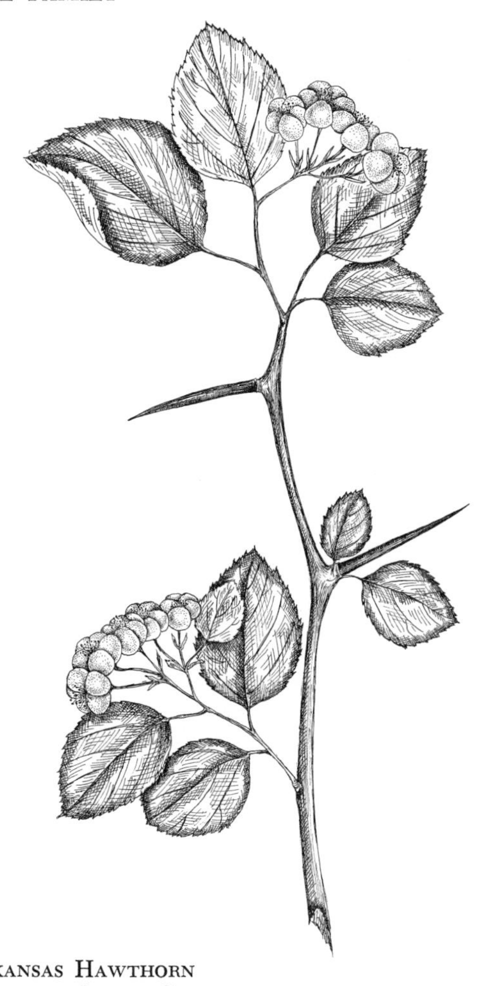

ARKANSAS HAWTHORN
Crataegus fastosa Sarg.

brown and pale-hairy, later grayish brown and glabrate but roughened by minute tubercles.

RANGE. In Missouri, Kentucky, and Arkansas. In Arkansas in Miller and Hempstead counties.

REMARKS. The genus name, *Crataegus,* is the classical name for the hawthorn. The species name, *verrucosa,* refers to the warty or minute tubercles of the twigs the first season.

ARKANSAS HAWTHORN

Crataegus fastosa Sarg.—Series **Punctatae** (11)

FIELD IDENTIFICATION. Tree attaining a height of 20 ft, with a trunk usually less than 12 in. in diameter, branches ascending to form an irregular crown.

FLOWERS. Opening in April in glabrous corymbs with slender pedicels, corolla white, 5-petaled, about ¾ in. across; bracts persistent, oblong-obovate or lanceolate, apex acute, margin glandular-serrate; calyx broadly obconic, 5-lobed, slender-acuminate at apex, margin glandular-serrate, glabrous externally, hairy within; stamens 20; anthers yellow; styles 5, woolly-hairy at base.

FRUIT. Maturing in October, clusters drooping, subglobose to short-oblong, orange-red, pale-dotted, about ⅜ in. in diameter; calyx deciduous with maturity; nutlets 3–5, ends narrowed, dorsally and obscurely ridged, about 5/16 in. long.

LEAVES. Simple, alternate, deciduous, oval to ovate, apex acute or rounded, base rounded to cuneate or some concave, margin coarsely serrate, blade length 1¾–2 in., width 1–2 in., when young pale-hairy and tomentose beneath, at maturity leathery and firm, upper surface shiny, yellowish green and glabrous; lower surface paler with conspicuous primary veins 3–5 to each side; leaves on young shoots often lobed; petioles densely hairy but eventually less so.

TWIGS. Slender, green to reddish brown later, shiny, armed with stout, lustrous, chestnut-brown spines 1½–2 in. long.

BARK. Broken into dark brown or black scales.

RANGE. Rather rare, known from near Fulton in southwestern Arkansas.

REMARKS. The genus name, *Crataegus,* is the classical name for the hawthorn. The meaning of the species name, *fastosa,* is obscure; possibly *fastuosa,* meaning "proud" or "stately," was intended, from the large size.

BRAZORIA HAWTHORN

Crataegus brazoria Sarg.—Series **Punctatae** (11)

FIELD IDENTIFICATION. Tree to 25 ft and 6–10 in. in diameter, the branches ascending to form a rounded crown.

FLOWERS. Opening in March, in hairy 7–9-flowered corymbs, pedicels slender, bearing white 5-petaled corollas; calyx-tube obconic, pale-hairy; lobes 5, oblong to linear, acuminate, margin entire or glandular-serrate, hairy externally and within; stamens 20, anthers red; styles 5, basally hairy.

FRUIT. Maturing September–October, corymbs lax and few-fruited, subglobose to short-oblong, yellow, some pale-dotted, length ⅓–½ in., flesh yellow and thin; nutlets 5, about ¼ in. long, dorsally rounded and grooved.

LEAVES. Simple, alternate, deciduous, oval to obovate, apex acute or acuminate, base cuneate or gradually narrowed, margin glandular-serrate except near the base, young leaves reddish and pale-hairy; older leaves dark green, shining and glabrous, lower surface paler, blade length 2–2½ in., width 1¼–1½ in.; leaves on young shoots usually larger, broad-ovate to

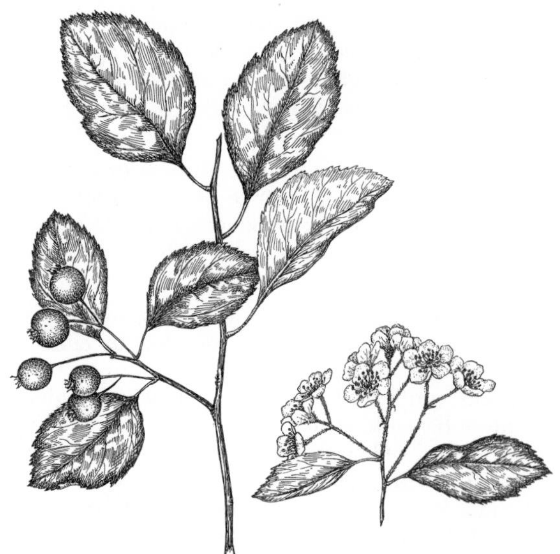

BRAZORIA HAWTHORN
Crataegus brazoria Sarg.

oblong, base rounded or broadly cuneate; petioles ⅓–¾ in., tomentose at first, glabrous later.

TWIGS. When young pale-hairy and reddish brown, later puberulous or almost glabrous, reddish brown to gray, unarmed, or with a few slender gray spines.

RANGE. Bottom lands of the Brazos River, Brazoria and Matagorda counties in Texas.

REMARKS. The genus name, *Crataegus*, is the classical name for the hawthorn. The species name, *brazoria*, is given for the town of Brazoria, Texas.

FIRE-BERRY HAWTHORN

Crataegus chrysocarpa Ashe—Series **Rotundifoliae** (12)

FIELD IDENTIFICATION. Stout, thorny, intricately branched shrub, or rarely a small tree to 18 ft.

FLOWERS. In May, borne in loose villose corymbs; petals 5, white, rounded, corolla ⅜–¾ in. wide when spread; stamens 5–10, anthers white or pale yellow; calyx-tube slightly villose, the 5 sepals lanceolate, nearly entire, glandular or finely glandular-serrate on margin, usually glabrous externally.

FRUIT. Borne August–October, subglobose or short-oblong, about ⅜ in. thick, dull or dark red (rarely golden yellow or orange); flesh thin, soft and yellow; nutlets 3–4, rarely 2.

LEAVES. Simple, alternate, deciduous, elliptic to oval or suborbicular, apex acute, base broadly cuneate to almost truncate, margin serrate with gland-tipped teeth, also often with 3–9 pairs of triangular and lateral lobes, blade length 1½–2½ in., width ¾–1¾ in., when young with short appressed hairs above, later yellowish green and more glabrate with impressed veins, lower surface more or less villous and paler, firm; petiole ⅜–¾ in., slender, flexuous, glabrous or slightly villous and glandular.

TWIGS. Stout, young ones villous, older ones glabrous, reddish brown to gray, thorns numerous ½–2¾ in. long.

RANGE. In thickets on rocky grounds along streams at altitudes of 5,000–6,000 ft. New Mexico; northward to South Dakota and Saskatchewan, and eastward to Iowa, Pennsylvania, New York, the New England states and southeastern Canada.

VARIETIES. A number of varieties are known as follows:

Eastern Fire-berry Hawthorn, *C. chrysocarpa* var. *phoenicea* Palmer, has all parts glabrous except appressed hairs on the upper side of young leaves. Pennsylvania and Michigan, northward to Nova Scotia and west to Saskatchewan.

Bicknell Fire-berry Hawthorn, *C. chrysocarpa* var. *bicknelli* (Egglest.) Palmer, has flowers in slightly villous corymbs and pink anthers; leaves averaging slightly larger, more deeply divided, lobes often reflexed at apex, glabrous except on the upper surface while young. Known from Massachusetts.

Hairy Fire-berry Hawthorn, *C. chrysocarpa* var.

FIRE-BERRY HAWTHORN
Crataegus chrysocarpa Ashe

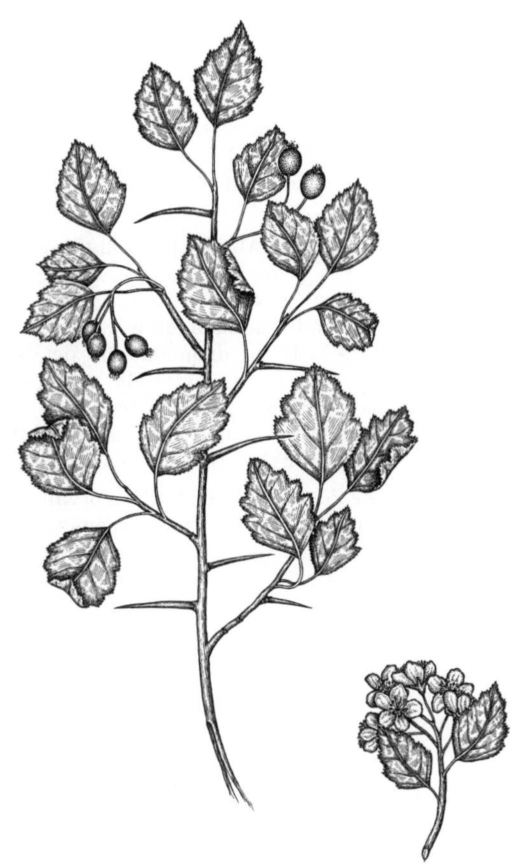

CERRO HAWTHORN
Crataegus erythropoda Ashe

caesariata (Sarg.) Palmer, has the flowering corymbs sparsely villous; fruit oblong to obovate, ⅜–⅝ in. long, to ⅜ in. thick, slightly puberulous at apices when young; nutlets 2–3. Known from New York.

Red-leaf Fire-berry Hawthorn, *C. chrysocarpa* forma *rubescens* (Sarg.) Palmer, has the foliage dark red when young and occasionally throughout the season. Known from Quebec.

REMARKS. The genus name, *Crataegus*, refers to the strong wood, and the species name, *chrysocarpa*, means "golden-fruited," because of the occasionally golden or yellow fruit.

CERRO HAWTHORN

Crataegus erythropoda Ashe—Series **Rotundifoliae** (12)

FIELD IDENTIFICATION. Small Western shrub or tree 6–15 ft. The branches are spreading and the trunk is short with a rough bark.

FLOWERS. Appearing April-May. Borne in 5–10-flowered corymbs, pedicels ¼–½ in. long, corolla ⅝–¾ in. across; petals 5, white, spreading, rounded, ¼–⅓ in. broad, reticulate-veined; calyx urn-shaped, 5-lobed, lobes longer than the calyx-tube, acuminate and gland-margined; stamens 1–8 (usually 5–8), anthers large, purple or rose-colored; styles usually 5, ovary 2–5-celled.

FRUIT. Pome cherry-red becoming black, brown, or mahogany-colored; seeds usually 5, large, bony, somewhat dissimilar, the pulp scanty.

LEAVES. Simple, alternate, deciduous, broadly elliptic-ovate, apex acute to acuminate, base abruptly contracted, margins with fine gland-tipped serrations and occasionally shallowly lobed, blade length 1¼–2 in., thin; upper surface olive-green to dark green, at first thinly hairy, later glabrous; lower surface pale, pubescent, more so on the veins, older leaves almost glabrous, veins conspicuous; petioles slender, glabrous, ¼–1 in. long, green to reddish.

TWIGS. Stout, reddish brown, later gray, lenticels small and pale; thorns numerous, stout, straight or curved, ¾–2 in. long.

RANGE. Along stream banks, in canyons, in the yellow pine belt at altitudes of 5,000–7,000 ft. New Mexico in the Sandia Mountains and adjacent Arizona; northward to Montana, Idaho, and Washington, and eastward to Missouri.

REMARKS. The genus name, *Crataegus*, is from a Greek word meaning "strong," with reference to the wood. The species name, *erythropoda*, means "red-footed," perhaps for the reddish leaf petioles. It is also known under the vernacular name of Manzana de Puya Larga. Sometimes listed under the name of *C. cerronis* A. Nels.

LARGE-SEEDED HAWTHORN

Crataegus macrosperma Ashe—Series **Tenuifoliae** (13)

FIELD IDENTIFICATION. Small tree to 25 ft, with wide-spreading branches to form an oval or rounded crown.

FLOWERS. Appearing when the leaves are nearly grown, corymbs 4–9-flowered, nearly simple, flowers on slender pedicels, white, 5-petaled, about ⅗ in. wide; calyx with 5 lanceolate divisions, ⅛–⅕ in. long, persistent and colored with the fruit; stamens 5–10, usually 5, the base of the stout filaments persistent and coloring with the fruits; styles 3–4.

FRUIT. Falling the last of September or October with slender pedicels, borne in small clusters, body ½–⅘ in. in diameter, dark bright red, glabrous or somewhat glaucous, flesh thick and mealy; nutlets 3–5, grooved and ridged dorsally, ⅕–⅓ in. long, ⅕ in. thick dorsoventrally, the lateral faces plane.

LEAVES. Simple, alternate, deciduous, deltoid or broadly oval, apex obtuse; base subcordate or on vigor-

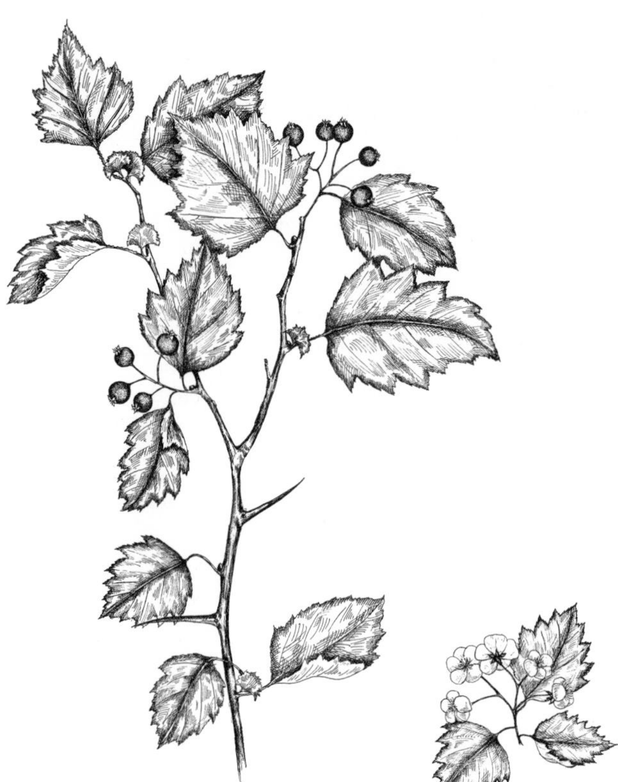

Large-seeded Hawthorn
Crataegus macrosperma Ashe

ous shoots cordate with a narrow sinus, varying greatly in size even on the same twig, blades 1¼–2⅓ in. long, ¾–2 in. wide, finely but sharply serrate to the base, doubly serrate or with 3–5 pairs of shallow lobes above, the serrations acutely gland-tipped, 4–6 pairs of prominent veins; petioles ⅜–¾ in. long, generally short and less than one third the length of the blade, channeled, narrow-margined, at least above, and with a few glands near the base of the leaf.

TWIGS. Young rather thick, reddish brown to purplish, sparingly glaucous-glabrous, marked with a few small grayish lenticels; winter buds rather large, oval or globular, dark reddish brown, the obtuse or rounded scales glabrous; older twigs gray; thorns numerous, short, very stout, ⅜–1¼ in. long, dark reddish brown or purplish to nearly black.

BARK. Grayish brown, broken into small oblong scales.

RANGE. Known from an isolated location in Pope County, Arkansas; elsewhere from Georgia, north to Newfoundland, and west to Minnesota.

REMARKS. The genus name, *Crataegus*, is from a Greek word referring to the strong wood. The species name, *macrosperma,* refers to the large seeds. Also known as Roan Mountain Haw or Variable Thorn. Closely related to *C. coccinea* L.

Shrub Hawthorn

Crataegus iracunda Beadle—Series **Silvicolae** (14)

FIELD IDENTIFICATION. Large shrub with many stems or a slender tree to 25 ft and 6–8 in. in diameter, the gray-spined branches ascending or spreading into a round-topped crown.

FLOWERS. Borne April–May, appearing when the leaves are small; bracts and bractlets reddish, linear, acuminate, glandular; corymbs simple, 3–7-flowered; peduncles glabrous and glandular or glandless; corolla white, 5-petaled, ½–¾ in. in diameter, the petals oval to obovate, veiny; calyx-tube narrowly obconic, the sepals 5, acuminate, glabrous, glandular, entire or serrate; stamens about ⅛ in. long, 10 in number, anthers large and dark purple; styles 3–5, hairy at the base.

FRUIT. Maturing September–October, corymbs erect and few-fruited, subglobose, ⅓–⅖ in. in diameter, green to red, flesh firm, thin, yellow; sepals falling away generally before maturity of fruit; nutlets 3–5, about ¼ in. long, hypostyle ⅛–⅙ in. long.

LEAVES. Broadly ovate to deltoid, apex acute or acuminate, base truncate, rounded or broadly cuneate, margin sharply and doubly serrate above the middle, some teeth short-acute and some long-acuminate, sometimes cut into short acute lobes as well as teeth, young leaves reddish and soft-pubescent; when older thin, yellowish green, upper surface smooth or slightly

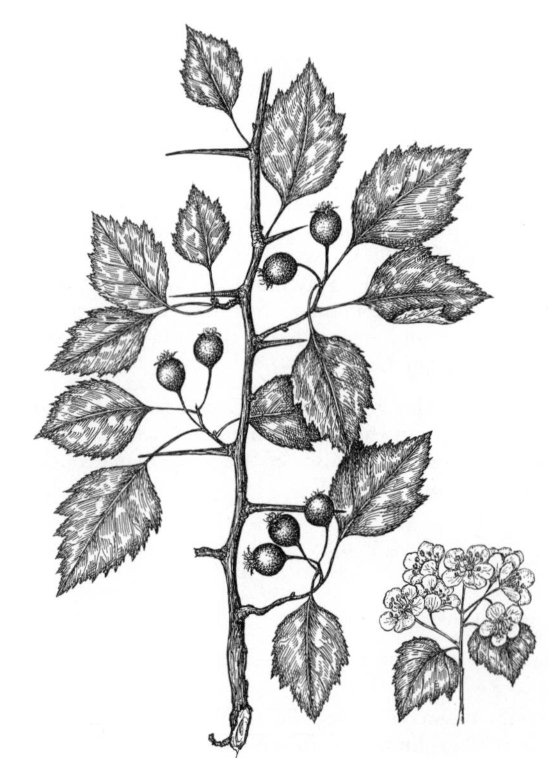

Shrub Hawthorn
Crataegus iracunda Beadle

scabrous, lower surface paler and slightly puberulent along the midrib especially, blade length ½–2½ in., width ½–2¼ in.; petiole ⅓–1 in., slender, glabrous, glandular.

TWIGS. Slender, green to reddish brown, eventually gray, white-pubescent at first, glabrous later; spines slender, straight or slightly curved, lustrous dark brown or gray on old branches, length ¾–2 in.

RANGE. Usually in moist flat woods. Louisiana (St. Tammany and Caldwell parishes), eastward to Georgia and north to Pennsylvania.

SYNONYMS AND VARIETY. It was formerly listed under the names of *C. drymophila* Sarg. and *C. silvicola* Beadle.

E. J. Palmer has separated a Southern variation under the name of *C. iracunda* var. *silvicola* (Beadle) Palmer, differing by having leaves deltoid or broad-ovate, with margins more finely serrate and with fewer lobes.

REMARKS. The genus name, *Crataegus*, is the classical Greek name for hawthorn. The species name, *iracunda*, means "irascible," "resentful," or "touchy," the significance of which is unknown.

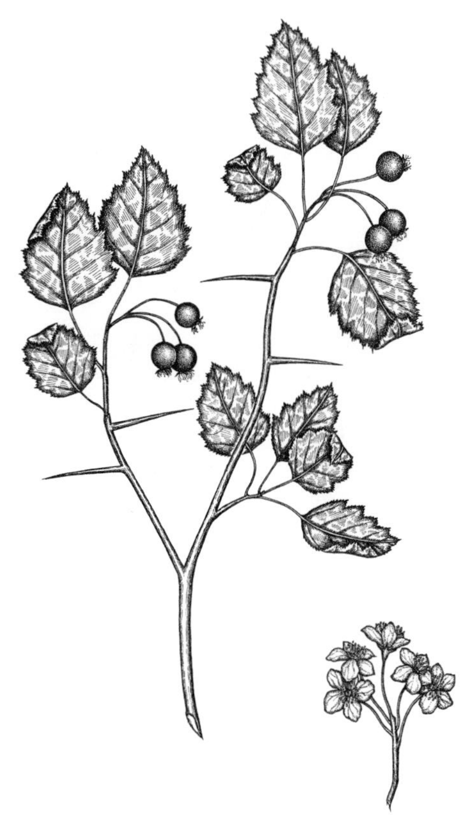

FROSTED HAWTHORN
Crataegus pruinosa (Wendl.) K. Koch

FROSTED HAWTHORN

Crataegus pruinosa (Wendl.) K. Koch—Series **Pruinosae** (15)

FIELD IDENTIFICATION. Divaricately branched shrub or tree, attaining a height of 20 ft and a diameter 4–8 in. The ascending and horizontal branches shape into an open irregular crown.

FLOWERS. Opening in May, corymbs glabrous and few-flowered, pedicels slender, bearing white 5-petaled corollas ¾–1 in. across; calyx-tube broadly obconic, smooth and glabrous, 5-lobed, the lobes acuminate-pointed and glandular-serrulate below; stamens 20, anthers rose-colored; styles 5, basally tomentose.

FRUIT. Borne in October, corymbs lax, few-fruited, pedicels green or red; body of fruit somewhat angled at first but at maturity mostly subglobose, ½–⅝ in. broad, dark red and often glaucous, shiny, often dotted, flesh yellow; nutlets 5, base rounded, apex acute, deeply grooved dorsally, about ¼ in. long; calyx conspicuous but the erect lobes usually deciduous.

LEAVES. Simple, alternate, deciduous, elliptic, apex acute, base cuneate, margin doubly serrate and also short-lobed; mature leaves firm and leathery, dark green and often with a bloom above, lower surface paler with 3–4 pairs of secondary veins, blade length 1–1½ in., width ¾–1 in.; leaves on young shoots broad-ovate, usually larger and with deeper lobes; petioles slender green or reddish, somewhat winged by the leaf base, ½–1 in. long.

TWIGS. Slender, glabrous, reddish brown, with stout, straight brown thorns ½–1½ in. long.

RANGE. Northern Arkansas; north to Newfoundland, and west to Ontario and Wisconsin.

REMARKS. The genus name, *Crataegus*, is the classical name for the hawthorn. The species name, *pruinosa*, means "frosty" and refers to the waxy bloom of the fruit. It is also known under the name of the Waxy-fruit Thorn. Some authors list the plant as *C. patrum* Sarg.

GATTINGER HAWTHORN

Crataegus gattingeri Ashe—Series **Pruinosae** (15)

FIELD IDENTIFICATION. Slender, spiny shrub or small tree to 25 ft.

FLOWERS. In May, corymbs few-flowered, pedicels slender and glabrous, ⅜–¾ in. long; petals 5, white, rounded; calyx-lobes 5, short, triangular, glabrous; stamens 20, anthers pink.

FRUIT. In October, dark dull red, sparingly pruinose, globular, ⅓–½ in. thick, flesh thin and dry, generally capped by the stalked calyx-lobes, persistent until after the foliage has fallen; nutlets 3–5.

GATTINGER HAWTHORN
Crataegus gattingeri Ashe

LEAVES. Simple, alternate, deciduous, oblong to ovate or deltoid, ¾–2½ in. long, ¾–2 in. wide, apex attenuate, base rounded to truncate or subcordate, margin finely serrate and some with 3–5 acute lateral lobes as well, terminal lobes sometimes elongate, surfaces glabrous, upper surface dark green, paler beneath; petiole slender, roughened above with 1–2 pairs of glands.

TWIGS. Slender, flexuous, glabrous, dark purplish brown to dark gray, sparingly glaucous, armed with numerous thorns 1–1½ in. long.

RANGE. Arkansas and Tennessee, northward to Missouri, Indiana, West Virginia, and Pennsylvania.

VARIETY. Rigid Gattinger Hawthorn, *C. gattingeri* var. *rigida* (Ashe) Palmer, is a variety with stout rigid branches, thorns stout, ⅜–⅝ in. long; flowers ½–⅝ in. wide. In Indiana and Kentucky.

REMARKS. The genus name, *Crataegus*, is the classical name. The species name, *gattingeri*, is in honor of Augustine Gattinger (1825–1903), Tennessee botanist of German birth. The type specimen was collected at Nashville by Gattinger in 1880. The original description was published by Ashe (1900). W. W. Ashe remarks: "The species probably has its nearest relative in *C. erythrocarpa* of the Atlantic coast, from which it is separated by its entirely different foliage and somewhat larger and more fleshy fruit."

MISSOURI HAWTHORN

Crataegus disjuncta Sarg.—Series **Pruinosae** (15)

FIELD IDENTIFICATION. Rather slender tree to 20 ft with slender branches and a loose open crown.

FLOWERS. Appearing April–May in glabrous 3–6-flowered corymbs with stout pedicels; bracts and bractlets glandular, dropping early; petals 5, white, rounded, when open the flower about ⅔ in. across; stamens 10, anthers pink or red; styles 4–5, tomentose at the base; calyx-tube obconic, glabrous, the 5 sepals elongate, acuminate, glabrous, margin entire or obscurely glandular-serrate.

FRUIT. Clusters loose and drooping, body green to reddish, glaucous, subglobose, or sometimes angled, ½–¾ in. in diameter, calyx-lobes erect or spreading, deciduous early; flesh yellowish or greenish, thin, mealy; nutlets 4, dorsally grooved, ends rounded, about ¼ in. long.

LEAVES. Simple, alternate, deciduous, blade broad-ovate, apex acute or acuminate, base entire and rounded to cuneate or somewhat concave, margin doubly glandular-serrate, usually with spreading acuminate lobes above the middle, blade length 2⅓–3 in., width 2¼–2½ in., upper surface dark green, glabrous, lower surface paler, midrib yellow, primary veins 4–5 pairs; petiole ½–1⅓ in. long, blade gradually narrowed downward into a short petiolar wing.

TWIGS. Stout, divaricate, green to reddish brown,

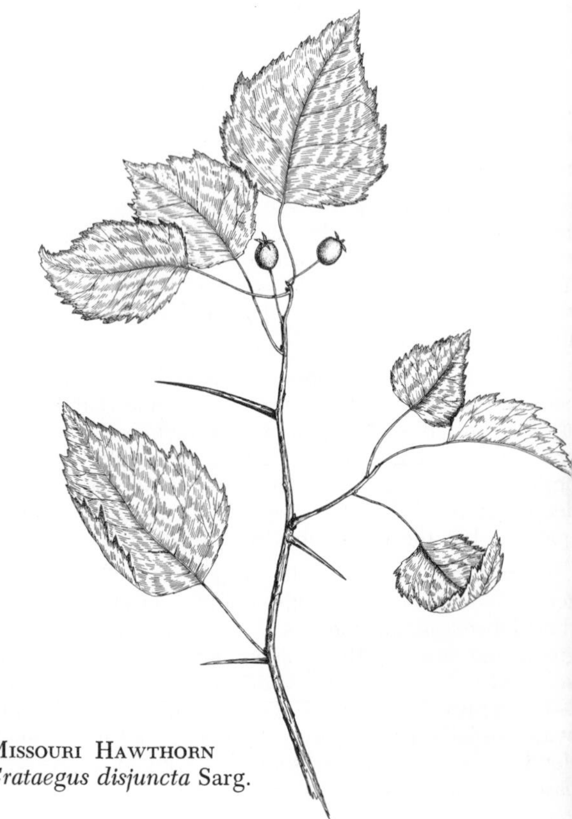

MISSOURI HAWTHORN
Crataegus disjuncta Sarg.

later grayish; spines stout, gray to purple, shiny, 1–3 in. long.

RANGE. Northern Arkansas, western Kentucky, and Missouri.

REMARKS. The genus name, *Crataegus,* is from an ancient Greek word referring to strong wood. The species name, *disjuncta,* refers to the "distinct" appearance of the tree as compared with other hawthorns, with large leaves and few flowers in a compact cluster.

Mackenzie Hawthorn

Crataegus mackenzii Sarg.—Series **Pruinosae** (15)

FIELD IDENTIFICATION. Shrub 6–8 ft, with slender, intricately branched, very spiny stems.

FLOWERS. Opening in April, borne in narrow, 5–6-flowered glabrous corymbs with stout pedicels; lower portion of the peduncle elongate from the upper leaf axils; bracts linear to oblong or obovate, pinkish, glandular; calyx-tube narrowly obconic, the 5 lobes short, gradually acuminate, margin entire or glandular-serrate; corolla ⅝–¾ in. wide, petals 5, white, rounded; stamens 20, anthers pink; styles 3–5.

FRUIT. Maturing in October in few-fruited clusters; pedicels stout, elongate, erect or spreading; fruit body subglobose to short-oblong, or often somewhat broader than high, occasionally somewhat angled, green at first, red later, pruinose, ⅜–½ in. in diameter; fruiting calyx prominent, tube short; cavity deep, wide, broad and tomentose in the bottom; lobes 5, spreading, apices deciduous from the mature fruit; flesh thin, dry, mealy, hard; nutlets 3–5, apices gradually narrowed and acute, base rounded, dorsally rounded and grooved, length about ⅕ in., width ⅙ in.

LEAVES. Simple, alternate, deciduous, ovate to deltoid, apex acute, base rounded to abruptly cuneate or truncate; margin serrate, often doubly so, with straight glandular teeth, also often with 3–4 pairs of narrow, acuminate lateral lobes; young leaves reddish, slightly hairy, later glabrous and firm; upper surface dark bluish green, smooth and shiny; lower surface paler, midrib slender and yellowish, primary veins thin, blade length 2–2⅓ in., width 1¾–2 in.; petioles slender, ⅝–¾ in. long, often rose-colored later, sometimes glandular near the apex; leaves on young vigorous shoots thicker, mostly coarse-serrate, more deeply lobed, 2–2½ in. long and wide.

TWIGS. Stout, zigzag, somewhat angled, first orange-green, later light to dark brown, armed with spines which are numerous, slender, straight, purple, lustrous, 1½–2¾ in. long.

RANGE. In thickets, open woods, or rock barrens; Arkansas and Oklahoma, north to Iowa, Missouri, and Kentucky.

VARIETIES. Bracted Mackenzie Hawthorn, *C. mackenzii* var. *bracteata* (Sarg.) Palmer, has bracts and leaves mostly ovate, rarely unlobed. Also listed as *C. bracteata* Sarg. Southern Missouri, Oklahoma, and Arkansas.

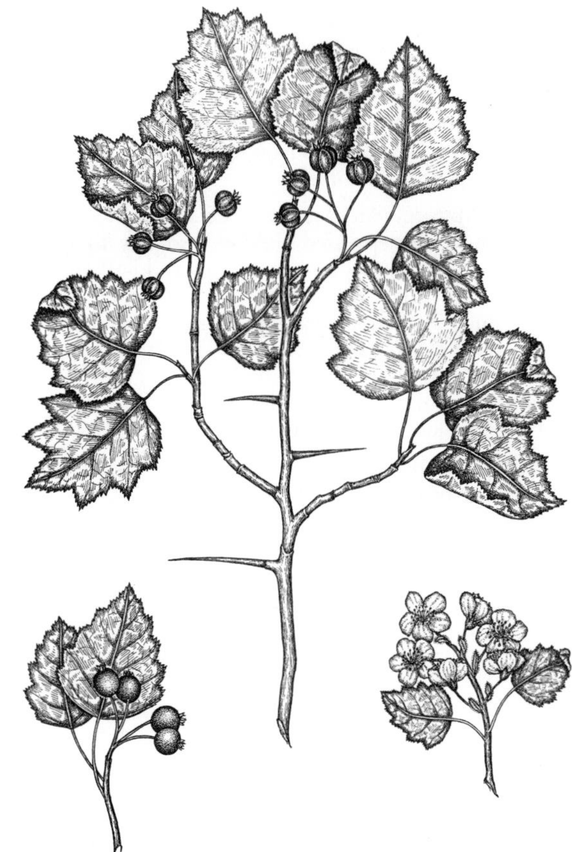

Mackenzie Hawthorn
Crataegus mackenzii Sarg.

Rough Mackenzie Hawthorn, *C. mackenzii* var. *aspera,* Palmer, has leaves short-hairy when young, somewhat pubescent along the veins later; inflorescence more or less pubescent. Also listed as *C. aspera* Sarg. and *C. decorata* Sarg. Known from southern Missouri.

REMARKS. The genus name, *Crataegus,* is from a Greek word for hard wood. The species name, *mackenzii,* honors Kenneth Kent Mackenzie (1877–1934), American botanist.

Flat-fruit Hawthorn

Crataegus platycarpa Sarg.—Series **Pruinosae** (15)

FIELD IDENTIFICATION. Tree to 23 ft, with a trunk 2–6 in. in diameter.

FLOWERS. Opening in April when the leaves are about half grown, borne on short glabrous pedicels in small compact 4–5-flowered corymbs; bracts and bractlets reddish, linear to obovate, glandular-serrate, persistent until the flowers open; petals 5, white, rounded; calyx-tube broadly obconic, glabrous, the 5 sepals wide, acuminate, entire or occasionally minutely dentate, glabrous, reflexed after anthesis; stamens 20; anthers large, bright red; styles 3–5, surrounded at the base by a narrow ring of pale tomentum.

FRUIT. Maturing in October, in short pediceled 1–3-fruited clusters, depressed-globose, broader than high, base somewhat angled, deep orange-red, about ⅜ in. in diameter; calyx prominent, with a broad shallow cavity, and 5 small, spreading or incurved, persistent sepals; flesh thick, yellow, becoming soft or succulent; nutlets 3–5, base acute, apex broader and rounded, dorsally irregular ridged, with a high narrow ridge, about ¼ in. long and ⅕ in. wide.

LEAVES. Simple, alternate, deciduous, when opening yellowish green above, pale and furnished below with small axillary tufts of white hairs; when mature thin, dark yellowish green, glabrous and shiny above, lower surface paler and nearly glabrous, blades 2–2⅓ in. long and 1½–1¾ in. wide, with slender midribs and thin primary veins; ovate, apex acuminate, base abruptly concave-cuneate or rounded and entire, margin sharply serrate above, with straight glandular teeth and divided into 4 or 5 pairs of small acuminate spreading lobes; petioles slender, narrow winged margin at apex, slightly villose when young, becoming glabrous, sparingly glandular, with usually deciduous glands, ⅝–1¼ in.; leaves on young shoots usually rounded or truncate at the broad base, more coarsely serrate and more deeply lobed, 2¼–2¾ in. long and nearly as wide.

TWIGS. Nearly straight, slender, glabrous, young ones light green tinged with red, becoming light brown, shiny, and with small pale lenticels becoming gray; spines numerous, stout, slightly curved, purplish, lustrous, 1–2 in. long.

RANGE. Dry ridges between swamps; northeastern Arkansas, Missouri, Indiana, and Illinois.

REMARKS. The genus name, *Crataegus*, is an old name referring to the hard wood. The species name, *platycarpa*, refers to the broad, depressed-globose fruit.

Downy Hawthorn

Crataegus mollis (Torr. & Gray) Scheele—
Series **Molles** (16)

FIELD IDENTIFICATION. Tree attaining a height of 40 ft and a diameter of 12–18 in., branches stout and spreading to form a round-topped crown.

FLOWERS. Opening April–May, corymbs many-flowered and broad, pedicels densely hairy; bracts and bractlets conspicuous; corolla white, 5-petaled, about 1 in. across; calyx-tube narrowly obconic, densely tomentose, lobes 5, linear to lanceolate, apex acuminate, margin serrate and red-glandular, externally villose, tomentose within; stamens 20, anthers large, light yellow, styles 4–5, basally hoary-tomentose.

FRUIT. Maturing in September, corymbs hairy, few-fruited, drooping, subglobose or short-oblong, terminally rounded, somewhat pubescent, scarlet, some dark dotted, ⅔–1 in. in diameter; flesh yellow, dry, mealy; nutlets 4–5, light brown, dorsally rounded and lightly ridged, about ¼ in. long; fruiting calyx-lobes deciduous when fruit half grown, at first erect, incurved, and hairy.

LEAVES. Simple, alternate, deciduous, broad-ovate, apex acute, base rounded or cordate, margin doubly serrate, and also with 4–5 lateral acute or rounded lobes on each side; mature leaves with upper surfaces dark yellowish green; lower surface paler and pubescent, particularly along the midrib, lateral veins 4–5 pairs, length and width of blades 3–4 in.; leaves on young shoots larger, lobes and basal sinus deeper; petioles ½–1¼ in. long, stout, densely hairy at first, later pubescent, some with minute dark glands.

TWIGS. Stout, densely white-villose at first, when older becoming glabrous, gray, bearing stout, shiny, chestnut-brown spines 1–2 in. long.

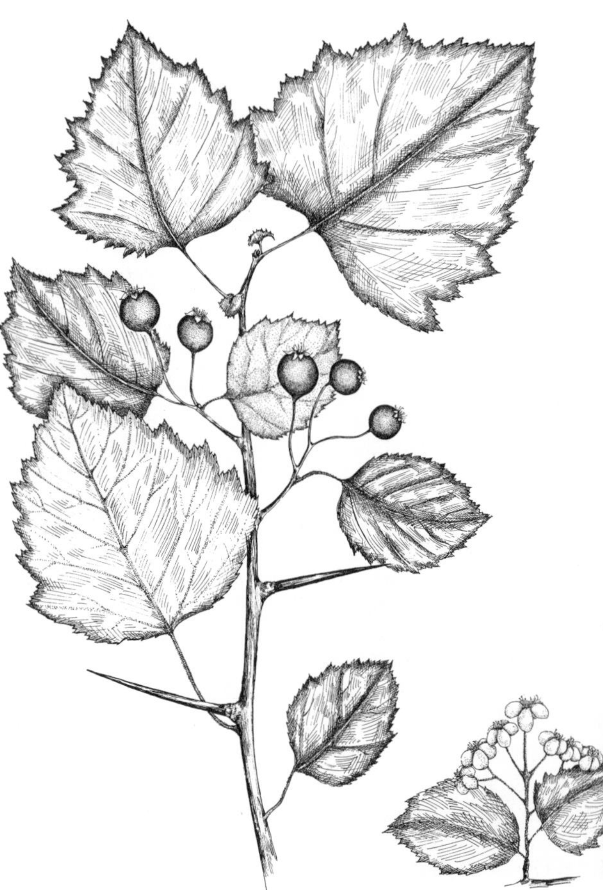

Downy Hawthorn
Crataegus mollis (Torr. & Gray) Scheele

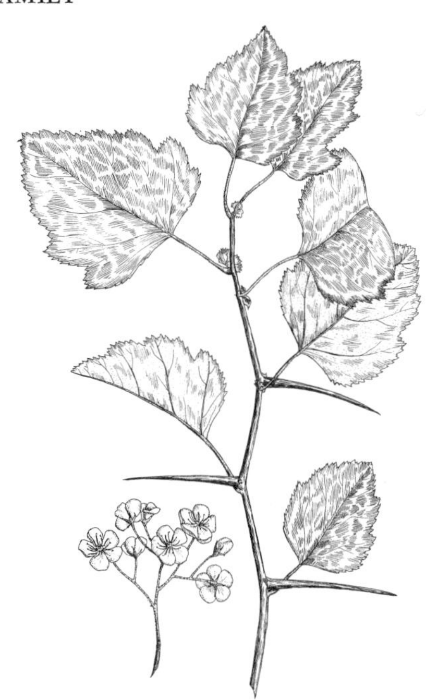

SHINY HAWTHORN
Crataegus limaria Sarg.

RANGE. In rich bottom lands. In Texas in Dallas, Grayson, and Lamar counties, and Oklahoma; east to Alabama, north to Ontario, and west to Minnesota and South Dakota.

SYNONYM. Downy Hawthorn has also been listed in the literature under the names of *C. arkansanum* Sarg., *C. lasiantha* Sarg., and *C. gravida* Beadle. Vernacular names in use are Red Haw and Downy Thorn.

REMARKS. The genus name, *Crataegus,* is the classical name for hawthorn, and the species name, *mollis,* refers to the soft-hairy foliage.

SHINY HAWTHORN

Crataegus limaria Sarg.—Series **Molles** (16)

FIELD IDENTIFICATION. Tree 20–30 ft, with a trunk 6–12 in. in diameter, the branches forming an irregular crown.

FLOWERS. March–April, borne in white-hairy corymbs of 12–20 flowers; petals 5, white, rounded, when spread the flower to 1 in. in diameter; stamens 15–18, anthers white; styles 3–5, hairy at base; calyx-tube obconic, white-hairy, the 5 sepals narrow, acuminate at apex, margin cut into glandular serrations.

FRUIT. Maturing September–October, few in a cluster, pedicels hairy, body ovoid to ellipsoid or short-oblong, base truncate or rounded, apex rounded, red, sometimes with pale spots, hairy at first, later glabrous, ½–⅗ in. in diameter; calyx with the 5 sepals hairy, reddish and persistent; flesh yellowish, mealy, insipid; nutlets 3–5. obscurely grooved dorsally, ends rounded, ⅕–¼ in. long.

LEAVES. Simple, alternate, young ones white-hairy above and densely tomentose beneath; mature leaves with upper surfaces light green and roughened, lower surface paler and tomentose, blade ovate, margin singly or doubly serrate (or 3–4-lobed on each margin), apex acute, base rounded to cordate or concave-cuneate, length 2½–3 in., width 1½–2 in.; petioles ¾–1½ in., slender, villose.

TWIGS. Young ones yellow to orange-brown and densely white-hairy, older ones brown to gray and eventually glabrous, spines purple to gray, sharp, slender, straight or slightly curved.

RANGE. Bottom lands of the Guadalupe, Cibolo, and San Antonio rivers in Texas. In Arkansas on the Red River near Fulton in Hempstead County. Also in southeastern Oklahoma.

SYNONYM. The tree is listed by some authors under the name of *C. mackensenii* Sarg.

REMARKS. The genus name, *Crataegus,* is from an ancient Greek word referring to the strength of the wood. The species name, *limaria,* refers to the lustrous, or shiny, fruits.

BLUE-LEAF HAWTHORN

Crataegus lanuginosa Sarg.—Series **Molles** (16)

FIELD IDENTIFICATION. Tree to 25 ft, with stout erect branches and an irregular crown. The bluish green leaves and very long slender spines are distinguishing features.

FLOWERS. Opening in April, corymbs many-flowered, villose with long pale hairs; corolla white, 5-petaled, about ¾ in. in diameter, borne on short stout pedicels; bracts and bractlets conspicuous, glandular-serrate, deciduous early; calyx-tube broadly obconic, hairy, 5-lobed, the lobes short, apex acute, minutely stipitate-glandular, externally villose, less villose on the inner surface; stamens 20, anthers pink or rose-colored; styles 5, basally white-hairy.

FRUIT. Maturing in October, corymbs few-fruited, pedicels short, erect, and tomentose, body subglobose to short-oblong, rounded and puberulent terminally, about ½ in. in diameter, dark crimson; flesh orange, thin, dry; nutlets 5, flattened, dorsally rounded and ridged, about ¼ in. long; fruiting calyx with the lobes persistent, erect, spreading or incurved, green or reddish.

LEAVES. Simple, alternate, deciduous, blade length 1½–2 in., width 1–1½ in., ovate to oval, apex acute or rounded, base broadly cuneate or rounded, margin sharply glandular-serrate but entire toward the base,

BLUE-LEAF HAWTHORN
Crataegus lanuginosa Sarg.

firm; upper surface distinctly bluish green, shiny, and scabrate; lower surface yellowish green and densely hairy, with 3–5 pairs of secondary veins; leaves on young shoots broad-ovate, base rounded or truncate, coarsely glandular-serrate, usually larger than the normal leaves; petioles ½–¾ in., stout, tomentose.

TWIGS. Stout, divaricate, green and densely hairy at first, later reddish brown or gray and pubescent, bearing lustrous, purple to gray, straight, slender spines ¾–3½ in. in length.

RANGE. Eastern Oklahoma, central Arkansas, Missouri, and Kansas.

SYNONYM. It is also listed under the name of *C. dasyphylla* Sarg. in some literature.

REMARKS. The genus name, *Crataegus,* is the classical name for the hawthorn, and the species name, *lanuginosa,* refers to the soft hairs on young parts. Another vernacular name is Woolly Thorn.

GREGG HAWTHORN

Crataegus greggiana Egglest.—Series **Molles** (16)

FIELD IDENTIFICATION. Shrub 9–12 ft, with a stem covered with thin scaly bark.

FLOWERS. Opening early in April, about ⅜ in. in diameter on slender villose pedicels in small compact 5–7-flowered corymbs, densely villose like the narrow

obconic calyx-tube, calyx-lobes slender, acuminate, en tire or minutely and irregularly glandular-serrate, pu bescent on the outer surface, densely villose on the in ner surface, deciduous from the ripe fruit; stamen 5–10, anthers yellow; styles 3–5.

FRUIT. Ripening early in October on more or les densely villose erect pedicels in small clusters, globose bright red, ⅜–½ in. in diameter, slightly pubescent wit a ring of white hairs surrounding the slightly enlarge calyx (or hairless in some forms), composed of a shor tube and a deep narrow cavity pointed in the bottom flesh dry and thick; nutlets 3–5, gradually narrowe and rounded at the ends, slightly ridged on the back ⅙–⅕ in. long and ⅛–⅙ in. wide, the broad conspicuou hypostyle extending to below the middle.

LEAVES. Ovate, acute and short-pointed at apex, con cave-cuneate at base, on weak shoots slightly lobed more deeply so on leading shoots; the lobes short, acu minate, usually occurring only above the middle, wit deep double serrations, often extending nearly to th leaflet base; the teeth slender, acuminate, gland-tipped expanding blades covered above with short pale hairs pale and villose below, especially along the midri and primary veins; mature blades thin, dark dull gree and scabrate on the upper surface, pale and nearl glabrous on the lower surface; blades 1½–2 in. lon and 1¼–1½ in. wide, with a slender slightly villos pale yellow midrib and primary veins; petioles slender slightly wing-margined at the apex by the decurren base of the blade, thickly covered early in the seaso

GREGG HAWTHORN
Crataegus greggiana Egglest.

with matted pale hairs, becoming pubescent, ¼–⅗ in. in length; stipules large, foliaceous, coarsely and sharply serrate.

TWIGS. Slender, nearly straight, covered when elongating with matted pale hairs, light orange-brown and slightly hairy at the end of the first season, ashy gray and glabrous the following year, armed with numerous, straight, gray spines 1¼–2 in. long.

RANGE. In Texas in Uvalde, Menard, Real, and Kerr counties; southward into Coahuila and Nuevo León, Mexico.

REMARKS. The genus name, *Crataegus*, is the classical name of the hawthorn, and the species name, *greggiana*, honors Josiah Gregg (1806–1850), early American botanical explorer of the West. This species has also been listed in the literature as *C. uvaldensis* Sarg.

SHORT-LEAF HAWTHORN

Crataegus brachyphylla Sarg.—Series **Molles** (16)

FIELD IDENTIFICATION. Tree 18–21 ft, with a trunk 2–7 in. in diameter, the slender spreading branches forming an often irregular head.

FLOWERS. About ⅝ in. in diameter, petals 5, white, rounded, appearing in the early part of April when the leaves are more than half grown, in small compact 5–8-flowered corymbs, densely covered, like the slender pedicels and narrow obconic calyx-tube, with long matted snow-white hairs; calyx-lobes narrow, long-acuminate, laciniately glandular-serrate, thickly covered with white hairs; stamens 20, anthers deep rose-colored.

FRUIT. Ripening early in September, on slightly villose pedicels, in erect clusters, subglobose, dull dark red, ⅖–½ in. in diameter, with thin flesh, the calyx little enlarged, with a deep narrow cavity pointed inward; nutlets usually 3, acute at base, rounded at the broader apex, only slightly ridged on the back, ¼–⅓ in. long and ⅛–⅙ in. wide, the broad hypostyle extending to the middle.

LEAVES. Broad-ovate, acute or rounded at apex, truncate or rounded at the wide base, coarsely, often doubly, serrate, with straight acuminate teeth, covered when they unfold with short hairs; below with long white matted hairs persistent during the season; at maturity thin, yellowish green and glabrous on the upper surface, blades 2–2¾ in. long and 2–2⅓ in. wide, with a slender midrib and primary veins; petioles slender, thickly covered with matted white hairs early in the season, becoming glabrous or nearly glabrous before autumn, ¾–1¼ in.; leaves on vigorous shoots rounded at apex, cordate at the broad base, slightly and irregularly lobed laterally, coarsely double-serrate and to 2¼–3 in. long and wide, with petioles ⅔–1¼ in. long.

TWIGS. Slender, nearly straight, at first densely white hairy, later reddish brown to gray and shiny; spineless, or some twigs armed with occasional straight or slightly curved spines 1¼–1½ in. long.

SHORT-LEAF HAWTHORN
Crataegus brachyphylla Sarg.

RANGE. Dry gravelly ridges or woods. Near Fulton, Hempstead County, Arkansas. Reported also from eastern Texas.

HYBRID. *C.* × *notha* is a hybrid between *C. brachyphylla* and *C. marshallii.*

REMARKS. The genus name, *Crataegus*, is the classical name of the hawthorn. The species name, *brachyphylla*, means "short-leaf." Sargent (1921) remarks: "This is one of the most distinct species of the Molles group, differing from the other described species in its comparatively small leaves without lobes except on vigorous shoots, small flowers in small few-flowered corymbs, and small fruit. It is unusual to find a tree of this group growing on gravelly hills."

VIBURNUM HAWTHORN

Crataegus viburnifolia Sarg.—Series **Molles** (16)

FIELD IDENTIFICATION. Tree attaining a height of 35 ft and a diameter of 12 in. The large spreading or ascending branches shaped into an irregular crown.

Viburnum Hawthorn
Crataegus viburnifolia Sarg.

FLOWERS. Opening in March, corymbs 5–12-flowered, spreading, on slender, densely tomentose pedicels; corolla white, 5-petaled, about ¾ in. across; bracts and bractlets large, leaflike, lanceolate to spatulate, obscurely serrate terminally; calyx-tube narrowly obconic, densely white-hairy, 5-lobed, lobes long-acuminate, margin laciniately glandular-serrate, moderately villose externally, inwardly densely villose; stamens 20; anthers white; styles 4 or 5.

FRUIT. Maturing in October, corymbs lax and few-fruited with somewhat hairy or semiglabrous pedicels, bright yellow, subglobose, ⅔–¾ in. in diameter; flesh yellow and soft; nutlets 4 or 5, ends rounded, dorsally ridged and grooved, length about ¼ in.

LEAVES. Simple, alternate, deciduous, elliptic to ovate or oval, apex acute to rounded, base cuneate or concave, margin doubly serrate with glandular teeth, entire toward the base, some leaves with a few short acute lobes; young leaves white-hairy above and densely hoary tomentose beneath; mature blades 2½–3½ in. long, 2–2½ in. wide, leathery, upper surface dark green, shiny, and roughened, lower surface paler and hairy; petioles densely tomentose at first, later almost glabrous, ⅖–1½ in. long.

TWIGS. Unarmed or with a few stout gray spines, hoary-tomentose at first, brown to gray, shiny and glabrous finally.

RANGE. In low wet woods, bottom lands of the Brazos River in Texas. Near Brazoria, Columbia, Sweeney, and Wharton, Texas. Specimens examined at the Missouri Botanical Garden Herbarium: B. F. Bush, No. 1219, common in woods at Columbia, Texas; E. J. Palmer, No. 4984, low woods, Colorado River, Wharton County, Texas; E. J. Palmer, No. 6636, low woods, Wharton County, Texas.

REMARKS. The genus name, *Crataegus*, is the classical name for hawthorn, and the species name, *viburnifolia*, refers to the *Viburnum*-like leaves.

Noel Hawthorn

Crataegus noelensis Sarg.—Series **Molles** (16)

FIELD IDENTIFICATION. Tree to 25 ft, with a trunk to 6 in. in diameter, the spreading branches forming a broad, flat or round-topped crown.

FLOWERS. Opening in April, borne in short-pediceled, white-hairy, 5–10-flowered corymbs; petals 5, white, rounded, the entire corolla when spread ¾–1 in. in diameter; calyx-tube narrowly obconic, white-hairy, the 5 sepals slender, long-acuminate, minutely glandular-serrate, slightly villose; stamens usually 10, sometimes 5, anthers rose-colored; styles 3–5, surrounded at base by a broad ring of pale tomentum.

FRUIT. Maturing in September, on slender drooping pubescent pedicels, subglobose, orange-red, about ⅜ in. in diameter, the calyx prominent with a short tube and sepals spreading or closely appressed; flesh thin, soft

Noel Hawthorn
Crataegus noelensis Sarg.

and yellow; nutlets 3–5, base rounded, apex narrowed and rounded, dorsally slightly grooved, ⅕–¼ in. long.

LEAVES. Simple, alternate, deciduous, young leaves covered above with short white hairs and densely villose-pubescent below, mature leaves dark yellowish green, upper surface smooth and glabrous, lower surface villose-pubescent, with a prominent midrib and thin conspicuous primary veins, ovate to oval, apex acute, acuminate or rarely rounded, base acutely or broadly cuneate, margin coarsely double-serrate with straight teeth; petiole slender, apex slightly wing-margined, hoary-tomentose at first, glabrous later, ¼–1¼ in. long; leaves on young shoots broadly ovate, apex acuminate, base broad, rounded or cuneate, more coarsely serrate, usually laterally lobed with short broad acuminate lobes, blades often 3½ in. long and 2½–3 in. wide.

TWIGS. Stout, divaricate, at first with matted white hairs, reddish brown, finally pubescent and gray, armed with few or many slender, straight, purple, shiny spines ¾–2¾ in. long, sometimes persistent and compound on old trunks.

RANGE. In southern Missouri and northern Arkansas.

REMARKS. The genus name, *Crataegus,* is an ancient name and refers to the hard wood. The species name, *noelensis,* is for Noel, Missouri, where this tree was discovered.

Texas Hawthorn

Crataegus texana Buckl.—Series **Molles** (16)

FIELD IDENTIFICATION. Shrub or broad, round-topped tree attaining a height of 30 ft.

FLOWERS. Borne in many-flowered hairy corymbs; pedicels and hypanthium villose; bracts villose; petals 5, oblong-obovate and acute, white, rounded, spreading, attached to the disk margin; perianth about ¾ in. in diameter; calyx-tube obconic with 5 lobes; lobes acuminate, villose, glandular-serrate; stamens 20, with red anthers; styles 5, surrounded by tomentum at base.

FRUIT. Ripening in October in drooping clusters; peduncles tomentose or glabrous; pome large, ½–1 in. long, bright red, dotted, obovate or short-oblong, rounded at the ends, with persistent pubescence, but glabrous at maturity; calyx persistent and enlarged; flesh soft, yellow, sweet, edible; nutlets 5, ¼–⅓ in. long, smooth or slightly grooved.

LEAVES. Simple, alternate, deciduous, ovate, blades 3–4 in. long, 2½–3 in. wide; apex acute or rounded; base truncate or cuneate; rarely subcordate, margins glandular serrate and incisely lobed into 8–10 divisions; dark green, shiny and mostly glabrous above; paler and pubescent beneath, especially along the veins; petioles stout, grooved, ½–¾ in. long, tomentose at first but almost glabrous later.

TEXAS HAWTHORN
Crataegus texana Buckl.

TWIGS. Green and hairy at first, reddish brown to gray and more glabrous later; spines lustrous brown, shiny, thin, almost straight, some 2 in. long; some trees with no thorns.

BARK. Scaly, reddish brown to gray.

WOOD. Tough, hard, heavy, not large enough to be commercially valuable.

RANGE. The Texas coast region, west to the Navidad River. In Jackson, Victoria, Gonzales, Bexar, Matagorda, Hardin, Galveston, Harris, and Jefferson counties.

REMARKS. The genus name, *Crataegus,* is an ancient Greek word meaning "strong," with reference to the tough wood, and the species name, *texana,* refers to the state of Texas as the habitat of the tree.

Turkey Hawthorn

Crataegus invisa Sarg.—Series **Molles** (16)

FIELD IDENTIFICATION. Tree attaining a height of 30 ft and a tall trunk 8–10 in. in diameter. Branches stout and spreading into a wide, open irregular crown.

FLOWERS. Opening in March, corymbs 6–12-flowered, broad, pedicels densely white-hairy; corolla white, 5-petaled, about ¾ in. across; calyx-tube densely white-hairy, lobes 5, apex acuminate, margin laciniately glandular-serrate, externally hairy, less so within; stamens 20, anthers white; styles 3–5, basally white-hairy.

FRUIT. Maturing in October, corymbs few-fruited,

TURKEY HAWTHORN
Crataegus invisa Sarg.

spreading or erect on pubescent or semiglabrous pedicels, body short-oblong, terminally puberulent, red to orange, often pale-dotted, about ½ in. in diameter; flesh yellow, scant, dry; nutlets 3–5, rounded at the ends, faintly grooved dorsally, ⅕–¼ in. long; calyx on mature fruit green or reddish, mostly hairy.

LEAVES. Simple, alternate, deciduous, oval to ovate, apex acute or acuminate, base cuneate or rounded, coarsely serrate or some with shallow acuminate or acute lobes above, blades 2½–3 in. long, 2–2½ in. wide; young leaves densely villose above at first; lower surface with long white hairs; mature leaves thin, yellowish green, roughened and shiny above, lower surface hairy; leaves on young shoots larger, broad-ovate, apex acuminate, base broadly cuneate, margins more coarsely and deeply toothed and lobed; petioles 1½–2 in. long, very villose at first, later much less so.

TWIGS. Stout, at first densely tomentose, gray to brown, later somewhat pubescent to glabrous, dark gray, spineless, or with scattered, slender, straight, brown spines 1–1¼ in. long. Bark on older branches and trunk dark brown, broken into small, tight scales.

RANGE. In woods of bottom lands; in Arkansas in Hempstead County, Oklahoma in Choctaw County, and Texas in San Augustine County.

HYBRID AND SYNONYMS. *C.* × *lacera* is a hybrid between *C. invisa* and *C. marshallii*. It has also been listed in the literature under the names of *C. limaria* Sarg and *C. berlandieri* Sarg.

REMARKS. The genus name, *Crataegus,* is the classical name for hawthorn. The species name, *invisa,* meaning "hateful" or "hostile," is not of certain application, but perhaps refers to the dreaded or hated thorns.

SPREADING HAWTHORN

Crataegus dispessa Ashe—Series **Molles** (16)

FIELD IDENTIFICATION. Tree 15–18 ft, with an open irregular crown. Bark brown to gray, divided by shallow fissures and low ridges. Larger branches slightly scaly.

FLOWERS. Borne in lax, 5–10-flowered, pubescent corymbs; corolla to 1 in. in diameter, petals 5, white, rounded; stamens 20, anthers cream-white or pale yellow; styles 5; calyx 5-lobed, the divisions long-acuminate, margin entire or toothed, densely pubescent on the surface.

SPREADING HAWTHORN
Crataegus dispessa Ashe

Dallas Hawthorn
Crataegus dallasiana Sarg.

FRUIT. Short-oblong or nearly globose, ⅜–½ in. thick, flesh thin and firm; nutlets 4–5.

LEAVES. Simple, alternate, deciduous, elliptic to oval, apex acute to obtuse, base broadly cuneate, margin finely serrate or slightly lobed on vigorous shoots, young leaves short-pilose above but later glabrous, permanently pubescent on the veins beneath, firm, at maturity yellowish green; petioles pubescent, one fourth to one half as long as the blades.

TWIGS. Young ones slender, mostly ascending, densely pubescent, green to reddish brown, later gray and glabrous, spineless or nearly so.

RANGE. In Arkansas and Missouri.

REMARKS. The genus name, *Crataegus*, is an ancient classical name which refers to the hard wood. The species name, *dispessa*, meaning "spread out," is for the form of the tree.

Dallas Hawthorn

Crataegus dallasiana Sarg.—Series **Molles** (16)

FIELD IDENTIFICATION. Tree attaining a height of 25 ft and a diameter of 4–10 in. The branches ascending or erect to produce an irregularly shaped crown.

FLOWERS. Opening in April, corymbs many-flowered with slender, very hairy pedicels; corolla white, 5-petaled, each about ⅝ in. across; calyx-tube densely hairy; lobes 5, hairy, long-acuminate, margin with a few glandular serrations; stamens 20, anthers pink; styles 5.

FRUIT. Maturing in July, corymbs few-fruited, pedicels pubescent and mostly erect, shape of fruit globose, dark red, ⅜–½ in. in diameter; nutlets 5, ends acute, body flattened, dorsally grooved or ridged, ¼–5/16 in. long; calyx persistent, conspicuous, lobes spreading and green or reddish.

LEAVES. Simple, alternate, deciduous, blades 1¾–2½ in. long, 1¼–1½ in. wide, oblong to ovate, apex acuminate or rounded, base cuneate to concave, margin doubly serrate except at base, sometimes shallowly lobed above; young leaves densely villose and tomentose; at maturity yellowish green, upper surface shiny and glabrous, lower surface paler and pubescent along the veins; petioles ¼–⅔ in. long, slender, tomentose at first but glabrous later, somewhat winged toward the blade.

TWIGS. Slender, rather divaricate, at first densely tomentose and reddish brown, later gray and shiny, bearing slender spines 1¼–2¼ in. long.

RANGE. Northeastern Texas and adjacent Oklahoma.

REMARKS. The genus name, *Crataegus*, is the classical name, and the species name, *dallasiana*, is for Dallas County, Texas, where it occurs along bottom-land streams.

Kansas Hawthorn

Crataegus coccinioides Ashe—Series **Dilatatae** (17)

FIELD IDENTIFICATION. Tree attaining a height of 20 ft, with a trunk to 12 in. in diameter, the stout branches developing into a broad rounded crown.

FLOWERS. Opening in May, corymbs 4–8-flowered, glabrous or slightly hairy; corolla 5-petaled, white, about ¾ in. in diameter; bracts and bractlets oblong to obovate, margin serrate and conspicuously red-glandular; calyx-tube broadly obconic, smooth and glabrous, lobes gradually acute at apex, margin coarsely glandular-serrate; stamens 20, anthers rose-colored; styles 5, basally tomentose.

FRUIT. Maturing in October, corymbs erect and few-fruited, body subglobose and terminally flattened, sometimes slightly angled, dark red, shiny, pale-dotted, about ¾ in. long and wide; flesh yellow or reddish, thick and firm; nutlets 5, pale, apex acute, base rounded, dorsally rounded and slightly ridged, about ⅓ in. long.

LEAVES. Simple, alternate, deciduous, broad-ovate, apex acute, base rounded or truncate, margin doubly serrate and some with short, acute lobes toward the apex, mature leaves rather thin but firm, upper surface dull dark green, paler beneath with a yellowish midrib, blade length 2½–3 in., width 2–2½ in., in autumn red to orange; leaves on young shoots larger and some semicordate; petioles ½–1 in. long, hairy at first but glabrous later, green or reddish, some with stalked red glands above.

TWIGS. Stout, nearly straight, glabrous, shiny, brown, often with reddish purple shiny spines 1–2 in. long.

KANSAS HAWTHORN
Crataegus coccinioides Ashe

RANGE. Eastern Oklahoma, Arkansas, Missouri, Kansas, and Illinois.

SYNONYMS. It is also listed in some literature under the synonyms of *C. callicarpa* Sarg. and *C. speciosa* Sarg. Another vernacular name for Kansas Hawthorn is Eggert Thorn.

REMARKS. The genus name, *Crataegus*, is the classical name of the hawthorn. The species name, *coccinioides,* alludes to this plant's resemblance to Scarlet Hawthorn, *C. coccinea.*

RIVER HAWTHORN

Crataegus rivularis Nutt.—Series **Douglasianae** (18)

FIELD IDENTIFICATION. Small Western tree 9–20 ft, with erect, ascending branches forming a narrow open head.

FLOWERS. Opening in May, borne in compact glabrous corymbs on long slender pedicels; corolla about ½ in. in diameter, petals 5, white, rounded; calyx-tube broadly obconic, slightly hairy at first but glabrous later, 5-lobed; lobes slender, entire or minutely glandular, glabrous externally, hairy within, sometimes reddish; stamens 10–20, anthers rose-colored; styles usually 5.

FRUIT. Ripening in September, corymbs drooping, few-fruited, long-pediceled; fruit body short-oblong, ends full and rounded, dark red to almost black, lustrous, usually white-dotted, length ⅓–½ in.; flesh yellow, thin, dry, mealy; nutlets 3–5, about ¼ in. long, apices narrowed or rounded, dorsally ridged, ventral cavities broad and shallow; calyx persistent, rather closely appressed, outer surface reddish and some slightly hairy below.

LEAVES. Simple, alternate, deciduous, lanceolate to narrowly oblong-obovate or elliptic, apices acute or abruptly acuminate, base gradually narrowed and cuneate to concave, entire, margin irregularly and crenately serrate above, with glandular teeth, rarely lobed; young leaves reddish and pale-hairy; mature leaves rather dull green, thin, smooth and glabrous above, paler and yellowish green beneath, blade length 1½–2 in., about ¾ in. wide, some twice as long as wide; midrib yellow, slender, primary veins obscure and 3–4 pairs; petiole about ½ in. long, slender, slightly winged above, hairy at first but reddish and glabrous later; young leaves almost round, coarsely toothed, some slightly incised-lobed, leathery, often to 3 in. long and 2 in. wide, petiole broadly winged.

TWIGS. Slender, reddish brown, lustrous, glabrous, lenticels numerous and pale, spineless, or spines straight, slender, blackish, ¼–1¼ in. long, bark on older stems or trunk dark brown and scaly.

RIVER HAWTHORN
Crataegus rivularis Nutt.

RANGE. At altitudes of 3,000–8,500 ft, borders of streams. Arizona, New Mexico, and northwestern Texas; northward to Idaho and Wyoming.

REMARKS. The genus name, *Crataegus*, is the classical name for the hawthorn, and the species name, *rivularis,* refers to its preference for the moist banks of small rivulets or streams.

PEAR HAWTHORN

Crataegus calpodendron (Ehrh.) Medic.—
Series **Macracanthae** (19)

FIELD IDENTIFICATION. Shrub or small tree attaining a height of 15–18 ft, the scaly branches thorny or thornless.

FLOWERS. In many-flowered villose or tomentose corymbs; petals 5, white, rounded, corolla ⅜–⅝ in. wide when spread; stamens about 20, anthers pink or rarely white; calyx-tube pubescent, the 5 lobes glandular-serrate or pectinate.

FRUIT. Oblong or obovoid, occasionally subglobose, about ⅓ in. thick, at first pubescent, later glabrous, bright red or orange-red; flesh thin, sweet, and succulent; nutlets 2–3, deeply pitted on the inner surface.

LEAVES. Simple, alternate, deciduous, blade oblong-elliptic or rhombic, margin coarsely serrate, except near the base, often with 3–5 pairs of irregular lateral lobes above the middle (sometimes lobed only on the vegetative shoots), young leaves short-villous above; when mature, pubescent beneath, dull yellowish green, firm, veins impressed above; petioles stout, ⅜–⅝ in. long, sometimes wing-margined nearly to the base.

TWIGS. Young ones tomentose, reddish brown, later glabrous and gray, bark on old trunks and limbs furrowed and thick; spineless, or spines ½–2 in. long, stout, slender, sharp, straight or slightly curved.

RANGE. The species and its varieties occur in open woods and thickets, often along small rocky streams. East Texas and Arkansas; east to Georgia, north to New York and Ontario, and west to Minnesota.

VARIETIES. A number of natural varieties have been segregated as follows:

Hispid Pear Hawthorn, *C. calpodendron* var. *hispida* (Sarg.) Palmer, has leaves more sharply serrate, with narrow acuminate teeth and small acuminate, spreading lateral lobes, hispidulous beneath at maturity; fruit obovoid or short-oblong, pubescent when young. From southern Missouri and southeastern Kansas.

Small-fruit Pear Hawthorn, *C. calpodendron* var. *microcarpa* (Chapm.) Palmer, has veins of leaves more crowded and impressed above; subglobose fruit ⅕–⅓ in. thick. Many intermediate forms connect it with the species. From Missouri to North Carolina and Virginia.

Globose-fruit Pear Hawthorn, *C. calpodendron* var. *globosa* (Sarg.) Palmer, has leaves obscurely lobed or

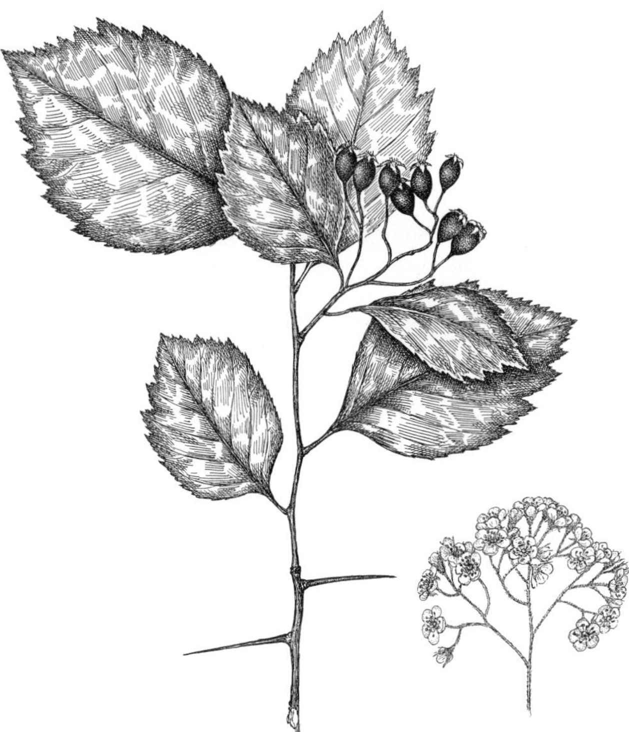

PEAR HAWTHORN
Crataegus calpodendron (Ehrh.) Medic.

unlobed; flowers fewer and in more compact clusters; fruit subglobose, about ⅓ in. thick with thin, dry, hard flesh. East Texas, Arkansas, Missouri, Kansas, and Kentucky.

Soft Pear Hawthorn, *C. calpodendron* var. *mollicula* (Sarg.) Palmer, is a shrub 6–9 ft high; leaves obscurely lobed or undivided, velvety-pubescent beneath; flowers ½–⅝ in. wide, with 10 or fewer stamens. In Arkansas and adjacent Missouri.

HYBRID. *C* × *incaedua* Sarg. is a hybrid of *C. calpodendron* and *C. collina.*

REMARKS. The genus name, *Crataegus*, refers to the hard wood of some species. The species name, *calpodendron,* means "urn-tree" and refers to the shape of the fruit.

CARROLL HAWTHORN

Crataegus carrollensis Sarg.—Series **Macracanthae** (19)

FIELD IDENTIFICATION. Stout shrub 6–9 ft, covered with scaly bark. Branches thick, erect, pale gray.

FLOWERS. In many-flowered corymbs, densely covered with matted white hairs, with conspicuous oblong-obovate to linear-lanceolate bracts and bractlets; calyx-tube narrow-obconic, villose, the lobes narrow-acuminate, laciniately glandular-serrate, sparingly villose or glabrous on the outer surface, glabrous on the inner surface; stamens 15–20; anthers pale yellow; styles 2–3; petals 5, white, rounded.

FRUIT. Ripening in October, on stout glabrous pedicels, ellipsoidal, dark red, the calyx with a short tube, spreading and reflexed lobes and a broad shallow cavity pointed in from the apex, ¼–⅓ in. long, ⅕–¼ in. wide; nutlets 2 or 3, acute at apex, rounded at base, slightly grooved on the back, penetrated on the inner face by short grooves, ⅕ in. long and ⅛ in. wide, with a narrow hypostyle extending to the middle.

LEAVES. Ovate to rarely obovate, acuminate at apex, abruptly or gradually narrowed and concave-cuneate at base, slightly and irregularly divided usually only above the middle, and coarsely, often deeply, serrate with straight gland-tipped teeth; more than half grown when the flowers open and then covered above with short white hairs and slightly villose below, especially on the midrib and veins; and at maturity thin, yellowish green, smooth, lustrous and glabrous above, still villose below, 2½–4 in. long and 2½–2¾ in. wide, with a stout midrib and slender primary veins; petioles stout, wing-margined toward the apex by the decurrent leaf blade, densely villous early in the season, becoming pubescent in the autumn, ¾–1 in. long.

TWIGS. Stout, yellowish green, young ones sparingly pale-hairy, later reddish brown and glabrous; spines numerous, straight, stout, shiny, chestnut-brown, 1–2½ in. long, some branched later and persistent.

RANGE. Rocky hillsides near Eureka Springs, Carroll County, Arkansas.

REMARKS. The genus name, *Crataegus*, is the classical Greek name of the hawthorn, and the species name, *carrollensis*, is for the type locality.

Hot Springs Hawthorn

Crataegus thermopegaea Palmer—
Series **Macracanthae** (19)

FIELD IDENTIFICATION. An intricately branched, very spiny shrub 3–5 ft.

FLOWERS. Opening very late in the season, in few-flowered, compact, densely villous corymbs with glandular deciduous bracts, and 5 serrate calyx-lobes; petals 5, white, rounded, spreading; stamens 10 or more; anthers deep red in bud.

FRUIT. Subglobose, ¼–⅖ in. in diameter, with a small, nearly sessile calyx and erect or spreading calyx-lobes, which are usually deciduous from the ripe fruit, bright scarlet, with yellow flesh, becoming succulent when mature; nutlets 2–3, broad and rounded at the ends, broadly ridged on the dorsal and shallowly pitted on the ventral surface.

LEAVES. Ovate in outline, obscurely lobed and sharply serrate, with broad acuminate teeth, acute or slightly acuminate at the apex, appressed-villous above and hirsute on the lower surface with short pale hairs, thin but firm at maturity, 1–1¾ in. long, ¾–1⅓ in. broad, or larger on vigorous shoots, where the base is also sometimes truncate or subcordate.

TWIGS. Slender, at first reddish brown and slightly pubescent, later gray and glabrous; spines ¾–1¾ in. long, straight, lustrous, gray to black.

RANGE. In thickets, on rocky banks, and along small upland streams in the vicinity of Hot Springs, Garland County, Arkansas.

REMARKS. The genus name, *Crataegus*, is the classical name of the hawthorn. The species name, *thermopegaea*, refers to its growth near Hot Springs. The above account is cited from E. J. Palmer's original description (Palmer, 1926, p. 124). Palmer also remarks: "This species differs strikingly in habit and general appearance from any other member of the group with which I am acquainted, resembling more some of the *Intricatae*, to which group I was inclined to assign it before examining the ripe fruit and nutlet."

Hybrid Hawthorn

Crataegus × *notha* Sarg. (*C. marshallii* Egglest. × *C. brachyphylla* Sarg.)—Series **Microcarpae** (2)

FIELD IDENTIFICATION. Tree to 21 ft, with a diameter of 6–8 in. The pale thin bark separating into small flakelike scales. The pale gray stout branches forming an open irregular head.

FLOWERS. Opening in March, corymbs compact, 12–15-flowered, on slender densely hoary-villose pedicels; petals 5, white, rounded, the corolla when spread ⅝–⅞ in. wide; calyx-tube narrow-obconic, densely villose, the 5 sepals gradually narrowed from the base, apex acuminate, glandular, margin laciniately serrate, slightly villose; stamens 20, anthers deep rose colored, styles 2–4.

FRUIT. Maturing in September, not abundant, on glabrous pedicels, ovoid, bright scarlet, ⅖ in. long and ¼–⅓ in. wide, with soft succulent flesh, the calyx enlarged with a deep narrow cavity; nutlets usually 4, acute at the ends, rounded and occasionally slightly ridged on the back, about ¼ in. long and ⅙ in. wide.

LEAVES. Simple, alternate, deciduous, when young matted with white hairs, when mature thin, upper surface glabrous, lower surface pubescent, 1¼–1½ in. long, 1–1¼ in. wide, with a thin midrib and slender primary veins occasionally running to the sinuses as well as to the points of the lobes, blades broad-ovate, apex acute, base abruptly cuneate, rounded or truncate, margin coarsely and sharply double-serrate, usually only above the middle, and slightly and irregularly lobed; petioles slender, densely villose early in the season, becoming nearly glabrous, ⅝–1⅓ in.; leaves on vigorous

HOT SPRINGS HAWTHORN
Crataegus thermopegaea Palmer

shoots truncate or subcordate at base, more coarsely serrate, usually 3-lobed by deep narrow lateral lobes pointed in the bottom, and to 2 in. long and 2–2½ in. wide, with petioles usually about 1 in. in length.

TWIGS. Slender, slightly zigzag, at first with matted white hairs, later glabrous and reddish brown and eventually gray to brown; unarmed, or armed with slender, straight, brown spines to 1¾ in.

RANGE. Dry gravelly hills, about five miles northwest of Fulton, Arkansas.

REMARKS. The genus name, *Crataegus*, is an ancient name referring to the hard wood. The species name, *notha*, means "false" or "bastard," referring to the hybrid origin of this species.

Sargent (1921), remarks as follows:

"Mr. [E. J.] Palmer, who has watched this tree for several years, suggests that it is a hybrid between *C. apiifolia* Michaux [*C. marshallii*] and *C. brachyphylla* Sargent, both of which are growing with it. The bark of the trunk is that of *C. apiifolia* and the fact that a primary vein sometimes extends to the base of a sinus of a leaf, the character by which the Microcarpae Group is best distinguished, also indicates its relationship with *C. apiifolia*. From that species it differs in its larger more pubescent only occasionally lobed leaves often cuneate at base, in its more pubescent corymbs of larger flowers with a more pubescent calyx, and in its larger fruit. The pubescence of *C. notha*, although less dense is in character and persistency that of *C. brachyphylla*; the larger rarely lobed leaves, the larger flowers and fruit and the nearly unarmed branches may also be due to the influence of that species. Five individuals are now known, one a solitary tree and the others in a group. They all grow in the immediate vicinity of their supposed parents. If *C. notha* is a hybrid, and there seems to be good reason for the belief, it is a plant of unusual interest, showing as it would the possibility of crossing species of two as distinct groups of the genus as are now recognized."

CAROLINA CHERRY-LAUREL

Prunus caroliniana (Mill.) Ait.

FIELD IDENTIFICATION. Evergreen tree attaining a height of 40 ft, with alternate, leathery, entire or toothed leaves.

FLOWERS. In short, axillary racemes; flowers perfect, white, pedicels about ½ in. long with acuminate bracts; calyx-tube narrow-obconic, 5-lobed; lobes suborbicular,

CAROLINA CHERRY-LAUREL
Prunus caroliniana (Mill.) Ait.

reflexed, deciduous; petals 5, white, boat-shaped, erect, smaller than the sepals; stamens numerous, exserted, orange, longer than the petals; filaments distinct; ovary sessile, small, 1-celled; style simple and slender; stigma club-shaped; ovules 2.

FRUIT. Drupe conic-ovoid or oval, abruptly pointed, black, lustrous, persistent; about ½ in. long, skin thick, flesh thin and dry, not edible; seed ovoid, acute, rounded at base, about ½ in. long, dorsal groove prominent.

LEAVES. Alternate, simple, persistent, coriaceous, shiny, margin entire or a few with remote spine-tipped teeth, oblong-lanceolate, apex acute or acuminate, mucronate, base wedge-shaped, glabrous and dark green above, paler beneath, 2–4½ in. long, ¾–1½ in. wide, aromatic when crushed, taste bitter because of prussic acid; petioles stout, about ⅓ in. long; stipules lanceolate.

TWIGS. Slender, glabrous, green to red or grayish brown.

BARK. Gray, thin, smooth, or later irregularly roughened, marked with blotches.

WOOD. Reddish brown, hard, heavy, strong, close-grained.

RANGE. In Texas, extending from the valley of the Guadalupe River, eastward through eastern Texas and Louisiana to Florida, northward to the Carolinas.

REMARKS. The genus name, *Prunus*, is the classical name. The species name, *caroliniana*, refers to the Carolina region. Vernacular names are Wild Peach, Cherry-laurel, Carolina Cherry, and Mock-orange. The leaves contain prussic acid which is injurious to livestock. A number of birds feed on the seeds. The tree is widely cultivated as a popular ornamental, and can be trained into hedges.

Mahaleb Cherry

Prunus mahaleb L.

FIELD IDENTIFICATION. Usually a small slender tree with fragrant foliage and flowers. Branches short and lateral to form an open, irregular or somewhat rounded crown. Maximum height 30 ft and diameter 8–10 in., often crooked or inclined. A species of Eurasian origin introduced to America principally for grafting stocks for garden cherries.

FLOWERS. April–May, fragrant, unfolding with the leaves on short lateral branches of the season, corymbs 4–10-flowered, some pedicels with small leaflike bracts; calyx imbricate in the bud, inferior, deciduous, the tube obconic, 5-lobed; the lobes obtuse; petals 5, white, spreading, inserted on the calyx, corolla about ⅝ in. broad; stamens numerous, inserted with the petals, distinct, filaments filiform; style simple, stigma small, ovary 1-celled and 2-ovuled.

FRUIT. Drupe maturing in July, ¼–⅜ in. long, globose to ovoid, reddish black, flesh thin and bitter, stone slightly flattened, seed suspended, endosperm none.

LEAVES. Simple, alternate, deciduous, ovate to round-cordate or orbicular, apex acute to obtuse, base rounded or slightly cordate, margin crenate-dentate and glandular between the teeth, blade length 1–3 in., width ¾–1 in., firm, deep-green, smooth, glabrous on both sides; petioles slender, terete, ¼–¾ in., puberulent at first, glabrous later.

TWIGS. Pubescent and green when young, from brown to gray and glabrous later, somewhat angled.

BARK. Gray to brown, rough with low erratic ridges and appressed scales, aromatic when bruised.

WOOD. Reddish brown, close-grained, heavy, hard, fragrant.

RANGE. Dry, well-drained soil in the sun. Native of Eurasia, escaped to roadsides and fields. Locally through the cherry-growing states, particularly in New York, Pennsylvania, Delaware, Missouri, and Arkansas; also in Ontario. Also cultivated in the Pacific Northwest.

VARIETIES AND FORMS. The following horticultural varieties and forms are known:

Mahaleb Cherry
Prunus mahaleb L.

Sand Cherry
Prunus pumila L.

P. mahaleb forma *chrysocarpa* Nichols is a form with yellow fruit.

P. mahaleb forma *variegata* Hort. is a form with white-edged leaves.

P. mahaleb forma *pendula* Dipp. is a form with drooping branches.

P. mahaleb forma *compacta* Hort. is a form with a compact compressed head.

P. mahaleb var. *albo-marginata* Dipp. has white-edged leaves.

P. mahaleb var. *globosa* Dieck has a rounded head.

P. mahaleb var. *monstrosa* Kirchn. has very short, thick branches.

P. mahaleb var. *cupiana* Fiori & Paol is smaller than usual; leaves much smaller, ½–1 in. long; peduncle short, 3–6-flowered, and flowers smaller.

P. mahaleb var. *transylvanica* Schur. has numerous small flowers with reflexed sepals.

REMARKS. *Prunus* is the ancient name of the plum, and the species name, *mahaleb*, is an ancient Persian name. The tree is also known as the Perfumed Cherry or St. Lucie Cherry, because of the aromatic flowers and wood. It is used for making cabinets, smoking pipes, and walking sticks. Oil from the seeds is used in perfumery and by the Arabs as a remedy for bladder disorders. The fruit yields a violet dye and also a fermented liquor which is used in Eurasia. Because of its hardiness, the stock is widely used for grafting other cherries. It is little planted for ornament in the United States.

Sand Cherry

Prunus pumila L.

FIELD IDENTIFICATION. Shrub of Northern distribution, rarely found in our area. Size varying 6 in.–6 ft, in favorable localities erect with strongly ascending branches. In other situations spreading or creeping with erect branches. Where stems touch the ground often sucker-forming.

FLOWERS. April–May, with or before the leaves, in 2–6-flowered umbellate, lateral clusters (mostly 2–3); pedicels filiform, erect or ascending; corolla ⅓–½ in. across, white; petals 5, inserted on the calyx, spreading, ovate to obovate; stamens 15–20, inserted with the petals, long-exserted, filaments filiform; pistil 1, ovary 1-celled and 2-ovuled, style simple, stigma small and capitate; calyx inferior, free from the ovary, tube obconic, glabrous; sepals 5, each about 1/12 in. long, margins glandular-serrate, often reddish.

FRUIT. Drupe ripe in August, ¼–½ in. in diameter, subglobose or slightly elongate, dark red or nearly black, without bloom, flesh thin and acid; pit solitary, ovoid to oblong, base rounded, apex apiculate, length ⅓–⅖ in., ⅙–¼ in. wide, dorsally ridged.

LEAVES. Mostly erect, simple, alternate or clustered, deciduous, elliptic to narrowly oblanceolate or spatulate, apex acute to acuminate, base narrowed to cuneate, margin appressed-serrate and gland-tipped, some leaves less serrate or entire toward the base, blade length 1½–3 in., width to ¾ in., upper surface deep-green, glabrous above and below or nearly so when mature, usually much paler beneath; stipules filiform, setose to fimbriate or glandular-serrate, deciduous; petioles ⅛–⅝ in., with or without glands near the base of blade.

RANGE. On moist sandy, gravelly, or calcareous shores. Indiana; east to New Jersey, northward to New Brunswick, and west through St. Lawrence basin to Minnesota and Wisconsin. Frequent along Lake Michigan and upper Atlantic coast. Listed by Moore in *A Check-list of the Ligneous Flora of Arkansas,* but probably quite rare in that state.

VARIETIES. Two major varieties have been segregated as follows:

Sprawling Sand Cherry, *P. pumila* var. *depressa* (Pursh) Bean, has a smaller fruit with a globose-ellipsoid stone; leaves often obtusely spatulate, or at least less acuminate, thinner, teeth more crenate, more glaucous beneath, veins less prominent, stipules less fimbriate. On beaches, Quebec to Massachusetts and Ontario.

BLACK CHERRY
Prunus serotina Ehrh.

Appalachian Sand Cherry, *P. pumila* var. *susquehanae* (Willd.) Jaeg., is more erect and to 3 or 4 ft, flowers about ⅓ in. across; fruit about ¼ in. in diameter; leaves mostly short-obovate to spatulate or elliptic-ovate, strongly serrate with appressed teeth toward the apex. Wet woodlands and bogs; Minnesota, Pennsylvania, Wisconsin, and Manitoba.

REMARKS. The genus name, *Prunus,* is the ancient Latin name for "plum," and the species name, *pumila,* refers to the dwarf size. Vernacular names are Dwarf Plum and Peach Plum. The cleaned seeds average 4,000 per lb. Dormancy is broken in part by stratification in moist sand for 90 days or more at 41°F. The seeds are eaten by a number of species of birds and casually browsed by white-tailed deer.

BLACK CHERRY

Prunus serotina Ehrh.

FIELD IDENTIFICATION. Tree to 100 ft, with oval-oblong or lanceolate leaves which are finely callous-serrate on the margin.

FLOWERS. March–June, racemes in spring with immature but nearly expanded leaves; individual flowers white, about ¼ in. in diameter, borne on slender pedicels; petals 5, obovate; stamens numerous, in 3 ranks; stigmas flattened; calyx-tube saucer-shaped, the lobes short, ovate-oblong, acute.

FRUIT. Ripening June–October, drupe borne in racemes, thin-skinned, black when mature, juicy, bittersweet, ⅓–½ in. in diameter; stone oblong-ovoid, about ⅓ in. long. The commercial seed-bearing age is about 10 years, with 25 to 75 as the most prolific and 100 as the maximum. Good crops are borne almost annually.

LEAVES. Simple, alternate, finely callous-serrate, apex acuminate, base cuneate, oval-oblong or oblong-lanceolate, firm, dark green and lustrous above, glabrous or hairy on midrib beneath, blades 2–6 in. long, 1–2 in. broad, with one or more red glands at base, tasting of hydrocyanic acid.

BARK. Reddish brown, gray or white, striped horizontally with gray to black lenticels, smooth when young, broken into small plates later, bitter to the taste.

WOOD. Rich reddish brown, sapwood whitish or light brown, heavy, moderately hard, strong, bends well, shock resistance high, works well, finishes smoothly, glues well, seasons well, shrinks moderately, moderately free from checking and warping, weighing about 36 lb per cu ft, taking a beautiful polish.

RANGE. Texas, Oklahoma, Arkansas, and Louisiana, eastward to Florida, north to Nova Scotia, west to North Dakota, Nebraska, and Kansas, and southward in Mexico.

PROPAGATION. The fruit may be gathered by shaking or flailing onto canvas or cloth. The seeds are then extracted by running through a hammer mill, or macerator, and floating off the residue. Some prefer to dry the entire fruit with pulp before storing or planting. The seed averages about 4,800 cleaned seeds per lb, with a commercial purity of 99 per cent and a soundness of 96 per cent. Pretreatment of seeds is done by stratification in moist sand or peat for 90–120 days at 41°F. Another method is to soak 30 minutes in H_2SO_4. A combination of these two methods is recommended, soaking in the acid first. Seed should be sown using about 25 seeds per linear ft, in drills 8–12 in. apart, covered with porous soil, and mulched until germination. Bird and rodent protection is also needed. Seedlings are sometimes attacked by a leaf-spot disease caused by the fungus *Coccomyces lutescens.* Spray with a 4–6–50 Bordeaux mixture. The fungus *Bacterium prunii,* also causing leaf-spot, can be controlled in the same manner. Powdery mildew, *Podosphaera oxyacanthae,* may be controlled with sulphur dust. Young trees seem to like partial shade. Budding or grafting methods are used for horticultural varieties.

VARIETIES. A number of natural varieties of Black Cherry have been described:

Shiny Black Cherry, *P. serotina* var. *cartilaginea* (Lebm.) Jaeg., has longer shiny leaves. Known in cultivation.

Underblue Black Cherry, *P. serotina* var. *montana* (Small) Britt., has leaves broadly elliptic and glaucescent beneath. Known from Alabama to Virginia.

Willow-oak Black Cherry, *P. serotina* var. *phelloides* Schwer., has lanceolate leaves. Known in cultivation.

Escarpment Black Cherry, *P. serotina* var. *eximia* (Small) Little, has flowers 35–60 on the rachis, pedicels ⅛–⅕ in. (or ⅙–¼ in. in fruit); the 5 petals 1/10–⅙ in. long or wide; calyx segments 1/25–1/15 in. long; drupe about ⅜ in. across; leaves 2–3½ in. long, 1¼–1¾ in. wide, margin with 4–6 teeth every ⅜ in., surfaces glabrous or some with axillary red tufts of hairs; petioles ⅜–¾ in. (to 1¼ in. sometimes). In Texas in the valley of the Colorado River, in San Saba and Burnet counties; south to Comal and Medina counties; west to the south fork of the Llano River in Kimble County and to the west fork of the Nueces River in Kinney County.

Alabama Black Cherry, *P. serotina* var. *alabamensis* (Mohr) Little, attains a height of 30 ft with tomentose twigs; flowers about ⅓ in. across borne in pubescent racemes 3¼–6 in. long; fruit eventually black, subglobose, about ⅓ in. across; leaves ovate to elliptic to oblong-ovate, 2⅓–5 in. long, apex short-acuminate, base somewhat obtuse, rounded or slightly narrowed, margin serrate with callous teeth; surfaces glabrate above and finely pubescent beneath. Known from eastern Georgia to northeastern Alabama, and south to northwestern Florida.

Capulin Black Cherry, *P. serotina* var. *salicifolia* (H.B.K.) Koehne, is listed in the literature as a native from central Mexico to Guatemala. It is cultivated in South America and in the United States, and has several horticultural forms such as the Fern-leaf (*asplenifolia*), with deeply and irregularly cut leaves; the Pyramid (*pyramidalis*), with a pyramidal shape; and the Weeping (*pendula*), with drooping branches.

ESCARPMENT BLACK CHERRY
Prunus serotina var. *eximia* (Small) Little

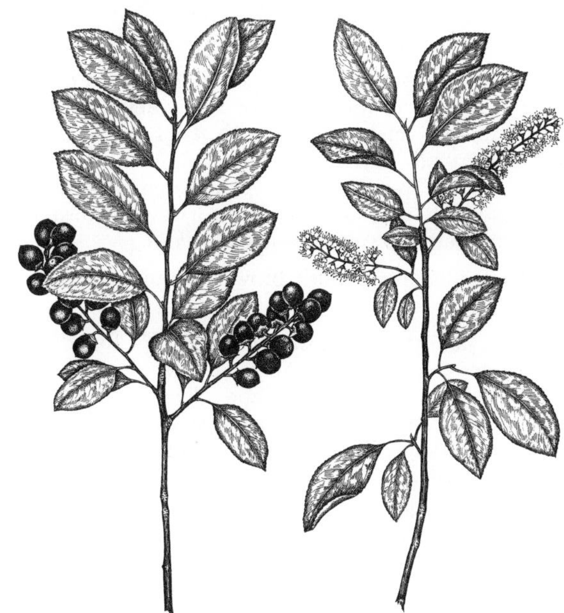

SOUTHWESTERN CHOKE CHERRY
Prunus virens (Woot. & Standl.) Shreve

REMARKS. *Prunus* is the ancient classical name, and the species name, *serotina,* means "late-flowering." It is also known under the vernacular names of Mountain Black Cherry and Rum Cherry. The bark is used medicinally as a cough remedy. The fruit is used as a basic flavoring extract, and is also eaten raw by man. It is eaten by a wide variety of wildlife, including 33 birds, raccoon, opossum, squirrel, bear, and rabbit. The foliage is considered to be poisonous to livestock. The wood is used for furniture, cabinetmaking, printer's blocks, veneer, patterns, panels, interior trim, handles, woodenware, toys, and scientific instruments. The tree has been cultivated for ornament since 1629.

SOUTHWESTERN CHOKE CHERRY

Prunus virens (Woot. & Standl.) Shreve

FIELD IDENTIFICATION. Partially evergreen shrub or small tree to 30 ft, the spreading branches forming a dense wide crown.

FLOWERS. Borne April–May, numerous in elongate glabrous racemes (or slightly hairy at base), in the leaf axils; flowering branches about 4 in. long; pedicels ⅛–⅙ in. long; calyx-lobes 5, 1/25–1/12 in. long; petals 5, white, about ⅛ in. long and wide, broadly obovate; stamens numerous, inserted on the calyx; pistil 1, ovary 1-celled, 2-ovuled.

FRUIT. Maturing August–September, sometimes abundant, pedicels ⅙–⅕ in., drupe about ⅜ in. across, purplish black, glabrous, fleshy, sweet and edible but astringent; seed solitary and bony.

LEAVES. Simple, alternate, persistent, blade ovate to oblong or elliptic, apex acute or rarely acuminate, base acute, margin finely serrate with 4–11 teeth (usually 5–7) every ⅜ in., blade length 1¾–2¾ in., width ¾–1⅓ in., leaves on some plants sometimes smaller, mostly glabrous, or less often a few reddish hairs beneath along the veins; petioles ⅕–⅓ in.

TWIGS. Slender, reddish brown or tan to gray later, usually glabrous.

RANGE. Southwestern Choke Cherry is found along mountain streams or canyons, at altitudes of 4,500–6,000 ft. Trans-Pecos Texas and New Mexico to Arizona; southward in Mexico in the states of Chihuahua, Sonora, Durango, Coahuila, San Luis Potosí, Guanajuato, and Jalisco.

VARIETIES. Gila Choke Cherry, *P. virens* var. *rufula* (Woot. & Standl.) Sarg., has the young twigs, leaves, and occasionally the inflorescence, thickly rufous-cinereous or hirsutulous. However, the degree of hairiness varies greatly, and there seems to be considerable intergrading with the species. It usually is found at altitudes above 4,000 ft in New Mexico and Arizona; Sonora and Chihuahua, Mexico.

Another closely related form was at one time named Parks Choke Cherry, *P. parksii* Cory (Cory, 1943a, pp. 325–327). Some authors now consider the name to be a synonym of *P. virens*. This plant also has a tendency to resemble *P. virens* var. *rufula*. It was first collected by V. L. Cory in the Carrizo Sands area near the Bexar-Wilson county line, southeast of San Antonio, Texas.

Southwestern Choke Cherry has also been given the name of *P. serotina* subsp. *virens* (Woot. & Standl.) McVaugh.

REMARKS. The genus name, *Prunus*, is the classical name of a plum-tree of Europe. The species name, *virens*, refers to the half-evergreen leaves. The foliage is good winter browse for stock. Bear and other animals eat the fruit. Southwestern Choke Cherry was introduced into cultivation in 1916.

COMMON CHOKE CHERRY

Prunus virginiana L.

FIELD IDENTIFICATION. Large shrub or small tree to 30 ft, with erect or horizontal branches.

FLOWERS. April–July, in short, dense, cylindric racemes 3–6 in. long; flowers ¼–⅓ in. in diameter, borne on slender, glabrous pedicels from the axils of early-deciduous bracts; corolla small, white, 5-merous; petals rounded, short-clawed; filaments glabrous; style thick, short, with orbicular stigmas; calyx-tube 5-lobed, lobes short, obtuse at apex, glandular-laciniate on the margin.

FRUIT. Ripening July–September, cherry ¼–⅓ in. thick, globose, lustrous, dark red, scarlet, or nearly black; skin thick, flesh juicy, acidulous and astringent, barely edible; stone oblong-ovoid, one suture ridged, the other suture acute. Good crops are borne almost annually.

COMMON CHOKE CHERRY
Prunus virginiana L.

LEAVES. Alternate, simple, deciduous, thin, blades ¾–4 in. long, ½–2 in. wide, oval to oblong or obovate, abruptly acuminate or acute at the apex, rounded to cuneate or somewhat cordate at the base, sharply serrate on the margin, dark green and lustrous above, paler on the lower surface, sometimes pubescent on veins, turning yellow in autumn, strong odor when crushed; petioles slender, ½–1 in., 2 glands at the apex.

TWIGS. Reddish brown to orange-brown, glabrous, slender, lenticels pale.

BARK. Dark gray to brown or blackish; old trees somewhat irregularly fissured into small scales with paler excrescences, smoother and tighter when young, inner bark ill-scented.

WOOD. Light brown, sapwood lighter, close-grained, moderately strong, hard, heavy, weighing 36 lb per cu ft.

RANGE. The species and its varieties rather widespread. Texas, New Mexico, Oklahoma, Arkansas, and Louisiana; eastward to Georgia, northward to Maine and Newfoundland, and west to British Columbia, Washington, Oregon, and California.

PROPAGATION. The fruits are gathered from the ground or flailed onto canvas. The seed may be separated from the pulp by running through a macerator or hammer mill with water and floating or skimming off the pulp. The seed can be dried after extraction before storing or sowing. The seed stored for 1 year at 26°F. showed little loss of viability, but higher temperatures resulted in increased mortality. It was also found that seed which fermented during extraction germinated sooner and better than seed from fresh drupes after 1 year storage. One hundred lb of fruit yield 15–20 lb of seed. The cleaned seed averages 5,800 seeds per lb, with a commercial purity of 97 per cent and a soundness of 94 per cent. Presowing treatment is recommended by stratification for 90–160 days at 41°F. The seed germinates in stratification if held too long. The pretreated seeds may be sown in spring in drills spaced 8–12 in. apart, with about 25 seeds per linear ft. The beds should be mulched until germination begins. A fungus, *Coccomyces lutescens,* causes a leaf-spot disease. Another leaf-spot disease is *Bacterium prunii.* Both of these may be controlled by a 4-6-50 Bordeaux mixture. Dusting with sulphur will help control the powdery mildew of the fungus *Podosphaera oxyacanthae.*

VARIETIES AND FORMS. Some authorities use the name *P. nana* Du Roi for the above species, and the name *P. virginiana* (L.) Mill. for *P. serotina* Ehrh.

Amber Choke Cherry, *P. virginiana* var. *leucocarpa* Wats., has amber-colored fruit. Distribution in cultivation since 1889.

Dwarf Choke Cherry, *P. virginiana* var. *nana,* is a low form. Widely distributed in cultivation.

Black Choke Cherry, *P. virginiana* var. *melanocarpa* (A. Nels.) Sarg., has smaller, thicker leaves; fruit darker, usually black; petioles glandless. New Mexico; westward to California and northward to North Dakota and British Columbia.

Western Choke Cherry, *P. virginiana* var. *demissa* (Torr. & Gray) Torr., has leaves usually rounded or subcordate at base and pubescent beneath, petioles glandular; branchlets and inflorescence glabrous or pubescent. From California northward into Washington. *P. virginiana* forma *pachyrrhachis* (Koehne) Sarg. is a form of *P. virginiana* var. *demissa,* with leaves broad-cuneate or rounded at base, twigs and rachis stout and pubescent. In Mexico.

Broad-leaf Choke Cherry, *P. virginiana* forma *duerinckii* (Martens) Zab., has leaves broadly elliptic. Distributed in cultivation.

Yellow Choke Cherry, *P. virginiana* forma *xanthocarpa* Sarg., is a yellow-fruited form related to *P. virginiana* var. *melanocarpa.* Distributed in cultivation since 1912.

REMARKS. The genus name, *Prunus,* is the classical name, and the species name, *virginiana,* refers to the state of Virginia. Vernacular names are Wild Black Cherry, Cabinet Cherry, Rum Cherry, Whiskey Cherry, Black Chokeberry, California Chokeberry, Eastern Chokeberry, Eastern Choke Cherry, Western Choke Cherry, and Caupulin. The tree is sometimes planted for ornament and for erosion control. It has been in cultivation since 1724. It has a tendency to form thickets of considerable extent from root sprouts. The fruit is used to make jellies and jams, and is eaten by at least 40 species of birds, and browsed by black bear and cottontail. The bark is sometimes used as a flavoring agent in cough syrup.

Bitter Cherry

Prunus emarginata (Dougl.) Walp.

FIELD IDENTIFICATION. Usually a shrub of high altitudes, 3–12 ft high. In more favorable locations a tree to 40 ft, with a trunk 8–14 in. in diameter. The slender upright or spreading branches form a symmetrical oblong crown. Sometimes forming dense thickets.

FLOWERS. April–June, when the leaves are about half expanded; corymbs or racemes 1–1½ in. long, 3–12-flowered, dull white or greenish, each flower ⅜–½ in. across; pedicels slender, glabrous or pubescent, ¼–⅖ in. long, bracts at base foliaceous, glabrous and glandular-serrate; calyx-tube obconic, glabrous or puberulous; sepals 5, shorter than the calyx-tube, about 1/12 in. long, apex obtuse or rounded, surfaces slightly pubescent, margin entire or slightly glandular, reflexed at maturity; petals 5, about ¼ in. long, apex rounded,

Bitter Cherry
Prunus emarginata (Dougl.) Walp.

notched or obovate, short-clawed at base; stamens numerous; pistil 1, style elongate, ovary 2-ovuled.

FRUIT. Maturing June–August, corymb pedicels to 2 in. long, drupe globose or oval, 1/4–1/2 in. long, red at first but eventually black, skin thick and translucent; flesh thin, bitter, and astringent; stone ovoid or ellipsoid, turgid, ends pointed, ventral suture ridged and grooved, dorsal suture rounded and faintly grooved.

LEAVES. Simple, alternate, often clustered, deciduous, obovate to oblong or oblanceolate to oval, margin finely crenulate-serrulate, the teeth glandular, apex obtuse to rounded or sometimes emarginate, more rarely acute, base cuneate, blade length 1–3 in., width 1/3–1 1/2 in., upper surface dark green and glabrous, lower surface paler and pubescent to eventually glabrous; petioles pubescent, 1/8–2/5 in. long, with 1–2 large dark glands at the apex; stipules lanceolate, apex acuminate, margin glandular-serrate above, deciduous.

TWIGS. Slender, young ones pubescent, reddish brown, later glabrous and red to purplish or eventually gray, lenticels large and pale.

BARK. On trunk light brown to gray, smooth, marked with horizontal yellowish or orange lenticels.

WOOD. Sapwood pale, heartwood brown or greenish, close-grained, soft, brittle.

RANGE. Moist slopes or stream banks in sun or shade, at altitudes of 5,000–9,000 ft. New Mexico; west through Arizona to California and northward to British Columbia and Montana.

VARIETIES. Several varieties of Bitter Cherry have been segregated by botanists.

P. emarginata var. *crenulata* (Greene) Kearney & Peebles has leaves elliptic or oblanceolate, apex mostly acute, mostly glabrous beneath. It occurs in New Mexico and Arizona.

REMARKS. The genus name, *Prunus*, is the ancient classical name, and the species name, *emarginata*, refers to the leaves sometimes being shallowly notched at the apex. However, the petals are also sometimes emarginate. Vernacular names are Quinine Cherry, Wild Mountain Cherry, and Plum-leaf Cherry. It was first introduced in cultivation in 1918.

TEXAS ALMOND CHERRY

Prunus glandulosa (Hook.) Torr. & Gray

FIELD IDENTIFICATION. Little-known shrub with crooked branches and subevergreen leaves, attaining a height of 2–5 ft. Fruit a velvety drupe to 3/4 in. long.

FLOWERS. Very small (3/16–3/8 in. across), from lateral buds in spring before or with the leaves, solitary or paired, perfect; pedicels pubescent, averaging about 1/8 in.; petals 5, white or pink, imbricate, spreading, rounded to obovate, inserted in the throat of the hypanthium; stamens numerous, filaments filiform, distinct; pistil solitary, ovary sessile, 1-celled, with 2 pendulous ovules; style simple and terminal; calyx about 1/8 in. long or less; with 5 imbricate sepals which are oblong, obtuse, densely hairy, glandular-toothed, and as long as the hypanthium.

TEXAS ALMOND CHERRY
Prunus glandulosa (Hook.) Torr. & Gray

FRUIT. Ripening April–May, 1/2–3/4 in. in diameter, subglobose to ovoid, one end rounded, the other short-pointed, slightly flattened, velvety-hairy, dry, thin-fleshed; stone solitary, ovoid, bony, with a membranous testa on the seed.

LEAVES. Simple, alternate or fascicled, blade 3/4–1 1/4 in. long, oblong to ovate-oblong or elliptic, apex obtuse, base acute or gradually narrowed, margin very glandular-serrate, surfaces dull grayish green and densely tomentose, especially beneath, the hairs short and curved; veins delicate and impressed above, more conspicuous beneath; petioles to 1/4 in., pubescent.

TWIGS. Slender, divaricate, flexuous, stiff, sometimes slightly spiniferous, when young reddish brown, densely tomentose, when older gray and glabrous.

RANGE. In sandy or sandy-loam soils on prairies in east-central, central, and southern Texas.

REMARKS. The genus name, *Prunus*, is the classical Latin name of the plum-tree, and the species name, *glandulosa*, refers to the glands of the leaf margins and calyx. The plant is little known, and has no particular economic value outside of furnishing a small amount of forage for sheep and goats. It has been listed under the name of *P. hookeri* Schneid. in some literature.

Desert Peach-brush

Prunus fasciculata (Torr.) Gray

FIELD IDENTIFICATION. Western thicket-forming shrub 2–9 ft. Sometimes shaped like a small tree. Branches rather dense, spreading, and spinescent.

FLOWERS. Borne March–May, mostly 1-sexed, with staminate and pistillate on different plants, solitary in the leaf axils or 2–5 together, 1/4–1/3 in. across, sessile or very short-pediceled, greenish white; calyx yellowish, free from the ovary; sepals 5, short-acuminate; petals 5, 1/12–1/8 in. long, oblanceolate to obovate, distinct, spreading; stamens numerous, exserted, borne on the edge of the recepto-calyx; pistil 1, ovary superior and 1-celled.

FRUIT. Maturing April–May, drupe very short-peduncled or almost sessile, subglobose to ovoid, apex shortly acute to obtuse-pointed, densely light brown hairy, about 5/8 in. long, the dry exocarp splitting on one side, flesh thin and dry; stone solitary, bony, smooth, narrowly winged on the ventral side.

LEAVES. Simple, deciduous, alternate or clustered on short stubby branches, blades 1/4–3/4 in. long, 1/8–1/4 in. wide, linear-spatulate, apex obtuse to acute, base tapering to a very short petiole or sessile, margin entire or with 1–3 remote teeth on a side, surfaces grayish green and finely pubescent, all veins except main vein obscure.

TWIGS. Short, stiff, spinescent, young ones semilustrous, puberulent and brown to gray, later dull gray and glabrous.

RANGE. Dry sunny sites on hillsides, mesas, or slopes, usually in sandy or gravelly soils, at altitudes of 2,500–6,500 ft. California and Arizona; northward to Nevada, Utah, and Colorado. To be expected in northern and western New Mexico, and perhaps in Trans-Pecos Texas.

VARIETY. Punctate Desert Peach-brush, *P. fasciculata* var. *punctata* Jepson, has young shoots pubescent and glabrous, glandular-punctate leaves. It occurs in Santa Barbara County, California.

REMARKS. The genus name, *Prunus,* is the classical Latin name of the plum tree of Europe. The species name, *fasciculata,* refers to the fascicled or clustered leaves of the reduced lateral spurs. The plant has also been listed in the literature under the name of *Emplectocladus fasciculatus* Torr., *Amygdalus fasciculata* Gray, and *Lycium spencerae* MacBr. The plant was introduced into cultivation in 1881. It furnishes some limited browse for sheep and goats. In some areas it is locally known as Desert Almond.

Desert Peach-brush
Prunus fasciculata (Torr.) Gray

Small-flower Peach-brush

Prunus minutiflora Engelm.

FIELD IDENTIFICATION. An intricately branched shrub 1–3 ft, sometimes forming dense thickets. Leaves small and leathery. Fruit a small velvety drupe less than 1/2 in. long.

FLOWERS. Solitary or 2–4-flowered corymbs, in March, lateral, scaly buds in early spring with the leaves, very short-pediceled or almost sessile, about 1/8 in. long; dioecious, the staminate plants more numerous than the pistillate; calyx campanulate, about 3/16 in. long, glabrous or puberulent; sepals 5, imbricate, deciduous, short and broadly triangular, apex obtuse to acute, about 1/25 in. long; petals 5, white, imbricate, spreading, inserted in the hypanthium throat, oval to obovate, base clawed, margin crenulate, 1/16–1/8 in. long; stamens numerous, inserted with the petals, filaments filiform and distinct; ovary sessile, 1-celled, 2-ovuled, style simple and terminal.

FRUIT. Maturing May–June. Drupe ovoid to globose-ovoid, 1/3–1/2 in. long, velvety-hairy, thin-fleshed, shell thin and brittle; stone bony, ovoid, 3/8–1/2 in. long, ends rounded or short-pointed, somewhat flattened, smooth, ridged on one side but grooved on the other, pale brown, seed solitary.

SMALL-FLOWER PEACH-BRUSH
Prunus minutiflora Engelm.

LEAVES. Simple, alternate, or clustered on short lateral branchlets, blades oblong to elliptic or obovate, blade length 3/8–3/4 in., apex obtuse to rounded, less often acute, cuneately narrowed into short petioles, margin usually entire or sometimes with coarse glandular teeth, leathery; upper surface dull green and glabrous, lower surface paler and minutely puberulent; petiole 1/16–3/16 in., puberulent to glabrous.

TWIGS. Slender, when young brown to gray with dense white tomentum, later gray and glabrous.

RANGE. Dry hills and slopes, mostly on limestone soils. On prairies of central and western Texas. From the Colorado River westward to beyond the Pecos. In Mexico in Chihuahua.

REMARKS. The genus name, *Prunus,* is the classical Latin name of a European plum-tree. The species name, *minutiflora,* refers to the tiny flowers. The plant has also been listed in the literature under the names of *Emplectocladus minutiflorus* (Engelm.) Gray and *Amygdalus minutiflora* (Engelm.) Wight.

HAVARD PLUM

Prunus havardii (Wight) Mason

FIELD IDENTIFICATION. Rather large, Western, rigidly-branched, spinescent shrub.

FLOWERS. May–June, less than 1/4 in. long, sessile and usually solitary at the base of the fascicled leaves; bud scales reddish brown, ovate to oval, ciliate, 1/25–1/16 in. long; petals 5, white, oval to obovate, short-clawed, about 1/16 in. long, externally puberulent to glabrate; stamens numerous, some slightly exserted; style included, ovary short; recepto-calyx campanulate, longer than the petals, 5-lobed, the lobes short-ovate or triangular, obtuse to acute, about 1/25 in. long, one third to one fourth the length of the recepto-calyx.

FRUIT. Drupe in July, sessile, to 1/4 in. long or less, subglobose to ovoid, blunt-pointed or rounded, puberulent to more glabrate later; stone solitary, ovate, bony.

LEAVES. Fascicled on short lateral spurs, 1/4–3/4 in. long, usually less than 1/2 in. wide, blade obovate to spatulate or oblanceolate, apex rounded to obtuse and usually toothed, base cuneate and usually entire, petiole pubescent, very short (1/25–3/16 in. long) or leaf almost sessile, leaf surface dull green and glabrous above with obscure impressed veins, lower surface reticulate-veiny and sometimes with minute scattered hairs or glabrous.

TWIGS. Slender, stiff, somewhat spine-tipped, usually with many short, almost lateral, side branchlets, at first gray or brownish and minutely puberulent, later light to dark gray and glabrous.

RANGE. In the mountains of Trans-Pecos Texas and adjacent Mexico. Observed by author as a shrub to

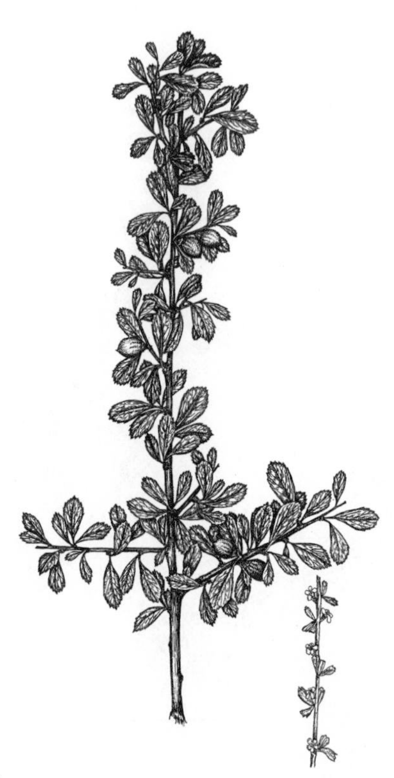

HAVARD PLUM
Prunus havardii (Wight) Mason

OKLAHOMA PLUM
Prunus gracilis Engelm. & Gray

5 ft in Oak Creek Canyon, Chisos Mountains, Brewster County, Texas.

REMARKS. The genus name, *Prunus*, is the ancient name, and the species name, *havardii*, honors Valery Havard, who made extensive collections along the Rio Grande River. An account of his work is given by Winkler (1915).

OKLAHOMA PLUM

Prunus gracilis Engelm. & Gray

FIELD IDENTIFICATION. Straggling thicket-forming shrub 1–15 ft.

FLOWERS. Opening in March, usually before the leaves, borne in sessile lateral umbels, 2–4 in a cluster; pedicels slender, pubescent ⅓–⅖ in. long; corolla ¼–⅓ in. broad; petals 5, white, rounded, imbricate, spreading, inserted in the throat of the hypanthium; stamens numerous, inserted with the petals, filaments filiform and distinct; ovary sessile, 1-celled, ovules 2 and side by side, pendulous; style simple and terminal; calyx with 5 imbricate sepals which are spreading, deciduous, ovate to ovate-lanceolate, obtuse to acute, entire or denticulate, and finely pubescent.

FRUIT. Drupe maturing June–August, subglobose or ovoid to somewhat pointed at the ends, to ⅝ in. in diameter, usually red with a slight bloom, pulpy; stone oval, indehiscent, obtuse at ends, bony, nearly smooth, slightly flattened; seed with membranous testa.

LEAVES. Simple, alternate or fascicled, deciduous, length 1–2 in., elliptic to oval or ovate, margin finely and sharply serrate with appressed teeth, apex acute or obtuse, base gradually narrowed, thickish, upper surface slightly pubescent to glabrous when mature, lower surface reticulate-veined and densely pubescent; winter buds with imbricate scales; petioles short, pubescent, glandless.

TWIGS. Slender, slightly divaricate, soft-pubescent, reddish brown, later gray and glabrous.

RANGE. On dry sandy soils in the sun. North Texas, Oklahoma, and western Arkansas, to Tennessee and Kansas.

VARIETIES, HYBRIDS, AND SYNONYMS. A hybrid between Oklahoma Plum, *P. gracilis,* and Chickasaw Plum, *P. angustifolia,* has been given the name of *P.* × *slavinii* Palmer.

It is also thought that Beach Plum, *P. maritima,* is a close relative of Oklahoma Plum.

A plant described under the name of Texas Plum, *P. venulosa,* is apparently a hybrid between Oklahoma Plum and Reverchon Plum, *P. reverchonii.*

Oklahoma Plum has been listed in the literature under the names of *P. normalis* Rydb. and *P. chicasa* var. *normalis* Torr. & Gray.

REMARKS. The genus name, *Prunus,* is the ancient Latin name, and the species name, *gracilis,* refers to the slender branches. The plant was introduced into cultivation in 1916. It is susceptible to black-knot fungus disease on the limbs and twigs. The fruit is edible but not of particularly good quality. Also known as the Sour Plum.

MURRAY PLUM

Prunus murrayana Palmer

FIELD IDENTIFICATION. Shrub attaining a height of 3–6 ft, or possibly becoming larger in protected situations, intricately branched or sometimes growing with erect stems and few ascending branches.

FLOWERS. Borne March–April with the leaves in simple 1–5-flowered umbels; pedicels pubescent, slender, ¼–½ in. long; flowers averaging ¼–½ in. in diameter when fully expanded; petals 5, white, oval or obovate, short-clawed, exceeding the numerous stamens; calyx 5-lobed, sepals oblong-lanceolate, apex obtuse, outer sur-

MURRAY PLUM
Prunus murrayana Palmer

face densely hispid-pubescent, inwardly less pubescent to glabrous, margin glandular-hairy.

FRUIT. Not seen.

LEAVES. Simple, alternate, ovate to elliptic or lanceolate, apex acute or acuminate, base rounded or cuneate, margin finely and evenly glandular-serrate (12–14 serrations to the ½ in.), thin but firm at maturity, upper surface scabrate, lower surface pilose-pubescent with short stiff hairs; petioles ⅕–⅖ in., densely pubescent, eglandular; stipules linear, ¼–⅖ in. long, glandular-serrate.

TWIGS. Slender, rather rigid, rarely spiny, green to brown and densely pubescent the first season, becoming light to dark gray, or retaining some of the pubescence the second season.

RANGE. Type specimen No. 33424 collected April 21, 1928, on rocky banks of a ravine, near the head of Big Aguja Canyon, Fowlkes Ranch, Davis Mountains, Texas. Type specimen No. 34562 collected from near Fort Davis, June 13, 1928. Specimens deposited in the United States National Herbarium.

REMARKS. The genus name, *Prunus*, is the classical name. The species name, *murrayana*, is for Andrew Murray (1812–1878), English botanist. The following comments concerning the plant are made by Palmer (1929, p. 38):

"This plant, which was the only wild plum seen in the Davis Mountains region, was found growing on steep rocky banks of a stream near the head of the Big Aguja Canyon in the Davis Mountains of Texas. It was also found in a little dry canyon off the Limpia Canyon, near Fort Davis. The plants at the first collection were in flower, while those collected at the latter show only the mature leaves.

"Although I have not seen the fruit, this species is so distinct in the character of its inflorescence and in the pubescence from that of any of the other plums with which I am acquainted that I venture to describe it as new. It is perhaps most closely related to *P. rivularis* Scheele, which it resembles in the habit of growth, but from which it is well distinguished by the characters mentioned in the above description."

It is to be noted that the artist made the illustration from the specimen with flowers and immature leaves, because the mature leaf specimen collected by Palmer was not available.

CHICKASAW PLUM

Prunus angustifolia Marsh.

FIELD IDENTIFICATION. Twiggy shrub forming dense thickets, or a short-trunked, irregularly-branched tree to 25 ft.

FLOWERS. March–April, in lateral 2–4-flowered umbels borne before the leaves on slender, glabrous pedicels ¼–½ in.; corolla white, about ⅓ in. across; petals 5, obovate, rounded at apex, somewhat clawed at base; calyx-tube campanulate, glabrous, 5-lobed; lobes ovate, obtuse, ciliate, pubescent within; stamens usually 15–20, filaments free with oval anthers; ovary 1-celled.

FRUIT. May–July, drupe globose, ½–¾ in. in diameter, red or yellow, yellow-dotted, lustrous, little bloom if any, skin thin with juicy, edible, subacid flesh; stone ovoid to oblong, about ½ in. long, rounded, somewhat grooved on the dorsal suture, rugose and turgid.

LEAVES. Alternate, simple, deciduous, 1–2 in. long, ⅓–⅔ in. wide, lanceolate or oblong-lanceolate, acuminate to acute and apiculate at the apex, rounded or broadly cuneate at the base, troughlike, glabrous and lustrous green above, paler and glabrous or pubescent beneath, sharply serrate with small glandular teeth; petioles slender, glabrous or puberulous, ¼–½ in. long, glandless or with 2 red glands near the apex.

TWIGS. Reddish brown, lustrous, hairy at first but

glabrous later, slender, zigzag, often with spinescent spurlike lateral divisions; lenticels horizontal, orange-colored.

BARK. Reddish brown to dark gray, scales thin and appressed; lenticels horizontal and prominent.

WOOD. Reddish brown, sapwood lighter, rather soft, not strong, fairly heavy, weighing 43 lb per cu ft.

RANGE. Thought to be originally native in Texas and Oklahoma but now rather rare in a wild state. Arkansas and Louisiana; eastward to Florida, northward to New Jersey, and west to Illinois.

PROPAGATION. The ripe fruit may be gathered by hand or by threshing or shaking the tree. The seeds are extracted by means of a macerator or hammer mill in water and the pulp floated off. If stored at low temperatures the seeds will remain viable for a year or so, and often longer. The cleaned seed averages about 1,060 seeds per lb, with a commercial purity of 96 per cent and a soundness of 96 per cent. Stratification is done in moist sand for about 150 days at 41°F. The seeds are sown in drills 8–12 in. apart and about ½ in. deep. One third to one half of the viable seed sown produce usable seedlings.

CHICKASAW PLUM
Prunus angustifolia Marsh.

VARIETIES. A number of varieties and pomological forms have been developed. Big Chickasaw Plum, *P. angustifolia* var. *varians* Wight & Hedrick, is larger than the species and has longer pedicels, and stones more pointed at the apex. It produces early yellow fruit. Some of its commercial forms are the Transparent, Emerson, Coletta, Clark, and African.

Sand Chickasaw Plum, *P. angustifolia* var. *watsonii* (Sarg.) Waugh, has rather rigid stems, leaves smaller and firm; fruit small, thin-skinned, red. Several horticultural forms from it are the Strawberry, Welcome, Red Panhandle, Yellow Panhandle, and Purple Panhandle.

The form El Paso is a hybrid of *P. angustifolia* × *P. munsoniana*. The form Junior Bruce is a hybrid of *P. angustifolia* × *P. salicina*.

REMARKS. The genus name, *Prunus*, is the classical name for the European plum, and the species name, *angustifolia*, refers to the narrow foliage. Often called Mountain Cherry in some localities. Seldom found in a wild state, but most often around dwellings. It is sometimes used in shelter-belt planting.

REVERCHON HOG PLUM

Prunus reverchonii Sarg.

FIELD IDENTIFICATION. Thicket-forming shrub attaining a height of 2–6 ft. Fruit yellow with a red cheek.

FLOWERS. Borne March–April, with or before the leaves, solitary or in clusters of 2–4; pedicels glabrous, ¼–⅖ in. long; flowers perfect, usually less than ½ in. across; petals 5, white, imbricate, spreading, obovate to oblong-obovate, somewhat clawed; inserted in the throat of the hypanthium; stamens numerous, filaments filiform, distinct; ovary sessile, 1-celled, ovules 2 and side by side, pendulous; style simple and terminal; calyx-lobes 5, shorter than the tube, imbricate, oblong to oblong-ovate, glandular-ciliate, sparingly hairy, erect or spreading.

FRUIT. Maturing July–September, drupe globose or subglobose, ⅝–⅞ in. in diameter, usually yellow with a red or orange cheek (rarely entirely red), with little or no bloom, pulpy; stone oblong, ends pointed, smooth or slightly reticulate, seed solitary with a membranous testa.

LEAVES. Simple, alternate or clustered on lateral spurs, deciduous, blades lanceolate to ovate-lanceolate, length 2–3 in., apex acuminate, base cuneate or rounded, margin glandular-serrate, strongly folded, surfaces rather densely pubescent when young, later the upper surface green and glabrous, lower surface paler and slightly pubescent; petiole ¼–½ in., slightly pubescent to glabrous, glandless or with 2–4 glands at apex.

REVERCHON HOG PLUM
Prunus reverchonii Sarg.

TWIGS. Slender, younger ones chestnut-brown to gray and glabrous, older ones light to dark gray.

RANGE. On well-drained moist sites in the sun, grows in limestone or other soils. In central and northern Texas, Oklahoma, Arkansas, and Louisiana.

REMARKS. The genus name, *Prunus,* is the classical Latin name of a European plum. The species name, *reverchonii,* honors Julien Reverchon (1837–1905), Texas plant collector of French birth. Reverchon Hog Plum was introduced into cultivation about 1916. The fruit is of rather poor quality, but is edible. Plants evidently can withstand considerable drought. Some individuals have been found which suggest hybridization with Mexican Plum, *P. mexicana.* L. H. Bailey (1917, Vol. 5, p. 2829) notes that it is closely related to Creek Plum, *P. rivularis,* and may only be a variation of that species, but is segregated by having more diffuse branching and less slender stems, very trough-like leaves, later ripening fruit, and a more pointed stone.

CREEK PLUM

Prunus rivularis Scheele

FIELD IDENTIFICATION. Slender-stemmed, thicket-forming shrub 3–8 ft, bearing bright red drupes.

FLOWERS. With or before the leaves in spring (March) from scaly buds on slender glabrous pedicels ⅓–½ in. long, usually in umbel-like clusters of 2–5 flowers or sometimes solitary; corolla about ½ in. wide; petals 5, imbricate, spreading, white, obovate-orbicular or oblong-obovate, short-clawed, inserted in the throat of the hypanthium; stamens numerous, inserted with the petals, filaments filiform and distinct; ovary sessile, 1-celled, the 2 ovules side by side and pendulous, style simple and terminal; calyx with 5 sepals, imbricate, as long as the tube, ovate to oblong, acuminate, glandular, slightly pubescent, somewhat deciduous or reflexed with age.

FRUIT. Drupes maturing in June, subglobose to short-oblong, ½–⅘ in. long, bright red, lustrous, without bloom or nearly so, drying black, exocarp pulpy; stone large, smooth, flattened, indehiscent; seed solitary, testa membranous.

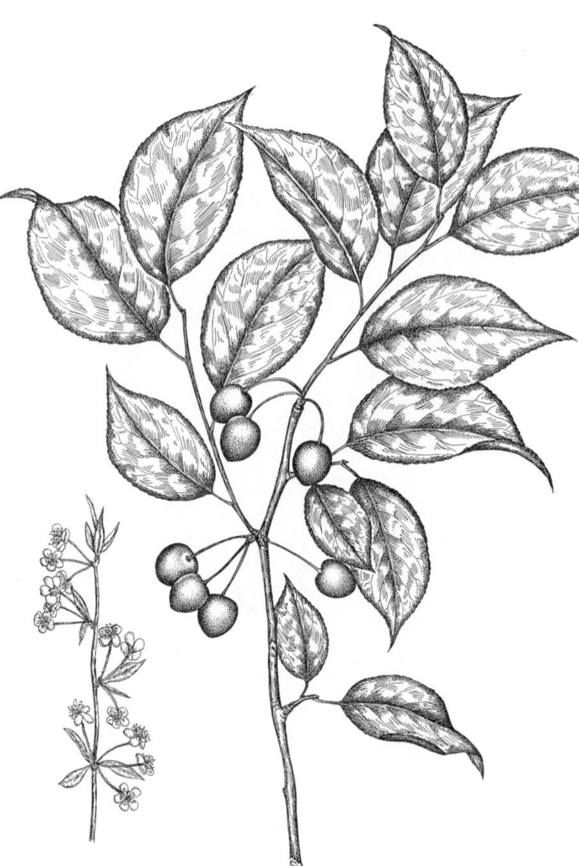

CREEK PLUM
Prunus rivularis Scheele

LEAVES. Simple, deciduous, alternate, or clustered on hort lateral spurs, blade ovate to oblong or sometimes obovate, apex acuminate, base rounded, margin glandular-serrate, length 1–3 in., upper surface olive-green and glabrous; lower surface slightly paler and sparingly pubescent, veins delicate and slender; petiole to ½ in. long, glabrous, green or reddish, glandless or with 1 or 2 glands.

TWIGS. Slender, reddish or gray, lustrous, glabrous, somewhat angled, later light to dark gray.

RANGE. Along streams in sunny sites. River valleys of Texas, along the Colorado, Guadalupe, and Leona rivers.

REMARKS. The genus name, *Prunus,* is the classical Latin name of a European plum. The species name, *rivularis,* refers to its preference for stream banks. The plant was first introduced into cultivation in about 1917. The fruit is about the size of a large cherry and is shiny cherry-red. It is poorly flavored, but was sometimes eaten by the Indians.

FLATWOODS PLUM
Prunus umbellata Ell.

FLATWOODS PLUM

Prunus umbellata Ell.

FIELD IDENTIFICATION. Shrub or small twiggy, flat-topped tree rarely over 20 ft.

FLOWERS. In 2–4-flowered umbels; individual flowers white, ½–¾ in. broad, borne on slender glabrous pedicels about ½ in. long; petals 5, rounded, clawed at base; stamens 15–20, filaments slender; ovary 1-celled; calyx obconic, glabrous or puberulent, 5-parted; sepals ovate, obtuse or acute at the apex, puberulent on the exterior, pubescent within.

FRUIT. Drupes on stems ½–1 in. long, about ½ in. in diameter, globose, no basal depression, usually dark purple with a bloom, sometimes red or yellow, skin tough; stone oval or subglobose, about ½ in. long, acute at ends, slightly roughened, ridged on the ventral suture and grooved on the dorsal.

LEAVES. Alternate, simple, deciduous, blades hardly over 2½ in. long, oblong, lanceolate, or occasionally oval, obtuse or acute at apex, rounded, cordate, or broadly cuneate at base, glandular-serrulate on margin, glabrous above, pubescent beneath along the midrib, firm; petiole ⅕–⅓ in. long, glabrous or pubescent.

TWIGS. Slender, spurlike, dark, reddish brown, lustrous, pubescent at first but later glabrous, more or less thorny.

BARK. Dark brown, scales small, persistent, appressed.

WOOD. Reddish brown, lighter sapwood, close-grained, heavy, hard.

RANGE. Usually in sandy soil; Texas, southern Arkansas, and Louisiana; eastward to Florida and northward on the coastal plain to southern North Carolina.

PROPAGATION. The fruit is picked from the ground, or the trees shaken and the fruits gathered in canvas or cloth. The pulp is removed by running through a hammer mill and floating off the pulp. Germination is hastened by stratification for 90 days at 41°F. in moist sand or peat. The planting is done in drills spaced 8–12 in. apart, at about 1 in. deep, using about 18 seed per linear ft. The seed beds should be mulched until germination begins and protected from birds and rodents. Fungus leaf-spot diseases can be controlled by spraying with a 4–6–50 Bordeaux mixture.

VARIETIES. *P. umbellata* var. *injucunda* (Small) Sarg., is a variety with elliptic to oblong leaves, acute or acuminate, broad-cuneate at the base, pubescent beneath; stone usually pointed. Occurs from North Carolina to Alabama and Mississippi.

P. umbellata var. *tarda* (Sarg.) Wight has elliptic-oblong or obovate leaves, acute to short-acuminate, cuneate at base, pubescent or glabrous above, tomentose beneath; petioles tomentose. Occurs from southern Arkansas to Texas, Louisiana, and Mississippi.

REMARKS. The genus name is the classical name for a plum of Europe, and the species name, *umbellata,* refers to the umbellate flower clusters. Some of the vernacular names are Hog Plum, Sloe, and Black Sloe. The fruit is used in considerable quantities for jellies and preserves.

WILD-GOOSE PLUM

Prunus munsoniana Wight & Hedrick

FIELD IDENTIFICATION. Thicket-forming shrub or small round-topped tree to 25 ft.

FLOWERS. In 2–4-flowered corymbs; pedicels slender, glabrous, ⅔–1 in., bearing flowers ½–⅗ in. in diameter; calyx-tube obconic, with 5 sepals which are ovate-oblong, acute or obtuse, glandular on the margin, glabrous or pubescent; petals about ¼ in. long, white, oblong-obovate, abruptly contracted into a short claw, entire or somewhat erose; stamens usually 15–20, filaments with oval anthers; ovary 1-celled, with 2 ovules, style terminal.

FRUIT. Drupe globose to oval, about ¾ in. long, red or yellow, white-dotted, bloom light, skin thin, flesh yellow and juicy; stone oval, pointed at the apex, truncate at the base, grooved on the sutures, roughened.

LEAVES. Alternate, simple, deciduous, blades 2½–4 in. long, ¾–1¼ in. wide, lanceolate to oblong-lanceolate, acute or acuminate at the apex, cuneate or rounded at the base, finely glandular-serrate on the margin, bright lustrous green above, sparingly pubescent, especially along the veins beneath; petioles slender, glabrous or pubescent, biglandular at the apex.

WILD-GOOSE PLUM
Prunus munsoniana Wight & Hedrick

TWIGS. Reddish brown, shiny, glabrous; lenticels pal and numerous.

BARK. Reddish or chestnut-brown, thin, smooth.

RANGE. Texas, Oklahoma, Arkansas, and Louisiana north to Kansas, Kentucky, and Illinois.

VARIETIES. Sixty-seven named horticultural forms ar known. The most important are Texas Belle, Newman Osage, Robinson, Schley, and Sophie; and hybrid such as El Paso, *P. angustifolia* × *P. munsoniana;* Nana *P. salicina* × *P. munsoniana;* and Wild Goose, *P. mun soniana* × *P. americana.*

REMARKS. The genus name, *Prunus,* is the classica name of a European plum, and the species name, *mun soniana,* refers to T. V. Munson (1823–1913), Amer ican botanist. The tree is grown both for ornament an for its fruit.

MEXICAN PLUM

Prunus mexicana Wats.

FIELD IDENTIFICATION. Shrub or small tree to 25 ft with an irregular open crown.

FLOWERS. White, ¾–1 in. in diameter, borne on slen der glabrous pedicels in 2–4-flowered umbels; petals 5 ovate-oblong, rounded, narrowed into a claw, entire or crenulate, pubescent, much longer than the calyx lobes; stamens 15–20; style elongate, ovary 1-celled calyx-tube obconic, puberulous on the exterior, tomen tose within, 5-lobed; lobes ovate to oblong, entire o serrate, ciliate and glandular on the margin, about as long as the tube.

FRUIT. Drupe subglobose to short-oblong, dark pur plish red with a bloom, 1¼–1⅓ in. in diameter; flesh juicy, of varying palatability; stone ovoid to oval, dorsa edge ridged, ventral edge grooved, smooth, turgid.

LEAVES. Alternate, simple, deciduous, thickish, blades 1¾–3½ in. long, 1–2 in. wide, ovate to elliptic or obo vate, abruptly acuminate at the apex, cuneate or round ed at the base; singly or doubly serrate with apiculate teeth; upper surface yellowish green, glabrous, shiny hairy below especially on the veins; prominently retic ulate-veined both above and below; petioles stout pubescent, hardly over ⅗ in. long, glandular at the apex.

TWIGS. Slender, stiff, glabrous, or pubescent early shiny, grayish brown.

BARK. Gray to black, exfoliating in platelike scales when young, when older rough and deeply furrowed.

RANGE. Texas, Louisiana, Arkansas, and Oklahoma north to Missouri, Tennessee, and Kentucky and southward in northeastern Mexico.

VARIETIES. A number of varieties have been described, but recent authors tend to consider most of

Mexican Plum
Prunus mexicana Wats.

them as synonyms of *P. mexicana* rather than separate varieties. Reticulate Leaf Plum, *P. mexicana* var. *reticulata* Sarg., and *P. mexicana* var. *tenuifolia* Sarg., near Larissa, Texas, are considered in that category.

Other varieties are Fulton Mexican Plum, *P. mexicana* var. *fultonensis* Sarg., and Polyandra Mexican Plum, *P. mexicana* var. *polyandra* (Sarg.) Sarg.

REMARKS. The genus name, *Prunus,* is the ancient classical name for a plum of Europe, and the species name, *mexicana,* refers to this species' Southwestern distribution. It is sometimes known as Big-tree Plum, because of the fact that it is treelike, and does not sucker to form thickets. It is rather drought-resistant, and has been used as a grafting stock.

Hortulan Plum

Prunus hortulana Bailey

FIELD IDENTIFICATION. Many-stemmed shrub or small tree to 30 ft, with a broad round-topped crown.

FLOWERS. Maturing late March–early May. Borne in 2–3-flowered umbels on slender, puberulous or glabrous pedicels 1/5–1/2 in. long; flower white, 2/3–1 in. in diameter; petals 5, oval-oblong, rounded at the apex, long-clawed, entire or erose; stamens numerous; ovary short; calyx-tube obconic, glabrous, 5-lobed; lobes glabrous, oblong-ovate, acute or obtuse, glandular-ciliate, pubescent, about as long as the tube.

FRUIT. Maturing August–October. Drupe globose to ellipsoid, 3/4–1 in. long, red or yellow-red, white-dotted, lustrous, little or no bloom, thin-skinned; stone turgid, reticulate, 2/3–3/4 in. long, rounded or short-pointed at the apex, rounded or truncate at the base, grooved on the dorsal suture.

LEAVES. Alternate, simple, deciduous, blades 4–6 in. long, 1–1½ in. wide, oblong-obovate to oblong-oval, acuminate at the apex, cuneate or rounded at the base; upper surface glabrous, dark green, lustrous; lower surface glabrous or slightly pubescent; finely glandular-serrate on the margin; petiole slender, orange-colored, 1–1½ in. long, glandular.

TWIGS. Dark reddish brown, stout, rigid, glabrous, or pubescent, occasionally somewhat spinescent.

BARK. Dark or light brown, exfoliating into large thin plates.

RANGE. Texas, Oklahoma, Arkansas, and Louisiana; northward to Missouri, Iowa, Kansas, Kentucky, and Tennessee.

VARIETIES AND FORMS. One of the best known varieties is Miner Hortulan Plum, *P. hortulana* var. *mineri* Bailey, with darker, duller, veiny leaves and more

Hortulan Plum
Prunus hortulana Bailey

scaly bark. Forms with close relationships are the Langsdon, Clinton, and Forest Rose.

Missouri Hortulan Plum, *P. hortulana* var. *pubens* Sarg., is a variety with somewhat pubescent leaves, petioles, and twigs.

Other horticultural forms are World Beater, Irby, Maquoketa, Golden Beauty, Cumberland, Neptune, Wayland, Moreman, and Sucker State. Reagan is a hybrid of *P. hortulana* and *P. americana.*

Hortulan Plum is considered by some to be the most beautiful American plum. It is known mostly from horticultural forms, being rather rare in a wild state. About 34 forms are known.

REMARKS. The genus name, *Prunus,* is the ancient Latin name, and the species name, *hortulana,* means "of gardens" and refers to its value in horticulture.

American Plum

Prunus americana Marsh.

FIELD IDENTIFICATION. Shrub propagating by root sprouts to form thickets, or a tree to 35 ft, with spreading, pendulous, more or less thorny branches.

FLOWERS. March–May, borne in 2–5-flowered umbels on slender, glabrous pedicels ⅓–⅔ in. long; flowers ¾–1¼ in. broad, perfect, white; petals 5, narrowly obovate, rounded or erose at the apex, narrowed into a claw which is red at base, about ½ in. long; stamens about 20, in numerous rows; pistil with a 1-celled ovary; calyx-tube obconic, glabrous or pubescent, red or greenish; lobes of calyx 5, lanceolate, acute or obtuse at the apex, entire or toothed toward the apex, glabrous or puberulous on the exterior, usually pubescent within.

FRUIT. Ripening June–October, variable in size, usually ¾–1 in. long, subglobose, red or sometimes yellow, somewhat glaucous or not at all, conspicuously marked with pale dots; skin tough; flesh yellow and succulent, varying in flavor, sometimes hardly edible, bittersweet, acid, very suitable for preserves; stone oval, rounded at apex, somewhat narrowed at base, smooth to slightly rugose, turgid, somewhat compressed, obscurely ridged on one suture and grooved on the other.

LEAVES. Alternate, simple, deciduous, 2½–4 in. long, 1½–2 in. wide, ovate to obovate or oblong, acuminate at apex, rounded or cuneate at the base; sharply and singly or doubly serrate on the margin; thick, firm, more or less rugose; dark green and glabrous above; paler and reticulate-veined, and glabrous or slightly pubescent along the veins beneath; petioles ½–¾ in. long, slender, glabrous or puberulous, eglandular or glandular at the apex.

TWIGS. Slender, glabrous, green to orange to reddish brown; lateral branchlets spurlike, sometimes spinescent; lenticels circular, raised, minute.

BARK. Dark brown to reddish, breaking into thin, long, persistent plates.

American Plum
Prunus americana Marsh.

WOOD. Dark brown, with lighter colored sapwood, close-grained, strong, hard, weighing 45 lb per cu ft of no commercial importance because of small size of trunks.

RANGE. Texas, New Mexico, Oklahoma, Arkansas, and Louisiana; eastward to Florida, northward to Massachusetts, New York, and southern Ontario, and west to the Rocky Mountains.

PROPAGATION. The ripe fruit may be gathered by hand or by threshing or shaking the tree. The seeds are extracted by means of a hammer mill or macerator in water and the pulp floated off. After drying they can be stored in open vessels at room temperature for 2–4 years with little loss in viability. The cleaned seed averages about 840 seeds per lb, with 96 per cent purity and 94 per cent soundness. For spring planting stratification is done in moist sand or peat for 150 days at 41° F. The seeds are sown in drills 8–12 in. apart and ½–1 in. deep. The seedlings can be protected from brown rot, caused by the fungus *Monilinia fruticola,* by applying 3–4–50 Bordeaux mixture and by burning diseased leaves and tips.

FORMS AND HYBRIDS. Names officially listed for horticultural clones and hybrids of *P. americana* by Kelsey and Dayton (1942):

Aitken (Haz., 1896)
Americana, *P. americana* × *P. salicina* (Burb., 1898)
Apple, *P. americana* × *P. salicina* (Burb., 1898)

Arkansas
Brackett (Terry, 1900)
Brittlewood (T. Will., 1897)
Cheney
Climax, *P. americana* × *P. salicina*
Desoto (Tup., 1853)
Downing (Ter., 1882)
Elliott, *P. americana* × *P. salicina*
Emerald, *P. americana* × *P. salicina*
Excelsior, *P. americana* × *P. salicina*
Flickinger, *P. americana* × *P. salicina*
Forest Garden (Hare, 1862)
Gaviota, *P. americana* × *P. salicina*
Golden, *P. americana* × *P. salicina*
Golden Beauty, *P. americana* × *P. salicina*
Gonzalez, *P. americana* × *P. salicina*
Hammer (Ter., 1888)
Handska, *P. americana* × *P. simonii*
Hawkeye (Terry, 1882)
Inkpa, *P. americana* × *P. simonii*
Juicy, *P. americana* × *P. simonii*
Koga, *P. americana* × *P. simonii*
Kahinta, *P. americana* × *P. simonii*
Laire
Mankato
Milton (Ter., 1885)
Mina
Miner (Dodd, 1814)
Monitor, *P. americana* × *P. salicina*
Morewan (Wayl., 1881)
Mound, *P. americana* × *P. salicina*
Neverfail
New Ulm (Hei., 1890)
Ocheeda (Hardo., 1872)
Omaha, *P. americana* × *P. salicina*
Oren (Oren, 1878)
Poole Pride (Kroh, 1885)
Pottawattomie
Radisson, *P. americana* × *P. salicina*
Reagan, *P. americana* × *P. hortulana*
Redwing, *P. americana* × *P. salicina*
Rollingstone
Shiro, *P. americana* & Mixed Hybrid (Burb., 1899)
Six Weeks, *P. salicina* × *P. americana*
Stevens
Stoddard (Baker, 1875)
Surprise, *P. americana* × *P. hortulana* (Penn., 1882)
Tawena, *P. salicina* × *P. americana*
Terrell, *P. americana* & Japanese Sp. (Ter., 1895)
Terry
Toka, *P. simonii* × *P. americana*
Tokata, *P. americana* × *P. simonii*
Tonka, *P. salicina* × *P. americana*
Valley River
Waneta, *P. salicina* × *P. americana*
Wayhard (Way., 1875)
Weaver (Wea., 1873)
Whitaker
Wildgoose, *P. munsoniana* × *P. americana* (McCa., 1820)
Wilson
Winnesboro
Winona, *P. salicina* × *P. americana*
Wolf (Wolf, 1852)
Wood (J. Wood, 1852)
Wyant (Wyant, 1866)

The following forms are also thought to be of *P. americana* Marsh. origin, but records are obscure: Blackhawk, Cherokee, Craig, Gaylord, Newton, Itasca, and Minnetonka.

Inch Plum, *P. americana* var. *lanata* Sudw., was formerly classified as a variety of American Plum, but has recently been given the status of a species. Apparently there are numerous intergrades between the two. Inch Plum has leaves which are permanently soft-pubescent beneath, as well as the calyx-lobes, pedicels, and twigs.

REMARKS. The genus name, *Prunus,* is the classical name for a European plum, and the species name, *americana,* is of obvious meaning. Other vernacular names are Horse Plum, Hog Plum, Goose Plum, Native Plum, River Plum, Wild Yellow Plum, Red Plum, and Thorn Plum. The fruit makes excellent jellies and preserves, or may be eaten raw or cooked. The fruit is also eaten by many species of birds. Many horticultural varieties have been developed. It is probably the best fruit plum for the Middle West and North.

INCH PLUM

Prunus americana var. *lanata* Sudw.

FIELD IDENTIFICATION. Thicket-forming shrub spreading by suckers, or a small slender tree to 30 ft, with a well-defined trunk.

FLOWERS. In 2–5-flowered umbels on hairy or glabrous pedicels ½–⅔ in. long; corolla about ¾ in. in diameter, with 5 petals which are oblong-oval, rounded, long-clawed, and about ¼ in. long; stamens about 25, filaments filiform; calyx-tube obconic, sepals acuminate, entire, ciliate or serrulate toward the apex, pubescent, reddish; ovary 1-celled.

FRUIT. Borne on drooping, glabrous pedicels, subglobose or ellipsoid, deep red with a bloom, about 1 in. long, flesh juicy; stone oblong, flattened, rounded at base, acute and apiculate at apex, dorsally ridged, somewhat grooved ventrally.

LEAVES. Alternate, simple, deciduous, blades 2¼–4½ in. long, about 2½ in. wide, ovate to oblong-obovate, acuminate at the apex, cuneate or rounded at the base, singly or doubly apiculate-toothed, thin; upper surface yellowish green, glabrous; lower surface paler and densely pubescent; petioles slender, pubescent, ½–⅔ in. long, glandless or occasionally having a gland at the apex.

TWIGS. Light yellowish green, slender, unarmed, pubescent, or glabrous later.

BARK. Pale gray to reddish brown, exfoliating in thin scales.

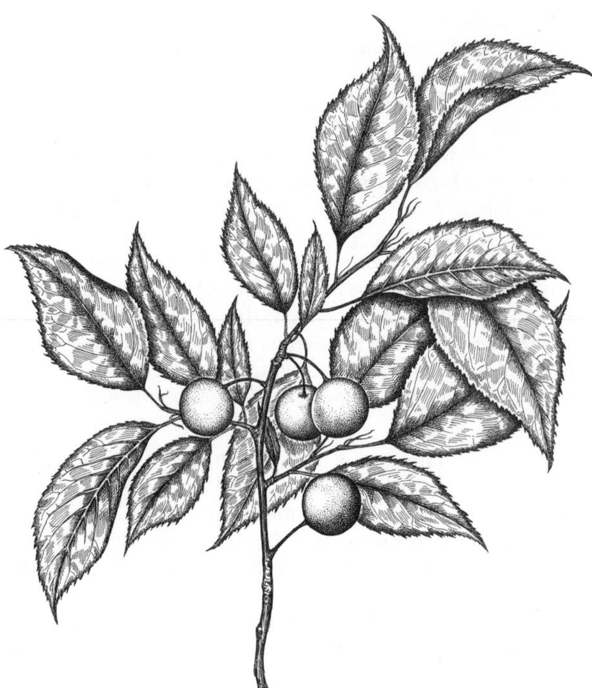

Inch Plum
Prunus americana var. *lanata* Sudw.

WOOD. Dark to light brown, heavy, hard, close-grained.

RANGE. Texas, Oklahoma, Arkansas, and Louisiana; north to Missouri, Indiana, and Iowa.

REMARKS. The genus name, *Prunus*, is the classical name for a plum of Europe; the species name, *americana*, distinguishes this as an American plum; and the variety name, *lanata*, refers to the woolly hairs of the leaves.

Peach

Prunus persica Batsch

FIELD IDENTIFICATION. Tree attaining a height of 24 ft, with a rounded crown and spreading branches.

FLOWERS. Usually expanding before the leaves March–May, subsessile, solitary or 2 together, fragrant, perfect, 1–2 in. across; petals 5, pink, spreading, rounded; calyx with the 5 sepals externally pubescent; stamens 20–30, filaments usually colored like the petals, exserted, slender, distinct; ovary tomentulose, 1-celled and sessile, the 2 ovules pendulous; pistil solitary with a simple terminal style.

FRUIT. Drupe maturing July–October, subglobose, grooved on one side, velvety-tomentose, 2–3⅓ in. in diameter, fleshy, separating in halves at the sutures; stone elliptic to ovoid-elliptic, usually pointed at distal end, deeply pitted and furrowed, very hard; fruit of escaped trees usually harder and smaller.

LEAVES. Numerous, conduplicate in the bud, almond-scented, impregnated with prussic acid, simple, alternate, some appearing clustered, elliptic-lanceolate to oblong-lanceolate, broadest at the middle or slightly above the middle, 3–6 in. long, apex long-acuminate, base varying from acute to acuminate or broad-cuneate, margin serrate or serrulate, surfaces glabrous and lustrous, bright green, thin; petioles glandular, ⅜–⅝ in.

RANGE. A native of China, but extensively cultivated from Texas eastward to Florida and northward to New York and southern Ontario. Sometimes escaped from cultivation in the southeastern United States.

FORMS, VARIETIES, AND HYBRIDS. The following ornamentals have been listed by Rehder (1940, pp. 463–464):

P. persica forma *alba* Schneid. has white flowers.

P. persica forma *duplex* (West) Rehd. has pink, double flowers.

P. persica forma *camelliaeflora* (Vanh.) Dipp. has deep red, semidouble flowers.

P. persica forma *albo-plena* Schneid. has white, semidouble flowers.

P. persica forma *rubro-plena* Schneid. has red, semidouble flowers.

P. persica forma *magnifica* Schneid. has bright red, double flowers.

P. persica forma *versicolor* (Vanh.) Voss has semidouble, white, red, and striped flowers on some plants.

P. persica forma *pyramidalis* Dippel has a narrow pyramidal habit.

Peach
Prunus persica Batsch

P. persica forma *pendula* Dippel has pendulous branches.

The numerous forms cultivated for their fruit may be classed under the following:

P. persica forma *scleropersica* (Reichb.) Voss has flesh adhering to the stone, and is known as the Clingstone Peach.

P. persica forma *aganopersica* (Reichb.) Voss has flesh separating from the stone, and is known as the Freestone Peach.

P. persica var. *nectarina* (Ait.) Maxim. is known as the Nectarine Peach. It has smaller, glabrous fruit and leaves more strongly serrate. This variety has two forms: *P. persica* forma *scleronucipersica* (Schuebl. & Mart.) Rehd. has the stone adhering to the flesh, and *P. persica* forma *aganonucipersica* (Schuebl. & Mart.) Rehder has the stone separating from the flesh.

P. persica var. *compressa* (Loud.) Bean is known as Flat Peach and has a fruit much flattened above, broader than high; stone small and irregular.

P. persica × *P. salicina* Lindl. is a hybrid plum-peach. The leaves are oblong, broad-cuneate at base; flowers pink and stalked; style longer than stamens; ovary rudimentary and hairy.

A cross of *P. persica* × *P. munsoniana* Wight & Hedr. has also been recorded.

REMARKS. The genus name, *Prunus*, is the classical Latin name of a plum of Europe. The species name, *persica*, means "Persian," and is also an old generic name for peach. The genus name of *Amygdalus* L. is used by some authors.

COMMON PEAR
Pyrus communis L.

COMMON PEAR

Pyrus communis L.

FIELD IDENTIFICATION. Tree generally pyramidal and upright, living to an old age. Attaining a height of 75 ft and 2–3 ft in diameter. The branches are stiff, upright, and sometimes thorny.

FLOWERS. March–May, with or before the leaves; cymes simple, terminal, 4–12- flowered, borne on short twigs of the preceding year; pedicels ½–2 in., pubescent at first, glabrous later; corolla white or pink, 1–2 in. broad, 5-petaled, petals broad-oblong, rounded, short-clawed; stamens numerous, exserted; anthers yellow, small, 2-celled, sacs longitudinally dehiscent; styles 5, distinct to the base, stigma small, ovules 2 in each cavity, cavities as many as the styles; disk cushion-like; calyx urn-shaped, the 5 acute lobes about as long as the tube.

FRUIT. Ripening July–October, pome pear-shaped, tapering to the base, in the wild form about 2 in. long, much longer in cultivated forms, yellow to reddish, flesh with abundant grit cells; seeds small, smooth, brown to black, endosperm none, cotyledons fleshy, the pome consisting of the thickened calyx-tube and receptacle enclosing the carpels.

LEAVES. Simple, alternate, deciduous, usually on short lateral spurs, ovate to elliptic or obovate, margin finely serrate to entire, apex acute or acuminate, sometimes abruptly so, base usually rounded, blade length 1½–4 in., young leaves downy and ciliate, mature leaves lustrous, dark green to olive-green above, lower surface paler, both sides glabrous or nearly so; petioles 1¼–3 in., as long as the blades or longer.

TWIGS. Somewhat pubescent at first, glabrous later, reddish brown to gray or black.

WOOD. Reddish brown, hard, fine-grained, 51 lb per cu ft.

RANGE. Common Pear is a native of Europe and Asia, and escapes cultivation in some areas in the Southwest and elsewhere in North America. It is often cultivated in Texas, Oklahoma, Arkansas, and Louisiana.

PROPAGATION. The fruit is usually collected by shaking down when mature. By macerating with water, the pulp is separated and floated off. About 100 lb of fruit yield a lb of cleaned seed, which averages about 15,000 seeds per lf. Good seed has a 98 per cent purity and a 96 per cent soundness. Viability of seed may be maintained 2–3 years or more by storage at room temperature. Seed dormancy may be overcome somewhat by stratification for 60–90 days at 32°–45° F. The stratified seed is sown in spring in drills and covered about ½ in. with soil. The seedlings are somewhat subject to mildew. Horticultural varieties are usually propagated by budding and grafting.

VARIETIES AND HORTICULTURAL FORMS. *P. communis* var. *pyrulaster* L. has globose fruit ⅝–¾ in. across; flowers 1 in. across; leaves rounded, strongly serrate, glabrous when young; usually thorny.

P. communis var. *cordata* (Desv.) Schneid. has globose to turbinate fruit about ½ in. across; flowers smaller; leaves ovate to orbicular, subcordate, ¾–1⅓ in. across; a spiny shrub of Great Britain and France.

P. communis var. *sativa* DC. is generally considered the parent of most cultivated forms; large thornless tree; fruit large, juicy and variously formed; flowers 1–2 in. across; leaves larger.

P. communis var. *longipes* Henry is an Algerian variety with leaves ovate, acuminate, subcordate, 2 in. long, 1 in. wide, fruit globose, about ½ in. in diameter; calyx deciduous.

P. communis var. *mariana* Willk. is a Spanish variety; tree small, leaves ovate, base rounded, 1 in. long, petiole long and slender; fruit globose, about ½ in. in diameter; calyx persistent.

The following garden varieties of *P. communis* are known:

P. communis var. *trilobata* Hort. has lobed leaves.
P. communis var. *heterophylla* Hort. has cut leaves.
P. communis var. *variegata* Hort. has variegated leaves.

The following horticultural forms are listed by Kelsey and Dayton (1942):

Andre Desportes (LeRoy, 1854)
Anjou
Ansault (LeRoy, 1863)
Barseck (Moore, 1890)
Bartlett (Stain, 1845)
Beirachmitt
Belle Lucrative
Bessemianka
Beurre Bosc (Van M., 1807)
Beurre Clairgeau (Clai., 1830)
Beurre D'Arenberg
Beurre Dejonghe (Dej., 1852)
Beurre Diel (Men., 1805)
Beurre Giffard (Gif., 1825)
Beurre Hardy (Bon., 1820)
Beurre Superfin (Gou., 1837)
Bloodgood
Bordeaux
Brandywine (Harv., 1820)
Buffum (Buf., 1828)
Cayuga (N. Y. S., 1920)
Clapp Favorite (T. Cla., 1860)
Clyde (N. Y. S., 1932)
Colonel Wilder (Fox, 1870)
Columbia (Cos., 1813)
Conference (Rev., 1894)
Covert (N. Y. S., 1935)
Dana Hovey (Dana, 1854)
Dearborn
Dorset (L. Cla., 1895)
Douglas (Ayer, 1897)
Doyenne Boussock (Van M., 1819)
Doyenne du Comice
Dr. Guyot
Duchesse D'Angouleme
Early Ely (Ely, 1906)
Early Harvest
Early Seckel (N. Y. S., 1935)
Easter Buerre (Van M., 1823)
Elizabeth (Van M., 1836)
Enie
Ewart
Fame
Feagan
Flemish Beauty (Van M., 1810)
Forelle
Fox (Fox, 1875)
Frederick Clapp (L. Cla., 1870)
Garber (Garb., 1880)
Glou Marceau (M. Hard., 1750)
Gorham (N. Y. S., 1923)
Guyat (Balt., 1870)
Henderson Special
Hood (Hood, 1911)
Howell (T. Howe., 1829)
Idaho (Mulk, 1867)
Jargonnelle
Josephine de Molines (Esp., 1830)
Kieffer (Kie., 1863)
Koance
Lawrence
Lawson
Le Conte (LeC., 1850)
Lincoln
Lincoln Careless
Louise Borne de Jersey (De Lo., 1780)
Madeleine
Magnolia
Marguerite
Marie Louise (Duq., 1809)
Mendel
Menie
Ming
Minie
Moe
Mount Vernon (Walk., 1868)
Onondaga
Ontario (Sun., 1856)
Osband
Ovid (N. Y. S., 1931)
Parker
Passe Colmar (A. Hard., 1758)
Patten
P. Barry (Fox, 1873)
Phileson
Pineapple (Stuc.)
Pitmaston (J. Will., 1841)
Pound (Uve., 1690)
President Drouard (Oli., 1886)

Pulteney (N. Y. S., 1925)
Pushkin (Hans., 1919)
Reeder (Reed., 1855)
Reidinger Bartlett
Riehl Best (Riehl, 1870)
Roosevelt
Roasney (Woo., 1881)
Russett Bartlett (Rut., 1893)
Rutter (Rut., 1860)
Seckel
Sheldon (Shel., 1815)
Souvenir D'Esperen
Souvenir Du Congress (Morel, 1852)
Sudduth (Cons., 1820)
Summer Doyenne (Van M., 1800)
Swanegg
Tait Dropmore
Tait No. 1
Tait No. 2
Tyson (Tyson, 1794)
Urbaniste (DeC., 1786)
Vermont Beauty (Mac., 1880)
Vicar of Wakefield
White Doyenne
Wilder Early (Green, 1884)
Willard (N. Y. S., 1931)
Winter Bartlett
Winter Nelis (Nelis, 1830)
Worden Seckel (Wor., 1881)

REMARKS. The genus name, *Pyrus,* is the classical name for the pear-tree, and the species name, *communis,* means "common." The tree is sometimes used for shelter-belt planting and for wildlife food.

SOUTHERN CRAB-APPLE
Pyrus angustifolia Ait.

SOUTHERN CRAB-APPLE

Pyrus angustifolia Ait.

FIELD IDENTIFICATION. Usually a small tree with rigid, thorny branches forming a broad open head. More rarely larger, and to 35 ft, with a diameter of 10 in.

FLOWERS. In simple, terminal, 3–5-flowered cymes, very fragrant; pedicels ¾–1 in. long, slender, puberulous at first, glabrous later; flowers perfect, about 1 in. in diameter; petals 5, obovate to oblong, clawed at base, about ¼ in. wide, white to pink; stamens shorter than the petals, about 20 in 3 series; ovary 5-celled, carpels leathery, styles 5, united and hairy below; calyx 5-lobed, lobes about as long as the tube, glabrous externally, tomentose within.

FRUIT. Pome depressed-globose, ¾–1 in. in diameter, often broader than long, waxy, pale yellowish green, fragrant, flesh acid and sour.

LEAVES. Simple, alternate, deciduous, or persistent southward, blade length 1–3 in., width ½–2 in., elliptic to oblong on flower spurs and crenately serrate (on young shoots oval to obovate and with lobes as well as serrations), apex obtuse to acute or rounded and often apiculate, base narrowed and cuneate, firm and leathery, dull green, hairy when young, but glabrous with age, midrib sometimes remaining pubescent; petiole slender, about ¾–1 in., green to reddish, villose at first, glabrous later; stipules linear, about ⅓ in. long, reddish.

TWIGS. Stout, smooth, light brown to reddish brown, pubescent at first, older twigs glabrous, lenticels scattered, orange-colored; buds small, about 1/16 in. long, dark brown, obtuse, the 4 outer scales imbricate and pubescent; leaf scars very narrow, bundle scars 3, pith homogeneous.

BARK. Dark reddish brown to gray, fissures rather deep, ridges narrow, separating into small, platelike scales.

WOOD. Heartwood reddish brown, sapwood yellow, close-grained, hard, heavy, weighing about 43 lb per cu ft.

RANGE. Woods, thickets, riverbanks. In Oklahoma, Arkansas, and Louisiana; eastward to Florida, northward into Illinois, and west to Missouri and Kansas.

PROPAGATION. The seeds are obtained by running fruit through a macerator and screening to float off the pulp. Seed may be stored for a year or two in dry sealed containers at temperatures above freezing. For spring planting, the seed is stratified for 85 days at 41°–46° F. in moist sand. Sowing is usually done in

PRAIRIE CRAB-APPLE
Pyrus ioensis (Wood) Bailey

drills, followed by a cover of about ¼ in. soil. A light clay-loam is generally considered best. The plant was first introduced into cultivation in 1750.

VARIATIONS. Drooping Southern Crab-apple, *P. angustifolia* forma *pendula,* is a form with slender pendulous branches. Known from cultivation.

Double-rosy Southern Crab-apple, *P. angustifolia* forma *rosea-plena,* is a form with pink double flowers. Known from cultivation.

Southern Crab-apple is considered by some authorities to be only a variety of the Wild Sweet Crab-apple, *P. coronaria* L.

REMARKS. The genus name, *Pyrus,* is the classical name for the Pear-tree, and the species name, *angustifolia,* refers to the narrow leaves. Also known under the vernacular names of Narrow-leaf Crab-apple and Wild Crab-apple. The wood of Southern Crab-apple is used for levers, tools, and small woodenware objects. The acid, sour fruit is made into preserves and cider. It is eaten by many species of birds and animals, including the bobwhite quail, blue jay, cardinal, ruffed grouse, prairie chicken, skunk, opossum, raccoon, cottontail, and red and gray fox.

PRAIRIE CRAB-APPLE

Pyrus ioensis (Wood) Bailey

FIELD IDENTIFICATION. Tree attaining a height of 28 ft and a trunk diameter of 18 in., the numerous rigid, crooked branches forming a rounded spreading crown.

FLOWERS. Borne April–June, fragrant, in 2–5-flowered clusters on very hairy pedicels 1–1½ in.; calyx-lobes 5, lanceolate-acuminate, longer than the tube, densely white-tomentose; petals 5, white or pink, about ½ in. wide, obovate, base narrowed into a slender claw; stamens numerous, shorter than the petals; styles 5, joined below and white-hairy.

FRUIT. Maturing September–October, peduncles ¾–1½ in. long, globose, somewhat depressed, apical and basal depressions shallow, greenish to yellow, sometimes with minute, yellow dots, surface waxy and greasy to the touch, length ¾–1¼ in., width ¾–1½ in., flesh sour and astringent.

LEAVES. Simple, alternate or clustered, deciduous, length and width variable on weak and vigorous shoots, blades 1½–5 in. long, ¾–4 in. wide, elliptic to oblong or obovate-oblong, apex acute or obtuse to rounded, base cuneate or rounded, margin singly or doubly crenate-serrate, or on some deeply lobed as well; at maturity coriaceous, dark green, lustrous and glabrous above, lower surface varying from almost glabrous to densely white-tomentose, turning yellow in autumn; petioles slender, at first with hoary-white tomentum, becoming pubescent or glabrous later.

TWIGS. Reddish brown to gray, densely tomentose at first but with age less so, finally glabrous, lenticels small and pale; twigs often set with numerous short lateral shoots bearing thorns terminally; winter buds small, obtuse, pubescent.

BARK. Reddish brown to dark gray, about ⅓ in. thick, scales small, narrow, persistent.

RANGE. In Texas, Oklahoma, Arkansas, and Louisiana; eastward to Alabama and north to Minnesota.

PROPAGATION. Good crops of fruit occur about every 2 years. The fruit is gathered from the ground or from the trees. The seeds are separated from the pulp by maceration in water and then dried. Storage is most satisfactory in sealed dry containers at temperatures just above freezing. The cleaned seed averages about 30,000 seeds per lb, with a commercial purity of 86 per cent and a soundness of 94 per cent. For breaking dormancy, stratification at 41° F. for 60–85 days is helpful. Freshly gathered seed may be sown in fall, or stratified seed in spring in drills and covered about ¼ in. with nursery soil.

VARIETIES, FORMS, AND HYBRIDS. A number of variations have been distinguished as follows:

Palmer's Crab-apple, *P. ioensis* var. *palmeri* (Rehd.) Bailey, differs chiefly by smaller leaves, more oblong and thinly pubescent, rounded at apex, and those of flower shoots not as lobed. The common variety of Missouri, Arkansas, and eastern Oklahoma.

Texas Crab-apple, *P. ioensis* var. *texana* (Rehd.) Bailey, differs by having smaller and broader leaves, slightly lobed or lobeless, persistently densely villous; very intricately branched and thicket-forming. A variety of the Texas Edwards Plateau (Blanco, Kendall, and Kerr counties).

Bush's Crab-apple, *P. ioensis* var. *bushii* (Rehd.) Bailey, differs by having oblong-lanceolate, acute, glabrescent, less deeply lobed leaves. A variety of southern Missouri, which may also occur in northern Arkansas.

Louisiana Crab-apple, *P. ioensis* var. *creniserrata* (Rehd.) Bailey, has leaves elliptic-ovate to ovate, nearly entire or crenately serrate (on vigorous shoots doubly serrate), pubescent below; petioles densely hairy; branches villose at fruit, slender, spineless. Near Pineville, Rapides Parish, and Crowley, Acadia Parish, Louisiana.

Bechtel Crab-apple, *P. ioensis* forma *plena* (Schneid.) Rehd., is a form with large, rose-colored double flowers. Known from cultivation since 1886.

Fringe-petal Crab-apple, *P. ioensis* forma *fimbriata* Slavin., is a form with petals double and fringed. Known from cultivation.

Soulard Crab-apple, *P.* × *soulardii* (Bailey) Britt., is a hybrid of *P. ioensis* × *P. malus*.

REMARKS. The genus name, *Pyrus*, is the classical name for the Pear-tree, and the species name, *ioensis*, refers to the state of Iowa where it was first described. Other vernacular names are Iowa Crab, Prairie Crab, and Western Crab-apple. The flesh of this crab-apple is sour and inedible, but is sometimes used for making vinegar. It has been cultivated for ornament since 1885. It is of considerable value as food for wildlife, the fruit known to be eaten by at least 20 species of birds and mammals, including bobwhite quail, ruffed grouse, ring-necked pheasant, gray and red fox, skunk, opossum, raccoon, cottontail, woodchuck, red squirrel, and fox squirrel.

LOUISIANA CRAB-APPLE
Pyrus ioensis var. *creniserrata* (Rehd.) Bailey

COMMON APPLE

Pyrus malus L.

FIELD IDENTIFICATION. Cultivated tree 15–50 ft, with a trunk diameter of 2–3 ft. The trunk also rather short, and the spreading branches forming a rounded head. Wild trees often with short spinescent branchlets.

FLOWERS. Usually appearing with the leaves April–May in close terminal cymes; inflorescence on stout tomentose pedicels ⅜–2 in. long; petals 5, white or pink, obovate or rounded, clawed at the base, when spread the corolla ½–3 in. broad; stamens numerous, anthers yellow; styles 5, commonly pubescent below the middle; calyx-like hypanthium urceolate, tomentose, with 5 sepals.

FRUIT. Pome September–November, variable in size and quality, 1⅓–4 in. in diameter, green to yellow or red, depressed-globose or elongate, base hollowed, fleshy, sweet or sour; the 5 papery locules imbedded in the flesh, mostly 2-seeded in each cavity, calyx remaining persistent.

LEAVES. Alternate, elliptic to oblong or ovate, base rounded to broad cuneate or cordate, apex obtuse or abruptly pointed to acuminate, margin finely serrate, blade length 1–4 in., width ¾–3 in.; upper surface dull green, glabrous or nearly so; lower surface paler and pubescent or soft-tomentose, both sides pubescent when young; petioles stout, ⅜–1¼ in.

TWIGS. Stout, young ones tomentose, older ones reddish brown, glabrous and lustrous, some branchlets short and spurlike.

WOOD. Reddish brown, close-grained, hard, weighing 50 lb per cu ft, sometimes used in turnery.

RANGE. Occasionally found in a naturalized condition in Arkansas and Oklahoma, but seemingly not at home farther south. Abundantly cultivated in the northern United States and Canada. A native of Eurasia. Escaping from cultivation and frequently naturalized about old fields, clearings, and abandoned home sites. Does best on a clay-loam soil, but will grow under a wide variety of soils and conditions in a moderate temperate climate.

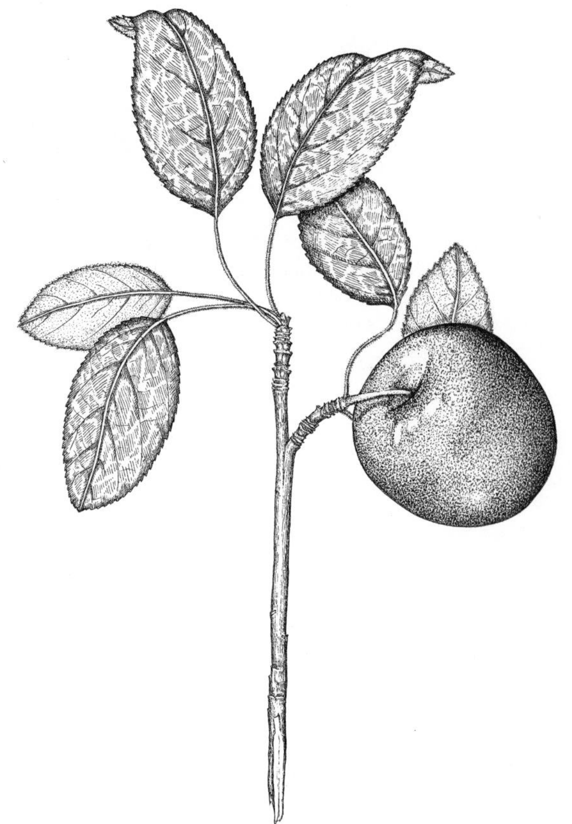

COMMON APPLE
Pyrus malus L.

PROPAGATION. Good fruit crops are obtained every 2 or more years. The seed is obtained by running the fruit through a macerator with water and screening off the pulp. The seed is then dried and, for storage up to 3 years, is stored in sealed containers just above freezing temperatures. It takes about 100 lb of fruit to yield ½–¾ lb of seed. The seed averages about 20,000 seeds per lb, with a commercial purity of 90 per cent and a soundness of 85 per cent. Stratification is generally in moist acid or neutral soil at 41°F. for 75 days. The seeds are sown in spring in drills, although fall sowing may also be practiced, and the seed covered with about ¼ in. soil. Most of the commercial apple varieties are propagated by budding or grafting onto hardy rootstocks. The fruit of seedling trees is generally not true to that of the parents and inferior in quality.

VARIETIES, HYBRIDS, AND CLONES. The cultivated apples are mostly considered to have *P. malus* as the original parent. However, some are hybrids with *P. sylvestris, P. coronaria, P. prunifolia, P. angustifolia,* and *P. ioensis.* Over 400 cultivated forms of apple are known. The following are some of the principal horticultural clones:

Double-pink *(translucens),* Elise Rathke, Niedzwetskyana, Weeping, Aldenham *(aldenhamensis),* Eley *(eleyi),* Lemoine *(lemoinei),* Weeping *(pendula),* and Purple Common Apple *(atrosanguinea* × *malus).*

REMARKS. The genus name, *Pyrus,* is the classical name of the pear-tree. The species name, *malus,* is the classical name of the apple. It is also known under the vernacular names of Pommier, Scarb-tree, Wilding-tree, and Crab-tree. The fruit is known to be eaten by at least 15 species of birds, including ring-necked pheasant and mourning dove, and is also eaten by opossum, raccoon, gray fox, and white-tailed deer.

RED CHOKEBERRY

Pyrus arbutifolia (L.) L.f.

FIELD IDENTIFICATION. Deciduous, swamp-loving shrub to 12 ft. Sometimes the young shoots overtop the compound flower clusters.

FLOWERS. Borne March–May, in terminal compound cymes ¾–2½ in. wide; axillary branches short, persistently villous, 9–20-flowered, sometimes overtopped by the sterile shoots; flowers small, white to pink, ⅓–½ in. broad; calyx urn-shaped, sepals 5, ovate to triangular, apex acute to obtuse, usually glandular, tomentose; petals 5, ⅙–¼ in. long, spreading, obovate to oval, concave, apex rounded, base short-clawed; stamens numerous (about 20), exserted, filaments shorter than

RED CHOKEBERRY
Pyrus arbutifolia (L.) L. f.

ARIZONA MOUNTAIN-ASH
Sorbus dumosa Greene

Dormancy may be broken by stratification in moist peat for 90 days at 33°–41°F. About 10,000 usable plants are obtained from a lb of seed. Red Chokeberry may also be propagated by suckers or layers, or by greenwood cuttings under glass.

VARIETIES. Some authors segregate varieties and horticultural forms on the basis of dwarfism, or large-leaved or large-fruited individuals. Since these variations are connected by intermediate forms, and since they are largely produced by habitat, they are not listed here.

About 3 species of Chokeberry are known in North America, and all are rather closely related. The genus is listed by some authors under the name of *Aronia*.

REMARKS. The genus name, *Pyrus*, is the classical name of the pear-tree. The species name, *arbutifolia*, refers to the *Arbutus*-like leaves. Also known under the vernacular names of Choke-pear and Dogberry. The fruit is a valuable wildlife food in fall and winter, being eaten by at least 13 species of birds, including bobwhite quail, ruffed grouse, ring-necked pheasant, and cedar waxwing. It has been cultivated since 1700. Red Chokeberry is somewhat subject to blight and borer attacks, withstands city smoke, and the leaves are tardily deciduous. It could be more extensively cultivated for its attractive flowers, brilliant fruit, and colorful autumn leaves.

the petals, anthers reddish or purplish; ovary woolly above, styles 5 and united at base, persistent.

FRUIT. Pome ripening September–October, globose or short pear-shaped, ⅙–¼ in. in diameter, conspicuously bright red at maturity, hairy at first, glabrous later, long persistent, carpels leathery; seeds 1–5, some usually abortive.

LEAVES. Convolute in the bud, simple, alternate, deciduous, oval to elliptic or oblong to obovate; apex obtuse or acute to short-acuminate and apiculate; base cuneate or narrowed; margin serrulate-crenulate, the teeth rounded, incurved, and glandular, blade 1–3 in. long; upper surface usually glabrous, midrib sometimes glandular; lower surface densely gray-tomentose; petioles ⅛–⅖ in., semiglabrous to tomentose; stipules narrow, early deciduous. Leaves turning red in autumn.

TWIGS. Brown to gray, persistently tomentose-hairy, older glabrate.

RANGE. Wet woods and swamps. East Texas, Oklahoma, Arkansas, and Louisiana; eastward to Florida, northward to Nova Scotia, and west to Minnesota.

PROPAGATION. The pomes may be picked by hand and dried without removing the pulp before planting. Or if desired, the seed may be cleaned by rubbing on a screen and floating off the pulp. One hundred lb of fruit yield 4–32 lb of seed, there being a high proportion of abortive seed. The seed averages about 256,000 seeds per lb, with a soundness of 70 per cent.

ARIZONA MOUNTAIN-ASH

Sorbus dumosa Greene

FIELD IDENTIFICATION. Shrub 3–15 ft, with slender branches.

FLOWERS. Maturing in June, inflorescences of terminal or axillary, rounded, compound cymes 1½–2 in. broad (rarely to 4 in.). Pedicels white-villous, individual flowers ¼–⅓ in. across; petals 5, white, spreading; stamens numerous; pistil 1, styles 3 and distinct, stigmas capitate and small; ovary inferior, usually 3-celled, ovules 2 in each cavity; calyx densely white-villous, the 5 teeth acute.

FRUIT. Fruiting peduncles with scattered silky hairs or eventually almost glabrous, pome red, berry-like, ⅓–⅖ in. in diameter.

LEAVES. Rachis densely white-hairy, leaves 3–6 in. long, odd-pinnately compound of 9–14 leaflets, each ¾–1¾ in. long, lanceolate to oblong-lanceolate, apex acute, base rounded or broadly cuneate, margin sharply serrate, upper surface dull green and glabrous, lower surface paler with pubescent veins or glabrous; petiolules very short or blade sessile; winter buds white-villous.

TWIGS. Young ones tan to reddish and densely white-hairy, older ones reddish brown or gray and glabrous.

RANGE. In sandy or gravelly moist soils at altitudes

of 7,500–10,000 ft. Often in coniferous forests. New Mexico and Arizona.

REMARKS. The genus name, *Sorbus,* is the ancient name of the Sorb-apple or Service-tree. The species name, *dumosa,* refers to its shrubby aspect. The plant is known to be browsed by sheep. The fruit is eaten by many birds and mammals. The attractive foliage and colorful fruit give it possibilities for use as an ornamental.

Greene Mountain-ash

Sorbus scopulina Greene

FIELD IDENTIFICATION. Rather stout Western shrub 3–15 ft.

FLOWERS. In June, in terminal, flat-topped, compound cymes 3½–6 in. wide, 15–40-flowered; calyx-tube turbinate, sepals 5 and 1/12–1/10 in. long; petals 5, spreading, white, oval, short-clawed, 1/6–1/4 in. long; stamens about 20; ovary inferior, adnate to the calyx-tube; styles 3–5, distinct, woolly at base, cells of ovary as many as styles, each with 2 ovules.

FRUIT. Maturing July–December, pome small, globose, 1/4–2/5 in. in diameter, red to orange, bitter, carpels papery-walled, cells 1–2-seeded.

LEAVES. Alternate, pinnately compound, deciduous, rachis glabrous or slightly pilose, length 4–10 in.; leaflets 11–15, length 1½–2¾ in., oblong-lanceolate to elliptic, margin sharply and doubly serrate, apex acute to acuminate, base cuneate to rounded, glabrous on both sides.

TWIGS. Sparingly villose at first, later glabrous to reddish brown or tan.

RANGE. Sandy or gravelly soil at altitudes of 6,000–10,000 ft. New Mexico in the Sacramento, Sangre de Cristo, Zuni, Manzano, and Tunitcha mountains; west to Arizona and north to Alberta and British Columbia.

REMARKS. The genus name, *Sorbus,* is the classical Latin name of the European species, and the species name, *scopulina,* means "rocklike," perhaps in reference to the habitat. Greene Mountain-ash is usually propagated by seeds. Livestock browse it usually only in stress, but it is much browsed by sheep in some areas. The seeds are known to be eaten by 5 species of birds, including the dusky grouse.

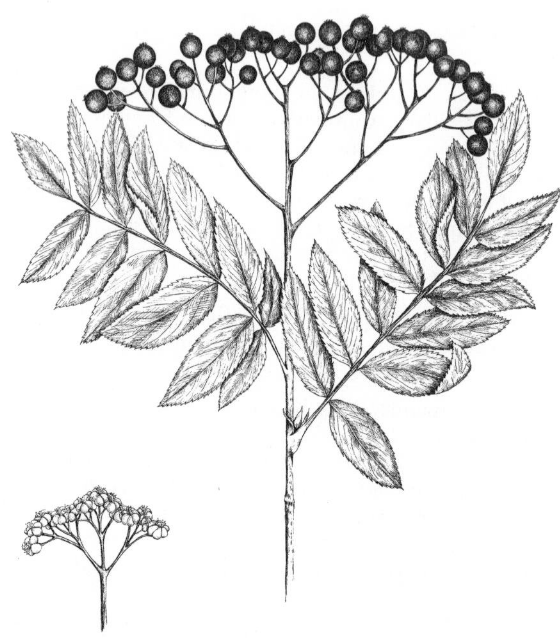

Greene Mountain-ash
Sorbus scopulina Greene

Saskatoon Service-berry

Amelanchier alnifolia Nutt.

FIELD IDENTIFICATION. Western thicket-forming shrub in unfavorable situations sometimes prostrate or dwarfed. On favorable sites sometimes a small tree to 6 or 12 ft.

FLOWERS. May–July, racemes 5–15-flowered, erect, 1¼–2½ in. long, dense; pedicels 1/5–3/5 in. long, conspicuously hairy; corolla white, fragrant, showy; petals 5, oval or oblanceolate, 1/4–1/2 in. long, 1/12–2/5 in. wide, ciliolate, basally hairy, claw very short; stamens about 20, with glabrous filaments 1/25–1/12 in. long; sepals 5, deltoid to lanceolate, 1/10–1/8 in. long, hairy within, reflexed later; styles 4–5, united at base, 1/20–1/10 in. long, ovary tomentose at apex.

FRUIT. Maturing July–August, good crops every year, pedicels glabrous, 1/5–3/5 in. long, fruit body globose to obpyriform, 2/5–3/5 in. in diameter, 10-celled normally, at maturity purple to black, with a bloom, sweet and juicy; seeds about 10, asymmetrical, oval, flattened, brown, smooth, 1/6–1/5 in. long, about 1/8 in. wide.

LEAVES. Simple, alternate, deciduous, dark green, oval to suborbicular in outline, apex obtuse to rounded or truncate, base rounded to truncate or subcordate, margins coarsely dentate to entire or entire toward the base, teeth coarse and somewhat incurved; young leaves tomentose, but later glabrous above and below or slightly hairy on the veins beneath, blade length 3/4–2 in., width 5/8–1⅔ in.; lateral veins conspicuous, rather parallel, curving, veins 8–13, intermediate veinlets obscure; stipules soon deciduous, linear, 1/4–3/4 in. long, hairy; petioles 1/3–3/4 in. long, hairy at first, glabrous later.

TWIGS. Young twigs silky-hairy, later glabrous, bark reddish brown to dark gray; winter buds conical, acute, dark brown, hairy, 1/8–1/4 in. long.

RANGE. In woods or thickets, along streams, in can-

Saskatoon Service-berry
Amelanchier alnifolia Nutt.

yons. Northwestern New Mexico; west to California and north to North Dakota, western Canada, and Alaska.

PROPAGATION. About 100 lb of fruit yield 2 lb of seed. The seed averages about 82,000 seeds per lb, with a soundness of 74 per cent. Maceration in water separates the seeds, and the pulp is floated off. The numerous abortive seeds are separated by a fanning mill or by screening. Storage of dry seed in a sealed container at 41°F. will maintain virility for at least a year. Dormancy may be broken by stratification in sand or peat at 35°–37° F. for 180 days. Nonstratified seed are sown in the fall, stratified in spring. Some seeds do not germinate until the second spring. Sowing is generally done in drills, with about 25 viable seed per linear ft and covered with about ¼ in. of firmed soil. During the first year half shade is furnished the seedlings. Plants may also be grown from cuttings or from suckers.

REMARKS. The genus name, *Amelanchier,* is derived from the French name of a European species, and the species name, *alnifolia,* means "alder-leaved." Also known under the vernacular names of June-berry, Western Shadbush, Alder-leaf Service-berry, Pacific Service-berry, and Common Service-berry.

The plant is considered a valuable browse for livestock, being especially palatable when young. Also a valuable food for wildlife, being eaten by California mule deer, black-tailed deer, plains white-tailed deer, blue grouse, Columbian grouse, sooty grouse, Richardson's grouse, mountain quail, and various chipmunks and ground squirrels.

Shadblow Service-berry

Amelanchier arborea (Michx. f.) Fern.

FIELD IDENTIFICATION. Slender shrub or small round-topped tree seldom over 25 ft.

FLOWERS. March–May, racemes 3–7 in. long, rather dense, erect or nodding, silky-hairy, fragrant, 6–12-flowered; calyx 5-cleft, campanulate, glabrous or hairy, sepals triangular-ovate; petals 5, white, elliptic to obovate, ½–1 in. long; stamens about 20; ovary 5-celled, terminating in 2–5 styles with broad stigmas.

FRUIT. June–July, on long pedicels, subglobose, ¼–½ in. in diameter, dry, reddish purple, tasteless or sweetish; seeds small and numerous (4–10, some abortive), usually dispersed by birds and animals.

LEAVES. Alternate, simple, deciduous, oval to oblong or obovate, acute or acuminate at the apex, rounded or cordate at the base, sharply and finely serrate on the margin, glabrous or nearly so above, paler and pu-

Shadblow Service-berry
Amelanchier arborea (Michx. f.) Fern.

bescent beneath or finally glabrous, blades 2–5 in. long, 1–2 in. wide; petioles 1½–2 in., slender, hairy at first but glabrous later.

TWIGS. Reddish brown to black, slender, rather crooked, somewhat hairy when young but glabrous later; lenticels numerous and pale.

BARK. Gray to black, thin, smooth, in age becoming shallowly fissured with scaly longitudinal ridges.

WOOD. Brown, close-grained, hard, strong, tough, elastic, weighing 49 lbs per cu ft.

RANGE. Oklahoma, Arkansas, Louisiana, and northeast Texas; eastward to Florida and northward to Quebec, Ontario, and Newfoundland.

VARIETY. Alabama Shadblow Service-berry, *A. arborea* var. *alabamensis* (Britt.) Jones, is a small tree with the summit of the ovary tomentose. It occurs in Alabama and Arkansas.

PROPAGATION. It has been cultivated since 1623. The seeds shoud be gathered immediately when ripe before being eaten by birds. They should be dried on shallow screens and macerated to remove the pulp. They are then dried again and winnowed to dispose of other debris and abortive seeds. The seed averages 80,000 seeds per lb, with a commercial purity of 93 per cent and a soundness of 82 per cent.

The germinative capacity averages about 50 per cent. The dormancy may be broken by stratification at 33°–41°F. for 90–120 days for planting in spring. Some planters report better results by immediate fall planting of cleaned seed, followed by mulching. Planting should be in drills of 25 seeds per linear ft, covered by firmed soil about ¼ in. Half shade is required. Plants 2–3 years old are transplanted. Propagation is also accomplished by cuttings in fall, or from suckers in spring.

REMARKS. The genus name, *Amelanchier,* is derived from the French name of a European species. The species name, *arborea,* refers to the treelike character of this species. Vernacular names are Boxwood, Billberry, June-plum, Indian-cherry, Swamp Shadbush, Indian-pear, Juice-pear, Sugar-pear, Plum-pear, and Berry-pear. It is occasionally cultivated in gardens for the showy white flowers. Dwarf plants are often found growing in sterile ground. The berries may be eaten uncooked or made into pies. Shadblow Serviceberry is a valuable wildlife plant, its fruit being eaten by at least 35 species of birds, and its foliage browsed by cottontail and white-tailed deer. The wood is sometimes used for making handles.

DWARF SERVICE-BERRY

Amelanchier pumila Nutt.

FIELD IDENTIFICATION. Shrub 3–9 ft, distinguished from other species by being completely glabrous, even on young parts.

FLOWERS. May–June, borne in 4–8-flowered racemes ¾–1¾ in. long, erect or ascending; pedicels glabrous ¼–½ in long; hypanthium glabrous, glaucous, campanulate, ⅛–⅙ in. long; sepals 5, either linear or triangular-lanceolate, apex acuminate, about ⅛ in. long later recurved, surfaces glabrous; petals 5, white, oval apex acute or obtuse, surfaces glabrous, length ⅓–½ in., width ⅛–⅙ in.; stamens 12–15, filaments glabrous, about 1/25–1/12 in. long; styles 4–5 united at base, 1/25–1/12 in. long, ovary completely glabrous at apex.

FRUIT. Somewhat depressed-globose, about ⅓–⅖ in. in diameter, dark purple, glabrous, juicy; seeds ⅙–⅕ in. long, brown.

LEAVES. Appearing with the flowers, oval to suborbicular, apex obtuse to truncate and sometimes mucronate, base rounded to subcordate or truncate, upper two thirds of blade coarsely toothed on margin, somewhat coriaceous, surfaces glabrous, upper surface deep green, lower surface pale and glaucescent; stipules glabrous, shorter than the blade; stipels linear, glabrous, soon deciduous.

TWIGS. Gray to reddish brown, glabrous throughout.

RANGE. Mountain slopes at altitudes of 6,500–10,000

DWARF SERVICE-BERRY
Amelanchier pumila Nutt.

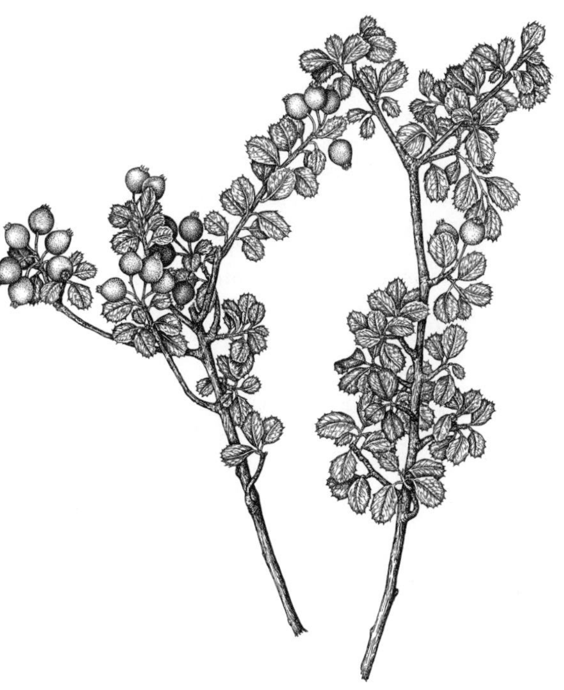

Big Bend Service-berry
Amelanchier denticulata (H. B. K.) Koch

t; in northern New Mexico, west to California and orth to Washington, Idaho, and Montana.

SYNONYMS. The synonomy of this plant is confusing, nd more research is needed to determine the relationhip of the following names: *A. glabra* Greene, *A. polycarpa* Greene, *A. goldmani* Woot. & Standl., *A. alnifolia* var. *pumila* A. Nelson, *A. canadensis* var. *pumila* Nutt.

The author has followed the treatment of George Neville Jones (1946, pp. 81–82).

REMARKS. The genus name, *Amelanchier*, is the French name of a European species, and the species name, *pumila*, refers to the low habit of the plant.

Big Bend Service-berry

Amelanchier denticulata (H. B. K.) Koch

FIELD IDENTIFICATION. Western densely branched shrub to 10 ft, or sometimes a small tree.

FLOWERS. Borne February–March, on short branches of the year, appearing as the leaves unfold or shortly before; pedicels reddish, pubescent, ¼–½ in. long, bracteate at base or some also bracteate near the middle, bracts linear to lanceolate, pubescent, scarious, ciliate, early deciduous; racemes much shortened to appear umbelliform, flowers usually 2–6, about ⅜–½ in. broad, perfect, regular; petals 5, white, rather showy, rounded, about ¼ in. long; stamens numerous, short, inserted on the rim of the calyx, filaments subulate; styles 2–5; calyx ⅛–3/16 in. long, campanulate, glabrous or nearly so on the surface, 5-lobed, lobes triangular to ovate, margins densely-hairy, persistent.

FRUIT. Maturing May–June, pome globose or ovoid, ⅓–½ in. in diameter, reddish to purplish or black, glabrous, remains of calyx and filaments persistent and often reflexed on fruit apex; locules 1-seeded, seeds small, smooth, dark brown, without endosperm.

LEAVES. Simple, deciduous, alternate or apparently clustered, often on short lateral scaly branchlets, oval to reniform or ovate, apex rounded or truncate, base rounded or broadly cuneate, margin entire, or sharply and remotely toothed mostly above the middle, blades ⅜–1 in. long or wide; upper surface olive-green, pubescent with scattered hairs or almost glabrous; lower surface grayish with very fine dense tomentum, veins inconspicuous; short-petiolate.

TWIGS. Slender, straight, terete, at first densely gray-tomentose, later gray to brown and glabrous; stipules linear and caducous; winter buds conspicuous, sessile, solitary, imbricate-scaled; leaf scars semilunar; pith pale and continuous.

RANGE. Chisos Mountains of Brewster County, Texas; south to Coahuila and Chihuahua, Mexico.

REMARKS. The genus name, *Amelanchier*, is from the French name of a European species. The species name, *denticulata*, refers to the small teeth of the leaves. Known in Mexico under the names of Membrillito, Membrillo, Cimarrón, Tlaxisqui, Tlaxistle, and Madronillo. The fruit is sometimes eaten by Indians. The wood is hard and ring-porous. The flexible stems are made by the Mexicans into canes known as "varitas de apizaco."

Utah Service-berry

Amelanchier utahensis Koehne

FIELD IDENTIFICATION. Bushy shrub, or sometimes a small much-branched tree to 25 ft.

FLOWERS. April–June, borne in racemes ¾–1¼ in. long, 3–6-flowered; pedicels and rachis usually hairy, or rarely glabrous; flower buds white to reddish, small; flowers perfect; petals 5, white, linear to oblanceolate, ¼–½ in. long; stamens 10–15, filaments smooth, 1/25–1/12 in. long; summit of ovary tomentose or rarely glabrous; styles 3–4 (occasionally 2 or 5), united below, glabrous, 1/12–⅛ in. long; hypanthium campanulate, ⅛–⅙ in. in diameter, glabrous to tomentose, sepals 5, linear to lanceolate, acuminate at apex, hairy or more rarely glabrous, about ⅛ in. long.

FRUIT. Pome in racemes 2–3 in. long; pedicels green to red, white-hairy or semiglabrous later, ¼–¾ in. long; pome puberulent when young, later bluish black

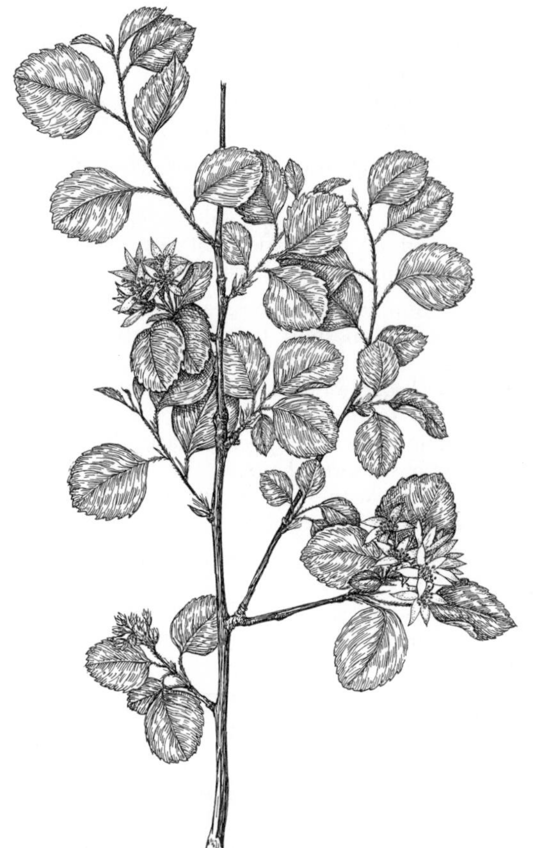

Utah Service-berry
Amelanchier utahensis Koehne

to purplish and glaucous, juicy, subglobose or obovoid, ¼–⅜ in. in diameter, crowned by the 5 erect or spreading hairy calyx-lobes (fruit sometimes drying on bush before maturing and then pale brown and leathery); seeds 4–6, ⅛–¼ in. long, about ⅛ in. wide, brown, ellipsoid, ends obtuse, dorsally rounded, other surfaces plane.

Leaves. Alternate, deciduous, blades ¼–1⅓ in. long, ¼–1 in. wide, oval to ovate or obovate; apex rounded to truncate (more rarely acute); base rounded to truncate or subcordate; margin acutely serrate-dentate, mostly above the middle, or rarely almost entire; upper surface finely hairy to glabrous and glaucescent, subcoriaceous; lateral veins delicate, 11–13 pairs; midrib yellow or reddish; lower surface paler, glabrous to puberulent; petiole ¼–¾ in. long, glabrous to white-hairy, slender, reddish or green; stipules linear, pubescent.

Twigs. Rigid, slender, reddish brown to gray, young densely villous, older more glabrous, lenticels small and scattered, winter buds glabrous or hairy.

Wood. Heavy, hard, rather strong, stems not large enough for commercial use except in making small woodenware articles.

Range. Dry canyons, rock slopes, and mountainsides, at altitudes of 4,000–8,000 ft. In Texas in the Guadalupe Mountains and in New Mexico; west to California and north to Washington, Idaho, and Montana.

Propagation. Utah Service-berry is a desirable ornamental because of the racemes of white or pink flowers in spring and the bluish purple fruit in the fall. The fruit is picked by hand and spread out to dry. The seed may be removed from the dry fruit by rubbing on screens and fanning away the debris, or from the fresh fruit by macerating in water and floating away the pulp. The seeds may be sown immediately after gathering and mulched over for cold protection during the winter, or may be planted in the spring after stratification in moist sand at 41° F. for 180 days or more. Seed have been stored for 2 years at 41° F. in dry sealed containers with only a small loss in viability. The cleaned seed numbers about 80,000 seeds per lb, with an average purity of 91 per cent and a soundness of 80 per cent. The seeds are generally sown in drills, using 20–30 seeds per linear ft. Partial shade is required for the seedlings which are planted out in 2–3 years. Propagation is also practiced with cuttings taken in fall or spring, or by suckers or grafting.

Remarks. The genus name, *Amelanchier*, is from the French name of a European species, and the species name, *utahensis*, is for the state of Utah. Also known by the vernacular names of Red Service-berry, Shadbush, Western June-berry, and Pemican-bush. It is a common and very variable shrub through the mountain and desert regions of western United States. Reflective of its great variability are the many botanical names given, but these have recently been reduced to synonyms by George Neville Jones (1946, pp. 90–93).

The fruit may be eaten raw, and it was also made into bread by the Indians. It was often made into a paste and mixed with cornmeal or jerked dry meat to make the "pemican" of frontier travelers. The fruit was also cooked in puddings and pies by the settlers. Birds and ground squirrels are fond of the ripe fruit and get much before it is ripe. The foliage is grazed by black-tailed deer, sheep, goats, and cattle.

Squaw-apple

Peraphyllum ramosissimum Nutt.

Field identification. Western shrub 3–10 ft, bearing upright, rigid, intricate branches.

Flowers. Appearing with the leaves April–May, solitary or in 2–3-flowered erect umbel-like racemes; individual flowers about ⅝ in. across, rose-colored to white, perfect, regular; petals 5, orbicular to obovate; disk rose-colored; stamens 20; styles 2–3, free, slightly longer than the stamens; ovary wholly inferior, 2–4-celled, each cell with 2 ovules divided by a false partition; calyx adnate to the ovary, the 5 lobes triangular, entire, pubescent within, reflexed, and persistent.

Fruit. Pome maturing July–September, pendulous,

⅓–½ in. thick, yellow or with a reddish or brownish cheek, globose, calyx-lobes persistent at the apex, each cell with 1 seed, fleshy, bitter.

LEAVES. Simple, alternate, clustered at the ends of short stubby branches, deciduous, narrowly oblong to oblanceolate, base cuneate, apex mostly acute but occasionally obtuse to rounded with a glandular tip, margin with brownish glandular-tipped teeth, or entire; young leaves silky-hairy at first but later glabrous, sessile or short-petioled; stipules short, minute and caducous.

TWIGS. Intricately branched, rigid, tomentose at first, eventually glabrous and gray; buds small, acute, pubescent.

RANGE. Arid soils on hills and sunny slopes at altitudes of 5,000–8,000 ft. Northwestern New Mexico and adjacent Arizona; northward to eastern Oregon.

PROPAGATION. Introduced into cultivation in 1870, but not much used. The growth is slow and the plant blossoms only after several years. Suitable for rocky slopes: Propagated by seeds or layers, or is grafted on Hawthorn or Service-berry. The fruit resembles small apples. Does not fruit regularly, but usually abundant when it does. Its value for livestock browse is doubtful.

REMARKS. The genus name, *Peraphyllum,* is from the Greek words *pera* ("excessively") and *phyllon* ("leaf"), referring to the abundant leaves. The species name, *ramosissimum,* refers to the intricate branching.

SQUAW-APPLE
Peraphyllum ramosissimum Nutt.

NARROW-LEAF FIRETHORN
Pyracantha angustifolia (Franch.) Schneid.

NARROW-LEAF FIRETHORN

Pyracantha angustifolia (Franch.) Schneid.

FIELD IDENTIFICATION. Cultivated half-evergreen shrub with diffusely spreading, irregular, spiny branches, or sometimes a tree to 20 ft. Occasionally almost prostrate in form.

FLOWERS. April–May, corymbs axillary, pubescent, many-flowered, ½–3 in. broad; calyx 5-lobed, lobes short, about 1/16 in. long, broadly obtuse to acute, white-hairy; margins thin and whitened, glandular or glandless, ciliate; corolla about ⅜ in. across, white, 5-petaled; petals spreading, suborbicular, narrowed to a broad base; stamens exserted, numerous, spreading, filaments white, anthers yellow; carpels 5, spreading, free on the ventral side, on the dorsal side partially connate with the calyx-tube.

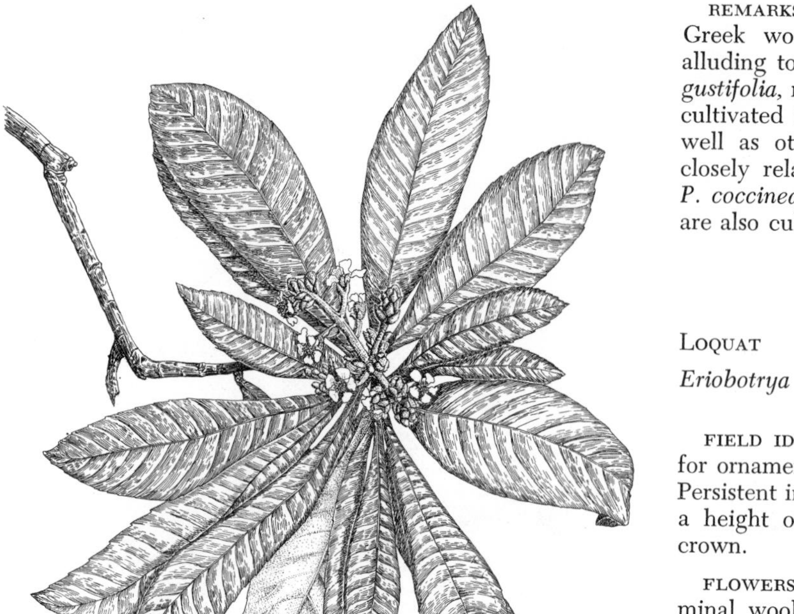

LOQUAT
Eriobotrya japonica Lindl.

FRUIT. Pomes persistent in the winter, numerous, red to orange, ⅕–⅜ in. across, depressed-globose, fleshy, calyx remnants persistent; seeds 5, about ⅛ in. long, black, lustrous, 2 surfaces plane, dorsal surface rounded, one end rounded, the other abruptly apiculate-pointed.

LEAVES. Simple, alternate, or somewhat clustered on short, lateral, spinescent branches, half-evergreen, leathery, oval-oblong to oblanceolate, apex obtuse to rounded or notched, base cuneate, magin entire, length ½–1¾ in.; upper surface dark green, shiny, apparently smooth (but often wtih scattered, white, cobwebby hairs under magnification); lower surface much paler, glabrous, or with a few white hairs along the main vein, obscurely reticulate-veined; petioles 1/16–¼ in., pubescent to glabrous; stipules minute, caducous; buds small and pubescent.

TWIGS. Green to grayish with dense fine pubescence; secondary twigs almost at right angles, and usually short and spiniferous; spines straight, ⅛–½ in. long, often pubescent at the base with apex reddish brown and more glabrous.

RANGE. Cultivated in gardens in Texas, Louisiana, Arkansas, Oklahoma, and elsewhere, sometimes escaping. Hardy as far north as Massachusetts. A native of southwest China.

PROPAGATION. Firethorns are propagated by seeds, layers, or cuttings or ripened wood under glass. It was introduced into cultivation about 1895.

REMARKS. The genus name, *Pyracantha*, is from the Greek words *pyr* ("fire") and *akanthos* ("thorn"), alluding to the bright red fruit. The species name, *angustifolia*, refers to the narrow leaves. The plant is often cultivated for ornament. Robin and cedar waxwing, as well as other birds, eagerly devour the fruit. Other closely related species, such as the Scarlet Firethorn, *P. coccinea* Roem., with crenate more glabrous leaves, are also cultivated extensively.

LOQUAT

Eriobotrya japonica Lindl.

FIELD IDENTIFICATION. Evergreen tree often planted for ornament in the warmer parts of the United States. Persistent in abandoned gardens for many years. Attains a height of 25 ft, with an open but rather rounded crown.

FLOWERS. August–November, fragrant, borne in terminal woolly panicles 4–7½ in. long, buds with conspicuously rusty-woolly tomentum, flowers about ½ in. across; calyx-lobes 5, ⅛–¼ in. long, acute, densely rusty-woolly; petals 5, white, oval to suborbicular, short-clawed; stamens 20; styles 2–5, connate below, ovary inferior, 2–5-celled, cells 2-ovuled.

FRUIT. In spring, pome edible, small, pear-shaped or spherical, yellow, 1½–3 in. long, endocarp thin; seeds large, ovoid, solitary or a few.

LEAVES. Simple, alternate, sessile or very short-petioled, crowded terminally and whorled to give a rosette-like appearance, rather stiff and firm, large and ornamental, 4–12 in. long, oval to oblong or obovate, margin with remote slender teeth but entire toward the base; upper surface with veins deeply impressed, dark green, lustrous, at maturity glabrous; lower surface much paler and densely feltlike wtih rusty tomentum.

TWIGS. Stout, green to brown, densely woolly-tomentose, when older brown to dark gray and glabrous; leaf scars large, half-round, bundle scars 3.

RANGE. A native of China and Japan, much planted for ornament in Texas, Louisiana, and other Gulf states. Grown as a pot plant in the North.

PROPAGATION. The tree may be propagated by seeds and is planted in orchards 20–24 ft apart. Individual trees are often planted in gardens for the ornamental leaves. For best results in fruit yield, improved varieties are bud-grafted on seedling stock. *E. japonica* var. *variegata* Hort. is a variety with variegated white, pale green, or dark green leaves.

REMARKS. The genus name, *Eriobotrya*, is the Greek word for "woolly cluster," referring to the hairy panicles. The species name, *japonica*, is for Japan, its original home. It is also known as Japanese-plum and China-plum. The fruit may be eaten raw, or prepared as jelly, jam, pies, and preserves. It has an agreeable acid flavor.

GOPHER-APPLE

Geobalanus oblongifolius (Michx.) Small

FIELD IDENTIFICATION. Shrub 1–4 ft, with slender, upright, simple branches from horizontal underground stems. Often growing in small patches or thickets.

FLOWERS. In spring, greenish white, the slender terminal panicles 1–3 in. long and ¾–2 in. wide, peduncle and pedicels scurfy-hairy; pedicels 1/16–3/16 in. long; individual flowers ⅛–3/16 in. across, perfect; sepals 5, triangular-ovate, acute to obtuse, scurfy-hairy; petals 5, white, oblong or oblong-obovate, 1/12–⅛ in. long, obtuse, pubescent within near the base; stamens numerous, borne on the edge of a collar; ovary 1-celled, inferior, sessile, glabrous; style filiform, entire.

FRUIT. Drupe ovoid to obovoid, apex rounded or obtuse, glabrous, dark brown, pulpy, ½–1¼ in. long, ⅓–½ in. wide; stone solitary, terete, no endosperm.

LEAVES. Alternate, blades 1¼–5 in. long, ½–1¼ in. wide, oblanceolate or oblong, apex obtuse or rounded and often mucronate, base attenuate, margin entire or undulate; thin and leathery, upper surface glabrous and lustrous green, prominently reticulate-veiny, lower surface paler; leaves almost sessile or on petioles ⅛–¼ in.

TWIGS. Brown to gray or black, glabrous, slender, upright from horizontal underground stems.

RANGE. Mostly in sandy pinelands. Reported from Washington Parish in Louisiana; eastward to Florida and Georgia.

GOPHER-APPLE
Geobalanus oblongifolius (Michx.) Small

REMARKS. The genus name, *Geobalanus,* means "earth-apple," and the species name, *oblongifolius,* refers to the oblong leaves. Also known under the vernacular names of Ground-oak or Ground-plum. Gopher-apple was formerly described under the name *Chrysobalanus oblongifolius* Michx.

BUSH CINQUEFOIL

Potentilla fruticosa L.

FIELD IDENTIFICATION. Low, much-branched, half-evergreen shrub ½–4 ft. The very leafy stems are erect or ascending, or occasionally sprawling in some forms. Rather variable in pubescence and foliage characters, and hence segregated into many varieties and horticultural forms.

FLOWERS. Maturing June–September, solitary or densely corymbose, large, showy, yellow, terminal, bisexual, ⅔–1⅓ in. across; hypanthium saucer-shaped; petals 5, nearly orbicular, longer than the sepals and bractlets; sepals 5, ovate, acuminate, alternate with the 5 bractlets which are lanceolate and hairy; stamens 15–25; pistils numerous, hairy, borne on a prolongation of the floral axis, styles lateral and filiform.

FRUIT. Disk and receptacle long-hairy, carpel body angular and hairy; achene about 1/15 in. long, densely hairy.

LEAVES. Some numerous, alternate, pinnately compound, 3–7-foliate, sometimes the terminal 3 confluent at apex, blade oblong to linear-oblong or oblanceolate, margin entire and often revolute, length ½–1 in., surfaces with appressed silky hairs; stipules scarious, petioles ⅛–¼ in.

TWIGS. Slender, at first silky-hairy and reddish brown to gray, later brown with shreddy bark.

RANGE. Around the world in subarctic regions, usually in wet alpine meadows or calcareous bogs at altitudes of 6,000–10,000 ft. In New Mexico in the Sangre de Cristo and Mogollon mountains; northward in the Rocky Mountains to Alaska. Also in the Sierra Nevada in California. Across the northern United States and Canada to Labrador, extending southward to New Jersey, Illinois, and Minnesota. Also in northern Europe and Asia.

HORTICULTURAL FORMS. It is a popular ornamental plant in various countries, and a number of horticultural varieties have been listed as follows:

Chinese Bush Cinquefoil, *P. fruticosa* var. *albicans* Rehd. & Wils.

Dahurian Bush Cinquefoil, *P. fruticosa* var. *dahurica* (Nestl.) Ser.

Big-flower Bush Cinquefoil, *P. fruticosa* var *grandiflora* Willd.

Manchurian Bush Cinquefoil, *P. fruticosa* var. *manchurica* Maxim.

Bush Cinquefoil
Potentilla fruticosa L.

Creamy Bush Cinquefoil, *P. fruticosa* var. *ochroleuca* (Spaeth) Bean
Little-leaf Bush Cinquefoil, *P. fruticosa* var. *parvifolia* Wats.
Dwarf Bush Cinquefoil, *P. fruticosa* var. *pumila* Hook. f.
Purdom Bush Cinquefoil, *P. fruticosa* var. *purdomii* Rehd.
Pyrenees Bush Cinquefoil, *P. fruticosa* var. *pyrenaica* Willd.
Himalaya Bush Cinquefoil, *P. fruticosa* var. *rigida* (Lehm.)
Slim-leaf Bush Cinquefoil, *P. fruticosa* var. *tenuiloba* Ser.
Veitch Bush Cinquefoil, *P. fruticosa* var. *veitchii* (Wils.) Bean
Vilmorin Bush Cinquefoil, *P. fruticosa* var. *vilmoriniana* Komar.

REMARKS. The genus name, *Potentilla*, is from the Latin word *potens*, referring to the "powerful" medicinal value of some species. The species name, *fruticosa*, means "shrubby." Some of the vernacular names of the plant are Golden Hardhack, Prairie-weed, and Shrubby Cinquefoil. The plant is occasionally browsed by white-tailed deer. It is considered to be important browse for sheep and goats in the Southwest, but inferior forage for cattle. It is used somewhat for erosion control.

Antelope Bitter-brush

Purshia tridentata (Pursh) DC.

FIELD IDENTIFICATION. Western shrub 2–10 ft. Rather intricately branched, erect with spreading branches or sometimes prostrate.

FLOWERS. April–August, solitary, terminating short branchlets, subsessile, perfect; calyx-tube funnelform, 1/12–1/8 in. long, tomentose and sometimes glandular, sepals 5, ovate to oblong; mature open flower about 1/3 in. across; petals 5, 1/5–1/3 in. long, spatulate or obovate, yellow, longer than the calyx-lobes; stamens 20–25; exserted on the calyx-tube margin; style stout and beaklike, pesistent.

FRUIT. Maturing July–September, achene 1/3–1/2 in. long, fusiform, acuminate at apex, pubescent, exceeding the calyx in length.

LEAVES. Alternate, but crowded on most shoots to appear fascicled, deciduous, 1/5–1 in. long, cuneate or wedge-shaped, apex tridentate, stiff, margins entire but revolute, somewhat hairy to glabrous above, lower surface densely white-tomentose, stipules small.

Antelope Bitter-brush
Purshia tridentata (Pursh) DC.

TWIGS. With many short, stubby, spurlike branchlets, brown or gray, pubescent at first but glabrous later, buds small and scaly.

RANGE. Found on sunny hillsides or slopes in sandy or clay soils at altitudes of 3,500–9,000 ft. In New Mexico in the Tunitcha Mountains; west through the Carrizo Mountains in Arizona to California and Colorado, and north to Montana, Idaho, and British Columbia.

PROPAGATION. For propagation the fruit of Antelope Bitter-brush may be collected when ripe, August–September, and beat in a sack for separation of the seeds. Seed stored at 35°–40° F. in sealed containers has been known to maintain vitality for 3 years. About 20,000 seeds are in a lb of seed, purity averaging about 77 per cent and the soundness 82 per cent. Germination is low unless the seed is stratified for 2 or 3 months to break dormancy before planting.

VARIETY. One variety has been classified as Glandular Antelope Bitter-brush, *P. tridentata* var. *glandulosa* Curran, which is more or less evergreen, with the smaller leaves usually 3-dentate, but sometimes 5-dentate, at apex, almost glabrous and glandular-punctate.

REMARKS. The genus name, *Purshia,* is in honor of J. T. Pursh (1774–1820), who was born in Germany, collected plants in North America, and wrote *Flora Americae Septentrionalis.* The species name, *tridentata,* refers to the 3-toothed leaves. The plant is also known under the name of Desert Bitter-brush. Cultivated for ornament only rarely, Antelope Bitter-brush is considered to be a very important browse plant for cattle and sheep, but is usually not eaten by horses. It is important also as a food for wildlife, being eaten by mule deer, pica, and a number of species of squirrels and chipmunks.

True Mountain Mahogany

Cercocarpus montanus Raf.

FIELD IDENTIFICATION. Western shrub or small tree to 12 ft, with upright or spreading branches. The present taxonomical treatment follows that of Martin (1950, pp. 91–111).

FLOWERS. Often crowded on short spurlike branchlets, solitary or fascicled; pedicels to ⅛ in., to ⅓ in. long in fruit; floral tube ⅕–⅓ in. long, elongate, cylindrical, spreading-villous or appressed-silky, usually somewhat enlarged below and at apex abruptly widened into a campanulate deciduous limb; sepals 5, ⅛–⅕ in. wide; corolla absent; stamens 22–44, filaments distinct, inserted in 2–3 rows, anthers hairy, emarginate at both ends, affixed dorsally above the base; pistil solitary, inserted in the bottom of the floral tube; ovary cylindric-fusiform, sessile; style terminal, elongate, exserted, silky-plumose; ovules solitary, ascending, affixed above the middle.

FRUIT. From ⅓–⅖ in. long, appressed silky-hairy,

True Mountain Mahogany
Cercocarpus montanus Raf.

cylindric-fusiform, about one third exserted from the floral tube at maturity; tail 1¾–2½ in. long, densely spreading, silky-plumose, testa membranaceous, seed cylindric; cotyledons linear, elongate, hypocotyl very short.

LEAVES. Simple, alternate, or somewhat fascicled on short spurlike branchlets, blade ovate to oval or obovate (in some varieties lanceolate, oblanceolate, or elliptic), apex acute to rounded, base usually cuneate, margin usually with coarse ovate to triangular teeth but sometimes entire, length ⅝–1¼ in., to ¾ in. wide; thin but firm, upper surface green to grayish green above, lower surface lighter and varying from glabrous to short villous or rather densely appressed-silky; lateral veins 3–6, parallel or somewhat curved toward the margins, prominent to less noticeable; petiole to ⅓ in. long; stipules adnate to petiole at base, lanceolate to ovate, acute, brown, scarious.

TWIGS. Rather stout, rigid, terete, roughened by leaf scars, often with short lateral spurs, gray to brown, pubescent, later roughened and fissured on old branches or trunk; wood hard and dark colored.

RANGE. Dry rocky bluffs or mountainsides at altitudes of 3,500–9,000 ft. In New Mexico in the Zuni, Jemez, Sandia, and Sangre de Cristo mountains; west to Arizona and northward to Wyoming and South Dakota.

VARIETIES. Silver Mountain Mahogany, *C. montanus*

var. *argenteus* (Rydb.) F. L. Martin, is a shrub 4–15 ft, with ascending branches; flowers June–July, floral tube 1/5–1/8 in. long (in fruit to 2/5 in. long), appressed-silky; calyx-limb 1/8–2/5 in. wide; pedicel 1/25 in. or less in length in flower (in fruit 1/8–1/6 in. long); leaves oblanceolate, narrowly obovate to narrowly elliptic; margin set with short, broad, triangular teeth which are apiculate-tipped or entire, length 3/4–1 1/2 in., width 3/8–3/4 in., surfaces white, appressed-silky, rarely sparsely villous or nearly glabrate, veins 5–6. In New Mexico and Texas. In Texas in Hutcheson, Randall, Lubbock, Presidio, Brewster, Pecos, Edwards, Kimble, Real, and Uvalde counties. In New Mexico in Sandoval, Santa Fe, San Miguel, Lincoln, and Otero counties. Rocky hillsides, canyons, mountains at altitudes of 4,000–8,500 ft. (Listed elsewhere as *C. argenteus* Rydb.)

Shaggy Mountain Mahogany, *C. montanus* Raf. var. *paucidentatus* (Wats.) F. L. Martin, is a shrub 3–15 ft; flowers May–November, floral tube 1/8–1/4 in. long (in fruit 1/8–1/3 in. long), spreading villous or appressed-silky; calyx-limb 1/12–1/5 in. wide; pedicel very short in flower, in fruit 1/12–1/6 in. long; tail of fruit 5/8–2 in. long; leaves lanceolate, oblanceolate or narrowly obovate, margin entire or with 3–5 short teeth at apex, short-villous, appressed-silky or glabrate, lateral veins 3–5 (rarely 6), length 3/8–3/4 in. (rarely to 1 1/4 in.), width 1/4–1/3 in. In Arizona, New Mexico, Texas, and Mexico. In New Mexico south to the Mexican border. In Texas in El Paso and Jeff Davis counties. In Mexico in Sonora, Chihuahua, Coahuila, San Luis Potosí, and Hidalgo. Rocky hills at altitudes of 4,500–8,500 ft.

SYNONYMS. The following names have been used for this plant in the literature: *C. breviflorus* A. Gray, *C. paucidentatus* (Wats.) Britt., *C. eximius* (C. K. Schneid.) Rydb., *C. parvifolius* Nutt. var. *paucidentatus* Wats., *C. parvifolius* Nutt. var. *breviflorus* Cov.

REMARKS. The genus name, *Cercocarpus*, refers to the long tail of the fruit. The species name, *montanus*, is for the mountainous habitat.

APACHE PLUME

Fallugia paradoxa (D. Don) Endl.

FIELD IDENTIFICATION. Straggling, clump-forming, Western shrub to 8 ft. Conspicuous because of the white, 5-petaled, roselike flowers about 1 in. across, and by the feathery balls of plumose, reddish-tinged achenes.

FLOWERS. June–August, solitary on long naked, slender peduncles, or corymbose; flowers perfect or rarely polygamous; 3/4–1 in. across; inflorescential leaves reduced to bracts; petals 5, white, oblong to obovate or oval; calyx cupulate, about 1/2 in. across, lobes ovate, apex long-acuminate or trifid; bractlets 5, linear-lanceolate or bifid, alternate with the calyx-lobes; stamens numerous, set on the rim of the hypanthium; pistils numerous on a conical torus, pubescent; styles slender, hairy.

APACHE PLUME
Fallugia paradoxa (D. Don) Endl.

FRUIT. Borne in fluffy, reddish clusters; achenes numerous, obovoid-fusiform, about 1/8 in. long, tipped with the persistent tail-like, reddish-tinged styles which are 1–2 in. long; styles conspicuous in fruit because of the plumose reddish hairs; seed hairy, solitary.

LEAVES. Alternate or fascicled, somewhat clustered on short spurs, thick, cuneate-obovate or spatulate, deeply lobed with 3–7 long, entire finger-like lobes, margin revolute, apex obtuse; attenuate into a slender base, undersurface white-tomentose, 1/4–1/2 in. long.

TWIGS. Slender, white, pubescent to glabrous, gray to brown when older.

BARK. Shreddy, white at first, dark gray to brown later.

RANGE. Along dry arroyos of deserts, or on rocky or gravelly slopes. Central, west, and northwest Texas. Uusually at altitudes of 3,000–8,000 ft. Abundant in the foothills of the Chisos Mountains in Brewster County. Also in New Mexico; west through Arizona to California and north to Colorado, Utah, and Nevada. In Mexico in Chihuahua, Durango, and Coahuila.

PROPAGATION. Seed averages about 420,000 seeds per lb, germinating 19–65 per cent in 4–10 days after sowing, seed sown broadcast on a prepared bed and covered 1/4 in. by fine loam or sand.

REMARKS. The genus name, *Fallugia*, is in honor of Virgilio Fallugi, an Italian botanical writer of the seventeenth century, and the species name, *paradoxa*, refers to the plant's paradoxical resemblance to the rose. Another vernacular name in use is Ponil. The name Apache Plume comes from the resemblance of the fruit to the Apache headdress. It grows rapidly when moisture is available and is used ornamentally because of the conspicuous, decorative, plumelike fruit. It is used to some extent for erosion control. Although of only low to fair palatability, it is grazed by cattle and goats, and in some areas is considered an important browse plant. The Indians used bundles of the twigs for brooms, and older stems for arrow shafts. The Hopi of Arizona used an infusion of the leaves as a supposed stimulant for hair growth.

Stansbury Cliffrose

Cowania stansburiana Torr.

FIELD IDENTIFICATION. Western shrub, much-branched, spreading, evergreen. Usually only 1–6 ft, but under favorable conditions a small tree to 25 ft.

FLOWERS. April–June, resinous and strong-smelling, bisexual or sometimes only staminate, solitary at the ends of small lateral branches; pedicels about ¼ in. long or less and glandular; corolla ½–¾ in. broad; petals 5, spreading, distinct, ¼–⅙ in. long, broadly ovate to obovate, white or yellowish; stamens numerous, inserted on the rim of the calyx-tube in 2 rows; pistils 5–10, distinct, densely villous; ovaries superior, 1-celled; calyx-tube ⅙–¼ in. long, hemispheric to turbinate; sepals 5, persistent, broadly ovate, densely glandular-tomentose externally, less so or glabrous within.

FRUIT. Mature in October, 5–10 per flower, achenes ⅛–⅓ in. long, narrowly oblong, striate-ribbed, glabrous at maturity, coriaceous, partly included in the calyx, the styles persistent, tail-like, silvery-hairy, several times the length of the achene, ½–2 in. long, achene germination 10–80 per cent.

LEAVES. Crowded, evergreen, simple, alternate or clustered, ¼–½ in. long, outline obovate to narrowly spatulate or cuneate, pinnately divided into 3–9 lobes, the lobes oblong to linear, or some toothed or lobed again, margins revolute, thick and coriaceous, mature leaves green-glabrate and glandular-dotted above, lower surface more or less white-tomentose, some with scattered glands; petioles absent or very short and margined.

TWIGS. Erect, stiff, green to reddish brown, puberulent, glandular, later becoming gray to black, with shreddy bark.

RANGE. On sunny sites in dry soils of slopes, mesas, or washes at altitudes of 3,500–8,500 ft. New Mexico; west through Arizona to California and north to Colorado, Utah, and Nevada.

VARIETY AND SYNONYM. A variety found in San Bernardino County, California, is known as the Doubtful Cliffrose, *C. stansburiana* var. *dubia* Brandg., having 2–3 pistils with short tails which are densely hairy but not plumose.

The plant is also listed in the literature under the names of *C. mexicana* D. Don and *C. mexicana* var. *stansburiana* (Torr.) Jepson.

REMARKS. The genus name, *Cowania*, is in honor of James Cowan, an English merchant who introduced many Peruvian and Mexican plants into England. The species name, *stansburiana*, is for the botanist Stansbury, who collected the type specimen on Stansbury Island, Great Salt Lake, Utah. It is known locally under the vernacular name of Quinine-bush. The Hopi Indians are reported to have used it as an emetic and as a wash for wounds and sores. The wood was also used for making arrow shafts. Some of the fibers used in the sandals, rope, and clothing of the ancient Basketmaker Indians of the Southwest were from the shreddy bark of this plant. Although not often grown in gardens, it was first cultivated in 1904, and could be used in rock gardens in the areas where it is native. The white flowers and plumose fruits are rather showy. It is an important browse plant for cattle and sheep, and is the staple food for mule deer in some areas. It is reported that it will stand grazing to 65 per cent of mass, but beyond that amount the plant does not recover readily.

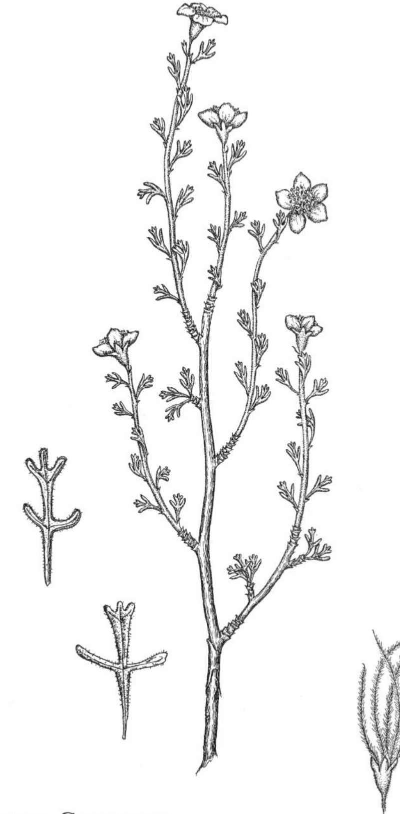

Stansbury Cliffrose
Cowania stansburiana Torr.

Heath Cliffrose

Cowania ericaefolia Torr.

FIELD IDENTIFICATION. Evergreen shrub to 3 ft, with much-branched stems and very small heathlike leaves.

FLOWERS. Often on short lateral branches, pedicels ⅓–⅔ in., often stipitate glandular; flowers showy, solitary, terminal; petals 5, white to yellowish, ¼–⅜ in. long, oblong-obovate; calyx spreading, broadly 5-lobed, lobes ovate to deltoid, apex acute or short-acuminate, densely hairy; stamens numerous, perigynous; styles elongate and conspicuously hairy.

FRUIT. Achenes 5–12, densely hairy, narrowly oblong, striate, tipped by the much-elongated, reddish brown, densely silvery-hairy style ½–1½ in. long.

LEAVES. In clusters along the stem, often on very short brown-scaly spurs, heathlike, very small, ⅙–¼ in. long, linear, often falcate, apex cuspidate, margins entire, stongly revolute with the curled portions sometimes touching beneath, upper surface glabrous or nearly so, lower surface densely white-hairy.

TWIGS. Stout, irregularly branched, brown to light or dark gray, sometimes almost black, bark separating into small, thin scales.

RANGE. Usually on dry, rocky, limestone soils in full sun. Brewster County, Texas, and adjacent Mexico.

REMARKS. The genus name, *Cowania,* honors John Cowan, English merchant, who introduced many Peruvian and Mexican plants into England. The species name, *ericaefolia,* refers to the *Erica*-like (heathlike) leaves. Heath Cliffrose is not to be confused with Apache Plume, *Fallugia paradoxa,* which has small, lobed leaves.

Heath Cliffrose
Cowania ericaefolia Torr.

Mountain Ninebark
Physocarpus monogynus (Torr.) Coult.

Mountain Ninebark

Physocarpus monogynus (Torr.) Coult.

FIELD IDENTIFICATION. Western shrub to 3 ft, with diffuse branches, the thin bark exfoliating in strips.

FLOWERS. May–July, corymbs terminal; pedicels short, usually only sparingly pubescent; bracts lanceolate and caducous; individual flowers small, regular, perfect, white or pink, ⅜–½ in. across; calyx campanulate, 5-lobed, lobes about ⅛ in. long, valvate, stellate-hairy but sometimes only thinly so; petals 5, spreading, suborbicular, ⅛–⅕ in. long; stamens numerous (20–40), inserted on the calyx-disk; carpels 1–3 (usually 2), turgid, united about half their length, styles divergent, stigmas capitate.

FRUIT. Follicles usually 2, mostly united to above the middle, divergent, beaked, inflated, membranous, pubescent, dehiscent along both sutures; seeds usually 1–3, yellowish to brown, lustrous.

LEAVES. Simple, alternate, deciduous, ovate to rounded, some cordate, margin palmately 3–5-lobed, the lobes rather rounded or acute and incised-serrate, usually more or less glabrous on both sides, blade length ½–2½ in.; petioles about ¼ in. or longer; stipules membranous and deciduous.

TWIGS. Slender, spreading, young ones sparingly stellate-hairy, older ones glabrous.

RANGE. Well-drained rocky sites in the sun. At altitudes of 4,000–10,000 ft, often in pine or spruce forests. In Trans-Pecos Texas in the Guadalupe Mountains and in New Mexico in the Sacramento, Black, Mogollon, Sandia, and Sangre de Cristo mountains; west to Arizona and north to South Dakota.

SYNONYMS. The plant is also listed in the literature under the names of *Spiraea monogynus* Torr., *Neillia torreyi* Wats., *P. torreyi* Maxim., and *Opulaster monogynus* (Torr.) Kuntze.

REMARKS. The genus name, *Physocarpus*, is from the Greek words *physa* ("bladder") and *karpos* ("fruit"), referring to the inflated fruit. The species name, *monogynus*, refers to a solitary pistil. The umbel-like corymbs of white flowers make Mountain Ninebark worthy of cultivation. It was first cultivated in 1879. Mountain Ninebark may be propagated from seed or green or hardwood cuttings. It is distinguished from Common Ninebark, *P. opulifolius* (L.) Maxim., by the fewer flowers in the corymbs and the pubescent carpels usually in twos.

COMMON NINEBARK

Physocarpus opulifolius (L.) Maxim.

FIELD IDENTIFICATION. Shrub 3–10 ft, with wide-spreading, graceful, recurved branches, and bark peeling off in conspicuous thin strips.

FLOWERS. May–July, corymbs peduncled, terminal, rounded, many-flowered, 1–2 in. broad; pedicels ½–⅔ in., slender, glabrous or slightly pubescent; individual flowers ⅓–⅖ in. wide, white or pinkish; calyx campanulate, glabrous or varying to somewhat pubescent, divided into 5 sepals; sepals 1/12–⅙ in. long, valvate, ovate; stamens inserted with the petals, numerous (30–40); styles 5, spreading, alternating with the calyx-lobes; stigma terminal, depressed-capitate; petals 5, spreading.

FRUIT. September–October, follicles in clusters of 3–5 (usually 3), length ¼–½ in., ovoid, obliquely subulate-tipped, inflated, papery, glabrous, shiny, reddish, 2-valved, dehiscent along both sutures; 2–4-seeded, seeds rounded to obovoid, crustaceous, lustrous, yellowish to pale brown, endosperm abundant.

LEAVES. Simple, alternate, deciduous, ovate-orbicular, obtusely or acutely 3–5-lobed, the lobes irregularly crenate-dentate, base broadly cuneate to truncate or cordate, blade length 1–3 in. (longer on vigorous shoots); upper surface dark green and glabrous; mostly glabrous beneath; petioles ⅜–1 in. long; stipules caducous.

TWIGS. Brown to yellowish, glabrous, bark peeling off in thin strips.

RANGE. The species is usually found in well-drained soils of sandy or rocky places, from Kansas southward to Georgia and north to Quebec. Only the Illinois Ninebark variety is found in our area, in Arkansas.

PROPAGATION. The plant is hardy and is often cultivated for its dark green leaves, almost globose heads of white flowers, and reddish fruit. It is propagated by seeds or hardwood cuttings. The seeds are eaten by a number of species of birds.

VARIETIES AND HORTICULTURAL FORMS. Illinois Ninebark, *P. opulifolius* var. *intermedius* (Rydb.) Robins., is a very closely related variety with narrower leaves; calyx and follicles stellate-pubescent. Connected with the species by intermediate forms. Found in Michigan, Illinois, and South Dakota; south to Alabama and Arkansas.

Gold-leaf Ninebark, *P. opulifolius* forma *aureus* Hort., has yellow to golden bronze leaves.

COMMON NINEBARK
Physocarpus opulifolius (L.) Maxim.

Dwarf Ninebark, *P. opulifolius* forma *nanus* Kirchn., has a dwarf stature with smaller, less-lobed, dark green leaves.

SYNONYMS. Some authors list the plant under the names of *Opulaster opulifolius* Kuntze, *Spiraea opulifolia* L., and *Neillia opulifolia* Brew. & Wats.

REMARKS. The genus name, *Physocarpus*, is from the Greek words *physa* ("bladder") and *karpos* ("fruit"), referring to the inflated capsules. The species name, *opulifolius*, means "having leaves of Opulus (the Cranberry-tree)."

Bush Rock-spirea

Holodiscus dumosus (Nutt.) Heller

FIELD IDENTIFICATION. Western shrub to 3 ft, often forming thickets, bearing a few erect stems and many spreading branches.

FLOWERS. Very showy, panicles terminal, ovoid, compound, diffuse, tomentose, length 2–8 in., width 1¾–4 in.; individual flowers small, white to pink; calyx-tube cup-shaped, sepals 5, valvate in the bud, 1/15–1/12 in. long, triangular-ovate; petals 5, oval to elliptic, short-clawed, little longer than the sepals; stamens 15–20, as long as or somewhat longer than the petals; disk entire; pistils 5, distinct, superior, styles terminal.

Bush Rock-spirea
Holodiscus dumosus (Nutt.) Heller

FRUIT. Follicles 5, small, distinct, beaked, membranous-woody, tardily dehiscent or not at all; achenes short-stipitate and hairy, broad-oblong.

LEAVES. Simple, alternate, deciduous, elliptic to ovate or obovate, apex obtuse to acute, base cuneate into a short-margined petiole, margin lobed and also doubly toothed with the teeth rounded to ovate and short-mucronate; upper surface smooth or pubescent; lower surface varying from pubescent to white-tomentose; length ¾–2 in.

TWIGS. Hairy at first, more glabrous later, dark red to reddish brown to gray, epidermis exfoliating in strips; buds small, ovoid, scales pubescent.

RANGE. Dry, well-drained, sunny sites of canyons and mountainsides, at altitudes of 3,000–10,500 ft. In Trans-Pecos Texas in the Chisos and Guadalupe mountains. North in New Mexico, Arizona, Colorado, Utah, and Wyoming. Also in Baja California and Chihuahua, Mexico.

VARIETIES. In Arizona apparently the species is represented by 2 varieties (Kearney and Peebles, 1951, p. 375) as follows:

H. dumosus var. *australis* (Nutt.) Heller is a shrub 6–9 ft, with usually ample inflorescences and leaves to 2 in., the lower surface white with dense sericeous-tomentose pubescence. It is found mostly in southern Arizona.

H. dumosus var. *glabrescens* (Greenman) Heller has leaves about as wide as long, green, and merely loosely villous beneath. It is found in Navajo and Coconino counties in Arizona. This may be the plant listed under the name of *Sericotheca schaffneri* Rydb. by Tidestrom and Kittell (1941, pp. 233–234).

REMARKS. The genus name, *Holodiscus*, is from the Greek words *holos* ("entire") and *diskos* ("disk"), referring to the entire disk of the flower. The species name, *dumosus*, is from the Latin noun *dumus*, which means "bramble." It is also known under the vernacular names of Cream-bush, Foam-bush, Ocean-spray, and Mountain-spray. It was first cultivated for ornament in 1879, and is propagated by seeds, layers, or greenwood cuttings under glass. The fruit is reported to be eaten by the Tewa Indians of New Mexico.

Black-bush

Coleogyne ramosissima Torr.

FIELD IDENTIFICATION. Western shrub 1½–6 ft, with the diffuse branches rigid and often spinescent.

FLOWERS. March–May, terminating short branchlets, solitary, subtended by 1–2 pairs of 3-lobed bractlets;

BLACK-BUSH
Coleogyne ramosissima Torr.

corolla absent; calyx-tube very short, coriaceous; sepals 4, the 2 outer lanceolate to oblong, the inner ones more ovate, acute, mucronate, green to yellowish or purplish, margin scarious, ⅕–⅓ in. long, persistent; stamens numerous, filaments filiform, inserted upon the base of an elongate sheathing tube ⅙–⅕ in. long that encloses the ovary; tube apex toothed, externally glabrous, densely white-hairy within; pistil 1, ovary 1-celled, ovule 1; style lateral, filiform, bent, hairy.

FRUIT. Achene ⅕–⅛ in. long, somewhat compressed, brown, glabrous, apex incurved, base hairy.

LEAVES. Crowded in opposite fascicles, deciduous, coriaceous, linear to oblanceolate, margin entire, length ⅕–⅗ in., width 1/25–⅕ in., upper surface flattened, lower surface with longitudinal ridges, hairy.

TWIGS. Opposite, short, rigid, divaricate, somewhat spinescent, gray at first but later becoming black.

RANGE. In well-drained gravelly or sandy soils in the sun at altitudes of 4,000–5,500 ft. New Mexico; west to California and north to Colorado and Utah.

REMARKS. The genus name, *Coleogyne,* is a Greek word meaning "sheath" or "scabbard," perhaps referring to the tubular sheath which encloses the ovary. The species name, *ramosissima,* means "much-branched." The shrub provides some browse for cattle and sheep in winter and can generally withstand heavy cropping.

HARDHACK SPIREA

Spiraea tomentosa L.

FIELD IDENTIFICATION. Shrub, simple or sparingly branched, attaining a height of 4 ft.

FLOWERS. July–September, borne in spirelike panicles 2–8 in. long, branches tomentose, the small flowers

HARDHACK SPIREA
Spiraea tomentosa L.

NARROW-LEAF VAUQUELINIA
Vauquelinia angustifolia Rydb.

crowded in spikelike racemes; flowers perfect, white or pink, 1/8–1/6 in. across; calyx turbinate, tomentose, tube about 1/25 in. long; sepals about the length of the tube, triangular, acute, reflexed; petals 5, wide-spreading, about 1/15 in. across, orbicular to ovate; stamens numerous, 15 or more, somewhat longer than the petals, inserted on the calyx tube between the disk and the sepals; pistils 5, distinct, alternate with the sepals, style terminal.

FRUIT. August–December, follicles 5, firm, about 1/15 in. long, densely tomentose, tips spreading, dehiscent along the ventral suture; seeds 4–7, minute, oblong, endosperm none, testa dull.

LEAVES. Simple, deciduous, ovate to oblong or lanceolate, blade length 1 1/4–2 1/2 in., width 3/8–1 1/4 in., apex acute to obtuse, base acute to rounded or cuneate, margin unequally and often doubly serrate, leathery, upper surface dark green and puberulent and sometimes rugulose, lower surface densely white or rusty-tomentose; petioles tomentose, 1/25–1/8 in. long.

RANGE. Generally in peaty or wet sandy soil with an acid or sour composition. Arkansas; eastward to Georgia, north to New Brunswick, and west to Minnesota.

VARIETIES. White Hardhack Spirea, *S. tomentosa* var. *alba* Macbr., has white flowers.

Rose Hardhack Spirea, *S. tomentosa* var. *rosea* (Raf.) Fern., has fewer flowers in the panicle and is usually found in the western part of the species' distribution, but intergrading forms occur.

REMARKS. The genus name, *Spiraea*, is from the Greek word *speira* ("wreath" or "band"), and the species name, *tomentosa*, refers to the dense tomentum of the leaves and stems. It also bears the common names of Steeple-bush and Thé du Canada. The plant is useful for ornamental planting in mass effects in low grounds. It was introduced into cultivation in 1736.

NARROW-LEAF VAUQUELINIA

Vauquelinia angustifolia Rydb.

FIELD IDENTIFICATION. Evergreen shrub or small tree to 30 ft, with an irregular form and contorted branches.

FLOWERS. Small, in terminal, rather flattened, panicles or corymbs; perianth 3/16–1/4 in. across; petals 5, elliptic to rounded, white, about 1/8 in. long; stamens numerous (10–25); sepals 5, green, triangular-ovate, acute at apex, glabrous or hairy.

FRUIT. Capsule densely hairy, later somewhat glabrous and woody with 5 follicles coherent at the base, 2 seeds in each capsule.

LEAVES. Simple, alternate, evergreen, coriaceous, linear, or linear-lanceolate, often falcate, remotely and coarsely toothed, acute or acuminate and apiculate at the apex, base gradually narrowed or attenuate to an almost winged petiole, 2–7 1/2 in. long, 1/4 in. broad or less, dark green and finely reticulate-veined, with the main vein conspicuously yellowish green, paler and pubescent or glabrous beneath; petiole glabrous or puberulent; stipules small and deciduous.

BARK. Dark brown to black, on old trunks curling loose in thin flakes.

WOOD. Brown to reddish, close-grained, heavy, hard.

RANGE. Occasionally in dry canyons in the Chisos Mountains in Brewster County, Texas, at altitudes of 5,300–6,500 ft; southward in Mexico from Coahuila and Chihuahua to Hidalgo.

REMARKS. The genus name, *Vauquelinia*, is in honor of Louis Nicholas Vauquelin (1763–1829), a French chemist. The species name, *angustifolia*, refers to the narrow leaves. Other vernacular names are Guauyul, Guayule, Palo Prieto, Arbol Prieto, Palo Verde, and Rosewood. The tree is usually too small to have commercial timber value, but the wood and bark are occasionally used for dyeing skins in Mexico.

TORREY VAUQUELINIA

Vauquelinia californica (Torr.) Sarg.

FIELD IDENTIFICATION. Evergreen Western shrub, or sometimes a tree to 20 ft, diameter of 6 in. The branches upright, rigid, and contorted.

FLOWERS. In June, borne in terminal, compound, tomentose corymbs 2–3 in. across; pedicels slender and bibracteolate; flowers about ¼ in. in diameter; petals 5, white, oblong or rounded, becoming reflexed and persistent; inner circle of the disk pilose; stamens 15–25, inserted in series, those of the outer row opposite the petals; filaments persistent, exserted, subulate; anthers versatile and extrorse; pistils 5, coherent at the base with styles dilated into capitate stigmas; calyx persistent; short-turbinate, 5-lobed, lobes ovate, obtuse or acute, erect, persistent, silky-hairy.

FRUIT. Maturing in August, persistent over the winter; capsule about ¼ in. long, ovoid, woody; petals reddish and coherent at base; carpels adherent below and dehiscent dorsally at maturity; seeds 2 in each cell, ascending, compressed, about 1/12 in. long; seed coat membranous expanding into a terminal, thin wing.

LEAVES. Simple, alternate, coriaceous, narrowly lanceolate, apex mostly acuminate or acute, base cuneate or slightly rounded, margin minutely serrate, blade length 1½–3 in., width ¼–½ in.; young leaf surfaces densely tomentose below and puberulous above, mature leaves leathery with upper surface glabrous and somewhat puberulent beneath; veins thin with reticulate veinlets; petioles thick, ⅓–½ in. long; stipules minute, acute, deciduous.

TWIGS. Slender, terete, younger ones reddish brown and tomentose, later light brown to gray and glabrous; leaf scars large and elevated; winter buds axillary, reddish brown, minute, acuminate, pubescent.

TUFTED ROCKMAT
Petrophytum caespitosum (Nutt.) Rydb.

TORREY VAUQUELINIA
Vauquelinia californica (Torr.) Sarg.

BARK. Dark reddish brown, 1/16–⅛ in. thick, scales small, platelike, persistent.

WOOD. Reddish brown, close-grained, heavy, hard.

RANGE. Dry rocky hillsides, canyons, or grassy slopes at altitudes of 2,500–5,000 ft. In the mountains of southwestern New Mexico; west through Arizona to California, and in Sonora, Mexico.

REMARKS. The genus name, *Vauquelinia,* honors the French chemist Louis Nicholas Vauquelin (1763–1829). The species name, *californica,* refers to the state of California, where this species is found. Another vernacular name is Arizona Rosewood.

TUFTED ROCKMAT

Petrophytum caespitosum (Nutt.) Rydb.

FIELD IDENTIFICATION. Very low woody-tufted perennials, usually less than 1 ft, forming flat tufts from crevices of rock.

FLOWERS. May–July, racemes solitary or branched at base; peduncles leafy, ¾–5 in. long with bractlike subulate leaves; pedicels short, flowers small, white, regular; calyx persistent, sepals 5, valvate, about 1/15 in. long, ovate-lanceolate, apex acute, silky-hairy; petals 5, white, imbricate, spatulate, apex obtuse, length about 1/12 in.; stamens numerous (about 20), ultimately exserted, arising from the calyx; pistils 5, styles slender, exserted, glabrous but hairy at base; ovary superior, not enclosed in the calyx-tube at maturity, disk entire.

FRUIT. Follicles about 1/12 in. long, 3–5, leathery, dehiscent on both sutures; seeds 2–4, linear.

LEAVES. Alternate, but very crowded and appearing fascicled on short tufted branches, linear-spatulate to elliptic or oblanceolate, apex obtuse and mucronulate, length ¼–½ in., width 1/12–1/6 in., surfaces densely silky-hairy, 1-nerved; leaves of the flowering scapes scattered and narrower.

RANGE. Often in barren limestone rock crevices at altitudes of 3,000–9,500 ft. In western Texas, particularly in the Trans-Pecos mountains. In the Del Norte Mountains, 15 miles east of Alpine; Bissett Hill, Glass Mountains; Hueco Pass, El Paso County; Van Horn Mountains, Culberson County; Altuda Mountain, 10 miles east of Alpine; 15 miles east of Leakey, Real County. In New Mexico in the Guadalupe and Big Hatchet mountains. Also in Arizona and California, north to Montana and South Dakota.

PROPAGATION. Rock-mat Spirea is suitable for rock gardens in sunny sites and has been cultivated since 1900. It is propagated by seeds and divisions.

SYNONYMS AND VARIETY. The plant has been listed in the literature under the scientific names of *P. acuminatum* Rydb., *Luetka caespitosa* Kuntze, *Spiraea caespitosa* Nutt., and *Eriogynia caespitosa* Wats.

Tall Rock-mat Spirea, *P. caespitosum* var. *elatior* Wats., is a variety with longer peduncles frequently branched and longer inflorescential bracts.

REMARKS. The genus name, *Petrophytum*, is from the Greek words *petros* ("rock") and *phyton* ("plant"). The species name, *caespitosum*, refers to its growing in tufts.

PRICKLY ROSE
Rosa acicularis Lindl.

PRICKLY ROSE

Rosa acicularis Lindl.

FIELD IDENTIFICATION. Shrub with stems to 3 ft, or more rarely to 8 ft.

FLOWERS. May–June, 1–3 on short lateral shoots from the old wood (shoots usually less than 2¾ in. long); pedicels ¾–1⅝ in. long, glabrous or slightly glandular-hispid; hypanthium glabrous, pear-shaped or elliptic; sepals 5, lanceolate, caudate-acuminate, about ¾ in. long, dorsally more or less pubescent and glandular, inner surface tomentose as well as the margin; petals 5, ¾–1 in. long, pink, obcordate; stamens numerous, inserted at the base of the fleshy hypanthium; styles not exserted, distinct, persistent.

FRUIT. Hips (mature hypanthia) pyriform, ⅜–⅝ in. broad, ⅝–¾ in. long, usually necked; sepals erect and persistent; achenes bony, numerous, inserted in the bottom and on the sides of the hypanthium.

LEAVES. Alternate, pinnate-compound; leaflets 3–9 (usually 5, rarely 9), broadly elliptic to narrowly oblong or oval, length 1½–2 in., apex and base usually acute, margins simply or doubly serrate but entire toward the base, upper surface dull and glabrous, lower surface paler and densely fine-pubescent (in the American form sometimes glandular-granuliferous beneath); stipules adnate, broad, pubescent to glandular-granuliferous, margins glandular-ciliate, about ⅝ in. long.

STEMS. Densely and weakly prickly, or bristly, or both; infrastipular prickles usually absent but bristles present.

RANGE. Plains, hills, slopes, usually at altitudes of 4,500–10,000 ft. A cosmopolitan species. In North America from New Mexico and Colorado northward to Alaska, eastward to New York; also in Siberia.

HYBRIDS AND VARIETIES. The following hybrids are listed in the literature:

R. acicularis × *R. blanda*, *R. acicularis* × *R. carolina*, *R. acicularis* × *R. fendleri*, *R. acicularis* × *R. muriculata*, and *R. acicularis* × *R. nutkana*.

Bourgeau Prickly Rose, *R. aciculata* var. *bourgeauiana* (Crep.) Crep., is a variety with leaflets broadly elliptic, margins doubly glandular-serrate, lower surface glandular and pubescent; flowers often to 2½ in. across; fruit usually subglobose with a very short neck. This variety seems to intergrade with the species in New Mexico and Arizona.

Engelmann Prickly Rose, *R. aciculata* var. *engelmannii* (Wats.) Crep., has the flowering branches usually bristly; leaflets double glandular-serrate, lower surface glabrous and glandular; fruit smaller, ellipsoid or pear-shaped. From Colorado northward to North Dakota and Montana. Introduced into cultivation in 1891.

REMARKS. The genus name, *Rosa,* is the Latin name used by the ancients, and the species name, *acicularis,* refers to the numerous prickles.

Arizona Rose

Rosa arizonica Rydb.

FIELD IDENTIFICATION. Much-branched shrubs usually less than 3 ft.

FLOWERS. May–July, floral branches to 4 in. long and flowers mostly solitary; pedicels glabrous, ¼–½ in. long; hypanthium globose and glabrous; sepals 5, broadly lanceolate, apex caudate-attenuate, length ⅜–⅝ in., margins tomentose, dorsally glabrous; petals 5, showy, obovate, pink, ¾ in. long or less; stamens numerous, inserted on an annular disk in the hypanthium; styles not exserted, distinct, persistent; stamens numerous.

Arizona Rose
Rosa arizonica Rydb.

FRUIT. Hips (mature hypanthia) about ⅓ in. broad, glabrous, subglobose to globose, constricted below the throat; achenes buried in the pulp.

LEAVES. Alternate, pinnately compound; rachis and petioles finely pubescent; leaflets mostly 5, occasionally 7, broadly oval, apex acute or obtuse, margin coarsely toothed, averaging about ¾ in. long, glabrous, bright green, finely puberulent and somewhat granuliferous beneath; stipules adnate, about ⅜ in. long, finely puberulent.

STEMS. Brown or white, bark flaking off, armed with small curved prickles ⅛–⅕ in. long.

RANGE. Usually along streams in pinewoods at altitudes of 4,000–9,000 ft. In New Mexico and Arizona.

VARIETY. Granular Arizona Rose, *R. arizonica* var. *granulifera* (Rydb.) Kearney & Peebles, has doubly serrate leaves which are very granuliferous beneath, and with more stipitate glands on the sepals.

REMARKS. The genus name, *Rosa,* is the ancient Latin name, and the species name, *arizonica,* refers to the distribution of this species in the state of Arizona.

Arkansas Rose

Rosa arkansana Porter

FIELD IDENTIFICATION. A subshrub with erect stems and attaining a height of 1½ ft.

FLOWERS. May–August, terminal, few or more in corymbs; peduncles ⅜–¾ in. long, glabrous; hypanthium subglobose; sepals 5, lanceolate, caudate-acuminate, ⅜–⅝ in. long; petals 5, obcordate, ¾–1 in. long; stamens numerous and exserted; styles persistent, distinct, not exserted.

FRUIT. Hips (mature hypanthia) ½–⅗ in. broad, subglobose to plump-ellipsoid; sepals persistent, erect or ascending; achenes bony, inserted on the bottom and sides of the hypanthium.

LEAVES. Alternate, pinnately compound; rachis and petiole glabrous or somewhat pubescent; leaflets 9–11, elliptic, terminally and basally acute, margin with coarse sharp teeth, length ⅝–2½ in., upper surface glabrous and lustrous, lower surface glabrous or with a few hairs on the veins; stipules adnate, ⅜–1 in. long, dilated, dorsally glabrous or sometimes glandular-granuliferous, margins glandular-toothed and ciliate.

STEMS. Usually simple and erect; very bristly and prickly; dies back in freezing weather.

Arkansas Rose
Rosa arkansana Porter

RANGE. Rocky slopes, thickets, and dry prairies at altitudes of 3,500–9,000 ft. In Texas and New Mexico; north through Colorado and Kansas to Wisconsin and Minnesota.

REMARKS. The genus name, *Rosa,* is the ancient Latin name. The species name, *arkansana,* refers to the Arkansas River, on the banks of which, near Cañon City, Colorado, the plant was first found.

Macartney Rose

Rosa bracteata Wendl.

FIELD IDENTIFICATION. Evergreen shrub, half climbing or trailing, and forming dense, impenetrable mounds to 20 ft.

FLOWERS. April–July (or sporadically all summer), solitary or several together, usually terminal, large, white, 5-petaled, 2–3 in. across; petals broadly obovate, apex notched, about 1 in. long, wide-spreading; stamens numerous, disposed in a circle about the pistil and inserted near the edge of the disk, about ⅓ in. long, or shorter; pistil sessile or nearly so, rounded, depressed, about ¼ in. wide, style distinct; calyx large in fruit, becoming fleshy, 1–1⅔ in. across, sepals 5, persistent, ovate to lanceolate, strongly revolute toward the acuminate apex and with scattered (almost spinose) teeth, densely pubescent especially on the margin and lower surface; bracts large, laciniately cut, pubescent.

FRUIT. Hips large, ⅔–1 in. long, subglobose or pyriform, orange-red, densely tomentose, subtended by a number of large foliaceous, hairy, laciniately lobed bracts about ½ in. long; achenes long, numerous, enclosed in the pulpy hypanthium, seed pendulous.

LEAVES. Odd-pinnately compound, 2–4 in. long, leaflets 5–11 (mostly 7); rachis glabrous or pubescent; petiolules about ⅛ in. long; leaflets leathery, oblong to elliptic or obovate; apex rounded to retuse or truncate and often mucronulate; base cuneate or rounded; margin crenate-serrulate; upper surface dark green, lustrous and glabrous or puberulent; lower surface paler and glabrous or puberulent, ½–1 in. long, ¼–½ in. wide; stipules with pectinate lobes.

STEMS. Spreading, stiff, divaricate, densely hairy; thorns paired at the nodes, recurved, large, stout, reddish brown, ¼–½ in. long; often thickly set with smaller internodal thorns.

BARK. Reddish brown to gray, tight, smooth, thorny.

RANGE. Forming clumps and mounds on prairies. A native of China and Formosa but escaping cultivation in Texas, Oklahoma, Arkansas, and Louisiana; east to Florida and north to Virginia.

HYBRID. Both the Cherokee Rose and Macartney Rose are natives of Asia and were introduced at an early date. A hybrid of the two produces the rose known as Marie Leonida, *R. alba* var. *odorata,* bearing double, white flowers. It was introduced into cultivation in 1832.

REMARKS. The genus name, *Rosa,* is the ancient

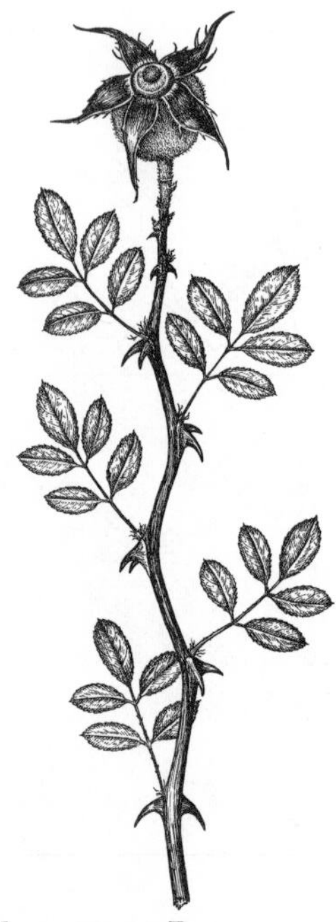

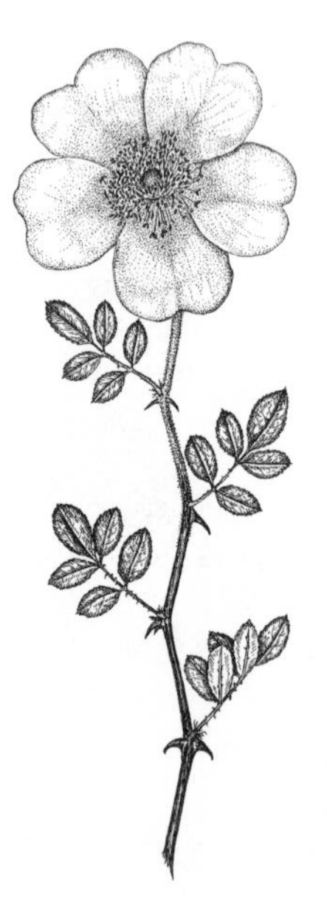

Macartney Rose
Rosa bracteata Wendl.

atin name, and the species name, *bracteata,* refers) the bracts of the flower. Vernacular names are 'rairie Rose, Evergreen Rose, Cherokee Rose, Hedge tose, Wild Rose, and Rosa Blanda. First introduced ıto cultivation in 1893. In frontier times it was planted s a windbreak for cattle on the Gulf Coast prairies, ut it is now reproducing so prolifically as to become pest in some areas. The plant has some value for ank-planting in erosion control. Its value as a wildife food has not been fully determined, but it serves s a dense protective cover for small birds and mamıals. Macartney Rose is often called Cherokee Rose ı some localities but true Cherokee Rose, *R. laevigata* Iichx., has 3 leaflets.

CAROLINA ROSE
Rosa carolina L.

AROLINA ROSE
osa carolina L.

FIELD IDENTIFICATION. Slender suckering rose, with imple or little-branched stems, attaining a height of in.–3 ft.

FLOWERS. May–July, mostly solitary, perfect, 1⅜–2¼ ı. broad; pedicels ⅜–1¼ in. long, glandular-hispid; ypanthium stipitate-glandular or smooth; petals 5, preading, pink, obovate or obcordate; stamens nuıerous, inserted near the orifice of the hypanthium, varies hairy; sepals 5, attenuate into linear apices, vith narrow foliaceous appendages, dorsally glanduar, tomentose internally, ⅝–¾ in. long.

FRUIT. Hips (mature hypanthia) red, about ⅓ in. ıroad, subglobose or appressed, orifice constricted, landular-hispid; achenes obovoid, ⅙–⅕ in. long, atached mostly to the bottom of the hypanthium; sepals eflexed and soon deciduous.

LEAVES. Altenate, pinnately compound; petioles and achis sparingly pubescent or glabrous, rarely with a ew prickles; leaflets 3–9 (usually 5), rather variable ı shape, elliptic to oblong or ovate-lanceolate (more arely oval or oblanceolate), firm, apex obtuse or acute, nargin sharply and coarsely serrate mostly above the niddle, length of blade ⅝–1⅝ in., upper surface glarous and dull or semilustrous, lower surface somewhat ıaler, and pubescent on veins, or glabrate; stipules omewhat folded, ⅕–⅝ in. long, 1/50–1/12 in. broad, adıate portion linear; free portion lanceolate and acuninate, glabrous or somewhat pubescent, margin glanlular-dentate or entire, ascending or spreading.

STEMS. Borne individually from stolons, rounded, /12–¼ in. thick; prickles scattered and short (some inernodes without prickles); infrastipular prickles slenler, straight or nearly so, solitary or paired, divergent ır somewhat reflexed, ⅙–⅖ in long. Branches usually ıot bristly, either with infrastipular prickles or someimes unarmed.

RANGE. Dry woods, thickets, and rocky or sandy banks. In Texas, Arkansas, and Louisiana; eastward to Florida, northward to Vermont, and westward to Michigan, Wisconsin, Minnesota, and Nebraska.

VARIETIES AND HYBRIDS. Hairy Carolina Rose, *R. carolina* var. *villosa* (Best) Rehd., has 5–7 leaflets which are pubescent beneath. From Louisiana and Oklahoma eastward to Georgia and northward to Minnesota and New Hampshire.

Three-lobe Carolina Rose, *R. carolina* var. *triloba* Rehd., has 3-lobed petals. Known in cultivation.

Large-flower Carolina Rose, *R. carolina* var. *grandiflora* (Baker) Rehd., has usually 7 leaflets which are obovate or oval, obtuse or subacute, sublustrous; flowers large, 1⅝–2¾ in. broad. From Arkansas eastward to Georgia and northward to Michigan, Wisconsin, Iowa, and Ontario.

Rehder Carolina Rose, *R. carolina* var. *glandulosa* Farwell, has more glands on the hypanthium, pedicels, and rachis; leaflets also more or less glandular-toothed. Northern Mexico; eastward to Florida and northward to Ontario and Massachusetts.

A form of *R. carolina* var. *villosa,* with white flowers, has been named *R. carolina* forma *alba.*

The following hybrids are known: *R. carolina* × *R. acicularoides; R. carolina* × *R. acicularis; R. carolina* × *R. palustris; R. carolina* × *R. virginiana;* and *R. carolina* × *R. suffulta.*

REMARKS. The genus name, *Rosa,* is the ancient Latin name. The species name, *carolina,* refers to the Carolinas, where the plant is found.

DEMAREE ROSE

Rosa demareei E. J. Palmer

FIELD IDENTIFICATION. Shrub attaining a height of 2–4 ft.

FLOWERS. May–July, solitary or rarely 2–3 together, subtended by 1–3 foliate bracts broader than the stipules; pedicels and hypanthium glabrous; flowers ¾–2 in. in diameter; petals 5, obovate to obcordate, pink; sepals 5, narrowly lanceolate, dilated at apex, ⅝–¾ in. long, margins usually entire, inner surface densely pubescent, outer surface glandular-hispid; stamens numerous; styles 12–14.

FRUIT. Not seen.

LEAVES. Alternate, pinnately compound, rachis and petioles densely pubescent and usually armed with weak prickles; leaflets 5–9 (usually 7), obovate, base narrowed or cuneate, apex rounded, blade length ⅕–1 in., width ⅙–⅗ in., margin coarsely and simply serrate, thin but firm, upper surface bright green and sparingly pilose above, lower surface much paler and copiously pubescent; stipules adnate, narrow, free portions lanceolate and erect.

STEMS. Reddish brown and often glaucous or pruinose, usually hispid with straight, slender spines ⅕–⅖ in. long.

DEMAREE ROSE
Rosa demareei E. J. Palmer

RANGE. In higher elevations of the Davis Mountains in Jeff Davis County, Texas. On moist bank and above springs in the upper Limpia and Madera canyons. On the northwest talus slopes of Mount Livermore at an altitude of 6,900 ft, growing under aspens.

REMARKS. The genus name, *Rosa,* is the ancient Latin name. The species name, *demareei,* is in honor of the Arkansas botanist Delzie Demaree, who accompanied E. J. Palmer when the plant was first discovered.

Palmer makes the following comments about the plant (1929, pp. 36–37):

"Although the mature fruit is unknown, this species, judging by the small number of styles, appears to belong to the *Gymnocarpae* group, and it is probably most nearly related to *Rosa gymnocarpa* Nutt. However, it differs markedly from that species in its obovate serrate leaflets with shorter petioles, in the character of the sepals, the enlarged-based spines, and especially in the broad stipules and modified bracts subtending the flowers. This last character appears to be quite constant and serves to distinguish our plant from related species."

EGLANTINE ROSE

Rosa eglanteria L.

FIELD IDENTIFICATION. Dense, stout shrub with stems branched, attaining a height of 10 ft.

FLOWERS. May–July, perfect, 1–4 together; pedicels ⅜–1¼ in. long, glandular-hispid, with foliaceous bracts at the base; hypanthium pear-shaped or broadly ellipsoid, orifice abruptly narrowed, base usually gradually narrowed, smooth or with a few bristles; sepals 5, ⅗–¾ in. long, subpersistent, reflexed, lanceolate, caudate-attenuate, subpectinate, dorsally glandular-hispid; petals 5, pink, ⅗–⅘ in. long; stamens numerous, spreading, inserted on the thickened margin of the hypanthium; styles somewhat exserted, pubescent.

FRUIT. Hips (mature hypanthia) ½–⅗ in. long, ⅖–½ in. thick, orange or scarlet; sepals spreading and tardily deciduous; achenes bony, inserted in the bottom and sides of the fleshy hypanthium.

LEAVES. Alternate, pinnately compound, 5–7-foliate (more rarely 9-); petioles and rachis glandular-hispid, sometimes pubescent, prickles absent or numerous; blades elliptic to oval or suborbicular, apex obtuse or acute (occasionally rounded), base rounded or obtuse, ⅜–1¼ in. long, margin doubly serrate with glandular teeth, both surfaces glandular-scurfy, but particularly densely pilose and resinous beneath; stipules ⅗–⅘ in. long, very glandular and pubescent below, margin glandular-ciliate.

STEMS. Branches slender, armed with prickles ¼–½ in. long, broad-based, curved, flattened, also sometimes with additional slender prickles.

Eglantine Rose
Rosa eglanteria L.

RANGE. Along roadsides, in clearings and thickets. A native of Europe, escaped from cultivation and naturalized in the United States. In Texas, Arkansas, Oklahoma, and Louisiana; eastward to Georgia, northward to Nova Scotia, and westward to British Columbia and the Pacific Coast states.

REMARKS. The genus name, *Rosa,* is the ancient Latin name. The species name, *eglanteria,* is the old English and French name latinized. The plant is also known as the Sweet-brier Rose. The shrub is handsome and hardy and grows compactly.

Leafy Rose

Rosa foliolosa Nutt.

FIELD IDENTIFICATION. Shrub with stems 6–20 in.

FLOWERS. May–June, usually solitary, or a few together; pedicels short, 1/12–2/5 in. long, sparsely glandular-hispid; hypanthium subglobose; flowers averaging about 1 5/8 in. across; sepals 5, dorsally glandular-hispid, inner surface tomentose, lanceolate and caudate-attenuate, length about 5/8 in., margins entire or a few with subulate lobes; petals 5, rose color, obcordate, about 3/4 in. long; stamens numerous; styles distinct, not exserted.

FRUIT. Hips (mature hypanthia) subglobose, smooth or glandular, about 1/3 in. broad; sepals spreading and early deciduous; achenes numerous, bony, in the bottom of the hypanthium.

LEAVES. Alternate, pinnate-compound; rachis and petiole glabrous or nearly so, often sparingly bristly and glandular-hispid; free portion of petiole very short; leaflets 5–11 (usually 9, rarely 11), oblong to linear-oblong or oblanceolate, apex and base acute, margins finely serrate, rarely glandular-ciliate, upper surface glabrous and lustrous, lower surface glabrous or pubescent on the veins, length 3/8–1 1/4 in.; stipules adnate, mostly narrow, glabrous or slightly pubescent, sometimes glandular-dentate, 3/8–5/8 in. long.

STEMS. Usually 4–8 in. long, prickles absent, or if present 1/12–1/5 in. long, straight but usually somewhat reflexed.

RANGE. On prairies of Texas, Oklahoma, and Arkansas.

VARIETY. A variety with white flowers has been named *R. foliolosa* var. *alba* Birdwell.

REMARKS. The genus name, *Rosa,* is the ancient Latin name, and the species name, *foliolosa,* refers to the many leaflets.

Pale-leaf Rose

Rosa hypoleuca Woot. & Standl.

FIELD IDENTIFICATION. Shrub attaining a height of 3 ft.

FLOWERS. Floral branches 4–8 in. long and usually armed with small prickles, corymbose at the ends of

Leafy Rose
Rosa foliolosa Nutt.

PALE-LEAF ROSE
Rosa hypoleuca Woot. & Standl.

the ascending branches; peduncles ⅜–¾ in. long, glabrous; hypanthium glabrous; sepals 5, narrowly lanceolate, caudate-acuminate, about ⅗ in. long, margin entire or with a few subulate lobes and tomentulose, also tomentulose on the inner side, apices glandular; petals 5, obovate, ⅗–⅘ in. long, notched, bright pink; stamens numerous; styles persistent, distinct, not exserted.

FRUIT. Hips (mature hypanthia) globose, glabrous, sepals persistent and erect.

LEAVES. Alternate, pinnate-compound, petioles and rachis glabrous or with a few stalked glands; leaflets usually 9, elliptic to oblong or oval, apex usually obtuse, base rounded, margin glandular-serrate, thin, length ⅜–1 in., upper surface dull and dark green, lower surface much whitened; stipules about ⅜ in. long, adnate, margins glandular-denticulate, apex acute, surfaces glabrous or glandular-pruinose.

STEMS. Reddish brown to purplish brown, usually glaucous when young; spines dense, ⅛–⅓ in. long, very slender, straight, terete.

RANGE. Type specimen collected near Kingston, New Mexico; also known from Arizona.

REMARKS. The genus name, *Rosa,* is the ancient Latin name, and the species name, *hypoleuca,* refers to the pale undersurface of the leaflets. Pale-leaf Rose is closely related to *R. woodsii* var. *fendleri* (Crep.) Rydb., but differs in the perfectly glabrous leaflets which are strongly glaucous beneath.

CHEROKEE ROSE
Rosa laevigata Michx.

FIELD IDENTIFICATION. Prickly evergreen shrub, trailing or high-climbing in habit, with stems 4–15 ft or more long.

FLOWERS. Borne in spring or summer, mostly in June, fragrant, solitary, perfect; peduncle and hypanthium strongly hispid; sepals 5, borne on the edge of the hypanthium, often somewhat glandular-bristly, more or less foliaceous, entire or toothed, reflexed or erect at maturity, bracts absent; petals 5, sessile at the base of the hypanthium, spreading, deciduous, cuneate-obcordate, 1¼–1½ in. long, white or more rarely pink; stamens numerous; styles distinct, lateral, included; ovule solitary and pendulous.

FRUIT. Mature hips (hypanthia) pyriform or obovoid, contracted to a stipelike base, bristly, 1⅓–1½ in. long; achenes enclosed in the pulpy fruit.

LEAVES. Alternate, pinnately compound, with 3 (rarely 5) leaflets; blades ovate-lanceolate or elliptic-ovate, margin finely but sharply serrate, evergreen,

CHEROKEE ROSE
Rosa laevigata Michx.

leathery, glabrous, lustrous, dark green above, under surface paler and reticulate, petioluled; stipules small and entire, almost free from the petiole.

STEMS. Slender, green, long-arching or trailing, glabrous; prickles stout, recurved, more or less flattened, rarely with a few bristles.

RANGE. Waste places, roadsides, banks, and thickets. A native of China, Formosa, and Japan. Often cultivated and sometimes escaping cultivation in the lower Gulf Coast states. From Texas eastward to Florida and northward to North Carolina.

HYBRIDS AND SYNONYMS. Cherokee Rose has also been described under the names *R. cherokeensis* Don, *R. sinica* Lindl., *R. ternata* Poir., *R. nivea* DC., and *R. camellia* Hort.

R. × *fortuneana* is a hybrid of *R. laevigata* and *R. banksiae*.

Anemone Rose is a hybrid with a tea rose and produces large, single, pink flowers.

REMARKS. The genus name, *Rosa*, is the ancient Latin name, and the species name, *laevigata*, refers to the smooth leaflets. This rose is sometimes confused with the Macartney Rose, but it is easily distinguished by the fewer leaflets.

SMALL-FLOWER ROSE
Rosa micrantha Sm.

SMALL-FLOWER ROSE

Rosa micrantha Sm.

FIELD IDENTIFICATION. Shrub with branched, climbing, or reclining stems attaining a height of 3–6 ft.

FLOWERS. May–July, usually 1–4 together, leafy-bracted; pedicels glandular-hispid, ⅜–¾ in. long; hypanthium narrowly elliptic, tapering at the apex and base; sepals 5, ½–¾ in. long, lanceolate, caudate-attenuate, dorsally glandular-hispid, tomentose within, margin more or less lobed; petals 5, ⅖–⅗ in. long, obcordate, rose-colored; styles distinct and glabrous; stamens numerous, inserted into the edge of the hollow disk which lines the receptacle.

FRUIT. Hips (mature hypanthia) elliptic, ⅗–⅘ in. long, ⅓–⅖ in. thick, fleshy; achenes inserted in the bottom and on the sides of the pulpy hypanthium.

LEAVES. Alternate, pinnately compound, 5–7-foliolate; petioles and rachis pubescent and glandular-hispid, rarely prickly; leaflets broadly ovate to elliptic or sometimes obovate, apex short-acuminate, base rounded or narrowed, length ⅜–1¼ in., margin doubly glandular-serrate, upper surface more or less pubescent, lower surface densely glandular-pruinose and pubescent; stipules adnate, ⅜–⅝ in. long, dilated above, lower surface glandular-pubescent, margins glandular-ciliate.

STEMS. Terete, prickles ⅕–⅖ in. long, rather uniform, hooked, flattened.

RANGE. A native of Europe. Sometimes escaping cultivation about old gardens or in roadside thickets. From Texas to South Carolina, New York, Massachusetts, and other scattered provinces.

REMARKS. The genus name, *Rosa*, is the ancient Latin name. The species name, *micrantha*, refers to the small flowers. This species is closely related to *R. eglanteria* L., there being a range of intergrading forms. Usually the Small-flower Rose has prickles of more uniform size; leaflets 5–7, more ovate, acuminate, nearly glandless above; and flowers smaller, paler, and not strongly fragrant.

NOOTKA ROSE

Rosa nutkana Presl.

FIELD IDENTIFICATION. Shrub with much-branched, stout, erect stems, and attaining a height of 2½–5 ft.

FLOWERS. June–July, floral branches mostly lateral, 1¼–4 in. long, from the old wood, usually with small infrastipular spines; pedicels ¾–1¼ in. long, more or less glandular-hispid; flowers mostly solitary but some-

Nootka Rose
Rosa nutkana Presl.

times 2–4 in a loose corymb; hypanthium glabrous; sepals 5, lanceolate, caudate-acuminate (sometimes with foliaceous appendages), margins villous and glandular-ciliate, dorsally usually glabrous or rarely glandular, tomentose within, ¾–1¼ in. long; petals 5, pink to white, broadly obcordate, length ¾–1½ in.; stamens numerous; styles not exserted.

FRUIT. Hips (mature hypanthia) globose, neckless, ⅗–¾ in. in diameter; achenes bony and numerous; sepals persistent and ascending.

LEAVES. Alternate, pinnately compound; rachis and petiole glandular-puberulent; leaflets 5–9 (usually 7), shape variable, broadly elliptic to oblong or ovate, apex acute to obtuse, base acute or obtuse to rounded, margin with double-serrate glandular teeth, upper surface dark green and glabrous, lower surface paler and glandular-puberulent, length ⅝–2 in.; stipules adnate, ⅜–¾ in. long, dilated above, glabrous, margin glandular-dentate, free portion ovate and acute.

STEMS. Surviving all but the severest winters, dark brown; less vigorous stems not usually bristly or prickly, if so only at the base, but infrastipular prickles present; on more vigorous stems prickles generally paired, straight, somewhat flattened below, ¼–½ in. long.

RANGE. Widespread in western United States and Canada, usually at altitudes of 3,500–10,000 ft. in our province. New Mexico, northward through the Rocky Mountains to Alberta and west to Alaska, British Columbia, Washington, and Oregon.

HYBRIDS, VARIETY, AND SYNONYMS. The following hybrids have been recognized: *R. nutkana* × *acicularis; R. nutkana* × *R. bourgeauiana; R. nutkana* × *R. gymnocarpa; R. nutkana* × *R. pisocarpa;* and *R. nutkana* × *R. pyrifera.*

A variety with the leaflets coarsely serrate and a glandular-hispid receptacle has been named Hispid Nootka Rose, *R. nutkana* var. *hispida* Fern. It ranges from British Columbia to Utah.

Synonyms of *R. nutkana* are *R. aciculata* Cockerell, *R. macdougali* Holz., *R. melina* Greene, *R. oreophila* Rydb., *R. pandorana* Greene, *R. underwoodi* Greene, *R. manca* Greene, and *R. bakeri* Rydb.

REMARKS. The genus name, *Rosa,* is the ancient Latin name, and the species name, *nutkana,* refers to Nootka Sound, British Columbia, where this species was discovered. Nootka Rose apparently has the largest flowers of any of the wild Western roses.

Palmer Rose

Rosa palmeri Rydb.

FIELD IDENTIFICATION. Shrub attaining a height of 20 in. or more.

FLOWERS. Floral branches usually with few prickles;

Palmer Rose
Rosa palmeri Rydb.

peduncles glandular-hispid, ⅜–¾ in. long; hypanthium glandular-hispid; sepals 5, ¾–1 in. long, caudate-attenuate, glandular-hispid; petals 5, obcordate, about ¾ in. long; stamens numerous; style persistent, not exserted.

FRUIT. Hypanthia (hips) at maturity about ½ in. thick, globose and glandular-hispid; sepals reflexed.

LEAVES. Alternate, pinnately compound; rachis and petioles more or less glandular-hispid; leaflets 5–9 (mostly 9), lanceolate or oblong (sometimes oblanceolate on floral branches), margin regularly serrate, length ¾–1⅝ in., upper surface dark green, lower surface paler and usually pubescent, especially along the veins; stipules ⅝–¾ in. long, puberulent, margin entire, apex lanceolate.

STEMS. Usually densely bristly and weakly prickly, somewhat glandular; prickles slender, terete, somewhat reflexed.

RANGE. Known from Texas, Arkansas, Oklahoma, and Missouri.

REMARKS. The genus name, *Rosa,* is the ancient Latin name. The species name, *palmeri,* is in honor of the botanist Ernest J. Palmer.

Swamp Rose

Rosa palustris Marsh.

FIELD IDENTIFICATION. Shrub with erect stems sometimes attaining a height of 1½–6 ft.

FLOWERS. June–August, usually corymbose, sometimes solitary, leafy-bracted; pedicels short, ⅜–¾ in. (rarely 1¼ in. long), usually glandular-hispid; hypanthia subglobose; sepals 5, narrowly lanceolate, caudate-attenuate, ¾–1 in. long, sometimes with foliaceous tips, glandular-hispid on the back, tomentose within and on the margins; petals 5, obcordate, rose-colored, ⅝–¾ in. long; styles distinct, numerous, not exserted, persistent; stamens numerous.

FRUIT. Hips (mature hypanthia) subglobose or somewhat depressed, glandular-hispid, usually acute at base, when mature ⅖–½ in. broad; sepals reflexed or spreading and soon deciduous.

LEAVES. Alternate, pinnately compound; rachis and petiole pubescent, rarely prickly, not glandular or, rarely, slightly so; free portion of the petiole above the stipules very short, ⅛–⅓ in. long; leaflets 7, rarely 9, dull and dark green, glabrous or nearly so above, paler and more or less finely appressed-puberulent beneath, at least on the veins, short-petiolate, lance-elliptic or oblanceolate, on vigorous shoots sometimes elliptic, usually acute at both ends, ¾–2½ in. long, finely and closely serrulate, with nonglandular teeth; stipules adnate, ¾–1¼ in. long, usually narrow, only the upper somewhat dilated, somewhat inrolled, usually pubescent beneath, somewhat glandular-denticulate or lobed on the margins, the free portion lanceolate or subulate, usually somewhat spreading.

SWAMP ROSE
Rosa palustris Marsh.

STEMS. Rather erect, sometimes reddish, glabrous, terete, armed with strong but rather short, more or less curved prickles, which are usually paired, flattened at the base, ⅙–¼ in. long.

RANGE. Along streams and marshes. Louisiana eastward to Florida, northward to Nova Scotia, and westward to Minnesota.

HYBRIDS. The following hybrids are known: *R. palustris* × *R. carolina; R. palustris* × *R. johannensis; R. palustris* × *R. nitida; R. palustris* × *R. virginiana;* and *R. palustris* × *R. serrulata.*

REMARKS. The genus name, *Rosa,* is the ancient Latin name. The species name, *palustris,* refers to the occurrence of the plant in marshy ground.

Field Rose

Rosa rudiuscula Greene

FIELD IDENTIFICATION. Shrub usually with a simple stem attaining a height of 3 ft.

FLOWERS. Borne in corymbs; pedicels about ¼–⅜ in. long, densely glandular-hispid; hypanthium globose; sepals 5, shape lanceolate, apex caudate-attenuate, dorsally glandular, margin mostly entire; petals 5, pink, obovate or obcordate; stamens numerous; styles distinct, persistent, not exserted.

FRUIT. Hips (mature hypanthia) globose, glandular-hispid, about ½ in. thick, red; sepals reflexed, deciduous later.

LEAVES. Alternate, pinnately compound, rachis and petiole densely pubescent and glandular; leaflets 5–7, rather crowded, subcoriaceous, margin coarsely serrate, apex and base acute, length ¾–1¼ in., upper surface dark green and lustrous, lower surface paler and densely pubescent; stipules adnate, narrow, about ⅜ in. long, margin mostly entire, dorsally pubescent and glandular.

STEM. Rather stout, gray, terete, bristly; prickles small, ⅛–⅕ in. long, very slender.

RANGE. Oklahoma to Missouri and Iowa.

REMARKS. The genus name, *Rosa,* is the ancient Latin name, and the species name, *rudiuscula,* refers to the growth of this species in old fields or waste places.

Prairie Rose

Rosa setigera Michx.

FIELD IDENTIFICATION. Shrub with high-climbing, trailing or leaning stems reaching a length of 6–15 ft.

FLOWERS. May–July, numerous, corymbose, perfect; pedicels glandular-hispid; hypanthium globose to ellipsoid; sepals 5, about ⅗ in. long, lanceolate, apex acuminate, dorsally glandular-hispid, inner surface tomentulose; petals 5, ¾–1¼ in. long, pink or white; stamens numerous, inserted on the thickened margin of the hypanthium; styles exserted, glabrous, united into a column about equaling the stamens, stigmas thickened.

FRUIT. Red, ⅓–½ in. long, globose to ellipsoid, usually glandular-hispid, contracted at the apex, fleshy, enclosing the bony achenes; sepals eventually reflexed and deciduous.

LEAVES. Alternate, pinnately compound; petioles and rachis glandular-hispid and often with a few prickles; terminal leaflet long-petioluled, lateral ones almost sessile; leaflets 3 or sometimes 5 or 7 on vigorous stems, lanceolate to oblong or ovate-oblong, apex acute to acuminate or rounded, base acute or more rarely subcordate, margin sharply serrate, length 1½–4 in., upper surface dark green and lustrous, lower surface paler and glabrous, veins sometimes glandular; stipules adnate, ⅜–¾ in. long, glabrous, entire, ciliate and sometimes glandular as well, free portion spreading, lanceolate, acuminate.

STEMS. Slender, green to reddish, glabrous; prickles scattered, curved, broad-based, ⅙–⅓ in. long.

RANGE. In thickets and along fence rows. Texas, Arkansas, and Louisiana; eastward to Florida, northward to New York, and westward to Indiana, Missouri, and Kansas.

VARIETIES. Spineless Prairie Rose, *R. setigera* var. *inermis* Palmer & Steyerm., has no prickles. Known in cultivation since 1923.

Fuzzy Prairie Rose, *R. setigera* var. *tomentosa* Torr. & Gray, has leaflets dull above and tomentose beneath. Known in cultivation since 1800.

REMARKS. The genus name, *Rosa,* is the ancient Latin name, and the species name, *setigera,* means "bristle-bearing," but this seems to be a misnomer.

Wonder Rose

Rosa mirifica Greene

FIELD IDENTIFICATION. Much-branched shrub with stems 16–24 in.

FLOWERS. June–September, solitary, terminal; peduncles about ⅜ in. long, hypanthium globose; sepals 5, lanceolate, about ¾ in. long, caudate-acuminate, usually with foliaceous tips, the outer more or less lobed, prickly on the backs, ciliate on the margins, and tomentose within; petals 5, deep rose-purple, 1–1¼ in. long; pistils numerous, styles not exserted, distinct, persistent; stamens numerous.

FRUIT. Hips (mature hypanthia) globose, about ⅗ in. broad, covered with numerous prickles; achenes numerous, bony, in the bottom of the hypanthium.

PRAIRIE ROSE
Rosa setigera Michx.

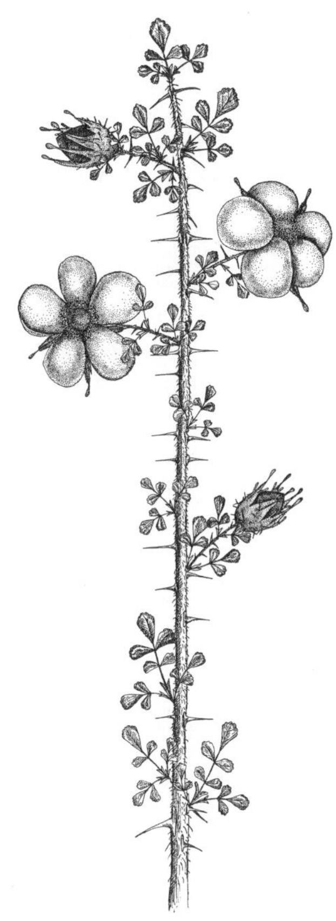

Wonder Rose
Rosa mirifica Greene

LEAVES. Alternate, pinnately compound; rachis and petiole mostly glabrate; leaflets usually 5 (or 3 on weak shoots), cuneate-obovate, apex rounded, with 7–10 rounded teeth above the middle, pilose, or glabrate, 1/3–4/5 in. long; stipules adnate, 1/4–2/5 in. long, often glandular-dentate, the free portion oblong and spreading.

STEM. Much-branched, prickles and bristles numerous and 1/12–2/5 in. long, more rarely glandular-hispid but not lepidote-stellate as in *R. stellata* Woot.

RANGE. Known from the White and Sacramento mountains, in grass and pinyon belts of New Mexico.

REMARKS. The genus name, *Rosa*, is the ancient Latin name. The species name, *mirifica,* means "wonderful," or "outstanding," and refers to the flowers.

Star Rose

Rosa stellata Woot.

FIELD IDENTIFICATION. Shrub with stems attaining a height of 16–24 in.

FLOWERS. June–September, borne terminally and solitary on stout pedicels 3/8 in. long; hypanthium globose; sepals 5, about 5/8 in. long, lanceolate, apex caudate-acuminate, margins entire or serrulate with glandular-ciliate lobes, dorsally prickly, inner surface tomentose; petals 5, broadly obovate, 3/4–1 in. long, rose-purple; pistils rather few; styles distinct and not exserted; stamens numerous.

FRUIT. Hips (mature hypanthia) globose, about 3/8 in. in diameter, glabrate or finely pubescent and covered with numerous straight prickles; achenes numerous, bony, inserted in the bottom of the hypanthium.

LEAVES. Alternate, pinnately compound, 3/8–5/8 in. long, 3-foliate; petioles finely pilose, leaflets cuneate to obovate, 1/5–1/3 in. long and about as broad, apex rounded to truncate, toothed toward the apex, the teeth glandular or glandless; surfaces finely hairy; stipules 1/5–1/3 in. long, oblong, adnate about half their length, entire or few-toothed, spreading at the free apex, pubescent.

STEM. Much-branched, often with stellate hairs; prickles 1/12–1/3 in. long, straight or slightly curved; sometimes bristly.

RANGE. Dry rocky places at altitudes of 6,500 ft. Known from western Texas, New Mexico, and Arizona.

Star Rose
Rosa stellata Woot.

REMARKS. The genus name, *Rosa*, is the ancient Latin name. The species name, *stellata*, refers to the stellate trichomes. The plant is evidently closely related to *R. mirifica* Greene, but the latter has no stellate hairs.

SEMITOOTHED ROSE

Rosa subserrulata Rydb.

FIELD IDENTIFICATION. Shrub with slender stems 24–32 in.

FLOWERS. Flowering branches very bristly and prickly and glandular-hispid; flowers solitary, borne on glandular-hispid pedicels ⅜–¾ in. long; hypanthium globose; sepals 5, lanceolate, apex caudate-attenuate, about ⅝ in. long, dorsally glandular-hispid, inwardly tomentose, margin entire or with a few subulate lobes; petals 5, rose-colored, about ¾ in. long, obcordate; styles distinct, persistent, not exserted.

FRUIT. Hips (mature hypanthia) globose, more or less glandular-hispid, ⅓–⅖ in. broad; mature sepals reflexed and soon deciduous; achenes bony, in the bottom of the hypanthium.

LEAVES. Alternate, pinnately compound; rachis and petiole more or less glandular-hispid and bristly; free portion of petioles 1¼–2 in. long; leaflets 3–7, lance-elliptic to oblanceolate or obovate, apex and base acute, thin, margin sharply serrate with lanceolate teeth which are glandular and often ciliate; stipules with the upper portion narrowly lanceolate, margin glandular-ciliate, surfaces glabrous or pubescent.

STEMS. Slender and usually densely bristly, infrastipular prickles slender, straight, divergent, averaging ⅙–⅕ in. long.

RANGE. In Texas, Arkansas, and Missouri.

REMARKS. The genus name, *Rosa*, is the ancient Latin name; and the species name, *subserrulata*, refers to the few subulate lobes of the sepals.

SEMITOOTHED ROSE
Rosa subserrulata Rydb.

STANDLEY ROSE

Rosa standleyi Rydb.

FIELD IDENTIFICATION. Shrub attaining a height of about 3 ft.

FLOWERS. Floral branches usually less than 4 in. long and often without prickles; pedicels glabrous, ⅜–¾ in. long; flowers usually solitary and perfect; petals 5, white or pink, rounded or obcordate; sepals 5, lanceolate, caudate-attenuate, puberulent but hardly glandular; stamens numerous, inserted on the thickened margin of the hypanthium; stigma thickened.

FRUIT. Mature hypanthia ellipsoid-elongate, with a distinct neck, dark purple; about ⅓ in. thick and ½ in. long; achenes bony, inserted on the bottom and sides of the hypanthium; sepals erect and persistent.

LEAVES. Alternate, pinnately compound; rachis and petioles from puberulent to glandular-puberulent; leaflets 5–7, oval, ⅜–1 in. long, upper surface dark green, glabrous, lower surface paler, puberulent and glandular-granuliferous, margins finely serrate; stipules adnate, about ⅜ in. long, dorsally puberulent and glandular-pruinose, margin glandular-ciliate.

STEMS. Gray to brown, branched; prickles infrastipular, usually paired, straight, weak.

RANGE. Type specimen collected near Pecos, New Mexico.

REMARKS. The genus name, *Rosa*, is the ancient Latin name. The species name, *standleyi*, is in honor of the American botanist Paul C. Standley.

SUNSHINE ROSE

Rosa suffulta Greene

FIELD IDENTIFICATION. A low subshrub hardly over 20 in., with erect and usually simple densely bristly stems.

SUNSHINE ROSE
Rosa suffulta Greene

FLOWERS. May–August, usually corymbose and terminating semiherbaceous stems of the year, or sometimes on short lateral branches on the older stems; pedicels usually glabrous, ⅜–¾ in. long; hypanthium globose, usually glabrous or occasionally bristly; sepals 5, lanceolate, apex caudate-attenuate, entire or sometimes lobed, length ⅜–⅝ in., inner surface tomentose, dorsally somewhat glandular; petals 5, pink, ¾–1¼ in. long, usually obcordate, spreading at anthesis; stamens numerous, inserted on the thickened margins of the hypanthium; styles somewhat exserted, stigma thickened; ovaries numerous, inserted on the bottom and sides of the hypanthium.

FRUIT. Hips (mature hypanthia) subglobose, about ⅜ in. broad, contracted at the mouth, pulpy; achenes bony, inserted in the bottom and on the lower sides of the hypanthium; sepals persistent on the fruit.

LEAVES. Alternate, pinnately compound; rachis and petioles finely pubescent (rarely glandular or bristly); leaflets 9–11 or occasionally 7, obovate or oblong-obovate, apex varying from obtuse to rounded, base acute or cuneate, length ⅝–1⅝ in., surfaces light green and usually densely pubescent, but upper surface eventually more glabrate, margin simply serrate; stipules adnate, mostly dilated, margins either entire or glandular-dentate, rather densely pubescent, occasionally somewhat glandular dorsally, length ⅝–¾ in.

STEMS. Erect and simple from a rootstock, often dying back to the base, light to dark green, usually densely prickly and bristly.

RANGE. On prairies and plains of Texas and New Mexico; northward to Alberta and Manitoba and eastward to Wisconsin, Illinois, and Indiana.

VARIETY, HYBRID, AND SYNONYM. White Sunshine Rose, *R. suffulta* var. *alba* Rehd., is a cultivated variety with white flowers.

R. suffulta is known to hybridize with *R. carolina* to form a large population on the prairies of the Middle West.

R. suffulta has also been described as a variety under the name of *R. arkansana* var. *suffulta* (Greene) Cockerell.

REMARKS. The genus name, *Rosa*, is the ancient Latin name. The species name, *suffulta*, means "propped up from beneath," because of the supernumerary reduced leaflets between the stipular lobes.

TOMENTOSE ROSE

Rosa tomentosa Sm.

FIELD IDENTIFICATION. An escaped shrub with a stem 3–6 ft.

FLOWERS. June–July, solitary or a few together; pedicels often glandular-hispid; hypanthium globose; sepals

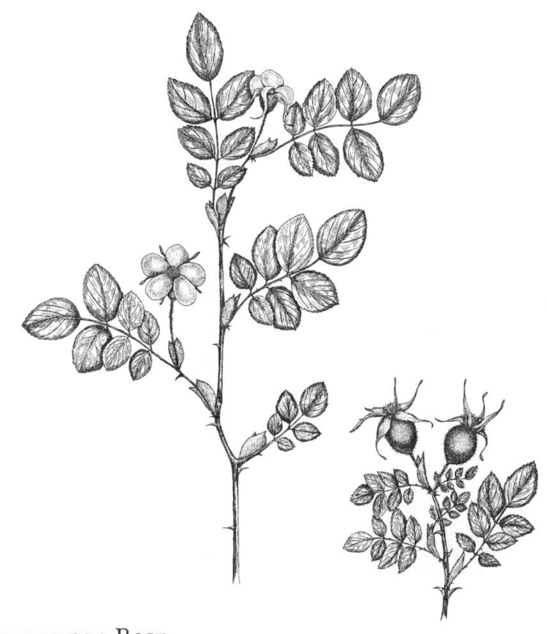

TOMENTOSE ROSE
Rosa tomentosa Sm.

5, lanceolate, glandular on the margins and dorsally, about ¾ in. long, apex acuminate, margin more or less lobed; petals 5, rose-colored, occasionally white, ⅔–⅘ in. long; stamens numerous.

FRUIT. Hips (mature hypanthia) globose or somewhat ellipsoid, base often glandular-hispid; sepals tardily deciduous before the ripening of the fruit.

LEAVES. Alternate, pinnately compound; rachis and petiole tomentulose and glandular-hispid, sometimes prickly; leaflets 5–7, oval or ovate, ¾–1⅝ in. long, base rounded, apex acute or short-acuminate, margin doubly glandular-serrate, upper surface finely appressed-pubescent, lower surface villous and glandular-pruinose; stipules broad, ⅜–⅝ in. long, apices abruptly slender-tipped, upper surface glabrous, pubescent beneath, margins glandular-ciliate.

STEMS. Often elongate and drooping, prickles paired, conic-subulate, straight or slightly curved, base somewhat flattened, ⅕–⅓ in. long; infrastipular prickles similar.

RANGE. In rich soil of thickets. A native of Europe, escaped from cultivation in the United States, at Big Sandy, Texas, and in North Carolina.

REMARKS. The genus name, *Rosa*, is the ancient Latin name, and the species name, *tomentosa*, refers to the hairy petiole and rachis.

Texarkana Rose

Rosa texarkana Rydb.

FIELD IDENTIFICATION. Shrub attaining a height of 3 ft.

FLOWERS. Borne in corymbs, on somewhat glandular-hispid pedicels ⅜–⅝ in. long; hypanthium globose; sepals 5, lanceolate, caudate-attenuate, dorsally glandular; petals 5, white, obcordate; stamens numerous, inserted at the base of the fleshy hypanthium.

FRUIT. Hips (mature hypanthia) ⅖–½ in. thick, globose, sparingly glandular; achenes bony, inserted in the bottom of the hypanthium.

LEAVES. Alternate, pinnately compound, rachis and petiole glandular-hispid and somewhat prickly; elliptic to oval, apex and base acute, margins serrate, length ⅜–1⅝ in., upper surface dark green, lustrous and glabrous, lower surface paler and glabrous to puberulent; stipules adnate, about ⅜ in. long, glabrous, apex lanceolate and somewhat spreading, marginal teeth with stipitate glands.

STEM. Slender, prickles ⅕–⅓ in. long, straight, slender, terete; also with conspicuous bristles.

RANGE. Northeast Texas, southwestern Arkansas, and northern Mississippi.

REMARKS. The genus name, *Rosa*, is the ancient Latin name. The species name, *texarkana*, refers to the type locality near Texarkana, Texas.

Trelease Rose

Rosa treleasei Rydb.

FIELD IDENTIFICATION. Shrub with a slender ster attaining a height of 3 ft.

FLOWERS. Borne in corymbs, pedicels somewha glandular-hispid, ⅜–⅝ in. long; hypanthium globose sepals 5, lanceolate, dorsally glandular, apex caudate attenuate; petals 5, obovate or obcordate; stamens nu merous.

FRUIT. Hips (mature hypanthia) globose or slightl depressed; achenes inserted in the bottom of the hy panthium; sepals soon deciduous.

LEAVES. Alternate, pinnately compound, rachis an petiole copiously glandular-hispid and somewhat prick ly; leaflets subcoriaceous, ½–1¾ in. long, elliptic o oval, apices and base acute, margin sharply serrate upper surface glabrous, dark green, lustrous; lowe surface paler and varying from glabrous to puberulent stipules adnate, about ⅜ in. long, apex free, lanceolate more or less spreading, margins with stipitate-glan dular teeth.

STEM. Slender, both prickly and bristly; prickle ⅕–⅓ in. long, slender, straight, terete.

RANGE. Known from Alden Bridge, Louisiana. Th type specimen was collected by William Trelease an is deposited in the Missouri Botanical Garden Herbar ium.

REMARKS. The genus name, *Rosa*, is the ancien Latin name. The species name, *treleasei*, is in hono of the botanist William Trelease.

Virginia Rose

Rosa virginiana Mill.

FIELD IDENTIFICATION. A shrub with stems attaining a height of 6 ft.

FLOWERS. Appearing in June in the Southwest. Usu ally borne in corymbs on glandular-hispid pedicel ⅜–1¼ in. long; hypanthium depressed-globose; sepal 5, narrowly lanceolate, caudate-acuminate, dorsally glandular-hispid, length ¾–1 in.; petals 5, rose-colored ¾–1⅓ in. long, obcordate; stamens numerous, insertec on the fleshy disk which lines the hypanthium; style not exserted, distinct and persistent.

FRUIT. Hips (mature hypanthia) red, depressed globose, ⅖–⅗ in. broad; sepals spreading or ascending soon deciduous; achenes obovoid, bony, tan to brown eventually ⅛–⅙ in. broad, attached in the bottom o the hypanthium.

LEAVES. Alternate, pinnately compound; petiole anc rachis usually prickly and sometimes glandular-hispic as well (free portion of the petiole 1⁄12–⅓ in. long) leaflets 5–11 (usually 7–9), firm to subcoriaceous, ellip tic to lance-elliptic or obovate, apex and base usually

Virginia Rose
Rosa virginiana Mill.

acute, or sometimes rounded at base, margin with coarse ascending teeth usually along the upper three fourths; upper surface dark green, lustrous and glabrous; lower surface duller and glabrous or with pubescent veins, length ¾–2½ in.; stipules ⅝–1¼ in. long, dilated toward the summit with free part lanceolate, surfaces glabrous or slightly pubescent, margin glandular-dentate.

STEM. Rather stout, glabrous, terete; prickles ⅓–½ in. long, straight or slightly curved but often reflexed, base flattened; bristles only on young shoots; internodal prickles few or absent.

RANGE. On banks, thickets, swamps, and shores. Arkansas, east to Alabama, north to Newfoundland, and west to Ontario.

HYBRIDS AND FORM. The following hybrids have been recognized:

R. virginiana × *R. bicknellii; R. virginiana* × *R. blanda; R. virginiana* × *R. carolina; R. virginiana* × *R. dasistema; R. virginiana* × *R. lyonii; R. virginiana* × *R. nitida; R. virginiana* × *R. palustris; R. virginiana* × *R. serrulata.*

R. virginiana forma *nanella* (Rydb.) Fern. is a dwarf compact form 4–16 in. high; leaflets ⅜–¾ in. long and ⅙–½ in. broad.

REMARKS. The genus name, *Rosa,* is the ancient Latin name. The species name, *virginiana,* refers to the state of Virginia, where this species was first found.

Woods' Rose
Rosa woodsii Lindl.

FIELD IDENTIFICATION. A shrub with stems attaining a height of 1½–6 ft, and not dying back to the base each winter.

FLOWERS. Borne June–August, floral branches lateral from the old wood, 4–8 in. long; flowers solitary or corymbose; pedicels ⅜–¾ in. long, glabrous; hypanthium globose; sepals 5, lanceolate, caudate-attenuate, about ⅗ in. long, dorsally glabrous or slightly glandular, margins and inner surface tomentose; petals 5, pink, ⅜–1 in. long; styles persistent, distinct, not exserted; stamens numerous.

FRUIT. Hips (mature hypanthia) globose or rarely ellipsoid, glabrous, in fruit ⅓–⅖ in. thick.

LEAVES. Alternate, pinnately compound; rachis and petiole glabrous, occasionally with a few prickles or stalked glands; leaflets 5–11 (usually 5–7); obovate or oval to elliptic, base cuneate, distinctly petioluled, margin serrate toward the apex but entire toward base, surfaces glabrous above, glabrous or somewhat puberulent and often glaucous beneath, ⅜–1¼ in. long; stipules adnate, somewhat dilated above, ⅜–⅝ in. long, glabrous, usually glandless, rarely slightly glandular-

Woods' Rose
Rosa woodsii Lindl.

Fendler Rose
Rosa woodsii var. *fendleri* (Crep.) Rydb.

pruinose, margin entire or toothed, free apex lanceolate to ovate.

STEMS. Reddish brown or gray, terete, glabrous; prickles rather numerous, straight or slightly curved, ⅙–⅓ in. long; infrastipular prickles present or absent.

RANGE. At altitudes of 3,500–10,000 ft, western Texas, west through New Mexico to Arizona, and northward to North Dakota and British Columbia.

VARIETY. Fendler Rose, *R. woodsii* var. *fendleri* (Crep.) Rydb., is a variety with straight slender prickles; stipules and petioles glandular; leaflets usually doubly glandular-serrate; flowers usually corymbose, sometimes solitary; usually taller and more glandless than the species. This variety intergrades with the species and is not recognized by some authors. It is a more eastern variety of Missouri to Minnesota, western Ontario, and British Columbia.

HYBRID. *R. woodsii* is known to hybridize with *R. gymnocarpa*.

REMARKS. The genus name, *Rosa*, is the ancient Latin name; and the species name, *woodsii*, is in honor of Joseph Woods (1776–1864), English student of the genus.

Genus Rubus

It is not the purpose of this author to present a monograph on the genus *Rubus*. The large number of illustrations and descriptions which would be required to give adequate coverage to the southwestern species would call for the work of a specialist who had spent many years studying the group.

The most exhaustive work on *Rubus* is that of the late L. H. Bailey in *Gentes Herbarum*, Volume V (1941–1945). This splendid contribution was published in ten illustrated fascicles, each with complete treatments of the species.

Permission has been granted by the L. H. Bailey Hortorium at Cornell University, through the kindness of Dr. George H. M. Lawrence, to use or adapt Bailey's material on the southwestern species. The original line drawings were prepared by Florence Mekeel Lambeth, onetime staff artist at the Hortorium, but were retouched by Sara Kahlden Arendale of Houston, Texas, to conform to the style of other illustrative material in this volume. Much of the text and keys has been adapted freely, and all direct quotations are so indicated.

The author is grateful to the L. H. Bailey Hortorium at Cornell University, for its co-operation and for granting permission to use this material.

McVaugh Blackberry

Rubus macvaughii L. H. Bailey

FIELD IDENTIFICATION. Bramble very prickly and glandular-hairy. The primocanes usually upright, but floricanes much-branched and scrambling.

FLOWERS. From 1–3 on short lateral shoots; pedicels prickly but mostly glandless, petals 5, white rounded; stamens numerous; calyx-lobes 5, ribbed, divaricate to deflexed, pubescent, usually glandless.

FRUIT. Large, about ¾ in., short-oblong, standing above the associated foliage.

FLORICANES AND LEAVES. Floricanes much-branched and scrambling, very prickly; leaves small, mostly 3-foliolate to simple, the ovate to lanceolate blades on the flowering parts glabrous above and pubescent beneath, closely serrate-dentate, ¾–1¼ in. long.

PRIMOCANES AND LEAVES. Axis of primocanes angled heavily armed with straight, whitish, broad-based prickles ⅛–¼ in. long, interspersed with setae and long gland-bearing hairs, these elements persisting on the floricanes; primocane leaves with long strongly armed petioles and petiolules; leaflets elliptic to oval short-pointed, 2¼–3¼ in. long, 1¼–2 in. broad, glabrous on upper face, soft-pubescent on lower face

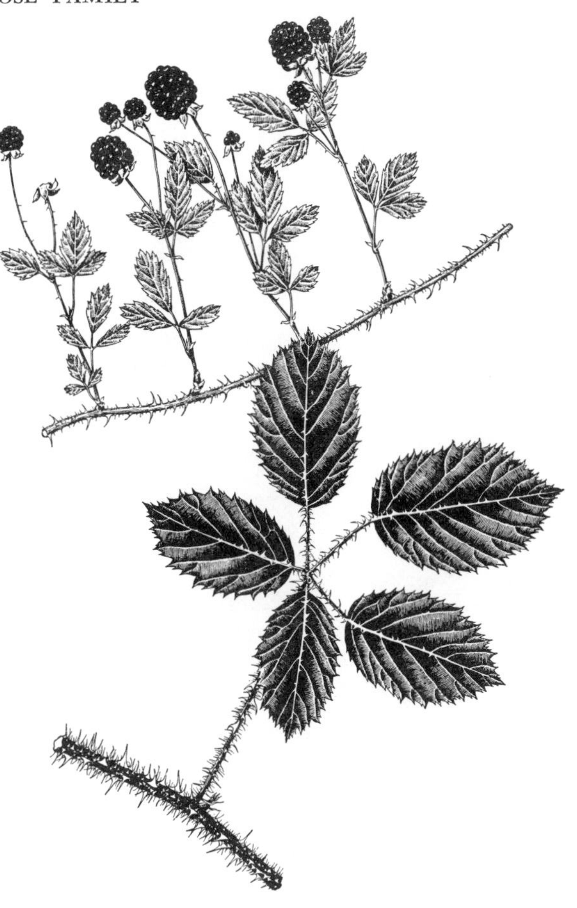

McVaugh Blackberry
Rubus macvaughii L. H. Bailey

prominently prickly on midrib underneath, margins irregularly serrate-dentate.

RANGE. In flat pinewoods, one mile northeast of Latex, Panola County, in northeast Texas.

REMARKS. The genus name, *Rubus,* is the old Roman name. The species name, *macvaughii,* is in honor of Rogers McVaugh, botanist at the University of Michigan.

Languid Blackberry

Rubus lassus L. H. Bailey

FIELD IDENTIFICATION. Bramble slender and weak. Canes erect at first but later arching over bushes, or diffuse, apparently tip-rooting, sparsely prickly, glandless.

FLOWERS. In May, borne in 1–4 clusters on ascending prickly pedicels that extend beyond the associated leafage; petals 5, white, rounded; calyx-lobes 5, pubescent, distinctly pointed but usually not elongate, soon reflexed; stamens numerous.

FRUIT. In May, small, black, globular to short-oblong, of few drupelets.

FLORICANES AND LEAVES. Sometimes crooked, sparsely prickly or bearing straight prickles ⅛–¼ in. long; leaves 3-foliolate to simple, small, somewhat puberulent underneath; leaflets narrowly elliptic and acute, 1¼–2¼ in. long, and less than ¾ in. broad, or those at base of floral shoots cuneate and nearly or quite obtuse.

PRIMOCANES AND LEAVES. Erect early in the season or nearly so, later arching; leaves 5-foliolate, thin, glabrous except slightly hairy on veins underneath, petioles somewhat prickly, very short; leaflets elliptic to oblong with more or less parallel sides, 3–3½ in. long, 1¼–1⅝ in. broad, briefly and not acutely pointed, margins rather bluntly serrate-dentate.

RANGE. In densely shaded situations in flood-plain woods. Four and one half miles northeast of Marshall, Harrison County, Texas.

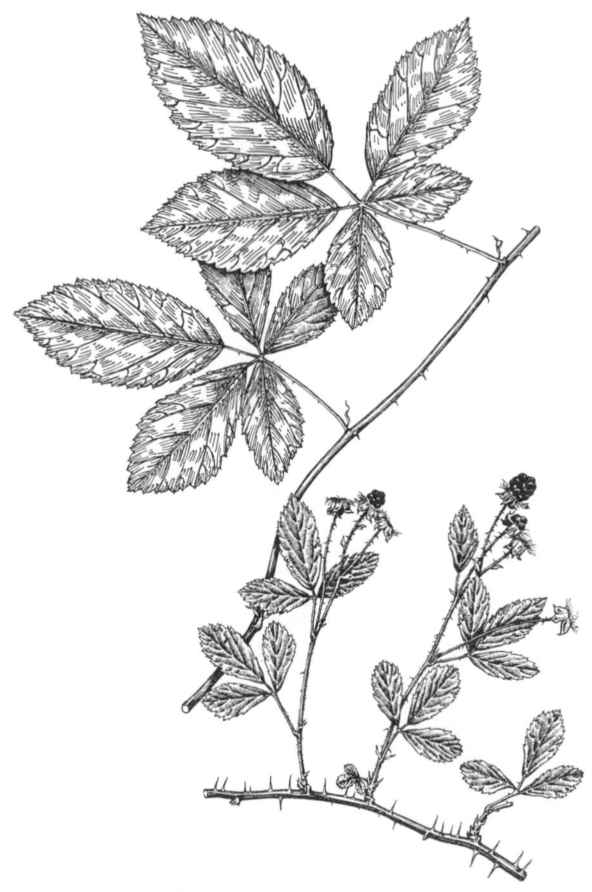

Languid Blackberry
Rubus lassus L. H. Bailey

REMARKS. The genus name, *Rubus,* is the old Roman name. The species name, *lassus,* refers to the languid character of the arching branches.

BARBED BLACKBERRY

Rubus uncus L. H. Bailey

FIELD IDENTIFICATION. Bramble very spiny on canes, petioles, and pedicels. Slender, prostrate, arching over bushes and herbs, probably tip-rooting, without glands.

FLOWERS. From 3–5 in a corymb, about equaling the leafage; corolla with 5 white spreading petals; calyx-lobes 5, soon divergent or reflexed, pubescent, acute but not elongate. Inflorescential leaves usually 3-foliolate or simple, blades very narrow, 1¼–1⅝ in. long and usually less than ¾ in. broad, often nearly lanceolate.

PRIMOCANES AND LEAVES. Primocanes terete, glabrous, bearing numerous scattered, nearly straight broad-based prickles ⅛–⅕ in. long; leaves 5-foliolate, firm, glabrous on upper face, more or less lightly pubescent on lower face, petiole and petiolules stout, somewhat pilose, stoutly hook-armed; leaflets long-elliptic, main ones about 3¼ in. long and about 1¼–1⅝ in. broad, long-acute or on some leaves nearly blunt, aciculate on midrib underneath, acutely double-serrate.

BARBED BLACKBERRY
Rubus uncus L. H. Bailey

ABUNDANT BLACKBERRY
Rubus largus L. H. Bailey

RANGE. Roadside bank six miles south of Buna, Jasper County, Texas.

REMARKS. The genus name, *Rubus,* is the old Roman name. The species name, *uncus,* means "barbed," with reference to the spines on the canes.

The following comments are made by L. H. Bailey (1941–1945):

"This species and *R. lassus* probably belong in the *R. apogaeus* complex, a group that extends from Alabama to eastern Louisiana and which may be represented in modified forms in eastern Texas by McVaugh collections in the Hortorium *Rubus* herbarium. Straight *R. apogaeus* is a plant with down-bending or overarching aspect rather than a flat or long trailer, with more pronounced lateral branches, pedicels short and hardly ascendate and much less armed, primocane aves more pubescent underneath. It will not be un- ›rstood until good collections are available."

BUNDANT BLACKBERRY

ubus largus L. H. Bailey

FIELD IDENTIFICATION. A large glandless bramble. he canes 6–9 ft long, at first upright, later arching 'er bushes and becoming prostrate, probably some- mes tip-rooting.

FLOWERS. Usually borne singly or 2–5 in axils of in- ›rescential leaves and little exceeding the leaves, ped- els stout, axillary, more or less aculeate; petals 5, hite, rounded; stamens numerous; calyx-lobes 5, ›reading or becoming reflexed, not elongate, pubes- ›nt.

FRUIT. Small, black, nearly globular, ¼–⅜ in. across.

FLORICANES AND LEAVES. Floricanes bearing many ›wering laterals, leaves 3-foliolate, soft-pubescent un- ›rneath; leaflets ovate to oblong, acute to obtuse, /4–2½ in. long.

PRIMOCANES AND LEAVES. Primocane axes stout, rongly angled, glabrous unless at tips, branched, ›aring few, scattered, broad-based prickles ⅙–¼ in. ng; leaves large, 5-foliolate, glabrous above, soft- ıbescent on lower face, petiole stout and armed; leaf- ts elliptic-acute with long point, sometimes broadest or above the middle, 4–5½ in. long, 2½–3¼ broad, ›uleate on midrib underneath, margins acutely serrate- ›ntate.

RANGE. Moist shady spots on hillsides in oak woods, ne miles northwest of Jacksboro, Jack County, north- n Texas.

REMARKS. The genus name, *Rubus*, is the old Roman ıme. The species name, *largus*, which means "plenti- ıl," is for the abundance of the species.

RONG-THORN BLACKBERRY

ubus valentulus L. H. Bailey

FIELD IDENTIFICATION. Bramble erect, glandless, well- 'med, to 6 ft. Canes sprawling.

FLOWERS. Usually 3–4 in each small cluster on side ıoots, pedicels prickly; petals 5, white, rounded; sta- ens numerous; calyx-lobes 5, soon reflexed, scantily ıbescent, point of lobes sharp and short.

FRUIT. Few, in an open broad cluster equaling or ›mewhat exceeding the leaves, nearly or quite globu- r, ⅜ in. or more in diameter, sweet, drupelets large ıd pulpy.

FLORICANES AND LEAVES. Floricanes arching or partially decumbent, axis angled, retaining armature, prickles about ¾ in. or more asunder, nearly or quite straight, ⅛–⅙ in. long; leaves 3-foliolate or simple in the inflorescence, leaflets not exceeding ¾–1¼ in. long.

PRIMOCANE LEAVES. Usually 5-foliolate, firm, glabrous on upper face, soft-pubescent on lower face, petiole armed with hooks, stipules small and very thin; leaflets ovate to elliptic, usually attenuate, 2⅓–3¼ in. long, 1¼–2 in. broad, margins coarsely and acutely serrate or almost dentate, midrib on under surface of larger leaflets more or less armed, lateral petiolules very short.

RANGE. Thickets along roads in sandy-gravelly soil. Short distance north of Collinsville, Grayson County, northeastern Texas.

REMARKS. The genus name, *Rubus*, is the old Roman name. The species name, *valentulus*, refers to the strong prickles.

R. valentulus appears to be related to *R. pensilvanicus*, but leaflets of the latter are much more attenuate and more coarsely serrate, pedicels of fruit elon-

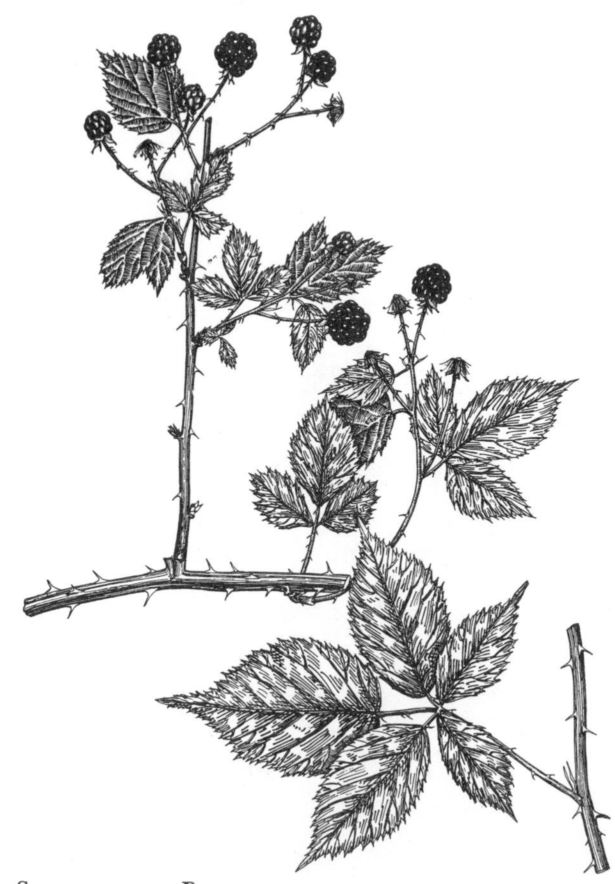

STRONG-THORN BLACKBERRY
Rubus valentulus L. H. Bailey

gate and divaricate, inflorescential leaflets broader, inflorescence apparently less evenly or regularly distributed.

Texas Blackberry

Rubus texanus L. H. Bailey

FIELD IDENTIFICATION. Shrub to 6 ft, erect, stiff, prickly, and glandless.

FLOWERS. Opening in May on short lateral shoots or spurs, on 2–10 short pedicels; petals 5, white rounded, stamens numerous.

FRUIT. Maturing May–June, black, in a close cluster or head, apparently mostly small and seedy but under good conditions nearly ¾ in. long, sweet, palatable; calyx-lobes 5, short, closely reflexed under the fruit.

FLORICANES AND LEAVES. Floricanes branched, often crooked and crabbed, axis angled; leaves 3-foliolate, those on the floral shoots very small; leaflets more or less obovate or cuneate, only briefly acute, ¾–1½ i long.

Texas Blackberry
Rubus texanus L. H. Bailey

PRIMOCANES AND LEAVES. Strongly angled, closel pubescent, with straight or slightly bent broad-base scattered prickles ⅕–⅓ in. long, stipules very narrow inconspicuous; leaves 5-foliolate, glabrous on uppe face, soft gray-pubescent underneath, petiole stout an armed; leaflets long-elliptic to nearly ovate, most c them with a narrowed base because the broadest pa is commonly above the middle, 2¾–3½ in. long, 1¼–1 in. broad at widest part, margins variously and acutel double-serrate.

RANGE. Low grounds, swamp margins, edges o woods. In Texas as follows: ten miles north of Pari in Lamar County; four miles east of Mineola in Woo County; six and one half miles north of Jasper i Jasper County; three miles west of Orange in Orang County; two and one half miles south of Arcola, For Bend County; two miles east of Navasota, Grime County; eleven and one half miles northeast of Bas trop, Bastrop County; five miles west of Angleton i Brazoria County.

REMARKS. The genus name, *Rubus*, is the old Roma name. The species name, *texanus*, refers to the stat where it grows.

Southern Dewberry

Rubus trivialis Michx.

FIELD IDENTIFICATION. A trailer appearing in numerous guises, rooting at tips of canes and branches an sometimes at nodes. Canes generally bearing reddis glandular hairs as well as small prickles.

FLOWERS. Opening March–April. Exceedingly divers in size, commonly ½–⅗ in. across, but sometimes twic these dimensions, on solitary, erect, armed, seldo glandiferous pedicels; petals 5, mostly broad contiguou or overlapping, very obtuse, white to pinkish; calyx lobes 5, soon reflexing, glandiferous; stamens numerous

FRUIT. Maturing June–July. Usually oblong, blac at maturity, large-seeded, perhaps ¼–⅜ in. long, bu sometimes ¾–1¼ in. long and then sweet, juicy, an attractive.

FLORICANE LEAVES. Usually 3-foliolate; leaflets widely variable in size and shape, glabrous, margins strongl serrate to dentate, petioles armed or nude, somewha glandular-hairy.

PRIMOCANES AND LEAVES. Primocanes often mor than 3 ft long, at first somewhat ascending but soo prostrate, becoming woody, branching, bearing many short, broad-based, often hooked prickles, variousl reddish glandular-hairy, particularly toward the growing tip, but sometimes glabrous on older parts; primocane leaves glabrous, 5-foliolate, firm or coriaceous, petiole stout and prickly and often glandular-hairy; leaflets narrow-elliptic to narrow-ovate, usually twice

SOUTHERN DEWBERRY
Rubus trivialis Michx.

or more longer than broad, more or less narrowed at base and not subcordate, sharply and coarsely serrate to dentate, apex acute, acuminate, or in many guises nearly obtuse, 1¾–2¾ in. long; floricanes flat on ground in open bare fields, but sometimes scrambling in bushes and fences, retaining the prickles and sometimes the glandular hairs, many of the 5-foliolate old leaves still persisting.

RANGE. Commonly a weedy plant in regions of lower elevation, mostly on sandy land but ranging into other habitats. In Texas as far south and west as Colorado, Gonzales, and Edwards counties. In Louisiana, Oklahoma, and Arkansas; eastward to Florida, northward to Pennsylvania, and west to Missouri.

VARIETIES AND HORTICULTURAL FORMS. Southern Dewberry is very variable as to size of flower, leaf, amount of glandulosity, and in exposure sensitivity. However, the most constant variety seems to be *R. trivialis* var. *seorsus* L. H. Bailey. It has leaflets broader than in *R. trivialis* to very broad (particularly those below apex of primocanes), often 1⅝ in. broad and 2½ in. long or 1¼ in. broad and 1⅝ in. long, broad or subcordate at base, nearly obtuse, obtusely serrate or dentate. In Crowder, Pittsburg County, and Muskogee, Muskogee County, Oklahoma.

Many large-fruited productive races have been introduced to cultivation, most of all from the wild, as pomological dewberries. The White Dewberry is a nearly colorless kind. Others are Advance, Bauer, Driskill, Eight Ells, Extra, Gregg, Houston, Howard, Lime Kiln, Long Branch, Lost Ball, McDonald (another McDonald is *R. velox*), Manatee, Muchee Grandee, Race Track, Rockledge, Rogers, and San Jacinto.

Nessberry is the result of a hand cross between *trivialis (rubrisetus)* pistillate and the red raspberry Brilliant.

REMARKS. The genus name, *Rubus*, is the old Roman name. The species name, *trivialis*, means "ordinary."

RIO GRANDE DEWBERRY

Rubus riograndis L. H. Bailey

FIELD IDENTIFICATION. Vinelike, long-running, and presumably rooting at tips, making mats or mounds 2 ft or more deep.

RIO GRANDE DEWBERRY
Rubus riograndis L. H. Bailey

FLOWERS. Appearing February–April, borne on 1–3 ascending glandular-hairy and weakly prickly pedicels on short lateral shoots that often bear leaves practically up to the flower; corolla about 1 in. across; petals 5, very broad and contiguous and very soon reflexed, glandless, stamens numerous.

FRUIT. Maturing May–June, black, oblong, ⅖–⅗ in. long, seedy.

FLORICANES AND LEAVES. Floricanes (or the slender branches) retaining the prickles and some of the glandular-hairiness; true floricane leaves small, 3-foliate, often soft-pubescent underneath; leaflets elliptic or oblong, 1½–2 in. long or less, obtusely notched and serrate to almost crenate, apex mostly obtuse.

PRIMOCANES AND LEAVES. Primocanes at first erect, shaggy with long gland-tipped hairs and curved prickles ⅕ in. long or less, soon arching over and becoming prostrate and repent, to 3 ft or more; primocane leaves 5-foliate, usually soft-pubescent to the touch underneath, soft and thin, petiole and petiolules glandular-hispid and prickly; leaflets (or at least the odd or central one) distinctly broad, ovate, broad-ovate, oval, oblong, 2–2¾ in. long and the 3 upper ones in the leaf perhaps two thirds as broad, the margins unevenly dentate with sinuses ⅛–⅕ in. deep and the immediate dentations also serrate, apex short-acute to obtuse, base of the center or upper ones broad to subcordate.

RANGE. In dry open places and woods. In Texas at Brownsville in the lower Rio Grande Valley area. Also known from Freestone, Kendall, Cooke, Kerr, and Blanco counties.

REMARKS. The genus name, *Rubus,* is the old Roman name. The species name, *riograndis,* refers to its habitat near the Rio Grande River in Texas.

MISSISSIPPI DEWBERRY
Rubus mississippianus, L. H. Bailey

MISSISSIPPI DEWBERRY

Rubus mississippianus L. H. Bailey

FIELD IDENTIFICATION. Stiffish low trailing bramble, 3–4 ft long, rooting at tip. Sometimes making a low mound 12–20 in. deep.

FLOWERS. Borne in March (Bexar County, Texas). On single and often elongate prickly, practically leafless, pedicels from short lateral spurs, often 1¼ in. and more across, but on weak plants perhaps one third that size, the 5 broad contiguous petals obtuse; calyx lightly pubescent, the 5 lobes soon reflexed; stamens numerous.

FRUIT. Short-oblong to nearly globular, black, about ⅗ in. long, seedy.

FLORICANES AND LEAVES. True floricane leaflets borne mostly close to the axis rather than on long branches, the leaves small and 3-foliate; leaflets ¾–1¼ in. or less long, oblong to indifferently obovate, obtuse or short-pointed, either acutely or bluntly serrate.

PRIMOCANES AND LEAVES. Primocane foliage mostly persisting until the following year and then lucid and bronzy, therefore in evidence even when the collector takes only the floricanes; primocane axis slender, wire-like, bearing scattered small hooked prickles, glabrous and glandless (very occasionally with glandiferous pilus), mostly flat on the ground; leaves of primocanes 5-foliolate, essentially glabrous, firm; leaflets very narrow, sometimes lanceolate (2 in. long and ⅝ in. broad), but often lance-ovate to oblong-lanceolate and perhaps 3¼–3½ in. long and 1¼–1½ in. broad, typically very acute but near the base of canes often obtuse, tapering to very narrow base, margins coarsely serrate to dentate.

RANGE. Banks and open land. Mississippi, Alabama, and northwestern Florida. To be expected in Louisiana and east Texas.

REMARKS. The genus name, *Rubus,* is the old Roman name. The species name, *mississippianus,* refers to the plant's habitat in Mississippi.

Shiny Dewberry
Rubus lucidus Rydb.

FIELD IDENTIFICATION. Essentially a trailer, without glands on any part, eventually rooting at tips, but making tangled low clumps or mounds.

FLOWERS. Flowers 3–5 on conspicuous long, hook-armed, glandless and nearly glabrous ascending pedicels; corolla medium to large in expansion, sometimes 1¼ in. across; petals 5, broad, very obtuse and contiguous; calyx-lobes 5, broad, short-pointed, pubescent, soon reflexing and then likely to appear narrow and acuminate because of infolding; stamens numerous.

FRUITS. Oblong, rather dry, black, ⅖–⅗ in. long.

FLORICANES AND LEAVES. Floricanes mostly prostrate, retaining the stout armature, often branched, producing stiffish leafy floral shoots 4–8 in. high, some of the upper leaves likely to be simple and broad; true floricane leaflets ovate to broad-ovate, 3 to a leaf, firm to coriaceous, either coarsely serrate-dentate or finely and unevenly serrate.

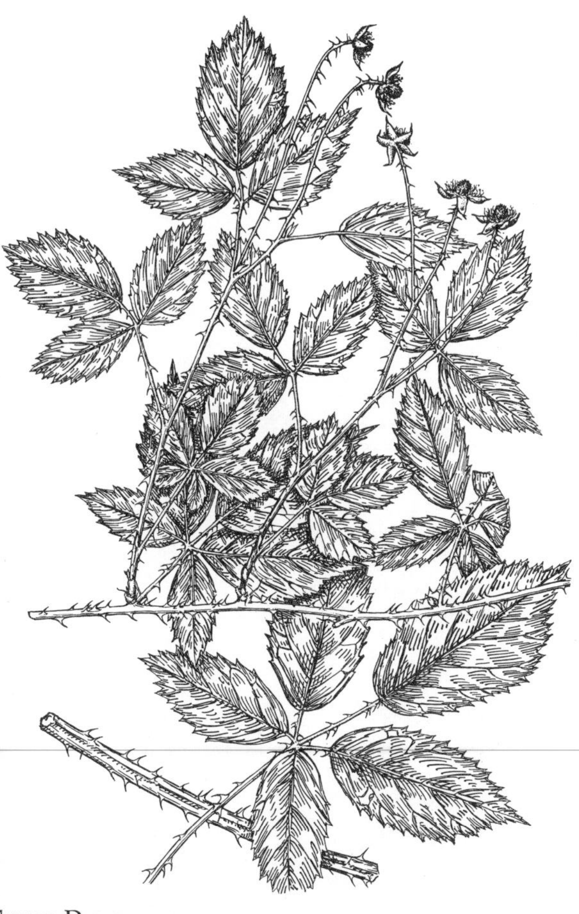

Shiny Dewberry
Rubus lucidus Rydb.

PRIMOCANES AND LEAVES. Primocane leaves often persisting over winter and therefore remaining on the floricanes and likely to be confused with the regular floricane foliage; primocanes soon becoming woody, at first erect or ascending but soon decumbent and extending 3–6 ft, axis of growing tips bearing many slender hooked prickles which become stout, broad-based, and formidable as the cane matures, commonly lying ⅜–¾ in. apart; leaves of primocanes firm, mostly lucid, glabrous except for a fine unnoticeable pubescence along veins or under surface, 5-foliate, stout petioles provided with strong hooked prickles; leaflets various, oblong-elliptic, elliptic, irregularly oval, somewhat obovate, differing even on same plant, odd or central one 2–3¼ in. long and 1¼–1¾ in. broad, either abruptly or acuminately pointed, margins strongly and almost simply serrate-dentate to finely and sharply serrate, base of odd leaflet usually broad and sometimes subcordate.

RANGE. In Pinellas and Hillsborough counties, Florida. Formerly recorded from Louisiana, Mississippi, and South Carolina, but more study is needed to determine its exact range.

REMARKS. The genus name, *Rubus*, is the old Roman name. The species name, *lucidus*, refers to the shining upper leaf surface.

Hurtful Dewberry
Rubus sons L. H. Bailey

FIELD IDENTIFICATION. Ascending with much-branched woody canes that make close tangles or clumps 3 ft or more high, the canes bending over and taking root.

FLOWERS. Opening May–June, borne on short upright floral branches well above the foliage on the long canes, mostly 2–6 on rather short pedicels, some of which may be armed; corolla about ¾ in. or somewhat more across, petals 5, rounded and contiguous; calyx thinly pubescent, glandless, the lobes with short points; stamens numerous.

FRUIT. Maturing late June–September, purplish black, oblong, to ½ in. long, small-seeded.

FLORICANES AND LEAVES. Floricanes retaining the armature even on the slender branches; leaves on floricanes small, 3-foliate; leaflets prevailingly not more than 1¼ in. long on the flowering parts, often less than one half as broad as long, oblong to elliptic, acute or obtuse, the serratures close.

PRIMOCANES AND LEAVES. Primocanes terete (at least when mature), glabrous, glandless, bearing broad-based woody hooked spines ⅙–⅕ in. long, perhaps 3–5 to each ⅜ in. of stem, the main cane becoming ⅜–1 in. in thickness; leaves of primocanes thin, usually not coriaceous or persistent, glabrous above but sometimes scantily pubescent beneath, 5-foliate, the long petiole sharply armed; leaflets elliptic to oblong-lanceo-

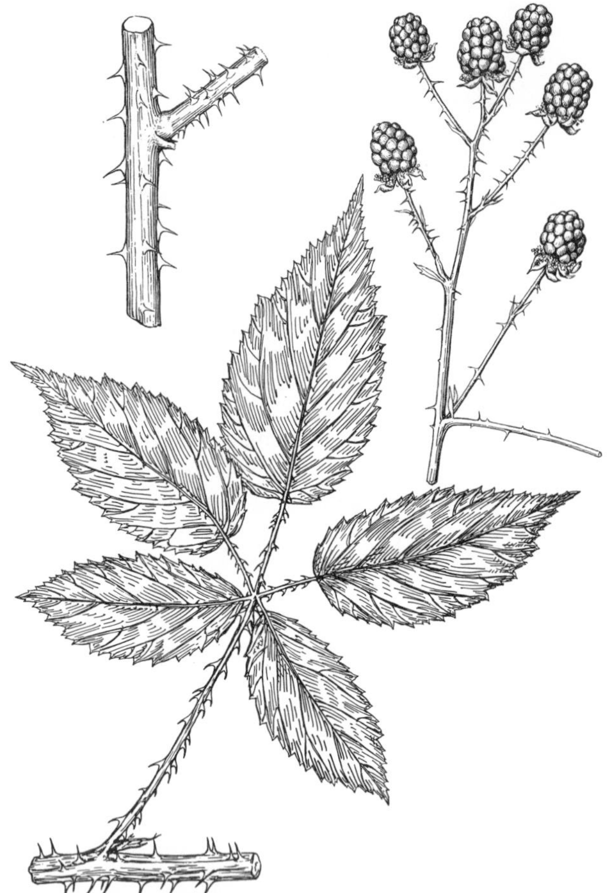

HURTFUL DEWBERRY
Rubus sons L. H. Bailey

late or sometimes ovate, long-petiolulate, markedly acuminate, the odd or central one 2¾–3½ in. long and one half as broad, and the base often cordate, although all of them are frequently narrowed to base, margins irregularly serrate or sharply jagged.

RANGE. Varying habitat, from flat dry lands to swamps, sometimes clambering on fences, but when alone and vigorous making dense formidable mounds or tangled heaps 5 ft high and 10 ft across. Gulf region in Louisiana from Baton Rouge southward, common about New Orleans, eastward to northwestern Florida.

REMARKS. The genus name, *Rubus,* is the Roman name. The species name, *sons,* means "criminal" or "guilty," alluding to the formidable spines.

FALLING DEWBERRY

Rubus apogaeus L. H. Bailey

FIELD IDENTIFICATION. Tumbled shrub, glandless throughout, the tangled loose mounds perhaps 2 ft deep, with few upstanding shoots, the general direction of all the growths bending toward the ground, the arms or branches 6–9 ft long and lying flat and rooting at ends.

FLOWERS. Flowers in June, 6–7 or fewer on short floral shoots, the pedicels not elongate (¾ in. or less), pubescent, unarmed or provided with a few little hooks; corolla often 1 in. across at expansion but sometimes only one half that size, petals 5, rounded or obtuse and more or less contiguous; calyx of 5 sepals; stamens numerous.

FRUIT. Maturing in July, deep violet, oblong, ⅜–½ in. long, seedy.

FLORICANES AND LEAVES. Floricanes retaining the straight armature which has now become stouter; leaves soft-pubescent underneath, small, 3-foliate, the leaflets elliptic or narrower, acute to obtuse, ¾ in. or less long, obtusely serrate.

PRIMOCANES AND LEAVES. Primocanes glabrous except at tip where they are somewhat pubescent, provided with weak essentially straight flat-based prickles, ⅛–⅕ in. long and spaced ⅕–⅖ in. apart; leaves of primocanes 5-foliolate, petioles long and armed, practically gla-

FALLING DEWBERRY
Rubus apogaeus L. H. Bailey

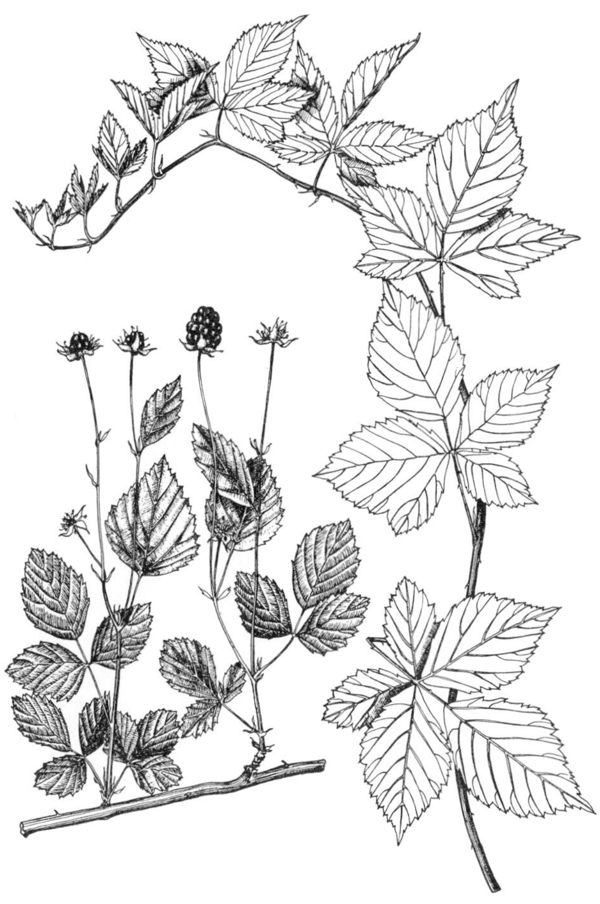

Kindred Dewberry
Rubus neonefrens L. H. Bailey

brous and somewhat lucid above, soft-pubescent to the finger underneath, not persistent; leaflets narrow, variable, those near top of cane oblong-lanceolate and long-acuminate (perhaps 2½ in. long and one half as broad) the main ones broader and short-pointed and at least the central or odd one broad or subcordate at base, all margins sharply and unequally serrate.

RANGE. Dry upland fields. East Texas, Louisiana, Mississippi, and Alabama.

REMARKS. The genus name, *Rubus*, is the Roman name. The species name, *apogaeus*, refers to the downward growth of the shoots.

Kindred Dewberry

Rubus neonefrens L. H. Bailey

FIELD IDENTIFICATION. Bramble stout, trailing, tip-rooting, glandless, and practically glabrous.

FLOWERS. From 3–5 on long ascending, nearly or quite glabrous, and frequently aculeate, pedicels; petals 5, white, rounded; stamens numerous; calyx-lobes 5, somewhat pubescent, attenuate.

FRUIT. Nearly globular, black, ⅖–⅗ in. in diameter.

FLORICANE LEAVES. Ovate to elliptic, larger ones 1½ in. long and ¾–1¼ in. broad, generally short-acute.

PRIMOCANES AND LEAVES. Axis of primocanes terete, ⅙–⅕ in. thick, bearing scattered short curved prickles only 1/25–1/12 in. long, stipules very large; primocane leaves 5-foliolate, glabrous on upper surface and perhaps very faintly pubescent along veins underneath, margins dentate and the nearly obtuse teeth often serrate, the teeth often ⅙–⅕ in. long, margins not ciliate; leaflets various but commonly ovate-acute or elliptic-acuminate, sometimes very broad at middle and abruptly short-pointed, odd or central one often 2¾–3¼ in. long and 2 or more in. broad, usually all of them narrowed to base, but sometimes the odd one approximately subcordate.

RANGE. On dry land, near Crowder, Pittsburg County, eastern Oklahoma. This, or a very similar species, grows in Cass County, Texas.

REMARKS. The genus name, *Rubus*, is the old Roman name. The species name, *neonefrens*, refers to the alliance of this species with *R. nefrens*.

Whiplash Dewberry

Rubus flagellaris Willd.

FIELD IDENTIFICATION. Plant vigorous, very strong, flat-trailing, or, when primocanes are young, making a tangled colony 20 in. deep but mainly close to the ground. Canes long-running, tip-rooting, and glandless.

FLOWERS. Borne May–June, inflorescence ascending, the long slender, frequently aciculate pedicels 1–3 from each axil, the whole floral shoot bearing 1–6 rather large flowers that stand well above the foliage; corolla ¾–1¼ in. broad when fully expanded; petals 5, white, rounded, obtuse, separated; calyx-lobes 5, narrow, short-pointed, pubescent, reflexed in anthesis.

FRUIT. Maturing June–August, black, large, or at least conspicuous, variable on different plants of a locale, globose to short-oblong, about ¼ in. thick, often attractively edible when fully ripe, drupelets rather large.

FLORICANE LEAVES. Usually 3-foliolate and much smaller than on primocanes, mostly elliptic rather than ovate and narrowed to base, or sometimes even nearly obovate, terminal one frequently 2–2½ in. long, but mostly shorter, sharply double-serrate, short-pointed; on strong floral shoots the uppermost leaves often simple and broader, on new growth often almost as broad as long and subcordate.

PRIMOCANES AND LEAVES. Primocanes often at first semierect but soon prostrate, branching, extending 6–9 ft; axis terete, glabrous, ⅙–⅕ in. thick, but very thin toward the long lashlike and almost leafless ends, prick-

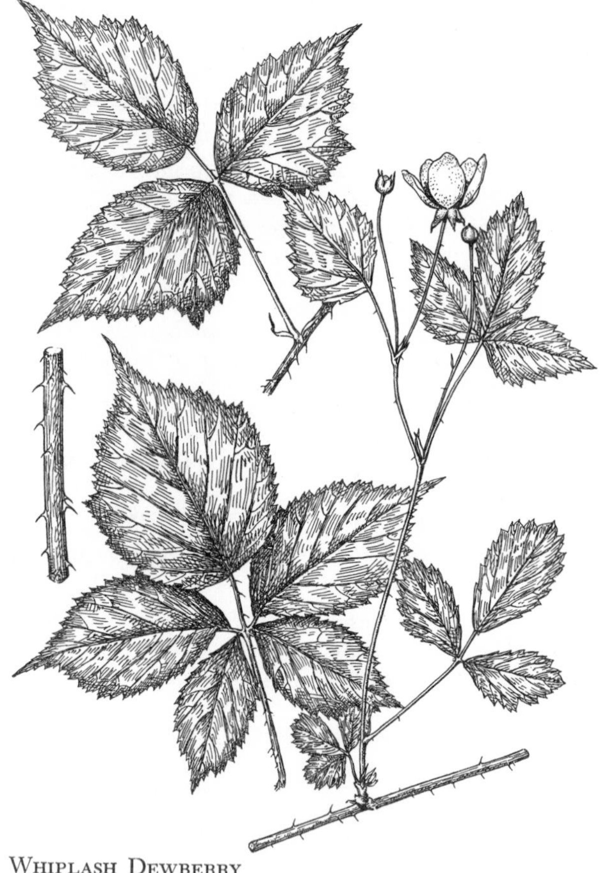

WHIPLASH DEWBERRY
Rubus flagellaris Willd.

les stout-hooked and very sharp, although short and seldom exceeding 1/8 in., or perhaps 1/6 in. in length, usually only 6–10 of them on 3/4 in. length of stem; leaves of primocanes 3–5-foliolate, firm, glabrous or essentially so on upper surface, thinly pubescent on under surface; petiole armed and either thinly pilose or glabrous; leaflets ovate or the lateral ones elliptic to rhomboid and mostly tapering to base, terminal or central one broad-based and sometimes subcordate, 2 1/3–3 1/4 in. long and 1 1/2 or more in. broad, tapering rather abruptly to a prominent narrow apex, margins strongly and doubly sharp-serrate.

RANGE. An abundant dewberry in fields, along banks and woods, woodsides, and on dry upper lands. Arkansas, Oklahoma, Texas, and Louisiana; east to Georgia, north to New Brunswick, and west to Minnesota.

REMARKS. The genus name, *Rubus*, is the old Roman name. The species name, *flagellaris*, refers to its whiplash-like habit. A full discussion of *R. flagellaris* is given by L. H. Bailey (1941–1945, pp. 244–250). He states (p. 246) that the plant occurs "in many perplexing manifestations, some of which will sometime be separated as taxonomic varieties or as species."

BOLL DEWBERRY

Rubus bollianus L. H. Bailey

FIELD IDENTIFICATION. Bramble stout, long-trailing to 15 ft, glandless, essentially glabrous. Prickles few, scattered, bent, but not hooked unless at the floricane stage, 1/25–1/12 in. long.

FLOWERS. Usually 2–3 to a floral shoot but sometimes twice as many, on ascending stout, pubescent, non-aculeate pedicels 3/8–1 5/8 in. long; petals 5, white, narrow and well separated; calyx-lobes 5, very pubescent or tomentose, lobes not extended into long points; stamens numerous.

FLORICANE LEAVES. Ample but much smaller, ovate but narrowed to base, 1 1/2–2 1/2 in. long, 1 1/4–1 5/8 in. broad, short-pointed, dentate, somewhat pubescent underneath.

PRIMOCANE LEAVES. Thin, 3–5-foliolate, glabrous on upper surface, slightly pubescent on veins underneath, margins dentate or strongly serrate-dentate; petiole sparingly pilose near base, armed; leaflets ovate, acuminate, somewhat shouldered or abrupt toward apex as in *R. flagellaris*, base sometimes subcordate, 3 1/4–4 in. long, 2–2 3/4 in. broad.

RANGE. In sandy post oak woods northwest of Grapevine, Tarrant County, and 4 miles north of Bonham, Fannin County, Texas.

REMARKS. The genus name, *Rubus*, is the old Roman name. The species name, *bollianus*, is in honor of Jacob Boll (1828–1880), Swiss naturalist, associate of Agassiz, who went to Texas in 1869 and lived in Dallas for the remainder of his life. He made important contributions on the zoology, botany, and geology of the Southwest.

L. H. Bailey (1941–1945, p. 250) comments as follows concerning *R. bollianus*:

"Texan ally of *flagellaris* but a larger, stronger plant, less armed but prickles on floricanes broad-based, leafage more abundant, leaflets larger and broader and very thin, serratures larger (serrate-dentate) and less acute, lateral ribs less pronounced on under surface of primocane leaves, floricane leaflets broader and dentate rather than serrate, flowers on stouter and relatively shorter pedicels, calyx tomentose or at least more pubescent."

NESS DEWBERRY

Rubus nessianus L. H. Bailey

FIELD IDENTIFICATION. Stout, much-branched, clambering glandless bramble with inflorescence of the Flagellares, probably at length tip-rooting, the diffuse, more or less prostrate canes sometimes 18 ft.

FLOWERS. Usually 2–5, on long ascending, very prickly pubescent, distinct pedicels; corolla 1 5/8 in. or more across, showy; petals 5, white, very broad and rounded;

stamens numerous; calyx pubescent, the 5 lobes broad and short-acuminate.

FLORICANE LEAVES. Leaflets elliptic, or the upper ones ovate and simple, 1⅝–2½ in. long, short-acute to almost obtuse or the odd one perhaps 2¾–3¼ in. long, glabrous on upper face, more or less thinly pubescent on veins underneath.

PRIMOCANES AND LEAVES. Primocane axis ⅙–⅕ in. thick, nearly or quite terete, glabrous, closely provided with stout, hard, bent, broad-based prickles ⅛–⅙ in. long; leaves 3-foliolate, probably sometimes 5-foliolate, coriaceous and sometimes persistent, margins sharply and unevenly serrate, practically glabrous on both surfaces; petiole abundantly hooked and more or less thinly pilose-pubescent; leaflets elliptic, shortly or abruptly acuminate, odd one 2¾–3¼ in. long and 1¼–1⅝ or more in. broad, narrowed to base.

RANGE. In eastern Texas; Navasota bottoms near Bryan, Brazos County; north of Saratoga, Hardin County; in pineland or red sand, east of Mineola, Wood County.

REMARKS. The genus name, *Rubus*, is the old Roman name. L. H. Bailey (1941–1945, p. 258) remarks as follows concerning the species name:

NESS DEWBERRY
Rubus nessianus L. H. Bailey

RAPID-GROWING DEWBERRY
Rubus velox L. H. Bailey

"It is satisfaction to dedicate this striking shrub to the memory of Helge Ness, born in Norway, November 4, 1861, died in Texas, December 30, 1928, Chief of the Division of Botany in the Texas Experiment Station. He originated the Nessberry . . . cross between Eubatus and Idaeobatus; also hybrid oaks, one of which is recently reported by Flory & Brison in Bulletin 612, Texas Agricultural Experiment Station."

RAPID-GROWING DEWBERRY

Rubus velox L. H. Bailey

FIELD IDENTIFICATION. Sprawling glandless Southwestern bramble with long prostrate or decumbent canes that tardily root at the tip, the inflorescence ascending.

FLOWERS. Maturing April–May, few on long pedicels from leaf axils, in an ascending cluster, bracts and stipules prominent, pedicels pubescent and sometimes bearing small remote prickles; corolla ¾ in. or more across, with 5 obtuse overlapping petals which are white and rounded; calyx-lobes 5, narrow and long, often nearly equaling the petals.

FRUIT. Maturing from late May–July, black, large, ¾ in. or more in length.

FLORICANE LEAVES. Usually 3-foliolate or those associated with the flowers simple, all elliptic-oblong and for the most part coarsely serrate or toothed, pubescent beneath.

PRIMOCANES AND LEAVES. Primocanes robust, becoming 6 ft or more long, more or less rooting at tips, prostrate or strongly arching, making slender vine-like branches, the leafy floral shoots erect, glandless, strong, angled, glabrous at or near the tip, bearing strong hooked prickles at about 6 to each ¾ in. of stem, but with some spineless intervals; leaves 5-foliolate for the most part, glabrous or practically so on upper surface, pubescent underneath, side veins close together and very sharply serrate; leaflets oblong or elliptic, acuminate to a prominent point, odd or terminal one 3¼–4 in. long and about one half as broad; petiole and terminal petiolule strong, pubescent and armed with hooks.

RANGE. L. H. Bailey (1941–1945, p. 260) comments on the range as follows:

"Texas, the range unknown. I have taken the plant on dry land in the wild in Wharton County, southeastern Texas. The plant was first known to me in the McDonald blackberry or dewberry, which was discovered near McKinney, Collin County, in northeastern Texas by the late E. W. Kirkpatrick on the farm of Mr. McDonald and introduced to cultivation early in this century. It was stated when the species was described that I expected it would be found in the wild when the Rubi of the Southwest were collected. To this species belongs the Haupt, discovered in the wild in Wharton County by W. W. Haupt, who died in 1907, also probably the Soft Core, Sonderegger, Sorsby May, Spaulding, all of Texas origin."

Collections at the University of Texas Herbarium are from Kerr, Freestone, Denton, Brazos, Montgomery, and Fannin counties.

REMARKS. The genus name, *Rubus*, is the old Roman name. The species name, *velox*, is for the swift growth of the plant.

WESTERN DEWBERRY

Rubus occidualis L. H. Bailey

FIELD IDENTIFICATION. Very leafy plant, canes long-extending and branching, rooting at tips; glandless throughout, pubescent.

FLOWERS. Maturing May–early June, ascending among the foliage of pilose-pubescent and aciculate floral shoots, few of them conspicuously overtopping the leafage; corolla usually very large, often 2 in. across at expansion; petals 5, white, broad, obtuse, overlapping at base; calyx pubescent, the 5 lobes prominently narrow-pointed, reflexed at anthesis; stamens numerous.

WESTERN DEWBERRY
Rubus occidualis L. H. Bailey

FRUIT. Short-oblong, black, ⅜–⅝ in. across.

FLORICANE LEAVES. Usually 3-foliolate or the upper ones on flowering shoots simple, much smaller than those on primocanes; leaflets ovate to elliptic, central one 1⅝–2 in. long and mostly short-pointed.

PRIMOCANES AND LEAVES. Primocanes at first erect or strongly arching, eventually prostrate and tangled, axis terete, glabrous or sparsely pilose on younger parts, stipules prominent and spreading; leaves 3–5-foliolate, thinly appressed-pilose on upper surface, soft-pubescent underneath, margins closely and doubly serrate-dentate but with tendency for teeth to become almost obtuse in certain cases; terminal or central leaflets ovate, taper-pointed, 3¼–4 in. long and 2¼–2¾ in. broad, broad to subcordate or cordate at base; lateral leaflets narrower, elliptic or elliptic-ovate, often asymmetric or indefinitely lobed.

RANGE. Dry lands in old fields, woods, hills, prairies, and hedgerows. Oklahoma, Missouri, Kansas, and Iowa.

REMARKS. The genus name, *Rubus,* is the old Roman name. The species name, *occidualis,* refers to the Western distribution of the species.

BOUNTIFUL DEWBERRY

Rubus almus L. H. Bailey

FIELD IDENTIFICATION. Very vigorous long-running, pubescent, glandless dewberry. Prostrate, repent, making dense masses of foliage, the canes extending to 6 ft or more, eventually tip-rooting.

FLOWERS. Few to several on each floral shoot, one or two from an axil, on stout ascending pilose and aculeate pedicels; corolla about 1¼ in. across; petals 5, white, rounded, broad, contiguous; calyx large, heavily pubescent, lobes 5 and prominently pointed; stamens numerous.

FRUIT. Oblong, about 1¼ in. long, firm although juicy, of good eating quality, the calyx-lobes at length spreading.

BOUNTIFUL DEWBERRY
Rubus almus L. H. Bailey

FLORICANE LEAVES. Mostly 3-foliolate or the upper ones on floral shoots simple, elliptic to ovate-pointed, sharply cut-toothed, pubescent on under surface.

PRIMOCANES AND LEAVES. Primocanes terete or nearly so, glabrous except near tips; prickles scattered, nearly straight, even if somewhat declined, not broad-based or very stout, ⅛–⅙ in. long; leaves mostly 5–7-foliolate, soft-pubescent underneath, the pilose and grooved petiole prickly; leaflets 2¾–3¼ in. long and one half or more as broad, mostly elliptic-ovate and narrowed at base although the terminal one is sometimes broad-ovate and subcordate, narrowly pointed, margins narrowly cut-toothed.

VARIETIES AND FORMS. *R. almus* is considered to be represented by the Mayes and its associates in horticulture. This question has been discussed more fully by Bailey (1941–1945, pp. 266–268).

"There is an unarmed phasis of *R. almus,* represented in the Thornless or Foster Thornless Dewberry, a pomological variety, said to have originated in Cooke County, north-central Texas, presumably as a sport in cultivation." (Bailey, 1941–1945, p. 268)

RANGE. Presumably native in northeastern Texas, but not recorded with certainty. Reported to have been found in the wild at Pilot Point, Denton County, northeastern Texas.

REMARKS. The genus name, *Rubus,* is the old Roman name. The species name, *almus,* means "nourishing."

HARMLESS DEWBERRY

Rubus nefrens L. H. Bailey

FIELD IDENTIFICATION. Bramble running, slender, glabrous, glandless, not heavily foliaged, the long thin ends rooting freely.

FLOWERS. Appearing May–June, small, on short, erect, leafy shoots from arching or prostrate floricanes, 1 or 2 to each shoot, about ¾ in. across or sometimes less, on very slender, usually nude and lightly pilose pedicels; petals 5, white, rounded, broad, overlapping; stamens numerous; calyx pubescent, the 5 very broad short-pointed lobes reflexed in anthesis.

FRUIT. Maturing mid-June–July, small, nearly globular to short-oblong, about ⅝ in. long.

FLORICANE LEAVES. Small, 3-foliolate, leaflets elliptic to oblong, hardly ovate, short-pointed or many of them obtuse, serrations small and not extending to base, the odd or central one on floral shoots usually ¾–2 in. long.

PRIMOCANES AND LEAVES. Primocane axis terete, thin, 1/12–⅛ in. thick, nearly unarmed, but frequently bearing very small curved prickles 1/25–1/12 in. long at considerable intervals; leaves small and firm, 3-foliolate or infrequently but imperfectly lobed, or 5-foliolate, margins closely but not very sharply serrate, glabrous or only minutely pubescent on veins underneath, margins usually ciliate, tapering both ways from middle, sometimes tending to be obovate, central or odd one 1½–2 in. long and about one half as broad, rather abruptly pointed, lateral leaflets broader and usually lobed on outer side; petiole sparsely aculeate.

RANGE. Banks and rocky woods, on dry land; eastern Oklahoma, northward to western Missouri and Kansas.

REMARKS. The genus name, *Rubus*, is the old Roman name. The species name, *nefrens* ("unable to bite"), refers to the relatively harmless prickles.

ARIZONA DEWBERRY

Rubus arizonensis Focke

FIELD IDENTIFICATION. Bramble very prickly, glandless, procumbent and trailing.

FLOWERS. From 1–6 on each floral shoot, on rather short ascending pilose pedicels, ⅝–¾ in. across; petals 5, white, short, obtuse and separate; stamens numerous; calyx pubescent, the 5 sepals acute, spreading, nearly or quite as long as petals.

HARMLESS DEWBERRY
Rubus nefrens L. H. Bailey

ARIZONA DEWBERRY
Rubus arizonensis Focke

FRUIT. Small, of few drupelets.

FLORICANE LEAVES. Usually 3-foliolate or the upper ones on flowering shoots simple, small, gray, ⅜–1¼ in. long, elliptic to oblong, obtuse or only briefly acute, closely but not narrowly serrate, very thinly pubescent on veins underneath.

PRIMOCANES AND LEAVES. Axes, petioles, and pedicels conspicuously armed with broad-based, very sharp prickles ⅛–⅙ in. long, axis thin and woody the second year; leaves 3–5-foliolate, thinly pilose on upper surface, pubescent underneath, the margins finely and sharply double-serrate; leaflets elliptic or nearly ovate, 1⅝–2½ in. long and more than one half as broad, pointed but hardly acuminate, the terminal one prominently stalked.

RANGE. Texas in the Chisos Mountains in Brewster County, westward through New Mexico to Arizona; in Mexico in Durango.

REMARKS. The genus name, *Rubus,* is the old Roman name. The species name, *arizonensis,* refers to the state of Arizona.

Aboriginal Dewberry

Rubus aboriginum Rydb.

FIELD IDENTIFICATION. Bramble running and tip-rooting, but often nearly erect on hard land in dry years, although short and low. Foliage glandless.

FLOWERS. In April, usually 1–4 on short floral shoots that are furnished with small 3-foliolate or sometimes simple leaves sharply serrate and mostly narrowed to base; pedicels slender, short-pilose, scantily if at all armed; corolla ¾–1 in. across, petals 5, white, obtuse, contiguous; stamens numerous; calyx small, pubescent, lobes 5 and soon reflexed.

FRUIT. Ripening in June, oblong, ⅗ in. or more long, black.

PRIMOCANES AND LEAVES. Primocanes at first erect, soon branching, glandless throughout, living tips rooting until late in the season; axis nearly or quite terete, lightly pilose when young but becoming glabrous the second year, bearing few scattered, hooked, short, broad-based prickles 1/12–⅛ in. long; leaves on old or developed primocanes 5-foliolate but on younger more vigorous shoots often 3-foliolate, softly but thinly pubescent both surfaces and on the under surface even until dropping in winter; leaflets ovate, elliptic to oval, briefly pointed, sharply double-serrate, lateral ones tapered to base, terminal one 2¾–3½ in. long and three fourths as broad, mostly rather broad and subcordate at base; petiole and strong terminal petiolule stoutly armed and sparsely pubescent.

RANGE. Big Sandy, Upshur County, northeastern Texas. Also reported from isolated localities in Oklahoma, Missouri, and Kansas.

REMARKS. The genus name, *Rubus*, is the old Roman name. The species name, *aboriginum*, means "aborigine" or "ancestor," probably because the plant was named from an early collection.

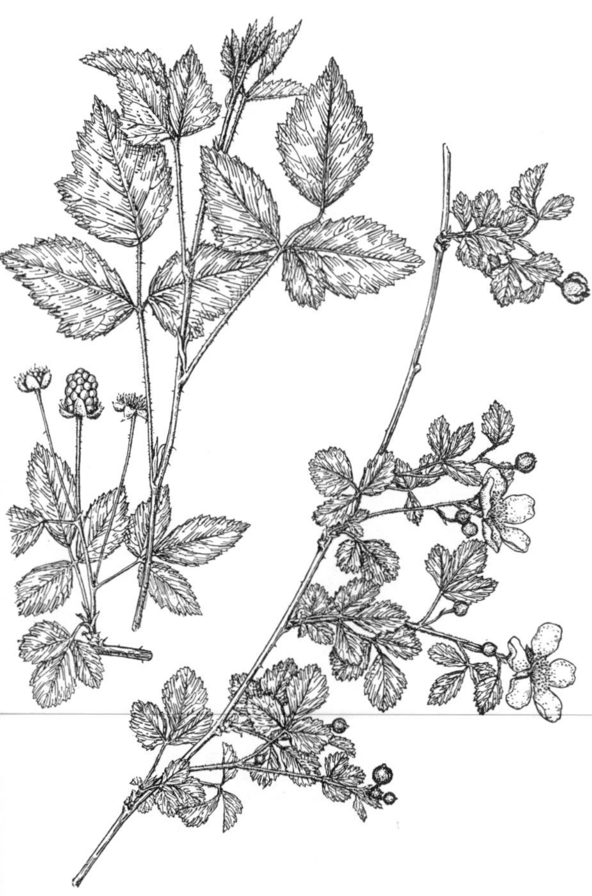

Aboriginal Dewberry
Rubus aboriginum Rydb.

Enslen Dewberry

Rubus enslenii Tratt.

FIELD IDENTIFICATION. Bramble weak, thin-leaved, scarcely woody, small, growing as single plants, erect and curving, and at length more or less trailing and tip-rooting. Glandless, unless sometimes on floral rachis and pedicels (or even on petioles and young axes), young growth erect 20–28 in. high but usually soon becoming somewhat or completely prostrate and 3 ft and more long, but in some populations remaining essentially ascending or semierect and the tips not reaching the ground.

FLOWERS. Appearing April–June. Commonly 1 to a short floral shoot, standing above the adjacent foliage and conspicuously white, frequently 1½ in. across, but for the most part smaller; petals 5, obtuse, separate; pedicels erect, thinly or stiffly pubescent and either nude or aciculate, in some populations glandless and in others conspicuously gland-bearing; stamens numerous; calyx pubescent, sometimes conspicuously reflexed at anthesis; lobes 5, narrow, slender-pointed.

FRUIT. Maturing June–August, black, short-oblong, seedy, ¼–⅜ in. or more long.

FLORICANE LEAVES. Inflorescential leaves commonly simple, ovate, often broad at base, 1⅝–2 in. long and 1¼–1⅝ in. broad; leaflets of floricane leaves (except those of parcifronds) oblong to elliptic to oval or ovate, sometimes obovate, mostly coarsely and bluntly toothed or serrate.

PRIMOCANES AND LEAVES. Axis thin, glabrous, becoming more or less woody the second year but usually appearing to be semiherbaceous, sometimes as thin as strings on the ground with nodes far apart; prickles very few and sometimes none, small and weak, not more than 1/25–1/12 in. long; leaves of primocanes various, 3-foliolate, lateral leaflets sometimes lobed, scattered-pilose on upper surface, thinly pubescent to nearly soft-pubescent underneath, margins usually acutely serrate on young canes but becoming coarsely and almost obtusely serrate or even dentate on subsequent parts

Enslen Dewberry
Rubus enslenii Tratt.

as the canes mature; petiole pilose, often spiculose and sometimes bearing a few glands.

RANGE. Mostly in dry woods. Kentucky, North and South Carolina, Virginia, and District of Columbia. Reported, but not with certainty, from Louisiana.

REMARKS. The genus name, *Rubus*, is the old Roman name. The species name, *enslenii*, is in honor of A. Enslen, nineteenth-century botanical explorer who collected in the southeastern states.

Clair Brown Dewberry

Rubus clairbrownii L. H. Bailey

FIELD IDENTIFICATION. More or less bushy trailer, with primocanes erect to decumbent. Glandless, with spines few and small, growing tips tomentose-pubescent.

FLOWERS. Mostly solitary to each shoot, but sometimes 2, small, on erect pubescent pedicels; petals 5, white, rounded; stamens numerous; calyx pubescen the 5 sharp-tipped sepals not prolonged; regular flor shoots 2–2⅜ in. long or less.

FRUIT. Globular, black, mostly ⅜ in. across.

FLORICANES AND LEAVES. Running 3–15 ft, not roo ing at tips, glandless throughout; leaflets small unles on parcifronds, ovate to obovate, mostly 1⅝ in. long c less, variously serrate or dentate, pubescent or sof pubescent underneath, apex mostly acute but som times blunt.

PRIMOCANES AND LEAVES. Axes terete, glabrous excep near tip, either nearly nude or provided with sma weak prickles ⅜–¾ in. apart; primocane leaves relativ ly large, mostly 5-foliolate, soft-pubescent underneat scantily short-pilose on upper surface, the large one coarsely and almost obtusely, but in other cases quit serrate; petiole pubescent and armed, terminal petiolu extended; leaflets ovate-acuminate, to 3¼ in. long an 2 in. broad, or the lateral ones elliptic or side-lobed.

RANGE. Cut-over pine hills near Pine Grove, S Helena Parish, Louisiana.

REMARKS. The genus name, *Rubus*, is the old Roma

Clair Brown Dewberry
Rubus clairbrownii L. H. Bailey

)eceiving Dewberry
ubus frustratus L. H. Bailey

ame. The species name, *clairbrownii,* honors Clair A. Brown, professor of botany at Louisiana State Uni-ersity.

)eceiving Dewberry

ubus frustratus L. H. Bailey

field identification. Bramble trailing and diffuse ut prostrate. Dull green or grayish, arching canes root-ng at the tips, glandless, leafage not heavy or profuse.

flowers. Appearing June–July, mostly 3–4 near top f floral shoots and 1 or 2 others in lower axils on lender, thinly pubescent and aciculate erect pedicels; orolla about 3/4 in. across, seldom 1 1/4 in.; petals 5, vhite, rounded, narrow at base and separate; stamens umerous; calyx pubescent, lobes 5 and narrow-pointed.

fruit. Maturing July–August, black, about 5/8 in. cross, of fine quality.

floricane leaves. Mostly 3-foliolate or some of hose on floral axes simple, mostly elliptic to narrow-vate, all except lowermost on the floral shoots short-acute, somewhat pubescent on veins beneath, main terminal or odd leaflet about 1 1/2 in. long and one half as broad, margins coarsely and acutely serrate or dentate.

primocanes and leaves. Primocanes at first erect but low-arching or horizontal, by end of season, axis glabrous unless at tip, more or less furrowed when dry, 1/6–1/5 in. thick, beset with sharp curved but hardly broad-based spines 1/8–1/6 in. long and 12–16 or fewer of them to 3/4 in. of stem; primocane leaves 3-5-foliolate, glabrous or with few scattered appressed hairs on upper surface, lightly pubescent beneath, unfolding leaves at tip soft-pubescent; petiole and terminal petiolule thinly pilose and sparsely aciculate; terminal or odd leaflet broad-ovate, base often subcordate, abruptly narrow-pointed, 1 3/8–1 3/4 in. long and 2–2 3/8 in. broad, lateral leaflets elliptic and often rhomboid or lobed at base, all margins doubly and rather coarsely serrate-dentate.

range. Prairies and old fields. Near Coyle, Logan County, central Oklahoma. Also from isolated localities in Missouri, and Kansas.

remarks. The genus name, *Rubus,* is the old Roman name. The species name, *frustratus,* applies to this apparently widespread, variable, and "deceiving" major species.

Southern Blackberry

Rubus austrinus L. H. Bailey

field identification. A glandless trailer with strong terete primocanes that bear scattered broad-based straight or bent prickles 1/12–1/8 in. long.

flowers. Solitary on each short lateral pedicel, elongate and equaling the foliage, mostly aculeate; petals 5, white, obovate to oval; stamens numerous.

fruit. Globular, 3/8–1/2 in. in diameter, of a few loosely coherent drupelets; calyx-lobes about 3/8 in. or less in diameter.

leaves. Primocane leaves 3–5-foliolate, glabrous on both faces, petiole and petiolule sparsely prickly; leaflets ovate to broad ovate, 2–2 1/3 in. long, 1 3/4–2 in. broad, truncate to perhaps subcordate at base, apex abruptly pointed, margins obtusely to subacutely dentate; floricane leaves 3-foliolate or 3-lobate, glabrous; leaflets elliptic-ovate, 3/8–1 1/3 in. long, 3/8–3/4 in. broad, sometimes somewhat cuneate but mostly broad at base, obtuse or nearly so at apex, margins irregularly semi-acutely serrate.

range. On sandy roadsides at edge of woods on low ground. Northern Texas, six and one-half miles south of Aubrey, Denton County.

remarks. The genus name, *Rubus,* is the old Roman name. The species name, *austrinus,* refers to this species'

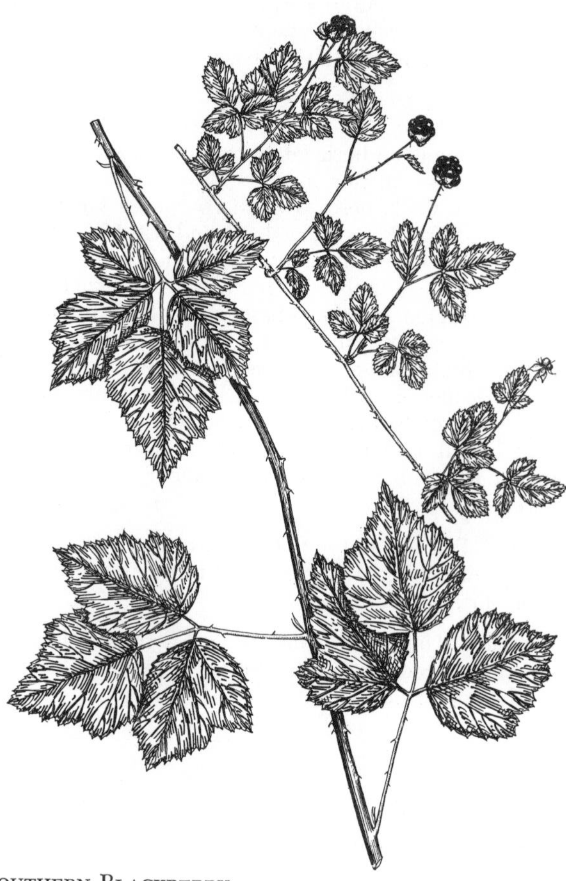

SOUTHERN BLACKBERRY
Rubus austrinus L. H. Bailey

Southern distribution. The plant lacks the common cuneate floricane leaflets of the otherwise similar *R. flagellaris* and is far out of range of that species. Additional distinguishing characteristics of the latter species include floricane leaflets blunter with more obtuse serratures, flowers solitary, pedicels shorter or stouter, fruit smaller and fewer drupelets, primocane leaflets with blunter dentations.

BAILEY BLACKBERRY

Rubus baileyanus Britt.

FIELD IDENTIFICATION. Bramble slender, glandless, low, tip-rooting, bright green.

FLOWERS. Usually 1–3, on very slender, nearly or quite glabrous, erect pedicels ¾–1⅝ in. long that usually slightly exceed the foliage; corolla ¾–1¼ in. across at expansion; petals 5, white, rather narrow and well-separated; stamens numerous; calyx-lobes 5, apiculate but not extended, reflexed, pubescent.

FRUIT. When well-grown and mature to ¾ in. long, broad-oblong, glossy black, with 15–25 fleshy succulent drupelets, sitting in an open cup of spreading-erect tomentose, briefly apiculate calyx-lobes.

FLORICANE LEAVES. Smaller, 3-foliolate or the inflorescential leaf solitary and somewhat lobed, elliptic to ovate, variable, sometimes semitruncate at base but often tapered, apex short but acute, margins closely and acutely serrate.

PRIMOCANES AND LEAVES. Primocanes terete, glabrous, very sparsely armed at maturity with little hooked prickles 1/25–1/12 in. long that sometimes stand as much as ¾–1⅝ in. apart; leaves thin, nearly or quite glabrous, those of primocanes 3-foliolate or infrequently 5-foliolate, petiole and terminal petiolule minutely aciculate; stipules prominent, not very narrow, thinly acute; leaflets ovate or the lateral ones elliptic, broadest at or below the middle, terminal one 1⅜–3¼ in. or 3½ in. long and 1⅝–2⅜ in. broad, and more or less subcordate, narrow apex beyond the main teeth ¼–⅜ in. long, margins sharply double-serrate.

RANGE. In wooded uplands, meadows, and open places, sometimes in dryish margins of swamps, often

BAILEY BLACKBERRY
Rubus baileyanus Britt.

nearly covered by other vegetation. Range not definitely known. Illinois, Michigan, Ohio, Pennsylvania, Maryland, New York, and Massachusetts. Recorded from Oklahoma and Arkansas by some authors.

REMARKS. The genus name, *Rubus,* is the old Roman name. The species name, *baileyanus,* is in honor of L. H. Bailey, who makes the following remarks concerning the plant (Bailey, 1941–1945):

"Likely to be confused with *R. flagellaris* which is a stouter more prickly plant with firmer leaves, shouldered primocane leaflets mostly not cordate and often narrowed to base, not narrow-pointed at apex, central floricane leaflets often broadest at middle or above or tending to obovate forms but sometimes simple and then perhaps broad to semicordate with notched and jagged margins, all leaves bluntly rather than acutely and openly serrate. Recent and more abundant collections will enable us to expand the definition of this species."

Lundell Dewberry

Rubus lundelliorum L. H. Bailey

FIELD IDENTIFICATION. Bramble slender, glandless, tip-rooting, with reclining more or less running canes to 15 ft long.

FLOWERS. About 5–10 on short, very slender, pubescent, naked or slightly divergent pedicels, forming a somewhat racemiform cluster; corolla 1¼ in. or more across, sometimes pinkish, the 5 petals separate and either very narrow or broad and rounded; stamens numerous; calyx closely pubescent, the 5 lobes narrow and slender-pointed.

FLORICANES AND LEAVES. Often 3 leaflets which are narrowly elliptic-acuminate to lanceolate; sometimes only a single narrow leaf at the inflorescence, or the odd or terminal leaflet long-cuneate to base and the apex hardly acute; floricanes bearing strong curved spines ⅛–⅕ in. long at intervals of ⅜–1¼ in.

PRIMOCANES AND LEAVES. Primocane axis slender, terete, glabrous, sometimes unarmed but usually the floricane with strong spines; leaves 5-foliolate, thin, practically glabrous on upper surface, pubescent to soft-pubescent underneath; petiole slightly pubescent, usually bearing a few very small hooks; stipules large, long-pointed, foliaceous; leaflets narrowly elliptic-acuminate to lanceolate, long-pointed, 2¼–3¼ in. long, and ¾–1⅝ in. broad, narrowed at base, side-ribs strongly ascending and prominent beneath.

RANGE. Eastern Texas, in mixed forest about seven miles north of Sour Lake, Hardin County; on bank of Neches River, Polk County; in bog nineteen miles southeast of Athens, Henderson County.

REMARKS. The genus name, *Rubus,* is the old Roman name, and the species name, *lundelliorum,* is in honor of C. L. and Amelia A. Lundell.

Sand Blackberry

Rubus cuneifolius Pursh

FIELD IDENTIFICATION. Bramble 3 ft high or less, sometimes taller southward, stiffly erect, harsh and recurved, spiny; roots shallow and horizontal.

FLOWERS. Appearing May–mid-July, flowers 3–5 in a little open erect cluster, ⅜–1¼ in. across; borne on pubescent slightly armed or unarmed pedicels; petals 5, white obtuse, well separated, opening wide; calyx-lobes tomentose and reflexed soon after flowering; stamens numerous.

FRUIT. Maturing July–September, globular or short-oblong, ¼–⅖ in. long, black, dry, and seedy, but usually palatable.

FLORICANES AND LEAVES. Floricanes erect and bearing short side branches; leaves similar to those of primocanes but much smaller, often less than ¾ in. long.

PRIMOCANES AND LEAVES. Primocanes usually angled when young but becoming terete and stiff, rather closely armed with stout prickles ⅛–¼ in. long, young axes

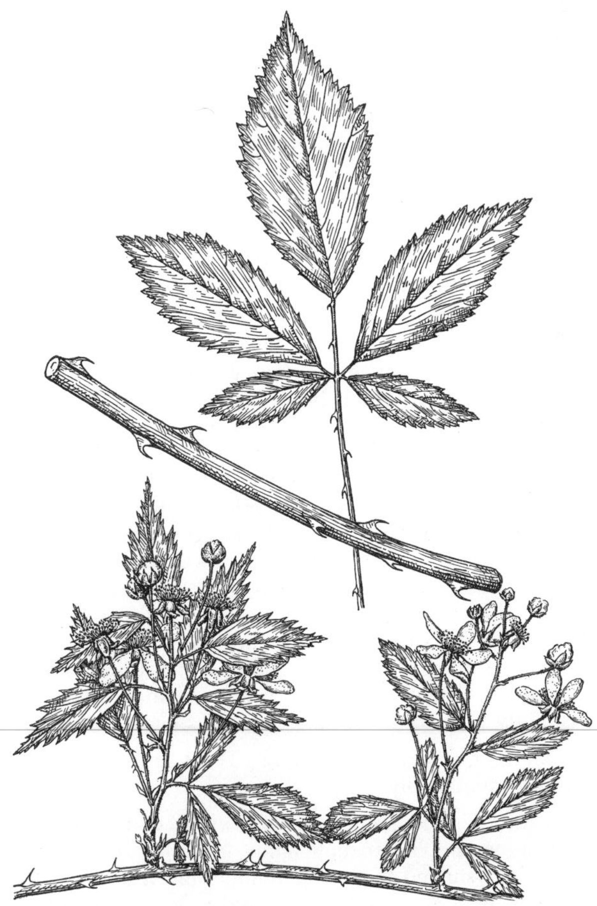

Lundell Dewberry
Rubus lundelliorum L. H. Bailey

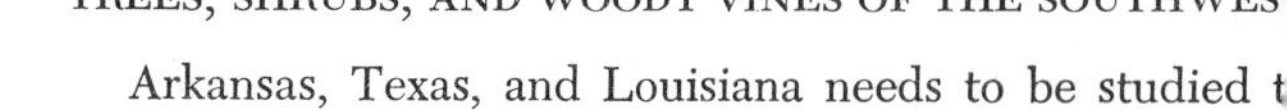

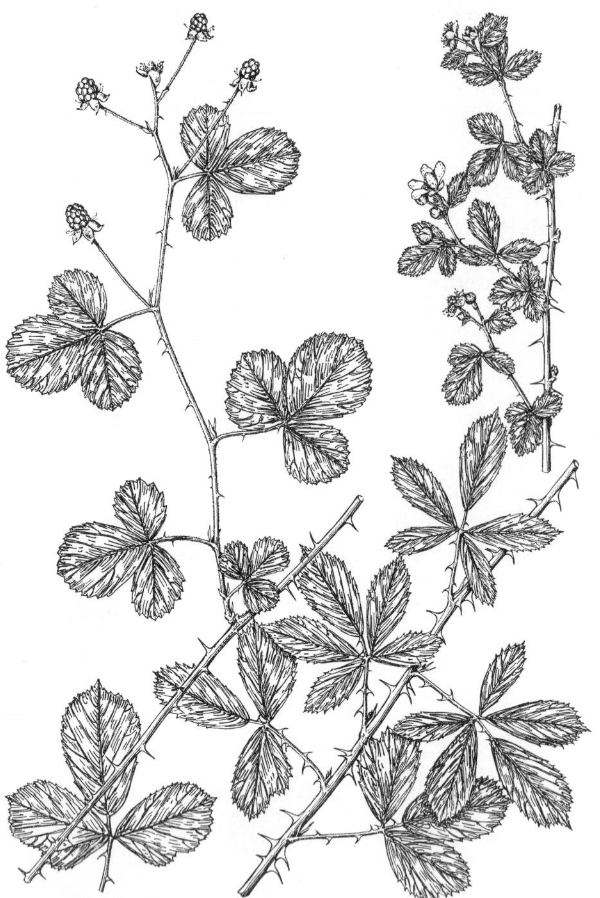

SAND BLACKBERRY
Rubus cuneifolius Pursh
Left, var. *subellipticus;* right, var. *austrifer*

gray-pubescent; leaves dark green and apparently glabrous on upper surface but actually bearing scattered short appressed hairs, markedly gray and pannose-tomentose underneath, those of primocanes commonly 3-foliate; petiole stout, pubescent, 1¼–2 in. long, commonly bearing stout recurved prickles; terminal leaflet stalked, the blade prominently cuneate-obovate, ¾–2 in. long and 1⅜ in. or less broad toward the apex, the end rounded or almost truncate to the brief central point, closely irregular serrate on the upper half or two thirds, but margin entire and likely to be revolute on the tapering lower one half or one third; lateral primocane leaflets sessile or nearly so and often less distinctly obovate in outline, usually shorter.

RANGE. Listed from Arkansas, Texas, and Louisiana, but only tentatively.

REMARKS. The genus name, *Rubus,* is the ancient Roman name. The species name, *cuneifolius,* refers to the wedge-shaped leaflets. A number of varieties of the Sand Blackberry have been assigned but most appear to be well out of the range of the present work and are therefore not listed here. Much more material from Arkansas, Texas, and Louisiana needs to be studied t determine the true status of the Sand Blackberry, eve as a species of our area.

EVERBEARING BLACKBERRY

Rubus serissimus L. H. Bailey

FIELD IDENTIFICATION. Plant prickly, stiff, glandles 2–4 ft. Canes sometimes 6 ft long and arching to th ground, but apparently not tip-rooting. Late bloomin and late fruiting.

FLOWERS. Several to many in close almost compac leafy racemes, the very short pedicels pubescent an prickly; petals 5, small, white, broad, obtuse; calyx lobes 5, short-pointed, tomentose; stamens numerous

FRUIT. Calyx either flaring or deflexed in fruit, i close clusters, body of fruit globular at maturity, intens black, with large drupes ⅗–⅘ in. in diameter eithe way.

FLORICANES AND LEAVES. Floricanes sometimes touch

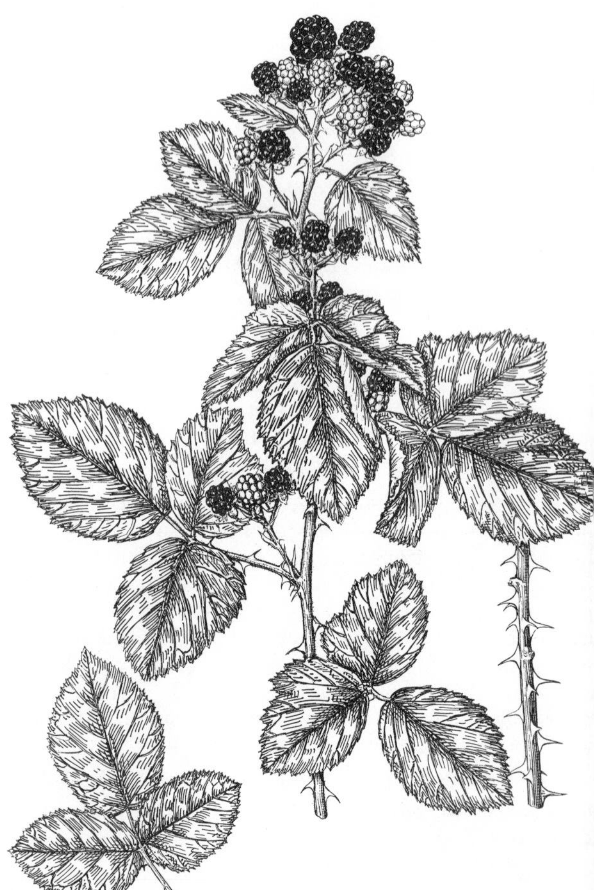

EVERBEARING BLACKBERRY
Rubus serissimus L. H. Bailey

ng the ground; leaflets similar to those of primocanes, very short-pointed to nearly obtuse or rounded.

PRIMOCANES AND LEAVES. Primocanes angled and pilose, bearing perhaps 8–12 nearly or quite straight flat-based prickles, these of different sizes, the largest ones being ⅕–⅓ in. long; primocane leaves 3-foliate, not large, the hirsute petioles and petiolules bearing strong recurved prickles; leaflets elliptic to nearly orbicular, 1¼–2 in. long, narrowed to base but not cuneate, short-pointed and not subtruncate, margins closely and doubly sharp-serrate, upper surface lightly pilose to glabrescent, under surface gray felty-tomentose.

RANGE. Cultivated, not yet collected in the wild, and nativity unknown.

REMARKS. The genus name, *Rubus*, is the old Roman name. The species name, *serissimus*, means very late flowering and fruiting. L. H. Bailey (1941–1945, p. 445) makes the following comment:

"I first met this berry in the Erwin Nursery, Denton, Texas, and herbarium material has been supplied by the active cooperation of J. W. Erwin, the proprietor; it is now growing in my garden. It is called Everbearing Blackberry, inasmuch as it has the habit of ripening its fruit three or four weeks later than other pomological varieties. . . .

"In Oklahoma it is said to be known as the Scott, from the late Mr. Scott of Ardmore who grew it in considerable quantity; also distributed as Hirschiberry, named for Mr. A. G. Hirschi of Oklahoma City. I have not seen these Oklahoma berries."

ALLEGHENY BLACKBERRY
Rubus allegheniensis Porter *ex* Bailey

ALLEGHENY BLACKBERRY

Rubus allegheniensis Porter *ex* Bailey

FIELD IDENTIFICATION. An erect or upright vigorous highbush blackberry. On good land to 3 ft, much less in hard dry fields. Taking many forms and guises in different locations. A composite plant, difficult to typify.

FLOWERS. Borne May–July, clusters racemiform, elongate, 3¼–10 in. long, extending beyond the foliage but provided with small leafage at base, pubescent and decidedly glandular, central axis continuing to the terminal short-pediceled flower, glandular pedicels ¾–1¼ in. long and markedly spreading to divaricate; flowers 12 to perhaps 30 on typical long clusters (but on outlying plants perhaps as few as 6), ¾ in. and somewhat more across, widely opening; petals 5, white, narrow and noticeably separate; calyx pubescent and glandular, the 5 lobes acute to acuminate.

FRUIT. Abundant, deep-violet, oblong conic, and ¾ in. and more long on vigorous plants, with many (50–70 or more) small drupelets, but almost globular on poorer plants, the flavor characteristically spicy and sweet.

FLORICANES AND LEAVES. Floricanes mostly retaining strong scattered prickles, sometimes pendent with fruit but normally erect; leaves of floricanes much like those on primocanes except smaller and leaflets prevailingly 3, foliage of floral shoots of short-pointed or even obtuse small leaflets.

PRIMOCANES AND LEAVES. Primocanes mostly angled but terete in less robust forms, sometimes partially zigzag node to node, bearing scattered sharp, straight or curved, broad-based prickles ⅛–⅕ in. long, glabrous for the most part except toward the top and sometimes with indefinite and variable glands; primocane leaflets mostly 5, with the upper 3 conspicuously stalked; petiole and petiolules prickly, pubescent, and glandular; blades typically oblong or oblong-acuminate to ovate-acuminate, 3¼–5 in. long (or longer on very vigorous canes) and nearly or quite one half as broad, the terminal one or three often subcordate, point almost caudate at times, rather closely serrate, thinly hairy-pubescent above and very soft-pubescent beneath, with characteristic yellowish green color.

RANGE. Dry places from lowlands to hills and mountains, open places in woodlands, along roadsides, on old fields, fence rows, and clearings. In Arkansas, Missouri,

Big Highbush Blackberry *Rubus alumnus* L. H. Bailey

and Tennessee; northward from the mountains of North Carolina, to Nova Scotia and Quebec.

REMARKS. The genus name, *Rubus*, is the old Roman name. The species name, *allegheniensis*, refers to its distribution in the Allegheny region. L. H. Bailey (1941–1945, p. 513) makes the following comment about the variability of the plant:

"Perhaps none of our highbush blackberries has so many disguises. In dry hard pastures it may reach only 2 feet high, but sometimes but 10–18 inches, yet laden with fruit. The expansive sprawly leaves may be reduced to short petiolules and abruptly pointed leaflets. Always the pubescent soft foliage remains, usually the glandular character, and the racemiform prominent clusters. Frequently the species makes large close colonies, particularly in old fields of light soil."

A number of varieties are known but are confined as far as known to the northeastern states.

Big Highbush Blackberry

Rubus alumnus L. H. Bailey

FIELD IDENTIFICATION. Tall shrub, 3–6 ft, with soft gray-pubescent leaves.

FLOWERS. Cluster short and little extending beyond the foliage, hardly racemiform, with 12 or fewer flowers on long pubescent and glandular pedicels, the lower more remote; bracts noticeable and often toothed; flowers large, to 1⅝ in. across; petals 5, white, broad, rounded, and overlapping; calyx 5-lobed, thinly pubescent and short-glandular; stamens numerous.

FRUIT. Large, to ¾ in. long or more, and about one half as thick, palatable.

FLORICANES AND LEAVES. Floricane leaflets usually 3, or reduced to 1 in the inflorescence, similar to those on primocanes except smaller and mostly less pointed, serratures often more obtuse, petiole pubescent and glandular.

PRIMOCANES AND LEAVES. Primocanes angled and usually furrowed, thinly pubescent and imperfectly glandular at least above, the stout prickles ⅕–¼ in. long and ¾–2 in. apart; leaves large, grayish, thinly hirsute above, soft-pubescent underneath, leaflets commonly 5; petiole and petiolules pubescent and usually gland-bearing, armed with stout hooks; terminal leaflet ovate and cordate to subcordate, 4–5½ in. long, short-pointed, the laterals elliptic and tapering either way, coarsely but not very sharply serrate.

RANGE. On prairies, bluffs, open and rocky woods. Arkansas, Missouri, Nebraska, and perhaps Wisconsin.

REMARKS. The genus name, *Rubus*, is the old Roman name. The species name, *alumnus*, is from the Latin, meaning a "nursling" or "foster child," but the connection is obscure.

Louisiana Blackberry

Rubus louisianus Berger

FIELD IDENTIFICATION. Glandless shrub, erect or arching, branching, with strongly furrowed canes, and straight or medium-stout scattered prickles. Usually 3–6 ft.

FLOWERS. Flowers borne April–May, large, in open leafy clusters among the foliage; petals 5, white, rounded; calyx-lobes 5, pubescent externally, tomentose inside, ovate-deltoid.

FRUIT. Borne in May, whitish, cylindric, with numerous small drupelets, sweetish; racemes pubescent, pedicels 7–8, erect, pubescent, and with a few hooked prickles.

FLORICANE LEAVES. Usually 3-foliolate, cuneate at the base, upper ones simple ovate-deltoid, short-stalked.

PRIMOCANE LEAVES. Leaflets 5, narrowly lanceolate or oblanceolate, rather long-pointed and somewhat narrower toward the base, dull green above, paler and pubescent underneath, rather regularly and sharply simple- or double-serrate; terminal leaflet larger, longer pointed and longer stalked, rounded or even subcordate at base, middle ones shortly stalked and lower ones sessile, all more or less acute at base; petiole rather stout, like the petiolules pubescent or villous and with scattered hooked prickles which extend to the midveins; stipules subulate, ciliate.

CANES. Erect or arching, sharply angled and deeply furrowed, downy when young, glabrous later on, green or greenish brown along the angles, with scattered middle-sized or large straight or hooked prickles.

RANGE. In Texas in Wood, Newton, and Shelby counties. In Louisiana in Washington, St. Helena, Palmetto, East Feliciana, and Jefferson parishes; also in Alabama; northward to Kentucky, Virginia, and Maryland.

REMARKS. The genus name, *Rubus,* is the old Roman name. The species name, *louisianus,* refers to the state of Louisiana. This species is possibly the basis of the cultivated blackberry known as the "Crystal White," or also possibly Burbank's "Iceberg" variety.

LAWLESS BLACKBERRY
Rubus exlex L. H. Bailey

LOUISIANA BLACKBERRY
Rubus louisianus Berger

LAWLESS BLACKBERRY

Rubus exlex L. H. Bailey

FIELD IDENTIFICATION. Shrub diffuse or erect and to 3 ft or more tall. Canes sulcate, older ones often erect or slightly arched, floriferous canes sometimes arching or repent in fruit, but apparently not tip-rooting. Plant glandless.

FLOWERS. Two to four in short clusters that may slightly exceed the foliage when in fruit; pedicels glandless and short pubescent, becoming divaricate and sometimes much elongate in fruit as if ascending; petals 5, white; stamens numerous.

FRUIT. Globular when well developed, to ⅗ in. or

more in diameter, drupelets many, berry black, red, or reddish when ripe, sometimes separating from calyx like a raspberry; calyx small, soon reflexed and inconspicuous under the fruit.

FLORICANES AND LEAVES. Prickles straight or bent on floricanes; leaves 3-foliolate, pubescent underneath; leaflets small, narrow-oval to lanceolate, 1¼–2 in. long, ⅜–¾ in. broad, short-acute to acuminate.

PRIMOCANES AND LEAVES. Erect and arching, prickles numerous and straight, 1/12–⅙ in. long; leaves many, firm, 5-foliolate, dark and essentially glabrous on upper face, thickly pubescent and gray on under face, margins finely and closely serrate; petiole stout, pubescent, and armed; leaflets oblong-acuminate or elliptic-acuminate, 2⅓–2¾ in. long, ¾–1¼ in. broad, base narrow.

RANGE. In an open field, East Baton Rouge, Louisiana.

REMARKS. The genus name, *Rubus*, is the ancient Roman name. The species name, *exlex* ("outside the law"), refers to the numerous prickles. The following comment has been made by L. H. Bailey (1941–1945, p. 639):

"This berry has something the look of *R. sons* (Verotriviales) but apparently is not tip-rooting, is much less prickly, primocane leaves soft-pubescent underneath, flowers fewer in scantier clusters that do not stand above the leaves, foliage probably not persistent over winter, serration much more minute, fruit more globular and pulpy."

DEMAREE BLACKBERRY
Rubus demareanus L. H. Bailey

DEMAREE BLACKBERRY

Rubus demareanus L. H. Bailey

FIELD IDENTIFICATION. Bramble to 3 ft on sterile sites or to 9 ft in swampy places. Canes to 12 ft or more long lodging in trees on creek bottoms, erect and stiff or in the robust occurrences arching over until tips may touch the ground.

FLOWERS. In short leafless lateral racemiform clusters, usually with 8–10 flowers or 1–2 separated flowers at base; pedicels ⅜–¾ in. long and unarmed but hairy-pubescent; corolla small, ⅝–2 in. across; petals 5, narrow; calyx-lobes small, narrow, apiculate; stamens numerous.

PRIMOCANES AND LEAVES. Terete, not very stout even when long, commonly not more than ¼–⅖ in. thick; prickles small and widely spaced, straight or nearly so, not much exceeding ⅕–¼ in. long, practically absent on some canes; leaves small, thin, 5-foliolate, nearly or quite glabrous on upper surface at maturity, soft-pubescent underneath, central leaflet and sometimes the upper lateral ones prominently petioluled, petiole bearing weak, more or less hooked, prickles; leaflets broadly lanceolate to nearly ovate-lanceolate, central or terminal one 2¾–3⅔ in. long and 1¼–1⅝ in. broad and sometimes abrupt at base, acuminate to apex, finely sharp-serrate.

RANGE. In southeastern Arkansas in Drew and Ashley counties.

REMARKS. The genus name, *Rubus*, is the old Roman name. The species name, *demareanus*, is in honor of Delzie Demaree of the Arkansas Agricultural and Mechanical College.

L. H. Bailey (1941–1945, p. 640) remarks on the species as follows:

"One of the *argutus* associates, differing from that species in the greater stature under prime conditions, less armature, less sharply and deeply toothed leaf margins, more racemiform clusters on shorter lateral branchlets, mostly smaller flowers, floral leaves or leaflets not tending to obovate or oblanceolate shapes."

PERSISTENT BLACKBERRY

Rubus persistens Rydb.

FIELD IDENTIFICATION. Shrub 3–6 ft tall, canes rather upright, entangled, and glandless. Branches numerous, horizontal or arching, but not trailing or tip-rooting; sometimes making close clumps.

FLOWERS. Flowers showy, 5–8, in small, short, racemiform clusters little if at all exceeding the foliage; the brief pedicels nude or carrying a few prickles; inflorescential leaflets or simple leaves hardly broader than long and most of them obtuse or only short-acute, not really acuminate; corolla 1–1¼ in. across, petals 5, obovate, obtuse, and soon more or less reflexed; calyx-lobes 5, oblong, not attenuate, but with a brief point.

FLORICANES AND LEAVES. Prickles becoming hooked, broad-based and about ⅛ in. long; leaflets small and nearly glabrous underneath at maturity, most of them nearly obtuse.

PRIMOCANES AND LEAVES. Prickles numerous and becoming hooked on old canes; canes essentially glabrous except leaves at first thinly pubescent underneath but only sparsely pilose on veins at full anthesis, more or less shining above; large primocane axes angled but branches essentially terete; primocane leaves not large, 3–5-foliolate, at first pilose above but becoming glabrous, petiole more or less armed, margin irregularly and mostly sharp-serrate; leaflets broad-lanceolate to ovate-lanceolate, terminal one 2¾–3½ in. long when full grown, 1¼–1¾ in. broad.

PERSISTENT BLACKBERRY
Rubus persistens Rydb.

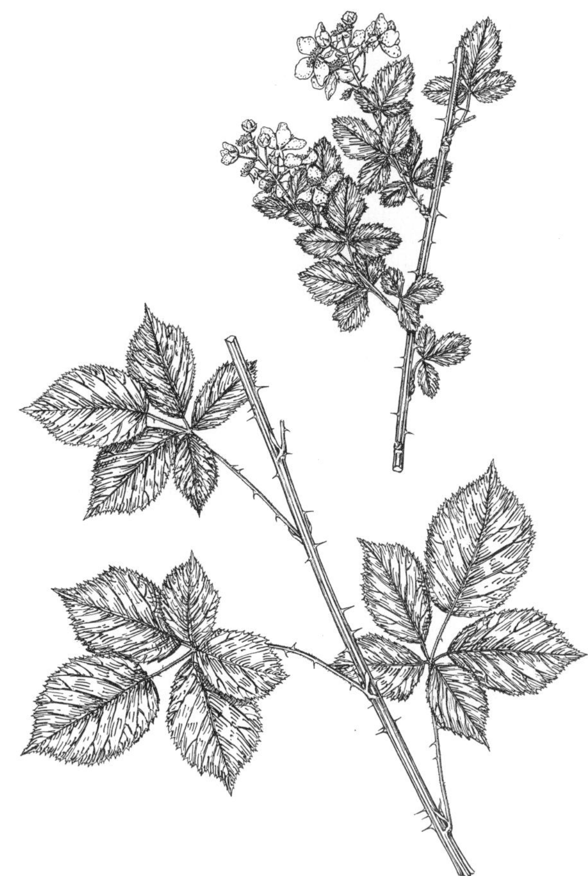

SOFT-LEAF BLACKBERRY
Rubus mollior L. H. Bailey

RANGE. Now considered distributed from southern Mississippi to western Florida. Recorded by some authors as from Louisiana and east Texas, but not without reservations. A single Texas specimen, deposited in the Herbarium of the University of Texas, was collected at Houston, Harris County.

REMARKS. The genus name, *Rubus,* is the old Roman name. The species name, *persistens,* is for the persistent habit of the plant.

SOFT-LEAF BLACKBERRY

Rubus mollior L. H. Bailey

FIELD IDENTIFICATION. Shrub often making large stands. Erect, stiff, branching, 3–4½ ft, glandless. Prickles few, but large at maturity and hooked.

FLOWERS. About 8–10 in a racemiform cluster with continuing axis and provided with small, short, obtuse or briefly acute leaflets; pedicels glandless, hirsute and unarmed; corolla ¾–1¼ in. across; petals 5, spatulate, obtuse and well separated; calyx-lobes 5, pubescent-tomentose outside, broad, short-pointed, becoming reflexed; stamens numerous.

FRUIT. Short-oblong, not described in its mature stage.

PRIMOCANES AND LEAVES. Primocane axis somewhat furrowed and bearing scattered prickles that become ⅕–¼ in. long, broad-based, and curved; leaves rather large and heavy, 5-foliolate, becoming glabrous on upper face but remaining soft-pubescent underneath; margins irregularly serrate-dentate.

FLORICANES. Axis nearly or quite terete, the strong prickles straight or hooked.

RANGE. Dry lands in Arkansas in Washington County near Fayetteville; also Newton and Conway counties. In eastern Oklahoma in Pittsburg County.

REMARKS. The genus name, *Rubus,* is the old Roman name. The species name, *mollior,* refers to the soft-pubescent leaves.

Field Blackberry

Rubus arvensis L. H. Bailey

FIELD IDENTIFICATION. An erect, prickly, common field blackberry. Attaining a height of 6 ft or more, with age forming large clumps and often occupying extensive areas, becoming much-branched and tangled, variable in size, habit, and leaf shape.

FLOWERS. In June, 8–12 on short pubescent, usually armed, divaricate, glandless pedicels, in small raceme-like clusters with central axis about equaling or a little exceeding the associated leafage, the bracts small but conspicuous; corolla about ¾ in. across; petals 5, separate, obtuse or emarginate; calyx-lobes 5, very pubescent, acuminate, later reflexed; stamens numerous.

FRUIT. May–June, deep-violet to black, small to medium in size, oblong to globular, pedicels usually elongated to beyond the foliage, seedy, ⅗ in. or less long or broad; stamen remnants persistent and calyx-lobes reflexed.

FLORICANES AND LEAVES. Sometimes at length falling or bending with weight, terete; prickles prominent but not very closely placed, curved, ⅛–⅕ in. long; leaves of floricanes conspicuously smaller, 3-foliolate or in the clusters simple, nearly or quite glabrous above, pubescent underneath, serratures often coarse, leaflets narrowly oblong-acute, ¾–1⅔ in. long, ⅜–⅝ in. broad, but sometimes much broader and nearly obtuse on the same plant, rather closely placed on the flowering parts.

PRIMOCANES AND LEAVES. Commonly angled or sulcate; prickles prominent, curved, not conspicuously broad-based; leaves 5-foliolate, sparsely pubescent on upper face, velvety-pubescent on lower face, margins irregularly but closely sharp-serrate; petiole long and armed with hooks; leaflets oblong or ovate-oblong, acuminate, 2¾–3½ in. long, 1¼–1⅝ in. broad (often much narrower on new growing tips), base of central one obtuse and often subcordate.

Field Blackberry
Rubus arvensis L. H. Bailey

RANGE. A prevailing highbush blackberry in old fields, abandoned waste lands, cut-over areas, along fences and woods margins, swamp borders, sometimes transplanted to gardens. In Texas in Smith, Harris, Montgomery, Bowie, Red River, Jefferson, and Freestone counties. In Louisiana in Washington, St. Tammany, Baton Rouge, Orleans, Jefferson, Terrebonne, Iberia, Rapides, Caldwell, Lincoln, and Webster parishes.

REMARKS. The genus name, *Rubus,* is the old Roman name. The species name, *arvensis,* is for the "old field" locations where it grows.

Abundant-flower Blackberry

Rubus abundiflorus L. H. Bailey

FIELD IDENTIFICATION. A strong briar making large, dense, tall clumps or colonies. The mounds sometimes to 9 ft, but much smaller on poor terrain.

FLOWERS. Flowers mostly 4–6, rather large under

good conditions, being perhaps 1¼ in., but often much smaller in broad clusters closely subtended by the many very small leaves; pedicels short and ascending, sometimes bearing a few short prickles; petals 5, narrow, well separated, obtuse; calyx-lobes 5, small, short-pointed, not reflexed at anthesis; stamens numerous.

FLORICANES AND LEAVES. Floricane with prickles mostly curved; leaves similar to those of primocanes but much smaller, those on the inflorescence simple or 3 leaflets ⅜–1¼ in. long, soft pubescent underneath, narrow.

PRIMOCANES AND LEAVES. Prickles on primocanes ⅜–1¼ in. apart, straight or nearly so, ⅙–⅕ in. long, somewhat expanded at base, becoming stout and curved or hooked on floricanes; leaves 3–5-foliolate, thinly pilose above, at least when young, pubescent underneath and midrib often armed, margins rather coarsely and unevenly double-serrate; petiole bearing hooks; leaflets lanceolate or elliptic to oblong and acuminate, rather small, tapering to base, 2–2¾ in. long, about ¾ in. broad.

RANGE. Dry and semimoist land, fence rows. Louisiana, eastward to Florida.

ABUNDANT-FLOWER BLACKBERRY
Rubus abundiflorus L. H. Bailey

WITHDRAWN BLACKBERRY
Rubus summotus L. H. Bailey

REMARKS. The genus name, *Rubus*, is the ancient Roman name. The species name, *abundiflorus*, means "abundant-flowered." L. H. Bailey (1941–1945, p. 705) makes the following comment:

"This briar, *Rubus abundiflorus*, is probably one of the major species of the southeastern United States, even though not yet generally recognized. Its habit of forming dense clumps when undisturbed in old fields and margins is a distinguishing quality. From *R. penetrans* it is known at once also by the finger-soft pubescence of the under surface of the leaves, much shorter and broader primocane leaves and leaflets; from other species by the narrower primocane leaflets and particularly by the smaller flowers that are not covered by larger broad ovate dull leaves and leaflets."

WITHDRAWN BLACKBERRY

Rubus summotus L. H. Bailey

FIELD IDENTIFICATION. Bramble erect, to 6 ft, glandless; with strongly angled drooping canes, apparently not tip-rooting.

FLOWERS. Usually 5–9 on ascending slender, thinly pubescent, prevailingly unarmed pedicels, making an open more or less corymbiform cluster that about equals the foliage; corolla ¾ in. or more across, petals narrow and separate, more or less pinkish; calyx-lobes 5, oblong, acute, soon reflexing; stamens numerous.

FLORICANES AND LEAVES. Main floricane leaves similar to those on primocanes, but those on floral branches narrow and more or less obtuse, tapered to base, serratures mostly obtuse, 1¼–1⅝ in. or less long.

PRIMOCANES AND LEAVES. Prickles scattered, straight on young primocanes but broad-based, more or less curved, on floricanes, ⅛–⅙ in. long; leaves of primocanes 3-foliolate (perhaps 5-foliolate on vigorous canes), light-colored, glabrous or soon glabrous on both surfaces, margins coarsely but not sharply serrate-dentate, petiole glabrous and scantily armed; leaflets (young) oblong to elliptic or sometimes narrow-ovate, short-pointed to nearly obtuse, base broad or narrowed, 2–2⅓ in. long, nearly or quite 1¼ in. broad.

RANGE. Uplands near Blanchard, McClain County, central Oklahoma.

REMARKS. The genus name, *Rubus*, is the old Roman name. The species name, *summotus*, means "withdrawn," because this species has been segregated from *R. oklahomus*.

Leaning Blackberry

Rubus saepescandens L. H. Bailey

FIELD IDENTIFICATION. Bramble dependent, scandent on shrubs or tree branches. The very long furrowed canes sometimes 50 ft, only remotely prickly.

FLOWERS. Ascending, rather few (6–10), in a broad corymbiform cluster that usually somewhat exceeds the accompanying foliage, borne on short lateral leafy spurs; pedicels slender, ascending or divaricate, about ¾ in. long at anthesis, pilose, weakly armed, glandless, sometimes forked; corolla about ¾ in. across, petals 5, white, narrow, obtuse, and well separated.

FLORICANES AND LEAVES. Floricane leaves 3-foliolate or the upper floral ones simple, blades oblong, ovate-oblong, obovate, short-pointed, 1½–2¾ in. long, ⅜–1¾ in. broad.

PRIMOCANES AND LEAVES. Primocanes very thin for the great length, usually less than ⅜ in. thick, commonly with one or two strong ridges, conspicuously glabrous, bearing a few straight flat-based prickles ¾–1¼ in. and more apart and $1/12$–⅛ in. long; primocane leaves mostly 5-foliolate, nearly or quite glabrous on upper face, soft-pubescent underneath, margins finely irregular-serrate, petiole armed; leaflets lance-oblong to elliptic, acuminate, 2¾–3½ in. long, 1½–2 in. broad, central one broad at base.

RANGE. Eastern Texas and southern Louisiana. In Texas near Port Arthur and Beaumont in Jefferson County. Louisiana near Cypremont Point, Iberia Parish.

REMARKS. The genus name, *Rubus*, is the ancient Roman name. The species name, *saepescandens*, means "often scandent."

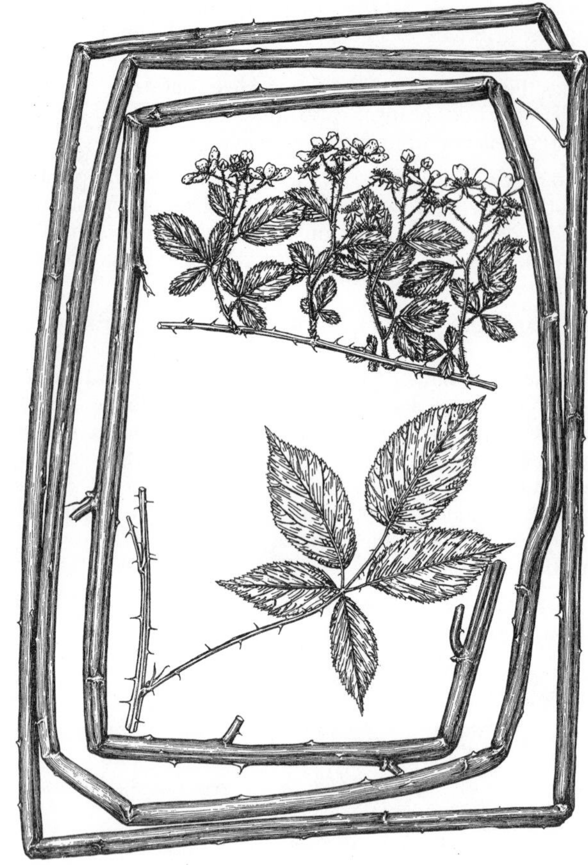

Leaning Blackberry
Rubus saepescandens L. H. Bailey

Highbush Blackberry

Rubus ostryifolius Rydb.

FIELD IDENTIFICATION. Bramble ascending to nearly erect, usually weak, some of the less prickly canes declined or even nearly horizontal, and very slender as if belonging to a trailer, but these parts not repent nor tip-rooting. Plant standing or mounding to 4½ ft.

FLOWERS. Borne April–May, large, 5–10 of them in a loose cluster about as broad as long and equaling or briefly exceeding the leaves; pedicels sometimes with very weak glandular hairs; corolla 1¼–1⅝ in. broad, the 5 petals white, obtuse, widely spreading; stamens very prominent; calyx-lobes 5, narrowly acuminate, widely spreading under the fruit.

FRUIT. Maturing May–June, black, short-oblong, small, loosely seeded, not pulpy, ⅖–½ in. long.

FLORICANES AND LEAVES. Floricane nearly or quite terete, prickles few and scattered; leaves usually 3-foliolate, sparsely pilose to glabrous on upper face, soft pubescent underneath, leaflets elliptic, commonly tapering both ways, varying from short-pointed to almost obtuse, 2¾–3¼ in. long, 1⅝–2 in. broad, finely serrate.

PRIMOCANES AND LEAVES. Sulcate to almost terete, prickles few and scattered but becoming ⅛–⅙ in. long with expanded base; leaves 3–5-foliolate, pilose but becoming nearly glabrous above, soft-pubescent to the touch underneath when young, less so with age; petiole and petiolules weakly and sparsely armed and pubescent; leaflets narrowly ovate-acuminate, 3½–4½ in. long, 2–2½ in. broad, point of the 3 upper ones narrow and projected, base sometimes subcordate, margins finely, acutely, and almost evenly serrate.

RANGE. Described by Bailey as in woodsy or bushy habitat near the sea at Atlantic Highlands, Monmouth County, east central New Jersey. Recorded by some authors from Arkansas, but perhaps this location should be checked in view of more recent study.

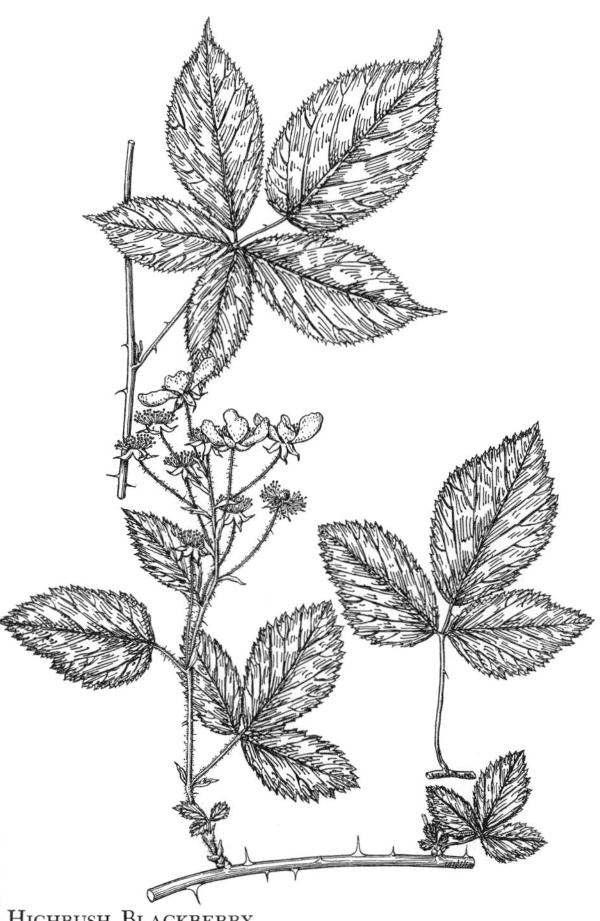

HIGHBUSH BLACKBERRY
Rubus ostryifolius Rydb.

OKLAHOMA BLACKBERRY
Rubus oklahomus L. H. Bailey

REMARKS. The genus name, *Rubus*, is the old Roman name. The species name, *ostryifolius*, is for the resemblance of the leaves to those of the Hornbeam-tree *(Ostrya)*. A full account of the taxonomy of this species is given by Bailey (1941–1945, pp. 726–729).

OKLAHOMA BLACKBERRY

Rubus oklahomus L. H. Bailey

FIELD IDENTIFICATION. A very prickly shrub, canes upright but arching, glandless, to about 3 ft. Branches long, slender, weak, horizontal or depressed, some of which probably take root at the tip.

FLOWERS. Few, usually 3–5, projecting on short laterals that are leafy at base, pedicels pubescent, ascending and armed; corolla large, about 1¼ in. across, the 5 petals very broad and rounded; calyx-lobes 5, very broad, apiculate or sometimes with foliaceous tips, becoming reflexed; stamens numerous.

FLORICANE LEAVES. Leaflets much smaller than those

on primocanes but otherwise similar, upper ones in flower cluster perhaps simple.

PRIMOCANES AND LEAVES. Canes terete for the most part at maturity, but sometimes angled; prickles straight or bent, ⅛–¼ in. long, broad-based, branchlets and petioles also armed; leaves 3–5-foliolate, thinly pubescent on upper face, more evidently pubescent and gray underneath, margins obtusely apiculate; leaflets broadly oval or the lower pair rhomboid, obtuse or nearly so at apex, rounded or expanded at base, about 2¾ in. long and 1⅝ in. or more broad.

RANGE. Low places, near Guthrie, Logan County, Oklahoma.

REMARKS. The genus name, *Rubus,* refers to an old Roman name. The species name, *oklahomus,* is for the state of Oklahoma.

Branched Blackberry
Rubus ramifer L. H. Bailey

FIELD IDENTIFICATION. Upright, much-branched and intertangled bramble to 3 ft. The slender, glandless branches with stout hooked prickles on the frondiferous and floriferous branches.

FLOWERS. In May, flowers few (4–5), on ascending or erect, pubescent, glandless pedicels, very little exceeding the associated leaves; corolla 3 in. or less broad at full expansion, white or pink; petals 5, obtuse; stamens numerous; calyx-lobes 5, small, finally completely reflexed.

FRUIT. Globular to oblong, black, on elongate pedicels about ⅗ in. long, many-seeded, the drupes small.

FLORICANES AND LEAVES. Prickles hooked and broad-based on the floricane parts; leaves of floricanes small (unless on parcifronds), 3-foliolate, leaflets of flowering branchlets variable, ⅜–1¼ in. long, ¼–⅝ in. broad, oval to oblong to lanceolate to obovate, long-acute to obtuse on the same plant, mostly glabrous above, pubescent beneath.

PRIMOCANES AND LEAVES. Prickles straight or hooked on the primocanes and ⅛–⅕ in. long; leaves of primocanes 3–5-foliolate, thinly pubescent to glabrous above, soft-pubescent underneath, margins rather coarsely and unevenly dentate-serrate; leaflets oblong to ovate-oblong, to somewhat obovate, 3¼–4 in. long, by 1½–2 in. broad, sharply or abruptly acuminate, some of them more or less subcordate at base.

RANGE. Near Huntington in Angelina County, eastern Texas. Also 5½ miles south of Carthage in Panola County.

REMARKS. The genus name, *Rubus,* refers to an old Roman name. The species name, *ramifer,* is for the branch-bearing habit.

Fruitful Blackberry
Rubus fructifer L. H. Bailey

FIELD IDENTIFICATION. An erect to arching, glandless shrub, to 3 ft or more. Prickles rather numerous and medium in size.

FLOWERS. From 3–5 on short lateral leafy-based shoots, on upright or ascending pubescent, mostly weakly armed, pedicels in ascendate clusters that somewhat exceed the accompanying foliage; petals 5, white; stamens numerous.

FRUIT. Globular, fleshy, juicy, black, ⅜–½ in. thick; drupelets few and large; calyx-lobes broad, spreading or even ascending under the fruit.

FLORICANES AND LEAVES. Fruiting canes often horizontal from weight but not trailing, somewhat angled, prickles usually many, straight or somewhat slanting, broad based, ⅙–⅕ in. long; prickles and 3-foliolate leaves similar to those on primocanes.

PRIMOCANES AND LEAVES. Somewhat angled, prickles straight to somewhat slanting and broad-based; leaves

Branched Blackberry
Rubus ramifer L. H. Bailey

FRUITFUL BLACKBERRY
Rubus fructifer L. H. Bailey

3–5-foliolate, nearly or practically glabrous on upper face, thick-pubescent on lower face, margin rather closely and not coarsely dentate-serrate; petiole stout, pubescent, and usually armed; leaflets oblong, elliptic-ovate or narrowly ovate, broad and the central one perhaps subcordate at base, apex short-pointed; terminal or central leaflet 2–2½ in. long and 1¼–1⅔ in. broad.

RANGE. Known from uplands near McCurtain, Haskell counties, eastern Oklahoma.

REMARKS. The genus name, *Rubus,* is the old Roman name. The species name, *fructifer,* refers to this species' fruitful character.

DISCERNIBLE BLACKBERRY

Rubus scibilis L. H. Bailey

FIELD IDENTIFICATION. Plant usually low, under 3 ft, glandless, leaflets small, prickles small and straight. Canes erect, but curving or arching the second year, and terete.

FLOWERS. Few (3–8), in an open ascendate cluster, little if at all surpassing the adjacent leaves, on slender, pubescent, unarmed and often spreading pedicels; corolla ¾ in. or less broad, the 5 white petals broad, obtuse and overlapping; calyx-lobes 5, briefly acute, reflexing in full anthesis; stamens numerous.

FLORICANES AND LEAVES. Floricane prickles scattered, straight, slanting, not broad-based, ⅛–⅙ in. long; leaves 3-foliolate, smaller, more gradually acute, elliptic, more finely and acutely serrate.

PRIMOCANES AND LEAVES. Prickles the same as on floricanes; primocane leaves 5-foliolate, sometimes 3-foliolate, soon glabrous on upper surface, pubescent on lower face, margins acutely serrate-dentate, petiole scantily pubescent to glabrous and weakly armed; leaflets (young) oblong-ovate to indifferently obovate, 2⅓–2¾ in. long, 1¼ in. and somewhat more broad, abruptly pointed, either broad or narrow at base.

RANGE. Known from uplands near Ochelata, Washington County, northeastern Oklahoma.

REMARKS. The genus name, *Rubus,* is the old Roman

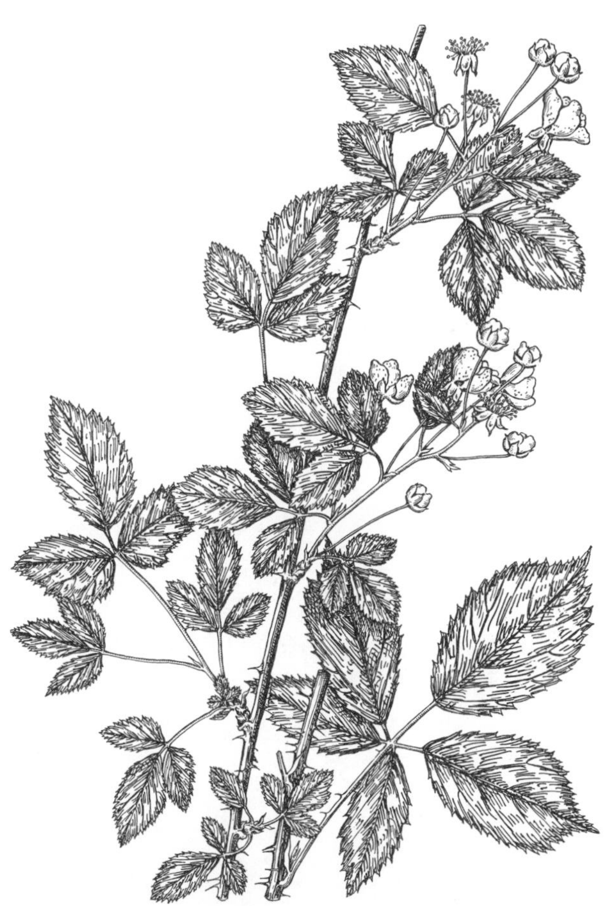

DISCERNIBLE BLACKBERRY
Rubus scibilis L. H. Bailey

name. The species name, *scibilis,* means "discernible," referring to the fact that this species is now distinguished from *R. oklahomus* Bailey.

Bush's Blackberry

Rubus bushii L. H. Bailey

FIELD IDENTIFICATION. A prickly, leafy, strictly erect shrub, low, usually less than 3 ft, from a horizontal root.

FLOWERS. Large, 3–6 per axis, ascending, pubescent, pedicels nude, forming a broad (not racemiform) short cluster among the leaves; other flower-bearing axes may be axillary on long leafy parcifronds; corolla variable, ¾–1⅝ in. across, petals 5, obtuse and separate; calyx-lobes 5, prominently acuminate, soon reflexed; stamens numerous.

PRIMOCANES AND LEAVES. Prickles on primocanes straight or bent, ⅛–⅕ in. long, many of them persisting on floricanes; leaves 3-5-foliolate, thick, slightly pilos above the midrib, densely soft-pubescent underneath margins closely and sharply serrate, veins prominent petiole pubescent and armed with hooks; leaflets ellip tic-ovate to oblong, abruptly acute, terminal one mor or less subcordate at base, 3–4 in. long, 1½–2½ in broad.

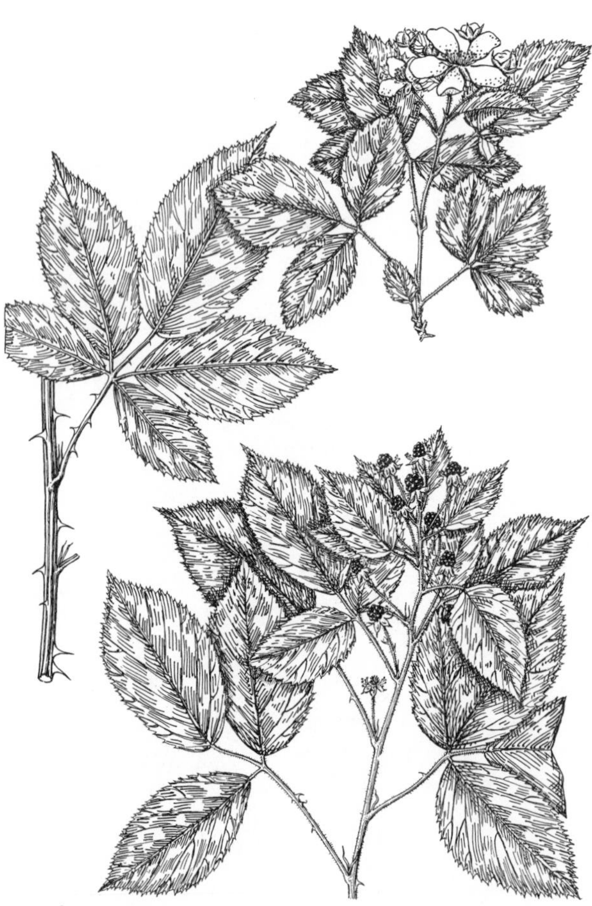

Bush's Blackberry
Rubus bushii L. H. Bailey

FLORICANES AND LEAVES. Leaflets mostly oval to ob long or the terminal one more or less obovate, coarsel serrate, 1¼–2 in. long; leaves associated with inflores cence usually simple and not very small.

RANGE. On rocky prairies near Webb City, Jaspe County, Missouri. Also recorded from Arkansas by som authors.

REMARKS. The genus name, *Rubus,* is the old Roma name. The species name, *bushii,* is in honor of B. F Bush, botanist of Missouri.

Pure Blackberry

Rubus putus L. H. Bailey

FIELD IDENTIFICATION. Bush prickly, diffuse o arched, upstanding, much-branched, usually 3–5½ ft

FLOWERS. April–May, mostly exceeded by the as sociated leafage, in more or less short ascending cluster which are not numerous (3–5); pedicels short, pubes cent, glandless, unarmed, ascending; corolla upwar of ¾–1 in. broad at full anthesis; petals 5, white, broa and rounded.

FLORICANES AND LEAVES. Floricanes usually furrowe at first, prickles ⅜–1¼ in. apart; leaves small, 3-foliolate or some of them simple, blades oval to ovate or broa ovate, 1¼–2 in. long, variable in width to 1⅔ in., soo glabrous above, decidedly pubescent underneath, most ly short-acute, either tapering or truncate at base strongly dentate-serrate or somewhat notched.

PRIMOCANES AND LEAVES. Canes mostly terete th second year; prickles not very numerous, mostly ⅜–1¼ in. apart on floricanes, 1/12–⅛ in. long, curved; leave of primocanes (young) 3–5-foliolate, thinly pubescen above, soft-pubescent underneath, margins conspicu ously doubly and acutely dentate-serrate, petiole wit curved or hooked prickles, or straight on young primo canes; leaflets ovate to ovate-oblong, 2½–2¾ in. long 1¼–2 in. broad, ribs brownish underneath and promi nent, short-acuminate, central one more or less subcor date.

RANGE. On dry ground, in northeastern and centra Texas. At Pilot Point in Denton County, and near Lak Lytle, Taylor County; also in Johnson and Tarran counties.

REMARKS. The genus name, *Rubus,* is the old Roma name. The species name, *putus,* is for the "pure" o unmixed strain. The plant is apparently a scant thoug conspicuous bloomer.

HARP-TOOTH BLACKBERRY

ubus argutus Link

FIELD IDENTIFICATION. Bramble erect or upright, but nore or less diffusely branching, and 3–4½ ft tall. The anes with scattered straight but not very large prickles.

FLOWERS. In April, usually 5–12 in a short somewhat acemiform cluster terminating brief leafy lateral shoots r on shoots from short branches of canes; corolla usually not exceeding 1 in. across, but variable in size on difering terrains; petals 5, white, rounded; stamens nunerous; calyx pubescent, lobes 5, acute but not elongate.

FRUIT. May–June, about ¼–⅔ in. in length, conic or arrow-oblong, black, seedy.

PRIMOCANES AND LEAVES. Primocanes slender, flexible, he axes thin, sometimes falling and prostrate; prickles cattered, straight, ⅕ in. or less long, not conspicuously road-based; primocane axis only obscurely angled vhen dry; leaves relatively small, 3-5-foliolate, dull reen above, lightly pubescent on veins underneath, even less so on intervening surface, and the midrib sparingly or not at all aculeate; petiole slender, hairy with relatively few (sometimes many) hooked prickles; leaflets broad-lanceolate to lanceolate, acuminate-pointed and 2½–3½ in. long, 1¼–1⅝ in. broad, base of leaflets typically tapering, but sometimes somewhat broadened, margins irregularly, doubly, and acutely serrate, the teeth 1/12–⅙ in. long.

SHARP-TOOTH BLACKBERRY
Rubus argutus Link

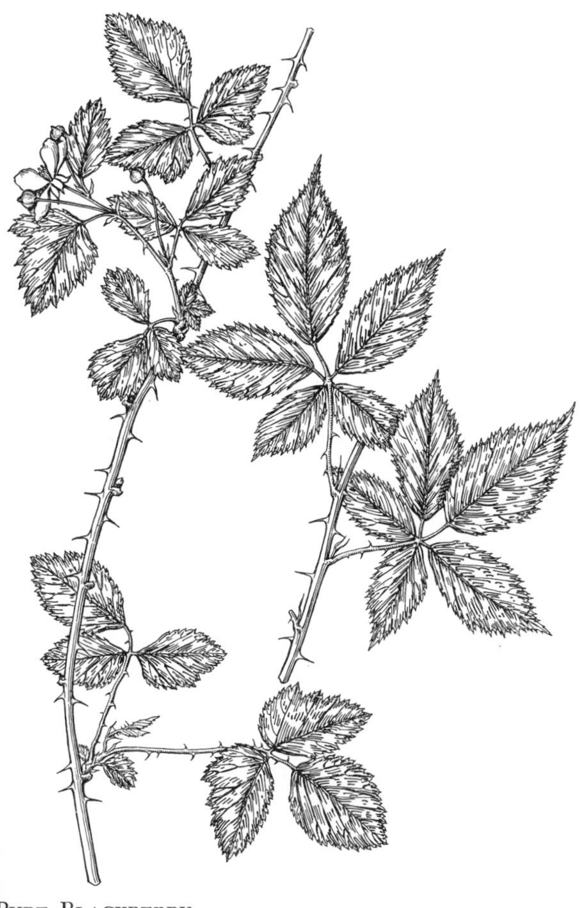

PURE BLACKBERRY
Rubus putus L. H. Bailey

RANGE. Dry and wooded areas of Alabama and Georgia, northward to Kentucky and Virginia. It has been recorded by some authors for Louisiana and Arkansas. L. H. Bailey (1941–1945, pp. 620–622) comments as follows:

"Once our writers accepted *Rubus argutus,* for the whole eastern country, even from Nova Scotia to Kansas. We now know that such distribution in species of blackberries is at once open to question. Recent opinion has been restricting this species to the southeastern states."

REMARKS. The genus name, *Rubus,* is the old Roman name. The species name, *argutus,* means "sharp-toothed," in reference to the leaf serrations. L. H. Bailey

(1941–1945, p. 620) remarks on the species as follows:

"A prominent mark of *Rubus argutus* is the argute or very acute serrations of the leaf margins, stressed by Link 'foliolis oblongis acuminatis duplicato-argute serratis.' It is largely on this primary character that I have arrived at my estimate of the species."

Ozark Blackberry

Rubus ozarkensis L. H. Bailey

FIELD IDENTIFICATION. An erect glandless shrub 3–6 ft. The prickles neither numerous nor conspicuous.

FLOWERS. Appearing May–July, usually 4–10 in open somewhat cyme-like clusters covered in the foliage or somewhat surpassing it; pedicels ascending or spreading, ¾–1¼ in. long, unarmed, pubescent; corolla small, not greatly exceeding ⅜–¼ in. broad, petals 5, white, rounded, well separated; stamens numerous; calyx-lobes broad, acuminate, reflexed.

Ozark Blackberry
Rubus ozarkensis L. H. Bailey

FRUIT. Maturing July–September, black, small, short-oblong, ¼–⅜ in. long, the many small drupelets seedy.

FLORICANES AND LEAVES. Floricanes sometimes arching with weight, bearing bent or slanting prickles ⅛–⅙ in. long; leaves 3-foliolate, leaflets similar in shape to primocane leaflets, 2¼–2¾ in. long, ¾–1¼ in. broad, prevailingly acute, but sometimes those in the inflorescence obovate and obtuse.

PRIMOCANES AND LEAVES. Canes angled or terete, prickles few, ⅜–¾ in. or more apart, rather straight; leaves ample, 5-foliolate, somewhat pubescent above, densely so underneath, acutely double-serrate, petiole pilose-pubescent and armed; leaflets ovate-oblong, lowest pair nearly broad-lanceolate, rather abruptly acuminate, mostly narrowed to base except the central one, which may be broadly rounded and 2¾–4½ in. long and half as broad or more.

RANGE. On low grounds and in woods of the Ozark Mountain region near Fayetteville, Washington County, Arkansas; and in isolated localities in the Ozark Mountains of Missouri.

REMARKS. The genus name, *Rubus*, is the old Roman name. The species name, *ozarkensis*, refers to the occurrence of this species in the Ozark Mountains.

Himalaya Blackberry

Rubus procerus P. J. Muell.

FIELD IDENTIFICATION. Robust, sprawling more or less evergreen glandless bramble, much-branched, making great mounds or banks and some of the canes perhaps standing 9 ft tall, others arching on the ground and frequently taking root at the tip.

FLOWERS. Appearing in July or even later, inflorescence a large, long terminal cluster with branches in lower axils, peduncles and pedicels cano-tomentose and prickly; flowers white or rose, ¾–1 in. or more across, the 5 petals broad; calyx-lobes 5, broad, cano-tomentose, conspicuously pointed, soon reflexed, stamens numerous.

FRUIT. Fruit late as compared with native blackberries, variable, ripening over a considerable period, globose to hemispherical, shiny black, to ¾ in. long, drupelets large and succulent, not very numerous.

FLORICANE LEAVES. Usually 3–5-foliolate, smaller than the primocanes, leaflets narrow and more gradually acuminate, most of them angustifoliate or subangustifoliate rather than expanded at base, margins coarsely double-serrate.

PRIMOCANES AND LEAVES. Primocanes pilose-pubescent and becoming nearly glabrous with age, very strongly angled and furrowed, bearing well-spaced, heavy, very broad-based straight or somewhat curved prickles ¼–⅖ in. long; primocane leaves large, 5-folio-

Himalaya Blackberry
Rubus procerus P. J. Muell.

ate, glabrous on upper face at maturity, cano-pubescent to cano-tomentose on lower face; petiole and long petiolule bearing heavy hooked prickles; leaflets large and broad, terminal leaflets roundish to broad-oblong, often with more or less parallel margins up to middle and with tendency to obovate outline, 4–4¾ in. long, 3–4 in. broad, abruptly contracted to short narrow apex or acumen, broad at base and truncate to subcordate; lateral leaflets of similar description except smaller and less broad with tendency to more pronounced obovate outlines; margins of primocane leaflets coarsely and unequally and acutely serrate-dentate to a height of 1/12–⅛ in. with a prominent projection or tooth terminating the main lateral veins so that the ambitus is irregular to almost jagged rather continuous.

RANGE. A native of western Europe. Probably escaped from cultivation in Oklahoma. Naturalized in California, Nevada, Oregon, Washington, and perhaps elsewhere.

REMARKS. The genus name, *Rubus*, is the old Roman name. The species name, *procerus*, refers to the elongate canes. A difficult species to circumscribe owing to its cultivated origin.

Twice-leafed Blackberry

Rubus bifrons Vest

FIELD IDENTIFICATION. A strong, low-arching glandless bramble making canes 6–9 ft or more long with apex lying or creeping on the ground, and sometimes tip-rooting, the clump or colony standing 3–4½ ft.

FLOWERS. Inflorescence May–June, narrow and long with a few branches in lower axils, peduncles and pedicels cano-tomentose, bearing straightish conspicuous prickles; corolla white or rose, ¾–2 in. across, petals 5, narrow; calyx-lobes 5, medium-broad, long-pointed, becoming reflexed; stamens numerous.

FRUIT. June–July, about ⅝ in. thick, black, drupelets not numerous and loosely placed.

FLORICANE LEAVES. Mostly 3-foliolate, leaflets similar to those of primocanes.

PRIMOCANES AND LEAVES. Primocane axis somewhat angled to almost terete, striate, scantily pilose or pubescent; prickles irregularly scattered and of different lengths, ⅙–⅓ in. long, moderately broad-based, nearly straight to somewhat hooked; primocane leaves 5-foli-

Twice-leafed Blackberry
Rubus bifrons Vest

ate, persistent, not very large, glabrous above, brownish-canescent or tomentose underneath; petiole and petiolules armed with short strong hooks; leaflets ovate-elliptic to oblong-elliptic or somewhat rhombic, 3–3½ in. long, 2–2½ in. broad, the short apex narrowed gradually or in some cases almost abruptly, mostly narrowed to base but sometimes the terminal leaflet nearly subcordate, margins finely and acutely but somewhat irregularly serrate to depth of 1/25–1/12 in. long, ambitus practically continuous.

RANGE. A native of Europe, naturalized in North America. East Texas, Louisiana, and Arkansas; east to Alabama, northward through North Carolina and Virginia to Rhode Island.

REMARKS. The genus name, *Rubus,* is the old Roman name. The species name, *bifrons,* refers to the often rather dimorphic leaves.

AMERICAN RED RASPBERRY
Rubus strigosus Michx.

FIELD IDENTIFICATION. A multifarious slender-growing species, inconstant in dimensions, armature, and glandulosity. Slender and erect, 4 in.–6 ft. Usually narrow-leaved, mostly glandiferous on some or all parts, usually strigose or bristly, but not armed with short wide-based prickles. Leaves thin, gray-pubescent to subtomentose underneath. Fruiting canes often spreading and decumbent by age and weight.

FLOWERS. Appearing late May–July, flowers single in lower axils but mostly 4–7 near the top, hardly racemiform or umbelloid, ⅜–½ in. broad when expanded, pedicels slender and commonly strigose or hairy and glandiferous; calyx more or less glandular, the 5–7 lobes narrow and pointed and soon reflexing, petals 5, narrow-oblong and widely separated, more or less obtuse, erect and about equaling the head of numerous erect stamens.

FRUIT. Maturing late June–October, more or less conic and variously pubescent before succulence, but becoming as broad as long or even broader, sometimes lobed, forming a loose aggregation of few red or rarely whitish drupelets that are likely to fall separately rather than in a firm cone or cap, the remaining torus or core very broad and low-conic; seeds or achenes oblong or somewhat reniform, much pitted when mature and dry.

FLORICANES AND LEAVES. Leaves on floricanes similar to those on primocanes except for reduced size, and those in inflorescence perhaps simple although prevailingly 3-foliate, and very occasionally 5-foliate.

PRIMOCANES AND LEAVES. Primocane axis 1/6–¼ in. in diameter, bark tawny or glaucous, flaking from the base, the canes usually conspicuously decorticating the second year, axis sometimes unarmed, in other situations or cases bearing slender bristles or stiff sharp-pointed hairs 1/25–1/5 in. long and not broad-based (strigose), often interspersed with stalked glands, frequently minutely grayish-pubescent underneath the armature; leaves of primocanes thin in texture or substance, grayish-pubescent underneath, mostly glabrous on upper surface, 3–5-foliate, when 3-foliate then the terminal leaflet commonly not deeply lobed, petiole for the most part strigose and thinly glandiferous; leaflets on primocanes lance-ovate to long-ovate, usually acuminate and narrowly pointed, for the most part narrowed or rounded at base, or if cordate then only narrowly so, terminal leaflet usually 3–4 in. long, 2–2½ in. broad, margins prominently and acutely double-serrate, apices apiculate.

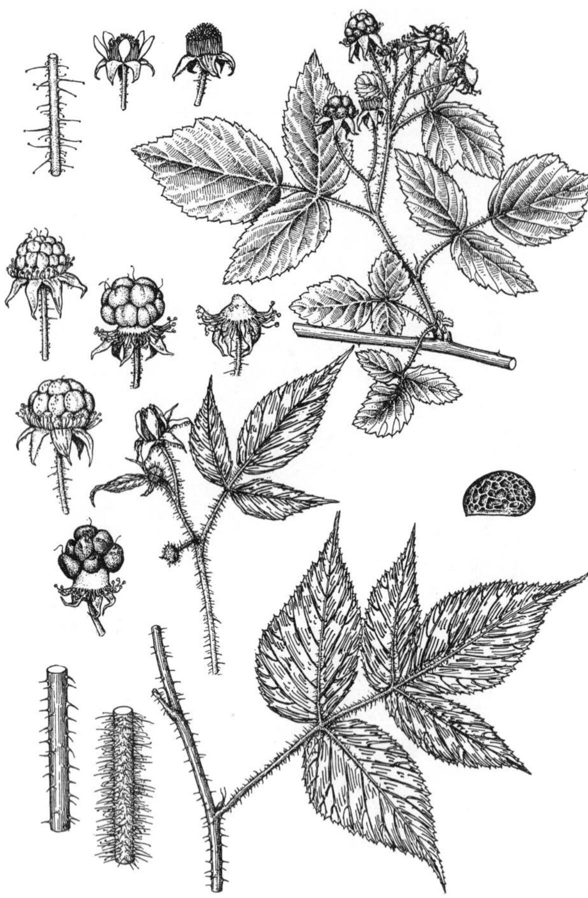

AMERICAN RED RASPBERRY
Rubus strigosus Michx.

RANGE. A cosmopolitan shrub in North America on many lands and in diverse exposures, usually in upper or dry locations. In New Mexico; west to California, northeastward to North Carolina, and north to Canada and Alaska.

REMARKS. The genus name, *Rubus,* is the old Roman name. The species name, *strigosus,* refers to the stiff bristles. As a reflection of the great variability in this species, many varieties have been designated, but these occur generally north of the area covered by the present volume.

Alice Raspberry

Rubus aliceae L. H. Bailey

FIELD IDENTIFICATION. Bramble erect, stiff, glandiferous, armed with prickles rather than bristles.

FLOWERS. Usually 5–8, near top of cane in a close cluster, with pedicels so short that it looks compact, the flowers subtended by foliage; pedicels armed like the petioles; petals 5, small, apparently white, much exceeded by the 5 extended very narrow calyx-lobes which are at first erect and ⅖–½ in. high, but becoming divaricate and the calyx then ¾ in. across, calyx-base and the lobes pubescent, armed, and glandiferous.

CANES. Terete, finely pubescent, with a mixed armament of (a) very short elevations, (b) pimples, (c) conspicuous broad-based straight prickles 1/12–⅛ in. long, (d) slender bristle-like prickles and gland-tipped hairs.

LEAVES. Usually 3–5-foliolate, glabrous and strongly side-veined on upper face, gray-tomentose underneath with side-veins upstanding, sometimes armed and midrib conspicuously short-armed; leaflets small, 1½–2 in. long, ¾–1¼ in. broad, ovate-acuminate, base broad or subcordate, very closely and evenly serrate, lateral leaflets sessile and terminal leaflet nearly or quite sessile or petiolule ⅙–⅓ in. long, petiole short, stout, and armed like the cane.

RANGE. In New Mexico in Santa Fe Canyon, near Santa Fe city, growing with *Pinus flexilis*.

REMARKS. The genus name, *Rubus*, is the old Roman name. The species name, *aliceae*, is in honor of Alice Eastwood, who collected the type plant.

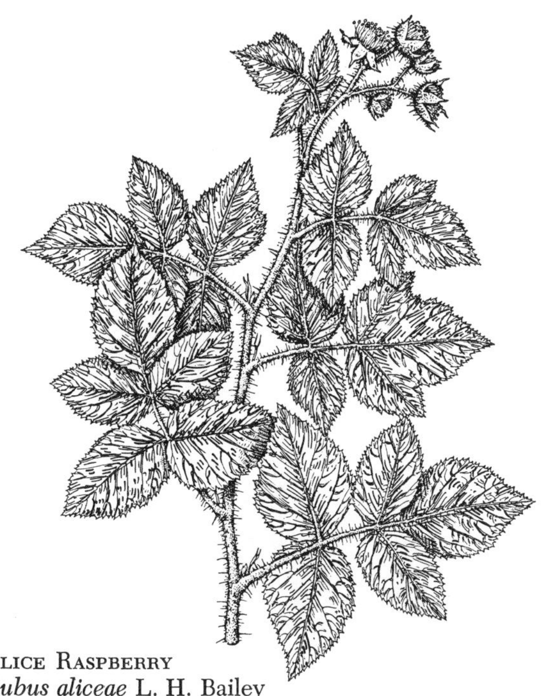

Alice Raspberry
Rubus aliceae L. H. Bailey

Blackcap Raspberry

Rubus occidentalis L.

FIELD IDENTIFICATION. Common native and cultivated shrub with tip-rooting stems and 3-foliolate leaves. Plant upright and attaining a height of 3–7½ ft.

FLOWERS. Maturing April–May, 3–7 rather close together, terminating lateral shoots, and perhaps others in lower contiguous axils on longer pedicels, but the inflorescence variable, all pedicels usually bearing stout hooked prickles but no glands; flower buds conic, canescent, ⅙–¼ in. across before expansion, the filiform calyx-tips forming a narrow erect apex; corolla ½–⅗ in. across in full expansion; petals 5, very narrow, obtusish or emarginate, at first erect and about equaling the compact upright stamens, soon spreading and then falling; calyx-lobes 5, prominent, attenuate-pointed; stamens numerous.

FRUIT. Maturing June–August, firm with many compactly placed drupelets, mostly depressed-globular but sometimes as long as broad, ⅗ in. or less broad, bloom black but varying sometimes to amber, yellowish (in vars. *pallidus* and *flavobaccus*), and purplish, aromatic, conspicuously subtended by the long-pointed divaricate calyx-lobes; drupelets nearly or quite glabrous at maturity; seeds curved-oblong, pitted, about 1/12 in. long.

PRIMOCANES AND LEAVES. Primocanes arching over at end of season and strongly rooting at tip, axis of cane terete and conspicuously pruinose, the bloom more or less wiping off with the finger, often disappearing with age, the smooth bark then glossy; prickles at base of new canes and on shoots arising from the rooted tips often many, small, straight, and close together, but on main part of primocane and floricane scattered or remote (sometimes lacking), stout, flattened, and hooked, ⅙–¼ in. long; leaves 3-foliolate or upper ones in flowering clusters perhaps simple, lower ones on primocanes sometimes nearly 5-foliolate, petiole long and usually bearing short, stout, curved prickles, terminal leaflet prominently stalked but laterals nearly or quite sessile; blades of leaflets broad-ovate to lance-ovate, terminal one subcordate to cordate and often as much as 5 in. long and 4 in. broad, but commonly 2¾–3¼ in. long, tapering to a narrow acute apex, margins sharply and unequally double-serrate, lateral leaflets more or less lobed; leaflets on flowering laterals smaller and narrower, perhaps 2½–3¼ in. long and 1½–2½ in. broad; under surfaces (unless in shade) prominently and closely white-canescent or white-tomentose.

RANGE. Woods borders, fields, fence-rows, copses, mostly on higher grounds. Oklahoma, Arkansas, Louisiana, and New Mexico; eastward to Georgia, north to New Brunswick, and west to Minnesota and Colorado.

HORTICULTURAL FORMS. Blackcap Raspberry has long been cultivated in many pomological variations; Bailey (1941–1945, p. 883) lists the following: Abundance, Ada, Black Pearl, Bristol, Buckeye, Caroline, Champion, Cumberland, Diamond (Black Diamond), Dundee,

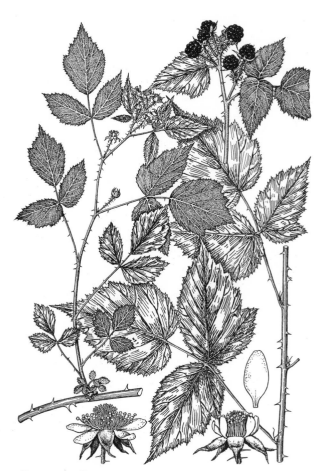

Blackcap Raspberry
Rubus occidentalis L.

Eureka, Evans, Gladstone, Gregg, Hilborn, Hoosier, Johnson, Kansas, Mid West, Munger, Naples, New Logan, Plum Farmer, Potomac, Royal (Royal Purple), Shaffer (Shaffer's Colossal), Sodus, Souhegan, Stone Fort, Tennessee Autumn, and Winfield.

REMARKS. The genus name, *Rubus,* is the old Roman name. The species name, *occidentalis,* means "western." The wild fruit, according to Bailey (1941–1945, p. 883), "is a source of much fruit of high edible quality in both the fresh state and in prepared jams and other products."

Whitebark Raspberry

Rubus leucodermis Dougl.

FIELD IDENTIFICATION. Differing from the Eastern species *(R. idaeus)* in the narrower and usually smaller, more finely and deeply dentate leaflets, much more prickly pedicels and floral shoots and abundantly armed primocanes, fruit mostly larger and purplish or often reddish. Attaining a height of 3–9 ft and upright.

FLOWERS. Flowers April–July, sometimes solitary, but usually 2–10 in compact terminal clusters, a few of them in contiguous axils, ¼–⅖ in. across, petals 5, narrow and at first upright, then deflexed and soon caducous; calyx-lobes 5, becoming strongly deflexed; stamens numerous.

FRUIT. Maturing July–September, firm, dark purple to almost black, sometimes yellowish red, depressed-globose, ⅗ in. or less across and usually less in height, gray-tomentose until nearly full grown, drupelets canescent at maturity; seed short-reniform, about 1/12 in. long, only obscurely pitted or rugose.

PRIMOCANES AND LEAVES. Vigorous, much-branched, arching over, taking root at tips, canes with a heavy whitish bloom that may more or less disappear with age; prickles many, those on main axis nearly or quite straight, ⅙–¼ in. long at maturity and not much expanded at base, those on floral shoots and pedicels 1/12–⅛ in. long, curved and mostly very broad-based; primocane leaves 3-foliolate to more or less imperfectly 5-foliolate or even 7-foliolate, white-canescent to tomentose underneath, thin, green and nearly or quite glabrous on upper face or thinly pilose along ribs; leaflets

Whitebark Raspberry
Rubus leucodermis Dougl.

ovate to narrow-ovate, sometimes almost lanceolate, terminal one long-stalked, mostly narrowly pointed, 2¾–3½ in. long and 1½–2 in. broad near the truncate or subcordate base, margins irregularly but sharply double-serrate and sometimes almost lobate, petiole and petiolules with small curved prickles.

RANGE. Reported from New Mexico, but usually northward; from Utah and Nevada to Montana and British Columbia.

RELATED SPECIES. L. H. Bailey (1941–1945, p. 884) makes the following comment:

"This species is close to *R. occidentalis* but has a separate geographic range, old canes are conspicuously white- or gray-glaucous, prickles on canes more numerous and not likely to be strongly curved, primocane leaflets with more attenuate narrower end and often a less expanded base, petioles and pedicels more stoutly and broadly prickly, fruit probably more uniformly purplish and bloomy."

REMARKS. The genus name, *Rubus*, is the old Roman name. The species name, *leucodermis*, refers to the bloomy cane epidermis.

BERNARDINO RASPBERRY

Rubus bernardinus Fedde

FIELD IDENTIFICATION. A diffuse, much-branched, more or less mounding bramble, attaining a height of 2–6 ft, but with recurved or decumbent canes.

FLOWERS. Axis of floral branchlets glaucous and sparingly glandular-hairy; petioles and pedicels sparsely glandular-hairy; flowers 2–5, at near end of floral shoot, not exceeding the accompanying foliage, ⅖–⅗ in. across, white, the 5 petals rounded; calyx-lobes 5, separated well to base, prickly on lower part, narrow and very long slender-pointed; stamens numerous.

FRUIT. Described as hemispheric, ½ in. high, with pubescent drupelets.

LEAVES. Foliage only slightly pubescent to practically glabrous on upper face, white-tomentose underneath, the leaves 3-foliolate, the petiole, terminal petiolule, and ribs underneath bearing many stout hooked prickles; leaflets ovate to oblong, 1¼–2 in. long and nearly as broad, base broad to almost truncate, apex short and acute, margins conspicuously double-serrate.

CANES. The second year stout, glabrous, purple-tinted and with conspicuous bloom; prickles stout, broad-based, nearly or quite straight, well separated, about ⅕ in. long.

RANGE. In southern California in San Bernardino County. Also recorded from New Mexico, but not clear as to whether wild or an escape from cultivation.

REMARKS. The genus name, *Rubus*, is the old Roman name. The species name, *bernardinus*, refers to the San Bernardino Mountains in California.

WESTERN THIMBLE RASPBERRY

Rubus parviflorus Nutt.

FIELD IDENTIFICATION. Shrub erect or upright, leafy, 3–6 ft, unarmed. Much like *R. odoratus* except usually less glandular and smoother. Flowers white and fewer in a cluster, leaves for the most part less deeply lobed and less limp.

FLOWERS. Maturing June–July, from 2–9 in a rather close cymiform cluster not exceeding the foliage, on short variously glandular-hairy pedicels; corolla white, 1–2⅓ in. across, with 5 flaring broad obtuse petals; calyx glandular or glandless, the lobes produced into 5 linear points and finally reflexed or spreading; stamens numerous.

FRUIT. Maturing August–September, red, mostly broader than long, ⅝–¾ in. thick, firm, the pubescent drupelets cohering, very edible in its eastern area of good rainfall, but less palatable westward.

CANES. Old canes becoming glabrous and more or less flaking or shredding; primocane axis glabrous or sparsely glandular-hairy.

LEAVES. Palmate, widely ovate-cordate, broadly and commonly shallowly 3–5-lobed, 5–12 in. broad, nearly or quite glabrous on upper face, sinuses mostly about one third depth of blade, lobes acute, margins closely and unevenly serrate to dentate, petiole with glandiferous hairs of different lengths on different plants.

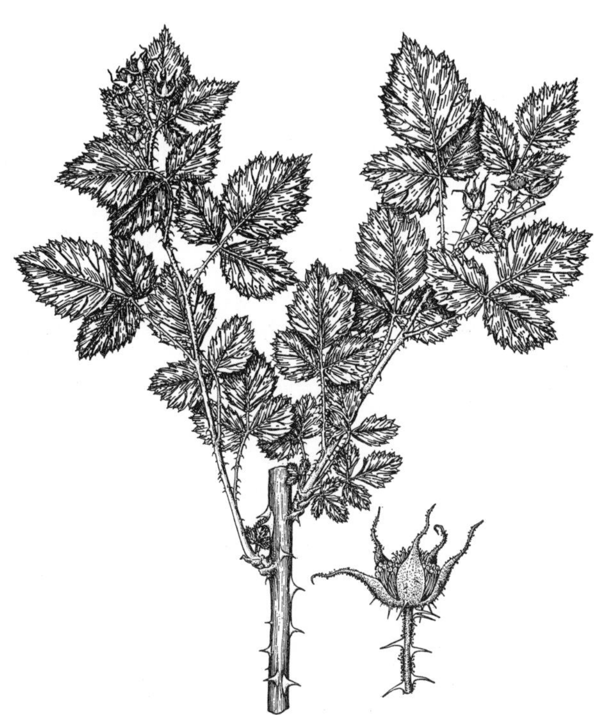

BERNARDINO RASPBERRY
Rubus bernardinus Fedde

Western Thimble Raspberry
Rubus parviflorus Nutt.

RANGE. Woodlands, copses, groves, borders, and open places. Mountains of New Mexico; northward to western Ontario, west to Alaska, British Columbia, and the Pacific Coast states, and southward in the mountains of Mexico.

VARIETIES AND FORMS. A great many varieties and forms have been described and much confusion exists concerning them. The following new varieties are presented by Fernald (1935a). Fernald's key to these is condensed and abbreviated for the present record:

Glands of pedicels all or nearly all stipitate.
- Calyx-lobes densely long-villous with pale trichomes hiding the glands: leaf blades velvety underneath .... var. *velutinus*
- Calyx-lobes not long-villous.
  - Pedicels and peduncles bearing glands that are very unequal, longest often 1–2 mm.
    - Lower surface of leaves soft to touch, pubescence divergent and abundant .... var. *hypomalacus*
    - Lower surface of leaves not obviously soft to touch, glabrate to glabrous .... var. *heteradenius*
  - Pedicels and peduncles bearing mostly subequal and short glands seldom more than .50 mm. long.
    - Lower surface of leaves soft to touch .... var. *bifarius*
    - Lower surface glabrous or soon glabrate .... var. *grandiflorus*

Glands of pedicels all or nearly all sessile, subsessile, or even wanting .... var. *scopulorum*

REMARKS. The genus name, *Rubus*, is the old Roman name. The species name, *parviflorus*, meaning "small-flowered," seems misapplied as the flowers are not small; the reference may be to the small inflorescences. An attractive shrub, fragrant from its glandular parts and frequently planted.

Boulder Raspberry

Rubus deliciosus Torr.

FIELD IDENTIFICATION. A showy plant with rose-like flowers, unarmed, prostrate or clambering.

FLOWERS. May–June; solitary terminating lateral short leaf-bearing branchlets, abundant and showy, 1½–2⅜ in. across, white; petals 5, obtuse, flaring or reflexed; calyx pubescent and sometimes semiglandiferous, the 5 lobes prolonged into narrow and sometimes laciniate ends; stamens numerous.

FRUIT. Maturing July–September, dark purple, hemispheric, about ½ in. or more thick, usually of indifferent edible quality.

FLORICANES AND LEAVES. Floricane leaves simple, mostly reniform, breadth equaling or exceeding the length, entire in outline or sometimes shallowly lobed above the middle, 1¼–2 in. broad, usually cordate at base, apex obtuse, margins shallowly and acutely dentate, essentially glabrous on upper face and more or less pubescent on lower face.

PRIMOCANES AND LEAVES. Canes sometimes 3 ft and more long; primocanes reddish, finely and loosely hairy-pubescent; primocane leaves simple, shallowly trilobate, broad-ovate in outline, 2–2½ in. long and broad, width nearly or quite equaling the length, broadly or narrowly cordate at base, more or less acute at apex, margins sharply but not deeply serrate, glabrous on upper face or soon becoming so, thinly pubescent underneath but at length glabrate, scarious stipules soon caducous.

RANGE. In rocky areas and among bushes in the Rocky Mountain region. New Mexico, adjacent Oklahoma, Colorado, and Wyoming.

REMARKS. The genus name, *Rubus*, is the old Roman name. The species name, *deliciosus*, was given because of the reportedly large delicious fruit, but this is not verified by subsequent authors. Sometimes planted as an ornamental.

New Mexican Raspberry

Rubus neomexicanus Gray

FIELD IDENTIFICATION. Shrub 3–7 ft, apparently not decumbent or trailing, glandless or nearly so, unarmed.

FLOWERS. One or 2 on short, leafy lateral branches, sometimes as much as 1½–2 in. across, white; petals 5, large, white, broad and very obtuse, wide-spreading at full anthesis; calyx pubescent, glandless or perhaps sometimes thinly glandiferous, lobes 5, broad, but extended to long narrow ends; stamens numerous.

FRUIT. Red, about ⅝ in. thick.

LEAVES. Floricane leaves round-cordate in outline or sometimes broader than long, 1½–3½ in. long, shallowly 3-lobed, or frequently sub–5-lobate with irregular periphery, base narrowly or broadly cordate, apex mostly acute even though the point may be short, margin irregularly sharp-serrate or sharp-dentate, upper surface thinly pubescent or pilose, under surface mostly soft-pubescent with spreading short hairs on ribs and veins; petiole with divergent hairs.

New Mexican Raspberry
Rubus neomexicanus Gray

TWIGS. Slender and glabrous, unarmed, old bark flaking off.

RANGE. In high elevations in canyons. New Mexico, Arizona, and Sonora, Mexico.

REMARKS. The genus name, *Rubus*, is the old Roman name. The species name, *neomexicanus*, refers to the state of New Mexico.

Boulder Raspberry
Rubus deliciosus Torr.

Similar Raspberry

Rubus exrubicundus L. H. Bailey

FIELD IDENTIFICATION. Plant upright or ascending, much-branched, to 3 ft. Usually unarmed and somewhat glandular.

FLOWERS. Solitary on short leafy lateral axes, rather large, white, ¾–1¼ in. across at expansion; petals 5, oblong and broad; calyx pubescent and glandiferous, the 5 long slender tips may equal the petals in length; peduncles pubescent and glandiferous; stamens numerous.

FRUIT. Small, dry, and red.

LEAVES. On floricanes broad-ovate and cordate, small, 1¼–1⅝ in. long and broad, obscurely shallow-trilobate with rounded apices, apex obtuse or only briefly acute, margins irregular in outline with many obtuse or abruptly acuminate teeth, basal sinus often broad, essentially glabrous above, somewhat closely pubescent along the veins underneath; petiole equaling the blades or shorter, closely pubescent.

TWIGS. Numerous, slender, divaricate, reddish and finely pubescent on floricanes.

RANGE. On faces of cliffs and in deep canyons. New Mexico in the Organ Mountains, west to Arizona.

REMARKS. The genus name, *Rubus,* is the old Roman name. The species name, *exrubicundus,* refers to its synonomy with *R. rubicundus* Tides. This plant has smaller leaves than those of the closely related *R. mexicanus* that are nearly glabrous, and pubescence of branches and petioles appressed rather than spreading-hairy; peduncles and calyx glandular.

SIMILAR RASPBERRY
Rubus exrubicundus L. H. Bailey

LEGUME FAMILY (*Leguminosae*)

Eastern Prairie Acacia
Acacia angustissima (Mill.) Kuntze var. *hirta* (Nutt.) Robins.

Eastern Prairie Acacia

Acacia angustissima (Mill.) Kuntze var. *hirta* (Nutt.) Robins.

FIELD IDENTIFICATION. Thornless acacia, mostly herbaceous in the North, but in the South tending to become a perennial with a woody base. Rarely over 3 ft, with delicate fernlike leaves.

FLOWERS. May–August, in white heads about ½ in. in diameter (more rarely pinkish), axillary or terminal, the heads disposed in a racemose or paniculate manner; calyx green, less than 1/25 in. long, lobes 4–5, deltoid, acute; petals 4–5, greenish, distinct but joined at the base, 1/12–⅛ in. long; stamens numerous (50–100 or more), exserted, filaments ¼–⅓ in. long, white to pink; ovary stipitate, style long-filiform.

FRUIT. Maturing July–August, linear to oblong, 2–3 in. long, ¼–⅖ in. wide, cuneate at apex, long-stipitate, margin straight or constricted, thin, flat, dry, glabrous, reddish brown, dehiscent; seeds oblong, mottled-gray or brown to black, ⅛–⅙ in. long, 1/12–⅛ in. wide.

LEAVES. Bipinnate, pinnae averaging 6–14 pairs; leaflets averaging 20–33 pairs (some plants with more numerous pinnae and leaflets); leaflets ⅛–¼ in. long, less than 1/25 in. broad, linear to oblong or elliptic, apex acute, surface glabrous or ciliate; stipules scalelike, ciliate, 1/12–⅛ in. long.

STEMS. Long, slender, deeply grooved, showing varying degrees of hairiness.

RELATED FORMS AND VARIETIES. The variety described above is closely related to *A. angustissima* (Mill.) Kuntze. It and its variable forms constitute an intergrading taxonomical complex. The type locality of the species is Campeche, Mexico, and it is distributed through Veracruz, Yucatán, Guatemala, and northward through Mexico into Texas. In its typical form it normally has from 12–25 pairs of pinnae and 30–75 pairs of leaflets. However, it is very variable, and in the southwestern United States shows all manner of gradations in the number of pinnae and leaflets, and in glabrous or hairy characters. Because of these numerous variations a large number of names have been applied to the species and its varieties.

According to the present classification status of these plants in the southern and southwestern United States, there seem to be two general categories, with inter-

TEXAS PRAIRIE ACACIA
Acacia texensis Torr. & Gray

mediate forms passing into the Mexican species. One of these varieties is known as Eastern Prairie Acacia, described above. It occurs in New Mexico, Arizona, Texas, Oklahoma, Arkansas, and Louisiana; eastward to Florida, northward into Missouri, and southward into Mexico.

The other variation, which is apparently distinct enough to merit recognition, is Texas Prairie Acacia, *A. texensis* Torr. & Gray. Some authors assign to it the name of *A. angustissima* (Mill.) Kuntze var. *cuspidata* (Schlect.) Benson. The plant is more or less glabrous with a few hairs on the twigs, and the legume is acuminate. It usually has 2–5 pairs of pinnae and 7–22 pairs of leaflets which are obtuse at the apex. However, the number of leaflets and pinnae are variable. Texas Prairie Acacia is seemingly confined to the hills and mountains of west Texas, New Mexico, Arizona; and Mexico in Chihuahua, Zacatecas, and Puebla.

For a detailed discussion of this problem the reader is referred to Benson (1943a, pp. 237–238).

REMARKS. The genus name, *Acacia,* is an ancient word meaning "a hard, sharp point," with reference to the spinescent stipules in many species. The species name, *angustissima,* refers to the very narrow leaflets, and the variety name, *hirta,* refers to the hirtellous (hairy) character. A form with glabrous stems is also known. Vernacular names are Fern Acacia, Thornless Acacia, Prairie Acacia, Huajilla, Guajillo, Xoax, Timbe, Cantemo, and Pala de Pulque. The legumes are sometimes eaten by cattle, and the seeds by bobwhite quail. The plant has ornamental possibilities and can be propagated by root divisions.

BLACK-BRUSH ACACIA

Acacia rigidula Benth.

FIELD IDENTIFICATION. Stiff, thorny shrub with many stems from the base, attaining a height of 15 ft. Often forming impenetrable thickets in west and southwest Texas.

FLOWERS. Heads April–May, axillary, white or light yellow, fragrant, arranged in oblong and densely flowered spikes; individual flowers small, sessile, calyx 4–5-toothed, corolla of 4–5 petals, stamens numerous, exserted, distinct.

FRUIT. Legume 2–3½ in. long, $\frac{3}{16}$–¼ in. wide, linear, often falcate, apex acuminate, base attenuate and stipitate, constricted between the seeds, flattened, red-

BLACK-BRUSH ACACIA
Acacia rigidula Benth.

dish brown to black, striate and puberulent, 2-valved and dehiscent; seeds about ¼ in. long, ends rounded or obtuse, lustrous, dark green to brown, compressed.

LEAVES. Bipinnate, rachis grooved and somewhat hairy, about ⅜–1 in. long, pinnae 1–2 pairs, leaflets 2–4 pairs (or rarely 5 pairs), ⅙–½ in. long and ¼–⅓ in. wide, sessile or nearly so, elliptic to oblong, oblique, apex rounded and mucronate or sometimes notched, base unsymmetric, firm; surface lustrous, dark green, glabrous, nerves conspicuous.

TWIGS. Divaricate, gray to reddish, glabrous, thorns usually paired at the nodes, straight or slightly curved, slender, pale, ⅓–1 in. long.

BARK. Light to dark gray, smooth, tight.

RANGE. Widespread over Texas in areas west and southwest of the Guadalupe River. Collected by the author in Bexar, Val Verde, and Brewster counties. In Mexico in the states of Tamaulipas, Nuevo León, Chihuahua, San Luis Potosí, and Jalisco.

SYNONYM. This plant is sometimes listed under the name of *A. amentacea* DC. However, the latter is a closely related species of central Mexico, with a more pubescent petiole and pinnae with only 2 pairs of leaflets.

REMARKS. The genus name, *Acacia,* is an old word meaning "a hard, sharp point," with reference to the spinescent stipules. The species name, *rigidula,* refers to the stiff or rigid branches. Vernacular names used are Chaparro Prieto, Catclaw, and Gavia. The plant has good possibilities for use as an ornamental in dry sandy or limestone areas. It could be used also as an erosion-control plant, and the flowers are a source of honey.

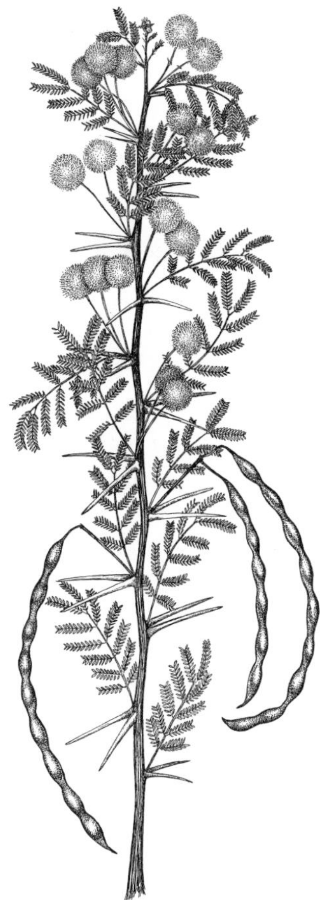

MESCAT ACACIA
Acacia constricta Benth.

MESCAT ACACIA

Acacia constricta Benth.

FIELD IDENTIFICATION. Spiny shrub to 18 ft, the slender spines in pairs at the nodes.

FLOWERS. Flowering date variable, borne on slender, puberulent peduncles ¾–1¾ in. long, flowers small, sessile, yellow, fragrant, in dense heads, ¼–⅓ in. in diameter; calyx campanulate, 4–5-lobed, lobes ciliate; petals 4–5, yellow, united into a tube below, 1/12–⅛ in. long; stamens numerous (30–40), exserted, filaments yellow and about ⅙ in. long, anthers yellow; ovary stalked.

FRUIT. July–September, legume linear, slender, falcate or straight, 2–4¾ in. long, 1/12–⅙ in. wide, somewhat constricted between the seeds; apex acute or mucronate; base narrowed and stipelike; surface reddish brown to black, glabrous, glandular or nonglandular, 2-valved, readily dehiscent; seeds in one row, lenticular, about ⅕ in. long and 1/12 in. broad, black or gray-mottled, smooth, germination about 45 per cent.

LEAVES. Alternate, or appearing clustered at the nodes, rachis pubescent, 1–2 in. long, bipinnate; pinnae 3–9 pairs (usually 3–7 pairs), petiole gland-bearing; leaflets 6–16 pairs (usually 6–12 pairs), oblong to linear, obtuse, thick, apparently nerveless, glabrous or puberulent, 1/16–⅙ in. long, 1/25–1/12 in. wide, somewhat viscid or nonviscid.

TWIGS. Reddish brown to dark brown or black, pubescent or glabrous, sometimes glandular; spines in pairs at the nodes, ¼–1½ in. long, straight or slightly curved, slender, sharp, white to ashy or brown. Plant sometimes spineless or nearly so.

BARK. Dark brown to black or reddish, smooth or roughened by small scales.

RANGE. In Texas, New Mexico, Arizona, and Mexico. Generally in dry sandy or caliche soils at altitudes of 1,500–6,500 ft. Scattered along the Rio Grande River drainage in west Texas, but mostly west of the Pecos River and beyond. Abundant in southern New Mexico and Arizona. In Mexico in the states of Sonora, Tamaulipas, Puebla, Zacatecas, Chihuahua, and San Luis Potosí.

VARIETIES. Spineless Mescat Acacia, *A. constricta*

Viscid Acacia
Acacia vernicosa Standl.

var. *paucispina* Woot. & Standl., is a variety with very few or no thorns and leaves which are less glandular and somewhat larger in size, and is found at higher elevations. The plant is usually of larger stature.

Viscid Acacia, *A. vernicosa* Standl., is a closely related species with 1–4 pairs of pinnae, each bearing 3–9 pairs of leaflets, characterized by being very sticky-viscid and glandular. The legume is usually less than 2½ in. long and is also sticky-viscid. The calyx is glandular-dentate and not ciliate as in the Mescat Acacia. However, where the Viscid and Mescat Acacia occur together, intermediate forms between the two are found.

REMARKS. The genus name, *Acacia,* means "a hard, sharp point," referring to the spinescent stipules of some species. The species name, *constricta,* refers to the constricted legume. Vernacular names are White-thorn Acacia, All-thorn Acacia, Huisache, Gigantillo, Vara Prieta, Chaparro Prieto, and Largancillo.

The seeds of Mescat Acacia serve as food for various species of birds including the Gambel's and Mearns' quail. Jack rabbits occasionally nibble the foliage. Cattle eat the unpalatable legumes when nothing else is available. The Indians of Arizona and New Mexico made a coarse meal known as "pinole" out of the legumes. The flowers are an important source of desert honey.

Viscid Acacia

Acacia vernicosa Standl.

FIELD IDENTIFICATION. Thorny shrub 2–6 ft, with many erect or spreading stems from the base, and very glandular-viscid throughout.

FLOWERS. Borne May–June, or sporadically after rains; inflorescence capitate, axillary on peduncles ⅜–⅞ in. long; heads fragrant, yellow, globose, dense; individual flowers small and crowded; stamens numerous, distinct, exserted, yellow; bracts and the 4–5 calyx teeth glandular or rarely ciliate; petals 4–5, yellow.

FRUIT. Borne on slender peduncles, legume 1–2¾ in. long, ⅛–⅕ in. wide, curved, more or less constricted between the seeds, flattened, dehiscent; the 2 valves very viscid-glandular, thin, convex, brown, lustrous; seeds ⅙–¼ in. long, oval-oblong, lenticular, gray- or black-spotted.

LEAVES. Very small, very viscid, hardly ½–1 in. long, bipinnate, pinnae 1–4 pairs (usually 1–2); leaflets 3–9 pairs (usually 5–7), oval-oblong, 1/25–⅛ in. long, thickish, fleshy, glabrous, apex rounded, venation obscure; petiole ⅛–⅓ in. long, usually bearing a cup-shaped gland near the apex, but sometimes glandless.

TWIGS. Gray or reddish, glabrous or nearly so, slightly zigzag; spines on each side of the petiole bases divergent, often paired, straight or slightly curved, slender, white or gray, ⅛–1 in. long.

BARK. Gray to black, but on some stems distinctly reddish-tinged, tight, shedding in small scales on older wood.

RANGE. On gravelly hillsides, or in dry caliche soils at altitudes of 2,000–5,000 ft. Trans-Pecos Texas, New Mexico, and Arizona. In Mexico in Zacatecas, Sonora, Chihuahua, and Puebla.

REMARKS. The genus name, *Acacia,* refers to the hard sharp points of the spinescent stipules of some species. The species name, *vernicosa,* means "shiny" or "varnished," with reference to the lustrous fruit or twigs. Viscid Acacia is very closely related to Mescat Acacia, *A. constricta* Benth. In fact, some authorities consider it to be only a more glandular-viscid variety of the latter. Although intergrading forms can be found, the typical representatives of each display sufficient differences to make a separation into species a valid proposal.

Schaffner Acacia

Acacia schaffneri (S. Wats.) Hermann

FIELD IDENTIFICATION. Usually a spiny, spreading shrub, with numerous stems from the base. More rarely a small tree to 20 ft, and 5–6 in. in diameter, with an irregular crown.

FLOWERS. In spring, the heads borne on solitary or clustered puberulent peduncles ⅓–1 in. long; bracts 2, minute, hairy; flower heads yellow, fragrant, globose, ¼–⅜ in. in diameter; calyx minute, shorter than the corolla, 5-lobed, lobes puberulous; corolla 5-parted, puberulous; stamens numerous, exserted, distinct, twice as long as the corolla; ovary pubescent.

FRUIT. Legume 2¾–5 in. long, about ¼ in. wide, elongate, linear, almost round but slightly compressed, somewhat constricted between the seeds, dark reddish brown to black, velvety-pubescent, tardily dehiscent, somewhat pulpy; seeds in one row, about ¼ in. long, obovoid, compressed, dark reddish brown, shiny.

Schaffner Acacia
Acacia schaffneri (S. Wats.) Hermann

LEAVES. Bipinnate, rachis slender and puberulent, short-petiolate, usually less than 1¼ in. long; pinnae 2–5 pairs (usually 3–4); leaflets 10–15 pairs, each leaflet linear to oblong, somewhat curved, apex mucronulate, base subsessile, surface light green and glabrous, 1/25–⅙ in. long.

TWIGS. Reddish brown or gray, slender, somewhat angled, glabrous or pubescent; lenticels numerous and minute; spines ¼–¾ in. long, terete, puberulent to glabrous.

BARK. Dark brown to black, deeply furrowed.

RANGE. Widespread in Texas, Mexico, the West Indies, South America, and the Galápagos Islands. In Texas from the valley of the Cibolo River to Eagle Pass on the Rio Grande. In Mexico in the states of Nuevo León, Chihuahua, Durango, Hidalgo, Puebla, Colima, and San Luis Potosí.

REMARKS. The genus name, *Acacia*, means "sharp-pointed" and refers to the spinescent stipules. The species name, *schaffneri*, is for Wilhelm Schaffner, pharmaceutical dentist from Darmstadt. He settled in Mexico City in 1856. Also known under the vernacular names of Huisache Chino and Huisachillo. Schaffner Acacia resembles Sweet Acacia, *A. farnesiana*, somewhat, but the legume of the former is much narrower and elongate. Schaffner Acacia is listed in recent literature as a synonym of Twisted Acacia, *A. tortuosa* (L.) Willd. At one time the two names described different species.

Schott Acacia

Acacia schottii Torr.

FIELD IDENTIFICATION. Glabrous, thorny or thornless shrub to 3 ft.

FLOWERS. In heads, May–September, borne on axillary peduncles ¾–1⅓ in. long; flowers sessile in a globose yellow head ¼–⅜ in. in diameter; stamens numerous, exserted, short, distinct.

FRUIT. Legume short-stipitate, 2–3 in. long, about ¼ in. broad, contorted or curled into a semicircle or circle, constricted between the seeds; valves coriaceous, glabrous, roughened and somewhat glandular, dehiscent; seeds 6–9, oblong, lenticular.

LEAVES. Fascicled at the nodes, bipinnate, ½–2 in. long, petiole gland-bearing or eglandular, petiole and rachis puberulent; pinnae 1 pair, leaflets 3–7 pairs, linear to filiform or subterete, thickish, ⅙–¼ in. long.

TWIGS. Gray to brown, pale and glabrous, smooth, bearing slender stipular spines, or sometimes spineless.

BARK. Smooth, light to dark gray.

RANGE. Known from Brewster County, Texas, and the Mexican state of Chihuahua. The type locality was the Canyon of the San Carlos in Texas. It has been

SCHOTT ACACIA
Acacia schottii Torr.

collected by the author in Brewster County, Texas, on Terlingua Creek near Study Butte.

REMARKS. The genus name, *Acacia*, means "point" or "thorn," in reference to the spinescent stipules of various species. The species name, *schottii*, is in honor of Arthur Schott, who served under Major William H. Emory in making a vegetational and geological survey of the United States–Mexico boundary in 1854.

SWEET ACACIA

Acacia farnesiana (L.) Willd.

FIELD IDENTIFICATION. Shrub with many stems from the base, or a thorny tree to 30 ft, and 18 in. in diameter. With either a flat top in coastal specimens, or round-topped with pendulous branches.

FLOWERS. February–March, very fragrant, solitary, or 2–5 heads together on puberulous peduncles 1–1½ in. long; bracts 2, minute; heads about ⅔ in. in diameter; stamens numerous, yellow, with distinct filaments (about 20), much longer than the corolla; corolla tubular-funnelform, shallowly 5-lobed, lobes as high as broad, about 1/12 in. long; calyx about half as long as corolla-lobes, somewhat hairy; ovary short-stipitate and hairy, style filiform.

FRUIT. Legume persistent, tardily dehiscent, cylindric oblong, thick, woody, stout, straight or curved, 2–3 in. long, ½–⅔ in. broad, short-pointed, reddish brown to purple or black, partitions thin and papery, pulp pithy seeds in solitary compartments, transverse, ovoid brown, shining, flattened on one side, about ¼ in. long peduncles stout and short.

LEAVES. Pinnately compound, alternate, deciduous 1–4 in. long; pinnae 2–8 pairs; leaflets 10–25 pairs linear, apex acute or obtuse, or with a minute mucro base inequilateral, length about 1/12–¼ in., width about 1/25 in., sessile or short-petioled, bright green and glabrous or puberulent.

TWIGS. Ascending, pendulous or horizontal, slender terete, striate, angled, glabrous or puberulent, armed with paired straight, rigid, stipular spines to 1 1/12 in. long.

WOOD. Reddish brown, sapwood paler, hard, heavy durable, close-grained, specific gravity 0.83.

RANGE. Cultivated in tropical and semitropical regions of both hemispheres. In Texas and Louisiana; eastward to Florida, northward in New Mexico and Arizona southward on the Texas coast through Mexico, through Central America and northern South America to the Guianas.

MEDICINAL USES. Standley (1920–1926, p. 379) reports that "An ointment made from the flowers is used in Mexico as a remedy for headache, and their infusion for dyspepsia. The green fruit is very astringent, and a decoction is employed for dysentery, inflammation of the skin and mucous membrane, etc. Seler reports even that in San Luis Potosí a decoction of the roots is employed as a supposed remedy for tuberculosis. The pulverized dried leaves are sometimes applied as a dressing to wounds."

VARIETIES. Sweet Acacia apparently occurs in two forms in Texas as a result of environmental factors Plants on the Texas coast often with many stems from the base, more shrubby, tops flattened, branches ascending or somewhat horizontal (these accentuated by distortion because of prevailing Gulf winds). Trees in inland areas often with a more solitary trunk, rounded head and pendulous branches, more arborescent, sometimes to 30 ft, reaching large size, particularly in the neighborhood of Uvalde and Brackettville, Texas.

REMARKS. The genus name, *Acacia*, refers to the hard sharp spinescent stipules of some species. The species name, *farnesiana*, honors Cardinal Odoardo Farnese (1573–1626) of Rome. This species was the first introduced to his gardens in 1611. Our species is cultivated in France under the name of Cassie, but is usually known in Texas as Huisache; however, because

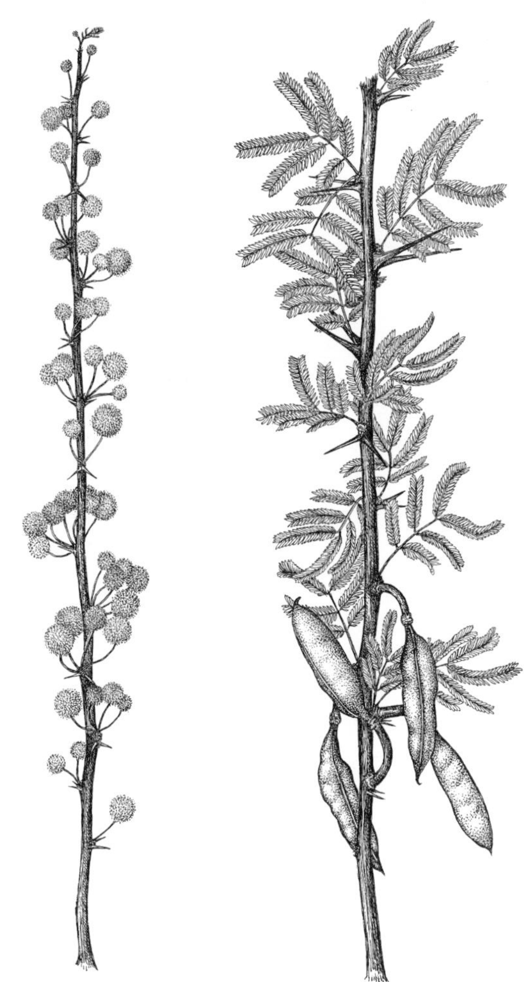

·WEET ACACIA
Acacia farnesiana (L.) Willd.

f confusion with other acacias, it seems best to apply he name of Sweet Acacia as used in Kelsey and Dayon (1942). Vernacular names in the United States and atin America are Acacia-catclaw, Honey-ball, Opopaax, Yellow-opopanax, Popinach, Hinsach, Binorama, 'inorama, Huisache, (from the Nahuatl *huitz-axin*), uisache Yondino, Guisache, Huisache de la Semilla, Iuixachin, Uisatsin, Xkantiriz, Matitas, Finisache, Bihi, spiño, Aroma, Zubin, Zubin-ché, Gabia, Gavia, Subin, roma Amarilla, Espiño Blanco, Cacheto de aroma, uji Cimarrón, Pelá, Uña de Cabra, and Espinillo.

Sweet Acacia is commonly cultivated as a garden rnamental in tropical countries. The wood is used for nany purposes, including posts, agricultural instrunents, and woodenware articles. It is considered a good vinter forage plant and a desirable honey plant in emiarid areas, being resistant to drought and heat. The ark and fruit are used for tanning, dying, and inknaking. Glue from the young pods is used to mend ottery.

BERLANDIER ACACIA
Acacia berlandieri Benth.

FIELD IDENTIFICATION. Spreading shrub, with many stems from the base, or sometimes a small tree to 15 ft. The gray to white branches are thornless or nearly so.

FLOWERS. Blooming November–March, in heads on axillary, solitary or clustered, pubescent peduncles arranged in racemes; the heads white, dense, fragrant, globose, or short-spicate, 3/8 in. or more in diameter; calyx 5-lobed, pubescent, lobes valvate; corolla 5-parted, pubescent; stamens numerous, exserted, distinct, anthers small; ovary densely white-hairy.

FRUIT. Legume matures June–July, 4–6 in. long, 1/2–1 in. wide, linear to oblong, flat, thin, firm, straight or somewhat curved, apex obtuse or apiculate, base stipitate, margins somewhat thickened, surface whitish when young, velvety-tomentose when older, valves dehiscent; seeds 5–10, about 3/8 in. long and broad, one margin straighter than the other, compressed, dark brown, lustrous.

LEAVES. Delicate and almost fernlike in appearance, bipinnate, 4–6 in. long; pinnae 5–9 pairs (sometimes to 18 pairs); leaflets 30–50 pairs (sometimes to 90

BERLANDIER ACACIA
Acacia berlandieri Benth.

pairs); leaflets ⅙–¼ in. long, linear to oblong, oblique, acute, prominently nerved, at first tomentose but more glabrate later; petiolar gland sessile; stipules small, early deciduous.

TWIGS. Gray to white, tomentose when young, when older glabrous, unarmed or prickly.

BARK. Gray, on older stems with shallow fissures and broad, flat ridges.

RANGE. In sandy or limestone soils from the Nueces River Valley westward and southward into Mexico. Abundant along the lower Rio Grande drainage. In Mexico in the states of Nuevo León, Tamaulipas, Querétaro, and Veracruz.

REMARKS. The genus name, *Acacia,* refers to the spinescent stipules of some species, and the species name, *berlandieri,* is for Luis Berlandier, a Belgian botanist who explored the United States–Mexico boundary for the Mexican government in 1828. Vernacular names are Thornless Catclaw, Mimosa Catclaw, Round-flowered Catclaw, Guajillo, Huajilla, and Matoral. This species is a very famous honey plant, producing a clear, white, excellent flavored honey. The wood is sometimes used for fuel locally, or for making handles and small woodenware articles. It is a desirable plant for cultivation both as a specimen plant or hedge plant in the areas suited to it. The fernlike leaves and globose white flower heads are very attractive. It can be propagated by seeds or by young seedlings which come up at the base of the old plant, but these are deep-rooted and hard to transplant.

Catclaw Acacia

Acacia greggii Gray

FIELD IDENTIFICATION. Thorny, thicket-forming shrub or small tree to 30 ft, and 12 in. in diameter. The numerous slender, spreading, thorny branches are almost impenetrable.

FLOWERS. Usually April–October; spikes 1¼–2½ in. long, about ½ in. in diameter, dense, oblong, creamy yellow, fragrant, peduncle usually about one half the length of the spike, sometimes a number of spikes clustered together at the ends of the twigs; calyx green, about 1/12 in. long (half as long as the petals), obscurely 5-toothed, puberulous on the outer surface; petals 5, ⅙–⅛ in. long, greenish, yellowish, and hairy on the margins; stamens numerous, exserted, about ¼ in. long, filaments pale yellow; ovary long-stalked, hairy.

FRUIT. Persistent from July through the winter, legume 2–5½ in. long, ½–¾ in. wide, compressed, straight, curved, or often curling and contorted, constricted between the seeds, apex acute or rounded, sometimes mucronulate, base obliquely narrowed into a short stalk, margins thickened, light brown or reddish, reticulate-veined, valves thin and membranous;

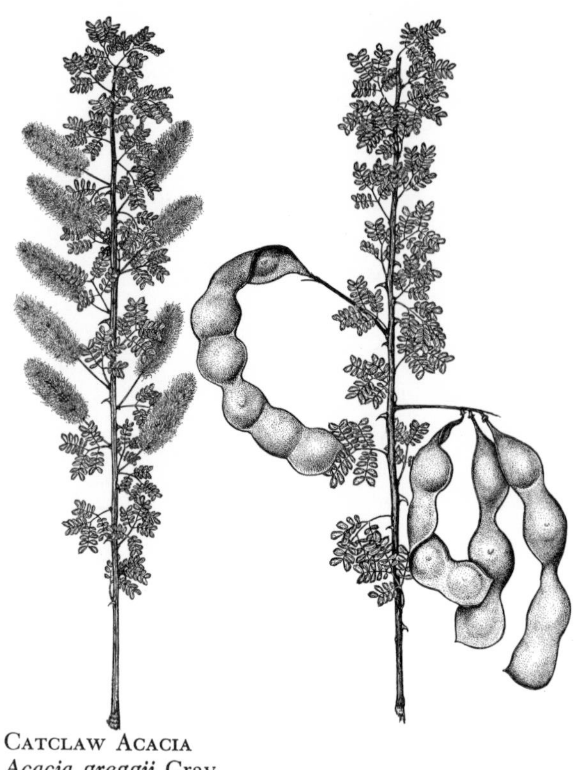

Catclaw Acacia
Acacia greggii Gray

seeds dark brown, shiny, almost orbicular, compressed, ¼–⅓ in. long, germination about 60 per cent.

LEAVES. Bipinnate, 1–3 in. long; pinnae short-stalked, 1–3 pairs, petiole with a minute brown gland near the middle; leaflets 3–7 (usually 3–5) pairs, obovate to oblong; apex rounded, obtuse or truncate, apiculate; base unequally contracted into a short petiolule; blade surface 2–3-nerved, lightly reticulate-veined, pubescent, length ⅙–¼ in.

TWIGS. Pale brown to reddish or gray, pubescent or glabrous; spines ⅛–¼ in. long, dark brown or gray, stout, recurved, flat at base, infrastipular.

BARK. Gray to black, about ⅛ in. thick, furrowed, the surface separating into small thin scales on older trunks.

WOOD. Brown to reddish, sapwood light yellow, close-grained, hard, heavy, strong, durable.

RANGE. At altitudes of 1,000–5,000 ft, on dry gravelly mesas, sides of canyons, and banks of arroyos. In west Texas there are two more or less disjunct ranges. The more southerly of these extends southward from Bexar County to Willacy County; the other from Taylor County southwestward, with heavy concentration in the Big Bend area. Also in New Mexico, Arizona, and Colorado; north to Nevada and Utah, west to California, and south into the Mexican states of Coahuila, Chihuahua, Sonora, and Baja California.

REMARKS. The genus name, *Acacia,* means "hard-pointed" and refers to the spinescent stipules of some of the species. The species name, *greggii*, honors Josiah Gregg (1806–1850), early botanist who collected in the Southwest and northern Mexico. Vernacular names used are Devil Claws, Texas Mimosa, Paradise Flower, Gregg Acacia, Long-flowered Acacia, Huajilla, Chaparral, Gatuña, and Uña de Gato.

Catclaw Acacia often grows in almost impenetrable thickets, furnishing shelter for various birds and mammals. The seeds are eaten by scaled quail, and the leaves are nibbled by jack rabbit when other food is scarce. The wood is used for fuel, small household articles, and singletrees. Cattle browse on the young foliage when grass is scarce, but is not palatable. It apparently stands heavy grazing well and is drought resistant. The fragrant yellow flowers furnish an excellent bee food, and honey from it is of light yellow and of good flavor. The Indians of west Texas, New Mexico, and Arizona made a meal, known as "pinole," from the legumes, which was eaten in the form of mush or cakes. It is also reported that the lac insect, *Tachardia lacca* Kerr., feeds on the sap and exudes the substance from its body. This substance is used as commercial lac. However, the infestation does not appear to be abundant enough to make the gathering of the lac commercially profitable on Catclaw Acacia.

Wright Acacia

Acacia wrightii Benth.

FIELD IDENTIFICATION. Spiny shrub, or sometimes a tree attaining a height of 30 ft, and to 1 ft in diameter. The glabrous, spreading branches form a wide irregular-shaped crown.

FLOWERS. March–May, or at odd times after rains, in cylindric spikes ¾–1½ in. long which are sometimes interrupted; peduncles slender, glabrous or pubescent, solitary or clustered; individual flower pedicels slender, subtended by minute caducous bracts; calyx minutely 5-toothed, pubescent on the outer surface; petals spatulate, slightly united at base, margin ciliate; stamens exserted, about ¼ in. long; ovary hairy, long-stalked.

FRUIT. Ripe June–September, legume often abundant and conspicuous, borne on peduncles ¾–2 in. long; legume 2–4 in. long, 1–1¼ in. wide; margin thick, straight or irregularly contracted or curved; apex rounded and usually short-pointed; base round or oblique and short-stipitate; valves much flattened, papery, thin, finely reticulate-veined, glabrous; seeds compressed, narrow-obovoid, light brown, sometimes marked with oval depressions, length about ¼ in.

LEAVES. Solitary or fascicled, 1–2 in. long, petiole and rachis pubescent, petiole ¼–1⅓ in. long, sometimes with a solitary gland near the apex; pinnae 1–3 pairs, each pinna with 2–6 pairs of leaflets obovate to oblong, apex obliquely rounded to obtuse or retuse, often apiculate; base sessile or short-petiolulate; length ¼–⅚ in.; surface 2–3-nerved and reticulate-veined, glabrous or pubescent, paler green on the lower surface.

TWIGS. Gray to brown or yellowish, mostly glabrous and somewhat striately angled; spineless, or armed with brown, recurved, flattened, sharp pointed infrastipular spines about ¼ in. long.

BARK. Rather thin, gray to brown, divided into broad ridges and shallow fissures, the ridges separating into thin, narrow scales on old trunks.

WOOD. Pale brown to reddish brown, sapwood white to yellow, close-grained, heavy, hard.

RANGE. On dry, rocky, prairie soils. In Texas from the valley of the Guadalupe River westward. Assuming tree habit and of greatest abundance in the vicinity of Brackettville, Uvalde, Sabinal, and Montell. In Mexico in Sonora, Tamaulipas, and Nuevo León.

REMARKS. The genus name, *Acacia*, refers to the sharp, pointed spinescent stipules of some species. The species name, *wrightii*, is in honor of Charles Wright, who collected plants in Texas in 1847–1848 while with the military forces at the Western forts. Wright Acacia can be transplanted or grown from seed and makes an excellent ornamental tree in the drier areas of the state. A few horticultural varieties have been developed. The legume is borne abundantly and is conspicuous. The spikes of yellow flowers make a good bee food, and the wood is sometimes used for fuel and posts. Vernacular names are Tree Catclaw, Texas Catclaw, Uña de Gato, and Negra. Wright Acacia can be distinguished from Catclaw Acacia, *A. greggii*, by the wider legume and leaflets which are twice as large.

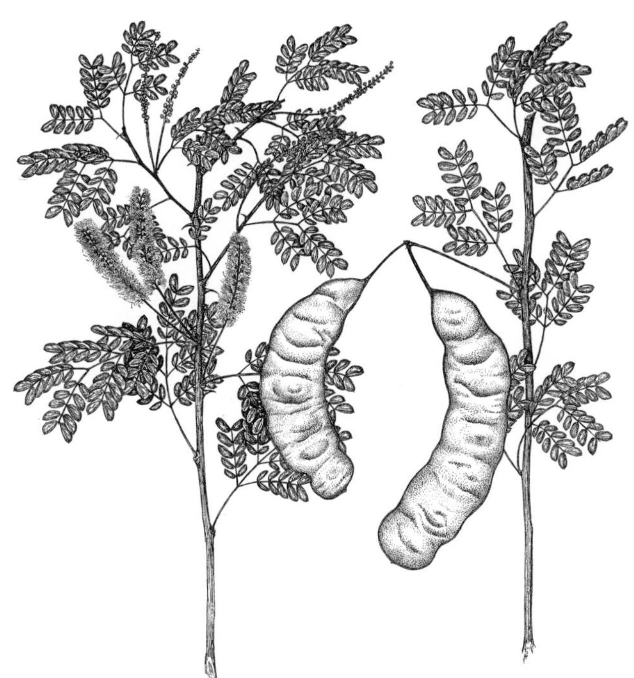

Wright Acacia
Acacia wrightii Benth.

Roemer Acacia

Acacia roemeriana Scheele

FIELD IDENTIFICATION. Prickly, round-topped shrub with many spreading branches. More rarely a small tree to 15 ft or more, with a maximum trunk diameter to 16 in.

FLOWERS. Heads mostly axillary; borne on slender, glabrous peduncles ⅓–1⅓ in. long; flowers white to pale greenish yellow, ¼–½ in. in diameter; calyx small, 5-lobed; corolla 5-lobed; stamens numerous, exserted, distinct, anthers small; ovary stalked.

FRUIT. Legume oblong to linear, 2–5 in. long, ¾–1¼ in. broad, straight or somewhat curved, compressed, leathery, margin entire to lobed or constricted, apex obtuse, base rounded to cuneate and stipitate, surface red or pink at maturity, glabrous and prominently nerved on the edge of the valves.

LEAVES. Bipinnate, 1⅓–4 in. long, pinnae 1–3 pairs, leaflets 3–8 pairs, each leaflet ⅓–⅗ in. long, oblong to oval or cuneate, oblique, obtuse to retuse and often apiculate, glabrous or nearly so, veins prominent beneath.

TWIGS. Gray to brown, smooth, with short curved prickles.

BARK. Gray to brown, smooth on young trunks, breaking into thin, small scales on old trunks.

RANGE. In dry, limestone soil, or on gravelly bluffs or banks. From the valley of the Colorado River south and west to El Paso County, Texas. The type specimen was collected at Austin. Abundant in the vicinity of New Braunfels, San Antonio, Langtry, and Del Rio, Texas. To altitudes of 4,500 ft in the foothills of the Chisos Mountains in Brewster County, Texas. In the lower Rio Grande Valley in Cameron and Hidalgo counties. Also in southern New Mexico near Carlsbad and in Mexico in the states of Coahuila, Chihuahua, and Baja California.

RELATED SPECIES. A plant listed in the literature as *A. malacophylla* Benth. *ex* Gray and *Senegalia malacophylla* (Benth. *ex* Gray) Britt. & Rose bears such close resemblance to *A. roemeriana* that it is doubtfully specific and appears to be only a more pubescent form. It is reported from the uplands of the Leona, western Texas. More field work should be done to determine its exact status.

REMARKS. The genus name, *Acacia*, refers to the sharp-pointed, spinescent stipules of some species. The species name, *roemeriana*, honors Ferdinand Roemer, a German geologist and naturalist who collected specimens, 1845–1847, in the vicinity of New Braunfels, Texas. Vernacular names for the plant are Round-flowered Acacia and Round-flowered Catclaw. It is sometimes planted for ornament, and is an important source of honey.

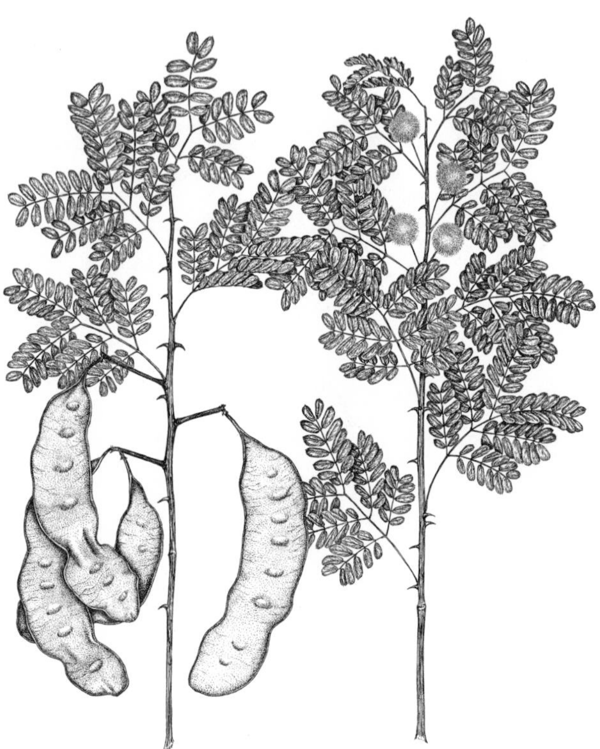

Roemer Acacia
Acacia roemeriana Scheele

Silk-tree Albizia

Albizia julibrissin (Willd.) Durazz.

FIELD IDENTIFICATION. Cultivated tree attaining a height of 40 ft, with a flat top and widely spreading branches, often broader than high.

FLOWERS. May–August, on the upper ends of the branches, in axillary, tassel-like, capitate clusters on slender, pubescent peduncles ½–2 in. long; heads 1½–2 in. broad, flowers perfect; stamens numerous, conspicuous, long-exserted, filamentous, 1–1½ in. long, pink distally, whitish proximally, united at base; pistils considerably longer than stamens, style white or pink, filiform, stigma minute, ovary short-stalked; corolla tubular, greenish, pubescent, ⅛–¼ in. long, 5-lobed, lobes oblong or ovate, obtuse or acute; calyx tubular, $\frac{1}{12}$–⅛ in. long, shallowly 5-lobed, pubescent.

FRUIT. Legume on a pubescent peduncle 1–2½ in. long, oblong to linear, 5–8 in. long, ¾–1 in. wide, margin straight or wavy, apex gradually or abruptly narrowed into a long spinose point, base cuneate; valves thin, flattened, papery, yellowish brown, not separating on margin, without partitions between the seeds; seeds brown, lustrous, flattened, rounded or obtuse at the ends, about ¼ in. long, averaging 11,000 seeds per lb, one fourth to one third germinating.

LEAVES. Alternate, deciduous, twice-pinnately com-

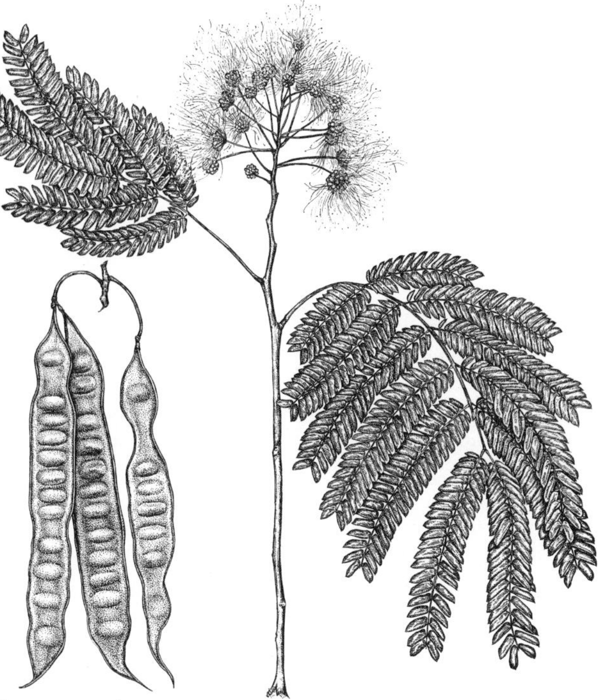

SILK-TREE ALBIZIA
Albizia julibrissin (Willd.) Durazz.

pound, 10–15 in. long; rachis green to brown, smooth or striate, pubescent, often glandular near the base; pinnae 2–6 in. long, 5–12 pairs, rachilla pubescent; leaflets 8–30 pairs per pinna, sessile or nearly so, ¼–⅜ in. long, oblong, slightly falcate, oblique; margin entire, ciliate, straighter on one side than the other; apex half-rounded and mucronulate; base truncate or rounded, attached to the rachilla on one side; main vein parallel to and close to the margin on one side; glabrous to puberulent.

TWIGS. Green to brown or gray, somewhat angular or rounded, glabrous, lenticels small but numerous.

BARK. Smooth, tight, blotched gray, sometimes brownish on young trunks or limbs.

RANGE. A native of Asia, cultivated from Washington, D.C., Philadelphia, and Indianapolis; and Maryland, south to Florida and west into Texas. Commonly planted on the streets of Houston and other Gulf Coast cities.

VARIETIES. Hardy Silk-tree Albizia, *Albizia julibrissin* var. *rosea* (Carr.) Mouillef, is a variety with brighter pink flowers, more dwarf in habit, and somewhat hardier. The color of the flowers seems to be rather variable, ranging from yellowish white to dark red in the species. Abyssinia Silk-tree Albizia, *Albizia julibrissin* var. *mollis* Benth., has leaves broader and densely pubescent. It is from the Himalayas and Abyssinia.

REMARKS. The genus, *Albizia,* was named after F. Degli Albizzi, an Italian nobleman and naturalist. The tree is often called Mimosa in the Southern states. Silk-tree was first introduced into cultivation in 1745 and is considered a top ornamental for the South, being somewhat drouth-resistant and fairly free from disease and insects. However, it has been reported that specimens in Georgia and the Carolinas have been very susceptible to the killing attacks of the root fungus *Fusorium perniciosum.* The tree does not seem to escape cultivation and reproduce readily. The seeds may furnish a limited amount of food for birds, squirrels, and other wildlife. The wood is used for cabinetmaking in its native Asiatic home.

FALSE-MESQUITE CALLIANDRA

Calliandra eriophylla Benth.

FIELD IDENTIFICATION. Low densely branched shrub ½–3 ft, or almost prostrate. Spreading by means of underground stems, the branches and petioles pubescent. Resembling some of the *Mimosa* species, but separated from them by its lack of spines, much longer globose heads of purple flowers, velvety silky-hairy legumes.

FLOWERS. March–May, in large, loose, globose, purple or white heads to 2 in. in diameter; calyx reddish

FALSE-MESQUITE CALLIANDRA
Calliandra eriophylla Benth.

purple, about 1/12 in. long, appressed-hairy; corolla reddish purple, 1/5–1/4 in. long, gamopetalous, elongate-tubular, appressed-hairy; stamens 20 or more, long-exserted, 3/5–1 in. long, filaments united at base, reddish purple, anthers minute and purple.

FRUIT. Maturing June–August, legume solitary or several together, lanceolate to oblanceolate, apex slender-apiculate, tapering from the middle to the base, straight or slightly curved, length 1–3 in., width 1/4–1/5 in., surfaces densely silky-hairy; margins strongly thickened and less hairy than the surface; valves dehiscent longitudinally from apex to base; seeds about 1/4 in. long or less, gray to black, smooth, obovate, germination about 64 per cent.

LEAVES. Bipinnately compound, pinnae 1–4 pairs, each 1/4–1/2 in. long; leaflets 6–12 pairs, imbricate, ovate to oblong, base asymmetrical, length 1/12–1/8 in., width 1/25–1/5 in., thickish, appressed-hairy and reticulate on the surfaces, later more glabrous; stipules setaceous, minute.

TWIGS. Slender, rigid, often divaricate, gray, densely appressed pubescent.

RANGE. Limestone hills, dry gravelly slopes and mesas at altitudes of 2,000–5,000 ft. Trans-Pecos Texas, New Mexico, Arizona, and California. In Mexico from Sonora and Coahuila to Puebla.

RELATED SPECIES. *C. eriophylla* is closely related to, or may be a variation of, *C. conferta* Benth., which has a very wide distribution in Mexico. Since the latter has such an extensive variation, and apparent integration of forms, it seems best to consider all variables in this complex as *C. eriophylla.*

REMARKS. The genus name, *Calliandra,* means "beautiful stamens," with reference to the purple color. The species name, *eriophylla,* is the Greek for "woolly leaf." The plant is also known under the vernacular names of Fairy Duster, Pink Mimosa, Pink-flowered Acacia, Huajillo, Plumita, and Gavia. It is considered to be a valuable browse for livestock and deer, being tolerant of grazing. The seed is eaten by bobwhite, scaled, Gamble's, and Mearns's quail.

RAYADO BUNDLE-FLOWER

Desmanthus virgatus (L.) Willd.

FIELD IDENTIFICATION. Erect or ascending perennial suffrutescent plant 1½–6 ft, the slender branchlets glabrous or sparingly pubescent. Usually herbaceous above, but sometimes woody near the base. A very variable plant represented in Texas by two varieties as described.

FLOWERS. Flowering April–November. Borne in several-flowered axillary peduncled heads; peduncles glabrous, 3/4–2 in. long; flowers green to white, sessile, perfect; calyx campanulate, teeth 5 and short; petals 5, distinct, valvate, about 1/8 in. long, blades spatulate; stamens 10, filaments distinct, exserted; ovary nearly sessile, ovules numerous.

FRUIT. Legumes usually several in a cluster straight or slightly curved, 2½–3½ in. long, 1/8–1/6 in. wide, flat, short-pointed at the apex, not constricted between the seeds, 2-valved, the valves somewhat coriaceous.

LEAVES. Bipinnate, 1½–3¼ in. long; petioles 1/5–2/5 in. long; pinnae 1–7 pairs, with an oblong or elliptic gland 1/25–1/12 in. long between the lowest pair, or rarely below an accessory pair of stipules; leaflets 10–20 pairs, linear to linear-oblong, 1/6–1/3 in. long, glabrous, acute or apiculate; stipules mostly filiform.

RANGE. The species is very variable and is distributed over a wide area including Florida, the West Indies, and central and tropical South America; and in Mexico in Coahuila, Nuevo León, Tamaulipas, San Luis Potosí, Morelos, Oaxaca, and Yucatán. Also in Old World tropics.

VARIETY. Turner Bundle-flower, *D. virgatus* var. *glandulosus* Turner, is described in Turner, (1950b) as follows: "Plant erect to ascending, suffrutescent, perennial, with a deep tap root, 8 in.–2½ ft tall; stems simple below, sparingly branched above; leaves 1½–4½ in. long; pinnae 3–7 pairs, a large oblong gland between the lower pair of pinnae and usually 1–4 glands between the upper pairs of pinnae; stipules not conspicuously auricled at base; resembling *D. virgatus* (typical) in all

TURNER BUNDLE-FLOWER
Desmanthus virgatus var. *glandulosus* Turner

Illinois Bundle-flower
Desmanthus illinoensis (Michx.) MacM.

respects except for the large orbicular glands located between the upper pinnae. . . . Exceptional specimens are found which show only the lower petiolar gland present." Known from Val Verde, Terrell, and Brewster counties in Texas. Also in Coahuila, Mexico.

Prostrate Bundle-flower, *D. virgatus* var. *depressus* (Humboldt & Bonpland *ex* Willd.) Turner, is distinguished from *D. virgatus* by the prostrate (rarely ascending) stems and the consistently smaller petiolar glands at base of lower pinnae only. A rather abundant coastal plant from the Rio Grande plain and northeast along the Texas coast to Harris County.

REMARKS. The genus name, *Desmanthus*, is from the Greek words *desme* ("a bundle") and *anthos* ("flower"). The species name, *virgatus*, is for the virgate or twiggy stems.

Illinois Bundle-flower

Desmanthus illinoensis (Michx.) MacM.

FIELD IDENTIFICATION. Plant herbaceous but somewhat bushy in form. Stems ascending or erect, glabrous or puberulent, 1–3½ ft.

FLOWERS. Flowering June–September. Peduncles ascending, 1¼–2¾ in. long, puberulent to glabrous later; heads ½–1¼ in. in diameter, greenish white, globose; individual flowers perfect, regular; calyx gamosepalous, campanulate, minutely 5-toothed, about 1/15 in. long; stamens 5, long-exserted, filaments distinct; petals 5, distinct, 1/12–⅛ in. long; ovary nearly sessile, ovules numerous.

FRUIT. Borne in a dense subglobose head, ⅝–1½ in. in diameter, legumes oblong or lanceolate, strongly curved, thin, flattened, margin entire or wavy, brownish, glabrous, 2-valved; seeds 2–6, impressed, ⅛–⅕ in. long, nearly as wide.

LEAVES. Bipinnately compound, petiole and rachis puberulent to glabrous; stipules filiform, setaceous, ¼–⅖ in. long; leaves 2½–4 in. long; pinnae 6–14 pairs, ¾–2 in. long, sometimes with a minute gland between them or the lower pair only, or absent; leaflets 20–30 pairs per pinna, lanceolate to linear, acute, glabrous or ciliate, 1/12–⅕ in. long.

STEMS. Green to brown, striate-angled, puberulent to glabrous later.

RANGE. On prairies, pastures, and riverbanks. New Mexico, Oklahoma, Arkansas, Texas, and Louisiana; eastward to Florida, northward to Ohio, and west through Ontario, Minnesota, and North Dakota to California.

REMARKS. The genus name, *Desmanthus*, is from the Greek words *desme* ("a bundle") and *anthos* ("a flower"), referring to the dense flower heads. The species name, *illinoensis*, is for the state of Illinois. It is also listed in some literature under the name of *Acuan illinoensis* (Michx.) Kuntze. Common names are Prairie Mimosa and Prickle Weed. It is only rarely cultivated for ornament.

Gregg Lead-tree

Leucaena greggii Wats.

FIELD IDENTIFICATION. Unarmed shrub or small tree attaining a height of 20 ft, with a trunk diameter of 3–5 in.

FLOWERS. Borne on slender, pubescent, solitary or clustered, peduncles 1–3 in. long; involucre of 2–3 lobed bracts; flowers small, white, dense, sessile, in globose heads ¾–1 in. in diameter; calyx-tube minutely 5-toothed, glabrous except for the pubescent teeth; petals 5, longer than the calyx, narrowly spatulate, glabrous or pubescent, each about ⅕ in. long; stamens 10, free, exserted, about ⅓ in. long; filaments filiform, anthers oblong, versatile; ovary stipitate, style slender, stigma minute and terminal.

FRUIT. Legume narrowly linear, flattened, 6–8 in. long, ⅓–½ in. wide, apex with a subulate beak ½–2 in. long, base narrowed into a short, stout stipe, cinereous-pubescent when young, at maturity glabrous, margins narrow and thickened; valves impressed between the seeds, seeds compressed and transverse.

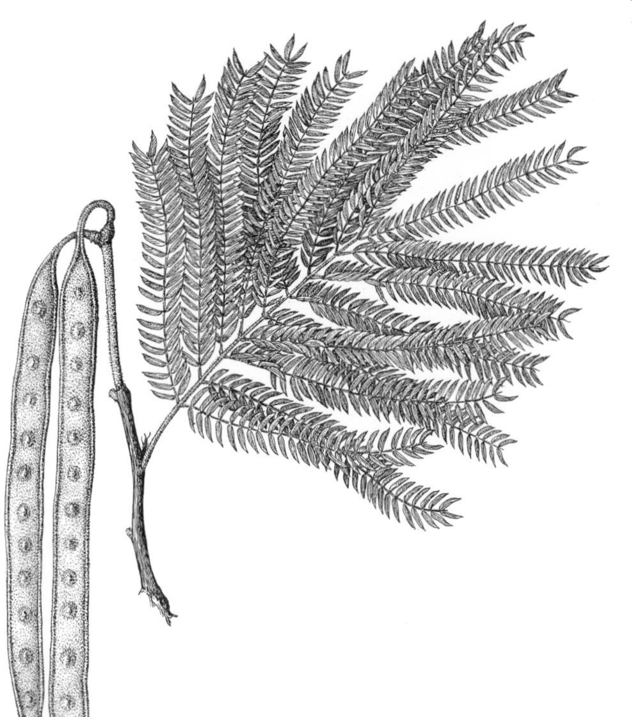

GREGG LEAD-TREE
Leucaena greggii Wats.

LEAVES. Bipinnate, 5–9 in. long and broad, rachis with a solitary, subcylindric gland between each pair of pinnae; pinnae 10–18, short-stalked; leaflets 15–30 pairs, lanceolate to linear, often somewhat falcate, apex acute or acuminate, base rounded on proximal side, almost straight on the distal side; grayish green, puberulent to glabrous, midvein nearly central and lateral veins obscure, sessile or short petiolulate, blade length ¼–⅓ in., width about ⅛ in.; petioles pubescent, ¾–1⅓ in. long; stipules persistent, rigid, acuminate, ⅓–½ in. long.

TWIGS. Reddish brown, stout, somewhat divaricate, pubescent at first, later glabrous, lenticels pale and numerous.

BARK. Dark brown, with low ridges which break into small, appressed scales on old trunks.

WOOD. Reddish brown, sapwood paler, close-grained, heavy, hard, specific gravity 0.92.

RANGE. Dry well-drained soil on the edges of rocky ravines. In Texas confined to a limited area from the upper San Saba to the Devil's River. Southward into Chihuahua, Coahuila, and Nuevo León, Mexico.

REMARKS. The genus name, *Leucaena,* refers to the white flowers of several species, and the species name, *greggii,* is in honor of Josiah Gregg (1806–1850). The Gregg Lead-tree is generally too small to be of much value for wood.

GOLDEN-BALL LEAD-TREE

Leucaena retusa Benth.

FIELD IDENTIFICATION. Shrub or small tree to 25 ft, and 8 in. in diameter.

FLOWERS. April–October, borne on stout, single or fascicled, axillary peduncles 1½–3 in. long and subtended by 2 villose bracts; bracts ⅙–¼ in. long, apex subulate; flower heads yellow, globose, dense, ¾–1¼ in. in diameter; calyx tubular, minutely 5-toothed, membranous, $\frac{1}{12}$–⅛ in. long; petals 5, free, narrowly oblong, barely longer than the calyx; stamens 10, distinct, free, exserted, ¼–⅓ in. long; filaments filiform, anthers glabrous and oblong; ovary stipitate, style slender, stigma minute and terminal.

FRUIT. Generally mature in August, solitary or clustered, borne on stout puberulous peduncles 1½–5 in. long; pods 3–10 in. long, ⅓–½ in. wide, stipitate, narrowly linear, flattened, somewhat constricted on the thickened margin, apex acute or acuminate, base

GOLDEN-BALL LEAD-TREE
Leucaena retusa Benth.

cuneate; rigid, thin, papery, glabrous; light to dark brown; seeds numerous, obliquely transverse, compressed, about ⅓ in. long and ¼ in. wide; seed coat thin, crustaceous, brown and lustrous.

LEAVES. Alternate, bipinnate, stipellate, 3–8 in. long, 4–5 in. wide, petiole and rachis slender; pinnae 2–5 pairs (usually 3–4), long-stalked and distant; leaflets 3–8 pairs (usually 3–6 per pinna), short-stalked; oblong to elliptic, or the upper obovate, basally asymmetric; apex rounded or obtuse to retuse, mucronulate; base rounded or cuneate; margin entire; length ⅓–1 in., width ⅓–½ in.; surface glabrous or pubescent, reticulate-veined, bluish green, thin; glands usually solitary and elevated, borne between the pinnae or leaflets; stipules ovate to lanceolate, subulate-tipped, ⅖–⅗ in. long.

TWIGS. Slender, grooved, brown to reddish, pubescent or puberulous, lenticels numerous and oval.

BARK. Light gray to brown, broken into small thin scales on old trunks, smooth on young branches.

RANGE. Mostly on dry, well-drained, rocky, limestone soils in central Texas and the Trans-Pecos area. Also in New Mexico and Mexico.

REMARKS. The genus name, *Leucaena,* refers to the white flowers of some species, and the species name, *retusa,* refers to rounded, sometimes concave-tipped, leaflets. It is apparently not abundant enough to warrant the application of other vernacular names except that of Mimosa and Wahoo-tree. Much browsed by cattle. Worthy of cultivation in the central and west Texas limestone areas because of the attractive flowers, fruit, and leaves.

WHITE POPINAC LEAD-TREE
Leucaena glauca (L.) Benth.

WHITE POPINAC LEAD-TREE
Leucaena glauca (L.) Benth.

FIELD IDENTIFICATION. Shrub or tree to 30 ft, and 4–6 in. in diameter. Branches unarmed, numerous and spreading, crown irregularly rounded, trunks short.

FLOWERS. Appearing after 3–4 years of age, borne usually on 1–3 axillary or terminal pubescent peduncles, ¾–1⅓ in. long; heads globose, dense, white or pink, ⅝–1¼ in. in diameter, the florets small, sessile, and perfect. Calyx 1/25–1/12 in. long, obconic, pubescent, teeth 5, short, minute, obtuse; petals 5, erect, distinct, linear-spatulate, ⅙–⅕ in. long or about twice as long as the calyx; stamens 10, distinct, exserted, about three times as long as the petals; ovary stipitate, pubescent, ovules numerous, style filiform, stigma minute.

FRUIT. Legume 4–9 in. long, about ⅝ in. wide, linear to oblong, obtuse to acute or abruptly pointed, base tapering, surface reddish brown, flat, finely pubescent, membranous; seeds 16–20, transverse, ovate, flattened, about ⅜ in. long, mottled-gray to black.

LEAVES. Alternate, bipinnate, 4–8 in. long, petiole 1¼–2½ in. long and glandular or glandless; pinnae 3–10 pairs; leaflets 10–20 pairs, oblique, linear-oblong or lanceolate, apex acute to obtuse, margin entire, thin, length ⅓–⅗ in., light green above, paler below.

TWIGS. Brown, slender, puberulent to glabrous, divaricate, lenticels numerous; older bark dull, reddish brown, with flattened overlapping ridges.

WOOD. Light brown, close-grained, hard, moderately heavy.

RANGE. Mostly in sandy soil of semitropical and tropical areas. Naturalized and occasionally escaping cultivation in southern Florida and the Keys. Also the lower Rio Grande Valley area of Texas in Cameron County between Brownsville and Boca Chica.

REMARKS. The genus name, *Leucaena,* is from the Greek word *leukos* ("to whiten"), referring to the flower colors. The species name, *glauca,* refers to the glaucous or bloom-covered foliage. Known in American tropics under the name of Jumby-bean, Xamim, Uaxi, Guacis, Aroma Blanca, Hediondilla, and Granolino. According to legend, if horses, mules, or pigs eat the plant their hair will fall out. However, goats and cattle are said not to be affected. The seeds are made into necklaces and bracelets, and are sometimes cooked and eaten with rice. It is also reported that the roots have emmenagogic and abortive properties. The plant may be cultivated in subtropical areas, and is planted as hedges in the Old World tropics.

BERLANDIER MIMOSA
Mimosa berlandieri Gray

GREAT LEAD-TREE

Leucaena pulverulenta (Schlecht.) Benth.

FIELD IDENTIFICATION. Tree attaining a height of 55 ft and a diameter of 2 ft. The smooth, gray to brown branches spread into a broad rounded crown.

FLOWERS. White, sweet-scented, perfect, in dense globose sessile heads about ½ in. in diameter, the slender peduncles ¾–1¾ in. long and fascicled in the axils; calyx 1/25–1/12 in. long, campanulate, 5-lobed, lobes obtuse; petals 5, distinct, linear to spatulate, acute, three to four times as long as the calyx, pubescent with appressed hairs; stamens 10, long-exserted, much longer than the petals; style filiform, ovary hairy, ovules numerous.

FRUIT. Legume on a stipe ⅜–1¾ in. long, often borne in pairs, 4–14 in. long, ⅔–¾ in. wide, flat, thin, glabrous, coriaceous, 2-valved, margins thickened, apex with a short point; seeds transversely arranged in the legume, 3/16—5/16 in. long.

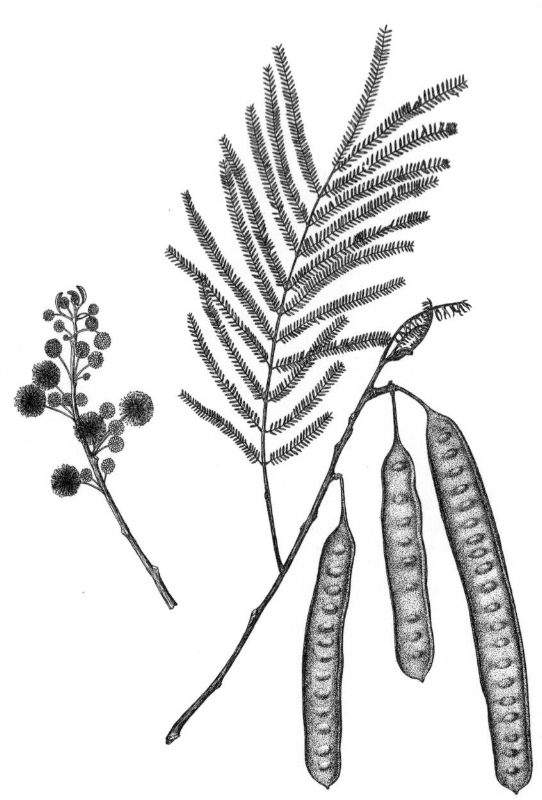

GREAT LEAD-TREE
Leucaena pulverulenta (Schlecht.) Benth.

LEAVES. Alternate, bipinnately compound, length 4–10 in., width 3–4 in.; petiole long, slender, often bearing a dark, round gland between the base and lowest pair of pinnae; pinnae 28–36, short-stalked; leaflets 15–60 pairs, linear, oblique, acute or obtuse, glabrous or puberulous, ⅛–¼ in. long, usually less than 1/12 in. broad, petiolules short or sessile, stipules minute and caducous.

TWIGS. White-tomentose at first, later almost glabrous or puberulous, pale brown, lenticels numerous, longitudinal.

BARK. Gray to cinnamon-brown, smoothish, later broken into short narrow scales.

WOOD. Dark brown, sapwood yellowish, close-grained, heavy, hard, specific gravity 0.67.

RANGE. Along streams and resacas. In Texas in Cameron, Kleburg, Willacy, Hidalgo, and Starr counties. In Mexico in the states of Nuevo León and Tamaulipas to Veracruz.

REMARKS. The genus name, *Leucaena,* is from the Greek *"leukos,"* meaning "white," in reference to the flowers. The species name, *pulverulenta,* is for the apparently dusty stems and flower parts. It is also known under the names of Mexican Lead-tree, Quiebra-hacha, and Tepeguaje. The wood is occasionally used for lumber and fence posts. The Great Lead-tree is a handsome tree with lacy foliage, and is often planted as a yard and street tree in the lower Rio Grande Valley area of Texas. It is very common in the city of Brownsville, Texas.

Berlandier Mimosa

Mimosa berlandieri Gray

FIELD IDENTIFICATION. Shrub to 9 ft, armed with stiff, straight spines.

FLOWERS. Peduncles slender, setose, ¾–2 in. long, axillary; flower heads pink, globose, perfect; calyx and corolla 4–5-parted; stamens 8–10, filaments distinct, exserted; ovary sessile or nearly so.

FRUIT. A narrowly oblong to linear legume, 1½–3 in. long, ⅓–½ in. wide, straight or somewhat curved, densely hispid, 8–15-jointed and -seeded, joints often falling away to leave the rim intact, base short-stipitate, apex rounded to obtuse and mucronate.

LEAVES. From 1¼–4 in. long, rachis and petiole pubescent; with or without spines; pinnae 4–6 pairs, 1–1½ in. long; leaflets many pairs (40–80 per pinna), linear to oblong, apex acute, glabrate or pubescent, ⅙–¼ in. long.

TWIGS. Puberulent to pubescent, little if at all setose, bearing stiff and straight, or slightly recurved, spines.

BARK. Reddish brown, smooth, tight, lenticels numerous.

RANGE. Low grounds on the edges of water courses or resacas. Mostly in the lower Rio Grande Valley of Texas. Collected by the author at Boca Chica in Cameron County, and near Port Lavaca, Texas. In Mexico in Tamaulipas, San Luis Potosí, Nayarit, Sinaloa, Chiapas, and Tabasco.

REMARKS. The genus name, *Mimosa,* comes from the word *mimos* ("to mimic"), referring to the animal-like rapidity of the closure of the leaves of some species. The species name, *berlandieri,* honors Luis Berlandier, an early Belgian botanist, who was sent as a naturalist by the Mexican government with the Mier Teran Expedition to determine the character of the United States–Mexico boundary. Other vernacular names are Zarza, Chaven, and Espina de Vaca. Some authorities list this species under the name of *M. pigra* L.

Catclaw Mimosa
Mimosa biuncifera Benth.

Catclaw Mimosa

Mimosa biuncifera Benth.

FIELD IDENTIFICATION. Spiny, thicket-forming shrub to 8 ft.

FLOWERS. Peduncles short, slender, pubescent; flowers in globose heads which are pale pink to whitish and fragrant; calyx and corolla pubescent, each 5-lobed; stamens 10, exserted, distinct, ovary sessile or nearly so.

FRUIT. Matures in September, legume linear, puberulent or glabrate, curved or straight, ¾–1½ in. long, ⅛–⅙ in. wide, constricted between the seeds, margins prickly or unarmed.

LEAVES. Bipinnate, leaves with petioles ⅝–1¾ in. long, internodes of rachis about ⅕ in. long or less; pinnae crowded, 4–10 pairs; leaflets 5–12 pairs, glabrous or pubescent, obtuse, linear to oblong, 1/25–⅙ in. long.

TWIGS. Angled, flexuous, pubescent; spines stout, recurved, flattened at base, pubescent or glabrous, solitary or paired.

RANGE. On dry hills and mesas, central and west Texas; westward through southern New Mexico to Arizona, south in Mexico in Nuevo León, Tamaulipas, Durango, Coahuila, Puebla, San Luis Potosí, and Chihuahua.

REMARKS. The genus name, *Mimosa,* is from the Greek word, *mimos* ("to mimic"), with reference to the animal-like rapidity of leaf movements. The species name, *biuncifera,* refers to the stout, paired thorns. Vernacular names are Uña de Gato, Gatura, Wait-a-bit, and Wait-a-minute. This species has possibilities as an erosion-control plant, and also for use as an ornamental hedge. It is browsed by deer, and by livestock during stress periods. The flowers are a good source of honey, and the seeds are eaten by Gambel's and scaled quail.

Lindheimer Mimosa

Mimosa lindheimeri Gray

FIELD IDENTIFICATION. An erect, low, very prickly shrub.

FLOWERS. In globose, pink heads, on puberulent peduncles ¼–½ in. long; calyx and corolla each 4–5-lobed and glabrous; stamens 10, exserted, filaments distinct; ovary sessile or nearly so.

FRUIT. Legume linear to oblong, straight or falcate, ¾–1½ in. long, ⅙–¼ in. wide, constricted between the seeds or unconstricted, 3–4-jointed, margin sparingly prickly.

LEAVES. Bipinnate, small, pinnae 2–7 pairs (usually 5–7 pairs); leaflets 7–12 pairs per pinna, leaflets oblong, 1/25–1/16 in. long, glabrous or barely puberulent, sessile or very short petiolulate, stipules setaceous.

Lindheimer Mimosa
Mimosa lindheimeri Gray

TWIGS. Divaricate, glabrous or nearly so; prickles solitary or paired, recurved, flattened, infrastipular.

RANGE. Foothills and dry canyons at altitudes of 3,000–6,000 ft. From the Edwards Plateau of central Texas and westward through New Mexico to Arizona. The type specimen was collected at New Braunfels, Texas, but the species has since been collected in Bexar, Val Verde, Jeff Davis, and Brewster counties, and is also known in Mexico in Coahuila, San Luis Potosí, Puebla, and Michoacán.

RELATED SPECIES. Lindheimer Mimosa is closely related to the Catclaw Mimosa, *M. biuncifera* Benth. In fact, by some it is assigned the status of a variety of Catclaw Mimosa as *M. biuncifera* Benth. var. *lindheimeri* (Gray) Robins. Also the name once applied as *M. biuncifera* Benth. var. *glabrata* Gray may have an application here. The principal differences between Catclaw Mimosa and Lindheimer Mimosa are the more glabrous flowers and leaves, and fewer pinnae on an average in the latter. However, intergrading forms are found.

REMARKS. The genus name, *Mimosa,* is from the Greek word *mimos* ("to mimic"), because of the animal-like rapidity of the movements of the leaves of certain species. The species name, *lindheimeri,* honors Ferdinand Lindheimer (1801–1879), a native of Germany who resided 30 years in Texas. Most of his plants were reported on by Asa Gray and George Engelmann.

Warnock Mimosa

Mimosa warnockii Turner

FIELD IDENTIFICATION. Prickly shrub attaining a height of 3 ft, growing in a low moundlike or depressed manner.

FLOWERS. Borne in globose heads; peduncles about ⅝ in. long or shorter; calyx and corolla small, 4–5-lobed, densely white pubescent; stamens 8–10; pistil stipitate.

FRUIT. Legume ¾–1¼ in. long, 1/12–⅛ in. wide, falcate, margins with straight or curved spines, margins usually straight but sometimes constricted between the seeds, surfaces short-villous, valves chartaceous and separating completely from the margin at maturity.

LEAVES. Twice-pinnate, pinnae 3–8 pairs (rarely to 10 pairs); leaflets 7–12 pairs per pinna, oblong to linear, apex obtuse, glabrous or puberulent, 1/25–⅕ in. long.

TWIGS. Flexuous, curved downward, slender, puberulent, prickles solitary or paired, rather connate, straight or slightly recurved.

RANGE. Trans-Pecos Texas in Jeff Davis County; west through New Mexico to Arizona and Sonora, Mexico.

RELATED SPECIES AND SYNONYMS. The plant is closely related to *M. biuncifera*. It has been listed under the names of *M. biuncifera* var. *flexuosa* Robins., *M. flexu-*

WARNOCK MIMOSA
Mimosa warnockii Turner

osa Benth., and *Mimosopsis flexuosa* (Benth.) Britt. & Rose.

REMARKS. The genus name, *Mimosa,* is from the word *mimos* ("to mimic"), with reference to the quick, animal-like movement of the leaves of some species. The species name, *warnockii,* is in honor of Dr. Barton Warnock, professor of botany at Sul Ross State College, Alpine, Texas.

EMORY MIMOSA

Mimosa emoryana Benth.

FIELD IDENTIFICATION. Low shrub with numerous spiny branches.

FLOWERS. Axillary, on densely hairy peduncles ½–1 in. long; flowers clustered in globose, pink heads about ½ in. wide or less; calyx densely hairy, about 1/25 in. long, 5-lobed; corolla tubular, about 1/16–⅛ in. long, densely hairy, 5-lobed, lobes ovate to oblong; stamens 10, conspicuously exserted, distinct, ovary sessile or nearly so.

FRUIT. Legume ¾–1½ in. long, ⅙–¼ in. wide, linear to oblong, often curved, apex acute and spiniferous, base obtuse to rounded, margin thickened, usually somewhat constricted, spiny or entire, surface setose with spines and also softly pubescent, at maturity reddish brown; seeds ⅛–3/16 in. long, black, compressed, oval or slightly beaked at one end and rounded at the other.

LEAVES. Twice-pinnately compound, usually less than ⅝ in. long; rachis of the leaf densely pubescent, with or without tiny prickles; pinnae 1–3 pairs, each pair about ¼–⅜ in. long; leaflets 3–7 pairs per pinna, often half-overlapping each other, 1/12–⅕ in. long, oblong, oblique, apex obtuse and minutely mucronate, base rounded and unequal, surfaces grayish green and densely soft-hairy on both sides.

TWIGS. Slender, somewhat grooved, younger ones reddish brown and densely puberulent, older stems brown to gray or black, glabrous; spines infrastipular and stipular, mostly curved, a few straight, flattened at base, brown to gray, on young stems puberulent, on older stems glabrous, usually less than ¼ in. long.

RANGE. In Trans-Pecos Texas and northern Mexico in Chihuahua and Durango at altitudes of 2,000–4,000 ft.

REMARKS. The genus name, *Mimosa,* is from the Latin *mimos* ("to mimic"), in reference to the movement of the leaflets. The species name, *emoryana,* honors W. H. Emory, who was in charge of making a government survey of the United States–Mexico boundary in 1854.

EMORY MIMOSA
Mimosa emoryana Benth.

Wherry Mimosa

Mimosa wherryana (Britt.) Standl.

FIELD IDENTIFICATION. Prickly shrub 2–3½ ft.

FLOWERS. May–September, on nearly filiform, axillary peduncles, ¼–¾ in. long, puberulent to glabrous; flowers numerous, borne in globose heads, glabrous or the calyx-teeth ciliolate, calyx and corolla small, 4–5-lobed; stamens 8–10.

FRUIT. A narrowly oblong legume, straight or somewhat falcate, flattened, glabrate, ½–1 in. long, ⅕–¼ in. wide, both margins usually with sharp, slender prickles (more rarely unarmed); valves chartaceous, the margins separating at maturity.

LEAVES. Twice-pinnate, ½–¾ in. long, rachis puberulent or short-pubescent; pinnae 1–3 pairs; leaflets 3–6 pairs per pinna, oblong, obtuse or rounded, 1/25–1/12 in. long, puberulent or glabrous.

TWIGS. Slender, zigzag, at first puberulent but later becoming glabrous, gray to dark brown or black; prickles small, mostly solitary, recurved.

RANGE. Southern Texas, type from Arroyo del Tigre, between Roma and Zapata, Texas. Also collected at Rio Grande City, Texas, by B. C. Tharp, his 5 sheets dated September 9, 1929, deposited in the University of Texas Herbarium. Evidently of a very limited range in southmost Texas and adjacent Tamaulipas, Mexico.

REMARKS. The genus name, *Mimosa,* is from the Greek word *mimos* ("to mimic"), suggesting the rapid animal-like movement of the leaves of some species. The species name, *wherryana,* is for the American botanist Edgar Theodore Wherry (b. 1904).

Wherry Mimosa
Mimosa wherryana (Britt.) Standl.

Pringle Mimosa

Mimosa pringlei Wats.

FIELD IDENTIFICATION. Rare Western, small to large densely branched shrub with short recurved spines.

FLOWERS. Rather numerous along the stems, axillary usually with the small leaves and spines at the base peduncles slender, ⅖–¾ in. long, puberulent; flowers in globose heads, white or pink, the heads about ½ in across; calyx very small, about 1/25 in. long, 4–5-lobed puberulent; corolla 4–5-lobed, lobes oblong to linear 1/16–⅛ in. long; pubescent stamens 8–10, long-exserted many grouped together to form the head.

FRUIT. Legume ¾–1¼ in. long, ¼–⅓ in. wide, straight or curved, linear, base acute, apex acute or obtuse margins unarmed or with a few scattered spines, not deeply segmented, flattened, glabrous, valves chartaceous, dehiscent at maturity; seeds about 3/16 in. long dark brown, flattened, ovoid to oval, sometimes bluntly beaked, asymmetrical.

LEAVES. Twice-pinnately compound, ¼–¾ in. long rachis pubescent, channeled, sometimes with tiny prickles; pinnae 1 pair, each about ⅜ in. long; leaflets 2–4 pairs per pinna, each 1/12–⅙ in. long, short-oblong to oval, oblique, apex rounded to obtuse, base unequally rounded, margin entire, surfaces appressed pubescent with minute hairs.

TWIGS. Slender, somewhat zigzag, when young reddish brown and puberulent, when older brown to dark gray or black, armed with small paired or solitary reddish brown to gray spines.

RANGE. In the state of Chihuahua, Mexico, and reported in adjacent Texas. A sheet, which was examined by the author, is deposited in the Missouri Botanical Garden Herbarium from C. G. Pringle, No. 724, year 1886, and gives the habitat as "cool slopes on rocky hills near Chihuahua City, Chihuahua, Mexico."

REMARKS. The genus name, *Mimosa,* is from the Greek word *mimos* ("to mimic"), referring to the sensitive leaf movement of animal-like speed in some species. The species name, *pringlei,* honors C. G. Pringle, who collected plants in the 1800's in Mexico and the southwestern United States.

Fragrant Mimosa

Mimosa borealis Gray

FIELD IDENTIFICATION. Prickly, much-branched, rigid glabrous to puberulent shrub to 6 ft.

FLOWERS. Peduncles slender, to ¾ in. long; flowers in heads, showy, pink, glabrous; calyx 5-lobed, corolla 5-lobed; stamens 10, distinct, exserted, anthers yellow ovary sessile or nearly so.

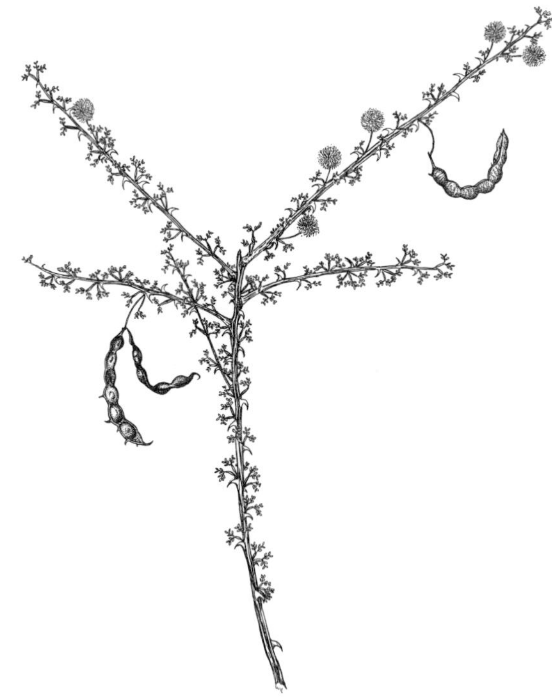

PRINGLE MIMOSA
Mimosa pringlei Wats.

FRUIT. Legume linear to oblong, stipitate, 1–2 in. long, 1/4–1/3 in. wide, falcate, more or less constricted between the 2–7 seeds, glabrous, flat, margin usually unarmed or with a few stout prickles, jointed, breaking away, partitions persistent.

LEAVES. Bipinnate, glabrous or puberulent, pinnae 1–3 pairs; leaflets 3–8 pairs per pinna, oblong to oval, 1/12–1/4 in. long, 1-3-nerved, apex obtuse to rounded, petiolules less than 1/25 in. long, internodes of petiole about 3/8 in. long; rachis exceeding upper pinnae axes as a small spur.

STEMS. Slender, brittle, glabrous or puberulent; prickles straight or apically curved, mostly solitary and stout; a few extra infrastipular ones sometimes present.

RANGE. Gravelly and limestone hills. Oklahoma, New Mexico, and Texas. In Texas collected by the author in Bexar, Travis, Kerr, and Brewster counties. The type locality is on the rocky banks of the Pedernales River in Texas. Also in northern Mexico.

SYNONYMS. Because of its variability a number of other names have been applied to it, such as *M. fragrans* Gray, *M. texana* Small, and *M. borealis* var. *texana* Gray.

REMARKS. The genus name, *Mimosa*, is from the word *mimos* ("to mimic"), referring to the quick animal-like movement of the leaves of some species. The species name, *borealis*, refers to its northward range into Oklahoma. Vernacular names are Pink Mimosa, Catclaw Mimosa, and Sensitive Mimosa. A beautiful plant when in full flower with a possibility for ornamental use. It is sometimes grazed by livestock but suffers from overgrazing.

VELVET-POD MIMOSA
Mimosa dysocarpa Benth.

FIELD IDENTIFICATION. Velvety shrub to 6 ft, with large, flat prickles.

FLOWERS. In summer, arranged in axillary, pink, fragrant, showy, spikes 3/8–1 in. long and about 1/2 in. in diameter; calyx small, pubescent, purple, the 5 teeth minute, 1/25–1/12 in. long; corolla 5-lobed, pubescent; stamens 10, distinct, filaments purplish pink, about 1/3 in. long.

FRUIT. A legume maturing in August, 1–2 in. long, 1/8–1/4 in. wide, linear, often falcate, more or less constricted between the seeds, acuminate, velvety, reddish brown-tomentose, valves jointed, segments separating at maturity, flattened but thickish; seeds brown, angular, obtuse or acute, 1/12–1/8 in. long.

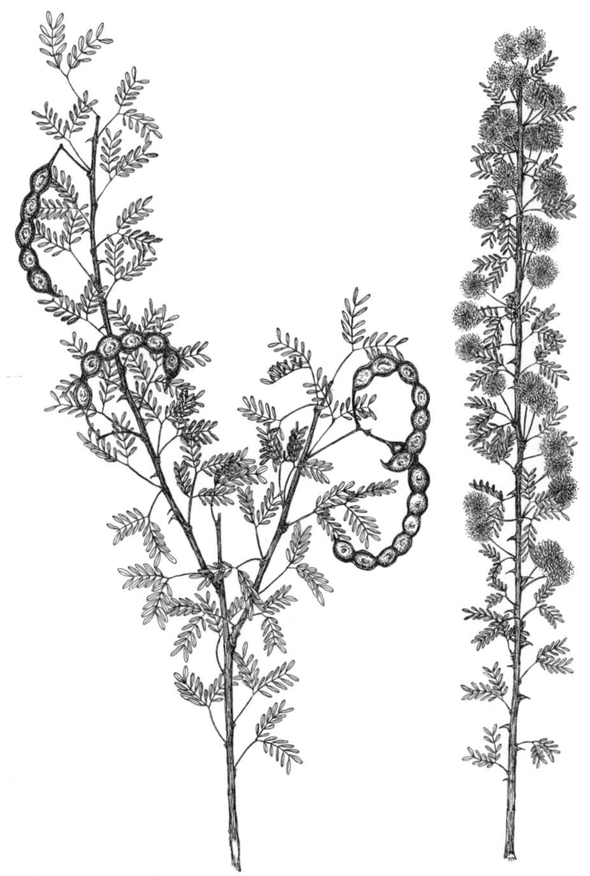

FRAGRANT MIMOSA
Mimosa borealis Gray

VELVET-POD MIMOSA
Mimosa dysocarpa Benth.

LEAVES. Bipinnate, some small prickles on leaf axes; pinnae 5–12 pairs; leaflets 6–16 pairs per pinna, crowded, leaflets oblong to linear, acute, pubescent above and below, ⅛–¼ in. long, less than half as wide.

TWIGS. Brown to yellowish, villous, very prickly; prickles recurved, stout, yellowish, ⅕–⅓ in. long, distributed irregularly.

RANGE. In Trans-Pecos Texas, through New Mexico to Arizona, at altitudes of 2,500–6,500 ft; south in Mexico in Durango, Chihuahua, and Sonora.

VARIETY. A variety known as Wright Velvet-pod Mimosa, *M. dysocarpa* Benth. var. *wrightii* (Gray) Kearney & Peebles, has narrow pods and is mostly unarmed, with the leaves more glabrous. However, there are numerous intergrading forms between this variety and the species.

REMARKS. The genus name, *Mimosa,* refers to the word *mimos* ("to mimic"), the quick movements of the leaves of some Mimosa species mimicking the movement of animals. The species name, *dysocarpa,* refers to the legume but the exact meaning is obscure. The seeds of Velvet-pod Mimosa are eaten by Gambel's and scaled quail, and the foliage is browsed somewhat by livestock. The plant has ornamental possibilities within its zone.

GRAHAM MIMOSA

Mimosa grahamii Gray

FIELD IDENTIFICATION. Spreading or decumbent shrub 1–2 ft. The branches somewhat angled, glabrous and prickly.

FLOWERS. April–July, heads white, globose, ½–¾ in. wide, individual flowers small and sessile; corolla about ⅙ in. long, 4–5-lobed, reddish; stamens distinct, white, about ⅓ in. long, anthers yellow; calyx small, 4–5-lobed.

FRUIT. Legume ⅜–1½ in. long, ⅕–⅓ in. wide, linear-oblong, flattened, margins straight or nearly so, prickly or entire, apex rounded and often apiculate, surface glabrous, valves dehiscent and chartaceous, segments 1-seeded.

LEAVES. Alternate, 2–4 in. long, twice-pinnately compound, petiole about ⅜ in. long, rachis prickly; pinnae 4–8 pairs and well separated (about ¼ in.); leaflets 8–15 pairs, blades oblong, length ⅙–¼ in., width 1/25–1/1[illegible] in., apex acute, distinctly veined.

GRAHAM MIMOSA
Mimosa grahamii Gray

Vine Mimosa
Mimosa malacophylla Gray

TWIGS. Young ones reddish brown, pubescent, later glabrous, grooved and ridged; prickles slender, broadened at base, straight or nearly so.

RANGE. Dry hillsides or mesas at altitudes of 3,500–6,000 ft. New Mexico to Arizona; south in the Mexican states of Sonora and Chihuahua.

VARIETY. A variety with abundant pubescence of flowers, leaves, and fruit has been classified as Lemmon Mimosa, *M. grahamii* var. *lemmonii* Kearney & Peebles.

REMARKS. The genus name, *Mimosa,* is from the word *mimos* ("to mimic"), and refers to the sensitive leaves of some species.

Vine Mimosa

Mimosa malacophylla Gray

FIELD IDENTIFICATION. A vinelike, decumbent or climbing shrub with stems 6–12 ft long, bearing recurved prickles.

FLOWERS. White, borne in terminal, condensed, oblong spikes on peduncles ⅓–1¼ in. long, corolla 5-lobed, pubescent; calyx 5-lobed, minute, pubescent; stamens 10; ovary sessile or nearly so.

FRUIT. Maturing August–September, 2–3 in. long, ⅓–⅖ in. wide, curved, long-stipitate, glabrous, shiny, veiny, 6–8-jointed and -seeded, joints sometimes falling out to leave rim of legume.

LEAVES. Petiole and rachis short-prickly; pinnae 3–7 pairs (usually 3–5); leaflets 3–8 pairs (usually 3–6), leaflets ⅓–⅗ in. long, oval to oblong or obovate, apex rounded to obtuse or acute, often apiculate, pubescent on both sides.

STEMS. Decumbent or climbing, soft-woody, tomentose, armed with many short, reflexed prickles.

RANGE. The lower Rio Grande Valley area of Texas, south into adjacent Tamaulipas and Nuevo León, Mexico.

VARIETY. A form with merely pubescent stems and branches, and larger, glabrous leaves is Smooth Vine Mimosa, *M. malacophylla* Gray var. *glabrata* Benth. At one time it was listed in the literature under the species name of Wooton Mimosa, *M. wootonii* Standl. However, it is so similar to *M. malacophylla,* except for more glabrous characters, that a varietal standing seems preferable. The type specimen comes from Hacienda Buena Vista in Tamaulipas, Mexico.

REMARKS. The genus name, *Mimosa,* is from the word *mimos* ("to mimic"), and refers to the rapidly moving

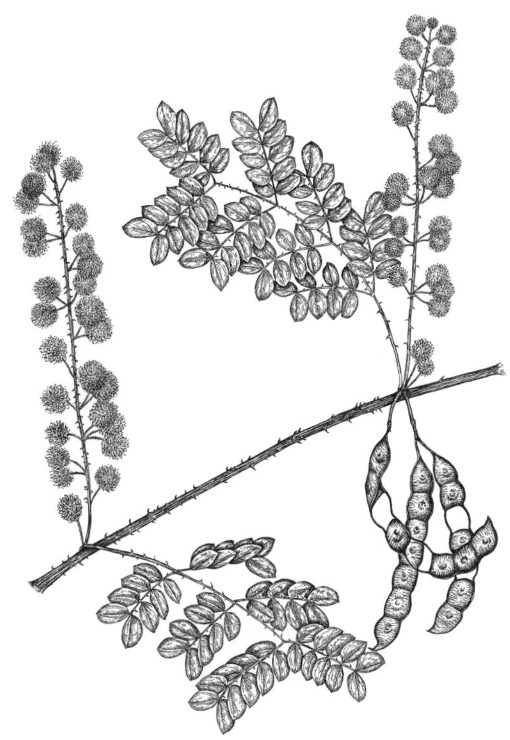

Smooth Vine Mimosa
Mimosa malacophylla Gray var. *glabrata* Benth.

leaves of some species mimicking animal motility. The species name, *malacophylla*, is from the Greek *malacos*, meaning "soft," and probably refers to the soft leaf pubescence. Vernacular names are Raspa-huevos and Raspilla.

Ebony Apes-earring

Pithecellobium flexicaule (Benth.) Coult.

FIELD IDENTIFICATION. Spiny, evergreen shrub or tree. Sometimes attaining a height of 40 ft, and forming a rounded spreading head with a trunk to 3 ft in diameter.

FLOWERS. June–August, borne in fascicled spikes, perfect, fragrant; peduncles ⅓–⅔ in. long, stout, pubescent; flower spikes ⅔–1½ in. long, cylindric, dense or interrupted; corolla about ⅛ in. long, much longer than the calyx, puberulous externally, 5-lobed, lobes longer than the tube; stamens numerous, long-exserted, filaments united below; calyx campanulate, 5-lobed, about 1/25 in. long; ovary glabrous, sessile.

FRUIT. Legume 4–6 in. long, 1–1¼ in. wide, dark brown to black, straight or somewhat curved, oblique at base, rounded or short-pointed at apex, narrowly oblong, flattened, thick, stout, woody, very hard, persistent for a year or more, valves very tardily dehiscent; seeds transverse in the legume and separated by thin tissue, bean-shaped, reddish brown, about ½ in. long and ¼ in. wide, seed coat thick and crustaceous.

LEAVES. Persistent, alternate, or clustered at the nodes, twice-pinnately compound, 1½–2 in. long and 2–3 in. broad, composed of 1–3 pairs of pinnae with the lower pair shortest; leaflets 3–5 pairs per pinna, obliquely elliptic, oval or obovate, obtuse or rounded at the apex, base unequal-sided, ⅕–½ in. long; thin or semicoriaceous; upper surface dark green, glabrous and lustrous, lower surface paler and finely reticulate-veined; petiolules almost absent or very short; petioles ⅓–1 in. long, puberulous, often glandular at the middle or apex.

TWIGS. Conspicuously flexuous, stout, green to reddish brown and pubescent at first, later becoming light gray and glabrous; spines usually in pairs at the nodes, persistent, straight or nearly so, ¼–½ in. long, brown to black or gray.

WOOD. Dark red to purple or brown, sapwood yellow, close-grained, very durable, hard, heavy, specific gravity 1.04.

RANGE. Texas and Mexico. In Texas from the shores of Matagorda Bay to the lower Rio Grande area. Abundant in Cameron County. Commonly planted on the streets of Brownsville, Texas. In Mexico in the states of Tamaulipas and Nuevo León, also Baja California.

REMARKS. The genus name, *Pithecellobium*, is from the Greek and means "sweet-pulp," with reference to the sugary pulp of some species. The species name, *flexicaule*, refers to the flexuous branches. Ebony Apes-earring is also known as Texas Ebony. Known in Mexico under the name of Ebano. The wood is used for posts, fuel, and cabinetmaking. It is often planted as a shade tree, being ornamental in bloom, and is considered by some to be the most valuable native tree in the lower Rio Grande Valley. The seeds are eaten by the Mexican people, and are boiled when green and roasted when ripe. The thick shells have been used as a coffee substitute.

Ebony Apes-earring
Pithecellobium flexicaule (Benth.) Coult.

Apes-earring

Pithecellobium pallens (Benth.) Standl.

FIELD IDENTIFICATION. Spiny shrub or small tree to 30 ft, occasionally 5–6 in. in diameter. Irregular or spreading with slender branches.

FLOWERS. May–August, in pubescent panicles 2–6 in. long; individual pedicels ⅓–1 in. long, pubescent, from lanceolate, acute, pubescent bracts; flowers white, in semiglobose heads, ⅔–¾ in. in diameter; calyx minute, 1/25 in. long, pubescent, 5-lobed, much shorter than the petals; corolla green, 1/16–⅛ in. long, tubular, puberulent, lobes 5, oblong-lanceolate, acute; stamens numerous, exserted, ½–⅗ in. long, persistent.

FRUIT. July–August, persistent, peduncles ½–2 in. long; pod glabrous or pubescent, linear-oblong, 2–5 in. long, about ½ in. wide, straight, apex abruptly acute or acuminate, stipitate, thin and somewhat membranous, sutures thickened, reticulate-veined, reddish brown, dehiscent at maturity into 2 papery valves; seeds ovoid or orbicular, about ¼ in. long, flattened, lustrous, dark brown to black, seed coat thin.

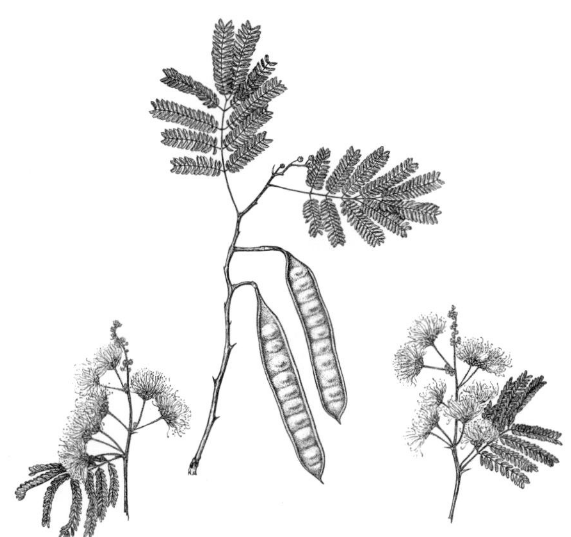

APES-EARRING
Pithecellobium pallens (Benth.) Standl.

LEAVES. Pinnately compound, 3–6 in. long, usually 3–6 pairs of pinnae; leaflets 7–20 pairs, oblong-linear, straight or slightly curved, ⅛–⅓ in. long, obtuse or acute, asymmetrical at base and apex, puberulent, slightly revolute; rachis slender, 1½–3 in. long, pubescent, bearing a small, cup-shaped gland near the middle, base with 2 short, slender spines; petioles of pinnae 1–2 in. long, petiolules about 1/25 in. long or less, pubescent.

TWIGS. Green to brown or gray, striate, pubescent or glabrous later, set with pairs of short, straight, slender spines, usually ¼–½ in. long; lenticels white and minute.

BARK. Gray to reddish, smooth, thin, breaking into small, thin flakes, sometimes bearing a few spines.

WOOD. Dark reddish brown, close-grained, hard, heavy.

RANGE. Usually on alluvial soils of stream bottoms, or on the edges of water holes. Known from Cameron, Willacy, Hidalgo, and Starr counties in Texas. Also cultivated in other counties in southwest Texas. Specimens collected by the author at Brownsville, San Benito, Harlingen, and Mission. Largest and most abundant in Mexico, extending into the states of Nuevo León, Tamaulipas, Coahuila, and San Luis Potosí. The type specimen was first collected near Monterrey, Nuevo León.

REMARKS. The genus name, *Pithecellobium*, refers to the sweetish fruit pulp of some species, and the species name, *pallens*, refers to the pale leaves. Some authorities list it under the name of *P. brevifolium* Benth. Vernacular names are Huajillo, Tenaza, Guajilla, and Mimosa-bush. The flowers are good bee food, and the foliage is sometimes browsed by sheep and goats in winter. The wood is occasionally used for small woodenware articles.

HONEY MESQUITE
Prosopis juliflora var. *glandulosa* (Torr.) Cockerell

FIELD IDENTIFICATION. Thorny shrub or small tree to 30 ft, with crooked, drooping branches and a rounded crown. Trunk dividing into branches a short distance above the ground. Root system radially spreading and deep.

FLOWERS. May–September, perfect, borne in yellowish green, cylindric, axillary, pedunculate spikelike racemes; calyx bell-shaped, minute, 5-lobed, with lobes triangular; corolla greenish white, small, with 5 linear petals which are erect, pubescent within, and much longer than the calyx-tube; stamens 10, exserted, with filiform filaments and oblong anthers; ovary stipitate, pubescent; style filiform; stigma small.

FRUIT. August–September, legumes in loose clusters, 4–9 in. long, somewhat flattened, glabrous, linear, straight or curved, somewhat constricted between the seeds, indehiscent; seeds oblong, compressed, shiny, brown, set in spongy tissue.

LEAVES. Alternate, deciduous, long-petioled, bipinnately compound of 2 (rarely 3 or 4) pairs of pinnae; with 12–20 leaflets; leaflets smooth, dark green, linear, acute or obtuse at apex, about 2 in. long and ¼ in. wide.

TWIGS. Zigzag, armed with stout straight spines to 2 in. long, or sometimes spineless.

BARK. Rough, reddish brown, with shallow fissures and thick scales.

HONEY MESQUITE
Prosopis juliflora var. *glandulosa* (Torr.) Cockerell

WOOD. Reddish brown, heavy, hard, durable, close-grained, sapwood yellow.

RANGE. Honey Mesquite is distributed from Kansas and New Mexico east into Oklahoma and Arkansas, and across Texas into Louisiana.

VARIETIES. The typical species of mesquite, *P. juliflora* (Swartz) DC., is considered as centering in distribution in Mexico and Central and South America. In Texas three of its varieties are represented. Besides the variety described above, which is the common variety of central, south, and east Texas, two Western varieties occur.

Velvet Mesquite, *P. juliflora* var. *velutina* (Woot.) Sarg., is a variety with short-pubescent foliage and twigs, found in west Texas, New Mexico, and Arizona.

Torrey Mesquite, *P. juliflora* var. *torreyana* (L.) Benson, is a small-leaved species of west and south Texas, New Mexico, Arizona, Nevada, California, and Mexico. In Texas it occurs along the Rio Grande, and then along the Gulf to the vicinity of Corpus Christi.

PROPAGATION. The pods can be gathered from the ground and stored in a dry place. The cleaned seed average about 14,000 seeds per lb. The purity and soundness varies from 75 to 90 per cent. Many seeds contain weevil larvae; the seeds should be fumigated before storage. To increase permeability of the seed coat the seeds should be soaked in sulphuric acid 15–30 minutes or placed in boiling water and allowed to cool as they soak for 24 hours. The germinative capacity averages about 89 per cent.

TORREY MESQUITE
Prosopis juliflora var. *torreyana* (L.) Benson

REMARKS. The genus name, *Prosopis,* is the old Greek name for the burdock. The species name, *juliflora,* refers to *julus,* or catkin-like inflorescence. The variety name, *glandulosa,* is for the glandular anther connectives of the flowers. Mesquite is often shrubby, forming thickets in dry areas and taking over grasslands rapidly. It readily sprouts from the stump, is very deep-rooted, and is not easily damaged by disease or insects. The wood is used for charcoal, fuel, furniture, building blocks, crossties, and posts, and is often the only wood available in regions in which it grows. Mesquite foliage and pods are eaten by livestock, and the seeds pass through the digestive tracts and grow where they fall. This fact has been responsible for much of its rapid distribution. The seeds are also considered an important wildlife food, being eaten by Gambel's quail, scaled quail, white-winged dove, rock squirrel, ground squirrel, coyote, skunk, jack rabbit, and white-tailed and mule deer. Mesquite beans played an important part in the diet of the Southwestern Indian. The legumes contain as high as 30 per cent sugar, and a meal known as "pinole" was prepared from them and made into bread. Fermentation of the meal also produced an intoxicating beverage. Exudation of gum from the branches and trunk offers possibilities as a substitute for gum arabic and is used locally to make candy. Honey Mesquite also produces a black dye and a cement for mending pottery.

VELVET MESQUITE
Prosopis juliflora var. *velutina* (Woot.) Sarg.

FREMONT SCREW-BEAN

Prosopis pubescens Benth.

FIELD IDENTIFICATION. Western tree attaining a height of 30 ft and a diameter of 1 ft. Often only shrubby with long, slender branches.

FLOWERS. Borne in the leaf axils and apparently attached to the petiole, but the insertions are distinct, spikes greenish white, cylindrical, linear, 1½–3 in. long, ¼–½ in. wide; individual flowers sessile; calyx 1/25–1/16 in. long, teeth 5, very minute, pubescent to tomentose; petals united, 1/12–⅛ in. long, inner side hairy, two to four times as long as the calyx; stamens exserted, filaments ⅙–⅕ in. long; style exserted, white-hairy basally.

FRUIT. Clustered on a rather stout peduncle, sessile or nearly so, yellow to brown, 1–2 in. long, ⅕–¼ in. in diameter, hairy at first, smoother later, tightly spiraled like a coil spring; seeds 1/16–1/12 in. long, ovoid to obovoid, asymmetrical, tan.

LEAVES. Deciduous, bipinnately compound, 1½–3 in. long, petioles ⅓–⅔ in.; rachis exceeding the outermost pinnae as a weak spine; stipules deciduous or spinescent; pinnae 1–2 pairs, 1½–2 in. long, leaflets 5–8 pairs per pinna, mostly sessile, oblong or slightly falcate, acute or apiculate, asymmetrical at base, puberulent, reticulate-veined, ¼–⅔ in. long.

TWIGS. Slender, reddish brown, often gray-striate; thorns stipular, solitary or paired, sharp, white, straight, ⅓–½ in. long.

BARK. Thick, gray to reddish brown, young branches smooth but thorny; old bark dark brown to black, flaking off into long shaggy strips.

WOOD. Light brown, sapwood lighter, close-grained, very hard, brittle, not strong, specific gravity about 0.76.

RANGE. In West Texas, New Mexico, Arizona, and California; north to Colorado, Utah, and Nevada. In Texas generally on alluvial bottom lands of the upper Rio Grande River and its tributaries; from Uvalde County west to El Paso County. Also in the Mexican states of Chihuahua and Sonora.

REMARKS. The genus name, *Prosopis*, is the old Greek name for the burdock. The species name, *pubescens*, refers to the hairy fruit. Also known under the names of Tornillo, Screw-pod Mesquite, and Twisted Bean. The powdered root bark was used to treat wounds by the Pima Indians of Arizona. The ground meal from the beans was made into pinole bread, and when steeped in water furnished a cooling drink. A crude syrup was obtained by boiling down the beans, which are also sweet enough to be chewed without preparation. The beans are also eaten by the roadrunner and by Gambel's and Mearns's quail. The wood is used locally for fuel, tool handles, hut building, and posts. It is rather durable in contact with the soil.

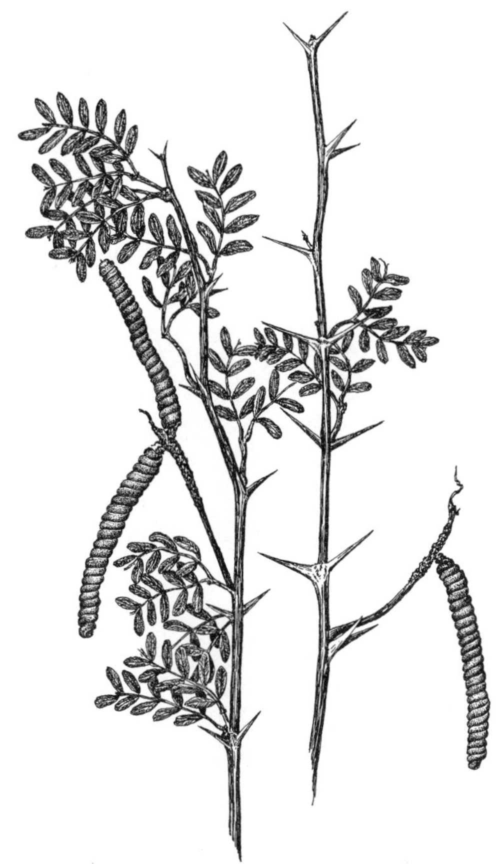

FREMONT SCREW-BEAN
Prosopis pubescens Benth.

DWARF SCREW-BEAN

Prosopis reptans Benth.

FIELD IDENTIFICATION. Low undershrub seldom over 2 ft, stems slender and wiry with long gray thorns.

FLOWERS. April–September, yellow, borne on pubescent peduncles ⅜–1¼ in. long, in capitate inflorescences, the globose heads ⅜–½ in. in diameter; individual flowers, perfect, puberulent to glabrous; calyx valvate, turbinate, about 1/12 in. long, 5-toothed; petals linear, ⅙–⅕ in. long, about twice as long as the calyx; stamens 10, slightly longer than the corolla, distinct, short, anthers bearing a deciduous gland; style filiform, stigma small and terminal, ovules numerous.

FRUIT. Legumes ½–1¾ in. long, ⅕–⅓ in. thick, yellow to reddish brown, tightly and spirally coiled, cylindric, puberulent to glabrous, woody externally, papery internally, indehiscent, with thick partitions between the seeds, no endosperm.

LEAVES. Alternate, bipinnate, petioles short; pinnae 1–2 pairs, rachilla puberulent; leaflets 5–12 pairs per pinna, sessile or nearly so, apex obtuse or rounded, base asymmetrical, subcoriaceous, length 1/25–⅛ in., surface puberulent or glabrous, margin entire and ciliate,

Dwarf Screw-bean
Prosopis reptans Benth.

main vein toward one side of the leaflet; stipules spinescent.

TWIGS. Very slender and wiry, brown to gray, puberulent when young, glabrous later; spines in pairs at the nodes, very slender, straight or slightly curved, ⅜–1 in. long.

RANGE. In moist soil along the lower Rio Grande River in Texas. Collected by the author at Boca Chica, near the mouth of the Rio Grande in Cameron County. Also in Mexico in Tamaulipas and Nuevo León.

REMARKS. The genus name, *Prosopis,* is the old Greek name for the burdock. The species name, *reptans,* means "creeping." The plant is usually very low and is easily overlooked. The oddly twisted legume is a good mark of identification.

Lead-plant Amorpha

Amorpha canescens Pursh

FIELD IDENTIFICATION. An erect or ascending shrub 1–3½ ft, leafy to the base, more or less densely gray-canescent throughout.

FLOWERS. June–July, inflorescence in slender, dense, crowded, short-peduncled, terminal racemes, often with subsidiary lateral ones below the terminal ones; racemes 1¼–10 in. long with a densely villous rachis; flowers very short-pediceled; calyx-tube obconic, slightly oblique, about ⅕ in. long, villous-canescent, 5-lobed; lobes lanceolate, nearly equal, about as long as the tube, 1/12–1/10 in. long; standard petal solitary (other 4 petals absent), bluish purple, erect, incurved, obovate-cuneate, apex truncate or sinuate, about ⅕ in. long, folded about the stamens; stamens 10, filaments exserted, united at base only; ovary short, 2-ovuled, style slender and bearded, stigma terminal.

FRUIT. Ripening August–September, legume ⅙–⅕ in. long, about 1/12 in. broad, nearly straight dorsally, apex rounder and broader with a long ascending beak, somewhat compressed, densely villous-canescent, glandular-dotted, almost indehiscent; seed solitary, oblong, brown, shining, about ⅛ in. long.

LEAVES. Odd-pinnately compound, spreading, 1½–5 in. long, rachis sometimes curved, densely villous; leaflets 13–49 or rarely more, sessile or nearly so, crowded or overlapping, elliptic to oblong or ovate, base usually rounded or truncate, apex rounded to obtuse or acute and usually mucronate, margin entire, length ⅓–⅗ in., width ⅙–⅖ in., upper surface appressed-pubescent, lower surface densely villous-canescent; lower leaves generally shortest and closest to the stem; petioles 1/25–1/12 in. long, stipules linear-setaceous.

STEMS. Erect or ascending, angled, densely villous-canescent, usually with few, if any, lateral branches.

Lead-plant Amorpha
Amorpha canescens Pursh

Mountain-indigo Amorpha
Amorpha glabra Desf.

RANGE. Sandy woods and stream banks. Texas, New Mexico, Oklahoma, Arkansas, and Louisiana; northward to Missouri, Iowa, Nebraska, Wisconsin, Illinois, Michigan, Minnesota, Indiana, North Dakota, and South Dakota.

PROPAGATION. The pods may be gathered by hand when they turn brown and may be spread out to dry. They may be planted as is, or the seed removed by flailing in a sack, but this is difficult to do. The ripe fruit averages 96,000 pods per lb, and the cleaned seed averages 296,000 seeds per lb, with a commercial soundness of 99 per cent. The seeds may be soaked in hot water prior to planting, or for spring planting stratified in sand for 40 days at 86°F. Germination averages about 30 per cent. One lb of seed will produce about 22,000 usable plants. The sowing is done in rows at a depth of about 3/16 in.

Smooth Lead-plant Amorpha, *A. canescens* var. *glabrata* Gray, is a variety in East Texas with branches and foliage only sparsely pubescent, or glabrate, and leaves greener. However, it is connected with the species by intergrading forms. Smooth Lead-plant Amorpha has been cultivated since 1883, and occurs from Illinois, Missouri, and New Mexico to Texas. The type locality is eastern Texas.

REMARKS. The genus name, *Amorpha*, means "without form," or here "without papilionate form," referring to the solitary petal. The species name, *canescens*, refers to the dense canescent hairs. Also known under the vernacular names of Shoe-strings and Spice-bush. The name in use refers to the leaden hue of the leaves. The plant has been cultivated as an ornamental in the United States and Europe since 1812. It is well adapted to sandy or gravelly soil in sunny places, and does well in rock gardens or as a border plant. It is also useful as game food and soil cover, but is rarely grazed by livestock.

Mountain-indigo Amorpha

Amorpha glabra Desf.

FIELD IDENTIFICATION. Stout shrub 3–6 ft, with parts glabrous or sparingly pubescent.

FLOWERS. Spikelike racemes single or clustered, somewhat loosely flowered, 4–8 in. long, glabrous; calyx campanulate, slightly oblique, 1/8–1/6 in. long, glabrous, margin ciliate, 5-lobed, the lobes very shallow or obsolete in some, when present acutish or rounded; standard petal bright purple (other petals absent), orbicular-spatulate, short-clawed, erect, folded around the stamens, 1/4–1/3 in. long; stamens 10, exserted, united at base only; style slender, straight, stigma terminal, ovary 2-celled.

FRUIT. Legume 1/4–1/3 in. long, obliquely ovate, dorsally straight, flattened, rounded on the ventral edge, apex with a short, erect or deflexed beak, surface light brown, glandular-dotted toward the apex; 1–2-seeded, seeds about 1/4 in. long, elliptic, brown.

LEAVES. Alternate, odd-pinnately compound, often spreading, 4–8 in. long, leaflets 9–21; blades oval-elliptic to ovate-elliptic or slightly obovate, 5/8–1 1/2 in. long, 1/3–1 in. wide, apex obtuse to rounded or often emarginate, base rounded or obtuse or subcordate, margin entire, thin, upper surface dark green and glabrous, lower surface paler, glabrous, glandular-dotted, veins inconspicuous; petiolules 1/12–1/8 in. long, glabrous.

TWIGS. Young ones slender, green, straw-colored, older ones reddish brown to gray and glabrous.

RANGE. Moist soil of riverbanks or prairies. Arkansas, Oklahoma, Alabama, Tennessee, Georgia, and North Carolina.

REMARKS. The genus name, *Amorpha*, means "without form," referring to the solitary petal. The species name, *glabra*, refers to the hairless or glabrous character of the plant.

Panicled Amorpha

Amorpha paniculata Torr. & Gray

FIELD IDENTIFICATION. Shrub to 7 ft, with grooved, densely hairy stems.

FLOWERS. June–July, panicles 6–15 in. long, slender,

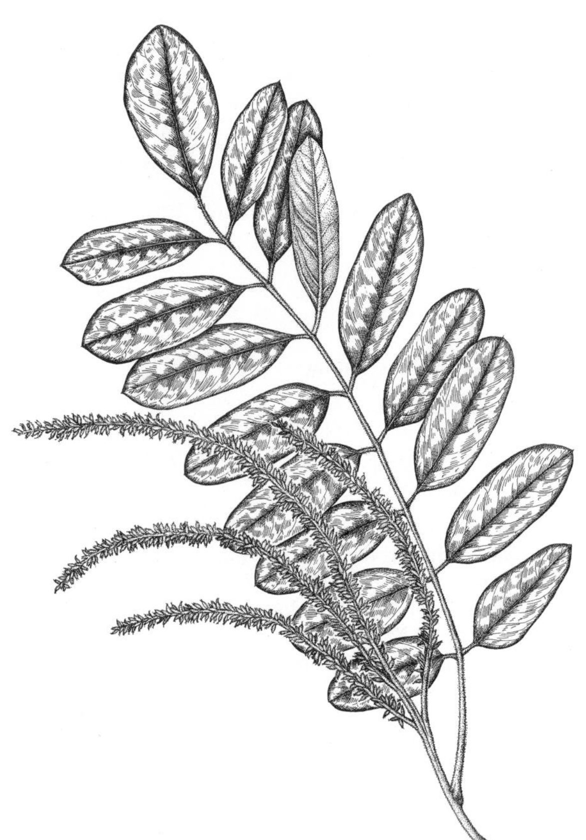

PANICLED AMORPHA
Amorpha paniculata Torr. & Gray

erect, tomentose; pedicels 1/25–1/16 in. long, tomentose; calyx subcampanulate, 5-lobed, densely hairy, 1/16–1/8 in. long; lobes slender, lanceolate, about half as long as the tube; standard petal (only corolla element present) dark purple, concave, obovate, clawed, about 1/16 in. long, folded around the stamens, wing and keel petals absent; stamens 10, exserted, anthers orange-colored, short-oblong, hardly over 1/25 in. long.

FRUIT. Pod 1/5–1/3 in. long, curved dorsally, pubescent, gland dots large and resinous.

LEAVES. Alternate, odd-pinnately compound, 6–9 in. long, 11–19 leaflets; rachis and petiole densely tomentose, angular; leaflets 1½–3 in. long, ¾–1¼ in. broad, mostly elliptic-oblong; apex rounded or emarginate, sometimes minutely mucronulate; base rounded or broadly cuneate; margin entire, slightly revolute; leathery; upper surface dark green, lustrous, glabrous or hairy along veins, veins impressed; lower surface dull, pale green, densely gray-tomentose, main veins prominent and nearly parallel; leaflet surface pellucid-punctate, the scattered dots seen when holding the leaflet up to the light; petiole tomentose, 1/8–1/4 in. long.

STEMS. Stems densely tomentose, grooved, forking above into long slender, erect panicles.

RANGE. Eastern Texas, Louisiana, southwestern Arkansas, and southern Oklahoma. First found and officially recorded from the state of Arkansas.

REMARKS. The genus name, *Amorpha,* is from the Greek work *amorphos,* meaning "deformed," with reference to the flower which bears only one petal. The species name, *paniculata,* refers to the structure of the inflorescence. Another name is False Indigo or Wild Indigo.

CALIFORNIA AMORPHA

Amorpha californica Nutt.

FIELD IDENTIFICATION. Slender shrub 3–9 ft. Foliage and branches pubescent, bearing prickle-like glands.

FLOWERS. May–July, in erect spikelike racemes 2–8 in. long; calyx turbinate, 1/5–1/4 in. long, slightly oblique, densely pubescent, with or without glands; calyx-lobes 5, lanceolate, as long as the tube or nearly so; standard petal solitary, reddish purple, obovate-cuneate, erect, incurved, about 1/5 in. long; stamens 10, exserted, filaments united at base; style slender, stigma terminal, ovary superior, 1-celled.

CALIFORNIA AMORPHA
Amorpha californica Nutt.

FRUIT. Legume ¼–⅓ in. long, about ⅛ in. wide, oblique, flattened, apex rounder and broader, curved dorsally, pubescent and much glandular-dotted, indehiscent or tardily so, 1–2-seeded, seeds elongate.

LEAVES. Alternate, ascending, 4–10 in. long, odd-pinnately compound of 11–17 leaflets, blades ⅜–1¼ in. long, ⅕–¾ in. wide, broadly oval to elliptic or oblong, apex obtuse to retuse or rounded, often somewhat emarginate and mucronate, base rounded or subcordate, at maturity thin and firm, margin entire, upper surface dull green and scabrate or glabrous, lower surface softly hairy and conspicuously glandular-dotted; petioles ⅜–1 in., pubescent, sometimes with sessile or stalked glands; stipules membranous, lanceolate to oblong.

RANGE. In moist wooded canyons in western New Mexico, Arizona, and California.

VARIETIES. Napa Mountain Amorpha, *A. californica* var. *napensis* Jepson, has racemes to 1–1½ in. long; calyx nearly glandless and glabrous, with minute lobes; rachis glandless; leaflets sessile and subglabrous. Found in California on Mount Howell in the Napa Range.

Hispid California Amorpha, *A. californica* var. *hispidula* (Greene) Palmer, is almost glabrous throughout with shorter calyx-lobes. It is found in Monterey County, California.

REMARKS. The genus name, *Amorpha,* means deformed," because of the solitary petal. The species name, *californica,* is for the state of California where this species is found. Also known under the vernacular names of Mock Locust and California Indigo-bush.

California Amorpha is distinguished from other amorphas by the glanduliferous, pubescent branches, petioles, and leaves, and by the calyx-lobes being as long as, or almost as long as, the tube.

Indigo-bush Amorpha

Amorpha fruticosa L.

FIELD IDENTIFICATION. Shrub often clumped, branching, to 18 ft. Very variable in size, shape, number of leaflets, degree of hairiness, and size of fruit.

FLOWERS. May–June, racemes spikelike, clustered (2–many), erect, closely flowered, 2½–6 in. long; pedicels $\frac{1}{25}$–$\frac{1}{12}$ in., erect; calyx $\frac{1}{12}$–⅛ in. long, turbinate, slightly oblique, glabrous to slightly pubescent, somewhat glandular, 5-lobed, margins ciliate, calyx-lobes all much shorter than the tube, upper lobes obtuse to rounded, lower triangular and acute; corolla of a single petal (the banner), erect, incurved, clawed, oval to obovate, apex emarginate, length about ¼ in., purplish blue (rarely pale blue or white); folded around the stamens; stamens 10, exserted, united at base only; ovary 2-ovuled, style slender and bearded, stigma terminal.

FRUIT. Maturing in August, legume ¼–⅓ in. long, $\frac{1}{12}$–⅛ in. wide, curved dorsally, conspicuously glandular-punctate and glabrous, somewhat compressed, rounded or broader at the apex, 1–2-seeded; seed about ⅙ in. long, oblanceolate-oblong, curved at one end, brown and glossy.

LEAVES. Odd-pinnately compound, usually ascending, length 2½–10 in.; leaflets 9–27, oval or elliptic to oblong, apex obtuse to rounded or abruptly pointed and mucronate, base rounded or cuneate, margin entire, length ⅝–1½ in., width ⅓–⅘ in.; upper surface dull green and glabrous; lower surface paler and glabrous or finely pubescent, glandular-dotted, densely hairy below when young; petioles ¾–2 in., glabrous or finely pilose; petiolules pubescent, about $\frac{1}{12}$ in. long.

TWIGS. Brown or gray, sparingly pilose or with age glabrous.

RANGE. Moist woods or stream banks, often in calcareous soil. Texas, New Mexico, Oklahoma, Arkansas, Louisiana, Arizona, and California; northward to Iowa, Ohio, Minnesota, Massachusetts, and New Hampshire.

PROPAGATION. The fruit of *A. fruticosa* and its varieties may be gathered by hand from the plants when brown. They may be dried and planted as is, or the seed may be extracted by flailing in a sack. The fruits number about 52,000 per lb. The cleaned commercial seed has a purity of 97 per cent and a soundness of 73 per cent. Germination is hastened by stratification in moist sand for 20 days at 86°F. The germination rate is 40–70 per cent. The seed is generally sown in rows about $\frac{3}{16}$ in. deep. A lb of seed produces 1,000–5,600 usable plants. Propagation may also be effected by cuttings, layers, or suckers.

REMARKS. The genus name, *Amorpha,* means "without form," and refers to the solitary petal, instead of the normal 5. The species name, *fruticosa,* is for the shrubby character of the plant. Also known under the vernacular names of False Indigo and Bastard Indigo. It is a very variable species and many forms and varieties have been separated from it. It has been cultivated since 1724 and is useful for landscape planting, erosion control, and game food. Bobwhite quail eat the seeds. Livestock seldom eat it, and it is reported as poisonous to them. The plant is rarely attacked by insects or disease. A monograph of the genus has been written by Palmer (1931).

The following forms and varieties are listed by Palmer, but the differences are sometimes indistinct, and more intensive study is needed to clarify the relationship.

White Indigo-bush Amorpha, *A. fruticosa* forma *albiflora* Shelden, has white flowers (cult.).

Drooping Indigo-bush Amorpha, *A. fruticosa* forma *pendula* Schneid., has slender recurved branches (cult.).

Crinkle-leaf Indigo-bush Amorpha, *A. fruticosa* forma *crispa* Schneid., has crisped leaf margins (cult.).

Ground Indigo-bush Amorpha, *A. fruticosa* forma *humilis* Palmer, is a dwarf form usually freezing back each winter (cult.).

Midwest Indigo-bush Amorpha
Amorpha fruticosa var. *angustifolia* Pursh

Cerulean Indigo-bush Amorpha, *A. fruticosa* forma *coerulea* Palmer, is a form with pale blue flowers.

Variegated Indigo-bush Amorpha, *A. fruticosa* forma *aureovariegata* Schwer., is a form with variegated foliage.

Midwest Indigo-bush Amorpha, *A. fruticosa* var. *angustifolia* Pursh, is a variety with leaflets narrower and elliptic, base narrowed and cuneate, pubescence more abundant and appressed. This appears to be the common variety from Wisconsin and Minnesota to Saskatchewan, and southward to Kansas, Texas, and northern Mexico. In Texas it has been recorded from Nolan, Brown, Armstrong, Val Verde, San Saba, Refugio, Chambers, Dallas, Kerr, and Harris counties. It is often cultivated in American and European gardens. It also has a glabrous form.

Emarginate Indigo-bush Amorpha, *A. fruticosa* var. *emarginata* Pursh, differs by having leaves larger, oval to ovate, apex blunt or emarginate. This variety appears intermediate between *A. fruticosa* and *A. fruticosa* var. *croceolanata*, somewhat resembling the latter in shape and size of leaflet, but differing in its gray, less copious pubescence. Recorded from Missouri, Mississippi, Tennessee, and Arkansas.

Tennessee Indigo-bush Amorpha, *A. fruticosa* var. *tennesseensis* Palmer, has more numerous oblong leaflets and slightly curved, or nearly straight, pods. It occurs from North Carolina south to Florida and west to Kansas, Oklahoma, and Texas.

Oblong-leaf Indigo-bush Amorpha, *A. fruticosa* var. *oblongifolia* Palmer, is an erect shrub 6–9 ft. Leaves 4¼–10 in. long; petioles ¾–1¼ in.; leaflets 21–41, oblong or narrowly oblong, ¾–2 in. long, 3/16–⅝ in. wide, rounded at both ends or slightly emarginate at the mucronate apex, and rarely abruptly narrowed at base, thin but firm, dark green and glossy above, much paler or sometimes slightly glaucous, black-punctate and glabrous or sparsely pubescent beneath; petiolules slender, 1/10–1/12 in. long. Inflorescence of a few or several erect spikes 2–8 in. long; calyx glabrous, its lobes much shorter than the tube; the two upper ones low and rounded, the lower ones broadly-triangular. Pod ¼–⅖ in. long, ⅛ in. wide, somewhat curved dorsally, with a short, erect beak, glandular-dotted. Known from Missouri, Arkansas, and Texas.

Yellow-wool Indigo-bush Amorpha, *A. fruticosa* var. *croceolanata* Mouillef., has leaves oblong to ovate-oblong, 1–2⅓ in. long, ends rounded, at first yellow-tomentose, villous at maturity. This is considered by some authors to be a species instead of a variety of *A. fruticosa*, but it appears to be so closely related to the latter by intergrading forms that it is here maintained as a variety.

Dwarf Indigo-bush Amorpha

Amorpha nana Nutt.

FIELD IDENTIFICATION. Low and bushy shrub, usually about 1 ft, but occasionally to 3 ft. All parts glabrous or with a few scattered hairs on the petiolules at first.

FLOWERS. May–June, fragrant, borne in terminal, usually solitary, densely flowered racemes 1¼–4 in. long; corolla of a solitary standard petal, purple, about ⅙ in. long, cuneate-obovate, clawed, erect, wrapped around the stamens and style; stamens 10, exserted, united at base; style slender, bearded, stigma terminal, ovary 2-ovuled; calyx turbinate, slightly oblique, glabrous, about ⅛ in. long, 5-lobed, the lobes lanceolate, acuminate, about half as long as the tube or more.

FRUIT. Maturing in July, legume about ⅕ in. long, 1/10–⅛ in. broad, obliquely ovate, flattened, dorsally straight, beak short, surface roughened and densely punctate-dotted, indehiscent, the 1 seed lustrous and about ⅙ in. long.

LEAVES. Alternate, 1¼–4 in. long, odd-pinnately compound, crowded, numerous; rachis grooved on one side, often with scattered glands, petioles ⅓–⅖ in. long; leaflets 13–19 (more rarely to 31), oblong to elliptic, or oval to slightly obovate, apex rounded to emarginate and mucronate, base rounded to abruptly narrowed or cuneate, blade length ⅓–¾ in., width ⅙–⅓ in., firm, slightly reticulate, margin entire, green and glabrous above, lower surface slightly paler and conspicuously glandular-dotted; petiolules 1/25–1/16 in., glabrous or slightly hairy; stipels in pairs, 1/12–⅙ in. long, soft-spinescent.

Dwarf Indigo-bush Amorpha
Amorpha nana Nutt.

TWIGS. Slender, finely grooved, glabrous, reddish brown to gray, glandular-warty.

RANGE. On dry sunny exposures of hills and prairies of the grass, pinyon, and yellow pine belts at altitudes of 4,500–7,500 ft. New Mexico, northward to Manitoba and Saskatchewan.

PROPAGATION. When mature the small legumes are stripped from the branches by hand and spread out to dry. The seeds need not be separated from the hull. The seed averages about 60,000 seeds per lb, with a soundness of about 85 per cent. Because of an impermeable seed coat there is a tendency toward persistent dormancy, which can be partially broken by stratification over winter or by soaking in concentrated sulphuric acid 5–10 minutes. The seed may be sown in fall or after stratification in spring at a depth of 3⁄16–1⁄4 in. Dwarf Indigo-bush may also be grown from greenwood or hardwood cuttings or by layers and suckers.

REMARKS. The genus name, *Amorpha,* means "without form," referring to the solitary petal, and the species name, *nana,* signifies the dwarf size. Some writers list it under the name of *A. microphylla* Pursh. It is also known locally as Fragrant Dwarf Indigo and Lead Plant. It has been cultivated for ornament since 1811 and is grown both in the United States and Europe. It is also of use in erosion control and as food for wildlife.

Georgia Amorpha
Amorpha nitens Boynton

FIELD IDENTIFICATION. Branching shrub 3–9 ft, with smooth and angled branches.

FLOWERS. Borne in solitary racemes or with a few shorter ones subtending, 2–9½ in. long, rachis glabrous or almost so; calyx narrowly conic, slightly oblique, about ⅛ in. long, striate, glabrous except the 5 ciliate lobes which are shorter than the tube; two upper lobes about 1⁄50 in. long, rounded or obtuse; lower lobes ovate, acute or obtuse; corolla of only a solitary blue or purple banner petal which is erect, incurved, cuneate, clawed and folded around the stamens; stamens 10, exserted, filaments united at base; ovary 2-ovuled, style slender, bearded, stigma terminal.

FRUIT. Legume ¼–⅓ in. long, narrow, nearly glandless, oblique, compressed, apex broader, indehiscent; seeds 1–2, elongate and curved.

LEAVES. Odd-pinnately compound, rachis glabrous or nearly so, 4–10 in. long, ascending; leaflets 7–19, oblong to oblong-ovate, ends rounded or obtuse and often slightly notched, margin entire, length ¾–2½ in., width ⅜–¾ in., thin, glandular-dotted; upper surface glabrous and lustrous; lower surface glabrous, or sparingly pubescent, and duller and paler; petioles ⅜–⅝ in., glabrous or nearly so, stipules setaceous.

RANGE. Georgia Amorpha occurs from Georgia to Alabama, north to Illinois, and in the Ozark region of western Arkansas.

VARIETY. Smooth Georgia Amorpha, *A. nitens* var.

Georgia Amorpha
Amorpha nitens Boynton

Desert-indigo Amorpha
Amorpha occidentalis Abrams

leucodermis Palmer, is a variety with glaucescent leaves found in Georgia.

REMARKS. The genus name, *Amorpha,* means "formless," referring to the flower's solitary petal, and the species name, *nitens,* is for shining or lustrous leaves. Some botanists consider that this species is so closely related to the Indigo-bush Amorpha, *A. fruticosa,* that it should be maintained only as a variety of it.

Desert-indigo Amorpha

Amorpha occidentalis Abrams

FIELD IDENTIFICATION. Shrub 6–9 ft, with leaves and twigs pubescent.

FLOWERS. Borne in spikelike, usually solitary, slender, pubescent racemes 4–9 in. long (sometimes with additional shorter racemes at the base); pedicels about 1/15 in. long; calyx ⅛–⅙ in. long, turbinate, 5-lobed, nearly glabrous or tube slightly hairy, lobes densely hairy and shorter than the tube; two upper lobes broad and obtuse, lower three triangular, acute or obtuse; banner petal of corolla dark blue, about ⅕ in. long, erect, incurved, obovate to cuneate, short-clawed, folded around the stamens; stamens 10, exserted, united at base only; ovary 2-ovuled, style slender, stigma terminal.

FRUIT. Legume ¼–⅓ in. long, slightly curved dorsally, oblique, apex broadened above, somewhat compressed, glabrous and conspicuously glandular-dotted, indehiscent, 1–2-seeded.

LEAVES. Alternate, odd-pinnately compound, ascending, rachis sparingly strigose, leaflets 11–27, not crowded, oblong to elliptic or oval, apex rounded or rarely abruptly pointed, mucronate, base rounded or acute, margin entire, length ⅝–1¾ in., width ¼–⅝ in., thin but firm, upper surface dark green; lower surface paler and more or less hairy, at least along the veins; petiolules about 1/12 in., hairy; petioles ⅝–1¼ in., sparingly hairy.

RANGE. The species is found from western Texas to Arizona, New Mexico, Wyoming, and California, and in Mexico in Chihuahua and Sonora.

VARIETIES. *A. occidentalis* var. *emarginata* Palmer has leaflets truncate or retuse and glabrous to subglabrous. It occurs in Arizona and California.

A. occidentalis var. *arizonica* Palmer has looser more copious pubescence on all parts. It occurs in Arizona and New Mexico.

REMARKS. The genus name, *Amorpha,* meaning "without form," refers to the solitary banner petal of the flower; the species name, *occidentalis,* means "western."

Stone Mountain Amorpha

Amorpha virgata Small

FIELD IDENTIFICATION. Shrub 3–6 ft, with 1–several stems, branched near the summit.

FLOWERS. The spikelike racemes mostly clustered, 3¼–8 in. long; calyx turbinate, slightly oblique, about ⅛ in. long, sparingly pubescent, 5-lobed, lobes all much shorter than the tube, glabrous except the ciliate lobe margins; upper lobes rounded or obtuse, lower 3 triangular and acute; corolla with a solitary blue or purple banner petal, erect, incurved, clawed, folded around the stamens; stamens 10, filaments exserted, united only at base; ovary 2-ovuled, style slender, stigma terminal.

FRUIT. Legume ¼–⅓ in. long, ⅛–⅙ in. broad, dorsally straight, somewhat compressed, oblique, apex rounder and broader, glabrous, with a few, large glandular dots, indehiscent, 1–2-seeded; seed about ⅙ in. long, oblong, brown, lustrous, somewhat curved at apex.

LEAVES. Alternate, 3¼–8 in. long; leaflets 9–19, elliptic to oblong, often twice at long as broad, ¾–2 in. long, ⅜–¾ in. broad; apex rounded or obtuse, sometimes notched, often apiculate; base obtuse to subcordate, margin entire; upper surface dark green and glabrous; lower surface paler, hairy, and strongly veined, glandular-dotted; petioles ½–1 in., sparsely hairy with the rachis; petiolules 1/16–¼ in., sparsely hairy; stipules setaceous.

TWIGS. Slender, green to light or dark brown, hairy at first, glabrous later.

RANGE. Ozark area of Arkansas and Oklahoma; eastward to Georgia, north Florida and Tennessee.

REMARKS. The genus name, *Amorpha* ("without form"), refers to the solitary banner petal, and the species name, *virgata,* refers to the branching habit.

TEXAS INDIGO-BUSH AMORPHA
Amorpha texana Buckl.

TEXAS INIGO-BUSH AMORPHA

Amorpha texana Buckl.

FIELD IDENTIFICATION. Shrub 3–9 ft, with spreading branches.

FLOWERS. Racemes 1–4, at the end of a stem, 2–6 in. long, loosely flowered on a pubescent rachis; pedicels about 1/25 in. long; calyx narrowly campanulate, 1/8–1/5 in. long, pubescent and glandular, 5-lobed; calyx-lobes much shorter than the tube, upper lobes rounded or obtuse, lower triangular to short-lanceolate, apex acute; the single petal standard blue, erect, incurved, clawed, suborbicular-spatulate, about 1/4 in. long, folded around the stamens; stamens 10, exserted, filaments united at base only; ovary 2-ovuled, style slender, stigma terminal.

FRUIT. Legume 1/4–1/3 in. long, about 1/8 in. wide, nearly straight dorsally, conspicuously glandular-dotted, oblique, somewhat compressed, rounded and broader at apex, almost indehiscent, 1–2-seeded.

LEAVES. Odd-pinnately compound, 4–7 in. long, spreading, leaflets 7–15, length 5/8–1 5/8 in., width 5/8–1 3/16 in., broadly-oval to oblong, ends mostly rounded and often notched, margin entire, subcoriaceous and reticulate; upper surface dark green, lustrous and glabrous; lower surface paler, more or less pubescent and glandular-dotted; petiolules 1/8–1/5 in., usually pubescent; petioles 3/8–3/4 in., pubescent.

RANGE. The species is confined to central and southwestern Texas in the limestone areas. It was first collected on the Pedernales River. Also found in Kerr and Comal counties.

VARIETY. A variety with foliage and stems glabrous to subglabrous is known as *A. texana* var. *glabrescens* Palmer, and has been found at New Braunfels, Bandera, and Medina Lake, Texas. This plant is very closely related to a plant described as *A. laevigata* Nutt.

REMARKS. The genus name, *Amorpha,* means "without form," referring to the solitary petal of the corolla. The species name, *texana,* refers to its being found in Texas.

STONE MOUNTAIN AMORPHA
Amorpha virgata Small

SMOOTH AMORPHA

Amorpha laevigata (Nutt.) Torr. & Gray

FIELD IDENTIFICATION. Glabrous shrub 3–6½ ft.

FLOWERS. Borne in 1–3 or more slender racemes 5–12 in. long; pedicels 1/25–1/8 in. long; calyx about 1/8 in. long, turbinate, slightly oblique, glabrous, glandular-dotted, 5-lobed; the 2 upper lobes obtuse or rounded, the lower 3 triangular and acute; corolla consisting of a standard petal only, purplish, reniform, about ¼ in. long or broad, erect, incurved; stamens 10, filaments united at base only, exserted; ovary 2-ovuled, style slender, stigma terminal.

FRUIT. Legume about ⅕ in. long, glandular-dotted, nearly straight dorsally, oblique, somewhat flattened, apex broadened, indehiscent.

LEAVES. Ascending, odd-pinnately compound, 4–8 in. long; leaflets 7–21, oblong or obovate, ⅜–¾ in. long, apex rounded or emarginate, base acute or cuneate, margin entire, surfaces glabrous or nearly so, conspicuously punctate and reticulate beneath; petioles slender, ⅜–¾ in. long; petiolules conspicuously glandular, 1/12–⅛ in. long, stipules setaceous.

RANGE. Moist rich soil along streams. Northeast Texas, Oklahoma, Arkansas, and Louisiana.

REMARKS. The genus name, *Amorpha,* means "without form," referring to the solitary petal. The species name, *laevigata,* refers to the smooth glabrous leaves. This species has been considered to be a synonym or a variety of *A. texana* by some authors.

SMOOTH AMORPHA
Amorpha laevigata (Nutt.) Torr. & Gray

SHORT-FRUIT AMORPHA

Amorpha brachycarpa Palmer

FIELD IDENTIFICATION. Rather slender shrub to 3 ft. Basal stems few, erect or ascending, grooved, somewhat pubescent to glabrous.

FLOWERS. Panicles 4–10 in. long, with many slender branches; flowers crowded on pedicels 1/25–1/15 in.; calyx ⅙–⅕ in. long, turbinate, angled, glabrous except the ciliate lobes which are lanceolate and acuminate at the apex, upper 2 lobes shorter than the tube, the lower lobes longer or as long as the tube; standard petal solitary, violet-blue, obovate, apex truncate or slightly emarginate, somewhat folded about the glabrous, exserted stamens and style.

FRUIT. Legume hardly longer than the calyx-lobes, ⅙–⅕ in. long, ⅛–⅙ in. wide, much flattened, obliquely obovate, dorsally almost straight, apex with a persistent style and curved beak, margins slightly thickened, dark brown and resinous-dotted, 1-seeded.

LEAVES. Numerous, alternate, deciduous, odd-pinnately compound, 2¾–5 in. long, rachis slender, glabrous or nearly so, channeled above; leaflets 21–45, sometimes crowded or overlapping, ⅓–⅗ in. long, ⅙–⅓ in. wide, oblong, symmetrical or slightly oblique, margin entire, apex rounded or slightly emarginate and mucronate, base rounded or subcordate, terminal leaflet sometimes smaller, rounder, and more deeply emarginate at the apex, leaflets thin and firm, usually glabrous, or the veins beneath and margins with a few scattered hairs, veins conspicuous beneath; petiolules about 1/25 in. long, glabrous; stipules linear-subulate, 1/25–1/15 in. long.

RANGE. In northeastern Arkansas and the Ozark region of Missouri.

REMARKS. The genus name, *Amorpha,* means "without form" and refers to the solitary petal. The species name, *brachycarpa,* is for the short fruit. Palmer (1931 p. 172) remarks as follows:

"This little species of *Amorpha* resembles *A. nana* in size, habit and foliage. However, it is well distinguished from that species by the large, many-branched panicle, and by the relatively short tube and long teeth of the calyx, the latter about equalling the broad pod with its strongly reflexed beak."

BUSH'S AMORPHA

Amorpha bushii Rydb.

FIELD IDENTIFICATION. Shrub 3–6 ft. Branches sparsely hairy when young but glabrous with age.

FLOWERS. Racemes slender and lax, 1–3, 4–12 in. long; pedicels about 1/25 in. long, sparingly pubescent; calyx about ⅛–⅙ in. long, turbinate, slightly oblique, sparingly hairy or almost glabrous, 5-lobed; the 2 upper lobes indistinct and somewhat rounded; the 3 lower ones triangular and acute; corolla composed of the standard petal only, purple, obovate-cuneate, erect, incurved, short-clawed, folded around the stamens; stamens 10, exserted, united at base only; ovary 2-ovuled, style slender, stigma terminal.

FRUIT. Legume about ¼ in. long and ⅛ in. broad, oblique, dorsally straight, glabrous and somewhat glandular-dotted, somewhat flattened, rounded and broader at apex, 1–2-seeded.

LEAVES. Strongly ascending, rachis finely hairy, 8–12 in. long, odd-pinnately compound; leaflets 11–25, lance-oblong or ovate-oblong, ¾–2 in. long, ⅜–¾ in. wide, margin entire, apex rounded, retuse or emarginate, base rounded or acutish, upper surface dull green and glabrous or short-hairy, lower surface paler, soft-hairy, and glandular-dotted; petioles 1¼–1⅝ in. finely hairy, petiolules 1/25–1/12 in. long.

RANGE. Moist soil along streams. Louisiana, near Alexandria, eastward to Florida.

REMARKS. The genus name, *Amorpha*, means "without form," referring to the solitary petal. The species name, *bushii*, is in honor of Benjamin Franklin Bush (1858–1937), botanist of Missouri.

American Yellowwood

Cladrastis lutea K. Koch

FIELD IDENTIFICATION. Tree attaining a height of 60 ft, with a diameter of 4 ft, the slender branches forming a wide head.

FLOWERS. Usually May–June, panicles loose, open, 10–20 in. long, 5–12 in. broad, drooping, many-flowered; pedicels slender, puberulous, often glaucous, bibracteolate; bracts lanceolate, scarious, early deciduous; individual flowers white, fragrant, irregular, perfect; calyx nearly tubular, base oblique and obconic, puberulous, 5-toothed; teeth nearly equal, obtuse or rounded, the 2 upper ones somewhat united; disk adnate to the calyx-tube within; corolla papilionaceous, about 1 in. long, standard nearly orbicular, entire or slightly notched, reflexed above, as long as the oblong wing petals or slightly longer, somewhat eared at the base, marked with a yellow blotch; keel petals free, oblong, obtuse, straight or slightly incurved, base biauriculate or somewhat subcordate; stamens 10, free, filaments threadlike, glabrous, anthers versatile; ovary short-stipitate, linear, red, pale-hairy; style slender, incurved, subulate; stigma terminal, minute; ovules numerous, superimposed.

FRUIT. Legume in August–September, short-stalked, narrowly linear-elliptic, length 3–4 in., flattened, upper margin somewhat thickened, apex with remnants of persistent style, tardily dehiscent, valves 2, thin, not pulpy between the seeds; seeds 2–6, oblong, flattened, seed coat dark brown, membranous; embryo large, cotyledons fleshy, oblong, flat.

LEAVES. Odd-pinnately compound, 8–12 in. long, deciduous, alternate, composed of 5–11 leaflets; leaflets 2–4 in. long, 1½–3½ in. wide, alternate, oval to suborbicular or ovate to elliptic, apex abruptly short-pointed, base cuneate, margin entire, young blades finely pubescent, later glabrous and dark yellowish green above, paler beneath, veins conspicuous, leaflets conspicuously yellow in the fall; petiole hollow and enlarged at base to enclose the bud.

TWIGS. Slender, terete, divaricate, light brown to reddish brown, smooth, shiny, pubescent at first, glabrous with age; lenticels numerous, dark colored; leaf scars large, raised.

BARK. Trunk bark ⅛–¼ in. thick, smooth, gray to light brown, somewhat resembling that of beech.

WOOD. With distinct but not prominent growth rings, tasteless, odorless, yellow to brown or yellow streaked with brown, sapwood nearly white, surface smooth, close-grained, hard, strong, weighing 39 lb per cu ft.

RANGE. Rich limestone soils of stream banks or cliffs. Rather rare and local. Northeastern Oklahoma and central Arkansas. Most abundant in Missouri, Alabama, Kentucky, Tennessee, Indiana, and Illinois.

American Yellowwood
Cladrastis lutea K. Koch

PROPAGATION. Good seed crops are borne on alternate years. The pods are collected at maturity by hand-picking or shaking into a canvas. The seeds are separated by beating in a sack and fanned to separate the chaff. The seed averages 11,300–14,600 seeds per lb. The average purity is 82 per cent and soundness 67 per cent. The seed may be stored over winter in sealed containers. For spring planting, the seed may also be stratified in moist sand or peat at 41°F. for 90 days. Alternatively, the seed may be scarified and stored for 30 days, or treated with sulphuric acid for 30–60 minutes before planting. The planting is usually done in beds and rows 8–12 in. apart and covered about ¼ in. with firmed soil. Fall-sown and untreated seeds should be mulched and screened over winter to protect from birds and rodents. Propagation is also practiced by root cuttings dug in the fall and kept moist and cool until spring.

REMARKS. The genus name, *Cladrastis,* is from the Greek and means "brittle-branched." The species name, *lutea,* refers to the yellow wood. Other vernacular names are Yellow-tree, Fustic-tree, Gopherwood, and Virgilia. The wood is used for fuel and gunstocks, and yields a clear yellow dye. The tree is often planted for ornament in the United States and Europe. It is hardy north to New England and Ontario, and was first introduced into cultivation in 1812.

Texas Kidneywood
Eysenhardtia texana Scheele

FIELD IDENTIFICATION. Irregularly shaped shrub usually less than 8 ft, growing on calcareous soils. Parts unpleasantly scented when bruised.

FLOWERS. Mostly in May, or intermittently after rains; racemes terminal, 1¼–4½ in. long; bracts ovate-lanceolate, brown, pubescent, 1/25–1/12 in. long, calyx about 1/12 in. long, hairy, glandular, unequally lobed, anterior lobes longer than posterior, split more deeply on the posterior side, apex obtuse or acute, lobes less than 1/25 in. long; petals 5, inserted on the hypanthium, white, nearly equal, distinct, concave, membranous, much longer than the calyx, standard petal somewhat broader and rounded to slightly emarginate; style slender, 1/12–⅛ in. long, pubescent, slightly hooked at the apex, bearing a reddish brown, stout gland (these stipel-like structures referred to by some authors as small apiculate pustules); stigma large and capitate; ovary sessile with 2–4 ovules; stamens 10, 9 of them united half their length, the other free.

FRUIT. Maturing in September, indehiscent, on pedicels 1/25 in. long or less, pod linear-oblong, ⅕–⅓ in. long, 1/25–⅛ in. wide, thickened, green to brown, glabrous, glandular-punctate, upwardly falcate and ascending; seed about ⅛ in. long, lanceolate to obovoid, brown, smooth, thickened, hilum near the distal end, germination about 50 per cent.

Texas Kidneywood
Eysenhardtia texana Scheele

LEAVES. Alternate, odd-pinnately compound, rachis grooved and glandular, leaves 1¼–2⅓ in. long; leaflets 15–31, ⅕–⅜ in. long, oblong, margin entire, apex rounded or notched, base rounded or obcordate; upper surface dull green, puberulent, obscurely reticulate, lower surface paler, puberulent and minutely punctate; stipules 1/12–⅛ in. long, brown, subulate; stipels subulate, persistent, less than 1/25 in. long.

STEMS. Slender, younger ones appressed-hairy; older gray and smooth, broken into thin, elongate plates later.

RANGE. On dry hills and in canyons, central and southwestern Texas. The type locality is at New Braunfels, Texas. Specimens have been collected by the author at Langtry, Kerrville, and Waring, Texas. Also known from Coahuila and Tamaulipas, Mexico.

RELATED SPECIES. A plant once described as *E. angustifolia* Pennell is now thought to be only a form of *E. texana,* with larger leaves. It occurs mostly west of the Pecos River in Texas.

REMARKS. The genus name, *Eysenhardtia,* honors Karl Wilhelm Eysenhardt (1794–1825), professor of botany at the University of Königsberg. The species name, *texana,* refers to the state of Texas, its natural habitat. It is also known as Rock Brush in the Texas hill country. The name Kidneywood is derived from the fact that closely related species are used in the treatment of renal disorders. The wood has been used to make dyes and is fluorescent in water.

Kidneywood

Eysenhardtia polystachya (Ort.) Sarg.

FIELD IDENTIFICATION. Western shrub with a number of stems from the base, or in more favorable situations, a small slender branched tree to 20 ft, with a diameter of 6–8 in.

FLOWERS. Maturing in May, borne in pubescent axillary spikes 3–6 in. long, pedicels slender and pubescent, bracts subulate and caducous, 1/25–1/12 in. long; calyx-tube campanulate, 1/12–1/8 in. long, many-ribbed, conspicuously glandular-dotted, pubescent, 5-toothed or split slightly more deeply on the posterior side, teeth about 1/50 in. long; erect, nearly equal; corolla white, petals erect, free, nearly equal, oblong-spatulate, apex rounded, clawed; standard concave, slightly broader than the wing and keel petals; disk cupuliform, adnate to the base of the calyx; stamens 10, inserted with the petals, 1 stamen free, superior and shorter than the rest, anthers oblong; style long and slender, geniculate and glandular near the apex; stigma inwardly oblique; ovary with 2–3 ovules attached to the inner angle.

FRUIT. Legume on pedicels about 1/25 in. long, pendent, oblong to oblanceolate or linear, sometimes falcate, about 1/2 in. long, 1/8–1/5 in. wide, tipped with the style remnants, thin, flattened, margins somewhat thickened, greenish brown, glabrous, indehiscent; seeds 1–2, compressed, light reddish brown, oblong-reniform, 1/5–1/4 in. long, seed coat coriaceous, embryo filling the seed cavity.

LEAVES. Alternate, odd-pinnately compound, 4–5 in. long, rachis pubescent and grooved on the upper side, stipules small, leaflets 10–23 pairs, each leaflet about 1/8–4/5 in. long, oval to oblong, apex rounded or slightly emarginate; dull green and puberulous or glabrous above; lower surface paler, pubescent, dotted with brown glands, somewhat reticulate-veined; petiolules stout; the stipel-like structures may be referred to as small, apiculate pustules.

TWIGS. Slender, terete, gray-pubescent at first, later reddish brown to gray and glabrous, with many glandular excrescences.

BARK. Light gray, about 1/16 in. thick, broken into plate-like scales, exfoliating into thin strips.

WOOD. Heartwood light reddish brown, sapwood yellow, close-grained, heavy, hard, dense, specific gravity 0.87.

RANGE. Arid slopes and dry ridges at altitudes of 3,500–5,000 ft. In Arizona and New Mexico; southward in Mexico in the states of Sonora, Chihuahua, Tamaulipas, and Oaxaca.

MEDICINAL USES. The following properties of the plant are described by Standley (1920–1926, p. 444): "This plant has long been known in Mexico because of the peculiar properties of the wood. An infusion of the heartwood in water has at first a golden-yellow color which soon deepens to orange. When held in a glass vial against a black background it will exhibit a beautiful peacock-blue fluorescence. The wood was well known in Europe as early as the 16th century, where it was called 'lignum nephriticum' because of its supposed diuretic properties. . . .

"The foliage of the tree is aromatic and the flowers are fragrant. In Mexico drinking troughs made from the wood are used for watering fowls, or a piece of the wood is put in their drinking water to ward off diseases. Palmer reports that in Sonora a decoction of the wood is given as a refreshing drink to fever patients. The wood is used in some localities for kidney and bladder affections. It also gives a yellowish brown dye."

REMARKS. The genus name *Eysenhardtia* honors Karl Wilhelm Eysenhardt (1794–1825), professor of botany at the University of Königsberg. The species name, *polystachya*, means "many-branched." The plant has also been listed under the name of *E. orthocarpa* (Gray) Wats. Vernacular names used in Latin American countries are Rosilla, Palo Cuate, Palo Dulce, Coatl, Coate, Cuate, Palo Dulce Blanco, Taray, Vara Dulce, Varaduz, Leña Nefrítica, and Tlapalezpatli. The plant is grazed by livestock to only a minor extent.

KIDNEYWOOD
Eysenhardtia polystachya (Ort.) Sarg.

KENTUCKY COFFEE-TREE
Gymnocladus dioica (L.) K. Koch

FIELD IDENTIFICATION. Tree attaining a height of 100 ft with a diameter of 4 ft. Branches stout and blunt, forming a narrow, round-topped crown.

FLOWERS. Appearing May–June, after the leaves, in loosely flowered terminal racemes; staminate and pistillate racemes separate on the same tree, or on different trees; raceme of staminate flower 3–5 in. long, on stout pedicels, flowers greenish white; calyx ⅜–⅔ in. long, elongate-tubular, hairy internally and externally, 5-lobed, lobes linear-lanceolate, acuminate at apex, ⅕–¼ in. long, shorter than the petals; petals 5, oblong, acute, equal, keeled externally, hairy on both sides, inserted on summit of calyx-tube, spreading or reflexed; stamens 10, those opposite the petals shortest, filaments filiform and hairy; anthers orange, oblong, introrse, opening lengthwise; stamens smaller and sterile in the pistillate flower, ovary rudimentary or none in the staminate flowers; pistillate raceme to 1 ft long, flowers on pedicels two to six times as long as those of the staminate flowers; ovary sessile or nearly so, superior, simple, hairy, 1-celled, 4–8-ovuled, ovules pendulous; style short, erect, oblique, lobes 2, inwardly stigmatic; pedicels of flowers from lanceolate, scarious, early deciduous bracts.

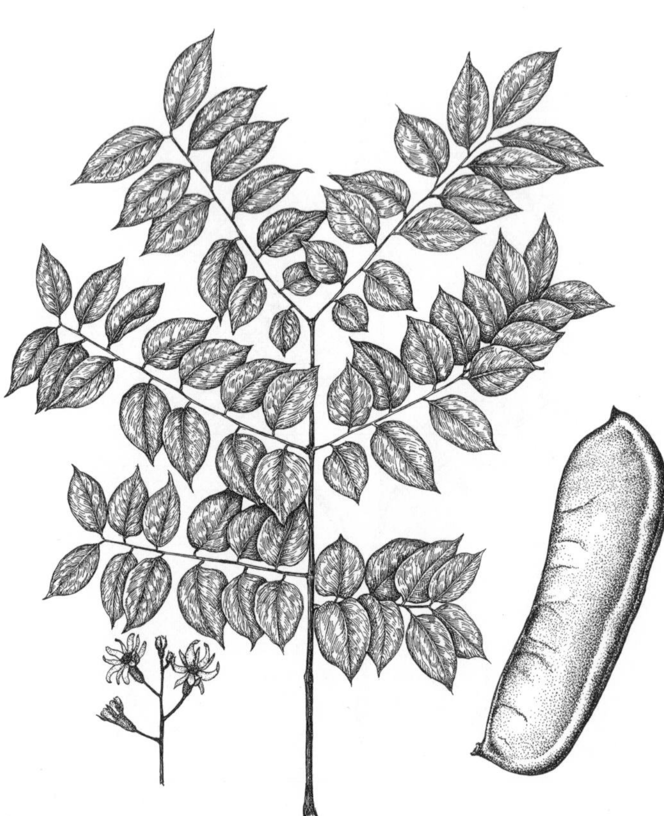

KENTUCKY COFFEE-TREE
Gymnocladus dioica (L.) K. Koch

FRUIT. Legume persistent during the winter and indehiscent, on peduncles 1–2 in. long, oblong, straight or falcate, flat, thick, woody, hard, reddish brown, sometimes glaucous, 4–10 in. long, 1–2 in. wide, green to brown, pulpy between the seeds; seeds several, ovoid to obovoid, lenticular, flattened, ⅜–¾ in. long, dark brown, bony, hard-shelled, endosperm thin.

LEAVES. Opening late in the spring, early deciduous in the fall, no terminal bud, alternate, twice-pinnately compound, 1–3 ft long, 1–2 ft wide; rachis long, slender, swollen at base; pinnae 5–13, each pinna with 7–14 leaflets which are ¾–3 in. long, 1–1½ in. wide, ovate to oval or elliptic; apex acute or acuminate, sometimes mucronate; base cuneate, rounded or truncate; upper surface dark green and glabrous or slightly hairy on midrib, veins obscure; lower surface paler and glabrous or pubescent; petiolules about 1/25 in. long, hairy to glabrous later; stipules about ⅓ in. long, lanceolate, apex glandular-serrate; the lower pair of pinnae sometimes with single leaflets; leaflets pink to bronze-green and tomentose at first, at maturity mostly glabrous, turning yellow before falling.

TWIGS. Short, blunt, contorted, pubescent at first and reddish, later glabrous and brownish to gray; leaf scars large, conspicuous, cordate; lenticels numerous, orange-colored; pith large, salmon-colored.

BARK. Gray to brown, furrowed, roughened by small, persistent scales.

WOOD. Reddish brown, sapwood yellowish white, heavy, straight-grained, moderately strong and durable, weighing 43 lb per cu ft, taking a high polish.

RANGE. Nowhere common. In low, rich woods, mostly west of the Appalachians and north of the Gulf Coast plain. Oklahoma, Arkansas, and a few records from north Texas; east to West Virginia, north to New York and Ontario, and west to Missouri and Nebraska.

MEDICINAL USES. The green pulp of the fruit has been used in homeopathic medical practice.

PROPAGATION. The seed averages 200–300 seeds per lb, with a high but rather irregular and slow germination rate. The pods may be gathered by threshing the limbs, and may be softened by macerating in water to remove the seeds. The seed is very hard and must be soaked 2–4 hours in concentrated sulphuric acid with a thorough washing in water afterward. If this treatment is not used, a soaking for 2–5 minutes in hot water at 185°F. will help. The germination rate averages 75 per cent. Seeds should be sown in spring in rows 24 in. apart, using a dozen seeds per linear foot, and covering 1–2 in. with firmed soil. Seedlings 1–2 years old may be set out.

The tree is planted as an ornamental in the North. It is also propagated by cuttings. The coarse thick twigs and compound foliage give a sturdy effect. Kentucky Coffee-tree was first introduced into cultivation about 1748, and it is often planted in Europe.

REMARKS. The genus name, *Gymnocladus*, means "naked-branched," with reference to the thick, blunt, naked branches. The species name, *dioica*, refers to the dioecious character of the flowers. Also known under the vernacular names of Chicot, Luck-bean, Coffee-nut, Kentucky-mahogany, American Coffee-tree, Nicor-tree, and Stump-tree.

The raw seeds are poisonous as they contain saponin and a toxalbumin very similar to ricin, the poisonous principle of the castor-bean plant. Roasting removes this poison, and the seeds were experimented with as a substitute for coffee in the Civil War period. The leaves are reported to be poisonous to cattle, but cases of such poisoning are infrequent. The wood is used for posts, crossties, furniture, fuel, cabinetmaking, interior finish, and construction.

Common Honey-locust
Gleditsia triacanthos L.

FIELD IDENTIFICATION. Tree to 100 ft, with a thorny trunk and branches and a loose, open crown.

FLOWERS. May–June, perfect or imperfect, borne in axillary, dense, green racemes; racemes of the staminate flowers often clustered, pubescent, 2–5 in. long; calyx campanulate; lobes of calyx 5, elliptic-lanceolate, spreading, hairy, acute; petals 4–5, longer than calyx-lobes, erect, oval, white; stamens 10, inserted on the calyx-tube, anthers green, pistil rudimentary or absent in the staminate flower; pistillate racemes 2–3 in. long, slender, few-flowered, usually solitary; pistil tomentose, ovary almost sessile, style short, stigma oblique, ovules 2 or many; stamens much smaller and abortive in pistillate flower.

Common Honey-locust
Gleditsia triacanthos L.

FRUIT. Ripening September–October, legume ½–1½ ft long, ½–1½ in. wide, borne on short peduncles, usually in twos or threes, dark brown, shiny, flattened, often twisted, falcate or straight, coriaceous, pulp succulent and sweetish between the seeds, occasional sterile legumes are seedless and pulpless, some trees with rather small legumes, others with large legumes, and some with a mixture of small, intermediate, and large; seeds about ⅓ in. long, oval, hard, compressed. The minimum commercial seed-bearing age is 10 years and the maximum 100 years.

LEAVES. Alternate, deciduous, once- or twice-pinnate, (the twice-pinnate usually on more vigorous shoots) 5–10 in. long; pinnae 4–8 pairs; leaflets 15–30, alternate or opposite, almost sessile, ¾–2 in. long, ½–1 in. wide, oblong-lanceolate, rounded or acute at apex, cuneate or rounded or inequilateral at base, crenulate or entire on margin, dark green and lustrous above, paler and often pubescent beneath; rachis and petioles pubescent.

TWIGS. Greenish or reddish brown, lustrous, stout, armed with solitary or 3-branched thorns which are rigid, sharp, straight, shiny, purplish brown.

BARK. Grayish brown to black, on older trees with fissures deep and narrow, separating into scaly ridges, often bearing heavy simple to multibranched thorns.

WOOD. Brown to reddish, sapwood yellowish, coarse-grained, strong in bending and end compression, stiff, highly shock resistant, tools well, splits rather easy, shrinks little, does not glue satisfactorily, rather hard, moderately durable, weighing 42 lb per cu ft.

RANGE. In moist fertile soil. Texas, Oklahoma, Arkansas, and Louisiana; eastward to Florida, northward to Pennsylvania and New York, and west to Nebraska.

PROPAGATION. The pods are gathered from the ground or picked when brown. After drying the seeds may be separated from the pod by flailing in a sack. The chaff may be removed by a fanning mill or by water flotation. Cleaned seed averages about 2,800 seeds per lb, with a commercial purity of 95 per cent and a soundness of 98 per cent. The seed may be stored at a temperature of 32°–45°F. in sealed containers. There are several methods recommended for pretreatment of seeds before planting as follows: (1) Soak for a short period in water at 185°–195°F.; they should then be planted immediately. (2) Scarify the seeds before planting. (3) Soak in concentrated commercial sulphuric acid at 60°–80°F. for 1–2 hours, then wash acid from seed before planting. The seeds are planted in rows 6–10 in. apart, with about 12 seeds per linear ft. A covering of ¾–1 in. of soil is adequate, and the soil should be kept moist until germination. Propagation is also done with cuttings. Sometimes other species or varieties are grafted on seedlings of Common Honey-locust.

VARIETIES, FORMS, AND HYBRIDS. A number of horticultural forms have been developed. Thornless Honey-locust, *G. triacanthos* var. *inermis* Willd., is thornless, or nearly so, and slender in habit.

Bushy Honey-locust, *G. triacanthos* var. *elegantissima* (Grosdemange) Rehd., is unarmed and densely bushy, with smaller leaflets.

Bujot Honey-locust, *G. triacanthos* var. *bujotii* (Neum.) Rehd., has slender pendulous branches and narrower leaflets.

Dwarf Honey-locust, *G. triacanthos* var. *nana* (Loud.) Henry, is a small compact shrub or tree with short and broad dark green leaflets.

Two other varieties of uncertain status are the Millwood and Smith honey-locusts.

Texas Honey-locust, *G.* × *texana* Sarg., is considered by some botanists to be a hybrid of *G. aquatica* Marsh. and *G. triacanthos* L. However, this matter is open to question because the latter is so variable that intergrading forms are found which fit the description of the hybrid. This tree is known to occur from Mississippi to east Texas and north to Arkansas and southwestern Indiana. The hybrid question is discussed by Sargent (1922).

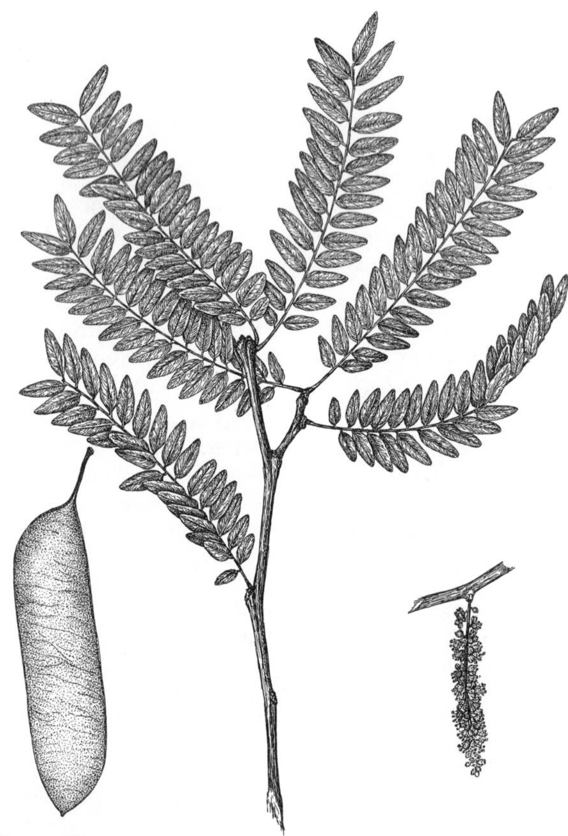

TEXAS HONEY-LOCUST
Gleditsia × *texana* Sargent *(G. triacanthos* × *G. aquatica)* A specimen showing a legume with straight margins from Brazoria County, Texas

TEXAS HONEY-LOCUST
Gleditsia × *texana* Sargent *(G. triacanthos* × *G. aquatica)* A typical specimen from Brazoria County, Texas

REMARKS. The genus name, *Gleditsia*, is a contraction and is in honor of Johann Gottlieb Gleditsch, an eighteenth-century German botanist. The species name, *triacanthos*, refers to the commonly 3-branched thorns. Other vernacular names are Honey-shucks, Sweet Locust, Thorn-tree, Thorny Locust, Three-thorned-acacia, and Sweet-bean. It has been known in cultivation since 1700, and is often planted for ornament, particularly the thornless form. The tree has few diseases or insect pests. The wood is used for farm implements, fuel, lumber, posts, vehicles, furniture, and railroad crossties. The flowers are reported to be a good bee food, and the Indians ate the fleshy sweet pulp of the young pods, older pods turning bitter. The pods are also eaten by cattle, white-tailed deer, gray squirrel, and cottontail.

TEXAS HONEY-LOCUST

Gleditsia × *texana* Sargent *(G. triacanthos* × *G. aquatica)*

FIELD IDENTIFICATION. Thornless, or thorny, tree attaining a height of 50–120 ft, with a diameter of ½–2½ ft. The branches are erect, ascending or spreading, and the bark is usually smooth and pale.

FLOWERS. Late in April, borne in racemes 3–4 in. ng, orange-yellow; calyx-lobes ovate, acute, villous, mewhat shorter than the petals, stamens exserted.

FRUIT. Legume straight or falcate, compressed, with r without pulp, apex rounded or short-pointed, base road and abruptly rounded or pointed, thin-walled, ark chestnut-brown, puberulent, margin only slight- thickened, length 3–14 in., width ¾–1½ in.; seeds umerous, oval, flattened, dark brown, shiny, about ½ . long or less.

LEAVES. Once- or twice-compound, puberulous, rachis ender and puberulous to glabrous later, 12–22-foliate; aflets oblong-ovate, often somewhat falcate, apex ounded or acute and apiculate, base obliquely round- d, margin obscurely crenulate-serrate, thick and firm, pper surface dark green and lustrous, paler on the wer surface, ½–1 in. long, short-petiolulate.

RANGE. Besides the Brazos River area, trees which re similar have been reported from other locations Texas. Some have been found along the Red River ear Shreveport, Louisiana, at Yazoo City, Mississippi, nd near Skelton, Gibson County, Indiana.

REMARKS. The genus name, *Gleditsia,* is in honor of ohann Gottlieb Gleditsch, an eighteenth century Ger- nan botanist. The hybrid name, *texana,* refers to its ccurrence in Texas in the bottom lands of the Brazos River.

This is a tree of rather uncertain relationship. It was irst described as a species under the name of *G. texana* Sarg. (Sargent, 1901).

Sargent found the grove of trees to which he gave he name of *G. texana* growing near Brazoria in the alley of the lower Brazos River in Texas. The author as examined numerous specimens in the Brazoria area nd has found so many intergrades with the common Honey-locust, *G. triacanthos* L., that it is hard to determine the typical form to which Sargent applied the name. Some trees were spineless, and others displayed all degrees of spinescence; some bore legumes over 1 ft long, and some only 3–5 in. long, and some trees produced fruit both small and large. The amount of fleshy tissue between the seeds appears to be variable also. In fact, the great variability of the trees of the Brazoria area is the rule rather than the exception.

Sargent later noted that some trees with these variable characters were scattered, and some were found growing in the vicinity of both *G. triacanthos* and *G. aquatica.* It was therefore suggested that the variable trees might possibly be hybrids of these two species, and the name of *G.* × *texana* Sarg. was given by Sargent (1922).

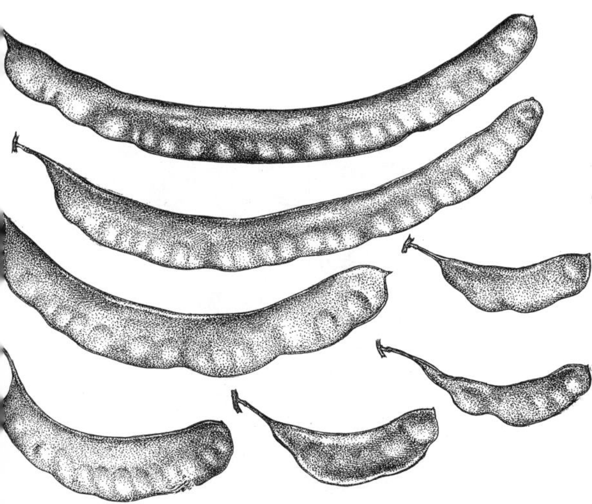

TEXAS HONEY-LOCUST
Gleditsia × *texana* Sargent *(G. triacanthos* × *G. aquatica)* Legumes showing variation in size and shape from different trees in Brazoria County, Texas

WATER LOCUST

Gleditsia aquatica Marsh.

FIELD IDENTIFICATION. Tree of deep swamps with contorted, spine-bearing branches, attaining a height of 60 ft.

FLOWERS. In May or June, perfect or imperfect, borne in slender, narrow, green, axillary racemes about 3–4 in. long; pedicels short; calyx bell-shaped, 3–5-lobed, pubescent; lobes as long as petals, narrow, acute or obtuse, somewhat pilose; petals 3–5, equal, sessile, inserted on rim of calyx-tube; stamens 3–10, distinct, exserted, filament hairy below, erect, filiform, anthers large and green, abortive and swollen in the pistillate flower; ovary nearly sessile, glabrous, rudimentary in staminate flowers, stigma smaller and oblique.

FRUIT. September–December, legume long-stipitate, oval-elliptic, oblique, brown, lustrous, thin, flat, pulpless, 1–3½ in. long; seeds 1–3, orange-brown, suborbicular, about ½ in. wide.

LEAVES. Pinnate or bipinnate, 5–10 in. long; pinnae 3–8 pairs; leaflets 5–12 pairs, ovate-oblong, obscurely crenulate or entire, rounded or emarginate at the apex, oblique at base, dark green and shining above, paler beneath, ½–1 in. long, ¼–½ in. wide; petiolules slightly hairy; petioles pubescent and grooved.

TWIGS. Gray or reddish, often with knobby spurs and corky lenticels; spines dark red and shiny, simple or compound, to 5 in. long, or spines sometimes absent.

BARK. Dull gray or reddish brown, firm, fissures narrow with small flat intervening scales, rough with small corky tubercles.

WOOD. Light reddish brown, sapwood yellow, strong, heavy, hard, close-grained, weighing 45 lb per cu ft.

RANGE. In swampland; east Texas, Oklahoma, Arkansas, and Louisiana; eastward to Florida and northward to South Carolina, Illinois, and Indiana.

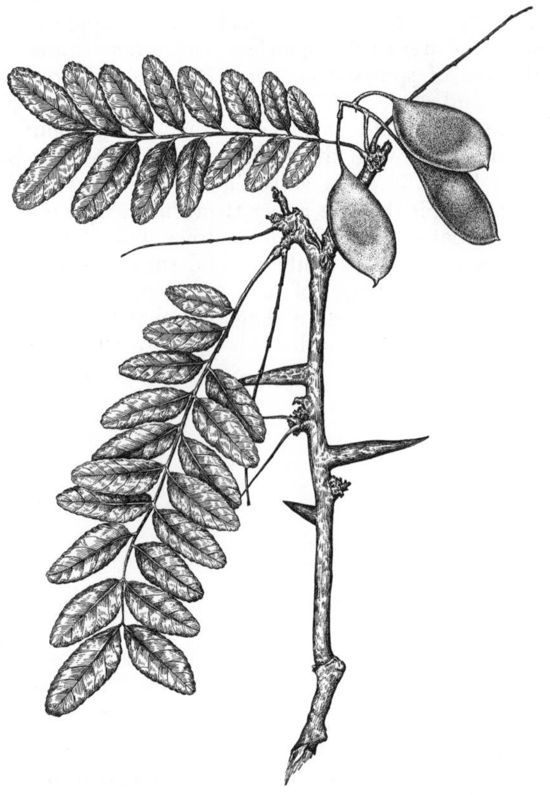

WATER LOCUST
Gleditsia aquatica Marsh.

PROPAGATION. In order to obtain the seeds, the brown and dried pods are hand-picked and threshed or flailed. A bath for 1–2 hours in sulfuric acid makes the seed coats more permeable, but must be followed by a thorough washing in water. Mechanical scarifiers can also be used with success. The seeds can also be soaked in water until swollen. Pretreated seeds are drilled in rows 6 to 10 in. apart and covered with about ½ in. of soil. Spring planting is better than fall planting. From 10 to 15 seeds are planted per linear ft. The germinative average is about 75 per cent. The Water Locust seedlings can stand much wetter soil than the other species of locust.

HYBRID. *G. aquatica* is known to hybridize with *G. triacanthos,* to produce *G.* × *texana* Sargent.

REMARKS. The genus name, *Gleditsia,* is a contraction and is named in honor of Johann Gottlieb Gleditsch, an eighteenth-century German botanist, and the species name, *aquatica,* refers to the plant's swampy habitat. Other vernacular names are Swamp Locust and Thorny Acacia. This tree is generally separated from Common Honey-locust, which it resembles, by its generally smaller leaves, slenderer thorns, and by the much smaller pulpless fruit. The wood has little commercial value except locally for posts and fuel.

TEXAS PALOVERDE
Cercidium texanum Gray

FIELD IDENTIFICATION. Spiny, green-barked shrub or small tree to 25 ft. Usually with a short crooked or leaning trunk, or with semiprostrate lower branches.

FLOWERS. Borne in clusters or short racemes; pedicels solitary, reddish, pubescent, ¼–½ in. long; corolla yellow, of 5 imbricate petals, each about ½ in. long, ovate to elliptic, apex acute to obtuse or rounded, margin frilled and crisped. One petal abruptly contracted into a long claw at base and red-spotted; stamens 10, filaments distinct, red, glabrous; anthers yellow, opening lengthwise; pistils equaling or slightly exceeding the stamens, filiform, hairy; calyx valvate, sepals 5, linear to oblong, puberulent to pubescent, reflexed.

FRUIT. Legume 1–2½ in. long, about ¼ in. wide, flattened, apex long-apiculate, margin straight or somewhat constricted between the seeds; surfaces light brown, at first finely hairy but later glabrate; seeds usually 1–4, about ¼ in. long, dark brown, shiny, flattened, short-oblong, ends rounded.

LEAVES. Early deciduous, leaving bare twigs and branches; leaves twice-pinnately compound, ½–¾ in. long; petiole puberulent, about ¼ in. long; leaflets 1–3 pairs (usually 1–2 per pinna); blades very short petioluled or sessile, short-oblong or narrowly obovate, some-

TEXAS PALOVERDE
Cercidium texanum Gray

what broader toward the rounded apex, ⅛–¼ in. long, grayish green with closely appressed pubescence.

TWIGS. Grayish green or dark green, zigzag, very thorny, smooth, finely grooved, densely appressed pubescent, later more glabrous; spines white to green or pale brown, straight or slightly recurved, sharp, pubescent or glabrous, ¼–½ in. long.

BARK. Thin, smooth, light or dark green.

RANGE. Usually on limestone soils of flats or gentle slopes. Abundant in the vicinity of Brackettville, Del Rio, and Langtry, Texas. In Mexico in Nuevo León, Tamaulipas, and Coahuila.

RELATED SPECIES. Texas Paloverde is closely related to Border Paloverde, *C. macrum* I. M. Johnston. Generally Texas Paloverde has fewer pinnae on an average, and a hairy ovary, but intermediate forms between the two are found where the ranges overlap in the lower Rio Grande River area.

REMARKS. The genus name, *Cercidium*, is from the Greek *kerkidion* ("a weaver's comb"), to which the fruit has a fancied resemblance. The species name, *texanum*, is for the state of Texas.

BORDER PALOVERDE
Cercidium macrum I. M. Johnston

BORDER PALOVERDE

Cercidium macrum I. M. Johnston

FIELD IDENTIFICATION. Green-barked spiny tree attaining a height of 25 ft, its branches crooked and sometimes close to the ground and horizontal.

FLOWERS. Borne in short racemes ¾–1¼ in. long; pedicels ⅛–½ in. long, puberulent to glabrous; corolla yellow, petals 5, ovate or elliptic, apex obtuse, margin crisped; one petal red-spotted, long-clawed, larger than the other petals; stamens 10, filaments distinct and hairy toward the base; pistil elongate, glabrous, calyx of 5 sepals, lanceolate or oblong, glabrous or nearly so, ⅛–³⁄₁₆ in. long, reflexed.

FRUIT. Pedicels ½–1½ in. long, legume 1–2½ in. long, ¼–½ in. wide, flattened, apex abruptly obtuse or acute to slender-mucronulate, margin straight or slightly undulate, dark brown, glabrous at maturity; seeds 1–5, flattened, shiny, dark brown, oval or short-oblong, with ends rounded.

LEAVES. Often early deciduous or releafing after rains, twice-pinnate, ¾–1 in. long; petioles puberulent to glabrous, about ¼ in. long; leaflets usually 2–3 pairs (more rarely 1 pair per pinna); blades oblong, apex mostly rounded or some semitruncate and slightly notched, base narrowed, margin entire, surfaces olive-green to dark green, ⅛–¼ in. long, young ones minutely puberulent, later glabrous; petiolules very short, about ¹⁄₂₅ in. long.

TWIGS. Zigzag, young ones light green, eventually dark green or dark brown on older branches, smooth, finely grooved; spines averaging about ¼ in. long, green to brown or black, slender, sharp, straight or slightly curved.

BARK. Thin, smooth, dark green to brown on old trunks.

RANGE. Either on sandy loams or clay soils. Valley of the lower Rio Grande River, Cameron, Willacy and Hidalgo counties in Texas, and in adjacent Mexico.

RELATED SPECIES. Border Paloverde is closely related to Texas Paloverde, *C. texanum* Gray, which grows in the Del Rio area. However, the latter has a hairy pistil, more pubescent leaves and twigs, and usually fewer leaflets (1–2 pairs) on an average. These characters are not always clearly distinguishable, especially where the two ranges approach each other and overlap.

REMARKS. The genus name, *Cercidium*, is from Greek, *kerkidion* ("a weaver's comb"), to which the fruit has a fancied resemblance. The species name, *macrum*, means "large," but the application is obscure.

BLUE PALOVERDE

Cercidium floridum Benth.

FIELD IDENTIFICATION. Tree usually less than 18 ft, with a low wide head. The trunk is usually short and crooked, and the bright green branches are spiny-armed.

FLOWERS. Usually borne April–July, or sporadically

BLUE PALOVERDE
Cercidium floridum Benth.

after rains, in few-flowered axillary racemes 2–4½ in. long (or solitary); pedicels jointed; bracts minute and early deciduous; flowers ½–¾ in. broad, petals 5, orbicular or short-oblong, clawed at base, bright yellow; the upper petal usually broader and longer-clawed than the others and usually glandular at the base; stamens 10, exserted, inserted on the disk margin, filaments filiform with the upper filament enlarged at base; ovary short-stalked, inserted at the base of the calyx-tube, style slender and involute with a minute terminal stigma; calyx ⅕–¼ in. long, glabrous, the 5 lobes equal, acute and reflexed at maturity, the petals twice as long.

FRUIT. Legume maturing in July, linear-oblong, glabrous, flattened, 1½–3 in. long, about ½ in. broad, margins thickened, straight or somewhat constricted between the seeds, apex acute, tardily dehiscent, the 2 valves papery and enclosing 1–8 seeds (usually 2–3).

LEAVES. Early deciduous, the tree appearing leafless or nearly so the greater part of the year, the green twigs and branches carrying on photosynthesis. Pinnae usually 1 pair; rachis pubescent and usually not spinulose, ¼–¾ in. long; each pinna with 2–4 pairs of leaflets; leaflets ovate to obovate, ⅛–⅜ in. long, apex obtuse or retuse, pale green, pubescent at first, more glabrous later.

TWIGS. Dull green to yellowish green, slender, tortuous; spines axillary, slender, straight, sometimes to 1 in. long; older branches dull green and smooth.

BARK. On old trunks near the base green to light brown, often with short, ridgelike excrescences.

WOOD. Light in weight, soft, close-grained, yellowish or greenish, sapwood paler.

RANGE. Trans-Pecos Texas, New Mexico, Arizona and northern Mexico.

RELATED SPECIES. It is rather difficult to separate this tree taxonomically from *C. texanum* Gray and *C. macrum* I. M. Johnston. There seems to be an overlap of *C. floridum* and *C. texanum* in Trans-Pecos Texas, and in turn *C. macrum* overlaps range with C. *texanum* in the Texas lower Rio Grande River area. More field work is needed to ascertain whether the three species should be maintained separately.

REMARKS. The genus name, *Cercidium*, is from the ancient Greek and signifies the resemblance of the legume to a weaver's instrument. The species name, *floridum*, refers to the profuse flowering habit of this very drought-resistant tree. A palatable meal is prepared from the ground pods and seeds by the Mexicans and Indians. Cattle browse the fruit also. The hard wood is sometimes used for fuel.

JERUSALEM-THORN

Parkinsonia aculeata L.

FIELD IDENTIFICATION. Green-barked, thorny shrub or tree to 36 ft. Branches slender, spiny, spreading, often pendulous to form a rounded head.

FLOWERS. Fragrant, borne in spring or throughout the summer especially after rains. Racemes 5–6 in. long, axillary, solitary or fascicled, pedicels ⅓–½ in. long; petals 5, imbricate in the bud, yellow, about ½ in. long, spreading, oval, clawed at the base, margin erose or entire; one petal larger that the others and bearing a gland at base, becoming red-dotted or orange with age; stamens 10, shorter than the petals, filaments distinct and hairy below, anthers yellow to reddish, opening lengthwise; ovary pubescent; calyx glabrous or nearly so, tube short, lobes oblong and reflexed, longer than the tube.

FRUIT. On pedicels ½–¾ in. long; a linear legume, 2–4 in. long, brown to orange or reddish, puberulent or glabrous, leathery, ends attenuate, constricted between the seeds, swollen portions almost terete, constrictions flattened; seeds 1–8, about ⅓ in. long, oblong, seed coat green to brown.

LEAVES. Bipinnate or rarely pinnate, petioles short, alternate or fascicled, linear, 8–16 in. long, rachis flat and winged; leaflets numerous (25–30 pairs), remote, linear to oblanceolate, about ⅓ in. long or less, inequilateral, dropping away early and leaving the per-

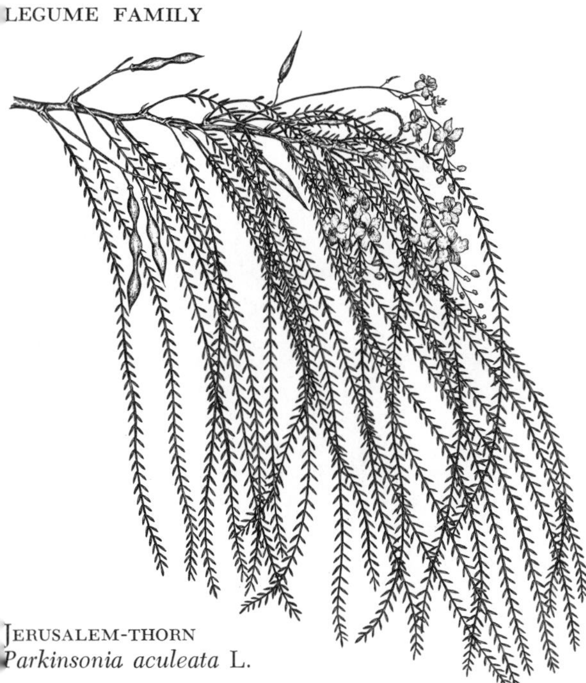

JERUSALEM-THORN
Parkinsonia aculeata L.

sistent, flat, naked, photosynthetic rachis, petiolules slender; stipules spinescent; rachis terminating with a weak spine.

TWIGS. Somewhat divaricate, green to yellowish, puberulent to glabrous; later gray to brown or orange; spines green, brown or black, to 1 in. long.

BARK. Thin, smooth, green, later brown to reddish with small scales on old trunks.

WOOD. Light brown to yellow, hard, heavy, close-grained, specific gravity 0.61.

RANGE. Usually in moist sandy soils, but resistant to saline situations. In the southern half of Texas; west through New Mexico to Arizona and southward through Mexico to northern South America. Probably an escape from cultivation in California, Florida, and the West Indies.

REMARKS. The genus name, *Parkinsonia*, refers to John Parkinson, an English botanical author. The species name, *aculeata*, refers to the spines. Other vernacular names are Horsebean, Cloth-of-Gold, Crown-of-thorns, Barbados Fence Flowers, Paloverde, Lluvia de Oro, Cambrón, Espilla, Espinillo, Espinillo de España, Junco, Junco Marion, Palo de Rayo, Flor de Rayo, Yabo, Calentano, Espino Real de España, Acacia de Aguijote, Guichi-belle, Mesquite Extranjero, Guacopano, Retama de Cerda, and Retama. A grove of trees is known as a "retamal" by the Mexican people. The tree is a rapid grower and is generally free of diseases and insects. It is often grown for ornament and hedges. The wood is occasionally used as fuel and was formerly used for papermaking. Standley (1920–1926, p. 428) reports that an infusion of the leaves is used in tropical America as a febrifuge, for diabetes, for epilepsy, and as a sudorific and abortifacient. The leaves and pods are eaten by horses, cattle, and deer, particularly in times of stress. The Indians formerly pounded the seeds into flour to make bread. The flowers are sometimes important as a bee food.

COW-FOOT BAUHINIA
Bauhinia mexicana Vog.

FIELD IDENTIFICATION. Cultivated tree with a rounded top, attaining a height of 25 ft.

FLOWERS. Racemes densely many-flowered, 1-2½ in. long, all parts densely soft-hairy. Individual flowers on slender peduncles ⅓–¾ in. long (average ½); calyx spathelike, green striped, hairy, ¼–⅜ in. long, splitting on one side and divided at the apex into 3 acuminate, subulate sepals; petals 5, white, delicate, almost equal in size, ½–⅝ in. long, lanceolate, apex long-acuminate, base cuneately narrowed into a long claw, margin somewhat undulate on the plane surface; fertile stamens usually only 1 or 2, almost as long as the pistil, balance of androecium composed of shorter, antherless staminodia, filaments white; pistil long-exserted, style filiform, longer than the stamens.

FRUIT. Borne on a slender peduncle, ¾–2½ in. long, ⅜–½ in. wide, linear, apex apiculate, base cuneate, surface flattened, dark brown to black, glabrous at maturity; seeds ¼–⅜ in. long, ovoid, flattened, obtuse at one end, rounded at the other, lustrous, green to brown.

LEAVES. Alternate, conduplicate, blade length 2–2½ in., width 1–1½ in., apex deeply cleft to form 2 acute

COW-FOOT BAUHINIA
Bauhinia mexicana Vog.

terminal lobes resembling a cloven hoof, base truncate to rounded or semicordate, margin entire; upper surface dull green, veins impressed, soft-puberulent; lower surface paler, palmately veined, soft-pubescent; petioles ¼–½ in. long, densely soft-pubescent.

TWIGS. Young ones green and soft-pubescent, older ones gray to brownish, glabrous.

BARK. Gray, separated into narrow, short and deep fissures between narrow confluent ridges.

RANGE. Occasionally cultivated for ornament in the lower Rio Grande Valley area of Texas. Native in Mexico and also escaping cultivation there.

REMARKS. The genus name, *Bauhinia,* is in honor of John and Caspar Bauhin, sixteenth-century herbalists. The species name, *mexicana,* is for Mexico, the tree's native habitat. Some vernacular names in use are Hierba de la Vaca, Pata-Vaca, and Timbe. At the time of this writing Robert Runyon of Brownsville, Texas, has a fine specimen in the yard of his home at 612 Saint Charles Street. It blooms profusely in May.

The bauhinias are tropical plants numbering over 200 species. A few are cultivated in southern Florida, southern California, and along the Texas coast. They grow best from seed and generally bloom in three or four years. Some species may be propagated from suckers. They are seldom grown in greenhouses in the United States. The bark of some species is used for dyes or ropemaking. A few have edible seeds, and some are used as fish poison. In tropical America various parts of the plants are employed in domestic medicine for liver affections.

Purple Bauhinia

Bauhinia purpurea L.

FIELD IDENTIFICATION. A small cultivated tree 15–25 ft, with an irregular rounded crown, the bark smooth, reddish brown, tan, or gray.

FLOWERS. Often abundant in winter or spring, peduncles 1–2½ in. long, green or purplish, grooved, puberulent to glabrate; flowers solitary or a few together in a loose cyme, large, about 3½ in. across when spread, 5-petaled, purple (one petal somewhat striped or mottled darker purple), oval to obovate, apex rounded, base abruptly narrowed into a short claw; stamens 5, purple, long-exserted, upcurved, 1 in. long or more; anthers about 3/16 in. long; style as long as the stamens, or longer, stigma minute; calyx ½–1 in. long, broadly ovate to oval, coherent into one piece at first, splitting later into several segments, longitudinally ribbed, green to reddish.

FRUIT. Legume gradually widened from a stout grooved peduncle 1–3 in. long, linear to oblong, flattened, 4–9 in. long, margin entire or undulate, apex acute, sometimes abruptly so and long-pointed, surfaces smooth and reticulate veiny, at maturity brown;

Purple Bauhinia
Bauhinia purpurea L.

seeds several, very thin, round, about ⅜ in. across, ta to brown.

LEAVES. Simple, alternate, deciduous, 2–3¾ in. lon or wide, oval, deeply cleft at apex and base, the apical lobes rounded, margin entire, surfaces dull gree and glabrous, slightly paler beneath, palmately veined petioles stout, 1–1½ in. long, puberulent at first, gla brous later.

TWIGS. Young ones green and puberulent, older one reddish brown and glabrous.

RANGE. Does best in well-drained soil. Cultivated i gardens along the coast from Texas to Florida. A na tive of India, China, and Burma.

PROPAGATION. A very beautiful tree for gardens an blooms readily in 3 or 4 years. It is known to stand temperature of 26° in Florida, but is generally ver tender and is subject to freezing on the Texas an Louisiana coasts. It grows readily from seeds but cut tings root only with some difficulty.

VARIETY. A variety with white flowers has bee named *B. purpurea* var. *alba* (Buch-Ham.).

REMARKS. The genus name, *Bauhinia,* is in honor o John and Caspar Bauhin, sixteenth-century herbalists an allusion suggested by the twin lobes of the leaves The species name, *purpurea,* is for the purple flowers The tree is sometimes called the Orchid-tree because o the flowers' superficial resemblance to an orchid flower

Anacacho Bauhinia

Bauhinia congesta Britt. & Rose

FIELD IDENTIFICATION. Western shrub or small tree, leaves bifoliate, each leaflet sessile, oblique.

FLOWERS. Borne March–April, racemes 1–2½ in. long, showy, pedicels ¼–½ in. long, densely short-hairy; flowers ¾–1 in. across when open; calyx to ½ in. long, tubular at first, splitting into 3–5 linear segments with short subulate teeth, densely hairy; petals 5, white, oblong to elliptic, apex obtuse to rounded, base gradually or abruptly long-clawed; style elongate, linear, stigma small and capitate; a single stamen elongate and fertile, the balance much shorter and apparently sterile, all very hairy at the base.

FRUIT. Borne August–September. Legume on a peduncle ¼–¾ in. long, oblong to linear, length 1¼–3½ in., width ¼–⅜ in., base cuneate or narrowed gradually, apex abruptly long- or short-pointed, the point sometimes curved, flattened or twisted, glabrous, dark brown at maturity, dehiscent into 2 valves, 1–4-seeded; seeds black, oval to short-oblong, flattened, ¼–5/16 in. long.

LEAVES. Alternate, divided to the base into 2 leaflets, occasionally simple, entire or bifid, leaflets ⅜–1¼ in. long, ¼–⅞ in. wide, sessile, asymmetrical, inner margin straighter than the outer, margins calloused and slightly wavy; apices rounded, upper surface dull green to semilustrous with scattered short stiff hairs, later almost glabrous; lower surface varying from very fine to brown-tomentose; especially on the palmate veins; veinlets reticulate; stipel about 1/25 in. long, tomentose; sometimes with 1 or 2 glands at the base of the tomentose petiole which is ¼–⅝ in. long.

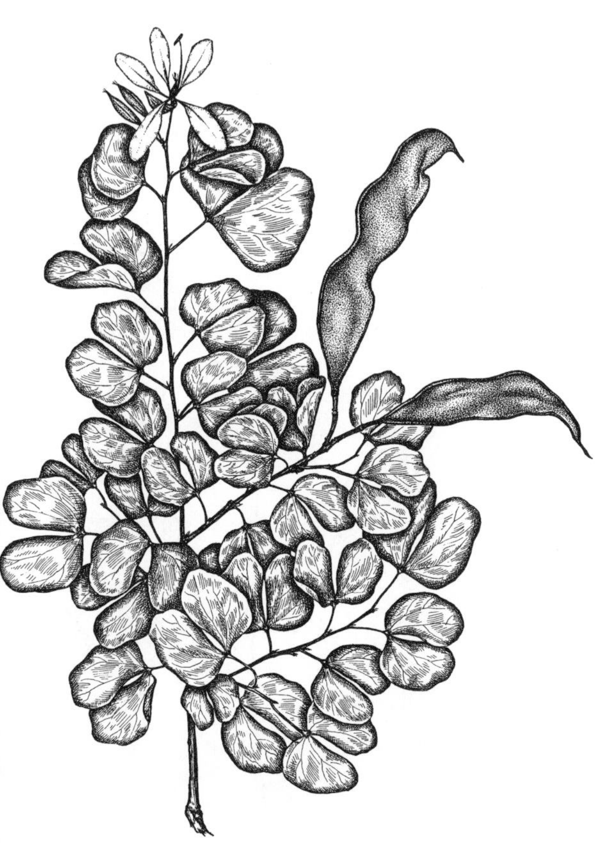

Anacacho Bauhinia
Bauhinia congesta Britt. & Rose

TWIGS. Slender, at first finely but densely brown-tomentose with short, stiff, curved hairs, later becoming brown to gray and glabrous; lenticels small, pale, scattered.

RANGE. Known in Texas only from the Anacacho Hills west of Uvalde. Type specimen examined from the United States National Herbarium, Edward Palmer, No. 285, north of Monclova, Coahuila, Mexico.

REMARKS. The genus name, *Bauhinia,* honors John and Caspar Bauhin, sixteenth-century herbalists, and the species name, *congesta,* refers to the crowded flowers.

Wislizenus Senna

Cassia wislizenii Gray

FIELD IDENTIFICATION. Much-branched, spreading shrub to 9 ft.

FLOWERS. Racemes 1–4 in. long or wide, terminal or axillary, loose, pubescent, many-flowered; pedicels slender, pubescent, ¼–1¼ in. long; flowers ¾–1½ in. across when open; petals 5, imbricate, yellow, somewhat unequal in size, ¼–⅝ in. long, oblong to oval or elliptic, apex rounded, base sessile or short-clawed; stamens 10, upper 3 abortive and minute, anthers large, opening by apical pores; pistil linear, curved, long-exserted; calyx 5-lobed, lobes rounded or obtuse.

FRUIT. Legume 2½–6 in. long, about ¼ in. broad or less, linear, straight or falcate, apex abruptly acute to acuminate and apiculate, base cuneate, surface flattened and sunken between the thickened sutures, glabrous, lustrous as if varnished, dehiscent into 2 thin valves; seeds brown, flattened, squarish to quadrangular, lustrous, about ⅛ in. long; fruiting peduncles ½–2 in. long, slender, pubescent.

LEAVES. Alternate or fasciculate, ½–1½ in. long, abruptly pinnately compound, rachis pubescent; leaflets 1–6 pairs (usually 3–4), petiolules about 1/25 in. long, pubescent; blades obovate to oval; apex rounded to truncate and apiculate (more rarely, acute), sometimes notched; base broadly cuneate and often inequilateral, thickish; upper surface dull green and glabrous to puberulent; lower surface with prominent thick veins, semiglabrous to puberulent, length ⅛–¼ in.; stipules filiform, pubescent.

TWIGS. Very leafy, green to brown or black, when young densely pubescent, when older glabrous and often lustrous, slender, rigid, slightly striate; lenticels

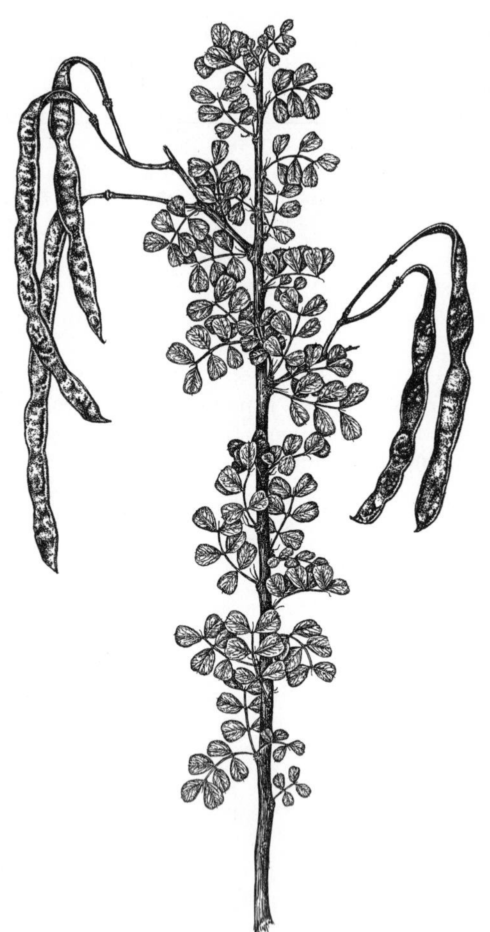

WISLIZENUS SENNA
Cassia wislizenii Gray

scattered, pale and minute; bark on old trunks roughened.

RANGE. Dry rocky soil to an elevation of 5,000 ft, in West Texas (Presidio County), New Mexico, Arizona, and south in Mexico in Chihuahua, Coahuila, and Zacatecas.

REMARKS. The genus name, *Cassia,* is the ancient name, and the species name, *wislizenii,* is for the botanist Friedrich Adolph Wislizenus. An attractive shrub in flower, sometimes cultivated in yards in the area where it grows.

FLOWERING SENNA

Cassia corymbosa Lam.

FIELD IDENTIFICATION. Cultivated shrub or small tree to 12 ft, with a trunk 2–5 in. in diameter. The numerous elongate, erect or spreading branches form a rounded crown.

FLOWERS. Blooms in late summer. Borne in corymbose panicles from the leaf axils; primary peduncle glabrous, 1–2½ in. long; secondary flower-bearing pedicels to ¾ in. long; petals 5, bright yellow, nearly equal, obovate, ⅜–½ in. long; stamens 7, 3 of them long exserted, upcurved, anthers large with oblique cup-like apices; 4 of the stamens more erect with anthers shorter; staminodia 2–3, short and flattened somewhat; style elongate, exserted, upcurved, short white-hairy; calyx of 5 sepals, green or yellowish, unequal in shape, narrowly elliptic to ovate, ¼–⅜ in. long.

FRUIT. Legume stout, thick, 1¾–6 in. long, sides convex and sutures furrowed, indehiscent.

LEAVES. Evenly pinnately compound, rachis averaging about 2 in. long, mostly glabrous; leaflets 4–6, 1–[illegible] in. long, terminal pair longest, lanceolate to linear-elliptic or elliptic-lanceolate, apex long-acute or acuminate, base rounded and often asymmetrical, margin entire, upper surface dull green and glabrous, lower surface paler and glabrous; petiolules ¹⁄₁₆–⅛ in. long, glabrous or with a few sparse white hairs, or a few dark, stalked glands in the axils.

TWIGS. Young ones elongate, slender, green, glabrous; older ones brown and glabrous.

BARK. On older branches and trunk smooth, dark brown to light or dark gray, marked with numerous pale, linear, horizontal lenticels.

FLOWERING SENNA
Cassia corymbosa Lam.

LINDHEIMER SENNA
Cassia lindheimeriana Scheele

RANGE. A native of Argentina, cultivated in gardens in many cities along the Gulf Coast. Occasionally escaping cultivation, or persistent about abandoned gardens of eastern Texas and Louisiana; eastward to Tennessee, Georgia, and Florida.

REMARKS. The genus name, *Cassia,* is the ancient name of an aromatic plant. The species name, *corymbosa,* is for the flower arrangement. It is listed by some authors under the name of *Adipera corymbosa* Britt. & Rose. This garden shrub is gaining in popularity because of its dense foliage, bright yellow flowers, and freedom from disease.

LINDHEIMER SENNA

Cassia lindheimeriana Scheele

FIELD IDENTIFICATION. Essentially a robust velvety-hairy herb, but becoming bushy in appearance and attaining a height of 1–3 ft. The stems solitary or several branching from the base.

FLOWERS. Appearing June–September; corymbs terminal, leafy, loosely many-flowered; petals 5, nearly equal, yellow, with slender brown veins, apices rounded or obtuse, ½–⅗ in. long; calyx hairy, the 5 lobes ovate-lanceolate, obtuse; stamens 10, the upper 3 abortive, anthers red or pale brown, opening by terminal pores; ovules numerous.

FRUIT. Legume broadly linear, straight or slightly curved, thickened-coriaceous, somewhat compressed, apex obtuse to acute or mucronate, marginal sutures thickened, surfaces sparingly hairy, length 1½–2½ in., width ⅙–¼ in., very tardily dehiscent, many-seeded.

LEAVES. Pinnately compound, leaflets 8–16 pairs, rather spreading, oval to elliptic or oblong, apex obtuse to rounded, often mucronate, base unequal-sided but rounded, blade length 1–1½ in., width about ⅝ in., surfaces lustrous with shiny velvety-silky hairs, especially beneath; with or without a stipitate-setiform gland between the pairs of leaflets; stipules linear, membranaceous, early deciduous.

STEMS. Green to brown or straw-colored, strongly angled, softly and densely hairy, stout but essentially herbaceous; the plant large, robust and shrubby in appearance.

RANGE. On dry mesas or hillsides to an altitude of 5,700 ft. Mostly on limestone soils. From the Colorado River through central and western Texas, New Mexico, and Arizona. In Mexico in Nuevo León.

REMARKS. The genus name, *Cassia,* is an ancient name of some aromatic plant. The species name, *lindheimeriana,* is in honor of Ferdinand Lindheimer (1801–1879), a native of Germany who resided 30 years in Texas. Most of the plants he collected were reported on by Asa Gray and George Engelmann. An attractive plant for its yellow flowers and gray leaves, worthy of cultivation.

PARADISE POINCIANA

Caesalpinia gilliesii Wall. *ex* Hook.

FIELD IDENTIFICATION. Shrub or small tree rarely over 8 ft. Branches green, light brown or reddish, glandular-pubescent, malodorous.

FLOWERS. Blooming most of the summer, borne in terminal, showy, open racemes; flowers ¾–1½ in. in diameter, solitary; pedicels ½–1½ in., yellowish and densely brown-glandular, some glands stalked; sepals larger than the petals, yellow, one noticeably incurved and hooded, imbricate, oblong-elliptic, somewhat narrower than the inner obovate or oval petals; petals 1–1½ in. long; stamens 10, conspicuously red, curved, long-exserted; stigma small, discoid.

FRUIT. Pod maturing at lower part of racemes as flowers open at apex, hence flowers and mature fruit often found on the same plant; pedicels stout, ¾–1½ in. long, often with stalked glands; pod flat, oblong-oblanceolate, somewhat falcate or straight, apex acuminate or abruptly acute; light tan or reddish brown when mature, 2–3½ in. long, ½–¾ in. wide; at dehiscence seeds ejected forcibly and fly out a considerable distance.

LEAVES. Bipinnate, 3–5 in. long; primary pinnae of 6–12 pairs, each about ¾–1 in. long; leaflets 5–9 pairs per pinna, each leaflet ⅛–3/16 in. long; oblong, apex rounded or obtuse, base asymmetrical; petiole, petiolules and rachis glabrous or nearly so.

STEMS. All parts of plant malodorous; older stems green, tan or reddish, glandular-pubescent; younger twigs thickly brown-glandular, some glands stalked.

RANGE. A native of Argentina, escaping cultivation and growing in a wild state in west Texas, New Mexico, and Arizona; also in Florida. Southward in Mexico.

REMARKS. The genus name, *Caesalpinia,* was named for Andreas Caesalpinus, chief physician to Pope Clement VIII. The species name, *gilliesii,* is for its discoverer, John Gillies (1747–1836). Another vernacular name is Bird-of-paradise.

Flower-fence Poinciana

Caesalpinia pulcherrima (L.) Swartz

FIELD IDENTIFICATION. An attractive tropical shrub or irregularly branched small tree to 15 ft. The large beautiful flowers have 5 petals which are red, orange, yellow, or mottled.

FLOWERS. In upright racemes 5–20 in. long, pedicels ⅓–3 in. long and glabrous; sepals 5, about ⅗ in. long, imbricate, petaloid, oblong to oblanceolate, apex rounded, much shorter than the petals (one sepal larger than the others, concave and overlapping the others in the bud); corolla of 5 unequal petals ¾–1¼ in. long, the standard petal generally the longest with a revolute, almost tubular claw, the other petals with shorter claws, blades flabellate, margin erose and crisped, variously colored red, orange, yellow, or blotched; stamens 10, long-exserted, up-curved, filaments filiform, red, distinct, about 1⅗ in. long, anthers opening lengthwise; style long-exserted, filiform, stigma minute.

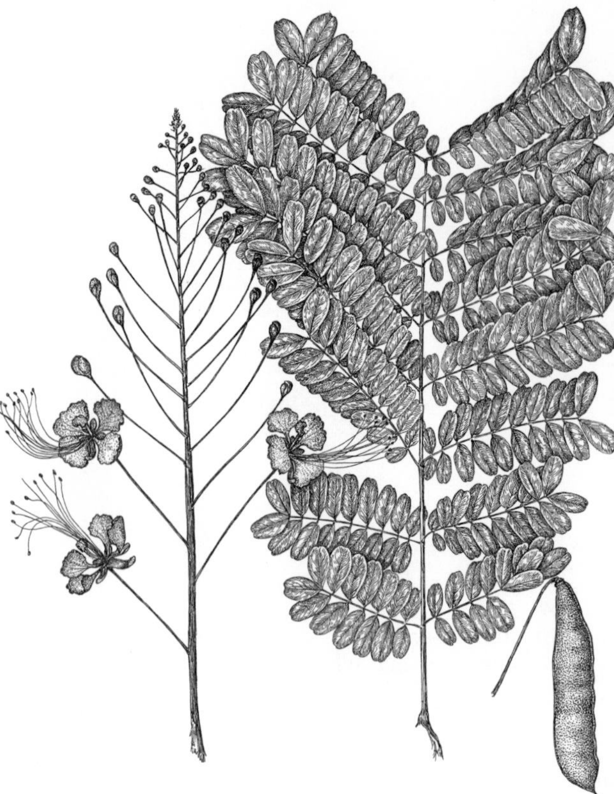

Flower-fence Poinciana
Caesalpinia pulcherrima (L.) Swartz

Paradise Poinciana
Caesalpinia gilliesii Wall. *ex* Hook.

FRUIT. Legume 3–5 in. long, ½–¾ in. wide, broadly linear, greatly flattened, one margin straight, the other margin slightly undulate, base inequilateral, apex abruptly one-sided into an apiculate point, surface dark brown or black when mature; seeds 5–9, flattened, impressed on the flattened surface of the legume; peduncle 2–3 in. long, glabrous.

LEAVES. Spreading, bipinnate, 8–15 in long, rachis green and glabrous or brownish, pinnae 3–9 pairs, each about 3½–6 in. long; leaflets 7–12 pairs per pinna, oblong to elliptic or cuneate, apex rounded or slightly notched with a minute mucro, base rounded and inequilateral, length ⅗–1 in., width ⅕–½ in., upper surface light green, glabrous, veins obscure; lower surface paler green, glabrous, delicately reticulate-veined; petiolules 1/25–1/16 in. long, glabrous or slightly puberulent.

TWIGS. Green, becoming dark brown or black with age, glabrate to slightly glaucous, finely grooved.

RANGE. In sandy soils, widely distributed and naturalized in tropical areas. In Texas in the lower Rio Grande River area. Often planted in gardens in Brownsville, Texas, sometimes escaping cultivation to grow on the edges of old resacas. Also in southern Florida, southern California, and Mexico; southward to West Indies and continental tropical America.

VARIETY. A variety with yellow flowers is known as *C. pulcherrima* var. *flava* Hort. The Flower-fence Poinciana is a popular plant in gardens in frostless areas. It is usually grown from seeds, or occasionally greenwood cuttings.

REMARKS. The genus name, *Caesalpinia,* is in honor of Andreas Caesalpinus, chief physician to Pope Clement VIII. The species name, *pulcherrima,* means "very beautiful." Other vernacular names are Barbados-flower, Barbados-Pride, Flame-tree, and Dwarf Poinciana.

Mexican Poinciana

Caesalpinia mexicana (Gray) Britt. & Rose

FIELD IDENTIFICATION. Small, open-branched tree rarely over 30 ft.

FLOWERS. Strongly fragrant, racemes 3–6 in. long, rachis green and puberulent; individual flower pedicels ¾–1½ in. long, green and glabrous or puberulent; calyx 5-lobed, lobes unequal, lower lobe reflexed and longer than the others (¼–⅜ in. long), other lobes ovate to oval, concave, apex acute or obtuse, margin ciliate, surface puberulent, length about ¼ in.; petals 5, short-clawed, claws glandular-pubescent, yellow, the standard with a reddish spot toward the base; standard oval to obovate, apex slightly notched, width about ½ in.; wing petals about the same size as the standard, oval, rounded, entire (occasionally notched); keel petals about ⅜ in. long, oval or somewhat asymmetrical; stamens 10, upcurved, ½–⅝ in. long, densely viscid-hairy or glandular toward the base; pistil simple, slender, about the same length as the stamens, stigma minute.

Mexican Poinciana
Caesalpinia mexicana (Gray) Britt. & Rose

LEAVES. Rachis slender, green to brown, glabrous; leaves 4–6 in. long, twice-pinnately compound with 2–3 lateral pairs of pinnae, and one terminal pinna, each about 2–3 in. long; leaflets 4–5 pairs per pinna, oblong to ovate, asymmetrical, apex rounded or obtuse, base rounded and oblique, essentially glabrous and dull green above, paler beneath; petiolules 1/25–1/16 in. long, glabrous or fine-hairy.

FRUIT. Legume light brown at maturity, puberulent and glandular-dotted, 1¾–2¾ in. long, oblong, somewhat curved, apex upturned, acute with a short point; valves thin, leathery, tough; seeds ¼–⅜ in. long, light brown, compressed, lenticular, lustrous, oval to obovate, rounded to truncate at base, apex rounded or obtuse; legume peduncles ¾–1½ in. long, yellowish green, puberulent or some glandular.

BARK. Smooth, mottled light or dark gray on young branches and twigs; older trunks with a few fissures.

RANGE. Cultivated, occasionally escaping along old resacas in the lower Rio Grande Valley of Texas. Cultivated on the streets and in parks in Brownsville. More common in northern Mexico.

PROPAGATION. Poincianas are easily grown from seed planted in flats with light sandy soil. When a few inches high they may be moved to pots, and then planted out after the first year or two, their growth being rapid. The seedlings are given partial shade at first.

REMARKS. The genus name, *Caesalpinia,* is in honor of Andreas Caesalpinus, chief physician to Pope Clement VIII. The species name, *mexicana,* refers to Mexico, the home of this species. Known under the Spanish name of Tabachin del Monte.

Flamboyant-tree

Delonix regia (Boj.) Raf.

FIELD IDENTIFICATION. Very beautiful tree, widely cultivated in the tropical regions of both hemispheres, attaining a height of 40 ft, with a flattened or rounded crown and horizontal or drooping branches.

FLOWERS. Borne in corymbose racemes on stout pedicels; sepals 5, about 1 in. long, much longer than the tube; corolla showy, red to orange or spotted, 3–4 in. across; petals 5, 2–3 in. long (1 petal usually longer than the others), blades imbricate, broader than long, flabellate toward the apex, long-clawed toward

FLAMBOYANT-TREE
Delonix regia (Boj.) Raf.

the base, margin undulate and crisped; stamens 10, long-exserted, declined, filaments distinct, densely silky-hairy or glandular toward the base, anthers opening lengthwise; pistil long-exserted, filiform, silky-hairy toward the base, thickened into a linear ovary on one side, stigma at apex minute.

FRUIT. Legume elongate-linear, ½–2 ft long, flattened, woody, valves 2, seeds several.

LEAVES. Bipinnately compound, 1–2 ft long, rachis green to reddish brown and pubescent distally, more glabrous proximally; large leaves with 20–50 pinnae which are 4–10 in. long; leaflets very numerous, blades linear to oblong, ¼–⅜ in. long, 1/12–⅙ in. wide, margin entire, slightly revolute, apex rounded, base inequilateral, lateral veins obscure; upper surface dull green, finely puberulent along the midrib especially; lower surface paler, pubescent with fine white hairs, especially on the midrib; petiolules 1/25 in. long, pubescent.

RANGE. Thought to be a native of Madagascar, escaping cultivation throughout its cultivated range. Southern Florida and the Keys and in southern California; in most tropical regions. Cultivated in gardens, but freezing back, in Brownsville, Texas.

REMARKS. The genus name, *Delonix,* is from the Greek and refers to the "long-clawed" petals. The species name, *regia,* pertains to the royal beauty of the tree in bloom. Other vernacular names are Framboyan, Royal Poinciana, Flame-tree, and Peacock-flower. The Flamboyant-tree is often confused with the tropical American poincianas. The propagation methods are the same as those of the poincianas.

BAGPOD

Sesbania vesicaria (Jacq.) Ell.

FIELD IDENTIFICATION. Widely branched herbaceous plant attaining a height of 12 ft. The old dead stems persistent with dry remnants of the legumes. Although not truly a woody plant it has a superficial resemblance to the Rattlebox, and the description is included for comparative purposes.

FLOWERS. Flowers August–September. Flowering peduncles 2–5 in. long, glabrous or sparingly white-hairy, lower one half to one third flowered; pedicels green to reddish, pubescent, ¼–⅜ in. long; corolla bonnet-shaped; standard ⅜–½ in. across, orange-red, yellow-spotted at base with a red margin, apex broadly rounded and notched; wing petals 2, red, about ¼ in. long, about as long as the keel, oblong, slightly falcate, clawed on one side of the base, apex slightly narrowed and rounded; keel petals joined below, falcate into an upward curved beak, enclosing the stamens; stamens all joined into a column except one; column also enclosing the curved style which is inwardly stigmatic; ovary flattened, elongate, stalked; calyx oblique, shallowly 5-toothed, white-hairy.

FRUIT. Legume on slender peduncles 3–6 in. long, body 1½–3 in. long, ⅜–½ in. wide, ellipsoid to oval, apex and base long acuminate, margin straight or constricted between the seeds; surface pale green, glabrous, smooth, inflated and bladdery, dehiscent, with outer and inner walls; seeds 1–2 (rarely 3), ¼–⅜ in. long, greenish to dull dark brown; oblong, with one end rounded, and bluntly curved into a short hook, other

BAGPOD
Sesbania vesicaria (Jacq.) Ell.

nd rounded and obtuse, hilum scar elongate, about two hirds the length of the seed.

LEAVES. Alternate, pinnately compound without terninal leaflet, 5–9 in. long; rachis green and glabrous; eaflets 10–27 pairs, ½–1 in. long, ¼–⅜ in. wide, obong, apex rounded or truncate, delicately apiculate, nargin entire, surfaces dull green and glabrous above, aler and glabrous or slightly white-hairy beneath.

TWIGS. Green, smooth, glabrous, herbaceous, lower art of main stem brownish and herbaceous or slightly voody.

FORM. A form from Florida has been given the name f *G. vesicaria* var. *atrorubium* (Nash) Brooks; it has purplish black corolla.

RANGE. In low moist soil. Texas, Oklahoma, Arkansas, nd Louisiana; eastward to Florida and northward to he Carolinas, mostly near the coast.

REMARKS. The genus name, *Sesbania,* is the Latinized ersion of the old Adansonian name, *Sesban,* which is resumably of Arabic origin. The species name, *vesiaria,* refers to the bladder-like legume.

)RUMMOND RATTLEBOX

esbania drummondii (Rydb.) Cory

FIELD IDENTIFICATION. A short-lived shrub of low, vet grounds to 15 ft, with many stems from the base, r sometimes a small tree to 20 ft.

FLOWERS. Racemes 2–6 in. long, shorter than the eaves, showy, loosely flowered, slender, axillary peuncles slender upper one third to one half flowered, he perfect bonnet-shaped flowers on slender pedicels –½ in. long; standard petal larger than the others, bout as broad as long, ½–¾ in. long, orbicular, notched t the apex, base clawed, yellow and streaked with red; ing petals oblong, obtuse, yellow, ½–¾ in. long; keel etals ½–¾ in. long; stamens 10, 9 united and one ree; pistil with a slender style and glabrous, stalked vary; calyx minute, campanulate, somewhat 2-lipped ith 5 short, acute, apiculate lobes.

FRUIT. Pod 4-sided and 4-winged, 2½–3½ in. long, n a stipe ⅜–⅝ in. long, persistent, leathery, indehisent, constricted somewhat between the seeds; the everal seeds separated by partitions, rattling when ry.

LEAVES. Folding in the hot sun, even-pinnately comound, 5–8 in. long, leaflets 12–60 (usually 13–40); lades narrowly oblong or elliptic, ⅜–1 in. long, base uneate, apex rounded and mucronate; upper surface ull green, glabrous, and inconspicuously veined; lower urface paler and glabrous or somewhat glaucescent eneath; petiolules about 1/16 in. long.

STEMS. When young green and smooth, when older ght brown; bark tight, on older stems separating into mall, thin scales.

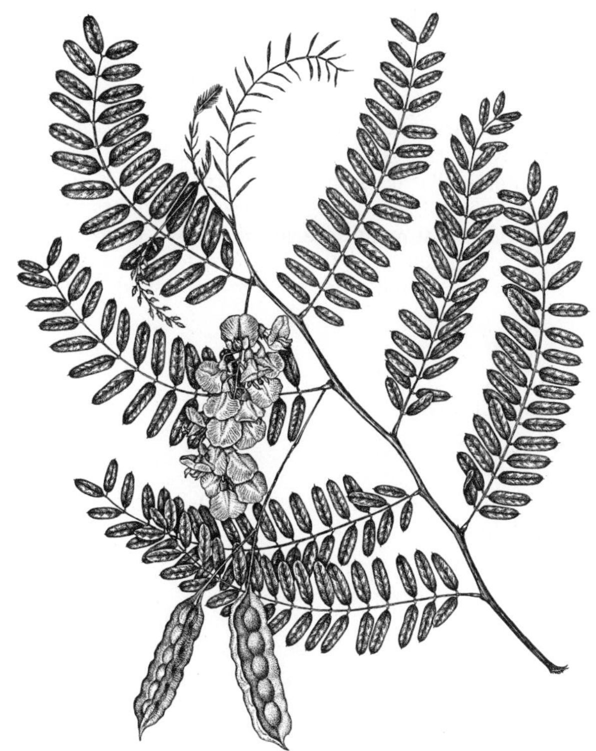

DRUMMOND RATTLEBOX
Sesbania drummondii (Rydb.) Cory

WOOD. Green to yellowish or white, pith large, of no commercial value.

RANGE. In low wet places. Arkansas, Texas and Louisiana; eastward to Florida and southward into Mexico.

RELATED SPECIES AND SYNONYMS. A South American species, Brazil Rattlebox, *S. punicea* Benth., is very ornamental with reddish purple flowers. Glorypea Rattlebox, *S. tripetii* Hort. *ex* F. T. Hubbard, is a species from Argentina with orange-red flowers and a scarlet standard petal yellow-spotted at the base.

Some authors list Drummond Rattlebox under the names of *Daubentonia drummondii* Rydb., *D. texana* Pierce, *D. cavanillesii* (S. Wats.) Standl., *D. longifolia* DC., or under the genus name of *Sesbania.*

REMARKS. The genus name, *Sesbania,* is the Latinized version of the old Adansonian name, *Sesban,* which has presumably an Arabic origin. The species name *drummondii,* is from Thomas Drummond, an English botanist, who collected in Texas, 1833–1834. He collected extensively in the vicinity of Galveston, Texas, where the plant is abundant. Other common names for the Drummond Rattlebox are Siene Bean, Rattle-bush, Rattle Bean, Coffee Bean, and Senna. The seeds of the Drummond Rattlebox are poisonous to livestock, especially sheep and goats, and death may follow if a quantity is eaten. The symptoms are diarrhea, extreme weakness, and lethargy from one to two days after eating. It is reported that the seeds were used as a substitute for coffee during the Civil War period. However, because of their toxic properties this seems to be open to question. The extent to which this toxicity is reduced by boiling has not been determined. Until suitable studies are made, its use for human consumption seems unwise and is not recommended by the author.

Drummond Rattlebox has some ornamental value for planting along the edges of streams or lakes. Some closely related tropical species are now being introduced into cultivation in the Gulf Coast states and seem to do well, occasionally escaping to grow wild.

Brazil Rattlebox

Sesbania punicea Benth.

FIELD IDENTIFICATION. Shrub to 10 ft, with an open rounded head.

FLOWERS. Borne April–May, in densely flowered racemes 2½–5 in. long; peduncles slender, drooping, glabrous or with a few white hairs; individual flower pedicels ¼–½ in., very slender; flowers bonnet-shaped, standard orange-red to purplish, orbicular, ½–¾ in. broad; keel petals yellow to red, upward curved, ½–¾ in. long, apex acute, base gradually narrowed into a claw, keel enveloping the staminal column; wing petals 2, orange-red to yellow, about the same length as the keel petals, oblong, almost straight, apex truncate or rounded; stamens 10 (9 united and 1 solitary), the solitary one attached below the others; pistil slender, falcate, as long as, or slightly longer than the stamens; stigma minute, capitate; calyx 3/16–⅛ in. long, green to reddish, campanulate, glabrous, flattened, lower side with 3 minute, shallow teeth, upper side with a central notch and 2 minute teeth on the side.

Brazil Rattlebox
Sesbania punicea Benth.

FRUIT. Legume 2–3 in. long, linear, 4-sided and 4-winged, glabrous, apex long-attenuate, base stipitate.

LEAVES. Even-pinnately compound, 3½–6 in. long, petiole glabrous or with a few fine white hairs; leaflets 12–40, linear to narrowly oblong, apex rounded to truncate and apiculate, base gradually narrowed, pale green and glabrous above, lower surface glabrous or slightly glaucous, finely reticulate-veined under magnification; petiolules 1/16–⅛ in. long, fine-hairy.

TWIGS. Slender, green to reddish brown, younger ones with fine white, scattered hairs, older ones reddish brown to gray and glabrous.

RANGE. In moist, fertile grounds, old fields, fence rows, and abandoned gardens. A native of Brazil, escaping cultivation from Florida to east Texas.

REMARKS. The genus name, *Sesbania*, is the Latinized version of the old Adansonian name, *Sesban*, which is presumably of Arabic origin. The species name, *punicea*, refers to the puniceous, or bright red, flowers. Other names are Sesban, Locust, and Red Siene Bean. A very handsome shrub in flower with good horticultural uses on the Gulf Coast plain area.

Hemp Sesbania

Sesbania exaltata (Raf.) Cory

FIELD IDENTIFICATION. Fast-growing, green-stemmed annual to 12 ft, but the dead stems woody with persistent dry legumes.

FLOWERS. Racemes shorter than the leaves, axillary, 1–6-flowered; pedicels ⅕–⅖ in. long, flowers ⅝–¾ in. long, bonnet-shaped, yellow, sometimes purple-spotted, standard petal erect or reflexed, ⅝ in. wide or less, suborbicular, sometimes notched, short-clawed; wings almost straight, base short-clawed with a small blunt angle; keel petals ⅖–½ in. long, oblanceolate, blunt, strongly curved, base with an angular lobe; stamens 10 (9 joined by their filaments and 1 free); ovary elongate, stipitate, many-ovuled; style glabrous, stigma small; calyx campanulate, 5-lobed, lobes shorter than the tube, equal, triangular, subulate.

FRUIT. Legume linear, slender, somewhat curved, 6–12 in. long, ⅛–⅙ in. wide, apex tipped by the subulate style, base subtended by the persistent calyx

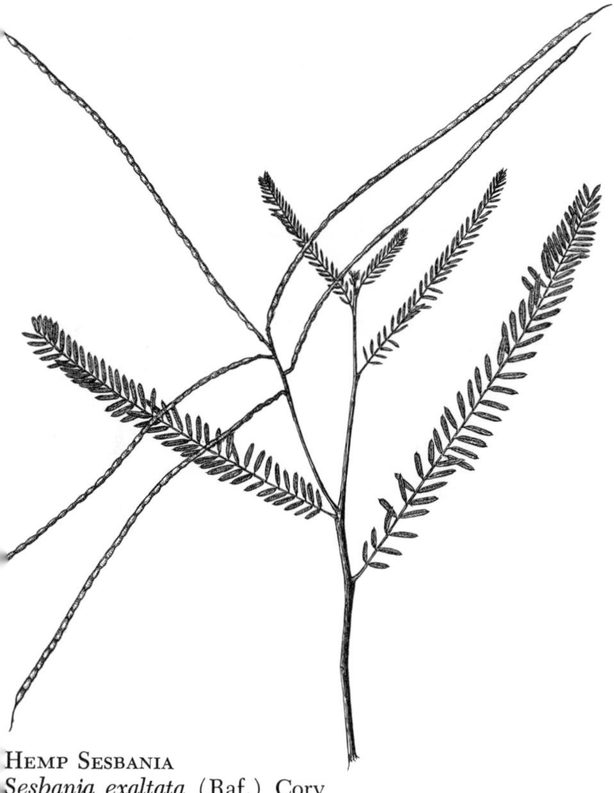

Hemp Sesbania
Sesbania exaltata (Raf.) Cory

sides knotted by the transverse partitions of the seeds; seeds numerous, about ⅛ in. long, oblong, terete, ends rounded, dark brown or mottled-brown, lustrous, hilum scar small.

LEAVES. Numerous, alternate, even-pinnately compound, length 4¾–16 in.; leaflets 30–70, blades oblong to linear or elliptic, length ⅜–1¼ in., margin entire, apex obtuse or rounded, mucronate; surfaces pale green, glabrous or glaucescent, thin, stipules fugacious.

RANGE. Usually in low moist grounds or along streams. Mostly on the coastal plain or adjacent areas. Texas, Oklahoma, Arkansas, and Louisiana; west through New Mexico to Arizona, eastward to Florida, and northward to South Carolina.

REMARKS. The genus name, *Sesbania,* is the Arabic name. The species name, *exaltata,* means "tall or lofty." Other vernacular names are Zacate de Agua, Bequilla, Siene Weed, Coffee Bean, and Florida Coffee Bean. It is listed by some authorities under the scientific name of *S. macrocarpa.*

The Indians of Mexico and Arizona made nets and fishlines from the long, strong fibers of the stems. The seeds are known to be eaten by a number of species of birds, including the bobwhite quail. When growing in the water, the submerged portion of the stem develops a thickened spongy aerenchyma.

Calderona Krameria

Krameria ramosissima (Gray) Wats.

FIELD IDENTIFICATION. Shrub usually under 2 ft, diffusely branched and rigid, sometimes almost prostrate. All parts, except the older branchlets, covered with dense appressed hairs.

FLOWERS. April–May, solitary on axillary bracted peduncles, maroon-purple, about ½ in. across or less, partially spreading at anthesis; sepals 5, petal-like, densely hairy; petals 5, unequal, the 3 upper long-clawed, the lower shorter, sessile, and fleshy; stamens 4, free above, anthers 2-celled, opening obliquely near the apex.

FRUIT. Fruiting peduncle to ½ in. long, densely appressed-hairy. Pod about ⅜ in. long, globose or ovoid, some slightly flattened, often with a reddish, raised, longitudinal ridge, densely silky-pubescent; bearing very short (1/25 in. long or less) slender, sharp, red, nonbarbed spines; indehiscent, 1-seeded.

LEAVES. Alternate or fascicled in the axils, simple, 1/12–¼ in. long, linear to lanceolate, acute or acuminate, margin entire, densely gray-tomentose.

TWIGS. Numerous, spreading, slender but stiff, younger ones densely tomentose with appressed hairs, older ones gray and glabrous, many more or less thorny.

RANGE. Sunny sites on limestone hillsides of southern and western Texas. Principally along the Rio Grande River from the vicinity of Del Rio southward. Also in Mexico in Tamaulipas, Coahuila, and Nuevo León.

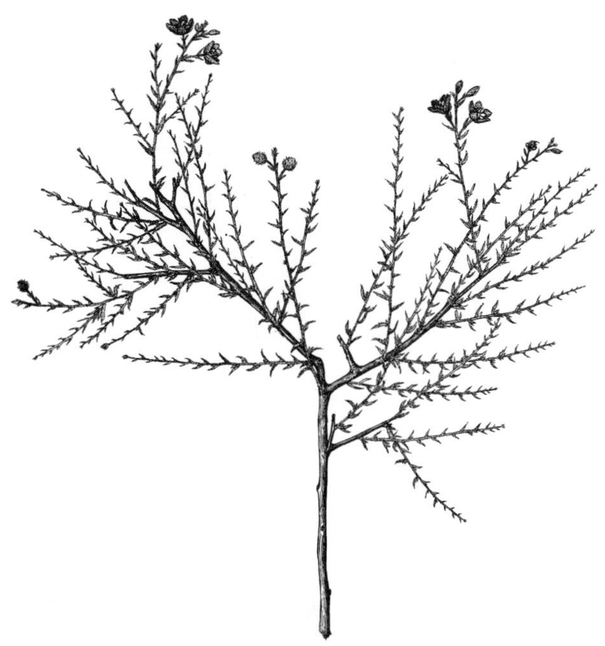

Calderona Krameria
Krameria ramosissima (Gray) Wats.

REMARKS. The genus name, *Krameria,* is in honor of John Henry Kramer of Austria, a botanist of the eighteenth century. The species name, *ramosissima,* means "much-branched."

LITTLE-LEAF KRAMERIA

Krameria parvifolia Benth.

FIELD IDENTIFICATION. Low, intricately branched shrub to 2 ft. The young parts are densely covered with long gray hairs.

FLOWERS. Appearing April–May, axillary, borne on silky peduncles 3/8–5/8 in. long, often with leaflike bracts; corolla very irregular, showy, purplish, about 2/5 in. in diameter; sepals 1/5–1/3 in. long, ascending, oblong to oblanceolate; petals 5, smaller than the sepals, about 1/8 in. long, the upper 3 oblong to obovate, more or less united toward the base into a short claw, the remaining petals reduced to nearly orbicular fleshy glands; stamens 3–4, more or less united.

FRUIT. Pod globose or nearly so, silky-hairy and with slender barbed spines, indehiscent, thick-walled, 1-seeded.

LITTLE-LEAF KRAMERIA
Krameria parvifolia Benth.

STICKY LITTLE-LEAF KRAMERIA
Krameria parvifolia Benth. var. *glandulosa* (Rose & Painter) Macbride

LEAVES. Simple, alternate, linear to oblanceolate, sessile, entire, 1/8–3/5 in. long, about 1/25 in. broad, or more rarely to 1/12 in. broad, gray with dense long hairs.

TWIGS. Slender, stiff, irregular, younger ones densely gray-hairy, older ones light to dark gray or brown to black, eventually thorny.

RANGE. Dry rocky slopes and plains, at altitudes of 500–5,000 ft. In western Texas and New Mexico; north to Utah and Nevada, westward through Arizona into California, and southward in Sonora, Mexico.

REMARKS. The genus name, *Krameria,* is in honor of J. H. Kramer, Austrian army physician and botanist of the eighteenth century. The species name, *parvifolia,* refers to the small leaves. Other vernacular names are Pima Ratany and Range Ratany. Indians once used a decoction of the plant as an eyewash and remedy for sores. It was also the source of a red or brown dye.

Several varieties of Little-leaf Krameria have been distinguished:

Sticky Little-leaf Krameria, *K. parvifolia* Benth. var. *glandulosa* (Rose & Painter) Macbride, has stipitate glands on flower pedicels and sometimes on leaves.

Spiny Little-leaf Krameria, *K. parvifolia* var. *imparata* (Benth.) Macbride, has no glands but denser spines of the fruit.

Gray's Krameria

Krameria grayi Rose & Painter

FIELD IDENTIFICATION. Thorny shrub to 2 ft, erect or semiprostrate, intricately branched.

FLOWERS. April–September, solitary, on axillary silky-tomentose peduncles which surpass the leaves, and bear 2 bracts at or above the middle; flowers about ½ in. across, perfect, showy, very irregular; sepals 5, pink to purple, unequal, lanceolate to linear, acuminate, surface puberulent, margin ciliate; petals 5, reddish purple, ⅛–¼ in. long, unequal; 3 upper petals distinct, spatulate, long-clawed, claws slender and yellow; 2 lower petals short, sessile, fleshy, scalelike; stamens 4, in unequal pairs, united below, anthers opening by oblique pores at the apex; stigma linear; ovary ovoid, hairy, spiny, 1-celled, 1–2-ovuled.

FRUIT. Pod subglobose, ridged on the side, about ⅓ in. long, coriaceous, indehiscent, densely woolly, style remnant persistent, covered with slender prickles each with a cluster of 3–4 barbs at the apex, 1-seeded.

LEAVES. Simple, alternate, linear to lanceolate, sessile at base, margin entire, surface densely silky-hairy, ⅛–⅗ in. long.

STEMS. Slender, thorny, when young whitish and densely gray-tomentose, when older gray and smooth.

RANGE. Dry soil of barren hillsides and desert areas at altitudes of 1,000–4,000 ft. In Trans-Pecos Texas and New Mexico; northward to Colorado and Nevada, westward into southern California, and southward into Mexico in the states of Chihuahua and Coahuila.

REMARKS. The genus name, *Krameria*, honors John Henry Kramer, eighteenth-century botanist of Austria. The species name, *grayi*, is for Asa Gray, American botanist. Also it is known under the vernacular names of Chacate, White Ratany, and Crimson Beak. The plant is very drought-resistant and is considered to be good browse for cattle in Arizona. The root bark is reported to be used by the Mexicans to make yellow or reddish brown dye. The Pima Indians used an infusion of the roots as a remedy for sores. Some of the closely related South American species are listed in official drug catalogues. Gray's Krameria, and others of the American Southwest, are reported as being partially parasitic on the roots of other desert plants.

Gray's Krameria
Krameria grayi Rose & Painter

Wislizenus Dalea

Dalea wislizenii Gray

FIELD IDENTIFICATION. Suffrutescent or subshrubby plant to 2 ft, with several slender erect stems.

FLOWERS. Borne in ovoid to short-cylindric spikes with the purple to reddish flowers crowded at the ends; petals inserted below the middle of the staminal tube; calyx-lobes subulate-setaceous, long, and plumose; fruit not seen. Corolla papilionaceous; stamens 9–10, monadelphous.

LEAVES. Usually less than 1 in. long, ascending; leaflets 11–21, each ⅛–¼ in. long, blade oblong or oblanceolate, apex abruptly acute or obtuse, sometimes apiculate, base acute or gradually narrowed, surfaces plane or involute, with dense long, silky hairs below and above, lower surface with scattered dark glands.

TWIGS. Slender, erect, reddish brown, densely villous, more glabrous and woody near the base.

RANGE. In mountain woods or on rocky hills at altitudes of 3,000–5,000 ft. New Mexico and Arizona; southward into Mexico.

VARIETIES. *D. wislizenii* var. *sanctaecrucis* (Rydb.) Kearney & Peebles has spikes mainly on the end of elongate branches, but the leaves and twigs are glabrous or sparsely pubescent.

D. wislizenii var. *sessilis* Gray has the spikes mostly at the ends of short lateral branches.

REMARKS. The genus name, *Dalea*, is in honor of Samuel Dale (1657–1739), an English botanist. The species name, *wislizenii,* is in honor of Friedrich Adolph Wislizenus (1810–1889), German-born physician of St. Louis, who collected plants in Mexico and the southwestern United States, 1846–1847.

FEATHER DALEA

Dalea formosa Torr.

FIELD IDENTIFICATION. Shrub small, usually under 3 ft, crooked, divaricately branched, glabrous except for the inflorescence.

FLOWERS. April–October, spikes capitate or sub-capitate, on short peduncles, 2–10-flowered, rather loose; bracts deciduous, ovate, glabrous dorsally and silky-villous on the margins; calyx-tube campanulate, 1/8–1/6 in. long, ribbed and glandular, lobes 5, 1/6–2/5 in. long, subulate, densely long-hairy; corolla papilionaceous, petals purple, blade of banner 1/10–1/6 in. long, claw about as long, sometimes yellowish; blade of wings 1/6–1/5 in. long; blade of keel 1/5–1/4 in. long, claws shorter and inserted below the middle of the staminal tube; stamens 9–10, monadelphous.

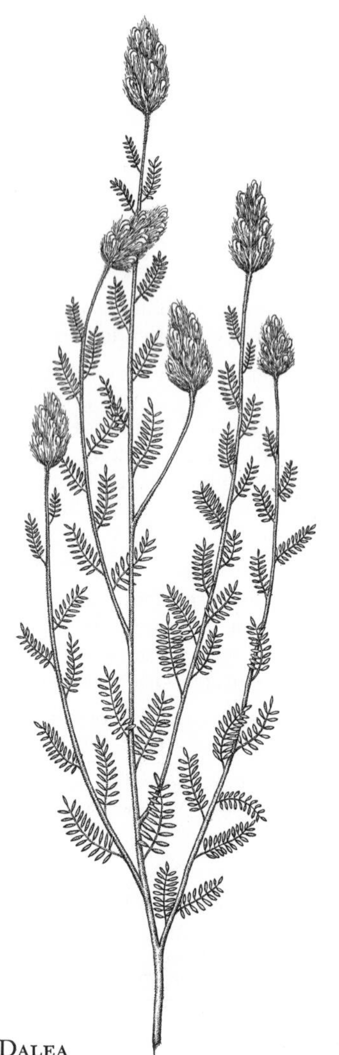

WISLIZENUS DALEA
Dalea wislizenii Gray

FEATHER DALEA
Dalea formosa Torr.

FRUIT. Legume small, flat, hairy, glandular-dotted, indehiscent, included in the calyx, 1–2-seeded.

LEAVES. Odd-pinnately compound, less than 2/5 in. long, leaflets 5–11, 1/25–1/6 in. long, thick, revolute or conduplicate, spatulate or cuneate-oblong, apex retuse or rounded, surfaces glabrous and glandular-dotted, sessile or very short petiolulate.

TWIGS. Divaricately branched, rigid, gray to brown, glabrous.

RANGE. Dry soil of hills at altitudes of 2,000–6,500 ft. In Texas from the Colorado River westward, New Mexico, Arizona, and Colorado; in Mexico in Chihuahua and Sonora.

REMARKS. The genus name, *Dalea,* honors Samuel Dale (1657–1739), English botanist, and the species name, *formosa,* refers to the handsome flowers. The plant is of little importance to livestock but is readily eaten by deer.

Broom Dalea

Dalea scoparia Heller

FIELD IDENTIFICATION. Shrub broomlike, 1–4 ft, stems slender, naked below, densely canescent-hairy and glandular-punctate.

FLOWERS. Borne on terminal and lateral peduncles ¼–2¾ in. long; spikes subglobose or oblong, ½–⅗ in. long, few-flowered; bracts deciduous, lanceolate, 1/12 in. long; calyx-tube turbinate, about 1/10 in. long, 10-ribbed, a row of 2–4 glands in each interval, white-villous; lobes 5, triangular, about 1/25 in. long, each with a dark gland at apex; corolla dark blue, petals clawed, inserted at the base of the staminal tube; banner petal broadly oval, base subcordate, apex retuse or rounded, claw about 1/10 in. long; wing petals about ⅕ in. long, oblong, basal lobe rounded; keel petals obliquely obovate; stamens 9–10, monadelphous, united high into a tube.

FRUIT. Legume about ⅛ in. long, obliquely obovate, turgid, densely pubescent and glandular-dotted, enclosed by the calyx, indehiscent, few-seeded.

LEAVES. Simple, alternate, linear to linear-spatulate; apex obtuse, rounded or slightly notched, length less than ¾ in., surfaces with appressed hairs and glandular-dotted, lower leaves sometimes with 3 linear leaflets.

TWIGS. Slender, green to tan or yellowish, densely hairy, conspicuously set with translucent yellow or brown glands, ends of twigs pointed but hardly spinescent, older twigs glabrous.

RANGE. On sand dunes, or in sandy bottom lands, Trans-Pecos Texas (known only from El Paso County), New Mexico, and southern Arizona. In Mexico in Chihuahua and Coahuila.

REMARKS. The genus name, *Dalea,* honors Samuel Dale (1657–1739), English botanist. The species name, *scoparia,* refers to the broomlike stems. The plant is very drought-resistant and is recommended as a sand binder.

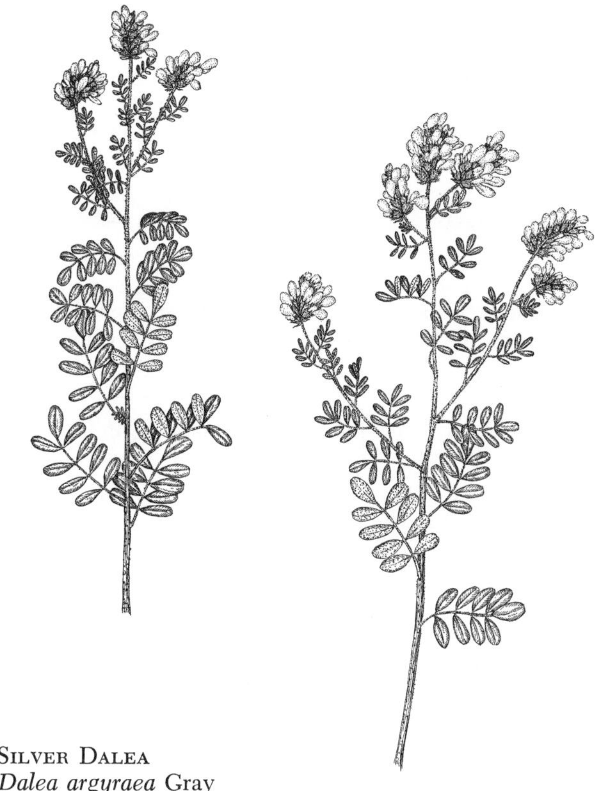

Silver Dalea
Dalea argyraea Gray

Broom Dalea
Dalea scoparia Heller

Silver Dalea

Dalea argyraea Gray

FIELD IDENTIFICATION. Plant shrubby, stems 1–2 ft, stout, erect, corymbosely branched above, with dense silvery-canescent hairs.

FLOWERS. Spikes ⅜–1½ in. long, spheroid to oblong, densely flowered, flowers perfect; bracts ovate, acuminate, hairy, about 1/10 in. long; calyx longer than the bracts, ribs obscure because of the dense silky-villous covering; lobes 5, lanceolate, somewhat shorter than the tube; banner petal purple or yellow, broad, long-clawed; wing and keel petals longer than the standard, their claws adnate to the staminal tube; stamens 9–10, monadelphous; ovules 2–3 in each cavity.

FRUIT. Legume small, included in the calyx, ovate, flat, membranous, silky-villous, indehiscent, 1–2-seeded.

LEAVES. Alternate, odd-pinnately compound, ½–1¼ in. long, leaflets 7–13 (mostly 7–9), oblong to obovate, apex rounded to obtuse, base gradually narrowed, margin entire, surfaces densely silky-hairy and glandular-dotted; petiolules about 1/25 in. long or less, densely hairy.

TWIGS. Leafy, densely and finely hairy, glandular-tuberculate, green to yellow or brown to grayish later.

RANGE. Dry soil in the sun. In Texas from the Nueces River westward, New Mexico, and Mexico in Nuevo León and Chihuahua.

REMARKS. The genus name, *Dalea,* honors Samuel Dale (1657–1739), English botanist, and the species name, *argyraea,* refers to the silvery hairs of the foliage.

BLACK DALEA

Dalea frutescens (Gray) Vail

FIELD IDENTIFICATION. Western shrub 1–3 ft, with smooth, slender, widely branching stems.

FLOWERS. Spikes paniculate, terminal, ⅓–½ in. long, subglobose to oblong, dense; bracts leathery, ovate, obtuse or abruptly pointed, scarcely exceeding the calyx; calyx smooth, 10-ribbed, glandular-dotted, teeth of calyx very short and triangular, villous-margined; corolla purple, papilionaceous, staminal column bearing 4 petals above the middle; standard petal broad, cordate, free, long-clawed; stamens 9–10, monadelphous; style subulate; ovules 2–3.

FRUIT. Legume membranous, ovate to broadly obovate, glabrous, glandular, flattened, included in the calyx, 1–2-seeded.

LEAVES. Alternate, ⅜–¾ in. long, odd-pinnately compound of 13–17 leaflets; each leaflet 1/12–¼ in. long, obovate to cuneate or obcordate, apex retuse or notched, surfaces grayish green, glabrous, conspicuously glandular beneath; petioles 1/25 in. long or less.

TWIGS. Very slender, glabrous, some sparingly glandular-dotted, green to reddish brown, striate, older stems gray to brown.

RANGE. Dry soil in the sun on rocky hills or plains. In Texas from the Colorado River to the Rio Grande and in southeastern New Mexico. In Mexico in Chihuahua, Coahuila, and Nuevo León.

REMARKS. The genus name, *Dalea,* honors Samuel Dale (1657–1739), English botanist and writer on pharmacology. The species name, *frutescens,* refers to the shrubby character of the plant. It is rather drought-resistant and somewhat browsed by livestock.

LOOSE-FLOWERED DALEA

Dalea frutescens var. *laxa* (Rydb.) Turner

FIELD IDENTIFICATION. Shrubby plant attaining a height of 3 ft.

FLOWERS. Borne in lax, slender spikes ¾–1¾ in. long; peduncles terminating the branches and about ⅜ in. long; bracts early deciduous, ovate, glabrous, conspicuously glandular-dotted; calyx glabrous with a ciliolate margin, calyx-tube about ⅛ in. long, conspicuously 10-ribbed, with a row of oblong, yellow glands in each interval, teeth broadly triangular with a short subulate tip, about 1/50 in. long and very broad, the sinuses between them rounded; petals with a cordate banner about ⅙ in. long, the claw about ⅛ in. long; wings and keel petals inserted near the middle of the staminal tube; wing blades about ⅕ in. long, keel about ⅕ in. long.

FRUIT. Not seen.

LEAVES. About ½–¾ in. long, spreading; leaflets more

BLACK DALEA
Dalea frutescens (Gray) Vail

Thyrsus Dalea
Dalea thyrsiflora Gray

or less cuneate, 1/6–1/5 in. long, slightly retuse, glabrous, glandular-dotted beneath; rachis slightly margined, on the upper side glandular-punctate; stipels glandlike and acute.

STEMS. Slender, terete, glabrous, straw-colored.

RANGE. Known from Texas; the type was collected near the Devils River, Texas.

REMARKS. The genus name, *Dalea*, is in honor of Samuel Dale (1657–1739), an English botanist. The species name, *frutescens*, refers to the shrubby character of the plant, and the variety name, *laxa*, to the lax or loosely-flowered spikes. The variety is very similar to the species except for an extension of the inflorescence.

Thyrsus Dalea

Dalea thyrsiflora Gray

FIELD IDENTIFICATION. Shrub attaining a height of 10 ft. The branches herbaceous, but the lower part of the stem woody.

FLOWERS. Very numerous, on short, slender, open branches; forming axillary or terminal, short-pedunculate, dense, thyrsoid panicles (appearing spikelike); flowers often scattered on the lower part of the rachis, or a few glomerate in the axils of the uppermost and reduced leaves; corolla exserted about one third or one fourth its length beyond the calyx, yellow to brownish purple; bracts ovate, pubescent, about equaling the sessile calyx-tube; calyx-tube pubescent, strongly 10-ribbed, conspicuously glandular-dotted, 1/10–1/8 in. long; lobes of calyx filiform, plumose, the lower about 1/5 in. long and hooked at the apex, the rest about 1/6 in. long; corolla papilionaceous; stamens joined into a staminal column.

FRUIT. Legume cinereous-pubescent, short, barely exceeding or enclosed by the persistent calyx.

LEAVES. Pinnately compound, 3/4–2 in. long, rachis short-villous, leaflets 3–5 (occasionally to 9 or upper leaves 1-2-foliate), oblong to oval or slightly obovate, apex rounded to retuse or occasionally notched, base usually rounded or occasionally semitruncate, margin entire, upper surface rather pale, dull, and pubescent; lower surface densely pubescent, copiously dotted with dark glands, length 1/8–3/8 in.; stipules subulate, villous, 1/8–1/5 in. long. Sometimes the primary stem bears smaller leaves than the branches.

STEMS. Young ones brown, densely villous, some with scattered dark glands; older ones brown to gray and more glabrous.

RANGE. Dry plains on clay or sandy loam soil, often along arroyo banks. In Texas in Cameron County near Brownsville; southward in Mexico through Nuevo León, Tamaulipas, and Chihuahua, to Guatemala.

REMARKS. The genus name, *Dalea*, is in honor of Samuel Dale (1657–1739), English botanist. The species name, *thyrsiflora*, refers to the thyrsus-like inflorescence.

Eastern Redbud

Cercis canadensis L.

FIELD IDENTIFICATION. Shrub or small tree to 40 ft, trunk usually straight, branching usually 5–9 ft from the ground, top broadly rounded or flattened. Distinctly ornamental in spring with small, clustered, rose-purple flowers covering the bare branches before the leaves.

FLOWERS. March–May, before the leaves, in clusters of 2–8, on pedicels 1/4–3/4 in. long; flowers 1/4–2/5 in. long, perfect, imperfectly papilionaceous, rose-purple, petals 5, standard smaller than the wings; keel petals not united, large; stamens 10, shorter than the petals; ovary pubescent, short, stipitate, style curved; calyx campanulate, 5-toothed.

FRUIT. September–October, persistent on the branches, often abundant, peduncles divaricate and reflexed, 1/3–3/5 in. long; legume 2–4 in. long, about 1/2 in. wide or less, tapering at both ends, flat, leathery, reddish

Eastern Redbud
Cercis canadensis L.

brown, upper suture with a somewhat winged margin; valves 2, thin, reticulate-veined; seeds several, oblong, flattened, ⅙–⅕ in. long.

LEAVES. Simple, alternate, deciduous, 2–6 in. long, 1¼–6 in. broad, ovate to cordate or reniform, apex usually abruptly obtuse or acute, base cordate or subtruncate, margin entire, palmately 7–9-veined; upper surface dull green and glabrous; lower surface paler and glabrous or somewhat hairy along the veins, membranous at first but firm (not coriaceous) later; petioles 1¼–5 in. long, essentially glabrous, stipules caducous.

TWIGS. Slender, glabrous, somewhat divaricate, brown to gray, when young lustrous, when older dull.

BARK. Reddish brown to gray, thin and smooth when young, older ones with elongate fissures separating long, narrow plates with small scales, lenticels numerous.

WOOD. Reddish brown, sapwood yellowish, close-grained, hard, weak, weighing 30 lb per cu ft.

RANGE. Eastern Redbud is found in rich soil along streams or in bottom lands from central Texas, Oklahoma, Arkansas, and Louisiana; eastward to Florida, northward to Connecticut and Ontario, and west to Michigan, Missouri, Nebraska, and Kansas.

Texas Redbud has been collected on limestone areas in Texas in Val Verde, Kerr, Austin, Comal, Erath, Brown, Dallas, and Hood counties. It is known in Oklahoma near Turner Falls Park in the Arbuckle Mountains, in Platt National Park at Antelope Springs, and in Mexico in Nuevo León and Tamaulipas.

Mexican Redbud has been collected on limestone areas in Texas in Brown, Dallas, Terrell, Nolan, Brewster, and Austin counties. It is the prevalent form in Trans-Pecos Texas, and occurs in Mexico in Coahuila, Nuevo León, and San Luis Potosí.

For a review of the Texas and Mexican Redbud varieties see Hopkins (1942).

MEDICINAL USES. Redbud bark has some medicinal value. In the form of a fluid extract it is an active astringent, and is used in the treatment of dysentery.

PROPAGATION. Five-year-old trees begin to bear legumes, and the maximum age is about 75 years. Good crops generally occur on alternate years. The legumes are picked by hand when they turn brown, or may be flailed into sacks and spread out to dry. The seeds are removed by threshing and the chaff screened or fanned off. About 30 lb of seed may be obtained from 100 lb of fruit. The seed averages about 14,000 seeds per lb, with a commercial purity of 90 per cent and a soundness of 85 per cent. Dry storage in sealed containers at 41°F. is adequate. Seed dormancy may be broken by soaking in sulphuric acid followed by 60 days stratification in moist sand at 41°F. The seed should be planted in drills in a rich, loamy seedbed in April or May and covered about ¼ in. The soil should be kept moist during germination. If sown in winter a leaf mulch should be used. Transplanting is usually done in

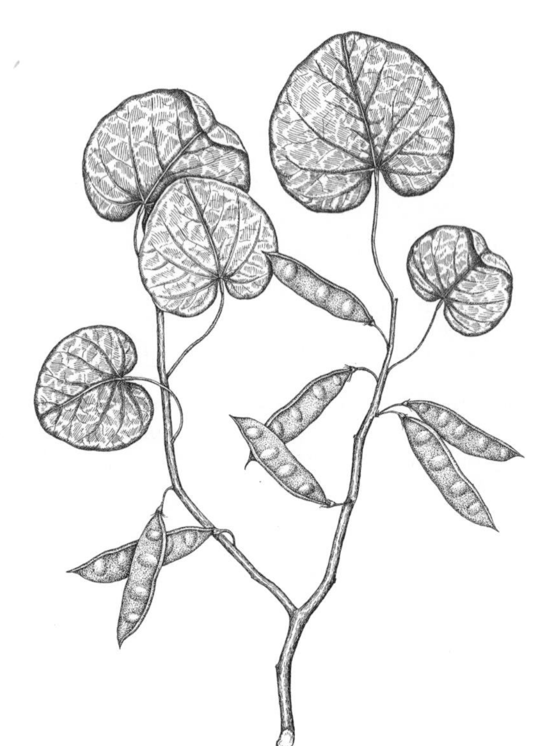

Texas Redbud
Cercis canadensis var. *texensis* (Wats.) Hopkins

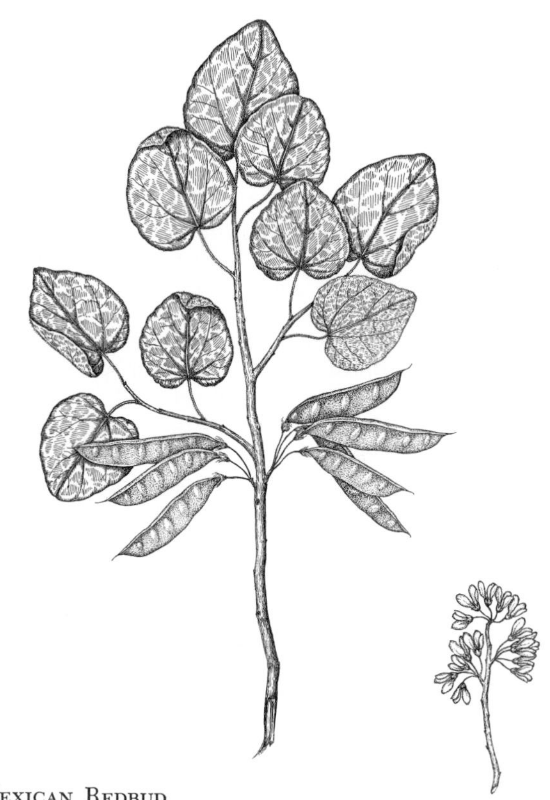

MEXICAN REDBUD
Cercis canadensis var. *mexicana* (Rose) Hopkins

rich, moist sandy loam. Propagation is also practiced by layering and by greenwood cuttings from forced plants in early spring.

VARIETIES AND FORMS. A number of varieties and horticultural forms are cultivated as follows:

White Eastern Redbud, *C. canadensis* forma *alba* Rehd., has white flowers.

Double Eastern Redbud, *C. canadensis* forma *plena* Sudw., has some stamens changed to petals.

Smooth Eastern Redbud, *C. canadensis* forma *glabrifolia* Fern., has glabrous leaves.

Hairy Eastern Redbud, *C. canadensis* var. *pubescens* Pursh, has more or less pubescent leaves.

Two native varieties of Redbud of Texas, Oklahoma, and Mexico are Texas Redbud, *C. canadensis* var. *texensis* (S. Wats.) Hopkins, and Mexican Redbud, *C. canadensis* var. *mexicana* (Rose) Hopkins.

Both the Texas and Mexican varieties differ from the Eastern Redbud by having conspicuously shiny and stiffly coriaceous leaves (leaves of the Eastern Redbud are dull green and not distinctly coriaceous). The leaf shapes of the species and the 2 varieties are rather similar, varying from cordate with an acute apex to reniform or rounded with the apex obtuse or emarginate.

The Texas and Mexican varieties are separated from each other on the dubious characters of the amount of hairs present on various parts. Texas Redbud has pedicels, young branchlets, and leaves glabrous or nearly so. Mexican Redbud has pedicels and young branchlets densely woolly-tomentose and leaves slightly so. These distinctions are so close that some botanists consider the Mexican Redbud to be only a hairy form of the Texas Redbud, instead of a distinct variety of the Eastern Redbud. However, since forms closely resembling *C. canadensis* var. *mexicana,* have been found in northeast Texas, it is more likely that the latter is derived from *C. canadensis.*

REMARKS. The genus name, *Cercis,* is the ancient name of the closely related Judas-tree of Europe and Asia. According to tradition Judas hung himself from a branch of the tree. The species name, *canadensis,* literally means "of Canada," where it is rather uncommon. Or perhaps the Linnean name refers to northeastern North America, before political boundaries were set up. The tree is a very handsome ornamental and has been in cultivation since 1641. It is reported that the acid flowers are sometimes pickled for use in salad, and in Mexico they are fried. The seeds are eaten by a number of species of birds and the foliage browsed by the white-tailed deer. Eastern Redbud also has some value as a source of honey.

CALIFORNIA REDBUD
Cercis occidentalis Torr.

FIELD IDENTIFICATION. An attractive shrub with several stems from the base, or a small tree seldom over 20 ft with smooth gray bark.

FLOWERS. Borne February–April, appearing before the leaves in scattered, sessile umbels of 2–5 flowers on the old wood, purplish red, irregular; pedicels ¼–½ in. long, glabrous; calyx campanulate, 5-lobed, ¼–⅜ in. broad; corolla about ⅓ in. long, petals 5 and distinct (not completely papilionaceous), about ½ in. broad, banner smaller than the wing petals; keel petals larger than the wing petals, not united; stamens 10, distinct; ovary superior, 1-celled, oblong, glabrous.

FRUIT. Legumes 1½–4 in. long, ½–⅝ in. wide, oblong to oblanceolate, flat and thin, dull red when mature, thin-walled, margins broad, not constricted between the several seeds.

LEAVES. Simple, alternate, deciduous, 2–3½ in. broad, heart-shaped or subreniform, apex rounded or retuse, base cordate, margin entire, somewhat palmately veined, both sides smooth, glabrous and glossy; petioles ⅖–1 in. long, glabrous; bud scales ciliate.

WOOD. Yellowish brown, close-grained, hard, specific gravity about 0.70.

RANGE. California Redbud is found in moist soil along streams or on rocky slopes at altitudes of 500–6,000 ft. In New Mexico; west to California and north to Utah and Nevada.

CALIFORNIA REDBUD
Cercis occidentalis Torr.

VARIETIES. A variety has been given the name of *C. occidentalis* var. *texensis* Wats. with smaller and more reniform leaves and legumes ⅖–⅗ in. broad. However, this form is best referred to under the name of *C. canadensis* var. *texensis* (S. Wats.) Hopkins (see description of *C. canadensis* for this variety). It occurs in Texas, New Mexico, and Mexico. Another variety of the California Redbud has been listed as *C. occidentalis* var. *orbiculata* (Greene) Tidestrom. It is characterized by having the sinuses of the leaves closed instead of open as in the typical form. However, this appears to be a dubious separation because intergrades are found with it and the species. It has been recorded as occurring in Diamond Valley, Utah, and in Arizona. Outside of glabrous and pubescent characters, the distinction between the varieties of Eastern Redbud and California Redbud is not very great.

REMARKS. The genus name, *Cercis,* is the ancient name of a closely related tree of Asia. The species name, *occidentalis,* refers to its western distribution. It is also known under the vernacular name of Pata de Vaca. It is reported that the flowers are fried and eaten in the state of San Luis Potosí, Mexico. The bark is used in domestic medicine as a mild astringent in the treatment of chronic diarrhea and dysentery. The plant is very attractive in flower in the spring and is worthy of cultivation in gardens.

CAMELTHORN

Alhagi camelorum Fisch.

FIELD IDENTIFICATION. Very rare shrub 1–2 ft, of Asiatic origin. It is very intricately branched and spiniferous.

FLOWERS. Maturing May–July, numerous, borne on short branchlets which terminate in spines; pedicels 1/16–⅛ in. long; calyx glabrous, about ⅛ in. long or less, irregular, crowned by 5, short, obtuse to rounded lobes; corolla purplish pink to yellow, about ⅓ in. long, irregular; banner and wing petals folded and flattened at first, later more spreading and the banner reflexed; banner and wing petals about the same length, or the wing petals slightly longer; reproductive organs bisexual; stamens 10, diadelphous; ovary superior, 1-celled.

FRUIT. Legume ⅛–½ in. long, linear, constricted into 1–3 segments which are often globular, 1–3-seeded.

LEAVES. Simple, alternate, deciduous, blade ½–⅞ in. long, elliptic to oblong-lanceolate, margin entire, apex rounded to obtuse or acute, surfaces glabrous, veins inconspicuous; petiole about 1/16 in. long.

TWIGS. Very spiniferous, smooth, green, glabrous; spines ½–1½ in. long, formed by the attenuated apices of lateral branches.

RANGE. Usually along streams or canals, or on rocky

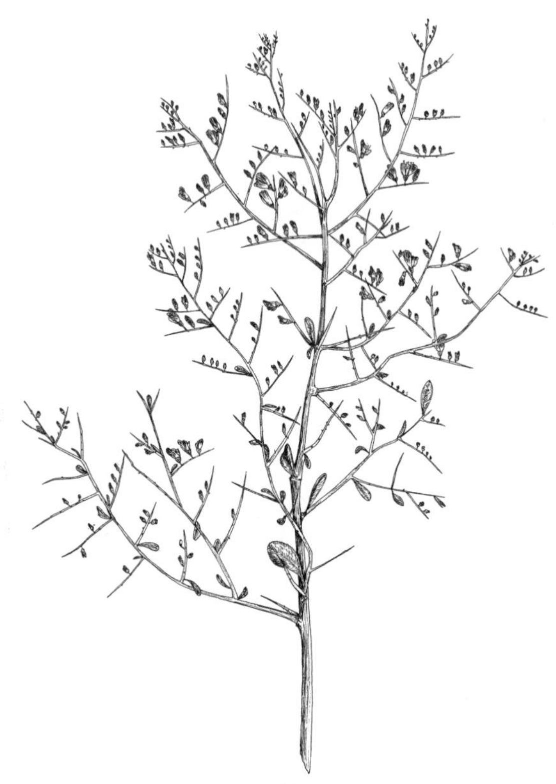

CAMELTHORN
Alhagi camelorum Fisch.

TEXAS BRONGNIARTIA
Brongniartia minutifolia Wats.

hillsides. In New Mexico; in Arizona in Navajo, Coconino, and Maricopa counties; also in California. It is a native of Asia.

REMARKS. *Alhagi* is the Mauritanian name. The species name, *camelorum*, means "of the camels." It is thought to be introduced into our southwestern states through impure alfalfa seed, or in packages of date cuttings. Considered to be of great value as browse in desert regions of Asia. The long spreading roots make it difficult to eradicate when once started.

TEXAS BRONGNIARTIA

Brongniartia minutifolia Wats.

FIELD IDENTIFICATION. Rare Western shrub. Erect, unarmed, delicately much-branched, 1–4 ft.

FLOWERS. Solitary on naked, glabrous, peduncles ¼–1½ in. long; rather conspicuous, yellowish or flesh-colored, ¼–⅜ in. long, standard petal broad, wings narrow and free; calyx obscurely glandular-dotted, lobes longer than the broadened base, ⅛–3⁄16 in. long, lanceolate, acuminate; stamens numerous, upcurved, persistent after the petals have dried and shriveled; style elongate, upcurved.

FRUIT. Legume ½–1½ in. long, about ⅜ in. wide, oblanceolate to obovate, apex abruptly rounded or obtuse with a slender mucronate point, base attenuate into a stipe, unequal-sided, sutures somewhat thickened, flattened, glabrous, yellowish green or brownish at maturity; seeds 1–3, light brown, flattened, about ¼ in. long, oval to short-oblong, hilum end oblique.

LEAVES. Odd-pinnately compound with a slender rachis ½–2 in. long, stipules oval to linear, herbaceous; leaflets numerous, 10–20 pairs, very small, 1⁄12–⅛ in. long, hardly 1⁄25 in. wide, linear, margin often revolute, surfaces mostly glabrous or slightly puberulent.

TWIGS. Slender, delicate, somewhat zigzag, finely striate, glaucous-green or tan at first, older twigs reddish brown with bark peeling in thin strips.

RANGE. Altitudes of 3,000–3,500 ft on dry alkaline soil. Collected by the author on the road between Hot Springs and Castalon, between the Chisos Mountains and the Rio Grande River, Brewster County, Texas. Known only from Brewster County.

REMARKS. The genus name, *Brongniartia,* honors Adolphe T. Brongniart, French botanist. The species name, *minutifolia,* refers to the very small leaves.

TEXAS BABY-BONNETS

Coursetia axillaris Coult. & Rose

FIELD IDENTIFICATION. A rare, densely branched shrub or small tree.

FLOWERS. Axillary, solitary, or in few-flowered racemes, borne on pubescent peduncles ⅙–⅓ in. long; calyx 5-lobed, hairy, the lobes almost equal, long-acuminate, as long as the campanulate tube or longer, ⅙–¼ in. long; flowers perfect, papilionaceous, white or yellowish, or sometimes with a pink cast, ⅜–½ in. long or broad; standard petal orbicular or reniform, reflexed; wings free and keel petals curved; style simple, free, incurved, very hairy above the middle, stigma capitate, ovary sessile with several ovules; stamens 10.

FRUIT. Legume reddish brown, glabrous at maturity, ¾–1½ in. long, about ¼ in. wide, narrow, linear, flat, thin, margin constricted-wavy, base broadened into a stipe, apices mostly rounded or obtuse to acute with the slender style remnant often persistent, bivalvate; seeds 2–6, light to dark brown, oval to short-oblong, rounded, flattened, thin, usually ⅛ in. long or less.

LEAVES. Alternate, pinnately compound, rachis surpassing uppermost pair of leaflets, ⅜–2 in. long; leaflets 3–5 pairs, ⅛–½ in. long, obovate to short-oblong or lower pairs often smaller and orbicular, apices mostly rounded or slightly truncate, minutely mucronate, base rounded or abruptly narrowed, surface rather dull green, reticulate, glabrous or slightly hairy above, lower surface conspicuously reticulate, pubescent along the veins especially; petiolules 1⁄25–1⁄16 in. long, densely hairy; stipules minute and setaceous.

TEXAS BABY-BONNETS
Coursetia axillaris Coult. & Rose

TWIGS. Slender, younger ones densely gray-tomentose, older ones pubescent to glabrous, light to dark gray.

RANGE. In Texas in the lower Rio Grande Valley, near San Diego, on a small rise known as La Lomita Mission Hill. Also near La Joya and Hidalgo, Texas. In Mexico in Tamaulipas.

REMARKS. The genus name, *Coursetia,* is in honor of Dumont de Courset, author of *Botaniste Cultivateur* (Paris: 1802). The species name, *axillaris,* refers to the axillary flowers.

BOYKIN CLUSTERPEA

Dioclea multiflora (Torr. & Gray) Mohr

FIELD IDENTIFICATION. Plant pubescent, with climbing or trailing semiwoody stems, attaining a length of 3–12 ft.

FLOWERS. Borne in raceme-like panicles which are shorter than the subtending leaves, pedicels 1/25–1/8 in. long, perfect; corolla purple, the standard suborbicular, reflexed, narrowed and auricled at the base, 1/2–3/5 in. long; wings free from the shorter, incurved obtuse keel; stamens 10, 9 joined and 1 free; ovary sessile or nearly so, ovules several; calyx-lobes unequal, the upper 2 united, the lateral smaller, the lower longer and lanceolate, all lobes longer than the tube.

FRUIT. Legume oblong or elliptic, flat, apex abruptly pointed, length 2–2 3/8 in., 2-valved.

LEAVES. Alternate, leaflets 3, blades 2–6 in. across oval to suborbicular or reniform, apex abruptly acuminate, base acute to subcordate, membranous, upper surface glabrate, lower pubescent to glabrate.

RANGE. Edges of ravines or on riverbanks. Louisiana and Arkansas, eastward to Kentucky and Georgia.

REMARKS. The genus name, *Dioclea,* is in honor of Diocles, considered by the ancients as second only to Hippocrates in his knowledge of plants. The species name, *multiflora,* refers to the numerous flowers.

WESTERN CORAL BEAN

Erythrina flabelliformis Kearney

FIELD IDENTIFICATION. Western shrub, often with barren spiny branches, or sometimes a small tree to 15 ft.

FLOWERS. In spring before the leaves or sometimes with the leaves; calyx about 1/3 in. long and 1/5–1/4 in. in diameter, teeth minute or obscure, body cylindric, lower side with a gland near the apex; corolla red, 1–2 in. long, about 1/4 in. wide; banner petal elongate, straight, strongly keeled, wing and keel petals small, barely as long as the calyx; stamens 10, one stamen solitary, as

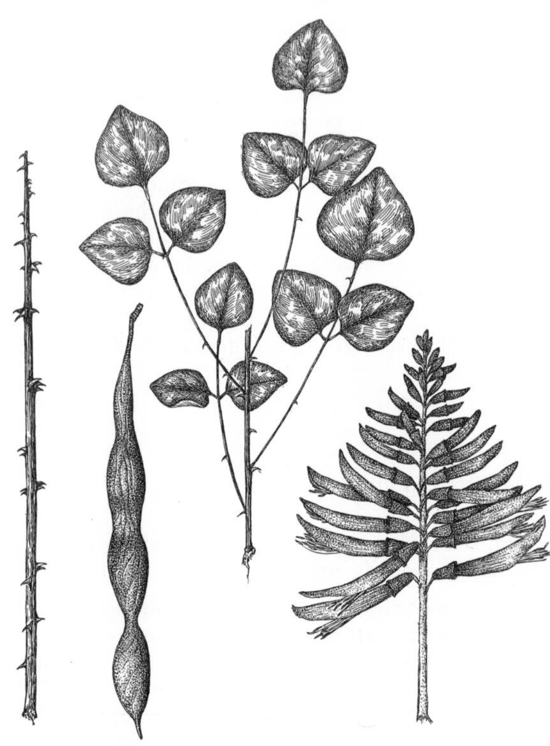

WESTERN CORAL BEAN
Erythrina flabelliformis Kearney

long as the banner petal, slightly exserted; pistil elongate.

FRUIT. Legume 4–10 in. long, ½–¾ in. wide, margin usually somewhat constricted between the seeds, at first densely short-pubescent, later more glabrous, brown, rounded or somewhat flattened, walls thick; seeds bright red, showy, ellipsoidal, about ½ in. long and ⅖ in. wide, often persistent in the dehiscent legume, poisonous.

LEAVES. Alternate, deciduous, trifoliate, the 3 leaflets with a triangular to broadly ovate shape with rounded angles, apex obtuse or acute, base truncate or broadly rounded, length 1–3 in., width 1½–3½ in., petiole often with small prickles; petiolules reddish brown, pubescent, those of terminal leaflet much longer.

TWIGS. Stout, soft-pubescent at first, straw-colored to light brown or gray, with scattered, solitary or paired, curved or straight spines; when older dark brown to gray, finely grooved.

RANGE. Dry rocky sites of canyon slopes or washes. In the desert mountains. Southwestern New Mexico, southeastern Arizona, and Mexico in the states of Baja California, Sonora, Chihuahua, and south to Michoacán.

REMARKS. The genus name, *Erythrina*, denotes the red flowers, and the species name, *flabelliformis*, is for the fan-shaped leaflets. It is also known under the vernacular names of Indian Bean, Coral-tree, and Chilicote. The soft wood is sometimes used for making corks and the red beans for necklaces. All parts of the plants are somewhat poisonous, especially the beans, which contain a powerful alkaloid. The seed germination is 20–40 per cent.

Eastern Coral Bean
Erythrina herbacea L.

FIELD IDENTIFICATION. Usually a shrub with many slender, spreading stems from the base, or more rarely a tree to 25 ft, with a trunk diameter of 10 in.

FLOWERS. Borne April–June in narrow, leafless spikes 8–13 in. long; individual flowers on short, slender, glabrous or pubescent pedicels 1/25–½ in. long; bracts linear-lanceolate, variable in size, mostly about ⅛ in. long and 1/25 in. broad, bracteoles smaller; lower flowers fading as upper ones open; calyx campanulate, chartaceous, dark red, somewhat oblique, entire or shallowly lobed, glabrous to pubescent, about ¼ in. long; corolla scarlet, showy, tubular, closed, perfect; standard narrow-oblanceolate to elliptic, falcate, apex rounded, base cuneate-clawed, slightly smaller than the keel petals, 1⅜–2¼ in. long, 1/25–⅜ in. broad; wing petals slightly longer than the calyx and larger than the keel petals; keel petals acuminate at apex, clawed at base, ¼–½ in. long and about 1/12–⅕ in. wide; stamens 10 (9 together and 1 separate), 1 3/16–1¾ in. long; pistil 1–1¾ in. long;

Eastern Coral Bean
Erythrina herbacea L.

ovary pubescent, stipitate, 1-celled, style solitary and subulate, incurved, naked; stigmas small, terminal; ovules numerous.

FRUIT. Legume on stout peduncles about 1 in. long, subligneous, green to dark brown or black, linear, often falcate, slightly flattened, strongly constricted between the seeds, 2–4 in. long, apex stiff-apiculate, base stipitate and about ¾ in. long; valves 2, thin, widely dehiscent to expose 5–10 (rarely 1–2) beans which are scarlet, lustrous, hard, bony, about ¼ in. long, and persistent by the basal hilum to the valve; hilum oblong, dark, about ⅛ in. long.

LEAVES. Persistent below (inflorescences usually leafless), alternate or clustered at the nodes, 6–8 in. long, with a slender petiole and rachis occasionally armed with small, flattened, recurved prickles; leaves 3-foliate, leaflets thin, entire, deltoid to hastate; base concave, cuneate or truncate, 3-nerved; apex acuminate or acute; lateral lobes broad and rounded, much shorter than the elongate, terminal lobe; dull green to yellowish green, smooth and glabrous, 2¼–3½ in. long, 1 1/12–2¼ in. wide; petiolules slender, glabrous or pubescent, 1/25–⅓ in. long; stipels minute, glandlike.

STEMS. Slender, terete, green to reddish brown, smooth, finely grooved, commonly bearing scattered, stout, broad, recurved thorns; older bark whitened, thick, soft, furrows vertical; herbaceous and freezing down in northern areas, tending to be woody and perennial, or arborescent, in subtropical latitudes; roots often large, thick and tuberous.

RANGE. Eastern Coral Bean is usually found in sandy soils from Texas and Louisiana to Florida, northward to North Carolina, and south in Mexico in the states of Tamaulipas, Nuevo León, and San Luis Potosí. Eastern Coral Bean is replaced in southwestern New Mexico and Arizona by Western Coral Bean, *E. flabelliformis* Kearney.

MEDICINAL USES. The poison beans are used in Mexico to kill rats and dogs, and contain the alkaloid erythroidine. This powerful poison is remarkably similar in action to that of curare, having strong effects on the motor nerves. Fish poison has also been made from the beans. The complex group of alkaloids peculiar to the *Erythrinas* of both the old and new world has been investigated by Folkers and Koniuszy (1940).

PROPAGATION. Many species of *Erythrina* (herbaceous and woody) are grown in gardens. Woody species are grown from cuttings of well-ripened wood, and the herbaceous species from cuttings or shoots from the old roots or divisions of the rootstock. All can be cultivated from seed. They apparently do best in Florida and California, although some species fail to bloom in California because of insufficient summer heat.

The best systematic treatment of the genus *Erythrina* is to be found in a monograph by B. A. Krukoff (1939).

VARIETIES, HYBRIDS, AND HORTICULTURAL FORMS. Eastern Coral Bean seems to assume several forms in response to climatic influences. In northern latitudes it is usually herbaceous, freezing down each year, but springing up from the roots. Along the Texas coast woody stems are developed which sometimes assume the proportions of trunks in the lower Rio Grande Valley. The herbaceous form and the arborescent form seem to be similar in all other respects, but some botanists have separated the two and given the arborescent form the name of *E. arborea* (Chapm.) Small or *E. herbacea* var. *arborea* Chapm.

Cockspur Coral Bean, *E. crista-galli,* a horticultural species treated elsewhere, is known to hybridize with *E. herbacea,* producing the beautiful hybrid *E.* × *bidwellii* Lindl.

REMARKS. The genus name, *Erythrina,* refers to the color of the flowers, and the species name, *herbacea,* to the herbaceous character of the plant in temperate regions. Also known under the vernacular English, Spanish, and Indian names of Cardinal Spear, Cherokee Bean, Red Cardinal-flower, Colorín, Corolillo, Patol, Pitos, Chilicote, Zampantle, Zumpantle, Tzampantle, Tzan-pan-cuohuitl, Cozquelite, Purenchegua, Pureque, Tzinacanquahuitl, Chijol, Chocolin, Pichoco, Jiquimite, Iguimite, Peonia, Chotza, Demthy, and Macayxtli.

Various species of *Erythrina* are planted in Mexico for ornament, as hedges, or for coffee or cocoa shade. The bark is said to yield a yellow dye. The beans are strung into necklaces and used by the Mexican Indians in a game called "patol." The wood of *Erythrina* is used in Mexico for carvings of miniature statues, for corks and tobacco boxes.

COCKSPUR CORAL BEAN
Erythrina crista-galli L.

FIELD IDENTIFICATION. Cultivated prickly shrub, c small tree to 15 ft, with a trunk to 7 in. in diamete The stout elongate branches form a rounded head.

FLOWERS. Borne in a large terminal panicle compose of many clusters of 1–7 flowers; pedicels ½–2 in. lon glabrous, green to reddish brown; calyx ⅜–½ in. lon reddish brown, somewhat flattened, shallowly 2-lippe lower lip with an apiculate appendage, upper lip round ed or notched; corolla red, composed of a large standar petal and smaller wing and keel petals; standard fre or nearly so, 1–2 in. long, conspicuous, erect, spreadin or half-folded, oval to elliptic or obovate, apex rounde to acute; wing petals small, ¼–⅓ in. long, sometime hidden by the calyx; keel petals 1–2 in. long, falcat clawless, androecium included within the envelopin keels; stamens 10, exserted, 9 together and united near to the summit, one stamen solitary and shorter tha the others; pistil slender, acuminate, falcate, flattene reddish, barely exserted, slightly shorter than the stamens, stigma minute, ovary short-stipitate.

FRUIT. Legume 3–10 in. long, cylindric; seeds vary ing from yellow to orange or red.

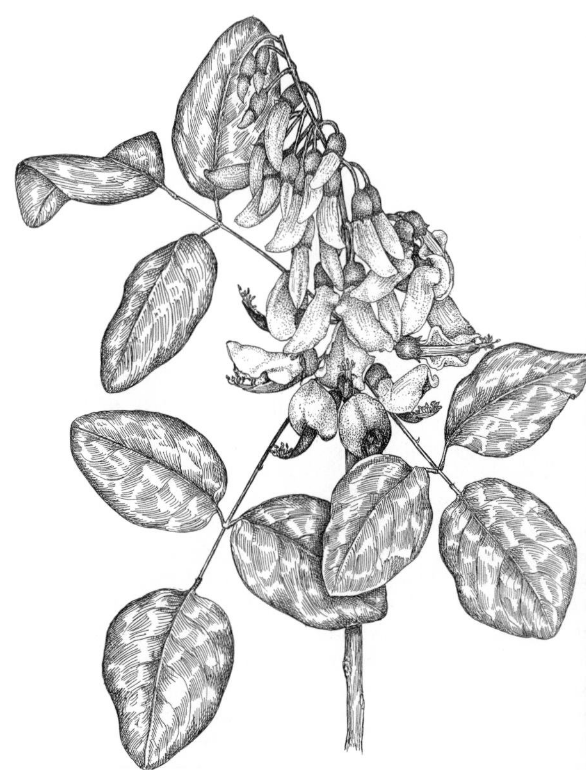

COCKSPUR CORAL BEAN
Erythrina crista-galli L.

LEAVES. Alternate, 3-foliate, 6–12 in. long (including he petiole); leaflets 2–5½ in. long, blades elliptic to val or ovate, margin entire, apex acute or obtuse, base ounded or broadly cuneate, blade often partly folded, urfaces dull green and glabrous; petioles slender, somewhat angular and finely grooved, often spiny, green to eddish brown; petiolules of terminal leaflet of 2 divi-ions; distal part down to the glandular stipules green, hickened, shorter than the proximal portion; petiolules f the lateral leaflets ¼–⅔ in. long, about the same ength as the distal portion of the terminal petiolule.

TWIGS. Elongate, stout, usually unbranched, green to eddish brown, glabrous, angled and finely grooved; pines ¼–½ in. long, green to brown, stout, flat, straight r slightly curved, at the petiole bases or scattered.

BARK. Green on young branches, older ones tan to rown or gray with narrow fissures and rather wide lat ridges.

WOOD. Soft, light, used for corks or carvings.

RANGE. A native of Brazil, cultivated and occasionally scaping in the Gulf states. In coastal Texas and Louisi-na, east to Florida.

MEDICINAL USES. All species of *Erythrina* beans pos-ess a number of powerful alkaloids. These include rythramine, erythraline, ersodine, erysopine, erysocine, nd erysovine, according to Folkers and Koniuszy 1940). Certain of these alkaloids produce a curare-ike paralyzing action when administered intravenously. No clinical use has been made of them, but further tudy is needed since their properties are remarkably imilar to those of curare. The sap of this and other pecies is used in tropical countries as a fish poison.

PROPAGATION. Cockspur Coral Bean is fairly hardy n the Gulf states if grown in a protected situation. In orthern areas the thick roots are taken up and stored luring the winter. It is a very handsome and desirable rnamental plant. About 50 species of *Erythrina* are nown, most from tropical regions. The herbaceous pecies are propagated by root divisions and woody nes from cuttings of growing wood. Older plants prout readily when cut back. Plants may be success-ully started from seeds also. The plants are often used o shade young coffee trees in South America.

FORM. *E. crista-galli* forma *compacta* Bull. is a form vith rich crimson flowers and a more compact habit. Some forms have variegated leaves.

REMARKS. The genus name, *Erythrina,* is the Greek name for red, referring to the flowers, and the species name, *crista-galli,* is the Latin for cockspur, referring to he spines of the plant. It is sometimes listed in the iterature under the name of *Micropteryx crista-galli* L.) Walp. Some of the vernacular names in use are Fireman's Cap, Common Coral-tree, Crybaby-tree, Drag-n's Teeth, and Immortelle. The seeds are sometimes nade into necklaces. The wood is carved into small ob-ects. The flowers seem to vary considerably in shades f red and density of the racemes.

Johnston Genistidium

Genistidium dumosa I. M. Johnston

FIELD IDENTIFICATION. Shrubby plant to 2 ft. Branches numerous from the base, dense and intricate, ascending or erect, slender and rigid.

FLOWERS. Usually solitary and axillary toward the upper part of the branches, or racemose and loosely flowered; peduncles 1/25–⅛ in. long with 2 subulate bracts 1/15–1/12 in. long; corolla papilionaceous, ⅛–3/16 in. long, yellow, sometimes with greenish spots; standard petal suborbicular, reflexed, clawed, slightly puberulent along the median line or glabrous; wings yellow, oblong-lunate, glabrous, about ⅙ in. long and 1/35 in. broad, base auricled; keel petals 1/25–1/12 in. long, yellowish to white, lunate, obtuse, base auricled; stamens ⅕–¼ in. long, somewhat hairy below, the vexilum stamen free; ovary hairy, subsessile, 4–6-ovulate; style incurved, hairy, subulate, the stigma capitate and minute; calyx campanulate, tube 1/12–⅛ in. long, slightly oblique, lobes cuneate and subulate.

FRUIT. Legume ¼–1 in. long, about ⅙ in. wide, linear, straight, apex rounded or obtuse, often abruptly apicu-

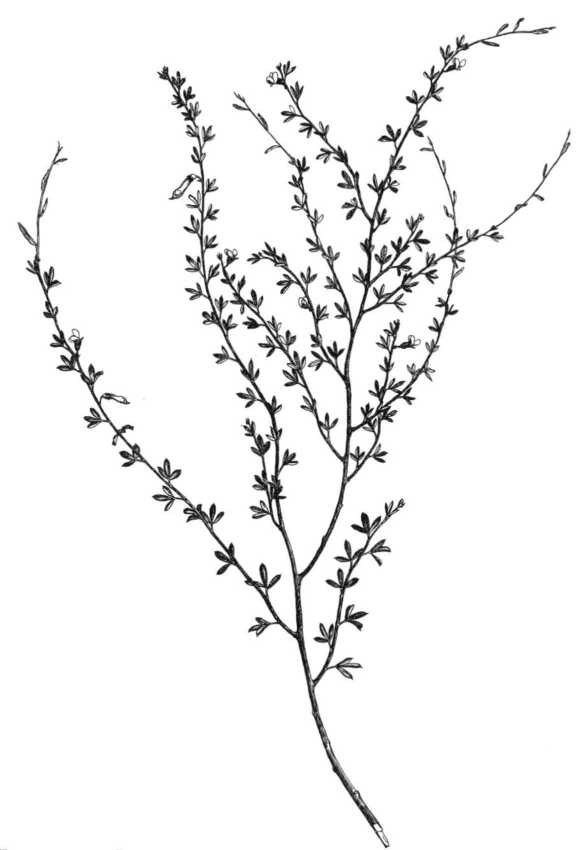

Johnston Genistidium
Genistidium dumosa I. M. Johnston

Chisos Hoffmanseggia
Hoffmanseggia melanosticta Gray

late, sutures thickened, bivalvate; seed suborbicular, flattened.

LEAVES. Numerous, alternate, trifoliate, or the upper unifoliate, 1/5–3/4 in. long, 1/16–3/16 in. wide, narrowly spatulate to oblanceolate or elliptic (or linear in the unifoliate ones and to 1 in. long), margin entire, apex acute to obtuse or rounded and mucronulate, base gradually narrowed, surface dull green with minute, white, appressed hairs, obscurely nerved, firm; the 2 lateral leaflets subsessile or short-petiolulate; the terminal leaflet often longer and with a petiolule slightly longer than the lateral ones; petiole 1/25–1/6 in. long, densely white-hairy and sometimes tufted at the base; petiolules also white hairy; stipules somewhat persistent, rigid, subulate, 1/25–1/15 in. long.

TWIGS. Slender, young ones greenish gray because of the minute white pubescence, distinctly 8–15-ribbed; older ones darker or olive-green, glabrous, and more woody at the base.

RANGE. Very rare in Texas. The author has seen only 4 plants. Between Terlingua and Lajitas, on the edge of rocky road cuts, near the Rio Grande River, in Brewster County. I. M. Johnston has described it as being frequent along the summit of the cliffs of volcanic tuff at San Antonio de los Alamos in Coahuila, Mexico. A 200-mile gap exists between the Mexico and Texas locations.

REMARKS. The genus name, *Genistidium*, denotes its resemblance to the genus *Genista* of the Old World. The species name, *dumosa*, refers to its bushy habit.

Chisos Hoffmanseggia

Hoffmanseggia melanosticta Gray

FIELD IDENTIFICATION. Erect, rigid, frutescent plant 8 in.–2 ft. Younger parts densely villose with black sessile glands.

FLOWERS. Racemes terminal or lateral, elongate, 2–6 in. or more long, loosely many-flowered; calyx densely dotted with brown pyriform glands; petals with a few scattered hairs on margin and veins, vexillúm densely villose below, with minute glands above; style with large setose projections and black sessile glands below; stamens 10, free.

FRUIT. Legumes borne on straight or suberect pedicels, broadly falcate, about 1 in. long and 1/3 in. wide, base rounded or gradually narrowed, muricate, the short murications stellate-pilose at apex, 2–3-seeded; seeds oblong to obovate, about 1/5 in. long, angular, greenish.

LEAVES. Scattered, 1–2½ in. long, usually with 2–4 pairs of pinnae and a terminal pinna; pinnae abrupt with a villous mucronate rachis; leaflets 3–5 pairs, outline elliptic, oblique, obtuse or retuse, very short-petiolulate, surfaces of leaflets villous and black-punctate beneath.

RANGE. Chisos Hoffmanseggia is evidently a rather rare species. Known from the Chisos Mountains in Brewster County, Texas. Also it was collected by G. C. Nealley and C. C. Parry in the valley of the Rio Grande below Donna, Texas. Also known from Mexico near Monterrey, Rinconada, and Buena Vista, Nuevo León.

VARIETIES. Parry Hoffmanseggia, *H. melanosticta* var. *parryi* Fisher, is more slender, not as woody or rigid; leaves larger and spreading; pinnae 3–5; leaflets longer and never more than 3 pairs; raceme of a few larger, pendent flowers; sepals less glandular; petals with no villi on margins or veins, vexillum glandless; ovary long with parallel sides; legume rhombic, ends acute, 1¼ in. long and ¼–3/8 in. wide, pendent on curved pedicels; seeds 3–4, elliptic-obovate, not angular, about ¼ in. long and 1/6 in. wide, brownish. Known from Mexico, below San Carlos on the Rio Grande. The only known American localities are in southern Brewster County, Texas.

REMARKS. The genus name, *Hoffmanseggia*, is in honor of Joh. Centurius (Count Von Hoffmansegg), a writer on plants of Portugal, who was born in 1766. The species name, *melanosticta*, refers to the black glands.

Anil Indigo

Indigofera suffruticosa Mill.

FIELD IDENTIFICATION. Sparingly branched perennial plant attaining a height of 1½–6 ft. The stems are rather erect and herbaceous above, but become woody near the base.

FLOWERS. From summer to fall, borne in axillary many-flowered racemes; pedicels pubescent, about 1/25 in. long, finally recurved; flowers perfect, calyx 1/25–1/12 in. long, densely pubescent, lobes about as long as the tube, deltoid to triangular-lanceolate, oblique, nearly equal to each other or the upper shorter; corolla reddish; standard petal broad, 1/6–1/5 in. long, persistent; wings slightly longer than the standard; keel petals scimitar-shaped; stamens 10, anthers alike; ovary sessile or nearly so, becoming curved, ovules numerous.

FRUIT. Legume becoming glabrous or nearly so, thickened at the sutures, linear, curved, apex with a long, slender beak; seeds 1/25–1/16 in. long, flattened and angled.

LEAVES. Pinnately compound, leaflets subsessile, 7–15 blades oblong to elliptic or oblanceolate to obovate, length 1/3–1¼ in., apex mucronate, base acute, upper surface sparingly hairy at first but becoming glabrous or nearly so, lower surface retaining the minute hairs longer.

Anil Indigo
Indigofera suffruticosa Mill.

STEMS. Slender, grooved, young ones green or grayish with minute appressed hairs, older stems gray to brown becoming glabrous.

RANGE. Usually in sandy soil of pinelands or old fields near the coast. From eastern and coastal Texas eastward to Florida. Known to the author from San Patricio, Karnes, Liberty, and Hardin counties in Texas. Widely distributed in coastal regions throughout the American tropics, including Mexico, Central and South America, and the West Indies; and also in the Old World tropics. However, some doubt exists as to its origin, it possibly being Asiatic rather than American.

RELATED SPECIES. The true indigo of commerce is the product of *I. tinctoria* L. of Asia, but indigo was also made from the West Indian species *I. suffruticosa* Mill. The latter was introduced into the United States, and grown commercially for dyes at one time, but coal-tar products replaced the dye derivative, and it is now seldom used.

I. suffruticosa Mill. is distinguished from the closely related *I. lindheimeriana* Scheele by the more glabrous upper surface of the leaflets and the lack of a swollen reddish knob at the base of the legume.

REMARKS. The genus name, *Indigofera,* is for the yield of indigo from some species. The species name, *suffruticosa,* refers to its suffrutescent manner of growth. *I. anil* L. is a synonym of *I. suffruticosa* Mill. according to some authors.

Lindheimer Indigo

Indigofera lindheimeriana Scheele

FIELD IDENTIFICATION. Erect perennial plant 2–6 ft. The stems herbaceous above but becoming somewhat woody near the base. The whole plant appearing gray or whitish because of the dense appressed hairs.

FLOWERS. Borne June–August. Axillary, in slender racemes ½–3 in. long; rachis slender, covered with minute, dense appressed, white to brown hairs; individual flower pedicels slender, recurved later, 1/25–1/12 in. long, very hairy; flowers small, perfect, ¼–1/3 in. long, green to white or reddish; calyx less than half the length of the corolla, 1/25–1/12 in. long, 4–5-lobed, lobes somewhat oblique and unequal, triangular to subulate-lanceolate, as long as the tube or longer, densely hairy; corolla composed of standard, wings, and keel petals; standard broad and rounded, blades sessile and clawed, persistent, almost completely enveloping the other petals except for the slightly longer wings; wings narrowed at base; keel petals erect, falcate, clawless, broadened toward the apex, deciduous; stamens 10, included; style slender, longer than the stamens, up-

Lindheimer Indigo
Indigofera lindheimeriana Scheele

curved; stigma minute and capitate; ovary sessile or nearly so, ovules numerous.

FRUIT. Legumes abundantly borne in spikelike racemes 1½–3 in. long; body of fruit linear, falcate, ½–¾ in. long, about ⅛ in. wide, apex apiculate, a swollen reddish knob often present at the base, sutures thickened, surfaces with fine, white, appressed hairs; seeds solitary in each compartment, about $\frac{1}{16}$ in. long, flattened and angled, dark green to brown or black.

LEAVES. Alternate, odd-pinnately compound, 2–4 in. long, rachis densely white-hairy; leaflets 7–15, short-petiolulate, firm, ⅓–¾ in. long; oblong to elliptic or oval, apex acute to obtuse or rounded, mucronulate, base cuneate or rounded, margin entire; surfaces densely strigose with white hairs and thus appearing pale gray.

STEMS. Young ones green or grayish green, herbaceous, somewhat angled or striate, white-hairy; older stems woody near the base, reddish brown to gray, striate, glabrous, sometimes roughened by lenticels; wood greenish white.

RANGE. Usually in dry limestone soils of central and western Texas and adjacent Mexico. Seen by the author in Kerr, Uvalde, Terrell, and Val Verde counties in Texas. Perhaps also in other counties. In Mexico known from Coahuila.

RELATED SPECIES. Lindheimer Indigo is closely related to Anil Indigo, but is distinguished by the more glabrous leaflets and lack of the reddish knob at the legume base of the latter species. The true indigo of commerce is obtained from *I. tinctoria* L. of Asia.

REMARKS. The genus name, *Indigofera*, is for the yield of indigo of some species. The species name *lindheimeriana*, is in honor of Ferdinand Lindheimer (1801–1879), German-born botanist and newspaper editor of New Braunfels, Texas.

Dune Broom
Parryella filifolia Torr. & Gray

FIELD IDENTIFICATION. Low, broomlike, much branched shrubs attaining a height of 3 ft. The herbage glandular-punctate and glabrous or sparingly strigose hairy.

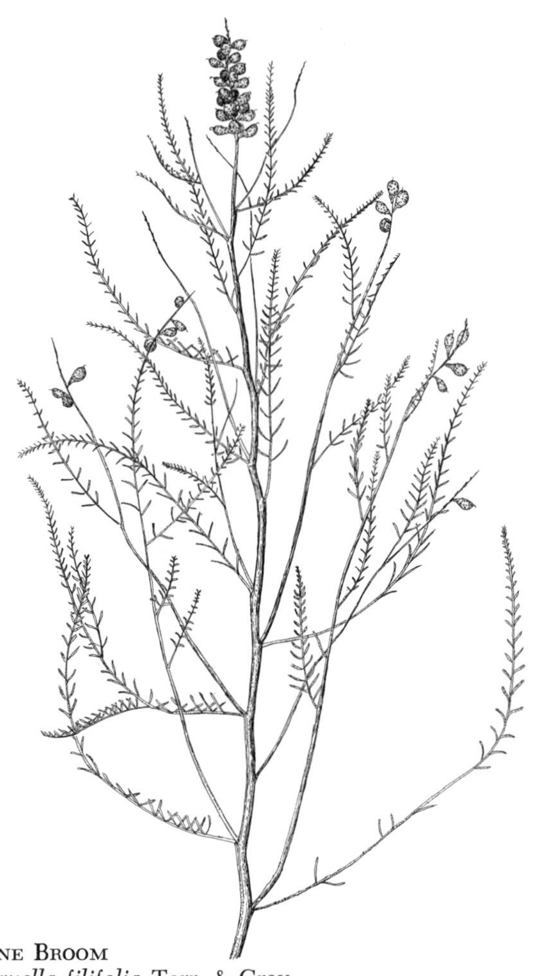

Dune Broom
Parryella filifolia Torr. & Gray

FLOWERS. On lax terminal racemes ¾–5 in. long; bracts minute and deciduous; pedicels 1/25–1/12 in. long; flowers about ¼ in. long or less, yellowish green; petals absent; stamens 9–10, filaments exserted, twice as long as the calyx, and inserted at its base, anthers uniform; ovary 2-ovuled, style thick, stigma glandlike and lateral; calyx turbinate, ⅙–⅛ in. long, 5-toothed, teeth triangular, short and equal, tube of calyx 10-ribbed and glabrous, and usually glandular-dotted externally.

FRUIT. Pods about ¼ in. long, conspicuously glandular-dotted, obliquely obovoid, short-beaked, indehiscent, 1-seeded; seed oval, somewhat compressed.

LEAVES. Ascending, odd-pinnately compound, 1–7 in. long; leaflets 11–45, linear-filiform to very narrowly elliptic, ⅕–⅗ in. long, 1/25–1/12 in. wide, often involute, strigose or glabrate, or sometimes glandular-punctate; stipules minute, subulate, deciduous.

TWIGS. Slender, broomlike, green to reddish brown, later brown to ashy-gray, glabrous or sparingly strigose, roughened by minute, somewhat raised, glands.

RANGE. Sandy rolling hillsides at altitudes of 4,400–6,000 ft in New Mexico and northern Arizona.

REMARKS. The genus name, *Parryella,* honors C. C. Parry, who collected plants, 1854–1858, for the United States–Mexico boundary survey. The species name, *filifolia,* refers to the filiform leaves. The Hopi Indians use the slender twigs for making brooms and baskets. The plant is also used by them as a cure for toothache and as an insecticide.

BLACK LOCUST
Robinia pseudo-acacia L.

BLACK LOCUST
Robinia pseudo-acacia L.

FIELD IDENTIFICATION. Spiny tree attaining a height of 100 ft, with a trunk diameter of 30 in. Trees rapid-growing, reaching maturity in 30–40 years.

FLOWERS. May–June, attractive, fragrant, in loose, pendent racemes 4–5 in. long; individual flowers perfect, bonnet-shaped, about 1 in. long; corolla white, petals 5; standard petal obcordate, rounded, a yellow blotch often on inner surface; wing petals 2, free; keel petals incurved, obtuse, united below; stamens 10, in 2 groups, 9 united and one free, the group forming a tube; pistil superior, ovary oblong, style hairy and reflexed, stigma small; ovules numerous; calyx 5-lobed, lower lobe longest and acuminate; pedicels about ½ in. long.

FRUIT. A legume, ripe September–October; brown, flattened, oblong-linear, straight or slightly curved, 2–5 in. long, about ½ in. wide, 2 valved, persistent; peduncle short and thick; seeds 4–8, hard, flat, mottled-brown, kidney-shaped, seed dispersed September–April from the persistent legume.

LEAVES. Pinnately compound, alternate, deciduous, 8–14 in. long; leaflets 7–19, each one ½–2 in. long and ½–1 in. wide, sessile or short-stalked; rounded at both ends or sometimes wedge-shaped at base; ovate-oblong or oval, entire, mature leaves bluish green and glabrous above, usually paler and glabrous except on veins beneath; young leaves silvery-hairy, mature leaves turning yellow in autumn; stipules becoming straight or slightly curved spines; leaves folding on dark days or in the evenings.

TWIGS. Stout, zigzag, brittle, greenish brown, glabrous, somewhat angular, stipules modified into sharp spines.

BARK. Gray to reddish brown, ½–1½ in. thick, rough, ridged, deeply furrowed, sometimes twisted, inner bark pale yellow, thorns paired and scattered.

WOOD. Greenish yellow or light brown, sapwood white and narrow, durable, hard, heavy, strong, stiff, close-grained, weighing about 45 lb per cu ft, is shock-resistant, shows little shrinkage, is very resistant to decay, but is difficult to hand-tool.

RANGE. Prefers deep, well-drained calcareous soil, probably originally native to the high lands of the Piedmont Plateau and Appalachian Mountains. Georgia; west to Oklahoma and Arkansas and northeast to New York. In other areas probably introduced and escaped from cultivation. Not native to Texas, but persistent along fence rows, abandoned fields, and old home sites.

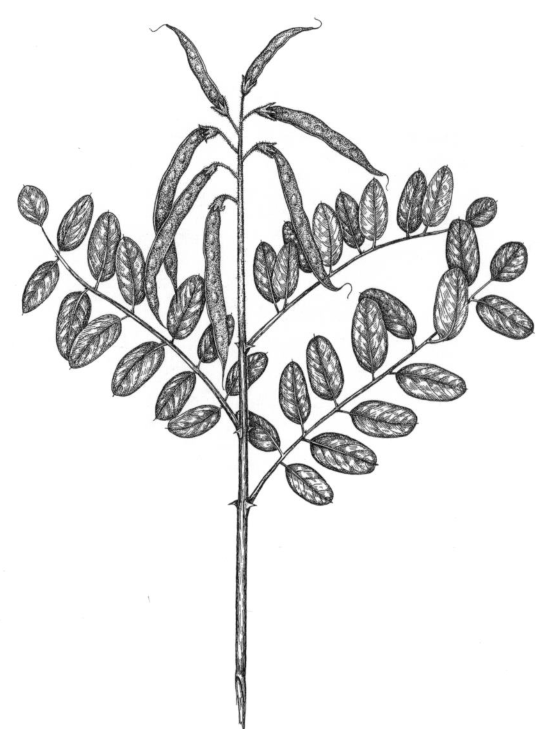

Rusby Locust
Robinia rusbyi Woot. & Standl.

Now commonly distributed by nurserymen in the Gulf Coast states.

PROPAGATION. The cleaned seed averages about 24,000 seeds per lb. Pods should be collected before ripening and threshed or flailed in August. Chaff and light seed can be removed by floating in water. Seed will retain viability for 10 years or more if placed in a closed container at 32°–40°F. Germination may be hastened by scarification, by immersing in sulphuric acid, or by soaking in hot water. Germination averages about 68 per cent and good seedlings about 25 per cent.

In nursery practice the seeds are planted in drills in well-drained loamy or sandy soil 9–12 in. apart, May–June. Beds should be kept moist, but not wet, and no mulch or shade is required. Damping-off may be prevented by dusting the seeds with a fungicide before planting. The tree is very susceptible to the attacks of the locust borer (*Cyllene robiniae*), and the tree has been exterminated in some areas. The chlorophyll-eating leaf-miner insect also causes minor damage.

VARIETIES AND CLONES. Kelsey and Dayton (1942) list the following clones, some of which may ultimately prove to be varieties or forms:
Besson *(bessoniana)*
Coluteoides
Crinkle-leaf *(crispa)*
Cut-leaf *(dissecta)*
Decaisne *(decaisneans)*
Dependens
Erecta
Golden *(aurea)*
Linearis
Little-leaf *(microphylla; R. pseudo-acacia angustifolia)*
Myrtifolia
One-leaf *(unifoliola; R. pseudo-acacia monophylla)*
Perpetual *(semperflorens; R. semperflorens)*
Purple-leaf *(purpurea)*
Pyramid *(pyramidalis)*
Rehder *(rehderi)*
Rozynsky *(rozynskyana)*
Shipmast *(rectissima)*
Stricta
Thornless *(inermis; spectabilis)*
Twisted *(tortuosa)*
Ulrich *(ulriciana)*
Umbrella *(umbraculifera)*
Weeping *(pendula)*

REMARKS. The genus name, *Robinia*, is in honor of Jean and Vespasian Robin, herbalists to Henry IV of France, and the species name, *pseudo-acacia* ("false-acacia"), refers to its resemblance to an acacia. Other vernacular names are White Locust, Yellow Locust, Red Locust, Red-flowering Locust, Green Locust, Honey Locust, Silver Locust, Post Locust, Pea-flower Locust, Silver-chain, and False-acacia Locust. The wood is highly resistant to decay.

Rusby Locust

Robinia rusbyi Woot. & Standl.

FIELD IDENTIFICATION. Spiny shrub or small tree.

FLOWERS. Rather large, crowded in dense axillary or terminal peduncled racemes to 4 in. long; pedicels about 1/4 in. long, densely pubescent and also with scattered larger glandular hairs; calyx 1/3–1/2 in. long; lobes 5, lanceolate, acuminate, ciliate, densely brownish-tomentose, also with stout thick hairs which are glandular-tipped or glandless; corolla about 3/4 in. long, papilionaceous, standard petal broad and reflexed; stamens 10 (9 united, 1 free); style elongate.

FRUIT. Legume 2–3 1/4 in. long, oblong-linear, flattened, sutures prominent, margin straight or somewhat undulate, apex acute or acuminate, often tipped with the style remnants, surfaces dark brown, glabrous to the eye but under magnification with scattered, minute, stiff, white hairs; dehiscent, seeds numerous.

LEAVES. Leaves alternate, odd-pinnately compound, 5–8 in. long, rachis puberulent with the 11–17 leaflets rather widely spaced, blades 3/4–1 1/2 in. long, oblong-elliptic to oval, apex rounded to obtuse, often mucronulate, base rounded or broadly cuneate, margin entire, upper surface dull dark green and glabrous, lower surface paler, slightly puberulent to glabrous; petiolules 1/8–1/4 in. long, puberulent.

TWIGS. Young ones reddish brown and puberulent,

older ones gray and glabrous; petiolar spines paired or solitary, curved or straight, usually short, brown.

RANGE. Usually in moist soil of slopes or along streams. In New Mexico in the Mogollon and Burro mountains. Type locality 15 miles east of Mogollon.

REMARKS. The genus name, *Robinia,* is in honor of Jean Robin (1550–1629) and his son Vespasian Robin (1579–1662), herbalists to kings of France. The species name, *rusbyi,* is in honor of H. H. Rusby, who collected in eastern Arizona in 1881, and who made large collections in 1883 in Yavapai County and on the San Francisco Peaks. This species is closely related to *R. neomexicana* and further study may show it to be a glabrous variety of it.

New Mexico Locust

Robinia neomexicana Gray

FIELD IDENTIFICATION. Spiny shrub or tree to 24 ft, with a trunk to 4 in. in diameter. Thicket-forming and sprouting freely from stumps and roots.

FLOWERS. April–August in axils of the leaves of the current year; racemes, large, showy, pendent, 2–4 in. long, all parts pubescent or glandular-hispid, the hairs straight and glanduliferous; calyx slightly 2-lipped, tube campanulate, ⅓–⅖ in. long, the teeth triangular; corolla rose-colored, ⅘–1 in. long, papilionaceous; standard large and rounded, reflexed; stamens 10 (9 united into a tube, the other free); style elongate.

FRUIT. Legume maturing September–October, 2–4 in. long, ⅜–⅖ in. wide, flat, thin, glandular-hispid, reticulate, thickened on one edge, 2-valved, 3–8-seeded.

LEAVES. Alternate, deciduous, odd-pinnately compound of 13–21 leaflets, blades elliptic-lanceolate, apex obtuse to rounded and mucronate, margin entire, length ⅜–1½ in., finely strigillose on both sides.

TWIGS. Young ones puberulent and glandular, older ones reddish brown to gray and more glabrous; prickles straight or subrecurved, sharp and stout.

RANGE. Generally in moist soil along streams in the sun. At altitudes of 4,000–8,500 ft in the conifer belt. Trans-Pecos Texas; west through New Mexico and Arizona, north to Colorado, Utah, and Nevada.

PROPAGATION. The seed is usually dispersed August–December. They should be collected before the pods open by stripping from the branches and beating in a bag. They are then spread out to dry, and the debris removed by fanning. About 20 lb of seed are yielded from 100 lb of fruit. Commercial seed has a soundness of 96 per cent and a purity of 98 per cent. If stored in a cool dry place they retain viability 1–4 years. For pretreatment the seed may be soaked in hot water at 212°F. or less 1–5 minutes, followed by soaking in water at room temperatures for 10 hours. The seeds may be sown in well-drained, fertile, sandy loam, in drills in spring 9–12 in. apart, and covered ¼ in. with soil. No mulch or shade is required.

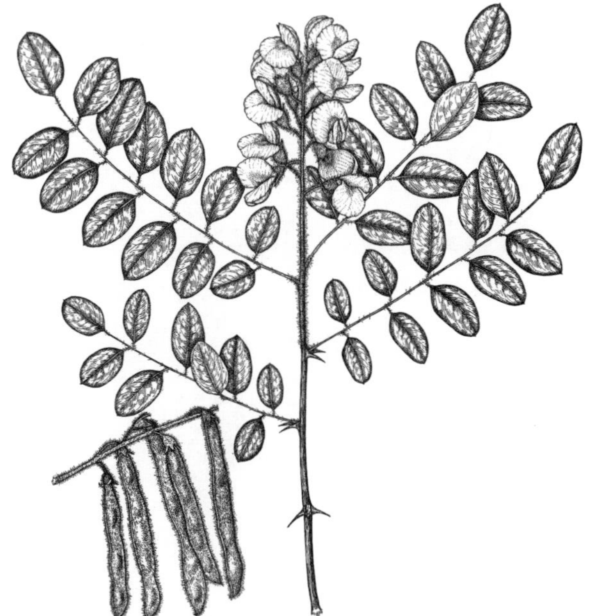

New Mexico Locust
Robinia neomexicana Gray

VARIETY. A variety has been named *R. neomexicana* var. *luxurians* Dieck, which has 15–21 oblong leaflets. Flowers rose or white; pods glandular-hairy. It has been found in Colorado, Utah, and New Mexico. However, this variety apparently intergrades considerably with the species and seems hardly worthy of segregation as a variety.

REMARKS. The genus name, *Robinia,* is in honor of Jean Robin (1550–1629) and his son Vespasian Robin (1579–1662), herbalists to kings of France. The species name, *neomexicana,* refers to the state of New Mexico where it grows. It is sometimes known under the name of Uña de Gato. It is reported to be used as a remedy for rheumatism by the Hopi Indians. The tree is valuable for erosion control. It is an important browse for goats, and is sometimes eaten by cattle and horses. It is also eaten by Gambel's quail, mountain sheep, mule deer, black-tailed deer, chipmunk, and porcupine. The tree has been planted for ornament in the United States and Europe since 1881.

Rose-acacia Locust

Robinia hispida L.

FIELD IDENTIFICATION. Shrub 3–18 ft, spreading by stoloniferous roots. Branches numerous, erect and diffuse; twigs, petioles, and rachises bristly with reddish hairs to ⅕ in. long.

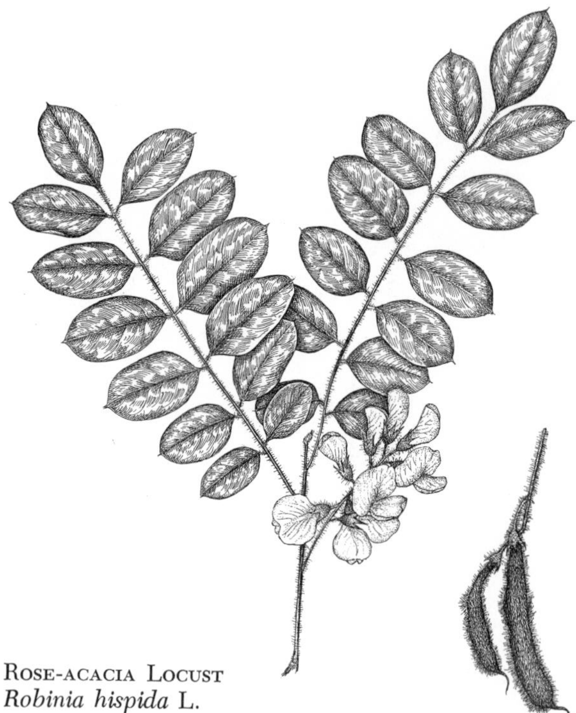

ROSE-ACACIA LOCUST
Robinia hispida L.

FLOWERS. May–June, racemes axillary, pendent, lax and open, 3-9-flowered, pedicels ¼–½ in. long, bristly-hairy; calyx short-hispid, somewhat 2-lipped, the upper 2 somewhat united, lateral lobes broadly-lanceolate and acuminate; corolla pink, bonnet-shaped, ⅔–1¼ in. long, bearing a banner petal, 2 wing petals, and 2 keel petals; standard petal about 1 in. wide, reflexed, about the same length as the wings and keel; wing petals oblong and base auricled; keel petals curved, obtuse, base with a rounded auricle; stamens diadelphous, ovary stalked and densely glandular-hairy.

FRUIT. Legume 2–3¼ in. long, linear, flattened, often terminated by the persistent style, margined along the upper suture, conspicuously and densely reddish brown hispid, several-seeded, tardily 2-valved.

LEAVES. Odd-pinnately compound, 5–9 in. long; leaflets 7–15, length ¾–2 in., width ⅝–1¾ in., suborbicular, ovate or oblong, ends rounded or broadly cuneate, apex mucronate, petiolules ⅛–¼ in. long and bristly-hairy, thin, light green to dark green, upper surface glabrous, lower surface glabrous or with occasional scattered hairs.

TWIGS. Young ones brown and densely reddish hispid; older ones gray to brown, striate and glabrous. The amount of hispidum varies considerably.

RANGE. Dry slopes, woods, and thickets. Oklahoma, eastward to Georgia and Virginia. Mostly north of the Gulf Coast plain. Often escaping from cultivation in more northern areas.

VARIETIES. Smooth Rose-acacia Locust, *R. hispida* var. *macrophylla* DC., has branchlets and petioles without bristles or nearly so; the leaflets and flower are somewhat larger.

Fruitful Rose-acacia Locust, *R. hispida* var. *fertilis* (Ashe) Clausen, has been described as a variety with numerous flowers ¾–1 in. long; leaves elliptic, acute or acutish and ⅔–1 in. long.

R. hispida also hybridizes with *R. pseudo-acacia* to produce *R.* × *margaretta* Ashe. Besides the name of Rose-acacia Locust the species is also known under the vernacular names of Honey Locust, Bristly Locust, and Moss Locust.

REMARKS. The genus name, *Robinia,* honors Jean Robin (1550–1629) and his son, Vespasian Robin (1579–1662), who first cultivated the Locust-tree in Europe. The species name, *hispida,* refers to the dense bristly hairs of foliage and inflorescence.

MESCAL-BEAN SOPHORA

Sophora secundiflora (Ortega) Lag.

FIELD IDENTIFICATION. Evergreen shrub, or sometimes a tree to 35 ft, with a narrow top, upright branches and velvety twigs.

FLOWERS. With the young leaves March–April; racemes densely flowered, terminal, 2–4¾ in. long, showy, violet, fragrant; pedicels ¼–⅖ in. long, subtended by subulate bracts about ½ in. long; calyx campanulate, ⅓–⅖ in. long, oblique, the 2 upper teeth almost united throughout, the lower 3 teeth triangular to ovate; corolla bonnet-shaped, violet; standard petal erect, broad, suborbicular to ovate, crisped, notched, ⅗–⅔ in. long, somewhat spotted at the base within; wing and keel petals oblong to obovate; stamens 10; ovary white-hairy.

FRUIT. Pod in September, on pedicels ¼–1 in. long, woody, hard, oblong, terete, densely brown-pubescent, somewhat constricted between the seeds, apex abruptly prolonged by persistent style remnants, indehiscent, 1–5 in. long, about ¾ in. wide, walls about ¼ in. thick; seeds red, 1–8 (usually 3–4), about ½ in. long, globose to oblong, often flattened at one end, hard, bony; hilum small, pale and about ⅛ in. long.

LEAVES. Odd-pinnately compound, 4–6 in. long, rachis grooved above; leaflets 5–13 (usually 5–9), persistent, elliptic-oblong or oval; apex rounded or obtuse, notched, or mucronulate; base gradually narrowed; margin entire; upper surface lustrous, leathery, reticulate, dark green, hairy when young but glabrous later; lower surface paler, glabrous or puberulent; 1–2½ in. long, ½–1½ in. wide; petioles stout, puberulous, sometimes leaflets sessile or nearly so.

TWIGS. With fine velvety tomentum at first, later becoming glabrous, or nearly so, green to orange-brown.

BARK. Dark gray to black, broken into shallow fissures with narrow flattened ridges and thin scales.

WOOD. Heartwood orange to red; sapwood yellow, hard, heavy, close-grained, specific gravity about 0.98, said to yield a yellow dye, otherwise of no commercial value.

RANGE. Usually on limestone soils in central, southern, and western Texas, New Mexico, and northern Mexico. In Texas from the shores of Matagorda Bay, almost at sea level, west into the Chisos and Davis mountains to altitudes of 5,000 ft. Frequent on the limestone hills around Austin, Texas. In Mexico from Nuevo León to San Luis Potosí.

REMARKS. The genus name, *Sophora*, is from the Arabic name, *Sophero*, and the species name, *secundiflora*, refers to the one-sided inflorescence. Other vernacular names are Texas Mountain-laurel, Frigolito, Frijollito, Frijolillo, Coral Bean, Big-drunk Bean, and Colorín.

Although called Mountain-laurel in Texas this plant is not a member of the Laurel Family (*Lauraceae*). However, neither is the so-called Mountain-laurel of the Eastern states, which is *Kalmia latifolia*, a heath. The true laurel, or Poet's Laurel, *Laurus nobilis*, is a native of Asia and south Europe. American representatives of the true Laurel Family native to our area are the Red Bay, *Persea borbonia*, and the Sassafras, *Sassafras albidum*.

The persistent, shiny leaves of Mescal-bean give a lustrous effect when seen in mass at a distance. The beautiful violet flowers are very fragrant, in fact offensively so to some people. A volatile liquid alkaloid, sophorine, which is identical with cytisine, exists in many species of *Sophora*, including *S. secundiflora*, *S. tomentosa*, and *S. sericea*. Other sophoras, including *S. angustifolia*, *S. alopecuroides*, and *S. pachycarpa*, contain the alkaloids matrine and sophocarpine.

MESCAL-BEAN SOPHORA
Sophora secundiflora (Ortega) Lag.

NECKLACE-POD SOPHORA
Sophora tomentosa L.

The narcotic properties of the red seeds of Mescal-bean Sophora were well known to the Indians. A powder from them, in very small amounts, was mixed with the beverage mescal to produce intoxication, delirium, excitement, and finally a long sleep. The seeds are poisonous to both humans and livestock. They were frequently used as a trade article by the Indians in the form of necklaces. The red seeds of *Erythrina herbacea* were also used for this purpose, and the two plants are often confused, but *Erythrina* has 3 leaflets only, and the pod is dehiscent at maturity. Mescal-bean is rather difficult to transplant, but this can be done if sufficient calcium is in the soil. The shrub is rather slow-growing.

NECKLACE-POD SOPHORA

Sophora tomentosa L.

FIELD IDENTIFICATION. Evergreen shrub to 9 ft.

FLOWERS. June–October, borne in elongate, erect, tomentose, racemes 4–16 in. long; corolla yellowish white, ½–1 in. long; standard clawed, blade ovate to rounded, to ½ in. broad and somewhat longer, notched at apex; keel and wing petals narrowly elliptic, auriculate at base on one side; calyx-tube campanulate, ⅕–½ in. long, obscurely 5-lobed; stamens 10, filaments distinct; ovary narrow, short-stalked, style incurved and subulate.

FRUIT. Pod 2–5 in. long, long-stalked, torulose, densely tomentose, strongly constricted between the seeds; seeds brown, smooth, lustrous, rounded or somewhat flattened on the sides, about 3/16 in. long.

LEAVES. Alternate, deciduous, 4–8 in. long, pinnately compound of 13-21 leaflets, rachis densely tomentose; leaflets elliptic, oblong or oval, 1–2½ in. long, acute to rounded at the apex, leathery, at first densely tomentose but later becoming glabrate above; petiolules tomentose, ¼–½ in. long.

TWIGS. Old twigs green to light brown, young twigs and all other parts of the season densely tomentose.

RANGE. Necklace-pod Sophora has a world-wide distribution along tropical seashores of both hemispheres. In North America it is found on coastal dunes of Baja California, Florida, and south Texas. The specimen for the drawing was collected by the author at Rockport, Texas, in sand close to the bay front. Also found at Boca Chica beach below Brownsville in Cameron County.

MEDICINAL USES. The seeds are poisonous and contain a powerful alkaloid. The plant is reported to have diuretic, sudorific, and purgative properties and is used in the West Indies as a remedy for venereal disease. In Ceylon the roots and leaves are used by the natives as a remedy for cholera.

RELATED SPECIES. Necklace-pod Sophora is sometimes confused with the Silky Sophora, S. *sericea* Nutt., but the latter species is a low perennial herb with silky hairs, oblong to obovate small leaflets ¼–⅗ in. long, and white flowers, and occurs in a widespread area from Texas to New Mexico, Arizona, Dakota, Wyoming, Kansas, and Nebraska.

REMARKS. The genus name, *Sophora,* is from the Arabic name, *Sophero,* which is applied to some tree of the same family, and the species name, *tomentosa,* refers to the woolly hairs of the young parts of the plant. In Mexico the plant is known as "Tambalisa."

Texas Sophora

Sophora affinis Torr. & Gray

FIELD IDENTIFICATION. A shrub or small tree to 25 ft, with spreading branches and a rounded head.

FLOWERS. In June, arranged in simple axillary racemes 2–6 in. long; pedicels slender, finely tomentose, about ½ in. long, subtended by small deciduous bracts; corolla bonnet-shaped, about ½ in. long, pink to white, somewhat fragrant; petals short-clawed, standard large, nearly orbicular, somewhat notched, reflexed, ⅜–⅝ in. long; wing and keel petals ovate to oblong and auriculate at base; stamens 10; ovary hairy and stipitate, stigma capitate; calyx campanulate, short, about ¼ in. long, abruptly narrowed at the base, obscurely 5-toothed with teeth triangular to ovate, somewhat pubescent.

FRUIT. Fruiting peduncle 2–4 in. long; legume ½–3½ in. long, abruptly constricted between the seeds, terete, black, often hairy, especially on the strictures, coriaceous, indehiscent, tipped with the prolonged style remnants, flesh thin, persistent; seeds 1–8 (mostly 4–8), ovoid or oval, seed coat brown. Constrictions between the seeds give the pod a beadlike appearance.

LEAVES. Alternate, odd-pinnately compound, deciduous; rachis lightly tomentose, 3–9 in. long, leaflets 9–19 (usually 13–15); leaflets elliptical or oval, margin entire; apex obtuse, acute, retuse and mucronulate; base rounded or cuneate, contracted into short stout petiolules ⅛–1/16 in. long; upper surface dark green to yellowish green, lustrous, glabrous or slightly hairy, lower surface slightly paler and pubescent; thin and soft; ¾–1½ in. long, about ½ in. wide; young leaves hoary-pubescent.

TWIGS. Slender, green to brown or streaked with lighter brown, nearly glabrous or puberulent, somewhat divaricate, somewhat swollen at the nodes.

TEXAS SOPHORA
Sophora affinis Torr. & Gray

KENTUCKY WISTERIA
Wisteria macrostachya Nutt.

BARK. Gray to reddish brown, broken into small, thin, oblong scales.

WOOD. Light red, sapwood yellow, heavy, strong, and hard.

RANGE. Usually on limestone soils of northwestern Louisiana and southwestern Oklahoma, down through central Texas. Occurs at Dallas, Kerrville, Austin, and San Antonio, Texas. Often in small groves on hillsides or along streams.

REMARKS. The genus name, *Sophora*, is from the Arabic word *Sophero*, which was applied to some tree of the same family, and the species name, *affinis*, means "related to." Another vernacular name is Eve's Necklace. The plant could be more extensively cultivated for ornament.

KENTUCKY WISTERIA

Wisteria macrostachya Nutt.

FIELD IDENTIFICATION. A twining, woody vine reaching a length of 30 ft, with a lower stem diameter of 1½ in.

FLOWERS. March–July, racemes slender, terminal on short lateral branches, pendent, 8–12 in. long, 1½–2 in. wide, loosely flowered, peduncles green and hairy; flower pedicels ¼–½ in. long, densely white-hairy and sometimes glandular; corolla papilionaceous, perfect; standard petal reflexed at maturity but enclosing the other petals before opening, orbicular, about ½ in. across, varying in color from white through blue-lilac or purple, with 2 basal appendages; wing petals about ⅜ in. long, oblong and falcate, the apex rounded, the base auricled with a linear spur about as long as the claw, white to purple; keel petals joined somewhat, about ⅜ in. long, upcurved, purplish, base auricled with slender claws; stamens 10 (9 united and 1 free), upcurved; pistil upcurved with the stamens and slightly longer; calyx campanulate, purple, densely hairy, slightly 2-lipped, upper lip with 2 short lobes; lower lip 3-lobed, lobes acuminate (the central lobe longest), densely hairy; ovary stalked, ovules numerous.

FRUIT. A legume maturing in October, stipitate, linear, 2–5 in. long, flattened, constricted between the seeds, reddish brown, rather obtuse, coriaceous, turgid; seeds large, oblong-cylindric, black, lustrous.

LEAVES. Alternate, 5–18 in. long, odd-pinnately compound of 5–11 leaflets; leaflets ovate, elliptic or oval, apex acute or acuminate, base rounded or slightly cordate, margin entire and slightly revolute, ciliate, length ¾–2½ in., width ½–1 in.; upper surface dull green or semilustrous, veins impressed, glabrous or nearly so; lower surface paler, bearing white hairs, especially along the veins; petiolules 1/16–3/16 in. long, densely white-hairy.

STEMS. Twining and climbing high on supports, young stems green, densely white-hairy; older stems glabrous, reddish brown to gray, becoming stout and branched.

RANGE. Kentucky Wisteria is usually found in low wet woods, or on edges of swamps in Louisiana and Texas, north into Missouri and Tennessee. Intermediate forms between it and American Wisteria are found.

VARIATIONS. A variety of Kentucky Wisteria, which is often cultivated, is the Lilac-pink Kentucky Wisteria, *W. macrostachya* var. *albolilacina* (Dipp.) Rehd.

Also, a rather similar species to Kentucky Wisteria is American Wisteria, *Wisteria frutescens* (L.) Poir. It has a shorter calyx, with the lower lobe shorter than the tube and lower spur of the wings shorter than the claw. Also the flower raceme is somewhat smaller than that of Kentucky Wisteria. It is found from east Texas to Florida and north to Virginia.

A number of showy Asiatic species of *Wisteria*, with numerous varieties and forms, are cultivated in gardens. The two most common ones are Japanese Wisteria, *W. floribunda* (Willd.) DC., and Chinese Wisteria, *W. sinensis* (Sims) Sweet. White, pink, blue, and purple forms of these are known horticulturally.

REMARKS. The genus, *Wisteria*, is named for Caspar Wistar (1761–1818), anatomist of Philadelphia. The species name, *macrostachya*, refers to the large racemes of flowers. Also known in some localities under the names of Virgins-bower and Kidney-bean-vine. The vine has been cultivated since 1855. Wistarin is a poisonous crystalline glycoside which has been obtained from the plant. However, it is recorded that the pioneers used the fresh flowers in salads and mixed them with batter to make fritters.

American Wisteria

Wisteria frutescens (L.) Poir.

FIELD IDENTIFICATION. A climbing woody vine growing to a length of 40 ft or more, with a stem several inches in diameter.

FLOWERS. April–June, borne before the leaves are fully mature; racemes 2–7 in. long, terminal on short lateral shoots, compact; pedicels 1/6–1/4 in. long, pubescent, more or less glandular; calyx campanulate, finely pubescent, sometimes also with clavate glands; somewhat 2-lipped, irregularly toothed, the upper lip of calyx about a third as long as the tube; corolla lilac-purple, 1/2–3/4 in. long; standard large, rotund, clawed, reflexed, bearing 2 small appendages at base; wings oblong to obovate, falcate, with one short and one slender auricle at the base; keel incurved, obtuse, auriculate at base; stamens 10, diadelphous; ovary stalked, elongate.

FRUIT. Maturing September–November, legume 1 1/4–4 in. long, linear, somewhat flattened, knobby, dehiscent into 2 coriaceous valves, seeds large.

LEAVES. Odd-pinnately compound to 1 ft long, of 9–15 leaflets, rachis and petiolules pubescent, blades

American Wisteria
Wisteria frutescens (L.) Poir.

ovate to oblong or lanceolate, apex acute or acuminate, base rounded, margin entire, length 1 1/2–2 1/2 in.; upper surface dark green and glabrous, lower surfaces paler and pubescent to glabrous.

RANGE. Low moist soil in woods or on riverbanks. Eastern Texas, Arkansas, and Louisiana; eastward to Florida and northward to Virginia.

VARIETIES AND RELATED SPECIES. A variety with white flowers has been given the name of *W. frutescens* var. *nivea* Lescuyer.

Kentucky Wisteria, *W. macrostachya* Nutt., is very closely related to American Wisteria, except that American Wisteria has calyx-lobes much shorter than the tube, and the racemes are usually shorter.

REMARKS. The genus name, *Wisteria*, honors Caspar Wistar (1761–1818), distinguished American physician and scientist of Philadelphia. The species name, *frutescens*, refers to the shrubby or woody character. It is also known under the names of Woody Wisteria, Kidney-bean-tree, and Virgins-bower. It could be more extensively cultivated for ornament. It was first planted in 1724.

Chinese Wisteria

Wisteria sinensis (Sims) Sweet

FIELD IDENTIFICATION. Climbing vine, in age with a stout, crooked, woody trunk becoming several inches in diameter. By repeated severe pruning it can be induced to assume a shrub or small tree form.

FLOWERS. April–May, and also with occasional flowers in late summer or fall; racemes pendent, terminal on axillary spurlike branches, 4–15 in. long or even longer, peduncle densely puberulent, pedicels 1/4–5/8 in. long and puberulent; calyx 3/16 in. long, puberulent, lavender or blue, 2-lipped, upper lip about 1/8 in. long, with a solitary acute or obtuse lobe, margin ciliate; lower lip 3-toothed, the teeth about 1/16 in. long, acuminate, ciliate; corolla violet to purple or blue; banner petal oval, 1/2–5/8 in. wide, erect, reflexed, sometimes with a reddish or purplish spot near the base; wings 2, about 1/2 in. long, oblong, asymmetrical (upper margin straight, lower curved); base auriculate with the longer claw on the bottom side; keel petals 2, slightly shorter than the claw or as long, closely folded, upcurved, the lower basal portion with a long claw; stamens enclosed by the keel petals, white (joined together, 1 free), anthers minute; style upcurved, filiform, about as long as the stamens, stigma minute.

FRUIT. Legume stipitate, 4–6 in. long, linear or sometimes widened toward the upper half, densely velvety-hairy, flattened, apex acute to acuminate and sometimes abruptly so, base gradually narrowed, 1–3-seeded.

LEAVES. Odd-pinnately compound, 6–12 in. long, rachis minutely puberulent to more glabrate later; leaflets 7–13, oblong to elliptic, apex acute to short-acuminate and sometimes abruptly so; base broadly cuneate to rounded, margin entire; upper surface semilustrous, glabrous, or sometimes puberulent on the impressed main veins; lower surface slightly paler with scattered pale appressed hairs or glabrous; petiolules ⅛–3/16 in. long; terminal petiolule ⅓–¾ in. long.

TWIGS. Young stems green and puberulent, older ones glabrous and gray to brown, the main stem often contorted.

RANGE. Prefers deep rich soil. A native of China and often planted for ornament in cities of the Texas and Louisiana coast. Sometimes escaping and growing along bayou banks, margins of woods, or fence rows.

PROPAGATION. Ripened wood cuttings may be grown under glass. Root grafting, or cuttings of roots, may also be used. During the summer layering is sometimes practiced. Propagation by seed usually gives good results, but for horticultural forms grafting on other stock is the best method. If the plant is pruned back continually to a height of 6–8 ft the main stem will eventually support the plant, and a crooked-trunked shrub or small tree will develop. Roots few but deeply penetrating.

HORTICULTURAL FORMS. White Chinese Wisteria, *W. sinensis* forma *alba* Lindl., is a white-flowered form introduced into cultivation in 1846.

REMARKS. The genus name, *Wisteria,* is in honor of Caspar Wistar (1761–1818), professor of anatomy at the University of Pennsylvania. The species name, *sinensis,* denotes its Chinese origin.

CHINESE WISTERIA
Wisteria sinensis (Sims) Sweet

CALTROP FAMILY (*Zygophyllaceae*)

Coville Creosote-bush
Larrea tridentata (DC.) Coville

FIELD IDENTIFICATION. Evergreen, aromatic shrub attaining a height of 11 ft.

FLOWERS. Axillary, solitary, perfect; petals 5, yellow, twisted on edge, ⅓–⅖ in. long; sepals 5, silky, yellowish, deciduous early; stamens 10, filaments winged; carpels 1-seeded, indehiscent.

FRUIT. Capsule small, globose, densely white-villous, 5-celled.

LEAVES. Bifoliolate; leaflets 2, small, opposite, divaricate, strongly falcate, united at base, pointed at apex, oblong to obovate, ⅕–⅖ in. long, thick, dark green to yellowish green, evergreen, sticky with resin, strong-scented.

TWIGS. Brown, flexuous, nodes conspicuous, giving a jointed appearance, lenticels small and pale.

BARK. Dark gray to black, older branches roughened by small scales.

RANGE. Usually in soils underlain with hardpan. Western Texas and New Mexico; north to Utah and Nevada, west to California, and south in Mexico to Durango and Querétaro. A widespread desert shrub of the most arid regions of the Southwest.

MEDICINAL USES. Extractions from the leaf were formerly used as an antiseptic dressing for cuts, bruises, and sores of both domestic stock and human beings. It is also employed as a treatment for rheumatism, venereal diseases, tuberculosis, and intestinal disorders, and used as an emetic.

REMARKS. The genus name, *Larrea*, is in honor of J. A. de Larrea, Spanish promoter of science. The genus was described by Cavanilles in an article published in 1800. The meaning of the species name, *tridentata*, is

Coville Creosote-bush
Larrea tridentata (DC.) Coville

vague, but it may refer to the 3-angled seeds of the 5-parted fruit. Vernacular names of Spanish or Indian origin are Cabonadera, Gobernadora, Hediondilla, Gaumis, Hedionda, and Falsa Alcaparra. Also known as Grease-wood in some regions of the Southwest. This is a very abundant xerophytic shrub of the Southwestern deserts. It is poisonous to sheep, is not eaten by cattle, but is consumed by various small mammals and antelope. It is reported that the buds are pickled in vinegar, and are eaten by the Mexican Indians. A crustaceous lac is often deposited on the branches by a scale insect, this lac is used by the Indians for cementing pottery and dyeing leather. The exudations of gum contain 17.27 per cent resin and 11.71 per cent mucilage and allied substances.

Texas Porlieria

Porlieria angustifolia (Engelm.) Gray

FIELD IDENTIFICATION. Shrub or tree to 21 ft, often in clumps, scrubby, compact, stiff, evergreen, the branches thick and stubby.

FLOWERS. Terminal, clustered or solitary, violet or purple, ⅓–¾ in. across, fragrant, attractive; petals 5, short-clawed, elliptic, apex often notched; sepals 5, shorter than petals, suborbicular, concave; stamens 10, as long as the petals, with scale-like appendages at the base; anthers yellow; ovary densely villous, 2–5-celled.

FRUIT. Capsule heart-shaped, mostly 2-lobed, somewhat winged on margin, apex abruptly attenuate-apiculate, surface reticulate, ⅓–⅔ in. broad; seeds 2 (sometimes 1 or 3), beanlike, large, shiny, red, yellow, or orange.

LEAVES. Often half-folded in the heat of the day, opposite or alternate, abruptly pinnate, rachis pubescent, composed of 4–8 pairs of leaflets; leaflets sessile or nearly so, about ⅔ in. long or less, entire, linear to oblong, leathery, lustrous, reticulate, dark green, acute and apiculate at the apex, oblique at the base, persistent; stipules persistent, somewhat spiniferous.

TWIGS. Short, stout, stiff, knotty, and gray.

BARK. Gray to black, broken into rough scales on old trunks and branches.

RANGE. From central through western and southwestern Texas; Mexico in Coahuila to Tamaulipas.

REMARKS. The genus name, *Porlieria,* is in honor of Porlier de Baxamar, Spanish patron of botany. The species name, *angustifolia,* refers to the narrow leaflets. Vernacular names for the plant are Soap-bush, Guayacán, and Guajacum. Texas Porlieria has possibilities as an ornamental. It is a good honey plant, and the wood is used for fence posts in some localities. The bark of the roots is sold on the Mexican markets as a soap for washing woolens. Extracts of the root are used to treat rheumatism and venereal disease, and they are also used as a sudorific.

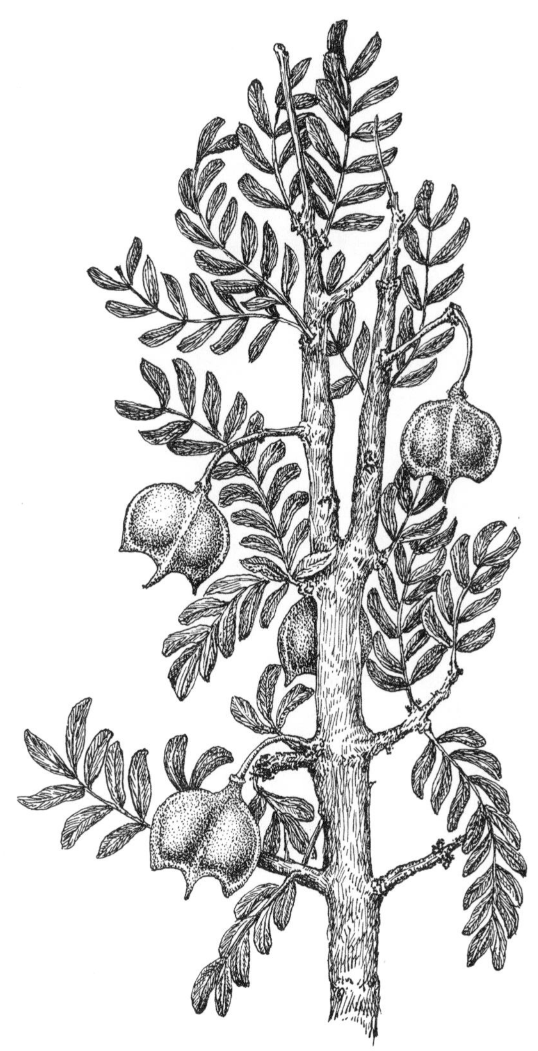

Texas Porlieria
Porlieria angustifolia (Engelm.) Gray

RUE FAMILY (*Rutaceae*)

MOUNTAIN TORCHWOOD AMYRIS

Amyris madrensis Wats.

FIELD IDENTIFICATION. Slender irregularly branched tree hardly over 25 ft, with all parts citrus-scented when bruised.

FLOWERS. Small, perfect, panicles axillary, 1¼–3 in. long, glandular-pubescent; individual pedicels 1/25–½ in. long, glandular-pubescent; calyx 4-lobed, lobes triangular, acute, very glandular; petals 4, imbricate, white to greenish or brown, ovate to oval, concave, clawed, margin erose, hardly over 1/16 in. long, glandular-dotted; stamens 8, sometimes fewer, inserted on the disk, filaments filiform, anthers opening lengthwise; ovary ovoid, 2-celled, style terminal, stigma capitate, glandular-hairy.

FRUIT. Drupe ¼–⅜ in. long, subglobose to ovoid, often glandular-pitted; seed solitary, testa membranous; peduncle ⅛–1 in. long, glandular-pubescent.

LEAVES. Aromatic when bruised, pinnately compound, 2–3½ in. long, rachis bumpy glandular-pubescent; leaflets 5–9 pairs with petiolules 1/25–⅛ in. long, glandular-hairy, elliptic or oval, often asymmetrical, apex obtuse to acute or rounded, base cuneate, margin entire to crenate and somewhat revolute, length ½–¾ in., width ¼–⅜ in., leathery, veins obscure; upper surface olive-green, lustrous, glabrous, minutely pellucid glandular-dotted; lower surface paler and finely pubescent, often with minute glands.

TWIGS. Younger ones green to gray, very glandular and densely fine-hairy; older ones gray to brown, glabrous, slender; lenticels numerous, linear-elongate, horizontal. Bark on trunk mottled light to dark gray, tight and smooth.

RANGE. A rare tree, in Texas known from the lower Rio Grande Valley area at Olmito in Cameron County. Also in Mexico in the state of Nuevo León and San Luis Potosí.

REMARKS. The genus name, *Amyris,* refers to the balsamic properties of the genus. The species name, *madrensis,* refers to its environment in the Sierra Madre Mountains of Mexico.

MOUNTAIN TORCHWOOD AMYRIS
Amyris madrensis Wats.

TEXAS TORCHWOOD AMYRIS

Amyris texana (Buckl.) P. Wilson

FIELD IDENTIFICATION. Aromatic shrub rarely over 9 ft, densely branched.

FLOWERS. In small corymbose-panicles, 1–3 in. long and 1–1½ in. wide, pedicels about ⅛ in. long, glanduar; individual flowers small, hardly over 1/16–3/16 in. vide, perfect, greenish white; petals 4, short-oblong o obovate, apex slightly mucronulate, 1/25–1/12 in. long; tamens 8, inserted on the disk, filaments filiform, anhers opening lengthwise; pistil gradually narrowed to n asymmetrical stigma, ovary 1-celled.

FRUIT. Drupe on clavate peduncles about 1/8–3/4 in. ong, subglobose, about 1/4 in. in diameter, glandular-otted, seed solitary.

LEAVES. Aromatic, alternate, trifoliate, 1–1½ in. long; eaflets elliptic to ovate, margin irregularly crenate-oothed, apex obtuse or rounded, sometimes obscurely otched, base rounded or broadly cuneate, leathery and rm, upper and lower surface glabrous and dull green, eins inconspicuous and covered with glandular dots; etiolule of terminal leaflet nearly sessile or to 1/4 n. long; petiolule of lateral leaflets 1/25–1/8 in. long; etiole 1/4–3/4 in. long, glandular.

BARK. Mottled gray, roughened by lenticels.

WOOD. Close-grained, hard, heavy, somewhat resinus and aromatic, used as fuel but hardly large enough or other purposes.

RANGE. Coastal Texas from Matagorda to Brownsville. Jsually on shell banks a short distance from tide waer. Scattered through the lower Rio Grande Valley rea. Also known in the Mexican states of Nuevo eón and Tamaulipas.

REMARKS. The genus name, *Amyris*, refers to the alsamic properties of the genus, and the species name, *exana*, is for the state of Texas, its original habitat. t is also known in Mexico under the name of Chapotillo. he leaves are glandular-dotted and aromatic when crushed, which character is peculiar to other members of the Rue Family. The name Torchwood was given because the wood ignites easily. It is usually a very low undershrub and, mixed with other dense chaparral shrubs, generally escapes notice.

EXAS TORCHWOOD AMYRIS
myris texana (Buckl.) P. Wilson

STARLEAF MEXICAN-ORANGE
Choisya dumosa (Torr.) Gray

STARLEAF MEXICAN-ORANGE

Choisya dumosa (Torr.) Gray

FIELD IDENTIFICATION. Low, aromatic, much-branched shrub to 6 ft. Recognized by the palmately 5–13-foliate, glandular-knotty leaves.

FLOWERS. Perfect, solitary, or in 2–4-flowered axillary cymes; pedicels 2/5–4/5 in. long, appressed-hairy, subopposite, 2 bracts below the middle; bracts scale-like, ovate, ciliate, glandular, appressed-hairy, 1/12–1/8 in. long or less; petals 4–5, white, obscurely glandular, broadly oblanceolate, about 1/3 in. long, about 1/6 in. broad; stamens 8–10, filaments unequal, in 2 series, outer series longest, ovary 2–5-lobed, 1/12–1/8 in. long, hairy, usually only 2 cells maturing, style about 1/25 in. long, stigma capitate and lobed; sepals 4–5, deciduous, ovate, obtuse, ciliate, glandular-punctate, bractlike.

FRUIT. On solitary or branched peduncles ½–1 in. long, pubescent, green to brown; follicle of 2–5 divergent carpels (becoming 2-celled by abortion); carpels dehiscent down the ventral suture, and dorsally as far as the dorsal horn; each carpel ovoid, about ¼ in. long, asymmetrical, rounded on the outer surface, apex with a straight or curved mucronulate protuberance, leathery, flattened, hairy, glandular-pitted; styles persistent and apiculate; seeds 1–2 in a carpel, somewhat oval, dark brown, shiny, longitudinally striate-reticulate, the concave side with a deep pit-like hilum.

LEAVES. Persistent, usually crowded toward the ends of the branches, opposite or subopposite, aromatic when bruised, palmately compound of 5–13 sessile, barely articulated leaflets; leaflets ⅓–2 in. long, 1⁄12–3⁄16 in. wide, linear, olive-green, leathery, at first sparsely pubescent-hairy, later glabrous or scurfy along the midrib, margin coarsely dentate or only undulate and conspicuously glandular-knotty, apex obtuse, truncate or rounded to glandular-emarginate, basally somewhat narrowed; petiole ½–1⅕ in. long, as long as the leaflets or shorter, with or without knotty glands, upper surface scurfy and flattened.

TWIGS. Rather slender, young ones green and densely or scantily pubescent, older ones mottled gray and glabrous, often warty with persistent raised glands.

BARK. Mottled gray to black, smooth or roughened toward the base.

RANGE. In west Texas, New Mexico, Arizona, and Mexico. In Texas in the pinyon belt of the Davis and Guadalupe mountains at altitudes of 3,000–7,000 ft. Also in El Paso County. In Mexico in the states of Chihuahua and Coahuila.

VARIETIES. Two varieties of Starleaf Mexican-orange have been described from Arizona as Arizona Mexican-orange, *C. dumosa* var. *arizonica* (Standl.) Benson, and Hairy Mexican-orange, *C. dumosa* (Torr.) Gray var. *mollis* (Standl.) Benson. These vary in degree of pubescence, number of glands, and size of flowers, and have leaflets 3–5 instead of 5–13 as in the species. For further data on these varieties see Benson and Darrow (1945, pp. 213–214).

REMARKS. The genus name honors J. D. Choisy, Swiss botanist (1799–1859). The species name, *dumosa*, refers to the bushy habit. Some authorities list the plant under the name of *Astrophyllum dumosum* Torr. Vernacular names are Fragrant Starleaf, Zorilla, and Sorilla. This rather odd and beautiful shrub could possibly be cultivated for ornament in the Western highlands. It is not eaten by livestock, but there is no conclusive evidence that it is poisonous. Like most members of the Rue Family it has aromatic glands.

Sour Orange

Citrus aurantium L.

FIELD IDENTIFICATION. Cultivated shrub or small tree, sometimes attaining a height of 30 ft. The regular branches form a rounded top, and the twigs often bear long flexible spines.

FLOWERS. Axillary, large, single or clustered, whit in the bud, very fragrant, perfect; petals 5, thick, stra shaped, imbricate; sepals 5, united below; stamen 20–24, inserted around a disk, polyadelphous; styl united and deciduous, containing as many tubes a there are cells in the globular ovary.

FRUIT. Of a globose or subglobose type, slightly fla tened at the tip and base, dark green to dark orang or reddish, 2¾–3¼ in. in diameter, rind thick, surfac rough, pulp sour and bitter, with a hollow core whe fully ripe, segments 10–12; seeds oval-cuneate, fla tened, white inside, endosperm lacking.

LEAVES. Simple, persistent, 3–4 in. long, ovate to el liptic, apex obtuse to acute or more or less acuminat base broadly rounded to cuneate, margin obscurel crenate or entire, dull green above, paler beneath, gla brous; petiole ¾–1¼ in. long, rather broadly winge ½–⅝ in. long, narrowing to the base.

TWIGS. Slender, stiff, green to brown, flattened to ward the nodes, often with single slender spines i the axils, or with stout spines to 2–3 in. long on rapidl growing shoots.

RANGE. Cultivated in south Texas and Florida, es caping cultivation in some areas. A native of southeas Asia. Adapted to heavy moist soils of good depth, bu will grow on many soil types.

Sour Orange
Citrus aurantium L.

VARIETIES AND HYBRIDS. The following hybrids are ormed between Sour Orange and other citrus trees.

Citradia is a hybrid of Sour Orange and Trifoliate Orange.

Lemonime is a hybrid of Sour Orange and Lemon.

Limequat is a hybrid of Sour Orange and Kumquat.

Eremoradia is a hybrid of Sour Orange and Desert Lime (*Eremocitrus glauca*).

The following key to the varieties of cultivated range has been proposed by H. J. Webber and Batcheor (1943, p. 491):

A. Fruits with acid (sour) pulp.
  B. Tree of standard size, 15–30 ft. high, commonly taller than broad, thorny; leaves large, broadly lanceolate, pointed; petioles broadly wing-margined; fruits large, seedy; odor typical .... Normal Citrus Aurantium
  Varieties: (1) *African;* (2) *Brazilian;* (3) *Rubidoux;* (4) *Standard.*
  BB. Tree medium to small, 8–15 ft high, spreading, broader than tall; thornless or nearly so; seeds few or none; leaves medium large; petioles medium wing-margined. .... Aberrant Group
    C. Fruits large, seedy; calyx large and fleshy.
    Varieties (5) *Daidai.*
    CC. Fruits medium to large, with few seeds; calyx of normal size, with thin tips that soon dry out.
      D. Seeds none or very few; leaves medium large, pointed at apex; odor like typical sour orange. Varieties: (6) *Goleta.*
      DD. Seeds few, 3 to 10; leaves of medium size, crowded on stem; apex rounded or blunt-pointed; very aromatic, similar to Bergamot Orange.
      Varieties: (7) *Bouquet*
  BBB. Tree small, 6 to 10 ft high, upright, conical, usually taller than broad; thornless; leaves very small, lanceolate, pointed, crowded; fruit small .... Myrtifolia Group
  Varieties: (8) *Chinotto.*
AA. Fruits with sweet pulp, otherwise like typical Sour Orange (B) .... Bittersweet Group
Varieties: (9) *Bittersweet;* (10) *Paraguay.*

REMARKS. The genus name, *Citrus,* is from the Greek word for "citron." The species name, *aurantium,* is the word for "orange." Also known under the vernacular ames of Seville Orange, Bittersweet Orange, and Bigaade. Sour Orange is extensively cultivated near Seville, pain, and shipped abroad for making orange marmalade. The fruit is too bitter to be eaten raw. The eel is sometimes candied and yields an essential oil. The petals yield a valuable perfume, oil of neroli, roduced principally in southern France and the Italian Riviera. The plant is able to withstand more cold than nost of the other citrus. It is grown on a world-wide commercial scale in semitropical or tropical areas for a grafting stock for other citrus. It is valuable for its resistance to the foot-rot disease, but is sometimes severely attacked by the scab fungus, *Elsinoe fawcettii.* Sour Orange is a very distinct species and reproduces fairly true to seed. It has maintained its identity from the Sweet Orange for 800 years.

LIME

Citrus aurantifolia (Christm.) Swingle

FIELD IDENTIFICATION. Small cultivated citrus tree with a thick crown, and rather irregular branches which bear short, stiff, very sharp spines.

FLOWERS. In short axillary lax racemes of 2–7 flowers (rarely single); calyx cupulate, 4–5-lobed; petals 4–5, white, ⅓–½ in. long, ⅛–⅙ in. wide; stamens 20–25; ovary depressed-globose, clearly delineated from the deciduous style; stigma depressed-globose.

FRUIT. Borne mostly in summer, but to some extent at other seasons, oval to subglobose, sometimes with small apical papilla, greenish yellow when ripe, diameter 1¼–2½ in.; peel prominently glandular-dotted, very thin; segments 9–12, pulp abundant, greenish, very acid; seeds small, oval, white within.

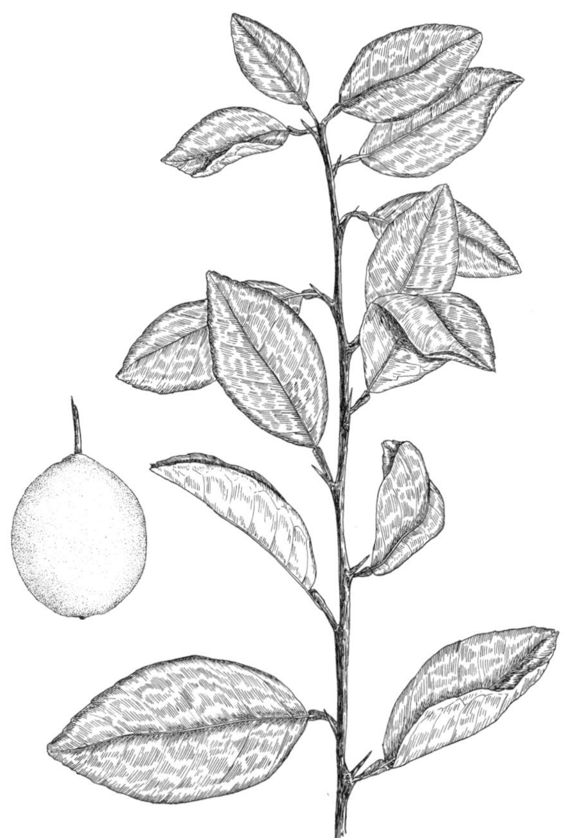

LIME
Citrus aurantifolia (Christm.) Swingle

LEAVES. Alternate, elliptic-ovate or oblong-ovate, apex obtuse or some almost rounded, base rounded, margins crenulate, blade length 2–3 in., surfaces rather pale green, minutely glandular-dotted; petioles narrowly winged, spatulate, thorn at the base.

TWIGS. Bright green to dark green, angular and flattened at the nodes, glabrous, with short, sharp spines.

RANGE. Probably native to the East Indian Archipelago. Now found in nearly all tropical countries. Cultivated in southern Florida and the Keys. Not successfully cultivated in Texas.

HYBRIDS AND VARIETIES. The following hybrids of the lime are known:

Lemonime is a hybrid of *C. limon* and *C. aurantifolia.*

Limequat is a hybrid of *C. aurantifolia* and one or more *Fortunella* Spp.

The following key to varieties of lime is proposed by Webber and Batchelor (1943, p. 620).

A. Fruit pulp very acid; chalazal spot dark reddish-brown .... Acid Limes
  B. Rind adherence tight, difficult to separate from the segments.
    C. Fruit small, diam. usually 1–2 in.; tree small and bushy; branches fine; leaves small .... True Limes: Mexican Group
      D. Branches thorny
        Varieties: (1) *Everglade;* (2) *Mexican;* (3) *Palmetto.*
      DD. Branches thornless or nearly so
        Varieties: (4) *Yung.*
    CC. Fruits large, diam. usually 2–2½ in. or more; tree large; branches coarse, semithorny to thornless; leaves large .... Large-fruited Limes: Tahiti Group
      Varieties: (5) *Bearss;* (6) *Pond;* (7) *Tahiti.*
  BB. Rind adherence loose, separating easily from the segments .... Mandarin-lime Group
    C. Fruits deep orange to reddish in color.
      Varieties: (8) *Rangpur.*
    CC. Fruits light orange to yellow in color
      Varieties: (9) *Kusaie.*
AA. Fruit pulp sweet, lacking acid; chalazal spot yellowish .... Sweet Lime Group
  B. Fruit large, diam. mainly 2 in. or over; chalazal spot yellowish; new twig growth light green.
    Varieties: (10) *Palestine;* (11) *Sweet.*
  BB. Fruits small, diam. mainly about 1½ in.; color of chalazal spot doubtful; new twig growth purplish.
    Varieties: (12) *Otaheite.*

REMARKS. The genus name, *Citrus,* is the Greek word for citron, and the species name, *aurantifolia,* means "orange-leaved." Lime is the most sensitive to cold of any cultivated citrus. Generally it cannot stand temperatures below 28°F. It can be grown from seed, and can also be propagated on stocks of sweet orange, rough lemon, sour orange, and grapefruit. Limes almost grow wild with a minimum of attention in cultivation, fertilizing, and pruning. The fruit does not keep well because of the thin skin, and is not processed like lemons. It is picked green and shipped and is principally used in limeade or as an ingredient in other beverages.

PUMMELO

Citrus grandis (L.) Osbeck

FIELD IDENTIFICATION. Cultivated tree grown in the United States mostly for ornament. Tree rather large for a citrus, spiny, round-topped. Distinctive because of the pubescent angular twigs, very large broadly winged leaves, large flowers and huge pale yellow fruit.

FLOWERS. Borne in axillary or subterminal clusters or singly, very large; sepals and petals 5; stamens 20–25, anthers linear; ovary globose, distinct from the style.

FRUIT. Very large, to 8 in. or more in diameter, varying in shape in different varieties, subglobose to oblate-spheroid or subpyriform; peel thick; pulp vesicles large, easily separating, membranes peeling easily from pulp; seeds numerous, large, thick, margined.

LEAVES. Large, 3–9 in. long, oval or elliptic, apex

PUMMELO
Citrus grandis (L.) Osbeck

blunt, base broadly rounded, sometimes subcordate, margin crenulate or entire, olive-green to dark green, veins obscure above, more conspicuous but delicate beneath; petioles very broadly winged, more or less cordate, usually pubescent.

TWIGS. Rather stout, angled, flattened at nodes, light to dark green, pubescent, older more glabrous, spines usually straight, stiff, sharp-pointed, ¼–2½ in. long.

RANGE. Native to Polynesia and Malaysia. Highly prized in southern China, Indo-China, and Siam. Widely cultivated in tropical areas over the world. Seldom grown in America.

REMARKS. The genus name, *Citrus*, is from the Greek word for "citron," and the species name, *grandis*, indicates the large size of the fruit. It is known in the Far East as the Pampelmous (Pummelo). Also, it is named the Shaddock from Captain Shaddock, who is said to have brought it from Barbados prior to 1707. Pummelo is seldom grown in North America except as a novelty. It is often grown and highly prized in southeast Asia. The flesh is less juicy and separates into segments much more easily than the grapefruit. Pummelos contain a slightly higher content of Vitamin C than grapefruit or oranges. Both pummelo and grapefruit contain a bitter glucoside known as "naringin." Some writers consider the pummelo to be the progenitor of the grapefruit. Both white- and pink-fruited varieties are known. They may be generally classified on the basis of fruit shape. Some are elongate and necked, others neckless and globose or flattened. The necked group includes the Kao-phuang, Siam, and sometimes the Moanalua. The neckless forms include the Pandan Wangi, Pandan Bener, Kao Panne, and Thesca.

LEMON

Citrus limon (L.) Burm. f.

FIELD IDENTIFICATION. Cultivated spreading shrub or small tree, with long irregular branches. Branches armed with stout short and stiff spines, or sometimes spineless.

FLOWERS. Rather large, solitary, or a few together in the leaf axils, perfect; sepals united at base; petals 5, reddish or purplish tinted in the bud, when opening white above and purplish beneath, waxy, fleshy; stamens 20–40, filaments inserted on an annular or cup-shaped disk; ovary superior, glabrous, tapering into the thick, deciduous, united styles, ovules several in each cavity.

FRUIT. Oval to oblong, apical papilla low and broad, color lemon-yellow when ripe, length 3–5 in., width 2–3 in.; peel rather thick, prominently glandular-dotted; pulp abundant and acid, 8–10 segments; seeds sometimes few or none, small, ovoid, pointed, smooth, white within.

LEAVES. Young leaves sometimes reddish, later pale

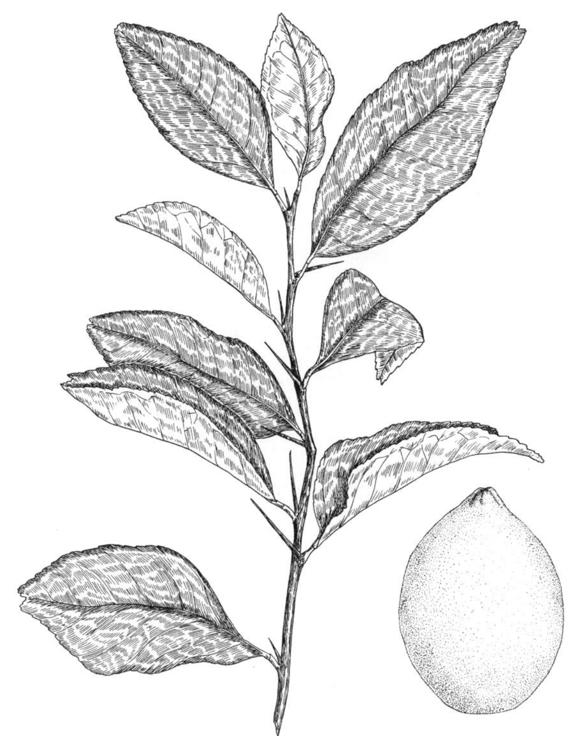

LEMON
Citrus limon (L.) Burm. f.

green, length 2–4½ in., elongate-ovate to elliptic or oval, apex acute to acuminate or obtuse, margin crenate or serrate, petioles wingless or sometimes narrowly margined, articulating with both the blade and the twig.

RANGE. Cultivated in southern Florida and the Florida Keys, southern California, and southern Texas. Escaping cultivation in frost-free areas. First introduced to the New World on the Island of Haiti. Origin not known with certainty. Probably from northern Burma and southern China. Brought from Palestine in about the thirteenth century.

VARIETIES AND HYBRIDS. The following hybrids of Lemon are recognized:

Lemandarin is a hybrid of *C. limon* and *C. reticulata.*
Lemonime is a hybrid of *C. limon* and *C. aurantifolia.*
Lemorange is a hybrid of *C. limon* and *C. sinensis.*
Eremolemon is a hybrid of *C. limon* and *Eremocitrus glauca.*

The following key to the varieties of the lemon has been proposed by H. J. Webber (Webber and Batchelor, 1943, pp. 594–595).

A. Fruit pulp very acid *(the acid lemons)*
  B. Fruit color light lemon-yellow; shape like typical lemon, elliptical, more or less drawn out at each end, nippled and necked.
    C. Tree open. spreading; branches and twigs few, coarse and rigid; foliage open; leaves dark green, rounded at apex or short-pointed . . . . Eureka Group

D. Tree and branches thornless.
Varieties: (1) *Eureka;* (2) *Genoa;* (3) *Wheatley.*

DD. Tree more or less thorny.
Varieties: (4) *Amerfo;* (5) *Belair;* (6) *Villa Franca.*

CC. Tree dense, upright; twigs and branches many, comparatively slender; foliage dense; leaves light-green, long-pointed .... Lisbon Group
Varieties: (7) *Bonnie Brae;* (8) *Kennedy;* (9) *Lisbon;* (10) *Messer;* (11) *Messina;* (12) *Sicily.*

BB. Fruit color light orange-yellow; shape more compressed and rounded and less drawn out at ends into nipple or neck .... Anomalous Group

C. Rind smooth and leathery
Varieties: (13) *Meyer.*
Rind rough and bumpy

CC. Varieties: (14) *Cuban;* (15) *Ponderosa;* (16) *Rough.*

AA. Fruit pulp sweet, lacking acid .... Sweet Lemon Group

B. Fruit the shape and size of typical lemons, nippled and necked.
Varieties: (17) *Dorsapo.*

BB. Fruit small, compressed, nearly spherical or obovate, not necked, and with low nipple mainly sunken in a deep apical depression.
Varieties: (18) *Millsweet.*

Varieties not included above: (19) *Everbearing,* (20) *Palestine Sweet;* (21) *Perrine;* (22) *Armstrong.*

REMARKS. The genus name, *Citrus,* is from the Greek word meaning "citron," and the species name, *limon,* is from the word for lemon. The lemon is very sensitive to cold. It is less hardy than the orange, but hardier than the lime. It is also less resistant to certain diseases than the orange or grapefruit. A fungus disease, Anthracnose, causes brown spots on the fruit, particularly during curing and after shipment. The lemon is gathered while green in color and about 2½ in. in diameter and ripened in curing houses.

In Florida the lemon is often grown on Sour Orange or Rough Lemon stock, according to soil and moisture conditions. It is sometimes grown on *Poncirus trifoliata* stock. The stock is usually budded some distance from the ground. The plantings are made 20–25 ft apart. Careful pruning is required to keep the trees low and compact. Lemon contains vitamins C, A, B, and P. The juice of lemon is much used for lemonade, the peel in cooking, and the oil extracted from it in cooking, perfumery, and art.

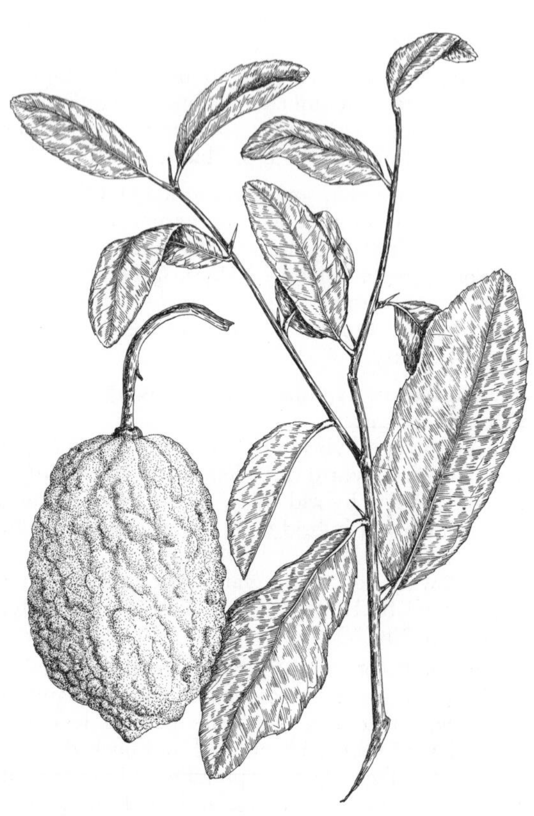

CITRON
Citrus medica L.

CITRON

Citrus medica L.

FIELD IDENTIFICATION. Cultivated citrus forming a shrub or small tree with long irregular branches. The twigs bearing short, stout, stiff thorns.

FLOWERS. Under favorable conditions more or less continually throughout the year, usually borne in short racemes in the leaf axils, large, reddish in the bud; when opening, large, white above, reddish purple beneath; petals 5, imbricate, thick, strap-shaped; stamens numerous, 30–40 or more, cohering toward the base; ovary large, swollen, cylindrical, gradually tapering into a thick, somewhat persistent style, 10–13-loculed (some flowers with an abortive pistil).

FRUIT. Shape lemon-like, oval to oblong, but much larger, yellow when ripe, rather fragrant; peel very thick, often rough and bumpy; flesh with small segments, vesicles pale greenish, acid or sweetish; seeds small, numerous, 9–10, pointed at base, smooth, ⅙–⅕ in. long and ⅛–⅙ in. wide.

LEAVES. Rather pale green, 3–7 in. long, 1–2 in. wide, ovate to elliptic or ovate to lanceolate, base cuneate to rounded, apex obtuse to rounded, margins serrate, surfaces glabrous, veins obscure above, more conspicuous beneath; petioles wingless or narrowly winged, not clearly articulated with the leaf base.

TWIGS. Rather dark green, younger angled, older ones becoming cylindrical, glabrous; spines axillary, short, solitary, stiff, stout.

RANGE. Sparingly grown in Florida and Texas. Sometimes escaping cultivation, especially on the Florida Keys. Introduced into the United States in 1894 by David Fairchild for the U. S. Department of Agriculture. Widely cultivated in tropical and subtropical regions. Origin not certain, possibly Arabia, China, or India.

PROPAGATION. Citron is more sensitive to cold than most citrus. It is usually propagated by cuttings, but it can be grafted on Rough Lemon or other stock. It is also used in propagating oranges in Spain. Methods of care are about the same as those for lemons.

VARIETIES. The sacred Jewish Citron, known as "Etrog" and used at the Feast of the Tabernacles, has fruit which is small, greenish yellow, ellipsoid to fusiform, externally rough, apex nippled. It is mostly grown on the Island of Corfu.

Another variety, known as the Fingered Citron, *C. medica* var. *sarcodactylis* (Noot.) Swingle, is unusual in having long finger-like lobes and very scanty pulp. It is sometimes grown as an ornamental.

The following key to the varieties of citron is by H. J. Webber (Webber and Batchelor, 1943, pp. 633–634):

A. Pulp sweet .... Sweet Citron Group
  Varieties: (1) *Corsican.*
AA. Pulp sour (the acid citrons)
  B. Fruits large, diam. 4–5½ in., height 5–7 in. .... Acid Citron Group
    C. Surface smooth or comparatively so
      D. Fruits somewhat lobed
        Varieties: (2) *Diamante.*
      DD. Fruits scarcely lobed
        Varieties: (3) *Earle.*
    CC. Surface rough and bumpy.
      Varieties: (4) *Cedressa.*
  BB. Fruits small, diameter 2–3 in., height 2¼–4 in. .... Anomalous Group
    C. Style persistent
      Varieties: (5) *Etrog.*
    CC. Style only partially persistent
      Varieties: (6) *China.*

REMARKS. The genus name, *Citrus*, is the Greek word for citron. The species name, *medica*, is for its once common medicinal use, especially by the ancient Greeks and Romans. It was probably introduced to the Mediterranean region about 300 B.C. The ancients considered it to be efficacious for curing rheumatism and sore mouth, but thought it to be inedible. Today the fruit is usually shipped in brine to the United States and Europe to be candied as a confection or used in cooking. Citron water is used also in flavoring liquors and Vermouth and for medical purposes.

GRAPEFRUIT

Citrus paradisi Macf.

FIELD IDENTIFICATION. Cultivated tree, round-topped, regularly branched, with dense foliage, and spineless, or with slender flexible spines.

FLOWERS. Axillary, large, white, borne singly or in small clusters; petals 5; calyx 5-lobed; stamens 20–25, with large linear anthers; ovary globose, sharply delineated from the style.

FRUIT. Very large, 4–6 in. in diameter, globose to oblate-spheroid or broadly obovate, smooth, pale lemon-yellow when ripe; peel ¼–½ in. thick, white and pithy within with numerous oil glands; segments 11–14, closed at the center, axis solid or semihollow; flesh fine-grained, acid, sweet, white, or reddish in some varieties; seeds usually numerous but fewer in some, variable in size (¼–⅝ in. long), flattened, whitened, asymmetrical, one end acute-pointed, the other rounded, polyembryonic.

LEAVES. Larger than those of Sweet Orange, elliptic to oblong or oval to ovate, apex obtuse, base broadly rounded or cuneate, margin crenate-serrate but often entire toward the base, length 2–5 in.; surfaces dark glossy-green, glabrous or nearly so; petiole ⅓–⅘ in. long, broadly winged at the leaf base; wing oblanceolate to obovate, apex rounded and touching the broadly rounded leaf base.

TWIGS. Green to brown, flattened and angular, often broadened at the nodes, puberulent at first but glabrous later.

GRAPEFRUIT
Citrus paradisi Macf.

RANGE. Introduced into Florida in the sixteenth century by the Spaniards. Now cultivated in Texas, New Mexico, Arizona, the West Indies, and the Orient. Origin uncertain; some propose that it originated in the West Indies, because no wild specimens have been found in Asia.

PROPAGATION. The budded trees bear in 3–4 years. Full bearing is reached in about 10 years. Trees are planted 18–25 ft apart. The seedlings are good grafting stock. Propagation is usually by budding because of variations. The varieties differ in susceptibility to scab, foot rot, and citrus canker. Commercial growing is usually below the frost line. The young plants are tender and the older ones more resistant.

HORTICULTURAL VARIETIES AND HYBRIDS. Grapefruit can be crossed with Tangerine, *C. reticulata,* to produce a fruit known as the Tangelo.

Another hybrid, between the grapefruit and the Trifoliate Orange, *Poncirus trifoliata,* produces the Citrumelo.

The following horticultural forms are listed by Kelsey and Dayton (1942, p. 253): (Imp. = Improved; Unc. = Uncolored; RC. = Red-colored):

Aurantium Unc., Orange; Cecily Unc.; Clayson Unc.; Conner; Davis; Desota Unc.; Duncan Unc.; Excelsior Unc.; Foster RC., Foster Pink-flesh; Hall Unc., Halls Silver Cluster; Henninger RC., Henningers Ruby; Imperial Unc.; Indian River Unc., Standard; Jordan Unc.; Josselyn Unc.; Leonardy Unc.; Manville Unc., Manville Imperial; Marsh Unc.; May Unc., Mays; McCarty Unc., Indian River; McKinley Unc.; Nectar Unc., Duarte Seedling; Pernambuco Unc.; Poorman Unc., Indian Pomelo; Royal Unc.; Thompson RC., Pink Marsh; Triumph Unc.; Walters Unc.; Webb RC., Webbs Redblush Seedless.

REMARKS. The genus name, *Citrus,* is from the Greek name, *citron,* first applied to the Citron tree *(C. medica).* The species name, *paradisi,* refers to the bountiful fruit. It is closely related to the Pummelo, *C. grandis,* and some botanists consider it as a variety of the Pummelo. The name Grapefruit was given because of the fruit's being borne in grapelike clusters of 3–12.

SWEET ORANGE

Citrus sinensis (L.) Osbeck

FIELD IDENTIFICATION. Cultivated medium-sized, round-topped tree, the branches regular and spiniferous or spineless.

FLOWERS. Medium-sized (smaller than those of Sour Orange), white in the bud, axillary, solitary, or in short racemes; petals 5, white, thick; stamens 20–25; ovary subglobose, distinct from the style; calyx 5-lobed.

FRUIT. Berry oval to flattened or subglobose, peel thin and tight, not bitter to taste, central axis solid, pulp sweet, segments 10–13; seeds variable in number or size, sometimes wanting, cuneate-ovoid, surfaces plane, edges margined, white within.

LEAVES. Oblong-ovate, 2–4½ in. long, apex obtuse to acute, or sometimes short-acuminate, base rounded, margin obscurely toothed or entire, light green to olive-green or dark green, lower surface paler, venation rather obscure; petiole narrowly winged, articulated with blade and twig.

TWIGS. Younger ones light green, older ones dark green to brown, angled, flattened at the nodes; spines slender, rather flexible, axillary, sometimes absent.

RANGE. Cultivated in Texas, California, and Florida. Considered to be a native of southeastern Asia, but it has been cultivated so long that it is no longer known in a wild state.

HYBRIDS AND VARIETIES. Sweet Orange hybridizes as follows:

Tangor is a hybrid of *C. sinensis* and *C. reticulata.*

Citrange is a hybrid of *C. sinensis* and *Poncirus trifoliata.*

Citrangor is a hybrid of *C. sinensis* and a cross of *C. sinensis* and *Poncirus trifoliata.*

Citrangequat is a hybrid of *C. sinensis* and a cross of *Poncirus trifoliata* and *Fortunella* Spp.

Citrangedin is a hybrid of a cross of *C. sinensis* and *Poncirus trifoliata* and a cross of *C. reticulata* var. *austera* and *Fortunella* Spp.

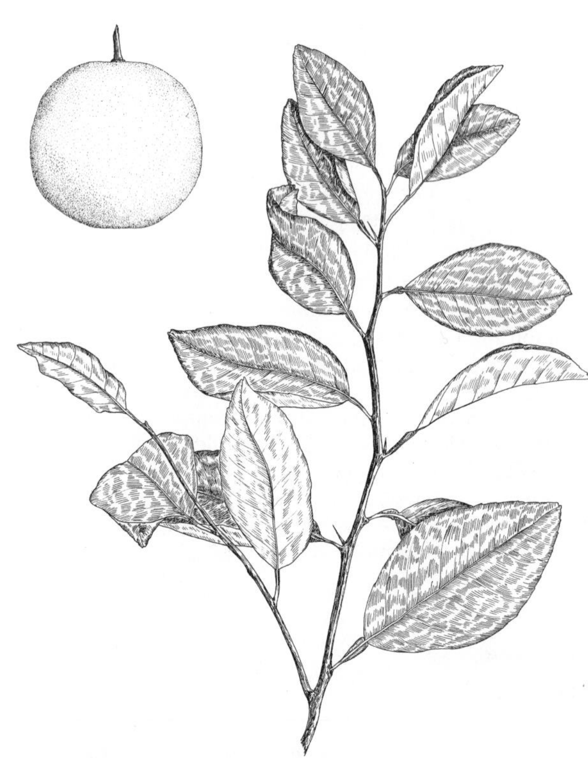

SWEET ORANGE
Citrus sinensis (L.) Osbeck

Citrangeremos is a hybrid of *C. sinensis, Poncirus trifoliata,* and *Eremocitrus glauca.*

The following key to varieties of Sweet Orange is quoted from H. J. Webber (Webber and Batchelor, 1943, pp. 502–503):

A. Fruits normal, not navels or bloods .... Normal Fruit Group
- B. Early-maturing varieties
  Varieties: Boone, Dillar, Early Oblong, Enterprise, Hamlin, Parson Brown, Sweet Seville, Centennial, Foster, Hicks, Trovita.
- BB. Midseason-maturing varieties
  Varieties: Acme, Arcadia, Conner, Circassian, Dummitt, Exquisite, Homosassa, Indian River, Jaffa, Joppa, Koethen, Majorca, Madam Vinous, Magnum Bonum, Marquis, Mays, Nonpareil, Old Vini, Osceola, Pineapple, Paperrind, Prata, Selecta, Shamouti, Star Calyx, Stark, Weldon, Whittaker, White.
- BBB. Late-maturing varieties
  Varieties: Bessie, Brazilian, Drake Star, Du Roi, Lamb Summer, Lue Gim Gong, Mediterranean Sweet, Maltese Oval, Valencia, Armstrong.

AA. Navel oranges (fruits that normally develop navels), no bloods .... Navel Fruit Group
- B. Glandular layer thick; oil glands large; oil abundant; primary glands with surface strongly to medium-depressed; secondary and tertiary glands with surface even or convex.
  - C. Navels mainly open, 50–100 per cent.
    Varieties: Australian, Carter, Golden Nugget, Texas.
  - CC. Navels largely closed, only 50 per cent or less open.
    Varieties: Washington, Surprise, Double Imperial.
- BB. Glandular layer thin; oil glands medium or small in size; oil little or medium; primary glands with surface medium-depressed; secondary and tertiary glands with surface even or depressed.
  - C. Navels mainly open, 50 per cent or more.
    Varieties: Thomson
  - CC. Navels mainly closed, less than 50 per cent open.
    Varieties:Navelencia, Buckeye.

Navel varieties not included above: Egyptian, Militensis, Parsons, Robertson, Sustain, Bahianinha, Summernavel.

AAA. Blood Oranges
Varieties: Egyptian Blood, Maltese, Ruby, Sanford, Saul, St. Michael.

REMARKS. The genus name, *Citrus,* is from the Greek word for citron, and the species name, *sinensis,* means "of China." Sweet Orange is rather tender, not as hardy as the Sour Orange, but much more cold-resistant than lemon or lime. It is grafted on stocks as are other citrus. The fruit is eaten fresh; the juice is used fresh or is canned; the peel is candied; or the fruit is used for making marmalade. An oil is also pressed from the peel. Far more Sweet Orange is grown than any other citrus. By the planting of early and late varieties a 12-month shipping season is possible in California. The two generally used are the Washington Navel and the Valencia.

MANDARIN ORANGE

Citrus reticulata Blanco

FIELD IDENTIFICATION. Cultivated small evergreen tree with slender branches.

FLOWERS. Small, borne singly or in clusters, white, perfect, regular; sepals 5, united into a cup-like calyx; petals 5, deciduous, strap-shaped, thick, imbricated in the bud; stamens numerous, cohering toward base in a few bundles, inserted around a disk; styles united and deciduous; ovary superior, several-celled.

FRUIT. Depressed-globose or subglobose, orange-yellow to reddish orange, smooth, variable in size, about 2¼ in. in diameter; peel thin and loose; flesh sweet, divisions 10–14 and loosely separating; seeds whitish, variable in size, numerous, about ¼ in. long, beaked, obovoid to ellipsoid, greenish within, endosperm absent, cotyledons fleshy.

MANDARIN ORANGE
Citrus reticulata Blanco

LEAVES. Persistent, simple, alternate, varying from narrowly to broadly lanceolate to elliptic, apex obtuse, some abruptly so, base cuneate, margin crenate-toothed, length 2¾–5 in., width ¾–1¾ in., glabrous on both sides, upper surface dull, dark green, lower surface paler with slender whitened veins more conspicuous; petiole with a wing very gradually narrowed to a short distance above the base.

TWIGS. Slender, dull green, glabrous, somewhat erratic in growth, flattened and ridged, especially widened toward the nodes.

RANGE. Cultivated in Texas, Florida, and California. Native to the Philippines, southeastern Asia.

HYBRIDS AND RELATED VARIETIES. The following hybrids are known:

Citrandarin is a hybrid of Mandarin and Trifoliate Orange.

Ichandarin is a hybrid of Mandarin and Ichang Bitter Orange, *C. ichangensis* Swingle.

Lemandarin is a hybrid of Mandarin and the Lemon.

Tangor is a hybrid of Mandarin and Sweet Orange.

Kumandarin is a hybrid of Mandarin and Kumquat.

Tangelo is a hybrid of Mandarin and Grapefruit.

Mandarin, *C. reticulata* Blanco, does not represent a pure stable species, but has many closely related citrus within its morphological boundaries.

The following key to groups of the Mandarin Orange was compiled by Webber (Webber and Batchelor, 1943, pp. 547–548):

I. Fruit sweet, high in sugar and very low in acid content.
  - A. Tree with few, rather coarse branches; leaves, flowers, and fruits comparatively large; petals 9⁄16–13⁄16 in. long.
    - B. Tree with upright branches; leaf petioles short, ¼–½ in.; petals reflexed, long, 9⁄16–11⁄16 in.; fruits medium large, seedy; rind thick, rough, bumpy; embryos white or slightly yellowish ...... King Group
      Varieties: (1) *King*.
    - BB. Tree with spreading branches; leaf petioles long, ½–1 in., petals only occasionally reflexed, ¾–13⁄16 in.; fruits medium in size, mainly seedless; rind smooth, thin; embryos light green .... Satsuma Group
      Varieties: (2) *Owari;* (3) *Wase;* (4) *Ikeda;* (5) *Zairai;* (6) *Mikado*.
  - AA. Tree with numerous, fine branches, bushy; leaves, flowers, and fruit small to medium; petals only occasionally reflexed, short, 7⁄16–½ in.; fruits seedy; embryos light green.
    - B. Fruits yellow to orange .... Mandarin Group
      Varieties: (7) *Emperor;* (8) *Oneco;* (9) *Willow-leaf*.
    - BB. Fruits deep orange to reddish............................ Tangerine Group
      - C. Fruits medium in size
        Varieties: (10) *Beauty;* (11) *Clementine;* (12) *Dancy;* (13) *Kinneloa;* (14) *Ponkan;* (15) *Trimble;* (16) *Weshart;* etc.
      - CC. Fruits small in size
        Varieties: (17) *Cleopatra;* (18) *Kinokuni*.

II. Fruit sour, low in sugar and high in acid content.
  - A. Tree with spreading, globose top; fruits averaging about 2⅛–2½ in. in diameter, closely resembling those of the mandarin in size and shape .... Mandarin-lime Group
    Varieties: *Rangpur* and *Kusaie* (both described as mandarin-limes).
  - AA. Tree with erect, columnar top; fruits much smaller, averaging about 1¼ in. in diameter.
    Varieties (19) *Calamondin* .... Mitis Group

REMARKS. The genus name, *Citrus*, is from the Greek word for citron. The species name, *reticulata*, refers to the fibrous inner peel and connective tissue. The loose rind makes it easy to peel.

CALAMONDIN

Citrus reticulata Blanco *sensu* Webber

FIELD IDENTIFICATION. Upright citrus tree attaining a height of 15–25 ft, with rather dense foliage and branches only slightly thorny. The following description is given by H. J. Webber (Webber and Batchelor, 1943, pp. 563–564):

FLOWERS. Similar to that of the Mandarin, calyx 5-tipped, small, persistent, frequently with a small areolar area.

FRUIT. Oblate to spherical, sometimes oblique, 1–1¼ in. long and 1–1⅝ in. in diameter, base flattened, even, or only slightly depressed, color orange, smooth and glossy, very finely pitted; rind thin, 1⁄16–⅛ in., loose, easily removed, tender and sweet; oil glands numerous, oil abundant, odor pleasant; segments 7–10; pulp orange-colored, tender, juicy, acid; seeds small, obovoid, plump, 0–5; chalazal spot brown; polyembryonic.

LEAVES. Similar to those of the tangerine, broadly oval, lighter green beneath, comparatively small, blades usually 1½–3 in. long; petioles short, ⅜–½ in. long, narrowly winged.

RANGE. Considered to be indigenous to the Philippine Islands. Cultivated in Texas, Florida, and California.

REMARKS. This citrus, once listed under the species name *C. mitis* Blanco, has been variously interpreted by pomologists in recent years. Walter T. Swingle (1942, pp. 25–26) regarded the Calamondin as a hybrid between *C. reticulata* var. *austera* Swingle and some species of *Fortunella*. Herbert J. Webber (Webber and Batchelor, 1943, p. 564), holding the view that distributional and temporal factors in the native ranges of the presumed parent taxa do not favor the explanation of hybridity, suggests that the Calamondin may have arisen by mutation from some mandarin type.

CALAMONDIN
Citrus reticulata Blanco *sensu* Webber

Known in Hawaii as Chinese Orange, and introduced to Florida as Panama Orange. In Texas it gives promise of becoming a good ornamental in coastal areas, developing a tall shapely, columnar form. The fruits are used to make marmalade and an acid drink. The tree apparently does not do so well in California. It is considered to be quite cold-resistant for a citrus, generally as hardy as the Satsuma, and developing a pronounced, deeply penetrating taproot.

KUMQUAT

Fortunella Spp.

FIELD IDENTIFICATION. Cultivated shrubs, sometimes persisting about abandoned fields or gardens. Dwarf and evergreen, some of the larger varieties to 15 ft, 8 ft wide, forming a dense symmetrical head. Some of the forms and varieties are so morphologically similar as to be scarcely distinguishable botanically.

FLOWERS. Often blooming several times during the year; singly or in clusters of 3–4; pedicels ¼–½ in. long, glabrous; flowers axillary, perfect, about ½ in. across, white, sweet-scented; petals 5, white, oblong, acute to obtuse, thick, waxy; calyx minutely and shallowly lobed; the numerous terminal anthers with very short filaments abruptly expanding into flattened erect appendages; stigma cavernous, style columnar, ovary globose and seated on a basal disc, 2 ovules in a cell.

FRUIT. Variable in size and shape according to the variety, oval or slightly longer than broad, ¾–1½ in. long, green to yellow or orange, skin thin to thick, tough; seeds 2–8, ⅜–½ in. long, somewhat flattened, ovoid to ellipsoid, rather unsymmetrical, ridged on one side and with a short flattened point, green in cross section.

LEAVES. Simple, alternate, evergreen, oblong to elliptic, apex acute to obtuse, base narrowly or broadly cuneate to rounded, margin entire, length 2–3 in., width ¾–1 in., firm and leathery, upper surface dull, dark green and glabrous, lower surface paler; petiole ¼–½ in. long, broadening into a narrow wing at the leaf base, sometimes with a short, sharp thorn at the base, or absent.

TWIGS. Young ones glabrous, dark green, flattened and grooved, older ones more terete and green to gray or brownish.

RANGE. Native of Asia, various species are cultivated

KUMQUAT
Fortunella Spp.

along the Gulf Coast from Florida to Texas and California.

PROPAGATION. Kumquat is desirable for cultivation because of the dwarf form, dark green leaves, and conspicuous orange fruit. It can be planted singly, or for hedges 6 ft apart, in rows 15 ft apart. It is ordinarily planted 10–15 ft apart. The plant should be fertilized heavily and pruned rather severely when fruit is removed for more abundant and larger fruits. It can be shield-grafted on almost any citrus stock, preferably *Poncirus trifoliata,* for hardiness.

HYBRIDS. The following hybrids of Kumquat with other citrus are known:

Citrangequat is a hybrid of the Citrange and *Fortunella* Spp. Varieties of this cross are known as Thomasville, Sinton, and Telfair. Citrange itself is a hybrid of *C. sinensis* and *Poncirus trifoliata.*

Citrumquat is a hybrid of *Poncirus trifoliata* (Trifoliate Orange) and *Fortunella* Spp.

Kumandarin is a hybrid of *C. reticulata* (Mandarin) and *Fortunella* Spp.

Limequat is a hybrid of *C. aurantifolia* (Lime) and *Fortunella* Spp.

Orangequat is a hybrid of *C. nobilis unshiu* (Satsuma) and *Fortunella* Spp.

The following Kumquats are cultivated:

Golden Hongkong Kumquat, *F. hindsi chinton;* Hongkong Kumquat, *F. hindsi* Swingle; Malay Kumquat, *F. polyandra* (Ridl.) Tan.; Meiwa Kumquat, *F. crassifolia* Swingle; Nagami Kumquat, *F. margarita* Swingle; and Marumi Kumquat, *F. japonica* Swingle.

REMARKS. The genus name, *Fortunella,* honors Robert Fortune, collector for the Royal Horticultural Society, London, who collected the first specimens in 1846 in the provinces of Foo-Choo-Foo, Chusan, and Mingpo, China. The fruit is in much demand for marmalade and jelly and is preserved in crystalline form. When eaten raw it is sweet, granular, and pleasantly acid.

Trifoliate Orange

Poncirus trifoliata (L.) Raf.

FIELD IDENTIFICATION. Green, aromatic, spiny tree to 30 ft, with stiff, flattened branches.

FLOWERS. Borne on the bare branches of old wood in spring, axillary, subsessile, white, spreading, 1½–2 in. across, perfect; petals 5, flat, thin, oblong-obovate to spatulate, at first imbricate but later spreading, base clawed, much longer than the sepals; sepals ovate-elliptic; stamens 8–10 free; ovary pubescent, 6–8-celled, ovules in 2 rows; style stout and short.

FRUIT. Berry September–October, yellow, aromatic, densely downy, globose, 1½–2 in. across; pulp thin, acid, rather sour; seeds numerous, large, taking up the greater part of the berry, ¼–½ in. long, flattened, obovate, rounded at one end and acute at the other, white to brown, smooth, often ridged or grooved on one side.

LEAVES. Trifoliate, about 3½ in. long or less; petiole broadly winged, ⅓–1 in. long or less; leaflets 3, aromatic, elliptic, oblong, oval or obovate; margin crenate-toothed; apex obtuse, rounded or emarginate; base often cuneate or rounded (in lateral leaflets often asymmetrical); sessile or very short-petioled; terminal leaflet 1–2½ in. long, laterals usually smaller; olive-green, glabrous, with pellucid dots above; lower surface paler, glabrous or puberulent along the veins.

TWIGS. Dark green, glabrous, divaricate, conspicuously flattened at the nodes to flare out into heavy, green, sharp, flattened spines ⅓–2¾ in. long.

BARK. Green or brown-streaked, smooth, often with thorns.

RANGE. Frequently planted for ornament and hedges from Texas to Florida, escaping cultivation in some areas. Hardy as far north as Washington, D. C. Native of China and Korea. Often planted for ornament in Japan.

HYBRIDS. Trifoliate Orange is often hybridized with other citrus to produce the following:

Cicitrange is a hybrid of the Citrange and *P. trifoliata.* (The Citrange itself is a hybrid of *C. sinensis* [Sweet Orange] and *P. trifoliata.*)

Citradia is a hybrid of *C. aurantium* (Sour Orange) and *P. trifoliata.*

Trifoliate Orange
Poncirus trifoliata (L.) Raf.

Citrandarin is a hybrid of *C. reticulata* (Mandarin) and *P. trifoliata.*

Citrange is a hybrid of *C. sinensis* (Sweet Orange) and *P. trifoliata.*

Citrumelo is a hybrid of *C. paradisi* (Grapefruit) and *P. trifoliata.*

Citrumquat is a hybrid of various *Fortunella* Spp. (Kumquats) and *P. trifoliata.*

Besides the uses of the Trifoliate Orange as a hybrid parent, it is often used as a stock to improve hardiness in other citrus particularly in Sweet and Satsuma oranges.

A well-known variety of Trifoliate Orange is Flying Dragon Trifoliate Orange, *Poncirus trifoliata* var. *monstruosa.*

REMARKS. The genus name, *Poncirus,* is from the French word *poncire,* a kind of citrus, and the species name, *trifoliata,* refers to the three leaflets. Vernacular names are Bitter Orange and Limoncito. Oil from the fruit has a rancid flavor and the pulp is aromatic.

LIMEBERRY

Triphasia trifolia P. Wilson

FIELD IDENTIFICATION. Cultivated shrub or small spiny tree 3–5 ft.

FLOWERS. Borne on peduncles ⅛–⅙ in. long, axillary, solitary or 2–3 together, fragrant, perfect; calyx-lobes 3, green, acute, 1/25–1/12 in. long; petals 3–4, thick, narrow, white, ½–⅗ in. long; stamens 6, free, filaments slender, glabrous, shorter than the corolla; style slender, elongate, shorter than the corolla, stigma 3-lobed and capitate; ovary ovoid to fusiform, 3–4-celled, ovules solitary; disk annular.

FRUIT. Resembling a miniature orange, ovate to ellipsoid or subglobose, sometimes apiculate, ⅖–⅗ in. long, dull reddish or orange, skin thick with many small oil glands; pulp mucilaginous and pulpy with a spicy taste; seeds 1–3 immersed in the flesh.

LEAVES. Three-foliate, short-peduncled or sessile, lateral leaflets usually smaller than the terminal one, blades ovate to elliptic or oval, ⅝–2 in. long, margin shallowly crenate or entire, apex retuse or obtuse to rounded, sometimes notched, base cuneate; a pair of straight sharp spines usually at the petiole base; petioles ⅛–⅕ in.; petiolules very short or absent.

TWIGS. Slender, light to dark green, terete, glabrous.

RANGE. Cultivated along the Gulf Coast from Florida westward to Texas, occasionally escaping cultivation in subtropical areas to persist in old fields and about abandoned gardens. The type specimen was collected in Java.

REMARKS. The genus name, *Triphasia,* is from the Greek word *triplex,* meaning "three-parted." The species name, *trifolia,* refers to the three leaflets. The plant is attractive when flowering and fruiting. It is sometimes planted for lawn specimens or for hedges, but suffers in freezes. Apparently it can stand considerable salinity in the soil. It is usually propagated from seed. Marmalade is made from the juicy, spicy fruit.

RUNYON ESENBECKIA

Esenbeckia runyoni Morton

FIELD IDENTIFICATION. Very rare small tree, known only from 4 specimens in the lower Rio Grande Valley. It has a rounded top and attains a height of 15 ft.

FLOWERS. In dense, pubescent panicles, hardly over 3 in. long, often blooming twice a year; flower peduncles pubescent, 1/12–⅛ in. long, subulate bracts at mid-center; corolla cream-white, about ⅓ in. broad; petals 5, oval to ovate, ⅙–⅛ in. long, glabrous, glandular dorsally; sepals triangular, pubescent, ciliolate, apex rounded, about 1/25 in. long; stamens 5, alternate with the petals; filaments about 1/12 in. long, glabrous; anthers oval, about 1/25 in. long; ovary 5-lobed, on a fleshy disk 1/12–⅛ in. broad; style short, glabrous, hardly over 1/25 in. long, stigma capitate.

LIMEBERRY
Triphasia trifolia P. Wilson

Runyon Esenbeckia
Esenbeckia runyoni Morton

FRUIT. A woody capsule borne on stout, pubescent peduncles 1½–3½ in. long, deeply 5-lobed (rarely 3–4-lobed), green at first, roughly rugose-tuberculate, about 1 in. long and 1½ in. wide, dehiscent into 5 carpels, each carpel with a 2-winged seed coat which twists upon drying to forcibly discharge the seeds a considerable distance; seeds ⅖–⅓ in. long, brown, rounded and asymmetrical, surmounted by a recurved beak.

LEAVES. Beautiful, laurel-like, leathery, trifoliate; blade 3–8 in. long, petiole pubescent to glabrous, 1–3 in. long; leaflets with petiolules ⅓–¼ in. long, blades 2–5 in. long, 1–2½ in. wide; elliptic-oblong, entire and occasionally wavy on the edge of the plane surface; apex obtuse, acute or rounded; base cuneate; upper surface dark green, lustrous, essentially glabrous; lower surface the same color or slightly paler, glabrous or somewhat pubescent along the veins.

TWIGS. Young twigs green and pubescent, older ones gray and glabrous.

BARK. Gray to black, smooth, tight.

RANGE. Known only from 4 trees on the banks of the Resaca Vieja, 3 miles northeast of Los Fresnos, Texas. Type specimen deposited in the United States National Herbarium, Sheet No. 1,438,940.

REMARKS. The genus name, *Esenbeckia,* commemorates Christian Gottfried Nees von Esenbeck (1776–1858), German botanist. The species name, *runyoni,* honors Robert Runyon, well-known Texas botanist and formerly mayor of Brownsville, Texas. It was first noticed on April 15, 1929, near Los Fresnos, Texas, by Harvey Stiles, who gave fruiting herbarium specimens to Runyon. Runyon visited the locality on July 8, 1929, and obtained flowering specimens. Both sets of specimens were sent to C. V. Morton at the United States National Herbarium, who described the tree as a new species (Morton, 1930).

Runyon has grown a beautiful specimen from seed in his garden at 812 St. Charles Street in Brownsville. The tree was observed to have many young seedlings springing up under it when last seen by the author in June of 1950.

Baretta

Helietta parvifolia (Gray) Benth.

FIELD IDENTIFICATION. Slender evergreen, thicket-forming shrub or small tree to 25 ft. Branches erect to form an irregular small crown. All parts pungently aromatic.

FLOWERS. Borne April–May in axillary panicled cymes 1–2½ in. long and 1–1½ in. wide, the slender primary pedicels pubescent, secondary pedicels about ¼ in. long, also pubescent; bracts very small, falling away early; corolla ⅛–³⁄₁₆ in. wide, petals 4, greenish white, about ⅛ in. long, ovate to oblong; apex acute

Baretta
Helietta parvifolia (Gray) Benth.

) obtuse, entire or slightly erose, four to five times s long as the calyx-lobes; calyx 4–lobed, lobes ovate, cute, puberulent; stamens 4, shorter than the petals, bout 1/25 in. long, filaments slightly flattened and glarous; pistil with united styles and stigma 3–4-lobed, vary depressed-ovoid, glandular, 3–4-lobed.

FRUIT. Maturing in October, oblong, of 3–4 dry carels which are winged on the back and indehiscent; vings about ⅜ in. long, about ¼ in. wide, rigid, broadvate, somewhat curved, rounded at the apex, deeply eticulate-veined.

LEAVES. Evergreen, mostly opposite, strongly aroıatic when crushed, trifoliate; terminal leaflet much ırger than the lateral ones, ½–2 in. long, about ½ in. vide, elliptic-oblong to narrowly obovate, sessile, entire n margin, or slightly crenulate, apex obtuse, rounded r slightly notched, base long-attenuate into a slightly ubescent and grooved petiole; lateral leaflets sessile, sually less than 1 in. long, oblong to elliptic or oboate, apex obtuse or slightly notched, base cuneate and essile; firm, leathery, puberulent at first but glabrous ıter, yellow green and lustrous above, paler beneath, ninute glandular-punctate, lateral veins inconspicuous, ungently aromatic.

TWIGS. Slender, terete, gray, brittle, pubescent when oung, glabrous later.

BARK. Gray, smooth and tight on young branches nd trunks, later gray or brownish and breaking into mall, thin scales on old trunks.

WOOD. Orange-brown to a whitish sapwood, closerained, hard, heavy, specific gravity about 0.88.

RANGE. On dry soil in extreme southern Texas (Hilalgo and Starr counties). In thickets near Rio Grande City. In Mexico from Nuevo León, Tamaulipas, and Coahuila to Querétaro.

REMARKS. The genus name, *Helietta,* is in honor of he French physician Lewis Theodore Helie (1804–867). The species name, *parvifolia,* refers to the small eaves. A thicket of trees is known as a "barretal" in outhwest Texas and Mexico.

Common Hop-tree
Ptelea trifoliata L.

FIELD IDENTIFICATION. Usually a rounded shrub, but occasionally a small tree to 25 ft. Leaves divided into 3 leaflets which are very unpleasantly scented. This s an extremely variable species in both form and habit of growth, and in the size and shape of the leaflets. Numerous varieties, forms, and species have been segregated by various authors. The entire genus *Ptelea* should be revised.

FLOWERS. March–July, polygamous, in terminal cymes; flowers small, borne on slender pedicels ¼–1½ in. long; calyx-lobes 4–5, obtuse, pubescent; petals 4–5, greenish white, oblong, somewhat puberulent, exceeding the calyx-lobes; stamens 4–5, alternating with the petals, filaments hairy, anthers ovoid; pistillate flowers with a raised pistil and abortive anthers; style short, sometimes glandular-dotted, stigma 2–3-lobed, ovary puberulous.

FRUIT. Ripening August–September, samaras borne in drooping clusters on slender reflexed pedicels; reticulate-veined, membranous, compressed, thin, wafer-like, suborbicular or obovate, ¾–1 in. across, persistent in winter, unpleasantly scented, 2–3-celled; seeds 2–3, oblong-ovoid, acute, leathery, reddish brown, about ⅓ in. long.

LEAVES. Alternate or opposite, trifoliate; leaflets sessile or nearly so, entire or finely serrulate, ovate-oblong, acute or acuminate, wedge-shaped at base with lateral leaflets often oblique, more or less pubescent and glandular below (in *P. trifoliata* var. *mollis* densely woolly beneath), darker green and lustrous above, unpleasantly scented, 4–6 in. long, 2–4 in. wide; petiole stout, base swollen, pubescent.

TWIGS. Slender, green, yellowish or reddish brown, pubescent, unpleasantly scented when bruised.

BARK. Thin, smooth, light to dark gray or brown, numerous excrescences, bitter to the taste.

WOOD. Yellowish brown, close-grained, hard, heavy, weighing about 51 lb per cu ft.

ROOT. Taste bitter, pungent, slightly acid, aromatic.

RANGE. Common Hop-tree, with its many varieties and closely related species, is distributed over a rather wide territory, growing in various types of soil; Texas, Louisiana, Arkansas, Oklahoma, and New Mexico; eastward

Common Hop-tree
Ptelea trifoliata L.

COMMON HOP-TREE
Showing leaf variation on the same plant

Five-leaf Common Hop-tree, *P. trifoliata* var. *pent phylla* (Moench.) DC., has 3–5 narrower leaflets.

Smooth Common Hop-tree, *P. trifoliata* var. *pube cens* Pursh, has branches, petioles, and inflorescen glabrous or nearly so, and leaflets somewhat pubesce beneath.

Golden Common Hop-tree, *P. trifoliata* var. *aure* Behnsch *ex* Hartwig, is a variety with yellow leave It is considered by some as only a form instead of variety.

Pyramid Common Hop-tree, *P. trifoliata* var. *fastigiat* Bean, has a pyramidal form with upright branches. is considered by some as only a horticultural form i stead of a variety.

At least 8 species of *Ptelea* have been described i the literature. However, the differences between som of the species appear to be indistinct. The specie with the most constant segregative characters are as fo lows:

Baldwin Hop-tree, *P. baldwinii* Torr., has somewha larger flowers in smaller corymbs; fruit 3/8–5/8 in. acros and apiculate; leaves 1–2½ in. long, narrowly ellipti ovate or oblong-ovate, pubescent beneath; branche dark brown. It occurs from central and southern Te to Florida, westward into Arizona and California, northward into Colorado, Utah, Nebraska, Minnesota, Michigan, Illinois, and continuing on east to Ontario, New York and Quebec.

MEDICINAL USES. A tincture of the root bark is occasionally used medicinally in tropical countries as a remedy for dyspepsia and as a mild tonic. Some Western physicians used it as a bitter stomachic. It contains the alkaloid berberine.

PROPAGATION. The green samaras may be picked by hand from the branches and spread out to dry. The cleaned seed averages 8,000–12,000 seeds per lb. Good commercial seed averages 98 per cent purity and 90 per cent soundness. If stored in sealed containers at 41°F. the vitality can be maintained for 16 months or longer. Pretreatment before planting is done by stratifying over the winter in sand or peat at 41°F. for 3–4 months. Planting should be done in good porous soil with medium moisture. Mulching for temperature control and partial shade for young seedlings helps. Spring sowing without stratification sometimes is practiced. The plant may also be propagated by budding, grafting, or layering.

The plant has some value as food for wild game. It is occasionally planted for ornament and is used in shelter-belt planting. It was first introduced into cultivation in 1724.

VARIETIES AND RELATED SPECIES. Woolly Common Hop-tree, *P. trifoliata* var. *mollis* Torr. & Gray, has branches, inflorescence, and leaflets rather densely tomentose. It appears to be the common variety in Texas and Louisiana.

BALDWIN HOP-TREE
Ptelea baldwinii Torr.

; to New Mexico, Colorado, Utah, northern California, ıd Arizona; also in Mexico.

Yellow Hop-tree, *P. lutescens* Greene, of Arizona, has hite, shiny, glabrous branches; leaflets 1¼–2½ in. ng, lanceolate, margin serrulate and surface minutely landular-dotted; fruit about ¾ in. across.

Shiny Hop-tree, *P. nitens* Greene, of Colorado and klahoma, has light yellowish brown smooth branches; uit about ⅝ in. across; leaves 1¼–2 in. long, rhombic lanceolate, apex acute to acuminate, margin slightly enate, surfaces light green, strongly glandular-dotted, ightly hairy or glabrous beneath.

Southwest Hop-tree, *P. polydenia* Greene, of west exas, New Mexico, Arizona, and Oklahoma, has leafts elliptic to ovate, apex acute or obtuse, margin crenate, leathery, surface bright green, strongly glandular-otted and short-hairy beneath; branches dark brown, ubescent to heavily tomentose.

Other species listed in the literature are difficult to istinguish from the species and varieties listed above. hey are *P. antoniana* Greene, *P. betulaefolia* Greene, *P. onticola* Greene, and *P. rhombifolia* Greene.

REMARKS. The genus name, *Ptelea*, is the classical ıme for the elm. The species name, *trifoliata*, refers the three leaflets. Vernacular names are Three-leaf op-tree, Shrubby-trefoil, Swamp Dogwood, Wafer-sh, Skunk-bush, Potatochip-tree, Quinine-tree, Ague-ark, Pickaway-anise, Prairie-grub, Cola de Zorillo, and Vingseed. All parts of the plant emit a disagreeable lor. The fruit was once used as a substitute for hops beer brewing.

EXAS DESERT-RUE

hamnosma texana (Gray) Torr.

FIELD IDENTIFICATION. Tufted plant with many wiry ems from the base, attaining a height of 8–16 in. he upper part of plant herbaceous but somewhat oody near the base.

FLOWERS. Appearing March–June, in loose bracted ıcemes along the upper part of the stems, solitary and orne on naked pedicels 1/25–1/12 in. long; flowers perfect, gular, campanulate, ⅛–⅕ in. long, less than ¼ in. cross, half-closed, yellowish to purplish; petals 4, ect, elliptic-oblong to semioval, obtuse, sometimes landular-dotted; stamens 8, shorter than the petals, aments glabrous and subulate, borne at the base of cup-shaped lobed disk; styles united, stigmas capitate, vary 2-celled and 2-lobed, ovules 5–6 in each cavity; alyx small, about 1/12 in. broad, with 4 obtuse sepals.

FRUIT. Capsule more or less stipitate, ⅕–¼ in. broad long, 2-lobed to near the middle, glandular-dotted, athery, 2-celled, opening at apex; seeds usually 4–6, ss than 1/12 in. long, broad, flat, testa tuberculate.

LEAVES. Numerous, scattered, deciduous, simple, al-rnate, linear to almost filiform, sessile or nearly so, argin entire, surfaces glaucous-green and glandular-otted, length ⅕–¾ in., strong-scented.

TEXAS DESERT-RUE
Thamnosma texana (Gray) Torr.

STEMS. Tufted at the base, when young grayish green and glandular-dotted, when older brown and somewhat woody toward the base.

RANGE. Usually growing singly or a few together on dry, poor, rocky soils at altitudes 2,000–4,500 ft. From central Texas through western Texas, New Mexico, Arizona, Colorado, and northern Mexico.

REMARKS. The genus name, *Thamnosma*, is from the Greek *thamnos* ("bush") and *osme* ("odor"), for the strong-scented foliage; the species name, *texana*, is for the state of Texas where it grows. It is also known under the vernacular names of Turpentine Bloom, because of the aromatic odor, or Dutchman's Breeches, for the oddly shaped fruit.

COMMON PRICKLY-ASH

Zanthoxylum americanum Mill.

FIELD IDENTIFICATION. Erect, much-branched, thicket-forming shrub or a small tree to 25 ft, with a diameter of 6 in.

FLOWERS. April–May, borne before the leaves on

wood of the previous year, in axillary cymes on slender pubescent pedicels, polygamo-dioecious, inconspicuous, very small, hardly over 1/12–1/10 in. broad, yellowish green, sometimes with minute reddish glands; calyx absent; petals 5, apex rounded, truncate or somewhat toothed; stamens 5, exserted, alternate with the petals; pedicels of pistillate flowers 1/25–1/4 in. long, shorter in the staminate; pistils 3–5, styles slender, distinct; ovules 2, collateral, pendulous, embryo straight, cotyledons broad.

FRUIT. Follicle maturing July–September, short-stipitate, green to reddish brown, strongly aromatic, subglobose to ellipsoid, 1/5–1/6 in. long, firm and fleshy, surface pitted, dehiscent at apex into 2 valves, the 1–2 seeds attached to a funiculus; seed about 1/8 in. long, seed coat black, smooth, shiny, crustaceous.

LEAVES. Alternate or somewhat clustered, 4–12 in. long, odd-pinnately compound of 5–11 leaflets, rachis pubescent or glabrous later; lateral leaflets sessile or nearly so, terminal leaflet short-petiolate and slightly winged by the leaf base, leaflets elliptic to ovate or oblong, apex mostly obtuse and slightly notched, base asymmetrical (except terminal one), margin crenate or some entire, length 3/4–3 in., width 3/8–1 1/2 in.; upper surface dull deep-green with veins impressed and pubescent at first but glabrous later; lower surface paler and pubescent, especially along the veins, lateral venation delicate and obscure; leaf rachis sometimes wit prickles on the back or unarmed.

COMMON PRICKLY-ASH
Zanthoxylum americanum Mill.

TWIGS. Smooth, reddish brown to gray, pubescent a first but glabrous later, unarmed, or with paired pe sistent spines 1/5–3/5 in. long, usually much flattened a the base and dark reddish to gray; bark of trun smooth, gray to brown.

WOOD. Light brown, soft, not strong, weighing 3 lb per cu ft.

RANGE. Open woods, rocky banks, or thickets; north eastern Oklahoma, Arkansas, Kentucky, and Georgia north to Missouri, Virginia, Minnesota, Nebraska, Kan sas, North Dakota, Ontario, and Quebec.

MEDICINAL USES. The bark and fruit are sometime used in the drug trade. The official drug is the drie bark known in extract form as "fluidextractum xantho xyli." The action of the drug is attributed to a pungen resin. It is irritant, producing a sense of heat in th stomach when swallowed and a tendency to diaphoresis It has been used as an internal remedy in rheumatism as a gastro-intestinal stimulant in flatulence and diar rhea, and as a masticatory in toothache, but at presen is rarely prescribed. The dose is 15–30 grains (1– gm.). Hercules-club Prickly-ash, *Z. clava-herculis* L has similar properties. (Wood and Osol, 1943, pp 1223–24)

PROPAGATION. The fruit is gathered by hand as soo as it begins to open and is spread out to dry. Th seeds are easily separated from the capsule by screen ing. Cleaned seed averages about 30,000 seeds per lb with a soundness of 85 per cent. It retains vitality stored at 41°F. for a year or two. Because of embryo dormancy, germination is hastened by stratification in sand for 120 days at 41°F. Sowing in fall immediately after gathering is sometimes practiced, as well as prop agation from root cuttings and suckers.

REMARKS. The genus name, *Zanthoxylum*, is from th Greek, meaning "yellow" and "wood," referring to th wood color. The species name, *americanum*, is for th tree's native home. Vernacular names in use are Angel ica-tree, Northern Prickly-ash, Toothache, Suterberry and Pepperwood. It is occasionally planted for orna ment and was introduced into cultivation about 1740 It has some use as a honey plant and as food for wild life, being eaten by bobwhite quail, red-eyed vireo pheasant, cottontail, and eastern chipmunk.

HERCULES-CLUB PRICKLY-ASH

Zanthoxylum clava-herculis L.

FIELD IDENTIFICATION. Small tree with a broad rounded crown, easily recognized by the corky-base prickles on the trunk and branches.

FLOWERS. Dioecious, greenish white, in large termina cymes; petals 4–5, oblong-ovate, obtuse, 1/8–1/4 in. long stamens 4–5, filaments slender and exserted, longe

HERCULES-CLUB PRICKLY-ASH *Zanthoxylum clava-herculis* L.

than the petals; pistils 2–3, styles short, with a 2-lobed stigma; calyx of 4–5 ovate or ovate-lanceolate, obtuse sepals.

FRUIT. Follicles 2–5 together, globose-obovoid, brownish, rough, pitted, apiculate, ⅙–¼ in. long, 2-valved; seed solitary, wrinkled, black, shining, persistent outside of follicle after dehiscence.

LEAVES. Alternate, 5–15 in. long, odd-pinnately compound of 5–19 leaflets, ½–4½ in. long, subsessile, sometimes falcate, ovate or ovate-lanceolate, acute or acuminate at apex, somewhat oblique and cuneate at base, crenate-serrulate, leathery, glabrous, lustrous above and more or less hairy below, spicy and dotted with pellucid glands, bitter-aromatic, stinging the mouth when chewed; petioles somewhat spiny, stout, hairy or glabrous.

TWIGS. Brown to gray, stout, hairy at first, glabrous later, often somewhat glandular, spinescent.

WOOD. Light brown or yellow, light, soft, close-grained, weighing 31 lb per cu ft.

BARK. Light gray, thin, covered with conspicuous, conelike corky tubercles.

RANGE. Texas, Louisiana, Oklahoma, and Arkansas; eastward to Florida and northward to Virginia.

MEDICINAL USES. It is sometimes used as a treatment for rheumatism. It has also been used as a gastrointestinal stimulant in flatulence and diarrhea, and as a masticatory in toothache, but at present it is little used.

REMARKS. The genus name, *Zanthoxylum,* comes from an erroneous rendering of the Greek word *xanthos* ("yellow") plus *xylon* ("wood"); the species name, *clava-herculis,* means "club of Hercules" and refers to the trunk's thorny character. Vernacular names are Toothache, Sea Ash, Pepperwood, Prickly Yellowwood, Yellow Prickly-ash, Tongue-bush, Rabbit Gum, Wild Orange, Sting Tongue, Tear Blanket, Pillenterry, and Wait-a-bit. A number of species of birds eat the fruit.

TEXAS HERCULES-CLUB PRICKLY-ASH

Zanthoxylum clava-herculis L. var. *fruticosum* Gray

FIELD IDENTIFICATION. Thorny shrub 3–15 ft, aromatic in all parts.

FLOWERS. Borne in early spring, dioecious, in cymes ⅓–2 in. long, 1–1½ in. wide; pedicels 1/12–⅛ in. long; sepals 5, minute, linear to subulate, acute; petals 5, greenish, elliptic, concave, about 1/12 in. long; stamens 5, exserted on filiform filaments about as long as the petals or shorter, but shorter than the anthers, wanting or rudimentary in the pistillate flowers; pistils 2–3, ovary sessile, style short, stigma entire or slightly 2-lobed.

Texas Hercules-club Prickly-ash
Zanthoxylum clava-herculis L. var. *fruticosum* Gray

FRUIT. Capsule borne in clusters ⅓–2 in. long on red, pubescent pedicels ¼–½ in., body of fruit subglobose, asymmetrical, glandular-dotted, apiculate, green at first, reddish brown later, about ¼ in. long, splitting into valves; seed black, shiny, obliquely ovoid, persistent to one valve.

LEAVES. Odd-pinnately compound, 1½–2½ in. long (more rarely to 4 in.); rachis red, pubescent, bearing reddish brown, straight, sharp thorns to ¼ in. long; leaflets 3–7 (usually 5), ½–1½ in. long, elliptic to oblong or oval, leathery, aromatic, glandular-dotted especially on the margin, crinkled, crenate, apex obtuse, base cuneate, lustrous above, dull beneath; lateral leaflets short-petioluled or sessile, the petiolule of the terminal leaflet longer.

TWIGS. Young ones greenish, pubescent; older ones gray and armed with stout, straight, gray or brown spines to ½ in. long.

BARK. Smooth, mottled light to dark gray; spines straight, or slightly curved, to 1 in. long, not built up on conspicuous corky bases as in *Z. clava-herculis.*

RANGE. On sandy or gravelly soil of central west Texas. Sandy areas south of San Antonio, also between Utopia and Tarpley. Reported from Arkansas.

REMARKS. The genus name, *Zanthoxylum,* comes from an erroneous rendering of the Greek word *xanthos* ("yellow") plus *xylon* ("wood"); the species name, *clava-herculis,* means "club of Hercules"; and the variety name, *fruticosum,* is for the shrubby habit.

Lime Prickly-ash

Zanthoxylum fagara (L.) Sarg.

FIELD IDENTIFICATION. Aromatic, thorny, evergreen shrub, or a tree to 30 ft, with a diameter of 10 in. The trunk is usually inclined, and the crown rounded or spreading with intricate branches.

FLOWERS. Borne mostly March–June on branchlets of the previous year, generally appearing after the leaves; staminate and pistillate on different trees; cymes axillary, small, cylindrical; bracts minute, ovate, obtuse, deciduous; flowers on short pedicels, small, hypogynous; sepals 4, membranous, yellowish green, triangular to ovate, acute, shorter than the petals; petals 4, yellowish

Lime Prickly-ash
Zanthoxylum fagara (L.) Sarg.

SMALL PRICKLY-ASH
Zanthoxylum parvum Shinners

green, oblong to ovate, concave, 1/12–1/8 in. long, about twice as long as the sepals; stamens 4, alternate with the petals, filaments slender and exserted; pistils 2 (rudimentary in staminate flower), style slender, stigmas capitate and obliquely spreading.

FRUIT. Ripening in September, follicle globose or obovoid, 1/8–1/4 in. long, rusty brown, pitted, warty, rugose, thin-walled, splitting at maturity; seed persistently attached to the follicle lip, solitary, subglobose, black, smooth, shiny, hilum rather conspicuous, cotyledons oval and foliaceous.

LEAVES. Alternate, evergreen, odd-pinnately compound, 3–4 in. long, with 5–13 leaflets (usually 7–9) on a broadly winged rachis; leaflets opposite, 1/3–1 1/4 in. long, sessile or nearly so, oval to obovate, apex rounded or notched, base cuneate or rounded, margin bluntly crenulate-toothed mostly above the middle, slightly revolute, coriaceous, young leaves often bronze-green; older with upper surface bright green, shiny and glandular-punctate, lower surface somewhat paler, aromatic and bitter to taste because of pellucid glands; petiole winged and jointed.

TWIGS. Numerous, zigzag, slender, smooth, gray to brown, glabrous to puberulent, set with hooked stipular spines; buds minute, globular, dark brown, woolly; leaf scars deltoid, vascular bundles with 3 scars, pith white and not interrupted.

BARK. Gray, thin, smooth or with warty excrescences; older trunks with small, thin, appressed scales, aromatic.

WOOD. Reddish brown, sapwood yellowish, compact, hard, heavy, close-grained, specific gravity about 0.74.

RANGE. Southern Florida and southwestern and coastal Texas. Extending eastward along the Texas coast to Harris and Galveston counties. Abundant in the Texas lower Rio Grande Valley area. In Mexico in the states of Tamaulipas, Sonora, Veracruz, Yucatán, Chiapas, and Baja California.

REMARKS. The genus name, *Zanthoxylum,* is from the Greek and means "yellow wood." The species name, *fagara,* is the old generic name. Vernacular names in use in various countries are Colima, Limoncillo, Uña de Gato, Palo Mulato, Espino, Espino Rubial, Corriosa, Tomeguín, Xic-ché, and Wild Lime. It is generally known as "Colima" to the Mexican people of the Texas Rio Grande Valley area. The plant has long been used medicinally in the Latin-American countries. Various extracts of the bark and leaves are taken as a sudorific and nerve tonic. The powdered bark and leaves are used as a condiment, and are also said to produce a yellow dye.

SMALL PRICKLY-ASH

Zanthoxylum parvum Shinners

FIELD IDENTIFICATION. Shrub to 4 1/2 ft, with spines erect or curved and 1/5–1/2 in. long at the nodes. Evidently closely related to *Z. americanum.*

FLOWERS. Inflorescence sessile or nearly so, umbellate, before the leaves mature. Flowers 2–12, pistillate pediceled, 1/12–1/6 in. long, scarcely pilose; sepals absent, petals 4, elliptic to ovate-oblong, 1/15–1/12 in. long, green, apices reddish-bearded; ovary thick, stipitate; carpels 2–4, connate, short tomentose; styles slender, becoming twisted, almost connate, at length free; stigmas short and subclavate; fruit not seen.

LEAVES. Leaflets 7–9, broad elliptic to ovate or lanceolate, apex obtuse, margin crenulate, base broadly to narrowly cuneate, both sides pilose, 1/4–1/2 in. long, sessile or nearly so.

TWIGS. Young ones reddish brown with a few long gray hairs, later dark gray to black and glabrous, rather erratic or crooked, stiff, irregularly roughened by leaf scars and nodes; spines either solitary or in pairs, at the nodes, mostly straight, slender, sharp-pointed, reddish brown to dark gray.

RANGE. The holotype was collected in rocky igneous

hills above Limpia Creek near Wild Rose Pass, about 15 miles northeast of Fort Davis, Jeff Davis County, Texas, by Rogers McVaugh, No. 7890, April 10, 1947 (in the herbarium of Southern Methodist University). High slopes under north-facing cliffs, branch canyon to east, just above pass. The specimens show flowers with very young leaves.

Dr. Lloyd H. Shinners, of Southern Methodist University, also makes the following comments concerning the plant:

"Evidently representing the same plant in sterile specimen in leaf, also from the Davis Mountains. 'Frequent in dense oak shinnery along trail to Tricky Gap, Buffalo Trail Scout Range, 5,500 [ft.],' *Barton H. Warnock & B. L. Turner* [sheet No.] *8089*, August 8, 1948 (SMU). The plant is evidently rare and of erratic flowering habits. Dr. Warnock informs me that Sul Ross State College Herbarium has only three specimens possibly belonging here, all sterile." (Shinners, 1956, p. 19)

REMARKS. The genus name, *Zanthoxylum*, is from the Greek *xanthos* ("yellow") and *xylon* ("wood"). The species name, *parvum*, refers to the shrub's small size.

QUASSIA FAMILY (*Simarubaceae*)

TREE-OF-HEAVEN
Ailanthus altissima (Mill.) Swingle

FIELD IDENTIFICATION. Cultivated tree attaining a height of 100 ft, and a diameter of 3 ft. Handsome and rapid-growing with a symmetrical open head and stout branches.

FLOWERS. April–May, borne in clusters of 1–5 in large, loose, terminal panicles 6–12 in.; pedicels subtended by small bracts or none; staminate and pistillate panicles on different plants or polygamous; flowers small, ⅕–⅓ in. across, yellowish green; staminate flowers unpleasantly scented; calyx regular, sepals 5, valvate in the bud, oval to oblong, spreading, ⅛–⅙ in. long, villous near the base, inserted at the base of the small 10-lobed disk; stamens 10 (in perfect flowers), staminate flowers with or without a rudimentary pistil; pistillate flowers smaller than staminate with 2 or 3 imperfect stamens or none; ovary deeply 2–5-cleft, the lobes flat and cuneate, ovules solitary in each cavity.

FRUIT. September–October, in persistent clusters of 1–5, samara linear-elliptic; ½–1½ in. long; flattened, thin, membranous, veiny, dry, twisted at the apex, sometimes curved, notched on one side, brownish red, the single flattened seed in the center, albumen thin.

LEAVES. Alternate, deciduous, odd-pinnately compound, length 8 in.–2½ ft, rachis pubescent or glabrous, unpleasantly odorous when bruised; leaflets 11–41; petiolules ⅛–⅓ in., pubescent or glabrous, leaflets ovate to oblong or lanceolate, sometimes asymmetrical, apex acute or acuminate, base cordate or truncate and often oblique, margin entire except for 2–4 coarse, glandular teeth at the base, length 2–5 in., upper surface dull dark green, glabrous or slightly hairy, lower surface paler and glabrous or with a few hairs on the veins; petiole swollen at base.

TWIGS. Coarse, blunt, yellowish orange or brown, younger ones pubescent, older ones glabrous, leaf scars large and conspicuous, pith reddish.

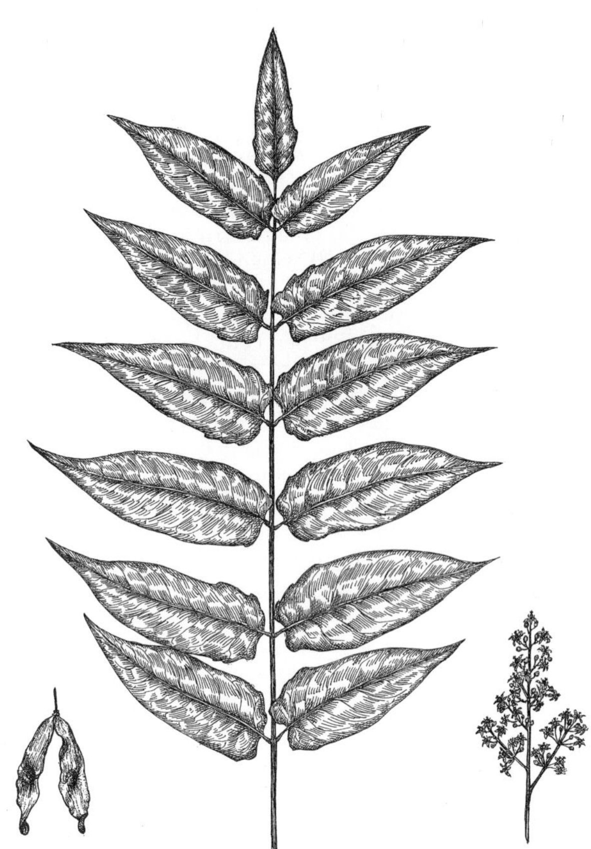

TREE-OF-HEAVEN
Ailanthus altissima (Mill.) Swingle

BARK. Pale grayish brown, fissures shallow.

WOOD. Pale yellowish brown, medium hard, not durable, weak, coarse, open-grained, said to make fairly good fuel, and sometimes used in cabinet work.

RANGE. A native of China, cultivated for ornament in the United States, and sometimes escaping cultivation in our area. Very hardy, seemingly growing well under adverse conditions of dust, smoke, and poor soil. Often found in waste places, trash heaps, vacant lots, cracks of pavement, crowded against buildings, and other situations. Does best on light, moist soils. In Texas, New Mexico, Oklahoma, Arkansas, and Louisiana; east to Florida and north to Massachusetts. Also cultivated westward throughout the interior to the Pacific Coast.

MEDICINAL USES. In China the powdered bark has been used successfully against tapeworm, and is also a popular remedy for dysentery. The powdered bark is greenish yellow, of strong nauseating odor, and bitter to the taste. When chewed it is said to produce nausea, vertigo, weakness, shivering, and cold sweats. It contains a bitter, fluorescent substance known as "ailanthin" and also a saponin.

PROPAGATION. Tree-of-heaven is often grown for its handsome foliage and clusters of reddish fruit. It appears to thrive under very unfavorable conditions of poor soils, dust, and smoke. For this reason it is often used for urban street planting. In most cases the female plant is used because the male flowers emit a disagreeable odor, which is said to cause catarrhal trouble. The seed is usually picked by hand and spread out to dry. It is fanned or flailed to remove the wings. From 30–90 lb of clean seed may be obtained from 100 lb of fruit. The seed averages 14,600 winged seeds per lb (some low lots 12,700 and some high 16,500), or 17,000 wingless. The germination is about 60 per cent and the average purity is 88 per cent. Seeds keep better if dry-sealed at a temperature of 40°F. Germination is hastened by stratification in moist sand for 60 days at 41°F. The seed may be sown in spring in drills, or broadcast, and covered with about ½ in. of firmed soil. About 26 per cent of the seedlings are usable. Propagation by root cuttings is sometimes practiced. Very vigorous shoots can be obtained by severely cutting back each year after the plant is 4 years old.

VARIETIES AND SYNONYM. The following varieties are listed by Rehder (1940):

Red-fruit Tree-of-heaven, *A. altissima* var. *erythrocarpa* (Carr.) Rehd., has leaves darker green above and more glabrous beneath; fruit bright red. Cultivated since 1867.

Weeping Tree-of-heaven, *A. altissima* var. *pendulifolia* (Carr.) Rehd., has large drooping leaves. Cultivated since 1900.

Szechwan Tree-of-heaven, *A. altissima* var. *sutchuensis* Rehd. & Wilson, has branchlets reddish brown, glabrous, lustrous; petioles purplish and glabrous; leaflets not ciliate, cuneate at base, at least on older plants; fruit about 2 in. long. Cultivated since 1895.

Besides the above, about nine other species are native and cultivated in central and southern Asia and Australia.

A common synonym is *A. glandulosa.*

REMARKS. The genus name, *Ailanthus,* is from a Chinese name, *Ailanto,* meaning "tree-of-heaven," and referring to its height. The species name, *altissima,* means "very tall." Vernacular names for the tree are Copal, Tree-of-the-Gods, Chinese Sumac, Heavenward-tree, False Varnish-tree, and Devil's Walkingstick.

The tree was introduced into the United States from China by William Hamilton in 1784. It is a very rapid grower, spreads freely from suckers and by seed, is rather free of insects but is easily storm-damaged. It was once used in erosion control work in the dune areas of the Black Sea, and was planted for timber in New Zealand. The seeds are known to be eaten by a number of birds, including the pine grosbeak and crossbill, and occasionally browsed by white-tailed deer. The tree is sometimes planted in China as host to a species of silkworm, *Attacus cynthia,* which produces a coarse inferior silk.

ALLTHORN CASTELA

Castela texana (Torr. & Gray) Rose

FIELD IDENTIFICATION. Densely branched, spinose shrub 3–10 ft.

FLOWERS. Polygamo-dioecious, axillary, solitary or clustered, sessile or nearly so, small, about ⅛ in. long, red or orange; petals 4, narrowly obovate; sepals 4, distinct or nearly so; stamens 8, hairy, on a disk; ovary 4-lobed.

FRUIT. Drupe solitary or in clusters of 2–4, brilliant red, sessile or short-stipitate, round, flattened, ¼–⅓ in. long; seed solitary, rugose, stony.

LEAVES. Alternate, often fascicled at the nodes, simple, entire and revolute, linear, oblong, or narrowly obovate, apex obtuse or mucronulate, base gradually narrowed, thick, shiny green above, densely silvery pubescent beneath, blades ¼–1 in. long; short-petioled or nearly sessile, petioles pubescent.

TWIGS. Mottled grayish white, divaricate, rigid, young

ALLTHORN CASTELA
Castela texana (Torr. & Gray) Rose

twigs grayish green, glabrous, ending in slender, sharp, straight spines which give rise to numerous lateral spines. Shrub thicket-forming.

RANGE. Central, southwestern, and western Texas. On rocky banks of the Rio Grande near Alamo, Hidalgo County; shores of the Gulf near Riviera, Kleberg County; also in the vicinity of Austin and San Antonio. In Mexico in Coahuila and San Luis Potosí.

MEDICINAL USES. It is a popular medicinal plant among the Mexican people. Extracts of the bark are used as a remedy for intestinal disturbances, skin diseases, fever, yellow jaundice, and dysentery; also as a tonic. In the treatment of amoebic dysentery it has been found that a fluidacetextract in the proportion of one part in a million is sufficient to render *Entamoeba histolytica* immobile.

REMARKS. The genus name, *Castela,* is in honor of Pierre Louis Richard Castel (1758–1832), French naturalist and poet. The species name, *texana,* refers to the state of Texas, where it grows. Vernacular names are Bisbirinda, Amargosa, and Goat-bush. The Spanish name Amargosa denotes the bark's bitter character.

Stewart Crucifixion-thorn

Holacantha stewartii C. H. Muller

Stewart Crucifixion-thorn
Holacantha stewartii C. H. Muller

FIELD IDENTIFICATION. Low shrub 5–24 in., usually broader than high, procumbent or somewhat ascending. Soon leafless; the stems, and stout spines, divaricately branched.

FLOWERS. Dioecious, sepals 6, ovate, acute, pubescent, about 1/25 in. long; petals 6, concave, fleshy, narrow, dorsally pubescent, about 1/6 in. long, 1/25–1/12 in. broad; stamens usually 12, filaments 1/15–1/12 in., broadened at base, hairy; pistillate calyx similar to the staminate; pistillate corolla not seen; staminodes similar to the functional filaments; carpels distinct, stigmas sessile and fused.

FRUIT. Drupes persistent for 1–2 years, individually sessile, but all divaricate from a central, short, densely pubescent peduncle about 1/8 in.; lenticular-ovate, green to tan or brown, flattened, apices obtuse or acute, somewhat asymmetrical, ventral margins obtusely ridged, 1/3–2/5 in. long, 1/5–1/4 in. broad, glabrous, nearly smooth, 1-seeded.

LEAVES. Buds small in axils of the spines, somewhat pubescent; leaves early-deciduous, blades 1/5–1/3 in. long, 1/12–1/10 in. broad, oblong, ends acute, base sessile, at first red and densely hirsute, later green and sparsely hirsute.

TWIGS. Forming a mass of divaricate thorns, grayish green to olive-green, branches terete, younger ones pubescent or appressed-sericeous, when older more glabrate; spines 1–4¾ in. long, 1/10–1/8 in. thick, tips subulate and brown.

RANGE. Deep heavy silt flats or rocky arroyos. In Texas known only from Brewster County—Big Bend National Park, 1.7 miles west of Sublett Ranch. In Coahuila, Mexico, Sierra de las Cruces, gulch in limestone hills 0.5 miles north of Santa Elena. In Zacatecas, Mexico, banks of arroyos in foothills, Hacienda de Cedros.

REMARKS. The genus name, *Holacantha,* is from the Greek words meaning "wholly" and "thorn," alluding to the spiny branches throughout. The species name, *stewartii,* is in honor of Robert M. Stewart of Santa Elena, Coahuila, Mexico.

C. H. Muller (1942) gives the following account of this new species:

"This species is the second described in this rare and striking genus, the first being *Holacantha emoryi* which is confined to Arizona and California. From that species *Holacantha stewartii* differs in its low sprawling habit, the sparse pubescence of the stems (comparing with the densely short-tomentose stems of *Holacantha emoryi*), the persistence of papillaelike hair bases, and its usually markedly acute and ventrally ridged carpels. The great discrepancy in the ranges of the two endemics further attests to their distinctness. . . . Its procumbent habit serves to bind the soil and form hillocks down the sides of which branches sprawl."

MAHOGANY FAMILY (*Meliaceae*)

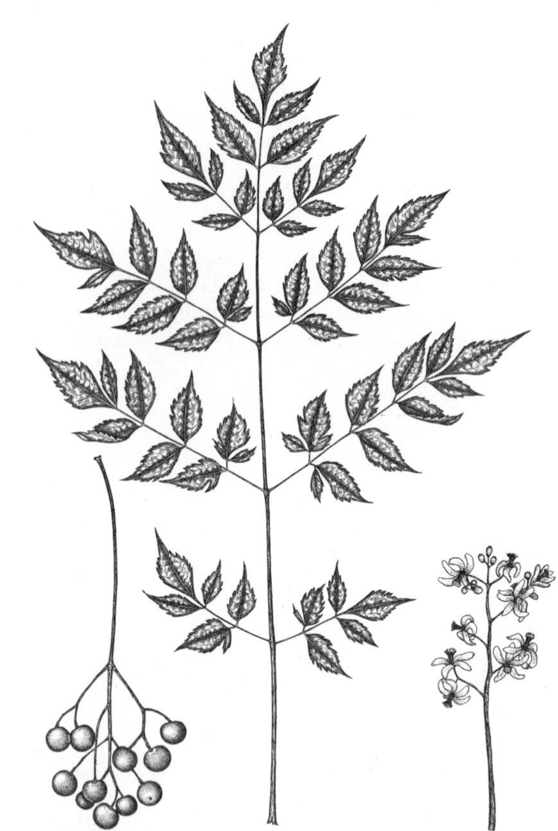

China-berry
Melia azedarach L.

China-berry
Melia azedarach L.

FIELD IDENTIFICATION. Tree to 45 ft; rounded crown; twice-compound leaves sometimes 2 ft long.

FLOWERS. March–May, panicles 4–6 in., loose, open, fragrant, showy, individual flowers about ½ in. across; sepals 5–6, lobes acute; petals 5–6, purplish, oblanceolate to narrow-oblong, obtuse; staminal tube with 10–12 stamens and sagittate anthers; ovary 5-celled, style elongate, stigma 3–6-lobed.

FRUIT. Ripening September–October, persistent, drupe ½–¾ in. in diameter, subglobose, coriaceous, fleshy, translucent, smooth, yellow, indehiscent, borne in conspicuous drooping clusters; stone ridged, seeds 3–5, smooth, black, ellipsoid, asymmetrical, acute or obtuse at ends, dispersed by birds or mammals.

LEAVES. Large, alternate, deciduous, twice-compound, to 25 in., long-petioled; leaflets numerous, ovate-elliptic, serrate or lobed, some entire, acute at apex, 1¼–2 in. long, mostly glabrous.

WOOD. Color variegated, durable, somewhat brittle.

RANGE. A native of Asia, introduced into the United States as an ornamental, and escaping to grow wild over a wide area. Texas; east to Florida, and north to Oklahoma, Arkansas, and North Carolina.

PROPAGATION. The fruit is collected by hand in the fall and can be planted as is, or the pulp can be removed by a macerator and by screening. The fruit averages about 640 seeds per lb. The seed can be dry-stored at least a year and retain its viability. It is not dormant and can be planted either in fall or spring. The average germinative capacity is about 65 per cent, and germination occurs about 3 weeks after planting. Nursery beds are prepared for planting in drills with the seeds sown 2–3 in. apart and covered with 1 in. of soil firmed down. One-year-old plants can be transplanted. Often necessary to top and root-prune old stock. Also propagated by cuttings and root suckers.

REMARKS. The genus name, *Melia*, is an old Greek name. The species name, *azedarach*, is from a Persian word meaning "noble-tree." Vernacular names are China-tree, Bead-tree, Indian-lilac, and Pride-of-India. The fruit is eaten by birds and swine, but if fermented it sometimes has a toxic effect. The fruit pulp is also used as an insect repellent and vermifuge. In some countries the hard seeds are made into rosaries. The wood was formerly used in cabinet work. China-berry grows rapidly, is rather free of insects, but cannot stand excessive droughts or much cold below zero. It is desirable as a shade tree and was formerly much planted in the South. The northern limit of hardiness is probably Virginia. Besides the species, there occurs an umbrella-shaped variety known as the Texas Umbrella China-berry, *M. azedarach* forma *umbraculiformis* Berckm., which was reported to be found by botanists originally near San Jacinto Battlefield, Houston, Texas.

MALPIGHIA FAMILY (*Malpighiaceae*)

Slender Janusia

Janusia gracilis Gray

FIELD IDENTIFICATION. Low, slender, twining or trailing perennial. The stems diffuse and woody toward the base.

FLOWERS. Flowers small, reddish yellow, petals 5. In axillary or terminal cluster, peduncles slender, usually with dichotomously minute bractlets toward the center; flowers either normal or abnormal; normal entire to denticulate, abruptly contracted into long claws; calyx with 8–10 glands; style 3-angled, ovaries 3; abnormal flowers often with rudimentary petals, no glands on calyx, style absent and 2 ovaries; stamens usually 5–6, perfect, or some with no anthers.

FRUIT. Samaras in clusters of 2–3, green to brownish, 1/3–1/2 in. long; wings divergent, asymmetrical and oblique, oblong to broad-lanceolate, surfaces with long appressed hairs or glabrous.

LEAVES. Simple, opposite, linear to lanceolate, apex and base mostly acute, margins entire, some with 2–3 toothlike glands at the base, both surfaces silky with long appressed hairs affixed by the middle, blade length 1/2–1 in., width 1/12–3/16 in., short-petioled.

TWIGS. Reddish brown with dense white appressed hairs; later dark brown to grayish and more glabrous, breaking into very small scales on old wood.

RANGE. Dry hills, canyons, or hillsides at altitudes of 1,000–5,000 ft. In Texas from the Colorado River westward into New Mexico, Arizona, and northern Mexico.

REMARKS. The genus name, *Janusia,* is in honor of Janus, the old Roman God of 2 faces (perhaps referring to the dimorphous flowers). The species name, *gracilis,* refers to the slender stems.

Slender Janusia
Janusia gracilis Gray

Barbados-cherry Malpighia

Malpighia glabra L.

FIELD IDENTIFICATION. Erect shrub with many slender stems from the base, attaining a height of 9 ft.

FLOWERS. In short, axillary umbels or cymes, 3–7-flowered, pedicels about 1/4 in., slender, articulate, vis-

BARBADOS-CHERRY MALPIGHIA
Malpighia glabra L.

cid, green to reddish-spotted, glabrous, or a few white appressed hairs, bracteolate; flowers perfect, regular, corolla pink, ½–¾ in. wide, slightly fragrant; sepals 5, ovate or oblong, acute, erect and incurved at maturity, viscid, green to reddish, 1/16–1/12 in. long, each with 1–2 large thick, oblong glands at the base; petals 5, pink, some larger than the others, apex oval or orbicular with a crisped and dentate margin, abruptly narrowed below into a long slender claw; stamens 10, erect, viscid, filaments pink, monadelphous; ovary 3-celled, styles 3, distinct.

FRUIT. Drupe red, about ⅓ in. in diameter, fleshy, acid, edible, skin thin, 3-celled; stones angled, crested on the back, indehiscent, sometimes abortive and asymmetrical.

LEAVES. Opposite, ovate or elliptic to lanceolate, apex acute, or acuminate, base cuneate or rounded, margin entire, slightly revolute, blade length 1–3 in., upper surface dull green and glabrous, lower surface glabrous or with scattered long white appressed hairs, veins obscure; petioles very short, 1/25–1/18 in., or leaf almost sessile, glabrous or hairy.

TWIGS. Slender, many stems from the base; young twigs green with long white hairs; older ones glabrous, brownish, or light to dark gray, roughened by abundant lenticels.

RANGE. In southmost Texas, in Cameron and Hidalgo counties, elsewhere only in cultivation. In Mexico in Nuevo León, Tamaulipas, Tabasco, and Yucatán. Also in Central and South America and the West Indies.

REMARKS. The genus name, *Malpighia*, is in honor of Marcello Malpighi, naturalist of Pisa (1628–1693). The species name, *glabra*, refers to the plant's glabrous character, but this is not distinctive because white hairs are occasionally present. The plant is sometimes known in cultivation under the name of Wild Crape-myrtle because of its superficial resemblance to Crape-myrtle. Many names are applied to the plant in tropical countries such as Manyonita, Cerezo de Jamaica, Cerezo de Castillo, Palo de Gallina, Escobillo, Chia, Arrayancito, Xocat, and Xochtatl. The fruit is edible and is sometimes made into preserves. The bark, known medicinally as Nance bark, is sometimes used as an astringent and febrifuge. It is reported as containing as high as

NARROW-LEAFED THRYALLIS
Thryallis angustifolia (Benth.) Kuntze

26.2 per cent tannin. The plant is a very desirable small shrub for gardens and may be propagated by seeds or from cuttings.

Narrow-leafed Thryallis

Thryallis angustifolia (Benth.) Kuntze

FIELD IDENTIFICATION. Western plant with slender tufted stems from the base, attaining a height of 1½ ft. Perhaps best considered as only herbaceous, but occasionally somewhat woody near the base.

FLOWERS. Through the summer, blossoming up the stalk in loose terminal racemes; pedicels bracteolate, usually less than ½ in., appressed-hairy; flowers about ½ in. across, perfect, regular, brilliant yellowish orange to red; petals 5, about ¼ in. long, oblong, apex rounded, base abruptly slender-clawed, with a dark midvein; stamens 10, exserted anthers about as long as broad; styles 3, 3/16–¼ in., stigmas very small; ovary 3-celled with solitary suspended ovules, embryo hooked; calyx with 5 lanceolate, reddish sepals about ⅛ in. long.

FRUIT. Capsule borne on short threadlike pedicels, about ⅙ in., 3-celled, the carpels separating and dehiscent.

LEAVES. Simple, opposite, linear to lanceolate or oblong, occasionally the lower oval, usually acute to acuminate, occasionally obtuse, base attenuate, margin entire, 2 minute glands at or near the base, blade length ½–1½ in., surfaces dull green and glabrous or nearly so when mature (young with a few medifixed hairs); petioles very short, or leaf sessile, glabrous or with a few white hairs; stipules linear, glabrous or hairy.

STEMS. Young slender, tufted below, greenish, with minute appressed white hairs; older stems reddish brown to black and slightly woody near the base.

RANGE. Dry rocky hillsides and prairies. In Texas from the Colorado River westward to the Rio Grande, also in Mexico in Sonora and Tamaulipas.

REMARKS. The genus name, *Thryallis,* is the old Greek name, and the species name, *angustifolia,* refers to the narrow leaves. Although some species of *Thryallis* are cultivated for ornament this species is probably too small and insignificant to be of value.

MILKWORT FAMILY (*Polygalaceae*)

Spiny Polygala

Polygala subspinosa Wats.

FIELD IDENTIFICATION. Western undershrub to 1 ft. Stems with slender spinose tips, branched above, woody toward the base.

FLOWERS. Racemes loosely flowered, bracts small and scarious, pedicels later reflexed and shorter than the flowers; flowers bisexual, perfect, very irregular; sepals 5, very unequal, lateral ones petaloid, naked or ciliate; petals yellow or purple, usually 3, unequal, united below into a dorsally cleft tube, the lower one (keel) united at base with the staminal tube and crested with a saccate process, petals 1/4–1/2 in. long; stamens usually 8, filaments united into a tube below, anthers opening by pores; style solitary, linear, long-curved, dilated above; ovary superior, 2-celled, with 2 united carpels.

FRUIT. Capsule 1/5–1/4 in. long, orbicular, emarginate, short-stipitate, flattened, membranous, dehiscent, 2-celled, 2-seeded, seeds with a caruncle at the hilum.

LEAVES. Simple, alternate, sessile or nearly so, scattered, obovate to elliptic, sometimes oblong or oblanceolate, apex acute to obtuse, base attenuate, blades 1/8–1/4 in. broad, 3/8–1 1/2 in. long, surfaces grayish green, puberulent to glabrate, veins rather obscure.

TWIGS. Erratic, slender, ending in spines, grayish green and puberulent at first, later gray to brown, glabrous, and surface breaking into small thin scales.

RANGE. Plains and hillsides to an elevation of 5,000 ft. New Mexico; west to California, and north to Colorado, Utah, and Nevada.

REMARKS. The genus name, *Polygala,* is from the Greek *polys* ("much") and *gala* ("milk"), referring to the milky sap, or to the fact that some species were formerly used to control the flow of milk. The species name, *subspinosa,* is for the weakly spine-tipped branchlets.

Spiny Polygala
Polygala subspinosa Wats.

Purple Polygala

Polygala macradenia Gray

FIELD IDENTIFICATION. Perennial plant, 5–12 in., with numerous stems from a woody base.

FLOWERS. April–July, short-peduncled, solitary or

aired, small, axillary, about ¼ in. long, puberulous; epals 5, the lateral ones large and petaloid; petals 3, nited below, wings purple and pubescent, obovate, bout ⅕ in. long, keel greenish yellow; stamens 6–8, nthers 1-celled with a terminal pore; style 1, ovary -celled, ovules solitary.

FRUIT. Capsules ovate, membranous, pubescent, glanular-dotted, flattened at right angles to the partition; eed very hairy, with an apical caruncle.

LEAVES. Numerous, simple, entire, blades ⅙–¼ in. ong, linear to oblong-lanceolate, apex obtuse, base ounded or broadly cuneate, surfaces densely canesent-puberulous, glandular-dotted, veins obscure; petiles very short and canescent-pubescent, or blade sesile.

TWIGS. Young ones slender, numerous, green to tan r brown, densely canescent-puberulous; older ones rown to gray, glabrous, flaking.

RANGE. On dry, rocky hills or mesas at altitudes of ,500–4,500 ft. Western Texas, New Mexico, and Arizoa; also in Coahuila, Mexico.

REMARKS. The genus name, *Polygala*, is from the reek word *polys* ("much") and *gala* ("milk"), some pecies formerly being used to control the flow of milk. The species name, *macradenia*, is translated "with large glands." The plant is browsed only occasionally.

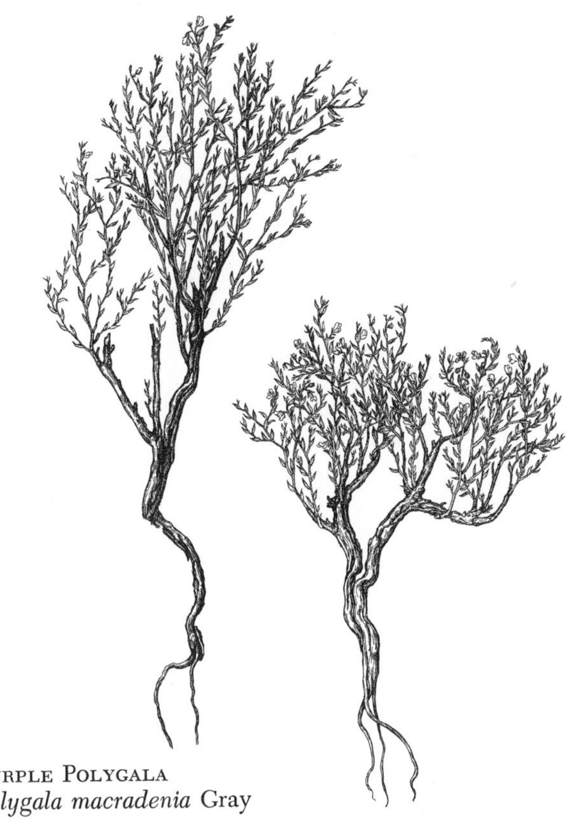

PURPLE POLYGALA
Polygala macradenia Gray

EUPHORBIA FAMILY (*Euphorbiaceae*)

VASEY ADELIA

Adelia vaseyi (Coulter) Pax & Hoffm.

FIELD IDENTIFICATION. Shrub attaining a height of 11 ft, with many long, slender upright stems from the base.

FLOWERS. February–April, dioecious; staminate borne in alternate clusters at the tomentose nodes; petals absent; calyx-tube short, spreading into 4–6 (usually 5) lobes, lobes oblong-lanceolate, acute, 1/16–1/8 in. long, white-hairy; stamens 8–15, exserted, filaments about 1/8 in. long, translucent-viscid, anthers ovoid-oblong; pistillate flowers terminal or axillary, solitary and long-pediceled, 3-lobed, capped by the persistent 3-lobed, laciniately cut styles.

FRUIT. Capsule, April–May, borne on long, slender, puberulent, solitary pedicels 1/3–1 1/4 in. long, pendent from the leaf axils at the nodes; body of fruit 3-lobed, depressed, to 1/2 in. wide and 1/4 in. high, puberulent and granular, more glabrous later; seeds 3, one in each cavity, round to broad-ovoid.

LEAVES. Scant, alternate or clustered at the remote, short, wart-like nodes; spatulate, cuneate or obovate, 1/2–1 1/2 in. long, 3/16–1/2 in. wide, entire; apex mostly rounded, some truncate or notched; base gradually narrowed and becoming sessile or very short-petiolate at the tomentose node; medium-green and puberulent above when young, glabrous later; lower surface paler, puberulent, minutely punctate, 3 veins rather conspicuous, the 2 lateral ones close to the margin.

TWIGS. Long, slender, rather rigid, gray, smooth, glabrous or pubescent when young; lateral branches few, mostly set with short, spurlike nodes at intervals.

BARK. Gray to brown, smooth and tight above, sometimes broken into thin, small scales on old stems.

RANGE. In dry soil in Cameron, Willacy, Hidalgo, and Starr counties, Texas. Near the mouth of the Rio Grande on bluffs at Boca Chica. Near Mission, Texas on La Lomita Hill. Also in Tamaulipas, Mexico.

VASEY ADELIA
Adelia vaseyi (Coulter) Pax & Hoffm.

REMARKS. The genus name, *Adelia*, is from the Gree[k] and refers to the small, obscure flowers. The specie[s] name, *vaseyi*, is in honor of George R. Vasey (1822-1893), botanist for the U.S. Department of Agricul-ture.

TUNG-OIL-TREE

Aleurites fordii Hemsl.

FIELD IDENTIFICATION. Tree introduced into cultivation in the Gulf states, attaining a height of 30 ft, with a rounded top and stout branches.

FLOWERS. Borne in spring, monoecious, in large conspicuous, terminal cymes; pedicels ½–5 in. long; petals 5, oblong to oblanceolate, sometimes orbicular-ovate, white, streaked with red at the base, apex rounded or more rarely truncate, length ¾–1½ in., width about ½ in.; sepals 2–3, valvate, reddish, about ¼ in. long, ovate to rounded; stamens 8–10, red, the inner row monadelphous; anthers attached terminally, splitting into 2 segments; ovary 4-celled, 1 ovule in each cell.

FRUIT. Drupe ripe in September, 1¼–3 in. in diameter, large, subglobose, smooth, seeds thick-shelled, oily, poisonous.

LEAVES. Alternate, large, ovate-cordate, margin entire or 3–5-lobed, apex acute or acuminate, base rounded or semicordate, palmately-veined, upper surface dull green with fine brown hairs, lower surface paler, veins heavy, sparingly hairy or glabrous; petiole long, stout, somewhat flattened, glabrous or puberulent; 2 red, lustrous glands at the petiole apex.

BARK. Light to dark gray, smooth.

TWIGS. Stout, glabrous, dark gray to brown, somewhat swollen at the nodes and terminally; leaf scars semiorbicular or oblong, lenticels numerous.

RANGE. On dry, thin soil; drought-resistant. A native of Central Asia. Grown for its oily nut in East Texas, Louisiana, Mississippi, Alabama, and Florida.

TUNG-OIL-TREE
Aleurites fordii Hemsl.

PROPAGATION. A good tree will yield 20–50 lb of nuts with about 24 per cent oil. Production age begins at about 5 years. Seedlings are transplanted when 1–2 ft. They may be grown from hardwood cuttings. The plant is also easily grown from seed which sprout in 4–5 weeks.

REMARKS. The genus name, *Aleurites,* is from the Greek and means "farinose" or "floury." The species name, *fordii,* is in honor of C. Ford, former superintendent of the Hongkong Botanic Garden. It is also known as China Wood-oil-tree and Candlenut-tree. This species is more hardy than other plants of the genus *Aleurites* in the Gulf Coast states area. Oil from the fruit is valuable as an ingredient of paint, varnish, soap, linoleums, etc. In Asia the oil is used as a fuel and for treating woodenware and cloth. The oil is pressed from the seeds after roasting.

MISSOURI MAIDEN-BUSH

Andrachne phyllanthoides (Nutt.) Muell. Arg.

FIELD IDENTIFICATION. An upright or straggling diffusely branched subshrub 1–3 ft.

FLOWERS. Mostly July–August, minute, greenish, axillary, monoecious, on glabrous threadlike pedicels ¼–⅚ in. long; staminate solitary or a few together, pistillate solitary; calyx segments 5–6, 1/12–⅙ in. long, spreading, oblong-obovate; petals 5–6, staminate ones slightly shorter than the sepals, greenish yellow, oblong to obovate, 3–5-toothed; pistillate rudimentary, broadly obovate and entire or obsolete; disk circular; stamens 5–6, filaments slender and distinct, not exceeding the sepals; ovary globose, 3-lobed and 3-celled; styles 3, stout, divergent, 2-parted; ovules 2 in each cell.

FRUIT. Capsule depressed-globose, ¼–⅓ in. in diameter, dry, subtended by the enlarged sepals, 6-seeded; seed somewhat falcate, roughened, endosperm fleshy, embryo curved.

LEAVES. Numerous, simple, alternate, deciduous, stipulate, broadly obovate to elliptic or oval, apex obtuse to retuse or rounded, often mucronulate, base rounded or broadly cuneate, margin entire, length ⅓–⅞ in., ⅖–⅗ in. wide, membranous, under surface yellowish green and essentially glabrous, lower surface paler with veins finely reticulate; short-petiolate or practically sessile.

TWIGS. Leafy, ascending, simple, slender and wiry, glabrous, or minutely puberulent, reddish brown to light brown or grayish later, shiny; buds ovate, scales ciliate.

RANGE. Dry rocky soil of ledges, arroyos, and barrens. Northern, central, and western Texas, Oklahoma, Arkansas, and Missouri.

VARIETY. Reverchon Maiden-bush, *A. phyllanthoides* var. *reverchoni* (Coult.) Pax & Hoffm., is a variety with leaves more pubescent, thickened, pale, about as broad

Missouri Maiden-bush
Andrachne phyllanthoides (Nutt.) Muell. Arg.

as long or orbicular; apex truncate to retuse or rounded and mucronate, base often subcordate. It is found in Trans-Pecos Texas.

REMARKS. The genus name, *Andrachne,* is the classical Greek name for the Purslane. The species name, *phyllanthoides,* indicates that it resembles the Phyllanthus plant. It is also listed under the name of *Savia phyllanthoides* (Nutt.) Pax & Hoffm. It has been cultivated since 1899 and is sometimes known under the vernacular name of Northern Andrachne.

Johnston Bernardia

Bernardia obovata Johnston

FIELD IDENTIFICATION. Green irregularly branched shrub to 3 ft.

FLOWERS. Dioecious, staminate axillary, rachis 1/8–1/6 in. long; bracts 1–4, triangular to cordate, about 1/25 in. long; 5–8-flowered, generally clothed with stellate hairs; stamens 5–8, usually 6, filaments about 1/35 in. long, glabrate, sometimes rose-colored, anthers as long as the filament or slightly longer, apices of the connective brownish red; disk minute, glandular, reddish; pistillate flowers usually solitary and nearly sessile, pear-shaped or subglobose, generally stellate-hairy, 3–4-lobed at anthesis; sepals 3, ovate, cordate, 1/25–1/15 in. long, ciliate; bracts 2–3, short-ovate, disk circular, inconspicuously toothed; ovary 2-ovuled, 2-celled, compressed, densely stellate-pubescent; styles 2, bilobed, lobes subulate, 1/25–1/15 in. long, recurved, entire.

FRUIT. Capsule solitary or 2-lobed, 1/4–1/3 in. high, 1/ in. thick, inconspicuously pubescent; seed about 1/5 in. high, facial axis plane, dorsal sides convex, 2-celled.

LEAVES. Simple, alternate, obovate to subelliptic, some widest above the middle, apex obtuse to rounded, base broadly acute, margin sparsely few-toothed, upper surface dull green with obscure veins and minute stellate hairs; lower surfaces paler with prominent veins 2–4 branched and heavily stellate-tomentose; petiole 1/25–1/12 in. long, densely stellate-hairy; stipules thick, stiff, oblique, 1/15–1/25 in. long.

TWIGS. Young ones green to brown with soft stellate glandular hairs, older ones soon glabrous and light to dark gray or brown.

RANGE. Igneous stony soils or sandstone regions. In

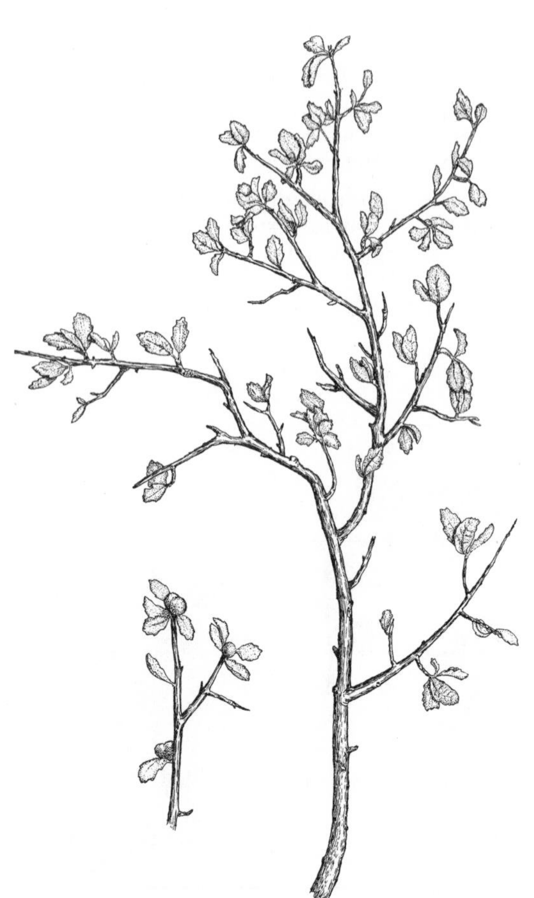

Johnston Bernardia
Bernardia obovata Johnston

SOUTHWEST BERNARDIA
Bernardia myricaefolia (Scheele) Wats.

Trans-Pecos Texas in Culberson, Hudspeth, and Brewster counties; also in Chihuahua, Mexico.

REMARKS. The genus name, *Bernardia,* is in honor of P. F. Bernard, a French botanist, and the species name, *obovata,* refers to the leaf shape. This species is distinguished from *B. myricaefolia* by its less elongate leaves, its sparser indument, thicker stipules, glabrescent male flowers, fewer stamens, smaller less hairy bilocular fruit, and simple style branches. Also *B. myricaefolia* is a plant of central and southern Texas, and *B. obovata* is found in the Trans-Pecos Texas area.

SOUTHWEST BERNARDIA

Bernardia myricaefolia (Scheele) Wats.

FIELD IDENTIFICATION. Shrub to 10 ft, densely branched, young parts copiously hairy with starlike tomentum.

FLOWERS. Small and inconspicuous, staminate and pistillate separate on the same plant or on different plants; staminate flowers axillary, 2–several in the axils of each bract of the short racemose panicles which are ⅜–¾ in. long; individual flower pedicels 1/25–1/12 in. long, hairy; petals absent; staminate calyx 3–4-lobed (rarely, 5-lobed), lobes oval to obovate or oblong, margin ciliate, apex rounded or truncate; stamens 3–8, alternate with minute glands, filaments distinct and erect, anthers globose; pistillate flowers terminal, solitary or paired, sessile, calyx 3–9-parted, hairy; styles 3, warty, stigma 2-lobed; ovary tomentulose, 3-celled and 3-ovuled, ovules solitary.

FRUIT. Capsule globular, somewhat depressed, usually 3-lobed, lobes rounded, ⅓–½ in. in diameter, densely pale stellate-tomentose; seeds about ⅕ in. long, ovoid to globose, grayish brown, dull, smooth.

LEAVES. Simple, alternate, or clustered, persistent; shape very variable, oblong to elliptic or ovate, orbicular or obovate; apex acute to obtuse or rounded, base cuneate to cordate or rounded, margin repand-dentate, length ⅜–2 in., upper surface dull green and minutely papillose with a few scattered simple or stellate hairs, veins obscure; lower surface paler and densely stellate-hairy, veins prominent; leaves sessile or on short petioles 1/25–⅛ in. long, hairy; stipules about equaling the petioles, acuminate, and fleshy.

TWIGS. Slender, many shortened; when young green to gray and densely stellate-hairy, when older light or dark gray and glabrous.

RANGE. Dry rocky slopes and canyons of central and southern Texas. The type locality is at New Braunfels, Texas. Plants in the Trans-Pecos area of Texas may be referred to under the name of *B. obovata* Johnston.

REMARKS. The genus name, *Bernardia,* is for P. F. Bernard, a French botanist, and the species name, *myricaefolia,* denotes its resemblance to a myrtle leaf. Some vernacular names in use are Myrtle-croton, Oreja de Ratón, and Palo de Tarugo. The plant is very drought-resistant and is browsed by cattle in time of stress.

SONORA CROTON

Croton sonoreae Torr.

FIELD IDENTIFICATION. Much-branched shrub attaining a height of 6 ft. Younger parts covered with a dense stellate pubescence.

FLOWERS. Racemes ¾–3⅓ in. long, terminal, slender, and loose, sometimes only the staminate flowers are present, but if pistillate flowers are also present in the same raceme, they are generally beneath the staminate; staminate calyx 5-parted, pubescent to almost glabrous; pedicels pubescent; petals 5, oblong, about as long as the calyx or slightly longer, ciliate; glands large and rounded; stamens 11–18, filaments exserted, slender or nearly so; bracts short, setaceous; pistillate flowers solitary or 2–4, very short-pediceled, sepals deltoid and glabrous or nearly so, bracts small and glabrous; petals

Sonora Croton
Croton sonoreae Torr.

5, linear-subulate, about equaling the sepals (petals sometimes absent); gland annular; ovary stellate-hairy, 3-chambered, ovules solitary; styles 3, bifid, slender, glabrous or slightly stellate-hairy.

FRUIT. Capsule 3-lobed, spherical to slightly oblong with scattered stellate hairs, more glabrescent when old, about ¼ or more in. long; 1 seed in each chamber, about ⅕ in. long, brown to gray, lenticular-ellipsoidal.

LEAVES. Simple, alternate, deciduous, oval-oblong, or ovate-lanceolate, apex acute or acuminate, base rounded, margins subentire, length ¾–2 in., upper surface thinly stellate-pubescent or almost glabrous with age, lower surface stellate-pubescent, especially on the veins; petioles ⅕–½ in. long, densely hairy.

TWIGS. Pale brown to gray, stellate-hairy, but glabrous on the older wood near the base.

RANGE. Dry rocky mesas and ravines at altitudes of 1,700–3,000 ft. Known from southern Arizona, and from Sonora to Oaxaca, Mexico. Reported from Texas by John M. Coulter (1891–1894, p. 398).

REMARKS. The genus name, *Croton,* is from the Greek word for "tick," because of its resemblance in shape. The species name, *sonoreae,* is for the state of Sonora, Mexico, where it grows most abundantly.

Fragrant Croton

Croton suaveolens Torr.

FIELD IDENTIFICATION. Stout much-branched, fragrant shrub, usually 1–2 ft. Herbage and young stems densely and coarsely stellate-tomentose with brownish or yellowish hairs.

FLOWERS. Monoecious or dioecious; staminate racemes ¾–1⅝ in. long, stout, closely flowered; bracts simple or variously lobed or parted, glandular; pedicels ⅙–⅓ in. long, densely stellate-tomentose; sepals 5, ovate to lanceolate, apex acute; petals 5, equaling the sepals, oblong, densely ciliate; stamens 12–15; receptacle tomentose and glandular; pistillate flowers below the staminate when in mixed inflorescence, solitary or a few together, short-pediceled or almost sessile, sepals oblong and acute at apex; petals absent or rudimentary in the form of subulate glands; disk glands 5; styles 3, 2-parted, ⅛–¼ in. long, terete or flattened.

FRUIT. Capsule globose, about ¼ in. long, 3-chambered; seeds about ⅕ in. long.

LEAVES. Simple, alternate, deciduous, thickish, blade elliptic to oblong, or ovate, apex obtuse and mucronate, base rounded or narrowed, margin entire, both surfaces stellate-tomentose, the lower very densely so, upper surface darker with the stellate hairs more minute, blade length ¾–1⅔ in., width ⅜–1 in.; petiole short, ⅕–⅗ in.

Fragrant Croton
Croton suaveolens Torr.

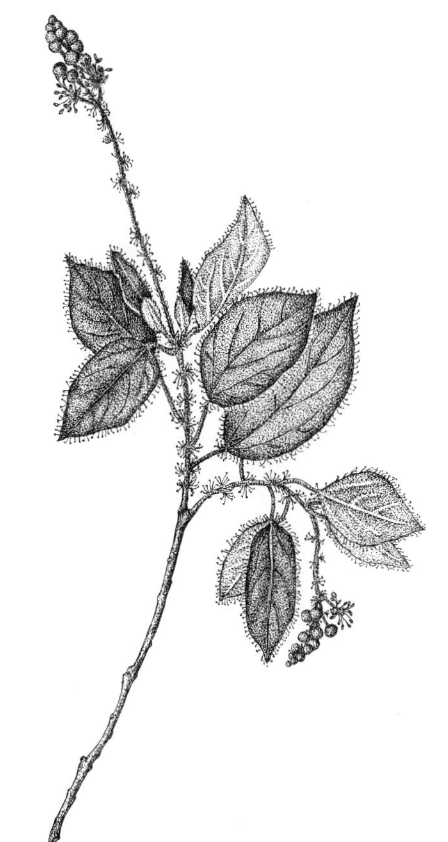

Mexican Croton
Croton ciliato-glandulosus Ortega

stipules cylindric-glandular, usually buried in the thick stellate tomentum.

TWIGS. Young ones densely stellate-tomentose with brown to yellowish hairs; older ones light brown to gray, smooth and glabrous.

RANGE. Ledges and clefts of rock on dry slopes of hills. In Texas at Fort Davis, Jeff Davis County, and in Val Verde County. In Mexico in Chihuahua, Coahuila, and San Luis Potosí.

REMARKS. The genus name, *Croton*, is from the Greek word for "tick." The species name, *suaveolens*, is for the fragrant herbage. It is known as "Encinillo" in Mexico, and is used in baths during convalescence from fevers.

Mexican Croton

Croton ciliato-glandulosus Ortega

FIELD IDENTIFICATION. An aromatic shrub attaining a height of 7 ft. Easily identified by the long, gland-tipped hairs on the leaf margins and in the leaf axils.

FLOWERS. Staminate flowers usually on the upper part of the stem, solitary, or a number together; pedicels rather short and densely tomentose; sepals 5, externally tomentose; petals 5, shorter or longer than the sepals, base ciliate; stamens numerous, exserted, filaments glabrous; pistillate flowers on the lower part of the stem; pedicels short, tomentose or sometimes glandular-hairy also; sepals 5, densely tomentose externally, glabrous within, margins ciliate-glandular; petals absent; ovary with 3 cavities, densely tomentose, ovules solitary; styles 3, each one 3–4-branched, stellate-pubescent at base.

FRUIT. Capsule globose, 3-lobed, densely stellate-pubescent or tomentose, about ¼ in. long; seeds 2–3, ⅕–¼ in. long, brown to tan or gray, lenticular, flattened, with a transverse caruncle.

LEAVES. Simple, alternate, deciduous, scented, blades ¾–4 in. long, ovate to ovate-lanceolate or heart-shaped, apex acute or acuminate, base rounded to cordate or semitruncate, margins entire but very noticeably ciliate or ciliate-glandular; surfaces dull green or grayish, lower surface densely tomentose, smoother above, with a few scattered hairs; petioles elongate, but usually shorter than the blades, densely pubescent or glandular-hairy, especially in the axils; stipules elongate, divided into conspicuous filiform segments.

TWIGS. Young ones densely tomentose and glandular-hairy, grayish green; older ones tan to brown or gray and glabrous.

RANGE. Mountainsides of the Mexican desert plateau area at altitudes of 3,000–4,500 ft. In Texas 3½ miles southwest of Roma in Starr County. In Arizona in Santa Cruz County. In Mexico in Tamaulipas, Nuevo León, Coahuila, and Chihuahua; southward to Guatemala and Honduras; also in Cuba.

REMARKS. The genus name, *Croton*, is the Greek name for "tick." The species name, *ciliato-glandulosus*, refers to the conspicuous glandular hairs of the leaf-margins. A number of vernacular names for the plant are used in Mexico including Solimán, Ciega-vista, Palillo, Chirca, Hierba de la Cruz, Enchiladora, Solimán Blanco, Canelillo, Picosa, and Xonaxl. The leaves are reported to be used in Mexico as a purgative and febrifuge.

Torrey Croton

Croton torreyanus Muell. Arg.

FIELD IDENTIFICATION. A whitish slender shrub 3–6 ft. Branches slender, straight, bearing dense stellate tomentum.

FLOWERS. Borne in racemes, staminate racemes ¾–1⅔ in. long, many-flowered; bracts ovate; sepals oblong, acute to obtuse; petals oblanceolate or elliptic, base somewhat ciliate; gland enlarged; stamens 12–15; pistillate flowers at base of raceme, pedicels short; sepals ovate, apices acute or acuminate; petals rudimentary,

TORREY CROTON
Croton torreyanus Muell. Arg.

subulate, densely tomentose; ovary densely stellate-tomentose, styles 3-parted, tomentose toward base.

FRUIT. Capsule subglobose to oblong, densely stellate-tomentose, about ¼ in. long, 3-lobed, apex depressed, 2–3-seeded; seeds about ⅙ in. broad, biconvex, caruncle oblong and stipitate.

LEAVES. Simple, alternate, deciduous, oblong to subelliptic, apex acute or obtuse and mucronulate, base acute, blades 1¼–2 in. long, width ⅗–1 in., dark green and tomentose above, lower surfaces pale and densely stellate-tomentose, 3-nerved at base; petioles ¼–⅖ in. long; stipules subulate, 1/12–⅛ in. long, deciduous.

TWIGS. Young ones densely and softly gray or brownish stellate-tomentose; older ones gray, smooth and glabrous.

RANGE. Rocky ravines and hillsides in dry gravelly soil; western Texas in Val Verde, Live Oak, and Hidalgo counties; New Mexico; in Mexico in Nuevo León, Tamaulipas, and Coahuila.

REMARKS. The genus name, *Croton*, is the Greek name for "tick." The species name, *torreyanus*, is in honor of John Torrey (1796–1873), American botanist of Columbia University.

BUSH CROTON

Croton fruticulosus Engelm.

FIELD IDENTIFICATION. Aromatic shrub attaining a height of 2–6 ft. Branches slender, the leaves densely gray or yellow stellate-hairy. Very variable in habit and leaf size as a result of environment.

FLOWERS. Monoecious, less often dioecious, in racemes 1–5 in. long; staminate racemes loose and interrupted; pedicels ⅛–⅙ in. long, stellate-hairy; bracts small and subulate; sepals 5, oval, acute at apex; petals 5, longer than the sepals, oblong to spatulate, ciliate; glands oval; stamens 9–10, exserted, glabrous but hairy at the base; pistillate flowers 2–5, nearly sessile; sepals 5, stellate-tomentose externally, oblong to oblanceolate; petals rudimentary or none; gland annular and lobed; styles 3, divided nearly to the base, about ⅕ in. long.

FRUIT. Capsule ⅕–¼ in. long, globose, depressed, somewhat 3-lobed, stigma remains persistent, finely and shortly stellate-puberulent; seeds oval, ⅙–⅕ in. long, caruncle oblong.

LEAVES. Simple, alternate, deciduous, blade 1¼–3¼ in. long, ¾–1½ in. wide, ovate to ovate-lanceolate, apex acute to acuminate, base rounded to cordate or truncate, margin entire or remotely and minutely denticulate-glandular, upper surface smooth and green to puberulent, lower surface soft-tomentose with dense, minute, stellate, spreading hairs; petioles ⅜–1½ in. long, or one third to one half as long as the blade.

TWIGS. Young ones grayish green to yellowish, with

BUSH CROTON
Croton fruticulosus Engelm.

Beach Croton
Croton punctatus Jacq.

dense stellate tomentum; older ones light to dark brown or gray and glabrous.

RANGE. Usually on limestone hills, bluffs, canyons, or rocky ravines at altitudes of 1,000–4,500 ft. In central and western Texas, New Mexico, and Arizona; south in Mexico to Nuevo León, Chihuahua, and Sonora.

VARIETY. A variety with long-acuminate leaves, almost glabrous above, yellow cinereous-tomentose beneath, has been given the name of *C. fruticulosus* var. *fuscescens* Muell. However, since intermediate forms between it and the species are found, the differences are not well defined.

REMARKS. The genus name, *Croton*, is the Greek name for "tick." The species name, *fruticulosus*, refers to its low shrubby habit. It also has the vernacular names of Encinilla and Hierba Loca.

Beach Croton

Croton punctatus Jacq.

FIELD IDENTIFICATION. Plant 6 in.–4 ft high, forming rounded mounds of herbage on sandy beaches. Upper stems herbaceous, lower stems semiwoody, spreading by underground stolons.

FLOWERS. Staminate and pistillate flowers on different plants, borne in terminal, few-flowered, densely stellate-hairy racemes; petals absent; sepals 5, triangular to oblong, apex obtuse, free for about two thirds to one half their length, often with scattered reddish glands; staminate flowers densely clustered distally, but proximally on short peduncles ½ in. long or less, stamens about 12, barely exserted, filaments pubescent; pistillate flowers 1–3, peduncles ¼–2 in. long, flattened; styles bifid, split into the linear segments either near the base or apex, ovary 3-celled, ovules 1 in each cavity.

FRUIT. Capsule subglobose, somewhat depressed, ⅕–⅓ in. long, 3-lobed (rarely, 2- or 4-lobed), lobes rounded, densely stellate-hairy; seed about ¼ in. long, ovoid to ellipsoid, light brown to somewhat mottled-gray, lined on one side, hilum scar rather large and terminal.

LEAVES. Alternate (or smaller leaves numerous around the flower clusters), oval to oblong or ovate to elliptic, blade length ¼–2½ in., width ⅜–1¼ in., apex rounded to obtuse; base rounded, truncate or semi-cordate; margin entire or occasionally undulate, often with reddish glands; upper surface pale green and densely stellate-hairy; lower surface silvery white and densely stellate-hairy, midrib often with reddish glands; petiole as long as the blade or shorter, the main vein sometimes decurrent one third to one half its length below; upper side flattened or grooved, densely gray to brownish stellate-hairy.

TWIGS. Flattened and grooved, petioles decurrent down the twig, young and old twigs gray to reddish brown, stellate-hairy and glandular, stems more terete toward the base.

RANGE. Sandy soils of sea beaches. Coastal Texas and Louisiana; eastward to Florida and northward to South Carolina.

REMARKS. The genus name, *Croton*, is reported to be the Greek name of the "tick," because of the seed's appearance. The species name, *punctatus*, refers to the punctate glands. Other local names are Silver-leaf Croton, Beach Tea, and Sand Croton. Of small economic value except as an erosion-control plant in sands.

Silver-leaf Croton

Croton argyranthemus Michx.

FIELD IDENTIFICATION. Plant 12–24 in. tall, the greater part herbaceous but somewhat suffrutescent in habit. Foliage strong-scented and conspicuously silvery-scaly.

FLOWERS. Racemes greenish, staminate and pistillate borne separately, ¾–2 in. long; pedicels 1/16–3/16 in. long, silvery-scaly; staminate flowers usually 10–15 and topping the raceme; petals oblong to elliptic, scaly externally; sepals 5, lanceolate, acute; stamens usually 10, filaments somewhat enlarged at the base; pistillate flowers solitary or few, pediceled or subsessile; sepals usually 5–7, partly united below; petals absent or rudimentary; ovary 3-celled, ovule solitary in each cell, stigma 2–3-lobed and sometimes with silvery scales; disk 5-lobed or almost entire.

FRUIT. Capsule oval-globose, 3-lobed, apex depressed,

Silver-leaf Croton
Croton argyranthemus Michx.

about ⅕ in. long, dehiscent into 3 2-valved carpels; seed solitary in each carpel, biconvex, about ⅙ in. long.

LEAVES. Simple, alternate, persistent, variously shaped, elliptic to lanceolate or obovate to oval, upper leaves usually narrower, margin entire, base gradually narrowed, apex obtuse or acute, blades ⅜–2 in. long, upper surface scaly or glabrate, lower surface clothed with conspicuous silvery scales, often with a central mass of rusty brown scales; petioles one fourth to one half as long as the blade.

STEMS. Usually divaricately branched toward the upper third of the plant, silvery and brown-dotted, scaly, later reddish brown and less scaly below, essentially herbaceous, but plant somewhat suffrutescent in habit.

RANGE. Usually in dry sandy soils of pinelands. Texas and New Mexico; eastward to Georgia and Florida.

REMARKS. The genus name, *Croton*, is the Greek word for "tick," referring to the fact that the seeds of some plants of this family resemble a tick. The species name, *argyranthemus*, refers to the silvery scales of the flowers.

New Mexican Croton

Croton neomexicanus Muell. Arg.

FIELD IDENTIFICATION. Strong-scented, slender, erect plant attaining a height of 1–2 ft. The several stems arise from a woody base and are often dichotomously branched. The plant is more or less silvery throughout as a result of its dense covering of argenteous scales.

FLOWERS. Dioecious, staminate ones often numerous, racemes 1½–3½ in. long, pedicels ⅛–⅙ in. long; petals absent; sepals 5, equal, relatively thin, lanceolate; glands large; stamens 10–12, filaments pubescent; pistillate racemes ¾–1⅓ in. long, flowers solitary or 2–3 together, pedicels about ¼ the length of the fruit; petals absent; sepals 5, ovate-oblong, apex obtuse; bracts ovate and minute; styles 3, palmately 2–4-cleft, ovary 3-celled, 1 ovule in each cavity.

FRUIT. Capsule ⅕–¼ in. broad, globose or somewhat 3-lobed, dehiscent into 3 2-valved carpels, each carpel 1-seeded; seeds ⅙–⅕ in. long, oval.

LEAVES. Simple, alternate, deciduous, narrowly oblong to lanceolate or elliptic, margin entire or repand, apex obtuse or rounded, base gradually narrowed, blade

New Mexican Croton
Croton neomexicanus Muell. Arg.

length 3/8–2 5/8 in., width 1/5–3/5 in., lower surfaces densely covered with radiate silvery scales, upper surface more green and less scaly; petioles one fourth to one half the length of the leaves.

TWIGS. Slender, silvery or brownish, with peltate scales and brownish glands; older ones gray to brown and more glabrous, woody near the base.

RANGE. Low rocky hills, mesas, and plains at altitudes of 4000–5,800 ft. Western Texas and New Mexico; Mexico in Nuevo León and Chihuahua.

REMARKS. The genus name, *Croton,* is the Greek name for "tick," perhaps referring to the shape of the seed. The species name, *neomexicanus,* is for the state of New Mexico, where it grows. In Mexico the root bark is reported to be used as a purgative.

Leather-weed Croton

Croton corymbulosus Engelm.

FIELD IDENTIFICATION. Suffruticose perennial 1/2–2 ft high, stems simple and somewhat woody near the base but corymbosely branched above. All parts silvery-hairy.

FLOWERS. Usually in loose corymbs 3/4–1 1/2 in. long and 3/8–1 1/3 in. broad, sometimes on peduncles 2–4 in. long, but usually much shorter; inflorescence mixed sexually, sometimes with only staminate on a plant, sometimes with only pistillate, or sometimes with both staminate and pistillate; pedicels of staminate flowers usually about 1/6–1/4 in. long; bracts setaceous, 1/25–1/12 in. long; calyx composed of 5 oblong, obtuse sepals; petals lanceolate to spatulate, densely ciliate; stamens 16–18, filaments pubescent near the base, 1/5–1/3 in. long; pedicels of pistillate flowers 1/4–1/2 in. or more rarely to 2 1/2 in. long; petals absent or rudimentary; sepals 5–6; disk glandular, 5-lobed; styles 3, split to below the middle, slender, 1/8–1/3 in. long.

FRUIT. Capsule oval or oblong, 1/6–1/3 in. long, stellate-scaly, calyx about one third as long; seeds oblong, 1/8–1/5 in. long, apices truncate, caruncles prominent.

LEAVES. Simple, alternate, blade oblong to oval or elliptic, 3/4–1 3/4 in. long, apex acute or mucronulate, base rounded or narrowed, margin entire, upper surface, greenish gray with closely appressed stellate pubescence or scales, paler below, but pubescence denser and looser, veins impressed above but conspicuous beneath the blade; petioles pubescent, usually about one half as long as the blades or more; stipules deciduous, small, subulate to foliaceous.

STEMS. Shrubby, with many erect, simple stems from a woody base, corymbosely branched above, younger ones densely silvery gray or brownish pubescent, older ones brown and less pubescent.

RANGE. Usually on sandy mesas or dry rocky slopes at altitudes of 2,000–6,000 ft. In Trans-Pecos Texas and the Organ Mountains of New Mexico to Arizona and adjacent Coahuila and Chihuahua, Mexico.

Leather-weed Croton
Croton corymbulosus Engelm.

REMARKS. The genus name, *Croton,* is the Greek word for "tick," because of the fact that the seeds of some plants of this genus resemble an engorged tick. The species name, *corymbulosus,* refers to the corymbulose flowering habit. It is reported that the Indians make tea from the foliage.

Cortes Croton

Croton cortesianus Kunth

FIELD IDENTIFICATION. Shrub 2–9 ft. The branches are usually 2–3-branched with the younger parts stellate-tomentose.

FLOWERS. Staminate racemes slender, interrupted, 4–8 in. long, flowers short-pediceled, bracts broad; petals oblong, ciliate; stamens 12–16; pistillate spike about 1 1/2 in. long, many-flowered, dense, flowers sessile; petals rudimentary or linear; glandular disk 5-

Cortes Croton
Croton cortesianus Kunth

lobed; calyx-lobes narrowly triangular or lanceolate, acuminate; ovary hispid-stellate, styles 3, 2-parted, about 1/5 in. long.

FRUIT. Capsule globose or shallowly lobed, densely hairy, 1/4–1/3 in. long, apex somewhat flattened; seeds about 1/5 in. long.

LEAVES. Simple, alternate, deciduous, oblong to lanceolate, margin entire or denticulate, apex acute or acuminate, base acute to rounded or subcordate, length 1–4 in., upper surface green and subglabrous, lower surface gray with stellate pubescence; petioles 1/5–3/4 in.

TWIGS. Slender, young ones yellowish green or light brown, densely tomentose; older ones tan to light brown, less tomentose, eventually glabrous.

RANGE. Southern and western Texas, Cameron and Hidalgo counties. In Mexico from Nuevo León and Tamaulipas to San Luis Potosí, Sinaloa, Campeche, and Chiapas.

REMARKS. The genus name, *Croton*, is the Greek name for "tick," referring perhaps to the shape of the seed. It is thought that the species name, *cortesianus*, was given in honor of the Spaniard Hernando Cortés, explorer of Mexico, but there is some doubt about the matter. The plant is known in Mexico under the various local names of Palillo, Pozual, Puzual, Ek-balam, and Pinolillo. It is used in domestic medicine as a caustic in the treatment of skin diseases.

Berlandier Croton

Croton humilis L.

FIELD IDENTIFICATION. Strong-scented, slender shrub attaining a height of 1–3 ft, with the young branches viscid-pubescent.

FLOWERS. Monoecious or dioecious, when monoecious the staminate flowers uppermost; staminate racemes slender, 2–2 3/4 in. long, individual flowers 1/8–1/6 in. broad, pedicels as long as the flower or longer; petals spatulate, ciliate below; sepals oval, about as long as the petals; stamens 15–35 (sometimes the staminate racemes contain 2–4 scattered pistillate flowers); pistillate racemes 3/4–1 in. long, usually 2–6-flowered, pedicels about 1/12 in. long or occasionally longer; petals absent, or rudiments subulate and about 1/25 in. long; sepals about 1/6 in. long, oblong-spatulate, apex acute, glands short-stipitate or sessile; styles 3, 4-parted or twice 2-parted, 1/8–1/5 in. long; ovary mostly 3-celled, 1 ovule in each cavity.

FRUIT. Capsule globose, 1/6–1/5 in. long, 3-lobed; 1 seed in each carpel, oval, flattened at the oblong caruncle.

Berlandier Croton
Croton humilis L.

LEAVES. Simple, alternate, deciduous, ovate to oblong or lanceolate, apex abruptly acute to acuminate, base rounded to subcordate, margin entire or repand; upper surface tomentose, especially when young; lower surface paler, pubescent, or often almost glabrous when mature; margin with minute glands at vein terminations.

TWIGS. Slender, young ones densely stellate-pubescent and viscid; older ones light brown to pale gray and glabrous.

RANGE. In dry well-drained sandy soils of coastal regions. Texas and Louisiana, eastward to Florida. Also from Nuevo León to Yucatán, Mexico; also the West Indies.

REMARKS. The genus name, *Croton*, is the Greek name for "tick." The species name, *humilis*, refers to its low growth. In Mexico the plant is known under the names of Ycaban and Icaban. It is also listed under the name of *C. berlandieri* Torr. by some authors. Florida plants seem to vary from Texas specimens in the number of stamens.

Wax Euphorbia

Euphorbia antisyphilitica Zucc.

FIELD IDENTIFICATION. Shrub under 3 ft. Stems rodlike, usually procumbent at base; branches numerous, solitary, erect, subterete, and leafless.

FLOWERS. Subtending bracts scalelike, blackish, triangular, leathery; flowers unisexual; involucre cuplike, puberulent, solitary, and axillary, a few forming a loose terminal spike; involucre ⅛–3/16 in. across, bearing 5 oblong glands with white, concave, toothed appendages (appendages petal-like, perianth absent); stamens 1 in the staminate flower.

FRUIT. Capsule on pendent peduncles ⅝ in. long or less, subglobose, ¼–⅓ in. in diameter, obtusely 2–4-lobed (mostly 3-lobed), glabrous, style persistent; seeds white, 3–4-angled, wrinkled.

LEAVES. Early deciduous and seldom seen, linear, about ⅔ in. long, rigid, recurved.

STEMS. Erect, usually only branched toward base, simple, rodlike, pale green, puberulent or glabrous, subterete, ¼–⅜ in. in diameter.

RANGE. Gravelly limestone hills of the Texas Big Bend area. In Mexico in Nuevo León, Coahuila, Durango, Hidalgo, Puebla, Zacatecas, and San Luis Potosí.

REMARKS. The genus name, *Euphorbia*, is an old classical name. The species name, *antisyphilitica*, was given because of its use in treating venereal disease. In Mexico the stems are boiled to obtain wax, which has many uses, including candlemaking, soap, ointments, sealing wax, phonograph records, insulation material, shoe polish, floor polish, waterproofing, and lubricants. The plant also has purgative properties. It is sometimes grown as a succulent in gardens. In Mexico it is known as "Candelilla" (little candle).

Common Poinsettia

Poinsettia pulcherrima Graham

FIELD IDENTIFICATION. Cultivated shrub or small tree to 24 ft, with many long stems from the base or trunk solitary. Very handsome with its brilliant red leaflike bracts in a crown below the small flowers.

FLOWERS. Unisexual, inflorescence cymose and inconspicuous, greenish yellow to reddish; involucre cuplike, 4–5-lobed, bearing a large fleshy yellow gland; 1 stamen in the staminate flower; perianth absent.

FRUIT. An exserted capsule with 3 rounded lobes, seed narrowed forward.

LEAVES. Simple, alternate, ovate-elliptic to lanceolate or panduriform, margin entire or sinuate-toothed or lobed, blade length 3–6 in., essentially glabrous at maturity; upper leaves narrower and more entire and graduating into brilliant red leaflike bracts below the flowers; stipules minute; petioles long.

STEMS. Solitary, or often many from the base, elongate, often arching, green to tan or brown.

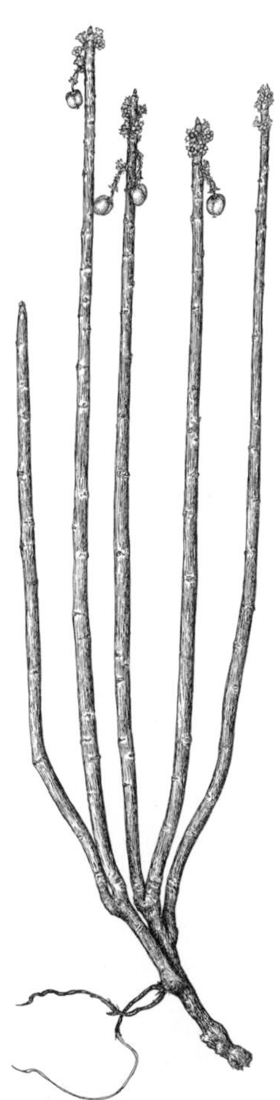

WAX EUPHORBIA
Euphorbia antisyphilitica Zucc.

Common Poinsettia
Poinsettia pulcherrima Graham

RANGE. A native of tropical America, occurring in Mexico in Jalisco, Veracruz, and Oaxaca. Introduced in the United States and grown for ornament. Subject to freezes in the Gulf states, but sometimes persisting in protected places. Grown as a pot plant in the North.

MEDICINAL USES. Standley (1920–1926, p. 600) reports that "in Mexico a decoction of the bracts is sometimes taken by nursing women to increase the flow of milk, but the practice is said to be dangerous. The leaves are applied as poultices for erysipelas and various cutaneous affections."

VARIETIES. Poinsettias can readily be grown from cuttings, and a number of horticultural forms have been developed. White Poinsettia, *P. pulcherrima* forma *alba*, has white bracts.

Double Poinsettia, *P. pulcherrima* forma *plenissima*, has all or most of the flowers turned into red leaflike bracts.

REMARKS. The genus name, *Poinsettia*, is in honor of Joel R. Poinsett (1779–1851), botanist of South Carolina. The species name, *pulcherrima*, refers to the beautiful bracts. This plant is known under many vernacular names in tropical America where it is native, including Flor de Fuego, Flor de Pascua, Flor de Santa Catarina, Santa Catarina, Catalina, Pano Holandés, Bandera, Bebeta, Pastora, Pastores, Pascuas, Flor de Nochebuena, and Cuitlaxochiti. The stems yield a milky juice which is said to have been used by the Indians to remove hair from the skin. The bark yields a red coloring principle and the bracts a scarlet dye.

Castor-bean

Ricinus communis L.

FIELD IDENTIFICATION. Mostly a stout, large herb to 15 ft in temperate regions, but in the tropics to 40 ft, soft-woody.

FLOWERS. In stout, pyramidal, terminal panicles on jointed peduncles, flowers monoecious, greenish, numerous, small, petals absent; pistillate flowers disposed above the staminate; staminate short-pediceled, calyx of 3–5 sepals which are valvate, oval, concave, reflexed, and purplish; stamens numerous and crowded, filaments much-branched; pistillate flowers long-pediceled, calyx caducous, sepals narrowly lanceolate; styles 3, red, plumose, linear, each branch 2-cleft, united at base; ovary 3-sided and rounded, 3-celled and 3-ovuled.

FRUIT. Maturing August–September, capsule ½–1 in. in diameter, glaucous, subglobose or 3-lobed, usually soft-spiny or sometimes smooth, suddenly dehiscent into 3 2-valved carpels, 1 seed in each carpel; seeds variously colored, lustrous, smooth, black to gray or brown and mottled, seed coat crustaceous, endosperm white, highly oily, sweetish to the taste, easily becoming rancid, caruncle large.

LEAVES. Alternate, conspicuously large, blades ⅓–1½ ft broad, nearly orbicular in outline, palmately 6–11-

Castor-bean
Ricinus communis L.

lobed, the lobes acute or acuminate, also irregularly serrate, smooth, glabrous or glaucous on both sides, petioles stout, very variable, but commonly approximating blade length.

STEMS. Of vigorous growth, erect, usually branching, hollow, smooth, glaucous, green to reddish or purplish.

RANGE. Cultivated and escaped to waste places, old fields, abandoned home sites; Texas and Louisiana; eastward to Florida and northward to New Jersey; also in southern California and southward to tropical America. Considered by some authorities as a native of Africa, by others as from India. Now grown on a commercial scale in some of the Gulf states.

MEDICINAL PROPERTIES. The following description of the seed and its properties is outlined by Wood and Osol (1943, pp. 779–781):

"A decoction or poultice of the leaves is sometimes used as a local galactagogue, and an infusion has been given internally for a similar purpose.

. . . .

"The seeds are active poisons; three have produced fatal gastro-enteritis in the adult. Their poisonous action . . . is due to an albumose called *ricin*. . . . It is so actively poisonous that 0.002 milligram is sufficient to kill a rabbit. The toxic symptoms, which frequently do not come on for several hours after the ingestion of the poison, are due primarily to the intensely irritant action of the substance, and consist of diarrhea, with suppression of the urine, and frequently jaundice, probably from irritation of the mucous membranes and biliary ducts. Müller (*Arch. Exp. Path. Pharmak.*, Vol. 42, 1899) states that the poison also has a direct action upon the medulla, leading to the fall of blood pressure and lessened respiratory activity.

. . . .

"The most important constituent of the castor beans is the fixed oil, of which they yield about 50 per cent. The cake left after the expression of the oil, known as *castor pomace*, contains a nitrogenous crystallizable body which differs from the alkaloids in not forming salts with acids. This was first described by Tuson in 1864 and named by him *ricinine*. Spaeth (*Ber. Dtsch. Chem. Ges.*, 58:2124, 1925) has synthesized ricinine; it is soluble in water or chloroform but only sparingly in alcohol. A typical *zymogen* is also present. It is soluble in fats and in a mixture of fats and ethyl ether, and is activated by acids.

. . . .

"The expression [of oil] may be hot or cold. The seeds are first decorticated by being passed between properly adjusted rollers and the kernels are separated from the husks by an air blast. The medicinal oil is expressed cold, or with the kernels not warmer than 50°C. as above this temperature ricinine is dissolved. The residue is sometimes expressed a second or even a third time, with elevation of the temperature. These later fractions of oil are used for industrial or soap-making purposes. The press cake . . . is unfit for feeding animals as it contains the poisonous constituents of the seed but it finds some use as a fertilizer, and also in the manufacture of tiles and certain plastics. About 40 per cent of oil is obtained by expression, 33 per cent in the first pressing and the balance in the later pressings.

"The albuminous principles which may have accompanied the oil are removed by steaming, which also destroys any of the enzyme that might be present and it is finally filtered.

. . . .

"The cathartic effects of castor oil are due to its ricinoleic acid which is split off in the intestinal tract. This, by gentle irritation of the mucous membrane, excites active peristalsis in the upper bowel and probably also somewhat increases the intestinal secretions, although the main action appears to be due to the increased peristalsis. Because of the thoroughness with which it empties the enteric canal there is on the following day an insufficient accumulation of intestinal contents and hence constipation is liable to follow. Because it cleans out the whole alimentary system, and that with relatively small degree of irritation, it is especially valuable in conditions of irritation or inflammation of the bowels, as colic or enteritis; on account of the tendency to subsequent constipation it is not to be recommended as a habitual laxative. It is frequently resorted to in cases of pregnant and puerperal women, and is an excellent cathartic for children. Infants usually require a larger relative dose than adults. Externally it is useful for its emollient effects in seborrhea and other skin diseases, having the advantage over other oils of being soluble in alcohol."

The official preparation of Castor Oil is known as Oleum Ricini Aromaticum, *N. F.*

HORTICULTURAL VARIETIES AND FORMS. Castor-bean is very variable as to stature and leaf and fruit characters. The following are the principal segregates.

African Castor-bean, *R. communis* var. *africanus*, has very large, green leaves.

Bourbon Castor-bean, *R. communis* var. *borboniensis*, has red stems and glaucous leaves (includes form *arboreus*).

Gibson Castor-bean, *R. communis* var. *gibsoni*, has small stature, with dark red, metallic foliage.

Cambodia Castor-bean, *R. communis* var. *cambodgensis*, has dark green foliage and stamens.

Cut-leaf Castor-bean, *R. communis* var. *laciniatus*, has lobes deeply cut.

Big-fruit Castor-bean, *R. communis* var. *macrocarpus*, has purplish red foliage and large fruit.

Big-leaf Castor-bean, *R. communis* var. *macrophyllus*, has large, purplish red leaves. This variety also includes the following forms:

Philippine (*philippinensis*), Purple (*purpureus*), purplish-red, Red (*sanguineus*), red leaves, and Zanzibar (*zanzibarensis*), bright green leaves with white veins.

Hybrid Castor-bean, *R. communis* var. *hybridus* (including the form *palermo*), a large form with dark, very glaucous foliage.

REMARKS. The genus name, *Ricinus*, is the classical

name of the tick, which the seed resembles in shape. The species name, *communis*, means "common." Other common names are Castor-oil-plant and Palma Christi. The plant is not subject to attack by insects or disease. The large, colorful foliage is useful as screens and for tropical effects. Propagation is generally practiced by planting the seeds in pots in fall and then planting out in May. The seeds yield the castor oil of commerce. The oil is known in French as *huile de ricin,* in German as *Castarol,* in Italian as *olio di ricino,* and in Spanish as *aceite de castor*. The oil is reported in legend as keeping away moles and malaria.

CHINESE TALLOW-TREE

Sapium sebiferum (L.) Roxb.

FIELD IDENTIFICATION. Small cultivated tree with a rounded crown, attaining a height of 30 ft.

FLOWERS. Male and female together on a yellowish green terminal spike, the pistillate below the staminate; calyx of staminate flowers 2–3-lobed, the lobes imbricate; petals absent; stamens 2–3, filaments free, anthers opening lengthwise; pistillate calyx of 2–3 sepals; ovary 1–3-celled, styles 2–3.

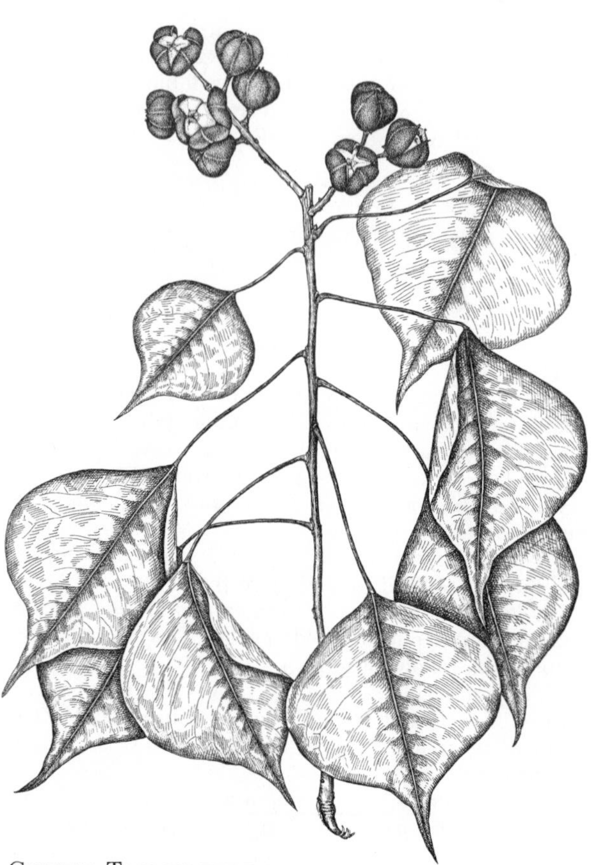

CHINESE TALLOW-TREE
Sapium sebiferum (L.) Roxb.

FRUIT. Capsule 3-lobed, lobes rounded externally and flattened against each other, ⅓–½ in. in diameter, dehiscent by the valves of the capsule falling away to expose the 3 white seeds; seed solitary, crustaceous.

LEAVES. Alternate, or rarely opposite, entire, rhombic-ovate, abruptly long-acuminate, base broadly cuneate, 1–3½ in. long, 1–3 in. broad, widest across the middle, deep green and glabrous above, paler below, turning deep red in winter; petiole slender, usually shorter than the blade, 2 glands at the apex.

TWIGS. Young twigs slender and green, older grayish brown and marked with numerous small lenticels.

BARK. Brownish gray, broken into appressed ridges, fissures shallow.

RANGE. Cultivated for ornament and sometimes escaping. Texas, Oklahoma, Arkansas, and Louisiana, eastward to Florida, along the Atlantic coast to South Carolina. A native of China.

REMARKS. The genus name, *Sapium,* was given by Pliny to a resinous pine, and the species name, *sebiferum,* refers to the vegetable tallow, or wax, of the fruit. Another vernacular name is Vegetable Tallow-tree. The milky sap is poisonous, but the tree is cultivated in China for the wax of the seed covering which is used for soap, candles, and cloth-dressing. The wax was formerly imported into the United States, but mineral waxes have almost entirely taken its place. The tree is easily propagated by seeds and cuttings, and is attractive with white seeds and red leaves in the fall.

SEBASTIAN-BUSH

Sebastiana ligustrina (Michx.) Muell. Arg.

FIELD IDENTIFICATION. Erect, irregularly branched shrub 3–12 ft.

FLOWERS. Borne April–August in inconspicuous, slender, yellowish green terminal spikes which are shorter than the leaves, or rarely over 2 in. long. The upper part of the spike bears the staminate flowers and the lower part the pistillate; staminate flowers on short pedicels hardly over 1/16 in. long, subtended basally by bracts which are reddish green, ovate, acute or acuminate, entire or with a few subulate teeth; sepals 3, ovate, obtuse, acute or acuminate, minute; stamens 3, exserted or barely so, united at the base; anthers cordate, opening lengthwise; pistillate flowers below the staminate, sepals 3, style parted into 2–3 reflexed stigmas, ovary 3-celled or by abortion fewer.

FRUIT. Borne on peduncles ⅛–½ in. long, capsule 3-lobed, or 2-lobed by abortion, depressed, ¼–⅜ in. across, about ¼ in. high; seeds ⅛–⅙ in. long, testa

SEBASTIAN-BUSH
Sebastiana ligustrina (Michx.) Muell. Arg.

thin, smooth, crustaceous, grayish brown, seed body solitary, subglobose, green to yellow or brown.

LEAVES. Half-evergreen, in autumn yellow, orange, or red, alternate, blades ¼–3 in. long, ½–1½ in. wide, firm; elliptic to oblong or oval; apex acute or short-acuminate; base cuneate; margin entire, sometimes undulate on the edge of the plane surface; upper surface dull green and glabrous with inconspicuous veins; lower surface paler, glabrous or rarely pubescent toward the base on the main vein, main vein raised; petiole ⅛–⅓ in., green or reddish, pubescent or glabrous, grooved above.

TWIGS. Slender, erect or spreading, younger ones green to brown, glabrous or nearly so, finely sulcate; older twigs gray to black, smooth and glabrous.

RANGE. In sandy swamps or on stream banks of the coastal plain. Texas and Louisiana, eastward to Florida.

REMARKS. The genus name, *Sebastiana,* is in honor of Antonio Sebastiani, a writer on Roman plants 100 years ago. The species name, *ligustrina,* refers to its privet-like character. Other vernacular names are Candleberry and False Gum. This plant has flowers which resemble the Chinese Tallow-tree, but are smaller, and the leaves are also brilliantly colored in the fall. It is half-evergreen in moderate winters but deciduous in severe ones, and merits some use as an ornamental plant.

CORKWOOD STILLINGIA
Stillingia aquatica Chapm.

FIELD IDENTIFICATION. Plant 2–7 ft, stems stout and glabrous, bearing an umbrella-like top, often standing in shallow water.

FLOWERS. Spikes as long as the leaves or shorter, stout, terminal; flowers small, most often yellow, monoecious, petals absent, bracts 1/25–1/15 in. high, glands smaller than the bracts; staminate flowers above the pistillate, several together in the axils of the bractlets; calyx about 1/25 in. long, slightly 2–3-lobed; stamens 2–3, exserted, free; pistillate flowers at the base of the spike, solitary in the axils of the lower bractlets, calyx 3-lobed; ovary 2–3-celled, with a solitary ovule in each cavity, style stout.

FRUIT. Capsule about ⅜ in. broad, smooth, 2–3-lobed, separating at maturity into 2 or 3 segments; seeds subglobose, ⅛–⅙ in. in diameter, rugose or reticulate.

LEAVES. Simple, alternate, linear to linear-oblong, somewhat broader above or below the middle, apex acute, margin finely serrate, base narrowed into short,

CORKWOOD STILLINGIA
Stillingia aquatica Chapm.

slender petioles or blade sessile, upper leaves sometimes yellowish.

RANGE. Swamps and wet pinelands. Southeastern Louisiana; eastward to Florida and northward to South Carolina.

REMARKS. The genus name, *Stillingia,* honors Benjamin Stillingfleet, English botanist of the eighteenth century. The species name, *aquatica,* refers to the fact that the plant often grows in the water. Vernacular names are Queen's Delight and Queen's Root. The plant is rather short-lived, and the wood is lighter than cork. The seed is known to be eaten by a number of species of birds, including the bobwhite quail.

CASSAVA

Manihot carthaginensis (Jacq.) Muell.

FIELD IDENTIFICATION. Large shrub or small tree to 25 ft, and 6 in. in diameter, with irregular branches forming a rounded crown; rhizomes fleshy.

FLOWERS. Axillary, solitary, or a few in short clusters; pedicels ⅓–¾ in. long; staminate and pistillate flowers borne separately; petals absent; calyx 5-parted, green with two reddish blotches on each sepal within, about ¾ in. across at anthesis; staminate with the sepals united about halfway to the base, stamens 10, slightly shorter than the sepals, the filiform filaments arising from between the lobes of a basal orange-colored dis pistillate flowers with the 5 oblong-ovate sepals fre to the base or nearly so; ovary dark green, subglobos smooth, immersed in a thick entire orange-colored dis ovules 1 in each of the 3 cells; stigmas sessile, yellowis green, 3-lobed, rounded and fimbriate on margin.

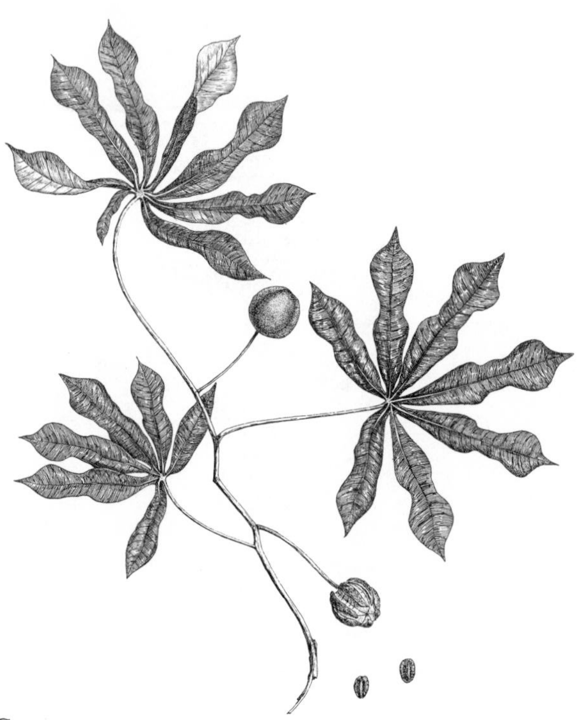

CASSAVA
Manihot carthaginensis (Jacq.) Muell.

FRUIT. Capsule subglobose, or slightly 3-lobed, ½– in. long, pale green, smooth, glabrous, 3-celled, seed somewhat flattened, ends obtuse or rounded, ⅜–½ i long; peduncles stout, ½–2½ in. long, glabrous, ape with a persistent orange disk.

LEAVES. Alternate, palmately parted into 6–9 seg ments (rarely, more or fewer), each segment 3–8 i long, oblanceolate, lanceolate or narrowly elliptic, ma gin entire or sinuate-wavy, apex acute or acuminat sometimes abruptly so, base gradually narrowed, uppe surface dark green, glabrous and semilustrous, lowe surface much paler and glabrous, venation yellowis and delicate; petiole glabrous, light green, shorter c longer than the leaf segments.

TWIGS. Young ones slender, smooth, glabrous, pal green, rather fleshy; older ones brownish and muc more woody.

BARK. On trunk rather smooth, light to dark brow marked with numerous horizontal, pale, linear to ob long lenticels.

RANGE. Cultivated in increasing amount the last fe years along the Gulf Coast from Texas and Louisian eastward to Florida. Sometimes escaping cultivatio about old gardens. Also grown in southern New Mexi co, Arizona, and California. A native of Venezuela Colombia and Brazil; also countries of Central America

REMARKS. The genus name, *Manihot,* is the nativ Brazilian name. It is also known under the names o Yuguilla, Yuca de Monte, Xcache, and Cuadrado. Th tree is a desirable ornamental because of the attractiv segmented leaves. The seeds have emetic and purgativ properties. The fleshy rhizomes are sometimes use as cassava, a starch derived after the poisonous aci is removed. However, most commercial cassava, fro which tapioca is obtained, comes from the rhizome of the closely related *M. utilissima.*

BLOOD-OF-CHRIST

Jatropha cardiophylla (Torr.) Muell.

FIELD IDENTIFICATION. Western shrub with flexibl stems to 3 ft, from running rootstocks.

FLOWERS. Borne in terminal or lateral cymes, monoe cious or dioecious; staminate cymes terminal on shor spurlike branches; calyx 5-lobed, about one third a long as the corolla, lobes ovate to deltoid, obtuse o acute (some lobes glandular-toothed in the pistillat flower); corolla tubular, about ¼ in. long, separatin into 5 short petals which are reflexed with maturity somewhat hairy deeply within; stamens 8–10, filaments

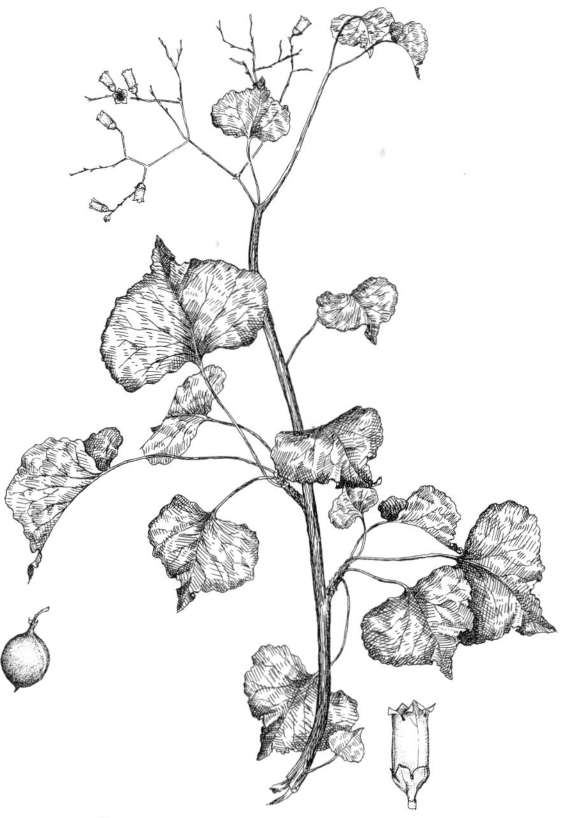

BLOOD-OF-CHRIST
Jatropha cardiophylla (Torr.) Muell.

united below, anthers in fives in 2 series, glands 5; pistillate flowers pediceled, usually 1–3 on short spurs; style entire, thickened, stigma lobed; ovary 1–3-celled, ovules solitary, embryo fleshy.

FRUIT. Globose or slightly flattened, greenish brown, somewhat ridged, abruptly short-pointed, ½ in. or less in diameter, indehiscent; seeds globose to ellipsoid, about ⅜ in. long, hard-carunculate.

LEAVES. Simple, alternate, or crowded on short spurs, cordate-deltoid, apex acute or abruptly acuminate, margin crenate with teeth rounded and often glandular, blade ½–2¾ in., also about as wide, upper surface lustrous green, lower surface somewhat paler with delicate veins, glabrous; petioles ¼–2 in., slender, glabrous.

TWIGS. Slender, straight or divaricate, very flexible, reddish brown, glabrous, younger parts sometimes glutinous, often with leaves and flowers on short, stubby, lateral spurs, twigs wrinkling considerably on drying.

RANGE. Dry rocky and sandy soils at altitudes of 2,000–3,000 ft. Southern Arizona and southern New Mexico to Sonora, Mexico.

REMARKS. The genus name, *Jatropha*, is of Greek derivation, meaning "I eat a remedy," and the species name, *cardiophylla*, refers to the heart-shaped leaf. It is also known under the name of Limber-bush because of the flexible stems, and Sangre-de-drago, because of the reddish root sap, which is used by the Indians for a red dye. The sap is also a coagulant and is used to staunch blood flow of wounds. The plant contains tannin and a small percentage of rubber.

RUBBER-PLANT

Jatropha dioica Sessé

FIELD IDENTIFICATION. Plant forming colonies from underground runners, usually less than 3 ft high (rarely, to as much as 15 ft). Stems succulent, but with some woody tissue, terete, very flexible and tough, can be tied in knots without breaking. Sap watery, yellowish, or reddish when bruised. Side branches often short and stubby.

FLOWERS. July–August, monoecious, usually flowering toward the apex of the stem or on short, stubby branches and cushioned by scalelike bracts. Flowers solitary or fascicled on pedicels 1⁄16–⅜ in. long; corolla white or pink, 3⁄16–¼ in. long, surpassing the calyx, semiglabrate to pubescent, tubular, 5-lobed; lobes shorter than the tube, reflexed, ovate to oblong, obtuse; calyx about one third as long as the corolla (about one half as long in the pistillate flower); sepals 5, lanceolate-ovate, apex acute or acuminate, reddish; stamens 8–10 (in 2 whorls of 4 or 5 each), barely exserted, united below, anthers oblong, introrse; ovary acute, ovules solitary in each cavity; styles thick, unequally 2-lobed, united into a column.

FRUIT. Capsule about ¾ in. long, globose-ellipsoid, somewhat asymmetrical, depressed, apiculate, base oblique, pubescent or glabrous, tardily dehiscent, style persistent; separating into 2–3 2-valved carpels with 2–3 seeds which are oblong to oval and less than ⅝ in. long, seed coat thin and papery.

LEAVES. Very variable, even on the same plant. Simple, alternate or fascicled on short, spurlike branches, glabrous and coriaceous, obovate-obcordate to spatulate, sometimes linear, margin entire, apex obtuse to rounded or occasionally 2–3-lobed or notched, length ⅜–2¾ in., petiole about 1⁄25 in. long or leaf sessile.

STEMS. Very flexible and rubbery, glabrous, reddish brown to gray and thin-barked, sometimes wrinkled.

RANGE. The two varieties are usually found on dry slopes, mesas, and rocky limestone bluffs. The Sessile-flower Rubber-plant is the most common form in the Southwest, and the Grass-leaf Rubber-plant is mostly confined to the Trans-Pecos Texas region.

MEDICINAL USES. According to Standley (1920–1926, p. 638): "The juice has astringent properties and is used for hardening the gums, for skin eruptions, sores, dysentery, hemorrhoids, and venereal diseases, to prepare a gargle for sore throat, as a wash to restore and give luster to hair, and to remove stains from teeth. The roots are also chewed to relieve toothache."

RUBBER-PLANT
Jatropha dioica Sessé

VARIETIES AND SYNONYMS. Some authors use the names *J. spathulata* Muell. Arg. or *Mozzina spathulata* Orteg. for the plant. McVaugh (1945, pp. 36–39) has given reasons for the use of the name *dioica,* and has also described two varieties as follows:

Grass-leaf Rubber-plant, *J. dioica* var. *graminea* McVaugh, has leaves narrow-oblanceolate to linear, two and one half to seventeen times as long as wide. Mostly $\frac{1}{25}$–$\frac{1}{6}$ in. wide. Margin often with lobes $\frac{3}{8}$–$\frac{5}{8}$ in. long (rarely, $1\frac{1}{2}$ in.). Apparently confined to the Big Bend area of Texas in Brewster and Presidio counties. In Coahuila, Chihuahua, Zacatecas, and San Luis Potosí, Mexico.

Sessile-flower Rubber-plant, *J. dioica* var. *sessiliflora* (Hook.) McVaugh, has leaves oblanceolate to obovate, two and one half to six times as long as wide, mostly $\frac{1}{4}$–$\frac{2}{5}$ in. wide (on shoots exceptionally $1\frac{1}{4}$ in. wide to $1\frac{5}{8}$ in. long); width of lobed blades across the widest points of the lobes $\frac{5}{8}$–$1\frac{3}{8}$ in., the central lobe $\frac{3}{8}$–$\frac{5}{8}$ in. wide. Widespread, on the limestone hills of central Texas, westward and southward. In Texas in Bexar, Blanco, Atascosa, Nueces, Val Verde, and Cameron counties. In New Mexico and Arizona. In Mexico in Tamaulipas, Nuevo León, Coahuila, San Luis Potosí, Zacatecas, Durango, Guanajuato, Querétaro, Hidalgo, Distrito Federal, Puebla, and Oaxaca.

The author has found Texas plants which present intermediate characters between the two varieties of McVaugh. However, McVaugh points out that in certain areas in Mexico intermediates do occur, and more material is needed to determine the exact status of the variations.

REMARKS. The genus name, *Jatropha,* is from the Greek, referring to its use in medicine. The species name, *dioica,* refers to the dioecious flowers of some species, but apparently not this species. It is known under many vernacular names in the Latin-American countries including Sangregrado, Sangregado, Sangre de Grado, Tecote Prieto, Tocote Prieto, Toxote Prieto, Telondillo, Tlapalezpatli, Piñón del Cerro, Coatli, Torte Amarillo, and Drago. The plant yields a dark red dye, but it is reported to be injurious to cloth. The flexible stems are used for whips or withes.

SMOOTH DITAXIS
Ditaxis laevis (Muell.) Heller

FIELD IDENTIFICATION. Perennial plant to 1 ft or more. Stems slender, much-branched, glabrous, woody at the base.

FLOWERS. Borne in 4–5-flowered spikelike racemes usually shorter than the leaves; staminate flowers with the 5 sepals linear-lanceolate; petals lanceolate, alternate with the sepals; stamens usually 5, sometimes more, united into a column; pistillate flowers with sepals lanceolate, the petals shorter; glands filiform, often lobed; styles 3, each 2-lobed and dilated above, ovary glabrous.

FRUIT. Capsule globular, depressed, 3-lobed, apiculate, rugose.

LEAVES. Simple, alternate, oblong to obovate or lanceolate-spatulate, about 1 in. long, apex acute, narrowed into a short petiole.

RANGE. Dry plains of the creosote-bush belt. West Texas, New Mexico, and Mexico.

REMARKS. The genus name, *Ditaxis,* is from the Greek word meaning "double-ranked," in allusion to the stamens. The species name, *laevis,* refers to the plant's smooth, glabrous character.

SUMAC FAMILY (*Anacardiaceae*)

AMERICAN SMOKE-TREE
Cotinus obovatus Raf.

FIELD IDENTIFICATION. Tree with slender, spreading branches attaining a height of 35 ft, with a diameter of 12–14 in. Most often only a straggling shrub.

FLOWERS. April–May, regular, dioecious or rarely polygamo-dioecious, greenish yellow, borne in loose, few-flowered terminal panicles 5–6 in. long and 2½–3 in. broad; pedicels slender, becoming conspicuously glandular-villose and purplish; subtending bracts linear or spatulate; corolla about ⅛ in. across, petals 5, oblong, acute, crisped, deciduous, about twice as long as the calyx-lobes and alternate with them; calyx-lobes 5, persistent, ovate-lanceolate, obtuse, disk at the base; stamens 5, inserted under the disk, included within the corolla, shorter than the petals and alternate with them, abortive or absent in the pistillate flower; ovary 1-celled, sessile, obovoid, oblique, compressed, rudimentary in the staminate flower; styles 3, short and spreading, stigmas large.

FRUIT. Drupelets ⅛–¼ in. long, peduncles 2–3 in. long, slender pedicels 1½–2 in. long, conspicuously purple or brown glandular-hairy; fruit body kidney-shaped or oblong-oblique, flattened, glabrous, veiny, pale brown, style remnants persistent; skin thin, dry, chartaceous; stone thick and bony.

LEAVES. Simple, alternate, deciduous, most abundant distally on the twigs, 1½–6 in. long, 2–3½ in. wide, elliptic-oval or obovate; apex rounded or obtuse or slightly emarginate; base broadly cuneate or rounded; margin entire or somewhat undulate, surface olive-green and glabrous to puberulent above, paler and pubescent below, veins conspicuous; petiole ¼–2 in., yellowish green to reddish, glabrous or pubescent, blade of leaf sometimes slightly decurrent on petiole.

TWIGS. Slender, young ones green to reddish or purple, lenticels small, abundant, pale; older ones gray, smooth, leaf scars large and conspicuous.

AMERICAN SMOKE-TREE
Cotinus obovatus Raf.

BARK. Gray to black, roughly breaking into thin, oblong, small scales.

WOOD. Orange to yellow, sapwood cream-white, coarse-grained, soft, light, weighing about 40 lb per cu ft.

RANGE. On rocky limestone hills of Texas, Oklahoma, Arkansas, Missouri, Alabama, Tennessee, and Kentucky. Nowhere very abundant or widespread. In Texas mostly on rocky banks of the upper Guadalupe and Medina rivers. In Kendall, Kerr, and Bandera counties. Near

Kerrville; between Utopia and Tarpley; also at Spanish Pass.

REMARKS. The genus name, *Cotinus,* is the classical name for the Wild Olive, and the species name, *obovatus,* refers to the leaf shape.

The wood yields a yellow dye and was much used for that purpose during the Civil War period. The wood is very durable in contact with the soil and is used for fence posts. The tree easily sprouts from the stump. The brilliant orange and red colors of the leaves in autumn make it a worthwhile ornamental, but the purple flower pedicels, which give it the name of Smoke-tree, are not as showy as those of the European Smoke-tree, *C. coggygria* Scop. American Smoke-tree has obovate, larger, thinner, basally cuneate leaves, a less showy fruiting pedicel, and lacks the white leaf margins of the European Smoke-tree. However, the two resemble each other closely, and are somewhat difficult to distinguish under cultivation.

TEXAS PISTACHE

Pistacia texana Swingle

FIELD IDENTIFICATION. Large shrub or small tree, with a number of trunks from the base. Attaining a height of 15–30 ft, with a spread of 15–30, rarely 36 ft. The larger trunks 8–10 in., rarely 12–14 in., in diameter. Forming compact rounded clumps. Young foliage in spring a beautiful red.

FLOWERS. Pistillate with or before the new leaves, loosely paniculate, glabrous, 1½–2¾ in. long; flowers small, subtended by a small reddish ciliate bract and 2 bractlets; perianth absent; ovary ovate or sub-globose; styles 3, 2 shorter ones with 2-lobed stigmas, the longer one with a 3-lobed stigma; staminate in compact panicles ¾–1½ in. long, more crowded than the pistillate, the subtending bracts reddish, anthers very evident, reddish yellow.

FRUIT. Oval to lens-shaped, reddish brown, finally glaucescent, ⅕–¼ in. long, ⅙–⅕ in. broad, 1/15–⅛ in. thick.

LEAVES. Persistent, odd-pinnately compound, 2–4 in. long, 1–2½ in. broad, petiole ⅖–⅗ in., or sometimes ⅘ in. on staminate trees, flattened and very narrowly winged; rachis very narrowly winged, puberulent above; leaflets 4–9 pairs (usually 5–8 pairs), opposite or subopposite, thin, reticulate-veined, ⅓–1 in. long, ⅕–⅓ in. broad, young leaves reddish, lanceolate, acute, at maturity leaves broadly rounded, spatulate or nearly so, apex mucronate, base deltoid or subcuneiform; lateral leaflets usually somewhat falcate and inequilateral; midrib near the twig side, upper surface dark green and sparingly pubescent along the midrib, lower surface pale green and glabrous, margin entire; lateral leaflets almost sessile, terminal leaflet with a petiolule ⅙–¼ in. long.

TWIGS. Slender, slightly pubescent, at first reddish, later brown to gray; buds small, puberulent, 1/15–1/10 in. long.

TEXAS PISTACHE
Pistacia texana Swingle

WOOD. Heartwood yellowish brown, sapwood pale yellowish, tough, strong, compact, fine-grained, weighing 60 lb per cu ft.

RANGE. In rocky limestone stream beds, or on limestone cliffs, sometimes in alluvial soils. Along the Pecos River near its junction with the Rio Grande in southwestern Texas and adjacent Mexico. Type locality near Hinojose Spring, Rio Grande Valley, near the mouth of the Pecos River, about 20 miles west of Comstock, Val Verde County, Texas.

REMARKS. The genus name, *Pistacia,* is from the Greek *pistake* or *pistakia* ("pistache"), and ultimately from the ancient Persian *pistah* ("pistache nut"). The species name, *texana,* refers to the state of Texas. This plant has been confused with the closely related *P. mexicana* H.B.K. of Mexico. Swingle (1920, p. 108) gave the following comparison: "This new species, *Pistacia texana,* differs from *P. mexicana* H.B.K. in having smaller leaves with fewer leaflets (4–9, usually 5–8 pairs, instead of 8–18, usually 12–16 pairs), which are more or less spatulate, broader and more obtuse at the tip, not so markedly mucronate and more or less curved. The young twigs more or less pubescent and have smaller and less pubescent flower-buds and bracts than in *P. mexicana.* The mature fruits of *P. texana* are dark reddish brown, slightly glaucescent rather than glaucous and purplish black, as in *P. mexicana.* The trunks of the trees of *P. texana* are much branched near the ground, while *P. mexicana* often (perhaps always) has a single trunk. The smaller branches are rough grayish brown whereas those of *P. mexicana* are smooth and often light brownish gray, almost silvery."

The tree has possibilities as an ornamental because of its dark, evergreen mature foliage and red new foliage in spring. It is highly intolerant of shade and self-prunes very rapidly.

FRAGRANT SUMAC
Rhus aromatica Ait.

FRAGRANT SUMAC
Rhus aromatica Ait.

FIELD IDENTIFICATION. Thicket-forming, ascending or prostrate, irregularly-branched shrub to 8 ft.

FLOWERS. Polygamo-dioecious, borne in terminal spikelike clusters, about 1½–2⅓ in. long and ¾–1⅓ in. broad, before the leaves; subtended by bracts which are ovate, pubescent, ciliate, about 1/25 in. long; individual flowers yellowish green, sessile or nearly so; sepals 5, persistent, ovate, obtuse, glabrous, 1/12–1/25 in. long; petals 5, ovate-oblong, obtuse, glabrous, 1/25–1/12 in. long, nearly erect; stamens 5, shorter than the petals, exserted; filaments subulate, as long as the anthers or longer; pistil sessile, ovary pubescent, styles 3.

FRUIT. Drupe persistent, July–August, subglobose, red, hairy with simple or glandular hairs, ⅕–⅓ in. long.

LEAVES. Alternate, deciduous, aromatic, 3-foliate, 1–4 in. long, leathery, glabrous or with varying degrees of pubescence, turning brilliant colors in autumn; petiole about 1 in. long but varying in length; leaflets sessile or nearly so, lateral ones oval, ovate or obovate, margin usually crenate-dentate (sometimes incised) toward the apex, entire toward the base; base rounded or truncate; inequilateral, pubescent or glabrous, ¾–1¾ in. long; ⅝–1⅓ in. wide, terminal leaflet short-stalked, ovate to obovate, apex acute or acuminate, base cuneate, usually pubescent, margin crenate-dentate, 1½–3 in. long, ¾–2½ in. wide.

TWIGS. Slender, brown, pubescent to glabrous later.

RANGE. Woods, hills, sand dunes, rocky soil. Eastern Texas, Oklahoma, Arkansas, and Louisiana; eastward to Florida, northward to Vermont, and westward to Minnesota, Michigan, and Missouri.

VARIETIES AND SYNONYMS. *R. aromatica* has been listed under the names of *R. canadensis* Marsh., *R. crenata* (Mill.) Rydb., and *Schmaltzia aromatica* Desv. Because of the fact that this plant is so variable a number of varieties and forms have been named as follows:

R. aromatica var. *serotina* (Greene) Rehd. has the terminal leaflet 1–2½ in. long, flabelliform-obovate, base cuneate, apex rounded, lower surface slightly pilose to glabrate, plant habit upright and to 6 ft. Flowers expanding with leaves. Known in northeast Texas, Oklahoma, and Arkansas; north to Nebraska and Illinois.

R. aromatica var. *illinoensis* (Greene) Rehd. has tomentulose branches and leaves velvety-tomentose beneath and appressed-pubescent above, terminal leaflet elliptic to rhombic-ovate, apex acutish, base subcuneate or tapering equally, length 1¼–3½ in.; flowers usually expanding before the leaves. It occurs in Missouri and Illinois.

R. aromatica var. *arenaria* (Greene) Fern. has a low, depressed habit of growth, leaflets with a velvety-tomentose lower surface, terminal one ¾–1½ in. long. In sandy soil in Ohio, Indiana, and Illinois.

R. aromatica var. *pilosissima* (Engelm.) Shinners, which was once known under the name of *R. trilobata* var. *pilosissima* Engler., has the young growth and leaflets very densely tomentose, round-ovate, crenate-dentate toward apex, entire toward the cuneate base; fruit is somewhat larger than in the species. It occurs in Trans-Pecos Texas and the Panhandle areas, also in New Mexico, Arizona, and California; and in Mexico in the states of Chihuahua, Durango, and Jalisco.

R. aromatica var. *flabelliformis* Shinners is described in full elsewhere in this book. It was formerly known as *R. trilobata* Nutt.

R. aromatica var. *laciniata* is a horticultural variety with deeply lobed leaves.

REMARKS. The genus name, *Rhus*, is the ancient

Latin name, and the species name, *aromatica*, refers to the aromatic leaves. It is also known as the Sweet-scented Sumac and Three-leaf Sumac. The fruit of Fragrant Sumac is eaten by numerous species of birds including wild turkey, ruffed grouse, and flicker; also eaten by raccoon, opossum, and white-tailed deer.

Skunk-bush Sumac

Rhus aromatica Ait. var. *flabelliformis* Shinners

FIELD IDENTIFICATION. An offensive-scented shrub with slender, spreading, crooked branches, attaining a height of 12 ft.

FLOWERS. March–April, numerous, small, borne before the leaves in terminal spikes; bracts 1/25–1/12 in. long, triangular, margin ciliate, upper surface glabrous, lower surface pubescent; individual flower pedicels 1/12–1/8 in. long; sepals 5, persistent, triangular-lanceolate to oblong, apex obtuse to rounded, margin ciliate, glabrous, 1/25–1/12 in. long, less than 1/25 in. broad; petals 5, greenish white, obovate or oblong, obtuse, glabrous above, somewhat hairy beneath, 1/12–1/8 in. long, anthers oval, less than 1/25 in. long, stamens 5, filaments as long as the sepals; pistil short, stigmas short.

FRUIT. Drupe August–September, persistent, 1/5–1/4 in. long, red, subglobose, with glandular or simple short hairs; seed lenticular, about 1/5 in. long or less, broader than long, somewhat roughened.

Skunk-bush Sumac
Rhus aromatica Ait. var. *flabelliformis* Shinners

LEAVES. Strongly pungent when crushed, alternate deciduous, trifoliate, 3/4–2 in. long; leaflets cuneate obovate, oval or spatulate; firm, sessile or nearly so dark green and glabrous above, paler and pubescent beneath; terminal leaflet cuneate, larger than the others, 3-lobed with smaller lobes or crenate-toothed, apex obtuse or rounded, base cuneate, about 3/4 in. long; lateral leaflets smaller than the terminal, obovate to oval, about 3/8 in. long, also 3-lobed, but often much less so; petiole 1/4–1/3 in., pubescent, reddish. Leaves brilliantly colored in autumn.

TWIGS. Slender, gray to reddish brown, puberulent at first, glabrous later.

RANGE. Often in limestone outcrops. Central, northern, western, and southwestern Texas, New Mexico, Arizona, and California. In Mexico in Nuevo León, Chihuahua, Coahuila, Tamaulipas, Durango, Puebla, Hidalgo, San Luis Potosí, and Colima.

PROPAGATION. Skunk-bush Sumac is drought-resistant, thicket-forming, and deep-rooted. It is sometimes planted for erosion control and shelter-belt purposes in the prairie plains region. It can be propagated by root cuttings. The seed averages about 20,300 seeds per lb. In nursery practice, the seeds may be sown in fall and should be stratified for 60 days at 34°–40°F. and treated with concentrated sulphuric acid for 1 hour. They are sown at a depth of 1/2 in., with 25 seeds per linear ft. Germination takes 10–30 days. The seedlings are very susceptible to damping-off.

REMARKS. The genus name, *Rhus*, is the ancient Latin name, and the variety name, *flabelliformis*, refers to the fan-shaped leaves. This plant was formerly known under the name of *R. trilobata* Nutt., but has been changed by Shinners (1951b). Other vernacular names are Ill-scented Sumac, Three-leaf Sumac, Lemonade Sumac, Squaw-bush, Quail-bush, Agrillo, and Lemita. The acid fruit is eaten, and a cooling drink is made by steeping it in water. The slender twigs are mixed with willow in basket weaving by the Indians, and also produce a yellow dye. The fruit is known to be eaten by 25 species of birds, especially the gallinaceous birds such as various species of quail, grouse, prairie chicken, sage-hen, and pheasant. It is browsed by mountain sheep and more rarely by deer. Its value for livestock browse seems to vary according to locality.

New Mexico Evergreen Sumac

Rhus choriophylla Woot. & Standl.

FIELD IDENTIFICATION. Low, western shrub with few branches and pinnately compound leaves, with 3–5(7) leaflets. Closely related to the Evergreen Sumac, but the inflorescences are more lateral and numerous, and leaflets larger, more glabrous, and fewer in number. Intergrading forms are to be found.

New Mexico Evergreen Sumac
Rhus choriophylla Woot. & Standl.

FLOWERS. Mostly July–August, inflorescence finely pubescent, 2 in. long or broad, petals 5, white, oval to ovate, length about 1/8 in., width about 1/16 in., glabrous or slightly hairy within; filaments shorter than the sepals, anthers about 1/25 in. long; sepals deltoid, apex rounded, length about 1/12 in., width 1/25–1/12 in., glabrous or nearly so, margin ciliate, glandular-hairy; bracts ovate, apex acuminate, length 1/8 in., width 1/25 in.; bracteoles persistent, about 1/12 in. long and broad, apex rounded, pubescent, and margin ciliate with glandular hairs.

FRUIT. About 1/4 in. long, subglobose, reddish or brownish pubescent with simple or glandular hairs; seed yellow to brown, oval to short-oblong, flattened, ends rounded, 1/8–3/16 in. long.

LEAVES. Pinnately compound, rachis 1–2½ in. long, glabrous or puberulent; leaflets 3–5 (rarely, 7), ovate, apex acute, base obtuse to subcuneate or often asymmetrical, margin entire or slightly revolute; upper surface dark green and dull, delicately veined; lower surface much paler, glabrous or slightly puberulent; terminal leaflet 1–2 in. long and 3/8–1 in. broad, coriaceous, petiole 1/8–1/4 in. long, glabrous or puberulent; lateral leaflets 1–2 in. long, and 3/8–1¼ in. broad, petiolules about 1/12 in. long.

TWIGS. Younger ones greenish brown, densely short-pubescent; when older gray to reddish brown, more glabrous.

BARK. Gray to brown, gray and smooth on branches; on trunks broken into thin, small, gray flakes.

RANGE. Trans-Pecos Texas, southwestern New Mexico, and Arizona. In Mexico in the state of Sonora.

REMARKS. The genus name, *Rhus*, is the ancient Latin name, and the species name, *choriophylla*, means "many-leaved."

Evergreen Sumac

Rhus sempervirens Scheele

FIELD IDENTIFICATION. Western evergreen shrub to 12 ft, forming rounded clumps. Branches spreading and often the lower ones touching the ground.

FLOWERS. Appearing irregularly during the summer after rains, borne in terminal or axillary, thyrsoid panicles 3/4–2 in. long, shorter than the leaves; panicles subtended by persistent bracts which are ovate, acute to rounded at the apex, 1/25–1/12 in. long, upper surface glabrous or nearly so, lower surface hairy, margin ciliate; sepals 5, oval, ovate or triangular, obtuse at apex, about 1/12 in. long, glabrous above and somewhat hairy beneath, the ciliate marginal hairs simple or glandular; petals 5, white or greenish, oblong-ovate to obovate, obtuse at apex, base obtuse or truncate, 1/8–1/6 in. long and 1/12 in. broad, glabrous above and hairy beneath; stamens 5, filaments as long as the sepals with

Evergreen Sumac
Rhus sempervirens Scheele

FLAME-LEAF SUMAC *Rhus copallina* L.

anthers about 1/25 in. long and broad; pistil with a slightly lobed stigma.

FRUIT. Maturing usually by September, subglobose, lenticular, oblique, 1/3–2/5 in. long, covered with red hairs which are simple and glandular; seed lenticular, broader than long, smooth.

LEAVES. Alternate, evergreen, 2–5½ in. long, odd-pinnately compound of 5–9 leaflets on a softly pubescent rachis, oblong, oval, ovate or lanceolate, apex acute, base cuneate, margin entire or subrevolute, leathery; upper surface dark green, lustrous and glabrous to puberulent; lower surface paler dull green with simple or glandular hairs; lateral leaflets ½–1½ in. long, 3/8–¾ in. broad, petiolules about 1/12 in. long; segments of rachis between leaflets ¼–½ in. long; terminal leaflet ¾–1½ in. long, 3/8–¾ in. wide, long-petiolulate; petioles about 5/8 in. long. Leaves turning red, yellow, or brown in autumn.

TWIGS. Brown to gray, slender, stiff, young puberulent, older glabrous.

BARK. Gray to brown, rough with small loose scales.

RANGE. In central and western Texas, New Mexico, and Mexico. On rocky hillsides, cliffs, and slopes at altitudes of 2,000–7,500 ft. In Mexico in the states of Nuevo León, Coahuila, Chihuahua, San Luis Potosí, Zacatecas, Durango, and Hidalgo.

RELATED SPECIES. A closely related species of Arizona is *R. choriophylla* Woot. & Standl., which has longer, more glabrous leaves and axillary inflorescence. However, some authors feel that it is only a geographical variation of *R. sempervirens*.

REMARKS. The genus name, *Rhus*, is the ancient Latin name, and the species name, *sempervirens*, refers to the evergreen habit. It has also been listed by some authors under the name of *R. virens* Lindh. Vernacular names in English, Spanish, and Indian dialects are Tobacco Sumac, Capulín, Lambrisco, Lantrioco, Ayume, Kinnikinnick, and Tamaichia. Tamaichia was the name given by the Comanche Indians, who gathered the leaves in the fall, sun-cured them, and mixed them with tobacco for smoking. It is also reported that the leaves were used in domestic medicine for relieving asthma. The acid, red fruit steeped in water makes a cooling drink. The red fruit and evergreen, lustrous leaves suit the plant for use in horticulture. The propagation methods of Evergreen Sumac seed are the same as those for Smooth Sumac.

FLAME-LEAF SUMAC

Rhus copallina L.

FIELD IDENTIFICATION. Slender-branched shrub or small tree to 25 ft.

FLOWERS. Polygamo-dioecious, about 1/8 in. across, borne in a densely pubescent, terminal thyrse about 4¾ in. long and 4 in. broad; pedicels about 1/12 in. long, pubescent; bracts very small, lanceolate, about 1/12 in. long; sepals 5, deltoid, pubescent, glandular-ciliate, 1/12–1/8 in. long, about 1/25 in. broad; stamens 5, anthers lanceolate; pistil 1, sessile, ovary pubescent, stigmas 3, styles 3; disk annular. Petals greenish white, 1/12–1/10 in. long, 1/25 in. broad, glabrous externally, a few hairs on the inner side, margin ciliate and glandular, deciduous.

FRUIT. In compact panicles, erect or drooping, persistent; drupe subglobose, flattened, red, glandular-hairy, 1/6–1/8 in. in diameter; seeds solitary, smooth.

LEAVES. Deciduous, alternate, pinnate, 5–12 in. long, rachis pubescent, broadly winged; leaflets 7–17, subsessile, inequilateral, elliptic or ovate to lanceolate, acute or acuminate at the apex, asymmetrical and obtuse, or rounded to subcuneate at the base, entire or few-toothed, lustrous, glabrous to pubescent above, hairy and glandular beneath; lateral leaflets sessile, 1–3½ in. long, ½–1¼ in. broad, terminal leaflet petiolulate or sessile.

TWIGS. Green to reddish brown, pubescent at first, glabrous later; lenticels dark.

BARK. Thick, greenish brown, excrescences circular, scales thin.

WOOD. Light brown to greenish, coarse-grained, soft, weighing 32 lb per cu ft, sometimes used for small posts.

RANGE. Moist soil in shade or sun. Texas, Oklahoma, Arkansas, and Louisiana; eastward to Georgia, northward to New Hampshire, and west to Michigan and Missouri.

VARIETIES. Flame-leaf Sumac is generally replaced in central Texas by the Prairie Flame-leaf Sumac, *R. copallina* L. var. *lanceolata* Gray, which has narrower and more falcate leaves, larger clusters of fruit, and a more treelike rounded form. It is thought that the two may hybridize in their overlapping areas.

White Flame-leaf Sumac, *R. copallina* var. *leucantha* (Jacq.) DC., is a variety with white flowers found near New Braunfels, Texas.

A variety with 5–13 broader oblong to narrow-ovate leaflets has been described from Oklahoma, Texas, Arkansas, and Louisiana as *R. copallina* L. var. *latifolia* Engl., but other authors have relegated it to the status of a synonym of the species.

REMARKS. The genus name, *Rhus,* is the ancient Latin name, and the species name, *copallina,* means "copal gum." Vernacular names for this plant are Mountain Sumac, Smooth Sumac, Black Sumac, Shining Sumac, Dwarf Sumac, Upland Sumac, and Winged Sumac.

The bark and leaves contain tannin and are used in the tanning industry. The crushed acrid fruit of this and other species was added to drinking water by the Indians to make it more palatable.

According to stomach records, the fruit has been eaten by at least twenty species of birds, and white-tailed deer occasionally browse it. It is propagated by seed in a manner similar to that of Smooth Sumac. It is conspicuous in fall because of the brilliant red leaves.

PRAIRIE FLAME-LEAF SUMAC

Rhus copallina L. var. *lanceolata* Gray

FIELD IDENTIFICATION. Clumpy shrub or small tree to 30 ft, with a rounded top.

FLOWERS. Borne in terminal panicles 4–6 in. long and 2–3 in. wide, peduncle pubescent, secondary pedicels about 1/6 in. long or flowers almost sessile; individual flowers yellowish green to white, about 1/8 in. long; bracts deciduous, ovate, 1/25–1/12 in. long, apex rounded, glabrous above, hairy below, margin ciliate; sepals 5, erect, ovate-triangular, acute or obtuse, 1/25–1/12 in. long, glabrous above, hairy beneath, margin ciliate with glandular or simple hairs; petals 5, yellowish green to white, about 1/8 in. long and 1/25 in. wide, oblong-ovate, obtuse, reflexed, smooth above, hairy beneath, margin ciliate; stamens 5, exserted, filaments elongate; anthers lanceolate, yellow, conspicuous, about 1/25 in. long and wide; ovary 1-celled, stigmas 3, styles 3.

FRUIT. In terminal showy panicles, drupe about 3/16 in. long, subglobose, flattened, dark red, glandular-hairy, the stigma persisting; seeds about 1/8 in. long and 1/12 in. broad, smooth, oval to obovate.

LEAVES. Alternate, 5–9 in. long, odd-pinnately compound of 9–21 leaflets; rachis between the leaflets narrowly winged, green to reddish, pubescent; leaflets sessile or short petioluled, 1–3 in. long, 1/4–1/2 in. wide, lanceolate to linear, falcate, apex acuminate, base cuneate or rounded, asymmetrical, margin entire or with coarse teeth, thin, subrevolute; upper surface dark green, lustrous, glabrous; lower surface duller, paler, pubescent or glandular-hairy, veins rather conspicuous; petiole 1–1½ in. long.

TWIGS. Slender, all young parts hairy, green to red; older twigs gray, glabrous, lenticels small.

BARK. Gray to brown, smooth when young, with small scales when older, excrescences lenticular, numerous, horizontal.

PRAIRIE FLAME-LEAF SUMAC
Rhus copallina L. var. *lanceolata* Gray

RANGE. Usually on dry and rocky soil of the Texas Edwards Plateau area; north into Oklahoma, northwest into New Mexico, and south into Mexico. In Mexico in the states of Coahuila, Puebla, and Tamaulipas.

REMARKS. The genus name, *Rhus*, is the ancient Latin name; the species name, *copallina*, means "gum copal," and the variety name, *lanceolata*, refers to the lanceolate-shaped leaflets. Vernacular names are Lance-leaf Sumac, Tree Sumac, Limestone Sumac, Mountain Sumac, Black Sumac, and Prairie Shining Sumac.

Prairie Flame-leaf Sumac is considered by some authorities to be a distinct species instead of a variety of the Flame-leaf Sumac, *R. copallina*. However, later classification takes into account the considerable variation of the Flame-leaf Sumac and has assigned the Prairie Flame-leaf Sumac as a variety of it. The Prairie Flame-leaf Sumac, in its most typical form, seems most abundant on the dry stony hills of central and west Texas. It has longer panicles of flowers and larger fruit than the Flame-leaf Sumac. The leaflets tend to be narrower, entire or remotely serrate, and falcate, and the rachis wing narrower. The habit approaches that of a small rounded tree instead of a straggling shrub.

The leaves of both the Flame-leaf and Prairie Flame-leaf Sumac contain considerable amounts of tannin and have been used as a substitute for oak bark in tanning leather. The acrid drupes, when crushed in water, produce a cooling drink known as "Sumac-ade" or "Rhus-ade." The drupes also produce a black dye for woolen goods. Because of the brilliant fall coloring of its red, purple, and orange leaves, Prairie Flame-leaf Sumac is being adapted to horticulture.

It has considerable wildlife value, the drupe being eaten particularly by the gallinaceous birds, such as various species of quail, grouse, prairie chicken, and ring-necked pheasant. It is also browsed by white-tailed deer and mule deer. The method of propagation by seed is similar to that of the Smooth Sumac.

STAGHORN SUMAC

Rhus typhina Torner

FIELD IDENTIFICATION. An irregularly branched, stout shrub with a rounded or flat-topped head, forming thickets by root suckers. The upright contorted branches are conspicuously velvety-hairy. Sometimes in favorable locations becoming a tree with a short, inclined trunk, a height to 40 ft, and a diameter of 3–14 in.

FLOWERS. Usually June–July, but opening gradually. Borne in dense, dioecious, terminal, greenish yellow panicles. Staminate panicles longest, usually maturing before the pistillate, and 7–12 in. long and 4–6 in. broad; panicles with acuminate bracts ½–2 in. long and early deciduous; individual flower pedicels slender with bracts 1/25–1/12 in. long, acute, hairy; petals 5, imbricated in the bud, inserted under the edge of the red, flattened disk, 1/12–⅛ in. long, linear to elliptic or

STAGHORN SUMAC
Rhus typhina Torner

oblanceolate, finally reflexed in the staminate flower, in the pistillate flower remaining erect and slightly hooded at the apex; calyx small, sepals 5, 1/25–1/12 in. long, deltoid to lanceolate, acuminate or acute, hairy beneath, ciliate; stamens 5, exserted in the staminate flowers, abortive in the pistillate flowers, filaments slender with large, orange-colored anthers 1/25–1/12 in. long; pistil ovoid, hairy; styles 3, with 3 short, fleshy capitate stigmas.

FRUIT. Ripening June–September, persistent during the winter, in dense, terminal, conical panicles 6–8 in. long and 2–4 in. wide; drupes ⅙–⅛ in. long and broad, depressed-globose, clothed with long, velvety, crimson hairs; flesh thin and dry, acrid to the taste; seed green to pale brown, somewhat oblique, slightly flattened, smooth, bony, no albumen.

LEAVES. Alternate, deciduous, odd-pinnately compound, 5–24 in. long; rachis green to reddish, with soft velvety hairs, segments ¼–⅝ in. long; petiole stout, 2¼–4 in. long, hairy, enlarged at base; leaflets 9–31, opposite or subopposite, middle pairs largest, laterals sessile or nearly so, terminal long-petiolulate, elliptic to lanceolate or ovate, sometimes falcate, apex acute or acuminate, base rounded or semicordate and slightly unequal, margin remotely serrate and subrevolute (laciniate in some varieties), length 1½–6 in., width ⅓–1¾ in., thin, veins conspicuous, upper surface light to dark green and dull to semilustrous with a few hairs or glabrous, lower surface whitish, and hairy on the veins, in fall turning brilliant shades of red, yellow, and orange.

TWIGS. Stout, brittle, brown to orange; younger with soft, brown to black, long, velvety hairs; older twigs glabrous; lenticels numerous, conspicuous, orange-brown; buds tan, conic, brown-hairy; leaf scars horseshoe-shaped, bundle scars in clusters of 3; pith round, orange-brown; sap sticky, milky, turning black on exposure.

BARK. Dark brown to gray, on young trees smooth, on older ones dark brown and separating into small squarish scales; lenticels elongate and horizontal, bark rich in tannin.

WOOD. Orange to yellow, streaked with green or brown, sapwood whitish, light, brittle, soft, coarse-grained, weighing 27 lb per cu ft.

RANGE. Roadsides, thickets, old fields, hillsides, and dry or gravelly slopes, to 2000 feet. Not definitely known in a wild state in our area, but cultivated in Louisiana. Native from Mississippi eastward to Florida, abundant northward through the Appalachian Mountains and throughout the eastern states to Quebec, westward to Ontario, Minnesota, and Iowa.

PROPAGATION. The fruiting panicles are hand-picked when ripe. Panicles picked early may require spreading out to dry. When dry seeds may be separated from the panicles by beating in a sack. The seed may be cleaned by macerating in water and floating off the pulp and bad seed. However, cleaning the seed is not necessary. The seed with pulp averages about 30,000 seeds per lb, and the cleaned seed about 53,300 seeds per lb, with commercial seed having a purity average of 89 per cent and a soundness of 71 per cent. Noncommercial seed is sometimes infected with the larvae of the chalcid fly.

Seed may be stored over winter without special treatment. However, for longer periods, storage in sealed containers at 41°F. is recommended. Germination is delayed because of the impermeable seed coat, but may be hastened by soaking seed in concentrated sulphuric acid for 60–80 minutes. The germination capacity is 40 per cent. Seed may be sown in the spring at a depth of one half in. with 25 viable seeds to the linear ft. Germination is complete in about 30 days with pretreated seed. The seedlings are susceptible to damping-off. The plant may also be propagated by root cuttings.

VARIATIONS. A variety of Staghorn Sumac which is often cultivated is Cut-leaf Staghorn Sumac, *R. typhina* var. *laciniata* Wood. It has leaflets and bracts deeply and laciniately toothed. A form of this variety with contorted bracts and pinnately dissected leaves has been given the name of Shred-leaf Staghorn Sumac, *R. typhina* forma *dissecta* Rehd.

Another variation of Staghorn Sumac was formerly known as Green Staghorn Sumac, *R. typhina* forma *viridiflora* Duhamel, but the differences between it and the species are so slight that the name is now considered a synonym.

REMARKS. The genus name, *Rhus*, is the ancient Greek and Latin name of Sicilian Sumac, *R. coriaria* L. The species name, *typhina*, refers to the hairy cattail-like twigs. Vernacular names are Hairy Sumac, Velvet Sumac, American Sumac, Virginia Sumac, and Vinegar-tree. The name Staghorn Sumac refers to the forked velvety twigs, which resemble a stag's young antlers. The wood is rarely used for cabinets or for small articles. The young hollow stems were formerly used for maple sap taps. In frontier times ink was made by boiling the leaves. The root, bark, and leaves are rich in tannin. The crushed acid fruits, steeped in water, make a cooling drink which was much used by the Indians. The seeds are known to be eaten by 94 species of birds, including mourning dove, bobwhite quail, ring-necked pheasant, and ruffed grouse; also by skunk, white-tailed deer, cottontail, moose, and opossum. The plant often reproduces by root suckers and hence is resistant to grazing. It is sometimes planted for erosion control and for ornament and has been cultivated since 1629.

Smooth Sumac

Rhus glabra L.

FIELD IDENTIFICATION. Thicket-forming shrub or small tree attaining a height of 20 ft. Leaves pinnate, bearing 11–31 elliptic or lanceolate, sharply serrate leaflets.

FLOWERS. June–August, in a terminal thyrse, 5–9 in. long; bracts narrowly lanceolate, about 1/25 in. long; calyx of 5 lanceolate sepals, each about 1/12 in. long; petals 5, white, spreading, lanceolate, 1/6 in. long or less; stamens 5; pistil 1, ovary 1-seeded, stigmas 3.

FRUIT. Ripening September–October, drupe subglobose, about 1/6 in. long, covered with short, red-velvety hairs; 1-seeded, stone smooth, seed dispersed by birds and mammals.

LEAVES. Alternate, pinnately compound, 11–31 leaflets which are elliptic, lanceolate, oblong or ovate, acuminate at the apex; rounded, subcordate, cuneate or oblique at the base, sharply serrate, usually dark green above, lighter to conspicuously white beneath; lateral ones sessile or almost so, 2½–4¾ in. long, ½–1¼ in. broad; terminal one sessile or petiolulate, 2–3¾ in. long, ½–1½ in. broad.

WOOD. Orange, soft, brittle.

RANGE. In moist, rich soil. Texas, New Mexico, Oklahoma, Arkansas, and Louisiana; eastward to Florida, northward to Quebec, and westward to British Columbia, Washington, Oregon, Utah, Colorado, and Missouri.

MEDICINAL USES. The Indians used the plant in many ways. The crushed acid fruits were added to water

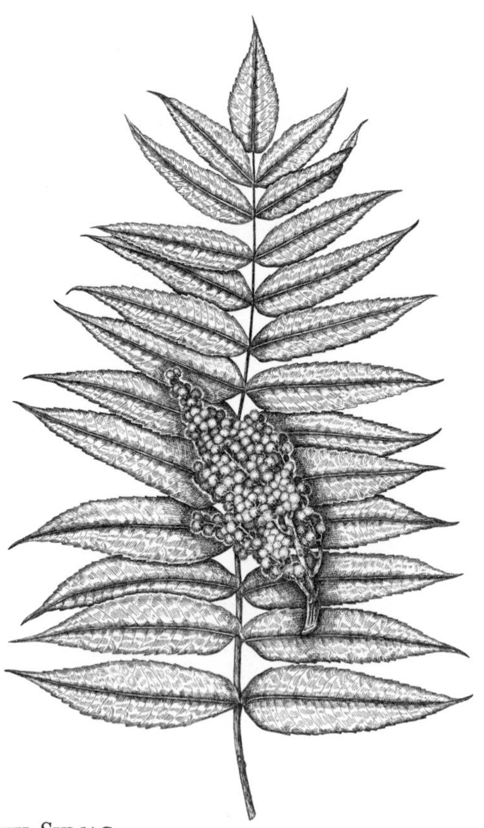

SMOOTH SUMAC
Rhus glabra L.

R. glabra var. *cismontana* (Greene) Daniels has 11–13 leaflets, and the fruiting thyrse pyramidal and compact.

R. glabra forma *flavescens* (D. M. Andr.) Rehd. is a form of the latter with yellow fruit.

R. glabra var. *borealis* Britt. has branches short-pilose and pubescent and occurs mostly east of the Rocky Mountains.

REMARKS. The genus name, *Rhus,* is the ancient Latin name, and the species name, *glabra,* refers to the plant's smoothness. Vernacular names are Scarlet Sumac, Red Sumac, White Sumac, Shoe-make, Vinegartree, Senhalanac, Pennsylvania Sumac, Upland Sumac, and Sleek Sumac. The leaves are reported to have been mixed with tobacco and smoked. The twigs, leaves, and roots contain tannin and were used for staining and dyeing. Smooth Sumac is now used extensively for ornament. Its red clusters of fruit and long graceful leaves which turn brilliant colors in the autumn are its attractive features. It should be more extensively cultivated for ornament. Records show that the date of earliest cultivation was 1620. It is occasionally planted for erosion control and used for shelterbelt planting in the prairie states. Thirty-two species of birds are known to feed on it. Wild turkey, bobwhite, cottontail, and white-tailed deer eat it eagerly.

to freshen it. Various concoctions of the bark, twigs, leaves, and flowers were used medicinally as astringents, to stop bleeding, and for renal disorders.

PROPAGATION. The seed is generally collected by hand and spread out to dry. When dry it is rubbed from the cluster by hand or on a screen. It is not necessary to remove all the pulp, but it can be removed by macerating in water and floating off the residue. The seeds have been stored in sealed containers at 41°F., and have retained their viability for 2½ years. The clean seed averages about 68,600 seeds per lb, and commercial seed averages a purity of 80–94 per cent and a soundness of 73 per cent. It is sometimes infested with the larvae of the chalcid flies. Because of a natural dormancy, the seed must be pretreated to hasten germination. The best method is to soak them in concentrated sulphuric acid 1–1½ hours for fall sowing, or followed by stratification and spring sowing. Germination is usually completed in about a month. The seed is planted in nursery beds using about 25 viable seeds per linear ft and covering with firmed soil about ½ in. deep. The various sumac species can also be propagated by root cuttings.

VARIETIES. A number of varieties of Smooth Sumac are recognized:

Cut-leaf Smooth Sumac, *R. glabra* var. *laciniata* Carr., has pinnately dissected leaves.

LITTLE-LEAF SUMAC

Rhus microphylla Engelm.

FIELD IDENTIFICATION. Clump-forming, intricately branched shrub attaining a height of 15 ft. Branches crooked, stiff, almost spinescent.

FLOWERS. Borne in dense, crowded, stiff, compound spikes before the leaves appear, spikes ½–2 in. long and broad, bracts hardly over 1/25 in. long and slightly broader, somewhat glandular-hairy, ciliate, persistent; pedicels about 1/25 in. long or shorter; flowers about 3/16 in. across, bracteolate; sepals 5, rounded to deltoid, concave, acute, ciliate, glabrous, about 1/25 in. long; petals 5, imbricate, greenish white, rounded to ovate, ciliate, glabrous on one side, hairy on the other; stamens 5, filaments exceeding the sepals; anthers about 1/25 in. long or less; disk 1/25–1/12 in. broad, lobed; styles united or nearly so.

FRUIT. Drupe May–July, subglobose to ovoid, reddish orange, glandular-hairy, 1/5–¼ in. long, 1-celled.

LEAVES. Small, deciduous, odd-pinnately compound, ½–1½ in. long, composed of 5–9 leaflets with a winged rachis; petiole about ⅛ in. long, leaflets oblong, elliptic, obovate or ovate, sessile; apex obtuse, acute, often mucronate; base cuneate, somewhat asymmetrical; margin entire, slightly revolute; thin, dull green and hairy above, somewhat paler and hairy beneath; lateral leaflets 3/16–⅜ in. long, 1/12–⅙ in. broad; terminal leaflet 3/16–⅝ in. long, ⅛–⅙ in. broad.

TWIGS. Almost spinescent, stiff, crooked, roughened, puberulent at first to glabrous later, lenticels prominent.

BARK. Dark gray to black, smooth when young, broken into small scales with age.

RANGE. On dry, rocky hillsides or gravelly mesas at altitudes of 2,000–6,000 ft. In Texas, New Mexico, and Arizona. Widespread in the western half of Texas. Also in the Mexican states of Chihuahua, Sonora, Coahuila, Durango, Nuevo León, San Luis Potosí, Zacatecas, and Baja California.

REMARKS. The genus name, *Rhus*, is the old Latin name, and the species name, *microphylla*, refers to the small leaves. Vernacular names are Winged Sumac, Small-leaf Sumac, Correosa, Agrito, and Agrillo. The reddish orange, hairy drupe is edible, but sour. It is eaten by a few species of ground squirrels and chipmunks. The leaves are rather poor browse for livestock, but are sometimes browsed by mule deer and Sonora deer. The method of propagation by seed is similar to that of Smooth Sumac.

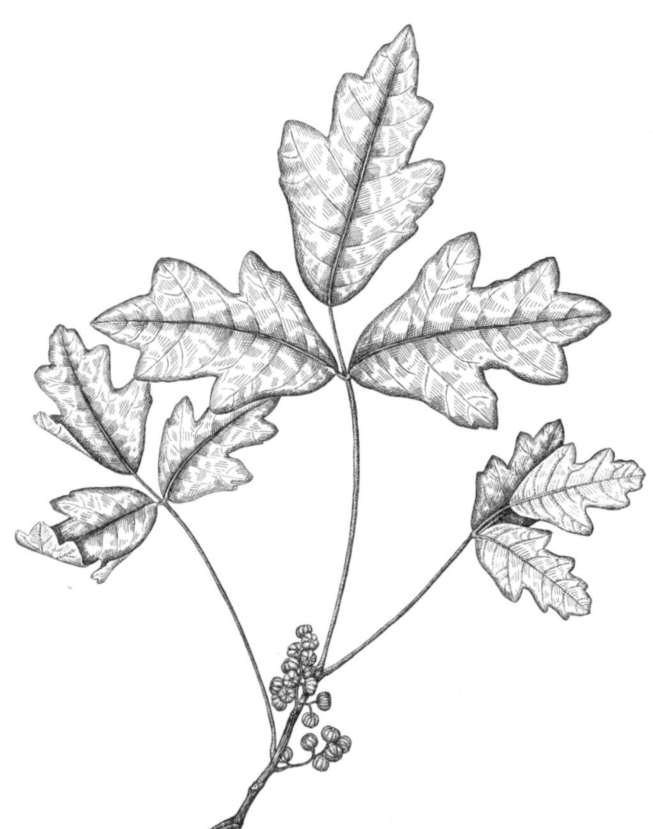

POISON-OAK
Toxicodendron quercifolium (Michx.) Greene

LITTLE-LEAF SUMAC
Rhus microphylla Engelm.

POISON-OAK

Toxicodendron quercifolium (Michx.) Greene

FIELD IDENTIFICATION. Poisonous, low-branching shrub to 3 ft. Leaves 3-foliate, densely tomentose, leathery, 3–7-lobed (oaklike) on the margin, rhombic-ovate.

FLOWERS. Borne in densely flowered, greenish yellow lateral panicles 1–3 in. long; panicles subtended by bracts which are deciduous, deltoid-lanceolate, glabrate, ciliate, less than 1/25 in. long; sepals 5, about 1/25 in. long, deltoid-ovate to oblong, apex obtuse, glabrate; petals 5, about 1/8 in. long, 1/25–1/12 in. broad, oblanceolate to oblong, obtuse at apex, veined, glabrous; stamens 5, erect, filaments about 1/25 in. long, linear-subulate, about as long as the anthers.

FRUIT. Cream-white, depressed-globose, pubescent when young, smooth and papillose when mature; seed about 1/8 in. long and 1/6 in. broad.

LEAVES. Trifoliate, long-petiolate, rhombic-ovate to obovate, margin irregularly lobate-dentate; lobes 3–7, blunt or subacute; apex rounded to obtuse or more rarely acute or acuminate; base obtuse or cuneate, leathery and firm; upper surface dark green and glabrous or pubescent; lower surface pale green and densely velvety-pubescent; terminal leaflet 2–3½ in. long, 1½–2¾ in. wide, petiolule about ¾ in. long; lateral leaflets smaller, asymmetrical, tending to be entire above and 3–7-lobed below; petiolules about ⅕ in. long, or absent.

TWIGS. Slender, densely pubescent, green to brown.

RANGE. Sandy soils of coniferous or mixed forests. Texas, Arkansas, Oklahoma, and Louisiana; eastward to Florida and north to Tennessee, Missouri, South Carolina, New Jersey, and Maryland.

REMARKS. The genus name, *Toxicodendron*, is from the Greek and means "poison tree," and the species name, *quercifolium*, refers to the oaklike appearance of the leaves. Also known locally as Poison-wood, Poison-weed, Scratch-ivy, Poison-ivy, and Poison Sumac.

Poison-oak and Poison-ivy *(T. radicans)* are very close kin and some botanists believe that one is only a variation of the other. However, the differences appear as rather constant factors to this author. Poison-oak leaves are much more leathery than Poison-ivy; also they are often more abundantly lobed and very densely pubescent beneath. The twigs are also densely pubescent. Poison-oak is inclined to be more upright and shrubby, whereas Poison-ivy tends to climb by aerial rootlets. They both contain the same poisonous principle, toxicodendrol, and the treatment for poisoning from either is the same. The description of Poison-ivy gives a detailed account of the remedies for the poison of both Poison-ivy and Poison-oak.

COMMON POISON-IVY
Toxicodendron radicans (L.) Kuntze

COMMON POISON-IVY

Toxicodendron radicans (L.) Kuntze

FIELD IDENTIFICATION. Poisonous suberect shrub, or woody-stemmed vine, which is very variable in habit of growth. The 3-foliate leaves vary greatly in shape and size and may be entire, coarsely-toothed, or deeply lobed on the margin.

FLOWERS. Borne in axillary panicles 1–4 in. long, small, greenish white, fragrant; bracts deciduous, about 1/25 in. long, deltoid, glabrate, margin ciliate; sepals 5, about 1/25 in. long, deltoid-ovate, obtuse, glabrous; petals 5, ⅛–⅙ in. long and 1/25 in. wide, oblanceolate to elliptic-ovate, glabrous; stamens 5, erect, filaments linear-subulate, 1/25–1/12 in. long, anthers ovate-lanceolate; ovary glabrous, styles short.

FRUIT. Drupes persistent, clustered, ⅛–¼ in. in diameter, dull white, waxy, glabrous or rarely sparsely pubescent, seed striate.

LEAVES. Trifoliate (rarely, 5-foliate), very variable in shape and size, long-petiolate; mostly oval to lanceolate; margin entire, dentate, or lobed; apex acute to acuminate; base cuneate to rounded; membranous, dull green and glabrous above, paler and with varying degrees of pubescence beneath; terminal leaflet 1⅓–8 in. long, ½–5 in. broad, petiolule ⅜–1¾ in.; lateral leaflets inequilateral, 1¼–6¾ in. long, ½–4 in. wide; petiolules 1/25–⅕ in., or leaflets sometimes sessile.

TWIGS AND STEMS. Climbing by aerial rootlets with disklike suckers which adhere to support. Stem often woody and attaining a diameter of 6 in. Sometimes climbing to the top of very tall trees and displaying a greater density of foliage than the tree itself. If no support is present, often assuming a suberect, rounded shape with smaller leaflets. The slender, spreading twigs are pubescent to glabrous.

RANGE. Poison-ivy, taken in the broad sense with all its variability, is widely distributed in the United States, Canada, Mexico, and the West Indies. In the southwestern states in Texas, New Mexico, Oklahoma, Arkansas, and Louisiana; eastward to Florida and northward to Maine on the east and to Washington and Oregon on the west. In Mexico in the states of Chihuahua, Durango, Jalisco, Mexico D. F., Michoacán, Nuevo León, Oaxaca, San Luis Potosí, Sonora, Tamaulipas, Veracruz, and Yucatán.

POISONOUS PROPERTIES AND MEDICINAL USES. Poison-ivy and its varieties contain a nonvolatile, poisonous oil known as "toxicodendrol" or "urushiol." Upon contact with the skin this oil produces blisters and eruptions accompanied by intense itching and burning. Apparently some people are not as susceptible as others to this Rhus dermatitis poisoning. However, this immunity does not appear to be a constant factor. People have been known suddenly to become affected who had previously come into contact with it for years with no harmful results. It is also believed that at certain seasons secretions of the oil are more abundant on the leaves and stems, and contact at that time of the year produces a double dose, so to speak. It is also claimed by some that the plant need not be touched, because the pollen grains floating on the wind may also contact a person. However, according to certain research done by Muenscher (*Chem. Abs.*, 24:5108, 1930; cited by Wood and Osol, 1943, p. 1498) it is stated that neither the pollen grains nor hairs contain the poison, but that the poison may adhere to the hairs and be carried by air currents. Also, there appears to be no conclusive evidence to prove that smoke from burning Poison-ivy may produce poisoning.

The remedies for Poison-ivy, both in folklore and professional medicinal practice, appear to be very numerous and varied. As many as forty so-called "sure-cures" have been suggested in various publications. Perhaps the best authority from the professional medical point of view is Wood and Osol (1943). The following quotation from this publication (pp. 1498–1499) gives a detailed account of the remedies: "The best treatment of Rhus dermatitis is the prophylactic one, to remove the irritant resin before it has set up the inflammation. Strong alcohol is an efficient solvent of the resin but unless very carefully employed it is likely to spread over a larger area. Probably the best method of removing the resin is with a thick lather of laundry soap; toilet soap is decidedly less efficient for this purpose. Various chemicals have been asserted to destroy the irritant properties of poison ivy. These are mostly in three groups: the salts of the heavy metals, the oxidizing agents and certain sulfur compounds. Among the heavy metals lead, iron and copper all produce precipitates with the juice but according to Gisvold (*J. Amer. Pharm. Ass.*, 30:17, 1941) these precipitates are almost as active as the original poison. However, there is considerable clinical evidence of protecting action of ferric chloride or sulphate if applied soon after exposure. Among the oxidants perhaps the most generally satisfactory is a 5 per cent solution of potassium permanganate; but hydrogen peroxide, benzoyl peroxide and sodium perborate all have their protagonists. Sodium perborate has recently received much commendation, but Shelmire (*J. Amer. Med. Ass.*, 116:681, 1941) asserts that Poison-ivy leaves soaked in a saturated solution of perborate for two weeks were not detoxified, and that an ointment of sodium perborate is no better than an ointment of zinc oxide. Of the sulfur compounds the most frequently employed is a sodium thiosulfate in 10 per cent solution. To relieve the itching and burning probably the best remedy is a 2 per cent solution of phenol.

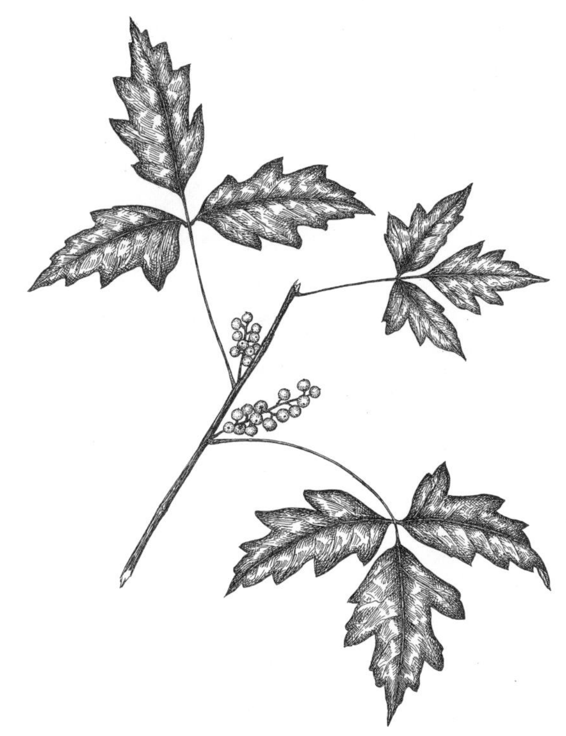

Poison-ivy (variety)
Toxicodendron radicans (L.) Kuntze var. *verrucosa* (Scheele) Barkley

"Attempts have been made to establish an immunity to poison ivy by the internal use—either by mouth or by injection—of various extracts of offending species of Rhus [or Toxicodendron]. There is an old superstition that eating the leaves of Poison-ivy induces an immunity but Silvers (*J. Amer. Med. Ass.*, 116:2257, 1941) reports a case in which violent inflammation of the mouth, pharynx and anus was caused by trial of this method. Although Spain and Cooke (*J. Immunology*, 13:93, 1927) claim to have produced a satisfactory immunity and several clinicians have reported favorable results, both in preventing and relieving attacks (see Strickler, *J. Amer. Med. Ass.*, 80:1588, 1923), the value of the measure is still *sub judice*. There are on the market several preparations for this purpose under various trade names as *Ivyol, Toxol, Poison Ivy Toxol, Rhus Venenata Antigen, Rhus Tox. Antigen, Poison Oak Extract,* etc.

"The poisonous Rhus *[Toxicodendron]* when taken internally appears to be possessed of narcotic besides irritant properties; vomiting, drowsiness, stupor, dilated pupils, convulsive movements, delirium, and fever have been present in poisoning by them. The tincture of Rhus has been used by homeopathic practitioners in the treatment of subacute and chronic rheumatism, and the practice has found imitation. Some years ago H. C. Wood made extended trials of the remedy, using a homeopathic tincture obtained from a homeopathic pharmacy, in various doses (homeopathic, small, and

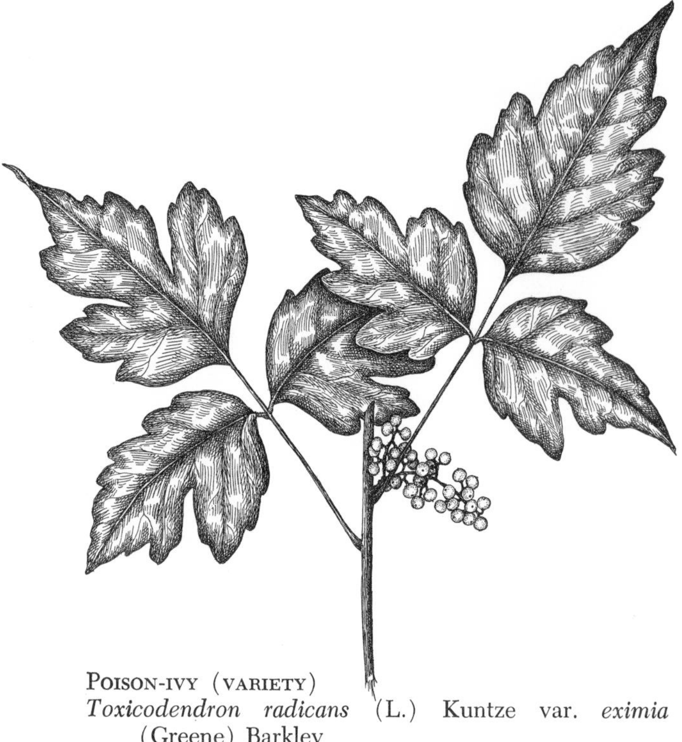

POISON-IVY (VARIETY)
Toxicodendron radicans (L.) Kuntze var. *eximia* (Greene) Barkley

large), upon a large number of cases of various types of rheumatism, in the Philadelphia Hospital, but he was not able to perceive that the patients progressed more rapidly when taking it than when they were simply nursed."

CLOSELY RELATED SPECIES, FORMS, VARIETIES, AND SYNONYMS. Some authorities place Poison-ivy under the genus name of *Rhus,* instead of *Toxicodendron.* However, there are sufficient practical, and taxonomic, reasons for maintaining it under the latter terminology and separating it, and also the Poison-oak and Poison Sumac, from the rest of the sumacs. Its smooth, white, waxy fruits, poisonous properties, style character, and axillary inflorescence are distinctive.

Poison-ivy varies considerably in shape and size of leaflets, size of panicle, and shrubby or vinelike habit. For this reason considerable change and adjustment has been made in the application of names by authors. There is at present a tendency to relegate the many races to varieties, instead of distinct species. Perhaps the most conclusive discussion of these varieties is to be found in Barkley (1937, pp. 420–436).

Of these varieties discussed by Barkley, two have been listed as occurring in Texas. *T. radicans* (L.) Kuntze var. *verrucosa* (Scheele) Barkley is described as a small shrub with trifoliate leaves, which are mostly glabrate except on the venation, rhomboid-ovate, apex acute to acuminate, margin regularly and deeply incised-dentate. It is known to occur in Texas and Oklahoma. In Texas near Montell, Brownwood, New Braunfels, Granbury, Fort Worth, Kerrville, Blanco, Kendalia, San Marcos, Fredericksburg, Austin, and Spanish Pass.

Another variety listed by Barkley is *T. radicans* (L.) Kuntze var. *eximia* (Greene) Barkley. This variety is a vine with leaves broadly ovate, puberulent, apex acuminate and often trilobate, the lobes rounded or rarely incised-dentate; fruit barely pubescent. It is known to occur in Mexico and Texas. In Texas in Bexar, Brewster, Maverick, Tarrant, Uvalde, and Val Verde counties. In Mexico in Durango and Nuevo León.

A third variety is listed as *T. radicans* (L.) Kuntze var. *littoralis* (Mearns) Barkley n. comb. It is a shrub or vine with trifoliate leaves; leaflets often broadly ovate, acute, usually entire but rarely crenate-dentate, usually subcordate at base; fruits distinctly pubescent. Found in Oklahoma and Florida; north to Indiana, Pennsylvania, New Jersey, Connecticut, and Maine.

T. radicans (L.) Kuntze var. *vulgaris* (Michx.) DC. has leaves membranous, conspicuously toothed or shallowly lobed. When the leaf surfaces and petioles are glabrous it is considered to be forma *vulgaris*; when the lower leaf surfaces are pubescent, but the petioles glabrous, it is considered to be forma *intercursa* Fern. With both leaf surfaces and petioles glabrous it is forma *negundo* (Greene) Fern. The variety and forms are found from Texas to Florida, north to Virginia, Ohio, Indiana, and Illinois.

T. radicans (L.) Kuntze var. *rydbergii* (Small) Rehd. has woody stems about 2–48 in. above the creeping bases, simple branched or spreading, aerial rootlets generally absent; leaflets subcoriaceous to mem-

FIVE-LEAF POISON-IVY
Toxicodendron radicans (L.) Kuntze
A form with 5 leaflets instead of 3

branaceous, margin dentate, undulate or rarely entire, terminal leaflet 2–6 in. long, broadly ovate to suborbicular, apex abruptly acuminate, venation often pubescent. Trans-Pecos Texas, Oklahoma, New Mexico, and Arizona; northward to Kansas, Minnesota, Michigan, West Virginia, New York, Nova Scotia, and British Columbia.

A form of Poison-ivy with 5 leaflets, instead of 3, has been collected by the author in Harris County, along Bray's Bayou, in Houston, Texas, and also in the San Jacinto River bottom lands near Sheldon, Texas. It is considered only a form, and not a variety. Some vines have 3- or 5-foliate leaflets and some have only 5-foliate. In all other characteristics it resembles Common Poison-ivy.

REMARKS. The genus name, *Toxicodendron*, is from the ancient Greek and means "poison tree." The species name, *radicans*, refers to its climbing habit. Also known under the vernacular names of Three-leaf Ivy, Climbing Ivy, Poison-oak, Mercury, Black Mercury-vine, Markry, Mark-weed, Picry, Hiedra, Mala Mujer, Mexye, Guadalagua, Hincha Huevos, Bemberecua, Gnaua, Hiedra Mala, and Hiedra Maligna.

The fruit is known to be eaten by at least 75 species of birds, particularly the gallinaceous birds such as the wild turkey, bobwhite quail, ruffed grouse, sharp-tailed grouse, and ring-necked pheasant. The leaves turn brilliant shades of red, yellow, and orange in autumn. The leaves bear a striking resemblance to Box-elder Maple leaves, but Box-elder is an upright shrub or tree. The poisonous milky sap turns black on exposure and stains black. It is sometimes used for making varnish.

Poison Sumac

Toxicodendron vernix (L.) Kuntze

FIELD IDENTIFICATION. Poisonous shrub or small tree to 25 ft, with pinnate leaves bearing 7–13 obovate, oblong-ovate, or oval leaflets.

FLOWERS. Polygamous, borne in axillary panicles 3–8 in. long; bracts lanceolate, 1/25 in. long or less; sepals 5, acute, about 1/25 in. long; petals 5, green, linear to oblanceolate, 1/12–1/8 in. long; stamens 5, filaments 1/25–1/12 in. long, anthers orange; styles short, stigmas 3, ovary glabrous.

FRUIT. Drupe subglobose, flattened, greenish white or gray, glabrous, 1/16–1/4 in. broad; seed flattened, striate, bony.

LEAVES. Alternate, deciduous, pinnate, 5–15 in. long; leaflets 7–13, obovate, oblong-ovate or oval, apex short-acuminate, base cuneate or rounded, entire, smooth and shining above, more or less pubescent beneath, 2½–4 in. long, 1–2 in. wide; lateral leaflets with short petiolules; petiolule of terminal leaflet longer; petioles stout and reddish.

TWIGS. Brown to orange, stout, glabrous, lenticels small and numerous.

Poison Sumac
Toxicodendron vernix (L.) Kuntze

WOOD. Yellowish brown, soft, tough, coarse-grained, weighing 27 lb per cu ft.

BARK. Gray to brown, thin, smooth, lenticels horizontal.

RANGE. In eastern Texas and Louisiana; eastward to Florida and northward to Minnesota, Ontario, New York, and Rhode Island.

REMARKS. The genus name, *Toxicodendron*, is from the ancient Greek and means "poison tree." The species name, *vernix*, means "varnish," erroneously referring to the Japanese Lacquer-tree. All parts of the plant are poisonous and produce an intense skin irritation known as "Rhus dermatitis." The tincture of *Rhus*, or *Toxicodendron*, has been used by homeopathic practitioners in the treatment of subacute and chronic rheumatism. (See description of Common Poison-ivy for detailed account of remedies and medicinal uses for Rhus dermatitis.) However, beneficial results of such treatment are in doubt. It is considered to be even more poisonous than Poison-ivy and Poison-oak. Vernacular names for the plant are Poison Elder, Poison Dogwood, Swamp Sumac, Poison-wood, and Poison-tree. It was formerly listed under the scientific name of *Rhus vernix* L., but it is sufficiently different to be placed in the genus *Toxicodendron*. The sap is used to make a high-grade varnish, and concoctions of the leaves have been used medicinally by the Indians. Fifteen species of birds and the cottontail feed upon it. Poison-ivy and Poison-oak are close kin to Poison Sumac, but they are vines or low shrubs and have only 3 coarsely toothed leaflets instead of 7–13 entire leaflets.

California Pepper-tree

Schinus molle L.

FIELD IDENTIFICATION. Cultivated evergreen tree attaining a height of 50 ft, with a spreading, rounded crown and graceful, pendulous branches.

FLOWERS. Panicles axillary to the terminal leaves, 4–6 in. long, minutely puberulent; bracts deltoid, ciliate, puberulent, early-deciduous; flowers small, numerous, yellowish white, less than ¼ in. across; calyx short, 5-lobed, lobes semiorbicular, glabrous or sparsely hairy; petals 5, imbricate, about 1/12 in. long, ovate to oblong, truncate, glabrous; disk annular, thick, saucer-shaped, with as many lobes as there are stamens and alternating with them; stamens 10, alternating with the petals, filaments somewhat enlarged toward the base, anthers oblong; pistil tricarpellate, styles 3 and free above, stigmas 3; ovary 1-celled, ovule solitary and suspended.

FRUIT. Drupe globose, ⅕–⅓ in. in diameter, pinkish, exocarp leathery, lustrous, mesocarp thin and resinous and coherent to the bony endocarp.

LEAVES. Alternate, pinnately compound, 4–12 in. long, rachis segments ⅙–½ in. long, glabrous; leaflets 19–41, alternate or opposite, ⅓–2⅖ in. long, firm, linear to lanceolate, apex acuminate or acute and often curved, base cuneate to obtuse, margin entire or obscurely serrulate, surfaces glabrous or puberulent.

TWIGS. Numerous, slender, brownish, glabrous or puberulent.

RANGE. Often in dry, sandy soil. Abundantly cultivated in southern California and occasionally in southwest Texas. Self-propagating in suitable areas. A native of Peru.

MEDICINAL USES. "The Peppertree is used in local medicine. The powdered bark or its decoction is used as a remedy for swollen feet and as a purgative in domestic animals; it is reported to have astringent and balsamic properties. The gum [known as Jesuits' Balm] which exudes from the trunk is bluish white, acrid, and bitter, and burns with a pleasant odor. It is often chewed as chewing gum, and is said to have purgative and vulnerary properties. It is applied in Mexico in the form of an emulsion to the eyes to hinder the development of cataracts, and is used for genito-urinary and venereal diseases. The leaves are chewed to harden the gums and to heal ulcers of the mouth. The fruit has been used by European physicians as a substitute for cubeb in the treatment of gonorrhea, and a syrup prepared with it is used in Mexico for bronchitis." (Standley, 1920–1926, p. 662)

The leaves contain an unidentified glycoside which has an emmenagogue action.

California Pepper-tree
Schinus molle L.

REMARKS. The genus name, *Schinus,* is the Greek name for the Mastic-tree, *Pistacia lentiscus,* for the mastic-like juice of some species. The species name *molle,* means "pliable" or "flexible," with reference to the branches. However, the use of the word "molle" is rather vague in this case, the word generally meaning "soft-hairy." Vernacular names are Peruvian Mastic-tree, Árbol del Perú, Perú, Pirul, Molle, Pimienta de América, Pimiento de California, Pimiento, Pimentero, Pelorquahuitl, Capoloatte, Capalquahuitl, Tizacthunni and Ttzacthumi.

An excellent and handsome shade tree, but not desirable in some areas because of the black scale which also attacks citrus orchards. The following account of its uses in Mexico is given by Standley (1920–1926, p. 662):

"The wood is useful for various purposes and the bark for tanning skins. When fragments of the leaves are placed in water, they execute quick jerking movements due to the sudden discharge of oil which they contain. The fruit contains a volatile oil and has a flavor resembling that of a mixture of fennel and pepper. [The oil is composed of beta-phellandrene, pinene, trausterpine and carvacrol.] The seeds are sometimes used to adulterate pepper. In Mexico the fruit is ground and mixed with *atole* or other substances to form beverages. An intoxicating liquor, known as 'copalocle' or 'copalote' is formed by fermenting the fruit with pulque for one or two days."

TITI FAMILY (*Cyrillaceae*)

AMERICAN CYRILLA

Cyrilla racemiflora L.

FIELD IDENTIFICATION. Swamp-loving shrub or small tree to 30 ft, with a short trunk and spreading irregular branches.

FLOWERS. In spring, racemes axillary, slender, erect or nodding, 4–6 in.; individual flowers on pedicels 1/12–1/6 in., small, white, fragrant, perfect, regular; petals 5, lanceolate-oblong, acute, about 1/8 in. long, with nectar glands at base; stamens 5, alternate with the petals, anthers oval; ovary sessile, ovoid, style short and thick with a 2-lobed spreading stigma; calyx minute, of 5 sepals which are ovate-lanceolate, acute, about 1/12 in.

FRUIT. Capsule in late summer, small, about 1/16–1/8 in. long, ovoid-conical, yellowish brown, dry, remnants of the style at apex, persistent, indehiscent, 2-celled with 2 seeds in each cell; seeds minute, dry, light brown.

LEAVES. Alternate, often clustered at the twig tips, petioles 1/8–1 in., half-evergreen, entire on margin, oblong-obovate to elliptic, obtuse or acute at apex, cuneate at base, dark green and lustrous above, paler and reticulate-veined beneath, 2–4 in. long, 1/4–1 in. wide.

TWIGS. Slender, smooth, shiny, brown to gray.

BARK. Thin, close, pale reddish brown or gray, scales small and thin, lower bark spongy.

WOOD. Reddish brown, close-grained, heavy, hard, not strong.

RANGE. In swamps on the Atlantic and Gulf coastal plains from eastern Texas eastward to Florida, and north to Virginia. Also in the West Indies, Mexico, and South America.

VARIETY. Small-leaf Cyrilla, *C. racemiflora* var. *parvifolia* Raf., is a variety with smaller leaves and racemes and grows from Florida to Louisiana.

AMERICAN CYRILLA
Cyrilla racemiflora L.

REMARKS. The genus name, *Cyrilla,* is in honor of Cyrillo Dominico, professor of medicine at Naples (1734–1799), and the species name, *racemiflora,* refers to the racemose flowers. Vernacular names are Black Titi, Burnwood Bark, Leatherwood, and He-huckleberry. The wood has no commercial value, but the flowers are considered good bee food. The spongy lower bark is reported to have been used formerly as a styptic. The leaves are apparently deciduous earlier, and turn various

SMALL-LEAF CYRILLA
Cyrilla racemiflora var. *parvifolia* Raf.

brilliant colors in the autumn in the northern part of the plant's range, but in the South they remain more green and persistent. American Cyrilla is occasionally cultivated for its attractive leaves and slender white racemes of flowers. It can be propagated by seeds and cuttings.

BUCKWHEAT-TREE

Cliftonia monophylla (Lam.) Britt.

FIELD IDENTIFICATION. Thicket-forming shrub with numerous irregular branches, or a tree to 30 ft, with a trunk 6–18 in. in diameter. The trunk is often inclined and contorted with a narrow crown at the top.

FLOWERS. February–April, borne terminally in nodding, or erect, racemes which attain a length of 1–3½ in.; pedicels slender, 1/12–⅛ in., subtended by conspicuous, red, membranous, acuminate bracts, which are deciduous before the flowers open; flowers about ⅓ in. across, small, fragrant, perfect; petals 5–8 (usually 5), lobes equal or nearly so, ovate, apex rounded or obtuse; stamens 10, included, 5 shorter than the others, filaments with a lateral flattened appendage near the middle, narrowed distally; ovary 2–4-angled and 2–4-celled, surrounded by a disk at the base, ovules 2 in each cell; style absent, stigma obscurely 2–4-lobed.

FRUIT. Ripening August–September, capsule oval to ovoid or oblong, brown, dry, body spongy, 2–4-winged, wings thin, notched at ends; style terminally persistent, length about ¼ in., 2–4-celled; seeds solitary in each cell, 1/16–⅛ in. long, terete, tapering to the ends, suspended, cotyledons very short.

LEAVES. Evergreen, simple, alternate, blade length 1½–2 in., width ½–1 in., elliptic to oblong or lanceolate, apex rounded or slightly notched, base cuneate, margin entire, firm and leathery; upper surface dark green and lustrous; lower surface paler and glandular-punctate.

TWIGS. Slender, rigid, terete, glabrous, reddish brown; winter buds acuminate, usually with 2 brown scales; leaf scars conspicuous, shield-shaped; pith conspicuous.

BUCKWHEAT-TREE
Cliftonia monophylla (Lam.) Britt.

BARK. Dark reddish brown, young branches thin and scaly, on older trunks deeply furrowed and broken into short, broad scales about ¼ in. thick.

WOOD. Reddish brown, lighter-colored sapwood, close-grained, fairly hard, weak, brittle.

RANGE. Damp, sandy or peaty, acid soil, usually in the coastal pine belts. Southeastern Louisiana; eastward to Florida and north to South Carolina.

REMARKS. The genus name, *Cliftonia*, is in honor of Francis Clifton (d. 1736), an English physician. The species name, *monophylla*, means "one leaf." This name was given because of the fact that it was formerly classified with *Ptelea*, which is a 3-leaflet genus. Also known under the vernacular names of Titi, Black Titi, and Ironwood. The fruit resembles that of a buckwheat, and hence the common name of the plant. The foliage is occasionally browsed by livestock, and the flowers are valuable for honey. The pithy twigs are used for pipe stems. The plant could be more extensively cultivated for its early delicate, fragrant flowers and fruit. It was first cultivated in 1806, and may be propagated by seeds and cuttings of half-ripened wood under glass. The wood is used for fuel locally.

HOLLY FAMILY (*Aquifoliaceae*)

Yaupon Holly
Ilex vomitoria Ait.

FIELD IDENTIFICATION. An evergreen, thicket-forming shrub with many stems from the base, or a tree to 25 ft, and a diameter of 12 in. The crown is low, dense, and rounded.

FLOWERS. April–May on branchlets of the previous year, solitary or fascicled in the leaf axils, polygamo-dioecious, pedicels slender, 1/25–1/6 in. long; staminate glabrous, 2–3-flowered; pistillate puberulent, 1–2-flowered; calyx-lobes 4–5, glabrous, ovate or rounded, obtuse, about 1/25 in. long; corolla white, petals 4–5, united at base, elliptic-oblong, 1/12–1/8 in. long, about 1/12 in. wide; stamens 4–5, almost as long as the petals, anthers oblong and cordate; staminodia of the pistillate flowers shorter than the petals; ovary ovoid, 1/25–1/12 in. long, 1–4-celled, stigma flattened and capitate.

FRUIT. Often in abundance, drupe shiny red, semi-translucent, subglobose, about 1/4 in. long, often crowned by the persistent stigma, nutlets usually 4, to 1/6 in. long, obtuse, prominently ribbed.

LEAVES. Evergreen, simple, alternate, elliptic-oblong to oval, margin crenate and teeth minutely mucronulate, apex obtuse or rounded and sometimes minutely emarginate and mucronulate, base obtuse or rounded, thick and coriaceous; upper surface dark lustrous green and glabrous, veins obscure; lower surface paler, glabrous, or with a few hairs on veins; petioles 1/25–1/4 in. long, grooved, glabrous or puberulent. Leaves vary considerably in size and shape on different plants.

TWIGS. Gray to brown, terete, stout, rigid, crooked, short, glabrous or puberulent; winter buds minute, obtuse, scales brown.

BARK. Averaging 1/16–1/8 in. thick, brownish to mottled gray or almost black, tight, smooth except for lenticels, or on old trunks eventually breaking into thin, small scales.

Yaupon Holly
Ilex vomitoria Ait.

WOOD. White, heavy, hard, strong, close-grained, weighing about 46 lb per cu ft, sometimes used for turnery, inlay work, or woodenware.

RANGE. Low moist woods, mostly near the coast. Texas, Oklahoma, Arkansas, and Louisiana. Evidently reaching its largest size in the east Texas bottom lands. Eastward to Florida and northward to Virginia.

MEDICINAL USES. A medicinal tea was formerly prepared by the Indians from the leaves, which contain caffeine and also possess emetic and purgative qualities. The brew was known as the "Black Drink" and was ceremoniously taken once a year. Perhaps other plants were used to adulterate it also. An analysis by Venable (1885, p. 390) showed a content of 7.39 tannic acid and 0.27 caffeine. Cabeza de Vaca, the early Spanish explorer, saw the Indians drinking it in what is now east Texas, between the years 1528 and 1536. He described the matter at length in the report of his travels published in Spain in 1542.

PROPAGATION. Yaupon Holly is often cultivated for ornament because of its bright red drupes and evergreen leaves. Because of its tight compact foliage and tolerance of shearing, it is used for hedges or for foundation planting. Dwarf forms have been developed. The seeds may be sown in flats with equal parts of peat, soil, or sand, covered about ¼ in. and kept moist by spreading a leaf mulch or by sprinkling. Because of a somewhat dormant embryo, all do not germinate at once, so it is a good plan to hold the flat over at least a second year for the slower germinating seeds.

For cuttings, a flat at least 6 in. deep is desirable, and a mixture of 1 part peat and 3 parts sand used, although sand alone will suffice. Dipping the base in any good chemical stimulant often helps. Cuttings are inserted in the growing media with the leaves resting on the surface and the slanted lower ends about 3 in. below the surface. A greenhouse temperature of 65°–75°F. for 60–90 days should start growth. At first syringing 3 or 4 times a day is necessary, and later only once or twice a day. Root cuttings about ⅜ in. in diameter can also be used, secured in autumn and planted with only the tips exposed. Stocks for budding or grafting are usually grown from seeds or cuttings. Yaupon Holly can be grown well on American Holly, *I. opaca,* stocks.

Yaupon Holly is subject to certain diseases such as the pitted scale, *Asterolecarium puteanum* Russell, which makes conical pits on the twigs. Tea scale, *Fiorinia theae* Green, which also attacks camellias, forms cottony patches on the underside of leaves. Spraying for scale should be done about the time the new growth starts in spring to destroy the eggs. Parthion has been found effective against scale.

REMARKS. The genus name, *Ilex,* is the ancient name for the Holly Oak, and the species name, *vomitoria,* refers to its use as a medicine. Local names in use are Cassena, Cassine, Cassio-berry-bush, Evergreen Cassena, Yapon, Yopan, Youpon, Emetic Holly, Evergreen Holly, South-sea-tea, Carolina-tea, Appalachian-tea, Yopon del Indio, Chocolate del Indio, Indian Black-drink, Christmas-berry.

The fruit is known to be eaten by at least 7 species of birds. In regions where it grows it is often used as a holiday decoration. Although yellow-fruited plants have been given variety names from time to time, they do not apparently reproduce true to color and are therefore to be regarded only as unstable forms.

AMERICAN HOLLY

Ilex opaca Ait.

FIELD IDENTIFICATION. An evergreen tree to 70 ft, with short, crooked branches and a rounded or pyramidal crown.

FLOWERS. April–June, in short-stalked, axillary, cymose clusters; polygamo-dioecious, staminate in 3–9-flowered cymes, small, white, petals 4–6; stamens 4–6, alternating with the petals; pistillate flowers solitary or 2–3 together; ovary 4–8-celled, style short with broad stigmas; pistil rudimentary in the staminate flowers; calyx 4–6-lobed, lobes acute, ciliate; peduncles with 2 bracts.

FRUIT. Maturing November–December, spherical or ellipsoid, mostly red, more rarely yellow or orange, ¼–½ in. long, nutlets prominently ribbed.

LEAVES. Variable in shape, size, and spines, simple, alternate, persistent, ovate to oblong or oval, flattened, keeled or twisted, stiff and coriaceous, margins wavy, set with sharp, stiff, flat or divaricate spines, sometimes spineless; apex acute, spinose; base cuneate or rounded; upper surface dark to light green, lustrous or dull, paler and glabrous to somewhat puberulous beneath, length 2–4 in., width 1–1½ in.; petioles short, stout, grooved, sometimes puberulent; stipules minute, deltoid, acute.

TWIGS. Stout, green to light brown or gray, glabrous or puberulous.

AMERICAN HOLLY
Ilex opaca Ait.

BARK. Light or dark gray, often roughened by small protuberances.

WOOD. White or brownish, tough, close-grained, shock resistance high, works with tools well, shrinks considerably, checks or warps badly unless properly seasoned, not durable under exposure, rather heavy, specific gravity when dry about 0.61. It is used for cabinets, interior finish, novelties, handles, fixtures, and scientific instruments.

RANGE. Texas, Oklahoma, Arkansas, and Louisiana; eastward to Florida, north to Massachusetts and New York, and westward through Pennsylvania, Ohio, Indiana, and Illinois to Missouri.

PROPAGATION. The fruit may be gathered by hand, and the pulp removed by running through a hammer mill and then allowed to ferment in warm water. The pulp is then removed by rubbing on a screen and floating off the pulp. There are about 35,000 seeds per lb. They are dried before storage or planting. The seed may be pretreated by stratifying in layers of sand or soil. Germination does not occur in all seeds the first season, but plantings can be held over for 2–3 years. All hollies can be grown from cuttings in damp sand or peat. Terminal twigs taken in late July–August are suitable. They should be 3–5 in. long, with 2–5 leaves left on the upper part, and planted in the sand up to the leaves. Rooting takes 60–90 days.

American Holly can also be grafted, and budded, by the use of diverse methods. Seedling stocks make good grafting because of a superior root development. Many hollies, and varieties, can be grafted on American Holly stock, including Dune Holly, Chinese Holly, Dahoon Holly, Yaupon Holly, and Myrtle-leaf Holly.

The staminate trees should be planted close to the pistillate trees. In nature there is usually a ratio of about one pistillate tree to four or more staminate trees. However, the sex cannot be distinguished until blooming age, which is 5–12 years. Sometimes individual pistillate or staminate trees do not bloom simultaneously and pollination does not take place. To offset this, sometimes a cut blooming staminate branch can be kept fresh in a can of water for a few days and hung in a pistillate tree to permit the bees to transfer the pollen. Sometimes it is possible to get a self-fertilizing tree by grafting staminate branches in the top of pistillate trees.

American Holly, as well as other species, should be transplanted during the dormant period, usually November–March. Small plants may be dug bare-rooted, but larger ones should be balled and burlapped. If bare-rooted, the roots should be immediately wrapped in wet Spanish-moss or excelsior, and kept wet until planted.

In planting, the hole should be about 1 ft wider and deeper than the root system. Water poured into the hole with the dirt will keep the roots moist. Also leave a shallow basin around the tree to retain water later. A good watering once a week is usually sufficient. A 2 in. mulch of leaves, straw, sawdust, or other material helps maintain root moisture. Fertilizers of various kinds can be used, such as stable manure or commercial balanced fertilizer. It may be worked into the ground with a hoe, or placed in holes 2 ft apart outwardly according to the diameter of the crown. Commercial fertilizer may be applied at about 2 lb for each in. of diameter of a tree under 6 in., and 3 lb for each in. of diameter over 6 in. applied annually.

American Holly can be used for yard, street, and park planting, or for hedges. The best forms of hedges are the Hedge Holly, Christmas Hedge, and Clark.

INSECT ENEMIES AND DISEASES. Hollies are subject to damage by certain insects and fungi as follows:

Leaf-mining insects, such as the European Holly leaf miner, *Phytomyza ilicis* Curt., and American Holly leaf miner, *Phytomyza ilicicola* Loew, do damage to leaves and may be controlled by spraying with DDT (2 lb of 50 per cent wettable powder in 10 gallons of water) when the flies emerge.

Spittle bugs, *Monecphora bicincta* Say, damage the young shoots, and the leaves may drop off or be distorted or stunted. Control is with DDT, the same as that of the leaf miner, except that a number of hatchings require spraying at intervals.

Scale insects, of which there are many, can be controlled by spraying the new growth in spring with a miscible oil spray such as Florida Volck, or parthion.

Red mite, *Paratetraanychus ilicis* McGregor, discolors the leaves by feeding on the juices. A good control is parthion, which is sold under different trade names. Dusting at intervals with sulphur also helps.

Black mold, *Englerulaster orbicularis* (Berk. & Curt.) Hohn, is a fungus which causes black spots on the lower leaf surfaces. Dusting the surfaces with a copper fungicide will assist in control.

VARIETIES, FORMS AND HYBRIDS. Three natural varieties have been described in the literature. However, two of these are rather weakly established taxonomically.

Whole-leaf American Holly, *I. opaca* var. *subintegra* Weatherby, is set apart as being spineless. Many intermediates are found between the spiny and less spiny forms; in fact, some branches on trees bear spineless (except apical spine), and other branches spiny leaves.

Yellow-fruit American Holly, *I. opaca* var. *xanthocarpa* Rehd., is segregated as having yellow fruit instead of red. Since yellow is a recessive trait in nearly all red-fruited hollies, and since plants from yellow fruit most often produce trees with red fruit, the character is not a dependable one.

Florida Dune Holly, *I. opaca* var. *arenicola* (Ashe) Ashe, is a more constant variety confined to the scrub dune lands of Florida. Its habit of growing in the deep sand, erect, fastigiate character, lanceolate or oblanceolate leaves with forward pointing teeth, and oval, shallowly grooved nutlets give it sufficient segregative characters. It was formerly known as *I. cumulicola* Small.

Topel Holly, *I.* × *attenuata* Ashe, is considered to be a hybrid of *I. opaca* and *I. cassine*. It has long spiny-pointed leaves and occurs in South Carolina and northwestern Florida.

Hume (1953, pp. 49–66) has listed the following horticultural forms of American Holly. The most dependable ones are italicized, and the name of the propagator follows the plant name:

Aalto (Wheeler), Amy (Wheeler), Arden (Nearing), Baker (Hume), Bountiful (Dilatush), Brooks (O'-Rourke), Canary (Dilatush), Cape Cod (Dilatush), *Cardinal* (Dilatush), Cheerful (Hohman), Christmas Hedge (Dilatush), Christmas Spray (Dilatush), Clark (White), *Cronenberry* (Thrasher), *Cumberland* (Wolf), Cup Leaf (Dilatush), Delia (White), Delia Bradley (Hohman), Dorothy (White), *East Palatka* (Hume), *Eleanor* (Wolf), Elizabeth (Wheeler), *Emily* (Wheeler), Faroge (White), Freeman (Wheeler), Goldie (White), *Griscom* (White), Hedgeholly (Basley), *Howard* (Hume), Hume #2 (Hume), Isaish (White), Judge Brown (Conners), Katz (Hume), Lady Alice (Wolf), Lake City (Hume), Laura (White), Mae (White), Makepeace (White), Manantico (Wolf), Manig (White), Marion (Hume), Maurice River (Wolf), Merry Christmas (Dilatush), Miss Helen (McLean), Mrs. Santa (Nearing), *Natole* (Wheeler), October Glow (Hohman), Old Faithful (Dilatush), *Old Heavyberry* (Dilatush), Old Leatherleaf (Dilatush), Osa (White), Perfection (Dilatush), Sallie (White), Savannah (Robertson), Silica King (Wolf), *Slim Jane* (Wolf), Slim Jim (White), St. Ann (Wheeler), St. Mary (Wheeler), Taber #3 (Hume).

REMARKS. The genus name, *Ilex,* is the old name for Holly Oak, and the species name, *opaca,* refers to the dull green leaf. Vernacular names are Yule Holly, Christmas Holly, and White Holly. The foliage and fruit are often used for holiday decorations and are sometimes browsed by cattle. At least 18 species of birds eat the fruit.

DAHOON HOLLY
Ilex cassine L.

DAHOON HOLLY
Ilex cassine L.

FIELD IDENTIFICATION. An evergreen, red-fruited holly, forming a shrub with slender, ascending branches, or a tree to 25 ft, with a diameter of 12–18 in. Usually with a low broad, rounded crown.

FLOWERS. Flowers polygamo-dioecious, usually on new growth in umbel-like clusters, numerous, small, white; on hairy peduncles to 1 in. long with acute, scarious bractlets, pedicels to 1/4 in., mostly 2–9-flowered in the staminate, corolla 1/6–1/5 in. broad; petals 4–6 (usually 4), obovate to elliptic, apex obtuse, united at base, about 1/10 in. long; calyx glabrous, 1/15–1/12 in. in diameter, lobes 4–6 (usually 4), ovate-triangular, margin ciliate-erose, apex acute or acuminate; stamens shorter than the petals, as many as the petals and alternate with them, filaments distinct; pistillate umbels usually 3-flowered, ovary superior, style very short; staminodia in pistillate flowers shorter than the petals, anthers small and abortive; ovary in pistillate flowers conical, about 1/12 in. long, 4-celled, stigma capitate.

FRUIT. Ripening September–March, often abundant, persistent until the following spring, globose, 1/5–1/4 in. in diameter, bright to dull red or yellow, flesh yellow and mealy, solitary or in clusters of twos or threes; nutlets 4, bony, pale brown, prominently few-ribbed on the back and sides, base rounded, apex acute, length about 1/6 in.

LEAVES. Evergreen, leathery, flat, simple, alternate, blade length 1 1/3–4 in., width 1/2–1 1/2 in., oblanceolate to oblong-obovate; apex acute to obtuse and mucronate to rarely rounded or occasionally emarginate; base gradually narrowed or cuneate; margin entire and revolute, or remotely and sharply toothed above the middle; upper surface at maturity dark green, glabrous, or a few hairs along the impressed midrib; lower surface paler and glabrous or slightly hairy, primary veins slender and conspicuous; petioles to 1/2 in., stout, thickened at base, canaliculate, glabrate or sparingly hairy.

TWIGS. Slender, pale, and rather densely pubescent the first 2 or 3 years, later dark brown and glabrous with scattered lenticels; winter buds very small, acute, pubescent; leaf scars crescent-shaped, bundle scar solitary, pith smooth and continuous.

BARK. Thin, gray and smooth at first, darker later and roughened by lenticels.

WOOD. Pale brown, close-grained, light, soft, not strong, weighing 30 lb to the cu ft, easily worked.

RANGE. Swamplands, in sandy, acid soil, often in pine barrens. Louisiana; eastward to Florida and Georgia, and north to Virginia. Probably erroneously recorded from Texas.

VARIETIES. The seeds germinate easily, or plants may be grown from cuttings or grafted on *I. opaca* stock. *I. cassine* also hybridizes with *I. opaca* to produce the Topel Holly (*Ilex* × *attenuata* Ashe), known from South Carolina and northwestern Florida.

It is also known to hybridize with Myrtle Dahoon Holly, *I. myrtifolia*. Dahoon Holly is very variable in leaf size, shape, and glossiness, size and amount of fruit, and habit of growth.

It is subject to attack by black mold, *Englerulaster orbicularis* (Berk. & Curt.) Hohn, which can be controlled by spraying beneath the leaves with a copper fungicide.

Alabama Dahoon Holly, *I. cassine* var. *angustifolia*, is a variety with narrow, elongate leaves, common in southern Alabama.

I. cassine forma *glencassine* Hume is a horticultural form densely branched and with a rounded head; leaves elliptic, length 4 in., width to 1½ in., apex obtuse and minutely mucronate, base narrowed, margin entire or nearly so, glossy green above, lighter beneath; petiole reddish or purplish above, ⅜–⅝ in. long; fruit globose, red, about ¼ in. long, densely clustered; pedicels about ⅓ in.; twigs dark.

REMARKS. The genus name, *Ilex*, is the ancient name for a European oak, and the species name, *cassine*, is from the Indian name, *Cassena*, which name was applied by them to a drink prepared from the Yaupon, *I. vomitoria*. Although the leaves of *I. cassine* were sometimes mixed with those of *I. vomitoria*, the former contains no caffeine. Vernacular names in use are Christmas-berry, Yaupon, and Henderson-wood. The fruit is eaten by about 10 species of birds and by the raccoon. The berries are much used during the Christmas season, and the plant has been cultivated since 1726. It is growing in demand as an ornamental plant.

MYRTLE DAHOON HOLLY
Ilex myrtifolia Walt.

MYRTLE DAHOON HOLLY

Ilex myrtifolia Walt.

FIELD IDENTIFICATION. An evergreen, red-fruited holly. Sometimes a small shrub with numerous crooked rigid stems, or a small tree to 23 ft, and a diameter of 10 in. Crown generally broad and compact.

FLOWERS. In May, borne on the new growth in simple or compound, short-peduncled cymes; staminate and pistillate on different trees or on the same tree; flowers numerous, small, inconspicuous; staminate cymes 3–9-flowered, corolla about ⅕ in. wide, petals usually 4, white, oblong-elliptic, united at base; stamens generally as many as the petals, alternate with them, filaments distinct and shorter than the petals; style of staminate flower reduced in size or abortive; calyx glabrous, 1/25–1/15 in. wide, lobes ovate-triangular; pistillate flowers 1–3 on a peduncle, ovary about 1/12 in. long, superior, 4-celled, style very short, stigma capitate, stamens usually reduced in size, in pistillate flower the pistil larger.

FRUIT. Maturing in late fall, persistent over winter, globose, about ¼ in. in diameter, red to orange or yellow, flesh thin and yellow; nutlets 4, ribbed; pedicels of fruit about ⅕ in.

LEAVES. Crowded, evergreen, simple, alternate, leathery and rigid, average length of blades ⅜–1½ in., width ⅕–⅓ in. (some plants with leaves to 2¾ in. long and 1¼ in. broad), apex acute to obtuse, mucronate-spined, base obtuse to acute or rounded, margin entire and revolute, occasionally serrate toward the apex with a few minute spinescent teeth, linear to lanceolate or oblong; upper surface dark green and glabrous or puberulent along the midrib; lower surface paler and glabrous or slightly puberulent, midrib conspicuous, primary veins generally obscure; petioles puberulent, usually less than ⅛ in. long, but sometimes to ½ in.

TWIGS. Slender, rigid, smooth, pale, puberulent at first, glabrate later.

BARK. Pale gray or almost white, roughened by warty protuberances.

RANGE. On the coastal plain, wet acid soil around cypress ponds and swamps, and in pine barrens. Probably erroneously recorded from Texas, except occasionally in cultivation. Louisiana; eastward to Florida and northward to North Carolina.

PROPAGATION. Considered by some gardeners to be our most beautiful native holly. It will grow readily from seed. When transplanted it should be cut back. It is known to hybridize with the Dahoon Holly, *I. cassine,* and can be grafted on American Holly, *I. opaca,* stock, and thus grown on drier soil.

VARIETY. A yellow-fruited form has been described as *I. myrtifolia* forma *lowii* Blake, but since most hollies produce yellow-fruited forms this is not a stable characteristic.

REMARKS. The genus name, *Ilex,* is the classical name for a European oak, and the species name, *myrtifolia,* refers to the small, myrtle-like leaves. Common names are Small-leaf Dahoon, Myrtle Holly, and Cypress Holly. Myrtle Dahoon Holly leaves contain no caffeine.

INKBERRY HOLLY

Ilex glabra (L.) Gray

FIELD IDENTIFICATION. Shrub to 12 ft, much-branched, evergreen, thicket-forming as a result of underground stolons.

FLOWERS. February–July, polygamo-dioecious, axillary, solitary or a few together on new growth; staminate flowers of 3 or more together on puberulent pedicels 1/25–⅕ in.; pistillate flowers often solitary or 2–3 together, pedicels shorter than those of the staminate flowers; corolla ⅕–¼ in. wide; petals 5–8, white, deciduous, united below, broadly elliptic or suborbicular, apex obtuse or rounded; about ⅛ in. long; stamens 4–6, shorter than petals, filaments erect and subulate, anthers introrse, 2-celled, opening longitudinally (stamens of pistillate flower usually abortive); ovary superior, free, about 1/12 in. long, 2–6-celled, stigma discoid, elevated (ovary in staminate flowers usually abortive).

FRUIT. Maturing in late autumn, persistent, solitary or 2–3 together, black, globose, ⅕–⅓ in. in diameter, stigma persistent; nutlets crustaceous, 5–8, each 1-seeded, about ⅙ in. long, black, flat.

LEAVES. Crowded, evergreen, simple, alternate, blade length ⅝–2⅓ in., width ¼–1¼ in., oblanceolate to obovate or elliptic to oval, apex obtuse to rounded and mucronulate, base acute or cuneate, margin entire or with 1–3 blunt low teeth on each side toward the apex, flat and coriaceous; upper surface lustrous dark green and glabrous, veins obscure; lower surface pale and punctate, semiglabrous or midvein puberulent; petioles 1/12–⅖ in., finely puberulent.

TWIGS. Slender, upright, terete or angled, young ones green and finely puberulent, older ones gray to black and glabrous.

RANGE. Generally near the coast in sandy, acid bogs of pinelands or prairies. Louisiana; eastward to Florida and Georgia and northward to Massachusetts. Most abundant in Florida.

PROPAGATION. Within its range, Inkberry is the most abundant of all American hollies. It is rather free from disease and insects, and sprouts back after fires. It has been cultivated since 1759, and is often planted for ornament, either as border plants or in hedges, and is also used as a Christmas decoration. In cultivation it seems to stand cold as far north as New England. The fruit may be hand-picked or flailed into a canvas. It is spread out to dry or the pulp removed by running through a macerator with water, and the debris floated off. Five lb of seed is yielded by 100 lb of fruit. Although the seeds germinate a little faster than some hollies, stratification in moist sand for 30 days at 80°F. is helpful. Even then some of the seeds germinate the second year. The seeds are sown in drills and covered

INKBERRY HOLLY
Ilex glabra (L.) Gray

with about ⅛ in. of rich, moist, sandy loam. Half-shade is required at first for the seedlings and they are planted out in 2–3 years.

Inkberry Holly can also be propagated from cuttings of ripened wood grown under glass, or it is sometimes grafted or budded on *I. opaca* stock. Because of the fact that it can also reproduce by underground stolons these divisions are sometimes used. Root cuttings about ⅜ in. in diameter can also be obtained and planted in flats so that only the tips are exposed. Another method is to root-prune the plant on one side, let it remain a year, and then ball and burlap the roots and replant. It helps to prune the top back when using this method. Clumps of plants obtained in a wild state seem to be either all male or all female.

VARIETY. A variety of Inkberry Holly has been named *I. glabra* var. *austrina* Ashe and is reported as having smaller, broader leaves than the species. However, the species is so variable that this appears to be a rather weak segregation. Forms with red berries have also been reported.

REMARKS. The genus name, *Ilex,* is the ancient name for the Holly Oak, *Quercus ilex,* of Europe. The species name, *glabra,* refers to the smooth glabrous leaves. The vernacular names are Gallberry, because of the bitter taste, and Inkberry which refers to the black fruit. Also known as Appalachian Tea, Canadian Winterberry, Evergreen Winterberry, and Dye-leaves. Inkberry is considered to be a good source of honey. The fruit is eaten by at least 15 species of birds, including bobwhite quail and wild turkey, and the leaves are browsed by marsh rabbit. The plant contains no caffeine.

TALL INKBERRY HOLLY
Ilex coriacea (Pursh) Chapm.

TALL INKBERRY HOLLY

Ilex coriacea (Pursh) Chapm.

FIELD IDENTIFICATION. Mostly an evergreen shrub, but sometimes a small tree to 15 ft.

FLOWERS. Polygamo-dioecious, solitary or clustered; staminate corolla about ¼ in. wide; petals 4–9, united below, white, oblong-elliptic, erose on the margin; stamens 4–6, almost as long as petals, anthers opening lengthwise, staminodia about one half as long as petals; calyx smooth with triangular and acute lobes; pistillate flowers inclined to be solitary, pedicel puberulent; pistil with a depressed ovary, thick short style, and discoid stigma.

FRUIT. Drupe globose, black, lustrous, ¼–⅓ in. long, tipped by the persistent stigma; nutlets 4–9, compressed, smooth, about ⅙ in. long.

LEAVES. Alternate, simple, evergreen, dark, shining, stiff, obovate, oblanceolate or elliptic, acute or short-acuminate and spinescent at the apex, rounded or acute at the base; entire on margin or with a few spinescent teeth; upper surface dark green and glabrous; punctate or sometimes glaucous, glabrous or puberulent beneath; petioles short and puberulent, about ¼ in., pubescent.

TWIGS. Slender, elongate, green to gray, glabrous or puberulent, lenticels numerous, small, round.

BARK. Mottled gray or brown, smooth and tight.

RANGE. In acid, sandy, low woods. Eastern Texas, eastward to Florida. In Texas in Hardin, Jasper, Newton, and Montgomery counties, and possibly others.

REMARKS. The genus name, *Ilex,* is the ancient name of the Holly Oak, and the species name, *coriacea,* refers to the thick, leathery leaves. Vernacular names are Shining Inkberry, Large Gallberry, Baygall-bush.

GEORGIA HOLLY

Ilex longipes Chapm.

FIELD IDENTIFICATION. Wide-spreading shrub or tree to 23 ft. Distinguished by elongate peduncles of the fruit. Leaves deciduous, or semipersistent southward.

FLOWERS. Inflorescences axillary, staminate and pistillate usually on different trees; staminate flowers solitary or fascicled; pistillate flowers usually solitary; pedicels ⅓–1¼ in., slender and glabrous, bearing small, 4-parted flowers; calyx about 1/12 in. in diameter, smooth, lobes of calyx triangular, acute at apex, denticulate on the margin; petals united below, elliptic; stamens about as long as the petals, but staminodia somewhat shorter; normal pistillate ovary ovoid, 4-celled, hardly over ⅛ in. long, stigma capitate.

FRUIT. Drupe red, globose, lustrous, ⅓–⅖ in. long, borne on a stalk ⅓–1¼ in.; nutlets 4, inconspicuously ribbed and striate.

LEAVES. Alternate, simple, persistent, thickish, blades ⅔–2½ in. long, ⅓–1⅓ in. wide, elliptic to elliptic-obovate or oval, acute or obtuse at the apex, minutely mucronulate, cuneate at the base; crenate-serrate on the margin, teeth often with minute bristles; glabrous and dark green above, paler beneath; petioles slender, ⅛–½ in. long, pubescent.

TWIGS. Slender, glabrous, round, green at first, gray with maturity.

RANGE. In low sandy woods. Eastern Texas in Newton, Jasper, Trinity, and San Jacinto counties, and possibly others; eastward to Florida and northward to North Carolina and Tennessee.

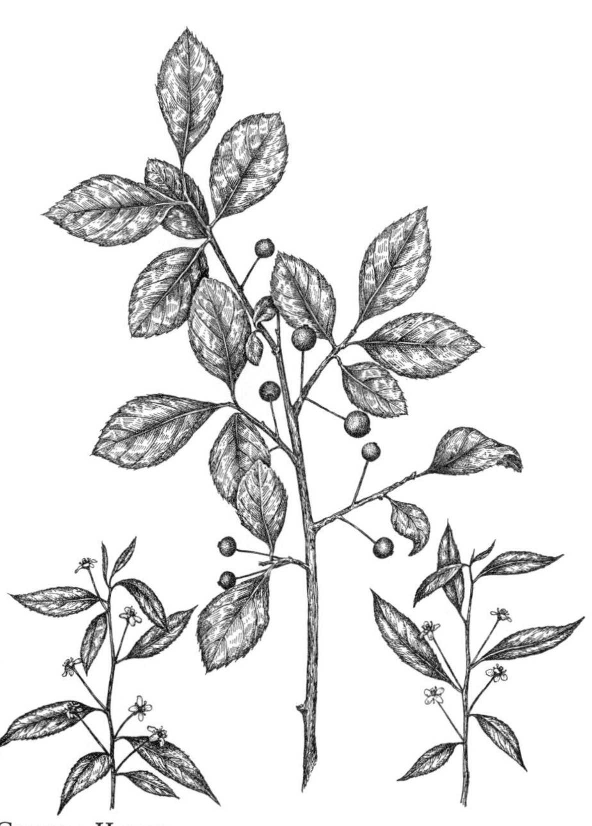

GEORGIA HOLLY
Ilex longipes Chapm.

CAROLINA HOLLY
Ilex ambigua (Michx.) Chapm.

VARIETY. Downy Georgia Holly, *I. longipes* var. *hirsuta* Lundell, has shorter fruiting pedicels, denser and coarser indument, and smaller leaves. Known from Harris, Madison, Newton, Polk, San Jacinto, Trinity, and Walker counties in Texas.

REMARKS. The genus name, *Ilex*, is the ancient classical name, and the species name, *longipes*, refers to the long-stemmed drupes, which is a conspicuous feature. Other names in common use are Chapman's Holly, Long-stem Holly, and Yaupon Holly. The plant has no economic importance other than the decorative value of the fruit.

CAROLINA HOLLY

Ilex ambigua (Michx.) Chapm.

FIELD IDENTIFICATION. Commonly a shrub, but sometimes a tree to 18 ft, with irregular branches and a rounded crown.

FLOWERS. In spring, borne in axillary inflorescences, staminate and pistillate usually on different trees, staminate usually fascicled but pistillate mostly solitary; pedicels smooth, slender, 1/25–⅙ in.; calyx smooth 1/25–1/12 in. in diameter, 4–5-lobed; lobes ovate, apiculate at the apex, somewhat hairy on the margin; corolla small, white, 4–5-parted; petals elliptic, ciliolate, hardly over ⅛ in. long, and longer than the stamens and staminodia;

ovary ovoid, 4–5-celled, hardly over 1/8 in. long with a capitate stigma.

FRUIT. Drupe in late summer on very short pedicels, dark red, translucent, subglobose, hardly over 1/8–1/4 in. long; nutlets 4–5, brown, striate, mostly less than 1/5 in. long.

LEAVES. Alternate, simple, deciduous, thin, oval-ovate or elliptic-lanceolate, acute or acuminate at the apex, rounded or cuneate at the base, crenate-serrulate on the margin but the lower third often entire, 3/4–3 in. long; mostly glabrous above and below, but sometimes with pubescence along the veins; petioles slender, often pubescent, 1/3–1/2 in.

TWIGS. Rather slender, smooth, brown to purple, with prominent lenticels.

BARK. Dark brown, sometimes black, lustrous, smooth, flaking later.

RANGE. In low sandy woods; eastern Texas, Louisiana, Arkansas, and Oklahoma; eastward to Florida and northward to North Carolina.

REMARKS. The genus name, *Ilex,* is the ancient classical name, and the species name, *ambigua,* refers to its vague similarity to other species. Vernacular names are Sand Holly and Possum Holly.

POSSUM-HAW HOLLY
Ilex decidua Walt.

POSSUM-HAW HOLLY

Ilex decidua Walt.

FIELD IDENTIFICATION. Usually a shrub with a spreading open crown, but sometimes a tree to 30 ft, with an inclined trunk to 10 in. in diameter.

FLOWERS. Borne March–May with the leaves, polygamo-dioecious, solitary or fascicled, borne on slender pedicels; pedicels of staminate flowers to 1/2 in. long, that of the pistillate flower generally shorter than the staminate; petals 4–6, white, united at the base, oblong to elliptic, 1/8–1/6 in. long; calyx-lobes 4–6, lobes ovate to triangular, acute or obtuse, entire to denticulate, sometimes ciliolate; stamens 4–6, anthers oblong and cordate, fertile stamens as long as the petals or shorter; staminodia of pistillate flowers shorter than the petals; ovary ovoid, about 1/12 in. long, 4-celled, stigma large, capitate and sessile.

FRUIT. Ripe in early autumn, persistent on branches most of the winter after leaves are shed, drupe globose or depressed-globose, orange-red, 1/4–1/3 in. in diameter, solitary or 2–3 together; nutlets usually 4, crustaceous, ovate or lunate, longitudinally ribbed, up to 1/5 in. long.

LEAVES. Simple, deciduous, alternate or often fascicled on short lateral spurs, obovate to spatulate or oblong, margin crenate-serrate with gland-tipped teeth; apex acute to obtuse, rounded or emarginate; base cuneate or attenuate, membranous at first but firm later; upper surface dark green and glabrous or with a few hairs, main vein impressed; lower surface paler and glabrous or pubescent on ribs, blade length 2–3 in., width 1/2–1 1/2 in.; petiole slender, grooved, glabrous to densely puberulent, length 1/12–1/2 in.; stipules filiform, deciduous.

TWIGS. Elongate, slender, often with many short spur-like lateral twigs, light to dark gray, glabrous or puberulent, lightly lenticellate, leaf scars lunate, buds small and obtuse.

BARK. Smooth, thin, mottled gray to brown, sometimes with numerous warty protuberances.

WOOD. White, close-grained, weighing 46 lb per cu ft, of no commercial value.

RANGE. In rich, moist soil, usually along streams or in swamps. Texas, Louisiana, Oklahoma, and Arkansas; eastward to Florida, and northward to Tennessee, Kentucky, Indiana, Illinois, Kansas, and Missouri.

PROPAGATION. The fruit can be picked off by hand in autumn and spread out to dry, or run through a macerator with water, and the pulp floated off. The seeds may then be stratified in moist sand or peat for spring planting. However, germination may be prolonged 1–3 years because of excessive dormancy of the embryo and a fairly hard seed coat. The seed is sown broad-ast or in drills, and covered with about ⅛ in. of soil. The seedlings require partial shade at first. Propagation nay be more successfully practiced by means of cuttings, layering, or grafting.

VARIETY. Curtis Possum-haw Holly, *I. decidua* var. *curtisii* Fern., is a closely related plant from Florida with eaves ⅓–⅔ in. long and small fruit about ¼ in. in diameter.

REMARKS. The genus name, *Ilex*, is the ancient name of the Holly Oak, and the species name, *decidua*, refers to the autumn-shed leaves. Local names are Deciduous Holly, Meadow Holly, Prairie Holly, Welk Holly, Bearberry, and Winterberry. It is occasionally planted for ornament, and is attractive in winter because of the persistent orange-red drupes. It is sometimes mistaken for a hawthorn in fruit, and possums are fond of it, hence the name, Possum-haw. At least 9 species of birds are known to feed upon the fruit, including the bobwhite quail.

Common Winterberry Holly

Ilex verticillata (L.) Gray

FIELD IDENTIFICATION. Deciduous, red-fruited holly forming a shrub or small rounded tree to 25 ft.

FLOWERS. Borne May–June, pedicels 2-bracted, shorter than the subtending petioles, staminate and pistillate pedicels about equal in length; flowers small, inconspicuous, dioecious or polygamo-dioecious; staminate in short peduncled cymes of 2–10 flowers; staminate corolla about ¼ in. across; petals 4–8, imbricate in bud, greenish white, oblong or rounded, apex obtuse, margin entire or minutely erose, united at base; stamens usually the same number as the petals and alternate with them, shorter than the petals and adnate to them at base; filaments distinct and erect, anthers introrse and longitudinally dehiscent; pistil usually reduced in size in the staminate flower; pistillate flowers 1–3 in a cluster, about ¼ in. across, pistil enlarged, compound, stigma broad and nearly sessile, stamens smaller and abortive; calyx-lobes imbricate, usually the same number as the petals, ovate-triangular, apex obtuse or acute, calyx about ⅒ in. wide in the staminate flower.

FRUIT. Ripening September–October, persistent in winter, nearly sessile, globose, ¼–⅓ in. in diameter, lustrous red to orange or yellow; nutlets usually 3 to 6, testa membranous, lunate, bony, smooth dorsally, embryo minute, albumen fleshy.

LEAVES. Simple, alternate, deciduous, texture varying from thin to firm-coriaceous, generally about twice as long as wide, length of blade 1½–4 in., width ⅔–2 in., ovate to elliptic or lance-oblong, margins sharply serrate, with short appressed teeth, apex acuminate or acute, base acute to cuneate or less often rounded, upper surface glabrous or nearly so; lower surface usually pubescent and strongly veined, rarely completely glabrous; petiole ⅕–⅖ in. long, channeled above, pubescent to glabrate. Leaves turning black in autumn.

TWIGS. Gray to reddish brown, smooth at first but usually roughened by warty lenticels later, glabrous or slightly pubescent.

RANGE. Common Winterberry has a very wide distribution. It can stand lower temperatures than any other native American holly. In swamps and wet woods.

Reported from southeastern Louisiana and Arkansas; eastward to Georgia and Florida, northward to Connecticut and Massachusetts, and westward to Michigan and Minnesota.

MEDICINAL USES. Common Winterberry has been used in domestic medicine, the fruit being bittersweet and acid to the taste. It and the bark were used as an astringent, internally for diarrhea, and locally in ulcerated skin lesions.

PROPAGATION. The drupes may be picked by hand in autumn, or flailed into a canvas. They may be spread out to dry and planted immediately, or they may be run through a macerator with water, and the pulp floated off. After drying, the seed may be stored in dry, sealed containers, or stratified for spring planting. The cleaned seed yields about 15 lb of seed per 100

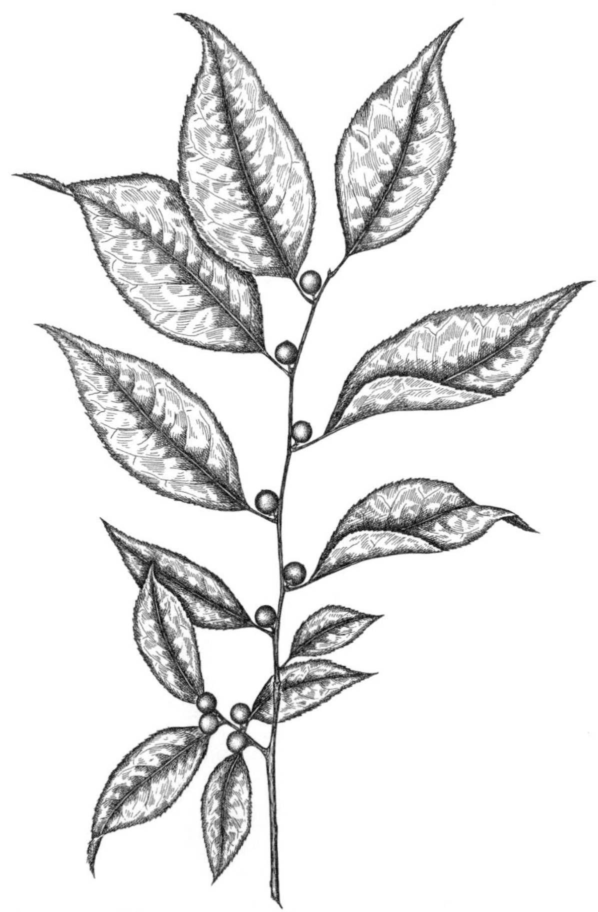

Common Winterberry Holly
Ilex verticillata (L.) Gray

lb of fruit, and averages about 92,000 seeds per lb. The commercial purity is 96 per cent and the soundness 75 per cent. Recommendations to break dormancy are stratification in moist sand for 60 days at 68°F., followed by 60 days at 41°F.; following this, the seed should germinate satisfactorily in sand flats in 60 days. Some seed seems to maintain dormancy until the second year. The planting may also be done broadcast or in drills, and covered ⅛ in. with rich, moist, well-drained soil. Transplanting should be done when dormant, but Common Winterberry Holly is easier to transplant than some of the other species. Propagation is also practiced by cuttings, layers, and grafting.

VARIETIES AND HORTICULTURAL FORMS. Common Winterberry is very variable in leaf shape, serration, and pubescence. A number of these variations have been recognized by botanists, but because of intergrading some are difficult to separate precisely.

Yellow-fruit Common Winterberry, *I. verticillata* forma *chrysocarpa* Robins., is a form with yellow-to-orange fruit. Introduced into cultivation in 1885.

Round-leaf Common Winterberry, *I. verticillata* var. *cylophylla* Robins., is a variety with leaves ¾–1⅔ in. long (about one to two and a half times as long as broad), crowded terminally, generally broadest above the middle or suborbicular, apex rounded and slightly apiculate, firm, glabrate above and pubescent below; densely flowered, cymes almost sessile. Cultivated since 1934. Found in Massachusetts, Michigan, and Illinois.

Plum-leaf Common Winterberry, *I. verticillata* var. *padifolia* (Willd.) Torr. & Gray, has leaves ⅞ in. or more wide, lower surface slightly to densely pubescent, somewhat resembling plum leaves. Found from New Jersey to Ontario and Minnesota, southward into Missouri and Mississippi.

Upright Common Winterberry, *I. verticillata* var. *fastigiata* (Bick.) Fern., is a dense shrub with ascending branches; leaves numerous, crowded, lanceolate to oblong, ¾–1⅔ in. long, less than ⅘ in. wide usually (mostly three to four times as long as wide), broadest near or below the middle, apex acute to subacuminate, base cuneate, margin sharply serrate, lower surface glabrous or slightly hairy on the veins. Found nearly always in wet sands near the sea. Connecticut to Massachusetts and Long Island.

Narrow-leaf Common Winterberry, *I. verticillata* var. *tenuifolia* (Torr.) Wats., has been listed as a variety with leaves obovate, thin, large and pellucid-puncticulate under the glass; fruit solitary. However, it seems hardly distinct from the species, and is not recognized by some botanists.

Several horticultural forms of this variation have been listed as the Prolific *(polycarpa)*, Round-leaf *(cyclophylla)*, and Yellow-fruit *(chrysocarpa)*.

A species at one time listed as *I. broxensis* Britt. is now considered to be the same as *I. verticillata* (L.) Gray.

REMARKS. The genus name, *Ilex*, is the classical name for a European oak, and the species name, *verticillata*, refers to the axillary clusters of flowers. Also known under the vernacular names of Striped Alder, False Alder, Black Alder, Fever-bush, and Virginia Winterberry. The fruit is valuable as a wildlife food, being eaten by at least 20 species of birds including bobwhite quail, ruffed grouse, sharp-tailed grouse, and ring-necked pheasant. The bright red fruit is also used for winter dry bouquets. It is more persistent on the branches in the southern parts of its distribution. Common Winterberry is the most popular of the red-fruited, deciduous hollies. It has been cultivated since 1736, and is used for ornamental planting in moist places. It is rather free of insects and disease.

MOUNTAIN WINTERBERRY HOLLY
Ilex montana Torr. & Gray

MOUNTAIN WINTERBERRY HOLLY

Ilex montana Torr. & Gray

FIELD IDENTIFICATION. Low spreading shrub or tree to 40 ft, and a trunk diameter of 10–12 in. Crown narrow and pyramidal.

FLOWERS. Dioecious or polygamous, borne usually in June when leaves are half-grown, on slender pedicels ⅛–⅓ (sometimes ½) in., staminate clustered, pistillate mostly solitary, crowded on lateral spurlike branchlets of the previous year, or solitary on branchlets of the current year; staminate calyx 1/8–1/6 in. wide, sepals 4–6 (mostly 4), triangular, acute, glabrous, ciliate; staminate corolla 1/6–1/5 in. wide; petals 4–6, white, obovate, apex obtuse; stamens as many as the petals, with shorter, distinct filaments; ovary superior, stigma broad and flat; staminate flowers with abortive pistils, and pistillate flowers with reduced stamens.

FRUIT. Ripening September–November, fruit somewhat larger than most hollies, 1/3–1/2 in. in diameter, globose, scarlet to orange-red or rarely yellow; nutlets about 1/4 in. long, prominently ribbed on the back and sides; peduncles 1/12–1/4 in., generally shorter than the petioles.

LEAVES. Simple, alternate, deciduous, membranous, ovate to oblong-lanceolate, apex acuminate or acute, base cuneate or rounded, margin sharply serrate with minute, glandular incurved teeth, length 2–6 in., width 1/2–2½ in.; upper surfaces dark green, and glabrous or slightly hairy along the prominently impressed arcuate veins; lower surface paler and glabrous or slightly pubescent; leaves early deciduous, turning yellow in autumn; petioles slender, 1/3–3/4 in.

TWIGS. Slender, more or less zigzag, reddish brown at first, gray later, glabrous; winter buds ovoid to subglobose, about 1/8 in. long, scales brown, keeled, apiculate.

BARK. Light brown to gray, about 1/16 in. thick, roughened by numerous lenticels.

WOOD. Creamy white, close-grained, hard, heavy, and strong.

RANGE. Usually on well-drained, wooded mountain slopes. In Louisiana recorded from West Feliciana and Winn parishes. Eastward to Florida and Georgia, northward to Tennessee, Pennsylvania, and New York, most often along the Appalachian Mountains.

PROPAGATION. The seeds may be gathered by hand in the fall or flailed into sacks. They may be dried and planted immediately, or depulped by running through a macerator with water, and the pulp floated off. They may then be dried and stored in sealed containers or stratified in moist sand for spring planting. The seed may be sown broadcast or in drills. Fall sowing should be mulched over the winter. Some seeds may not germinate until the second or third year. The cleaned seed averages about 35,000 seeds per lb, with a commercial purity of 90 per cent and a soundness of 84 per cent.

VARIETY. Beadle Mountain Winterberry, *I. montana* var. *beadlei* Ashe, is a variety of the North Carolina mountains, and has leaves ovate to elliptic-oblong, paler and velvety-hairy below, apex scarcely acuminate, also larger berries and a villous calyx.

A Japanese plant has also been named as a variety, *I. montana* var. *macropoda* (Miq.) Fern., but this remote relationship is in doubt.

I. montana Torr. & Gray is also listed in the literature as *I. monticola* Gray.

REMARKS. The genus name, *Ilex*, is the ancient name for a European oak, and the species name, *montana*, refers to its mountain habitat. Also known by the vernacular names of Large-leaf Holly, Mountain Holly, and Hulver Holly. The leaves are sometimes browsed by white-tailed deer, and the fruit eaten by a number of species of birds. It has been in cultivation since 1870.

SERVICE-BERRY HOLLY

Ilex amelanchier Curt.

FIELD IDENTIFICATION. Deciduous, red-fruited holly attaining a height of 7 ft. Distinguished from other hollies by the elliptic to lanceolate leaves, which are so shallowly toothed on the margin that they appear entire. Also distinguished by the peduncles or pedicels of the flowers, which surpass the leaf petioles in length. It is evidently closely related to Georgia Holly, *I. longipes* Chapm.

FLOWERS. Polygamo-dioecious, small, in axillary 6–9-flowered clusters in the staminate; usually solitary in the pistillate; pedicels or peduncles equalling or mostly surpassing the subtending petioles; petals white, 4–8, slightly united at base; stamens as many as the petals and alternate with them, barely adnate to the corolla-tube below; filaments distinct, anthers introrse, longitudinally dehiscent; ovary 4–8-celled, ovules solitary in each cell, suspended; pistil 1 and style very short; calyx 4–6-parted, sepals ciliate and persistent.

SERVICE-BERRY HOLLY
Ilex amelanchier Curt.

FRUIT. A berry-like drupe, mostly solitary, $\frac{1}{3}$–$\frac{2}{5}$ in. in diameter, globose, dull red or less often orange or yellow; nutlets 1-seeded, hard, bony, ribbed on the back.

LEAVES. Simple, alternate, deciduous, firm, length 2–4 in., elliptic to lanceolate or oblong, more rarely oval, base rounded or broadly-cuneate, apex acute to obtuse or short-acuminate; margin serrulate with very obscure teeth, apparently entire unless closely examined; upper surface glabrous and dull green, finely reticulate; lower surface more or less soft-pubescent. Twigs also slightly pubescent.

RANGE. Usually in sandy swamps and wet woods on the coastal plain. In Louisiana recorded only from Washington and St. Tammany parishes; northward in scattered locations to South Carolina and southeastern Virginia.

REMARKS. The genus name, *Ilex*, is the classical name of the Holly Oak of Europe. The species name *amelanchier*, refers to its superficial resemblance to a service-berry *(Amelanchier)*. It has been cultivated since 1880. This species is of uncertain relationship and should be more closely studied. It is also listed in the literature under the name of *I. dubia* (G. Don) B. S. P. It is distinguished from the Common Winterberry Holly, *I. verticillata*, and the Mountain Winterberry Holly *I. montana*, by the long flower pedicels and obscure serrations of the leaf.

STAFF-TREE FAMILY (*Celastraceae*)

AMERICAN BITTER-SWEET
Celastrus scandens L.

FIELD IDENTIFICATION. Diffuse woody vine, climbing to heights of 20 ft or trailing on the ground. Thicket-forming by underground stolons.

FLOWERS. May–June, in terminal panicles or racemes 2–4 in. long, many-flowered; flowers polygamo-dioecious, inconspicuous, small, greenish white; calyx about ⅓ in. wide, with 4–5 sepals; corolla of 5 petals, petals much longer than the calyx, spreading or reflexed, inserted under the disk, ovate to oval or oblong, margin erose; disk cup-shaped, 5-lobed; stamens 5, shorter than the petals, borne at the sinuses of the disk; style short, thick, stigma 3-lobed; ovary free, inserted on the disk, 2–4-lobed, 2–4-celled, ovules 1–2 in each cell, enclosed in a fleshy crimson aril, embryo straight, endosperm fleshy, cotyledons flat.

FRUIT. Ripening August–October, capsule borne in drooping clusters, subglobose, ⅓–⅜ in. in diameter, orange-yellow, leathery, finely wrinkled, loculicidally dehiscent into 3 valves, each valve 3-celled, each cell exposing 1–2 globose seeds with a conspicuous crimson aril, persistent and showy in autumn.

LEAVES. Simple, alternate, deciduous, membranous, blade length 2–4 in., width 1–2 in., ovate to oval or elliptic to lanceolate, apex acute to acuminate, base acute or rounded, margin entire or crenate-serrate, both surfaces glabrous; petioles ⅓–⅘ in. long; stipules minute and slender. Foliage occasionally variegated.

STEMS. Diffuse, green to gray or brown, pith solid and white.

RANGE. In many types of soil, thickets, woods, fence rows, and along streams. Central Texas, Oklahoma, and Arkansas; eastward through Louisiana to Georgia, northward to Quebec, and west to Ontario, Minnesota, South Dakota, and Nebraska.

AMERICAN BITTER-SWEET
Celastrus scandens L.

MEDICINAL USES. The root bark, according to Wood and Osol (1943, p. 1308), "has been used in chronic affections of the liver and in secondary syphilis, and is said to be emetic, diaphoretic, and alterative. Gisvold (*J. Amer. Pharm. Assoc.*, 29:432, 1940) has separated from the bark of the root . . . a red pigment which he calls *celastrol*. It is an alkyl-substituted naphthoquinone. Wayne isolated . . . white minute crystals, to which he gave the name *celastrine*. According to Ugolino Mosso (*Rivista Clinica*, 1891), celastrine arrests the frog's heart in systole, does not affect markedly the blood pressure in mammals, and influences the respiration only through its action on the vagus [nerve]. The stimulant action is especially manifest on the brain and is not followed by a secondary depression. There is marked, persistent elevation of the temperature."

PROPAGATION. The vine is valuable for covering trellis work, trees, or rocks and walls. It grows in almost any type of soil and in sunny or shaded positions. It has been in cultivation since 1736. It may be propagated by seeds sown in fall or stratified for spring planting, by root cuttings or layers, or by the freely produced suckers. The cuttings may be obtained from soft or mature wood. Good seed crops are borne nearly every year. The seeds are collected as soon as the capsules open and are spread out to dry for 2–3 weeks. The valves may be removed by flailing or by running through a hammer mill with water. Three to eight seeds are obtained from each fruit, with about 20,000 seeds per lb. About 3,000 usable plants are usually obtained from a lb of seed. The commercial purity is about 98 per cent, with a soundness of 85 per cent. The seed may be fall-sown, or stratified for spring planting in moist sand or peat for 3 months at 41°F.

REMARKS. The genus name, *Celastrus,* is from the ancient Greek name *kelastrus,* and the species name, *scandens,* refers to its climbing habit. Known also by the vernacular names of False Bitter-sweet, Shrubby Bitter-sweet, Climbing Bitter-sweet, Staff-vine, Fever-twig, Gnome's-gold, Waxwork, Roxbury-waxwork, Jacob's Ladder, and Climbing-orangeroot.

The plant sometimes injures trees by constriction. Fruit persistent and ornamental in winter because of the scarlet aril. Sometimes used for indoor floral decorations. The fruit is known to be eaten by at least 15 species of birds, including ruffed grouse and bob-white quail. Consumed by cottontail and fox squirrels. Considered poisonous to horses but avoided by them.

STRAWBERRY-BUSH

Euonymus americana L.

FIELD IDENTIFICATION. Shrub to 6 ft, erect or sometimes creeping or trailing, the slender branches stiffly divergent.

FLOWERS. May–June, cymes mostly 1–3-flowered, peduncles 1/5–4/5 in. long, pedicels 1/12–1/6 in. long; flowers perfect, regular, 3/8–1/2 in. broad; calyx short, flattened, 1/5–1/3 in. long, sepals 5, rounded; corolla with 5 petals which are spreading, separated, ovate to suborbicular with undulate margins, greenish to greenish purple or reddish; stamens 5, alternate with the petals, arising from the edge of the disk, filaments very short, anthers about 1/25 in. wide, sacs 2 and divergent; disk flat, inconspicuously 3–5-lobed; stigma confluent with the disk, ovary sessile, 3–5-celled and 3–5-valved later.

FRUIT. September–October, capsule depressed-globose, 3–5-lobed, fleshy, bearing warty tubercles, pink, about 5/8 in. thick, valves dehiscent loculicidally to expose 1–2 seeds; seeds 1/6–1/4 in. long, elliptic, flattened, with a conspicuous orange-red aril.

LEAVES. Persistent southward, simple, opposite, sessile or petioles to 1/8 in.; blade length 3/4–4 in., width 1/2–1 in., oblong to lanceolate or broadly oval, apex acute or acuminate, base narrowed or occasionally rounded to subcordate, margin obscurely crenate-serrate, upper surface glabrous and dark green, lower surface glabrous or slightly hairy on the veins beneath.

TWIGS. Divaricate, sometimes spreading and horizontal, green to brown or gray, smooth, somewhat 4-angled, sometimes rooting at the nodes if in contact with the soil, winter buds imbricate; wood light colored, close-grained, tough.

RANGE. In rich moist woods, east Texas, Oklahoma, Arkansas, and Louisiana; eastward to Florida, northward to New York and west through Pennsylvania, Indiana, and Illinois to Nebraska.

VARIETY. Narrow-leaf Strawberry-bush, *E. americana* var. *angustifolius* Dipp., has lanceolate to linear-lanceolate leaves and is half-evergreen southward.

PROPAGATION. For propagation purposes the seed is hand picked in fall and spread out to dry. The seeds may be extracted from the hull by flailing in a sack, and the chaff removed by winnowing. The pulpy aril may be removed by rubbing on a large mesh wire screen, but care must be taken not to injure the seed. The seed averages about 30,000 seeds per lb, and in most species of *Euonymus,* respond to stratification at 32°–50°F. for 90–120 days prior to spring sowing. Germination averages 30–70 per cent. Propagation can also be practiced by cuttings of ripened wood in fall, by layers, or by grafting and budding.

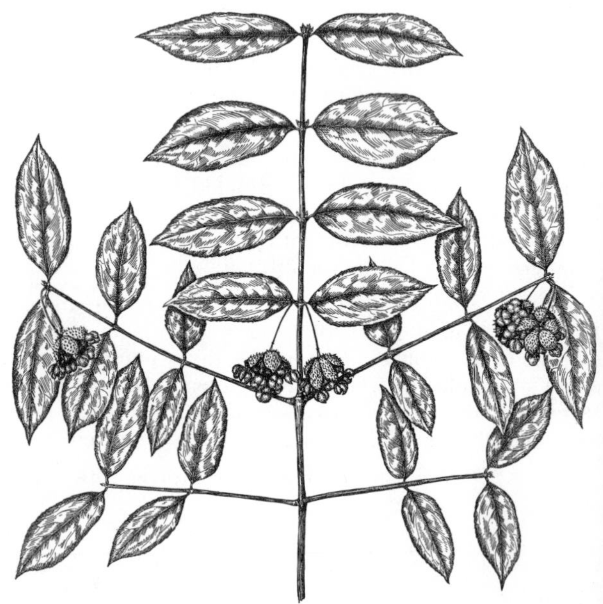

STRAWBERRY-BUSH
Euonymus americana L.

EASTERN WAHOO
Euonymus atropurpureus Jacq.

REMARKS. The genus name, *Euonymus*, is the Greek name meaning "true name," and the meaning of the species name, *americana*, is obvious. Also known under the vernacular names of Bursting-heart, Fish-wood, and Burning-bush. Strawberry-bush was introduced into cultivation in 1697. It is a handsome shrub because of the dark green leaves and pink warty fruit which opens to display the orange-red seeds. The fruit is eaten by a number of species of birds.

EASTERN WAHOO

Euonymus atropurpureus Jacq.

FIELD IDENTIFICATION. Usually a shrub, but sometimes a small tree to 25 ft, with spreading branches and an irregular crown.

FLOWERS. May–July, borne in 7–15-flowered, axillary, trichotomous cymes; peduncles slender, 1–2 in. long, with individual perfect flowers about ½ in. wide; petals 4, purple, obovate, undulate or obscurely toothed, borne on the edge of a 4-angled disk; stamens 4, short, with 2-celled purple anthers; ovary 4-celled, style short, stigma depressed.

FRUIT. September–October, capsule deeply 3–4-lobed, smooth, about ½ in. across, persistent on long peduncles; valves purple or red, splitting open to expose brown seeds about ¼ in. long enclosed by a scarlet seed coat.

LEAVES. Opposite, petioled, deciduous, 2–5 in. long, 1–2 in. wide, ovate-elliptic, acuminate or acute at the apex, acute or cuneate at the base, finely crenate-serrate on the margin, bright green above, pale and puberulent beneath; petioles ½–1 in. long.

TWIGS. Slender, somewhat 4-angled, purplish green to brownish later; lenticels pale and prominent.

BARK. Smooth, thin, gray, with minute scales.

WOOD. Almost white, or tinged with yellow or orange, close-grained, heavy, hard, tough, weighing 41 lb per cu ft.

RANGE. Eastern Texas and Arkansas; east to northern Alabama, north to New York, and west to Ontario, Montana, Nebraska, and Kansas.

MEDICINAL USES. Powdered *Euonymus* bark is known to have a powerful action on the heart. At one time it was used as a diuretic in dropsy, but today it is employed only as a cathartic.

The name "Wahoo" was given by the Indians, who used the powdered inner bark in uterine trouble.

PROPAGATION. The tree has been cultivated since 1756. The seeds may be collected by hand in late summer and spread out to dry for a few days. The hulls may be removed by beating them in a bag or by rubbing through coarse mesh wire. The seed averages about 11,800 seeds per lb, with a purity of 75 per cent, soundness 83 per cent, and germination 5–30 per cent. Spring sowing of seed stratified about 3 months at 30° to 50°F. is best. It appears to grow in a variety of soils. The plant can also be propagated by cuttings, budding, and grafting.

REMARKS. The genus name, *Euonymus*, is a translation of an ancient Greek term meaning "true name," and the species name, *atropurpureus*, refers to the purple flowers and fruit. Vernacular names are Spindle-tree, Burning-bush, Bleeding-heart, Arrowwood, Indian-arrow, Bitter-oak, and Strawberry-tree. The tree is sometimes used as an ornament because of its beautiful scarlet fruit in autumn, although it is subject to scale and fungus diseases. It is known to be purgative to livestock. The seeds are eaten by a number of species of birds.

PLAINS GREASE-BUSH

Forsellesia planitierum Ensign

FIELD IDENTIFICATION. Irregularly branched grayish green, spinescent shrub. Rather low or sometimes matted.

FLOWERS. Borne May–July, axillary, bisexual, regular, on slender peduncles 1/25–1/8 in. long, with small scarious bracts at the base; sepals 5, ovate, 1/15–1/12 in. long, about 1/25 in. wide; petals 5, white, distinct, oblanceolate, 1/6–1/4 in. long, 1/15–1/12 in. wide, much longer than

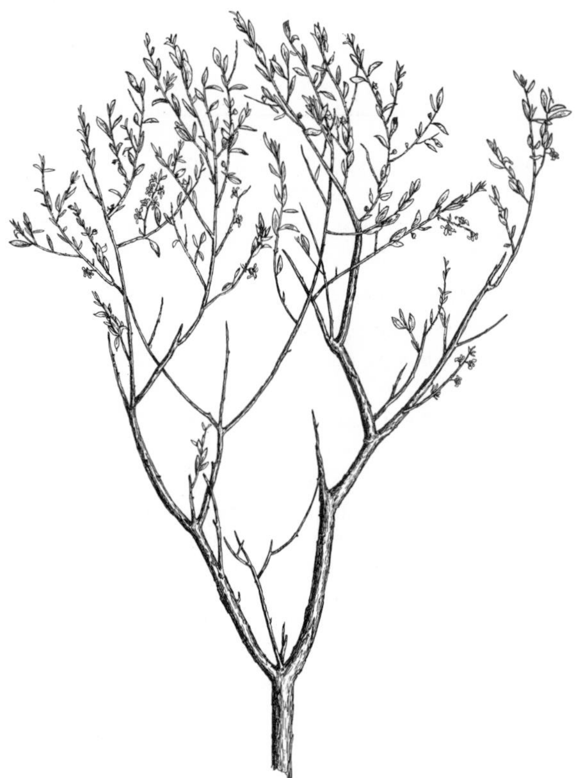

PLAINS GREASE-BUSH
Forsellesia planitierum Ensign

the sepals; stamens 8, unequal in length; disk fleshy and crenately lobed; ovary superior, 1-celled and 1–2-ovuled.

FRUIT. Maturing in August, follicles ovoid, 1/6–1/5 in. long, asymmetrical, coriaceous, striate, opening along the ventral suture.

LEAVES. Simple, alternate, deciduous, lanceolate, apex acute, base cuneately narrowed, margin entire, blades 1/4–1/2 in. long, 1/12–1/6 in. wide, surfaces pubescent, veins obscure above but thickened beneath, thin; petioles less than 1/25 in.; stipules subulate, 1/50–1/16 in. long, conspicuously black.

TWIGS. Slender, grayish green, puberulent, angled, nodes with decurrent lines; older twigs light to dark gray with thin scaly bark.

RANGE. Usually on rocky calcareous slopes. In the Panhandle region of Texas and in Oklahoma. In Texas in Randall, Armstrong (Gamble Ranch), Lubbock, and Marion counties. In Oklahoma 4 miles north of Kenton, Cimarron County.

REMARKS. The genus name, *Forsellesia,* is in honor of James Henry Forselles, a Swedish mining engineer and botanical writer of the last century. The species name, *planitierum,* refers to its habitat on the plains. It is distinguished from *F. spinescens* and *F. texana* by the presence of stipules and by broader petals. Also it is less spinescent than *F. spinescens.*

SPINY GREASE-BUSH

Forsellesia spinescens (Gray) Greene

FIELD IDENTIFICATION. Shrub to 3 ft, branches green, intricately branched, weakly spinescent.

FLOWERS. February–May, solitary or cymose in the leaf axils on pedicels 1/16–1/5 in. long, with reduced scarious bracts at base; flowers small, inconspicuous, regular, perfect; sepals 5, ovate, acute or obtuse, hyaline-margined; petals 5, deciduous, white, narrowly oblanceolate or ligulate, 1/8–1/3 in. long, much longer than the sepals; stamens 5–10, exserted, inserted below a fleshy, crenate-lobed disk; style short, ovary 1-celled, 1–2-ovuled.

FRUIT. Follicle about 1/5 in. long, ovoid or subglobose, coriaceous, dry, asymmetrical, longitudinally striate, brown, slightly glutinous; dehiscent along the ventral suture; seed solitary, shiny, dark brown, asymmetrical, beaked, with 2 lateral circular grooves.

LEAVES. Simple, alternate, 1/8–1/2 in. long, 1/16–1/8 in. wide, lanceolate to oblanceolate or obovate, apex obtuse to acute or rounded and mucronate, base attenuate, grayish green to glaucous, glabrous to puberulent, venation of upper surface obscure, lower surface with 3–5 veins more prominent, early-deciduous, short-petiolate, stipules absent.

TWIGS. Young ones slender, green, stiff, angled, decurrent from the nodes, glabrous and glaucous; older stems gray to brown with bark exfoliating in narrow thin strips, terminally and weakly spinescent.

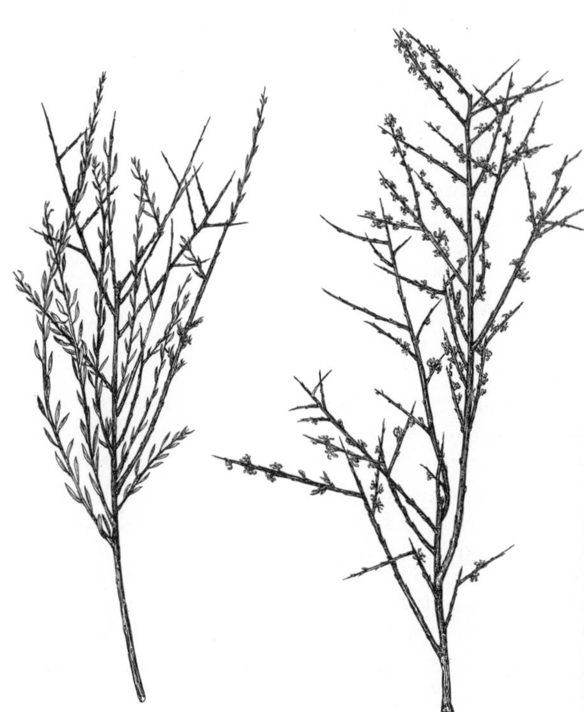

SPINY GREASE-BUSH
Forsellesia spinescens (Gray) Greene

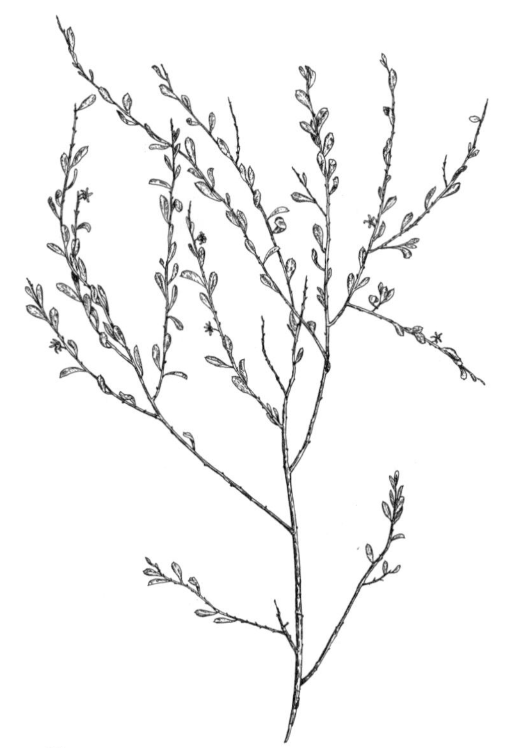

Texas Grease-bush
Forsellesia texensis Ensign

RANGE. On rocky slopes of dry limestone hills in the pinyon-juniper belts of the West. In Texas in the Trans-Pecos and Panhandle areas. The type specimen from near Frontera, Texas. In New Mexico in the San Andres and the Organ mountains, and on the Llano Estacado. Westward in Arizona and California, and northward to Colorado, Utah, Washington, and Oregon. In Mexico in northern Chihuahua.

REMARKS. The genus name, *Forsellesia,* is in honor of James Henry Forselles, a Swedish mining engineer and botanical writer of the last century. The species name, *spinescens,* refers to the spinescent twigs. The plant is known to be browsed by sheep and mule deer.

Texas Grease-bush

Forsellesia texensis Ensign

FIELD IDENTIFICATION. Low to medium-sized, intricately branched, somewhat spinescent or spineless shrub. Branches angled, light to dark gray.

FLOWERS. Axillary, bisexual, regular, borne on slender pedicels with small scarious bracts at the base; sepals 5, entire, ovate, hyaline-margined, about 1/12 in. wide; stamens 7–9, equal; petals 5, white, narrowly oblanceolate, distinct, much longer than the sepals, inserted under the edges of a crenate-lobed disk; ovary superior, 1-celled, 1–2-ovuled, attenuate to the stigma.

FRUIT. Mature follicle broadly ovoid, 1/6–1/5 in. long, asymmetrical, leathery, striate, dehiscent at the ventral suture.

LEAVES. Simple, alternate, deciduous, broadly oblanceolate, apex acute and mucronate, margins entire and thickened, length of blades 3/8–3/4 in., width 1/8–1/5 in., surfaces pubescent to glabrous, veins thickened below; petioles about 1/25 in.

TWIGS. Slender, greenish, angled, often with decurrent lines at the nodes, spinescent or spineless.

RANGE. Dry limestone soils. Known from a chalk bluff on the Nueces River, near Montell, Uvalde County, Texas.

REMARKS. The genus name, *Forsellesia,* is in honor of James Henry Forselles, a Swedish mining engineer and botanical writer of the last century. The species name, *texensis,* is for the state of Texas where it grows. Texas Grease-bush is distinguished from the other species by the broad oblanceolate leaves with thickened margins and lack of stipules.

Guttapercha Mayten

Maytenus phyllanthoides Benth.

FIELD IDENTIFICATION. Creeping evergreen shrub or a small crooked tree rarely over 20 ft.

FLOWERS. Axillary, solitary or fascicled, peduncles 1/16–1/8 in. long, hardly over 3/16 in. across; petals 5, green to white, triangular, apex obtuse or acute; stamens 5, alternate with the petals, borne under the margin of a 5-angled viscid disk, shorter than the petals, filaments about 1/25 in. long, anthers ovoid-cordate; pistil thick, short, immersed in the disk; stigmas 3–4, sessile or nearly so, ovary 3–4-celled; calyx of 5 persistent sepals, apex obtuse or rounded, reddish, shorter than the petals.

FRUIT. Ripening in November in Texas, capsules on short axillary peduncles 1/16–1/8 in. long, ovoid to ellipsoid, 3/8–1/2 in. long and about 1/4 in. broad, abruptly mucronulate at apex, base abruptly narrowed, 3–4-angled; dehiscent into 2–4 thin, elliptic to oval, recurved valves at maturity; seeds 2–4, bony, white to brown, ellipsoid to oblong, apices acute; dorsal face rounded, ventral face plane, about 3/16 in. long, covered with a conspicuous, red, loose, fleshy aril, calyx-lobes 4–5, minute, acute, about 1/25 in. long, pointed downward.

LEAVES. Alternate, simple, persistent, stiff, leathery, oval to oblong or elliptic, 3/4–1 3/4 in. long, 1/2–1 in. wide; apex rounded, obtuse, or notched; base rounded to cuneate; margin entire or undulate; surfaces dull grayish green and glabrous with obscure veins; petioles stout, short, mostly less than 1/4 in. long; stipules very small, caducous.

GUTTAPERCHA MAYTEN
Maytenus phyllanthoides Benth.

TWIGS. Slender, gray, glabrous, or when younger slightly puberulent, obscurely striate, sometimes prostrate, runner-like, and rooting at the nodes, or erect on larger plants.

BARK. Mottled gray to brown, smooth, thin, rougher on old bark near the base of the trunk.

RANGE. Usually on sandy bluffs on tidal beaches. On the coasts of Texas and Florida, also Baja California; south to Sonora, Puebla, and Yucatán in Mexico.

MEDICINAL USES. The leaves have been used as remedies for scurvy and toothache, though without official medical approval. The gum, which is sometimes substituted for guttapercha, is used to bind splints for broken limbs, for insulating material, and in golf balls. The true guttapercha gum comes from trees of the Sapotaceae Family, notably *Palaquium gutta* Burck.

REMARKS. The genus name, *Maytenus,* is a name for a Chilean species, and the species name, *phyllanthoides,* means "*Phyllanthus*-like," or like the phyllanthus plant. It is known also under the vernacular names of Leatherleaf, Mangle, Mangle Dulce, Mangle Aguabola, and Aguabola. The wood is sometimes used for fuel in tropical America, but in Texas the plant is only a small shrub.

GREGG MORTONIA
Mortonia greggii Gray

FIELD IDENTIFICATION. Low branching shrub with erect pubescent stems.

FLOWERS. Borne in terminal, spikelike racemes; petals 5, inserted under the disk margin, white, oval, apex obtuse, margin somewhat erose, length about 1/12 in.; stamens 5, inserted in the lobes of the disk, filaments slender; ovary ovoid, imperfectly 5-celled, ovules 2 in each cell, style long, stigma 5-lobed; calyx glabrous, sepals 5, oval, apex acute, margin scarious, length about 1/12 in.; bracts acute.

FRUIT. Capsule 1/6–1/5 in. long, oblong, thick-walled, slightly grooved, indehiscent; seed solitary and erect.

LEAVES. Crowded, alternate, evergreen, blade linear to linear-spatulate or oblong, 3/8–1 in. long, apex obtuse or acute, mucronate, margin entire, base narrowed, surfaces glabrous when mature; rather thick; petiole hardly over 1/25 in. or blade sessile.

RANGE. On limestone hills in southwestern Texas. In Mexico in Coahuila.

REMARKS. The genus name, *Mortonia,* is in honor of C. V. Morton, American botanist. The species name, *greggii,* is for Josiah Gregg (1806–1850), early botanist who collected in the Southwest and in northern Mexico. Known under the vernacular name of "Afinador" in Mexico.

GREGG MORTONIA
Mortonia greggii Gray

Texas Mortonia

Mortonia scabrella Gray

FIELD IDENTIFICATION. Western shrub to 6 ft, erect, very leafy, branches stiff and divaricate.

FLOWERS. Borne at the ends of the branches March–September in narrowly cymose-paniculate clusters ¼–1½ in. long. Individual flowers ⅛–⅙ in. across, perfect, regular; petals 5, imbricate, inserted below the disk, white, rounded, about 1/12 in. long; calyx obconic, 5-lobed, lobes acute at apex, hairy; stamens 5, alternate with the petals, inserted on the margin of the broad disk, anthers subglobose and mucronulate; style short, 5-lobed; ovary 5-celled, surrounded by the disk, producing a 1-seeded fruit.

FRUIT. Capsule 1/12–¼ in. long, about 1/12 in. in diameter, short-oblong, cylindric, apex rounded with stigma remnants persistent, dry, hard, indehiscent, 1-seeded.

LEAVES. Simple, alternate, crowded, small, ⅛–⅓ in. long, ⅛–¼ in. wide, persistent, oblong to elliptic or oval, apex rounded to obtuse or acute, base cuneate and sessile or short-petiolate, margin thickened and revolute, thick and scabrous, yellowish green, veins obscure except midrib.

TWIGS. Very leafy, slender, stiff, divaricate, often spine-tipped, young twigs greenish yellow, somewhat pubescent; older glabrous, bark gray, tight and smooth.

RANGE. Dry limestone hillsides and canyons, at altitudes of 2,000–5,000 ft, western Texas, New Mexico, Arizona, and Mexico. In Mexico in Chihuahua and Sonora.

Myrtle Pachystima
Pachystima myrsinites (Pursh) Raf.

Texas Mortonia
Mortonia scabrella Gray

VARIETY. A variety with larger leaves to ⅗ in. long has been described as *M. scabrella* var. *utahensis* Coville, and is found in Utah, Arizona, and southern California. A plant formerly described as *M. sempervirens* Gray appears to be a synonym of *M. scabrella* Gray.

REMARKS. The genus name, *Mortonia,* is in honor of C. V. Morton, American botanist, and the species name, *scabrella,* refers to the rough, scabrous leaves. The shrub has no particular commercial value except for erosion control. It is not browsed by livestock.

Myrtle Pachystima

Pachystima myrsinites (Pursh) Raf.

FIELD IDENTIFICATION. Evergreen shrub attaining a height of 1–3 ft. Rather densely branched and often depressed or creeping.

FLOWERS. April–July, very small, solitary or in few-flowered, axillary cymes; pedicels with paired minute bracts; flowers regular and perfect; calyx-tube short, sepals 4, imbricate and rounded, about 1/25 in. long or less, green; petals 4, about twice as long as the sepals, imbricate, greenish to reddish brown or purplish, inserted beneath the disk in the calyx; stamens 4, alternate with the petals, filaments twice as long as the anthers, inserted on the disk; ovary half-inferior, 2-celled, immersed in the disk, style very short, stigma obscurely 2-lobed.

FRUIT. Maturing July–September, capsule 1/5–1/3 in. long, ovoid to oblong or obovoid, leathery, opening by 2 valves, seeds 1–2, invested by a white, lacerated, membranaceous aril.

LEAVES. Simple, opposite, persistent, oval to elliptical or oblanceolate, apex rounded, base cuneate or rounded, blade length 1/4–1 1/2 in., width 1/8–1/2 in., margin entire or often serrulate on the upper one half or two thirds of the blade, also somewhat revolute, surface dull or semilustrous, smooth and glabrous, firm and thickened, veins obscure; petioles 1/25–1/8 in.; buds ovoid, with 1–2 pairs of outer scales; stipules minute and deciduous.

TWIGS. Slender but rather stiff, brown to gray, essentially glabrous but somewhat roughened, 4-angled above but more terete below.

RANGE. In well-drained soil from the pinyon zone upward to the subalpine, at altitudes of 6,000–10,000 ft. New Mexico; west to California and north to Alberta and British Columbia.

REMARKS. The genus name, *Pachystima,* is from *pachus* ("thick") and *stigma* ("stigma"), referring to the thickened stigma. The species name, *myrsinites,* refers to the genus *Myrsine,* which it resembles. It is also known under the vernacular names of Myrtle-bush, Myrtle-boxleaf, Mountain-lover, False-box, and Oregon-boxwood. The shrub has been cultivated since 1879 and is used ornamentally in rock gardens, evergreen borders, or for ground cover. It is propagated by seeds, layers, and cuttings under glass. It is also useful to wildlife, being eaten by white-tailed deer, black-tailed deer, mountain sheep, and grouse. It is occasionally browsed by livestock.

DESERT-YAUPON

Schaefferia cuneifolia Gray

FIELD IDENTIFICATION. Densely branched, rigid, evergreen shrub attaining a height of 4 ft.

FLOWERS. Dioecious, axillary, sessile or subsessile, inconspicuous, greenish; petals 4, oblong, obtuse; calyx 4-parted, lobes suborbicular; stamens 4, anthers oval and 2-celled; ovary 2-celled, stigma 2-parted.

DESERT-YAUPON
Schaefferia cuneifolia Gray

FRUIT. Drupe red, subglobose, flattened, grooved, about 1/5 in. long, dry, 2-celled and 2-seeded.

LEAVES. Alternate or fasciculate, on short lateral spurs, simple, sessile or subsessile, cuneate to obovate or oval, apex rounded or notched, base gradually narrowed, entire or rarely crenate toward the apex, 1/4–1/2 in. long, 1/12–1/3 in. wide, coriaceous, pale green and glabrous, obscurely reticulate-veined.

BARK. Gray, tight, smooth.

TWIGS. Green to gray, divaricate, rigid, glabrous or puberulent, almost spinescent.

RANGE. Western and southwestern Texas, also northern Mexico. Along the Rio Grande at La Joya, Hidalgo County; at Langtry, Val Verde County; and in Big Bend National Park, Brewster County. In Mexico in the state of Coahuila.

REMARKS. The genus name, *Schaefferia,* is dedicated to the German naturalist Jakob Christian Schaeffer, and the species name, *cuneifolia,* refers to the cuneate leaves. Known in Mexico under the name of "Capul," where the roots are reported to be used as a remedy for venereal disease.

BLADDERNUT FAMILY (*Staphyleaceae*)

American Bladdernut
Staphylea trifolia L.

FIELD IDENTIFICATION. Deciduous upright shrub or small tree to 25 ft, with striped twigs.

FLOWERS. In May, in terminal or axillary drooping racemes 2–4 in. long; pedicels ⅓–¾ in.; flowers bisexual, regular, perfect, small, white, about ⅓ in. long; petals 5, white, somewhat longer than the sepals, spatulate, obtuse, pubescent at base; calyx of 5 sepals, greenish white, erect, ⅓–⅖ in. long, imbricate, lanceolate to oblong; stamens 5, exserted, inserted beneath the disk edge; filaments erect, pubescent below the middle, about as long as the petals, anthers apiculate; style exserted, composed of 3 carpels which are united in the axis; ovary pubescent, 3-celled; stigma capitate.

FRUIT. July–September, solitary or a few together on pedicels ⅓–¾ in., capsule bladder-like, 1¼–2½ in. long, 3-lobed, obovoid, net-veined, green to brown, opening at the apiculate apex; seeds 1–4, ⅕–3⁄16 in. long, rounded, somewhat flattened, yellowish brown, hard, lustrous.

LEAVES. Opposite, trifoliate-compound, the 3 leaflets elliptic, ovate or obovate, 1½–4 in. long, margin sharply serrulate; apex acuminate, acute or obtuse; base obtuse or cuneate, lateral leaflets symmetrical, upper surface bright green and pubescent on the veins; lower surface slightly paler and pubescent; petiolules pubescent, that of the terminal leaflet ½–1½ in., stipules deciduous, linear, ⅕–½ in.

TWIGS. Green to brown or gray, often striped, smooth, curved, ascending.

RANGE. Usually in moist, rich soil in the shade. Oklahoma, Arkansas, Missouri, and Georgia; north to South Carolina, Minnesota, Quebec, and Ontario.

REMARKS. The genus name, *Staphylea*, is from the Greek word meaning "cluster of grapes," in reference to the drooping raceme-like clusters of flowers. The species name, *trifolia*, denotes the 3 leaflets. Bladdernut can be propagated by cuttings, layers, suckers, and seeds. It is sometimes cultivated for ornament. Most of the species cultivated in gardens are of Asiatic origin.

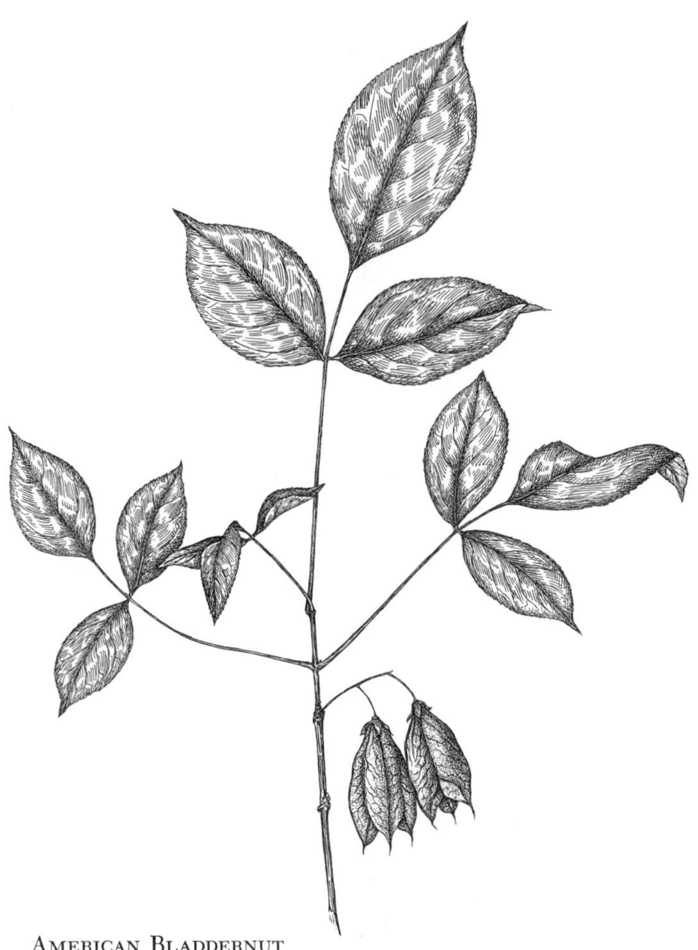

American Bladdernut
Staphylea trifolia L.

Dwarf American Bladdernut, *S. trifolia* var. *pauciflora* Zab., is low and suckering in habit. The flowers are in short racemes, and the glabrous leaflets are smaller and broader. It was first cultivated in 1888.

MAPLE FAMILY (*Aceraceae*)

Red Maple
Acer rubrum L.

Red Maple
Acer rubrum L.

FIELD IDENTIFICATION. Beautiful tree attaining a height of 100 ft, with a narrow, rounded crown.

FLOWERS. Borne in early spring before the leaves, in staminate and pistillate axillary fascicles on the same tree, or on different trees, red to yellowish green; petals 5, same length as the calyx-lobes, linear-oblong; stamens 5–8, anthers red; pistil short, ovary glabrous; styles 2, stout and spreading; calyx campanulate, 5-lobed.

FRUIT. Ripening March–June, borne on slender, drooping pedicels 2–4 in.; composed of a pair of winged, red, yellowish green or brown, flattened samaras, with a seed at the base; wings spreading, ½–1 in. long.

LEAVES. Opposite, simple, deciduous, blades ovate to oval, 2–6 in. long, 3–5-lobed; lobes acuminate or acute, irregularly serrate, sinuses angular; base truncate or somewhat cordate; bright green and glabrous above, whitish beneath; turning beautiful shades of yellow, orange, or red in autumn; petioles slender, 2–4 in., green to red.

TWIGS. Slender, glabrous, reddish, lenticels pale.

BARK. Light gray and smooth at first, becoming darker, furrowed and flaky on old trunks.

WOOD. Light reddish brown, close-grained, hard, heavy, weak, weighing 38 lb per cu ft.

RANGE. In Texas, Oklahoma, Arkansas, and Louisiana; eastward to Florida, northward to Newfoundland, and west to Ontario, Minnesota, Wisconsin, and Missouri.

PROPAGATION. Red Maple has been cultivated since 1656. The clean seed averages about 22,800 seeds per lb, with a commercial purity of 92 per cent. The seed seems to vary in its germinating period, some germinate soon after falling, and others are dormant until the following spring. To break dormancy, stratification in sand at 41°F. for 60–75 days is practiced. Another method is to soak the seed in cold running water for about 5 days. Nursery germination usually averages about 35–80 per cent. The beds are planted in drills spaced about 1 ft apart, and about 20 seeds planted per ft. They are covered with ¼ to 1 in. of firmed soil and mulched. If seed is stratified for spring planting the stratification period should not be long or germination will occur.

DRUMMOND RED MAPLE
Acer rubrum var. *drummondii* (Hook. & Arn.) Sarg.

The tree is short-lived, grows rapidly, and is subject to rot and fire damage. However, it is a popular ornamental tree for its red flowers and fruit in spring and its brilliantly colored leaves in fall.

VARIETIES. Red Maple is subject to considerable variation, and some of the more constant varieties are as follows:

Drummond Red Maple, *A. rubrum* var. *drummondii* (Hook. & Arn.) Sarg., is perhaps more common in east Texas than the species, with 3–5-lobed leaves which are generally broader than long, woolly white-hairy beneath, and the scarlet fruit is somewhat larger than in the species.

Trident Red Maple, *A. rubrum* var. *tridens* Wood, has smaller 3-lobed leaves, obovate, narrowed and rounded, and sparingly toothed or entire below the 3 short lobes; dark green above, white-tomentose beneath; flowers often yellowish green.

Pale-flower Red Maple, *A. rubrum* var. *pallidiflorum* Pax, has yellow flowers. A number of horticultural clones, which reproduce asexually, have been developed. The most popular are the Column *(columnare)*, Dwarf *(globosum)*, and Schlesinger *(schlesingeri)*.

REMARKS. The genus name, *Acer*, is the ancient Celt name, and the species name, *rubrum*, may refer to the red flowers, fruit, or autumn leaves. Vernacular names are Scarlet Maple, Shoe-peg Maple, Swamp Maple, Soft Maple, and Hard Maple. The wood is used for furniture, turnery, fuel, gunstocks, and woodenware. The seeds are eaten by squirrel and chipmunk, and the foliage browsed by cottontail and white-tailed deer.

SUGAR MAPLE

Acer saccharum Marsh.

FIELD IDENTIFICATION. Large, broad, round-topped tree attaining a height of 100 ft, and a diameter of 2½ ft.

FLOWERS. Borne March–May, polygamous, pedicels slender, hairy, ¾–3 in. long, together forming a greenish yellow corymb; petals absent; calyx campanulate, 5-lobed, lobes obtuse and hairy; stamens 7–8, filaments slender and glabrous in the staminate flower, and shorter in the pistillate; ovary pale green, hairy, with 2 long-exserted stigmas.

FRUIT. Ripening September–October, borne in clusters; samara double, reddish brown, 1–1½ in. long, winged; wings ¼–½ in. wide, flat, thin, parallel or angle of divergence exceeding 45°; seeds at the samara base smooth, red, about ¼ in. long, easily wind-borne because of the attached wing, good seed crops every 3–7 years.

LEAVES. Simple, opposite, deciduous, blades 3–6 in. long and as wide, usually 5-lobed, but occasionally 3-lobed; lobes entire or irregularly toothed or lobed, apex acuminate; base cordate, rounded, or truncate; upper surface at maturity dark green, paler and whitish to glaucous beneath; leaf turning beautiful shades of yel-

TRIDENT RED MAPLE
Acer rubrum var. *tridens* Wood

SUGAR MAPLE
Acer saccharum Marsh.

low, orange, or red in autumn; petioles slender, glabrous, 1½–3 in.

TWIGS. Slender, lustrous, glabrous, green at first, reddish brown later; lenticels conspicuous, pale, oblong; buds conical, reddish brown, ⅛–⅕ in. long.

BARK. Light gray to brown, smooth when young, on older trees darker and fissured with irregular scaly plates.

WOOD. Reddish brown, sapwood white, close-grained, straight- or curly-grained, hard, strong, tough, stiff, odorless, strong in bending and endwise compression, shock resistance high, tools well, stays smooth under abrasion, takes a high polish, stains well, holds nails well, fair in gluing, dries easily, shrinks moderately, not durable when exposed to decay conditions.

RANGE. Almost the entire eastern half of the United States; west to east Texas, Arkansas, Oklahoma, Louisiana, Kansas, Nebraska, North Dakota, and South Dakota. In Canada in Manitoba, Ontario, Quebec, New Brunswick, and Newfoundland.

PROPAGATION. Sugar Maple has been cultivated since 1753, and is a popular ornamental tree. The fruit is picked by hand or flailed into canvas. It may be dried and stored as is, or de-winged by threshing. The seed must not dry out completely because viability is lost. The seed will keep a year in sealed containers at 36°–40°F. Cleaned seed averages about 6,100 seeds per lb, with a commercial purity of 94 per cent and a soundness of 52 per cent. Germination is delayed because of embryo dormancy, which can be overcome for spring planting by stratification in moist sand or peat at 36°–41°F. for 60–90 days, or by soaking in running water for about 5 days. The nursery germination capacity is about 15 per cent. Stratification should not be prolonged because of early germination. The seeds may be sown in drills about 1 ft apart, 20 seeds to the ft. The seedlings should be given some shade, and are sometimes susceptible to top wilt caused by *Verticillium*. One-year seedlings can be planted out in fertile soil. Propagation is also practiced by grafting and budding or by cuttings.

VARIETIES AND FORMS. A closely related species is Florida Sugar Maple, *A. barbatum* Michx. It occurs with the species in some areas, and many intergrading forms are found. Subsequent investigation may prove that Florida Sugar Maple is a Southern variety of the Northern Sugar Maple.

At one time the maple of the limestone regions of central Texas and Oklahoma was considered a variety of the Northern Sugar Maple and was listed as Uvalde Sugar Maple, *A. saccharum* var. *sinuosum* (Rehd.) Rousseau. Later investigators have pointed out (Little, 1944) that it has a closer affinity with the Western *A. grandidentatum*, and should therefore be known as Uvalde Big-tooth Maple, *A. grandidentatum* var. *sinuosum* (Rehd.) Little.

According to Rehder (1940, p. 573), the following horticultural clones and varieties should be recognized for Sugar Maple:

Blue Sugar Maple, *A. saccharum* forma *glaucum* (Schmidt) Pax, has leaves glaucous beneath.

Conic Sugar Maple, *A. saccharum* forma *conicum* Fern., has ascending branches forming a dense conical head.

Sentry Sugar Maple, *A. saccharum* forma *monumentale* (Temple) Rehd., has a columnar habit and ascending branches.

Schneck Sugar Maple, *A. saccharum* var. *schneckii* Rehd., is a variety with leaves densely villous on the veins beneath, petioles villous, rarely glabrous.

Rugel Sugar Maple, *A. saccharum* var. *rugelii* (Pax) Rehd., has leaves 3-lobed, generally broader than long, with entire lobes, glaucous, or somewhat pale green beneath, and leathery.

REMARKS. The genus name, *Acer*, is the ancient name of a maple of Europe, and the species name, *saccharum*, refers to the sugar content of the sap. Vernacular names are Hard Maple, Black Maple, and Sweet Maple. This tree is considered by many to be the most valuable hardwood tree in America. The wood is used for furniture, shoe lasts, turnery, pegs, shipbuilding, fuel, crossties, veneer, athletic equipment, musical instruments, boxes, woodenware, handles, shuttles, toys, and general millwork. Individual trees produce the bird's-eye and curly grains used in furniture. The wood is also rich in potash, and the sap is a source of the sucrose-rich maple sugar which is produced mostly in the Northern States and Canada. It is usually not refined but owes its flavor to impurities in the crude product.

Florida Sugar Maple

Acer barbatum Michx.

FIELD IDENTIFICATION. Tree over 50 ft, with small spreading branches and rounded crown.

FLOWERS. Polygamo-dioecious, borne on long puberulent pedicels in yellowish green corymbs; petals none; stamens 7–8, exserted; ovary 2-lobed, 2-celled; styles 2, inwardly stigmatic; calyx campanulate, 5-lobed, lobes puberulent.

FRUIT. Two green to reddish winged samaras joined at the seeded base and divergent-winged toward the apices, hairy at first but becoming glabrous later, hardly over ¾ in. long, flattened; seed solitary, lustrous, green.

LEAVES. Opposite, simple, deciduous, blades 1½–3 in. long, 3–5-lobed; lobes short, acute or obtuse, entire, or lobes very coarsely toothed, truncate or subcordate or rounded at the base; dark green and shiny above, glaucous and pubescent beneath, turning red to yellow in autumn; petiole 1½–3 in., pubescent at first, glabrous later.

Florida Sugar Maple
Acer barbatum Michx.

TWIGS. Slender, light reddish brown, glabrous.

BARK. Whitish, close, smooth when young, becoming darker and shallow-fissured later.

WOOD. Hard, strong, light, light reddish brown, individual trees sometimes found with a curly grain which makes the bird's-eye maple of commerce.

RANGE. Moist rich soil; Texas, Oklahoma, Arkansas, and Louisiana; east to Florida and north to Missouri and Virginia.

SYNONYMS AND VARIETIES. It appears likely that this tree is really only a southern variety of the Northern Sugar Maple, *A. saccharum* Marsh. (*A. saccharophorum* K. Koch), instead of a distinct species. In fact, at one time it did hold the rank of a variety. It appears that the principal differences are: the smaller size of the Florida Sugar Maple tree, the smaller leaves with short acute lobes, the smaller samaras, and a more whitish bark. However, these characters are not always constant, and many intergrades are found between the typical Florida Sugar Maple and the Northern Sugar Maple in east Texas.

Some botanists list Florida Sugar Maple under the scientific name of *A. floridanum* (Chapm.) Pax, but it is now shown by Fernald (1945a, pp. 156–160) that the name *A. barbatum* Michx. has preference.

REMARKS. The genus name, *Acer*, is from the Celt, and means "hard," in reference to the wood. The species name, *barbatum*, meaning "bearded," is from the beard in the flower.

The wood of Florida Sugar Maple is used for furniture, flooring, and shoe lasts. It is also widely known as a source of maple sugar and syrup. The average tree produces 3–4 lb of sugar, and it takes about 15 quarts of sap for 1 lb of sugar. The early settlers were taught the process of maple sugar making by the Indians. The seasons are apparently not distinct enough for a proper sap flow from Southern trees. The method of propagating by seed is similar to that of Northern Sugar Maple.

Chalk Maple

Acer leucoderme Small

FIELD IDENTIFICATION. Shrub or tree 20–40 ft, with a diameter of 4–18 in., often crooked. The branches are rather slender, forming a round-topped crown.

FLOWERS. In April, in small, nearly sessile corymbs, yellow, ⅛–¼ in. long, pedicels slender and glabrous; corolla none; calyx campanulate, about 1/12 in. long, lobes rounded to obtuse, ciliate, glabrous or slightly hairy; stamens 7–8, filaments hairy, longer than the calyx (in the pistillate shorter); ovary hairy, style elongate with divergent lobes.

FRUIT. Maturing in September, samaras conspicuous,

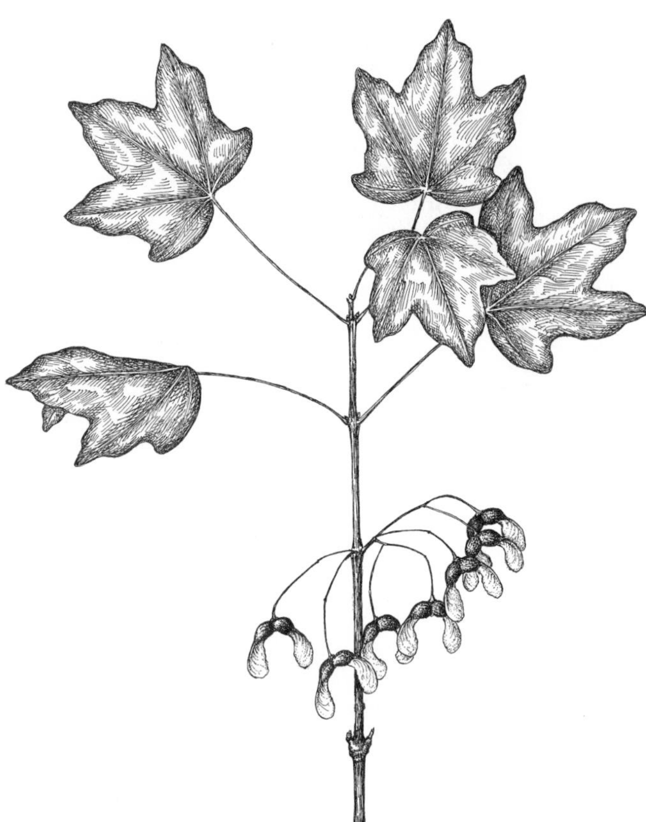

CHALK MAPLE
Acer leucoderme Small

reddish brown, drooping, at first with pale hairs, later glabrous; wings parallel or some wide-spreading toward maturity, oblong to elliptic or spatulate, ⅜–¾ in. long; seeds smooth, about ¼ in. long.

LEAVES. Appearing with the flowers, orbicular in outline, or some blades broader than long, 2–3½ in. across (some to 6 in.), base truncate or slightly cordate, margin divided into 3–5 acute or acuminate lobes (lobes at base often obtuse), lobes often dentate or sinuate, when young with pale hairs; older leaves with upper surface dark green to yellowish green, glabrous or nearly so, lower surface velvety-pubescent, green to occasionally reddish, prominently veined, scarlet in autumn; petioles slender, glabrous, 1–2 in. long.

TWIGS. Slender, glabrous, green to reddish, lustrous, later reddish brown to gray; lenticels numerous, oblong, gray; winter buds ovoid, acute, about 1/16 in. long, dark brown, inner scales red later.

BARK. On young stems and large branches smooth, light to dark gray or grayish brown, trunk very pale gray with narrow ridges, separated by deep fissures, and with closely appressed scales.

RANGE. In moist soil of rocky riverbanks, edges of the Gulf Coast plain and Piedmont Plateau. Southeastern Oklahoma, southern Arkansas, and western Louisiana; eastward through northern Alabama to Florida, and north to North Carolina and Tennessee.

REMARKS. The genus name, *Acer*, is the ancient Celt name, and the species name, *leucoderme*, refers to the very white bark. It is sometimes planted as a shade tree in the South and has been cultivated since 1900. Although closely related to Sugar Maple, *A. saccharum* Marsh., and Florida Sugar Maple, *A. barbatum* Michx., Chalk Maple may be distinguished by the green leaves (not glaucous) which are velvety-pubescent beneath, and by the very white bark.

SILVER MAPLE

Acer saccharinum L.

FIELD IDENTIFICATION. Tree to 100 ft, with a rounded crown and slender spreading branches.

FLOWERS. Before the leaves in spring, from imbricate involucres in sessile, or short-stalked, axillary clusters, greenish yellow, polygamo-dioecious; no petals; stamens 3–7, exserted; filaments slender with red anthers; pistil short, with a pubescent ovary and 2 spreading stigmatic styles; calyx obscurely 5-lobed, narrower in the staminate than in the pistillate flower, greenish yellow, pubescent.

FRUIT. Ripe when the leaves are almost mature, and

SILVER MAPLE
Acer saccharinum L.

endent on slender pedicels, composed of 2 samaras ith nutlets at the base and thin, widely divergent ings; samaras rather large, 1½–2 in. long, hairy at rst but glabrous later, flattened, reddish brown or reen, wrinkled, thin, falcate, dispersed April–June, ood seed crops nearly every year.

LEAVES. Opposite, simple, deciduous, blades 6–7 in. ong, borne on red, drooping petioles about 4 in., truncate or somewhat cordate at the base, 5-lobed; lobes eep, narrow, acuminate, variously toothed and cut, nuses acute or rounded; upper surface pale green, ower surface silvery white, turning yellow in autumn.

TWIGS. Slender, brittle, shiny, reddish brown.

BARK. At first smooth and gray, later breaking into oose flakes.

WOOD. Pale brown, close-grained, somewhat brittle, ven-textured, moderately strong, easily worked, decaying rapidly on exposure, weighing 32 lb per cu ft.

RANGE. East Texas, Oklahoma, Arkansas, and Louisiana; eastward to Florida, north to New Brunswick, nd west to Ontario, the Dakotas, Nebraska, and Kanas.

PROPAGATION. Silver Maple has been cultivated since 725. The cleaned seed averages about 1,400 seeds er lb, with a commercial seed purity of 97 per cent nd a soundness of 97 per cent. The seed should not e allowed to dry out too much. If the moisture content falls below 30–34 per cent the seed loses viability. However, it has been known to retain its viability for year stored cold and moist at about 32°–50°. The eed often germinates naturally immediately after it alls and has no dormancy. It can be sown immediately n mulched beds in the fall. It has a nursery germination rate of about 18 per cent. One-year-old seedings can be transplanted. Maple seed is planted in rills about a ft apart, with 20 seeds per ft. They re covered with ¼–1 in. of firmed soil and mulched. The seedlings may be attacked occasionally by a *Verticillium* wilt. Propagation can also be done by hardwood uttings.

VARIETIES AND FORMS. A number of horticultural lones have been developed such as Cut-leaf *(laciniaum)*, Pyramid *(pyramidale)*, Skinners *(skinneri)*, Trefoil *tripartitum)*, Weeping (*pendulum*), Wier (*wieri*), and Yellow Bronze *(lutescens)*.

REMARKS. *Acer* is the ancient Celt name, and the pecies name, *saccharinum*, refers to the sweet sap. Vernacular names are White Maple, Soft Maple, River Maple, Creek Maple, and Swamp Maple. The tree is rittle and subject to wind damage. It is also rather usceptible to insect and fungus diseases. Although hort-lived it is a rapid-grower, and is often planted for rnament and occasionally for shelter-belt planting and tream-bank protection. Sugar is made from the sap ccasionally, and the wood is made into flooring and urniture, and used for fuel. The fruit is eaten by number of species of birds, and by squirrel and chipmunk.

BLACK MAPLE

Acer nigrum Michx.

FIELD IDENTIFICATION. Tree attaining a height of 120 ft and a diameter of 3–4 ft. The stout, spreading or erect, branches form an ovoid, or broad and rounded, crown.

FLOWERS. Appearing in May when the leaves are about half-grown, borne in nearly sessile corymbs which are many-flowered, pendent, and hairy; pedicels 2–3 in.; individual flowers greenish yellow, about ¼ in. long; staminate and pistillate flowers on different trees, on the same tree, or sometimes in the same cluster; calyx broadly campanulate, 5-lobed, hairy below; corolla absent; stamens 7–8, filaments slender and glabrous, in the staminate flower much longer than the calyx, but in the pistillate flower shorter than the calyx; ovary lobed, long-hairy.

FRUIT. Ripening in autumn, 2-winged, glabrous, reddish brown, wings ½–1 in. long, convergent or widespreading, seeds smooth, about ¼ in. long.

LEAVES. Opposite, blades usually 5–6 in. long or wide, base cordate with the sinus often closed by the overlapping of the basal lobes; margin with 3–5 lobes which are acuminate, or abruptly and acutely pointed; margins of the lobes entire or undulate, or sometimes with short lobules; young leaves densely tomentose at first; at maturity thickish, dull, dark green and glabrous above, lower surface yellowish green and soft-hairy, in autumn drooping and yellow; petioles stout,

BLACK MAPLE
Acer nigrum Michx.

usually densely tomentose but less so or almost glabrous later, 3–5 in., base swollen and almost enclosing the buds; supernumerary leaves ¼–1½ in. long, deltoid and dentate.

TWIGS. Stout, lustrous, green to orange or brown to gray; lenticels pale and oblong; young generally with pale, caducous hairs, glabrous later. Buds ovoid, acute, about ⅛ in. long; scales acute, reddish brown, densely pubescent, inner scales yellowish and puberulous.

BARK. On young branches thin, smooth, pale gray, later becoming deeply furrowed and dark.

WOOD. Hard, heavy, strong, weighing about 44 lb per cu ft.

RANGE. Southwestern Arkansas and Louisiana; eastward to Georgia, north to Ontario, and west to Minnesota, Wisconsin, Michigan, and Missouri.

VARIETIES. It is closely related to the Sugar Maple, *A. saccharum*, and at one time was classified as a variety of it, but is generally distinguished by the orange color of the twigs, and by the drooping leaves which are dark green above and green-pubescent beneath.

A variety has been described as Palmer Black Maple, *A. nigrum* var. *palmeri* Sarg., which has 3 lobes, with the terminal one broad and long-acuminate, and the leaf base rounded or slightly cordate. It is rare and local, occurring in Johnson County, Illinois; Putnam and Lawrence counties, Indiana; and Clark, Jackson, and Dunklin counties, Missouri.

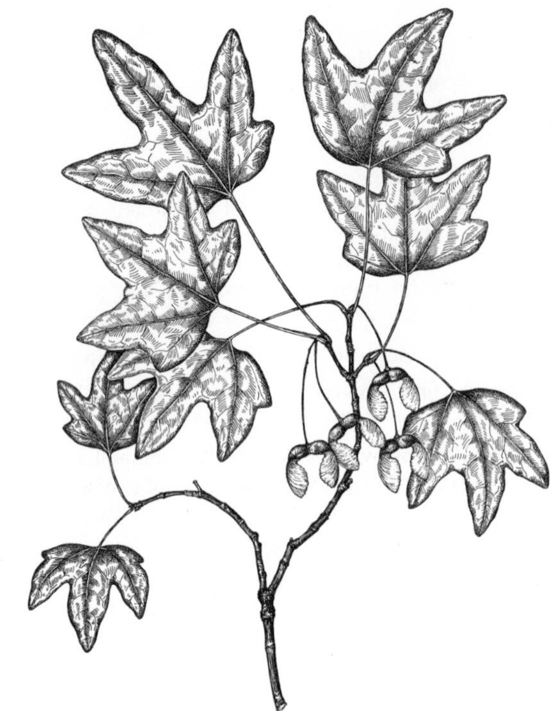

UVALDE BIG-TOOTH MAPLE
Acer grandidentatum var. *sinuosum* (Rehd.) Little

BIG-TOOTH MAPLE
Acer grandidentatum Nutt.

REMARKS. The genus name, *Acer*, is the classical name of the maple, and the species name, *nigrum*, refers to the dark bark or leaves. The tree is also known under the names of Black Sugar Maple, Sugar Maple, Hard Maple, and Rock Maple. It is long-lived and rather free of insects and disease. It produces maple sugar; the seeds are eaten by birds and the foliage browsed by many species of mammals. It is often cultivated for ornament, and in the fall is very attractive because of the scarlet, orange, and yellow leaves.

BIG-TOOTH MAPLE

Acer grandidentatum Nutt.

FIELD IDENTIFICATION. Western shrub or tree to 50 ft, and a diameter of 8–10 in. The branches rather stout and usually erect to form an open rounded crown.

FLOWERS. Appearing with the leaves April–May, axillary, corymbs short-stalked or almost sessile, hairy, yellow, few-flowered; pedicels slender, drooping, hairy; calyx persistent, campanulate, about ¼ in. long, lobes broad, rounded, and pale-hairy; corolla absent; stamens 6–8, elongate in the staminate flower; ovary superior, glabrous, 2-carpellate, 2-lobed, and 2-celled.

FRUIT. A double samara, mature in September, green

or rose-colored; wings ¾–1¼ in. long, usually divaricate, glabrous or nearly so, reticulate; seed reddish brown, smooth, about ¼ in. long, calyx persistent.

LEAVES. Simple, alternate, deciduous, 2–5 in. wide, 3–5-lobed, the lobes acute or obtuse and usually sinuate-dentate, more rarely entire, sinuses broad and rounded, base of leaf cordate or truncate, coriaceous; upper surface dark green, lustrous and glabrous; lower surface paler, glaucescent, and somewhat pubescent, turning red or yellow in autumn; petiole stout, 1¼–2 in. long.

TWIGS. Slender, reddish to brown, gray later; lenticels pale, small, numerous; leaf scars narrow, almost circular; buds reddish brown, about 1/16 in. long, scales ovate to obovate, puberulous and ciliate.

BARK. Thin, dark brown to gray with narrow fissures and flat ridges separating into platelike scales.

WOOD. Heart brown, sapwood nearly white, and wide, heavy, hard, close-grained, often used for fuel.

RANGE. Valleys, canyons, banks of mountain streams, at altitudes of 4,000–6,000 ft.

In Trans-Pecos Texas in the Guadalupe, Davis, and Chisos mountains. In New Mexico in the Organ, White, and Sacramento mountains. In Oklahoma in the Wichita Mountains; also in Arizona, north to Idaho, Utah, Colorado, and Wyoming.

VARIETIES. Uvalde Big-tooth Maple, *A. grandidentatum* var. *sinuosum* (Rehd.) Little, is a variety with leaf lobes longer and more sinuate. However, this form of leaf is not constant, sometimes appearing on certain branches, or on individual trees, and the rest of the leaves more closely resembling the species. At one time this variety was considered as being more closely related to Eastern Sugar Maple, *A. saccharum* var. *sinuosum* (Rehd.) Sarg. (Sargent, 1919). Later references assigned it a variety of the Big-tooth Maple (Little, 1944). Uvalde Big-tooth Maple occurs on the limestone regions of the Texas Edwards Plateau. Collected by the author on limestone bluffs 6½ miles northeast of Vanderpool, Texas.

Another variety has been described as Southwestern Big-tooth Maple, *A. grandidentatum* var. *brachypterum* (Woot. & Standl.) E. J. Palmer (Palmer, 1929), smaller samaras, about ⅗ in. long; leaf lobes much shorter and few-toothed or entire; also more densely pubescent beneath. It occurs in southeastern Arizona and southern New Mexico. It appears to intergrade with the species and may be only a xerophytic form of it.

Florida Sugar Maple, *A. barbatum* Michx., of the Southeastern states is also considered as being closely related to Big-tooth Maple.

REMARKS. The genus name, *Acer*, is the classical name of maple; the species name, *grandidentatum*, refers to the large teeth or lobes on the leaf margin. It is reported that the sap is sometimes used for sugar making, and the foliage occasionally browsed by deer and livestock.

Rocky Mountain Maple

Acer glabrum Torr.

FIELD IDENTIFICATION. Western shrub or small tree to 30 ft, and 1 ft in diameter. The small upright branches form an irregular crown, and the bark is smooth and gray.

FLOWERS. Appearing May–July in branching, 6–15-flowered corymbs about 1 in. long, either with staminate and pistillate on different trees, or on the same tree; individual flowers greenish yellow, ⅛–¼ in. across, long-pedunculate; sepals about 8, oblong, obtuse, petal-like; petals greenish yellow; stamens 7–8, glabrous, unequal, very short or rudimentary in the pistillate flower; ovary obtusely lobed, glabrous, rudimentary or absent in the staminate flowers; style deeply divided into 2 stigmatic lobes.

FRUIT. Maturing August–September, samaras several in a cluster, 2-winged; wings ½–1½ in. long, nearly parallel or spreading 60°–90°, tan or reddish, glabrous; seed ovoid, about ¼ in. long, chestnut brown.

LEAVES. Opposite, simple, or in some forms palmately compound, blade 1–3 in. long or broad, outline of blade rounded or cordate-reniform, palmately 3–5-lobed, lobes unequally and sharply toothed, sinuses acute, base subcordate to truncate or broadly cuneate (the forms and variations with the sinuses extending to the midrib to form a 3-foliate leaf have been given

Rocky Mountain Maple
Acer glabrum Torr.

varietal standing by some botanists); upper surface rather dark green, lustrous and glabrous; lower surface paler or glaucescent, thin, turning red or yellow in autumn; petioles slender, grooved, often reddish, variable in length.

TWIGS. Slender, green to brown or reddish brown, glabrous; winter buds about ⅛ in. long, red to yellow, tomentose within.

BARK. Reddish brown, thin, smooth.

WOOD. Heart light brown, sapwood thicker and paler, heavy, hard, close-grained.

RANGE. Usually in poor, well-drained soils in the sun, in canyons and on mountainsides, at altitudes of 4,000–9,000 ft. Usually in the pinyon, yellow pine, and spruce belts. New Mexico, Arizona, and southern California; northward to British Columbia and Alberta.

VARIETIES. Rehder (1940, p. 579) lists the following varieties for Rocky Mountain Maple:

Douglas Rocky Mountain Maple, *A. glabrum* var. *douglasi* (Hook.) Dipp., has 3-lobed leaves, with lobes short-acuminate, often lobulate, terminal one broad-ovate with acuminate teeth, base subcordate; fruit to 1 in. long with broader wings.

Three-leaf Rocky Mountain Maple, *A. glabrum* var. *tripartitum* (Nutt.) Pax, has smaller, usually 3-foliate, leaves. It is also listed under the name of *A. glabrum* var. *neomexicanum* (Greene) Kearney & Peebles, but some botanists separate the two, principally on the size of the leaflets.

Rose-fruit Rocky Mountain Maple, *A. glabrum* var. *rhodocarpum* Schwer., has bright red fruit when fully ripe.

REMARKS. The genus name, *Acer,* is the classical name of the maple, and the species name, *glabrum*, refers to the smooth character of the leaves and fruit. It has also appeared in the literature under the names of *A. tripartitum* Nutt., *A. neomexicanus* Greene, and *A. bernardinum* Abrams. It has also been given the vernacular names of Dwarf Maple, Mountain Maple, and Sierra Maple. The tree was introduced into cultivation about 1882. It is sometimes planted for erosion control, and is browsed by mule and white-tailed deer, or by mountain sheep occasionally. The seed averages about 16,000 seeds per lb. They have an embryo dormancy and should be stratified in sand for 90 days at 41°F., and given light scarification, before planting.

Box-elder Maple
Acer negundo L.

FIELD IDENTIFICATION. Tree attaining a height of 75 ft, with a broad rounded crown.

FLOWERS. March–May, dioecious, small, greenish, drooping, on slender stalks; staminate in fascicles 1–2 in. long; no petals; stamens 4–6, exserted, filaments hairy with linear anthers; calyx campanulate, hairy, lobed; pistillate flowers in narrow racemes; ovary pube cent, style separating into 2 elongate, stigmatic lob calyx of pistillate flower narrowly 5-lobed.

Box-elder Maple
Acer negundo L.

FRUIT. Ripening August–October, borne in early su mer in drooping clusters 6–8 in. long; samara doub greenish, minutely pubescent or glabrous, 1–2 in. lon wings divergent to about 90°, straight or falcate, th and reticulate; seeds at base of wings, solitary, smoo reddish brown, about ½ in. long.

LEAVES. Deciduous, opposite, 6–15 in. long, odd-pi nately compound of 3–7 (sometimes 9) leaflets; leafl short-stalked, 2–4 in. long, 1½–3 in. wide, ovate-ellip or oval-obovate; margin irregularly serrate or lob mostly above the middle; acute or acuminate at ap rounded, cuneate or cordate at the base, sometim unsymmetrical; light green, glabrous or slightly pub cent above, paler and pubescent beneath, especially axils of veins; petioles glabrous.

TWIGS. Slender, smooth, shiny, green to purpl green.

BARK. Young bark green, smooth, thin, later pale gr to brown, divided into narrow rounded ridges wi short scales and shallow fissures.

WOOD. Whitish, light, soft, not strong, close-grain weighing 27 lb per cu ft.

RANGE. Texas, Oklahoma, Arkansas, and Louisian eastward to Florida, northward to New Brunswick, a west to Ontario, Michigan, Minnesota, and Nebrask

PROPAGATION. Box-elder Maple was first cultivat in 1688. The seeds are gathered from the ground when floating on water. They may be cleaned by han by screening, or by fanning. The cleaned seeds avera about 11,800 seeds per lb, and have a commerc purity of 92 per cent and a soundness of 62 per ce

The dormant embryo causes delayed germination. Pretreatment is applied by stratification in sand at 41°F. for 90 days, or an alternative treatment of soaking for 2 weeks in cold running water. Stratified seed are planted in spring, but stratification should not be for too long or germination will occur. The germinative capacity is on the average about 33 per cent. Seeds are drilled about a ft apart, with about 20 seeds per ft. The seeds are covered with about ¼ in. of firmed nursery soil and mulched. Partial shade is beneficial.

VARIETIES. Rehder (1940, p. 586) has listed a number of varieties as follows:

A. negundo var. *variegatum* Jacq. has leaves with white margins.

A. negundo var. *aureo-marginatum* Schwerin has leaves with yellow margins.

A. negundo var. *pseudo-californicum* Schwerin has vigorous growth and green branches.

A. negundo var. *violaceum* (Kirchn.) Jaeg. has twigs purplish or violet with a glaucous bloom, leaflets 3–11, usually 5–7, pubescent beneath.

A. negundo var. *interius* (Britt.) Sarg. has twigs with short pale pubescence, rarely nearly glabrous; leaflets 3, slender-stalked, glabrous beneath or villous on midrib.

A. negundo var. *californicum* (Torr. & Gray) Sarg. has twigs hoary-tomentose; leaflets 3, coarsely serrate or nearly entire, pubescent above, tomentose beneath when young, later densely pubescent; fruit puberulous or glabrous, the nutlet not constricted at base.

A. negundo var. *texanum* Pax has twigs pale-tomentose; leaves 3-foliate, tomentose beneath when young, later glabrous; fruit puberulous, constricted into a short stipe.

The variety *A. negundo* var. *texanum* Pax (*Rulac texana* [Pax] Small) occurs in east Texas with the species, and is found west to the San Antonio River.

REMARKS. The genus name, *Acer*, is from an old Celt word, and the application of the species name, *negundo*, is obscure. The synonyms *Negundo aceroides* Moench and *Rulac negundo* (L.) A. S. Hitchcock are also used by some writers. Vernacular names are Maple-ash, Ashleaf Maple, Water-ash, Sugar Maple, Red River Maple, Black Maple, and Manitoba Maple. The tree is easily transplanted when young, grows rapidly, is short-lived, and is easily damaged by rot, insects, storm, and fire. It is widely planted as a quick-growing ornamental tree, and has been extensively used in shelter-belt planting in the prairie states. The wood is used for paper pulp, cooperage, woodenware, interior finish, and cheap furniture. The seeds are eaten by many species of birds and squirrels. It is occasionally tapped for its sugary sap, which is inferior to that of Sugar Maple.

BUCKEYE FAMILY (*Hippocastanaceae*)

Texas Buckeye
Aesculus arguta Buckl.

Texas Buckeye
Aesculus arguta Buckl.

FIELD IDENTIFICATION. Usually a shrub, but under favorable conditions a tree to 35 ft. The branches are stout and the crown is rounded to oblong.

FLOWERS. Inflorescence a dense, yellow panicle, borne after the leaves, 4–8 in. long, 2–3½ in. broad, primary peduncle of panicle densely brown-tomentose, secondary racemes ¾–½ in. long, 3–15-flowered, tomentose; individual flower pedicels ⅛–¼ in., densely tomentose; calyx campanulate, ⅕–¼ in. long, brown-tomentose, 4–5-lobed above, lobes imbricate in the bud, lobes unequal, apices obtuse to rounded or truncate; petals 4, pale yellow, some reddish at the base, ⅓–¾ in. long, upright, parallel, clawed, thin, deciduous, densely hairy, margin ciliate; upper pair oval to broadly oblong; lateral pair elongate, oblong to spatulate, long-clawed, apex rounded to truncate or notched; disk hypogynous, annular, depressed; stamens 7–8, inserted on the disk, long-exserted, upcurved, unequal, filiform, hairy, longer than the petals; anthers yellow, ellipsoid, introrse, 2-celled, opening longitudinally; style 1, slender, elongate, curved; stigma terminal and entire; ovary 3-celled, sessile, cells often 1-ovuled by abortion.

FRUIT. Maturing May–June, peduncles stout, hairy at first, glabrate later; capsule ¾–1¾ in. in diameter, subglobose to obovoid or asymmetrically lobed, armed with stout warts or prickles or occasionally smooth, light brown, dehiscent into 2–3 valves at maturity; seeds usually solitary and rounded, if 2, usually flattened by pressure, coriaceous, smooth, lustrous, brown, ⅝–¾ in. in diameter.

LEAVES. Deciduous, opposite, palmately compound of 7–9 leaflets (often 7), sessile or nearly so, narrowly elliptic to lanceolate, or more rarely obovate, apex long-acuminate (less often broader and abruptly acuminate), base attenuate to narrowly cuneate, margin finely serrate or occasionally incised above the middle, blade length 2½–5 in., width ½–2 in.; upper surface olive-green, lustrous, a few fine hairs mostly along the veins; lower surface paler and more pubescent; petioles slender, 3–5 in., grooved, woolly-hairy.

BARK. Gray to black; fissures narrow, short and irregular; ridges broken into small short, rough scales.

TWIGS. Tough, stout, terete, the young ones green and glabrous to somewhat hairy, the older ones gray to reddish brown, lenticels small; leaf scars lunate or horseshoe-shaped, fibrovascular marks 3–6.

RANGE. Texas Buckeye is found on limestone or granite soils in the Edwards Plateau area of Texas, north to southern Oklahoma, and in Missouri.

REMARKS. The genus name, *Aesculus,* is the old name for a European mast-bearing tree, and the species name, *arguta,* means "sharp-toothed," perhaps referring to the foliage. Texas Buckeye is closely related to Ohio Buckeye, *A. glabra* Willd., and at one time was listed as a variety of it under the name of *A. glabra* var. *arguta* (Buckl.) Robinson. Although listed only as a shrub in most literature, Texas Buckeye becomes a tree to 35 ft and 18 in. in diameter on the Texas Edwards Plateau. The leaves and flower parts of Texas Buckeye appear to be only lightly pubescent on some specimens and heavily tomentose on others.

Woolly Buckeye

Aesculus discolor Pursh

FIELD IDENTIFICATION. Usually a shrub, but occasionally a tree to 25 ft, with stout branches and an open crown.

FLOWERS. Borne April–May in terminal, rather narrow, pubescent panicles 6–8 in.; individual flowers borne on pubescent pedicels ¾–1 in.; petals 4–5 (usually 4), unequal, clawed, red and yellow, ¾–1½ in. long, connivent, shorter than the stamens, oblong to obovate, minutely glandular or glandless on margins; stamens 5–8, longer than the petals, filaments hairy; calyx tubular, pubescent externally, tomentose within, 5-lobed.

FRUIT. Fruiting peduncle light brown, stout, thickened at the base, about 1 in. long or longer; capsule solitary or several in a cluster, 1½–2½ in. long, subglobose or obovoid, often asymmetrical and abruptly pointed, light brown, smooth, leathery, 3-valved, 3-celled, but often by abortion 1-celled and 1-seeded; seeds large, light brown, variously shaped, testa thin.

Woolly Buckeye
Aesculus discolor Pursh

Yellow Woolly Buckeye
Aesculus discolor var. *flavescens* Sarg.

LEAVES. Deciduous, opposite, composed of palmately compound leaves bearing 5 leaflets; leaflets 3–7 in. long, elliptic to oblong-obovate, apex acuminate, base narrowed, margin serrate, upper surface dark green and lustrous, lower surface pubescent to tomentose; petioles usually 4–6 in., grooved, tomentose to pubescent or almost glabrous; leaf scars cordate, about ¼ in. long and wide.

TWIGS. Stout, green and pubescent at first, becoming gray to light brown and glabrous later, lenticels small and abundant.

BARK. Gray to light brown, smooth, pale, thin.

RANGE. Woolly Buckeye is found in Texas mostly on the Edwards Plateau, where it is represented by several color variations; also in Oklahoma, Arkansas, and Louisiana; eastward through Alabama to Georgia and northward to Missouri.

VARIETIES. The following varieties have been named in connection with Woolly Buckeye:

Scarlet Woolly Buckeye, *A. discolor* var. *mollis* Sarg., has scarlet flowers, and dense tomentum on the lower leaf surface; occurs mostly as a shrub in central and east Texas, and into Louisiana.

Yellow Woolly Buckeye, *A. discolor* var. *flavescens* Sarg., has yellow flowers, and occurs as a shrub in a limited area near San Marcos, New Braunfels, Boerne, Comfort, and Kerrville, Texas. It is occasionally found growing with Scarlet Woolly Buckeye, and some botanists consider it only a color form of the latter.

Shrubby Woolly Buckeye, *A. discolor* var. *koehnei* Rehd., has red and yellow flowers, and is placed as an intermediate cultivated form between Scarlet Woolly Buckeye and Yellow Woolly Buckeye.

The closely related Red Buckeye, *A. pavia* L., a shrub occurring from east Texas eastward to Georgia, varies from Scarlet Woolly Buckeye mostly in the leaves being glabrous, or somewhat pubescent, instead of tomentose. The exact relationship between *A. pavia* and *A. discolor* has never been clearly defined botanically. Some botanists feel that the latter is only a hairy Western variation of the more glabrous Eastern plant. However, the author feels that more study is needed before a change in name is made.

The other buckeyes which resemble each other are Texas Buckeye, *A. arguta* Buckl., and White-bark Ohio Buckeye, *A. glabra* var. *leucodermis* Sarg. Both have capsules with prickles, but Texas Buckeye has 7–9 leaflets and White-bark Ohio Buckeye usually 5 leaflets. They both have yellow flowers and become trees.

REMARKS. The genus name, *Aesculus*, is the ancient name for a mast-bearing tree, and the species name, *discolor*, refers to the vari-colored flowers.

OHIO BUCKEYE
Aesculus glabra Willd.

OHIO BUCKEYE

Aesculus glabra Willd.

FIELD IDENTIFICATION. Tree attaining a height of 75 ft and a diameter of 2 ft. The branches are rather thick, and the crown broad and rounded.

FLOWERS. March–May, panicles terminal, narrow, loose, pubescent, 4–8 in. long, 2–3 in. wide; the pedicels 4–6-flowered; flowers yellowish green, ½–¾ in. long, petals 4, parallel, almost equal in length, proximal ends clawed and often hairy at the base; the lateral pair broad-ovate or oblong; the upper pair much narrower, oblong-spatulate, yellow (occasionally red-striped); stamens usually 7 or 8, long-exserted, filaments curved upward and pubescent, anthers orange-colored; ovary pubescent and prickly; calyx campanulate, 4–5-lobed above.

FRUIT. Capsule ripening September–October, on stout pedicels ½–1 in., irregularly subglobose or obovoid, capsule ¾–2¼ in. long, pale brown, leathery, roughened by warty tubercules and prickles, valves thick, 3-celled and 3-seeded, but often by abortion 1-celled and 1-seeded; seeds 1–1½ in. wide, lustrous brown, sometimes depressed, hilum scar large, round, and pale; cotyledons thick and fleshy.

LEAVES. Opposite, on petioles 4–6 in., palmately compound of 5 leaflets (or more in the varieties); leaflets sessile or nearly so, 4–6 in. long, 1½–2½ in. broad, elliptic to oblong or oval to obovate, apex mostly acuminate, base cuneate, margin finely serrate, upper surface glabrous, lower surface paler and glabrous to pubescent on the veins, turning yellow in the autumn, fetid when crushed.

TWIGS. Reddish brown to gray, pubescent at first but glabrous later, lenticels orange-colored, bud scales keeled and apiculate.

BARK. Young bark dark brown and smoother, older gray and broken into plates roughened by small, numerous scales, odor fetid (almost white in *A. glabra* var. *leucodermis* Sarg.).

WOOD. Whitish, fine-grained, weighing 28 lb per cu ft, weak in bending and endwise compression, durability low, shock resistance low, shrinks moderately, hand tools easily, but machining properties rather low.

RANGE. In moist, rich soil of woodlands or riverbanks. Oklahoma, Arkansas, northern Mississippi, northern Alabama, and northeastern Texas; eastward to West Virginia, north to Pennsylvania, and west to Michigan, Iowa, and Kansas.

MEDICINAL USES. There is an old legend that the seeds will prevent rheumatism if carried in the pocket.

The bark of some of the species of *Aesculus* contains a glycoside, aesculin, which is used in a 4 per cent solution as an ointment to protect skin from sunburn. It has the property of absorbing ultraviolet rays, which are then gradually given off.

PROPAGATION. The capsules are hand-picked, or shaken into canvas when they turn yellowish brown. After a short period of drying the large seeds can be separated from the hull. Excessive drying will dry out the seed and prevent germination. If the seeds are not planted at once they should be stratified in moist sand at 41°F. for 120 days for spring planting. The seeds should be planted in moist, rich soil in nursery rows. Propagation is also practiced by side-grafting or budding, or by semiripe wood under glass.

CLOSELY RELATED VARIETIES AND FORMS. Pale Ohio Buckeye, *A. glabra* var. *pallida* (Willd.) Scheele, is a form with leaves permanently pubescent beneath. It has been cultivated since 1809.

Sargent Ohio Buckeye, *A. glabra* var. *sargentii* Rehd., has 6–7 (rarely 8) leaflets which are obovate to lanceolate, acuminate at the apex and finely pubescent beneath. It has been in cultivation since 1922, and occurs naturally in Ohio, Iowa, Oklahoma, and Arkansas.

Oklahoma Buckeye, *A. glabra* var. *monticola* Sarg., has 6–7 leaflets, and leaves glabrous or puberulent on veins beneath; the fruit is subglobose, ⅝–¾ in. long, and usually 1-seeded. It occurs in Oklahoma, and was introduced into cultivation about 1922.

White-bark Ohio Buckeye, *A. glabra* var. *leucodermis* Sarg., has an almost white bark with leaflets pale-green and glaucous beneath. It was brought into cultivation about 1901.

A closely related species is Texas Buckeye, *A. arguta* Buckl., with 7–9 narrowly lanceolate leaflets which are usually glabrous at maturity. It has been cultivated since 1909.

Bush's Buckeye, *A.* × *bushii* Schneid., is a hybrid of *A. glabra* and *A. pavia,* and is known to occur in Arkansas and Mississippi. It is named for Benjamin Franklin Bush (1858–1937), botanist of Missouri.

REMARKS. The genus name, *Aesculus,* is the ancient name for a European mast-bearing tree. The species name, *glabra,* refers to the smooth leaves. Other local names are Fetid Buckeye, Stinking Buckeye, and American Horse-chestnut. The tree is sometimes planted for ornament in the eastern United States and in Europe. It has been in cultivation since 1809. It is short-lived; the young shoots are poisonous to cattle; and hogs are poisoned by the seed. The wood is used for fuel, paper pulp, artificial limbs, splints, woodenware, boxes, crates, toys, furniture, veneer for trunks, drawing boards, and occasionally for lumber.

Red Buckeye
Aesculus pavia L.

FIELD IDENTIFICATION. Shrub with an inclined stem, or more rarely a tree attaining a height of 28 ft and a diameter of 10 in. The crown is usually dense and the branches short, crooked, and ascending.

FLOWERS. March–May, panicles narrow to ovoid, erect, pubescent, 4–8 in. long; 1–numerous-flowered, pedicels slender, ¼–½ in.; flowers red, ¾–1½ in. long; calyx tubular, ⅜–⅝ in. long, dark red, tube about five times as long as the rounded lobes; petals 4, red, connivent, ⅝–1 in. long, oblong to obovate, apex rounded, base narrowed into a claw, limbs of the 2 superior petals shorter than the 2 lateral pairs; stamens about equaling the longest petals, exserted, usually 8 in number, filaments filiform and villous below; ovary sessile, villous, 3-celled (or by abortion fewer), style slender.

FRUIT. Capsule 1–2 in. in diameter, subglobose or obovoid, light brown, smooth but finely pitted, dehiscent into 2–3 valves; seeds 1–3, rounded, or flattened by pressure against each other, lustrous, light to dark brown, about 1 in. in diameter.

LEAVES. Deciduous, opposite, palmately compound of 5 leaflets (rarely, 3 or 7), leaflets oblong to elliptic or oval to obovate, apex acute to short-acuminate, base gradually narrowed, margin coarsely serrate, leaflet length 3–6 in., width 1–1½ in., firm; upper surface lustrous, dark green, glabrous except a few hairs on the veins; lower surface paler, almost glabrous to densely tomentose; petiole nearly glabrous or with varying degrees of hairiness, red, 3–7 in.

BARK. Gray to brown, smooth on young branches, on old trunks roughened into short plates which flake off in small, thin scales.

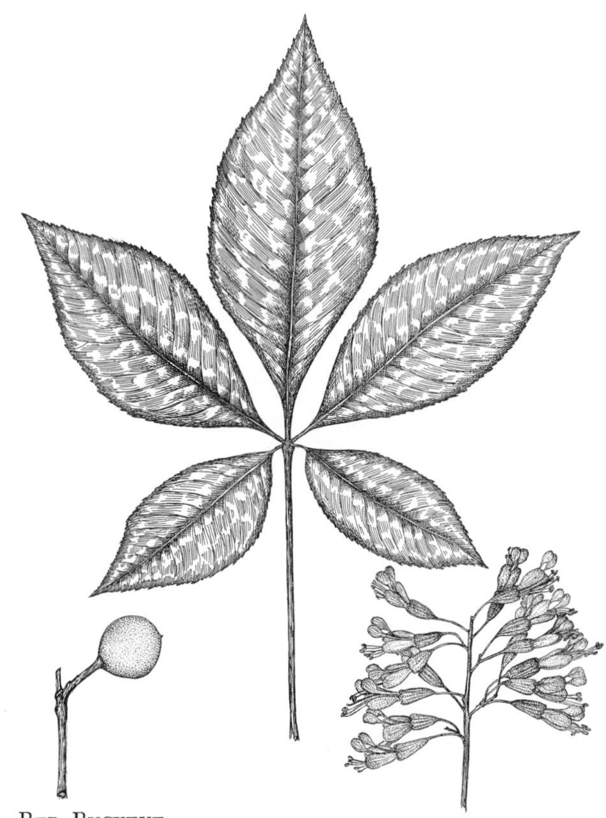

RED BUCKEYE
Aesculus pavia L.

TWIGS. Green to gray or brown, crooked, stout, smooth; lenticels pale brown to orange; leaf scars large and conspicuous, with 3 fibrovascular bundles.

RANGE. Mostly along streams of the coastal plains. East and central Texas to the Edwards Plateau, Oklahoma, and Arkansas; eastward through Louisiana to Florida, north to southern Illinois, and west to southeast Missouri.

VARIETIES AND FORMS. Prostrate Red Buckeye, *A. pavia* forma *humilis* Mouillef, is a low, often prostrate, shrub 2–4 ft, with coarsely and unequally serrate leaves slightly pubescent beneath.

Wine-red Buckeye, *A. pavia* var. *atrosanguinea* Rehd., has dark red flowers.

Cut-leaf Red Buckeye, *A. pavia* var. *sublaciniata* Wats., has narrower oblong, deeply serrate leaves and dark flowers. Forms with variegated leaves are also known.

Red Buckeye, *A. pavia* L., is also known to hybridize with Ohio Buckeye, *A. glabra* Willd., to produce Bush's Buckeye, *A.* × *bushii* Schneid., named after B. F. Bush (1858–1937), a botanist of Missouri.

Some botanists have questioned the segregation of Red Buckeye, *A. pavia* L., and Woolly Buckeye, *A. discolor* Pursh, as separate species, maintaining that Woolly Buckeye is only a more hairy Western variation of the Red Buckeye of the East. Pending more conclusive research the writer prefers to maintain the two species as distinct. A monographic study is needed to determine the exact placement of the numerous variations, not only of hairiness, but also of flower color and distribution of the forms.

REMARKS. The genus name, *Aesculus,* is the ancient name for an old mast-bearing tree, and the species name, *pavia,* honors Peter Paaw (d. 1617), of Leyden. Some vernacular names in use for the plant are Scarlet Buckeye, Woolly Buckeye, Firecracker-plant, and Fish-poison–bush. It is reported that the powdered bark is used in domestic medicine for toothache and ulcers, the roots for washing clothes, and the crushed fruit for fish poison.

SOAPBERRY FAMILY (*Sapindaceae*)

WESTERN SOAPBERRY

Sapindus drummondii Hook. & Arn.

FIELD IDENTIFICATION. Tree attaining a height of 50 ft, with a diameter of 1–2 ft. The branches are usually erect to form a rounded crown.

FLOWERS. May–June, in large, showy panicles 5–10 in. long and 5–6 in. wide; perianth about 1/25 in. across; petals 4–5, obovate, rounded, white; sepals 4–5, acute, concave, ciliate on margin, shorter than the petals; stamens usually 8, inserted on the disk; style single, slender, with a 2–4-lobed stigma; ovary 3-lobed and 3-celled, each cell containing 1 ovule.

FRUIT. September–October, globular, fleshy, from white to yellowish or blackish, translucent, persistent and shriveled; seed 1, obovoid, dark brown, the other 2 seeds seeming to atrophy.

LEAVES. Short-petiolate, deciduous, alternate, 5–18 in. long, abruptly pinnate; leaflets 4–11 pairs, 1½–4 in. long, ½–¾ in. wide, falcate, lanceolate, acuminate at the apex, asymmetrical at base, veiny, yellowish green, glabrous above, soft pubescent or glabrous beneath.

BARK. Gray to reddish, divided into narrow plates that break into small reddish scales.

TWIGS. Yellowish green to gray, pubescent to glabrous, lenticels small.

WOOD. Light brown or yellowish, close-grained, hard, strong, weighing 51 lb per cu ft.

RANGE. In moist soils along streams; Texas, New Mexico, Arizona, Oklahoma, and Arkansas; eastward into Louisiana, north to Missouri and Kansas, and south into Mexico.

PROPAGATION. For planting the fruit can be hand-picked or flailed. The pulp is removed from the seed by maceration in water. The seed numbers about 1,160 seeds per lb, and tests about 77 per cent sound. The germination rate averages about 31 per cent. The thick seed coat makes it necessary either to stratify for 60–90 days, or treat with sulphuric acid for 2–3 hours. Either treatment should precede spring sowing. The seeds are sown at a depth of ¾ in. and the soil well-firmed above. Propagation may also be practiced with hard-wood cuttings planted in spring.

WESTERN SOAPBERRY
Sapindus drummondii Hook. & Arn.

REMARKS. The generic name is from *sapo* ("soap") and *Indus* ("Indies"), referring to the fact that some of the West Indian species are used for soap. The species name, *drummondii*, is in honor of the botanist Thomas Drummond. Vernacular names are Amole de Bolita, Tehuistle, Palo Blanco, Jaboncillo, Wild China-tree, and Indian Soap-plant.

The fruit of this tree, and related species, contains the poisonous substance saponin, which produces a good lather in water. It is used in Mexico as a laundry soap. The fruit is also used medicinally as a remedy for renal disorders, rheumatism, and fevers. Buttons and necklaces are made from the seeds. The wood is of little value except for making baskets and frames, and as fuel. Soapberry makes a desirable shade tree, and could be more extensively planted for ornament. It has been cultivated since 1900, and is sometimes used for shelter-belt planting.

Short-fruited Serjania-vine

Serjania brachycarpa Gray

FIELD IDENTIFICATION. Slender, somewhat hairy, scandent vine, usually herbaceous but sometimes woody near the base. Attaining a length of 14 ft.

FLOWERS. Panicles axillary, thyrsoid, about 1 in. long, or longer at maturity; peduncles as long as the panicles, or longer; tendrils inflorescential, flowers polygamo-dioecious, small, yellow, irregular; sepals 5, imbricate, outer ones smaller, white-tomentose; petals 4, 1/12–1/8 in. long, scales of upper petals large, bicorniculate-bifid; stamens 8, filaments united near base and inserted on disk; ovary 3-celled, with 2 small and 2 large glands, styles united to above the middle, ovary 1-celled.

Short-fruited Serjania-vine
Serjania brachycarpa Gray

Incised Serjania-vine
Serjania incisa Torr.

FRUIT. Composed of 3 united winged samaras, 3/8–5/8 in. long and broad, ovate and cordate at base, seed-bearing portion toward the apex, lenticular, surface obscurely veined, puberulent or glabrate, wings 1/8–1/5 in. broad.

LEAVES. Alternate, compound, biternate; leaflets ovate to lanceolate or rhombic, apex acute to obtuse or mucronulate, base attenuate, margin remotely and coarsely toothed, thin, densely tomentose beneath, upper surface brown-hairy, length 3/8–1½ in.

STEMS. Herbaceous, or somewhat woody near the base, younger hairy and older glabrate, 5-angulate with conspicuous striations.

RANGE. Known in Texas from the lower Rio Grande Valley area. In Mexico in Tamaulipas, Nuevo León, and southward.

REMARKS. The genus name, *Serjania*, is in honor of the French botanist Paul Serjeant, and the species name, *brachycarpa*, refers to the short fruit. This slender vine superficially resembles certain species of *Clematis*, and also has the general aspects of the vine *Cardiospermum*. Short-fruited Serjania is distinguished from Incised Serjania by the densely white-tomentose sepals and lenticular fruit which is deeply cordate at base.

Incised Serjania-vine

Serjania incisa Torr.

FIELD IDENTIFICATION. Slender, delicate vine 3–6 ft long, and climbing by paired inflorescential tendrils; usually herbaceous, but with older stems partially woody close to the base.

FLOWERS. Peduncles axillary, elongate, slender, grooved, minutely white-hairy, 1–4 in.; the racemose panicle 1–1¼ in. long with inflorescence on the upper part, usually with paired, coiled tendrils at the base of the panicle; small flowers hardly over ⅛–³⁄₁₆ in. broad, on individual pedicels ⅛–¼ in. long, yellowish white, irregular, polygamous; sepals 5, free and connate, somewhat imbricate, outer ones smaller, glabrate to puberulent; petals 4, appendaged; disk undulate with 4 glands (2 large and 2 small); stamens 8, filaments united near the base, inserted on the disk; styles united to above the middle, ovary 3-celled, 1 ovule in each cavity.

FRUIT. Composed of 3 indehiscent samaras, winged, broadest toward the apex, attached by their backs, obovate-oblong, ½–1¾ in. long, ⅜–¾ in. wide, glabrate to pubescent, subacute or rounded at apex; base rounded, obtuse or truncate, often undulate; cells inflated, more or less produced beyond the axis, longer than broad, wings ⅕–⅖ in. wide, seed portion reticulate-veined; 1 ovule in each cavity.

LEAVES. Alternate, pinnate or bipinnate, 1–3 in. long, rachis minutely white-hairy, often winged, usually with 2–4 pairs of 3-foliate pinnae; lowest pinnae often 5-foliate; leaflets ovate to rhombic, margin incised-dentate or irregularly lobed, apex acute to acuminate and often apiculate, base broadly or narrowly cuneate into a winged petiole, or sessile; surface pubescent with stiff, white hairs, ⅜–1¼ in.

STEMS. Delicate, slender, elongate, sulcate, glabrate or hairy, herbaceous, or woody near the base.

RANGE. In western and southwestern Texas, mostly confined to the lower Rio Grande Valley area. Also in Coahuila, Mexico.

REMARKS. The genus name, *Serjania*, is in honor of Paul Serjeant, a French botanist, and the species name, *incisa*, refers to the incised leaves.

Urvillea-vine
Urvillea ulmacea H. B. K.

Urvillea-vine

Urvillea ulmacea H. B. K.

FIELD IDENTIFICATION. Tomentose scandent vine.

FLOWERS. Borne in long-peduncled racemes 1–4 in., sometimes longer than the leaves, pedicels jointed; tendrils modified inflorescential branches; sepals 4–5, imbricate, oblong or obovate, the outer 2 shortest; petals 4–5, about as long as the inner sepals, crisped, concave; stamens 8, as long as the petals, declined, inserted on the disk; disk 1-sided, produced into 4 glands opposite the 2 smaller petals; ovary 3-celled, styles united to above the middle; ovules 1 in each cavity, attached by the middle of the axis.

FRUIT. Composed of 3 samaras attached dorsally, each with a seed centrally fixed, with thin wings; the entire fruit obovate to elliptic or oval in outline, narrowed at each end, ¾–1 in. in length, surfaces pubescent or glabrous; seeds solitary in each cell, smooth, black, shining.

LEAVES. Alternate, numerous, petioled; leaflets 3, blades ⅜–3 in. long, ovate to deltoid or oval, apex acute or acuminate, margin doubly serrate or often incised, sometimes lobate, upper surface glabrous or thinly hairy, lower surface pubescent or tomentose.

RANGE. On plains and prairies, southwestern Texas; and Mexico from Nuevo León to Yucatán, Oaxaca, and Nayarit.

REMARKS. The genus name, *Urvillea*, is in honor of Capt. Dumont D'Urville, French botanist and naval officer. The species name, *ulmacea,* refers to the elmlike leaflets. In Yucatán it is known under the Mayan name of Apaac. It is sometimes cultivated in gardens in Mexico.

Mexican-buckeye

Ungnadia speciosa Endl.

FIELD IDENTIFICATION. Western shrub or tree to 30 ft, and 10 in. in trunk diameter. Branches small and upright to spreading with an irregularly shaped crown.

FLOWERS. Borne in pubescent fascicles in spring, appearing with the leaves, or just before them; polygamous and irregular, about 1 in. across, fragrant; petals usually 4 or occasionally 5, rose-colored, deciduous, obovate, erect, clawed, margin crenulate, somewhat tomentose, with a tuft of fleshy hairs; disk 1-sided, oblique, tongue-shaped, connate with the ovary base; calyx campanulate, 5-lobed, lobes oblong-lanceolate; stamens 7–10, unequal, exserted, inserted on edge of disk, filaments filiform and pink, anthers oblong and red; ovary ovoid, stipitate, hairy, 3-celled, rudimentary in the staminate flower, ovules 2; style subulate, elongate, filiform, slightly upcurved; stigma terminal, minute.

FRUIT. Capsule stipitate, broad-ovoid, crowned by the style remnants, about 2 in. broad, leathery, roughened, reddish brown, 3-celled and 3-valved, dehiscent in October while still on the trees, but hardly opening wide enough to release the large seeds; seeds usually solitary because of the abortion of the others, about ½ in. in diameter, round, smooth, shining, black or brown, hilum scar broad.

LEAVES. Odd-pinnately compound, alternate, deciduous, 5–12 in. long, leaflets 5–7 (rarely, 3), ovate to lanceolate, apex acuminate, base rounded or cuneate, margin crenate-serrate, 3–5 in. long, 1½–2 in. wide, rather leathery, upper surface dark green, lustrous and glabrous, lower surface paler and pubescent to glabrous; petiole 2–6 in.; petiolule of terminal leaflet ¼–1 in.; lateral leaflets sessile or very short-petioluled.

TWIGS. Buds imbricate, ovate to almost globose, leaf scars large and obcordate, twigs slender, terete, brown to orange and pubescent at first, later reddish brown and glabrous.

BARK. Mottled gray to brown, thin, tight, smooth, broken into shallow fissures on old trunks.

WOOD. Reddish brown, sapwood lighter, brittle, soft, close-grained.

RANGE. Usually in limestone soils of stream banks,

Mexican-buckeye
Ungnadia speciosa Endl.

moist canyons, or on bluffs. In Texas, New Mexico, and Mexico. In Texas of greatest abundance west of the Brazos River. A few trees found as far east as Harris County, Texas. In Mexico in the states of Nuevo León, Coahuila, and Chihuahua.

REMARKS. The genus name, *Ungnadia,* is in honor of Baron Ungnad, ambassador of Emperor Rudolph II.

The species name, *speciosa,* means showy, with reference to the flowers. Also known under the vernacular names of Monillo, Texas-buckeye, Spanish-buckeye, New Mexican-buckeye, False-buckeye, and Canyon-buckeye. The tree should be grown more for ornament, being very beautiful in the spring. The flowers resemble Redbud or Peach blossoms at a distance. It is also a source of honey. The sweet seeds are poisonous to human beings. Seemingly a few may be eaten with impunity, but a number cause stomach disturbances. The leaves and fruit may also cause some minor poisoning to livestock, but they are seldom browsed except in times of stress. Children in west Texas sometimes use the round seeds for marbles.

BUCKTHORN FAMILY (*Rhamnaceae*)

TEXAS ADOLPHIA

Adolphia infesta Meissn.

FIELD IDENTIFICATION. Densely green-thorny shrub 2–6½ ft.

FLOWERS. January–April, in fasciculate axillary clusters of 1–4 flowers; pedicels pubescent, mostly less than ⅛ in. long; individual flowers ⅛–3/16 in. wide, greenish white; calyx-tube broad, pubescent, minutely punctate, 5-lobed; lobes more glabrous, thin, scarious, triangular to ovate, obtuse to acute, straw-colored to brownish, reflexed at maturity; petals 5, white, 1/25–1/16 in. long, much narrower than the calyx-lobes, distinct, hooded, abruptly contracted into a long slender claw; stamens 5, opposite the petals, exserted somewhat at anthesis; ovary 2–4-celled, ovules solitary.

FRUIT. Capsule small, subglobose, beaked by the persistent style, dry, coriaceous, 3-celled; nutlets 3, perforated at base.

LEAVES. Leaflets, or the leaves early deciduous, simple, opposite, 1/12–⅖ in. long, pubescent on both surfaces, dull green, linear to sublanceolate, entire or slightly serrulate, 1-nerved, short-petioled; stipules small, persistent, acute.

TWIGS. Slender, green, opposite, puberulent to pubescent; striate with the dilated decurrent petiole bases; usually ending in straight, or slightly curved, slender, green, sharp-pointed spines.

RANGE. Foothills or mountains of Trans-Pecos Texas. Also in Mexico in Baja California, Chihuahua, Zacatecas, Hidalgo, and Oaxaca.

REMARKS. The genus name, *Adolphia,* honors Adolphe T. Brongniart, French botanist, author of a monograph on the Rhamnaceae Family. The species name, *infesta,* refers to the "unsafe" pernicious spines. Also known under the names of Junco or Allthorn, but these names also apply to *Koeberlinia spinosa,* which this shrub somewhat resembles.

TEXAS ADOLPHIA
Adolphia infesta Meissn.

Alabama Supplejack

Berchemia scandens (Hill) Trel.

FIELD IDENTIFICATION. High-climbing, large, woody, twining vine with small parallel-veined leaves.

FLOWERS. March–June, in loose terminal panicles, perfect or polygamous, greenish yellow; petals 5, about as long as the sepals, obovate, acute, hooded, clawless; calyx glabrous, the 5 sepals ovate-triangular, acute or acuminate; stamens 5, equal to the petals or shorter, to which they lie opposite, filaments filiform; style short, 2 united stigmas, ovary 2-celled and half covered by a thick flat disk.

FRUIT. July–October, drupe oblong, ovoid or ellipsoid, slightly flattened, fleshy, bluish black, ¼–⅓ in. long, containing a 2-celled crustaceous stone.

LEAVES. Alternate, dark green, glabrous, with veins impressed above, conspicuously parallel-veined beneath, leathery, shiny, oblong, ovate or elliptical; apex acute or acuminate; base cuneate or rounded; margin entire, undulate or slightly serrulate, 1½–3 in. long and ¾–1½ in. broad; petioles about ¼ in., glabrous or puberulent.

STEMS. Much-branched, glabrous, terete, slender, green and gray-streaked to brown and finely grooved when young, reddish brown or black when older, twining, pliant, strong.

RANGE. In moist rich soil of eastern and southern Texas, Oklahoma, Arkansas, and Louisiana; eastward to Florida and north to Virginia, Kentucky, and Missouri.

Alabama Supplejack
Berchemia scandens (Hill) Trel.

REMARKS. The genus name, *Berchemia,* is in honor of Berthout van Berchem, eighteenth century Dutch botanist. The species name, *scandens,* refers to the climbing or leaning habit. Another vernacular name is Rattan-vine in allusion to the strong, pliant stems, which are much used in making wickerware. The strong twining stems often girdle and in time kill good-sized trees. The astringent drupe stains purple. Seventeen species of birds are known to feed upon it, including the turkey, bobwhite quail, and mallard duck.

Jersey-tea Ceanothus

Ceanothus americanus L.

FIELD IDENTIFICATION. Shrub to 3 ft. Branches spreading, herbaceous above, woody toward the base.

FLOWERS. Borne in axillary, short-cylindric to ovoid panicles (sometimes branched), usually 2–5 in. long, ¾–1¼ in. in diameter; peduncles 2–10 in. long, elongate, naked, usually subtended by 2–3 bracts; calyx campanulate, lobes 5, incurved, cohering below the disk and ovary; petals 5, hooded, usually notched, clawed, longer than the calyx and inserted under the disk; stamens 4–5, filaments filiform and elongate; ovary immersed in the disk and adnate to it, base 3-lobed; style short, 3-cleft.

FRUIT. Drupe depressed-obovoid, ⅕–¼ in. long, 3-lobed, dry, dehiscent into 3 carpels; nutlets 3, one nutlet in each locule, seed coat smooth.

LEAVES. Simple, alternate, ovate to ovate-oblong, apex acute to obtuse or acuminate, base cordate or rounded to broadly cuneate, margin serrate, distinctly 3-ribbed, upper surface green and glabrous, somewhat gray-hairy below, blades 2–4 in. long, width 1–2⅓ in.; petioles about ½ in. long.

RANGE. Prairies, woodlands, barrens. In Texas, Arkansas, and Louisiana; eastward to Florida and northward to Quebec, Ontario, and Manitoba.

VARIETIES. A variety known as the Pitcher Jersey-tea Ceanothus, *C. americanus* var. *pitcheri* Torr. & Gray, has leaves broadly oblong-ovate or oval, apices rounded or obtuse, upper surface pilose, lower surface soft-velvety pubescent, branches densely villose. It occurs from Texas, Oklahoma, and Louisiana; eastward to Florida and northward to Indiana, Illinois, Iowa, and Kansas.

Georgia Ceanothus, *C. americanus* var. *intermedius* (L.) Trel., is a variety described in full by the author in this manual following this description.

MEDICINAL USES. The official medicinal uses are given by Wood and Osol (1943, p. 1308):

"The root is astringent and imparts a red color to water. Ceanothus contains *ceanothic acid* and also three alkaloids, one named *ceanothine,* the others unnamed (see Clark, *Amer. J. Pharm.,* 100:240, 1928, and Julian, *J. Amer. Chem. Soc.,* 60:77, 1938). Bertho and Won Sang Liang *(Archiv. der Pharmacie,* 271:273, 1933)

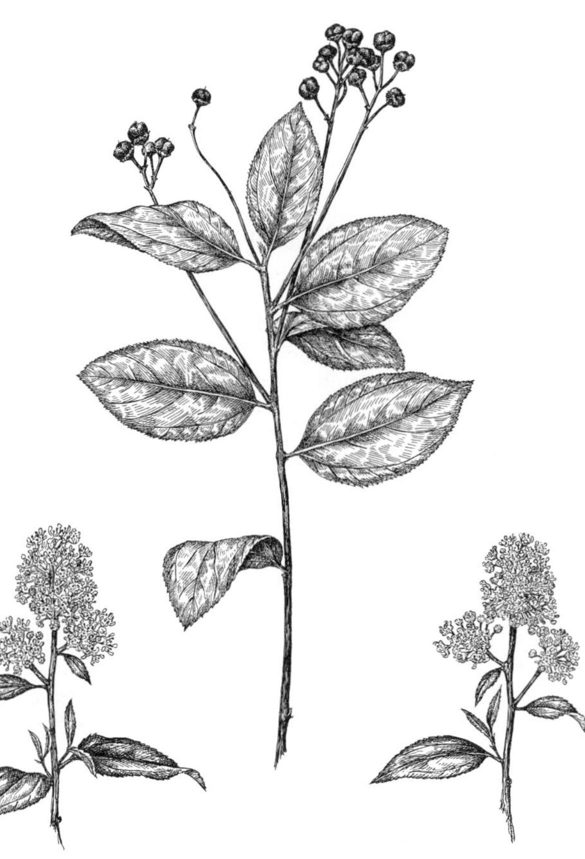

ERSEY-TEA CEANOTHUS
Ceanothus americanus L.

olated 1 per cent of ceanothine from *C. americanus*, nd found the purified alkaloid to have the formula $C_{29}H_{36}N_4O_4$, and to have a melting point of 227°–28°C.

"Ceanothus was at one time used in syphilis but here is no reason to believe it of value. Within more ecent years a proprietary tincture of Ceanothus has een used to considerable extent for the purpose of ncreasing the coagulability of the blood; especially or the prevention of hemorrhage from surgical operations. Goot *(J. Pharmacol. Exper. Therapeutics*, 30:275, 928) finds that the mixed alkaloids cause a fall of he blood pressure and a definite decrease in the coagulation time of the blood, and a number of clinicians ave reported more or less favorable results on the action of this drug. But in an elaborate study, undertaken or the Council of Pharmacy and Chemistry, Tewksury and Connery (*J. Amer. Med. Ass.*, 20:1287, 1931) vere unable to note any material change in the coagulaion time and they criticize the methods used by some f the earlier investigators. *Ceanothyn* is a proprietary incture of Ceanothus which is recommended in *doses* f 4 fluidrachms (16 cc.) at intervals of 30 minutes, nore or less, as may be required."

REMARKS. The genus name, *Ceanothus*, is from an obscure name given by Theophrastus, probably not for this genus. The species name, *americanus*, refers to its growth in North America. It is also known under the names of Wild Snowball, Spangles, Walpole-tea, Wild-pepper, Mountain-sweet, Red-root, and Mountain-tea Bohea. The leaves were considered to be the best substitute for tea during the American Revolution.

GEORGIA CEANOTHUS

Ceanothus americanus var. *intermedius* (L.) Trel.

FIELD IDENTIFICATION. An erect shrub 1–3½ ft, usually much-branched and finely pubescent.

FLOWERS. In spring, the common peduncles elongate, panicles ⅜–1 in. long, densely flowered; sepals 5, triangular, acute, membranous, incurved, about 1/25 in. long, deciduous; petals 5, about 1/15 in. long, inserted under a disk, white, blade long-clawed, strongly hooded; stamens 5, exserted, filaments filiform, anthers emarginate; ovary immersed in the disk, 3-celled; style short, united below, stigmas 3.

GEORGIA CEANOTHUS
Ceanothus americanus var. *intermedius* (L.) Trel.

FENDLER CEANOTHUS
Ceanothus fendleri Gray

FRUIT. Drupe ⅙–⅕ in. in diameter, 3-lobed, epicarp thin, separating into 3 crustaceous nutlets, testa smooth, endosperm fleshy.

LEAVES. Simple, alternate, deciduous, ovate to elliptic, ⅜–1¼ in. long, apex obtuse or acute, base rounded or broadly cuneate, margin finely serrate, rather thick; upper surface dull green and glabrous, or with scattered hairs mostly along the 3 impressed veins; lower surface paler with rather dense silky-appressed hairs, especially along the main veins, veinlets reticulate; petioles densely hairy, 1⁄16–⅛ in.; stipules minute and caducous.

TWIGS. Rather leafy, slender, brown, with appressed or spreading long hairs, when older light to dark brown or gray and glabrous.

RANGE. Pine woods areas of Louisiana; eastward to Florida, also Tennessee and Georgia.

REMARKS. The genus name, *Ceanothus,* is the ancient Greek name. The variety name, *intermedius,* refers to its close relationship to other species.

FENDLER CEANOTHUS

Ceanothus fendleri Gray

FIELD IDENTIFICATION. Thicket-forming, spinescent shrub to 3 ft. Usually erect but sometimes procumbent at high altitudes.

FLOWERS. June–August, in simple, terminal racemes ¾–1¼ in. long. Usually at the end of lateral leafy branchlets; calyx-lobes 5, petaloid, white, disk filling the hypanthium; petals 5, hooded and clawed; stamens 5, exserted; style elongate, mostly 3-lobed, ovary immersed in the disk.

FRUIT. Capsule August–September, obovoid, dry, 3-lobed, ⅛–3⁄16 in. in diameter, reddish brown; calyx cuplike, shallowly lobed, covering the lower third of fruit; peduncle ⅛–¼ in. long; seed germination not less than 16 per cent.

LEAVES. Simple, alternate, narrowly to broadly elliptic (rarely, suborbicular), entire or nearly so; apex acute or obtuse, base cuneate, ⅓–1 in. long, 3⁄16–¼ in. wide, grayish green, thickish, pubescent above, white-tomentose beneath, veins prominent, stipules thin and caducous, petiole short.

TWIGS. Densely tomentose at first, more glabrate later, gray to black or reddish, rigid, terete; spines straight, sharp, slender, ⅓–1 in. long.

RANGE. At altitudes of 3,000–10,000 ft. In the Trans-Pecos Texas area to New Mexico and Arizona; north to Wyoming and South Dakota.

VARIETY. *C. fendleri* Gray var. *venosus* Trel. has thicker, prominently veined leaves which are sericeous above. However, it is not very distinct, and intergrading forms are numerous.

REMARKS. The genus name, *Ceanothus,* is an ancient Greek name. The species name, *fendleri,* honors August Fendler (1813–1883), an early German botanical explorer of New Mexico.

This plant is sometimes known under the name of Buck-brush, and is an important browse for deer and cattle over its range. It is also eaten by porcupine and nibbled by jack rabbits.

Fendler Ceanothus has been cultivated since 1893 and is considered the most hardy of the Western species. It is used most in rock gardens and sandbank plantings.

DESERT CEANOTHUS

Ceanothus greggii Gray

FIELD IDENTIFICATION. Half-evergreen, leafy shrub with a rather dense, rounded form and short, rigid, divaricate branches.

FLOWERS. March–June, axillary or terminal; cymes often barely surpassing the leaves; borne on puberulent, glandular or glandless pedicels mostly less than ¼ in long; individual flowers small, perfect, 5-merous; sepals 5, petaloid, white; petals 5, white or bluish, apex hooded and base long-clawed; stamens 5, exserted filaments slender, disk annular; pistil exserted, stigma 2–3-cleft; ovary partly adnate to, and immersed in, the disk.

FRUIT. In July, drupe about ⅛ in. across, green to reddish brown, subglobose to somewhat 3-lobed with 3 warty protuberances above the middle, the petaloid sepals and elongate pistil persistent, separating when dry into 3 1-seeded nutlets.

LEAVES. Opposite, or clustered at intervals along the stem, blades hardly over ¼–⅗ in. long, about ⅛ in. wide, rather thick, elliptic to somewhat obovate; apex rounded to obtuse or retuse; base rounded or cuneate, margin entire, or rarely with a few small teeth; upper surface dark green, minutely hairy and granular; lower surface with scattered hairs and patches of fine white tomentum, granular; petiole 1/25–1/16 in. long, pubescent; stipules reddish brown, white-hairy, thick, ovate to lanceolate.

TWIGS. Gray, or almost white with tomentum, rigid, short, divaricate.

RANGE. On gravelly slopes, or in rocky canyons at altitudes of 2,000–5,600 ft. Trans-Pecos Texas and New Mexico; west to southern California and north to Utah and Colorado. In Mexico in the states of Chihuahua, Coahuila, San Luis Potosí, Hidalgo, Puebla, and Oaxaca.

CUP-LEAF DESERT CEANOTHUS
Ceanothus greggii var. *perplexans* (Trel.) Jeps.

DESERT CEANOTHUS
Ceanothus greggii Gray

VARIETIES. Cup-leaf Desert Ceanothus, *C. greggii* var. *perplexans* (Trel.) Jeps., is a variety with yellowish green, more serrate leaves, which are broadly elliptic to obovate and about ¾ in. long. Some botanists consider it distinct enough to carry the species name of *C. perplexans* Trel. Known from west Texas, New Mexico, Arizona, Utah, Nevada, and California.

Another variety with short, broad leaves has been described as *C. greggii* var. *orbicularis* Kelso. Known from California.

A third variety listed as Mohave Desert Ceanothus, *C. greggii* var. *vestitus* Greene, is thought to be a distinct California species rather than a variety. Many intergrading specimens are found.

REMARKS. *Ceanothus* is an ancient Greek name, and the species name, *greggii*, is in honor of the botanist Gregg. Also known by the name of Gregg Ceanothus, New Jersey-tea, Indian-tea, Red-root, and Mountain-lilac. However, some of the names should more properly apply to the related species *C. herbacea* Raf., which was used as a substitute for tea in Civil War days, and to *C. americanus* L. which was similarly used. Red dyes are obtained from the roots of several species. In Mexico, the flowers of *Ceanothus*, when rubbed in water, give a cleansing lather for washing clothing. Also the plants have been used in domestic medicine as a treatment for syphilis and are said to possess purgative properties. Most species can be propagated by seeds sown in spring, softwood cuttings, or from mature cuttings in autumn grown in a greenhouse. The species grows best in a light, well-drained soil. Desert Ceanothus has possibilities for use as a hedge plant in the Western areas.

Inland Ceanothus

Ceanothus herbacea Raf.

FIELD IDENTIFICATION. An upright shrub attaining a height of 3 ft, with slender, upright, puberulent branches.

FLOWERS. April–July, panicles short, some corymbose, terminating leafy shoots of the season; peduncles ¼–2 in. long; flowers perfect, regular, perigynous; petals 5, spreading, hooded, on slender claws longer than the calyx; stamens 5, exserted; ovary solitary, immersed in the annular disk, style 3-lobed; calyx with 5 short, incurved lobes.

FRUIT. Maturing September–November, capsules ⅙–⅕ in. long, dry, 3-lobed, coriaceous, eventually loculicidally dehiscent into 3 carpels.

LEAVES. Simple, alternate, deciduous, blades ¾–2½ in. long, ⅜–1 in. broad, narrowly elliptic or elliptic-lanceolate, margin crenate-serrate with glandular-tipped teeth, apex acute or obtuse, base cuneate to rounded, 3-nerved at base, upper surface dull green and semi-lustrous, lower surface paler and glabrous or somewhat pubescent along the veins; petioles ⅛–½ in., puberulent to glabrous later.

TWIGS. Slender, young ones green to brown or straw-colored, somewhat pubescent; older ones brown to dark gray or almost black and glabrous.

RANGE. Usually in sandy or rocky calcareous areas. Central Texas, Louisiana, Oklahoma, and Arkansas; east to Florida, north to Quebec, and west to Illinois, Minnesota, and South Dakota.

VARIETY. A variety has also been named *C. herbacea* var. *pubescens* (Torr. & Gray) Shinners and has leaves lightly to densely pubescent, smaller and narrower, with veins more prominent. It is rather rare in Texas, coming south to Kerr County, but more common in Oklahoma, Arkansas, Colorado, Kansas, and Nebraska; north and east to Ontario, overlapping with the species in some areas.

The species and variety afford some good browse for livestock in certain areas. The seeds are eaten by bobwhite quail and other birds. Various species of *Ceanothus* are propagated for gardens. The seeds are sown in spring, or cuttings of mature wood are taken in autumn and started in a cold frame or greenhouse. They may also be increased by layers.

REMARKS. The genus name, *Ceanothus,* is an ancient Greek name of obscure meaning. The species name, *herbacea,* refers to the leafy branchlets. Other vernacular names used include Jersey-tea and Red-root. Some authors list this plant under the name of *C. ovatus* Desf. For a full discussion of this name change, see Shinners (1951a).

Inland Ceanothus
Ceanothus herbacea Raf.

Mogollon Ceanothus

Ceanothus integerrimus Hook. & Arn.

FIELD IDENTIFICATION. Much-branched Western shrub attaining a height of 2–12 ft, rather handsome with small white flowers in panicles.

FLOWERS. May–July, panicles 2–6 in. long, 1–4 in. wide, terminal, many-flowered, leafless or nearly so; flowers very small, perfect; petals white (or pink or bluish in some variations) hooded, long-clawed; stamens 5, exserted; ovary immersed in the disk, 3-lobed, 3-celled, style 3-cleft; sepals 5, calyx adnate by disk to the lower part of ovary.

FRUIT. Maturing June–August, capsule globose to triangular, 3-celled, ¼ in. wide or less, apex slightly depressed, viscid, with 3 very small lateral crests, the elastic cell walls forcibly discharging the seed, separating into 3 sections.

LEAVES. Dense, simple, alternate, rather persistent,

Mogollon Ceanothus
Ceanothus integerrimus Hook. & Arn.

outline broadly elliptic to ovate to nearly oblong, apex acute to obtuse, base rounded, margin entire (in some variations minutely toothed near the apex), blades 1–3 in. long, width ⅜–1½ in., 3-veined from the base; upper surface light green to puberulent or glabrous; lower surface somewhat paler and sparsely pilose on the veins; stipules early-deciduous; petiole about ½ in. long or less.

TWIGS. Slender, green to brown or yellowish, obscurely puberulent to glabrous.

RANGE. On dry sites in the sun at altitudes of 2,000–6,700 ft. New Mexico in the Mogollon Mountains; west through Arizona to California and north to Oregon.

PROPAGATION. The plant withstands grazing well, is rapid-growing, and sprouts after cutting or fires. The seed averages 65,000 seeds per lb. They have both an embryo and seed coat dormancy, and can be stratified in moist sand for spring planting; 85 per cent germination of seed can be obtained by heating to 176°F. in water, cooling and stratifying 3½ months at 36°. About 68 per cent germination can be obtained by boiling for one minute and stratifying for 3 months. Propagation is also practiced with cuttings of mature wood under glass in autumn, or softwood cuttings in early spring. Layering is also practiced.

VARIETIES. A number of segregates have been proposed because of the variability of the plant, but many of these intergrade. The most distinct forms are:

C. integerrimus var. *peduncularis* Jepson, with simple narrow inflorescences 3–5 in. long. Known in northern California and adjacent Oregon.

C. integerrimus var. *puberulus* Abrams, with inflorescence broad and compound; leaves pubescent to puberulent above. In California.

C. integerrimus var. *californicus* (Kell.) Benson, with leaves glabrous above, inflorescence compound. Known from California.

REMARKS. The genus name, *Ceanothus*, is the ancient Greek name, and the species name, *integerrimus*, is for the very entire leaves. It is listed in some literature under the name of *C. mogollonicus* Greene and *C. myrianthus* Greene. Vernacular names applied in some localities are Deer-brush, Sweet-birch, Chaquira, and Chaquirilla.

The plant is very valuable browse for livestock and mule deer. The seeds are eaten by a number of species of birds and the bark is nibbled by porcupine. Bees make a good honey from the flowers. The Southwestern Indians are reported to use the bark in making a tonic and the flowers for a lather in water.

Texas Colubrina

Colubrina texensis (Torr. & Gray) Gray

FIELD IDENTIFICATION. Thicket-forming shrub rarely over 15 ft, with light gray divaricate twigs.

FLOWERS. April–May, borne in axillary, subsessile clusters; perfect, tomentose, greenish yellow, less than ⅓ in. across; petals 5, hooded and clawed; calyx 5-lobed, persistent on the fruit, lobes triangular-ovate; stamens 5, inserted below the disk, opposite the petals, filaments filiform; ovary 3-celled, immersed in the disk, styles 3, stigma obtuse.

FRUIT. Pedicels recurved, ¼–⅓ in. long, drupes borne at the twig nodes, ovate to subglobose, tomentose at first, later glabrous, dry, crustaceous, brown or black, about ⅓ in. in diameter, style persistent to form a beak, separating into 2–3 nutlets; seeds one in each partition, about 3/16 in. long, rounded on the back, angled on the other 2 surfaces, dark brown, shiny, smooth.

LEAVES. Simple, alternate or clustered, grayish green, blades ½–1 in. long, ovate, obovate or elliptic, densely hairy at first and glabrous later; 3-nerved, margin denticulate and ciliate; apex rounded, sometimes apiculate; base cuneate, rounded, truncate or subcordate; petioles ⅛–⅜ in. or less, hairy, reddish.

TWIGS. Slender, ashy-gray, noticeably divergent, scarcely spiniferous, densely white-tomentose at first but glabrous later.

BARK. Gray, smooth, close, cracked into small, short scales later.

TEXAS COLUBRINA
Colubrina texensis (Torr. & Gray) Gray

RANGE. Central, western, and southwestern Texas and New Mexico; Mexico in the states of Nuevo León, Coahuila, and Tamaulipas.

REMARKS. The genus name, *Colubrina*, is from *coluber* ("a serpent"), perhaps for the twisting, divaricate branches, or for the sinuate grooves on the stems of some species. The species name, *texensis*, refers to the state of Texas. A vernacular name is Hog-plum. The dark brown or black drupes are persistent.

LOTEBUSH CONDALIA

Condalia obtusifolia (Hook.) Weberb.

FIELD IDENTIFICATION. Stiff, spiny, much-branched shrub with grayish green grooved twigs.

FLOWERS. Inconspicuous, in clustered umbels; pedicels $\frac{1}{25}$–$\frac{1}{12}$ in. long, pubescent or villous; corolla small, green, 5-parted; petals 5, hooded and clawed, shorter than the sepals; sepals 5, triangular, acute, soft-hairy; stamens 5, inserted on the edge of the disk, opposite the petals; filaments subulate with anther sacs opening lengthwise; style partly 2-cleft with a basal 2–3-celled ovary immersed in a flattened, obscurely 5-lobed disk.

FRUIT. In June, drupe globular, $\frac{1}{3}$–$\frac{2}{5}$ in. in diameter, black, fleshy, not palatable; stone solitary, hard, with a thin, membranous testa.

LEAVES. Alternate, green, thin, firm, glabrous or puberulent, blades $\frac{1}{2}$–$1\frac{1}{4}$ in. long, elliptic, ovate, or narrowly oblong; apex obtuse, acute, retuse or occasionally emarginate; margin entire to coarsely serrate; base narrowed and 3-veined; petioles one third to one fourth as long as the blade.

TWIGS. Divaricate, grayish green, grooved, glaucous; spines stout, straight or nearly so, green or brown, to 3 in. long.

BARK. Smooth, light or dark gray.

RANGE. Widespread in central, southern, and western Texas. At altitudes of 5,000 ft in the Chisos Mountains in Brewster County to almost sea level at the mouth of the Rio Grande in Cameron County. On the central Texas limestone plateau. Also in Arizona, New Mexico, and northern Mexico.

SYNONYMS. The question has been posed by Cory (1947, pp. 130–131) as to whether *C. obtusifolia*

LOTEBUSH CONDALIA
Condalia obtusifolia (Hook.) Weberb.

(Hook.) Weberb. and *C. lycioides* (Gray) Weberb. are really separate species. He says that they are the same 'in young growth or in new growth of old plants the foliage is typically that of *C. obtusifolia,* and very frequently the young growth contrasts markedly with the more abundant older growth of the same plant which has foliage typically that of *C. lycioides.* A plant with the linear, oblong or elliptic leaves characteristic of *C. lycioides* will, upon being cut off at the surface of the ground, send up vigorous sprouts with orbicular leaves as much as ¾ in. in diameter. At and toward the eastern limit of its range in Texas the plant more commonly has foliage typical of *C. obtusifolia,* while a few hundred miles farther west the more common type is that of *C. lycioides.* Complete intergradation of the two extremes, however, is evidence of their conspecifity. Specimens verifying this conclusion have been deposited at the Arnold Arboretum, Harvard University. The common species occurring from central Texas westward to southern California and south into northern Mexico should be known, therefore, as *Condalia obtusifolia* (Hook.) Weberb."

The present author feels, in view of Cory's conclusion, that one of the forms might perhaps be segregated as a variety of the other, but a more extensive study should be made over a much wider area before any conclusion of a synonymous status is applied. Meanwhile, the two are maintained as separate species.

REMARKS. The genus name, *Condalia,* is in honor of Antonio Condal, a Spanish physician, and the species name, *obtusifolia,* refers to the obtuse apex of the leaf. Vernacular names are Texas Buckthorn, Lote-bush, Chaparral, Chaparro Prieto, and Abrojo. The mealy drupe is edible, but not tasty. It is eaten by gray fox, raccoon, and various birds. The roots are used as a soap substitute, and as a treatment for wounds and sores of domestic animals.

SOUTHWESTERN CONDALIA
Condalia lycioides (Gray) Weberb.

SOUTHWESTERN CONDALIA

Condalia lycioides (Gray) Weberb.

FIELD IDENTIFICATION. Rigid, spinose shrub to 9 ft, with divaricate, grayish green branchlets.

FLOWERS. Solitary, or in axillary umbellate clusters, on pedicels about ⅛ in. long, perianth about ⅛ in. wide, inconspicuous, perfect; calyx of 5 triangular sepals about 1/12 in. long; petals 5, white, smaller than sepals and alternate with them, linear to spatulate, about 1/25 in. long; stamens 5, opposite the petals on the disk margin; pistil of 2 or 3 united carpels, stigma 2- or 3-lobed.

FRUIT. Drupe dark blue, glaucous, globose or slightly oblong, ¼–⅓ in. long, pulp thin, edible; seed ovoid, acute at one end and rounded at the other, somewhat rugose, about 3/16 in. long.

LEAVES. Simple, alternate, short-petioled, coriaceous, ovate or oblong-elliptic, obtuse or rounded or emarginate at the apex, cuneate or rounded or subcordate at the base, mostly entire, but occasionally serrate, glabrate, or pubescent in *C. lycioides* var. *canescens* (Gray) Trel.; pale green, ⅕–¾ in. long, about ⅕–¼ in. wide, often prematurely deciduous.

TWIGS. Divaricate, grayish green, glaucous, pubescent, striate, ending in stout, rigid, straight or decurved spines.

RANGE. On dry hills and desert flats, ascending to an altitude of 5,000 ft; extreme west Texas, through New Mexico and southern Arizona to southern California; southward into Mexico in the states of Chihuahua, Nuevo León, San Luis Potosí, and Zacatecas.

SYNONYMS. The reader is referred to the description of *C. obtusifolia* for a discussion of relationships between it and *C. lycioides.*

REMARKS. The genus name, *Condalia,* is in honor of Antonio Condal, a Spanish physician, and the species name, *lycioides,* refers to the lycium-like foliage. Various vernacular names used in Mexico and the southwestern United States are Lote-bush, Crucillo, Garrapata, Barchatas, and Garambullo. The dark blue fruit is edible, and the bark of the root is used in Mexico as a soap substitute. It is also reported that a medicine made from the root was used as an eye treatment by the Indians. The fruit is eaten by various species of birds, particularly the Gambel's and scaled quail.

Knife-leaf Condalia
Condalia spathulata Gray

FIELD IDENTIFICATION. Western shrub with very spiny, grayish green, divaricate branches, attaining a height of 10 ft.

FLOWERS. Solitary or a few together in the leaf axils, short-peduncled, perfect, greenish, hardly over 1/8 in. broad; petals absent; sepals 5, persistent, spreading, green, triangular, acute; stamens 5, incurved, shorter than the sepals and alternating with them, attached to the edge of the 5-angled disk; ovary superior; pistil solitary, stout, stigma 2–3-lobed.

FRUIT. On short pedicels, drupe black, subglobose or obovoid, apiculate, to 3/8 in. in diameter, fleshy, juicy, thin-skinned, stains purple, edible; seed solitary, about 3/16 in. long, ovoid, truncate and paler at the hilum end, obtuse at the other end, somewhat rugose.

LEAVES. Persistent, alternate or fascicled, 1/5–1/2 in. long, 1/25–1/12 in. wide, spatulate to obovate, cuneate, entire, apex varying from rounded to obtuse or acute and mucronulate, glabrous or pubescent, very short-petioled.

Knife-leaf Condalia
Condalia spathulata Gray

TWIGS. Stiff, divaricate, ending in gray or reddish brown spines, older parts grayish green to reddish, young parts puberulent, but glabrous later.

BARK. Dark gray to black, smooth, or with small thin scales.

RANGE. Knife-leaf Condalia is found on dry hillsides and along stony arroyos from the Edwards Plateau into southwest and west Texas. It is a common species in the Chisos Mountains of Brewster County, ascending to altitudes of 5,500 ft. It is found in New Mexico and Arizona; west into southern California. Southward it enters the Mexican states of Sonora, Zacatecas, Tamaulipas, and Coahuila.

RELATED SPECIES. Another closely related plant is Mexican Condalia, *C. mexicana* Schlecht. It occurs in west and southwest Texas, Arizona, and New Mexico; south into the Mexican states of Chihuahua, Coahuila, San Luis Potosí, Querétaro, Hidalgo, and Puebla. It appears to vary from Knife-leaf Condalia in minor characters such as inconspicuous leaf venation, subsessile fruit pedicels, and stouter branchlets.

REMARKS. The genus name, *Condalia*, is for Antonio Condal, Spanish physician of the eighteenth century, and the species name, *spathulata*, refers to the spatulate leaves. It is known under the vernacular names of Mexican Crucillo, Squaw-bush, Chamis, Abrojo, and Tecomblate. The fruit is eaten by the Gambel quail, and the young shoots are occasionally browsed by stock and game.

The leaves and fruit of Knife-leaf Condalia are smaller than those of the Bluewood Condalia, but the two species resemble one another in many other respects. It seems to replace Bluewood Condalia in west Texas, but the ranges overlap in south central Texas.

Bluewood Condalia
Condalia obovata Hook.

FIELD IDENTIFICATION. Thicket-forming spinescent shrub or tree to 30 ft, with a diameter of 8 in. The branches are rigid and divaricate.

FLOWERS. Solitary, or 2–4 in axillary clusters, sessile, or on short pedicels about 1/16 in. long; calyx 1/16–1/8 in. broad, glabrous or nearly so; sepals 5, green, spreading, triangular, acute, persistent; disk fleshy, flat, somewhat 5-angled; petals absent; stamens 5, shorter than the sepals, incurved, inserted on the disk margin; ovary superior, 1-celled, or sometimes imperfectly 2–3-celled, styles stout and short, stigma 3-lobed.

FRUIT. Ripening at intervals during the summer, drupe black at maturity, shiny, smooth, subglobose, somewhat flattened at apex, 1/4–1/3 in. in diameter, thin-skinned, fleshy, sweet, juice purple; seed solitary, ovoid to globose, flattened, acute at one end and truncate at the other, light brown, crustaceous, about 1/8 in. long.

Bluewood Condalia
Condalia obovata Hook.

LEAVES. Alternate, or fascicled on short, spinescent branches, obovate to broadly spatulate, margin entire, apex rounded to retuse or truncate and mucronate, base attenuate or cuneate, leathery, ⅓–1½ in. long, ⅓–½ in. wide; light green and lustrous, pubescent at first, glabrous later, paler beneath, midrib prominent, subsessile or short-petioled.

TWIGS. Divaricate, ending in slender, reddish or gray thorns, green to brown or gray, finely velvety-pubescent at first but glabrous later.

BARK. Smooth, pale gray to brown or reddish on branches and young trunks. Old trunks with narrow, flat ridges, deep furrows, and breaking into small, thin scales.

WOOD. Light red, sapwood yellow, close-grained, heavy, hard, dense, specific gravity 1.20.

RANGE. Dry soil, central, southern, and western Texas. On the Texas coast from Matagorda County to Cameron County, frequent along the lower Rio Grande River. In central Texas in greatest abundance on the limestone plateau area and west to the Pecos River. Less common west of the Pecos and north into the Panhandle area of Texas. Also in Mexico in Nuevo León and Tamaulipas.

VARIETY. A variety has been described as the Edwards Bluewood Condalia, *C. obovata* var. *edwardsiana* V. L. Cory (Cory, 1947, pp. 128–129). It differs from the typical species in its longer and narrower light colored leaves, which are spatulate instead of obovate. At the type locality the tallest shrubs were 3.1 m. high (about 9 ft), and 2.5 m. (about 7½ ft) was the average. The type locality was 29 airline miles northwest of Rocksprings, Edwards County, Texas, at an altitude of approximately 2400 ft.

REMARKS. The genus name, *Condalia,* is in honor of Antonio Condal, Spanish physician of the eighteenth century, and the species name, *obovata,* refers to the obovate leaves. Other vernacular names are Brazil, Logwood, Bluewood, Purple Haw, Capulín, Capul Negro, and Chaparral. The wood yields a blue dye and is sometimes used as fuel. The flower pollen serves as bee food, and the black fruit makes good jelly, but it is difficult to gather because of the thorns. Birds eagerly devour the fruit. The bushes often form impenetrable thickets.

Mexican Bluewood Condalia

Condalia mexicana Schlecht.

FIELD IDENTIFICATION. Large, spinose shrub with divaricate, rigid, stout branchlets.

FLOWERS. Axillary, umbels sessile or nearly so, perfect, greenish; petals absent; calyx 5-lobed, enclosing the disks; stamens 5; style 2- or 3-lobed; ovules solitary in each carpel.

Mexican Bluewood Condalia
Condalia mexicana Schlecht.

FRUIT. Drupe sessile or nearly so, black, ellipsoid or ovoid; 1/4 in. long or less, short-beaked; stone solitary.

LEAVES. Short-petiolate, spatulate, obovate, or oblanceolate, obtuse or rounded at the apex, cuneate at the base, entire or denticulate, glabrate or pubescent, 1/5–1/2 in. long, 1/6–1/4 in. wide, lateral veins inconspicuous.

RANGE. Well-drained stony hillsides in extreme west Texas, New Mexico, and Arizona. In Mexico in the states of Chihuahua, Coahuila, San Luis Potosí, Querétaro, Hidalgo, and Puebla.

REMARKS. The genus name, *Condalia,* is for Antonio Condal, Spanish physician. Another vernacular name is Bindo. This shrub is closely related to *C. spathulata* and *C. obovata.* It differs in minor characters such as stouter branches, inconspicuous leaf venation, and subsessile leaf petioles. It should probably be relegated to a varietal standing instead of that of a species.

REED'S GREEN CONDALIA
Condalia viridis Johnston var. *reedii* Cory

GREEN CONDALIA

Condalia viridis I. M. Johnston

FIELD IDENTIFICATION. Shrub 3–9 ft, with spreading spinescent branches.

FLOWERS. Axillary, solitary or paired, glabrous, 5-merous; hypanthium disklike, glabrous, about 1/15 in. in diameter; 5 lobes triangular, about 1/15 in. long, half-crested within and above; petals absent; stamens 1/25–1/15 in. long, affixed within and below the calyx bowl; ovary glabrous, stigma rather obscurely 2-lobed.

GREEN CONDALIA
Condalia viridis I. M. Johnston

FRUIT. Drupe black, globose, about 1/5 in. long; seed ellipsoid, slightly less than 1/5 in. long.

LEAVES. Bright green, oblanceolate to oblong-obovate blades 1/6–3/4 in. long, 1/12–1/4 in. wide; somewhat wider above the middle, gradually attenuated toward the base into a petiole 1/25–1/12 in., apex obtuse to rounded and mucronulate, midrib slender, lateral veins 2–3 and easily seen; surfaces bright green, glabrous; young leaves sparsely pubescent to glabrous later; stipules about 1/25 in. long, triangular, persistent, margin ciliate.

TWIGS. Young ones minutely hispid, at maturity glabrous, bark grayish.

RANGE. Green Condalia is found along gravelly washes or dry stream beds. In Texas in Val Verde, Hudspeth, and Brewster counties. In the latter county, between Burro Mesa and the Chisos Mountains. In Mexico in Chihuahua.

VARIETY. Reed's Green Condalia, *C. viridis* Johnston var. *reedii* Cory, has been described as differing from the species in being (on the average) twice as tall and having twice its spread, and with leaves about twice as large. The type locality is along the Frio River, Garner State Park, Uvalde County, Texas, at altitudes of 1,500–1,600 ft.

REMARKS. The genus name, *Condalia,* is in honor of Antonio Condal, Spanish physician of the eighteenth century. The species name, *viridis,* refers to the bright green leaves.

Ivan Johnston makes the following remarks (Johnston, 1939, p. 235):

"Though some of the Texas specimens have been accepted as forms of *C. obovata* Hook., this plant is evidently most closely related to *C. mexicana* Schlecht. From the latter species it differs in its glabrous ovary, calyx and mature leaves, and in the green rather than reddish brown upper face of the leaves. It occurs in a region north of the known range of *C. mexicana.* From *C. obovata,* a species of central and southern Texas and adjacent Mexico, *C. viridis* differs in its smaller bushy statue, its much smaller leaves, green on both surfaces, and its more westerly occurrence."

Common Jujube
Ziziphus jujuba Lam.

FIELD IDENTIFICATION. Tree to 50 ft, with a diameter of 10 in. The short trunk supports slender ascending branches formed into a rounded head.

FLOWERS. Borne March–May in the axils of the leaves, solitary or a few together on glabrous pedicels 1/16–3/16 in. long; flowers perfect, 1/16–1/8 in. across, yellowish green; calyx campanulate, spreading, 5-lobed, lobes ovate-triangular, acute, keeled within; petals 5, hooded, clawed, 1/25–1/12 in. long, much smaller than the sepals and alternating with them; stamens 5, as long as the petals or shorter and opposite to them; ovary 2–4-loculed, style 2-parted.

FRUIT. Ripe July–November, slender-pediceled, drupe very variable in size and shape, subglobose to oblong, 1/2–1 in. long, green at first, turning yellowish, reddish brown or black at maturity, pulp yellow, sweet, shriveling later, acidulous; 1–3-celled; seeds usually 2, deeply furrowed, oblong, reddish brown to gray, apices pointed, about 3/4 in. long.

LEAVES. Alternate on short thickened spurlike twigs, often somewhat fascicled, ovate to oblong or lanceolate; apex obtuse; base rounded or broadly cuneate, sometimes inequilateral; margin shallowly toothed, some teeth minutely mucronulate; distinctly 3-nerved at base; upper surface dark waxy green and glabrous; lower surface paler and the 3 nerves more conspicuous, glabrous to pubescent, stipules spinescent.

TWIGS. Stout, green to gray or black, some nodes thickened, lateral branchlets thickened and leaves often fascicled on them. Spinelike stipules 1/8–3/4 in. long, straight or curved.

BARK. Mottled gray or black, smooth on younger branches, on older branches and trunks roughly furrowed and peeling in loose shaggy strips.

RANGE. Grows on most soils, except very heavy clays or swampy ground. Considered to be a native of Syria. Widely distributed in the warmer parts of Europe, south Asia, Africa, and Australia. Cultivated in North America in Florida, California, and the Gulf Coast states. First introduced into America from Europe by Robert Chisholm in 1837 and planted in Beaufort, North Carolina.

Common Jujube
Ziziphus jujuba Lam.

MEDICINAL USES. An extract of the fruit is sold as a tonic under the name of Hsuan Tsao Ren. It is also used in domestic medicine in the form of pectoral pastes, tablets, and syrups. However, since the medicinal quality of the demulcent is due to the gum arabic and sugar, the Jujube constituent is considered of little importance. The drug commonly sold under the name often contains none of the fruit. (See Wood and Osol, 1943, p. 1584.)

PROPAGATION. Usually a good crop of fruit is borne each year. It is hand-picked, soaked in water, run through a macerator, and the residue floated off. The seeds are then dried for storage or prepared for planting. The seed averages about 25 lb of seed per 100 lb of fruit. The average number of seeds is about 750 per lb, with about 99 per cent purity and 65 per cent soundness. Storage is usually in sealed containers at 41°F. No data are available as to the viability under storage.

The dormant embryo and hard seed coat necessitate considerable pretreatment. Scarification for 6 hours and stratification in moist sand for 60–90 days at 41°F. seems to be the best method. The stratified seed may be sown in spring in drills about 1 in. deep in light soils. About one third of the seeds produce viable seedlings. One- or two-year-old plants can be planted out. It can also be propagated by offsets from well-established trees.

VARIETY. *Ziziphus jujuba* var. *inermis* is a thornless variety.

REMARKS. The genus name, *Ziziphus,* is from an ancient Greek name derived from the Persian *zizafun.* The species name, *jujuba,* is the French common name, derived from the Arabic. The tree is also known as the Chinese Date. It is popular with the Chinese and as many as 400 varieties have been cultivated by them. The fruit exhibits great variety in shape, size, and color, sometimes becoming as large as a hen egg. It is processed with sugar and honey and is sold in Chinese shops.

In India the wood is used for fuel and small timber, and the leaves for cattle fodder and as food for the Tasar silkworm and lac insect. In Europe it has long been used as a table dessert and dry winter sweetmeat. The Common Jujube has also been used in shelter-belt planting and for wildlife food.

HUMBOLDT COYOTILLO

Karwinskia humboldtiana (R. & S.) Zucc.

FIELD IDENTIFICATION. Shrub or small tree attaining a height of 24 ft.

FLOWERS. In axillary, sessile or short-pedunculate, few-flowered cymes ⅓–½ in. long, persistent, perfect; petals 5, hooded, clawed; calyx of 5 sepals, glabrous, about ⅛ in. broad, sepals triangular, acute, keeled within; stamens 5, inserted on the edge of a disk, longer than the petals, filaments subulate; styles united except at apex, stigmas obtuse, ovary immersed by the disk and 2–3-celled.

FRUIT. In October, peduncle one third to one half as long as the drupe, subglobose, brown to black at maturity, apiculate, ¼–⅜ in. long; stone solitary, ovoid, smooth, grooved on one side, about ¼ in. long, poisonous.

LEAVES. Opposite or nearly so, ovate, elliptic or oblong; apex acute, obtuse or mucronate; base cuneate or rounded, ¾–1¾ in. long, ½–¾ in. wide, margin entire or undulate, dark lustrous green above, firm, glabrous or puberulent, paler beneath, conspicuously pinnate-veined and marked with black spots and longitudinal marking on veins and young twigs; petioles slender, 1–3 in.

TWIGS. Gray to reddish brown, smooth, glabrous or puberulent; lenticels abundant, white, oval to oblong.

BARK. Gray, smooth, tight.

WOOD. Hard, strong, tough, but of no commercial value.

RANGE. Dry plains and prairies in the western and southwestern parts of Texas. Abundant near the mouth of the Pecos River and near the mouth of the Rio Grande in Cameron County. In Mexico in the states of Tamaulipas, Veracruz, Yucatán, Oaxaca, and Baja California, and south into Central America.

HUMBOLDT COYOTILLO
Karwinskia humboldtiana (R. & S.) Zucc.

REMARKS. The genus name, *Karwinskia,* is in honor of Wilhelm Friedrich von Karwinski, a Bavarian botanist, who collected plants in Mexico in 1826. The species name, *humboldtiana,* honors Alexander von Humboldt (1769–1859), a Prussian naturalist who explored South America and Mexico 1799–1804.

In Mexico this plant is known under many local names such as Tullidora, Capulincillo, Capulincillo Cimarrón, Capulín, Palo Negrito, Margarita, Cacachila, China, Cacohila Silvestre, Frutillo Negrito, Cochila, and Margarita del Cero. The oily seeds are poisonous, and when eaten cause paralysis in the limbs of human beings and domestic animals. It is reported that a decoction of the leaves and roots is less poisonous and is used locally in Mexico for fevers. The plant is easily propagated by root divisions.

Javelina-brush
Microrhamnus ericoides Gray

FIELD IDENTIFICATION. Spiny, densely branched evergreen shrub to 4½ ft. Usually low, sprawling, and irregular in shape.

FLOWERS. Small, solitary, pediceled, yellowish; petals 4–5, hooded and clawed; disk lining the 5-lobed calyx-tube; stamens 4–5, alternate with the lobes of the calyx.

FRUIT. Pedicel 1/12–1/8 in. long; drupe ovoid-ellipsoid, reddish brown to black, 1/5–1/3 in. long, apex abruptly pointed, often with persistent style remnants; stone solitary, 1–3-celled.

LEAVES. Alternate, fascicled, persistent, dark green, minute, 1/12–1/4 in. long, heathlike, linear-oblong, acute at apex, margins entire but strongly revolute, the enclosed grooves of lower surface short-tomentose; stipules broadly triangular, ciliate.

TWIGS. Stout, short, gray, often ending in straight slender spines. Leaves clustered on short spurs.

BARK. Gray, smooth above, somewhat roughened by small scales on old stems.

RANGE. Western and southwestern Texas, mostly west of the Pecos River, Brewster County at Terlingua, Guadalupe Mountains at Pine Springs, and Hueco Mountains near El Paso; also in Mexico.

Javelina-brush
Microrhamnus ericoides Gray

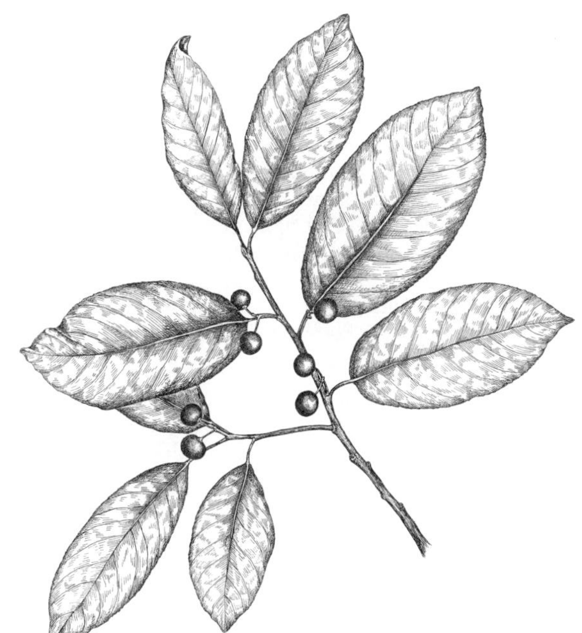
Carolina Buckthorn
Rhamnus caroliniana Walt.

REMARKS. The genus name, *Microrhamnus*, refers to the small size of the buckthorn-like plant, and the species name, *ericoides*, is for the tiny heathlike leaves. Other vernacular names are Little Buckthorn, Abrojo, and Tecomblate. This shrub is easily recognized by its small enrolled leaves. It might possibly be confused with *Castela*, but the latter genus has larger leaves and brilliant red, oval to flattened fruit.

Carolina Buckthorn
Rhamnus caroliniana Walt.

FIELD IDENTIFICATION. Shrub or small tree attaining a height of 35 ft, with a diameter of 8 in.

FLOWERS. May–June, borne solitary or 2–10 in peduncled umbels, peduncles to 2/5 in. long or often absent, pedicels 1/8–1/4 in.; flowers perfect, small, greenish yellow; petals 5, each about 1/25 in. long or broad, apex broad and notched, base acute, concave; stamens 5, included, anthers and filaments less than 1/25 in. long; style equaling the calyx-tube, stigma 3-lobed, ovary glabrous and 3-celled; calyx-tube campanulate, about 1/12 in. long, 1/8 in. wide at apex, the 5 sepals glabrous and triangular, apices acuminate.

FRUIT. Drupes August–October, persistent, sweet, spherical, 1/3–2/5 in. in diameter, red at first, at maturity black and lustrous, 3-seeded (occasionally 2–4-seeded); seeds 1/5–1/4 in. long, reddish brown, rounded almost equally at apex and base, rounded dorsally, inner side with a triangular ridge from the apex to notch at base.

LEAVES. Abundant, scattered along the branches, simple, alternate, deciduous, elliptic to broadly oblong, apex acute or acuminate, base cuneate to acute or rounded, sometimes inequilateral, margin indistinctly serrulate or subentire, rather thin, prominently parallel-veined; upper surface bright green, smooth and lustrous, pubescent to glabrous; lower surface velvety pubescent to only puberulent or glabrous, length of blade 2–6 in., width 1–2 in., turning yellow in the fall; petiole slender, ⅖–⅗ in., widened at base, glabrous or pubescent.

TWIGS. Slender, young ones green to reddish, later gray; pubescent at first, glabrous later; sometimes terminating in a cluster of very small folded leaves.

BARK. Gray to brown, sometimes blotched, smoothish, furrows shallow.

WOOD. Light brown, sapwood yellow, close-grained, fairly hard, rather weak, weighing 34 lb per cu ft.

RANGE. Most often in low grounds in eastern, central, and western Texas as far west as the Pecos River, Arkansas, Oklahoma, and Louisiana; eastward to Florida, northward to North Carolina, and west to Missouri.

REMARKS. The genus name, *Rhamnus*, is from an ancient Greek word. The species name, *caroliniana*, refers to the state of South Carolina where it grows. Other vernacular names are Yellow Buckthorn, Indian-cherry, Bog-birch, Alder-leaf Buckthorn, and Polecat-tree. The fruit is eaten by several species of birds, especially the catbird. The shrub appears susceptible to the crown rust of oats. The handsome leaves and fruit make the tree a good ornamental possibility. It is apparently adjustable to both moderately acid and alkaline soil types.

LANCE-LEAF BUCKTHORN
Rhamnus lanceolata Pursh

LANCE-LEAF BUCKTHORN

Rhamnus lanceolata Pursh

FIELD IDENTIFICATION. Small to large shrub to 9 ft, with an erect and widely branched habit.

FLOWERS. Appearing with the leaves April–May, solitary or in axillary clusters of 2–4 on pedicels 1/12–¼ in. long, dioecious, greenish, 1/12–⅛ in. broad; petals 4–5, suborbicular or obcordate, concave, apex deeply notched, about half as long as the sepals, smaller in the pistillate flowers, inserted on the margin of the disk; stamens 4–5, as long as the petals, filaments short, inserted on the edge of the disk, anthers obtuse; ovary 2–4-celled, glabrous, nearly free, styles united below, bifid above (styles usually short and included in the staminate flower, longer and exserted in the pistillate); calyx-tube campanulate, about ⅛ in. broad; sepals 4–5, triangular-ovate, acute, about as long as the tube.

FRUIT. Ripening June–September, drupe ⅕–⅓ in. thick, obovoid to globose, black, 2-seeded; seeds ⅕–¼ in. long, cartilaginous, deeply grooved dorsally, endosperm fleshy.

LEAVES. Appearing with the flowers, abundant, deciduous, simple, alternate, blades 1–3½ in. long, ½–1 in. wide, elliptic to lanceolate or oblong, rarely oval, apex acute to acuminate or obtuse on floral branches, base cuneate, obtuse or rounded, margin finely serrulate with incurved glandular teeth, upper surface glabrous or nearly so, lower surface lighter green and more or less pubescent, especially pubescent on the 6–9 pairs of parallel, but inconspicuous, veins; petiole ⅕–⅖ in. long, pubescent to glabrous.

TWIGS. Slender, reddish brown and pubescent at first, later gray, glabrous, and smooth.

RANGE. The species is generally found in moist calcareous soils or on banks or hillsides. Eastern Texas, Arkansas, and Alabama; north to Pennsylvania and west to South Dakota, Illinois, and Missouri.

VARIETY. Smooth Lance-leaf Buckthorn, *R. lanceolata* var. *glabrata* Gl., is a variety with leaves and twigs glabrous or nearly so. It occurs in Ohio, Kentucky, Nebraska, Kansas, and Oklahoma.

REMARKS. The genus name, *Rhamnus*, is the ancient Greek name, and the species name, *lanceolata*, refers to the lance-shaped leaves. The plant is of no great economic importance. The fruit is eaten by at least 5 species of birds. It is occasionally cultivated for ornament. The seeds are sown in the fall, or stratified for spring planting, and the plant is also propagated by cuttings or grafting. Lance-leaf Buckthorn is apparently hardy as far north as Massachusetts.

Smith Buckthorn

Rhamnus smithii Greene

FIELD IDENTIFICATION. An upright, rounded, densely leafy shrub 3–15 ft.

FLOWERS. Dioecious, small, greenish, maturing April–May, solitary or in 2–3-flowered umbels in the axils of the leaves; petals 4 (sometimes absent), about 1/25 in. long, borne on the margin of a disk, deeply notched; stamens 4, inserted on the disk margin; ovary almost free, sometimes abortive, 2–4-celled, style small and short, stigma 2-parted; calyx campanulate, 1/25–1/12 in. long with 4 sepals; pedicels 1/12–1/5 in. long.

Smith Buckthorn
Rhamnus smithii Greene

FRUIT. Maturing June–August, drupe 1/4–1/3 in. long, slightly longer than broad, black, subglobose, dry, dehiscent by a longitudinal slit; nutlets usually 2, apex rounded, base narrowed, surface finely reticulate, exterior with a groove about 1/25 in. wide.

LEAVES. Appearing with the flowers from mixed buds, deciduous, opposite, alternate or fascicled, usually less than 1¼ in. long (rarely to 2¾ in.), elliptic to oblong or ovate to lanceolate, apex obtuse to acute, base rounded to cuneate or obtuse, margin serrulate to crenulate with glandular teeth, firm and thickish, surfaces pubescent with straight hairs or later glabrous, lower surface paler and green to yellowish green; petioles pubescent, 1/8–2/5 in.

TWIGS. Slender, yellowish to gray or brown, younger ones pubescent, older ones smooth and glabrous, often lustrous; bud scales membranaceous, light brown, about ¼ in. long or less, margins fimbriate, apex obtuse, base truncate.

RANGE. Open hillsides, high mountain canyons, stream banks, at altitudes of 5,000–7,500 ft. In Trans-Pecos Texas in the Chisos and Davis mountains. In New Mexico in the Guadalupe, Sierra Blanca, and Sacramento mountains. Also in southern and western Colorado.

VARIETIES. The plant is listed by some authors under the synonymous name of *R. fasciculata* Greene. However, others accept this name only in a varietal category as the Clustered Smith Buckthorn, *R. smithii* var. *fasciculata* (Greene) C. B. Wolf. The following key separates the species and variety in some instances.

Blades 1¼–2¾ in. long, apex acute, surface glabrous, green beneath, margin crenate ...... *R. smithii.*

Blades 1–1⅞ in. long, apex obtuse, surfaces pubescent, yellowish brown beneath, margin serrate .... *R. smithii* var. *fasciculata.*

Since the above variety occurs in the same ranges as the species, and since intermediate forms connect the two, the segregation is sometimes dubious. *R. smithii* is also closely related to *R. lanceolata* Pursh, and further study may prove it to be a xerophytic variation of it. For further references on the *Rhamnus* of the Southwest the author recommends the monograph by Wolf (1938).

REMARKS. The genus name, *Rhamnus*, is the ancient Greek name. The species name, *smithii*, is for B. H. Smith, who collected it at Pagosa Springs, Colorado. Smith Buckthorn was first introduced into cultivation in 1905. It is susceptible to crown rust of oats.

Birch-leaf Buckthorn

Rhamnus betulaefolia Greene

FIELD IDENTIFICATION. Upright rounded shrub, or sometimes a small tree to 20 ft, often branched near the base.

FLOWERS. In June, borne after the leaves appear, in

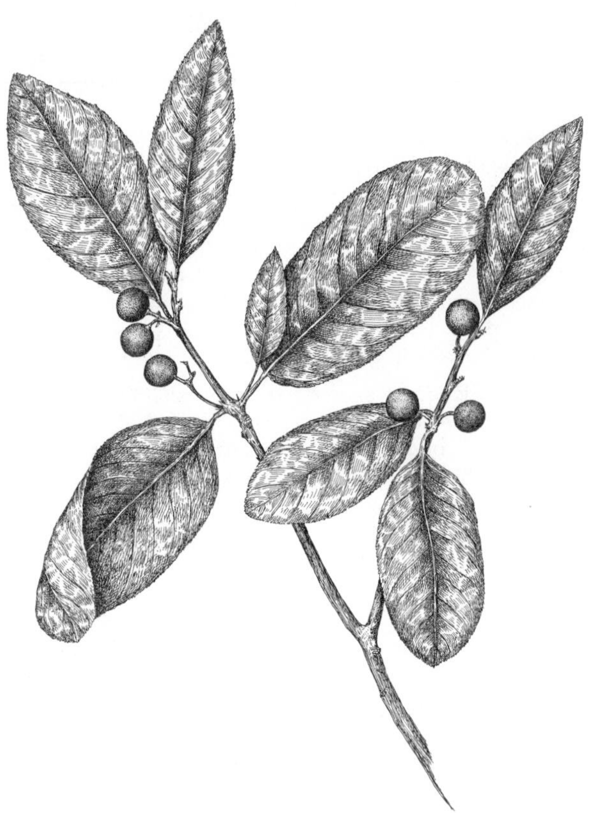

Birch-leaf Buckthorn
Rhamnus betulaefolia Greene

peduncled, pubescent, axillary umbels; peduncles ⅙–1⅕ in. long, pedicels to ⅖ in. long, both elongate in fruit; petals 4–5 (or sometimes absent), small, greenish, about 1/25 in. long, notched, base narrowed, inserted on the disk; stamens 4–5, inserted on the edge of the disk, filaments about 1/25 in. long; ovary essentially superior, 2–3-celled, style equaling the calyx-tube, stigma 2–3-lobed; calyx-tube funnelform, about ⅛ in. long, lobes 5, triangular.

FRUIT. Maturing July–October, drupe subglobose or somewhat depressed, about ⅓ in. in diameter, apex rounded, base narrowed, inner side with a triangular ridge, black to dark purple, glabrous and lustrous, flesh dry, 3-seeded; seeds dorsally rounded, plane on the 2 inner surfaces, peduncles ⅛–⅓ in. long, pubescent.

LEAVES. Simple, alternate, deciduous, elliptic to oblong, apex rounded to obtuse or acute, base rounded or broadly cuneate, margins serrulate to subcrenate, or rarely entire, blades 2–6 in. long, 1–1½ in. wide, rather thin, upper surface bright green, glabrous and lustrous, lower surface paler and pubescent or glabrate, veins pinnately 10–11 pairs; petioles slender, pubescent, ⅕–⅖ in. long.

TWIGS. Rather leafy, sometimes with embryonic leaves at the apex, young twigs pubescent and green to reddish, older ones and trunk gray to dark and glabrous.

RANGE. Usually in moist canyons at altitudes of 4,000–7,500 ft. In Trans-Pecos Texas, New Mexico, and Arizona; north to Utah and Nevada; southward to Mexico in Sonora, Chihuahua, Durango, Tamaulipas, and Nuevo León.

VARIETY. Obovate-leaf Buckthorn, *R. betulaefolia* var. *obovata* Kearney & Peebles, is a variety with obovate leaves and occurs in northern Arizona, Utah, and Nevada.

REMARKS. The genus name, *Rhamnus,* is the ancient Greek name, and the species name, *betulaefolia,* refers to the birchlike leaves. The young foliage is browsed by mule deer, and the seeds are eaten by a number of species of birds. The seed averages about 4,000 seeds per lb. They may be sown in fall or stratified for spring planting, or the plant reproduced by cuttings or grafts. Birch-leaf Buckthorn is closely related to the more northerly species, Cascara Buckthorn, *R. purshiana* DC., from which the drug cascara sagrada is obtained.

California Buckthorn

Rhamnus californica Esch.

FIELD IDENTIFICATION. Western shrub, or more rarely a small tree to 18 ft. Very variable in leaf size, shape, and amount of pubescence, and hence separated into a number of varieties.

FLOWERS. April–June, in peduncled axillary cymes, peduncles ¼–1 in. long, pedicels ⅛–½ in. long, pubescent; flowers bisexual or unisexual, inconspicuous, greenish; calyx free from the ovary, sepals 5; petals 4–5; or absent, inserted on the margins of the disk; stamens 4–5; pistils 1, style not exserted, stigma 2–3-lobed, ovary 2–3-celled.

FRUIT. Maturing June–October, drupe berry-like, depressed-globose, ¼–⅜ in. in diameter, green to red, at maturity black, 2–3-seeded.

LEAVES. Evergreen or tardily deciduous, alternate, blades 1–3 in. long, ½–1 in. wide, elliptic to oblong, apex obtuse to acute, base rounded or broad-cuneate, margin serrulate to entire, sometimes revolute; lateral veins 8–12 pairs, curving somewhat upward; dark green and glabrous above, lower surface paler, sometimes yellowish green and glabrous or pubescent (densely so in some varieties); petioles ¼–⅗ in. long, glabrous to pubescent.

TWIGS. Grayish brown or reddish, pubescent to glabrous, bark yellow within, bitter to the taste.

RANGE. In canyons and rocky sheltered ravines, generally at altitudes of 4,000–7,500 ft. New Mexico; west to California and north to Washington and Colorado.

In New Mexico represented by *R. californica* var. *ursina* (Greene) Wolf.

PROPAGATION. The plant was first cultivated in 1871. The fruit is gathered by hand immediately upon ripening. It is run through a macerator with water and the pulp floated off. The seed is spread out in flats to dry before storing. Seed germination is slow, and stratification in moist sand at 41°F. for 90 or more days is helpful. Unstratified seed is sown in fall immediately when gathered, or stratified in spring, in drills about 8 in. apart and covered about ½ in. with soil. Propagation can also be practiced by layers, cuttings, or grafting.

VARIETIES. McMinn (1951, pp. 327–328) has given the following key to the varieties of *R. californica:*

Leaf-blades glabrous on both surfaces or only slightly puberulent beneath.
  Leaf-blades dark green above, paler beneath, serrulate or entire . . . . *R. californica*
  Leaf-blades yellowish-green, usually entire but sometimes serrulate . . . .
  . . *R. californica* var. *occidentalis* (Howell) Jepson.
Leaf-blades finely white-tomentose, white-velvety, or silvery beneath.
  Leaf-margins entire or with blunt teeth; the blades pubescent with short white hairs.
    Leaf-blades broadly elliptical, very thick and leathery, finely white-tomentulose on both surfaces . . . . *R. californica* var. *crassifolia* Jepson.
    Leaf-blades narrowly elliptical, thinner, finely tomentulose beneath with short, dense hairs, nearly glabrous above . . . .
    . . . . *R. californica* var. *tomentella* Brew. & Wats.
  Leaf-margins sharply serrulate; the blades whitish beneath with short dense hairs and intermingled longer hairs.
    Leaf-margins dentate . . . .
    . . . . *R. californica* var. *viridula* Jepson
    Leaf-margins serrate or rarely entire; mountains of the Mohave Desert . . . .
    . . . . *R. californica* var. *ursina* (Greene) Wolf.

CALIFORNIA BUCKTHORN
Rhamnus californica Esch.

REMARKS. *Rhamnus* is the ancient name of the Buckthorn, and the species name, *californica,* refers to the state of California where it frequently occurs. It is also known under the names of California Coffee-berry, Coast Coffee-berry, and Pigeon-berry. The fruit is eaten by at least 7 species of birds, including the band-tailed pigeon, also eaten by the black bear and California mule deer. It has no value as forage for cattle and is poor to fair for sheep.

It is of some importance as a honey plant. The plant is sometimes confused with, and the bark used for, the official drug cascara sagrada, which comes from *R. purshiana* DC.

WRIGHT SAGERETIA

Sageretia wrightii Wats.

FIELD IDENTIFICATION. Slender, straggling, spreading shrub 2–12 ft. The branches rather rigid, spiny, and divaricate.

FLOWERS. March–September, borne in scattered glomerules on leafy, slender, loosely branched axillary or terminal spikes; petals 5, white, hooded and clawed, inserted on the calyx; calyx valvate, obconic, shallow, 5-lobed, a disk lining the calyx; stamens 5; ovary superior, style short and 3-lobed.

FRUIT. Drupe almost black, fleshy, juicy; nutlets 3, coriaceous, indehiscent.

LEAVES. Opposite or alternate, elliptic to obovate or oblong, cuneate, margins entire or serrulate, blades ⅓–1 in. long, pinnately veined, tomentulose when young, soon glabrous and shiny, short-petiolate.

RANGE. Dry canyons among rocks at altitudes of 3,000–5,000 ft. In Texas known from the Davis and Chinati mountains of the Trans-Pecos region. In New Mexico and Arizona, southward into Mexico from Sonora to Jalisco.

REMARKS. The genus name, *Sageretia,* is in honor of

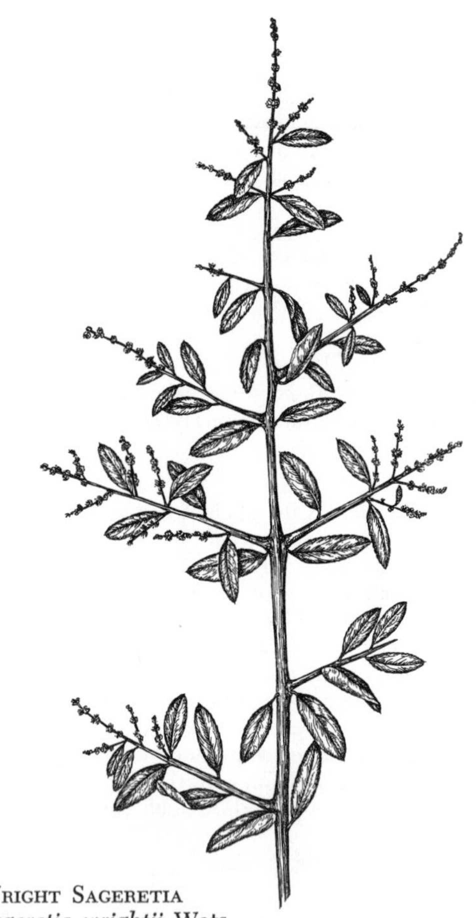

Wright Sageretia
Sageretia wrightii Wats.

Augustin Sageret, French botanist (1763–1851). The species name, *wrightii,* is in honor of Charles Wright, botanist and surveyor for the United States–Mexico boundary survey in 1851.

Gulfcoast Sageretia

Sageretia minutiflora (Michx.) Trel.

FIELD IDENTIFICATION. Shrub, trailing or straggling, sometimes vinelike, branches spinescent and diffuse.

FLOWERS. April–May, fragrant, spikelike or occasionally panicled, ⅜–3 in. long; individual flowers perfect, minute, about 1/10 in. across; calyx about 1/12 in. broad; sepals triangular, acute, keeled within, white-hairy externally; petals 5, about half as long as the sepals, white, concave, ovate to orbicular, emarginate, short-clawed; stamens 5, about as long as the petals, filaments distinct, anthers versatile and 2-celled; ovary superior, immersed in the disk, 2–3-celled, ovules solitary, style short with 3 stigmas.

FRUIT. Drupe ¼–⅓ in. in diameter, subglobose, pulpy, when dry some becoming shriveled and somewhat flattened, indehiscent, containing 1–3 nutlets which are rounded in outline, flattened on one side, the other side sometimes ridged, endosperm thin.

LEAVES. Simple, alternate or opposite, ovate to elliptic or orbicular, blades ⅜–2½ in. long, apex acute to short-acuminate and sometimes submucronate, base rounded to broadly cuneate or subcordate, margin serrulate, surfaces with minute cobwebby hairs when young, later more glabrous and lustrous, rather finely reticulate-veined; petioles 1/16–⅓ in. long, mostly with minute white cobwebby hairs; winter buds small, with several outer scales.

TWIGS. Slender, with lateral stiff spine-tipped branchlets, at first with white cobwebby hairs, but later glabrous, reddish brown to gray.

RANGE. Not known definitely in Louisiana, but to be looked for on calcareous bluffs and hammocks near the eastern state line. Mississippi, eastward to Florida, north on the coastal plain to North Carolina.

REMARKS. The genus name, *Sageretia,* honors Augustin Sageret, French botanist. The species name, *minutiflora,* refers to the small flowers. The plant is propagated by seeds and cuttings.

Gulfcoast Sageretia
Sageretia minutiflora (Michx.) Trel.

GRAPE FAMILY (*Vitaceae*)

HEART-LEAF AMPELOPSIS

Ampelopsis cordata Michx.

FIELD IDENTIFICATION. Glabrous vine climbing by forked tendrils.

FLOWERS. Borne May–June, in flat-topped loose greenish cymes at the nodes, 1–3½ in. broad; peduncles slender and often acting as tendrils; calyx small, barely lobed; petals 5, 1/12–⅛ in. long, elliptic, separate, spreading; stamens 5, exserted; disk cup-shaped, as high as the ovary, nearly free; ovary 2-celled, ovules 2 in each cavity, style subulate.

FRUIT. Borne on slender, glabrous pedicels ⅓–½ in. long, mature August–November, subglobose, somewhat flattened, ⅛–⅓ in. in diameter, bluish green or reddish, flesh dry and thin, not edible; seeds 1–3, nearly as broad as long, about ⅕ in. in diameter, dorsally smooth and rounded, granular, raphe narrow.

LEAVES. Alternate, broadly ovate or triangular-ovate (rarely 3-lobed); margin with remote, coarse, apiculate teeth; apex acute or acuminate, base truncate or semicordate; rather thin; upper surface olive-green, rather dull, glabrous or with a few scattered white hairs; palmately veined; lower surface paler and glabrous or with a few scattered white hairs especially on the veins, length of blade 2–5 in.; petiole ½–4 in., glabrous or with a few white hairs.

STEMS. Young stems green, grooved or angular, glabrous or slightly pubescent; older stems gray to light brown, smooth except for numerous oval, warty lenticels; tendrils from the nodes, each opposite a leaf, stout, forked terminally.

RANGE. Texas, Oklahoma, Arkansas, and Louisiana; eastward to Florida, northward to Ohio, Illinois and Nebraska, and south into Mexico.

REMARKS. The genus name, *Ampelopsis,* is from the Greek and means "vinelike," and the species name, *cordata,* refers to the cordate leaf. Another vernacular name is Simple-leaf Ampelopsis, probably in comparison with the Pepper-vine, *A. arborea,* which has pinnate leaves. Heart-leaf Ampelopsis is often cultivated for ornament and is rather rapid-growing. The seeds number about 9,600 seeds per lb, and are usually stratified before sowing. A number of species of birds consume the fruit including the cardinal, bobwhite quail, flicker, brown thrasher, and wood thrush.

HEART-LEAF AMPELOPSIS
Ampelopsis cordata Michx.

Pepper-vine

Ampelopsis arborea (L.) Koehne

FIELD IDENTIFICATION. Rather slender vine climbing by forked tendrils.

FLOWERS. June–July, borne in slender peduncled cymes ¾–2½ in. across; peduncles ⅓–⅔ in., glabrous or pubescent; pedicels solitary or branched, puberulent, ⅛–3⁄16 in.; flowers small, inconspicuous; petals 5, greenish white, spreading, ovate to ovate-oblong, acute at apex, margin ciliate, 1⁄16–1⁄12 in. long; calyx shallowly 5-lobed, lobes low and rounded; disk thick, shallowly crenulate-undulate; stamens 5, about 1⁄25 in., exserted, erect, attached to the disk margin, anthers introrse; pistil simple and rather depressed, stigma solitary, ovary 2-celled, ovules usually 2 in each cavity.

FRUIT. A few in a cymose cluster, subglobose, slightly flattened, often apiculate, at first green to pink or bluish, at maturity shiny black, often warty-dotted, ¼–⅜ in. long, juicy, flesh thin, not edible; seeds 1–4, about ⅛ in. long, green to brown, dorsal surface smooth and round, ventral surface angled.

LEAVES. Alternate, bipinnate, or lowest tripinnate, 3–8 in. long; leaflets ovate to rhombic-ovate, ½–1½ in. long, margin coarsely toothed, incised, or lobed; apex acute or acuminate; base cuneate or slightly cordate; rather thin; upper surface dark green, glabrous, or with a few scattered hairs; lower surface lighter green, glabrous, or with scattered white hairs especially along the veins; young leaves and shoots sometimes conspicuously reddish or bronze; petioles shorter than the blades, pubescent, sometimes with reddish blotches at the base of petioles and petiolules.

STEMS. Erect, ascending or bushy, tendrils present or absent; young stems green to reddish, glabrous or white-hairy; older stems tan to reddish brown, rounded or angular, sometimes roughened by oval, warty lenticels.

Pepper-vine
Ampelopsis arborea (L.) Koehne

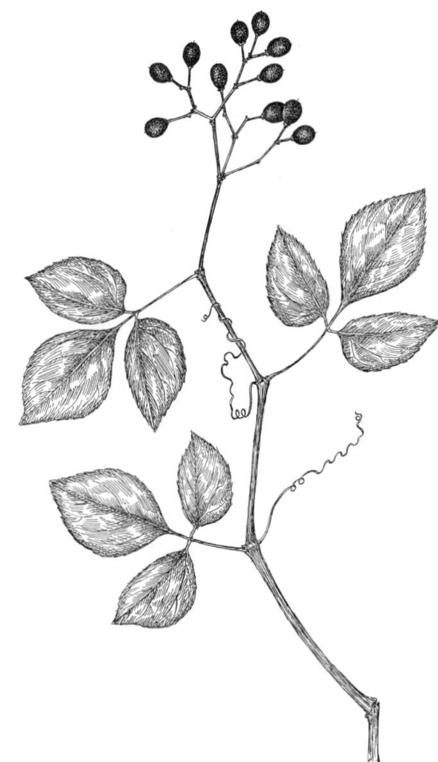

Mexican Ampelopsis
Ampelopsis mexicana Rose

RANGE. Rich moist soil. Texas, Oklahoma, Arkansas, and Louisiana; eastward to Florida, northward to Virginia, and west to Missouri.

REMARKS. The genus name, *Ampelopsis,* is from the Greek and means "vinelike," and the species name, *arborea,* refers to its climbing habit in the trees. Vernacular names are Pinnate-leaved Ampelopsis, Cow Vine, and Wild-sarsaparilla. It is sometimes confused with Poison-ivy, but Poison-ivy and Poison-oak have once-pinnate, trifoliate leaves.

Mexican Ampelopsis

Ampelopsis mexicana Rose

FIELD IDENTIFICATION. Large vine climbing by means of coiled tendrils, and somewhat woody toward the base.

FLOWERS. Borne in long-pedunculate loose cymes; individual flowers small, perfect; petals 5, valvate in the bud, distinct and spreading later, disk cupular; stamens 5, opposite the petals; calyx minute.

FRUIT. Black when dry, subglobose or somewhat obovoid, about ¼ in. long, with persistent stigma, flesh tough; seeds 1–3, obovoid to ellipsoid, brown, bony.

LEAVES. Alternate, composed of 3 leaflets, lateral ones unsymmetrical and smaller than the terminal one, 1⅓–3 in. long, ovate to elliptic, apex acute to short-acuminate, base broadly cuneate, upper and lower surface finely pubescent with short, white, curved hairs especially on the main veins, veinlets reticulate, margin serrate with low slender teeth; petioles 1–2 in., densely hairy; petiolules ⅛–¼ in., hairy.

STEMS. Somewhat divaricate, tendril-bearing at the nodes and opposite the leaves, glabrous, glaucescent, somewhat angled, brown to gray, bark slipping away easily.

RANGE. In sandy soils, mostly in Mexico in the states of Sinaloa and Guerrero. Rarely extending into southwest Texas, or possibly only escaping from cultivation there.

REMARKS. The genus name, *Ampelopsis,* is from the Greek *ampelos* ("vine") and *opsis* ("appearance"). The species name, *mexicana,* refers to its Mexican habitat.

IVY TREEBINE
Cissus incisa Desmoul.

IVY TREEBINE

Cissus incisa Desmoul.

FIELD IDENTIFICATION. Vine mostly herbaceous or occasionally semiwoody toward the base. Climbing by tendrils and identified by the thick, fleshy, trifoliate leaves, which have a disagreeable odor when bruised.

FLOWERS. Borne in 3–5-rayed, compound, umbellate clusters 1–2 in. wide; main peduncle ½–1 in., rays ⅓–½ in., individual flower pedicels 1/16–3/16 in., recurved, all parts glabrous or slightly puberulent; receptacle obconic, calyx minutely toothed or truncate; petals 4, hardly over 1/16 in. long, greenish, oblong, obtuse, erect or spreading, flattened or concave; stamens 4, erect, inserted on the margins of a 4-lobed disk; pistil erect, columnar, gradually narrowed to the apex.

FRUIT. On recurved, green to reddish brown pedicels ⅓–1 in.; berry subglobose to broad-ovoid, ¼–⅜ in. long, shiny black, sometimes apiculate; seed ovoid, acute at one end and rounded at the other, about ¼ in. long, flesh thin, not edible.

LEAVES. Alternate or opposite, trifoliate (or deeply 3-parted), very fetid when crushed, ovate to elliptic or obovate; apex obtuse, acute or rounded; base cuneate; margin coarsely toothed, or terminal leaflets 3-lobed, and lateral leaflets 2-lobed and asymmetrical, length ¾–3 in.; thick and fleshy; surfaces dull green and glabrous, or with a few minute white hairs, veins obscure; petiole ⅕–½ in., leaflets sessile or nearly so.

STEMS. Green to reddish, angled, when younger herbaceous, when older brown to gray, often warty with some woody fiber near the base of old stems, climbing by coiled nodal tendrils.

RANGE. Texas, Oklahoma, Arkansas, and Louisiana; eastward to Florida and northward to Missouri and Kansas.

REMARKS. The genus name, *Cissus,* is the Greek name for ivy, and the species name, *incisa,* refers to the incised leaves.

WATERWITHE TREEBINE

Cissus sicyoides L.

FIELD IDENTIFICATION. Fleshy high-climbing vine, mostly herbaceous, or somewhat woody toward the base.

FLOWERS. Clusters umbel-like cymes, loose and open, peduncles and pedicels of small flowers hairy; perfect or polygamous, greenish white; petals 4, distinct, expanding after anthesis; disk cuplike, 4-lobed; stamens 4, inserted on the disk margin; style subulate, ovary 2-celled and adnate to the base of disk, ovules 2 in each cavity. Flowers sometimes enlarged and modified by attacks of the fungus *Ustilago cissi.*

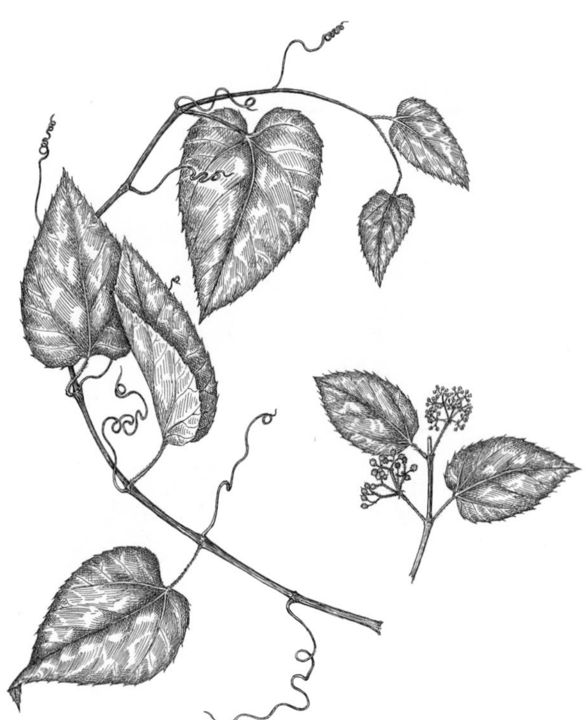

WATERWITHE TREEBINE
Cissus sicyoides L.

FRUIT. Berries about ⅜ in. in diameter, subglobose or obovoid, apiculate, black at maturity, flesh thin, not edible; seeds 1–2, ⅙–⅕ in. long, somewhat angled, base acute.

LEAVES. Simple, alternate, deciduous, thick and succulent, ovate to oblong-ovate, length ¾–6 in., apex acute to acuminate, base rounded to truncate or cordate, margins bristle-toothed, pale green, upper surface glabrous, lower surface pubescent and usually more glabrous with age, but very variable in pubescence; petioles ⅜–1¾ in., pubescent.

STEMS. High-climbing, green to brown or red, tuberculate or smooth, terete or striate and flattened, young parts often pubescent, older glabrous, tendrils opposite the leaves.

RANGE. Peninsular Florida, southwest Texas, Mexico, West Indies, Central America, and northern South America. In Mexico from Sonora to Tamaulipas, Yucatán, and Chiapas.

REMARKS. The genus name, *Cissus,* is the Greek name for ivy. The species name, *sicyoides,* denotes its resemblance to the cucumber vine. The plant seems to be very variable in leaf size and pubescence. In Florida, because of its presence near the coast, it is sometimes known as Marine Ivy. In Texas, in the lower Rio Grande Valley area, it is known under the local name of Bejuco Loco. Standley (1920–1926, p. 731) has listed the following names which are used in Mexico and other counties of Central and South America: Hierba del Buey, Tripa de Zopilote, Bejuco Loco, Tab kanil, Vid Silvestre, Tripas de Judas, Tumba-vaqueros Molonqui, Temecatl, Tripa de Vaca, Iasu, Bejuco Iasu Bejuco Comemano, Bejuco Castro, Bejuco Chirriador Uvilla, Ubí, Caro, and Bejuco de Caro.

In some tropical countries the fiber of the stems i used for twine and basketmaking. The macerated leave are also used for washing clothes because of their lath ery exudation. The leaves are also applied to sores, an a decoction of the stems is used for the treatment o rheumatism. The fruit is said to yield a blue dye.

THICKET CREEPER

Parthenocissus vitacea (Knerr) A. S. Hitchc.

FIELD IDENTIFICATION. Woody vine climbing by slen der-tipped, few-branched tendrils with or without ad hesive disks.

FLOWERS. Appearing June–July, mostly in panicle opposite the leaves, the two branches divergent an producing a broad rounded cluster, pedicels ⅛–¼ in. reddish; calyx shallow and obscurely 5-lobed; petals 5 spreading at anthesis, concave, thick; stamens 5, ex serted, filaments slender, anthers introrse; ovary 2

THICKET CREEPER
Parthenocissus vitacea (Knerr) A. S. Hitchc.

celled, sessile, narrowed into a minute stigma, ovules 2 in each cavity, erect and anatropous.

FRUIT. Maturing August–October, berry subglobose, bluish black, sometimes with a slight bloom, ⅓–⅖ in. long, thin-fleshed; seeds 1–4, angled, bony.

LEAVES. Alternate, palmately compound with 5 leaflets, narrowly to broadly elliptic to oblong, apex acute to acuminate, base cuneate or gradually narrowed, margin coarsely and sharply toothed, glossy green above, paler and reticulate-veined beneath, glabrous (or pubescent in some forms); petiolules reddish brown, channeled, ¼–¾ in.; petioles channeled, 1½–5½ in.; stipules deciduous.

STEMS. When young reddish to brownish, glabrous, somewhat grooved, later brown to gray, the branched tendrils usually without disks, leaf scars large and circular.

RANGE. Well-drained soil in sun or shade, at altitudes of 3,000–7,000 ft; Texas and New Mexico; northeast to Nova Scotia and west to Manitoba, Wyoming, and California.

VARIATIONS. Large-leaf Thicket Creeper, *P. vitacea* forma *macrophylla* (Louche) Rehd., has leaflets to 8 in. long.

Cut-leaf Thicket Creeper, *P. vitacea* forma *laciniata* (Planch.) Rehd., has smaller, narrower leaflets, deeply incised or with flaring teeth.

Doubtful Thicket Creeper, *P. vitacea* forma *dubia* (Rehd.) Rehd., has leaves ranging from pubescent to almost glabrous.

REMARKS. The genus name, *Parthenocissus*, is from the Greek words *parthenos* ("virgin") and *kissos* ("ivy"), and the species name, *vitacea*, refers to the grapelike tendrils. It has also been listed in the literature under the names of *P. inserta* (Kerner) K. Fritsch and *P. quinquefolia* (L.) Planch. var. *vitacea* Knerr. Vernacular names are Five-leaf Ivy and Woodbine. The plant is useful in cultivation for covering rocks, arbors, or bushes, and for the yellow or red leaves in fall. It has been cultivated since 1800. The fruit is eaten by many species of birds. A vigorous, rapid-growing, rather drought-resistant vine.

Virginia Creeper

Parthenocissus quinquefolia (L.) Planch.

FIELD IDENTIFICATION. Climbing or trailing vine with tendrils sometimes provided with adhesive tips.

FLOWERS. Borne in panicles of compound cymes, greenish, perfect or polygamo-monoecious; calyx small, 5-lobed; petals 5, spreading, concave; stamens 5, exserted; ovary 2-celled, sessile, ovules 2 in each cavity; style short and thick.

FRUIT. Ripe in the fall, cymes spreading, erect; peduncles red, slender, 3–6 in. long; pedicels red, ¼–½ in.; berries in clusters of 1–5, greenish blue, subglobose, slightly flattened, 1/16–⅓ in. in diameter; seeds 2–4, about ⅛ in. long, round to short-oblong, dorsal side rounded and half-grooved by the raphe, ventral side 2-angled.

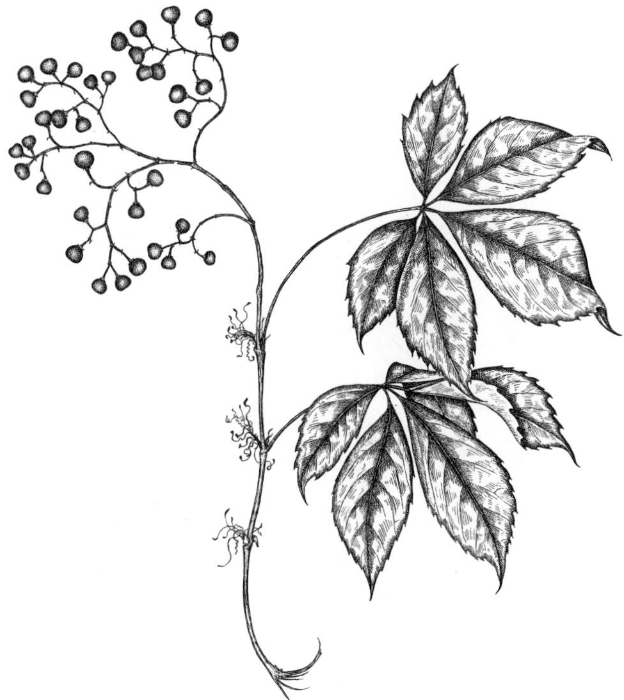

VIRGINIA CREEPER
Parthenocissus quinquefolia (L.) Planch.

LEAVES. Palmately-compound of 5 leaflets (rarely, 7). Leaflets elliptic, oblanceolate, obovate or oval, lateral ones often inequilateral; margins coarsely serrate-dentate except near the base and apex; apex acute or acuminate; base gradually narrowed into the petiolule; upper surface dull dark green, essentially glabrous; lower surface paler green and glabrous, sometimes glaucescent; blade 2–6 in. long, 1–3 in. broad; petiolules slender, ¼–½ in., gradually expanding into the blade; petiole slender, elongate, 3 in.–1 ft, flattened, grooved.

STEMS. Climbing by means of slender tendrils which sometimes are provided with terminal, adhering expansions; old stems brown, somewhat roughened; young stems green to red or brown, glabrous, flattened, somewhat finely grooved. The vines may climb tree trunks or walls, clamber over objects, trail on the ground, or sometimes are supported by aerial rootlets.

RANGE. Widespread in the United States, Canada, Cuba, and the Bahamas. In Texas, Oklahoma, Arkansas, and Louisiana; east to Florida, and northward throughout eastern North America; also in the Mexican states of Nuevo León, Veracruz, Hidalgo, and Michoacán.

VARIETIES AND FORMS. Rehder (1940, p. 619) has listed the following:

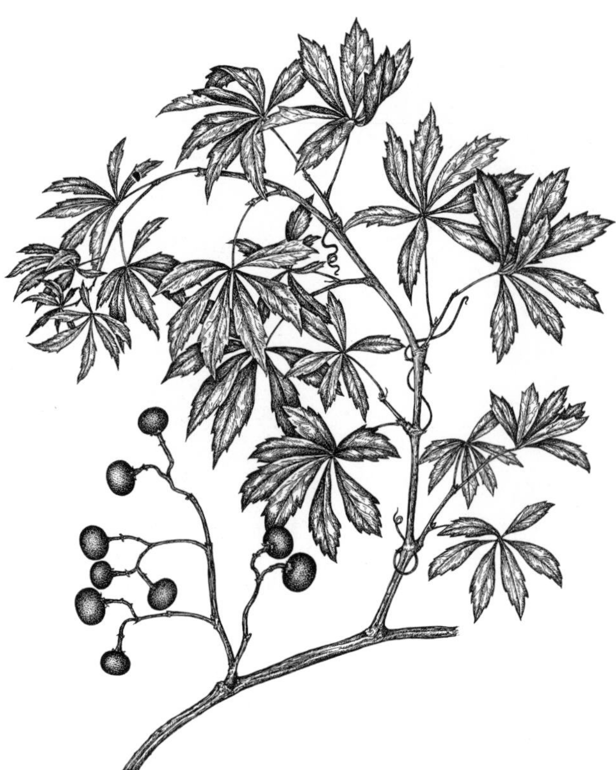

SEVEN-LEAF CREEPER
Parthenocissus heptaphylla (Buckl.) Britt.

Engelmann Virginia Creeper, *P. quinquefolia* forma *engelmannii* (Graebn.) Rehd., has smaller leaves than the species.

Red-twig Virginia Creeper, *P. quinquefolia* var. *hirsuta* (Pursh) Planch., has leaflets, twigs and inflorescence soft-pubescent.

Little-leaf Virginia Creeper, *P. quinquefolia* var. *minor* (Graebn.) Rehd., has small, oval to orbicular-ovate leaflets, rounded at base, on stalks ¼–⅓ in. long.

Wall Virginia Creeper, *P. quinquefolia* var. *murorum* (Focke) Rehd., has shorter tendrils, 8–12 divisions, leaflets usually broader, outer ones broad-ovate, rounded at base.

St. Paul Virginia Creeper, *P. quinquefolia* var. *saintpaulii* (Graebn.) Rehd., has young branches finely pubescent, tendrils with 8–12 divisions; leaflets oblong-obovate, cuneate, short-stalked, pubescent beneath, sharply serrate with usually flaring teeth.

REMARKS. The genus name, *Parthenocissus,* is from two Greek words, *parthenos* ("virgin") and *kissos* ("ivy"), and the species name, *quinquefolia,* refers to the 5 leaflets. Vernacular names for the vine are Woodbine, Five-finger Ivy, Five-finger Creeper, Five-leaf Ivy, and American Ivy. The vine is often cultivated for ornament because of its glossy palmate leaflets, which become red in autumn. The bark has been used in domestic medicine as a tonic, expectorant, and remedy for dropsy.

SEVEN-LEAF CREEPER

Parthenocissus heptaphylla (Buckl.) Britt.

FIELD IDENTIFICATION. Glabrous woody vine climbing to heights of 30 ft by means of tendrils with sucker-like disks.

FLOWERS. In spring in loose-clustered, greenish corymbs 1¾–3⅓ in. broad, often pendulous; pedicels ⅛–½ in., glabrous and reddish; flowers small, perfect or polygamo-monoecious; calyx pediceled, entire or very obscurely lobed, shallow, thin; petals 5, spreading, thickened, concave; stamens 5, exserted, introrse, anthers oblong; ovary sessile, style stout and conical.

FRUIT. Berry clustered, bluish black, subglobose, ⅓–½ in. in diameter, pulp scant and inedible; seeds 3–4, about ⅙ in. long.

LEAVES. Alternate, digitately compound of 5–9 (usually 7) leaflets, leaflets 1–3 in. long, width ⅓–1 in., oblanceolate to oblong-lanceolate, apex acuminate or acute; base sessile or very short-petioled and cuneately narrowed, margin coarsely and sharply serrate at least above the middle, dark green above and paler beneath, glabrous; petioles ⅓–3 in., reddish brown, glabrous; petiolules short or leaflets sessile.

STEMS. When young reddish brown, when older brown to gray, somewhat angled or rounded, tendrils long and forking with sucker-like disks.

RANGE. In rocky or sandy soil usually in or near mountains in southern or western Texas.

REMARKS. The genus name, *Parthenocissus,* is from the Greek *parthenos* ("virgin") and *kissos* ("ivy"). The species name, *heptaphylla,* refers to the 7 leaflets. Also known under the vernacular names of Texas Woodbine and Southwest Ivy. It climbs on fences, trees, shrubbery, and walls. The leaves turn attractive shades of red or orange in fall. Some botanists classify the plant as only a variation of the Virginia Creeper.

SUMMER GRAPE

Vitis aestivalis Michx.

FIELD IDENTIFICATION. Vigorous vine, high-climbing by tendrils. Readily identified by the reddish hairy undersurface of the usually 3–5-lobed leaves.

FLOWERS. May–July, dioecious, peduncle ½–3 in.; panicle 2–6 in., slender, loose, cylindrical, interrupted, fragrant; calyx very short; petals 5, coherent, early-deciduous; disk hypogynous, with 5 nectariferous glands alternating with 5 stamens which are reflexed in the pistillate flower; style short and thick, ovules 2 in each cavity.

FRUIT. September–October, persistent, numerous, ⅕–½ in. in diameter, globose, dark blue to black with a thin bloom, not musky, variable in quality, perhaps a result of hybridization, either dry and astringent or sweet and juicy; seeds 2–4, large for the size of fruit, pyriform, base contracted, 1/5–1/4 in. long, 1/6–1/5 in. broad.

LEAVES. Simple, alternate, cordate-ovate to subrotund, blades 2–8 in. long or broad, margin of some irregularly toothed and unlobed, others shallowly to deeply 3–5-lobed, lobes acute and sinuses acute or rounded; basal lobes with narrow or broad sinuses; apex acute or often broadly triangular; thin at first but thickish later; when young, surface densely reddish-tomentose; later smooth, dull and glabrate or pubescent above; lower surface varying from densely rusty-tomentose to barely loose-hairy and glaucous (the more glaucous form has been treated as a separate variety); petiole one half to two thirds the length of the blade, cobwebby-tomentose at first but later glabrous or somewhat pubescent.

TWIGS. Reddish brown, woolly at first but soon glabrous, terete, internodes medium to short, nodes swollen, 1/12–1/8 in. thick, pith interrupted at the nodes by a biconcave diaphragm; bark loose and shreddy.

RANGE. Warm sandy soil, dry woods, thickets, and along roadsides. Oklahoma, Arkansas, Texas, and Louisiana; eastward to Florida, northward to New Hampshire, and west to Wisconsin and Kansas. This species occurs east of the region occupied by *V. lincecomii* and south of the region where *V. bicolor* grows.

PROPAGATION. The plant is fairly resistant to cold and drought, and also quite resistant to mildew and rot. It is about 75 per cent resistant to plant lice of the genus *Phylloxera*. Cultivation is mostly from hardwood cuttings or layering, with about 50 per cent of the cuttings taking root.

VARIETIES, FORMS, AND HYBRIDS. Silver-leaf Summer Grape, *V. aestivalis* var. *argentifolia* (Munson) Fern., has leaves conspicuously glaucous or whitened beneath, and glabrate or slightly hairy; twigs and petioles glabrous and glaucous. It intergrades with the species.

Bourquin Summer Grape, *V. aestivalis* var. *bourquiniana* (Munson) Bailey, is a variety which has been developed in cultivation since 1847. Leaves slightly rusty-pubescent beneath and gray or brownish-colored; berries large and juicy, black or amber-colored.

V. aestivalis and *V. aestivalis* var. *bourquiniana* are known to develop many horticultural forms in combination with *V. champinii, V. labrusca, V. lincecomii, V. rupestris, V. vinifera,* and *V. vulpina.* See Kelsey and Dayton (1942).

REMARKS. The genus name, *Vitis*, is the classical Latin name, and the species name, *aestivalis,* refers to the summer blooming. Also known under the vernacular names of Pigeon Grape and Bunch Grape. The fruit is known to be eaten by many species of birds and mammals, including cardinal, cowbird, bobwhite quail, ruffed grouse, wild turkey, and white-tailed deer. Fine jellies and wine are made from the grape. It resembles the northern *V. bicolor* somewhat but differs chiefly in having a rusty, cottony, thin wool on lower surface of the often deeply lobed leaves. The foliage is also often more dense, and the fruit in more open clusters.

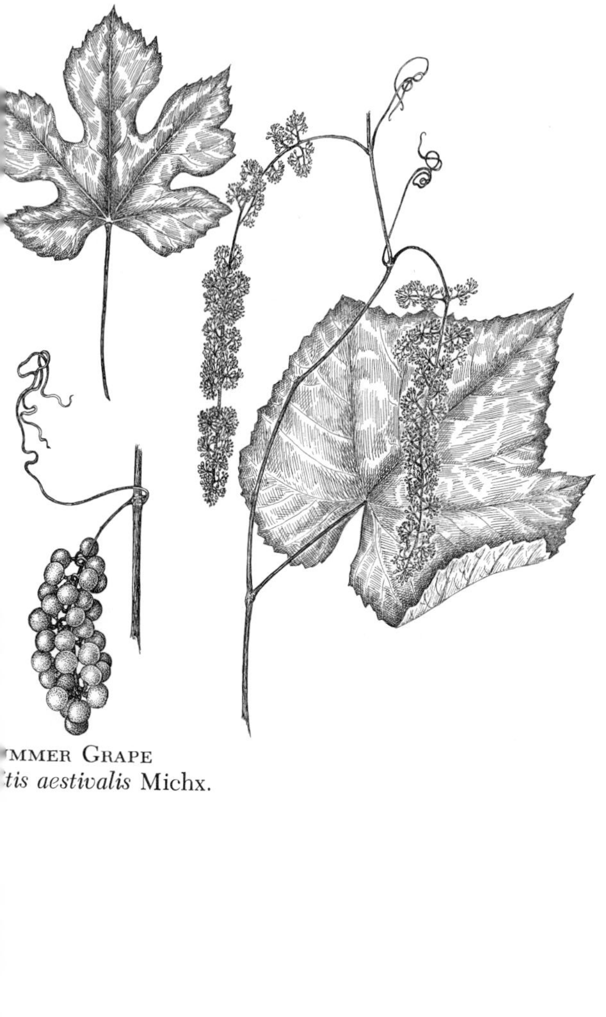

SUMMER GRAPE
Vitis aestivalis Michx.

CANYON GRAPE

Vitis arizonica Engelm.

FIELD IDENTIFICATION. Vine, scarcely climbing, much-branched, weak, slender, tapering rapidly in diameter from base of stem to the apices.

FLOWERS. Clusters 2–4 in. long on an average, peduncle and rachis cottony, pedicels about 1/8 in., slender; fertile flowers with 5 strongly recurved stamens; disk distinct, lobed; ovary globose; style short, thick, stigma broad; stamens of sterile flowers ascending and slender; petals 5, coherent at apex.

FRUIT. Maturing July–August, in clusters usually shorter than the leaves, globose or ovate, about 2/5 in. thick, black, slightly glaucous, very sweet; seeds 1–3, usually 2, oval to broadly obcordate, 1/6–1/5 in. long, 1/8–1/6 in. broad, raphe a rather thin line on the keel, chalaza rounded or short-oblong.

LEAVES. Generally cordate-ovate, length or width 2–4 in., basal sinus usually broad but occasionally narrowly inverted U-shaped, apex triangular, margin sometimes acutely lobed or shouldered, teeth 1/12–1/6 in. high, acute to rarely right-angled and mucronate-pointed, both surfaces of the leaves floccose or covered with gray-cottony hairs, either thin or thickened; stipules 1/12–1/8 in.,

Canyon Grape
Vitis arizonica Engelm.

thinly cottony; petiole dark red when young, about half as long as the blade is wide, somewhat grooved or striate, usually pubescent and cottony.

STEMS. Young stems gray-cottony, angled, obscurely striate; mature bark dark, almost black, cracked after the first year and separating into fragile, thin, non-fibrous plates; mature wood very dense and hard; nodes scarcely enlarged, mostly straight, diaphragm about 1/12 in. thick or nearly lacking; internodes very short, 3/4–2 1/2 in.; tendrils mostly once- or twice-forked, about the same length as leaves, cottony, weakly attached to support, or deciduous the first year if not clinging.

RANGE. Ravines and gulches at altitudes of 2,000–7,500 ft. In parts of western Texas (Uvalde and Jeff Davis counties); at Chloride, New Mexico; near Phoenix and Prescott, Arizona.

PROPAGATION. Canyon Grape endures cold and drought well, but does not endure excessive moisture, which may cause mildew and rot. It grows easily from cuttings, with seed germination from early to late spring; foliation and inflorescence medium season, ripening of fruit medium to late, exfoliation late.

VARIETIES. Smooth Canyon Grape, *V. arizonica* var. *glabra* Munson, has leaves larger than the species, shiny smooth, nearly glabrous. It occurs in the same area as the species.

Galvin Canyon Grape, *V. arizonica* var. *galvini* Munson, has leaves larger than the species, more ser rate or lobed and bright green, with larger fruit. I Chihuahua, Mexico.

REMARKS. The genus name, *Vitis*, is the classica name. The species name, *arizonica*, is for the state o Arizona. The fruit is eaten by many kinds of birds an mammals including Gambel's quail, scaled quail Mearns's quail, skunk, fox, coyote, and bear; also mucl eaten by captive deer. Pueblo Indians, who formerl cultivated this vine, eat the fruit, either fresh or dried Sometimes it is made into preserves or grape juice. Th plant has some value as an erosion control.

Winter Grape

Vitis berlandieri Planch.

FIELD IDENTIFICATION. Vine rather stocky, branched moderate- or high-climbing.

FLOWERS. Clusters 3/4–3 1/4 in. (fruiting clusters ofte longer), sometimes with a false tendril which become a secondary cluster; rachis once- or twice-compound pubescent or cottony, pale green; pedicels 1/5–1/4 in. slender, enlarged at receptacle; calyx absent; petals 5 cohering at summit, separating at base; stamens i

Smooth Canyon Grape
Vitis arizonica var. *glabra* Munson

WINTER GRAPE
Vitis berlandieri Planch.

fertile flowers 5, recurved and bent laterally; ovary ovate, style slender, stigma solitary and small; staminate flowers with stamens long, slender and ascending.

FRUIT. Ripening August–October, rachis sometimes to 6 in., but generally much shorter, berries ⅕–⅓ in. in diameter, globular, black or purple, sometimes red, glaucous coating either light or heavy (far more than in *V. cinerea*—a good distinction); skin thin; pulp juicy, sweet if allowed to hang till frost; seeds variable, 1–3 (usually 1), about ⅕ in. long or broad, globular or broadly ovoid, grayish coffee-colored; beak very small and short; raphe (a fine thread) generally invisible; chalaza usually flat or depressed, ovate or rounded.

LEAVES. Blade cordate-ovate to cordate-orbicular, usually 2–4 in. wide or longer (more rarely to 5–6 in.), often broader than long, basal sinus narrow or broad, either U- or V-shaped; shoulder lobes (if any) usually not prominent, apex of blade short and acutely tapering; marginal teeth small and short-apiculate; venation of 3–4 pairs of opposite rather prominent ribs; young leaves pinkish or green, thinly covered with short delicate hairs giving an ashy appearance; at maturity upper surface smooth and dark glossy green; lower surface with short delicate, cottony hairs (especially on the venation); dense and leathery; leaves from ground shoots of old wood usually 3–5-lobed, with rounded lateral sinuses; stipules ⅛–⅙ in.; petiole green or reddish, velvety-tomentose or with gray-cottony hairs, distinctly striate, groove narrow and shallow.

STEMS. Distinctly angled, covered with dull ashy pubescence, persistent the first year, or also with thin cottony hairs; bark dull brown, finely striate or angled; nodes slightly swollen, much bent; diaphragm $\frac{1}{12}$–⅙ in. in thickness, nearly plane; internodes usually 2–4 in. long (more rarely 6–7 in.); tendrils once- or twice-forked, at first cottony and pubescent, later smooth, striate, red or green; wood rather soft, appearing porous and with distinct rays in sectional view.

RANGE. Commonly on calcareous soil, but sometimes on well-drained sandy soil in the sun. Rather abundant on limestone hillsides, hilltops, and creek bottoms of central and southwestern Texas. West of the Brazos River to the Rio Grande, and into Mexico from Coahuila to Veracruz.

PROPAGATION. Winter Grape is cold- and drought-resistant, resistant to black rot and *Phylloxera,* but subject to mildew. The vines are long-lived, seed germination slow, and about 40 per cent of cuttings striking root; roots hard, little branched, penetrating deeply.

HYBRIDS AND RELATED SPECIES. This grape offers a fine basis for a group of hybrids to secure a combination of properties for very limy soils. It naturally hybridizes with *V. candicans, V. rupestris,* and toward the Rio Grande with *V. arizonica.* The best varieties are prolific, the fruit of pure quality and free of rot. Although closely allied to *V. cinerea,* it is easily distinguished from that species by its lighter, green-glossy foliage, with few ribs, its usually few seeds, its smaller, more compact, short-peduncled clusters bearing generally larger, bloomier berries.

REMARKS. The genus name, *Vitis,* is the classical name. The species name, *berlandieri,* is in honor of the Swiss botanist, Luis Berlandier, who was sent as naturalist of the Mier Teran Expedition to determine the character of the country along the proposed United States–Mexico boundary in 1828. Also known under the vernacular names of Little-mountain Grape, Fall Grape, Spanish Grape, Uña Cimarrona.

MUSTANG GRAPE

Vitis candicans Engelm.

FIELD IDENTIFICATION. Vine vigorous, high-climbing, rampant. Old vines sometimes attaining a length of 40 ft or more, and a trunk diameter in excess of 6 in. A striking character is the inverted-saucer shape of the leaf blade, which is borne on the petiole in such a manner as to provide a shingled canopy effect when the foliage is dense.

FLOWERS. Pistillate cluster usually small, 2–3 in. long and nearly equally forked, the 5 petals of pistillate flowers crimson in the bud and coherent at the apex; stamens 5, very short, reflexed and curved laterally; pistil large with a thick and very short style appearing nearly sessile, ovary globose; staminate cluster much larger and commonly forked or with one main axis; peduncle

Mustang Grape
Vitis candicans Engelm.

short, woolly or floccose; stamens 5, medium strong, with large anthers and abundant pollen.

FRUIT. Maturing June–August, generally persistent until late autumn, berries 3–12 in a cluster, ½–⅞ in. in diameter, globular, black (very rarely red, and still more rarely white), bloom none, more or less punctate; skin thick and tough, possessed of a fiery pungency which, unless the skin is carefully removed, irritates the mouth; pulp tough, sometimes transparent, of a pleasant to rather insipid sweetish taste; a jelly-like juice, sometimes red, occurs between skin and pulp; berry separating from the pedicel with difficulty, the pedicel always drawing a white or crimson core from the berry; seeds 2–4, ⅙–⅓ in. long and nearly as broad, ovoid when solitary, rounded; beak like that of unparched corn, with a brownish tinge; ridge distinctly grooved, raphe a thin line on keel, chalaza small and depressed.

LEAVES. Blade 2½–5 in. long, and 2½–6 in. wide; cordate-ovate to reniform-ovate, apparently broad for its length; apex usually triangular-pointed; basal sinus broadly V-shaped or truncate; margin entire to sinuate, or obscurely shallow-toothed, or 3–5-lobed, sinuses broad and rounded, apex of lobes at acute right angles; venation of 5–7 opposite or subopposite pairs of ribs; lower surface of leaves of nearly snow-white felty tomentum, becoming dull ashy-colored; upper surface of leaves thinly covered with whitish cobwebby hairs when young, which become flocculent, and are shed leaving a dull dark green color; leaves on ground shoots of old roots 3–9-lobed; stipules broad, blunt, crimson or pink tomentose; petioles 2–3 in. long with indistinct groove densely woolly.

STEMS. More or less angled when young, densely covered with dense to cobwebby whitish or yellowish wool, becoming floccose late in the season, persistent until the second year; bark on mature annual wood dull grayish brown, with warty blisters; bark on old wood finely checked and fibrous; nodes moderately enlarged, little bent, diaphragm ⅛–⅕ in. or more thick nearly plane; internodes 2–6 (even 10) in. long; pith medium, dark brown; tendrils nearly always forked twice, strong, clinging well, woolly when young, later becoming smooth, pale brown, finely striate.

RANGE. Of greatest size in the bottom lands of limy cretaceous hills of southwestern Texas; also in central, northern, and eastern Texas, and western Louisiana. From the Arkansas River in Oklahoma southward in western Arkansas.

PROPAGATION. Vigorous climbing vine, enduring great drought and heat, generally free from disease subject to attack of the leaf folder (*Desmia funeralis*). Hard to grow from cuttings, about 75 per cent resistant to *Phylloxera*, long-lived, seedlings at first feeble, later vigorous.

HYBRIDS. Mustang Grape hybridizes with *V. longii*, *V. vulpina*, *V. cinerea*, *V. lincecomii*, *V. berlandieri*, *V. monticola*, and *V. rupestris*. In France many artificial hybrids have been produced between Mustang Grape and *V. vinifera*. Two well-known horticultural hybrids are Elvicand, which combines with *V. candicans*, *V. labrusca*, and *V. vulpina;* and Valhallah, which combines with *V. candicans*, *V. bourquiniana*, *V. labrusca*, *V. vinifera*, and *V. vulpina*.

V. candicans var. *diversa* Bailey has leaves glossy above and duller gray beneath with the tomentum less dense, shoots apparently soon shedding the wool; berries much smaller, oblong or pyriform, milder in quality, lacking the excessive harshness of *V. candicans;* seeds ⅕–¼ in. long. In Gonzales County, southwestern Texas, perhaps also from Bexar and Brazos counties.

REMARKS. The genus name, *Vitis*, is the classical name, and the species name, *candicans*, refers to the white-woolly hairs of the leaves.

Champin's Grape

Vitis champinii Planch.

FIELD IDENTIFICATION. Robust upright vine climbing to a height of 40–50 ft.

FLOWERS. Pistillate cluster 1½–4 in., frequently shouldered; peduncles generally as long as the cluster, floccose nearly until fruit maturity; pedicels smooth, 1/20–1/16 in.; petals 5, coherent above, parting below; staminate cluster 3–6 in., shouldered, greenish, tomentose; pedicels 1/12–1/10 in.; pistillate with the 5 stamens short-recurved, pollen impotent, pistil thick with short thick style and large stigma; staminate flower stamens medium ascending, pollen very abundant.

FRUIT. Maturing in August, globose, ⅖–¾ in. thick, black, with little or no pruinose bloom when ripe; skin rather thin, tough, heavily pigmented, hardly pungent, pulp generally juicy, deficient in sugar but of a rich, agreeable flavor; seeds 1–3, ⅙–¼ in. long, ⅐–⅙ in. broad, obovoid when solitary; beak short and blunt; raphe extending from beak as a very slender imbedded thread and passing over the seed apex into the elliptic, sunken, or slightly elevated chalaza.

LEAVES. Blade 2¼–4¾ in. wide, broadly cordate to nearly reniform; basal sinus broad, inverted V-shaped; apex short-triangular and acute; margins rarely lobed but sometimes with shoulders, marginal teeth irregular, dentate, obtuse, with small mucro; venation arising generally from about 3 pairs of ribs; upper surface of blade at first thinly tomentose, becoming dark glossy green when mature; lower surface much paler, thin-woolly, becoming nearly devoid of tomentum at maturity but not glossy; stipules ⅛–⅕ in. long or wide, subovate, thinly tomentose; petiole 1–2 in. long, stout cylindrical, obscurely grooved on the upper side, thin-woolly.

STEMS. Young ones red to green, white-tomentose, strongly striate or angled, becoming cylindrical; older ones becoming dull reddish brown, bark rather finely checked and flaking off easily; nodes swollen, diaphragm 1/16–⅛ in. long and nearly plane; internodes 1½–4 in.; pith somewhat thicker in diameter than surrounding annual wood; tendrils abundant, strong, 3–5 in., once-forked, thinly tomentose.

RANGE. Central southern Texas in the following counties: Johnson, Llano, Lampasas, Bell, Mills, Mason, Gillespie, San Saba, Tom Green, Coryell, Schleicher, and Val Verde. Usually in the same regions as *V. berlandieri,* but less in the bottom lands.

PROPAGATION. Vine stocky and vigorous, grows easily from cuttings. Fairly resistant to *Phylloxera,* resistant to Anaheim disease, resistant to cold and drought; staminate vines commonly fewer than pistillate ones in the wild. This species has been cultivated since 1880.

VARIETIES AND HYBRIDS. Champin's Grape takes and carries grafted varieties well. It is used as a graft stock for *V. vinifera* in dry, adobe, drought soils, and is useful for hybridizing other varieties on the same type of soil. Also it hybridizes with *V. rupestris, V. monticola,* and *V. berlandieri.* Champin's Grape is possibly itself a hybrid of *V. candicans, V. rupestris,* and *V. berlandieri.*

The following horticultural clones are known involving *V. labrusca, V. vinifera, V. bourquiniana,* and *V. champinii:*

Champanel *(cha., lab., vin.)* (Mun., 1893)
Ladano *(bou., cha., lab., vin.)* (Mun., 1902)
Lufato *(bou., cha., lab., vin.)* (Mun., 1902)
Nitodal *(bou., cha., lab., vin.)* (Mun., 1902).

REMARKS. The genus name, *Vitis,* is the classical name, and the species name, *champinii,* honors M. Champin, French viticulturist.

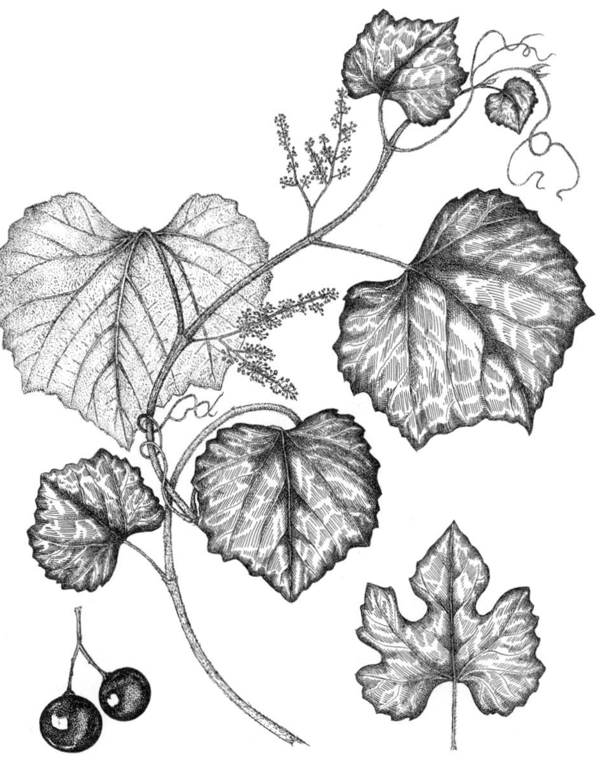

CHAMPIN'S GRAPE
Vitis champinii Planch.

SWEET WINTER GRAPE

Vitis cinerea Engelm.

FIELD IDENTIFICATION. Lax, high-climbing vine attaining large size.

FLOWERS. Usually May–June, in rather open, canescent or gray-floccose thyrses; pistillate clusters 4–4¾ in. long; peduncle elongate, striate, usually with a simple false tendril at the lowest node, this sometimes being modified as a small secondary cluster; rachis usually twice-divided, cottony or pubescent; pedicels numerous, slender, warty; calyx obsolete; petals 5, cohering at apex; stamens of the pistillate flowers short, recurved and bent laterally, pistil large; staminate flowers with stamens slender and ascending; style short; stigma solitary and minute; ovary 2-celled.

FRUIT. Berries numerous, ripening August–November, ⅙–⅓ in. in diameter, globose, black or purple, with a slight bloom or none, skin tough; pulp tender, juicy, initially acid, finally sweet, not pungent; seeds 1–3, usually 3, ⅙–⅕ in. long, ovoid when solitary, grayish purple, beak small and sharp; raphe prominent and

SWEET WINTER GRAPE
Vitis cinerea Engelm.

about filling the groove, extending prominently from the beak to the chalaza.

LEAVES. Blade 4–8 in. long or broad, thin, subrotund to cordate; basal sinus deep and narrow or broad, base lobes generally short and obtuse; apex long-triangular-pointed; margin unlobed, or with 2–4 more or less prolonged lobes, marginal teeth irregularly and finely to coarsely serrate; side ribs 3 rather prominent, conspicuous pairs; upper surface at first with thick, white, cobwebby tomentum, diminishing at maturity, rugose, dark green; lower surface permanently densely felty and gray-tomentose; stipules 1/12–1/6 in. long, more or less hairy; petiole shorter than the blade and pubescent or floccose.

STEMS. Young ones angled, densely white-felty, at first green, later pale brown to gray under the floccose coating; bark at first finely striate, later flaking into fine fibrous ridges; nodes usually enlarged, the diaphragm commonly 1/8–1/6 in. thick, often thicker, biconcave; tendril once-forked, or more often twice-forked, very long cottony, smooth with maturity.

RANGE. In moist alluvial soil along streams, thickets, and bottom lands. Texas, Oklahoma, Arkansas, and Louisiana; eastward to Florida and northward to Missouri, Indiana, Wisconsin, and Illinois.

PROPAGATION. Germination from early spring to summer, moderately cold-resistant, resistant to both drought and excess water, rather free of disease, long-lived, and difficult to grow from cuttings in the pure species.

HYBRIDS AND VARIETIES. Sweet Winter Grape hybridizes with *V. lincecomii, V. cordifolia, V. rupestris, V. candicans, V. aestivalis,* and *V. labrusca.*

V. cinerea var. *floridana* Munson varies by having young tips always red, rusty-tomentose, and the lower surface of the leaves rusty-cinereous, instead of light cineraceous, as in most Western and Northern forms. It is found in Florida and southern Georgia, proceeding westward along the Gulf Coast and northward, grading into the species.

V. cinerea var. *canescens* Bailey has shorter clusters with berries and seeds slightly larger; leaves more regularly cordate-ovate and less frequently sharp-lobed and without the long deltoid apex; lower surface less floccose and prominently soft-pubescent or canescent at maturity. Known from near Austin, Texas, as well as in central Missouri and southern Illinois.

REMARKS. The genus name, *Vitis*, is the classical name, and the species name, *cinerea*, refers to the gray tomentum of the leaves. Also known under the vernacular names of Ashy Grape, Gray Grape, Gray-bark Grape, Pigeon Grape, and Downy Grape; it is eaten by many species of birds and mammals.

RIVERBANK GRAPE

Vitis cordifolia Lam.

FIELD IDENTIFICATION. Very vigorous vine attaining a large size. Old trunks with a girth of 5 ft have been reported. The dark green leaves cover trees or bushes with a canopy-like covering.

FLOWERS. Pistillate clusters medium to large, open and loose, simple or broadly shouldered; peduncle strong, 1/2–2/3 the length of rachis, bracted above the middle, smooth or pubescent; rachis simple or forked, branches short, bractless or bracted, yellowish green, or bronze green in some forms; pedicels 1/4–1/3 in., slender, little enlarged at receptacle, thinly warty; stamens 5, very short, reflexed, curved laterally, having scant weak pollen; disk prominently 5-lobed, orange-yellow; ovary small and subconical; style short, slender, pointed; staminate cluster much larger than pistillate (to 4 3/4 in. long); rachis compound, often dividing into several nearly equal short branches, each 3–12-flowered; pedicels whitish, very slender, 1/6–1/4 in.; stamens (in staminate flower) usually 5, 1/10–1/6 in., ascending, pollen abundant; petals 5, coherent above and separated below.

FRUIT. Berries ripening August–October, persistent, 1/8–2/5 in. in diameter, globular or slightly oblate, sometimes shining but mostly dull black, rarely glaucous; skin thick or rigid, and possessed of a biting pungency until well frosted, but then becoming sweet; epidermal pigment violet, pulp scanty; seeds 1/6–1/5 in. long, nearly as broad, obovate or subglobose, dark brown; beak short, sub-bilobed, blunt or acute, raphe extending from the extremity of beak prominently over top of seed; ventral depression short and shallow; chalaza circular, small but well marked.

LEAVES. Simple, alternate, deciduous, blade length 2–6 in., width about two thirds that of the length, outline cordate-ovate or in some triangular at the apex; basal sinus narrow or broad and moderately deep, basal lobes sometimes closed or overlapping; shoulder and apex acute or taper-pointed (rarely long lance-pointed in the most acute-toothed form); margin in the most typical form not lobed, but shouldered or rarely lobed in some with the blade averaging broader; marginal teeth small to large, irregular, mostly acute, rarely obtuse, apiculate; venation from usually 7 pairs of subopposite ribs which are densely and stiffly pubescent, also often with white pubescent tufts in the forks; blade surfaces at first slightly hairy, some plants with young leaves green, others violet or bronze-green, and some with both; mature leaves with a smooth glossy green upper surface; lower surface much paler, varying in pubescence but mostly glabrate; thick and leathery. Leaves on ground shoots from old wood generally 3–5-lobed with rounded sinuses; stipules ciliate, soon shedding; petiole averaging over one half the width of blade, striate, with a shallow groove above, stiffly pubescent.

RIVERBANK GRAPE
Vitis cordifolia Lam.

STEMS. Young ones a pale yellowish green or bronze-red, surfaces with fawn-colored pubescence; on old stems the bark but slightly adhering, separating in broad, nearly regular plates which can be easily separated when the wood is but a year old, of a drab or hazel color, with darker and lighter markings, making them in nearly every instance appear mottled, deeply striate; surfaces smooth, except in some plants producing a rasplike feeling when the finger is passed over them; without any pruinose bloom near the nodes (or rarely with a trace); annual wood rather soft and tough; nodes little enlarged, slightly bent, often pubescent; diaphragm thick, ⅛–⅕ in. or more, biconcave; internodes long to very long, 3–7 in. or more; tendrils usually once-forked, rarely twice, intermittent, finely striate, smooth; pith large at lower end, rather dark brown.

RANGE. Grows well in alluvial moist bottom lands and also vigorously in strong limy soils. In eastern and central Texas, central and eastern Oklahoma, Arkansas, and Louisiana; eastward to Georgia and Florida, northward to Virginia, and westward to Illinois, Kentucky, and Missouri.

PROPAGATION. Seed germination usually early, foliage and wood maturing very early (about the first of our native species), leaves turning yellow before frost, falling and exposing the fruit for ripening (in other forms the leaves persist very late); vigor great; endures cold, mildew, and rot well. Its hybrids with *V. rupestris* have been found eminently resistant to *Phylloxera* by Prof. A. Millardet. Cuttings grow with difficulty. The immense size and great age it may attain suit it well as graft-stock, though the difficulty of rooting the cuttings makes it less desirable than the more easily rooted hybrids of *V. rupestris* and *V. vulpina.*

HYBRIDS AND VARIETIES. Numerous natural hybrids of this species with *V. rupestris, V. vulpina, V. lincecomii,* and *V. cinerea* were found by H. Jaeger in southwest Missouri; with *V. cinerea* and *V. lincecomii* by T. V. Munson in northern Texas; with *V. coriacea, V. simpsoni, V. aestivalis,* and rarely even with *V. munsoniana* by J. H. Simpson in Florida; and it has been obtained from Virginia and Georgia hybridized with *V. aestivalis, V. labrusca,* and *V. cinerea.*

REMARKS. The genus name, *Vitis,* is the classical name. The species name, *cordifolia,* refers to the cordate leaves. Also known under the vernacular names of Frost Grape and Sour Winter Grape in Texas, Fox Grape in Illinois, and Possum Grape in Virginia, North Carolina, and South Carolina.

DOAN GRAPE

Vitis doaniana Munson

FIELD IDENTIFICATION. Vine climbing vigorously to 15–30 ft when among trees, but bushy and ascending when without support.

FLOWERS. Clusters 1¼–3 in. long, peduncles 2–4 in. long, compact-shouldered; staminate much longer than the pistillate, branches and divisions more or less woolly. Fertile flowers with 5 short, recurved stamens, small anthers, nonvirile pollen; pistils short and thick, style very short, stigma broad; staminate with 5 stamens ascending, medium in length, anthers large, bearing abundant pollen; calyx essentially absent; petals 5, coherent above, separating at base.

FRUIT. Maturing July–October, berries persistent, about ½ in. in diameter, globose, black, bloom heavy and white; skin tough and pungent; quality good though a little pulpy; seeds 2–4, about ¼ in. long, ⅕ in. wide; beak thick, conical, blunt, slightly bilobed at tip; raphe sunken or invisible at notch, becoming prominent as it approaches the chalaza which is moderately elevated and narrowly ovate.

LEAVES. Average length of blade 4 in., average width 4⅓ in.; basal sinus ½–1 in. deep, margin usually 3-lobed (ground-shoot leaves often 5-lobed), apex of lobes usually acute, more rarely taper-pointed; teeth irregular, obtuse or acute, mucronate; venation pubescent with 6 pairs of ribs, mature blades with upper surface dull and dark green, lower surface with an ashy-bluish tomentum (varying to cottony or floccose on some geographical forms); stipules ⅛–¼ in. long, pinkish, usually hairy; petiole about one third to one half the length of the blade, striate, grooved, cottony or pubescent.

STEMS. Young stems covered with white cottony hairs which become flocculent at maturity; nodes little-bulged, diaphragm medium to thin; internodes short; tendrils once- or twice-forked, medium to strong, woolly or hairy when young; old bark persistent, finely checked and fibrous; mature wood pale grayish brown.

RANGE. On sandy loam or well-drained calcareous soils in the sun. From Greer County, Oklahoma, through the Panhandle of Texas; also in New Mexico to the Pecos River. Headwaters of the Brazos and Colorado rivers in Texas. Abundant in Donley County, Texas, west of Clarendon. Possibly from Crosby, Harris County, Texas.

PROPAGATION. Seemingly an abundant bearer, of promise as a species for dry climates. Rather rapid-growing, vigorous, remaining bushy if failing to find support; roots fibrous, hard, spreading, penetrating; resistant to drought, cold, and disease; 60 per cent resistant to *Phylloxera;* 60 per cent of cuttings take root.

REMARKS. The genus name, *Vitis,* is the classical name, and the species name, *doaniana,* is in honor of J. Doan of Wilbarger County, Texas, who made fine wine from the fruit which had been gathered in Greer County, Oklahoma.

DOAN GRAPE
Vitis doaniana Munson

HELLER GRAPE

Vitis helleri Small

FIELD IDENTIFICATION. Vine somewhat resembling a round-leaved, smooth scallop-toothed Berlandier Grape, but with glossy foliage something like the Riverbank Grape. The following description follows that of Bailey (1934, pp. 223–224).

FLOWERS. Clusters 2½–3¾ in., and perhaps longer in fruit, branched, main divisions long and narrow, at least in staminate inflorescences; pedicels prominent; stamens 5, in staminate flowers long and erect; flower petals 5, adherent above, separating at base.

FRUIT. In July, berries about ¼–⅖ in. thick; seeds rather large, broad, ⅕–¼ in. long and nearly as wide, glossy, beak abruptly very short, raphe little evident on the keel, but conspicuous although not elevated at the ridge and to the small but prominent chalaza.

LEAVES. Blades with a circular aspect due to breadth mostly greater (3–5¼ in.) than length and either a short triangular apex or none (sometimes a leaf or two on a branch more or less pointed), sides lacking tendency to bear a shoulder or short lobe as in Berlandier Grape, rather thin, strong side ribs, 2–4 pairs and spreading, basal sinus mostly deep and narrow to open, margins with broad, usually bluntish teeth except for the small apiculus terminating each nerve; both surfaces later

HELLER GRAPE
Vitis helleri Small

slender, stigma small; staminate panicle looser, compound, blooming a few days earlier than the pistillate panicle, with long and erect stamens, anthers 2-celled; ovary 1, ovules 1–2 in each cavity.

FRUIT. Maturing August–October, fruit cluster often as broad as long, usually bearing less than 20 berries, which commonly drop singly when ripe; berries globose to oblate, ⅓–1 in. in diameter, color varying from dull black to brownish purple or yellowish (rarely red or greenish white); skin thick and tough, usually with only a thin bloom, pungent and musky to the taste; pulp tough, slippery, acid; seeds 2–6, erect, ⅕–⅓ in. long, strongly angled, apex notched, brown to purple, beak short, thick, generally distinct; raphe a thin line in groove until near its passage over top of seed, then small and slightly elevated to base of beak; chalaza mostly a circular depression.

LEAVES. Alternate, each one opposite a forked tendril or a flower cluster; thick and heavy; ovate to cordate-ovate or quadrate-orbicular, blades varying 3–8 in. long or broad (average about 4 in.); margin varying from entire or dentate to unevenly scallop-toothed with short mucronate teeth, or with 3–5 lobes; the lobes acute to obtuse at apex, many with shallow narrow sinuses; smooth or with fine pubescence on veins underneath and perhaps remains of webs, undersurface conspicuously glossy; stipules ⅛–⅙ in.; petioles shorter than blades and glabrous.

STEMS. Young growth light cottony, soon becoming glabrous, angled and striate, short-jointed, new unfolding tip perhaps brownish.

RANGE. Southern Texas–Kerr County (type locality); Ottine, Gonzales County; and Del Rio, Val Verde County. The plant has been confused with Champin's Grape, as if a smooth, glossy form; and its relationship is yet to be determined.

REMARKS. The genus name, *Vitis,* is the classical name. The species name, *helleri,* is in honor of the botanist A. Arthur Heller.

FOX GRAPE
Vitis labrusca L.

FIELD IDENTIFICATION. Strong-growing vine with abundant foliage, trailing or climbing to 40 ft. Twigs, tendrils, petioles, and lower leaf surfaces densely rusty-tomentose.

FLOWERS. May–June, pistillate panicles 2–4 in., generally short-peduncled; pedicels ⅛–¼ in., enlarged somewhat at apex, more or less warty; stamens 5, short-recurved and bent laterally; ovary small, ovate, style

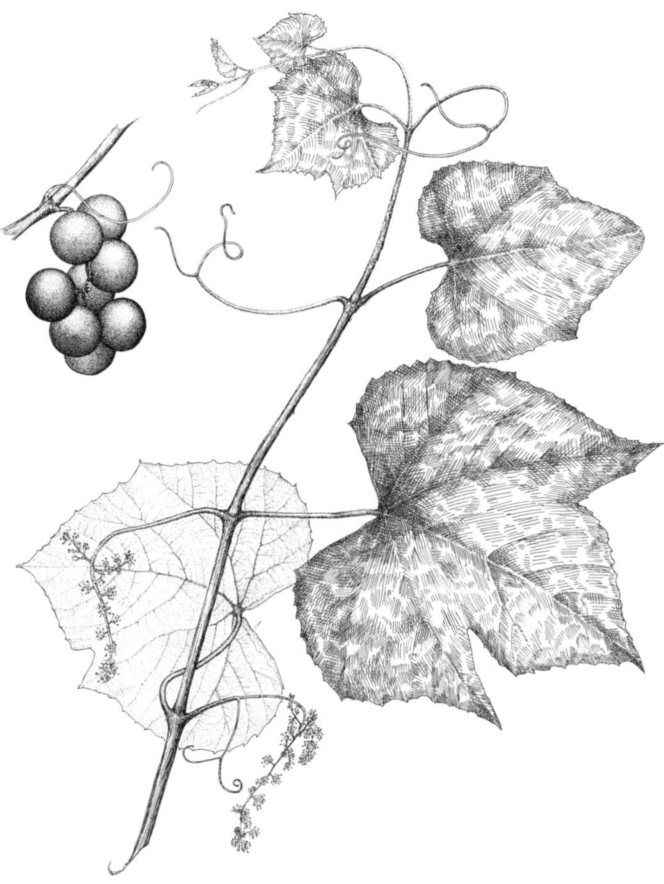

FOX GRAPE
Vitis labrusca L.

basal sinus either broad or narrow; veins 6–8, subopposite; when young the upper surface densely appressed-hairy and buff to pink, later becoming floccose, eventually dull dark green and glabrous; lower surface reddish brown felty-hairy (or tawny gray on some leaves); petiole enlarged toward base, cottony and pubescent; stipules ⅛–⅙ in. long, ⅒–⅛ in. wide, membranaceous, delicately hairy.

STEMS. Cylindrical or obscurely angled, at first whitish to yellowish or with reddish brown tomentum, later shedding, but some with spinous pubescence on older stems; bark finely striate, color light to dark brown, surface shreddy the second year; nodes bent, moderately enlarged; diaphragm ⅛–⅕ in. thick, biconcave; internodes 2–6 (occasionally 8–10) in. long on young shoots; pith moderately and gradually enlarged below, light brown; tendrils arising at each node, 1–2-forked, medium to long-cottony at first, at maturity glabrous, striate, brown.

RANGE. In sandy, moist, alluvial soil. Infrequent in Arkansas; east to Georgia, and north to Massachusetts, Ohio, and Indiana.

PROPAGATION. Fox Grape withstands cold well but suffers from heat and drought; while not susceptible to downy mildew, it suffers from *Phylloxera.* Ripe fruit dropped singly. Although seeds rarely produce plants true to type, propagation is easily accomplished by means of cuttings.

VARIETIES AND HYBRIDS. Short-tooth Fox Grape, *V. labrusca* var. *subedentata* Fern., has leaves of fertile portions very densely felted with close tomentum; margins with poorly developed shoulders and low teeth. South Carolina to New York.

Fox Grape hybridizes with other grape species, or other hybrids, very readily. The following species most readily hybridize with it: *V. aestivalis, V. bourquiniana, V. candicans, V. champinii, V. cinerea, V. lincecomii, V. riparia, V. rotundifolia, V. rupestris, V. vinifera,* and *V. vulpina.*

A large list of horticultural Fox Grape hybrids may be consulted in Kelsey and Dayton (1942).

REMARKS. The genus name, *Vitis,* is the ancient Latin name, and the species name, *labrusca,* was the classical name of a wild vine. The word "fox" as applied to this grape originated from the old verb "to fox," meaning to intoxicate. The colonists from Europe were interested in the wine-making quality of New World grapes, more than in the fruit product. Some popular accounts variously ascribe the name "foxy" to the reddish brown hairs of the lower leaf surface resembling a fox's pelt, or to the lobed leaves looking like a fox's track, or to the fruit having a foxy pungent odor or taste. Also known under the names of Northern Fox Grape, Plum Grape, Northern Muscadine, Swamp Grape, and Wild Vine. It is distinguished from other grapes by the regularly occurring tendrils, prickles on stems of the season, densely rusty tomentum on lower leaf surface, sunken chalaza or seed, and the shallow fibrous roots. The fruit is eaten by a number of species of birds and mammals including the ruffed grouse, ring-necked pheasant, gray fox, striped skunk, Virginia opossum, and raccoon.

This has been the most important grape in the development of North American viticulture.

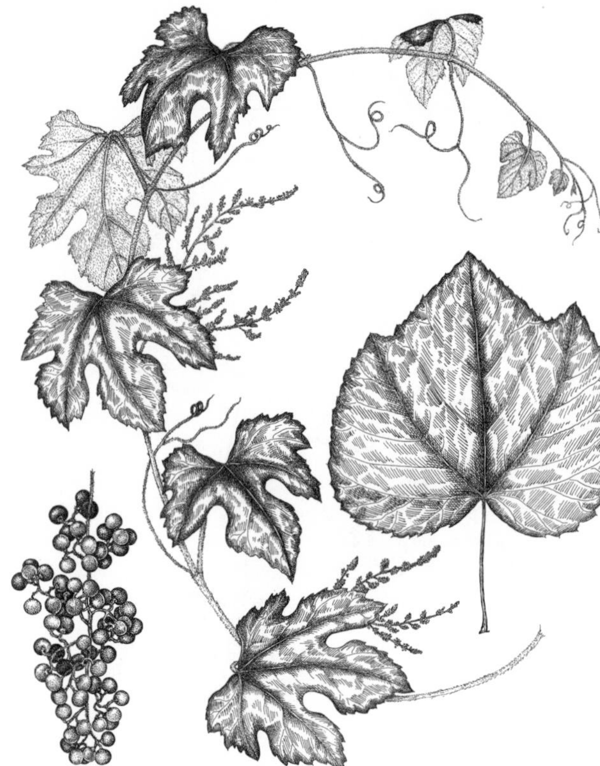

PINEWOODS GRAPE
Vitis lincecomii Buckl.

PINEWOODS GRAPE

Vitis lincecomii Buckl.

FIELD IDENTIFICATION. Vine vigorous, stout, branching moderately, and climbing to as high as 40 ft. Sometimes a shrub of diffuse form when no support is available.

FLOWERS. May–June, pistillate clusters 3–10 in. long (averaging 3–4 in.), shorter than the leaves, divisions simple; rachis bluish green, at first cottony, smooth later; pedicels ⅙–¼ in., enlarged above, warty; stamens in the pistillate flower weak, bending laterally, approximate to the enlarged 2-celled ovary, style short, stigma solitary; staminate panicles generally larger than pistillate; stamens 5, in staminate flower, strong, ascending; petals 5, cohering at summit, separating at base, falling early.

FRUIT. July–September, persistent or falling at once; fruiting rachis tomentose; berries ⅜–1 in. in diameter, spherical to nearly oblate, purplish black or black, with a thin bloom; skin thin, tough, dry; flesh very acid and astringent (in the variety *glauca* more juicy, tender, and vinous); seeds obovate-pyriform, ¼–⅓ in. long, ⅕–¼ in. broad, short-beaked or beakless, purplish brown; raphe usually a prominent keel on face and chalaza small and sunken lying in the groove at ridge, generally ovate, or narrowly elliptic.

LEAVES. Simple, alternate, length or width 3¼–8 in., nearly always broader than long, quadrate or broadly cordate-ovate, basal sinus generally broadly or narrowly V-shaped with doubly curved sides, sometimes rounded with limbs approaching or lapping; margin usually obscurely shouldered to 3–5- (or 7-) lobed; lateral sinuses rounded, occasionally with a large tooth at the base, the lobes more or less closing up or lapping across the sinuses; teeth of margin rather large and coarse, irregular, sometimes scalloped; venation of 6–8 pairs of prominent ribs, covered densely with more or less rusty cotton-like tomentum in the typical form of southern Texas and Louisiana, or with whitish tomentum in northern Texas and Missouri, becoming floccose at maturity in both. The young leaves pinkish to crimson and densely tomentose, mature leaves with the upper surface slightly wrinkled, soon losing its tomentum, becoming smooth, dark green; lower surface becoming rusty ash-colored, abundantly woolly (or in the variety *glauca*, bluish glaucous-green, with little wool); petioles tomentose, pubescence heavy at first but thinly so later, gray or tawny.

GLAUCOUS PINEWOODS GRAPE
Vitis lincecomii var. *glauca* Munson

STEMS. When young cylindrical, with whitish or light brown tomentum, soon becoming floccose, the tomentum disappearing on mature annual wood; older stems glabrous, dark reddish brown, pruinose, often bearing numerous, short, black, glandular prickles, especially near the nodes; nodes slightly enlarged, much bent; diaphragm $\frac{1}{16}$–⅛ in. thick, biconcave; internodes usually 2–4 in. long; pith slightly enlarged at lower end of diaphragm, pale brown; tendrils intermittent, rarely 3–5 in succession, twice-forked in the typical forms (ordinarily once-forked in var. *glauca*), slightly cottony in var. *glauca*, or smooth when young, to densely rusty-woolly in the species, at first finely striate, reddish or green, becoming same color as wood at maturity, strong, persistent.

RANGE. Usually in dry sandy soils of pine woods or post-oak lands. Central and eastern Texas, Louisiana, Oklahoma, and Arkansas; eastward to Mississippi, northward to Indiana, and westward to Missouri.

PROPAGATION. Fairly cold- and heat-resistant, very drought-resistant, susceptible to mildew and rot, about 75 per cent resistant to *Phylloxera;* about 50 per cent of cuttings take root (although reproduction from cuttings difficult); fertile vines generally incapable of self-fertilization; seeds germinate promptly.

VARIETIES AND HORTICULTURAL HYBRIDS. *V. lincecomii* var. *glauca* Munson has leaves glaucous below and veins rusty-pubescent. In Missouri, Texas, Arkansas, and Oklahoma. Very abundant in northern Texas.

V. lincecomii is known to hybridize with other hybrid grapes, with parentage from the following: *V. cordifolia, V. cinerea, V. rupestris, V. aestivalis, V. bourquiniana, V. candicans, V. vulpina, V. champinii, V. labrusca, V. riparia, V. rotundifolia,* and *V. vinifera.*

REMARKS. The genus name, *Vitis,* is the classical name. The species name, *lincecomii,* is in honor of Gideon Lincecum (1793–1874), early collector of Texas plants. Vernacular names for the plant are Post-oak Grape, Turkey Grape, and Big Summer Grape. The green fruit is sometimes eaten by domestic turkey.

LONG'S GRAPE

Vitis longii Prince

FIELD IDENTIFICATION. Often a stocky, erect, much-branched, vigorous shrub without support, or a small-to-large, very vigorous vine when support is available.

FLOWERS. Usually in May, fragrant; pistillate clusters 2–3 in. long, simple, very compact, peduncle very short, without false tendril, rachis forked, the shoulder nearly half as large as main cluster, woolly when young but becoming glabrous; pedicels short, thick, warty; staminate clusters simple, little if any larger than the fertile cluster; pistillate flowers with the 5 stamens short, weak, reflexed and curved laterally, pollen nonvirile; staminate flowers with the stamens long and strong, and with large anthers, pollen abun-

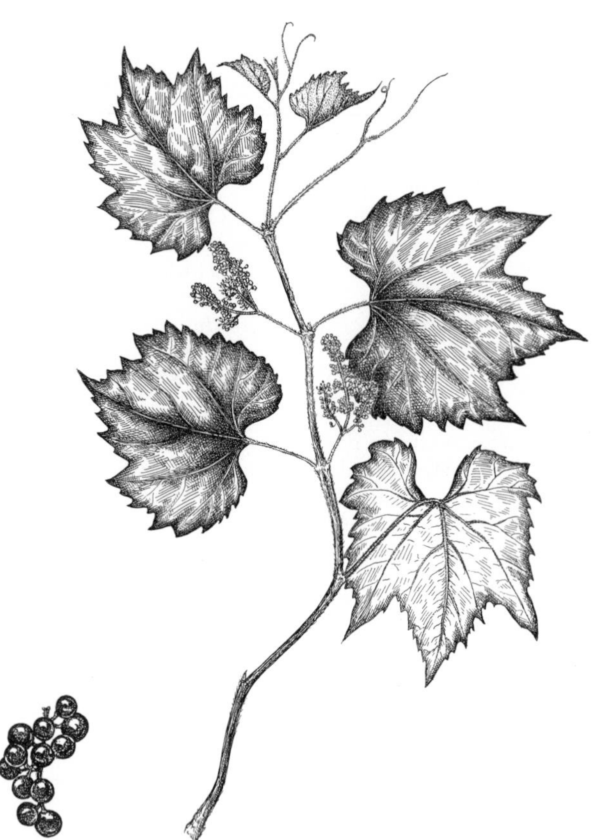

Long's Grape
Vitis longii Prince

dant and very efficient; calyx absent; the 5 petals cohering at summit and separating at base, falling early.

FRUIT. Berries maturing July–August, ⅓–½ in. in diameter, globular or oblate, black with a heavy bloom; skin thin; pulp tough, with copious sanguine pigment, juicy, at first somewhat sour but developing much sugar in full maturity; seeds pear-shaped, ⅕–¼ in. long, or broad, usually 2–3, nearly hemispherical when two, or globose-pyriform when one, reddish brown, apex acute (scarcely a beak); raphe begins as a distinct fine thread on the keel, extends in a slight depression to top of seed where it becomes less prominent or nearly obsolete as it passes through the groove extending from top of seed to small sunken chalaza.

LEAVES. Blades 3½–6 in. wide, 3–5 in. long, broadly cordate-ovate, nearly rotund, rarely slightly 3-lobed or subreniform, either shouldered or nonshouldered, apex taper-pointed or triangular, basal sinus broad or narrow, generally U-shaped; basal lobes spreading or approaching each other, sometimes lapping, teeth large, coarsely angular, usually acute, mucronate; venation of lower surface prominent, with 6–7 nearly always opposite pairs of pubescent and slightly cottony ribs (only one pair of divergent side ribs very prominent); upper surface appearing white when young, usually quite hairy, or some leaves slightly hairy, becoming appressed-cottony or floccose, eventually smooth, glossy, and dark green; lower surface smooth and much paler green than upper surface, the ribs usually retaining the pubescence; firm and leathery; lobes on ground shoots or old roots 3–5-lobed with lateral sinuses narrow and acute; entire and not lobed on seedlings of first year; stipules ⅕–¼ in. long or more, membranaceous, becoming pale brown and shedding when leaves are about half grown; petiole one half to three fourths the length of midrib, grooved moderately on upper side, striate; both pubescent and cottony, eventually becoming glabrous, pale to dark red, usually set at about a right angle with blade.

STEMS. Growing tips whitish or grayish with cottony hairs which become floccose and almost disappear the first year except about the nodes; young wood finely striate, in maturity becoming light grayish brown; on old wood the bark closely persisting for several years, then becoming fibrous and shedding in thin plates; nodes slightly or much enlarged, slightly bent, internodes short, 1–3 in.; pith small and light-colored; diaphragm plane and thin, 1/25–⅛ in. long; tendrils generally once-forked, rarely twice-forked, short, slender, soon caducous unless they engage a support, when they become very thickened, persist and cling with great strength; red, at first bearing cottony hairs.

RANGE. Along sandy banks and bluffs, nearly always in ravines and gulches tributary to the larger streams along the Red, Canadian, Cimarron, and Arkansas rivers, and through the Texas Panhandle; west into New Mexico, and north to Colorado, Oklahoma, and Kansas.

PROPAGATION. Though chiefly found in sandy soils Long's Grape thrives well on limy soils, and endures drought well. The wood resists great severity of cold. It is about 75 per cent resistant to *Phylloxera*. In wet seasons mildew and anthracnose attack young leaves of this species in the Texas Panhandle region (but not the var. *microsperma*) to some extent, especially in its less woolly forms. Black rot attacks it very little. Pollen of staminate plants is very prolific and prepotent in impressing its characteristics in hybridizing with other species. Pollen of the pistillate plants is very feeble (bearing vines rarely set fruit unless staminate vines grown near by). About 60 per cent of cuttings root. Owing to its great vigor, hardiness, and ease of growth, this species must stand in very high repute as a stock. As a direct producer of wine, it is not large enough in cluster, but offers unsurpassed vinous quality for combination in hybridization. The germination, foliation, inflorescence, and ripening of fruit are among the earliest in our species.

HYBRIDS AND VARIETY. A variety has been named *V. longii* var. *microsperma* (Munson) Bailey, which is represented by very vigorous plants, with smaller seeds, and which seems to be especially adapted to resisting drought. The wood is of a lighter straw color. It is found along the Red River in Grayson County, northeastern Texas.

Long's Grape is considered to be possibly a hybrid between *V. rupestris* and *V. arizonica,* but it occurs in sufficient abundance over a wide enough area to make its maintenance of a species a practical one, and it is so accepted by most botanists.

Long's Grape has been found naturally hybridized with *V. candicans, V. doaniana,* and sometimes in another form which bears some resemblance to *V. treleasei.*

REMARKS. The genus name, *Vitis,* is the classical name. The species name, *longii,* is in honor of the western explorer Stephen H. Long (1784–1864). Also known by the vernacular names of Maple-leaf Grape, Bush Grape, Sand Grape, Sugar Grape, and Woolly Riparia. The fruit is sometimes gathered to make pies and homemade wine in the Texas Panhandle area.

Sweet Mountain Grape

Vitis monticola Buckl.

FIELD IDENTIFICATION. Vine slender and climbing 10–30 ft in its native habitat, with unique foliage giving the plant an airy aspect.

FLOWERS. Fertile clusters 1½–3 in. long; peduncles 1–2 in., nearly equally forked, branches nearly half as long as the entire cluster and usually glabrous; fruiting peduncles of a dull, rusty brown, firm but brittle; pedicels ⅛–⅙ in., enlarging rapidly upward, very warty; staminate similar to pistillate but with more numerous flowers; pistillate flowers with 5 petals, the 5 stamens short and recurved; pistil small, short; style short and very slender, with a very small stigma; staminate flowers with stamens erect, straight, and short.

FRUIT. Berries 12–50 per cluster, about ¼–½ in. in diameter, spherical or oblate, nearly always black when fully ripe (red or white varieties very rarely found, but long before they are ripe they assume a rose or pink color, hence are described by some as red grapes), with moderate or little bloom; skin thin, becoming brown, with rough lenticular dots or lenticels; pulp scant, not very juicy, astringent, of a sweetish taste; seeds 1–4, ⅕–⅓ in. long by about ⅕ in. broad, globular when solitary, cinnamon-brown; beak short, thick, blunt; raphe obscure on keel but more prominent at beak, especially so as it passes the barely grooved top of seed, then expanding rapidly into the generally broad, prominent chalaza.

LEAVES. Blade diameter usually 2–3 in., or rarely to 4 in., usually somewhat broader than long, cordate-ovate to deltoid-ovate, outline continuous or with only indications of shoulders, or on certain ground shoots the leaf may be 3-lobed; basal lobes sometimes lapping a little with the basal sinuses commonly inverted U-shaped; teeth medium to large, usually a little acute or right angled and with a sharp or blunt mucro; ribs 3–4 pairs (more rarely to 7 pairs); young leaves pale reddish, then becoming bright green, the upper surface on mature leaves lustrous and glabrous, lower surface smooth and glabrous or sometimes with a little cotton remaining on the veins; thick but brittle; stipules 1/10–⅙ in. long, slightly hairy outside, green to pink or crimson, soon caducous; petiole 1–2 in., terete, faintly or not at all grooved, thin-hairy or cottony along obscure striae when young, becoming smooth or floccose with maturity and green to dark red.

STEMS. When young distinctly angled, at first thinly hairy, becoming terete, finely striate, glabrous, light reddish brown, outer bark often splitting freely in thin fibrous strips; nodes very little enlarged, nearly straight, diaphragm plane or nearly so, 1/12–⅛ in. thick; internodes 2–5 in.; pith small, meeting diaphragm abruptly; tendrils persistent, strong, intermittent, 1–2-forked, somewhat hairy or cottony, reddish brown when mature.

RANGE. Growing abundantly on top of limestone hills of central-southwest Texas in Coryell, Lampasas, Bell, Llano, Travis, Gillespie, and Uvalde counties, and to greater or lesser extent in adjoining counties.

PROPAGATION. Worthy of cultivation as an ornamental vine; seeds germinate slowly; the foliation, flowering, and ripening of fruit is late to very late; endures drought well, withstands cold, matures wood early. The vine always appears healthy, the foliage resisting

Sweet Mountain Grape
Vitis monticola Buckl.

mold well and persisting well into the fall. Black rot does not affect the fruit. The fruit is not particularly valuable owing to the smallness of the cluster, and the fact that the berry is dry and seedy. The vine is a valuable stock for grafting in dry climates and very limy soils. While its slenderness in the wild state was first thought to detract from its value, in cultivation it becomes quite stocky. Cuttings root with some difficulty.

HYBRIDS. It is known to hybridize with *V. candicans, V. berlandieri, V. rupestris,* and *V. arizonica.*

REMARKS. The genus name, *Vitis,* is the classical name. The species name, *monticola,* refers to its preference for mountain habitats.

Cat Grape

Vitis palmata Vahl

FIELD IDENTIFICATION. Beautiful, high-climbing, slender, delicate vine.

FLOWERS. May–July, clusters slender, loose, much-branched; pistillate clusters 4–6 in. long at full development, broad, conical, compound; peduncle and rachis smooth, yellowish green, sometimes thinly hairy; pedicels about 1/5 in., slender, warty near the summit; staminate cluster larger than pistillate and more compound; staminate flowers with the 5 stamens long, slender, erect, with small anthers; pistillate flowers with the 5 stamens short and recurved; pistil minute, style slender, stigma small and hemispherical; petals 5 in all flowers, connate (adjoined) at apex, free at base.

FRUIT. Berries September–October, persistent, 1/5–1/4 in. in diameter, spherical, bluish black, with little or no bloom; skin thick and rigid in breaking; possessed of much coloring matter; pulp juicy, without much pungency, sweet; seeds usually 1–2, largest of any species in proportion to size of berry, short, thick, 1/5–1/4 in. long, 1/6–1/5 in. broad; globose when solitary, hemispherical when in pairs; dark brown or chestnut; beak short, thick, poorly defined, lightly bilobed, raphe threadlike from beak to top of seed, then becoming obscure or invisible in a rather deep groove to the chalaza, which is usually moderately prominent and depressed.

LEAVES. Foliage variable, even on the same shoot, blade 2¾–4¾ in. long, about as wide, cordate or ovate in outline; basal sinus generally narrowly U-shaped; shoulders, or lobes, and apex acute to acuminate, margin either sharply shouldered or 3–5-lobed with narrow, acute or rounded sinuses, frequently many different forms found on the same plant; marginal teeth large, irregular, acute, with long mucronate points; venation with usually 6 subopposite pairs of prominent ribs and their divisions, upper surface dark glossy green with a little short pubescence on the ribs; lower surface smooth, much paler green, glabrous except for pubescence on ribs and tufts in the axillae; stipules 1/8–1/6 in. long, broad at base, apex rounded to obtuse, membranaceous, ciliate-margined; petiole attached to base at obtuse angle, red, length one half to two thirds the width of blade, slender, striate, with a deep narrow groove, pubescent, occasionally with a few cottony hairs; leaves on ground shoots deeply 3-lobed with rounded enlarging sinuses and additional side lobes, giving the blade a skeletonized appearance.

Cat Grape
Vitis palmata Vahl

STEMS. Young growth slender and herblike, with the tips naked or not leafy, thinly covered with whitish cottony hairs, angularly grooved, with a few scattered warts especially near the nodes; at maturity stems conspicuously red or chestnut, finely striate, shiny, smooth or slightly scabrous; bark on old wood separating in wide, thin plates; nodes little enlarged, slightly bent; diaphragm 1/8–1/5 in. thick, biconcave; pith small, slightly enlarged at the proximal end; tendrils persistent, intermittent, forked, strong, when young red and having a few cottony hairs.

RANGE. Low, wet, sandy woods, and margins of swamps and streams. Eastern Texas, Louisiana, Oklahoma, and Arkansas; northward to Illinois and Missouri.

PROPAGATION. Seed germination slow, growth the first year slender and feeble. Resistant to cold and drought when established. Resistant to *Phylloxera,* and otherwise disease-free. Inflorescence the latest to appear of any native species; fruit ripening very late; roots rather thick, abundant, finely divided near the surface, slightly wrinkled transversely, hard, and penetrating.

REMARKS. The genus name, *Vitis*, is the classical name, and the species name, *palmata*, refers to the palmately lobed leaves. Also known under the name of Cat-bird Grape. A beautiful ornamental vine suitable for trellis or arbor, somewhat resembling an *Ampelopsis*.

Muscadine Grape

Vitis rotundifolia Michx.

FIELD IDENTIFICATION. Slender, high-climbing vine lacking the shreddy bark of most grapes.

FLOWERS. Polygamo-dioecious, borne in panicles which are dense and short-branched; pistillate panicle hardly over 2 in. long, sometimes bearing a tendril branch; staminate panicle larger than the pistillate, with slender peduncles; calyx minute; corolla early-deciduous; petals 5, cohering in a cap and never expanding; stamens 5 or more, exserted, alternating with the lobes of the disk, in staminate flowers ascending and straight, in pistillate flowers short and recurved laterally; styles short-conical, stigma small and hemispheric; ovary mostly 2-celled, disk visible.

FRUIT. Maturing July–September, dropping singly as soon as ripe, borne in short, loose, globular clusters of 2–8 or more; peduncles glabrous or pubescent, ½–2 in.; pedicels ¼–⅓ in.; berries subglobose, purplish black, rarely yellowish bronze (or silvery green in the horticultural variety Scuppernong), with little or no bloom, often warty-dotted, ⅓–1 in. in diameter; skin tough; flesh tough, thick, musky, edible; seeds 1–4 (usually 2–3), ¼–⅜ in. long, about ¼ in. broad, oval-lenticular when 2, ellipsoidal when 1, ends rounded or obtuse, dorsal side rounded with a narrow raphe showing as a very thin line, ventral side flattened or 2-angled, with a small, narrow, depressed chalaza.

LEAVES. Alternate, oval to suborbicular or broadly cordate, margin seldom lobed, coarsely angular-dentate, teeth sometimes mucronate (25–35); apex acute or short-acuminate; base cordate with sinus generally narrow, rarely closed, sometimes nearly truncate; thick, firm, leathery; upper surface glabrous, dark green, lustrous, ribs in 5–6 pairs and impressed; lower surface slightly paler, glabrous or a few hairs on the nerves, or tufts at the vein angles; blade 1¾–5 in. long (average 2½ in.), 2–4 in. broad; petioles green or dull red, slender, flattened, narrowly to deeply grooved above, glabrous to pubescent along obscure striations; stipules minute, about 1/10 in. long, truncate.

STEMS. Climbing to as high as 100 ft, young stems green to reddish brown, angled, often minutely warty; thinly pubescent at first; stems sometimes producing aerial rootlets; old stems hard, terete or flattened, glabrous, reddish brown to grayish and warty, bark tight and not shreddy as in other grapes; tendrils simple, 4–6 in. long, absent from every third node; pith continuous through the nodes (pith apparently not typical, but seemingly a dense, green, cellular, nonfibrous wood).

RANGE. In high bottom lands where the land is well drained, and in moist shady situations. Eastern Texas, Oklahoma, Arkansas, and Louisiana; eastward to Florida, northward along the Atlantic coastal plain to Washington, D.C., and westward to Missouri and Kansas.

PROPAGATION. The vigor, long life, and freedom from disease of this species is unsurpassed in *Vitis*. It is never attacked by rot or mildew. The cuttings root with great difficulty, but layers of young wood root easily if twisted withelike at one point when covered.

HYBRIDS. *V. rotundifolia* has never been found naturally hybridized with any species other than *V. munsoniana*, so far as is known. It has been artificially hybridized with *V. vinifera* and with *V. rupestris*. The latter hybrid proved to be sickly and nonproductive, however, and resembled *V. rupestris* more than *V. rotundifolia*. A number of vigorous hybrids of the Scuppernong variety (a variety with silvery amber-green fruit) can be produced with hybrid varieties of the true bunch grapes.

Some of the horticultural varieties now recognized for the Muscadine Grape are: Eden (Guild), Flowers (Flo., 1819), James, Memory (Mem., 1868), Mish (Mish, 1846), and Scuppernong.

REMARKS. The genus name, *Vitis*, is the ancient Latin name, and the species name, *rotundifolia*, refers to the rounded leaves. Vernacular names are Southern Fox Grape, Bullace Grape, Bullet Grape, Bull Grape, and Scuppernong.

Muscadine Grape
Vitis rotundifolia Michx.

Red-shank Grape

Vitis rufotomentosa Small

FIELD IDENTIFICATION. Strong high-climbing vine with forking tendrils, recognized by its twigs and foliage being copiously red or reddish-tomentose.

FLOWERS. Borne in ample racemes, sterile ones 3¼–4¾ in. long in anthesis; peduncles much shorter than the racemes, red, felty-floccose; fertile racemes shorter than the sterile ones, stamens 5 and not strongly reflexed; flowers rather small, dioecious or polygamo-dioecious; calyx minute; petals 5, adhering at top like a cap, early-deciduous; disk hypogynous; stamens exserted and alternating with the lobes of the disk; style short, ovary 2-celled with 2 ovules in each cell.

FRUIT. Berries black, with little or no bloom, ⅕–¼ in. in diameter; seeds few, pear-shaped, gradually narrowed to the beak; raphe very prominent on keel and over ridge to the small but conspicuous circular chalaza.

LEAVES. Simple, alternate, blade 4–8 in. long, usually ovate to suborbicular or broadly cordate-ovate, as broad as long or broader, margin coarsely and sinuately toothed and usually with 2–4 lateral acute or obtuse lobes (on ground shoots 3–5-lobed and with deep rounded sinuses, basal sinus narrow or broad); base cordate; upper surface dull green with scattered webby hairs, lower finely webby with densely reddish brown tomentum especially on the veins; lateral veins 3–4, very conspicuous below leaf and red-rusty; stipules small and caducous; petioles short, 1⅓–2¾ in., stout, rusty-tomentose or shaggy.

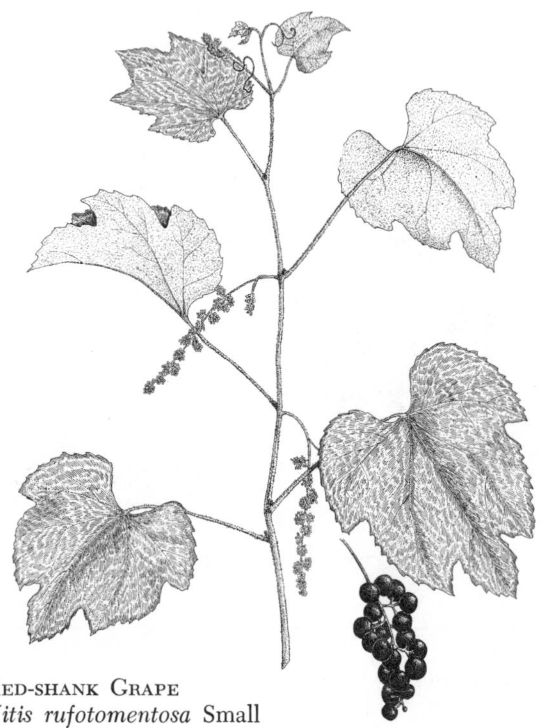

Red-shank Grape
Vitis rufotomentosa Small

STEMS. Rather stout, densely reddish woolly-tomentose; older woody, gray to reddish and more glabrous, shreddy; internodes short, diaphragm 1/12–⅕ in. long.

RANGE. In sandy soil from eastern Texas and Louisiana eastward into Florida. Specimen cited deposited at the Houston Museum of Natural History, George L. Fisher, B-1548J, Hemphill, Texas.

REMARKS. *Vitis* is the classical Latin name, and the species name, *rufotomentosa,* refers to the reddish brown tomentum of the foliage.

Sand Grape

Vitis rupestris Scheele

FIELD IDENTIFICATION. Vine prostrate or ascending, or merely reclining, scarcely climbing, rarely more than 4–8 ft. Branch ends very leafy because of the rapidly unfolding tips and the short internodes.

FLOWERS. Borne in May, pistillate clusters ⅓–2 in., simple or nearly so, seldom shouldered, the peduncle equal in length or shorter; rachis pale yellowish-green, glabrate or somewhat floccose; pedicels about ⅛ in., slender, thickened toward the receptacle; the 5 stamens of the pistillate flower recurved and bent laterally, producing nonvirile pollen (hence pistillate plants will not bear alone), ovary 2-celled and globose, style short and slender, stigma medium; staminate cluster usually considerably larger than the pistillate, forked; the 5 stamens slender and ascending; anthers large and bearing abundant virile pollen; petals 5 in both the staminate and pistillate flowers, cohering at apex, separating at base.

FRUIT. Borne late June–August, soon dropping, ¼–½ in. in diameter, globose or somewhat depressed, often somewhat doubled (like two berries coalesced), black, with or without bloom, skin very thin and tender; pulp tender and sweet; seeds 3–4, ⅙–⅕ in. long and nearly as broad; light chocolate-colored, beak rather short and acute; raphe slender, threadlike, extending over the notch (if notch is present) and continuing to the chalaza which is small and slightly elevated.

LEAVES. Width averaging 3–4 in., sometimes to 8 in., usually broader than long, outline reniform to reniform-ovate, folded upward or trough-shaped to expose the pale green lower surface; basal sinus very broad and often somewhat truncate, but usually terminating acutely at insertion of the petiole; margin with only indications of shoulders, teeth large, with the apices short taper-pointed and mucronate; venation rising mostly from 5 pairs of generally opposite ribs which are sometimes pubescent, or with tufts of hairs in the axils; upper surface light glaucous green and glabrous; lower surface glabrous and yellowish green; dense, somewhat

SAND GRAPE
Vitis rupestris Scheele

leathery but fragile; blades on ground shoots of old roots often a little lobed; stipules 1/6–1/5 in., lanceolate, membranaceous, crimson; petiole deeply and broadly grooved, distinctly striate, usually smooth or thinly pubescent along the striae, green to dark crimson.

STEMS. When young leafy to the tip, slightly 5–6-angled, becoming terete, finely striate when mature, red, eventually with bark a dark cinnamon color, growing darker with age, scaling off in broad nonfibrous plates after the second or third year; wood dense and not very hard; nodes slightly enlarged, nearly straight, diaphragm 1/16–1/8 in. thick and nearly plane; internodes 1/3 in.; pith rather large, enlarging downward in each internode toward diaphragm; tendrils thin, smooth, crimson, or sometimes tendrils absent.

RANGE. Sand hills or calcareous soils along streams or gulches, often at the head of ravines. In central and southwestern Texas to the Rio Grande, and in Arkansas along the Ozark ridge into Oklahoma; eastward to Tennessee and Kentucky, and northward to southern Missouri and Illinois.

PROPAGATION. Sand Grape succeeds in poor calcareous and sandy soils. The seed germinates readily, foliation is fairly early, and fruit ripening is early. This species has the faculty of bearing fruit on young shoots from 2–4-year-old wood, in case the last year's growth has been destroyed during the winter. However, it is generally resistant to cold weather.

The vigor and hardiness of Sand Grape are usually great where there is permanent moisture a few feet below the ground surface. The foliage is resistant to fungus and insect attacks. Although readily attacked by the speckled form of anthracnose no serious damage results. It resists black rot completely. Propagation is easily accomplished from cuttings, with about 80 per cent of them taking root. Pollen from the staminate plants is very prepotent in fertilizing and hybridizing with other species. The plant is sometimes used as a grafting stock in vineyards out of its natural range.

HYBRIDS AND FORMS. Sand Grape is known to hybridize with the following species, or with mixed hybrids from them:

V. aestivalis, *V. bourquiniana*, *V. candicans*, *V. champinii*, *V. cinerea*, *V. labrusca*, *V. lincecomii*, *V. riparia*, *V. rotundifolia*, *V. vinifera*, *V. vulpina*, *V. berlandieri*, *V. cordifolia*, *V. monticola*, *V. arizonica*, and *V. longii*.

Cut-leaf Grape, *V. rupestris* forma *dissecta* (Eggert) Fern., has laciniately toothed leaves with a strong tendency to irregular lobing.

REMARKS. The genus name, *Vitis*, is the classical name, and the species name, *rupestris*, means "of rocks," referring to this species' habitat preference of stony ground. Vernacular names are Sand Beach Grape, July Grape, Currant Grape, Rock Grape, and Bush Grape.

TRELEASE GRAPE

Vitis treleasei Munson

FIELD IDENTIFICATION. Much-branched shrubby plant, usually showing little tendency to climb. The foliage bright green and shining.

FLOWERS. Clusters small, averaging 2–3 in., staminate flowers with 5 ascending, slender stamens, pollen grains small; pistillate flowers with stamens recurved; disk distinct, lobed; ovary globose, style short and thick,

TRELEASE GRAPE
Vitis treleasei Munson

stigma broad; peduncle and rachis slender, smooth or with a few cottony hairs; peduncles 1/5–1/4 in., little warty, or not at all, enlarging moderately at summit.

FRUIT. Clusters sometimes to 4½ in., glabrous and somewhat branched, berries maturing in September, 1/4–1/3 in. in diameter, covered with thin bloom, globular, black, thin-skinned; juice red; pulp juicy, sweet, of very agreeable flavor; seeds 1–3, usually 2, 1/5 in. long, 1/6 in. broad, obcordate, of variable form; beak short, blunt; raphe a fine thread to chalaza, which is prominent and narrow.

LEAVES. Very broad-ovate to nearly reniform, blade 3–5 in. wide, mostly wider than long, apex triangular, basal sinus broad and often having shallow acute sinuses either side of central lobe, apex triangular, margin indistinctly 3-lobed, with teeth rather unequal and acute-pointed; ribs 5–6 nearly opposite pairs; both upper and lower surfaces smooth and lustrous, except for short pubescence along the ribs and their axils; petiole generally more than half as long as width of blade, grooved on upper side, faintly striate, smooth or thinly pubescent, bright red when young; stipules 1/12–1/8 in., thinly membranaceous.

STEMS. Pale when mature, the bark splitting after first year and separating into thin fibrous plates; nodes straight, little enlarged, if at all; diaphragm 1/16–1/6 in. thick, slightly biconcave; internodes short, 1–3 in.; tendrils small, generally shorter than the leaves and mostly once-forked, smooth or sparingly downy, deciduous first year, unless clasping some object, then gripping feebly.

RANGE. Ravines and gulches of southwestern Texas (Chisos Mountains), western New Mexico, and Arizona.

PROPAGATION. Foliation and inflorescence considerably later than *V. vulpina,* and somewhat earlier than in *V. arizonica;* ripening of fruit late, exfoliation late, endures cold well. It has little horticultural value, and is very sensitive to downy mildew and black rot.

RELATED SPECIES. It appears to stand midway between *V. vulpina* and *V. arizonica.* In its Eastern range it has affinities with *V. vulpina,* and in the West with *V. arizonica.*

REMARKS. The genus name, *Vitis,* is the classical Latin name. The species name, *treleasei,* was given by T. V. Munson for his botanist friend William Trelease. It is very closely related to *V. arizonica.*

Frost Grape

Vitis vulpina L.

FIELD IDENTIFICATION. Vigorous high-climbing vine, moderately branching, attaining a length of 10–30 ft or more; trunk sometimes as large as 1–2 ft thick.

FLOWERS. May–June, pistillate clusters small to medium, 3–5 in. long, including peduncle which is one

Frost Grape
Vitis vulpina L.

third to one half as long as the rachis; shoulder on fourth to one third as large as the main cluster; peduncle and rachis pubescent or smooth; pedicels from little divisions of rachis, or direct, 3/16–1/4 in., enlarging slightly toward summit; stamens of pistillate flower reclining and curving laterally, producing nonvirile pollen when the stamens are reflexed; staminate cluster much larger than the pistillate, very floriferous and fragrant, commonly prominently shouldered and branched staminate flowers with long and ascending stamens, with pollen very abundant and virile; petals 5, attached a the apex, spreading at base and deciduous.

FRUIT. Maturing September–October, persistent, berries 1/4–1/3 in., or rarely ½ in., in diameter, globular, blac (very rarely white), with a heavy pruinose bloom when ripe; skin thin; pulp juicy, rich in sugar and tartari acid, vinous-tasting when fully ripe, but until then exceedingly acid; seeds 2–4, about 1/6–1/4 in. long, 1/8–3/1 in. broad, coffee-colored; beak very short, poorly defined, acute, raphe scarcely visible from chalaza ti it reaches top of seed, thence becoming a well-define thread reaching the beak; chalaza itself narrowly elliptic and slightly elevated.

LEAVES. Blade 3–5 in. long, sometimes broader tha long, about one third folded toward the upper surface outline cordate-ovate with a long and sharp apex usually with a short lobe on either shoulder with an open sinus; lateral lobes rarely separated by narrow round sinuses; basal sinus broadly U-shaped, basal lobes rarely if ever lapping; teeth on margin large, irregular and acute, with smaller teeth between; venation moderately prominent from the 6–7 pairs of subopposite, pubescent ribs; the ribs also with axillary tufts of hairs, both surfaces of blade otherwise smooth; upper surface of blade rather dark green in Eastern forms or brighter green in Western forms; thicker in Western than in Eastern forms. Leaves on ground shoots generally 3-lobed; stipules large, 3⁄16–5⁄16 in. long, ½–⅔ as wide, membranaceous, becoming light brown and shedding when the leaf is about full size; petiole usually shorter than blade, cylindrical or slightly elevated on upper side, with a slight groove, the sides of groove pubescent even in otherwise smooth specimens, in some specimens pubescent all over; when young pale red or green, sometimes dark red.

STEMS. Young ones pubescent to glabrous later, terete, reddish, becoming clear green; bark of annual wood finely to obscurely striate when mature, of a dark crimson color in Eastern forms, or lighter in Western forms, cracking and easily separating into thin, weakly fibrous plates at close of first year and during second year; wood rather soft, probably the softest of any native species; nodes slightly swollen and little bent, with very thin diaphragms 1⁄25–1⁄12 in. thick; internodes 1–4 in., terete; pith rather large, meeting the diaphragm abruptly, light brown.

RANGE. Prefers moist but well-drained soil on loamy or sandy banks of streams or lakes. In New Mexico, Oklahoma, Arkansas, Texas, and Louisiana; eastward to Tennessee, northward to Virginia and Nova Scotia, and westward to Manitoba, Montana, and Colorado.

PROPAGATION. The vine was first cultivated in 1806. The seed germination is quick and vigorous, with the seed numbering about 14,500 seeds per lb. The foliage, inflorescence, and fruit mature very early, some varieties later by one or two weeks. Frost Grape is easily grown from cuttings, with about 85 per cent of them taking root, also easily grafted. It apparently endures cold better than any species. However, its ability to withstand severe summer heat and long drought is rather poor. The pollen is prepotent upon other species when blooming together, equally so with *V. rupestris.*

HYBRIDS. Owing to the resistance to *Phylloxera,* ease of growth from cuttings, and grafting, *V. vulpina* and *V. rupestris* are extensively used in Europe to graft favorite *vinifera* varieties, but they are not suited to the very limy soils of the Charente-Inférieure in which it is found that *V. berlandieri* succeeds best.

Frost Grape is known to hybridize with the following species, or mixed hybrids of them: *V. aestivalis, V. bourquiniana, V. candicans, V. champinii, V. cinerea, V. labrusca, V lincecomii, V. riparia, V. rotundifolia, V. rupestris, and V. vinifera.*

VARIETIES. The following natural varieties are listed by Bailey (1934, p. 233).

Dune Frost Grape, *V. vulpina* var. *syrticola* Fern. & Wiegand, has petioles and fully expanded blades, soft-pilose beneath to the touch. On sand dunes on the southeastern shore of Lake Michigan in Michigan and Indiana; probably in similar intermediate places, and perhaps of still wider distribution.

June Frost Grape, *V. vulpina* var. *praecox* Bailey, has flower clusters small, 2½ in. long or less, very early (April); berries many and small, ¼–⅓ in. thick or less, on short pedicels in clusters, with bloom, sweet, ripe in June; seeds small, sometimes nearly circular in outline aside from the very short abrupt beak, about ⅙ in. long, raphe little evident on the keel and in the groove; chalaza sunken, small, and little-developed. In the vicinity of St. Louis, Missouri, on both sides of the Mississippi River.

REMARKS. The genus name, *Vitis,* is the classical name, and the species name, *vulpina,* pertains to the fox, which eagerly devours the fruit. Vernacular names are Fox Grape, Winter Grape, Chicken Grape, Raccoon Grape, Riverbank Grape, and August Grape. The fruit can be made into preserves and wine. A number of birds and mammals also eat the fruit including the ruffed grouse, bobwhite quail, wild turkey, mourning dove, white-tailed deer, red fox, striped skunk, and cottontail.

LINDEN FAMILY (*Tiliaceae*)

American Linden

Tilia americana L.

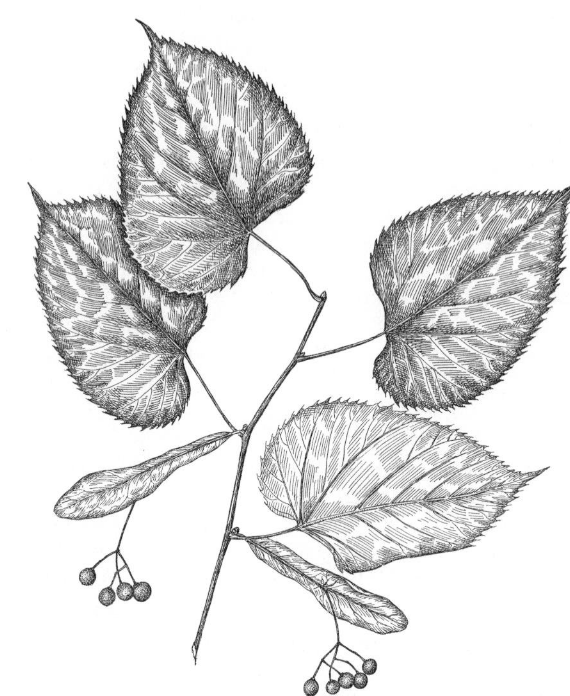

American Linden
Tilia americana L.

FIELD IDENTIFICATION. Tree known to attain a height of 130 ft, with a diameter of 4 ft. The branches are usually small and horizontal, or drooping, forming a broad, round-topped head. The root system is wide-spreading, and root sprouts often grow at the base of the trunk.

FLOWERS. May–June, perfect, borne in loose drooping cymes with pedicels slender and glabrous, 6–15-flowered; peduncle 1½–4 in., slender and glabrous, attached to a large foliaceous bract; bract 2–5 in. long, ¾–1½ in. wide, membranous, glabrous, strongly veined, narrowly oblong, apex rounded or obtuse, base narrowed, sessile or short-stalked; sepals 5, small, ovate, acuminate, pubescent within, puberulent externally, considerably shorter than the petals; petals 5, yellowish white, lanceolate, crenate, alternate with the sepals, a small scale alternating with each petal; stamens numerous, shorter than the petals; ovary 1, 5-celled, each cell 2-ovuled, ovules anatropous; style simple and tomentose.

FRUIT. Drupaceous, dry, persistent, globose to ovoid, apex rounded or pointed, ¼–⅓ in. long, densely brown-tomentose; 1–2-seeded.

LEAVES. Simple, alternate, deciduous, broad-ovate, apex acute or acuminate, base commonly unsymmetrical, obliquely cordate or almost truncate, margin coarsely serrate, individual teeth with slender, gland-tipped apices; thick and firm, upper surface dark green and glabrous, lower surface paler green and glabrous or some with axillary tufts of hairs, length of blades 5–6 in., width 3–4 in., prominently yellow-veined; petioles 1½–2 in. long, about one third the length of the blade.

TWIGS. Slender, somewhat divaricate, glabrous, smooth, green to brown, gray later; lenticels numerous, oblong; winter buds dark red, ovoid, about ¼ in. long, mucilaginous; leaf scars half-elliptic, stipule scars conspicuous.

BARK. Light brown to gray, about 1 in. thick on the trunk, the deep furrows separated by narrow, flat-topped, confluent ridges which shed small thin scales.

WOOD. Light brown to reddish brown, straight-grained, rather soft, moderately weak in bending and endwise compression, low in shock resistance, works well with hand tools, finishes smoothly, holds glue, paint, and lacquer well, resistant to splitting, holds nails poorly, seasons well but shrinks considerably, low in durability, weighing 28 lb per cu ft.

RANGE. Rich, moist soil of woods and bottom lands. Northeast Texas, Oklahoma, Arkansas, Missouri, Tennessee, Kentucky, and Georgia; north to Nebraska, Kansas, North Dakota, Minnesota, Illinois, Indiana, Michian, Pennsylvania, Maine, New Brunswick, Quebec, nd Ontario.

PROPAGATION. The winged bracts are removed from he fruit by flailing or rubbing. The hull may be re- oved by running through a coffee grinder. The seed verages about 5,000 seeds per lb, with a commercial urity of 99 per cent and a soundness of 81 per cent. 'he seed will keep for at least 2 years if stored dry t room temperature. Because of excessive seed dorman- y, due to a hard seed, the seed must be pretreated y soaking in concentrated sulphuric acid for 20 min- tes and stratified in moist sand or peat for 3–5 months t 30°–40°F. However, even with this treatment ger- nination may be irregular over 2–3 years. Stratified eeds are covered ¼–½ in. with soil when sown in the pring. The plant may also be propagated by layers or y budding or grafting.

HORTICULTURAL FORMS. Big-tooth American Linden, *. americana* forma *dentata* (Kirchn.) Rehd., is a form vith large leaves, and teeth very coarse and some oubly serrate.

Big-leaf American Linden, *T. americana* forma *mac- ophylla* (Bayer) V. Engl., has very large broad leaves, –8 in. long.

Grape-leaf American Linden, *T. americana* forma *mpelophylla* V. Engl., has leaves large, lobed, and oarsely and irregularly toothed with acuminate teeth.

Pyramidal American Linden, *T. americana* forma *astigiata* (Slavin) Rehd., has a narrow pyramidal form vith ascending branches.

REMARKS. *Tilia* is the classical Latin name, and the pecies name, *americana*, refers to its distribution. It as many vernacular names such as Bast-tree, Lin- ree, Lime-tree, Bee-tree, Blacklime-tree, White Lind, Vhitewood, Southern Lind, American Basswood, Yel- ow Basswood, Whistlewood, Spoonwood, Daddynuts, Monkeynuts, and Wickyup.

The American Linden has been cultivated for orna- nent since 1752. The tough inner bark was formerly sed for mat fiber and rope by the Indians. The lowers are valuable for bee pasture, and young trees urnish wildlife cover. The fruit is known to be eaten y a number of species of birds and rodents. The wood s valuable for paper pulp, woodenware, cheap furni- ure, plywood, veneers, panels, cooperage, boxes and rates, casks and coffins, handles, shades, blinds, fix- ures, appliances, excelsior, and millwork.

Carolina Linden

Tilia caroliniana Mill.

FIELD IDENTIFICATION. Large tree with an irregular, ounded top.

FLOWERS. Borne on slender peduncles, pubescent, in 3–15-flowered cymes and subtended by conspicuous papery bracts; bracts linear, elliptic to obovate, cuneate at the base, rounded or acute at apex, somewhat pubescent at first, becoming glabrous later, 4–5 in. long, about ⅘ in. wide, decurrent or almost so to the peduncle base; sepals 5, shorter than petals, ovate, acuminate, ciliate, brown-pubescent on the exterior, white-hairy within; petals 5, lanceolate, acuminate, somewhat longer than sepals; stamens many, with filaments forked at apex; staminodia about as long as sepals; ovary superior, 5-celled, with a slender style and 5-lobed stigma.

FRUIT. Nutlet subglobose to ellipsoid, apiculate, tomentose, or pubescent, rather small, about ⅛ in. in diameter, 1–3-seeded.

LEAVES. Alternate, simple, deciduous, blades 2⅓–4½ in. long, 1¾–3½ in. wide, broadly ovate, abruptly acuminate, base truncate or cordate and asymmetrical; margin coarsely dentate with glandular apiculate teeth; upper surface dark green, lustrous and glabrous at maturity, lower surface thinly tomentose with brownish fascicled hairs; leaves tomentose on both sides when young; petioles stout, glabrous or slightly pubescent, 1–1½ in.

TWIGS. Slender, reddish brown, pubescent at first, glabrous later.

BARK. Gray, with shallow fissures and flat ridges.

WOOD. Light-colored, soft, light, easily worked.

RANGE. Central and eastern Texas, southwestern Arkansas, and western Louisiana; eastward to Georgia and Florida, and north to North Carolina.

VARIATIONS. The species has a variety known as *T. caroliniana* var. *rhoophila* Sarg. which occurs in Ar-

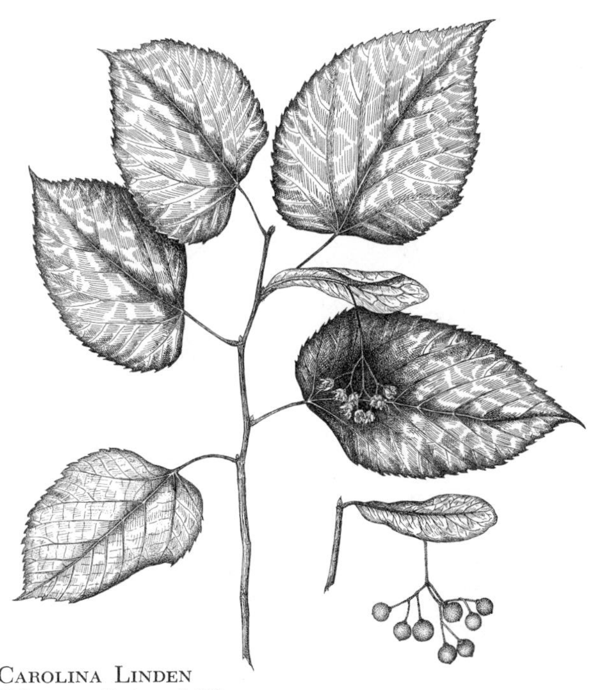

Carolina Linden
Tilia caroliniana Mill.

kansas, Louisiana, and Texas, west to the Guadalupe River. It is distinguished by pubescent twigs and winter buds, larger leaves and tomentose clusters of more numerous flowers.

The following Texas plants, which at one time were considered as separate species, cannot be separated by any constant characters and are hereby considered as synonyms of *T. caroliniana:*

T. texana Sarg., occurring in Brazos, Kendall, and Kerr counties; *T. phanera* Sarg., of south central Texas; and *T. phanera* var. *scabrida* Sarg., found in Blanco and Brazoria counties.

REMARKS. The genus name, *Tilia,* is the classical name, and the species name, *caroliniana,* refers to the states of Carolina. Vernacular names are Basswood, Lime-tree, Whitewood, and Bee Basswood. The tree is a rapid grower, and is often planted for ornament. The flowers make good honey, and the wood is used with other species for making interior finishing and woodenware.

Florida Linden

Tilia floridana Small

FIELD IDENTIFICATION. Tree attaining a height of 40 ft, with an irregular, rounded crown.

FLOWERS. Borne on tomentose pedicels from large membranous bracts in drooping cymes; bracts decurrent on the petiole, membranous, greenish, glabrous, linear to oblong or spatulate, often falcate, rounded at

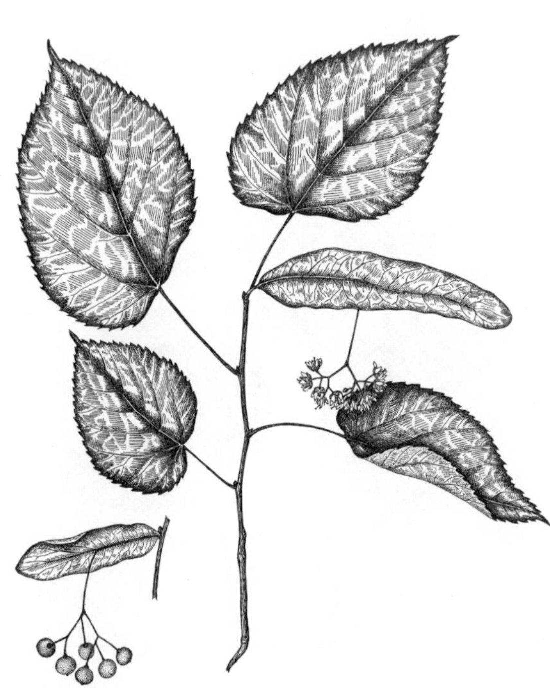

Florida Linden
Tilia floridana Small

the apex, 3–6 in. long, flowers in flattened cymes; sepal 5, ovate, acuminate at apex, tomentose, shorter than th petals; corolla about ¼ in. long, yellowish; petals 5 lanceolate; stamens numerous; staminodia often present oblong-obovate, shorter than the petals; ovary broad tomentose, 5-celled, style slender, with 5-lobed stigmas

FRUIT. Nutlet globose or ovoid, apiculate, tomentose about ¼ in. in diameter, usually 1–3-seeded, ripe in September.

LEAVES. Alternate, simple, deciduous, blades 3–5 in long, 2½–3½ in. wide; broad-ovate, acuminate at apex sometimes abruptly so; truncate, rounded, or cordate and unsymmetrical at the base; margin coarsely serrat with apiculate teeth; at maturity dark green and glabrous above, pale and sometimes silvery glaucous, o with axillary hairs beneath; young leaves tomentose petioles slender, usually glabrous, ¾–1 in.

TWIGS. Slender, reddish brown to yellow, tomentos or glabrous.

BARK. Gray, furrows shallow, ridges flat-topped, interlacing into small angular blocks.

WOOD. Light-colored, light, soft, easily worked.

RANGE. Central and eastern Texas, eastern Oklahoma Arkansas, and Louisiana; eastward to Florida and northward to North Carolina and Virginia. In Mexico in Nuevo León, Coahuila, and Chihuahua.

VARIETIES AND SYNONYMS. The form with silver bloom beneath the leaf has been classified as *T. floridana* var. *hypoleuca* Sarg.

Gulf Linden, *T. leucocarpa* Ashe, formerly describe as a separate, but closely related, species, has bee relegated to a synonym of *T. floridana* Small, since th two are connected by intermediate forms. Other forme Southwestern species reduced to synonyms of *T. floridana* Small are: *T. cocksii* Sarg., *T. alabamensis* Ashe and *T. nuda* Sarg.

REMARKS. The genus name, *Tilia,* is the classica name, and the species name, *floridana,* refers to th plant's Floridian habitat. Other names are Bee Linde and Florida Basswood. The tree is propagated b stratified seeds and by layering. It is a rapid grower and sometimes is planted for ornament. The nectar i a good bee food, and the wood is used along wit other species for interior trim, wood carving, and smal woodenware articles.

Bee-tree Linden

Tilia heterophylla Vent.

FIELD IDENTIFICATION. Tree attaining a height of 80 ft and a diameter of 2½ ft. The trunk is long, clear o branches for a considerable distance, and the spreading branches form an irregular head. Root system wide and deep.

FLOWERS. Borne May–July, in 8–20-flowered pubescent corymbs; pedicels slender, clavate, pubescent

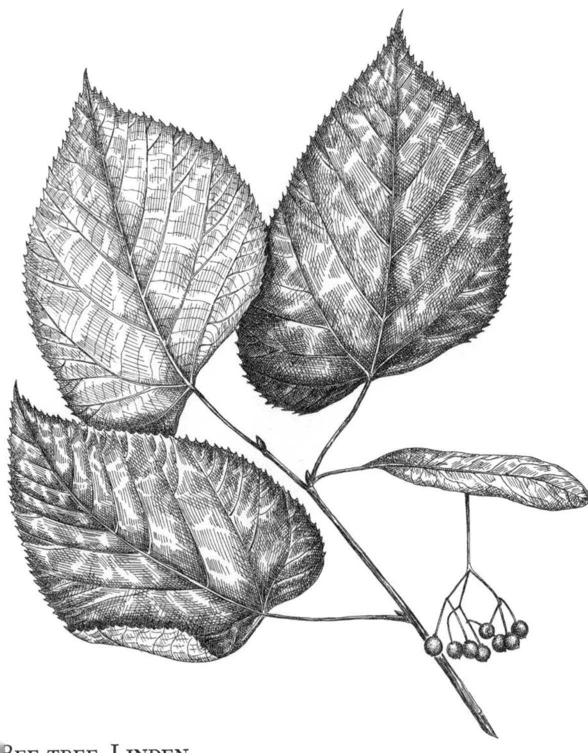

BEE-TREE LINDEN
Tilia heterophylla Vent.

petioles of bracts glabrous; bracts oblong, or narrowed and rounded at apex, base cuneate, pubescent above, tomentose beneath, later glabrous, length 4–6 in., width 1–1½ in.; flowers fragrant, nectariferous; sepals 5, distinct, acuminate, pubescent above, villose beneath, white-hairy at base; petals 5, imbricate in the bud, lanceolate, acuminate, longer than the sepals and alternate with them; stamens inserted on a short hypogynous receptacle, collected into 5 clusters; a petaloid spatulate scale placed opposite each petal; style erect, dilated at apex into 5 spreading stigmatic lobes, villous at base; ovules 2 in each cell.

FRUIT. August–September, nutlike, woody, ellipsoid or globose, apiculate at apex, ¼–⅓ in. long, brown-tomentose, 1-celled, 1–2-seeded; seeds obovoid, ascending, reddish brown, seed coat cartilaginous, embryo large.

LEAVES. Simple, alternate, deciduous, conduplicate in the buds, ovate, apex gradually narrowed and acuminate, margin finely apiculate and glandular-toothed, at maturity dark green and glabrous above, lower surface with white or brown tomentum, or with brown hair-tufts in the vein axils, blade length 3¼–5¼ in., width 2½–2¾ in.; petioles 1½–1¾ in., slender, glabrous or somewhat pubescent.

BARK. Gray to brown, rather thick and deeply furrowed on older trunks, ridges flattened with scaly surfaces.

TWIGS. Slender, terete, yellowish to reddish brown, leaf scars lunate, stipule scars prominent; winter buds ⅕–⅓ in. long, oblong to ovate, somewhat compressed, scales apically ciliate; axillary buds large, acute, compressed, scales imbricate; pith white and homogeneous; inner bark tough and fibrous.

WOOD. The heartwood is pale brown to reddish, and the sapwood white to brownish. It is straight-grained and has a specific gravity of 0.40, rather soft, weak in endwise compression, low in shock resistance, moderately weak in bending, works well, takes a smooth finish, takes paint well, glues well, has poor nail-holding ability, seasons well, and holds shape fairly well. The wood is used for piano keys, venetian blinds, handles, rollers, cabinets, laundry appliances, caskets and coffins, woodenware, crates, excelsior, cooperage, trunk panels, furniture, and all types of veneer.

RANGE. Northern Arkansas; eastward to Georgia and Florida, northward to New York, and west to Illinois and Missouri.

VARIETY. A variety known as Michaux Bee-tree Linden, *T. heterophylla* var. *michauxii* (Nutt.) Sarg., is seemingly more abundant than the species, and is found in Ohio, Indiana, Arkansas, Alabama, and Georgia. The leaves are ovate to oblong, apex acute or abruptly acuminate, base cordate or obliquely truncate, margin coarsely serrate, whitish-tomentose beneath, blade length 3–6 in.; peduncle pubescent at first but glabrous later; fruit ¼–⅓ in. in diameter.

REMARKS. The genus name, *Tilia*, is the classical Latin name, and the species name, *heterophylla*, means "various-leaved." Also known under the vernacular name of White Basswood. It has been in cultivation since 1800 and is rapid-growing and short-lived.

MALLOW FAMILY (*Malvaceae*)

Rio Grande Abutilon

Abutilon hypoleucum Gray

FIELD IDENTIFICATION. Plant to 4 ft, herbaceous above but somewhat lignescent toward the base.

FLOWERS. Perfect, mostly axillary, peduncles white-tomentose, usually solitary and exceeding the petioles (1–3 in. long); calyx of 5 valvate white-tomentose sepals, about as long as the petals, ovate to subcordate, apex acuminate to acute, some overlapping at base, ½–¾ in. long; petals 5, alternate with the sepals, convolute, distinct, obovate to oval, ⅔–1 in. long; stamens numerous, monadelphous in a column; anthers 1-celled, kidney-shaped; stigma terminal and capitate.

FRUIT. Carpels numerous, leathery, stiff, long-beaked, densely white-hairy, length ½–⅝ in., 2-valved, deciduous from the central axis at maturity, 3–9 seeded, tardily dehiscent; seeds heart-shaped, about 1/25 in. long, dark brown, puberulent, hilum large.

LEAVES. Simple, alternate, deciduous, broadly ovate to lance-ovate, apex acute to long-acuminate, base cordate, margin erose-serrate, blade length 1½–4½ in., upper surface dull green and scantily pubescent to almost glabrous except along the veins, lower surface velvety white-tomentose, palmately veined; petioles ½–2 in. long, pubescent.

TWIGS. Young stems slender, green, densely tomentose, later tan to reddish brown and glabrous.

RANGE. At New Braunfels, Texas, and southward along the lower Rio Grande. Collected by the author in the palm grove on Rabb Ranch below Brownsville, Cameron County, Texas. In Mexico in Nuevo León, Coahuila, San Luis Potosí, and Puebla.

REMARKS. The genus name, *Abutilon,* is from an Arabic name given by Avicenna. The species name, *hypoleucum,* refers to the dense white tomentum of the leaves. About 100 species of *Abutilon* are known, mostly from tropical regions. They are grown from seed, or from cuttings taken early for bedding.

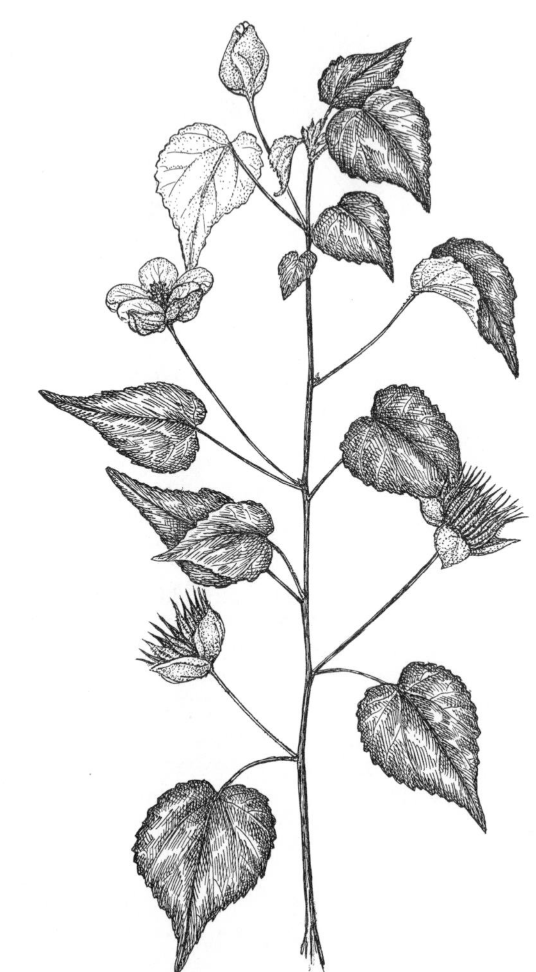

Rio Grande Abutilon
Abutilon hypoleucum Gray

Berlandier Abutilon

Abutilon lignosum (Cav.) Don

FIELD IDENTIFICATION. Velvety-pubescent perennial, sometimes woody near the base. The erect or ascending stems 2–6 ft.

FLOWERS. On axillary peduncles, solitary, or a few together, perfect, regular, yellow or orange; petals 5, distinct, convolute, apex rounded or nearly truncate, to 5/8 in. long, alternate with the sepals; calyx somewhat accrescent, valvate, with 5 sepals which are ovate, acuminate, 3/8–1/2 in. long, stellate-pubescent, united below; stamens numerous; filaments monadelphous, united at the base with the claws of the petals; anthers reniform, 1-celled; stigma terminal and capitate.

FRUIT. Capsule 2/5–1/2 in. long, of about 9 radially disposed, beaked carpels which tardily separate at maturity, shorter than the calyx, finely stellate-tomentose; seeds 2–several in each valve, finely pilose.

LEAVES. Simple, alternate, blade ovate or broadly cordate, apex acuminate or acute, margin irregularly crenate-dentate, blade length 3/4–6 in., upper surface dull green, somewhat pubescent or glabrate, lower surface paler, prominently veined and densely stellate-tomentose; petiole about one half as long as the blade.

Berlandier Abutilon
Abutilon lignosum (Cav.) Don

Mexican Bastardia
Bastardia viscosa (L.) H. B. K.

TWIGS. Young ones green to straw-colored or brown, densely stellate-tomentose; older ones brown to gray and less tomentose to glabrous and woody near the base.

RANGE. On dry prairies, in southwest Texas (Cameron County); and in Mexico in Sonora, Chihuahua, Jalisco, and Yucatán. Also in the West Indies.

REMARKS. The genus name, *Abutilon*, is of ancient Arabic origin, and the species name, *lignosum*, refers to the woody base. Some other vernacular names are Indian Mallow, Pelotazo, Pelatazo Blanco, and Colotahue. This plant resembles *A. hypoleucum* Gray, but the latter has leaves white-tomentose beneath, and flowers to 1 in. long, with carpels stellate-hirsute. The fiber is sometimes used in Mexico for twine or rope.

Mexican Bastardia

Bastardia viscosa (L.) H. B. K.

FIELD IDENTIFICATION. Mostly herbs to 3 ft, but occasionally somewhat woody toward the base. All parts more or less viscid-hairy.

FLOWERS. Peduncles axillary, usually 1-flowered, 1/2–1 1/2 in. long, tomentulose; calyx about 1/4 in. long, separated one half to three fourths its length into 5 sepals which are ovate to lanceolate, acuminate or acute and white-hairy externally; petals 5, yellow or orange-yellow, about ¼ in. long, rounded or obovate; stamens numerous, united to form a short column, carpels 5.

FRUIT. Capsule dry, depressed-globular, about ¼ in. in diameter, green at first, brown later; carpels 5, chartaceous, verticillate about the axis, dehiscent loculicidally; seeds 1 in each cell, about 1/16 in. long, flattened or 3-sided, margins rounded, surfaces puberulent.

LEAVES. Simple, alternate, deciduous, blade ¾–3 in. long, ovate-cordate, apex long-acuminate, base cordate, margin sinuate-dentate, upper surface dull green and finely tomentulose, lower surface paler and more densely tomentulose, palmately veined with veins more conspicuous beneath; petioles ¼–1½ in. long, tomentose.

STEMS. Slender, flexuous, angular or terete; young stems green, hairy with stiff, spreading, gray, and often glandular-viscid hairs; older stems light brown to gray, finely tomentulose to eventually glabrous.

RANGE. In Texas in the lower Rio Grande Valley area, in the vicinity of Brownsville, Cameron County; also in Mexico in Tamaulipas and Veracruz. Widespread in the West Indies and South America.

REMARKS. The genus name, *Bastardia,* is in honor of Toussaint Bastard (1784–1846), French botanist. The species name, *viscosa,* refers to the glandular-viscid hairs on the stems.

MOUNTAIN ROSE-MALLOW
Hibiscus cardiophyllus Gray

MOUNTAIN ROSE-MALLOW

Hibiscus cardiophyllus Gray

FIELD IDENTIFICATION. Woody-based perennial 1–2 ft, with densely stellate-hairy branches.

FLOWERS. Crimson, showy, perfect, on peduncles longer than the petioles; sepals 5, broadly lanceolate to oblong, densely hairy, the hairs simple or stellate, lower surface 3-veined, veins inconspicuous above; involucre of 8–11 bracts which are lanceolate, hairy, and ½–1 in. long; petals 5, spreading, brilliant crimson, 1–1¼ in. wide, obovate or somewhat oblique, apex rounded, conspicuously veined; stamens numerous, borne on the lower part of the staminal column which is ½–¾ in. long; filaments 3/16–¼ in. long, solitary or bifid, anthers small and yellow; pistil with 5 spreading styles above, stigmas capitate and about 1/16 in. broad; ovary 5-celled, sessile.

FRUIT. Capsule globose, depressed, ½–¾ in. in diameter, formed by the indurated sepals, glabrous or hairy along the sutures; bracts persistent.

LEAVES. Alternate, ovate to reniform, rarely somewhat lobed, blades 1–3 in. long, margin crenate-dentate, apex obtuse or acute, base cordate, 5–7-veined from the base; upper surface finely pubescent, velvety-gray or brown-hairy beneath; petioles stout; stipules persistent, subulate, curved.

RANGE. Usually on gravelly hillsides, but not abundant anywhere. Mostly along the breaks of the lower Rio Grande River. Collected by the author on a gravel hill known as La Lomita, near Mission, in Hidalgo County, Texas. In Mexico in Coahuila, Nuevo León, Tamaulipas, San Luis Potosí, and Puebla.

REMARKS. The genus name, *Hibiscus,* is the ancient name of the European Marsh-mallow, and the species name, *cardiophyllus,* refers to the heart-shaped leaves. It is known in southwest Texas under the Spanish name of Tulipa de Monte and Malva Rosa del Monte. The plant was first collected in Texas by Charles Wright in 1849. It has a very beautiful flower worthy of cultivation in rock gardens, and blooms throughout the summer, especially after rains.

DESERT ROSE-MALLOW

Hibiscus coulteri Harv.

FIELD IDENTIFICATION. Plant low and straggling, 4–12 in., herbaceous above but woody toward the base. All parts with appressed white stellate hairs.

FLOWERS. In continuous succession, peduncles 2–5 in.,

xillary, solitary; bractlets about 10, ½–¾ in. long, linear, etaceous, appressed stellate-hairy; calyx 5-lobed, lobes anceolate-acuminate, about one half as long as the etals; petals yellow, 1–1½ in. long, obovate to oval; taminal column with scattered anthers; stigmas 5, capiate, spreading.

FRUIT. Capsule globose, hairy to glabrate later, many-eeded, seeds densely woolly; pedicels usually disrticulating at maturity of fruit.

LEAVES. Alternate, blades ½–1½ in. in diameter, vate to oblong or lanceolate, margin irregularly toothed nd often also shallowly or deeply 3-cleft or 3-lobed, pex rounded to obtuse, base rounded to broadly cuneate; upper surface dull green with scattered stiff, white, ppressed hairs; lower surface slightly paler and also airy; petioles ¼–1¼ in. long, appressed white-hairy.

STEMS. Slender, green, strigose with forked hairs, when older gray to light brown and glabrous, woody bove the caudex.

RANGE. Rocky slopes and canyons at altitudes of 1,500–3,500 ft. Trans-Pecos Texas, New Mexico, and Arizona; southward into Mexico.

DESERT ROSE-MALLOW
Hibiscus coulteri Harv.

PALEFACE ROSE-MALLOW
Hibiscus denudatus Benth.

REMARKS. The genus name, *Hibiscus*, was the ancient name for the European Marsh-mallow. The species name, *coulteri*, is for John M. Coulter, author of *A Botany of Western Texas* (1891–1894).

PALEFACE ROSE-MALLOW

Hibiscus denudatus Benth.

FIELD IDENTIFICATION. Plant herbaceous above but subshrubby toward the base. Attaining a height of 1–2½ ft, much-branched and very tomentose.

FLOWERS. March–April, axillary, solitary, showy, large, regular; peduncles ¼–1 in. long, densely tomentose; bractlets 3–8, ⅛–⅓ in. long, linear, tomentose, inconspicuous, bristle-tipped, less than half to fully the length of the calyx-tube; calyx ⅓–⅗ in. long, 5-lobed, lobes lanceolate, tomentose; petals 5, about ½–1 in. long, obovate or oval, white to light purple or lavender, red or purplish at the base, sometimes with a narrow rose band; staminal column antheriferous for much of its length, but naked at the apex; stigmas 5, capitate and spreading.

FRUIT. Capsule shorter than the calyx, about 1/4 in. long, pubescent to glabrous, 5-celled, loculicidal, valves chartaceous; seeds numerous, hairy.

LEAVES. Simple, alternate, deciduous or semipersistent, blades 5/8–1 1/2 in. long, oval to ovate or oblong, apex rounded to obtuse, base rounded or occasionally cordate, margin sinuate to dentate, some entire toward the base, thickish, surfaces pale green to yellowish green, densely stellate-tomentose above and beneath; petioles 1/8–1/3 in., densely tomentose.

TWIGS. Slender, greenish yellow and densely tomentose at first, older wood toward the base gray and glabrous.

RANGE. On bluffs or mesas at altitudes of 2,000–7,000 ft. Mostly along the upper Rio Grande River in Texas; west through New Mexico and Arizona to California, south in Mexico to Baja California, Durango, Coahuila, and Chihuahua. Type from Magdalena Bay, Baja California.

VARIETY. Bracted Paleface Rose-mallow, *H. denudatus* var. *involucellatus* Gray, has a better developed and more persistent involucel and occurs mostly in saline or caliche soil.

REMARKS. The genus name, *Hibiscus,* was the ancient name used for the European Marsh-mallow. The application of the species name, *denudatus,* to this plant is obscure.

SHRUB-ALTHEA
Hibiscus syriacus L.

SHRUB-ALTHEA

Hibiscus syriacus L.

FIELD IDENTIFICATION. Much-branched shrub or small tree 3–18 ft. Often grown for ornament and developed into a large number of horticultural forms with single and double, variously colored flowers.

FLOWERS. Showy, perfect, solitary, axillary, on peduncles of variable length; bractlets subtending the calyx usually 5–7, 3/8–3/4 in. long, linear to linear-spatulate; calyx longer or shorter than the bractlets; sepals 5, triangular ovate or lanceolate, about as long as the tube; petals 5, very variable in color in the many horticultural forms, white, pink, lavender, rose, with a crimson or purplish blotch at the base, 1 3/4–3 in. long, rounded to obovate, margins sometimes undulate; staminal column prominent with numerous anthers below, and apex 5-parted with capitate stigmas; ovary sessile, 5-celled and loculicidally 5-valved.

FRUIT. Capsule oblong-ovoid, apex drawn tightly, 3/4–1 in. long, pubescent, more or less dry, larger than the calyx.

LEAVES. Alternate, triangular to rhombic-ovate or elliptic to oval, blades 1 1/2–4 3/4 in. long, margin variously crenate-toothed or notched with rounded or acutish teeth, usually more or less 3-lobed also; young leaves pubescent, when older becoming glabrous or nearly so palmately veined with 3 veins more conspicuous; petioles generally shorter than the blades; winter buds minute. Some forms with variegated leaves are known.

BARK. Gray to brown, somewhat roughened.

RANGE. Cultivated in gardens and occasionally escaping to roadsides, thickets, and woods. Texas, Oklahoma, Arkansas, and Louisiana; eastward to Florida and northward to Missouri, Ohio, and Massachusetts. Also in coastal regions of Canada.

HORTICULTURAL FORMS. Propagation is by seeds or by cuttings of ripened wood taken in the fall. Named forms are obtained by grafting on common seedling stock. The following horticultural forms are listed in Kelsey and Dayton (1942):

Admiral Dewey, Amaranthus, Amplest *(amplissimus)*, Anemone *(anemonaeflorus)*, Ardens, Banner, Bicolor, Boule de Feu, Coelestis, Comte de Hainault, Comte des Flandres, Double Precoce, Double White *(albus plenus)*, Duc de Brabant, Flesh *(incarnatus)*, Granville, Jeanne D'Arc, Lady Stanley, Leopard-leaf *(variegatus)*, Meehan *(meehani)*, Peony *(peoniflorus)*, Pulcherrimus, Purity

Purple *(purpureus semiplenus),* Rosy *(roseus),* Showy *(speciosus),* Snowdrift, Snowstorm *(totus-albus),* Souv. Charles Breton, Spot-leaf *(foliovariegatus),* W. R. Smith.

REMARKS. The genus name, *Hibiscus,* is the ancient name of the European Marsh-mallow, and the species name, *syriacus,* is for Syria, where it was once supposed to be native. However, more recent investigations prove it to be originally from China and India. It is also known under the vernacular names of Rose-of-Sharon and Rose-mallow. It was introduced into cultivation about 1600.

Wright Malvastrum

Malvastrum wrightii Gray

FIELD IDENTIFICATION. Perennial plant with stems mostly herbaceous, but sometimes woody near the base. Attaining a height of 1–2 ft.

FLOWERS. Perfect, mostly solitary, almost sessile in the axils; calyx densely scaly, sepals 5, partially united, triangular-ovate, ⅕–⅓ in. long, longer than the tube; involucel bractlets usually 3, foliaceous, ovate or lanceolate, sometimes cordate, adnate to the calyx-tube; petals 5, distinct, yellow, ½–⅗ in. long; stamen column bearing anthers at the summit; carpels as many as the styles, stigmas capitate, ovule solitary and ascending.

Wright Pavonia
Pavonia lasiopetala Scheele

FRUIT. Carpels 15–20, leathery, smooth, apex hirsute, 2-gibbous, ventrally 2-beaked, 1-celled, mostly dehiscent; seed kidney-shaped, solitary, filling the cell.

LEAVES. Alternate, blades suborbicular to ovate or ovate-oblong, apex obtuse, margin crenate-serrate, base rounded to subcordate or truncate; petioles one half as long as the blades or equaling them in length.

STEMS. Rigid, ascending, pubescence scalelike and stellate.

RANGE. In dry soil of central and western Texas, mostly west of the Colorado River. Often in mesquite thickets.

REMARKS. The genus name, *Malvastrum,* is the name derived from *Malva,* a closely related genus, and means "of the mallows." The species name, *wrightii,* is in honor of Charles Wright, botanist and surveyor of the United States–Mexico boundary survey in 1851.

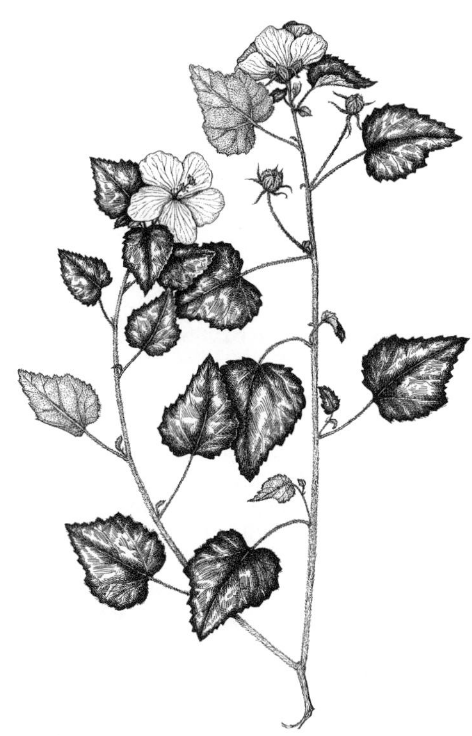

Wright Malvastrum
Malvastrum wrightii Gray

Wright Pavonia

Pavonia lasiopetala Scheele

FIELD IDENTIFICATION. Western plant 1½–4 ft, herbaceous above and slightly woody toward the base. The stems, petioles, and peduncles velvety-pubescent.

FLOWERS. Throughout the summer, peduncles as long as the petioles or longer, axillary or terminal; flowers solitary, showy, rose-colored, 1¼–1¾ in. in diameter,

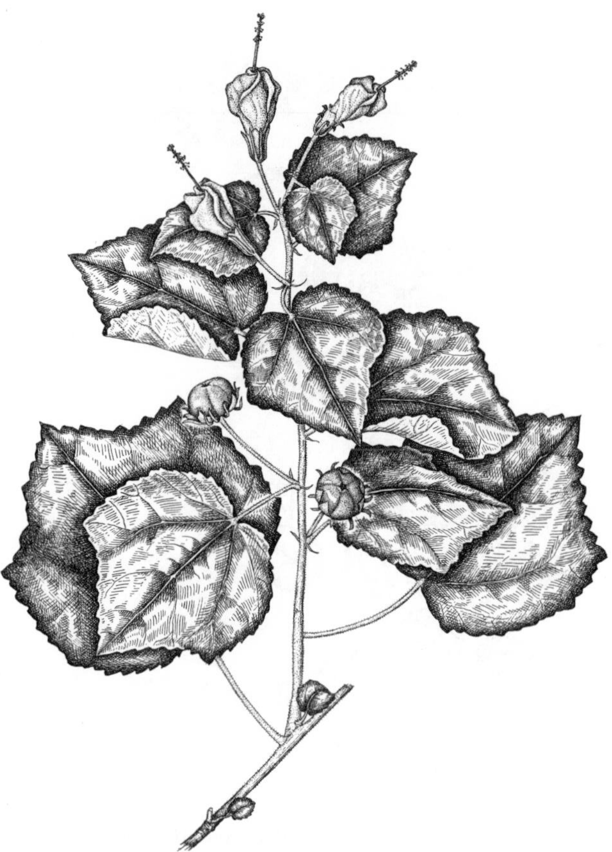

Drummond Wax-mallow
Malvaviscus drummondii Torr. & Gray

5-petaled; stamens numerous; styles 10 (twice as many as the carpels) stigmas capitate, ovules solitary, borne in 5 1-ovuled, crustaceous, obtuse carpels; calyx densely stellate-hairy, sepals 5, partially united, ovate, acuminate, 3–5-nerved; bractlets 5 or more, densely stellate-hairy, usually distinct, linear, ⅜–⅝ in. long.

FRUIT. Capsule dry, rough, ⅕–¼ in. long, deeply lobed into 5 loose carpels at maturity, each segment 1-seeded.

LEAVES. Alternate, stipulate, cordate-ovate, margin sharply and irregularly dentate or somewhat angulate, apex acute or obtuse, base cordate, blade length 1–2½ in.; upper surface dark green with veins obscure; lower surface paler, densely stellate-pubescent with 3 stout veins at base; petiole nearly as long as the blade or as long, stellate-pubescent.

TWIGS. Slender, green and whitened with stellate hairs, later brown and less hairy.

RANGE. Dry rocky woods or banks, southern, central, and western Texas. In Mexico in Coahuila, Nuevo León, and Durango.

REMARKS. The genus name, *Pavonia,* honors J. Pavon (d. 1844) joint author of Ruiz and Pavon, *Flora Peruviana et Chilensis*. The species name, *lasiopetala,* means "shaggy-petaled," with reference to the undulate petal margins. The plant has some ornamental value, but is little used as browse. The various species of pavonias can be propagated by cuttings in spring or summer, or by seeds.

Drummond Wax-mallow

Malvaviscus drummondii Torr. & Gray

FIELD IDENTIFICATION. Rather large, coarse, tomentose perennial to 10 ft, somewhat woody near the base.

FLOWERS. Solitary, perfect, red, on axillary, pubescent peduncles 1–2 in. long; petals 5, red, less than 1¼ in. long, inequilateral, auricled, not spreading, erect, connivent, twisted into a tube; pistillate column long-exserted, twisted, glandular, hairy, styles bifid to make 10, ovary 5-celled; stamens borne on the column beneath the stigmas; calyx about ½ in. long, tubular-campanulate, hairy, lobes 5, ovate to lanceolate, shorter than the tube; calyx subtended by 8–10 involucral bracts, linear to spatulate, reflexed, ⅕–¼ in. long.

FRUIT. August–September, drooping, baccate or berry-like, flattened, red, ⅜–¾ in. in diameter, mealy, fleshy, edible, separating gradually into 5 1-seeded carpels.

LEAVES. Simple, alternate, oval to ovate, margin coarsely toothed or sometimes angulately 3-lobed, apex acute or obtuse, base cordate, blades 2–4 in. long or broad, palmately veined; upper surface dull green and glabrous or puberulent; lower surface velvety-hairy and conspicuously veined; petioles pubescent, 1–3 in. long; stipules linear to subulate.

STEMS. Simple or branched, upright or half declining, green to brown, hairy, mostly herbaceous above, somewhat woody toward the base.

RANGE. Usually in sandy, low grounds along streams. On the coastal plain from Texas eastward to Florida; also in the West Indies and Mexico. In Mexico in the states of Tamaulipas, Veracruz, San Luis Potosí, and Yucatán.

REMARKS. The genus name, *Malvaviscus,* is from a Greek word meaning "sticky-mallow," and the species name, *drummondii,* is in honor of Thomas Drummond, who collected in Texas 1833–1834. Vernacular names are Texas Mallow, Mexican Apple, Red Mallow, May Apple, Manzanilla, and Wild Turk's Cap. This plant is worthy of cultivation, but for best results it should be pruned back after 2 years. It may be propagated by seeds and greenwood cuttings. The leaves are used as an emollient, and the flowers are used in Mexico in a decoction for inflammation of the digestive tract and as an emmenagogue. The small, apple-like fruit is edible, either raw or cooked, and has a mealy taste. It is also eaten by a number of species of birds and mammals. Livestock occasionally browse the leaves.

Mexican Wissadula

Wissadula amplissima (L.) R. E. Fries

FIELD IDENTIFICATION. An herbaceous plant to 3½ ft, but occasionally suffrutescent (in Texas). Recorded as a shrub in Mexico by Standley (1920–1926, p. 756).

FLOWERS. In axillary, several-flowered panicles on pubescent peduncles ½–3 in. long; bracts at base of peduncles filiform and pubescent; calyx bowl-shaped, spreading, pubescent, 5-lobed, lobes about as long as the calyx-tube, broadly ovate to triangular, acute to obtuse; corolla about ⅜ in. across, petals 5, ⅛–¼ in. long, distinct, obovate, yellow; stamens numerous, united to form a column; styles short, stigmas terminal, ovules 1–2.

FRUIT. A cluster of 4–5 carpels verticillate about an axis and forming a depressed head about ⅜ in. wide and ¼ in. high; carpels dry, obovate, chartaceous, smooth, abruptly apiculate at apex, laterally constricted and dehiscent, 3–5-seeded; seeds dark brown to black, obovoid to subglobose, about 1/16 in. long, glandular-dotted and with a few tufts of hairs.

LEAVES. Simple, alternate, deciduous, broadly heart-shaped, apex acuminate or acute, base cordate, margin entire or some slightly sinuate-toothed, blade length 2–6 in., upper surface dull green and minutely puberulent, lower surface much paler and soft with minute stellate hairs, palmately nerved; petioles pubescent, becoming much shorter than the blades toward the inflorescence.

TWIGS. Slender, green to brown, puberulent to pubescent or with dense stellate hairs.

RANGE. In dry soil, roadsides, thickets, and hillsides; western and southern Texas. Near Brownsville, Cameron County, Texas. In Mexico from Sinaloa to Tamaulipas, Morelos, and Oaxaca. Also in Central America, South America, and Africa.

SYNONYMS. Some other names listed for the Mexican Wissadula are *W. mucronulata* Gray and *Abutilon mucronulatum* Gray.

REMARKS. The genus name, *Wissadula,* is an East Indian name of uncertain origin. The species name, *amplissima,* refers to the "amplest" or best-developed flower panicles.

Mexican Wissadula
Wissadula amplissima (L.) R. E. Fries

CHOCOLATE FAMILY (*Sterculiaceae*)

Berlandier Ayenia
Ayenia berlandieri Wats.

FIELD IDENTIFICATION. Suffrutescent plant, sometimes quite woody below, attaining a height of 2–6 ft.

FLOWERS. Borne on pedicels mostly less than 1 in. long; perianth hardly over ⅛ in. across; sepals 5, densely white-hairy, oblong to lanceolate, obtuse or acute; petals 5, yellowish green, clawed, crisped, incurved, adnate to the stamen tube; anthers in the sinuses of the stamen tube; carpels 5.

FRUIT. On pedicels ½–1½ in. long, capsule 5-lobed, subglobose, somewhat flattened, about ⅜ in. in diameter, covered with soft spinules and short, stellate hairs, the 5 carpels bivalvate at maturity.

LEAVES. Mostly in clusters at the nodes, 1–3 in. long, ½–1½ in. wide, ovate, apex acuminate, base rounded to cordate, margin coarsely serrate; upper surface dark green, puberulent, somewhat papillose; lower surface paler, densely covered with fascicled white hairs; petioles ¼–1 in. long, densely hairy.

TWIGS. Young twigs slender, spreading, green, hairy; older twigs light brown, glabrous, striate.

RANGE. In Texas in the lower Rio Grande area, especially Cameron County. In Mexico from Tamaulipas south to Jalisco, Guerrero, and Morelos.

REMARKS. The genus name, *Ayenia,* is in honor of Duc d'Ayen. The species name, *berlandieri,* is in honor of J. L. Berlandier, a Swiss botanist who collected, 1827–1830, in Mexico.

Berlandier Ayenia
Ayenia berlandieri Wats.

Small-leaf Ayenia
Ayenia microphylla Gray

FIELD IDENTIFICATION. Low densely branched Western subshrub. Usually less than 1 ft high, with densely stellate-tomentose leaves, petioles, and stems.

FLOWERS. May–July, long-pedicellate, axillary, small, red to purplish; calyx small and inconspicuous, 5-lobed; petals 5, deltoid-reniform, hooded, apex more or less lobed and glandular, claw longer than the blade; staminal column short and funnelform; stamens 10, 5 fertile, the 5 sterile ones forming large, fleshy, crenate staminodia; anthers 2–3-celled, solitary in the sinuses of the stamen tube; stigma capitate, ovary very short-stipitate.

FRUIT. Capsule about 1/4 in. in diameter, densely woolly-tomentose, conspicuously tuberculate, 5-lobed, separating into 1-seeded carpels, these bivalvate; seeds about 1/25 in. long, oblong, one end gradually acute, the other end more or less truncate, dark brown, deeply wrinkled.

LEAVES. Simple, alternate, deciduous, 1/4–3/4 in. long, orbicular to ovate, broadest near the base and cordate to rounded; apex rounded or obtuse, margin coarsely crenate-dentate; surfaces dull green and stellate-tomentose, densely so beneath; petioles 1/16–1/4 in., densely tomentose.

TWIGS. Slender, the young ones green to gray or brown and densely tomentose, older twigs light gray and glabrous.

RANGE. Dry rocky or gravelly hillsides at altitudes of 2,000–3,500 ft. Trans-Pecos Texas through New Mexico to Arizona; south in Mexico to Chihuahua and Coahuila.

SMALL-LEAF AYENIA
Ayenia microphylla Gray

DWARF AYENIA
Ayenia pusilla L.

REMARKS. The genus name, *Ayenia,* is in honor of Duc d'Ayen. The species name, *microphylla,* refers to the small leaves.

DWARF AYENIA

Ayenia pusilla L.

FIELD IDENTIFICATION. Slender, often prostrate, Western plant. Essentially an herbaceous perennial, but the wiry stems sometimes woody toward the base.

FLOWERS. Small, perfect, axillary, long-pedicellate, solitary, or a few in a cluster, reddish purple; sepals 5, small and inconspicuous; petals 5, hooded, apex inflexed, long-clawed, adnate to the staminal tube, dorsally appendaged; stamens 10, connate at the base to form a slender pedicellate 5-lobed cup, with 5 of the stamens abortive into small, toothlike staminodia; anthers 2–3-celled, solitary in the sinuses of the stamen tube; stigma capitate, ovary on a slender stipe, with 3 parallel, ovoid cells.

FRUIT. Capsule long-stipitate, depressed-globose, ¼ in. in diameter, pubescent, sharply muricate, 5-lobed, each carpel 1-seeded and bivalvate.

LEAVES. Simple, alternate, very variable in form, lanceolate to suborbicular, ⅜–2 in. long, apex acute to acuminate or obtuse to rounded, margin dentate, light green and finely puberulent to glabrous above: lower surface slightly paler and puberulent, finely reticulate-veined; petioles ⅛–¼ in. long, pubescent.

TWIGS. Very slender, upright or prostrate, green, finely white-pubescent, becoming light to dark brown and glabrous near the base with age.

RANGE. Canyons and rocky banks at altitudes of 2,000–4,000 ft. Southern and western Texas; west through New Mexico to California and southward over a wide area in Mexico.

REMARKS. The genus name, *Ayenia,* is in honor of Duc d'Ayen. The species name, *pusilla,* refers to the dwarf stature.

Chinese Parasol-tree

Firmiana simplex W. F. Wight

FIELD IDENTIFICATION. Cultivated shrub or tree to 35 ft, with a smooth green bark and a rounded crown.

FLOWERS. In a pubescent, terminal panicle 4–12 in. long; flowers monoecious, small, greenish yellow; petals absent; calyx petal-like, sepals 5, valvate, colored, linear to narrowly oblong, reflexed, ⅓–⅖ in. long; stamens united in a column, bearing a head of 10–15 sessile anthers; carpels 5, nearly distinct above, each terminating into a peltate stigma, carpels 2¼–4 in. long at maturity.

Chinese Parasol-tree
Firmiana simplex W. F. Wight

FRUIT. A follicle, stipitate, leathery, carpels veiny, finely pubescent, distinct before maturity and spreading open into 5 leaflike bodies bearing 2–several seeds on their margins; seeds globular, pealike, about ¼ in. or less in diameter, albuminous; carpels rudimentary and free in the staminate flowers. A peculiar feature is the black or brown fluid which covers the fruit and is liberated when the follicle bursts.

LEAVES. Alternate, simple, blades 4–12 in. broad and long, cordate to orbicular, palmately 3–5-lobed, lobes entire and acuminate, dull green and glabrous or lightly pubescent above, lower surface glabrous to tomentulose; petiole glabrous or pubescent, usually 5–18 in. long.

TWIGS. Stout, smooth, grayish green, bark of trunk also smooth and green.

RANGE. A native of Japan and China. Cultivated in the United States and escaping to roadsides, woods, and thickets. From South Carolina to Florida and westward to Texas and California. Hardy as far north as Washington.

VARIETY. A variety with creamy-white, variegated leaves is known as *F. simplex* var. *variegata.* Two other species are cultivated in California.

REMARKS. The genus name, *Firmiana,* honors Karl Joseph von Firmian (1718–1782), at one time governor of Lombardy. The species name, *simplex,* refers to the lobed but simple leaves. It is also known as Varnish-tree, Phoenix-tree, and Bottle-tree. It is an excellent tree for lawn and shade and has been cultivated since 1757, being easily grown from seed.

Hermannia

Hermannia texana Gray

FIELD IDENTIFICATION. Densely tomentose herbaceous plant 1–4 ft, or sometimes shrubby and lignescent near the base.

FLOWERS. In axillary tomentose cymes, the peduncles ⅜–1 in. long, few-flowered; pedicels shorter than the peduncles, recurved; flowers small, perfect, dull red; petals 5, about ⅓ in. long, erect or spreading, obovate, claws slender and hollow; stamens 5, filaments joined at base to form a staminal tube, anthers pubescent, sagittate, 2-celled; ovary 5-celled, ovules numerous in each cavity; styles 5, nearly distinct, stigmas capitate; sepals 5, lanceolate or triangular-lanceolate, acuminate, about ⅛ in. long, densely woolly-tomentose.

FRUIT. Capsule globose or short-oblong, ⅖–¾ in. long, densely stellate-tomentose, cristate dorsally on the 5-valves, loculicidally dehiscent at maturity; seeds nu-

HERMANNIA
Hermannia texana Gray

merous, about 1/30 in. long, oval, cleft on one side, flattened, black, endosperm fleshy.

LEAVES. Simple, alternate, deciduous, oval to ovate or ovate-oblong, apex rounded, base cordate or truncate, margin coarsely dentate, blades 1–2½ in. long or wide, surfaces densely stellate tomentose-canescent, petioles ⅛–1 in., densely tomentose, stipules foliaceous.

TWIGS. Slender, erect or ascending, young ones green and densely tomentose, herbaceous; older ones less tomentose and eventually glabrous, dark gray and sometimes woody near the base.

RANGE. In dry rocky, often calcareous, soil of hillsides and prairies. In Texas from the Colorado River westward and southward into Mexico in Tamaulipas, Coahuila, Nuevo León, and San Luis Potosí.

REMARKS. The genus name, *Hermannia,* is in honor of the botanist Leyde Hermann, and the species name, *texana,* is for the state of Texas, where it grows.

PYRAMID-BUSH
Melochia pyramidata L.

FIELD IDENTIFICATION. Plant to 4½ ft, herbaceous above, somewhat woody toward the base. Stems diffuse, wide-spreading or arched, slender and elongate.

FLOWERS. Solitary or in few-flowered corymbs in the axils; peduncles slender, to 1 in. long, densely hairy, pedicels 1/16–⅛ in. long; calyx persistent, tube campanulate, about 1/16 in. long, divided into 5 long-acuminate lobes; lobes reddish-glandular down the center and margins, also with a few stiff, white, scattered hairs; corolla ⅛–¼ in. long, pink, petals 5, spatulate, flat, apex rounded, margin slightly ciliolate, base narrowed into yellowish orange claws below; stamens 5, exserted when the petals are spread, erect; filaments white, slender, glabrous, adnate to the petal bases; anthers oblong, yellow, 2-celled; stigmas 5, linear, at the apex of the hairy, ovoid-pointed ovary.

FRUIT. A short-pedunculate capsule, pyramidal, bladdery, less than ½ in. wide, apex acuminate, base truncate or broadly rounded; wings 5, thin, triangular, acuminate or acute, subglabrate or with scattered white hairs and reddish glands; capsule loculicidally 5-valvate, tardily dehiscent; seeds mostly 5, about 1/25 in. long, ovoid, somewhat asymmetrical, dull brown to black.

PYRAMID-BUSH
Melochia pyramidata L.

TWIGS. Slender, elongate, spreading, green to reddish brown, young subglabrate or white-hairy with simple or stellate hairs, older ones glabrous.

LEAVES. Simple, alternate, thin, elliptic to oblong or narrowly ovate to lanceolate, blade length 1–1½ in., width ½–¾ in., apex acute or short-acuminate, base rounded to broadly cuneate, margin coarsely toothed and minutely ciliolate or glabrate, upper surface with veins conspicuously impressed, dull green, glabrous, or a few hairs along the veins; lower surface with veins prominent, often with 2 lateral basal ones joining the main vein, glabrous or with a few stiff, white, simple or fascicled hairs along the veins; petioles ¼–½ in., clothed with white, simple or stellate hairs.

RANGE. In dry or rocky soil, southwest Texas and adjacent Mexico.

REMARKS. The genus name, *Melochia,* means "yellow honey." The species name, *pyramidata,* refers to the pyramid-like capsule. Known under the vernacular names of Broomwood, Malva Común, Malva Cimarrona, Bretónica, and Suponite. Occasionally grazed by livestock. Closely related to *M. tomentosa,* which has leaves densely tomentose.

WOOLLY PYRAMID-BUSH
Melochia tomentosa L.

WOOLLY PYRAMID-BUSH

Melochia tomentosa L.

FIELD IDENTIFICATION. Plant herbaceous above or woody below, 1–6 ft.

FLOWERS. In loose, several-flowered, terminal cymes, pedicels 1/12–1/5 in. long; petals 5, pink or violet, ⅓–½ in. long, spatulate, flat, withering early, convolute in the bud; stamens 5, filaments well united upward, adnate to the claws of the petals within; styles 5, distinct, stigmas capitate; ovary 5-celled, ovules 2 in each cavity; calyx-lobes 5, lanceolate to linear-lanceolate, joined at the base, stellate-tomentose or pubescent.

FRUIT. Fruiting stipe ⅛–¼ in. long, capsule pyramidal, ¼–⅓ in. long, long-beaked, 5-lobed, the lobes rounded or blunt-tipped at the base and thin, surface densely stellate-pubescent, 5-celled, loculicidally 5-valved, valves easily separating; seeds about 1/25 in. long, solitary in each cavity, ascending, endosperm fleshy, embryo straight, cotyledons broad.

LEAVES. Simple, alternate, ½–2½ in. long, slender-petioled, rhombic-ovate to oblong or linear-lanceolate, apex rounded to obtuse or acute, base rounded to truncate or subcordate, margin serrate or crenate, firm and leathery, surfaces conspicuously veined, whitish beneath with stellate-tomentose hairs.

RANGE. In dry, rocky places or shell mounds near the coast; southwestern Texas, southern Florida, nearly throughout Mexico, Central and northern South America, and the West Indies.

REMARKS. The genus name, *Melochia,* is from the Greek and means "yellow honey." The species name, *tomentosa,* refers to the densely hairy leaves. The plant is known in Central and South American countries under the vernacular names of Malva, Malvavisco, Bretónica, and Varita de San José.

Woolly Pyramid-bush can be distinguished from the rather similar Pyramid-bush, *M. pyramidata,* by the conspicuously whitish, densely tomentose leaves; the somewhat larger flowers with the filaments joined somewhat higher up; the fruit, which is more densely tomentose, and the lobes at the base being more rounded or blunt instead of acute. However, some authors believe that the 2 species are very closely related and may even hybridize, because forms are found which appear to be intermediate.

FLORIDA WALTHERIA

Waltheria americana L.

FIELD IDENTIFICATION. Stiff-shrubby or herbaceous plant 1–5 ft, usually erect, but sometimes spreading and decumbent.

FLOWERS. Borne in dense axillary clusters, varying from almost sessile to long-pedunculate; flowers small, perfect, yellow, sweet-scented; petals 5, convolute in the bud, persistent, 1/6–1/4 in. long, slightly longer than the sepals, spreading, spatulate, slender-clawed; stamens 5, filaments united below, anthers with 2 sacs; ovary 1-celled, sessile, ovules 2 in each cavity, style simple, stigma brushlike; calyx 1/6–1/5 in. long, sepals 5, subulate or subulate-lanceolate, as long as the ribbed tube or shorter, villous-hirsute; involucel of 3 villous bractlets.

FRUIT. A capsule 1/12–1/8 in. long, longer than thick, rounded and oblique, canescent, 1-celled, longitudinally 2-valved; seed solitary, ascending, endosperm fleshy, embryo straight.

LEAVES. Alternate, ovate to orbicular or oblong; apex obtuse to rounded; base obtuse, rounded or subcordate; margin dentate-serrate, length 1/3–3 1/3 in., surface dark to pale green, velvety-tomentose above, more densely so beneath; petioles stout, variable in length, tomentose; stipules narrow.

RANGE. In rocky or sandy soil. Southwest Texas, Arizona, Mexico, Florida Keys, West Indies, Central America, and the warmer regions of South America.

MEDICINAL USES. In Mexico the plant is used in domestic medicine as a febrifuge, antiseptic, antisyphilitic, and for the treatment of cutaneous diseases.

REMARKS. The genus name, *Waltheria*, is in honor of the American botanist Thomas Walter, and the species name, *americana*, refers to the plant's extensive distribution in the Americas. Vernacular names in use in Central and South America include Malva, Basora Prieta, Hierba del Pasmo, Malva del Monte, Malva Blanca, Malvavisco, and Bretónica. It is known to the Mexican people of southwest Texas as Hierba del Soldado ("soldier-weed"). It is said to be of some value as forage for cattle and is sometimes grown for ornament in Florida.

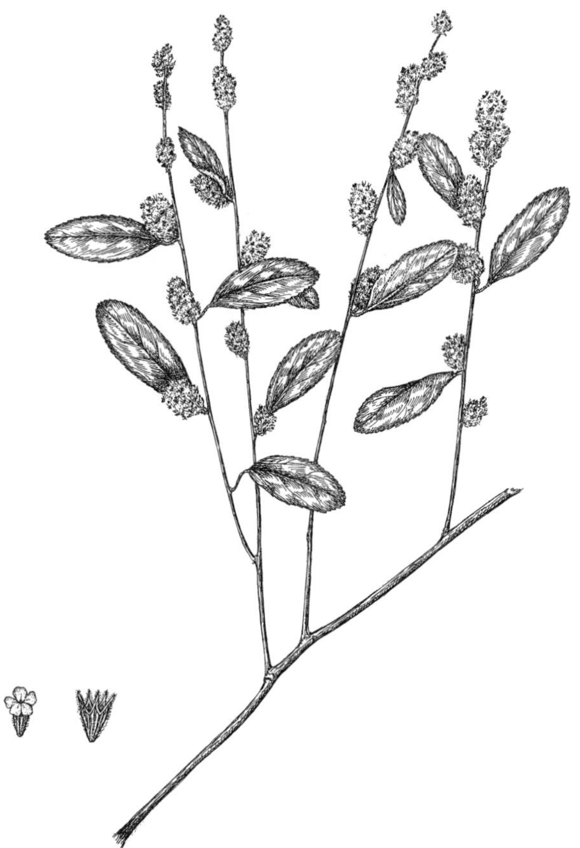

FLORIDA WALTHERIA
Waltheria americana L.

PRINGLE KIDNEY-PETAL

Nephropetalum pringlei Robins. & Greenm.

FIELD IDENTIFICATION. Small subshrub covered with stellate tomentum on young stems and leaves.

FLOWERS. Borne in axillary 2–3-flowered, umbelliform cymes about 1 in. long; peduncles and pedicels about equal in length; individual flowers greenish, about 1/12 in. in diameter; calyx with 5 deep ovate segments; petals 5, clawed, slightly adnate at base to the staminal cup, apex free; blade of petal small, reniform, with a deep basal sinus at the claw, concave, not appendaged or glandular; stamens 5, united into a short cup; anthers sessile or nearly so, extrorse; staminodia 5, alternate with the stamens and opposite the sepals, rounded and cucullate at summit; ovary sessile, globose, 5-celled; cells 2-ovuled; style short and terete with a capitate stigma.

FRUIT. Globose, set with numerous pubescent processes, ovules superposed on the axil placentae, mature seeds solitary by abortion in the cells.

LEAVES. Simple, alternate, ovate, margin crenate-dentate, apex acuminate to acute or obtuse, base cordate with a narrow sinus, blade length 3 1/2–5 in., width 1 1/2–3 in., palmately 7-nerved from the base, upper surface finely stellate-pubescent, lower surface paler and tomentulose; petioles canescent-tomentulose, about 1 1/2 in. long; stipules setaceous, deciduous, nearly 1/6 in. long.

TWIGS. Young stems stellate-tomentose, older ones terete and nearly glabrate.

RANGE. In Texas in the lower Rio Grande Valley area and adjacent Mexico. Collected by C. G. Pringle at Hidalgo, Texas, August 6, 1888 (Robinson and Greenman, 1896, p. 168).

REMARKS. The genus name, *Nephropetalum*, refers to the kidney-shaped petals. The species name, *pringlei*, is in honor of C. G. Pringle, botanical collector of Mexico and the southwestern United States. The plant is most closely related to the genera *Ayenia* and *Buettneria*, and differs from the former in its reniform petals and sessile ovary, and from the latter in the entire absence of the appendage of the petals as well as the very different habit.

CAMELLIA FAMILY (*Theaceae*)

Loblolly-bay Gordonia
Gordonia lasianthus (L.) Ellis

FIELD IDENTIFICATION. Shrub, or less often a large tree, to 60 ft, with a trunk diameter of 12–18 in. The stout, short, crooked branches form a cylindric or conical head.

FLOWERS. May–July, continuing to open for several months thereafter; pedicels from the axils of new growth, stout, green to reddish, glabrous or somewhat puberulent, ¾–3 in., sometimes with 3–4 minute bractlets; sepals 5, unequal, ovate to oval, rounded, concave, coriaceous, about ½ in. long, margin ciliate with long hairs, surface densely and finely hairy; petals 5, white, obovate, concave, apex rounded, base contracted, dorsal surface silky-puberulent, incurved, length 1¼–1½ in., width about 1 in., deciduous; stamens numerous, filaments short, united at base on the margin and top of a 5-lobed fleshy cup; ovary sessile, ovoid, pubescent; style short and stout; stigma fleshy, minutely and irregularly 5-lobed.

FRUIT. Capsule ripening in the fall, about ¾ in. long, woody, silky-hairy to almost glabrous, ovoid, long-pointed, remnants of style persistent, loculicidally dehiscent into 5 valves below the middle; seeds 10–40, small, winged, wing as long as the seed body.

LEAVES. Simple, alternate, persistent, leathery, thick, narrowly elliptic or oblanceolate, apex acute to obtuse, base cuneate, sometimes narrowed into a short wing, margin serrate with appressed teeth mostly above the middle, blade length 1½–6 in., surface smooth, light to dark green and lustrous above, lower surface paler and glabrous or pubescent; petiole stout, about ½ in., grooved above.

TWIGS. Gray to dark brown, rather stout, smooth, with a few scattered lenticels; buds ovoid to globose, leaf scars shield-shaped.

BARK. Gray to reddish brown, thick, roughened by numerous interlacing flat-topped ridges separated by narrow rough furrows.

Loblolly-bay Gordonia
Gordonia lasianthus (L.) Ellis

WOOD. Reddish, close-grained, light, soft.

RANGE. In acid, swampy soils of pinelands on the coastal plain. To be looked for in eastern Louisiana and also growing close to the state line of Louisiana and Mississippi, eastward to Florida, and northward to Virginia.

REMARKS. The genus name, *Gordonia,* is in honor of James Gordon (1728–1791) an English nurseryman and the species name, *lasianthus,* is the old name of the genus, meaning "hairy-flowered." Other vernacular names are Tan-bay, Red-bay, and Black-laurel. It is a good ornamental tree and could be more extensively cultivated. It is propagated by seeds or greenwood cuttings under glass. The bark was formerly used for tanning, and the wood used in small amounts for cabinet work.

Virginia Stewartia

Stewartia malacodendron L.

FIELD IDENTIFICATION. Shrub with upright branching stems, or a small tree to 20 ft.

FLOWERS. Borne April–July, axillary, showy, perfect, solitary or occasionally in pairs, peduncles very short; sepals 5, imbricate, broad-ovate to orbicular, apex acute or obtuse, about ⅜ in. long, surface pubescent, united at base and subtended by 1–2 bractlets; petals 5, spreading, concave, white, obovate, margin crenulate, length 1¼–2 in., somewhat silky-pubescent, united at base; stamens numerous, in 3–4 series; filaments purple, united at base into a short tube and hairy below; anthers blue, introrse, versatile; style 1, compound, stigma terminal and 5-lobed, surpassing the stamens; ovary superior, 5-celled, ovules 2 in each cavity.

FRUIT. Capsule depressed-globose, crustaceous, ½–¾ in. in diameter, angles very low, 5-celled, loculicidally dehiscent; seeds lenticular, shiny, thick and crustaceous, marginless; embryo straight.

LEAVES. Simple, alternate, deciduous, membranous, blades 2–4½ in. long, 1½–2 in. broad, oval to elliptic or ovate to obovate, apex short-acuminate or acute, base acute, margin finely and sharply serrate with mucronate teeth, mostly glabrous above, lower surface light green and pubescent; petioles ⅙–⅓ in., pubescent to glabrous.

TWIGS. Slender, light to dark brown or gray, smooth, pubescent at first, glabrous later.

RANGE. In well-drained, rich, deep soils of wooded banks or hillsides. Eastern Louisiana. Rarely found in Texas, but one colony known 15 miles northwest of Burkville on Little Cow Creek in Newton County. Eastward to Florida and northward to eastern Virginia.

PROPAGATION. Virginia Stewartia is a very desirable ornamental shrub worthy of cultivation for the large white flowers with purple stamens. It is tender north of Washington, D.C. It grows best in a mixture of loam or peat soil. Propagation is by seeds sown as soon as ripe, by layers, or by cuttings or half-ripened wood in late summer under glass.

Virginia Stewartia
Stewartia malacodendron L.

REMARKS. The genus name, *Stewartia*, is in honor of John Stuart (1713–1792), Earl of Bute and patron of botany. The species name, *malacodendron*, is from the Greek *malakos* ("soft") and *dendron* ("tree"), referring to the silky pubescence. Also known under the vernacular names of Silky Camellia and Round-fruited Stewartia.

ST. JOHN'S-WORT FAMILY (*Hypericaceae*)

Atlantic St. Peter's-wort

Ascyrum stans Michx.

FIELD IDENTIFICATION. Plant shrubby, stems simple or sparingly branched above the base, 1–3 ft tall. Flowers rather large and bright yellow.

FLOWERS. From summer to fall, pedicels ⅓–½ in. long, 2-bracted, solitary or in few-flowered cymes; sepals 4, imbricate, very unequal, outer ones ⅜–¾ in. long, nearly orbicular, apex rounded or some abruptly short-pointed, base cordate or rounded; inner sepals ¼–⅗ in. long, lanceolate, short-acuminate; petals 4, convolute in the bud, very deciduous, bright yellow, showy, obovate, longer than the sepals; styles 3–4, relatively short; stamens numerous, filaments distinct; ovary superior, 1-celled, with 2–4 parietal placentae, ovules numerous.

FRUIT. Capsule ⅓–⅖ in. long, 1-locular, 2–4-valved, included in the calyx, seed with no albumen.

LEAVES. Opposite, erect or ascending, broadly oblong to elliptic, occasionally obovate or oval, apex rounded or obtuse, base sessile, clasping, some cordate, blade length ⅔–1¾ in., width ⅜–¾ in., margin entire, surfaces rather pale green and glabrous with minute punctate dots, thickish.

TWIGS. Slender, light to dark brown, glabrous, flattened, 2-edged, older twigs more terete and bark often shedding in thin strips.

RANGE. Moist or dry sandy woods, meadows and barrens. Texas, Oklahoma, Arkansas, and Louisiana; eastward to Florida and north to Kentucky, Tennessee, New Jersey, Pennsylvania, and New York.

REMARKS. The genus name, *Ascyrum*, is the ancient Greek name of a related plant, and the species name, *stans*, means "standing upright." The plant has been cultivated since 1816. It is propagated from seed.

Atlantic St. Peter's-wort
Ascyrum stans Michx.

Flaxleaf St. Peter's-wort

Ascyrum linifolium Spach

FIELD IDENTIFICATION. Small shrub 6–15 in., with branches erect, slender, and densely leafy.

FLOWERS. Borne in spring, terminal, perfect, regular, hypogynous; petals 4, bright yellow, equal, oblique convolute; stamens numerous, filaments distinct and

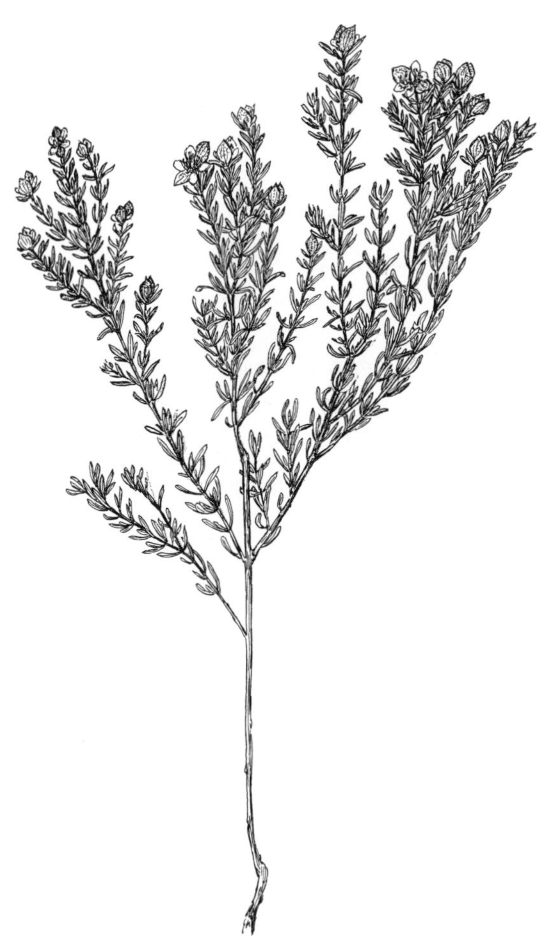

LAXLEAF ST. PETER'S-WORT
scyrum linifolium Spach

ongate, anthers short; ovary superior, compound, with -3 parietal placentae, ovules numerous; sepals 4, very nequal, inner ones very small and petaloid; outer ones rger, elliptic to ovate, ¼–⅓ in. long.

FRUIT. Capsule ⅕–⅓ in. long, inserted in the calyx, barely exserted at the tip, dehiscent, septicidal, many-eded.

LEAVES. Simple, opposite, or fascicled, rather dense, ariously shaped, linear-spatulate to oblanceolate or liptic to obovate-elliptic, apex acute, base sessile and asping, margin entire, blade length ¼–1½ in., surces glandular-punctate and glabrous.

TWIGS. Slender, erect, reddish brown, young ones omewhat angled, older ones somewhat fibrous barked.

RANGE. Sandy pinelands of Texas and Arkansas, east Florida and north to South Carolina. Also in the Vest Indies.

REMARKS. The genus name, *Ascyrum,* is from the reek *askuron,* and is the ancient name for a glabrous lant. The species name, *linifolium,* is for the flaxke leaves.

ST. ANDREW'S CROSS
Ascyrum hypericoides L.

FIELD IDENTIFICATION. Leafy, glabrous subshrub ½–3 ft, either branched at the base or branched above.

FLOWERS. June–September, terminal or axillary, on rather short, lateral, leafy stems; pedicels 1/12–¾ in., with 2 subulate bracts near the summit; sepals 4, in 2 sets, the inner small and inconspicuous, the outer 2 oval to ovate or elliptic, apex acute, base narrowed or occasionally cordate, 1/12–⅓ in. long, obscurely 3–5-nerved, conspicuously subtending the petals; petals 4, convolute in the bud, dropping away early, bright yellow, linear to oblong, apex acute, oblique, about equaling the outer sepals; stamens numerous, distinct, yellow; anthers oval; styles 2, short, persistent.

FRUIT. Capsule about ⅓ in. long, ovoid to spindle-shaped, flattened, striate, styles persistent, 1-celled, 2–4-valved, dehiscent at the placentae, enclosed by the persistent calyx, seeds numerous.

LEAVES. Opposite, or sometimes fascicled, the larger leaves often subtended by smaller ones, sessile, blade length ½–1½ in., width about ¼ in., apex obtuse to acute or rounded, base narrowed, margin entire, oblong to linear or oblanceolate, thin, surface bright green and glabrous, punctate.

TWIGS. Much-branched, slender, brown, flattened when young, the thin ridges shreddy.

RANGE. Usually in sandy soil. Texas, Oklahoma, Arkansas, and Louisiana; north to Massachusetts and west

ST. ANDREW'S CROSS
Ascyrum hypericoides L.

Naked St. John's-wort
Hypericum nudiflorum Michx.

to Nebraska. Also in the West Indies, Jamaica, and Cuba. In Mexico in the states of Veracruz, Oaxaca, and Chiapas.

VARIETIES. Several varieties of St. Andrew's Cross have been proposed, but the differences between them are not distinct, and intergrading forms often occur. *A. hypericoides* var. *multicaule* (Michx.) Fern. is a low and diffuse form with many stems from the base, and to 20 in. tall. The sepals are ⅓–⅖ in. long and the leaves oblanceolate. It occurs mostly in dry soil.

A. hypericoides var. *oblongifolium* (Spach) Fern. is to 32 in. tall, with erect stems branched above the middle. The sepals are ⅖–⅗ in. long and the leaves are oblanceolate. It seems to prefer moist soil.

REMARKS. The genus name, *Ascyrum*, is from the Greek *askuron*, and is the ancient name for a glabrous plant. The species name, *hypericoides*, is for its resemblance to certain species of *Hypericum*. It is known by the name of "Arrayanalla" in Puerto Rico. The plant is occasionally browsed by white-tailed deer. An extract of the leaves is used in home medicine as an astringent and resolutive. The seeds are reported to have purgative properties. It is sometimes cultivated for ornament, but is rather short-lived and should be protected from severe cold in the North. It is propagated by seeds and divisions. The flowers shatter easily.

Naked St. John's-wort

Hypericum nudiflorum Michx.

FIELD IDENTIFICATION. Branching shrub 1½–6½ ft.

FLOWERS. Appearing June–July, the open terminal cymes dichotomous and many-flowered, 1½–4½ in. broad; petals 5, ¼–⅖ in. long, oblique, obtuse or acute, convolute in the bud; stamens very numerous, filaments elongate; ovary 1-celled, divided by intruding placentae, styles 3, closely connivent at base, persistent; sepals 5, about equal, narrowly oblong, ⅛–⅕ in. long, usually less than one half as long as the petals; bracts minute.

FRUIT. Capsule conic-ovoid, ⅙–¼ in. long, ⅙–⅕ in. thick, splitting by dehiscence, styles persistent, 1-celled with 3 inwardly projecting placentae, seeds short and cylindric.

LEAVES. Simple, opposite, some leaves with other minute, subulate, bracted leaves in their axils, linear-oblong to narrowly elliptic to oval-lanceolate, blade length 1–2½ in., apex obtuse to rounded, base sessile or nearly so, somewhat clasping, margin entire, thin, pale green, minutely punctate.

TWIGS. Slender, upright, 2-angled, young ones greenish, older ones reddish brown, glabrous.

RANGE. In sandy swamps. East Texas and Louisiana, eastward to Florida and north to Tennessee, North Carolina, and Virginia.

REMARKS. The genus name, *Hypericum*, is from the Greek word *huperikon*, referring to a species of St. John's-wort. The species name, *nudiflorum*, refers to the fact that the inflorescence has only small subtending leaves.

Golden St. John's-wort

Hypericum aureum Bartr.

FIELD IDENTIFICATION. Shrubby perennial with wide branching, slightly winged stems, attaining a height of 1–4 ft.

FLOWERS. Very showy, sessile, usually solitary, or occasionally 2–3 in terminal or axillary cymes; sepals 5, very unequal, foliaceous, ⅖–⅗ in. long, broadened toward the apex; corolla golden yellow, 1–2 in. broad; petals 5, glabrous, oblique, ⅗–1 in. long; stamens numerous; filaments distinct and grouped at the base; ovary 1-celled; styles 3.

;OLDEN ST. JOHN'S-WORT
Iypericum aureum Bartr.

FRUIT. Capsules conic, ⅖–½ in. high, apex acumi-
ate, thick-walled, incompletely 3-celled by the pla-
ental intrusion; seeds numerous.

LEAVES. Opposite, blade leathery, elliptic to ovate-
lliptic, apex obtuse or mucronate, base narrowed into
very short petiole or sessile, margin entire, length
–2¾ in., lower surface glaucous, somewhat dotted.

RANGE. On cliffs or along streams, often in calcareous
oil. Eastern Texas, Louisiana, and Oklahoma; eastward
o Mississippi, Tennessee, and Georgia, and northward
o North Carolina.

REMARKS. The genus name, *Hypericum,* is from the
;reek word *huperikon,* referring to a species of St.
ohn's-wort; the species name, *aureum,* is for the golden
olor of the flowers. The plant is also known under the
ynonym of *H. frondosum* Michx.

;HRUBBY ST. JOHN'S-WORT
Iypericum prolificum L.

FIELD IDENTIFICATION. Shrub 1–6 ft high, usually
nuch branched at the base with the stems erect or as-
:ending.

FLOWERS. Appearing July–September, solitary or in twos or threes, forming narrow panicles or loose cymes from the upper axils; peduncles short; petals 5, ⅓–⅘ in. long, convolute, nearly equilateral, corolla when spread ¾–1 in. across; stamens numerous, filaments elongate, distinct; styles 3–4, elongate, stigmas minute; calyx of 5 sepals, unequal, shorter than the petals, ovate to obovate, ⅙–¼ in. long.

FRUIT. Capsule conic to narrowly ovoid, ⅓–⅖ in. long, apex acute, styles persistent, dehiscent, completely 3-celled, seeds numerous.

LEAVES. Opposite, sometimes with smaller ones in their axils, linear to oblong or narrowly elliptic, more rarely lanceolate, apex obtuse, often mucronate, base gradually narrowed into a short and often winged petiole, margin entire and slightly revolute, blade length ¾–3⅓ in., width ⅙–⅗ in., dull green above, lower surface paler.

TWIGS. Slender, glabrous, reddish brown at first and sharply 2-edged, older gray to straw-colored and less angled.

RANGE. Usually in calcareous rocky or sandy soil.

SHRUBBY ST. JOHN'S-WORT
Hypericum prolificum L.

Dense St. John's-wort
Hypericum densiflorum Pursh

Louisiana and Arkansas; north to New York and west to Ontario and Minnesota.

REMARKS. The genus name, *Hypericum,* is from the ancient Greek word *huperikon* referring to a species of St. John's-wort. The species name, *prolificum,* means abundant in number. It is known also as Broom-brush, Paint-brush, and Rock-rose.

Dense St. John's-wort

Hypericum densiflorum Pursh

FIELD IDENTIFICATION. Shrubby perennial to 6 ft, with many-branched, slender, winged stems.

FLOWERS. Numerous, borne in terminal, dense corymbs; sepals 5, unequal, elliptic to elliptic-oblong or lanceolate, 1/12–1/10 in. long; corolla smooth, bright yellow, ⅜–⅝ in. broad, with 5 obliquely pointed petals; stamens numerous, distinct, borne in groups; styles mostly 3, more or less connate at base.

FRUIT. Capsule ovoid, ⅙–¼ in. long, shallowly 3-lobed, 3-celled; styles persistent as a beak.

LEAVES. Often with smaller ones in their axils, blades linear to oblanceolate or oblong, blades ⅜–2 in. long, ⅕–⅓ in. wide, apex acute or obtuse, base narrowed into a short petiole, margin more or less revolute, more or less dotted.

RANGE. The species is found in east Texas, Oklahoma, Arkansas, and Louisiana; eastward to Florida, northward to South Carolina and west to Missouri.

VARIETY. The variety *H. densiflorum* var. *lobocarpum* (Gatt.) Svenson, has a deeply lobed capsule ⅕–⅓ in. long and most often 4–5-celled. It occurs in Louisiana, Oklahoma, Missouri, and North Carolina.

REMARKS. The genus name, *Hypericum,* is from the Greek word *huperikon,* referring to a species of St. John's-wort; the species name, *densiflorum,* is for the dense flowers.

Rockrose-leaved St. John's-wort

Hypericum cistifolium Lam.

FIELD IDENTIFICATION. Woody plant 1–3 ft, with yellow flowers and globose capsules.

FLOWERS. Usually in numerous, terminal or axillary small-flowered cymes; petals 5, convolute in the bud, cuneate to obovate, ⅕–⅓ in. long; stamens numerous,

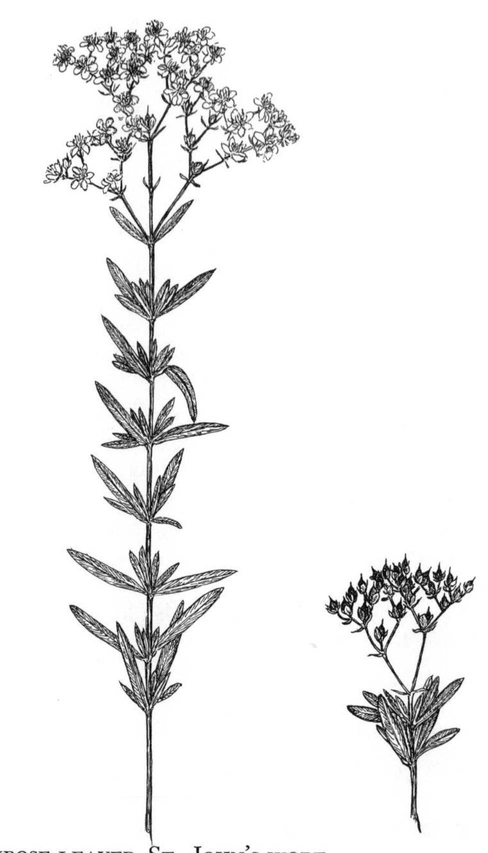

Rockrose-leaved St. John's-wort
Hypericum cistifolium Lam.

BEDSTRAW ST. JOHN'S-WORT
Iypericum galioides Lam.

ilaments elongate; styles coherent below, stigma minute; sepals 5, ovate to lanceolate, ⅛–⅙ in. long.

FRUIT. Capsule globose to globose-ovoid, ⅙–¼ in. ong, style remnants persistent, 1-celled, with parietal lacentae, seeds wrinkled.

LEAVES. Opposite, often with smaller ones in their xils, blade linear-elliptic to linear-lanceolate, or narowly elliptic, ⅝–3¼ in. long, margin entire, apex acute r obtuse, base sessile or nearly so and abruptly narowed, surfaces glabrous and minutely punctate.

TWIGS. Slender, erect, reddish brown, glabrous, oung ones generally flattened and 2-edged, older ones ess flattened.

RANGE. Rocky slopes and banks. Louisiana and Aransas; eastward to Florida and northwestward to Misouri, Iowa, and Kansas.

REMARKS. The genus name, *Hypericum*, is from the ncient Greek word *huperikon*, referring to a species f St. John's-wort. The species name, *cistifolium*, means 'leaves like the Rockrose." It is also known as the Round-podded St. John's-wort.

BEDSTRAW ST. JOHN'S-WORT

Hypericum galioides Lam.

FIELD IDENTIFICATION. Evergreen, branching, somewhat woody plant 1–6 ft.

FLOWERS. Appearing July–September, borne in axillary cymes which form panicles, short-pediceled, corolla ¼–½ in. broad; petals 5, convolute in the bud, obliquely pointed, base cuneately narrowed, ⅙–⅓ in. long; stamens numerous, distinct, filaments elongate; styles elongate; sepals linear, ⅛–⅙ in. long, foliaceous, slightly revolute, resembling the adjacent leaves.

FRUIT. Capsule conic, acute, ⅙–¼ in. long, dehiscent, styles persistent, incompletely 3-celled by the projecting placentae, seeds numerous.

LEAVES. Opposite, more or less spreading, linear-oblong or oblanceolate, with smaller ones clustered in the axils, apex obtuse or acute, base narrowed and sessile or very short petioled, margin entire, somewhat revolute, thickish, glabrous, blade length ½–2½ in., width 1/12–⅙ in.

TWIGS. Slender, reddish brown, glabrous, somewhat angled or older ones almost terete.

RANGE. In pineland swamps. Louisiana eastward to Florida and north to Tennessee, North Carolina, and Delaware.

REMARKS. The genus name, *Hypericum*, is from the Greek word *huperikon*, once applied to a species of St. John's-wort. The species name, *galioides*, means "like *Galium*," or appearing like a bedstraw plant.

SANDBUSH ST. JOHN'S-WORT

Hypericum fasciculatum Lam.

FIELD IDENTIFICATION. Tall evergreen shrub, under favorable conditions treelike and 3–20 ft in height.

FLOWERS. Through the spring and summer, borne in narrow loose panicles, or some inflorescences corymbose; petals 5, convolute, obliquely apiculate, about ⅓ in. long; stamens numerous, filaments elongate; styles long; sepals 5, about equal, linear, about as long as the petals or fully half as long, ⅛–⅙ in. long, 1/15–1/25 in. wide.

FRUIT. Capsule ovoid or conic-ovoid, ⅙–⅕ in. long, styles persistent, splitting by dehiscence, incompletely 3-celled by intrusion of the placentae.

LEAVES. Numerous, simple, fascicled, a group of smaller ones clustered in the axils of the larger ones, linear-filiform, fleshy, acute at apex, sessile at base, margin entire and revolute, length ⅜–¾ in.

TWIGS. Numerous, green to brownish, glabrous, sharply angled.

RANGE. Mostly in low, sandy, acid pinelands, often in shallow ponds where its presence indicates a hard

SANDBUSH ST. JOHN'S-WORT
Hypericum fasciculatum Lam.

sandy bottom. Eastern Texas, eastward through Louisiana to Florida.

REMARKS. The genus name, *Hypericum,* is from the ancient Greek word *huperikon,* which was applied to a species of St. John's-wort. The species name, *fasciculatum,* refers to the clustered (fascicled) leaves.

PINELAND ST. JOHN'S-WORT

Hypericum aspalathoides Willd.

FIELD IDENTIFICATION. Shrubby evergreen perennial plant ½–3 ft, with simple or much-branched stems.

FLOWERS. Borne in several terminal corymbose cymes, pedicels slender, ⅛–⅙ in.; sepals 5, essentially equal, less than one half as long as the petals, linear or nearly so, 1/12–⅛ in. long; petals convolute, 5, bright orange-yellow, ¼–⅓ in. long, oblique, smooth, glabrous; stamens numerous, distinct; styles rather long.

FRUIT. Capsule conic, ⅙–¼ in. long, acuminate upward, incompletely 3-celled by the intrusion of the placentae, seeds numerous.

LEAVES. Very numerous, opposite, or in conspicuous axillary clusters, often with smaller ones in the axis of the larger ones, linear-filiform or linear-subulate, obtuse, entire, revolute, sessile, ⅕–⅓ in. long, most less than ⅖ in. long.

TWIGS. Slender, reddish brown to dark brown, glabrous, flattened and sharply angled at first, later more terete and brown to gray.

RANGE. Pinelands and prairies from Florida to Louisiana, northward to North Carolina.

REMARKS. The genus name, *Hypericum,* is from the ancient Greek name *huperikon*. The species name, *aspalathoides,* refers to the heathlike *Aspalathus* plants of South Africa, which it resembles.

PINELAND ST. JOHN'S-WORT
Hypericum aspalathoides Willd.

FRANKENIA FAMILY (*Frankeniaceae*)

JAMES FRANKENIA
Frankenia jamesii Torr.

FIELD IDENTIFICATION. Low erect undershrub, usually less than 3 ft.

FLOWERS. Usually solitary, or occasionally in cymose, axillary clusters; small, white, sessile, perfect, regular, hypogynous; petals 5, blade 1/6–1/4 in. long, apex cuneiform, erose, base narrowed into a claw, claw bearing a basal crown; stamens 5–6, exserted, alternate with the petals; style 2–4-cleft, divisions filiform; ovary superior, 1-celled with 2–4 parietal placentae; calyx tubular, about 1/5 in., sepals persistent, about 1/25 in.

FRUIT. Capsule linear, small, 1-celled, included in the calyx; seeds few, oval to oblong.

LEAVES. Numerous, crowded, small, subsessile, opposite or fascicled, a membrane connecting the base of the subsessile petioles, fascicles usually 1/4–1/2 in. apart on the stems, blade 1/5–1/3 in. long, linear to filiform. Margin revolute, apex acute or obtuse, surface glabrous or nearly so, stipules absent.

TWIGS. Erect, somewhat fascicled, green to gray or brown, scabrous-puberulent.

RANGE. Altitudes to 5,500 ft, usually on alkaline or salt flats in west Texas, New Mexico, and Arizona; north to Colorado and Nevada, and south into Mexico.

REMARKS. The genus name, *Frankenia,* honors Johann Franke (latinized, Frankenius; 1590–1661), a Swedish botany professor at Uppsala. The species name, *jamesii,* is in honor of Erwin James, botanist and historian of Long's expedition to the Rocky Mountains in 1820. The plants are used only to a small extent in cultivation for ornament.

JAMES FRANKENIA
Frankenia jamesii Torr.

TAMARISK FAMILY (*Tamaricaceae*)

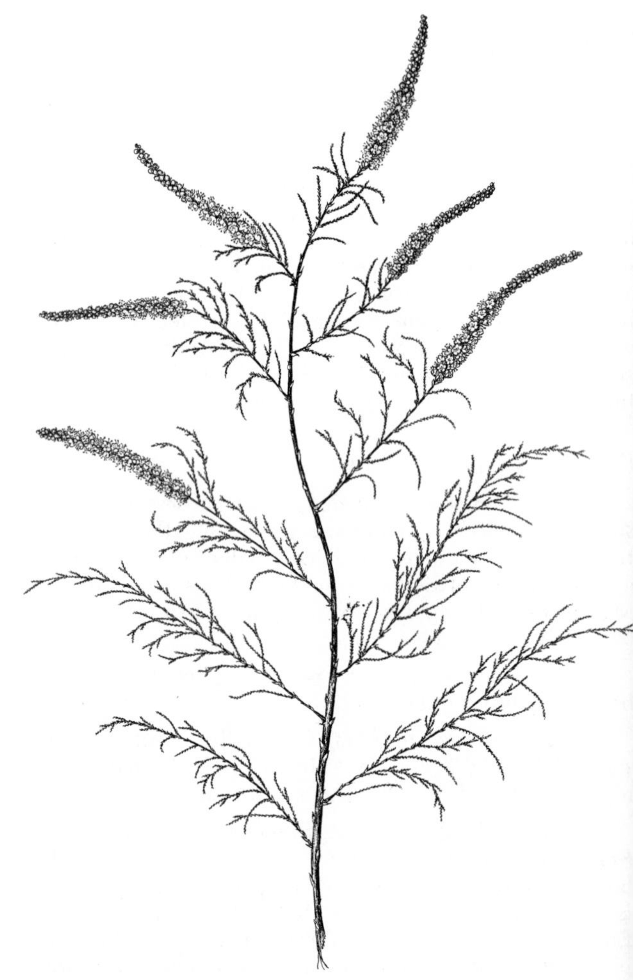

FRENCH TAMARISK
Tamarix gallica L.

FRENCH TAMARISK
Tamarix gallica L.

FIELD IDENTIFICATION. Shrub with contorted branches, or a tree to 30 ft, with a twisted trunk.

FLOWERS. Borne in summer on the current wood, in white or pink racemes, which are grouped to form terminal panicles of variable length; pedicels about 1/50 in. long; sepals 5, 1/25–1/12 in. long, ovate; corolla petals 5, 1/25–1/12 in. long, oblong, mostly deciduous from the mature fruit; stamens 5, filaments 1/12–1/10 in., enlarged toward the base and attached to the corners of the 5-angled disc, anthers mucronate, 2-celled; ovary 1/25–1/15 in. long, set on the disc; styles 3, about 1/50 in. long, clavate.

FRUIT. Capsule very small, 1/12–1/8 in. long, dehiscing into 3 parts; seeds numerous, minute, tufted with hairs at apex.

LEAVES. Foliage sparse, delicate, grayish green, scale-like, alternate, imbricate, 1/50–1/8 in. long, deltoid to lanceolate, acute to acuminate, entire and scarious, glabrous; bracts 1/25–1/15 in. long.

BRANCHES. Drooping and graceful, often sweeping the ground, young ones glabrous or glaucous, reddish to gray later.

WOOD. Light-colored, close-grained, takes a high polish, often twisted or knotty.

RANGE. French Tamarisk was introduced to the United States from Europe. It now grows as an escape from cultivation from Texas eastward to Florida, westward to California, and north to Arkansas and South Carolina.

PROPAGATION. It is widely cultivated for ornament, erosion control, and windbreaks, in Texas and the Southwest generally. Preferring moist soil and an open sunny situation, it has adjusted to the Gulf Coast prairie areas, having a high tolerance for saline or alkaline sites. It apparently does well in the coastal brackish, sandy marshes. It can readily be propagated from cuttings taken in early winter from hardwood of the previous summer.

RELATED SPECIES. *T. gallica* is very closely related to *T. pentandra* Pallas. Both bloom in summer and appear very similar except for the tiny flowers. *T. gallica* has the filaments broadened toward the base and confluent with angles of the disk, also the petals are usually deciduous from the fruits. *T. pentandra* Pallas has the filaments inserted between the lobes of the disk, and petals usually present on the fruit.

Another tamarisk planted in the Southwest is the Athol or Athel Tamarisk, *T. aphylla* (L.) Karst. It also has flowers borne in summer, and it can be recognized by its reduced sheathing leaves and bracts which give the stem a jointed appearance; stem leaves and floral parts are covered with punctate salt-secreting glands. It is native to North Africa, tropical Africa, and northern India.

REMARKS. The genus name, *Tamarix*, is the ancient name, probably with reference to the Tamaracine people of southern Europe, where the plant grew. The species name, *gallica*, refers to a Gallic tribe, who lived where the plant grew. Other names are Salt-cedar, Manna-bush, Athel, Eshel, Asul, Athul, and Atle. The low-sweeping branches make excellent wild-life cover.

FIVE-STAMEN TAMARISK
Tamarix pentandra Pallas

OCOTILLO FAMILY (*Fouquieriaceae*)

OCOTILLO
Fouquieria splendens Engelm.

OCOTILLO

Fouquieria splendens Engelm.

FIELD IDENTIFICATION. Spiny shrub with no definite trunk, but sending up numerous, erect, whiplike, slender stems 6–21 ft high from the base. The leaves are spatulate or oblanceolate, early deciduous, the plant appearing leafless most of the year.

FLOWERS. Appearing before the leaves, borne on short pedicels in narrow panicles or racemes 2–8 in.; corolla crimson, ½–1 in. long, perfect, regular, broadly tubular with a spreading, 5-lobed limb; lobes ovate to triangular, recurved; stamens 10–19, red, exserted, borne on the corolla-tube within; ovary of 3 septiform placentae, ovules 4–6 in each cell; sepals 5, reddish, ⅕–¼ in. long, apices somewhat rounded.

FRUIT. Capsule ovoid, ⅓–¾ in. long, imperfectly 3-celled; seeds numerous, flat, winged, wings often fringed.

LEAVES. Simple, solitary or fascicled, petiolate or sessile, spatulate, oblanceolate or oblong, entire on margin, thick, leathery, blades ½–2 in. long; numerous at first, or after rains, but falling away early to leave the plant leafless; spines formed by the indurate midribs, ⅔–1¼ in. long.

STEMS. Radiating from the base, slender, whiplike, conspicuously grooved and ridged, spines decurrent.

RANGE. Usually on dry, rocky hillsides, or desert flats; western and southern Texas, New Mexico, and westward through Arizona to California; north to Nevada; and south into Mexico. In Mexico in the states of Coahuila, Chihuahua, Sonora, Zacatecas, and Baja California.

REMARKS. The genus name, *Fouquieria,* is in honor of P. E. Fouquier of Paris. The species name, *splendens,* is for the showy flowers. Vernacular names for the plant are Coachwhip-cactus, Vine-cactus, Slimwood, Jacob's Staff, Candlewood, Albarda, Barda, Ocote, Ochotilla, and Ocotilla. The slender stems are use for walking sticks, and are planted close together t make fences and hut walls, often sprouting from the barren stem. The plant is also used ornamentally in cactu or succulent gardens for its unusual form and strikin crimson flowers. The flowers and seed pods are sometimes eaten by the Cahuilla Indians, and a beverage and also a cough medicine, are made from the flowers The bark contains gum, resin, and wax, and was formerly used for waxing leather. The powdered root is use for dressing wounds and swellings by the Apache Indians. The wood is hard, heavy, and burns easily Ocotillo is often confused with cactus by people wh see it for the first time.

FLACOURTIA FAMILY (*Flacourtiaceae*)

Mexican Xylosma

Xylosma blepharodes Lundell

FIELD IDENTIFICATION. An evergreen shrub, or an irregularly branched spiny tree, to 20 ft.

FLOWERS. Dioecious, very small, borne in axillary fascicles on pedicels 1/12–1/5 in.; petals absent; sepals 4–5, imbricate; stamens numerous, filaments free or nearly so, surrounded by a disk; ovary superior, styles 2–3.

FRUIT. In clusters at the nodes, peduncles 1/8–3/16 in., slender; fruit body subglobose, 1/5–1/4 in. in diameter, green when immature, red later, indehiscent; seeds usually 5, 1/16–1/8 in. long, white, translucent, flattened or angled at base.

LEAVES. Variable in size and shape, simple, alternate, elliptic to oblong, ovate to obovate, coriaceous; apex obtuse to acute or acuminate; base cuneate or rounded; margin coarsely and remotely toothed, or sometimes entire, somewhat revolute; blade length 1–2½ in., width ½–1¼ in., upper surface olive-green, glabrous, or pubescent along the veins; lower surface paler, glabrous or pubescent along the veins; petiole variable in length, usually short.

TWIGS. Slender, gray to brown, glabrous or pubescent, with a single, slender, straight spine arising from each node, about ¼ in.; bark smooth when immature, broken into small scales later.

RANGE. Lower Rio Grande Valley area of Texas, near Combes, Cameron County, Texas. A cultivated specimen grows in the yard of Robert Runyon, 812 St. Charles Street, Brownsville, Texas. Known in the Mexican states of Nuevo León, Veracruz, and Chiapas, and also in Guatemala. The type specimen was collected at Jalapa, Veracruz, Mexico.

REMARKS. The genus name, *Xylosma,* is from the Greek words *xylos* ("wood") and *osme* ("smell"). The species name, *blepharodes,* means "fringelike," but the reason for its application here is obscure. Vernacular names are Manzanillo, Coronilla, and Huichichiltemel. It is reported to be a local remedy for tuberculosis in Central American countries. A number of species are known from Mexico but are not often cultivated in the United States. They can be propagated from seed or from softwood cuttings; they do best under glass, being rather tender.

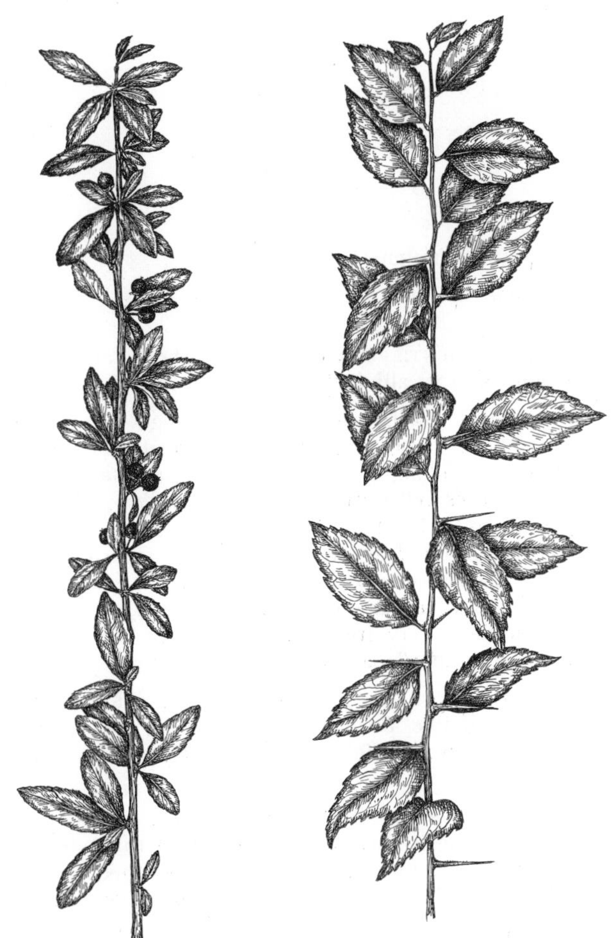

Mexican Xylosma
Xylosma blepharodes Lundell

TURNERA FAMILY (*Turneraceae*)

DAMIANA TURNERA
Turnera diffusa Willd.

DAMIANA TURNERA
Turnera diffusa Willd.

FIELD IDENTIFICATION. Diffuse shrub attaining a height of 6 ft, although usually less than 3 ft.

FLOWERS. Axillary, small, perfect, regular; calyx sessile, about 1/6 in. long, tubular, tomentose; sepals 5, imbricate, oblong or somewhat ovate, shorter than the tube, apices reflexed; corolla small, yellow, 5-petaled, petals 1/4–1/3 in. long, narrowly cuneate, sometimes twisted, inserted at the throat of the calyx-tube; stamens 5, inserted near the petals; filaments free and distinct, somewhat flattened; anthers 2-celled, sacs opening lengthwise; ovary free, 1-celled, ovules numerous, in 2 rows on the 3 parietal placentae.

FRUIT. Capsule ovoid, 1/6–1/4 in. long, dehiscent from apex to middle, 3-valved; seeds curved, with a rough testa; embryo straight in the copious, fleshy endosperm.

LEAVES. Simple, alternate or clustered, aromatic, narrowly oblong to spatulate, apex obtuse to acute, base short-cuneate, margin coarsely 2–10-toothed on each side, blade length 3/8–1 in., to 2/5 in. wide, thickish; veins prominent beneath and impressed above; upper surface smooth and light olive-green; lower surface whitish, varying from glabrous with a few hairs on the ribs to tomentose; buds gray with appressed pubescence; petioles very short, twigs yellow to reddish brown, pubescent when young, glabrous later.

RANGE. Western Texas, southern California, Mexico, West Indies, and South America.

MEDICINAL USES. Standley (1920–1926, p. 848) reports that "In Mexico the plant is often used as a substitute for Chinese tea and for flavoring liquors. It has a wide reputation as an aphrodisiac and [a decoction of the leaves] is administered also for dysentery, malaria, syphilitic diseases, pains in the stomach and intestines, dyspepsia, and even paralysis. Diuretic, astringent, tonic, expectorant, and laxative properties are ascribed to it. The plant was introduced into Europe about 1874 under the name damiana, and was for some time recommended for all types of renal and vesical diseases."

The name damiana is also applied to other plants which are sometimes used as adulterants.

The *National Formulary* (7th ed., quoted in Wood and Osol, 1943, p. 362) describes the drug damiana officially as follows: " 'Damiana . . . the dried[, powdered] leaf, contains not more than 15 per cent of the stems of the plants and not more than 3 per cent of other foreign organic matter, and yields not more than 4 per cent of acid-insoluble ash.' " Wood and Osol (1943, p. 362) note that Bolivia and Mexico are the chief sources of supply for damiana imported for commercial use. As to the supposed aphrodisiac properties of the drug, Wood and Osol (p. 362) make the following statement: "Although damiana has achieved some repute in the treatment of sexual impotence, it is worthy of note that it is always given in conjunction with strychnine, phosphorus, or some other stimulant. There is no convincing evidence of its remedial value and it is now rarely prescribed by physicians."

REMARKS. The genus name, *Turnera,* honors William Turner, English herbalist (d. 1568), and the species name, *diffusa,* refers to its spreading habit. Also known under the vernacular names of Hierba de Vemulo, Hierba de la Pastora, Xmisiboc, Hierba de Vendo, and Oreganillo. The plant is also listed in some literature as *T. aphrodisiaca* Ward.

PAPAYA FAMILY (*Caricaceae*)

PAPAYA
Carica papaya L.

PAPAYA

Carica papaya L.

FIELD IDENTIFICATION. Small, usually unbranched, milky-sapped tree 9–25 ft. The very large 7–9-lobed leaves are borne at the top of the tree only.

FLOWERS. The staminate and pistillate flowers are borne on separate trees, or more rarely both sexes are on the same tree. Mostly axillary, staminate in slender, long-peduncled panicles which are 4 in. long or more; corolla about 1–1¼ in. long, salverform, yellowish white to reddish; tube slender, dilated near the top, limb 5-lobed, the segments lanceolate to elliptic or oblong, barely one half as long as the tube; stamens 10, inserted in the corolla-throat (5 short-filamented and 5 sessile staminodia); anthers adnate to the filaments, erect, oblong-linear; calyx of staminate flowers 1/25–1/12 in. high, 5-lobed. Pistillate flowers much larger than the staminate, yellow to white, solitary, or in short-peduncled, few-flowered cymes; petals 5, distinct, linear to lanceolate, twisted, ¾–1 in. long; gynoecium a compound pistil, swollen below, style short, stigmas 5-lobed and sessile; ovules numerous, inserted in series on the 5 placentae; calyx of the pistillate flower 1/5–2/5 in. high, 5-lobed.

FRUIT. Usually November–June, but in tropical regions borne almost continually. A melon-like berry, borne near the top of the trunk on very short peduncles, usually 2–8 in. long, but rarely to 18 in. and with a weight of 20 lb, ovoid to oblong or subglobose, indehiscent with a thick rind; flesh firm, yellow or orange, taste cloyingly sweet and insipid, odor pungent; seeds numerous, subglobose, black, roughened.

LEAVES. At the top of the trunk, alternate, very large, blade 8–30 in. wide, suborbicular in outline, palmately 7–9-lobed, each lobe pinnately lobed, segments obtuse or acute, or the larger ones acuminate, upper surface dark green, lower surface paler or glaucous with prominent veins; petiole commonly as long as the blade o longer.

BARK. Green, gray, or purplish; leaf scars large and prominent.

WOOD. Exterior woody parts thin, pith large and porous, trunk often with a cavity within.

RANGE. Considered to be a native of tropical America and cultivated in the warm regions of both hemispheres. Escaped cultivation in southern Florida and the Keys. Cultivated in the lower Rio Grande area of Texas, and the fruit is often sold in the markets there.

OFFICIAL MEDICINAL PROPERTIES. The following properties are officially described for *C. papaya* (Wood and

Osol, 1943, p. 1467): "The milky juice of the papaw [Papaya] contains a proteolytic enzyme to which the name of *papaïn* and *papayotin* have been given. The enzyme has been isolated in crystalline form. . . . It is protein in character.

"The seeds of the [papaw] tree contain a glycoside, *caricin,* which resembles sinigrin. These seeds also contain the ferment, myrosin, and by the reaction of the two a volatile pungent body is produced, suggestive of oil of mustard in odor. An alkaloid, *carpaine,* . . . has been obtained from the leaves. . . . Physiologically, this alkaloid is said to slow the heart and depress the nervous system. . . . [It is also said to be] an active amebicide.

"Under the name of *papayotin, papaïn,* or *papoid* the dried juice of the *Carica Papaya* is put upon the market. As it occurs in commerce, papaïn is a grayish, fine powder, which in appearance, odor and taste strongly suggests pepsin. It varies greatly in its quality, some specimens being grossly adulterated with starch and almost inert. The best grades will render soluble more than 200 times their weight of coagulated egg albumin in alkaline media. It has been used . . . in dyspepsia and gastric catarrh, and as a local application for the destruction of false membranes, warts, tubercles, walls of old sinuses, and even of epithelioma. . . . It is stated that from time immemorial the fresh leaves of the Papaya have been used by the Indians to wrap meat in to make it tender . . . and as a dressing to foul wounds. According to Brunton, the *dose* is from five to ten grains (0.32–0.65 Gm.). Injected into the venous circulation it acts as a powerful poison, of which a single grain is sufficient to kill a rabbit."

NONOFFICIAL MEDICINAL USES. Standley (1920–1926, p. 852) notes the following: "Various medical properties are attributed to the papaya plant. The seeds and the milk from the roots are often employed as a vermicide, and the milk is applied to the skin to assist in the removal of chiggers. The infusion of the flowers is reported to have emmenagogue, febrifuge, and pectoral properties; a decoction of the leaves is employed as a remedy for asthma; and the juice is administered for indigestion. Grosourdy states that the juice of the ripe fruit was used as a cosmetic, to remove freckles.

"The Papaya is treated at length by Oviedo . . . who stated that in Hispaniola it was known as 'papaya,' but among the Spaniards of the mainland it was called 'higos de mastuerzo,' the latter name being given because the seeds had a pungent flavor like cress (mastuerzo). In Nicaragua, he states, the plant or fruit was called 'olocotón.' He claims also that the plants were not native to the West Indies, but were brought there by the Spaniards from the mainland, which may or may not be true. The plant is mentioned by all the early writers, and is described by Hernandez.

"Ramírez has described and illustrated a fruit known as 'papaya voladora,' which is presumably a form of this species. It is noteworthy in having peduncles as much as 11 in. long. The flowers of *Carica papaya* are usually dioecious, but occasionally both kinds of flowers are found upon the same plant."

REMARKS. The genus name, *Carica,* is from the Latin word for a dried fig, referring to the shape of the fruit. The species name, *papaya,* is thought to be from the Carib Indian name *ababai.* The *papaya* is used for food and medicine by people of the tropical and subtropical regions. The following excellent account is given by Standley (1920–1926, pp. 851–852):

"Known generally in Spanish-speaking countries as papaya, . . . the English names *Papaw* or *Pawpaw* are used, also 'tree-melon,' but Papaya is the preferable name. The following additional names are reported: Chick-put, Put, Papaya de los Pájaros, Papaya de Arbol, Papayero, Melón Zapote, Melón Chapote, Manón, Papaya Montes, Frute Bomba, Dzoosadzahuidium, and Lechosa. It is remarkable that no Nahuatl name is known for the plant.

"The Papaya is one of the best known of tropical American fruits. The fruits vary greatly in shape, size and quality; they sometimes attain a length of 18 in. and a weight of 20 lb. They resemble some forms of muskmelons, especially on the inside. The flesh is ¾–2 in. thick and orange-yellow or deep orange. The fruit is eaten like muskmelon or sliced and served with sugar and cream, made into salads or candied; made into preserves, pickles, jellies, pies, or sherbets; or sometimes cooked and eaten like a vegetable. The plants grow easily and rapidly from seeds and they bear fruit almost throughout the year. A confection is sometimes made by boiling the flowers in syrup.

"The fruit and other parts of the plant contain an abundant milky juice from which an enzyme, papain, resembling animal pepsin in its digestive action, has been separated. This product has become an article of commerce, being used for the treatment of dyspepsia and related affections, and also for clarifying beer. The digestive properties of the juice were well known to the original inhabitants of tropical America, like those of today, who often wrap meat in the leaves and leave it thus overnight, to make it tender. Sometimes leaves are boiled with meat for this purpose, but if too much papaya juice comes in contact with the meat, or for too long a time, the meat will fall apart in shreds. Indeed, it is even popularly believed that the plant is even more efficient, for it is said that if old hogs or poultry are fed on the leaves and fruit, their flesh will become tender, and if a piece of tough meat is hung among the leaves of the tree for a few hours it also will be made tender. This last property is attested by so eminent an authority as Heber Drury, who states that he proved it by experiment. The leaves are sometimes used in Mexico as a substitute in washing clothes."

CACTUS FAMILY (*Cactaceae*)

Barbwire Acanthocereus

Acanthocereus pentagonus (L.) Britt. & Rose

FIELD IDENTIFICATION. Cactus semierect or clambering 3–21 ft. Forming thickets by the arching, 3–5-angled stems, which root where they touch the ground.

FLOWERS. One at each areole, blooming at night, 5½–8 in. long, tubular, flaring into a funnelform throat with perianth segments numerous; outer segments shorter than the inner, green, narrowly lanceolate to linear; inner segments white, acuminate; tube withering and persistent on ripe fruit; areoles on tube and ovary with brown felt and slender spines; stamens numerous.

FRUIT. Berry oblong, red, rind thick, flesh red and edible, seeds numerous and black; cotyledons broadly ovate, ⅕–⅓ in. long, united at base.

STEMS. From 2–4½ in. in diameter, strongly 3–5-angled or rounded on old trunks; young growth sometimes 6–8-ribbed; exterior of stems pulpy and thick, interior hard and woody with a small pith; areoles 1⅓–2 in. distant; spines numerous, gray, acicular to subulate, the radial spines 6–7 (sometimes to 12), ⅜–1⅓ in. long or longer; central spines 1 or more, longer than the radials.

RANGE. Coastal Texas (from Corpus Christi southward), southern Florida, east coast of Mexico, Guatemala, Panama, Venezuela, Colombia, Cuba, and St. Thomas and St. Croix Islands. Collected by the author on clay mounds between Brownsville and Boca Chica Beach, Cameron County, Texas.

REMARKS. The genus name, *Acanthocereus*, is from the Greek meaning "thorn-cereus." The species name, *pentagonus*, refers to the predominantly 5-angled stems. Other vernacular names are Pitahaya, Pitahaya Naranjada, and Pitahaya Morada. The plant is very variable in size of branches, spines, and size of flowers.

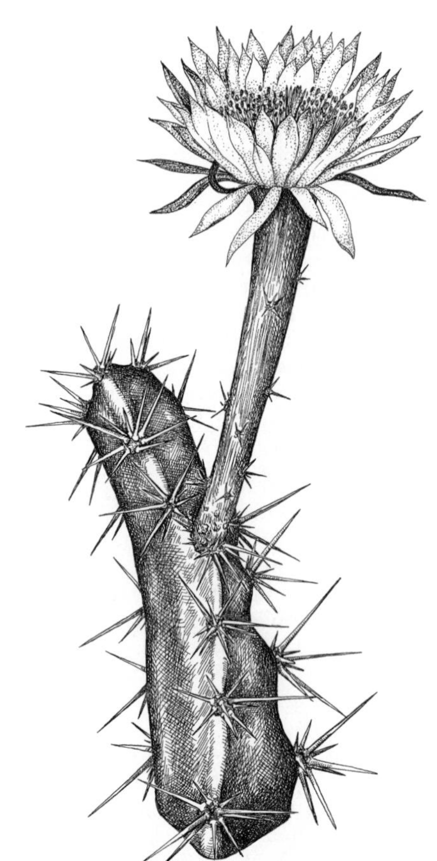

Barbwire Acanthocereus
Acanthocereus pentagonus (L.) Britt. & Rose

Deerhorn Cactus

Peniocereus greggii (Engelm.) Britt. & Rose

FIELD IDENTIFICATION. Cactus erect or sprawling, slender-stemmed, few-branched, 1–3 (rarely to 6) ft. Often mixed with other shrubs and easily overlooked. Root much enlarged, turnip-like.

FLOWERS. Usually May–June, very fragrant, blooming

DEERHORN CACTUS
Peniocereus greggii (Engelm.) Britt. & Rose

for only a night or two; tubular-funnelform, 4–8 in. long, tube slender and long, throat funnelform, 3½–5 in. in diameter, upper scales with long hairs, but lower scales spiny; perianth segments numerous, white, or the outer reddish, lanceolate, acute, to 1¾ in. long, somewhat reflexed in the outer, stamens numerous, filaments exserted and erect; style slender, stigma lobes about ⅜ in. long.

FRUIT. Berry terminal, spiny, red, ovoid, tuberculate, long-beaked, 2–4 in. long, 1–1½ in. in diameter, beak about 1 in. long, fleshy, edible; seeds dull black, roughened, hilum large and oblique.

STEMS. Usually rather slender, weak, rambling, less than 3 ft, or taller when partially supported, 3–6-angled (usually 4–5), spines clustered at areoles which are ¼–1 in. in diameter; spines all rather similar, appressed, dense, either white or dark colored; radials 11–13 (usually 6–9), centrals 1 to 2; young parts of stem often pubescent.

ROOT. Very large, turnip-shaped, 5–12 in. long and 2–8 in. in diameter, weighing 5–15 lb, rarely 50 lb or more.

RANGE. Dry alluvial soil at altitudes of 2,000–4,000 ft. In western Texas, New Mexico, Arizona, and Mexico. Collected by the author in Brewster County, Texas. In New Mexico in the Tortugas Mountains. In Mexico in Chihuahua, Sonora, and Zacatecas.

REMARKS. The genus name, *Peniocereus,* means "thread-cactus," with reference to the long hairs of the upper scales. The species name, *greggii,* honors Josiah Gregg (1806–1850), early American botanist who collected plants in the southwestern United States and Mexico. It is also known by the vernacular names of Night-blooming Cereus, Chaparral Cactus, Queen-of-the-night, Reina-de-la-noche, and Huevo de Venado.

The flowers are exceedingly fragrant and the plants can often be located at night by the fragrance. The seeds are eaten by birds, and pass through their digestive tracts, often as they roost in thickets. Hence the plant is commonly found in chaparral thickets. The stems are rather brittle and are easily broken and destroyed in the open.

It is also reported that the Mexicans sliced the turnip-like tubers and fried them in deep fat. The slices were also bound to the chest as a cure for chest colds.

WILCOX CACTUS
Wilcoxia poselgeri (Lemaire) Britt. & Rose

FIELD IDENTIFICATION. A rare and inconspicuous cactus attaining a height of 2–3 ft, with slender erect or reclining stems about ⅓ in. in diameter, each with 8–10

WILCOX CACTUS
Wilcoxia poselgeri (Lemaire) Britt. & Rose

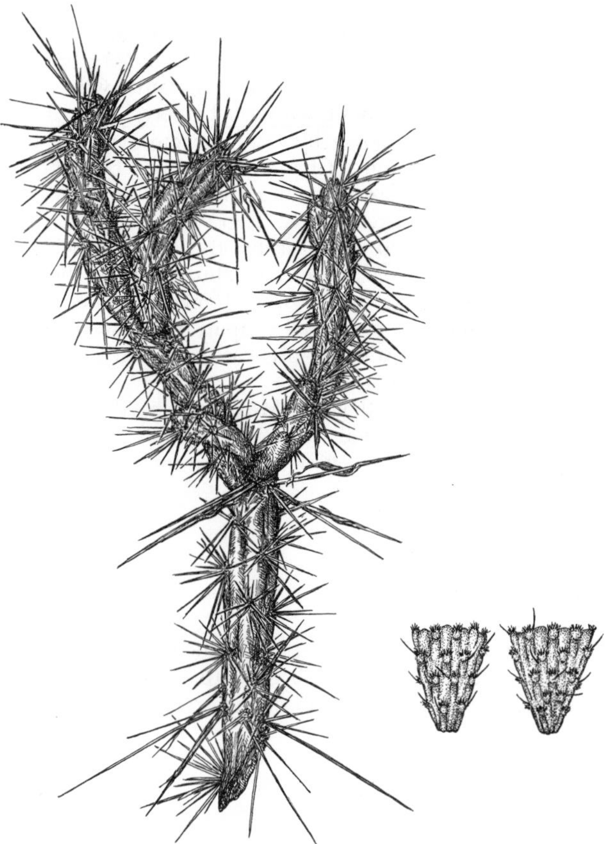

JEFF DAVIS CHOLLA
Opuntia davisii Engelm. & Bigel.

inconspicuous spiny ribs. Often growing entangled with other thorny shrubs in thickets and easily overlooked.

FLOWERS. Blooming March–May. Mostly terminal and solitary from an areole, some fragrant, but older ones scentless. Opening in the afternoon, but closing at night, for 5–9 days. The 1 in. long flower tube and the ovary clothed externally with long hairs and slender clustered spines which are reddish brown or black and 1/16–¼ in. long; perianth segments numerous, overlapping, 1–1½ in. long, ⅛–¼ in. wide, linear, acute to acuminate at apex, wide-spreading or recurved, pink with usually a darker strip down the center; stamens numerous; filaments slender, filiform, green below to white above, about ¼ in. long, erect around the style base; anthers ovoid to short-oblong, flattened, yellow to red, about 1/16 in. long; style greenish, stigmas 8–12, erect, linear, slender, green, apex obtuse or rounded and somewhat incurved, length about ⅜ in., exserted beyond the stamen ring.

FRUIT. Seeds about ⅓ in. long, rugose, pitted, black, aril large and basal.

STEMS. Green to gray, 1–3 ft high, slender, solitary or branched, ¼–⅓ in. thick, ribs 8–10 and inconspicuous; spine clusters dense, low, radial spines 9–12, appressed, delicate, gray to green, puberulent; central spines ascending or appressed, white to gray, often tipped black, about ⅜ in. long, stouter and longer than the radials; old stems sometimes spineless.

ROOTS. Near the surface of the ground, tuberous, fusiform, black, 1–4 in. long, usually several together.

RANGE. On dry gravelly or sandy hillsides. In southwest Texas and adjacent Coahuila, Mexico. In Texas from Hidalgo County to Laredo along the Rio Grande River.

REMARKS. The genus, *Wilcoxia*, was named after General Timothy E. Wilcox of the United States Army. The species name, *poselgeri*, honors H. Poselger, a German botanist who, in an article written in 1853, described the plant under the name of *Cereus tuberosus* Poselger. It is also known under the vernacular names of Sacasil and Lead Pencil Cactus. Rather rare, but also easily overlooked owing to its occurrence among thick chaparral. A 1-ft specimen was grown by the author in a pot and bloomed on March 16. The pink flowers were not fragrant and opened only a few hours in the afternoon for a period of 8 days. It is sometimes grafted on *Selenicereus pteranthus* and does well.

JEFF DAVIS CHOLLA

Opuntia davisii Engelm. & Bigel.

FIELD IDENTIFICATION. A branched cactus usually less than 2 ft. The yellowish spines make the plant conspicuous.

FLOWERS. Flowers usually less than 1½ in. long; petals yellow to olive-green, broadly obovate or oval, apex rounded and mucronate; areoles on the ovary large, usually set with a few spines.

FRUIT. Green to yellowish, about 1¼ in. long (including ovary), naked and somewhat tuberculate.

STEMS. Pale brown or gray, bark thin and glabrous; terminal joints slender, 2¼–3¼ in. long, about ⅜ in. in diameter, conspicuously tuberculate; spines dense, radiating, 6–12, slender, acicular, reddish brown, unequal in length, ¾–2 in. long, with papery and lustrous sheaths; glochids numerous and yellow.

RANGE. The type locality has been designated as the Upper Canadian, about Tucumcari Hills, near the Llano Estacado, western Texas, and eastern New Mexico.

REMARKS. The genus name, *Opuntia*, was the name given by Theophrastus for some other plant which grew near Opus, a town in ancient Greece. The species name, *davisii*, is in honor of Jefferson Davis, who was secretary of war when Lieutenant A. W. Whipple was United States Army commander of a Pacific railroad expedition from the Mississippi River to Los Angeles and made his report on the region.

ANDLE CHOLLA

)puntia kleiniae DC.

FIELD IDENTIFICATION. A slender cholla attaining a eight of 2–6½ ft, the branches cylindrical and an-ularly branched to form an open head. The older tems become woody.

FLOWERS. From ¾–1¼ in. in diameter, purple; petals umerous, obovate, apex rounded.

FRUIT. Long-persistent, red at maturity, obovoid, –1 in. long, about ½ in. in diameter, strongly tuber-led, not spiny, fleshy; seeds ⅙–⅕ in. long.

LEAVES. Deciduous, acute, about ⅗ in. long.

BRANCHES AND SPINES. The terminal branches about 2 in. in diameter, prominently tuberculate, tubercles 2–⅝ in. long, ⅛–³⁄₁₆ in. broad and about ⅛ in. high; nterior of branches woody with pith about one fourth f the total diameter; spines solitary or 2–4 from the reole base, usually less than 1 in. long; sheaths de-iduous from spines on older branches; glochids minute, ellow to brown.

RANGE. In limestone hills of desert areas at altitudes f 3,000–4,000 ft. Trans-Pecos Texas, west through New Mexico to Arizona, southward into central Mexico.

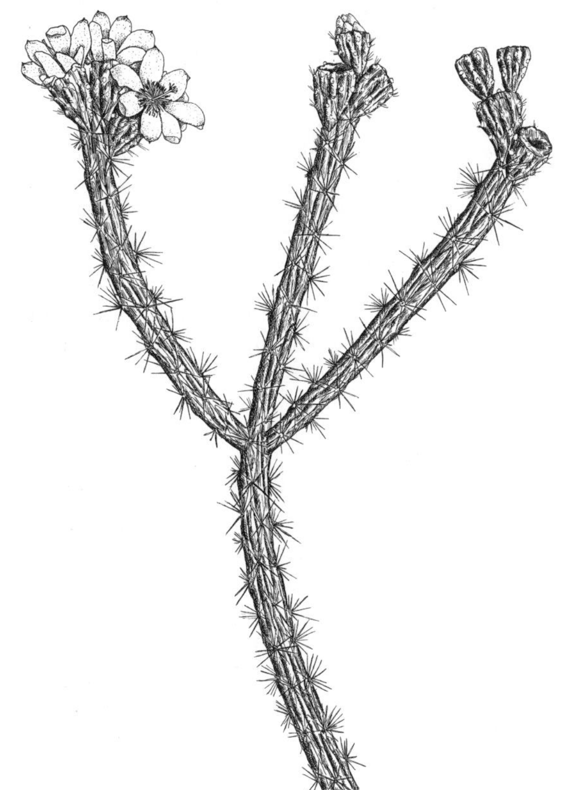

STAGHORN CHOLLA
Opuntia versicolor Engelm.

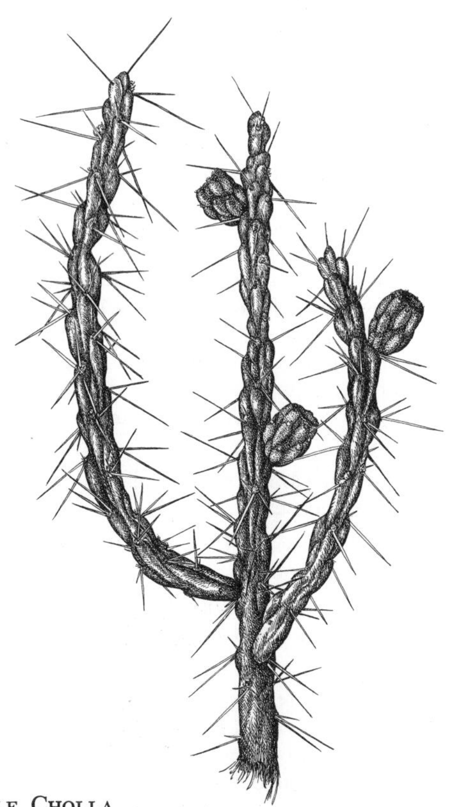

CANDLE CHOLLA
Opuntia kleiniae DC.

REMARKS. The genus name, *Opuntia,* was given by Theophrastus to another plant which grew near Opus, a town in ancient Greece. This cactus resembles *O. leptocaulis* DC., but has stouter, larger stems and purple flowers. However, *O. kleiniae* is very variable over a wide area and a typical form is difficult to ascertain.

STAGHORN CHOLLA

Opuntia versicolor Engelm.

FIELD IDENTIFICATION. Usually low and bushy, but becoming a shrub or small tree with an open, branched, rounded crown, attaining a height of 3–12 ft and about as broad as high. The purple joints are a good means of identification. Branches about ¾ in. thick, usually with 4 or more spines at each areole.

FLOWERS. Borne laterally from areoles of last year's branches, numerous, sessile, cup-like, 1¼–2 in. in diameter, color varied from orange to yellow, bronze, or red; corolla rotate; petals spreading, obovate, joined at the base; stamens numerous, shorter than the petals; anthers oblong; style cylindric, longer than the stamens, obclavate, branches inwardly stigmatic; ovary tuberculate, with large woolly glochids and deciduous bristles. Flowers sometimes produced from areoles of old fruit.

FRUIT. Persistent, usually spineless, succulent, at maturity pear-shaped or globose, green or purplish, 1–1½ in. long, about 1 in. in diameter; seeds embedded in the pulp, flattened, discoid, white, about ⅕ in. broad.

LEAVES. Small and scalelike, terete, subulate, deciduous soon, with axillary areoles.

JOINTS. Green or purplish; up to 1 ft long or more; glochids small, reddish brown; spines rather early deciduous, 5–11, ¼–1 in. long, purple or gray, the sheath pale white or gray.

STEMS. Much branched with branches cylindrical; tubercles about ¾ in. long, ¼ in. wide, 1/16–⅛ in. high; old stems contain tubular, woody, reticulate skeletons.

RANGE. Along arroyos, desert foothills, and upper mesas at altitudes of 2,000–3,000 ft. Southwestern New Mexico and southern Arizona; also in Sonora, Mexico.

REMARKS. The genus name, *Opuntia,* is reported to be a name given by Theophrastus for another plant growing near the ancient town of Opus in Greece. The species name, *versicolor,* refers to the varied color of the flowers. It is also known as the Tree Cholla. Staghorn Cholla is closely related to Tasajo Cholla and apparently intergrades with it.

Tasajo Cholla

Opuntia spinosior (Engelm.) Toumey

FIELD IDENTIFICATION. A cactus of treelike appearance attaining a height of 3–10 ft, and a spread of 3–10 ft, forming an open irregular head. The stems bear whorls of divaricate branches which usually spread at right angles.

FLOWERS. April–May, persisting for several days, 2–2½ in. across; petals usually 9–11, varying in color from pink to purple or occasionally yellow or whitish, obovate, narrowed to the base; sepals obovate; stamens numerous, red, sensitive; stigma 6–9-parted; ovary about 1 in. long, tuberculate, set with long white, easily deciduous bristles and small purplish leaves.

FRUIT. Usually clustered terminally, persistent sometimes for a year (sometimes proliferous), 1–1½ in. long, globose to obovoid or broadly oblong, apex cupped, about 1½ in. in diameter, at maturity yellow, at first the 20–30 tubercles very prominent, but much less so as the fruit matures; areoles small, covered with short bristles, with or without slender spines (if spiny then spines deciduous November–January); seeds orbicular or nearly so, white, smooth, ⅙–⅕ in. in diameter.

LEAVES. Deciduous in about a month, about 1 in. long, terete, tapering to the setulose apex.

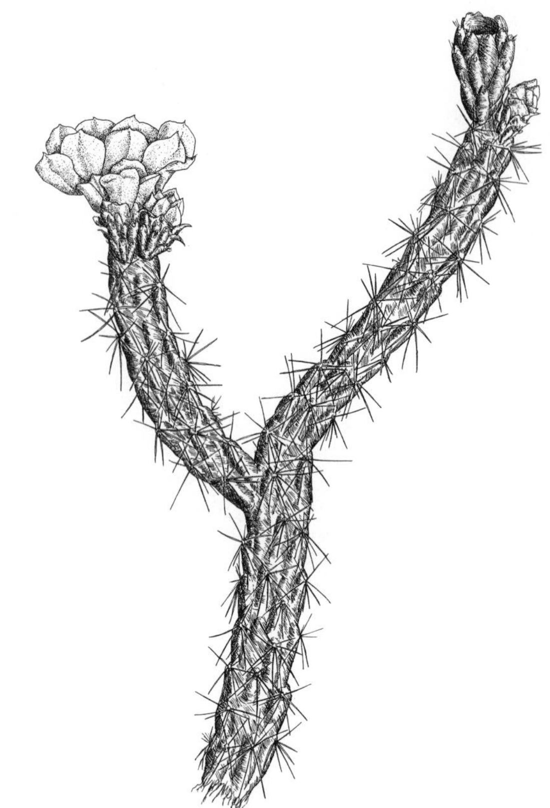

Tasajo Cholla
Opuntia spinosior (Engelm.) Toumey

BRANCHES AND AREOLAE. Branches usually 4–12 in. long, ¾–1 in. thick, cylindrical, surfaces dark green to purplish, becoming woody with age; tubercules rather numerous, longer than broad, more or less flattened laterally, each tubercule ⅜–¾ in. long; areolae oval, covered with a pale tomentum and bearing brown bristles; spines usually 6–12 (or on old branches as many as 25–50), gray to reddish brown, ½–¾ in. long, radiate-spreading, radial ones usually shortest; sheaths thin and shiny, deciduous the first year.

BARK. On the trunk and old branches about ¼ in. thick, usually spineless, brown to blackish, roughened by closely appressed scales.

WOOD. Reddish brown, soft, very reticulate, rays and annual rings conspicuous.

RANGE. At altitudes of 2,000–7,000 ft on dry desert flats or hillsides. New Mexico, Arizona, and in Mexico in Sonora.

REMARKS. The genus name, *Opuntia,* is from Opus, a town in ancient Greece. The species name, *spinosior,* refers to the very spiny nature of this plant. It is also known under the vernacular name of Chain Cholla. The wood is sometimes used to make small articles such as picture frames, canes, and light furniture. Cattle sometimes eat the fruit.

TESAJO CACTUS (TASAJILLO)

Opuntia leptocaulis DC.

FIELD IDENTIFICATION. Cactus thicket-forming, usually bushy, erect or sometimes reclining on other plants, ½–6½ ft. Trunk or main stem ½–4 in. in diameter.

FLOWERS. Borne mostly July–August, but sometimes at irregular periods, usually open during the day for several days and closing at night, small, inconspicuous, greenish yellow, ½–¾ in. in diameter; sepals ovate, apex acute or cuspidate; ovary obconic, areoles woolly, glochids brown; stamens numerous.

FRUIT. Small, globular to obovoid or clavate, usually ⅖–¾ in. long, red, or rarely yellow, turgid, somewhat fleshy and juicy; seeds ⅛–⅙ in. long, flattened, margins acute; fruit breaking off with pieces of the stems and rooting.

LEAVES. Green, awl-shaped, ½ in. long or less, acute, early deciduous.

STEMS. Varying shades of green, branches slender, ascending, cylindric, ¼–½ in. in diameter or sometimes larger, joints varying in length from 1 in. to 1 ft, fruiting ones spineless or spiny, often spreading at right angles, easily detached, areoles about ¼ in. apart; internal core of stems very woody and solid, bark green or brown to gray.

TESAJO CACTUS (TASAJILLO)
Opuntia leptocaulis DC.

SPINES. Very variable in abundance and length, solitary or 2–3 together at the areoles, ¾–2 in. long, very slender, white; sheaths of spines tight or loose, papery, yellow to white or brown; areoles with very short, white wool; glochids few and small.

RANGE. At elevations to 3,000 ft, in Texas, usually west of the Brazos River, west through New Mexico to California; south in Mexico to Puebla.

VARIATIONS. A short-spined variety has been described as *O. leptocaulis* var. *brevispina*. The species is also known to hybridize with the Walking-stick Cholla, *O. imbricata*. Tesajo might also be confused with the Candle Cholla, *O. kleiniae*, which has stouter stems, yellow fruit about 1½ in. in diameter, and a dull purple flower with greenish striations. Tesajo is very variable as to length of spines and character of spine sheaths. However, these variations seem to be connected by intergrading forms.

REMARKS. The genus name, *Opuntia*, is from the old name, Opus, a town in ancient Greece. The species name, *leptocaulis*, means "slender-stemmed." It is also known as the Rat-tail Cactus, Slender-stem Cactus, and Garambullo. The red fruit is very conspicuous in winter, and a yellow-fruited form is also known. The plant grows easily from cuttings or sections of the joints. The small, barbed spines stick into the flesh of man and livestock and are difficult to remove.

WALKING-STICK CHOLLA

Opuntia imbricata (Haworth) DC.

FIELD IDENTIFICATION. An arborescent cactus, with a short, woody trunk, and many erect candelabrum-like branches. Attaining a height of 9 ft and trunk diameter of 10 in. Sometimes forming dense thickets.

FLOWERS. Terminal, 1½–2½ in. long, 2–3 in. broad, petals purple, rounded to obovate, ovary tuberculate, upper areoles with a few bristles.

FRUIT. Yellow, dry, 1–1½ in. long, nearly hemispheric, tuberculate-cristate, sometimes falling off and producing new plants from the sprouting tubercles; seeds 1/12–⅙ in. in diameter.

JOINTS. Cylindric, ¾–1¼ in. in diameter, strongly and prominently tuberculate-crested, tubercles ¾–1 in. long, flattened laterally, very spiny.

LEAVES. Terete, ⅓–1 in. long, soon deciduous.

SPINES. Stellate-divaricate, 8–30, length ¾–1¼ in., barbed almost entirely, with papery, white to green, brown-tipped sheaths.

RANGE. Texas, north to Oklahoma and Kansas, south and west through New Mexico. Collected by the author in the Big Bend National Park, Brewster County, Texas. Sometimes in lava-rock soil of foothills.

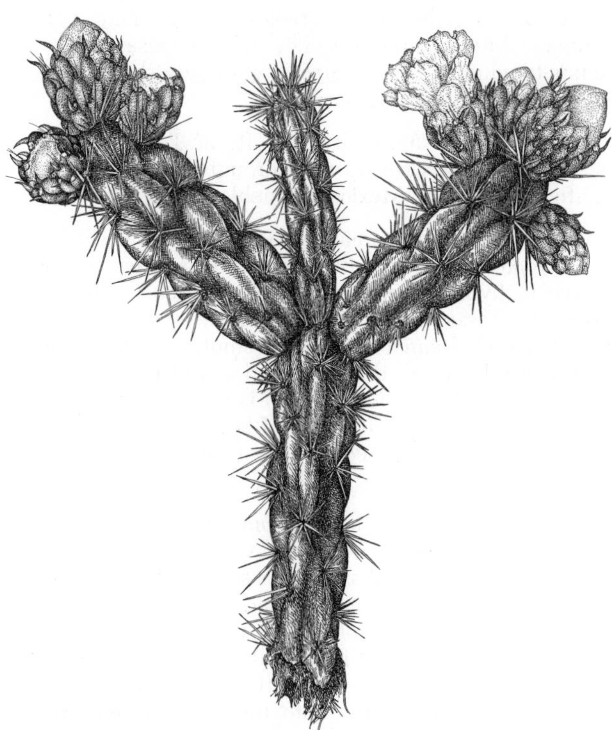

WALKING-STICK CHOLLA
Opuntia imbricata (Haworth) DC.

REMARKS. The genus name, *Opuntia,* is from Opus, a town in ancient Greece. The species name, *imbricata,* means "overlapping," with reference to the ridges of the joints. It is also known as Cholla, Tree Cactus, Cane Cactus, Velas de Coyote, Candelabrum Cactus, Devil's Rope, Coyote Prickly Pear, Cardenche, Tuna Juell, Goconoxtie, Estrara, Xoconochtli, Xoconostle, Joconoxtle, Joconostle, Tarajo, Coyonostle, Coyonoxtle, Coyonostli, Tuna Joconctla, and Tuna Huell. The plant covers a wide area, and variable races are known, but they seem to be connected by intermediate forms.

The stems with the tissue removed form a hollow cylinder with a framework of rhombic holes and meshes. Woody skeletons of this sort are made into odd-looking walking canes. The fruit is not known to be eaten by man or animal. Although the spines are sometimes burned off, and the stems fed to cattle, they are not very palatable.

Schulz and Runyon (1930) report that before the introduction of coal-tar dyes into Mexico the fruit had an important place in the arts. The fruit was gathered, chopped into small pieces, and boiled, the fiber and seed being filtered out and the extract used to dissolve and set cochineal dye. It is still used this way to a limited extent. Its mordanting property is doubtless due to the high concentration of acids and salts of organic acid present.

SONORA JUMPING CHOLLA
Opuntia fulgida Engelm.

FIELD IDENTIFICATION. A shrubby or treelike cactus attaining a height of 3–12 ft, about as wide as high. The woody trunk sometimes 3–12 in. in diameter, branching a short distance above the ground to form an irregular, somewhat flattened, crown. A distinctive character is the fruit, which grows in a pendulous cluster hanging one from the other.

FLOWERS. June–September, diurnal, borne laterally, sessile, rotate, borne either from last year's wood or from tubercles of the immature fruit, color very variable, ranging through white, pink, purple, magenta, saffron, or maroon, ¾–1½ in. broad when open; sepals 8–10, their margins crenulate; petals usually about 8, obovate or cuneate to obtuse, the inner ones more lanceolate and acute, margins crenate; stamens numerous, shorter than the petals, with oblong anthers; style cylindric, longer than the stamens; stigmas 5, erect, receptive surfaces inward.

FRUIT. Borne in long chains of 8–9 (to 14), usually proliferous, deciduous, or if persistent, then succeeding flowers arising from the tubercules, pear-shaped or oval, ¾–2 in. long, young ones tuberculate but later smooth, usually green in color; areoles rather large and tomentose, bearing numerous bristles, but essentially spineless or with only a few spines; seeds flattened, thin, angular, 1/12–1/6 in. in diameter, commonly absent.

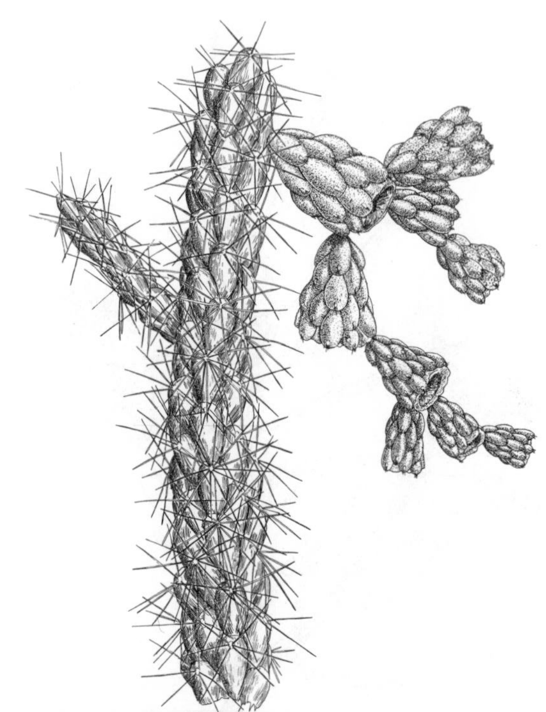

SONORA JUMPING CHOLLA
Opuntia fulgida Engelm.

LEAVES. Early-deciduous, ½–1 in. long, light green, gradually narrowed to acuminate apex.

BRANCHES. Light green, 1¼–2 in. in diameter, cylindric, set with tubercles ½–⅝ in. long, ¼ in. wide, and 3/16–¼ in. high; terminal joints 4–8 in. long, succulent and easily broken off.

AREOLES AND SPINES. Areoles with pale yellowish tomentum and pale slender bristles; spines 2–12, spreading, rather stout, acicular, ¾–1½ in. long, yellowish; sheaths loose, papery, straw-colored, persistent, the branches having a yellowish halo-like appearance owing to the color and density of the spines.

BARK. Brown to black on old branches and trunk, about ¼ in. thick, separating into large, loose scales, the spines absent, falling away easily.

WOOD. Yellow, light, hard, medullary rays broad, annual layer easily seen, pith thick, the woody skeletons developed the second or third year.

RANGE. Usually in sandy soils of flats, valleys, or mesas at altitudes of 1,000–4,000 ft. Rare in southwestern New Mexico, most abundant in Pima County, Arizona, southward in Mexico through adjacent Sonora to Sinaloa.

VARIETY. A closely related variety with intergrading forms is *O. fulgida* var. *mammillata* (Schott) Coulter. It attains a height of 2–4 ft, is more succulent than the species, has more prominent tubercles, and bears shorter spines in groups of 2–6. It is known from Arizona.

REMARKS. The genus name, *Opuntia,* is a name given by Theophrastus to some plant which grew near Opus in ancient Greece. The species name, *fulgida,* means "shining" and refers to the lustrous yellow spines of the branches. The name Jumping Cholla is given because the joints attach themselves readily to objects touching them. The juicy fruits are eaten by cattle. The seeds are rather difficult to germinate and propagation is generally accomplished more easily by rooting the terminal joints or the fruits themselves.

Dollar-joint Prickly Pear

Opuntia chlorotica Engelm. & Bigel.

FIELD IDENTIFICATION. An erect shrubby cactus attaining a height of 3–10 ft, and spreading to form a rounded head several feet across. Trunk short, strong, spine-matted, branching several times with 3–10 joints on a branch.

FLOWERS. About 2½ in. in diameter, oval or obovate, light yellow; stamens numerous.

FRUIT. Externally purple to gray, internally green, fleshy, spherical, usually spineless, 1–1½ in. in diameter.

JOINTS. Broadly obovate or oval, flat, 6–10 in. long, 5–8 in. broad, about ⅜ in. thick, yellowish green, spiny almost to the base.

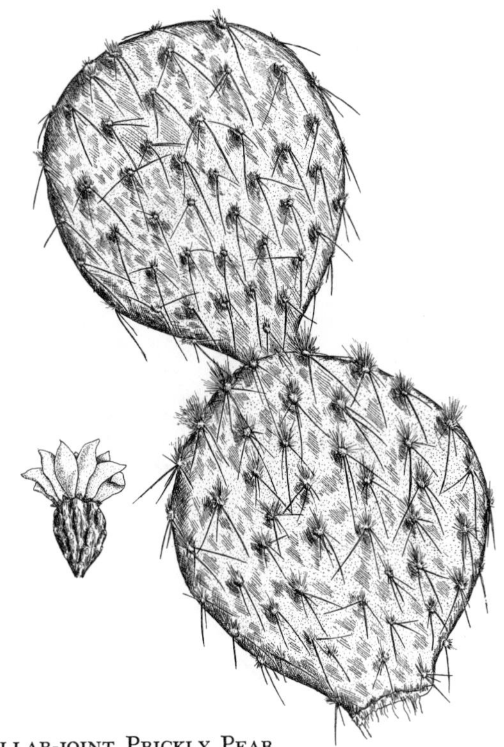

Dollar-joint Prickly Pear
Opuntia chlorotica Engelm. & Bigel.

AREOLES. About an inch or so apart; glochids mostly about 3/16 in. long; spines growing from most of the areoles except the very basal ones on some plants, usually turned downward, 1–6 (sometimes to 12), yellowish, averaging about 1 in. long, somewhat flexible, tapering, flattened.

RANGE. Rocky ledges, hills, and mountains, at altitudes of 2,500–6,000 ft. In New Mexico, Arizona, southwestern Utah, and California. Also in Baja California and Sonora, Mexico.

REMARKS. The genus name, *Opuntia,* is from Opus, a town in ancient Greece, in reference to some kind of plant which grew there. The species name, *chlorotica,* is for the greenish or yellow flowers and joints.

Engelmann Prickly Pear

Opuntia engelmannii Salm-Dyck

FIELD IDENTIFICATION. The common prickly pear of the Southwest. A bushy cactus to 6 ft, usually without a definite trunk.

FLOWERS. April–May, flowers about 3 in. in diameter, funnelform, yellow; petals numerous, oval to obovate,

ENGELMANN PRICKLY PEAR
Opuntia engelmannii Salm-Dyck

apex rounded to truncate with an abrupt point; sepals smaller than the petals and passing into them; stamens numerous; style elongate with lobed stigmas; receptacle bearing a few spines or small early-deciduous leaves.

FRUIT. Mostly obovoid, 1¼–3 in. long, 1¼–1½ in. in diameter, not spiny, red to dark purple, edible; seeds small, ⅛–⅙ in. broad.

LEAVES. Inconspicuous, about ⅗ in. long, subulate, early deciduous.

JOINTS. Oval to oblong or obovate, 7–12 in. long, 6–8 in. broad, thick, pale green; areoles 1¼–1½ in. apart, rather large.

SPINES. Mostly white or gray, sometimes with tips or bases reddish brown to black, 1–5 (usually 3–4); spines usually at all areoles, or the lower absent, usually variously spreading but not reflexed (often pointed downward); larger spines 1–2 in. long, stout, rigid, gradually tapering, flattened; glochids numerous, conspicuous, ⅕–½ in. long, brown with yellow tips.

RANGE. At elevations of 1,000–4,500 ft in arroyos, low hills, deserts, and grasslands. In Trans-Pecos Texas; west through New Mexico to Arizona, north to Nevada and Utah, and southward into Mexico in the states of Chihuahua, Durango, and Sonora.

REMARKS. The genus name, *Opuntia*, is from Opus, a town in Greece, in reference to some kind of plant which grew there. The species name, *engelmannii*, honors George Engelmann of St. Louis, who collected and named many Southwestern plants. Engelmann Prickly Pear is very variable in size, habit of growth, and size and abundance of joints and spines. A number of varieties have been described, but these all seem to have intergrading forms with the species. Engelmann Prickly Pear is also often confused with Lindheimer Prickly Pear, but the former has stouter, white to gray spines with reddish brown or black bases and tips, and the plant seldom has a distinct trunk. The juice is sometimes pressed from the fruit and boiled down with sugar to make syrup. It is also cooked and prepared in ways similar to those given under the Lindheimer Prickly Pear.

LINDHEIMER PRICKLY PEAR

Opuntia lindheimeri Engelm.

FIELD IDENTIFICATION. A thicket-forming cactus, heavy-bodied, with a definite cylindrical trunk, erect, or much lower and prostrate. Attaining a height of 3–12 ft.

FLOWERS. April–June, numerous, shallowly bowl-shaped, usually 1 to an areole; sepals and petals numerous, intergrading, hardly distinct, yellow to orange or red, a plant usually producing only 1 shade of flowers, oval to obovate or spatulate, apices rounded or abruptly short-pointed, length ½–2½ in.; stamens numerous, much shorter than the petals; ovary inferior, 1-celled, ovules numerous on thick, fleshy stalks, pla-

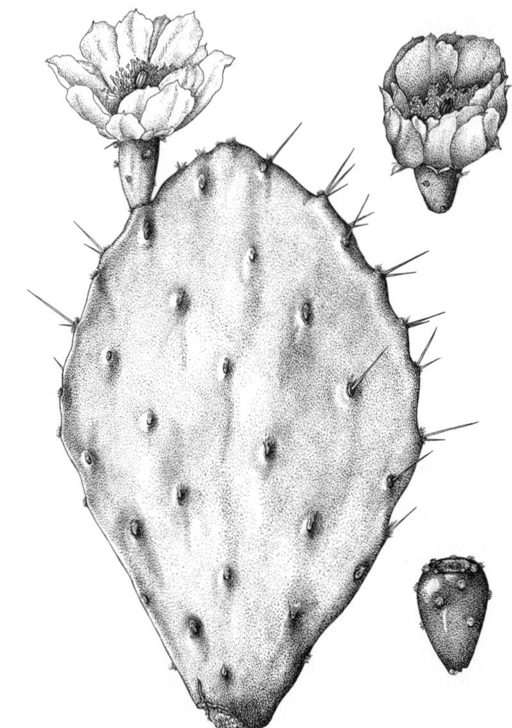

LINDHEIMER PRICKLY PEAR
Opuntia lindheimeri Engelm.

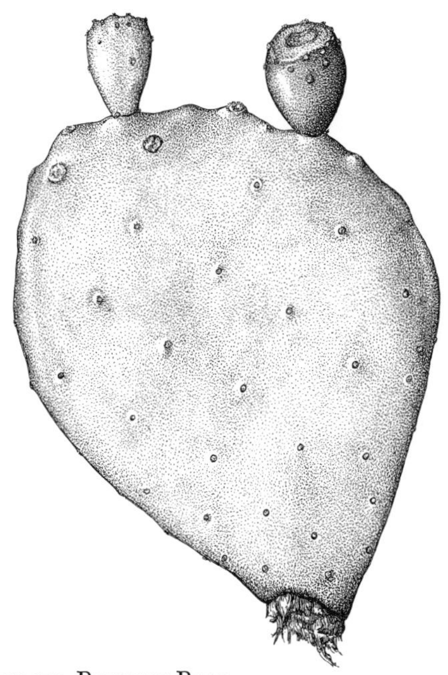

Lindheimer Prickly Pear
Opuntia lindheimeri Engelm.
Spineless form

centae parietal, withered perianth crowning the ovary and later crowning the fruit; style longer than the stamens, single, thick, stigma lobes short.

FRUIT. Ripening July–September, berry very variable in size and shape, clavate to oblong or globose, length ½–2½ in., red to purple, with scattered tufts of glochids; skin thin, rind thick, pulp juicy; seeds very numerous, about ⅛ in. long, flattened, curved, with a thick bony aril on the edge.

LEAVES. Very small, ⅛–⅙ in. long, pointed, flattened, early deciduous.

JOINTS. Green to bluish green, orbicular or obovate, or sometimes asymmetrical, length to 11 in., flat, waxy, succulent; set with areoles 1–2⅓ in. apart, which produce dense tufts of yellow to brown barbed, minute glochids less than 3/16 in. long; larger spines 1–6, usually 1–2, one erect or semierect, the others generally smaller and somewhat spreading, color of spines pale yellow to almost white, sometimes brown or black at base; some joints spineless or nearly so. Some forms are known which lack spines.

RANGE. From coastal southwestern Louisiana westward in drier regions of central Texas (not in east Texas woodlands). The type specimen was collected at New Braunfels, Texas. It is common around San Antonio, Corpus Christi, and Brownsville, Texas. It is not to be confused with *O. engelmannii* of Trans-Pecos Texas.

PROPAGATION. Opuntias are propagated easily by the joints. The joint is exposed to the sun a few days to callous the raw end, then laid flat on the surface of the ground. Roots develop from the lower side. The plants are subject to certain diseases. The shot-hole disease, caused by the fungus *Gloesporium lunatum,* turns areas on the joints brown or gray with black spore spots. Sunscald disease is caused by the organism *Hendersonia opuntiae* and causes light brown patches. Black sooty, spore spots are caused by *Perisporium wrightii.*

REMARKS. The genus name, *Opuntia,* is the Latinized name for the town Opus, in ancient Greece. The species name, *lindheimeri,* is in honor of Ferdinand Lindheimer, a German-born botanist, who collected extensively in Texas in 1836 and 1842.

The plant and its relatives are known under many names in Latin-American countries, such as Nochtli, Culhua, Cancanopa, Pacal, Potzotz, Toat, Pare, Caha, Xantha, and more commonly as Nopal. The fruit is known as "tuna."

The Indians and Mexican people formerly used the plant extensively for food. The fruits may be eaten raw or made into a preserve. The joints, when young and tender, are cooked and served with dressing and pepper, and are also made into candy. Syrup is made by boiling the ripe fruit and straining off the seed. The boiled and fermented juice is known as "colonche." A thick paste made by boiling down the juice is known as "melcocha." Queso de tuna (tuna cheese) is prepared from a pulp of the fruit seed. After evaporation, it is made into small cheeselike pieces. The Indians also believe that a tea made from the fruit will cure ailments caused by gallstones. Commercial alcohol has been made from the sap. The tender young joints are sometimes used as poultices to reduce inflammation. Juice of the joints is boiled with tallow in candlemaking to make the candles hard. The joints are made edible to cattle by burning off the spines. A number of animals and birds feed on the fruit. According to folklore, the coyote brushes the spines off the fruit with his tail before eating it.

The Nopal occupies a prominent place in the history and legend of Mexico. The Nopal and Caracara (Mexican Eagle) are the national emblems of Mexico. About 33 species of *Opuntia* are known in Texas, and about 250 species are found mostly in the southwestern United States, Mexico, and Central and South America.

Red-joint Prickly Pear

Opuntia macrocentra Engelm.

FIELD IDENTIFICATION. A bushy cactus with ascending branches 1½–3 ft. Joints 4–10, red to purple or bluish, bearing a very long, slender, brown or black, central spine from each of the uppermost areoles.

FLOWERS. From 2–4 in. broad, yellow, petals rounded or obovate; sepals ovate, apex acuminate; stamens numerous, filaments short; ovary ovate, areoles scattered, with brown glochids.

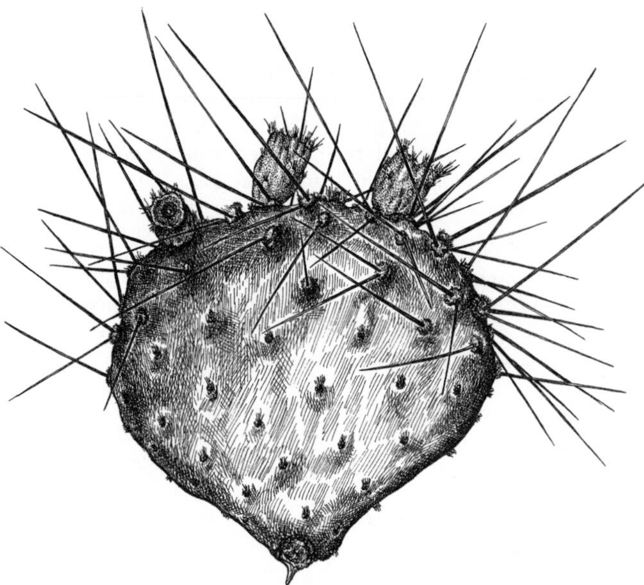

RED-JOINT PRICKLY PEAR
Opuntia macrocentra Engelm.

FRUIT. Ovoid, 1–2 in. long, about ¾ in. in diameter, purple to red, not spiny, fleshy, seeds numerous, ⅙–⅕ in. long.

JOINTS. Orbicular to obovate, or sometimes broader than long, 4–8 in. long, 3–9 in. broad, thin (less than ⅜ in.), often red to purple or bluish, usually with spines on the sides and along the margin at least from the upper areoles; areoles ¾–1 in. apart, glochids ⅛–¼ in. long.

SPINES. Central spine always present from upper areoles, usually 1–2, rarely 3 together, brown or black, sometimes white, erect or spreading, slender, 1½–5 in. long, round in cross section.

RANGE. Desert foothills, or gravelly flats, at altitudes of 2,000–5,000 ft. Western Texas to New Mexico and eastern Arizona; also Chihuahua, Mexico. Common near Fort Stockton, Texas. Collected by the author in Brewster County, Texas, between Hot Springs and the Ranger headquarters of Big Bend National Park.

REMARKS. The genus name, *Opuntia,* is from Opus, a town in ancient Greece. The species name, *macrocentra,* is for the large central spine. The bluish or reddish purple joints are very showy and make the plant worthy of cultivation.

COW-TONGUE PRICKLY PEAR
Opuntia linguiformis Griff.

FIELD IDENTIFICATION. Plant erect or semiprostrate. Joints glaucous bluish green when young, later yellow to brown, scaly and scurfy.

FLOWERS. Up to 3 in. across, yellow, petals numerous, obovate or obcordate, apex cuspidate, margin minutely serrulate; stamens numerous, white to greenish; stigma yellow, 7–9-parted.

FRUIT. Obovate to pear-shaped, reddish purple, length 1½–2½ in.; areoles small, rounded; spicules small, brown to yellowish; spines small, recurved, deciduous; seeds numerous, about ⅛ in. in diameter.

LEAVES. About ¼ in. long, apex cuspidate, circular in cross section, early deciduous.

JOINTS. Very variable, linear to narrowly ovate-oblong, 4–32 in. long, sometimes ovate or rounded; areoles ⅙–⅕ in. in diameter, subcircular; spicules yellowish brown and about ¼ in. long; spines yellow to reddish, flattened, angular, 2–6, one erect or downward-pointed, others shorter and spreading, length ⅓–2½ in.

RANGE. Apparently rare in a wild state, but rather commonly cultivated. The type was collected by David Griffiths near San Antonio, Texas (Griffiths, 1908, pp. 270–271).

REMARKS. The genus name, *Opuntia,* is the Latin name for some plant which grew near Opus, a town in ancient Greece. The species name, *linguiformis,* refers to the "tongue-shaped" joints. It is sometimes cultivated for ornament. Some botanists consider it to be a mutant variety of *O. Lindheimeri.*

BLIND PRICKLY PEAR
Opuntia rufida Engelm.

FIELD IDENTIFICATION. A more or less erect prickly pear, 8 in.–5 ft, with or without a definite trunk. Joints without conspicuous spines.

FLOWERS. Corolla rotate, 1½–2 in. long (including

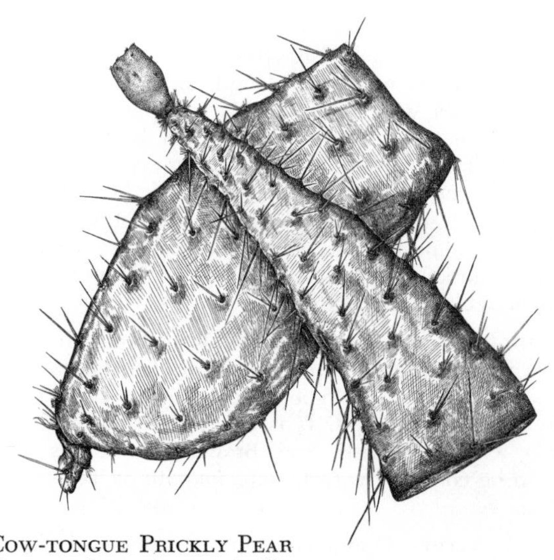

COW-TONGUE PRICKLY PEAR
Opuntia linguiformis Griff.

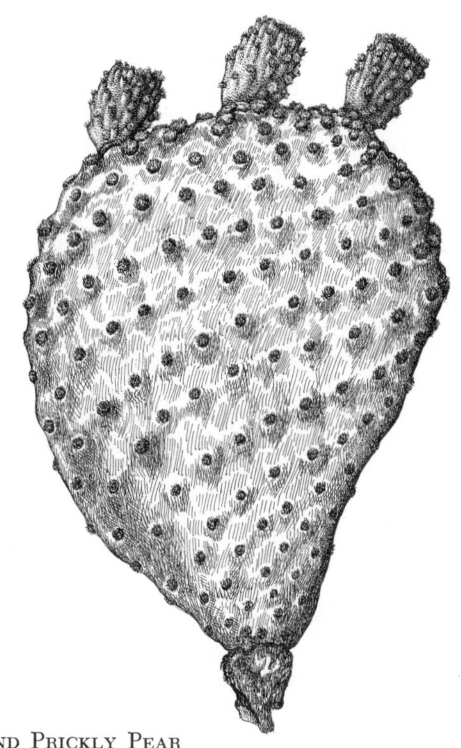

Blind Prickly Pear
Opuntia rufida Engelm.

the ovary); petals yellow to orange, obovate, ¾–1 in. long; stamens numerous, filaments greenish white, about ⅜ in. long; style about ⅝ in. long, thick; stigma lobes 5, deep green; ovary globular, ⅝ in. in diameter, areoles large.

FRUIT. Rounded or short-obovoid, smooth, red, length 2–3 in., width ¾–1 in.

JOINTS. Oval or nearly orbicular, 2½–10 in. in diameter, thick, flattened, dull grayish green, velvety-tomentose; areoles large, glochids numerous and brown.

LEAVES. Green or reddish, 2–2½ in. long, falling early.

RANGE. Trans-Pecos Texas and northern Mexico. In Brewster County between Hot Springs and the Chisos Mountains.

REMARKS. The genus name, *Opuntia*, is a name given by Theophrastus, but applies to some other plant which grew near Opus, a town in ancient Greece. The species, *rufida*, refers to the reddish brown glochids.

Notes on Other Species of Opuntia

The following species of *Opuntia* are known to occur in Texas, Louisiana, Arkansas, Oklahoma, and New Mexico. They are not described or illustrated because their low stature or lack of definite woody characters excludes them from the scope of the present volume. However, full descriptions and illustrations can be found in Britton and Rose (1919–1923, Vol. 1).

O. aciculata Griffiths. Low bushy plant, to 3 ft or more. Lower branches decumbent and sending up erect branches. Type locality, on high gravelly ground near Laredo, Texas.

O. allairei Griffiths. Low and spreading, often no spines, tuberous-rooted. Type locality, mouth of the Trinity River in Texas. Southern Texas and western Louisiana.

O. anahuacensis Griffiths. Reclining or prostrate, sometimes to 20 in. Type locality, Anahuac, Texas, at the mouth of the Trinity River.

O. arenaria Engelm. Stems prostrate. Type locality, sandy bottom lands of the Rio Grande near El Paso. Texas and southern New Mexico.

O. atrispina Griffiths. Low and spreading, sometimes up to 2 ft, and 6 ft across. Type locality, near Devils River in Texas. Abundant between Del Rio, Texas, and the Devils River, being one of the more common species in that region.

O. ballii Rose. Low and spreading. Type locality, near Pecos, Reeves County, west Texas.

O. clavata Engelm. Plant low, usually not over 6 in. Type locality, Albuquerque, New Mexico; central New Mexico.

O. fragilis (Nutt.) Haworth. Low and spreading, small and inconspicuous, but sometimes forming mounds 8 in. high in the center and 16 in. in diameter. New Mexico and Arizona; northward to Oregon, Washington, and British Columbia, eastward to Kansas and Wisconsin.

O. fuscoatra Engelm. Diffuse or prostrate. Type locality, sterile prairies west of Houston, Texas, and eastern Texas also.

O. grahami Engelm. Low mounds, often half-buried in the sand. Type locality, at El Paso, Texas. West Texas, New Mexico, and Mexico.

O. grandiflora Engelm. Plant low, branches somewhat ascending. Type locality, on the Brazos River, Texas; east Texas.

O. hystricina Engelm. & Bigel. Plant diffuse and spreading. Type locality, Little Colorado River and on San Francisco Mountains. New Mexico to Arizona and Nevada.

O. juniperina Britt. & Rose. Plant semiprocumbent. Type locality, on dry hills among junipers in vicinity of Cedar Hill, San Juan County, New Mexico.

O. macateei Britt. & Rose. Plants small and prostrate. Type locality, Rockport, Texas, collected by W. L. MacAtee.

O. mackensenii Rose. Plant low and spreading, roots tuberous. Type locality, near Kerrville, Kerr County, Texas.

O. macrorhiza Engelm. Plant low, nearly prostrate. Type locality, rocky places on the upper Guadalupe River of central Texas. Also in Dallas County, Texas. From Texas northward to Missouri and Kansas.

O. nemoralis Griffiths. Plant clump-forming, low, prostrate, to 12 in. Type locality, pine woods and fields about Longview, Texas.

O. phaeacantha Engelm. Plant low and usually prostrate. Type locality, about Sante Fe, New Mexico. Also in Texas, Arizona, and Chihuahua, Mexico.

O. polyacantha Haworth. Plant low and spreading. Type locality, arid situations on the plains of the Missouri. Texas, northwestern Oklahoma, New Mexico, and Arizona; northward to Utah, North Dakota, Washington, and Alberta.

O. pottsii Salm-Dyck. Plant low and spreading. Type locality, near Chihuahua City, Mexico. In Texas on the Rio Grande below El Paso, also in New Mexico.

O. schottii Engelm. Plant prostrate. Type locality, on the Rio Grande and Pecos rivers in Texas. From Hidalgo County along the Rio Grande River to the Pecos River. Abundant near Rio Grande City.

O. sphaerocarpa Engelm. & Bigel. Plant small and spreading. Known only from the type locality in the mountains near Albuquerque, New Mexico.

O. stenochila Engelm. Plant prostrate. Type locality, Canyon of Zuni, western New Mexico and Arizona.

O. stricta Haworth. Bush, low, spreading, sometimes forming large clumps, seldom over 32 in. Western Cuba and Florida, sometimes planted in southern Texas. Best known for its great spread over wide areas when introduced to Australia.

O. strigil Engelm. Plant suberect, to 24 in. Type locality, in crevices of limestone rock in western Texas. From the Pecos River to El Paso, Texas.

O. tardospina Griffiths. Plant low and spreading. Type locality, near Lampasas, Texas; in central and eastern Texas.

O. tenuispina Engelm. Plant low and spreading, to 12 in. Type locality, sand hills near El Paso, Texas. Southwestern Texas and adjacent parts of Mexico and New Mexico, apparently extending into Arizona.

O. tortispina Engelm. Plant prostrate and creeping. Type locality, on the Comanchica Plains near the Canadian River. Texas, New Mexico, Colorado, Kansas, South Dakota and Wisconsin.

O. trichophora (Engelm.) Britt. & Rose. Plant low and spreading to form small clumps 2–3½ ft in diameter. Type locality, mountains near Albuquerque, New Mexico; also in Texas and Oklahoma.

O. viridiflora Britt. & Rose. Plant spreading with stems 1–2 ft. Type locality, vicinity of Santa Fe, New Mexico.

O. whipplei Engelm. & Bigel. Plant low and much-branched. Type locality, about Zuni, New Mexico. In northern New Mexico and Arizona to southwestern Colorado, and probably southern Utah and southern California.

MEZERUM FAMILY (*Thymelaeaceae*)

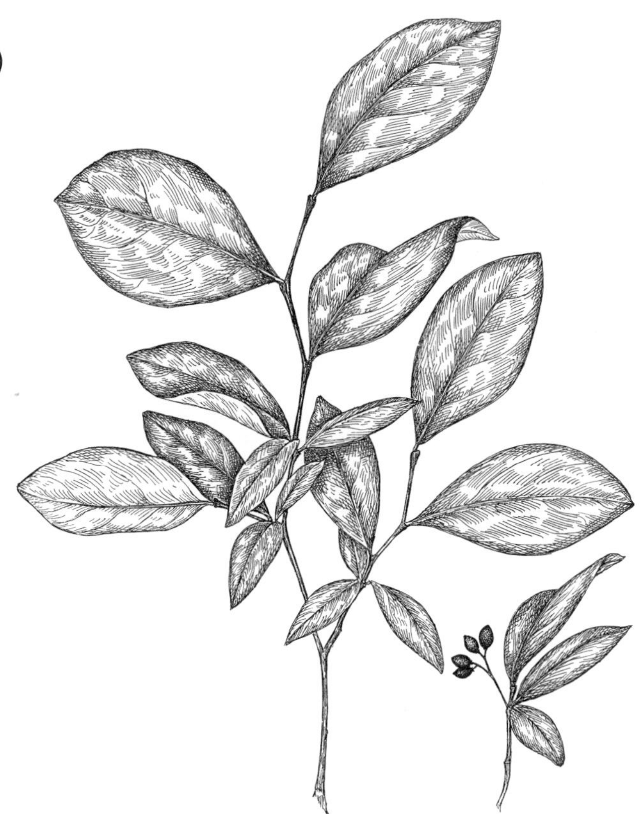

Atlantic Leatherwood
Dirca palustris L.

Atlantic Leatherwood
Dirca palustris L.

FIELD IDENTIFICATION. Widely branching shrub to 7 ft, with a trunk 4 in. in diameter. Sometimes miniature and treelike.

FLOWERS. March–April before the leaves, falling when the leaves expand, fairly abundant, from axillary buds with 3–4 scales, fascicles of 2–4 flowers, pedicels short or almost sessile; flowers ½–⅓ in. long, perfect, petals absent; calyx yellowish, narrowly funnel-form, somewhat constricted above the ovary and then cylindric-tubular, apex undulately and obscurely 4-toothed, teeth shorter than the tube; stamens 8, in 2 lengths, exserted about the middle; ovary nearly sessile, glabrous, free, with a single hanging ovule, style elongate, filiform, exserted, somewhat exceeding the stamens, stigma small and capitate.

FRUIT. Ripening May–June, sometimes hidden by the dense foliage, peduncles ⅛–⅖ in. long; drupe early-deciduous, red to orange, ellipsoid to oval or oblong, length ¼–⅖ in.; seed solitary, testa bony, dark brown and streaked lengthwise, anatropous, cotyledons fleshy.

LEAVES. Abundant, simple, alternate, deciduous, membranous, obovate to oval, margin entire, apex obtuse or subacute, base narrowed or rounded, length 2–4 in., width to 2¾ in.; surface pubescent when immature, at maturity upper surface light green, smooth and glabrous; lower surface glaucescent and pubescent to glabrate; petiole 1/25–⅕ in. long, base of petiole covering buds of the next season.

TWIGS. Yellow to green and glabrous, apparently socket-jointed by circular scars at the beginning of the annual growth; buds small, conical, enclosed by the petiole base; scales 3–4, oval to oblong, brown-pubescent, deciduous; bark mostly smooth and gray on old stems and roughened at the base of old trunks, very tough; wood soft, white, and brittle.

RANGE. In rich wet soil of woods or thickets. Louisiana, eastward to Florida, and north to Minnesota and New Brunswick.

REMARKS. The genus name, *Dirca*, is from Dirke, mythological name of a spring near Thebes. The species name, *palustris*, refers to the plant's habitat in low wet soil. Also known by the vernacular names of Wicopy, Wickup, Moose-wood, Rope-bark, Swamp-wood, Leather-bark, Leather-bush, and American Mezereon. The tough bark, being used for baskets and cordage by the Indians, has given it some of its names. It is also said to cause inflammation and pustulation on the skin of some people. Only 8 grains of the bark distillate taken internally causes vomiting and purging. The berries are reported to be narcotic. The plant was introduced into cultivation about 1750. It is a good ornamental plant. It grows slowly and is seemingly adjustable to many soil types. It is relatively free from insects and diseases, and when grown in the open develops a well-rounded head. The plant is easily propagated from seed or by layering.

OLEASTER FAMILY (*Eleagnaceae*)

Silver Buffalo-berry

Shepherdia argentea (Nutt.) Greene

FIELD IDENTIFICATION. Large, thorny, thicket-forming shrub to 18 ft.

FLOWERS. April–June, in short axillary racemes, small, short-peduncled, dioecious, petals absent; staminate perianth yellowish, 4-parted, lobes spreading; pistillate perianth oblong and tubular, limb 4-cleft, throat nearly closed by an 8-lobed disk; stamens 4–8, borne on the perianth; pistil simple, style slender, stigma 1-sided, ovary 1-celled; winter buds with 1–2 pairs of outer scales.

FRUIT. Often abundant July–August, drupelike, ellipsoid or oval, red or yellow, 1/6–1/4 in. long, not scurfy, fleshy and succulent, acid, sour, edible; seed solitary, smooth, shiny, ovate to ellipsoid, seed coat membranous.

LEAVES. Deciduous, opposite, blades 3/4–2 in. long, broadly to narrowly elliptic or oblong, apex obtuse, base cuneate, margin entire, silvery with brown scales beneath, but less so above; petioles 1/3 in. long or less, scurfy-silvery and brownish.

TWIGS. Spreading, often spiny, brown with scurfy scales, later silvery gray.

RANGE. Sandy banks of plains and canyons of the artemisia and pinyon belts at altitudes of 3,000–7,500 ft. New Mexico, Colorado, Utah, and Kansas; north to Saskatchewan and Alberta.

PROPAGATION. The cleaned seed averages 5–10 lb from 100 lb of fruit. The seed averages about 45,000 seeds per lb, with a commercial purity of 87 per cent and a soundness of 90 per cent. The seeds vary considerably in dormancy and should be stratified in moist sand at 41°F. for 60–90 days before planting in spring. About 30 seeds per linear ft per row are used.

REMARKS. The genus name, *Shepherdia*, is in honor of John Shepherd (1764–1836), at one time curator of Liverpool Botanic Garden. The species name, *argentea*, refers to the silvery scales.

Silver Buffalo-berry
Shepherdia argentea (Nutt.) Greene

The plant is sometimes used for hedges in the Northwest, but is rather hard to transplant from the wild. It is sometimes planted for shelter belts and erosion control, and is used in horticulture. The silvery gray leaves show off well in front of a dark green background. It has been cultivated since about 1818. It is very cold- and drought-hardy. The fruit is usually gathered before frost, and contains a considerable amount of pectin, which makes it highly suitable for conserves and jellies. Indians often cooked the fruit with buffalo meat. The seed is known to be eaten by at least 15 species of birds and by the porcupine and chipmunk. It has no value as livestock browse. A yellow-fruited horticultural form has been named S. *argentea* forma *xanthocarpa*. The largest specimen officially recorded grows in Malheur County, Oregon. It has a trunk circumference at 4½ ft high of 2 ft 7 in., with an over-all height of 18 ft and a spread of 18 ft.

Russett Buffalo-berry

Shepherdia canadensis (L.) Nutt.

FIELD IDENTIFICATION. Spreading shrub attaining a height of 3–8 ft.

FLOWERS. Usually April–June, short-racemose, individual flowers about ⅙ in. across, yellowish, petals absent; staminate flowers larger, 4-parted, the lobes spreading; pistillate flowers with an oblong, tubular perianth and a 4-parted limb; stamens 4–8, alternate with the lobes of a thick disk; style slender, stigmatose on one side; ovary 1-celled and 1-ovuled.

FRUIT. Maturing July–September, berry-like, ¼–⅓ in. long, oval to ellipsoidal, yellow or red, very juicy, insipid, bitter; seed ellipsoid, smooth, shiny, brown to black, sometimes with a longitudinal groove.

Russett Buffalo-berry
Shepherdia canadensis (L.) Nutt.

LEAVES. Opposite, oval to ovate or elliptic, blade length ½–2½ in., apex obtuse, base rounded or narrowed, margin entire, dull green above, lower surface silvery and rusty scurfy-scaly; petioles to ¼ in., scurfy-scaly.

TWIGS. Densely brown-scurfy at first, later gray to brown or black and glabrous.

RANGE. Usually on limestone rocks or ledges at altitudes of 5,000–12,000 ft, or on moist, open wooded slopes. New Mexico, Colorado, and Utah; north to Canada and eastward across the northern tier of states and southern Canada, to Newfoundland.

PROPAGATION. The fruit is usually abundant each year and is gathered by thrashing into a canvas. It is fanned to remove debris, and then run through a macerator and the pulp floated off. The seeds are then laid out thinly to dry. Since they have an embryo dormancy they should be stratified in moist sand at 41°F. for 60–90 days. The average germination capacity is 57 per cent. Sowing is done in spring with stratified seed, using about 30 seeds per linear ft, and covered about ¼ in. with soil. A form with yellow fruit is known as S. *canadensis* forma *xanthocarpa* Rehd.

REMARKS. The genus name, *Shepherdia*, is in honor of John Shepherd (1764–1836), English botanist. The species name, *canadensis*, refers to its growth in Canada. It is also known under the vernacular names of Canadian Buffalo-berry and Soapberry. The mashed berries produce a pink foam in water, which was sweetened and eaten by the Indians. The fruit was also made into preserves and jellies by white settlers; but is mostly considered too bitter and unpalatable. It is a good food for wildlife, being eaten by at least 12 species of birds, including various species of grouse and waxwings, and is also eaten by the Alpine chipmunk. White-tailed deer occasionally browse it, but it is considered unfit as browse for cattle. It is sometimes used in erosion control and has been cultivated since 1759.

Russian-olive

Elaeagnus angustifolius L.

FIELD IDENTIFICATION. Cultivated shrub or small tree to 25 ft, with thorny or thornless branches. Easily recognized by the lanceolate-to-oblong leaves which are conspicuously silvery and brown-dotted beneath.

FLOWERS. In June, axillary, scattered on the branches in clusters of 1–3, short-pediceled; flowers small, pale yellow or silvery, fragrant, perfect, regular; petals absent; perianth campanulate, 4-cleft, tube as long as the limb; stamens included, filaments short; style at base included by a tubular disk.

FRUIT. Maturing August–October, drupelike, on pe-

Russian-olive
Elaeagnus angustifolius L.

duncles about ¼ in.; fruit body ⅜–½ in. long, oval to ellipsoid, apex and base rounded or slightly flattened, yellow to tan, at first densely silvery or brown-scaly, later less so and becoming lustrous; flesh yellow, waxy, mealy, sweet; stone solitary, ellipsoid to oblong, about ⅜ in. long, brown, with darker longitudinal bands.

LEAVES. Simple, alternate, blades 2–3½ in. long, ⅜–¾ in. wide, lanceolate to oblong-lanceolate, margin entire, apex acute, base gradually narrowed or cuneate; upper surface bright green with veins obscure except main vein; lower surface densely silvery-scaly, veins more evident; petiole ⅕–½ in., silvery-scaly.

TWIGS. Slender, elongate, younger silvery-lepidote, older ones reddish brown, lenticels small and numerous; buds ovoid, ⅛–3/16 in. long, silvery-scaly.

RANGE. Tolerant of considerable amounts of salinity or alkalinity, ascending to an altitude of 5,500 ft. Cultivated in the United States, and occasionally escaping in New Mexico, Arizona, and western Texas. A native of southern Europe and western Asia to the Himalaya Mountains.

PROPAGATION. The fruit is gathered by hand in the fall, and the pulp separated by running through a macerator with water. Sometimes the dry fruit is used without removing the pulp. From 15–60 lb of seed is yielded by 100 lb of fruit. The average number of seeds per lb is 5,200. The commercial purity is 97 per cent and soundness 92 per cent. The naturally dormant seed may be induced to germinate by stratification in moist sand for 90 days at 41°F. The seeds may be sown in spring in drills with 40 viable seeds per linear ft and covered about ¾ in. The beds should be mulched to cover the soil and prevent rain spattering. Transplanting is usually done in the spring. Other methods of propagation include stem cuttings, root cuttings, layers, and grafting.

VARIETIES. Several varieties are known in cultivation as follows:

Oriental Russian-olive, *E. angustifolius* var. *orientalis* Dipp., is often spineless; leaves oblong to elliptic or oval, base rounded, stellate-hairy and scaly, upper surface glabrous later; petiole ⅙–¼ in., fruit to 1 in. long.

Spiny Russian-olive, *E. angustifolius* var. *spinosa* Schneid., has spiny branches; leaves elliptic to elliptic-oblong, blade length 1¼–2¾ in., base narrowed, scaly both above and below, but more so below; fruit globose to ellipsoid.

REMARKS. The genus name, *Elaeagnus*, is from the Greek word *elaia* ("olive") and the Latin *agnus* ("lamb"), probably with reference to the stellate pubescence on the undersides of the leaves. The species name, *angustifolius*, refers to the narrow leaves. Another vernacular name in use is Oleaster. The shrub is used in shelter-belt planting, and is ornamental because of the silvery leaves and decorative fruit. It also has some value as a honey plant. The fruit is edible and is also beneficial to wildlife, being eaten by many species of birds. The plant is rarely attacked by insects and is drought-resistant.

LOOSESTRIFE FAMILY (*Lythraceae*)

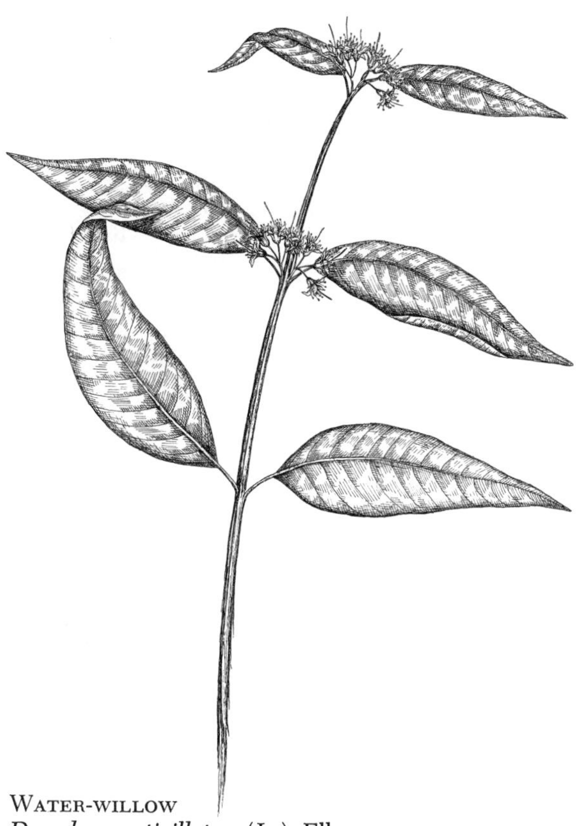

WATER-WILLOW
Decodon verticillatus (L.) Ell.

WATER-WILLOW

Decodon verticillatus (L.) Ell.

FIELD IDENTIFICATION. An aquatic plant, herbaceous above, often somewhat shrubby below, with elongate arching stems.

FLOWERS. Borne in axillary, peduncled cymes, peduncles short and stout, pedicels pubescent, 1/5–2/5 in.; sepals 5–7, about one half as long as the tube, alternating with subulate teeth at the sinuses; petals 5–7, purple to pink, about 1/3 in. long, lanceolate to ovate, ends acuminate, margin undulate-crisped; stamens 10, 5 short and 5 long, the longer exserted, filaments filiform; ovary 3–6-celled; stigma capitate; hypanthium campanulate, about 1/5 in. high, glabrous, ribbed.

FRUIT. Capsule subglobose, about 1/5 in. long, calyx remnants stiffly adherent on some, loculicidally 3–6-valved; seeds angled, 1/25–1/12 in. long.

LEAVES. Deciduous, simple, opposite or whorled, entire, blades 1¼–8 in. long, lanceolate to elliptic, apex acuminate, base gradually narrowed or broadly cuneate, dark green above, pale green beneath; petioles 1/16–½ in.

TWIGS. Recurved, 4–6-angled, tan to brown, arching tips sometimes rooting, submerged portions of stem and roots with spongy, reddish brown tissue.

RANGE. In swamps and ponds. Texas and Louisiana; eastward to Florida, northward to New York, Massachusetts, and New Hampshire, and westward to Illinois, Michigan, and Minnesota.

VARIETY. A variety with all parts glabrous is known as the Smooth Water-willow, *D. verticillatus* var. *laevigatus* Torr. & Gray.

REMARKS. The genus name, *Decodon,* is from the Greek *deca* ("ten") and *odon* ("tooth"), from the toothed calyx. The species name, *verticillatus,* refers to the whorled or verticillate leaves. Vernacular names are Swamp Loosestrife, Willow Herb, Water Oleander, and Peat-weed. It is desirable for colonizing on pond margins.

WILLOW-LEAF HEIMIA

Heimia salicifolia (H. B. K.) Link & Otto

FIELD IDENTIFICATION. Spreading, much-branched, glabrous shrub attaining a height of 9 ft.

FLOWERS. Axillary, on slender peduncles 1/25–3/16 in., bearing 2 oblong to spatulate, foliaceous bracts one half to fully as long as the campanulate calyx; corolla yellow, petals 5–7, spreading, obovate, yellow, ½–⅔ in. long, deciduous early; stamens 10–18 (generally 12), filaments long, slender, exserted, inserted on the calyx-tube; style slender, surpassing the stamens, ovary sessile and 4-celled; calyx campanulate, ⅕–⅓ in. long; sepals 6, triangular, acute, erect or converging, with subulate, hornlike appendages at base of the lobes.

FRUIT. Capsule obovoid, septicidal, dry, brown, ribbed, calyx-lobes closed above, seeds minute and clavate.

LEAVES. Blades opposite or whorled; linear to lanceolate or oblong; apex acute, acuminate or obtuse; base gradually narrowed to a short petiole or sessile, margin entire, blade length ½–2½ in. Upper surface bright green and glabrous; lower surface paler, glabrous or minutely papillose.

TWIGS. Young ones green to reddish brown, glabrate, somewhat angled or striate; older ones gray to brown, glabrous, sometimes shreddy; internodes ⅛–1¾ in. long.

RANGE. Usually along streams or resacas in southwest Texas. Specimens collected by the author on the banks of the Rio Grande River at Brownsville. In Mexico in Baja California and Coahuila, also Veracruz and Oaxaca. Southward into South America. Also in the west Indies.

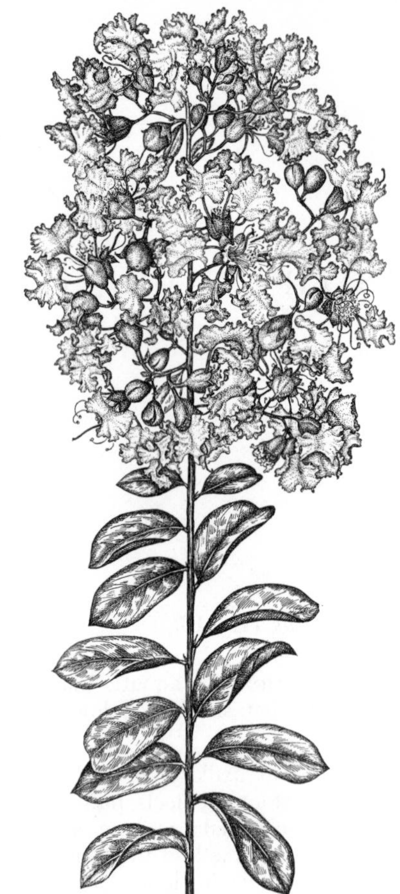

COMMON CRAPEMYRTLE
Lagerstroemia indica L.

WILLOW-LEAF HEIMIA
Heimia salicifolia (H. B. K.) Link & Otto

MEDICINAL USES. According to Standley (1920–1926, p. 1026), "The plant is much used locally in medicine, emetic, antisyphilitic, hemostatic, febrifuge, diuretic, laxative, vulnerary, sudorific, tonic and astringent properties being ascribed to it. It is employed most commonly for syphilitic affections. The leaves are said to contain 9 per cent of a bitter principle, nesine, and about 14 per cent of a resin, the latter being the active principle. If the juice or a decoction of the plant is taken internally it is said to produce a mild and pleasant intoxication, during which all objects appear to be yellow."

REMARKS. The genus name, *Heimia*, is in honor of Dr. Heim (Geheimerath), of Berlin, who died in 1834. The species name, *salicifolia,* refers to the willow-like leaves. The plant has many vernacular names in the countries of Central and South America such as Huachinal, Hauchinol, Hauchinoli, Hanchinol, Huauchinal, Hanchinoli, Hanchinal, Jarilla, Sinicuiche, Siniculche, Sinicuil, Granadillo, Escobella del Rio, Quiebra Yugo, and Quiebra Arado.

Common Crapemyrtle

Lagerstroemia indica L.

FIELD IDENTIFICATION. Commonly cultivated shrub or small tree to 35 ft, with a very smooth fluted trunk.

FLOWERS. In showy, terminal panicles 2½–8 in. long, pedicels and peduncles bracted; corolla 1–1½ in. across; petals 5–7, but usually 6, purple, pink, white, red, lavender, or blue; stamens numerous, elongate, some upcurved; calyx of 5–8 sepals, shorter than the hypanthium; ovary 3–6-celled, style long and curved, stigma capitate.

FRUIT. Capsule oval-globose, about ⅓ in. long; seeds with a winged apex.

LEAVES. Opposite or alternate, deciduous, obovate-oval, entire, acute or obtuse, broad-cuneate or rounded at base, blades ½–2 in. long, subsessile, glabrous and lustrous above, paler and glabrous or pilose along veins beneath.

TWIGS. Pale, glabrous, 4-angled.

BARK. Thin, exfoliating to expose a smooth, often convoluted, pale surface.

RANGE. A native of India. Cultivated extensively in Texas, Louisiana, Oklahoma, and Arkansas; eastward to Florida, and northward to Virginia. Also cultivated in California.

PROPAGATION. Common Crapemyrtle blooms when quite young. Old trees produce the most flowers when cut back and bloom almost continually during the summer. It can be grown from seed planted in boxes in autumn and kept moist until large enough to plant out. It is also grown from cuttings of ripe wood.

FORMS. Some of the common color forms of crapemyrtle are Dwarf *(nana)*, Dwarf Blue *(lavandula)*, Pink, Purple *(purpurea)*, Red *(magenta rubra)*, and White *(alba)*.

REMARKS. The genus name, *Lagerstroemia*, is in honor of Magnus Lagerstroem (1696–1759), a Swedish friend of Linnaeus, and the species name, *indica*, is for India. Another vernacular name is Ladies' Streamer.

POMEGRANATE FAMILY (*Punicaceae*)

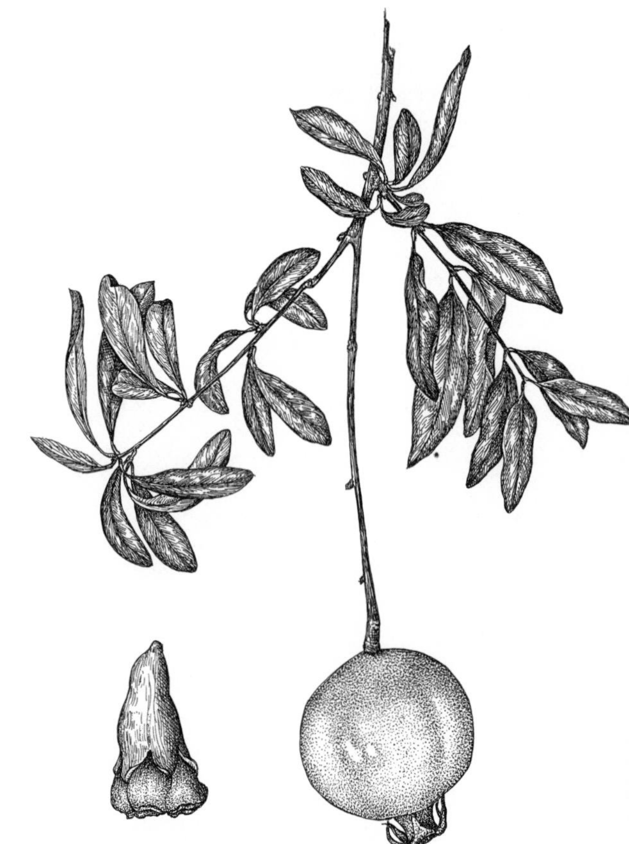

POMEGRANATE
Punica granatum L.

POMEGRANATE

Punica granatum L.

FIELD IDENTIFICATION. Cultivated, clumped shrub or tree to 25 ft. The branches erect or ascending to form an irregularly shaped crown.

FLOWERS. In simple, axillary or terminal racemes with 1–5 solitary short-peduncled flowers; flowers to 2 in. across, perfect, perigynous, showy, usually red (occasionally white or pink); calyx tubular to campanulate, later subglobose, persistent; 5–7-lobed; lobes valvate, ascending, fleshy, triangular to lanceolate, apex acute, stiffly persistent in fruit; petals 5–7, inserted on the upper part of the calyx between the lobes, imbricate, wrinkled, ⅝–1 in. long, suborbicular or obovate; stamens numerous, in several series, filaments filiform; style 1, stigma capitate; ovary inferior, embedded in the calyx-tube, comprising several compartments in 2 series, ovules numerous.

FRUIT. Berry maturing in September, pendulous, 2–4 in. in diameter, subglobose or depressed, crowned with the persistent calyx; rind thick, leathery, reddish yellow; pulp juicy, pink or red, acidulous; septa membranous, many-celled; seeds numerous, cotyledons convolute and auricled at base.

LEAVES. Simple, deciduous, alternate, opposite or clustered, blades ¾–3½ in. long, oval to oblong or elliptic to lanceolate, apex obtuse or acute, base attenuate into a short wing, margin entire, surface bright green; main vein prominent below, impressed above, other veins inconspicuous, glabrous on both surfaces; petiole ⅛–⅓ in., grooved above, green to red; winter buds with 2 pairs of outer scales.

TWIGS. Slender; younger ones green to reddish brown, somewhat striate or angular; older ones gray and more terete. Bark on older limbs and trunks gray to brown, smooth at first but eventually breaking into small, thin scales, shallowly reticulate.

WOOD. White to yellowish, hard, close-grained, specific gravity about 0.93.

RANGE. Old fields, waste places, and abandoned homesites. A native of Arabia, Persia, Bengal, China, and Japan. Hardy in the United States as far north as Washington, D.C., but does best in subtropical southern regions. Escaping cultivation in the Gulf Coast states.

MEDICINAL USES. Pomegranate has been much used

in home remedies as an astringent in dysentery and diarrhea and also as a gargle. The drug pelletierine tannate is obtained from the powdered bark of the root and stems. This drug (according to the *United States Pharmacopoeia*, 12th ed., quoted in Wood and Osol, 1943, p. 825) " 'is a mixture in varying proportions of the tannates of the several alkaloids obtained from pomegranate. . . . It contains an amount of the alkaloids equivalent to not less than 20 per cent as the hydrochloride.' " Wood and Osol (1943, p. 826) describe the uses of pelletierine as follows:

"Although the value of pomegranate in the treatment of tape-worm was known to the ancients and commented on by Dioscorides, the drug fell into disuse in Europe for several centuries until revived by Buchanan in 1805. The efficacy of its alkaloids as taeniacides has been abundantly confirmed, and it appears to be established that the tannate of the alkaloids is the most effective and the least dangerous form of the remedy—probably because its insolubility prevents its rapid absorption and enables it to come in prolonged contact with the worm.

. . . .

"The chief symptoms from overdoses of pelletierine are muscular weakness and sometimes sensations of dizziness. Large doses cause in addition mydriasis, amblyopia, vomiting, and diarrhea. The blindness seems to be due to inflammation of the optic nerve, which in occasionally reported cases has been permanent."

LEGENDS. Many legends have been handed down by Asiatic people concerning the pomegranate. Theophrastus described it three hundred years before the Christian era, and Pliny considered it one of the most valuable of ornamental and medicinal plants.

The many seeds of the pomegranate are supposed to be a symbol of fertility, and in Turkey a bride casts a ripe pomegranate to the ground. The number of seeds falling out indicates the number of children she will bear. Ancient legend also says that the pomegranate was the "tree of life" in the Garden of Eden, and from this belief it became the symbol of hope and eternal life in early Christian art. Early Hebrews used the design of the fruit and flower on the walls of temples and the robes of priests. The erect calyx-lobes were the inspiration for Solomon's crown and for all future crowns. A full account of the uses of the pomegranate in Biblical times is given by Moldenke and Moldenke (1952, pp. 190–191).

PROPAGATION. The plant can be propagated by seeds, but it is usually done by softwood cuttings under glass in summer, or by hardwood cuttings planted out in February. The shoots, with roots attached, can be set out also. For orchards the plantings should be set 12–18 ft apart. There is a highly prized variety in India which is seedless. Several double-flowered varieties, which do not bear fruit, are cultivated in the South and in greenhouses in the North.

VARIETIES AND FORMS. The following varieties and forms are cultivated:

Double Pomegranate, *P. granatum* forma *pleniflora* Hayne, has double scarlet flowers.

Pale Pomegranate, *P. granatum* forma *albescens* DC., is a white-flowered form.

Legrelle Pomegranate, *P. granatum* forma *legrellei* Vanh., is a form with double-striped red and yellowish white flowers.

Yellow Pomegranate, *P. granatum* var. *flavescens* Sweet, is a yellow-flowered variety.

Dwarf Pomegranate, *P. granatum* var. *nana* (L.) Pers., is a low variety with smaller linear-lanceolate leaves and smaller flowers and fruit.

REMARKS. The genus name, *Punica,* is from the Latin words for "Punic apple," and the species name, *granatum,* means "many-seeded apple." Known in the countries of Central and South America under the Spanish and Indian names of Granada de China, Granada Agria, Granado, Granada, Tzapyan, Yaga-zehi, and Yutnudidzi. The wood is sometimes used by engravers as a substitute for Boxwood *(Buxus).*

From the days of Solomon pomegranate has been used for making cooling drinks and sherbets; it is also eaten in its natural state. The astringent rind yields a black or red dye and has been used for tanning morocco leather and for ink. The fruit is sweet in the better varieties, and a spiced wine is made from the juice. The soft seeds are also eaten, sprinkled with sugar, or, when dried, as a confection.

MYRTLE FAMILY (*Myrtaceae*)

PINEAPPLE GUAVA
Feijoa sellowiana Berg.

PINEAPPLE GUAVA
Feijoa sellowiana Berg.

FIELD IDENTIFICATION. Cultivated shrub or small tree to 15 ft, the branches spreading to form a dense rounded head.

FLOWERS. In late spring, at the ends of branches, or axillary, solitary, 1–1½ in. in diameter, bisexual; petals 4–5, cupped, spreading, fleshy, white outside, reddish purple within; stamens numerous, long-exserted, tufted, dark red, about 1 in.

FRUIT. Maturing in early winter, berry oval to oblong, 1–2 in. long, ¾–1½ in. thick, dull green to reddish, flesh white-granular, translucent, flavor somewhat pineapple-like; fruit tipped by the irregular persistent calyx; seeds very small and numerous.

LEAVES. Opposite, or some alternate, blade 1–3 in. long, elliptic to oblong or oval, margin entire, apex rounded or obtuse, base broadly narrowed or cuneate; upper surface dark glossy green and puberulent at first to glabrous later; lower surface gray-white with soft, dense, fine tomentum; upper surface dark lustrous green; veins on the upper surface rather obscure and impressed, those of the lower surface raised and conspicuous; petiole ¼–½ in., somewhat channeled above, densely tomentose.

TWIGS. Young ones slender, densely gray-tomentose; older ones gray to reddish brown and glabrous; bark gray on older branches and trunk, separating into thin short scales to expose a reddish under surface.

RANGE. In Texas mostly grown as an ornamental shrub and seldom for fruit. Introduced into the United States in 1900. A native of Paraguay, Uruguay, and Argentina. Suitable for subtropical areas, but will stand some frost. Specimens grown on the University of Texas campus at Austin since 1936 have stood 15°F.

PROPAGATION. The fruit is gathered when it falls at maturity, and is stored in a cool dry place to ripen and soften. When properly stored, the seeds retain viability at least a year. The seeds may be germinated in sand or sawdust, but seedlings are often sensitive and die from damping-off. Germination requires about 3 weeks. The seedlings are usually transferred to small pots before planting out.

Named varieties are usually propagated by 4 in. cuttings taken from young wood and rooted in sand under glass. Sometimes layering is practiced by bending a branch and covering with moist soil 3–6 in. deep. Veneer and whip grafting are also done successfully in the greenhouse. Seedlings should start bearing at 3–5 years of age. However, some plants are apparently self-sterile; a number planted together will better insure fruit production. Ordinarily the plants are set 15–18 ft apart and pruned but little. The Pineapple Guava is rather drought-resistant, but plants do best if watered well and planted in a rich loamy soil. Fertilizing with manure is recommended. The plants are sometimes attacked by the black scale, *Saissetia oleae*, but usually not seriously.

FORMS. The following horticultural clones are recognized: Andre, Besson, Choicenna, Coolidge, Hehre, Pineapple, and Superba.

REMARKS. The genus name, *Feijoa*, is in honor of J. da Silva Feijo, of San Sebastián, Spain. The species name, *sellowiana*, is for Friedrich Sello (1789–1831), German traveler in South America. *Feijoa* is closely related to *Psidium*, but has seeds with albumen and stamens suberect in the bud. The fruit may be eaten raw, but is usually crystallized or made into jam or jelly.

Common Guava

Psidium guajava L.

FIELD IDENTIFICATION. Tropical shrub or small tree to 30 ft, with a trunk to 1 ft in diameter, the trunk often branching close to the ground.

FLOWERS. In spring, buds tomentose to glabrate, solitary, or in 2–3-flowered cymes; pedicels axillary, ¾–1½ in.; calyx tube oblong-ovate, slightly constricted above the ovary, closed before anthesis, splitting into 3–4 irregular segments ⅓–⅗ in. long, whitish and sparsely hairy within; petals 4–5, white, broadly oval to obovate, thin, delicate, ⅗–⅘ in. long, much longer than the sepals; stamens very numerous, borne on the disk, exserted, erect or spreading, about ½ in. long, anthers tan; stigma subcapitate and greenish; ovary inferior, ovules numerous in each cavity.

FRUIT. Maturing August–September, berry variously shaped, from globular to pyriform in certain varieties, crowned by the persistent calyx, ¾–4 in. long, pulp somewhat acid, flavor sweet, odor very musky; seeds numerous, rounded or flattened, yellowish, white, or pink.

LEAVES. Opposite, leathery, elliptic to oblong or oval, margin entire, apex obtuse to acute or rounded, base rounded or broadly cuneate, blade length 1½–6 in., width 1½–2½ in., at maturity glossy green above and glabrate with impressed veins, lower surface finely pubescent with veins prominent; petioles puberulent to glabrous, 1–2½ in.

TWIGS. Somewhat angled, tomentulose at first, later reddish brown to gray and glabrous; bark on young plants rather smooth, on old trees scaly.

WOOD. Brown to reddish, close-grained, hard, strong, elastic, specific gravity about 0.69, takes a good polish, durable on contact with the soil, but stems usually too small for commercial use.

RANGE. Native of tropical America and cultivated in subtropical areas in southern Texas, southern California, and Florida, where it sometimes escapes cultivation in Peninsular Florida and the Keys. Also in the West Indies, Mexico, and Central and South America. Naturalized in Asia and Africa.

Common Guava
Psidium guajava L.

MEDICINAL USES. A decoction of the flower buds is reported to be made into a remedy for diarrhea in Mexico. The leaves are considered to be a remedy for itch. It is also said that a decoction of the bark is taken for ulcers and stomach pains.

PROPAGATION. Guava seeds germinate well and are spread mostly by domestic animals. In its native habitat the plant grows rapidly and is hard to eradicate.

FORMS. Various horticultural forms have been developed with names based mostly on the quality of the fruit, such as the Sweet, Sour, Red, Redflesh, Lemon, Pear, Hawaiian, and Guinea.

REMARKS. The genus name, *Psidium*, is from the Greek word *psidion* ("pomegranate"). The species name, *guajava*, is from the Spanish word *guayabo*. Other names sometimes used in tropical American countries are Gauave, Gaiaba, Gujavabaum, Guayabo de Venado, Guayaba de China, Guayaba Colorado, Guayaba Peruana, Guayaba Perulera, Guayaba de Gusano, Guabaya Manzano, Xalxocoth, Posh, Posh-keip, Enandi, Poos, Poos-cuy, and Bayaba. A grove of these trees is often known as a "guaya-bales" in Mexico. It is also given the English names of Sand-apple and Sand-plum.

Common Guava is a very widely known tropical fruit. It is grown in a number of varieties for its fruit, which is very variable in size and palatability. The raw fruit is only mediocre in flavor, but is excellent when made into jelly, jam, paste, or confection. The fruit has a pronounced musky odor. The bark is used for tanning.

GINSENG FAMILY (*Araliaceae*)

Devil's Walkingstick
Aralia spinosa L.

Devil's Walkingstick

Aralia spinosa L.

FIELD IDENTIFICATION. Spiny, few-branched shrub, or slender, flat-topped tree to 35 ft. Recognized by very large, twice-pinnate leaves which are 3–4 ft long and 2–4 ft wide.

FLOWERS. July–August the large terminal panicle of flowers is very conspicuous and is divided into smaller umbels of individual flowers. Pedicels light yellow; corolla white, 1/8 in. across; petals 5, ovate, acute; stamens 5, alternate with petals; ovary 5-celled or abortive, styles distinct, sepals triangular.

FRUIT. September–October, drupe black, fleshy, juice purple, diameter about 1/4 in., flattened, 3–5 angled, style persistent; seed solitary, oblong, rounded at ends, flattened, crustaceous, brownish.

LEAVES. Alternate, compound, generally borne at top of trunk, blades 3–4 ft long, 2–4 ft wide, twice-pinnate; pinnae also pinnate with 5–6 pairs of lateral leaflets and a terminal leaflet; leaflets ovate, acute or acuminate, serrate or crenate, cuneate or rounded at base, dark green above, paler beneath, 1–4 in. long, tiny prickles often on midrib, yellow in autumn; petiole 18–20 in. long, clasping the base, prickly; stipules about 1 in. long, acute, ciliate.

BARK. Dark brown, fissures shallow, ridges irregular, armed with orange prickles, inner bark yellow, leaf scars abundant and conspicuous.

WOOD. Brown to yellow, weak, soft, light, close-grained.

RANGE. In rich, moist soil, edges of streams, woods and thickets. East Texas, Oklahoma, Arkansas, and Louisiana; east to Florida; north to New York, Indiana, and Iowa.

PROPAGATION. Shrub has been cultivated since 1688. Drupes contain 2–5 nutlets, each nutlet containing a solitary seed. Mature fruit should be picked by hand and macerated in water to remove the pulp and other matter. Cleaned seed averages about 100,000 seeds per lb, and has a commercial purity of about 90 per cent and a soundness of 86 per cent. For best results seed should be stratified at low temperatures before planting in spring. Tree may be transplanted by root cuttings.

REMARKS. Origin of the genus name, *Aralia*, is from the French-Canadian *Aralie*, the name appended to the original specimens sent to Tournefort by the Quebec physician, Sarrasin. The species name, *spinosa*, refers to the spiny trunk and branches. Vernacular names are Hercules Club, Angelica-tree, Prickly-ash, Prickly-elder, Pick-tree, Pigeon-tree, and Toothache-tree. The seeds are eaten by many birds and the leaves browsed by white-tailed deer. It is reported that the bark, roots, and berries are occasionally used in medicine. Shrub has high ornamental value and is often planted in Europe.

English Ivy

Hedera helix L.

FIELD IDENTIFICATION. Evergreen woody vine, climbing by aerial rootlets to a height of 90 ft, or creeping on the ground.

FLOWERS. Borne in terminal, globose, or somewhat paniculate umbels, usually several forming a raceme; peduncles slender and pubescent; individual flowers small, solitary, regular, perfect; calyx, pedicels, and tips of young branches with grayish white stellate hairs usually 5–6-rayed; petals 5, valvate, greenish, ovate to oblong, apex obtuse, length 1/12–1/8 in.; stamens 5, inserted at edge of the disk, filaments short and ascending, anthers ovate; ovary inferior, 5-celled, cells 1-ovuled, style connate into a short column.

FRUIT. Drupe berry-like, somewhat fleshy, black, globose or angled, about 1/4 in. in diameter; nutlets 2–5, papery or membranous.

LEAVES. Simple, alternate, entire, or on sterile branches often 3–5-lobed, ovate to rounded, blades 1½–4 in. long, apex acute to acuminate or obtuse, sometimes abruptly pointed, base rounded to truncate, coriaceous, upper surface dark green or blotched, often with whitish veins, sometimes lustrous, lower surface paler and dull or yellowish green, leaves on fertile branches almost always entire or somewhat smaller; petioles usually shorter than leaves, green to brown, glabrous.

TWIGS. Stems stiff and ropelike, green to brown, glabrous, with or without aerial rootlets.

RANGE. Cultivated in Texas, Oklahoma, Arkansas, and Louisiana, and throughout the United States except in desert areas. Often persisting, or escaping, about abandoned gardens, cemeteries, or buildings. Usually in rich moist soil. A native from Europe to the Caucasus, also North Africa.

MEDICINAL USES. The fresh, bitter leaves were once used in domestic medicine as a local application in skin diseases and old ulcers. Both leaves and berries are poisonous when eaten, and in some persons contact with the leaves causes dermatitis. The berries contain the glycosides hederin and hederagenin. Resin from the bark of old plants has been used in medicine under the name of "ivy gum." It was at one time used to plug carious teeth for the relief of toothache.

English Ivy
Hedera helix L.

PROPAGATION. Many garden forms and a number of geographical varieties are known. English Ivy has been cultivated in gardens in both hemispheres since ancient times. It is planted as a ground cover in shady places, as a wall cover, and for borders of shrubbery. The seeds usually germinate the second year. Also grown from cuttings and layers, and is sometimes grafted.

REMARKS. The genus name, *Hedera,* is the ancient Latin name of ivy. The species name, *helix,* is an old generic name meaning "twining."

DOGWOOD FAMILY (*Cornaceae*)

FLOWERING DOGWOOD
Cornus florida L.

FIELD IDENTIFICATION. Shrub or tree to 40 ft, with a straggling, spreading crown.

FLOWERS. March–June, perfect in terminal dense clusters; corolla tiny, about 1/8 in. wide, greenish white, tubular; petals 4, linear, acute; stamens 4; true flowers subtended by 4 large white or pink, obcordate, emarginate bracts 1¼–2½ in. long; calyx 4-lobed; ovary 2-celled, with a slender style.

FRUIT. Drupes clustered, conspicuous, bright red, ovoid, lustrous, ¼–½ in. long; seeds 1–2, channeled, ovoid; calyx and style persistent; fruit ripe September–October, dispersed by birds and mammals.

LEAVES. Petioled, simple, opposite, entire, or barely and minutely toothed, blades 3–5 in. long, 1½–2½ in. wide, oval to ovate or elliptic, acute or acuminate at the apex, somewhat cuneate at the base, often unequal at the base, shiny green and somewhat hairy above, much paler and pubescent below, heavily veined; petioles stout, grooved, about ¾ in.

TWIGS. Slender, yellowish green to reddish, pubescent to glabrous.

BARK. Grayish brown or black, broken into squarish rough checks.

WOOD. Brown or reddish, close-grained, strong, weighing 51 lb per cu ft.

RANGE. Oklahoma, Arkansas, Texas, and Louisiana; eastward to Florida, northward to Maine, and west to Minnesota and Ontario.

MEDICINAL USES. The dried bark of the root has been used medicinally in the form of a powder or a fluid extract as a remedy for intermittent fever. Its efficacy for this purpose is doubtful, and it is now used infrequently.

FLOWERING DOGWOOD
Cornus florida L.

PROPAGATION. Dogwood is extensively planted and does well under cultivation. It is one of our most handsome native ornamentals, and efforts should be made to prevent its extermination in the wild state. Formerly rather abundant, it is now becoming scarce in some sections. It was first cultivated in 1731. For planting, the seeds are usually hand-picked. The hulls can be extracted by maceration and washing away the pulp. The clean stones are then dried and stored. The seed averages about 4,500 seeds per lb, with an average commercial purity of 97 per cent and a soundness of

76 per cent. The seed has a dormant embryo, should be stratified in sand or peat for 100–130 days at 41°F. and planted in April or early May. The seeds are sown in drills of about 40 viable seeds per sq ft and covered with about ¼ in. of nursery soil. The germination rate is about 80 per cent in the nursery. It is reported that dogwood seedlings in nurseries of the Southeastern states are sometimes defoliated by a leaf spot caused by *Cercospora conicola*. An application of 4–6–50 Bordeaux mixture at 2-week intervals is helpful. Another common leaf trouble is believed caused by the leafhopper, *Graphocephala versuta*.

One-year seedlings can be transplanted to moist sites. Dogwood can also be propagated by layering, root cuttings, and divisions.

VARIETIES AND FORMS. Yellow-berry Flowering Dogwood, *C. florida* forma *xanthocarpa* Rehd., has yellow fruit.

Weeping Dogwood, *C. florida* forma *multibracteata* Rehd., has 6–8 large bracts and a number of small ones.

Pink Dogwood, *C. florida* var. *rubra* West., has pink floral bracts.

REMARKS. The genus name, *Cornus,* is a Latin word for "tough wood," and the species name, *florida,* refers to the showy petal-like bracts. Vernacular names are Arrowwood, Boxwood, Cornelius-tree, False Box, Nature's Mistake, Florida Dogwood, and White Cornel. The wood is used for small woodenware articles, tool handles, wheel hubs, and pulleys. The fruit of dogwood is eaten by at least 28 species of birds and ranks 21 on the list of quail-food plants of the Southeast. It is preferred food for wild turkey, also it is much eaten by squirrels and white-tailed deer. It is reported that Indians made a dye from the roots.

Stiff Dogwood

Cornus stricta Lam.

FIELD IDENTIFICATION. Shrub with stiff, upright irregular branches, or sometimes a small tree to 15 ft.

FLOWERS. Borne May–June in round-topped, rather open cymes 1¼–2½ in. broad; peduncles glabrous, 1–2¾ in. long; flowers perfect, calyx of 4 minute sepals; petals 4, small, white, oblong, valvate in bud, spreading in anthesis; stamens 4, filaments long and slender, anthers bluish and versatile; style elongate, stigma capitate and terminal; ovary 2-celled, with 1 ovule in each cell.

FRUIT. Drupe maturing August–October, subglobose, ⅕–¼ in. in diameter, pale blue; seed solitary, longer than broad, slightly furrowed.

LEAVES. Simple, opposite, blade lanceolate to elliptic or ovate-lanceolate, apex acuminate and often abruptly so, base narrowly to broadly cuneate or gradually tapering, blade length 1½–3½ in., one third to one half as wide, margin entire, green on both sides, lower surface

Stiff Dogwood
Cornus stricta Lam.

slightly paler, glabrous or sparingly puberulent; petiole ¼–1 in., glabrous or sparingly puberulent.

TWIGS. Young ones reddish, later greenish to brown or gray; older ones gray and glabrous.

RANGE. Wet woods, bottom lands in sun or shade. East Texas, Louisiana, and Arkansas; east to Florida, north to Virginia, and west to Missouri.

REMARKS. The genus name, *Cornus,* is from the Latin word for "horn," referring to the hard wood. The species name, *stricta,* refers to the stiff, upright branches. It is also known as Stiff Cornel and has been cultivated since 1758.

Redosier Dogwood

Cornus stolonifera Michx.

FIELD IDENTIFICATION. Stoloniferous, thicket-forming shrub with ascending, spreading or prostrate branches, attaining a height of 3–7½ ft.

FLOWERS. May–August, dull white, perfect, cyme puberulent, flat-topped and 1⅓–2 in. across; corolla ovoid in bud; calyx minutely 4-toothed; petals 4, oblong, spreading; stamens 4, filaments slender, anthers cream-colored; style slender, stigma terminal; disk usually red.

FRUIT. Maturing July–October, drupe white or lead-colored (rarely, bluish), globose, about ⅕ in. across; stone rounded at base, as broad as high or slightly broader; 2-locular and 2-seeded.

LEAVES. Opposite, blades ovate to oblong or lanceolate, blades 2⅓–4¾ in. long, veins in 5–7 pairs, apex acuminate, base rounded, margin entire, upper surface dark green and appressed-pubescent to glabrous later, lower surface glaucous and appressed-pilose to glabrous later; petioles ⅜–1 in.

TWIGS. Ascending or some prostrate, slender, red, smooth; pith white and large.

RANGE. Often in sandy, wet swamps or along streams, but somewhat tolerant of alkaline soil. At altitudes of 1,500–9,000 ft. New Mexico, Arizona, and California; northward to Canada and Alaska, eastward to Newfoundland; Kentucky, Virginia, and Nebraska.

PROPAGATION. One hundred lb of fruit yield 15–20 lb of seed. The cleaned stones average about 18,700 stones per lb, with a purity of 99 per cent and a soundness of 85 per cent. Most stones have a dormant embryo and some lots also show a hard seed coat. Usually stratification in sand or peat 90–120 days at 41°F. will suffice, or for lots with hard pericarps mechanical scarification plus 120 days' stratification at 41°F. Planting methods are the same as those for *C. alternifolia*.

FORMS, VARIETIES, AND HYBRID. Bailey Redosier Dogwood, *C. stolonifera* forma *baileyi* (Coult. & Evans) Rickett, has the pubescence of the leaves distinctly spreading instead of appressed.

Interior Redosier Dogwood, *C. stolonifera* forma *interior* (Rydb.) Rickett, has the pubescence of young stems and inflorescence densely tomentose.

Yellow-twig Redosier Dogwood, *C. stolonifera* var. *flaviramea* (Spaeth) Rehd., has yellow twigs.

Green-twig Redosier Dogwood, *C. stolonifera* var. *nitida* (Koehne) Schneid., has green branches in winter; leaves lustrous above, with 6–8 pairs of veins.

Colorado Redosier Dogwood, *C. stolonifera* var. *coloradensis* (Koehne) Schneid., has brownish red twigs; leaves smaller, less pale beneath; fruit bluish white; stone higher than broad.

Redosier Dogwood hybridizes with *C. rugosa* Lam., to produce *C.* × *slavini* Rehd., with purple branches; leaves intermediate in shape, more or less woolly beneath; fruit bluish, rarely white.

REMARKS. The genus name, *Cornus*, is from the Latin word *cornu* ("a horn"), referring to the hard wood. The species name, *stolonifera*, is for the stolons or runners by which the shrub forms thickets. Also known as Harts Rouges or Poison Dogwood. Called Redosier Dogwood because the bark resembles that of the Osier Willow. Seeds known to be eaten by at least 12 species of birds including ruffed grouse, sharp-tailed grouse, bobwhite quail, and Hungarian partridge. Foliage browsed by white-tailed deer, mule deer, cottontail, snowshoe hare, elk, and moose. The plant is valued in horticulture and was introduced into cultivation in 1656.

REDOSIER DOGWOOD
Cornus stolonifera Michx.

GRAY DOGWOOD

Cornus racemosa Lam.

FIELD IDENTIFICATION. Thicket-forming shrub to 7 ft. Stems much-branched, ascending, divergent and irregular, smooth, and light gray to brown.

FLOWERS. May–June, borne in open, paniculate, convex cymes 1¼–2½ in. high or broad; peduncles of cymes ⅓–1½ in. long, conspicuously red, appressed-pubescent to nearly glabrous; individual flowers perfect, small, white; petals 4, valvate in the bud, ⅛–⅙ in. long, oblong and spreading or recurved; stamens 4, borne on the disk margin, filaments slender; style 1, ovule single and anatropous; calyx-tube adherent to the ovary, limb minutely 4-toothed.

FRUIT. Drupe ripening July–November on red pedicels, rather persistent, white to gray or greenish, depressed-globose, ⅕–⅓ in. high, pulpy, 1–2-seeded; stone subglobose, apex and base rounded, sometimes shallowly furrowed, about ⅙ in. long and wide, embryo nearly as large as the albumen, cotyledons large and foliaceous.

LEAVES. Opposite, oblong-lanceolate, elliptic or narrowly ovate, apex acuminate, base acute or occasionally rounded, margin entire, blade length 1–4 in., width ½–1½ in., veins in 3–4 pairs, upper surface olive-

GRAY DOGWOOD
Cornus racemosa Lam.

green, lower surface glaucous and glabrous or slightly appressed pubescent; petioles ⅛–⅗ in., glabrous or appressed-pubescent.

TWIGS. Older ones gray and smooth or somewhat angled, younger ones brown to red, bark bitter, pith white to brown.

RANGE. Sandy or gravelly soil, roadsides and fence rows, thickets, and riverbanks. Arkansas and Oklahoma; east to Virginia, north to Maine, and west to Ontario.

PROPAGATION. The fruit is collected by hand from the plant when mature, and the seeds separated from the pulp by macerating in water or by use of a hammer mill. One hundred lb of fruit yield 18–25 lb of seed, averaging about 12,000 cleaned seed per lb. Commercial seed has a purity of 99 per cent and a soundness of 83 per cent. The seed has a hard pericarp and a dormant embryo. Pretreatment consists of stratification in sand or peat for 60 days at 68°–86°F., plus 120 days at 41°F. Another method of treatment is by soaking in sulphuric acid for 2 hours plus 120 days' stratification at 41°F. The shrub may also be propagated by cuttings.

REMARKS. The genus name, *Cornus,* refers to the hard wood, and the species name, *racemosa,* is for the raceme-like flowers (however, they are more cymose-paniculate in form). The shrub has also been known in the literature under the name of *C. paniculata* Gray, *C. femina* B. & B., *Svida femina* Small, and *Svida foemina* Rydb.

Gray Dogwood has been cultivated for ornament since 1758. It is persistent on unfavorable sites and endures city smoke. Its fruit is known to be eaten by at least 25 species of birds, including ruffed grouse, sharp-tailed grouse, ring-necked pheasant, and bobwhite quail. The wood is hard, heavy, and durable, but does not get large enough for commercial use.

PALE DOGWOOD

Cornus obliqua Raf.

FIELD IDENTIFICATION. Open irregularly branched shrub to 9 ft.

FLOWERS. May–July, borne on axillary, hairy peduncles 1–2½ in.; individual pedicels hairy, 1/16–¼ in. long, flowers in rounded cymes, perfect; calyx minutely 4-toothed; petals 4, valvate in the bud, ⅛–3/16 in. long, oblong, acuminate, spreading at anthesis; stamens 4, filaments long, slender and exserted; style solitary, slender, stigma terminal and capitate; ovary 2-celled, with 1 ovule in each cell; cotyledons large and foliaceous.

FRUIT. Drupe late May–October, persistent, blue, globose, 3/16–¼ in. in diameter, style remnant persistent, 2-locular and 2-seeded.

LEAVES. Numerous, opposite, deciduous, variable in size, blades mostly 1½–3¼ in. long, ½–1½ in. wide, ovate to lanceolate or oblong, apex acuminate or acute, base acute or narrowly cuneate, margin entire, upper surface smooth and green, lower surface whitish or glaucous and with minute appressed, white hairs, older more glabrous, lateral veins 3–5 on each side; petioles ⅛–⅓ in., pubescent.

PALE DOGWOOD
Cornus obliqua Raf.

TWIGS. Slender, light to dark brown, smooth; when young whitish with appressed hairs, when older glabrous.

RANGE. Stream banks, swamps, and wet thickets. Oklahoma, Arkansas, and Kentucky; northward to New Brunswick and west to North Dakota.

HYBRIDS AND RELATED SPECIES. Pale Dogwood is closely related to *C. amomum* Mill. and apparently hybridizes with it in some areas. It is also known to hybridize with *C. racemosa* Lam.

REMARKS. The genus name, *Cornus,* refers to the hard or hornlike wood. The species name, *obliqua,* refers to the unequal bases seen in some leaves.

Rough-leaf Dogwood

Cornus drummondii C. A. Meyer

FIELD IDENTIFICATION. Irregularly branched shrub or small spreading tree.

FLOWERS. May–August, perfect, yellowish white, borne in terminal, spreading, long-peduncled cymes 1–3 in. across; peduncles 1–2 in. long, pubescent; individual pedicels ⅛–¾ in. long, branched, glabrous or pubescent; corolla ⅛–3/16 in. across, short-tubular; petals 4, spreading, oblong-lanceolate, acute; calyx-teeth 4, minute, much shorter than the hypanthium; stamens 4, exserted, filaments slender, white, longer than the pistil; style simple, slender, with a terminal somewhat capitate stigma; ovary inferior, 2-celled; annular ring viscid and reddish.

FRUIT. Ripening August–October, drupe globular, about ¼ in. in diameter, white, style persistent, flesh thin; 1–2-seeded, seeds subglobose, slightly furrowed.

LEAVES. Simple, opposite, deciduous, blades 1–5 in. long, ½–2½ in. broad, conspicuously veined, ovate to lanceolate or oblong to elliptic, apex acute or acuminate, base rounded or cuneate, margin entire, plane surface somewhat undulate; upper surface olive-green and rather rough-pubescent above; lower surface paler, pubescent and veins prominent; petiole ⅕–¾ in. long, slender, rough-pubescent, green to reddish.

TWIGS. Young ones green and pubescent, older ones reddish brown and glabrous.

BARK. On young branches and trunks rather smooth, pale gray to brown. On old trunks gray with narrow ridges and fissures, scales small.

WOOD. Pale brown, with sapwood paler, heavy, hard, strong, durable, close-grained.

RANGE. Edges of thickets, streams, and fence rows. Central, southern, and eastern Texas, Oklahoma, Arkansas, and Louisiana; east to Alabama and northward to Ontario.

PROPAGATION. The fruit is collected by hand or shaken into a canvas. It can be planted immediately or the pulp removed by macerating in water or running through a hammer mill, floating off the debris. The clean seed averages 15,700 seeds per lb, with a commercial purity of 89 per cent and a soundness of 67 per cent. The seed may be mechanically scarified or stratified in moist sand or peat for 60 days at temperatures alternating diurnally 70°–85°F., followed by a longer period at much lower temperatures. The seeds are sown in drills, with about 40 seeds per sq ft, and covered with about ¼ in. soil. A mulch is used until germination begins. Seedlings may be protected partially from leaf-spot fungus diseases by applications of 4-6-50 Bordeaux mixture at about 2-week intervals. One-year-old stock can be transplanted. The plant can also be propagated by root cuttings, layering, and divisions, or by nearly ripened cuttings in frames in summer.

Rough-leaf Dogwood
Cornus drummondii C. A. Meyer

REMARKS. The genus name, *Cornus,* is from *cornu* ("a horn"), in reference to the hard wood, and the species name, *drummondii,* is in honor of Thomas Drummond (1780–1835), a Scottish botanical explorer. Vernacular names are Cornel Dogwood, Small-flower Dogwood, and White Cornel. The word "dogwood" comes from the fact that a decoction of the bark of *C. sanguinea* was used in England to wash mangy dogs. Rough-leaf Dogwood was formerly listed by some authorities under the name of *C. asperifolia* Michx., but hat name is no longer valid. See Rickett (1942, pp. 59–261).

Rough-leaf Dogwood is sometimes used in shelter-belt planting in the prairie-plains region. It has been nown in cultivation since 1836. The wood is used for mall woodenware articles, especially shuttle-blocks and harcoal. The fruit is known to be eaten by at least 40 pecies of birds, including bobwhite quail, wild turkey, nd prairie chicken.

Pagoda Dogwood

Cornus alternifolia L.

FIELD IDENTIFICATION. Shrub or small tree to 18 ft. Branches often in tierlike layers.

FLOWERS. May–July, inflorescence in open or flat-topped puberulous cymes, 1¼–2½ in. broad, perfect, epals minute or absent; petals 4, small, about ⅛ in. ong, valvate in the bud; stamens 4, filaments long and lender, anthers versatile; stigma capitate, style elongate; ovary 2-celled, 1 ovule in each cell.

FRUIT. July–September, borne on red pedicels, drupe bluish black, with a bloom, 2-celled but 1-seeded by abortion, pulp scant; stone solitary, ¼–⅓ in. across, subglobose, deeply pitted at the apex.

LEAVES. Mostly alternate, a few opposite, but often crowded terminally, blade either ovate, oblong, elliptic, or obovate, 2–4¾ in. long, ¾–2½ in. wide, apex abruptly acuminate, base cuneate or rarely rounded, margin entire, upper surface glabrous and yellowish green, lower surface paler and somewhat appressed-hairy, trichomes pale or reddish, lateral veins 4–6 on each side; petioles ⅓–2 in. long.

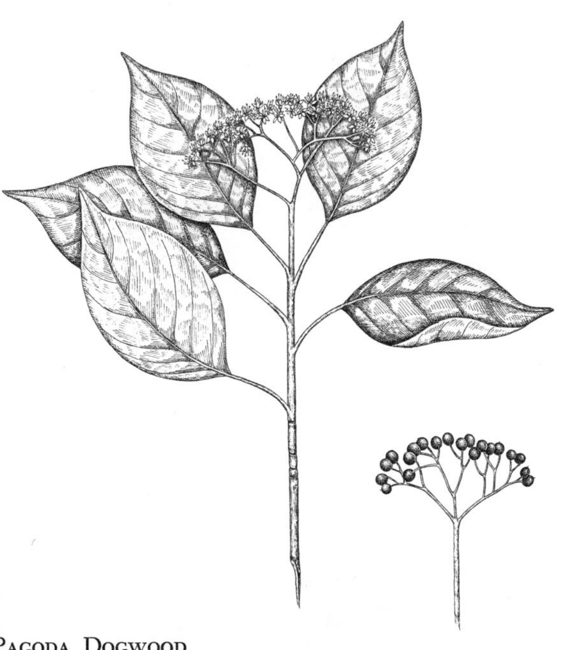

Pagoda Dogwood
Cornus alternifolia L.

TWIGS. Often horizontal or ascending, slender, smooth, glabrous, green, pith white and small.

RANGE. Rich woods, thickets, rocky slopes. Arkansas; eastward to Florida, northward to Nova Scotia and Newfoundland, and westward to Minnesota and Missouri.

PROPAGATION. The cleaned stones average about 8,000 per lb, with an average soundness of 74 per cent. The stones are usually extracted by macerating the fruit in water or running through a hammer mill, and the empty stones and pulp washed away. The stones may then be planted, or stratified, because of embryo dormancy and hard seed coat, in sand or peat for 60 days at 68°–86°F. plus 60 days at 41°F. Sometimes mechanical scarification or acid in combination with cold stratification helps. Ordinarily fruit having both embryo and hard pericarp dormancy will lay over until the second year before germination, but in some cases fruit collected somewhat green in July and sown immediately gives full germination the next spring.

The stones are generally sown in drills and covered with about ¼ in. soil. A mulch of straw or leaves is used until germination begins.

Pagoda Dogwood may also be propagated by root cuttings, layers, and divisions.

FORMS AND HYBRIDS. *C. alternifolia* forma *argentea* Temple has leaves variegated with white.

C. alternifolia forma *ochrocarpa* Rehd. has yellow fruit instead of blue.

Pagoda Dogwood hybridizes with the Redosier Dogwood, *C. stolonifera* Michx., to produce the Acadian Dogwood, *C.* × *acadiensis* Fern., with elliptic to oval, short-pointed, opposite leaves crowded in false whorls; cymes small; drupes blue and fleshy; branchlets purplish brown.

REMARKS. The genus name, *Cornus*, is from the Latin word meaning "horn," with reference to the hard wood. The species name, *alternifolia*, refers to the alternate leaves, unusual in this genus. It is also known under the vernacular names of Alternate-leaf Dogwood, Red-osier Dogwood, Osier Dogwood, Blue Dogwood, and Gray Dogwood. It was first introduced into cultivation in 1760. It is often used in ornamental planting. Wildlife feed upon the fruit, the seeds having been taken from the stomachs of 11 species of birds, including the ruffed grouse. Its leaves are browsed by the white-tailed deer and cottontail.

Silky Dogwood

Cornus amomum Mill.

FIELD IDENTIFICATION. Upright or spreading shrub attaining a height of 3–10 ft. The young twigs with dark brown pith.

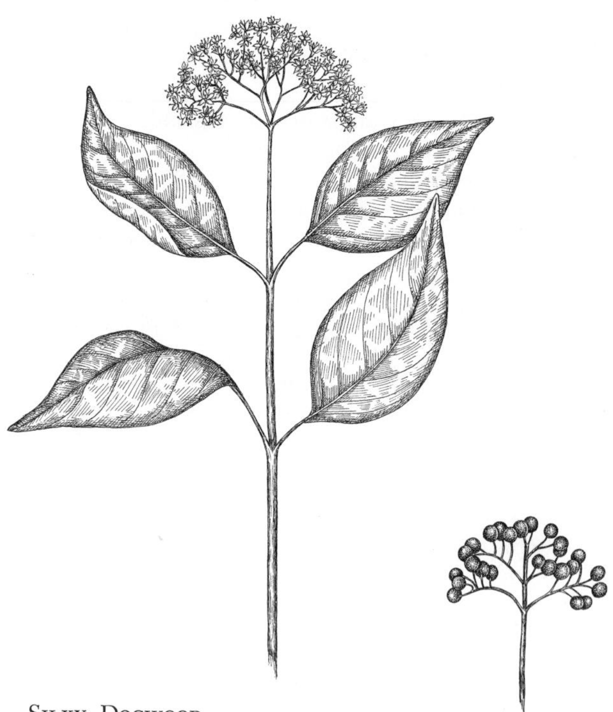

Silky Dogwood
Cornus amomum Mill.

FLOWERS. Borne May–July in rather open, flat or rounded cymes 1½–2¾ in. broad; peduncle 1¼–2 in. long, both it and the pedicels silky-pilose; calyx minutely 4-toothed, the teeth pubescent, lanceolate-linear, 1/25–1/12 in. long; petals 4, yellow-white, oblong, spreading, about ⅙ in. long; stamens 4, filaments slender; style slender, stigma terminal.

FRUIT. Borne August–October, drupe blue to bluish white, spherical, ¼–⅓ in. in diameter, 2-loculed; seed subglobose, irregularly furrowed.

LEAVES. Simple, opposite, blade 2½–4¾ in. long and about half as wide, rather thin, oval to ovate or broadly elliptic, apex acute to obtuse or short-acuminate, sometimes abruptly so, base broadly cuneate to rounded, margin entire, upper surface dark green and glabrous or slightly appressed-pubescent, lower surface lighter green with gray or reddish brown, flattened trichomes attached near their centers; petioles ⅜–¾ in. long, also hairy.

TWIGS. Slender; young ones reddish to brown, brown-hairy, older ones gray and glabrous, pith tawny.

RANGE. Usually in moist soil in the sun or shade. Texas, Arkansas, Louisiana, Tennessee, Alabama, Kentucky, Indiana, Illinois, and South Carolina.

REMARKS. The genus name, *Cornus,* is from the Latin word *cornu* ("a horn"), referring to the hard wood. The species name, *amomum,* is the Latin name of some fragrant, balsam-yielding shrub. A vernacular name for the plant is Kinnikinnick.

Water Tupelo

Nyssa aquatica L.

FIELD IDENTIFICATION. Large, semiaquatic tree wit a conspicuously swollen base, and attaining a height o 100 ft. Male and female flowers on different trees, o sometimes on the same tree.

FLOWERS. March–April, in axillary clusters, polygamo dioecious, appearing before or with the leaves; stami nate clusters capitate; peduncles slender and hairy bractlets linear and ciliate; petals 5–12, thick, oblong early deciduous; calyx cup-like, obscurely 5-toothed shorter than the petals; pistillate flowers solitary on slender peduncle; bractlets 2–4, oblong, ciliate; calyx tube exceeding the petals; style stout, reflexed, an revolute.

FRUIT. September–October, on drooping peduncle 3–4 in. long; drupe about 1 in. long, oblong-obovoid dark purple, skin thick, pale-dotted, flesh thin; ston obovoid, compressed, rounded or pointed, brown o white, channeled with about 10 sharp longitudina ridges.

LEAVES. Simple, alternate, deciduous, oblong-obovat or oval, apex acuminate or acute; base rounded, wedge-

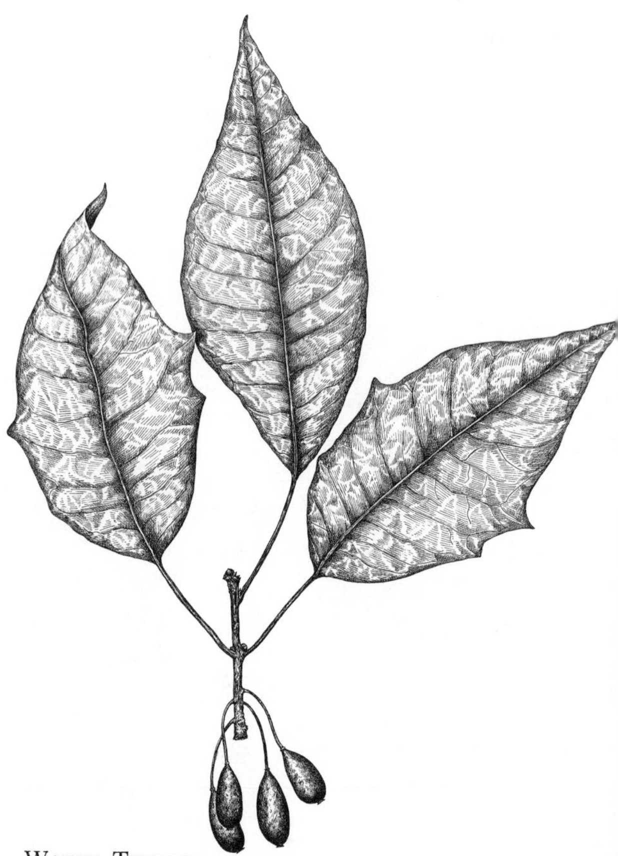

Water Tupelo
Nyssa aquatica L.

shaped or subcordate; margin entire or irregularly scallop-toothed, dark lustrous-green above, paler beneath; glabrous above, pubescent beneath, 5–10 in. long, 2–4 in. wide; petioles about 2 in. long, stout, grooved, pubescent.

TWIGS. Dark red, stout, tomentose, pithy.

BARK. Grayish brown, thin, with longitudinal small-scaled ridges.

WOOD. Light brown to white, soft, light, tough, close-grained, weighing 32 lb per cu ft. Wood of roots very light, soft and spongy.

RANGE. In swamps of Texas, Oklahoma, Arkansas, and Louisiana; eastward to Florida and northward to Virginia, Illinois, and Missouri.

PROPAGATION. The seed is gathered by hand and the pulp removed by macerating and rubbing on screens in water. The seed averages about 456 seeds per lb. They may be stored in moist sand at low temperature over the winter. Stratification at 30°–50°F. in moist sand for 60–90 days is adequate. Untreated seed can be sown in the fall, but stratified seed can be sown in the spring. Seeds are sown at the rate of 15 seeds per linear ft and covered with ½–1 in. firmed soil. The soil should not be allowed to dry out. Stratified seed has been found to give a germination of 38–50 per cent. Better stock is obtained after the first year by transplanting and root-pruning.

REMARKS. The genus name, *Nyssa,* means "water nymph," and the species name, *aquatica,* refers to the tree's habitat. Vernacular names are Tupelo Gum, Cotton Gum, Bay Poplar, Swamp Tupelo, Black Gum, Sour Gum, and Hornbeam. The fruit is eaten by at least 10 species of birds, and the wood is made into boxes, woodenware, and fruit crates. The wood of the light spongy roots is made into net floats and corks. The flowers have some value as a bee food.

BLACK TUPELO
Nyssa sylvatica Marsh.

BLACK TUPELO
Nyssa sylvatica Marsh.

FIELD IDENTIFICATION. Tree to 100 ft, with horizontal branches. Male and female flowers on separate trees, or sometimes together.

FLOWERS. In axillary clusters April–June; polygamo-dioecious, greenish; staminate flowers in long-peduncled capitate clusters; petals small, thick, ovate or oblong, rounded, erect, early deciduous; calyx 5-lobed, disc-like; stamens 5–12, exserted, and inserted on the calyx-disc below; pistillate flowers in slender-peduncled clusters of 2 or more; calyx like that of staminate flowers; bracts small but conspicuous and foliaceous; pistil 1–2-celled, style tubular, stigmas exserted.

FRUIT. Ripening September–October, drupelike, 1–3 in a cluster on long peduncles; bluish black, glaucous, about ½ in. long, ovoid, acid, bitter, flesh thin; stone solitary, ovoid to oblong, round or flattened, light brown, indistinctly 10–12-ribbed.

LEAVES. Simple, alternate, deciduous, entire, or with a few coarse remote teeth, 2–6 in. long, 1–3 in. wide; apex acute or acuminate; base cuneate or rounded; ovate or obovate to oval; thick, firm, lustrous green above, paler and hairy below; petiole about 1 in. long, villous-pubescent to glabrous.

BARK. Gray to brown or black, sometimes reddish-tinged; deeply fissured and broken into small irregularly shaped blocks.

WOOD. Tough, heavy, hard, light brown, grain close and twisted, hard to work, warping easily.

RANGE. In moist rich soils. Texas, Oklahoma, Arkansas, and Louisiana; east to Florida, north to Maine, and west to Michigan and Wisconsin.

PROPAGATION. The tree has been in cultivation at least since 1750. The seed is gathered by hand, and the pulp is removed by macerating and rubbing on screens in water. The cleaned seed averages about 3,300 seeds per lb, with a commercial soundness of 87 per cent. They may be stored in moist sand at low temperature over the winter. Stratification at 30°–50°F. in moist sand for 60–90 days is adequate. Untreated seed can be sown in the fall, stratified seed being sown in the spring. Seeds are sown using 25 seeds per linear ft and covered with ½–1 in. firmed soil. The soil should be kept moist or germination will be delayed. Stratified seed has been found to give a germination of 38–50 per cent. Better stock is obtained by root-pruning and transplanting the 1-year-old seedlings.

VARIETIES. A variety of Black Tupelo is Swamp Black Tupelo, *N. sylvatica* var. *biflora* (Walt.) Sarg. It closely resembles the species, but has narrower obtuse leaves and drupes usually in pairs, and the seeds have ridges and ribs which are more prominent. However, trees with characters intermediate to the species and the variety are often found, so the distinctions are not clear cut. Both trees are desirable for yard planting.

Other varieties of the Black Tupelo have been described by Fernald (1935b, pp. 433–437). Two of these are *N. sylvatica* var. *caroliniana* (Poir.) Fern., with rhombic-ovoid leaves tapering at either end, and *N. sylvatica* var. *dilatata* Fern., with oval leaves having a rounded apex. In what degree these variations apply to Southwest material is yet to be determined.

REMARKS. The genus name, *Nyssa,* means "water nymph," and the species name, *sylvatica,* refers to the wooded habitat. Vernacular names are Swamp-hornbeam, Yellow Gum, Snag-tree, Beetle-bung, Hornbeam, Hornpipe, Hornpine, Hornbine, Pepperridge, Bee Gum, and Sour Gum. The wood is used for veneer, plywood, railroad crossties, boxes, cooperage, pulp, woodenware, hubs, wharf piles, handles, and planing-mill products. Thirty-two species of birds eat the fruit. The foliage is browsed by black bear and white-tailed deer. The flowers serve as bee food.

SWAMP BLACK TUPELO
Nyssa sylvatica var. *biflora* (Walt.) Sarg.

VHITE-ALDER FAMILY (*Clethraceae*)

SUMMERSWEET CLETHRA
Clethra alnifolia L.

UMMERSWEET CLETHRA
'lethra alnifolia L.

FIELD IDENTIFICATION. Straggling shrub seldom more ıan 10 ft.

FLOWERS. July–September, racemes 3–8 in. long, ›rminal, erect, pubescent, fragrant, individual flowers erfect, on pedicels 1/25–1/8 in. long, subtended by narrow racts; sepals 5, imbricate in the bud, persistent, ob›ng to elliptic, obtuse to acute, canescent, 1/12–1/8 in. ›ng, when older longitudinally ridged; petals 5, imricate, white or pink, erect, obovate-oblong, hooded, pex often notched, 1/5–1/4 in. long; stamens 10, filaments ›ng and glabrous, anthers sagittate, inverted in anthesis, acs opening by pores; ovary superior, of 3 united arpels; style slender, glabrous to pubescent; stigma -lobed.

FRUIT. Capsule subglobose, enclosed in the calyx, rect, canescent, hardly more than 1/8 in. long, 3-celled, he valves 2-cleft at maturity.

LEAVES. Simple, alternate, deciduous, short-petioled, xstipulate, obovate to cuneate, apex obtuse, acute or hort-acuminate, base cuneate, margin sharply toothed ›ward the apex, surface prominently straight-veined, labrous or slightly pubescent, length 1–3 in.

TWIGS. Slender, ascending, young pubescent, older labrous, bark brown.

RANGE. In wet, sandy, acid soils in woods and wamps. East Texas and Louisiana eastward to Florida, nd north, mostly near the coast, to Maine.

REMARKS. The genus name, *Clethra*, is the Greek ame for "Alder," and the species name, *alnifolia*, means alder-leaved." Some vernacular names are White-alder, Vhite-bush, Spice-bush, and Sweet Pepper-bush. It vas introduced into cultivation in 1906.

Three varieties have been segregated from the species s follows:

Panicled Summersweet Clethra, *C. alnifolia* var. *paniculata* Arb., has leaves cuneate-lanceolate, fewer-toothed, green and glabrous on both sides, and flowers panicled.

Pink Summersweet Clethra, *C. alnifolia* var. *rosea* Rehd., has pink flowers.

Woolly Clethra, *C. alnifolia* var. *tomentosa* Michx., has larger leaves which are densely white-tomentose beneath.

About 30 species of *Clethra* are known, mostly natives of America, East Indies, and Madeira. A few hardy species are cultivated by seed germinated in sand, or by greenwood cuttings under glass. Some are propagated by layers and divisions of large plants. They are rarely attacked by insects or disease.

HEATH FAMILY (*Ericaceae*)

Arizona Madrone

Arbutus arizonica (Gray) Sarg.

FIELD IDENTIFICATION. Western evergreen tree attaining a height of 40–50 ft, and a diameter of 18–24 in. The terete smooth red branches with exfoliating bark form a compact rounded head.

FLOWERS. Appearing in May, in loose terminal racemes 1½–3 in. long, with persistent ovate bracts at their base; pedicels stout, clavate, hairy; calyx free from the ovary, imbricate in the bud, the 5 divisions ovate, acute, scarious, and persistent; corolla ¼–⅓ in. long, globular to ovoid, white to pink, 5-toothed, the teeth short, obtuse and recurved; stamens 10, filaments free, shorter than the corolla, subulate, dilated and hairy at base; anthers short, laterally compressed, dorsally 2-awned, the cells with an apical pore; ovary glabrous and glandular-roughened, immersed in the 10-lobed disk, 4–5-celled; style columnar, simple, exserted, with an obscurely 5-lobed stigma; ovules arising from a central placenta.

FRUIT. October–November, berry-like, dark orange to red, tessellate-warty, about ⅓ in. in diameter, flesh thin, sweet and dry, 5-celled; stone papery and incompletely developed; seeds numerous, small, compressed, puberulous; seed coat coriaceous, dark reddish brown, albumen copious, embryo clavate, radicle terete and erect.

LEAVES. Evergreen, simple, alternate, lanceolate to oblong, coriaceous, firm, apex acute to rounded and apiculate, base cuneate or rounded, margin entire or denticulate, length of blades 1½–3 in., ½–1 in. wide; at maturity light green and glabrous above, lower surface paler with a yellowish midrib and delicately reticulate veinlets; young leaves reddish and somewhat puberulent.

TWIGS. Stout, divergent, reddish brown, glaucous and pubescent at first, later with thin red scaly bark; winter

ARIZONA MADRONE
Arbutus arizonica (Gray) Sarg.

buds about ⅓ in. long, red, outer scales largest, acut and apiculate, ridged dorsally.

BARK. On branches dark red and smooth, but shec ding in long thin scales; older trunk bark with muc thicker squarish plates separated by longitudinal fis sures, with scales gray to white or reddish.

WOOD. Light reddish brown, sapwood lighter colorec close-grained, heavy, soft, brittle, specific gravity abou 0.71.

RANGE. On well-drained, gravelly, sunny sites a altitudes of 4,000–8,000 ft. In the Santa Catalina an Santa Rita mountains of southern Arizona. In Nev Mexico in the San Luis and Animas mountains. I Mexico in Sonora and Chihuahua, southward to Sa Luis Potosí and Jalisco.

REMARKS. The genus name, *Arbutus*, is the classical name of the south European species. The species name, *arizonica*, refers to its distribution in Arizona. The wood is sometimes used for charcoal and in the manufacture of gunpowder. It is rarely browsed except by goats.

Texas Madrone

Arbutus texana Buckl.

FIELD IDENTIFICATION. Tree hardly over 30 ft, and 1 ft in diameter, often only shrublike. Easily identified by the very smooth pink or white bark. The old bark scaling off in papery layers. Branches are usually crooked, stout, and spreading.

FLOWERS. February–March, borne in tomentose panicles about 2½ in. long; bracts scaly, ovate, acute, tomentose; calyx-lobes 5, ovate, acute, scarious; corolla ovoid-urceolate, white or pink, 5-lobed; lobes obtuse, reflexed; stamens 10, filaments subulate, swollen and hairy at base; anthers short, compressed, 2-awned on the back, cell opening by a terminal pore; ovary 5-celled, granular, smooth or hairy, sessile or nearly so, disk 10-lobed; style simple with obscurely 5-lobed stigmas.

FRUIT. Borne in clusters 2–3 in. long; body of fruit ¼–⅓ in. in diameter, subglobose, dark red to yellow, pubescent when young, waxy-granular, 5-celled; seeds numerous, white, small.

LEAVES. Simple, alternate, persistent, mostly borne distally on twigs, blades 1–3 in. long, ⅔–1 in. wide, oblong to elliptic, ovate or oval, apex obtuse to rounded, base rounded or broadly cuneate; thick and leathery; upper surface dark green and glabrous; lower surface paler or glaucous, or somewhat pubescent, especially along the veins; venation inconspicuous, except the yellow mid-vein; petioles stout, one third to one half as long as the leaf, green to reddish, pubescent to glabrous.

Texas Madrone
Arbutus texana Buckl.

BARK. Very conspicuous, younger bark very smooth, pink or white. Older bark on lower parts of old trees generally dark brown to gray or black, breaking into rather small, short scales, but exfoliating in thin papery layers to expose the new bark.

WOOD. Reddish brown, sapwood lighter, close-grained, hard, heavy, specific gravity about 0.75.

RANGE. On limestone or igneous hills and mountains. Central Texas, west Texas, and southeastern New Mexico. In Mexico in the states of Nuevo León, Chihuahua, Veracruz, Oaxaca, and Sinaloa; south into Guatemala.

PROPAGATION. Some of the species are horticulturally used. Propagation is by seeds, half-ripe cuttings, layering, budding, and grafting. In the South they generally require well-drained soil, in the sun. In the North they are confined to greenhouses.

RELATED SPECIES. Besides the Texas Madrone, closely related species are the Arizona Madrone, *A. arizonica* (Gray) Sarg., of Arizona and New Mexico, and the Pacific Madrone, *A. menziesii* Pursh, of the Pacific Coast. *A. menziesii* also occurs as far north as Prince George, British Columbia, and is cultivated in Juneau, Alaska. Other species of *Arbutus* are scattered through Central America, Mexico, Asia, and Europe.

REMARKS. The genus name, *Arbutus*, is the classical name of a European species, and the species name, *texana*, refers to the state of Texas. Also known under the vernacular names of Texas Arbutus, Madroño, Naked Indian, Lady Legs, Manzanita, and Nuzu-ndu. The wood is used for tools, handles, rollers, fuel, and charcoal for gunpowder. The bark and leaves are astringent, and are occasionally used medicinally in Mexico. The fruit is sweetish and is eaten by a number of species of birds. It is also browsed lightly by cattle and heavily by goats.

Point-leaf Manzanita

Arctostaphylos pungens H. B. K.

FIELD IDENTIFICATION. Shrub of the Southwestern mountains, thicket-forming, erect or spreading, 1–10 ft. Sometimes rooting where branches touch the ground.

FLOWERS. January–March, in short, congested, simple or branched, spicoid racemes; pedicels glabrous to subtomentose; flowers small, nodding, white or pink, about ¼ in. long; bracts short-deltoid, pubescent or tomentose, thick, firm, 1/12–⅛ in. long; rachises often distinctly thickened and club-shaped at tip; corolla cylindric-urceolate, 5-lobed at the summit, lobes small and recurved; stamens 10 (rarely, 8), included; filaments

POINT-LEAF MANZANITA
Arctostaphylos pungens H. B. K.

dilated at base, with 2 recurved dorsal awns, opening by rounded, terminal pores; ovary superior, glabrous, a 10-lobed hypogynous disk at the base, 4–10-celled, a single ovule in each cell, style slender; calyx 5-lobed, lobes glabrous or ciliolate.

FRUIT. Maturing July–April, berry-like, 1/5–1/3 in. in diameter, depressed-globose, smooth, dark-brown or reddish, glabrous, pulp soft; nutlets several, ridged on the back.

LEAVES. Evergreen, simple, alternate, leathery, blade length 1/2–1 1/4 in., width 3/8–5/8 in., oblong, elliptic, lanceolate or ovate, less often obovate or oblanceolate, apex acute and mucrocuspidate, base rounded or subcuneate, surfaces pale and glaucescent-green, finely tomentulose when young, glabrous with age; petioles about 1/8 in. long.

STEMS. Rigid, at first white-tomentulose; later smooth, reddish brown and flaky; buds small, ovoid, outer scales few.

WOOD. Dark brown, hard, smooth, brittle.

RANGE. Rocky mesas and dry slopes, usually at altitudes of 3,000–8,000 ft. In western Texas, New Mexico, Arizona, California, Utah, and Nevada; south in Mexico in Chihuahua, Coahuila, Veracruz, and Oaxaca.

PROPAGATION. By seeds sown in spring or fall in well-drained soil. Germination may be hastened by soaking 2–3 hours in concentrated sulphuric acid before planting. The seeds are sown in drills and covered with 1/4 in. of soil. Germination averages about 35 per cent. Propagation by cuttings of half-ripened wood under glass is also practiced, as is budding or veneer-grafting. If propagation is by layers, 2 years are required for development.

REMARKS. The genus name is from the Greek *Arkto* ("a bear") and *staphylo* ("a grape cluster"), referring to the fact that bears feed on the clustered fruit. The species name, *pungens*, refers to the pointed leaf. Many vernacular names are applied to the plant in the United States and Mexico, such as Bear-berry, Mexican Manzanita, Kinnikinnick, Manzanilla, Pingüica, Palo de Pingüica, Manzana, Tnu-ndido, Gayuba del País, Tepezquite, Tepeizquitl, Tepesquisuckil, Pinquiqua, and Leña Colorada. The fruit is often sold in the markets in Mexico and is made into a delicious jelly. It is also recorded as being eaten by sooty grouse, northern hooded skunk, black-tailed deer, blue grouse, Mearns's quail, Gambel's quail, bear, pigeon, fox, and coyote. The leaves are occasionally browsed by goats. In Mexico, the leaves and fruit are used as household remedies in dropsy, bronchitis, venereal diseases, and other affections.

Point-leaf Manzanita may be distinguished from other species by the enlarged club-shaped rachises, the absence of glandular-hairs on young parts, and a glabrous ovary. It has been cultivated since 1840.

BEAR-BERRY MANZANITA

Arctostaphylos uva-ursi (L.) Spreng.

FIELD IDENTIFICATION. Western evergreen shrub, much-branched, depressed or sometimes prostrate; branches terminally erect, to 6–8 in.; bark thin and exfoliating.

FLOWERS. April–July, in short, dense terminal racemes; pedicels reflexed, 1/8–1/6 in., glabrous or pubescent; corollas white or pink, small, nodding, urceolate, 1/6–1/4 in. long, the 5 lobes short and recurved; stamens 10, included, dilated and pubescent below; anthers short, subglobose, reddish, with 2 recurved dorsal sacs opening by terminal pores; disk 10-lobed and hypogynous; ovary conic-ovoid, glabrous, 5-celled, ovules solitary in the cavity, style slender; sepals about 1/25 in., broad-ovate, acute.

FRUIT. Drupe berry-like, maturing July–October, somewhat persistent, globose to oblate-spheroid, 1/4–1/3 in. in diameter, bright red, lustrous, smooth; nutlets 5, bony, rugose, dorsally ridged, coherent into a separable stone; flesh dry, insipid, and tasteless when raw, but reported to be edible when cooked.

LEAVES. Simple, alternate, persistent, evergreen, leathery, blade length 1/2–1 1/4 in., width 1/5–2/5 in., obovate to oblong or oval, base cuneate or narrowed, apex obtuse, retuse or rounded, margins entire and revolute, upper surface bright green, glabrous and lustrous; paler and glabrous beneath or pubescent on the veins; petioles 1/12–1/6 in., often pubescent.

BARK. On twigs and branches dark reddish brown, thin, exfoliating in conspicuous flakes, at first tomentulose or viscid-glandular but glabrous later; winter buds small, ovoid, with few outer scales.

RANGE. Dry, well-drained soil in sun or shade, at altitudes of 5,000–10,000 ft, from the pine belt upward to the spruce belt. In New Mexico in the Tunitcha and Sangre de Cristo mountains; west to California, north through Colorado, Utah, Nevada, Oregon, and Washington to Alaska; on the east from Pennsylvania, Illinois, and Virginia north to Newfoundland. Also in Europe and Asia.

MEDICINAL USES. It was once used by the settlers for treating urinary disorders.

PROPAGATION. The fruit may be collected from July to the following spring. It is hand-picked, run through a macerator, and the pulp floated off. The cleaned seed averages about 25,000 seeds per lb, with a commercial purity of 99 per cent and a soundness of 76 per cent. Because of slow germination, the seeds should be soaked in concentrated sulphuric acid, or stratified in moist sand at 77°F. for 60 days, plus an additional 60-day period at 40°. Sowing is in drills in early summer, the seeds being covered with 1/4 in. of soil. Germination occurs the following spring. Mulching beds over winter is required. Propagation can also be done by cuttings taken in late summer and rooted under glass.

VARIATIONS. Because of the great variability of Bear-berry Manzanita three segregations have been recognized by Fernald and MacBride (1914).

1. Branchlets minutely tomentulose, commonly somewhat viscid, soon losing all hairs—*A. uva-ursi* (L.) Spreng.
2. Branchlets with dense canescent tomentum, this persistent for several years, not viscid—*A. uva-ursi* var. *coactilis* Fern. & MacBr.
3. Branchlets viscid-villous, commonly intermixed with black stipitate glands—*A. uva-ursi* var. *adenotricha* (Fern.) MacBr.

REMARKS. The genus name, *Arctostaphylos*, is from the Greek *Arkto* ("bear") and *staphylo* ("bunch of grapes"), and the species name *uva-ursi*, of Latin origin, is also translated *uva* ("grape") and *ursi* ("bear"). Also known under the vernacular names of Sand-berry, Sierra Bear-berry, and Kinnikinnick. The wood is hard and smooth. The plant has been cultivated since 1800, and is considered to be a valuable mat-covering for rocky slopes and sandy banks. Known to be eaten by at least 18 species of birds, including ruffed grouse, dusky grouse, Richardson's grouse, spruce grouse, sharp-tailed grouse, and wild turkey; also browsed by mountain sheep, black-tailed deer, and white-tailed deer. It is of little value to livestock.

BEAR-BERRY MANZANITA
Arctostaphylos uva-ursi (L.) Spreng.

COAST LEUCOTHOË

Leucothoë axillaris (Lam.) D. Don

FIELD IDENTIFICATION. Evergreen shrub to 6 ft, with spreading, usually recurved, branches.

FLOWERS. February–May, in axillary racemes 3/4–2 3/4 in. long, sessile, simple or compound, shorter than the leaves, densely many-flowered, one-sided; bracts 1–2, broad-ovate, concave, 1/12–1/8 in. long; pedicels 1/25–1/6 in. long, minutely pubescent; calyx white, petal-like, saucer-shaped, 1/6–1/5 in. broad; sepals 5, imbricate, about 1/12 in. long, obtuse or rounded; corolla greenish in bud, white later, cylindric-ovoid, 1/5–1/4 in. long; lobes 5, erect or spreading, very short and rounded, contracted near the apex; stamens 10, included; filaments puberulent, flat, very narrowly triangular, anthers oblong, dorsifixed, prominently bimucronate; pollen sacs separate distally, each opening by a terminal pore; disk 10-lobed, ovary subglobose, 5-celled, with numerous ovules, style slender, stigma depressed and 5-lobed.

FRUIT. Capsule globose-depressed, 1/4–1/3 in. long, loculicidally 5-valved, valves membranous and entire; seeds numerous, minute, flat, shiny, angular, reticulate.

Coast Leucothoë
Leucothoë axillaris (Lam.) D. Don

LEAVES. Alternate, persistent, leathery, blade length 2–6 in., width ½–1½ in., elliptic to oblong or lanceolate to oval, apex acute or short-acuminate, base acute to rounded, upper surface dark green and glabrous, lower surface paler and sparingly pubescent, margin very short-spiny toothed or often entire toward the base; petiole 3/16–⅜ in. long, pubescent.

TWIGS. Spreading, mostly weak or recurved, terete, puberulent when young, glabrous later, pith solid.

RANGE. Acid woods and swamps near the coast. Southeastern Louisiana; also eastward to Florida and northward to Virginia.

REMARKS. The genus name, *Leucothoë,* is for Leucothoë, daughter of Orchamus, king of Babylonia. The species name, *axillaris,* refers to the axillary inflorescence. The plant is sometimes planted for ornament and has been in cultivation since 1765.

Long Leucothoë

Leucothoë elongata Small

FIELD IDENTIFICATION. Branching shrub 3–9 ft.

FLOWERS. Racemes one-sided, 4–8 in. long; pedicels 1/25–⅛ in.; corolla white to pink, about ⅓ in. long, cylindric, somewhat constricted at the throat, with 5 short, often recurved lobes; stamens 10, included, filaments adnate to base of corolla-tube, anthers short with subulate awns, opening by terminal pores, disk 10-lobed; ovary 5-celled, ovules numerous; calyx with 1–2 bracts at base which are lanceolate and acuminate; sepals 5, ⅛–⅙ in., lanceolate, often narrowly so, acuminate, ciliolate, puberulent, sepals about one half as long as the corolla.

FRUIT. Peduncles 1/16–⅛ in., puberulent, with 1–2 acuminate bracts persistent at apex, capsule ⅛–⅙ in., spheroidal, depressed, much shorter than the sepals, loculicidally 5-valved; styles persistent.

LEAVES. Rather crowded, deciduous, alternate, blades ¾–2 in. long, often with smaller ones near base of larger more terminal ones, blades elliptic to elliptic-lanceolate or oblong-oblanceolate, margin sharply and finely serrate, apex acute or acuminate, base gradually or abruptly narrowed into a short pubescent petiole, upper surface dull green and glabrous; lower surface paler and puberulent, especially along the veins, young leaves puberulent.

TWIGS. Young ones slender, rigid, terete, brown, pubescent; older ones gray and glabrous.

RANGE. Moist sandy soil in the sun, often in acid swamps. Louisiana; eastward to Florida and northward to Virginia.

REMARKS. The genus name, *Leucothoë,* is for the daughter of Orchamus, king of Babylonia. The species name, *elongata,* refers to the long racemes. This species differs from the closely related *L. racemosa* Gray chiefly by the longer sepals.

Long Leucothoë
Leucothoë elongata Small

Sweet-bells Leucothoë

Leucothoë racemosa (L.) Gray

FIELD IDENTIFICATION. Shrub 3–12 ft, widely branching, erect or ascending, glabrous to puberulent.

FLOWERS. April–June, racemes terminating the twigs of the previous year, usually longer than the leaves, solitary or clustered, 1¼–4 in. long, simple or branched, one-sided; pedicels 1/25–⅛ in.; calyx saucer-shaped, subtended by 1–2 bracts, glabrous; sepals 5, imbricate, 1/12–⅛ in., triangular-lanceolate, apex acute, ciliolate, less than one half as long as the corolla; corolla white or pink, cylindric, ⅓–⅖ in. long, constricted at throat; lobes 5, ovate, short, recurved; stamens 10, included, filaments glabrous, flat, narrowly triangular, adnate to base of the corolla-tube; anthers much shorter than filaments, with 4 long, subulate awns at apex, opening by terminal pores; disk 10-lobed; ovary subglobose, 5-celled, ovules numerous, style slender, stigma truncate.

FRUIT. Capsule depressed-globose, ⅙–⅕ in. broad, loculicidally 5-valved, surpassing the persistent calyx; seeds minute, irregular.

LEAVES. Alternate, deciduous, firm, blade length ¾–3 in., width ½–1 in., oblong to elliptic or ovate, apex acute or shortly acuminate, margin finely toothed, glabrous above or a few hairs, pubescent beneath, at least on the veins; petioles 1/25–⅛ in.

TWIGS. Erect or divergent, puberulent at first, glabrous later, pith lamellate.

RANGE. Low woods and acid swamps. Southeastern Louisiana in the pinelands; also eastward to Florida and northward to Massachusetts.

REMARKS. The genus name, *Leucothoë,* is for the daughter of Orchamus, king of Babylonia. The species name, *racemosa,* refers to the flower racemes. Vernacular names are White-osier and Pepper-bush. The plant is also listed in the literature as *Eubotrys racemosa* (L.) Nutt. It has been in cultivation since 1736. The leaves are reported as being poisonous to livestock, and a powder of the leaves and bark is considered to be a powerful errhine.

Sweet-bells Leucothoë
Leucothoë racemosa (L.) Gray

Stagger-bush Lyonia

Lyonia mariana (L.) D. Don

FIELD IDENTIFICATION. Shrub 3–6 ft, with erect, mostly glabrous, black-dotted branches.

FLOWERS. May–July, in nodding lateral umbels, forming elongate, compound racemes along leafless branchlets of the previous year; pedicels ⅜–⅝ in. long, bracteolate at the base, flowers perfect; calyx often puberulent, sepals 5, valvate in the bud, soon spreading, distinct, ⅙–⅖ in. long, lanceolate or narrowly oblong, apex acute to acuminate, firm in age, deciduous with the leaves; corolla nodding, white or pinkish, ovoid-cylindric, length ⅖–½ in., ⅕–¼ in. thick, lobes 5, mostly spreading; stamens 10, included, much shorter than tube, filaments pubescent, adnate and dilated at base, appendaged near apex; anthers dorsifixed, opening by 2 terminal pores; disk 10-lobed; style columnar, stigma truncate, ovary 5-celled, ovules numerous in each cavity.

FRUIT. Capsule ovoid-pyramidal, ¼–⅖ in. long, truncate at the contracted apex, 5-angled, sutures thickened, loculicidally dehiscent; seeds numerous, club-shaped, about 1/25 in. long, testa smooth and membranous.

LEAVES. Alternate, deciduous, blades membranous, oblong to oval or elliptic, or slightly broadest above the middle, blades ¾–3 in. long, margin entire, slightly revolute, apex obtuse and apiculate, base narrowed or obtuse, upper surface glabrous, lower surface sparingly pubescent and sometimes black-dotted beneath; petioles 1/12–⅙ in. long.

TWIGS. Erect or nearly so, when young pubescent, when older glabrous, sometimes black-dotted.

RANGE. Sandy pinelands or acid prairies. East Texas and Louisiana; also eastward to Florida and northward to Rhode Island.

REMARKS. The genus name, *Lyonia,* honors John Lyon (d. 1818), American botanical explorer. The species

STAGGER-BUSH LYONIA
Lyonia mariana (L.) D. Don

name, *mariana,* means "of Maryland." The plant was introduced into cultivation about 1736. Also known by the vernacular names of Sorrel-tree and Wicks. The foliage is said to be poisonous to lambs and calves.

HE-HUCKLEBERRY LYONIA

Lyonia ligustrina (L.) Britt.

FIELD IDENTIFICATION. Branching shrub 3–12 ft, with more or less pubescent, entire, oblong to elliptic leaves.

FLOWERS. May–July, panicles elongate with umbel-like 2–6-flowered clusters, the panicle naked (except in some varieties in which the panicles are leafy-bracted); pedicels single or clustered, 1/25–1/4 in. long, pubescent; calyx 1/8–1/6 in. broad, sepals 5, spreading, triangular-ovate, 1/25–1/12 in. long, apex acute; corolla white, subglobose to ovoid-globose, 1/8–1/5 in. long, lobes 5, short; stamens 8–12 (usually 10), included, much shorter than the corolla-tube; filaments flat, incurved, pubescent, anthers opening by 2 terminal pores; disk 8–12-lobed; style columnar, stigma truncate, ovary 4–6-celled (usually 5), ovules numerous in each cavity, pendulous.

FRUIT. Capsule dry, globose-depressed, 1/8–1/6 in. long, much longer than the persistent sepals, the 5 sutures thickened and riblike, loculicidally dehiscent into 5 valves.

LEAVES. Deciduous, alternate, firm, blades oblong to elliptic or obovate, blade length 1 1/4–3 1/2 in., apex acute or abruptly acuminate, base narrowed into a petiole 1/12–1/6 in. long, margin entire or obscurely serrulate, upper surface usually glabrous, lower surface pubescent or glabrate with age.

TWIGS. Terete, spreading widely, pubescent when young, glabrous later.

RANGE. In sandy soils from Arkansas and Oklahoma eastward to Florida, and north to Quebec.

PROPAGATION. It may be propagated by seeds sown in sand and sphagnum moss in flats in the fall, and transplanted in spring. It is also propagated by divisions or cuttings. It has been cultivated since 1748.

VARIETIES. Downy He-huckleberry Lyonia, *L. ligustrina* var. *pubescens* Rehd., has leaves densely soft-pubescent, thick, reticulate, entire or nearly so on the margin.

Bracted He-huckleberry Lyonia, *L. ligustrina* var. *foliosiflora* Michx., differs from the species by having panicles copiously leafy-bracted and flowers less crowd-

HE-HUCKLEBERRY LYONIA
Lyonia ligustrina (L.) Britt.

FETTER-BUSH LYONIA
Lyonia lucida (Lam.) K. Koch

ed; corolla white or pinkish with sepals ovate or half-orbicular; foliage glabrous or sparingly pubescent; distinctly serrulate. Usually on the coastal plain from Florida to southern Virginia, and westward into east Texas. Intermediate forms occur between this variety and the species.

REMARKS. The genus name, *Lyonia,* is in honor of John Lyon, American botanical explorer, who died in Asheville, North Carolina, in 1818. The species name, *ligustrina,* denotes its resemblance to privet, or *Ligustrum.* Also known under the vernacular names of Male-berry, Pepper-bush, Seedy-buckberry, White-wood, and White-alder.

FETTER-BUSH LYONIA

Lyonia lucida (Lam.) K. Koch

FIELD IDENTIFICATION. Shrub 2–6 ft, with glabrous, drooping, angled branches.

FLOWERS. April–May, borne in axillary, nodding and spreading umbels of 3–10 flowers, pedicels 1/8–1/3 in. long; calyx persistent, segments 5, essentially distinct, ovate to lanceolate, apex acuminate, rigid, soon spreading, 1/8–1/5 in. long, glabrous or minutely pubescent, green to purplish; corolla nodding, ovoid-conic, cylindric, narrowed at throat, white to pink, about 3/8 in. long, 5-lobed, lobes short and reflexed; stamens 10, included, filaments glabrous, slender above the slightly dilated base, 2-spurred; style columnar, stigma truncate; ovary 5-celled, ovules numerous in each cavity; disk 10-lobed.

FRUIT. Capsule ovoid-globose, rounded at apex, 5-lobed, sutures thick, woody, 1/8–1/6 in. long; seeds club-shaped, testa smooth and membranous.

LEAVES. Simple, alternate, blades 1 1/2–3 1/4 in. long, 1/2–1 1/4 in. wide, variable in shape, oblong to elliptic or lanceolate, or oval to obovate, persistent, stiff, leathery, glabrous; upper surface dark green and lustrous; paler below and black-dotted; margin thickened, entire, revolute, bordered by an extra nerve; apex acute or acuminate, sometimes abruptly so, base cuneate; petioles about 1/16–3/16 in.

TWIGS. When young reddish brown, when older gray; slender, ascending or erect, leafy, acutely angled, sparingly black-dotted.

RANGE. Low, wet, sandy, acid woods, on the coastal plain. In Louisiana on Pushepetappa Creek near Varnado in Washington Parish; also in other areas in southeastern Louisiana; eastward to Florida and northward to Virginia.

VARIETY. Red Fetter-bush, *L. lucida* forma *rubra* (Lodd.) Rehd., is a form with red flowers.

REMARKS. The genus name, *Lyonia,* is in honor of John Lyon (d. Asheville, N.C., 1818), who introduced many American plants to England. The species name, *lucida,* refers to the lustrous leaves. Other local names are Pipe-stem and Stagger-bush. It is occasionally grown in gardens and has been in cultivation since 1765.

SOUR-WOOD

Oxydendrum arboreum (L.) DC.

FIELD IDENTIFICATION. Tree to 70 ft, diameter of 20 in. Branches spreading and pendulous with a short, oblong or rounded crown.

FLOWERS. June–August, fragrant, in terminal or axillary panicles of one-sided, puberulent racemes, 6–12 in. long, erect or curving, slender, loose, one-sided, many-flowered, axis and short pedicels canescent; pedicels 1/8–2/5 in., drooping in flower, 2-bracteolate at or above the middle; bracts linear, acute, caducous; sepals 5, imbricate in the bud, almost distinct, persistent, ovate to lanceolate, acute to acuminate, 1/25–1/12 in., outer surface puberulous; corolla conic or ovate-cylindric, narrowed at throat, white, waxy, puberulous, tardily expanding, hypogynous, 1/4–1/3 in. long, 5-lobed; lobes minute, reflexed, ovate, obtuse or acute, mucronate; stamens 10, included, about as long as corolla, filaments pubescent, wider than the linear anthers, anther sacs opening by large chinks; ovary broad-ovoid, pubescent, 5-celled; ovules numerous, albumen fleshy; style columnar, thick, slightly exserted, stigma small, simple, capitate; disk thin, obscurely 10-toothed.

SOUR-WOOD
Oxydendrum arboreum (L.) DC.

FRUIT. Capsule maturing September–October, ovoid-pyramidal, small, dry, ¼–½ in. long, style persistent at apex, canescent, woody, 5-valved, dehiscent loculicidally, 5-celled, the empty capsules persistent, pedicels curved; seeds numerous, brown, oblong, ⅛ in. long, ascending or erect, elongate, extending into long, slender points, seed coat reticulate.

LEAVES. Revolute in the bud, bronze-green at first, red in the fall, acid to the taste, simple, alternate, deciduous, blades 4–7 in. long, 1–3 in. wide, elliptic, oblong, oval or lanceolate, apex acute to acuminate, base narrowly or broadly cuneate, margin sharply fine-toothed, or entire toward base, upper surface shiny dark green and glabrous, lower surface paler, glabrous, or slightly hairy on veins, finely reticulate-veined, thin; petioles slender, glabrous or slightly hairy, ¼–1 in., mostly ¼–½.

TWIGS. Slender, smooth, glabrous, terete, green to brown or reddish; when older gray; lenticels conspicuous, small, numerous; leaf scars elevated, half-round to triangular.

BARK. Gray to reddish brown, smooth above; old bark with rather short, deep fissures; ridges broad, rounded and confluent with small, thick, appressed scales.

WOOD. Light reddish brown, sapwood yellowish brown or pinkish brown, hard, close-grained, strong in bending, moderately strong in shock resistance and endwise compression, shrinks considerably, takes a fine polish, weighs 46 lb per cu ft.

RANGE. Well-drained rich soil of slopes and ridges. Eastern Louisiana (Washington and West Feliciana parishes) and Arkansas; eastward to Kentucky and South Carolina, and north to Pennsylvania, Ohio, Indiana, and Illinois.

MEDICINAL USES. The leaves are acid and are reported to be tonic, refrigerant, and diuretic. It was deemed by eclectic physicians to be useful in the treatment of dropsy and fevers, and was administered as either a semisolid or fluid extract.

PROPAGATION. Sour-wood was first cultivated in 1747. It is occasionally planted for its small cup-shaped flowers and red leaves in fall. The fruit is collected from the trees in the fall. The seed may be freed from the capsule by beating in a bag, and by screening and fanning. The seeds are very small, averaging about 5,500,000 cleaned seeds per lb, with about 96 per cent soundness and 33 per cent purity. No particular preplanting treatment is necessary. The seeds are germinated in acid sand or peat, in small flats or pans in the greenhouse, and later transplanted to cold frames, and then to nursery beds in the spring of the second year. The cut trees sprout readily from the stump.

REMARKS. The genus name, *Oxydendrum*, is Latinized from the Greek, and is translated "sour-tree." The species name, *arboreum*, also means "tree." Vernacular names are Sorrel-tree, Sour-gum, Elk-tree, and Titi. The flowers produce a good honey, and the wood is used for fuel, tool handles, machinery bearings, sled runners, and small articles. The leaves are occasionally browsed by white-tailed deer.

HAIRY HUCKLEBERRY

Gaylussacia hirtella (Ait.) Klotzsch

FIELD IDENTIFICATION. Shrub ½–5 ft, often with underground stems. All parts usually clothed with silvery, bristly-hispid, glandular hairs.

FLOWERS. Branches of inflorescence spreading, pubescent with short, glandular or glandless hairs; racemes many-flowered on previous year's growth; bracts and bractlets large and persistent; calyx hairy, about ¼ in. broad; sepals 5, triangular, apex acute or short-acuminate, about as long as the hypanthium; corolla white or pink, broadly campanulate, about ⅓ in. long; lobes 5, ovate to triangular, apex recurved, margin revolute; stamens 10, about equaling the corolla; filaments distinct, more or less winged; anthers longer than filaments, sacs terminating in tubes; disk present; ovary 10-celled, ovules solitary in each cavity.

FRUIT. Drupe ⅓–⅖ in. in diameter, black, lustrous, pubescent, 10-celled, nutlets 10; seeds solitary, bony, flattened, testa very thin.

Hairy Huckleberry
Gaylussacia hirtella (Ait.) Klotzsch

LEAVES. Simple, alternate, blades 1¼–2⅓ in. long, oblanceolate to spatulate or elliptic, margin entire and glandular-ciliate, apex acute to obtuse and apiculate, base narrowed into a short petiole, both surfaces minutely pubescent when young, later sparingly hispid or glabrate above, more hispid beneath, texture thickened and veiny with age, tardily deciduous.

RANGE. In sandy, usually acid wet soil, from Louisiana eastward to Florida.

SYNONYMS. Sometimes listed in the literature under the scientific name of *Lasiococcus moisieri* Small, or as a variety under *G. dumosa* (*G. dumosa* var. *hirtella* [Ait. f.] Gray). However, the author is of the opinion that it should be maintained as a species.

REMARKS. The genus name, *Gaylussacia*, honors J. L. Gay-Lussac (1778–1850), a French chemist. The species name, *hirtella*, refers to the small bristly hairs. Also known under the names of Gopherberry and Dwarf Huckleberry.

Blue Huckleberry

Gaylussacia frondosa (L.) Torr. & Gray

FIELD IDENTIFICATION. A freely branched slender shrub 4–9 ft.

FLOWERS. April–June, on twigs of previous year; racemes usually surpassing the leaves, slender, lax, open, to 3 in. long, few-flowered; bracts minute, linear-oblong to elliptic, early deciduous; pedicels to 1 in., glabrous or sparsely pubescent, glands absent, or a few sessile ones; calyx 5-lobed, lobes about one fifth as long as corolla-tube, broadly deltoid; corolla tubular-campanulate, white to greenish purple, ⅛–⅕ in. long, about two thirds as thick; lobes 5, very short, erect or slightly reflexed; stamens 10, included, about ⅛ in. long; filaments glabrous, narrowed upward into long tubules which are longer than the filaments; anthers 2-celled, opening by a slit at the ends; ovary inferior, 10-celled.

FRUIT. Maturing July–September, drupe dull dark blue, glaucous, about ⅓ in. in diameter, fleshy, sweet, juicy, edible, calyx persistent, containing 10 seed-like nutlets.

LEAVES. Simple, alternate, deciduous, elliptic to oblong or oblong-obovate, apex obtuse to retuse, base gradually narrowed, margin entire, blades 1¼–2½ in. long, firm, upper surface pale green; lower surface finely pubescent, glaucous, often with minute resinous glands; short-petioled.

TWIGS. Reddish brown or green to gray, glabrous or merely puberulent; winter buds ovoid, with about 3 outer scales.

Blue Huckleberry
Gaylussacia frondosa (L.) Torr. & Gray

RANGE. Moist woodlands; Louisiana; eastward to Florida, north to Massachusetts and New Hampshire, and west to Ohio.

VARIETY. A variety has been described as Dwarf Blue Huckleberry, *G. frondosa* var. *tomentosa* Gray. It has a dwarf habit and is more densely hairy.

REMARKS. The genus name, *Gaylussacia,* honors J. L. Gay-Lussac (1778–1850), a French chemist, and the species name, *frondosa,* means "leafy." The plant is listed in some literature under the name of *Decachaena frondosa* Small. Vernacular names are Dangleberry, Blue Tangle, Tangleberry, and Blue Tangleberry.

Blue Huckleberry is eaten by a number of species of birds, including cedar waxwing, bobwhite quail, and ruffed grouse.

DWARF HUCKLEBERRY

Gaylussacia dumosa (Andr.) Torr. & Gray

FIELD IDENTIFICATION. Low shrub 1–2 ft, sending up erect, solitary or tufted branches from underground stems. All parts of the plant glandular-pubescent.

FLOWERS. May–June, in axillary, rather dense, drooping racemes; bracts foliaceous, persistent, oval to oblong, pubescent or glandular, ⅕–½ in. long; calyx glandular-hairy, about ⅕ in. broad; sepals 5, triangular-ovate, sometimes unequal, acute, glandular-ciliat corolla campanulate, ⅕–¼ in. long, white or pin waxy; lobes 5, broadly-ovate, more or less recurve or revolute; stamens 10, usually included, ⅙–⅕ in.; fila ments distinct, pubescent, anthers longer than the fila ments, with tuberular appendages slightly longer tha the sacs, opening by terminal pores; disk present, ovar 10-celled, ovules solitary in each cavity, pendulous.

FRUIT. June–October, drupe globose, black, some what pubescent, ¼–⅓ in. in diameter, calyx-lobes pe sistent, taste watery and insipid, with 10 nutlets, eac 1-seeded; seeds solitary, bony, flattened, testa very thir

LEAVES. Deciduous, alternate, leathery, oval to obo vate or oblanceolate to spatulate, rarely linear-lanceo late, blade length ¾–1½ in., margin entire and rarel ciliate, apex obtuse and mucronate, base short-petiole or sessile, when young minutely pubescent and resinous glandular, later deep green above and lustrous wit age, lower surface paler and glandular-pubescent.

RANGE. Acid swamps and sandy or rocky soil. Loui siana, eastward to Florida, and northward to Newfound land.

VARIETY. A variety of the northeastern states, know as Bigelow Dwarf Huckleberry, *G. dumosa* var. *bige loviana* Fern., has conspicuous glandular bracts an slightly larger flowers.

Hairy Huckleberry, *G. hirtella* (Ait.) Klotzsch, i sometimes listed in the literature as a variety of Dwar Huckleberry, *G. dumosa* var. *hirtella* (Ait. f.) Gray However, it seems sufficiently distinct to maintain as species. Dwarf Huckleberry has been cultivated sinc 1774.

REMARKS. The genus name, *Gaylussacia,* is in hono of J. L. Gay-Lussac (1778–1850), French chemist. Th species name, *dumosa,* refers to its bushy habit. Als known under the vernacular names of Bush Huckle berry and Gopherberry.

DWARF HUCKLEBERRY
Gaylussacia dumosa (Andr.) Torr. & Gray

BLACK HUCKLEBERRY

Gaylussacia baccata (Wangh.) K. Koch

FIELD IDENTIFICATION. Rigid, upright, much-branche shrub attaining a height of 1–4 ft.

FLOWERS. May–June, short-paniculate or racemose from wood of previous season. Branches of the panicl finely pubescent and ⅜–1 in. long, sometimes resinous sticky when young. Pedicels 1/12–⅓ in. long, bractlet minute, reddish, deciduous; calyx about 1/12 in. broad glabrous but usually resinous; sepals 5, ovate to deltoid apex obtuse; corolla ⅕–¼ in. long, conic to cylindric pink or reddish, lobes 5, short, erect or recurved; sta mens 10, included, about ⅙ in. long, anther sacs pro longed into tubes attached by the back; style slightl exserted, ovary 10-celled, ovules solitary in each cavity

FRUIT. Maturing July–September, a berry-like drupe length ¼–⅖ in., black, shining, without a bloom, sweet

BLACK HUCKLEBERRY
Gaylussacia baccata (Wangh.) K. Koch

nutlets 10, bony, seeds solitary, flattened, testa very thin.

LEAVES. Simple, alternate, deciduous, firm, blade elliptic-oval or obovate-ovate, margin entire, apex obtuse or mucronate, base broadly cuneate, blade length ¾–1¾ in., upper surface yellowish green, glabrous or nearly so, lower surface paler, puberulent or glabrous and sometimes sticky-resinous, venation finely rugose, short-petioled.

TWIGS. When young green, puberulent and somewhat glandular-sticky, when older gray and glabrous.

RANGE. In sandy or rocky acid soils of swamps or thickets. Louisiana eastward to Florida, northward to Maine, and west to Manitoba and Iowa.

FORMS. Several forms of Black Huckleberry have been segregated. *G. baccata* forma *glaucocarpa* (Robins.) Mackenzie, has large blue fruit with a heavy bloom.

G. baccata forma *leucocarpa* (Porter) Fern., has a white or pinkish, somewhat translucent, fruit.

REMARKS. The genus name, *Gaylussacia,* is in honor of J. L. Gay-Lussac (1778–1850), a French chemist. The species name, *baccata,* refers to the baccate, or berry-like, fruit. Also known as the High-bush Huckleberry and Black Snap. The fruit is eaten by at least six species of birds, including ruffed grouse, sharp-tailed grouse, greater prairie chicken, bobwhite quail, wild turkey, and mourning dove.

FARKLEBERRY

Vaccinium arboreum Marsh.

FIELD IDENTIFICATION. Stiff-branched, evergreen shrub, or small crooked tree attaining a height of 30 ft.

FLOWERS. In axillary racemes, leafy-bracted, perfect; corolla about ½ in. long, bell-shaped, pendent, 5-lobed, white or pinkish; calyx 5-toothed; stamens 10, within the corolla, filaments hairy, anthers slender-tubed; ovary inferior, 5-celled, style filiform with a minute stigma.

FRUIT. About ⅓ in. in diameter, globose, black, shiny, sweet, mealy, dry, many-seeded, persistent, long-peduncled, ripening in winter.

LEAVES. Alternate, simple, 1–3 in. long, about 1 in. wide, oval to obovate or elliptic, entire or obscurely denticulate, mucronate and acute or rounded at apex, cuneate at base, veiny, leathery, glossy above, duller

FARKLEBERRY
Vaccinium arboreum Marsh.

green and slightly pubescent below, deciduous in the North, persistent in the South; petioles 1/25–3/16 in. long.

TWIGS. Slender, light brown to dark brown or grayish, divergent, glabrous or puberulous.

BARK. Gray or grayish brown, thin, smooth, ridges narrow and shredding into large plates.

WOOD. Brown to reddish brown, close-grained, hard, weighing 48 lb per cu ft.

RANGE. Eastern Texas, Oklahoma, Arkansas, and Louisiana; east to Florida, northward to Virginia, and west to Missouri.

MEDICINAL USES. The root bark, in the form of an extract, has been used for treatment of diarrhea.

VARIETY. Missouri Farkleberry, *V. arboreum* var. *glaucescens* (Greene) Sarg., has glaucescent leaves, glabrous or pubescent, with bracts of inflorescence larger. Louisiana, Texas, and Oklahoma; north to Kentucky, Missouri, and Illinois.

REMARKS. The genus name, *Vaccinium,* is the classical name for an Old World species, and the species name, *arboreum,* refers to the treelike habit, which is unusual for huckleberries. Vernacular names are Whortleberry, Sparkleberry, Tree Huckleberry, Gooseberry, and Winter Huckleberry. The edible fruit is eaten by a number of species of birds. The wood is used for making tool handles, and the bark for tanning leather.

Common Deerberry

Vaccinium stamineum L.

FIELD IDENTIFICATION. Irregularly branched shrub of sandy, acid soils, rarely more than 6 ft.

FLOWERS. April–June, borne in loose leafy racemes, pedicels 1/4–3/4 in. long, pubescent; calyx 1/8–3/16 in. long, pubescent, 5-lobed, lobes spreading, acuminate, ciliolate on the margin; corolla short-campanulate, 1/4–1/3 in. across, white, glabrous, spreading into 5 short, rounded, obtuse lobes, sometimes obscurely mucronulate; stamens 10, 3/16–1/4 in. long, conspicuously exserted beyond corolla and standing erect around pistil; filaments distinct, anthers 2-awned on the back, sacs prolonged into slender tubes; pistil straight, exserted beyond stamens; stigma minute; ovary inferior, 5-celled.

FRUIT. Borne May–September, berry globose, green or yellowish, about 2/5 in. in diameter, seeds few, acid to taste but hardly edible raw.

LEAVES. Deciduous, alternate, oblong to elliptic or oval; apex acute to short-acuminate; base cuneate or rounded; margin entire and ciliate, blades 3/4–2 3/4 in. long, 1/3–1 in. wide; upper surface light green, shiny, and mostly glabrous; lower surface duller and pubescent, veins conspicuous. The flowering branches often bear leaves considerably smaller in size than others.

Common Deerberry
Vaccinium stamineum L.

TWIGS. Slender, growth of current year green to reddish, pubescent; later green to brown and glabrous.

BARK. Gray to brown; rather smooth, tight.

RANGE. In moist, sandy, acid soils of east Texas, Louisiana, Oklahoma, and Arkansas; east to Georgia, north to Maine, and west to Ontario and Minnesota.

VARIETIES. Common Deerberry is a close relative to Southern Deerberry, *V. stamineum* var. *neglectum* (Small) Deam. The latter varies from the species principally by glabrous stems and leaves, which are somewhat glaucous beneath, and by glabrous sepals and hypanthium. The fruit is green to dark purple. Southern Deerberry occurs in New Jersey, West Virginia, Indiana, Oklahoma, Missouri, and Kansas.

Georgia Deerberry, *V. stamineum* var. *melanocarpum* Mohr, is a variety with berries bluish black, calyx densely tomentose, and leaves pubescent beneath. North Carolina to Missouri and Oklahoma.

Inland Deerberry, *V. stamineum* var. *interius* (Ashe) Palmer & Steyermark, has pubescent twigs, glabrous calyx, and leaves glabrous or slightly hairy on the veins beneath. Maryland to Kansas; south to Virginia, Arkansas, and Oklahoma.

Langlois Deerberry, *V. langloisii* (Greene) Cory, is also closely related to Common Deerberry.

REMARKS. The genus name, *Vaccinium,* is the ancient Latin name, and the species name, *stamineum,* refers to the prominent stamen structure. Vernacular names in use are Squaw Huckleberry and Buckberry. Recognized in flower by the open-campanulate corolla with conpicuous erect stamens. The mature green or yellowish fruit is also a mark of identification. The fruit is known to be eaten by at least 14 species of birds, including bobwhite quail and ruffed grouse, also by gray fox and white-tailed deer.

LANGLOIS DEERBERRY

Vaccinium langloisii (Greene) Cory (*Polycodium langloisii* Greene)

FIELD IDENTIFICATION. The Deerberry genus (*Polycodium* of some authors) constitutes a large interbreeding population with considerable variation over a wide area. Many of the variants have not been clearly defined taxonomically. The best attempt at classification was made by W. W. Ashe (1931b).

The author has never been positive about the identification of Langlois Deerberry as compared to Common Deerberry. It is apparent that a large series of specimens is needed, from a much wider area than the type locality at Covington, Louisiana, to determine the true status of the species. Until this can be done, the descriptions of Edward L. Greene and W. W. Ashe must suffice as follows:

Greene (1910–1912) offers the following note on *P. langloisii:*

"Branches slender, minutely and sparsely pubescent with short curved hairs, the pubescence permanent, at least during several seasons; leaves ample, thin, deep-green and almost alike as to color on both faces, of oval-elliptic outline, 2½–3½ in. long, 1½–1¾ in. wide, subsessile, narrow below the middle, yet ending abruptly and usually subcordate at base, the apex very acute: bracts of the short racemes small, oval, obtuse, rather strongly ciliate under a lens; corolla extremely short yet apparently mature; fruit unknown. Species very distinct by the foliage; known only as collected by the late Rev. A. B. Langlois, at Covington, Louisiana, 16 April, 1894."

Ashe (1925) notes with respect to *P. langloisii* Greene as follows:

"The description of this species is drawn from a flowering specimen only. In the general absence of important floral differences in this genus, determination of a species from a description based entirely on a flowering specimen may be doubtful. Specimens both in flower and fruit from a vicinity of Covington, Louisiana, (type locality) are so different from any recognized species and conform so closely to the original description that it is believed they represent Dr. Greene's plant, and that it should be considered valid. The mature foliage is dark yellowish green on both faces but paler below, glabrous above except the midrib which is finely pubescent, the lower surface sparingly pubescent with loose matted hairs especially along the midrib; the fruit is wine-colored sometimes nearly black and not or rarely glaucous; seed pitted. This seems to be the prevailing deerberry over much of Louisiana and southern Arkansas. The flowers described by Dr. Greene as small, are actually of medium sizes ¼–⅓ in. long, and purplish; calyx occasionally sparingly pubescent. It seems to be more closely related to *P. sericeum* having the same kind of pubescence."

GROUSE WHORTLEBERRY

Vaccinium scoparium Leiberg

FIELD IDENTIFICATION. Small Western undershrub 4–12 in., or under optimum conditions as high as 6 ft.

FLOWERS. June–July, corolla solitary, axillary, ⅛–⅙ in. long, globose-urceolate, minutely 5-toothed, white; stamens twice as many as the corolla-teeth, filaments glabrous, anthers included, 2-awned, opening by terminal slits; ovary 4–5-celled, crowned at top by the epigynous disk, inferior; style straight, filiform, stigma small.

FRUIT. Maturing July–September, berry-like, subglobose, about ⅕ in. across, red or wine-colored, sweet, edible.

LEAVES. Alternate, deciduous, ovate to oval or elliptic, blade length ⅙–⅝ in. (usually less than ½ in. long), margin serrulate to subentire, apex acute, base rounded or some broadly cuneate to truncate, thin; surfaces light

GROUSE WHORTLEBERRY
Vaccinium scoparium Leiberg

green, glabrous, shiny; conspicuous reticulate veins; petiole very short or leaf sessile.

TWIGS. Numerous, slender-ascending (broomlike), sharply angled, mostly glabrous in the grooves, young ones bright green, older ones reddish brown to grayish.

RANGE. At altitudes of 6,000–11,500 ft. Often in deep woods of spruce, yellow pine, or aspen. Usually in moist, well-drained soil in the shade. Northern New Mexico, west to mountains of California, north to British Columbia and Alberta.

REMARKS. The genus name, *Vaccinium,* is the ancient Latin name, and the species name, *scoparium,* refers to the broomlike branchlets. Also known under the name of Little-leaf Huckleberry. It is considered to be good food for wildlife, but an inferior browse for livestock. The plant has been cultivated since 1904. The edible fruit is of good flavor. It is very closely related to the black-fruited *V. myrtillus* L., and perhaps could be better segregated as a red-fruited variety of it.

Myrtle Whortleberry

Vaccinium myrtillus L.

FIELD IDENTIFICATION. Shrub of the Western mountains. Slender-branched, deciduous, seldom over 2 ft.

FLOWERS. Maturing in May, mostly solitary, perfect, short-pediceled, gamopetalous, globose-ovoid, pinkish, about ¼ in. long, shortly 5-lobed at apex; calyx with rim almost entire, adnate to ovary; stamens 10, included, anthers prolonged into tubes and opening by terminal pores; ovary inferior, crowned by an epigynous disk, style straight with a small stigma.

FRUIT. Maturing in July, bluish black to black, sometimes slightly glaucous, globose, ⅕–⅓ in. in diameter, sweet, edible, many-seeded.

LEAVES. Simple, alternate, deciduous, blades ⅜–1¼ in. long; ovate to elliptic, occasionally oval, oblanceolate or obovate; apex obtuse to acute or sometimes rounded; base somewhat rounded to broadly cuneate, or more rarely subcordate; margin minutely serrate; surfaces glabrous, but lower surface reticulate-veiny; petiole very short.

TWIGS. Yellowish to reddish or brown, glabrous or puberulent, acutely angled.

RANGE. In the spruce and subalpine belts at altitudes of 7,000–12,000 ft. In northern New Mexico, Arizona, and California; northward to British Columbia and Alberta. Also in Europe and northern Asia.

VARIETY. A white-fruited form has been given the name of *V. myrtillus* var. *leucocarpum* Dumort. It is thought by some botanists that *V. myrtillus* is an autopolyploid derived from *V. scoparium,* with possible subsequent slight genetic contamination from other species.

REMARKS. The genus name, *Vaccinium,* is the ancient Latin name, and the species name, *myrtillus,* refers to the myrtle-like foliage. The plant is sometimes described under the name of *V. oreophilum* Rydb.

Myrtle Whortleberry
Vaccinium myrtillus L.

Dwarf Blueberry

Vaccinium depressum Small

FIELD IDENTIFICATION. Dwarf depressed shrub hardly over 12 in.; branches slender, finely pubescent.

FLOWERS. Borne in leafy-bracted panicles, the bracts much smaller than the leaves; pedicels ¼–⅜ in., velvety pubescent with curved hairs or sometimes glandular; sepals 5, persistent, elongate, narrow, erect, obtuse or acute, pubescent; corolla white, open-campanulate, about ⅙–⅕ in. long; corolla-lobes 5, rounded, deeply divided; stamens 10, erect, ¼ in. or longer, longer than the corolla; filaments distinct; anthers conspicuously exserted, 2-awned on the back, sacs prolonged into slender tubes; ovary inferior and 5-celled.

FRUIT. Red to dark purple or black, subglobose, glan-

Ɔwarf Blueberry
/accinium depressum Small

ɩular or pubescent when young, glabrous at maturity, ⁄4–3⁄8 in. in diameter, few-seeded.

LEAVES. Simple, alternate, deciduous, elliptic to ob-ɒng, apex acute, base rounded, margin entire, blade ɐngth 3⁄4–2 1⁄3 in., upper surface dull green and pubes-ɛent, lower surface paler and velvety-pubescent with ɛattered glands, veins impressed above and conspicuous ɒeneath.

TWIGS. Young ones brown and pubescent, older ones ɭarker brown and glabrous.

RANGE. In sandy Longleaf Pine woods. Gulf Coast ɒlain from southern Louisiana to western Florida.

VARIETY. *V. depressum* var. *minus* Ashe has leaves ᴠhich are not always velvety-pubescent and the glan-ɭular hairs are absent or not abundant. The type speci-nen was originally collected in Rapides Parish, Loui-ıiana.

REMARKS. The genus name, *Vaccinium,* is an ancient ıame, thought to be from the Latin *vaccinus* ("of ɛows"). The species name, *depressum,* refers to the ɒlant's habit of growth.

High-bush Blueberry
Vaccinium corymbosum L.

FIELD IDENTIFICATION. Shrub 3–12 ft. Very variable in size and aspect. Sometimes several-stemmed and forming individual plants, at other times producing runners and forming dense colonies.

FLOWERS. Appearing when the leaves are half-grown, maturing May–June, borne in rather dense corymbs; calyx 5-lobed, the divisions glabrous or glaucous and acute; corolla cylindric to broadly urceolate or less often subcampanulate, white to pinkish, 1⁄4–1⁄2 in. long, 1⁄6–1⁄4 in. wide, apex with 5 short lobes; stamens 8–10, anthers awnless and included, opening by a terminal pore; style slender.

FRUIT. Maturing June–August, subglobose, blue with a bloom, or bluish black, dull, 1⁄5–1⁄2 in. in diameter, sweet and juicy, 4–5-locular.

LEAVES. Simple, alternate, deciduous, narrowly to broadly elliptic or ovate, apex acuminate to acute, margin entire and ciliate to sharply serrate in some varieties, blades 1 1⁄2–3 1⁄4 in. long, 1⁄3–3⁄4 in. wide, surfaces green to slightly glaucous or very glaucous, lower sur-

High-bush Blueberry
Vaccinium corymbosum L.

THICK-LEAF BLUEBERRY
Vaccinium fuscatum Ait.

face often pubescent, at least on veins. Leaves about half grown when the flowers appear.

TWIGS. Young ones slender, yellowish green, sometimes with warty excrescences, older ones brown to gray.

RANGE. Moist soils of bogs or low woods. Louisiana and Arkansas; eastward to Florida, north to Nova Scotia, and west to Ontario, Minnesota, and Indiana. W. H. Camp states that the Louisiana, Arkansas, and Florida segregates may possibly be of complex hybrid origin.

VARIETIES. The following varieties have been listed: Pale High-bush Blueberry, *V. corymbosum* var. *glabrum* (Gray) Camp, has glaucous, glabrous, and ciliate-serrulate leaves.

Entire-leaf High-bush Blueberry, *V. corymbosum* var. *caesariense* (Mackenzie) Camp, has glabrous entire leaves and white flowers.

White-fruit High-bush Blueberry, *V. corymbosum* forma *albiflorum* (Hook.) Camp, has white flowers and fruit, serrulate leaves.

The following species are closely related to *V. corymbosum* and are connected by intermediate forms of hybrid origin: *V. arkansanum, V. simulatum, V. australe,* and *V. marianum.* See Camp (1945, pp. 260–262).

REMARKS. The genus name, *Vaccinium,* is thought to be from the Latin *vaccinus* ("of cows"). The species name, *corymbosum,* refers to the corymbose, or clustered, flowers. Vernacular names are Whortleberry, Swamp Blueberry, and Northern High-bush Blueberry. The fruit has commercial value but flavor varies from mediocre to good. Wildlife such as the mourning dove, ruffed grouse, ring-necked pheasant, and cottontail feed on the fruit. The plant transplants well and suckers freely.

THICK-LEAF BLUEBERRY

Vaccinium fuscatum Ait.

FIELD IDENTIFICATION. Southeastern shrub 3–9 ft. Sometimes in small colonies, irregularly branched to form small crowns. Leaf blades thick-coriaceous.

FLOWERS. More or less fascicled; pedicels glabrous, to ¼ in.; scales about ⅛ in., reddish, lanceolate to oblong, acute to obtuse; corolla pink to reddish, ¼–⅓ in., about twice as long as thick, cylindric, urceolate, tipped by 5 obscure short lobes (about 1/25 in.) which are erect or recurved; calyx glabrous, sepals 5, ovate to deltoid or somewhat rounded, obtuse; stamens ¼–⅓ in., included; filaments pubescent, anthers with apical tubes; style filiform, stigma minute and capitate.

FRUIT. Berry subglobose, dark, dull or subglaucous, ¼–⅖ in. in diameter, usually of poor flavor.

LEAVES. Simple, alternate, firm and coriaceous, evergreen to semipersistent; lanceolate, narrowly to broadly elliptic, or ovate to oval; apex acute, base rounded or broadly cuneate, margin entire but sometimes subserrate or ciliate, blade length 1⅓–2 in., width ⅝–1 in., upper surface dark green, or occasionally subglaucous, lower surface varying from puberulent to pubescent on veins only, or glabrous; petiole very short, usually ⅛–3/ in., pubescent.

TWIGS. Slender, rather erratic, when young finely pubescent, light green to brown, when older brown and glabrous.

RANGE. Sandy flatwoods, low pinelands, cut-over areas, eastern Louisiana and Arkansas; eastward to Florida and Georgia.

HYBRIDS. The plant shows complex elements and the following crosses seem possible: *V. fuscatum* × *V. virgatum; V. fuscatum* × *V. myrsinites; V. fuscatum* × *V. australe;* and *V. fuscatum* × *V. arkansanum.*

REMARKS. The genus name, *Vaccinium,* is the ancient name, and the species name, *fuscatum,* refers to the brownish hairs. The plant is characterized by its erratic branching and leaves with subcoriaceous blades. It was introduced to the Kew Gardens in 1770 by William Young from North America.

LARGE-CLUSTER BLUEBERRY

Vaccinium amoenum Britt.

FIELD IDENTIFICATION. Shrub with spreading branches, and producing suckers to form rather dense clumps. Attaining a height of 3½–6½ ft and sometimes forming a distinct crown.

FLOWERS. In good-sized clusters, corolla ⅖–½ in. long, usually deep pink, broadly cylindro-urceolate; stamens included, anthers with apical tubes; calyx with the 5 sepals persistent, deltoid-ovate, acute.

FRUIT. Globular, black, shiny, ⅓–⅖ in. in diameter, thick-skinned, flavor insipid.

LEAVES. Simple, alternate, deciduous, obovate to narrowly elliptic or elliptic, apex acute or acuminate and sometimes abruptly so, base cuneate or gradually narrowed, blades 1½–2 in. long, margin sharply serrate, upper surface dark green or slightly glaucescent, lower surface pubescent along the veins or eventually glabrous, also conspicuously glandular; petiole very short and pubescent or blade sessile.

TWIGS. Slender, when young finely pubescent, reddish brown, later glabrous and brown or gray.

LARGE-CLUSTER BLUEBERRY
Vaccinium amoenum Britt.

PALE BLUERIDGE BLUEBERRY
Vaccinium pallidum Ait.

RANGE. Acid soil of uplands or open woods, often on the margins of swamps or streams. Eastern Texas, Louisiana, and Arkansas; eastward to Florida and north to South Carolina.

REMARKS. The genus name, *Vaccinium,* is the ancient name, and the species name, *amoenum,* meaning "pleasant" or "delightful," is an allusion to its delightful or charming habit and pretty flowers. It is also known as Tall Huckleberry. For a full discussion of this species, and its relationships, the reader is referred to Camp (1945).

PALE BLUERIDGE BLUEBERRY

Vaccinium pallidum Ait.

FIELD IDENTIFICATION. Shrub 1–3 ft, usually occurring in colonies of considerable extent. Controversial species said to be closely related to *V. vacillans* Torr. Divided into an Eastern and a Western variation.

FLOWERS. Corolla cylindraceous or cylindro-campanulate, greenish white or pink, ⅙–¼ in. long, the 5 lobes at apex short; stamens 8–10, anthers 2-celled, opening by a pore, awnless; style slender; calyx-lobes 5, acute.

FRUIT. Berry-like, subglobose, dark blue to black, glaucous in some forms, ⅕–⅓ in. in diameter, rather dry and of mediocre quality, 4–5-celled, many-seeded.

LEAVES. Simple, alternate, deciduous, elliptic to ovate, apex acute to acuminate, margin finely serrate, base broadly cuneate, thin but firm, blade length 1¼–2 in., width ⅝–1 in., upper surface dull green, lower surface paler or sometimes glaucous, glabrous, or somewhat puberulent on the venation; petiole very short and pubescent, or blade sessile.

TWIGS. Slender, at first finely pubescent but later glabrous, light to dark brown.

RANGE. Usually on rocky hillsides to altitudes of 3,500 ft. In Arkansas in the Ozark Mountains; east to northern Alabama and Georgia, north to New York, and west to Missouri.

REMARKS. The genus name, *Vaccinium*, is from the Latin *vaccinus* ("of cows"). The species name, *pallidum*, refers to the pale or glaucous leaves.

Blueridge Blueberry

Vaccinium vacillans Torr.

FIELD IDENTIFICATION. Low, stiffly branching, glabrous shrub, ½–3 ft high and often in extensive colonies. A very variable plant, often hybridizing with closely related species; difficult to segregate with exactitude.

Blueridge Blueberry
Vaccinium vacillans Torr.

FLOWERS. Flowering when leaves are partly expanded, racemes terminating branches or from the old axils; calyx 5-lobed, often reddish; corolla variable, ⅙–⅓ in. long, cylindro-urceolate to narrowly campanulate, the 5 lobes short and erect to reflexed, sometimes constricted at the throat, color varying from white to greenish or purplish, sometimes reddish tinged (especially in the bud); stamens 8–10, included, anthers 2-celled, with a terminal pore; style slender.

FRUIT. Berry maturing June–September, ⅕–⅓ in. in diameter, globose, dull dark blue to almost black, with a faint bloom, many-seeded.

LEAVES. Simple, alternate, deciduous, variable in shape, commonly elliptic to oval, varying to ovate, lanceolate, or even oblanceolate, blade length ¾–1¾ in., about two thirds as wide, apex obtuse to acute (more rarely apically rounded), margin entire to finely serrulate especially near the apex, pale green or glaucous, glabrous or finely pubescent, finely reticulate-veined beneath, firm; short-petioled.

TWIGS. Rather loosely arranged, stiff, green to brown, glabrous.

RANGE. Often in sandy soil in dry woods, old fields, and on rocky ledges. Arkansas; eastward through Mississippi to Alabama, northward to Nova Scotia, and west to Minnesota, Michigan, Illinois, and Missouri.

SYNONYMS AND HYBRIDS. A very variable species particularly because of hybridization. For a full discussion of these complexities see Camp (1945).

Known to hybridize with *V. angustifolia*, *V. caesariense*, *V. elliottii*, *V. pallidum*, and *V. tenellum*. Typical plants are reported as having glabrous leaves, those with pubescent leaves are considered as hybrids of *V. atrococcum* or *V. myrtilloides*.

Evidently very closely related to *V. pallidum* Ait., and considered by some botanists as a synonym of it. *V. vacillans* is also listed under the name of *Cyanococcus vacillans* (Rydb.) Small.

REMARKS. The genus name, *Vaccinium*, is the ancient Latin name, and the species name, *vacillans*, refers to the variable or inconstant nature of the plant. It is locally known under the vernacular names of Low Huckleberry, Sugar Huckleberry, Dwarf Dryland Blueberry, Low Blueberry, Early Sweet Blueberry.

Introduced into cultivation about 1884; recovers from fire damage well. Known to be eaten by gray fox, ruffed grouse, wild turkey, cottontail, and white-tailed deer.

Rabbit-eye Blueberry

Vaccinium virgatum Ait.

FIELD IDENTIFICATION. Shrub 1–3 ft, often found in fairly extensive colonies.

FLOWERS. Perfect, regular, clustered; sepals 5, persistent, broad, deltoid; corolla cylindro-urceolate, pink-tinged, ⅙–¼ in. long, apex with 5 short, erect or recurved lobes; stamens included, twice as many as the

Rabbit-eye Blueberry
Vaccinium virgatum Ait.

sepals; anthers awnless, with apical tubes opening by slits, ovary inferior and 4–5-celled.

FRUIT. Berry-like, black, lustrous, globose, ¼–⅖ in. in diameter, many-seeded, usually poor in flavor.

LEAVES. Simple, alternate, deciduous, spatulate to narrowly elliptic or oblanceolate, apex acute or acuminate, base gradually narrowed or broadly cuneate, margin sharply serrulate, blade length 1¼–2 in., width ⅜–¾ in., upper surface dull green or semilustrous, and lower surface glabrous, pubescent (especially along midvein) and rather conspicuously glandular, old leaves almost glabrous beneath; very short-petioled and pubescent, or blade sessile.

TWIGS. Slender; young ones pubescent or glandular-pubescent, green to reddish brown; older ones reddish brown to gray.

RANGE. Eastern Texas, Louisiana, Arkansas, and Oklahoma; eastward to Georgia and Florida.

RELATED SPECIES. *V. virgatum* var. *ozarkense* Ashe and *V. virgatum* var. *speciosum* Palmer have been suggested to be segregate individuals and populations of *V. virgatum* and *V. arkansanum*. See Camp (1945).

V. virgatum is known to hybridize with *V. australe*, *V. arkansanum*, and *V. myrsinites*.

REMARKS. The genus name, *Vaccinium*, is an ancient name, thought to be from the Latin *vaccinus* ("of cows"). The species name, *virgatum*, refers to the slender twigs. It is also listed under the vernacular name of Medium-cluster Blueberry.

Ground Blueberry

Vaccinium myrsinites Lam.

FIELD IDENTIFICATION. Evergreen, much-branched shrub, 9 in.–3 ft, often in colonies.

FLOWERS. In umbel-like clusters, perfect, regular; corolla cylindro-urceolate, ¼–⅓ in. long, white to deep pink, lobes 5, short, erect or reflexed; sepals 4–5, persistent, broadly ovate to deltoid; stamens 10, included, filaments pubescent, anthers about ⅕ in. with apical tubes; ovary 4–5-celled, inferior, styles united.

FRUIT. Black to subglaucous, globular, about ⅙ in. in diameter, juicy, of only fair quality, many-seeded.

LEAVES. Alternate, persistent, blade obovate to elliptic or spatulate to narrowly elliptic, margin entire or obscurely serrulate, surfaces either shining, dull, or subglaucous, lower surface conspicuously glandular with minute-stalked glands and sparingly pubescent, blade 5/16–⅘ in. long, ¼–⅜ in. wide.

RANGE. In acid soils of pinelands or prairies. Loui-

Ground Blueberry
Vaccinium myrsinites Lam.

siana, eastward to Florida and northward to South Carolina.

HYBRIDS, VARIETIES, SYNONYMS. The following quotation on the races of this species is from Camp (1945, pp. 221–222):

"*V. myrsinites* is an allopolyploidic species . . . derived out of hybrids between *V. tennellum* and *V. darrowi;* because of this there is a certain amount of segregation, the subglaucous 'darrowoid' phase being concentrated along the Florida gulf coast and the non-glaucous 'tenelloid' phase in northeastern Florida and in southern Georgia, where even a few partially deciduous plants are known. There are strong arguments for a pre-Pleistocene origin of this species.

"*V. myrsinites* hybridizes freely with all tetraploids of the area, those with *V. arkansanum* and *V. australe* yielding the most striking segregate and back-cross combinations because of the wide difference in plant characters. In some of these, intermediate characters may be evident, indicating a balance of genes; in others the height factor of *V. myrsinites* may be coupled with the general leaf-types of the others, or the reverse condition is found. The latter group is of particular interest because of its height (as much as 2 meters), its small evergreen leaves borne on gracefully drooping branches and abundant flowers, deep red in bud and becoming pink. *V. myrsinites* also hybridizes freely with *V. fuscatum,* the segregates and back-crosses adding to the difficulties of interpretation of the latter species.

. . . .

"It is strongly suspected that certain plants in parts of southern Georgia and northern Florida are the result of hybridization between *V. myrsinites* and *V. virgatum.* As in the foregoing species, plants grown under pot culture in the greenhouse sometimes have excessively lax branches. Such a plant led to the description of var. *decumbens.* Although the plant on which *V. nitidum* var. *decumbens* was based was said to come from South Carolina, no authentic records are known of such material from this state."

Synonyms for *V. myrsinites* are *V. nitidum* Andr., *V. nitidum* var. *decumbens* Sims, and *Cyanococcus myrsinites* (Lam.) Small.

REMARKS. The genus name, *Vaccinium,* is the Latin name, translated "of cows." The species name, *myrsinites,* was given because of its resemblance to certain species of the genus *Myrsine.* A vernacular name for the plant is Florida Evergreen Blueberry.

SMALL-CLUSTER BLUEBERRY

Vaccinium tenellum Ait.

FIELD IDENTIFICATION. Shrub 6 in.–1 ft, often growing in extensive colonies from horizontal underground stems. The slender ascending branches are copiously pubescent.

FLOWERS. Borne in dense, fasciculate racemes, perfect, regular; corolla narrowly urceolate or slender subcylindric, apex with 5 short lobes, length ⅙–¼ in., white to pink or red; stamens 10, included, filaments hairy, anthers 2-celled with apical pores; ovary 4-5-celled, style slender; sepals ovate to deltoid.

FRUIT. Maturing June–July, globose, black, shiny, sometimes with a bloom, ⅕–⅓ in. in diameter, many-seeded, usually of poor flavor and texture.

LEAVES. Alternate, deciduous, blades ¾–1¼ in. long, ¼–½ in. wide, spatulate to narrowly elliptic, apex acute, base tapering, margin sharply serrate, green, lower surface conspicuously glandular with minute, reddish, glandular hairs, pubescent along the midrib or glabrous above.

RANGE. Sandy pinelands, or in open forests or meadows, particularly common in areas formerly burned over, or on margins of swamps. Louisiana and Mississippi; eastward to northern Florida and Georgia, and northward to Virginia.

HYBRIDS, FORMS, SYNONYMS. *V. tenellum* hybridizes

SMALL-CLUSTER BLUEBERRY
Vaccinium tenellum Ait.

DARROW EVERGREEN BLUEBERRY
Vaccinium darrowi Camp

with *V. darrowi, V. elliottii, V. vacillans, V. caesariense,* and *V atrococcum.* Forms with narrowly elliptic or glabrous leaves are evidently hybrid strains, particularly with *V. darrowi.*

Synonyms are *V. galezans* Michx., *V. virgatum* var. *tenellum* Gray, and *Cyanococcus tenellus* Small.

REMARKS. The genus name, *Vaccinium,* is from an old Latin name meaning "of cows." The species name, *tenellum,* alludes to its delicate or tender appearance when compared with various other species. Also known under the vernacular name of Low Huckleberry.

DARROW EVERGREEN BLUEBERRY

Vaccinium darrowi Camp

FIELD IDENTIFICATION. Low shrub 8–18 in., urceolate corolla, blue fruit, and very small entire leaves. This account follows Camp (1945).

FLOWERS. In small axillary clusters, pedicels about 1/4 in. or less, glabrous; pedicel scales less than 1/8 in., glabrous or nearly so, ovate to lanceolate, concave; corolla urceolate, 1/5–1/4 in. long, pink to red; calyx about 1/12 in. long, reddish brown, glabrous or nearly so; sepals 5, broadly ovate to deltoid, less than half the length of the calyx body.

FRUIT. Blue, subglobose, 1/6–1/4 in. in diameter, of only fair flavor.

LEAVES. Evergreen, coriaceous, blade length 3/8–5/8 in., narrowly elliptic to spatulate, margin entire, surface glabrous above; lower surface glabrous or occasionally puberulent, especially along the midvein, nonglandular, usually very glaucous.

STEMS. Plant often in extensive colonies, stems solitary below, divided above into numerous, slender, delicate twigs; when young brown and pubescent, when older dark brown and glabrous.

RANGE. East Texas and eastward through Louisiana to Florida. Type locality New Orleans, Louisiana. Introduced into cultivation at Beltsville, Maryland, in 1940.

HYBRIDS. Very complex species, seldom genetically pure and therefore difficult to treat as a species. Known to hybridize with numerous other species. The following hybrids are mentioned by W. H. Camp:

V. darrowi × *V. elliottii* produces a hybrid with leaves 3/4–1 1/4 in. long, serrate to subserrate, glabrous to slightly puberulous, thin; plant 1 1/2–3 ft.

V. darrowi × *V. atrococcum* produces a hybrid with leaves entire, pubescent, usually very coriaceous.

V. myrsinites Lam. is considered to be an allopolyploid species derived from hybrids of the cross *V. darrowi* and *V. tenellum.* W. H. Camp points out that "these separate into two forms, the subglaucous 'darrowoid' phase being concentrated along the Florida Gulf, and the non-glaucous 'tenelloid' phase in northeastern Florida and southern Georgia, where even a few deciduous plants are known."

The form of *V. tenellum,* with narrowly elliptic and/or glabrous leaves—now fairly common, especially through the southernmost part of the range—seems to be the result of the flow of residual nonselective genes through the population following ancient hybridization, particularly with *V. darrowi.*

V. darrowi is also suspected of hybridizing with *V. caesariense* in north Florida.

REMARKS. The genus name, *Vaccinium,* is the ancient Latin name, and the species name, *darrowi,* honors George McMillan Darrow (b. 1889), who first brought the plant into cultivation. It is the only diploid evergreen blueberry in North America.

ELLIOTT BLUEBERRY

Vaccinium elliottii Chapm.

FIELD IDENTIFICATION. Shrub 3–12 ft, sometimes in colonies. Branches numerous, slender, horizontal or ascending, often forming a crown.

FLOWERS. Usually appearing before the leaves expand, more or less fascicled, corolla cylindric to narrowly urceolate, 1/4–1/3 in. (about twice as long as thick), apex with 5 short teeth, various shades of pink; stamens 10, included, 1/6–1/5 in., filaments pubescent, anthers 2-celled, with apical pores; ovary 4–5-celled, style slender; calyx with 5 deltoid sepals.

ELLIOTT BLUEBERRY
Vaccinium elliottii Chapm.

FRUIT. Black or dark bluish, dull to lustrous or occasionally glaucous, subglobose, ⅕–⅓ in. in diameter, sometimes even greater, flavor fair to poor, seeds many.

LEAVES. Deciduous, alternate, blade ⅝–1¼ in. long, ¼–⅝ in. wide, usually broadly elliptic, sometimes ovate to oval, apex acute or obtuse, base narrowed or rounded, margin entire or finely serrulate, rather thin; upper surface green, usually glabrous, but sometimes with the midrib pubescent, rather lustrous; lower surface finely pubescent to glabrous, nonglandular.

RANGE. In river-bottom lands subject to flooding, along ravines, open flatwoods, low pinelands, and thickets. Eastern Texas, Louisiana, and Arkansas; eastward to Florida and northward to Virginia.

HYBRIDS, FORMS, SYNONYMS. This complex species has been thoroughly investigated by W. H. Camp, from whom we quote the following (1945, pp. 229–230):

"The hybrids with *V. tenellum* are usually common and the segregates often migrate into areas where the parents are unable to persist; such plants have been described as *Cyanococcus cuthbertii* (*V. cuthbertii* [Small] Uphoff). In parts of Florida and Mississippi, *V. elliottii* × *V. darrowi* hybrids are known to be locally common. The segregates and back-crosses of this combination run the whole gamut of expected variation one extreme being a plant which looks like a coarse form of *V. darrowi*, the other extreme being an *elliottii* like plant, but with leaves which persist through the winter. While the leaves of *V. elliottii* appear to be basically thin-textured, green and shining, their surfaces non-glandular, the margins serrate, and while the plants bear black fruit, aberrancies may be noted. These sometimes may be traced to hybridization and gene-exchange of recent origin; or they may be residual, having persisted and having carried through the population of *V. elliottii* by means of gene-flow following former and possibly ancient hybrid combinations with *V. atrococcum, V. caesariense, V. tenellum, V. darrowi* and *V. vacillans.*

"The source of the more notable aberrancies seem to be the following: A fine puberulence over the whole leaf-surface—*V. atrococcum;* stiff pubescence limited to the midvein—*V. tenellum;* subcoriaceous and/or subpersistent leaves—*V. darrowi;* subglaucous leaves and fruit—*V. darrowi, V. caesariense* and *V. vacillans;* narrowly elliptic leaves, below average in length—*V. darrowi;* narrowly elliptic leaves, above average in length—*V. atrococcum* and *V. caesariense;* leaves bearing some gland-hairs—*V. tenellum;* and leaf margins subserrate to entire—all except *V. tenellum.* It will be apparent that this is only a partial list of the aberrant characters found in *V. elliottii* for such things as height of plant and type of colony, character of the twigs and branches, and shape and size of flowers etc., have been omitted, all of which would be of value in a complete analysis and assist materially in placing unusual individuals or local communities.

"A complete outline of such aberrancies has not been presented for the reason that it would occupy too much space in an already overlong paper; in the treatment of this species, as in the others, no more can be done than give brief mention of a few of the outstanding forms which are to be found following inter-specific gene exchanges.

"In addition, *V. elliottii,* at times seems to have been the vehicle for the transmission of genes between two other species. For example, in western Florida, where much of *V. elliottii* is puberulent (or even pubescent), and where *V. darrowi* and *V. atrococcum* appear to be disjunct, the puberulence of some of the material of *V. darrowi* seems to have been derived by means of hybridization with *V. elliottii* which, in turn, had received its puberulence by gene exchange with the pubescent *V. atrococcum.* Yet, in spite of all the aberrancies noted, throughout much of its range *V. elliottii* remains one of our better marked and more easily distinguished species."

V. elliottii is sometimes listed by authors under the name of *Cyanococcus elliottii* Small.

REMARKS. The genus name, *Vaccinium,* is from the Latin and is translated "of cows." The species name, *elliottii,* is in honor of Stephen Elliott (1771–1830), an early student of the flora of South Carolina. A vernacular name is Mayberry.

Arkansas High-bush Blueberry

Vaccinium arkansanum Ashe

FIELD IDENTIFICATION. Erect plant 3–12 ft, usually with several stems at base and crown-forming.

FLOWERS. Corolla varying from greenish to white, pink, or striped; cylindro-urceolate, 5-parted at apex, length ¼–⅓ in.; hypanthium green; ovary 4–5-celled; anthers not awned.

FRUIT. Diameter ⅓–⅖ in., globose, dull black to slightly glaucous, of poor quality, many-seeded.

LEAVES. Deciduous, blade elliptic to ovate, apex acute, margin entire, blades 1¼–1⅝ in. long, ¼–⅓ in. wide, rather dark green, glabrous or nearly so above, lower surface pubescent but not glandular.

RANGE. Margins of swamps or streams in sandy soil. Arkansas and northeastern Texas, westward to Florida. Type locality about 6 miles north of Hot Springs, Arkansas, on the Little Rock road.

HYBRIDS. The plant is thought to be a tetraploid, and hybridizes with other tetraploids such as *V. myrsinites* Lam., *V. virgatum* Ait., and *V. australe* Small. See discussion by Camp (1945, pp. 256–257).

REMARKS. The genus name, *Vaccinium*, is from the classical Latin name, *vaccinus*, meaning "of cows." The species name, *arkansanum*, refers to the state of Arkansas where it grows. The plant is also known by the name of Black High-bush Blueberry. Some plants reproduce by underground stolons.

Mountain-laurel

Kalmia latifolia L.

FIELD IDENTIFICATION. Thicket-forming, evergreen shrub or small tree to 36 ft, 18–20 in. in diameter. Trunk often short and gnarled. Branches often distorted to form a round-topped or asymmetrical, dense head.

FLOWERS. March–July in dense, terminal or lateral, many-flowered, showy, compound corymbs 3–6 in. across; pedicels green or reddish, ⅜–1½ in. long, erect, slender, clammy, and viscid-pubescent, with bracts and 2-bracteolate at base; flowers regular, perfect; calyx-lobes 5, elliptic to ovate, acute, pubescent or almost glabrous, viscid, green, thin, about 1/12 in. long; corolla saucer-shaped, tube short, limb 5-lobed, ⅘–1 in. wide, white or pink, often marked purple within; exterior viscid-pubescent, interior of corolla with 10 sacs, anthers held back in these sacs and elastically straightened to discharge the pollen by mechanical means (often by insect disturbance); stamens 10, filaments filiform, shorter than the corolla, ⅖–½ in. long, anthers short, attached near the top; opening by an apical oblong pore; ovary 5-celled, superior, glandular; style slender, exserted, persistent; stigma capitate; disk 10-lobed.

Mountain-laurel
Kalmia latifolia L.

FRUIT. Mature in September, persistent, capsule somewhat woody, ⅕–¼ in. long, globular, depressed at apex, glandular-hairy, dehiscent into 5 septicidal, crustaceous valves; style and calyx persistent; seeds numerous, minute, brown, oblong, terminally winged, disseminated by the valves splitting down the middle.

LEAVES. Persistent, simple, alternate or irregularly whorled, elliptic to oblong or oval, margin entire, ends acute or acuminate, blade length 2–5 in., width ½–¾ in., upper surface dark green and glabrous, lower surface paler, glabrous or pubescent, veins obscure except main vein; thick, coriaceous and rigid; petioles pubescent, ⅜–¾ in.

TWIGS. Green or reddish when young, glandular-hairy; older ones glabrous, terete, gray to reddish brown, often crooked and distorted, exfoliating in narrow thin flakes.

BARK. Reddish brown to dark gray, narrowly and deeply fissured longitudinally, about 1/16 in. thick.

WOOD. Reddish brown, hard, sapwood lighter color, heavy, strong, brittle, fine-grained, weighing 44 lb per cu ft.

RANGE. Dry rocky woods in sandy acid soil, or in low moist grounds of swamp margins. Eastern Louisiana (Washington Parish) northward to Ontario.

MEDICINAL USES. The leaves have been used medicinally, in internal treatment of diarrhea and syphilis, and in external treatment of skin diseases.

PROPAGATION. The capsules are picked by hand and flailed in a sack. The seed may be sown in boxes in early spring under glass. Germination may be speeded by exposing sown seeds to winter temperature in a cold frame for 2–3 months. Varieties are increased by side grafting in seedlings in the greenhouse, or by layers. It grows less readily from cuttings. The plants do best in acid, sandy or peaty loams. A mulch helps protect the roots from drying out in summer, or from frost in winter. The plant can be rather easily transplanted if soil conditions are right; can be readily forced in a greenhouse and makes a handsome pot plant. It is relatively free of insects and disease, and is often cultivated for ornament in the United States and in Europe.

VARIETIES. A number of varieties have been listed by Rehder (1940, pp. 726–727), as follows:

Smooth Mountain-laurel, *K. latifolia* var. *laevipes* Vern., has pedicels glabrous or nearly so.

Myrtle-leaf Mountain-laurel, *K. latifolia* var. *myrtifolia* Bosse, has small leaves, blades 1–2 in. long, deep green, plant slow-growing, forming a low dense bush.

Banded Mountain-laurel, *K. latifolia* var. *fuscata* Rehd., has interior of corolla with a broad, dark purplish brown band.

Obtuse Mountain-laurel, *K. latifolia* var. *obtusata* Rehd., has buds rounded at apex, leaves elliptic to oval, obtuse at ends, the shrub slow-growing and compact in habit.

Feathery Mountain-laurel, *K. latifolia* var. *polypetala* Nichols, has a corolla divided into 5 narrow petals which give a feathery appearance.

Pink Mountain-laurel, *K. latifolia* var. *rubra* Sweet, has deep pink flowers.

White Mountain-laurel, *K. latifolia* var. *alba* Bosse, has white flowers.

REMARKS. The genus name, *Kalmia,* honors Peter Kalm, a Swedish botanist, who collected, 1748–1751, in North America. The species name, *latifolia,* means "broad-leaved." Vernacular names are Big-leaf Ivy, Ivy-leaf-laurel, Spoonhunt, Spoonwood, American-laurel, Wood-laurel, Small-laurel, Poison-laurel, Broad-leaf Kalmia, Calico-bush, Clamoun, and Southern Mountain-laurel. The leaves and tops contain the poisonous glucosides arbutin and andromedotoxin. The leaves are poisonous to sheep or cattle; seem to be eaten by white-tailed deer in small quantities with impunity, but are toxic in large amounts. Seeds have been found in the stomach of ruffed grouse. Records also exist concerning human fatalities from Kalmia honey. The wood is used for making small woodenware articles, pipes, handles, and fuel.

TEXAS AZALEA

Rhododendron oblongifolium (Small) Millais

FIELD IDENTIFICATION. Shrub 1–5 ft with spreading stems.

FLOWERS. Flower buds fully formed in March after the leaves appear, ⅓–½ in. long, elongate-tapering; scales imbricate, oval to oblong, pubescent, apex ciliate, rounded and abruptly apiculate, reddish; pedicels ⅛–¼ in. long, glandular-hairy; corolla trumpet-shaped, white, viscid with glandular hairs, 1¼–1½ in. long, rather abruptly dilated into a 5-lobed limb; lobes reflexed, each ½–¾ in. long, lanceolate to oblong, apex acute or acuminate; stamens 5, exserted, filaments filiform, glabrous above, hairy below; anthers about ⅛ in. long with terminal pores; pistil considerably longer than stamens, filiform, stigma capitate; calyx ⅛–¼ in. long, 5-lobed, lobes ovate, obtuse, glandular-hairy.

FRUIT. Capsule persistent, elongate, about ¾ in. long, septicidally dehiscent into 5 long-acuminate, persistent, revolute, concave valves, thickly covered on the back with long, stiff, glandular hairs; peduncles ½–1 in. long, thickly glandular-hairy.

LEAVES. Alternate, deciduous, disposed in terminal radiate clusters, blades 1–2 in. long, ⅓–1 in. wide, oblong, elliptic or obovate, usually widest above the middle; margin entire, slightly revolute, ciliate with short white hairs; apex acute, base gradually narrowed; upper surface dull green and glabrous or somewhat hairy, lower surface paler and hairy along the veins, tending to be more glabrous with age; petioles ⅛–³⁄₁₆ in. long or almost absent, white-hairy.

STEMS. Gray to reddish brown, smooth, tight, usually erect and single-stemmed below, radiate-forking above.

TWIGS. Reddish brown, slender, rather smooth, divaricate, spreading, bearing terminal radiate clusters of leaves.

RANGE. In moist, sandy, acid soil; nowhere abundant. Texas, Arkansas, and perhaps adjacent Louisiana. Specimens collected by the author in San Augustine County Texas, in the Boykin Springs recreational area. Also said to be in Tom Green County.

TEXAS AZALEA
Rhododendron oblongifolium (Small) Millais

Piedmont Azalea
Rhododendron canescens (Michx.) G. Don

REMARKS. The genus name, *Rhododendron,* is an ancient name meaning "rose-tree," and the species name, *oblongifolium,* refers to the leaf shape. Also known under the vernacular names of White Azalea, Bush Azalea, Thicket Azalea, and White-bush Honeysuckle. Texas Azalea has white flowers and blooms after the leaves are mature, while the Piedmont Azalea has pink flowers and blooms before the leaves appear, or as they appear.

Piedmont Azalea

Rhododendron canescens (Michx.) G. Don

FIELD IDENTIFICATION. Shrub to 8 ft with a loose, open shape. Leaves tending to be clustered terminally in a rotate manner.

FLOWERS. March–April, pink, slightly fragrant, conspicuous, borne in terminal, whorl-like clusters before the leaves appear, or with the young leaves; pedicels ⅓–½ in. long, green or pink, finely glandular-hairy; calyx-lobes ovate, deltoid or rounded, ciliate, about 1/25 in. long; corolla pink (more rarely white), long-tubular, tube about 1 in. long, abruptly flaring into a 5-lobed limb ¾–1 in. across; lobes shorter than tube, pink or white, faintly grooved, ovate-oblong, apex acute or obtuse, reflexed, margin somewhat revolute; stamens 5, long-exserted, 1–1¼ in. long, white or pink, hairy at base, anthers about 1/16 in. long with terminal pores; pistil equal to or exceeding stamens in length; stigma capitate, about 1/16 in. long.

FRUIT. Capsule ellipsoid-cylindric, narrowed above, opening at apex, canescent-hairy, about ¾ in. long.

LEAVES. Alternate, deciduous, tending to be clustered terminally in a somewhat rotate manner, 1½–4 in. long, ¾–1¼ in. wide, oblong to oblanceolate or elliptic; apex acute, glandular-apiculate; base cuneate; margin set with small glandular teeth; firm and thickish; upper surface dark green, puberulent and with scattered, slender, glandular hairs; lower surface canescent with white hairs, especially along the midrib, and with scattered, glandular hairs.

STEMS. Spreading, slender, brown to mottled gray, rather smooth, tight.

TWIGS. Young twigs of the year brown, densely pubescent with tiny, stalked glands; smoother and glabrous when older.

RANGE. In sandy, acid pinelands from East Texas and Arkansas; eastward to Florida and north to Tennessee and North Carolina. In Texas collected near Saratoga, Kuntze, Jasper, Silsbee, Beaumont, Zavala, Woodville, and Livingston.

VARIATIONS. Piedmont Azalea has two variations and may itself be a variation.

White Piedmont Azalea, *R. canescens* var. *candidum* (Small) Rehd., has white flowers.

Smooth Piedmont Azalea, *R. canescens* forma *subglabrum* Rehd., has leaves glabrous beneath, except on margin and midrib.

Piedmont Azalea is also considered to be very closely related to the more northern Pinxterbloom Azalea, *R. nudiflorum* L. In fact, some botanists consider the former to be only a more pubescent southern variation of the northern plant.

REMARKS. The genus name, *Rhododendron,* is an ancient name, meaning "rose-tree," and the species name, *canescens,* refers to the plant hairs. Vernacular names are Honeysuckle Azalea, Wild Honeysuckle, Pinxter Flower, Early Azalea, and Hoary Azalea. At times the Piedmont Azalea bears a large, gall-like, green, translucent structure on the twigs. This body is edible and was used by early settlers for pickling. The structure is probably caused by bacterial action on the twig.

Hammock-sweet Azalea

Rhododendron serrulatum (Small) Millais

FIELD IDENTIFICATION. The southernmost American Azalea. Erect with irregular branching stems and growing to a height of 20 ft.

FLOWERS. Bud scales numerous and marked with a

dark band, maturing June–August after the leaves; borne in few or many terminal clusters; pedicels rather slender, ⅜–1 in. long, glandular-hirsute; calyx-lobes ovate to elliptic, long-ciliate, obtuse; corolla white, sweet-scented, tube slender, cylindric, slightly enlarged near base, expanded at the limb, 1¼–1¾ in. long, very viscid-glandular hairy; limb with 5 lobes, spreading, unequal, narrow, shorter than tube; stamens 5, conspicuously exserted, declined; anthers opening by terminal pores; style long-exserted, glabrous or slightly puberulent; ovary 5-celled, ovules numerous in each cavity.

FRUIT. Capsule elongate, slenderly ovoid to ellipsoid, ⅜–⅝ in. long, glandular-setose, 5-valved.

LEAVES. Clustered at ends of branches, simple, alternate, deciduous, thin, narrowly obovate to elliptic or oblanceolate, apex acute to obtuse, mucronate, base cuneate or gradually narrowed, margin serrulate-ciliate, blade length 1½–3½ in., width ⅝–1½ in., lustrous, green on both sides, finely reticulate, glabrous except for occasional scattered hairs on the midrib beneath, short-petioled.

TWIGS. When young reddish brown, pubescent, and with strigose hairs; later brown to gray and glabrous; winter buds with numerous glabrous scales (15–20 or more), ciliate, dark-banded, the inner especially with apices aristate-mucronate.

RANGE. Usually in sandy swamps, from southeastern Louisiana eastward into Lake County, Florida, and northward into central Georgia.

REMARKS. The genus name, *Rhododendron*, is from the Greek and means "rose-tree." The species name, *serrulatum* refers to the serrulate margin. At one time the plant was classified as *Azalea viscosa*, but was reported as a new species by John K. Small and later placed in the genus *Rhododendron* by Millais. It differs from *A. viscosa* by its strongly hairy, reddish brown twigs and banded buds.

HAMMOCK-SWEET AZALEA
Rhododendron serrulatum (Small) Millais

LEADWORT FAMILY (*Plumbaginaceae*)

CLIMBING PLUMBAGO
Plumbago scandens L.

FIELD IDENTIFICATION. Plant usually herbaceous, with semierect or scandent stems 1½–4½ ft long. Occasionally subwoody near the base.

FLOWERS. In spikelike racemes ¾–6 in. long, slender, many-flowered, terminal; individual flowers perfect and regular; corolla white to bluish, salverform; tube slender, limb 4–5-lobed; lobes spreading, ovate to oblong, obtuse to retuse; stamens 4–5, exserted, distinct, filaments dilated at base, anthers oblong-linear, blue; style single, divided above into 5 filiform branches, stigmatic within, ovary 1-celled; calyx tubular, tube about ⅜ in. long, 5-ribbed, covered with stalked, viscid hairs, apex 4–5-lobed, lobes erect.

FRUIT. Capsule membranous, adherent to the hardened calyx, about ⅓ in. long, linear, angled, beaked, seed solitary.

LEAVES. Simple, alternate, oblong or elliptic-lanceolate, apex acute or acuminate, base acute and sometimes clasping, margin entire or somewhat undulate, length ¾–4 in., surface glabrous and lustrous; petiole short, often with smaller leaves at base of larger ones.

STEMS. Erect, scandent or prostrate, herbaceous above, woody below, much branched, somewhat grooved, swollen at the nodes.

RANGE. Southwest Texas, Arizona, New Mexico, Florida, West Indies, and Central and South America. Collected by the author on banks of Rio Grande River in Cameron County, Texas.

MEDICINAL USES. Certain parts of the plant produce blisters when rubbed on the skin. Beggars are said to apply the leaves to raise sores upon the body for the purpose of exciting pity. The leaves and root when taken internally are poisonous, and a decoction was once used as an emetic. Various drugs have been prepared from it to treat skin diseases and toothache.

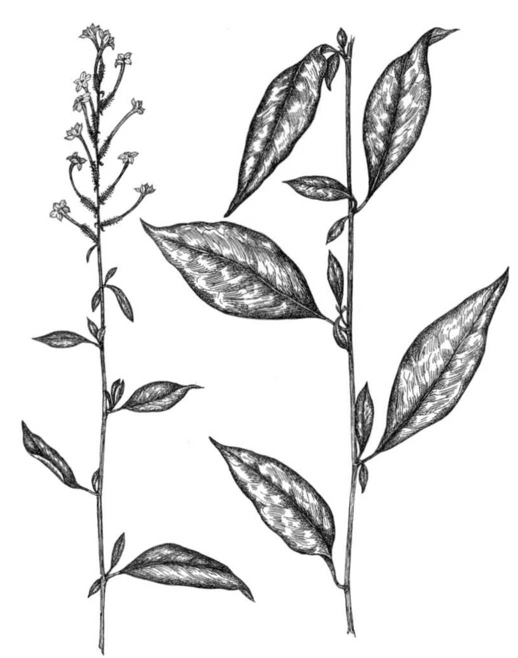

CLIMBING PLUMBAGO
Plumbago scandens L.

REMARKS. The genus name, *Plumbago*, may have been derived from several sources. The European species, *P. europaea* L., was once reported as being used as a remedy for an eye disease called "plumbum," or the name may have been given because of the lead color of the flowers. An English derivation of the Latin *plumbago* is "leadwort." The species name, *scandens*, refers to the plant's climbing habit.

About 12 species of *Plumbago* are known, mostly distributed in warm regions. A number are cultivated for ornament. The common plumbago of the gardens is *P. capensis* Thunb., of South Africa. *Plumbago* is well represented in the Mediterranean countries and South Africa. Many vernacular names have been applied in this country, and in Central and South America, to the plant as follows: Leadwort, Clammy Plumbago, Dentelaria, Hierba del Diablo, Hierba del Negro, Hierba de Alacrán, Lagán de Perro, Canutillo, Chapak, Pilillo, Turicua, Tlepatli, Tlalchichinolli, Embeles, Jazmín Azul, Beleza, Veleza Enredadera, Meladillo, Higuillo, Centella, Hierba del Monte, Pegajoso, and Guachochile.

SAPODILLA FAMILY (*Sapotaceae*)

WOOLLYBUCKET BUMELIA
Bumelia lanuginosa (Michx.) Pers.

FIELD IDENTIFICATION. Shrub or an irregularly shaped tree to 60 ft, with stiff, spinose branchlets.

FLOWERS. June–July, in small fascicles ¼–1½ in. across, pedicels hairy or subglabrous, 1/12–3/5 in. long; corolla white, petals 5, each 3-lobed, middle lobe longest, fragrant, ⅛–⅕ in. long, tube about 1/12 in. long; stamens 5, normal and fertile, also 5 sterile stamens (staminodia) which are deltoid-ovate, petaloid, and nearly equaling the corolla-tube; ovary 5-celled, hairy, style 1; calyx 5-lobed, hairy or nearly glabrous, 1/12–⅛ in. long, lobes suborbicular or ovate.

FRUIT. Berry September–October, borne on slender, drooping peduncles, subglobose or obovoid, ⅓–1 in. long, lustrous, black, fleshy; seed solitary, large, brown, rounded, scar small and nearly basal, ¼–½ in. long, no endosperm, cotyledons fleshy.

LEAVES. Alternate or clustered, often on short lateral spurs, oblong-obovate, elliptic or wedge-shaped, apex rounded or obtuse, base cuneate, margin entire, blade length 1–3 in., width ½–1 in., leathery, shiny green and smooth above, varying from rusty to white or gray-woolly beneath; petioles short, averaging about ½ in. long, tomentose.

TWIGS. Gray to reddish brown, zigzag, slender, stiff, spinose, hairy at first with gray, white, or rusty tomentum.

BARK. Dark brown or grayish, fissured and reticulate into narrow ridges with thickened scales.

WOOD. Yellow or brown, fairly hard, close-grained, weighing about 40 lb per cu ft.

RANGE. The species occurs in east Texas, Oklahoma, Arkansas, Louisiana, eastward to Florida, north to Kansas, Missouri, Illinois, and Virginia. In central and west Texas represented by its varieties.

WOOLLYBUCKET BUMELIA
Bumelia lanuginosa (Michx.) Pers.

PROPAGATION. Plant may be grown from seeds or cuttings. Fruit is generally hand-picked when ripe macerated in water, and the fleshy parts floated off The germinative capacity is about 50 per cent. Soaking the seed in sulphuric acid for 2 hours and washing in water before planting hastens germination. Seed may also be stratified in moist sand or peat in fall for spring sowing.

VARIETIES AND CLOSELY RELATED SPECIES. It is interesting to note that some botanical authors have split

3. *lanuginosa* into a number of varieties and forms ac-ording to color, density of hairs on the foliage, and lower parts. Robert Brown Clark (1942) considers hat the most typical representatives of the species *anuginosa* occur mostly east of the Mississippi River nd have tawny or rufous hairs. Also, that the same pecies occurring west of the Mississippi River, and aving gray or nearly white hairs, should be classified s the Oblong-leaf Woollybucket Bumelia, *B. lanuginosa* ar. *oblongifolia* (Nutt.) Clark.

A variety with very long, silvery white hairs, which occurs from Oklahoma and east Texas to Nuevo León, Mexico, usually near the coast, is classified as Gum Voollybucket Bumelia, *B. lanuginosa* var. *albicans* Sarg.

A variety from central and west Texas, New Mexico, Arizona, and Mexico, is classified as the Rigid Woolly-bucket Bumelia, *B. lanuginosa* var. *rigida* Gray. It has arrower leaves ⅜–2 in. long, hairs gray or nearly white, occasionally tawny at first, the older leaves be-coming less hairy, but the pedicels and sepals remaining airy. The branches are quite rigid and spinescent, and he plant is usually less than 15 ft tall. When this ariety becomes more glabrous on the leaves and flower arts it is classified by some authorities as Brazos Wool-ybucket Bumelia, *B. lanuginosa* var. *texana* Buckl., or y other authorities as a species under the name of *B. nonticola* Buckl. Since the latter variant is more com-non in central and west Texas, southwestern Okla-ioma, and adjacent Mexico, than the Rigid Woolly-bucket Bumelia, and seems to be more constant in its haracters, it is described as a species by the author vith the knowledge that it does have intergrades in ertain localities. For a review of this problem see Cronquist (1945).

REMARKS. The genus name, *Bumelia,* is the ancient Greek name for the European Ash, and the species name, *lanuginosa,* refers to the woolly hairs of the leaf. Vernacular names are Woolly-buckthorn, Woolly Bumelia, Gum Elastic, Gum Bumelia, Chittamwood, False-buckthorn, and Blackhaw. The black fruit is edi-ble, but not tasty, and produces stomach disturbances nd dizziness if eaten in quantity (at least this is the xperience of the author). Birds are very fond of the ruit; in fact, they get it as soon as it is barely ripe. The wood is used in small quantities for tool handles nd cabinetmaking. A gum is freely exuded from vounds on the trunk and branches. The tree has been n cultivation since 1806.

BUCKTHORN BUMELIA

Bumelia lycioides (L.) Gaertn.

FIELD IDENTIFICATION. Large shrub or small tree at-taining a height of 25 ft, with spreading branches usually armed with stout spines.

FLOWERS. In dense, many-flowered, axillary fascicles; pedicels glabrous, about ½ in. long; corolla white, about ⅕ in. wide, 5-lobed, lobes each 3-parted (or toothed), lanceolate; staminodia ovate, petal-like, denticulate, about one half as wide as the corolla; stamens 5, anthers sagittate; calyx 5-lobed, lobes oval or ovate; ovary 5-celled, ovoid, style thick and short.

FRUIT. Drupe oval-oblong, black, ¼–⅔ in. long, thin-skinned, pulpy, bittersweet; stone solitary, large, smooth, ovoid or obovoid, abruptly pointed at the apex, hilum large.

LEAVES. Usually clustered on short lateral spurs, ex-cept on vigorous growth, simple, alternate, deciduous, blades 2–6 in. long, ½–2 in. wide, entire on margin; acute or acuminate at apex, or rarely obtuse; base gradually tapering; bright green and glabrous above; paler, glabrous, or pubescent, and reticulate-veined be-neath; petiole slender, ⅓–1 in. long, pubescent at first but glabrous later.

TWIGS. Rather stout, thick, with lateral spurlike branchlets, unarmed, or armed with stout spines, pubes-cent at first but later glabrous, shiny, reddish brown to gray.

BARK. Smooth, thin, reddish brown to gray, scales small and thin.

WOOD. Brown to yellow, close-grained, heavy, hard, not strong, weighing 46 lbs per cu ft.

BUCKTHORN BUMELIA
Bumelia lycioides (L.) Gaertn.

RANGE. Eastern Texas and Arkansas, eastward to Florida and north to Kentucky, Indiana, and Illinois. Seemingly rather rare in east Texas.

REMARKS. The genus name, *Bumelia,* is the ancient name for a kind of ash-tree, and the species name, *lycioides,* applies to the *Lycium*-like fruit, which in turn is named for the country of Lycia. Vernacular names are Buckthorn, Mockorange, Ironwood, Chittamwood, and Gum Elastic. The leaves are browsed occasionally by livestock, and the fruit eaten by a number of species of birds. Various authors state that the fruit is edible, but the author has found it so bitter as to be unpalatable.

Brazos Bumelia

Bumelia monticola Buckl.

FIELD IDENTIFICATION. Spiniferous shrub, or small tree to 25 ft, with an irregular crown.

FLOWERS. May–June, pedicels 1/4–1/2 in. long, hairy at first, glabrous later, flowers in fascicles; calyx green, hairy, lobes ovate, margin ciliate, shorter than corolla-lobes; corolla short-campanulate, white; lobes 5, broad-ovate, rounded at apex, with a lanceolate appendage on each side at the base; stamens 5, filaments filiform, anthers sagittate; staminodia 5, petaloid, apex obtuse or rounded, margin erose or entire, longer than corolla-lobes; style elongate, simple, stigmatic at apex.

Brazos Bumelia
Bumelia monticola Buckl.

FRUIT. August–September, borne in fascicles on spurs 1/4–1/3 in. long, black, oblong, obovoid or subglobose, apex rounded or apiculate; seed hardly over 3/8 in. long, oblong, smooth, obtuse or rounded at apices, straight or somewhat asymmetrical, light brown to white.

LEAVES. Deciduous, fascicled on short spurs, blades 1–3 in. long, 1/3–1 1/4 in. wide, pubescent when young, glabrous at maturity, dark green, lustrous, slightly paler beneath, elliptic to oblong or spatulate, margin entire and barely revolute, apex obtuse, rounded or acute, base gradually narrowed, reticulate-veined; petiole 1/4–1/2 in. long, slightly pubescent but glabrous later.

TWIGS. Young twigs zigzag, reddish brown, smooth, the laterals often ending in stout, gray to reddish brown thorns; older twigs gray and smooth.

BARK. Gray, reddish brown beneath, broken into flat, narrow scales and shallow fissures.

WOOD. Brown to yellowish, sapwood lighter, hard, moderately strong.

RANGE. In Texas mostly west of the Brazos River on the Edwards Plateau limestone hills. In Kerr, Kendall, Comal, Real, Uvalde, Palo Verde, Val Verde, Pecos, Brewster, Crockett, Callahan, Coleman, Shackleford, Terrell, and Brown counties. Also in Oklahoma. In Mexico in Coahuila.

REMARKS. The genus name, *Bumelia,* is the ancient classical name for the ash-tree. The species name, *monticola,* refers to its habitat of hilly or mountainous regions. Vernacular names used are Gum-elastic, Chittamwood, Mountain-gum, and Gum-buckthorn.

Some botanists have listed Brazos Bumelia under the names of *B. texana* Buckl., *B. lanuginosa* var. *texana* Buckl., or *B. riograndis* Lundell.

Saffron-plum Bumelia

Bumelia angustifolia Nutt.

FIELD IDENTIFICATION. Spiny, divaricately branched shrub, or rarely a tree, to 27 ft, 8 in. in diameter.

FLOWERS. October–November, usually in small fascicles of 2–10 flowers on lateral, spiniferous twigs; pedicels 1/4–1/2 in. long, glabrous or slightly hairy; individual flowers fragrant, petals 5, 1/6–1/8 in. long, greenish white, oblong to ovate or suborbicular, concave; base bearing lateral, small, narrowly lanceolate lobes; also a set of 5 staminodia which are petal-like, elliptic to lanceolate or round-ovate, erect, erose-dentate on the margin, 1/12–1/8 in. long; stamens 5, anthers sagittate; ovary narrow-ovoid, hairy beneath but glabrous above, tyle 1/12–1/6 in. long, elongate; calyx glabrous and granlar, 1/25–1/16 in. long, sepals in 2 series, inner series sually 3, broadly rounded, outer series 2–3, smaller, ounded to obtuse.

FRUIT. Maturing April–June on the Texas coast, borne n peduncles 1/4–1/2 in. long, glabrous or nearly so; body f fruit black, 1/4–3/4 in. long, about 1/4 in. in diameter, blong or ellipsoid-cylindric, sometimes slightly wider istally, sometimes asymmetrical; apex of fruit truncate o rounded, with or without an apiculate style remnant; esh thick, tough, mucilaginous, sweet; seed 3/16–1/2 in. ong, elliptic to oblong, apices obtuse, acute, or roundd, somewhat flattened, seed coat white and translucent.

LEAVES. Persistent, alternate, or fascicled at nodes, ariable in size, mostly less than 1 in. long, but blades n vigorous shoots sometimes to 2 1/2 in. long, 1/4–1/2 in. road, oval to obovate or spatulate; margin entire; apex ounded, obtuse, truncate, or slightly notched; base radually attenuate, cuneate, or rarely rounded; firm nd leathery, finely and obscurely reticulate-veined, uper surface glabrous, dark green, and glossy; lower urface essentially glabrous or slightly puberulent when oung; petioles 1/16–3/16 in. long, glabrous or pulverulent, tout.

TWIGS. Crooked, gray to brown, glabrous to slightly ubescent, laterals often short and stiff, ending in long, lender, sharp spines.

BARK. Mottled gray to brown on young stems, smooth r finely furrowed; on old trunks with gray, short, quarish, flat ridges and narrow confluent furrows.

WOOD. Brownish orange, sapwood lighter, hard, eavy, close-grained, weak, specific gravity about 0.79.

RANGE. In southern and western Texas. Often on shell nounds near the Gulf of Mexico with other chaparral rowth. From Matagorda County to Cameron County. Also in peninsular Florida and the Keys. In Mexico in he states of Nuevo León, Tamaulipas, Veracruz, Oaxaa, and Chiapas. Also in Guatemala, El Salvador, Panana, Colombia, Venezuela, Cuba, and the Bahamas.

REMARKS. The genus name, *Bumelia*, is the classical ame for ash-tree, and the species name, *angustifolia*, efers to the narrow leaves. Some botanists refer this species under the scientific synonyms of *B. celastrina* H. B. K., *B. schottii* Britt., and *B. spiniflora* DC. Some of the vernacular names are Schott Bumelia, Antwood, Downward-plum, Bagne, Bebelamilla, Coma, Caimito, Hopuche, and Coma Resinera. The fruit is sometimes eaten in Mexico, and is also used as an aphrodisiac. The heartwood is occasionally used in cabinet work.

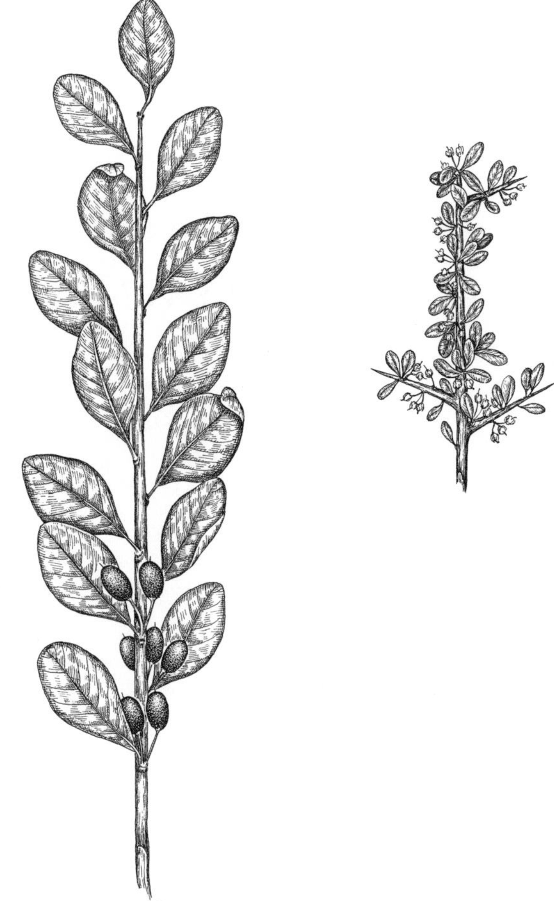

SAFFRON-PLUM BUMELIA
Bumelia angustifolia Nutt.

PERSIMMON FAMILY (*Ebenaceae*)

Common Persimmon

Diospyros virginiana L.

Common Persimmon
Diospyros virginiana L.

FIELD IDENTIFICATION. Tree generally less than 40 ft, rarely reaching 70–100 ft. Habit of growth variable, usually disposed to an upright or drooping type with rounded or conical crown. Branches spreading or at right angles. Twigs self-pruning or some breaking with heavy fruit to form an irregularly shaped tree.

FLOWERS. April–June, staminate and pistillate on separate trees; pollen light and powdery, spread by wind and insects; staminate in 2–3-flowered cymes, tubular, ⅓–½ in. long, greenish yellow; stamens usually 16; pistillate solitary, sessile or short-peduncled, about ¾ in. long or less, stamens 8, some stamens abortive and some fertile; ovary 8-celled, styles 4, 2-lobed at apex; corolla fragrant, 4–5-lobed, greenish yellow, thick, lobes recurved.

FRUIT. Berry persistent, variable as to season, locality, or individual tree, some early or some late (August–February). The very early or very late fruit generally smaller than fruit which ripens about when the leaves fall, seedless fruit also generally smaller; diameter generally ¾–1½ in., shape variable from subglobose or oblate to short-oblong; calyx thick, lobes ovate and recurved; color when mature yellow to orange or dark red, often with a glaucous bloom, flesh pale and translucent, astringent and puckery to taste when green; when ripe somewhat softer and sweet with a high sugar content; 4–8-seeded, seeds large, oblong, flat, leathery, wrinkled, dark brown, about ½ in. long; some trees seedless.

LEAVES. Deciduous, simple, alternate, entire, ovate-oblong to elliptic, apex acute or acuminate, base rounded, cuneate or subcordate, blade length 2–6 in., width 1–3 in., upper surface dark green and lustrous, lower surface paler and pubescent; petiole about 1 in. long or less, glabrous or pubescent.

BARK. Brown to black, fissures deep, ridges broke into rectangular checkered sections.

WOOD. Dark brown to black, sapwood lighter, fine grained, strong, hard.

RANGE. Thrives on almost any type of soil from sand to shales and mud bottomlands. Generally in the southeastern United States. Gulf states to Iowa and Connecticut. Seemingly the best zone is from Maryland Virginia, and Carolinas, westward through Missour and Arkansas. In Texas west to the valley of the Colorado River.

PROPAGATION. Transplanting of trees is a difficult matter because of the long taproot which requires carefu digging. If planting is to be done with seeds, they shoul be stratified immediately after gathering and soake 2–3 days before planting. They may be planted eithe in spring or fall in shallow drills in light soil with plent of humus, and covered to a depth of about ½ in. Roo cuttings 6–8 in. long and ⅓ in. in diameter can be use also, provided the ends are sealed with pitch or wa to prevent rot. Older twigs can be similarly used. The can be buried in sand in a nursery row until ready to plant.

Trees may also be grafted by chip-budding, cleft-grafting, or whip-grafting, using the same procedure as in other fruit trees. Nursery stock should be set about 6 in. apart and root-pruned each year with a spade or tree digger. Stock 1–2 years old may be transplanted, but this should be done in moist deep soil because of the deep root system. Trees can be set 20–25 ft apart and because of absence of surface roots vegetable crops can be grown between. If an orchard is desired, the land can be prepared as for other fruit trees. Although the trees are somewhat self-pruning, tops should be thinned to allow for better fruit production. Long twigs can be removed by clipping young growing tips. All cuts should be well painted to prevent decay.

VARIETIES. Fuzzy Common Persimmon, *D. virginiana* var. *pubescens* (Pursh) Dipp., is a variety with branches and leaves densely pubescent.

Oklahoma Common Persimmon, *D. virginiana* var. *platycarpa* Sarg., is a variety with leaves broader, rounded or cordate at the base, and pubescent beneath; fruit depressed-globose, 1½–2⅛ in. across, trees scattered through Missouri, Arkansas, and Oklahoma. A number of horticultural clones are also known. The following list was prepared by G. A. Zimmerman, H. P. Kelsey, and William A. Dayton (see Kelsey and Dayton, 1942, p. 260): Delman, Early Golden, Ford, Garretson, Glenwood, Glidewell, Goldengem, Harris, Hicks, Ida, Josephine, Lambert, Miller, Ruby, Silkyfine, Stout, Woodward, and Kawakami (the last thought to be a hybrid of *D. virginiana* × *D. kaki).*

DISEASES. Persimmon is subject to attack by the hickory twig girdler, *Oncideres cingulata* Say, which girdles the twigs after depositing eggs in them. The twig will die and drop off. All dead twigs should be gathered from the ground in June or July and burned. The eggs hatch in the twigs in 7–9 days and the pupa stage is 10–14 days before adult emergence.

In recent years a serious leaf-wilt disease has appeared, caused by the fungus, *Cephalosporum*. The upper leaves die first, followed by the lower leaves, and the wood shows fine, brownish black streaks. Pinkish spores appear between the bark and wood, and infected trees usually die. All diseased trees should be burned, and cuts or bruises painted on other trees to prevent entry by the wind-borne spores. No disease-resistant trees have as yet been found.

COOKING PERSIMMONS. Persimmons may be eaten uncooked when they become somewhat softened, but the green fruits are very astringent and cause a puckering of the mouth. Wine has also been made from the fermented fruit.

The following recipes and methods of cooking persimmons are quoted from Fletcher (1942):

"The suggestions given in this section for serving persimmons are based on tests made with fruit of the Miller, Josephine, Ruby, Hicks and Kawakami varieties grown at the Arlington Experiment Farm at Arlington, Va. Fruit so ripe that it had dropped from the trees was used. The different varieties were judged for quality and utilized in various ways. The Miller and Ruby were thought to have the best flavor, with less astringency and more body than the others. The Josephine and Hicks were less uniform in astringency. Some of the ripe fruits of each variety were very astringent, others were free from astringency. The Kawakami was considered less desirable than either the Miller or Ruby, as it was less sweet and had less flavor, and the flavor was not entirely pleasant. There seemed to be little difference in the quality of the tree-ripened fruit that had dropped to the ground, and the less mature fruit picked from the trees and ripened at room temperature before it was used.

"Before the advent of the white man, the Indians mixed the pulp with crushed corn and made it into a kind of bread. . . .

"The astringency of the persimmon pulp may be counteracted to some extent if soda is added to the pulp before it is heated. The quantity of soda to use varies with the degree of ripeness, and the variety of the fruit, but ¼ teaspoon of soda to 1 cup of the persimmon pulp is an average. Persimmon will discolor if cooked in utensils made of tin. Products made without additional acid will turn dark if they stand too long before cooking or serving."

RECIPES USING PERSIMMON PULP.

Baked Persimmon Pudding

2 cups pulp	½ teaspoon cinnamon
3 eggs	½ teaspoon nutmeg
1¾ cups milk	1½ cups sugar
2 cups sifted flour	3 tablespoons melted butter
½ teaspoon soda	
1 teaspoon salt	

Mix the persimmon pulp, beaten eggs, and milk. Sift the dry ingredients together and pour the liquid mixture into them. Stir in the melted butter and pour into a shallow greased pan to the depth of about 2 in. Bake for about 1 hour in a very moderate oven (300°–325°F.). When cold, cut into squares and serve with plain or whipped cream.

Steamed Persimmon Pudding

¼ cup butter	⅜ teaspoon soda
½ cup sugar	½ cup sifted flour
2 eggs	2 teaspoons baking powder
1½ cups pulp	¼ teaspoon salt

Cream the butter; add the sugar, beaten eggs, and persimmon pulp. Sift the remaining dry ingredients and combine with the first mixture. Pour into a greased mold, cover, and steam for 1½ hours.

Persimmon Custard

2 cups pulp	¼ teaspoon cinnamon
½ cup sugar	⅛ teaspoon nutmeg
½ teaspoon soda	1⁄16 teaspoon salt
2 egg yolks, beaten slightly	

Combine the ingredients, pour into a baking dish surrounded by hot water, and bake in a slow oven (250°–300°F.) for about 15 minutes.

Make a meringue, using two egg whites, ¼ cup of sugar, and 1⁄16 teaspoon of salt. Place on top of the custard and bake in a slow oven (250°–300°F.) for about 1 hour.

Persimmon Whip

1 cup pulp	¼ teaspoon salt
½ cup sugar	3 tablespoons lemon juice
5 egg whites	

Heat the persimmon pulp with the sugar. Fold the hot mixture into the stiffly beaten egg whites to which the salt has been added. Mix in the lemon juice, place in a baking dish surrounded by hot water, and bake in a very slow oven (225°–250°F.) for about 1 hour.

Persimmon Cornstarch Pudding

½ cup sugar	1½ cups boiling water
⅓ cup cornstarch	¼ teaspoon soda
⅛ teaspoon salt	1 cup pulp
½ cup cold water	1 tablespoon cinnamon
1 tablespoon lemon juice	

Mix the sugar, cornstarch, salt, and cold water thoroughly in a double boiler. Add the boiling water and stir until the mixture thickens. Add the soda to the persimmon pulp and combine with the cornstarch mixture. Cover and cook for 15 minutes. Add the cinnamon and lemon juice, pour into a mold, and chill. Serve plain or with cream.

Persimmon Cake

½ cup fat	¼ teaspoon soda
1 cup sugar	2 cups sifted flour
1 cup pulp	3 teaspoons baking powder
2 eggs	½ teaspoon salt

Cream the fat and sugar together, add the persimmon pulp and beaten eggs. Sift the dry ingredients and add to the liquid mixture. Beat well, pour into a greased pan, and bake in a moderate oven (325°–375°F.) for about 1 hour.

Persimmon Sherbet

1 cup water	Juice ½ lemon
1 cup sugar	⅛ teaspoon salt
2 cups pulp	1 egg white

Boil the water and sugar for 1 minute, and put aside. When cold, add the persimmon pulp, lemon juice, salt, and unbeaten egg white, and freeze with a mixture of 1 part salt to 4 to 6 parts of ice. Turn the crank slowly until the mixture is firm. Remove the dasher, pack the freezer with more ice and salt, and let the sherbet stand for an hour or more to ripen.

Preserved Persimmon Pulp

Mix equal amounts by measure of persimmon pulp and sugar and place in glass jars. Partially seal and process pint jars for 15 minutes in boiling-water bath. Seal and store in a cool place.

REMARKS. The genus name, *Diospyros,* is translated "fruit-of-the-gods," and the species name, *virginiana,* refers to the state of Virginia. Vernacular names are Jove's-fruit, Winter-plum, and Possum-wood. The fruit was known and appreciated by early settlers and explorers, being mentioned in writings of De Soto in 1539, Jan de Laet in 1558, and John Smith in the seventeenth century. The wood of Persimmon is used for handles and shoe lasts, but three fourths of the supply is made into golf clubs and shuttles. Its hardness, smoothness, and even texture make it particularly desirable for these purposes. The tree is suitable for erosion control on deeper soils because of its deep root system, but this same characteristic makes it difficult to transplant. Also, the rapid spread of a new leaf-wilt disease introduces a factor of caution before extensive plantings are made.

The fruit is eaten by at least 16 species of birds, also by the skunk, raccoon, opossum, gray and fox squirrel, and white-tailed deer. Fallen fruit is also useful in providing some forage for hogs, having a high carbohydrate content. The bark is known to have astringent medicinal properties.

TEXAS PERSIMMON

Diospyros texana Scheele

FIELD IDENTIFICATION. An intricately branched, smooth-barked shrub or tree up to 40 ft.

FLOWERS. Dioecious, small, solitary or in few-flowered clusters; corolla urn-shaped, greenish white, pubescent, about ⅓ in. long; lobes 5, spreading, suborbicular, often notched at apex; stamens 16, included, anthers glabrous; ovary sessile, pubescent, 4–8-celled, style united with narrow stigmas; calyx-lobes 5, ovate, obtuse, spreading or reflexed, thickened, pubescent.

TEXAS PERSIMMON
Diospyros texana Scheele

FRUIT. Depressed-globose, black, apiculate, pulp sweet when mature, astringent when green, about ¾–1 in. long; seeds 3–8, triangular, flattened, hard, shiny, about ⅓ in. long.

LEAVES. Persistent, alternate, leathery, entire, oblong or obovate; apex obtuse, retuse, rounded or emarginate, abruptly narrowed at base, 1–2 in. long, dark green, glabrous above or somewhat pubescent, tomentose below.

BARK. Very smooth, gray, thin layers flaking off.

WOOD. Heavy, black, compact, sapwood yellow, takes a high polish.

RANGE. In Texas and northern Mexico. In central and west Texas usually on rocky hills or the sides of ravines and canyons. Especially abundant in the Texas Edwards Plateau area. When near the coast usually on soils with lime composition because of marine shells. Probably reaching its easternmost limit in Harris County, Texas, near the coast. In Mexico in the states of Nuevo León, Coahuila, and Tamaulipas.

REMARKS. The genus name, *Diospyros,* is translated "fruit-of-the-gods," and the species name, *texana,* refers to the state of Texas. Vernacular names are Mexican Persimmon, Black Persimmon, Chapote, and Chapote Prieto. The fruit is somewhat smaller than that of the Common Persimmon, but is likewise sweet and juicy at maturity, and is eaten by many birds and mammals. The black juice is used to dye skins in Mexico, and the wood is used for tools, and engraving blocks. It was also used in a craft now little practiced—that of ornamenting wooden objects by burning designs into them with an iron.

SWEETLEAF FAMILY (*Symplocaceae*)

COMMON SWEETLEAF
Symplocos tinctoria L'Her.

COMMON SWEETLEAF
Symplocos tinctoria L'Her.

FIELD IDENTIFICATION. Semievergreen shrub or tree up to 35 ft tall and 9 in. in diameter. Branches are slender and upright, giving the tree a wide, loose-spreading appearance.

FLOWERS. Yellowish white, borne in axillary clusters, on branches of previous year, mostly March–May; scales orange-colored, ovate, acute, margin brown and ciliate; bracts 3, at base of each flower, oblong, apex obtuse or rounded, margin ciliate, one longer than others; calyx 5-lobed, oblong, sepals minute, ovate, partly united and shorter than the tube, dark green, puberulent; calyx adherent to ovary as far as sepals. Clusters of flowers 4–14, dense, sessile or short-peduncled; flower fragrant, deciduous, regular, perfect, epigynous; petals 5, obovate to spatulate, ¼–⅓ in. long, united at base; stamens numerous, conspicuously exserted, borne in clusters of 5, one cluster adhering to base of each petal, filaments slender, anthers orange-colored, 2-celled; ovary 3-celled, with 5 dark glands at apex and opposite calyx-lobes; style slender, gradually thickened toward apex and longer than corolla.

FRUIT. Drupe ripe in summer or early autumn, dry, orange-brown, about ¼–½ in. long, 1-celled; seed solitary, ovoid, pointed at ends; seed coat brown, thin, papery; calyx-lobes crowning the fruit.

LEAVES. Persistent in the South, revolute in the bud, drooping on upcurved twigs, alternate, simple, oblong to lanceolate or elliptic, apex acute or acuminate, base cuneate, margin obscurely and remotely serrate, or almost entire, teeth often with small, dark, caducous glands, leaf texture thickish and leathery; upper surface dark green, glabrous, and lustrous (sometimes yellowish green); lower surface paler and pubescent with arcuate veins and reticulate veinlets, blades 2–6 in. long, 1–2 in. wide, petioles ⅓–½ in. long, slightly winged, glabrous to pubescent.

TWIGS. Upcurved terminally, slender, terete, gray to reddish brown, pubescent at first, more glabrate later, lenticels scattered; leaf scars low, horizontal.

BARK. Gray to reddish, mostly smooth, but on old trunks has warty excrescences or a few narrow fissures.

WOOD. Light red to brown, sapwood almost white, close-grained, brittle, weak, soft, light, weighs about 33 lb per cu ft; no commercial value.

RANGE. Mostly on the Atlantic and Gulf Coast plain in low, rich grounds of river bottoms and bay flats. From eastern Texas, Oklahoma, Arkansas, and Louisiana, eastward to Florida and northward to Delaware.

VARIETIES. Two varieties of the Sweetleaf have recently been recognized as follows:

A hairy variety from the mountains of South Carolina has been named the Ashe Sweetleaf, S. *tinctoria* var. *ashei* Weatherby.

Dwarf Sweetleaf, S. *tinctoria* var. *pygmaea* Fern., is a variety hardly over 3½ ft high with leaves ¾–2½ in. long.

REMARKS. The genus name, *Symplocos*, is from the Greek and refers to the united stamens. The species name, *tinctoria*, is applied because a tincture, or dye, is made from the leaves and roots. The leaves and roots are also reported to have medicinal properties. Vernacular names are Yellow-wood, Wild-laurel, Florida-laurel, Dye-leaves, and Horse-sugar. The sweet leaves are greedily eaten by livestock. The seeds have also been found in the stomach of the phoebe.

STORAX FAMILY (*Styracaceae*)

CAROLINA SILVER-BELL
Halesia carolina L.

CAROLINA SILVER-BELL
Halesia carolina L.

FIELD IDENTIFICATION. Wide-spreading shrub or tree to 40 ft, diameter 8–18 in. Branches are spreading or erect to form a rounded or irregular crown.

FLOWERS. Borne March–May, axillary, in fascicles of 1–5; pedicels ½–¾ in. long, slender, drooping, villose; axillary bracts ovate, serrate, pubescent or semiglabrous, apex rounded, caducous; calyx obconic, 4-ribbed, glabrous to pubescent (or hoary-tomentose in var. *mollis*), adherent to ovary surface; teeth 4, small and triangular, ciliate; corolla epigynous, open campanulate, 4-lobed, narrowed below into a short tube, white, or more rarely slightly pink, about ¾ in. across; stamens 8–16, shorter than the corolla, inserted near base of tube and united into a ring, monadelphous for part of their length, filaments villose with a few white hairs, anthers linear-oblong; ovary inferior, ovules 4 in each cell, style gradually contracted into a stigmatic apex.

FRUIT. Drupe ripening in autumn, persistent, dry, bony, reddish brown at maturity, oblong to oblong-obovate, 1–2 in. long, ½–¾ in. in diameter, 4-winged; stone ellipsoid to somewhat obovoid, at base narrowed into a short stipe, at apex terminating into an elongate persistent style, ½–⅝ in. long, somewhat angled; seed solitary by abortion (sometimes 2–3), ends narrowed and rounded, dispersed by water, wind, or gravity.

LEAVES. Simple, alternate, deciduous, elliptic or oblong-obovate, apex usually acuminate, base rounded to cuneate or gradually narrowed, margin dentate; young leaves at first with upper surface densely stellate-pubescent above, lower surface with thick hoary tomentum; at maturity upper surface dark yellowish green and glabrous, lower surface pale and glabrous or slightly villose on the midrib and primary veins, blades 3–4 in. long, 1½–2 in. wide (on vigorous shoots 6–7 in.), texture thin, turning yellow in the fall; petiole slender, ¼–½ in. long, at first pubescent to tomentose but later semiglabrous.

TWIGS. Slender, terete, pithy, at first densely pubescent, later becoming slightly pubescent to glabrous, orange-brown to reddish brown eventually; leaf scars large and obcordate; winter buds axillary, ellipsoid to ovoid, ⅛ in. long; scales thick, broad-ovate, dark red, acute, puberulous.

BARK. Reddish brown, separating into closely appressed scales, slightly ridged, about ½ in. thick.

WOOD. Heart light reddish brown, sapwood lighter colored, soft, close-grained, weighing 35.07 lb per cu ft when dry.

RANGE. In rich well-drained soils of stream banks or wooded slopes, generally where protected. Oklahoma, Arkansas, northeast Texas, and northern Louisiana; eastward to northwest Florida, northward to the Virginias, and westward to Illinois.

PROPAGATION. The fruit is collected from trees in late fall and early winter. Wings should be removed to facilitate handling. No data are available on longevity in storage. One hundred fruits average 106–135 seeds. There are 1,200–2,500 dewinged seeds per lb. Untreated seeds will not germinate satisfactorily until the second year after planting. Dormancy is broken by stratifying the seed at high temperatures (56°–86°F.) for 60–120 or more days and then at low temperatures (33°–41°F.) for 60–90 days, although some lots germinate fairly well with cold stratification alone. The average germinative capacity on test lots is about 53 per cent for pretreated seeds.

The plants may also be propagated by layers, root cuttings, and greenwood cuttings. It does best in rich, well-drained soil in sheltered positions.

VARIETY. Suwanee Carolina Silver-bell, *H. carolina* var. *mollis* Sarg., has a densely hairy calyx and pedicels, and more tomentose leaves. It is the most common form in western Florida.

REMARKS. The genus name, *Halesia,* is in honor of Stephen Hales (1677–1761), English clergyman and author of *Vegetable Staticks.* The species name, *carolina,* refers to the states of Carolina where it grows. Vernacular names for the tree are Opossum-wood, Silver-bell, Snowdrop-tree, Calico-wood, Tisswood, Bell-tree, Olive-tree, and Chittimwood. The tree is rarely attacked by insect pests, but is easily storm-damaged. It is very ornamental and is often cultivated in the eastern United States, California, and central Europe. It was first introduced into cultivation in 1756.

TWO-WING SILVER-BELL
Halesia diptera Ellis

TWO-WING SILVER-BELL
Halesia diptera Ellis

FIELD IDENTIFICATION. Shrub or small tree to 30 ft, with diameter 3–12 in. Branches are slender and form a small, rounded crown.

FLOWERS. March–April, on branches of the previous year with the young leaves, borne in 2–6-flowered axillary fascicles; pedicels slender, hairy, 1½–2 in. long; flowers perfect, corolla white, drooping, broadly campanulate; tube short, about 1 in. long, nearly divided to the base into 4, spreading, oval to obovate lobes, puberulent, calyx-tube obconical, ⅛–⅙ in. long, pubescent; lobes 4, triangular, acuminate; stamens 8 or more, nearly as long as the corolla, filaments hairy; ovary 2- or rarely 4-celled, style slender, elongate, pubescent; bracts obovate, apex rounded or acute, puberulous.

FRUIT. Drupe oblong to ellipsoid, dry, flattened, 1–2 in. long, with 2 broad thin wings, the remaining angles sometimes with lesser wings, smooth, beak rather short; stone solitary, about ¾ in. long, ellipsoid, ridged, acuminate at both ends.

LEAVES. Buds obtuse, leaves simple, alternate, deciduous, ovate to oval or elliptic to obovate; apex rounded or acute to acuminate, sometimes abruptly so; base rounded or cuneate, margin with remote teeth, blades 3–4 in. long, 2–3 in. wide, upper surface light green and glabrous to pubescent, lower surface paler and soft-pubescent, veins pale and conspicuous; petioles ½–¾ in. long, light green, pubescent, slender.

TWIGS. Slender, gray to brown, lustrous; leaf scars cordate, large, raised; buds ovoid, obtuse, hairy; pith chambered.

BARK. Reddish brown to gray, fissures irregular; flakes small, thin, tight.

WOOD. Light brown, sapwood lighter, close-grained, light, soft, brittle, of little commercial value.

RANGE. In sandy, moist soil along streams or in bottomlands. Texas eastward to Florida. Evidently a Gulf Coast plain species, but extending north into Arkansas, Oklahoma, Tennessee, and South Carolina.

PROPAGATION. The seeds are collected by hand in fall, and may be stored without special treatment for spring planting. The wings can be broken off by a hammer mill. Cleaned seed averages about 1,500 seeds per lb. Two years is required for seed to germinate unless pretreated. Pretreatment consists of stratification in sand or peat at 56°–86°F. for 60–150 days, followed by 2–3 months at 33°–40°F. The average germinative capacity in test lots is 53 per cent. Germination requires 30–80 days for completion. The plant is also propagated by layers, root cuttings, or greenwood cuttings in spring or autumn.

REMARKS. The genus name, *Halesia,* is in honor of Stephen Hales, an English clergyman (1677–1761). The species name, *diptera,* refers to the 2-winged fruit. Vernacular names are Snowdrop-tree, Snow-bell, and Cowlicks. The tree is not subject to insect damage and, being easily damaged by storms, prefers a sheltered growing site. Gray and fox squirrels sometimes eat the fruit. Silver-bell flowers are attractive but unfortunately the tree is generally irregular in shape. It has been cultivated since 1758, and is occasionally grown in Europe.

FLORIDA SILVER-BELL

Halesia parviflora Michx.

FIELD IDENTIFICATION. Shrub, or sometimes a slender tree, 25–30 ft, 5–10 in. in diameter.

FLOWERS. Maturing March–April, appearing with the leaves, in short racemes, or in axillary clusters, on branchlets of the previous year; on densely pubescent pedicels which become glabrate later and are ⅓–⅖ in. long; flowers white, drooping, bell-shaped, perfect; corolla ⅓–½ in. long, the 4 lobes almost erect; calyx 1/10–⅛ in. long, densely tomentose, narrowly obconic, with 4 minute teeth; stamens 10–16, exserted, partly adnate to the corolla, anthers oblong and basi-fixed, filaments slightly hairy; ovary inferior and 3–5-celled, ovules 4 in each cavity, style elongate, stigma scarcely enlarged.

FRUIT. Maturing August–September, brown, capsule elongate, clavate, base long-stipitate, apex short and sharp-beaked, length ¾–1¾ in.; margin 4-winged, the wings equal or 2 wider than the others; seeds 1–3, ovoid, base abruptly narrowed to a short stipe, apex gradually narrowed, obscurely ridged.

FLORIDA SILVER-BELL
Halesia parviflora Michx.

LEAVES. Simple, alternate, deciduous, blades 2½–3¼ in. long, 1–1¼ in. wide, oblong-ovate to obovate or elliptic, apex abruptly acute or acuminate, base rounded or broadly cuneate, margin finely glandular-serrate, young leaves densely stellate-hairy, later pubescent, especially pubescent on veins beneath; mature leaves dull green and pubescent to glabrous above.

TWIGS. Slender, young ones pubescent and reddish brown, older ones gray and glabrous; winter buds obtuse.

BARK. On trunk dark brown to black, ridges rounded and separated by deep fissures.

RANGE. Usually on dry, rich, sandy uplands in partial shade or sun. Oklahoma, Arkansas, eastern Mississippi, Alabama, Georgia and northern Florida.

PROPAGATION. For propagation the seeds may be sown immediately or stratified for spring planting. This species can also be grown from layers, root cuttings, or cuttings of greenwood from forced plants.

VARIETIES. Florida Silver-bell, or Little Silver-bell, as it is sometimes known, is closely related to the Carolina Silver-bell, *H. carolina,* and is possibly only a variety of it with smaller leaves, flowers, and fruits.

REMARKS. The genus name, *Halesia,* is in honor of Stephen Hales (1677–1761), English clergyman and author of *Vegetable Staticks.* The species name, *parviflora,* is for the small flower

MOUNTAIN SILVER-BELL

Halesia monticola Sarg.

FIELD IDENTIFICATION. Tree attaining a height of 90 ft and diameter of 3 ft. Pyramidal when young, but the small erect branches forming a round-topped crown later.

FLOWERS. Borne in May, in clusters of 2–5 from bracts ½–¾ in. long, elliptic to obovate, apex acute, surface pubescent; pedicels of flowers ½–1 in. long, pubescent to glabrous; corolla about 1 in. across with 4 almost erect lobes; stamens 10–16, filaments hairy below, ovary 4-celled; calyx obconic, pubescent to glabrous.

FRUIT. Capsule 1¾–2 in. long, 1 in. wide, oblong-obovoid, base cuneate, 4-winged; stone ovoid-ellipsoid, short-stipitate, apex elongate, angled.

LEAVES. Simple, alternate, deciduous, elliptic to oblong or obovate, apex abruptly acuminate or acute, base cuneate or rounded, margin remotely toothed, blades 3–6 in. (rarely to 11 in.) long, 1½–3 in. wide, thin, white-hairy when young, upper surface dark green

American Snow-bell
Styrax americanum Lam.

and glabrous, lower surface paler and glabrous or somewhat pubescent, especially along the veins; petioles ½–¾ in. long, pubescent at first, glabrous later.

TWIGS. Slender, pubescent to glabrous, brown in varying shades, shiny; buds ellipsoid to ovoid, acuminate, flattened, scales about ⅓ in. long, lustrous, gibbous dorsally.

BARK. About ½–¾ in. thick, reddish brown, separating into larger, loose, platelike scales.

RANGE. Altitudes of 2,000–4,000 ft in the mountainous regions of Arkansas, Oklahoma, Tennessee, Georgia, and North Carolina.

VARIETIES, FORMS, AND SYNONYMS. Mountain Silverbell is considered by some botanists as not being sufficiently distinct for a species, and is given the varietal name of *H. carolina* var. *monticola* (Sarg.) Rehd., being separated only on the basis of larger leaves and flowers.

A variety has been segregated as Fuzzy Mountain Silver-bell, *H. monticola* var. *vestita* Sarg., with leaves often rounded at base and white-tomentose to pubescent later. Occurs in North Carolina and Arkansas. A color form, Pink Mountain Silver-bell, *H. monticola* forma *rosea* Sarg., is distinctive with rose-colored flowers.

REMARKS. The genus name, *Halesia,* honors Stephen Hales (1677–1761), an English clergyman. The species name, *monticola,* refers to the tree's mountainous habitat.

American Snow-bell

Styrax americanum Lam.

FIELD IDENTIFICATION. Widely branched shrub attaining a height of 3–9 ft.

FLOWERS. Borne May–June, fragrant, on short, lateral leafy branchlets; racemes axillary, short, subtended by small leafy bracts; flowers solitary or 2–7; pedicels drooping, 1/12–⅗ in. long, when young slightly glandular-pubescent, when older glabrous; calyx greenish, truncate, ⅛–⅙ in. high, lower half adherent to the ovary; lobes 5, small, broadly triangular, sometimes glandular, usually glabrous (pubescent in var. *pulverulentum*); corolla white, ⅖–⅗ in. long, rotate; petals 5, valvate in the bud, elliptic-oblong or lanceolate-oblong, acute to linear-tipped, slightly puberulent externally or glabrous; stamens 10, filaments flat, erect, pubescent below and adnate to the corolla-base; anthers bright yellow, elongate, erect, the sacs united, basi-fixed; ovary half-inferior, 3-celled or at length 1-celled by obliteration of the septa; ovules several in each cavity, ascending; stigma 3-toothed.

FRUIT. Drupe maturing September–October, persistent, dry crustaceous, subglobose or obovoid, about ¼–⅓ in. in diameter, finely tomentose, calyx persistent on the lower third, dehiscent into 3 thin valves; seed usually solitary, rarely 2–3, globular, erect, hard-coated.

LEAVES. Simple, alternate, deciduous, blades oval, elliptic, or oblong, sometimes ovate to obovate, length ¾–4 in., apex acute or short-acuminate, base usually acute, margin varying from entire to serrate or remotely toothed; upper surface dark green and usually glabrous; lower surface paler or somewhat puberulent to glabrous, pinnately veined; petioles glandular and slightly pubescent to semiglabrous, 1/12–⅙ in. long.

TWIGS. Slender, stellate-pubescent when young but glabrous later; bark thin, smooth, reddish brown to gray.

RANGE. Margins of swamps and streams in rich, moist soil. Arkansas, perhaps also in Texas, and Louisiana; eastward to Florida, northward to Virginia, and westward to Illinois and Missouri.

VARIETY. A variety has been segregated and named *S. americanum* var. *pulverulentum* (Michx.) Perkins, on the basis of the dense hairiness of the pedicels, calyxes, and petioles. However, it is sometimes difficult to separate the species and variety with certainty because many intermediate forms are found.

REMARKS. The genus name, *Styrax,* is the Greek name of the Old World tree producing storax. The species name, *americanum,* refers to its North American habitat. Also known under vernacular name of Spring Orange.

Downy American Snow-bell

Styrax americanum var. *pulverulentum* (Michx.) Perkins

FIELD IDENTIFICATION. Shrub of sandy lowlands to 12 ft tall and 3 in. in trunk diameter.

FLOWERS. March–April, borne in lateral, leafy-bracted, loosely flowered racemes from 1–4 in. long. Individual flowers 1–4, in the axils, pendent, fragrant. Pedicels ⅛–¼ in. long, scurfy-hairy. Corolla rotate, petals 5, reflexed, imbricate in the bud, white, about ¼ in. long, elliptic to oblong or elliptic to lanceolate, puberulent, apex acute or obtuse; stamens 10, adnate to the corolla-base, exserted; filaments white and threadlike; anthers nearly erect, elongate, about 1/12 in. long, linear, flattened, yellow, longer than filaments; pistil slender, filiform, exserted considerably beyond stamens; ovary nearly superior; calyx persistent, densely scurfy-hairy; shallowly 5-lobed, lobes apiculate.

FRUIT. Capsule subglobose, ¼–⅓ in. in diameter, puberulent, dry with a hard coat, 3-valvate at the summit; seed solitary.

Downy American Snow-bell
Styrax americanum var. *pulverulentum* (Michx.) Perkins

TWIGS. Young ones slender, elongate, green to gray or brown, densely scurfy-hairy; older ones dark gray to reddish brown, glabrous.

BARK. Rather smooth, dark gray to brown.

LEAVES. Alternate, deciduous, elliptic to oval or obovate to oblong, blades 1–2½ in. long, margin shallowly toothed, slightly revolute, apex acute to abruptly short-acuminate, base cuneate, upper surface dull green, veins impressed, with a few scattered hairs or almost glabrous, lower surface much paler and densely scurfy-hairy; petioles 1/16–¼ in. long, densely scurfy-hairy.

RANGE. Moist places in sandy soil; East Texas, Arkansas, and Louisiana; eastward to Florida and north to Virginia.

MEDICINAL USES. Two closely related foreign species, *S. benzoin* Dryander and *S. tonkinensis* (Pierre) Crab *ex* Hartwich, yield medicinal balsamic resins known in commerce as Sumatra benzoin and Siam benzoin, respectively. However, the true medicinal storax balsam (or copalm balsam) comes from the Oriental Sweetgum, *Liquidambar orientalis* Miller. Smaller quantities are also obtained from the Central American Sweetgum varieties of *L. styraciflua* var. *mexicana,* and *L. styraciflua* var. *macrophylla.*

SPECIES AND VARIETIES. Some authorities classify Downy American Snow-bell as a distinct species instead of a variety of American Snow-bell. However, the two are so similar that a varietal standing seems most fitting.

REMARKS. The variety name, *pulverulentum,* refers to the scurfy indument and hairs on the lower leaf surface. Also known by the vernacular names of Powdery Storax and Snow-bell in some localities. Has been in cultivation since 1794. May be propagated by seeds, layers, or by grafting.

Bigleaf Snow-bell

Styrax grandifolia Ait.

FIELD IDENTIFICATION. Broad shrub or small tree to 25 ft, 8 in. in diameter. The smooth, ascending branches form a round-topped crown.

FLOWERS. Appearing with leaves in March–April, fragrant, solitary, or in loose axillary clusters or racemes 3–6 in. long, from the axils of small caducous bracts which are leaflike only at the base of the racemes; pedicels about ¼ in. long, pubescent to canescent; corolla ¾–1 in. long, 5-petaled, petals longer than the tube, imbricate in the bud, oblong to elliptic, apex rounded or acute, exterior finely stellate-pubescent; calyx campanulate, ⅙–⅕ in. long, tomentose, more or less adnate to ovary base, margin truncate and obscurely 5-toothed, teeth triangular and acute; stamens 10, inserted at corolla-base, distinct, about as long as corolla, filaments flattened, hairy below the middle, anthers erect, elongate, yellow, cells parallel; ovary nearly superior, obo-

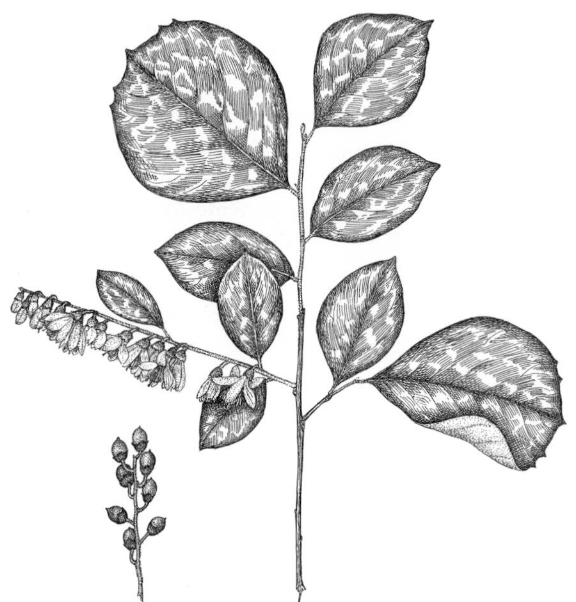

Bigleaf Snow-bell
Styrax grandifolia Ait.

void, tomentose, 3-celled (becoming 1–2-celled after anthesis), ovules 3–4 in a cell; style persistent, exserted, slender, filiform, glabrous.

FRUIT. Drupaceous, 1/4–1/3 in. in diameter, somewhat obovoid or subglobose, tomentose, tipped by style remnants, outer coat hard and crustaceous, indehiscent or only irregularly and partially split at apex, surrounded basally by the calyx; seed erect, large, obovoid or globose, dark brown, tightly enclosed by the calyx.

LEAVES. Involute in the bud, simple, alternate, deciduous, oval to obovate or elliptic, apex acute to short-acuminate or rounded, base cuneate or rounded, margin entire or remotely serrate, blades 2–6 in. long, 1–3 in. wide (longer on young shoots); when young very densely pale-tomentose and ciliate; when mature upper surface green and glabrous with impressed veins; lower surface densely and finely gray-tomentose; petioles 1/5–1/4 in. long, densely tomentose at first, eventually pubescent.

TWIGS. Slender, terete, light to dark brown, when young densely stellate-pubescent, later glabrous; axillary buds 1–3, superimposed, acute, scales scurfy-tomentose, about 1/8 in. long; pith small, rounded, homogeneous.

BARK. Dark reddish brown or grayish brown, close, smooth, 1/3–1/2 in. thick.

RANGE. In low, sandy, wet woods. Eastern Louisiana and Arkansas; eastward to Florida northward to Tennessee and Virginia.

REMARKS. The genus name, *Styrax,* is the ancient Greek name, first given to *S. officinalis,* which produces the resin storax of the drug trade. The species name, *grandifolia,* refers to the large leaves. Vernacular names are Coast Snow-bell, Storax, and Mock-orange. The fruit is known to be eaten by the wood duck. The shrub is sometimes grown for ornament, but is not hardy in the North. It is very closely related to the American Snow-bell, S. *americana* Lam., which has leaves glabrous beneath.

Sycamore-leaf Snow-bell

Styrax platanifolia Engelm.

FIELD IDENTIFICATION. Shrub to 12 ft, open, irregular crown and slender branchlets.

FLOWERS. In axillary, semidrooping racemes 1½–2½ in. long, pedicels 1/4–1/2 in. long, semiglabrous to pubescent, subtended by minute bracts; calyx, semiglabrous (or densely hairy in S. *platanifolia* var. *stellata*), shallowly 5-toothed; flowers perfect and regular; petals 5, oblong to elliptic or obovate, apex acute or obtuse, distinct, white; stamens 10, adnate to corolla-base, filaments flattened; anthers elongate, linear, erect, introrse, sacs united; ovary superior, 3-celled or later 1-celled by abortion, ovules several in a cavity, style slender and united.

FRUIT. Drupe maturing in June, borne singly or several in a cluster, peduncles 1/4–1/2 in. long, slightly pubescent (or densely tomentose in S. *platanifolia,* var. *stellata*); calyx tightly coherent to the fruiting base one fourth to one third the length, flaring widely from

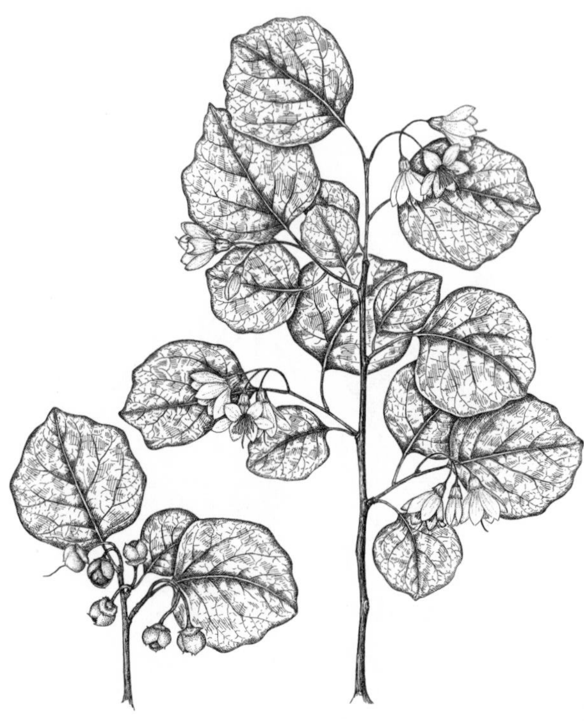

Sycamore-leaf Snow-bell
Styrax platanifolia Engelm.

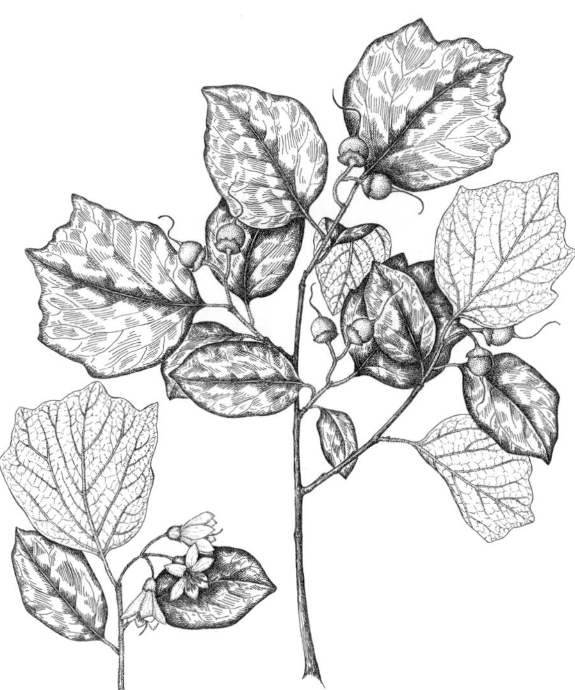

Hairy Sycamore-leaf Snow-bell
Styrax platanifolia var. *stellata* (Engelm.) Cory

the clavate peduncles, margin set with remote, abruptly pointed, subulate teeth; surface semiglabrous to densely tomentose; body of fruit subglobose or obovate, ¼–⅜ in. long, the style persistent and apiculate; pericarp tough, leathery, dehiscent, flesh thin; seeds 1–2, about ¼ in. long, testa thin, oval to obovate, rounded or obtuse at apex and base, a ridge running from the hilum down the side.

LEAVES. Simple, alternate, deciduous, blades 1½–3 in. long, about as broad, ovate or broadly heart-shaped, margin entire or some with short acute or obtuse lobes, apex acute to obtuse, base cordate or semitruncate; upper surface light to dark green, semilustrous to dull, semiglabrate or set with minute, scattered stellate hairs, veins numerous and finely reticulate; lower surface paler, reticulate veins more conspicuous than above, smooth or softly and densely stellate-hairy; petiole ¼–½ in. long, glabrate to stellate-hairy.

TWIGS. Slender, when young brown and pubescent; with age brown to gray and glabrous.

BARK. Gray to black, when young smooth, on old trunks near the base broken into small scales.

RANGE. Wooded rocky banks and ledges in central and western Texas—Spanish Pass, Kendall Co.; Enchanted Rock, Llano County; Little Blanco River, Blanco County; and Travis Peak, Travis County. The hairy variety, S. *platanifolia* var. *stellata* (Engelm.) Cory, has been collected 9 miles west of Boerne, and in Sabinal Canyon 6½ miles north of Vanderpool, Texas.

VARIETIES. Specimens vary from quite glabrous to very densely stellate-hairy. However, these are connected by intermediate forms with varying degrees of hairiness. Those with copious stellate tomentum have been relegated to the status of a variety known as Hairy Sycamore-leaf Snow-bell, S. *platanifolia* var. *stellata* (Engelm.) Cory.

The snow-bells may be propagated by seeds or layers. Some species are grafted on the closely related *Halesia carolina.*

REMARKS. The genus name, *Styrax,* is the ancient Greek name, and the species name, *platanifolia,* refers to the sycamore-like foliage.

Texas Snow-bell

Styrax texana Cory

FIELD IDENTIFICATION. Shrub to 15 ft, with slender and irregular branches; often appearing one-sided and unshapely because of its frequent occurrence on the faces of bluffs or cliffs.

FLOWERS. In spring, axillary, solitary, or in clusters of 2–5; peduncles ½–¾ in. long, finely tomentulose; individual flower pedicels ¼–½ in. long; calyx 3/16–¼ in. long, spreading at the apex, shallowly set with remote, minute teeth; corolla ½–¾ in. long, petals 5–6, distinct, white, elliptic to oblong, apex obtuse or acute; stamens

Texas Snow-bell
Styrax texana Cory

equaling the petals, or shorter, included when the petals are not reflexed; style filiform, sometimes exceeding the corolla.

FRUIT. Maturing August–September, peduncles finely tomentose, ¼–½ in. long, gradually expanded into the shallow, cup-shaped, minutely toothed calyx; fruit subglobose, about ⅜ in. long, green at first, brown-tomentose later, dehiscent into 2–3 valves; seed solitary, globose or slightly longer than wide, smooth, often with 1 or 2 shallow grooves on the side, dark lustrous-brown, about 5/16 in. long.

LEAVES. Simple, alternate, deciduous, blades 2–3 in. long, almost as broad, mostly oval or a few broadly elliptic, margin entire, base abruptly contracted into petiole, apex rounded or blunt-pointed, upper surface pale green and glabrous, veins delicate and impressed, lower surface conspicuously white with veins raised and more prominent; petiole ⅜–¾ in. long, green to reddish, grooved above, essentially glabrous.

TWIGS. Slender, when young reddish brown, older gray and glabrous, minutely white-scaly under the glass; bark of trunk smooth, light gray to dark gray.

RANGE. Rare and local, confined to limestone areas of the central Texas Edwards Plateau–Edwards County, on Polecat Creek 14½ miles southeast of Rocksprings, also on Cedar Creek; Real County, 3 miles north of Vance on Hackberry Creek near old post-office site.

REMARKS. The genus name, *Styrax*, is the ancient Greek name, and the species name, *texana*, refers to the state of Texas.

YOUNG'S SNOW-BELL
Styrax youngae Cory

YOUNG'S SNOW-BELL

Styrax youngae Cory

FIELD IDENTIFICATION. Shrub 6–9 ft. Very rare plant known only from the type collection at the University of Texas Herbarium. The following description of this specimen is given by Cory (1943b, p. 113):

FLOWERS. Appearing the middle of April, in racemose clusters of 3–7, peduncle stout, ⅙–⅘ in. long, coarsely stellate-pubescent, pedicels stout, ⅙–⅓ in. long, densely and coarsely stellate-pubescent; calyx campanulate, about ⅙ in. long and ⅙ in. broad, dark brown, densely stellate-pubescent, apex truncate, non-glandular, teeth inconspicuous; petals 5, white, ⅗–¾ in. long, narrowly elliptic, obtuse, densely stellate-puberulent; style stout, about ⅗ in. long; fruit not seen.

LEAVES. Smaller leaves orbicular, larger leaves elliptic, to 2 in. long and 1⅜ in. broad, subentire, more or less rounded at the apices and bases, or somewhat acute at the apices, thin; green above, but densely pubescent with coarse stellate hairs; tomentose below, but not silvery, with a very fine and dense indumentum which is beset with coarse stellate hairs; veins very prominent and straw-colored; short-petioled.

RANGE. Igneous soil at an altitude of 4,000 ft. In a canyon, Davis Mountains, Texas. Type collected May 12, 1914.

REMARKS. The genus name, *Styrax*, is the ancient Greek name. The species name, *youngae*, honors Mary S. Young, formerly with the University of Texas. V. L. Cory remarks, "It is the only collection of *Styrax* from the mountains of southwestern Texas . . . and it might possibly be a species of northern Mexico."

The author has compared the type specimen with specimens of *S. texana* in the University of Texas Herbarium and finds a very close resemblance. *S. youngae* has smaller, less conspicuously veined leaves than *S. texana*, and the flowers appear to be more numerous in the cluster (3–8). Both are similar in having a very dense whitish tomentum on the under surface of the leaves. However, since only one specimen of *S. youngae* exists, it is difficult to judge validity and status of this taxon, and until more material can be collected it seems best to maintain it as a separate species.

The above lack of certainty is brought about by precise collection data being absent from the collection sheet. This points up the need for better data on herbarium specimens in general. Many botanists are seemingly loath to spend the time needed to insure proper data on collections.

ASH FAMILY (*Oleaceae*)

White Fringe-tree
Chionanthus virginicus L.

FIELD IDENTIFICATION. Usually a shrub with crooked branches, but sometimes a tree to 35 ft, with a narrow, oblong crown.

FLOWERS. March–June, perfect or polygamous, in delicate drooping panicles 4–6 in. long; pedicels pubescent; bracts of panicles sessile, oval-oblong, leaflike; calyx small, 4-lobed, persistent, lobes ovate-lanceolate and acute, green, glabrous; petals 4–6, linear, acute, about 1 in. long, white with purple spots near the base, barely united at base, longer than the tube, fragrant; stamens 2; filaments short, adnate to corolla-tube; anthers 2, ovate, light yellow, subsessile; ovary ovoid, 2-celled; style short, thick, 2-lobed; staminate flowers sometimes with sterile pistils.

FRUIT. August–October, drupe borne in loose clusters; bracts leaflike, oval or short-oblong, some 2 inches long; drupe bluish black, glaucous, globose-oblong, ½–¾ in. long, pulp thin, 1-celled, 1–3-seeded; seeds about ⅓ in. long, ovoid, brown, somewhat reticulate. Plants 5–8 years old begin to produce seed.

LEAVES. Simple, opposite, deciduous, oval to oblong or obovate-lanceolate; apex obtuse, acute or acuminate; base wedge-shaped; margin entire or wavy; 4–8 in. long, 1–4 in. wide, dark green and glabrous above, paler below with hairs on veins; petioles ½–1 in. long, puberulent.

TWIGS. Stout, pubescent, light brown to orange, later gray.

BARK. Brown to gray, thin, close, appressed, broken into small thin scales.

WOOD. Light brown, sapwood lighter, hard, heavy, close-grained, weighing about 39 lb per cu ft.

RANGE. Oklahoma, Arkansas, Texas, and Louisiana; eastward to Florida and northward to Pennsylvania and New Jersey.

White Fringe-tree
Chionanthus virginicus L.

PROPAGATION. The fruit is collected for planting when it has turned dark in color. The pulp is removed by rubbing over a suitable screen and cleaned by washing. The seed averages about 1,800 seeds per lb, with a commercial purity of about 99 per cent and soundness of about 95 per cent. The potential germination is about 50 per cent. Seed dormancy may be broken somewhat by stratification in sand or peat at 41°F. for a year or so before spring planting. If planted in the fall, the cleaned seed should be planted in beds 8–12 in. apart, covered about ¼ in. with firmed soil, and covered with a straw or leaf mulch until after the last spring frost. Propagation is also done by layering, grafting, or budding.

VARIETY. *C. virginicus* var. *maritimus* Pursh is a variety with more pubescence on leaves and panicles.

REMARKS. The genus name, *Chionanthus,* is a combination of two Greek words meaning "snow flower," and the species name, *virginicus,* refers to the state of Virginia. Vernacular names are Flowering Ash, Old Man's Beard, Grandfather-graybeard, Snowflower-tree, Sunflower-tree, Poison Ash, White-fringe, Shavings, and Graybeard-tree. The bark has medicinal uses as a diuretic and fever remedy. The tree is cultivated to some extent for the fragile panicles of flowers in spring and for the dark green foliage. It has been cultivated since 1736. The staminate plants display the attractive delicate, drooping flowers but bear no fruit. The leaves are persistent in winter in the Gulf Coast area, but farther north fall after turning bright yellow.

Downy Forestiera

Forestiera pubescens Nutt.

FIELD IDENTIFICATION. Sometimes a small tree to 15 ft, and 5 in. in diameter, but usually only a straggling, irregularly shaped shrub.

FLOWERS. Polygamo-dioecious, appearing before the leaves in spring from branches of the preceding year; clusters lateral, from bracts which are obovate, 1/12–1/8 in. long, ciliate, densely pubescent; staminate fascicles greenish; sepals 4–6, small, early-deciduous; petals absent; stamens 2–5; pistillate clusters on slender pedicels of short spurs; ovary 2-celled, 2 ovules in each cell, style slender, stigma capitate or somewhat 2-lobed.

Downy Forestiera
Forestiera pubescens Nutt.

FRUIT. June–October, drupes pediceled, clustered, bluish black, glaucous, ellipsoid, 1/4–1/3 in. long, fleshy, 1-seeded; stone oblong to ellipsoid, ribbed.

LEAVES. Simple, opposite, deciduous, 1/2–1 3/4 in. long, varying from elliptic to oblong or oval, margin obscurely serrulate, apex obtuse or rounded, base cuneate or rounded; dull green and glabrous or slightly pubescent above; lower surface densely soft-pubescent; petioles short, yellowish green, pubescent.

TWIGS. Green to yellowish and pubescent when young, older ones light to dark gray and glabrous.

RANGE. Mostly in rich, moist soil along streams. New Mexico, Texas, and Oklahoma; eastward to Florida.

VARIETY AND SYNONYM. A variety (which is discussed immediately following) is *F. pubescens* Nutt. var. *glabrifolia* Shinners. It has been previously listed in the literature as *F. neomexicana* Gray. (See Shinners, 1950.) The plant listed as *F. neomexicana* var. *arizonica* Gray is to be considered synonymous with *F. pubescens,* and extends the range of the latter as far west as Arizona.

REMARKS. The genus name, *Forestiera,* honors the French physician and naturalist, Charles Le Forestier, and the species name, *pubescens,* refers to the soft-hairy leaves. Also known under the vernacular names of Devil's-elbow, Chaparral, Spring-herald, Spring-goldenglow, and Tanglewood. The shrub has no particular economic value but has been recommended for erosion control and wildlife cover. It has been cultivated since 1900. About 20 species of *Forestiera* are known, these being distributed in North America, West Indies, and Central to South America. The various species are propagated by cuttings and seeds, and some are rooted by layering.

New Mexico Forestiera

Forestiera pubescens Nutt. var. *glabrifolia* Shinners

FIELD IDENTIFICATION. Erect, spreading shrubs or small trees to 12 ft, often clumped at the base and with semispinescent branches.

FLOWERS. March–May before the leaves in the axils of the last year's leaves; flowers small, polygamo-dioecious, crowded, the dense, sessile clusters subtended by 4 small bracts; petals absent; stamens 2–4, anthers oblong and yellow; ovary superior, ovate, 2-celled, with 2 pendulous ovules in each cell; style slender, stigma somewhat 2-lobed; pistillate flowers with 2–4 sterile stamens.

FRUIT. Ripening June–September, drupe 1/5–1/3 in. long, bluish black, ovoid to ellipsoid, obtuse, 1-celled and 1-seeded; seeds bony, germinating 40–70 per cent.

NEW MEXICO FORESTIERA
Forestiera pubescens Nutt. var. *glabrifolia* Shinners

LEAVES. Simple, opposite, deciduous, spatulate-oblong to ovate-oblong, apex obtuse or short acuminate, base cuneate, margin minutely serrulate or sometimes entire toward the base, length ½–1¾ in., width ¼–¾ in., surfaces glabrous above and below, grayish green, membranous; petioles ⅛–¼ in. long, glabrous; smaller leaves sometimes fascicled at the base of the older ones.

TWIGS. Stiff, lateral ones often shortened, gray to whitened, smooth, glabrous.

RANGE. New Mexico Forestiera is found on hillsides or mesas, or in moist valleys, at altitudes of 3,000–7,000 ft, from the northern parts of Trans-Pecos Texas to New Mexico, Arizona, Colorado, Utah, and westward into California.

PROPAGATION. It has some ornamental value when mass-planted with other plants along streams. It may be propagated from cuttings, layers, or seeds.

SYNONYM AND RELATED FORM. New Mexico Forestiera is also listed in the literature as *F. neomexicana* Gray. *F. neomexicana* var. *arizonica* Gray is not accepted. Instead, the listed characters of *arizonica* appear to make it a synonym of *F. pubescens*.

REMARKS. The genus name, *Forestiera,* honors Charles Le Forestier (d. *circa* 1820), a French naturalist and physician, and the variety name, *glabrifolia,* refers to the smooth foliage. It is also known as Desert Olive and Palo Blanco. The Hopi Indians are said to have made digging-sticks of the branches.

WRIGHT FORESTIERA

Forestiera wrightiana C. L. Lundell

FIELD IDENTIFICATION. Shrub to 6 ft, with short lateral branches.

FLOWERS. Staminate flowers unknown; pistillate flowers lateral, fasciculate or borne in reduced racemes, usually glabrous, sometimes sparsely to densely hirtellous, pedicels 1/25–1/12 in. long; bracts elliptic, ciliate, 1/12–1/10 in. long; calyx small, usually 2–4-lobed, the lobes unequal; petals not evident; staminodia usually 4, reduced; ovary 2-celled, with 2 ovules in each cell; style slender.

FRUIT. Bluish black, glaucous, subglobose, ¼–⅖ in. long, depressed apically.

LEAVES. Opposite, crowded at the tips of the spur-like branchlets; petioles pubescent, slender, canaliculate; blade subcoriaceous, elliptic or ovate-elliptic, ½–2 in. long, ¾–1 in. wide, apex obtuse, rounded or acutish,

WRIGHT FORESTIERA
Forestiera wrightiana C. L. Lundell

base acute, slightly decurrent, margin serrulate, densely pilose beneath over the entire surface, pubescent above along the midvein, primary veins and veinlets slightly impressed above, the primary veins inconspicuous beneath.

TWIGS. Usually shortened, at first pubescent, later glabrous.

RANGE. The following localities are given by Lundell: Texas, Newton County, just off U.S. highway 190, in woods above Cow Creek, September 10, 1942, C. L. Lundell and S. W. Geiser, 11878 (type in the University of Michigan Herbarium, duplicate in the herbarium of Southern Methodist University); same locality and date, Lundell and Geiser, 11879; Harris County, Houston, in low woods, Sept. 18, 1915, E. J. Palmer, 8582.

REMARKS. The genus name, *Forestiera,* is in honor of Charles Le Forestier (d. *circa* 1820), French physician and naturalist at Saint-Quentin. The species name, *wrightiana,* is in honor of Charles Wright, early botanical collector of Texas. Cyrus Longworth Lundell, who described the species (Lundell *et al.,* 1943a), remarks: "*F. wrightiana,* a species closely allied to *F. ligustrina* (Michx.) Poir. . . . grows in the southeastern region where Wright as a young man made his first collections in the state."

The author has noted a very close resemblance between specimens of *F. wrightiana* with *F. pubescens* Nutt. also. More study is needed to determine whether the two plants are the same.

TEXAS FORESTIERA
Forestiera acuminata Poir.

TEXAS FORESTIERA

Forestiera acuminata Poir.

FIELD IDENTIFICATION. Straggling shrub or tree to 30 ft, growing in swampy ground.

FLOWERS. Dioecious or polygamous; staminate in dense green fascicles subtended by yellow bracts; calyx ring narrow, slightly lobed; petals none; stamens 4; filaments long, slender, erect; anthers oblong, yellow; ovary in staminate flowers abortive; pistillate flowers in short panicles, ¾–1¼ in. long; ovary ovoid with a slender style and 2-lobed stigma, stamens usually abortive or absent.

FRUIT. Drupe ovoid-oblong, purplish, apex acute, tipped with style remnants, base rounded, fleshy, dry, about 1 in. long, young fruit somewhat falcate; seed usually solitary, ridged, compressed, light brown, about ⅓ in. long, one side often flatter than the other.

LEAVES. Simple, opposite, deciduous, elliptical or oblong-ovate, acuminate at apex, cuneate at base, remotely serrulate above the middle, 2–4½ in. long, 1–2 in. wide, glabrous and yellowish green above, paler with occasional hairs on veins beneath; petioles slender, ¼–½ in., slightly winged by leaf bases.

TWIGS. Light brown, glabrous, slender, warty, with numerous lenticels, sometimes rooting on contact with the mud.

BARK. Dark brown, thin, close, slightly ridged.

WOOD. Yellowish brown, close-grained, light, weak, soft, weighing about 39 lb per cu ft.

RANGE. In swamps or bottom lands, Oklahoma, Arkansas, Texas, and Louisiana; eastward to Florida and northward to Tennessee, Indiana, Illinois, and Missouri.

VARIETY. A variety of Texas Forestiera has been given the name of *F. acuminata* var. *vestita* Palmer. It varies by having the leaves, petioles, and young branchlets more or less clothed with short straight pubescence, which is persistent to the end of the season, and even in some cases is found on the slender branchlets of the second season; on the typical form there is only a slight trace of pubescence on the petioles and veins of the younger leaves. Known from Miller, Hempstead, and Crawford counties, Arkansas; also near Richland, Rapides Parish, Louisiana.

REMARKS. The genus name, *Forestiera,* is in honor of the French physician and botanist, Charles Le Forestier, and the species name, *acuminata,* refers to the acuminate leaves. Another vernacular name is Swamp Privet. It has no particular economic use, except that the fruit is considered to be a good wild duck food.

ET-LEAF FORESTIERA

orestiera reticulata Torr.

FIELD IDENTIFICATION. A small to medium-sized hrub, with an irregular crown, and many small stems rising from the base; or more rarely a small spreading ree to 12 ft, with a single trunk.

FLOWERS. In short, crowded racemose clusters from he axils of last year's leaves; buds scaly, scales imricate and straw-colored; flowers dioecious, petals abent, calyx of 4 minute greenish sepals; stamens 2–4, xserted, anthers oblong; ovary 2-celled, ovules 2 in ach cell, becoming a 1-celled and 1-seeded drupe; tyle slender, stigma somewhat 2-lobed or capitate.

FRUIT. Borne in axillary, short clusters, from scales which are oval to ovate, acute to obtuse, ciliate, and 1/25–1/16 in. long; drupes on glabrous pedicels ⅛–3/16 in. ong, ovoid to ellipsoid or obovoid, asymmetrical, usu.lly less than ¼ in. long, tipped by the persistent style emnants, dark reddish-brown; 1-celled and 1-seeded.

LEAVES. Opposite, pairs ⅓–1 in. apart on the branchlets, length ½–1½ in., width ¼–⅘ in., coriaceous, ovate to short-oblong, a few oval; apex acute to obtuse, base rounded, margin entire, or with obscure, appressed teeth mostly toward the apex; upper surface dull green, essentially glabrous and finely reticulate; lower surface paler, glabrous, semiglaucous, reticulate-veiny and porulose-punctate (under a glass); petioles mostly less than ¼ in. long, glabrous or puberulent.

TWIGS. Slender, terete, rather straight; younger ones brown, glabrous or puberulent; older ones light to dark gray and glabrous; lenticels small, pale, and scattered.

RANGE. Dry hillsides and canyons. Texas and Mexico. In Texas on the Edwards Plateau, west to the Pecos River and perhaps beyond. Apparently nowhere abundant. Collected by Valery Havard at the mouth of the Pecos. Specimens examined by the author deposited in the Missouri Botanical Garden Herbarium by E. J. Palmer, No. 12974, from south and west slopes of high limestone hills at Montell, Uvalde County, Texas.

REMARKS. The genus name, *Forestiera,* honors Charles Le Forestier (d. *circa* 1820), a French physician and naturalist at Saint-Quentin and first botany teacher of Poiret. The species name, *reticulata,* refers to the netveined leaves. This shrub is not very well known.

NET-LEAF FORESTIERA
Forestiera reticulata Torr.

NARROW-LEAF FORESTIERA

Forestiera angustifolia Torr.

FIELD IDENTIFICATION. Evergreen, dense, stiff, intricately branched shrub. Sometimes a small tree to 25 ft, with a short, crooked trunk.

FLOWERS. Polygamo-dioecious, inconspicuous, greenish, in clusters from scaly bracts; sepals 4, minute, early-deciduous; petals absent; staminate flowers sessile or nearly so; bracts imbricate, oval to ovate, margin fimbricate, yellowish green, about 1/16 in. long; stamens 2–4 in a cluster, conspicuously exserted, erect or spreading, ⅛–3/16 in. long; anthers reddish brown, oblong, hardly over ⅛ in. long; pistil about ⅛ in. long, style slender and gradually swollen below into a 2-celled ovary, developing a stipelike base later, ovules 2 in each cavity.

FRUIT. Drupe short-peduncled, oblong-ovoid, somewhat falcate, acute, ¼–½ in. long, black, 1-seeded, edible but astringent.

LEAVES. On older twigs often clustered on short knotty spurs, on young shoots mostly opposite and more distant, persistent, linear to oblanceolate, apex obtuse, margin entire and somewhat revolute, leathery, light green and glabrous, veins obscure, somewhat porous, blade length ½–1¼ in., width ⅙–¼ in., sessile or nearly so.

TWIGS. Gray, slender, stiff, smooth, sometimes spines-

NARROW-LEAF FORESTIERA
Forestiera angustifolia Torr.

cent; bark of older branches and trunk smooth and gray.

RANGE. On dry, well-drained hillsides, or along stony arroyos in Texas and Mexico. In Texas in the central, western, and southern portions. Also following the coastal shell-banks (limy soil) along the Gulf as far east as Chambers County. Rare in Harris County, but found on Hog Island at Tabbs Bay Ferry and at La Porte and Seabrook close to the bayside. The coastal plant may be referable to *F. texana* upon more investigation. In Mexico in the states of Tamaulipas, Nuevo León, and Coahuila.

RELATED SPECIES. V. L. Cory has separated a form in Cameron and Hidalgo counties, Texas, under the name of *F. texana*. The reader is referred to that description for the comparative features.

REMARKS. The genus name, *Forestiera,* is in honor of Charles Le Forestier, a French naturalist and physician, and the species name, *angustifolia,* refers to the narrow, linear leaves. In Mexico it is known as Panalero and Chaparral Blanco. The fruit is eaten by a number of birds and mammals, including the scaled quail and gray fox. The plant may be propagated by seeds and layers. It has some possibility as an ornamental in close proximity to the Gulf, in saline or limy soil, where plants are subject to heavy buffeting by winds.

CAMERON FORESTIERA
Forestiera texana Cory

FIELD IDENTIFICATION. Small tree 7½–12 ft, with grayish bark and slender, not stiff, branchlets. Long thought to be only a southern Texas form of *F. angustifolia* but herewith considered a species as described by Cory (1944, p. 252):

FLOWERS. Pedicels mostly ⅙–⅕ in. long (flowers otherwise not described).

FRUIT. Oblong, usually curved, acute, tipped with a slender style which is 1/25 in. long or somewhat more, nutlet oblong, many-ribbed, truncate at base, usually curved, acute at apex, to ⅓ in. long and about 1/10 in. broad.

LEAVES. Chiefly opposite, scarcely fasciculate, occurring in pairs or frequently with the pairs doubled, oblong, elliptic, elliptic-oblong, rounded at the apex, cuneate to rounded at base, short-petioled to subsessile, to 2 in. long and about ⅓ in. broad, but mostly 1 in. long or less and ¼ in. or less broad, being much longer and broader late in the season than in early spring, the under surface porulose and conspicuously 1-nerved, both surfaces light green.

RANGE. The type specimen of Cameron Forestiera was collected 9 miles south of La Feria, Cameron County, Texas, April 4, 1938, Cory No. 28393, Arnold Arboretum, Harvard University. The isotype is in the Tracy Herbarium, Agricultural and Mechanical College of Texas. Other specimens seen by Cory were near Brownsville, Cameron County, also southwest of Donna,

CAMERON FORESTIERA
Forestiera texana Cory

Hidalgo County, and at La Joya. Cory states also that he has seen the plant growing only in Cameron and Hidalgo counties.

RELATED SPECIES AND VARIETY. In order to clearly differentiate between *F. texana* and *F. angustifolia,* contrasting parts are listed by Cory as follows:

F. texana: habit, moderately branched small tree; bark, grayish or pale; branchlets, elongate, slender, not stiff; foliage, pale green; leaves, averaging about ⅘ in. long and ⅙ in. broad, not fasciculate; pedicels, ⅙–⅕ in. long; nutlets, slender, acute, ¼–⅓ in. long and ⅒ in. broad.

F. angustifolia: habit, densely branched bush or shrub; bark, dark to almost black; branchlets, stout, short, stiff; foliage, dark green; leaves, average ⅖–⅗ in. long, comparatively narrow, fasciculate in clusters of 2–6; pedicels, 1/12 in. long or less; nutlets, stout, rounded, ¼ in. long and ⅛ in. broad.

Cory has also named a variety, *Palmer Forestiera, F. texana* var. *palmeri* Cory, which he relates as differing from the species in its denser foliage and its shorter, narrower, usually clustered leaves, averaging ⅗ in. long and less than ⅛ in. broad. Type from Val Verde County, Texas; also La Salle, Uvalde, and Live Oak counties.

REMARKS. The genus name, *Forestiera,* is in honor of Charles Le Forestier (d. *circa* 1820), French physician and naturalist at Saint-Quentin and first botany teacher of Poiret. The species name, *texana,* is for the state of Texas, its native home.

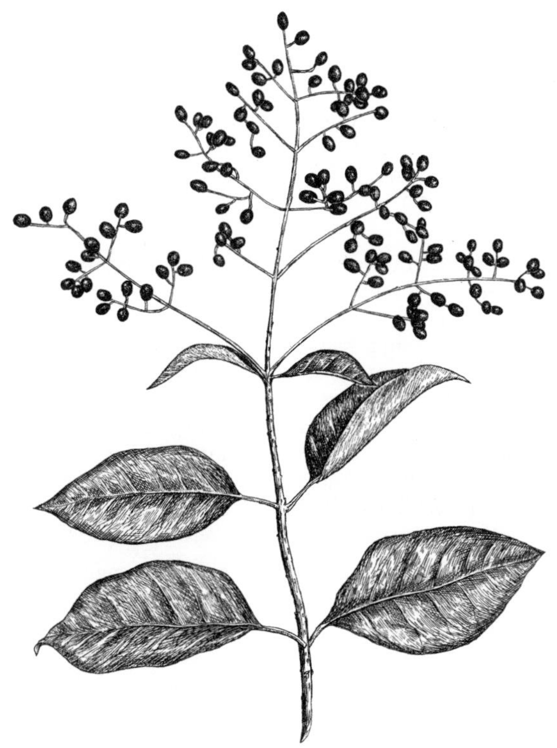

JAPANESE PRIVET
Ligustrum japonicum Thunb.

JAPANESE PRIVET
Ligustrum japonicum Thunb.

FIELD IDENTIFICATION. Shrub or small tree to 35 ft, trunks often inclined and clumped.

FLOWERS. June–August, in panicles which are terminal, broad, loosely flowered, 4–8 in. long and 2–6 in. wide; individual flowers perfect, sessile or nearly so. Calyx tubular, about 1/16 in. long, very shallowly 4-lobed, glabrous; corolla white, tubular, about ⅛ in. long, flared into 4 reflexed ovate to short-oblong, acute or obtuse lobes about 1/16 in. long; stamens 2, considerably exserted, filaments adnate to the corolla-tube below; anthers rather large, yellow, one end rounded or truncate, the other end cleft; pistil shorter than the stamens at anthesis, stigma 2-lobed.

FRUIT. Persistent, bluish black, somewhat glaucous, about ¼ in. long, oval to short-oblong, flesh thin; seed brown to black, solitary, somewhat rounded and rugose on one side, the other surfaces plane.

LEAVES. Evergreen or nearly so, opposite, 2–4½ in. long, elliptic to oblong or oval, apex acute to acuminate, often apiculate, base cuneate, margin entire; dark, dull green, semilustrous, and glabrous above; lower surface paler and glabrous with minute black-punctate dots; petioles ¼–¾ in. long, green to brown, glabrous, somewhat channeled above.

TWIGS. Younger ones green to brown, or gray later, essentially glabrous; lenticels small, brown, orbicular on younger branches, slitlike and horizontal on older branches and trunks.

RANGE. A native of Asia. Often cultivated in gardens in Texas and Louisiana, sometimes escaping to grow wild along roadsides, stream banks, or edges of woods.

PROPAGATION. The plant is propagated by seed sown in fall, by green or hardwood cuttings, and by grafting.

VARIETY AND HORTICULTURAL FORMS. A variety known as the Round-leaf Japanese Privet, *L. japonicum* var. *rotundifolium* Bl., is described as a compact shrub to 6 ft, with stiff, short branches, and leaves crowded, broad-ovate or sub-orbicular, 1¼–2½ in. long, margin obtuse or emarginate, surface dark green and shiny, often curved; panicle dense, 2–4 in. long; flowers sessile; fruit subglobose, about ⅕ in. across. Introduced into cultivation about 1860.

The following horticultural forms have been developed: Golden-tip *(aureifolium marginatum)*; Silver-leaf *(excelsum superbum)*; Variegated *(variegatum)*; Yellow-leaf *(aureifolium)*.

REMARKS. The genus name, *Ligustrum,* is the ancient name, and the species name, *japonicum,* refers to Japan, its native province. Vernacular names are Privet-berry and Prim. It is much planted for ornament, and the fruit is greedily eaten by the cedar waxwing and other birds.

Glossy Privet

Ligustrum lucidum Ait.

FIELD IDENTIFICATION. Handsome cultivated shrub or small tree to 30 ft. Trunk often inclined and solitary or clumped from the base, branches spreading to form a rounded top (or variously shaped in horticultural forms).

FLOWERS. In terminal or axillary panicles 2–8 in. long, often closely crowded by dense adjoining leaves; individual flowers sessile or short-peduncled; calyx about 1/16 in. long, minutely 4-lobed or almost truncate; corolla white, funnelform, tube about 1/8 in. long, flared above into 4 reflexed, short-oblong, obtuse to acute lobes about as long as the tube; stamens 2, adnate to the corolla-tube, exserted on filaments about as long as the corolla-tube; anthers large, yellow, oblong, cleft at one end; pistil usually inclined or barely exserted at anthesis, apex spatulate and somewhat 2-lobed.

FRUIT. Drupe subglobose to oblong or narrowly obovate, bluish black, 3/16–1/4 in. long, ripening October–November; seed linear to oblong, slightly granular, 1/8–3/16 in. long.

LEAVES. Opposite, oval to ovate or oblong to elliptic, apex obtuse to acute, base rounded to broadly cuneate; margin entire, somewhat revolute, flat or wavy on the plane surface; thick, stiff, and leathery; upper surface dark, shiny green, and glabrous, main vein yellowish green, other veins inconspicuous and obscure toward the margin; lower surface much paler green, glabrous, veins obscure, minutely and closely punctate (under the glass), length 1½–4 in.; petiole glabrous, grooved above, 1/4–3/4 in. long.

TWIGS. Green to brown, gray later, essentially glabrous, roughened by numerous, pale, orbicular or short-oblong lenticels.

BARK. Rather smooth, light to dark gray or almost black, branches gray to brownish, lenticels numerous.

RANGE. A native of China, Korea, and Japan, introduced into gardens in the Gulf Coast states and occasionally escaping cultivation. Commonly planted in parks and gardens in Texas and Louisiana.

FORMS. The following horticultural forms of Glossy Privet have been recorded by Kelsey and Dayton (1942):

Big-leaf *(macrophyllum)*, Black-leaf *(nigrifolium)*, Compact *(compactum)*, Crinkly-leaf *(recurvifolium)*, Graceful *(gracile)*, Griffings Waxleaf *(compactum)*, Pyramid *(pyramidale)*, Spreading *(repandens)*, Tricolor, Upright *(erectum)*, Yellow-leaf *(aureovariegatum)*.

REMARKS. The genus name, *Ligustrum*, is the ancient classical name, and the species name, *lucidum*, refers to the lucid, or shining, leaf. It has been cultivated since 1794. Glossy Privet is sometimes confused with the Japanese Privet, *L. japonicum*, but the former has denser, smaller, lustrous leaves and a shorter panicle which is often nestled in the upper leaves.

Glossy Privet
Ligustrum lucidum Ait.

Chinese Privet

Ligustrum sinense Lour.

FIELD IDENTIFICATION. Shrub or small tree to 20 ft, and 5 in. in trunk diameter. Trunks often clumped and inclined, branches slender and spreading.

FLOWERS. Inflorescence March–May, fragrant, perfect; borne in panicles which are terminal, narrow, elongate, 2–6 in. long and ½–3 in. wide; corolla white, tubular, limb about 3/8 in. across, 4-lobed, lobes spreading, oblong or ovate, acute; stamens 2, filaments adnate to the corolla-tube, exserted, longer than the prominent corolla-lobes; pistil shorter than the stamens, stigma spatulate, flattened; calyx campanulate, about 1/12 in. long, glabrous, shallowly 4-lobed, lobes acute; pedicels 1/16–1/8 in. long, pubescent.

FRUIT. Drupe bluish black, subglobose to oval or obovoid, seeds 1–2.

LEAVES. Opposite, oval to elliptic, apex rounded to obtuse or slightly notched, base cuneate or rounded, margin entire, length 1–2 in., width ½–1 in., main vein apparent but others obscure; upper surface dark green and semilustrous, glabrous or slightly pubescent along the main vein; lower surface paler, glabrous or slightly pubescent on the main vein; petiole 1/8–½ in., pubescent.

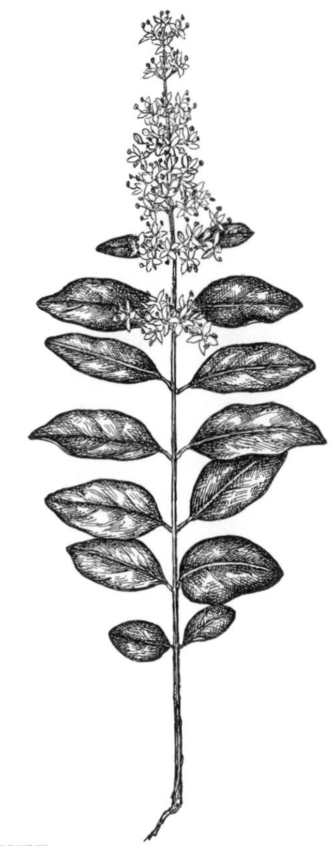

Chinese Privet
Ligustrum sinense Lour.

TWIGS. Slender, spreading, gray to brown, pubescent; older branches and trunk smooth, glabrous, and various shades of gray to brown, lenticels pale and scattered.

RANGE. A native of southeast Asia. Grown for ornament in Texas, Oklahoma, Arkansas, Louisiana, and elsewhere throughout the North Temperate Zone, sometimes escaping cultivation.

PROPAGATION. The species and varieties of Chinese Privet grow in many types of soil. The seeds may be sown in fall or stratified in sand or peat for spring sowing; some may not germinate until the second year. Propagation may also be practiced by cuttings of green wood or hardwood in summer under glass.

VARIETIES. Staunton Chinese Privet, *L. sinense* var. *stauntoni* Rehd., is not as tall as the species and has more spreading branches, leaves oval to ovate, usually obtuse, pubescent on the midrib beneath; panicle broader and looser.

Shining Chinese Privet, *L. sinense* var. *nitidum* Rehd., has branchlets puberulous or minutely pilose; leaves ovate-oblong to ovate-lanceolate, acute or acuminate, lustrous above.

Many-flower Chinese Privet, *L. sinense* var. *multiflorum* (Bowles) Bean, has abundant flowers and reddish brown anthers.

REMARKS. The genus name, *Ligustrum,* is the ancient classical name, and the species name, *sinense,* refers to its Chinese origin. It is a handsome plant, much cultivated for hedges and screens in the South. A number of varieties have been listed.

Quihoui Privet
Ligustrum quihoui Carr.

FIELD IDENTIFICATION. Shrub cultivated, slender, erect or spreading, to 10 ft.

FLOWERS. April–June, heavy-scented, borne in narrow racemes 2–8 in. long and ½–1 in. wide, lateral branches of inflorescence ¼–½ in. long, rather densely flowered from the axils of smaller leaves below, usually leafless above; calyx sessile or nearly so, puberulent, shallowly 4-toothed; corolla white, tubular, about ⅛ in. long, the tube as long as the 4 lobes of the limb or longer, lobes ovate, acute to obtuse; stamens 2, much exserted, anthers short-oblong, about 1/16 in. long; pistil much shorter than the stamens, included or slightly exserted, simple, erect, stigma slightly capitate.

FRUIT. Ripening September–November, bluish black, slightly glaucous, 3/16–¼ in. long, subglobose or slightly flattened; seeds 1–2, about 3/16 in. long, short-oblong to oval; when 2-seeded the outer surfaces are rounded and sculptured and the inner faces plane.

LEAVES. Opposite, simple, partly folded, ⅔–1½ in. long, 3/16–¼ in. wide, linear to narrowly oblong or elliptic, margin entire; apex obtuse, sometimes slightly notched, base gradually narrowed, upper surface dark green and glabrous; lower surface paler and duller green, glabrous or barely puberulent on the midrib; leaves sessile or short-petiolate, glabrous or puberulent.

TWIGS. Younger ones green and finely pubescent, older ones gray or pale brown and glabrous; bark gray, smooth, and with numerous pale lenticels.

RANGE. A native of China, cultivated in the Gulf Coast states, sometimes escaping.

REMARKS. The genus name, *Ligustrum,* is the ancient classical name. The species name, *quihoui,* honors Antoine Quihou, a French botanist who worked in the late nineteenth century. The plant has been cultivated since 1862. It is an attractive shrub with erect or spreading stems and late-flowering habit and grows well in the Houston area.

Rough Menodora
Menodora scabra Gray

FIELD IDENTIFICATION. Western perennial plant ½–2½ ft. The numerous erect stems are puberulent-scabrous, branching near the base, and leafy almost to the top. It is mostly herbaceous but sometimes woody near the base.

Rough Menodora
Menodora scabra Gray

FLOWERS. Maturing May–August, numerous, rather showy, in terminal cymose clusters, bisexual, peduncles erect, leaves of inflorescence smaller; calyx ⅙–⅖ in. long, lobes 7–15, linear, often unequal, persistent; corolla bright yellow, subrotate, ⅖–⅗ in. long and ⅗ in. across, the 5 obtuse lobes much longer than the tube; stamens 2–3, exserted on slender filiform filaments, anthers linear-oblong, emarginate; ovary superior, 2-celled, 2 seeds in each cell, style slender and stigma capitate.

FRUIT. Ripening September–October, capsule bispherical, inflated, thin-walled, ¼–⅓ in. long, ⅓–½ in. wide.

LEAVES. Simple, alternate above or opposite below, lanceolate to oblong, base sessile or nearly so, margin entire, apex acute to obtuse, length ⅓–1¼ in. (usually less than ¾), width ⅙–⅓ in., surfaces dull green and scabro-puberulent to almost glabrous, upper leaves smaller.

RANGE. Dry rocky mesas, desert grasslands, or in oak woodland at altitudes of 1,500–7,000 ft. In western Texas and New Mexico; west to California, north to Utah and Colorado, and south into Mexico in Chihuahua, San Luis Potosí, Durango, and Baja California.

PROPAGATION. Good seed crops usually occur every year. Clean seed will average 108,000 seeds per lb. In one sample, purity was 41 per cent and soundnes 98 per cent. The seed should be stored in a dry plac and will germinate freely without stratification.

VARIETIES. Broom Menodora, *M. scabra* var. *glabrescens* Gray, has glabrous leaves and often bears fewe sepals. It has been listed under the name of *M. scopari* Engelm.

Branched Rough Menodora, *M. scabra* var. *ramosissima* Steyerm., is a variety with longer, more wood stems.

Long-tube Rough Menodora, *M. scabra* var. *longitub* Steyerm., has a longer corolla-tube.

REMARKS. The genus name, *Menodora,* is from th Greek words *menos* ("force") and *daron* ("gift"), i reference to the force or strength it gave to animals The species name, *scabra,* is for the rough leaf surface, as a result of scabrid hairs. The plant is reporte to be of some value as livestock forage.

Showy Menodora

Menodora longiflora Gray

FIELD IDENTIFICATION. Western suffrutescent her 6–18 in. high. Stems numerous, simple, tufted, sometimes woody near the branching base.

Showy Menodora
Menodora longiflora Gray

FLOWERS. In terminal or axillary, few-flowered corym-ose clusters, or solitary; individual flowers large, howy, yellow, perfect; corolla salverform, the tube lender, 1½–2 in. long; lobes 5, oblong to ovate with nucronate-acuminate apices, ½–⅝ in. long; stamens sually 2, included, anthers nearly sessile on the throat, near, apiculate; ovary superior, 2-celled, style slen-ler, stigma 2-lobed; calyx pediceled, with 10 linear etaceous lobes to ½ in. long.

FRUIT. Capsule ⅜–⅝ in. across, 2-parted at or near he middle, the divisions subglobose and bladdery, 2-elled and 2-seeded.

LEAVES. Simple, opposite or sometimes alternate be-ow, alternate above, linear to lanceolate, occasionally -cleft below, apex acute to obtuse and sometimes nucronate, sessile or nearly so at the narrowed base, nargin entire and slightly revolute at maturity, rarely vith minute hairs on the margins, surfaces dull green or slightly paler beneath and essentially glabrous; main ein conspicuous beneath, other veins obscure.

STEMS. Ascending, herbaceous, often tufted near the voody base, young slightly hairy to glabrous and finely grooved, later glabrous and tan to brown.

RANGE. Dry hillsides in southern and western Texas and southeastern New Mexico; south to Coahuila, Mexico.

REMARKS. The genus name, *Menodora,* is from the Greek words *menos* ("force") and *daron* ("gift"), for he strength or force it gives animals. The species name, *ongiflora,* refers to the long-tubed corolla. The plant s showy in bloom and is worthy of cultivation.

DEVILWOOD OSMANTHUS

Osmanthus americana Benth. & Hook.

FIELD IDENTIFICATION. Shrub or tree to 50 ft, and 1 t in diameter, with a narrow-oblong crown.

FLOWERS. Staminate and pistillate flowers borne on different trees or on the same tree, disposed in sessile or short-pedicellate, 3–many-flowered axillary, dense cymes; bracts scaly, triangular, acute and finely pubes-cent; calyx prominently 4-lobed, lobes deltoid, acute, rigid; corolla with a short tube, ⅛–⅙ in. long, greenish white; lobes 4, imbricate, about as long as the tube, broad-ovate, rounded; stamens 2, filaments short and terete, attached to about the middle of the corolla-tube, slightly exserted or included; anthers notched, opening longitudinally; stamens small and rudimentary in the pistillate flower; ovary subglobose, abruptly con-tracted into a stout, columnar style and a capitate stigma, ovules 2. Stigma reduced in the staminate flowers.

FRUIT. Drupe ripening in September, oval to ovoid or obovoid, greenish yellow to purple, ⅖–1 in. in diam-eter, thin-skinned, flesh thin and succulent, bitter; stone ovoid, thin-walled, about ⅜ in. long; seed solitary, brown, striate, hard, bony.

DEVILWOOD OSMANTHUS
Osmanthus americana Benth. & Hook.

LEAVES. Simple, opposite, persistent, thick and leathery, narrowly elliptic to oblanceolate or lanceolate, margin revolute and entire, apex very variable, either obtuse or acute to rounded and notched, base gradually narrowed or wedge-shaped, length 2–6 in., width 1–2½ in., upper surface bright green, glabrous and shiny, paler below; petioles stout, to ¾ in. long.

TWIGS. Slender, reddish brown to gray, puberulous to glabrous, slightly angled or terete; lenticels pale and minute; leaf scars small, raised, orbicular; buds reddish brown, to ½ in. long, narrowly lanceolate, pu-berulous; pith white and homogeneous.

BARK. Gray to reddish brown, with age breaking in-to small thin, appressed scales.

WOOD. Dark brown, sapwood lighter, close-grained, hard, strong, specific gravity 0.81, difficult to work.

RANGE. In southeastern Louisiana (Washington Parish); eastward to Florida and northward to Vir-ginia. In Mexico in Veracruz and Oaxaca.

REMARKS. The genus name, *Osmanthus,* is from the Greek words *osme* ("odor") and *anthus* ("flowers"), referring to the fragrant flowers. The species name, *americana,* refers to its North American habitat. It is also known as American Wild-olive. It has some value as a garden plant and has been cultivated since 1758.

FRAGRANT ASH

Fraxinus cuspidata Torr.

FIELD IDENTIFICATION. Shrub or small slender tree to 20 ft, sometimes forming thickets. Branches slender, smooth, gray.

FLOWERS. April–May, fragrant, borne in loose, ter-minal, glabrous panicles 3–4 in. long; corolla about ⅔

Fragrant Ash
Fraxinus cuspidata Torr.

in. long, white, 4-petaled; petals linear-oblong, exceeding the anthers; anthers oblong, almost sessile; stigma 2-lobed, almost sessile, ovary 2-celled; calyx cup-shaped, about 1/16 in. long, teeth acute and apiculate.

FRUIT. In drooping panicles 2–5 in. long; samaras on slender, glabrous peduncles ½–1 in. long, wing and seed together ¾–1¼ in. long, flattened, oblong-linear or spatulate, rounded or obtuse or notched at the apex, pale green.

LEAVES. Compound, 3–5½ in. long, composed of 5–7 leaflets; leaflets delicate, lanceolate or narrowly ovate, long acuminate or cuspidate at the apex, cuneate at the base, margin sharply and remotely serrate, less so toward the apex and base, blades 1½–2½ in. long, ½–¾ in. wide; dark green, thin, glabrous above, paler beneath; long-petiolate, petiolules grooved above, slightly wing-margined, those of lateral leaflets ¼–½ in. long, that of the terminal leaflets ½–1 in. long.

TWIGS. Slender, younger ones green to brown and glabrous, older ones gray, lenticels small.

BARK. Gray, smooth, rather tight, on old trunks broken into small, short scales and irregular fissures.

RANGE. In dry well-drained soil at altitudes of 3,500–5,500 ft, on mountainsides of the grass, pinyon, and yellow pine belt. In Trans-Pecos Texas in rocky canyons of the Pecos, Devil's, and Rio Grande rivers. Also in the Chisos Mountains of Brewster County. In New Mexico and Arizona. In Mexico in Nuevo León, Chihuahua, and Coahuila.

VARIETY. Arizona Fragrant Ash, *F. cuspidata* Torr. var. *macropetala* Rehd., is a variety occurring in the Grand Canyon area in Arizona and differing by havi 3–7 broader, often ovate, entire leaflets, and larg flowers.

REMARKS. The genus name, *Fraxinus,* is the ancie Latin name, and the species name, *cuspidata,* refers the leaf apex. In Mexico it is known under the Spani name of Fresno. This small Western ash is unique having floral fragrance and petals. Deer and livesto occasionally browse the foliage.

Gregg Ash

Fraxinus greggii Gray

FIELD IDENTIFICATION. Western clump-forming shr or small tree to 25 ft.

FLOWERS. Panicles ½–¾ in. long, individual flowe perfect or 1-sexed, on slender pedicels ⅛–¼ in. lon springing from brown-pubescent, ovate, acumina bracts; petals absent; calyx campanulate, scarious; st mens solitary or paired, filaments longer than the caly anthers about ⅛ in. long; style short, stigmas with r flexed stigmatic lobes, ovary rounded, longer than t calyx.

FRUIT. Samaras ½–¾ in. long, on peduncles ⅛–3/16 i long; wing narrowly elliptic or oblong, apex rounde retuse or notched, often tipped by the style remnar extending at least part way down the terete seed.

LEAVES. Opposite, persistent, ¾–1½ in. long, od

Gregg Ash
Fraxinus greggii Gray

pinnately compound of 3 leaflets (rarely 5–7); leaflets sessile or nearly so, winged, ⅓–1 in. long, ⅛–¼ in. wide, terminal leaflet usually largest, spatulate, elliptic, oval or narrowly ovate, apex obtuse, rounded or notched, base gradually narrowed, margin entire or sparingly crenate-serrate, sometimes revolute, thick and coriaceous, veins inconspicuous, glabrous or puberulent, olive-green above, paler beneath; petioles distinctly winged.

TWIGS. Slender, young twigs dark green and puberulent; older twigs gray, glabrous or puberulent; lenticels small, rounded, raised, gray or brown.

BARK. Dark gray to black, mostly smooth, broken into thin scales on old trees.

WOOD. Hard, heavy, close-grained, brown, sapwood lighter, specific gravity 0.79.

RANGE. On dry rocky hillsides and arroyo banks. Trans-Pecos Texas, rather abundant near Del Rio and Langtry in Val Verde County. In Brewster County in the Chisos Mountains at altitudes of 4,000–7,800 ft. Also near Mount Locke in Jeff Davis County. In New Mexico and southern Arizona. In Mexico in Nuevo León, Tamaulipas, Coahuila, and Zacatecas.

REMARKS. *Fraxinus* is the ancient Latin name, and the species name, *greggii*, honors the botanist Josiah Gregg. Other vernacular names are Fresno, Escobillo, Barreta, and China. The wood is occasionally used for fuel, and the leafy twigs are used for crude brooms in Mexico.

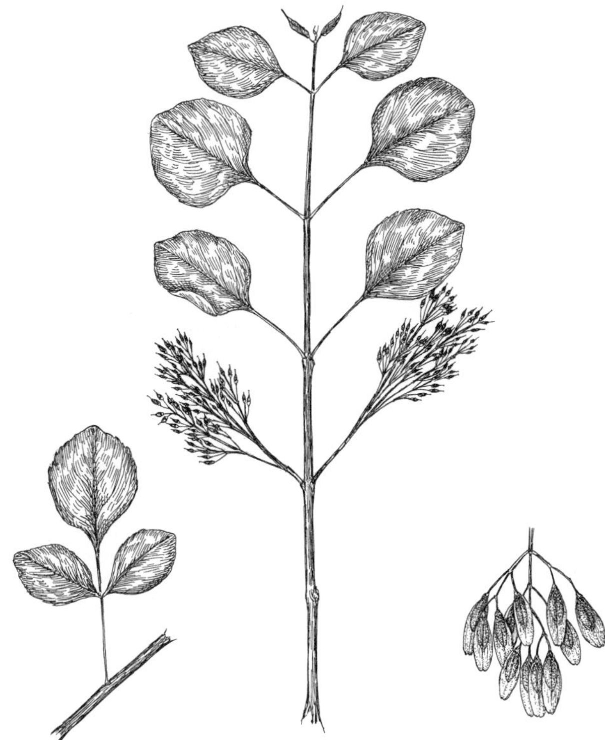

SINGLE-LEAF ASH
Fraxinus anomala Wats.

SINGLE-LEAF ASH

Fraxinus anomala Wats.

FIELD IDENTIFICATION. Low shrub or small tree to 30 ft, with a trunk diameter of 5–7 in. Branches numerous and spreading in the shrub form, and in the tree form short and crooked to form a round-topped crown.

FLOWERS. April–May, borne in short pubescent axillary panicles; bracts about ½ in. long, linear to lanceolate, thickly brown-tomentose; stamens 2, in some panicles aborted, the 2 forms sometimes in the same panicle; filaments slender, about the same length as the style; anthers yellowish orange, linear to oblong; ovary 2-celled, style united with a terminal 2-lobed stigma; corolla absent; calyx cup-shaped, with 4 very small teeth.

FRUIT. Samara narrowly to broadly-obovate, ½–¾ in. long and ⅓ in. wide, thin, flattened, striately nerved, surrounded by the wing; wing rounded to retuse or truncate, often deeply emarginate; seed solitary, compressed, apices gradually narrowed and rounded; pedicels about ¼ in. long or less.

LEAVES. Simple or compound, leaflet either solitary or 2–3-foliate, orbicular to ovate, apex rounded to acute or sometimes obcordate, base cuneate to cordate, margin varying from entire to somewhat crenate or serrate, mostly above the middle, blade length usually 1½–2 in., width 1–2 in. (usually smaller when 2–3-foliate), young ones with short whitish hairs on the upper surface and pubescent below, at maturity glabrous on both surfaces or nearly so, midrib conspicuous, but lateral veins rather obscure; petiole variable in length (½–2 in.), usually longer on the solitary leaflet, grooved, pubescent at first but glabrous later.

TWIGS. Usually 4-angled when young, later terete; young ones dark green to reddish brown and pubescent; older ones turning gray and glabrous; lenticels somewhat elevated and pale; leaf scars narrow and lunate; winter buds terminal, broad-ovoid, tomentose, ⅛–¼ in. long.

BARK. Reddish brown to dark brown, ridges narrow and bearing thin, closely appressed scales, fissures rather shallow.

WOOD. Heart light brown, sapwood lighter colored and thicker, close-grained, heavy, hard.

RANGE. Along streams in the sun at altitudes of 2,000–6,000 ft. Northwestern New Mexico, Colorado, Utah, and Nevada; westward into Arizona and California; also in northwestern Mexico.

VARIETY. Lowell Ash, *F. anomala* var. *lowelli* (Sarg.) Little, is a variety which becomes a small tree 21–24 ft, with deeply fissured reddish brown bark; branchlets quadrangular and winged, yellow to brown with brown cinereous hairs; leaflets 5, or varying 3–7, ovate to elliptic, apex acute or acuminate, more rarely obtuse, base cuneate, teeth on margin rather remote (sometimes the teeth are small, but on other specimens rather large and coarse), surfaces light green, mostly glabrous but on some slightly pubescent beneath; fruit maturing in July, in long glabrous panicles, oblong-elliptic to oblong-obovate, 1–1⅜ in. long, about one third as wide, apex rounded or occasionally emarginate, winged to the base. It is known to intergrade in regard to number of leaflets, their shape, and texture, with the species. Some authors have assigned it the rank of a species under the name of *F. lowelli* Sarg. However, the range of the species and the variety seem to be mostly separate. Sargent observed that the ash was somewhat intermediate between *F. quadrangulata* Michx. of the East and *F. anomala* Torr. Centering in Arizona in Coconino, Yavapai, and Mohave counties.

REMARKS. The genus name, *Fraxinus*, is the ancient Latin name. The species name, *anomala*, refers to the simple leaves in a genus characterized by compound leaves. Vernacular names are Dwarf Ash and Fresno. The seed averages about 22,000 seeds per lb, giving rise to about 5,000 usable plants per lb of seed. It provides at least poor, and sometimes fair to fairly good, browse for goats, sheep, and cattle.

Berlandier Ash

Fraxinus berlandieriana A. DC.

FIELD IDENTIFICATION. Small round-topped tree of Western distribution, seldom seen east of the Colorado River except in cultivation. Rarely over 30 ft.

FLOWERS. Dioecious, greenish, staminate and pistillate flowers on different trees; calyx of staminate flower obscurely 4-lobed; stamens 2, filaments short, anthers linear-oblong and opening laterally; calyx of pistillate flower campanulate, deeply cleft; ovary with a slender style and stigmas 2-lobed.

FRUIT. Ripening in May. Samara spatulate to oblong-obovate, 1–1½ in. long, about ¼ in. wide; wing acute or acuminate at apex, decurrent down the seed body almost to the base, set in a deeply lobed calyx; samaras sometimes 3-winged.

LEAVES. Deciduous, opposite, odd-pinnate, slender, petioled, 3–10 in. long; leaflets 3–5, petiolulate, elliptic, lanceolate or obovate, acuminate to acute at apex, cuneate or rounded at base, entire or remotely serrate, thickish, dark green and glabrous above, glabrous or a few axillary hairs beneath, 3–4 in. long, ½–1½ in. wide, petiolule of terminal leaflet longer than those of lateral leaflets. Leaflets fewer, smaller, more coarsely toothed, and more widely separated than those of White Ash or Green Ash.

TWIGS. Green, reddish or gray, with scattered lenticels, leaf scars small, raised, oval.

BARK. Gray or reddish, fissures shallow and ridges narrow.

Berlandier Ash
Fraxinus berlandieriana A. DC.

WOOD. Light brown, sapwood lighter, close-grained, light, soft.

RANGE. Moist canyons and stream banks. Central Texas to Trans-Pecos Texas; southward in Mexico in Coahuila, Durango, and Veracruz.

MEDICINAL USES. The bark contains the glucoside fraxin, which is used in Mexico as a tonic and febrifuge. A decoction of the leaves is locally used as a treatment for yellow fever, malaria, gout, and rheumatism. Folk tales recount that ash leaves are so offensive to rattlesnakes that they are seldom found near them, and hunters often thrust the leaves in their boots for protection.

REMARKS. The genus name, *Fraxinus*, is the ancient Latin name, and the species name, *berlandieriana*, is in honor of the Swiss botanist, Jean Louis Berlandier (1805–1851), who collected extensively in Mexico and Texas. Local names for the tree are Plumero, Fresno, and Mexican Ash. The wood has no particular commercial importance, but the tree is widely planted as an ornamental in western and southwestern Texas and Mexico.

Velvet Ash

Fraxinus velutina Torr.

FIELD IDENTIFICATION. Slender tree with a diameter to 18 in. and a height of 50 ft, but generally only half that size, the branches spreading and forming a rounded crown. Velvet Ash is extremely variable in its form, size of leaflets, fruit, and the amount of hairiness. A number of varieties are recognized, some occurring with the species and connected by intergrades.

FLOWERS. Usually March–May with the unfolding leaves; dioecious, borne in long, pubescent panicles on slender pedicels; calyx densely pubescent, cup-shaped, stamens short, anthers oblong and apiculate; ovary enveloped in the calyx, stigma lobes subsessile.

FRUIT. Ripening in September, clusters often abundantly fruited, oblong to obovate or elliptic, about ¾ in. long and ⅙ in. wide, wing with apex rounded to acute or emarginate, seed terete, many-rayed, about ½ in. long.

LEAVES. Rachis slender, grooved, pinnately compound, 4–5 in. long, composed of 3–5 leaflets, ovate to elliptic or obovate, apex acute, base cuneate to rounded, margin semientire to finely crenate-serrate above, thickened, when young densely tomentose, when older pale green and glabrous on the upper surface, tomentose beneath, midrib conspicuous, veins reticulate, length 1–1½ in., width ¾–1 in.; lateral petiolules to ⅙ in. long, terminal petiolule to ½ in. long.

TWIGS. Densely tomentose at first, later glabrous and gray; leaf scars large, obcordate; buds about ⅛ in. long, acute, scales 6 or more, ovate to linear, tomentose, ¼–½ in. long.

BARK. Gray to brown or reddish, ⅓–½ in. thick, fissures deep, ridges broad and flat, scales small and thin.

VELVET ASH
Fraxinus velutina Torr.

WOOD. Light brown, sapwood lighter, close-grained, rather soft, not strong, fairly heavy.

RANGE. Usually in mountain canyons at altitudes of 2,000–6,000 ft. Trans-Pecos Texas, New Mexico, Arizona, Utah, Nevada, and California. In Mexico in Baja California, Sonora, and Chihuahua.

PROPAGATION. The seeds are gathered when mature in late summer. The wing may be removed or seed planted with the wing attached. The cleaned seed averages about 20,600 seeds per lb, with a purity of about 92 per cent and a soundness of 20–74 per cent, depending on the lots. Different lots of seed vary greatly as to germination time. For spring planting, stratification over winter for 90 days at 41°F. or soaking at room temperature for 10–21 days will hasten germination. The seeds may be sown in drills or broadcast, with about 30 seeds per linear ft, and covered about ¼ in. with good nursery soil. Fall-sown seeds may be mulched with straw until germination starts in the spring.

VARIETIES. A number of segregates are generally recognized for Velvet Ash as follows:

Leather-leaf Velvet Ash, *F. velutina* var. *coriacea* (Wats.) Rehd., has thicker, more leathery, coarsely serrate leaves and branchlets pubescent to glabrate. It occurs mostly from Utah to California, and has been cultivated since 1900.

Smooth-leaf Velvet Ash, *F. velutina* var. *glabra* (Thornb.) Rehd., has leaves 3–7-foliate and glabrous with the twigs. It usually occurs with, and intergrades with, the species. It has been in cultivation since 1916.

Toumey Velvet Ash, *F. velutina* var. *toumeyi* (Britt.) Rehd., has leaflets 5–7, elliptic to lanceolate, 1¼–2¾ in. long, apex acuminate, base cuneate, margin finely toothed toward the apex, finely pubescent beneath, petiolules ⅛–⅖ in. long; samara ⅝–1 in. long, wing oblong to spatulate, about as long as the rounded seed or somewhat longer. In New Mexico, Arizona, and Mexico. Introduced into cultivation in 1891.

REMARKS. The genus name, *Fraxinus,* is the classical name of Ash, and the species name, *velutina,* refers to the velutinous (velvety) hairs of the leaves. Other names are Arizona Ash, Desert Ash, Smooth Oregon Ash, and Fresno. The wood is used for ax handles and in the manufacture of wagons. The tree is used in shelter-belt planting and in ornamental planting in arid regions of the Southwest. Velvet Ash has been in cultivation since 1900. It has also some use as a food for wildlife.

CAROLINA ASH
Fraxinus caroliniana Mill.

FIELD IDENTIFICATION. Tree of deep swamps, with small branches and rounded open head, to 40 ft.

FLOWERS. Dioecious, appearing in fasciculate panicles before the leaves, yellowish green; no petals; staminate

Carolina Ash
Fraxinus caroliniana Mill.

panicles dense; staminate calyx minute; stamens 2–4, with linear-oblong, apiculate anthers opening longitudinally; pistillate flowers in slender clusters about 2 in. long; pistillate calyx campanulate, deeply cleft, persistent; ovary globose, elongated into a forked style.

FRUIT. Samara oblong-obovate to elliptic, yellowish brown, flattened, thin, smooth, 1–3 in. long, ½–¾ in. broad; seed elliptic, flattened, surrounded by the broad wing; wing pinnately veined, midvein impressed, apex acute, rounded, or emarginate.

LEAVES. Deciduous, opposite, odd-pinnately compound of 5–9 (usually 7) leaflets, 7–12 in. long, petiole elongate; leaflets long-petiolulate, the blades 2–6 in. long, ½–3 in. broad, oblong-ovate, thick, acute, or acuminate at apex, rounded or cuneate at base, serrate or sometimes entire on margin; dark green, lustrous and glabrous above; paler and glabrous or pubescent beneath.

TWIGS. Slender, terete, green, pubescent at first, later brown to gray and glabrous.

BARK. Gray, often blotched, thin, smoothish, with small scales.

WOOD. Yellowish white, close-grained, soft, weak, weighing 22 lb per cu ft, not important commercially.

RANGE. Swamplands, eastern Texas, Arkansas, and Louisiana; eastward to Florida and northward to Washington, D.C., Virginia, and Missouri.

VARIETY. *F. caroliniana* var. *pubescens* (M. A. Curtis) Fern. is a variety having yellowish green leaves, wit more pubescence beneath than in the species.

REMARKS. The genus name, *Fraxinus,* is the ancien Latin name, and the species name, *caroliniana,* refer to the states of Carolina. Vernacular names are Popp Ash, Pop Ash, and Water Ash.

Green Ash

Fraxinus pennsylvanica var. *lanceolata* (Borkh.) Sarg

FIELD IDENTIFICATION. Spreading, round-topped tre attaining a height of 70 ft or more.

FLOWERS. Dioecious, borne in spring in slender pediceled, terminal, glabrous panicles; no petals; stam inate with a campanulate, obscurely toothed calyx stamens 2, composed of short, terete filaments an linear-oblong, greenish purple anthers; calyx of pistil late flowers deeply cleft; ovary 2–3-celled, style elon gate, with 2 green stigmatic lobes.

FRUIT. Samaras in panicles; samaras flat, 1–2 in long, ¼–⅓ in. wide, winged; wing decurrent down th side of seed body often past the middle, spatulate o oblanceolate; end of wing square, notched, rounde or acute; seed usually 1-celled, or rarely 2–3-celled.

LEAVES. Deciduous, opposite, odd-pinnately com pound, 8–12 in. long, rachis glabrous; leaflets 5–9 (usu

Green Ash
Fraxinus pennsylvanica var. *lanceolata* (Borkh.) Sarg

lly 7) ovate to oblong-lanceolate, acute or acuminate t apex, cuneate at base, entire or irregularly serrate n margin, lustrous green on both sides or somewhat aler beneath; glabrous above, usually glabrous below r with scant pubescence on veins, 2–6 in. long, 1–2 n. wide.

TWIGS. Gray, glabrous, terete; lenticels pale.

BARK. Brown, tight, ridges flattened, furrows shallow, cales thin and appressed.

WOOD. Light brown, sapwood lighter, coarse-grained, eavy, hard, strong, weighing 44 lb per cu ft.

RANGE. Texas, New Mexico, Oklahoma, Arkansas, nd Louisiana; eastward to Florida, northward to Nova cotia, and west to Manitoba, Montana, Wyoming, Colorado, and Kansas.

PROPAGATION. Green Ash is often planted as a shade ree and is also much used for shelter-belt planting in he prairie-plains states. It was first cultivated in 1823. The seeds may be hand-picked or obtained by pruners n fall. The stems may be removed by light flailing and he seeds spread out to dry. They are stored best in ealed containers and kept dry. The cleaned seed aver-ges about 17,300 seeds per lb, with a purity of 89 er cent and a soundness of 88 per cent. The seed may e planted in the fall or in early spring. If sown in pring, they may be stratified in sand or peat 60–90 lays at 41°F., or they may be water-soaked at room emperature 10–21 days. Planting is done in nursery ows in drills, or seeds may be broadcast. About ¼ in. oil is spread over them. If sown in the fall, the beds re covered with a mulch of leaves, straw, or burlap. Nursery germination averages 50–60 per cent.

Old trees are subject to the attack of a fungus, *Polyporus fraxinophilus,* which turns wood into a yellow ulp. The ash-leaf rust, *Aecidium fraxinii,* sometimes ttacks the leaves, but does little damage. It can be enerally controlled by spraying with Bordeaux mixture 4–6–50) at 2-week intervals. A 2 per cent solution f lime-sulphur is often used.

VARIETIES AND RELATED SPECIES. Green Ash is closely elated to Red Ash, *F. pennsylvanica* Marsh., from which it is distinguished by the lustrous, green, lanceo-ate, sharply serrate leaves and glabrous twigs. How-ver, there are numerous intergrading forms.

Aucuba-leaf Ash, *F. pennsylvanica* var. *aucubaefolia* K. Koch) Rehd., has yellow-mottled leaves.

Berlandier Ash, *F. berlandieriana* A. DC., is a closely elated species of southwest Texas and Mexico.

REMARKS. The genus name, *Fraxinus,* is the ancient Latin name for Ash-tree. The species name, *pennsyl-vanica,* refers to the state of Pennsylvania, and the arietal name, *lanceolata,* to the lanceolate leaflets. Vernacular names are Water Ash, River Ash, Red Ash, he Swamp Ash. The wood is not as desirable as that f White Ash but used for the same purposes: tool andles, furniture, interior finishing, cooperage, and wagons. A number of birds eat the seed and the foliage s browsed by white-tailed deer and cottontail.

TEXAS ASH

Fraxinus texensis (Gray) Sarg.

FIELD IDENTIFICATION. Tree to 50 ft, 2–3 ft in diameter, with a short trunk and contorted branches.

FLOWERS. With the leaves in March, in large glabrous panicles, buds with ovate, rounded, brown to orange-colored scales, from the axils of last year's leaves; staminate and pistillate panicles separate; staminate with petals absent and a minute 4-lobed calyx; stamens 2, filaments short, anthers linear-oblong and apiculate, purplish; pistillate calyx with 4 deep, acute lobes; ovary attenuate into a slender style.

FRUIT. Samara borne in compact panicles 2–3¾ in. long on slender pedicels ⅛–¼ in. long, fruit ½–1 in. long, 3/16–¼ in. wide, body rounded; wing terminal on the seed body or extending only slightly on the sides, apex rounded or notched (rarely with more than 1 wing).

LEAVES. Odd-pinnately compound, 5–8 in. long; leaflets 5 (more rarely 7); petiolules ¼–1½ in. long, slender, yellowish green; blades elliptic to oblong or ovate to obovate, apex acute, base broadly cuneate or rounded, margins obscurely serrate or entire toward the base, length 1–3 in., width ¾–2 in.; upper surface olive-green to dark green and glabrous; lower surface paler and often somewhat glaucous and glabrous, or with a few white hairs on the main vein.

TWIGS. Numerous, stout, terete, green to reddish

TEXAS ASH
Fraxinus texensis (Gray) Sarg.

brown or gray, younger slightly puberulous, older glabrous, smooth; lenticels scattered, oblong, pale; leaf scars large, raised, with conspicuous fibrovascular bundles; buds acute, ovate, apex rounded or truncate, brown to orange, densely hairy.

BARK. Gray to brown or black, ½–¾ in. thick, furrows deep, the wide ridges confluent to give a netlike appearance.

WOOD. Light brown, sapwood paler, strong, hard, heavy.

RANGE. From the Arbuckle Mountains of Oklahoma southward over the limestone Edwards Plateau of Texas. In Texas in Dallas, Tarrant, Travis, Bandera, Kerr, Edwards and Palo Pinto counties. Collected by the author between Utopia and Tarpley, Texas.

REMARKS. The genus name, *Fraxinus,* is the ancient Latin name, and the species name, *texensis,* refers to the state of Texas where it occurs. It is also known as Mountain Ash because of its growth on limestone hills.

The wood is used for fuel or flooring, but is hardly abundant enough to be of commercial importance. It is a handsome tree and should be more extensively grown for ornament. It was first cultivated in 1901. Texas Ash is closely related to White Ash, *F. americana,* and some botanists consider it as only a variety of the latter.

WHITE ASH
Fraxinus americana L.

WHITE ASH

Fraxinus americana L.

FIELD IDENTIFICATION. Tree attaining a height of 100 ft and a diameter of 3 ft. Records show that some trees have reached a height of 175 ft and a diameter of 5 to 6 ft, but such trees are no longer to be found. The general shape is rather narrow and rounded.

FLOWERS. Borne April–May, dioecious, with or before the leaves in staminate and pistillate panicles; staminate clusters short and dense; individual flowers minute, green to red, glabrous; no petals; calyx campanulate, 4-lobed; stamens 2–3, filaments short, anthers oblong-ovate and reddish; pistillate clusters about 2 in. long, slender, calyx deeply lobed; style split into 2 spreading, reddish purple stigmas.

FRUIT. Ripening August–September. Samaras in dense clusters often 6–8 in. long; seed body terete; wing slightly extending down the body of the seed, but usually not at all, oblong or spatulate; often notched at the end, thin, smooth, flat, yellow to brown, 1–2½ in. long, about ¼ in. wide.

LEAVES. Simple, opposite, deciduous, odd-pinnately compound, 8–13 in. long, leaflets 5–9, usually 7, ovate-lanceolate, acuminate or acute, rounded or cuneate at base, entire or crenulate-serrate on margin, dark lustrous green above, paler and whitish and glabrous or pubescent beneath, 3–5 in. long, 1½–3 in. wide; petiole glabrous.

TWIGS. Green to brown or gray, stout, smooth with pale lenticels.

BARK. Light gray to dark brown, ridges narrow and separated by deep fissures into interlacing patterns.

WOOD. Brown, sapwood lighter, close-grained, strong, hard, stiff, heavy, tough, weighing 41 lb per cu ft, seasons well, takes a good polish, moderately durable, shock resistant.

RANGE. Typical White Ash is distributed in Oklahoma, Arkansas, Texas, and Louisiana; eastward to Florida, northward to Nova Scotia, and west to Ontario, Minnesota, Michigan, and Nebraska.

PROPAGATION. A good crop of seeds is borne every 3–5 years, with light crops in intervening years. The minimum seed-bearing age is 20 years, with a maximum of 175 years. The seed may be gathered by hand, or cut off with pruning shears. It is spread out a few days to dry, and the stems removed by flailing in a sack. Dewinging is usually unnecessary, but is sometimes done. Cleaned seed averages about 10,000 seeds per lb, with a commercial purity of 92 per cent and a soundness of 80 per cent. The seeds are best stored at low temperatures with a low moisture content. For best success in planting, the seeds may be stratified in sand or peat for 60–90 days at 41°F. Some growers soak them in water 10–27 days at 70°F. before planting. The germinative capacity averages 38 per cent for 40–60 days with stratified seed. Seeds are sown at the rate of 25–30 per linear ft in drills or nursery rows and covered with a thin layer of soil. Beds sown in fall should be mulched with leaves, burlap, or straw. Sometimes seedlings are subject to leaf damage by the defoliating fungus, *Marsonia fraxinii*. Spraying with Bordeaux mixture at 2-week intervals is recommended for control.

VARIETIES. A number of varieties of White Ash are listed in the literature as follows:

Purple-fruit White Ash, *F. americana* var. *iodocarpa* Fern., has samaras tinged with purple.

Small-fruit White Ash, *F. americana* var. *microcarpa* Gray, is a shrubby form with small seeds about ½ in. long.

Leather-leaf White Ash, *F. americana* var. *subcoriacea* Sarg., has entire or slightly serrate and leathery leaves.

Walnut-leaf White Ash, *F. americana* var. *juglandifolia* (Lam.) Rehd., has leaflets more or less serrate, less lustrous above, less white beneath, and with varying amounts of pubescence.

Texas Ash, *F. texensis* Sarg., is a species considered to be closely related to White Ash and grows from Dallas, Texas, over the limestone hills to the Devil's River. It usually has only 5 leaflets which are elliptic to obovate, obtuse or acute and crenulate-serrate on the upper half of the leaf.

REMARKS. The genus name, *Fraxinus,* is the ancient Latin name, and the meaning of the species name, *americana,* is obvious. Also known as Small-seed White Ash, Cane Ash, Biltmore Ash, and Biltmore White Ash. It has been known in cultivation since 1724. It is an important timber tree and is widely planted as an ornamental. It is estimated that 45 per cent of all ash lumber used is from the White Ash. The center of production is now the lower Mississippi Valley. No differentiation is made in the lumber trade as to the species of ash, however the term "white ash" generally designates top quality ash. Ash wood is used for tanks, silos, toys, musical instruments, cabinets, refrigerators, millwork, sash, doors, frames, vehicle parts, farm utensils, woodenware, butter tubs, veneer, fuel, railroad cross ties, sporting goods, furniture, cooperage, handles, ships, boats, railroad cars, and frame parts of airplanes. It usually grows in association with other hardwoods in well-drained soils on slopes. It is valuable in small tracts for woodland management. Although sometimes used, it is not as valuable for shelter-belt planting as Green Ash. The fruit is known to be eaten by a number of birds, including the purple finch and pine grosbeak, and the foliage is browsed by rabbit, porcupine, and white-tailed deer.

BLUE ASH
Fraxinus quadrangulata Michx.

BLUE ASH

Fraxinus quadrangulata Michx.

FIELD IDENTIFICATION. Tree attaining a height of 80 ft and a diameter of 2–3 ft. The small branches spread into a slender or rounded crown. The root system is spreading and shallow, and the tree is rather rapid-growing.

FLOWERS. April–May, axillary, developing in axils of leaves of the previous year, perfect, dioecious or polygamous, panicles loosely flowered; stamens 2, anthers nearly sessile, linear-oblong, obtuse, dark purple; corolla absent; calyx obscure; ovary oblong-ovoid, 2-celled, 2 ovules in each cell; style short, dividing into 2 short, purplish lobes.

FRUIT. Maturing September–October, samara often twisted on its axis, linear-oblong to elliptic, apex often notched or some rounded to obtuse, surface faintly parallel-veined; length 1–2 in., width ¼–½ in., seed body flattened, extending more than half way to apex, surrounded by the wing.

LEAVES. Opposite, pinnately compound, deciduous, 8–12 in. long, rachis slender and glabrous or puberulent; leaflets 5–11 (mostly 7), lateral ones with petiolules ⅛–¼ in. long or almost sessile, terminal petiolule ⅜–⅝ in. long, oblong to lanceolate or ovate, apex long-acuminate, base rounded to cuneate and inequilateral, margin serrate with short incurved teeth; young leaves brown-tomentose, older ones firm and rather leathery, yellowish green to dull green, upper surface smooth and glabrous, lower surface paler and glabrous or slightly hairy along the midrib and veins, length 3–5 in., width 1–2 in., veins arcuate near the margin, leaves yellow in autumn.

TWIGS. Stout, reddish brown and pubescent at first, later glabrous and brown to light gray, conspicuously 4-sided with the angles sometimes slightly winged; lenticels pale and scattered; leaf scars large, U-shaped, bundle scars crescent-shaped; terminal buds ovoid, reddish brown, larger than the lateral buds, about ¼ in. long, with 2–3 pairs of scales.

BARK. Light brown to gray, irregularly fissured and broken into large plates with smaller thin scales.

WOOD. Yellowish brown, sapwood lighter, close-grained, weighing 47 lb per cu ft, hard, rather brittle, very durable.

RANGE. Usually in rather dry, calcareous soils of hills or bottom lands. Most abundant in the upper Mississippi and Ohio valleys. Arkansas, Oklahoma, and Alabama; northward to Minnesota and southern Ontario.

PROPAGATION. The minimum commercial seed-bearing age is 25 years, and the optimum 40–125 years. The seeds are picked by hand and spread out to dry. Dewinging is not necessary, but can be done if care is taken not to damage the seed. The seed averages about 7,000 seeds per lb, with an average purity of 72 per cent and a soundness of 85. Storage in ordinary containers at room temperature is satisfactory for 1 year, but for longer periods dry storage in sealed containers at low temperatures helps maintain vitality. Dormancy is of both seed coat and embryo types and can be broken by stratification in moist sand for 69–90 days at 68°–86°F., plus 120 days at 41°F. If not stratified the seed may be sown in spring or summer and beds mulched until the following spring when they germinate. The seeds are usually sown in drills 6–12 in. apart and covered ½–¾ in. with nursery soil.

REMARKS. The genus name, *Fraxinus*, is the classical name of the ash. The species name, *quadrangulata*, refers to the 4-angled twigs. It is also known under the name of Hoop Ash. The name Blue Ash refers to the fact that a blue dye is obtained by immersing the inner bark in water. The tree has been in cultivation since 1823. The wood is generally sold as White Ash, and is used for flooring and interior finishing.

Pumpkin Ash

Fraxinus tomentosa Michx.

FIELD IDENTIFICATION. A tree attaining a height of 120 ft. The trunk sometimes measuring 3 ft in diameter, but usually larger nearer the ground because of the much-swollen and buttressed base. Branches rather spreading to form a narrow open crown. The leaves are generally larger than those of most ashes, and the frui[t] is borne in dense clusters.

FLOWERS. April–May, dioecious, panicles elongate much-branched, pubescent; bracts oblong to obovate scarious; staminate calyx campanulate, very small minutely 4-toothed; stamens 2–3, filaments slender and elongate, anthers oblong and apiculate; pistillate calyx persistent, larger, ⅙–⅕ in. long (sometimes only ⅛) deeply lobed; ovary with an attenuate style.

FRUIT. Maturing September–October, clusters long and drooping, fruit variable in size and shape, abundant oblong to linear or spatulate, 2–3 in. long, to ½ in wide, straight or falcate; apex rounded or emarginate often apiculate, decurrent on the seed body nearly to the base or sometimes only one fourth to one half its length; body of seed terete or somewhat flattened striate.

LEAVES. Opposite, pinnately compound, deciduous 9–18 in. long, petiole tomentose; leaflets usually 7 sometimes 3 or 9, blade lanceolate to elliptic, margin entire to serrulate, apex acuminate, some abruptly so base cuneate to rounded and sometimes unsymmetric young leaflets with dense hoary tomentum, older leaflets dark yellowish green and almost glabrous above lower surface soft-pubescent, 5–10 in. long and 1½–

Pumpkin Ash
Fraxinus tomentosa Michx.

in. wide; petiolules 1/4–1/2 in. long (terminal one often longer), tomentose at first, glabrous later.

TWIGS. Young ones light gray, at first densely velvety-tomentose, later pubescent, and eventually glabrous on older twigs; lenticels large and pale; leaf scars large; terminal buds reddish brown, ovate, obtuse, pubescent.

BARK. Light gray, ridges flattened or rounded, with small appressed scales, furrows shallow; bark somewhat resembling that of White Ash.

RANGE. Swamps and bottom lands in the sun. Louisiana, eastward to northern Florida, northward to New York and up the Mississippi Valley to southern Illinois, Indiana, and Ohio.

VARIETY. Palmer Pumpkin Ash, *F. tomentosa* var. *ashei* Palmer, is a variety with twigs and leaves either smooth or hairy only along the midribs of the blade. From eastern Maryland southward to western Florida.

REMARKS. The genus name, *Fraxinus,* is the classical name of the ash. The species name, *tomentosa,* refers to the thick hairs on the young leaves. Some authors list the plant under the names of *F. profunda* Bush and *F. michauxii* Britt. However, the author follows the more recent nomenclature of Fernald (1938, pp. 450–452). Some vernacular names for the plant are Red Ash and Swell-butt Ash. Wood similar in appearance to that of White Ash but inferior in quality.

LOGANIA FAMILY (*Loganiaceae*)

EMORY-BUSH

Emorya suaveolens Torr.

FIELD IDENTIFICATION. Rare, much-branched shrub attaining a height of 6 ft, with all young parts densely tomentose.

FLOWERS. In narrow pedunculate panicles; calyx-tube oblong, about 5/16 in. long, tubular, densely stellate brown-tomentose, 4-lobed, lobes as long as the tube or slightly shorter, linear, subulate; corolla sweet-scented, greenish white or yellowish, 1–1½ in. long, long-tubular, gradually widened toward apex, densely hairy, 4-lobed; lobes about ⅛ in. long, ovate-triangular, obtuse; stamens 4, exserted, filaments filiform, inserted on middle of tube; style filiform and exserted longer than the stamens; stigma entire, small; ovary free from calyx, oblong-conical, base surrounded by a glandular ring.

FRUIT. Capsule globose to oblong, apiculate at apex, septicidal, 2-celled; valves 2-cleft at apex; seeds numerous, imbricate, testa loose and cristate at both ends.

LEAVES. Opposite, blades ½–2 in. long, ovate to deltoid or hastate, margin coarsely sinuate-dentate, revolute, apex obtuse to acute or rounded, base obtuse or truncate, upper surface dull green or yellowish green with scattered stellate hairs or almost glabrous, lower surface conspicuously white-tomentose, pinnately nerved; petioles ¼–½ in. long, densely tomentose, some narrowly margined by leaf almost to base.

STEMS. Slender, green, with dense stellate tomentum, later more glabrous and light brown to dark brown or gray, slightly striate.

RANGE. Trans-Pecos Texas and northern Mexico. Canyons of the Rio Grande below Presidio. Also in Nuevo León and Coahuila, Mexico. Specimen examined by the author collected by Ivan M. Johnston, No. 7215 from Saucilla, Coahuila, Mexico. Deposited in the United States National Herbarium.

EMORY-BUSH
Emorya suaveolens Torr.

REMARKS. The genus name, *Emorya*, honors W. H. Emory, United States commissioner to the United States–Mexico boundary survey. The species name, *suaveolens*, refers to the sweet scent.

CAROLINA JESSAMINE

Gelsemium sempervirens (L.) Ait.

FIELD IDENTIFICATION. Beautiful, slender, evergreen vine.

FLOWERS. February–April, borne on short, scaly-bracted pedicels, solitary or clustered in the leaf axils, very fragrant, funnelform, yellow, 1–1½ in. long, 5-lobed; calyx deeply 5-lobed, lobes ovate to oblong or elliptic, obtuse, ⅛–¼ in. long; subtended by a number of green, oblong to elliptic, acuminate bracts 1/16–⅛ in. long; stamens 5 with anthers linear-oblong, sagittate, almost as long as filaments which are attached below in corolla-throat; style slender, filiform, much longer than stamens, stigma lobes 3–4, linear; ovary oblong, 2-celled. Flowers dimorphous, occurring in forms with either exserted or included stamens.

FRUIT. Capsule elliptic, oblong, or ovoid, veiny, flattened, channeled and sutured, beaked, ⅓–⅔ in. long, septicidally dehiscent, 2-valved; seeds flat, oblique-oblong, winged, ⅕–¼ in. long, numerous.

LEAVES. Evergreen, opposite, short-petioled, pairs rather widely separated on stem, lanceolate or ovate, apex acuminate, base cuneate, entire, dark green, shiny, glabrous, blades 1–3 in. long, ½–1 in. wide, persistent.

CAROLINA JESSAMINE
Gelsemium sempervirens (L.) Ait.

TWIGS. Stem up to 20 ft, smooth, slender, green or reddish brown, twining and tangled, sometimes prostrate, but usually climbing.

RANGE. Generally in sandy moist soil of eastern and southern Texas, Arkansas, Oklahoma, and Louisiana; eastward to Florida and north to Virginia. Also Mexico in the states of Veracruz, Puebla, Oaxaca, and Chiapas. Also Guatemala.

MEDICINAL USES. The dried rhizome of Carolina Jessamine contains the poisonous alkaloid gelsemine, which is used medicinally. It is usually administered in the form of a tincture, and is reported to be efficacious in treatment of facial neuralgia, rheumatism, and gonorrhea. The drug should be administered only under advice of a physician because improper usage has resulted in vertigo, and even death. It is also poisonous to livestock.

REMARKS. The genus name, *Gelsemium,* refers to the Italian name for Jessamine, *gelsomino,* and the species name, *sempervirens,* is given because of its evergreen habit. Some vernacular names in use are Yellow Jessamine, Jasmine, Carolina Wild Woodbine, and Evening Trumpet Flower. This native vine is not to be confused with the cultivated jessamines and jasmines, which belong to the genera *Jasminum* and *Cestrum,* members of the Olive and Nightshade families, respectively. The seeds have been known to be eaten by bobwhite quail and the leaves by marsh rabbit. The poisonous effects on wildlife, if any, are unknown.

ESCOBILLA BUTTERFLY-BUSH

Buddleia scordioides H. B. K.

FIELD IDENTIFICATION. Aromatic western shrub to 3 ft, all parts ferruginous-tomentose.

FLOWERS. In dense sessile clusters ⅕–⅓ in. in diameter, a number together forming a globular head in axils of upper leaves; individual flowers small, crowded, perfect, regular; calyx campanulate, the 4 teeth hidden in dense brown tomentum; corolla whitish, densely pubescent, rotate-campanulate, the 4 lobes ovate to orbicular, spreading at anthesis; stamens 4, included, sessile or nearly so; styles 2, united; ovary 2-celled, ovules solitary.

FRUIT. Capsule oblong, septicidally dehiscent at the apex, bivalvate; seeds numerous and minute.

LEAVES. Opposite, blades ⅜–1⅓ in. long, ⅛–¼ in. wide, sessile or nearly so, shape narrowly oblong to cuneate-linear, apex obtuse, base gradually narrowed, margin coarsely crenate, upper surface rugose, with sunken veins and short velvety hairs, lower surface with denser brown tomentum.

TWIGS. Slender; young twigs with dense brown tomentum, older tan to light brown or gray, glabrous and often shreddy.

ESCOBILLA BUTTERFLY-BUSH
Buddleia scordioides H. B. K.

RANGE. On dry sunny open sites. Southwestern Texas, New Mexico, and in Mexico from Chihuahua to San Luis Potosí, Hidalgo and Mexico, D.F.

REMARKS. The genus name, *Buddleia,* honors Adam Buddle (1660–1715), an English botanist. The species name, *scordioides,* refers to a plant with a strong odor (*Teucrion scordium* L.). The plant is also known in Mexico under the names of Salvia, Hierba de las Escobas, and Golondrilla. A tea is made from the leaves as a remedy for indigestion. Considered to be fairly good browse for sheep, goats, and cattle. About 100 species of *Buddleia* are known in tropical and temperate regions of America, Asia, and South Africa.

TEXAS BUTTERFLY-BUSH

Buddleia racemosa Torr.

FIELD IDENTIFICATION. Low, loose-branching shrub to 3 ft, bearing slender racemes of small flowers in op posite, tomentose, pedunculate heads.

FLOWERS. In slender racemes 4–12 in. long, the op posite, globular, dense, tomentose heads of flowers 1/5– in. broad; corolla barely exceeding the tomentose calyx yellowish white, rotate-campanulate, 1/12–1/8 in. long tube straight and hairy above; limb 4-lobed, lobe broad, ovate to orbicular, spreading; stamens 4–5, in cluded, anthers sessile or nearly so on the corolla throat; ovary 2-celled, styles 2 and united; calyx smal campanulate, 4-lobed, densely tomentose, almost a long as the corolla.

FRUIT. Capsule oblong or globose, septicidally 2 valved, the valves 2-cleft at the apex and separating th united placentae.

LEAVES. Simple, opposite, short-petioled, oblong to ovate or lanceolate, margin crenate-dentate, apex obtus or acute, base truncate to cuneate or slightly hastate

TEXAS BUTTERFLY-BUSH
Buddleia racemosa Torr.

›lades 2–4 in. long, upper surface green and glabrous, ɔwer surface pale puberulent-canescent.

TWIGS. Stems loosely branching and glabrous, nearly erete, terminated by the virgate racemes of flowers which are often pendent on sheer cliff sides.

RANGE. Growing on edges of limestone cliffs, on rocky helves, or on banks of dry stony ravines. In central, vestern, and southwestern Texas.

PROPAGATION. The buddleias are propagated by seeds, reenwood cuttings, or hardwood cuttings taken in fall nd protected from frost during the winter.

VARIETIES. A variety of Texas Butterfly-bush with eaves brown canescent-tomentose beneath is *B. racemosa* var. *incana* Torr. Lindley Butterfly-bush, *B. lindeyana* Fort., is a Chinese species, often cultivated in ardens on the Gulf Coast.

REMARKS. The genus name, *Buddleia*, is dedicated o Adam Buddle (1660–1715), an English botanist, and the species name, *racemosa*, refers to the flower acemes. About 100 species of *Buddleia* are known rom America, Asia, and South Africa.

Woolly Butterfly-bush
Buddleia marrubiifolia Benth.

Woolly Butterfly-bush

Buddleia marrubiifolia Benth.

FIELD IDENTIFICATION. Western shrub to 3 ft, all parts densely brown-tomentose.

FLOWERS. In terminal, solitary, globose, dense heads ½–⅝ in. across; peduncles short but distinct; calyx minute, campanulate, densely tomentose, 4-toothed, teeth obtuse or rounded; corolla golden yellow to orange-red, rotate-campanulate, tube tomentose, limb 4-lobed; lobes imbricate, spreading at anthers, ovate to orbicular, less tomentose than tube or glabrous; anthers 4, included, sessile or nearly so, attached to middle of corolla-tube; stigma slightly exserted.

FRUIT. Capsule septicidal, bivalvate, seeds small and numerous.

LEAVES. Opposite, very short petioled, elliptic to oval or obovate, base cuneate or decurrent, apex obtuse or rounded, margin coarsely crenate or entire near base, blades ⅓–1¼ in. long, texture thick and velvety, surfaces with dense golden brown or brown tomentum, veins obscure above, coarse beneath.

TWIGS. Slender, young densely tomentose, older tan to gray, glabrous, often striate or shreddy.

RANGE. On well-drained sunny sites. Southern and western Texas along the Rio Grande and adjacent Mexico. In Chihuahua, Coahuila, Nuevo León, and Zacatecas.

REMARKS. The genus name, *Buddleia*, honors Adam Buddle (1660–1715), English botanist. The species name, *marrubiifolia*, refers to the woolly marrubium-like leaves. Also known by vernacular names of Saffron, Azafrán, Azafrán del Campo, Azafrancillo, Azafranillo, and Topocan. An infusion of the flowers is used to impart a yellow color to butter in Coahuila, Mexico. Also used as a bath for rheumatic disorders, and as an aperitive and diuretic. It is a minor source of browse for livestock. A number of Asiatic species of *Buddleia* are grown for ornament.

Humboldt Butterfly-bush

Buddleia humboldtiana R. & S.

FIELD IDENTIFICATION. Southwestern shrub or small tree to 18 ft. Young stems and foliage copiously brown-tomentose.

FLOWERS. Borne in large, ample, naked, brown-tomentose, terminal panicles. Clusters numerous with perfect small flowers hardly over ⅛ in. long; corolla rotate, short-campanulate, 4–5-lobed, the lobes imbricate, spreading in anthesis, rounded; throat hairy with the anthers sessile or nearly so; calyx-teeth 4, triangular, obtuse, ¼–⅓ the length of the campanulate tube, ovary

HUMBOLDT BUTTERFLY-BUSH
Buddleia humboldtiana R. & S.

free from calyx with fruiting capsule septicidally biva vate.

LEAVES. Opposite, narrowly lanceolate to oblong c ovate, apex acute or acuminate, base broadly acute c rounded, some slightly cordate, blades 2–6 in., margi entire or serrulate; upper surface dark green, whe young with scattered short brownish hairs, later almos glabrous; lower surface conspicuously net-veined wit fine brown stellate tomentum; petiole ¼–1¾ in. lon angular, finely brown-tomentose, sometimes slightl winged by the leaf base.

TWIGS. Angled, young finely brown-tomentose, late gray and more glabrous

RANGE. In southwest Texas. Also in Mexico in th states of Nuevo León, Chihuahua, San Luis Potos Chiapas, and Oaxaca.

MEDICINAL USES. Decoctions of the bark and roc are used as a diuretic, for uterine affections, and fc rheumatism.

REMARKS. The genus name, *Buddleia,* honors Adar Buddle, English botanist, and the species name, *hum boldtiana,* is in honor of Alexander von Humboldt. Th plant is known in Mexico under the names of Tepozá and Tepozán Blanco.

DOGBANE FAMILY (*Apocynaceae*)

ARIZONA COCKROACH-PLANT
Haplophyton crooksii L. Benson

FIELD IDENTIFICATION. Slender-stemmed plants to 2½ ft with green, herbaceous twigs, but woody toward the base.

FLOWERS. Terminal, solitary, or a few in a cymose cluster, showy, perfect; corolla yellow, salverform, short tube expanding into a 5-lobed limb, about 1 in. across; lobes about ½ in. long, longer than the tube, obovate or oval, delicately veined; stamens 5, alternate with the lobes, often with basal sterile appendages; pistils 2, ovaries 2 and separate, styles and stigmas united; calyx of 5 linear-subulate, pubescent sepals, about ¼ in. long.

FRUIT. A pair of linear, terete, slender follicles; seeds black, about ¼ in. long, linear, grooved, rugose, each end with a tuft of white, deciduous hairs.

LEAVES. Simple, alternate or opposite, entire on margin, lanceolate to ovate, apex acuminate, base rounded or cuneate; upper surface dull green and glabrous or pubescent; lower surface paler green with short, stiff, white, scattered hairs; blade length ⅗–1¼ in., width ⅙–⅖ in., petiole 1/25–1/16 in. long, pubescent.

STEMS. Slender, erect, on upper parts green, herbaceous, and pubescent; lower parts brown, glabrous, and somewhat woody.

RANGE. In Texas, New Mexico, Arizona, and adjacent Mexico. In Texas in Hudspeth and Brewster counties.

REMARKS. The genus name, *Haplophyton*, is from the Greek and means "single plant." The species name, *crooksii*, is named for D. M. Crooks, head of the division of drug and related plants, Bureau of Plant Industry, Washington, D.C. The plant was described by Benson (1943b, pp. 630–632). Some vernacular names of the plant are Hierba de la Cucaracha, Raíz de la Cucaracha, Atempatli, and Actimpatli. It is closely related to *H. cimicidum* A. DC. of Mexico and Central America. The Arizona form has smaller, more lanceolate leaves, and smaller fruit and seeds. Both plants contain a poison which is often used as a domestic insecticide. A decoction is mixed with molasses or cornmeal for a cockroach poison. Sometimes it is applied in the form of a lotion to the human body to kill parasites, or as a sweetened infusion for a mosquito and flea repellent.

ARIZONA COCKROACH-PLANT
Haplophyton crooksii L. Benson

Common Oleander

Nerium oleander L.

FIELD IDENTIFICATION. Cultivated, clumped shrub to 18 ft, 3–8 in. in diameter at the base.

FLOWERS. Odorless, blooming during summer in compound, terminal cymes; flowers variously colored, and often double; corolla-tube funnelform, dilated into a narrow-campanulate throat with crownlike appendages 3–5-toothed; limb salverform, 1½–3 in. across, 5-lobed; lobes convolute in the bud, obliquely apiculate, twisted to the right; stamens 5, alternating with corolla-lobes, filaments partly adnate to corolla-tube; anthers with 2 basal tails, apex long-attenuate, hairy, 2-celled; styles united, slender, stigma simple, ovary superior and 2-carpellate; calyx of 5 persistent sepals, imbricate in the bud, lanceolate, acuminate, ⅙–¼ in. long.

FRUIT. The 2 ovaries forming follicles, erect or nearly so, 4–8 in. long, seeds twisted.

LEAVES. Numerous, opposite, or in whorls of 3–4, linear to elliptic, margin entire and often whitened, revolute, apex and base acute or acuminate, firm and leathery, many-nerved; dark green and glabrous with a conspicuous yellowish green main vein above; paler beneath with numerous, delicate, almost parallel lateral veins.

Luckynut Thevetia
Thevetia nereifolia Juss.

Common Oleander
Nerium oleander L.

TWIGS. Erect or arching, young ones green, older ones light brown to gray; lenticels numerous, oval.

RANGE. Cultivated in gardens in Texas and Louisiana, sometimes escaping cultivation. A native of Asia, and widely distributed from the Mediterranean region to Japan. Cultivated throughout the tropics and subtropics.

PROPAGATION. In the South oleanders may be grown outside and require little care. The top may be pruned back every third year for best flowering. The ripened shoots can be rooted in a bottle of water and transplanted into pots to set out the following spring. They are subject to scale and mealy bugs, which can be sponged or hosed off. Occasionally it is also propagated by layers and seed.

FORMS AND RELATED SPECIES. Flowers of the oleander vary in colors of white, blush, copper, crimson, and dark purple; variegated and double forms are also found.

The following horticultural forms have been listed by Kelsey and Dayton (1942):

Cardinal, Comdr. Barthelemy, Dr. Golfin, Lilian Henderson White, Luteum, Mme. Peyre, Mme. Sarah Bernhardt, Mrs. Roeding, Professor Bodkin, Rosepink, Rubra, Shellpink, Single Red, Sister Agnes, Splendens.

A closely related species with an odor is the Fragrant leander, *N. odorum* Soland., with linear-lanceolate aves in threes, and rosy-pink flowers; segments of e crown 4–7, long and narrow; anther appendages rotruding. It has the same color variations as the Comon Oleander, with single and double flowers also.

REMARKS. The genus name, *Nerium*, is from the reek *neros* ("moist"), referring to places the wild lants grow. The species name, *oleander*, is from the atin, meaning "olive-like," referring to the leaves. The owers are poisonous if eaten by human beings, and e leaves have been known to kill cattle. They also ontain a small amount of rubber. Oleander has been sed for rat poison in Europe for many centuries. The mptoms of poisoning in human beings are abdominal ain, dilation of the pupil, vomiting, vertigo, insensibily, convulsive movements, small and slow pulse, and fatal cases epileptiform convulsions with coma endg in death. The erratic pulsation of the heart where eath has not followed has been pronounced, the pulse r 5 days remaining as low as 40 per minute. An inusion made from 4 ounces of the root is affirmed to ave taken life. The active principle of the plant is a lycoside, oleandrin, which hydrolyzes into a gitoxigen. (See Wood and Osol, 1943, p. 1454.)

UCKYNUT THEVETIA

hevetia nereifolia Juss.

FIELD IDENTIFICATION. Cultivated shrub attaining a eight of 8 ft, resembling the oleander in habit.

FLOWERS. Terminal few-flowered cymes; pedicels –1¾ in. long, glabrous to puberulent; flowers perfect nd regular; calyx-lobes 5, imbricate in the bud, spreadg, ⅕–⅓ in. long, lanceolate, long-acuminate, glandular glandless, puberulent; corolla saffron to yellow or hite, funnelform, fragrant, tube about ½–¾ in. long, ilated into 5 lobes above, lobes twisted, falcate, inquilateral, ascending; stamens 5, alternating with the orolla-lobes, converging toward the stigma; filaments artly adnate to corolla-tube; anthers 2-celled, small, nceolate; styles united, stigma simple and 2-pointed; vary superior, 2-lobed and 2-celled.

FRUIT. Drupe ¾–1½ in. broad, somewhat triangular nd broader than long, apex truncate with 2 small igma remnants, base abruptly narrowed, compressed, lack, hard, 2-carpellate and 2–4-seeded; seeds large, ngular, edges acute.

LEAVES. Simple, alternate, numerous, half-evergreen, ssile or nearly so, narrowly linear, margin entire and ightly revolute, gradually narrowed to the base and pex, blades 2½–6½ in. long, less than ½ in. wide, pper surface glossy green and glabrous; veins obscure, xcept main vein; lower surface paler, essentially glarous or slightly puberulent.

TWIGS. Slender, gray to brown, younger ones greenish and slightly puberulent, older ones glabrous; leaf scars half-moon–shaped.

WOOD. Soft, fibrous, specific gravity 0.80, of no commercial value.

RANGE. In sandy soil, old gardens, and waste places. Cultivated in gardens in southern Texas. Also in Florida, and escaping cultivation in southern Florida and the Keys. In Mexico in San Luis Potosí, Veracruz, Yucatán, Chiapas, and Guerrero; southward to South America. Also in the West Indies. Cultivated in the East Indies and Hawaii.

MEDICINAL USES. It is reported that in Yucatán, cotton soaked in the juice is placed in cavities of teeth to relieve toothache. The seeds have been reported to yield a glucoside, thevetine. A tincture of the bark is considered a powerful febrifuge, and in large doses a violent purgative and emetic. The fruit is used as a topical application in hemorrhoids. A number of cases of poisoning by the seeds of *Thevetia* have occurred. The symptoms have been repeated vomiting, a slow, very feeble pulse, delirium, convulsive movements, and coma.

REMARKS. The genus name, *Thevetia*, is in honor of Andre Thevet (1502–1590), a French monk, and the species name, *nereifolia*, is for its resemblance to the oleander.

Many vernacular names have been given the plant in tropical America, and in the East Indies and Hawaii, where it is cultivated, these being as follows: Chirca Campanilla, Acitz, Naranjo Amarillo, Yoyote, Yoyotli, Narisca Amarilla, Chilca, Cabollon, Cabalonga, Cobalonga, Amancay, Aje de Monte, Pepa de Cruz, Castaneto, Chilindrón, Lengua de Gato, Retama, Camanche, and Carvache. The fruits are known as "Huesos de Fraile" or "Codos de Fraile." Names in Hawaii are Kokilphil and Ahouai. English vernacular names are Trumpet-flower, Yellow-oleander, Lucky-seed, Be-still, and Friar's Elbow Bones.

Shrub is often cultivated for ornament outdoors in semitropical areas of the South, or under glass in the North, and is usually grown from cuttings. It can stand relatively light frosts, and, although the tops die, it will sprout back if the base is banked with sand.

CLIMBING STAR-JASMINE

Trachelospermum difforme (Walt.) Gray

FIELD IDENTIFICATION. Twining vine, usually only herbaceous, but sometimes woody close to base.

FLOWERS. In slender, cymose clusters in leaf axils, peduncles slender, ½–¾ in. long, bearing several pedicels with linear-filiform bracts at base; calyx-lobes 5, lanceolate-subulate, ⅛–⅙ in. long, pubescent; corolla funnelform, greenish yellow, 5-lobed; tube ⅕–¼ in. long, striped orange-red within, lobes shorter than tube, spreading, reflexed, ovate; stamens included, anthers sagittate and converging about stigma; disk 5-lobed, glandular; ovary with 2 carpels, ovules numerous in each carpel; style slender, stigma thickened toward the apex, ovoid.

FRUIT. Follicles 2, slender, elongate, 5–9 in. long, hardly more than 1/6 in. wide; seeds numerous, linear-oblong, tufted.

LEAVES. Opposite, deciduous, 1–3 in. long, 1/2–2 in. wide, lanceolate to broadly elliptic or oval, apex abruptly acute or acuminate, base cuneate to rounded, margin entire, texture thin; upper surface dark green and glabrous to puberulent along the veins; lower surface much paler green, pubescent; petioles 1/8–1/3 in. long, pubescent.

STEMS. Slender, twining, younger parts herbaceous, green to reddish brown and pubescent; at base glabrous and somewhat woody with age.

RANGE. Low rich grounds on the margins of thickets. Texas eastward to Florida, northward to Arkansas, Oklahoma, Missouri, Indiana and Delaware; southward into Mexico.

REMARKS. The genus name, *Trachelospermum*, means "neck-seeded" (but the seed of this species is neckless), and the species name, *difforme*, refers to the double fruit. Also known under vernacular names of Dogbane, Double-pod, and Southern Jasmine.

CLIMBING STAR-JASMINE
Trachelospermum difforme (Walt.) Gray

MILKWEED FAMILY (*Asclepiadaceae*)

GRECIAN SILK-VINE

Periploca graeca L.

FIELD IDENTIFICATION. Cultivated partly woody vine attaining a length of 40 ft. Stems slender, twining, glabrous; leaves handsome and deciduous.

FLOWERS. Borne July–August, in few- to many-flowered, long-peduncled, loose cymes; corolla to 1 in. across, rotate, 5-lobed; lobes obtuse, linear, purple, green-margined and pubescent within; crown with 5 slender, threadlike appendages, crown itself of 5 lobed scales adnate to the corolla; stamens 5, pollen grains in groups of 4; anthers merely connected at apex and pubescent; carpels 2, stigmas united; calyx-lobes 5, ovate to deltoid, glandular within.

FRUIT. Composed of 2 brownish follicles, generally united at apex, long-subulate, curved, nearly terete, glabrous, to 4 in.; seeds numerous and fusiform, about ⅓ in. long.

LEAVES. Opposite, membranous, elliptic or oblong to ovate, apex acute or acuminate, base rounded or obtuse, margin entire, blades 1¼–5 in. long, 1–2 in. broad, upper surface dark green and glossy, lower surface paler; petioles ⅕–½ in. long.

STEMS. With milky sap, young ones green and glabrous, older reddish brown to gray, twining and elongate, woody toward the base.

RANGE. In cultivated grounds or old gardens, or along fence rows. Cultivated in Texas, Louisiana, Oklahoma, eastward to Florida, north to Kansas, New York, and Massachusetts. A native of Europe in Syria and Grecian Islands region.

REMARKS. The genus name, *Periploca,* is from the Greek, and means "around" and "to twine," referring to its twining habit. The species name, *graeca,* means "Grecian." The plant is sometimes grown for ornament and is used on arbors, trellises, and trunks of trees. May be propagated by seeds, greenwood cuttings under glass, or layers.

GRECIAN SILK-VINE
Periploca graeca L.

MORNINGGLORY FAMILY (*Convolvulaceae*)

Tree Morningglory

Ipomoea fistulosa Mart.

FIELD IDENTIFICATION. Beautiful, subshrubby plant to 10 ft, with erect branches.

FLOWERS. June–September, borne in terminal, several-flowered, corymbose clusters from densely puberulent peduncles 2–5 in. long; individual flower pedicels ⅛–1¼ in. long, densely puberulent; flower buds finely white-tomentose; corolla pink, funnelform, limb 5-angled, banded longitudinally, exterior finely puberulent, length 2½–4 in.; flower parts perfect and regular; calyx of 5 sepals partially united below, sepals imbricate, unequal in size, ovate to subrotund, apex rounded or obtuse, finely pubescent, length ⅛–¼ in.; androecium of 5 straight stamens, partially adnate to corolla within and alternate with lobes; gynoecium of 2 united carpels, styles included.

FRUIT. Capsule 1/25–⅛ in. long, ovoid or subglobose, valvular and dehiscent; seeds shaggy, pubescent, 1–2 in each cavity.

LEAVES. Simple, alternate, deciduous, blade 3–6 in. long, ovate or occasionally sublanceolate, apex long-acuminate or acute, base cordate or semitruncate, margin entire or nearly so; upper surface dull green and glabrous or slightly puberulent along the veins; lower surface paler and more densely pubescent, or subglabrate when older, thin and chartaceous; petiole 1½–5 in. long, slender, glabrous to puberulent.

TWIGS. Upper stems herbaceous, soft, pubescent; lower woody, smooth, green to light or dark brown, glabrous.

RANGE. A native of Brazil, but escaping cultivation on the Gulf Coast plain in Florida and southwest Texas. One of the earliest plants introduced by the Spanish in the vicinity of Brownsville, Texas, and now growing spontaneously in waste places, about old residences, and on the edges of resacas.

Tree Morningglory
Ipomoea fistulosa Mart.

PROPAGATION. The plant may be propagated by several methods. Germination of the seed may be hastened by scarification, or by soaking the seed in water for several hours before planting. Cuttings may also be secured from well-ripened wood under glass, by layers, or by divisions of the rootstocks. In the fall, tops of plants may be cut off and the roots mulched heavily to carry the plant over the winter.

REMARKS. The genus name, *Ipomoea,* according to Linnaeus, is from *ips* ("bindweed") and *homois* ("like"), because of the resemblance to *Convolvulus.* The species name, *fistulosa,* refers to the fistulous, or tubular, flower. The Tree Morningglory has rather unusual growth habit in a genus predominated by vines.

BORAGE FAMILY (*Boraginaceae*)

GREGG COLDENIA

Coldenia greggii (Torr.) Gray

FIELD IDENTIFICATION. Low, erect, much-branched shrub rarely more than 2 ft with all parts densely canescent-hairy.

FLOWERS. Small, clustered in dense, short, terminal heads, white to pink, sessile; calyx campanulate, 5-lobed, lobes linear-subulate, conspicuously plumose with long hairs; corolla regular, short-funnelform or subrotate, longer than the calyx, 5-lobed; lobes imbricate, spreading, somewhat unequal, obovate or rounded, crenulate; stamens 5, alternate with lobes of the corolla, filaments short, equally or unequally inserted at base of corolla, anther sacs spreading; style 2-cleft; ovary superior, an obscure glandular ring at the base, obscurely 4-lobed and 4-celled, by abortion finally 1-celled and 1-seeded.

FRUIT. Nutlet dry, ovate to oblong, walls thin, 1-celled and 1-seeded.

LEAVES. Simple, alternate, deciduous, blades 1/6–1/3 in. long, ovate to oval, margin entire and revolute, veins obscure, surface densely canescent-tomentose; petioles 1/16–1/8 in. long, densely canescent.

TWIGS. Slender, young growth silvery to brown-tomentose, when older light gray to tan and glabrous, often with fibrous bark.

RANGE. Rocky hillsides and ravines, Trans-Pecos Texas, New Mexico, and Mexico in the states of Chihuahua, Coahuila, Durango, and Zacatecas.

REMARKS. The genus name, *Coldenia,* is in honor of Cadwallader Colden, at one time lieutenant governor of New York, and a correspondent of Linnaeus. The species name, *greggii,* honors the botanist Josiah Gregg (1806–1850), who collected in the southwestern United States and Mexico. The plant is ornamental due to the purplish, plumed flowers.

GREGG COLDENIA
Coldenia greggii (Torr.) Gray

ANACAHUITA

Cordia boissieri A. DC.

FIELD IDENTIFICATION. Tree with a spreading rounded top and stout branches. It rarely attains a height of 30 ft with a trunk up to 10 in. in diameter. Often it is only an irregularly branched shrub.

FLOWERS. April–June in terminal cymes, individual flowers sessile or short-pediceled; corolla showy, white with a yellow spot in the throat, trumpet-shaped,

ANACAHUITA
Cordia boissieri A. DC.

TWIGS. Gray to brown, puberulous to tomentose, leaf scars obcordate, somewhat elevated.

BARK. Dark gray or reddish, irregularly broken into broad, flat, confluent ridges which on old trunks may break into thin, elongate scales.

WOOD. Brown, sapwood lighter, close-grained, soft, specific gravity about 0.68.

RANGE. In dry soil of the lower Rio Grande Valley in Texas. In Cameron, Hidalgo, Starr, Zapata, and Willacy counties. In Mexico in the states of Nuevo León, Tamaulipas, Veracruz, Hidalgo, and San Luis Potosí.

MEDICINAL USES. The fruit, in the form of a jelly, is employed as a domestic remedy for coughs and colds in Mexico. The leaves are used as a remedy for rheumatism and bronchial disturbances. Although popular, the efficiency of these remedies is a matter of conjecture. It is not listed in the *United States Dispensatory* (Wood and Osol, 1943).

REMARKS. The genus name, *Cordia,* is in honor of the German pharmacist and botanist, Valerius Cordus (1515–1544), and the species name, *boissieri,* refers to the botanist Boissier. Vernacular names used are Flor de Anacahuita, Nacahuite, Nacahuitl, Nacagüita, Siricote, Trompillo, Rosca Vieja, and Cicuas de Cueramo. It is a handsome ornamental and is extensively planted in the lower Rio Grande Valley and northern Mexico. The wood has little economic value, but is occasionally made into woodenware objects and yokes. The fruit is edible and sweet, but excessive consumption is said to produce dizziness and intoxication. It is relished by livestock.

spreading into 5 lobes, 1–1½ in. long, 2–2½ in. wide, the tube about 1 in. long and puberulous; lobes of corolla about ½ in. long, rounded to broad-ovate, apex rounded to obtuse, margin ciliate and somewhat erose; stamens 5, of unequal lengths, basally attached to corolla-tube within, slightly exserted; anthers oblong, about ⅛ in. long, filaments filiform, yellowish green, glabrous; pistil of varying length, usually ⅓–½ as long as the stamens, narrowed to a slender apex and split into 2 stigmas, each stigma bifid; calyx tubular, about ½ in. long at anthesis, densely brown-tomentose, distinctly ribbed, the ribs ending in abruptly acuminate teeth 1/16–⅛ in. long, teeth usually 5.

FRUIT. Calyx greatly enlarged at fruit maturity with ribs connected by densely brown-tomentose sculptured reticulations. Drupe maturing July–September, about 1 in. long and ¾ in. wide, ovoid-oblong, white to yellowish green and brownish later, flesh thin, pulpy, sweet; seed solitary, ovoid, acute, hard, bony, white, ¼–¾ in. long.

LEAVES. Oblong to ovate, apex rounded or acute, base rounded or subcordate, margin entire or obscurely crenate-serrate, blades 3–5 in. long, 3–4 in. wide, thick and firm, velvety with soft brown tomentum, especially beneath; petioles 1–1½ in. long, stout, tomentose.

ANAQUA

Ehretia anacua (Mier & Berland.) Johnst.

FIELD IDENTIFICATION. Half-evergreen shrub or tree with a rounded head and attaining a height of 50 ft.

FLOWERS. March–April, but occasionally in fall after rains. Panicles 2–3 in. long, fragrant; bracts linear, acute, deciduous, about ¼ in. long; corolla small, white, about ¼ in. long, ⅓–½ in. across when open, tube short and campanulate, 5-lobed; lobes oval to ovate; calyx lobes 5, ovate to linear or lanceolate, acute at apex, almost as long as the corolla-tube; stamens 5, adnate to the tube within, filaments filiform; ovary oblong, style split into 2 capitate stigmas.

FRUIT. Drupe ¼–⅓ in. in diameter, globular, yellowish orange, juicy, sweet, edible, 2-celled and 2-seeded, stone separating into 2 bony nutlets rounded on the back and plane on the inner face, seeds terete and erect. A handsome tree when laden with the yellowish orange drupes.

LEAVES. Simple, alternate, half-evergreen, oval to oblong or elliptic, margin entire or coarsely serrate above the middle, apex acute and sometimes apiculate, base cuneate or rounded, leathery, stiff, upper surface very

ANAQUA
Ehretia anacua (Mier & Berland.) Johnst.

oughened by small, crowded tubercles, upper surface dull olive-green, glabrous or slightly pubescent, lower surface paler and pubescent, veins coarse, blade length 1–3½ in., width ¾–1½ in.; petiole stout, grooved, pubescent, ⅛–¼ in. long.

TWIGS. Slender, crooked, brown to gray, when young with pale or brownish hairs, older glabrous; leaf scars small, obcordate and depressed; lenticels pale and numerous.

BARK. Thick, reddish brown to gray or black, broken into narrow, flat-topped ridges and deep fissures, the platelike scales exfoliating into gray or reddish flakes.

WOOD. Light brown, sapwood lighter, close-grained, tough, heavy, hard, difficult to split, specific gravity about 0.64.

RANGE. Usually attaining its largest size in rich river valleys of central and south Texas. Only a shrub on poor, dry soil of hillsides. Southward into the Mexican states of Nuevo León, Tamaulipas, Coahuila, Guanajuato, and Veracruz. Occasionally as far east as Houston, Texas. Abundantly planted as shade trees in Victoria, Texas.

REMARKS. The genus name, *Ehretia,* is in honor of George Dionysius Ehret (1708–1770), a German botanical artist. The species name, *anacua,* is a latinization of the vernacular name Anaqua. A scientific synonym no longer valid is *E. elliptica* DC. Other vernacular names are Anacahuite, Knackaway, Nockaway, Sugarberry, Manzanita, and Manzanillo. The wood is used for posts, wheels, spokes, axles, yokes, and tool handles. A number of birds and mammals feed upon the fruit. Anaqua is a desirable tree for ornamental planting in its native region. It forms persistent clumps by root suckers, which can be used for transplanting. It is drought-resistant and comparatively free of disease. Of the 40 species of *Ehretia* known in the warm regions of the world, only a few are used in horticulture.

CROWDED HELIOTROPE

Heliotropium confertifolium Torr.

FIELD IDENTIFICATION. Perennial suffruticulose plant, but sometimes woody toward the base. The stems much-branched and hardly over 1 ft tall. All parts with a silky and silvery pubescence.

FLOWERS. Small, borne with leafy bracts toward ends of branches, not scorpoid; calyx sessile, lobes 5, linear-lanceolate, 1/12–1/10 in. long; corolla salverform, pale purple; tube cylindric, about as long as the calyx, silky-pubescent; limb angularly 5-lobed, about ⅙–⅕ in. broad; stamens 5, included, filaments short, adnate

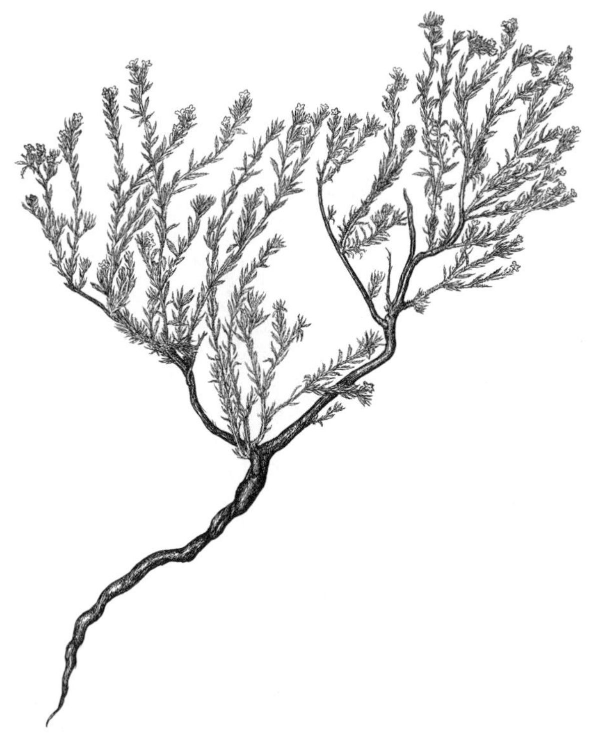

CROWDED HELIOTROPE
Heliotropium confertifolium Torr.

to the corolla-tube; anther sacs appendaged and puberulent; styles united and short; ovules pendulous.

FRUIT. Four-lobed, dry, separating into 4, 1-celled and 1-seeded nutlets.

LEAVES. Alternate, crowded, imbricate above, blades linear to narrowly oblong, 1/5–1/3 in. long, acute, revolute, surfaces equally whitened with silvery pubescence; short-petioled.

RANGE. In dry soil of southern and western Texas. In Mexico in Coahuila, Nuevo León, and San Luis Potosí.

REMARKS. The genus name, *Heliotropium,* is from the Greek words, *helios* ("the sun") and *tropium* ("to turn"), because the flowers were believed to turn with the sun. The species name, *confertifolium,* refers to the crowded leaves.

MEXICAN TOURNEFORTIA
Tournefortia volubilis L.

NARROW-LEAF HELIOTROPE

Heliotropium angustifolium Torr.

FIELD IDENTIFICATION. Plant slender, erect, much-branched from a distinctly woody base. All parts densely hairy with long appressed hairs.

FLOWERS. In slender, nearly straight, secund, bracteate or bractless spikes; individual pedicels about 1/25 in. long, very hairy; flowers very small, less than 1/8 in. across, perfect, yellowish white; calyx 5-lobed, white hairy, lobes partly united below, linear, acute; corolla gamopetalous, salverform, tube with dense, white, appressed hairs, 5-lobed, each lobe linear to lanceolate long-acuminate to acute, 1/16–1/10 in. long, hairy; stamens 5, adnate to the corolla, anthers nearly sessile, anther tips glabrous and mucronate; style short and united stigma annular.

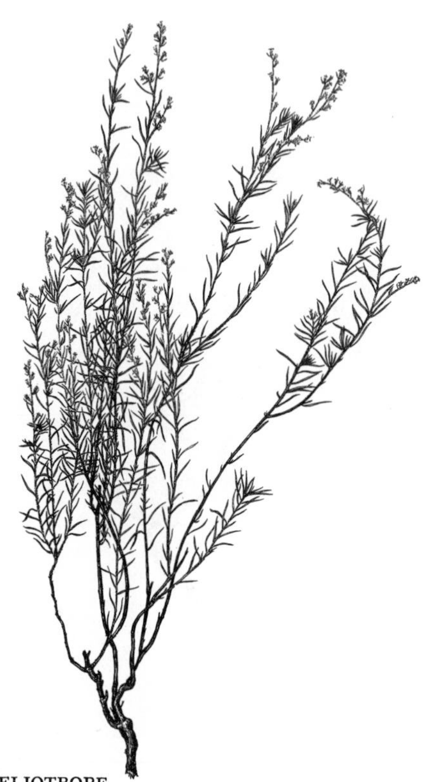

NARROW-LEAF HELIOTROPE
Heliotropium angustifolium Torr.

FRUIT. Capsule 4-lobed, separating into 4 nutlets 1-celled and 1-seeded.

LEAVES. Numerous, alternate, narrowly linear to almost filiform, apex acute, base sessile, margin often revolute, blades 1/4–3/4 in. long, about 1/12 in. wide whitened with stiff, appressed, white hairs.

TWIGS. Leafy, slender, rigid, numerous, when young green with dense, white, appressed hairs; older twigs brown to gray or black and glabrous toward the base.

RANGE. Southwestern and western Texas. In Mexico from Coahuila and Tamaulipas to San Luis Potosí.

REMARKS. The genus name, *Heliotropium,* is from the ancient names *helios* ("the sun") and *tropium* ("to turn"), because the flowers were believed to turn with the sun. The species name, *angustifolium,* refers to the narrow leaves.

Mexican Tournefortia

Tournefortia volubilis L.

FIELD IDENTIFICATION. Suberect shrub with pendent or scandent branches, or a climbing, twisting woody vine.

FLOWERS. Small, perfect, regular, spikes few or numerous, wide-spreading, lax, long, slender, somewhat scorpoid; corolla white, salverform, about 1/8 in. long, 5-lobed; lobes erect, lanceolate to subulate, acuminate, shorter than the tube; tube cylindric, somewhat contracted below the middle and at the throat; calyx about 1/25 in. long, nearly sessile, lobes triangular or lanceolate and acuminate; stamens included, inserted on the corolla, as many as the corolla lobes and alternate with them, anthers lanceolate; styles elongate, united, stigma-appendage long-conic; ovary of 2 carpels which are each 2-carpellate, ovules pendulous from lateral attachments.

FRUIT. Drupe small, somewhat depressed, 1/12–1/8 in. wide, 2–4-lobed, lobes black-spotted; nutlets 2–4, more rarely solitary.

LEAVES. Alternate, blade lanceolate to oblong-ovate, 1½–4 in. long, margin entire, apex acute or acuminate, base rounded or obtuse, sometimes oblique, surface glabrous or finely pubescent above, lower surface grayish sericeous to rusty pubescent, sometimes densely so; petiole shorter than blade.

RANGE. In sandy soils, near Brownsville, Cameron County, Texas, in groves of Texas Palm *(Sabal texana)* on Rabb Ranch. In Mexico in Tamaulipas, Baja California, southward into Sinaloa, Veracruz, Oaxaca, and Yucatán. Also in southern Florida, Central and South America, and the West Indies.

REMARKS. The genus name, *Tournefortia,* is in honor of James Pitton de Tournefort (1656–1708), one of the earliest systematic botanists. The species name, *volubilis,* refers to the scorpoid spikes.

VERBENA FAMILY (*Verbenaceae*)

HIGH-MASS
Aloysia wrightii (Gray) Heller

HIGH-MASS
Aloysia wrightii (Gray) Heller

FIELD IDENTIFICATION. Western shrub to 6 ft, slender-branched, hairy, aromatic.

FLOWERS. April–May, small, numerous, dense, crowded on paired axillary spikes ⅜–1⅝ in. long and longer than the leaves, peduncles short, 1/25–⅖ in. long, puberulent; flowers perfect, hypogynous, corolla white, cylindric, 2-lipped, tube about ⅛ in. long, limb about 1/12 in. wide, more or less puberulent outside; stamens 4, included, inserted near middle of tube in sets of 2, anther cells parallel; pistil 1, included, stigma obliquely capitate, ovary 2-celled, cells with 1 erect ovule; calyx tubular, densely hairy, about 1/12 in. long, 4-lobed, lobes small and nearly equal; bractlets lanceolate, acuminate, 1/25–1/12 in. long, densely puberulent.

FRUIT. Calyx splitting into 2 small, dry, thin-walled segments, the 2 nutlets minutely papillose, with no endosperm.

LEAVES. Numerous, small, simple, deciduous, aromatic, decussate-opposite, shape ovate or orbicular, apex rounded, base rounded to cuneate, margin set with low crenate-serrate teeth; upper surface with veins impressed, strigose-hairy and roughened; lower surface densely canescent-tomentose and glandular, blade ⅛–⅝ in. long, ⅛–½ in. wide; petiole 1/25–⅙ in. long or leaves subsessile, densely yellow- or gray-hairy.

TWIGS. Slender, brittle, divaricate, 4-angled, tan to gray or brown, pubescent, glabrous later, bark fibrous.

RANGE. Usually in rocky canyons, on stony slopes or arroyo banks at altitudes of 2,000–6,000 ft. In western and southern Texas, New Mexico, Arizona, Nevada, and California. In Mexico in Coahuila, Sonora, Durango, and Zacatecas.

REMARKS. The genus name, *Aloysia,* is in honor of Maria Louisa, wife of Charles IV of Spain. The species name, *wrightii,* honors Charles Wright, American botanist, who collected in Texas from 1847–1852. Vernacular names for the plant are Alta Misa, Vara Dulce, and Wright's Lippia.

Some books list the plant as *Lippia wrightii* Gray. The aloysias and lippias have been reviewed in monograph form by Moldenke (1942, pp. 53–55, 62–66).

It is considered to be a good honey plant, and furnishes fairly good browse for cattle.

Sweet-stem

Aloysia macrostachya (Torr.) Moldenke

FIELD IDENTIFICATION. Aromatic shrub seldom over 6 ft with densely short-pubescent, 4-angled twigs.

FLOWERS. Almost throughout the year and especially after rains, spikes axillary, erect or ascending, 2–6¼ in. long; peduncles white-hairy, glandular, slender, ⅜–1⅝ in. long; corolla pink, cylindric, hypocrateriform, 2-lipped, tube ⅙–⅕ in. long, limb about ⅙ in. wide; stamens 4, in pairs, included, inserted near middle of corolla-tube, anthers with 2 parallel cells; pistil 1, included; ovary 2-celled, each cell 1-ovuled; calyx tubular, ⅛–⅕ in. long, glandular, white-hairy, 4-lobed, lobes linear-subulate, barely over 1/25 in. long; bractlets densely hirsute, 1/12–⅛ in. long, lanceolate, apex acuminate.

FRUIT. Drupe small, dry, thin-walled, separating into 2 nutlets.

LEAVES. Numerous, often crowded, simple, deciduous, decussate-opposite, ovate, apex obtuse to rounded, base truncate or subcordate, margin crenate at least above middle; upper surface dark green, rough, glandular, densely strigose-hirsute, veins impressed; lower surface pale, canescent-tomentose and glandular, blades ¼–1¾ in. long, ⅕–⅞ in. wide; petioles ⅗–2 in. long, densely white-hairy.

Sweet-stem
Aloysia macrostachya (Torr.) Moldenke

TWIGS. Slender, erect or ascending, brittle, 4-angled; when young with dense, white, glandular, spreading hairs; older more glabrous and gray to brown, nodes annulate.

RANGE. On rocky hillsides and in dry gravelly arroyos in southern and western Texas. In Texas recorded from Webb, Duval, Zapata, Starr, Hidalgo and Cameron counties. In Mexico in the states of Tamaulipas, Coahuila, and San Luis Potosí.

REMARKS. The genus name, *Aloysia,* is in honor of Maria Louisa, wife of Charles IV of Spain. The species name, *macrostachya,* means "large-spiked." Plant is also listed in the literature under the name of *Lippia wrighti* var. *macrostachya* Torr., and *Lippia macrostachya* (Torr.) S. Wats. The aloysias have been described by Moldenke (1942, pp. 62–66). This species is also commonly known as Vara Dulce.

Privet Lippia

Lippia ligustrina (Lag.) Britt.

FIELD IDENTIFICATION. Shrub 3–14 ft, thicket-forming, much-branched, slender, aromatic.

FLOWERS. Blooming intermittently March–November, especially after rains, vanilla-scented, borne in erect or ascending, densely flowered, axillary spikes from ¾–2¾ in. long; peduncles ⅕–⅗ in. long, puberulent; rachis subfiliform, densely pubescent, bractlets deciduous, minute, about 1/25 in. long, puberulent, lanceolate, acuminate; corolla cylindric, 2-lipped, upper lip notched, lower larger and 3-lobed, white or violet, tube ⅛–⅙ in. long, limb about ⅛ in. wide, throat villous; stamens 4, included, 2 somewhat longer, attached to the corolla-tube; pistil 1, stigma obliquely capitate, ovary 2-celled, each cell with 1 ovule; calyx tubular, campanulate, 1/12–⅛ in. long, densely hairy and glandular, 4–5-ribbed, 4-lobed, lobes unequal, linear-subulate.

FRUIT. Drupe small, dry, containing 2 separate nutlets, enclosed in the calyx.

LEAVES. Aromatic, decussate-opposite, lanceolate to oblong or elliptic; apex obtuse or acute, often mucronulate or emarginate; base gradually narrowed; margin somewhat revolute and entire, or more rarely serrate, ⅛–1¼ in. long, 1/12–⅓ in. wide, coriaceous; upper surface scaberulous and fine-hairy; lower surface densely puberulent and glandular-punctate; veins obscure; petioles 1/25–⅛ in. long, leaves almost sessile.

TWIGS. Gray, slender, stiff, squarish, brittle, pulverulent, internodes ¼–1¼ in. long, somewhat spinescent terminally.

WOOD. White to yellow, weak, soft, brittle.

RANGE. The species is usually found on rocky, limestone soil at altitudes of 1,000–4,000 ft in southern, central, and western Texas; also in New Mexico, Arizona, and southern California. In Mexico in the states of

Privet Lippia
Lippia ligustrina (Lag.) Britt.

Nuevo León, Chihuahua, Zacatecas, Durango, Sonora, and Puebla. Also in South America.

VARIETY. Schulz Privet Lippia, *L. ligustrina* var. *schulzii* Standl. (*Aloysia ligustrina* var. *schulzii* [Standl.] Moldenke), is a variety which has leaves broadly elliptic, obovate or oblong-elliptic, with 2–8 teeth on the margin and hairs denser below. Known to occur in Brewster, Jeff Davis, and Cameron counties.

REMARKS. The genus name, *Lippia*, is in honor of Auguste Lippi (1678–1703), a French naturalist. The species name, *ligustrina*, is from the Latin *ligare* ("to tie or bind"), with reference to flexible character of the branches. Some authorities list the plant under the name of *Aloysia ligustrina* (Lag.) Small. Vernacular names are Bee-brush, White-brush, Bee-blossom, Agrito, Vara Blanca, Vara Dulce, Huele de Noche, Hierba Dulce, Cabradora, Jaboncillo, Jazmincillo, and Jasminillo. The leaves and flowers are used medicinally in Mexico as a remedy for diseases of the urinary tract. Plant is very fragrant, blooms profusely after rains, and is considered good bee food. Cattle find it palatable. Sometimes planted for ornament. Easily transplanted, and may be propagated from seed or cuttings.

Red-brush
Lippia graveolens H. B. K.

FIELD IDENTIFICATION. Western, aromatic, pubescent shrub, more rarely a small tree to 27 ft.

FLOWERS. Almost throughout the year, especially after rains, the 4–6, pubescent to villous peduncles shorter than the leaves and arising from their axils; flowering heads oblong, 1/6–1/2 in. long; corolla fragrant, yellowish white (yellow-centered), tube 1/8–1/4 in. long, limb somewhat 2-lipped and 4-parted; stamens 4, in pairs, included, inserted near the middle of the corolla-tube; style short, stigma obliquely capitate, ovary 2-celled; calyx 4-toothed, about 1/12 in. long; pubescent to villous; bracts imbricate in series of 4, folded, keeled, narrowly ovate or lanceolate, apex acute, surface tomentose and glandulose.

FRUIT. Drupe dry, splitting into 2, thin-walled pyrenes.

LEAVES. Simple, opposite, deciduous, ovate to oblong or elliptic, apex rounded to obtuse, base subcordate to rounded or acute, margin crenate with low teeth; upper surface pubescent to villous, hairs appressed;

Red-brush
Lippia graveolens H. B. K.

lower surface pubescent to strigose-tomentose and glandular, blades ⅓–2¾ in. long, ¼–1¼ in. wide; petioles slender, 1/12–⅘ in. long, densely puberulent.

TWIGS. Slender, spreading, terete or 4-angled above, with densely appressed or spreading hairs, glandular-punctate.

RANGE. On rocky slopes, in dry soil of arroyos and chaparral thickets. In southern, central, and western Texas southward through Mexico to Nicaragua.

MEDICINAL USES. Also known under the vernacular name of Orégano in Central America, where it is used as a tonic, stimulant, and expectorant, and also as a condiment.

REMARKS. The genus name, *Lippia*, honors Auguste Lippi (1678–1703), French naturalist. The species name, *graveolens*, means "strong smelling," with reference to the plant's aromatic scent. The plant is listed in some botanical publications under the names of *Lantana origonoides* Mort. & Gal., *Lippia berlandieri* Schau., and *Goniostachyum graveolens* (H. B. K.) Small, and it is also known under the vernacular name of Hierba Dulce. The lippias have been reviewed by Moldenke (1942, pp. 53–55).

WHITE-FLOWERED LIPPIA
Lippia alba (Mill.) N. E. Brown

WHITE-FLOWERED LIPPIA

Lippia alba (Mill.) N. E. Brown

FIELD IDENTIFICATION. Much-branched, aromatic shrub to 6 ft.

FLOWERS. Peduncles 1–2 at the nodes, as long as the leaf petioles or longer; borne in dense, capitate, or slightly elongate heads about ⅓–½ in. long; heads crowded by ovate to lanceolate bracts which are acuminate, villous, ciliate and 1/25–⅕ in. long; calyx 2-lobed, compressed and keeled, villous-hirsute, 1/25–1/12 in. long; corolla white (more rarely violet or purple), much longer than the calyx, tube cylindric, minutely hairy, ⅙–⅕ in. long; limb 2-lipped and 4-parted, lower lip much the largest; stamens 4, in pairs, inserted near middle of tube, anthers ovate, cells parallel; ovary 2-celled, each cell with 1 ovule.

FRUIT. Small, dry, enveloped by calyx, pericarp crustaceous, nutlets 2.

LEAVES. Opposite or ternate, blades ovate to oblong-ovate, ½–2¾ in. long, ⅜–1 in. wide, apex acute or obtuse, base cuneate, margin serrate but entire toward base; upper surface rugose-reticulated, veins impressed, finely hairy; lower surface densely pubescent or tomentose, petioles ⅛–½ in. long, densely hairy.

TWIGS. Elongate, slender, ascending; young twigs green to tan or brown with white to brownish hair, often finely striate, older twigs pale and glabrous.

RANGE. In southwest Texas and Mexico, along the lower Rio Grande drainage from Zapata to Brownsville. Reported also in Wharton County, Texas. Widespread in the tropics. In Mexico in the states of Tamaulipas, Veracruz, and Oaxaca. Also in Central America, South America, and the West Indies.

MEDICINAL USES. In tropical American countries a tea prepared from the leaves is used in domestic medicine as a sudorific, antispasmodic, stomachic, and emmenagogue.

REMARKS. The genus name, *Lippia*, honors Auguste Lippi (1678–1703), a French naturalist. The species name, *alba*, refers to the white flowers. Vernacular names used for the plant are Hierba Negra, Hierba del Negro, Mirto, Juanslama, Juanislama, Salvia, Sonora, Mastranto, Hierba Buena, Té del País, Té de Maceta, Té del Pan. This plant has been confused in the literature with *Lantana involucrata* L., not found in Texas.

BLACK MANGROVE

Avicennia nitida Jacq.

FIELD IDENTIFICATION. Shrub, or small tree, of sandy beaches with opposite, oblong or lanceolate, entire leaves.

FLOWERS. July, borne in axillary or terminal spikes; calyx campanulate, sepals 4–5, ovate, rounded or obtuse, densely tomentose; bracts 4–5, imbricate, obtuse or

Black Mangrove
Avicennia nitida Jacq.

acute, densely tomentose; corolla fragrant, campanulate-rotate, ½–¾ in. long, creamy-white, throat yellow, tube short and densely hairy; limb 4-lobed, spreading, lobes oval to obovate, densely white-hairy and ciliate, about 1/12 in. long, some lobes abruptly contracted at base; upper lobe somewhat broader, reflexed, cleft at apex; stamens 4, equal, exserted, glabrous or puberulent, filaments broadened and flattened toward base, partially adnate to corolla; anthers oblong, somewhat asymmetrical, opening below; pistil exserted, pubescent, gradually enlarged toward base; stigma 2-cleft, one lobe larger.

FRUIT. Capsule 2-valved, compressed, oblique, ovate or ellipsoid, very densely pubescent, beaked at the apex, ½–1½ in. long, ⅓–¾ in. wide.

LEAVES. Short-petiolate, entire, leathery, elliptic, oblong or lanceolate, 1½–6 in. long, ⅓–1¾ in. wide, obtuse at apex, acute or acuminate at base, shining above, densely pale-pubescent beneath.

BARK. Dark brown, thin, green when young.

WOOD. Dark brown, close-grained, hard.

RANGE. Sandy tidal flats and lagoons of the Texas coast. On the east end off Galveston Island near the stone breakwaters. East to Florida, south to Mexico, also the West Indies, and Central and South America. In South America it becomes a tree with curious aerating projections rising from superficial roots.

REMARKS. The genus name, *Avicennia*, is in honor of Abdallah Ibn Sina, Arabian philosopher and scientist. The species name, *nitida*, means "shining," and refers to the leaves. Also known as White Mangrove or Mangle Blanco. Standley (1920–1926, p. 1251) reports that the bark is used for tanning.

Texas Bouchea

Bouchea spathulata Torr.

FIELD IDENTIFICATION. Rare, very leafy, densely branched, low shrub.

FLOWERS. July–October, in elongate terminal spikes 1½–4 in. long, ⅝–¾ in. wide, spikes many-flowered and crowded by large, leaf-like bracts; bracts ¼–⅖ in. long, 1/12–⅙ in. wide, shape spatulate, apex acute, base gradually narrowed, pubescent; calyx tubular, ⅓–½ in. long, 1/12–⅛ in. wide, 5-ribbed and toothed, teeth unequal in size and subulate-mucronate; corolla white or purplish, salverform, ½–1 in. long, cylindric-tubular,

Texas Bouchea
Bouchea spathulata Torr.

somewhat curved, spreading into a 5-lobed limb at apex; 2 lobes somewhat shorter than the other 3, about 1/5 in. long; stamens 4, included, one pair inserted near mouth of tube and other pair nearer center of tube, filaments filiform and short, anthers ovate and 2-celled; pistil included, style curvate, stigma 2-lobed; one lobe erect, flattened and stigmatose, the other lobe abortive, minute, reflexed; ovary erect, oblong-conic, 2-celled, each cell 1-ovulate.

FRUIT. Achene-like, dry, linear, beaked, about 1/2 in. long or less, 1/25–1/12 in. wide, not exserted from the calyx, but at maturity dehiscent into 2 nutlets, each 1-celled and 1-seeded; nutlets somewhat ridged and hairy dorsally. Other surfaces smooth and glabrous or pubescent, beak smooth or hairy.

LEAVES. Usually densely clustered, borne opposite or in threes, sessile or nearly so, spatulate or obovate, 1/5–1 in. long, 1/12–1/4 in. wide, apex rounded to obtuse or acute, mucronate, base gradually narrowed, texture thick, surfaces similar in color, pubescent or rugose, upper leaves narrower and passing into the subtending bracts of dense inflorescence.

TWIGS. Light brown to dark brown, stout, young ones rather densely hairy, older more glabrate, leaf scars numerous and prominent, leaves densely clustered.

RANGE. Known in the United States from a few locations in Brewster County, Texas. In this area on dry rocky ridges and canyons in mountains east of Tornillo Creek in Big Bend National Park. Also on gravelly slopes above Boquillas Canyon. Herbarium specimens collected in the Tornillo Creek area by Valery Havard in 1883 were examined by the author. This sheet, numbered 252,007, is deposited in the herbarium of the Chicago Museum of Natural History.

REMARKS. The genus name, *Bouchea*, means "closed," and refers to the enclosed nutlets, and the species name, *spathulata*, refers to the large spathulate bracts. Because of the rarity of the plant no particular vernacular name has been applied.

Flaxleaf Bouchea

Bouchea linifolia Gray

FIELD IDENTIFICATION. Low leafy shrub with numerous erect, striate-angled branchlets. Seldom over 3 ft.

FLOWERS. Spikes terminal, erect, 2–5 in. long, 1/3–2/3 in. wide, rachis slender and ridged, pedicels absent or very short; calyx appressed to rachis, tubular, 2/5–3/5 in. long, about 1/8 in. wide, straight or slightly curved, prominently ribbed, unequally 5–6-toothed, 4 teeth about 1/16–1/8 in. long, subulate and erect, the other minute; bracts setaceous-subulate, 1/25–1/5 in. long, often with 2 minute bractlets on the sides at the base; corolla white or purple, hypocrateriform, tube narrow and cylindric, 1/4–3/4 in. long, glabrous without; limb spreading, irregularly 4–5-lobed, lobes obovate to oval, about 1/4 in. long; stamens 4, inserted within the corolla, filaments filiform, glabrous, about 1/25 in. long; anthers ovate, about 1/25 in. long, 2-celled; pistil single, compound, style filiform, stigma 2-lobed, one lobe larger than the other; ovary 2-celled, oblong, 1/25–1/12 in. long.

FLAXLEAF BOUCHEA
Bouchea linifolia Gray

FRUIT. Fruiting calyx is purplish to brown, ribbed, dry, linear, crowned with persistent calyx teeth and separates into 2 beaked nutlets; pedicels absent or very short; nutlets 1/3–1/2 in. long, faintly ridged, hairy, concave on one surface and faintly ridged on the other, beak short and pointed.

LEAVES. Numerous, decussate-opposite, sessile, linear to narrowly lanceolate, margin entire, sometimes revolute when older, apex acute and submucronate, base narrowed; surface glabrous and bright green above and below, veins obscure, blades 1/4–1 3/4 in. long, 1/8–1/6 in. wide.

TWIGS. Much branched, slender, erect, rigid, prominently ridged longitudinally, glabrate; younger green to brown, nodes annulate, older brown to gray and broken into small, short scales.

RANGE. Dry valleys and rocky hillsides in western Texas and Mexico. In Texas in Brewster, Presidio, Uvalde, Sutton, Kinney, and Val Verde counties. In Mexico in Coahuila and Chihuahua.

REMARKS. The genus name, *Bouchea*, means "closed," and refers to the enclosure of the nutlets by the calyx. The species name, *linifolia*, refers to the flaxlike leaves.

AMERICAN BEAUTY-BERRY
Callicarpa americana L.

FIELD IDENTIFICATION. Stellate-tomentose shrub to 9 ft.

FLOWERS. June–November, borne in axillary, many-flowered dichotomous cymes 1/3–1 1/2 in. long and multidichotomous; peduncle slender, 1/8–1/5 in. long; bractlets about 1/25 in. long, setaceous; corolla rose to pink or pale blue (rarely white), tubular, 1/12–1/6 in. long, 4-lobed; lobes equal, rounded or ovate, obtuse or apiculate at apex, 1/25–1/12 in. long; stamens 4, exserted, adnate to corolla-tube below; style slender, about 1/5 in. long, anthers oblong and yellow; stigma stout and capitate, ovary depressed and globular, 2-celled with 2 ovules in each cell; calyx campanulate, 1/25–1/12 in. long, granulose and puberulent, with 4 obscure teeth.

FRUIT. August–November, a berry-like drupe, borne in conspicuous expanded clusters in leaf axils, peduncles hardly over 1/4 in. long, rose to purple or violet to blue, globose, fleshy, 1/8–1/4 in. long, mealy, sweet; seeds 4, about 1/16 in. long, ovoid to ellipsoid, dorsally rounded, ventrally flattened, light brown and translucent at first; fruiting calyx very shallow, subtruncate or irregularly short-lobed.

LEAVES. Simple, aromatic, decussate-opposite at distant intervals on stem, slender-petioled, coarsely-serrate except near ends, ovate, oval, elliptic or oblong; acuminate, acute or obtuse at apex; cuneate at base; 3–9 in. long, 1 1/2–5 in. broad; dark green and glabrous or puberulent above, paler and stellate-tomentose beneath; petioles slender, 3/4–2 in. long, stellate-tomentose.

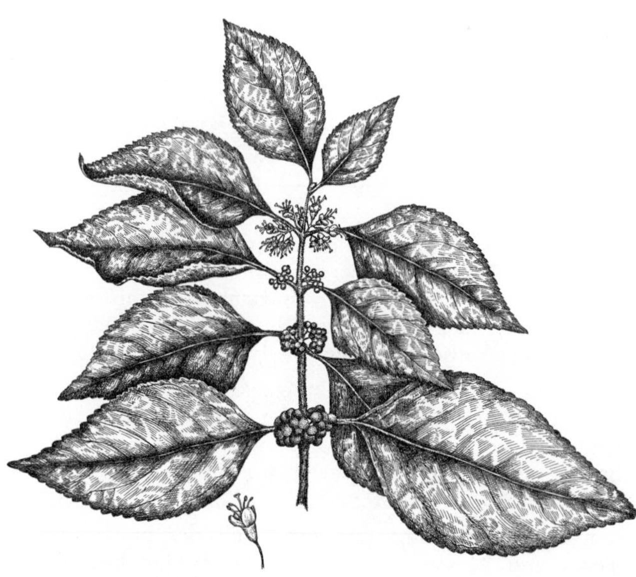

AMERICAN BEAUTY-BERRY
Callicarpa americana L.

BERLANDIER FIDDLEWOOD
Citharexylum berlandieri Robins.

TWIGS. Terete or 4-sided, slender, gray to reddish brown, stellate-tomentose, glabrous later; bark on old stems smooth, tight, or somewhat roughened with small, thin scales below.

RANGE. In rich woods and thickets, especially in pinelands of the coastal plain. Texas, Oklahoma, Arkansas, Louisiana, eastward to Florida, and northward to Virginia. Also in the West Indies.

VARIETY. A variety bearing white fruit, *C. americana* var. *lactea* F. J. Muller, is sometimes found growing with the species, and is similar except for fruit color.

REMARKS. The genus name is from the Greek word *kallos* ("beauty") and *karpos* ("fruit"). Generally known in cultivation under name of Beauty-berry. Other local names in use are Spanish-mulberry, French-mulberry, Bermuda-mulberry, Sour-bush, and Sow-berry. According to stomach records at least 10 species of birds feed upon the fruit, especially the bobwhite quail. Also known to be eaten by raccoon, opossum, and gray fox. Can be cultivated from seeds or cuttings.

3ERLANDIER FIDDLEWOOD

:itharexylum berlandieri Robins.

FIELD IDENTIFICATION. Crooked shrub, or small, ;narled tree to 27 ft.

FLOWERS. Almost throughout summer, fragrant, small, orne in densely flowered, upright racemes on short, ateral axillary twigs, 1/3–4 in. long, 1/3–1 in. wide; peduncles 1/6–1/2 in. long, or sometimes longer, densely short-hairy; pedicels slender, about 1/25 in. long, short-pubescent; calyx campanulate, 1/8–1/6 in. long, 1/12–1/8 in. wide, lensely puberulent, shallowly 5-lobed, ciliolate; corolla vhite, tubular, tube longer than calyx; limb 5-lobed, bout 1/6 in. wide, lobes somewhat irregular; throat ubescent; stamens 4, included, with an occasional bortive fifth stamen; filaments short, anthers ovate, rect, opening longitudinally; style included, stigma hort and bifid, ovary 4-celled.

FRUIT. Drupe yellow or red, subglobose, 2-lobed, bout 1/5–1/4 in. long, lustrous, glabrous, often apiculate t apex, fleshy, 2-celled and 2-seeded, pyrenes separated y a partition.

LEAVES. Opposite, deciduous, thin, blades 1–4 in. ong, 3/4–1 3/4 in. wide; lanceolate, oval, or ovate; apex ounded or obtuse, base rounded or broadly cuneate; pper surface light green and glabrous or slightly hairy long the veins; lower surface paler with soft, minute airs; margin entire or with a few coarse, broad teeth oward the apex; petiole about 1/25–2/5 in. long, pubesent, sometimes glandular at base.

TWIGS. Young twigs slender, green to brown or gray, triate, tetragonal, barely to densely pubescent; older ray, striate, semiterete.

BARK. Mottled gray, smooth, tight.

RANGE. In Texas confined to the valley of the lower Rio Grande River in Cameron, Hidalgo, and Willacy ounties. One mile from Boca Chica Beach below Brownsville, Texas. Also in Mexico in the states of Tamaulipas, Sinaloa, San Luis Potosí, and Veracruz.

REMARKS. The genus name, *Citharexylum*, is from the Greek *cithara*, a stringed musical instrument, and *xylon* ("wood"), because of the fact that musical instruments were made from the wood of some species. The species ame, *berlandieri*, is in honor of the Swiss botanist, Luis Berlandier, who explored the United States–Mexico boundary in 1828. Other names used in Mexico for the tree are Orcujuela, Orcajuela, Negrito, Revienta-Cabra, and Saúco Hediondo. Reported that the plant is used as a remedy for colds in Mexico.

MEXICAN FIDDLEWOOD

Citharexylum brachyanthum Gray

FIELD IDENTIFICATION. Shrub rarely more than 5 ft high, with stiff, irregular branches.

FLOWERS. Axillary or terminal on abbreviated branchlets, solitary or a few together; pedicels hardly over 1/12 in. long, slender, pubescent; corolla white, tubular, cylindric, irregularly 5-lobed; stamens 4 with a rudimentary 5th, filaments short; style included, stigma 2-lobed, carpels 2, each 2-seeded; calyx 5-toothed, about 1/6 in. long, 1/12–1/6 in. wide, shallowly cup-shaped, rim truncate, calyx-teeth short.

FRUIT. Drupe subglobose, 1/5–1/4 in. long, lustrous, glabrous, fleshy, style persistent; pyrenes 2-celled and 2-seeded.

LEAVES. Small, decussate-opposite or somewhat fascicled, sessile or nearly so, 1/12–1 in. long, 1/12–1/6 in. wide, entire, spatulate to obovate or oblanceolate, margin entire, apex acute to rounded, base cuneate, texture thin, surfaces densely pubescent.

TWIGS. Slender, irregular, laterally attenuated and ending in a blunt thorn, tetragonal to subterete, sometimes with corky ridges on the angles, internodes 1/5–1 in. long, gray to black, young densely pubescent, older glabrous.

RANGE. Dry rocky soil of valleys and hills, western Texas and Mexico. In Texas reported from Webb County; abundant in northern Mexico.

VARIETY. The Smooth Fiddlewood, *Citharexylum brachyanthum* var. *glabrum* C. L. Hitchc. & Moldenke, is a variety having glabrous leaves and twigs. Known from Hidalgo and Starr counties, Texas.

REMARKS. The genus name, *Citharexylum*, is from

MEXICAN FIDDLEWOOD
Citharexylum brachyanthum Gray

the Greek *cithara,* the name of a stringed musical instrument, and *xylon* ("wood"), because of the fact that musical instruments were made from wood of some species. The species name, *brachyanthum,* is from *brachy* ("short") and *anthos* ("flower").

Creeping Skyflower

Duranta repens L.

FIELD IDENTIFICATION. Spiny, or spineless, shrub or small tree to 18 ft with drooping or trailing branches.

FLOWERS. February–August, in long racemes, peduncles 2–5 in. long, glabrous or sparsely white-hairy; individual pedicels pubescent, 1/16–1/4 in. long; corolla tubular, 1/8–3/16 in. long, somewhat curved, greenish, puberulent, flaring into a salverform, slightly oblique, limb with 5 lobes; lobes violet or purple; lower lobe largest, about 1/4 in. long, oval to suborbicular or obovate, base rather abruptly contracted, margin somewhat crisped and ciliolate; 2 lateral lobes of limb 1/8–3/16 in. long, obovate, somewhat asymmetrical, margin crisped and ciliolate; 2 upper lobes smallest, 1/16–3/16 in. long, obovate to spatulate, margin crisped and ciliolate, center usually with a dark purple stripe or spots; throat of tube usually white and hairy; 4 stamens, 2 long and 2 short, adnate to the corolla-tube, included; stigma capitate, 3–4-lobed; calyx tubular, plicate, 3/16–1/4 in. long, green, pubescent or almost glabrous; sepals 5, 1/25–1/16 in. long, subulate.

FRUIT. In long, pendent, loose, heavily-fruited racemes, sometimes up to 1 ft long; pedicels 1/8–1/2 in. long, glabrous or nearly so; fruit body golden yellow, shiny, subglobose, formed by the accrescent calyx, sepals

Creeping Skyflower
Duranta repens L.

drawn together at the apex to form a short beak, flesh juicy; nutlets 4, 1/8 in. long or less, short-ellipsoid c oval, rounded and grooved on the dorsal surface, ventr surface flattened or angled and notched at one en brown.

LEAVES. Opposite, elliptic-oblong, ovate to obovat margin serrate along distal 2/3, or nearly entire; ape acute, obtuse or apiculate; base gradually narrowed t form a semiwing on the petiole; upper surface du green, glabrous, veins puberulent and impressed; lower surface paler green, glabrous or pubescent along th prominent veins; blades 2–4 in. long, 3/4–1 1/4 in. wid petiole short, pubescent, portion not winged by blad mostly less than 1/4 in. long.

STEMS. Young stems green to brown or gray, pube cent at first but glabrous later, 4-angled; older sten mostly light brown, rather smooth; spines paired at le bases, straight, slender, stiff, sharp, green or brow pubescent at first, glabrous later, 1/8–3/4 in. long.

RANGE. In Texas escaping from cultivation in th lower Rio Grande Valley area. In Mexico in the state of Baja California, Sinaloa, and Chiapas, to Puebl Veracruz, and Yucatán.

VARIETIES. A number of horticultural and natur varieties are known.

White Creeping Skyflower, *D. repens* var. *alb* (Masters) L. H. Bailey, has white flowers.

Variegated Creeping Skyflower, *D. repens* var. *varie gata* L. H. Bailey, has variegated leaves.

Large-flower Creeping Skyflower, *D. repens* va *grandiflora* Moldenke, has large flowers.

Small-leaf Creeping Skyflower, *D. repens* var. *mi crophylla* (Desf.) Moldenke, has small leaves.

Hairy Creeping Skyflower, *D. repens* var. *canescen* Moldenke, has densely canescent parts.

REMARKS. The genus name, *Duranta,* is in honor o Castor Durantes, physician and botanist in Rome wh died about 1500. The species name, *repens,* mean "creeping," perhaps with reference to the drooping o trailing branches. Vernacular names are Golden Dew drop, Violet Duranta, Pigeon-berry, Espina Blanca Xcambocoche, Adonis Blanco, Adonia Morado, Gar bancillo, Espino Negro, Celosa, Celosa Cimmarona Violetina, Espina de Paloma, Lluvia, Azota Caballo Lila, Cuenta de Oro, Pensamiento, Lora, Heliotropio and Chulada. This plant is often cultivated for orna mental purposes, and the fruit used as a domesti remedy for fevers.

Common Lantana

Lantana camara var. *mista* (L.) L. H. Bailey

FIELD IDENTIFICATION. Irregularly shaped shrub to 6 ft high, branches ascending or spreading.

FLOWERS. Borne on stout, axillary, hairy peduncle which are 3/4–3 1/4 in. long, longer or shorter than th

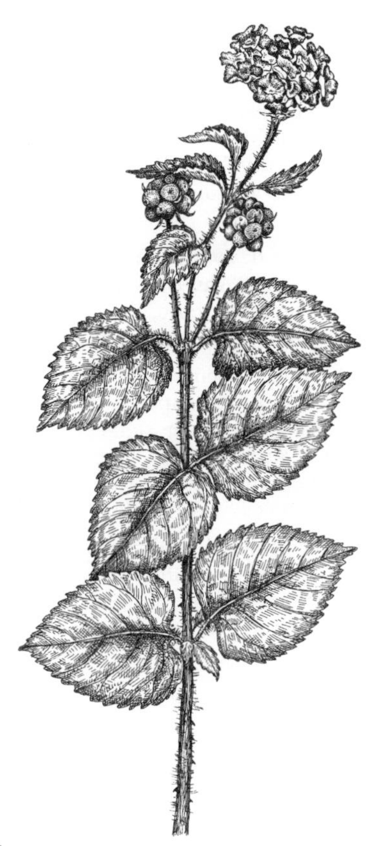

OMMON LANTANA
antana camara var. *mista* (L.) L. H. Bailey

ubtending leaves; densely many-flowered to produce a emispheric head ¾–1⅓ in. across; bractlets ⅙–⅓ in. ong, 1/25–1/15 in. wide, equal, oblong to lanceolate, apex cute, surfaces with long appressed hairs; calyx-tube bout ⅛ in. long, membranous; corolla-tube about ⅖ n. long, slightly expanded above, straight or barely urved, externally puberulent, limb irregularly 5-lobed; uter ring of corollas saffon to brick-red, inner ring ellow to orange later, conspicuous; stamens 4, didy-nous, filaments adnate to about middle of corolla-tube, nthers ovate with parallel cells; style short, stigma thick nd oblique; ovary 1-carpellate and 2-celled, ovules olitary in each cell.

FRUIT. Drupelike, ⅛–¼ in. in diameter, black, pulpy, vatery, nutlets 2, no endosperm.

LEAVES. Decussate-opposite, ovate to oblong, firm in exture, blades ¾–4¾ in. long, 1–2¾ in. wide, apex cute to short-acuminate or rarely obtuse, base abruptly ounded to acutely narrowed or subcuneate, margin renate-serrate; upper surface reticulate-veined, rough nd scabrous; lower surface densely short-pubescent, specially thick on the veins; petioles slender, ⅕–⅘ in. ong, with stiff spreading hairs or glandular-puberulent.

TWIGS. Varying from unarmed to scarcely or abun-lantly prickly, also with dense gray, spreading, rigid airs, older becoming glabrous, stems angled or ridged.

RANGE. Usually in sandy soil, but in gardens apparently thriving in many types of soil. Cultivated from Texas and Louisiana to Florida, escaping cultivation occasionally. Also in the West Indies, Central and South America, Asia, and Africa.

VARIETIES. Bailey (1917) lists the following varieties:

L. camara var. *nivea* Bailey has white flowers, the center becoming bluish.

L. camara var. *mutabilis* Bailey varies in color greatly—from white to yellowish, lilac, rose, or blue. Outer flowers open white and run through yellowish, rose, and lilac; the inner open yellowish.

L. camara var. *crocea* Bailey has flowers opening sulphur-yellow and changing to saffron.

L. camara var. *sanguinea* Bailey has flowers opening saffron-yellow; changing to brick-red.

REMARKS. The genus name, *Lantana,* is the old name once applied to the viburnum. The species name, *camara,* is derived from a Latin word meaning "to arch," in reference to the arching stems. Odor of the plant disagreeable when crushed. Easily rooted from softwood cuttings.

MEXICAN-MARJORAM

Lantana macropoda Torr.

FIELD IDENTIFICATION. Western, aromatic, hairy shrub attaining a height of 3 ft.

FLOWERS. Borne on solitary or opposite peduncles, axillary, often two opposite on a node, ⅜–4½ in. long (two to three times as long as the leaves), heads elongate or rounded, ⅜–¾ in. long, ⅜–⅝ in. wide, densely many-flowered; corolla hypocrateriform, white or pink, often yellow-centered, obscurely 2-lipped and 4–5-lobed, tube ⅙–⅕ in. long, limb about ⅛ in. wide; stamens 4, included, in sets of 2, anthers ovate and cells parallel; style short, ovary 1-carpellate and 2-celled, each cell with 1 ovule; calyx small, thin, membranous; bracts imbricate, ovate, acuminate, white-hairy, punctate.

FRUIT. Drupe with thin flesh, splitting into 2 1-celled pyrenes without endosperm.

LEAVES. Simple, deciduous, decussate-opposite, shape ovate to lanceolate or oblong, blades ⅜–1¾ in. long, ⅛–1¼ in. wide, apex acute, base more or less cuneate, upper surface white-hairy, lower surface hairy and glandular-punctate, veins terminating in the sinuses; margin sharply serrate; petioles slender, 1/12–⅖ in. long, white-hairy.

TWIGS. Slender, 4-angled, strigose with white or yellow hairs, glabrate later.

RANGE. Dry soil of stony hills, or along arroyos, roadsides, and waste places. From central Texas westward to Brewster and El Paso counties, southward to Cameron, Hidalgo, and Willacy counties. In Mexico in Chihuahua, Coahuila, Nuevo León, and Durango.

MEXICAN-MARJORAM
Lantana macropoda Torr.

REMARKS. The name, *Lantana,* is an old word once applied to the viburnum plant, and the species name, *macropoda,* is from *macro* ("large") and *poda* ("foot"), in reference to the elongate flower peduncles. This aromatic shrub is known along the Texas–Mexico border under the name of Mejorana, or Marjoram. The true Marjoram, however, is not this plant, but is *Origanum majorama,* a member of the Mint Family. The Texas species of *Lantana* have been reviewed in a monograph by Moldenke (1942, pp. 48–52).

TEXAS LANTANA

Lantana horrida H. B. K.

FIELD IDENTIFICATION. Perennial herb, or low wide-spreading shrub, sometimes to 6 ft, the whole plant with an unpleasantly pungent scent. Branches terminally ascending and bearing varicolored flowers of red, orange, or yellow.

FLOWERS. Blossoms throughout summer from axils of upper leaves. Borne in many-flowered rounded heads on 2 long, hairy peduncles from each node. Corolla red, orange, or yellow, tubular, ⅓–½ in. long; limb spreading and 4-lobed, densely pubescent; lobes rounc ed, revolute, 2 lobes larger than the others, one lob slightly apiculate; 4 stamens, 2 long and 2 short, ir cluded, inserted near middle of tube; anthers ovate style short, stigma thick and oblique, ovary 1-carpellate 2-celled; calyx hardly ¼ as long as corolla-tube; bract ⅕–⅓ in. long, linear to lanceolate or oblong; acute hairy.

FRUIT. Borne on hairy, somewhat grooved, peduncle 1–4 in. long; drupe August–September, on short re ceptacle, dark blue, globose, lustrous, fleshy; nutlet 2, obovate, about ⅕ in. long, hard, bony.

LEAVES. Simple, decussate-opposite, conduplicate margin coarsely serrate, shape ovate to broadly ovate apex acute, base truncate or semicordate, blade ⅓–1 in. long; upper surface dark green, rugose, scabrou and veiny; lower surface paler and hairy, especiall on the veins; petioles ¼–½ in. long, very rough hairy

STEMS. Lower sometimes almost prostrate but termi nally ascending, young stems 4-angled, green to red dish, very rough-scabrous and hairy (more rarely al most smooth). Some plants with prickly stems.

RANGE. Growing mostly in sandy soil in Texas, Loui

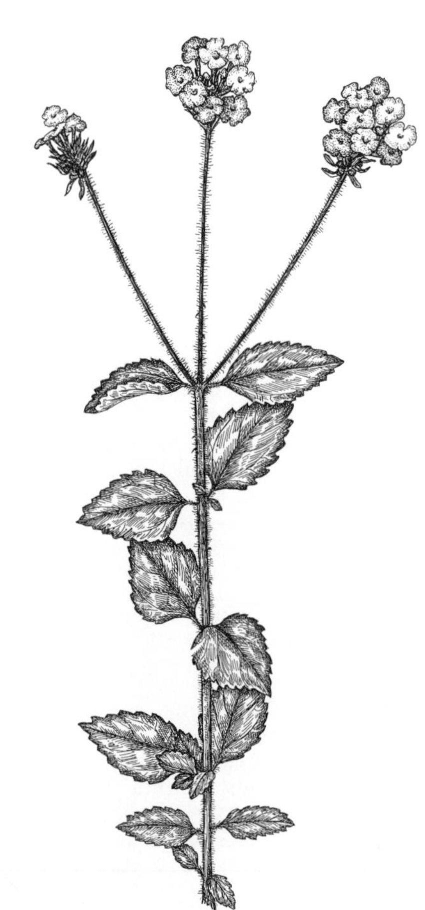

TEXAS LANTANA
Lantana horrida H. B. K.

Vhite-flowered Lantana
.antana citrosa (Small) Moldenke

iana, Mississippi, and Mexico. In Texas apparently nost abundant in coastal areas, but also scattered cross central Texas to the Rio Grande and north to Cook and Archer counties.

MEDICINAL USES. Plant is reported to be poisonous o cattle and sheep though usually not browsed by hem. It contains the alkaloid lantanine, which is an ctive antipyretic. Standley (1920–1926, p. 1249) reorts that "a decoction of the leaves is sometimes emloyed [in Mexico] . . . as a tonic for the stomach. In inaloa the plant is a favorite remedy for snake bites, a trong decoction of the leaves being taken internally and poultice of crushed leaves applied to the wound."

PROPAGATION. Lantana is an attractive shrub, despite he species name, *horrida,* blooming almost throughout he year. Valuable for planting on roadsides and in arks, and as background landscaping. Does best in ight, sandy soils, and is resistant to hot, dry weather. Aay be propagated by seeds, the long root, or cuttings.

SYNONYMS. There appears to be considerable confuion in the literature concerning use of the names *L. orrida* H. B. K. and *L. camara* L. Some authors have onsidered the two names as synonymous, and others ave separated the names as applying to distinct pecies. The author has recognized the name *L. horrida* H. B. K. as applying to the plant herewith described. This plant grows without cultivation in Texas, Louisiana, Mississippi, and Mexico. *L. camara* var. *mista* (L.) L. H. Bailey, which is similar, is confined to cultivation in the areas mentioned, and is native to Florida, Bermuda, West Indies, Mexico, and Central and South America. It differs from *L. horrida* H. B. K. by its larger leaves with more abundant closely appressed teeth. Numerous other species and horticultural forms of *Lantana* are cultivated in gardens. Most of these are introductions from tropical America.

REMARKS. Lantana is an old name which once applied to a viburnum; the species name, *horrida,* refers to the pungent, unpleasant odor of the crushed leaves. The following vernacular names have been recorded for Lantana in the United States and in the Latin-American countries: Bunch-berry, Calico-bush, Hierba de Cristo, Palabra de Mujer, Cinco Coloraditos, Santo Negrito, San Rafaelito, Venturosa Colorado, Filigrana, Comida de Paloma, Poley Cimarrón, Corioguillo, Sarrito, Corronchocho, Zarzamora, Confite, Confite Negro, Sonora Roja, Lampana, Alantana, Confituría, Matizadilla, Mora, Peonía Negra, Siete Colores, Xo-hexnoc, Flor de San Cayetano, Alfombrilla Hedionda, Orozuz del País, Zapotilla, Uña de Gato, Tres Colores, and Cinco Negritos.

White-flowered Lantana

Lantana citrosa (Small) Moldenke

FIELD IDENTIFICATION. Slender shrub attaining a height of 6 ft.

FLOWERS. Borne February–August on slender hairy axillary peduncles ⅜–2 in. long (shorter or longer than the leaves). Flower heads globose or oblong in shape, ⅙–½ in. long, ¼–½ in. wide; bractlets overlapping in series of fours, about ¼ in. long and ⅛ in. wide, or smaller, ovate, base rounded and sessile, apex acuminate, keeled on back, densely appressed-hairy; calyx small, about 1/25 in. long, flattened, membranous, obscurely 2-lobed; corolla-tube cylindric, slender, about ⅙ in. long, white, yellow in the throat and glabrous, exterior hairy; limb about ⅛ in. wide, spreading, 2-lipped, the 4 lobes rounded or truncate, lower and upper lobes longest; stamens 4, in 2 pairs, included and inserted near middle of corolla-throat; anthers ovate, cells parallel; ovary 1-carpellate and 2-celled, each cell 1-ovulate; style short, stigma thick.

FRUIT. Composed of a dry drupe about 1/25–1/12 in. in diameter, nutlets 2 each and 1-celled, flattened, no endosperm.

LEAVES. Opposite, larger ones often with small ones in their axils, varying from ovate to lanceolate or elliptic, length ⅓–2½ in., width ¼–1 in.; apex acute to acuminate; base attenuate, margin finely toothed but often entire toward the base; upper surface with veins impressed and densely canescent-hairy, lower surface pubescent and canescent; petioles 1/12–1/3 in. long, slender, densely canescent-hairy, sometimes margined by the attenuate leaf base.

TWIGS. Slender, squarish, pale gray to white, densely canescent with appressed hairs, older twigs more glabrate.

RANGE. Known in Texas only from Hidalgo and Cameron counties, but widely distributed in Mexico, Guatemala, Cuba, and southern Florida.

SYNONYMS AND RELATED SPECIES. This species has been erroneously recorded from Texas under the name of *L. involucrata* L.

L. citrosa is distinguished from the closely related *L. macropoda* Torr. by the peduncles being mostly shorter than the subtending leaves and the leaf blades finely crenate. Reported as being used in Mexico for its aromatic essential oil.

REMARKS. The genus name, *Lantana*, is an old name once applied to a viburnum. The species name, *citrosa*, refers to its citrus-like aromatic odor.

Lilac Chaste-tree

Vitex agnus-castus L.

FIELD IDENTIFICATION. Cultivated, aromatic tree to 30 ft, often with many trunks from base. Branches slender and spreading outward to form a wide, broad-topped crown.

FLOWERS. May–September, panicled spikes conspicuous, terminal, dense, puberulent to pulverulent, 4–12 in. long, 1/2–1 1/4 in. wide; flowers sessile or very short pediceled; corolla blue to purplish, 1/4–1/3 in. long, funnelform; tube slightly curved, densely white-pubescent above the calyx; limb 1/5–1/4 in. broad, slightly oblique, ciliate, white hairs at the limb sinuses, somewhat 2-lipped, 5-lobed, upper 2 lobes and lateral 2 lobes ovate and obtuse, lower lobe largest, obtuse to rounded; stamens 4, exserted, 2 sometimes longer than the others but not always, anthers with nearly parallel, arched or spreading sacs; stigma exserted, slender, 2-cleft; ovary 4-celled and 4-ovuled; calyx campanulate, 1/12–1/8 in. long, densely white-puberulent, irregularly and shallowly 5-toothed, teeth triangular and acute; bractlets and bracteoles linear-setaceous, 1/25–1/6 in. long.

FRUIT. Small, globular, brown to black, 1/8–1/6 in. long, the persistent calyx membranous; stone 4-celled, no endosperm.

LEAVES. Internodes 1 1/4–4 in. long, leaves decussate-opposite, deciduous, digitately compound of 3–9 (mostly 5–7) leaflets, (blades) 1 1/4–5 in., central one usually largest, linear to linear-elliptic or lanceolate, apex attenuate-acuminate; base gradually narrowed into a semiwing and channeled, sessile in the smaller leaves; margin entire or plane surface undulate-repand, texture thin; upper surface dull green and glabrous or

Lilac Chaste-tree
Vitex agnus-castus L.

minutely pulverulent; lower surface paler or almost whitened, puberulent, veins reticulate under magnification; petioles 1/2–3 in. long, densely grayish or reddish brown, puberulent to pulverulent or cinereous, resinous-granular.

TWIGS. Slender, elongate, quadrangular, green to reddish brown or gray, grayish puberulent and pulverulent, pith stout.

BARK. Smooth, light to dark gray on young branches, on old trunks gray with broad ridges and shallow fissures.

RANGE. Dry, sunny situations, in various types of soils. From China and India, widely cultivated in Europe and Asia. In the United States cultivated from Texas eastward to Florida and northward to North Carolina. Sometimes escaping cultivation.

PROPAGATION. The fruit may be hand-picked or flailed into a canvas. The seed are extracted by running hrough a macerator and floating off the hulls. Cleaned eed averages about 40,000 seeds per lb, with a commercial purity of 80 per cent and a soundness of 55 er cent. Pretreatment to break dormancy consists of tratification in moist sand or peat for 90 days at 41°F. eed are sown in loamy, rich soil and covered ¼ in. haste-tree may also be propagated by layers and by reenwood cuttings under glass. The tree was first ultivated in the year 1570.

VARIETIES. A number of horticultural varieties and orms have been developed such as:

V. agnus-castus var. *caerulea* Rehd., which has blue lowers;

V. agnus-castus var. *alba* West, which has white lowers;

V. agnus-castus var. *latifolia* (Mill.) Rehd., which has oblong-lanceolate leaves up to 1 in. broad, more hardy;

V. agnus-castus forma *rosea* Rehd., which has pink flowers;

V. agnus-castus var. *serrata,* which has serrate leaflets.

REMARKS. The genus name, *Vitex,* is the ancient Latin name. The species name, *agnus-castus,* is from *agnus* ("lamb") and *castus* ("pure, holy, or chaste"). Vernacular names are Monk's Pepper-tree, Wild Pepper, Indian Spice, Abraham's Balm, Hemp-tree, Sage-tree, Wild Lavender, Common Chaste-tree, True Chaste-tree, Tree of Chastity, Chaste Lamb-tree.

The seeds of the Chaste-tree are reported to be sedative. In Brazil a perfume is made from the flowers, and the aromatic leaves are used to spice food.

MINT FAMILY (*Labiatae*)

HAIRY ROSEMARY-MINT
Poliomintha mollis (Torr.) Gray

HAIRY ROSEMARY-MINT

Poliomintha mollis (Torr.) Gray

FIELD IDENTIFICATION. Slender Western plant 6–24 in. Stems essentially herbaceous, but occasionally somewhat woody close to base. All parts densely grayish-tomentose.

FLOWERS. Often abundant in axillary glomerules; corolla white to pink, 1/4–1/2 in. long; tube gradually broadened upward and with scattered white hairs; limb 2-lipped, upper lip more erect and emarginate, lower lip 3-lobed with middle lobe often emarginate; fertile stamens 2, parallel under the upper lip; style filiform, apex usually cleft; calyx 1/4–5/16 in. long, tube slender and terete, many-nerved, surface densely tomentose, apical teeth 5, sometimes unequal, reddish, about 1/25 in. long.

FRUIT. Nutlets small and elongate.

LEAVES. Deciduous, opposite, often with smaller ones in their axils, blade 3/8–1 in. long, ovate to oval or oblong, apex obtuse or acute, base rounded or gradually narrowed, margin entire, grayish green, densely stellate-tomentose, veins obscure above but more prominent beneath; petiole short, about 1/8 in. or less, tomentose.

STEMS. Very slender, mostly herbaceous, angled, green and densely tomentose, later from tan to brownish and glabrous close to base.

RANGE. Dry rocky hillsides and mountain canyons. In western Texas mostly west of the Pecos River. Also in adjacent Mexico.

REMARKS. The genus name, *Poliomintha,* is from the Greek word *polios* ("gray-white") and *mintha* ("mint"), referring to the hairs. The species name, *mollis,* refers to the softness of the hairs.

Smooth Rosemary-mint

Poliomintha glabrescens Gray

FIELD IDENTIFICATION. Low Western plant shrubby toward the base. All parts at first canescently puberulent but later almost glabrate.

FLOWERS. Numerous in axillary glomerules; corolla-tube up to ½ in. long, white and sometimes also purplish-speckled, loosely white-hairy; limb 2-lipped, upper lip erect and emarginate, lower lip 3-lobed and spreading, with the middle lobe often emarginate; tube pilose-annulate within; fertile stamens 2, sometimes slightly exserted, parallel under the upper lip; style slender, filiform, apex 2-cleft, sometimes slightly exserted; calyx up to ¼ in. long, terete and tubular, distinctly ribbed, obscurely glandular-dotted, densely and finely white-hairy, apex with 5 acute or acuminate erect teeth hardly more than 1/12 in. long.

FRUIT. Nutlets small, smooth, oblong.

LEAVES. Deciduous, opposite, often with smaller ones in their axils, blades ¼–¾ in. long, ⅛–3/16 in. wide, linear-oblong, margin entire, apex obtuse, base gradually narrowed, surfaces dull pale-green, veins obscure, conspicuously glandular-dotted, puberulent to glabrate; petiole 1/16–1/25 in. long, minutely puberulent, leaf sometimes almost sessile.

TWIGS. Slender, brown and finely puberulent, angled, older brown to tan and glabrous, distinctly angled.

RANGE. Known from western Texas and adjacent Mexico. Standley (1920–1926, p. 1271) states that the type specimen was collected at Soledad, southwest of Monclova, Coahuila, Mexico.

REMARKS. The genus name, *Poliomintha,* is from the Greek words *polios* ("gray-white") and *mintha* ("mint"), referring to the hairs of same species. The species name, *glabrescens,* signifies the nearly hairless character of the older stems.

Smooth Rosemary-mint
Poliomintha glabrescens Gray

Hoary Rosemary-mint

Poliomintha incana (Torr.) Gray

FIELD IDENTIFICATION. Western subshrub to 3 ft. All parts densely white-tomentose.

FLOWERS. May–September, in small axillary glomerules toward the ends of the branches; pedicels about 1/25 in. long and tomentose; corolla pale purplish or bluish, or white with purple dots, ⅖–½ in. long, tube with hairy rings in the throat; limb 2-lipped, upper lip erect and emarginate, lower lip 3-cleft and the middle lobe often emarginate; fertile stamens 2, attached above middle of corolla-tube and ascending parallel under the upper lip; style slender, cleft at apex; calyx about ¼ in. long, terete, 15-nerved, tube silky-hairy; teeth 5, minute, subequal, more or less connivent, calyx strongly annulate in the throat.

FRUIT. Nutlets small, smooth, oblong.

LEAVES. Deciduous, opposite, often with smaller ones in their axils, blades ¼–1 in. long, linear to oblong or lanceolate, margin entire, apex obtuse to acute, base gradually narrowed, surfaces finely and velvety tomentose on both sides, minutely punctate, veins obscure; leaf sessile or petioles very short.

TWIGS. Slender, greenish, densely silvery-tomentose, minutely punctate-dotted; older woody, brown, eventually glabrous.

RANGE. On dry, sunny, sandy sites; sometimes found on alkaline or gypsum soils. At altitudes of 4,000–5,000 ft. In Texas west of the Pecos River, and north in New Mexico and Arizona to Colorado and Utah.

REMARKS. The genus name, *Poliomintha,* is from the Greek words *polios* ("gray-white") and *mintha* ("mint"), referring to the hairs. The species name, *incana,* also

HOARY ROSEMARY-MINT
Poliomintha incana (Torr.) Gray

refers to the gray hairs. The plant is attractive and deserves more use in cultivation. It is reported that the Hopi Indians eat the leaves raw or boiled, sometimes drying them for winter food. The flowers are also used for seasoning.

GREASELEAF SALVIA

Salvia pinguefolia (Fern.) W. & S.

FIELD IDENTIFICATION. Aromatic Western shrub to 5 ft.

FLOWERS. Borne in terminal spikes 2–5 in. long, on pedicels about ⅛ in. long; corolla blue, bilabiate, ¼–⅜ in. long, upper lip saccate and erect, mostly including the 2 stamens, lower lip 3-lobed; style hairy, usually with the 2 filiform lobes slightly exserted; calyx bilabiate, widely flared at apex, surface canescent with minute simple appressed hairs and glandular-dotted, green or wine-colored.

FRUIT. Nutlets 4, ovoid, smooth, nestled in the calyx base.

LEAVES. Opposite, ovate to deltoid, base truncate or subcordate, but sometimes slightly extended on the petiole, apex acute or obtuse, margin crenate-toothed, surfaces varying in hairiness; on some finely pubescent above and densely white-canescent beneath, on others greener and puberulent, usually minutely glandular-dotted; petioles ⅛–¾ in. long, glandular-pubescent.

TWIGS. Slender, tetragonal, when young tan to brownish, densely canescent-tomentose; older glabrous, reddish brown to light brown or gray, irregularly shreddy.

RANGE. Dry mesas or hillsides at altitudes of 2,000–7,000 ft. Extreme west Texas, New Mexico, and Arizona; also Chihuahua, Mexico.

REMARKS. The genus name, *Salvia,* is from the Latin *salvus* ("to heal"), referring to its medicinal properties, and the species name, *pinguefolia,* means "grease-leaf" or "fat-leaf," the leaves having a slightly greasy feeling when rubbed between the fingers. The description in-

GREASELEAF SALVIA
Salvia pinguefolia (Fern.) W. & S.

Shrubby Blue Salvia
Salvia ballotaeflora Benth.

cludes also the species known at one time as S. *vinacea* W. & S., which was distinguished as having a wine-colored calyx. However, since the two seem to intergrade as to green or wine-colored calyx characters, the segregation seems too weak to warrant two specific names. Greaseleaf Salvia has also been described as a variety under the name of S. *ballotaeflora* var. *pinguefolia* Fern.

Shrubby Blue Salvia

Salvia ballotaeflora Benth.

FIELD IDENTIFICATION. Very aromatic, much-branched, square-stemmed shrub attaining a height of 6 ft.

FLOWERS. Generally throughout the summer, borne in short, simple racemes, ¾–1½ in. long, in axils of upper leaves; calyx ⅕–⅗ in. long, pendulous at maturity, campanulate, conspicuously ribbed, puberulent, 2-lipped, flaring into 3, nearly equal, broadly ovate, obtuse lobes; corolla bluish or purple, ¼–⅓ in. long, tube shorter than calyx, 2-lipped; lower lip about ¼ in. long, much longer than the upper, 3-lobed and spreading, middle lobe largest and notched at apex, glabrous; upper lip short, concave, hairy, paler blue than lower lip.

FRUIT. Sides of the fruiting calyx folded together at maturity, nutlets slightly more than 1/12 in. long.

LEAVES. Aromatic, simple, opposite, ovate to triangular or deltoid, apex acute or obtuse, base truncate or cuneate to subcordate, margin crenate, blades ⅓–1½ in. long; young leaves tomentulose-canescent, older leaves rugose and reticulate-veined, petioles slender.

STEMS. Square, brittle, pale.

BARK. Varying shades of gray to black, smooth on young stems.

RANGE. Usually in dry, rocky soils on limestone hillsides. Scattered through the Edwards limestone area of central Texas and westward, also New Mexico. In Mexico in the states of Tamaulipas, Nuevo León, Chihuahua, San Luis Potosí, and Zacatecas.

REMARKS. The genus name, *Salvia,* is from the Latin *salvus* ("to heal"), referring to medicinal qualities of the plants. The species name, *ballotaeflora,* means "having flowers of *Ballota.*" Vernacular names used in the Southwest are Shrubby Sage, Engorda-cabra, Mejorana de País, and Crespa. The dried aromatic leaves are used for flavoring meats and other foods. During drought periods the leaves are deciduous, but new leaves appear with the rains.

Royal Sage

Salvia regla Cav.

FIELD IDENTIFICATION. Shrub of west Texas mountains, attaining a height of 2–6 ft, often with many stems from the base.

FLOWERS. Very conspicuous, in loose terminal racemes; corolla scarlet, exserted 1–1¾ in. beyond the calyx, widely 2-lipped, upper lip saccate and somewhat hooded over the stamens, lower lip pendent and 3-lobed, middle lobe longest, exterior of corolla puberulent and often with golden orange minute glandular dots; stamens 2, slightly exserted or included, the lower fork of the connectives bearing an anther cell; style solitary, exserted slightly, usually longer than stamens, style with 2 linear lobes of unequal length; ovary deeply 4-lobed; calyx ¼–⅝ in. long, somewhat inflated and gradually enlarged upward, tinged with red, 2-lipped, one lip broadly obtuse or almost truncate, the other lip with two obtuse to acute or triangular teeth.

FRUIT. Nutlets 4, nestled in bottom of calyx-tube.

LEAVES. Opposite, broadly deltoid to ovate, apex acute to obtuse, base truncate or subcordate, margin coarsely crenate, blades ¾–2 in. long, upper surface dull-green with minute short scattered hairs, lower surfaces much paler, puberulent, and often dotted with minute golden orange glands; petioles ⅜–1 in. long, puberulent and scatteredly glandular-dotted.

STEMS. When young slender, squarish, reddish to brown, pubescent, sometimes glandular-dotted; older stems with brown to gray bark sometimes shreddy.

RANGE. A Mexican species evidently confined to the

Royal Sage
Salvia regla Cav.

mountains of western Texas and southward. In the Chisos Mountains in Brewster County, along Oak Creek Canyon and Green Gulch. In Mexico in Coahuila and Durango to Oaxaca. Type from Regla, Hidalgo, Mexico.

REMARKS. The genus name, *Salvia*, is from the Latin name used by Pliny, meaning "safe, unharmed," referring to the medicinal properties of some of the species. The species name, *regla*, means "a standard" or "model"; the plant is perhaps named after the town of Regla in Hidalgo, Mexico. Very showy shrub conspicuous on the mountainsides; worthy of cultivation.

Autumn Salvia

Salvia greggii Gray

FIELD IDENTIFICATION. Western shrub attaining a height of 3 ft.

FLOWERS. Throughout summer and into fall, borne in terminal, erect 3–12-flowered racemes; pedicels 1/8–1/4 in. long, glandular-puberulent; corolla exserted 3/4–1 1/4 in. beyond the calyx, red to purplish red, 2-lipped, somewhat inflated at the middle; upper lip saccate, lower lip 3-lobed with middle lobe longest; stamens 2, on short filaments, jointed with the elongate transverse connective; style barely exserted, hairy along the uppe side, split at apex into unequal filiform stigmas; caly 2/5–1/2 in. long, broadly campanulate, flared toward th apex, reddish-ribbed, puberulent-glandular, 2-lippe lower lip with 2 abruptly mucronate points, upper li acute or mucronate-pointed.

FRUIT. Nutlets 4, smooth, nestled at base of calyx.

LEAVES. Opposite or somewhat fascicled, narrowl oblong to linear or oblanceolate to oblong-obovate, ape obtuse to acute, base cuneate or gradually narrowe margin entire and glabrous or sometimes ciliate, blad length 3/8–1 1/4 in., width 3/16–3/8 in.; surfaces dull pale green and glabrous, or obscurely puberulent, obscurel glandular-punctate, 1-ribbed, almost veinless, coria ceous; petiole obscurely puberulent to glabrate, 1/16–1 in. long, or base winglike and leaf sessile.

TWIGS. Slender, squarish, glandular-puberulent, gree to reddish brown, later brown and roughened by loos scaly bark.

Autumn Salvia
Salvia greggii Gray

Parry Sage
Salvia parryi Gray

RANGE. On dry sunny sites in southern and western Texas and New Mexico. Also in Mexico in Coahuila, Sonora, and Durango.

FORMS. Several color forms are known. White Autumn Salvia, S. *greggii* forma *alba*, has white flowers. Pink Autumn Salvia, S. *greggii* forma *rosea*, has pink flowers. Both the species and its two forms are desirable ornamentals.

REMARKS. The genus name, *Salvia*, is from the Latin name used by Pliny, meaning "safe, unharmed," referring to the medicinal properties of some of the species. The species name, *greggii*, honors Josiah Gregg (1806–1850), early American explorer and botanist.

Parry Sage

Salvia parryi Gray

FIELD IDENTIFICATION. Plant mostly herbaceous and up to 2 ft, numerous stems from a woody base.

FLOWERS. Maturing April–August, in slender interrupted spikes 3–6 in. long, usually 4–6 in axillary whorls at short distances along the stem, the upper forming a more continuous spike; bracts oval to ovate, abruptly pointed, about 1/4 in. long, purple or yellowish, slightly tomentose; calyx 1/8–1/3 in. long, laterally compressed, 2-lipped, the divisions slender and acute, surface purple and densely woolly with branched hairs; corolla blue, 1/4–1/3 in. long, strongly 2-lipped; lower lip much larger and 3-lobed with middle lobe longest; upper lip smaller and hoodlike; stamens 2, connectives strongly developed; style exserted, bifid, arched under the hood.

FRUIT. Enclosed in the dry calyx, the 4 nutlets small and smooth.

LEAVES. Simple, opposite, sometimes with small leaves in axils of large ones, blade ovate to oblong or lanceolate, sometimes deltoid, margin crenulate, apex obtuse or acute, base narrowed to rounded, or almost truncate, blades 1/2–1 1/2 in. long, 1/4–3/4 in. wide, veins obscured by the dense white to gray or brown tomentum; petiole 1/4–3/8 in. long, densely tomentose.

TWIGS. Slender, erect, angled, young ones pale grayscurfy, older ones tan to brown and glabrous, woody near the base with bark somewhat fibrous.

RANGE. Rocky slopes and desert grasslands at altitudes of 3,500–5,000 ft. In southwestern New Mexico, Arizona, and Sonora, Mexico.

REMARKS. The genus name, *Salvia*, is the old classical name, and the species name, *parryi*, is in honor of C. C. Parry, a naturalist connected with the United States–Mexico boundary survey in the 1850's.

Branched Sage

Salvia ramosissima Fern.

FIELD IDENTIFICATION. Spreading plant with slender branches attaining a height of 8–18 in. Somewhat woody near the base.

FLOWERS. Corolla-tube purple to blue or rarely white, about 1/2–3/4 in. long, 2-lipped; lower lip 3-lobed and much longer than the upper lip, middle lobe largest and cleft; upper lip hooded; fertile stamens 2; style bifid, curved, slightly exserted under upper lip; calyx tubular, gradually widened above, 2-lipped, 1 lip deeply cleft and acutely pointed, surfaces green and conspicuously ribbed.

FRUIT. Formed by the dry adherent calyx which encloses 4 small, smooth nutlets.

LEAVES. Simple, opposite, sometimes with smaller ones in their axils, blade linear to oblong or elliptic, margin entire or minutely serrulate, older somewhat revolute, apex obtuse to acute or rounded, base gradually or abruptly narrowed into a short winged petiole, blade length 1/4–1 in., upper surface dull green, glabrous or nearly so, minutely punctate.

TWIGS. Young ones slender, green and puberulent,

Branched Sage
Salvia ramosissima Fern.

4-angled, older ones brown to gray and glabrous, sometimes fibrous.

RANGE. Canyons and dry hills of the grass or pinyon belt. Western Texas near Signal Peak, Guadalupe Mountains in Culberson County. New Mexico in the Organ Mountains. Also in Mexico.

REMARKS. The genus name, *Salvia*, is the old classical name, and the species name, *ramosissima*, refers to the many branches.

Bladder-sage

Salazaria mexicana Torr.

FIELD IDENTIFICATION. Western shrub 1–3½ ft with wide-spreading, subspinose branches.

FLOWERS. In few-flowered terminal racemes 2–4 in. long, generally in pairs from the axils of the miniature upper leaves; corolla bluish or purplish, tube white and puberulent; limb 2-lipped, rather short, lateral lobes more or less joined with the upper lip and enclos ing the stamens and style; stamens 4, paired; ovary 4 lobed, superior, style bifid; calyx pouch-shaped, abou ⅜ in. long, later becoming inflated.

FRUIT. Calyx at maturity bladdery and inflated, vesic ular and reticulated, ½–¾ in. in diameter, reddish enclosing 4 roughened nutlets.

LEAVES. Simple, opposite, remote, oblong to ellipti or lanceolate to ovate, apex obtuse or acute, base nar rowed, margin entire or few-toothed, blade length ¼–1½ in., width ⅛–¼ in., surface glabrate and obscurely 1–3 veined; petioles about ⅛ in. long.

TWIGS. Diffuse, slender, squarish, white, subspinose soft-canescent.

RANGE. Desert areas, rocky ravines, foothills and ar royos at elevations of 1,000–3,500 ft. Western Texas New Mexico, west to California, north to Nevada and Utah, south in Mexico to Coahuila and Chihuahua.

REMARKS. The genus name, *Salazaria*, is for Don José Salazar, Mexican commissioner for the United States–Mexico boundary survey. The species name *mexicana*, is given because it is abundant in certain parts of Mexico. A vernacular name for the species is Paperbag-bush, because of the inflated calyx. It is re ported to be of some value for livestock in arid areas.

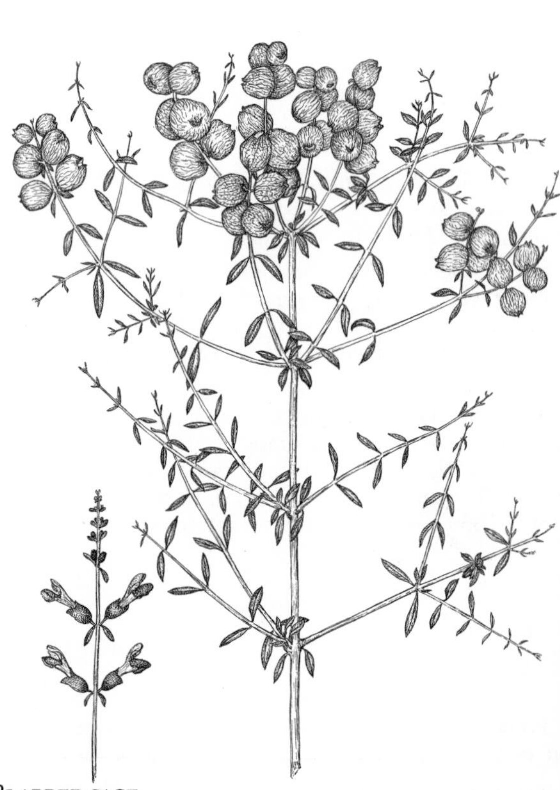

Bladder-sage
Salazaria mexicana Torr.

NIGHTSHADE FAMILY (*Solanaceae*)

BUSH PEPPER
Capsicum frutescens L.

FIELD IDENTIFICATION. Low undershrub to 5 ft, with a broad spreading top, somewhat woody toward the base.

FLOWERS. Solitary, borne on glabrous or slightly puberulent pedicels ½–1½ in. long; perianth about ⅜ in. wide; petals 5, white, oblong, apex acute, puberulent, about 3/16 in. long, reflexed later; stamens 5, 1/16–⅛ in. long, filaments short, hardly over 1/25 in. long; anthers oblong, erect, closely appressed to and surrounding the pistil, about 1/16 in. long; style simple, white, somewhat exceeding the stamens in length, stigma minute, ovary 2–3-celled; calyx very shallow and broadly lobed, lobes abruptly and minutely pointed at the apex, sometimes nearly truncate.

FRUIT. Borne on erect, green, clavate, glabrous peduncles ½–1¼ in. long. Fruit body variously shaped, oval, obovate, oblong or conate, green at first, orange to red later, ¼–⅓ in. long, glabrous, lustrous, fleshy, pungently hot to the taste; seeds numerous, flattened, oval, whitish, minutely beaked, about ⅛ in. wide.

LEAVES. Blade elliptic, ovate or oval, apex acute or acuminate, base broadly cuneate, 1–1½ in. long, upper surface rather dull green and glabrous, lower surface paler and glabrous, margin entire to serrulate; petioles ⅓–⅔ as long as the blade, glabrous or puberulent.

TWIGS. Crooked, smooth, slender, green to light brown.

RANGE. The Bush Pepper is distributed from Arizona, New Mexico, and Texas east to Florida, and south through the tropics.

MEDICINAL USES. Capsicum contains as its most important constituent an irritant, *capsaicin,* which was isolated in 1876, according to Wood and Osol (1943),

BUSH PEPPER
Capsicum frutescens L.

who describe the medicinal uses of capsicum as follows (pp. 257–258):

"Capsicum is a powerful local stimulant, producing, when swallowed, a sense of heat in the stomach, and a general glow over the body without any narcotic effect. It is much employed as a condiment, especially in hot climates, and is useful in cases of atony of the stomach or intestines. It is strongly contraindicated by the existence of gastric catarrh of the ordinary type, but in the chronically inflamed stomachs of persons of intemperate habits it frequently appears to do good.

It is also of value in certain cases of serious diarrhea, not dependent on true inflammation of the intestines.

"When applied to the skin in proper concentration solutions of capsicum produce at first a sensation of warmth, and, with more powerful solutions, later an almost intolerable burning. It differs from other local irritants in that there is practically no reddening of the skin even when there is very severe subjective sensation. In other words, while it exerts a strongly irritating effect upon the endings of the sensory nerves, it has very little action upon the capillary or other blood vessels. Therefore it does not cause blistering in strong solution. It is not proper to call capsicum a rubefacient because it does not produce reddening of the skin. It is, however, frequently added to counter-irritant applications. The sensation of warmth which it produces is often very grateful to the patient, but we are somewhat skeptical as to whether it exerts the same therapeutic value as the rubefacients. In the form of lozenges, it is sometime used for the treatment of relaxed uvula and other similar conditions of the pharynx. The tincture has been employed externally in the treatment of chilblains."

VARIETIES, FORMS, AND SYNONYMS. The history of the taxonomy of *Capsicum* is very complex. Parks and Cory (1936) list two species as native to Texas—*C. frutescens* L. and *C. baccatum* L. Small (1933, p. 1117) states that the principal difference between these two species is that the former has a toothed calyx and subglobose berry and the latter a truncate or barely undulate calyx and an ellipsoid to conic berry. These differences do not appear to be constant, however, and some botanists consider that the two are but the same plant, showing considerable variation.

The botanist Asa Gray (Irish, 1898, p. 55) believed that there are only two basic species of *Capsicum*—one which is herbaceous and annual or biennial, represented by *C. annuum*, and the other shrubby and perennial, represented by *C. frutescens*. Bailey (1925, pp. 659–660), on the other hand, believed that the annual forms have descended from the perennial *C. frutescens* and that all forms and varieties have it as a basic heritage. Working upon the basis of this opinion, the following varieties and forms are listed:

Red Pepper, *C. frutescens* forma *typicum*, has oblong to spherical fruit, usually red, about ¼–¾ in. long. (Including *C. baccatum* and *C. annuum.*)

Ornamental Bush Pepper, *C. frutescens* var. *abbreviatum* Fingh., has ovoid, rugose fruit up to 2 in. long, used for pickling.

Cherry Pepper, *C. frutescens* var. *cerasiforme* Irish, has erect or leaning spherical fruit up to one inch across, yellow or purple, very pungent.

Tabasco Pepper, *C. frutescens* var. *conoides* Irish, has conical to oblong-cylindrical fruit.

Cluster Pepper, *C. frutescens* var. *fasciculatum* Irish, has fruit very slender, up to 3 in. long, very pungent.

Sweetbell Pepper, *C. frutescens* var. *grossum* Sendt., is tall, stout; pepper oblong to bell-shaped, red to yellow, fleshy, swollen, sides furrowed, apex sunken, flavor mild.

Long Pepper, *C. frutescens* var. *longum* Sendt., has drooping, tapering fruit up to a ft long. It includes such forms as Long Red, Long Yellow, Chili, Cayenne, and others.

REMARKS. The genus name, *Capsicum*, is of vague origin, and the specific name, *frutescens*, refers to the frutescent, or shrubby, habit. Vernacular names are Chilipitin, Chillipiquin, Chili Pepper, Red Pepper, Hot Pepper, and Cayenne Pepper. The pungent berries are rubbed on sun-drying meat in the Southwest to keep flies away. A seasoning is also made by soaking the berries in vinegar. A domestic remedy for coughs and colds was formerly made by steeping the berries in boiling water and adding sugar. Wild turkeys eat the berries readily, but it is said that their eating too many makes the flesh unpalatable.

Night-blooming Cestrum

Cestrum nocturnum L.

FIELD IDENTIFICATION. Vigorous cultivated shrub to 12 ft. Branches glabrous, elongate, often flexuous, arching. The very fragrant flowers opening at night.

Night-blooming Cestrum
Cestrum nocturnum L.

FLOWERS. June–August, borne in axillary cymes or anicles 2–5 in. long, peduncles flattened and grooved; edicels 1/25–1/8 in. long, glabrous, slender, flattened, ometimes channeled; calyx tubular, glabrous, less than in. long; sepals 5, acute, about 1/25 in. long; corolla reen or yellowish, tubular, slender at the base and radually broadened at the throat, 3/4–1 in. long, apex ith 5 lobes about 1/8 in. long, ovate, acute to short-cuminate; ovary 2-celled, style slender, slightly exerted or included, stigma slightly 2-lobed; stamens 2–5 some abortive), included, attached to wall of the ibe within, appendaged near the point of insertion, nther sacs parallel.

FRUIT. Berry maturing July–August, 1/4–3/8 in. long, void to short-oblong or subglobose, white, flesh thin; eeds 1–2, black, oval to ovoid, flattened, about 1/8 in. ong, albuminous.

LEAVES. Simple, alternate, deciduous, very variable n size, blades 1½–6 in. long, the longest on vigorous oung shoots, elliptic to oblong or ovate to lanceolate, pex acute to acuminate, base rounded to broadly uneate, sometimes asymmetrical, margin entire, texture rm, dark green and glabrous above with veins impressed, lower surface paler, smooth and glabrous with eins more conspicuous; petioles glabrous, 3/16–1/2 in. ong.

TWIGS. Young ones slender, arching, glabrous, light o dark green; older ones gray to brownish with minute ale lenticels.

RANGE. A native of the West Indies. Cultivated in outhern Texas and Louisiana, eastward to Florida; in Mexico in Coahuila, Guerrero, Oaxaca, and Veracruz. Persistent about abandoned gardens or escaping in ome regions.

REMARKS. The genus name, *Cestrum,* is from an old Greek name of uncertain application, and the species ame, *nocturnum,* refers to the night-blooming habit. However, other species of this genus also bloom at ight. The fruit and juice are reported to be poisonous f taken internally. It has many vernacular names including Jessamine, Hule de Noche, Hierba Hedionda, Galán de Tarde, Galán de Noche, Dama de Noche, Reina de la Noche, Palo Hedionda, and Pipiloxihuitl.

Chilean Cestrum
Cestrum parqui L'Her.

FIELD IDENTIFICATION. Half-hardy upright, branching shrub attaining a height of 6 ft. Known mostly as a cultivated plant in the United States.

FLOWERS. Borne in axillary or terminal cymes forming panicles; calyx about 1/6 in. long, tube cylindric, the 5 lobes deltoid to triangular-ovate; corolla yellowish green, fragrant at night, tube slender and gradually dilated to the mouth, nearly cylindric, 3/4–1 in. long; corolla-lobes 1/8–1/6 in. long, spreading, acute or obtuse, apiculate; stamens 5, inserted about the middle of the tube, included, anthers globular, sacs parallel; style slender, stigma enlarged, ovary 2-celled, 2–6 ovules in each cell.

Chilean Cestrum
Cestrum parqui L'Her.

FRUIT. Berry globose, violet-brown, seeds 1–4, smooth, albuminous.

LEAVES. Simple, alternate, lanceolate to elliptic or oblong, margin entire, apex acute or acuminate, gradually narrowed or cuneate at base, 2–8 in. long, upper surface bright green, lower slightly paler, finely pubescent along the veins at first but later glabrous; petioles 1/5–2/5 in. long, finely pubescent.

TWIGS. Younger ones pale green to tan, angled, finely pubescent, older ones tan to light or dark brown and glabrous.

RANGE. Woods, thickets, or waste places; cultivated and persistent about old gardens. Perhaps not truly escaping cultivation except in Florida. From Texas and Louisiana coastal-area gardens eastward to Florida and Georgia. A native of Chile.

REMARKS. The genus name, *Cestrum,* is from the Greek name *kestron,* of uncertain application. It is also known under the name of Night-blooming Jessamine, but is not to be confused with *C. nocturnum,* which is also a night-bloomer. Another name is Willow-leaved Jessamine. It can be propagated by seeds, or by cuttings taken February–March and potted in sand.

Day Cestrum

Cestrum diurnum L.

FIELD IDENTIFICATION. Shrub attaining a height of 15 ft, but usually only 3–5 ft tall. The slender erect stems are branched and minutely pubescent on the younger parts.

FLOWERS. Borne on long stems in loose, mostly axillary, pedunculate racemes; individual flowers pediceled or sessile; calyx tube campanulate, about ⅛ in. long, somewhat pubescent or glabrous, the 5 lobes broadly ovate; corolla-tube white, ½–⅗ in. long, gradually widened to the apex, the 5 lobes reflexed, much shorter than the tube, rounded and about as wide as long; stamens 5, filaments adnate to above the middle of the corolla-tube, anthers globular with sacs parallel; stigma capitate, ovary 2-celled.

FRUIT. Berry subglobose to oblong, drying brown, about ¼ in. long, 2-celled; seeds solitary or a few, smooth.

LEAVES. Simple, alternate, persistent, rather thick in

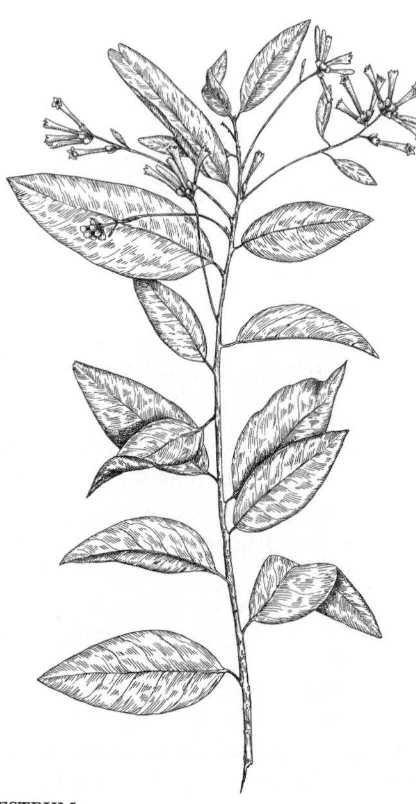

Day Cestrum
Cestrum diurnum L.

texture, oblong to elliptic, 2–5 in. long, margin entir apex short-acute or obtuse, base rounded or broad cuneate, surface glossy green above, slightly paler b neath, petioles mostly less than ⅖ in. long.

TWIGS. Slender, young ones green to brownish wi scattered pubescence, older ones brown to gray and gl brous.

RANGE. Cultivated in gardens, but occasionally e caping cultivation and persistent in sandy soil in woo and waste places. Southern Texas and the Florida Key Also in Mexico in Yucatán and Sinaloa. Native of th West Indies.

REMARKS. The genus name, *Cestrum,* is from th Greek name *Kestron,* originally applied to a plant of u certain identity. The species name, *diurnum,* refers t its fragrance by day, in contrast to certain other ce trums which are fragrant only at night. It is also know as Day-jessamine, Jaun de Noche and Galán de Día.

Anderson Wolfberry

Lycium andersonii Gray

FIELD IDENTIFICATION. Spiny scraggly or rounde shrub attaining a height of 1–9 ft. Branches numerou with the older ones often spineless.

FLOWERS. January–May, mostly axillary, solitary o in pairs; pedicels ⅛–⅖ in. long; calyx cup-shaped 1/25–⅛ in. long, 2-lipped or with 4–5 teeth which ar about ¼ as long as the tube, surfaces puberulent o glabrous, margins ciliate; corolla tubular-funnelform gradually widened upward, ¼–⅗ in. long; lobes o corolla 4–5, spreading, ovate, about ¼ as long as th tube, margin fimbriate or ciliate, color light purple lavender, or white; stamens exserted or as long as th corolla, filaments adnate partially, hairy near the base style slender, exserted, or as long as the corolla.

FRUIT. Berry ellipsoid to ovoid, ⅛–⅓ in. long, juicy fleshy, red; seeds numerous, about 1/12 in. long, brown.

LEAVES. Alternate or clustered, fleshy, ⅛–⅗ in. long, 1/25–⅛ in. wide, linear to spatulate, base gradually narrowed, apex acute or rounded, surfaces glabrous o younger ones rather scurfy; petioles averaging abou 1/16 in. long but variable on young shoots.

TWIGS. Somewhat flexuous, slender, grayish white to silvery or light brown; spines numerous, slender, sharp, ⅕–⅘ in. long.

RANGE. Gravelly washes, sandy or alkali flats up to an altitude of 5,000 ft. In New Mexico to California, north to Colorado, Nevada, and Utah. In Mexico in Sonora and Sinaloa.

VARIETIES. *L. andersonii* var. *wrightii* Gray, is a variety with small flowers ⅕–⅓ in. long, elliptic-spatulate leaves, and stigma 1/12–⅛ in. long. It is found in Arizona and Sonora, Mexico.

Anderson Wolfberry
Lycium andersonii Gray

with a few scattered hairs toward the base, 4–5-lobed, lobes reflexed at maturity, 1/6–1/3 the length of the tube; stamens 4–5, barely exserted, inserted by the filaments within the tube and hairy, anthers about 1/25 in. long; style slender, filiform, about as long as the stamens, stigma expanded, green, ovary bicarpellate and 2-celled.

FRUIT. Berry red, globose to ellipsoid, about 1/6 in. thick, 8–30-seeded; seeds irregular, some minutely pitted, embryo coiled, cotyledons long and slender.

LEAVES. Alternate or 2–3 in a cluster, linear to spatulate, 3/8–1 in. long, 1/16–1/8 in. wide, apex rounded to obtuse or acute, base attenuate, margin entire, surfaces gray, glabrous or puberulent.

TWIGS. Slender, crooked, often drooping, gray to reddish, spineless or with slender gray to brown spines, glabrous to pubescent.

RANGE. The species is found up to an altitude of 3,000 ft in Texas, New Mexico, Arizona, and Mexico. In Texas in the counties of Maverick, El Paso, Bexar, Tom Green, Val Verde, Cameron, Presidio, Jackson, Zapata, Brewster, Uvalde, and Starr. In Mexico in Coahuila, Nuevo León, and Hidalgo.

VARIETIES AND FORM. Two varieties and a form have been described as follows:

L. andersonii forma *deserticola* C. L. Hitchcock, has leaves 4/5–1 2/5 in. long and 1/25–1/12 in. broad.

REMARKS. The genus name, *Lycium,* is the Greek name of the country of Lycia. The species name, *andersonii,* is for Edgar Anderson, American botanist. Other vernacular names for the plant are Anderson Desert-thorn and Squawberry. It can be propagated by seed, suckers, layers, and hardwood cuttings. The fruit is eaten by Gambel's quail and the flowers attract the black-chinned hummingbird.

Berlandier Wolfberry

Lycium berlandieri Dunal

FIELD IDENTIFICATION. Shrub spinose, glabrous or pubescent, sparingly branched, spreading or reclining, attaining a height of 7 ft.

FLOWERS. Axillary, solitary or several together, on filiform pedicels 1/8–4/5 in. long; calyx campanulate, 1/25–1/12 in. long or wide, glabrous to pubescent, 3–5-lobed, lobes about 1/3 as long as the tube; corolla blue to lavender (more rarely white), campanulate but much flared toward the apex and constricted at the base, length 1/6–1/3 in., mostly glabrous or occasionally

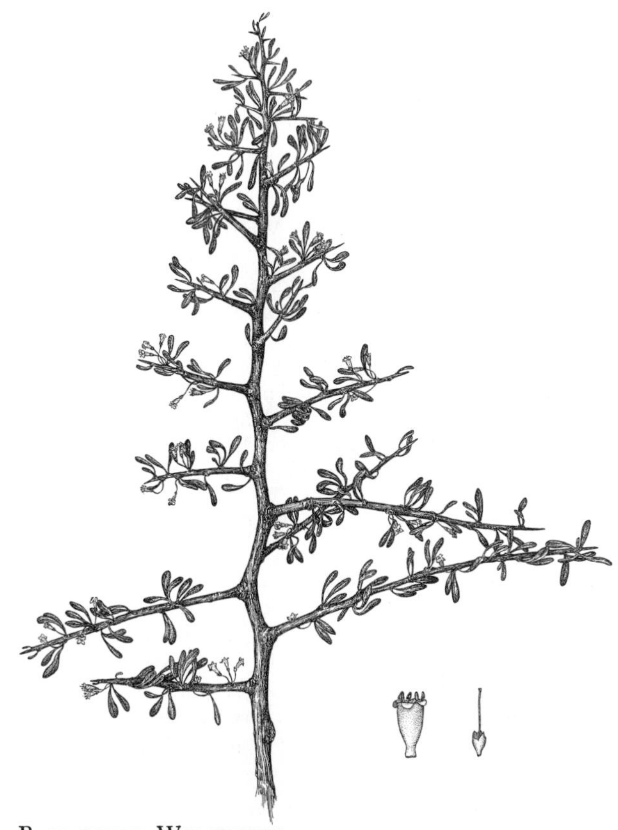

Berlandier Wolfberry
Lycium berlandieri Dunal

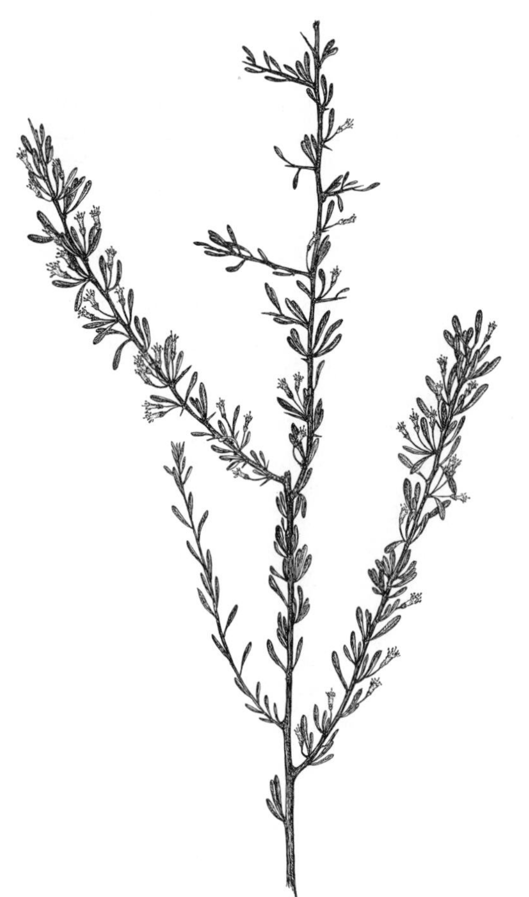

SHORTLOBED BERLANDIER WOLFBERRY
Lycium berlandieri var. *brevilobum* C. L. Hitchcock

The Smallflower Berlandier Wolfberry, *L. berlandieri* forma *parviflorum* (Gray) C. L. Hitchcock, is a form with smaller flowers, and stouter and more leafy branches. It occurs in Arizona and Mexico.

The Longstyle Berlandier Wolfberry, *L. berlandieri* var. *longistylum* C. L. Hitchcock, has leaves ⅓–1 in. long and 1/25–⅛ in. wide; stamens 1/25–⅛ in. longer than the corolla-lobes and style as long as the stamens or longer; corolla much broadened toward the apex, the 5 lobes ⅓–½ as long as the tube. It is found in southern Arizona.

The Shortlobed Berlandier Wolfberry, *L. berlandieri* var. *brevilobum* C. L. Hitchcock, is a variety with the lobes of the corolla about 1/25 in. long and not reflexed, with slightly exserted stamens; corolla-tube funnelform and flared at the top; pedicels 1/12–¼ in. long; leaves ⅛–⅗ in. long, 1/25–1/12 in. wide, in clusters of 2–5; branches usually spineless, straight and glabrous. In Texas recorded from Terlingua Creek, Brewster County; Presidio del Norte, Presidio County; also in Zacatecas, Mexico.

REMARKS. The genus name, *Lycium*, is for the country of Lycia, and the species name, *berlandieri*, honors the Swiss botanist, Luis Berlandier, who was sent by the Mexican government to determine the character of the United States and Mexican boundary in 1828. It is also known under the vernacular name of Cilindrillo. Berlandier Wolfberry may be grown from hardwood cuttings, suckers, layers, and seeds.

CAROLINA WOLFBERRY

Lycium carolinianum Walt.

FIELD IDENTIFICATION. Thorny, widely-branched, or semitrailing shrub of saline soil. Attaining a height of 6 ft with alternate or seemingly fascicled, thick, grayish green leaves.

FLOWERS. Axillary, solitary or clustered, rotate-campanulate, on glabrous peduncles about ¼–1¼ in. long, lavender to purple with darker purple streaks within; corolla-tube flaring into a 4-lobed limb about ⅜–⅝ in. across; lobes equal to the tube or longer or shorter, 3/16–5/16 in. long, ovate to oblong, apex obtuse to rounded; calyx 3/16–¼ in. long, 4-lobed, glabrous, cuplike, lobes ovate and acute, 1/12–⅙ in. long; stamens 4–5, exserted ¼ in. or more, filaments filiform, white or purplish, adnate near middle of tube, woolly-hairy below; anthers about 1/25–1/12 in. long, opening lengthwise;

CAROLINA WOLFBERRY
Lycium carolinianum Walt.

istil nearly same length as stamens, style filiform, tigma small and capitate, ovary 2-celled.

FRUIT. Berry 3/8–1/2 in. long, ovoid to short-oblong, eddish orange, fleshy; seeds numerous, flattened, round-d to almost triangular.

LEAVES. Alternate or fascicled at the nodes, sessile r nearly so, 1/2–1 1/2 in. long, 1/8–3/16 in. wide, widest bove the middle, linear to narrowly spatulate or semi-erete, base gradually narrowed, apex rounded to ob-use, fleshy, thick, grayish green and somewhat glauces-ent, veins obscure.

TWIGS. Rigid or elongate-arching or trailing, smooth, labrous, younger parts somewhat succulent; lateral pines slender, sharp, stiff, gray to brown, averaging bout 3/8 in. long, terminating lateral shoots.

RANGE. Saline soils, brackish marshes, shell mounds, andy lagoons, mostly on the coast. Texas, Louisiana, astward to Florida, and northward to North Carolina. n Mexico in Tamaulipas, Hidalgo, Michoacán, Sinaloa, nd Baja California; also in the West Indies.

PROPAGATION. The plant may be propagated by eeds, suckers, or hardwood cuttings. It is sometimes ultivated, as are several other species of *Lycium*, for rnamental hedges or specimen plants. It grows well n sand, and can stand in water for considerable periods, et is also resistant to drought and apparently well uited to cultivation in semidesert areas.

VARIETY. The Large-flowered Wolfberry, *L. carolin-anum* var. *quadrifidum* (Moc. & Sessé *ex* Dunal) . L. Hitchcock, was once distinguished as a variety ecause of the larger leaves, flowers, and more spines, s well as a more western distribution. However, these haracters are not constant, there being many inter-rades. Therefore, segregation of the variety does not ppear justifiable.

REMARKS. The genus name, *Lycium*, is for the coun-ry of Lycia. The species name, *carolinianum*, is for he states of Carolina. Vernacular names are Matri-nony-vine, Box-thorn, and Christmas-berry. The red erry, which resembles a tomato, was formerly eaten aw or dried by the Indians. It also is consumed by vild fowl. The leaves of a related species, *L. europeum* ., are used in Europe as a cooked green.

ATRIMONY-VINE

ycium halimifolium L.

FIELD IDENTIFICATION. Spiny or spineless shrub, cul-ivated erect or with recurved and drooping branches, ometimes trailing or vinelike, attaining a length of –25 ft. Hardly over 3–6 ft when erect.

FLOWERS. Axillary, solitary or in clusters of 2–4, orne on filiform pedicels 3/8–4/5 in. long; calyx cam-anulate, persistent, about 1/10 in. long, glabrous, 3–5-obed, lobes about 1/2 as long as the tube; corolla greenish to purplish, rotate-campanulate, tube 1/8–1/3 in. long, gradually widened toward the apex, lobes 4–5, imbricated, ovate to oblong, 1/6–1/5 in. long; stamens 5, adnate to the middle of the corolla-tube, filaments filiform, subequal, somewhat hairy above point of adnation, anthers oval, about 1/25 in. long, cells longitudinally dehiscent; style slender, equaling or exceeding the stamens, stigma dilated, ovary 2-celled.

MATRIMONY-VINE
Lycium halimifolium L.

FRUIT. Berry in August–May, orange to red, ovoid to short-oblong, length 2/5–3/5 in., bicarpellate and 2-celled; seeds 10–20, shape irregular, embryo coiled, cotyledons long and slender.

LEAVES. Alternate or fascicled, shape various, from oval to ovate or elliptic and lanceolate to rarely spatulate, blades 1/2–2 1/3 in. long, width 1/4–1 in., apex obtuse or acute, base attenuate to a petiole 1/8–2/5 in. long or leaf sessile, margin entire, texture thick and firm, surfaces grayish green and glabrous.

TWIGS. Slender, flexible, arching or recurved, somewhat angled, smooth, gray to tan, spiny or spineless; spines slender, sharp, 3/8–1/2 in. long.

RANGE. Usually in thickets, along fence rows, and waste places. A native of Europe and adjacent Asia, cultivated in the United States and escaping cultivation in various provinces. Also in the West Indies and parts of Mexico. In Louisiana near Lake Charles; in Leflore County, Oklahoma; near Santa Fe, New Mexico.

Reported as escaping in Georgia, Kansas, Minnesota, California, and Ontario, Canada.

VARIETIES. Several varieties are known in horticulture:

The Lanceleaf Matrimony-vine, *L. halimifolium* var. *lanceolatum* (Poir.) Schneid., has lanceolate leaves and ellipsoid fruit.

The Roundfruit Matrimony-vine, *L. halimifolium* var. *subglobosum* (Dunal) Schneid., has subglobose fruit, lanceolate leaves, and a dwarf habit.

REMARKS. The genus name, *Lycium,* refers to the country of Lycia, and the species name, *halimifolium,* is for the leaves resembling *Halimus.* Also known by the local names of Matrimony-vine, Box-thorn, Bastard-jessamine, Jasmine, and Jackson-vine. A plant commonly cultivated in the United States, and propagated by means of hardwood cuttings, suckers, layers, or seeds. It is sometimes used for hedges or arbors. It is reported that in Europe the young leaves are cooked and eaten. A tincture of the leaves is also considered of value as an antispasmodic in whooping cough.

PALE WOLFBERRY
Lycium pallidum Miers

PALE WOLFBERRY

Lycium pallidum Miers

FIELD IDENTIFICATION. Shrub thicket-forming, spreading or upright, densely branched, spiny, 3–6 ft tall.

FLOWERS. In April–May, axillary, solitary or paired, on slender pedicels about as long as the calyx; calyx campanulate, 1/5–1/3 in. long, glaucous-green, 5-lobed, lobes ovate to lanceolate or elliptic, apex acute, about as long as the tube, glabrous to pubescent; corolla-tube 1/2–4/5 in. long, funnelform, expanding from base to tip from 1/12–1/4 in., color greenish yellow or veins purple, 5-lobed, lobes oval to rhombic, from 1/5–1/3 as long as the tube, margins ciliolate; stamens 5, slightly exserted, adnate within to the corolla-tube and hairy, anthers about 1/25 in. long; style filiform, slender, varying in length from about equal to the tube to 1/6–1/5 longer than the tube, shortly 2-lobed, ovary bicarpellate, 2-celled.

FRUIT. Berry red, glaucous, subglobose to ovoid, about 3/8 in. in diameter, juicy, 20–50-seeded; seeds about 1/25 in. long, embryo coiled, cotyledons slender.

LEAVES. Alternate or fascicled, oblong to spatulate or broadly elliptic, apex acute, obtuse or rounded, base attenuate into a short petiole, margin entire, length 3/8–2 in., width 1/8–3/5 in., texture fleshy, color glaucous-green, midrib and primaries apparent.

TWIGS. Flexuous, gray, yellow, reddish, or purplish, pubescent or glabrous later, spineless or spines 1/8–3/8 in. long, slender and sharp.

RANGE. Pale Wolfberry is found at altitudes of 3,000–7,000 ft in western Texas, New Mexico, Arizona, Colorado, Utah, and southern California; also in Mexico in Sonora, Chihuahua, Zacatecas, and San Lui Potosí.

VARIETY. A variety is *L. pallidum* var. *oligospermur* C. L. Hitchcock, of California and Nevada. It has fruit 1/4–1/3 in. in diameter and 4–8 seeds; corolla seldor over 4/5 in. long, filaments hairy most of their length leaves 2/5–4/5 in. long, 1/8–1/5 in. broad.

REMARKS. The genus name, *Lycium,* is for the cour try of Lycia, and the species name, *pallidum,* refers t the pale flowers and leaves. The berries are eaten b the Arizona Indians, and are also consumed by bird The plant is occasionally cultivated for ornament an is propagated from hardwood cuttings, suckers, layer and seeds. It has been cultivated since 1878. It sprout back from the base when cut down. In some area it has value as a winter browse for livestock. Easil identified by its large pale leaves which are large than most lyciums, and large funnelform flowers.

HAIRY WOLFBERRY

Lycium puberulum Gray

FIELD IDENTIFICATION. Small, sparingly branche spiny shrub to 3½ ft.

FLOWERS. Axillary, 1–2 in the leaf fascicles, pedice about 1/12 in. long or flowers sessile; calyx campanulat 1/6–1/4 in. long, densely pubescent, 5-lobed, lobes ob long to ovate, as long as or much longer than the calyx tube; corolla tubular-campanulate, tubular portion white, 1/3–1/2 in. long, base about 1/15 in. wide, apex about 1/8 in. wide, glabrous externally; lobes 5, yellowish green, 1/5–1/4 as long as the tube, reflexed, ovate-triangular and acute; stamens 5, equal to the corolla-tube or shorter, adnate to the tube within, glabrate, tube densely hairy toward the base; anthers affixed to the connectives near the middle, about 1/25 in. long, cells longitudinally dehiscent; style long and slender, about as long as the filaments, included.

FRUIT. In June, red, rather dry, bicarpellate, 2-celled, 1–2 seeds in each cell, seeds irregular in shape, embryo coiled, cotyledons slender.

LEAVES. Alternate or 2–6 together, elliptic to oblong or obovate, length 1/5–3/5 in., width about 1/3 as broad as long, apex rounded, sessile or subsessile, densely short-pubescent, not fleshy, midnerve evident.

TWIGS. Usually slender and divergent, green to gray or purplish, young pubescent, older glabrous, usually with slender spines 1/5–2/5 in. long.

RANGE. In the larrea belt along the Rio Grande River in southern Texas. At Presidio in Presidio County, also in Brewster County in the foothills of the Chisos Mountains. Type collected at El Paso by Charles Wright.

REMARKS. The genus name, *Lycium,* is for the country of Lycia, and the species name, *puberulum,* refers to the lightly hairy flowers and twigs.

AIRY WOLFBERRY
cium puberulum Gray

TORREY WOLFBERRY
Lycium torreyi Gray

TORREY WOLFBERRY

Lycium torreyi Gray

FIELD IDENTIFICATION. Heavily spined to spineless shrub divaricately branched, 3–6 ft.

FLOWERS. April–June, in axillary fascicles of 1–3 flowers on pedicels 1/5–4/5 in. long, bisexual and regular; corolla greenish white to lavender, funnelform, slightly expanded from the base toward the apex, 1/3–3/5 in. long, exterior glabrous or nearly so, hairy within, apex of tube flared into 4–5 lobes; lobes rounded or oval, minutely white-ciliate on the margin, 1/5–1/4 as long as the tube; stamens 5, barely exserted, adnate to about the middle of the tube, filaments woolly at base, anthers about

1/25 in. long, opening lengthwise; style slender, slightly longer than the stamens and exserted 1/12–1/8 in. beyond the tube, stigma capitate, ovary superior, 2-celled; calyx cup-shaped, 1/12–1/5 in. long, about 1/12 in. wide, glabrous or puberulent, 5-lobed; lobes deltoid, ciliate, 1/4–1/2 as long as the tube.

FRUIT. Berry ripening June–September, bright red, ovoid, 1/5–2/5 in. long, fleshy, juicy, 8–30-seeded, edible but not palatable.

LEAVES. Simple, alternate or fascicled, shape varying from linear to elliptic or spatulate to oblanceolate, apex obtuse or acute, base attenuate, margin entire, texture thickish, surfaces glabrous and veins obscure except main vein, length 3/8–2 in., width 1/25–5/8 in.; petiole 1/12–1/8 in. long.

STEMS. Erect or spreading, densely leafy, divaricate, spineless, or more often with stout spines 1/5–1/2 in. long, green to gray.

RANGE. Along streams, on alkali flats, sometimes thicket-forming, usually below an altitude of 3,000 ft. The Rio Grande River drainage of Texas, also in New Mexico, Arizona, southern California, south in Mexico to Chihuahua, Sonora, and Hidalgo.

REMARKS. The genus name, *Lycium,* is for the country of Lycia, in Asia Minor, and the species name, *torreyi,* honors John Torrey (1796–1873), American botanist of Columbia University. The plant is also known by the vernacular names of Desert-thorn, Rabbit-thorn, Box-thorn, Squaw-berry, Tomatillo, and Garambullo.

TREE TOBACCO
Nicotiana glauca Graham

TREE TOBACCO

Nicotiana glauca Graham

FIELD IDENTIFICATION. Shrub to 18 ft; stems slender; all young parts glabrous and glaucous.

FLOWERS. April–May, disposed in loose, long, terminal panicles; pedicels about 2/5 in. long or less; flower bisexual, regular; calyx 1/2–3/5 in. long, lobes 5, unequal, lanceolate to triangular; corolla nearly tubular, green to yellowish, slightly enlarged above the calyx and somewhat constricted at the throat, soft-pubescent externally, 1–2 in. long, limb about 2/5 in. wide; lobes 5, small, shorter than the tube, erect, ovate to lanceolate, acute to acuminate; stamens 5, included, inserted on the corolla-tube, filaments slender, anther sacs opening lengthwise; stigma capitate, ovary superior and 2-celled.

FRUIT. Pedicels incurved in fruit, capsule narrowly ovoid to elliptic or oblong, 2/5–1/2 in. long, dehiscent at apex, 2-celled; seeds numerous, minute.

LEAVES. Simple, alternate, evergreen or nearly so, shape ovate to elliptic-ovate or subcordate, 2 in. to 1 ft long, 1–3 in. wide, apex obtuse or acute, base narrowed to a petiole 1–2 in. long, margin entire or undulate, bluish green with a bloom on both surfaces.

STEMS. Slender, elongate, herbaceous above, soft woody below, glabrous and covered with a white bloom.

RANGE. Ditches, stream banks, roadsides, waste places. A native of Argentina. Cultivated and escaping in the southern states from Florida westward through Texas, New Mexico and Arizona to California. Also in Mexico and the West Indies.

MEDICINAL USES. The plant is reported to contain an alkaloid, anabasine, which is efficacious in killing aphids. The leaves are often applied as poultices in domestic medicine to relieve pain, especially headaches. It is also reported as poisonous if taken internally.

REMARKS. The genus name, *Nicotiana,* is for Jean Nicot, French diplomat, and the species name, *glauca,* refers to the glaucous ("bloom") covering of the leaves and stems. The plant is known under many vernacular names such as Sacred Mustard, Tronadora, Tabaco Amarillo, Conetón, Virginio, Gigante, Tabaquillo, Mostaza Montés, Don Juan, Lengua de Buey, Mar huana, Arbol de Tabaco, Tabaco Cimarrón, Buena Moza, Tepozan Extranjero, Gretana, Tabacón, Tacote, Palo Virgen, Marquiana, and Hierba del Gigante. It is grown from seed.

Mullein Nightshade

Solanum verbascifolium L.

FIELD IDENTIFICATION. Tree to 30 ft, 6 in. in diameter, but usually only an irregular, open flat-topped shrub. Young branchlets, leaves, and flowering cymes are densely set with stellate velvety hairs.

FLOWERS. Blooming intermittently throughout the year in stellate-hairy, many-flowered, flat-topped cymes on peduncles 1–4 in. long; peduncle branched into a number of stout pedicels ¼–¾ in. long; corolla white, rotate, ½–⅝ in. across, the 5 lobes ovate to oval, acute at the apex, and somewhat ciliate on the margin; stamens exserted, adnate near the throat of corolla-tube, anthers as long as the filaments or longer, oblong, introrse, 2-celled; ovary pubescent, ovules numerous; calyx ⅜–¼ in. wide, densely stellate-hairy, the 5 lobes ovate to triangular, acute at apex.

FRUIT. Calyx stellate-hairy; berry oblong to globose, green at first, yellow at maturity, young with scattered stellate hairs, older more glabrous, about ½–¾ in. in diameter, rind thick, pulp juicy; seeds very numerous, about 1/25 in. long, orbicular to obovoid, lenticular, flattened, white to yellow or brownish.

Mullein Nightshade
Solanum verbascifolium L.

LEAVES. Alternate, ovate to elliptic or oblong, apex acute or acuminate, base cuneate or rounded, margin entire or slightly undulate, length 4–12 in., width 1–3 in., texture thin, yellowish green and stellate-hairy above, paler and more densely hairy beneath, midrib prominent; petiole ½–1¼ in. long, densely stellate-hairy.

TWIGS. Younger stellate-hairy, older gray to brown, smooth, with tight bark.

BARK. Smooth, thin, green to brown or gray.

RANGE. Usually in sandy soil, in the United States in southern Florida and the Keys, southern California, and in the lower Texas Rio Grande Valley. In Texas along the edges of old resacas at Olmito in Cameron County. Also in Mexico, Central and South America, West Indies, tropics of the Old World, and in southeastern China. Considered by some authorities as naturalized rather than native in the United States.

REMARKS. The genus name, *Solanum,* is from the Latin *solamen,* meaning "solace or quieting," probably with reference to some of the medicinal virtues of related species. The species name, *verbascifolium,* refers to the resemblance of the leaves to those of *Verbascum* (mullein). It is also known under the vernacular name of Potato-tree. It is reported that the soft velvety leaves are used for cleaning dishes. They are also heated and applied to the forehead to relieve headache and are applied as a poultice to ulcers and boils.

The plant has also been listed under the following names in Central and South America: Salvadora, Sacamanteca, Guardolobo, Xtuhny, Tom-poop, Xaxox, Zoza, Sosa, Hierba de San Pedro, Hoja de Manteca, Galantea, Friegaplato, Berenjera, Tabaco Cimarrón, Preadedera Hedionda, Prendedera Macho, Pendejera Macho, Berenjena de Paloma, Berenjena Cimarrona, Tabacón Pelado, and Tapabayote.

Many well-known food and ornamental plants belong to the genus *Solanum,* including the Irish Potato, Eggplant, Jerusalem-cherry, and Bittersweet. Tomato and pepper are in closely related genera. Some members of the group possess toxic properties, as is true of the various nightshades.

Texas Nightshade

Solanum triquetrum Cav.

FIELD IDENTIFICATION. Green vinelike perennial, mostly herbaceous above but somewhat woody toward the base.

FLOWERS. Borne in few-flowered, loose, umbellate cymes on club-shaped pedicels; calyx 5-lobed, tube turbinate and glabrous, lobes triangular-ovate, acute; corolla about ⅝ in. broad, white to violet or purple, rotate, deeply 5-lobed, lobes oblong to lanceolate, apex ciliolate; stamens 5, filaments short, anthers yellow, narrow, about 1/25 in. broad, sacs opening by 2 terminal pores or chinks; ovary 2-celled.

FRUIT. Berries red, globose, about 1/3–2/5 in. in diameter, 2-celled, seeds flattened.

LEAVES. Simple, alternate, deciduous, length 3/4–2 1/2 in., ovate to lanceolate or deltoid, entire, or some 3-lobed toward the truncate or cordate base, apex acute to long acuminate, surfaces deep-green; petioles barely 1/2 as long as the blades or shorter.

STEMS. Glabrous, slender, vinelike, spreading or reclining on other plants, 1–5 ft long, herbaceous above, lignescent toward the base.

RANGE. Often on fences or intertwined with other plants. Low grounds in southern and western Texas. Also in Nuevo León, Coahuila, and San Luis Potosí, Mexico.

REMARKS. The genus name, *Solanum*, is the classical name for the nightshade. The species name, *triquetrum*, means "3-angled," and refers to the triangular leaves. It is also known as White Nightshade, Climbing Nightshade, and Hierba Mora. It is worthy of cultivation.

TEXAS NIGHTSHADE
Solanum triquetrum Cav.

IGWORT FAMILY (*Scrophulariaceae*)

/RIGHT FOXGLOVE
rachystigma wrightii (Gray) Pennell

FIELD IDENTIFICATION. Perennial herb 1–2 ft high, mewhat lignescent near the base. The slender stems e erect and scabrous-puberulent.

FLOWERS. August–September, in loose racemes from e axils of the smaller upper leaves; pedicels about –½ in. long, scabrous-puberulent; corolla ½–1 in. ng, yellow or orange-yellow, open-funnelform, slight- irregular, the lobes spreading, externally puberulent glabrate, glabrous within; stamens 4, included, fila- ents pubescent, nearly as long as the corolla-tube or nger, anthers 2-celled; ovary superior, style slender, igma capitate; calyx about ⅛ in. long, 5 sepals about 25 in. long and acute at apex.

FRUIT. Capsule ¼–⅜ in. long, partly exserted, apex ith 2 acuminate reflexed beaks, longitudinally dehis- ent; seeds numerous and small.

LEAVES. Opposite, alternate, or in whorls of 3, linear, ntire on margin, often revolute, surfaces with short, iff white backward-pointing hairs.

TWIGS. Young ones slender and scabrous-hairy, green brownish; older ones brown to gray and somewhat oody near the base.

RANGE. Mountain sides of the pinyon and yellow pine elt at altitudes of 5,000–7,500 ft. In Arizona; in New Iexico in the San Luis Mountains; also in Sonora, Iexico.

REMARKS. The genus name, *Brachystigma,* refers to he short stigma. The species name, *wrightii,* is in honor f Charles Wright, who collected plants in the South- vest 1847–1848 while with the military forces of the Vestern forts.

WRIGHT FOXGLOVE
Brachystigma wrightii (Gray) Pennell

Texas Silverleaf

Leucophyllum frutescens (Berl.) Johnst.

FIELD IDENTIFICATION. Western shrub to 10 ft, conspicuous because of the ashy-gray leaves.

FLOWERS. Axillary, short-peduncled, barely campanulate, 2/5–1 in. long, pale violet to purple, or pink to white (rare), externally and internally hairy; limb 4/5–1 in. across, 2-lipped and 5-lobed, lobes spreading and rounded; calyx tomentose, sepals 5, lanceolate to oblong, 1/5–1/4 in. long; stamens 4 (rarely 5) in pairs, included, anthers distally narrowed, cells about 1/25 in. long; styles united, ovary 2-celled, ovules numerous.

FRUIT. Capsule 2-valved, 1/6–1/5 in. long, seeds numerous, strongly rugose.

LEAVES. Simple, alternate, or clustered, sessile or short-petioled, obovate, elliptic or oblong; apex obtuse or rounded, base acute, margin entire, surface white, densely tomentose, midrib evident.

TWIGS. Young ones white, tomentose; older gray to brown.

BARK. Grayish black on old trunks and roughened with small scales.

RANGE. Texas Silverleaf is distributed in central,

Texas Silverleaf
Leucophyllum frutescens (Berl.) Johnst.

Lesser Texas Silverleaf
Leucophyllum minus Gray

western, and southwestern Texas, New Mexico, and i Mexico in Coahuila, Nuevo León, and Tamaulipas.

VARIETIES, FORMS, SYNONYMS, CLOSELY RELATE SPECIES. Texas Silverleaf is sometimes listed under th name of *L. texana* Benth., but the name *L. frutescer* (Berl.) Johnst. has precedence.

White-flower Texas Silverleaf, *L. frutescens* form *alba* E. U. Clover, is a form with white flowers whic has been found in Starr County, Texas.

A horticultural variety with abundant flowers ha been listed as *L. frutescens* var. *floribunda*.

Another with glaucous leaves has been listed as *L frutescens* var. *glaucum*.

A closely related species has been listed by Penne (1941) as the Lesser Texas Silverleaf, *L. minus* Gray It differs from Texas Silverleaf by having a corolla wit a narrow throat (funnelform), usually less than 1/2 in long, all the lobes somewhat hairy within, but the up per lobes much less hairy than the lower; anther cell usually less than 1/25 in. long, not distally narrowed leaves cuneate-spatulate, longer than wide, closely to mentose, 1/4–3/8 in. long; old branches persistent an semispinescent. Distributed in the mountains of Trans Pecos Texas in Brewster County at Santiago Peak also east of the Chisos Mountains, in the Persimmo

Gap area between Marathon and Big Bend National Park, near Sierra Blanca in Hudspeth County, and near Fort Stockton in Pecos County. Occurs in southern New Mexico and in Mexico in the states of Chihuahua, Nuevo León, and Zacatecas. A form of the Lesser Texas Silverleaf has been described as *L. minus* forma *argentea* (Gray) O. E. Sperry, and is reported from Brewster County, Texas. The leaves are silvery-canescent and less spatulate than the species.

A new species has also been described by Pennell as the Violet Texas Silverleaf, *L. violaceum* Pennell, as follows (1941, pp. 295–296): "A shrub 1–4 feet tall, stiffly and intricately branched, branches leafy to the apex, young growth tomentose-silvery or faintly yellowish. Leaf blades ¼–¾ in. long, obovate-oval to nearly circular, midrib obscure, cuneate to nearly or quite sessile at the base. Pedicels 1/25–⅛ in. long. Sepals ⅛–⅕ in. long, lanceolate to ovate, obtusish, in fruit incurved around the capsule. Corolla ⅖–¾ in. long, hortense-violet, throat purplish-violet, anteriorly pale and with orange-yellow blotches, externally pubescent, internally hirsute-pubescent on all lobes. Anther cells each about 1/25 in. long or less, oblong, rounded, not tapering distally."

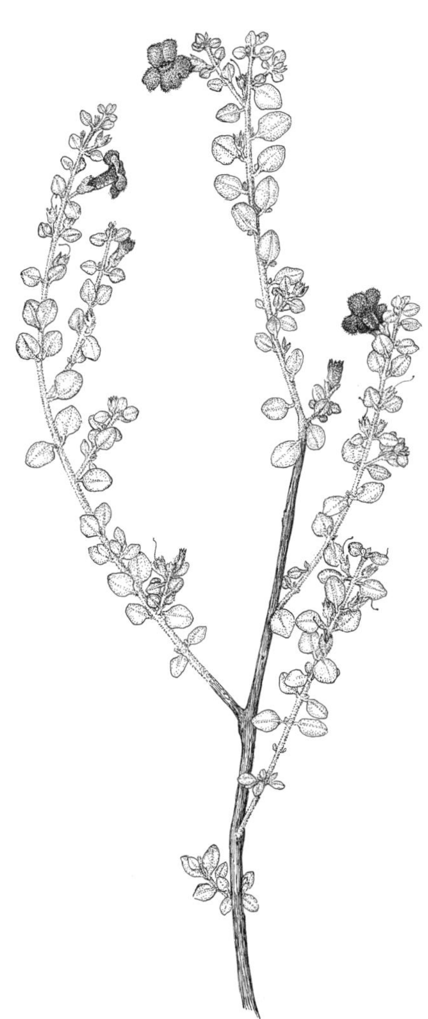

Violet Texas Silverleaf
Leucophyllum violaceum Pennell

In the Big Bend area of Texas, in Brewster County at Glen Springs to the base of the Chisos Mountains, and on the Hot Springs road. Type specimen in flower collected at latter locality by B. H. Warnock, Sheet No. 1124, and deposited in the United States National Herbarium.

Remarks. The genus name, *Leucophyllum*, refers to the grayish white leaves, and the species name, *frutescens*, designates the plant's shrubby habit. The Spanish vernacular name, Cenizo, is much used in the Southwest and means "ash-like" with reference to the pale leaves. In some localities it is known as Barometer-bush, because of the fact that it blooms profusely after rains. Other vernacular names are Ash-bush, Wild Lilac, Purple Sage, Senisa, Cenicilla, Palo Cenizo, and Hierba del Cenizo. This beautiful plant is often cultivated for ornament in the Southwest because of the attractive white foliage and violet to purple flowers. In Texas it is used along the highways for culvert and hedge plantings. It is used by the Mexican Indians for the treatment of chills and fever.

Baccharis-leaf Pentstemon

Pentstemon baccharifolius Hook.

Field identification. Much-branched undershrub rarely more than 2 ft with herbaceous twigs, and woody toward the base of the stems.

Flowers. In loose, glandular racemes; corolla deep red, showy, perfect, straight-tubular, expanded toward the apex, 1–1¼ in. long, 2-lipped, upper lip 2-lobed, lower lip 3-lobed; stamens 5, the fifth naked and sterile, all attached to the corolla-tube; anthers glabrous, 2-celled; style simple; calyx 5-parted, glandular-pubescent; pedicel glandular-hairy.

Fruit. Capsule flattened, 2-lobed at maturity, each lobe with 2 attenuate, spinose tips, interior 2-celled and many-seeded, seeds less than twice as long as wide; fruiting calyx almost as long as the capsule, glandular-pubescent.

Leaves. Opposite, oblong to elliptic, sessile or nearly so, 1–2 in. long, often smaller; entire, or rigidly dentate, especially toward the apex; apex obtuse; base acute or cuneate; texture thick-coriaceous, glabrous or finely puberulent; upper leaves reduced to small ovate scales.

Twigs. Herbaceous above, glabrous or puberulent, distinctly woody toward the base.

Bark. Green to gray, smooth above, breaking into small scales toward the base.

Range. In the Texas Trans-Pecos area, at Boquillas Canyon in Brewster County, on the rocky banks of the Rio Grande at Langtry, in Val Verde County, and

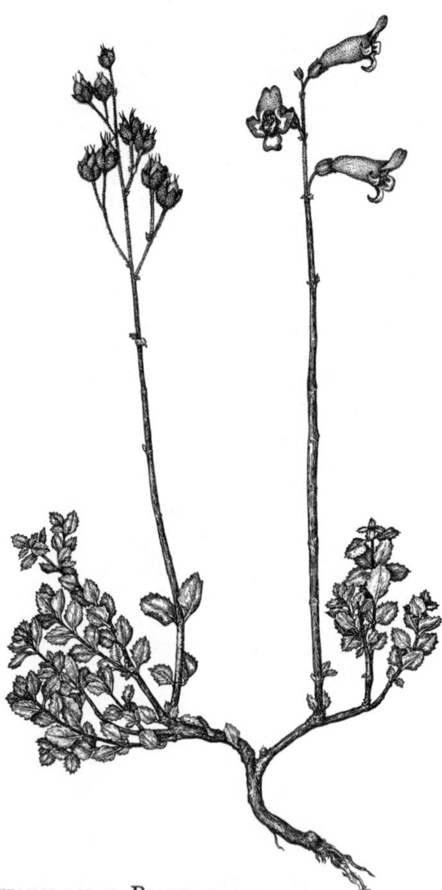

Baccharis-leaf Pentstemon
Pentstemon baccharifolius Hook.

at Sanderson in Terrell County. Also in adjacent Mexico south to the state of San Luis Potosí.

REMARKS. The genus name, *Pentstemon,* is from the Greek and refers to the 5 stamens, one of which is often sterile. The species name, *baccharifolius,* refers to the *Baccharis*-like leaves. Vernacular names are Red Pentstemon, Rock Pentstemon, Limestone Beardtongue. The plant grows erect or semiprostrate from the crevices of seemingly soilless cracks in limestone ledges. Apparently it would be very suitable for rock gardens.

Thurber Pentstemon

Pentstemon thurberi Torr.

FIELD IDENTIFICATION. Slender plant ½–2½ ft tall, usually much-branched near the woody base.

FLOWERS. Borne in open axillary panicles, corolla bluish purple or pink, funnelform, ⅖–¾ in. long, somewhat 2-lipped, the upper lip 2-lobed, the lower 3-cleft, pubescent at base of lower lip, limb scarcely oblique; fertile stamens 4, exserted, arching and in pairs; the sterile filament glabrous, attached to uppe side of corolla near base of the throat, anthers 2-cellec calyx with 5 ovate to short-lanceolate, acute sepals, eac about ⅛ in. long.

FRUIT. Capsule ovoid, style persistent, about 3/16 in long, septicidal, seeds numerous and angled.

LEAVES. Crowded, opposite, linear, entire, ape mucronate, sessile or semiclasping, grayish green.

TWIGS. Often branched near the base, young one grayish green and herbaceous, older brown to gray woody and roughened near the base.

RANGE. On sandy or rocky hillsides at altitudes o 2,000–4,500 ft. New Mexico, Arizona, and California

RELATED SPECIES. *P. thurberi* is closely related t *P. ambiguus*. The former may be distinguished by it bluish purple corolla which is slightly shorter, and limb pubescent only at base of lower lip.

REMARKS. The genus name, *Pentstemon,* refers to the 5 stamens (1 is sterile). The species name, *thurberi* is in honor of George Thurber, a naturalist attached to the United States–Mexico boundary survey in 1850.

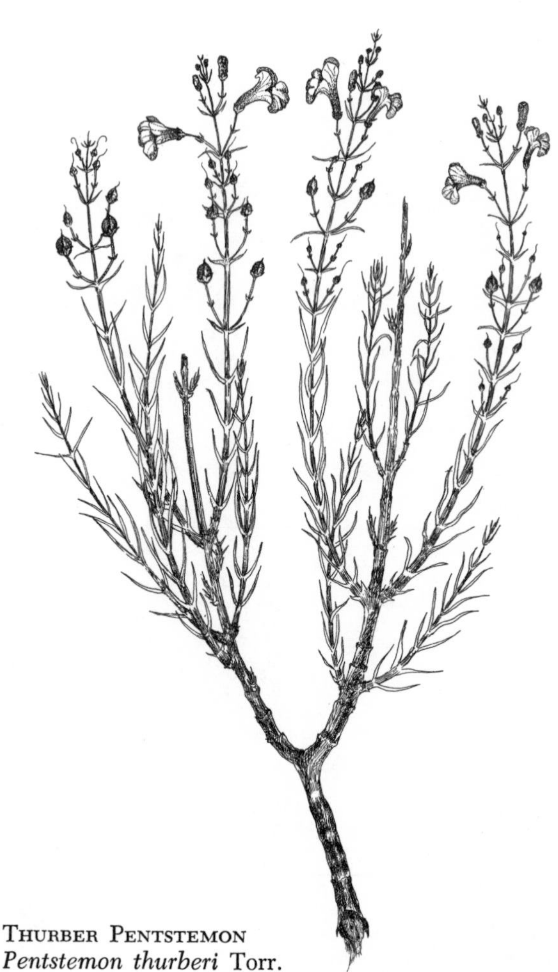

Thurber Pentstemon
Pentstemon thurberi Torr.

GILIA PENTSTEMON

Pentstemon ambiguus Nutt.

FIELD IDENTIFICATION. Plant herbaceous above, with simple, slender, tufted, glabrous stems 4–20 in.

FLOWERS. Borne in loose panicles, corolla rose-colored to white later, tube ⅗–1 in. long, slender, decurved, slightly widened at the rotate, oblique limb; limb barely 2-lipped, the 5 lobes orbicular to oval, densely puberulent on the sides of the orifice; stamens 5, included, 4 parallel in pairs and fertile, the fifth sterile or sometimes with a rudimentary anther; style filiform with a small capitate stigma; calyx campanulate, ⅛–⅙ in. long, glabrous, 5-parted, lobes ovate to ovate-lanceolate, apices acute or acuminate, margin scarious.

FRUIT. Capsule ovoid, acute, seeds numerous, angled.

LEAVES. Opposite, filiform to linear, entire, upper floral-leaves smaller, surfaces glabrous, scabrid-puberulent in some forms.

RANGE. On sandy mesas or grasslands of the Great Plains at altitudes of 4,000–6,500 ft. In Texas in the Panhandle area and adjacent New Mexico, Colorado, and Utah, north to Kansas and west to California.

REMARKS. The genus name, *Pentstemon*, refers to the 5 stamens, and the species name, *ambiguus*, means "vague, of uncertain relationship." The plant has also been listed in the literature under the name of *Leiostemon ambiguus* (Torr.) Greene. It is also known to the Hopi Indians under the name of Cow Tobacco. Recent study has assigned a glabrous form of the plant with the name of *P. ambiguus* subsp. *laevissimus* Keck.

GILIA PENTSTEMON
Pentstemon ambiguus Nutt.

TRUMPET-CREEPER FAMILY (*Bignoniaceae*)

Cross-vine

Bignonia capreolata L.

FIELD IDENTIFICATION. Woody, semievergreen vine, climbing by forked tendrils and bearing opposite, entire, 2-foliate leaves.

FLOWERS. April–June in axillary, 2–5-flowered clusters, very showy on pedicels ½–2 in. long; corolla campanulate, about 2 in. long, rounded or ovate, somewhat irregularly 5-lobed or 2-lipped; red-orange externally, yellow internally; calyx campanulate, ⅕–⅓ in. long, shallowly 5-toothed, membranous; stamens 4, included, sometimes showing an abortive fifth, filaments hairy at the base; stigma 2-lobed.

FRUIT. Capsule linear, 4–7 in. long, flattened, leathery, dehiscent, valves 1-nerved, 2-celled; seeds flattened, transversely winged, elliptic to oblong, ends rounded, ⅜–½ in. long, numerous.

LEAVES. Opposite, compound, 2-foliate, rarely 1-foliate, tendril-bearing, often with smaller axillary leaves; leaflets entire, ovate, oblong or lanceolate; apex acute, acuminate or obtuse; base heart-shaped and auricled; blades 2–7 in. long, ¾–1½ in. wide; dark green and glabrous or slightly puberulent above; paler, veiny, and glabrous or puberulent beneath; petiolules ¼–⅓ in. long, rachis ½–¾ in. long. The tendrils, which are leaf structures, bear terminal, flattened, disklike appendages for clinging on vertical surfaces.

STEMS. Young ones green, glabrous or pubescent, finely grooved, climbing by means of leaf tendrils as high as 70 ft.

BARK. Brown to gray on old stems, thinly flaking.

RANGE. In rich, moist woods and swamps, eastern and southern Texas, Oklahoma, Arkansas, Louisiana, eastward to Florida, north to New Jersey, and westward to Illinois and Ohio.

Cross-vine
Bignonia capreolata L.

FORM. Redpurple Cross-vine, *B. capreolata* forma *atrosanguinea* Hook., is a form with showy, large, reddish purple flowers and longer, narrower leaves.

REMARKS. The genus name, *Bignonia,* is in honor of Abbé Jean Paul Bignon, court librarian to Louis XV. The species name, *capreolata,* means "winding" or "twining." Vernacular names are Tendriled Trumpet-creeper, and Quarter-vine. The sections of stems are smoked like cigars in some localities and given the name Smoke-vine. This is one of our most beautiful native vines and should be more extensively cultivated.

[CO]MMON TRUMPET-CREEPER

[Bi]gnonia radicans L.

FIELD IDENTIFICATION. Woody vine, climbing tall [tr]ees by means of aerial rootlets, or creeping prostrate [on] the ground.

FLOWERS. All summer, in 2–9-flowered corymbs, on [st]out pedicels 3/16–3/4 in. long; corolla tubular-funnel[fo]rm, almost equally 5-lobed, lobes suborbicular or reni[fo]rm, limb 1–1½ in. broad; tube orange-red, 2–3½ in. [lo]ng; 4 stamens, 2 long and 2 short; filaments curved [an]d yellowish white; anthers large, yellow, bifid, about [1/1]6 in. long; pistil solitary, as long as the stamens or [lo]nger, stigma spatulate, thin, flattened; calyx tubular-[ca]mpanulate, 3/4–1 in. long, the 5 lobes triangular-ovate [an]d nearly equal, about 1/4 in. long.

FRUIT. Capsule stout, fusiform, leathery, 2–6 in. long, [na]rrowed toward the ends, slightly flattened, somewhat [ri]dged by the valve margins, persistent, splitting at [m]aturity, partitioned at right angles to the concave [va]lves, 2-celled; seeds transversely thin-winged.

LEAVES. Opposite, odd-pinnate, 8–15 in. long; leaf[le]ts 7–13, ovate, oval or elliptic, acute or acuminate at [th]e apex, narrowed at the base, and somewhat asym[m]etrical, coarsely serrate blades 3/4–3 in. long, 1/2–2 in. [w]ide; olive-green, lustrous and glabrous above, veins [im]pressed; lower surface glabrous, paler, veins distinct; [p]etiolules wing-margined somewhat by the leaf bases, about 1/4 in. long; rachis grooved above, rounded below, smooth, green to reddish brown.

STEMS. Green to reddish, glabrous, rather stiff, climbing by aerial rootlets to great heights or creeping on the ground. Often covering fences, stumps, and old buildings.

RANGE. Along streams, fences, thickets, in moist soil; Texas, Louisiana, Oklahoma, and Arkansas, east to Florida, north to New Jersey and Pennsylvania, and west to Missouri.

FORMS AND SYNONYM. A number of horticultural forms have been developed, such as the Dwarf *(minor)*, Early *(praecox)*, Purple *(atropurpurea)*, Showy *(speciosa)*, and Yellow *(flava)*.

Sometimes the name *Campsis* is used as a synonym for *Bignonia*, to separate those members of the Linnaean genus *Bignonia* which lack tendrils *(B. radicans* is one of these).

REMARKS. The genus name, *Bignonia*, is in honor of Abbé Jean Paul Bignon, librarian to Louis XV, and the species name, *radicans*, refers to its aerial rooting habit. Some of its vernacular names are Trumpet-vine, Trumpet-ash, Trumpet-flower, Devil's-shoestring, Foxglove-vine, and Cow-itch. Though nonpoisonous it is often confused with Poison-ivy, but Poison-ivy has only 3 leaflets and small greenish flowers. Trumpet-creeper is the state flower of Kentucky, and while native in the South, is often cultivated in the North.

[C]OMMON TRUMPET-CREEPER
[B]ignonia radicans L.

SOUTHERN CATALPA

Catalpa bignonioides Walt.

FIELD IDENTIFICATION. Tree to 60 ft and 2–4 ft in diameter. Trunk short, crown broad and rounded, branches stout and brittle.

FLOWERS. May–July, borne in large, erect, broad-pyramidal panicles up to 10 in. long; irregular, perfect, showy and attractive; calyx deeply 2-lipped, about 1/2 in. long, glabrous, purplish, lips broad-concave, abruptly pointed; corolla white, bell-shaped, tube expanded into throat, 1½–2 in. long, about 1½ in. wide, channeled and keeled on lower side; limb oblique, 2-lipped, upper lip 2-lobed, lower lip 3-lobed, lobes crisped; interior of corolla with 2 rows of yellow and purple spots; fertile stamens 2 (rarely, 4), but some sterile and rudimentary; filaments filiform with spreading linear-oblong anthers; ovary sessile, 2-celled; style filiform with 2 exserted stigmas.

FRUIT. Ripening in October, capsule beanlike, linear, cylindrical, woody, 8–18 in. long, 3/8–1/2 in. wide, thin-walled, pointed, tardily dehiscent into 2 valves; seeds numerous, flat, about 1 in. long and 1/4 in. wide, apices winged and fringed with white hairs.

LEAVES. Simple, opposite or whorled, deciduous, ovate, apex abruptly acuminate or acute, base subcordate or truncate, entire, sometimes lobed, blades 5–12

SOUTHERN CATALPA *Catalpa bignonioides* Walt.

in. long, upper surface glabrous and light green, lower surface paler and pubescent, strong-scented; petioles stout, terete, shorter than the blade, 5–6 in. long, pubescent.

TWIGS. Stout, brittle, greenish purple to grayish brown, lustrous; lenticels large and pale; leaf scars suborbicular, bundle scars about 10; buds large, subglobose; pith large, white.

BARK. Thin, light brown to gray, divided into narrow scales.

WOOD. Grayish brown to lavender-tinged, sapwood lighter, coarse and straight-grained, no characteristic odor or taste, very durable, weighs about 28 lbs per cu ft, shrinks little, weak in endwise compression, soft, moderately high in shock resistance, weak in bending.

RANGE. Thought to be native from Florida and Georgia, westward into Louisiana. Doubtful whether native in Texas, Oklahoma, and Arkansas, though planted for many years and escaping cultivation in those states and elsewhere.

MEDICINAL USES. Seeds contain a bitter glycoside, catalpin, which is reported to be used for asthmatic, cardiac and antispasmodic purposes. Large doses of the seed are said to produce weak pulse, nausea, and vomiting. Some people develop a skin allergy when handling the flowers.

PROPAGATION. The following method of handling seed is recommended by the U.S. Department of Agr culture: Good seed are borne every 2–3 years and th commercial seed-bearing age is about 20 years. Colle tion of the pods is made by hand as soon as they tur brown; the pods can be separated from the seeds b flailing in sacks. Cleaned seeds average about 20,00 per lb, with a commercial purity of 60 per cent and soundness of 96 per cent. Seeds may be stored for year or two in dry form in sealed containers at 50°F No previous treatment is required to hasten germina tion. Sowing is done in late spring in drills, using abou 30 seeds per linear ft, covered with about ⅛ in. wit rich, moist soil. Germination averages about 85 per cen Seedlings are subject to nematode infestations.

FORMS. A number of horticultural forms are know
C. bignonioides forma *nana* Bur. is a dwarf form
C. bignonioides forma *aurea* Bur. has yellow leave
C. bignonioides forma *koehnei* Dode has yello leaves with green blotches.

REMARKS. The genus name, *Catalpa*, is the Ame ican Indian name, and the species name, *bignonioide* refers to its flowers resembling the Bignonia-vine. Abb Jean Paul Bignon was the court librarian to Louis XV Vernacular names for the tree are Candle-tree, Ciga tree, Smoking-bean, Bean-tree, Catawba, Indian-bea and Indian-cigar. The wood of Southern Catalpa used for posts, poles, rails, crossties, interior finish, an cabinet work.

Northern Catalpa

Catalpa speciosa Warder

FIELD IDENTIFICATION. Tree usually less than 50 ft, with a narrow or broad head; rarely up to 120 ft and 4½ ft in diameter.

FLOWERS. May–June, borne in panicles 5–8 in. long and broad, terminal, relatively few-flowered, attractive, fragrant; pedicels slender, glabrous, green to purplish, 1–3-bracteate; calyx closed in the bud, ⅖–½ in. long, green to purplish, subglabrate, or pubescent, splitting irregularly into 2 broad-ovate, entire, apiculate lobes; corolla campanulate, about 2 in. long and 2½ in. wide, limb oblique and 5-lobed, posterior 2-lobed, anterior 3-lobed, lobes all spreading and margins undulate, externally white, throat marked by rows of yellow blotches with pale purplish spots spreading on lips; stamens 2, filaments flattened, anthers oblong and introrse, staminodia 3, minute or absent; ovary 2-celled, style filiform with 2 stigmatic lobes; ovules in numerous series on the central placenta.

FRUIT. Capsule maturing in October, persistent, solitary or 2–3 together, 8–20 in. long, ⅜–½ in. in diameter near the middle, tapering to either end, shape elongate-linear, straight or curved, terete, light brown, wall thick, splitting into 2, loculicidally dehiscent concave valves, seeds numerous, inserted in 2–4 ranks, free from the capsule wall, about 1 in. long and ⅓ in. wide, oblong, compressed, papery-thin, light brown to gray, veined longitudinally, notched at base of seed and ends fringed with hairs.

LEAVES. Involute in the bud, simple, opposite or in threes, deciduous, blades 6–12 in. long, 4–8 in. wide, broadly ovate to oval, apex long-acuminate, base cordate or somewhat rounded, margin mostly entire or with 1–2 lateral teeth, surfaces when young with whitish or brownish tomentum, later dark green and smooth above, paler and pubescent beneath, ribs rather prominent and sometimes with dark axillary glands; petioles 4–6 in. long, stout, terete.

TWIGS. Robust, terete, green to reddish brown or purplish, at first somewhat glaucous and hairy, later more glabrous; lenticels numerous; leaf scars conspicuous, large, raised, each with a circular row of fibrovascular scars; buds large, scales imbricate, brown, ovate to obovate, acute or apiculate at apex, pubescent.

BARK. Dark gray to brown or reddish, ¾–1 in. thick on the trunk, surface broken into irregular fissures and scaly flat-topped ridges.

WOOD. Heart light brown, sapwood nearly white, coarse-grained, not strong, soft, weighs 26 lb per cu ft, very durable in contact with the soil, moderately weak in bending, weak in endwise compression, moderately high in shock resistance.

RANGE. In woods in deep, moist, rich soil. East Texas, Louisiana, Oklahoma, and Arkansas, north to Missouri, Tennessee, Kentucky, Indiana, and Illinois. Naturalized in Virginia, Kansas, and Ohio.

PROPAGATION. Minimum commercial seed-bearing age about 20 years. Good crops borne about every 2 years. Cleaned seed averages about 21,000 seeds per lb, with a purity of 90 per cent and a soundness of 90 per cent. Germination averages about 80 per cent. Seeds may be planted when gathered, but yield a higher germination rate when planted the following spring. Pretreatment by stratification not of particular benefit. Planting is in drills at the rate of about 30 seeds per linear ft and covered with about ⅛ in. of rich moist soil. About 5,000 usable plants per lb of seed are usually obtained. Seedlings are sometimes attacked by nematodes and powdery mildew occurs on the leaves.

REMARKS. The genus name, *Catalpa,* is the American Indian name. The species name, *speciosa,* is given for the large showy flowers. It is known under the vernacular names of Larger Indian-bean, Bois-plant, Indian-cigar, Smoking-bean, Cigar-tree, Shawnee-wood, Hardy Catalpa, Western Catalpa, and Catawba-tree. The wood is used for fence posts, rails, poles, and occasionally for furniture and interior finish. The tree seems to coppice easily, is rapid-growing, short-lived, easily injured by storms, insects, and fungi. It is subject to the attack of the catalpa sphinx, which defoliates the tree. Northern Catalpa is distinguished from Southern Catalpa primarily by the smaller, fewer-flowered lax panicle of the former; by the sparser, paler spots on the corolla-throat, and the lobes flatter, with the limb up to 2–2⅓ in. long; also by the leaves having a less fetid odor and the pods thicker walls.

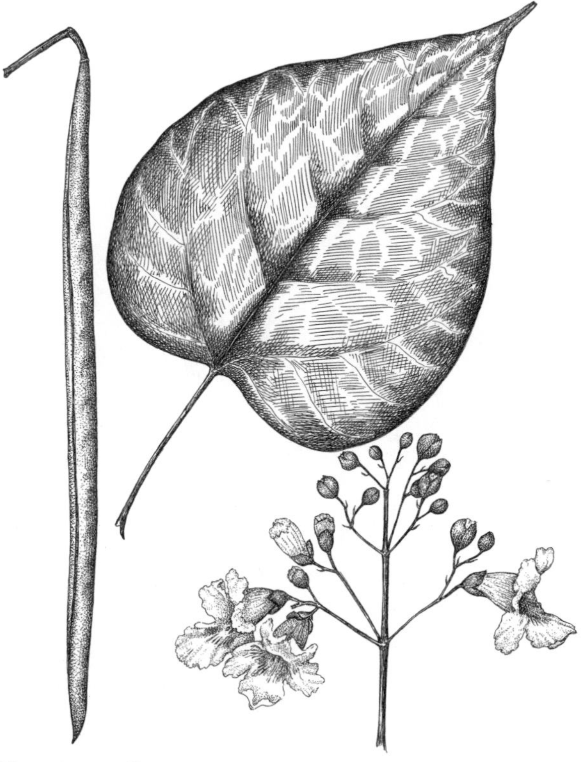

Northern Catalpa
Catalpa speciosa Warder

Hardy Yellow Trumpet

Tecoma stans (L.) H. B. K. var. *angustata* Rehd.

FIELD IDENTIFICATION. Irregular shrub or small tree to 24 ft high.

FLOWERS. Panicles or racemes 3–17-flowered, 3–5 in. long; pedicels 1/12–1/2 in. long; corolla bright yellow, funnelform-campanulate, 1–2 in. long; flared into a spreading limb about 3/4 in. across, 5-lobed and slightly 2-lipped; lower lip 3-lobed, middle lobe the longest; upper lip 2-lobed, all lobes rounded; throat with faint dark streaks of orange; 4 stamens, 2 long and 2 short (sometimes with a fifth stamen bearing an abortive anther), attached to the corolla-throat within, pubescent; anthers about 1/4 in. across when spread in anthesis, hairy; pistil thinly spatulate but 2-cleft; ovary with numerous ovules borne in 2 rows in each cavity; calyx about 1/8–1/4 in. long, tubular, 5-lobed; lobes triangular, acute or acuminate, glabrous.

FRUIT. Capsule 4–6 in. long and 1/4 in. wide, linear, flattened, acute, valves leathery, loculicidally dehiscent, pedicel about 1/2 in. long; seeds about 3/16 in. long, flat and winged.

LEAVES. Opposite, odd-pinnate, 4–8 in. long; leaflets 5–13 (usually 7–9) sessile or nearly so, lanceolate to elliptic or linear, acuminate at apex, cuneate but rather asymmetrical at base, margin sharply and coarsely serrate, surface usually glabrous and olive-green above, paler and glabrous or pubescent beneath (pubescent to tomentose in the Velvety Yellow Trumpet, *T. stans* var. *velutina*); petiole about 1/4–1/2 in. long, winged and grooved by the continuing leaf base.

Hardy Yellow Trumpet
Tecoma stans (L.) H. B. K. var. *angustata* Rehd.

TWIGS. Young ones green, slender, older ones brown to gray, striate, lenticels elongate.

RANGE. Well-drained dry soil in full sun. The species, *stans,* grows in western and southern Texas, New Mexico, Arizona, and Florida. Southward it is widely distributed through the countries of Central and South America, and the West Indies.

MEDICINAL USES. Standley (1920–1926, p. 1319) reports that in Veracruz, Mexico, a decoction of various parts of the plant is administered for stomach cramps, and in other localities as a remedy for diabetes. In Guadalajara the roots are much used in domestic medicine and are considered to have diuretic, tonic, antisyphilitic, and anthelmintic properties. A native beer is prepared from the root.

VARIETIES. It appears that the variety found in Texas and adjacent northern Mexico is the Hardy Yellow Trumpet as described above. It is apparently more cold-resistant, and has narrower, incised-serrate leaflets.

Velvety Yellow Trumpet, *T. stans* var. *velutina,* has leaves copiously pubescent or tomentose beneath. It does not appear to be stable because of its many intergrades into the glabrous forms.

Elderleaf Yellow Trumpet, *T. stans* var. *sambucifolia,* has elder-like leaves.

REMARKS. The genus name, *Tecoma,* is from an Indian name meaning "pot-tree." The species name, *stans,* signifies the upright habit of this plant. English vernacular names are Yellow-bells, Trumpet Flower, and Yellow-elder. In Mexico, Central and South America, and the West Indies, there are many local names as follows: Retamo, Retama, Tronodor, Tronodora, Trompetilla, Trumpeta, Gloria, Kanlo, Xkanlol, Guie-bicki, Tulosuchil, Palo de Arco, Flor de San Pedro, Corneta Amarilla, Nixtamaxochitl, Nextamalxochill, Borla de San Pedro, Hierba de San Nicolás, Flor Amarilla, Roble Amarillo, Miñona, Mazarca, Huachacata, Ichulili, Saúco Amarillo, Ruibarba, Copete, Sardinillo, Fresno, Chirlobirlos, Paulo Huesa, Tache, Tosto, Candillo, Garrocha, Garanguay Amarillo, Guaram-guaran, San Andrés, Morchucha, and Tagualaishte. The nectar is reported to be good bee food. The Indians made bows from the wood. The plant is often cultivated for ornament.

Desert-willow

Chilopsis linearis (Cav.) Sweet

FIELD IDENTIFICATION. Shrub or slender tree to 30 ft, with trunks usually leaning.

FLOWERS. Mostly May–June, but blooming sporadically after rains in other months, showy, perfect, in short

DESERT-WILLOW
Chilopsis linearis (Cav.) Sweet

panicles 2–4 in. long; corolla funnelform-campanulate, slightly oblique, 1–1½ in. long, 5-lobate, lobes suborbicular and undulate on the margins, disposed in 2 lips; lower lip 3-lobed, dark pink or purple with the central lobe longest; upper lip 2-lobed and pink, throat white, yellow, or streaked purple (corolla variable in color shades from white or purple); 4 stamens, 2 long and 2 short, included, adnate to the wall of the corolla within; filaments filiform, glabrous; anthers oblong, cells divergent at maturity; a solitary staminodium, shorter than the stamens, which is also included; pistil simple, usually longer than the stamens; ovary 2-celled, glabrous, lobes flattened, ovate, rounded; ovules numerous; calyx splitting into 2 lips, lips about ¼ in. long, ovate, concave, acute, thin, papery, pubescent or glabrous.

FRUIT. Capsule borne on stout peduncles ½–1 in. long, linear, 4–12 in. long, about ¼ in. thick, subterete, striate, 2-valved, apiculate at the apex, persistent; seeds numerous, compressed, oblong, about ⅓ in. long, in 2 ranks, extended into wings with long fimbriate white hairs.

LEAVES. Deciduous, opposite or alternate, linear to lanceolate, entire, thin, 3-nerved, 4–12 in. long, average length 3–5 in., ¼–⅓ in. wide, attenuate long-pointed at the ends, pubescent or glabrous, rather pale green on both sides, sometimes viscid-sticky; petiole short or none, almost winged by the leaf base.

TWIGS. Slender, green the first year, somewhat pubescent later, gray to reddish brown and glabrous.

BARK. Smooth and brown on young trunks, dark brown to black later and breaking into broad ridges with small scales, fissures irregular and rather deep.

WOOD. Dark brown with lighter streaks, coarse-grained, soft, weak, rather durable in contact with the soil, specific gravity 0.59.

RANGE. The species, *linearis,* and its varieties, grow along arid desert washes or dry arroyos from Texas north into New Mexico, west to Arizona and California, south into the Mexican states of Nuevo León, Tamaulipas, Zacatecas, Chihuahua, Sonora, and Durango; and Baja California.

VARIATIONS. Fosberg (1936) has separated two varieties of the Desert-willow. One of these is referred to as the Western Desert-willow, *C. linearis* var. *arcuata* Fosberg, with curved, glabrous leaves about 1⁄12–⅙ in. wide. This variety is found from New Mexico north to Nevada, west to California, and south to Sonora, Mexico.

The other variety is known as the Sticky Desert-willow, *C. linearis* var. *glutinosa* (Engelm.) Fosberg. It is reported as having straight, viscid leaves ¼–⅓ in. wide and is listed as occurring in Texas, New Mexico, and Mexico.

A white-flowered horticultural form, *C. linearis* forma *alba,* is popular.

The trees of the Texas Big Bend area appear to fit Fosberg's desciption of the Sticky Desert-willow, but those of the northern Trans-Pecos area show wide foliage variations, the extremes being viscid and nonviscid, straight and curved, and wider or narrower leaves. The intergrades between the two varieties are rather numerous.

REMARKS. The genus name, *Chilopsis,* is from the Greek word *cheilos* ("lip") and *opsis* ("likeness"), with reference to the corolla lips. The species name, *linearis,* refers to the narrow leaves. Also known under the vernacular names of Flowering-willow, Willowleaf-catalpa, and Flor de Mimbre. These names arise from the fact that the flowers resemble those of the Catalpa-tree and the leaves are willow-like in appearance.

The wood is used for fence posts and fuel, and baskets are woven from the twigs. It is reported that in Mexico a decoction of the flowers is used for coughs and bronchial disturbances. The flowers also make good honey. The foliage is unpalatable to livestock and is eaten only under stress. Various birds consume the winged seeds, which average about 75,000 per lb. Only about half the seeds are viable, and according to United States Forest Service data only about 4,000 usable plants can be obtained from a lb of seed. The tree grows readily from cuttings and is being planted extensively for ornament.

Royal Paulownia

Paulownia imperialis Sieb. & Zucc.

FIELD IDENTIFICATION. Cultivated tree to 50 ft, with a diameter of 4 ft. The stout spreading branches develop into a rounded or flattened head.

FLOWERS. April–May, borne before the leaves in terminal panicles up to 10 in. long with densely tomentose pedicels; corolla 1½–2½ in. long, irregularly shaped, pale violet to blue, puberulent, corolla-tube slightly curved and spreading at the apex into an oblique, 5-lobed limb; upper corolla-lobes reflexed, lower corolla-lobes spreading, lobes unequal and shorter than the tube; corolla-throat broad and 2-ridged with spots or yellowish stripes; 4 stamens, 2 long and 2 short, included, anther sacs divergent and pendent; style slender, slightly thickened distally, stigmas distinct, plate-like, inwardly stigmatic; calyx campanulate, ⅖–⅗ in. long, nearly regular; sepals 5, united half their length, ovate to rounded, obtuse, thick, tomentose.

FRUIT. Capsule persistent, conspicuous in winter, length 1¼–1¾ in., woody, broadly ovoid, acute, beaked, longitudinally dehiscent into 2 valves; seeds numerous, about 2,000 per capsule, small, flat, striate, broadly winged.

LEAVES. Deciduous, opposite, blades 5–12 in. long, cordate to suborbicular, margin entire or 3-lobed, apex acuminate, base heart-shaped; young leaves densely short-stellate hairy, somewhat sticky, older ones pubescent above, tomentose beneath; petioles about 3–8 in. long, terete, pubescent.

TWIGS. Stout, smooth; bark thin and flaky on older branches, winter buds with several outer scales, pith segmented.

WOOD. Brown to purplish, sapwood thin, light, soft, easily worked, has a satiny finish.

RANGE. Persistent about old gardens and waste places; Texas, Louisiana, Oklahoma, and Arkansas. A native of central China. Escaping to roadsides and open woods throughout almost all the eastern United States, north to New York.

PROPAGATION. The tree may be propagated by seeds sown in spring or by root or greenwood cuttings under glass. In northern areas the buds freeze, but if the plant is cut back to the ground, it will send out vigorous sprouts. However, if cut back very often the plant is weakened.

VARIETIES. Woolly Royal Paulownia, *P. imperialis* var. *lanata* Schneid., has leaves densely woolly-tomentose beneath, and the calyx tomentose with acute lobes.

Pale Royal Paulownia, *P. imperialis* var. *pallida* Dode, has pale violet to whitish flowers and dull green leaves.

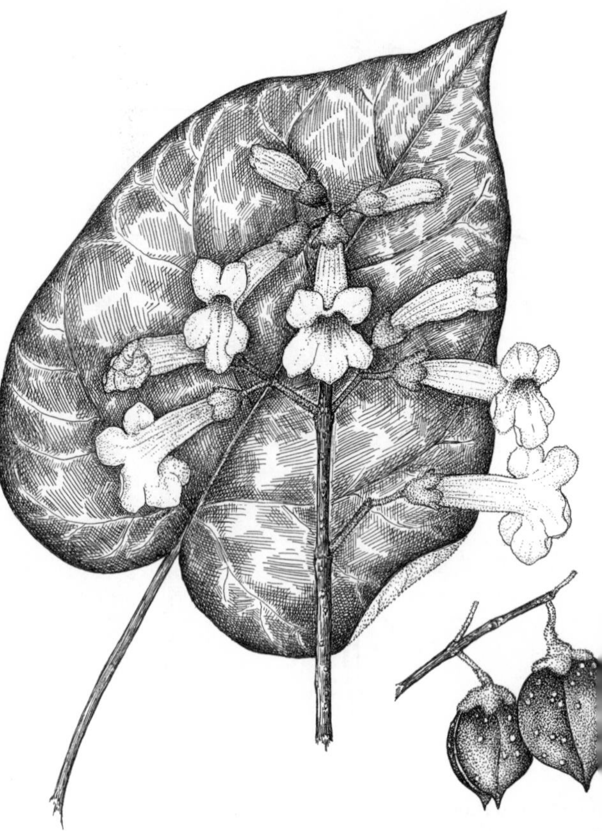

Royal Paulownia
Paulownia imperialis Sieb. & Zucc.

FAMILY CHARACTERS. Botanists differ about the family placement of the Royal Paulownia. The presence of the endosperm suggests Scrophulariaceae, while the arboreal habit and large terminal inflorescence favor the Bignoniaceae.

REMARKS. The genus name, *Paulownia,* is in honor of Anna Paulownia (1795–1865), Princess of the Netherlands. The species name, *imperialis,* refers to the royal name. It is also known under the names of Empress-tree, Princess-tree, Cotton-tree, and Karri-tree. The flowers are delightfully fragrant and the large catalpa-like leaves are conspicuous.

CANTHUS FAMILY (*Acanthaceae*)

RIGHT ANISACANTH

ıisacanthus wrightii (Torr.) Gray

FIELD IDENTIFICATION. Western shrub 2–4 ft high d irregularly branched.

FLOWERS. Borne singly, or in slender, naked or small-acted, loosely paniculate, one-sided racemes; short-dicellate or sub-sessile; bracts and bracteoles lanceo-te-acuminate, 1/12–1/5 in. long, puberulent; calyx 1/12–1/3 . long, sessile, glabrate to glandular-puberulent, lobes ovate to lanceolate or oblong-lanceolate, acute; co-lla-tube 1¼–1¾ in. long, red to orange, elongated, raight or curved, longer than the lobes, puberulent, lated above into 2 lips; upper lip lobe narrowly ovate, otuse and entire; lower lip lobe 3-parted; stamens 2, long as the corolla or longer, filaments adnate to ove the middle of the tube, anthers 2-celled, cells rallel and contiguous; ovary 2-celled, 2 ovules in ch cavity; style filamentous, slender, persistent, as ng as the stamens or longer.

FRUIT. Capsules on a stipelike base as long as the ody or longer, about 3/5–¾ in. long, about ¼ in. thick, avate, flattened, striate, acute to obtuse at the apex; eds 2–4, flat, wrinkled, orbicular, about 1/5 in. in ameter.

LEAVES. Opposite, lanceolate to oblong or ovate, apex cute to acuminate, base rounded or abruptly narrowed, argin entire, length ¾–2 in., surfaces olive-green, uberulent or glabrate; petioles pubescent, ⅛–⅜ in. ng.

TWIGS. Slender, green to tan or gray, often puberu-nt when young, or pubescent in 2 lines, glabrous later.

RANGE. In rich soil in thickets, southern and western exas (Bexar, Uvalde, and Kinney counties). In Mexico Tamaulipas, Nuevo León, Coahuila, and Michoacán.

MEDICINAL USES. Various parts of the plant are em-loyed by Mexican Indians as a remedy for colic.

WRIGHT ANISACANTH
Anisacanthus wrightii (Torr.) Gray

VARIETY AND RELATED SPECIES. A closely related species is *A. thurberi* (Torr.) Gray.

Shortlobe Wright Anisacanth, *A. wrightii* var. *breviloba* Hagen, has narrow subsessile leaves ½–1¾ in. long, ⅛–⅙ in. broad, with glabrate surfaces; calyx ⅙–1/5 in. long; tube several times longer than the triangular acute lobes. Occurs in northern Coahuila, Mexico.

REMARKS. The genus name, *Anisacanthus*, is from *anise* ("unsymmetrical") and *akanthos* ("thorn"). The species name, *wrightii*, is in honor of Charles Wright (1811–1886), botanist who collected in Texas and the Southwest. A vernacular name for the plant in Mexico is Muicle. Sometimes eaten by livestock.

THURBER ANISACANTH

Anisacanthus thurberi (Torr.) Gray

FIELD IDENTIFICATION. Western shrub 1½–8 ft high with conspicuous gray or white, shreddy, exfoliating bark.

FLOWERS. In spring, or sporadically during the year; borne singly or in 2–5-flowered, leafy, axillary panicles; pedicels glandular, ⅛–⅓ in. long; bracts and bracteoles lanceolate to linear, acute, about 1/12–⅖ in. long, 1/25–⅛ in. broad, puberulent; calyx of 5 linear-attenuate sepals united only at the base, about ⅜ in. long, glandular-puberulent, longer than the stipe of the fruit or sometimes as long as the body; corolla 1⅜–1⅝ in. long, usually orange-red, but sometimes yellow to orange or bluish, puberulent, perfect but more or less irregular, tubular-funnelform, bilabiate and 4-lobed, anterior lip 3-lobed, posterior entire or slightly emarginate, lobes shorter than the tube, or as long, lobes ¾–1⅜ in. long, narrow, recurved; the tube slender, widening at the throat, stamens 2, inserted below the sinuses of the anterior lip, long exserted, filaments about ¾ in. long;

THURBER ANISACANTH
Anisacanthus thurberi (Torr.) Gray

anthers symmetrical, parallel and contiguous, 2-celle stigma minute, style long-exserted, ovary superior.

FRUIT. Stipe usually shorter than the clavate or su ovoid capsule which is glabrous, shiny, flattened, abo ¼ in. in diameter, valves longitudinally dehiscent, celled; seeds flattened, 2 in a cell, slightly tubercula about ⅕ in. in diameter.

LEAVES. Simple, opposite, but fore-shortened and a pearing fascicled, lanceolate to oblong or ovate-lance late, margin entire and ciliate, apex acute or acumina base gradually narrowed, length ⅓–2⅓ in., width ⅛ in., surfaces rather dull, olive-green, sparsely puber lent or glabrous later; petioles 1/25–⅕ in. long, or le sometimes sessile.

TWIGS. Young ones terete, minutely striate, tan brownish, minutely pubescent or with a few hairs opposite lines; older ones gray to white and glabrou exfoliating in shreddy thin strips.

RANGE. Dry soil in the sun, mesas, canyons, hillside arroyo banks at altitudes of 2,500–5,500 ft. In Ne Mexico in Luna, Hidalgo, and Grant counties. Also r ported from the mountains of western Texas. In ce tral Arizona, and southward into Sonora, Mexico.

REMARKS. The genus name, *Anisacanthus,* is fro *anise* ("unsymmetrical") and *akanthos* ("thorn"). T species name, *thurberi,* honors the early botanist Geor Thurber, who collected in Arizona. It is also known u der the vernacular names of Desert-honeysuckle, Tap rosa, and Chuparosa. Palatable to cattle and shee and sometimes grazed closely.

DWARF ANISACANTH

Anisacanthus insignis Gray

FIELD IDENTIFICATION. Slender vinelike shrub attai ing a height of 7 ft. Older branches whitish, shreddin into fibrous strips.

FLOWERS. In lateral, short, loose, racemose cluster borne on glandular pedicels ⅛–⅓ in. long; bracts an bractlets 1/12–⅖ in. long, obovate to ovate or ellipti puberulent-glandular; corolla tubular, curved, 1½–2 i long, reddish orange or salmon-colored, minutely pube cent, tube bilabiate into 1 lip which is 3-lobed and entire lip; lobes recurved, linear, longer than the tub stamens 2, exserted, filaments orange-red and inserte at base of central anterior lobe, anthers 2-celled an about ⅛ in. long; style filiform, slender, equal to slightly exceeding the stamens, stigma simple and lobed; disc at base of ovary about 1/25 in. high; caly ¼–⅖ in. long, puberulent, stipitate-glandular, 5-lobe lobes linear and separating almost to the base.

FRUIT. Capsule about ¾ in. long, club-shaped, born on a stipe as long as the body or longer, flattened, ligh green, often with dark green stripes, glabrous and shin 2-celled; seeds 2–4, often with 1–2 aborted, flattene

Dwarf Anisacanth
Anisacanthus insignis Gray

discoid, short-pointed at one end, brown, more or less tuberculate, about 1/5 in. across.

LEAVES. Opposite, linear-lanceolate to ovate, somewhat acuminate, margin entire, 1–2¾ in. long, ⅜–1¼ in. wide, surfaces puberulent, pilose or tomentose, base with tomentose petioles ⅛–⅖ in. long.

TWIGS. Younger ones green to reddish, glabrous or puberulent in lines, finely grooved; older ones whitened and fibrous.

RANGE. The species is reported to be confined to Mexico from Coahuila and Chihuahua to Durango. However, intergrades are found between the species reported in Mexico and the variety reported in Brewster County, Texas.

VARIETY. Narrowleaf Dwarf Anisacanth, *A. insignis* var. *linearis* Hagen, has been described as having the lower leaves, and those of the flowering branches, sessile; leaves generally narrower, linear to linear-lanceolate, ⅜–⅔ in. long, about 1/15 in. broad, surfaces essentially glabrous. Known from the Chisos Mountains, Brewster County, Texas, and Coahuila, Mexico.

REMARKS. The genus name, *Anisacanthus,* refers to a combination of words, *anise* ("unsymmetrical") and *akanthos* ("thorn"). The species name, *insignis,* means "striking" or "remarkable." This is a showy shrub having possibilities for ornamental use. The foliage is sometimes browsed by deer, antelope, and livestock.

Heath Carlowrightia

Carlowrightia linearifolia (Torr.) Gray

FIELD IDENTIFICATION. Western plant, herbaceous above but somewhat woody toward the base. Usually not over 1 ft in height.

FLOWERS. August–September, in narrow loose panicles, pedicels about ¼ in. long; calyx 4–5-parted, segments linear; corolla imbricate in the bud, purple or streaked pink and purple, almost rotate, tube shorter than lobes which are ⅕–⅜ in. long; stamens 2, inserted on the tube, filaments nearly equaling the corolla-lobes; anthers sagittate, cells parallel and contiguous.

FRUIT. Capsule set on a long slender stipe, ovoid, flattened, apex abruptly acute or acuminate, ½–⅝ in. long, 2-celled, elastically dehiscent, 2–4-seeded; seed flat and minutely scabrous.

LEAVES. Simple, opposite, uppermost passing into leaflike bracts and bractlets of the inflorescence, ⅓–3 in. long, filiform to narrowly linear, entire, glabrous or puberulent, yellowish green, only main vein apparent below.

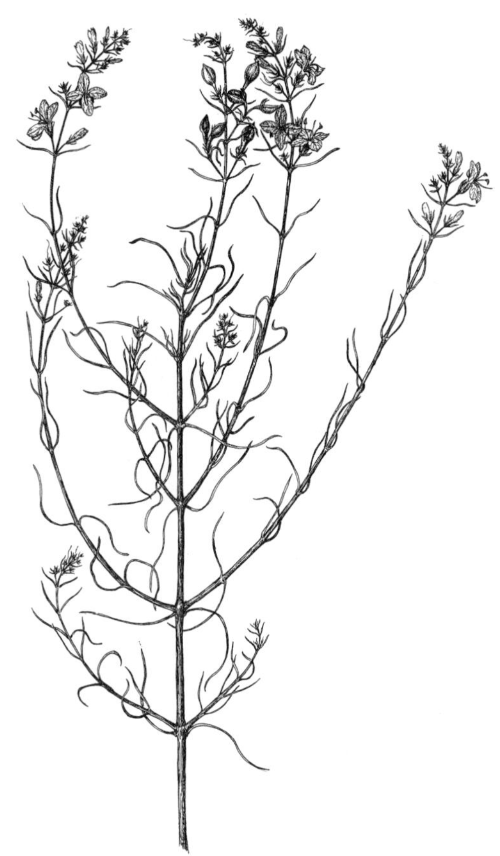

Heath Carlowrightia
Carlowrightia linearifolia (Torr.) Gray

TWIGS. Slender, mostly opposite; young green, puberulent and finely striate; older gray and glabrous.

RANGE. Dry sites in open sun at altitudes of 2,500–3,000 ft. Trans-Pecos Texas, New Mexico, Arizona, and Mexico.

REMARKS. The genus name, *Carlowrightia,* honors Charles Wright, botanist and surgeon attached to the United States–Mexico boundary commission in 1851. The species name, *linearifolia,* refers to the very narrow leaves.

ARIZONA CARLOWRIGHTIA

Carlowrightia arizonica Gray

FIELD IDENTIFICATION. Spreading shrub to 2 ft, woody toward the base. Upper cauline leaves small and bractlike.

FLOWERS. In interrupted elongate spikes on filiform branchlets, sometimes in a few-flowered panicle; corolla imbricate in the bud, white, tube slender, narrow, not dilated at throat; limb 4-lobed, about twice as long as the tube, upper lobe yellow-spotted and purplish; stamens 2, anthers parallel, 2-celled, much shorter than the filaments; stigma minute and capitate, ovary superior; calyx deeply cleft.

PARRY RUELLIA
Ruellia parryi Gray

ARIZONA CARLOWRIGHTIA
Carlowrightia arizonica Gray

FRUIT. Capsule borne on a slender clavate stipe often longer than the capsule body, flattened, apex abruptly acuminate, elastically dehiscent, 2-celled; seeds 2–4, flat, minutely scabrous.

LEAVES. Simple, opposite or nearly so, lanceolate to ovate-lanceolate or oblong, margin entire, apex acute to acuminate, base very short-petiolate or sessile and cuneate to rounded, blade length ¼–1¼ in., about ¼ in. wide or less, light green and finely canescent, only the main vein prominent.

TWIGS. Numerous, slender, young ones green to brownish and minutely puberulent, later gray, glabrous, and woody near the base.

RANGE. Dry stony hillsides, or rocky mesas, at altitudes of 2,500–4,500 ft. In western New Mexico, western Texas, Arizona, and Mexico from Sonora to Sinaloa.

REMARKS. The genus name, *Carlowrightia*, is in honor of Charles Wright, botanist and surgeon attached to the United States–Mexico boundary commission in 1851. The species name, *arizonica*, is for the state of Arizona. The plant is browsed by cattle and sheep.

Parry Ruellia

Ruellia parryi Gray

FIELD IDENTIFICATION. Low much-branched subshrub, usually less than 2 ft.

FLOWERS. Large, ¾–1 in. long, axillary, mostly solitary, white or blue, perfect, corolla funnelform, slender; limb 5-lobed, spreading, almost regular, about ½ in. wide, lobes rounded; stamens 4, included and inserted on the tube, anthers parallel; ovary 2-celled, styles recurved above; calyx 5-parted, segments linear.

FRUIT. Capsule 2-celled, pointed, ellipsoid to oblong, dehiscent longitudinally; ovules 3–10 in each cavity, seeds flattened.

LEAVES. Simple, opposite, about 1 in. or more long, petiolate, texture thick and stiff, pubescent on both surfaces, margin ciliate, oblong to elliptic or obovate, apex acute or obtuse, base cuneate or gradually narrowed.

RANGE. Western and southwestern Texas, also southern New Mexico and northeastern Mexico.

REMARKS. The genus name, *Ruellia*, from the French herbalist, Jean Ruelle (1474–1537), and the species name, *parryi*, from the botanist, C. C. Parry, who was a member of the United States–Mexico boundary survey commission in 1854.

MADDER FAMILY (*Rubiaceae*)

SCARLET BOUVARDIA
Bouvardia ternifolia (Cav.) Schlecht.

SCARLET BOUVARDIA

Bouvardia ternifolia (Cav.) Schlecht.

FIELD IDENTIFICATION. Attractive Western shrub attaining a height of 4 ft.

FLOWERS. Cymes terminal, few-flowered; corolla red, slender, tubular, ¾–1½ in. long, with 4–5 short lobes; stamens 4–5, inserted in the corolla-throat; style slender, exserted, with 2 stigmas; calyx campanulate, of 4 persistent, linear sepals.

FRUIT. On peduncles ¼–½ in. long; 2-celled ovary forming a globose-didymous, coriaceous capsule about 3/16–¼ in. long, enclosed one half to three fourths its length in the calyx; seeds numerous, compressed, winged.

LEAVES. In threes, fours, or pairs above, 1–2 in. long, short-petiolate or sessile, ovate-oblong to lanceolate-oblong, scabrous, hirtellous, apically acute or long-attenuate.

TWIGS. Gray to white, divaricate, older stems same color.

RANGE. Southwestern and western Texas. Abundant in some canyons of the Chisos Mountains in Big Bend National Park. Also in Arizona and New Mexico. In Mexico in the states of Sonora to Coahuila, Veracruz, and Oaxaca.

MEDICINAL USES. In Mexico and other parts of tropical America, roots are used as a remedy for dysentery, hydrophobia, heat exhaustion, and as a preventive for excessive bleeding.

REMARKS. The genus name, *Bouvardia,* is dedicated to Charles Bouvard, at one time director of the Paris Garden of Plants, and the species name, *ternifolia,* refers to the ternate leaf arrangement. Standley (1920–1926, p. 1362) lists a number of Spanish and Indian names in use for the plant in Mexico, and Central America, such as Trompetilla, Doncellita, Mirto, Tlacoxochitl, Tlacosuchiu, Contrayerba, Hierba del Pasmo, and Hierba del Indio. It is often cultivated.

SMOOTH BOUVARDIA

Bouvardia glaberrima Engelm.

FIELD IDENTIFICATION. Suffrutescent or shrubby plant attaining a height of about 3 ft. Upper stem herbaceous, the lower woody.

FLOWERS. Perfect, regular or nearly so, borne in terminal, few-flowered cymes; corolla narrowly tubular, gradually widened toward the apex, ¾–1¼ in. long, red, occasionally white or pink; lobes 4–5, valvate in the bud, erect or somewhat spreading, 1/16–⅛ in. long; calyx-tube adnate to the ovary, lobes 4, persistent, subulate, ⅛–¼ in. long; stamens distinct, borne on corolla alternate with lobes, dimorphic in length; style 1, dimorphic like the stamens, stigmas 2, obtuse, ovary inferior and 2-celled.

FRUIT. Capsule didymous, subglobose, coriaceous, separating into 2 indehiscent 1-seeded carpels; seeds flat, peltate, winged, imbricate on the placenta.

LEAVES. Usually in whorls of 3, occasionally 2–4, lanceolate to ovate-lanceolate, 1–3¼ in. long, apex acute or acuminate, base gradually narrowed into a short petiole or sessile, margin entire, surfaces glabrous above, somewhat hispidulous or glabrous beneath.

STEMS. Branched, slender, glabrous; bark pale at first, brown later.

RANGE. On dry shady hillsides and canyon sides at altitudes of 3,000–9,000 ft. Southern New Mexico and Arizona, and Mexico in Sonora, Chihuahua, and Durango.

REMARKS. Genus name, *Bouvardia,* is in honor of Charles Bouvard, at one time director of the Paris Garden of Plants. The species name, *glaberrima,* refers to the smooth character of the plant. It is also listed under the name of *B. ovata* Gray. The plant offers possibilities as a garden plant because of the attractive flowers.

SMOOTH BOUVARDIA
Bouvardia glaberrima Engelm.

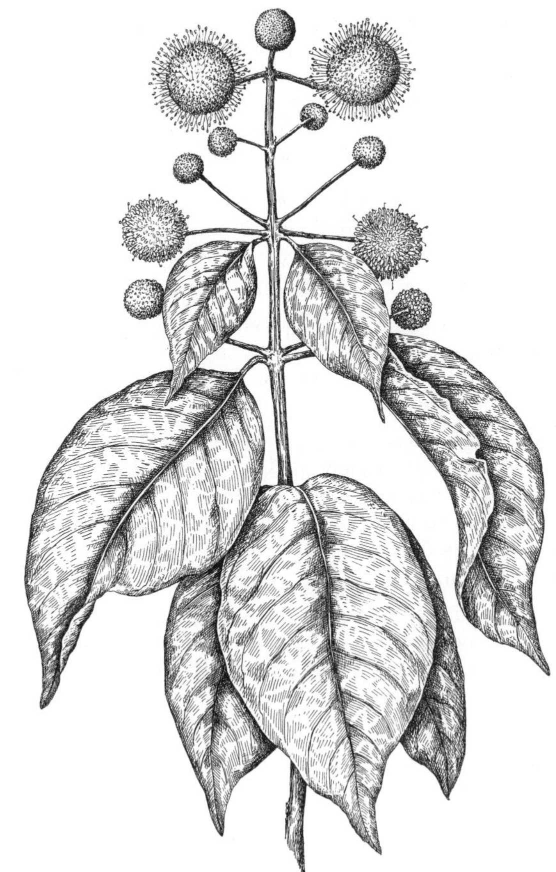

COMMON BUTTON-BUSH
Cephalanthus occidentalis L.

COMMON BUTTON-BUSH

Cephalanthus occidentalis L.

FIELD IDENTIFICATION. Shrub or small tree to 18 ft, growing in low areas, often swollen at the base.

FLOWERS. June–September. Borne on peduncles 1–3 in. long, white, sessile, clustered in globular heads 1–1½ in. in diameter; corolla ¼–½ in. long, tubular with 4 short, ovate, spreading lobes; stamens 4, inserted in corolla-throat, anthers oblong; style slender, exserted; stigma capitate; ovary 2-celled; calyx-tube obpyramidal with 4–5 rounded sepals.

FRUIT. September–October. Round cluster of reddish brown nutlets; nutlets dry, obpyramidal, ¼–⅓ in. long.

LEAVES. Opposite, or in whorls of 3, simple, deciduous, short-petioled, ovate or lanceolate-oblong, acuminate or acute at apex, rounded or narrow at base, entire; blades 2–8 in. long, 1–3 in. wide, dark green and glabrous above or somewhat hairy beneath; petioles glabrous, stout, ½–¾ in. long; stipules small, triangular.

TWIGS. Dark reddish brown, lustrous, glaucous when young; lenticels pale and elongate.

BARK. Thin, smooth, gray to brown, later with flattened ridges and deep fissures.

RANGE. New Mexico, Oklahoma, Texas, Arkansas, and Louisiana; eastward to Florida, and throughout North America from southern Canada to the West Indies, also in eastern Asia.

MEDICINAL USES. Bark has been used medicinally in the treatment of bronchial, skin, and venereal diseases; also used as a laxative and as a fever remedy. The bark should, however, be used with caution because it contains the poison "cephalanthin," which may cause paralysis, convulsions, and vomiting.

PROPAGATION. For planting, the small seeds should be gathered before the heads break up. Each fruit breaks into 1–4 1-seeded nutlets. The seed averages about 134,000 seeds per lb. The germination rate is low. Cuttings planted in moist sandy soil do well.

VARIETIES. Southern Button-bush, *C. occidentalis* var. *pubescens* Raf., is a variety with pubescent leaves and twigs.

Narrow-leaf Button-bush, *C. occidentalis* var. *angustifolius* Dippel, has oblong-lanceolate leaves usually in threes.

REMARKS. The Greek genus name, *Cephalanthus*, means "head-flower," and the species name, *occidentalis*, means "western." Vernacular names for the shrub are Spanish Pincushion, River-brush, Swampwood, Button-willow, Crane-willow, Little-snowball, Pinball, Box, Button-wood, Pond-dogwood, Uvero, and Crouper-brush. It is frequently cultivated as an ornamental shrub and provides good bee food. According to stomach records the nutlet is eaten by at least 25 species of birds, mostly water birds. The wood is of no economic value.

David Milkberry

Chiococca alba (L.) Hitchc.

FIELD IDENTIFICATION. Erect, spreading, or reclining shrub.

FLOWERS. Borne in slender, axillary racemes or panicles 1½–4 in. long; pedicels 1/16–3/16 in. long; hypanthium 1/25–1/12 in. long; corolla white or yellow, ¼–⅓ in. long, short-funnelform, the 5 short lobes ovate to lanceolate and spreading to reflexed; stamens 5, free from corolla-tube, united around base of style, anthers linear; styles united, elongate, filiform, exserted, ovary 2–3-celled, disk cushion-like; calyx persistent, 5-lobed, lobes ovate or triangular and shorter than the tube.

FRUIT. Drupe white, subglobose, flattened, leathery about ¼ in. in diameter; carpels 2, each 1-seeded endosperm fleshy.

LEAVES. Opposite, 1–2¾ in. long, lanceolate to elliptic or ovate, margin entire, apex acute or short acuminate, base acute, texture leathery, surfaces glabrous and lustrous or somewhat pubescent beneath petioles ⅛–⅜ in. long, stipules persistent.

RANGE. Southwestern Texas, Mexico, Florida, West Indies, and tropical South America. In Mexico in Baja California, Sonora, Tamaulipas, Veracruz, Yucatán, and Chiapas. In Texas in the lower Rio Grande Valley area in Cameron County, where it is difficult to ascertain whether it is native or an escape from cultivation.

MEDICINAL USES. Formerly much used in Central and tropical South America and Europe as a medicine, being considered especially efficacious in the treatment of dropsy, venereal disease, and rheumatism. Poultices were also made from the leaves.

The following excerpt from Wood and Osol (1943, pp. 1295–1296) gives the official status of the drug, which is listed as cahinca, David's Root, Snowberry, *Radex Caincae*, and cainca: "This medicine attracted at one time considerable attention. The name of *cahinca*,

David Milkberry
Chiococca alba (L.) Hitchc.

Fascicled Bluet
Houstonia fasciculata Gray

or *cainca,* was adopted from the language of the Brazilian Indians. The Portuguese of Brazil called the medicine *raiz pretta,* or black root. It is derived from *Chiococca racemosa* L. *[C. alba]* (Family *Rubiaceae*), and perhaps other species of the genus. The virtues of the root reside almost exclusively in its bark. A glycoside, *caincin,* has been extracted from cahinca by Costa (*Rev. Brazil Med. Pharm.,* 8:124, 1932) and the pharmacology of the drug has been studied. The drug is used as a diuretic and purgative.

"Cahinca is an emeto-cathartic capable of producing serious gastro-intestinal irritation. In Brazil it has long been used by the natives as a remedy for the bites of serpents, also in rheumatism and in dropsy. *Dose,* of the powdered bark, from twenty grains to a drachm (1.3–3.9 Gm.). For further information see *U. S. Dispensatory* 20th ed., p. 1295."

REMARKS. The genus name, *Chiococca,* is derived from the words *Chio* ("snow") and *cocca* ("seed"), refering to the white drupe. The species name, *alba,* also means "white." It is occasionally listed in the literature under the name of *C. racemosa.* Vernacular names used within the range of this plant are Snowberry, Milkberry, David's Root, Cahinca, Cainea, Canica, Caninara, Aceitillo, Perlilla, Madreselva, Lágrimas de San Pedro, Lágrimas de María, Oreja de Ratón, Suelda, Consuelda, Bejuco de Berac, Bejuco de Berraco, and Xcanchac-che. The plant is an attractive shrub and could be more extensively cultivated.

Fascicled Bluet

Houstonia fasciculata Gray

FIELD IDENTIFICATION. Small Western shrub usually under 18 in. high, with rigid, spreading, puberulent branches.

FLOWERS. In small cymes ½–1 in. across, or some solitary; flowers short-pediceled, perfect, or some dimorphous, white, about ⅛–⅙ in. long, corolla broadly funnelform, tube longer than the 4 valvate lobes; calyx minute, persistent, 4-lobed, lobes triangular, 1/25–1/16 in. long; stamens 4; style single, stigmas 2 and narrow, ovary 2-celled.

FRUIT. Capsule top-shaped, barely 1/12 in. long, about one third free, more or less 2-lobed, loculicidally dehiscent; seeds 4 or 5 in each cell, elongate-oblong.

LEAVES. Opposite or fascicled in whorls, linear, subulate, margin entire and revolute, ⅙–½ in. long, rigid, glabrate or scabrid.

TWIGS. Young green to tan or light brown, angled, hirtellous, rigid, divaricate; older terete, gray to black, slender.

RANGE. On dry sunny sites. In Trans-Pecos Texas, New Mexico, and Mexico in Chihuahua and Coahuila.

REMARKS. The genus name, *Houstonia,* honors William Houston (1695–1733), who collected in tropical America. The species name, *fasciculata,* refers to the fascicled leaves.

Texas Randia

Randia aculeata L.

FIELD IDENTIFICATION. Rigid, thorny shrub attaining a height of 12 ft.

FLOWERS. Axillary, solitary or few in a cluster, on puberulent pedicels ⅛–¼ in. long, perfect; calyx about 1/16 in. long, 4–5-lobed, lobes about 1/25 in. long, ovate to almost subulate; corolla short-tubular, flared into a 5-lobed spreading limb, white above, greenish white below; lobes about ⅛ in. long, short-oblong, elliptic or ovate, apex rounded to obtuse and mucronulate; stamens 5, adnate to the corolla-throat, short; anthers barely exserted from the hairy throat, linear; pistil about 1/25 in. long, slightly exserted, 2-lobed, flattened, granular.

FRUIT. Scattered along the stems, sessile or nearly so, globose, smooth, greenish when young, black when ripe, ¼–½ in. long; seed solitary, black, disklike, round, compressed, thin on the edges, center with a short ridge on each side, ⅛–3/16 in. wide.

LEAVES. Clustered at the nodes but essentially opposite, ½–1½ in. long, broadly-obovate, oval or suborbicular; apex rounded to truncate and often obcordate; base cuneate or gradually narrowed and continuing down the petiole; margin entire and ciliate, slightly wavy occasionally; texture leathery, papillose, pubescent on both sides but at maturity more glabrous above, with hairs mostly on the veins beneath, light to dark green, veins obscure above but more obvious beneath; petioles pubescent, semiwinged by the acuminate leaf base.

TWIGS. Young twigs densely tomentose, later glabrous, smooth, gray; conspicuously armed with opposite, stiff, stout, slender thorns ¼–¾ in. long.

BARK. Mottled gray, smooth, tight.

RANGE. South Texas along the lower Rio Grande River, also southern Florida and the West Indies. In Mexico in the states of Sinaloa, Tamaulipas, Oaxaca, and Veracruz; south through Panama to Colombia.

CLOSELY RELATED SPECIES. Diverse opinions exist as to whether the Texas Randia should be classified as *R. aculeata* L. or *R. rhagocarpa* Standl. Key characters and descriptions as shown by Standley (1920–1926, pp. 1372–1376) indicate characters either of which may fit the Texas plant. For example, the Texas plant has green young fruit and black mature fruit, a thin to thickened pericarp, and pubescent to glabrous leaves. Perhaps one of the species should be reduced to a pubescent variation of the other. Further study is needed to determine the exact relationship of the two. At least 26 species of Randia occur in Mexico.

REMARKS. The genus name, *Randia,* is in honor of Isaac Rand, and the species name, *aculeata,* refers to the spines. *R. mitis* L. is a common synonym. Vernacular names in use in Central America and northern South America are Crucilla, Cruceto, Crucecilla de la Costa, Crucete, Tintillo, Escambron, Cambron, Palo de Catarro, Maiz Tostado, Agalla de Costa, Yamaguey, Yamoguey de Costa, Pitajoni Bravo, Pitajoni Espinoso, Espino Cruz, and Papachilla.

The crosslike paired thorns are a conspicuous feature supporting the name of Crucilla ("little cross") among the Mexican people of south Texas and adjacent Mexico. The plant is also reported to be used in domestic medicine as a remedy for dysentery and the ripe fruit is said to produce a blue dye.

TEXAS RANDIA
Randia aculeata L.

IONEYSUCKLE FAMILY (*Caprifoliaceae*)

LACK-BEAD ELDER
ambucus melanocarpa Gray

FIELD IDENTIFICATION. Shrub 3–12 ft, or rarely a mall tree to 30 ft.

FLOWERS. Borne May–August in terminal compound onvex cymes, often somewhat pyramidal with the nain axis, continuous beyond its lower branches, 2–3 in. r more broad; corolla small, yellowish white; petals 5, qual, imbricate, rotate at maturity or saucer-shaped, ypanthium turbinate; sepals 5, minute; stamens 5, dnate to the base of the corolla, anthers opening by lefts; ovary 3–5-celled, inferior, stigmas 3.

FRUIT. Maturing July–September, berry-like, black t maturity, not glaucous, ⅛–¼ in. in diameter, somewhat juicy; nutlets 3–5, 1-seeded, rugose.

LEAVES. Opposite, deciduous, pinnately compound f 5 (sometimes 7–9) leaflets, blades 2–6 in. long, anceolate to oblong-ovate, margin sharply serrate, apex cute or acuminate, sometimes abruptly so; base broady cuneate to rounded or often asymmetrical, upper urface dark green and glabrous at maturity, lower surace scurfy puberulent or sparsely villous; petiolules hort.

TWIGS. Young ones green to tan or reddish brown, curfy-pubescent, somewhat angled; wood soft; pith vhite to light brown later.

RANGE. Rocky, moist stream banks, or canyons in he conifer belt at altitudes of 5,000–10,000 ft. New Mexico; west to California and north to Alberta, British Columbia, and southern Alaska.

Black-bead Elder is listed by some authors as a variety of Red-fruited Elder, S. *racemosa* var. *melanocarpa* (Gray) McMinn. S. *racemosa* itself is considered o be a cosmopolitan species of western America, Europe, and Asia.

A form of the Black-bead Elder is S. *melanocarpa* forma *fuerstenbergii* Schwer., with reddish brown fruit.

BLACK-BEAD ELDER
Sambucus melanocarpa Gray

REMARKS. The genus name, *Sambucus,* is the classical Latin name, and the species name, *melanocarpa,* refers to the black fruit. Also known as the Black Elderberry and Mountain Elder. It was first cultivated in 1891 and furnishes some minor forage for livestock.

AMERICAN ELDER

Sambucus canadensis L.

FIELD IDENTIFICATION. Stoloniferous shrub with many stems from the base, or under favorable conditions a tree to 30 ft. Stems thinly woody with a large white pith.

FLOWERS. Borne May–July in conspicuous, large, terminal, convex or flattened cymes, sometimes as much as 10 in. across. Peduncles and pedicels striate, green at first, reddish later. Corolla white, ⅕–¼ in. wide, tube short and expanding into 5 lobes; lobes ovate to oblong, rounded, about ⅛ in. long; calyx minute, 5-lobed; stamens 5, exserted, inserted at the base of the corolla; filaments slender, white, about ⅛ in. long, anthers oblong, yellow; style short, depressed, 3-parted; ovary inferior, 4-celled and 1-seeded.

FRUIT. Drupe berry-like, deep purple or black, subglobose, ⅙–¼ in. in diameter, bittersweet, 4-celled, seed roughened.

LEAVES. Deciduous, opposite, 4–12 in. long, odd-pinnately compound of 5–11 (usually 5–7) leaflets; rachis glabrous or pubescent; leaflets elliptic to lanceolate or ovate to oval, apex acute or acuminate, base rounded or broadly cuneate, margin sharply serrate, blades 2–6 in. long, width 1–2 in.; petiolules ⅛–¼ in. long, pubescent; upper surface lustrous, bright green and glabrous; lower surface paler, barely or copiously pubescent.

STEMS. Smooth to angular or grooved, green to red when young, older stems reddish, yellowish or gray, bark sometimes with warty protuberances, nodes sometimes enlarged; external woody layer thin, pith large and white.

RANGE. In rich moist soil, along streams, low places, fence rows. In Oklahoma, Arkansas, Texas, and Louisiana; eastward to Florida and Georgia, northward to Nova Scotia, and westward to Kansas and Manitoba.

AMERICAN ELDER
Sambucus canadensis L.

MEDICINAL USES. In domestic medicine, the leaves have been used as a poultice for sores and tumors. The flowers have been used as a diaphoretic, diuretic, febrifuge, and as an alternative for rheumatism and syphilis. The drupes, bark, and leaves have been used as a laxative. Despite its wide use in domestic medicine, it appears doubtful that the plant possesses the extensive curative properties ascribed to it. For the official properties of the drug the reader is referred to Wood and Osol (1943, pp. 946–947).

PROPAGATION. American Elder and its numerous varieties are often cultivated for the large, attractive cymes of flowers and fruit and variable leaves. It may be propagated by seeds, cuttings of hard and soft wood and suckers. Seeds are produced from the plant about 3 years after sowing and average 175,000–468,000 per lb.

VARIETIES, HORTICULTURAL FORMS, AND CLOSELY RELATED SPECIES. Rehder (1940, p. 828), has listed the following segregates for American Elder:

Red-fruit American Elder, S. *canadensis* forma *rubra* Palmer & Steyerm., is a form with bright red fruit.

Large-leaf American Elder, S. *canadensis* forma *maxima* (Hesse) Schwer., is a form of vigorous growth, large leaves, and cymes to 15 in. across.

Green-fruit American Elder, S. *canadensis* forma *chlorocarpa* Rehd., is a form with green fruit and pale green leaves.

Gold-leaf American Elder, S. *canadensis* var. *aurea* Cowell, is a variety with golden yellow leaves and red fruit.

Cut-leaf American Elder, S. *canadensis* var. *acutiloba* Ellw. & Barry, is a variety with dissected leaflets, the lower pinnatifid, the upper serrate and narrow. This variety is listed by some authors as S. *canadensis* var. *laciniata* Cowell.

Hairy-leaf American Elder, S. *canadensis* var. *submollis* Rehd., is a variety with grayish green, softly pubescent leaflets. This seems to be the most widespread of the natural varieties, occuring with the species over a wide area.

Gulf American Elder, S. *simpsoni* Rehd., has been reported as the Common Elder of the Gulf states from Florida to Louisiana. It has been segregated on the basis of having mostly 5 leaflets, with the lower pair pinnately 2–3 foliate, and by a flat flower cyme. The author has examined a large number of plants from Texas to Florida and has found numerous intergrades with it and the American Elder. The number of leaflets, and pinnate lower leaflets, can be found to vary on the same plant. Also the character of a flat or convex cyme is a variable feature. Additional study should be made to determine if the Gulf American Elder should be maintained as a species or reduced to a varietal rank, or to a synonym of the American Elder.

REMARKS. The genus name, *Sambucus*, is the classical Latin name, and the species name, *canadensis*, refers to Canada, where the plant grows at its most northern limit. It is also known under the local names of Elderberry, Common Elder, Sweet Elder, Pie Elder, and Elder-blow. The various parts of the plant have been ed for food or medicine in domestic practice. The ıit is made into pies, wines, and jellies. The flowers e used to flavor candies and jellies, and the Indians ade a drink by soaking them in water. The dried ıves have been used as an insecticide, and the bark preparing a black dye. The stems, with the pith re- oved, were formerly used as drains in tapping maple gar, and children make whistles, flutes, and pop- ns from them. The plant has considerable value as wildlife food, being eaten by about 45 species of rds, especially the gallinaceous birds such as the ıail, pheasant, and prairie chicken. It is also browsed white-tailed deer and is considered hightly palatable livestock; however, its use by livestock should be vestigated further. The writer has noted that in some calities it is browsed, but in other localities the cattle fuse to touch it even under stress conditions.

MEXICAN ELDER
Sambucus mexicana Presl.

EXICAN ELDER

mbucus mexicana Presl.

FIELD IDENTIFICATION. Shrub or half-evergreen tree 30 ft, with a diameter of 6–18 in. Trunk often much ickened at the base.

FLOWERS. Mostly April–June, but sometimes at other asons after rains; cymes flat or depressed, occasional- somewhat rounded, 4–8 in. broad, the axis seldom ntinuous; calyx minutely 5-sepaled; corolla yellowish hite, ⅛–⅙ in. broad, tube short; petals 5, rotate, equal, stinct; stamens 5, adnate to the corolla-base, filaments ort, anthers opening extrorsely by clefts; stigmas 3, yles short, ovary 3-celled, ovule solitary in each cavity d pendulous.

FRUIT. Berry-like drupe about ¼ in. in diameter, void to subglobose, black, more rarely slightly glau- us, when glaucous having a bluish appearance; seeds ongate and somewhat flattened, usually 1–3 in a upe.

LEAVES. Opposite, odd-pinnately compound of 3–5 aflets (more rarely 7), the lower ones rarely 3-parted, chis densely pubescent; leaflets elliptic to narrowly long or ovate, occasionally somewhat obovate, apex ort-acuminate, margins finely serrate, mostly less than in. long, pale green, thick and leathery, rather varia- e in hairiness on the surfaces, some plants almost gla- ous, others with varying degrees of pubescence to ensely so at the other extreme.

TWIGS. Slender, green to reddish brown, when young riate, when older less so, pithy.

BARK. Gray, thickened, on older trunks furrowed and aly.

WOOD. Brownish, soft, coarse-grained, specific gravity out 0.46.

RANGE. Along low places, ditches, and streams, at titudes of 1,000–4,000 ft in the desert or desert grassland. In Texas and New Mexico, west to southern California; also in Mexico.

MEDICINAL USES. The flowers, fruit, and bark are used medicinally and for food. The flowers serve as a gentle excitant and sudorific. The fruit is a diaphoretic and has been used as an alternative in treating rheumatism and syphilis. The inner bark is a hydragogue cathartic and in large doses is emetic. It has been employed for dropsy and epilepsy.

PROPAGATION. It is sometimes planted for ornament in western Texas and in Mexico. It is grown from seeds, which germinate readily, by greenwood or hardwood cuttings, or by root cuttings or suckers.

VARIETIES AND SYNONYMS. Mexican Elder has recently been described as a variety of the Blueberry Elder under the name of S. *caerulea* var. *mexicana* (Presl.) Benson, by Benson (1943b). Mexican Elder has also been described under the names of S. *canadensis* L. var. *mexicana* Sarg., S. *caerulea* var. *arizonica* Sarg., and S. *glauca* Nutt. var. *arizonica* Sarg. *ex* Jepson.

REMARKS. The genus name, *Sambucus*, is the classical Latin name, and the species name, *mexicana*, refers to Mexico. It is known in the Latin-American countries under the names of Saúco, Azumiatl, Cubemba, Cumdumba, Xumetl, Uttzirza, Bixhumi, Yutnucate, Shiilsh, and Coyopa. The fruit is also of considerable value for wildlife, being eaten by at least a dozen species of birds. The leaves are browsed by mule deer and have some value as livestock browse in winter. Dyes from the stems are used for coloring baskets. The fruit makes excellent wines and pies, and is often dried by the Indians and stored for future use.

Scarlet Elder

Sambucus pubens Michx.

FIELD IDENTIFICATION. Shrub or small tree to 24 ft, with puberulent or smooth branches.

FLOWERS. April–July, inflorescence a pyramidal, thyrsoid cyme, longer than broad, to 4 in. long, central axis extended beyond the lowest branches of the cyme, peduncle ¾–2½ in. long, puberulent to glabrous; flowers ⅛–⅙ in. across, rotate; petals 5, rounded, spreading; stamens 5, exserted, anthers oblong and extrorse; style short, stigmas 3, ovary 3-celled, 1-ovule in each cell.

FRUIT. Persistent, maturing June–November, drupe

Scarlet Elder
Sambucus pubens Michx.

broadly ovate, about ⅕ in. in diameter, red, juice y lowish; nutlets 3, less than ⅛ in. long, minutely roug ened.

LEAVES. Opposite, pinnately compound of 5–7 le lets, blades ovate-lanceolate to narrowly oblong, rare oval, apex acuminate, base narrowed or rounded a some inequilateral, margin sharply serrate, length 2½ in., upper surface dark green and glabrous, lower s face soft-pubescent, but glabrous later; petiole a petiolules puberulent; terminal petiolule ⅖–1 in. lon lateral ones very short or blade sessile; stipules minu or absent.

TWIGS. Young ones pubescent, gray to reddish yellowish brown, glabrous later; bark on older ste with warty excrescences; pith brown.

RANGE. Along streams or on shaded hillsides altitudes of 7,500–12,000 ft. New Mexico; east Georgia, north to Newfoundland, and west to Alas and the montane regions of western North America.

PROPAGATION. The ripe fruit is run through macerator and the pulp floated off. After drying, t seed may be stored 1–2 years in sealed containers 41°F. One hundred lb of fruit yield about 4 lb seed, with about 286,000 seeds per lb. The commerc purity is about 98 per cent and the soundness 97 p cent. There is both a seed coat and an embryo d mancy, and the seed can be stratified in moist sa 30–60 days at 68°–86°F., plus 90–150 days at 41° Germination is sometimes not completed until the f lowing spring. The seed can be planted in drills wi 40 viable seeds per linear ft and covered ¼ in. wi soil. The seedlings should be given half shade. Scarl Elder, like most elders, may also be propagated fro cuttings.

VARIETIES. The following varieties and forms ha been segregated:

Smooth Scarlet Elder, S. *pubens* forma *calva* Ver has glabrous twigs and leaves.

Cut-leaf Scarlet Elder, S. *pubens* forma *dissec* (Britt.) Vern., has dissected leaves.

Pearl Scarlet Elder, S. *pubens* forma *leucocar* (Torr. & Gray) Vern., has white fruit.

Yellow-fruit Scarlet Elder, S. *pubens* forma *xanth* *carpa* (Cockerell) Vern., has yellow fruit ⅙–¼ in. diameter.

Rose-flower Scarlet Elder, S. *pubens* forma *rosaeflo* (Carr.) Dansereau, has cymes panicle-like from ovo to convex or pyramidal; flowers yellowish white pink, ill-scented.

Tree Scarlet Elder, S. *pubens* var. *arborescens* Dipp has a treelike growth.

Amber Scarlet Elder, S. *pubens* forma *chrysocar* Eames & Godfr., has golden yellow fruit.

REMARKS. The genus name, *Sambucus*, is possib the classical Latin name. The species name, *puben* refers to the soft hairs of the leaves and pedicels. is also known under the vernacular names of Re rried Elder, Stinking Elder, and Soreau Rouge. The ant is closely related to European Red Elder, S. cemosa L., and is considered by some authors to be e same. The plant is valuable as an ornamental and wildlife food, being fed upon by at least 23 species birds, moose, red squirrel, and white-tailed deer.

.UEBERRY ELDER

ımbucus caerulea Raf.

FIELD IDENTIFICATION. Often a broad shrub with ımerous spreading stems, or occasionally a tree to -40 ft, with a diameter of 8–18 in. Branches rather out and spreading into a round-topped head. Twigs ıd leaves glabrous or nearly so. A Western Elder mewhat resembling the American Elder, but more osely branched and taller with bluish fruit. Mexican lder is more pubescent on the leaves, and fruit black only slightly glaucous.

FLOWERS. April–August, cymes 5-rayed, 4–10 in. oad, mostly flattened or occasionally convex, rays ongate and compound, axis seldom extending beyond e lowest branches; bracts and bractlets early decidu-ıs, green, linear, acute; flower buds globose, glaucous reddish; calyx reddish brown, lobes 5, small, acute, arious; corolla yellowish white, rotate, about ⅛–⅙ in. diameter; petals 5, oblong, apex rounded, as long as e stamens; stamens 5, inserted at the base of the corolla, filaments short, anthers extrorse; style short, 3-cleft, ovary 3-celled, ovules solitary and pendulous.

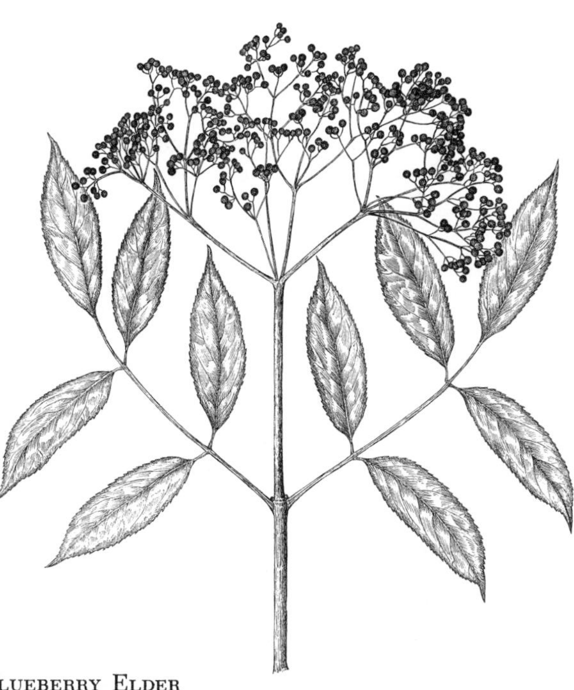

LUEBERRY ELDER
ımbucus caerulea Raf.

FRUIT. August–September, often abundant, conspicuous, a berry-like drupe, subglobose, ¼–⅓ in. in diameter, black, but appearing blue because of a heavy mealy bloom, flesh sweet, juicy, edible; seeds 1–3.

LEAVES. Odd-pinnately compound, deciduous, 5–10 in. long; petiole stout, grooved; leaflets 5–9 (rarely 3), ovate to narrow-oblong or lanceolate, long acuminate, base rounded or unequally cuneate, margin rather coarsely and sharply serrate, blade 1–6 in. long, ⅓–2 in. wide; lower leaflets sometimes 3-parted or pinnate; upper surface yellowish green and glabrous, lower surface paler, at first somewhat hairy, later glabrous, firm, veins inconspicuous; petiolules slender, ¼–2 in. long, those of the terminal leaflets longest; stipels 1/16–½ in. long, early deciduous, linear to lanceolate or ovate, apex acute to obtuse, margin entire to serrate; stipels sometimes absent.

TWIGS. Stout, smooth, somewhat angled, lustrous, green to reddish brown, pubescent at first, but glabrous later; leaf scars large, deltoid, bundles conspicuous; pith white to brown; buds axillary, scales 2–3 pairs, brown, ovate.

BARK. Dark brown to reddish, fissures rather deep and irregular, scales appressed, squarish, small.

WOOD. Yellowish brown to dark brown, sapwood lighter, coarse-grained, soft, weak, durable in contact with the soil, specific gravity 0.50.

RANGE. Moist places along creeks or in canyons at altitudes of 2,000–8,000 ft. In Trans-Pecos Texas and New Mexico; west to southern California, north to British Columbia and Alberta, and south in Mexico to Sonora and Chihuahua.

VARIETIES AND SYNONYMS. Blueberry Elder has been listed under the scientific synonyms of S. *glauca* Nutt., S. *neomexicana* Woot., S. *glauca* var. *neomexicana* (Woot.) A. Nels., and S. *intermedia* Carr. var. *neomexicana* (Woot.) Schwer.

The two varieties most generally recognized are:

Velvet-leaf Blueberry Elder, S. *caerulea* var. *velutina* (Dur. and Hilg.) Schwer., which has leaves and twigs densely short-, white-pubescent. It occurs in California and has been listed as S. *californica* K. Koch.

New Mexico Blueberry Elder, S. *caerulea* var. *neomexicana* (Woot.) Rehd., has 3–5 leaflets, narrow lanceolate, grayish green and slightly pubescent beneath. It is sometimes listed as S. *intermedia* Carr. It occurs in New Mexico and Arizona.

REMARKS. The genus name, *Sambucus*, is the classical Latin name. The species name, *caerulea*, refers to the bluish fruit. The plant sprouts readily from the base and grows rapidly. It was introduced into cultivation before 1850. It serves as food for many forms of wildlife. The foliage is browsed by mule deer, and the fruit eaten by at least 12 species of birds, including the Gambel's, valley, and California quail, and the ring-necked pheasant.

Canyon Elder

Sambucus vestita Woot. & Standl.

FIELD IDENTIFICATION. Shrub to 9 ft, with numerous stout stems from the base. Young branches minutely and densely velvety-pubescent.

FLOWERS. Borne in open flat-topped cymes 4–8 in. across; rays numerous, slender, pubescent, the axis not continuous; flowers hardly over 1/8–1/6 in. in diameter, white, rotate; petals 5, imbricate, rounded; stamens 5, adnate to the corolla-base, anthers opening transversely by clefts; ovary 3–5-celled.

FRUIT. Drupes abundant, black, glaucous, about 1/5 in. in diameter, 1-seeded.

LEAVES. Opposite, pinnately compound, leaflets large, the blades 3–6 in. long, lanceolate or narrowly so, apex long-attenuate, base rounded and often oblique, margin with coarse, sharp, acute teeth, thin, rather pale green, puberulent along the veins above, surface puberulent beneath; petioles and petiolules densely and finely pubescent.

TWIGS. Young ones green, angled, densely pubescent; older ones tan to brown and glabrous; wood soft; pith large.

RANGE. Along streams in canyons at altitudes of 6,000–9,000 ft. In New Mexico known from the Mogollon, Organ, and Black mountains; also in Gila County, Arizona.

REMARKS. The genus name, *Sambucus,* is the classical Latin name. The species name, *vestita,* refers to the velvety-hairy stems and leaves. This species is apparently closely related to S. *caerulea* Raf. *(S. neomexicana),* but it has smaller flowers, more pubescent branches, and a clump of simple shoots instead of a well-developed trunk.

Mountain Snowberry

Symphoricarpos oreophilus Gray

FIELD IDENTIFICATION. An erect, deciduous shrub to 3½ ft. Branches slender and spreading with roots stoloniferous.

FLOWERS. June–July, axillary or in few-flowered terminal spikes, small, perfect, rose-colored; corolla tubular-funnelform, slender, ½ in. long or less, symmetrical, glabrous to somewhat pilose; lobes 4 or 5, slightly spreading, 1/12–1/8 in. long; tube with 5 basal nectaries within; stamens 4 or 5, inserted on the corolla, anthers about 1/12 in. long or about as long as the filaments; style solitary, glabrous, about 1/8 in. long, stigma capitate; ovary inferior, 4-celled, 2 cells with fertile ovules, the other 2 with abortive ovules; calyx glabrous, lobes deltoid, 1/50–1/25 in. long; bracts oval, acute, glabrous, about 1/25 in. long.

Mountain Snowberry
Symphoricarpos oreophilus Gray

FRUIT. August–September, a berry-like drupe, whi ovoid to ellipsoid, 1/3–2/5 in. long; nutlets lanceoloi flattened, ends obtuse or acute, 1/5–1/4 in. long, 1/12– in. wide, endosperm present, embryo minute.

LEAVES. Rather variable in size and lobing, sho petiolate, simple, opposite, oval, glabrous or nearly s glaucous-green above, pale green below with fi prominent veins, thin, margin entire to dentate, ap acute, obtuse or rounded, blade length 1/3–1 in., wid 1/3–2/3 in., petiole 1/12–1/8 in. long; leaf buds ovoid a acute.

TWIGS. Young twigs glabrous, older ones brown ar shreddy.

RANGE. At altitudes of 5,000–9,000 ft, in west Texa New Mexico, Arizona, Utah, Nevada, and Sonora, Me ico.

REMARKS. The genus name, *Symphoricarpos,* is fro the Greek *sympherein* ("to bear together") and *karp* ("fruit"), referring to the clustered drupes. The speci name, *oreophilus,* meaning "mountain-loving," refe to its growth in high, rocky places. The plant is browse by livestock and deer, and the seeds are eaten b chipmunks, ground squirrels, and many species grouse, pheasant, and quail. It may be propagated fro seed and from softwood or hardwood cuttings and h been cultivated since 1894. The tubular-funnelfor corolla, flattened, acute, lanceoloid nutlets, and glabro leaves and twigs serve to segregate it from other speci within its area.

COMMON SNOWBERRY

Symphoricarpos albus (L.) Blake

FIELD IDENTIFICATION. Thicket-forming shrub, ½–3 ft, stems slender, loosely ascending, some branches curved or declined and occasionally rooting.

FLOWERS. Borne May–July, in few-flowered interrupted spikes or spikelike raceme, short-pediceled; bracts lanceolate; bractlets deltoid; calyx irregularly 5-toothed, glabrous to ciliate; corolla pink, campanulate, base somewhat gibbous, length ⅕–⅓ in.; lobes about half as long as the tube (one lobe slightly larger), 1/12–⅛ in. long, densely hairy within; stamens slightly shorter than the corolla; anthers 1/25–1/12 in. long; style 1/12–⅛ in. long, glabrous, stigma capitate; ovary 4-celled, 2 cells with abortive ovules, 2 perfect.

FRUIT. Maturing August–September, persistent, solitary or paired in the axils of the upper leaves, white, depressed-globose, ¼–⅖ in. in diameter, not edible; nutlets 2, light tan, ends rounded, one surface plane, the other convex, length ⅙–⅕ in.

LEAVES. Simple, opposite, membranous, broad to narrowly oval or ovate, blade length ⅜–1½ in., width ⅓–1 in., base somewhat narrowed or rounded, apex obtuse or acute, occasionally apiculate, margin entire or some sinuate, ciliate when young, upper surface sparsely puberulent to glabrous later, lower surface pale or glaucous, puberulent to pubescent, especially on the veins; petioles 1/12–⅛ in. long, sparsely hairy.

TWIGS. Slender, puberulent, but later glabrous, older branches with thin, gray, shreddy bark; winter buds with about 2 pairs of ciliate or puberulent scales.

RANGE. Dry rocky soil, calcareous barrens, gravelly banks. New Mexico, northward through Rocky Mountains to Alberta and Saskatchewan, east to Quebec, New England states, New York, and Pennsylvania, and west to British Columbia, Washington, and Oregon.

PROPAGATION. The fruit is collected from the branches by hand after maturing in the fall, and the seed is removed by maceration in water. If sun-dried, it can be stored in a sealed container at 41°F. for 2 years or more. The dried fruit averages about 18,000 seeds per lb, and cleaned seed about 75,000 seeds per lb, with an average commercial purity of 90 per cent and a soundness of 75 per cent. Dormancy is caused by a hard seed coat and immature embryo, which can be broken by stratification in sand or peat for 90–120 days at 77°F., plus 180 days at 41°F. In spring, sow about 30 viable stratified seed per linear ft of nursery row and cover to a depth of ¼ in. Nursery germination is 50–70 per cent.

VARIETIES. The variations of Common Snowberry fall into three general segregates as follows:

Garden Common Snowberry, *S. albus* var. *laevigatus* (Vern.) Blake (*S. racemosus*), is often cultivated. The twigs are usually glabrous beneath; flowers generally in more elongate terminal racemes; fruit ⅜–½ in. in diameter; nutlets ⅙–¼ in. long; leaves usually larger than in the species; habit to 6 ft. Introduced into cultivation about 1806. From Quebec southwestward.

Broad-leaf Common Snowberry, *S. albus* forma *ovatus* (Spaeth) Rehd., is a form of *S. albus* var. *laevigatus*, with broadly-ovate, bluish green leaves, base rounded or nearly truncate. Cultivated since 1888.

Dwarf Common Snowberry, *S. albus* var. *pauciflorus* (Gray) Blake, is apparently not cultivated; stature dwarf; leaves smaller, grayish green, pubescent beneath; flowers terminal, 1–3. In New Mexico.

REMARKS. The genus name, *Symphoricarpos*, is from the Greek *sympherein* ("to bear together") and *karpos* ("fruit"), referring to the clustered fruit. The species name, *albus*, means "white," and also refers to the fruit. It is rather important as a forage plant in winter in the Western states, but cannot stand overgrazing. It is used in ornamental planting and has been cultivated since 1806. It is also used for erosion control and for wildlife food and cover, known to be eaten by at least 20 species of birds, including ruffed grouse, sharp-tailed grouse, prairie sharp-tailed grouse, Richardson's grouse, Hungarian partridge, ring-necked pheasant, and California and bobwhite quail. It is occasionally eaten by mule deer.

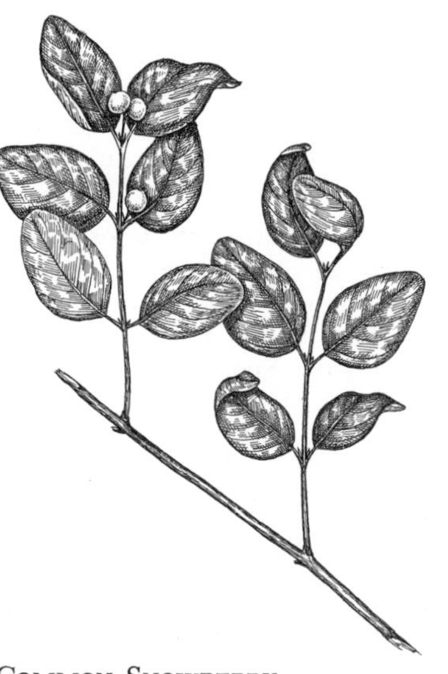

COMMON SNOWBERRY
Symphoricarpos albus (L.) Blake

INDIAN-CURRANT SNOWBERRY

Symphoricarpos orbiculatus Moench

FIELD IDENTIFICATION. Hardy, slender, erect or ascending, thicket-forming shrub spreading by stolons. Usually 2–6 ft, with brown, shreddy bark.

INDIAN-CURRANT SNOWBERRY
Symphoricarpos orbiculatus Moench

FLOWERS. June–August, borne in many-flowered, dense, axillary or terminal spikes; individual flowers sessile or nearly so, greenish white, sometimes pinkish, campanulate, 1/8–1/6 in. long, somewhat hairy within, 4-lobed, lobes about 1/16 in. long or as long as the tube, perfect; stamens 4–5, included, adnate to the throat of the corolla-tube, anthers about 1/12 in. long, shorter than the filaments; ovary inferior, 4-celled, only 2 seeds maturing; style included, about 1/12 in. long, thickened at base, hairy, stigma capitate; calyx-teeth 5, triangular, persistent on the fruit.

FRUIT. September–October, often prolific, persistent through most of the winter, in dense axillary clusters, the berry-like drupe pink to coral-red, glaucous, subglobose to ellipsoid, 1/6–1/4 in. thick, the beak about 1/25 in. long, fleshy; nutlets 2, oval, flattened, 1/12–1/8 in. long, about 1/12 in. wide, obtuse.

LEAVES. Numerous, tardily deciduous, opposite, oval to ovate, or orbicular to elliptic; margin entire or somewhat undulate on the plane surface; apex acute, obtuse or rounded; base rounded or cuneate; upper surface dull green, glabrous or slightly puberulent; lower surface paler, glabrous to pubescent or glaucescent, veins more prominent; leaves red in autumn; blades 1/2–1 1/2 in. wide; petioles 1/12–1/16 in. long, puberulent.

TWIGS. Young twigs slender, brown, puberulent to pubescent, older twigs gray with shreddy bark.

RANGE. Oklahoma, Arkansas, Texas, and Louisiana; east to Florida, north to New England, and west to Nebraska and Colorado.

PROPAGATION. The following instructions are given for cultivation of the 3 most useful species of *Symphoricarpos* for erosion control, wildlife food, and ornamental planting, including the Common Snowberry, *S. albus* (L.) Blake, the Western Snowberry, *S. occidentalis* Hook., and the Indian-currant Snowberry, *S. orbiculatus* Moench (U.S.D.A., 1948, p. 347):

"[In nursery and field practice] seed should be sown in seedbeds or nursery rows in the spring and these kept mulched until the following spring, at which time germination will occur. Sulphuric acid treatment followed by either fall sowing or stratification over winter followed by spring sowing, should give good results the first spring. Fresh seed of *Symphoricarpos orbiculatus* can sometimes be fall-sown or stratified over winter prior to spring sowing with complete germination taking place the first season. About 30 viable seeds should be sown per linear ft of nursery row and covered to a depth of about one-fourth inch. [In] nursery germination [the average is:] *S. albus,* 50–70 per cent; *S. orbiculatus,* 90 per cent. *Symphoricarpos* can also be propagated from stem and root cuttings."

VARIETY, FORM, AND HYBRID. Goldnet Indian-currant, *S. orbiculatus* var. *variegatus* (Cowell) Schneid. is a variety with leaves variegated with yellow.

White-fruit Indian-currant, *S. orbiculatus* forma *leucocarpos* (D. M. Andrews) Rehd., is a form with greenish yellow flowers and white fruit.

Indian-currant Snowberry, *S. orbiculatus* Moench and Pink Snowberry, *S. microphyllus* H. B. K., hybridize to produce the Chenault Snowberry, *S.* × *chenaulti* Rehd. It differs from Indian-currant Snowberry by having smaller, more pubescent leaves, a tubular corolla and fruit both red and white blotched.

REMARKS. The genus name, *Symphoricarpos,* comes from the Greek *sympherein* ("to bear together") and *karpos* ("fruit"). The species name, *orbiculatus* refers to the rounded leaf. It is also known under the vernacular names of Coralberry, Snapberry, Buckberry Wolfberry, Waxberry, and Turkey-bush. It is often planted for erosion control on barren, sterile soils and has considerable wildlife value. It is known to be eaten by 12 species of birds, including the cardinal, bobwhite quail, wild turkey, greater prairie chicken, ruffed grouse, sharp-tailed grouse, and ring-necked pheasant and is also browsed by white-tailed deer. The plant is desirable for cultivation for its attractive red fruit and leaves in autumn, being first introduced into horticulture in 1727.

LMER SNOWBERRY

mphoricarpos palmeri G. N. Jones

FIELD IDENTIFICATION. Trailing, deciduous shrub at-ining a length of 3–9 ft, with erect or ascending anches.

FLOWERS. June–July, often abundant, short-pediceled, litary or several together; corolla pink, tubular-fun-lform, perfect, 1/3–1/2 in. long (averaging 3/8 in.), 5-bed, lobes less than half the length of the tube; tube iry within; stamens 5, inserted on the corolla; anthers –1/8 in. long, style glabrous, 1/12–1/6 in. long, stigma pitate; ovary inferior, 4-celled, 2 cells with abortive ules, the other 2 cells with fertile single ovules in ch; calyx regularly 5-toothed, tube somewhat con-icted below the teeth, glabrous or puberulent, 1/16–1/8 long; bracts lanceolate and puberulent, bractlets oval ovate and puberulent.

FRUIT. Berry-like drupe, white, ellipsoid, 1/4–1/3 in. ng; nutlets 2, oval, white, apex rounded, base acute, ngth 1/6–1/5 in., 1/12–1/8 in. wide, flattened, plano-convex.

LEAVES. Simple, opposite, short-petiolate, oval to ate, 3/8–3/4 in. long, 1/5–3/4 in. wide; apex rounded to tuse or acute, sometimes apiculate; base cuneate to unded; margin entire or crenate to lobed; often ciliate; ely reticulate-veined, finely hairy above and below, darker green above than below; petiole stout, 1/25–1/8 in. long, rather densely hairy. Leaves very variable in size and shape on younger and older shoots.

ALMER SNOWBERRY
ymphoricarpos palmeri G. N. Jones

TWIGS. Young twigs densely hairy, with short, curved hairs; older twigs gray, ridged and shreddy.

RANGE. Rocky ledges and moist stony banks at altitudes of 6,500–8,500 ft. In west Texas, New Mexico, Arizona, and southern Colorado.

REMARKS. The genus name, *Symphoricarpos*, is from the Greek *sympherein* ("to bear together") and *karpos* ("fruit"), referring to the clustered drupes. The species name, *palmeri*, is in honor of Ernest Jesse Palmer, botanist of the Arnold Arboretum. The plant was described by George Neville Jones (1940). It may be propagated by seed or softwood or hardwood cuttings.

ROUND-LEAF SNOWBERRY

Symphoricarpos rotundifolius Gray

FIELD IDENTIFICATION. Slender, straggling shrub usually less than 3 ft.

FLOWERS. May–June, axillary, sessile or nearly so; corolla pink, tubular-funnelform, symmetrical, 1/3–2/5 in. long, 5-lobed; lobes about 1/12 in. long, somewhat spreading or erect; stamens 5, inserted on the corolla, anthers about as long as the free portion of the filament or sometimes shorter; style solitary, glabrous, 1/8–1/6 in. long; base of tube with 5 small nectaries, hairy within; ovary inferior, 4-celled, 2 cells with perfect ovules, the other 2 with abortive ovules; calyx campanulate, lobes 1/25 in. long.

FRUIT. July–September, a berry-like drupe, white, ovoid to ellipsoid, about 3/8 in. long, 1/4–1/3 in. wide; nutlets 2, about 1/5 in. long and 1/8 in. wide, oval, flattened, striate, acute or obtuse at ends, endosperm present, embryo minute.

LEAVES. Simple, opposite, petioles 1/25–1/8 in. long, densely short-hairy; rotund-oval to ovate, apex rounded to obtuse, margin entire or dentate to lobate, blade length 3/8–1 1/4 in., width 1/4–1 in., grayish green, softly pubescent above and below, but more so below.

TWIGS. Lightly or densely pubescent with straight spreading hairs, older stems glabrous and shreddy.

RANGE. Mountain canyons and rocky slopes at altitudes of 4,000–10,000 ft in west Texas, New Mexico, Arizona, Colorado, and in Mexico in northern Sonora.

REMARKS. The genus name, *Symphoricarpos*, is from the Greek *sympherein* ("to bear together") and *karpos* ("fruit"), referring to the clustered drupes. The species name, *rotundifolius*, refers to the round leaves. This highly ornamental plant has been cultivated since 1896 and may be propagated by seeds and by softwood or hardwood cuttings. It has considerable food value for

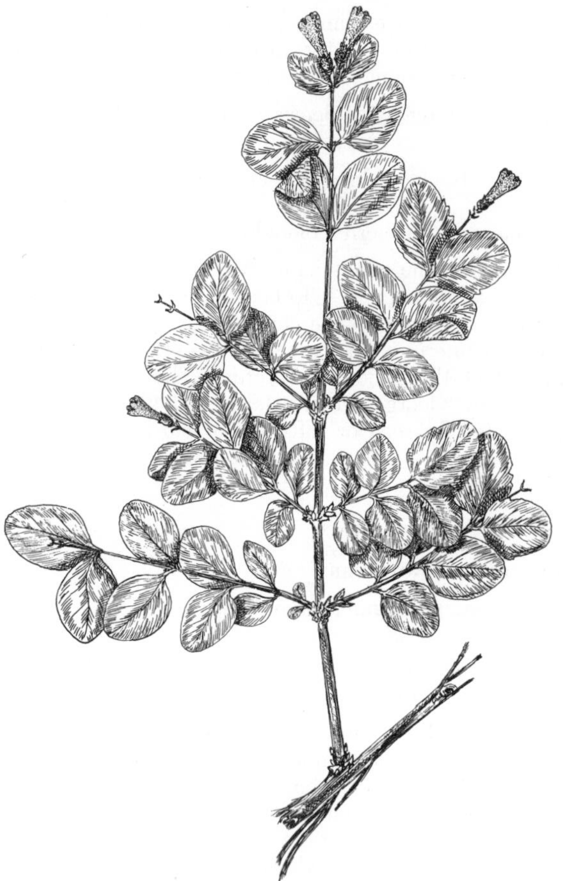

Round-leaf Snowberry
Symphoricarpos rotundifolius Gray

tary, about 1/5–1/3 in. long, hairy above the midd stigma capitate; ovary inferior, 4-celled, 2 cells w single, well-developed ovules, the other 2 cells w ovules abortive; calyx 5-lobed, lobes deltoid, about in. long, glabrous and glaucous or puberulent.

FRUIT. Berry-like drupe, white, ellipsoid, 1/3–2/5 long; nutlets 2, about 1/5 in. long and 1/8 in. wide, ba acute, endosperm present, embryo minute.

LEAVES. Simple, opposite, deciduous, blades 1/4–3/5 long, 1/12–2/5 in. wide, lanceolate to oblanceolate or ov apex acute to obtuse, base cuneate to rounded, marg entire, surfaces glabrous or slightly hairy and glaucou green; petioles 1/25–1/8 in. long, glabrous to hairy.

TWIGS. Young twigs glabrous or sparsely hairy, old ones whitish to gray, fibrous and shreddy.

RANGE. Rocky foothills and canyons at altitudes 4,000–8,000 ft. West Texas and New Mexico; west southeastern California and north to eastern Oregon.

REMARKS. The genus name, *Symphoricarpos,* is fro the Greek *sympherein* ("to bear together") and *karp* ("fruit"), referring to the clustered drupes. The speci name, *longiflorus,* refers to the long-tubed corolla. T plant may be propagated by seeds and softwood hardwood cuttings. This plant and other related speci are browsed by livestock and deer, and the seeds eat by a number of species of birds, especially the ga

wildlife. It and other Southwest species of *Symphoricarpos* are browsed by cattle, sheep, and deer. The seeds are also eaten by many species of birds, especially the grouse, pheasant, and quail, as well as ground squirrels and chipmunks. The Round-leaf Snowberry may be distinguished from other similar species by the dense, villous, spreading pubescence of the young twigs, and by the tubular-funnelform corolla which is about 2/5 in. long.

Longflower Snowberry

Symphoricarpos longiflorus Gray

FIELD IDENTIFICATION. Low, spreading, deciduous shrub with somewhat drooping branchlets.

FLOWERS. May–June, solitary, paired, or in few-flowered racemes; corolla pink, fragrant, salverform, about 1/2 in. long or less, symmetrical; tube narrow, much longer than the 5 spreading lobes, glabrous externally and internally, with a solitary nectary; stamens 4 or 5, filaments barely or wholly inserted on the corolla, anthers sessile or nearly so, 1/12–1/8 in. long; style soli-

Longflower Snowberry
Symphoricarpos longiflorus Gray

WESTERN SNOWBERRY
Symphoricarpos occidentalis Hook.

linaceous birds such as the grouse, pheasant, and quail. The Rocky Mountain pica and various ground squirrels and chipmunks also eat the seeds. Distinguished from other species by the salverform corolla, solitary basal nectary, the pilose style ⅕–⅓ in. long, and the oblanceolate and glaucous leaves.

WESTERN SNOWBERRY

Symphoricarpos occidentalis Hook.

FIELD IDENTIFICATION. Shrub 1–4 ft, stoloniferous, thicket-forming, erect, more or less branching.

FLOWERS. Borne May–July, in terminal or axillary, simple or divided, crowded, spikelike clusters, each cluster ⅜–1 in. long; bracts and bractlets oval and ciliate; calyx sessile, margin with 5 ovate, ciliate, minute teeth; corolla funnelform-campanulate, 4–5-lobed to the middle or beyond, ¼–⅓ in. long, pink, hairy within, lobes obtuse, ⅛–⅙ in. long (as long as the tube or slightly longer); stamens inserted on the corolla basally, slightly exserted terminally, anthers about half as long as the filaments; style filiform, glabrous or hairy near the middle, about twice as long as the corolla-tube and exserted beyond the stamens; stigma yellow and capitate; ovary superior, 4-celled, only 2 with single ovules.

FRUIT. Drupe berry-like, ripening August–September, persistent in winter, subglobose, greenish white, ¼–⅓ in. in diameter, beaked by the persistent calyx-lobes; nutlets 2, about ⅙ in. long, about 1/12 in. wide, oval, smooth, plano-convex, light tan, obtuse at the ends, endosperm fleshy, embryo minute.

LEAVES. Simple, opposite, deciduous, blades 1–4¼ in. long, ⅝–2¾ in. wide, elliptic to oval or ovate, apex obtuse to acute and apiculate, base cuneate or rounded, margin entire or crenate-sinuate, thickish; upper surface dull, dark green, glabrous or short-hairy along the veins; lower surface grayish green, finely pubescent or almost glabrous; petioles ⅙–⅖ in., pubescent.

TWIGS. Slender, rather stiff and upright, reddish brown and puberulent at first, later gray with shreddy bark; winter buds with 2 pairs of opposite scales.

RANGE. Dry rocky soil of bluffs, prairies and plains. Oklahoma and New Mexico; north to Ontario and west to British Columbia.

PROPAGATION. The fruit is collected by hand after ripening and is spread out to dry. The pulp is separated from the seed by macerating in water. One hundred lb of fruit yields 5–10 lb of seed, which average 73,000 seeds per lb. The seed has a commercial purity of 90 per cent and an average soundness of 83 per cent. The seed may be stored in dry sealed containers at 41°F. for 2 years with only moderate loss in viability. Seed dormancy may be broken by soaking in sulphuric acid for 75 minutes and then stratifying for 180 days at 41°F. Sowing is in nursery rows, with 30 seeds per linear ft and covered about ¼ in. with soil. A mulch is used until germination the following spring. Propagation is also practiced with stem and root cuttings.

VARIETY. A variety known as Thin-leaf Western Snowberry, S. *occidentalis* var. *heyeri* Dieck, has thinner leaves, more obscurely veined, and shorter stamens and style. It occurs in Colorado.

REMARKS. The genus name, *Symphoricarpos*, is from the Greek and means "fruit borne together", with reference to the clustered berry-like drupes. The species name, *occidentalis*, refers to the westerly distribution of the plant. The Western Snowberry has been in cultivation since 1880. It is valued as a bee food, for erosion control, as a good forage for cattle, and for ornamental planting. It is known to be eaten by at least 15 species of birds, including the greater prairie chicken. It is also known as Buck-brush and Western Wolfberry.

WESTERN WHITE HONEYSUCKLE

Lonicera albiflora Torr. & Gray

FIELD IDENTIFICATION. Scandent shrub or partly twining vine to 9 ft.

FLOWERS. In spring, in a terminal cluster, corolla funnel-form, 2-lipped and -lobed, the lips nearly as long as the tube, white to yellowish, ½–⅔ in. long, glabrous on the exterior, tube slightly gibbous-pubes-

WESTERN WHITE HONEYSUCKLE
Lonicera albiflora Torr. & Gray

cent within; lobes of the lip obtuse, lower lip spreading and reflexed; stamens 5, inserted near the summit of the corolla-tube, alternate with the lobes, filaments glabrous or nearly so; style pubescent, ovary 2–3-celled.

FRUIT. October–November, bracts less than half as long as the ovary, berry globose, pubescent, ⅕–⅖ in. long, few-seeded.

LEAVES. Opposite, short-petiolate or upper pair perfoliate, margin entire, broadly oval to orbicular or obovate, apex acute to obtuse or rounded, firm, olive-green and glabrous above, paler and glabrous and glaucous beneath.

STEMS. Scandent or twining, brown to gray or whitened, glabrous and shreddy.

RANGE. In central and western Texas, north into Oklahoma, west into New Mexico and Arizona. Also in the Mexican states of Chihuahua, Coahuila, and Sonora.

VARIETY. Hairy White Honeysuckle, *L. albiflora* var. *dumosa* Gray, is a variety with hairy stems and leaves.

REMARKS. The genus name, *Lonicera*, is for Adam Lonitzer (1528–1586), a German herbalist, and the species name, *albiflora*, refers to the white flowers. The fruit of the species and its variety are said to be used in domestic medicine as an emetic and cathartic.

UTAH HONEYSUCKLE

Lonicera utahensis Wats.

FIELD IDENTIFICATION. Small or large, erect, clump-forming shrub to 5 ft, the branches slender and spreading.

FLOWERS. April–June, axillary, sessile, paired, at the ends of solitary peduncles; bracts 2, minute, glabrous, subulate (sometimes absent); calyx minute, corolla showy, yellow or ochroleucous, or sometimes purple-tinged, fragrant, tubular-funnelform, apex slightly 2-lipped, with 5 nearly equal teeth, tube ½–1 in. long, gibbous at base, pubescent within; stamens 5; style slender, stigma capitate, slightly exserted, ovary inferior and 2–3-celled.

FRUIT. Berry maturing June–August, orange to yellow or red, globular, about ¼ in. in diameter, borne in 2 divergent pairs.

UTAH HONEYSUCKLE
Lonicera utahensis Wats.

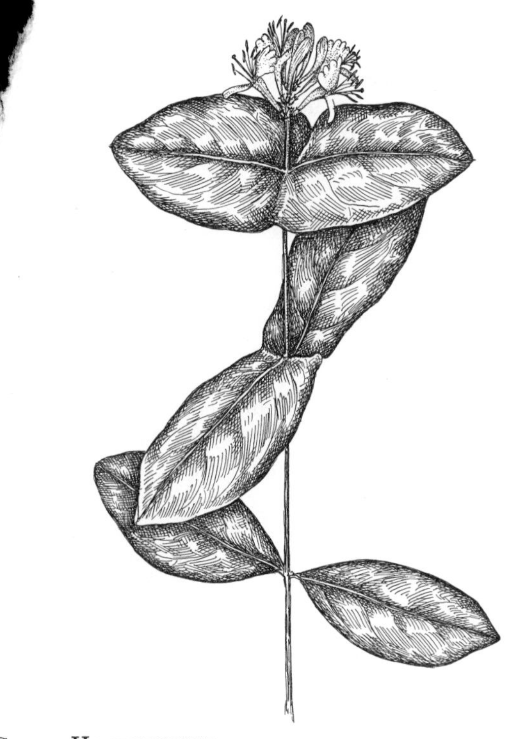

Grape Honeysuckle
Lonicera prolifera (Kirchn.) Rehd.

LEAVES. Simple, opposite, deciduous, blade oval to broad-ovate or elliptic-oblong, ends rounded, or apex sometimes obtuse, margin entire, rather thin, glabrous or nearly so, reticulate-venulose, length ¾–2¼ in., width ½–1¼ in., short-petioled.

TWIGS. Slender, spreading, glabrous, reddish brown to straw-colored, older ones sometimes fibrous.

RANGE. Slopes and canyons of the conifer belt at altitudes of 8,000–11,000 ft. New Mexico, west to California and north to Alberta and British Columbia.

REMARKS. The genus name, *Lonicera,* is in honor of Adam Lonitzer (1528–1586), a German physician and botanist. The species name, *utahensis,* refers to the state of Utah where it grows. The plant is also known as Fly Honeysuckle. It has some small value as livestock browse.

Grape Honeysuckle

Lonicera prolifera (Kirchn.) Rehd.

FIELD IDENTIFICATION. Woody twining vine 4–12 ft, or sometimes semibushy when no support is present. Stems and leaves whitened with a glaucous bloom and glabrous, or more or less puberulent on the lower leaf surfaces.

FLOWERS. May–June, terminal, showy, fragrant; in short-peduncled spikes of 2–6 whorls of 3, or nearly sessile; corolla pale-yellow, tubular, tube slightly swollen at base, gradually enlarged above, externally glabrous, internally hairy, about ⅝–1 in. long; apex of tube strongly spreading into two divergent lips, one lip with a single narrow lobe, the other lip shortly 4-lobed; stamens 5, exserted, inserted on the corolla-tube, anthers linear; style slender, stigma capitate; ovary inferior, 2–3-celled with several ovules in each cell and pendulous; calyx with 5 short teeth.

FRUIT. Berry maturing July–October, about ⅓ in. in diameter, subglobose, red to orange-red, fleshy; seeds several, ⅙–⅕ in. long, ⅛–⅙ in. wide, oval and flat, endosperm fleshy, embryo terete.

LEAVES. Simple, opposite, 1–4 pairs; upper pairs often joined together to produce a more or less orbicular disk, rounded to retuse at the ends, margin entire, glaucous on the upper surface; lower leaves broadly oval to obovate, sessile or nearly so at base, apex obtuse to rounded or emarginate, margin entire, length 1½–4 in.; smooth and dark green above or sometimes glaucous; lower surface glaucous, or more or less puberulent to pubescent, or nearly glabrous; leaves on sterile shoots with petioles to ⅝ in. long.

STEMS. Slender, twining, green to tan or reddish brown, later grayish, finely grooved.

RANGE. On banks, fence rows, rocky woods, and thickets. Oklahoma, Arkansas, and Mississippi; east to Tennessee and Indiana, north to Minnesota and Ontario, and west to Kansas.

PROPAGATION. Grape Honeysuckle has been cultivated since 1840. It is propagated by seeds sown in fall, or stratified for spring sowing, or by cuttings of ripened wood. It seems to be susceptible to attack by green aphis.

VARIETY. Smooth Grape Honeysuckle, *L. prolifera* var. *glabra* Gl., has leaves glabrous beneath. It occurs in Arkansas and southern Missouri.

REMARKS. The genus name, *Lonicera,* honors Adam Lonitzer (1528–1586), a German herbalist. The species name, *prolifera,* means "abundant growth." This plant has also been listed in the literature under the name of *L. sullivantii* Gray.

Arizona Honeysuckle

Lonicera arizonica Rehd.

FIELD IDENTIFICATION. Woody vine, stiff, trailing or scandent.

FLOWERS. In terminal, pedunculate spikes; flowers often in whorls of 2–4 and sessile; corolla trumpet-shaped, ¾–1½ in. long, tube red and glabrous externally, orange internally, limb short and only slightly 2-lipped, ⅕–¼ as long as the tube; stamens 5, deeply inserted and adnate to the corolla-tube, anthers slightly exserted; style glabrous, ovary 2–3-celled, ovules numerous; calyx teeth 5, short.

Arizona Honeysuckle
Lonicera arizonica Rehd.

FRUIT. Berrylike, globose, red, fleshy, few-seeded.

LEAVES. Simple, opposite, deciduous, oval to ovate or oblong to elliptic; apex obtuse, acute or rounded; length to 2⅔ in., margin entire, sometimes glandular ciliate, upper surface green and glabrous, lower surface glaucous, petiole ¼–½ in. Upper leaves sessile and connate-perfoliate, lower separated and short-petiolate, no stipules.

STEMS. Trailing, scandent, or semi-twining, gray to brown, bark shreddy.

RANGE. In Trans-Pecos Texas, New Mexico, Arizona, and Utah.

PROPAGATION. The fruit is gathered by hand as soon as ripe and spread out to dry. The seed may be extracted by macerating in water and floating off the pulp. Immediate fall planting is best, but stratification to break seed dormancy is needed if not planted until spring. Viability in storage is lost rapidly after one year in sealed containers at atmospheric temperatures. In planting, the seed is covered with rich, moist soil about ¼ in. Arizona Honeysuckle is also propagated by greenwood cuttings under glass.

RELATED SPECIES. Arizona Honeysuckle is closely related to Eastern Trumpet Honeysuckle, *L. sempervirens* L., and to *L. ciliosa* Poir. of California, Nevada, and Utah.

REMARKS. The genus name, *Lonicera*, is for the man botanist Adam Lonitzer (1528–1586), and species name, *arizonica*, is for the state of Arizo The fruit is eaten by birds, squirrels, and chipmunk but is considered inedible for human beings because o its cathartic effects. It has been cultivated since 1900.

Bearberry Honeysuckle

Lonicera involucrata (Richards) Banks

FIELD IDENTIFICATION. Western shrub, thicket-forming by suckers, erect, ascending, or sometimes reclining on other shrubs. Root system shallow and fibrous.

FLOWERS. June–July, inflorescence in pairs (rarely threes) on axillary, clavate, glabrous peduncles ¾–1¾ in. long; bracts oval to cordate, foliaceous, large, in 2 sets, (2 outer and 4 inner) or united in pairs, green to reddish or purple, ⅜–¾ in. long, finally reflexed, the 2 corollas embraced at base by the bracts; bractlets large, pubescent, exceeding the ovary; corollas 2 together, ⅖–½ in. long, yellow, sessile, ascending, funnelform, saccate at base; lobes 5, nearly equal and scarcely spreading, about half as long as the tube, surfaces glandular-pubescent; stamens 5, inserted on the corolla within, terminally exserted somewhat or included; style slightly longer than the stamens, slender, glabrous, stigma capitate; ovary 2–3-celled, ovules numerous in each cavity, pendulous.

Bearberry Honeysuckle
Lonicera involucrata (Richards) Banks

FRUIT. August–September, berry in separate pairs, bout ⅓ in. in diameter, globose or oval, purplish black, hiny, fleshy, taste disagreeable; seeds few, with fleshy ndosperm and a terete embryo, seed averages about 26,000 seeds per lb; berry surrounded at base by the eflexed bracts.

LEAVES. Simple, opposite, less often ternate, ascending, ovate to obovate or elliptic, length 2½–5 in., apex hort-acuminate or acute, base tapering to cuneate or ounded into a short petiole, margin entire, upper surace green and glabrous, lower paler and pubescent on eins beneath, midrib stout.

TWIGS. Glabrous, somewhat 4-angled, more rounded nd bark shreddy when older.

RANGE. Cold calcareous soils of moist woods and mountain meadows at altitudes of sea level–10,500 ft. n New Mexico in the Tunitcha and Sangre de Cristo mountains; and in Oklahoma, north and east to Quebec and west to British Columbia, Washington, Oregon, and California.

VARIETY. A variety and a horticultural form are known. Yellow Bearberry Honeysuckle, *L. involucrata* var. *flavescens* (Dipp.) Rehd., has leaves 2¾–5 in. long, oblong to ovate, light green, glabrous or nearly so; corolla gibbous, not saccate at base. In Oregon, Wyoming, and Utah.

FORM. Orange Bearberry Honeysuckle, *L. involucrata* forma *serotina* Rehd., has nearly glabrous leaves; corolla ¾ in. long, orange-yellow, flushed scarlet; bracts not reflexed in fruit. In Colorado, cultivated since 1903.

REMARKS. The genus name, *Lonicera,* honors Adam Lonitzer (1528–1586), German botanist, and the species name, *involucrata,* refers to the large involucral bracts. It is also known under the vernacular names of Twinberry Honeysuckle, Fly Honeysuckle, Black Twinberry, Pigeonberry, and Inkberry. The fruit of Bearberry Honeysuckle is known to be eaten by at least 6 species of birds and visited by the blue-throated and Rivoli hummingbirds. It is only rarely grazed by livestock. It should be more extensively cultivated for ornament.

Japanese Honeysuckle

Lonicera japonica Thunb.

FIELD IDENTIFICATION. Woody, shreddy-barked, twining, or trailing vine often to 18 ft.

FLOWERS. June–August, in pairs from leafy bracts; corolla ½–1½ in. long, tubular, slender, white, pink or later yellow, pubescent, distinctly 2-lipped; 1 lip 4-lobed, the other lip 1-lobed, both lips recurved; stamens and style exserted; stamens 5, adnate to the corolla-tube; style slender, stigma capitate, ovary 2-celled; calyx of 5 slender, subulate sepals.

Japanese Honeysuckle
Lonicera japonica Thunb.

FRUIT. Berries ripe September–March, subglobose, ⅙–¼ in. long, black, pulpy, several-seeded.

LEAVES. Opposite, half-evergreen, short-petioled, 1–3 in. long, entire, and often ciliolate, pubescent, ovate, orbicular-ovate or oblong; apex acute, obtuse, or short-acuminate; base rounded or truncate.

RANGE. Low moist ground along streams, fence rows, and borders of woods. Texas, Oklahoma, Arkansas, and Louisiana; eastward to Florida and north to Massachusetts; also in Oregon and Washington. Introduced into America from Asia and freely escaping cultivation especially in the Southern states.

VARIETIES. Many varieties are known to the floral trade as follows: Purple Japanese Honeysuckle, *L. japonica* var. *chinensis;* Hall's Japanese Honeysuckle, *L. japonica* var. *halliana* (Dippel) Nicholson; Creeping Japanese Honeysuckle, *L. japonica* var. *repens.* A popular horticultural form of the last-mentioned is the Yellownet *(aureoreticulata).*

REMARKS. The genus name, *Lonicera,* is for Adam Lonitzer (1528–1586), a German herbalist, and the species name, *japonica,* denotes the plant's Asiatic origin. Vernacular names are Southern Honeysuckle, White Honeysuckle, and Chinese Honeysuckle. This is a very popular vine in the Southern states, where it is cultivated for its penetrating fragrance and showy flowers. It reproduces itself so abundantly, however, that it is now becoming a troublesome weed.

At least 5 species of birds are known to eat the fruit, and the vine is often used as a cover for landscaping unsightly objects.

Yellow Honeysuckle

Lonicera flava Sims

FIELD IDENTIFICATION. Woody trailing climber with glabrous branches.

FLOWERS. April–May, conspicuous, bright orange-yellow, borne in heads, or interrupted leafless spikes, mostly in terminal 3-flowered cymules, producing a 6-flowered whorl; corolla ¾–1¼ in. long, tube cylindric, slender, barely enlarged at the limb, glabrous externally, glabrous or slightly pubescent within, longer than the lobes; limb irregularly 2-lipped, one lip narrow and with a solitary lobe, the upper lip broad and 4-lobed; stamens 5, long-exserted, glabrous, inserted on the corolla-tube; style slender, elongate, exserted, gl brous, stigma capitate; ovary inferior, 2–3-celled, se eral pendulous ovules in each cell; calyx with 5 sho teeth.

Yellow Honeysuckle
Lonicera flava Sims

FRUIT. Ripening August–September, crowded in se sile or short-peduncled heads, berries subglobose, re to orange-red, about ¼ in. in diameter, fleshy, man seeded, endosperm fleshy.

LEAVES. Simple, opposite, the upper pairs often co nate to form an oval to subrotund disk, disk ofte longer than broad, apices rounded to obtuse; leave below the disk elliptic to obovate or suborbicular, apic acute to obtuse, base narrowed or rounded, sessile nearly so, margin entire, upper surface bright gree and glabrous, lower surface grayish green.

STEMS. Slender, twining, glabrous, green to tan brown, later gray and shreddy.

RANGE. Rocky woods and bluffs. Oklahoma and A kansas; east to Georgia, north to North Carolina, an west to Missouri.

REMARKS. The genus name, *Lonicera,* honors Ada Lonitzer (1528–1586), a German herbalist. The speci name, *flava,* refers to the yellow color of the flower This handsome species has been cultivated since 181 and may be propagated by seeds sown in fall, by cu tings of ripened wood, or by greenwood cuttings u der glass in summer.

Trumpet Honeysuckle

Lonicera sempervirens L.

FIELD IDENTIFICATION. Smooth, twining, evergreen shrubby vine 3–18 ft.

FLOWERS. Borne in sessile whorls, which are include in terminal, spikelike panicles; corolla scentless, red o yellow, slender trumpet-shaped tube slightly dilate above, 1½–2 in. long, glabrous or puberulent, ape spreading into 5 slightly irregular lobes; lobes ⅛–³⁄₁₆ in long, broad-ovate, rounded or obtuse at the apex; sta mens 5, exserted about ¼ in. or more, filaments re and attached to the wall of the corolla-tube within anthers linear, about ⅛ in. long; style filiform, glabrous elongate, equal to the stamens or almost as long, stigm capitate.

FRUIT. Berries in clusters September–October, sub globose, red or orange, several-seeded; calyx-teeth per sistent.

LEAVES. Upper pairs of leaves joined together to be come connate-perfoliate; lower leaves petioled, entire on margin, oval, oblong or lanceolate, obtuse or apicu late at the apex, leathery, dark green and shiny above glaucous or slightly pubescent beneath, 2–3 in. long petioles about ⅛ in.

STEMS. Twining, green to red, glabrous, older stems

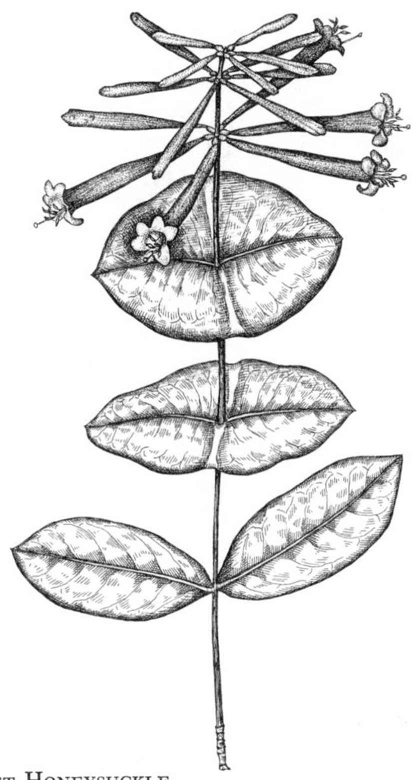

TRUMPET HONEYSUCKLE
Lonicera sempervirens L.

gray and fibrous when mature; lenticels numerous, small, black.

RANGE. Texas, Oklahoma, Arkansas, and Louisiana; eastward to Florida and north to Maine, Missouri, and Nebraska.

VARIETY AND FORMS. Hairy Trumpet Honeysuckle, *L. sempervirens* L. var. *hirsutula* Rehd., is a variety with a stem which is short-pilose and glandular or glabrous; leaves ovate and ovate-oblong to oval, obtuse at the apex, sparingly pilose above, especially near the margin, puberulous beneath, finely ciliate, 1½–3½ in. long, upper pair connate into an elliptic disk, often constricted in the middle; petioles, bractlets, and calyx glandular, or calyx glabrous and ovaries glandular; corolla hairy and glandular, about 1⅓ in. long.

The three most commonly grown horticultural forms are: *minor,* which is a smaller Southern form; *superba,* with scarlet flowers; and *sulphurea,* which is yellow-flowered.

REMARKS. The genus name, *Lonicera,* is for Adam Lonitzer (1528–1586), a German botanist, and the species name, *sempervirens,* means "ever-living," because of its evergreen habit. Vernacular names are Woodbine, Scarlet Trumpet, Honeysuckle, Red Honeysuckle, and Red Woodbine. The fruit is eaten by a number of species of birds, including the cardinal and purple finch. The vine is often cultivated for its beautiful flowers.

RUSTY BLACKHAW VIBURNUM
Viburnum rufidulum Raf.

FIELD IDENTIFICATION. An irregularly branched shrub or tree to 40 ft, with opposite, finely-serrate, shiny leaves.

FLOWERS. In flat cymes 2–6 in. across with 3–4 stout rays and minute subulate bracts and bractlets; corolla small, ¼–⅓ in. in diameter, regular; petals 5, rounded orbicular or oblong, white; stamens 5, attached to the corolla, exserted, anthers oblong and introrse; pistil with style absent and stigmas 1–3, sessile on the ovary, ovary 3-celled, only 1 cell maturing.

FRUIT. Ripe July–October, in drooping clusters, drupes ⅓–½ in. long, oblong to obovoid, bluish black, glaucous; seed solitary, flattened, oval to ovate, ridged toward one end.

LEAVES. Simple, opposite, deciduous, or half-evergreen southward, dark green, leathery, shiny above, paler below with red hairs on veins, margin finely serrate, elliptic to obovate or oval, apex rounded to acute or obtuse, base cuneate or rounded, 1½–4 in. long, 1–2½ in. broad, petiole grooved, wing-margined, clothed with red hairs, length ½–¾ in.

BARK. Rather rough, ridges narrow and rounded, fissures narrow, breaking into dark reddish brown or black squarish plates.

RUSTY BLACKHAW VIBURNUM
Viburnum rufidulum Raf.

TWIGS. Young ones gray with reddish hairs, older ones reddish brown, more glabrous.

WOOD. Fine-grained, hard, heavy, strong, with a disagreeable odor.

RANGE. In river-bottom lands or dry uplands. Oklahoma, Arkansas, Texas, and Louisiana; eastward to Florida, northward to Virginia, and west to Kansas.

MEDICINAL USES. The official drug Fluidextractum Viburni Prunifolii can be made from the dried bark of the roots or stems of either *Viburnum rufidulum* Raf. or *V. prunifolium* L. See section MEDICINAL USES under description of the latter plant.

PROPAGATION. The ripe fruits are picked by hand and either dried on a screen or run through a macerator with water and the pulp floated off. The dried seed may be stored in sealed containers at 41°F. for 2 years without loss of viability. Stratification for 180 days or more at 70°F. is helpful. Even stratification for as long as a year is recommended for better results because of the slowness of germination. The seed is sown usually in early fall in fertile, well-drained soil, and in drills 8–12 in. apart. Partial shade is recommended for the seedlings. Rusty Blackhaw Viburnum may also be propagated by cuttings of green or mature wood, or by layers.

VARIETIES. Margarette Rusty Blackhaw Viburnum, *V. rufidulum* Raf. var. *margarettae* Ashe, differs from the species by having very broad, often suborbicular leaves, somewhat cordate at base, and densely glaucous fruit ½–¾ in. long. First collected by W. W. Ashe at Groveton in eastern Texas where it is the common form.

Florida Rusty Blackhaw Viburnum, *V. rufidulum* var. *floridanum* Ashe, differs from the species mostly by the oblong leaves, and fruit ⅖–½ in. long and scarcely glaucous. It was first collected by W. W. Ashe in Walton County, Florida, in 1923.

REMARKS. The genus name, *Viburnum*, is the classical name of the Wayfaring-tree, *V. lantana* L., of Eurasia, which is often cultivated. The species name, *rufidulum*, refers to the rufous-red hairs on young parts. Vernacular names are Rusty Nanny-berry, Southern Nanny-berry, Blackhaw, Southern Blackhaw. The tree is worthy of cultivation because of the lustrous leaves, cymes of white flowers in April, and bluish black fruit in October. The wood has no value.

BLACKHAW VIBURNUM
Viburnum prunifolium L.

FIELD IDENTIFICATION. Shrub or small tree to 28 ft, with a diameter of 10 in., the stiff spreading branches forming an irregular crown.

FLOWERS. April–June, with the leaves or slightly before, cymes sessile or nearly so, compound, round-topped, white, 3–4-rayed, 2–4 in. wide; individual flowers ⅙–¼ in. wide; calyx 5-toothed; corolla-tube short, deeply 5-lobed, rotate and spreading, lobes suborbicular; stamens 5, filaments slender, exserted, inserted on the corolla-tube; anthers oblong, introrse; style absent; stigmas 3, minute and sessile; ovary 3-celled, 1 ovule usually maturing.

BLACKHAW VIBURNUM
Viburnum prunifolium L.

FRUIT. August–October, drupe ⅓–½ in. long, ellipsoid to subglobose, bluish black, glaucous or only slightly so, flesh thin and dry but sweet; seed solitary, flat, oval or elliptic, or one side slightly convex, anatropous, embryo small, albumen fleshy.

LEAVES. Simple, opposite, deciduous, 1–2¾ in. long, oval to ovate or obovate, or oblong to elliptic; apex acute to obtuse or rounded; base obtuse to rounded, margin finely serrulate; upper surface dull green (not lustrous), smooth, lower surface paler, membranaceous but later subcoriaceous; petiole green to red, slender, ¼–¾ in. long, not winged or only slightly so; buds slender, short-pointed, with small reddish fascicled hairs and whitish crystals, scales involute.

TWIGS. Slender, rigid, green to reddish or brown, some with short lateral spurs.

BARK. Gray to brown with narrow rounded ridges broken into short sections.

WOOD. Reddish brown, hard, weighs 52 lb per cu ft, of no commercial importance.

RANGE. Thickets, roadsides, borders of woods, and along streams. Oklahoma, Arkansas, Texas, and Loui-na; east to Florida, north to Connecticut, and west Michigan and Kansas.

MEDICINAL USES. A drug prepared from the powdered rk of the root or stem is officially known as Fluidex-ctum Viburni Prunifolii; at one time it was used in e treatment of uterine colic and as a general anti-asmodic. Evans and associates (*J. Pharm. Exper. ierapeutics*, 75:174, 1942; cited in Wood and Osol, 43, p. 1222) have separated a pharmacognostically entified sample of *V. prunifolium* and found a yellow ystalline glycoside which had a relaxing action on e uterine muscle. Munch (1940) has reported some periments on human beings, both from oral ad-inistration and local application, which indicate that e drug lowers both the contractility and the tonus the uterus. Since the bark is often adulterated, how-er, the findings appear in doubt, and the question hether or not *Viburnum* is therapeutically useful can-)t be positively answered.

PROPAGATION. Good crops of seed are borne every -2 years. The fruit may be picked by hand and spread it to dry, or run through a macerator with water and e pulp floated off. The seed may be dried and stored r 1–2 years in sealed containers a little above freezing. About 20–50 lb of seed are yielded from 100 lb of fruit. The seed averages about 5,000 per lb, with a purity of 98 per cent and a soundness of 95 per cent. The seeds are slow to germinate and dormancy may be broken by stratification in moist sand or peat for 150–200 days at 68°–86°F., plus 30–45 days at 41°F. The sowing is in drills 8–12 in. apart in spring or late summer, and covered about ½ in. with soil. The seed beds may be mulched with straw or leaves over the winter. The plant is also propagated by greenwood or hardwood cuttings or by layers.

Warnock Viburnum
Viburnum australe C. V. Morton

VARIETIES. Small-fruit Blackhaw Viburnum, *V. prunifolium* var. *globosum* Nash, is a variety with small globose fruit about ¼ in. in diameter. It occurs in New Jersey and Pennsylvania.

Bush's Blackhaw Viburnum, *V. prunifolium* var. *bushii* (Ashe) Palmer & Steyerm., has leaves membranaceous at maturity and oblong to lanceolate, their petioles somewhat winged. From southern Illinois to Arkansas.

Blackhaw Viburnum differs from Rusty Blackhaw Viburnum by the lack of rufescent tomentum on young leaves, petioles, and stems, and by the fact that the leaves lack luster and are less coriaceous.

REMARKS. The genus name, *Viburnum*, was the Latin name applied to the Wayfaring-tree, but is of uncertain origin. The species name, *prunifolium*, means "plum-leaved." It also is known under the vernacular names of Sheep-berry, Nanny-berry, Sweet-haw, Sweet-sloe, Stag-bush, and Arrow-wood. The plant has been cultivated for ornament since 1727. The fruit is sweet and edible and is known to be eaten by 12 species of birds, including bobwhite quail, and also by the gray fox and white-tailed deer.

Warnock Viburnum

Viburnum australe C. V. Morton

FIELD IDENTIFICATION. Rare shrub 4–6 ft, with angled, glabrous, grayish branches.

FLOWERS. Opening May–June, borne in twice-compound convex cymes to 2⅜ in. wide and 1¼ in. long, the primary rays 6–8 and densely glandular; peduncle to 1⅜ in. long, conspicuously glandular, also with a few simple or stellate hairs; bracts at base of inflorescence linear, to ⅝ in. long and 1/25–1/12 in. broad, glandular and somewhat pubescent with simple or stellate hairs, margin sparingly ciliate; calyx-lobes 5, about 1/25 in. long, broadly obtuse or acute, margins ciliate with long simple hairs, glabrous or slightly glandular dorsally; corolla 1/7–1/6 in. long, white, petals 5, oval to obovate, glabrous; stamens 5, slightly exserted, inserted at the base of corolla-tube, filaments glabrous and about 1/6 in. long; style glabrous, short, thick, stigmas 3; ovary inferior, adnate to the calyx-tube, 1-celled, ovule solitary.

FRUIT. Maturing in July, drupe much flattened, about

⅜ in. long, ⅓ in. wide and ⅛ in. thick, apex obtuse or rounded, or with a very short knobby point, base rounded, one side 3-sulcate with the central groove very slight, the lateral deeper, the other side 2-sulcate.

LEAVES. Deciduous, opposite, broadly ovate to oval, length to 2⅜ in., width to 1¾ in., apex acute or abruptly short-acuminate, base cordate to subcordate or rounded, margin with large coarse dentations except near the base, also ciliate with white hairs, upper surface somewhat strigose, especially along the veins, lower surface glandular and hairy on the veins and veinlets, and especially densely hairy in the venation axils; petioles grooved above, to ½ in. long, densely hairy and glandular; stipules about ⅕ in. long, glandular and hispidulous, linear, borne often 1/25–1/12 in. above the base of the petiole, but not always so.

TWIGS. Slender, angled, young ones reddish brown, somewhat glandular, older ones brown to pale gray and glabrous.

RANGE. Local in the mountains of west Texas and in Nuevo León and Coahuila, Mexico.

This shrub has been discovered growing in the United States for the first time by Barton H. Warnock of Sul Ross State College. He kindly lent the author herbarium specimens for examination. His sheet No. 6490, July 20, 1947, is a fruiting specimen collected in igneous soil at an elevation of 6,000 ft, on north slopes of Timber Mountain above Madera Springs, Davis

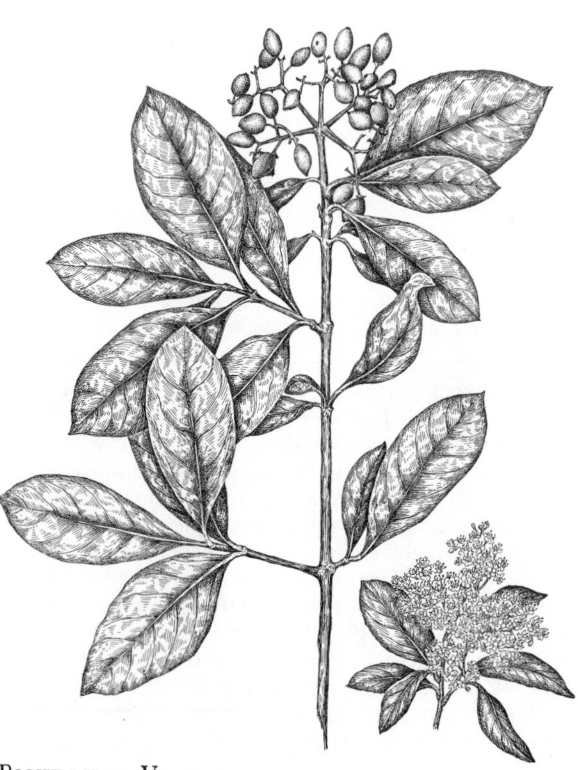

POSSUM-HAW VIBURNUM
Viburnum nudum L.

Mountains, Jeff Davis County, Texas. His sheets N 9067 *a* and 9067 *b* from the same locality, dated M 14, 1950, show mature flowers.

Besides the locations listed above the followi herbarium specimens from Mexico were examined:

(1) Type collected in the Sierra Madre abo Monterrey, Nuevo León, Mexico, at an altitude of ab 2,700 ft, April 25, 1906, by C. G. Pringle. No. 4622 in the National Herbarium.

(2) At General Cepida, Coahuila, Nelson's sh No. 6725, April 20, 1902. In the National Herbariu

(3) In the Caracol Mountains, 21 miles southe of Monclova, Coahuila, October, 1898, by E. J. Palm No. 388. In the Gray and National herbariums.

REMARKS. The genus name, *Viburnum,* is the classi Latin name of a tree of Eurasia. The species nan *australe,* refers to the plant's southern distribution. T species was first described by C. V. Morton, botan at the United States National Herbarium. He consider the plant to be closely related to *V. affine,* but distir enough for species rank because of its more deep cut calyx-lobes and the calyx-tube with large, cc spicuously stipitate glands, and because of the ha on both veins and veinlets beneath.

POSSUM-HAW VIBURNUM

Viburnum nudum L.

FIELD IDENTIFICATION. Irregularly branched shrub small tree, rarely over 20 ft.

FLOWERS. April–June, white, perfect, borne in fl or round-topped cymes 2–4½ in. across, prima peduncles ½–1 in. long, often with ovate, acute brac pedicels ¼–⅓ in. long; corolla about 3/16 in. broad, lobed; lobes spreading or reflexed, broad-ovate rounded, apex obtuse to rounded; stamens 5, exsert about ⅛ in. beyond the corolla; filaments white, fi form, with short-oblong anthers; pistil short, include at anthesis; calyx about 1/16 in. long, greenish, shor tubular, lobes short-ovate, obtuse or acute.

FRUIT. Mature in autumn, in cymes 2–4½ in. acros pink at first but glaucous-blue later, about ¼ in. diameter, subglobose to short-oblong or somewhat fla tened, wrinkling early, fleshy; seed black, compresse rounded, rugose, ridged down the middle, apices wi abrupt points; fruiting peduncles with rusty-brow scales, finely grooved, ½–1 in. long, rebranched in short pedicels about ¼ in. long.

LEAVES. Opposite, leathery, shiny, variable in siz shape, and scurfiness, oblong to elliptic or oval, blad 2–5 in. long, 1–2 in. broad, margin entire or obscure serrulate; apex acute, obtuse, acuminate, rounded abruptly pointed; base rounded or cuneate into a na row wing on petiole; olive-green to dark green an lustrous above; lower surface paler and with rust brown scales; petioles slender, somewhat winged, pu bescent and brown-scaly, about ½ in. long; buds us

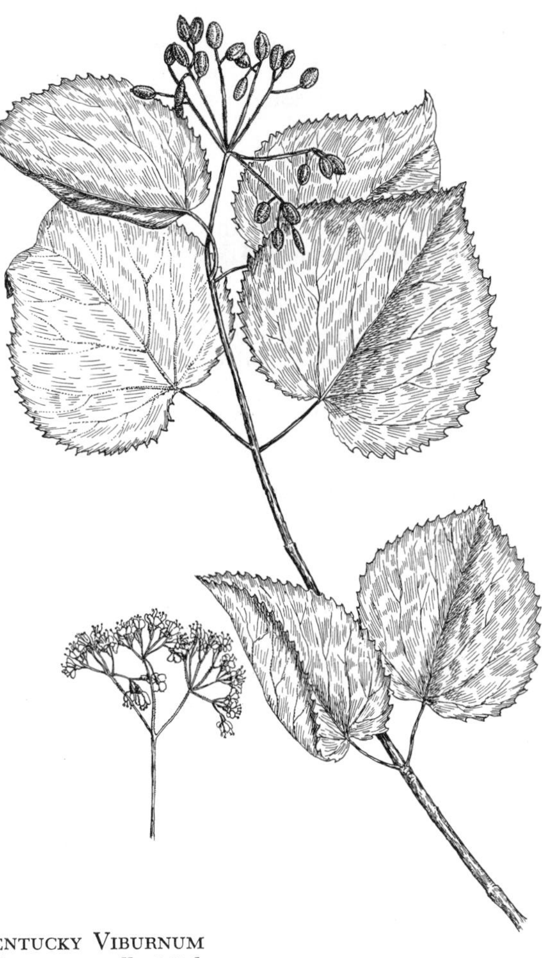

ENTUCKY VIBURNUM
iburnum molle Michx.

ly acuminate, ½–¾ in. long, with reddish brown ales.

BARK. Gray to brown, rather smooth.

TWIGS. Young ones gray to reddish brown or green, strous, finely grooved, rusty-brown, scaly; older twigs abrous.

RANGE. In sandy, acid swamps, usually in pinelands. ast Texas, Louisiana, Oklahoma, and Arkansas; eastard to Florida and north to Tennessee and Kentucky.

VARIETY. Shining Possum-haw Viburnum, *V. nudum* ır. *angustifolium* Torr. & Gray, is a variety with leaves ırrowly elliptic and sometimes less than 2 in. long and in. wide.

REMARKS. Viburnum is the classical name of a Eurian tree, and the species name, *nudum,* refers to the ther naked stems with remote leaves. Vernacular ımes in use are Withe-rod, Smooth Withe-rod, Naked /ithe-rod, Bilberry, Nanny-berry, and Swamp-haw iburnum. The sweet but unpalatable fruit is consumed y a number of species of birds, including the bobhite quail.

KENTUCKY VIBURNUM

Viburnum molle Michx.

FIELD IDENTIFICATION. Shrub 9–12 ft, with gray exfoliating bark.

FLOWERS. May–June, numerous, on short lateral or terminal branches; cymes 2–3⅓ in. broad, 4–7-rayed, peduncle 1¼–2 in. long, puberulent, often glandular; flowers perfect, corolla spreading, white, about ¼ in. across, 5-petaled; stamens 5, exserted, anthers oblong and introrse; stigmas 3, minute, sessile; ovary 3-celled at first, but only 1 cell maturing; ovule solitary.

FRUIT. Drupe maturing August–October, ellipsoid, bluish black, about ⅜ in. long, flattened, ends obtuse; pulp soft; stone solitary, crustaceous, somewhat grooved.

LEAVES. Simple, opposite, deciduous, suborbicular to broadly ovate, base cordate or truncate, apex short acute or acuminate sometimes abruptly so, margin coarsely dentate, length 2–5 in. and about as wide, upper surface dark green and glabrous or slightly pilose, lower surface pale and soft-pubescent (sometimes with reddish-glandular, simple or fascicled hairs); petioles slender, ⅝–2 in. long, glabrous or glandular-hairy; stipules small, early-deciduous, linear-filiform.

TWIGS. Young ones glabrous or sparingly glandular-pilose, older ones gray to black, or reddish brown after exfoliation.

RANGE. Often in calcareous soils on rocky hills or bluffs. Arkansas and Missouri, northeastward to Pennsylvania.

FORM. A form with the lower leaf surfaces glabrous, except on the primary veins, has been given the name of *V. molle* forma *leiophyllum* Rehd.

REMARKS. The genus name, *Viburnum,* is the ancient Latin name. The species name, *molle,* refers to the soft undersurface of the leaves. It is sometimes known as the Soft-leaf Viburnum. The seeds are eaten by a number of species of birds.

OZARK VIBURNUM

Viburnum ozarkense Ashe

FIELD IDENTIFICATION. Shrub 3–7 ft, with wandlike or dichotomously forked stems, usually several together. The following description follows that of Ashe (1928, pp. 463–464).

FLOWERS. About the middle of May when the leaves are nearly grown. Inflorescence a 5–7-rayed compound cyme, 1⅝–2½ in. wide, 1–1¼ in. high, and subtended by linear-lanceolate, ciliate bracts, ⅜–½ in. long, often equaling in length the rays and partly persistent; raylets with bracts 3/16–½ in. long, erect or ascending, largely persistent and much overtopping the flowers, which are white, about ⅓ in. across, the hairy corolla

Ozark Viburnum
Viburnum ozarkense Ashe

having deep rounded lobes; calyx-teeth short and ciliate; stamens about ⅗ in. long, anthers yellow.

FRUIT. Ripening the last of August. Bluish black, ⅓–½ in. long; seeds about two thirds the length of the fruit; flattened on 1 face and with 2 shallow lateral grooves, convex on the other with 2 deep lateral grooves separated by a prominent ridge.

LEAVES. Opposite, blades ovate or broadly ovate, 3¼–4⅜ in. long and 1⅜–3¾ in. wide, with a broad shallow sinus at base and abruptly taper-pointed at the apex, usually with 5–7 pairs of prominent veins, those above the much-branched basal pair rarely branching more than twice, with distant but shallow sinuate teeth, soon glabrous above, ciliate on margin and pubescent over entire lower surface, but more so on the veins, with short simple hairs; petioles ¾–1½ in. long, channeled, but at length seeming terete from incurving of the edges, granular-glandular and ciliate, with 1–2 pairs of linear ciliate deciduous stipules about ⅜ in. long near the base.

TWIGS. Obtusely angled, glabrous, grayish brown, becoming gray the second year, and marked by few small lenticels; pith solid, occupying about one third of the diameter of the winter twig of the year; winter buds short, broadly conical, ⅛–⅙ in. long, of about pairs of opposite scales, the outer short and glabrou the inner pubescent, ciliate, more or less granular-gla dular and accrescent in estivation.

RANGE. Frequent along streams, particularly on sanc soils in Stone County, Arkansas. Type from Sylamor

REMARKS. The genus name, *Viburnum*, is the classic name of a Eurasian tree, and the species name, *ozar ense*, refers to its location in Arkansas. This species a parently has been confused with *V. molle* Michx. Missouri, Kentucky, and Illinois, from which it is se arated by having prevailingly oval or broadly ova leaves of relatively longer and narrower outline, thos of *V. molle* being nearly orbicular in outline an abruptly acuminate; by its coarser venation, 5–7 pai of prominent veins, *V. molle* prevailingly having 8–1 by its coarser serration, shorter petioles, longer pe duncles, glandular pubescence, and bracted cymes.

Rafinesque Viburnum

Viburnum rafinesquianum Schultes

FIELD IDENTIFICATION. Rather loose straggling shru to 6 ft.

FLOWERS. May–June, borne in cymes ⅝–2½ in. broad individual flowers small and numerous, $\frac{3}{16}$–¼ in. acros corolla white, 5-lobed, lobes rounded, spreading; caly tube much longer than the 5 short broadly-acute lobes stamens 5, exserted, ovary subglobose to ovoid, stigma short.

FRUIT. July–October, ellipsoid to subglobose, dar purple, ¼–⅖ in. broad, pulp soft; stone crustaceou brown, flattened, margins rounded, faces somewha sulcate.

LEAVES. Opposite, 1–3 in. long, ovate to oblong, ape acute or acuminate, base cordate to subcordate or round ed, subsessile, or petioles to ¼ in. long; some leave with linear-subulate stipules; margins coarsely dentat upper surface dull green, glabrous or slightly hairy lower surface soft and densely hairy; veins rathe straight and conspicuous.

TWIGS. At first light to dark brown and pubescent t glabrous, eventually gray and glabrous; winter bud with 2 pairs of outer scales.

RANGE. The species is usually found in calcareou soil of dry slopes or open woods in Arkansas, Missour and Kentucky, northward into Quebec and Manitoba

VARIETY. *V. rafinesquianum* var. *affine* (Bush House is a variety described as having leaves glabrou or pilose on nerves beneath. It is known to occur i Arkansas, north into Virginia, Minnesota, and Ontario

REMARKS. The genus name, *Viburnum*, is the classica name of a Eurasian tree. The species name, *rafines quianum*, honors its discoverer, Constantine Samue

RAFINESQUE VIBURNUM
Viburnum rafinesquianum Schultes

Rafinesque (1783–1840). It is also listed in some literature as *V. pubescens* (Ait.) Pursh and *V. affine* var. *hypomalacum* Blake. The vernacular names of Hairy Nanny-berry and Downy Arrow-wood are used in some areas.

MAPLE-LEAF VIBURNUM

Viburnum acerifolium L.

FIELD IDENTIFICATION. Shrub attaining a height of 2–6 ft, sometimes forming thickets. Branches erect or ascending and rather slender.

FLOWERS. Maturing May–August, in 3–7-rayed cymes which are convex or flattened and ¾–3½ in. wide; peduncles slender, ⅝–2¾ in. long, pubescent with simple or stellate hairs; corolla small, ⅙–¼ in. wide, creamy-white, 5-lobed; stamens 5, long-exserted, filaments ⅛–⅙ in. long, inserted on the corolla-tube, anthers oblong and introrse; ovary 3-celled at first, usually only 1 ovule maturing, style none, stigma sessile on the ovary summit; calyx with the 5 sepals reniform and about 1/50 in. long.

FRUIT. Maturing July–October, drupes persistent, reddish to purplish black at maturity, about ⅓ in. long, globose to oblong or ellipsoid, flesh thin and purplish; stone lenticular, flattened, crustaceous, 3-grooved on one side and 2-grooved on the other, endosperm fleshy, embryo minute.

LEAVES. Simple, opposite, deciduous, blades suborbicular to ovate, margin coarsely dentate and often also with 3 acute to acuminate lobes, base rounded or subcordate, length or width 1¼–5 in., venation palmately 3-ribbed, surfaces usually dull green and pubescent on both sides, lower surface more densely so and often conspicuously dotted, sometimes with tufted axillary hairs, thin; petioles ⅜–1 in., slender, pubescent, often stipulate at base.

TWIGS. Slender, tan to reddish brown, finely stellate-pubescent, later gray and glabrous.

RANGE. Dry sandy or rocky woods. Arkansas, Texas, and Louisiana; eastward to Florida, northward to Quebec, and west to Ontario and Minnesota.

PROPAGATION. The fruit is collected when ripe and spread out to dry to prevent heating. The fruit may be planted as is, or the seed removed by macerating in water and floating off the pulp. If stored, the dry seed can be kept in a sealed jar at 41°F. for as long as 2 years. About 30 lb of seed are yielded from 100 lb of fruit, with an average of 13,200 seeds per lb.

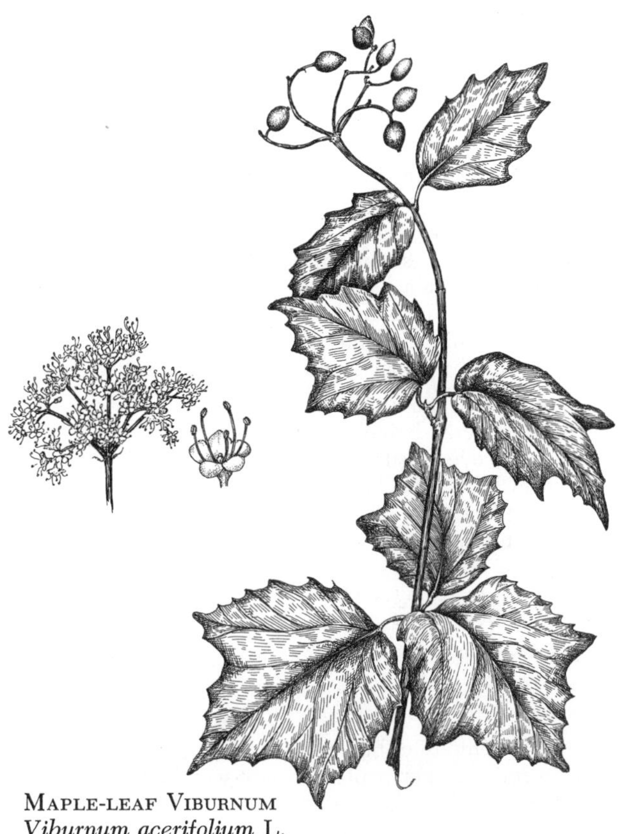

MAPLE-LEAF VIBURNUM
Viburnum acerifolium L.

They have a soundness of 96 per cent. Dormancy may be partially broken by stratification in moist sand or peat for 180 days at 68°–86°F., plus 60–120 days at 41°F. Planting is usually done in drills 8–12 in. apart and covered ½ in. with soil. A mulch of straw is removed after germination begins. The beds are protected from birds or rodents by screens. Propagation may also be practiced by greenwood or hardwood cuttings or from layers.

VARIETIES AND FORMS. White-fruit Maple-leaf Viburnum has been named *V. acerifolium* forma *eburneum* House.

Pink-flower Maple-leaf Viburnum is *V. acerifolium* forma *collinsii* Rouleau.

Smooth Maple-leaf Viburnum, which is glabrous except along the veins, has been named *V. acerifolium*, var. *glabrescens* Rehd.

Short-lobe Maple-leaf Viburnum, *V. acerifolium* var. *ovatum* Rehd., has leaves with short lobes.

REMARKS. The genus name, *Viburnum*, is the classical name of a Eurasian tree, and the species name, *acerifolium*, refers to the maple-like foliage. It is also known under the vernacular names of Dockmackie, Arrow-wood, Possum-haw, Squash-berry, and Guelder-rose. The plant has been cultivated since 1736. The creamy-white cymes of flowers in summer and dark purple to crimson foliage in fall make it attractive. The foliage endures city smoke well. The fruit is eaten by at least 4 species of birds, including ruffed grouse, and the leaves are sometimes browsed by white-tailed deer and cottontail.

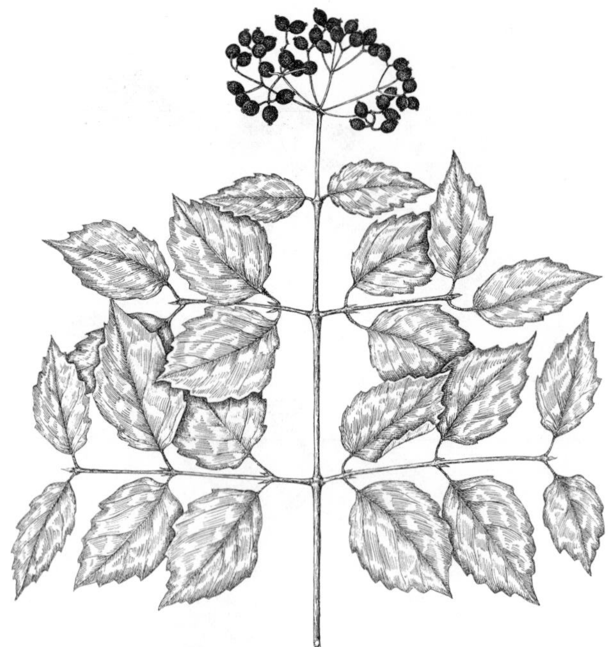

ARROW-WOOD VIBURNUM
Viburnum dentatum L.

ARROW-WOOD VIBURNUM

Viburnum dentatum L.

FIELD IDENTIFICATION. Shrub attaining a height of 3–15 ft. Branches slender, elongate, ascending or semipendent. Main stem either solitary or numerous from a clumped base. Very variable in size and shape of leaves and cymes and amount of pubescence, and hence divided into a number of varieties by authors.

FLOWERS. Maturing June–August, cymes terminal or on short lateral branches, flower parts more or less pubescent; cyme usually 5–7-rayed, 1¼–4½ in. broad; peduncle 1⅛–2⅜ in. long; corolla small, white, regular, spreading, rotate, 5-lobed; stamens 5, exserted; anthers oblong, introrse; style very short or mostly absent, the 3 stigmas minute and sessile; ovary 3-celled, only 1 cell with a maturing ovule.

FRUIT. Drupe ripening August–November, bluish black, subglobose to ovoid, pulp soft, ⅕–⅖ in. long; stone solitary, crustaceous, ellipsoid to ovoid, deeply grooved on the ventral side.

LEAVES. Opposite, blade 1–4½ in. long, ovate-lanceolate to rotund, apex acute or short-acuminate (in some forms rounded), base rounded to subcordate, margin serrate to dentate, the teeth rather triangular; usually thick and firm, especially in sun-exposed plants of dry situations; upper surface glabrous or nearly so, or with stellate pubescence; veins 5–11 pairs, straight or nearly so, conspicuous beneath the leaf blade; petioles very slender, ⅓–1¼ in. long, glabrous to stellate-pubescent.

TWIGS. Slender, elongate, straight or arching, young ones pubescent, older ones glabrous, bark gray to grayish brown or reddish brown.

RANGE. Arrow-wood Viburnum and its forms occur in many types of soil, but mostly in moist sandy lands. Arkansas, eastern Texas, and Louisiana; eastward to Florida and northward to Massachusetts.

VARIETIES. This species is very variable and many of the named varieties are connected by intermediate forms. Probably the most constant variety within our range is *V. dentatum* var. *scabrellum* Torr. & Gray, which has the hypanthium and corolla hirsute or setose. It is found in east Texas and Louisiana, eastward to Florida. A number of other varieties are found north of our area and have been listed as *V. dentatum* var. *lucidum* Ait., *V. dentatum* var. *venosum* (Britt.) Gl., *V. dentatum* var. *deamii* (Rehd.) Fern., and *V. dentatum* var. *indianense* (Rehd.) Gl. These are described by Gleason (1952, p. 295).

REMARKS. The genus name, *Viburnum*, is the classical name. The species name, *dentatum*, refers to the coarsely toothed margins of the leaf. It is also known under the vernacular names of Southern Arrowwood, Mealy-tree, Withe-rod, and Withe-wood. Indians were known to have made arrows from the straight stems.

COMPOSITE FAMILY (*Compositae*)

Bigelow Sage-brush

Artemisia bigelovii Gray

FIELD IDENTIFICATION. Shrub 18 in. or less, with many slender, erect flowering stems. Whole plant mildly and pleasantly fragrant.

FLOWERS. September–October, panicles spikelike, 3–8 in. long, ⅜–1½ in. broad; ray and disk flowers (more rarely ray or disk separate), sessile, several heads together; involucre 1/12–1/7 in. high, 1/12–1/10 in. broad; bracts 8–12, variable in shape, ovate to oblong, apex obtuse, outer thicker, margins thin; ray flowers 1–2, or absent; corolla 1/25–1/15 in. long, cylindric, narrowed toward the apex, teeth minute; disk flowers 2 (more rarely 1 or 3), fertile, funnelform, 5-toothed, 1/12–⅛ in. long, glabrous; styles of 2 forms, those of the ray flowers exserted and bifid, in disk flowers included and the 2 divisions erose-truncate.

FRUIT. Achenes of ray and disk flowers ellipsoid to oblong, glabrous, 5-ribbed.

LEAVES. Sessile or very short-petioled, linear to linear-cuneate or narrowly elliptic, length ⅜–⅘ in., width 1/12–⅙ in.; margin entire except many acutely 3-toothed at the apex, the balance acute; surfaces silvery canescent-hairy.

TWIGS. Numerous, spreading toward the base, younger silvery-canescent, those bearing the inflorescence slender and erect.

RANGE. Mostly on stony grounds of juniper and pinyon savannas at altitudes of 2,000–7,000 ft, forming dominant societies where not crowded out by sod grasses. Type from the upper Canadian River in Texas. New Mexico, Arizona, Colorado, Utah, and California.

REMARKS. The genus name, *Artemisia,* is for the wife of Mausolus, king of Caria. The species name, *bigelovii,* honors J. M. Bigelow, a botanist who participated in the United States–Mexico boundary survey in 1850.

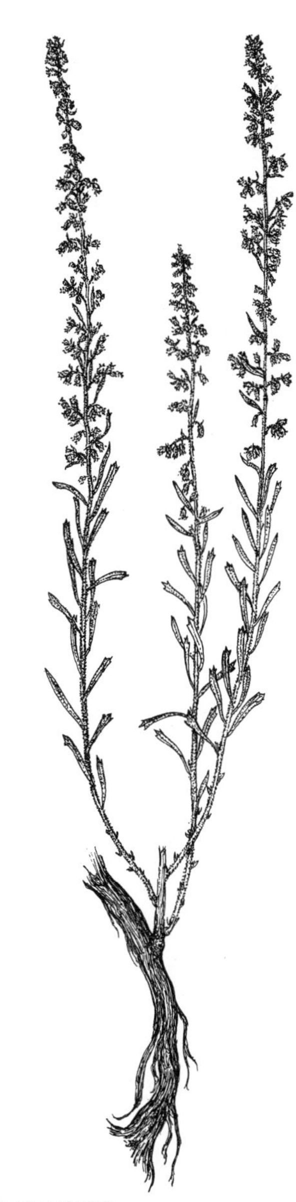

Bigelow Sage-brush
Artemisia bigelovii Gray

Another local name is Flat Sage-brush. The plant is not sufficiently abundant to be of widespread grazing benefit, but is very palatable to livestock, especially goats. It is very closely related to the larger and more northerly distributed Big Sage-brush, *A. tridentata* Nutt.

Silver Sage-brush

Artemisia cana Pursh

FIELD IDENTIFICATION. Shrub sometimes forming extensive colonies. The stems much branched to form a rounded bush 1½–5 ft.

FLOWERS. Borne in a leafy panicle 6–12 in. long and ¾–2½ in. wide; heads in glomerules, erect and sessile; involucre campanulate, ⅙–⅕ in. high, ⅛–⅙ in. broad; bracts 8–15, the outer series orbicular or narrower above, the inner series elliptic-spatulate, obtuse, margin scarious, surfaces canescent or tomentose; ray flowers absent; disk flowers 6–20, fertile; corolla tubular-funnelform, 1/12–⅛ in. long, 5-toothed at apex, surface resinous-glandular; style branches with lacerate margins.

FRUIT. Achenes cylindric-turbinate, truncate at apex, 4–5-ribbed, granuliferous.

LEAVES. Blade linear, apex acute, base sessile, margin entire or with 1–2 irregular teeth or lobes, length ¾–2 in., width 1/25–⅙ in., surface of leaves canescent, less so with age.

TWIGS. Young ones green to straw-colored and finely canescent or tomentose; older ones brown to gray and glabrous, bark fibrous.

RANGE. Dry soils of the Great Plains area. New Mexico, California, Colorado, Utah, Wyoming, North Dakota, Nevada, Oregon, and Alberta.

REMARKS. The genus name, *Artemisia,* is in honor of Artemisia, wife of Mausolus. The species name, *cana,* refers to the silvery-gray or hoary pubescence. The plant is also known under the name of White Sage-brush or Hoary Sage-brush. It is cultivated in England for the silvery leaves and stems. It is considered to be an important browse for sheep, cattle, and horses.

Silver Sage-brush
Artemisia cana Pursh

Fringed Sage-brush

Artemisia frigida Willd.

FIELD IDENTIFICATION. Plant semierect or decumbent and mat-forming, mostly herbaceous but sometimes decidedly woody toward the base. Seldom over 2 ft, foliage with a pleasant fragrance.

FLOWERS. On narrow racemose branches 4–12 in. long and ⅜–4 in. broad; heads sessile, or short-peduncled, nodding; involucre rounded, 1/12–⅛ in. high, ⅙–⅕ in. broad; bracts 11–19, long-hairy, outer linear, inner lanceolate to ovate, margins pale and thin; receptacle densely long-hairy; ray flowers 10–18, fertile, corolla narrowed above, about 1/25 in. long; disk flowers 25–50, fertile, corolla funnelform, granular, 1/16–1/12 in. long, 5-toothed; ray flowers with obtuse style branches, but truncate and fringed in the disk flowers, pollen smooth, 3-lobed.

FRUIT. Achenes almost cylindric, base narrowed, apex truncate to rounded, surface barely ribbed or smooth, glabrous.

LEAVES. Crowded below, but essentially alternate, short-petiolate or sessile, outline rounded, 1/16–⅜ in. long, dissected three to five times into linear to narrowly oblanceolate, acute segments, surfaces white-hairy; base of petiole with simple or 3-parted stipular-like structures.

TWIGS. Branched from the base, semierect or decumbent and mat-forming; herbaceous upper stems very leafy and silky-hairy; lower ones brown to gray and more glabrous.

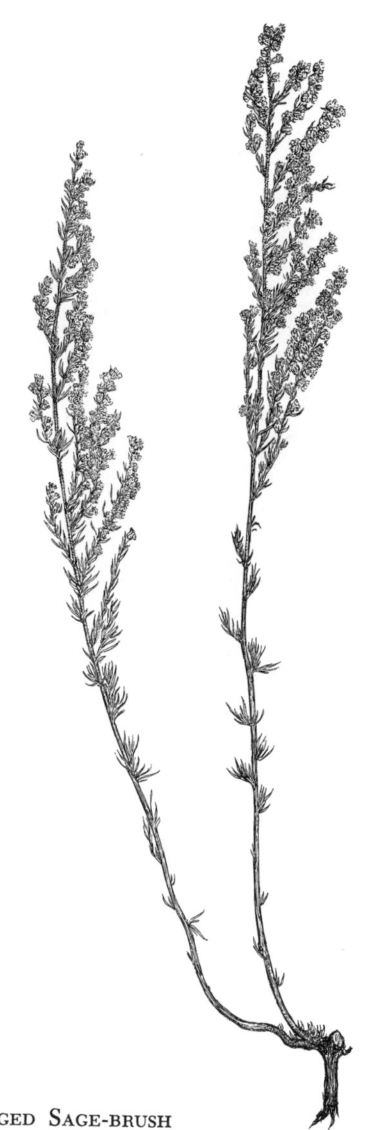

FRINGED SAGE-BRUSH
Artemisia frigida Willd.

RANGE. On dry, stony soil to an altitude of 7,000 ft. Widely distributed in western Texas, New Mexico, Arizona, Colorado, Utah, North and South Dakota, Nebraska, and Idaho; northward into Canada, Alaska, and northeast Asia. Introduced into Nova Scotia and New Jersey.

MEDICINAL USES. It was also formerly much used in pioneer days in the Western states as a diuretic and mild cathartic. The leaves contain 0.26–0.41 per cent of a fragrant essential oil which contains borneol camphor and cineole, these possessing valuable antiseptic properties. The pollen causes some hay fever and is used in remedial extracts.

REMARKS. The genus name, *Artemisia,* is for the wife of Mausolus, king of Caria. The species name, *frigida,* is of uncertain application, perhaps referring to its habitat in high, cold mountain valleys. Other vernacular names in use are Estafiata, Sierra Salvia, Rocky Mountain Sage, Wild Sage, Colorado Sage, Pasture Sage-brush, and Mountain Worm-wood. It is reported that the Indians roasted the leaves with sweet corn to flavor it. The plant ranks high as a fall and spring forage for livestock in the Western mountain region. It is also of value as a food for wildlife; the seeds are eaten by sage grouse, and the plant winter-grazed by elk and bighorn sheep. It is an excellent indicator of overgrazing.

SAND SAGE-BRUSH
Artemisia filifolia Torr.

FIELD IDENTIFICATION. Rounded, aromatic shrub, freely branched, usually less than 4 ft high.

FLOWERS. Borne in dense leafy panicles 3–12 in. long and ⅜–2 in. wide; heads small, peduncles short and stiffly pendent; involucre 1/25–1/12 in. long or broad;

SAND SAGE-BRUSH
Artemisia filifolia Torr.

bracts 5–9, densely silky-hairy, outer short and indurate; inner thin, broadly elliptic, apex obtuse; receptacle glabrous and smooth; ray flowers 2–3, fertile, corolla tubular, about 1/25 in. long, anthers longer than the filaments; disk flowers 1–6, sterile, corolla funnelform, 1/25–1/16 in. long, 5-lobed; style 1/25–1/16 in. long, apex fused or bifid, stigmatic margins erose.

FRUIT. Achenes ellipsoid, faintly or conspicuously 4–5-ribbed, surface glabrous, pappus absent, fruits of disk flower often abortive, fruiting heads often gall-forming.

LEAVES. Alternate, sessile, often fascicled, blade length 1½–3 in., width 1/25 in., upper leaves usually entire, lower often divided into threadlike divisions, silky-hairy, incurved or ascending.

TWIGS. Slender, striate, pubescent to canescent, older twigs glabrous, dark gray to black.

RANGE. An indicator of sandy soils, to an altitude of 6,000 ft. In Texas abundant in the lower Panhandle sandy lands, also in New Mexico, Arizona, Nevada, Colorado, Utah, and Chihuahua, Mexico.

REMARKS. The genus name, *Artemisia,* is for the wife of Mausolus, king of Caria. The species name, *filifolia,* refers to the threadlike leaves. It is also known as Sand Worm-wood. In Mexico a decoction of the leaves is used for intestinal worms and affections of the stomach. The plant is palatable to livestock, but is not grazed when grasses are present. The seed germinates 40–50 per cent.

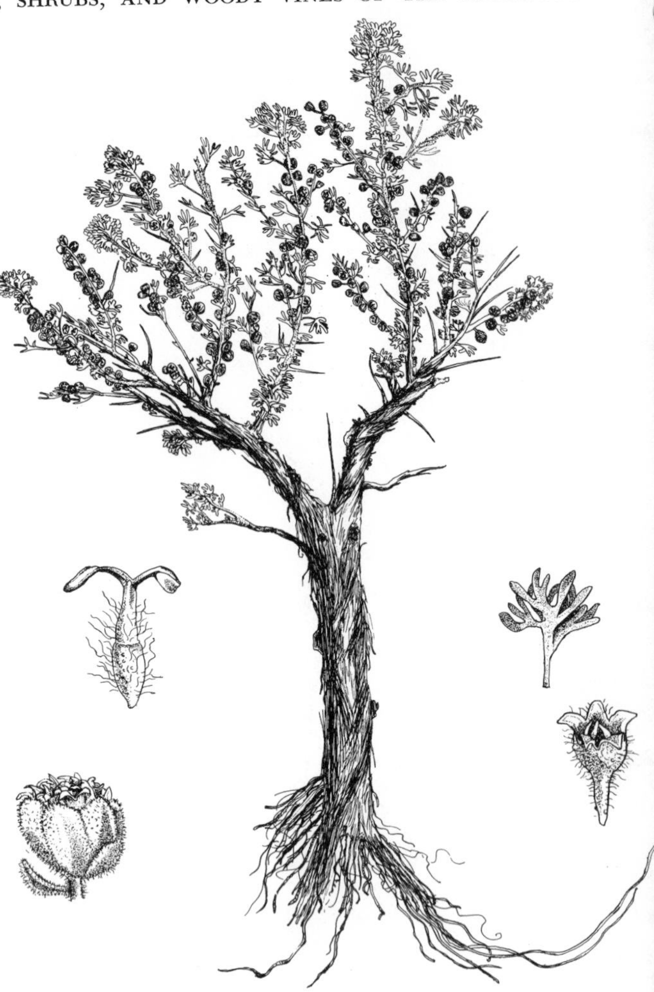

BUD SAGE-BRUSH
Artemisia spinescens (DC.) Eaton

BUD SAGE-BRUSH

Artemisia spinescens (DC.) Eaton

FIELD IDENTIFICATION. Western shrub 1–2 ft, rounded, much-branched, rigid, spinescent. Parts of the current season densely white-tomentose, strongly odorous when bruised. Root system widespread.

FLOWERS. Maturing March–June, borne on short leafy-bracted branchlets which later become slender, naked, spinescent twigs; racemes composed of small, glomerate, nodding, discoid, subsessile or short-pedunculate heads; involucre broadly turbinate, 1/8–1/6 in. broad; bracts 4–8, broadly obovate to orbicular, apex obtuse, base cuneate, thickened, margins scarious, surfaces densely hairy; receptacle naked; ray flowers 2–6, no stamens present but fertile ones; the shorter corollas are hairy, about 1/25 in. long, 2–3-toothed; disk flowers 5–13, sterile, corolla hairy, ventricose-campanulate, 5-toothed, 1/12–1/8 in. long; style of disk flower undivided at the expanded and radiately penicillate summit.

FRUIT. Achenes from ray flowers only, oblong-obovate or ellipsoid, densely long-hairy.

LEAVES. Simple, alternate, outline fan-shaped, 3–5-parted into linear-spatulate lobes or these again divided, upper leaves usually less divided, length ¼–¾ in.; width ¼–⅝ in.

TWIGS. Older ones much branched from the base, thickened and rigid, bark brown and fibrous; younger ones ascending, white-tomentose to short-hairy, those of inflorescence often becoming spine-tipped after the flowers have fallen.

RANGE. On dry sunny sites of plains and slopes, in alkaline soils, at altitudes of 4,000–8,500 ft; New Mexico, west to California and north to Montana, Idaho, and Oregon.

REMARKS. The genus name, *Artemisia,* is in memory of the wife of Mausolus. The species name, *spinescens,* refers to the spine-tipped flowering twigs. The name of Bud Sage-brush comes from the early development of foliage before other shrubs in some areas. It is considered to be very valuable as early browse for sheep in Utah, Nevada, and eastern California. Occasionally thought to be poisonous for calves when not browsed with other plants. The pollen is known to cause hay fever in the Great Basin area. The plant is resistant to drought or overgrazing. It is unique among artemisias for its spinescent twigs, which become lignified and hence more or less persistent, and it is subject to little variation over a wide area.

Big Sage-brush

Artemisia tridentata Nutt.

FIELD IDENTIFICATION. The most widely distributed shrub of the western United States. Varying in height, 1½–10 ft, or in favorable situations even becoming a small tree. Often dwarfed or prostrate as a result of cropping by cattle. Stem mostly erect with ascending branches. Young parts silvery-canescent, aromatic and bitter to the taste.

FLOWERS. Borne in panicles 4–10 in. long and ¾–4 in. wide, sometimes spikelike, often leafy-bracted; heads numerous, sessile or nearly so, erect or slightly drooping; involucre ovoid or campanulate, ⅛–⅕ in. high, 1/25–⅙ in. wide; bracts 8–18, the outer ones short, orbicular to ovate, oval, or somewhat narrowed; inner bracts oblong to elliptic or spatulate, margin scarious, canescent to glabrous; ray flowers absent, disk flowers 3–12 (rarely more in the subspecies), fertile; corolla funnelform, 5-toothed, 1/12–⅛ in. long, resinous; styles with 2 flattened apical divisions which are enlarged and erose.

Big Sage-brush
Artemisia tridentata Nutt.

FRUIT. Achenes cylindric-turbinate, border raised, 4–5-ribbed, resinous-granuliferous.

LEAVES. Very leafy, sessile or slightly petioled, cuneate or flabelliform, narrowed at base, apex rounded or truncate and 3–7-toothed (usually 3), sometimes the teeth deeply cleft, surfaces finely silvery-canescent; ⅜–1¾ in. long, 1/12–⅕ in. wide; upper leaves linear, acute, mostly entire.

TWIGS. Slender, stiff, at first gray or white with dense tomentum, later gray to tan or brownish to black, and separating into small scales or fibrous.

RANGE. Dry and stony soils usually in deep soil pockets. Very widely distributed in the West, especially on the arid plains of the Great Basin, but up to timber line in the mountains. Also British Columbia and northern Mexico, Texas, New Mexico, Arizona, Colorado, California, North Dakota, Montana, Wyoming, and Washington.

MEDICINAL USES. The pollen is known to cause hay fever, and an extract of it is used in treatment of that malady. In domestic medicine, the plant is used as a diaphoretic, antiperiodic, or laxative.

CLOSELY RELATED TAXA. Because of its wide distribution, *A. tridentata* varies considerably, and the following segregates have been named by Hall and Clements (1923):

The typical species is tall with broad inflorescence; heads medium, 1/12–1/10 in. broad; 4–6-flowered; involucre canescent; leaves 3-dentate.

A. tridentata parishii Gray, is tall, heads small; 4–6-flowered; inflorescence various; achenes villous.

A. tridentata nova Nelson, is low; heads small, about 1/12 in. broad, 3–6-flowered; involucre greenish; leaves dentate.

A. tridentata trifida Nutt. is low; heads small, 1/12–1/10 in. broad, 5–8-flowered; involucre canescent; leaves trifid.

A. tridentata arbuscula Nutt., is low; heads large, 1/10–⅛ in. broad, 5–9-flowered; leaves less than ⅝ in. long.

A. tridentata rothrocki Gray, is of medium height; heads large, ⅛–⅙ in. broad, 6–20-flowered; leaves over ⅝ in. long.

A. tridentata bolanderi Gray, is low; heads large, 1/10–⅛ in. broad, 8–15-flowered; tomentum loose and white.

REMARKS. The genus name, *Artemisia*, is for the wife of Mausolus, king of Caria. The species name, *tridentata*, refers to the 3-toothed leaves. It is also known as Black Sage. The plant is of fairly rapid growth and is an indicator of soils largely free of alkalinity. Root sprouts are sometimes formed. The germination of the seed is 35 per cent or less. The plant has been cultivated since 1881, but has never been a popular ornamental.

The wood makes a quick hot fire. The plant is a range browse of first importance, especially in autumn and winter, and carries large numbers of cattle. It is apparently not much eaten by horses, and is persistent under close grazing by sheep. The brittle branches are sometimes used for thatch or temporary sheds. The Cahuilla Indians of California ground the seeds into meal which was made into a kind of pinole. Many kinds of birds and mammals eat the seed or browse upon the plant, including the dusky grouse, sharp-tailed grouse, sage grouse, black-tailed deer, mule deer, and bighorn sheep.

Arizona Baccharis

Baccharis thesioides H. B. K.

FIELD IDENTIFICATION. Shrubby dioecious plant to 6 ft.

FLOWERS. August–September, borne in numerous, rounded, terminal, paniculate cymes; heads 1/6–1/4 in. high, discoid, numerous; bracts of involucre imbricate, graduated, 1/25–1/8 in. long, linear to narrowly oblong, pale, scarious-margined; staminate heads with hermaphrodite infertile flowers, corollas tubular, 5-toothe pistillate heads composed entirely of fertile tubula filiform flowers.

FRUIT. Achenes about 1/16 in. long, glabrous or gla dular, ribbed; pappus in the pistillate flowers 1/8–1/6 i long, of copious, persistent, capillary bristles; papp in the staminate flowers of stiffer, shorter, scabrou clavellate bristles.

LEAVES. Simple, alternate, deciduous, length 3/4–3 in., width 1/15–1/3 in., varying from linear-lanceolate t linear-oblanceolate, apex obtuse to acute, or rounde base gradually narrowed into a short petiole, marg sharply spinose-serrulate, dull green above and beneat porous, often glutinous, 1–3-nerved.

TWIGS. Slender, erect or ascending, angular grooved, glabrous and glandular, green at first, old bark brown.

RANGE. On rocky slopes at altitudes of 4,000–8,00 ft. Western Texas, southern New Mexico, and Arizon In Mexico in Sonora and Jalisco.

REMARKS. The genus name, *Baccharis*, is from th Greek word *bakcharis*, but the name applied to som other plant. The species name, *thesioides*, is from th Greek *thes* ("divine"), but the application is obscur

Arizona Baccharis
Baccharis thesioides H. B. K.

Narrow-leaf Baccharis

Baccharis angustifolia Michx.

FIELD IDENTIFICATION. Shrub attaining a height o 1–12 ft, with numerous resinous branches.

FLOWERS. Maturing in autumn, dioecious; head numerous, solitary on short peduncles or 2–4 i peduncled clusters, the whole broadly paniculate; in volucres 1/6–1/5 in. long, campanulate, many-flowered bracts in several series, ovate to oblong-lanceolate, in ner elliptic, obtuse, those of the staminate involucr fewer in number than those of the pistillate; receptacl flat and naked; staminate corolla tubular, with 5 lanceo late lobes; pistillate corolla filiform, stigmas slender.

FRUIT. Achene 1/25–1/15 in. long, flattened, 10-ribbed pistillate pappus copious, white, soft, fine, flaccid, twic as long as the involucre; staminate pappus considerabl shorter than that of the pistillate.

LEAVES. Numerous, simple, alternate, narrowly linea 3/8–3 1/4 in. long, margin entire or nearly so, apex acute base sessile; some lower leaves broadly lanceolate an more serrate; surface glabrous and some glutinous.

TWIGS. Young ones green to tan and glutinous, olde ones brown to gray and less glutinous.

RANGE. Low places in coastal marshes and brackis swamps, or sometimes along streams in eastern an south central Texas. Louisiana eastward to Florida an north to North Carolina.

NARROW-LEAF BACCHARIS
Baccharis angustifolia Michx.

REMARKS. The genus name, *Baccharis*, is from the Greek word *bakcharis*, but the name was probably applied to some other plant. The species name, *angustifolia*, refers to the narrow leaves.

SEEPWILLOW BACCHARIS

Baccharis glutinosa Pursh

FIELD IDENTIFICATION. Western shrub of stream banks, forming thickets in clumps, 3–12 ft, with wandlike stems from the base, roots wide and spreading.

FLOWERS. Borne February–May, in staminate and pistillate flowers on different plants; peduncles terminal, ½–3 in. long; flowers yellow, cymose-paniculate, discoid, ray flowers absent; pistillate heads with pappus 1/5–1/4 in. long in fruit, corolla tubular-filiform; staminate heads 1/8–1/5 in. long, corolla tubular, 5-lobed, infertile, anthers united; involucre imbricate in several series; bracts thin, chartaceous, scarious-margined, yellowish, outer bracts ovate, inner bracts oblong-lanceolate.

FRUIT. Achene 4–5-ribbed longitudinally, pappus of white capillary bristles, scanty and twisted in the staminate plant; more abundant in the pistillate, about 3/16 in. long, rigid, minutely scabrous.

LEAVES. Willow-like, simple, alternate, linear to lanceolate, apex acuminate, gradually narrowed into a short petiole at base, margin entire or with coarse remote teeth, length 2½–6 in.; surfaces bright green, glabrous, lustrous, sticky-glutinous, conspicuously 3-nerved.

TWIGS. Herbaceous above, woody below, wandlike, slender, erect, leafy, green to brown, striate to smooth.

RANGE. Along streams, west Texas, New Mexico, Arizona, California, Mexico, and southward to Chile. From almost sea level to an altitude of 5,000 ft.

REMARKS. The genus name, *Baccharis*, is from the

SEEPWILLOW BACCHARIS
Baccharis glutinosa Pursh

Greek word *bakcharis*, but was applied to some other plant. The species name, *glutinosa*, refers to the sticky glutinous leaves. Also known by the common names of Sticky Baccharis, Water Wally, and Batamote. The plant has been recommended for erosion control along streams. The Indians prepared an eyewash from the resinous leaves. It has no forage value. Cuttings from the long, wandlike stems readily strike root.

Broom Baccharis

Baccharis sarothroides Gray

FIELD IDENTIFICATION. Much-branched, broomlike, evergreen shrub 3–9 ft. The leaves are few and small, and the flowers are borne on separate staminate and pistillate plants.

FLOWERS. Borne August–September, heads discoid, mostly solitary, on elongate, nearly leafless branches forming a dense panicle; receptacle not chaffy, flattened; involucral bracts much imbricate, acute, oval to ovate, scarious-margined; staminate heads broadly obconical, about 1/6 in. long, with tubular, 5-toothed corollas, infertile; pistillate heads cylindrical, composed of tubular filiform flowers about 3/8 in. long.

FRUIT. Achenes about 1/16 in. long, 10-nerved, glabrous; pappus of fertile flowers 1/4–1/3 (to 1/2) in. long; capillary bristles abundant, soft, lustrous, white to brownish; pappus of staminate flowers scant and shorter.

LEAVES. Few, remote, simple, alternate, 1/4–1 in. long, sometimes scalelike, linear, margin entire and sometimes revolute, apex obtuse to acute, base sessile, surface glabrous, some resinous, 1-nerved.

TWIGS. Slender, tufted, resinous, glabrous, strongly striate, green to straw-colored or reddish brown.

RANGE. Sandy or gravelly washes of low hills or desert grasslands at altitudes of 1,000–5,000 ft. New Mexico, Arizona, California, and adjacent Mexico.

REMARKS. The genus name, *Baccharis*, is from the Greek word *bakcharis*, but was applied to some other plant. The species name, *sarothroides*, means "*Sarothra*-like," referring to a genus in the St. John's-wort Family. It is also known under the vernacular names of Greasewood, Desert-bloom, Rosin-brush, and Groundsel. The plant is sometimes used as an ornamental and is planted for erosion control. It is not palatable to livestock.

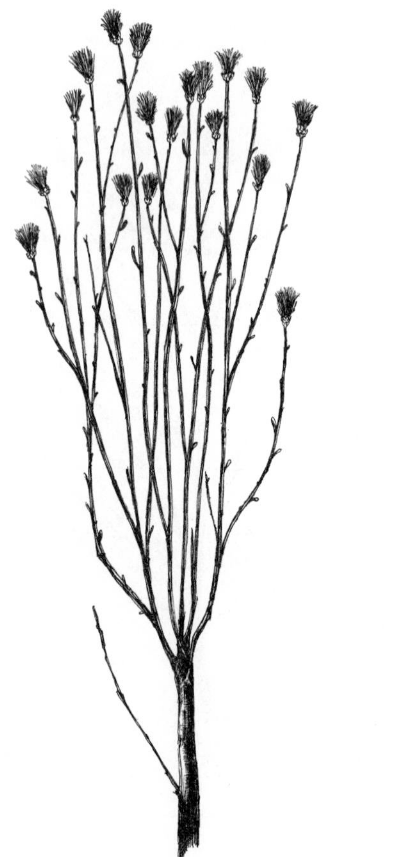

Broom Baccharis
Baccharis sarothroides Gray

Bigelow Baccharis

Baccharis bigelovii Gray

FIELD IDENTIFICATION. Glutinous perennial plant 1–2 ft. Stems slender and loosely branched near the woody base.

FLOWERS. Dioecious, heads discoid, about 1/6 in. high, borne in cymose panicles; bracts of involucre much imbricate, oblong to linear, firm, scarious margined, costa green and prominent; receptacle flat and naked; pistillate flowers tubular-filiform; staminate flowers hermaphroditic, tubular, 5-toothed, infertile.

FRUIT. Achenes small, brown, flattened, 4–5-nerved; pappus scanty, not elongate, not exceeding the styles at maturity.

LEAVES. Simple, alternate, cuneate-oblong or oblong-lanceolate, margin rather coarsely and unequally spinulose-toothed on the upper half or two thirds, apex obtuse, base cuneate, sessile, or with a very short and channeled petiole, length 1/2–1 1/2 in., width 1/6–2/5 in., stiff, upper surface dull green, glabrous and glutinous, lower surface paler, 1-nerved or obscurely 3-nerved.

TWIGS. Young ones green, herbaceous, glutinous, striate-angled; older ones tan to brown, less glutinous and woody toward the base.

RANGE. Often in coniferous forests at altitudes of 4,500–6,000 ft in western Texas, New Mexico, and Arizona; in Mexico in Sonora and Coahuila.

REMARKS. The genus name, *Baccharis*, is from the Greek word *bakcharis*, but the name was probably

BIGELOW BACCHARIS
Baccharis bigelovii Gray

applied by the Greeks to some other plant. The species name, *bigelovii,* is in honor of J. M. Bigelow, one of the naturalists of the first United States–Mexico boundary survey in 1850.

TEXAS BACCHARIS

Baccharis texana (Torr. & Gray) Gray

FIELD IDENTIFICATION. Plant 8–36 in. Stems simple, rigid, branched, woody near the base, herbaceous and leafy to the top.

FLOWERS. Heads dioecious, discoid, few, corymbose; receptacle flat, naked, pitted; involucre campanulate, bracts imbricate, in several series; staminate involucre about 1/5 in. high with bracts oblong-lanceolate to linear; pistillate involucre 1/3–2/5 in. high with bracts lanceolate to linear, apex acute, margin scarious and ciliate, dorsal rib thickened and greenish; outer bracts sometimes slightly glandular; corollas of the staminate flowers tubular and 5-lobed, anthers entire; corollas of the pistillate flowers filiform, stigmas slender.

FRUIT. Achene about 1/6 in. long, more or less 8-ribbed, flattened, glandular-scabridulous or glabrous in some; pappus of capillary bristles, that of the pistillate flower at least twice as long as the involucre, in several series, tan, spreading, thick, soft; pappus of the staminate plant short and tortuous.

LEAVES. Simple, alternate, linear to linear-oblanceolate, apex acuminate or acute, margin entire, base sessile, length 3/8–2 in., width 1/25–1/8 in., firm, surfaces glabrous and pale green, midrib impressed above and prominent beneath.

TWIGS. Slender, numerous, glabrous, green to tan on young ones, light to dark brown and roughened on old wood at the base.

RANGE. Often in colonies on dry prairies of southern and western Texas. Also in adjacent Tamaulipas, Mexico.

REMARKS. The genus name, *Baccharis,* is from the Greek word *bakcharis,* but was probably a name given for some other shrub. The species name, *texana,* refers to the state of Texas where it grows.

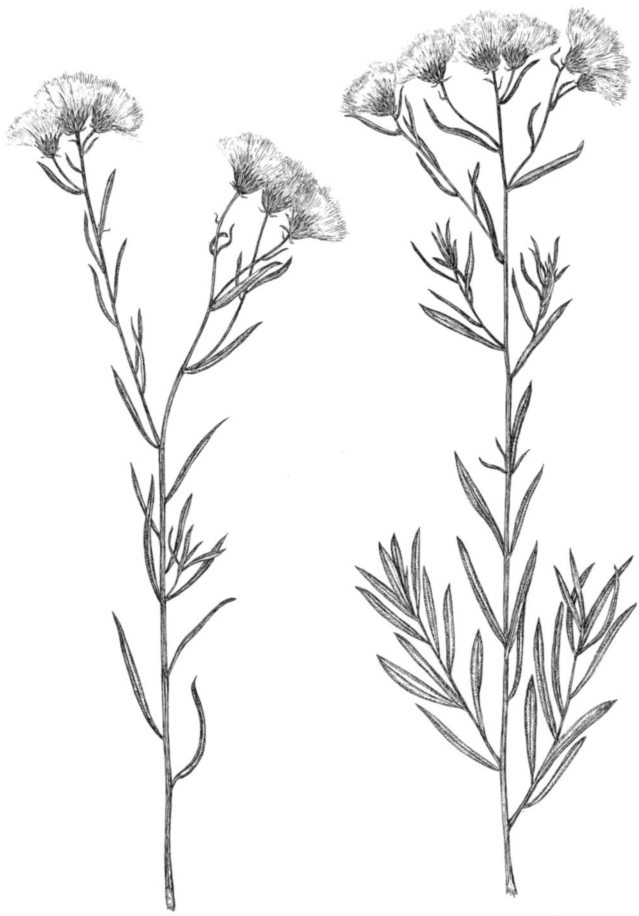

TEXAS BACCHARIS
Baccharis texana (Torr. & Gray) Gray

Willow Baccharis

Baccharis salicina Torr. & Gray

FIELD IDENTIFICATION. Much-branched shrub attaining a height of 3–6 ft. The branches are ascending, glabrous and glutinous.

FLOWERS. May–July, dioecious, borne in paniculate clusters; heads hemispheric, ⅕–⅖ in. broad; involucre campanulate, about ¼ in. high; bracts imbricate in several series, ovate to lanceolate, obtuse or subacute, the outer ones shorter; receptacle flat and naked; pistillate flower corollas slender, style branches smooth and exserted; staminate flower corollas tubular and 5-lobed, style branches rudimentary.

FRUIT. Mature in August, pappus a single series of almost white capillary bristles, copious and soft; that of the staminate flowers short; achenes more or less compressed and ribbed.

LEAVES. Simple, alternate, deciduous, not crowded generally, 1–2 in. long, ⅙–½ in. wide, lanceolate to oblong, apex mostly obtuse, base cuneate or gradually narrowed and subsessile, margin entire or repand-dentate with remote teeth, more or less 3-nerved.

Willow Baccharis
Baccharis salicina Torr. & Gray

Emory Baccharis
Baccharis emoryi Gray

TWIGS. Slender, striate-angled, young ones green to tan and glabrous, older ones light to dark brown.

RANGE. Usually in moist soil in the sun at altitudes of 1,000–5,500 ft. Often in the artemisia and pinyon belts. In central and western Texas, New Mexico, Colorado, and Kansas.

REMARKS. The genus name, *Baccharis*, is from the Greek word *bakcharis*, but the name originally applied to some other plant. The species name, *salicina*, refers to the slender, willow-like leaves. Willow Baccharis resembles the Narrow-leaf Baccharis somewhat, but the latter has linear to linear-lanceolate, mostly acute leaves.

Emory Baccharis

Baccharis emoryi Gray

FIELD IDENTIFICATION. Shrub attaining a height of 3–12 ft. Rather loosely branched and erect with evergreen leaves.

FLOWERS. Borne June–September, dioecious; heads borne in loose, sparingly leafy panicles; peduncles short and naked or flowers sessile; involucre 1/6–1/3 in. high; involucral bracts closely imbricate, outer ones ovate or oval, inner narrower and thinner, apices acute or obtuse; pistillate heads cylindric, composed entirely of tubular-filiform corollas; staminate of tubular, 5-toothed infertile corollas about 1/6 in. long.

FRUIT. Achene small, glabrous, 8–10-nerved; pistillate pappus with the capillary bristles white, shiny, copious, 3/8–1/2 in. long or more, surpassing the styles; pappus in staminate flowers shorter and bearded at the apex.

LEAVES. Simple, alternate, persistent, oblanceolate to oblong or sometimes ovate or obovate, margin on older leaves coarsely toothed on the lower half or third, upper leaves sometimes linear and entire, apices acute or obtuse, base gradually narrowed, length 3/4–1 1/2 in., width 1/8–3/4 in., 3-nerved at base; petiole short or leaf sessile.

TWIGS. Young ones green, striate-angled, glabrous, older tan to brownish.

RANGE. Moist soil along streams at altitudes of 500–5,000 ft. Western Texas, New Mexico, Arizona, California, Colorado, Nevada, and Utah.

REMARKS. The genus name, *Baccharis,* is from the Greek word *bakcharis,* but the name was originally applied to some other plant. The species name, *emoryi,* is in honor of Major Wm. H. Emory, who first collected it on the Gila River, Arizona, in 1846. It is sometimes known under the name of Water-willow. In Arizona it is grazed to some extent by livestock in summer and fall.

HAVARD BACCHARIS

Baccharis havardii Gray

FIELD IDENTIFICATION. Shrub 1–1 1/2 ft with copiously branched slender stems.

FLOWERS. Heads loosely paniculate, dioecious, about 1/6 in. long; involucral bracts imbricate, ovate to oblong, margins scarious, apex acute or obtuse, costa green, lower bracts shortest; pistillate flowers slender and threadlike; staminate tubular and 5-lobed.

FRUIT. Achene 5-nerved, pappus slightly elongate, copious, tan, very short and rigid in the staminate flower.

LEAVES. Alternate, sometimes crowded and clustered, very variable in length (1/8–1 in. long) usually about 1/4 in., linear-oblanceolate, some falcate, margin with irregular, remote, flaring teeth, or some laciniate-pinnatifid, apex acute or acuminate, sessile or nearly so; rather thickened, wavy on the plane surface, densely glandular-punctate; upper leaves usually smaller, narrowly linear, entire or 2–3-toothed.

TWIGS. Young ones slender, green to tan or light brown, rather densely glandular; older ones darker brown and striate-angled, bark fibrous.

RANGE. Dry rocky hillsides in the Chisos, Davis, and Guadalupe mountains of Texas. Perhaps also in adjacent New Mexico and Mexico.

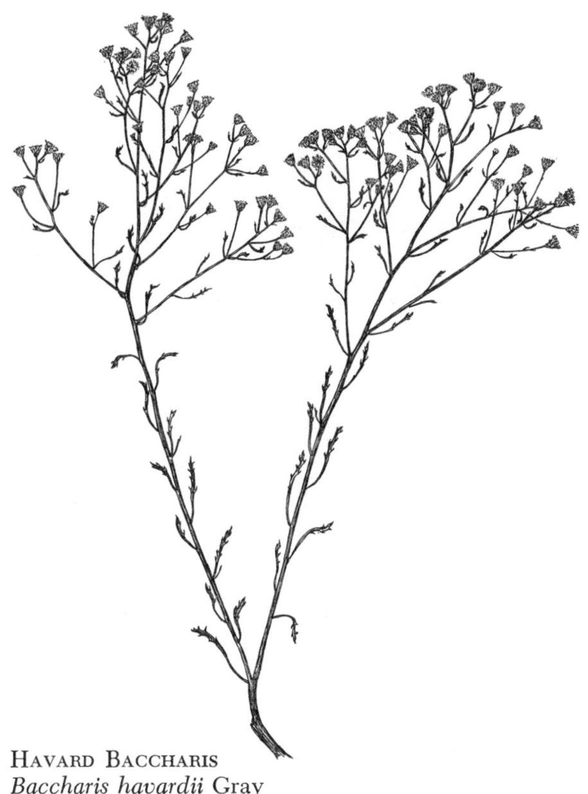

HAVARD BACCHARIS
Baccharis havardii Gray

REMARKS. The genus name, *Baccharis,* is from the Greek word *bakcharis,* but the name was probably applied to some other plant. The species name, *havardii,* is in honor of Valery Havard, an early botanist who collected extensively along the valley of the Rio Grande and adjacent territory.

YERBA-DE-PASMO BACCHARIS

Baccharis pteronioides (DC.) Gray

FIELD IDENTIFICATION. Low, diffusely branched Western shrub. The small leaves generally clustered.

FLOWERS. Dioecious, sessile or on short peduncles 1/16–1/8 in. long; heads generally solitary on short branchlets and arranged racemosely in the axils of clustered leaves. Heads 1/8–1/3 in. long and broad, much imbricate, bracts 1/16–1/8 in. long, linear to oblong, acute, puberulent, margins white-scarious, midline green. Heads of the staminate flowers average smaller than the pistillate, with shorter, puberulent, scarious-margined bracts; pappus short and twisted, mixed with the numerous florets. Pistillate flowers larger in fruit, with abundant, lustrous, straw-colored pappus borne in a single series.

FRUIT. Achene 8–10-nerved, linear, obtuse, 1/16–1/25

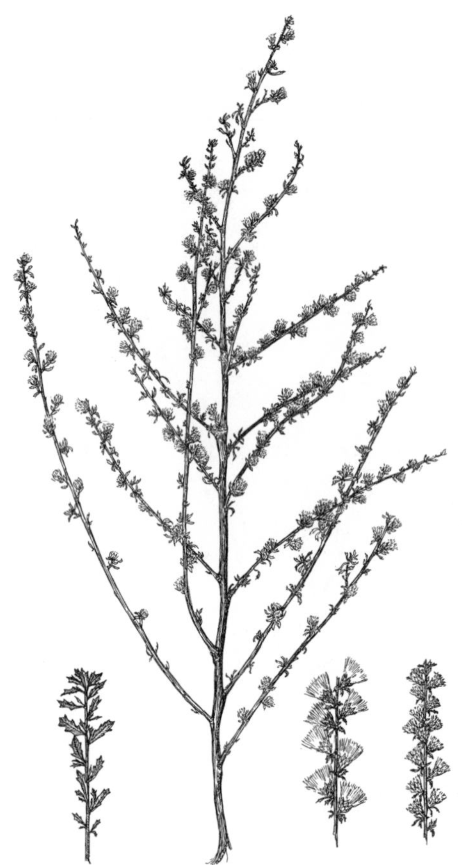

YERBA-DE-PASMO BACCHARIS
Baccharis pteronioides (DC.) Gray

in. long, slightly hairy, somewhat glandular and scabridulous; pappus tawny.

LEAVES. Rather early-deciduous, fasciculate, simple, alternate, small, ⅕–½ in. long, ⅛–⅕ in. wide, varying from linear to lanceolate and entire on the flowering branches to cuneate or obovate and coarsely toothed on the lower branches. Teeth irregularly disposed, usually toward the upper half of the leaf, flaring or recurved, sometimes 1 side of leaf with fewer lobes than the other, sometimes with several terminal lobes forming an asymmetrical apex, leaf surface pitted with small resin glands.

TWIGS. Diffusely branched, tan to white, or grayish on older stems, somewhat striate-angled or fibrous, roughened by small resin glands.

RANGE. Dry soil of hills in open sun at altitudes of 3,000–5,000 ft in the Upper Sonoran zone. In Texas in the Chisos Mountains, Brewster County; in Limpia Canyon, Jeff Davis County; and in upper McKittrick Canyon, Guadalupe Mountains, Culberson County. In New Mexico in the Organ Mountains. In Mexico in Puebla, Chihuahua, and Tamaulipas.

REMARKS. The genus name, *Baccharis*, is from the Greek word *bakcharis*. The species name, *pteronioides*, refers to the "winged" or pappus-bearing achenes. The Spanish name Yerba-de-pasmo means "chill weed," an infusion of the leaves being used as a chill tonic. In Coahuila, Mexico, the dried, powdered leaves were used for treating sores. In tropical America other names for the plant are Tepopote, Tepopotl, Jaral Blanco, Boshi, and Popotillo. The plant is reported to be poisonous to sheep and is grazed by cattle only when little else is available.

EASTERN BACCHARIS
Baccharis halimifolia L.

FIELD IDENTIFICATION. Branching shrub 3–10 ft. The branches rather numerous to form a rounded crown.

FLOWERS. Borne August–November, in dioecious, pyramidal or nearly globose panicles; heads in clusters of 1–5 generally; involucre campanulate, ⅛–¼ in. long, the bracts oblong to ovate in several series, appressed, imbricate, glutinous, obtuse to acute and greenish at the apices (bracts of the pistillate flowers lanceolate and acute); receptacle flat and naked; pistillate flower

EASTERN BACCHARIS
Baccharis halimifolia L.

with a slender filiform corolla; staminate flower corolla tubular and 5-lobed, ovary abortive.

FRUIT. Maturing November–December, achene more or less flattened and ribbed; pappus of pistillate flowers white, copious, ¼–⅓ in. long, composed of a series of capillary bristles; pappus of the staminate flowers short and scanty.

LEAVES. Simple, alternate, tardily deciduous, inflorescence leaves smaller and narrower, blade length 1–3 in., width ¼–2 in., varying from oblanceolate or elliptic to obovate or deltoid-obovate, margin entire or with a few remote, coarse, angled teeth, apex obtuse or acute, base cuneiform and sessile or short-petiolate.

TWIGS. Slender, glabrous or slightly scurfy, sometimes glutinous, green to tan or light brown and grooved, older ones darker brown or gray.

RANGE. Open woods, low prairies, or margins of salt marshes or rivers in coastal areas. Texas, Louisiana, Oklahoma, and Arkansas; eastward to Florida. Also Mexico, Cuba, and the Bahamas.

REMARKS. The genus name, *Baccharis,* is from the Greek word *bakcharis,* but the name was evidently applied to some other plant. The species name, *halimifolia,* refers to the *Halimus*-like leaves, with which there is a marked similarity to those of the orach plant, at one time known under the name *Halimus.* It is also known under the vernacular names of Groundsel-tree, Cottonseed-tree, Consumption-weed, Sea Myrtle, and Ploughman's Spikenard. The white pappus is very conspicuous in autumn, and the seeds are dispersed by the wind. The symmetrical growth of the plant gives it some ornamental value. Narrow-leaved forms of this plant have been often referred to under the name of *B. neglecta* Britt., but this name is of doubtful status.

WRIGHT BACCHARIS

Baccharis wrightii Gray

FIELD IDENTIFICATION. Plant mostly herbaceous from a woody base. Attaining a height of 1–3 ft, nearly erect, slender and much-branched, glabrous.

FLOWERS. Borne April–July, heads usually solitary and terminating slender branches, ⅓–½ in. high, rays absent; involucre of the staminate heads about ¼ in. high, pistillate somewhat larger; bracts imbricate in several series, lanceolate, apices acuminate or acute, margins scarious, dorsal vein green; corolla of the pistillate flower slender, that of the staminate 5-lobed and tubular; style branches narrow.

FRUIT. Achene 5–10-nerved, glandular and scabrous, flattened; pappus in several series of bristles, ⅜ in. long or more; in the pistillate flowers copious and soft, tawny or purplish-colored; in the staminate flowers short and tortuous.

LEAVES. Simple, alternate, deciduous, sparse, linear, to oblanceolate, upper ones linear-subulate, apex acute, base gradually narrowed and sessile, margin entire, 1-nerved on the surface, length 1/12–¾ in.

WRIGHT BACCHARIS
Baccharis wrightii Gray

TWIGS. Slender, smooth, glabrous, younger ones pale green to tan, striate-angled, brown and woody near the base when older.

RANGE. In dry, often saline, soil at altitudes of 4,000–6,000 ft. Western Texas, New Mexico, Arizona, Colorado, and Kansas. In Mexico in Chihuahua and Durango.

REMARKS. The genus name, *Baccharis,* is from the Greek, but the name was originally applied to some other plant. The species name, *wrightii,* is in honor of Charles Wright, who collected plants, 1847–1848, in Texas.

RUSH BEBBIA

Bebbia juncea (Benth.) Greene

FIELD IDENTIFICATION. Low, strong-scented, Western plant usually under 3 ft, with green, slender, rushlike stems. Mostly herbaceous, but old plants somewhat woody toward the base.

FLOWERS. Flowering throughout the summer. Heads terminal, solitary or a few together at the ends of long, slender, naked stems, ⅜–¾ in. in diameter. Ray flowers absent, disk yellow, 20–30-flowered; involucre campanulate, strongly graduate, ⅕–⅓ in. high; bracts overlapping in 2–3 series, the outer series green with scari-

Rush Bebbia
Bebbia juncea (Benth.) Greene

ous, ciliate, whitened margins, ovate to oblong, canescent-pubescent, apex obtuse to rounded or acute; the inner series of bracts thinner, whitened, subcareous; proper tube of the corolla glandular-hairy.

FRUIT. Achenes somewhat flattened, turbinate, hirsute, faintly 5-nerved; pappus of 15–20 plumose bristles in a single series, more than twice as long as the achene.

LEAVES. Absent, or remote, upper alternate, lower often opposite, linear to linear-lanceolate, entire or few toothed, grayish green, fleshy, ⅜–1½ in. long and ⅙ in. broad, glabrous or hairy with the hairs stout, appressed, broadened and somewhat tuberulate at the base.

TWIGS. Young ones green, slender, elongate, erect, glabrous or hairy; older twigs roughened, straw-colored or brown.

RANGE. In dry, well-drained soil of the larrea belt to an altitude of 4,000 ft. In Trans-Pecos Texas, New Mexico, Arizona, and Colorado; west to California and south into the Mexican states of Sinaloa and Sonora.

VARIETY. A common variety is Sweet Rush Bebbia, *B. juncea* (Benth.) Greene var. *aspera* Greene. It has roughened, more or less densely hispidulous, stems and leaves, and bracts which are more acute or acuminate, and is a desert form.

REMARKS. The genus name, *Bebbia,* is in honor of Michael Shuck Bebb, American botanist. The species name, *juncea,* refers to the rushlike, slender branches.

Rough Brickellia

Brickellia scabra (Gray) A. Nels.

FIELD IDENTIFICATION. Slender undershrub 1–4 ft. Twigs wiry and ascending, numerous and clump-forming. The upper branchlets bearing many reduced bract-like leaves.

FLOWERS. Yellowish to purple, 10–15 in a head, heads terminal and on short lateral branches; ray flowers none; receptacle flat and naked; corolla slender, 5-toothed at apex; involucres campanulate, about ⅓ in. high; bracts imbricate, firm, green, apices spreading glandular-viscid, lower ones herbaceous and sometimes passing into the leaves of the branchlets.

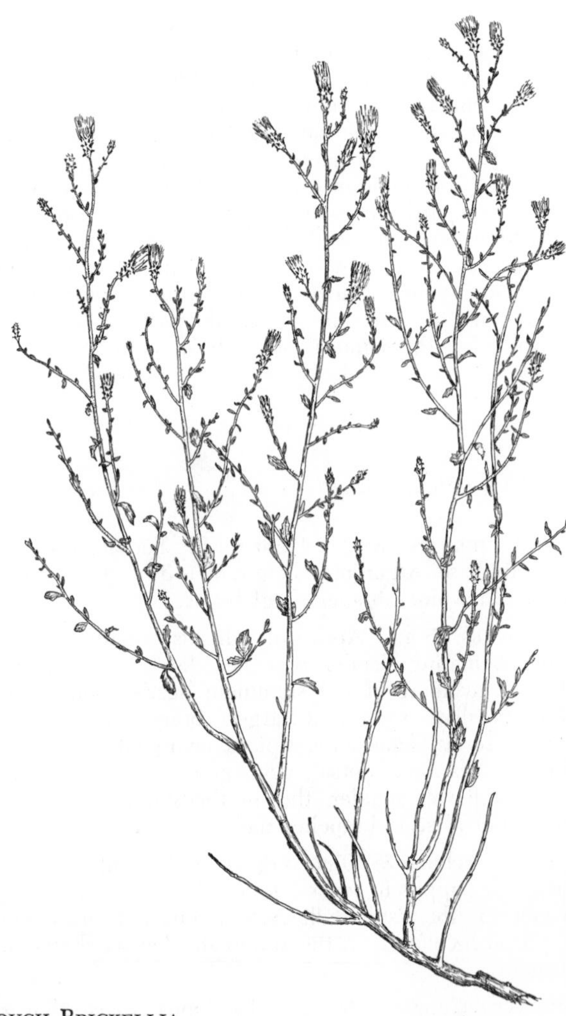

Rough Brickellia
Brickellia scabra (Gray) A. Nels.

COULTER BRICKELLIA
Brickellia coulteri Gray

FRUIT. Achenes about ⅛ in. long, 10-ribbed, the angles scabrous or finely hispidulous; pappus composed of a single series of whitish capillary bristles.

LEAVES. Simple, alternate, ⅕–⅗ in. long, often much reduced on the flowering branchlets, blade ovate to lanceolate or oblong, apex obtuse to acute, sometimes apiculate, base broadly cuneate to subcordate, margin entire or sparingly dentate, surface somewhat scabrous; petioles to ⅛ in. long or leaf sessile.

TWIGS. Very slender, paniculately much-branched, young ones straw-colored to ashy, glandular-puberulent or scabrous-puberulent; older ones woody near base, dark brown to gray and glabrous.

RANGE. In sandy or rocky soils of dry canyons or foothills at altitudes of 4,500–7,500 ft. New Mexico and Arizona; northward to Colorado, Utah, Nevada, Wyoming, and Oregon.

REMARKS. The genus name, *Brickellia,* is in honor of John Brickell (1749–1809), a native of Ireland, longtime resident of Georgia, and a distinguished botanical scholar. The species name, *scabra,* refers to the stiff rough hairs of the twigs and leaves.

COULTER BRICKELLIA

Brickellia coulteri Gray

FIELD IDENTIFICATION. Brittle slightly woody plant with opposite branches, attaining a height of 1–3 ft.

FLOWERS. Heads erect, but later lax, cymose-paniculate, on slender peduncles ¼–1 in. long, bearing 15–17 flowers; bracts of the involucre about 21, narrowly lanceolate to linear, apex acute to acuminate, 4-striate, apices of the external ones somewhat lax, outer dorsal surface glabrous to puberulent; corolla yellowish or greenish white or sometimes purple-tinged, about ⅓ in. long, limb 5-lobed.

FRUIT. Achenes about ⅛ in. long, yellowish to brown with short gray hairs; pappus with 30–40 bristles about ⅕ in. long.

LEAVES. Opposite, blade length ⅝–1½ in., ⅜–¼ in. wide, deltoid to rhombic-ovate, apex acute to acuminate, base hastate or 1–3-dentate or more rarely subentire, entire or toothed toward the apex, both sides rather sparsely-stiff pubescent or pulverulent; slender-petioled.

STEMS. Ascending, opposite, gray to white or greenish, glandular-puberulent or only puberulent when young, older ones more glabrous.

RANGE. Rocky hills at altitudes of 2,000–4,000 ft. Western Texas, New Mexico, Arizona, Baja California, and Mexico.

REMARKS. The genus name, *Brickellia,* is in honor of John Brickell (1749–1809), a botanist of Georgia. The species name, *coulteri,* is for John M. Coulter, author of *Botany of Western Texas* (1891–1894).

SCALY BRICKELLIA

Brickellia squamulosa Gray

FIELD IDENTIFICATION. Low Western shrub 16–24 in. The stems are erect, simple and fastigiately branched above. An unusual feature is the small scalelike leaves of the flowering branches.

FLOWERS. Maturing May–September, borne in dense, narrow, or pyramidal panicles; heads sessile or subsessile, ⅓–½ in. high, 6–12-flowered; corolla white or pinkish, tubular, gradually wider toward the apex, lobes 5 and short; stigma bifid and exserted; involucre scales seriate, imbricate, pale green; outer series of scales at base of involucre shorter and more crowded; inner series oblong to linear, thickened, apex obtuse or rounded, surface puberulent and obscurely 3-nerved.

FRUIT. Achenes about ⅛ in. long, 10-ribbed, glabrous or hispid above; pappus bristles straw-colored, numerous, fine, length unequal.

LEAVES. Those of the stem filiform to linear, ¼–2¾ in. long, margin entire, sessile, 3–5-nerved; leaves of

Scaly Brickellia
Brickellia squamulosa Gray

branchlets abundant, clustered, some in sets of four, short and scalelike, mostly deciduous, oblong, obtuse at apex, somewhat puberulent to glabrous; the scalelike leaves said to be capable of reproducing the plant.

twigs. Slender, young ones green to straw-colored or brownish, minutely puberulent or glabrous; older ones brown to gray and glabrous, finely grooved, woody near the base; flowering branchlets somewhat appressed to the main stem in some specimens examined.

range. Dry soil in the sun at altitudes of 4,000–6,000 ft. Mountain valleys of the pinyon and yellow pine belts in southwestern New Mexico and adjacent Arizona, and in Mexico in Chihuahua, San Luis Potosí, and Guanajuato.

remarks. The genus name, *Brickellia,* is in honor of John Brickell (1749–1809), a native of Ireland, long-time resident of Georgia, and distinguished botanical scholar. The species name, *squamulosa,* refers to the numerous, scalelike leaves. In cultivation the brickellias are usually reproduced by means of cuttings.

Small Brickellia

Brickellia parvula Gray

field identification. Subshrub attaining a height of 5–12 in. The scabrous-puberulent, slender stems with ascending leaves.

flowers. Corymbs terminating in 3–10 heads, pedicels slender, usually to 1½ in. long; heads erect, usually solitary, about ½ in. tall and ⅕–¼ in. in diameter, each about 12-flowered; involucral bracts few, about 14, loosely imbricate, 3–5-striate, narrowly mucronulate, dorsal surface pubescent; corolla ⅕–¼ in. long, style branches clavate.

fruit. Achenes hirsute, ribbed; pappus bristles obscurely barbellate.

leaves. Mostly opposite, ovate, acute at apex, base subtruncate and subsessile, margin coarsely few-toothed, blade length ⅓–⅗ in., width ¼–⅖ in., both surfaces green and scabrous-puberulent; upper leaves smaller and sparser.

twigs. Slender, curved, ascending, leafy.

range. Trans-Pecos Texas along the Limpia River. Not common.

remarks. The genus name, *Brickellia,* is in honor of John Brickell (1749–1809), a botanist of Georgia. The species name, *parvula,* refers to the small size.

Small Brickellia
Brickellia parvula Gray

BACCHARIS-LEAF BRICKELLIA
Brickellia baccharidea Gray

FIELD IDENTIFICATION. Shrubby plant to 3 ft, with leafy ascending branches.

FLOWERS. September–November, in rather dense terminal or axillary clusters; heads ⅜–½ in. long, about ⅙ in. in diameter, many-flowered (15–18), short-pediceled; involucre bracts green to straw-colored or brown, 3–4-striate, narrowly oblong or lanceolate, mostly obtuse at apex, subglabrous to villose above, sometimes resinous, about 1/25 in. broad.

FRUIT. Achenes 1/12–1/10 in. long, pale, 10-ribbed, finely hairy; pappus bristles 15–25, white, about ¼ in. long, scarcely scabrid.

LEAVES. Usually alternate, rhombic-ovate, coriaceous, blades ⅜–1⅝ in. long, ⅜–¾ in. wide, margin coarsely laciniate-dentate toward the apex (usually with 2–5 large teeth per side), base entire and cuneate, apex obtuse to acute; surfaces subglabrous, rather glossy, somewhat glandular-dotted or viscid, lower surface reticulate-veined.

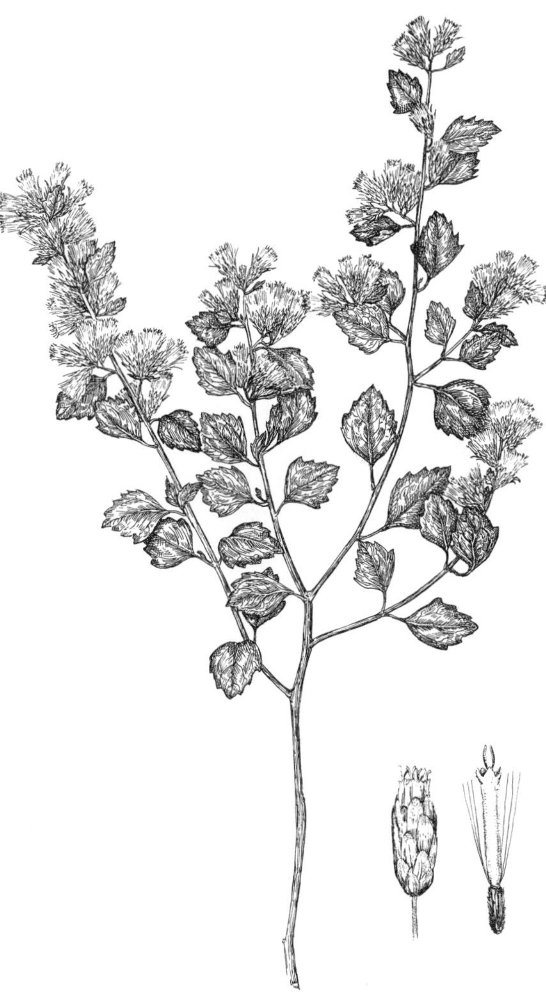

BACCHARIS-LEAF BRICKELLIA
Brickellia baccharidea Gray

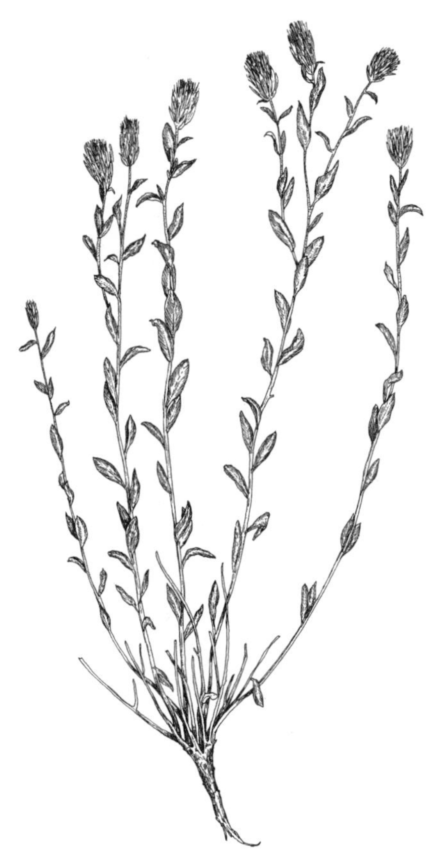

NARROWLEAF MOHAVE BRICKELLIA
Brickellia oblongifolia var. *linifolia* (D. C. Eaton) Robins.

TWIGS. Slender, flexuous, tan to gray or brown, finely puberulent at first, scabrous, glabrous, and bark exfoliating later.

RANGE. In limestone areas of hills or desert-grassland areas. At altitudes of 500–5,000 ft. In the mountains of Trans-Pecos Texas. In New Mexico in the Organ Mountains. In Arizona in Yuma and Pima counties. In Mexico in Chihuahua, Coahuila, and San Luis Potosí.

REMARKS. The genus name, *Brickellia,* is in honor of John Brickell (1749–1809), a botanist of Georgia. The species name, *baccharidea,* denotes its resemblance to a *Baccharis* plant.

NARROWLEAF MOHAVE BRICKELLIA

Brickellia oblongifolia var. *linifolia* (D. C. Eaton) Robins.

FIELD IDENTIFICATION. Round-topped bush 8–16 in., often in dense clumps.

FLOWERS. April–June, heads discoid, solitary, or a

few on elongate peduncles in a terminal cyme, 20–50-flowered, white to greenish; receptacle naked; ray flowers absent; anthers united; style branches exserted and club-shaped; involucre ⅖–⅗ in. high, the bracts in several series, imbricate, dry, thin, striate, outer ovate, inner linear to oblong-lanceolate, apex acute and appressed.

FRUIT. Achene ⅙–¼ in. long, 10-ribbed, pappus composed of a single series of numerous barbellate capillary bristles.

LEAVES. Numerous, simple, alternate, ½–1 in. long, ovate-oblong to linear, margin entire or with 1–2 teeth, apex obtuse or acute, base cuneate or gradually narrowed, sessile or very short-petiolate, semi-clasping, surfaces puberulent and often minutely glandular, obscurely 3-nerved.

TWIGS. Leafy, slender, erect or ascending, young ones pale green to straw-colored or white, glandular-puberulent and obscurely 3-nerved; narrowed to a sessile base.

RANGE. Usually on sandy banks of hills or plains, or in dry rocky places at altitudes of 4,500–9,000 ft. New Mexico, Arizona, California, Colorado, Utah, and Nevada.

REMARKS. The genus name, *Brickellia,* is in honor of John Brickell (1749–1809), a botanist of Georgia. The species name, *oblongifolia,* refers to the oblong leaves, and the variety name, *linifolia,* is for their flaxlike appearance.

MOHAVE BRICKELLIA

Brickellia oblongifolia Nutt.

FIELD IDENTIFICATION. Densely clumped plant 8–16 in. Mostly herbaceous, but sometimes woody near the base. Simple or corymbosely branched at the summit.

FLOWERS. Greenish to cream or purplish, heads large, terminating elongate leafy branches; receptacle flat and naked; ray flowers absent; disk flowers 40–50, corolla slender, 5-toothed at apex, anthers united, style branches clavate and exserted; involucre ⅖–⅗ in. high, bracts imbricated in several series, outer ones ovate, inner linear to oblong-lanceolate, apices acute and appressed.

FRUIT. Achenes with rows of minute glandular bristles along the ribs, ⅕–¼ in. long; pappus whitish, composed of a single series of bristles.

LEAVES. Simple, alternate, blade elliptic-lanceolate or elliptic-ovate or oblong, apex obtuse to acute, length ½–1½ in., margin usually entire, surfaces more or less puberulent and obscurely 3-nerved; narrowed to a sessile base.

TWIGS. Simple, erect, glandular-puberulent, younger ones greenish straw-colored, older ones woody near the base and grayish brown.

RANGE. Arid sandy soils of foothills and canyons at altitudes of 4,000–9,000 ft. New Mexico, Arizona, and California; north into Colorado, Utah, Nevada, Oregon, Washington, and British Columbia.

REMARKS. The genus name, *Brickellia,* is in honor of John Brickell (1749–1809), a native of Ireland, longtime resident of Georgia, and distinguished botanical scholar. The species name, *oblongifolia,* is for the oblong leaves. *B. oblongifolia* var. *linifolia* (D. C. Eaton) Robins., is a variety with the achenes hispidulous and the glands few or absent.

CALIFORNIA BRICKELLIA

Brickellia californica (Torr. & Gray) Gray

FIELD IDENTIFICATION. Straggling, rounded, aromatic shrub attaining a height of 1½–3½ ft. The numerous stems from a woody base.

FLOWERS. June–November, heads about ⅗ in. high, numerous, discoid, solitary in the axils or often cymose clustered or in a leafy panicle; secondary lateral branches almost spicate with slender pedicels, 8–18-flowered (average 15); receptacle glabrous; involucre bracts graduated, narrow-oblong, subacute to almost round at apex, 3–6-striate, green to straw-colored or purple-tinged, external ones gradually shortened, the outermost oval; corolla ¼–⅓ in. long, yellowish white, sometimes purplish.

FRUIT. Achenes about ⅛ in. long, ribbed, pappus

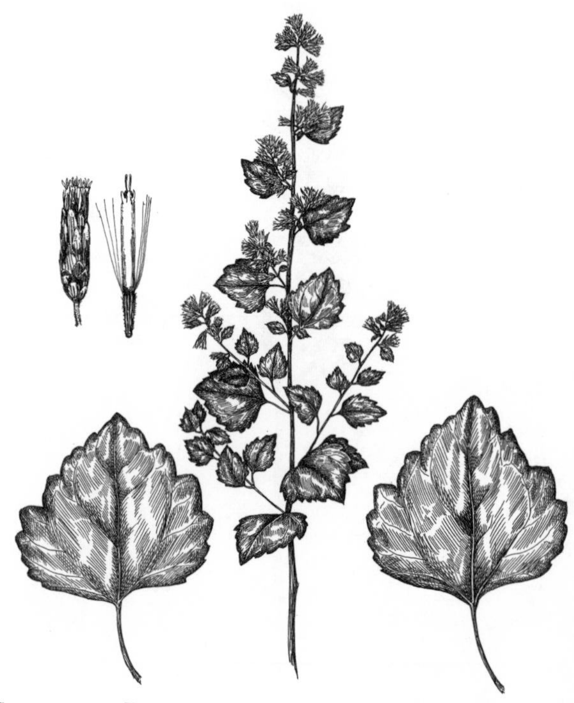

CALIFORNIA BRICKELLIA
Brickellia californica (Torr. & Gray) Gray

VEINY BRICKELLIA
Brickellia venosa (W. & S.) Robins.

bristles numerous, white or sordid to purplish or brown, somewhat scabrid.

LEAVES. Usually alternate, deltoid-ovate or nearly reniform, apex mostly obtuse, base subtruncate to broadly cordate, margin dentate to crenate-serrate, surfaces with short, rough, grayish pubescence, 3-nerved, ⅜–2 in. long and wide; petioles ⅙–⅘ in. long.

TWIGS. Rather rounded, short-pubescent, at first pale, later purple or brown to gray and glabrous.

RANGE. Usually in dry gravelly stream beds or rocky hillsides at altitudes of 3,000–7,500 ft. Western Texas, New Mexico, Arizona, Colorado, and California. In Mexico in Sonora, Chihuahua, and Baja California.

VARIETY. A variety known as *B. californica* var. *desertorum* (Cov.) Parish, has ovate leaves, and heads ⅓–⅖ in.

REMARKS. The genus name, *Brickellia,* is in honor of John Brickell (1749–1809), a botanist of Georgia. The species name, *californica,* is for California, where it was first collected by Douglas.

VEINY BRICKELLIA

Brickellia venosa (W. & S.) Robins.

FIELD IDENTIFICATION. Plant 2–3 ft, with gray-pulverulent to tomentose stems, woody near the base.

FLOWERS. September–October, panicles loose, branches rather naked or with a few scattered leaves; heads about ½ in. high and ¼ in. wide, averaging about 24 flowers; pedicels slender, erect, as long as the heads or longer; involucre subcampanulate-cylindric, bracts averaging about 25, imbricate, purplish, 6–7-ribbed, lower ones shorter and rounded, upper ones ovate to lanceolate, apices obtuse to retuse, often abruptly mucronulate, surfaces pubescent; receptacle glabrous; corolla-tube about ¼ in. long, yellowish or purplish, limb short and 5-lobed; stigma bifid and exserted.

FRUIT. Achene 10-ribbed, sericeous, bristles about 38.

LEAVES. Opposite, or some alternate, blade length 1–3½ in., width 1/12–⅖ in., narrowly oblong to linear, apex obtuse, margin entire or sometimes crenate near the middle, base rounded and entire, thin but firm, reticulate-veined, olive-green, short-petiolate or sessile.

TWIGS. Young ones slender, straw-colored to brownish, densely pubescent or tomentose, older ones brown and glabrous.

RANGE. Usually in dry calcareous soil in mountainous areas at altitudes of 4,000–5,500 ft. New Mexico, Arizona, and Mexico in Chihuahua and Sonora.

REMARKS. The genus name, *Brickellia,* is in honor of John Brickell (1749–1809), a native of Ireland, longtime resident of Georgia, and distinguished botanical scholar. The species name, *venosa,* refers to the veiny leaves.

CUTLEAF BRICKELLIA

Brickellia laciniata Gray

FIELD IDENTIFICATION. Shrub ½–2 ft, with pale gray, fibrous bark. The upper part subfastigiately branched.

FLOWERS. Panicle elongate-pyramidal, heads subracemose, ½ in. tall and 1/12–⅛ in. in diameter, 8–12-flowered; pedicels 1/25–⅛ in. long; involucral bracts about 13, green to straw-colored, 4-striate, narrowly oblong, somewhat obtuse; corolla about ¼ in. long, green to yellowish green or whitish, the limb sometimes purplish-tinged.

FRUIT. Achene pale gray, ribbed, villose, 1/25–⅛ in. long; pappus bristles hardly barbellate, about ⅕ in. long.

LEAVES. Mostly alternate, to 1¼ in. long (rarely more), ovate-cuneate, apex obtuse or acute, base cuneate to truncate, margin laciniately toothed or lobed, obscurely veined, glabrous or slightly puberulous or scabrous, membranous, petiolate.

RANGE. In Trans-Pecos Texas and New Mexico; in

Cutleaf Brickellia
Brickellia laciniata Gray

Mexico in Nuevo León, Coahuila, Chihuahua, and San Luis Potosí.

REMARKS. The genus name, *Brickellia,* is in honor of John Brickell (1749–1809), a botanist of Georgia. The species name, *laciniata,* refers to the laciniate leaves.

Bigelow-bush

Carphochaete bigelovii Gray

FIELD IDENTIFICATION. Plant herbaceous above, but somewhat woody toward the base, usually 8–24 in., with fasciculate glabrous branches.

FLOWERS. March–July, mostly terminating very leafy short branches, solitary or clustered, sessile or short-peduncled; heads discoid, 4–6-flowered, cylindrical, perfect, disk flowers only; involucre rose-colored, narrowly cylindric, about 1 in. high; bracts narrowly oblong, imbricate, unequal, acute to acuminate, sparsely puberulent to glabrous, densely glandular-punctate.

FRUIT. Achenes slender, 8–10-ribbed, awn-bearing scales of the pappus 11–14, linear-attenuate, subulate, scabrous, rigid, alternating with 1–5 small, short, nearly nerveless, pointless ones.

LEAVES. Opposite, or apparently clustered because of smaller axillary ones, sessile, 1–3-nerved, 3/5–1 1/2 in. long, 1/12–1/5 in. wide, linear to spatulate-oblong or elliptic, apex acute or obtuse, surfaces glabrous or nearly so and densely glandular-punctate.

TWIGS. Slender, brittle, reddish brown, young ones somewhat puberulent and glandular, older ones glabrous and dark brown to black.

RANGE. Rocky hillsides at altitudes of 3,500–6,000 ft. In Trans-Pecos Texas, New Mexico, Arizona, and Mexico.

REMARKS. The genus name, *Carphochaete,* refers to

Bigelow-bush
Carphochaete bigelovii Gray

SOUTHWEST RABBIT-BRUSH
Chrysothamnus pulchellus (Gray) Greene

the bristly pappus. The species name, *bigelovii*, honors J. M. Bigelow, naturalist connected with the United States–Mexico boundary survey in 1850. The plant is sometimes browsed by livestock and deer.

SOUTHWEST RABBIT-BRUSH

Chrysothamnus pulchellus (Gray) Greene

FIELD IDENTIFICATION. Densely branched Western shrub usually less than 3½ ft.

FLOWERS. Borne in corymbose clusters or loose terminal cymes; involucre ⅖–½ in. long; bracts 20–30, concave, keeled, stiff, glabrous, apex attenuate-pointed and green; flowers 3–5; corolla tubular, length ⅖–⅗ in., glabrous or granular, expanded into lobes which are less than 1/12 in. long, lanceolate and semierect; anthers lanceolate, acute, less than 1/25 in. long; style bifid, long-exserted, stigmas half the length of the style branches.

FRUIT. Prismatic to ellipsoid, 4-angled and ribbed, length ¼–⅓ in., glabrous to puberulent; pappus longer than the tubular corolla, soft, brownish.

LEAVES. Alternate, sessile, filiform to linear or oblong, flattened or revolute, entire, apex mucronate, blade length ⅜–1½ in., width 1/25–1/12 in.; main vein apparent, other veins obscure; surfaces green and glabrous, but varying to puberulous or scabrous.

TWIGS. Green to gray or white, short, brittle, leafy, striate, glabrous, older bark gray to brown.

RANGE. Southwest Rabbit-brush is often found in sandy soil to an altitude of 5,500 ft. In the southern Rocky Mountain area of Trans-Pecos Texas, New Mexico, Arizona, Colorado, Utah, and Kansas.

VARIETIES. The following varieties are generally recognized: Bailey Southwest Rabbit-brush, *C. pulchellus* var. *baileyi* (Gray & Greene) Woot. & Standl., is a shrub usually less than 2 ft, with linear to linear-oblong leaves and margins minutely ciliolate; bracts abruptly acuminate and bristle-pointed. Occurring in Texas, New Mexico, and Kansas. In Texas near Mustang Springs.

Tall Rabbit-brush, *C. pulchellus* var. *elatior* (Gray & Greene) Standl., is a slender shrub to 3½ ft, with leaves linear and surfaces densely puberulent; peduncles puberulous, involucre ⅓–½ in. long; bracts abruptly acuminate.

REMARKS. The genus name, *Chrysothamnus*, is from the Greek words *chrysos* ("gold") and *thamnos* ("shrub"), referring to the golden-yellow flowers. The species name, *pulchellus*, means "beautiful." The plant has only small forage value. It is reported that the Southwestern Indians made arrows and wickerwork from this and related species, and also obtained dyes from the inner bark.

RUBBER RABBIT-BRUSH

Chrysothamnus nauseosus (Pallas) Britt.

FIELD IDENTIFICATION. Western shrub 9 in.–9 ft in height, bearing several erect stems from the base to form a rounded clump; root system deep.

FLOWERS. Usually August–December, cymes rounded or aggregated in thyrses; involucre ¼–½ in. high; bracts in imbricate, vertical ranks, 20–25, lanceolate, apex obtuse or acute, firm; flowers 5–6, corolla tubular-funnelform, gradually narrowed below, ⅓–½ in. long, glabrous or pubescent, 5-lobed; lobes less than 1/25 in. long, erect or spreading, glabrous or sparingly pubescent; anther tips lanceolate, acute, less than 1/25 in. long; style branches long-exserted, stigmatic portion as long as the forked appendages or shorter.

FRUIT. Achenes linear, narrowed to one end, 5-angled, ⅕–¼ in. long, varying from villous to subglabrous; pappus copious, rather rigid, mostly shorter than the corolla, but some slightly longer, dull white or tawny.

LEAVES. Alternate, sessile, filiform to broadly linear, apex acute, blade length ¾–2¾ in., width 1/25–⅕ in., 1–3-nerved, densely hairy to subglabrous.

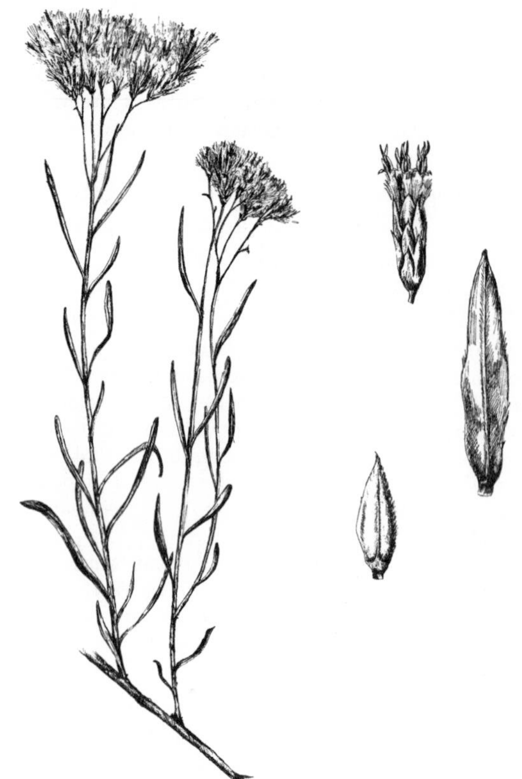

RUBBER RABBIT-BRUSH
Chrysothamnus nauseosus (Pallas) Britt.

TWIGS. Erect, flexible, moderately leafy, often densely felty-tomentose, striate beneath the tomentum; main stems brown to gray and fibrous.

RANGE. Rubber Rabbit-brush and its numerous variations occur from western Texas northward through New Mexico, Colorado, the Dakotas, Washington, British Columbia and Alberta, Canada; westward to California, and southward into Mexico.

PROPAGATION. For propagation purposes the heads can be collected October–November and the seed removed by flailing in bags. The seed averages about 330,000 seeds per lb, with many sterile. The seeds are usually sown in the fall without pretreatment. Germination averages 20–40 per cent. Established plants transplant well and can be divided by the crowns. After cuttings, sprouts grow from the base. It is a subclimax dominant of the sage-brush.

VARIETIES AND FORMS. This species represents a very composite entity of many varieties, subspecies, and races; even a large number of species have been separated from it by different authors. These segregates have been evaluated by Hall and Clements (1923). Only those occurring within the area with which the present work is concerned are mentioned here. These are the following:

C. nauseosus var. *latisquameus* Gray has leafy erect or ascending twigs, striations obscured by the tomentum; leaves ¾–2 in. long and 1/25 in. wide; flowers in a rounded compound cyme; involucre about ⅓ in. high; bracts in 5 vertical rows, keeled, obtuse, inner glabrous, outer tomentulose; corolla about ⅓ in. long; tube short, pubescent or glabrous; lobes ovate or short-lanceolate, 1/25 in. long or less, erect; style-appendage longer than the stigmatic portion; achenes densely pubescent. Texas in Hudspeth County; New Mexico in Grant County; Mexico at Santa Cruz, Sonora.

C. nauseosus var. *bigelovii* Gray is a shrub 1–3½ ft, with leafy or leafless erect branches, green to yellowish, striate, tomentum smooth; leaves linear to filiform, ⅜–1 in. long, 1/25 in. or less wide, 1-nerved, younger ones tomentulose; flowers in loose, terminal cymes; involucre ⅖–½ in. long; bracts 4–5 in a row, acute to acuminate, tomentulose and ciliate; corolla ⅓–⅖ in. long; tube glabrous or nearly so; lobes ovate, erect or spreading, averaging about 1/25 in. long; style appendages longer than the stigmatic part; achene glabrous, 5-nerved. In Texas, New Mexico, Arizona, and Colorado.

REMARKS. The genus name, *Chrysothamnus*, is from the Greek words *chrysos* ("gold") and *thamnus* ("shrub"), with reference to the golden yellow flowers. The species name, *nauseosus*, refers to the somewhat pungent odor of the plant.

DWARF RABBIT-BRUSH
Chrysothamnus depressus Nutt.

The plant is useful in erosion control and is a good honey producer. It is browsed by the moose, elk, and mule deer in more northern areas, but is seemingly of small value as forage for cattle over most of its range. If fed in concentrated form, or to sickly animals, it is reported to be poisonous. It contains a small amount of rubber, but not enough to be of economic importance. Some plants from alkali sites have produced to 6.5 per cent, but the average for good varieties is about 2.83.

Dwarf Rabbit-brush

Chrysothamnus depressus Nutt.

FIELD IDENTIFICATION. Shrub or undershrub 4–18 in. Usually dense and irregularly branched, clump-forming from a decumbent woody base.

FLOWERS. Borne in small terminal cymes; involucre ⅓–½ in. long; bracts 18–25, in vertical ranks, lanceolate, gradually acuminate to a setaceous tip, keeled, with rough-puberulent minute hairs; corolla tubular-funnelform, gradually widened above, ¼–⅓ in. long, glabrous, the 5 lobes lanceolate and 1/25–1/10 in. long; style branches exserted, usually shorter than the stigmatic portion; anthers lanceolate, acute, about 1/50 in. long.

FRUIT. Achenes, angled or 8-ribbed, about ⅕ in. long, glabrous or pubescent; pappus somewhat exceeding the corolla, brownish, soft.

LEAVES. Erect or ascending, rigid, oblanceolate to lanceolate or spatulate, apex acute and mucronate, ¼–¾ in. long, 1/25–⅙ in. wide, 1-veined, minutely puberulent.

TWIGS. Leafy up to the fasciculate cymes, short, numerous, brittle, striate, cinereous or obscurely scabro-pubescent, rather pale, older stems gray or brown.

RANGE. Dry rocky slopes or cliff edges, plains, or hills. New Mexico, Colorado, Arizona, California, Utah, and Nevada.

REMARKS. The genus name, *Chrysothamnus,* is from the Greek words *chrysos* ("gold") and *thamnus* ("shrub"), meaning the "golden-flowered shrub." The species name, *depressus,* refers to its low decumbent habit of growth. The plant is closely browsed by sheep but it is of little value as forage.

Greene's Rabbit-brush

Chrysothamnus greenei (Gray) Greene

FIELD IDENTIFICATION. Bushy, much-branched shrub 3–9 ft.

FLOWERS. Borne August–September in terminal flattened cymes; involucres ⅕–⅙ in.; bracts imbricate, 15–20, oblong, apex acute, sometimes abruptly so; corollas about 5, white to greenish, tubular, about ⅙ in. long, the 5 lobes 1/30–1/15 in. long, spreading, glabrous, lanceolate; style branches exserted, stigmatic portion longer than the appendage.

Greene's Rabbit-brush
Chrysothamnus greenei (Gray) Greene

FRUIT. Achenes about ⅛ in. long, densely hairy; pappus stiff and scant, white, as long as the corolla or shorter.

LEAVES. Simple, alternate, blade length ⅜–1½ in., width 1/25–1/15 in., filiform to linear, apex acute, rather stiff, 1-nerved, viscid, scabrous, ciliate.

TWIGS. Erect, crowded, glabrous, young ones green, older ones white and lustrous.

RANGE. Sandy or rocky alkaline plains or hills. New Mexico, Colorado, Arizona, Nevada, and Utah.

VARIETY. A variety known as the Thread-leaf Rabbit-brush, *C. greenei* forma *filifolius* Rydb., is a form with leaves mostly less than ¾ in. long and less than 1/25 in. wide. It occupies the same range as the species.

REMARKS. The genus name, *Chrysothamnus,* is from the Greek words *chrysos* ("gold") and *thamnus*

Vasey Rabbit-brush
Chrysothamnus vaseyi (Gray) Greene

("shrub"), meaning "golden-flowered shrub." The species name, *greenei,* is in honor of Edward L. Greene (1842–1915), American botanist who collected in the Southwest.

Vasey Rabbit-brush

Chrysothamnus vaseyi (Gray) Greene

FIELD IDENTIFICATION. Low undershrub rarely more than 1 ft, assuming a rounded form from a single stem, or sometimes multi-stemmed.

FLOWERS. Maturing August–September, heads borne in small cymes; involucre about $\frac{1}{4}$ in. high; bracts 13–15, oblong, apex obtuse, slightly keeled, margin scarious, ciliolate, somewhat erose, surface glabrous and viscid, outer bracts greenish near the apex; flowers 5–8; corolla $\frac{1}{5}$–$\frac{1}{4}$ in. long, tubular-funnelform, gradually widened toward the apex, glabrous; corolla lobes 5, spreading, $\frac{1}{15}$–$\frac{1}{12}$ in. long; anthers lanceolate; style branches slightly exserted, appendage shorter than the stigmatic portion or more rarely as long.

FRUIT. Achenes terete-turbinate, about $\frac{1}{5}$ in. long, usually with 10 striations, glabrous or slightly pubescent near the apex; pappus scanty, shorter than the corolla, dull white, smooth, lustrous.

LEAVES. Simple, alternate, narrowly oblanceolate to linear, flattened or twisted, apex acute or obtuse, base sessile or nearly so, blade length $\frac{3}{8}$–1 in., width $\frac{1}{25}$–$\frac{1}{10}$ in., 1-nerved, firm, surface glabrous and resinous.

TWIGS. Erect or ascending, brittle, green to white and glabrous.

RANGE. Grassy mountain valleys and hillsides at altitudes of 6,000–8,500 ft. New Mexico, Colorado, Utah, and Wyoming.

REMARKS. The genus name, *Chrysothamnus,* is from the Greek words *chrysos* ("gold") and *thamnos* ("shrub"), for the brilliant golden yellow flowers. The species name, *vaseyi,* is in honor of George Vasey (1822–1893) an eminent agrostologist of the U.S. Department of Agriculture. The plant apparently has little value as a browse for livestock.

Parry Rabbit-brush

Chrysothamnus parryi (Gray) Rose

FIELD IDENTIFICATION. Small shrub generally less than 2 ft, sometimes diffuse in dwarf forms.

FLOWERS. Racemes terminal and leafy; involucre

Parry Rabbit-brush
Chrysothamnus parryi (Gray) Rose

⅖–⅗ in. long, bracts in vertical ranks, 8–20, lanceolate, apex acute, stiff, puberulent; heads with 4–20 flowers; corolla ⅓–½ in. long, tubular-funnelform, the 5 lobes 1/35–1/10 in. long, erect, glabrous or pubescent; anthers 1/35–1/25 in. long, linear-lanceolate; styles long-exserted, the stigmatic portion much shorter than the appendage.

FRUIT. Achenes ⅕–¼ in. long, 4-angled, somewhat wider toward the apex, densely appressed-hairy; pappus white or tan, soft, shorter or longer than the corolla.

LEAVES. Linear to narrowly spatulate, acute and mucronate at apex, base gradually narrowed, blade length ⅜–3¼ in., width 1/35–⅓ in., 1–3-nerved, surfaces green to grayish and viscid-glandular to tomentose.

TWIGS. Leafy, ascending, flexible, tomentose, older ones brown, more glabrous, and sometimes fibrous.

RANGE. Rocky hills, mountainsides, or in dry grasslands. New Mexico, north to Nebraska and Wyoming and west to California.

VARIETIES. The plant is very variable as to flower and leaf characters, and at least 10 variations have been listed by Hall and Clements (1923, pp. 198–208).

C. parryi var. *attenuatus* Jones is a variant with flowers 5–7 in a head; bracts 13–15, with slender straight apices; upper leaves narrowly linear, with the upper ones seldom overtopping the inflorescence, green and somewhat viscid, not tomentulose but pubescent or almost glabrous. Known from New Mexico, Arizona, Colorado, Utah, Idaho, and Nevada.

REMARKS. The genus name, *Chrysothamnus,* is from the Greek words *chrysos* ("gold"), and *thamnus* ("shrub"), meaning "golden-flowered shrub." The species name, *parryi,* is in honor of C. C. Parry, one of the naturalists of the United States–Mexico boundary survey in the early 1850's. The plant is occasionally browsed by livestock during times of stress.

DOUGLAS RABBIT-BRUSH

Chrysothamnus viscidiflorus (Hook.) Nutt.

FIELD IDENTIFICATION. Very variable round-topped shrub attaining a height of ½–8 ft., usually branched near the base and with fibrous brown bark.

FLOWERS. Borne in terminal cymes; involucre ⅕–⅓ in., with 10–15 bracts which are linear to oblong to lanceolate, apex obtuse or acute, thickened, surface glabrous or puberulent, sometimes with a greenish thickened spot near the apex; flowers 4–6; corolla 5-lobed, tubular-funnelform, gradually expanded toward the apex, ⅙–¼ in. long, glabrous or somewhat viscid, corolla-lobes 1/25–1/12 in. long, erect or recurved; anthers lanceolate, acute; styles short-exserted, stigmatic portion longer than the lanceolate appendage.

FRUIT. Achenes ⅛–⅙ in. long, narrowed to base, 5-angled, usually densely villous-hairy; pappus rigid, whitish, as long as the corolla or longer.

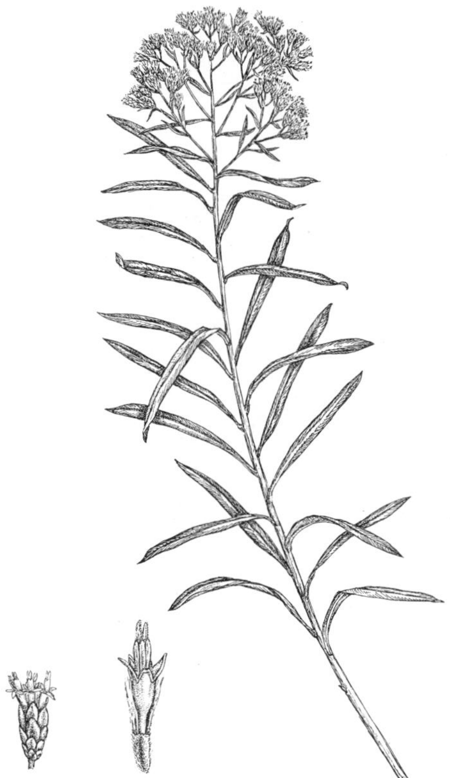

DOUGLAS RABBIT-BRUSH
Chrysothamnus viscidiflorus (Hook.) Nutt.

LEAVES. Alternate, linear to oblong or lanceolate, sometimes twisted, apex acute or obtuse, blade length ¾–2½ in., width 1/25–⅖ in., nerves 1–5, surfaces glabrous or puberulent, viscid, margins scabrous.

TWIGS. Mostly erect, rather leafy, brittle, striate, greenish or whitish, glabrous or puberulent.

RANGE. The species and its varieties are widespread in the western United States. West Texas and New Mexico; northward to North Dakota and west to British Columbia, Washington, Oregon, and California.

REMARKS. The genus name, *Chrysothamnus,* refers to the golden-flowered branches. The species name, *viscidiflorus,* is for the viscid exudations on the flowers and other parts of the plant. The plant is very variable as to size, flowers, and leaves. For the subspecies and intergrading forms, the reader is referred to Hall and Clements (1923, pp. 181–186).

MEXICAN AGERATUM

Ageratum corymbosum Zucc.

FIELD IDENTIFICATION. An erect or ascending perennial to 3 ft, occasionally somewhat woody toward the base.

FLOWERS. Light purple, in densely set corymbs or

MEXICAN AGERATUM
Ageratum corymbosum Zucc.

short racemes, ¾–1½ in. across with 5–10 heads; flower pedicels ¼–⅓ in. long, pubescent, some with a solitary linear bract and 2 filiform-linear bracts ⅛–¼ in. long at the base of the corymbs; heads about ¼ in. in diameter; involucre campanulate, 2–3-seriate, bracts numerous, linear, stiff, erect, grooved, ⅛–3/16 in. long, densely hairy, ciliate, basal ones fewer and shorter; florets 1/25–1/12 in. long, lobed at apex, stamens exserted, disk flowers numerous.

FRUIT. Achene 5-angled, pappus a paleaceous crown.

LEAVES. Very variable in shape, either ovate, oval or lanceolate, opposite, in remote pairs 1½–3 in. apart, blades ¾–4 in. long, ½–¾ in. wide, apex acuminate to acute or obtuse, base broadly cuneate, margin with the upper two thirds with a few coarse blunt teeth, lower third entire, upper surface dull green and pubescent, lower surface densely brown-tomentose and reticulate-veiny; petiole 1/16–¼ in., densely tomentose; leaves often with small ones at their base.

TWIGS. Green to brown, young ones densely pubescent, older ones more glabrous.

RANGE. On high rocky slopes, western Texas, New Mexico, and Arizona; in Mexico from Sonora to Zacatecas.

REMARKS. The genus name, *Ageratum,* is from a Greek word meaning "ageless," and the species name, *corymbosum,* refers to the corymbose arrangement of heads.

BLUE EUPATORIUM

Eupatorium azureum DC.

FIELD IDENTIFICATION. Shrub with spreading tomentulose branches.

FLOWERS. Borne in small, dense, terminal corymbs, 40–70-flowered; heads azure-blue, slender-pediceled, about ⅜ in. in diameter; involucre campanulate; bracts herbaceous, in series, imbricate, striate, mostly acute; style branches elongate, exserted, thickened toward the apex; receptacle flattened and glabrous.

FRUIT. Achene usually 5-angled, pappus of one series of capillary bristles.

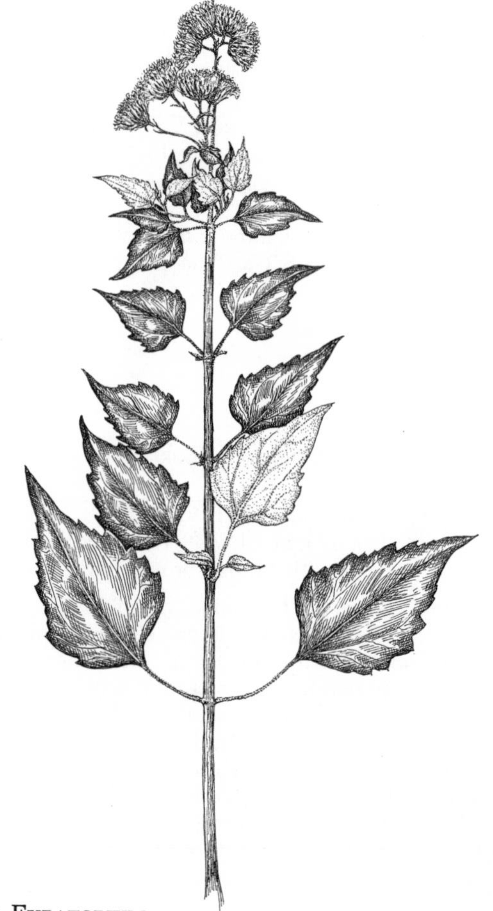

BLUE EUPATORIUM
Eupatorium azureum DC.

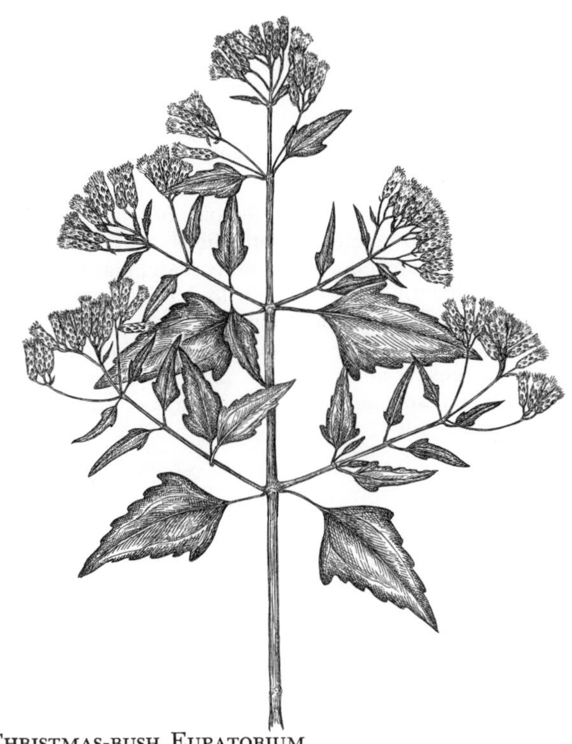

CHRISTMAS-BUSH EUPATORIUM
Eupatorium conyzoides Vahl

LEAVES. Opposite, those of the branches smaller and narrower than those of the stem, deltoid-ovate, apex acute or short-acuminate, base cuneate to truncate or subcordate, margin coarsely crenate-dentate, except the upper third and base, blade length 1½–2½ in., width 1¼–1½ in., lower surface paler and pubescent to gray tomentose, upper surface more glabrate; palmate-veined at or near the base; petioles ¼–1¼ in., pubescent.

RANGE. In caliche soil, southwest Texas; in Mexico in Tamaulipas, Nuevo León, and San Luis Potosí.

REMARKS. The genus name, *Eupatorium,* is for Mithridates Eupator (132–63 B.C.), who is said to have used a closely related plant in medicine. The species name, *azureum,* means "sky-blue," for the color of the flower. The plant is reported to be used in Mexico for astringent poultices.

CHRISTMAS-BUSH EUPATORIUM

Eupatorium conyzoides Vahl

FIELD IDENTIFICATION. Vigorous shrubby plant attaining a height of 3–10 ft. The habit is erect or reclining, and the branches are herbaceous above and somewhat woody near the base.

FLOWERS. Borne October–November, rather showy, abundant, fragrant, pale blue or white; heads in flattish corymbs, numerous and open, 15–25-flowered; ray flowers absent; receptacle flat and naked; involucre cylindric, bracts closely imbricate and appressed, coriaceous, striate; corolla tubular, 5-toothed; style branches exserted, thickened toward the apex.

FRUIT. Achene 5-angled, truncate; pappus in 1 series, slender, capillary bristles.

LEAVES. Variable in size, opposite, rhombic-ovate to triangular or ovate-lanceolate, apex acuminate, base mostly cuneate, margins sparsely and acutely toothed or more rarely entire (basal teeth sometimes enlarged and hastate); 3–5-nerved; surfaces varying from puberulent to pubescent or even tomentose; blades 1–5 in. long; petioles ¼–1 in., pubescent to tomentose.

STEMS. Slender, elongate, striate, green to brown or gray, pubescent, older ones more glabrous.

RANGE. In rocky or clay soils. In Texas along the Rio Grande River. In Jim Wells, Kleberg, Nueces, Hidalgo, and Cameron counties. Widespread in Mexico and tropical America.

REMARKS. The genus name, *Eupatorium,* is for Mithridates Eupator (132–63 B.C.), who is said to have used a related plant for medicine. Some vernacular names used for the plant in Latin-American countries are Xtokabal, Cihuapatli, Ciguapazle, Crucito, Santa María, Crucito Olorosa, Garrapato, and Varejón de Caballo. It is reported that the leaves are used as an emmenagogue in Mexico.

GOLDENROD EUPATORIUM

Eupatorium solidaginifolium Gray

FIELD IDENTIFICATION. Suffruticose plant attaining a height of 1–2 ft, but southward in Mexico becoming a much-branched shrub.

FLOWERS. May–October, mostly white, thyrsoid-paniculate, 2–8 in. long, leafy; involucral bracts linear to lanceolate, acute, inner ones ⅙–⅕ in. long, 8–15; heads sessile or subsessile on short terminal branches, 3–6-flowered; receptacle naked; corolla tubular, 5-toothed; style branches elongate, exserted, thickened toward the apex.

FRUIT. Achenes 5-angled, truncate, pubescent; pappus a single row of slender capillary bristles.

LEAVES. Opposite, oblong to narrowly ovate-lanceolate, apex acute or acuminate, strongly palmately nerved at or near the rounded or subcordate base, margin entire or sometimes obscurely and remotely dentate, blade length ¾–2½ in., or longer on lower stems of some plants, surfaces glabrate or puberulent to pubescent along the veins, especially beneath; petioles ⅙–⅖ in.

STEMS. Slender, green to light brown, striate, puberulent, when older brown and glabrous.

RANGE. Usually on dry hills in calcareous soil at

GOLDENROD EUPATORIUM
Eupatorium solidaginifolium Gray

altitudes of 3,000–5,500 ft. Trans-Pecos Texas, New Mexico, and Arizona; in Mexico in Chihuahua, Coahuila, and Durango.

REMARKS. The genus name, *Eupatorium,* is for Mithridates Eupator (132–63 B.C.), who is said to have used a related plant in medicine. The species name, *solidaginifolium,* refers to the goldenrod-like leaves.

AGERATUM EUPATORIUM

Eupatorium ageratifolium DC.

FIELD IDENTIFICATION. Shrub attaining a height of 1–6 ft, with slender spreading branches. Very variable in leaf form and degree of pubescence.

FLOWERS. Borne in convex or flattened corymbs, white or purplish, fragrant; receptacle flat and naked; involucre 1/6–1/5 in. high, bracts in 1–2 series, narrowly lanceolate to linear, obtuse or acute, green to purple; heads numerous, about 2/5 in. long, well-pediceled, 10–30-flowered; rays absent; corolla white, tubular, 5-lobed; stigmas elongate, exserted, thickened toward the apex.

FRUIT. Achenes 1/12–1/8 in. long, 5-angled, truncate; pappus a single row of white or rosy capillary bristles.

LEAVES. Opposite, blade deltoid or rhombic-ovate, apex obtuse to acute or sometimes acuminate; base rounded, truncate, cuneate or subcordate; margin coarsely and rather obtusely dentate to obscurely serrate or subentire; average blade length 1–3 in. (sometimes to 5 in.), width 5/8–1 5/8 in. (sometimes to 3 in.), firm and subcoriaceous; 3-nerved at the base, surfaces minutely puberulent (especially along the veins beneath) or glabrous; petioles 1/4–1 in.

STEMS. Slender, terete, striate, green to brown or reddish, puberulent; older ones gray to brown and glabrous.

RANGE. Rocky ravines, limestone ledges, and edges of bluffs, central and western Texas. In Mexico in Coahuila, Nuevo León, Tamaulipas, San Luis Potosí, and Veracruz; also in Cuba and the Bahamas.

VARIETY. Variety with all parts covered with a dense

AGERATUM EUPATORIUM
Eupatorium ageratifolium DC.

WRIGHT EUPATORIUM
Eupatorium wrightii Gray

fine pubescence has been named *E. ageratifolium* var. *acuminatum* Coult. It is known from the Texas coast at Point Isabel.

REMARKS. The genus name, *Eupatorium*, is for Mithridates Eupator (132–63 B.C.), who is said to have used a related plant in medicine. The species name, *ageratifolium*, refers to the *Ageratum*-like leaves. Also listed under the name of *E. havanense* H. B. K. Other vernacular names are Thoroughwort, Boneset, and Mist Flower.

WRIGHT EUPATORIUM

Eupatorium wrightii Gray

FIELD IDENTIFICATION. Shrubby, spreading, puberulent plant with leafy branches, attaining a height of 1–2 ft.

FLOWERS. September–October, white or purplish; corymbs small but numerous; involucre campanulate, ⅛–⅕ in. high, scarcely as long as the flowers; bracts in 1–2 series, green or purplish, oblong-lanceolate, acute, obscurely 3-nerved; ray flowers absent; heads ¼–⅓ in. long, about 12-flowered; corollas tubular and 5-toothed; receptacle flat and naked; style branches elongate, exserted, thickened toward the apex.

FRUIT. Achenes 5-angled, truncate; pappus a single row of many slender hairy bristles.

LEAVES. Opposite, small, blade length to 1¼ in. (average length ½–⅗ in.), upper leaves much smaller, ovate to oval, apex obtuse or subacute to rounded, base rounded or subcordate, margin entire or obscurely few-toothed, membranous to chartaceous or thickish, surfaces scabrous, abruptly contracted into a short-winged petiole.

STEMS. Much-branched, slender, terete, light brown to gray, puberulent, older ones gray and glabrous.

RANGE. On limestone hills and slopes at altitudes of 5,000–6,000 ft. In Trans-Pecos Texas in the Guadalupe, Davis, and Chisos mountains. In Mexico in Chihuahua, Coahuila, and San Luis Potosí.

REMARKS. The genus name, *Eupatorium*, is for Mithridates Eupator (132–63 B.C.), who is said to have used a related plant in medicine. The species name, *wrightii*, honors Charles Wright, who collected plants in Texas in 1847 and 1848, and who, in 1851, was sent as botanist and surveyor on the United States–Mexico boundary survey.

AMERICAN TAR-BUSH

Flourensia cernua DC.

FIELD IDENTIFICATION. Shrub to 6 ft, much-branched, leafy, erect or procumbent, resinous-viscid, with a hop-like odor.

FLOWERS. Borne in corymbs or panicles; peduncles short, curved, with 1–3 ovate to elliptic bracts at apex; heads about ⅜ in. high, discoid, yellow, nodding, solitary in the upper leaf axils; rays absent, disk subturbinate, 12–20-flowered, disk corollas about ⅛ in. long, resinous-dotted; involucre 2–4-seriate, bracts subequal, erect, lanceolate-oblong, acutish, resinous, the outer sometimes foliaceous and graduating into leaves; receptacle flat with scarious chaff folded around the achenes.

FRUIT. Achenes about ¼ in. long and 1/12 in. wide, oblong-cuneate, very villous, callous-margined, laterally flattened, thickened, apex truncate; awns 2–4, unequal, subulate, ciliolate, borne from the summit or with smaller ones between.

LEAVES. Simple, alternate, persistent, elliptic to oblong or ovate to oval, margin entire, apex acute and often mucronulate, base cuneate or acute, thickish, upper surface green and glabrous, somewhat resinous, lower surface paler and glabrous to puberulent, veins mostly obscure, but occasionally 1–2 pairs of lateral veins conspicuous, blade length ⅔–2 in., width ¼–½ in., bitter, odor aromatic and hoplike; petioles ⅛–1 in., puberulent.

TWIGS. Young ones slender, resinous, light brown to gray, pubescent; older ones gray to whitish, glabrous, somewhat striate.

RANGE. In dry soil of valleys, mesas, and flats to an

AMERICAN TAR-BUSH
Flourensia cernua DC.

altitude of 5,000 ft. Western Texas, New Mexico, and Arizona. In Mexico in Sonora, Chihuahua, Coahuila, Durango, San Luis Potosí, Zacatecas, and Mexico, D.F.

MEDICINAL USES. In Mexico, a decoction is made from the leaves and flower heads as a remedy for indigestion, and is also used for female ailments. The plant is considered to be unpalatable to livestock.

REMARKS. The meaning of the genus name, *Flourensia,* is not clear. The species name, *cernua,* refers to the drooping flower heads. It is also known under the vernacular names of Varnish-bush, Hojasen, and Hojase.

BROOM SNAKEWEED

Gutierrezia sarothrae Britt. & Rusby

FIELD IDENTIFICATION. Plant bushy, herbaceous above and woody toward the base, attaining a height of 4–16 in. The branches numerous, erect, and redivided into slender branchlets.

FLOWERS. June–October, inflorescence cymose-paniculate, heads numerous, yellow, radiate, about 1/8 in. thick, clavate-oblong; receptacle small and naked; disk and ray flowers usually 3–8 each; disk flowers fertile, with well-developed ovaries; involucre turbinate, about 1/12 in. thick, with thick, narrow, green-tipped bracts.

FRUIT. Achenes oblong, sericeous-pubescent, pappus of numerous, chaffy, oblong to linear scales.

LEAVES. Numerous, alternate, 3/8–4 in. long (mostly 3/8–1 1/2 in.), 1/25–1/12 in. wide, linear-filiform, entire, revolute when dry, puberulent to scaberulous.

STEMS. Very slender, green and herbaceous above, lower brown to gray, smooth or scaly and woody.

RANGE. The species occurs on arid rocky plains at altitudes of 2,800–7,000 ft. Western Texas, New Mexico, Arizona, California, Utah, Montana, Idaho, Nevada, Kansas, and Saskatchewan.

REMARKS. The genus name, *Gutierrezia,* is the name of a noble Spanish family, and the species name, *sarothrae,* is from the Greek word for "broom." The bushy plant is sometimes used for brooms by the Mexican and Indian people of the Southwest. Other names in common use are Broom-weed, Sheep-weed, Yellow-weed, Snake-weed, Hierba de Víbora, Hierba de San Nicolás, and Cayaye. A decoction of the plant is reported to be used in New Mexico as an emmenogogue and for gastric disturbances. In times of stress it is eaten in

BROOM SNAKEWEED
Gutierrezia sarothrae Britt. & Rusby

TICKY SNAKEWEED
utierrezia lucida Greene

nited amounts by sheep and horses. It is an indicator overgrazed land.

Thread-leaf Snakeweed, *G. sarothrae* (Pursh) Britt. ır. *microcephala* (DC.) Benson, is a variety with flow- heads averaging only 1/10 in. across; ray flowers usu- ly 4–5 and disc flowers 1–4. It occurs in Idaho, rizona, New Mexico, and west Texas, southward to oahuila.

TICKY SNAKEWEED

utierrezia lucida Greene

FIELD IDENTIFICATION. Woody-based perennial ½–2 with many slender branched stems from the base or ar it.

FLOWERS. June–October, crowded terminally, in 2–5- wered sessile glomerules; heads narrowly cylindric, out 1/25 in. thick, yellow, radiate; rays 1–2, usually stillate and fertile; disk flowers usually 1 or 2, usually rile, corolla 5-lobed, anthers united; bracts charta- ous, scarious-margined, tips greenish, imbricate, usu- ly 2 larger ones enclosing the ray and disk flowers, and e others small; ovary minute and sterile in disk wer.

FRUIT. Achenes small, oblong, pappus of several narrow scales.

LEAVES. Simple, alternate, yellowish green, narrowly linear to filiform, ½–2½ in. long, 1/25–1/12 in. wide, margin entire, 1-nerved, glabrous to puberulent, some sticky-glutinous, resin dots minute.

STEMS. Slender, numerous, often woody toward the base, green to dark brown or grayish.

RANGE. Common on dry stony mesas or on desert grasslands at altitudes of 1,200–7,000 ft. In western Texas, New Mexico, Arizona, southern California, Colorado, Nevada, and northern Mexico.

REMARKS. The genus name, *Gutierrezia*, is the name of a noble Spanish family, and the species name, *lucida*, may refer to the shiny, glutinous leaf surface. The plant is of no particular economic value. Sheep and goats usually avoid it, and it is considered to be poisonous to them if eaten in quantity.

NAKED-SEED WEED

Selloa glutinosa Spreng.

FIELD IDENTIFICATION. Perennial plant 12–48 in. Stems slender, wandlike, glabrous and glutinous, herbaceous above and woody near the base. Closely resembling *Gutierrezia*, but distinguished from it by the lack of pappus and minute size of the ray flowers.

FLOWERS. Borne in yellow, corymbose, terminal clusters ¾–2 in. across; heads numerous, crowded, radiate; ray flowers several (5–9), usually about 6, ligules no longer than the disk corolla, fertile; disk flowers about 6, perfect, sometimes a few fertile; anthers obtuse at base; stigmas flattened; receptacle small and naked, involucre ⅛–⅙ in. high, closely imbricate; bracts few, subcoriaceous, margins narrow and scarious, concave, obtuse, green to pale tan.

FRUIT. Achenes oblong or nearly so, about 1/25 in. long, puberulous or almost glabrous, pappus absent.

LEAVES. Simple, alternate, sometimes with smaller ones fascicled in their axils, linear to oblanceolate or lanceolate, ⅓–3½ in. long, ⅛–⅖ in. wide, margin entire, apex acute, base sessile, some 3-nerved, glabrous and glutinous, punctate-dotted.

TWIGS. Slender, simple, branched near the base; young ones green and glandular-glutinous; older ones tan to brown, less glutinous and woody near the base.

RANGE. Rocky soils of dry hillsides and arid grasslands at altitudes of 2,000–6,000 ft. Western Texas, New Mexico, and Arizona; in Mexico from Chihuahua to Chiapas.

MEDICINAL USES. It is reported that in Mexico a decoction of the plant is used as a household remedy for diarrhea, and a solution of the gum is used externally as a remedy for rheumatism and ulcers.

Naked-seed Weed
Selloa glutinosa Spreng.

remarks. The genus name, *Selloa,* is in honor of Friedrich Sellow (earlier spelled "Sello") (1789–1831), German botanist and collector in Brazil. The species name, *glutinosa,* refers to the glutinose, or sticky, exudation found on the foliage and inflorescence. The plant is listed by some authors under the name *Gymnosperma corymbosa* DC. It is known in Latin-American countries under the names of Jarilla, Moto, Mariquita, Tatatencho, Motita, Cola de Zorra, Xonequitl, Hierba Pegajosa, Jucu Ndede, Zazal, Pegajosa, and Escobilla.

Chomonque

Gochnatia hypoleuca Gray

field identification. Rigid branching shrub to 6–8 ft, or rarely a small tree with twigs pale and white-tomentose.

flowers. In white heads in leafy, paniculate fascicles crowded toward the tips of the branches; heads ⅓–⅖ in. long, discoid; involucres cylindric-campanulate, 4–7-flowered; phyllaries in series, graduate, imbricate, dry and indurate, ovate to oblong-elliptic, apex obtuse to acute or obtusely-acuminate; flowers perfect, corolla regular, white, similar, deeply 5-cleft into linear, revolute lobes; anthers long-caudate at base, the auricles tailed; stigmas short, round to obtuse or nearly truncate at apex.

fruit. Achenes about ⅛ in. long, oblong or nearly so, obscurely ribbed; bristles of pappus rigid, scabrous and capillary.

leaves. Simple, alternate, deciduous, elliptic to lanceolate, or oblong to oval, apex acute, base narrowed to a very short petiole, or subsessile, upper surface bright green and glabrous at maturity, lower surface finely white-tomentose, leathery, length ¾–2 in.; twigs and young leaves white-tomentose.

range. On dry hillsides, southwestern and western Texas, from the Nueces River westward. In Mexico from Nuevo León and Coahuila to Querétaro.

remarks. The genus name, *Gochnatia,* is for Caroli

Chomonque
Gochnatia hypoleuca Gray

Dune Sumpweed
Iva imbricata Walt.

Gochnat, Argentine botanist, who wrote on the Composite Family. The species name, *hypoleuca,* refers to the white under surface of the leaf. Known also under the Mexican name of Ocotillo, but this name is better used with the unrelated *Fouquieria splendens.*

Dune Sumpweed

Iva imbricata Walt.

FIELD IDENTIFICATION. Perennial glabrous plant of coastal sand dunes or rocky shores. Stem 2–10 ft, erect or decumbent, branched, somewhat fleshy above but woody toward the base.

FLOWERS. August–October, spicate or racemose, terminating leafy stems with the upper leaves reduced in size (bractlike), and subtending the heads; heads short-pedunculate or almost sessile, solitary, drooping or partly so, greenish; involucre campanulate, subtended by 6–9 bracts which are 2-seriate, imbricate, foliaceous, suborbicular to oblong, obtuse or retuse at the apices, ⅛–⅕ in. long; peripheral flowers 2–4, fertile, tube truncate, 5-lobed; disk flowers more numerous, mixed with spatulate chaff, perfect, not fruit-producing, corolla ⅙–⅕ in. long, funnelform, sparingly glandular; lobes 5, ovate, acute, recurved; anthers yellow, entire at base, with mucronate appendages; stigmas 2, simple, dilated above.

FRUIT. Achenes about 1/12 in. long, obovoid, slightly flattened, glabrous or slightly glandular, pappus absent.

LEAVES. Alternate or opposite, numerous, somewhat fleshy and glabrous, sessile or obscurely petiolate, linear-elliptic to spatulate-oblong, margin entire or with a few remote teeth on the margin, apex acute to obtuse and some mucronate, base gradually narrowed, length ¼–2 in., width ⅙–⅓ in., upper ones much reduced and bractlike among the flowers, dull green, both surfaces glabrous and obscurely 1–3-veined.

TWIGS. Erect or somewhat decumbent, stout, glabrous, somewhat grooved longitudinally, green to brown.

RANGE. Coastal sand dunes or rocky shores. Virginia, south to Florida and westward to Louisiana. Also in Cuba and the Bahamas.

REMARKS. The genus name, *Iva,* is an old name for a medicinal plant, *Ajuga iva.* The species name, *imbricata,* refers to the imbricate bracts of the receptacle. The foliage has a heavy odor.

Big-leaf Sumpweed

Iva frutescens L.

FIELD IDENTIFICATION. Shrub of saline marshes to 12 ft, with many stems from the base.

FLOWERS. July–August, inconspicuous, in terminal racemes 3–8 in. long, individual pedicels 1/16–⅛ in. long; flower heads axillary, drooping, depressed-hemispheric, greenish, about 3/16 in. across, subtended by 4–5 oval to obovate, obtuse, veiny bracts hardly more than 1/16 in. long; disk corollas perfect, not fruit-producing, tubular, 1/12–⅛ in. long; lobes 5, ovate, acute; stamens 5, anthers entire at base with mucronate appendages; marginal pistillate flowers about 5; stigmas simple, exserted, dilated above.

FRUIT. Achenes obovate, cuneate, compressed, 1/12–⅛ in. long, glabrous or puberulent, pappus absent.

LEAVES. Opposite, lanceolate-oblong to linear, 2–6 in. long, ½–2 in. wide; thick and fleshy, 3-veined at base; margin serrate with large, coarse, remote teeth or none toward the apex and base; apex acute or acuminate; base cuneate, sometimes abruptly so; both sides pale

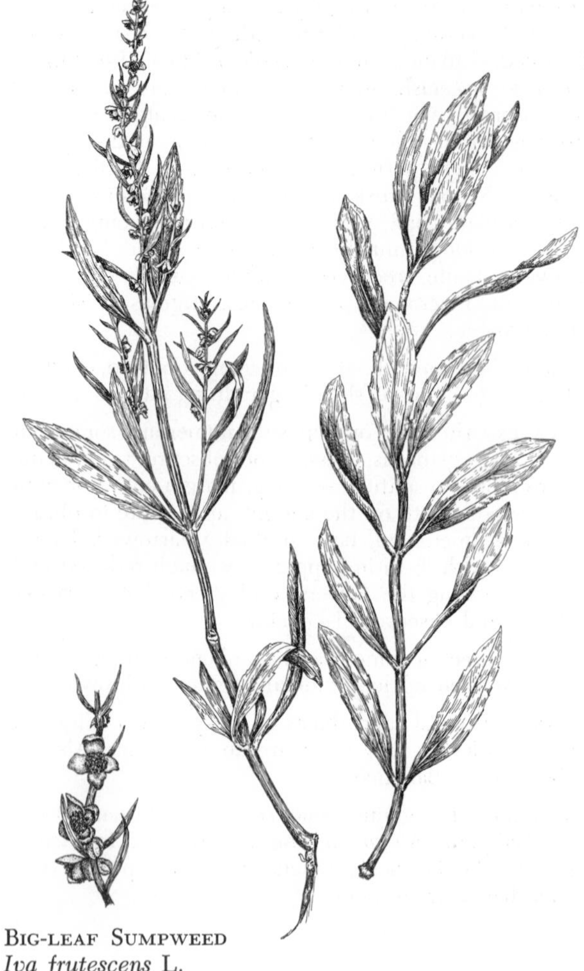

Big-leaf Sumpweed
Iva frutescens L.

green, slightly viscid or strigose (porulose under the glass); petiole grooved on upper side, ¼–⅓ in. long, stout.

TWIGS. Rather herbaceous above, more woody toward the base, opposite, rather remote, leafy toward the apex, slender, gray to light brown, smooth, lenticels rather abundant.

RANGE. On the coast in saline or brackish marshes or on sea beaches. Common on the Louisiana and upper Texas coast; eastward to Florida and northward to Massachusetts.

VARIETY. Northern Sumpweed, *I. frutescens* var. *oraria* (Bart.) Fern. & Griscom, is a variety hardly over 3 ft, with elliptic-ovate or linear leaves, obtuse or acute; heads ⅕–¼ in. across with 5, rarely 6, bracts. Known from Massachusetts to Maryland.

REMARKS. The genus name, *Iva,* is a medieval name for a medicinal plant, and the species name, *frutescens,* refers to the shrubby character. Vernacular names are High Water Shrub, Salt Bush, Salt Marsh Elder, Jesuit Bark, and False Jesuit Bark.

Prickly Oxytenia

Oxytenia acerosa Nutt.

FIELD IDENTIFICATION. Slender-stemmed erect perennial 3–6 ft, somewhat woody at the base. The leaves narrow and threadlike.

FLOWERS. Borne July–September, the panicles dense with many, small, whitish, discoid heads; inner flowers 10–20, sterile, staminate, with white corollas and distinct anthers, pistil rudimentary, receptacle convex; pistillate of about 5 outer flowers without corollas; involucre ⅙–⅕ in. wide; bracts 5, distinct, ovate, acuminate, canescent.

FRUIT. Achenes about 1/12 in. long, obovoid, long-villous, pappus absent or reduced to a minute scale, areola large.

LEAVES. Alternate, upper leaves often entire, lower

Prickly Oxytenia
Oxytenia acerosa Nutt.

leaves pinnately parted into 3–5 long filiform lobes, each about 1/25 in. wide.

TWIGS. Erect, green to gray-strigose; especially near the ends, leafless and rushlike or sometimes with many leaves; woody parts at base gray to brown.

RANGE. On dry plains or in valleys, often in alkaline soil, at altitudes of 3,500–6,500 ft. New Mexico, Arizona, and California; north into Colorado, Utah, and Nevada.

REMARKS. The genus name, *Oxytenia,* is from the Greek and refers to the pointed leaves, meaning literally "pointed thread." The species name, *acerosa,* means "needle-shaped," also referring to the leaves. A vernacular name is Copperweed. It is reported to be poisonous to stock, but is usually not browsed.

Arrow-weed Pluchea

Pluchea sericea (Nutt.) Cov.

FIELD IDENTIFICATION. Western shrub 3–12 ft, with numerous, erect, willow-like branches. Sometimes forming pure stands in alluvial soil along streams.

FLOWERS. May–June, in small, terminal, cymosely glomerate discoid heads, pink or purple; receptacle flat and naked; involucre campanulate, pubescent, imbricate in 2–3 series; outer involucral bracts purplish, ovate, obtuse or acute, margins scarious and fimbriate, coriaceous; inner bracts thinner, linear, white to straw-colored, deciduous; corollas all tubular, those in the center bisexual, but often sterile, with 5-cleft tubes; marginal corollas pistillate and fruit-producing, with a slender, truncate, or 2–3-toothed tube; anthers sagittate at the base with caudate auricles.

FRUIT. Achenes small, 4–5-ribbed, glabrous, not beaked, with a single series of capillary bristles, pappus of sterile flowers with bristles more rigid and apices thickened.

LEAVES. Numerous, simple, alternate, odor unpleasant, length 1/2–1 3/4 in., width 1/12–1/4 in., narrowly elliptic or linear-lanceolate, apex acute, base tapering and sessile, margin entire, both surfaces with silvery-white, appressed pubescence, pinnately veined.

TWIGS. Slender, pale, obscurely nerved, densely silvery appressed-hairy, the numerous erect twigs almost willow-like.

RANGE. Usually below an altitude of 3,000 ft, in sandy or saline soil along water courses. Trans-Pecos Texas, New Mexico, Arizona, Utah, and California. In Mexico in Sonora, Chihuahua, and Baja California.

MEDICINAL USES. An infusion of the leaves was employed by the Pima Indians as a remedy for sore eyes.

REMARKS. The genus name, *Pluchea,* honors N. A. Pluche, a Parisian naturalist of the eighteenth century. The species name, *sericea,* refers to the silky, closely appressed hairs. It is also known under the vernacular names of Arrow-wood, March Fleabane, Cachimilla, and Cachanilla. The flowers are considered to be a good bee food. The foliage is browsed by deer and livestock. The slender twigs were used by the Indians for arrows, bird cages, storage bins, baskets, and wattle for roofs and walls of huts.

Arrow-weed Pluchea
Pluchea sericea (Nutt.) Cov.

Slender Poreleaf

Porophyllum gracile Benth.

FIELD IDENTIFICATION. Slender woody-based perennial 1–3 ft.

FLOWERS. Maturing April–May, borne in loose cymes; heads straw-colored to whitish with purple streaks, discoid, 5/8–3/4 in. high; phyllaries 5, 1/12 in. wide or less, 1-seriate, oblong, obtuse, scarious-margined, marked with conspicuous linear oil glands; receptacle small and naked; ray flowers absent.

FRUIT. Achenes 1/3–2/5 in. long, slender, elongate, apex

Slender Poreleaf
Porophyllum gracile Benth.

attenuate; pappus of numerous capillary, scabrous bristles.

LEAVES. Opposite or alternate, ⅜–2½ in. long, about 1/15 in. wide, filiform to linear, sessile, minutely porose, odorous when crushed.

TWIGS. Slender, striate, green to gray and glaucous at first; older wood toward the base with bark gray to brown and glabrous.

RANGE. On arid plains, dry rocky slopes, or canyons at altitudes to 3,500 ft. Western Texas, New Mexico, Arizona, California, Nevada, Baja California, and Sonora.

MEDICINAL USES. A bitter decoction of the leaves is sometimes used by the Indians and Mexicans for intestinal affections.

REMARKS. The genus name, *Porophyllum,* is from the Greek words *poros* ("a passage"), and *phyllon* ("leaf"), with reference to the oil glands of the phyllaries. Th species name, *gracile,* refers to the plant's slender habi It is also known under the vernacular name of Yerb del Venado. It is said to be browsed occasionally b deer and cattle.

Shrubby Poreleaf

Porophyllum scoparium Gray

FIELD IDENTIFICATION. Western shrubby-based perennial with numerous slender rushlike branches.

FLOWERS. In pedunculate, solitary, discoid, yellowis heads; receptacle small and naked, ray flowers absent involucre campanulate, bracts 7–9, ¼–⅜ in. long, 1/12 in wide or less, broadly lanceolate, 1-seriate, equal an free, green, with 1 or 2 stripes of black oil glands.

FRUIT. Heads when spread in the fruiting stage ½–¾ in. across and ⅝–½ in. high; achenes slender, elongate pappus composed of long, copious, free, tawny-brown scabrous bristles.

LEAVES. Opposite or alternate, sessile, linear-filiform

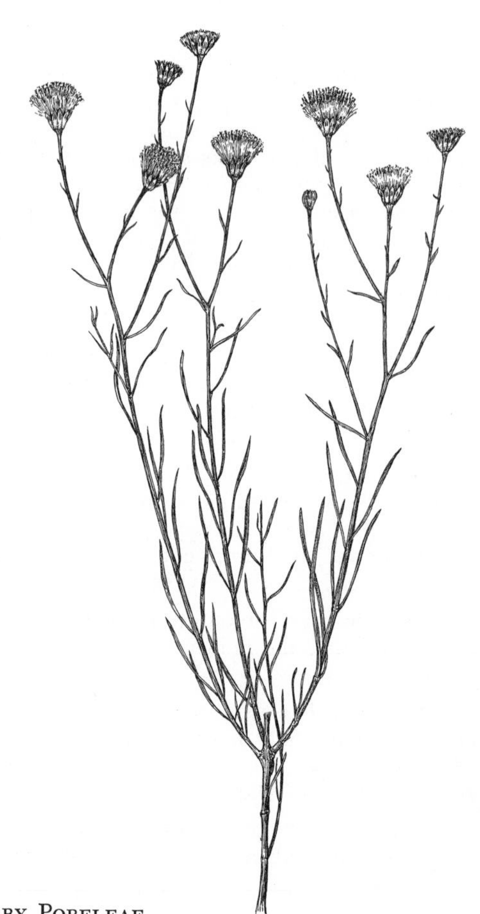

Shrubby Poreleaf
Porophyllum scoparium Gray

mucronate, entire, thick, firm, ⅜–2½ in. long, about 1/25 in. wide.

TWIGS. Numerous, rushlike, slender, green or somewhat glaucous at first, later brown to black or gray.

RANGE. Rocky banks and plains of southwestern Texas, New Mexico, and Mexico in the states of Chihuahua and Coahuila.

REMARKS. The genus name, *Porophyllum*, means "pore-leaf" and refers to the small pores of the phyllaries. The species name, *scoparium*, denotes the broomlike twigs. The plant is also known under the names of Hierba del Venado, Jarilla, and Pomerillo. In Mexico it has been used by the natives as a remedy for fevers, rheumatism, and affections of the stomach and intestines.

Willow-leaf Stevia

Stevia salicifolia var. *stenophylla* (Gray) Robins.

FIELD IDENTIFICATION. Rigid-stemmed subshrub 8–16 in., somewhat woody toward the base.

FLOWERS. Borne in flattish, terminal, crowded panicles ¾–3 in. wide; heads small, perfect, white or pinkish, 3–5-flowered; involucre narrowly cylindric, ⅙–⅓ in. high; bracts 5–6, mostly equal, linear, acute, appressed, rather rigid, sometimes glutinous; receptacle naked; corolla tubular and 5-toothed.

FRUIT. Achenes slender, linear, 5-angled, sparsely hispidulous; pappus 1–3-awned or with a few squamellae.

LEAVES. Numerous, opposite, smaller ones often clustered in their axils, blade linear-lanceolate, margin entire or with a few remote teeth, apex acute or acuminate, base narrowed and short-petiolate or subsessile, length 1¼–4 in., upper surface dull green and glabrous, lower surface slightly paler, puberulent to glabrous, leaves sometimes a little glutinous.

STEMS. Slender, stiff, pubescent, reddish brown, later glabrous and dark brown or gray to black.

RANGE. On prairies in southwestern Texas, southward in Mexico to Chihuahua and Querétaro.

REMARKS. The genus name, *Stevia*, honors the Russian Christian von Steven (1781–1863). The species name, *salicifolia*, refers to the willow-like leaves, and the variety name, *stenophylla*, alludes to their narrowness.

Willow-leaf Stevia
Stevia salicifolia var. *stenophylla* (Gray) Robins.

Narrow-leaf Horse-brush

Tetradymia filifolia Greene

FIELD IDENTIFICATION. Plant low, straggling, irregularly branched, woody near the base.

FLOWERS. In narrow yellow heads, involucre cylindric, ⅓–⅖ in. long; bearing 4–6 bracts which are imbricate, ¼–⅜ in. long, oblong to linear, apex obtuse or rounded, surface densely gray-tomentose; ray flowers absent; receptacle flat; corolla-tube ¼–⅜ in. long, somewhat longer than the 5 linear to narrowly lanceolate lobes; stigma bifid, segments linear, exserted beyond the corolla-lobes at maturity.

FRUIT. Achene to ⅛ in. long, terete, 5-nerved, narrowed somewhat toward one end, dark colored, glabrous; pappus of numerous soft, capillary bristles to ½ in. long.

LEAVES. Alternate, crowded, filiform to linear, ½–1¾ in. long, apex spinose, surfaces densely and finely gray-tomentose.

TWIGS. Young ones numerous, slender, crowded toward the apex of branches, densely gray-tomentose; older ones gray and glabrous and roughened by the leaf scars, flaking away to expose the brownish inner bark.

RANGE. Known only from the type locality on Round Mountain, near Tularosa, New Mexico.

NARROW-LEAF HORSE-BRUSH
Tetradymia filifolia Greene

REMARKS. The genus name, *Tetradymia*, is from the Greek *tetradymos* ("fourfold") and refers to the 4-flowered heads, and the species name, *filifolia*, is for the filiform, or threadlike, leaves.

GRAY HORSE-BRUSH

Tetradymia canescens DC.

FIELD IDENTIFICATION. Western shrub 1–3 ft, herbaceous above but woody below. Spreading-assurgent, stiffly much-branched.

FLOWERS. June–August, terminal, short-peduncled, the compact cymes with 5–20 heads; heads discoid, bisexual, yellow, 4-flowered, rays absent; corolla-lobes linear, spreading, often longer than the elongate slender tube; anthers exserted, base sagittate, tips triangular-lanceolate; style branches flat and obtuse; receptacle flat, small, naked; involucre cylindric, ¼–⅖ in. long; bracts 4, thickened, imbricate, equal, linear to oblong, acute to obtuse, ridged dorsally, concave, margins thin and scarious, densely gray-tomentose.

FRUIT. Achenes linear to slightly obovoid, terete, faintly 5-nerved, about ¼ in. long or less, densely silky-hairy, crowned with numerous white, soft, minutely-scabrous bristles.

LEAVES. Rather persistent, simple, alternate, permanently and densely woolly-canescent, linear to lanceolate or narrowly oblanceolate, sessile or nearly so, margin entire, apex acute and callous-tipped, midrib evident, lateral nerves obscure, length ¼–1¼ in., sometimes crowded.

TWIGS. Short and rather stout, branching above, at first silvery-canescent-tomentose; older glabrous, gray and shedding in fibrous strips to expose the brownish inner bark.

RANGE. On dry sunny sites of barren plains and rocky or sandy soils. New Mexico, west to California and north to Montana and British Columbia.

MEDICINAL USES. Hopi Indians are reported to use the leaves and roots as a tonic and for uterine disorders.

VARIETY. *T. canescens* var. *inermis* (Nutt.) Gray is a form with shorter, broader leaves and involucres, but many intergrading forms occur.

REMARKS. The genus name, *Tetradymia*, is from the Greek word *tetradymos* ("fourfold"), referring to the 4-flowered heads. The species name, *canescens*, means "gray-hairy." The plant is considered to be poor or worthless livestock browse.

LONG-STALK GREEN-THREAD

Thelesperma longipes Gray

FIELD IDENTIFICATION. Strong-scented Western perennial plant to 24 in. Herbaceous above, but woody toward the much-branched and leafy base.

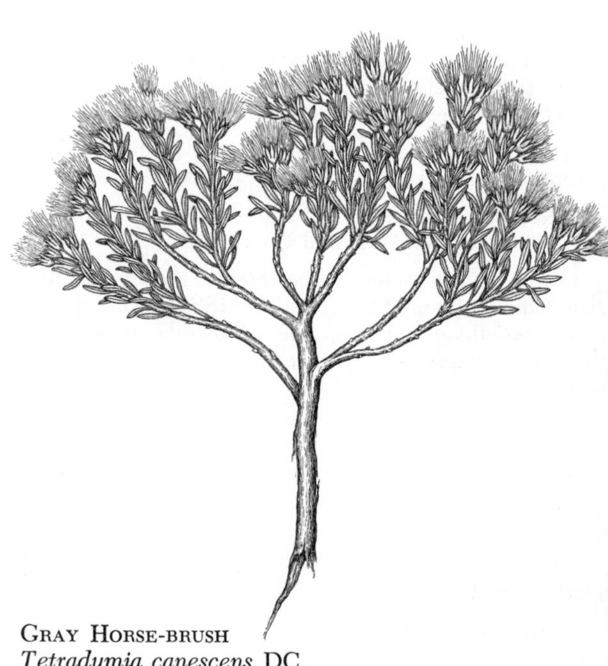

GRAY HORSE-BRUSH
Tetradymia canescens DC.

Long-stalk Green-thread
Thelesperma longipes Gray

FLOWERS. June–September, borne on simple, elongate, filiform, finely grooved peduncles to 10 in. long, and much exceeding the leaf stems; heads discoid, yellow, ¼–⅖ in. wide; involucre double, the outer phyllaries small, herbaceous, lance-shaped, less than 1/16 in. long; inner submembranous, dark, joined at about the middle, margin thin and scarious, about ⅛–3/16 in. long; heads rayless; lobes of disk corolla yellow, ovate to oblong, decidedly shorter than the cylindraceous throat; receptacle flat with white-scarious deciduous chaff.

FRUIT. Achene 1/12–⅙ in. long, oblong to linear, curved, thickened, muricate, almost terete, pappus absent (rarely with 2 very short teeth).

LEAVES. Opposite, crowded at the base of the stems, 1–3 in. long, entire or parted into 3–5 filiform divisions no wider than the rachis, glabrous, entire, dull green.

STEMS. Numerous, slender, leafy below, glabrous, reddish brown, finely grooved, woody only toward the base.

RANGE. Dry hills and mesas, slopes and canyons at altitudes of 5,000–6,000 ft. In western Texas in the Guadalupe Mountains and perhaps elsewhere. Collected by the author in McKittrick Canyon, near Pine Springs, Texas. In New Mexico in the Capitan, Organ, White, Tortugas, and Guadalupe mountains; also in southern Arizona. In Mexico in Coahuila, Nuevo León, and San Luis Potosí.

REMARKS. The genus name, *Thelesperma,* is from the Greek words *thela* ("nipple") and *sperma* ("seed"), on account of the papillose achenes. The species name, *longipes,* refers to the elongate flower peduncles. Vernacular names in use are Cota and Hierba de San Nicolás. It is reported to be used as a tea in Mexico, giving a reddish tinge when boiled.

Texas Varilla

Varilla texana Gray

FIELD IDENTIFICATION. Glabrous and frutescent plant to 2 ft, much branched and leafy near the base.

FLOWERS. Heads small, discoid, several in corymbose cymes, or more often borne on long naked, or minutely bracteate, peduncles 3–6 in. long. Involucres short and broad, several-flowered, with the bracts few, linear-lanceolate or subulate, about ⅕ in. long, appressed-imbricate, usually striate, indurate. Receptacle even-

Texas Varilla
Varilla texana Gray

tually oblong or conical; ray flowers absent; disk flowers fertile and perfect; corollas yellow and tubular, with bractlets like those of the involucre; stigmas with obtuse or apiculate conic tips; disk rounded, much surpassing the involucre.

FRUIT. Achenes narrow linear-oblong, subcylindric, 8–15 nerved, pappus absent.

LEAVES. Opposite or nearly alternate, sessile, linear, terete, apex obtuse, very thick and succulent, length ½–1½ in., width 1/25–1/12 in., margin entire.

RANGE. From the Nueces River drainage to the Rio Grande. Also in adjacent Mexico and in New Mexico near El Paso. Often in saline soil.

REMARKS. The genus name, *Varilla*, is derived from a Spanish name meaning "little stick." The species name refers to the state of Texas, where it is most abundant.

JIMMY GOLDENWEED

Haplopappus heterophyllus (Gray) Blake

FIELD IDENTIFICATION. Plant perennial, woody toward the base, stems numerous, slender, ascending, strict, leafy in entirety, 1½–2½ ft.

FLOWERS. June–September, borne in crowded cymes 2–6 in. across, heads numerous, discoid, yellow; involucre 1/6–1/5 in. high, turbinate; bracts in several series, erect, oblong to lanceolate, apex obtuse or acute, glabrous or slightly pubescent; ray flowers absent; disk flowers 7–16, slender, tubular, 1/5–¼ in. long, glabrous or slightly puberulent, lobes 1/25–1/12 in. long, erect, linear to lanceolate, acute; branches of the style somewhat exserted, 1/15–1/12 in. long, appendages ovate and acute, less than one half as long as the stigmas.

FRUIT. Achene obconic, about 1/12 in. long, silky-hairy; pappus composed of a single row of brownish capillary bristles about as long as the corolla.

LEAVES. Alternate, simple, upper ones linear-lanceolate to oblanceolate, entire, acute, base narrowed, lower leaves sometimes remotely serrate or laciniate-dentate, blade length ¾–2½ in., width 1/25–1/5 in., only the midvein apparent, pale green, glabrous, or in some variations hairy; sometimes with secondary clustered leaves in the axils of the others.

STEMS. Gray to white, glabrous and lustrous, older stems dark gray and fibrous, wood hard and brittle.

RANGE. Dry hills and plains to an altitude of 6,000 ft. A roadside weed in some irrigated areas. In Trans-Pecos Texas, New Mexico, eastern Colorado, Arizona, and Mexico.

REMARKS. The genus name, *Haplopappus*, is from the Greek words *haploos* ("simple") and *pappos* ("down"), referring to the simple pappus hairs. The species name, *heterophyllus*, refers to the variable leaves. It is not generally grazed by cattle, and when eaten produces trembles, which is transferable through the milk to human beings.

JIMMY GOLDENWEED
Haplopappus heterophyllus (Gray) Blake

BURRO GOLDENWEED

Haplopappus hartwegii (Gray) Blake

FIELD IDENTIFICATION. Western woody subshrub with numerous branches from the base.

FLOWERS. Borne in racemose cymes; involucre 1/8–1/5 in. long and broad, turbinate; involucre composed of overlapping bracts which are oblong, stiff, glabrous, granular to resinous, margin thin, and apex obtuse and resinous; ray flowers absent, disk flowers 10–20; corolla 1/6–¼ in. long, tube slender and inflated above, apex almost closed, glabrous or puberulent; lobes of corolla lanceolate, acute, erect, glabrous to puberulent; style branches 1/25–1/15 in. long, barely exserted.

FRUIT. Achenes 1/8–1/6 in. long, quadrangular, slightly narrowed toward the base, silky-hairy; pappus abundant, brown, unequal in length.

LEAVES. Alternate, blades ⅝–1¾ in. long, 1/6–⅔ in.

wide, oblong to spatulate-oblong, cut into 5–11 linear remote lobes, lobes mucronulate; rather thick, hairy, or scabrid, pitted resin dots often conspicuous, smaller leaves sometimes clustered in the axils.

TWIGS. Slender, pale gray to brown or tan, hard, brittle, obscurely striate, scabrid.

BARK. Brown or gray, breaking longitudinally on old woody portions.

RANGE. Rather widespread in the desert, Lower Sonoran zone of the Southwest. Rather tolerant of alkaline soils. At altitudes of 1,500–4,000 ft. Western Texas to Arizona; in Mexico in Sonora, Jalisco, San Luis Potosí, and Puebla.

RELATED SPECIES AND SYNONYMS. This species has also been listed in the literature under the names of *Aplopappus fruticosa* Blake, *Bigelovia coronopifolia* Gray, *Isocoma fruticosa* Rose & Standl., *Linosyris coronopifolia* Gray, *Isocoma hartwegii* (Gray) Greene, and *Bigelovia hartwegii* Gray.

It is also closely related to *H. venetus* (H. B. K.) Blake, but is separated by the more deeply cleft corollas and definitely lobed leaves.

BURRO GOLDENWEED
Haplopappus hartwegii (Gray) Blake

LARCH-LEAF GOLDENWEED
Haplopappus laricifolius Gray

REMARKS. The genus name, *Haplopappus*, is from the Greek and means "simple pappus," with reference to the hair of the achenes. The species name, *hartwegii*, is for Karl Hartweg, German collector for the Horticulture Society of London (1836–1843 and 1845–1848). He collected in Ecuador.

LARCH-LEAF GOLDENWEED

Haplopappus laricifolius Gray

FIELD IDENTIFICATION. Shrub 1½–3½ ft, erect, flat-topped, fastigiately branched, densely leafy, conspicuously resinous-punctate and viscid.

FLOWERS. September–November; heads small, numerous, crowded, terminating short, slender, leafy branches, in irregular dense cymes; involucre ⅛–⅖ in. long, turbinate; usually less than 20 bracts, imbricate, appressed, in 2–3 series, various sizes, linear to lanceolate, apex acute, margins white and scarious, center brownish, glabrous or puberulous; ray flowers conspicuous, bright yellow, ligules 3–6, length ⅙–⅕ in., width 1/15–1/12 in.; disk flowers 8–16, corolla funnelform, tube slender, length ⅕–¼ in., glabrous or puberulent, 5-lobed; lobes spreading, lanceolate, apex acute, glabrous, length about 1/25 in.; style branches long-exserted, stigmatic portion shorter than the basal portion.

FRUIT. Achene slender, apex truncate, slightly 4-angled, length 1/8–1/6 in.; pappus about the same length as the corolla, silky, appressed, light brown.

LEAVES. Crowded, alternate, linear-acerose or linear-spatulate, rigid, apex acute, margin entire, blade length 3/8–3/4 in., width 1/25–1/12 in., midrib apparent, surface glabrous and glutinous to viscid, resinous-punctate; sometimes with small leaves fascicled in the axils of the larger ones.

TWIGS. Rather short, leafy, glabrous, some striate, resinous-punctate; older branches 1/4–3/4 in. thick, green to light or dark tan, sometimes darker.

RANGE. Upper desert and desert grasslands, dry rocky hills and canyons at altitudes of 3,000–6,000 ft. Type locality at Guadalupe Pass, New Mexico. In Texas in the Franklin and Guadalupe mountains; also in Mexico in the states of Chihuahua and Sonora.

REMARKS. The genus name, *Haplopappus*, is from the Greek words *haploos* ("simple") and *pappos* ("down"), with reference to the simple pappus hairs. The species name, *laricifolius* ("larch-leaved"), refers to the thick leaves resembling those of the larch. A very handsome plant in bloom with its crown of yellow flowers. It is highly resinous with a turpentine-like odor which gives it the name of Turpentine Bush in some localities. It has no value as forage for livestock. Experiments show that its sap contains a small amount of rubber, averaging about 2.01 per cent.

THREE-FLOWER GOLDENWEED
Haplopappus trianthus Blake

THREE-FLOWER GOLDENWEED

Haplopappus trianthus Blake

FIELD IDENTIFICATION. Low shrub 1/2–2 ft, with erect or ascending branches.

FLOWERS. Borne in loose open terminal panicles to 4 in. long and 3 in. wide, the upper leaves becoming small and bractlike and subtending the peduncles or pedicels; heads slightly glutinous, 1/4–3/8 in. wide; true bracts 1/16–3/16 in. long, imbricate, numerous, the outer ones shorter than the inner, straw-colored, linear; apices acute to obtuse, thickened and glandular-darkened, margins thin and scarious, some margins ciliate, dorsally somewhat convex; disk flowers 2–4, usually 3, about 3/16 in. long; corolla-lobes 5, linear-lanceolate, acuminate, one fourth to one third as long as the tube; style bifid, stigmas filiform, as long as or longer than the corolla-lobes, stamens erect.

FRUIT. Achene about 1/8 in. long, straw-colored, densely clothed with long appressed hairs, crown bearing a ring of tawny, scabrous, spreading pappus hairs.

LEAVES. Simple, alternate, deciduous, ascending, blades 1/8–2/3 in. long, 1/25–1/15 in. wide, linear, thickened, fleshy, almost terete, entire, revolute, grooved beneath, veins vague, apex obtuse or rounded, surface glabrous often with smaller leaves in the axils of the longer ones.

STEMS. Slender, erect or ascending, young ones greenish, slightly glutinous and often with particles of sand or soil adhering; older stems glabrous, woody toward the base, light brown to gray.

RANGE. Apparently known only from the Brewster County area of Texas. Collected by Omer E. Sperry 3 miles north of Lajitas in Brewster County. Frequent along road to Terlingua. Infrequent on limestone wastelands 1 mile beyond Tornillo Flats toward the Chisos Mountains. Also in Big Bend National Park.

REMARKS. The genus name, *Haplopappus*, is from the Greek *haploos* ("simple") and *pappos* ("down"), referring to the simple pappus hairs. The species name, *trianthus*, means "3-flowered."

DRUMMOND GOLDENWEED

Haplopappus drummondii (Torr. & Gray) Blake

FIELD IDENTIFICATION. Plant to 20 in., herbaceous above, woody toward the base, branches leafy, erect or partly ascending, herbage resinous.

FLOWERS. June–September, borne in terminal, rounded cymes 1–2½ in. across, composed of peduncled discoid heads; involucre turbinate, ¼–⅓ in. high, ⅙–⅕ in. broad; bracts 4–5-seriate, imbricate, erect, linear-oblong to lanceolate, base coriaceous; apex acute or obtuse, thickened, greenish, mostly glabrous; ray flowers absent; disk flowers 14–35, corolla ¼–⅓ in. long, tubular, smooth or roughened, 5-lobed; lobes 1/25–1/15 in. long, erect, lanceolate, acute, glabrous; style branches 1/25–1/12 in. long, somewhat exserted, the ovate, acute appendage one third to one half as long as the stigmatic portion.

FRUIT. Achenes slender, prismatic, silky-hairy; pappus bristles copious, brownish, about as long as the corolla.

LEAVES. Simple, alternate, linear to narrowly spatulate, mostly entire, or rarely the lower with a few divergent lobes, apex acute, base narrowed, blade length 1¼–2 in., width 1/12–⅕ in., main vein usually apparent, others obscure, thickened, surfaces pale green and glabrous or resinous-granular.

TWIGS. White to gray, darker on older stems, glabrous or granular, stiff and brittle.

DRUMMOND GOLDENWEED
Haplopappus drummondii (Torr. & Gray) Blake

MEXICAN GOLDENWEED
Haplopappus gymnocephalus DC.

RANGE. In alkaline soil, at altitudes of 15 ft on the Texas coast to 5,000 ft in the mountain regions of the Southwest. In New Mexico, Arizona, Utah, Colorado, and Mexico. In Mexico in Coahuila and Nuevo León.

REMARKS. The genus name, *Haplopappus,* is from the Greek words *haploos* ("simple") and *pappos* ("down"), referring to the simple hairs of the pappus ring. The species name, *drummondii,* honors Thomas Drummond (1780–1835), a Scotch botanical explorer, who discovered it. It is distinguished from other closely related species by the absence of secondary fascicled leaves.

MEXICAN GOLDENWEED

Haplopappus gymnocephalus DC.

FIELD IDENTIFICATION. Western shrubby-based perennial of arid areas attaining a height of 8–18 in., with ascending stems.

FLOWERS. Maturing June–October, borne in 1 or 2 heads on leafy-bracted peduncles; involucre ¼–⅓ in. high, hemispheric-depressed; bracts numerous, imbricate, unequal, linear to lanceolate, acuminate, rigid, often awned, scabrous; ray flowers numerous (to 40 or more), pink, ⅖–⅗ in. long; disk corolla ⅕–¼ in. long, puberulent; lobes 5, ovate, about 1/35 in. long, puberulent to glabrous; styles 1/25–1/12 in. long, appendage longer than the stigma, very slender.

FRUIT. Achenes 1/12–⅒ in. long, short-turbinate, densely hairy; disk pappus brown, the bristles rigid and unequal.

LEAVES. Ascending, oblong, margin serrate with bristle-tipped teeth, apex round to obtuse, base half clasping, blade length ⅝–1⅓ in., width ⅙–¼ in., midrib prominent, other veins obscure, except occasionally with two parallel veins, surface scabrid and glandular, upper leaves smaller.

STEMS. Ascending, pale gray and scabrid-glandular, later darker, sometimes striate from the leaf bases.

RANGE. On dry hillsides at altitudes of 3,000–7,500 ft. Western Texas and New Mexico, southward in Mexico to Mexico City.

REMARKS. The genus name, *Haplopappus,* is from the Greek words *haploos* ("simple") and *pappos* ("down"), referring to the simple hairs of the pappus ring. The species name, *gymnocephalus,* means "naked head." This is a species of difficult classification, seemingly closely related to *H. spinulosus* and *H. phyllocephalus.* For a detailed discussion see Hall (1928, pp. 65–68).

THRIFT GOLDENWEED

Haplopappus armerioides (Nutt.) Gray

FIELD IDENTIFICATION. Dense, tufted, low subshrub 4–8 in. Branches with few leaves and borne on a short woody stem.

FLOWERS. Maturing May–July, heads scattered and solitary on elongate peduncles; involucre campanulate, ⅖–½ in. high; bracts seriate, imbricate, pale, leathery, glabrous, slightly glutinous, outer ones oval and obtuse, inner ones longest and oblong; disk flowers 15–30, ¼–⅖ in. long, gradually narrowed to the base, lobes 5, ovate to lanceolate; style exserted, 1/12–⅙ in. long; ray flowers 8–15, ligules ⅖–½ in. long, about ⅙ in. wide.

FRUIT. Achenes about ⅙ in. long, flattened, pappus about as long as the corollas, soft, white, silky, rather abundant.

LEAVES. Mostly basal, erect or nearly so, linear to spatulate, base gradually narrowed, apex sharply acute, margin entire, blade length 1½–3⅓ in., width ⅛–⅓ in., rigid, 3-nerved, surfaces glabrous, somewhat glutinous. Occasionally with a few small leaves on the elongate peduncles or on the lower part of flowering stem.

TWIGS. Reduced to short branches on a woody caudex, upper ones reddish brown and tomentose to glabrous, older ones gray to brown, glabrous, and slightly woody.

RANGE. Mountain valleys, dry hills, and plains of the Rocky Mountain region. In New Mexico, Arizona, Colorado, Utah, Nebraska, and Montana.

REMARKS. The genus name, *Haplopappus,* is from the Greek words *haploos* ("simple") and *pappos* ("down") referring to the simple pappus hairs. The species name *armerioides,* is for its resemblance to the genus *Armeria* which also grows in tufts.

NUTTALL GOLDENWEED

Haplopappus nuttallii Torr. & Gray

FIELD IDENTIFICATION. Perennial herb or partially woody plant seldom over 1 ft. Sometimes erect, at other times depressed and densely tufted.

FLOWERS. May–October; heads discoid, solitary or cymose and borne on bracted peduncles ¼–2 in. long, involucre campanulate, ¼–⅖ in. long, ⅓–⅖ in. broad, bracts usually in 3 series, few in number, imbricate, lanceolate to linear-oblong, mostly appressed, apex acute and sharply pointed, surfaces glandular-hispid; ray flowers absent; disk corolla gradually widened toward the apex, ¼–⅓ in. long; lobes 5, lanceolate, puberulent, about 1/25 in. long; style branches barely exserted, stigmatic portion as long as the appendage or shorter.

FRUIT. Achene 1/12–⅛ in. long, turbinate-prismatic, densely hairy; pappus reddish brown, as long as the corolla or shorter.

LEAVES. Simple, alternate, erect or ascending, oblong to spatulate, margin serrate to dentate with white bristle-tipped teeth, apex obtuse, base sessile or semiclasping, blade length ⅗–1½ in., width ⅙–⅖ in., main vein evident, 2 lateral veins obscure, surfaces canescent to glandular-hispid, later almost glabrate.

TWIGS. Faintly striate, young ones tomentose, older ones glabrous, gray to white, woody areas near base darker.

RANGE. On poor arid soils of hills and mesas at altitudes of 3,000–8,000 ft. New Mexico and Arizona, north to Saskatchewan and Alberta.

REMARKS. The genus name, *Haplopappus,* is from the Greek words, *haploos* ("simple") and *pappos* ("down"), referring to the simple pappus hairs. The species name, *nuttallii,* is in honor of Thomas Nuttall (1786–1859), English and American botanist and ornithologist. It is reported that the Hopi Indians prepare a cough medicine from the roots.

MEXICAN TRIXIS

Trixis radialis (L.) Kuntze

FIELD IDENTIFICATION. Shrub 3–4½ ft, much-branched, glabrous or somewhat pubescent.

FLOWERS. Borne in many-headed, somewhat leafy, corymbose panicles; heads usually 8–12-flowered, crowded, ⅝–¾ in.; outer bracts linear or lance-linear, usually half or two thirds as long as the inner; inner

Mexican Trixis
Trixis radialis (L.) Kuntze

bracts 8–10; flowers all fertile, yellow; corolla bilabiate, outer lip 3-toothed, inner 2-cleft; anthers caudate at base.

FRUIT. Achene subcylindric, papillose; pappus bristles numerous, brownish.

LEAVES. Simple, alternate, elliptic to lanceolate or oval, blade length 1⅓–4 in., width ⅜–1½ in., margin entire or remotely dentate, apex acuminate to acute and mucronate, base sessile or petiole ⅛–⅜ in., usually glabrous but occasionally sparingly pubescent on the main veins beneath, finely reticulate-veiny.

TWIGS. Slender, young ones glabrous or sparingly appressed-pubescent, finely grooved, straw-colored to brown.

RANGE. In Texas in the lower Rio Grande Valley area (Cameron County). In Mexico in Tamaulipas near Matamoros, southward to Yucatán, Guatemala, and Panama. Also in the West Indies and South America.

REMARKS. The genus name, *Trixis,* is the classical name and refers to the 3-cleft corolla. The species name, *radialis,* is for the radiate, or rayed, flower head. It is known in Central American countries under the names of Tokaban, Tokabal, Plumella, Hierba del Aire, Falsa Arnica, Juan de Calle, Arnica de Monte, Chucha, Diente de León, Palo de Santa María, San Pedro, Santo Domingo, Tulan Verde, and Carmen.

American Trixis

Trixis californica Kellogg

FIELD IDENTIFICATION. Western plant, herbaceous above and woody toward the base, 1–3 ft.

FLOWERS. February–June, borne in corymbose heads at the ends of short branchlets; heads 8–12-flowered, yellow, clustered or solitary, all bisexual and fertile; involucre not imbricate, bracts in 2 series, the outer leaflike, linear to elliptic; the inner 8–12, linear, acuminate, firm, light colored, about ½ in. long, thickened at the base; corollas 2-lipped, lower lip 3-toothed, upper lip 2-cleft; receptacle hairy.

FRUIT. Achenes slender, narrowed to the apex, sometimes beaked, glandular, densely hairy; pappus soft, of numerous straw-colored capillary bristles.

American Trixis
Trixis californica Kellogg

LEAVES. Simple, alternate, sessile, rather rigid, lanceolate to oblong, apex acute or obtuse, base narrowed, margin entire or remotely denticulate, blade length 1–4¼ in., width ¼–1⅓ in.; both surfaces green, lower densely puberulent and glandular.

RANGE. Rocky hills and canyons to an altitude of 4,500 ft. Trans-Pecos Texas, west to California and south into Mexico in Sonora, Coahuila, Zacatecas, and San Luis Potosí.

REMARKS. The genus name, *Trixis*, is the Greek word for "threefold," referring to the 3-cleft lower lip of the corolla. The species name, *californica*, is for the state of California. The plant is browsed occasionally by cattle in winter.

Bushy Sea-oxeye

Borrichia frutescens (L.) DC.

FIELD IDENTIFICATION. Small fleshy, whitened shrub to 4 ft, of saline marshes and sandy shores.

FLOWERS. Throughout the year, but usually April–September, borne on solitary stiff, stout peduncles, or sometimes a few together; heads of both ray and disk flowers; rays 15–25, short, each about ⅜ in. long, pistillate and fertile; disk yellowish brown, corolla perfect and fertile, cylindric, funnelform, 5-toothed, tube short, anthers dark-colored; style branches slightly flattened and hairy.

Bushy Sea-oxeye
Borrichia frutescens (L.) DC.

FRUIT. Involucre hemispheric, about ⅝ in. broad, the imbricate bracts spreading, leathery, chaffy, ovate, those of the receptacle rigid, lanceolate and spiniferous; achenes wedge-shaped, 3–4-angled, pappus crown short, 4-toothed. The spiny, persistent, chaffy heads are conspicuous at maturity.

LEAVES. Erect or ascending, simple, opposite, thick-fleshy, blade length 1–3 in., width ⅙–¾ in., spatulate to obovate or lanceolate, apex acute to obtuse and mucronulate, base tapering or somewhat connate, margin entire or often toothed toward the base, surface finely whitish-canescent above and below.

STEMS. Sparingly branched, stout, rounded, grooved, whitened with silky hairs or more glabrous on older stems.

RANGE. Salt marshes, sandy shores, sea beaches, and saline prairies. Near the coast from Texas eastward to Florida and northward to Virginia. Also in the West Indies and Bermuda. In Mexico in Tamaulipas, Veracruz, and San Luis Potosí.

REMARKS. The genus name, *Borrichia*, honors Ole Borch (Latinized Olaus Borrichius), a Danish botanist (1626–1690). The species name, *frutescens*, refers to the shrubby habit.

Parish Goldeneye

Viguiera deltoidea var. *parishii* (Greene) Vasey & Rose

FIELD IDENTIFICATION. Much-branched shrub to 2½ ft.

FLOWERS. Mostly April–June, solitary on long, terminal, rough-hairy peduncles; composed of both ray and disk flowers; rays bright yellow, spreading, ½–1 in. long, linear to oblong, apex acute, surface lightly parallel-veined, sterile; disk flowers numerous and fertile, stigmas with slender tips; involucral bracts much-imbricate, ⅜ in. long or less, linear, apex acute to obtuse, clothed with short, stiff, white hairs.

FRUIT. Achenes ⅛–3/16 in. long, oblong, flattened, white-hairy, pappus persistent and composed of 1–2 slender awns often as long as the achene body, also with short scales at the base of the awns.

LEAVES. Opposite below and often alternate above, deltoid-ovate with a broad cuneate or almost truncate base, apex acute or acuminate, margin entire or coarsely-toothed, blade length ⅝–1½ in., width ⅝–¾ in., stiff and rigid, some half-folded, surface dull green and harshly roughened by stiff hairs, veins reticulate, with several main veins often at the base; petioles 1/16–¼ in., white hispid-hairy.

TWIGS. Slender, erect or ascending, tan to brown, somewhat finely grooved, harshly stiff-hairy, older stems less hairy, brown to grayish.

RANGE. In rocky arroyos, canyons, and deserts at

PARISH GOLDENEYE
Viguiera deltoidea var. *parishii* (Greene) Vasey & Rose

altitudes of 1,000–3,500 ft. Trans-Pecos Texas, New Mexico, and Arizona; west to California and north to Nevada.

REMARKS. The genus name, *Viguiera*, honors L. G. A. Viguier, a physician and botanist of Montpellier, France. The species name, *deltoidea*, refers to the deltoid shape of the leaves, and the variety name, *parishii*, is for W. F. Parish, who collected around Camp Lowell, near Tucson, Arizona, in the early 1880's. It is also known as the Desert Sunflower.

SKELETON-LEAF GOLDENEYE

Viguiera stenoloba Blake

FIELD IDENTIFICATION. Western shrub much branched a short distance above the base. Often with a rounded top, attaining a height of 4 ft.

FLOWERS. Solitary on long naked peduncles 8 in. or less; disk 1/4–1/3 in.; involucral bracts in 3 series, bracts ovate-lanceolate, apex abruptly linear, base strongly indurate and ribbed, surface pale and strigose-hairy; rays yellow, 10–12, 1/3–3/5 in. long, 1/12–1/6 in. wide; disk corolla puberulous with teeth reflexed, 1/8–1/6 in.

FRUIT. Achene 1/8–1/6 in. long, glabrous, substriate, subquadrangular, pappus none.

LEAVES. Alternate or opposite, sometimes a few linear leaves present, but generally ovate and divided almost to the center into 3–7 lobes; lobes linear-lanceolate, entire or few-toothed, attenuate, 1/25–1/5 in. wide; dull green and strigose-tuberculate above, canescently-strigose below, over-all length 1–2½ in. or less, width 1½ in. or less.

STEMS. Slender, gray, hairy or glabrous, nodes and leaf scars somewhat prominent; older stems with narrow ridges and shallow fissures.

RANGE. On rocky ground of the Pecos and Rio Grande river basins. In Trans-Pecos Texas, in Presidio, Brewster, El Paso, and Cameron counties. In New Mexico in Doña Ana County. In Mexico in Chihuahua, Coahuila, Nuevo León, and Tamaulipas.

REMARKS. The genus name, *Viguiera*, honors L. G. A. Viguier, a physician and botanist of Montpellier, France; the species name, *stenoloba*, refers to the narrowly lobed leaves. The plant is occasionally browsed by livestock in time of stress.

DEVIL-WEED ASTER

Aster spinosus Benth.

FIELD IDENTIFICATION. Plant 1½–6 ft, herbaceous above, woody at the base, branchlets very erect, slender, green, and glabrous.

SKELETON-LEAF GOLDENEYE
Viguiera stenoloba Blake

Devil-weed Aster
Aster spinosus Benth.

FLOWERS. September–January, heads mostly solitary on slender paniculate branches and axillary to the minute leaves; involucre turbinate, 1/6–1/5 in. high; bracts imbricate, appressed, green, ovate to lanceolate or linear, apex subulate-attenuate or acute, margins slightly white-scarious, outer surface rounded, somewhat viscid, length 1/25–1/8 in. long; ray flowers numerous, ligules white, oblong to linear, apex obtuse or slightly notched, base gradually narrowed, styles white, segments elongate; disk flowers somewhat exserted, yellowish, 5-lobed, lobes acute and spreading, the stigma much-exserted, yellow, and bifid at apex.

FRUIT. Achene linear, flattened, brown, glabrous, slightly viscid, less than 1/25 in. long; apex truncate, bearing a collar of long, white, capillary bristles forming a copious pappus.

LEAVES. Sometimes absent, when present upper alternate and inconspicuous, lower linear to linear-spatulate, 3/4–1 1/2 in. long, those of the twigs reduced to subulate scales, some with subulate soft spines in or above their axils.

TWIGS. Young erect, very slender, lithe, green and glabrous or slightly glaucous; older green to brown or gray and striate.

RANGE. Sand flats and banks of rivers and canals, on the Texas coast at an altitude of 15 ft, to an altitude of 1,300 ft in west Texas, New Mexico, Arizona, California, Colorado, Utah, and Nevada; in northern Mexico south to Guatemala; also in Costa Rica.

REMARKS. The genus name, *Aster,* is from the Greek word *astere* ("a star"), referring to the starlike flowers. The species name, *spinosus,* refers to the soft spines. The plant is sometimes useful for erosion control.

Palmer Aster

Aster palmeri Gray

FIELD IDENTIFICATION. Shrubby, erect, divaricately branched aster to 6 ft, or partially decumbent.

FLOWERS. Heads terminal, on bracteate peduncles, turbinate, about 1/4 in. long; involucral bracts numerous, green to reddish, imbricate, appressed, oblong to lanceolate, acute or obtuse at apex, margin scarious-hairy and

Palmer Aster
Aster palmeri Gray

granulose, usually less than ⅛ in. long; rays 8–10, short, white, about $1/12$ in. long, or sometimes absent.

FRUIT. Achenes with pappus ⅛–⅓ in. long.

LEAVES. Numerous, small, less than ¾ in. long, $1/25$–⅛ in. wide, entire, linear, sessile, thickened, granulose or somewhat viscid.

TWIGS. Numerous, slender, divaricate, clustered toward the upper part of main stem, green to gray or brown, puberulent, later glabrous and irregularly grooved or striate.

BARK. On old stems gray to brown and rather deeply furrowed and fibrous.

RANGE. Mostly on shell banks of the Texas Gulf Coast from Matagorda to Cameron counties, also along the Rio Grande in Hidalgo and Starr counties. The type specimen was collected in the Corpus Christi Bay area. In Mexico in the states of Tamaulipas, Nuevo León, and San Luis Potosí.

REMARKS. The genus name, *Aster*, is from the Greek, referring to the starlike radiate heads of flowers, and the species name, *palmeri*, is in honor of Ernest J. Palmer, American botanist.

DAMIANITA

Chrysactinia mexicana Gray

FIELD IDENTIFICATION. Evergreen undershrub, resinous-aromatic, seldom over 2 ft, much-branched.

FLOWERS. Solitary, terminal, borne on slender, upright pedicels ¼–¾ in. long; involucre ⅙–⅕ in. high; bracts numerous, somewhat reflexed, linear-lanceolate; apex obtuse or rounded, granular, each bract with a round gland on the back near the apex; ray flowers 8–10, $3/16$–¼ in. long, golden-yellow, linear; disk flowers 5-lobed, yellow.

FRUIT. Achenes linear, rather short, with abundant, capillary, scabrous bristles.

LEAVES. Alternate, heathlike, mostly less than ⅓ in. long, linear or filiform, apex cuspidate-mucronate, margin entire, grooved above, rounded below and set with large round oil glands, aromatic when crushed.

TWIGS. Slender, gray to black, crowded toward the apex of the branches.

BARK. Gray to black, roughened with small irregular scales.

RANGE. Western Texas, New Mexico, and Mexico. In Texas in the Guadalupe, Davis, and Chisos mountains, and perhaps other mountain ranges of the Trans-Pecos. In Mexico in the states of Tamaulipas and Chihuahua, south to Veracruz and the state of Mexico.

MEDICINAL USES. All parts of the plant were used medicinally by the Indians for fever, rheumatism, and as a diuretic, sudorific, antispasmodic, and aphrodisiac.

DAMIANITA
Chrysactinia mexicana Gray

This plant is not to be confused with the official drug damiana, which is prepared from *Turnera diffusa* Willd. & Schultes.

REMARKS. The genus name, *Chrysactinia*, refers to the golden-yellow color of the rays. The species name, *mexicana*, is for the plant's predominantly Mexican range. Known under the vernacular names of False Damiana, Mariola, Romerillo, Garanona, San Nicolás, Calanca, and Yeyepaxtle.

CLAPPIA

Clappia suaedaefolia Gray

FIELD IDENTIFICATION. Plants 8–24 in., mostly fleshy and herbaceous, but much branched and somewhat woody near the base.

FLOWERS. On thickened, elongate, leafless peduncles 2–5 in. long; heads solitary, radiate, yellow, ¾–1⅓ in. wide; involucral bracts imbricate in 2–4 series, oval to oblong, obtuse, leathery, striate, thin-margined; receptacle naked, convex with 12–15 fruiting rays; disk flow-

CLAPPIA
Clappia suaedaefolia Gray

ers perfect, fruit-producing; anthers entire at base; stigmas of disk with triangular appendages.

FRUIT. Achenes about 1/8 in. long, oblong-linear, terete, 8–10-ribbed, sparsely hairy; pappus bristles 20–35, longer than the achene, unequal, rigid, scabrous.

LEAVES. Alternate, simple or 3–5-parted below, sessile, blades 3/4–2 1/2 in. long, 1/25–1/8 in. wide, filiform to linear, fleshy, terete or flattened, entire; pale green.

TWIGS. Slender, green, finely grooved, rather fleshy, glabrous; when older light tan to brown.

RANGE. Sandy soils of southern and western Texas, New Mexico, and Mexico.

REMARKS. The genus name, *Clappia,* is in honor of A. Clapp, author of a synopsis of medicinal plants of the United States. The species name, *suaedaefolia,* is for *Suaeda* (Chenopodiaceae), genus of Old World halophytic herbs with thickened fleshy leaves.

PRICKLY-LEAF DOGWEED

Dyssodia acerosa DC.

FIELD IDENTIFICATION. Subshrubby plant with numerous stems woody toward the base. Hardly more than 9 in. or less, with very small needle-like leaves set with conspicuous glands.

FLOWERS. Heads solitary, subsessile, yellow, radiate, 1/4–1/3 in. high and about 5/8 in. wide, rays oblong; involucre calyculate, subtended by bractlets; phyllaries oblong, equal, 1–2-seriate, glandular-dotted.

FRUIT. Achene linear, striate, each with 5–20 scales and divided above into 3–7 capillary bristles.

LEAVES. Mostly opposite, sessile, 1/8–3/4 in. long, less than 1/25 in. wide, filiform to linear, rigid, needle-like, acerose, entire, marked with translucent glands, often with shorter axillary leaves in their axes.

STEMS. Numerous, divaricate, somewhat glandular or hirtellous, when older tan to gray and glabrous.

RANGE. Dry rocky slopes and mesas at altitudes of 3,500–6,000 ft. Western Texas, New Mexico, and Arizona; in Mexico in Sonora to Coahuila, south to Zacatecas, Hidalgo, and San Luis Potosí.

REMARKS. The genus name, *Dyssodia,* is from the Greek *dysodia* ("ill smell"). The species name, *acerosa,*

PRICKLY-LEAF DOGWEED
Dyssodia acerosa DC.

Bush Encelia
Encelia frutescens Gray

refers to the sharp-pointed leaves. It is also known under the vernacular names of Hierba de San Nicolás and Contrayerba.

Bush Encelia

Encelia frutescens Gray

FIELD IDENTIFICATION. Western shrub 2-4 ft. Stems much branched and covered with dense, short, very rough pubescence. Flower heads solitary on long naked scabrous peduncles.

FLOWERS. Heads discoid, 3/8–1 in. broad, disk yellow, involucre 1/4–2/5 in. high, scales unequal, linear-lanceolate to ovate-acuminate, 3-rowed, hispid-scabrous; rays usually absent, but if present to 12, with apex 3-lobed, about 1/3 in. long; disk corollas with tube and limb hairy and sometimes glandular; pales about 3/8 in. long.

FRUIT. Achenes black, obcordate, 1/4–1/3 in. long, 1/12–1/8 in. wide, edges hairy, sides pubescent or almost glabrous, awnless, or with 1–2 minute hairy awns.

LEAVES. Simple, alternate, oblong to ovate, apex obtuse to acute, base cuneate or truncate, blade length 3/8–1 1/4 in., width 3/16–5/8 in., surfaces roughened by the tuberculate bases of scattered white hairs, 3-nerved at base, lateral veins obscure; petioles 1/16–1/4 in. long, hairy and scabrous.

STEMS. Slender, elongate, tan to brown or whitened and roughened by tuberculate hairs.

RANGE. The species occurs in New Mexico, Arizona, and California.

VARIETIES. Virgin River Encelia, *E. frutescens* var. *virginensis* (A. Nels.) Blake, has the outer involucral scales linear-lanceolate; rays nearly always present; leaves broadly ovate, cinereous-scabrous with a fine glandular pubescence intermixed with stouter tuberculate-based hairs. In Utah and Arizona.

Acton Encelia, *E. frutescens* var. *actoni* (Elmer) Blake, has involucral scales mostly ovate-acuminate; rays nearly always present; leaves ovate, base cuneate or truncate, very rarely toothed, whitened with a rather soft fine pubescence; pubescence of stem and peduncles also softer than the typical form. Known from Nevada, California, and Arizona.

REMARKS. The genus name, *Encelia,* honors Christopher Ensel, who wrote a book on oak galls in 1577. The species name, *frutescens,* refers to the frutescent or shrubby stems of the plant.

New Mexico Encelia

Encelia scaposa Gray

FIELD IDENTIFICATION. A plant perhaps better classified as an herbaceous perennial, but occasionally somewhat lignescent at the base. Simple and erect and to 18 in., the solitary-headed scape devoid of leaves, but sometimes with two linear bracts.

FLOWERS. Scape puberulent, striate, whitish; head solitary, terminal, almost 1/2 in. high and 3/4 in. wide, not including the rays; scales linear-lanceolate, subequal, somewhat 2-rowed, dorsally hispid; rays yellow, somewhat pubescent, about 3/5 in. long and to 1/4 in. wide; disk corolla about 1/5 in. long, 5-toothed, tube short, pubescent; pales about 2/5 in. long, 5–9-nerved, pubescent above.

FRUIT. Achenes about 1/5 in. long, flattened, margins ciliate, apex with 2 chaffy awns.

LEAVES. Lower ones somewhat scalelike, upper ones long and clustered but really alternate, linear, blade length 1–3 1/2 in., width 1/25–1/6 in., apex attenuate, base subsessile, margin entire, rough-pubescent on both sides.

RANGE. Seemingly a rather rare species. Recorded

New Mexico Encelia
Encelia scaposa Gray

from western Texas, and in New Mexico between the Mimbres and the Rio Grande rivers.

REMARKS. The genus name, *Encelia,* is in honor of Christopher Ensel, who wrote a book on oak galls in 1577. The species name, *scaposa,* refers to the scapose flower heads.

Single-whorl Burro-brush

Hymenoclea monogyna Torr. & Gray

FIELD IDENTIFICATION. Shrub 3–6 ft, thicket-forming, much-branched, slender, minutely canescent or glabrous.

FLOWERS. Numerous, profuse, heads small, unisexual, scattered or glomerate-paniculate, the two sexes sometimes intermixed in the same leaf axil, monoecious or dioecious; staminate heads with a flattish 4–6-lobed involucre; pistillate flower solitary, involucre ovoid to fusiform, indurate, beaked, lower part winged.

FRUIT. The fruiting involucre about 1/6 in. long, wings 7–12 in a single series, whorled, cuneate to orbicular or obovate, thin, scarious, silvery-white, 1/25–1/12 in. long.

LEAVES. Numerous, simple, alternate, 1–3 in. lon linear-filiform, entire, lower ones sometimes parted i to linear-filiform divisions, achenes turgid.

RANGE. Sandy stream banks, washes, and arroyo usually at altitudes of 2,000–4,000 ft. Trans-Pecos Te as, New Mexico, Arizona, and southern California. I Mexico in Sonora, Sinaloa, and Chihuahua.

REMARKS. The genus name, *Hymenoclea,* is from th Greek *hymen* ("membrane") and *kleio* ("to close") in reference to the winged bracts. The species nam *monogyna,* refers to the solitary pistillate flower. Th plant has some value for erosion control, but is usuall avoided by livestock.

Mariola

Parthenium incanum H. B. K.

FIELD IDENTIFICATION. Shrub to 3 ft, densel branched, twigs grayish green with dense gray tomen tum.

FLOWERS. In flat-topped, long-peduncled paniculat cymes; heads numerous, small, white, radiate; ray flow ers fertile, with 5 ligules which are erect, short, broa

Single-whorl Burro-brush
Hymenoclea monogyna Torr. & Gray

Mariola
Parthenium incanum H. B. K.

and obcordate, producing achenes at the base; disk flowers sterile, the involucre of 2 ranks of dry, appressed, short ovate to oval scales.

FRUIT. Achene small, flattened, conical, margin callous, crowded by the persistent ray corolla, and adnate at base at first to the pales of the two apposed disk flowers.

LEAVES. Alternate, obovate in outline, length 3/8–3/4 in., width 3/5–5/8 in., variously pinnatifid into 3–7 lobes which are oblong, rounded or obtuse, surfaces gray-cinereous-tomentulose but less so above.

RANGE. On dry hills, gravelly mesas, and desert grasslands. At altitudes of 3,000–6,000 ft, in western Texas, Arizona, New Mexico, and Mexico.

Collected by the author in Big Bend National Park in Brewster County, Texas. Also known from Hudspeth, Presidio, Pecos, and El Paso counties. In New Mexico in the Organ, White, and Tortugas mountains; also in Arizona. In Mexico in Sonora and Coahuila, south to Hidalgo.

REMARKS. The genus name, *Parthenium*, is the ancient name, and the species name, *incanum*, refers to the gray hairs of the plant. Vernacular names are Hembra de Guayule and Tananini. It is sometimes called Guayule in error, but the name should apply more properly to *P. argentatum* Gray. Mariola also contains rubber, but in much less amount than in the Guayule. It is only slightly browsed by livestock.

Guayule

Parthenium argentatum Gray

FIELD IDENTIFICATION. Shrub to 3 ft. Branches numerous and erect, with conspicuous silvery-canescent hairs on leaves, young stems, and cymes.

FLOWERS. Small, borne on naked long-peduncled cymes; flower heads about ¼ in. wide or less, radiate, whitish; the fertile ray flowers 5, projecting beyond the disk with broad, obcordate ligules; involucre hemispheric with sterile disk flowers, chaff of 2 ranks of short ovate or rounded scales.

FRUIT. Achenes conical, compressed, surrounded by a callous margin, with a pappus of 2 persistent, chaffy scales.

LEAVES. Alternate, crowded or scattered, lanceolate

Guayule
Parthenium argentatum Gray

to ovate or spatulate, margin entire or with a few acute teeth or 2–7 lateral lobes, apex acute or acuminate, base acute or gradually narrowed, blade length ½–1¾ in., width 3/16–1 in., surfaces with silvery-canescent hairs; petioles ¼–1 in.

RANGE. Native to the United States only in the Big Bend area of Trans-Pecos Texas, also in adjacent Mexico.

REMARKS. The genus name, *Parthenium,* is the ancient name, and the species name, *argentatum,* refers to the silvery foliage. Known by many local names in Mexico, such as Hierba del Hule, Hierba Blanca, Hierba Ceniza, Hule, Copallin, Afinador, Xihuite, Jihuite, and Tatanini.

This is the well-known Rubber Plant of the Southwest. Rubber is not taken directly from the shrub, but the stem is cut, and various mechanical and chemical processes are used to remove the rubber from the stem. The first attempt at commercial exploitation began in Zacatecas, Mexico, in 1892. Between July 1, 1905, and July 1, 1909, 32,000,000 lb of guayule rubber were imported into the United States. However, because of the competition of East India rubber, the industry declined. With the advent of World War II, it was revived and commercial plantings were extensively made in the southwestern United States. A number of improved varieties have been developed through selection, but the industry is still considered only an emergency source of rubber.

WHITE-STEM PAPERFLOWER
Psilostrophe cooperi (Gray) Kuntze

WHITE-STEM PAPERFLOWER

Psilostrophe cooperi (Gray) Kuntze

FIELD IDENTIFICATION. Western, rounded, clump-forming shrub seldom over 2 ft. The stems are loosely and corymbosely branched. It is recognized by the white tomentose stems and contrasting yellow, paper-like corollas of the ligulate flowers.

FLOWERS. April–October, borne in long-peduncled, radiate, solitary or cymose heads ⅜–1¼ in. across, bearing both ray and disk flowers; involucre cylindric-campanulate, ¼–⅓ in. long, about ⅕ in. wide; receptacle small and naked; bracts 15–30, in 3 series, outer stiff, inner soft, connivent; ray flowers papery, 4–8 (usually 5), yellow, ½–¾ in. long, ⅖–⅗ in. wide, apex 3-toothed; disk flowers 4–12, bisexual, regular, the 5 lobes short and glandular.

FRUIT. Achene slender, terete, glabrous or nearly so, obscurely striate, pappus of 4–6 oblong, obtuse, hyaline scales.

LEAVES. Simple, alternate, blade length ⅜–2 in., width 1/25–⅙ in., sessile at base, linear or linear-lanceolate, mostly entire, the lower occasionally pinnatifid, white-tomentose, later green and more glabrous.

STEMS. Erect, slender, simple or corymbosely branched above, covered with dense white tomentum, less so and woody near the base.

RANGE. Usually in alluvial soils of deserts or desert-grasslands at altitudes of 2,000–5,000 ft. From southwestern New Mexico westward through Arizona to southern California and northward in Colorado, Utah and Nevada.

REMARKS. The genus name, *Psilostrophe,* is from the Greek, *psilos* ("bare") and *strophe* ("to turn"), with reference to the plant's becoming less hairy on old stems. The species name, *cooperi,* honors J. G. Cooper distinguished ornithologist, who collected plants near Fort Mohave in the 1860's. The yellow flowers are rather showy and attractive, giving the plant possibilities as a garden ornamental.

THREAD-LEAF GROUNDSEL

Senecio longilobus Benth.

FIELD IDENTIFICATION. An erect, usually branched plant attaining a height of 1–3½ ft. The branches rather diffuse, permanently tomentose, herbaceous above and often woody near the base, leafy to the summit.

FLOWERS. Few or numerous, corymbose, yellow, heads ⅜–⅝ in. high, broadly campanulate and radiate, involucre essentially uni-seriate with 13–21 equal, narrow, herbaceous phyllaries; calyculate with a few small inconspicuous bracteoles; ray flowers 7–13, linear, elon-

THREAD-LEAF GROUNDSEL
Senecio longilobus Benth.

gate, fertile; disk flowers more numerous, hermaphrodite and fertile; style branches in the disk flowers spreading, recurved, and truncate.

FRUIT. Achenes strigose-canescent, ribbed; pappus of soft, white, capillary bristles.

LEAVES. Alternate, filiform or pinnatilobate with 5–9 linear to filiform divisions which are often unequal, apices obtuse, margin entire and revolute.

RANGE. Dry plains, sage-brush areas, grasslands, at altitudes of 2,000–7,500 ft. In western Texas, New Mexico, Arizona, Colorado, and Utah. In Mexico in Chihuahua and Coahuila.

REMARKS. The genus name, *Senecio*, is the Latin name of a plant, from *senex* ("old man"), alluding to the white hairs of many species. The species name, *longilobus*, refers to the long lobes of the leaves. It is listed by some authors under the name of S. *douglasii* DC. var. *longilobus* (Benth.) Benson, and S. *filifolius* Nutt. The plant, especially the young growth, is considered poisonous to livestock; however, they seldom graze it except in times of stress. The Indians are reported to use the plant medicinally.

ZALUZANIA

Zaluzania grayana Robins. & Greenm.

FIELD IDENTIFICATION. Branched suffrutescent perennial 1–2½ ft. The branches are herbaceous above but sometimes woody near the base of the stem.

FLOWERS. Terminal or axillary, loosely clustered, corymbose; peduncles to 3 in. long, finely grooved, with scattered white hairs; heads yellow, radiate; receptacle conic; involucre hemispheric, bracts appressed, in about 2 series, canescent, oblong-lanceolate, acute, 1/16–1/8 in. long; ray and disk flowers fertile; rays linear to oblong, slightly notched at apex; tubular flowers of disk crowded, about 1/8 in. long or less, canescent-hairy, lobes 5, acute, about 1/25 in. long.

FRUIT. Achenes about 1/12 in. long, flattened or angled, lower end truncate, the upper end obtuse; those of the rays with a few short deciduous scales.

LEAVES. Alternate or opposite, ovate, often 3–5-lobed,

ZALUZANIA
Zaluzania grayana Robins. & Greenm.

Orange Zexmenia
Zexmenia hispida Gray

the lobes again slightly lobed or toothed, apex acute, base cordate or truncate and sometimes the blade slightly extending down the petiole, blade length 1¼–3 in., width ¾–2½ in., upper surface dull green with scattered stiff, curved or straight appressed hairs to glabrate; lower surface paler green and stiff hairs more numerous, especially on the veins; petiole ¼–1 in., grooved, pubescent.

TWIGS. Slender, finely grooved, reddish brown to straw-colored, puberulous; older ones glabrous toward the base, dark brown, roughened by longitudinal grooves.

RANGE. In New Mexico in the San Luis Mountains, Arizona, and Mexico in Chihuahua.

REMARKS. The genus name, *Zaluzania,* is in honor of the Polish botanist Adam Zaluziansky von Zaluzian. The species name, *grayana,* honors the American botanist Asa Gray.

Orange Zexmenia

Zexmenia hispida Gray

FIELD IDENTIFICATION. Perennial plant attaining a height of 3 ft. Herbaceous above but somewhat woody near the base. The slender branched stems and leaves very roughly strigose-hispid to the touch.

FLOWERS. Peduncles elongate, heads terminal, radiate, rather showy, solitary or in a cyme of 3; involucre 2–3-seriate, cylindric to campanulate, outer bracts lanceolate and loose, inner bracts oblong to ovate with acute or acuminate apices; ray flowers with 7–9 orange yellow ligules about ⅜ in. long; corolla-lobes puberulent-ciliolate.

FRUIT. Achenes very variable in size and shape, but usually obcordate; ray flowers with achenes 3-angled or 2–3-winged; disk flowers with achenes broadly 2-winged or with 2 upward auricles; pappus separated from the body of the achene by a constriction surrounded by a row of several short scales and 1–3 small awns.

LEAVES. Simple, opposite or alternate (sometimes with small leaves in the axils), rhombic to lanceolate or ovate to lanceolate, apex acute or acuminate, base acute or cuneate and sessile or nearly so, margin sparingly dentate or subhastately lobed, both sides roughly strigose.

TWIGS. Branched at the base, green to brown, strigose or strigose-hispid, herbaceous above, more or less woody near the base.

RANGE. Dry soil of hillsides or arroyo banks. Southwestern and western Texas. In Mexico in Tamaulipas, San Luis Potosí, Hidalgo, Veracruz, and the state of Mexico.

VARIETY. A variety of the Orange Zexmenia has been named Branched Orange Zexmenia, *Z. hispida* var. *ramosissima* Greenman, with stems 6–9 ft and much-branched peduncles. However, it is connected with the species by intermediate forms. It occurs in Yucatán, Mexico.

REMARKS. The genus name, *Zexmenia,* is an altered spelling for Ximenes. Francisco Ximenes (d. 1620) was a Dominican friar who collected plants in Mexico. He also edited the works of Francisco Hernandez. The species name, *hispida,* refers to the roughly strigose hairs of the leaves and stems.

Shorthorn Zexmenia

Zexmenia brevifolia Gray

FIELD IDENTIFICATION. Usually a perennial herb, but older specimens becoming quite woody and much-branched near the base, attaining a height of 3 ft.

FLOWERS. Borne solitary and terminally on naked peduncles; heads medium-sized, yellow, radiate; rays

5–9, pistillate and fertile; involucre ¼–⅖ in. high, bracts subequal or graduate, in 3–4 series, outer series broadest and mostly spatulate or obovate, surfaces strigose, apices herbaceous and spreading; disk flowers several, perfect, fertile; corollas yellow; stigmas of the disk with acute pubescent appendages.

FRUIT. Achenes of the rays 3-angled; those of the disk flattened, oblong or obovate, at maturity with callous wings, and often with persistent scales between the pappus of 1–3 awns.

LEAVES. Alternate or opposite, sometimes with smaller ones in their axils, blade length ¼–½ in., occasionally to 1 in., width ⅛–½ in., ovate to oval or suborbicular, apex obtuse or rounded, base rounded to subcordate, margin entire or occasionally with a few coarse teeth, surfaces very rough to the touch with strigose-scabrous whitened hairs; petioles 1/16–⅜ in. long, strigose.

TWIGS. Slender, light brown to gray, young ones strigose and scabrous, rough to the touch; older ones smooth, gray, and glabrous.

RANGE. In caliche or limestone soils of rocky banks, canyon slopes, and crevices of rock. In southwestern Texas and New Mexico. In Mexico in the states of Chihuahua and Coahuila, south to Zacatecas and San Luis Potosí.

REMARKS. The genus name, *Zexmenia*, is an altered spelling for Ximenes. Francisco Ximenes (d. 1620) was a Dominican friar who collected plants in Mexico. The species name, *brevifolia*, refers to the small short leaves.

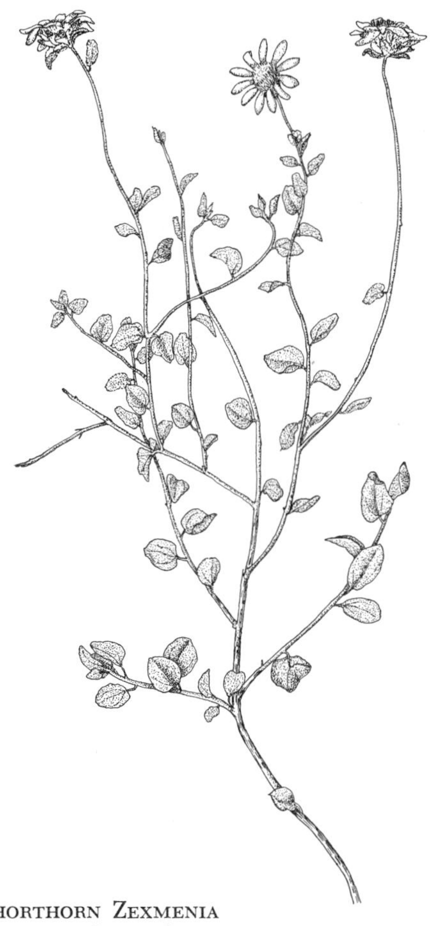

SHORTHORN ZEXMENIA
Zexmenia brevifolia Gray

GREGG ALKALI-BUSH

Haploesthes greggii Gray

FIELD IDENTIFICATION. Plant a perennial, fleshy above, and woody below, attaining a height of 1–2 ft. The stems are usually branched near the base.

FLOWERS. Borne in few-flowered corymbs or cymes, the heads radiate and borne on slender naked peduncles; involucres ⅛–1/7 in. high; bracts few (usually 4–5), imbricate, many-nerved, oval to orbicular, apex rounded, margin thin, green or yellowish; receptacle flat and naked; ray flowers few, pistillate, fertile, ligules pale yellow and 1/12–⅙ in. long; disk flowers perfect, fertile, stigmas capitate–truncate, anthers united.

FRUIT. Achenes about 1/12 in. long, linear, terete, 10-ribbed, glabrous or hispidulous; pappus of slender, rigid, scabrous, white bristles.

LEAVES. Opposite, fleshy, connate and sheathing at the base, linear to filiform, ¾–1¾ in. long, margin entire, apex acute, 1-nerved.

STEMS. Leafy, branched at the base, smooth, glabrous, striate, pale green.

RANGE. Usually in dry saline or alkaline soil. Western Texas, New Mexico, Oklahoma, Arizona, Colorado, and Kansas. In Mexico in the state of Coahuila.

REMARKS. The genus name, *Haploesthes*, means "simple garment" and refers to the unusually few phyllaries. The species name, *greggii*, is in honor of the botanist Josiah Gregg (1806–1850), botanical explorer of the southwestern United States.

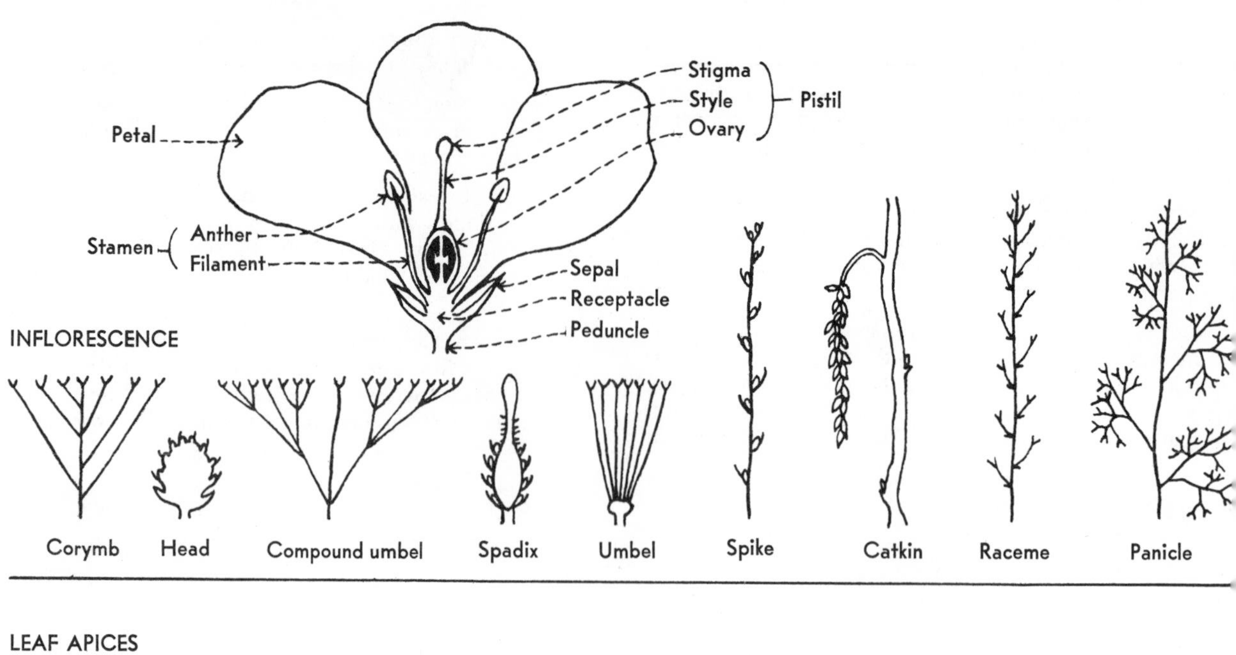
Petal
Stamen
Anther
Filament
Stigma
Style
Ovary
Pistil
Sepal
Receptacle
Peduncle
INFLORESCENCE
Corymb
Head
Compound umbel
Spadix
Umbel
Spike
Catkin
Raceme
Panicle

LEAF APICES

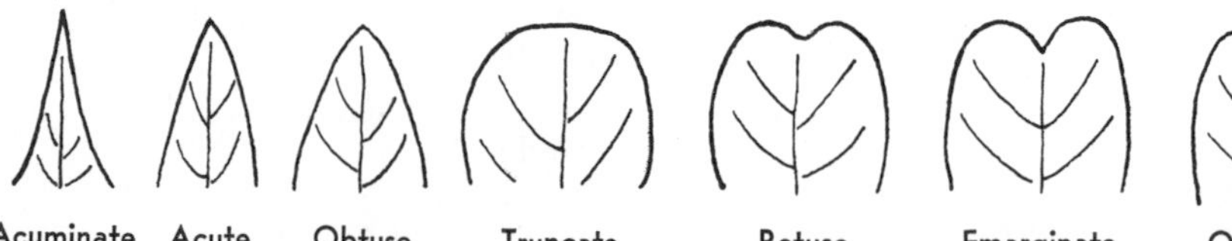
Acuminate
Acute
Obtuse
Truncate
Retuse
Emarginate
Obcordate
Cuspidate
Mucronate

LEAF MARGINS

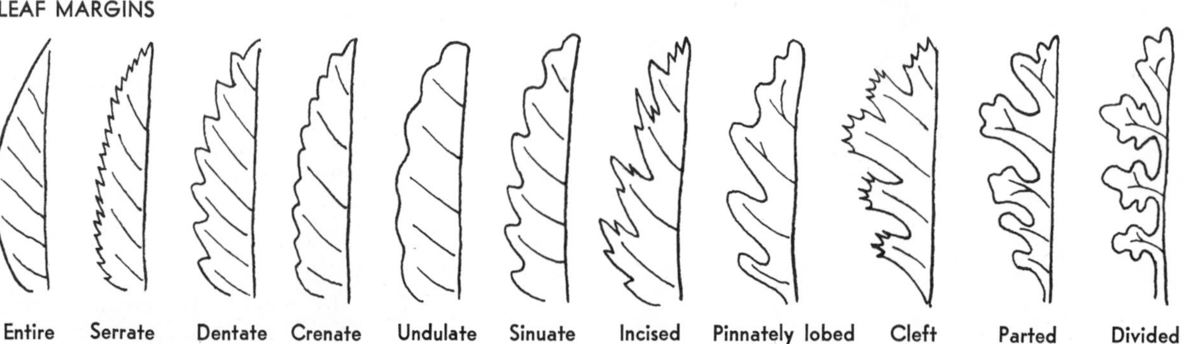
Entire
Serrate
Dentate
Crenate
Undulate
Sinuate
Incised
Pinnately lobed
Cleft
Parted
Divided

LEAF FORMS

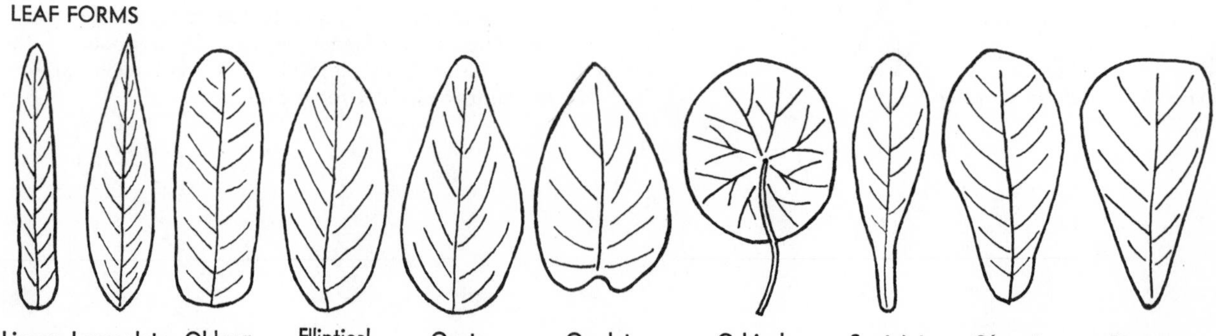
Linear
Lanceolate
Oblong
Elliptical
Ovate
Cordate
Orbicular
Spatulate
Obovate
Cuneate

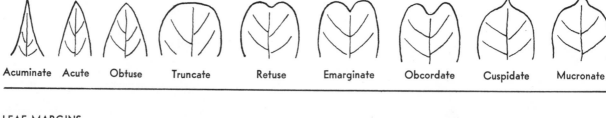

GLOSSARY

A- (prefix). Without, lacking.
Abaxial. On the dorsal side, away from the axis.
Aberrant. Not normal, atypical.
Abortive. Imperfectly developed.
Abruptly acuminate. Suddenly pointed from a rounded or truncate apex.
Abruptly pinnate. A pinnate leaf ending with a pair of leaflets; even-pinnate.
Acaulescent. Without a stem aboveground.
Accrescent. Increasing in size with age.
Accumbent. Lying against and face to face.
Acerose. Needle-like.
Achene. A small, dry, hard, one-seeded, indehiscent fruit.
Acicular. Bristle-like.
Acidulous. Slightly acid or bitter.
Acorn. The fruit of oaks.
Acotyledonous. Without cotyledons.
Acrid. Bitter or sharp tasting, usually referring to the fruit or sap.
Acropetal. Borne in succession toward the apex, as in certain inflorescences.
Acuminate. Referring to an acute apex whose sides are concave.
Acute. Terminating with a sharp angle.
Adherent. Referring to two dissimilar parts or organs which touch but are not fused.
Adnate. Grown together; organically united.
Adventive. A recent, perhaps temporary introduction, not as yet naturalized, or barely so.
Aerenchyma. Spongy respiratory tissue in stems of many aquatic plants, characterized by large intercellular spaces.
Aerial. Parts above the ground or water.
Aggregate. Referring to a type of fruit with a cluster of ripened coherent ovaries.
Alkaloid. An organic base produced in some plants, sometimes with medicinal or poisonous properties.
Allergic. Subject to irritation by foreign substances.
Alpine. At high elevations; above the tree line.
Alternate. Placed singly at different levels on the axis.
Alternation of generations. Alternate sexual and asexual generations.
Alveola. Surface cavity of carpel or seed.
Ament. A catkin or scaly-bracted, often pendulous, spike of naked or reduced flowers.
Amplexicaul. Clasping the stem.
Ampliate. Expanded or enlarged.
Anastomosing. Netted; said of leaf blades with cross veins forming a network.
Anatropous. Referring to an ovule's position when the micropyle is close to the point of funiculus attachment.
Androecium. Stamens, in the collective sense.
Androgynous. Bearing staminate and pistillate flowers in the same inflorescence.
Androphore. Supporting stalk of a group of stamens.
Angulate. Angled.
Annulus. A ring-shaped part or organ.
Anterior. Away from the axis; the front; toward a subtending leaf or bract.
Anther. The polleniferous part of a stamen.
Anther cells. The actual chambers or locules of an anther.
Anther sac. Pollen sac of an anther.
Antheriferous. Having anthers.
Anthesis. Time of expansion of a flower; pollination time.
Anthocarp. A fruit condition in which at least a portion of the perianth is united with the ovary wall itself.
Antrorse. Directed upward.
Apetalous. Lacking petals.
Apex. The top or termination of a part or organ.
Apiculate. Ending in a short pointed tip.
Apomixis. Reproduction without sexual union.
Appendages. Various subsidiary or secondary outgrowths.
Appressed. Closely and flatly pressed against.
Approximate. Adjacent or close.
Aquatic. Growing in water.
Arachnoid. Bearing weak, tangled, cobwebby hairs.

Arboreal. Referring to trees, treelike.
Arborescent. Like trees in size and growth habit.
Arcuate. Somewhat curved.
Arenicolous. Sand loving.
Areole. Areola (pl. *Areolae*). An open space or island formed by anastomosing veins in a foliar organ; spine-bearing area of cactus.
Aril. An appendage or complete additional covering of the seed, arising as an outgrowth from hilum or funicle.
Aristate. With a stiff, bristle-like appendage.
Articulated. Jointed and cleanly separating to leave a scar.
Ascending. Growing upward, but not erect.
Asexual. Without sex, reproduction without sexual union, such as by cuttings, buds, bulbs, etc.
Assurgent. Ascending.
Asymmetric. Not symmetrical; with no plane of symmetry.
Atypical. Not typical.
Attenuate. Slenderly tapering.
Auricle. Earlike attachment, such as at the base of some leaves or petals.
Auriculate. Furnished with auricles.
Autophytic. Having chlorophyll, self-sustaining.
Axial. Referring to the axis.
Axil. The upper angle formed by a leaf with the stem, or veins with other veins.
Axile. Situated on the axis.
Axillary. Situated in an axil.
Axis. The center line of any organ; the main stem.

Baccate. Berry-like; bearing berries.
Banner. Upper or posterior petal of a papilionaceous flower.
Barb. A bristle-like hooked hair or projection.
Bark. External tissues of woody plants, especially the dead corky layer external to the cortex or bast in trees.
Basal. At or near the base.
Beak. A narrow pointed outgrowth of a fruit, a petal, etc.
Berry. A fruit in which the ovary becomes a fleshy or pulpy mass enclosing one or more seeds, as is seen in the tomato.
Bi- (prefix). Two or two-parted.
Bifid. Cleft into two lobes.
Biglandular. With two glands.
Bilabiate. Two-lipped, as in some irregular corollas.
Bipinnate. Twice pinnate.
Bisexual. Having both sex organs in the same flower.
Biternate. Twice ternate.
Blade. The expanded portion of a leaf.
Brackish. Partially saline.
Bract. A more or less modified leaf subtending a flower or belonging to a cluster of flowers.
Bracteole. A small bract or bractlet.
Bractlet. The bract of a pedicel.
Bud. The rudimentary state of a shoot; an unexpanded flower.
Bud Scales. Modified leaves protecting a bud, oft dry, resinous or viscid.

Caducous. Falling early.
Caliche. A hard calcareous soil.
Callose. Bearing a callus.
Callus. A thickened or hardened protuberance region.
Calyculus. Usually in the Compositae, referring to th short outer sepal-like phyllaries of an involucre.
Calyx. The outer perianth whorl of the flower.
Campanulate. Bell-shaped.
Canaliculate. Having one or a few prominent longitu dinal grooves.
Cancellate. With cross hatching or latticed ridges.
Canescent. Hoary with a gray pubescence.
Capillary. Hairlike.
Capitate. In a head, or headlike.
Capsule. A dry dehiscent fruit composed of more tha one carpel.
Carina. A dorsal ridge or keel.
Carpel. A simple pistil, or one member of a compoun pistil; the ovule-bearing portion of a flower, believe to be of foliar origin.
Cartilaginous. Firm or tough tissue, suggestive animal cartilage.
Caruncle. A seed protuberance; a small hard aril.
Catkin. A delicate, scaly-bracted, usually pendulou spike of flowers; an ament.
Caudate. Long-attenuate; tail-like.
Caudex. Enlarged basal part of stems, or combine root and stem.
Caulescent. Having an evident stem, as contrasted t acaulescent.
Cauline. Borne on the stem.
Centrifugal Inflorescence. Flowers developing fro the center outward; a determinate type such as cyme.
Centripetal Inflorescence. Flowers developing from th outer edge toward the center; an indeterminate typ such as a corymb.
Cespitose, caespitose. In tufts.
Chaff. A small scale, as is found on the receptacle c many Compositae.
Chaparral. A type of low scrub, commonly with dens twiggy, thorny habit and evergreen leaves.
Chartaceous. Like old parchment; papery.
Ciliate. Bearing marginal hairs.
Ciliolate. Bearing short marginal hairs.
Cinereous. Ashy-gray.
Circinate. Referring to a leaf that is coiled from th tip toward the base, the lower surface outermost.
Cirrhus. A tendril.
Clambering. Leaning on other plants or objects, no self-supporting.
Clasping. Enveloping the stem partially or wholly a the base, as leaf bases, bracts, stipules, etc.
Clavate. Club-shaped.
Claw. Narrowed petiole-like base of a petal or sepal
Cleft. Incised or cut nearly or quite to the middle.

Cleistogamous. Said of flowers that are self-fertilized without expanding; modified or reduced flowers.
Climber. Plant seeking support by twining or by tendrils.
Coalescent. Grown together, as in similar parts.
Coat. Covering of a seed.
Coherent. Having similar parts in close contact but not fused.
Column. A body formed by the union of parts, such as a union of stamens to form the staminal column in Malvaceae.
Coma. A hair tuft, usually at the apex of seeds.
Commisure. A line of coherence, such as where two carpels are joined.
Compound Leaf. A leaf with the blade divided into two or more leaflets.
Compressed. Flattened.
Concavo-convex. Convex on one side and concave on the other.
Concolorous. The same in color.
Conduplicate. Folded lengthwise.
Cone. An inflorescence of flowers or fruit with overlapping scales.
Confluent. Gradually passing into each other.
Congested. Crowded, as flowers in a dense inflorescence.
Conic. Cone-shaped.
Coniferous. Cone-bearing.
Connate. Said of similar parts which are united, at least at the base.
Connective. Extension of the filament of a stamen connecting the anther lobes.
Connivent. Arched inward so that the tips meet.
Consimilar. Alike.
Constricted. Narrowed between wider portions.
Contiguous. Adjacent or adjoining similar or dissimilar parts, not fused.
Contorted. Twisted.
Contracted. Narrowed or shortened.
Convergent. Tending toward a single point, as with leaf veins approaching each other.
Convolute. Rolled together, as with petals in buds of certain plants.
Copious. Abundant, plentiful.
Cordate. Heart-shaped, as in some leaves.
Coriaceous. Leathery in texture.
Corneous. Horny, as in margins of some leaves.
Corniculate. Hornlike in appearance.
Cornute. Spurred or horned.
Corolla. The inner perianth whorl of a flower.
Corona. A crownlike structure on the corolla and the stamens.
Coroniform. Crown-shaped, applied to the pappus of certain Compositae.
Corrugate. Strongly wrinkled.
Corymb. A flat-topped or convex open flower cluster, with its marginal flowers opening first.
Costa. A thickened vein or midrib.
Cotyledon. Plant embryo leaf, usually rich in stored food.
Creeping. Referred to a trailing shoot which strikes root along most of its length.
Crenate. Dentate with the teeth much rounded.
Crenulate. The diminutive of crenate.
Crest. A crown or elevated ridge, often entirely or partly of hair.
Crispate. Crisped or crumpled.
Cross-pollination. Transfer of pollen from flower to flower.
Crown. Usually referring to the branches and foliage of a tree; or thickened bases of stems.
Cruciform. Shaped like a cross.
Crustaceous. Dry and brittle.
Cucullate. Hood-shaped.
Cuneate. Wedge-shaped.
Cupulate. Cup-shaped, as the involucre of an acorn.
Cuspidate. Having a sharp rigid point.
Cyathium. A specialized form of inflorescence of some of the Euphorbias.
Cylindric. Elongate and circular in cross section.
Cylindroid. Like a cylinder but elliptic in cross section.
Cymbiform. Boat-shaped.
Cyme. A usually broad and flattish inflorescence with its central or terminal flowers blooming earliest.
Cymose. Cymelike.
Cymule. A small cyme.

Deciduous. Not persistent, not evergreen.
Declined. Turned downward or outward, as in some stamens.
Decompound. Several times divided, as in repeatedly compound leaves and inflorescences.
Decumbent. Reclining, but ascending at the apices.
Decurrent. Extensions downward, as of some petioles or leaves along the stem.
Decurved. Curved downward.
Decussate. Leaves opposite, with each pair at a right angle to the pair above or below.
Deflexed. Bent downward.
Dehiscent. Opening regularly by valves.
Deltoid. Triangular.
Dentate. Toothed, specifically when teeth have sharp points and spreading bases.
Denticulate. Minutely toothed.
Denudate. Becoming bare.
Depressed. Flattened from above.
Dermatitis. Irritation or inflammation of the skin.
Determinate. Having fixed limits, as in an inflorescential axis ending with a bud; also referring to a cymose, or centrifugal, inflorescence.
Diadelphous. With stamens in two groups.
Dichotomous. Forked in pairs.
Didymous. Twice, two-lobed, in pairs.
Didynamous. With four stamens in two pairs of differing length.
Diffuse. Spreading.
Digitate. Said of a compound leaf in which all the leaflets arise from one point.
Dilated. Expanded or enlarged.
Dimidiate. Reduced to one half, often by abortion.

Dimorphic. Having two forms.
Dioecious. Unisexual, with the two kinds of flowers on separate plants.
Diploid. Having the somatic number of chromosomes; twice as many as in the germ cells after reduction.
Disarticulate. Breaking at a joint.
Disc, disk. Outgrowth of the receptacle or hypanthium within the perianth, often composed of fused nectaries; central portion of the head in Compositae.
Disciform. Disk-shaped.
Discoid. Referring to a rayless head of flowers, as in some Compositae.
Discolored. Usually of two colors, as in leaf surfaces.
Dissected. Deeply cut or divided into many narrow segments, as in some leaves.
Distal. The apex of an organ, the part most distant from the axis.
Distended. Swollen.
Distichous. In two vertical ranks.
Distinct. Separate, not united.
Diurnal. Opening during the day.
Divaricate. Widely spreading.
Divergent. Spreading apart.
Divided. Lobed to near the base.
Division. Segment of a parted or divided leaf.
Dorsal. Upon or relating to the back or outer surface of an organ.
Dorsifixed. Attached to the dorsal side.
Drupe. A fleshy or pulpy fruit in which the inner portion is hard and stony, enclosing the seed.
Drupelet. A small drupe.

E- (prefix). Without or lacking a structure or organ.
Echinate. Prickly.
Ecology. Science dealing with plants in relation to their environment.
Eglandular. Without glands.
Ellipsoid. Referring to the geometric figure obtained by rotating an ellipse on its longer axis.
Elliptic. Of the form of an ellipse.
Elongate. Lengthened, drawn out.
Emarginate. Referring to a notch, usually of a leaf apex.
Embryo. A young plant, still enclosed in the seed.
Endemic. Known only in a limited geographical area.
Endocarp. Inner layer of the pericarp of the ovary or fruit.
Endosperm. Part of the seed outside the embryo; the albumen or stored food.
Ensiform. Sword-shaped.
Entire. Without toothing or divisions.
Ephemeral. Lasting for only a short period.
Epidermis. The superficial layer of cells.
Epigaeous, epigeous. Growing on the ground, or close to the ground.
Epigynous. Attached to or borne on the pistil, as in some stamens or petals.
Erose. Irregularly toothed.
Estipulate. Exstipulate.
Exalbuminous. Without albumen or endosperm, the embryo filling the seed instead.
Excavated. Referring to some pitted or channeled seeds.
Excurrent. Prolongation of nerves into awns or mucros of a leaf or fruit body; also the prolongation of the main stem or axis of a plant in certain conifers.
Excurved. Curved outward.
Exfoliating. Referring to bark separating into strips or flakes.
Exocarp. The outer layer of pericarp.
Exogenous. Growth by increase of tissue beneath the expanding bark; the cambium region in dicotyledonous woody plants; as opposed to endogenous growth by internal multiplication of tissues in monocotyledonous plants.
Expanded. Spreading, opened to the greatest extent.
Explanate. Flattened out.
Exserted. Projecting beyond an envelope, as stamens from a corolla.
Exstipulate. Lacking stipules.
Extra-axillary. Being near the axil but not truly axillary, as in some flowers.
Extrafloral. Outside the flower, as extrafloral nectaries.
Extrorse. Directed outward.
Exudate. An excretion of wax, gum, sap, etc.

Faceted. Usually applied to seeds with several plane surfaces.
Facial. On the plane surface or face rather than the margin.
Falcate. Sickle- or scythe-shaped.
Farinose. Covered with mealy or floury particles.
Fascicle. A close bundle or cluster.
Fastigiate. Having closely set erect branches.
Faveolate. Honeycombed.
Feather-veined. With secondary veins branching from the main vein.
Fertile. Applied to flowers with pistils capable of producing seeds, or to stamens with functional pollen.
Fetid. Malodorous.
Fibrilla. A very small fringe.
Fibrillose. Having fine fibers, as the leaf margins of some Yuccas.
Fibrous. Bearing a resemblance to fibers, or possessing fibers, as in fibrous roots.
Filament. The part of a stamen which supports the anther.
Filamentose. Having threadlike structures; thread-like.
Filiferous. Bearing threads.
Filiform. Thread-shaped.
Fimbriate. Fringed, frayed at the ends or on the margin.
Flabellate. Fan-shaped.
Flaccid. Lax, limp, flabby.
Flexuose, flexuous. Zigzag, or bent in an alternating manner.
Floccose. Clothed with tufts of soft woolly hairs.
Floral. Referring to flowers.
Floret. A small flower, as in the disk flowers of Compositae.
Floriferous. Having flowers, usually abundantly.

Floristic. Referring to the aggregate aspects of the vegetation, as to species, abundance and distribution in a geographical sense.

Fluted. Regularly marked by alternating ridges and groovelike depressions.

Foliaceous. Leaflike in texture and appearance.

Foliate. Referring to leaves as opposed to leaflets.

Foliolate. Referring to the leaflets in a compound leaf, such as bifoliolate or trifoliolate.

Follicle. A dry fruit opening along the single suture, the product of a simple pistil.

Fruit. Seed-bearing part of a plant.

Frutescent. Shrubby.

Fugacious. Referring to early-deciduous parts, such as petals or sepals of certain plants.

Fulvous. Tawny, dull yellow.

Functional. Able to produce normally.

Funiculus. The stalk attaching ovule to ovary wall or placenta.

Funnelform. Shaped like a funnel.

Fuscous. Dark brownish gray.

Fusiform. Spindle-shaped.

Gamopetalous. Referring to the petals being more or less united.

Gamosepalous. Referring to the sepals being more or less united.

Geminate. In pairs, twins.

Gene. A unit of inheritance, which occupies a fixed place on a chromosome.

Geniculate. Bent like a knee joint.

Gibbous. Swollen or inflated on one side.

Glabrate. Somewhat glabrous or becoming glabrous.

Glabrous. Smooth; pubescence or hairs absent.

Gland. A secreting protuberance or appendage.

Glandular. Bearing glands, or of the nature of a gland.

Glaucous. Covered with a white, waxy bloom.

Globose. Globular, spherical.

Glochid. A hair or minute prickle, often with retrorse or hooklike projections, as in some cacti.

Glomerate. In a headlike or crowded inflorescence.

Glomerule. A small headlike inflorescence.

Glutinous. Gluelike, sticky.

Graduated. Rows or series of bracts, or phyllaries, of different size, as on the involucre of some Compositae.

Granules. Small particles on the surface of a plant, such as resinous granules, wax granules, etc.

Gynoecium, gynecium. The total female element of a flower.

Gynophore. The stalk of a pistil.

Habitat. The environment of a plant.

Halophyte. A plant tolerant of saline conditions.

Haploid. Possessing half of the diploid number of chromosomes, as in the germ cells after the reduction division.

Hastate. Halberd-shaped; sagittate, but with the basal lobes more or less at right angles.

Head. A spherical or flat-topped inflorescence of sessile or nearly sessile flowers borne on a common receptacle.

Heartwood. The oldest wood, inclosing the pith; the hard central, often deeply colored, portion of a tree trunk.

Hermaphrodite. Bisexual.

Heterogamous. Having two or more kinds of flowers with respect to the distribution of sex organs.

Heteromorphic. Unlike in form or size, as sometimes the length of stamens or pistils on different plants of the same species.

Heterostylous. Having styles of different length or character.

Hexamerous. With the parts in sixes.

Hilum. The scar of a seed, marking the point of attachment.

Hip. Fruit in *Rosa,* composed of swollen hypanthium bearing achenes within.

Hirsute. Covered with rather coarse or stiff hairs.

Hirsutulous. Finely or minutely hirsute.

Hispid. Beset with rigid or bristly hairs or with bristles.

Hispidulous. Minutely hispid.

Hoary. Densely grayish-white pubescent.

Homochromous. Of uniform color.

Homogeneous. Of the same kind, uniform.

Homomorphic. Of uniform size and shape.

Hood. A concave organ, usually referring to certain petals.

Hyaline. Very thin and translucent.

Hybrid. Product of genetically dissimilar parents.

Hybridization. The production of a hybrid.

Hydrophilous. Referring to a tendency to grow in water.

Hydrophyte. An aquatic plant.

Hypanthium. Upward or outward extension of receptacle derived from fused basal portions of perianth and androecium.

Hypogynous. Inserted beneath the gynoecium, but free from it.

Imbricate. Overlapping.

Immersed. Submerged.

Imparipinnate. Pinnate with a single terminal leaflet; odd-pinnate.

Imperfect. Diclinous; lacking functional stamens or pistils.

Implicate. Twisted together or interwoven.

Impressed. Lying below the general surface; as, impressed veins.

Incanous. Hoary with whitish pubescence.

Incised. Cut sharply, irregularly, and more or less deeply.

Included. Not protruding, not exserted.

Incomplete. Descriptive of flowers in which one or more perianth whorls are wanting.

Incurved. Curved inward.

Indehiscent. Not opening.

Indeterminate. Applied to inflorescences in which the flowers open progressively from the base upward.

Indigenous. Native to an area.
Indument. A covering of hairs or wool.
Induplicate. Having the edges folded together.
Indurate. Hardened.
Inequilateral. Asymmetrical.
Inferior. Usually referring to an ovary being adnate to and appearing as if below the calyx.
Infertile. Not fertile or viable.
Inflated. Bladder-like.
Inflexed. Bent inward.
Inflorescence. A flower cluster; the disposition of flowers.
Inframedial. Below the middle, but not at the base.
Infrastaminal. Below the stamens.
Inodorous. Lacking odor.
Inserted. Attached to or growing out of.
Inter- (prefix). Between.
Intercostal. Between the ribs, veins, or nerves.
Internode. That portion of stem lying between two successive nodes.
Interrupted. Referring to an inflorescence with sterile intervals, mostly of varying length, between the flowers.
Intricate. Densely tangled, as in some branches.
Introflexion. State of being inflexed.
Introrse. Turned inward; facing the axis.
Intrusion. Protruding or projecting inward.
Inverted. Reversed, opposite to the normal direction.
Investing. Enclosing or surrounding.
Involucel. A secondary involucre, subtending a secondary division of an inflorescence as in Umbelliferae.
Involucral. Belonging to an involucre.
Involucre. One or more series of bracts, surrounding a flower cluster or a single flower.
Involute. With edges rolled inward.
Irregular. Said of flowers that are not bilaterally symmetrical.

Joint. A node; a unit of a segmented stem as in *Opuntia.*

Keel. The united pair of petals in a papilionaceous flower; a central or dorsal ridge.

Laciniate. Cut into lobes separated by deep, narrow, irregular incisions.
Lanate, lanuginose. Covered with matted hairs or wool.
Lanceolate. Shaped like a lance-head.
Lanulose. Short-woolly.
Leaf. The usually thin and expanded organ borne laterally on the stem; in a strict sense inclusive of the blade, petiole, and stipules; but in common practice referred to the blade only.
Leaflet. A single division of a compound leaf.
Legume. A dry fruit, the product of a simple unicarpellate pistil, usually dehiscing along 2 lines of suture.
Lenticels. Corky growths on young bark.
Lenticular. Having the shape of a double convex len
Lepidote. Covered with minute scales or scurf.
Liana. A climbing woody plant.
Ligneous. Woody.
Ligulate. Tongue-shaped.
Ligule. Usually referring to an expanded ray flow in the Compositae.
Limb. The ultimate or uppermost extension of gamopetalous corolla or calyx; distinct from the tub or throat.
Linear. Long and narrow.
Lingulate. Tongue-shaped.
Lip. One of the (usually two) divisions of an un equally divided corolla or calyx.
Littoral. Growing near the sea.
Lobe. Any segment or division of an organ.
Lobed. Divided into lobes, or having lobes.
Lobulate. With small lobes.
Locule. A compartment of an anther, ovary, or fruit.
Loculicidal. Longitudinally dehiscent dorsally, midwa between the septa.
Loment. A fruit of the Leguminosae usually constricte between the seeds, the one-seeded indehiscent por tions breaking loose.
Longitudinal. Lengthwise.
Lunate. Crescent-shaped.
Lyrate. Lyre-shaped, or a pinnatifid form with th terminal lobe usually longest.

Marcescent. Persistent after withering.
Marginal. On or pertaining to the margin of a plan organ.
Mealy. Farinose or floury.
Medial, median. Referring to or at the middle.
Membranaceous, membranous. Thin, rather soft, an more or less translucent.
Meniscoid. Concavo-convex, like a watch crystal.
Mericarp. A one-seeded indehiscent carpel of a fru whose carpels separate at maturity.
-Merous (suffix). Referring to the number of parts as 4-merous.
Mesophyte. Plant requiring a moderate amount o water.
Microphyllous. Small-leaved.
Midnerve. The central vein or rib.
Midrib. The main rib of a leaf.
Monadelphous. Union of all stamens by their filaments
Moniliform. Necklace-like.
Monochasial. Applied to a cymose inflorescence wit one main axis.
Monochromatic. Of one color.
Monoecious. With stamens and pistils in separate flow ers, but on the same plant.
Monopodial. Applied to a stem having a single con tinuous axis.
Montane. Growing in the mountains.
Mottled. Spotted or blotched.
Mucilaginous. Sticky and moist.
Mucro. A short and small abrupt tip.
Mucronate. Tipped with a mucro.

Mucronulate. Diminutive of mucronate.
Multicipital. With several or numerous stems from a single caudex or taproot.
Multifoliolate. Referring to a compound leaf with numerous leaflets.
Multiple Fruit. One that results from the aggregation of ripened ovaries into one mass.
Muricate. With surfaces bearing hard sharp tubercles.
Muriculate. Diminutive of muricate.

Naked. Lacking a customary part or organ; flowers without a perianth, a receptacle without chaff, a stem without leaves, etc.
Nectariferous. Producing nectar.
Nectary. A gland, or an organ containing the gland, which secretes nectar.
Nerve. An unbranched vein.
Neuter. Referring to flowers lacking stamens or pistils; or in some flowers of the Compositae having an ovary but lacking style or stigmas.
Nigrescent. Turning black.
Nocturnal. Night-blooming.
Nodding. Arching downward.
Node. The joint of a twig; usually a point bearing a leaf or leaflike structure.
Nut. A hard, indehiscent, one-celled and one-seeded fruit.

Ob- (prefix). Inversely, upside down, as an obovate leaf (inversely ovate), etc.
Obcordate. Inverted heart-shaped.
Oblanceolate. Lanceolate with the broadest part toward the apex.
Oblique. Slanting; with unequal sides.
Oblong. Longer than broad and with nearly parallel sides.
Obovate. Inverted ovate.
Obovoid. Appearing as an inverted egg.
Obsolete. Not evident; very rudimentary; vestigial.
Obtuse. Blunt.
Ochroleucous. Yellowish-white.
Ocrea. A tubular sheath formed by the union of a pair of stipules.
Odd-pinnate. A pinnate leaf with a single terminal leaflet.
Opposite. Opposed to each other, such as two opposite leaves at a node.
Orbicular. Circular.
Organ. A part of a plant with a definite function, as a leaf, stamen, etc.
Oval. Broadly elliptical.
Ovary. The part of the pistil that contains the ovules, or seeds after fertilization.
Ovate. Egg-shaped, with the broader end closer to the stem.
Ovoid. Referring to a solid object of ovate or oval outline.
Ovulate. Bearing ovules; referring to the number of ovules.
Ovule. The body which after fertilization becomes the seed.

Palea. A chaffy bract on the receptacle of Compositae.
Pallid. Pale or light colored.
Palmate. Said of a leaf radiately lobed or divided.
Palmatifid. Palmately cleft or lobed.
Palmatisect. Palmately divided.
Palustrine. Growing in wet ground.
Panicle. A branched raceme.
Pannose. Feltlike, covered with closely interwoven hairs.
Papilionaceous. Descriptive of the flowers of certain legumes having a standard, wings, and keel petals; resembling a sweet pea.
Papilla. A small soft protuberance on leaf surfaces, etc.
Papillate. Bearing papillae.
Pappus. Modified calyx of certain Compositae, often elaborate in the fruit, consisting of hairs, bristles, awns or scales.
Parcifrond. A long leafy usually sterile shoot of a floricane, arising from below the ordinary floral branches.
Parietal. Attached to the wall within the ovary.
Parted. Cleft nearly to the base.
Parthenogenetic. Asexual development of an egg, without fertilization by a male sex cell.
Parti-colored. Variegated.
Pectinate. With narrow, toothlike divisions.
Pedate. Palmate, with the lateral lobes or divisions again divided.
Pedicel. The stalk of a single flower in a cluster.
Pedicellate. Borne on a pedicel.
Peduncle. Primary flower stalk supporting either a cluster or a solitary flower.
Pedunculate. Borne on a peduncle.
Pellucid. Transparent or nearly so.
Peltate. Referring to a leaf blade in which the petiole is attached to the lower surface, instead of on the margin as in the garden nasturtium.
Pendent. Hanging or drooping.
Pendulous. More or less hanging.
Pentagonal. Five-angled.
Pentamerous. Of five parts, as a flower with five petals.
Perennial. Usually living more than two years.
Perfect. Referring to a flower having both pistil and stamens.
Perfoliate. A sessile leaf whose base passes around the stem.
Perforate. Pierced with holes; dotted with translucent openings which resemble holes.
Perianth. The floral envelope, usually consisting of distinct calyx and corolla.
Pericarp. The outer wall of an ovary after fertilization, hence of the fruit.
Perigynous. Referring to petals and stamens whose bases surround the pistil, or pistils, and are borne on the margin of the hypanthium, as in some Rosaceae.
Persistent. Said of leaves that are evergreen, and of

flower parts and fruits that remain attached to the plant for protracted lengths of time.
Petal. The unit of the corolla.
Petaloid. Resembling a petal.
Petiole. The stalk of a leaf.
Petiolar. Relating to or borne on the petiole, as a petiolar gland.
Petiolate. Having a petiole.
Petiolulate. Having a petiolule.
Petiolule. The footstalk of a leaflet.
Phyllary. One of the bracts of the involucre in Compositae.
Pilose. Hairy, especially with long soft hairs.
Pinnae. The primary divisions of a pinnate leaf.
Pinnate. Descriptive of compound leaves with the leaflets arranged on opposite sides along a common rachis.
Pinnatifid. Pinnately cleft, the clefts not extending to the midrib.
Pinnatilobate. Pinnately lobed.
Pinnatisect. Pinnately divided, the clefts extending to the midrib.
Pinnule. A leaflet or ultimate segment of a pinnately decompound leaf.
Pistil. The seed-bearing organ of the flower, consisting of ovary, stigma, and style.
Pistillate. Provided with pistils, but lacking functional stamens.
Pith. The central tissue of a (usually exogenous) stem, composed of thin-walled cells.
Placenta. The structure in an ovary to which the ovules are attached.
Plano-convex. Flat on one side, convex on the other.
Pleiochasium. A compound cyme with more than two branches at each division.
Plicate. Folded into plaits.
Plumose. Having fine hairs on each side like the plume of a feather.
Pluriseriate. In many series.
Pod. Any dry and dehiscent fruit.
Pollen. The male germ cells, contained in the anther.
Pollen sac. Pollen-bearing chamber of the anther.
Pollination. Deposition of pollen upon the stigma.
Polygamo-dioecious. Essentially dioecious, but with some flowers of other sex or perfect flowers on same individual.
Polygamous. With both perfect and unisexual flowers on the same individual plant or on different individuals of the same species.
Polygonal. With several sides or angles.
Polymorphic. With a number of forms, as, a very variable species.
Polyploid. With chromosome number in the somatic nuclei greater than the diploid number, sometimes accompanied by increased size and vigor of the plant.
Pome. A kind of a fleshy fruit of which the apple is the typical form.
Pore. A small opening, as found in certain anthers or fruits.
Porrect. Reaching or extending perpendicular to th surface, as with spines of some cacti.
Posterior. Toward the rear, behind, toward the axi
Prickle. A rigid, straight or hooked, outgrowth of th bark or epidermal tissue.
Procumbent. Lying on the ground.
Prominent. Higher than the adjacent surface, as i prominent veins.
Prophyllum. A bracteole.
Prostrate. Flat on the ground.
Proximal. Near the base or axis, the opposite of dista
Pruinose. Covered with a white bloom or powder wax.
Puberulent. Very slightly pubescent.
Pubescence. A covering of short hairs.
Pubescent. Covered with hairs, especially if short, so and downlike.
Pulverulent. Powdered, appearing as if covered b minute grains of dust.
Punctate. Dotted with depressions or translucent i ternal glands or colored spots.
Pungent. Sharply and stiffly pointed; bitter or hot t the taste.
Pyriform. Pear-shaped.

Quadrangular. Four-angled.
Quadrate. Square.

Raceme. A simple indeterminate inflorescence pediceled flowers upon a common more or le elongated axis.
Racemose. Resembling a raceme; in racemes.
Rachis. The axis of a compound leaf or of an ir florescence.
Radial. Developing around a central axis.
Raphe. In anatropous ovules, the ridge formed by th fusion of a portion of the funicle to the ovule coa ridge often persistent and prominent in seeds.
Ray. Usually referring to the more or less strap-shape flowers in the head of Compositae.
Receptacle. The expanded portion of the axis th bears the floral organs.
Reclining. Bending or curving toward the ground.
Recurved. Curved downward or backward.
Reduced. Not normally or fully developed in size.
Reflexed. Abruptly bent or turned downward.
Regular. Uniform in shape or distribution of part radically symmetrical.
Remote. Referring to leaves or flowers which are di tant or scattered on the stem.
Reniform. Kidney-shaped.
Repand. Having a somewhat undulating margin.
Repent. Prostrate, with the creeping stems rooting the nodes.
Reticulate. Forming a network.
Retrorse. Turned downward or backward.
Retuse. Rounded and shallowly notched at the apex.
Revolute. Rolled or turned downward or toward th under surface.

Rhizome. A prostrate stem under the ground, rooting at the nodes and bearing buds on nodes.
Rhombic. With equal sides forming oblique angles.
Rib. A primary vein.
Rigid. Stiff.
Rosette. A cluster of basal leaves appearing radially arranged.
Rostrate. With a beaklike point.
Rotate. Wheel-shaped, usually referring to a flattened, short-tubed, sympetalous corolla.
Rotund. Rounded in shape.
Rudiment. A vestige, a much reduced organ, often nonfunctional.
Rufous. Reddish brown.
Rugose. Wrinkled.
Runcinate. Sharply incised pinnately, the incisions retrorse.
Runner. A stolon.

Sagittate. Shaped like an arrowhead.
Salient. Conspicuously projecting, as salient teeth or prominent ribs.
Salverform. Referring to a sympetalous corolla, with the limb at right angles to the tube.
Samara. An indehiscent winged fruit as in the maples and elms.
Scaberulous. Minutely scabrous.
Scabrous. Rough to the touch.
Scale. A leaf much reduced in size, or an epidermal outgrowth.
Scandent. Climbing.
Scape. A flower-bearing stem rising from the ground, the leaves either basal or scattered on the scape and reduced to bracts.
Scarious. Thin and dry.
Schizocarp. Referring to a septicidally dehiscent fruit with one-seeded carpels.
Sclerotic. Hardened, stony.
Scurf. Small scales, usually borne on a leaf surface or on stems.
Secondary. The second division, as in branches or leaf veins.
Secund. One-sided.
Seed. A ripened ovule.
Segment. Part of a compound leaf or other organ, especially if the parts are alike.
Self-pollination. Pollination within the flower.
Semi- (prefix). Approximately half; partly or nearly.
Sepal. A unit of the calyx.
Septicidal. Splitting or dehiscing through the partitions.
Septum (pl. *septa*). A partition or crosswall.
Seriate. Arranged in rows or whorls.
Sericeous. Bearing straight silky hairs.
Serrate. Having sharp teeth pointing forward.
Serrulate. Serrate with small fine teeth.
Sessile. Without stalk of any kind.
Seta (pl. *setae*). A bristle.
Sheath. A tubular structure, such as a leaf base enclosing the stem.
Shoot. A young branch.
Shrub. A woody plant with a number of stems from the base, usually smaller than a tree.
Silicle. A silique broader than long.
Silique. A long capsular fruit typical of the Cruciferae, the two valves separating at maturity.
Silky. Covered with close-pressed soft and straight hairs.
Simple. In one piece or unit, not compound.
Sinuate. Deeply or strongly wavy.
Sinus. The cleft or recess between two lobes.
Solitary. Alone, single.
Somatic. Referring to the body, or to all cells except the germ cells.
Spatulate. Gradually narrowed downward from a rounded summit.
Spherical. Round.
Spike. A racemose inflorescence with the flowers sessile or nearly so.
Spine. Modification of a stipule, petiole, or branch to form a hard, woody, sharp-pointed structure.
Spinescent. Ending in a spine.
Spiniferous. Bearing spines.
Spinulose. With small spines.
Spur. A tubular or sac-like projection of the corolla or calyx, sometimes nectar-bearing; a short, compact twig with little or no internodal development.
Squamella. A small scale, usually as in the pappus of some Compositae.
Squamose. Bearing scales.
Squamulose. Diminutive of squamose.
Stamen. The pollen-bearing organ, usually two or more in each flower.
Staminal column. A column or tube formed by the coalescence of the filaments of the stamens.
Staminate. Bearing stamens but lacking pistils.
Staminodium (pl. *staminodia*). A sterile stamen, often reduced or otherwise modified.
Standard. Usually referring to the upper or posterior petal of a papilionaceous corolla; a banner.
Stellate. Having the shape of a star; starlike.
Stem. Main axis of the plant.
Sterile. Unproductive.
Stigma. That part of the pistil which receives the pollen.
Stigmatic. Relating to the stigma.
Stipe. A stalklike support of a pistil or of a carpel.
Stipel. The stipule of a leaflet in compound leaves.
Stipitate. Having a stipe.
Stipulate. Having stipules.
Stipule. An appendage at the base of a petiole or on each side of its insertion.
Stolon. A branch or shoot given off at the summit of a root.
Stoma. A minute orifice in the epidermis of a leaf.
Stone. The hard endocarp of a drupe.
Striate. With fine grooves, ridges, or lines.
Strict. Erect and straight.
Strigillose. Diminutive of strigose.
Strigose. Bearing hairs which are usually stiff, straight, and appressed.

Strobile. An inflorescence or cone with imbricate bracts or scales.
Style. Upward extension of the ovary terminating with the stigma.
Sub- (prefix). Under; below; less than.
Subshrub. Perennial plant with lower portions of stems woody and persistent.
Subtend. Adjacent to an organ, under or supporting; referring to a bract or scale below a flower.
Subulate. Awl-shaped.
Succulent. Juicy.
Sucker. A stem originating from the roots or lower stem.
Suffrutescent. Slightly shrubby or woody at the base.
Suffruticose. Referring to a slightly shrubby plant; especially applied to low subshrubs.
Sulcate. Grooved or furrowed.
Superior. Above or over; applied to an ovary when free from and positioned above the perianth.
Supra- (prefix). Above.
Suture. Line of dehiscence; a groove denoting a natural union.
Symmetric. Divisible into equal and like parts; referring to a regular flower having the same number of parts in whorl or series.
Sympetalous. Petals more or less united.
Sympodial. Development by simultaneous branching rather than by only a main continuous axis.

Taproot. The main or primary root.
Taxonomy. The science of classification.
Tendril. Usually a slender organ for climbing, formed by modification of leaf, branch, inflorescence, etc.
Terete. Circular in transverse section.
Terminal. At the end, summit, or apex.
Ternate. Divided into three parts.
Testa. The outer seed coat.
Tetragonal. Four-angled.
Tetrahedral. Four-sided.
Theca. Anther sac.
Thorn. A sharp-pointed modified branch.
Throat. The orifice of a sympetalous or gamopetalous corolla.
Thyrse. A shortened panicle with the main axis indeterminate and the lateral flower clusters cymose; loosely, a compact panicle.
Tomentellous. Diminutive of tomentose.
Tomentose. Densely hairy with matted wool.
Tomentulose. Slightly pubescent with matted wool.
Toothed. Having teeth; serrate.
Tortuous. Zigzag or bent in various directions.
Torulose. Cylindrical and constricted at intervals.
Torus. The receptacle, or thickened terminal axis of a flower head, especially in the Compositae.
Toxic. Poisonous.
Trailing. Growing prostrate but not rooting at the nodes.
Translucent. Transmitting light but not transparent.
Transpiration. Passage of water vapor outward, mostly through the stomata.
Transverse. A section taken at right angles to the longitudinal axis.
Tri- (prefix). In three parts, as tri-lobate (3-lobed) trifid (3-cleft).
Trichome. A hair arising from an epidermal cell.
Trichotomous. Three-branched or forked.
Truncate. Ending abruptly, as if cut off transversely.
Trunk. The main stem or axis of a tree below the branches.
Tube. A hollow cylindric organ, such as the lower part of a sympetalous calyx or corolla, the upper part usually expanding into a limb.
Tuber. A thickened underground stem usually for food storage and bearing buds.
Tubercle. A small tuber-like body.
Tumid. Inflated or swollen.
Turbinate. Top-shaped.
Turgid. Swollen.
Twining. Climbing by means of the main stem or branches winding around an object.

Ultimate. The last or final part in a train of progression, as the ultimate division of an organ.
Umbel. An inflorescence with numerous pedicels springing from the end of the peduncle, as with the ribs of an umbrella.
Umbilicate. Depressed in the center.
Umbo. A central raised area or hump.
Uncinate. Scythe-shaped, sometimes referring to hooked hairs or prickles.
Undershrub. A low plant generally woody only near the base.
Undulate. With a wavy surface or margin.
Unequally Pinnate. Pinnate with an odd terminal leaflet.
Uni- (prefix). Solitary or one only; such as unifoliate (with one leaflet) or unilocular (with one cell).
Unilateral. One-sided.
Uniseriate. In one series, or in one row or circle.
Unisexual. Of one sex, either staminate or pistillate only.
Urceolate. Pitcher-shaped, usually with a flaring mouth and constricted neck.
Utricle. An achene-like fruit, but with a thin, loose outer seed covering.

Vallecula (pl. *valleculae*). A channel or groove between the ridges on various organs, such as on stems or fruits.
Valve. One of the pieces into which a capsule splits.
Vascular. Referring to the conductive tissue in the stems or leaves.
Vascular Bundle. A bundle or group of vascular tubes or ducts.
Vegetative. Nonreproductive, as contrasted to floral.
Veins. Ramifications or threads of fibrovascular tissue in a leaf, or other flat organ.
Velutinous. Covered with dense velvety hairs.
Venation. A system of veins.
Venose. Veiny.

Ventral. Belonging to the anterior or inner face of an organ; the opposite of dorsal.
Ventricose. Asymmetrically swollen.
Vernation. Arrangement of leaves in a bud.
Verrucose. Covered with wartlike excrescences.
Versatile. Swinging free, usually referring to an anther attached above its base to a filament.
Verticil (adj. *verticillate*). A whorl of more than two similar parts at a node; leaves, stems, etc.
Verticillate. Disposed in a whorl.
Vesicle (adj. *vesicular*). A small inflated or bladder-like structure.
Vespertine. Opening in the evening.
Vestigial. A rudimentary, usually nonfunctioning, or underdeveloped organ.
Villosulous. Diminutive of villous.
Villous, villose. Bearing long soft hairs.
Virgate. Straight and wandlike.
Viscid. Glutinous or sticky.
Viviparous. Precocious development, such as the germination of seeds or buds while still attached to the parent plant.

Whorl. Cyclic arrangement of like parts.
Wing. Any membranous or thin expansion bordering or surrounding an organ.
Woolly. Clothed with long and tortuous or matted hairs.

Xeric. Characterized by aridity.
Xerophilous. Drought-resistant.
Xerophyte. A desert plant or plant growing under xeric conditions.

ACKNOWLEDGMENTS

It is indeed difficult to give adequate thanks to the many gracious persons and institutions that have given assistance in the completion of this project. A large number have aided by financing the botanical drawings. Some have assisted by supplying botanical specimens as loans or gifts. Others have helped by giving permission to quote from their publications. Many friends have offered kind words of encouragement, stimulating my zeal. In fact, this book may be considered to be a public undertaking, the author being only the catalyst who bonded together the facts. Because they are so many, it is impossible to do more than list the people whose gifts of time, counsel, and financial resources have made this book possible, but my gratitude to each is far from perfunctory. Any possible omissions occurred only through oversight, and the author apologizes for this inadvertence.

ILLUSTRATIONS

Most of the illustrations were prepared by Sarah Kahlden Arendale. Mrs. Arendale, working diligently with the author for a period of seventeen years, completed 1,200 drawings. Credit should go to her for both the quantity and the superior quality of her work.

About 50 drawings were prepared by other artists. Among these were Mrs. Miriam Monteabaro, Miss Annabelle Peck, Miss Barbara Pell, Miss Alice Sayers, Don Smith, and Barry Tinkler.

The *Rubus* drawings were adapted from L. H. Bailey's monograph on *Rubus*. See the special acknowledgments concerning *Rubus* in the chapter on *Rosaceae.*

PATRONS

Sustaining Patrons

Mr. and Mrs. W. L. Anderson
Mrs. Herman Brown
Miss Nina Cullinan
Mrs. Earle C. Douglas
Mr. and Mrs. Arthur Lefevre, Jr.
Mr. and Mrs. T. S. Maffitt, Jr.
Mr. and Mrs. A. J. Wray
Camellia Garden Club
Garden Club of Houston
River Oaks Garden Club
Southern Garden Club

Contributing Patrons

Allen, Mrs. N. B.
Allen, Mrs. N. N.
Amerman, Mrs. A. E.
Anderson, Mr. and Mrs. Ben M.
Anderson, Mrs. Thomas D.
Anderson, Mr. and Mrs. W. L.
Angly, Mr. and Mrs. Maurice
Arnold, Aileen Lovejoy
Arnold, Mrs. Alphonse
Arnold, Mrs. H. K.

Baker, Mr. and Mrs. James A., Jr.
Baker, Mrs. Malcolm G.
Bateman, Mr. and Mrs. Dupuy, Jr.
Benton, Mr. and Mrs. F. Fox
Berry, Mr. and Mrs. Albert *(Drawings given in the name of Mr. and Mrs. R. J. Jackson)*
Berry, Phyllis Edna
Bessell, Mrs. Alfred, Jr.
Best, Dr. and Mrs. Paul
Best, Paul W., Jr.
Blaine, Mrs. Robert M.
Bloxsom, Mr. and Mrs. Dan E.
Boice, Mrs. Mary
Boone, Mrs. James C.
Boswell, Mrs. E. R.
Bosworth, Mrs. L. S.
Boyles, Mrs. Edward S.

Bradley, Mrs. Palmer
Briscoe, Mrs. Birdsall P.
Broun, Mrs. Conway
Brown, Mr. and Mrs. George R.
Brown, Mrs. Herman
Bruce, Mrs. George S., Jr.
Bruhl, Adalene Wellborn
Bruhl, Dr. and Mrs. Charles Kennedy
Bruhl, Charles K., Jr.
Bryan, Mrs. Guy M.
Burnett, Mrs. T. J. *(Drawing of Laredo Mahonia)*

Camellia Garden Club
Carl, Mrs. Noble
Carlton, Clara
Carter, Mrs. Victor
Cashin, Mrs. D'Arcy M. *(Mimosa drawings)*
Casperson, Mrs. Andrew
Cline, Mabel *(Wisteria drawing)*
Conner, Billy
Conner, Cora
Conner, Mr. and Mrs. E. H.
Cook, Cecil N.
Cook, Florence Elliott
Cook, Mrs. Raymond
Copley, Mr. and Mrs. George N.
Crain, Mrs. W. O.
Cravens, Mary Patricia
Cravens, Mrs. Rorick
Cravens, Rorick, Jr.
Cub Scouts, Pack 505, St. Vincent de Paul School
Cullen, Mrs. Roy G.
Cullinan, Mrs. Craig
Cullinan, Mrs. Frank
Cullinan, Miss Nina
Cummings, Mrs. Hatch W., Jr.

Day, Mr. and Mrs. Stephen D.
Dentler, George H.
Dickson, Dr. and Mrs. J. Charles
Dobbs, Dr. and Mrs. E. W.
Douglas, Mrs. Earle C.
Drouet, Mrs. E. N.
Dudley, Mrs. Ray L.

Eastham, Mrs. Clarence S.
Edwards, Mrs. John C.
Edwards, John Charles, Jr.
Edwards, Robert James
Eglin, Mr. and Mrs. Ted
Elles, Dr. Norma B.

Farish, Mrs. W. S.
Favrot, Lawrence H.
Fay, Mrs. Albert
Fay, Katherine Bel
Fay, Marion
Fay, Spencer
Filkins, Mrs. Alice
Flaitz, Mr. and Mrs. J. M.
Flude, Cecile
Flude, Mr. and Mrs. John W. *(Mr. Flude's request, Mississippi trees; Mrs. Flude's request, Louisiana trees)*
Flude, Susan
Fondren, Mrs. W. W.
Foster, Dr. and Mrs. John H.
Fuqua, Mr. and Mrs. C. T.
Francis, Mrs. W. H., Jr.

Garwood, Mrs. St. John
Gershinowitz, Mary
Goldston, Mrs. W. J.
Goss, Dr. and Mrs. J. M.
Griffith, Mr. and Mrs. James P. S.
Gunn, Mr. and Mrs. Ralph Ellis
Guthrie, Dr. and Mrs. Thomas H., Jr.

Haltom, B. D.
Hamill, Mrs. Allen, Jr.
Hamill, Carol
Hamilton, Mrs. Arthur
Hamilton, Mrs. Carlos R.
Hamman, Mr. and Mrs. John, Jr.
Hancock, Mrs. William T.
Heitmann, Mrs. F. A.
Helm, Mrs. Shirley
Henderson, Agnes Raleigh
Henderson, Mr. and Mrs. Homer
Henderson, Mrs. Ralph
Hirsch, General and Mrs. Maurice
Hobby, Mrs. Oveta Culp
Hogg, Miss Ima
Howard, Mrs. Philo
Howe, Dorothy Knox
Howe, Harris Milton
Howe, Mrs. Knox B.
Hunt, Mrs. W. S.

Jacomini, Mr. and Mrs. Victor
Johnson, Mr. and Mrs. John M.
Jones, Franklin C.
Jones, Mrs. Jesse H.
Judd, Mrs. Ardon

Kelley, Allie Autry
Kelley, Mrs. Edward
Kelly, Maureen
Kelly, Mr. and Mrs. T. F. P.
Kelly, Thomas F. W.
Kelsey, Dr. and Mrs. Mavis P.
Kempner, Mrs. I. H., Jr.
Kirkland, W. A.

Knipp, Mrs. Ernest A.
Kuldell, Mrs. R. C.

La Bauve, Mrs. E. C.
Lefevre, Mr. and Mrs. Arthur, Jr.
Lykes, Mrs. James M.

McAshan, Mrs. Harris
McAshan, Mrs. James E.
McAshan, Mrs. S. M., Jr.
Maffitt, Peter C.
Maffitt, Mrs. T. S., Jr.
Maffitt, Thomas F.
Maher, Mr. and Mrs. John F.
Manion, Mrs. Kenneth
Marechal, Mrs. Greer
Maresh, John
Maresh, Mrs. R. E.
Mars, Mr. and Mrs. Forrest E.
Marshall, Mr. and Mrs. Whitfield H.
Massingill, Mrs. Darris Jennings
Mavor, Mr. and Mrs. James E.
Meyers, Mrs. John H.
Meynier, Dr. and Mrs. M. J., Jr.
Michaux, Frank W.
Moore, Perryman S.
Moore, Mrs. Preston

Nelms, Mrs. Agnese Carter
Neuhaus, Mr. and Mrs. Ralph
Neuhaus, Ralph Sells
Neuhaus, Sally Jane

Olsen, Olaf La Cour
Owen, Carol
Owen, Jane
Owen, Mrs. Kenneth Dale

Pan-American Round Table of Houston
P.E.O. Sisterhood, Chapter A.D.
Pearson, Agnes
Pearson, Betsey
Pearson, Mr. and Mrs. Edward G.
Pearson, Hannah
Peden, Mrs. E. A.
Perry, Mrs. Hally Bryan
Pillot, Mrs. Norman V.
Pollard, Jack C.
Pressler, Mrs. H. P., Jr.
Pressler, Townes G.

Ragan, Mrs. Cooper K.
Riesen, Miss Alberta J.
Risien, Mr. and Mrs. Raymond
Roberts, Mrs. Herbert
Robertson, Mrs. C. J.
Rogers, Mrs. Sam D.
Rouse, Mr. and Mrs. Edward

Safford, Mrs. Winifred
Sander, Charles H.
Schumacher, Dr. and Mrs. Paul
Schwartz, Mr. and Mrs. Andrew K.
Schwing, Mr. and Mrs. L. Sanford
Schwing, Lilla Anne Bryant
Schwing, Walter Edward
Scurry, Mrs. Tom
Second National Bank of Houston
Sewall, Mrs. Campbell
Sharp, Mrs. Dudley
Sharrar, Mr. and Mrs. Lee M.
Sherman, Mr. and Mrs. W. A.
Smith, Mrs. Forrest B. *(Drawings given in the name of Peter C. Maffitt and Thomas F. Maffitt)*
Smith, Mrs. Frank C.
Smith, Mr. and Mrs. Lloyd
Smith, Mr. and Mrs. R. E.
Smith, Sandra
Smith, Sharon
Southern Garden Club
Staub, Mr. and Mrs. John
Stilwell, Miss Leota
Stone, Mrs. Lee
Symonds, Mr. and Mrs. Gardiner

Taylor, Mrs. C. M.
Taylor, Miss Kathleen
Teas, Edward
Thompson, Almeria
Thompson, Mr. and Mrs. Ben F.
Thompson, Helen
Thompson, Kay
Torrens, Mrs. W. B.
Townes, Mrs. E. E., Sr.
Townes, Mrs. Edgar, Jr.
Townes, Edgar Goss
Townes, Mary Louise

Underwood, David F.
Underwood, Mr. and Mrs. Milton R.
Underwood, Peter M.

Vaughan, Dr. and Mrs. Luther M.

Waggaman, Adele
Waggaman, Louise
Walker, Mrs. W. L.
Walne, Mrs. Walter H.
Wells, Mr. and Mrs. Damon
West, Mrs. Gordon
West, Mrs. R. C.
White, Mr. and Mrs. Lewis N.

Wier, Mrs. Robert W.
Wilkerson, Mrs. Edward A.
Willborn, Mr. and Mrs. J. E.
Williams, Mrs. Herbert E.
Williams, Mrs. John B.
Williams, Lawrence H.
Williams, Mrs. T. Walter *(Drawing given in the name of Francita Stuart)*
Wilson, Mr. and Mrs. E. C.
Wilson, Mrs. Wallace D.
Wintermann, Mr. and Mrs. David
Wirtz, Mr. and Mrs. L. M.
Wortham, Diana
Wortham, Mrs. Gus
Wortham, Lyndall
Wray, Mrs. A. J.
Wray, Lucie

Younger, Mrs. A. M.
Younger, Gordon
Younger, Lee

TECHNICAL ASSISTANCE

Those institutions and individuals who, by loan or by gift, made available plant specimens or scientific aid

The Academy of Natural Sciences of Philadelphia
The Arnold Arboretum
The Bailey Hortorium of Cornell University
Brooklyn Botanical Garden
California Academy of Science
Chicago Natural History Museum
Colorado A. & M. College
Florida Agricultural Experiment Station
Gray Herbarium of Harvard University
Houston Museum of Natural History
Louisiana State University
Missouri Botanical Garden
The National Arboretum
New York Botanical Garden
Oklahoma A. & M. College
Rancho Santa Ana Botanic Garden
Sam Houston State Teachers College
Santa Barbara Botanic Garden
Southern Methodist University
Sul Ross State College
Texas A. & M. College Agricultural Experiment Station
Texas Forest Service
Tracy Herbarium of Texas A. & M. College
Tulane University
United States Department of Agriculture
United States Forest Service
United States National Herbarium

Cory, V. L.
Fisher, George L.
Greulach, Dr. V. A.
Heiser, J. M., Jr.
Hilliard, R. E.
Irwin, Dr. Howard
Lowrey, Lynn
Marsters, L. G., Jr.
Miner, Mrs. Edna
Palmer, Ernest J. *(See the special acknowledgments concerning* Crataegus *in the chapter on Rosaceae)*
Parks, H. B.
Runyon, Robert
Shinners, Dr. Lloyd
Smith, E. C.
Tharp, Dr. B. C.
Turner, Dr. B. L.
Warner, Dr. S. R.
Warnock, Dr. Barton
Wilkinson, Leroy
Willborn, Mr. and Mrs. J. E.
Winkler, Paul

University of Arkansas
University of California
University of Colorado
University of Houston
University of New Mexico
University of Notre Dame
University of Oklahoma
University of Texas
University of Utah
University of Washington
University of Wyoming

BIBLIOGRAPHY

Those publications which have been of unusually great assistance and from which references or quotes were often made

Bailey, L. H. 1917. *The Standard Cyclopedia of Horticulture,* 2d ed. 6 vols. New York: The Macmillan Company.

Camp, W. H. 1945. The North American Blueberries with notes on other groups of Vacciniaceae, *Brittonia,* **5** (3): 203–275 (March, 1945).

Kelsey, Horton P., and William A. Dayton. 1942. *Standardized Plant Names,* 2d ed. Harrisburg, Pa.: J. Horace McFarland Co.

Little, Elbert L., Jr. 1953. *Check List of Native and Naturalized Trees of the United States (Including Alaska).* U.S. Dept. Agric., Forest Service, Agric. Handbook No. 41.

Muller, Cornelius H. 1951. The Oaks of Texas, *Contr. Texas Res. Foundation,* **1** (3): 21–311 (September 15, 1951).

Munson, T. V. 1909. *Foundations of American Grape Culture.* New York: Orange Judd Company.

Shinners, Lloyd H. 1951. *Yucca freemanii,* a new species from northeastern Texas, *Field & Lab.,* **19** (4): 168–171 (October, 1951).

Standley, Paul C. 1920–1926. *Trees and Shrubs of Mexico. Contr. U.S. Natl. Herbarium,* Vol. 23. Washington, D.C.: Government Printing Office.

U.S.D.A. 1948. *Woody-Plant Seed Manual.* Prepared by the Forest Service. U.S. Dept. Agric., Misc. Pub. No. 654. Material under the heading, Propagation, has been abstracted from the *Woody-Plant Seed Manual* (U.S.D.A., 1948) unless acknowledgment of another source is given.

Webber, Herbert John, and Leon Dexter Batchelor. 1943. *The Citrus Industry,* Vol. 1. Berkeley and Los Angeles: University of California Press.

Wood, Horatio C., and Arthur Osol. 1943. *The Dispensatory of the United States of America,* 23d ed. Philadelphia: J. B. Lippincott Co.

BIBLIOGRAPHY

Abrams, Leroy. 1940–1959. *Illustrated Flora of the Pacific States.* 4 vols. Palo Alto, Calif.: Stanford University Press.

Adams, J. E. 1940. A systematic study of the genus *Arctostaphylos* Adans., *Jour. Elisha Mitchell Sci. Soc.*, **56** (1): 1–62 (June, 1940).

American Rose Society. 1931. *What Every Rose-Grower Should Know.* Harrisburg, Pa.: American Rose Society.

Apgar, Austin Craig. 1910. *Ornamental Shrubs of the United States.* New York: American Book Company.

Ashe, W. W. 1899. New east American species of *Crataegus, Jour. Elisha Mitchell Sci. Soc.*, **16** (2): 70–79 (July–December, 1899).

———. 1900. Some east American species of *Crataegus, Jour. Elisha Mitchell Sci. Soc.*, **17** (1): 4–20.

———. 1925. Notes on woody plants, *Charleston Museum Quart.*, **1** (2): 28–32 (Second Quarter, 1925).

———. 1928. Notes on Southeastern woody plants, *Bull. Torr. Bot. Club,* **55** (7): 463–466 (November, 1928).

———. 1931a. Notes on *Magnolia* and other woody plants, *Torreya,* **31** (2): 37–41 (March–April, 1931).

———. 1931b. Polycodium, *Jour. Elisha Mitchell Sci. Soc.* **46** (2): 196–213 (June, 1931).

Bailey, L. H. 1917. *The Standard Cyclopedia of Horticulture,* 2d ed. 6 vols. New York: The Macmillan Company.

———. 1925. *Manual of Cultivated Plants.* New York: The Macmillan Company.

———. 1934. *The Species of Grapes Peculiar to North America. Gentes Herbarum,* Vol. 3, Fasc. 4 (pp. 149–241). Ithaca: The Bailey Hortorium of the New York State College of Agriculture at Cornell University.

———. 1941–1945. *The Genus* Rubus *in North America. Gentes Herbarum,* Vol. 5. Ithaca: The Bailey Hortorium of the New York State College of Agriculture at Cornell University.

Bailey, L. H., and Ethel Zoe Bailey. 1941. *Hortus Second: A Concise Dictionary of Gardening, General Horticulture and Cultivated Plants in North America.* New York: The Macmillan Company.

Ball, Carleton R. 1938. New varieties and combinations in *Salix, Jour. Washington Acad. Sci.,* **28** (10): 443–452 (October 15, 1938).

———. 1948. *Salix petiolaris* J. E. Smith: American, not British, *Bull. Torr. Bot. Club,* **75** (2): 178–187 (March, 1948).

———. 1950. New combinations in Southwestern *Salix, Jour. Washington Acad. Sci.,* **40** (10): 324–335 (October 15, 1950).

Barber, Harry Lee. 1884. *Menispermum canadense, Amer. Jour. Pharm.,* **56** (8): 401–404 (August, 1884).

Barkley, Fred A. 1937. A monographic study of *Rhus* and its immediate allies in North and Central America, including the West Indies, *Ann. Missouri Bot. Gard.,* **24** (3): 265–498 (September, 1937).

———. 1944. *Schinus* L., *Brittonia,* **5** (2): 160–198 (September, 1944).

Batson, F. S. 1942. *An Illustrated Guide to Identification and Landscape Uses of Mississippi Native Shrubs.* Miss. Agric. Exp. Sta., Bull. 369 (July, 1942).

Beadle, C. D. 1899. Studies in *Crataegus* (I), *Bot. Gaz.,* **28** (6): 405–417 (December, 1899).

———. 1900. Studies in *Crataegus* (II), *Bot. Gaz.,* **30** (5): 335–346 (November, 1900).

———. 1902. New species of thorns from the southeastern states. II. *(Crataegus pyracanthoides), Biltmore Bot. Studies,* **1** (2): 136–137 (April 30, 1902).

Benson, Lyman. 1940. *The Cacti of Arizona.* Biol. Sci. Bull. No. 5. Univ. Ariz. Bull., Vol. 11, No. 1 (January 1, 1940).

———. 1941. The Mesquites and Screw-beans of the United States, *Amer. Jour. Bot.,* **28** (9): 748–754 (November, 1941).

———. 1943a. Revisions of status of Southwestern desert trees and shrubs (I), *Amer. Jour. Bot.,* **30** (3): 230–240 (March, 1943).

———. 1943b. Revisions of status of Southwestern

desert trees and shrubs (II), *Amer. Jour. Bot.*, **30** (8): 630–632 (October, 1943).

Benson, Lyman, and Robert A. Darrow. 1945. *A Manual of Southwestern Desert Trees and Shrubs.* Biol. Sci. Bull. No. 6. Univ. Ariz. Bull., Vol. 15, No. 2 (April, 1944 [Distributed August 6, 1945]).

———. 1954. *The Trees and Shrubs of the Southwestern Deserts.* Tucson: The University of Arizona.

Bergen, Joseph Y. 1901. *Foundations of Botany.* Boston: Ginn & Co.

Bertho, Alfred, and Wor Sang Liang. 1933. Notiz über ein Alkaloid aus *Ceanothus americanus, Archiv der Pharmacie,* **271**: 273–276.

Billings, W. D. 1936. A bud and twig key to the southeastern arborescent oaks, *Jour. Forestry,* **34** (5): 475–476 (May, 1936).

Bilsing, S. W. 1927. *Studies on the Biology of the Pecan Nut Case Bearer.* Texas Agric. Exp. Sta. Bull. No. 347 (April, 1927).

Bishop, G. Norman. 1940. *Native Trees of Georgia.* Athens: University of Georgia School of Forestry.

Blake, S. F. 1918. *A Revision of the Genus* Viguiera, *Contr. Gray Herbarium,* n.s., No. 54.

———. 1920. A preliminary revision of the North American and West Indian avocados, *Jour. Washington Acad. Sci.,* **10** (1): 9–21 (January 4, 1920).

———. 1921. Revisions of the genera *Acanthospermum, Flourensia, Oyedaea,* and *Tithonia, Contr. U.S. Natl. Herbarium,* **20** (10): 383–436.

Bogusch, E. R. 1931. A new variety of *Koberlinia, Torreya,* **31** (3): 73–74 (May–June, 1931).

Bomhard, Miriam L. 1935. *Sabal louisiana,* the correct name for the polymorphic Palmetto of Louisiana, *Jour. Washington Acad. Sci.,* **25** (1): 35–44 (January 15, 1935).

Britton, Nathaniel Lord. 1926. The swamp cypresses, *Jour. N.Y. Bot. Gard.,* **27** (321): 205–207 (September, 1926).

Britton, Nathaniel Lord, and Hon. Addison Brown. 1913. *An Illustrated Flora of the Northern United States, Canada, and the British Possessions,* 2d ed., rev. and enl. 3 vols. New York: Charles Scribner & Sons.

Britton, Nathaniel Lord, and J. N. Rose. 1919–1923. *The Cactaceae.* 4 vols. Washington, D.C.: The Carnegie Institution of Washington, Publ. No. 248.

———. 1928a. (Rosales:) Mimosaceae, *N. Amer. Flora,* **23** (1): 1–76 (February 11, 1928).

———. 1928b. (Rosales:) Mimosaceae, *N. Amer. Flora,* **23** (2): 77–136 (September 25, 1928).

———. 1928c. (Rosales:) Mimosaceae, *N. Amer. Flora,* **23** (3): 137–194 (December 20, 1928).

Britton, Nathaniel Lord, and John Adolph Shafer. 1908. *North American Trees.* New York: Henry Holt & Co.

Brown, Clair A. 1943. Vegetation and lake level correlations at Catahoula Lake, Louisiana, *Geog. Rev.,* **33** (3): 435–445 (July, 1943).

———. 1945. *Louisiana Trees and Shrubs.* Baton Rouge: Louisiana Forestry Commission, Bull. No. 1.

———. 1956. *Commercial Trees of Louisiana.* Baton Rouge: Louisiana Forestry Commission.

Brown, H. B. 1921. *Trees of New York State, Native and Naturalized.* Technical Publication No. 15. Syracuse: The New York State College of Forestry, Syracuse University.

Brown, H. P., A. J. Panshin, and C. C. Forsaith. 1949–1952. *Textbook of Wood Technology.* 2 vols. New York: McGraw-Hill Book Co.

Brown, Nelson Courtlandt. 1950. *Forest Products.* New York: John Wiley & Sons.

Buckley, S. B. 1860. Descriptions of several new species of plants, *Proc. Acad. Nat. Sci.,* Philadelphia, **12**: 443–445.

Camp, W. H. 1942a. A survey of the American species of *Vaccinium* subgenus Euvaccinium, *Brittonia,* **4** (2): 205–247 (September, 1942).

———. 1942b. The *Crataegus* problem, *Castanea,* **7** (4–5): 51–55 (April–May, 1942).

———. 1942c. Ecological problems and species concepts in *Crataegus, Ecology,* **23** (3): 368–369 (July, 1942).

———. 1945. The North American Blueberries with notes on other groups of Vacciniaceae, *Brittonia,* **5** (3): 203–275 (March, 1945).

———. 1951. Biosystematy, *Brittonia,* **7** (3): 113–127 (January, 1951).

———. (ed.). 1955. *Taxonomic Index.* **18** (6). [Organ of Amer. Soc. Plant Taxonomists. Dept. of Botany, Univ. of Connecticut, Storrs, Conn.]

Camp, W. H., H. W. Rickett, and C. A. Weatherby. 1947. International rules of botanical nomenclature, *Brittonia,* **6** (1): 1–120 (April, 1947). [Also issued separately, pub. by the New York Bot. Garden in co-operation with the Amer. Soc. Plant Taxonomists.]

Castetter, Edward F., Willis H. Bell, and Alvin R. Grove. 1938. *The Early Utilization and the Distribution of* Agave *in the American Southwest (Ethnobiological Studies in the American Southwest, VI).* Univ. of N. Mex. Bull. No. 335, Biol. Ser. Vol. 5, No. 4.

Chapman, A. W. 1897. *Flora of the Southern United States,* 3d ed. New York: American Book Co.

Clark, A. H. 1928. The alkaloids of *Ceanothus americanus.* II. Extraction of the alkaloids, *Amer. Jour. Pharm.,* **100** (4): 240–242 (April, 1928).

Clark, Ora M. 1934. *The Cacti of Oklahoma. Proc. Okla. Acad. Sci.,* Vol. 14. Norman: University of Oklahoma.

Clark, Robert Brown. 1942. A revision of the genus *Bumelia* in the United States, *Ann. Missouri Bot. Garden,* **29** (3): 155–182 (September, 1942).

Clements, Frederic Edward, and Edith Schwartz Clements. 1914. *Rocky Mountain Flowers: An Illustrated Guide for Plant-lovers and Plant-users.* New York: H. W. Wilson Co.

Clepper, Henry E. 1934. *Hemlock, the State Tree of Pennsylvania.* Harrisburg: Pa. Dept. Forests and Waters, Bull. No. 52.

Clover, Elzada U. 1937a. Vegetational survey of the lower Rio Grande Valley (I, II), *Madroño,* **4** (2): 41–66 (April 7, 1937).

———. 1937b. Vegetational survey of the lower Rio

Grande Valley, Texas (III), *Madroño,* **4** (3): 77–100 (July 1, 1937).

Clute, Willard N. 1940. *American Plant Names,* 3d ed. Indianapolis: W. N. Clute & Co.

Coker, William Chambers. 1944. The woody smilaxes of the United States, *Jour. Elisha Mitchell Sci. Soc.,* **60** (1): 27–69 (August, 1944).

Coker, William Chambers, and Henry Roland Totten. 1945. *Trees of the Southeastern States,* 3d ed. Chapel Hill: University of North Carolina Press.

Collingwood, G. H., and Warren D. Brush. 1955. *Knowing Your Trees,* rev. ed. Washington, D.C.: American Forestry Association.

Connery, Joseph E., and Margaret H. Tewksbury. 1931. Studies on ceanothyn as a blood coagulant in man, *Jour. Amer. Pharmaceut. Assoc.,* **20** (12): 1287–1290 (December, 1931).

Cory, V. L. 1936. Three junipers of western Texas, *Rhodora,* **38** (449): 182–187 (May, 1936).

———. 1943a. The Wild Cherry of the Carrizo Sands of Texas, *Rhodora,* **45** (536): 325–327 (August, 1943).

———. 1943b. The genus *Styrax* in central and western Texas, *Madroño,* **7** (4): 110–115 (October 14, 1943).

———. 1944. *Forestiera* in southern and southwestern Texas, *Madroño,* **7** (8): 252–255 (October 30, 1944).

———. 1947. Two new varieties of *Condalia* from Texas, *Madroño,* **9** (4): 128–131 (July 17, 1947).

Cory, V. L., and H. B. Parks. 1937. *Catalogue of the Flora of the State of Texas.* Texas Agric. Exp. Sta., Bull. No. 550 (July, 1937).

Coulter, John M. 1891–1894. *Botany of Western Texas: A Manual of the Phanerogams and Pteridophytes of Western Texas. Contr. U.S. Natl. Herbarium,* Vol. 2.
No. 1, pp. 1–152, June 27, 1891
No. 2, pp. 153–346, July 1, 1892
No. 3, pp. 347–588, May 10, 1894

Coulter, John M., and Aven Nelson. 1909. *New Manual of Botany of the Central Rocky Mountains.* New York: American Book Company.

Croizat, Leon. 1942. A study of *Manihot* in North America, *Jour. Arnold Arb.,* **23** (2): 216–225 (April, 1942).

Cronquist, Arthur. 1944. Studies in the Simaroubaceae. IV. Resumé of the American genera, *Brittonia,* **5** (2): 128–147 (September, 1944).

———. 1945. Studies in the Sapotaceae. III. *Dipholis* and *Bumelia, Jour. Arnold Arb.,* **26** (4): 435–471 (October, 1945).

———. 1946. Studies in the Sapotaceae. II. A survey of the North American genera, *Lloydia,* **9** (4): 241–292 (December, 1946).

Cutler, Hugh C. 1939. Monograph of the North American species of the genus *Ephedra, Ann. Missouri Bot. Gard.,* **26** (4): 373–427 (November, 1939).

Davis, Helen Burns. 1936. *Life and Work of Cyrus Guernsey Pringle.* Burlington: University of Vermont.

Davis, William C., Ernest Wright, and Carl Hartley. 1942. *Diseases of Forest-Tree Nursery Stock.* Federal Security Agency Civilian Conservation Corps, Forestry Publ. No. 9 (February, 1942).

Deam, Charles Clemon. 1924. *Shrubs of Indiana.* Indiana Dept. of Conservation, Publ. No. 44.

———. 1931. *Trees of Indiana,* 2d rev. ed. Indiana Dept. Conservation, Publ. No. 13.

Dormon, Caroline. 1941. *Forest Trees of Louisiana and How to Know Them.* Baton Rouge: La. Dept. Conservation, Bull. No. 15.

Dyal, Sarah C. 1936. A Key to the species of oaks of eastern North America based on foliage and twig characters, *Rhodora,* **38** (446): 53–63 (February, 1936).

Eastwood, Alice. 1934. A revision of *Arctostaphylos, Leaflets West. Bot.,* **1**: 105–127.

Eggleston, W. W. 1909. The *Crataegi* of Mexico and Central America, *Bull. Torr. Bot. Club,* **36** (9): 501–514 (September, 1909).

Emory, William Hemsley. 1859. *Report of the United States and Mexican Boundary Survey,* Vol. 2, Pt. 1. Botany of the Boundary: Introduction, by C. C. Parry; Botany, by John Torrey; Cactaceae, by George Engelmann. House of Representatives Exec. Doc. 135; Serial No. 862.

Engelmann, George. 1882. The black-fruited *Crataegi* and a new species, *Bot. Gaz.,* **7** (11): 127–129 (November, 1882).

Ensign, Margaret. 1942. A revision of the celastraceous genus *Forsellesia (Glossopetalon), Amer. Midl. Nat.,* **27** (2): 501–511 (March, 1942).

Erickson, Ralph O. 1943. Taxonomy of *Clematis* section Viorna, *Ann. Missouri Bot. Gard.,* **30** (1): 1–62 (February, 1943).

Erlanson, Eileen Whitehead. 1934. Experimental data for a revision of the North American Wild Roses, *Bot. Gaz.,* **96** (2): 197–259 (December, 1934).

Evans, William E., Jr., William G. Harne, and John C. Krantz, Jr. 1942. A uterine principle from *Viburnum prunifolium, Jour. Pharmacol. and Exper. Therapeutics,* **75** (2): 174–177 (June, 1942).

Fassett, Norman C. 1943. The validity of *Juniperus virginiana* var. *crebra, Amer. Jour. Bot.,* **30** (7): 469–477 (July, 1943).

———. 1944. *Juniperus virginiana, J. horizontalis* and *J. scopulorum.* I. The specific characters, *Bull. Torr. Bot. Club,* **71** (4): 410–418 (July, 1944).

Featherly, Henry Ira. 1954. *Taxonomic Terminology of the Higher Plants.* Ames: Iowa State College Press.

Ferguson, A. M. 1901. *Crotons* of the United States, *Missouri Bot. Gard.,* 12th Ann. Rep., pp. 33–72.

Fernald, Merritt Lyndon. 1908. *Gray's Manual of Botany,* 7th ed. New York: American Book Company.

———. 1935a. Critical plants of the upper Great Lakes region of Ontario and Michigan, *Rhodora,* **37** (440): 272–301 (August, 1935).

———. 1935b. Midsummer vascular plants of southeastern Virginia, *Rhodora,* **37** (442): 423–454 (December, 1935).

———. 1936. Plants from the outer coastal plain of

Virginia, *Rhodora,* **38** (456): 414–452 (December, 1936).

———. 1938. Noteworthy plants of southeastern Virginia, *Rhodora,* **40** (479): 434–459 (November, 1938).

———. 1943a. Fruit of *Dirca palustris, Rhodora,* **45** (532): 117–119 (April, 1943).

———. 1943b. Virginian botanizing under restrictions, *Rhodora,* **45** (538): 357–413 (October, 1943).

———. 1944. Overlooked species, transfers and novelties in the flora of eastern North America (concluded), *Rhodora,* **46** (542): 32–58 (February, 1944).

———. 1945a. Botanical specialties of the Seward Forest and adjacent areas of southeastern Virginia, *Rhodora,* **47** (557): 149–182 (May, 1945).

———. 1945b. Botanical specialties of the Seward Forest and adjacent areas of southeastern Virginia (concluded), *Rhodora,* **47** (558): 191–204 (June, 1945).

———. 1945c. Some North American Corylaceae (Betulaceae). I. Notes on *Betula* in eastern North America, *Rhodora,* **47** (562): 303–329 (October, 1945).

———. 1945d. Some North American Corylaceae (Betulaceae). II. Eastern North American Representatives of *Alnus incana, Rhodora,* **47** (563): 333–361 (November, 1945).

———. 1950. *Gray's Manual of Botany,* 8th ed. New York: American Book Company.

Fernald, Merritt Lyndon, and Alfred Charles Kinsey. 1943. *Edible Wild Plants of Eastern North America* (Gray Herbarium of Harvard University, special publ.). Cornwall-on-Hudson, N.Y.: Idlewild Press.

Fernald, Merritt Lyndon, and J. Francis MacBride. 1914. The North American variations of *Arctostaphylos uva-ursi, Rhodora,* **16** (192): 211–213 (December, 1914).

Fisher, E. M. 1892. Revision of the North American species of *Hoffmanseggia, Contr. U.S. Natl. Herbarium,* **1** (5): 143–150.

Fletcher, W. F. 1942. *The Native Persimmon,* rev. ed. U.S. Dept. Agric., Farmers' Bull. No. 685.

Flory, W. S., Jr., and F. R. Brison. 1942. *Propagation of a Rapid Growing Semi-evergreen Hybrid Oak.* Texas Agric. Exp. Sta. Bull. No. 612 (May, 1942).

Folkers, Karl, and Frank Koniuszy. 1940. *Erythrina* alkaloids. VIII. Studies on the constitution of erythramine and erythraline, *Jour. Amer. Chem. Soc.,* **62** (7): 1673–1677 (July, 1940).

Fosberg, F. Raymond. 1936. Varieties of the Desert-Willow, *Chilopsis linearis, Madroño,* **3** (8): 362–366 (October 28, 1936).

Frye, Theodore C., and George B. Rigg. 1914. *Elementary Flora of the Northwest.* New York: American Book Company.

Geiser, S. W. 1936. A century of scientific exploration in Texas. I. 1820–1880, *Field & Lab.,* **4** (2): 41–55 (April, 1936).

———. 1939. A century of scientific exploration in Texas. IB. 1820–1880, *Field & Lab.,* **7** (1): 29–52 (January, 1939).

———. 1944. Dr. David Porter Smythe, an early Texan botanist, *Field & Lab.,* **12** (1): 10–16 (January, 1944).

Gleason, Henry A. 1952. *New Britton and Brown Illustrated Flora of the Northeastern United States and Adjacent Canada.* 3 vols. New York: New York Botanical Garden.

Gooding, Leslie N. 1940. *Willows in Region* 8. U.S. Dept. Agric., Soil Conservation Service Regional Bull. No. 65 (mimeographed).

Gray, Asa. 1858. *Botany for Young People and Common Schools: How Plants Grow.* New York: Ivison, Blakeman, Taylor, & Company.

Greene, Edward L. 1906. The genus *Ptelea* in the western and southwestern United States and Mexico, *Contr. U.S. Natl. Herbarium,* **10** (2): 49–78.

———. 1910–1912. Notes on *Polycodium langloisii, Leaflets of Botanical Observation and Criticism,* **2**: 226–227.

Greenman, J. M. 1914. Descriptions of North American Senecioneae, *Ann. Missouri Bot. Gard.,* **1** (3): 263–290 (September, 1914).

———. 1915–1918. Monograph of the North and Central American species of the genus *Senecio, Ann. Missouri Bot. Gard.,* **2** (3): 573–626 (September, 1915); **3** (1): 85–194 (February, 1916); **4** (1): 15–36 (February, 1917); **5** (1): 37–108 (February, 1918); (not completed).

Griffiths, David. 1908. Illustrated studies in the genus *Opuntia, Missouri Bot. Gard. 19th Ann. Rep.,* pp. 259–272.

Groot, James T. 1927. The pharmacology of *Ceanothus americanus.* I. Preliminary studies: Hemodynamics and the effects on coagulation, *Jour. Pharmacol. and Exper. Therapeutics,* **30** (4): 275–291 (February, 1927).

Gunderson, Alfred, and Arthur H. Graves. 1942. Trees in the Brooklyn Botanic Garden, *Brooklyn Bot. Gard. Rec.,* **31** (1, [January, 1942]).

Hagen, Stanley Harlan. 1941. A revision of the North American species of *Anisacanthus, Ann. Missouri Bot. Gard.,* **28** (4): 385–408 (November, 1941).

Hall, Harvey M. 1928. *The Genus* Haplopappus: *A Phylogenetic Study in the Compositae.* Washington, D.C.: Carnegie Institution of Washington, Publ. No. 389.

Hall, Harvey M., and Frederic E. Clements. 1923. *The Phylogenetic Method in Taxonomy: The North American Species of* Artemisia, Chrysothamnus *and* Atriplex. Washington, D.C.: Carnegie Institution of Washington, Publ. No. 826.

Hamilton, Clyde C. 1948. *Insects and Spider Mites Attacking Holly.* Washington, D.C.: Holly Soc. Amer., Bull. No. 2 (March, 1948).

Hanson, Herbert C. 1920. *Key to the Malvaceous Plants in Texas.* Texas Agric. Exp. Sta. Circ. No. 22 (April 30, 1920).

Harbison, T. G. 1928. Notes on the genus *Hydrangea,*

Amer. Midl. Nat., **11** (5): 255–257 (September, 1928).

———. 1931. *Symplocos tinctoria, ashei:* a new dye-bush from the Southern mountains, *Jour. Elisha Mitchell Sci. Soc.*, **46** (2): 218–220 (June, 1931).

Harlow, William M. 1942. *Trees of the Eastern United States and Canada, Their Woodcraft and Wildlife Uses.* New York: McGraw-Hill Book Co.

Harlow, William M., and Ellwood S. Harrar. 1950. *Textbook of Dendrology, Covering the Important Forest Trees of the United States and Canada,* 3d ed. New York: McGraw-Hill Book Co.

Harper, Francis. 1943. *Quercus incana* Bartram, *Bartonia,* **22**: 3.

Harper, Roland M. 1926. The cedar glades of Middle Tennessee, *Ecology,* **7** (1): 48–54 (January, 1926).

———. 1928. *Catalogue of the Trees, Shrubs and Vines of Alabama* (Economic botany of Alabama, pt. 2). Ala. Geol. Survey, Monograph 9.

———. 1943. *Forests of Alabama.* Ala. Geol. Survey, Monograph 10.

Harrar, Ellwood S., and J. George Harrar. 1946. *Guide to Southern Trees* (Whittlesey House Field Guide series). New York: Whittlesey House, McGraw-Hill Book Co.

Harrington, Harold D. 1954. *Manual of the Plants of Colorado.* Denver, Colo.: Sage Books.

Havard, Valery. 1885. A report on the flora of western and southern Texas, *Proc. of the U.S. Natl. Mus.,* **8**: 449–553.

Heiser, Charles B., Jr. 1944. Monograph of *Psilostrophe, Ann. Missouri Bot. Gard.,* **31** (3): 279–300 (September, 1944).

Hinckley, L. C. 1944. The vegetation of the Mount Livermore area of Texas, *Amer. Midl. Nat.,* **32** (1): 236–250 (January, 1944).

Hitchcock, Charles Leo. 1932. A monographic study of the genus *Lycium* of the Western Hemisphere, *Ann. Missouri Bot. Gard.,* **19** (2–3): 179–374 (April–September, 1932).

———. 1943. The xerophilous species of *Philadelphus* in southwestern North America, *Madroño,* **7** (2): 35–36 (April 28, 1943).

Holly Society of America. 1957. *Handbook of Hollies.* Special issue, *Natl. Horticultural Magazine,* Vol. 36 (January, 1957).

Hopkins, Milton. 1942. *Cercis* in North America, *Rhodora,* **44** (522): 193–211 (June, 1942).

Hopkins, Milton, and Umaldy Theodore Waterfall. 1943. Notes on Oklahoma plants, *Rhodora,* **45** (532): 113–117 (April, 1943).

Hough, Romeyn Beck. 1947. *Handbook of the Trees of the Northern States and Canada East of the Rocky Mountains.* New York: The Macmillan Company.

Howell, John Thomas. 1945. Studies in Rosaceae, tribe Potentilleae. Ia. Reconsideration of the genus *Purpusia, Leaflets West. Bot.,* **4**: 171–175.

Hoyt, Roland Stewart. 1938. *Check Lists for the Ornamental Plants of Subtropical Regions.* Los Angeles, Calif.: Livingston Press.

Hume, H. Harold. 1953. *Hollies.* New York: The Macmillan Company.

Hutchinson, John. 1926. *Key to the Families of the Dicotyledons.* London: Macmillan and Company, Ltd.

Irish, H. C. 1898. A revision of the genus *Capsicum* with especial reference to garden varieties, *Missouri Bot. Gard.,* 9th Ann. Rep., pp. 53–110.

Jepson, Willis Linn. 1901. *A Flora of Western Middle California.* Berkeley, Calif.: Encina Publishing Co. [2d ed., San Francisco: Cunningham, Curtiss & Welch, 1911].

———. 1909. *The Trees of California.* San Francisco: Cunningham, Curtis & Welch [2d ed., Berkeley: Sather Gate Bookshop].

———. 1909–1922. *A Flora of California,* Vol. 1, Pts. 1–7. Berkeley: Associated Students Store, University of California. [Pts. 1–4, San Francisco: Cunningham, Curtiss & Welch; Pts. 4–5, San Francisco: H. S. Crocker Co.].

Johnston, Ivan Murray. 1924a. Expedition of the California Academy of Sciences to the Gulf of California in 1921. The botany (the vascular plants), *Proc. Calif. Acad. Sci.,* 4th ser., **12** (30): 951–1218 (May 31, 1924).

———. 1924b. Taxonomic records concerning American spermatophytes, 1. *Parkinsonia* and *Cercidium, Contr. Gray Herbarium,* n.s., No. 70, pp. 61–68.

———. 1937. Studies in the Boraginaceae (XII). 1. *Trigonotis* in southwestern China. 2. Novelties and critical notes, *Jour. Arnold Arb.,* **18** (1): 1–25 (January, 1937).

———. 1938. Some undescribed species from Mexico and Guatemala, *Jour. Arnold Arb.,* **19** (2): 117–128 (April, 1938).

———. 1939. New phanerogams from Mexico, Jour. Arnold *Arb.,* **20** (2): 234–240 (April, 1939).

———. 1940. New phanerogams from Mexico (III), *Jour. Arnold Arb.,* **21** (2): 253–265 (April, 1940).

———. 1941a. New phanerogams from Mexico (IV), *Jour. Arnold Arb.,* **22** (1): 110–124 (January, 1941).

———. 1941b. Gypsophily among Mexican desert plants, *Jour. Arnold Arb.,* **22** (2): 145–170 (April, 1941).

———. 1943a. Noteworthy species from Mexico and adjacent United States (I), *Jour. Arnold Arb.,* **24** (2): 227–236 (April, 1943).

———. 1943b. Publication dates for the botanical parts of the Pacific Railroad Reports, *Jour. Arnold Arb.,* **24** (2): 237–242 (April, 1943).

———. 1943c. Plants of Coahuila, Eastern Chihuahua, and adjoining Zacatecas and Durango (I), *Jour. Arnold Arb.,* **24** (3): 306–359 (July, 1943).

———. 1944a. Plants of Coahuila, Eastern Chihuahua and adjoining Zacatecas and Durango (III), *Jour. Arnold Arb.,* **25** (1): 43–83 (January, 1944).

———. 1944b. Plants of Coahuila, Eastern Chihuahua and adjoining Zacatecas and Durango (IV), *Jour. Arnold Arb.,* **25** (2): 133–182 (April, 1944).

———. 1944c. Plants of Coahuila, Eastern Chihuahua, and adjoining Zacatecas and Durango (V), *Jour. Arnold Arb.,* **25** (4): 431–453 (October, 1944).

———. 1948. Noteworthy species from Mexico and adjacent United States (II), *Jour. Arnold Arb.*, **29** (2): 193–197 (April, 1948).

———. 1950. Noteworthy species from Mexico and adjacent United States (III), *Jour. Arnold Arb.*, **31** (2): 188–195 (April, 1950).

Jones, D. L., Frank Gaines, and R. E. Karper. 1932. *Trees and Shrubs in Northwest Texas.* Texas Agric. Exp. Sta., Bull. No. 447 (April, 1932).

Jones, George Neville. 1939. A synopsis of the North American species of *Sorbus, Jour. Arnold Arb.*, **20** (1): 1–43 (January, 1939).

———. 1940. A monograph of the genus *Symphoricarpos, Jour. Arnold Arb.*, **21** (2): 201–252 (April, 1940).

———. 1946. *American Species of* Amelanchier. *(Ill. Biol. Monographs,* Vol. 20, No. 2) Urbana: University of Illinois Press.

Jones, W. W. 1905. A revision of the genus *Zexmenia, Proc. Amer. Acad. Arts and Sci.*, **41** (7): 143–167 (June, 1905).

Julian, Percy L., Josef Pikl, and Ray Dawson, 1938. Constituents of *Ceanothus americanus.* I. Ceanothic acid, *Jour. Amer. Chem. Soc.*, **60** (1): 77–79 (January, 1938).

Kearney, Thomas H., and Robert H. Peebles. 1942. *Flowering Plants and Ferns of Arizona.* U.S. Dept. Agric., Misc. Pub. No. 423.

———. 1951. *Arizona Flora.* Berkeley: University of California Press.

Keller, Allan C. 1942. *Acer glabrum* and its varieties, *Amer. Midl. Nat.*, **27** (2): 491–500 (March, 1942).

Kelsey, Horton P., and William A. Dayton. 1942. *Standardized Plant Names,* 2d ed. Harrisburg, Pa.: J. Horace McFarland Co.

King, Eleanor Anthony. 1941. *Bible Plants for American Gardens.* New York: The Macmillan Company.

Krukoff, B. A. 1939. The American species of *Erythrina, Brittonia,* **3** (2): 205–337 (October, 1939).

———. 1941. Supplementary notes on the American species of *Erythrina* (I), *Amer. Jour. Bot.*, **28** (8): 683–691 (October, 1941).

———. 1943. Supplementary notes on the American species of *Erythrina* (II), *Bull. Torr. Bot. Club,* **70** (6): 633–637 (November, 1943).

Kumlien, Loraine L. 1946. *The Friendly Evergreens.* Dundee, Ill.: D. Hill Nursery Co.; New York: Rinehart & Company, Inc. (reissue, 1954).

Kurz, Herman. 1942. *Florida Dunes and Scrub, Vegetation and Geology.* Fla. Geol. Survey, Geol. Bull. No. 23. Tallahassee: State of Florida, Dept. of Conservation.

Lanjouw, Joseph, and F. A. Stafleu. 1954. *Index Herbariorum,* Pt. 2: Collectors–First Instalment, A-D. *Regnum Vegetabile,* Vol. 2. Utrecht: International Bureau for Plant Taxonomy and Nomenclature.

———. 1956. *Index Herbariorum,* Pt. 1: The Herbaria of the World, 3d ed. *Regnum Vegetabile,* Vol. 6. Utrecht: International Bureau for Plant Taxonomy and Nomenclature.

———. 1957. *Index Herbariorum,* Pt. 2: Collectors–Second Instalment, E-H. *Regnum Vegetabile,* Vol. 9. Utrecht: International Bureau for Plant Taxonomy and Nomenclature.

Lawrence, George H. M., and Arnold E. Schulze. 1942. *Hederae Cultorum* (The Cultivated *Hederas*). *Gentes Herbarum,* Vol. 6, Fasc. 3 (pp. 105–173). Ithaca: The Bailey Hortorium of the New York State College of Agriculture at Cornell University.

Lee, Frederic P. (ed.). 1952. *The Azalea Handbook.* Washington, D.C.: American Horticultural Society.

Lewis, Isaac M. 1915. *The Trees of Texas: An Illustrated Manual of the Native and Introduced Trees of the State.* Bull. Univ. Texas 1915, No. 22 (April 15, 1915).

Ley, Arline. 1943. A taxonomic revision of the genus *Holodiscus* (Rosaceae), *Bull. Torr. Bot. Club,* **70** (3): 275–288 (May, 1943).

Lindsay, T. S. 1923. *Plant Names.* London: The Sheldon Press; New York: The Macmillan Company.

Little, Elbert L., Jr. 1944. *Acer grandidentatum* in Oklahoma, *Rhodora,* **46** (551): 445–450 (November, 1944).

———. 1948. Older names for two western species of *Juniperus, Leaflets West. Bot.*, **5**: 125–132.

———. 1950. *Southwestern Trees: A Guide to the Native Species of New Mexico and Arizona.* U.S. Dept. Agric., Agric. Handbook No. 9.

———. 1951. *Key to Southwestern Trees* [Supplement to Little, 1950]. U.S. Forest Service, Southwestern Forest and Range Exp. Sta., Res. Rep. No. 8 (September, 1951).

———. 1953a. *Check List of Native and Naturalized Trees of the United States (Including Alaska).* U.S. Dept. Agric., Forest Service, Agric. Handbook No. 41.

———. 1953b. Five varietal transfers of United States trees, *Phytologia,* **4** (5): 305–310 (August, 1953).

Little, Elbert L., Jr., and Keith W. Dorman. 1952. Slash Pine *(Pinus elliottii),* its nomenclature and varieties, *Jour. Forestry,* **50** (12): 918–923 (December, 1952).

Longyear, B. O. 1927. *Trees and Shrubs of the Rocky Mountain Region.* New York: G. P. Putnam's Sons.

Lounsberry, Alice, and Mrs. Ellis Rowan. 1901. *Southern Wild Flowers and Trees.* New York: Frederick A. Stokes Co.

Lundell, Cyrus Longworth, *et al.* 1943a. New vascular plants from Texas, Mexico, and Central America, *Amer. Midl. Nat.*, **29** (2): 469–492 (March, 1943).

———. 1943b. Anacardiaceae, Cyrillaceae, Aquifoliaceae, Convolvulaceae, genus *Cuscuta. Flora of Texas,* Vol. 3, Pt. 2. Dallas: Southern Methodist University Press.

———. 1944. Palmaceae, Bromeliaceae, Onagraceae. *Flora of Texas,* Vol. 3, Pt. 4. Dallas: Southern Methodist University Press.

McClintock, Elizabeth. 1951. Studies in California plants. III. The Tamarisks, *Jour. Calif. Hort. Soc.*, **12**: 76–83.

McDougall, W. B., and Omer E. Sperry. 1951. *Plants*

of Big Bend National Park. Washington, D.C.: National Park Service.

McKelvey, Susan Delano. 1938. *Yuccas of the Southwestern United States,* Vol. 1. Cambridge, Mass.: Arnold Arboretum of Harvard University.

———. 1947. *Yuccas of the Southwestern United States,* Vol. 2. Cambridge, Mass.: Arnold Arboretum of Harvard University.

Mackenzie, Kenneth Kent, and Benjamin F. Bush. 1902. *Manual of the Flora of Jackson County, Missouri*. Kansas City, Mo.

McMinn, Howard E. 1942. A Systematic Study of the Genus *Ceanothus, Ceanothus,* by Maunsell Van Rensselaer and Howard E. McMinn. Santa Barbara, Calif.: Santa Barbara Botanic Garden.

———. 1951. *An Illustrated Manual of California Shrubs*. Berkeley: University of California Press.

McNair, James B. 1923. *Rhus dermatitis . . . Its Pathology and Chemotherapy*. Chicago: University of Chicago Press.

McVaugh, Rogers. 1943. The status of certain anomalous native crabapples in eastern United States, *Bull. Torr. Bot. Club,* **70** (4): 418–429 (July, 1943).

———. 1945. The *Jatrophas* of Cervantes and of the Sessé & Mociño Herbarium, *Bull. Torr. Bot. Club.,* **72** (1): 31–41 (January, 1945).

———. 1951. A revision of the North American Black Cherries (*Prunus serotina* Ehrh. and relatives), *Brittonia,* **7** (4): 279–315 (December, 1951).

Manning, Wayne E. 1949. The genus *Carya* in Mexico, *Jour. Arnold Arb.,* **30** (4): 425–432 (October, 1949).

Marshall, Rush P., and Alma M. Waterman. 1948. *Common Diseases of Important Shade Trees*. U.S. Dept. Agric., Farmers' Bull. No. 1987.

Martin, Floyd L. 1950. A revision of *Cercocarpus, Brittonia,* **7** (2): 91–111 (March, 1950).

Martínez, Maximino. 1936. *Plantas útiles de México,* 2d ed. México, D.F.: Ediciones Botas.

———. 1937. *Catálogo de nombres vulgares y científicos de plantas mexicanas*. México, D.F.: Ediciones Botas.

Mathews, F. Schuyler. 1915. *Field Book of American Trees and Shrubs*. New York: G. P. Putnam's Sons.

Mattoon, Wilbur R. 1915. *The Southern Cypress*. U.S. Dept. Agric., Bull. No. 272.

———. 1926. *Loblolly Pine Primer*. U.S. Dept. Agric., Farmers' Bull. No. 1517.

———. 1936. *Forest Trees and Forest Regions of the United States*. U.S. Dept. Agric., Misc. Publ. No. 217.

———. 1939. *Slash Pine,* rev. ed. U.S. Dept. Agric., Farmers' Bull. No. 1256 (Superseded by Pomeroy and Cooper, 1956).

———. 1940a. *Longleaf Pine Primer,* rev. ed. U.S. Dept. Agric., Farmers' Bull. No. 1486 (Superseded by Muntz, 1954).

———. 1940b. *Shortleaf Pine,* rev. ed. U.S. Dept. Agric., Bull. No. 1671.

Mattoon, Wilbur R., and C. B. Webster. 1928. *Forest Trees of Texas: How to Know Them*. College Station: Texas Forest Service, Bull. No. 20.

Medsger, Oliver Perry. 1939. *Edible Wild Plants*. New York: The Macmillan Company.

Merrill, E. D. 1945. In defense of the validity of William Bartram's binomials, *Bartonia,* **23**: 10–35.

Metz, Sister Mary Clare. 1934. *A Flora of Bexar County, Texas*. Washington, D.C.: The Catholic University of America Press.

Mohr, Charles, 1901. *Plant Life of Alabama*. Prepared in co-operation with the Geological Survey of Alabama. *Contr. U.S. Natl. Herbarium,* Vol. 6.

Moldenke, Harold N. 1942. Eriocaulaceae, Avicenniaceae, Verbenaceae. *Flora of Texas,* Vol. 3, Pt. 1. Dallas: Southern Methodist University Press.

Moldenke, Harold N., and Alma L. Moldenke. 1952. *Plants of the Bible*. Waltham, Mass.: Chronica Botanica Company.

Morrison, R. C. 1933. *Planting and Care of Trees and Shrubs*. Fort Worth, Texas: Board of Park Commissioners (rev. 1942).

Morton, C. V. 1930. A new species of *Esenbeckia* from Texas, *Jour. Washington Acad. Sci.,* **20** (7): 135–136 (April 4, 1930).

———. 1933. The Mexican and Central American species of *Viburnum, Contr. U.S. Natl. Herbarium,* **26** (7): 339–366.

———. 1941. Notes on Juniperus, *Rhodora,* **43** (512): 344–348 (August, 1941).

———. 1944. Taxonomic studies of tropical American plants, *Contr. U.S. Natl. Herbarium,* **29** (1): 86.

———. 1945. Mexican phanerogams described by M. E. Jones, *Contr. U.S. Natl. Herbarium,* **29** (2): 87–116.

Mulford, A. Isabel. 1896. A study of the agaves of the United States, *Missouri Bot. Gard.,* 7th Ann. Rep., pp. 47–100.

Mulford, Furman Lloyd. 1929. *Transplanting Trees and Shrubs*. U.S. Dept. Agric., Farmers' Bull. No. 1591.

Muller, Cornelius H. [*sometimes* Mueller]. 1934. Some new oaks from western Texas, *Torreya,* **34** (5): 119–122 (September–October, 1934).

———. 1936a. Studies in the oaks of the mountains of northeastern Mexico, *Jour. Arnold Arb.,* **17** (3): 160–179 (July, 1936).

———. 1936b. New and noteworthy trees in Texas and Mexico, *Bull. Torr. Bot. Club,* **63** (3): 147–155 (March, 1936).

———. 1937a. Studies in Mexican and Central American plants, *Amer. Midl. Nat.,* **18** (5): 842–855 (September, 1937).

———. 1937b. Vegetation in Chisos Mountains, Texas, *Trans. Texas Acad. Sci.,* **20**: 5–31.

———. 1938. Further studies in Southwestern Oaks, *Amer. Midl. Nat.,* **19** (3): 582–588 (May, 1938).

———. 1940a. Oaks of Trans-Pecos Texas, *Amer. Midl. Nat.,* **24** (3): 703–728 (November, 1940).

———. 1940b. A revision of *Choisya, Amer. Midl. Nat.,* **24** (3): 729–742 (November, 1940).

———. 1941. The Holacanthoid plants of North America, *Madroño,* **6** (3): 128–132 (October 15, 1941).

———. 1951. The oaks of Texas, *Contr. Texas Res. Foundation,* **1** (3): 21–311 (September 15, 1951).

Munch, James C. 1940. The uterine sedative action of *Viburnums* (VIII), *Pharmaceut. Archives,* **11** (3): 33–37 (May, 1940).

Munson, T. V. 1909. *Foundations of American Grape Culture.* New York: Orange Judd Company.

Muntz, H. H. 1954. *How to Grow Longleaf Pine.* U.S. Dept. Agric., Farmers' Bull. No. 2061.

Murrill, William Alphonse. 1945. *A Guide to Florida Plants.* Gainesville, Fla.: Published by the author.

———. 1946. *Familiar Trees.* Gainesville, Fla.: Published by the author.

Nelson, Ruth A. 1950. The genus *Vitis* in Oklahoma, *Proc. Okla. Acad. Sci.,* **31**: 20–23.

Osol, Arthur, and George E. Farrar, Jr. 1955. *The Dispensatory of the United States of America,* 25th ed. Philadelphia: J. B. Lippincott Co.

Palmer, Ernest J. 1925a. Synopsis of North American *Crataegi, Jour. Arnold Arb.,* **6** (1): 5–128 (January, 1925).

———. 1925b. Is *Quercus arkansana* a hybrid? *Jour. Arnold Arb.,* **6** (3): 195–200.

———. 1926. The ligneous flora of the Hot Springs National Park and vicinity, *Jour. Arnold Arb.,* **7**: 104–135.

———. 1929. The ligneous flora of the Davis Mountains, Texas, *Jour. Arnold Arb.,* **10** (1): 8–45 (January, 1929).

———. 1931. Conspectus of the genus *Amorpha, Jour. Arnold Arb.,* **12** (3): 157–197 (July, 1931).

———. 1932. The *Crataegus* problem, *Jour. Arnold Arb.,* **13** (3): 342–362 (July, 1932).

———. 1942. The Red Oak Complex in the United States, *Amer. Midl. Nat.,* **27** (3): 732–740 (May, 1942).

———. 1943a. *Quercus prinus* Linnaeus, *Amer. Midl. Nat.,* **29** (3): 783–784 (May, 1943).

———. 1943b. The species concept in *Crataegus, Chron. Bot.,* **7** (8): 373–375 (December, 1943).

———. 1945. *Quercus durandii* and its allies, *Amer. Midl. Nat.,* **33** (2): 514–519 (March, 1945).

———. 1946. *Crataegus* in the northeastern and central United States and adjacent Canada, *Brittonia,* **5** (5): 471–490 (September, 1946).

———. 1948. Hybrid oaks of North America, *Jour. Arnold Arb.,* **29** (1): 1–48 (January, 1948).

Palmer, Ernest J., and Julian A. Steyermark. 1935. An annotated catalogue of the flowering plants of Missouri, *Ann. Missouri Bot. Gard.,* **22** (3): 375–758 (September, 1935).

Parks, H. B. 1937. *Valuable Plants Native to Texas.* Texas Agric. Exp. Sta., Bull. No. 551 (August, 1937).

Parks, H. B., V. L. Cory, *et al.* 1936. The Fauna and Flora of the Big Thicket Area. Issued through the co-operation of Sam Houston State Teachers College, the Texas Agricultural Experiment Station, the East Texas Big Thicket Association, the Forest Committee, and the Texas Academy of Science.

Patraw, Pauline Mead. 1951. *Flowers of the Southwest Mesas* (drawings by Jeanne Russel Janish). Santa Fe, N. Mex.: Southwestern Monuments Association Popular Series No. 5.

Pennell, F. W. 1919. [Rosales:] Fabaceae: *Eysenhardtia, N. Amer. Flora,* **24** (1): 34–40 (April 25 1919).

———. 1935. The Scrophulariaceae of Temperate Eastern North America. *Monograph Acad. Nat. Sci.* Philadelphia, **1**: 1–650.

———. 1941. Scrophulariaceae of Trans-Pecos Texas *Proc. Acad. Nat. Sci.,* Philadelphia, **92**: 289–308.

Perkins, Janet Russell. 1907. The Leguminosae of Porto Rico. *Contr. U.S. Natl. Herbarium,* **10** (4): 133–220

Piper, Charles V. 1906. *Flora of the State of Washington. Contr. U.S. Natl. Herbarium,* Vol. 11.

Piper, Charles V., and R. Kent Beattie. 1914. *Flora of Southeastern Washington and Adjacent Idaho.* Lancaster, Pa.: Press of the New Era Printing Co.

———. 1915. *Flora of the Northwest Coast.* Lancaster Pa.: Press of the New Era Printing Co.

Pomeroy, Kenneth B., and Robert W. Cooper. 1956 *Growing Slash Pine.* U.S. Dept. Agric., Farmers Bull. No. 2103.

Putnam, J. A., and Henry Bull. 1932. *The Trees of the Bottomlands of the Mississippi River Delta Region.* U.S. Forest Service, Southern Forest Exp Sta., Occasional Paper No. 27.

Reed, E. L. 1935. A new species of *Ephedra* from western Texas, *Bull. Torr. Bot. Club,* **62** (1): 43 (January, 1935).

Rehder, Alfred. 1903. Synopsis of the genus *Lonicera, Missouri Bot. Gard.,* 14th Ann. Rep., pp. 27–232.

———. 1917. The genus *Fraxinus* in New Mexico and Arizona, *Proc. Amer. Acad. Arts and Sci.,* **53** (2): 199–212 (October, 1917).

———. 1940. *Manual of Cultivated Trees and Shrubs Hardy in North America, Exclusive of the Subtropical and Warmer Temperate Regions,* 2d ed. New York: The Macmillan Company.

Rehder, Alfred, Ernest J. Palmer, and Leon Croizat. 1938. Seven binomials proposed as nomina ambigua, *Jour. Arnold Arb.,* **19** (3): 282–290 (July, 1938).

Rickett, Harold William. 1936. Forms of *Crataegus pruinosa, Bot. Gaz.,* **97** (4): 780–793 (June, 1936).

———. 1942. *Cornus asperifolia* and its relatives, *Amer. Midl. Nat.,* **27** (1): 259–261 (January, 1942).

———. 1944a. *Cornus stolonifera* and *Cornus occidentalis, Brittonia,* **5** (2): 149–159 (September, 1944).

———. 1944b. Legitimacy of names in Bartram's "Travels," *Rhodora,* **46** (551): 385–391 (November, 1944).

Robinson, Benjamin Lincoln. 1898. Revision of the North American and Mexican species of Mimosa, *Contr. Gray Herbarium,* n.s., No. 13; *Proc. Amer. Acad. Arts and Sci.,* **33** (17): 303–331 (May, 1898).

———. 1917. *A Monograph of the Genus* Brickellia. *Memoirs Gray Herbarium,* Vol. 1. Cambridge, Mass.: Harvard University Press.

Robinson, Benjamin Lincoln, and J. M. Greenman. 1896. A new genus of Sterculiaceae and some other noteworthy plants, *Bot. Gaz.*, **22** (2): 168–170 (August, 1896).

———. 1899. I. Revision of the genus *Gymnolomia*. II. Supplementary notes upon *Calea, Tridax* and *Mikania, Contr. Gray Herbarium*, n.s., No. 17; *Proc. Boston Soc. Nat. Hist.*, **29** (5): 87–108 (August, 1899).

———. 1904. Revision of the Mexican and Central American species of *Trixis, Proc. Amer. Acad. Arts and Sci.*, **40** (1): 6–14 (July, 1904).

Rollins, Reed C. 1939. The cruciferous genus *Stanleya, Lloydia*, **2** (2): 109–127 (June, 1939).

Runyon, Robert. 1947. *Vernacular Name of Plants Indigenous to the Lower Rio Grande Valley of Texas.* Brownsville, Texas: Brownsville News Publishing Co.

Rydberg, Per Axel. 1906. *Flora of Colorado.* Agric. Exp. Sta. Colo. Agric. Coll., Bull. No. 100.

———. 1913. Rosaceae *(Pars), N. Amer. Flora,* **22** (5): 389–480 (December 23, 1913).

———. 1918. Rosaceae *(Conclusio), N. Amer. Flora,* **22** (6): 481–533 (December 30, 1918).

———. 1919. [Rosales:] Fabaceae: Psoraleae, *N. Amer. Flora,* **24** (1): 1–64 (April 25, 1919).

———. 1932. *Flora of the Prairies and Plains of Central North America.* New York: New York Botanical Garden.

Safford, William Edwin. 1909. Cactaceae of Northeastern and Central Mexico, Together with a Synopsis of the Principal Mexican Genera. *Ann. Rep. Smithson. Instn.*, **1908**: 525–563.

St. John, Harold. 1942. Nomenclatorial changes in *Glossopetalon, Proc. Biol. Soc.*, Washington, **55**: 109–112.

Sargent, Charles Sprague. 1889. Notes upon some North American trees (VIII), *Garden and Forest,* **2** (80): 423–424 (September 4, 1889).

———. 1891–1902. *The Silva of North America: A Description of the Trees Which Grow Naturally in North America Exclusive of Mexico.* 14 vols. Boston and New York: Houghton Mifflin Co.

———. 1901. New or little known North American trees (II), *Bot. Gaz.*, **31** (1): 1–16 (January, 1901).

———. 1905–1913. *Trees and Shrubs: Illustrations of New or Little Known Plants, Prepared Chiefly from Material in the Arnold Arboretum of Harvard University.* 2 vols. Boston and New York: Houghton Mifflin Co.

———. 1908. *Crataegus* in Missouri, *Missouri Bot. Gard.*, 19th Ann. Rep., pp. 35–126.

———. 1917. Botanical activities of Percival Lowell, *Rhodora,* **19** (218): 21–24 (February, 1917).

———. 1918. Notes on North American trees. I. *Quercus, Bot. Gaz.*, **65** (5): 423–459 (May, 1918).

———. 1919. Notes on North American Trees (IV), *Bot. Gaz.*, **67** (3): 208–242 (March, 1919).

———. 1920. Notes on North American Trees. VI. *Hamamelis.* Crataegus, *Jour. Arnold Arb.*, **1** (4): 245–254 (April, 1920).

———. 1921. Notes on North American trees. IX. New species and varieties of *Crataegus, Jour. Arnold Arb.*, **3** (1): 1–11 (July, 1921).

———. 1922. Notes on North American Trees. X. New species and varieties of *Crataegus, Jour. Arnold Arb.*, **3** (4): 182–206 (October, 1922).

———. 1923. Notes on North American trees, XI. New species of *Crataegus, Jour. Arnold Arb.*, **4** (2): 99–107 (April, 1923).

———. 1925. Notes on *Crataegus–Crataegus padifolia* var. *incarnata, Jour. Arnold Arb.*, **6** (1): 1–5 (January, 1925).

———. 1933. *Manual of the Trees of North America (exclusive of Mexico).* Boston and New York: Houghton Mifflin Co.

Sax, Karl. 1921. The origin and relationships of the Pomoideae, *Jour. Arnold Arb.*, **12** (1): 3–21 (January, 1921).

Schulz, Ellen D. 1928. *Texas Wild Flowers.* New York: Laidlaw Brothers, Inc.

Schulz, Ellen D., and Robert Runyon. 1930. *Texas Cacti: A Popular and Scientific Account of the Cacti Native of Texas. Proc. Texas Acad. Sci.*, Vol. 14.

Shinners, Lloyd H. 1950. *Forestiera pubescens* Nutt. var. *glabrifolia* Shinners, *Field & Lab.*, **18** (2): 99–100 (April, 1950).

———. 1951a. *Ceanothus herbacea* Raf. for *C. ovatus:* a correction of name, *Field & Lab.*, **19** (1): 33–34 (January, 1951).

———. 1951b. *Rhus aromatica* Ait. var. *flabelliformis* Shinners, var. nov., *Field & Lab.*, **19** (2): 86–87 (April, 1951).

———. 1951c. *Yucca freemanii,* a new species from northeastern Texas, *Field & Lab.*, **19** (4): 168–171 (October, 1951).

———. 1956. *Zanthoxylum parvum* (Rutaceae), a new species from Trans-Pecos Texas, *Field & Lab.*, **24** (1): 19–20 (January, 1956).

Small, John K. 1913. *Flora of the Southeastern United States,* 2d ed. New York: Published by the author.

———. 1923. The Cabbage-tree–*Sabal palmetto, Jour. N.Y. Bot. Gard.*, **24** (284): 145–158 (August, 1923).

———. 1926. The Saw-Palmetto–*Serenoa repens, Jour. N.Y. Bot. Gard.*, **27** (321): 193–202 (September, 1926).

———. 1927. The Palmetto-Palm–*Sabal texana, Jour. N.Y. Bot. Gard.*, **28** (330): 132–143 (June, 1927).

———. 1930. Chronicle of the palms of the continental United States, *Jour. N.Y. Bot. Gard.*, **31** (363): 57–66 (March, 1930).

———. 1931a. The Cypress, southern remnant of a northern fossil type, *Jour. N.Y. Bot. Gard.*, **32** (378): 125–135 (June, 1931).

———. 1931b. Palms of the continental United States, *Sci. Monthly,* **32**: 240–255 (March, 1931).

———. 1932. *Tamala littoralis, Addisonia,* **17** (3): 45–46 (October 28, 1932).

———. 1933. *Manual of the Southeastern Flora.* New York: Published by the author.

———. 1935. Chronicle of the cacti of eastern North America, *Jour. N.Y. Bot. Gard.*, **36** (422): 25–36 (February, 1935).

Sperry, Omer E. 1938. *A Check List of the Ferns, Gymnosperms, and Flowering Plants of the Proposed*

Big Bend National Park. Sul Ross State Teachers College Bull., Vol. 19, No. 4 (December 1, 1938).

———. 1941. *Additions to the Check List of Plants of the Proposed Big Bend National Park Area.* Sul Ross State Teachers College Bull., Vol. 22, No. 1 (March 1, 1941).

Stafleu, F. A., and Joseph Lanjouw. 1954. *The Genève Conference on Botanical Nomenclature and Genera Plantarum,* January 25–30, 1954. *Regnum Vegetabile,* Vol. 5. Utrecht: International Bureau for Plant Taxonomy and Nomenclature.

Standley, Paul C. 1909. The Allionaceae of the United States with notes on Mexican species, *Contr. U.S. Natl. Herbarium,* **12** (8): 303–389.

———. 1920–1926. *Trees and Shrubs of Mexico. Contr. U.S. Natl. Herbarium,* Vol. 23. Washington, D.C.: Government Printing Office.
Pt. 1, pp. 1–170, October 11, 1920
Pt. 2, pp. 171–515, July 14, 1922
Pt. 3, pp. 517–848, July 18, 1923
Pt. 4, pp. 849–1312, December 31, 1924
Pt. 5, pp. 1313–1721, November 15, 1926.

———. 1940a. *Studies of American Plants* (IX), *Field Museum of Nat. Hist. Publ.,* Bot. Series, Vol. 22, No. 1.

———. 1940b. *Studies of American Plants* (X), *Field Museum of Nat. Hist. Publ.,* Bot. Series, Vol. 22, No. 2.

Sterrett, W. D. 1915. *The Ashes: Their Characteristics and Management.* U.S. Dept. Agric., Bull. No. 299.

Steyermark, Julian A. 1932. A revision of the genus *Menodora, Ann. Missouri Bot. Gard.,* **19** (1): 87–160 (February, 1932).

———. 1940. *Spring Flora of Missouri.* St. Louis: Missouri Botanical Garden; Chicago: Field Museum of Natural History.

Steyermark, Julian A., and John Adam Moore. 1933. Report of a botanical expedition into the mountains of western Texas, *Ann. Missouri Bot. Gard.,* **20** (4): 791–806 (November, 1933).

Stillwell, Norma. 1939. *Key and Guide to Native Trees, Shrubs, and Woody Vines of Dallas County, Texas.* Dallas: Boyd Printing Co.

Stockwell, William Palmer, and Lucretia Breazeale. 1933. *Arizona Cacti.* Biol. Sci. Bull. No. 1, Univ. Ariz. Bull. Vol. 4, No. 3 (April 1, 1933).

Stokes, Susan G. 1936. The Genus *Eriogonum: A Preliminary Study Based on Geographic Distribution.* San Francisco: J. H. Neblett Pressroom.

Sudworth, George B. 1908. *Forest Trees of the Pacific Slope.* U.S. Forest Service.

———. 1915. *The Cypress and Juniper Trees of the Rocky Mountain Region.* U.S. Dept. Agric., Bull. No. 207.

———. 1916. *The Spruce and Balsam Fir Trees of the Rocky Mountain Region.* U.S. Dept. Agric., Bull. No. 327.

———. 1927. *Check List of the Forest Trees of the United States, Their Names and Ranges,* rev. ed. U.S. Dept. Agric., Misc. Cir. 92.

———. 1934. *Poplars, Principal Tree Willows and Walnuts of the Rocky Mountain Region.* U.S. Dep Agric., Tech. Bull. No. 420.

Svenson, H. K. 1939. *Quercus rubra* once more *Rhodora,* **41** (491): 521–524 (November, 1939).

Swingle, Walter T. 1920. A new species of *Pistaci* native to southwestern Texas, *P. texana, Jour. Arnol Arb.,* **2** (2): 105–110 (October, 1920).

———. 1942. Three new varieties and two new com binations in *Citrus* and related genera of the Orang subfamily, *Jour. Washington Acad. Sci.,* **32** (1) 24–26 (January 15, 1942).

Texas Forest Service. 1946. *Forest Trees of Texas: Ho to Know Them.* Prepared by the Texas Forest Serv ice, A.&M. College, W. E. White, Director. Col lege Station: Texas Forestry Association.

Tharp, Benjamin Carroll. 1926. *Structure of Texa Vegetation East of the 98th Meridian.* Univ. Texa Bull. No. 2606 (February 8, 1926).

———. 1939. *The Vegetation of Texas.* Houston: An son Jones Press, for the Texas Academy of Science

Thornber, J. J. 1911. *Native Cacti as Emergency Forag Plants,* Univ. Ariz. Agric. Exp. Sta., Bull. No. 67.

Tidestrom, Ivar. 1925. *Flora of Utah and Nevad Contr. U.S. Natl. Herbarium,* Vol. 25.

Tidestrom, Ivar, and Sister Teresita Kittell. 1941. *Flora of Arizona and New Mexico.* Washington D.C.: The Catholic University of America Press.

Trelease, William. 1902. The Yucceae, *Missouri Bo Gard.,* 13th Ann. Rep., pp. 27–133.

———. 1911a. The desert group Nolineae, *Proc Amer. Philos. Soc.,* **50**: 405–443.

———. 1911b. Revision of the *Agaves* of the grou Applantae, *Missouri Bot. Gard.,* 22d Ann. Rep., pp 85–97.

———. 1916. The Genus *Phoradendron:* A Mono graphic Revision. Urbana: University of Illinois.

———. 1924. *The American Oaks. Natl. Acad. Sc Mem.,* Vol. 20.

Turner, B. L. 1950a. *Hoffmanseggia parryi* (E. M Fisher) Turner, comb. nov., *Field & Lab.,* **18** (1) 46 (January, 1950).

———. 1950b. Vegetative key to Texas *Desmanthu* (Leguminosae) and similar genera, *Field & Lab* **18** (2): 54–65 (April, 1950).

———. 1950c. Mexican species of *Desmanthu* (Leguminosae), *Field & Lab.,* **18** (3): 119–13 (June, 1950).

Turner, Lewis M. 1937. *Trees of Arkansas.* Univ. Ark Agric. Ext. Service, Ext. Circ. No. 180.

U.S.D.A. 1948. *Woody-Plant Seed Manual.* Prepare by the Forest Service. U.S. Dept. Agric., Misc. Pub No. 654.

Van Dersal, William R. 1938. *Native Woody Plants o the United States, Their Erosion-control and Wild life Value.* U.S. Dept. Agric., Misc. Pub. No. 303.

Van Eseltine, G. P. 1933a. *Notes on the Species o Apples. I. The American Crabapples.* N.Y. Stat Agric. Exp. Sta., Tech. Bull. No. 208.

———. 1933b. *Notes on the Species of Apples. II The Japanese Flowering Crabapples of the Siebold*

Group and Their Hybrids. N.Y. State Agric. Exp. Sta., Tech. Bull. No. 214.

———. 1934. *Ornamental Apples and Crabapples.* N.Y. State Agric. Exp. Sta., Circ. No. 139.

Van Melle, P. J. 1950. *Juniperus utahensis* Lemm., *Phytologia,* **3** (6): 299–300 (April, 1950).

———. 1952. *Juniperus texensis* sp. nov.—West-Texas Juniper in relation to *J. monosperma, J. ashei, et al., Phytologia,* **4** (1): 26–35 (March, 1952).

Van Rensselaer, Maunsell, and Howard E. McMinn. 1942. *Ceanothus.* Santa Barbara, Calif.: Santa Barbara Bot. Garden.
Contents: *Ceanothus* for gardens, parks, and roadsides, by Maunsell Van Rensselaer;
A systematic study of the genus *Ceanothus,* by Howard E. McMinn.

Venable, F. P. 1885. Analysis of the leaves of *Ilex cassine, Amer. Jour. Pharm.,* **57** (8): 389–390 (August, 1885).

Vines, Robert A. 1943. A proposed manual of the woody plants of Texas, *Trans. Texas Acad. Sci.,* **26**: 52–53.

———. 1953. *Native East Texas Trees.* Houston: The Adco Press.

Vines, Robert A., and V. A. Greulach. 1940. *A Key to the Trees of the Houston Region.* Houston: University of Houston Press.

Waterfall, Umaldy Theodore. 1952. *A Catalogue of the Flora of Oklahoma.* Stillwater: Oklahoma Research Foundation.

Weaver, Howard E. (ed.). 1952. *A Manual of Forestry, with Special Reference to Forestry Problems in East Texas.* College Station: Texas Forest Service, Bull. No. 45.

Webber, Herbert John, and Leon Dexter Batchelor. 1943. *The Citrus Industry,* Vol. 1. Berkeley and Los Angeles: University of California Press.

Webber, John Milton. 1945. The Navajo Yucca, a new species from New Mexico, *Madroño,* **8** (4): 105–110 (November 5, 1945).

West, Erdman, and Lilian E. Arnold. 1956. *The Native Trees of Florida,* rev. ed. Gainesville: University of Florida Press.

Wiggins, Ira L. 1934. A report on several species of *Lycium* from the Southwestern deserts, *Contr. Dudley Herbarium,* **1** (6): 197–206 (November 15, 1934).

———. 1940. Taxonomic notes on the genus *Dalea* Juss. and related genera as represented in the Sonoran Desert, *Contr. Dudley Herbarium,* **3** (2): 41–64 (July 10, 1940).

———. 1942. *Acacia angustissima* (Mill.) Kuntze and its near relatives, *Contr. Dudley Herbarium,* **3** (7): 227–239 (February 25, 1942).

Willis, John Christopher. 1931. *A Dictionary of the Flowering Plants and Ferns,* 6th ed., rev. (Cambridge Biological Series). London: Cambridge University Press.

Winkler, Charles Herman. 1915. *The Botany of Texas: An Account of Botanical Investigations in Texas and Adjoining Territory.* Bull. Univ. Texas, 1915, No. 18 (March 25, 1915).

Wolf, Carl B. 1938. *The North American Species of* Rhamnus. Monographs of *Rancho Santa Ana Bot. Gard.,* Bot. Ser., No. 1.

Wood, Horatio C., and Arthur Osol. 1943. *The Dispensatory of the United States of America,* 23d ed. Philadelphia: J. B. Lippincott Co. *N.B.* (A 25th edition of the Dispensatory, completely revised and reset, appeared in 1955 [see Osol and Farrar, 1955], when Vines's work was substantially completed. Most of the material quoted or abstracted herein from the 23d edition, however, has been found to have been retained substantially unchanged in the 25th edition, or appears there in abridged form.—*Ed.*)

Wooton, E. O. 1913. *Trees and Shrubs of New Mexico.* N. Mex. Agric. Exp. Sta., Bull. No. 87 (June, 1913).

Wooton, E. O., and Paul C. Standley. 1915. *The Flora of New Mexico. Contr. U.S. Natl. Herbarium,* Vol. 19.

Young, Mary Sophie. 1920. *The Seed Plants, Ferns, and Fern Allies of the Austin Region.* Univ. Texas Bull. No. 2065 (November 20, 1920).

Youngken, Heber Wilkinson. 1919. The Comparative Morphology, Taxonomy and Distribution of the Myricaceae of the Eastern United States. *Contr. Bot. Laboratory of the Univ. Pa.,* **4** (2): 339–400.

Zimmerman, G. A. 1941. Hybrids of the American Papaw, *Jour. Hered.,* **32** (3): 83–91 (March, 1941).

INDEX

The number of the page on which the illustration of a species is found appears in parentheses. It is usually, but not always, the same as the number of the page where the species description is found. Other numbers indicate pages on which the name is discussed in relation to other species.

INDEX OF COMMON NAMES

INDEX OF SCIENTIFIC NAMES